McGRAW-HILL ENCYCLOPEDIA OF ELECTRONICS AND COMPUTERS

McGRAW-HILL

ENCYCLOPEDIA OF ELECTRONICS AND COMPUTERS

SYBIL P. PARKER

Editor in Chief

McGraw-Hill Book Company

New York   St. Louis   San Francisco

Auckland   Bogotá   Guatemala   Hamburg
Johannesburg   Lisbon   London   Madrid   Mexico
Montreal   New Delhi   Panama   Paris   San Juan
São Paulo   Singapore   Sydney   Tokyo   Toronto

McGRAW-HILL ENCYCLOPEDIA OF ELECTRONICS
AND COMPUTERS
Copyright © 1984 by McGraw-Hill, Inc.

1234567890 KPKP 892109876543

ISBN-0-07-045487-6

Library of Congress Cataloging in Publication Data

McGraw-Hill encyclopedia of electronics and computers.

   "Much of the material in this volume has been published previously in the McGraw-Hill encyclopedia of science & technology, fifth edition" — T.p. verso
   Bibliography: p.
   Includes index.
   1. Electronics — Dictionaries.   2. Computers — Dictionaries.
I. Parker, Sybil P.   II. McGraw-Hill Book Company.
III. McGraw-Hill encyclopedia of science & technology (5th ed.)
TK7804.M43   1984   621.381′C3′21   83-9897
ISBN 0-07-045487-6

PREFACE

Electronics, which is properly a part of electrical engineering, had its beginning in 1883 with the invention of the electron tube by Thomas Edison. This device dominated the field for more than a half century, until 1947 when the transistor was invented. This was an outstanding event in the history of electronics, and what followed was an explosive growth of the industry coupled with an impressive variety of applications. Among these, computers have been the principal benefactors.

The purpose of this Encyclopedia is to thoroughly explore the discipline of electronics and the manifold applications of electronic devices, with particular emphasis on computer science and engineering. To accomplish this in one volume, the scope of coverage and the selection of individual subjects have been restricted to only those topics considered essential by the editors.

In choosing entries for the Encyclopedia recognition has been given to three basic areas: (1) the scientific principles of concern to electronics, that is, the conduction and control of electricity flowing through semiconducting materials, vacuum, or gases; (2) the science and technology of electronic devices; and (3) the applications of these devices. The result is a volume that effectively fulfills its purpose. The contents include articles on such important topics as integrated circuits, electrical conduction in gases, diodes, communications, optical fibers, computer storage technology, control systems, semiconductors, programming languages, consumer electronics, lasers, artificial intelligence, and holes in solids.

Many of the articles have been taken from the *McGraw-Hill Encyclopedia of Science and Technology* (5th ed., 1982) — some with minor updates and revisions — while others were written exclusively for this volume. In all, there are 477 articles by 272 contributors, all recognized authorities in academia and industry. The organization of the Encyclopedia is alphabetical. Each article is signed, and there is a complete list of contributors with their affiliations at the back of the book.

Access to the information in the Encyclopedia is enhanced by the copious use of cross-references as well as by a detailed and analytical index, where subjects are listed both individually and categorically. In addition, most articles have a bibliography giving references in which the reader can explore a given subject in greater depth. Illustrations complement the text with additional information.

While there are a number of dictionaries and encyclopedias available on computers, this is the only one-volume encyclopedia to provide comprehensive, authoritative, and up-to-date coverage of electronics — as a science and an industry. It is a "first-stop" entry into the world of electronics. As such, it will be a useful reference for computer professionals and nonspecialists alike. Librarians, writers, business people, students, technicians, and consumers will find this Encyclopedia indispensable.

Sybil P. Parker
EDITOR IN CHIEF

McGRAW-HILL ENCYCLOPEDIA OF ELECTRONICS AND COMPUTERS

A-Z

Abacus

The earliest known computing instrument, consisting of beads (counters) strung on wires mounted in a frame (Fig. 1); technically the term also covers the earlier counting tables and the grooved tablets of the Romans. The abacus is still used in parts of the world.

The Latin word *abacus* means simply "a flat surface" and is derived from the Greek word *abax*. The ancient Hebrew word *abaq* means "dust," and there are references in the literature which indicate that sanded surfaces were used for reckoning. Stones were probably used as counters on the first counting tables, and the Latin word *calculus* means "little stone." The Romans also had counters of ivory, metal, and glass, while in the Orient pieces of bamboo were used.

The earliest known example of a counting board is the Greek Salamis marble tablet. Its exact date is unknown, but the Greek historian Herodotus (485–425 B.C.) gave examples of mathematical problems, and these have been worked out by using an abacus of the Salamis type.

The Romans also had a hand abacus, examples of which survive. This was a bronze tablet with grooves in which small spherical counters could slide, remarkably like the abacus which came into use later in the Middle and Far East. Although this type of abacus is described in the Chinese literature of the 2d century, it was not until the 13th century that its use became widespread in China. The Chinese abacus had one more counter in each section than the Roman hand abacus, and the counters were mounted on wires in a frame rather than in grooves. It was introduced into Japan in the 15th century, and by the 19th century a modified version had come into wide use there. In China the abacus is called the *suan-pan*, and in Japan the *soroban* (Fig. 2). The Chinese character *suan* means "to calculate" and is formed by the characters for two hands holding a counting board made of bamboo. In the Middle East, still another version of the abacus is used, which in the Soviet Union is called the *schoty*.

By the 19th century the counting-table type of abacus had essentially gone out of use in the Western world, and there the beads-on-shaft type had never been widely used. Conversely, the 19th and 20th centuries, with their economic development, brought an upsurge in the use of the latter type of abacus in the Middle and Far East. By the end of World War II, abacus operation was included in the curriculum of the fourth and upper grades in Japan, and licenses were issued for proficiency in its use. Teachers prefer it over the pocket electronic calculator as it forces students to grasp fundamental mathematical concepts. In the Soviet Union and China, as in Japan, the abacus is still widely used, but it remains to be seen whether it will survive competition from the pocket electronic calculator, whose use is continually becoming more widespread. *See* CALCULATORS.

[VELMA R. HUSKEY]

Bibliography: N. S. Kagisho, *Soroban, the Japanese Abacus, Its Use and Practice*, 1967; K. Menninger, *Number Words and Number Symbols*, 1969; J. M. Pullan, *The History of the Abacus*, 1968.

Abstract data types

Mathematical models which may be used to capture the essentials of a problem domain in order to translate it into a computer program.

Motivation. It is generally understood that people join a line by standing at the end of the line and are serviced from its front. Stepping into the middle of a line violates the rules of line etiquette, as does servicing someone from the middle of the line.

To describe a particular line of people, it suffices to keep track of the first and last person as the line moves. Then, it will always be clear behind whom a new person must stand and who must be served next. Such a description of a line is an example of an abstract data type (ADT). It is a model for representing some sort of structure, with rules for making changes to that structure. The rules may be expressed as operators that perform specific functions.

ABACUS

Fig. 1. Type of abacus now in use.

ABACUS

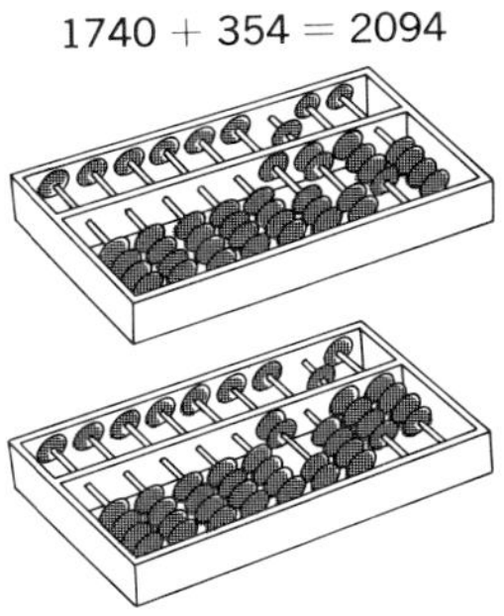

1740 + 354 = 2094

Fig. 2. Addition with a soroban.

In the field of computer programming, lines are usually called queues. For a person to enter a queue would be to perform an insertion; to leave the front of a queue would be a deletion. When the rules of line etiquette are broken, that is, an individual enters into or is serviced from the middle, the queue becomes a list. A list is a queue where insertions and deletions may be made at any point.

There are many examples of abstract data types, but this formalism is rarely useful in everyday life. However, an ADT queue would be quite helpful in representing a customer line as a computer program. The queue could be used to express certain business operations in terms of a well-understood model for representing information. This would allow the programmer to simplify the problem to that of implementing the queue in a computer language and then using the queue to represent a business function. Techniques for implementing queues are readily available.

Abstract data types allow the use of previously discovered programming techniques. Another benefit of ADTs is that they allow a programmer to encapsulate the rules for altering a model within a set of procedures, and ignore these details when using the given ADT. In other words, the programmer concerned with modeling a business function could forget exactly how an insertion or deletion is made when procedures that perform these functions are used.

ADTs and data structures. An ADT is indeed abstract, in that it must be translated into a programming language. This is done by using program data structures, which are formed from the primitive abstract data types supported directly by the given language. The programmer must also implement operators (for example, insert and delete in the queue example) that follow the rules for manipulating the ADT. In essence, then, ADTs are conceptualizations built of simple data structures and augmented with high-level operators. When an ADT is used as the model of information in a specific application, an instance of the ADT is said to be defined.

For example, an instance of a queue used to model a business function could be implemented as a sequential set of text strings in main memory.

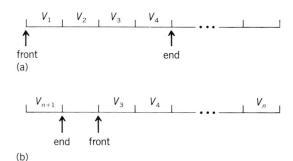

Fig. 1. A queue (a) with four values entered, and markers for the front and end of the queue, and (b) with values 1 and 2 removed from the front, and the queue wrapped around so that value $n + 1$, the last value, is situated where value 1 used to be.

This implementation is illustrated in Fig. 1. Each text string contains the name of a person in line. The rules for manipulating the queue could be implemented by keeping track (in two variables) of the location of the first and last position in the queue. In Fig. 1a, the queue is shown with four values entered, and markers for the front and end of the queue. If insertions and deletions cause the first position in the queue to migrate to the end of the allocated queue space, the queue will wrap around on itself. When the last position in the queue is used, the next new value will be inserted in the first position. In Fig. 1b, values 1 and 2 have been removed from the front, and value 3 is now first. The queue has wrapped around, and value $n + 1$, the last value, is situated where value 1 used to be.

Classification of data types. Abstract data types may be grouped according to the ways they arrange information. Typical abstract data types are discussed below; with each one, particular classes of problems that are best solved using the given ADT are discussed. First, linearly organized data types represent information as a one-dimensional series of elements. Second, nonlinear modeling techniques arrange this information so that one element may be logically followed by more than one other element. (This corresponds to a customer line where there is more than one person directly behind any specific person.) Third, an unordered data type assumes no specific order of the elements of the type. (This corresponds to customers waiting in a chaotic group.) The first three data types are examples of models used to represent information in the fast-access, main memory of a computer. Other data types are oriented toward mass storage devices like magnetic tapes and disks.

Before more interesting ADTs may be considered, however, it is necessary to discuss the primitive ADTs that are typically used to form more complicated data types.

Primitive ADTs. There are a number of simple abstract types that are typically implemented directly in a high-level programming language. These include integers and real numbers (with appropriate arithmetic operators), booleans (with appropriate logical operators), text strings (often a programming language provides no special operators for manipulating strings), and pointers. An instance of a pointer data type has as its value a logical address of the instance of another ADT. The ADT pointer is usually implemented as an integer containing the main memory address of some data item. The primitive ADTs generally serve as implementations of the elements of a more sophisticated data type. For example, in the queue example, the data type text string was suggested as a model of the underlying implementation. *See* PROGRAMMING LANGUAGES.

Queues, lists, and stacks. Queues, lists, and stacks are one-dimensional; what varies are the operators used to manipulate each ADT. A queue needs only two operators: one to insert and one to delete. To use a list, it is also necessary to have operators that may delete any element within the list or insert an element anywhere in the list. This may be done by an operator that steps through the list, locating the elements to insert or delete one by one, or by an operator that locates an element to insert or delete in the list according to some means of identification. A list is a useful ADT for keeping track of sorted information in a situation where insertions and deletions in the middle are common. As an example, a personal telephone directory could be viewed as a list.

A list is clearly a queue with somewhat more generalized data operations. Varying the accessing methods of an ADT usually implies a change in the underlying data structure as well; for example, in a list it is necessary to be able to make insertions and deletions in the middle, not just at the ends. Thus, a list could be represented as a data structure which consists of a set of variables arranged in pairs. These variables would be instances of two different primitive ADTs of the programming language. In each pair, one variable, of type text string, would describe the value of an element (for example, in the telephone directory, a person's name and phone number) and the other, of type pointer, would indicate the next element pair in the queue (Fig. 2).

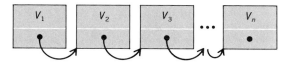

Fig. 2. Implementation of a list; arrows are used to indicate pointer variables.

A stack is essentially a queue in which insertions and deletions are both made at the front. Thus, the last element inserted is the first deleted. An insertion is commonly called a push and a deletion is called a pop. Stacks are used when modeling tasks that must be interrupted and then completed. An uncompleted task is pushed onto the stack when it is interrupted, and popped off when it is resumed. Several tasks may be pushed on top of each other; they are popped in reverse order.

Trees and graphs. A tree is an ADT where each element may be logically followed by two or more other elements. These are called its children. Each element in a tree may, as always, be of some other abstract data type and is called a node. The element that points to a node is called its parent. A tree may typically be accessed at any node; thus a tree in which each node has at most one child is a list.

A search of, an addition to, or a deletion from a tree is usually based upon an assigned order within the tree. For example, suppose that the tree consists of a number of nodes, each of which may point to two others. (This structure is called a binary tree.) Besides the two pointers, each node would contain a value or set of values. The tree may be ordered such that the left child of any node has a value less than the parent that points to it, and the right child has a larger one (Fig. 3). Trees are often used to model complex decision spaces; each node represents a process with some num-

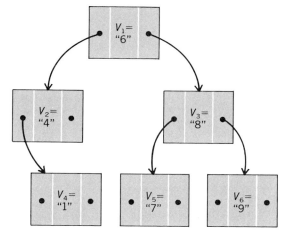

Fig. 3. Implementation of a sorted binary tree. The values V_1, V_2, and so forth are keys that are used to order the tree.

ber of possible outcomes. *See* DECISION THEORY.

A graph (Fig. 4) is in essence a generalized tree: each node may have more than one parent. In some graphs, there is no parent/child dominance; a path between two nodes (called a link) may be traversed in either direction. Further, in many graphs there may be any number of links emanating from a given node. Graphs are often used to model communication and transportation networks. For example, a graph could represent the American road network, with each node representing a city and each link representing a road. Common operations on a graph are to add a link of a specified weight between two nodes or to find the shortest path between two nodes. *See* GRAPH THEORY.

Sets. A set is a data structure that models some group of related concepts. Just as with mathematical sets, the elements of the ADT set are not ordered. For example, an ADT set called colors may consist of the various colors of the rainbow. The set may be represented within a program as the

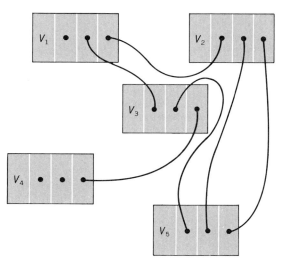

Fig. 4. ADT graph.

set of character strings: red, yellow, and so forth.

Set operators usually perform insertions and deletions, form set unions and intersections, compare two sets for equality, test to see if an element is in a set, and so forth. Sets are commonly used to solve algebraic problems. For example, a compiler typically contains a module called a lexical analyzer which breaks the input program into identifiable pieces. One of the primary tasks of a lexical analyzer is to judge the legality of each character. For example, when the analyzer encounters a "[" or an "a" it might check to see if the character is an element of a set that contains all the legal characters of the language.

Files and data bases. A file is collection of identically formatted records, each containing some number of fields. Each field value is an atomic value, consisting of an integer, character string, real number, and so forth. Unlike the previously mentioned abstract data types, files typically model data stored on an external medium (magnetic disk, tape, and so forth).

Files are accessed either sequentially or randomly, meaning either that one is forced to step sequentially through the file until the desired record is located (for example, a tape file) or that a given record may be located with one logical search operation (for example, a disk). Operators on sequential files usually locate the first record in the file and then step to the next sequential record. Random files also provide operators for locating a record or set of records, given some value or set of values for one or more of the fields. (There is a clear parallel between queues and lists as main memory models and sequential and random files as mass storage models.)

Files are the backbone of external storage structures. They are used whenever main memory is insufficient in size to contain all the information required to solve a programming problem. Often, a file is constructed to consist of a series of records where each one will be transformed into an element in another data structure as the file is read into main memory. Thus, a number of records in a file may be read in and placed in a queue for processing, and when the queue is empty another group of records is read in.

A data base is an extension of the ADT file. The difference between a data base and a file is that a data base is a complex system containing many files and rules for interrelating the data on various files. For example, a typical data base operator is MERGE, which takes two files that share a common field format and forms one file containing the elements of both, sorting them in some manner. *See* DATA-BASE MANAGEMENT SYSTEMS.

Relation to algorithms. Abstract data types have been discussed as conceptualizations of typical program data structures. Algorithms are techniques for solving programming problems which use the data structures and programs that implement ADTs. A sound knowledge of ADTs and a familiarity with common implementation techniques provide the background necessary to simplify the derivation of many algorithms. *See* ALGORITHMS.

[ROGER KING]

Bibliography: A. V. Aho, J. E. Hopcroft, and J. D. Ullman, *Data Structures and Algorithms,* 1983; T. A. Standish, *Data Structure Techniques,* 1980.

Acceptor atom

An impurity atom in a semiconductor which can accept or take up one or more electrons from the crystal and become negatively charged. An atom which substitutes for a regular atom of the material but has one less valence electron may be expected to be an acceptor atom. For example, atoms of aluminum, gallium, or indium are acceptors in germanium and silicon, and atoms of antimony and bismuth are acceptors in tellurium crystals. Acceptor atoms tend to increase the number of holes (positive charge carries) in the semiconductor. The energy gained when an electron is taken up by an acceptor atom from the valence band of the crystal is the ionization energy of the atom. *See* DONOR ATOM; SEMICONDUCTOR.

[H. Y. FAN]

Admittance

The reciprocal of the impedance of an electric circuit. Admittance is expressed in the unit mho, coined from the inverse spelling of ohm, the unit of impedance. Admittance is used primarily in computations of parallel alternating-current circuits. *See* ALTERNATING-CURRENT CIRCUIT THEORY.

By using admittance Y, current I can be expressed as $I = EY$, where E is the voltage across the impedance Z. In terms of complex quantities

$$Y = \frac{1}{Z} = \frac{1}{R \pm jX} = \frac{R}{R^2 + X^2} \mp j\frac{X}{R^2 + X^2} = G \pm jB$$

where R is the total circuit resistance, X the total circuit reactance, G the conductance, and B the susceptance. However, these are not simple conductance $(1/R)$ and susceptance $(1/X)$. As seen from the equation, both G and B are combinations of resistance and reactance. *See* CONDUCTANCE; SUSCEPTANCE.

[BURTIS L. ROBERTSON]

Algorithm

A precise formulation of a method for doing something. In computers, an algorithm is usually a collection of procedural steps or instructions organized and designed so that computer processing results in the solution of a specific problem.

Algorithms play an important role in computers. Donald Knuth has suggested that science may be defined as knowledge which is understood well enough to be taught to a computer. In this view, the concept of an algorithm or computer program furnishes an extremely useful test for the depth of knowledge about any particular subject, and the process of going from an art to a science involves learning how to construct an algorithm.

Algorithms are employed to accomplish specific tasks using data and instructions when applying computers. The task may be well definable in either mathematical or nonmathematical terms; it may be either logical or heuristic, and either simple or complex; and it may be either computational or data-processable, or involve sensing and control. In any case, the task must be definable. Then an algorithm can be devised and specified for a computer to perform the task.

Algorithm **5**

Properties. Algorithms are further characterized by several properties. Either the data set over which the algorithm will operate or the process of how the computer is to get access to the data must be specifiable. The process required to be performed can be defined with a finite set of operations or actions together with a unique starting point. The sequence of steps, the tree, the list, or the network describing the process is mappable. This, however, does not imply that the path through these steps is known, since in some cases the data or prior process steps will dictate the actual path. That is, for classical computers it must be possible to "program" the algorithm. The algorithm process must terminate, either with the task completed or with some kind of indication that the task (problem) is unsolvable.

Algorithms define the method of operation for performing a task. However, for each task to be performed, there usually are many different mappable methods for the computer to execute it — but they are not of equal desirability. From the above definitions and characterizations it follows that any computer program that does its intended task is also an algorithm. But to be practical an algorithm must perform its task within the time and memory capacity constraints of the system.

Hardware implementation. The so-called silicon revolution has spawned the emergence and growth of smart machines. These devices have embedded microcomputerlike logic which gives them some degree of adaptability to their environment and functionality well beyond that of their dumb forerunners. In such a role, and others, the microcomputer logic becomes an algorithm, the hardware embodiment of a task process. *See* EMBEDDED SYSTEMS; MICROPROCESSOR.

As each generation of computers has emerged, a growing amount of software (programs and algorithms) has been cast in hardware, resulting in so-called hard software. Most computers in the pre-first generation (the late 1940s) did not have sequenceable and combinational primitive instructions, such as multiply and divide, wired in. In the first generation of computers (the early 1950s) these primitives were wired into the hardware. The second computer generation — (the 1960s) saw the hardware implementation of computational algorithms, such as hardware algorithms for indexing, floating point, and trigonometric and square root functions. With the third computer generation (the 1970s) came hard software language/control algorithms, such as executive control, I/O (input/output), HLL (High Level Language), microprocessor, and system hardware algorithms. Fourth-generation computers (the 1980s) have hard software application algorithms, such as hardware in the form of peopleware primitives, profession (for example, management) algorithms, accounting primitives (for example, payroll), and MIS and courseware primitives. General-system hardware algorithms, are expected after 1985, and institutional, inference, artificial intelligence, and robotic hardware algorithms in the 1990s.

Most future algorithms for hard software will either (1) be incorporated as part of the hardware architecture of computers or "calculator" devices; (2) be cast as an optional adjunct for attachment to a computer system, memory, calculator, or information appliance to make it smarter; or (3) become a stand-alone, special-purpose machine — for example, a "payroll machine" or an "electronic file cabinet." In the late 1980s, smart people/information appliances could well become the major interface to computers, data bases, information bases, and knowledge-based systems driven by hardware algorithms. *See* DIGITAL COMPUTER.

Artificial intelligence. Human capabilities are extremely limited when unaided by technology. In the past, technological advances amplified human abilities for such activities as performing mechanical operations, sensing light and sound, resisting diseases, coping with weather, performing arithmetic operations, and processing data. A new era is now beginning for amplifying perception, reasoning, decision making, inventing, creating, thinking, and other mental activities. This new era results from the acquisition of knowledge for constructing artificial intelligence expert machines. Such machines are designed to imitate the brain's reasoning by storing expert knowledge for the purpose of raising the productivity of the minds which apply them. Advances in expert systems are putting society at the brink of massive application of artificial intelligence. However, such artificial intelligence expert systems require processes that are codifiable algorithmically.

Attempts to automate reasoning. For more than 2000 years, philosophers, mathematicians, logicians, and scientists have been trying to formalize the rules of inference and mathematical reasoning in order that knowledge could be produced and checked automatically. Various algorithmic approaches have been attempted. Euclid's geometric axioms are early examples of such a step toward automatizing the process. In 1677, G. W. Leibniz put forward the beginnings of an idea for a "calculus of reasoning." He imagined such a calculus for reasoning processes for all inquiries, including grammar, mathematics, physiology, politics, theogy, philosophy, and the arts of discovery. It was over 200 years before F. L. G. Frege and others advanced formal logic to the point where it seemingly would be adequate for all reasoning in pure mathematics alone.

Bertrand Russell discovered that Frege's system was inconsistent and suffered from a serious pitfall in that it allowed mathematicians to prove false results as well as true ones. Many, including Russell and A. N. Whitehead, devised logical reasoning systems which at first appeared to avoid such fatal flaws. In the 1920s, D. Hilbert tried to prove that these systems enabled mathematicians to prove the whole truth of mathematics, and nothing but the truth.

In the 1930's, K. Gödel, A. M. Turing, A. Church, and S. C. Kleene showed that these attempts at achieving a calculus of reasoning were doomed to failure — and that such systems were impossible, for either machines or humans. They proved that there could not exist any logical system whose proofs could be checked mechanically and were both complete and consistent, even for a small fragment of mathematics, let alone the other fields of logical inquiry. This does not mean that humans have powers of reasoning which machines can never possess. It implies that neither machines nor humans can have an automatic process for proving logical statements.

Thus, it seems that as artificial intelligence becomes an ingredient in machines and robots, they will, like humans, generate in some instances educated opinions rather than absolute truths. At least this will often be the case when reasoning and discourse are involved. It should be possible to design and program machines to imitate human heuristic methods of proof and reasoning even if mechanically automatic methods will never exist. Artificial intelligence programs to prove the correctness (within certain bounds) of equations and some problem statements (programs and algorithms) have been developed.

Expert systems. Considerable progress has been made in devising programs (that is, algorithms) that make inferences from knowledge bases. Heretofore computers were known for their ability to speed up and take the routine drudgery out of mathematical calculations. Now they are beginning to do likewise for mathematical proofs and logical inference. The difference is that it will be necessary in many cases to be satisfied with opinionated electrical assistance rather than absolute facts. Working within this unavoidable limitation, researchers are beginning to formalize electronic advisers, known as expert systems. Already these systems answer questions, perform artificial intelligence tasks, and explain the logical and heuristic process that they used to reach their conclusions, opinions, or recommended courses of action. In the process, they behave in general like an electronic equivalent of a professional (human) adviser. They consist of knowledge bases containing data, information, expert knowledge, and logical reasoning processes in algorithm form. The reasoning processes are imitations of human processes for learning, deductive logic, inferences, discovery, and the like. Knowledge bases are produced from extensive study of the knowledge of human experts by systematically extracting and programming this knowledge into computer systems. Such expert knowledge is much more than facts and rules. It includes less tangible and less codifiable factors—such as opinions and judgments (based upon logical reasoning), and educated guesses—as well as factual information and well formulatable logic rules for reasoning. That is, knowledge-based expert systems contain a simulation of human experts in the form of both logical and heuristic processes.

Therefore, whether knowledge is explicit and logical, heuristic or fuzzy, electronic expert systems require algorithms for their characterization, storage, and inference processing. *See* ARTIFICIAL INTELLIGENCE; COMPUTER; DIGITAL COMPUTER PROGRAMMING; EXPERT SYSTEMS.

[EARL C. JOSEPH]
Bibliography: *Collected Algorithms from ACM*, vols. 1–3, and updates.

Alternating current

Electric current that reverses direction periodically, usually many times per second. Electrical energy is ordinarily generated by a public or a private utility organization and provided to a customer, whether industrial or domestic, as alternating current.

One complete period, with current flow first in one direction and then in the other, is called a cycle, and 60 cycles per second (60 hertz, or Hz) is the customary frequency of alternation in the United States and in all of North America. In Europe and in many other parts of the world, 50 Hz is the standard frequency. On aircraft a higher frequency, often 400 Hz, is used to make possible lighter electrical machines.

When the term alternating current is used as an adjective, it is commonly abbreviated to ac, as in ac motor. Similarly, direct current as an adjective is abbreviated dc.

Advantages. The voltage of an alternating current can be changed by a transformer. This simple, inexpensive, static device permits generation of electric power at moderate voltage, efficient transmission for many miles at high voltage, and distribution and consumption at a conveniently low voltage. With direct (unidirectional) current it is not possible to use a transformer to change voltage. On a few power lines, electric energy is transmitted for great distances as direct current, but the electric energy is generated as alternating current, transformed to a high voltage, then rectified to direct current and transmitted, then changed back to alternating current by an inverter, to be transformed down to a lower voltage for distribution and use.

In addition to permitting efficient transmission of energy, alternating current provides advantages in the design of generators and motors, and for some purposes gives better operating characteristics. Certain devices involving chokes and transformers could be operated only with difficulty, if at all, on direct current. Also, the operation of large switches (called circuit breakers) is facilitated because the instantaneous value of alternating current automatically becomes zero twice in each cycle and an opening circuit breaker need not interrupt the current but only prevent current from starting again after its instant of zero value.

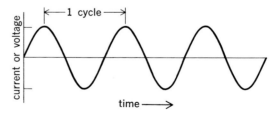

Fig. 1.   Diagram of sinusoidal alternating current.

Sinusoidal form. Alternating current is shown diagrammatically in Fig. 1. Time is measured horizontally (beginning at any arbitrary moment) and the current at each instant is measured vertically. In this diagram it is assumed that the current is alternating sinusoidally; that is, the current i is described by Eq. (1), where I_m is the maximum in-

$$i = I_m \sin 2\pi ft \qquad (1)$$

stantaneous current, f is the frequency in cycles per second (hertz), and t is the time in seconds.

A sinusoidal form of current, or voltage, is usually approximated on practical power systems because the sinusoidal form results in less expensive construction and greater efficiency of operation of electric generators, transformers, motors, and other machines.

Measurement. Quantities commonly measured by ac meters and instruments are energy, power, voltage, and current. Other quantities less commonly measured are reactive volt-amperes, power factor, frequency, and demand (of energy during a given interval such as 15 min).

Energy is measured on a watt-hour meter. There is usually such a meter where an electric line enters a customer's premises. The meter may be single-phase (usual in residences) or three-phase (customary in industrial installations), and it displays on a register of dials the energy that has passed, to date, to the system beyond the meter. The customer frequently pays for energy consumed according to the reading of such a meter.

Power is measured on a wattmeter. Since power is the rate of consumption of energy, the reading of the wattmeter is proportional to the rate of increase of the reading of a watt-hour meter. The same relation is expressed by saying that the reading of the watt-hour meter, which measures energy, is the integral (through time) of the reading of the wattmeter, which measured power. A wattmeter usually measures power in a single-phase circuit, although three-phase wattmeters are sometimes used.

Current is measured by an ammeter. Current is one component of power, the others being voltage and power factor, as in Eq. (5). With unidirectional (direct) current, the amount of current is the rate of flow of electricity; it is proportional to the number of electrons passing a specified cross section of a wire per second. This is likewise the definition of current at each instant of an alternating-current cycle, as current varies from a maximum in one direction to zero and then to a maximum in the other direction (Fig. 1.) An oscilloscope will indicate instantaneous current, but instantaneous current is not often useful. A dc (d'Arsonval-type) ammeter will measure average current, but this is useless in an ac circuit, for the average of sinusoidal current is zero. A useful measure of alternating current is found in the ability of the current to do work, and the amount of current is correspondingly defined as the square root of the average of the square of instantaneous current, the average being taken over an integer number of cycles. This value is known as the root-mean-square (rms) or effective current. It is measured in amperes. It is a useful measure for current of any frequency. The rms value of direct current is identical with its dc value. The rms value of sinusoidally alternating current is $I_m/\sqrt{2}$, where I_m is the maximum instantaneous current. See Fig. 1 and Eq. (1). *See* AMMETER; OSCILLOSCOPE.

Voltage is measured by a voltmeter. Voltage is the electrical pressure. It is measured between one point and another in an electric circuit, often between the two wires of the circuit. As with current, instantaneous voltage in an ac circuit reverses each half cycle and the average of sinusoidal voltage is zero. Therefore the root-mean-square (rms) or effective value of voltage is used in ac systems. The rms value of sinusoidally alternating voltage is $V_m/\sqrt{2}$, where V_m is the maximum intantaneous voltage. This rms voltage, together with rms current and the circuit power factor, is used to compute electrical power, as in Eqs. (4) and (5). *See* VOLTMETER.

The ordinary voltmeter is connected by wires to the two points between which voltage is to be measured, and voltage is proportional to the current that results through a very high electrical resistance within the voltmeter itself. The voltmeter, actuated by this current, is calibrated in volts.

Phase difference. Phase difference is a measure of the fraction of a cycle by which one sinusoidally alternating quantity leads or lags another.

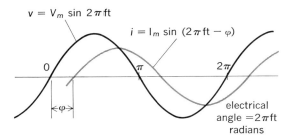

Fig. 2. The phase angle φ.

Figure 2 shows a voltage v which is described in Eq. (2) and a current i which is described in Eq. (3).

$$v = V_m \sin 2\pi f t \qquad (2)$$

$$i = I_m \sin (2\pi f t - \varphi) \qquad (3)$$

The angle φ is called the phase difference between the voltage and the current; this current is said to lag (behind this voltage) by the angle φ. It would be equally correct to say that the voltage leads the current by the phase angle φ. Phase difference can be expressed as a fraction of a cycle or in degrees of angle, or as in Eq. (3), in radians of angle, with corresponding minor changes in the equations.

If there is no phase difference, and $\varphi = 0$, voltage and current are in phase. If the phase difference is a quarter cycle, and $\varphi = \pm 90$ degrees, the quantities are in quadrature.

Power factor. Power factor is defined in terms of the phase angle. If the rms value of sinusoidal current from a power source to a load is I and the rms value of sinusoidal voltage between the two wires connecting the power source to the load is V, the average power P passing from the source to the load is shown as Eq. (4). The cosine of the phase

$$P = V I \cos \varphi \qquad (4)$$

angle, $\cos \varphi$, is called the power factor. Thus the rms voltage, the rms current, and the power factor are the components of power.

The foregoing definition of power factor has meaning only if voltage and current are sinusoidal. Whether they are sinusoidal or not, average power, rms voltage, and rms current can be measured, and a value for power factor is implicit in Eq. (5). This gives a definition of power factor when V and I are not sinusoidal, but such a value for power factor has limited use.

$$P = VI \text{ (power factor)} \quad (5)$$

If voltage and current are in phase (and of the same waveform), power factor equals 1. If voltage and current are out of phase, power factor is less than 1. If voltage and current are sinusoidal and in quadrature, power factor equals zero.

The phase angle and power factor of voltage and current in a circuit that supplies a load are determined by the load. Thus a load of pure resistance, as an electric heater, has unity power factor. An inductive load, such as an induction motor, has a power factor less than 1 and the current lags behind the applied voltage. A capacitive load, such as a bank of capacitors, also has a power factor less than 1, but the current leads the voltage, and the phase angle φ is a negative angle.

If a load that draws lagging current (such as an induction motor) and a load that draws leading current (such as a bank of capacitors) are both connected to a source of electric power, the power factor of the two loads together can be higher than that of either one alone, and the current to the combined loads may have a smaller phase angle from the applied voltage than would currents to either of the two loads individually. Although power to the combined loads is equal to the arithmetic sum of power to the two individual loads, the total current will be less than the arithmetic sum of the two individual currents (and may, indeed, actually be less than either of the two individual currents alone). It is often practical to reduce the total incoming current by installing a bank of capacitors near an inductive load, and thus to reduce power lost in the incoming distribution lines and transformers, thereby improving efficiency.

Three-phase system. Three-phase systems are commonly used for generation, transmission, and distribution of electric power. A customer may be supplied with three-phase power, particularly if he uses a large amount of power or if he wishes to use three-phase loads. Small domestic customers are usually supplied with single-phase power.

A three-phase system is essentially the same as three ordinary single-phase systems (as in Fig. 2, for instance) with the three voltages of the three single-phase systems out of phase with each other by one-third of a cycle (120 degrees) as shown in Fig. 3. The three voltages may be written as Eqs. (6), (7), and (8), where $V_{an(\max)}$ is the maximum

$$v_{an} = V_{an(\max)} \sin 2\pi ft \quad (6)$$

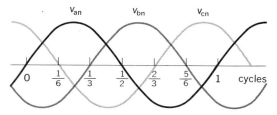

Fig. 3. Voltages of a balanced three-phase system.

$$v_{bn} = V_{bn(\max)} \sin 2\pi (ft - 1/3) \quad (7)$$
$$v_{cn} = V_{cn(\max)} \sin 2\pi (ft - 2/3) \quad (8)$$

value of voltage in phase an, and so on. The three-phase system is balanced if relation (9) holds,

$$V_{an(\max)} = V_{bn(\max)} = V_{cn(\max)} \quad (9)$$

and if the three phase angles are equal, 1/3 cycle each as shown.

If a three-phase system were actually three separate single-phase systems, there would be two wires between the generator and the load of each system, requiring a total of six wires. In fact, however, a single wire can be common to all three systems, so that it is only necessary to have three wires for a three-phase system (a, b, and c of Fig. 4)

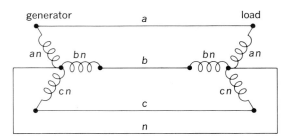

Fig. 4. Connections of a simple three-phase system.

plus a fourth wire n to serve as a common return or neutral conductor. On some systems the earth is used as the common or neutral conductor.

Each phase of a three-phase system carries current and conveys power and energy. If the three loads on the three phases of the three-phase system are equal and the voltages are balanced, then the currents are balanced also. Figure 2 can then apply to any one of the three phases. It will be recognized that the three currents in a balanced system are equal in rms (or maximum) value and that they are separated one from the other by phase angles of 1/3 cycle and 2/3 cycle. Thus the currents (in a balanced system) are themselves symmetrical, and Fig. 3 could be applied to line currents i_a, i_b, and i_c as well as to the three voltages indicated in the figure. Note, however, that the three currents will not necessarily be in phase with their respective voltages; the corresponding voltages and currents will be in phase with each other only if the load is pure resistance and the phase angle between voltage and current is zero; otherwise some such relation as that of Fig. 2 will apply to each phase.

It is significant that, if the three currents of a three-phase system are balanced, the sum of the three currents is zero at every instant. Thus if the three curves of Fig. 3 are taken to be the currents of a balanced system, it may be seen that the sum of the three curves at every instant is zero. This means that if the three currents are accurately balanced, current in the common conductor (n of Fig. 4) is always zero, and that conductor could theoretically be omitted entirely. In practice, the three currents are not usually exactly balanced, and either of two situations obtains. Either the common neutral wire n is used, in which case it carries little current (and may be of high resistance compared to the other three line wires), or else the common neutral wire n is not used, only three line wires being installed, and the three phase currents are thereby forced to add to zero even though this requirement results in some inbalance of phase voltages at the load.

It is also significant that the total instantaneous power from generator to load is constant (does not vary with time) in a balanced, sinusoidal, three-phase system. Power in a single-phase system that has current in phase with voltage is maximum when voltage and current are maximum and it is instantaneously zero when voltage and current are zero; if the current of the single-phase system is not in phase with the voltage, the power will reverse its direction of flow during part of each half cycle. But in a balanced three-phase system, regardless of phase angle, the flow of power is unvarying from instant to instant. This results in smoother operation and less vibration of motors and other ac devices.

Three-phase systems are almost universally used for large amounts of power. In addition to providing smooth flow of power, three-phase motors and generators are more economical than single-phase machines. Polyphase systems with two, four, or other numbers of phases are possible, but they are little used except when a large number of phases, such as 12, is desired for economical operation of a rectifier.

Power and information. Although this article has emphasized electric power, ac circuits are also used to convey information. An information circuit, such as telephone, radio, or control, employs varying voltage, current, waveform, frequency, and phase. Efficiency is often low, the chief requirement being to convey accurate information even though little of the transmitted power reaches the receiving end. For further consideration of the transmission of information *see* RADIO; TELEPHONE; WAVEFORM.

An ideal power circuit should provide the customer with electric energy always available at unchanging voltage of constant waveform and frequency, the amount of current being determined by the customer's load. High efficiency is greatly desired. *See* ALTERNATING-CURRENT CIRCUIT THEORY; CAPACITANCE; CIRCUIT (ELECTRICITY); ELECTRIC CURRENT; ELECTRIC FILTER; ELECTRICAL IMPEDANCE; ELECTRICAL RESISTANCE; INDUCTANCE; JOULE'S LAW; NETWORK THEORY; OHM'S LAW; RESONANCE (ALTERNATING-CURRENT CIRCUITS).

[H. H. SKILLING]

Alternating-current circuit theory

The mathematical description of conditions in an electric circuit when the circuit is driven by an alternating source or sources. *See* ALTERNATING CURRENT.

The alternating quantity is often assumed to be sinusoidal. With this assumption, an alternating voltage v may be described by Eq. (1), where t is

$$v = V_m \cos 2\pi f t \qquad (1)$$

time (seconds), f is frequency (in cycles per second, or hertz), and V_m is the maximum instantaneous value of the alternating voltage (volts). In some treatments of circuit theory the sine function is used, rather than the cosine function, as above, but this difference is immaterial because it does not affect the voltage, except as to an arbitrary time reference. Although a sine function of voltage is perhaps easier to visualize, the cosine function provides a readier graphical interpretation.

Phasors. It is usual to represent a sinusoidally varying quantity by a rotating line. In Fig. 1a a line of length V_m rotates at an angular speed ω where $\omega = 2\pi f$; the line therefore makes one revolution for each electrical cycle represented. The projection of the rotating line on the fixed horizontal axis (Fig. 1a) is $V_m \cos \omega t$, and because $\omega = 2\pi f$, this projection is identically equal to voltage v of Eq. (1). Only at one particular instant does Fig. 1a represent Eq. (1); the line in the figure must be visualized as rotating at synchronous speed.

At the instant when t was zero, the rotating line was horizontal and its projection was V_m, for at this instant $\cos \omega t = \cos 0 = 1$. To represent a voltage that is not zero at zero time, a phase angle θ is introduced. Such a voltage is shown by Eq. (2),

$$v = V_m \cos(2\pi f t + \theta) = V_m \cos(\omega t + \theta) \qquad (2)$$

where θ is in radians of angle; Fig. 1b shows a graphical representation of Eq. (2).

Such rotating lines are called phasors. (The term vector was formerly used, and sinor is employed by some authors.) Phasors can represent any sinusoidally varying quantities, such as alternating voltage or current. Thus Fig. 1c shows voltage v with maximum instantaneous value V_m, and current i of the same frequency with maximum instantaneous value I_m; the phase angle between v and i is φ. This angle φ remains unchanging, as the two phasors rotate with the same velocity ω.

In a phasor diagram such as Fig. 1c, it is understood that the phasors shown are all rotating together at the same speed. Hence, their lengths and their angles relative to each other are important, but their angles relative to a fixed horizontal axis are not significant; consequently, a fixed axis is usually not shown in the diagram. However, in drawing such a diagram, it is always necessary to assume an arbitrary angular position for one of the phasors, called the reference phasor, and to draw the others with correct relative angles.

A phasor diagram can be drawn with the measured length of a line proportional to the maximum instantaneous value of the voltage or current represented, as in Fig. 1, or with the measured length

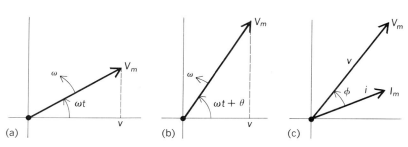

Fig. 1. Phasor representations. (a) Of voltage $v = V_m \cos \omega t$. (b) Of voltage $v = V_m \cos(\omega t + \theta)$. (c) Of voltage V_m and current I_m separated by phase angle ϕ.

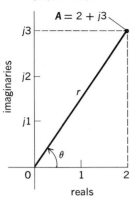

Fig. 2. Number $2+j3$ represented on the complex plane.

proportional to the root-mean-square (rms) value of that voltage or current. Because for sinusoidal quantities the rms value equals the maximum value divided by $\sqrt{2}$, either method can be used.

Complex notation. Common practice in all modern writing on alternating currents is to use a convention (introduced by C. P. Steinmetz in 1893) that assumes the phasors to lie in the mathematical complex plane. The actual instantaneous value of voltage or current is then the real component of the phasor. A brief discussion of the algebra of complex quantities is given below.

Algebra deals with numbers and with operations, such as addition and multiplication, performed on numbers. A number can be real (positive or negative), imaginary (positive or negative), or complex (the sum of a real number and an imaginary number). A complex number can be written in the form of Eq. (3), where symbol j indi-

$$Z = x + jy \qquad (3)$$

cates that the associated number is imaginary. Thus $3 + j4$ is a complex number of which the real component is 3 and the imaginary component is 4. (The symbol i is often used in place of j, but electrical engineers commonly use j, reserving i as a symbol for current.)

By definition, the product of two positive imaginary numbers is real and negative. Thus $(j5)(j7) = -35$. Most notably, $(j1)(j1) = -1$, or the relation is expressible as Eq. (4). By this definition the square

$$\sqrt{-1} = j1 \qquad (4)$$

root of a negative real number is defined and this is the basis of complex algebra.

In general, operations of complex algebra are those of the algebra of real numbers, with the additional rules shown below.

Equality. Complex numbers are equal if their real components are equal and their imaginary components are also equal.

Addition and subtraction. The sum of two complex numbers has a real component equal to the sum of the real components, and an imaginary component equal to the sum of the imaginary components. Thus, $(3 + j4) + (5 + j6) = (8 + j10)$. To subtract, change the signs of both components of the subtrahend and add.

Multiplication. The product of a real number and an imaginary number is imaginary; thus $3(j4) = j12$. The product of two positive imaginary numbers is real and negative (as stated above); thus $(j3)(j4) = -12$; but $(-j3)(j4) = +12$.

Complex numbers are multiplied by the rules of real algebra supplemented by the foregoing rules. These rules give rise to Eq. (5), and Eq. (6) is a numerical example.

$$(a + jb)(c + jd) = (ac - bd) + j(ad + bc) \qquad (5)$$

$$(3 + j4)(5 + j6) = -9 + j38 \qquad (6)$$

Conjugates. Complex numbers are conjugate if their real components are equal and their imaginary components are equal in magnitude but opposite in sign. Thus $(2 + j7)$ and $(2 - j7)$ are conjugate. An asterisk is commonly used to indicate the conjugate; thus A and A^* are conjugates. Equations (7) and (8) indicate that the sum of conjugates

$$(a + jb) + (a - jb) = 2a \qquad (7)$$

$$(a + jb)(a - jb) = a^2 + b^2 \qquad (8)$$

is real, and that the product of conjugates is real.

Division. Division is defined as the inverse of multiplication. It is possible to divide by the following technique. To divide $(5 + j10)$ by $(2 + j1)$, multiply each by the conjugate of the latter, Eq. (9).

$$\frac{5 + j10}{2 + j1} = \frac{(5 + j10)(2 - j1)}{(2 + j1)(2 - j1)}$$

$$= \frac{(10 + 10) + j(20 - 5)}{4 + 1} = 4 + j3 \qquad (9)$$

Thus a ratio can be evaluated by rationalization of the denominator. However, both multiplication and division of complex numbers are easier in the exponential or polar form, as given below.

Complex plane. Two axes are drawn at right angles to each other (Fig. 2), calibrated in equal divisions with the point of intersection (origin) being zero on each axis. All real numbers are represented by points on the horizontal axis, and all imaginary numbers by points on the vertical axis. Each point in the plane represents a complex number, its projection on the axis of reals being its real component and its projection on the axis of imaginaries being its imaginary component. Thus any complex number $A = a + jb$ is represented by a point in the plane; the point represented in Fig. 2 is $A = 2 + j3$.

Addition (and subtraction) are easily visualized in the complex plane. Following the foregoing rule for addition, real components are added and imaginary components are added. Figure 3 shows $A + B$ where $A = 2 + j3$ and $B = 4 + j1$; the sum, as shown by the parallelogram, is $6 + j4$.

A line marked r can be drawn from the origin to the point A, as in Fig. 2. Its length, called A, is $(a^2 + b^2)^{1/2}$, and it makes an angle θ with the real axis. By using trigonometry, Eq. (10) is obtained. In

$$A = a + jb = A(\cos\theta + j\sin\theta) \qquad (10)$$

the example of Fig. 2, $a = 2$, $b = 3$, $A = \sqrt{13}$, and θ is the angle whose tangent is 3/2, written mathematically as $\theta = \tan^{-1} 3/2$.

With this interpretation, another notation is used for complex numbers. This notation is shown in Eq. (11), where $a = A \cos\theta$, $b = A \sin\theta$, $A =$

$$A = a + jb = A\underline{/\theta} \qquad (11)$$

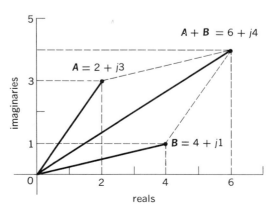

Fig. 3. Addition of complex numbers.

$(a^2 + b^2)^{1/2}$, and $\theta = \tan^{-1} b/a$. The notation using a and b is called rectangular, and that using A and θ is called trigonometric or polar.

Euler's theorem. Equations (12a) and (12b) show

$$\cos \theta = \tfrac{1}{2}(e^{j\theta} + e^{-j\theta}) \qquad (12a)$$

$$j \sin \theta = \tfrac{1}{2}(e^{j\theta} - e^{-j\theta}) \qquad (12b)$$

the relation of the sine and cosine functions to the exponential e, which is the base of natural logarithms ($e \approx 2.718$). By adding Eq. (12a) and (12b), Eq. (13) is obtained, which is Euler's theorem. By

$$e^{j\theta} = \cos \theta + j \sin \theta \qquad (13)$$

using Euler's theorem in Eq. (10), Eq. (14) is obtained, where A is any complex number and A and

$$A = Ae^{j\theta} \qquad (14)$$

θ are real with values found as in Eq. (11. The expression for A in Eq. (14) is the exponential form of a complex quantity.

Multiplication in exponential form. Multiplication and its inverse, division, are easier when complex numbers are expressed in exponential form. To multiply two complex numbers $A = Ae^{j\alpha}$ and $B = Be^{j\beta}$, the familiar rules of algebra give the relationships of Eq. (15). Hence the product of com-

$$A \cdot B = (Ae^{j\alpha})(Be^{j\beta}) = ABe^{j(\alpha + \beta)} \qquad (15)$$

plex numbers is the product of the magnitudes at the sum of the angles. In polar form the relationship is written as indicated in Eq. (16). Division follows the same rule. That is, as shown in Eq. (17),

$$A \cdot B = (AB)\,\underline{/\alpha + \beta} \qquad (16)$$

$$A/B = (A/B)\,\underline{/\alpha - \beta} \qquad (17)$$

magnitudes are divided and angles subtracted.

Multiplication (and division) are easily visualized in the complex plane. Equation (16) or, employing exponential notation, Eq. (15) can be illustrated, as shown in Fig. 4: the multiplication $(0.9\underline{/30°})(1.2\underline{/70°}) = 1.08\underline{/100°}$. Figure 4 could also illustrate the division $(1.08\underline{/100°})/(0.9\underline{/30°}) = 1.2\underline{/70°}$.

In accordance with custom, arrowheads have been added to radial lines in Fig. 4. Actually, only a point in the plane is meaningful, but the radial line and arrowhead are perhaps helpful graphically to

indicate the location of the point relative to the origin. A point at unit distance from the origin, at the tip of a radial line of unit length, as in Fig. 5a, that makes an angle θ with the horizontal axis of reals, represents a complex quantity of magnitude 1 and angle θ.

Similarly, a complex quantity of magnitude A and angle ωt is indicated by a radial line in Fig. 5b. Because t is time, the angle of the radial line increases with time. Thus the complex quantity $A\underline{/\omega t}$, which can also be written as $Ae^{j\omega t}$, is represented by a line of length A in the complex plane that revolves about the origin with angular velocity ω. Such a line represents an electrical phasor.

Powers and roots. The raising of complex quantities to powers follows the same rules as multiplication. Thus powers of A, where $A = A\underline{/\theta}$, are shown by Eq. (18). For instance, the square of a complex

$$A^n = A^n\underline{/n\theta} \qquad (18)$$

number is the square of the magnitude at twice the angle.

The most important powers are those of $j1$. The first four are tabulated for reference by Eqs. (19).

$$
\begin{array}{ll}
(j1)^2 = -1 & (jx)^2 = -x^2 \\
(j1)^3 = -j1 & (jx)^3 = -jx^3 \\
(j1)^4 = +1 & (jx)^4 = x^4 \\
(j1)^5 = j1 & (jx)^5 = jx^5
\end{array}
\qquad (19)
$$

A complex number has multiple roots; it has two distinct square roots, three distinct cube roots, and so on. Thus if $A = 8\underline{/60°}$, its cube roots are shown by Eq. (20). This is true because

$$(A)^{1/3} = 2\underline{/20°} \text{ or } 2\underline{/140°} \text{ or } 2\underline{/260°} \qquad (20)$$

$(2\underline{/140°})^3 = 8\underline{/420°}$, which is indistinguishable from $8\underline{/60°}$ if each is considered a point in the complex plane, because $420° - 360° = 60°$. In this way three distinct cube roots of a complex number can be computed.

The algebra of complex quantities is applied below to alternating electrical quantities as represented by phasors in the complex plane.

Alternating voltage and current. By using Euler's theorem, Eq. (13), one has the equivalency shown by Eq. (21), and Eq. (1) can be written as

$$e^{j\omega t} = \cos \omega t + j \sin \omega t \qquad (21)$$

Eq. (22), where $\mathscr{R}e$ means "the real component of."

$$v = V_m \mathscr{R}e\, e^{j\omega t} \qquad (22)$$

In Eq. (21) $\cos \omega t$ is identically the real component of $e^{j\omega t}$. With this notation, Eq. (2) can be written in the form of Eq. (23).

$$v = V_m \mathscr{R}e\, e^{j(\omega t + \theta)} = V_m \mathscr{R}e\, e^{j\omega t} e^{j\theta}$$
$$= \mathscr{R}e \sqrt{2}\,(Ve^{j\theta})\, e^{j\omega t} \qquad (23)$$

The quantity $Ve^{j\theta}$ is called the transform of voltage v, and is commonly symbolized by $\mathbf{V}$, as shown in Eq. (24), where V is the rms value of

$$\mathbf{V} = Ve^{j\theta} \qquad (24)$$

voltage, and θ is the phase angle of voltage (relative to an assumed reference), as in Eq. (2). By using this symbol, Eq. (23) becomes Eq. (25).

$$v = \mathscr{R}e \sqrt{2}\, \mathbf{V} e^{j\omega t} \qquad (25)$$

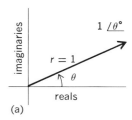

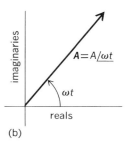

Fig. 5. Representations on the complex plane. (a) A line of unit length at angle θ. (b) A line revolving at rate ω that has reached at time t an angle ωt.

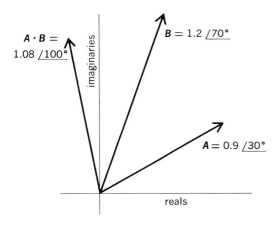

Fig. 4. Multiplication of complex numbers.

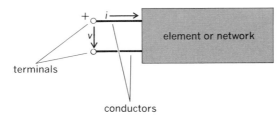

Fig. 6. Current *i* entering a network with voltage *v* across the terminals.

In this expression $\sqrt{2}\,\mathbf{V}e^{j\omega t}$ is a line revolving about the origin of the complex plane: It is the phasor of voltage *v*, as discussed in the first section of this article. (The phasor may be defined without $\sqrt{2}$ if preferred, but the definition given here is more usual.) Transform $\mathbf{V}$ is a complex quantity. It is not voltage *v*, although it is often loosely called a voltage in ac circuit theory.

As an alternative notation, with the same meaning, Eq. (24) can be written as Eq. (26).

$$\mathbf{V}=V\underline{/\theta} \qquad (26)$$

Parameters. There is a relation between the electric current in an element or network of elements and the voltage between the terminals of that element or network (Fig. 6). In the figure the arrow beside a conductor indicates the nominal positive direction of current *i*; the choice of direction is arbitrary, but all equations must be consistent with this arbitrary choice. The nominal positive direction of voltage is indicated in the figure in two alternative ways. (1) An arrow from the upper terminal to the lower terminal marks the direction in which positive voltage produces electric field in the space between the terminals and, hence, the direction in which there would be positive current in a resistive load, such as a voltmeter, from terminal to terminal. (2) Alternatively, the + beside the upper terminal indicates the same conditions. Both these conventions are used to indicate the nominal positive direction of voltage. (A third convention, sometimes seen, uses an arrow with the head pointing toward the + terminal, opposite to the arrow shown in Fig. 6.)

It is found by experiment and measurement that most actual networks, as in Fig. 6, can be treated theoretically as if they were made up of various combinations of resistance, capacitance, and inductance, including mutual inductance, and sources that are either independent of all currents and voltages (independent sources) or that are proportional to specified currents or voltages (controlled sources). For analysis an actual electrical device or network is idealized, being replaced by a model that behaves in a similar manner but that is composed of idealized parameters.

To define the parameters, the voltage-current relations, shown by Eqs. (27)–(29) and which are taken from the general theory of electricity, are used for ac circuit theory. Voltage across resistance *R* is shown by Eq. (27). Voltage across constant inductance *L* is shown by Eq. (28). Current

$$v=Ri \qquad (27)$$

stant inductance *L* is shown by Eq. (28). Current

$$v=L(di/dt) \qquad (28)$$

entering constant capacitance *C* is shown by Eq. (29). In these equations *R* is resistance (ohms); *L* is

$$i=C(dv/dt) \qquad (29)$$

inductance (henrys); *C* is capacitance (farads); *v* is voltage (volts); *i* is current (amperes) with nominal directions, as in Fig. 6; and *t* is time (seconds). For mutual inductance *see* COUPLED CIRCUITS.

Theorem. An important theorem relating to steady-state ac operation of networks of linear elements is based on the mathematical fact that the sum of sinusoidal functions is itself sinusoidal. The theorem is: Because a sinusoidal current through an element requires a sinusoidal voltage across that element, it follows that if voltage or current at any part of any linear network is sinusoidal, voltages and currents at every part of the network are sinusoidal with the same frequency. This theorem applies only to steady-state operation, after all transient components of current and voltage have died away.

Voltage and current phasors. Equation (30) is the same as Eq. (25) and is an expression for alternating voltage. Similarly, alternating current is expressed by Eq. (31). In these equations $\mathbf{V}$ and $\mathbf{I}$

$$v=\mathscr{R}e\sqrt{2}\,\mathbf{V}e^{j\omega t} \qquad (30)$$

$$i=\mathscr{R}e\sqrt{2}\,\mathbf{I}e^{j\omega t} \qquad (31)$$

are complex quantities, the transforms of voltage and current. When these expressions are inserted into Eq. (27), it is found that Eq. (30) expresses the relation between them across pure resistance, Eq. (32). This identity (true at all values of time and

$$\mathscr{R}e\sqrt{2}\,\mathbf{V}e^{j\omega t}=R\,\mathscr{R}e\sqrt{2}\,\mathbf{I}e^{j\omega t} \qquad (32)$$

with *R* constant) can be valid only if Eq. (33) is

$$\mathbf{V}=R\mathbf{I} \qquad (33)$$

true. Hence the transform or phasor for voltage across a purely resistive element is equal to the resistance of the element times the current transform or phasor (Fig. 7*a*).

Inductance. In the same way Eqs. (30) and (31) are inserted into Eq. (28), which relates to inductance, giving Eq. (34). The derivative of the real

$$\mathscr{R}e\sqrt{2}\,\mathbf{V}e^{j\omega t}=L\frac{d}{dt}(\mathscr{R}e\sqrt{2}\,\mathbf{I}e^{j\omega t}) \qquad (34)$$

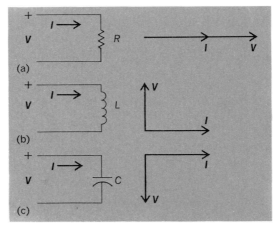

Fig. 7. Phasors. (*a*) For a resistive element. (*b*) For an inductive element. (*c*) For a capacitive element.

component is equal to the real component of the derivative: differentiating and dividing each side by $\sqrt{2}$ gives Eq. (35). This identity (to be true for all

$$\mathscr{R}e\,\mathbf{V}e^{j\omega t} = \mathscr{R}e\,j\omega L\mathbf{I}e^{j\omega t} \qquad (35)$$

t) requires that Eq. (36) hold true. Hence the pha-

$$\mathbf{V} = j\omega L\mathbf{I} \qquad (36)$$

sor for voltage across a purely inductive element is proportional to the current phasor, to the inductance of the element, and to the frequency of the current; furthermore, the voltage phasor leads the current phasor by 90° because of the right-angle relationship in the complex plane that is indicated by the symbol j (Fig. 7b).

Capacitance. Likewise, Eqs. (30) and (31) are inserted into Eq. (29), which relates to capacitance, giving Eqs. (37) and (38). This identity (to be

$$\mathscr{R}e\sqrt{2}\,\mathbf{I}e^{j\omega t} = C\frac{d}{dt}(\mathscr{R}e\sqrt{2}\,\mathbf{V}e^{j\omega t}) \qquad (37)$$

$$\mathscr{R}e\,\mathbf{I}e^{j\omega t} = \mathscr{R}e\,j\omega C\,\mathbf{V}e^{j\omega t} \qquad (38)$$

true for all t) requires that Eq. (39) hold true.

$$\mathbf{I} = j\omega C\mathbf{V} \qquad (39)$$

Hence the phasor for current to a purely capacitive element is proportional to the voltage phasor, to the capacitance, and to the frequency; furthermore, the current phasor leads the voltage phasor by 90° (Fig. 7c).

Impedance and admittance. Impedance of an element or network is the ratio of the phasor of applied voltage to the phasor of entering current (Fig. 6). (Impedance is not the ratio of instantaneous voltage to instantaneous current.) Admittance is the ratio of the current phasor to the voltage phasor. Thus impedance $\mathbf{Z}$ and admittance $\mathbf{Y}$ are defined by Eq. (40). Because $\mathbf{V}$ and $\mathbf{I}$ are com-

$$\mathbf{Z} = \frac{\mathbf{V}}{\mathbf{I}} \qquad \mathbf{Y} = \frac{\mathbf{I}}{\mathbf{V}} = \frac{1}{\mathbf{Z}} \qquad (40)$$

plex quantities, Z and Y are complex quantities also, but they are not phasors; that is, they do not represent sinusoidally varying quantities.

The foregoing paragraph defines self-impedance and self-admittance, because $\mathbf{V}$ and $\mathbf{I}$ are taken at the same pair of entering terminals. If $\mathbf{V}$ represents a voltage between any two points in a network and $\mathbf{I}$ represents current at some other point, Eq. (40) gives transfer impedance and transfer admittance between the respective locations. More generally, a transfer function can be a ratio of phasors of voltage to voltage, voltage to current, current to voltage, or current to current at different points in a network.

Real and imaginary parts of impedance are called resistance R and reactance X. Real and imaginary parts of admittance are called conductance G and susceptance B. Thus are obtained Eqs. (41) and (42).

$$\frac{V}{I} = \mathbf{Z} = R + jX = |\mathbf{Z}|\,(\cos\varphi + j\sin\varphi) \qquad (41)$$

$$\varphi = \tan^{-1}\frac{X}{R}$$

$$\frac{I}{V} = \mathbf{Y} = G + jB = |\mathbf{Y}|\,(\cos\theta + j\sin\theta) \qquad (42)$$

$$\theta = \tan^{-1}\frac{B}{G}$$

Equation (43) follows, and the components of $\mathbf{Y}$ are therefore expressed by Eqs. (44).

$$\mathbf{Y} = \frac{1}{\mathbf{Z}} = \frac{1}{R + jX} = \frac{R - jX}{R^2 + X^2} \qquad (43)$$

$$G = \frac{R}{R^2 + X^2} \qquad B = \frac{-X}{R^2 + X^2} \qquad (44)$$

Similarly, Eq. (45) follows, and its components are expressed by Eqs. (46).

$$\mathbf{Z} = \frac{1}{\mathbf{Y}} = \frac{1}{G + jB} = \frac{G - jB}{G^2 + B^2} \qquad (45)$$

$$R = \frac{G}{G^2 + B^2} \qquad X = \frac{-B}{G^2 + B^2} \qquad (46)$$

It should be observed that G is not the reciprocal of R unless X is zero, nor is R the reciprocal of G unless B is zero.

The unit of resistance, reactance, or impedance is the ohm. The unit of conductance, susceptance, or admittance is called the mho.

Series and parallel connections. Elements are said to be in series if the same current passes through them (Fig. 8a). Elements are said to be in parallel if the same voltage is across them (Fig. 8b).

The total impedance of elements or networks in series is the sum of the impedances of the several elements or networks. Thus Eq. (47) expresses the total impedance shown in Fig. 8a.

$$\mathbf{Z} = \mathbf{Z}_1 + \mathbf{Z}_2 + \mathbf{Z}_3 \qquad (47)$$

The total admittance of elements or networks in parallel is the sum of the admittances of the several elements or networks. Thus Eq. (48) expresses

$$\mathbf{Y} = \mathbf{Y}_1 + \mathbf{Y}_2 + \mathbf{Y}_3 \qquad (48)$$

the total admittance shown in Fig. 8b.

The impedance of several elements or networks in parallel (Fig. 8b) is found as shown by Eq. (49).

$$\mathbf{Z} = \frac{1}{\mathbf{Y}} = \frac{1}{\mathbf{Y}_1 + \mathbf{Y}_2 + \mathbf{Y}_3} = \frac{1}{1/\mathbf{Z}_1 + 1/\mathbf{Z}_2 + 1/\mathbf{Z}_3}$$

$$= \frac{\mathbf{Z}_1\mathbf{Z}_2\mathbf{Z}_3}{\mathbf{Z}_1\mathbf{Z}_2 + \mathbf{Z}_2\mathbf{Z}_3 + \mathbf{Z}_3\mathbf{Z}_1} \qquad (49)$$

An important special case is that of two elements or networks in parallel (Fig. 8c). If the impedances

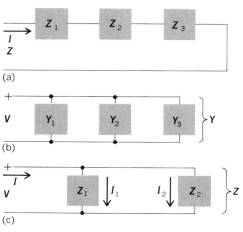

(a)

(b)

(c)

Fig. 8. Circuits of elements. (a) Elements in series. (b) Elements in parallel. (c) Two impedances in parallel.

are $\mathbf{Z}_1$ and $\mathbf{Z}_2$, the impedance of the two in parallel is shown by Eq. (50). Total current I divides (Fig.

$$\mathbf{Z} = \frac{\mathbf{Z}_1 \mathbf{Z}_2}{\mathbf{Z}_1 + \mathbf{Z}_2} \tag{50}$$

8c) as indicated by Eqs. (51) and (52). Equations

$$I_1 = \frac{V}{\mathbf{Z}_1} = I \frac{\mathbf{Z}}{\mathbf{Z}_1} = \frac{\mathbf{Z}_2}{\mathbf{Z}_1 + \mathbf{Z}_2} I \tag{51}$$

$$I_2 = \frac{\mathbf{Z}_1}{\mathbf{Z}_1 + \mathbf{Z}_2} I \tag{52}$$

(50), (51), and (52) are frequently used in ac circuit theory.

Power. The power entering an ac load at its two terminals is computed from the rms current into the load, the rms voltage between the terminals, and the power factor. If current and voltage vary sinusoidally with time, the power factor is $\cos \varphi$, where φ is the angle of phase difference between current and voltage. Thus the power, averaged through an integer number of cycles, is shown by Eq. (53).

$$P = VI \cos \varphi \tag{53}$$

Power is expressed usefully in several ways. For instance, the power to impedance $\mathbf{Z} = R + jX$ when the impedance carries current I is shown by Eq. (54). The final form of this equation is obtained by

$$P = (|\mathbf{Z}|I)I \cos \varphi = I^2 |\mathbf{Z}| \cos \varphi = I^2 R \tag{54}$$

noting that φ, the angle between voltage and current in Eq. (53), is also the angle of the impedance in Eq. (41), and therefore $|\mathbf{Z}| \cos \varphi = R$.

Another useful formula for power to a resistor, found from Eq. (54), is shown by Eq. (55), where V is the voltage across resistance alone.

$$P = I^2 R = \left|\frac{V}{R}\right|^2 R = \frac{V^2}{R} \tag{55}$$

Devices and models. Frequently the best model of a practical device is obtained from parameters in series. Thus a coil of wire is commonly represented by resistance and inductance in series (Fig. 9a), although in reality both resistance and inductance reside in the whole coil. This model of a coil using lumped parameters (Fig. 9a) is quite satisfactory if, in the actual wire, current is essentially the same from end to end. The model fails if appreciable current is carried by distributed capacitance between one part of the wire and another, or between part of the wire and ground, as may well happen at radio frequency.

A battery is usually represented as a source of constant, unidirectional voltage in series with resitance. An ac generator is represented as a source of voltage that varies sinusoidally with time, connected in series with resistance and inductance (Fig. 9b). (This model can be made more exact by using different values of L under different circumstances.) A medium-length transmission line (up to 30 mi long at 60 hertz) can be represented satisfactorily by a model using lumped resistance, inductance, and capacitance (Fig. 9c).

Impedance of a coil. The impedance $\mathbf{Z}$ of the model of a coil (Fig. 9a) is shown by Eq. (56), where

$$\mathbf{Z} = R + j\omega L \tag{56}$$

R is the resistance of the coil and ωL is its reactance. Resistance R is basically independent of frequency, although for accuracy it may be best to distinguish between the dc resistance and the slightly higher ac resistance, which takes also into account other losses, such as skin effect in the wire and core losses due to eddy currents or hysteresis if the coil has a ferromagnetic core.

Inductance L is nearly independent of frequency, except as L may be slightly reduced by skin effect. As a first approximation, R and L are usually considered to be independent of frequency.

Impedance of a generator. Figure 9b shows a simple model of an ac generator. There is an independent source of sinusoidal electromotive force e in series with R and L, and the output voltage is v.

By using transforms, the transform of the electromotive force is $\mathbf{E}$, that of the output voltage is $\mathbf{V}$, that of the current is $\mathbf{I}$, impedance is $\mathbf{Z} = R + j\omega L$, and the transform equivalent is shown by Eq. (57).

$$\mathbf{V} = \mathbf{E} - \mathbf{ZI} \tag{57}$$

Electromotive force appears in a circuit where energy of some other nature is transformed into electrical energy. In the generator, mechanical energy is changed into electrical energy. In a battery, chemical energy is changed into electrical energy. Voltage results from the presence of an electromotive force. Both voltage and electromotive force are measured in volts.

Symbols. Instantaneous values of electromotive force and voltage are represented by e and v, respectively, as in Eq. (1). For current (originally called intensity), i is used. The maximum instantaneous value of an alternating quantity is often distinguished as $V_{\max}$ or V_m, as in Eq. (1). An italic letter without subscript indicates the rms value, as V or I; Eq. (24) illustrates the use. A phasor or transform is boldface italic, as $\mathbf{V}$ or $\mathbf{I}$ in Eq. (57). The nominal positive direction of every current, voltage, and electromotive force, either direct or alternating, must be indicated on circuit diagrams.

Instantaneous power is usually p, and power averaged through one or more complete cycles or periods is P. Time is t. Numbers represented by symbols in this article are the ratio of circumference to diameter $\pi \approx 3.1416$, and the base of natural logarithms $e \approx 2.718$. (The numerical value of e is not required in this article; its importance comes

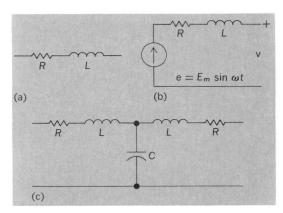

Fig. 9.   Models of circuit elements. (a) Model of a coil. (b) Model of an ac generator. (c) Model of medium-length single-phase transmission line.

from Euler's law, Eq. (13). Also, the unit imaginary quantity $j1 = \sqrt{-1}$ is used in ac circuit theory.

The complex number representing impedance is also bold-face italic, as **Z** in Eq. (57). When only the real component is referred to, an italic capital letter is used.

Graphical symbols for resistance R, inductance L, and capacitance C are used in Figs. 7 and 9. Figure 10 shows forms of the common symbols.

It is convenient to follow the convention adopted by some authors of using circles and narrow rectangles to indicate independent sources. A circle represents an independent voltage source, meaning that the source has an electromotive force within it, and hence a voltage between its terminals, that has always a specified value. In particular, the voltage across the source is independent of current through the source. The value of the independent electromotive force or voltage must be given. For example, the model of a storage battery may contain an independent source of 12 volts in series with some amount of resistance; a model of an ac generator (Fig. 9b) may contain an independent source of $e = 162 \sin \omega t$ volts in series with appropriate amounts of R and L.

An independent current source, represented by a narrow rectangle, has within it such an electromotive force that the current through the source always has the value specified for that current source. Thus an independent ac current source may have a current that is always $5 \sin \omega t$ amperes.

The specified value of an independent source is usually indicated on a diagram beside the symbol for the source.

It must be understood that an independent voltage source is not without current. Its voltage is specified and its current depends on the circuit in which the source is connected. Similarly, an independent current source is not without voltage. Its current is given and its voltage depends on the circuit to which the source is connected.

Ideal sources cannot exist alone in actuality, but they are useful in linear network models. An independent voltage source is a physical impossibility; if short-circuited, it would be required to develop an unlimited (infinite) current. Similarly, an independent current source would have to supply unlimited (infinite) voltage if its terminals were open-circuited.

A type of source used in models of electronic circuits (but not needed in this article) is a dependent or controlled source, the voltage or current of which depends on some other voltage or current elsewhere in the system. For example, the voltage in the collector circuit of a transistor model might be proportional to the emitter current.

Units. For electrical work the meter-kilogram-second (mks) system of mechanical units is extended by adopting a fundamental electrical unit. This unit is commonly either the coulomb or the ampere. The resulting mksc or mksa system is compatible (not requiring dimensional constants) and includes volts of potential or electromotive force, ohms of impedance, mhos of admittance, henrys of inductance, and farads of capacitance. The compatible unit of angle is the radian, and that of frequency (symbol ω) is the radian per second. (Frequency measured in cycles per sec-

ond, or hertz, is incompatible; the symbol is f and $\omega = 2\pi f$.)

Prefixes. It is common practice to use prefixes to indicate powers of 10. Thus 1 kilowatt means 10^3 watts or 1000 watts. Similarly 1 microampere means 10^{-6} ampere or one-millionth of an ampere. The prefixes are:

tera-	$= 10^{12}$	milli-	$= 10^{-3}$
giga-	$= 10^9$	micro-	$= 10^{-6}$
mega-	$= 10^6$	nano-	$= 10^{-9}$
kilo-	$= 10^3$	pico-	$= 10^{-12}$

Examples. The following examples illustrate alternating-current circuit theory. The first example concerns elements in series and division of voltage; the second example concerns elements in series and parallel, division of current, and computation of power.

Example 1. An incoming electric line provides power at 115 volts, 60 Hz. This is the amount of ac voltage as read on an ordinary voltmeter, and is therefore the rms value. Both voltage and frequency are essentially independent of the amount of current received from this line by the load that is about to be considered. The voltage, known to be alternating, is assumed to be sinusoidal.

Onto this line is connected a motor with 60-Hz impedance (at standstill) of $\mathbf{Z}_m = 1 + j5$ ohms in series, with a starting resistance of $\mathbf{Z}_s = 11$ ohms. The motor does not turn, being too heavily loaded. Compute the current that will flow steadily. Also, find voltage across the motor alone with this starting resistance in the circuit.

Solution. Draw Fig. 11 and start Fig. 12. Take voltage to be the reference phasor with zero angle, $\mathbf{V} = 115\underline{/0°}$. Obtain the total impedance **Z** by adding, as shown in Eq. (58a). The complex number $12 + j5$ can also be written in polar form, as shown by Eq. (58b). Using this polar form and dividing, it is found, as shown in Eq. (58c), that the steady current has an rms value of 8.85 amp, and that the sinusoidal wave of current lags (is delayed behind)

$$\mathbf{Z} = \mathbf{Z}_s + \mathbf{Z}_m = 11 + 1 + j5 = 12 + j5 \quad (58a)$$

$$\mathbf{Z} = 12 + j5 = 13\underline{/22.6°} \quad (58b)$$

$$\mathbf{I} = \frac{\mathbf{V}}{\mathbf{Z}} = \frac{115\underline{/0°}}{13\underline{/22.6°}} = 8.85\underline{/-22.6°} \quad (58c)$$

the sinusoidal wave of voltage by 22.6°. Voltage and current phasors are shown in Fig. 12a, and the corresponding waves are shown in Fig. 12b. Usually, phasors are shown but not waves (time functions of voltage and current); this is consistent with the usual practice of calling the phasor a current and leaving the reader to understand that the magnitude is the rms value of current and the

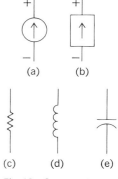

(a)   (b)

(c)   (d)   (e)

Fig. 10. Commonly used graphic circuit symbols. (a) Independent voltage source. (b) Independent current source. (c) Resistance. (d) Inductance. (e) Capacitance.

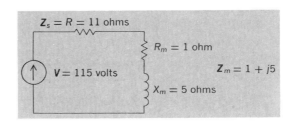

Fig. 11. Impedances in series: $\mathbf{Z}_s$ is starter resistance; R_m and X_m are model of motor with stalled rotor.

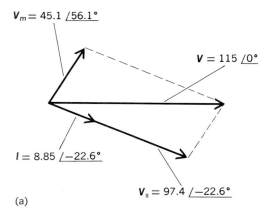

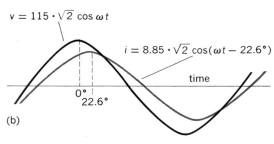

(a)

(b)

Fig. 12. Phasors and waves for circuit of Fig. 11.

angle is phase relative to an assumed reference.

In the present example, one is also required to find voltage across the motor with the starting resistance in the circuit. This voltage $\mathbf{V}_m$ is shown by Eq. (59). The phasor representing motor voltage is included in Fig. 12a.

$$
\begin{aligned}
\mathbf{V}_m &= \mathbf{Z}_m\mathbf{I} \\
&= (1+j5)(8.85\underline{/-22.6°}) \\
&= (5.10\underline{/78.7°})(8.85\underline{/-22.6°}) = 45.1\underline{/56.1°} \quad (59)
\end{aligned}
$$

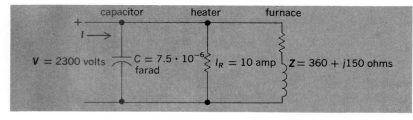

Fig. 13. Capacitor, heater, and furnace connected in parallel.

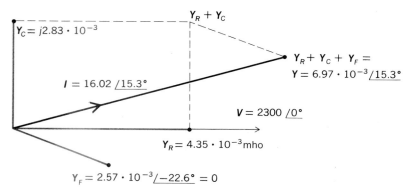

Fig. 14. Phasors for circuit of Fig. 13.

Check. As a check on the accuracy of the solution, if one computes voltage across the starting resistor and then adds this to the motor voltage, the resultant sum should be the line voltage. This computation is shown in Eqs. (60). The sum of the

$$
\begin{aligned}
\mathbf{V}_s &= \mathbf{Z}_s\mathbf{I} = 11(8.85\underline{/-22.6°}) = 97.4\underline{/-22.6°} \\
&= 89.9 - j37.4 \\
\mathbf{V}_m &= 25.1 + j37.4 \\
\mathbf{V} &= 115.0 + j0
\end{aligned}
\quad (60)
$$

voltage phasors is $115.0+j0$, which is the correct line voltage. Graphical addition is illustrated in Fig. 12a.

Example 2. An electric furnace of the induction type has impedance $\mathbf{Z}=360+j150$ ohms at its rated frequency of 60 Hz. It is connected to an incoming line that provides 2300 volts at this frequency. A purely resistive electric heater is also connected to the line (Fig. 13), and draws 10 amp at this voltage. To improve the power factor, management has connected to the incoming line a capacitor with 7.5 microfarads (μf) capacitance. Find the current supplied from the incoming line; also find total power supplied and power to each load.

Solution. Draw Fig. 13 and start Fig. 14. Take the voltage of the incoming line as the reference phasor, $2300\underline{/0°}$. Compute the admittance of each element of the network and add. Computations are shown in Eqs. (61).

For the capacitor,

$$
\begin{aligned}
\mathbf{Y}_C &= j\omega C = j2\pi(60)(7.5)10^{-6} \\
&= j(2.83)10^{-3} \text{ mho} \quad (61a)
\end{aligned}
$$

For the resistor,

$$
\mathbf{Y}_R = \frac{10}{2300} = (4.35)10^{-3} \text{ mho} \quad (61b)
$$

For the furnace,

$$
\begin{aligned}
\mathbf{Y}_F &= \frac{1}{360+j150} = \frac{1}{390\underline{/22.6°}} \\
&= (2.57)10^{-3}\underline{/-22.6°} \quad (61c)
\end{aligned}
$$

or $\mathbf{Y}_F = \dfrac{360-j150}{(360)^2+(150)^2} = (2.37-j0.986)10^{-3}$ mho

Total line current is then

$$
\begin{aligned}
\mathbf{I} &= \mathbf{VY} = 2300(j2.83+4.35+2.37-j0.986)10^{-3} \\
&= 2.30(6.72+j1.84) = 16.02\underline{/15.3°} \quad (61d)
\end{aligned}
$$

Power computations shown in Eqs. (62)–(64) are derived from earlier equations as indicated. Power from line by Eq. (53) is

$$
\begin{aligned}
P &= VI\cos\varphi = (2300)(16.02)\cos 15.3° \quad (62) \\
&= 35,520 \text{ watts}
\end{aligned}
$$

Power to resistor by Eq. (53) is

$$
P_R = (2300)(10) = 23,000 \text{ watts} \quad (63)
$$

Power to impedance of furnace by Eq. (54) is

$$
P_F = I^2R = \left(\frac{2300}{390}\right)^2 360 = 12,520 \text{ watts} \quad (64)
$$

As a check, no power is supplied to the capacitor, and the sum of power to the resistor, 23.0 kw, and power to the furnace, 12.52 kw, equals the incoming line power, 35.52 kw, which is correct. *See* COUPLED CIRCUITS; NETWORK THEORY;

RESONANCE (ALTERNATING-CURRENT CIRCUITS); Y-DELTA TRANSFORMATIONS.

[H. H. SKILLING]

Bibliography: H. H. Skilling, *Electric Networks*, 1974; H. H. Skilling, *Electro-Mechanics*, 1962; H. H. Skilling, *Fundamentals of Electric Waves*, 2d ed., 1947.

Amateur radio

Two-way radio communications by private individuals as a leisure-time activity. Amateur, or "ham," radio is defined in international treaty as a "service of self-training, intercommunications, and technical investigations carried on by amateurs; that is, by duly authorized persons interested in radio technique solely with a personal aim and without pecuniary interest." *See* RADIO.

Public service. More than 750,000 hams throughout the world translate these formal words into public service and private pleasure. With two-way radio equipment in their homes or cars they help a doctor in Africa consult with physicians at Duke University; they carry messages from a soldier overseas to his family in the States; they experiment with space communications through amateur-built Earth satellites (the Oscar series, ham-constructed and launched piggyback on United States government space shots at no additional cost to the taxpayers); they set up warning and relief networks during the hurricane season and handle communications after blizzards have torn down the phone lines; and they get acquainted with each other through casual "people-to-people" chats about the weather, the family, school or work, sports, and technical topics.

Requirements. There are about 275,000 amateur operators in the United States. Their licenses are issued by the Federal Communications Commission in five classes—novice, technician, general, advanced, and amateur extra. Knowledge of the international Morse code (at speeds varying from 5 words per minute for the novice to 20 words per minute for the amateur extra) and of radio theory and regulations (relatively simple for the novice; comprehensive enough to stump some professionals for the amateur extra) is required for a license of any class.

Privileges. With each succeeding license greater privileges are granted. Novices can use only code, in four frequency bands. Power is restricted to 75 watts input. Other classes are permitted to use 1000 watts input power. Any of the following modes may be used: code (CW), standard amplitude-modulated voice (AM), the newer single-sideband suppressed-carrier voice (SSB), frequency modulation (FM), radioteletype (RTTY), facsimile, television, or even pulse transmissions akin to radar. *See* AMPLITUDE-MODULATION RADIO; FACSIMILE; FREQUENCY-MODULATION RADIO; TELEVISION.

Each grade carries greater choice of frequencies, with the amateur extra class licensee being permitted to use authorized modes on all of the following frequency bands (expressed in millions of cycles per second, or megahertz, MHz):

1.8 – 2.0 MHz	21.0 – 21.45
3.5 – 4.0	28.0 – 29.7
7.0 – 7.3	50.0 – 54.0
14.0 – 14.35	144 – 148
220 – 225 MHz	5650 – 5925
420 – 450	10,000 – 10,500
1215 – 1300	24,000 – 24,250
3300 – 3500	Others above 48,000

Regulations. In addition to the international treaty, United States amateurs are governed by Part 97, Rules and Regulations of the Federal Communications Commission. Unlike most radio services, there is no requirement for "type-approved" equipment. Amateurs may design and build their equipment, may assemble it from kits, or purchase it ready to go. Although the latter course is now the most popular, according to a 1974 survey more than half of the ham antennas and a fifth of the ham transmitters were home-constructed. Amateurs are not hemmed in by precise technical parameters; instead, the quality of the signals transmitted on the air must be up to the state of the art.

Technical developments. This open policy toward amateur radio by the government has made it possible for amateurs to pioneer in several areas. In 1923, for instance, amateurs led the way to the short waves after an amateur in Connecticut talked with a ham in France on the previously ignored wavelength of 110 meters (m). Until that achievement the accepted theory had been that only wavelengths above 200 m were really useful.

By the early 1930s communications between points on opposite sides of the globe had become commonplace on wavelengths of 80 through 20 m (frequencies of 3.5 through 14.4 MHz). Many hams then turned their attention to higher frequencies. Again amateurs shattered previous conceptions, especially concerning "line-of-sight" limitations on communications at 56 MHz and above. These pioneers discovered and used means of radio propagation which reached far beyond the usual horizon: reflections from the aurora borealis, from meteor trails, from sporadic *E* layers (patches of ionized particles about 70 mi above the Earth), and recently from the Moon; bending of radio waves through layers of stable air; and a phenomenon still not thoroughly understood, transequatorial scatter, by which stations on one side of the Equator may communicate with stations on the other over distances of more than 1000 mi at times when such work would not otherwise be expected to succeed. This mode and several of the others were subjects of an International Geophysical Year study undertaken by amateurs through the American Radio Relay League.

Other technical developments by amateurs, first published in amateur journals, are superregeneration (1922), crystal control (1923), high *C* oscillator circuits (1928), single-signal superheterodyne receivers (1932), and *pi*-section antenna couplers (1934). Amateurs have led to the development of compact systems for SSB; for instance, the Air Force completed its change to SSB for long-range air-ground communications several years ahead of schedule by adapting equipment designed by and for amateurs. High-gain directional antennas and low-noise-figure receivers have been favorite technical projects of amateurs. Solid-state, integrated circuit technology has been picked up by amateurs, putting to new uses products designed by the industry.

The amateur service also contributes to the

state of the art by influencing the choice of careers of talented young people; about 50% of amateurs are employed in some phase of the electronics industry.

Associations. The developments just described required a clearinghouse, an information exchange, which is embodied in the amateurs' organization, the American Radio Relay League. The League, founded in 1914, is a nonprofit, nonstock association of radio amateurs in North America with headquarters at 225 Main Street, Newington, Conn. 06111. It publishes *QST* magazine each month (the name comes from an operating signal meaning "Calling all radio amateurs"), an annual *Radio Amateur's Handbook*, and 13 other publications covering various aspects of the hobby, ranging from beginner material to complex subjects. The League, which has about 110,000 members, serves as spokesman for amateurs in regulatory matters; it presents new technical developments and conducts operating contests and other activities to increase the hams' enjoyment of, and skill in, the hobby; it organizes networks to handle messages described earlier and coordinates emergency communications training. The League is also headquarters for the International Amateur Radio Union, composed of some 85 similar national radio societies.

[PERRY F. WILLIAMS]

Bibliography: American Radio Relay League, *Radio Amateurs Handbook*, published annually; and *Radio Amateurs License Manual*, 78th ed., 1980; L. Buckwalter, *Beginner's Guide to Ham Radio*, 1978; R. L. Shrader, *Electronic Communications*, 3d ed., 1975.

Ammeter

An electrical instrument for the measurement of electric current. In the usual indicating ammeter, an electromechanical system causes a pointer to traverse a calibrated scale, its position on the scale indicating the value of the current.

Several kinds of ammeter mechanism are used for different kinds of current, different applications, and differing degrees of accuracy.

Permanent-magnet movable-coil type. This type of ammeter is used universally for the measurement of direct currents. It is developed from the D'Arsonval movement. Figure 1a shows the general arrangement of the mechanism; Fig. 1b is a typical instrument. In the mechanism an external permanent magnet produces a uniform radial magnetic field across the air gap through which the coil rotates. With an air-gap flux density of B gauss, the torque produced by a current of I amperes through N turns in the rotating coil is shown in the equation below, where L is the length

$$T = \frac{B2RLIN}{10} = \frac{BAIN}{10} \quad \text{dyne-cm}$$

of the active conductor, R is the radius of action of those conductors, and A is the effective area of the moving coil.

It is important that the magnetic system be of such materials and proportions that a constant flux density is maintained in the air gap. For the older tungsten steel magnets, the ratio of the magnet length to air-gap length times the ratio of the air-gap cross section to magnet cross section should be more than 100 for a permanent magnetic

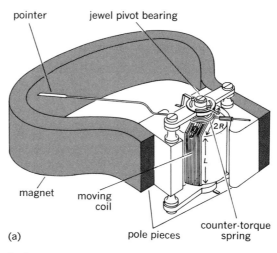

Fig. 1. Permanent-magnet movable-coil ammeter. (a) Mechanism. (b) Typical instrument. (*Weston Instruments, Inc.*)

system. For the Alnico permanent-magnet alloys with their high coercive force, however, this product can be as low as 10 for a stable system. The resulting short magnet can then be placed inside the rotating coil, as in the core-magnet system of Fig. 2. Flux densities typically range from 1500 gauss in small instruments to 5000 gauss in larger systems for special uses.

In Fig. 1 the coil is usually wound on a metal form and swung on highly polished steel or hard-alloy pivots between V-cup jewels or bearings of sapphire or hard glass. Bronze springs serve to carry the current in and out of the coil as well as to oppose the electrical torque, resulting in a deflection proportional to the coil current. The coil turns may vary from as few as 15 to as many as 5000; wire size, in turn, may be as small as 0.0006 in. in diameter for coils of 1000 turns or more.

Obviously, very weak control springs would allow for a normal deflection on the low torque produced by only a few microamperes in the coil. But the lower limit of useful torque is that which will overcome the residual bearing friction by a factor of several times the expected accuracy in terms of deflection. This sets a definite lower limit to a pivoted instrument, which is usually in the

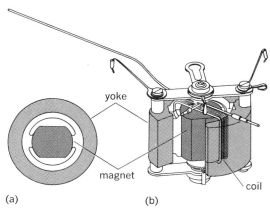

(a)   (b)

Fig. 2. Core-magnet version of permanent-magnet movable-coil mechanism. (*a*) Magnetic system. (*b*) Cutaway view. (*Weston Instruments, Inc.*)

order of $10-20$ μA full scale in a coil of several thousand ohms resistance.

The high limit of current through the coil is set by thermal limits; about 0.03 A is the maximum current in a moving coil. For higher values the instrument is shunted; that is, the bulk of the current is bypassed around the moving coil through a shunt, with only a definite fraction of the current passing through the moving coil itself.

Within the limits given, the permanent-magnet movable-coil ammeter for direct current has high torque and good accuracy, and is made in models varying from a 2-in.-diameter panel type to the large laboratory standards accurate within 0.1%.

Taut-band type. In a modification of ammeter design, the jeweled bearings and control springs are replaced by a taut metallic band rigidly held at the ends. The coil is firmly attached to the band. As the coil rotates in the magnetic field under the influence of the current which is being measured, the restoring torque is supplied by the twisting of the band. Thus, bearing friction is completely eliminated, and the resolution of the instrument thereby improved. This type of coil mounting is coming into wide use in applications where high resolution and sensitivity are desired.

Polarized-vane type. This instrument, shown in Fig. 3, is of only moderate accuracy. It is used in large numbers in battery chargers and automobiles

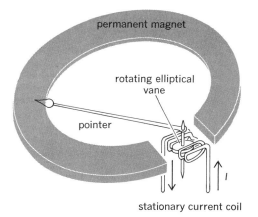

Fig. 3. Polarized-vane ammeter mechanism. (*Weston Instruments, Inc.*)

because of its low cost. Current through the small coil distorts the field of the circular permanent magnet; the iron vane aligns itself with the axis of the distorted field, the deflection being roughly proportional to the current. Reversed electrical polarity reverses the direction of motion; the instrument thus indicates current flowing into or out of a battery.

Electrodynamic type. This is also a movable-coil instrument, but the coil rotates in the magnetic field produced by a fixed coil. The instrument responds to alternating as well as direct current, and is thus a transfer instrument. Such instruments are precise and may be calibrated accurately and used for secondary standards, but the magnetic field of the fixed-coil system is relatively weak, 60 gauss being a common value. The controlling forces are thus much lower than in other types, and the electrodynamic ammeter is mainly

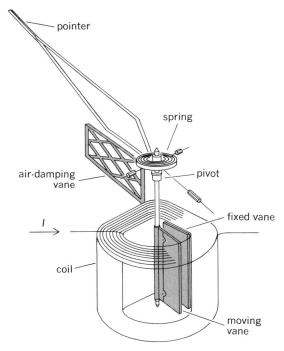

Fig. 4. Repulsion-vane or book-type ammeter mechanism. (*Weston Instruments, Inc.*)

used in the laboratory for calibrating the more rugged types.

Soft-iron type. These are widely used for alternating-current measurements. Figure 4 shows a repulsion-vane type with pointer, control spring, and damping vane on the moving element.

Requiring about 300 ampere-turns for full-scale deflection in the larger sizes, the common 5-A range will have a 60-turn coil of 0.06-in.-diameter copper wire, ample to carry even considerable overloads. The iron vanes tend to saturate magnetically on overload and thereby act as further overload protection. Actuating forces are high, and only the limitations in the iron vanes in responding exactly to the coil current limit the accuracy. The most accurate instruments of this type are accurate to within 0.25%.

On direct current, iron-vane instruments tend to

read low on increasing current and high on decreasing current because of hysteresis in the vanes. Iron-vane instruments should therefore always be calibrated on alternating current using an electrodynamometer standard.

Thermal type. Such ammeters function through the medium of the heat generated by the passage of electric current through a resistance element. Early hot-wire ammeters contained a platinum alloy or other resistance wire, heated by the current in question. The rise in temperature of the platinum caused it to expand, and its linear increase was amplified by a mechanical lever system to cause a pointer to travel across a scale calibrated in amperes. Although useful up to very high frequencies, hot-wire ammeters had high electrical losses, were difficult to compensate for variations in ambient temperature, and tended to burn out on moderate overloads. They are now completely obsolete, having been replaced by the thermocouple type developed about 1920.

Modern thermal ammeters consist of a thermal converter and a sensitive dc millivoltmeter. Figure 5 shows such a thermal converter mounted on the instrument terminals. The short and efficient heater is a platinum alloy wire or tube, to which is welded a thermocouple of two dissimilar metals. When the junction of the thermocouple is heated, a small voltage is generated which is proportional to the temperature rise; this in turn is applied to the millivolt mechanism to deflect its pointer. Typically, the heater is so proportioned as to have a temperature rise of 200°C on the desired full-scale current; about 10 mv is then available for the dc mechanism. Losses are represented by a potential drop through the heater of 200 mv at full-scale current.

The temperature rise of the heater is a function of the square of the current; thus the instrument scale follows a square law, compressed at the start and expanded at full scale.

Using a 1-mil heater wire, full scale represents about 1 A. Higher ranges are made with heavier heaters; losses increase correspondingly, with 60 A representing a practical top limit. For measuring currents that are under 1 A, the thermoelement is placed in vacuum to reduce heat losses; these are now useful as low as 1mA. A special bridge form of open thermoelement is also used in the 100-mA range.

Calibration of thermoammeters should be made on a low-frequency (such as 60 Hz) alternating current using appropriate standards. The instruments will then be within 2%, their typical accuracy, on frequencies as high as 50 MHz. On still higher frequencies they tend to read high because of excessive skin-effect losses in the heater.

In addition to the several kinds of ammeter described, there are several composite types widely used in communications and electronics for low values of current. Iron-vane instruments take considerable energy, about 1 watt for full scale; low-range thermal instruments are expensive and easily burned out. A rectifier-type ac milliammeter consisting of a small copper oxide or germanium bridge rectifier feeding a conventional dc milliammeter has low losses, good overload capacity, and adequate accuracy. It is widely used to monitor voice-frequency currents, with ranges available as low as 100 μA full scale. Similarly, vacuum-tube and transistor amplifying and rectifying systems

Kinds of ammeters

Type of mechanism	Kind of current	Application and accuracy*
Permanent-magnet movable-coil (D'Arsonval)	Direct	Very general: panel, switchboard, portable, laboratory instruments; accuracy 0.1–2%
Taut band	Direct	General, requiring high resolution and sensitivity; accuracy better than 0.1%
Polarized-iron vane	Direct	Battery charging, automobiles; accuracy only moderate
Electrodynamic†	Direct and alternating	Laboratory; high accuracy, 0.1%; portable testing, 0.25%
Soft-iron vane, repulsion, attraction, inclined-vane types	Alternating, low-frequency	Panel, switchboard, portable; accuracy 0.25–2%
Thermoelectric‡	Alternating, frequencies up to 100 MHz	Panel, switchboard, general testing; accuracy 0.5–3%
Miscellaneous composite types	Usually alternating	Communications, electronic circuit testing; accuracy 0.5–5%

*Accuracy stated as % maximum error of full-scale reading.

†Important as a transfer instrument from basic direct-current standards to use on alternating current.

‡Also may be used as a transfer standard, although usually from low-frequency alternating current via an electrodynamic ammeter to radio frequency.

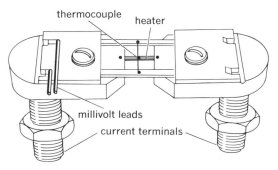

thermocouple   heater

millivolt leads

current terminals

Fig. 5.   Thermal converter of thermocouple ammeter. (*Weston Instruments, Inc.*)

used with sensitive dc microammeters further expand the coverage in range, frequency, and sensitivity to make available ammeters for most applications. Commercial ammeters are available with range 10^{-10} A full scale to 10^{-3} A with 14 overlapping ranges. The table summarizes the common types of ammeters. *See* CURRENT MEASUREMENT.

[JOHN H. MILLER/JESSE BEAMS]

Bibliography: W. D. Cooper, *Electronic Instrumentation and Measurement Techniques*, 2d ed., 1978; E. Frank, *Electrical Measurement Analysis*, 1959; F. K. Harris, *Electrical Measurements*, 1952; K. Henney (ed.), *Radio Engineering Handbook*, 5th ed., 1959; P. Kantrowitz, *Electronic Measurements*, 1979.

Amorphous solid

A rigid material whose structure lacks crystalline periodicity; that is, the pattern of its constituent atoms or molecules does not repeat periodically in three dimensions. In the present terminology amorphous and noncrystalline are synonymous. A solid is distinguished from its other amorphous counterparts (liquids and gases) by its viscosity: a material is considered solid (rigid) if its shear viscosity exceeds $10^{14.6}$ poise ($10^{13.6}$ Pa·s).

Preparation. Techniques commonly used to prepare amorphous solids include vapor deposition, electrodeposition, anodization, evaporation of a solvent (gel, glue), and chemical reaction (often oxidation) of a crystalline solid. None of these techniques involves the liquid state of the material. A distinctive class of amorphous solids consists of glasses, which are defined as amorphous solids obtained by cooling of the melt. Upon continued cooling below the crystalline melting point, a liquid either crystallizes with a discontinuous change in volume, viscosity, entropy, and internal energy, or (if the crystallization kinetics are slow enough and the quenching rate is fast enough) forms a glass with a continuous change in these properties. The glass transition temperature is defined as the temperature at which the fluid becomes solid (that is, the viscosity = $10^{14.6}$ poise = $10^{13.6}$ Pa·s) and is generally marked by a change in the thermal expansion coefficient and heat capacity. [Silicon dioxide (SiO_2) and germanium dioxide (GeO_2) are exceptions.] It is intuitively appealing to consider a glass to be both structurally and thermodynamically related to its liquid; such a connection is more tenuous for amorphous solids prepared by the other techniques.

Types of solids. Oxide glasses, generally the silicates, are the most familiar amorphous solids. However, as a state of matter, amorphous solids are much more widespread than just the oxide glasses. There are both organic (for example, polyethylene and some hard candies) and inorganic (for example, the silicates) amorphous solids. Examples of glass formers exist for each of the bonding types: covalent [As_2S_3], ionic [KNO_3 — $Ca(NO_3)_2$], metallic [Pd_4Si], van der Waal's [o-terphenyl], and hydrogen [$KHSO_4$]. Glasses can be prepared which span a broad range of physical properties. Dielectrics (for example, SiO_2) have very low electrical conductivity and are optically transparent, hard, and brittle. Semiconductors (for example, As_2SeTe_2) have intermediate electrical conductivities and are optically opaque and brittle.

Metallic glasses (for example, Pd_4Si) have high electrical and thermal conductivities, have metallic luster, and are ductile and strong.

Uses. The obvious uses for amorphous solids are as window glass, container glass, and the glassy polymers (plastics). Less widely recognized but nevertheless established technological uses include the dielectrics and protective coatings used in integrated circuits, and the active element in photocopying by xerography, which depends for its action upon photoconduction in an amorphous semiconductor. In optical communications a highly transparent dielectric glass in the form of a fiber is used as the transmission medium. In addition, metallic amorphous solids have been considered for uses that take advantage of their high strength, excellent corrosion resistance, extreme hardness and wear resistance, and unique magnetic properties. *See* OPTICAL COMMUNICATIONS.

Semiconductors. It is the changes in short-range order (on the scale of a localized electron), rather than the loss of long-range order alone, that have a profound effect on the properties of amorphous semiconductors. For example, the difference in resistivity between the crystalline and amorphous states for dielectrics and metals is always less than an order of magnitude and is generally less than a factor of 3. For semiconductors, however, resistivity changes of 10 orders of magnitude between the crystalline and amorphous states are not uncommon, and accompanying changes in optical properties can also be large.

Electronic structure. The model that has evolved for the electronic structure of an amorphous semiconductor is that the forbidden energy gap characteristic of the electronic states of a crystalline material is replaced in an amorphous semiconductor by a pseudogap. Within this pseudogap the density of states of the valence and conduction bands is sharply lower but tails off gradually and remains finite due to structural disorder (Fig. 1). The states in the tail region are localized; that is, their wave functions extend over small distances in contrast to the extended states that exist elsewhere in the energy spectrum. Because the localized states have low mobility (velocity per unit electric field), the extended states are separated by a mobility gap (Fig. 1) within which charge transport is markedly impeded. In each band, the energy at which the extended states meet the localized states is called the mobility edge. *See* BAND THEORY OF SOLIDS.

An ideal amorphous solid can be conceptually

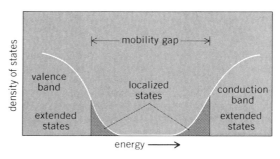

Fig. 1. Density of states versus energy for an amorphous semiconductor.

defined as having no unsatisfied bonds, a minimum of bond distortions (bond angles and lengths), and no internal surfaces associated with voids. Deviations from this ideality introduce localized states in the gap in addition to those in the band edge tails due to disorder alone. One important defect is called an unsatisfied, broken, or dangling bond. These dangling bonds create states deep in the gap which can act as recombination centers and markedly limit carrier lifetime and mobility. A large number of such states introduced, for example, during the deposition process will dominate the electrical properties.

Charge transport can occur by two mechanisms. The first is conduction of mobile extended-state carriers (analogous to that which occurs in crystalline semiconductors), for which the conductivity is proportional to exp $(-E_g/2kT)$, where E_g is the gap width, T is the absolute temperature, and k is Boltzmann's constant. The second mechanism is hopping of the localized carriers, for which the conductivity is proportional to exp $[-(T_0/T)^{1/4}]$, where T_0 is a constant (Mott's law). At low temperatures carriers hop from one localized trap to another, whereas at high temperatures they can be excited to the mobility edge.

Glassy chalcogenides. One class of amorphous semiconductors is the glassy chalcogenides, which contain one (or more) of the chalcogens sulfur, selenium, or tellurium as major constituents. These amorphous solids behave like intrinsic semiconductors, show no detectable unpaired spin states, and exhibit no doping effects. It is thought that essentially all atoms in these glasses assume a bonding configuration such that bonding requirements are satisfied; that is, the structure accommodates the coordination of any atom. These materials have application in switching and memory devices. *See* GLASS SWITCH.

Tetrahedrally bonded solids. Another group is the tetrahedrally bonded amorphous solids, such as amorphous silicon and germanium. These materials cannot be formed by quenching from the melt (that is, as glasses) but must be prepared by one of the deposition techniques mentioned above. An amorphous to crystalline transformation in these materials is irreversible.

When amorphous silicon (or germanium) is prepared by evaporation, not all bonding require-

ments are satisfied, so a large number of dangling bonds are introduced into the material. These dangling bonds are easily detected by spin resonance or low-temperature magnetic susceptibility and create states deep in the gap which limit the transport properties. The number of dangling bonds can be reduced by a thermal anneal below the crystallization temperature, but the number cannot be reduced sufficiently to permit doping.

Amorphous silicon prepared by the decomposition of silane (SiH_4) in a plasma has been found to have a significantly lower density of defect states within the gap, and consequently the carrier lifetimes are expected to be longer. This material can be doped p- or n-type with boron or phosphorus (as examples) by the addition of B_2H_6 or PH_3 to the SiH_4 during deposition. This permits exploration of possible devices based on doping, which are analogous to devices based on doping of crystalline silicon.

One reason plasma-deposited silicon has a significantly lower density of defect states within the gap is that the process codeposits large amounts of hydrogen (typically 5–30% of the atoms, depending upon deposition conditions), and this hydrogen is very effective at terminating dangling bonds (Fig. 2). Other possible dangling-bond terminators (for example, fluorine) have been explored.

The ability to reduce the number of states deep in the gap and to dope amorphous silicon led directly to the development of an amorphous silicon photovoltaic solar cell. Intense effort has been devoted to improving the efficiency of these cells to the 8% level thought to be required for large-scale application. The appeal of amorphous silicon is that it holds promise for low-cost, easily fabricated, large-area cells.

Amorphous silicon solar cells have been constructed in heterojunction, p-i-n-junction, and Schottky-barrier device configurations and have been introduced for use in calculators and watches. The optical properties of amorphous silicon provide a better match to the solar spectrum than do those for crystalline silicon, but the transport properties of the crystalline material are better. Experiments indicate that hole transport in the amorphous material is the limiting factor in the conversion efficiency. *See* SEMICONDUCTOR; SEMICONDUCTOR DIODE; SEMICONDUCTOR HETEROSTRUCTURES.

[BRIAN G. BAGLEY]

Bibliography: R. H. Doremus, *Glass Science*, 1973; N. F. Mott and E. A. Davis, *Electronic Processes in Non-Crystalline Materials*, 2d ed., 1979; J. Tauc (ed.), *Amorphous and Liquid Semi-conductors*, 1974.

Amplifier

A device capable of increasing the magnitude or power level of a physical quantity that is varying with time, without distorting the wave shape of the quantity. The great majority of amplifiers are electronic and depend upon transistors or vacuum tubes for their operation; such amplifiers will be discussed in this article. A small number of electronic amplifiers are magnetic amplifiers, while others take the form of rotating electrical machinery, such as the Amplidyne. A common nonelectrical amplifier is the hydraulic actuator, which is an amplifier of mechanical forces.

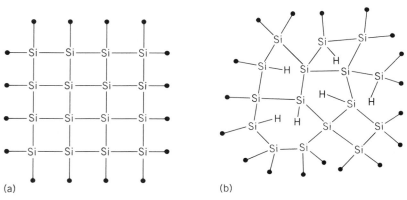

Fig. 2. Bonding of silicon. Bonds which continue the network are shown terminated by a dot. (a) Crystalline arrangement. (b) Amorphous structure with dangling bonds terminated by hydrogen.

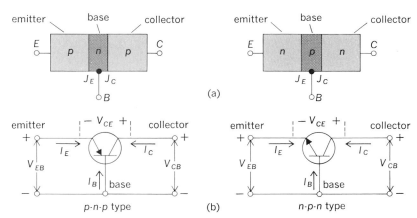

Fig. 1.   Two types of transistor. (*a*) The *p-n-p* and *n-p-n* types. The emitter (collector) junction is $J_E(J_c)$. (*b*) Circuit representation of the two types.

TRANSISTOR AMPLIFIERS

The transistor, since its discovery in 1947, has played an extremely important role as an active circuit element in amplifiers. The great majority of amplifiers employ transistors because of their low cost, long life, high reliability, better overall characteristics, and small size and because the physical nature of the transistor makes possible circuits that are not possible with vacuum tubes.

Basic principles. A junction transistor consists of a silicon (or germanium) crystal in which a layer of *n*-type silicon is sandwiched between two layers of *p*-type silicon. Alternatively, a transistor may consist of a layer of *p*-type between two layers of *n*-type material. In the former case, the transistor is referred to as a *p-n-p* transistor, and in the latter case as an *n-p-n* transistor. The semiconductor sandwich is extremely small and hermetically sealed against moisture inside a metal or plastic case. *See* TRANSISTOR.

The two types of transistor are represented in Fig. 1*a*. The representations employed when transistors are used as circuit elements are shown in Fig. 1*b*. The three portions of a transistor are known as emitter, base, and collector. The arrow on the emitter lead specifies the direction of current flow when the emitter-base junction is biased in the forward direction. In both cases, however, the emitter, base, and collector currents, I_E, I_B, and I_C, respectively, are assumed positive when the currents flow into the transistor. The symbols V_{EB}, V_{CB}, and V_{CE} are the emitter-base, collector-base, and collector-emitter voltages, respectively. (More specifically, V_{EB} represents the voltage drop from emitter to base.).

Potential distribution. The essential features of a transistor as an active circuit element are depicted in Fig. 2*a*. Here a *p-n-p* transistor is shown with voltage sources which serve to bias the emitter-base junction in the forward direction and the collector-base junction in the reverse direction. The variation of potential through an unbiased (open-circuited) transistor is shown in Fig. 2*b*. The potential variation through the biased transistor is indicated in Fig. 2*c*. The dashed curve applies to the case before the application of external biasing voltages, and the solid curve to the case after the application of biasing voltages. In the absence of applied voltage, the potential barriers at the junctions adjust themselves to the contact potential V_o (a few tenths of a volt) required so that no current flows across each junction. If now external potentials are applied, these voltages appear essentially across the junctions. Hence the forward biasing of the emitter-base junction lowers the emitter-base potential barrier by $|V_{EB}|$, whereas the reverse biasing of the collector-base junction increases the collector-base potential barrier by $|V_{CB}|$. The lowering of the emitter-base barrier permits the emitter current to increase, and holes are injected into the base region. The potential is constant across the base region (except for the small ohmic drop), and the injected holes diffuse across the *n*-type material to the collector-base junction. The holes which reach this junction fall down the potential barrier, and are therefore collected by the collector. *See* SEMICONDUCTOR.

A load resistor R_L is in series with the collector supply voltage V_{CC} of Fig. 2*a*. A small voltage change ΔV_i between emitter and base causes a relatively large emitter-current change ΔI_E. The symbol α defines that fraction of this current change which is collected and passes through R_L. The change in output voltage across the load resistor $\Delta V_o = \alpha R_L \Delta I_E$ may be many times the change in input voltage ΔV_i. Under these circumstances, the voltage amplification $A \equiv \Delta V_o/\Delta V_i$ will be greater than unity, and the transistor acts as an amplifier. If the dynamic resistance of the emitter junction is r_e, then $\Delta V_i = r_e \Delta I_E$, and Eq. (1) exists.

$$A \equiv \frac{\alpha R_L \Delta I_E}{r_e \Delta I_E} = \frac{\alpha R_L}{r_e} \qquad (1)$$

For example, if $r_e = 25\Omega$, $\alpha = 0.99$, and $R_L = 3000\Omega$, $A = 118$. This calculation is oversimplified, but in essence it is correct and gives a physical explanation of why the transistor acts as an amplifier. The transistor provides power gain as well as voltage amplification. From the foregoing explanation it is clear that current in the low-resistance input circuit is transferred to the high-resistance output circuit. The word transistor, which originated as a contraction of "transfer resistor," is based upon the preceding physical picture of the device.

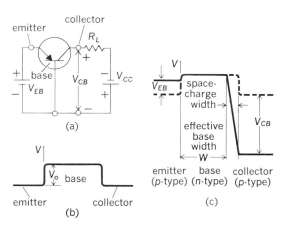

Fig. 2.   A *p-n-p* transistor with biasing voltages. (*a*) Circuit diagram. (*b*) Potential barriers at junction of the unbiased transistor. (*c*) Potential variation through the biased transistor.

AMPLIFIER

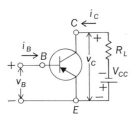

Fig. 3. Transistor common-emitter connection.

Current amplification as well as voltage amplification may be provided by the transistor if the emitter is used as a common, or grounded, terminal as shown in Fig. 3. In this configuration, the input signal voltage is applied between the base and emitter electrodes, as it is in the common-base arrangement of Fig. 2a, but the signal source needs only provide the base current Δi_b, which is small in comparison with the emitter signal current Δi_E. In order to show the relationship between the output signal current Δi_C and the input signal current Δi_B, one begins with the relationship for the common base amplifier, Eq. (2), where

$$\Delta i_C = -\alpha \Delta i_E \qquad (2)$$

α is defined as the ratio of Δi_C to Δi_E and the negative sign indicates that the current flows into one terminal while flowing out of the other terminal. Then, using Kirchhoff's current law, that the total current flowing into a junction is zero, or $\Delta i_B + \Delta i_C + \Delta i_E = 0$, Eq. (3) is obtained. Substituting this value for Δi_E into Eq. (2) results in Eq. (4). Finally, the explicit solution of Eq. (4) for Δi_C as a function of Δi_B is given by Eq. (5). Equation (5) shows that

$$\Delta i_E = -\Delta i_B - i_C \qquad (3)$$

$$\Delta i_C = \alpha(\Delta i_B + \Delta i_C) \qquad (4)$$

$$\Delta i_C = \frac{\alpha}{1-\alpha} \Delta i_B \qquad (5)$$

$\alpha/(1-\alpha)$ is the ratio of collector signal current to the base signal current. This ratio, or current amplification factor, is commonly known as either β or h_{fe}. Since values of α range from about 0.9 to 0.999, values of β range from about 10 to 1000.

The small input current of the common-emitter amplifier leads to a higher input resistance as compared with the common-base amplifier. In fact, the input resistance is increased by the factor $\beta + 1$, so the voltage amplification of the common-emitter amplifier is about the same as that of the common-base amplifier. However, the power amplifier is higher by the factor β because of its current amplification.

VACUUM-TUBE AMPLIFIERS

The basic electron-tube amplifier consists of a vacuum triode tube and associated circuitry, as shown in Fig. 4. The signal voltage to be amplified is applied to the input, or grid, circuit and the amplified signal appears across the load resistor R_L in the anode (plate), or output, circuit. A gain in voltage, current, and power is thereby provided. By selection of tubes and associated circuitry, these gains may be maximized.

Basic principles. In normal operation a small negative voltage called a bias voltage, V_{GG}, is applied to the grid of tube. This voltage establishes a definite plate-to-cathode current through the tube for a fixed load resistor R_L and high-potential plate-supply voltage V_{PP}. This plate current I_P is called the quiescent current since it is the zero-signal current. A dc voltage drop $I_P R_L$ appears across the output terminals because of the flow of the quiescent current through the load resistor. In the vacuum tube, electrons are emitted by the cathode and pass through the grid to the plate. The direction of "conventional" current is opposite to that of the electron, and therefore the plate current (conventional notation) flows from plate to cathode. *See* VACUUM TUBE.

If a small signal voltage is added to the grid bias voltage, the instantaneous grid voltage v_G is the algebraic sum of the bias voltage V_{GG} and the signal voltage v_G. As the instantaneous grid voltage changes, the plate current i_p changes in proportion to the grid voltage changes. The result is a net plate current i_p that is the algebraic sum of the quiescent plate current I_P and a varying or signal component of plate current e_v.

The passage of signal current e_v through the load resistor results in an output voltage i_p that is the product $i_p R_L$. The ratio of i_p to the grid signal voltage v_s is the voltage gain A of the amplifier.

If the root-mean-square (rms) value of signal current is I_o, the signal power delivered to the load is $I_o^2 R_L$ watts. For the typical voltage-amplifier vacuum tube the output power may be as high as 1 watt. For output tubes used in radio, television, and phonograph amplifiers the power output may run between 2 and 20 watts, with a few units having much greater output ratings for public address amplifiers and the like. Large power amplifiers used in broadcasting and commercial rf heating generators range up to 50 kW in power rating. Larger tubes are occasionally used.

Classifications. Amplifiers of either the transistor or tube variety may be classified in any one of several ways. One is by means of the coupling circuitry, such as an *RC*-coupled (resistance-capacitance-coupled) amplifier. Other categories include classification by purpose, such as hi-fi (high-fidelity audio amplifier), and mode of operation, such as class A. Some of the more common methods of classification follow with a brief description of the operating properties.

Classification by coupling methods. A great deal of information about an amplifier is conveyed by merely stating the interstage coupling method used. The more common types are listed here.

1. *RC*-coupled amplifier. This amplifier (Fig. 5) is so named because of the coupling capacitor C_b from the load resistor of stage A_1 to the input of the following stage, A_2. This capacitor blocks the dc collector or plate voltage from reaching the base or grid of the next stage. It is by far the commonest form of audio amplifier because of its simplicity and because it can be designed with nearly constant amplification over a wide frequency range. It is essentially an audio amplifier, although with high-frequency compensation the upper half-power frequency can be extended to several megahertz (MHz). The lower half-power frequency may be on the order of 20 Hz, and can be extended to lower frequencies by using low-frequency compensation.

2. Transformer-coupled amplifier. This amplifier is so named because two stages are coupled through a transformer. The dc collector or

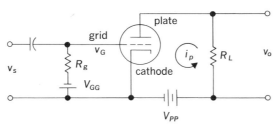

Fig. 4. Diagram of a basic vacuum-tube amplifier.

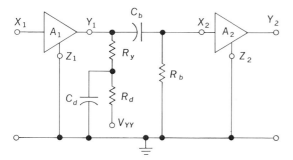

Fig. 5. Diagram illustrating the use of the decoupling filter R_d and C_d for compensation.

plate voltage of the transistor or tube is blocked from reaching the base or grid of the following stage by the transformer. Transformer coupling finds wide application in tuned amplifiers, such as rf and i-f amplifiers, where tuned air-core transformers or transformers with a core of powdered iron are used.

3. Direct-coupled amplifier. This form of amplifier, usually called a dc amplifier, allows the amplification of dc as well as ac signals. It has special circuitry, either in the coupling network or in the type of amplifier stages, which eliminates the need for the coupling capacitor. In direct-coupled amplifiers problems arise with drift produced by changes in temperature or aging of components. Therefore, special techniques must be employed to reduce the undesirable signals in the output to negligible values. This type of amplifier finds extensive application whenever very-low-frequency signals must be amplified, such as in certain medical instruments and servomechanisms. It also finds extensive application in the electronic differential analyzer (analog computer).

Classification by selectivity. An important feature of some amplifiers is their ability to amplify only those signals lying within a certain band of frequencies and to reject signals outside this band. Other amplifiers are designed to amplify an extremely wide range of frequencies.

1. Tuned amplifiers. A tuned amplifier is one which is designed to amplify signals in a frequency band that is centered on a chosen carrier frequency and not to amplify signals which lie outside this band. Tuned amplifiers form an important part of radio and television receivers in the rf and i-f amplifier sections. The signal picked up by the antenna and appearing at the terminals of the receiver is composed of the desired signal along with signals from other stations. The receiver is tuned to the carrier frequency of the desired signal, and the tuned amplifiers amplify the desired signal. A tuned amplifier is formed by using a collector or plate load impedance which is a resonant circuit. The gain is normally maximum at the resonant frequency and decreases for frequencies on either side of the resonant frequency. The resonant frequency of the amplifier (the frequency of maximum gain) may be selected by varying either the inductance or capacitance in the circuit. For an i-f amplifier, where the resonant frequency is constant at the intermediate frequency, the tuning for maximum gain is done at the time that the receiver is aligned and remains fixed, except for adjustments made necessary by aging of the components

or replacement of the transistors or tubes. On the other hand, the tuning of an rf amplifier in a radio receiver must be variable in order to receive stations transmitting at different carrier frequencies. The tuning is usually done with a variable capacitor. Cascaded amplifiers sometimes have different resonant frequencies for each stage in order to produce the required frequency response. Such amplifiers are called stagger-tuned amplifiers.

2. Untuned amplifiers. An untuned amplifier is one in which the coupling circuitry is not a tuned circuit. The audio amplifier and the direct-coupled amplifier are the most important of this type.

Classification by operating mode. The operating modes of vacuum tubes are prescribed by the designations class A, class B, and class C, with certain divisions among the classes. The distinction between the various classes is determined for a sinusoidal signal voltage applied to the input. The position of the quiescent point and the extent of the characteristic that is being used determine the method of operation. Whether the transistor or tube is operated as a class A, B, AB, or C amplifier is determined from the following definitions.

Class A. In this amplifier the operating point and the input signal are such that the current in the output circuit (in the collector or plate) flows at all times. A class A amplifier operates essentially over a linear portion of its characteristic.

Class B. This amplifier has the operating point at an extreme end of its characteristic, so that the quiescent power is very small. Hence either the quiescent current or the quiescent voltage is approximately zero. If the signal voltage is sinusoidal, amplification takes place for only one-half cycle. For example, if the quiescent output-circuit current is zero, this current will remain zero for one-half cycle.

Class AB. This amplifier operates between the two extremes defined for class A and class B. Hence the output signal is zero for part, but less than one-half, of an input sinusoidal signal cycle.

Class C. In this amplifier the operating point is chosen so that the output current (or voltage) is zero for more than one-half of an input sinusoidal signal cycle.

In the case of a vacuum-tube amplifier the suffix 1 may be added to the letter, or letters, of the class identification to denote that grid current does not flow during any part of the input cycle. The suffix 2 may be added to denote that grid current does flow during some part of the input cycle.

Classification by application. Amplifiers may be classified by the use for which they are designed.

1. Audio-frequency (af) amplifiers are intended to operate over the general range of about 20–20,000 Hz. Those capable of operating over this full range with a minimum of distortion are often referred to as high-fidelity (hi-fi) amplifiers if they are used with sound-reproducing equipment. Special-purpose amplifiers, such as those used in automatic control equipment, often operate in the audio-frequency range.

2. Radio-frequency (rf) amplifiers are used to amplify signals in the range of about 100–1,000,000 kHz. They are usually used as the first amplifier stage in selective radio receivers, including automobile receivers. Because of the additional cost they are usually not used in the average home receiver designed for local reception.

3. Intermediate-frequency (i-f) amplifiers are used in radio, radar, and television receivers. The majority of broadcast radio receivers employ i-f amplifiers operating at 445 kHz; some automobile receivers use 262 kHz. Communication receivers commonly use 1600-kHz i-f transformers. In frequency-modulated (FM) receivers an i-f of 10.7 MHz is used. Television receivers use 26- or 46-MHz i-f frequencies. Radar receivers commonly use 30- or 60-MHz i-f amplifiers.

Cascade amplifier. Since, in most applications, one stage of amplification does not provide enough gain, two or more stages are connected together (cascaded) to provide the required gain. For example, many radio receivers have two i-f amplifiers, and the more sensitive receivers have three stages. Similarly, there may be two stages of amplification, and possibly a preamplifier, preceding the power amplifier in an audio amplifier.

The gain of an amplifier is defined as the ratio of the output voltage to the input voltage for a sinusoidal input voltage. Under the usual method for analyzing ac circuits, the gain is a complex number indicating the magnitude of the gain and the phase angle by which the output voltage lags the input voltage in time. Since the gain is a function of frequency, because of the reactive circuit elements, the figure for the gain should be accompanied by the frequency at which it was determined. The practice is to quote the gain for the mid-band region of an audio amplifier or the gain at the resonant frequency for a tuned amplifier. The terms gain and amplification are synonymous, the former being more commonly used. The gain of a cascaded amplifier at any frequency is equal to the product of the gain of each stage. The gain of each stage must be measured or calculated when the stages are connected together, if the frequency at which the gain is being measured is above the mid-band range for an audio amplifier, or if the amplifier is a tuned amplifier. This is necessary because the input impedance of the following stage causes the gain of a stage to be different from what it would be if the following stage were not connected. This is important at high frequencies because the dominant component of the input is a shunt capacitor. If the frequency at which the measurement is made is in the midband range or lower for an audio amplifier, the reactance of the input capacitance is sufficiently high to allow it to be neglected, and the gain is the product of the gain of each stage, where the stage gain is measured before cascading. *See* VOLTAGE AMPLIFIER.

Signal response. An amplifier cannot exactly reproduce the waveform or frequency spectrum of its input signal. The amplifier's ability to faithfully reproduce a signal is measured by either its frequency response or its time response.

High-frequency compensation. As the frequency is increased above the upper limit of the mid-band region, the gain decreases because the reactance of the shunt capacitance is decreasing. High-frequency compensation is employed to increase the upper half-power frequency. One technique is to place a small inductance in series with the load resistor. The reactance of the inductance, and hence the effective amplifier load, increases with increasing frequency. This increased load tends to offset the effect of the decrease in reactance of the

shunt capacitance as the frequency increases. The result is that the value for the inductance can be chosen to extend the mid-band region of the compensated amplifier without introducing undesirable characteristics into the frequency response.

Low-frequency compensation. As the frequency is decreased below the lower limit of the mid-band region, the gain of an *RC*-coupled amplifier decreases because the reactance of the coupling capacitor is increasing.

The decrease in gain can be compensated for by using the compensation network shown in Fig. 5. Devices A_1 and A_2 may be transistors or vacuum tubes. A high resistance R_d in series with the collector or plate resistance R_y connects to the supply voltage. A large capacitance C_d is used to bypass R_d to ground. The components R_d and C_d are often used with a multistage amplifier as a decoupling filter to minimize the interactions between stages which result from the use of a common power supply. This same decoupling network compensates for the low-frequency distortion introduced by C_b. Thus, at high frequencies (in the mid-band region) C_d acts as a short circuit across R_d, and the gain of the amplifier is determined by R_y. At low frequencies, however, C_d becomes a large reactance, and the effective output-circuit resistance increases toward $R_y + R_d$. This increase in amplification tends to compensate for the loss in output due to the attenuation caused by the reactance of C_b, which increases as the frequency is reduced.

Pulse rise time. An amplifier may be called upon to increase the size of a signal whose wave shape is that of a narrow pulse. The definition for pulse rise time is the time required for the pulse to increase from 10% of its maximum value (assuming no overshoot) to 90% of its maximum value (see Fig. 6). The rise time of the pulse is a function of the frequency response of the amplifier. The rise time t_r is inversely proportional to bandwidth B; the constant of proportionality is taken as 0.35. Thus $\Delta\tau = 0.35/B$.

Bandwidth is inversely proportional to equivalent load resistance R_{eq} and shunt capacitance C_o of an amplifier stage; thus the expression for pulse rise time may be written as $t_r = 2.2R_{eq}C_o$.

Undesirable conditions. There are a number of undesirable conditions that may occur in amplifiers, produced by improper circuit design and by inherent limitations in the physical operation of the devices. In general, good circuit design can reduce all of the undesirable conditions, in-

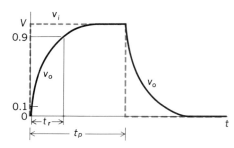

Fig. 6.   The output waveform v_o resulting when a pulse v_i is applied to an amplifier.

cluding those caused by physical limitations, to the point where they are not noticeable.

Distortion. If the output signal from an amplifier is not an exact replica of the input signal, distortion has occurred. In theory it is impossible for an amplifier to avoid introducing distortion. On the other hand, amplifiers have been designed in which the distortion is extremely small. Distortion is introduced by two factors in the amplifier. One factor is a nonlinear relation of the device between the input signal and the output signal. If the input signal is a sinusoidal signal, the output will be composed of harmonics of the input. The second factor is the frequency response of the amplifier. Since the gain of the amplifier is not the same for all frequencies, the amplification of each harmonic component of an input signal is not necessarily the same as that of the others. The output signal, since it is a superposition of the harmonic components, differs from the input signal. Distortion caused by nonlinearities in the amplifier and by the frequency response of the amplifier can be reduced by the proper use of feedback.

Noise. The noise encountered in amplifiers may, in general, be classified into thermal noise and shot noise. Thermal noise is caused by the random motion of electrons inside resistors, conductors, tubes, and transistors. It has the characteristic of having uniform power per unit bandwidth. Furthermore, the noise power is directly proportional to the temperature when expressed in degrees Kelvin. Shot noise is the name given to the noise generated in transistors or vacuum tubes by the random emission of holes or electrons from the emitter or cathode. The random emission produces minute fluctuations in the average value of collector or plate current. These fluctuations produce a small noise voltage of the same order of magnitude as thermal noise (a few microvolts).

To analyze the effect of these noise sources upon the circuit and to design the optimum circuit, noise generators are used, and the analysis proceeds in a straightforward manner. Although noise voltages can become bothersome in audio amplifiers, the most common situation where they must be considered is in the design of rf amplifiers, where quite often the signal to be amplified is not much larger than the rms value of the noise voltage. *See* ELECTRICAL NOISE.

Degeneration. This undesirable condition is the loss of gain through unintentional negative feedback. Two common sources of degeneration are improperly bypassed emitter or cathode resistors and screen dropping resistors. Ideally the emitter, cathode, and screen voltages should be constant with respect to the signal voltage. However, if the emitter bypassing is not effective, that is, if the signal current flowing through the emitter-to-ground impedance produces a varying voltage across the impedance, then the base-to-emitter voltage is less than the signal voltage by the amount of the voltage from emitter to ground. This is equivalent to reducing the gain of the amplifier stage. Degeneration can be reduced to a negligible value by having a bypass capacitor with sufficiently small reactance at the signal frequency.

Regeneration. This undesirable condition is an increase in gain through unintentional positive feedback. In its worst form, the amplifier becomes an oscillator. Regeneration can be caused by unwanted capacitive feedback coupling between certain stages. If the regeneration is just less than the amount necessary to cause oscillation, the frequency-response characteristic for the amplifier contains a region where the amplification is much greater than that in the mid-band region. This causes excessive phase and amplitude distortion of the signal and the possibility of lightly damped transient oscillations. *See* OSCILLATOR.

DIFFERENTIAL AMPLIFIERS

Amplifiers that use a single active device such as a vacuum tube or transistor cannot usually achieve high dc gain, good bias stability, and low noise simultaneously, because large emitter resistors must be used for high bias stability. These resistors restrict the dc gain but may be bypassed for high ac gain. The differential amplifier uses two active devices in an arrangement that overcomes this problem, as illustrated in the balanced transistor amplifier of Fig. 7. As the base of one is driven positive, the other is driven negative. If the transistors are matched and operate in their linear regions, the emitter-current increase of one transistor is equal to the emitter-current decrease of the other. Therefore, the current through the emitter resistor R_E and the voltage drop across R_E remain constant. Thus, R_E causes no degeneration and needs no bypassing. On the other hand, when signals having the same magnitude and polarity are applied to the two bases, the current change through both transistors must flow through R_E, and the resulting voltage change across R_E opposes the input voltage, causing degeneration. This is the same effect observed with an unbypassed emitter resistor in a conventional amplifier. This equal-polarity signal is known as a common-mode signal, and the voltage amplification of a differential transistor amplifier with a common-mode input signal is approximately R_C/R_E. Voltage or current changes due to temperature changes are common-mode signals. Therefore, best thermal stability is achieved when the common-mode voltage gain is very low. A figure of merit for the differential amplifier is the

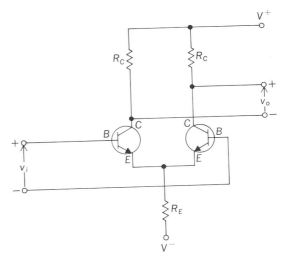

Fig. 7. A balanced differential transistor amplifier.

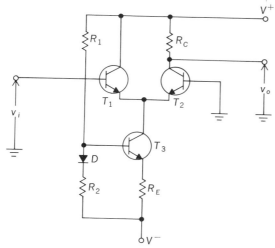

Fig. 8. An unbalanced differential, or emitter-coupled, amplifier.

ratio of differential voltage gain to the common-mode voltage gain. This ratio is known as the common-mode rejection ratio (CMRR). Since the differential gain is the same as that of a single amplifier using the same amplifying device and the same load resistance while the common-mode gain is inversely proportional to the emitter resistor R_E, the CMRR improves as R_E is increased. However, the value of R_E in Fig. 7 should not appreciably exceed the value of R_C in order to maintain suitable dynamic range for the amplifier.

The input signal of the differential amplifier may be single-ended, or referenced to ground, and the output signal may be either differential or single-ended, as shown in Fig. 8. Besides being unbalanced, an additional transistor T_3 is included in the emitter circuit of this differential amplifier. This transistor accurately controls the bias currents in the differential amplifier and therefore acts as a current source, or very high resistance, in the emitter circuit, thus providing a very high CMRR. In this circuit, the voltage across R_E is the same as the voltage across R_1. Therefore, the value of R_E controls the bias value of emitter current in all three transistors. The diode D provides temperature compensation. As the input signal goes positive, the forward bias, and hence the emitter current, of transistor T_1 is increased. This current increase cannot pass through transistor T_3 but opposes the emitter bias current of transistor T_2. Thus, the currents in transistor T_2 decrease by the same amount as the currents in transistor T_1 increase, as they did in the balanced amplifier.

When the desired output from the differential amplifier is unbalanced, as shown in Fig. 8, there is no need to include a load resistor in the collector circuit of both transistors. In fact, the omission of the load resistance from the collector circuit of T_1 increases the bandwidth of the amplifier considerably, because the amplifier then appears as an emitter-follower amplifier coupled to a common-base amplifier and both of these configurations may have much wider bandwidth than the common-emitter configuration. The arrangement shown in Fig. 8 is commonly known as an emitter-coupled amplifier. The bandwidth increase actually results from the reduction of the effective capacitance, known as Miller capacitance, in transistor T_1.

OPERATIONAL AMPLIFIERS

The term "operational amplifier" was coined by people in the analog computer field and was used to designate an amplifier with very high voltage gain, high input impedance, and low output impedance. This type of amplifier is used as the basic element in the performance of the basic mathematical operations such as addition, subtraction, multiplication, integration, differentiation, and logarithms. These operations can be performed with high accuracy, providing the amplifier has a voltage gain of the order of 100,000, an input impedance of the order of 1 megohm or higher, and an output impedance no greater than a few hundred ohms. Thus, the operational amplifier is a special type of cascade amplifier having these characteristics. See ANALOG COMPUTER.

Operational amplifiers, called op amps, are used extensively in the areas of electronic instrumentation and communication in addition to analog computing devices. These amplifiers are available in integrated-circuit, or solid-state-chip, form and come in a variety of bandwidths. The op amp is normally direct-coupled, so it may be used for either dc or ac applications and its gain may be accurately controlled by negative feedback. Thus, the op amp has become a universal amplifying device. The following applications illustrate the operating principles as well as some uses of the op amp. See INTEGRATED CIRCUITS.

Voltage amplifier. A voltage amplifier simply multiplies the input voltage by a constant, the voltage gain of the amplifier, which may be controlled by negative feedback as illustrated in Fig. 9a. The op amp, represented by the triangular symbol, is assumed to have very high voltage amplification a_v so the input voltage v_a to the amplifier is negligibly small, since $v_a = v_o/a_v$. Then, the current i provided by the driving source v_s is essentially given by Eq. (6). Since the amplifier input voltage v_a is very small and the input resistance of the amplifier is very high, the amplifier input current is negligible compared with the source current i. Therefore, essentially all of the current i must flow through the feedback resistor R_f. Then, since v_a is negligible compared with v_o, Eq. (7) follows. Substituting Eq. (6) into Eq. (7) and rearranging, Eq. (8) is obtained. But the voltage gain of the amplifier with feedback is $A_v = v_o/v_s$. Therefore Eq. (9) is valid.

$$i = \frac{v_s}{R_s} \qquad (6)$$

$$v_o = iR_f \qquad (7)$$

$$\frac{v_o}{v_s} = \frac{R_f}{R_s} \qquad (8)$$

$$A_v = \frac{R_f}{R_s} \qquad (9)$$

Thus the voltage amplification is controlled by the ratio of the feedback resistance R_f to the source resistance R_s. The resistance R_s may be the actual internal resistance of the driving source or it may include resistance external to the driving source, such as resistance needed to reduce the loading on the driving source. *See* VOLTAGE AMPLIFIER.

Most operational amplifiers may have their dc input and output voltages referenced to ground when the power supply provides balanced voltages V^+ and V^- with respect to ground, as indicated in Fig. 9a. However, when the op amp uses a transistor differential amplifier as the input stage, the base currents of the two input transistors must flow from ground through the external circuit resistances to both the inverting and noninverting inputs of the amplifier. When the iR drop to the inverting input differs from the iR drop to the noninverting input, a dc offset voltage appears between the input terminals. This voltage is amplified by the gain A_v of the amplifier and appears as a magnified dc voltage in the output. This unwanted voltage may be largely eliminated by placing a

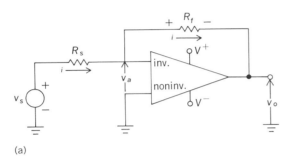

(a)

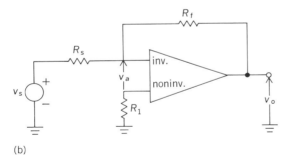

(b)

Fig. 9. An operational amplifier with feedback gain control. (a) With input referenced to ground. (b) With equalized resistances between the inputs and ground to reduce the input offset voltage.

resistor R_1 between the noninverting input and ground, as shown in Fig. 9b. This resistance should have a value equal to the parallel combination of R_s and R_f, assuming the two bias currents are equal. On the other hand, the value of R_1 may be adjusted to compensate for any difference, or offset, in the bias currents.

It can be seen from Fig. 9a that a voltage polarity reversal occurs between the input and output of the amplifier since the input voltage is applied to the inverting input terminal. Thus, the input is multiplied by a negative constant. When the voltage gain of such an amplifier is adjusted to one, or unity, the amplifier becomes a sign changer or simply an inverter.

Summing amplifier. Several different voltage sources may be connected to the input of an op amp, as shown in Fig. 10. Since both the input volt-

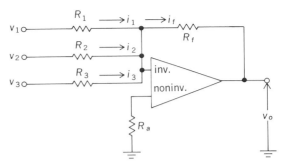

Fig. 10.   A summing amplifier, or summer.

age and the input current of the amplifier are negligible, i_f is given by Eq. (10). Also, $i_1 = v_1/R_1$, $i_2 = v_2/R_2$, $i_3 = v_3/R_3$, and $i_f = v_o/R_f$. Therefore, Eq. (11) follows. Solving Eq. (11) for v_o explicitly, Eq. (12) is obtained. When $R_f = R_1 = R_2 = R_3$, $v_o = v_1 + v_2 + v_3$ and the output voltage is simply the sum of the input voltages. However, as Eq. (12) shows, multiplication may be performed as well as addition when the resistance values are not equal. Again, the negative sum is obtained unless an inverter is used to restore the initial polarities. On the other hand, an inverter may be placed in an input lead to perform subtraction on that input.

$$i_1 + i_2 + i_3 = i_f \qquad (10)$$

$$\frac{v_1}{R_1} + \frac{v_2}{R_2} + \frac{v_3}{R_3} = \frac{v_o}{R_f} \qquad (11)$$

$$v_o = \frac{R_f}{R_1}v_1 + \frac{R_f}{R_2}v_2 + \frac{R_f}{R_3}v_3 \qquad (12)$$

Integrator. When the feedback resistor R_f is replaced by a capacitor C, as shown in Fig. 11, the op amp circuit becomes an integrator. Again $i_i = v_i/R$ and $i_i = i_f$. However, the current through a capacitor is proportional to the rate of change of voltage across it, as shown in Eq. (13). By solving this equation explicitly for v_o, Eq. (14) is obtained,

$$i_f = C\frac{dv_o}{dt} \qquad (13)$$

$$v_o = \frac{1}{C}\int i_f\,dt + k \qquad (14)$$

where k is the constant of integration or initial output voltage. When this initial voltage is assumed to be zero and the relationship $i_f = i_1 = v_i/R$

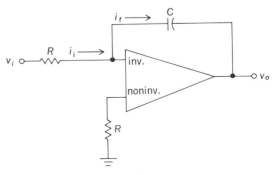

Fig. 11.   An integrating amplifier, or integrator.

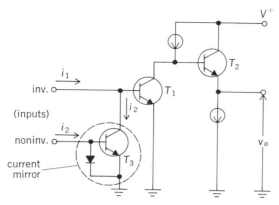

Fig. 12.   A simplified Norton amplifier circuit.

is substituted into Eq. (14), this equation becomes Eq. (15). Thus, the circuit is an integrator having a multiplying factor $1/RC$ which may be made equal to unity if desired.

$$v_o = \frac{1}{RC} \int v_i \, dt \qquad (15)$$

The positions of R and C could be exchanged in the circuit of Fig. 11 to produce a differentiating amplifier, or differentiator. However, the differentiator is seldom used because of its inherent high noise level.

Norton amplifier. The Norton amplifier is a type of operational amplifier that yields some cost benefits because of its simplicity. A simplified circuit diagram of this amplifier is given in Fig. 12. When i_2 is equal to zero, T_3 is nonconducting and transistor T_1 is a high-gain inverting amplifier with an emitter follower T_2 as its load. The current sources in the collector circuit of T_1 and the emitter circuit of T_2 provide unusually large load resistance for transistor T_1 and, therefore, this transistor may provide a voltage gain of several thousand. A general-purpose op amp needs a noninverting input as well as an inverting input so the current mirror, consisting of transistor T_3 and diode D, is added to the circuit to provide the additional input. The diode in combination with transistor T_3 provides a current gain very nearly equal to one. Thus, a current i_2 flowing into the noninverting input causes the same amount of current to flow into the collector of T_3. This current must be supplied by i_1, so the current actually flowing into the base of the high-gain transistor T_2 is $i_1 - i_2$. Since this base current is very small, typi-

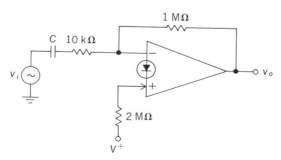

Fig. 13.   A Norton amplifier used as an ac amplifier.

cally a few nanoamperes, i_2 is approximately equal to i_1 under quiescent conditions. Both inputs are about 0.6 V above ground and the quiescent output voltage should be about $V+/2$. This amplifier is known as a Norton amplifier because it is driven by a differential current instead of a differential voltage and the concept of the current source was developed initially by E. L. Norton. The input currents are transformed to input voltages by placing fairly large resistances in series with the inputs.

The use of a Norton amplifier as a simple ac amplifier is illustrated in Fig. 13. Diode symbols have been added to the traditional triangular symbol to identify the Norton amplifier. The voltage gain of the amplifier is $1\mathrm{M}\Omega/10\mathrm{k}\Omega = 100$, the same as for a conventional op amp, because essentially all of the input signal current that flows through the 10-kΩ resistor also flows through the 1-MΩ feedback resistor. The amplifier is biased so $V_o = V+/2$, because the bias currents of the two inputs are approximately equal so $V_o/1\ \mathrm{M}\Omega \simeq V+/2$ MΩ so $V_o = V+/2$ for maximum dynamic range. The dc component in the output may be blocked from the actual load by a capacitor, if desired. The $V+$ bias source must be well filtered or the ripple voltage will introduce hum into the amplifier output. Hum is the word commonly used to describe the sound from a loudspeaker or headset that results from the power supply ripple.

The Norton amplifier is very inexpensive. Because of its simplicity, four amplifiers are available on a single low-cost chip. Also, the single-ended power supply is less expensive than the dual supply recommended for conventional op amps. Compared with a conventional op amp, however, the Norton amplifier has lower open-loop gain, lower input resistance, and higher output resistance. It also has a higher noise level because of the large values of resistance typically used in the input circuits.

[CHRISTOS C. HALKIAS; CHARLES L. ALLEY]

Bibliography: C. L. Alley and K. W. Atwood, *Electronic Engineering*, 3d ed., 1973; D. J. Comer, *Modern Electronic Circuit Design*, 1976; C. A. Holt, *Electronic Circuits: Digital and Analog*, 1978; D. L. Schilling and C. Belove, *Electronic Circuits: Discrete and Integrated*, 2d ed., 1979.

Amplitude (wave motion)

The maximum magnitude (value without regard to sign) of the disturbance of a wave. The term "disturbance" refers to that property of a wave which perturbs or alters its surroundings. It may mean, for example, the displacement of mechanical waves, the pressure variations of a sound wave, or the electric or magnetic field of light waves. Sometimes in older texts the word "amplitude" is used for the disturbance itself; in that case, amplitude as meant here is called peak amplitude. This is no longer common usage.

As an example, consider one-dimensional traveling waves in a linear, homogeneous medium. The wave disturbance y is a function of both a space coordinate x and time t. Frequently the disturbance may be expressed as $y(x,t) = f(x \pm vt)$, where v denotes the wave velocity. The plus or minus sign indicates the direction in which the wave moves, and the shape of the wave dictates the functional form symbolized by f. Then, the amplitude of the disturbance at some point x_0 is the maximum magnitude (that is, the maximum absolute value) achieved by f as time changes over the duration required for the wave to pass point x_0. A special case of this is the one-dimensional, simple harmonic wave $y(x,t) = A \sin [k(x \pm vt)]$, where k is a constant. The amplitude is A since the absolute maximum of the sine function is $+1$. The amplitude for such a wave is a constant. *See* HARMONIC MOTION; SINE WAVE.

If the medium which a wave disturbs dissipates the wave by some nonlinear behavior or other means, then the amplitude will, in general, depend upon position.   [S. A. WILLIAMS]

Amplitude modulation

The process or result of the process whereby the amplitude of a carrier wave is changed in accordance with a modulating wave. This broad definition includes applications using sinusoidal carriers, pulse carriers, or any other form of carrier, the amplitude factor of which changes in accordance with the modulating wave in any unique manner. *See* MODULATION.

Practical examples of amplitude modulation (AM) include AM radio broadcasting, single-sideband transmission systems, vestigial-sideband systems, frequency-division multiplexing, time-division multiplexing, phase-discrimination multiplexing, and reduced-carrier systems.

Amplitude modulation is also defined in a more restrictive sense to mean modulation in which the amplitude factor of a sine-wave carrier is linearly proportional to the modulating wave. AM radio broadcasting is a familiar example. At the radio transmitter the modulating wave is the audio-frequency program signal to be communicated; the modulated wave that is broadcast is a radio-frequency, amplitude-modulated sinusoid.

In AM the modulated wave is composed of the transmitted carrier, which conveys no information, plus the upper and lower sidebands, which (assuming the carrier frequency exceeds twice the top audio frequency) convey identical and therefore mutually redundant information. J. R. Carson in 1915 was the first to recognize that, under these conditions and assuming adequate knowledge of the carrier, either sideband alone would uniquely define the message. Apart from a scale factor, the spectrum of the upper sideband and lower sideband is the spectrum of the modulating wave displaced, respectively, without and with inversion by an amount equal to the carrier frequency.

For example, suppose the audio-frequency signal is a single-frequency tone, such as 1000 Hz, and the carrier frequency is 1,000,000 Hz; then the lower and upper sidebands will be a pair of single-frequency waves. The lower-sideband frequency will be 999,000 Hz, corresponding to the difference between the carrier and audio-signal frequencies; and the upper-sideband frequency will be 1,001,000 Hz, corresponding to the sum of the carrier and audio-signal frequencies. The amplitude of the signal appears in the amplitude of the sidebands. In practice, the modulating waveform will be more complex. A typical program signal might occupy a frequency band range of perhaps 100 Hz to 5000 Hz.

An important characteristic of AM, as illustrat-

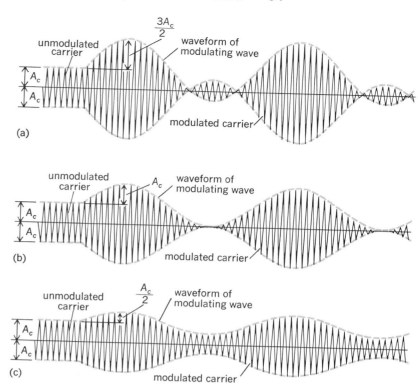

Fig. 1.   Amplitude modulation of a sine-wave carrier by a sine-wave signal, with (a) 50% overmodulation, (b) 100% modulation, and (c) 50% modulation. (*From H. S. Black, Modulation Theory, Van Nostrand, 1953*)

ed by Fig. 1, is that, apart from a scale factor and constant term, either the upper or lower envelope of the modulated wave is an exact replica of the modulating wave, provided two conditions are satisfied: first, that the carrier frequency exceeds twice the highest speech frequency to be transmitted; and second, that the carrier is not overmodulated.

Single-sideband modulation (SSB). This is modulation whereby the spectrum of the modulating wave (message to be transmitted) is shifted in frequency, either without or with inversion, by an amount equal to the carrier frequency. One of several ways to produce SSB is to suppress the carrier and one of the sidebands of an amplitude-modulated sinusoid. At the receiving end, in order to be able to recover the exact waveform of the original modulating wave, it is necessary to know the precise frequency and phase of the carrier. By saving bandwidth occupancy and signal power, SSB achieves two important objectives of any communication system.

SSB implies that the two sidebands, one wanted and the other unwanted, of an amplitude-modulated sinusoid can be separated. Actually, as the lowest significant frequency f_1 of the modulating wave is reduced, this becomes increasingly difficult. Clearly, the sidebands cannot be separated unless f_1 exceeds zero. This inability to handle relatively low frequencies constitutes one of the important disadvantages of SSB.

Another less obvious disadvantage is the delay inherent in the production of a single sideband. This unavoidable delay must exceed $1/8f_1$ and is a necessary consequence of virtually suppressing the unwanted sideband. In most practical applications not only will this unavoidable delay be 10 or more times longer than $1/8f_1$, but a similarly long delay is also encountered at the receiver. For some purposes delays of such magnitude would be objectionable.

Vestigial-sideband modulation (VSB). This is modulation whereby in effect the modulated wave to be transmitted is composed of one sideband plus a portion of the other adjoining the carrier. The carrier may or may not be transmitted. VSB is like SSB except in a restricted region around the carrier. The overall frequency response to the wanted sideband and to the vestigial sideband is so proportioned by networks that upon demodulation, preferably but not necessarily by a product demodulator, the original modulating wave will be recovered with adequate accuracy.

By thus transmitting a linearly distorted copy of both sidebands in a restricted frequency region above and below the carrier, the original modulating wave can now be permitted to contain significant components at extremely low frequencies, even approaching zero in the limit. By this means, at the cost of a modest increase in bandwidth occupancy, network requirements are greatly simplified. Furthermore, the low-frequency limitation and the inherent delay associated with SSB are avoided.

In standard television broadcasting in the continental United States the carrier is transmitted, envelope detection is used, and the vestigial sideband possesses a bandwidth one-sixth that of a full sideband.

Uses of AM in multiplexing. Multiplexing is the process of transmitting a number of independent messages over a common medium simultaneously. To multiplex, it is necessary to modulate. Two or more communication channels sharing a common propagation path may be multiplexed by arranging them along the frequency scale as in frequency division, by arranging them in time sequence as in time division, or by a process known as phase discrimination, in which there need be no separation of channels in either frequency or time.

Frequency-division multiplexing. When communication channels are multiplexed by frequency division, a different frequency band is allocated to each channel. Single-sideband carrier telephone systems are a good example. At the sending end, the spectrum of each channel input is translated by SSB to a different frequency band. For example, speech signals occupying a band of 300–3300 Hz might be translated to occupy a band range of 12,300–15,300 Hz corresponding to the upper sideband of a sinusoidal carrier, the frequency of which is 12,000 Hz. Another message channel might be transmitted as the upper sideband of a different carrier, the frequency of which might be 16,000 Hz.

At the receiving end, individual channels are separated by electric networks called filters, and each original message is recovered by demodulation. The modulated wave produced by SSB at the sending end becomes the modulating wave applied to the receiving demodulator.

When communication channels are multiplexed by frequency division, all channels may be busy simultaneously and continuously, but each uses only its allocated fraction of the total available frequency range.

Time-division multiplexing. When communication channels are multiplexed by time division, a number of messages is propagated over a common transmitting medium by allocating different time intervals in sequence for the transmission of each message. Figure 2 depicts a particularly simple example of a two-channel, time-division system. Ordinarily, the number of channels to be multiplexed would be considerably greater. Transmitting and receiving switches must be synchronized; time is of the essence in this system, and the problem is to switch at the right time.

On the theoretical side there is a certain basic, fundamental question which must always be answered about any time-division system. The question is: At what rate must each communication channel be connected to its common transmitting path? Today it is known from the sampling principle that for successful communication each channel must be momentarily connected to the common path at a rate that is in excess of twice the highest message frequency conveyed by that channel. *See* PULSE MODULATION.

Viewed broadly, amplitude modulation of pulse carriers generates the desired amplitude and phase relationships essential to time-division multiplexing. In addition, whereas each communication channel may use the entire available frequency band for transmitting its message, it may transmit only during its allocated fraction of the total time.

Phase-discrimination multiplexing. This type of multiplexing, like SSB, saves bandwidth, may save signal power, and, like AM and VSB, has the important advantage of freely transmitting extremely low modulating frequencies. Furthermore, each communication channel may utilize all of the available frequency range all of the time.

When n channels are multiplexed by phase discrimination, the modulating wave associated with each channel simultaneously amplitude-modulates $n/2$ carriers, with a different set of carrier phases provided for each channel. All sidebands are transmitted; $n/2$ carriers may or may not be transmitted. At the receiving end, with the aid of locally supplied carriers and an appropriate demodulation process, the n channels can be separated, assuming distortionless transmission, ideal modulators, and so on. Systems with odd numbers of channels can also be devised. Equality of sidebands and their exact phases account for the suppression of interchannel interference.

Day's system. This is a simple example of phase-discrimination multiplexing. Two sine-wave carriers of the same frequency but differing in phase by 90° are amplitude-modulated, each by a different message wave. The spectrum of each modulated sinusoid occupies the same frequency band. These modulated sinusoids are then added and propagated without distortion to a pair of demodulators. Quadrature carriers of correct frequency and phase are applied locally, one to each demodulator. Theoretically, a faithful copy of each message can be recovered.

Within the continental United States, for pur-

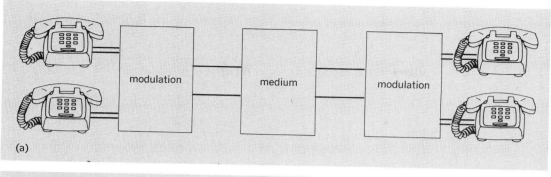

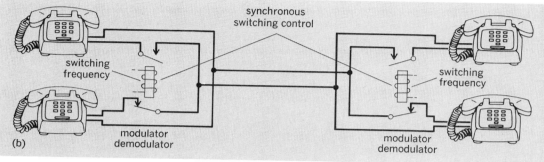

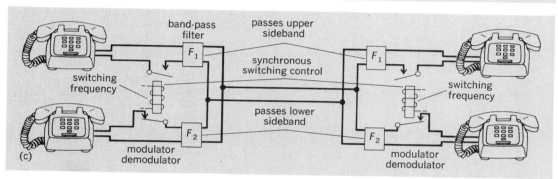

Fig. 2.   Two-channel, time-division carrier system and two-channel, frequency-division carrier system. (*a*) General diagram of two-channel carrier system. (*b*) Amplitude modulation, time division. (*c*) Single-sideband modulation, frequency division. (*From H. S. Black, Modulation Theory, Van Nostrand, 1953*)

poses of saving bandwidth, Day's system is used for multiplexing the two so-called color components associated with color tv broadcasting.

Modulator and demodulator. Many methods of modulating and demodulating are possible, and many kinds of modulators and demodulators are available for each method. *See* MODULATOR.

Fundamentally, since the sidebands of an amplitude-modulated sinusoid are generated by a multiplication of wave components which produces frequency components corresponding to the products, it is natural to envisage a product modulator having an output proportional to the product of two inputs: modulating wave and carrier. An ideal product modulator suppresses both modulating wave and carrier, transmitting only upper and lower sidebands. *See* AMPLITUDE MODULATOR.

At the receiving end, the original message may be recovered from either sideband or from both sidebands by a repetition of the original modulating process using a product modulator, commonly referred to as a product demodulator, followed by

a low-pass filter. Perfect recovery requires a locally applied demodulator carrier of the correct frequency and phase. For example, a reduced carrier system creates its correct demodulator carrier supply by transmitting only enough carrier to control the frequency and phase of a strong, locally generated carrier at the receiver. *See* AMPLITUDE-MODULATION DETECTOR.

SSB systems commonly generate their modulator and demodulator carriers locally with independent oscillators. For the high-quality reproduction of music, the permissible frequency difference between modulator and demodulator carriers associated with a particular channel is limited to about 1–2 Hz. For monaural telephony, frequency differences approaching 10 Hz are permissible.

Unbalanced square-law demodulators, rectifier-type demodulators, and envelope detectors are often used to demodulate AM. However, even though overmodulation (Fig. 1) is avoided, significant distortion may be introduced. In general, the amount of distortion introduced in this manner will depend upon the kind of demodulator or detector used, the amount of noise and distortion introduced prior to reception, and the percentage modulation. [HAROLD S. BLACK]

Bibliography: J. Betts, *Signal Processing, Modulation and Noise*, 1971; P. F. Panter, *Modulation, Noise and Spectral Analysis*, 1965; M. Schwartz, *Information, Transmission, Modulation and Noise*, 2d ed., 1970.

Amplitude-modulation detector

A device for the recovery of the modulating signal from an amplitude-modulated carrier. A detector, like a modulator, must employ a nonlinear device and therefore usually includes either a diode or a nonlinear amplifier. Detectors are sometimes known as demodulators. Several types of detector circuits are discussed in this article. *See* MODULATION.

Diode detectors. A simple diode detector circuit (Fig. 1) was frequently used as an entire receiver in the early days of radio; it was known as a crystal set. Its radio-frequency (rf) input signal, which was

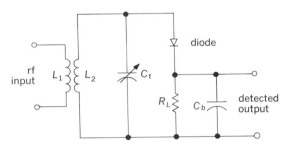

Fig. 1. Typical diode detector circuit.

obtained by attaching an antenna and a ground to the input terminals, induced a voltage in a secondary coil L_2 by magnetic induction. Capacitor C_t was adjusted to resonate with inductance L_2 at the desired input frequency. Because of the high resonant impedance of the tuned circuit, a voltage was induced in a narrow band of frequencies about the resonant frequency to a high enough level for satisfactory operation of the diode. The desired radio station was thus selected. *See* RADIO.

Modern diode detectors operate similarly. The average value of the modulated input voltage is zero because the negative half cycles have the same amplitude as the positive half cycles. Only high, or radio, frequencies, consisting of the carrier plus sideband frequencies, are present in this input signal; the original modulating frequency appears as the modulation envelope. The diode or rectifier recovers this envelope by allowing current to flow primarily in only one direction, thereby eliminating either the positive or negative half cycles (Fig. 2). Average current through the diode is not zero, but has a component which varies with the amplitude of the input signal and thus contains the original modulating frequencies. Theoretically, nonlinearity produces sum and difference frequencies, as well as harmonics, of the input frequencies. The difference between the carrier frequency and a sideband frequency is the modulating frequency that produced the sideband frequency during modulation. Thus amplitude modulation and detection are basically the same process. *See* RECTIFIER.

The diode current contains the carrier frequency and harmonics as well as the original modulating frequencies and a dc component produced by the rectification.

Conditions for high efficiency. Load resistor R_L is placed in series with the diode to produce an output voltage. Without capacitor C_b the voltage across the load resistor would have the same wave form as the current through the diode. However, capacitor C_b greatly reduces the magnitude of the high-frequency components and increases the desired modulating frequency components if the circuit components have the proper relationships. High-detection efficiency can be obtained if the following conditions are met:

1. Load resistance should be large in comparison with forward resistance of the diode so the peak voltage across the load will be almost equal to the peak voltage applied from the secondary of the coupling coil.

2. Reactance of bypass capacitor C_b should be small compared with the load resistance at the carrier frequency, but large compared with the load resistance at the highest modulating frequency.

3. The diode is fast enough so its charge storage time is short compared with the period of the input carrier frequency.

Under these conditions bypass capacitor C_b will charge quickly through the small forward resistance of the diode, and the capacitor voltage will closely follow the input voltage while the diode is conducting. However, the capacitor cannot discharge through the diode because of its unilateral characteristics and therefore must discharge through the load resistor. Because under the above conditions time constant $R_L C_b$ is long compared with the period of the carrier, the capacitor voltage almost follows the peaks of the input voltage (Fig. 2b). In most applications the ratios of carrier frequency to modulator frequency are much higher than that illustrated in Fig. 2. Therefore, the high-frequency sawtoothlike variations in output voltage are much smaller than those shown in Fig. 2b.

Typical values of load resistance and bypass capacitance for broadcast-frequency radio receivers are $R_L = 2000-10,000$ ohms for semiconductor diodes, and $0.1-0.5$ megohm for vacuum diodes; $C_b = .002-.01$ microfarad for semiconductor di-

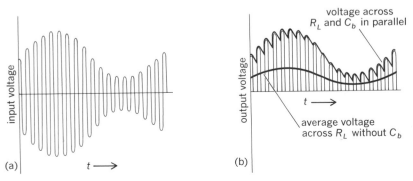

Fig. 2. Input and output waveforms of a detector. (a) Modulated rf input. (b) Detected output, with modulation envelope recovered by rectification and filtering.

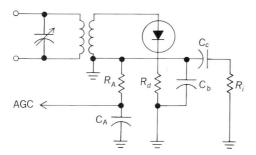

Fig. 3. Diode detector circuit with features usually found in a radio receiver.

odes, and 100–500 picofarads for vacuum diodes.

Special features. A diode detector in a modern radio receiver usually includes additional features and refinements (Fig. 3). The tuning capacitor usually appears on the primary side of the coupling coil in transistor-type receivers. The dc component of voltage across diode load resistor R_d is blocked by the ac coupling (or dc blocking) capacitor C_c, because this dc component would otherwise upset the dc operating potentials of the amplifier that normally follows the detector. The input resistance of this amplifier is represented by resistor R_i. The reactance of coupling capacitor C_c should be small in comparison with R_i at the lowest desired output frequency in order to avoid reduction of the output voltage.

The dc component of voltage across R_d is often used to automatically control the gain of the amplifiers that precede the detector in a radio receiver, thus making the output voltage relatively independent of the strength of the input signal to the receiver. This automatic gain control (AGC) voltage, which is proportional to the carrier voltage into the detector, must have proper polarity to reduce the gain of the controlled amplifiers. Because this polarity can be reversed by reversing the direction of the diode, the proper polarity is easily obtained. The modulating frequencies as well as the carrier frequencies are filtered from the AGC voltage by capacitor C_A and resistor R_A combination. *See* AUTOMATIC GAIN CONTROL (AGC).

Requirements for linearity. The diode detector just described is linear, the output voltage being linearly related to the waveform of the modulation envelope. This linearity is based on the requirement that the load resistance be large in comparison with the average forward resistance of the diode. Another requirement for linearity is that the ac load resistance be almost as large as the dc load resistance. In Fig. 3 the dc load resistance is R_d only if no direct current flows through AGC resistor R_A. However, ac or signal currents flow through R_A and R_i as well as R_d, so these three resistors in parallel constitute the ac load resistance R_L for the diode. Of course the high-frequency bypass C_b is in parallel with R_L and lowers the ac load impedance Z_L at high modulating frequencies. Therefore, the ac load impedance can be nearly as large as the dc load resistance only if R_i and R_A are both large in comparison with R_d, and if the reactance of bypass capacitor C_b is greater than R_d at the highest modulating frequency component of interest.

When the ac load resistance is much lower than the dc load resistance, distortion results (Fig. 4). This phenomenon, called negative clipping, is caused by the fact that the diode is kept from conducting for a number of carrier cycles. The load voltage then cannot follow the modulating voltage. Let I_D be the average diode current, which is also the average current through dc load resistance R_d. Bypass capacitor C_b (Fig. 3) causes voltage V_L across the load to follow the peaks of the carrier input, neglecting the forward voltage drop across the diode. The average, or dc, voltage V_L across the load is shown in Eq. (1). This voltage is also

$$V_L = I_D R_L \qquad (1)$$

equal to peak carrier voltage V_c, by using the approximations given above.

Assume that the carrier is sinusoidally modulated with a peak voltage variation V_m (Fig. 4a). Then this varying component is coupled through the coupling capacitors and appears across the total ac load resistance r_{ac} to produce a peak current variation as shown in Eq. (2).

$$I_p = V_m/r_{ac} \quad \text{or} \quad V_m = I_p r_{ac} \qquad (2)$$

But the largest possible decrease in diode current from the average value is the average current I_D. Then the maximum permissible voltage variation is as shown in Eq. (3). If the peak variation

$$V_{m(\max)} = I_D r_{ac} \qquad (3)$$

given by Eq. (2) exceeds the value given by Eq. (3), diode current remains zero for part of the modulation cycle, and the voltage across the load cannot follow the modulation voltage.

This is the negative clipping phenomenon, mentioned above. It is shown by Fig. 4b, in which average load voltage V_L and average diode current I_D are given by the solid line with slope $1/R_d$, known as the dc load line. Similarly, the relationship between the varying, or modulation, voltage V_m and the change in load current is given by the dashed line, with slope $1/r_{ac}$, known as the ac load line. Clipping begins where the ac load line intersects the $I_D = 0$ axis. *See* CLIPPING CIRCUIT.

A simple relationship exists between the maximum modulation index that will not produce clipping and the ratio of ac load resistance to dc load resistance. Because the modulation index is defined as $M = V_m/V_c$, Eq. (2) can be divided by Eq. (3) to obtain Eq. (4). Thus, if the modulation index

$$M_{(\max)} = \frac{V_{m(\max)}}{V_c} = \frac{I_D r_{ac}}{I_D R_d} = \frac{r_{ac}}{R_d} \qquad (4)$$

is 90%, distortionless detection or demodulation can be obtained only if $r_{ac} \geq 0.9 R_d$.

Square-law detectors. Another important detector is the square-law detector. This device has an output current proportional to the square of the input voltage (Fig. 5). Almost all diodes and electronic amplifying devices have this type of characteristic, basically, and linear operation is obtained only by careful selection of the combination of operating point, signal level, driving source resistance, and load resistance. For example, the diode provides linear detection only when the load resistance is large in comparison with the average

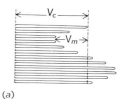

(a)

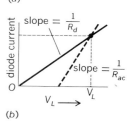

(b)

Fig. 4. Distortion. (a) Sinusoidal modulation of the carrier. (b) Graphical construction showing that distortion results when ac load resistance is much smaller than dc load resistance.

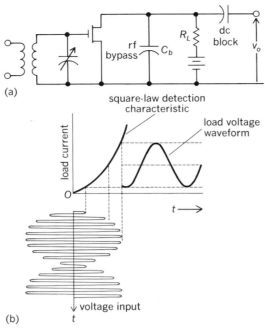

(a)

(b)

Fig. 5. Square-law detector and its operation. (a) Typical circuit. (b) Characteristics and operation.

forward resistance of the diode, as previously mentioned. On the other hand, the diode is a square-law detector if the load resistance is small in comparison with diode resistance. Amplification can be obtained in addition to square-law detection if an amplifying device, such as a transistor or vacuum tube, is used with bias such that the quiescent operating point is near zero current. This condition is obtained with zero bias in an enhancement-mode insulated-gate transistor circuit (Fig. 5).

The square-law detector distorts or changes the waveform of the modulation envelope, as shown in Fig. 5b. This distortion is sometimes desirable, as in single-sideband transmission, where one sideband has been eliminated. Single-frequency modulation then produces a nonsinusoidal modulation envelope, but the square-law detector produces a sinusoidal or single-frequency output. In other words, as can be shown mathematically, a square-law detector is necessary for distortionless single-sideband transmission. Square-law detectors are sometimes used for convenience in other applications where preservation of the modulation envelope waveform is not necessary or desirable. *See* DISTORTION (ELECTRONIC CIRCUITS).

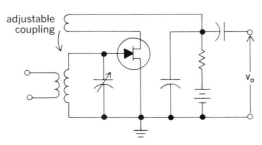

Fig. 6. Elementary regenerative detector.

Regenerative detectors. A high-gain narrow-band radio receiver can be achieved with a single amplifying detector by providing positive or regenerative feedback from the output to the input (Fig. 6). This arrangement is known as a regenerative detector. The feedback adjustment is critical and distortion is high because the feedback must maintain operation on the verge of oscillation.

The feedback adjustment can be made less critical and the gain increased if a signal from an oscillator is injected into the detector to periodically squelch its oscillations, which tend to build up slowly. This type of detector is known as a super regenerative detector. The oscillator frequently must be above the audible range, but much lower than the carrier frequency. *See* DETECTOR; OSCILLATOR.

Synchronous detectors. A four-quadrant multiplier, available as an integrated circuit, may be used as a demodulator or detector, as shown by the trigonometric identity in Eq. (5), where ω_c is

$$A \cos \omega_c t \left[\cos(\omega_c + \omega_m)t + \cos(\omega_c - \omega_m)t \right]$$
$$= A \left[\cos \omega_m t + \tfrac{1}{2} \cos(2\omega_c + \omega_m)t \right.$$
$$\left. + \tfrac{1}{2} \cos(2\omega_c - \omega_m)t \right] \quad (5)$$

the angular carrier frequency and ω_m is the angular modulating frequency. A filter in the output of the multiplier will eliminate the high-frequency terms on the right-hand side of the equation and leave only the modulating-frequency term $\cos \omega_m t$. This type of detector is known as a synchronous detector because the original carrier, without modulation, must be reinserted into the detector. In the multiplication above, the carrier multiplied only the upper and lower sidebands with no carrier, so suppressed carrier transmission was assumed. The multiplier performs equally well when the carrier is not suppressed. Then there is a carrier frequency term at each of the multiplier input terminals, but their product is a double carrier frequency term $2\omega_c$ and a dc, or constant, term, both of which are removed by the output filter. The unmodulated carrier frequency ω_c is usually generated in the receiver by either a local oscillator or a phase-lock loop. When double sideband transmission is used, these devices must be synchronized with the original carrier. *See* AMPLITUDE MODULATION; AMPLITUDE MODULATOR; PHASE-LOCKED LOOPS.   [CHARLES L. ALLEY]

Bibliography: C. Alley and K. Atwood, *Electronic Engineering*, 3d ed., 1973; D. J. Comer, *Modern Electronic Circuit Design*, 1976.

Amplitude-modulation radio

Radio communication employing amplitude modulation of a radio-frequency carrier wave as the means of conveying the desired intelligence. In amplitude modulation the amplitude of the carrier wave is made to vary corresponding to the fluctuations of a sound wave, television image, or other information to be conveyed. *See* AMPLITUDE MODULATION; RADIO.

Amplitude modulation (AM), the oldest and simplest form of modulation, is widely used for radio services. The most familiar of these is broadcasting; others include radiotelephony and radiotelegraphy, television picture transmission, and navigational aids. The essentials of these radio systems are discussed in this article.

Low frequency (long wave). European and Asian countries use frequencies in the range 150–255 kilohertz (kHz) for some broadcast services. An advantage of these frequencies is stable and relatively low-attenuation wave propagation. When not limited by atmospheric noise, large areas may be served by one station. In the United States these frequencies are reserved for navigational systems and so are not available for broadcasting.

Low-frequency (lf) broadcast antennas are omnidirective and radiate vertically polarized waves. Unless special means are used to reduce antenna selectivity, the highest program frequencies transmitted are substantially below 10,000 hertz (Hz).

Medium frequency. The frequencies in the range from 535 to 1605 kHz are reserved all over the world for AM (standard) broadcasting. In the Western Hemisphere this band is divided into channels at 10-kHz intervals, certain channels being designated as clear, regional, and local, according to the licensed coverage and class of service. The local channels are occupied by stations, usually of 250-watt output, servicing smaller localities. Many stations occupy a channel, but they are separated far enough to permit interference-free coverage in the local area. Fewer stations of higher power, but with greater distances between them, share the regional channels. A few clear channels are occupied by high-power stations (50,000-watt maximum output in the United States). These stations may have exclusive use of a channel, or may share it with another distant station.

Interference between co-channel regional stations and clear-channel stations is minimized by use of directive antennas, which suppress radiation toward other stations and direct it to main populated areas from a properly located station.

European medium-frequency (mf) broadcasting channels are assigned at 9-kHz intervals rather than the 10-kHz intervals used in the Western Hemisphere. This reduced spacing provides more channels within the mf band. The technique of directive antennas, which also provides more channels within a band by minimizing interference between stations, has not been used extensively in Europe.

Vertically polarized radiaton is used at medium and low frequencies propagated over the Earth's surface. There is also propagation of high-angle radiation via reflection from the ionosphere, a phenomenon that predominates at night but is relatively absent during daylight. This sky-wave propagation accounts for the familiar long-distance reception at night. At distances where the downcoming sky waves overlap the ground wave, fading and distortion of the signal occurs. Receivers in the ground-wave zone get stable signals day and night. In the overlap zone, daylight reception may be stable but night reception disturbed by fading. In the sky-wave zone at night, assuming no interference from other stations, satisfactory service may be received over long distances. Reception in this zone depends on atmospheric noise, which varies seasonally, and the state of the ionosphere, which varies greatly from night to night depending upon astronomical conditions that affect the upper atmosphere.

Individual AM broadcast stations transmit program frequencies ranging from 30 to 10,000 Hz with excellent fidelity. To obtain suitable tuning selectivity between channels, AM broadcast receivers may reproduce program frequencies only up to 4000 Hz, or even less, according to make, cost, and condition of the receiver.

High frequency (shortwave). Small bands of frequencies between 3000 and 5000 kHz are used in tropical areas of high atmospheric noise for regional broadcasting. This takes advantage of the lower atmospheric noise at these frequencies and permits service under conditions where medium frequencies have only severely limited coverage. Wave propagation day and night is by sky wave. Ground-wave coverage from such stations is usually negligible, since high-angle horizontally polarized radiation is used. Short-distance coverage by this mode is by ionospheric reflection of waves radiated almost vertically.

Long-distance international broadcasting uses high-power transmitters and directive (beam) antennas operating in the bands 5950–6200 kHz, 9500–9775 kHz, 11,700–11,975 kHz, 15,100–15,450 kHz, 17,700–17,900 kHz, 21,450–21,750 kHz, and 25,600–26,100 kHz. These bands are allocated throughout the world for this purpose, and the band used for any path depends upon ionospheric conditions which vary with direction, distance, hour, season, and the phase of the 11-year sunspot cycle. Typically, waves are propagated in low-angle beams by multiple reflections between ionosphere and Earth to cover transoceanic distances, and signals are often distorted during transmission. These bands are so crowded that a signal is seldom received without interference for any prolonged period. Reception from particular stations can be improved by the use of special directive receiving antennas.

The technical performance of high-frequency (hf) broadcast transmission systems is usually to the same standards employed for mf broadcasting, although propagation and interference conditions seldom make this evident to a listener.

AM telephony and telegraphy. The first radiotelephony was by means of amplitude modulation, and its use has continued with increasing importance. Radiotelephony refers to two-way voice communication. Amplitude modulation and a modified form called single-sideband are used almost exclusively for radiotelephony on frequencies below 30 megahertz (MHz). Above 30 MHz, frequency or phase modulation is used almost exclusively, a notable exception being 118–132 MHz, where amplitude modulation is used for all two-way vhf radiotelephony in aviation operations.

The least expensive method known for communicating by telephony over distances longer than a few tens of miles is by using the high frequencies of 3–30 MHz. Furthermore, since radio is the only way to communicate with ships and aircraft, hf AM radiotelephony has remained essential to these operations, except for short distances that can be covered from land stations using the very high frequencies. Therefore, hf radiotelephony has become established for a great variety of pioneering and exploration enterprises and private, public, and government services needing telephone communication, fixed or mobile, over substantial distances where there are no other ways to tele-phone. Because AM techniques are simple and inexpensive, this form of modulation has predomi-nated. The economic development of distant hin-terland areas depends greatly on hf AM radio te-lephony to the outside world.

Widespread use has led to serious crowding of the hf band. All governments are suspending the use of the hf band wherever it is technically and economically feasible to employ frequencies above 30 MHz using either direct transmission or radio repeater stations. The trend to single-sideband modulation also alleviates the pressure of conges-tion in the hf band.

Many of the radiotelephone systems in use oper-ate two ways on one frequency in simplex fashion, that is, all stations on the frequency are normally in a receiving status. Transmission, by press-to-talk (manual) or voice-operated carrier (automatic) switching from reception to transmission, occurs only while the sender is talking. Many stations can thus occupy one frequency provided the volume of traffic by each of the stations is small. Two-way telephony must be strictly sequential. This system is not adapted for connection to a normal two-wire telephone.

Full duplex radiotelephony, for interconnection with wire telephone systems, is essential for most public correspondence. This requires two frequen-cies, each available full time in one direction. Even so, typical fading of signals during propagation requires that voice-operated antiregeneration de-vices be used to maintain circuit stability. Talking must be sequential between speakers as there can be no interrupting, but the system will intercon-nect with conventional business or home tele-phones.

Amplitude-modulated telegraphy consists of interrupting a carrier wave in accordance with the Morse dot-dash code or codes used for the printing telegraph. Much of the radiotelegraph traffic of the world uses AM telegraphy, although there has been extensive conversion to frequency-shift (frequency-modulation) telegraphy since 1944, the latter being better adapted to automatic teleprint-ing operations. Radiotelegraph operations have been refined, speeded, and mechanized, but, un-der adverse noise and fading conditions, AM man-ual telegraphy between experienced operators is still more reliable. Most aviation and marine radi-otelegraphy uses AM manual methods.

Single-sideband (SSB) hf telephony. This is a modified form of amplitude modulation in which only one of the modulation sidebands is trans-mitted. In some systems the carrier is transmitted at a low level to act as a pilot frequency for the regeneration of a replacement carrier at the re-ceiver. Where other means are available for this purpose, the carrier may be completely sup-pressed. Where intercommunication between AM and SSB systems is desired, the full carrier may be transmitted with one sideband.

Since 1933 most transoceanic and interconti-nental telephony has been by single-sideband reduced-carrier radio transmission on frequencies between 4000 and 27,000 kHz. Since 1954 the use of SSB has expanded rapidly in replacing common AM telephony for military and many nonpublic

radio services. In time, SSB will gradually displace AM radiotelephony to reduce serious interference due to overcrowding of the radio spectrum. SSB transmission also is less affected by selective fad-ing in propagation.

Multiplexing, both multiple-voice channels or teleprinter channels included with voice, is ap-plied to SSB transmission. Teleprinting and data transmission by frequency multiplex using SSB radiotelephone equipment is increasing and dis-placing older methods of radiotelegraphy for fixed point-to-point government and common-carrier services.

Aviation and marine navigation aids. Ampli-tude-modulated radio has a dominant role in guid-ance and position location, especially in aviation, which is almost wholly under radio guidance. Radio is used in traffic control from point to point, in holding a position in a traffic pattern, and in approach and landing at airports. Distance meas-uring from a known point, radio position markers, runway localizers, and glide-path directors all use AM in some form, if only for coded identification of a facility.

Marine operations are not so dependent on radio facilities as are those of aviation, but almost every oceangoing ship makes use of direction finding, at least to determine its bearing from a radio station of known location. Special marine coastal beacon stations emit identified signals solely for direction-finding purposes. Certain navigational systems (Decca, loran) for aviation are also used by ships for continuous position indication and guidance.

Television broadcasting. Amplitude modula-tion is used everywhere for the broadcasting of the picture (video) portion of television. In England, France, and a few other places amplitude modula-tion is also used for the sound channel associated with the television picture, but frequency modula-tion is more commonly used for sound.

Countries of the Western Hemisphere, Japan, Philippines, Thailand, and Iran broadcast televi-sion video in an emission band of 4.25 MHz; The English video bandwidth is 3 MHz; the French sys-tem, 10 MHz. The rest of continental Europe (except the Soviet Union) use a bandwidth of 5.25 MHz. The carrier frequencies employed are be-tween 40 and 216 MHz, and 470 to 890 MHz. A channel allocation includes the spectrum needed for both sound and picture. Japan and Australia also use 88–108 MHz for television broadcasting.

The English and French systems employ posi-tive video modulation, in which white corresponds to higher amplitudes of the modulation envelope and black corresponds to lower amplitudes. All other established video broadcasting uses negative modulation, working in the opposite sense. Syn-chronizing pulses in the negative system are at maximum carrier amplitude. The dynamic range from black to white in the picture varies from 75 to 25% of maximum amplitude.

The upper-frequency portion of one video side-band is suppressed by filters so that its remaining vestige, together with the other complete side-band, is transmitted. This is called vestigial side-band transmission and avoids unnecessary spec-trum usage.

[EDMUND A. LAPORT]

Amplitude modulator

A device for amplitude-modulating a carrier signal. The carrier is usually the radio frequency (rf) in a communications system, but it may also be a carrier signal in a multichannel cable communication system, a telemetering system, a control system, or a data-collecting system. The modulator is usually a vacuum tube or semiconductor amplifier, the output power of which controls the output level of another amplifier which modulates the carrier. *See* AMPLITUDE MODULATION; MODULATION.

Basic requirements. The goal of a modulator is to vary the amplitude of a carrier in proportion to the modulating voltage. Departure from this linear relation results in modulation distortion. If the carrier being modulated is at a high power level, such as the rf in the final stage of a radio transmitter, the power output capability and efficiency of the modulator and the efficiency of the carrier amplifier become important.

The process of amplitude modulation produces new frequencies known as side frequencies or sidebands; they are the sum and difference of the carrier and modulating frequencies. Because new frequencies are produced only in nonlinear devices, the carrier amplifier in which modulation is to take place must be nonlinear, insofar as the carrier signal is concerned, and must therefore be either a class B or class C amplifier. This requirement will be discussed further when different types of modulators are considered.

The carrier frequency must not be changed by the modulating voltage. This type of change is known as frequency modulation (FM) and occurs when the modulating voltage is coupled into the oscillator that generates the carrier frequency. Frequency modulation is avoided in high-quality transmitters by including an amplifier known as a buffer between the oscillator and the amplifier being modulated. However, in some transmitters, such as very-high-frequency transmitters where amplification is difficult, the oscillator is actually modulated.

High-level modulation. The most frequently used method of amplitude modulation employed in radio communications is known as plate or collector modulation. As this name implies, the modulator varies the plate or collector voltage in order to accomplish the modulation (Fig. 1).

The vacuum-tube circuit in Fig. 1 will be used to illustrate the operating principles of the modulator and the modulated amplifier. The carrier signal is a medium-level signal obtained from an oscillator, usually through a buffer amplifier. The tuned input coupling coil eliminates dc and undesired harmonics from the carrier and also provides impedance transformation from the previous stage.

Capacitor C_1 and resistor R provide self-bias, usually class C, for the grid. This bias is obtained because the input signal drives the grid positive with respect to the cathode on positive peaks of the carrier. During these peaks the resulting grid current charges the grid side of C_1 negatively. During the remainder of the input cycle the grid is negative with respect to the cathode, and so the charge on C_1 can only leak off through R. But resistance R is so large that time constant RC_1 is long in comparison with the period of the input cycle, so that the charge on C_1 remains substantially constant. Also, the resistance R is large in comparison with the resistance between grid and cathode while the grid is forward-biased. The magnitude of bias voltage I_gR, where I_g is the average value of grid current, can be controlled by the choice of R. The *rf* choke L has high reactance at the carrier frequency and therefore reduces the signal currents through R and consequently reduces signal power loss.

The plate current flows only during about one-third of the carrier input cycle because of the high class C bias. The peak amplitude of each partial cycle of plate current depends on the instantaneous applied plate voltage. As shown by the schematic (Fig. 1a), the voltage is the sum of the plate supply voltage $+V_{pp}$ and the voltage developed by the modulator across the secondary of its output transformer. If the modulation signal is sinusoidal, the

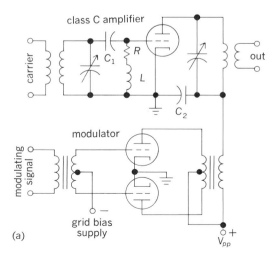

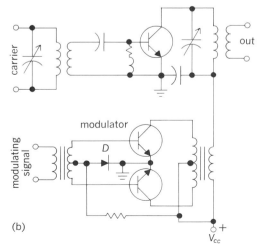

Fig. 1.   Schematics of basic high-level modulators. (a) Plate-modulated vacuum-tube amplifier with a vacuum-tube modulator. (b) Collector-modulated transistor amplifier with a solid-state modulator.

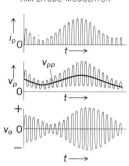

Fig. 2. Current and voltage waveforms in plate circuit of a plate-modulated amplifier.

resulting plate current i_p is as shown in Fig. 2. Plate voltage v_p varies in a continuous and nearly sinusoidal manner, because the parallel tuned plate circuit filters out harmonics in the nonsinusoidal plate-current pulses. Also, the plate current flows only during the part of the cycle when plate voltage is low. Because plate dissipation, or power lost at the plate, equals the product of plate current and plate voltage at any instant, plate dissipation is low compared with power output; therefore, plate circuit efficiency is usually about 85%. Plate voltage v_p varies almost sinusoidally about modulated plate supply voltage v_{pp}, as shown.

The tuned coupling circuit transfers the modulated rf which consists of the carrier and the side frequencies into the output, but eliminates the dc and the modulating frequencies. Therefore the output voltage v_o varies almost sinusoidally about the zero axis. The rf cycles are not quite sinusoidal because of the amplitude modulation, all harmonics of the carrier frequency are eliminated, but the carrier and the first-order sideband frequencies remain. *See* RADIO.

Carrier harmonics are most effectively removed by a high Q coupling circuit. However the coupling efficiency, which is the ratio of power output to plate circuit power, is high only when the coil Q is high compared with the circuit Q. Therefore typical coupling circuits are designed to have values of circuit Q around 12 to 15, whereas the coil Q may be in the order of 100 to 200. Then high efficiency is attained and the harmonics are not excessive.

Low-distortion modulation is easily attained by plate modulation, and high efficiency is attained in the modulated amplifier. However, the power output required from the modulator is high, being for 100% sinusoidal modulation one-half the power provided by the power supply to the rf or carrier power amplifier. This power relationship arises because the voltage induced into the secondary of

the modulation transformer alternately adds to and subtracts from power supply voltage V_{pp}; therefore the peak modulating voltage must equal power supply voltage V_{pp} in order to cancel this voltage and provide 100% modulation. Also, current I_p, which flows from the power supply through the modulation transformer secondary to the rf amplifier, must vary between zero and twice the average value for 100% modulation. Therefore the ac modulator power in the secondary of the modulation transformer is $P = V_{pp}I_p/2$, where the factor 2 is required to reduce peak values to root-mean-square (rms) values. Product $V_{pp}I_p$ is the power delivered by the power supply to the rf (or carrier) power amplifier.

The value of load resistance R_L which the rf amplifier presents to the secondary of the modulation transformer is calculated from the ratio of the peak voltage to peak current; it is $R_L = V_{pp}/I_p$.

As an example of plate modulation, consider a transmitter capable of radiating 1000 watts of carrier power. If the rf amplifier and coupling circuit combined are 80% efficient, the rf plate power input is 1250 watts, and the power output from the modulator must be 625 watts for 100% modulation. This additional power goes into the sidebands and increases the radiated power. If $V_{pp} = 1250$ volts and $I_p = 1.0$ amp, the load resistance presented to the modulator is 1250 ohms. High-power modulators are usually push-pull class B.

Collector modulation for a transistor amplifier is almost identical to plate modulation. The principal difference is in the comparatively low impedances of the transistor circuit resulting from the low voltages and high currents. Also class B and class C biases are more easily obtained because a transistor is biased beyond cutoff at zero bias; therefore, little or no additional bias is needed for class C operation. A few tenths of a volt forward bias is required for class B operation; diode D in Fig. 1

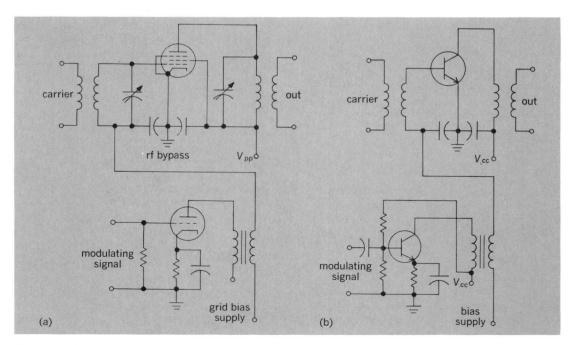

(a)   (b)

Fig. 3.   Schematics of basic low-level modulators. (a) Grid-modulated vacuum-tube amplifier with a vacuum-tube modulator. (b) Base-modulated transistor amplifier with a solid-state modulator.

provides this bias in the modulator; without this bias crossover distortion occurs.

Low-level modulation. Amplitude modulation can be produced by applying the modulating voltage to one of the grids of a vacuum tube or to the base of a transistor rf or carrier amplifier (Fig. 3). Modulation is accomplished by varying the bias of the grid or base (Fig. 4). The tube or transistor must be operated class B or class C, as in the plate-modulated amplifier. For the tube amplifier, grid bias, without modulation, is usually class C, or about twice cutoff. The rf or carrier signal amplitude is adjusted until signal peaks are half way between the maximum normally used for class C operation and cutoff. These points are shown as B, A, and C, respectively, in Fig. 4. The modulating voltage in the secondary of the modulation transformer alternately adds to and subtracts from the bias supply voltage, causing the carrier signal peaks to vary between the maximum permissible value V_A and cutoff V_C. The peak modulating voltage must therefore be $(V_A - V_C)/2$.

Power required for grid modulation is small compared with that required for plate modulation. However, the power output of the grid-modulated amplifier is only about one-fourth as great as that of a plate-modulated amplifier of the same type, because the amplifier only reaches its full power potential at the peaks of the modulation cycle; average power output is only one-fourth this value. Also, linear modulation is not easily achieved because the grid circuit appears as a high impedance to the modulator over that portion of the modulation cycle during which no grid current flows, but the impedance drops markedly at the positive peaks of the modulation cycle when grid current flows. Distortion can be made small by employing a modulator with a low output impedance compared

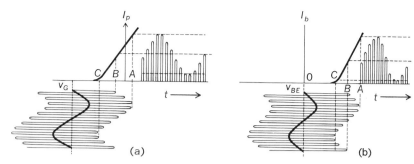

Fig. 4.   Voltage-current relationships in low-level modulators for (a) grid-modulated and (b) base-modulated amplifiers.

with the minimum impedance presented by the modulated amplifier. The grid-modulated amplifier also has a much lower efficiency than the plate-modulated amplifier.

The base-modulated transistor amplifier (Fig. 3b) operates much like the grid-modulated vacuum-tube amplifier. Power output at the peak of the modulation cycle may be limited by either the peak signal or the average power dissipation capabilities of the transistor, depending upon (1) the transistor used, (2) the potentials applied, and (3) the effectiveness of the heat sink. Linearity of the modulation can be controlled to a large extent by the source impedance of the driver and the modulator.

A transformer is not required to couple the modulator to the modulated amplifier. Resistance coupling can be used (Fig. 5). Other coupling systems can also be devised.

The screen grid, suppressor grid, or cathode of a vacuum tube can also be modulated, as can the emitter of a transistor, but the power required

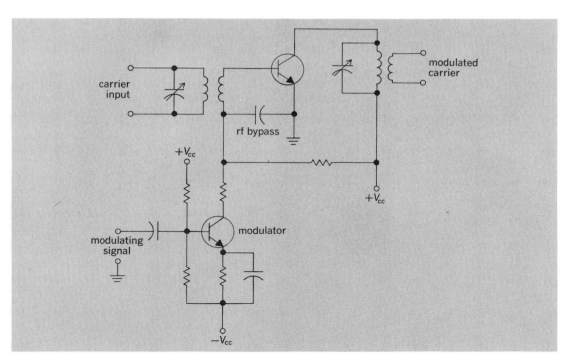

Fig. 5.   Schematic of a typical resistance-coupled base modulator.

from the modulator is somewhat higher than for grid or base modulation. Many different circuits can be devised to modulate these elements. Field-effect transistors can also be modulated in a manner very similar to a triode tube.

Figure 6 shows that a linear class A amplifier cannot be modulated. In such an amplifier the plate or collector current must flow at all times. Therefore the amplitude of the carrier does not change. The signal intended to modulate is only mixed with the carrier signal, and a tuned output circuit will pass only an unmodulated carrier. Thus cross modulation and other undesirable modulation effects can be avoided in a radio receiver by the use of linear or low-distortion amplifiers in stages where modulation or detection is not wanted.

Multiplier modulators. Low-level modulation can be accomplished by a four-quadrant multiplier, as shown by the trigonometric identity in Eq. (1), where A_m is the peak amplitude and ω_m

$$(A_m \cos \omega_m t)(A_c \cos \omega_c t)$$
$$= \frac{A_m A_c}{2}\left[\cos (\omega_c + \omega_m)t + \cos (\omega_c - \omega_m)t\right] \quad (1)$$

is the radian frequency of a sinusoidal modulating signal, and A_c is the peak amplitude and ω_c the radian frequency of a sinusoidal carrier. Observe that the multiplication of these two signals produces the two sidebands having frequencies $\omega_c + \omega_m$ and $\omega_c - \omega_m$. The output of the multiplier modulator contains neither the modulating frequency nor the carrier frequency but only the sidebands. This type of modulator is known as a suppressed carrier or balanced modulator. A four-quadrant multiplier which will handle either positive or negative voltages on both inputs is available as an inexpensive integrated circuit. The linearity of the modulator depends only on the accuracy of the multiplier. Distortion of 1% of less is typical. *See* INTEGRATED CIRCUITS.

Regular amplitude modulation, without carrier suppression, may be produced by adding an offset bias voltage to the modulating input of the four-quadrant multiplier, as shown by Eq. (2). This bias

$$(1 + A_m \cos \omega_m t)(A_c \cos \omega_c t) = A_c \cos \omega_c t$$
$$+ \frac{A_c A_m}{2}\left[\cos (\omega_c + \omega_m)t + \cos (\omega_c - \omega_m)t\right] \quad (2)$$

offset voltage may be obtained by merely adjusting the input offset adjustment, which is normally included in the multiplier, until the carrier voltage

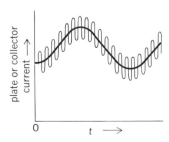

Fig. 6. Current in a linear class A amplifier, showing the absence of modulation.

output without modulation is one-half of the peak value obtained when the maximum modulating voltage is applied. The maximum modulation is then 100%. On the other hand, the multiplier acts as a balanced modulator when the input offset adjustment is such that no carrier signal appears in the output when the modulating input voltage is zero.   [CHARLES L. ALLEY]

Bibliography: C. Alley and K. Atwood, *Electronic Engineering*, 3d ed., 1973; D. J. Comer, *Modern Electronic Circuit Design*, 1976.

Analog computer

A computer or computational device in which the problem variables are represented as continuous, varying physical quantities. An analog computer implements a model of the system being studied. The physical form of the analog may be functionally similar to that of the system, but more often the analogy is based solely upon the mathematical equivalence of the interdependence of the computer variables and the variables in the physical system. *See* SIMULATION.

COMPUTER TYPES

Analog computers can be classified in two ways: in accordance with their use, and in accordance with the type of components used to assemble them. In terms of use, there are general-purpose and special-purpose analog computers. The general-purpose computer is designed so that it can be programmed to solve many kinds of problems or permit the development of different simulated models as needed (Fig. 1). The special-purpose computer has a fixed program with a few or no permitted adjustments; it is generally built into or appended to the physical system it serves (Fig. 2).

In terms of the components with which computers are made, there are mechanical, hydraulic, pneumatic, optical, electrical, and electronic analog computers serving in various applications. Because of its ease of programming, flexibility of operation, and repeatable results, the general-purpose electronic analog computer has come into wide use. Therefore this article deals specifically with features and applications of electronic analog computers. The flexibility of the electronic analog computer has allowed it to be augmented with interface channels to the electronic digital computer, so that during the 1960s a third type of general computer, the hybrid computer, came into being. The characteristics of this computer are also described in this article.

Digital equivalents. Because of the practical difficulties of scaling, interconnecting, and operating an electronic analog computer, the equivalent of its powerful problem-solving capability has been sought and realized in other forms. Since 1965 many digital computer programs have been developed which essentially duplicate the functions of analog computers via digital algorithms. Such programs as CSMP, DARE, SIMSCRIPT, CSSL, or CPSS are available for almost all widely used digital computers; these programs approximate the parallel, continuous-time operation of the electronic analog computer by stepwise-incrementing a solution "time" and repeatedly solving the programmed dynamic equations. Digital equivalents to the electronic analog computer are free from scaling requirements and are convenient to run; on the other hand, no interaction within the dynamic response is possible, the solution speed is slower by a factor of a hundred times or more, and little or no on-line model building and exploring is possible.

Digital multiprocessor analog system. Another type of analog computer is the digital multiprocessor analog system (Fig. 3), in which the relatively slow speeds of sequential digital increment calculations have been radically boosted through parallel processing. In this type of analog computer it is possible to retain the programming convenience and data storage of the digital computer while approximating the speed, interaction potential, and parallel computations of the traditional electronic analogs.

The digital multiprocessor analog computer typically utilizes several specially designed high-speed processors for the numerical integration functions, the data (or variable) memory distributions, the arithmetic functions, and the decision (logic and control) functions. All variables remain as fixed or floating-point digital data, accessible at all times for computational and operational needs.

The digital multiprocessor analog computer achieves an overall problem-solving efficiency comparable to the very best continuous electronic analog computers, at a substantially lower price. An example of such a computer, the model AD10 (Fig. 3a), can solve large, complex, and multivariate problems at very high speeds and with the advantages of all-digital hardware. Its computation system (Fig. 3b) is based on five parallel processors working in a special computer architecture designed for high-speed operation. The various elements are interconnected by means of a data bus (MULTIBUS). A highly interleaved data memory of up to 10^6 words serves the data and program storage functions. The five processors working in parallel are: the control processor (COP), which controls all operations; the arithmetic processor (ARP), which runs the numerical calculations; the decision processor (DEP), which executes the logic parts of the program; the memory address processor (MAP), which makes sure that all data are fetched and stored efficiently; and the numerical integration processor (NIP), which carries out the integration functions that are crucially important in an analog computer.

HISTORY

The slide rule, which originated in the 17th century, represents the first analog computing aid to become a common engineering tool. In it a mechanical position represents the problem variable, and the computations take the form of adding or subtracting linear positional displacements. *See* SLIDE RULE.

Beginning approximately in 1825, several mechanical integrating devices were developed for measuring the area under a curve, and in 1876 William Thomson (later Lord Kelvin) presented a complete description of a process for solving a general ordinary differential equation by analog means. In 1881 the integraph was introduced by C. V. Boys and Abdank-Abakanowicz for drawing the integral curve of a given curve. The mechanical analog computer was greatly improved during

Fig. 1. General-purpose electronic analog computer, model EAI-2000. (*Electronic Associates, Inc.*)

World War I; in the succeeding years it was applied primarily to naval gunfire systems.

Significant advancements were made by Vannevar Bush and his colleagues at the Massachusetts Institute of Technology when they developed the first large-scale general-purpose mechanical differential analyzer in the early 1930s. The success of

Fig. 2. Special-purpose pneumatic analog computer, model Foxboro 556. (*Foxboro Company*)

this machine led to the construction of similar machines in the United States and other countries and prompted the MIT group to construct a more elaborate machine, placed in operation in 1942.

The electronic analog computer had its origin in the dc network analyzer developed about 1925 by General Electric Company and Westinghouse Corporation; the dc network analyzer was followed in 1929 by the more versatile ac analyzer. The electronic differential analyzer, as it was called, had its fastest growth during World War II and soon displaced mechanical computers except for some special-purpose applications, such as jet-engine fuel controllers. C. A. Lovell and D. B. Parkinson of Bell Telephone Laboratories were responsible for the first published use of operational amplifiers as computer components. They used operational amplifiers in the real-time computer of the M9 antiaircraft-gun director built by Western Electric Company. J. B. Russell of Columbia University brought the circuits used in the M9 computer to the attention of J. R. Ragazzini, R. H. Randall, and F. A. Russell, who then built the first general-purpose electronic analog computer under contract with the National Defense Research Committee. This work led to the publication of the first article, in May 1947, describing the operational amplifier as a computer component. Meanwhile G. A. Philbrick had independently pioneered the use of high-gain direct-coupled amplifiers as components of fast-time electronic analog computers. The invention of the chopper-stabilized direct-coupled amplifier was primarily due to E. Goldberg of RCA Laboratories; it essentially eliminates the problem of drift in dc computation. In 1947 Reeves Instrument Corporation, under a Navy contract, developed a computer which was the forerunner of the present-day electronic analog computer.

DESCRIPTION AND USES

The typical modern general-purpose analog computer consists of a console containing a collection of operational amplifiers; computing elements, such as summing networks, integrator networks, attenuators, multipliers, and function generators; logic and interface units; control circuits; power supplies; a patch bay; and various meters and display devices. The patch bay is arranged to bring input and output terminals of all programmable devices to one location, where they can be conveniently interconnected by various patch cords and plugs to meet the requirements of a given problem. Prewired problem boards can be exchanged at the patch bay in a few seconds and new coefficients set up typically in less than a half hour. Extensive automatic electronic patching systems have been developed to permit fast setup, as well as remote and time-shared operation.

The analog computer basically represents an instrumentation of calculus, in that it is designed to solve ordinary differential equations. This capability lends itself to the implementation of simulated models of dynamic systems. The computer operates by generating voltages that behave like the physical or mathematical variables in the system under study. Each variable is represented as a continuously varying (or steady) voltage signal at the output of a programmed computational unit. Specific to the analog computer is the fact that individual circuits are used for each feature or equation being represented, so that all variables are generated simultaneously. Thus the analog computer is a parallel computer in which the configuration of the computational units allows direct interactions of the computed variables at all times during the solution of a problem.

(a)

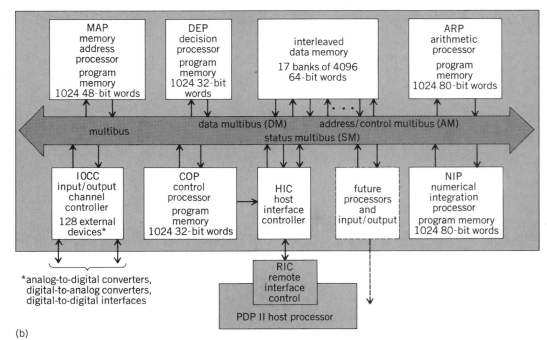

(b)

Fig. 3. Multiprocessor analog computer, model AD10. (*a*) Exterior. (*b*) Organization. (*Applied Dynamics International*)

An example of a general-purpose electronic analog computer, the model EAI-2000, is shown in Fig. 1. This analog/hybrid computer is controlled by digital means, and operator input/output is through a keyboard-equipped cathode-ray tube. Plug-in analog/digital converters and digital/analog multipliers quickly convert the computer for complete hybrid operation, discussed below. Computational speed is equivalent to 15,000,000 operations per second, or more.

Applications. The analog computer or its equivalent, the continuous system simulation by a digital computer, is employed in every area of science and technology. The range of applications is constantly increasing, from the social sciences, to economics, human relations, political problems and policy/decision-making, to ecology and environmental studies, medicine, health and welfare, up to the more traditional analog computer application areas: science, technology, and hardware and systems development and engineering.

The analog computer is used to gain better insight into research topics, to develop and test methods and equipment, to carry out the design of new processes, or to predict by trial-and-error methods the model responses of system applications beyond known boundaries as well as to improve the behavior of systems already in operation. Special-purpose analog computers are directly connected to or built into physical systems to serve such on-line tasks as immediate and continuous computation or data reduction, or as part of dynamic control. An example of this type is the pneumatic computer shown in Fig. 2. This computer uses standard ($3-15$ psi or $20-100$ kPa) air-pressure signals, a mechanical flexure construction, and a readily interchangeable switch plate to perform any one of the following functions: multiplication, division, squaring, and square-root extraction.

The analog computer serves especially well in two categories of systems engineering. In the first, the inductive process of model building, an analytical relation between variables is hypothesized to describe the physical system of interest. Forcing functions identical to those in the physical system can then be applied to the hypothetical model, enabling a comparison of the response of the model with that of the actual physical system. This will often indicate improvements and extensions that should be made in the model to fit it to the system better and thus make it more valid. Because of the ease with which parameter variations and model changes can be accomplished, the analog computer is useful in conducting many trial-and-error experiments on the model to obtain the best fit to the physical system.

In the second category, the deductive process of systems analysis, mathematical statements describing the physical system to be studied are necessary. These mathematical statements (equations) are often supplemented by graphical information as well as by logic statements. Also, validity ranges for parameters are established for the study. Experiments are then performed, varying the inputs and parameters that describe the system, to obtain (finally) a set of optimum responses for the system, to develop a better understanding of the intrinsic nature of the system by

studying input-output relations, or to lay a foundation for further investigations into the system.

Unique features. The unique features of the analog computer which are of value in science and technology are as follows:

1. Within the useful frequency bandwidth of the computational units and components, all programmed computations take place in parallel and are for practical purposes instantaneous. That is, there is no finite execution time associated with each mathematical operation, as is encountered with digital computer methods.

2. The dynamics of an analog model can be programmed for time expansion (slower than real system time), synchronous time (real time), or time compression (faster than real time).

3. The computer has a flexible addressing system so that almost every computed variable can be measured, viewed with vivid display instruments, and recorded at will.

4. One control mode of the computer, usually called HOLD, can freeze the dynamic response of a model to allow detailed examination of interrelationships at that instant in time, and then, after such study, the computing can be made to resume as though no stop had occurred.

5. By means of patch cords, plugs, switches, and adjustment knobs the analog computer program or model can be directly manipulated, especially during dynamic computation, and the resultant changes in responses observed and interpreted.

6. Because of the fast dynamic response of the analog computer, it is easy to implement implicit computation through the use of problem feedback. (This important and powerful mathematical quality of the analog computer is discussed more fully below.)

7. The computer can be used for on-line model building; that is, a computer model can be con-

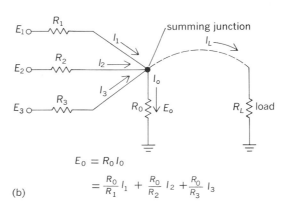

(a)

(b)

$$E_0 = R_0 I_0$$

$$= \frac{R_0}{R_1} I_1 + \frac{R_0}{R_2} I_2 + \frac{R_0}{R_3} I_3$$

Fig. 4. Addition of electric currents by using a passive network of resistors. (a) Individual voltages and resistors. (b) Voltages and corresponding currents summed into a common resistor.

structed in a step-by-step fashion directly at the console by interconnecting computational units on the basis of one-for-one analog representation of the real system elements. Then, by adjusting signal gains and attenuation parameters, dynamic behavior can be generated that corresponds to the desired response or is recognizable as that of the real system. This method allows a skillful person to create models when no rigorous mathematical equations for a system exist.

8. For those applications to which it is well suited, the analog computer operates at relatively low cost, thus affording the analyst ample opportunity to investigate, develop, and experiment within a broad range of parameters and functions.

COMPONENTS

Manipulations of the signals (voltages) in the analog computer are based upon the fundamental electrical properties associated with circuit components. The interrelation of voltages for representing mathematical functions is derived by combining currents at circuit nodes or junctions. *See* KIRCHHOFF'S LAWS OF ELECTRIC CIRCUITS; OHM'S LAW.

Linear computing units. The simplest arrangement of components for executing addition would be to impress the voltages to be added across individual resistors (Fig. 4a). The resistors would then be joined (Fig. 4b) to allow the currents to combine and to develop the output voltage across a final resistor. Use of this simple configuration of elements for computation is impractical, because of the undesirable interaction between inputs. A change in one input signal (voltage) causes a change in the current that flows through the input resistor; this changes the voltage at the input resistor junction, and the change secondarily causes a different current to flow in the other input resistors. The situation gets more interactive when another computing circuit is attached so that part of the summing current flows away from the summing junction. This interaction effect also prevents exact computing. If, in some way, each voltage to be summed could be made independent of the other voltages connected to the summing junction, and if the required current fed to other circuits could be obtained without loading the summing junction, then precise computation would be possible. *See* DIRECT-CURRENT CIRCUIT THEORY.

Function of operational amplifier. The electronic analog computer satisfies these needs by using high-gain (high-amplification) dc operational amplifiers. A symbol to represent a dc direct-coupled amplifier is shown in Fig. 5. According to convention, the rounded side represents the input to the amplifier, and the pointed end represents the amplifier output. A common reference level or ground exists between the amplifier input and output, and all voltages are measured with respect to it. The ground reference line is understood and is usually omitted from the symbol. The signal input network (consisting of summing resistors) connects to the inverting (or negative) amplifier input terminal. The noninverting (or positive) amplifier input terminal is normally connected to the reference ground. Generally the inverting input is called the summing junction (SJ) of the amplifier. Internal design of the amplifier is such that, if the signal at

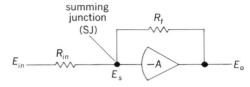

Fig. 6. High-gain amplifier which has been made into an operational amplifier through the inclusion of an input resistor and a feedback resistor tied together at the amplifier's summing junction (SJ).

the summing junction is positive with respect to ground, the amplifier output voltage is negative with respect to ground. The amplifier has an open-loop voltage gain of $-A$; therefore an input voltage of E_s results in an output voltage of $-AE_s$. Gain of a commercial computing amplifier is typically 10^8; thus, an input voltage of less than 1 μV can produce several volts at the amplifier output. *See* AMPLIFIER; DIRECT-COUPLED AMPLIFIER.

Because the operational amplifier thus inverts the signal (changes its sign or polarity), it lends itself to the use of negative feedback, whereby a portion of the output signal is returned to the input. This arrangement has the effect of lowering the net gain of the amplifier signal channel and of improving overall signal-to-noise ratio and increasing computational accuracy.

Circuit operation (Fig. 6) can be viewed in the following manner. A dc voltage E_{in} applied to input resistor R_{in} produces a summing junction voltage E_s. The voltage is amplified and appears at the amplifier output as voltage E_o, (equal to $-AE_s$, where A is voltage gain of the amplifier). Part of output voltage E_o returns through feedback resistor R_f to the summing junction. Because the returned or feedback voltage is of opposite (negative) polarity to the initial voltage at the summing junction it tends to reduce the magnitude of E_s, resulting in an overall input-output relationship that may be expressed as Eq. (1). In fact, the summing junction voltage E_s is so small that it is considered to be at practically zero, a condition called virtual ground.

$$\frac{E_o}{E_{in}} = \frac{-A}{A+1} = \frac{-1}{1+1/A} \simeq -1 \qquad A > 10^8 \qquad (1)$$

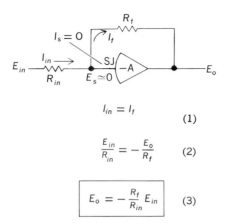

$$I_{in} = I_f \qquad (1)$$

$$\frac{E_{in}}{R_{in}} = -\frac{E_o}{R_f} \qquad (2)$$

$$\boxed{E_o = -\frac{R_f}{R_{in}} E_{in}} \qquad (3)$$

Fig. 7. Summing junction currents into an operational amplifier create the fixed gain function, determined by resistor values, as indicated in the box.

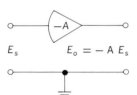

ANALOG COMPUTER

$$E_s \qquad E_o = -A\,E_s$$

Fig. 5. Symbol for a high-gain direct-current amplifier, with one inverting input referenced to ground.

To illustrate how the operational amplifier serves the needs of the computing network, consider the currents that flow and the corresponding algebraic expressions (Fig. 7). The operational amplifier is designed to have high input impedance (high resistance to the flow of current into or out of its input terminal); consequently the amplifier input current I_g can then be considered to be practically zero. The resulting current equation states that input current I_{in} is equal to feedback current I_f. Since the amplifier has a very high gain, the summing junction voltage is virtually zero. Voltage drop across R_{in} is thus equal to E_{in}; voltage drop across R_f is E_o. The equation in the box is the fundamental relationship for the amplifier. As long as the amplifier has such a high gain and requires a negligible current from the summing junction, the amplifier input and output voltages are related by the ratio of the two resistors and are thus not affected by the actual electronic construction of the amplifier. If several input resistors are connected to the same summing junction and voltages are applied to them (Fig. 8), then because the summing junction remains at practically zero potential, none of the inputs will interfere with the other inputs. Thus all inputs will exert independent and additive effects on the output.

Because amplifier gain has a negative sign, output voltage equals the negative sum of the input voltages, weighted according to the values of the individual resistors in relation to the feedback resistor, as shown in the box in Fig. 8.

When a computing circuit is connected to the amplifier output, a demand for load current is introduced. Such a required output load current must be supplied by the amplifier without a measurable change in its output voltage. That is, the operational amplifier must act as a voltage controller, supplying whatever current is required within limits, while maintaining the net voltage gain established by the mathematical ratio of its input and feedback elements. The operational amplifier and network is a linear network because once the input-feedback ratios are adjusted, signal channel gains remain constant (a straight-line function) during computation.

Accessories with operational amplifier. The input and feedback elements that can be used with an operational amplifier are not restricted to resistors alone. A reactive feedback component such as a capacitor may also be used, resulting in mathematical operations analogous to integration with respect to time. The capacitor is a device that accumulates electric charge. The voltage that exists across it is related to its electrical size (capacitance in farads) and the quantity of charge (in coulombs) it contains. Should more charge flow into it (an electric current is actually equivalent to charge per unit time), the voltage across the capacitor would rise. To state this concept mathematically: the capacitor integrates electric current (flow of charge) with respect to time. An operational amplifier with a feedback capacitor (Fig. 9) will thus generate an output voltage that is the time integral of its input voltage. *See* CAPACITANCE.

To solve equations with specific numerical values, many different circuit gains are needed. Although it is possible to provide almost any desired gain by selecting appropriate resistances for the input-feedback ratio of operational amplifier channels, each one usually remains fixed once it has been established. In practice the choice of gain is limited to a few integer values (amplifier gains such as 1, 5, 10) which are obtained with high accuracy, using matched resistors, on the general-purpose analog computer.

Another linear computing element, called the attenuator, also known as the potentiometer or pot, is used to adjust signal gains to desired fractional values. It thus allows parameters (coefficients in the problems) to be adjusted either manually or automatically, between or during computations. Traditionally, the attenuator is a three-terminal potentiometer consisting of a multiturn, resistive winding with a movable wiper (Fig. 10a). Output voltage is a fraction of the input, depending upon the position of the wiper (Fig. 10b). For manually adjusted potentiometers, a dial is mounted on the wiper shaft so that the actual pot position can be read out (Fig. 10c). Servo-set potentiometers are adjusted automatically, using individual dc motors in a measurement and control loop fashion; a desired attenuation coefficient value is entered via the console keyboard, and the corresponding potentiometer is set to that value in less than a second. The most modern attenuator is the digitally set one, in which the coefficient value is established via electronic multiplication techniques, making it possible for a digital computer (in a hybrid setup) to control at high speed the coefficients used in the analog computer program.

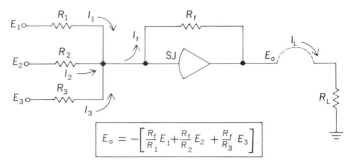

$$E_o = -\left[\frac{R_f}{R_1}E_1 + \frac{R_f}{R_2}E_2 + \frac{R_f}{R_3}E_3\right].$$

Fig. 8. Operational amplifier which has been made into a summer through the use of several input resistors connected to the summing junction. Output is equal to the inverted weighted sum of the inputs.

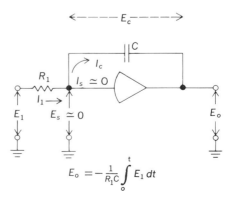

$$E_o = -\frac{1}{R_1 C}\int_0^t E_1\, dt$$

Fig. 9. Use of a capacitor as a feedback element to turn a high-gain amplifier into an integrator, with a time scale determined by the RC time constant.

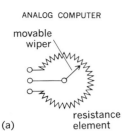

ANALOG COMPUTER

movable
wiper

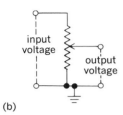

resistance
element

(a)

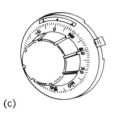

input
voltage

output
voltage

(b)

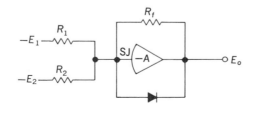

(c)

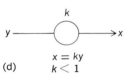

k

$y \longrightarrow \bigcirc \longrightarrow x$

$x = ky$
$k < 1$

(d)

Fig. 10. Attenuators.
(a) Traditional
potentiometer voltage
divider. (b) Circuit used
in analog computers. (c)
Typical manually
adjusted dial with
indication of mechanical
position of wiper in 10-
turn potentiometer.
(d) Circuit symbol for
attenuator with
coefficient k.

An in-line circle symbol is used to represent the attenuator in analog circuit diagrams (Fig. 10d). *See* POTENTIOMETER (VARIABLE RESISTOR).

Because the classical potentiometer type is a passive element rather than an active one like the operational amplifier, it can only provide signal attenuations of less than unity. When it is connected to other circuits, including amplifier inputs, some current is drawn through its wiper arm. Therefore, for greatest accuracy in setting, the attenuator must be adjusted while measuring the voltage at the wiper arm when it is under load and with a steady, known reference voltage applied to the potentiometer input. This procedure compensates for such slight loading voltage drops as occur when input resistors of value commensurable with the potentiometer resistance are connected to the wiper. Otherwise the result is a dial reading that is higher than the electrical attenuation. As will be described below, attenuators in conjunction with other elements allow the computer to be set up for any required problem value or analog model scale factor.

Nonlinear computing elements. In nature nearly all systems or processes exhibit nonlinear features in the form of constraints, discontinuities, and variable gains. If an analog computer is to be programmed and operated as a realistic system model, it must include devices with programmable nonlinear characteristics. The most versatile computing element for this service is the diode function generator. *See* SEMICONDUCTOR DIODE.

The diode operates as an electrical one-way check valve, allowing current to flow one way but not the other. Electrically, a diode has two states or conditions: the ON or conducting condition and the OFF or nonconducting condition. Due consideration should be given to the fact that a diode operates solely according to the voltage across it. A diode will assume the ON condition whenever the anode (the positive terminal) is more positive than the cathode. For example, if the anode is at -6 V and the cathode is at -8 V, the diode is in a conducting (ON) condition; if the anode is at $+9$ V and the cathode is at $+15$ V, the diode will be in a nonconducting (OFF) condition.

Within various limitations, the diode may be used in the input, feedback, and output circuits of operational amplifiers to simulate nonlinear features that are functions of the variable voltages being computed. To illustrate diode action, the output of an amplifier (Fig. 11) might represent the computer altitude of a space vehicle above the lunar surface. Upon descent, the constraint is encountered at touchdown that no negative signal (and thus no negative altitude, meaning "flying underground") be allowed to occur, because such conditions obviously make the system equations invalid. Should the amplifier output signal attempt to go negative, the cathode of the diode would assume a lower potential than the anode at the summing junction's virtual ground, placing the diode in the ON condition. The result is to shunt the feedback resistor so that the effective R_f becomes nearly zero. When the amplifier output signal once again goes positive, the diode switches to its OFF condition and no further limiting occurs.

Diodes, in addition to being used to implement simulations of unique nonlinearities such as limits,

backlash, and deadzones, can be combined with a network of resistors to produce particular nonlinear functions. A diode function generator (DFG) uses the one-way conducting property of diodes to selectively alter the resultant gain of an operational amplifier in accordance with a desired function of the input voltage. The amplifier input circuit of Fig. 12 is adjusted so that the breakpoint resistors R_b permit individual diodes to conduct at successively higher values of the input signal E_{in}. The resultant input-output gain is determined by the sum of the ratios of feedback resistor R_f and output increment (or slope) resistors R_s, as illustrated by the equation for point Q on the curvature in Fig. 13. Specific resistor values for the set of R_b's and R_s's can be calculated to generate many mathematical functions, such as $x, x^2, \log x, e^x, \cos x$, or $\sin x$. The DFG can be arranged to operate on either positive or negative input signals by choosing the orientation of diodes and reference voltage polarity. Further, by utilizing fully adjustable potentiometers for breakpoint and slope resistances, a variable diode function generator (VDFG) is established, permitting the programming of most analytic, arbitrary, and empirical functions, including functions with one or more inflections.

Multipliers. The principal device for multiplication of one variable by another (to yield a product of variables) is the electronic quarter-square multiplier. The mathematical function "multiplication" corresponds to a nonlinearity in which one signal input is acted on through a second input, such as a variable gain. Traditionally, an electromechanical device, the servo-multiplier, accomplished this by dynamically positioning the wiper arm of a potentiometer through a motor drive that responded to the second input signal. Servo-multipliers are now infrequently used in electronic analog computers. However, they represent a type of computing unit having relatively low cost, great versatility, and low signal bandwidth.

The all-electronic quarter-square multiplier achieves the multiplication function to frequencies of several thousand hertz by implementing an algebraic identity (Fig. 14). The squaring operations are accomplished with high accuracy by two DFG circuits, as described earlier. Typically, 80 or more properly oriented diodes are employed to permit the accurate generation of the square-law terms of all polarity combinations required for complete

$R_1 = R_2 = R_f$
$E_0 = E_1 + E_2$ when $E_1 + E_2 > 0$
$E_0 = 0$ when $E_1 + E_2 \leq 0$

Fig. 11. Circuit containing one-way conducting diode that limits the output of an amplifier to positive signals (since a negative output would draw a short-circuit current from the summing junction through the diode).

four-quadrant multiplication. The simplified symbols which are used in the circuit diagram are more fully discussed below.

Signal-controlled programming. An electronic comparator–analog switch combination provides the analog computer with decision-making and preprogrammed automatic signal-rerouting capabilities (Fig. 15). These are nonlinear functions preset to take place at some particular point in a calculation run and to cause a change in coefficients or signals, that is, a branching in mathematical procedure or a change in the control of the run. The technique consists of using a comparator with two or three inputs, generating a logic output signal (TRUE-FALSE) which depends on whether or not the sum of the input signals is greater than zero. Basically the comparator responds to its input analog signals and abruptly shifts its output from one logic level (FALSE) to the other (TRUE) whenever there is a polarity change to positive at its summing junction, that is, whenever the sum of the input voltage goes through zero. The comparator must switch its output rapidly because there is no intermediate value assumed in the transition. The comparator output signal m may be used in logic units (AND gates, OR gates, flip-flops, and counters) to achieve complex logic functions and automatic (programmed) control of computations. *See* LOGIC CIRCUITS.

The electronic analog switch is an ON-OFF input network used in connection with an operational amplifier. The network is ON, permitting an analog input signal to propagate through it to the associated amplifier when, and only when, a logic high (TRUE) signal is applied to its command input. If the command input is logic low (FALSE) or absent, the signal switch network remains in its disconnected or shut-off mode, and no analog input signal will be conducted through it.

One or more of the comparator inputs can be dynamically variable inputs, so that the comparison function itself is a function of the analog computation. The analog signal switch can be used to alter the configuration of computing circuits or to change signal gains by switching between pairs of attenuator potentiometers set to different coefficient values. In modern computers, each comparator also applies power to its indicator light bulb on the console front panel to indicate the state of the comparison function.

PROGRAMMING

To solve a problem using an analog computer, the problem solver goes through a procedure of general analysis, data preparation, analog circuit development, and patchboard programming. He or she may also make test runs of subprograms to examine partial-system dynamic responses before eventually running the full program to derive specific and final answers. The problem-solving procedure typically involves eight major steps, listed below:

1. The problem under study is described with a set of mathematical equations or, when that is not possible, the system configuration and the interrelations of component influences are defined in block-diagram form, with each block described in terms of black-box input-output relationships.

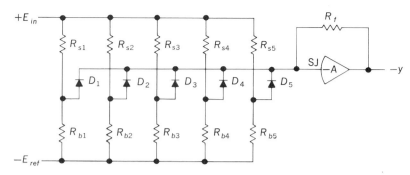

Fig. 12. Diode function generator with five diodes, each biased to a different voltage and contributing a specified current to the summing amplifier, to generate a desired function of the input signal.

2. Where necessary, the description of the system (equations or system block diagram) is rearranged in a form that may better suit the capabilities of the computer, that is, avoiding duplications or excessive numbers of computational units, or avoiding algebraic (nonintegrational) loops.

3. The assembled information is used to sketch out an analog circuit diagram which shows in detail how the computer could be programmed to handle the problem and achieve the objectives of the study.

4. System variables and parameters are then scaled to fall within the operational ranges of the computer. This may require revisions of the analog circuit diagram and choice of computational units.

5. The finalized circuit arrangement is patched on the computer problem board.

6. Numerical values are set up on the attenuators, the initial conditions of the entire system model established, and test values checked.

7. The computer is run to solve the equations or simulate the black boxes so that the resultant values or system responses can be obtained. This

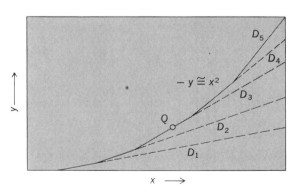

at $Q: -y = x \left[\dfrac{R_f}{R_{s1}} + \dfrac{R_f}{R_{s2}} + \dfrac{R_f}{R_{s3}} \right]$

D_1, D_2, and D_3 on

D_4, D_5 off

Fig. 13. Resultant function of diode function generator, made up of the sum of currents from the individual diodes; the desired function is approximated by a number of straight-line segments.

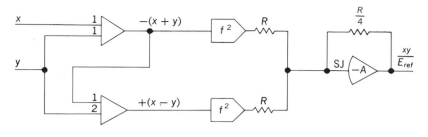

$$xy = 1/4[(x + y)^2 - (x - y)^2]$$
$$= 1/4(x^2 + 2xy + y^2 - x^2 + 2xy - y^2)$$
$$= 1/4(4xy) = xy$$

Fig. 14. Electronic quarter-square multiplier, which uses an algebraic identity to achieve the multiplication of variables.

gives the initial answers and the "feel" for the system.

8. Multiple runs are made to check the responses for specific sets of parameters and to explore the influences of problem (system) changes, as well as the behavior which results when the system configuration is driven with different forcing functions.

Symbols. A number of concepts and techniques are used to ease the process of programming an analog computer. For example, it usually is too tedious and cumbersome to show all of the wiring associated with the computing devices used in a program; therefore, simplified symbols are used to represent the various computer components (Fig. 16). These symbols are easy to draw, and they allow a complex program to be designed quickly, yet with sufficient clarity for unambiguous patching.

When an amplifier symbol is shown with a straight front, it is understood to have a standard feedback resistor; but when the front is curved, the feedback circuit is not included and must be explicitly programmed in the computer diagram. In the integrator, the constant of integration C is applied to a separate, special input terminal through a mode switching network of the integrator to provide for establishment of initial condi-

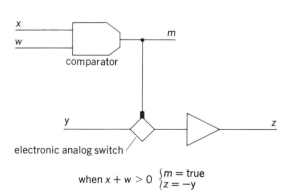

when $x + w > 0$ $\begin{cases} m = \text{true} \\ z = -y \end{cases}$

when $x + w \leq 0$ $\begin{cases} m = \text{false} \\ z = 0 \end{cases}$

Fig. 15. Comparator and electronic analog switch, combined to give signal and coefficient changes in the analog circuits based on the comparison results.

tions. Function-generating networks as such do not show a polarity reversal because they represent characterized impedances (resistor-diode arrays), but they are usually permanently connected to one or two summing amplifiers. The multiplier normally comes with an output amplifier; thus it provides an inverted output.

Representation of variables. When equations are to be solved or a system is to be modeled on the analog computer, all individual parameters and variables must be fitted (scaled) to the available three dimensions of the computer: voltage range, input gains, and computing time. The voltage range spans from negative reference voltage, through zero (ground), to positive voltage reference. The range on large multiconsole analog computers often extends from −100 to +100 V; on most small desktop or single-console-type computers the range is −10 to +10 V. Reference voltages are carefully established and rigidly regulated to provide a means of generating accurate input signals and to ensure a reliable base against which computed variables can be measured.

The zero point is usually referred to as high-quality ground; it is wired separately from the standard electronic chassis ground and power return ground. All computing signals are referenced to the high-quality ground. The precision of the computing gain of an amplifier is dependent on carefully matching its input and feedback resistors to provide the desired ratio with great accuracy. To avoid errors due to ambient temperature variations, these resistors as well as the feedback capacitors used in the integrators are either specially compensated against thermal errors, or they are physically maintained at a constant temperature in an oven mounted immediately behind the patch bay. By using multiturn potentiometers as attenuators, excellent resolution is obtained for coefficient adjustment. The maximum range of such coefficient adjustment, using the available input gains (including the combination of a potentiometer and an amplifier), may span approximately six decades (0.0001 to 100).

The time duration required for the dynamic response of a system model determines how long the computer must be programmed to run. A single run can be as short as 1 ms or as long as hours. The precision of the computational results is determined by how accurately all the computer circuits function and how well the operational amplifiers are stabilized against drift.

Just as the analog voltage range can be scaled to represent the range of any real system variable, so the analog time dimension can be used to represent a system dimension other than time. For example, the problem to be programmed might be the calculation of the temperature profile of a metal rod heated at one end and cooled at the other. The analog computer could be programmed to sweep continuously in time along the length of the rod, developing a varying analog voltage scaled to represent the temperature. Thus, the computer time dimension would correspond to the rod length dimension.

Inverse operations. The complementary nature of the input and feedback networks of the operational amplifier gives it one of its most flexible properties. If, for example, a squaring module in an amplifier input circuit is moved to the amplifier

ELEMENT	CIRCUIT	SYMBOL	FUNCTION
attenuator	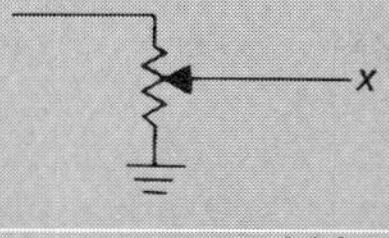	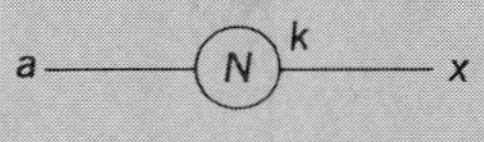 $a \xrightarrow{} (N)^{k} \xrightarrow{} x$	$x = ka$
inverter	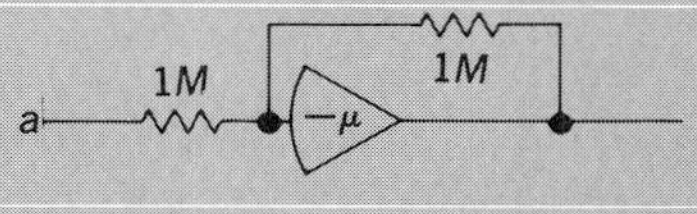 $a \xrightarrow{} 1M \quad -\mu \quad 1M$	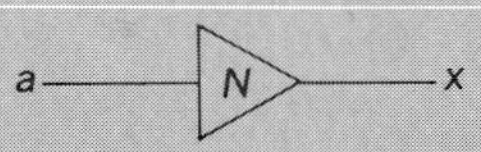 $a \xrightarrow{} N \xrightarrow{} x$	$x = -a$
summer	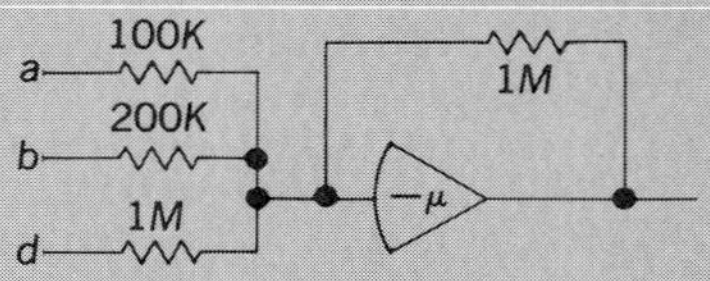 $a \xrightarrow{} 100K \quad b \xrightarrow{} 200K \quad 1M \quad d \xrightarrow{} 1M \quad -\mu$	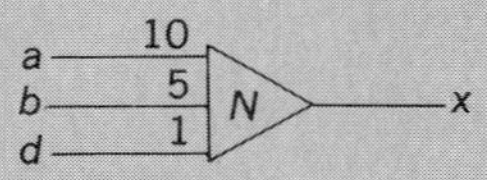 $\begin{array}{l} a \xrightarrow{10} \\ b \xrightarrow{5} \; N \\ d \xrightarrow{1} \end{array} \xrightarrow{} x$	$x = -(10a + 5b + d)$
high-gain amplifier	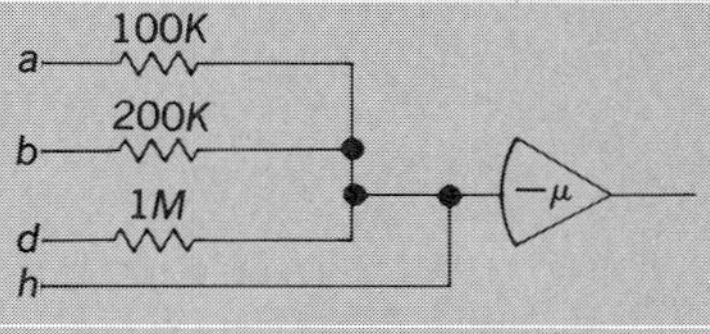 $a \xrightarrow{} 100K \quad b \xrightarrow{} 200K \quad d \xrightarrow{} 1M \quad h \quad -\mu$	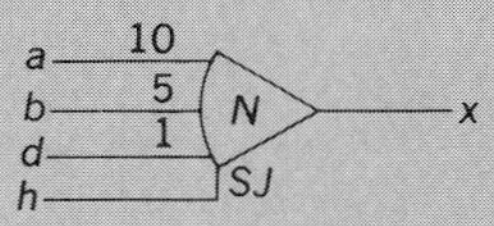 $\begin{array}{l} a \xrightarrow{10} \\ b \xrightarrow{5} \\ d \xrightarrow{1} \; N \\ h \xrightarrow{SJ} \end{array} \xrightarrow{} x$	$x = -f(10a + 5b + d + h\mu)$ f is defined when a feedback circuit is supplied
integrator	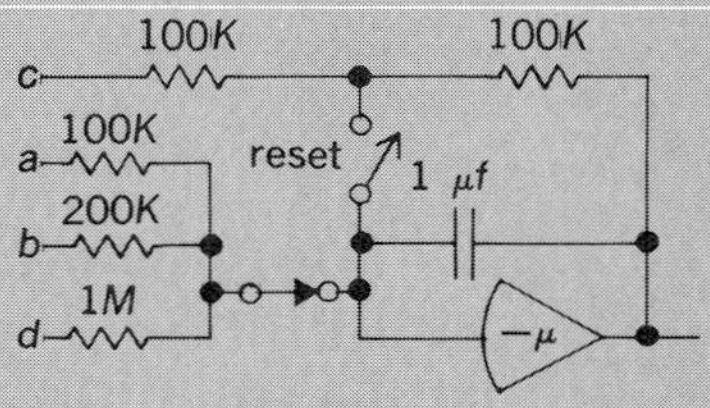 $c \xrightarrow{} 100K \quad 100K \quad a \xrightarrow{} 100K \quad \text{reset} \quad 1 \; \mu f \quad b \xrightarrow{} 200K \quad d \xrightarrow{} 1M \quad -\mu$	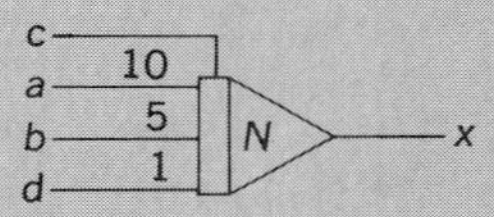 $\begin{array}{l} c \xrightarrow{10} \\ a \\ b \xrightarrow{5} \; N \\ d \xrightarrow{1} \end{array} \xrightarrow{} x$	$x = -\left[\int_0^t (10a + 5b + d)dt + C \right]$
fixed-function generator	see Figs. 12 and 13	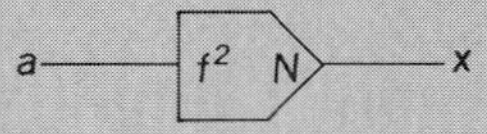 $a \xrightarrow{} \boxed{f^2 \; N} \xrightarrow{} x$	$x = a^2$
fixed-function generator	similar to above	 $a \xrightarrow{} \boxed{\log \; N} \xrightarrow{} x$	$x = \log a$
arbitrary function generator	similar to above but with adjustable R_s and R_b	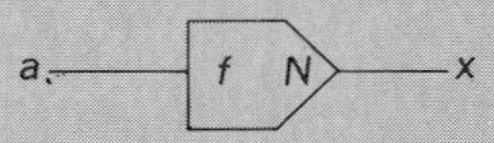 $a \xrightarrow{} \boxed{f \; N} \xrightarrow{} x$	$x = f(a)$
multiplier	see Figs. 14 and 19	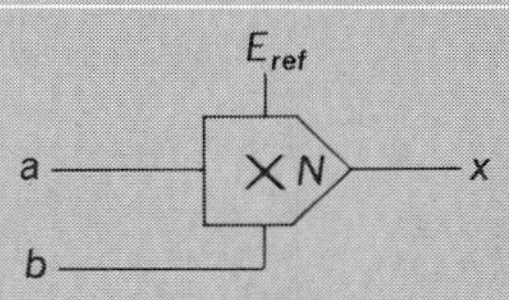 $\begin{array}{l} E_{ref} \\ a \xrightarrow{} \\ \quad \times N \xrightarrow{} x \\ b \xrightarrow{} \end{array}$	$x = -\dfrac{ab}{E_{ref}}$
comparator switch	see Fig. 15	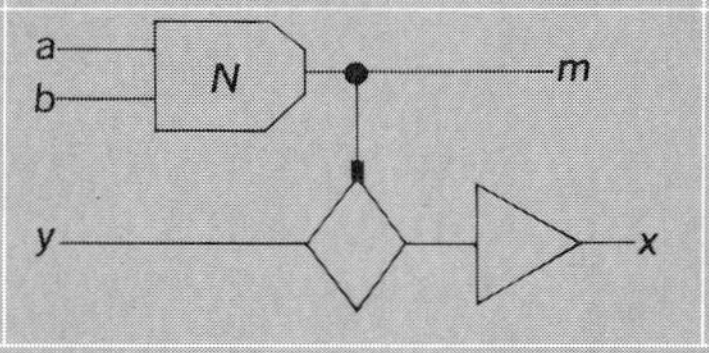 $\begin{array}{l} a \xrightarrow{} \\ b \xrightarrow{} N \xrightarrow{} m \\ y \xrightarrow{} \diamond \triangleright \xrightarrow{} x \end{array}$	when $(a + b) > 0 \begin{cases} m = \text{true} \\ x = -y \end{cases}$ when $(a + b) < 0 \begin{cases} m = \text{false} \\ x = 0 \end{cases}$
signal flow	analog signal (voltage)	mechanical contact 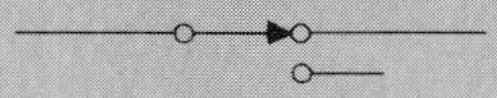	connections 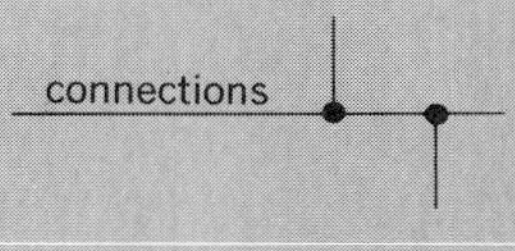 ● ● no connection 
notes	$M =$ megohms (10^6 ohms) $K =$ kilohms (10^3 ohms)	N is assignment number of element μ is open loop gain of amplifier	

Fig. 16. Symbols used in the programming of analog circuit diagrams.

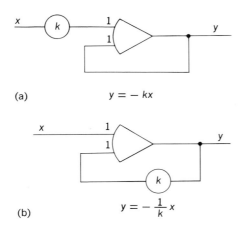

(a) $\qquad y = -kx$

(b) $\qquad y = -\dfrac{1}{k}x$

Fig. 17. Movement of an attenuator from the input to the feedback network of an operational amplifier to create the inverse function: the coefficient changes from k in a to the inverse, $1/k$ in b.

feedback circuit, the amplifier will compute the square root. In general, when a function is exchanged between the input and feedback positions of an operational amplifier, the inverse function is obtained. For example, the constant k in Fig. 17a becomes the reciprocal $1/k$ when relocated in the feedback circuit of Fig. 17b. Similarly, the log function exchange (Fig. 18a and b) provides the exponential function.

No special device is needed in an analog computer to divide one variable by another. Division is achieved by placing a multiplier in the feedback path of an amplifier, as shown in Fig. 19. (For the division to be stable, the denominator W in Fig. 19b must always be negative and larger than zero, so that there is no positive feedback loop around the amplifier.) Similarly, the derivative of a function can be obtained by using this same principle. Since differentiation is the inverse mathematical operation of integration, the feedback circuit in Fig. 20 incorporates an integrator to satisfy the requirement. The function of differentiation, due to its nature, exhibits higher gain at higher signal

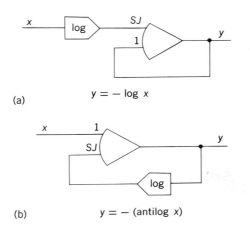

(a) $\qquad y = -\log x$

(b) $\qquad y = -(\text{antilog } x)$

Fig. 18. Movement of a logarithm function generator from the input to the feedback network of an operational amplifier to change the amplifier function from (a) log x to (b) antilog x (exp x).

frequencies, a characteristic which may often lead to excessive amplification of high-frequency circuit noise; therefore, the addition of a feedback attenuator, indicated with a broken line in Fig. 20, is often useful to reduce the gain above a selected high-frequency roll-off corner. *See* ELECTRIC FILTER.

Solution of algebraic equations. Programming an analog computer to solve an equation such as Eq. (2) can be described by referring to Fig. 21.

$$y = x^2 + 4x - 50 \qquad (2)$$

Basically, a circuit is devised to generate each term of the equation; the output of each is then summed algebraically by the final amplifier.

The first term results from a squaring function generator in the input circuit of amplifier 6 as defined previously in Fig. 12. The arithmetic for the second term is programmed by dividing the forward gain of amplifier 5 by its feedback gain, 20/5,

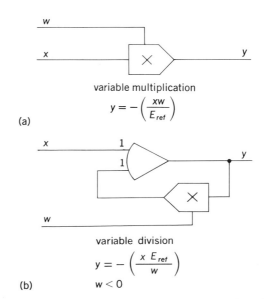

variable multiplication

(a) $\qquad y = -\left(\dfrac{xw}{E_{ref}}\right)$

variable division

(b) $\qquad y = -\left(\dfrac{x\,E_{ref}}{w}\right)$
$\qquad w < 0$

Fig. 19. Use of a multiplier (a) to create the variable multiplication function, and (b) in the feedback path of an amplifier to create the variable division function.

to yield a net gain of 4. The third term derives from the reference voltage by an adjustment of attenuator 2. Each circuit of the program operates in parallel with all others and instantaneously, so that y is explicitly and immediately developed; thus the circuit will handle either varying or steady-state values of the input signal x.

Because the analog computer has a limited operational range and therefore resolution, the desired practical ranges of the various terms in a problem must be scaled to fit the analog voltage range. This requirement corresponds somewhat to that of keeping track of the decimal point when using the slide rule. The circuit of Fig. 21b is scaled for a specific range such that a signal voltage V is related to the numerical value N by a scale factor α. The maximum value of 80 for x can be directly accommodated within the 100-V range of the computer. Accordingly, its scale factor α_1 is set at unity. But the corresponding maximum value for y

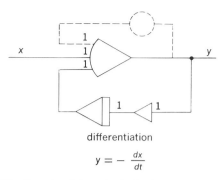

differentiation

$$y = -\frac{dx}{dt}$$

Fig. 20. Use of an integrator in the feedback path of an operational amplifier to achieve differentiation with respect to time, dx/dt.

would exceed the unscaled computer range, necessitating an assignment of 0.01 for α_9 (meaning the y signal will not rise above 66.70 V). All the signals collected at the summing junction of amplifier 3 must have scales that are compatible with the output variables; therefore α_2, α_5, and α_8 are matched to α_9. Taking into consideration the reference voltage and the choice of slope resistors used to establish the square function generator, a scale of 0.01 at amplifier 6 seems practical when a feedback gain of 10 is used. This scale factor relates directly with α_1^2 to yield the desired α_2 factor. To adjust the second-term circuit for α_5, the forward gain of amplifier 5 must be lowered by a decade to set α_3, and attenuator 8 was added to insert α_4 so that the product of α_3 and α_4 would produce the proper scale factor α_5. The necessary α_8 in the third circuit arises from the α_6 associated with a 100-V reference and a readjustment of attenuator 2, taking into account α_7 ($10^2 \times 10^{-4} = 10^{-2}$).

Use with calculus. Because of its fundamental ability to perform continuous integration, the analog computer is particularly well suited to solve ordinary differential equations, typically arising in the problems of calculus. Because of this ability, the analog computer has at times been referred to as the electronic differential analyzer. Significantly, the analog computer integrator will respond correctly to any simple or complex signal waveform presented at its input; it is not restricted to applications involving conventional classes of functions defined by analytic calculus.

The analog program for a third-order, nonlinear differential equation is illustrated in Fig. 22. The equation is nonlinear because the third term has a variable coefficient, $7x$, operating on the first derivative. As a result, this equation has no known analytic solution in closed form. Nevertheless, the analog computer does generate the solution for x by virtue of its instantaneous, parallel structure, its inherent capability to carry out continuous integrations, its ability to multiply variables, and its feature of problem feedback permitting implicit computation. The first step in the programming procedure is to algebraically rearrange the equation so that the highest-order derivative stands alone on the left side, as shown in the second expression in Fig. 22. The second step in the programming is to assume that a combination of signals representing the highest time derivative exists in the form of the summed inputs of an integrator

(this condition is justified below). Then, by integration, the next lower time derivative will appear at the output of this integrator. By using a cascade of integrators (composed of three integrators in this case), each derivative is developed in turn until the dependent variable finally is generated. Next, each term on the right side of the expression may now be assembled through separate analog components and circuits and then gathered at the input of the first integrator. Now, when the circuits are activated electronically, the signals that the expression equates as the highest derivative will exist in fact at the integrator, justifying the original assumption. The circuit configuration involves the feedback of problem variables; therefore it is said that an implicit rather than an explicit solution of the equation is being made.

The analog computer can only represent the independent variable as time (clock time or real time). That is, regardless of the nature of the independent dimension in the simulated system, all derivatives in the computer representation are taken with respect to time. Because the computer accommodates a practical but finite time span, time ranges must be scaled in a fashion similar to that for voltage amplitudes discussed earlier. Computer time T is related to real system time t by a scale factor β. Also, an integrator operates on a

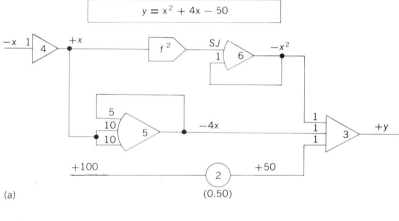

(a)

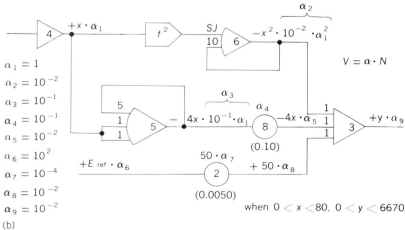

$\alpha_1 = 1$
$\alpha_2 = 10^{-2}$
$\alpha_3 = 10^{-1}$
$\alpha_4 = 10^{-1}$
$\alpha_5 = 10^{-2}$
$\alpha_6 = 10^2$
$\alpha_7 = 10^{-4}$
$\alpha_8 = 10^{-2}$
$\alpha_9 = 10^{-2}$

(b)

Fig. 21. Analog computer circuits for solving a problem in algebra. (a) Unscaled circuit reflecting the equation given in box. (b) Circuit scaled for computation in computer with 100-V range, with appropriate scale factors $\alpha_1 - \alpha_9$.

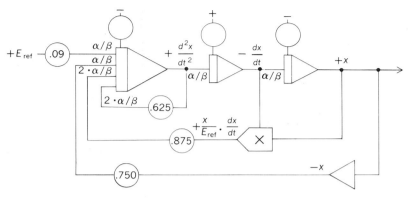

$$4\frac{d^3x}{dt^3} + 5\frac{d^2x}{dt^2} + 7x\frac{dx}{dt} - 3x + 36 = 0$$

$$-\frac{d^3x}{dt^3} = \frac{5}{4}\frac{d^2x}{dt^2} + \frac{7}{4}x\frac{dx}{dt} - \frac{3}{4}x + 9$$

Fig. 22. Analog computer circuit for solving a problem in calculus, with third-order differential equation shown in box.

(responses take place at speeds slower than real time); and when β is less than one, computer time is compressed (and responses occur faster than in real time). In arranging the time scales, it is mandatory that the β value be identical for all integrators within one interdependent system program. The potentiometer shown at the top of each integrator is used to preset the integrator to its proper initial condition (also called the constant of integration), as will be discussed later.

Use as simulator. The dynamic system in Fig. 23a is presented to illustrate how an analog computation can be built up block by block directly from the characteristics of a system. Each part of the automobile suspension system is treated as a black box containing a second-order response mechanism. These boxes are then strongly intercoupled to the main chassis member. The design objective would be to adjust the masses M_i, the spring constants K_i, and the dampers D_i to minimize x_1, the seat displacement that the driver experiences under various road conditions. Figure 23b shows the assembled analog computer model, with its variables labeled.

Finding a single combination of parameter settings is difficult, because at different road speeds the forcing functions at the front and rear wheels, F_2 and F_3, reflect a shift in frequency content and a difference in the excitation delay time from front to back. Once the building block model has been assembled, it is possible to see that a set of four

voltage with respect to time, so that the gain of an integrator is determined by the ratio α/β. When β is one, the computer calculates in real time; when β is greater than one, computer time is expanded

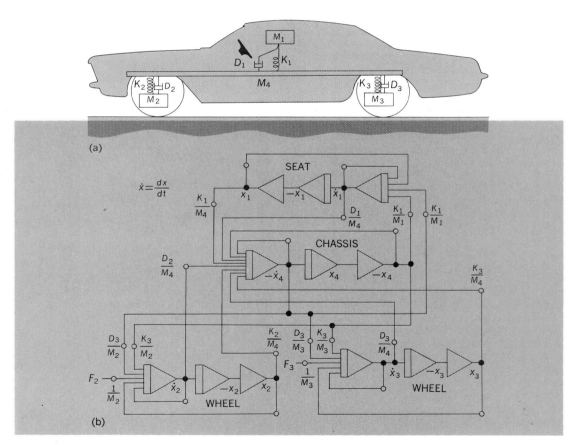

Fig. 23. Construction of an analog computer model. (a) Approximation of the suspension system of an automobile with damped second-order systems. (b) Diagram of corresponding analog computer model. (*From T. H. Truitt and A. E. Rogers, Basics of Analog Computers, John F. Rider, 1960*)

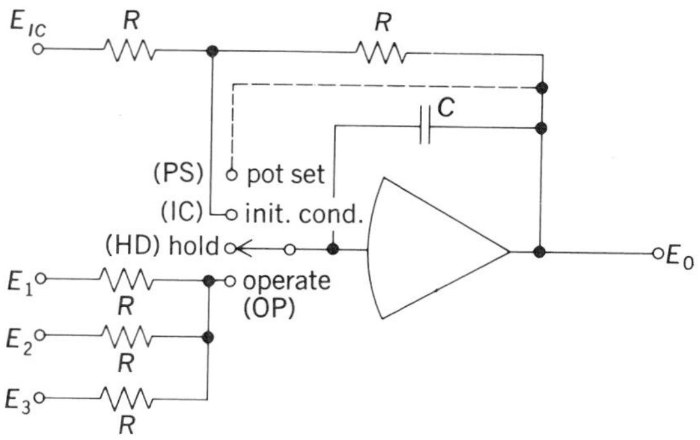

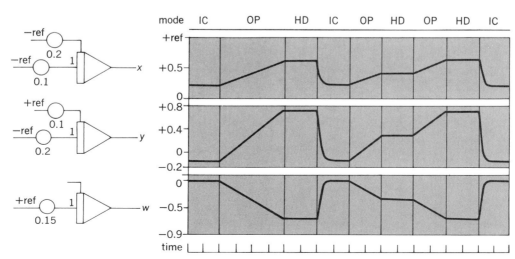

PS: $E_0 = 0$

IC: $E_0 = -E_{IC}$

OP: $E_0 = -E_{IC} - \dfrac{1}{RC}\displaystyle\int_0^t (E_1 + E_2 + E_3)\, dt$

HD: $E_0 = E_0$

Fig. 24. Integrator controls.

simultaneous differential equations could be extracted from the analog circuit configuration to represent the system mathematically.

OPERATION

After a program has been connected up on the patchboard in accordance with the foregoing programming procedures, the problem patchboard inserted in the patch bay, and the attenuators adjusted to their proper values, the analog computer is ready for a solution run. The mode of operation of the computer is governed by integrator controls, as shown in Fig. 24. The input to an integrator amplifier is connected to one of four possible points: potset (PS) for adjusting potentiometers, in which the integrator is effectively short-circuited; the reset or initial condition (IC) mode, in which the capacitor is charged up to the initial value E_{IC}; hold (HD), which is a stop and freeze mode mainly for readout purposes; and operate (OP) mode, in which the intended continuous integration with respect to time takes place. The switching be-

tween these modes is done with high-speed electronic analog signal switches (older computers use relay contacts) controlled from the computer console or via logic command signals connected on the patchboard. At the beginning of a run, the computer is put in the reset or initial condition mode, energizing a bus circuit so that the integrators assume their initial conditions. Next, the computer is switched to the operate mode, removing the initial condition inputs and connecting the integrand inputs to the integrators. The generated and connected input signals are now integrated by the integrator according to its programmed time scale.

A study of the recorded responses in Fig. 25 for the x, y, and w integrators reveals how each linear ramp function develops by the integration of its particular constant-voltage input signal. As the computer modes are exercised, each integrator behaves in accordance with its initial condition and the polarities and amplitudes of input signals imposed upon it. During the normal problem run, the integrators respond to rapidly varying, interacting signals, and many different dynamic results are generated.

In operating the computer, the user must be prepared to deal with many practical circumstances and to diagnose the troubles or errors that arise from the analog computer program. Mistakes in circuit patching, incorrect attenuator settings, wrong reference polarities, misapplication of ranges of computer units and display devices, and inappropriate time scale factor assignments may occur, as well as a host of details associated with the electronic properties of the computer. A smooth-running program, however, returns an extremely large amount of problem insight and experience, well worth the effort required to get the problem running.

HYBRID COMPUTERS

The accuracy of the calculations on a digital computer can often be increased through double precision techniques and more precise algorithms, but at the expense of extended solution time, due to the computer's serial nature of operation. Also,

Fig. 25. Integrator control: outputs from three integrators are recorded with respect to time, showing how they depend on the computer control modes, integration initial conditions, and integrand inputs.

Fig. 26. Analog-hybrid laboratory equipped with a HY-SHARE 600 system. This system can include up to six analog computer consoles and one or more digital processors for multiuser, multitask applications. (*Electronic Associates, Inc.*)

the more computational steps there are to be done, the longer the digital computer will take to do them. On the other hand, the basic solution speed is very rapid on the analog computer because of its parallel nature, but increasing problem complexity demands larger computer size. Thus, for the analog computer the time remains the same regardless of the complexity of the problem, but the size of the computer required grows with the problem.

Interaction between the user and the computer during the course of any calculation, with the ability to vary parameters during computer runs, is a highly desirable and insight-generating part of computer usage. This hands-on interaction with the computed responses is simple to achieve with analog computers. For digital computers, interaction usually takes place through a computer keyboard terminal, between runs, or in an on-line stop-

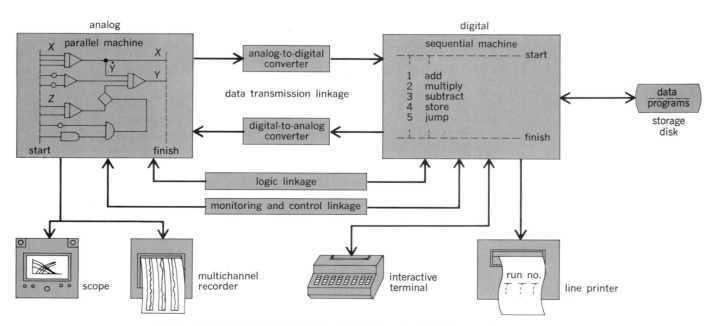

Fig. 27. Block diagram representation of a hybrid analog/digital computer.

go mode. An often-utilized system combines the speed and interaction possibilities of an analog computer with the accuracy and programming flexibility of a digital computer. This combination is specifically designed into the hybrid computer.

A modern analog-hybrid console is depicted in Fig. 26. It contains the typical analog components plus a second patchboard area to include programmable, parallel logic circuits using high-speed gates for the functions of AND, NAND, OR, and NOR, as well as flip-flops, registers, and counters. The mode switches in the integrators are interfaced with the digital computer to permit fast iterations of dynamic runs under digital computer control. As shown in Fig. 27, data flow in many ways and formats between the analog computer with its fast, parallel circuits and the digital computer with its sequential, logic-controlled programs. Special high-speed analog-to-digital and digital-to-analog converters translate between the continuous signal representations of variables in the analog domain and the numerical representations of the digital computer. Control and logic signals are more directly compatible and require only level and timing compatibility. *See* ANALOG-TO-DIGITAL CONVERTER; DIGITAL-TO-ANALOG CONVERTER.

The programming of hybrid models is a more complex challenge than described above, requiring the user to consider the parallel action of the analog computer interlaced with the step-by-step computations progression in the digital computer. For example, in simulating the mission of a space vehicle, the capsule control dynamics will typically be handled on the analog computer in continuous form, but interfaced with the digital computer, where the navigational trajectory is calculated. *See* COMPUTER.   [PER A. HOLST]

Bibliography: R. A. Collacott, *Simulators: International Guide*, 1973; P. A. Holst, *Computer Simulation 1951–1976: An Index to the Literature*, 1979; A. S. Jackson, *Analog Computation*, 1960; G. A. Korn and T. M. Korn, *Electronic Analog and Hybrid Computers*, 1971; Society for Computer Simulation, *Simulation*, monthly; R. E. Stephenson, *Computer Simulation for Engineers*, 1971.

Analog-to-digital converter

A device for converting the information contained in the value or magnitude of some characteristic of an input signal, compared to a standard or reference, to information in the form of discrete states of a signal, usually with numerical values assigned to the various combinations of discrete states of the signal.

Analog-to-digital (A/D) converters are used to transform analog information, such as audio signals or measurements of physical variables (for example, temperature, force, or shaft rotation) into a form suitable for digital handling, which might involve any of these operations: (1) processing by a computer or by logic circuits, including arithmetical operations, comparison, sorting, ordering, and code conversion, (2) storage until ready for further handling, (3) display in numerical or graphical form, and (4) transmission.

If a wide-range analog signal can be converted, with adequate frequency, to an appropriate number of two-level digits, or bits, the digital representation of the signal can be transmitted through a

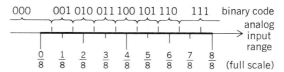

Fig. 1. A three-bit binary representation of a range of input signals.

noisy medium without relative degradation of the fine structure of the original signal. *See* COMPUTER GRAPHICS; DATA COMMUNICATIONS; DIGITAL COMPUTER.

Concepts and structure. Conversion involves quantizing and encoding. Quantizing means partitioning the analog signal range into a number of discrete quanta and determining to which quantum the input signal belongs. Encoding means assigning a unique digital code to each quantum and determining the code that corresponds to the input signal. The most common system is binary, in which there are 2^n quanta (where n is some whole number), numbered consecutively; the code is a set of n physical two-valued levels or bits (1 or 0) corresponding to the binary number associated with the signal quantum.

Figure 1 shows a typical three-bit binary representation of a range of input signals, partitioned into eight quanta. For example, a signal in the vicinity of 3/8 full scale (between 5/16 and 7/16) will be coded 011 (binary 3). *See* NUMBER SYSTEMS.

Conceptually, the conversion can be made to take place in any kind of medium: electrical, mechanical, fluid, optical, and so on (for example, shaft-rotation-to-optical); but by far the most commonly employed form of A/D converters comprises those devices that convert electrical voltages or currents to coded sets of binary electrical levels (for example, +5 V or 0 V) in simultaneous (parallel) or pulse-train (serial) form, as shown in Fig. 2. The serial output is not always made available.

The converter depicted in Fig. 2 converts the analog input to a five-digit "word." If the coding is binary, the first digit (most significant bit, abbreviated MSB) has a weight of 1/2 full scale, the second 1/4 full scale, and so on, down to the nth digit (least significant bit, abbreviated LSB), which has

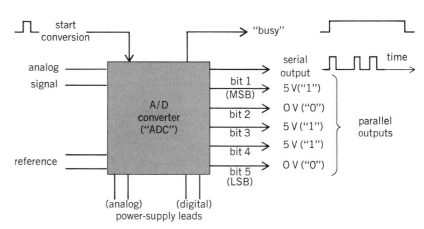

Fig. 2. Basic diagram of an analog-to-digital converter, showing parallel and serial (return-to-zero) output formats for code 10110.

Binary resolution equivalents*

Bit	2^{-n}	$1/2^n$ (fraction)	dB (decibels)	$1/2^n$ (decimal)	%	ppm (parts per million)
FS†	2^0	1	0	1.0	100	1,000,000
MSB‡	2^{-1}	1/2	−6	0.5	50	500,000
2	2^{-2}	1/4	−12	0.25	25	250,000
3	2^{-3}	1/8	−18.1	0.125	12.5	125,000
4	2^{-4}	1/16	−24.1	0.0625	6.2	62,500
5	2^{-5}	1/32	−30.1	0.03125	3.1	31,250
6	2^{-6}	1/64	−36.1	0.015625	1.6	15,625
7	2^{-7}	1/128	−42.1	0.007812	0.8	7,812
8	2^{-8}	1/256	−48.2	0.003906	0.4	3,906
9	2^{-9}	1/512	−54.2	0.001953	0.2	1,953
10	2^{-10}	1/1024	−60.2	0.0009766	0.1	977
11	2^{-11}	1/2048	−66.2	0.00048828	0.05	488
12	2^{-12}	1/4096	−72.2	0.00024414	0.024	244
13	2^{-13}	1/8192	−78.3	0.00012207	0.012	122
14	2^{-14}	1/16,384	−84.3	0.000061035	0.006	61
15	2^{-15}	1/32,768	−90.3	0.0000305176	0.003	31
16	2^{-16}	1/65,536	−96.3	0.0000152588	0.0015	15
17	2^{-17}	1/131,072	−102.3	0.00000762939	0.0008	7.6
18	2^{-18}	1/262,144	−108.4	0.000003814697	0.0004	3.8
19	2^{-19}	1/524,288	−114.4	0.000001907349	0.0002	1.9
20	2^{-20}	1/1,048,576	−120.4	0.0000009536743	0.0001	0.95

*From D. H. Sheingold (ed.), *Analog-Digital Conversion Handbook*, Analog Devices, Inc., 1972.
†Full scale.
‡Most significant bit.

a weight of 2^{-n} of full scale (1/32 in this example). Thus, for the output word shown, the analog input must be given approximately by the equation shown. The number of bits, n, characterizes the

$$\frac{16}{32} + \frac{0}{32} + \frac{4}{32} + \frac{2}{32} + \frac{0}{32} = \frac{22}{32} = \frac{11}{16}\ \text{FS (full scale)}$$

resolution of a converter. The table translates bits into other conventional measures of resolution in a binary system.

Figure 2 also shows a commonly used configuration of connections to an A/D converter: the analog signal and reference inputs; the parallel and serial digital outputs; the leads from the power supply, which provides the required energy for operation;

and two control leads—a start-conversion input and a status-indicating output (busy) that indicates when a conversion is in progress. The reference voltage or current is often developed within the converter.

Second in importance to the binary code and its many variations is the binary-coded decimal (BCD), which is used rather widely, especially when the encoded material is to be displayed in numerical form. In BCD, each digit of a radix-10 number is represented by a four-digit binary subgroup. For example, the BCD code for 379 is 0011 0111 1001. The output of the A/D converter used in digital panel meters is usually BCD.

Techniques. There are many techniques used for A/D conversion, ranging from simple voltage-

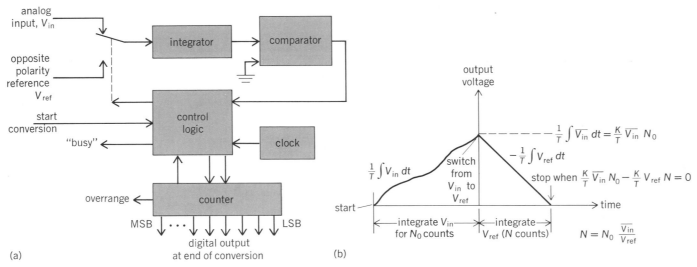

(a)

(b)

Fig. 3. Example of dual-slope conversion. (a) Block diagram of converter. (b) Integrator output. Here, k is a constant, and T is the RC (resistor-capacitor) time constant of the integrator.

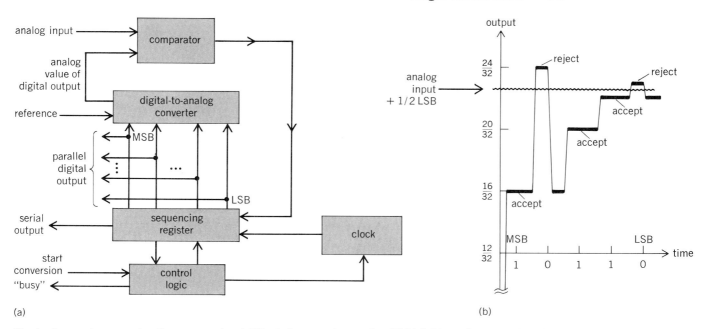

Fig. 4. Successive-approximations conversion. (*a*) Block diagram of converter. (*b*) Digital-to-analog converter output for the example in Fig. 2.

level comparators to sophisticated closed-loop systems, depending on the input level, output format, control features, and the desired speed, resolution, and accuracy. The two most popular techniques are dual-slope conversion and successive-approximations conversion.

Dual slope. Dual-slope converters have high resolution and low noise sensitivity; they operate at relatively low speeds—usually a few conversions per second. They are primarily used for direct dc measurements requiring digital readout; the technique is the basis of the most widely used approach to the design of digital panel meters.

Figure 3*a* is a simplified block diagram of a dual-slope converter. The input is integrated for a period of time determined by a clock-pulse generator and counter (Fig. 3*b*). The final value of the signal integral becomes the initial condition for integration of the reference in the opposite sense, while the clock output is counted. When the net integral is zero, the count stops. Since the integral "up" of the input over a fixed time (N_0 counts) is equal to the integral "down" of the fixed reference, the ratio of the number of counts of the variable period to that of the fixed period is equal to the ratio of the average value of the signal to the reference.

Successive approximations. Successive-approximations conversion is a high-speed technique used principally in data-acquisition and computer-interface systems. Figure 4*a* is a simplified block diagram of a successive-approximations converter. In a manner analogous to the operation of an apothecary's scale with a set of binary weights, the input is "weighed" against a set of successively smaller fractions of the reference, produced by a digital-to-analog (D/A) converter that reflects the number in the output register. *See* DIGITAL-TO-ANALOG CONVERTER.

First, the MSB is tried (1/2 full scale). If the signal is less than the MSB, the MSB code is returned to zero; if the signal is equal to or greater than the MSB, the MSB code is latched in the output regis-

ter (Fig. 4*b*). The second bit is tried (1/4 full scale). If the signal is less than 1/4 or 3/4, depending on the previous choice, bit 2 is set to zero; if the signal is equal to or greater than 1/4 or 3/4, bit 2 is retained in the output register. The third bit is tried (1/8 full scale). If the signal is less than 1/8, 3/8, 5/8, or 7/8, depending on previous choices. bit 3 is set to zero; otherwise, it is accepted. The trial continues until the contribution of the least-significant bit has been weighed and either accepted or rejected. The conversion is then complete. The digital code latched in the output register is the digital equivalent of the analog input signal.

Physical electronics. The earliest A/D converters were large rack-panel chassis-type modules using vacuum tubes, requiring about 1/25 m³ of space and many watts of power. Since then, they have become smaller in size and cost, evolving through circuit-board, encapsulated-module, and hybrid construction, with improved speed and resolution. Single-chip 12-bit A/D converters with the ability to interface with microprocessors are now available in small integrated-circuit packages. Integrated-circuit A/D converters, with 6-bit and better resolution and conversion rates to beyond 50 MHz, are also commercially available. *See* MICROPROCESSOR.   [DANIEL H. SHEINGOLD]

Bibliography: D. J. Dooley (ed.), *Data Conversion Integrated Circuits*, 1980; B. M. Gordon et al., *Data-Conversion Systems Digest*, 1977; *IEEE Transactions on Circuits and Systems*, special issue on analog/digital conversion, AS-25, July 1978; D. Sheingold (ed.), *Analog-Digital Conversion Notes*, 1977; E. Zuch (ed.), *Data Acquisition and Conversion Handbook* (1979).

Angle modulation

Modulation in which the angle (entire argument) of a sinusoidal carrier is the parameter changed by the modulating wave. This variation in angle may be related to the modulating wave in any predetermined unique manner. Frequency and phase modulation are particular forms of angle modulation. Often the term frequency modulation is used to connote angle modulation. For example, as the frequency of a sinusoidal signal input sweeps across the program frequency band, the output of a typical FM (frequency-modulation) broadcast transmitter varies from almost pure phase to almost pure frequency modulation, and in between the angle changes some other way. *See* FREQUENCY MODULATION; MODULATION; PHASE MODULATION. [HAROLD S. BLACK]

Antenna

A portion of a radio system especially designed to couple electromagnetic energy between free space and transmission lines. At a radio transmitter the antenna couples energy from the transmitter circuits into free space, thus serving to radiate the transmitter's output as radio waves. At the radio receiver the antenna couples energy from the radio waves into the receiver's circuits. The same antenna structure functions equally well for transmis-

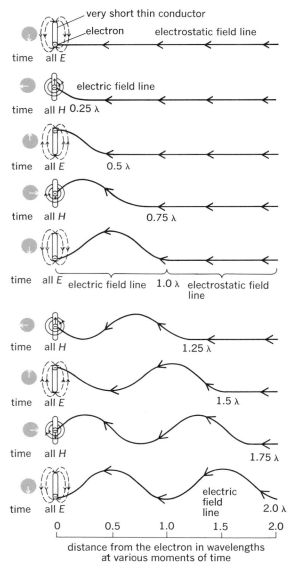

Fig. 1. Graphical representation of electric E and magnetic H fields of oscillating electron.

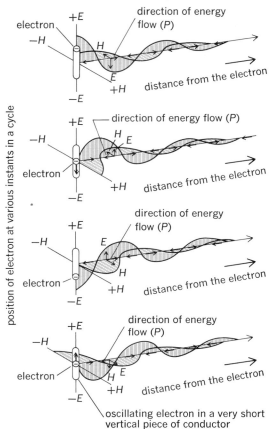

Fig. 2. Induction field of oscillating electron.

sion or reception. Because of this reciprocity, all antennas can serve for either transmission or reception; however, in describing antenna operation it is often more direct to describe the radiation action, with the understanding that reception is the reciprocal. *See* RECIPROCITY PRINCIPLE.

Despite this reciprocity antennas used for transmission may differ significantly from those used for reception because of the power levels at which they operate. An antenna for transmitting to a major portion of the world or to a space vehicle must handle large powers; the antenna for receiving from radar reflections or from communication satellites handles negligible power. For these reasons antennas are classified by use; they may also be classified by operating frequency. Antennas for very low frequencies may be a mile long, whereas antennas for very high frequencies may be only a few feet long. More basically, if antenna size is measured in units of the wavelength of the frequency at which the antenna operates, structures that differ greatly in size can be described in similar terms. In describing antennas here, their principal dimensions in wavelengths will be given.

Mechanism of radiation. For an antenna to radiate, it must contain charges that oscillate. Such a charge is the electron shown in Fig. 1. Because no other nearby charges oscillate to counteract the effect of the electron, its field extends infinitely into space. This field contains two components: an electric field E and a magnetic field H. Energy stored in the electric field when the electron is stationary equals energy in the magnetic field when the electron is moving most rapidly. Basically, the changing electric field induces an adjacent magnetic field, and the changing magnetic field induces an adjacent electric field. The result of this continuing energy exchange is radiation.

Figure 1 is an elementary graphical representation of the radiating electric field while a single electron oscillates on a short, thin conductor. At the top of the figure, one line of the electric field terminates at the electron. Successive drawings down the figure show this field line as the electron oscillates on the conductor. Motion of the electron displaces the line. This displacement travels along the field line, moving in the figure to the right, with the velocity of light. After the electron has completed one cycle of oscillation on its conductor, the initial displacement has traveled one wavelength λ from the conductor.

More rigorously, the oscillating electron has an induction field which decreases rapidly with distance (Fig. 2). Energy in this field flows alternately away from and toward the electron and never escapes into space. The magnetic component of this field is a maximum when the electric component is zero. In the radiation field, which extends indefinitely into space, energy flows only outward at all instants of time and all locations in space; magnetic and electric components of this field pass through zero together (Fig. 3).

Along with the oscillating negative charge, whose fields are shown in Figs. 1–3, is a positive charge. If all oscillating charges on a short, thin conductor, called a doublet, are considered, the

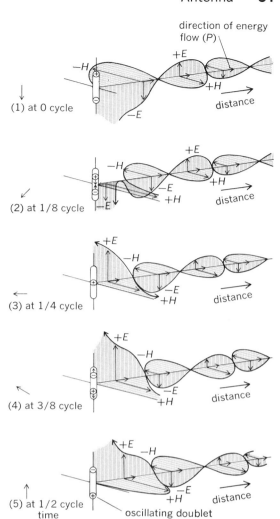

Fig. 4. Electromagnetic field near oscillating doublet.

total electromagnetic field is found to contain terms that decrease with distance in the ratios $1/r^3$ and $1/r^2$ in the induction field and $1/r$ in the radiation field. The real part of the resultant field near the doublet is shown in Fig. 4. The $1/r$ decrease in radiation field as it spreads out into space from the doublet is due to the greater area of the sphere that bounds it. See ELECTROMAGNETIC WAVE TRANSMISSION.

The elemental radiator can be either a short electric device or a small current device. If infinitesimally small, their far fields are as given in Fig. 5. Even if the radiators have maximum dimensions of 0.1 λ, the equations are useful approximations for short electric conductors and small current loops.

Pattern shape. As Fig. 5 shows, a small electric or magnetic dipole radiates no energy along its axis, the contour of constant energy being a toroid. This contour in space, called the radiation pattern, is usually the most basic requirement of an antenna. The purpose of a transmitting antenna is to direct power into a specified region; whereas the purpose of a receiving antenna is to accept signals from a specified direction. In the case of a vehicle, such as an automobile with a car radio, the receiving antenna needs a nondirectional pattern so that

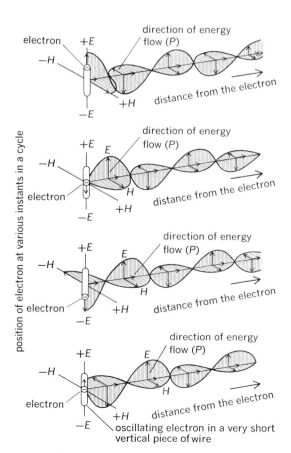

Fig. 3. Radiation field of oscillating electron.

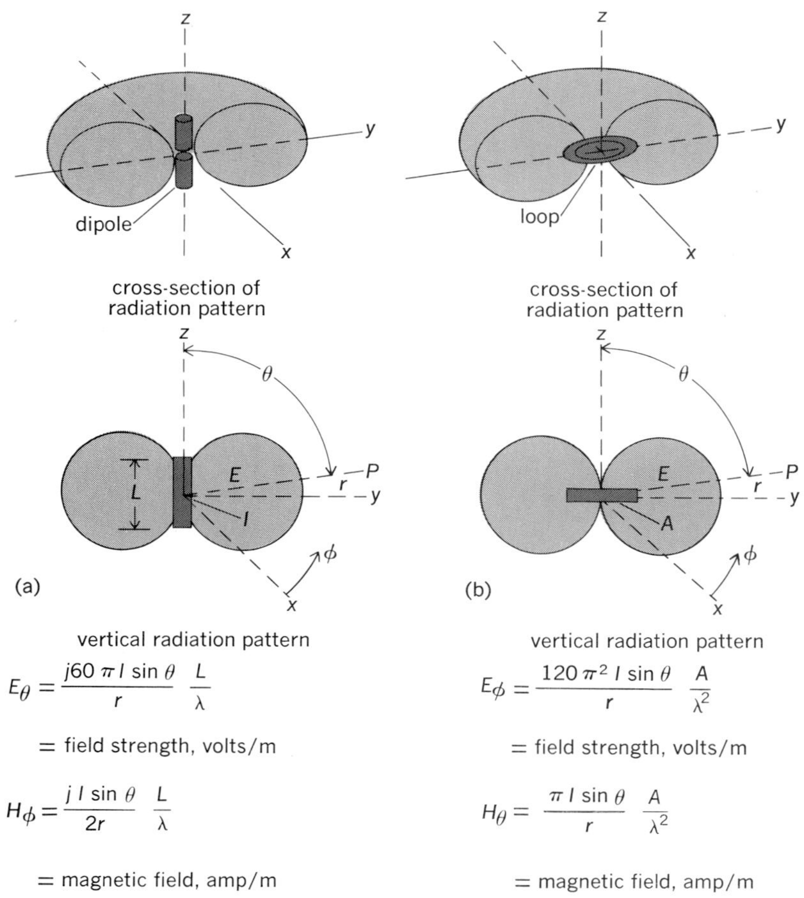

vertical radiation pattern

$$E_\theta = \frac{j60\,\pi\,I\,\sin\theta}{r}\,\frac{L}{\lambda}$$

= field strength, volts/m

$$H_\phi = \frac{j\,I\,\sin\theta}{2r}\,\frac{L}{\lambda}$$

= magnetic field, amp/m

vertical radiation pattern

$$E_\phi = \frac{120\,\pi^2\,I\,\sin\theta}{r}\,\frac{A}{\lambda^2}$$

= field strength, volts/m

$$H_\theta = \frac{\pi\,I\,\sin\theta}{r}\,\frac{A}{\lambda^2}$$

= magnetic field, amp/m

Fig. 5. Far-field patterns. (a) Small electric dipole. (b) Small magnetic dipole.

it can accept signals from variously located stations, and from any one station, as the automobile moves. The antenna of a broadcast station may be directional; for example, a station in a coastal city would have an antenna that concentrated most of the power over the populated land. The antenna for transmission to or from a communication satellite should have a narrow radiation pattern directed toward the satellite for efficient operation, preferably radiating essentially zero power in other directions to avoid interference. Each special application of radio waves has its unique requirements for radiation patterns. *See* RADAR.

For highly directional beam-type patterns, the concept of directivity is useful in the description of pattern shape. A simple approximation is based on beam area B at the half-power beam width of the pattern, that is, at the angles to the sides of the central beam axis where the radiated power is half the power radiated along the beam axis. Beam area B is defined as the solid angle through which all power would be radiated if the power were constant over the complete solid angle and at maximum level. For these hypothetical conditions directivity D is defined as the surface of the sphere (4π radians) divided by beam area B in radians, as shown in Eq. (1). If the beam is described as its width in degrees in two orthogonal directions, $D =$

$$D = \frac{4\pi}{B} \qquad (1)$$

$41,253/\theta\phi$, wheren θ and ϕ are the two beam widths. These approximations neglect power in secondary lobes and losses in the antenna.

Total radiated power P_r from an antenna can be computed by integrating the energy flow outward through an imaginary sphere surrounding the antenna system. For a hypothetical spherical radiation pattern surface which has a radius of r, the radiated power P_r is given by Eq. (2), where E_s is

$$P_r = 4\pi\,E_s^2\,r^2/R_c \qquad (2)$$

the field strength of the radiated electric field at the surface of the imaginary sphere, and R_c is the characteristic resistance of free space. If radiated power is 1 kw, field strength E_s at the surface of the imaginary sphere 1 mi in radius is 107.6 mv/m. This uniform spherical radiator has no directivity and hence no gain with a lossless antenna system. It is referred to as an isotropic radiator and is used as the standard for comparison of the gain of directional antennas.

By analogy one can think of a sphere of given volume as representing the total radiated energy. Field strength at every point on the surface of the sphere is the same as represented by the constant radius. If the sphere is deformed with its volume held constant, the distance from the origin to the surface is longer in some directions and shorter in others. These distances represent field strengths at their various locations on the surface. By the deformation, field strength is gained in a desired direction, although it is decreased in another direction. This antenna gain can be defined as the ratio of maximum field strength from the antenna of interest relative to the field strength that would be radiated from a lossless isotropic antenna fed with the same input power. In relation to directivity D, gain G can be expressed as in Eq. (3), where η is antenna efficiency.

$$G = \eta D \qquad (3)$$

Efficiency. Antenna efficiency η is defined as the radiated power divided by the input power to the antenna. Input power P_i supplies antenna system losses P_l and the radiated power P_r. This is shown by Eq. (4). Losses in an antenna system take place in the resistance of the transmission line to the antenna, the coupling network between line and antenna, antenna conductors, insulators,

$$\eta = \frac{P_r}{P_l + P_r} \qquad (4)$$

and the ground system. For electrically small antennas even small loss resistances can significantly reduce the efficiency. In the design of transmitting antennas, the heating effect of losses must also be considered together with losses due to corona.

Polarization. As shown in Fig. 4, the plane of the electric field depends on the direction in which the electric charge moves on the antenna. The electric field is in a plane orthogonal to the axis of a magnetic dipole (Fig. 5). This dependence of the plane of the radiated electromagnetic wave on the orientation and type of antenna is termed polarization. By combining fields from electric and magnetic dipoles that have a common center, the radiated field can be elliptically polarized; by control of the contribution from each dipole, any ellipticity from plane polarization to circular polarization can be produced. A receiving antenna requires the same polarization as the wave that it is to intercept.

Impedance. The input impedance to an antenna is the ratio at its terminals, where the transmission line and transmitter or receiver are connected, of voltage to current. If the antenna is tuned to resonance at the operating frequency, the input impedance will be a pure resistance; otherwise it will also have a reactance component. For simple antennas, input or self-impedance can be computed from a knowledge of radiation pattern and current distribution. For more complicated radiating structures, input impedance is usually measured, computing values serving only as a guide.

To visualize antenna impedance, consider a thin conductor transmission line with sinusoidal current distribution, as shown in Fig. 6; the lines can be opened out to form an antenna. Input impedance of the quarter-wave line in Fig. 6a will be a low resistance, because the transmission line is resonant and because the opposing currents prevent radiation. *See* TRANSMISSION LINES.

However, as the conductors are partially opened out in Fig. 6b, current distribution will remain substantially the same; the input impedance will still be resistive because of resonance, but its value will be larger because the opposing currents are farther apart and there is now an in-phase component. Finally, when the transmission line is completely opened out in Fig. 6c, input impedance is pure resistance because of resonance and a maximum of 73 ohms, because only the in-phase component of current remains. The result is a very practical free-space half-wave antenna often used as a standard reference antenna. If one-half of this antenna is operated above ground (Fig. 6d), one has the practical quarter-wave antenna, which is also used as a reference standard with 36.5 ohms of input resistance.

If the input terminals of the half-wave antenna (Fig. 6c) are moved from the center toward the end of the resonant antenna, the input impedance remains a pure resistance but can be increased from 73 ohms at the center to a high value near the end.

From Fig. 6d the electrical phenomena can be visualized. At the base of the antenna, current I is high while voltage E is low. Because impedance Z_{in} is the complex ratio of E/I, at the base of the antenna Z_{in} is low. If the feed point is moved up the antenna where current I is lower and voltage E is higher, the driving-point impedance becomes appreciably higher.

In addition to this effect on input impedance by the standing wave on the antenna, the power radiated into space affects input impedance. Although power radiates from the whole antenna, this radiated power enters at the feed point. Here, radiated power, like all electrical power, can be expressed as the product of in-phase components of voltage E_r and current I_r. The ratio E_r/I_r of these values defines a resistance R_r, which is termed radiation resistance. For a linear conductor in free space carrying uniform current over its length s, which is very short compared to the operating wavelength λ, resistance R in ohms is shown by Eq. (5).

$$R = 80\pi^2 s^2 / \lambda^2 \qquad (5)$$

As a component of a receiving circuit, the antenna may be represented by Thévenin's or Norton's equivalence theorems. In the former the representation is a simple series circuit containing impedance, and also containing voltage induced from the incoming wave. Useful power is calculated in terms of the resulting current in the resistance of the load circuit. *See* NETWORK THEORY; THEVENIN'S THEOREM (ELECTRIC NETWORKS).

If a current element of length s is placed in a field E parallel to it, the root-mean-square voltage induced is E_s. If the element is tuned, the available power P which it captures is shown by Eq. (6),

$$P = \frac{E^2 \lambda^2}{320\pi^2} \qquad (6)$$

and is independent of s. The average power per square meter P_{av} in the plane wave is shown in Eq. (7), which indicates that the effective area of the element is $3\lambda^2/8\pi$.

$$P_{av} = \frac{E^2}{120\pi} \qquad (7)$$

If the reactive component of antenna impedance is appreciable, techniques similar to those used in other transmission circuits are required to obtain maximum power; that is, the impedance of the connected circuit must be conjugate to that of the antenna. Reflections caused by a mismatched antenna can cause trouble at a distant point in receiv-

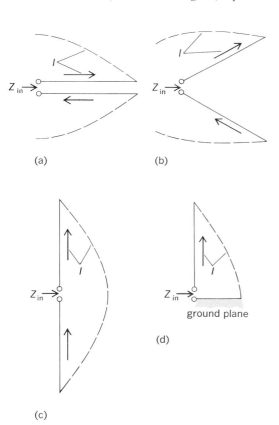

(a)

(b)

(c)

(d)

ground plane

Fig. 6. Antenna as an opened-out transmission line. (*a*) Standing wave on quarter-wave open circuit transmission line. (*b*) Standing wave on partially opened-out transmission line. (*c*) Standing wave on half-wave antenna. (*d*) Standing wave on quarter-wave antenna.

ing circuitry. The solution may be to obtain a better match or to employ a nonreciprocal device which discriminates against backward-traveling waves.

When more than one antenna is used in an array, input impedance is called a driving-point impedance because it contains self-impedance plus mutual-impedance terms due to current in other elements of the array. In an array of n elements, the driving-point impedance Z_k of the kth element can be written as in Eq. (8), where I_n is current in

$$Z_k = \frac{V_k}{I_k} = \frac{I_1}{I_k} Z_{k1} + \frac{I_2}{I_k} Z_{k2} + \cdots \\ + Z_{kk} + \cdots + \frac{I_n}{I_k} Z_{kn} \quad (8)$$

the nth element, Z_{kn} is mutual impedance between the kth element and the nth element, and V_k is the input voltage to the kth element. From a family of such equations it is possible, with a knowledge of impedances and currents, to solve a set of mesh circuit equations for all of the driving-point impedances. This information is necessary for proper design of a directional-antenna feeder system, such as is commonly used in the standard broadcast band.

Because of interaction between elements of an array, a distinction is made between a driven element to which a transmission line directly connects and a parasitic element that couples into the array only through its mutual electromagnetic coupling to driven elements. Driving-point impedance and radiation pattern are controlled by the tuning and spacing of parasitic elements relative to driven elements. If a parasitic element reduces radiation in its direction from a driven element, it is called a reflector; if it enhances radiation in its direction, it is called a director. *See* YAGI-UDA ANTENNA.

Any conductor, such as a guy wire, in the vicinity of a primary or driven antenna element may act as a parasitic element and affect antenna impedance and radiation pattern.

Bandwidth. Bandwidth of an antenna is related to its input impedance characteristics. Bandwidth may be limited by pattern shape, polarization characteristics, and impedance performance. Bandwidth is critically dependent on the value of Q; hence the larger the amount of stored reactive energy relative to radiated resistive energy, the less will be the bandwidth. *See* Q (ELECTRICITY).

For efficient operation of low standing-wave ratio at the antenna input terminals is necessary over the operating frequency range. Usually the input impedance will vary in both resistance and reactance over the desired frequency interval. It is, therefore, of interest to determine the bandwidth that can be achieved for a given standing-wave ratio. For a particular antenna the bandwidth that can be obtained by compensating networks has a theoretical limit; therefore, if a wider band is required, it becomes necessary to select antenna types with the inherent characteristic of wider bandwidth.

There are two principal approaches to frequency-independent antennas. The first is to shape the antenna so that is can be specified entirely by angles; hence when dimensions are expressed in wavelengths, they are the same at every frequency. Planar and conical equiangular spiral antennas adhere to this principle. The second principle depends upon complementary shapes so that if, for example, an antenna that is cut from a flat conducting sheet is exactly the same shape as the part removed, its input impedance

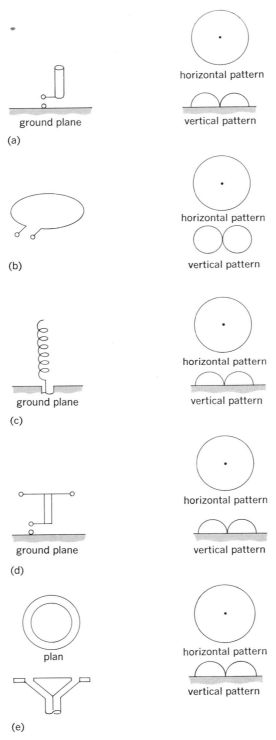

Fig. 7. Typical electrically small antennas. (*a*) Monopole. (*b*) Small loop. (*c*) Helical monopole. (*d*) Top loaded monopole. (*e*) Annular slot.

will be frequency-independent. Many log-periodic antennas employ this second principle, and before the structure shape changes very much, when measured in wavelengths, the structure repeats itself. By combining periodicity and angle concepts, antenna structures of very large bandwidths become feasible.

Electrically small antennas. Antennas whose mechanical dimensions are short compared to their operating wave-lengths are usually characterized by low radiation resistance and large reactance. This combination results in a high Q and consequently a narrow bandwidth. Current distribution on a short conductor is sinusoidal with zero current at the free end, but because the conductor is so short electrically, typically less than 30° of a sine wave, current distribution will be essentially linear. By end loading to give a constant current distribution, the radiation resistance is increased by 4 times, thus greatly improving the efficiency but not noticeably altering the pattern.

An end-fed monopole antenna (Fig. 7a) is common at very low frequencies for long-range communication, commercial broadcasting, and mobile use. The small loop antenna is used extensively for direction finding and navigation (Fig. 7b). Where height is a limiting factor at higher frequencies, the monopole height can be reduced by forming the conductor into a helical whip (Fig. 7c) or by top loading (Fig. 7d).

Where no height is permitted, a slot is used. Just as the magnetic loop is related to the electric dipole, so a slot in a conductive surface is related to a conductive wire in space. The conductive surface is usually the outside surface of a wave guide within which the radio energy travels. A slot in the wave guide allows the energy to radiate. Typically the slot is narrow and half a wavelength long. Configuration of the electric field radiated from the slot is the same as the magnetic field from a wire of like dimensions; thus one is the dual of the other (Fig. 7e). The slot may be fed by a transmission line connected across its narrow dimension, by a resonant cavity behind it, or (as mentioned above) from a wave guide. Because it is flush with a metallic surface, a slot antenna is advantageous in aircraft.

Resonant antennas. Dipoles and monopoles illustrate resonant antennas which have an approximate sinusoidal current distribution and a pure resistance at their input terminals. However, when the ratio of diameter to length is small, the input impedance varies rapidly and precludes their use as broadband antennas. The impedance bandwidth limitation can be mitigated by increasing the diameter to form a cylinder or by using conical conductors (Fig. 8).

Input resistance of a horizontal dipole varies as the antenna height increases. However, as the dipole is moved farther from the ground toward free space, its impedance tends toward 73 ohms. By folding the dipole this impedance can be increased by n^2, where n is the number of conductors; hence by using two conductors a 300-ohm line can be used to feed the antenna, or by using three conductors a 600-ohm line can be used.

Nonresonant antennas. Long-wire antennas, or traveling-wave antennas, are usually one or more wavelengths long and are untuned or nonresonant.

Horizontal single-conductor type. The radiation

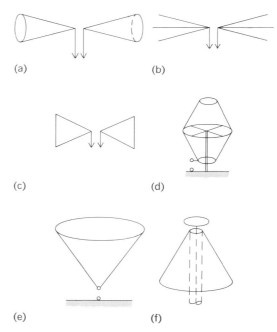

Fig. 8. Broadband dipole and monopole antennas. (*a*) Conical broadband dipole. (*b*) UHF-TV broadband dipole, built of fanned wires. (*c*) UHF-TV broadband dipole, built of flat sheet metal. (*d*) Conical monopole. (*e*) Inverted discone. (*f*) HF discone.

pattern of a long conductor in free space depends upon its length in wavelengths. A 0.5-λ conductor will radiate broadside, but as it is made longer the pattern splits, the major lobe coming closer to, but never reaching, the direction of the conductor. A cross section of the pattern is a figure of revolution around the conductor as an axis.

When a horizontal unterminated line with standing electric current waves of uniform amplitude is placed above perfect ground (Fig. 9a), it has a radiation pattern symmetrical about the center with its major lobes closest to the line in both directions. In the practical case with constant phase velocity and exponential variation in amplitude, the pattern lobes will be unsymmetrical with larger lobes toward the open end. The total number of lobes is the same as the number of 0.5 λ lengths of the conductor.

If the far end of the line is terminated (Fig. 9b), the antenna radiates only one major lobe in the direction the wave travels down the line to the termination.

Arrays of long-conductor type. Because a single long-conductor terminated antenna has rather high side lobes and a major lobe at an angle to the axis of the conductor, two long terminated conductors can be combined into a horizontal V array (Fig. 10a). The angle between the V array is determined by the length of the conductors and is usually chosen so that major lobes will add in phase.

The commonest type of long-conductor array is the rhombic antenna (Fig. 10b). Major lobes of four legs add in phase to form the resultant major lobe, while major lobes at right angles tend to cancel leaving only smaller side lobes. Because of its simplicity, low cost, and wide bandwidth, the rhombic array is widely used for both transmission and re-

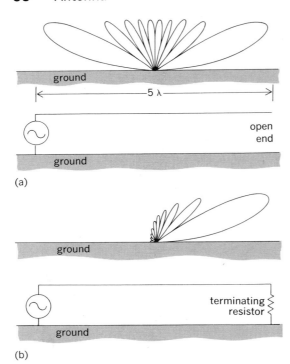

(a)

(b)

Fig 9. Long single-conductor antennas produce multiple-lobed patterns. (a) Unterminated conductor and pattern. (b) Terminated conductor and pattern.

ception where the side lobes can be tolerated and the real estate dimensions permit.

Dielectric type. Dielectric material, either as a solid rod or as a cylinder, is a wave guide. The wavelength inside a large, solid dielectric rod is less than the free-space wavelength; rather, the wavelength is this free-space length divided by the square root of the dielectric constant of the rod. If rod diameter is large compared to the wavelength inside it, most of the energy in the wave travels inside the dielectric. Even if the rod diameter is somewhat less than the wavelength inside the dielectric, the wave travels along the rod, some inside the dielectric and some outside. However, if the diameter is reduced below a half wavelength, the velocity increases to that of free space and the wave continues beyond the end of the rod into free space (Fig. 11). The major lobe is then in the direction of the rod. *See* WAVE GUIDE.

Frequency-independent type. Equiangular and log-periodic structures conform to both the angle condition and repetitive condition, hence should expand from a zero center out toward infinity. Actually, it is practical to start the structure at a radius corresponding to the shortest wavelength and to end at one representing the longest.

The antenna structure is fed at the inner radius with a coupling network that will not unduly influence the pattern shape or impedance over the operating frequency range. One method is to run a transmission line into the rear of the antenna, or long-wavelength end, then along one of the radiating elements to the feed point at the short-wavelength end, or within the transmission line that feeds the log-periodic elements.

The log-periodic dipole antenna can be used as a free-space antenna, or it can be placed over

ground to produce either horizontal or vertical polarization. A diversely polarized log-periodic antenna is obtained by using two coaxial transmission line feeds; it can generate orthogonal patterns independently, or by proper combination can produce any desired elliptically polarized pattern.

Direct-aperture type. When they are to be used at short wavelengths, antennas can be built as horns, mirrors, or lenses. Such antennas use conductors and dielectrics as surfaces or solids in contrast to the antennas described thus far, in which the conductors were used basically as discrete lines. *See* MICROWAVE OPTICS.

Horn radiators. Directivity of a horn increases with the size of the mouth of the horn; however, tolerances must be held close if the variations in magnitude and phase distribution are to be low enough to maintain high gain over a broad band. Conical and pyramidal horns are commonly used for high gain beams (Fig. 12a and b). Biconical horns (Fig. 12c) have an omnidirectional pattern with either vertical of horizontal polarization.

Luneberg lens. If in a dielectric sphere the index of refraction varies with distance from the center

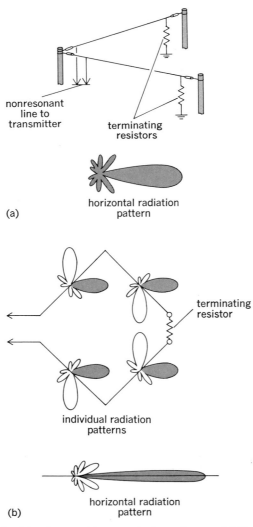

(a)

(b)

Fig. 10. Arrays of long-conductor antennas. (a) Horizontal terminated V. (b) Horizontal terminated rhombic.

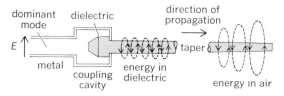

Fig. 11. Dielectric-rod antenna.

of the sphere, a plane wave falling on the surface of one hemisphere is focused at the center of the opposite hemisphere. Such an arrangement is called a Luneberg lens (Fig. 12*d*). By reciprocity, energy is fed into the lens at the focal point for transmitting; energy is removed from the lens at this point for receiving. Only the feed point need be moved around the lens to steer the pencil beam.

For high frequencies a horizontal conducting grid structure serves as a Luneberg lens antenna. Because it will focus rays from any azimuth direction, several high-frequency beams can be generated in different azimuth directions at the same time.

Reflector type. By using reflectors one can achieve high gain, modify patterns, and eliminate backward radiation. A low-gain dipole, a slot, or a horn, called the primary aperture, radiates toward a larger reflector called the secondary aperture. The large reflector further shapes the radiated wave to produce the desired pattern.

Plane sheet reflectors. One common plane reflector antenna consists of a dipole parallel to a flat conducting sheet (Fig. 13*a*). The perfectly conducting screen creates an image, and the forward pattern is the same as that of a two-element dipole array. If the screen is large enough, it prevents all back radiation. Usually the screen is limited in size and some radiation exists to the rear. Gain depends on spacing and can exceed 6 dB above a free-space dipole.

Corner reflectors. If two flat conducting sheets intersect along the *y* axis at an angle α, as shown in Fig. 13*b*, an effective directional antenna results when a dipole is placed a distance *S* from the corner. Gain depends on corner angle α, dipole spacing *S*, and antenna losses; gain can exceed 12 dB for $\alpha = 45°$. *See* CORNER REFLECTOR ANTENNA.

Parabolic reflectors. By placing an isotropic source at the focus of a parabola, the radiated wave will be reflected from the parabolic surface as a plane wave at the aperture plane (Fig. 13*c*). The physical size of the parabola determines the size of the aperture plane.

If the parabolic reflector is a surface of translation (Fig. 13*d*), an in-phase line source serves as the primary aperture, such as the two dipoles in the illustration. Beam width in the vertical, *z* direction is controlled by the parabola aperture, while the azimuth beam width, in the *y* direction, is controlled by the length of the cylindrical parabola.

If the parabolic reflector is a surface of rotation, it converts a spherical wave from an isotropic source at its focus to a uniform plane wave at its aperture. The presence of the primary antenna in the path of the reflected wave has two disadvantages: (1) The reflected wave modifies the primary radiator input impedance. (2) The primary radiator

obstructs the reflected wave over the central portion of the aperture (Fig. 13*e*).

To avoid both of these disadvantages, only a portion of the paraboloid can be used; the focus, where the primary aperture must be placed, is then to one side of the secondary aperture. For example, a primary aperture horn placed at one side illuminates the parabolic reflector, as shown in Fig. 13*f*. This type of antenna is commonly used in point-to-point microwave systems, because it is broad-band and has a very low noise level. Because of its high front-to-back ratio, the horn paraboloid reflector is used at the ground station for satellite communication.

Two-reflector antennas. A beam can be formed in limited space by a two-reflector system. The commonest two-reflector antenna, the Cassegrain system, consists of a large paraboloidal reflector. It is illuminated by a hyperbolic reflector, which in turn is illuminated by the primary feed (Fig. 14*a*). The Gregorian system (Fig. 14*b*) is similar, except that an elliptical reflector replaces the hyperbolic reflector and consequently must be placed farther out than the focal point of the paraboloid. This system is less compact than the Cassegrain system and for this reason is less popular.

In microwave relay systems one common practice is to place the paraboloid at ground level (Fig. 14*c*). The feed system is readily accessible, and long lengths of transmission line are not required. A flat sheet at the top of a tall tower redirects the beam over obstacles to the next antenna.

By using a portion of a hemisphere (Fig. 14*d*) illuminated by a feed located at focus *F*, the spherical reflecting surface causes the wave to be reflected off the ground to form a beam at an elevation angle θ. With this arrangement it is possible to

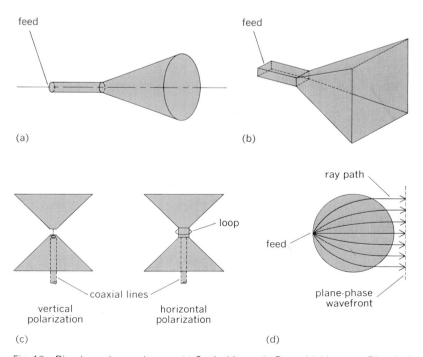

Fig. 12. Direct-aperture antennas. (*a*) Conical horn. (*b*) Pyramidal horn. (*c*) Biconical horns, electrically fed for vertical polarization or magnetically fed for horizontal polarization. (*d*) Luneberg lens with dielectric sphere between feed and plane-phase wavefront.

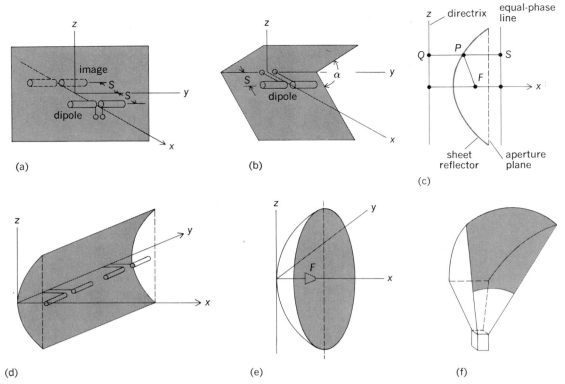

Fig. 13. Reflector antennas. (*a*) Plane sheet reflector. (*b*) Corner reflector. (*c*) Parabolic reflector. (*d*) Cy-

lindrical parabolic reflector. (*e*) Paraboloidal reflector. (*f*) Horn paraboloid reflector.

steer the beam both in elevation θ and in azimuth ϕ. Figure 15 shows such a shortwave multiplex steerable antenna system. Figure 16 shows the Arecibo instrument, a 1000-ft radio telescope antenna.

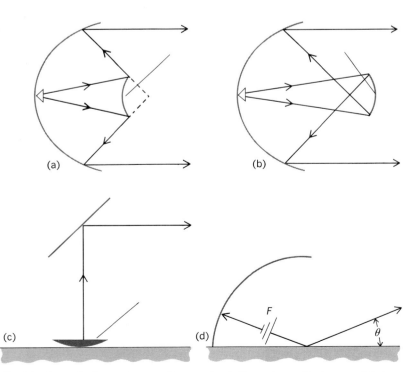

Fig. 14. Two-reflector systems. (*a*) Cassegrain. (*b*) Gregorian. (*c*) Periscope. (*d*) Spherical steerable system.

Array antennas. An array of antennas is an arrangement of several individual antennas so spaced and phased that their individual contributions add in the preferred direction and cancel in other directions. One practical objective is to increase the signal-to-noise ratio in the desired direction. Another objective may be to protect the service area of other radio stations, such as broadcast stations.

The simplest array consists of two antennas. The azimuth pattern is given by Eq. (9), where E is

$$E = E_1 \, \underline{/0°} + E_2 \, \underline{/S_2 \cos \phi + \psi_2} \qquad (9)$$

field strength for unit distance at azimuth angle ϕ, E_1 and E_2 are individual antenna nondirectional values of field strength, S_2 is electrical spacing from antenna 1 to antenna 2, and ψ_2 is electrical phasing of antenna 2 with respect to antenna 1.

The four variables that the simple two-antenna array provides make possible a wide variety of radiation patterns, from nearly uniform radiation in azimuth to a concentration of most of the energy into one hemisphere, or from energy in two or more equal lobes to radiation into symmetrical but unequal lobes.

In a directional antenna a feeder system is required for proper division of power and control of phase of the radiated fields. The feeder system is also called upon to match impedances between transmitter or transmission line and antenna.

For further control over the radiation pattern a preferred arrangement is the broadside array. In this array, antennas are placed in a line perpendicular to the bidirectional beam. Individual antenna currents are identical in magnitude and phase. The array can be made unidirectional by placing an

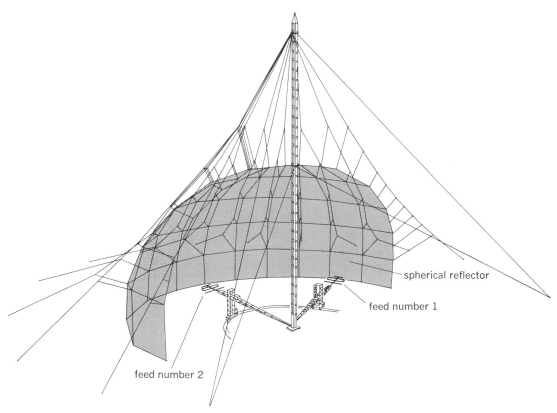

spherical reflector

feed number 1

feed number 2

Fig. 15. Superpower multiplex steerable shortwave antenna system with two feeds. Each feed is independently positionable to steer its beam and a spherical reflector of wire mesh which is accurately suspended in position from a tower.

identical array 90° to the rear and holding its phase at 90°.

The directivity of such a box array increases with the length or aperture of the array. Individual pairs of antennas in the broadside array must be spaced 217° for maximum gain with two pairs; spacing increases as more pairs are used.

Another popular arrangement is the in-line array, in which the current in each antenna is equal but the phase varies progressively to give an end-fire unidirectional pattern. The end-fire array, for a given number of elements, does not have as much gain as the broadside box array.

If antennas are stacked vertically in line to form a collinear array with the currents of each antenna in phase, a symmetrical pattern results in the azimuth plane. This type of array is used for commercial broadcasting to increase the nondirectional ground coverage for a given transmitter power.

Where more than two antennas are required to give the desired pattern shape, it is possible to use more antennas in the array. If a symmetrical pattern is satisfactory, antennas can be in line. If the arrays are not in line, the protection directions can be specified by the nulls of the individual pairs of antennas used in the array. When necessary the number of pairs of antennas can be increased to provide further control of the pattern shape.

[CARL E. SMITH]

Bibliography: American Radio Relay League, *ARRL Antenna Anthology*, 1979; American Radio Relay League, *ARRL Antenna Book*, 2d ed., 1980; H. Jasik, *Antenna Engineering Handbook*, 1961; M. T. Ma, *Theory and Application of Antenna Arrays*, 1974; W. V. Rusch and P. D. Potter, *Analysis of Reflector Antennas*, 1970; C. E. Smith, *Directional Antenna Patterns*, 2d ed., 1958; C. E. Smith, *Theory and Design of Directional Antennas*, 1959; C. E. Smith et al., *Log Periodic Antenna Design Handbook*, 1966; C. E. Smith and D. B. Hutton, *Standard Broadcast Antenna Systems*, 1968; C. H. Walter, *Traveling Wave Antennas*, 1970.

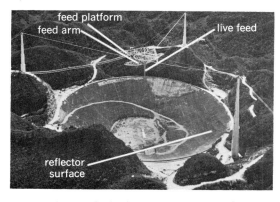

feed platform

feed arm

live feed

reflector surface

Fig. 16. Arecibo Radio Observatory antenna. Reflector is supported on a valley floor that has feed suspended from cables. Rotation of Earth serves to scan the reflector across the sky; reflector has 870-ft radius of curvature and circular aperture with 1000-ft diameter. Line feed is moved along feed arm to steer beam.

Antiresonance

The condition for which the impedance of a given electric, acoustic, or dynamic system is very high, approaching infinity. In an electric circuit consisting of a capacitor and a coil in parallel, antiresonance occurs when the alternating-current line voltage and the resultant current are in phase. Under these conditions the line current is very small because of the high impedance of the parallel circuit at antiresonance. The branch currents are almost equal in magnitude and opposite in phase.

The principle of antiresonance is used in wave traps, which are sometimes inserted in series with antennas of radio receivers to block the flow of alternating current at the frequency of an interfering station, while allowing other frequencies to pass. *See* RESONANCE (ALTERNATING-CURRENT CIRCUITS).   [JOHN MARKUS]

Arc discharge

A type of electrical conduction in gases characterized by high current density and low potential drop. It is closely related to the glow discharge but has a much lower potential drop in the cathode region, as well as greater current density. *See* GLOW DISCHARGE.

There are many arc devices, and they operate under a wide range of conditions. For example, an arc discharge may be sustained at either high pressure (of the order of atmospheres) or low pressure. The cathode may or may not be heated from an external source. Furthermore, the applied potential difference may be either direct or alternating. Numerous applications have been made of such devices, some having large commercial value. A few of these applications are illuminating devices, high-current rectifiers, high-current switches, welding devices, and ion sources for nuclear accelerators and thermonuclear devices.

Arc production. Although an arc may be initiated in several ways, it is instructive to consider the transition from a cold-cathode glow discharge (see illustration). In the normal condition, the glow partially covers the cathode, and the voltage drop remains nearly constant as the current is in-

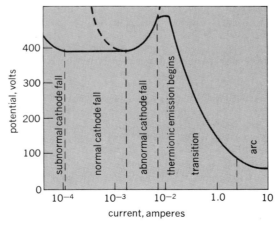

Transition from glow discharge to arc with increase of current. (*From L. B. Loeb, Basic Processes of Gaseous Electronics, University of California Press, 1955*)

creased. In this condition, the ionization is produced primarily by electron impact. As the current is increased, the cathode glow spreads out, eventually covering the cathode completely. Further increase in current can now be obtained only by an increase in the potential drop across the discharge. The cathode temperature increases in this process. This is the abnormal glow region. As the cathode temperature is increased further, thermionic emission becomes an important factor. At this point the discharge characteristic may acquire a negative slope; that is, a further increase in current may increase the cathode temperature, and the resulting thermionic emission, enough actually to reduce the potential drop. Unless the external resistance is sufficiently great to make the overall resistance positive, the discharge will change suddenly to the arc mode. Typical values in this region are a potential drop of a few tens of volts and a current ranging from amperes to thousands of amperes. For a more detailed discussion of the foregoing *see* ELECTRICAL CONDUCTION IN GASES.

Regions of an arc. There are three geometrical regions in an arc. These are the cathode fall, the anode rise, and the arc body.

Cathode fall. The cathode region is characterized by a high potential gradient. There will be a large positive ion current to the cathode, but the electron current from the cathode may be greater than this. It is not difficult to understand the large thermionic emission for a cathode of refractory metal having a very high melting temperature. However, for a metal with a low melting temperature, such as copper, the situation is not quite so clear. The best explanation appears to be that advanced first by J. J. Thomson and later by L. B. Loeb, who pointed out that the cathode must be viewed on a microscopic basis, and that the local temperature may be vastly greater than the macroscopic cathode temperature. Under these conditions it is not unreasonable to believe that there may be extensive thermionic emission. As in the glow discharge, the cathode is surrounded by a positive ion sheath, resulting in a large space charge.

Anode rise. The anode region requires considerable explanation. There is to this electrode a large electron current which may raise the anode to a temperature even greater than that of the cathode. Thus it is also a thermionic emitter, although the electrons are returned to the anode. Hence there is a large negative space charge and a resulting anode rise. Cooling the anode may result in a reduction in anode rise, indicating a decrease in thermionic emission. Again it is necessary to use the microscopic viewpoint.

Arc body. The main body of the arc is characterized by secondary ionization. This is predominantly a temperature effect. The electrons produce very little ionization by impact, because the electron energy is generally low. However, energy can be imparted to the gas molecules in the form of atomic and molecular excitation. With the many intermolecular collisions that occur, this energy is readily degraded to thermal energy of the gas. Thus a high temperature may be achieved and ionization occur by virtue of intermolecular collisions. This is expressed quantitatively by the following equation, derived by M. N. Saha. Here n_i and

$$\log_{10}\left(\frac{n_i{}^2}{n_0}\right) = -5040\,\frac{v_i}{T} + {}^3\!/_2\log_{10}(T) + 15.385$$

n_0 are the ionic and neutral densities, respectively, v_i the ionization potential in volts, and T the Kelvin temperature. This equation refers to the equilibrium condition.

It is difficult to make accurate and meaningful measurements because of the high temperature and large current. Thus there is much that is not understood or substantiated.

An interesting situation obtains in connection with an externally heated cathode. Here may exist an arc in which the potential drop is less than the ionization potential of the gas. Several factors are important in this case. First, there is a positive space charge in the arc body which results in a maximum potential there, and there is actually a potential drop from that point to the anode. Second, there may be many excited or metastable atoms or molecules in the discharge. It seems likely that ionization may take place in several steps rather than one. The most important mechanism, however, appears to be that for removal of electrons that have lost so much energy by inelastic collisions that they are trapped near the potential maximum. A fast electron may give up enough energy by coulomb interaction with one of these so that both electrons may reach the anode. If this were not so, neutralization would occur and the arc would be quenched.

Finally, it should be pointed out that in any arc there will be vaporization of the electrodes, and that the gas will have in it molecules of the electrode material. For this reason the electrode material may have a profound effect on the arc characteristics.

[GLENN H. MILLER]

Bibliography: P. Llewellyn-Jones, *The Glow Discharge and an Introduction to Plasma Physics*, 1966; L. B. Loeb, *Basic Processes of Gaseous Electronics*, 1955; F. A. Maxfield and R. R. Benedict, *Theory of Gaseous Conduction and Electronics*, 1941; J. Millman and S. Seely, *Electronics*, 2d ed., 1951.

Artificial intelligence

The subfield of computer science concerned with understanding the nature of intelligent action and constructing computer systems capable of such action. It embodies the dual motives of furthering basic scientific understanding and making computers more sophisticated in the service of humanity.

Many activities involve intelligent action—problem solving, perception, learning, symbolic activity, creativity, language, and so forth—and therein lie an immense diversity of phenomena. Scientific concern for these phenomena is shared by many fields, for example, psychology, linguistics, and philosophy of mind, in addition to artificial intelligence. This implies no contradiction, the phenomena being rich enough to support many starting points and diverse perspectives. The starting point for artificial intelligence is the capability of the computer to manipulate symbolic expressions that can represent all manner of things, including the programs of action of the computer itself.

The approach of artificial intelligence is largely experimental, although it contains some general principles and small patches of mathematical theory. The unit of experimental investigation is the computer program. New programs are created as probes to explore and test new ideas about how intelligent action might be attained. Artificial intelligence lives in a world of throw-away programs; once a program has yielded its bit of scientific evidence, it is often of little further interest.

Foundations. The foundations of artificial intelligence are divided into represenatation, problem-solving methods, architecture, and knowledge. The four together are necessary ingredients of any intelligent agent.

Representation. To work on a task, a computer must have an internal representation in its memory, for example, the symbolic description of a room for a moving robot, or a set of algebraic expressions for a program integrating mathematical functions. The representation also includes all the basic programs for testing and measuring the structure, plus all the programs for transforming the structure into another one in ways appropriate to the task.

Many representations may be used. The task may pass through a sequence of representations with successive processing, or through alternative representations to find a good one. The representation used for a task can make an immense amount of difference, turning a problem from impossible to trivial.

Problem-solving methods. Given the representation of a task, a method must be adopted that has some chance of accomplishing the task. Artificial intelligence has gradually built up a stock of relevant problem-solving methods (the so-called weak methods) that apply extremely generally: generate-and-test (a sequence of candidates is generated, each being tested for solutionhood); heuristic search (sequences of operations are tried to construct a path from an initial situation to the desired one); hill climbing (a measure of progress is used to guide each step); means-ends analysis (the difference between the desired situation and the present one is used to select the next step); operator subgoaling (the inability to take the desired next step leads to a subgoal of making the step feasible); planning by abstraction (the task is simplified, solved, and the solution used as a guide); and matching (the present situation is represented as a schema to be mapped into the desired situation by putting the two in correspondence). An important feature of all the weak methods is that they involve the process of search. One of the most important generalizations to arise in artificial intelligence is the ubiquity of search. It appears to underlie all intelligent action.

Architecture. A general intelligent agent has multiple means for representing tasks and dealing with them. Also required is an operating frame within which to select and carry out these activities. Often called the executive or control structure, it is best viewed as a total architecture (as in computer architecture), that is, a machine that provides data structures, operations on those data structures, memory for holding data structures, accessing operations for retrieving data structures from memory, a programming language for expressing integrated patterns of conditional operations, and an interpreter for carrying out programs.

Any digital computer provides an architecture, as does any programming language. Architectures are not all equivalent, and the scientific question is what architecture is appropriate for a general intelligent agent.

In artificial intelligence, this question has taken the form of determining what language is good for programming artificial intelligence systems. Although the question is seemingly about research tools, in reality its investigation has been a search for the properties that make intelligence possible. The main development has been that of list-processing languages, which embody a general homogeneous and flexible notion of symbols and symbolic structure. LISP, one of the early list-processing languages, has evolved until it functions as the common language for artificial intelligence programs. Work on so-called planner languages has added to the architecture four notions that previously had existed only separately: (1) the goal, a data base to hold all knowledge, avoiding ad hoc and be an associative focus for knowledge relevant to obtaining the desired situation; (2) a uniform data base to hold all knowledge, aboiding ad hoc encodings for each type of data; (3) pattern-directed invocation, or finding what method or process to use by matching rather than having to know its name; and (4) the incorporation of search into the fabric of the programming language, so that search can be used anywhere.

Knowledge. The basic paradigm of intelligent action is that of search through a space (called the problem space) for a goal situation. The search is combinatorial, each step offering several possibilities, leading to a cascading of possibilities in a branching tree. What keeps the search under control is knowledge, which tells how to choose or narrow the options at each step. Thus the fourth fundamental ingredient is how to represent knowledge in the memory of the system so it can be brought to bear on the search when relevant.

A general intelligent system will have immense amounts of knowledge. This implies a major problem, that of discovering the relevant knowledge as the solution attempt progresses. This is the second problem of search that inevitably attends an intelligent system. Unlike the combinatorial explosion characteristic of the problem space search, this one involves a fixed, though large, data base, whose structure can be carefully tailored by the architecture to make the search efficient. This problem of encoding and access constitutes the final ingredient of an intelligent system.

Examples. Three examples will provide some flavor of research in artificial intelligence.

Games. Games form a classical arena for artificial intelligence. They provide easily defined, isolated worlds which still permit the indefinite play of reason.

All the ingredients of an artificial intelligence program can be seen in game programs. A chess program must have a representation of the current chess position. Besides the bits that describe the position, there must be procedures for analyzing the position and for making the moves to create new legal positions. All these capabilities constitute the representation. Their efficiency is important; some special computing hardware has even been designed to make them faster.

The basic method used for chess is to generate the moves and test their worth. The test is a form of heuristic search that explores the consequences of a candidate move many potential moves into the future. This search is combinatorial and much too big to accomplish without the aid of knowledge. The main technique for bringing knowledge to bear is the evaluation function, which can be applied to a position to estimate directly its chance of being on a winning path. Many features of the position are calculated, each representing a bit of chess knowledge (for instance, queens are better than pawns). Such value functions are heuristic, only approximating a correct analysis. Evaluation turns the search into hill climbing. Not all the knowledge stems from the evaluation functions; many subtleties of the search allow great improvement.

The Northwestern chess program searches about 10^6 positions per move in tournament play (about 2.5 min per move). It has played in many official tournaments against human competition and has an official U.S. Chess Federation rating of a low expert. At fast play (5 min per game) it has beaten several grand masters and has a rating of about 2300, which is high master. The program's large search compared with that of humans (estimated at a few hundred positions at most) illustrates that exploiting the speed and reliability of computers can produce a distinctive style of problem solving.

A program for another game, backgammon, illustrates further the role of knowledge. In 1979 a program developed at Carnegie-Mellon University played an exhibition match with the world backgammon champion Luigi Villa, winning 7 points to 1. Since backgammon involves chance as well as skill, the program in fact is not the superior player, but it is definitely formidable.

The interest in the feat from an artificial intelligence point of view is twofold. First is the calibration against human talent under real conditions. Second is the mechanisms that enabled such play; these are the basic scientific contribution. In backgammon, the number of moves from a single position is so large that search is almost totally ineffective. Instead, the evaluation function must contain an immense amount of knowledge of the game. The success of the program rests in part on using separate evaluation functions during different regions of play and, moreover, adjusting the transitions between these different views so that they are smooth and continuous. Analysis reveals a general principle: discontinuous jumps in value (in crossing a boundary of two regions) inevitably lead to serious errors. The discovery of this principle, its testing by being embodied in a program, and its tentative support by the program's success form a typical example of scientific activity and progress in artificial intelligence.

Perception. Perception is the formation, from a sensory signal, of an internal representation suitable for intelligent processing. Though there are many types of sensory signals, computer perception has focused on vision and speech. Perception might seem to be distinct from intelligence, since it involves incident time-varying continuous energy distributions prior to interpretation in symbolic terms. However, all the same ingredients occur:

representation, search, architecture, and knowledge. Thus, perception by computers is a part of artificial intelligence.

Speech perception starts with the acoustic wave of a human utterance and proceeds to an internal representation of what the speech is about. A sequence of representations is used: the digitization of the acoustic wave into an array of intensities; the formation of a small set of parametric quantities that vary continuously with time (such as the intensities and frequencies of the formants, bands of resonant energy characteristic of speech); a sequence of phones (members of a finite alphabet of labels for characteristic sounds, analogous to letters); a sequence of words; a parsed sequence of words reflecting grammatical structure; and finally a semantic data structure representing the final meaning.

Speech perception is difficult, because the encoding of the intended utterance by the speaker is extremely convoluted, the acoustic wave at any point being influenced in complex ways by substantial context on either side. Added to this are multiple sources of noise and variability (for example, speakers differ and vary over time). To unravel this requires search, generating hypotheses about the identity of some element in the representation (for example, a phone or a word) and then testing how well it fits with the other hypothesized elements. Several possibilities can be hypothesized at each point (and at each level of representation), the whole becoming combinatorial. Thus, knowledge about the encoding and noise at each level must be brought to bear to control the search: knowledge of speech articulation, phonological regularities, lexical restrictions, grammar, the task, and so forth.

There are speech recognition programs that recognize continuous speech with high accuracy (a few percent utterance error) from several speakers with vocabularies of about 1000 words in environments that are not too noisy, within restricted grammatical and task contexts. These higher-level restrictions are extremely important at present, and recognition of freely spoken conversation or dictation is not yet in sight.

Expert programs. Extensive experience enables humans to exhibit expert performance in many tasks, even though no firm scientific or calculational base exists and the knowledge does not exist in any explicit form. So-called expert artificial intelligence programs attempt to exhibit equivalent performance by acquiring and incorporating the same knowledge that the human expert has. Many attempts to apply artificial intelligence to socially significant tasks take the form of expert programs. Even though the emphasis is on knowledge, all the standard ingredients are present: representation of the task, methods of manipulating the representation, an architecture to make processing easy, as well as the knowledge.

An example is MYCIN, developed at Stanford University, which makes judgments on the diagnosis of bacterial infection in patients and proposes courses of therapy with antibiotics. MYCIN operates as a consultant by interacting with a physician who knows the history of the patient. The knowledge of the program is encoded in a large number of if-then rules, for example, "If the site of the culture is blood, and the identity of the organisms may be pseudomonas, and the patient has ecthyma gangrenosum skin lesions, then there is strongly suggestive evidence (.8) that the identity of the organism is pseudomonas." There are several hundred such rules, which jointly encode what expert physicians know about the signs, symptoms, and causes of disease and the effects of drugs. The program uses these rules to search for a diagnosis, letting them guide what specific knowledge it must seek from the physician about the patient.

Obtaining and codifying such hitherto implicit expert knowledge is a critical aspect of developing artificial intelligence expert programs. That entire bodies of knowledge could be embodied as sets of active if-then rules, even though they did not yet form a coherent scientific theory, has been an important discovery.

A number of expert programs (including MYCIN) have shown performance at levels of quality equivalent to average practicing professionals (for example, average practicing physicians) on the restricted domains over which they operate, as shown by careful tests. Incorporation into ongoing practice, which has not yet generally occurred, depends on many other aspects besides the intrinsic intellectual quality (for example, ease of interacting, system costs). From a scientific viewpoint, the lessons in understanding the nature of knowledge and its organization are already apparent. These expert programs illustrate that there is no hard separation between pure and applied artificial intelligence; finding what is required for intelligent action in a complex applied area makes a significant contribution to basic knowledge.

Scope and implications. Research in artificial intelligence spreads out to explore the full range of intellectual tasks. The three areas above — games, speech, and medical diagnosis — are only a sample. Significant work has been done on puzzles and reasoning tasks, induction and concept identification, symbolic mathematics, theorem-proving in formal logic, natural language understanding, vision, robotics, chemistry, engineering analysis, and computer-program synthesis and verification, to name only the most prominent. As in any developing science, there occurs both a recurrence of the basic ideas (about representation, search, architecture, and knowledge) and the discovery of new mechanisms; both extend existing ideas and reveal limits.

Artificial intelligence has close ties with several surrounding fields. As part of computer science, it plays the role of expanding the intellectual sophistication of the tasks to which computers can be applied. Various subfields, once viewed as part of artificial intelligence, have become autonomous fields, most notably symbolic mathematics and program verification. The work on vision and speech shares its concern with signal processing with the field of pattern recognition in electrical engineering.

The relation of artificial intelligence to the study of cognitive and linguistic processes in humans is especially important. Human experimental psychology and linguistics underwent a transformation in the late 1950s, coincident with the birth of artificial intelligence, to an essentially symbolic and information-processing viewpoint. The sources were much broader than artificial intelligence, especially in linguistics, but the history of these three fields has been intertwined ever since. The relations with linguistics have been marked with controversy, but the relations with cognitive psychology (as human experimental psychology has come to be called) have been mostly symbiotic. The psychological theories of problem solving, long-term memory structure, and the organization of knowledge are essentially the theories from artificial intelligence.

Fascination with accomplishing mental activities by mechanical means has a long history; it became channeled into cybernetics, which had its advent in the 1940s. Cybernetics focused on the role of feedback mechanisms and on the analysis of purpose they made possible in mechanical terms. Artificial intelligence did not really begin until the emergence of the digital computer. This added to the notions of cybernetics the notions of symbolic systems and programmability; it shifted concern from the construction of simple adaptive circuits to experimentation with programs, which is still the hallmark of artificial intelligence. Many of the pioneers of digital computers and operational mathematics played some part in the prehistory of artificial intelligence: Claude Shannon, John von Neumann, Norbert Wiener, and above all Alan Turing. However, the start of the field is usually located in the last half of the 1950s, when a large amount of important work was done, associated with the names of John McCarthy, Marvin Minsky, Oliver Selfridge, Herbert Simon, and Allen Newell.

The major centers for artificial intelligence research are in the United States (primarily at Stanford, the Massachusetts Institute of Technology, and Carnegie-Mellon), but research exists in all major countries. Japan especially is doing vigorous work in artificial intelligence, as are Britain and the Soviet Union. *See* AUTOMATA THEORY; COMPUTER; DIGITAL COMPUTER; DIGITAL COMPUTER PROGRAMMING.

[ALLEN NEWELL]

Bibliography: N. Neilson, *Principles of Artificial Intelligence*, 1980; A. Newell and H. A. Simon, *Human Problem Solving*, 1972; E. Shortliffe, *Computer Based Medical Consultations: MYCIN*, 1976; P. Winston, *Artificial Intelligence*, 1977.

Attenuator

An arrangement of fixed and variable resistors mounted in a compact package, used to reduce the strength of a radio- or audio-frequency signal by a desired amount without causing appreciable distortion. The attenuator is so designed that its impedance matches that of the circuit in which it is connected, regardless of the amount of attenuation introduced. This characteristic is achieved by properly proportioning the resistance values of the series and shunt elements.

Attenuators are often calibrated in decibels to indicate the amount of attenuation that is introduced at each setting of the control. In this form, attenuators are widely used to control the output levels of signal generators, oscilloscope input levels, and audio levels in broadcast studios. The corresponding nonadjustable device is usually called a pad. *See* RESISTOR.   [JOHN MARKUS]

Audio amplifier

An electronic circuit for amplification of signals within, and in some cases above, the audible range. The term may mean either the complete amplifier, consisting of the voltage-amplifier and power-amplifier stages, or it may mean just one stage. In the usual case, the term implies that the amplifier is not capable of amplifying a direct-current signal; such an amplifier would more likely be called a direct-coupled or dc amplifier. *See* DIRECT-COUPLED AMPLIFIER.

The function of an audio amplifier is to amplify a weak electrical signal, such as from a microphone or a phono pickup, to a level capable of driving a loudspeaker at the desired output sound level. The sound power level may be small in such applications as a hearing aid.

Audio amplifiers may be constructed with vacuum tubes, but current technology favors transistors and integrated circuits. Transistor amplifiers generally use RC (resistor-capacitor) coupling between stages because of its simplicity, good frequency response, and low cost. When integrated circuits are used, the complete voltage-amplifying portion of the circuit may consist of one chip (containing several stages of amplification) and a few external resistors and capacitors. The voltage amplifier drives the power-amplifier section, composed of transistors that must be capable of supplying the desired power to the loudspeaker. For low-power applications there are integrated circuits that provide the necessary power gain and connect directly to the loudspeaker. *See* INTEGRATED CIRCUITS; TRANSISTOR; VOLTAGE AMPLIFIER.

The amplifier can be designed with a frequency response that is adequate to produce nearly distortionless amplification, particularly when negative feedback is used. Proper feedback can overcome most of the distortion introduced by the nonlinearities in the amplifier transistors or vacuum tubes, the power output transformer, and the loudspeaker, and the lack of a uniform frequency response of the amplifier and loudspeaker. High-fidelity amplifiers use feedback. *See* DISTORTION (ELECTRONIC CIRCUITS).

Audio amplifiers are capable of a power output ranging from a few watts to a few hundred watts, depending on the application.

The term af (audio-frequency) amplifier is occasionally used in place of audio amplifier. This is consistent with the designations of rf (radio-frequency) and i-f (intermediate-frequency) amplifiers. *See* AMPLIFIER.   [HAROLD F. KLOCK]

Bibliography: J. Lenk, *Manual for Integrated Circuit Users*, 1973; J. Millman, *Microelectronics*, 1979; D. Schilling and C. Belove, *Electronic Circuits: Discrete and Integrated*, 1979.

Audio delayer

A device for introducing delay in the audio signal in a sound-reproducing system. The loudspeaker-pipe-microphone combination, the magnetic tape recorder-reproducer, and the digital delayer are the most common.

Loudspeaker-pipe-microphone system. An audio delay system consisting of an equalizer, power amplifier, loudspeaker, pipe, microphone, and additional amplifiers is shown in Fig. 1. The

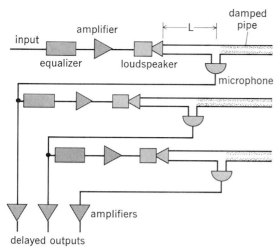

Fig. 1. Loudspeaker-pipe-microphone audio delay system.

audio delay is given by Eq. (1), where $D=$ delay in

$$D=\frac{L}{1.1} \qquad (1)$$

milliseconds, and $L=$ distance between the microphone and loudspeaker in feet. There is attenuation in the pipe, which increases with frequency. Amplitude compensation with respect to frequency is introduced in the input to the power amplifier to compensate for the frequency discrimination due to attenuation. The pipe beyond the microphone is damped with tufts of felt to absorb the sound waves beyond this point and thereby eliminate reflections at the end of the pipe. Delay of any practical value can be obtained by the assembly of appropriate units as shown in Fig. 1. *See* FRE-QUENCY-RESPONSE EQUALIZATION.

Magnetic tape recorder-reproducer system. An audio delay system consisting of an audio magnetic tape recorder and reproducer with multiple spaced heads is shown in Fig. 2. The delay is given by Eq. (2), where $D=$ delay in milliseconds,

$$D=\frac{1000L}{V} \qquad (2)$$

$V=$ speed of the magnetic tape in inches per second, and $L=$ distance between the record and reproduce heads in inches. The system provides a simple means for introducing delay; however, the objection is that the tape belt wears out after a few hours and must be replaced. Delay of any practical value can be obtained by appropriate arrangement of the record and reproduce heads. *See* MAGNETIC RECORDING.

Digital system. An audio delay system employing a digital delay system is shown in the schematic diagram of Fig. 3. The first element is a low-pass filter which eliminates all audio components above 12,000 Hz. A sampling switch operates at a frequency of 30,000 Hz. The sampled signal is converted into the proper digital format by the analog-to-digital converter and fed into the digital shift register serving as a memory. The shift register elements are connected so that each element transfers its digital word to the next stage and reads the word from the previous stage during the sampling interval. In this way each word is entered at the input of the memory and moves down the line relatively slowly. The output of the shift register is fed to the digital-to-analog converter and then to the

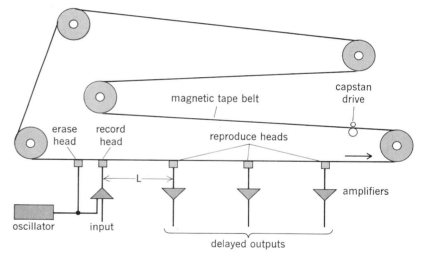

Fig. 2. Magnetic tape recorder-reproducer audio delay system.

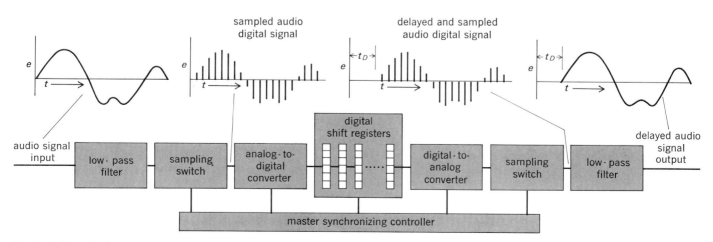

Fig. 3. Schematic diagram of digital audio delay system, with graphs of signal voltage *e* versus time *t* at various points along the transmission path of the signal.

sampling switch. The output of the sampling switch is shown in Fig. 3, with a delay indicated as t_D. The output of the sampling switch is smoothed by the low-pass filter. The audio signal output, also shown in the figure, is the same as the audio wave input but with a delay of t_D seconds. The master synchronous controller synchronizes and controls all the functions with respect to time. The storage requirements of the memory are given by Eq. (3),

$$W = \frac{1}{3}\left(\frac{S}{N}\right)f_M t_D \qquad (3)$$

where W = storage capacity in bits, S/N = signal-to-noise ratio in decibels, f_M = maximum audio frequency passed by the system in hertz, and t_D = delay time in seconds. The signal-to-noise ratio of the system is given by Eq. (4), where S/N = signal-to-

$$\frac{S}{N} = 20\log_{10}(2^n) \qquad (4)$$

noise ratio in decibels, and n = number of bits in the digital word. The advantage of the digital delay system is the all-solid-stage electronics with no mechanical moving parts. *See* ANALOG-TO-DIGITAL CONVERTER; DIGITAL-TO-ANALOG CONVERTER; ELECTRIC FILTER.

Applications. One of the major applications for audio delayers is in sound-reinforcement systems. Audio delay of up to 500 msec can be obtained with any of the three systems described above. A signal-to-noise ratio of 60 dB is attainable. In general, the audio frequency bandwidth is usually of the order of 12,000 Hz.

[HARRY F. OLSON]

Bibliography: H. F. Olson, *Modern Sound Reproduction*, 1972.

Audion

The name given in 1906 to the electronic amplifying tube by its inventor, Lee DeForest. Although initially copyrighted by DeForest as a trade name, the term was used generically to described grid-controlled amplifier tubes until the 1920s, when the appearance of multigrid tubes led to an expanded terminology (for example, triode, tetrode, pentode, hexode). *See* ELECTRON TUBE; VACUUM TUBE.

Although the term "audion" is merely of historical interest in modern electronics, the circumstances surrounding its invention are of continuing interest. The fact that electric current can be carried between a heated filament and a nearby metal plate, when both are enclosed in a rarefied gas or vacuum, had been discovered in 1883 by Thomas Edison, but the mechanism of the transfer of charge was a mystery at that time.

By 1900 J. J. Thompson had advanced the electron theory, which identified the quantified nature of electric current. It was then realized that the Edison effect could be explained by the emission of electrons from the hot filament and their subsequent collection by the metal plate. The current would not travel in the reverse direction since the cold plate was incapable of emitting electrons. This one-way passage gave rise to the use of a device, first developed in 1897 by J. Ambrose Fleming, which rectified alternating current, and acted as a detector of radio waves. At this stage the

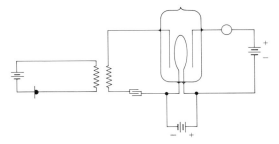

The DeForest audion circuit, as shown in his patent application. (*U.S. Patent no. 841,387, 1907*)

Fleming "valve" occupied the same position held by the semiconductor diode in the early 1940s, prior to the invention of the transistor. Neither device could amplify, and without amplification, the vast potential of electronic and semiconductor technology could not be realized at the time. *See* TRANSISTOR.

It was DeForest's seminal concept that a third electrode might be inserted into the Fleming valve and that changes in voltage applied to the third electrode would produce corresponding changes in the current passing between the heated filament and the metal plate. This changing current, passed through a resistance external to the device, would generate a magnified copy of the voltage applied to the device; that is, amplification would take place.

DeForest applied for his patent on the three-electrode amplifier on Oct. 25, 1906, and the following day presented his paper "The Audion—A New Receiver for Wireless Telegraphy" to the American Institute of Electrical Engineers. The patent (no. 841, 387) was issued on Jan. 15, 1907. In one form shown in the first patent (see illustration) the signal to be amplified is applied between the heated filament and one plate, and the output is taken from the filament and another plate, which is at the opposite side of the filament, a procedure not favorable to amplifying action in a highly evacuated device. However, reference is made in the patent to a "gaseous medium" within the tube, maintained in a condition of "molecular activity." It thus appears that the observed amplification resulted from changes in ionization, rather than from attraction and repulsion of the flow of electrons. In fact, the early audions were subject to wide variations of performance arising from uncontrolled factors in their design and construction (particularly the amount and nature of the gas content). Such tubes were known as "soft tubes" because of their relatively high gas content. It remained for others, notably Irving Langmuir, to understand the importance of high vacuum in obtaining reproducible characteristics in audions, which later became known as "hard tubes." A similar situation occurred in the early development of transistors, which exhibited erratic behavior and had short life-spans until the importance of hermetically sealing the transistor against intrusion of moisture was understood. *See* ELECTRONICS.

[DONALD G. FINK]

Bibliography: Lee DeForest, The audion: A new receiver for wireless telegraphy, *Trans. A.I.E.E.*, 25:735, 1907; U.S. Patents no. 841, 387 (1907) and no. 879, 532 (1908), issued to Lee DeForest.

Automata theory

In its most general sense, automata theory is concerned with ways in which inputs to a system combine with current states of a system to produce outputs from the system. The subject matter can be divided into (1) abstract theory, (2) structural theory, and (3) self-organizing and adaptive systems. These three categories will be briefly treated here. The abstract-theory discussion covers general definitions, Turing machines, and language theory. The structural category is concerned more with practical machine construction and covers the propositional calculus, switching algebra (or Boolean algebra), and information theory. The self-organizing discussion covers neural nets, artificial intelligence, game theory, and pattern recognition.

Abstract theory. An automaton can be characterized by three sets and two functions: the set A of all input signals, the set S of internal states of a system subjected to the inputs, and a set Z of all output signals (see illustration). At any given time the set of internal states S is determined from both a combination of the set of possible states and the set of possible input terms, so that $A \times S$ maps into S. Similarly, the output functions, or members of the set Z, are determined from system states and inputs, so that $A \times S$ maps into Z.

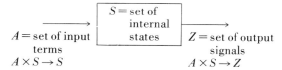

Representation of an automaton.

When the sets A and S and Z are all finite, the automata are referred to as finite automata. When one or more of A or S or Z is an infinite set, then the automata are called infinite automata.

Turing machines. The Turing machine basically consists of a computing element and a tape. A series of characters, $C_1, C_2, \ldots, C_n$, can be written on the tape so that each character occupies a finite space along the tape, or, say, one square. The computing element can assume a number of states, $S_1, S_2, \ldots, S_m$. The tape passes through the computing element one square at a time so that the machine operation can be described by the quadruple $S_i C_j C_k S_l$, which is read: When the machine is in state S_i and the character C_j is read, then replace the character C_j with character C_k and proceed to state S_l. The character C_k can be considered as being written over the character C_j. In addition, the instruction R or L may be given, calling for a movement of the tape one square to the right or left. Thus the command $S_i C_j R S_l$ is read: When the machine is in state S_i and character C_j is read, then move the tape one square to the right and proceed to state S_l.

Studies with Turing machines have been concerned with the consideration of those functions which are computable with this machine organization. Whether all functions that are computable can be computed on a Turing machine is generally regarded as unprovable. *See* COMPUTER.

Language theory. A grammar can be considered as a quadruple $G = (\xi, T, B, P)$ where ξ is the set of symbols available, T is a set of words on ξ, B is a particular beginning symbol ($B \epsilon \xi$), and P is a set of productions. Productions are of the form $axb \rightarrow ayb$, where a, b, x, y are arbitrary strings of symbols. Consider the following example: Let $\xi =$ (boy, jumped, the, S_0, S_1, S_2, S_3, S_4), $B = S_0$, and the elements of P be as follows:

1. $a S_0 b \rightarrow a S_4 S_2 b$
2. $a S_4 b \rightarrow a S_4 S_2 b$
3. $a S_2 b \rightarrow a$ jumped b
4. $a S_1 b \quad a$ boy b
5. $a S_3 b \quad a$ the b

With this defined grammar, $G = (\xi, T, B, P)$, it can be proved, as an example, that the structure "The boy jumped" is a valid construction:

1. S_0	$B = S_0$ beginning symbol
2. $S_4 S_2$	Rule 1 $a = \{0\}, b = \{0\}$
3. $S_3 S_1 S_2$	Rule 2 $a = \{0\}, b = S_2$
4. $S_3 S_1$ jumped	Rule 3 $a = S_3 S_1, b = \{0\}$
5. S_3 boy jumped	Rule 4 $a = S_3, b =$ jumped
6. The boy jumped	Rule 5 $a = \{0\}, b =$ boy jumped

Of particular interest is the application of the theory of grammars to push-down store automata, where symbols are available on the last-in, first-out basis. These grammars find use in the construction of higher-level programming languages such as ALGOL.

Language theory can be formalized to treat areas other than those of automatic language translation and compilation. The principle applied in the above example may be extended to define grammars capable of describing a variety of systems of interest to automata theory; the propositional calculus, nerve networks, sequential machines, and programming schemes are examples of such systems.

Structural theory. The propositional calculus concerns itself with the manipulation of linguistic statements. Rules of symbolic logic are employed to reduce compound statements into simplfied component statements. Of central interest is whether a particular statement is "true" or "false." For example, consider the following statements: "New York is a city" is true; "St. Louis is the mayor of New York" is false; "Cats are mammals" is true. Suppose now that the symbols X, Y, and Z are assigned to these statements, respectively. New statements can be made using the symbols. The new statement "Both X and Z" is true, since the first and last statements are both true. The statement "Both X and Y" is false, since the second statement is false. The statement "Either X of Y" is true, since at least one of them is true.

When the statements considered are limited to those that can unambiguously be said to be true (T) or false (F), or, more succinctly, when the sets A, S, and Z each have two members, $\{T, F\}$, then the language is referred to as switching algebra, switching theory, or Boolean algebra. The study of switching functions is of particular importance in computing machines where binary sequences are used to represent numbers. *See* BOOLEAN ALGEBRA; SWITCHING THEORY.

Information theory was principally originated by Claude Shannon in 1948–1949. Communication processes are concerned with the flow of information from a transmitter, through a transmission medium or channel, to a receiver. The word "information" is associated with received intelligence that is changing and unpredictable. For example, a well-worn story or familiar melody conveys little information, but when a text is changing in an unpredictable fashion so that the receiver is unable to predict what will come next, the information rate is said to be high.

The following measure of uncertainty, or entropy, or the average amount of information that is associated with a finite sample space, may be stated in the equation below. Here $H(x)$ is the uncer-

$$H(x) = -\sum_{i=1}^{h} p_i \log p_i$$

tainty of the finite space sample, i and h establish the bounds of the space to be examined, and p_i is the probability of occurrence of a given transmitted event. In communication problems a convenient unit of information is the bit, or binary character, and the log is taken to the binary base. *See* INFORMATION THEORY.

Self-organizing systems. There are two distinct research areas to which the term neural modeling has been applied. The intent of one area is to represent physiological phenomena by constructing systems whose building blocks have properties based on physiological correlates. The intent of the second area is to build generalized automata, whether or not they duplicate in exact detail any actual physiological functions.

In one scheme a neuron consists of a body, or soma, from which axons lead to one or more end bulbs. A nerve net is an arrangement of both "input" and "inner" neurons, where any given axon impinges on not more than one other cell body, and where each synapse is either excitatory or inhibitory. With these building blocks, one can perform such operations as logical conjunction and disjunction, and also work with delays.

Of major interest in these constructions is the matter of representability of events, as well as problems of synthesis and the fabrication of reliable organisms from unreliable components.

Artificial intelligence. Artificial intelligence generally relates to the attempt to use computer programming to model the behavioral aspects of human thinking, learning, and problem solving. To the programmer the computer represents a tool for manipulating symbols whose collective behavior can be abstracted from the details of a particular computer construction. While solutions to many problems can be found by exploring every possible outcome and selecting the appropriate alternative, in more complex situations such methodical searching is prohibitive. For example, tracking down every possible line of play in a chess game would involve analyzing 10^{120} positions. Hence, the programmer must discover methods of investigating only those alternatives which are most likely to result in early solutions.

It is this aspect that makes heuristic program-

ming interesting and significant. For these purposes the term intelligent behavior can be ascribed to those methods, techniques, or algorithms which decide what approach is to be tried next on the basis of a current analysis. *See* ARTIFICIAL INTELLIGENCE; DIGITAL COMPUTER PROGRAMMING.

Game theory. In its broadest interpretation game theory deals with the study of any conflict situation. Conflicts ranging from parlor games to athletic events to international wars share a common principle: participants do not have identical objectives or their ultimate interests are in conflict.

In ideal cases one can describe a game by organizing arrays of possible outcomes and associated rewards into payoff matrices. A player then selects those courses of action which constrain the possible outcomes to that set which is consistent with his goals; that is, he adopts a strategy.

Each player's preferences, as well as basic playing rules, are generally known to all players, although each player has limited control over the outcomes of his actions. Strategy choices may lead to eliminating an opponent entirely, or to more desirable ends, such as maximizing winnings per play without bankrupting the opponent, or employing mixed strategies, where one player may deviate from his own optimum strategy in order to keep the enemy from consistently pursuing his own optimum strategy.

Pattern recognition. The basic problem in pattern recognition is to construct algorithms capable of processing the range of input patterns of interest to yield a set of output patterns such that the occurrence of each input causes a uniquely identifiable output pattern to occur.

It is essentially a decoding process in which discriminations are made. In simpler cases tree methods of serial segmentation find value. Ideally, such measures enable one to divide remaining patterns at each decision node into two disjoint patterns, similar in principle to the basis for binary decoding. Another approach makes use of as many measures as possible that appear to be related to the desired discriminations. A search is then made for an appropriate differential weighting of these measures so as to obtain maximal spacing between patterns. This approach is sometimes referred to as factor analysis.

Other efforts have employed adaptive threshold logics, such as the perceptron and Adaline. These mechanisms use simple reinforcement signals to control structural modifications within the network. Such modifications are accomplished by altering the threshold weightings of individual logic elements. Thus, a multiple set of threshold logics have their relative weightings varied in an effort to generate a distinct code at the network outputs for each input pattern. It is not necessary for one to know exactly how a particular discrimination is made, just as one does not know how information flows in biological neural nets when an instructor programs a student to make a specific discrimination.

[NORMAN B. REILLY]

Bibliography: Z. Bavel, *Introduction to the Theory of Automata and Sequential Machines,* 1982; J. E. Hopcroft and J. D. Ullman, *Introduction to Automata Theory, Languages and Computation,* 1979; R. Y. Kain, *Automata Theory: Machines and Language,* 1981.

Automatic frequency control (AFC)

The automatic control of the intermediate frequency in a radio, television, or radar receiver, to correct for variations of the frequency of the transmitted carrier or the local oscillator. In high-fidelity broadcast receivers AFC keeps distortion due to detuning to a low figure. In the reception of long-haul telegraph signals, AFC reduces the error rate due to signal pulse distortion or interference from lower intensity signals in the same frequency band.

Single-sideband receivers receive signals which are transmitted with a carrier level that is reduced to as small a proportion as 5%, or less, of the sideband (intelligence) amplitude. Proper demodulation requires the generation locally of a carrier-frequency signal synchronized to the transmitted carrier by AFC. Since propagation at frequencies of 3–30 MHz is dependent upon reflections from the ionosphere, motion in this medium will speed up and retard the arrival of the wave, causing Doppler-effect frequency changes. Transmission to and from speeding aircraft will also suffer Doppler frequency drift. To reduce these effects, the carrier is transmitted for synchronization so that the frequency difference between the carrier and side frequency is maintained on reception.

AFC techniques are varied but are mainly of two types. One uses a discriminator to furnish a voltage whose magnitude and polarity are determined by the frequency change. This voltage is used to adjust the frequency of the local oscillator of the receiver, thereby keeping the intermediate frequency constant. The other uses two-polarity pulse accumulation which furnishes a dc potential proportional to frequency error.

To select only the carrier for AFC, very narrow bandwidths are utilized. As an example, a bandwidth of 30 Hz at 70% of maximum response is quite common. The response of the control circuits is usually designed to be slow and to be inactive below a determinable level of carrier input. These techniques reduce noise and interference as well as frequency-control capture by undesired adjacent carriers. *See* RADAR; RADIO RECEIVER; TELEVISION RECEIVER. [WALTER LYONS]

Automatic gain control (AGC)

The automatic maintenance of a nearly constant output level of an amplifying circuit by adjusting the amplification in inverse proportion to the input field strength, also called automatic volume control (AVC). Almost all radio receivers in use employ AGC. In broadcast receivers AGC makes it possible to receive incoming signals of widely varying strength, yet have the sound remain at nearly

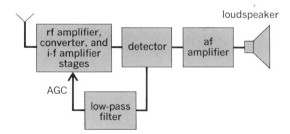

Fig. 1. Block diagram of broadcast receiver using AGC.

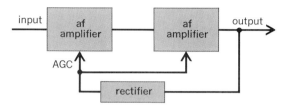

Fig. 2. Block diagram of AGC of an audio amplifier.

the same volume. In communications receivers a type of AGC circuit called a squelch circuit is used to prevent noise during periods of no transmission, such as in the reception of on-off keying, frequency-shift keying (FSK), and phone. AGC is also useful in accelerating the switching action between receivers in diversity connection. *See* RADIO RECEIVER.

AGC action depends on the characteristic, possessed by most electronic tubes and transistors, of adjustment of gain by the variation of the applied bias voltage. If the dc voltage applied to the control grid of a vacuum tube is made more negative, the amplification of that stage will be reduced.

In most broadcast receivers the AGC voltage is taken from the detector. This dc voltage, proportional to the average level of the carrier, adjusts the gain of the radio-frequency (rf) and intermediate-frequency (i-f) amplifiers and the converter, as shown in Fig. 1. AGC tends to keep the input signal to the audio-frequency (af) amplifier constant despite variations in rf signal strength. There are several modifications of this basic circuit.

Perfect AGC action would provide a constant output for all values of input signal strength. A slightly rising output characteristic with increased signal strength is generally desirable to facilitate proper tuning. The figure of merit applied to AGC action is given as the change in input required for a given output change. An example of a good figure of merit may be seen when an 80-dB change in input carrier signal results in an output change of no more than 3dB. This applies to the unmodulated carrier strength only, since the modulation of the carrier must always vary as the modulation of the transmitter.

AGC circuits are also used in dictation recording equipment, public address systems, and similar equipment where a constant output level is desirable. Figure 2 shows a typical block diagram for such equipment. *See* AMPLIFIER.

[WALTER LYONS]

Backward-wave tube

A type of microwave traveling-wave tube in which energy on a slow-wave circuit flows opposite in direction to the travel of electrons in a beam. Chief characteristics of backward-wave tubes are regenerative feedback produced by interaction of circuit and beam, and a wide range of electrical tuning, easily produced by changing the beam voltage. Such tubes are useful as voltage-tuned oscillators for signal generators, as power sources for quick tuning transmitters, and as local oscillators in receivers for systems that have quick tuning transmitters. If backward-wave tubes are operated as regenerative amplifiers, they are useful as narrowband amplifiers in wide-range rapidly tuned receivers.

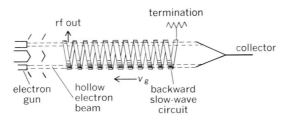

Fig. 1. O-type backward-wave oscillator (or O-carcinotron), which uses a helix as the slow-wave circuit and with a hollow cylindrical electron beam.

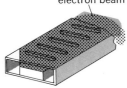

Fig. 2. Karp circuit, used at millimeter wavelengths.

O-type backward-wave oscillator.

An O-type backward-wave oscillator (or O-carcinotron) may be similar in appearance to a forward-wave traveling-wave tube. An electron gun produces an electron beam; this beam is focused longitudinally throughout the length of the tube. A slow-wave circuit interacts with the beam, and at the end of the tube a collector terminates the beam (Fig. 1). *See* TRAVELING-WAVE TUBE.

Energy in the slow-wave circuit travels from collector end toward the gun end of the tube. Hence, microwave energy is coupled out adjacent to the electron gun. At the collector end the slow-wave circuit is terminated with a matched load, usually internally, so that no microwave terminal is provided. For a tube in which the slow-wave circuit is a helix, the termination usually consists of lossy material sprayed on the collector end of the circuit.

The electron beam is usually hollow, the electrons thereby all being close to the helix. The electrons experience the strongest electric field from the microwave signal when they pass a gap between two helix turns. Electron velocity is adjusted so that any one electron experiences approximately the same phase of the microwave signal as it passes successive gaps. Hence, axial forces due to the microwave field cause some electrons to speed up and others to slow down as they travel past successive gaps. These accelerations and decelerations cause the electrons to bunch in the axial direction. Average velocity of a bunch is such that it drifts into a retarding electric field as it travels down the tube; thus the bunched beam transfers energy to the slow-wave circuit. This action provides continuous feedback along the tube, the beam providing a forward flow of energy and the circuit a backward flow.

To examine the synchronism condition for backward-wave interaction, consider two successive gaps of a helix. The time required for an electron to travel from the first gap to the second equals p/u_o, where p is helix pitch, and u_o is electron velocity. During this transit time, a microwave signal of frequency f on the helix will have changed its phase at any one gap by $2\pi fp/u_o$. In addition, at any instant the microwave fields at the two adjacent gaps differ in phase by $2\pi fp/v_g$, where v_g is the velocity at which microwave energy travels along the helix toward the gun. Thus, an electron in moving from the first to the second gap experiences a total field phase change of ϕ. This is defined in the equation below, where synchronism occurs when

$$2\pi fp/u_o + 2\pi fp/v_g = \phi$$

phase change ϕ is exactly one cycle or 2π radians. Frequency of oscillation f can be controlled by changing electron velocity u_o, which depends on the helix-to-cathode voltage.

As in any feedback oscillator, gain must exceed internal losses. To obtain this minimum required gain, beam current is raised above a value called the start-oscillation current. Normally, operation is at a current in the order of twice the start-oscillation current. Too much beam current would permit higher-order electronic modes that interfere with the desired mode.

Backward-wave oscillators with a helix for the slow-wave circuit typically develop outputs of 10–200 milliwatts, with efficiencies of a few percent or less. Frequency can be voltage-tuned typically over a 2:1 range. Another slow-wave circuit, used at millimeter wavelengths, is the Karp circuit (Fig. 2).

M-type backward-wave oscillator.

An M-type backward-wave oscillator (or M-carcinotron) is similar in principle to the O-type, except that focusing and interaction are through magnetic fields, as in magnetrons. In the M-type oscillator (Fig. 3), a transverse magnetic field and a static radial electric field between sole and backward-wave circuit structure confine the beam to the interaction space. Either the voltage connected to the sole or to the slow-wave circuit tunes the frequency. *See* MAGNETRON.

The commonest slow-wave circuit is an interdigital structure, consisting of an array of vertical bars alternating up and down. Efficiency of the M-type tube is considerably higher than that of the O-type tube, typical efficiencies being 20–30%. However, noise and spurious output power are also greater in the M-type tube. Continuous output powers of several hundred watts are typical.

Backward-wave amplifier.

When either type tube is operated with currents below the start-oscillation value, narrow-band regenerative amplification is obtained. Amplifier frequency is electrically tunable over a wide range by change of beam voltage. Input to a backward-wave amplifier is at

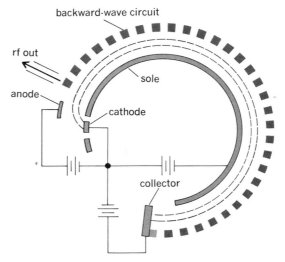

Fig. 3. M-type backward-wave oscillator with the beam focused by a static magnetic field directed into the plane of the page.

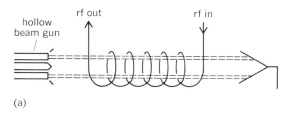

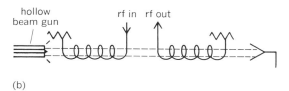

(a)

(b)

Fig. 4. Backward-wave amplifiers. (a) Single-circuit type. (b) Cascade type, used for increased stability.

the collector end and output is at the gun end (Fig. 4). Separation of the slow-wave circuit into two or more parts enhances stability. A cascade backward-wave amplifier has a stable gain in excess of 20 dB, a bandwidth of a fraction of a percent, a tuning range easily of 2:1, and a noise figure as low as 3.5 dB. *See* MICROWAVE TUBE.

[J. W. GEWARTOWSKI]

Bibliography: J. W. Gewartowski and H. A. Watson, *Principles of Electron Tubes*, 1965; A. L. Lance, *Introduction to Microwave Theory and Measurements*, 1964; S. Ligo, *Microwave Devices and Circuits*, 1980.

Band theory of solids

A quantum-mechanical theory of the motion of electrons in solids. Its name comes from the fact that it predicts certain restricted ranges, or bands, for the energies of electrons in solids.

Suppose that the atoms of a solid are separated from each other to such a distance that they do not interact. The energy levels of the electrons of this system will then be those characteristic of the individual free atoms, and thus many electrons will have the same energy. Now imagine the atoms being slowly brought closer together. As the distance between atoms is decreased, the electrons in the outer shells begin to interact with each other. This interaction alters their energy and, in fact, broadens the sharp energy level out into a range of possible energy levels called a band. One would expect the process of band formation to be well advanced for the outer, or valence, electrons at the observed interatomic distances in solids. Once the atomic levels have spread into bands, the valence electrons are not confined to individual atoms, but may jump from atom to atom with an ease that increases with the increasing width of the band.

Although energy bands exist in all solids, the term energy band is usually used in reference only to ordered substances, that is, those having well-defined crystal lattices. In such a case, an electron energy state can be classified according to its crystal momentum **p** or its electron wave vector **k** = **p**/$\hbar$ (where $\hbar$ is Planck's constant h divided by 2π). If the electrons were free, the energy of

an electron whose wave vector is **k** would be as shown in Eq. (1), where E_0 is the energy of the

$$E(\mathbf{k}) = E_0 + \hbar^2\mathbf{k}^2/2m_0 \qquad (1)$$

lowest state of a valence electron and m_0 is the electron mass. In a crystal, however, the electrons are not free because of the effect of the crystal binding and the forces exerted on them by the atoms; consequently, the relation $E(\mathbf{k})$ between energy and wave vector is more complicated. The statement of this relationship constitutes the description of an energy band.

A knowledge of the energy levels of electrons is of fundamental importance in computing electrical, magnetic, optical, or thermal properties of solids.

Allowed and forbidden bands. The bands of possible electron energy levels in a solid are called allowed energy bands. It often happens that there are also bands of energy levels which it is impossible for an electron to have in a given crystal. Such bands are called forbidden bands, or gaps. The allowed energy bands sometimes overlap and sometimes are separated by forbidden bands. The presence of a forbidden band, or energy gap, immediately above the occupied allowed states (such as the region A to B in the illustration) is the principal difference in the electronic structures of a semiconductor or insulator and a metal. In the first two substances there is a gap between the valence band or normally occupied states and the conduction band, which is normally unoccupied. In a metal there is no gap between occupied and unoccupied states.

The presence of a gap means that the electrons cannot easily be accelerated into higher energy states by an applied electric field. Thus, the substance cannot carry a current unless electrons are excited across the gap by thermal or optical means.

Effective mass. When an external electromagnetic field acts upon the electrons in a solid, the resultant motion is not what one would expect if the electrons were free. In fact, the electrons behave as though they were free but with a different mass, which is called the effective mass. This effective mass can be obtained from the dependence of electron energy on the wave vector, $E(\mathbf{k})$, in the following way.

Suppose there is a maximum or minimum of the function $E(\mathbf{k})$ at the point $\mathbf{k} = \mathbf{k}_0$. The function $E(\mathbf{k})$ can be expanded in a Taylor series about this point. For simplicity, assume that $E(\mathbf{k})$ is a function of the magnitude of **k** only, that is, is independent

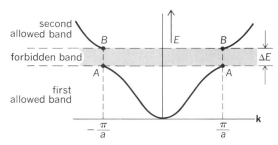

Electron energy E versus wave vector **k** for a monatomic linear lattice of lattice constant a. (*After C. Kittel, Introduction to Solid State Physics, 2d ed., Wiley, 1956*)

of the direction of $\mathbf{k}$. Then, by dropping terms higher than second order in the Taylor series, Eq. (2) results. By analogy with Eq. (1), a quantity

$$E(\mathbf{k}) = E(\mathbf{k}_0) + \tfrac{1}{2}(\mathbf{k} - \mathbf{k}_0)^2 \left(\frac{d^2E}{dk^2}\right)_{k_0} \qquad (2)$$

m^* with the dimensions of a mass can be defined by the relation in Eq. (3).

$$\hbar^2/m^* = \left(\frac{d^2E}{dk^2}\right)_{k_0} \qquad (3)$$

The quantity m^* is called the effective mass of electrons at $\mathbf{k}_0$. For many simple metals, the average effective mass is close to m_0, but smaller effective masses are not uncommon. In indium antimonide, a semiconductor, the effective mass of electrons in the conduction band is $0.013\ m_0$. In a semiclassical approximation, an electron in the solid responds to an external applied electric or magnetic field as though it were a particle of mass m^*. The equation of motion of an electron is shown in Eq. (4), where $\mathbf{v}$ is the electron velocity,

$$m^*\frac{d\mathbf{v}}{dt} = e(\mathbf{E} + \mathbf{v} \times \mathbf{B}) \qquad (4)$$

$\mathbf{E}$ the electric field, $\mathbf{B}$ the magnetic induction, and e the charge of the electron.

It may happen that the energy $E(\mathbf{k})$ does depend upon the direction of $\mathbf{k}$. In such a case, the effective mass is a tensor whose components are defined by Eq. (5). Equation (4) remains valid with

$$\hbar^2/m_{ij}^* = \left(\frac{\partial^2E}{\partial k_i\,\partial k_j}\right)_{k_0} \qquad (5)$$

a tensor m^*. Bismuth is an example of a metal in which the effective mass depends strongly on direction.

Transitions between states. Under external influences, such as irradiation, electrons can make transitions between states in the same band or in different bands. The interaction between the electrons and the vibrations of the crystal lattice can scatter the electrons in a given band with a substantial change in the electron momentum, but only a slight change in energy. This scattering is one of the principal causes of the electrical resistivity of metals.

An external electromagnetic field (for example, visible light) can cause transitions between different bands. In such a process, momentum must be conserved. Because the momentum of a photon $h\nu/c$ (where ν is the frequency of the light and c its velocity) is quite small, the momentum of the electron before and after collision is nearly the same. Such a transition is called vertical in reference to an energy band diagram. Conservation of energy must also hold in the transition, so absorption of light is possible only if there is an unoccupied state of energy $h\nu$ available at the same $\mathbf{k}$ as the initial state. These transitions are responsible for much of the absorption of light by semiconductors in the visible and near-infrared region of the spectrum. *See* HOLES IN SOLIDS; PHOTOEMISSION; SEMICONDUCTOR.

[JOSEPH CALLAWAY]

Bibliography: P. W. Anderson, *Concepts in Solids*, 1964; N. W. Ashcroft and N. D. Mermin, *Solid State Physics*, 1976; J. Callaway, *Quantum Theory of the Solid State*, 1974; M. Cardona, *Modulation Spectroscopy*, Suppl. 11 of *Solid State Phys.*, 1969; G. C. Fletcher, *The Electron Band Theory of Solids*, 1971; W. Jones and W. H. March, *Theoretical Solid State Physics*, 1973; C. Kittel, *Quantum Theory of Solids*, 1963; J. C. Slater, *Quantum Theory of Molecules and Solids*, vol. 2: *Symmetry and Energy Bands in Crystals*, 1965; J. M. Ziman, *Principles of the Theory of Solids*, 2d ed., 1972.

Bandwidth reduction

Techniques for reducing the high transmission rate required for digital transmission of television signals. These techniques, also known as bit-rate compression, involve elimination of redundant information from the signals.

Digital transmission. Digital transmission of signals is accomplished by sending on-off pulses, in contrast to analog transmission, where a continuous waveform of the signal, such as the picture or speech intensity, is transmitted. Digital transmission has a number of advantages: it offers flexibility; digital signals of different types can be easily multiplexed or encrypted; and transmission over long distances can be achieved with easy regeneration of the signal, without adding additional noise or distortion to the signal. Digitization of the analog signal is done by first sampling the signal and then representing each sample by a string of binary digits (bits), 1 or 0, specifying the on or off nature of the pulse. The transmission cost would be proportional to the product of the number of samples that are generated per second and the number of bits required to specify a sample. *See* ANALOG-TO-DIGITAL CONVERTER; PULSE MODULATION.

The television signal contains a sequence of "snapshots" (called television frames) taken from a scene at a rate of 30 times a second. Within each television frame, sampling is done horizontally to generate 525 scan lines, each of which contains approximately 500 samples (monochrome television); for color television a larger number of samples (between 700 and 900) is required. Each sample is represented by 8 bits, giving a total of about 64,000,000 bits per second for monochrome television and 80,000,000 to 100,000,000 bits per second for color television. This transmission rate is rather high, but it can be reduced by techniques of bandwidth compression. *See* TELEVISION; TELEVISION SCANNING.

Techniques. Bandwidth compression, or more correctly bit-rate reduction, is accomplished by eliminating from transmission redundant information normally contained in the signals. Picture signals contain a significant amount of redundancy. Although there are approximately 8,000,000 samples generated per second, their intensities are not independent of one another. For example, the intensity of one picture sample contains considerable information about the intensity of a sample that is next to it, either horizontally or vertically in the same television frame or at the same spatial location in the previous television frame. Thus, in the illustration, intensities of picture elements A, B, D, E, F, G are close to each other most of the time. Bandwidth reduction techniques have been devised to take advantage of the similarity of the intensities of spatially as well as temporally adjacent picture samples. These techniques can be generally classified into two categories: differential pulse-code modulation (DPCM), in which intensity differences are sent; and transform coding, in which linear combinations of intensities in a block of samples are taken and only some of the combinations are selected for transmission. Practical systems based on transform coding have not yet proliferated due to their complexity.

Conditional replenishment. One popular application of DPCM is conditional replenishment, in which each television frame is divided into two parts: one part which is practically the same as the previous frame, and the other part (called the moving area) which has changed since the previous frame. Two types of information are transmitted about the moving area: addresses specifying the location of the picture samples in the moving area, and information by which the intensities of the moving area picture samples can be reconstructed at the receiver. Comparison with the previous frame intensities requires storage of an entire television frame (about 2,000,000 bits for black and white television and 4,000,000 bits for color television), both at the transmitter and at the receiver.

Since the motion in a real television scene occurs randomly and in bursts, the amount of information about the moving area will change as a function of time. To transmit it over a channel, which works at a constant bit rate, the output of the encoder has to be smoothed out by storing it in a storage buffer prior to transmission. The encoded data enter the buffer at an irregular rate, but they exit the buffer and enter the channel at the constant bit rate of the channel. If the buffer gets nearly full, then certain samples are deleted from transmission, thereby reducing the resolution. Some of the strategies for deleting samples from transmission are: (1) transmitting 1 in every n ($n = 2, 4, \ldots$) samples along a scan line—this reduces horizontal resolution; (2) transmitting 1 in every n ($n = 2, 4, \ldots$) scan lines—this reduces vertical resolution; (3) transmitting 1 in every n frames—this reduces temporal resolution. The blur introduced by these strategies is less visible since it is generally introduced when there is large motion in the scene, in which case the human visual acuity is low.

In the last 5–10 years several improvements of the conditional replenishment technique have been made, resulting in commercially available encoders. For television signals generated from video conferences, where the camera motion is limited and the scenes do not contain a large amount of movement, reasonable-quality pictures can be obtained at a transmission rate of 3–6 megabits per second, which is a reduction over the uncompressed bit rate by a factor of about 10–20. For broadcast television, on the other hand, where there can be large changes from one frame to the next and where there is a stricter picture quality requirement, reductions in the bit rate can be made only by a factor of 3–4.

Motion-compensated coding schemes. In television scenes which contain moving objects, a more efficient encoding can be performed by estimating the motion of objects and then using the motion to compare intensities in successive frames which are spatially displaced by an amount equal to the motion of the object. Such schemes are called motion-compensated coding schemes. The operation of such a scheme is shown in the illustration. If a point on an object moves from location C

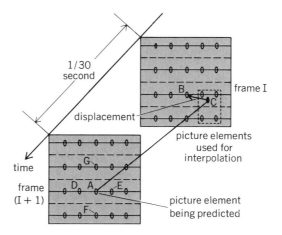

key: – – – scan line from alternate field

Diagram of the sampled nature of the television signal, showing the operation of a motion-compensated coding scheme.

to location B in a frame time, then, instead of comparing intensities A and B having the same spatial location, as in conditional replenishment, intensities A and C are compared. If the estimate of displacement is accurate, the intensity difference between A and C is much lower than the intensity difference between A and B, resulting in a lower amount of information to be transmitted. The intensity at point C is computed by interpolating from the intensities of the picture elements in the small box in frame I in the illustration. Procedures for estimating translations of moving objects from frame to frame by recursive adjustments have been devised recently. These are simple to implement in hardware. Computer simulations show that, for many video conferencing and broadcast types of scenes, the bit rate can be reduced by a factor of 1.5–3 over the conditional replenishment. The resulting bit rate for monochrome television signals from video conferencing scenes would then be 1–2 megabits per second and for broadcast television 10–20 megabits per second. Hardware implementation of these schemes is expected in the new few years with many modifications to improve the efficiency and decrease the complexity.

[ARUN N. NETRAVALI]

Bibliography: B. G. Haskell, F. W. Mounts, and J. C. Candy, *Proc. IEEE*, 60(7):792–800, July 1972; T. Ishiguro et al., *National Telecommunications Conference Record*, Dallas, pp. 6.4-1 to 6.4-5, November 1976; J. O. Limb, R. F. W. Pease, and K. A. Walsh, *Bell Sys. Tech. J.*, 53(6):1137–1173, August 1974; F. W. Mounts, *Bell Sys. Tech. J.*, 48(7):2545–2554, September 1969; A. N. Netravali and J. D. Robbins, *Bell Sys. Tech. J.*, 58(3):631–670, March 1979.

Bandwidth requirements

The difference between the limiting frequencies of a frequency band containing the useful components of a signal is called the bandwidth, expressed in the unit of frequency, hertz (Hz); the unit was formerly cycles per second (cps). Each communication system requires some optimum bandwidth for the satisfactory transmission of information.

Electrical communications involve the variation of some electrical quantity as a representation of the signal to be transmitted. In the case of simple wire telephony, where the speech sound waves cause variation of current in the line, the bandwidth required is the same as the desired bandwith of the signal to be reproduced, or about 3000 Hz. In wire telegraphy, the bandwidth required depends on the signaling speed used. For both wire and radio telegraphy, a reduction in the required bandwidth can be achieved by shaping the pulses to approach sine wave shape, in which latter case the bandwidth for wire telegraphy is equal to the signaling speed in bauds.

When the signal to be transmitted causes the variation of the amplitude, phase, frequency, or pulsing of a carrier wave, the process is known as modulation. Bandwidth requirements then depend on the frequency stability of the carrier wave, the character of the modulation process, and the bandwidth of the modulating signal. The minimum bandwidth possible is equal to the bandwidth of the modulating signal. The bandwidth of the modulating signal for a given message may, however, be altered by encoding the message so as to require either greater or less signal bandwidth than that of the information in its original form. *See* INFORMATION THEORY; MODULATION.

Overall bandwidth requirements for the transmission and reception circuits of a simple amplitude-modulation system may be determined from the frequency response specifications for the system. There will be a bandwidth, called a sideband, on each side of the radio carrier, which is equal to the modulating frequencies. For example, if transmission of voice frequencies up to 3000 Hz is required, the radio-frequency bandwidth, when both upper and lower sidebands are included, must be at least 6000 Hz. A single-sideband system, transmitting the upper sideband or the lower sideband only, occupies a bandwidth of about 3000 Hz.

In frequency-modulation systems, when the input signal-to-noise ratio exceeds a certain threshold value, the signal-to-noise power ratio is improved at the expense of increased bandwidth, relative to that of a smple amplitude-modulation system, in accordance with the equation below,

$$\frac{\text{FM signal/noise ratio}}{\text{AM signal/noise ratio}} = 3\frac{F^2}{B^2}$$

where F is the greatest frequency deviation of the carrier from the quiet or unmodulated condition and B is the highest modulation frequency. Other kinds of modulation systems, such as pulse modulation, also provide improved signal-to-noise ratio in exchange for greater required bandwidth, and have an improvement threshold.

Bandwidth requirements determine the amount of radio spectrum needed by communications systems. In allocating the spectrum to the many radio services and stations, their bandwidth requirements are taken into account.

Transmitted signal spectra can be computed when the waveform of the modulating signal and modulation parameters are known, by use of Fourier analysis for amplitude- and pulse-modulation systems and by use of Bessel functions for frequency-modulation systems.

Radio communications bandwidth requirements are characterized by reference (1) to the bandwidth of the transmitted signal (emission bandwidth), or (2) to the frequency response of the electrical circuits through which the signal passes, including transmitter and receiver. Emission bandwidth is usually stated in terms of the frequency band which contains a specified portion of the total signal power. For example, the International Radio Regulations, 1947, defines occupied bandwidth as "the frequency bandwidth such that, below its lower and above its upper frequency limits, the mean powers radiated are each equal to 0.5% of the total mean power radiated by a given emission."

Bandwidths necessary for various kinds of communications systems have been designated in Appendix 5 of the International Radio Regulations. The following list shows the approximate bandwidth necessary for several kinds of communications systems.

System	*Bandwidth, Hz*
Telegraphy	
100 words/min	170–400
Telephony	3,000
High-speed data transmission	
1000 bits/sec	2,000–3,000
AM radiotelephony	
Commercial	6,000
Broadcasting	8,000–20,000
FM radiotelephony	
Commercial	36,000
Broadcasting	180,000
Radiotelegraphy	
Frequency shift	600
AM facsimile	5,000
FM facsimile	25,000
Television, U.S.	
Broadcasting	6×10^6
Radio relay	25×10^6

For further information of specific systems *see* FACSIMILE; RADIO; TELEVISION.

[EDWARD W. ALLEN]

Bibliography: S. Guiasu, *Information Theory with New Applications*, 1977; D. H. Hamsher, *Communications System Engineering Handbook*, 1967; Howard W. Sams Engineering Staff, *Reference Data for Radio Engineers*, 6th ed., 1975; P. B. Johns and T. R. Rowbotham, *Communication System Analysis*, 1972.

Barium titanate

A dielectric crystalline material which exhibits the anomalous properties typical for ferroelectricity. Barium titanate, $BaTiO_3$, is a thoroughly studied important compound and is representative of the physically related materials crystallizing in the perovskite structure.

On lowering the temperature, cubic $BaTiO_3$ transforms to a tetragonal structure at about 130°C(Curie temperature), where the crystal spontaneously acquires an electric polarization (a permanent electric dipole moment). As the polarization can in general point in any one of the originally cubic crystal axes, the crystal consists, in the ferroelectric state, of regions (domains) which differ in the direction of the spontaneous polarization. An applied electric field can align the polarization in the whole crystal and can reverse the polarization direction whereby electric hysteresis occurs. At lower temperatures, 5 and −90°C, additional phase transitions take place. The spontaneous polarization successively changes its direction relative to the crystallographic axes and increases in absolute value. *See* FERROELECTRICS.

Many other compounds of the general formula ABO_3 have a perovskite-type structure and rather similar properties. As $BaTiO_3$ forms solid solutions with many of these compounds, its properties can be varied systematically over wide ranges. Such materials can be readily produced in a polycrystalline ceramic form which is fully adequate for many applications, some of which are discussed below.

Barium titanate has a high dielectric constant, typically 2000 for a ceramic sample at room temperature. Thus, capacitors of very small dimensions can be produced for electronics applications. As a further advantage, the temperature coefficient of capacitance can be varied over a range of positive and negative values by admixtures of other compounds, such as strontium titanate and lead titanate.

Barium titanate ceramics, when suitably doped with a trivalent oxide such as lanthanum oxide, for example, exhibit a variation of the electrical conductivity by a factor up to 10^6 in the vicinity of the Curie temperature. This effect can be used for sensitive temperature-control devices.

Ceramics of barium titanate and, in particular, of a mixed lead titanate zirconate can be poled by an electric field and retain a substantial remanent polarization. These materials then have a strong piezoelectric effect. In addition, they are mechanically rugged and rather insensitive to temperature and humidity. Therefore, these ceramics find wide applications in devices such as microphones, ultrasonic and underwater transducers, and spark generators. *See* PIEZOELECTRICITY.   [H. GRANICHER]

Bibliography: M. Deri, *Ferroelectric Ceramics*, 1969; M. E. Lines and A. M. Glass, *Principles and Applications of Ferroelectrics and Related Materials*, 1977; T. Mitsui, *An Introduction to the Physics of Ferroelectrics*, 1976.

Bias of transistors

The establishment of an operating point on the transistor volt-ampere characteristics by means of direct voltages and currents. Since the transistor is a three-terminal device, any one of the three terminals may be used as a common terminal to both input and output. In most transistor circuits the emitter is used as the common terminal, and this common emitter, or grounded emitter, is indicated in illustration *a*. If the transistor is to be used as a linear device, such as an audio amplifier, it must be biased to operate in the active region. In this region the collector is biased in the reverse direction and the emitter in the forward direction. The area in the common-emitter transistor characteristics to the right of the ordinate $V_{CE} = 0$ and above $I_C = 0$ is the active region. Two more biasing regions are of special interest for those cases in

which the transistor is intended to operate as a switch. These are the saturation and cutoff regions. The saturation region may be defined as the region where the collector current is independent of base current for given values of V_{CC} and R_L. Thus, the onset of saturation can be considered to take place at the knee of the common-emitter transistor curves. *See* AMPLIFIER.

In saturation the transistor current I_C is nominally V_{CC}/R_L. Since R_L is small, it may be necessary to keep V_{CC} correspondingly small in order to stay within the limitations imposed by the transistor on maximum-current and collector-power dissipation. In the cutoff region it is required that the emitter current I_E be zero, and to accomplish this it is necessary to reverse-bias the emitter junction so that the collector current is approximately equal to the reverse saturation current I_{CO}. A reverse-biasing voltage of the order of 0.1 volt across the emitter junction will ordinarily be adequate to cut off either a germanium or silicon transistor.

The particular method to be used in establishing an operating point on the transistor characteristics depends on whether the transistor is to operate in the active saturation or cutoff regions, on the application under consideration, on the thermal stability of the circuit, and on other factors.

Fixed-bias circuit. The operating point for the circuit of illustration *a* can be established by noting that the required current I_B is constant, independent of the quiescent collector current I_C, which is why this circuit is called the fixed-bias circuit. Transistor biasing circuits are frequently compared in terms of the value of the stability factor $S = \partial I_C/\partial I_{CO}$, which is the rate of change of collector current with respect to reverse saturation current. The smaller the value of S, the less likely the circuit will exhibit thermal runaway. S, as defined here, cannot be smaller than unity. Other stability factors are defined in terms of dc current gain h_{FE} as $\partial I_C/\partial h_{FE}$, and in terms of base-to-emitter voltage as $\partial I_C/\partial V_{BE}$. However, bias circuits with small values of S will also perform satisfactorily for transistors that have large variations of h_{FE} and V_{BE}. For the fixed-bias circuit it can be shown that $S = h_{FE} + 1$, and if $h_{FE} = 50$, then $S = 51$. Such a large value of S makes thermal runaway a definite possibility with this circuit.

Collector-to-base bias. An improvement in stability is obtained if the resistor R_B in part *a* of the figure is returned to the collector junction rather than to the battery terminal. Such a connection is shown in illustration *b*. In this bias circuit if I_C tends to increase (either because of a rise in temperature or because the transistor has been replaced by another), then V_{CE} decreases. Hence I_B also decreases and, as a consequence of this lowered bias current, the collector current is not allowed to increase as much as it would if fixed bias were used. The stability factor S is shown in Eq. (1). This value is smaller than $h_{FE} + 1$, which is the value obtained for the fixed-bias case.

$$S = \frac{h_{FE} + 1}{1 + h_{FE}R_L/(R_L + R_B)} \qquad (1)$$

Self-bias. If the load resistance R_L is very small, as in a transformer-coupled circuit, then the previous expression for S shows that there would be no improvement in the stabilization in the

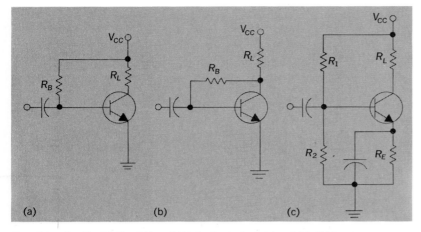

Transistor circuits. (*a*) Fixed-bias. (*b*) Collector-to-base bias. (*c*) Self-bias.

collector-to-base bias circuit over the fixed-bias circuit. A circuit that can be used even if there is zero dc resistance in series with the collector terminal is the self-biasing configuration of illustration *c*. The current in the resistance R_E in the emitter lead causes a voltage drop which is in the direction to reverse-bias the emitter junction. Since this junction must be forward-biased (for active region bias), the bleeder R_1-R_2 has been added to the circuit.

If I_C tends to increase, the current in R_E increases. As a consequence of the increase in voltage drop across R_E, the base current is decreased. Hence I_C will increase less than it would have had there been no self-biasing resistor R_E. The stabilization factor for the self-bias circuit is shown by Eq. (2), where $R_B = R_1 R_2/(R_1 + R_2)$. The

$$S = (1 + h_{FE}) \frac{1 + R_B/R_E}{1 + h_{FE} + R_B/R_E} \qquad (2)$$

smaller the value of R_B, the better the stabilization. Even if R_B approaches zero, the value of S cannot be reduced below unity.

In order to avoid the loss of signal gain because of the degeneration caused by R_E, this resistor is often bypassed by a very large capacitance, so that its reactance at the frequencies under consideration is very small.   [CHRISTOS C. HALKIAS]

Bibliography: J. Millman, *Microelectronics*, 1979.

Bit

In the pure binary numeration system, either of the digits 0 or 1. The term may be thought of as a contraction of binary digit. *See* NUMBER SYSTEMS.

In a binary notation, bits are used as two different characters. For example, in the American National Standard Code for Information Interchange (ASCII), a seven-bit coded character set, the seven bits 1000111 represent the letter G.

Bit is widely used as a synonym for binary element, a constituent element of data that takes either of two values or states (on-off, yes-no, zero-one, and so forth). The brains of animals and registers, memories, and other storage devices of digital computers and electronic calculators store bits (binary elements) as the smallest unit of information.

During a meeting in the late winter of 1943/1944

in Princeton, NJ, convened by John von Neumann and Norbert Wiener, engineers, physiologists, and mathematicians found that it was convenient to measure information in terms of numbers of yeses and noes and to call this unit of information a bit. Strictly speaking, the bit of von Neumann, Wiener, and their associates is the binary element and not the binary digit.

A byte is a string of bits (binary elements) operated on or treated as a unit and usually shorter than a word. An eight-bit byte comprises eight bits. In precise usage, *n*-bit bytes are called quartet, quintet, . . ., octet, and so forth. Byte is derived from bite.

A word is a string of bits (binary elements) that consists of two or more bytes. The terms halfword, fullword, and doubleword are also used.

A computer word is a word stored in one computer location, usually 16, 32, 36, 48, or 64 bits in length, depending on the design of the computer.

In microcomputers or in particular applications, a nybble is a string of bits (binary elements) operated on as a unit, larger than a bit, and smaller than a byte. Nybble is derived from nibble. *See* DIGITAL COMPUTER.

In information theory, bit is a synonym for the new, preferred term shannon, a unit of measure of information equal to the decision content of a set of two mutually exclusive events. For example, the decision content of a character set of eight characters equals three shannons (the logarithm of 8 to base 2). A shannon also is called an information content binary unit. A hartley is the information content decimal unit. These unit names were adopted by the International Organization for Standardization (ISO) in 1975, honoring C. E. Shannon and R. C. L. Hartley. *See* INFORMATION THEORY.   [HELMUT E. THIESS]

Bibliography: M. H. Weik, Jr., et al., *American National Dictionary of Information Processing*, 1977; N. Wiener, *Cybernetics*, 2d ed., 1961; N. Wiener, *I Am a Mathematician*, 1956.

Blocking oscillator

A relaxation oscillator that generates a short-time-duration pulse using a single transistor or vacuum tube and associated circuitry. The input and output are coupled together by a transformer in a regenerative feedback arrangement, with the out-

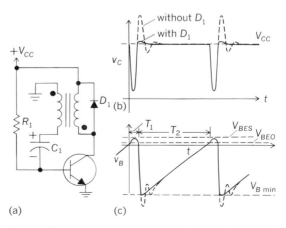

Fig. 1.   Typical free-running blocking oscillator. (*a*) Circuit. (*b*) Collector voltage. (*c*) Base voltage.

put current being periodically interrupted because of the regenerative action which drives the input far into the reverse-bias or cutoff region. The duration of the period between pulses, which coincides with the period during which the output current is blocked, is dependent upon the value of the RC time constant in the input circuit. The blocking oscillator may be made to operate in either a monostable or astable fashion. When operated as a monostable device, considerable control can be exercised over the shape and duration of the pulse, making the circuit useful as a pulse generator. *See* PULSE GENERATOR; RELAXATION OSCILLATOR.

Free-running astable type. A simple astable blocking oscillator using an *n-p-n* transistor is shown in Fig. 1. A field-effect transistor or vacuum tube may be used as an alternative. If one assumes initially that the transistor is forward-biased at cutoff level V_{BEO} of the input waveform such that base current is just beginning to flow, the regenerative action will increase the current up to a maximum value corresponding to saturation level V_{BES}, as shown. At some such level the output will no longer drive the input to higher current levels because of transformer losses and decreased transistor input impedance. Then, the base current reverses and the voltage drops to the level V_{BEO}. This portion of the waveform corresponds to the negative going collector pulse, as shown. During this conducting period T_1, the capacitor voltage changes by an amount given by Eq. (1), with polari-

$$V_{c1} = \int_0^{T_1} i_B dt \qquad (1)$$

ty as indicated. At the end of the conducting interval, the voltage across the transformer winding in series with the base collapses, and the base voltage drops to the negative value $V_{B\min}$, as shown, which is approximately equal in magnitude to the change in capacitor voltage during the previous conducting interval. At this time the voltage at the base rises toward the supply voltage V_{CC}, according to Eq. (2). Time T_2 is the time taken to

$$v_B = V_{B\min} + (V_{CC} - V_{B\min})(1 - e^{-t/(R_1 C_1)}) \qquad (2)$$

reach the point at which the transistor again conducts, and the entire process repeats itself.

Synchronized type. The blocking oscillator may be synchronized with external pulses occurring at a rate slightly faster than its own natural fre-

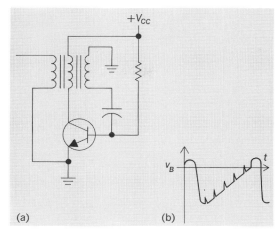

Fig. 3. The blocking oscillator as a frequency divider. (*a*) Circuit. (*b*) Base voltage.

quency in the same manner as multivibrators, as shown in Fig. 2. It may be used in the same manner as a frequency divider, as shown in Fig. 3. *See* MULTIVIBRATOR; SCALING CIRCUIT.

The blocking oscillator becomes a monostable device if resistance R_1 is returned to a voltage sufficiently far in the reverse-bias direction so that an external trigger must be applied before an action can be initiated. There is more freedom of control of the circuit elements for control of pulse duration in the monostable circuit than in the astable circuit, if it is desired to use the blocking oscillator as a pulse generator.

[GLENN M. GLASFORD]

Bibliography: B. Chance et al. (eds.), *Waveforms*, 1949; G. M. Glasford, *Fundamentals of Television Engineering*, 1955; J. Millman and H. Taub, *Pulse, Digital, and Switching Waveforms*, 1965; L. Strauss, *Wave Generation and Shaping*, 1970.

Boolean algebra

A branch of mathematics that was first developed systematically, because of its applications to logic, by the English mathematician George Boole, around 1850. Closely related are its applications to sets and probability.

Boolean algebra also underlies the theory of relations. A modern engineering application is to computer circuit design. *See* DIGITAL COMPUTER; SWITCHING THEORY.

Set-theoretic interpretation. Most basic is the use of Boolean algebra to describe combinations of the subsets of a given set I of elements; its basic operations are those of taking the intersection or common part $S \cap T$ of two such subsets S and T, their union or sum $S \cup T$, and the complement S' of any one such subset S. These operations satisfy many laws, including those shown in Eqs. (1), (2), and (3).

$$\begin{aligned} S \cap S = S \qquad S \cap T = T \cap S \\ S \cap (T \cap U) = (S \cap T) \cap U \end{aligned} \qquad (1)$$

$$\begin{aligned} S \cup S = S \qquad S \cup T = T \cup S \\ S \cup (T \cup U) = (S \cup T) \cup U \end{aligned} \qquad (2)$$

$$\begin{aligned} S \cap (T \cup U) = (S \cap T) \cup (S \cap U) \\ S \cup (T \cap U) = (S \cup T) \cap (S \cup U) \end{aligned} \qquad (3)$$

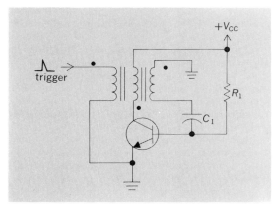

Fig. 2. Synchronized blocking oscillator.

If O denotes the empty set, and I is the set of all elements being considered, then the laws set forth in Eq. (4) are also fundamental. Since these laws

$$O \cap S = O \qquad O \cup S = S \qquad I \cap S = S$$
$$I \cup S = I \qquad S \cap S' = O \qquad S \cup S' = I \qquad (4)$$

are fundamental, all other algebraic laws of subset combination can be deduced from them.

In applying Boolean algebra to logic, Boole observed that combinations of properties under the common logical connectives *and*, *or*, and *not* also satisfy the laws specified above. These laws also hold for propositions or assertions, when combined by the same logical connectives. *See* LOGIC CIRCUITS.

Boole stressed the analogies between Boolean algebra and ordinary algebra. If $S \cap T$ is regarded as playing the role of st in ordinary algebra, $S \cup T$ that of $s + t$, O of 0, I of 1, and S' as corresponding to $1 - s$, the laws listed above illustrate many such analogies. However, as first clearly shown by Marshall Stone, the proper analogy is somewhat different. Specifically, the proper Boolean analog of $s + t$ is $(S' \cap T) \cup (S \cap T')$, so that the ordinary analog of $S \cup T$ is $s + t - st$. Using Stone's analogy, Boolean algebra refers to Boolean rings in which $s^2 = s$, a condition implying $s + s = 0$. *See* RING THEORY; SET THEORY.

Boolean algebra arises in other connections, as in the algebra of (binary) relations. Such relations $\rho, \sigma, \ldots$ refer to appropriate sets of elements I, $J, \ldots$. Any such ρ can be defined by describing the set of pairs (x,t), with x in I and y in J, that stand in the given relation—a fact symbolized $x \rho y$, just as its negation is written $x \rho' y$. Because of this set-theoretic interpretation, Boolean algebra obviously applies, with $x(\rho \cap \sigma)y$ meaning $x \rho y$ and $x \sigma y$, and $x(\rho \cup \sigma)y$ meaning $x \rho y$ or $x \sigma y$.

Abstract relationships. Before 1930, work on Boolean algebra dealt mainly with its postulate theory, and with the generalizations obtained by abandoning one or more postulates, such as $(p')' = p$ (Brouwerian logic). Since $a \cup b = (a' \cap b')'$, clearly one need consider $a \cap b$ and a' as undefined operations. In 1913 H. M. Sheffer showed one operation only $(a \mid b = a' \cap b')$ need be taken as undefined. In 1941 M. H. A. Newman developed a remarkable generalization which included Boolean algebras and Boolean rings. This generalization is based on the laws shown in Eqs. (5) and (6). From these assumptions, the idempotent, commutative, and associative laws (1) and (2) can be deduced.

$$a(b + c) = ab + ac \qquad (a + b)c = ac + bc \qquad (5)$$

$$a1 = 1 \qquad a + 0 = 0 + a = a \qquad aa' = 0 \qquad (6)$$
$$a + a' = 1$$

Such studies lead naturally to the concept of an abstract Boolean algebra, defined as a collection of symbols combined by operations satisfying the identities listed in formulas (1) to (4). Ordinarily, the phrase Boolean algebra refers to such an abstract Boolean algebra, and this convention is adopted here.

The class of finite (abstract) Boolean algebras is easily described. Each such algebra has, for some nonnegative integer n, exactly 2^n elements and is algebraically equivalent (isomorphic) to the algebra of all subsets of the set of numbers $1, \ldots, n$, under the operations of intersection, union, and complement. Further, if m symbols $a_1, \ldots, a_m$ are combined symbolically through abstract operations $\cap$, $\cup$, and $'$ assumed to satisfy the identities of Eqs. (1) to (4), one gets a finite Boolean algebra with 2^{2m} elements—the free Boolean algebra with m generators.

Infinite relationships. The theory of infinite Boolean algebras is much deeper; it indirectly involves the whole theory of sets. One important result is Stone's representation theorem. Let a field of sets be defined as any family of subsets of a given set I, which contains with any two sets S and T their intersection $S \cap T$, union $S \cup T$, and complements S', T'. Considered abstractly, any such field of sets obviously defines a Boolean algebra. Stone's theorem asserts that, conversely, any finite or infinite abstract Boolean algebra is isomorphic to a suitable field of sets. His proof is based on the concepts of ideal and prime ideal, concepts which have been intensively studied for their own sake. Because ideal theory in Boolean algebra may be subsumed under the ideal theory of rings (via the correspondence between Boolean algebras and Boolean rings mentioned earlier), it will not be discussed here. A special property of Boolean rings (algebras) is the fact that, in this case, any prime ideal is maximal.

The study of infinite Boolean algebras leads naturally to the consideration of such infinite distributive laws as those in Eqs. (7) and (7').

$$x \cap \left(\bigcup_B y_\beta \right) = \bigcup_B (x \cap y_\beta)$$
$$x \cup \left(\bigcap_B y_\beta \right) = \bigcap_B (x \cup y_\beta) \qquad (7)$$

$$\bigcap_C \left[\bigcup_{A\gamma} u_{\gamma,\alpha} \right] = \bigcup_F \left[\bigcap_C u_{\gamma,\phi(\gamma)} \right]$$
$$\bigcup_C \left[\bigcap_{A\gamma} u_{\gamma,\alpha} \right] = \bigcap_F \left[\bigcup_C u_{\gamma,\phi(\gamma)} \right] \qquad (7')$$

For finite sets B of indices $\beta = 1, \ldots, n$, if $\bigcup_B y\beta$ means $y_1 \cup \cdots \cup y_n$, and so on, the laws (7) and (7') follow by induction from (1) to (3). Also, if the symbols x, y_β, and so on, in (7) and (7') refer to subsets of a given space I, and if $\bigcup_B y_\beta$ and $\bigcap_B y_\beta$ refer to the union and intersection of all y_β in B, respectively, then (8) and (7') are statements of general laws of formal logic. However, they fail in most infinite Boolean algebras. This is shown by the following result of Alfred Tarski: If a Boolean algebra A satisfies the generalized distributive laws (7) and (7'), then it is isomorphic with the algebra of all subsets of a suitable space I. A related result is the theorem of L. Loomis (1947): Every σ-complete Boolean algebra is isomorphic with a σ-field of sets under countable intersection, countable union, and complement.

In general, such completely distributive Boolean algebras of subsets may be characterized by the properties of being complete and atomic. These properties may be defined roughly as the properties that (*a*) there exists a smallest element

$\bigcup_{B} y_{\beta}$ containing any given set B of elements y_{β}, and

(b) any element $y > 0$ contains an atom (or point) $p > 0$, such that $p > x > 0$ has no solution (from Euclid, "A point is that which has no parts"). Condition (b) is also implied by the "descending chain condition" of ideal theory.

Other forms. Nonatomic and incomplete Boolean algebras arise naturally in set theory. Thus, the algebra of measurable sets in the line or plane, ignoring sets of measure zero, is nonatomic but complete. The field of Borel sets of space is complete as regards countable families B of subsets S_{β}, but not for uncountable B. Analogous results hold for wide classes of other measure spaces and topological spaces, respectively. In any zero-dimensional compact space, the sets which are both open and closed (which "disconnect" the space) form a Boolean algebra; a fundamental result of Stone shows that the most general Boolean algebra can be obtained in this way.

Many other interesting facts about Boolean algebra are known. For instance, there is an obvious duality between the properties of $\cap$ and $\cup$ in the preceding discussion. However, so many such facts have natural generalizations to the wider context of lattice theory that the modern tendency is to consider Boolean algebra as it relates to such generalizations. For instance, the algebras of n-valued logic, intuitionist (Brouwerian) logic, and quantum logic are not Boolean algebras, but lattices of other types. The same is true of the closed sets in most topological spaces.

[GARRETT BIRKHOFF]

Bibliography: G. Birkhoff, *Lattice Theory*, 3d ed., Amer. Math. Soc. Colloq. Publ., vol. 25, 1967; G. Birkhoff and S. MacLane, *Survey of Modern Algebra*, 4th ed., 1977; G. Boole, *An Investigation of the Laws of Thought*, 1854; R. Sikorski, *Boolean Algebras*, 3d ed., 1969.

Breadboarding

Assembling an electronic circuit in the most convenient manner on a board or other flat surface, without regard for final locations of components, to prove the feasibility of the circuit and to facilitate changes when necessary. Standard breadboards for experimental work are made with mounting holes and terminals closely spaced at regular intervals, so that parts can be mounted and connected without drilling additional holes.

Printed-circuit boards having similar patterns of punched holes, with various combinations of holes connected together by printed wiring on each side, are often used for breadboarding when the final version is to be a printed circuit. *See* CIRCUIT (ELECTRONICS); PRINTED CIRCUIT.

[JOHN MARKUS]

Breakdown

A large, usually abrupt rise in electric current in the presence of a small increase in electric voltage. Breakdown may be intentional and controlled or it may be accidental. Lightning is the most familiar example of breakdown.

In a gas, such as the atmosphere, the potential gradient may become high enough to accelerate the naturally present ions to velocities that cause further ionization upon collision with atoms. If the region of ionization does not extend between oppositely charged electrodes, the process is corona discharge. If the region of ionization bridges the gap between electrodes, thereby breaking down the insulation provided by the gas, the process is ionization discharge. When controlled by the ballast of a fluorescent lamp, for example, the process converts electric power to light. In a gas tube it provides controlled rectification. *See* GAS TUBE.

In a solid, such as an insulator, when the electric field gradient exceeds 10^6 volts/cm, valence bonds between atoms are ruptured and current flows. Such a disruptive current heats the solid abruptly: the rate of local temperature rise may fracture the insulator, the high temperature may carbonize or otherwise decompose the insulation, or as occasionally happens when lightning strikes a tree, the heat may ignite the insulator.

In a semiconductor if the applied backward or reverse potential across a junction reaches a critical level, current increases rapidly with further rise in voltage. This avalanche characteristic is used for voltage regulation in the Zener diode. In a transistor the breakdown sets limits to the maximum instantaneous voltage that can safely be applied between collector and emitter. When the internal space charge extends from collector junction through the base to the emitter junction, even a voltage below the avalanche level can produce a short circuit, in which case the phenomenon is termed punch-through. *See* BREAKDOWN POTENTIAL; TRANSISTOR; ZENER DIODE.

[FRANK H. ROCKETT]

Breakdown potential

That potential difference in a gaseous discharge at which a sudden increase in its electrical conductivity takes place. An alternative term is sparking potential. *See* ELECTRIC SPARK.

In general, the breakdown potential is a complicated function which depends on such factors as electrode material, both macroscopic and microscopic electrode shape, chemical surface states of the electrodes, constitution and pressure of the gas, temperature, electrode separation distance, and time-dependence of applied potential difference. It is also dependent, at least to some extent, on the current in the initial conduction state. Because each of these factors plays a part in determining the breakdown potential, it is exceedingly difficult to obtain precise information about this potential. Invariably, several factors will change during the course of an experiment, perhaps masking changes in the effect presumably under investigation.

In certain cases the nature of the influence of one of the above factors may be readily determined. The presence of sharp points at the electrodes, and particularly at the cathode, produces a marked lowering of the breakdown potential. This is because the strong electric field around such a point results in intense ionization once an initial electron enters the region. In the mechanism proposed for the propagation of the spark through the gap, such intense ionization will result in a copious supply of photons which then produce ionization by the photoelectric effect in other parts of the gap. As an example of how conditions may change, however, the evaporation and sputtering of such a point will smooth it off. Thus the next spark will have a greater breakdown potential.

Because the easy production of electrons is an important condition, reduction of the surface work function will generally result in a reduction of the sparking potential, though this effect may be small and difficult to observe. Again from the standpoint of electron production, the ionization potential of the gas is important. Other effects may obscure this, however. For example, the addition of a slight impurity to either helium or neon will result in a decrease in the sparking potential. This is due to collisions of the second kind, in which an excited atom of the main gas may ionize an impurity atom with which it collides. In other cases the addition of a gas with a high density of molecular states may increase the breakdown potential. Here the excitation energy of an atom may be transferred to molecular excitation in a collision, thus reducing photon production.

For a discussion of the effects of pressure and electrode separation *see* BREAKDOWN; ELECTRICAL CONDUCTION IN GASES; SPARK GAP.

[GLENN H. MILLER]

Bibliography: A. E. Fitzgerald and D. E. Higginbotham, *Electrical and Electronic Engineering Fundamentals*, 1964.

Bridge circuit

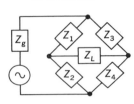

BRIDGE CIRCUIT

Schematic of bridge circuit. Z_g is the impedance of the seat of the electromotive force.

An electric network composed of four impedances forming a diamond-shaped network. These impedances, shown in the illustration as Z_1, Z_2, Z_3, and Z_4, may be combinations of resistors, capacitors, and inductors. Bridge circuits perform numerous functions, depending upon the type of circuit elements used in the arms of the bridge. For example, the bridge circuit can be used to determine the value of an unknown resistance. This is done by making the arm impedances resistances with one of them the unknown. The load impedance Z_L can be that of a measuring instrument, such as a galvanometer. The current through Z_L, and the voltage across it, will be zero when $Z_1 Z_4 = Z_2 Z_3$. With the proper combination of resistors and capacitors in the bridge arms, the voltage across Z_L will be zero for some frequency. Thus, the bridge acts as a frequency-measuring device. *See* FREQUENCY MEASUREMENT; WHEATSTONE BRIDGE.

The property of the bridge circuit that the voltage across Z_L will be zero for the proper arm impedances, and the proper frequency makes the circuit useful in an oscillator known as the Wien-bridge oscillator. It also has been used in the feedback path of some special amplifiers where its properties give the amplifier unusual characteristics. *See* OSCILLATOR.

[HAROLD F. KLOCK]

Bibliography: A. Budak, *Circuit Theory Fundamentals and Applications*, 1978; D. Tuttle, *Circuits*, 1977.

Brush discharge

A particular form of corona discharge characterized by strongly ionized streamers. It generally occurs at a field slightly less than that required for complete breakdown. The conditions for a brush discharge are as follows: The pressure should be about atmospheric; one or both of the electrodes should have a radius of curvature small compared

to the gap separation; and the gap itself should be large, at least of the order of a few centimeters. The ionized streamers take on a treelike form. The light originates from recombination and radiative transitions from excited states. *See* BREAKDOWN POTENTIAL; CORONA DISCHARGE.

[GLENN H. MILLER]

Calculators

Portable or desktop devices, primarily electronic, that are used to perform arithmetic and other numerical processing operations at the direction of an operator or a stored program.

Mechanical calculators. Early calculators were exclusively mechanical. An older mechanical calculator provides identical columns of numeral selection keys, one column for each digit of the operand range, with the units selection keys as the rightmost column. A numerical value is selected by depressing one or more keys, one in each column for which a nonzero value is desired. The selected operand is added to the accumulated sum or subtracted from the remaining difference either as the selection of value is made or as the operator takes some further action. The result is displayed in a mechanical accumulator register.

Mechanical calculators are best suited to addition and multiplication. Multiplication is accomplished by repeated addition of the multiplicand to form a series of partial sums. Additions for the units digit of the multiplier are made into the rightmost accumulator locations. A shift of one position to the left precedes the additions for each subsequent multiplier digit. Subtraction involves addition of the tens complement of the subtrahend. Division consists of repeated subtractions, beginning in the leftmost accumulator locations and shifting right by one location for each quotient digit to be determined. The operator must keep track of the location of any decimal point appearing in any calculation and must align decimal points for an addition or subtraction.

Improvement in mechanical calculators consisted of incorporating motors to effect additions upon key command and adding mechanisms to automatically control execution of the repeated additions needed in multiplication and division. Few mechanical units could perform any operations beyond the four basic arithmetic functions. Although some printing 10-key mechanical calculators are used in offices and small businesses, mechanical calculators have been almost completely replaced by quieter, more efficient, and more capable electronic units.

Electronic calculators. Calculators changed radically in the mid-1960s when transistorized models were developed to replace the mechanical units. The early electronic units were faster and quieter, but merely emulated their mechanical predecessors—only arithmetic operations were executed. As development continued, capabilities increased and sizes decreased. By 1975, inexpensive hand-held calculators employing large-scale integrated circuits as processing elements were in wide use. Arithmetic units costing less than $10 are available, more expensive models provide a wide range of operations. *See* ELECTRONICS.

The rapid evolution of the electronic calculator has been made possible by technological improvements in integrated circuits. An integrated circuit with an area of less than 0.5 in.^{2} (3 cm^2) can hold all the circuit elements needed to implement the algorithms and the control and timing functions of a calculator. Multiplication, for example, is still accomplished by multiple additions and shifts, but every step in the operation is under the direction of the integrated circuit. Forty or more algorithms may be held in memory, to be retrieved and used when requested by depression of a function key. A hand-held calculator may use a single integrated circuit to interpret key-switch closures, carry out the requested operation, and multiplex the individual digits of the result to the display. *See* INTEGRATED CIRCUITS.

Although most electronic calculators are similar with regard to basic operation, wide differences in functional capabilities exist between models. The user must evaluate the application and select a unit providing the needed features. Some of the differences in calculator capabilities are discussed in the following.

Display. Several display formats are prevalent. The simple fixed-point display, in which the decimal point remains in a fixed position and answers are automatically aligned on the decimal point by the calculating unit, appeared in early electronic models but is seldom used in more recent designs. The floating-point format, in which the decimal point can appear in any display location, is more flexible. The processor carries information on decimal-point location along with each operand, aligns decimal points of operands involved in additions and subtractions, and calculates decimal-point locations for results of all other arithmetic processes, passing this information to the display.

More elaborate calculators employ either scientific notation or engineering notation or both. In

Programmable calculator combined with printer. The calculator has an alphanumeric, liquid-crystal display. The thermal printer prints a variety of characters and is equipped with a plotting routine. (*Hewlett-Packard Co.*)

scientific notation, any displayed result consists of a mantissa and a characteristic, the characteristic being the power of 10 associated with the number. The characteristic may be positive or negative. The mantissa always has one digit, which may be a zero, to the left of the decimal point. For example, the displayed value 4.86750 07 corresponds to the floating-point number 4.86750 $\times$ 10^7, or 48,675,000. A display of 2.13447-02 represents 2.13447 $\times$ 10^{-2}, or 0.0213447. Some units capable of scientific notation routinely display results in floating-point form and automatically shift to scientific notation when the results are too large or too small to be accurately represented in floating-point form. The calculator definition of floating-point format does not conform to the same expression in computer terminology, where scientific notation is considered floating-point.

Engineering notation is similar to scientific notation with the restrictions that the characteristic must be a multiple of three and from one to three nonzero digits must appear to the left of the decimal point. The examples provided earlier in scientific notation would appear as 48.6750 06 and 21.3447-03 in an engineering notation display. Since engineers and scientists often express units in powers of 3, engineering notation facilitates the use of common engineering prefixes such as mega (10^6) and micro (10^{-6}).

The numbers of display digit locations vary with the calculator model and the mode of display. A floating-point display might have 8 digit positions and a sign location. Scientific and engineering formats usually occupy 12 or 14 display positions, with 8 or 10 digit locations for the mantissa, 2 for the characteristic, and 1 each for the sign of the mantissa and the sign of the characteristic. The operator can select the number of mantissa digits displayed and the display mode on some models.

Although the early electronic calculators used high-voltage gas discharge tubes for display, most hand-held units now utilize the more reliable and power-saving light-emitting diode (LED) displays, which are energized at a low voltage level and which can be readily multiplexed from the calculator integrated circuit. Since an LED display can account for most of the power consumed in a calculator, and therefore for battery drain, many units automatically remove power from the displays after a predetermined interval of inactivity, retaining the displayed value in semiconductor memory for recall. *See* LIGHT-EMITTING DIODE.

Some designs make use of liquid-crystal display (LCD) elements which avoid the current drain of LED displays. An LCD assembly provides a significant saving in power consumption over an LED display, but it is less easily viewed by the operator, particularly in a bright light environment. *See* LIQUID CRYSTALS.

A display usually provides an overflow or underflow indication when answers exceed the range of the calculator, or an error indication, such as a blinking display, if an illegal operation is requested or unacceptable results are generated. Most displays also indicate impending battery discharge. *See* ELECTRONIC DISPLAY.

Operand range. The range of operands that can be accepted, processed, and displayed by a calculator is a function of the display and the processing capabilities of the unit. An eight-digit floating-point display can exhibit numbers between .00000001 and 99,999,999, a 16-decade dynamic range. A 12-location scientific display can represent numbers from 1.0×10^{-99} to 9.9999999×10^{99}, a range of about 200 decades. If the available range is exceeded, the operator must reformulate or rescale the problem.

Entry notation. The user of a calculator must enter a sequence of operands and commands to direct the unit through the required processing. For entry, most calculators use either algebraic notation or reverse polish notation (RPN). With algebraic notation, the calculator accepts and processes data and commands in the order in which they would be written in an equation. If, for example, the operator wished to add 21 and 38 and divide the result by 14, the operator would enter, in sequence, the number 21, an add command (+), the number 38, a divide command (÷), the number 14, and a command to complete the calculation (=). Some calculators using algebraic notation must perform special operations or impose restrictions on the operator in order to handle the nesting of parentheses that occurs in more complex equations.

The RPN entry mode is more convenient for evaluation of complex expressions. This is also known as postfix notation because an operational command always follows the operand to which it applies. For the example provided previously, the operator would enter 21, an ENTER command, 38, an add command (+), 14 and a divide command (÷).

Storage registers. Availability of registers in which operands and intermediate results can be stored is of importance in many applications. If storage is inadequate, the operator may be forced to record intermediate results by hand and reenter them as they are needed for the calculation. Most calculators supply at least one memory register, to be exercised by store and recall keys. Some units contain stacks of three or four registers each; selection of one register in a stack is made by shifting the contents of the registers through the visual display until the desired location reaches the display. Any of a number of scientific calculators provides 10 or more storage locations. Storage and retrieval (into the displayed register) are done by activation of the store or recall key and a location-designating key, in sequence.

Speed. Calculators are designed to carry out computations previously performed by hand, the automation yielding speed and accuracy advantages. By computer standards, calculators are slow. A scientific calculator may require a second or more to execute one of its more complex operations. Execution time may also be dependent upon operand value since some operations involve iterative computations. The relatively slow speed of the calculator is seldom perceived as a handicap by the user.

Programmable calculators. Any calculator can be considered a computer in that it contains in nonvolatile memory a fixed set of algorithms for execution of the processing operations in its repertoire. These algorithms are available whenever power is applied to the unit. Programmable calculators are able to accept and act upon a higher level of programming that directs the units through sequences of processing steps without operator intervention other than operand entry. Programming may be done by the operator or, in advance, by the manufacturer, and storage of the program may be volatile or nonvolatile. Programming relieves the drudgery of repeated keystroke entry when a repetitive calculation must be made.

Programming by the operator on the unit to be used for the calculation is known as keystroke programming. The desired sequence of operations is entered into calculator memory by depression of keys as though a calculation were being made, but with the unit in program mode. Each keystroke enters an instruction into the semiconductor memory of the unit. When the unit is returned to the normal run mode, the program sequence can be initiated by the operator through depression of a special function key on the keyboard. Three or more special function keys may be provided; a number of programs can simultaneously reside in memory.

The semiconductor storage in a calculator is usually volatile—program information is lost when power is removed. Some units can accept magnetic tape cartridges or miniaturized magnetic cards, from which the volatile memory can be loaded and to which the memory contents can be written. The magnetic medium provides a means of saving and reentering a program written by the keystroke method. Manufacturers also offer preprogrammed tapes and cards for specialized calculations in a wide range of fields, such as surveying, statistics, financial decisions, mathematics, and circuit analysis. A tape or card can customize a general-purpose calculator to a particular application. Since the miniaturized cards are easily carried with a calculator, the nonvolatile storage capacity greatly improves the computational capacity, and hence the value of a hand-held calculator, especially for use in the field.

Memory of the complementary metal-oxide semiconductor (CMOS) type, which exhibits very low power consumption when storage or retrieval is not under way, has been incorporated into several models. These units prevent program loss by continuously maintaining power on the memory section. The combination of a CMOS memory and an LCD readout can provide long battery life and protection against program loss in a self-sufficient calculator unit.

Several programmable calculators accept plug-in elements such as memory extender modules, read-only memory modules for specialized computations, magnetic card readers, and printers. Programmer assistance in the form of prompting is provided in an alphanumeric display—a display capable of showing letters as well as numbers. A calculator with these capabilities could be likened to a numerical processing computer with peripheral devices. As the range of calculator expansion elements increases and calculation capabilities increase, the differences between calculators and computers will become even less distinct. *See* COMPUTER; DIGITAL COMPUTER.

Special-purpose calculators. Although any programmable calculator with sufficient capability can be specialized to a task by insertion of a program, some units are designed for particular applications. Calculators performing all the basic engineering, scientific, or statistical calculations are numerous. A financial calculator can carry out determinations of yields, compound interest, loan and mortgage amortization, depreciation, and many related items. Some calculators perform arithmetic in, and conversions between, the several number systems used in computer programming. A checkbook calculator maintains current balance in its memory. Several models serve as teaching aids by posing arithmetic problems to elementary students and evaluating answers. Calculatorlike units serve as teaching aids for spelling, in translation from one language to another, and in a variety of electronic games. The proliferation of calculator applications is expected to continue, fostered by the continuing improvements in the capabilities of large-scale integrated circuits. [W. W. MOYER]

Bibliography: J. E. Carter and D. Young, *Electronic Calculators*, 1981; G. Immerzeel and E. Ockenga, *The Calculator*, 1981.

Calibration

The process of determining, by measurement or by comparison with a standard, the correct value of each scale reading on a meter or the correct value of each setting of a control knob. In a radio receiver, for example, calibration would mean adjusting the tuned circuits in the oscillator to make the readings of the tuning dial correspond exactly to the frequencies of the incoming signals.

With measuring instruments, calibration generally involves adjusting the values of internal components so that the indication is correct at a specified number of points on the indicating scale and approximately correct between these points. With highly accurate instruments, calibration involves the preparation of a graph or table that gives the exact value corresponding to each line on the indicating scale.

[JOHN MARKUS]

Capacitance

The ratio of the charge q on one of the plates of a capacitor (there being an equal and opposite charge on the other plate) to the potential difference v between the plates; that is, capacitance (formerly called capacity) is $C = q/v$.

In general, a capacitor, often called a condenser, consists of two metal plates insulated from each other by a dielectric. The capacitance of a capacitor depends on the geometry of the plates and the kind of dielectric used, since these factors determine the charge which can be put on the plates by a unit potential difference existing between the plates.

For a capacitor of fixed geometry and with constant properties of the dielectric between its plates, C is a constant independent of q or v, since as v changes, q changes with it in the same proportion. This statement assumes that the dielectric strength is not exceeded and thus that dielectric breakdown does not occur. (If it does occur, the device is no longer a capacitor.) If either the geometry or dielectric properties, or both, of a capacitor change with time, C will change with time.

In an ideal capacitor, no conduction current flows between the plates. A real capacitor of good quality is the circuit equivalent of an ideal capacitor with a very high resistance in parallel or, in alternating-current (ac) circuits, of an ideal capacitor with a low resistance in series.

Properties of capacitors. One classification system for capacitors follows from the physical state of their dielectrics. For a discussion based on this classification *see* CAPACITOR.

Charging and discharging. These processes can occur for capacitors while the potential difference across the capacitor is changing if C is fixed; that is, q increases if v increases and q decreases if v decreases. If C and v both change with time, the rate of change of q with time is given by Eq. (1).

$$\frac{dq}{dt} = C\frac{dv}{dt} + v\frac{dC}{dt} \qquad (1)$$

Since the current i flowing in the wires leading to the capacitor plates is equal to dq/dt, Eq. (1) gives i in the wires. In many cases, C is constant so $i = C\,dv/dt$.

Energy of charged capacitor. This energy W_C is given by the formula $W_C = vq/2$, and is equal to the work the source must do in placing the charge on the capacitor. It is, in turn, the work the capacitor will do when it discharges.

Geometrical types. The geometry of a capacitor may take any one of several forms. The most common type is the parallel-plate capacitor whose capacitance C in farads is given in the ideal case by Eq. (2), where A is the area in square meters of

$$C = \frac{Ak\epsilon_0}{d} \qquad (2)$$

one of the plates, d is the distance in meters between the plates, ϵ_0 is the permittivity of empty space with the numerical value 8.85×10^{-12} coul²/newton-m², and k is the relative dielectric constant of the dielectric between the plates. The value of k is unity for empty space and almost unity for gases. For other dielectrics, k ranges in value from one to several hundred. In order for Eq. (2) to give a good value of C for an actual capacitor, d must be very small compared to the linear dimensions of either plate.

Each plate of a parallel-plate capacitor may be made up of many thin sheets of metal connected electrically with a corresponding number of thin sheets of metal making up the other plate. The sheets of metal and their intervening layers of dielectric are chosen and stacked in such a way that A will be large and d small without making the whole capacitor too bulky. The result is appreciable capacitance in a reasonable volume.

The spherical capacitor is made of two concentric metal spheres with a dielectric of relative dielectric constant k filling the space between the spheres. The capacitance in farads of such a capacitor is given by Eq. (3), where r_2 is the radius in

$$C = 4\pi k\epsilon_0 \frac{r_1 r_2}{r_2 - r_1} \qquad (3)$$

meters of the outer sphere and r_1 that of the inner sphere.

The cylindrical capacitor, as the name implies, is made of two concentric metal cylinders, each of length l in meters, with a dielectric filling the

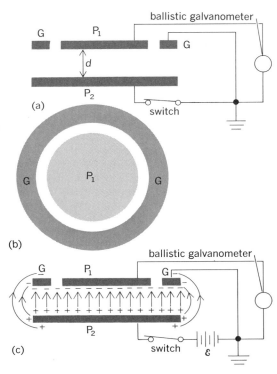

Parallel-plate capacitor P_1, P_2 with guard ring GG. Spacing between guard ring and plate is exaggerated. (*a*) Cross section with all parts of capacitor at ground potential. (*b*) Top view showing plate P_1 surrounded by guard ring. (*c*) Cross section of capacitor charged from battery whose electromotive force is $\mathscr{E}$.

space between the cylinders. If r_2 and r_1 are the radii in meters of the outer and inner cylinders, respectively, and l is very large compared to $r_2 - r_1$, the capacitance C in farads is given by Eq. (4), where k is the relative dielectric constant

$$C = \frac{2\pi k \epsilon_0 l}{\ln\,(r_2/r_1)} \qquad (4)$$

of the dielectric and $\ln\,(r_2/r_1)$ indicates the natural logarithm of the ratio r_2/r_1.

Guard ring. This is often used with a standard parallel-plate capacitor, as shown in the diagram, in order that Eq. (2) shall more accurately represent its capacitance. It is the fringing of the electric lines of force which makes Eq. (2) inaccurate for an actual capacitor and, as shown in Fig. 2c, the fringing is nearly all at the outside edge of the guard ring, and thus is not associated with the charge Q which is put onto plate P_1 while the capacitor is being charged. It is the charge Q that determines the deflection of the ballistic galvanometer during the charging process. Then Eq. (2) gives the correct value of C that is needed to relate Q to the potential difference $\mathscr{E}$ across the plates by the equation $Q = C\mathscr{E}$. Thus, with C known from Eq. (2) and $\mathscr{E}$ known from a potentiometer measurement, Q may be computed and the ballistic galvanometer calibrated. This illustrates one use of a standard capacitor with a guard ring.

Body capacitance. When a part of the human body, say the hand, is brought near a high-impedance network, the body serves as one plate of a capacitor and the adjacent part of the network as the other plate. This situation is the equivalent of a capacitor of very low capacitance, in parallel between that part of the network and ground, since the human body can usually be considered as being a grounded conductor. This capacitance is known as body capacitance and enters as a part of the distributed capacitance of the network. A high-impedance network must be well shielded in order to eliminate the variable and undesirable effects of body capacitance.   [RALPH P. WINCH]

Bibliography: Berkeley Physics Course, vol. 2: *Electricity and Magnetism*, 1970; R. Resnick and D. Halliday, *Physics*, 3d ed., 1977; F. W. Sears et al., *University Physics*, 5th ed., 1976.

Capacitance measurement

In a multiconductor system capacitances between each of the conductors may be defined. In general, these capacitances are functions of the total geometry, that is, the location of each and every conducting and dielectric body. When, as is usually true, one is interested in the capacitance between two conductors only, the presence of other conductors is an undesirable complication. It is then customary to distinguish between two-terminal and three-terminal capacitors and capacitance measurements. In the case of a two-terminal capacitor, either one of the conductors of primary interest essentially surrounds the other, in which case the capacitance between them is independent of the location of other bodies except in the vicinity of the terminals; or one accepts the somewhat indefinite contributions of the other conductors to the capacitance of interest.

A three-terminal capacitor consists of two active electrodes surrounded by a third, or shield, conductor. The direct capacitance between the two active electrodes is the capacitance of interest, and, when shielded leads are used, it is independent of the location of all other conductors except the shield. Only certain of the methods described are suitable for the measurement of three-terminal capacitors.

Every physically realizable capacitor has associated loss in the dielectric and in the metal electrodes. At a single frequency these are indistinguishable, and the capacitor may be represented by either a parallel or series combination of pure capacitance and pure resistance. The measurement of capacitance, then, in general involves the simultaneous measurement of or allowance for an associated resistive element.

Most capacitance measurements involve simply a comparison of the capacitor to be measured with a capacitor of known value. Methods which permit

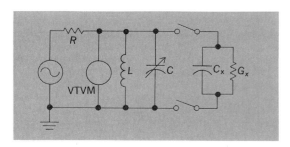

Fig. 1. Susceptance variation method.

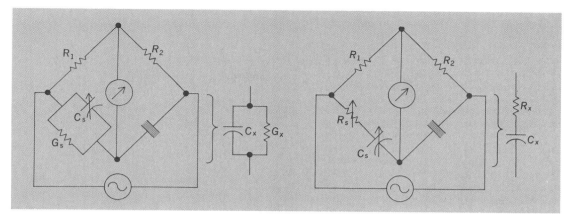

Fig. 2. Resistance-ratio bridges, in which two adjacent arms are resistors.

comparison of essentially equal capacitors by simple substitution of one for the other at the same point in a circuit are frequently possible and almost always preferable.

Resonance method. This method is suitable for two-terminal capacitance measurements when only moderate accuracy is required. The resonance method involves resonating an inductor with a calibrated variable capacitor which has a range of variation larger than the value of any capacitor to be measured. The parallel or series LC circuit is tuned to resonance by adjustment of the variable capacitor. The capacitor to be measured is then connected in parallel with the variable unit and the latter readjusted until resonance at the same frequency is again achieved. The value of the capacitor being measured is equal to the change in the variable capacitance standard.

A variety of techniques is available for determining when the resonance condition is achieved, some using series and some parallel resonance, but they differ significantly only in the way in which the source and detector are connected.

Susceptance variation. This method is a resonance technique capable of yielding both the effective parallel capacitance and conductance components of a capacitor (Fig. 1). The calibrated variable capacitor C is first adjusted to the value C_1 for which the vacuum-tube voltmeter (VTVM) indicates a maximum V_m. C is then increased to a new value C_2 for which the VTVM reads $V_m/\sqrt{2}$. Simple network theory shows that the equivalent parallel conductance of the LC network is given by Eq. (1).

$$G_{eq} = \omega(C_2 - C_1) \qquad (1)$$

If now the capacitor to be measured, represented by the parallel combination C_x and G_x, is connected across C and the process repeated yielding values C'_1 and G'_{eq}, the Eqs. (2) and (3) hold.

$$C_x = C'_1 - C_1 \qquad (2)$$
$$G_x = G'_{eq} - G_{eq} \qquad (3)$$

Alternate resonance techniques include use of a parallel LC circuit as the frequency-determining element in an oscillator or as one arm of a Wheatstone bridge whose other three arms are resistors. In the latter case balance of the bridge defines the resonant condition and also provides a measure of the equivalent resistive component.

Bridge methods. When high accuracies are required, bridge methods must be adopted. *See* BRIDGE CIRCUIT; WHEATSTONE BRIDGE.

Resistance-ratio bridges. These are Wheatstone-bridge configurations in which two adjacent arms are resistors, as indicated in Fig. 2. Application of the generalized Wheatstone bridge equation, Eq. (4), leads to Eqs. (5) and (6) for the parallel arrange-

$$Z_1 Z_x = Z_2 Z_s \quad \text{or} \quad Y_1 Y_x = Y_2 Y_s \qquad (4)$$

$$C_x = C_s \frac{R_1}{R_2} \qquad (5)$$

$$G_x = G_s \frac{R_1}{R_2} \qquad (6)$$

ment, and to Eqs. (7) and (8) for the series arrange-

$$C_x = C_s \frac{R_1}{R_2} \qquad (7)$$

$$R_x = R_s \frac{R_2}{R_1} \qquad (8)$$

ment. These are the conditions for bridge balance as indicated by zero voltage across the detector terminals.

Schering bridge. This bridge yields a measurement of the equivalent series-circuit representation of a capacitor (Fig. 3). The equations of bal-

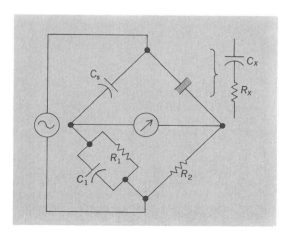

Fig. 3. The Schering bridge.

ance are written as Eqs. (9) and (10).

$$C_x = C_s \frac{R_1}{R_2} \qquad (9)$$

$$R_x = R_2 \frac{C_1}{C_s} \qquad (10)$$

The arrangement shown is used when it is desirable for high voltages to be applied to the capacitor under test. The relative impedances are usually such that most of the applied voltage appears across the capacitors and, if balance is effected by varying C_1 and R_1 or R_2, the operator need not approach the high-voltage elements.

In general with a given applied voltage, higher sensitivity is obtained if the source and detector connections are interchanged; this so-called conjugate bridge is more frequently used for measurements of high precision. Because variable capacitors are less subject to errors from associated residuals than variable resistors, balance is then usually effected by varying C_s and C_1; the standard capacitor C_s is assumed to be free from loss.

Wagner branch. The resistance-ratio and Schering bridges (Fig. 4) are useful for two-terminal capacitance measurements. Their use may be extended to three-terminal measurements and extended in accuracy and range by the introduction of shielding and the addition of the Wagner branch.

Balance is effected alternately by adjustment of the main bridge arms with the switch open, and by adjustment of the Wagner branch elements with the switch closed. The final balance may be con-

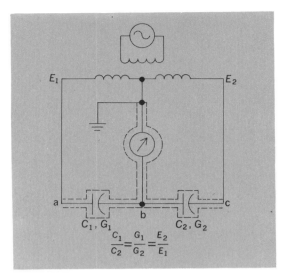

Fig. 5. Transformer ratio-arm bridge, used in capacitance measurements requiring high accuracy.

sidered to be that with the switch open. Capacitances to ground at both the A and C corners of the bridge then shunt the Wagner arms and do not affect the balance of the main bridge. Capacitances to ground at the B and D corners carry no current at balance and therefore cannot introduce error.

Transformer ratio-arm bridges. These are particularly suitable for three-terminal capacitance measurements. Of the variety of inductive and transformer ratio-arm bridges that have been developed, the one shown in Fig. 5 seems most suitable for capacitance measurements of high accuracy.

Capacitance to ground from b reduces sensitivity but does not affect the condition of balance. Capacitances to ground from a and c introduce loads on the transformer secondaries, but with proper transformer design the errors introduced may be kept negligible.

Audio-frequency three-winding transformers with secondary open-circuit voltage ratios equal to the turns ratios within 10 parts in 1,000,000 are quite readily constructed. With care in design 1:1 and 10:1 transformers with ratio errors of the order of 1 part in 10^7 are entirely feasible. The secondary-winding resistances and leakage inductances under conditions of use introduce departures from the open-circuit voltage ratio, and therefore this bridge circuit is most useful for the measurement of capacitors less than 1 microfarad at audio frequencies. Transformer design for radio-frequency use is more difficult. However, the bridge circuit is used with reduced accuracy at capacitance levels of the order of 100 picofarads and below, at frequencies up to 10 MHz.

Bridged-T and parallel-T networks. These complex bridges possess a significant advantage over four-arm bridges for medium-accuracy radio-frequency measurements of capacitance; the source and detector have a common point of connection which may be grounded (Fig. 6).

Use of the Y-delta network transformation permits reduction of either the bridged-T or the

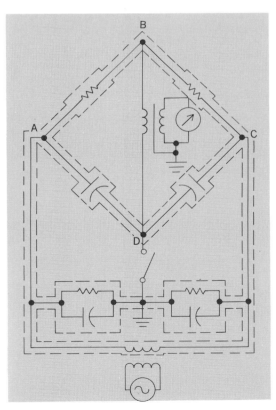

Fig. 4. Shielded resistance ratio-arm bridge with Wagner branch for three-terminal measurements.

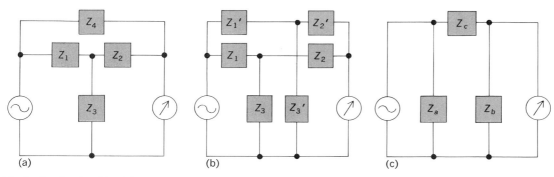

Fig. 6. T networks. (a) Bridged T. (b) Parallel T. (c) Equivalent circuit.

parallel-T to the equivalent circuit shown. Balance, or null indication of the detector, is achieved by variation of the network components until the equivalent impedance Z_c is infinite. In the application of the parallel-T network to capacitance measurements, as shown in Fig. 7, two balances are required, one with and the other without the unknown connected. The equivalent parallel components of the unknown admittance are obtained from the differences in the values of the two capacitors indicated.

Time-constant methods. If a direct voltage is suddenly applied to the series combination of a resistor and an initially discharged capacitor, the charge and the voltage on the capacitor increase exponentially toward their full magnitudes with a time constant equal in seconds to the product of the resistance in ohms and the capacitance in farads. Similarly, when a charged capacitor is discharged through a resistor, the charge and the voltage decay with the same time constant. A variety of methods are available for the measurement of capacitance by measurement of the time constant of charge or discharge through a known resistor.

Ballistic galvanometer method. This method (Fig. 8) compares the charge lost by the capacitor in a known time interval to the total charge. With the capacitor completely discharged the switch is closed, and the product of the galvanometer deflection and the multiplying factor of the shunt is proportional to the total charge Q_T. The switch is opened for a measured time interval Δt and again closed. The product of the galvanometer deflection

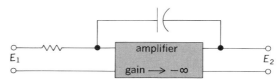

Fig. 8. Ballistic galvanometer method.

and the shunt factor is then proportional to the charge Q_a lost by the capacitor. Because Eq. (11) is valid, Eq. (12) holds.

$$Q_T - Q_a = Q_T e^{-\Delta t/RC} \qquad (11)$$

$$C = \frac{-\Delta t}{R \ln \left[(Q_T - Q_a)/Q_T \right]} \qquad (12)$$

Integrating operational amplifier. In this method the time required for the output voltage of an amplifier to increase to a value equal to the step-function input voltage is determined by an electronic voltage-comparison circuit and timer (Fig. 9). With the assumption of ideal characteris-

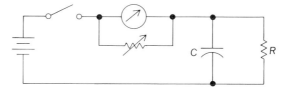

Fig. 9. Integrating operational amplifier.

tics for the amplifier, such as infinite gain without feedback, infinite input impedance, and zero output impedance, the measured time interval is equal to the product of the values of the known resistance and the capacitance being measured.

Distributed capacitance. An inductor is a particularly impure circuit component. In addition to the series resistance of the winding, distributed capacitance is always present from turn to turn and layer to layer, making the effective inductance a function of frequency. It is customary to assume that the effect of the distributed capacitance in an inductance coil may be represented by a single capacitor connected between the coil terminals. In some cases the value of this equivalent capacitor may be obtained by a determination of the self-resonant frequency of the coil; however, as a result

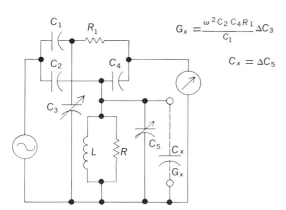

$$G_x = \frac{\omega^2 C_2 C_4 R_1}{C_1} \Delta C_3$$

$$C_x = \Delta C_5$$

Fig. 7. Parallel-T network to measure capacitance.

CAPACITANCE MEASUREMENT

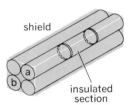

Fig. 10. Thompson-Lampard capacitor.

of the distributed nature of both the inductance and the capacitance, the coil may exhibit several resonance modes, and a self-resonance determination is then ambiguous.

A procedure that usually surmounts this difficulty is that of determining the resonant frequency for several settings of a variable capacitor connected in parallel with the coil. A plot of the capacitance of the observed auxiliary capacitor against the reciprocal of the square of the corresponding resonant frequency and extrapolated to infinite frequency then gives a value for the lumped equivalent of the distributed capacitance.

Capacitance standards. These are high-quality, stable capacitors of known capacitance against which other capacitors may be compared. A computable capacitance standard is a capacitor of such design and construction that its capacitance can be calculated from measured dimensions. Simple geometrics, such as concentric spheres, parallel circular plates, and coaxial circular cylinders, have been used successfully. With the two latter types, guard sections (which make the capacitors three-terminal) are required to define the edge effects sufficiently well for precise calculation. In the Kelvin guard-ring capacitor the island electrode is insulated from the guard ring by a very narrow gap. If, for purposes of calculation, the guard ring and island are assumed to be at the same potential, which is different from that of the base, the field is everywhere normal to the plates except in the immediate vicinity of the gap. It may be shown that to a first-order approximation the existence of a narrow gap increases the effective diameter of the island electrode by the width of the gap. If A is the effective area, d the plate separation, and ϵ the permittivity of the dielectric, the capacitance, in rationalized mks units, is $C = \epsilon A / d$.

The design of the Thompson-Lampard cylindrical cross-capacitor is such that the capacitance may be calculated from one length measurement only. One form consists of four insulated circularly cylindrical bars in a suitable shield (Fig. 10). Bar a has end sections insulated from the central section. The direct capacitance between bar b and the central section of bar a is a function only of the length L of that central section. Its value, in rationalized mks units, is given by Eq. (13).

$$C = \frac{\epsilon \ln 2}{\pi} L \qquad (13)$$

Laboratory capacitance standards of the order of 0.001 microfarad and lower in nominal value are usually parallel-plate air-dielectric capacitors. Capacitors from 0.001 to 1 microfarad in value are usually mica dielectric. The common construction is a stack of flat parallel conductors separated by thin mica sheets. The conducting plates may be evaporated films of silver on the faces of the mica sheets or sheets of metal foil clamped between the mica sheets. [F. RALPH KOTTER]

Bibliography: W. D. Cooper, *Electronic Instrumentation and Measurement Techniques*, 2d ed., 1978; F. K. Harris, *Electrical Measurements*, 1952; M. C. McGregor et al., New apparatus at National Bureau of Standards for absolute capacitance measurement, *IRE Trans. Instrum.*, 1–7: 253–261, 1958.

CAPACITOR

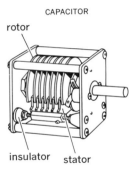

Fig. 1. Variable air capacitor.

Capacitor

A device for introducing capacitance into a circuit. In general, a capacitor consists of two metal plates insulated from each other by a dielectric. The capacitance of a capacitor depends primarily upon the shape and size of the capacitor and upon the relative dielectric constant of the medium between the plates. In vacuum, and approximately so in air and most gases, k, the dielectric constant, is unity. For other dielectrics, k ranges from one to several hundred. *See* CAPACITANCE.

Classification. One classification of capacitors comes from the physical state of their dielectrics. The dielectric may be a gas (or vacuum), a liquid, a solid, or a combination of these. Each of these classifications may be subdivided according to the specific dielectric used. Capacitors are also classified as fixed, adjustable, or variable.

Fixed capacitors. The capacitance of fixed capacitors remains unchanged, except for small variations (caused by temperature fluctuations or vibration, for example).

Adjustable capacitors. The capacitance of adjustable capacitors may be set at any one of several discrete values.

Variable capacitors. The capacitance of variable capacitors may be adjusted continuously and set at any value between minimum and maximum limits fixed by construction.

Trimmer capacitors. These are relatively small variable capacitors used in parallel with larger variable or fixed capacitors to permit exact adjustment of the capacitance of the parallel combination.

Air, gas, and vacuum types. Made in both the fixed and variable types, these capacitors are constructed with flat parallel metallic plates (or cylindrical concentric metallic plates) with air, gas, or vacuum as the dielectric between plates. Alternate plates are connected, with one or both sets supported by means of a solid insulating material such as glass, quartz, ceramic, or plastic. Figure 1 shows a variable air capacitor. Gas capacitors are similarly built but are enclosed in a leakproof case. If the gas capacitor is variable, the shaft supporting the movable plates, or rotor, is brought out through a pressure-tight insulated seal. Vacuum capacitors are of concentric-cylindrical construction and are enclosed in highly evacuated glass envelopes.

The purpose of a high vacuum, or a gas under pressure, is to increase the voltage breakdown value for a given plate spacing. For high-voltage applications when increasing the spacing between plates is undesirable, the breakdown voltage of air capacitors may be increased by rounding the edges of the plates. Air, gas, and vacuum capacitors are used in high-frequency circuits. Fixed and variable air capacitors incorporating special design are used as standards in electrical measurements.

Solid-dielectric types. These capacitors use one of several dielectrics such as a ceramic, mica, glass, or plastic film. Alternate plates of metal, or metallic foil, are stacked with the dielectric, or the dielectric may be metal-plated on both sides. The dielectric of the ceramic capacitor is a mixture of titanium oxide with other titanates. The dielectric

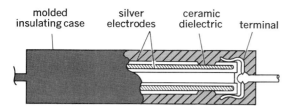

Fig. 2. Cutaway drawing of typical molded-case ceramic capacitor. (*K. Henney and C. Walsh, eds., Electronic Components Handbook, McGraw-Hill, 1957*)

constant may range from about six to several hundred, depending upon the composition of the mixture. Ceramic capacitors (Fig. 2) have a high capacitance-to-volume or -mass ratio and a stable temperature-capacitance characteristic, but are affected by vibration, aging, and overvoltage.

Mica types. These capacitors use thin rectangular sheets of mica as the dielectric. The dielectric constant of mica is in the range of 6−8. The electrodes are either thin sheets of metal foil stacked alternately with the mica sheets, or thin deposits of silver applied directly to the surface of the mica sheets. Mica capacitors are used chiefly in radio-frequency applications. They have a low dielectric loss at very high frequencies, good temperature, frequency, and aging characteristics and low power factor, but have a low capacitance-to-volume or mass ratio. They are made with dc voltage ratings from a few hundred to many thousand volts and with radio-frequency current ratings up to about 50 amps.

Plastic-film types. These capacitors use dielectrics such as polystyrene or Mylar with dielectric constants in the range of 2−6. These plastics may be used alone or as a laminate with paper. Construction is similar to paper capacitors. These capacitors are used for dc voltages up to about

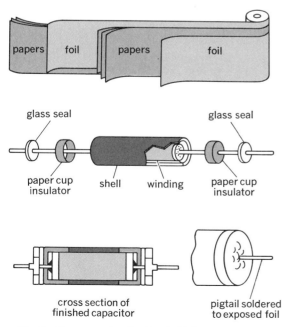

Fig. 3. Components of extended-foil paper capacitor. (*K. Henney and C. Walsh, eds., Electronic Components Handbook, McGraw-Hill, 1957*)

2000 volts and where very low dc leakage current is required.

Paper types. A dielectric of kraft paper usually impregnated with mineral oil, chlorinated naphthalene, or chlorinated diphenyl is used in paper capacitors. Paper (Fig. 3) and plastic-film capacitors are constructed by stacking, or forming into a roll, alternate layers of foil and dielectric. Paper capacitors have low cost per unit capacitance and moderate capacitance-to-volume ratio, but are pervious to moisture, unless hermetically sealed, and change capacitance with temperature.

Thick-film types. These capacitors are made by means of successive screen-printing and firing processes in the fabrication of certain types of microcircuits used in electronic computers and other electronic systems. They are formed, together with their connecting conductors and associated thick-film resistors, upon a ceramic substrate. Their characteristics and the materials used are similar to those of ceramic capacitors. *See* PRINTED CIRCUIT.

Liquid-dielectric types. The plate assemblies of these capacitors are mounted in a tank filled with a suitable oil or liquid dielectric. External connection is through insulators or bushings, and provision is made for expansion and contraction of the liquid during temperature changes. They have high capacity and low losses.

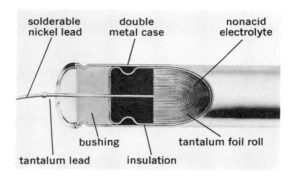

Fig. 4. Tantalum electrolytic capacitor showing cross section. (*General Electric Co.*)

Electrolytic types. A large capacitance-to-volume ratio and a low cost per microfarad of capacitance are chief advantages of electrolytic capacitors. These use aluminum or tantalum plates (Fig. 4) on which an oxide film forms and acts as the dielectric. The characteristics of the oxide film depend on the polarity of the applied voltage. Such a device is said to be polarized. Polarized electrolytic capacitors must be connected to the circuit correctly. They can be used only in circuits in which the dc component of voltage across the capacitor exceeds the crest value of the ac ripple.

Nonpolarized electrolytic capacitors can be constructed for use in ac circuits. In effect, they are two polarized capacitors placed in series with their polarities reversed. Nonpolarized capacitors have only one-half the capacitance of polarized capacitors for a given dc voltage and foil area.

Two important types of electrolytic capacitors are (1) the dry-type aluminum oxide film, and

(2) the solid-electrolyte tantalum capacitor. Type (1) is used widely as a filter and bypass capacitor in radio and television receivers, other electronic devices, and instruments. Type (2) is useful over a fairly wide range of temperature, is not affected by idleness as is type (1), can withstand vibration and shock, and affords greater capacitance for a given volume. The maximum voltage rating of type (2) is low, but it has applications such as in transistor circuitry.

[BURTIS L. ROBERTSON;
WILSON S. PRITCHETT]

Bibliography: D. G. Fink and H. W. Beaty (eds), *Standard Handbook for Electrical Engineers*, 11th ed., 1978; T. Jones, *Electronic Components Handbook*, 1978; W. F. Mullin, *ABCs of Capacitors*, 3d ed., 1978.

Carrier

Any wave to which modulation is subsequently applied. Commonly encountered examples are a sinusoidal wave, a recurrent series of pulses, and a direct current or voltage. *See* AMPLITUDE MODULATION; ANGLE MODULATION; FREQUENCY MODULATION; PHASE MODULATION. For a definition of pulse carrier *see* PULSE MODULATION; for a definition of subcarrier *see* MODULATION.

It is not always necessary or even desirable to transmit the carrier wave; when not transmitted, it is termed a suppressed carrier.

[HAROLD S. BLACK]

Cascode amplifier

A transistor amplifier consisting of a common-emitter stage in series with a common-base stage. The input resistance and current gain are nominally equal to the corresponding values for a single common-emitter stage (see illustration). The output resistance is approximately equal to the high output resistance of the common-base stage. The reverse open-circuit amplification parameter h_r is very much smaller for the cascode connection than for a single common-emitter stage. The small value of h_r makes this circuit particularly useful in tuned amplifier design where the collector load is replaced with a tuned circuit.

A vacuum tube version of the cascode amplifier is composed of a grounded cathode stage connected to a grounded grid stage. This circuit has the characteristics of a pentode, that is, a high amplification factor and a large plate resistance.

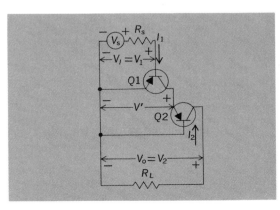

Cascode configuration. Supply voltages are not shown.

The cascode amplifier finds its principal application in the amplification of high-frequency signals of 25 MHz and higher. It is particularly useful in low-level input stages where the signal-to-noise ratio is a problem. One widespread use for the cascode circuit is in the tuner of a television receiver. In the fringe areas where the signal strength is small, an amplifier is needed which can amplify the signal the required amount without introducing an excessive amount of noise. *See* TELEVISION RECEIVER.

[CHRISTOS C. HALKIAS]

Bibliography: C. A. Holt, *Electronic Circuits: Digital and Analog*, 1978; J. Millman, *Microelectronics*, 1979.

Cathode-ray tube

An electron tube in which a beam of electrons can be focused to a small area and varied in position and intensity on a surface. In common usage, the term cathode-ray tube is usually reserved for devices in which the surface referred to is cathodoluminescent, that is, luminescent under electron bombardment. In these tubes the output information is presented in the form of a pattern of light. The character of this pattern is related to, and controlled by, one or more electrical signals applied to the cathode-ray tube as input information.

The most familiar form of the cathode-ray tube is the television picture tube, found in home television receivers. Cathode-ray tubes are also used in large quantities in measuring instruments such as oscilloscopes, which have become indispensable in laboratories devoted to experimental studies in the physical and biological sciences and in engineering. The oscilloscope has also been adopted as a vital tool in the fields of automobile, aircraft, and television-receiver maintenance and repair. For navigation the cathode-ray tube is the output device of radars. Large users of cathode-ray tubes are the military services, which employ these devices to display data to guide the employment of weapons systems.

Increasingly, cathode-ray tubes are finding use in input/output terminals of digital computers. Cathode-ray-tube terminals of this type have important advantages over the impact-type printers previously used. They can display a page of information much more quickly, using formats that are much less restrictive, and they do not deluge the user with masses of unwanted printed copy. These advantages are especially important in systems where a single computer services on a time-shared basis a number of terminals, each being used to conduct a user-computer dialog.

The cathode-ray tube is used in increasing numbers in the general field of printed communications, both as an adjunct to, and as a competitor of, traditional means of preparing "hard copy." In the first instance, cathode-ray tubes expose photographic masters that, in turn, are used to prepare offset printing plates. Where only a single copy is required, photosensitive paper is exposed by a cathode-ray tube to deliver printed copy in the business offices from an electronic communication link. *See* ELECTRONIC DISPLAY; OSCILLOSCOPE; STORAGE TUBE; TELEVISION RECEIVER.

The basic elements of a cathode-ray tube are the envelope, electron gun, and phosphor screen.

Envelope. This element serves as the vacuum enclosure, the substrate for the phosphor screen, and the support for the electron gun. The envelope of a cathode-ray tube is typically funnel-shaped; its small opening is stopped by the stem. This stem is essentially a disk of glass through which the metallic leads which support and carry applied voltages to the several elements of the gun are sealed. These leads are generally arrayed on one or more circles concentric with the tube axis. Usually the exhaust tubulation, through which the envelope is evacuated during processing, joins the envelope at the center of the stem. In most cases the stem leads are carried through and soldered to tubular pins in a plastic base cemented to the neck of the envelope in the immediate vicinity of the stem. This base serves to protect the stem and the tipped-off exhaust tubulation, as well as to provide means of contact between the tube socket and the elements of the tube itself.

The large end of the funnel is closed by a glass faceplate on the inside of which the phosphor screen is deposited. In most cases the faceplate is spherically shaped to resist the forces of atmospheric pressure. Often, however, an optical or near-optical flatness is desired for certain tube applications. In these cases the faceplate must be substantially thicker.

The envelopes of most types of cathode-ray tubes are glass, although occasionally ceramic or metal envelopes are used.

Some typical cathode-ray tubes are shown in Fig. 1. Cathode-ray tubes are made in a wide variety of sizes. Tubes having faceplate diameters from 1 to 22 in. (2.5–55 cm) are produced. Picture tubes having faceplates of rectangular outline are considered standard in the television industry. These faceplates range in diagonal dimension from 8 to 27 in. (20–67 cm). Rectangular-faceplate tubes are also fabricated in smaller sizes for use in instrumentation where space availability is restricted.

Electron gun. This electrode structure produces, controls, focuses, and deflects an electron beam. In some cases one or more of the functions enumerated above are performed by components not included in the gun or even in the cathode-ray tube itself. These components are invariably electromagnetic-field-producing devices; a common

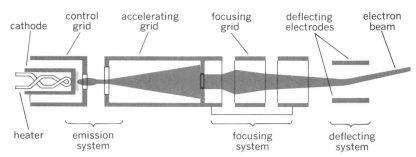

Fig. 2.   Simplified electron gun employing electrostatic focus and deflection.

example is a magnetic solenoid used for beam focusing. A typical electron gun used in a cathode-ray tube for oscillographic applications (Fig. 2) is functionally self-contained because no external components are required to perform beam-intensity control, focusing, or deflection.

Emission system. This part of an electron gun includes the thermionic cathode which is the source of electrons in the beam. The emitting material is generally a barium-calcium-strontium oxide coating deposited on the end of a deep-drawn nickel cup with cylindrical walls. The walls enclose a twisted tungsten wire coated with a refractory insulating material such as aluminum oxide. The passage of current through this wire generates sufficient heat, transmitted by conduction and radiation to the nickel cup, to maintain the oxide coating at emitting temperatures of the order of 1050°K. This heater-cathode structure is supported coaxially within, and insulated from, a cylindrical cup having a greater diameter, called the control grid. The planar section of the control-grid cup is spaced several thousandths of an inch from the cathode surface and has a circular aperture of diameter somewhat smaller than that of the cathode cylinder. Spaced some 0.02–0.25 in. (0.5–6.25 mm) from the control-grid plane is a second apertured disk, called the accelerating grid.

The voltage applied to the accelerating grid sets up a field which penetrates the opening in the control grid and determines the magnitude of the space-charge-limited current drawn from the cathode. As the control grid is made more negative in potential, the penetration of the accelerating field and, consequently, the beam current are reduced.

The electrostatic field in the emission system of a cathode-ray-tube electron gun is so shaped by the geometry of the several elements that electrons leaving the cathode perpendicularly enter trajectories which carry them across the axis of the system in the vicinity of the control-grid—accelerator-grid space. The location at which the electrons cross the axis is called the crossover point. Were it not for space-charge effects, nonhomogeneity of emission velocities, nonperpendicularity of emission directions, and certain imperfections in the shape of the field itself, the crossover would be a point in fact as well as in name. Because of these effects, unavoidably encountered in practice, the ideal crossover point is only approximated by the electronic equivalent of the optical disk of least confusion. Even so, the crossover represents the minimum beam cross section in the vicinity of the cathode surface. Be-

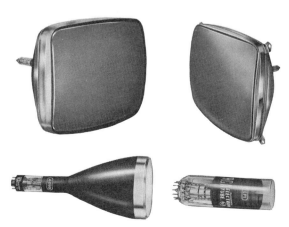

Fig. 1.   Typical cathode-ray tubes.

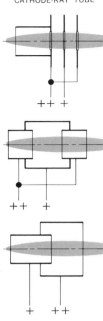

Fig. 3. Typical mechanical arrangements of electrostatic focus lenses.

yond this point, beam electrons enter the field-free drift space within the cylindrical electrode attached to the accelerating grid and travel along straight diverging trajectories toward the focusing system. *See* ELECTRON EMISSION.

Focusing system. After traversing the drift space, beam electrons enter an electrostatic field in which they experience forces which, in combination, redirect the electron trajectories toward the gun axis. The electrode structure to which voltages are applied to produce this field is known as the focusing system or main electron lens. The performance of such an electron lens can be described in terms familiar to the science of light optics. For example, the location of principal planes and focal points for a particular electrode configuration operated at a given set of voltage conditions can be deduced from the potential gradient along the lens axis. Typical electrostatic lenses are shown in Fig. 3. *See* ELECTRON LENS; ELECTRON OPTICS.

The voltages applied to the electron lens in a cathode-ray tube are adjusted to produce the smallest possible beam area at the phosphor screen. It is generally agreed that this condition exists when the object plane of the main lens coincides with the axial location of the crossover previously discussed.

Electron lenses are subject to aberrations very closely analogous to those encountered in glass-lens technology. Excluding those aberrations introduced inadvertently during fabrication, the most troublesome from the point of view of the cathode-ray-tube designer is spherical aberration. This aberration is usually minimized by making lens elements large in diameter compared with the beam or, conversely, by restricting the beam diameter by means of a stop or limiting aperture, as has been done in the gun of Fig. 2. It is obvious that the current available for excitation of the phosphor screen has been reduced by this approach.

Deflecting system. In the electron gun shown in Fig. 2, the deflecting system consists of a single pair of parallel electrodes mounted symmetrically about the axis of the gun. The electron beam entering and traversing the space between the electrodes does so at a constant axial velocity corresponding to the average potential of the electrodes. This potential is usually adjusted to be essentially that of the final element in the focusing system. Signals to be applied to the deflecting electrodes are first passed through a paraphase amplifier in which two components equal in voltage but opposite in polarity are generated. These two components can be applied to the opposing deflecting electrodes without affecting the average potential in the space between them. Under the influence of the applied deflection signals an electric field is set up between the electrodes. This field imparts an acceleration to the electron beam which is perpendicular to the direction of constant beam velocity. In the space between the electrodes, therefore, the beam travels in a parabolic path. After leaving this transverse field, the beam continues to the phosphor screen on a straight path tangent to the parabola. The assumption is made that no post-deflection acceleration is present. If the signals applied to the electrodes vary with time, a straight line is traced out upon the phosphor

screen. In a practical tube, a second pair of deflecting electrodes is mounted further along the gun axis at right angles to the first. These electrodes provide a second independent axis of deflection, orthogonal to the first.

The deflection produced by a pair of parallel electrodes is given by Eq. (1), where y is the

$$y = \frac{LbV_d}{2aV_o} \qquad (1)$$

deflection observed at the phosphor screen in any linear units, L is the distance from axial midpoint of electrodes to phosphor screen in the same linear units, b is the deflecting-electrode length in the same linear units, a is the deflecting-electrode spacing in the same linear units, V_d is the potential difference between electrodes in volts, and V_o is the beam acceleration potential in volts. *See* ELECTRON MOTION IN VACUUM.

The effectiveness of a pair of deflecting electrodes is often described in terms of its deflection factor, that is, the number of volts necessary to deflect the beam a unit distance (usually in inches) at the phosphor screen for a given accelerating potential (usually 1 kV).

Plane-parallel electrodes are rarely found in modern cathode-ray tubes. Overall improvements in performance are obtained by forming the electrodes so that spacing increases with distance in the direction of beam travel. The theoretically ideal curved electrode structure is approximated in practice by electrodes consisting of two or three planar sections, each tilted at a slight angle to the plane of its neighbor.

Deflection systems, particularly electrostatic systems, invariably distort to some degree the circular cross section of an electron beam which undergoes deflection. This distortion increases with angle of deflection. Taking into consideration all factors, the designer of practical cathode-ray tubes using electrostatic deflection rarely attempts to design for angles of deflection much greater than 15° with respect to the gun axis.

Electrostatic deflection is usually preferred over magnetic deflection for cathode-ray tubes intended for instrumentation applications because of the wider frequency range over which operation is readily possible. The bandpass of an electrostatic deflection system is limited in a practical sense by the amount of power the equipment designer can justifiably expend in achieving adequate display brightness and size. Some relief in this regard is obtained if deflection-structure reactances, both capacitive and inductive, are kept to a minimum. A straightforward way of minimizing reactances is to bring the deflecting-electrode leads directly to contacts sealed through the neck wall instead of through the base, as in cathode-ray tubes for less-critical applications. Laboratory oscilloscopes designed around conventional cathode-ray tubes and having useful bandpass to the order of 50 MHz are readily obtained. Even higher performance can be obtained if needed, although at increased instrument cost.

A more fundamental limitation to the high-frequency performance of deflection structures such as that discussed above is related to the time required for the beam to traverse the axial length of a pair of electrodes. It is understandable that a deflecting signal might well reverse in polarity during the electrode transit time, resulting in a completely erroneous indication at the phosphor screen. A practical solution to this problem, used in cathode-ray tubes for specialized instrumentation applications, is to use a traveling-wave structure, for instance a helix, as an electrostatic deflection element. In this case the speed of propagation of a signal along the deflection structure is made to match the velocity of an electron beam passing between the structure and a radio-frequency plane. Thus a particular beam electron traveling through the system "sees" what appears to be a constant field transverse to its initial motion and experiences a deflection which is truly indicative of the amplitude of the deflecting signal at the time the electron entered the deflection structure. Such deflection systems have been demonstrated to operate at gigahertz frequencies.

Many cathode-ray tubes, especially those intended for television or radar display service, employ magnetic deflection systems. An elementary system of this kind takes the form of a pair of coils, each wound on a rectangular form, positioned with major planes parallel on opposite sides of the neck of a cathode-ray tube. The magnetic field lines are thus perpendicular to the axis of the electron gun mounted therein. On entering this field, the beam electrons encounter forces which impart acceleration in the form of a change in direction. If the flux density is uniform, the path followed by the electrons is circular and lies in a plane perpendicular to the lines of flux. On leaving the deflection field, the beam travels in a straight-line path tangent to the circle just described. The deflection given an electron beam which traverses a field of uniform flux density limited to the space between two planes perpendicular to the initial direction of the beam motion, that is, the tube axis, is given the approximation (2), where y is the deflection ob-

$$y \approx \frac{LbB}{3.37\sqrt{V}} \qquad (2)$$

served at the phosphor screen in centimeters, L is the distance from axial midpoint of field to phosphor screen in centimeters, b is the axial extent of uniform field in centimeters, B is the flux density in gauss, and V is the beam acceleration potential in volts. By an appropriate arrangement of the windings of a magnetic-deflection coil pair, more commonly called a deflecting yoke, it is possible to obtain useful total deflection angles up to 110°.

Figure 4 shows a typical electron gun designed for use with magnetic deflection. This gun employs the so-called tetrode configuration, in contrast to the triode configuration characteristic of the all-electrostatic gun previously discussed. The addition of an accelerating element, grid no. 2, to be operated at several hundred volts potential with respect to the cathode serves two purposes. It isolates the performance of the emission system from fluctuations in the final accelerating-voltage source, and it serves as one element in an electron lens which prefocuses the beam, reducing its diameter in the vicinity of the main lens and the deflecting system. This gun is also designed for use with a magnetic main focus lens. This lens takes the form of a solenoid almost completely

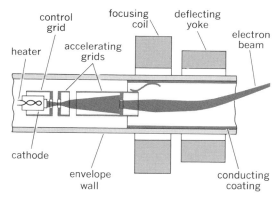

Fig. 4. Simplified electron gun with magnetic deflection.

encased in a suitable high-permeability shell.

As will be discussed in a following section, the light output from a cathode-ray tube increases with acceleration potential. The equipment designer, therefore, will usually specify the largest acceleration potential consistent with deflection voltage or power considerations as well as others, such as freedom from corona and personnel safety. The limitation imposed by deflection considerations is often severe in the case of electrostatic deflection, because required deflection voltages increase linearly with acceleration potential. In an attempt to defeat this limitation, the concept of postdeflection acceleration (PDA) was developed. In a tube employing PDA, the electron beam is not accelerated to its final velocity until after it has traversed the deflection fields. In this way the deflection factor in relation to the final phosphor screen potential is decreased and thereby made more favorable. The simplest practical approach to PDA is implemented by applying the envelope coatings in two sections separated by an insulating gap. This arrangement displays electron-optical lens properties in addition to the electron-acceleration function for which it is primarily intended. The gain in deflection factor occasioned by PDA is therefore offset to some extent by the converging action of the lens formed by the wall coatings. Furthermore, this lens introduces distortions to the pattern displayed on the phosphor screen, particularly at high ratios of postdeflection acceleration to acceleration voltage. These effects are minimized by using several stages of PDA, comprising a series of wall coatings, each operated at a progressively higher voltage. A modern approach is to replace a solid-envelope wall coating by a narrow ribbon of resistive material arranged in a continuous helical spiral extending from the neck coating in the vicinity of the deflecting electrodes to the region of the phosphor screen. The PDA voltage is applied across this spiral, which serves to set up a gradual gradient extending practically the entire distance from the gun to the screen.

PDA cathode-ray tubes suffer in image contrast by comparison with those using a single accelerating voltage. This situation arises because secondary electrons invariably produced in some degree by collision of the beam with deflecting electrodes or the edges of limiting apertures are attracted as a general spray to the phosphor screen by the postacceleration field.

Phosphor screen. This element converts electrical energy to visible radiation. Materials known as phosphors are said to be luminescent; that is, they are able to emit light at temperatures substantially below those which produce incandescence. A phosphor which is excited to luminescence by electron bombardment is described as cathodoluminescent. In the case of phosphors for cathode-ray tubes, luminescence invariably persists after cessation of excitation. Luminescence which continues for more than 10^{-8} sec after excitation is removed is called phosphorescence. Luminescence which is coincident in time with excitation is known as fluorescence.

Screen materials. A wide variety of materials displays the property of cathodoluminescence. Among these materials are nonmetallic elements, large classes of both organic and inorganic compounds, and many glasses. All the phosphors used in commercially fabricated cathode-ray tubes are inorganic, nonmetallic, crystalline materials (see table). Although some of these compounds are found in nature, considerations of reproducibility and uniformity of product dictate the use of phosphors synthesized under controlled conditions. Some nonmetallic crystalline materials, notably calcium tungstate, will emit useful quantities of radiation in the pure state. Most, however, display practical luminescence only when activated by an impurity. These impurities, deliberately introduced in amounts ranging from 1 part in 100,000 to 1 part in 100 play a profound role in determining the efficiency, color, and persistence of the emission obtained from a given phosphor.

Phosphors are fabricated by firing together the base material, the activator, and a flux (to promote crystallization) in a suitable vessel at high temperature. By careful control of the cooling process after firing, the phosphor powder can be made to crystallize in a form which lends itself readily to the usual methods of applying phosphor screens. These methods are spraying, dusting, pouring, or settling. In each case, except dusting, the phosphor is mixed with a vehicle appropriate to the method of application. For example, in the settling process, the phosphor powder is introduced into the envelope of the cathode-ray tube in a suspension of distilled water to which have been added an electrolyte and a gelling agent to promote adherence of the screen. After a period of hours, the phosphor has settled out of the suspension and the remaining liquid is slowly and carefully decanted. Drying of the screen by a gentle flow of warm air follows.

Luminous efficiency. Typical base materials and activators are shown in the table, in which the properties of commercially important cathode-ray-tube phosphors are listed.

One of the criteria by which a phosphor material is judged is its ability to convert electrical energy to useful radiation. In most cases, cathode-ray-tube screens are intended to be viewed by human observers. It is therefore appropriate to speak in terms of visible flux output. The luminous efficiencies of practically all commercially significant phosphors are 5–50 lumens/watt of exciting power, measured under conditions where the beam-bombarding voltage is about 8 kv and the beam-current density about 1.5 μa/cm². For a con-

stant current density, the light output from a phosphor varies as the nth power of the beam accelerating voltage, where n is usually 1.5–2.0. Light output increases with increasing bombarding-current density, exhibiting definite evidences of saturation at current-density levels which are characteristic of particular phosphors. For some phosphors, saturation effects are quite evident at the 10 μa/cm² level. Others display relatively undiminished luminous efficiencies at levels of several hundred μa/cm².

The color of the emission from a phosphor is governed both by the base material and by the activator. The spectral-energy distributions of most phosphors are fairly broad, although colors of high saturation are readily obtained. A wide range of hues is available; the tube designer is restricted in this matter only insofar as particular efficiencies and persistences must be achieved simultaneously with a given color.

Persistence. The law governing the decay of phosphorescence is either an exponential or a power law in nature, depending upon the phosphor under consideration. Usually, in both types of decay, simple expressions are not adequate for an accurate description of a phosphor's performance. Instead, a multiterm expression involving several constants is required. Phosphors are available for a wide range of applications, from flying-spot video-signal-generator service requiring fractional microsecond decay times to radar indicator service in which images are desired many seconds after screen excitation.

Secondary-electron emission. It is necessary, in the choice of a phosphor for a particular application, to be aware of its secondary-electron-emission properties. Even if penetration effects were ignored, it would be impossible to increase the light output of a phosphor arbitrarily by unrestricted increase of accelerating potential. Bombardment of a phosphor with electrons accelerated by voltages in excess of the potential at which the secondary-electron-emission ratio drops below unity, the so-called second crossover potential, will result in light output no greater than that obtained at second crossover potential. Because of the low conductivity of the phosphor screen, charges accumulate thereon, driving the potential of the phosphor to a stable equilibrium at second crossover potential. The potential at which equilibrium occurs can be increased by coating the phosphor particles with a less-than-monomolecular layer of some excellent secondary emitter such as magnesium oxide. *See* SECONDARY EMISSION.

A more complete solution to the problem is to coat the phosphor screen with a thin electron-transparent film of aluminum by vacuum-evaporation techniques. This process not only stabilizes the phosphor potential at the applied acceleration voltage, but also provides an increased light output because of the mirror effect of the highly reflective layer, which redirects radiation that would otherwise be lost in the interior of the envelope.

Depositing of screen. The fineness of detail in an image presented on a phosphor screen is limited to some extent by the granular nature of the screen. Both particle size and layer thickness affect performance in this regard. Substantial improvement may be obtained by the use of phosphor screens in

Characteristics of some cathodoluminescent materials

EIA* no.	Base material (activator)	Persistence category†	Fluorescent color‡	Phosphorescent color‡
P1	Zinc orthosilicate (Mn)	Medium	Yellowish-green	Yellowish-green
P2	Zinc sulfide (Ag + Cu)	Medium	Yellowish-green	Yellowish-green
P3	Zinc beryllium silicate (Mn)	Medium	Yellowish-orange	Yellowish-orange
P4	Zinc sulfide (Ag) + zinc cadmium sulfide (Ag)	Medium short	White	White
P4	Zinc sulfide (Ag) + zinc beryllium silicate (Mn)	Medium to medium short	White	White
P4	Calcium magnesium silicate (Ti) + zinc beryllium silicate (Mn)	Medium	White	White
P5	Calcium tungstate (W)	Medium short	Blue	Blue
P7	Zinc sulfide (Ag) on top of zinc cadmium sulfide (Cu)	Long	White	Greenish-yellow
P11	Zinc sulfide (Ag)	Medium short	Blue	Blue
P12	Zinc magnesium fluoride (Mn)	Long	Orange	Orange
P14	Zinc sulfide (Ag) on top of zinc sulfide (Cu)	Medium	Bluish-white	Orange
P15	Zinc oxide (Zn)	Short	Green	Green
		Very short	Ultraviolet	Ultraviolet
P16	Calcium magnesium silicate (Ce)	Very short	Bluish-purple	Bluish-purple
P17	Zinc oxide (?) on top of zinc cadmium sulfide (Cu)	Long	White	Yellow
P18	Calcium magnesium silicate (Ti) + zinc beryllium silicate (Mn)	Medium	White	White
P19	Zinc fluorosulfide (?)	Long	Orange	Orange
P20	Zinc cadmium sulfide (Ag)	Medium short	Yellow-green	Yellow-green
P21	Magnesium fluoride	Medium	Orange	Orange
P22§	⎧ Zinc sulfide (Ag)	Medium short	Purplish-blue	Purplish-blue
	⎨ Zinc cadmium sulfide (Ag)	Medium short	Greenish-yellow	Greenish-yellow
	⎩ Yttrium oxysulfide (Eu)	Medium short	Red	Red
P24	Zinc oxide (Zn)	Short	Green	Green
P25	Calcium silicate (Pb, Mn)	Medium	Orange	Orange
P26	Zinc fluoride (Mn)	Very long	Orange	Orange
P27	Zinc orthophosphate (Mn)	Medium	Reddish-orange	Reddish-orange
P28	Zinc sulfide (Ag, Cu)	Long	Yellow-green	Yellow-green
P31	Zinc sulfide (Cu)	Medium short	Green	Green
P32	Calcium magnesium silicate (Ti) and zinc cadmium sulfide (Cu)	Long	Purplish-blue	Yellowish-green
P33	Magnesium fluoride (Mn)	Very long	Orange	Orange
P34	Zinc sulfide (Pb, Cu)	Very long	Bluish-green	Yellow-green
P35	Zinc sulfide selinide (Ag)	Medium short	Green	Blue
P36	Zinc cadmium sulfide (Ag, Ni)	Very short	Yellow-green	Yellow-green
P37	Zinc sulfide (Ag, Ni)	Very short	Blue	Blue
P38	Zinc magnesium fluoride (Mn)	Very long	Orange	Orange
P39	Zinc silicate (Mn, As)	Long	Yellowish-green	Yellowish-green
P40	Zinc sulfide (Ag) and zinc cadmium sulfide (Cu)	Long	White	Yellowish-green

*Electronic Industries Association. For P17 and P19, activator material is not identified.

†Persistence categories are based upon time for radiant output to drop to 10% of initial level following interruption of excitation.

Very long	1 s or more	Medium short	10 μs to 1 ms
Long	100 ms to 1 s	Short	1 μs to 10 μs
Medium	1 ms to 100 ms	Very short	less than 1 μs

‡Colors listed here are in accordance with the Kelly system for color designation; see *J. Opt. Soc. Amer.*, 33:629, 1943.

§P22 shown is one of several combinations in common use.

the form of transparent films. Such films, which owe their transparency to their microcrystalline nature, may be prepared by one of several techniques, including simple evaporation of the powder material, evaporation of the powder followed by refiring at high temperature, or vapor-phase reaction of the constituents. In general, transparent phosphor films have displayed substantially lower efficiency than conventional powder screens. This reduced efficiency is explained in part by the loss of light trapped within the transparent film itself and "piped" to the edge of the screen. However, the advantages of high resolution, high contrast in ambient illumination, and thermal stability are important.

Cathode-ray-tube photography. It is sometimes necessary or desirable to photograph the pattern displayed on a cathode-ray tube. Users of oscilloscope instruments frequently resort to photography for purposes of record keeping. In some cases, such as in the study of highly transient phenomena, the photographic capture of a waveform alone permits its full analysis. Until the advent of high-performance video-tape recorders, many live television programs were recorded, for delayed broadcast, on motion-picture film exposed to pictures displayed on special high-quality cathode-ray tubes. In military fields of application, photographic records are frequently kept of the progress of operational or training missions as evidenced by the signals displayed on plan position indicators or fire-control radar indicators. In combination with the character-display cathode-ray tube, photographic techniques provide a rapid means of printing "hard" copy comprising the output of high-speed digital computers.

The technique of photographing oscilloscope patterns is readily mastered because instrument manufacturers offer cameras especially adapted for the purpose. These cameras often are designed to use Polaroid Land film, which has found wide acceptance among oscilloscope users.

For more specialized photographic applications, care is taken to match cathode-ray-tube characteristics (especially phosphor color and persistence) with those of the photographic medium. Because of its high actinic brightness, the blue phosphor designated P11 is widely used. In choosing a film for cathode-ray-tube photography, the ASA Index should be used only as a rough guide. Preliminary exposure experiments are always recommended.

Special-purpose tubes. Although each type of cathode-ray tube is built for a particular application, the features discussed thus far differ relatively little from use to use. In cathode-ray tubes built for specialized applications, however, unique tube characteristics often require distinctive features.

Multigun tubes. Cathode-ray tubes employing multiple independent electron guns are often required for use in oscilloscopes for the simultaneous display of several concurrent signals. Tubes having two guns are common; some have been produced with as many as ten guns. For color television, three guns, one to excite each of the three primary color phosphors, are built into the tube.

Flat cathode-ray tubes. A novel development in the cathode-ray-tube field is the so-called flat picture tube. This device may take the form shown in Fig. 5. The electron beam from the gun in the appended neck is first deflected essentially 90° in a vertical direction, and then 90° in a direction perpendicular to both previous directions. Thus the beam is directed to the phosphor screen, sometimes transparent, deposited on one of the large walls of the envelope. The opposite large wall bears conductive strips (sometimes transparent) extending parallel with its long axis. The voltage pulse trains which set up the final 90° deflecting field are applied to these strips. The initial 90° deflecting field is established by applying appropriately timed voltage pulse trains to a series of U-shaped elements situated astride the gun axis.

Optically ported cathode-ray tubes. In some applications it is desirable to relate the information pattern created on the phosphor screen of a cathode-ray tube to a relatively fixed background pattern. For this purpose cathode-ray tubes are sometimes built with an optical port located in the funnel portion of the envelope. This port allows rear projection of a background image upon the

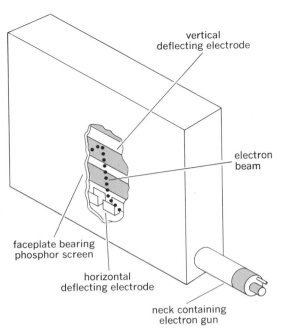

Fig. 5. Schematic diagram of flat cathode-ray tube.

phosphor screen. Alternatively, a photographic camera can be focused through the port and a record made of the images appearing on the phosphor screen.

Character-display cathode-ray tubes. As was mentioned above, the use of cathode-ray tubes in computer input/output terminals is increasing rapidly. Most commonly, the information to be displayed in these cases is in the form of alphanumeric messages. There are a wide variety of electronic techniques available for generating alphanumeric characters. In many instances these techniques require as a display device a cathode-ray tube of conventional design, with due attention to proper choice of electron gun and phosphor screen. However, in some techniques, the combined requirements for high speed and low power consumption require a cathode-ray tube with dual deflection systems. An electromagnetic deflection yoke is used to position the electron beam at the correct location for a particular character. A set of relatively insensitive deflection plates which are positioned downbeam from the yoke then causes the beam to trace out the outline of the character at high speed.

A character-producing tube is used in some alphanumeric display systems. It employs an electron gun of essentially conventional design to "illuminate," one at a time, characters etched in a thin metal stencil mounted in the envelope neck perpendicular to the tube axis. The electron beam issuing from the stencil, having assumed a cross section corresponding to the character being illuminated, is deflected to an appropriate position on the phosphor screen, on which a luminous image of the selected character then appears.

Printing cathode-ray tubes. In the simplest printing systems, the image on the face of a conventional cathode-ray tube, usually designed to have a very small luminous spot and an optically flat faceplate, is optically projected onto the photosensitive medium page by page. Higher optical coupling efficiency, important where speed of printing must be maximum, can be obtained if the cathode-ray tube faceplate is made of an array of optical fibers, fused together so as to be vacuum-tight. Such a device permits contact printing on the photographic medium. Frequently, to reduce device cost, the fiber optics faceplate takes the shape of a narrow rectangle and the printing is done line by line by moving the paper continuously past the faceplate. Cathode-ray tubes are being made in which the fiber optics assembly is replaced by a window of glass or mica several thousandths of an inch thick. The width of the window is restricted to ensure its integrity under the forces of atmospheric pressure. Printing cathode-ray tubes are available in which fine wires, suitably insulated from each other, are brought through the faceplate. These wires conduct charges deposited by the electron beam directly to charge-sensitive paper, which is then passed through a suitable toner material to develop the final printed image.

Color cathode-ray tubes. Television-type color cathode-ray tubes are generally not used in specialized display systems because the dot or line structure does not permit an image of sufficiently high resolution. This limitation is overcome in a color cathode-ray tube of the penetration type. In this device two or more layers of phosphor, each emitting radiance of a different color, are deposited on the inside of the faceplate. If the electron gun is operated at an appropriate overall voltage, only the phosphor layer closest to the gun is excited. At higher voltages a further layer absorbs most of the beam energy and thus its characteristic color dominates the display. By controlling the beam accelerating voltage, therefore, the display can include a range of colors.

[M. DUFFIELD HARSH]

Bibliography: D. G. Fink (ed.), *Electronic Engineer's Handbook*, 1975; S. Sherr, *Electronic Displays*, 1979.

Cathode rays

The name given to the electrons originating at the cathodes of gaseous discharge devices. The term has now been extended to include low-pressure devices such as cathode-ray tubes. Furthermore, cathode rays are now used to designate electron beams originating from thermionic cathodes, whereas the term was formerly applied only to cold-cathode devices. *See* CATHODE-RAY TUBE.

The basis for the nomenclature is purely historical. The first outward evidence was fluorescence from the glass walls of cold-cathode discharge tubes. This fluorescence appeared as the pressure was reduced to the region where the mean free path became greater than the tube dimensions. At these pressures, the gas in the tube no longer emitted an appreciable amount of light. It was ascertained that the wall fluorescence had its origin in rays of particles coming from the cathode. Furthermore, it was demonstrated that these particles traveled in approximately straight lines. If an object was interposed between the cathode and the wall, the fluorescence disappeared in the optically shadowed region of the wall. In 1897 J. J. Thomson was able to show, using electric and magnetic fields, that the particles were negatively charged.

In his experimental arrangement, he eliminated all but a very narrow beam of these rays. By producing an electrostatic deflection of the beam and then counteracting this with a magnetic deflection, he was able to determine the charge-mass ratio e/m. This was found to be the same as that of the electron, and resulted in the identification of these particles.

The electrons produced at the cathode were the result of positive-ion bombardment. Thus the presence of gas in the tube was necessary for the production of the cathode rays. The descendants of this first type of cathode-ray tube include fluorescent devices such as oscilloscopes and television tubes. The major changes have been the introduction of the thermionic cathode and the addition of a very efficient fluorescent screen. Because gas is no longer necessary for electron production, the tubes are evacuated, and the scattering of the electron beam is reduced. An even more advanced application of a cathode-ray device is the electron microscope. Here use is made of diverging rays of electrons to obtain tremendous magnification. The science of electron and ion optics has been advanced to a high degree, so that such devices have become useful, reliable, and efficient. *See* ELECTRON LENS; OSCILLOSCOPE.

It is interesting and appropriate to note that Thomson's measurements also form the basis for modern β-ray spectroscopy. In this case e/m is known, so that either an electric or magnetic field may be used to determine the energy of the electrons which emanate from various radioactive materials, such as β-rays.

An additional application of a cathode-ray device is an x-ray tube. Here a beam of electrons is accelerated in an evacuated tube and allowed to strike a target. The major part of the electron energy is dissipated as heat in the target. A small part of the energy, however, comes off as short-wavelength electromagnetic radiation. *See* ELECTRICAL CONDUCTION IN GASES; ELECTRON EMISSION. [GLENN H. MILLER]

Bibliography: F. K. Richtmyer et al., *Introduction to Modern Physics*, 6th ed., 1969; M. R. Wehr. J. A. Richards, and T. W. Adair, *Physics of the Atom*, 3d ed., 1978.

Cathodoluminescence

A luminescence resulting from the bombardment of a substance with an electron beam. The major application of cathodoluminescence is in cathode-ray oscilloscopes and television picture tubes. In these devices a thin layer of luminescent powder is evenly deposited on the transparent glass face plate of a cathode-ray tube. The electron beam originating in the cathode, after undergoing acceleration, focusing, and deflection by various electrodes in the tube, impinges on the luminescent coating. The resulting emission is normally observed through the glass face plate of the tube, that is, from the unbombarded side of the coating. Depending on the type of signal information to be displayed on the tube, phosphors or phosphor blends of various emission colors and persistences are used.

The brightness B of a phosphor under cathode-ray bombardment depends linearly on the current density j for sufficiently small values of j, but saturation effects set in at higher current densities. B depends on the bombarding voltage according to the empirical relation $B = kf(j)(V - V_0)^q$, where k is a constant of the face plate material, $f(j)$ represents the dependence of brightness on current density, V is the accelerating voltage, V_0 is a so-called dead voltage, and q is a constant for the particular phosphor, the values for different phosphors lying between 1 and 3. The efficiency of conversion of the electron-beam energy into light energy is at most about 20%, considerably less than the best efficiencies found with photoluminescence. This relative inefficiency is due to the complex mechanism of excitation by the electron beam, which loses its energy by interaction with the ions of the crystal lattice and consequently dissipates a large fraction of its energy before reaching a luminescent center.

An important requirement for a good cathodoluminescent phosphor is that it have good secondary electron emission properties; otherwise it will charge up negatively and decrease the effective potential of the bombarding beam. Phosphors in cathode-ray tubes are subject to deterioration by prolonged action of the electron beam and by bombardment by residual gas ions that remain in the tube or are generated at the electrodes or tube walls. *See* SECONDARY EMISSION.

[CLIFFORD C. KLICK; JAMES H. SCHULMAN]

Cavity resonator

An enclosure with conducting walls surrounding a dielectric medium, capable of resonating with electromagnetic waves. It is used mainly as a reservoir of electromagnetic oscillations at microwave or ultra-high frequencies. The conducting walls are usually made of metal such as copper or silver, while the dielectric medium is usually air or vacuum. The enclosure has at least one aperture for coupling the electromagnetic energy between the inside and outside of the cavity.

Cavity resonance. For an electromagnetic wave to exist inside a cavity, the electric and magnetic field intensities must be solutions of Maxwell's equations, and at the same time they must satisfy the boundary conditions on the inside surfaces of the cavity walls. For all practical purposes it can be assumed that the wall conductivity is so high that the tangential component of the electric field vanishes everywhere along the inside walls. This restriction requires that only waves of certain discrete frequencies be allowed in the cavity. Thus, when the frequency of an exciting wave is varied, the cavity does not extract or contain any energy

from the wave unless the exciting frequency happens to be one of the allowed values. When the latter condition holds, then the cavity will resonate strongly with the exciting wave by absorbing a part or all of the incoming energy. The cavity is thus acting as a resonator.

The values of the resonant frequencies of a cavity depend on the size, shape, dielectric material, and the mode of operation (TE or TM modes). The values can be calculated with high precision for cavities of regular shapes (rectangular, circular cylindrical, and so on), or they can be directly calibrated by measuring the resonant frequencies with respect to a frequency standard. Whenever there is a choice, it is customary to use a cavity at its lowest resonant frequency.

Cavity quality factor. For simplicity of discussion, assume the cavity has a single resonant frequency f_R. Suppose a field probe like a small loop is placed inside the cavity which can measure the field intensity by an external detecting circuit (Fig. 1) while drawing negligibly small power from the cavity. The current I in the meter can be taken to be proportional to the square of the magnetic field intensity. As the exciting frequency f is varied, resonance spreads over a narrow band of frequencies instead of being a sharp line at f_R (Fig. 2). As a measure of resonance width, a qual-

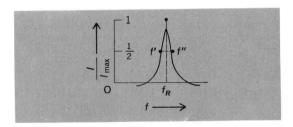

Fig. 2.   Typical resonance curve of a cavity.

ity factor Q_L is defined by Eq. (1), where $f'' - f'$ is

$$\frac{1}{Q_L} = \frac{f'' - f'}{f_R} \tag{1}$$

the frequency width of resonance at $I/I_{max} = 1/2$ point.

It can be shown that the quality factor has the physical significance given by Eq. (2).

$$\frac{1}{Q_L} = \frac{\text{energy lost by cavity per radian of oscillation}}{\text{energy stored in cavity}} \tag{2}$$

Combining Eqs. (1) and (2), it is seen that, for a constant energy storage in the cavity, the smaller the energy loss, the sharper the resonance. A cavity which has a small energy loss is commonly known as a high-Q cavity. Except for special requirements, a cavity's Q is usually preferred to be as high as possible.

The energy loss in a cavity can be divided into the internal loss due to energy dissipation on the cavity walls and in a "lossy" dielectric material (if any), and the external loss of energy through the coupling aperture, sometimes called the window. Application of Eq. (2), leads to Eq. (3), where Q_i is

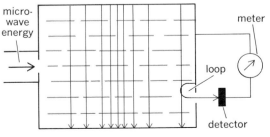

Fig. 1.   A resonant cavity with a field-measuring probe. Solid lines indicate the electric field and dashed lines indicate the magnetic field.

$$\frac{1}{Q_L}=\frac{1}{Q_i}+\frac{1}{Q_e} \tag{3}$$

known as the internal, or unloaded, Q; Q_e the external, or window, Q; and Q_L the loaded Q. Since the ratio Q_e/Q_i can be directly determined by measuring the standing-wave ratio (caused by incident and reflected waves) external to the cavity, the individual quality factors can all be evaluated. In particular, if $Q_i=Q_e$, then the cavity becomes nonreflective at f_R with respect to the incident wave and is said to be matched, a condition often desired in practice.

Forms of resonators. Various forms of cavity resonators have been used for microwave or ultrahigh frequency application. A rectangular cavity is particularly simple for construction and analysis. For instance, a short section of wave guide closed by two end conducting plates forms a usable cavity. The resonant frequency of a rectangular cavity (Fig. 3) can be determined by Eq. (4), where n, m, l

$$f_{nml}=v_0\sqrt{\left(\frac{n}{2a}\right)^2+\left(\frac{m}{2b}\right)^2+\left(\frac{l}{2c}\right)^2} \tag{4}$$

are integers (only one of them can be zero for a TE mode and none can be zero for a TM mode), v_0 is the velocity of propagation in the dielectric medium (substantially the velocity of light for air or vacuum), and a, b, c are the dimensions of the cavity. Loaded Qs for rectangular cavities are typically of the order of 5000–10,000.

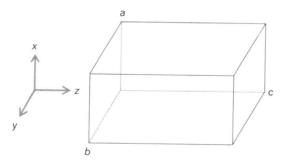

Fig. 3. A rectangular cavity (coupling hole omitted).

Circular cylindrical cavities are much used, particularly the tunable ones, for wavemeter application (Fig. 4). Here the movable cavity wall is in the form of a circular piston. Lack of intimate contact between the piston and the cylindrical cavity wall is either nonessential in some cases (where no current flows across the gap), or can be compensated for by a choke arrangement which artificially creates a nodal zone for the current at the gap. Wavemeters can be made into accurate direct-reading instruments for measuring wavelengths. In other areas of application, circular cylindrical cavities are particularly desirable for obtaining higher Qs (for instance with a TE_{01l} mode) than is possible with a rectangular cavity. *See* WAVEMETER.

Coaxial cavity resonators are often used as wavemeters, particularly for lower than microwave frequencies. The so-called reentrant cavities are sometimes used for certain special purposes.

Since cavity resonators are essentially fre-

quency-selective circuit elements for microwaves, they are very useful as components in a microwave filter network. In a radar set, a resonant cavity can be used in a so-called TR box, where a gas-filled cavity breaks down to a short-circuit impedance to protect the receiver from damage when the high-powered transmitter starts working. In high-energy linear accelerators, resonant cavities are used to accelerate charged particles in intense high-frequency electric fields. *See* KLYSTRON; RADAR.   [C. K. JEN]

Bibliography: H. A. Atwater, *Introduction to Microwave Theory*, 1962; S. Liao, *Microwave Devices and Circuits*, 1980.

Character recognition

The technology of using machines to automatically identify human-readable symbols, most often alphanumeric characters, and then to express their identities in machine-readable codes. This operation of transforming numbers and letters into a form directly suitable for electronic data processing is an important method of introducing information into computing systems. *See* DATA-PROCESSING SYSTEMS.

Character recognition machines, sometimes called character readers, print readers, scanners, or reading machines, automatically convert printed alphanumeric characters or symbols into a machine-readable code at high speeds. The output of the character recognition machine may be temporarily stored on magnetic disks or magnetic tape. Alternatively, the recognition system may be operated on line with the data processor (Fig. 1). The first commercial application of character recognition was in the banking industry, which adopted magnetic ink character recognition (MICR). In most applications, the printed or typed characters are sensed optically; this process is called optical character recognition (OCR).

Optical mark reading (OMR) refers to the simpler technology of optically sensing marks. The information being read is encoded as a series of marks such as lines or filled-in boxes on a test

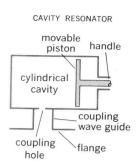

CAVITY RESONATOR

Fig. 4. A circular cylindrical cavity used as a wavemeter.

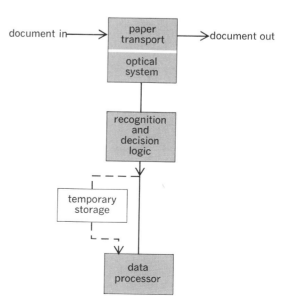

Fig. 1. Diagram of optical character recognition system.

answer sheet, or as some special pattern such as the Universal Product Code (UPC) which has been adopted by the retail food industry. The UPC uses a predetermined and reproducible standard pattern which corresponds to each number in a 10-digit code. A relatively simple optical scanner detects the pattern and decodes the number.

There are various manual alternatives to automatic character recognition equipment for data entry. These include keypunch, key tape, and key disk systems, in which the data are entered through a keyboard and are recorded on punched cards, magnetic tape, or rotating magnetic disks, respectively. Keyboard terminals are another approach to on-line data entry. Manual keystroking is a tedious and costly operation, and it is therefore advantageous to use automatic character recognition equipment whenever the form of the information to be entered is consistent with the capabilities of the character recognition system. Limited-vocabulary isolated-word voice recognition equipment is now available for data entry. However, voice input is also a slow process and would not be competitive with OCR in situations where the data already exist in printed or typed form.

PATTERN RECOGNITION

The technology of automatically recognizing complex patterns is being developed in the context of numerous military and civilian applications such as: the analysis of aerial and satellite images obtained with visible, infrared, radar, and multispectral sensors, for the detection and classification of military objects, terrain types, weather patterns, and land use patterns; identification of individuals or medical conditions using fingerprint, palmprint, blood cell, tissue cell, chromosome, and x-ray images; processing of voice, sonar, electrocardiogram, electroencephalogram, seismic, and other waveforms to detect and classify "signatures" of different events or conditions; automatic detection of flaws in sheet glass, bottles, textiles, paper, printed circuit boards, and integrated circuit masks; recognition of industrial parts for automated machining, handling, inspection, and assembly; recognition of alphanumeric characters and special symbols for automatic processing of bank checks, sale and inventory documents, postal addresses, and for text input to computers or reading for the blind; and analysis, clustering, and classification of survey and experimental data obtained in diverse disciplines.

OPTICAL CHARACTER RECOGNITION

Of all the above application areas, OCR is the single most important commercial application of pattern recognition technology.

Functional systems. The technologies used in OCR systems include optical, electronic, mechanical, and computer techniques. In general an OCR system has the following functional systems: input; transport; scanner; preprocessor; feature extraction and classification logic; output.

Transport. The problems of feeding, transporting, and handling paper, especially at high speeds, have caused much difficulty in some applications of OCR, for example, in postal address reading for mail sorting. Turnaround documents, such as credit card slips, have tight constraints on size and quality of paper and so are much easier to transport at high speeds than mail in which size, thickness, and paper quality have only recently become subject to some constraints. Page readers may have to handle only a couple of documents per second, but each page may contain up to 2000 characters. And in some applications, for example, the reading of sales tags with a hand-held "wand," motion of the paper is completely avoided.

Scanning. The scanner converts reflected or transmitted light into an electric signal which is then digitized by an analog-to-digital (A/D) converter. Although simpler to design, transmitted light scanning requires the additional step of making a (film) transparency of every image to be read, and hence is less suitable than reflected light scanning for high-volume applications unless the documents already exist in microfilm form or are routinely microfilmed for ease of storage. *See* ANALOG-TO-DIGITAL CONVERTER.

One categorization of optical scanners is whether they employ a flying-spot or a flying-aperture principle. In the first case a spot of light sequentially illuminates successive portions of the material to be read, and all the reflected or transmitted light is collected by a detector. An example is the cathode-ray-tube (CRT) flying-spot scanner, where the CRT beam may be moved to sequentially strike the phosphor on the tube face, and the glow of the spot on the phosphor screen illuminates the corresponding spot on the document. In flying-aperture devices the entire document is flooded with light, but light is collected sequentially spot by spot from the illuminated image. An example is Vidicon scanners in which a document is flooded with light from an ordinary light source, and the reflected or transmitted light impinges upon the photoconductive target of the Vidicon. The image on the surface causes a variation in the local charge concentration, which is converted into a video signal by sequentially scanning the photoconductive surface with an electron beam. *See* CATHODE-RAY TUBE; TELEVISION CAMERA TUBE.

Mechanical scanners, television cameras, CRT flying-spot scanners, solid-state linear and two-dimensional array scanners, and electrooptical scanners which use a laser as the source of illumination are the main scanning techniques which have been employed in commercial OCR systems.

Mechanical motion of the document is usually used to scan successive lines of print. Earlier OCR systems also used a rotating or oscillating mirror or a prism to scan across each line of print, and to collect the signal from a single vertical column in a character or form an entire character using a photocell array.

In solid-state scanners, instead of mechanically or electronically scanning a single beam across the region of interest, the region is sampled in small, discrete, adjacent areas by electronically switching between elements in the array. Flying-spot devices use linear or two-dimensional light-emitting-diode (LED) arrays. Flying-aperture devices use arrays of photodiodes or phototransistors. *See* LIGHT-EMITTING DIODE; PHOTODIODE; PHOTOTRANSISTOR.

The most commonly used scan pattern is a raster scan in which the flying spot or flying aperture

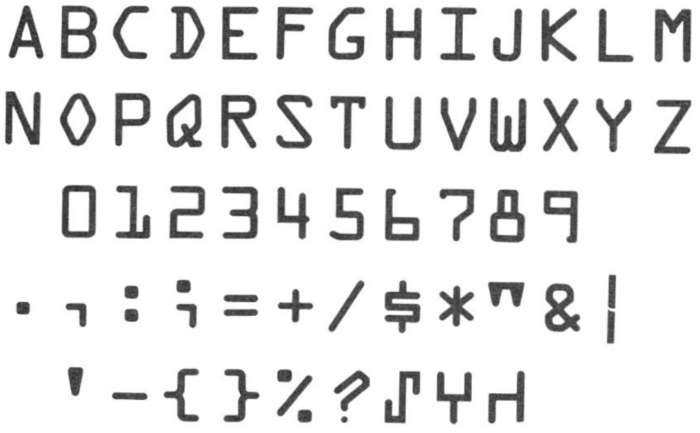

ABCDEFGHIJKLM
NOPQRSTUVWXYZ
0123456789
• ¬ : ; = + / $ * ″ & |
' - { } % ? ♩ Ч Ꮰ

Fig. 2. Character set (ISO-A) approved by the American Standards Association for use in optical character recognition applications.

sequentially scans the character area by using a sawtooth pattern. A line-following pattern has been used in some hand-print readers. Because of microprocessors, completely programmable scanners are economically feasible, making it easier to rescan rejected characters, scan blank areas at low resolution for increased throughput, and perform various preprocessing functions. *See* MICRO-PROCESSOR.

Preprocessing. Preprocessing refers to line finding, character location and isolation, normalization and centering, and related processing functions that may be needed prior to feature extraction and classification. The nature and degree of preprocessing needed depend on whether the material being read consists of stylized fonts, typescript, typeset text, or hand-printed characters.

Stylized font characters such as the ISO-A or OCR-A (Fig. 2) and ISO-B or OCR-B (Fig. 3) character sets have well-defined and closely controlled formats and line spacing. Documents using stylized font characters usually also have special symbols to guide the scanner to each field of data; special ink invisible to the optical scanner is used to print material not to be read by the OCR. Contrasted with these well-defined formats which are easy to handle automatically, are the format control problems created when reading general typeset material involving text interspersed with illustrations, tables, and so forth. In such situations, current commercial OCRs require human direction of the scanner to the appropriate data field.

Procedures available for line-finding include adaptive line-following algorithms which compensate for baseline drift in the lines of print. For situations where characters are not uniformly well spaced or in which easily detectable boundaries do not occur where expected, character segmentation may involve a scanning aperture smaller than that used for subsequent classification. Various heuristic procedures are used to separate touching characters, eliminate noise such as isolated dots, and smooth out gaps or breaks in line segments.

The extent of rotation and skew correction, character segmentation, size normalization, centering, and noise elimination achievable on individual characters prior to classification determines how sophisticated the feature extraction and algorithms must be for a given application. Many commercial OCRs use simple template matching

classification logic, which gives adequate performance only if variations of the above type have been essentially eliminated.

Feature extraction and classification. When simple template matching is inadequate, recognition is achieved by extracting distinctive features and using them in a decision logic to classify the characters. Decision logics are designed by using statistics of features obtained from sets of learning samples representative of the applications for which the OCR is intended.

Most high-speed commercial OCRs use special hardware for preprocessing, feature extraction, and classification, although there exists at least one high-performance, high-speed postal address reader which uses general multiprocessors and software recognition. Optical correlation, resistor summing networks, and parallel digital logic circuits represent some of the ways in which character and feature templates and weighted masks have been implemented in hardware. Centering of the character being scanned within a recognition "window," referred to as registration, may be done by shifting the digitized character through a discrete number of successive positions in a one- or two- dimensional shift register. Approaches to segmenting a line of print into individual characters include comparing successive vertical scans to give an explicit segmentation of the entire line or alternatively looking for peaks in the output of the classifier to implicitly segment each character.

Element design. The design of different elements of an OCR, that is, the transducer, the preprocessing, features, and classification logic, is determined by the type of characters and quality of material to be read. Some aspects which affect the design, such as line and character spacing, character-size and stroke-width variability, number of character classes, and differentiability of most similar characters, are a function of the character fonts used. Independent of the character fonts are other aspects, such as the reflectance of the paper, quality of the printing, and format of the document. Various standards by the American National Standards Institute (ANSI) contain specifications and recommendations for the paper, print quality, format rules, and measuring techniques for various classes of OCR applications.

OCR APPLICATIONS

The classes of OCR applications range from highly controlled stylized fonts to unconstrained multifont typewritten, printed, and handwritten characters encountered by postal address readers. Important elements in rating the performance of an OCR are the error rate, the throughput, the number of fonts capable of being recognized, and the cost, size, and reliability of the machine. The error rate has two components: the undetected substitution rate and the reject rate, that is, the proportion of characters not classified into any character class.

Stylized font characters. Stylized characters are designed to make automatic recognition easier. Two widely used special fonts are the ISO-A (Fig. 2), which was the first character set standardized by the International Standards Organization, and ISO-B (Fig. 3), originally promoted in Europe as being more natural looking. ANSI

CHARACTER RECOGNITION

1234567890
ABCDEFGHIJKLM
NOPQRSTUVWXYZ
abcdefghijklm
nopqrstuvwxyz
★+-=/.,:;"'_
?!()<>[]%#&a^
¤£$¦\ ¥■ ——

Fig. 3. ISO-B character set for international usage.

0 I 2 3 4 5 6 7 8 9 A B C D E F G H I J
K L M N O P Q R S T U V W X Y Z + – . ,
ƀ † " # $ % ¢ ´ ( ) * / : ; < = > ? @ \
^ _ [ ] ¡ À Ē Ī Ō Ū Á É Í Ó Ú Ã È Ì Ò Ù
Â Ê Î Ô Û Å Ø Ç Ñ ß £ ¥ ↑ ↕

Fig. 4. Proposed standard character set for hand-printed OCR applications.

has also proposed a standard for hand-printed characters (Fig. 4). Stylized fonts are used extensively on turnaround documents such as utility bills and gasoline credit card slips (Fig. 5). A common example of stylized font character recognition is the magnetic ink character recognition (MICR) system which has been used on bank checks since the late 1950s. This special font, adopted as a standard by the American Banking Association, is called E-13B (Fig. 6). Error rates for single-font stylized character readers can be about 1 substitution in 200,000 characters and 1 reject in 20,000.

Single-stylized-font OCR wands, that is, hand-held readers, are being used more and more in inventory and point-of-sale applications. User-programmable hand-held wands in which a microprocessor can be programmed through the wand by reading special codes in the stylized character set have been introduced.

Postal address readers. At the opposite end of the scale from stylized font characters is the variability of character fonts used in postal address readers. About 24 large postal address readers sort mail in large cities in the United States, and additional OCR and bar code readers are to be installed. Together with the introduction of a nine-digit business zip code and special incentives for large-volume business mailers, this equipment should improve service for large-volume mailers. High-volume business mailers already participate in a special "red tag" program in which batches of mail suitable for automatic reading and sorting are sorted and tagged before being sent to the post office for automatic mail sorting.

In the machine reading of outgoing mail one can make use of relationships between the city, state, and zip code on the last line of the address. Thus

better recognition and missort rates can be expected on outgoing mail than on incoming mail. Performance figures on the postal address reader SARI show that the state of the art of postal address reading is well advanced. On a sample of 1,250,487 live mail pieces, a true throughput of 32,078 letters per hour was obtained with a recognition rate of 97% and a missort rate of 0.01% on outgoing mail, and a recognition rate of 93% and missort rate 0.46% on incoming mail. On incoming mail the missort rate figure includes missorting of street numbers. The volume of mail is such that even when a substantial percentage is rejected by the OCR and diverted to subsequent manual sorting, automatic sorting of the rest can still be cost-effective.

The reading of hand-printed zip codes has been successfully accomplished in Japan by requiring that the numerals be carefully written in boxes preprinted on a standard location on the envelope. Such constraints avoid the problem of segmenting the characters and help humans adapt to the char-

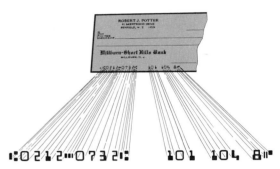

Fig. 6. E-13B font in magnetic ink on the bottom of a typical bank check to permit electronic processing.

acteristics suitable for machine readability of hand-printed numerals. Preprinted boxes on envelopes are likely to be adopted by many postal services. To encourage experimentation and comparative evaluation of approaches to automatic reading of handwritten zip codes, the U.S. Postal Service has prepared two data bases. The first, CONSCRIPT, consists of multilevel digitized data tapes of the video scan from five-digit postal zip codes and six-digit numeric codes from constrained handwritten OCR forms. The second, ZIPSCRIPT, consists of multilevel digitized tapes of zip codes selected from samples or envelopes collected at dead-letter mail offices.

Typewritten and typeset characters. Many data-entry and word-processing applications are aided by the automatic reading of material typed in one of the various popular typewriter fonts. Performance figures similar to those quoted for stylized font characters may be obtained with certain typewriter fonts, that is, modified Courier, when typed on a specially aligned typewriter using carbon film ribbon, high-quality paper, and well-specified format conventions. In general, with standard typefaces, error rates are likely to be one or two orders of magnitude higher, depending on the typeface and print quality.

The large number of styles and the segmenta-

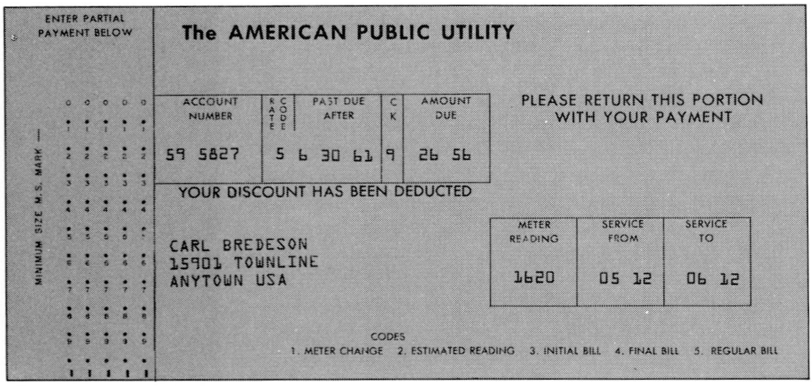

Fig. 5. Typical document for optical character recognition systems use.

tion of variable-width characters make the automatic reading of typeset text much more difficult than typewritten material. There are larger number of classes of characters in each font, many symbols, and combinations of characters called ligatures, for example, fi, ffi, and fl. OCR performance on variable-pitch typeset material is much lower than on typewritten and stylized OCR fonts; 1 in 100 characters may be incorrectly classified.

A major application for automatic reading of typeset text is computerized information retrieval. A highly useful application of mixed-font OCR is a commercially available reading machine for the blind. By coupling the reader to a speech synthesizer, the OCR's output is voiced at an adjustable rate between 100 and 300 words per minute. The human listener soon learns to make sense of the output even when a high fraction of the characters is incorrectly recognized.

Hand-printed characters. A number of manufacturers offer OCRs for constrained hand-printed numerics, and a few offer readers for constrained alphanumeric hand-printed characters. State payroll tax forms, driver's license applications, magazine subscription renewal forms, and other short forms with boxes and directions on how to print the characters give evidence of the business use of constrained hand-printed OCR. The recognition performance depends, among other things, on the number of writers, the number of different character classes, and the training of the writers.

The problem of recognizing unconstrained hand-printed characters is a challenging one and is similar to the problem of classifying reasonably good-quality machine-printed characters from an unlimited number of fonts all mixed together. The variability of characters printed by the same individual is large, with an even larger variability encountered between different individuals. Performance figures of the order of 95–98% correct recognition for alphanumerics and 99.5% correct recognition for numerals only, have been found for an "untrained but motivated population." Providing feedback and display of rejected and misclassified characters to the writers can result in a much lower error rate. A commercially available OCR for reading unconstrained numerals handwritten on bank checks is in operation.

Cursive writing. On-line recognition of cursive writing by using features of the stylus motion appears to lead to the future possibility of devices for signature verification, and is being studied by a number of investigators. Non-real-time approaches to automatic recognition of cursive alphanumeric script have not gone beyond the laboratory investigation phase.

Different alphabets. In principle, OCR can be performed on characters of any alphabet belonging to any language. Work has been reported on OCR for Cyrillic, Devanagri (Hindi), Arabic, Chinese, and Katakana alphabets. One OCR recognizes Russian capital letters, digits, punctuation marks, and arithmetic and other symbols—53 characters in all. Chinese or Japanese character recognition is a very special problem in which the number of possible categories is very large—tens of thousands. But the variability in each character is very small, for the font as known in English does not exist. Thus a particular Chinese character in a newspaper or a book does not vary much from sample to sample. Moreover, the problems of character separation in Chinese or Japanese are easier to resolve than in English since a character usually has a definite size and constant pitch.

PROSPECTS

The increasing use of microprocessor technology is aimed at developing low-cost, decentralized OCR devices which can handle a wide variety of fonts and formats. Combinations of word-processing and OCR equipment, and OCR and facsimile store and forward communications are likely to lead to many new applications for OCR. *See* COMPUTER; DIGITAL COMPUTER; MICROPROCESSOR; WORD PROCESSING.

[LAVEEN N. KANAL]

Bibliography: L. N. Kanal, Patterns in pattern recognition, 1968–1974, *IEEE Trans. Inform. Theory*, pp. 697–722, November 1974; L. N. Kanal (ed.), *Pattern Recognition*, 1968; G. Nagy, Engineering considerations in optical character recognition, *Proc. COMPCON*, San Francisco, IEEE Cat. no. 8OCH1491-OC, pp. 402–406, February 1980; C. Y. Suen, Automatic recognition of hand-printed characters: The state of the art, *Proc. IEEE*, 68:469–487, April 1980; J. R. Ullman, *Pattern Recognition Techniques*, 1973.

Characteristic curve

A curve that shows the relationship between two changing values. A typical characteristic curve is that which shows how the changes in the control-grid voltage of an electron tube affect the anode current. When three variables are involved, a family of characteristic curves is frequently drawn, with each curve representing one value of the third variable. Other common examples of curves include speed-torque characteristics of motors, frequency-gain characteristics of amplifiers, and voltage-temperature characteristics of thermocouples. The curves are produced by making a series of measurements and plotting the results.

With appropriate electronic circuits the measurements can now be made automatically and fast enough so that the entire characteristic curve is visible on the screen of a cathode-ray oscilloscope. *See* OSCILLOSCOPE.

[JOHN MARKUS]

Charge-coupled devices

Semiconductor devices wherein minority charge is stored in a spatially defined depletion region (potential well) at the surface of a semiconductor, and is moved about the surface by transferring this charge to similar adjacent wells. The formation of the potential well is controlled by the manipulation of voltage applied to surface electrodes. Since a potential well represents a nonequilibrium state, it will fill with minority charge from normal thermal generation. Thus a charge-coupled device (CCD) must be continuously clocked or refreshed to maintain its usefulness. In general, the potential wells are strung together as shift registers. Charge is injected or generated at various input ports and then transferred to an output detector. By appropriate design to minimize the dispersive effects associated with the charge-transfer process, well-defined charge packets can be moved over relatively long distances through thousands of transfers.

Control of charge motion. There are several methods of controlling the charge motion, all of which rely upon providing a lower potential for the charge in the desired direction. When an electrode is placed in close proximity to a semiconductor surface, the electrode's potential can control the near-surface potential within the semiconductor. The basis for this control is the same as for metal oxide semiconductor (MOS) transistor action. If closely spaced electrodes are at different voltages, they will form potential wells of different depths. Free charge will move from the region of higher potential to the one of lower potential. Figure 1 shows a case where by alternating the voltage on three electrodes in proper phase, a charge packet can be moved to the right. Figure 2 illustrates another scheme whereby an asymmetry built into a well can direct the charge in a given direction. Asymmetries of this type are easily created by using implanted ion layers or varying dielectric thickness. The three-phase structure shown in Fig. 1 has the ability to reverse the charge direction by a change in electrical phase.

Transfer efficiency. An important property of a charge-coupled device is its ability to transfer almost all of the charge from one well to the next. Without this feature, charge packets would be quickly distorted and lose their identity. This ability to transfer charge is measured as transfer efficiency, which must be very good for the structure to be useful in long registers. Values greater than 99.9% per transfer are not uncommon. This means that only 10% of the original charge is lost after 100 transfers.

Figure 3 shows a close-up of a stage of a charge-coupled device. Several mechanisms influence the transfer of charge from one well to the next. Initial self-induced drift acts to separate the charge. This

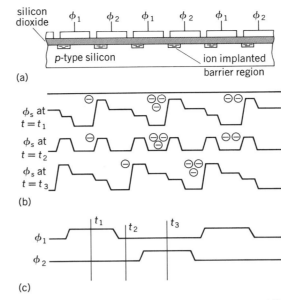

(a)

(b)

(c)

Fig. 2. Operation of two-phase charge-coupled shift register. (a) Cross section of register through channel. (b) Profile of surface potential ϕ_s for three different clock time intervals t_1, t_2, and t_3. (c) Voltage waveforms ϕ_1 and ϕ_2 for the two-phase clocks.

repulsion of like charge is the dominant transfer mechanism for large signals, and is effective for the first 99% or so of the charge. Near the edge of the transfer electrode the potential gradient creates a field which sweeps the charge onward. Thermal diffusion accounts for the transfer of the remaining charge.

Electron traps. It would appear that, given sufficient time, almost all of the charge could be transferred. Two other mechanisms are at work to counter this. Within the silicon and at its surface are sites that can act as electron traps. This is especially true at the surface where numerous surface states exist. These traps collect charge when exposed to a large charge packet and then slowly release it during later cycles of small charge packets. By stringing a large number of empty charge packets together, the traps can be emptied completely. When the first packet containing charge arrives, it could be completely consumed, recharging the traps.

Other than process steps that minimize the trap density, there are two methods to alleviate the severity of this problem. The first is never to allow a series of completely empty charge packets to occur. Instead of an empty charge packet, a minimum charge quantity is always present. This charge, called a "fat zero," can be 10–20% of the well capacity. Under some conditions, this can reduce the trapping effects to tolerable levels. The second approach is to use a channel for the charge whose potential minimum does not occur at the surface. This is called a buried channel device, as opposed to a surface channel device. Since the charge is located within the bulk silicon below the surface, it is not exposed to the surface-state traps. Bulk traps remain, but because of their low density they are almost insignificant.

The process to shift the channel potential mini-

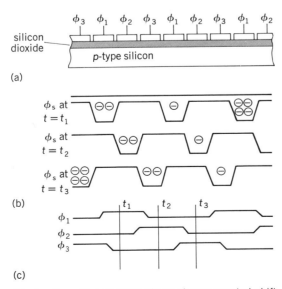

(a)

(b)

(c)

Fig. 1. Operation of three-phase charge-coupled shift register. (a) Cross section of register through channel. (b) Profile of surface potential ϕ_s for three different clock time intervals t_1, t_2, and t_3. (c) Voltage waveforms ϕ_1, ϕ_2, and ϕ_3 for the three-phase clocks.

mum below the surface can be either an ion-implant layer or an epitaxy layer combined with an implant layer. Because the charge is further removed from the surface control electrode, the maximum charge density has been reduced by a factor of 2 to 3. For the same reason, the fringing fields are greater and charge transfer can be much faster. The charge also moves with bulk mobility rather than the lower surface mobility, further enhancing performance. Fractional charge losses of as low as 5×10^{-5} at clock frequencies greater than 100 MHz have been reported for buried-channel charge-coupled device structures. This means that such a structure could transfer 10^6 electrons from one well to another in less than 10 ns, leaving fewer than 50 electrons behind. *See* ION IMPLANTATION.

Localized potential minima. In addition to electron-trapping sites, there is another reason that charge can be left behind. Where the two adjacent surface electrodes come together, there is necessarily a gap. This gap is usually quite small, and sometimes may even be covered by one of the electrodes. However, the gap represents a region of poorly controlled surface potential. It is possible, under some conditions, for the transition from one potential well to the next to have perturbations in it. These perturbations, or "glitches," can trap charge. They represent a localized minipotential well within the larger one. Charge trapped by the glitch will be left behind after a transfer. If the glitch remained filled, continued clocking of the structure would not represent a problem. However, the size of the glitch may change, depending upon electrode clock amplitude. Also, charge is released over the glitch as thermionic emission over a barrier potential. Thus a glitch can be emptied by a series of empty charge packets to trap the next packet-containing charge. Many times, glitches occur at input or output ports without detriment to the structure's operation since their effect may be small. However, if a glitch occurs within the repetitive register, the multiplication effect can have a serious impact on overall performance.

Lifetime. A second important property of a charge-coupled device register is its lifetime. When the surface electrode is clocked high, the potential within the semiconductor also increases. Majority charge is swept away, leaving behind a depletion layer. If the potential is taken sufficiently high, the surface goes into deep depletion until an inversion layer is formed and adequate minority charge collected to satisfy the field requirements. The time it takes for minority charge to fill the well is the measure of well lifetime. The major sources of unwanted charge are: thermal diffusion of substrate minority charge to the edge of the depletion region, where it is collected in the well; electron-hole pair generation within the depletion region; and the emission of minority charge by traps. Surface-channel charge-coupled devices usually have better lifetime, since surface-state trap emission is suppressed and the depletion regions are usually smaller.

Input and output ports. Once adequate transfer efficiency and sufficient lifetime have been achieved, input and output ports must be established. The port structure of a charge-coupled device depends upon its application. Analog registers

require linear inputs and outputs with good dynamic range. Digital registers require precisely metered and detected ones and zeros. Imaging devices use photoemission at the register sites as input with analog-output amplifiers.

The most common input is shown in Fig. 4. This input structure relies upon a diode as a source of minority charge. Whenever the diode potential drops below the threshold potential of the adjacent gate, an inversion layer is formed beneath the gate. By pulsing the diode potential low or the adjacent gate potential high, charge can be injected into the register. To prevent the charge from returning to the diode, the next gate in the register immediately collects it. The amount of charge injected can be controlled by adjusting the channel potential difference between the first two gates.

Figure 4 also shows a common output circuit. The output diode is reset at the start of each cycle to the reset level. During the second part of the cycle, charge from the charge-coupled device register is dumped on the floating diode, causing its voltage to change. The diode is connected to a gate which can be used in an amplifier configuration. The example in Fig. 4 shows a source follower out-

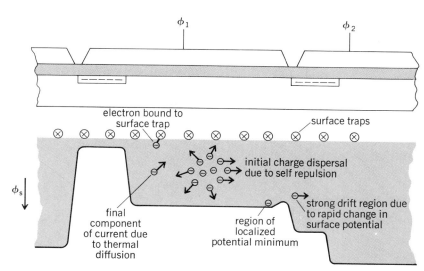

Fig. 3. Mechanisms determining charge-transfer rate. The three terms contributing to charge transfer include self-induced drift, channel drift fields, and thermal diffusion. Charge is left behind due to electron traps and localized potential minima.

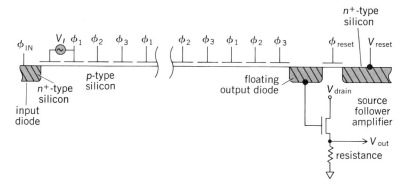

Fig. 4. Charge-coupled register showing analog input and output ports.

put. For digital applications, the output goes to a regenerative circuit with sufficient gain to produce full-level signals at the output. *See* COMPUTER STORAGE TECHNOLOGY; INTEGRATED CIRCUITS; SEMICONDUCTOR.

[MARK R. GUIDRY]

Choke (electricity)

An inductor used in a low-pass filter in which the useful output is a direct current, as in the filter of a power rectifier. At usual power frequencies a filter choke has an iron core, with an air gap to minimize variation of inductance with direct current. In a choke-input filter (a filter in which the first element is an inductance) some economy is sometimes afforded by designing the choke so that its inductance is allowed to reduce because of saturation at high currents. Such a choke is called a swinging choke. *See* ELECTRONIC POWER SUPPLY.

[WILBUR R. LE PAGE]

Chopper

Usually an electromechanical component that synchronously switches a signal circuit. By extension, the term chopper is also applied to photoelectric, electronic, and transistor circuits adapted to perform the same function as a mechanical chopper. An older related device is an optical shutter arranged to sequentially pass and interrupt, or chop, a light beam.

Basically a chopper serves as a suppressed carrier square-wave modulator. Such a component is shown schematically in the diagram. An electromagnet driven by a source of alternating current (typically 60 or 400 Hz, although higher

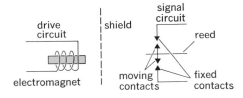

Basic chopper structure.

frequencies may be used) sets a reed into vibration. The reed carries a moving contact that alternately contacts one and the other of two fixed contacts in a signal circuit. Thus the signal (typically in the order of 1 milliwatt or less at 1 volt) is periodically interrupted.

A usual chopper application is the inversion of a slowly varying signal into a rapidly varying one, the latter being the more readily amplified. An electrostatic shield and other design precautions minimize coupling between drive and signal circuits. A frame supports the parts, usually with some degree of vibration isolation.

[FRANK H. ROCKETT]

Chopping

The act of interrupting an electric current, beam of light, or beam of infrared radiation at regular intervals. This can be accomplished mechanically by rotating fan blades in the path of a beam or by placing a vibrating mirror in the path of the beam to deflect it away from its intended source at regular intervals.

A current can be chopped with an electromagnetic vibrator having contacts on its moving armature. A current can also be chopped electronically by passing it through a multivibrator or other switching circuit.

Chopping is generally used to change a direct-current signal into an alternating-current signal that can more readily be amplified. Control systems for guided missiles make extensive use of chopping. *See* MULTIVIBRATOR.

[JOHN MARKUS]

Circuit (electricity)

A general term referring to a system or part of a system of conducting parts and their interconnections through which an electric current is intended to flow. A circuit is made up of active and passive elements or parts and their interconnecting conducting paths. The active elements are the sources of electric energy for the circuit; they may be batteries, direct-current generators, or alternating-current generators. The passive elements are resistors, inductors, and capacitors. The electric circuit is described by a circuit diagram or map showing the active and passive elements and their connecting conducting paths.

Devices with an individual physical identity such as amplifiers, transistors, loudspeakers, and generators, are often represented by equivalent circuits for purposes of analysis. These equivalent circuits are made up of the basic passive and active elements listed above.

Electric circuits are used to transmit power as in high-voltage power lines and transformers or in low-voltage distribution circuits in factories and homes; to convert energy from or to its electrical form as in motors, generators, microphones, loudspeakers, and lamps; to communicate information as in telephone, telegraph, radio, and television systems; to process and store data and make logical decisions as in computers; and to form systems for automatic control of equipment.

Electric circuit theory. This includes the study of all aspects of electric circuits, including analysis, design, and application. In electric circuit theory the fundamental quantities are the potential differences (voltages) in volts between various points, the electric currents in amperes flowing in the several paths, and the parameters in ohms or mhos which describe the passive elements. Other important circuit quantities such as power, energy, and time constants may be calculated from the fundamental variables. For a discussion of these parameters *see* ADMITTANCE; CONDUCTANCE; ELECTRICAL IMPEDANCE; ELECTRICAL RESISTANCE; REACTANCE; SUSCEPTANCE.

Electric circuit theory is an extensive subject and is often divided into special topics. Division into topics may be made on the basis of how the voltages and currents in the circuit vary with time; examples are direct-current, alternating-current, nonsinusoidal, digital, and transient circuit theory. Another method of classifying circuits is by the arrangement or configuration of the electric current paths; examples are series circuits, parallel circuits, series-parallel circuits, networks, coupled

CIRCUIT (ELECTRICITY)

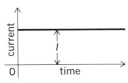

Fig. 1.   Direct current.

circuits, open circuits, and short circuits. Circuit theory can also be divided into special topics according to the physical devices forming the circuit, or the application and use of the circuit. Examples are power, communication, electronic, solid-state, integrated, computer, and control circuits. *See* CIRCUIT (ELECTRONICS); NEGATIVE-RESISTANCE CIRCUITS.

Direct-current circuits. In dc circuits the voltages and currents are constant in magnitude and do not vary with time (Fig. 1). Sources of direct current are batteries, dc generators, and rectifiers. Resistors are the principal passive element. For a discussion of direct-current circuits *see* DIRECT-CURRENT CIRCUIT THEORY.

Magnetic circuits. Magnetic circuits are similar to electric circuits in their analysis and are often included in the general topic of circuit theory. Magnetic circuits are used in electromagnets, relays, magnetic brakes and clutches, computer memory devices, and many other devices.

Alternating-current circuits. In ac circuits the voltage and current periodically reverse direction with time. The time for one complete variation is known as the period. The number of periods in 1 sec is the frequency in cycles per second. A cycle per second has recently been named a hertz (in honor of Heinrich Rudolf Hertz's work on electromagnetic waves).

Most often the term ac circuit refers to sinusoidal variations. For example, the alternating current in Fig. 2 may be expressed by $i = I_m \sin \omega t$. Sinusoidal sources are ac generators and various types of electronic and solid-state oscillators; passive circuit elements include inductors and capacitors as well as resistors. The analysis of ac circuits requires a study of the phase relations between voltages and currents as well as their magnitudes. Complex numbers are often used for this purpose. For a detailed discussion *see* ALTERNATING-CURRENT CIRCUIT THEORY.

Nonsinusoidal waveforms. These voltage and current variations vary with time but not sinusoidally (Fig. 3). Such nonsinusoidal variations are usually caused by nonlinear devices, such as saturated magnetic circuits, electron tubes, and transistors. Circuits with nonsinusoidal waveforms are analyzed by breaking the waveform into a series of sinusoidal waves of different frequencies known as a Fourier series. Each frequency component is analyzed by ac circuit techniques. Results are combined by the principle of superposition to give the total response. *See* NONSINUSOIDAL WAVEFORM.

Electric transients. Transient voltage and current variations last for a short length of time and do not repeat continuously (Fig. 4). Transients occur when a change is made in the circuit, such as opening or closing a switch, or when a change is made in one of the sources or elements. For a discussion of dc and ac transients *see* ELECTRIC TRANSIENT.

Series circuits. In a series circuit all the components or elements are connected end to end and carry the same current, as shown in Fig. 5. *See* SERIES CIRCUIT.

Parallel circuits. Parallel circuits are connected so that each component of the circuit has the same potential difference (voltage) across its terminals, as shown in Fig. 6. *See* PARALLEL CIRCUIT.

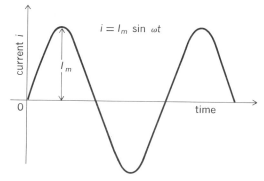

Fig. 2. Alternating current.

Series-parallel circuits. In a series-parallel circuit some of the components or elements are connected in parallel, and one or more of these parallel combinations are in series with other components of the circuit, as shown in Fig. 7.

Electric network. This is another term for electric circuit, but it is often reserved for the electric circuit that is more complicated than a simple series or parallel combination. A three-mesh electric network is shown in Fig. 8. *See* NETWORK THEORY.

Coupled circuits. A circuit is said to be coupled if two or more parts are related to each other through some common element. The coupling may be by means of a conducting path of resistors or capacitors or by a common magnetic linkage (inductive coupling), as shown in Fig. 9. *See* COUPLED CIRCUITS.

CIRCUIT (ELECTRICITY)

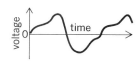

Fig. 3. Nonsinusoidal voltage wave.

CIRCUIT (ELECTRICITY)

Fig. 4. Transient electric current.

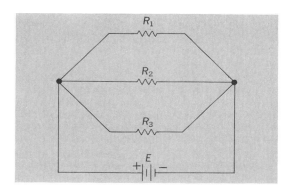

Fig. 5. Series circuit.

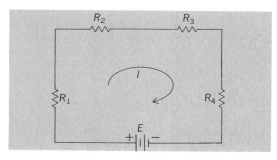

Fig. 6. Parallel circuit.

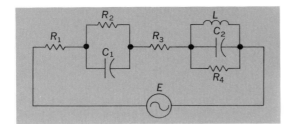

Fig. 7. Series-parallel circuit.

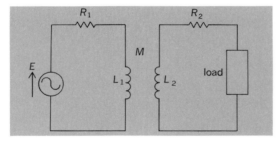

Fig. 8. A three-mesh electric network.

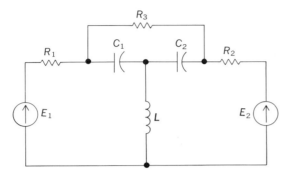

Fig. 9. Inductively coupled circuit.

Open circuit. An open circuit is a condition in an electric circuit in which there is no path for current flow between two points that are normally connected. *See* OPEN CIRCUIT.

Short circuit. This term applies to the existence of a zero-impedance path between two points of an electric circuit.

Integrated circuit. The integrated circuit is a recent development in which the entire circuit is contained in a single piece of semiconductor material. Sometimes the term is also applied to circuits made up of deposited thin films on an insulating substrate. *See* INTEGRATED CIRCUITS.

[CLARENCE F. GOODHEART]

Bibliography: R. Boylestad, *Introductory Circuit Analysis,* 3d ed., 1977; E. Brenner and M. Javid, *Analysis of Electric Circuits,* 2d ed., 1967; P. Chirlian, *Electronic Circuits: Physical Principles, Analysis, and Design,* 1971; A. E. Fitzgerald et al., *Basic Electrical Engineering,* 4th ed., 1975; W. Hayt, Jr., and J. E. Kemmerly, *Engineering Circuit Analysis,* 3d ed., 1978; W. W. Lewis and C. F. Goodheart, *Basic Electric Circuit Theory,* 1957; R. E. Scott, *Elements of Linear Circuits,* 1965; R. Smith, *Circuits, Devices, and Systems,* 1966.

Circuit (electronics)

An electric circuit in which the equilibrium of electrons in some of the components is upset by a means other than an applied voltage. The means by which electron equilibrium is upset leads to further designation of electronic circuits. In circuits with thermionic tubes, electron equilibrium is upset by thermal emission. In solid-state circuits, electron equilibrium is upset by intentionally introduced irregularities of crystals.

Electronic circuits are also classified by their functions. *See* AMPLIFIER; DETECTOR; DIGITAL COUNTER; ELECTRONIC POWER SUPPLY; MODULATOR; OSCILLATOR; SWEEP GENERATOR; SWITCHING CIRCUIT; WAVE-SHAPING CIRCUITS.

Electronic circuits can consist of discrete elements exclusively, coupled integrated circuits exclusively, or combinations of both kinds (Fig. 1). Integrated circuits can consist of groups of lumped components such as resistors, capacitors, and active devices, or of exotic combinations of passive and active components intended to have properties not normally available in lumped discrete-element networks. *See* INTEGRATED CIRCUITS.

Electronic circuits find application in all branches of industry and in the home, both for entertainment equipment and increasingly for control. Because of their low power dissipation and fast response, they are excellent control circuits. Computers, communication systems, and navigation systems use many types of electronic circuits. *See* COMPUTER; ELECTRICAL COMMUNICATIONS.

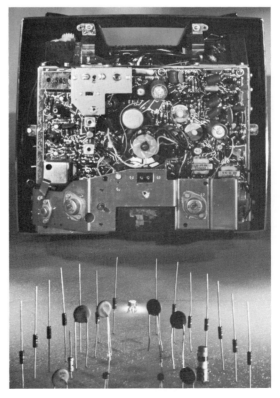

Fig. 1. Television receiver circuit is a combination of technologies. Printed wiring connects discrete components and integrated circuits. A single integrated circuit, center foreground, replaces all discrete components at either side. (*Radio Corporation of America*)

Electronic components. For an electric circuit to be classed as an electronic circuit, some of its components must be able to switch currents or voltages without mechanical switches, or to amplify or control voltages or currents without mechanical command or other nonelectrical command. Such devices as electron tubes, transistors, magnetic amplifiers, diodes, quartz crystals, resistors, capacitors, inductors, transformers, and ferrite devices are electronic components; tube sockets, interconnecting wires, and chassis, for example, are not, although they form a part of the electronic circuit. For a discussion of typical components *see* CAPACITOR: ELECTRON TUBE: FERRITE DEVICES: INDUCTOR: PIEZOELECTRIC CRYSTAL: RESISTOR: SEMICONDUCTOR: SEMICONDUCTOR DIODE: TRANSFORMER: TRANSISTOR.

Typically, electronic components have the capability of dissipating, controlling, or storing electric energy, or some combination of these functions. In addition, the current and voltage applied to the component may bear a relatively linear relation to each other (as in a resistor), a nonlinear relation as a function of current and voltage (as in a diode, biopolar transistor, field-effect transistor, electron tube, or saturable reactor), a relation that is a function of frequency (as in an inductor, capacitor, or transformer), a time-controlled switching function (as in a thyratron, ignitron, unijunction transistor, or a silicon-controlled rectifier), or a combination of the relations. Current and voltage may be inversely related. *See* NEGATIVE-RESISTANCE CIRCUITS.

Although a component may be intended to introduce only one or two of the voltage-current relations or characteristics into a given circuit, certain undesired or parasitic characteristics are also introduced. Because of the wide range of power levels at which electronic components must function, from less than 10^{-25} watt to 10^7 watts and higher, and the wide range of frequencies, from direct current to 10^{11} hertz (Hz), the parasitic characteristics are of great importance in circuit behavior. Noise generated in the components is important in low-power circuits. *See* ELECTRICAL NOISE.

Some classes of electronic components, such as those for military, marine, or airborne service, are packaged to protect them from environmental conditions, whereas ordinary components are given little more than electrically insulating mechanical protection. Specially packaged components that are used in equipment required to give reliable operation must also be used conservatively, that is, derated in accordance with environmental conditions to fractions of their rated values of voltage, current, and power dissipation. Determination of deratings is a function of circuit design, and is discussed later in this article.

Complex electronic systems often use assemblies of components interconnected as subassemblies. The most elaborate subassemblies, called integrated circuits, are collections of active and passive components formed by deposition, diffusion, or epitaxial growth on a common poorly conducting or insulating substrate (Fig. 2). They have many practical advantages compared with discrete-component circuits (including those of the welded cord-wood variety), and several disadvantages. Probably the most important advantage is the extremely small segment, or chip, required to

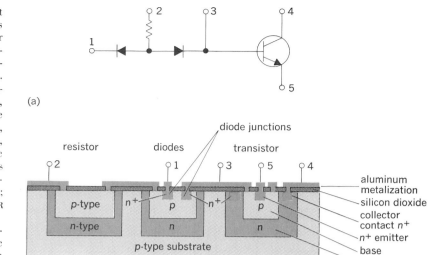

(a)

(b)

Fig. 2.   Integrated circuit combines active and passive components on a single chip. (*a*) Schematic of discrete circuit. (*b*) Cross section of integrated equivalent circuit, with corresponding terminals numbered. (*From J. Millman and C. C. Halkias, Electronic Devices and Circuits, McGraw-Hill, 1967*)

support a relatively complex circuit. For example, an operational amplifier having a gain in the tens of thousands or more may often be housed in a single transistor can whose volume is on the order of 0.1 in.[3]. Previous equivalent tube-type designs typically required more than 10 in.[3] of space.

Signal transmission times among integrated circuit (IC) components can be less by as much as two or more orders of magnitude compared with those of corresponding tube circuits because lead lengths can be reduced from many inches to as little as a few thousandths of an inch in a typical integrated circuit. These short lead lengths make possible circuit analysis based on lumped parameters both for operation at higher frequencies and for faster switching rates. As a consequence, a computer with the computation capability of the original electronic digital computer, the ENIAC, but based on integrated circuits, would require one or two racks of equipment instead of approximately 50 racks for the ENIAC, and it would have at least 100 times the speed of the ENIAC (Fig. 3). In addition, power requirements would be in the hundreds of watts, instead of in the kilowatts.

Electronic circuit design. The design of electronic circuits is the process of establishing appropriate electrical and environmental conditions in a network to achieve a required reaction to a given stimulus. In the synthesis of design, both the form of the network as a whole and the kinds and sizes of the parts must be selected to assure proper operation on demand from the stimulus.

Design of an electronic circuit is also a process of compromise in that it is necessary to establish a steady-state electrical environment in which the required small-signal behavior can be achieved. Even reliable operation of a switching circuit depends, in the final analysis, on small-signal characteristics for its action in transferring between its rest states. In particular, switching between states occurs only if the instantaneous gain of the circuit as a whole exceeds unity. As a result, efficient circuit design depends on the engineering quality of

Fig. 3. ENIAC, first large-scale electronic digital computer, contained 18,000 vacuum tubes. Photograph shows some of the computer racks with their patch panels for programming. (*Ballistic Research Laboratories*)

specifications developed for the component characteristics, both nominal values and tolerances, and the ingenuity of the designer in evolving the circuit configuration.

In addition to electrical factors, such other environmental factors as mechanical and thermal must all be considered. Importance of these other factors is a function of (1) use of the circuit, (2) complexity of the system in which it operates, and (3) reliability of operation that is required.

Even simple circuits may use components having relatively wide electrical tolerances. For this reason it is important that the designer determine the relative importance of each design factor involved and that he know which factors can be specified most precisely and used most effectively in the design procedure. Some design procedures evolve through experience, but other procedures can be developed only after intensive study of both the physical fundamentals of specific devices and the practical requirements for their reliable operation.

Component specifications. It is a design tautology that an electronic circuit can be only as good as the information, electrical and otherwise, on which it is based. For example, a designer's understanding of the nonlinear characteristics of devices, particularly active nonlinear devices, is usually gained through extensive applicational experience. Because such experience can be ap-

plied only in a limited way to other devices, development of reliable electronic circuits has been restricted, in great degree, to linear operation. Even the best circuit engineers who have extremely able circuit technicians working with them usually use active devices with which they are well acquainted.

More fundamental description of active devices can aid circuit designers in carrying over experience from one device to another. The description can be based on more specific identification of a device in terms of precise physical properties or can be based on more inclusive classification of device production characteristics by statistical methods.

Careful analysis of properties of known active devices in the circuit configurations ordinarily encountered shows that there is a set of device specifications that can lead to efficient design and useful reliability. With these specifications, the most important parameters of devices can typically be specified accurately in terms of a prime device variable, such as a current that is directly controlled by the designer. For electron tubes, bipolar transistors, and field-effect transistors, the most important parameter is the Fermi constant Λ, which has the value of 39 mhos per ampere at room temperature. The reciprocal of 39 mhos per ampere is 26 millivolts or 26 ohm-milliamperes.

Traditional treatments of parameter relations

and their applications to circuit design are based on the form of the specifications used by manufacturers to present data on resistors, capacitors, inductors, transistors, tubes, and similar devices. For entertainment applications, this convention, although not ideal, is adequate. With more stringent requirements placed on circuits for computers, controllers, and communication systems, designs based on these data have not always proved to be fully adequate. Two solutions have been found for this dilemma, one based on the choice of parameters that measure the primary properties of the device in question in terms of physical coefficients, and the other on use of statistical techniques for circuit design. The latter procedure helps to minimize the difficulties resulting from wide tolerances because the use of statistics is effective when the theoretically selected statistical distribution is reasonably similar to the distribution encountered with practical devices. The statistical procedure has become practical with the advent of large-scale high-speed computers.

Examination of parameter data on active devices in particular shows that the tolerances on some parameters must be wide, whereas other parameters can be more precisely controlled. Statistical procedures are required when design is based on such wide-tolerance derived parameters as transistor current gain, or β, which is the ratio of one current to the difference between two small changes in current. Fundamental parameters, such as the Fermi constant, $\Lambda = q/kT$, where q is electron charge, k is Boltzmann's constant, and T is absolute temperature, are truly independent of the methods by which they are measured. Use of fundamental parameters becomes increasingly important as integrated circuits encompass more elements.

Development and use of integrated circuit assemblies as fundamental building blocks simplify and at the same time complicate the design of electronic systems. Design is simplified insofar as a single mass-produced integrated circuit assembly can often replace a complex circuit constructed from discrete components. Design is complicated by the fact that this mass-produced integrated circuit consists of a complex of nonlinear elements which, although deposited in a specific configuration, differ from unit to unit. As a consequence, the transfer function, measuring the ratio of the circuit response to its input signal, differs for each unit from the required transfer function.

These differences limit the designer when he models or simulates a circuit during its development. Applicability of the model depends on two chief criteria: (1) conditions of operation in the model need to duplicate those to which published specifications apply; and (2) there is a unique and close correspondence between observable behavior in the model and published specifications. Implicit in these criteria is the need that published specifications for components intended to be used in circuits meaningfully describe characteristics and tolerances which are fundamental to circuit operation.

From the above, it follows that specifications placed on components, subassemblies such as integrated circuits, and circuit assemblies are important in the design of electronic circuits.

Preparation and selection of specifications strongly influence use of a component or a subassembly; reliability and maintenance of the finished circuit, both of which contribute to its long-term costs, also depend on the adequacy of component specifications.

Circuit specifications. Normally, characteristics of an existing or a required circuit are expressed as a performance specification. This specification states, often in considerable detail, performance limitations, leaving the choice of structural and circuit details that will meet these limits to the circuit designer. Such a specification typically includes, for an amplifier, bandwidth between half-power (3-dB) points (the lower and upper frequency limits at which power gain is half of the nominal value), overall power gain (typically given in decibels), power output (in watts), and permitted distortion or intermixing limit (expressed in a form such as harmonic distortion or intermodulation distortion). Other electrical characteristics important for the application at hand will be part of the specification. In addition, because electronic circuits are expected to perform satisfactorily over a wide range of environmental temperatures, the specifications include the permissible change in characteristics of the circuit over the temperature range.

Mechanical, chemical, atmospheric, and possibly other design requirements must also be met. Circuits may be required to withstand sinusoidal vibration and noise excitation, to undergo accelerations, and sometimes to sustain impulse loading without failing. Assemblies must withstand the chemical action of their own encapsulants and also the action of corrosive atmospheres at alternately elevated and depressed temperatures. A series of tests exposing the circuit to these environmental conditions is often specified to establish minimum requirements for acceptance. These tests reflect special environmental applications such as marine, military, aircraft, or space use. Such qualification tests are designed to measure compliance of production circuits to performance limits under electrical or environmental conditions that experience has shown may lead to difficulty or failure. Because new problems appear as rapidly as new techniques, such tests can only represent a starting point for evaluation of reliability.

Reliability considerations. An electronic circuit may or may not function as intended during its expected life. The probability that it will function as intended is its reliability. Reliability of a circuit depends, in turn, on the reliability of each element or component as operated in the circuit in question. Consequently, reliable circuits are made from reliable components used under conditions that provide an additional margin of safety beyond the safe limits specified by the manufacturer.

Mechanical reliability of an electronic circuit depends largely on the number of interconnecting wires, the mechanical support provided to components, and the manner in which connections are made and connectors used. Mechanical design of parts, particularly connectors and relays, can also be important. Low reliability can be expected on interconnection wiring, particularly when soldering is involved. If such wiring is replaced with welded connections or, better still, with vacuum-deposited wiring, significantly improved operation generally can be achieved. With large-scale integration, external interconnections may be as few as 2% of those in a circuit wired between discrete components. The potential improvement in reliability is self-evident.

Disadvantages of integrated circuits are their low voltage and power ratings and their highly nonlinear parasitic leakage paths, which can exist in the semiconductor material. Typical integrated circuits should be operated at supply voltages less than 12–20 volts. Their power dissipation capability depends on the amount of heat-conducting and -radiating material in the integrated circuit and the paths by which the heat developed on the chip can be conducted away or radiated. Unfortunately, a design which achieves low thermal resistance (high capability for conducting heat away from the chip) also tends to have high parasitic capacitances.

The nature of integrated circuits is such that parasitic diodes cannot be avoided in their structures, although their effects can be minimized, particularly on thin-film integrated circuits. Parasitic coupling effects often can be minimized by proper use of inverting barriers. *See* INTEGRATED CIRCUITS.

Component ratings. Circuit components are rated to indicate the conditions below which they can probably provide a specified kind of service without failure and above which they cannot function without excessive chance of failure. In reality, probability of failure is usually a continuous function of operating conditions. When a circuit must be designed to achieve a specified reliability and must simultaneously meet a conflicting requirement, a rating curve is preferable to the step-function rating inherent in tabulated values. An example of conflicting requirements is the need for high reliability in airborne circuits in combination with small size and low inertia.

Ratings are normally a function of operating environment. For example, a pair of resistors may be able to dissipate 5 watts of power each in a given environment; yet both might burn out under the same environment if they were mounted adjacent to one another. Common ratings are power dissipation, peak voltage or peak inverse voltage, peak pulse current or maximum average current, and insulation resistance or dielectric strength. Power dissipation is a sensitive function of such factors as heat radiation, convection, and conduction; it should be specified in terms of these factors.

The maximum voltage a component can withstand normally is a function of the temperature of the component, and may be a function of humidity, ambient air pressure, and similar factors. It may also be a function of the method of use. For example, a capacitor used in electronic flash unit can be intermittently charged to a voltage significantly higher than the safe voltage at which it can be steadily operated as a filter component in a high-current power supply.

Environmental factors. Environmental conditions under which any electronic circuit operates must be considered. Already mentioned is the importance of temperature; it may range between −45 and +65°C ambient (−50 and +150°F). Equipment may encounter environmental conditions of up to 100% humidity with salt air conditions, air with high dust content, and atmospheric pressures from below sea level to interplanetary space.

Steps commonly taken to protect equipment against such adverse environments include selection of specially designed components or subassemblies, use of fungus-resistant insulating varnishes, drying and filtration of air, circulation of cooling air, and maintenance of a positive air pressure differential between the container housing the circuits and the external environment. Protection of circuits for field use requires specialized engineering experience.

Stabilization of circuits. The stability of a circuit is a measure of the ranges of environmental and operating conditions over which performance remains within specified tolerances. Because characteristics of electronic circuits drift to a greater or lesser extent, they can change with both environment and with time. Control of these changes is simplified by designing circuits in terms of parameters having the least significant stability problems and in ways that minimize the effects of the remaining parameters.

Typical circuits using electron tubes or transistors are often stabilized by control of the steady-state current flowing in the output circuit because forward admittances can generally be specified with considerable precision in terms of instantaneous output current. When necessary, feedback or degeneration, achieved by introduction of output signal into the input of an amplifier, helps stabilize amplifier characteristics. Preaging, whereby a circuit operates in a prescribed environment for a predetermined time prior to testing or use, may cause it to complete most of its drifting so that, thereafter, operation remains adequately stable.

Electronic circuit tests. Tests normally made on electronic circuits may be classified into (1) engineering design (worst case) tests, (2) production limit tests, (3) functional tests, and (4) failure (or repair) tests. Engineering design tests are performed to verify that the engineering design results in the required performance with production-tolerance components and that it will meet specifications under all conditions likely to be encountered in use. Production limit tests verify that tolerance limits established in design are met by production equipment. Limits may not be met because of changes in components resulting from procurement difficulties. Functional tests serve to assure that the unit continues to function within tolerances in the field-use environment. Failure tests are used to locate failure and diagnose its cause. The number, kind, and severity of the tests applied depend on the nature and intended function of the equipment.

Electronic equipment intended for home use (radios, televisions, hi-fi equipment, tape recorders, intercoms, and so on) being sold in a highly competitive market may only be tested to ensure initial operation in such an environment. Equipment intended for a radio or a television station or for use as part of a computer or an industrial control system will be tested much more thoroughly, both for the relatively moderate environmental conditions it normally must survive and for stability during expected service life. Field equipment, particularly for military use, and other equipment on whose continued operation many lives may depend, are given extremely rigorous performance and environmental tests.

Engineering design tests. After a trial design has been established, a laboratory model is constructed and tested to measure characteristics of the circuit. During these tests, values of components are varied, supply voltages are varied, and a variety of active devices or integrated circuits representing the full expected range of acceptable units is substituted within the model. The extent of these tests depends on the application and the number of units to be made and on the innovations in earlier equipment of similar nature.

Tests on the engineering model are not necessarily conclusive, however. For example, a circuit may function as intended in the presence of parasitic conductances and capacitances existing in the model; yet it may not function in the prototype because these parasitic conductances and capacitances are significantly different from those in the engineering design. As a consequence, a series of engineering design tests is made on the final prototype model to determine whether production problems are likely to occur. Any significant change in test limits between engineering and prototype models, unless explainable in terms of known changes in the design or structure, may be cause for a further study. The prototype will be given the full spectrum of tests; for military equip-

ment this includes temperature, cycling, humidity, salt spray, fungus, vibration, and mechanical shock, as required. Life tests to determine the mean-time-before-failure (MTBF) can also be applied, either normally or under overstress conditions as indicated in the performance specifications.

Production line tests. The more complex the circuit, the more extensive will be the testing, more in a geometric progression than in an arithmetic one. For simple circuits, direct comparison between a standard circuit and each production circuit may be sufficient. Such tests have the advantage that they can be conducted by self-stepping test equipment, the test equipment can be adapted to a new version of the circuit simply by replacing the standard circuit with the new one, and test data can be related directly to specific features of the circuit. For complex circuits, simply verifying the correctness of all wiring may require such automatic test procedures (Fig. 4).

When circuits are to be assembled into larger assemblies, production tests screen circuits before they become parts of the larger assemblies. For such equipment as computers or guidance systems, a hierarchy of such in-line test aids in ensuring that each subassembly has been properly manufactured before it is built into the next higher assembly.

Choice of circuit boundaries may be influenced

Fig. 4. Magnetic tape, or perforated paper tape, programs this automatic test equipment that checks out communication equipment. The computer-controlled tester prints out a description of any malfunction it detects. (*Radio Corporation of America*)

as strongly by production test requirements as by circuit functions. That is, if packaging of a multiplicity of circuits into manufacturable subassemblies divides circuits into units that can be tested independently of each other, both production testing and field failure testing can give results directly related to the units under test. As a consequence, manufacturing can be more economically controlled, and field troubleshooting and maintenance can be more quickly completed.

Production tests may also call for a small percentage of production assemblies to be put through full engineering design final tests. This procedure gives further assurance that significant unexpected changes have not occurred to cause production tests to become meaningless.

Statistical test methods. Because of the importance of failure-free operation during such activities as aircraft or space flights, accurately predictive test methods are essential. In the Department of Defense, a special set of production-limit tests has been developed for use in evaluating electronic circuits that must meet stringent requirements. These are based on statistical methods, because 100% testing is neither wise nor economical even if the tests can be nondestructive.

For statistical methods to be applicable, all circuits of a given type (the test population) must share common characteristics. Consequently, the first step in preparing for statistical tests is to sort out those factors that cannot be handled statistically and to test them separately. Tests may show that dissipation or voltage breakdown failures are occurring with one or two elements of a circuit. Such failures must then be treated separately. Any predominant cause of failure must be treated separately; only those causes of failure not occurring abnormally often can be treated statistically.

The Department of Defense, in particular, maintains a list of components meeting specific tests which it has found important in military applications. Such a list, called a qualified parts list (QPL), provides a basis for parts selection for an electronic circuit to which statistical methods could be applied. The fact that a component has a QPL designation does not necessarily assure that it will be satisfactory for a given application. The circuit designer should make certain that his method of application is compatible with all conditions for which the part was tested. In a circuit so designed, all parts can be expected to withstand the test conditions equally.

Another condition needed before statistical theory can be applied in electronics is the basis for defining acceptance. Too high an acceptance level raises the producer's risk, which in effect measures the probability that many circuits in a production lot will be satisfactory but that the test samples will fail. Too low a level increases the user's risk, which measures the probability that the user will accept some defective merchandise. These risks contribute to the determination of the acceptable quality level (AQL). Sufficient samples are then tested from a production lot to measure this level to the required confidence.

Reliability. Many specifications now contain reliability clauses, requiring the contractor to estimate the expected mean-time-between-failure and to demonstrate compliance with the estimate. A

test for circuit characteristic can be useful if data are properly used. The question is, what constitutes a failure? In field equipment designed to be operated by an untrained, or relatively untrained, individual, any adjustment required by the equipment could represent a failure, whereas if the field equipment is meant to be operated by a trained technician thoroughly versed in its operation, only an unexpected interruption of operation might represent a failure.

Functional tests. In such equipment as computers, radars, military electronic systems, and television networks, it is considered good practice to provide functional tests which measure the conformance of the equipment to the overall system specifications just prior to its use. This kind of test is particularly important in satellite and missile tracking systems and in electronics associated with expensive experiments. Typically, monitors serve this function for television stations. In addition, a variety of general-purpose test equipment is usable for function and repair tests.

Failure tests. No electronic equipment or system can be made to be failure-free. As a consequence, it is necessary to have test equipment capable of diagnosing the causes of failure so that repairs may be made. Self-contained and easily accessible equipment such as radio and television receivers can often be tested adequately with a volt-ohmmeter. This kind of testing is simplified through the use of schematics having supplementary tables of test-point voltages and resistances.

Failures of plug-in components or circuit assemblies are usually found relatively easily by substitution of a unit known to operate properly for the one suspected to be the cause of failure.

With complex industrial and military equipment, construction is usually based on replaceable subassemblies, each such subassembly having test points for diagnosis. These test points serve, first, for location of the defective subassembly within the equipment and, second, for localization of the failure to a component or integrated circuit of the subassembly.   [KEATS A. PULLEN]

Bibliography: P. M. Chirlian, *Electronic Circuits: Physical Principles, Analysis, and Design*, 1971; C. F. Coombs, Jr., *Printed Circuits Handbook*, 2d ed., 1979; C. A. Holt, *Electronic Circuits: Digital and Analog*, 1978; D. L. Schilling and C. Belove, *Electronic Circuits: Discrete and Integrated*, 2d ed., 1979.

Clamping circuit

An electronic circuit that effectively functions as a switch to connect a signal point to a fixed-reference voltage or current level, either at a specific time interval or at some prescribed amplitude level of the signal itself. The circuit is also called, simply, a clamp. Clamps are frequently used to reset the starting level of periodic waveforms, such as sweep generators, and to establish a direct-current (dc) level in a signal which may have lost its dc component because of capacitance coupling. They are also used in many frequency and voltage comparison circuits and in control circuits.

The basic elements of a clamping circuit are shown in Fig. 1. The clamp between terminals A and B is shown as a switch S in series with reference voltage V_R, having a large resistance r_R when

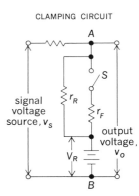

CLAMPING CIRCUIT

Fig. 1. Clamp elements.

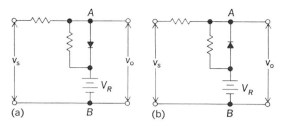

Fig. 2.   Single diode clamp circuits. (a) Voltage-amplitude-controlled clamp. (b) Unidirectional clamp.

the switch is open and a small resistance r_F when the switch is closed.

Voltage-amplitude-controlled clamp. In Fig. 2a the diode functions as a clamp whenever the potential at point A starts to rise above V_R; then the diode is in its forward-biased condition and acts as a very low resistance. Such a clamp is referred to as a voltage-amplitude-controlled, or continuous, clamp. The circuit of Fig. 2b is similar, except that the polarity is reversed. Point A is connected to V_R through a low resistance whenever the potential at point A drops below V_R. These clamps function only when a fixed polarity of signal with respect to V_R appears at point A, and are therefore referred to as unidirectional clamps.

Keyed, or synchronous, clamp. If a clamp functions at specific time intervals controllable from separate voltage or current sources rather than from the signal itself, it is a keyed, or synchronous, clamp, and the sources, usually pulses, used to actuate the clamping action are called keying, or clamping, pulses. Such a keyed clamp may be unidirectional but more often it is a bidirectional clamp, defined as a clamp that functions at the prescribed time irrespective of the polarity of the signal source at the time the keying pulses are applied.

A bidirectional clamp using four diodes with two opposite-polarity keying pulses is shown in Fig. 3. The peak of the negative pulse is established at a level negative with respect to V_R, causing D_2 to conduct heavily when the pulse is applied, while that of the positive-going pulse is above V_R, causing D_4 to be forward-biased and to conduct heavily. If, at the same time, point A is above V_R, D_1 conducts and point A is clamped to V_R, while if it is below V_R, D_3 conducts and point A is clamped to V_R. When point A is exactly at V_R, all diodes are forward-biased and conducting. Between clamping pulse intervals, all of the diodes are reverse-biased, and the clamp is open. A unidirectional keyed clamp results if either the left or right pair of diodes, together with their keying pulses, is eliminated.

In form, the bidirectional clamp is similar to the essential element of the diode transmission gate and the sample-and-hold circuit, and the voltage-controlled diode clamp is the main element of the limiter. *See* GATE CIRCUIT; LIMITER CIRCUIT.

DC restorer. A dc restorer is a clamp circuit used to establish a dc reference level in a signal without modifying to any important degree the waveform of the signal itself. For this function to be performed, the signal must have a self-contained reference level which repeats at periodic intervals. A typical example of such a waveform is that comprising the video signal of the television system. *See* TELEVISION.

An example of a simple waveform, assumed to be periodic at intervals of T, applied to a single-diode voltage-controlled clamp is shown in Fig. 4. Restoration of the peak of the waveform to the level of V_R depends upon current flowing through the capacitance when the input level is above V_R and the diode is forward-biased. The capacitance charges in the direction shown during such successive peaks, and the peak level is finally established at V_R. The circuit does not operate if the peak of the waveform at point A is not above the level of V_R initially. The reverse time constant $[R_1 + R_2 r_R/(R_2 + r_R)]C$ must be long compared to the period T, while the forward time constant $(R_1 + r_F)C$ must be very short, with $(R_1 + r_F)$ as small as possible. A bidirectional keyed clamp is required to function as a dc restorer on the waveform shown in Fig. 5. A bidirectional keyed clamp thus establishes the recurrent reference level regardless of the signal polarity between pulses.

Triode clamp. A three-terminal, high-impedance device such as a field effect transistor (FET) may be used as a keyed clamp to eliminate the need for balanced keying pulses. One common form of such a clamp is shown in Fig. 6a using an n-channel

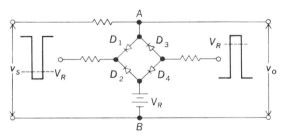

Fig. 3.   Four-diode bidirectional keyed clamp. D= diode.

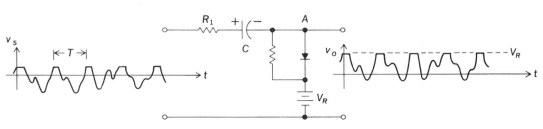

Fig. 4.   Diode clamp as dc restorer.

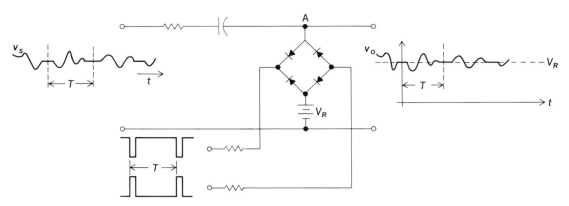

Fig. 5. Keyed clamp as dc restorer.

junction FET having the low voltage characteristics shown in Fig. 6b.

When no keying pulse is applied, diode D-1 conducts and the gate of the FET is held sufficiently negative with respect to the source that its drain current is cut off; hence, the drain-source terminals represent essentially an open circuit. When the keying pulse is applied, D-1 cuts off and sufficient gate current flows through R to cause the drain-source path through the FET to be a very low resistance, thus essentially short-circuiting point A to V_R. The junction FET is a symmetrical device

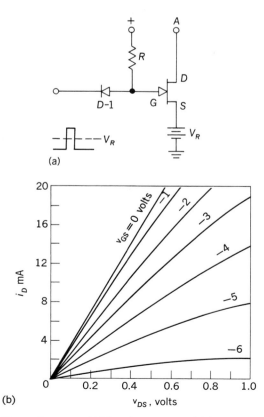

(a)

(b)

Fig. 6. Example of bidirectional clamp which uses complementary field-effect transistors. (a) Circuit. (b) The low-voltage characteristics of a low-resistance field-effect transistor of the n-channel type.

with the source and drain interchangeable; therefore the clamp will function whether A is positive or negative with respect to V_R before the clamp pulse is applied.

[GLENN M. GLASFORD]

Bibliography: G. M. Glasford, *Fundamentals of Television Engineering*, 1955; J. Millman and H. Taub, *Pulse, Digital, and Switching Waveforms*, 1965; J. Millman, *Microelectronics*, 1979.

Clipping circuit

An electronic circuit that prevents transmission of any portion of an electrical signal exceeding a prescribed amplitude. The clipping circuit operates by effectively disconnecting the transmission path for the portion of the signal to be clipped.

Diode. A diode may be used as a switch to perform the clipping action. The direct-coupled circuit of Fig. 1 represents a typical example. A time-varying signal voltage v_s is applied to the input of the network containing a diode having terminals p and n. The diode functions as a very low resistance when p is positive and as a very high resistance when p is negative. Thus, if any portion of the waveform at p lies below the common reference or ground level, that portion will be removed, or clipped, and will not appear across output load resistance R_L, as illustrated by the three separate examples of output waveforms for load voltage v_L. If bias voltage V_B is positive, the upper plot is typical; if it is zero, the upper half periods, as shown by the center waveform, will appear; and if it is negative, more than half of the peak-to-peak amplitude will be clipped, as shown by the lower waveform. The output waveforms are for a sinusoidal input as shown. In the plots of typical outputs, when $V_B \neq 0$, the dashed lines are the zero axis of this sinusoidal waveform from which the output is clipped. That is, the action of bias voltage V_B is to displace the zero axis of the signal waveform from the zero axis of the output voltage.

In the circuit shown in Fig. 1, when the diode is nonconducting, assuming infinite diode resistance, the voltage v_p at p is shown by Eq. (1), and when it is conducting, assuming zero diode resistance, it is shown by Eq. (2). The value of v_p from both of

$$v_p = \frac{R_2}{R_1 + R_2} v_s + \frac{R_1}{R_1 + R_2} V_B \qquad (1)$$

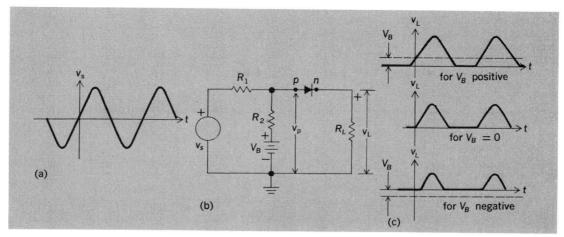

Fig. 1. An example of series diode clipping. (*a*) Waveform of time-varying input voltage. (*b*) Circuit diagram.

(*c*) Waveform of output voltage v_L when V_B is positive, $V_B = 0$, and V_B is negative.

$$v_p = \frac{R_2 v_s + R_1 V_B}{(R_1 + R_2)\left[1 + \dfrac{R_1 R_2}{R_L(R_1 + R_2)}\right]} \qquad (2)$$

these equations must be the same at the time of transition from reverse to forward bias of the diode. Thus, the value of v that, when substituted into Eqs. (1) and (2), gives equal values of v_p is the transition signal voltage $v_{s(\text{trans})}$. The value of this voltage is indicated by Eq. (3), which allows appro-

$$v_{s(\text{trans})} = -\frac{R_1}{R_2} V_B \qquad (3)$$

priate circuit constants to be selected for a desired clipping level. The nonclipped waveform is attenuated by the divider network, which for a perfect diode is given by $v_L = v_p$ during signal transmission, in which case Eq. (2) applies for v_L. Clipping of the opposite polarity can be obtained by reversing the connections to the diode and the polarity of the bias voltage V_B.

Bidirectional clipping can be achieved by connecting two diodes as in Fig. 2. The circuit contains two bias voltages and a resistance network. Bias voltage V_B is positive relative to point b, and bias voltage V_A is negative relative to point b. In a region below a level determined by these voltages, network R_1, R_2, and R_4, and diode D_1, input voltage v_s will produce no change in output voltage v_L; thus these parts of the circuit set the lower clipping level. For signals v_s above this level, both diodes can conduct up to a level at which the voltage at a rises

above a value determined primarily by V_B, V_A, R_3, and R_L. When v_s is above that value, D_2 becomes nonconductive. This action sets the upper clipping level. By proper choice of circuit values, clipping can be made approximately symmetrical about the axis, as shown by the output waveform v_L. Such a bidirectional clipper is sometimes referred to as a slicer, since it allows transmission of only a center section of a waveform. *See* DIODE.

Triode clipper. Clipping action may be achieved by using bipolar or field-effect transistors. To do so the bias at the input is set at such a level that output current cannot flow during a portion of the amplitude excursion of the input voltage or current waveform. A particular example, using a *p*-channel insulated-gate field-effect transistor, is shown in Fig. 3. Advantages of this particular device are that it operates normally with the gate forward-biased and that, with a single dc voltage supply in the source lead as shown, both the input and output voltages can be referenced to a common reference or ground potential. As indicated by load line R_L on the set of characteristic curves, the gate is set at a dc voltage determined by R_1, R_2, and V_{SS}, so that at the output the axis of the signal voltage waveform is near drain-current cutoff. Operation is along the load line determined by R_L. Excursions of the output waveform with respect to V_{DS} as a reference, plotted directly below the set of characteristic curves, correspond to the actual output voltage waveform with respect to the common reference seen plotted to the right of the circuit.

This particular circuit can also function as a lim-

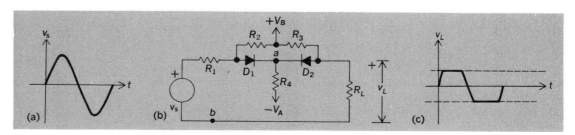

Fig. 2. Two-diode clipping. (*a*) Waveform of time-varying signal. (*b*) Circuit diagram. (*c*) Clipped waveform.

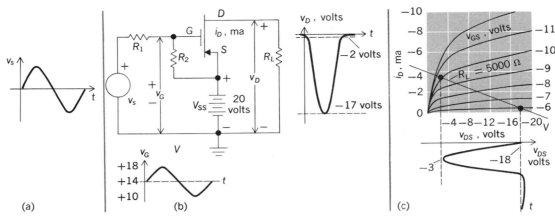

Fig. 3. Triode clipper using insulated gate transistor. (a) Input waveform v_s. (b) Circuit diagram with waveform of voltage v_G from gate G to ground, and one of voltage v_D across output load resistance R_L. (c) Relation of output waveform to transistor characteristic curves as determined by load line for an R_L of 5000 ohms in circuit.

iter if the input amplitude is increased, or if the input bias point is shifted such that the output-circuit saturation level is exceeded at very low values of V_{DS}. Likewise, clipping can be achieved using bipolar transistors by shifting their bias levels from the limiting conditions to values such that they have sufficient reverse bias to prevent output current from flowing for a portion of the amplitude excursion of the input voltage or current waveform. Slicing, as described above for the diodes, can be accomplished by a combination of limiting and clipping actions if the bias is adjusted properly and the input amplitude is sufficiently large. See LIMITER CIRCUIT; TRANSISTOR.

[GLENN M. GLASFORD]

Bibliography: G. M. Glasford, *Fundamentals of Television Engineering*, 1955; J. Millman and H. Taub, *Pulse, Digital, and Switching Waveforms*, 1965; J. Millman, *Microelectronics*, 1979.

Closed-circuit television

A video communication system in which the signal is transmitted from the point of origin only to those specific receivers that have access to it by previous arrangement. Closed-circuit television enables visual information to be transmitted, often in parallel with audio information, to a selected audience or between selected groups of people. This private nature of closed-circuit television distinguishes it from broadcast television, in which all receivers in a given area have access to the broadcast signals. See TELEVISION.

Applications. The number of closed-circuit television installations is counted in the hundreds of thousands and far exceeds the number of television broadcast stations. These installations employ a wide variety of system arrangements tailored to the even wider variety of uses to which closed-circuit television is put. This article describes some of these applications.

Monitoring. Closed-circuit television is used for a large number of different types of monitoring applications in which the remote viewing of an ongoing process or activity at a location provides economic, safety, or educational advantages. Examples include the monitoring of hazardous situations such as the input to a rock crusher, the interior of a furnace, or the operation of a process

in a hostile environment. Closed-circuit television is used to monitor a number of remote locations from a centralized point, for example, to observe traffic flow in a city, for control; to observe critical locations in a bank or plant, for security reasons; or to read critical gages and instruments, for process control. It can be used to monitor microscopic or telescopic images, for greater ease of viewing; for example, closed-circuit television is used in hospital operating rooms so that new surgical techniques can be demonstrated to a large group of people. Often, a video tape recording is an integral part of the monitoring system, permitting later viewing or analysis.

Educational applications. A growing number of educational institutions and corporations are using closed-circuit television systems. A professor can lecture from a centralized classroom to students in a number of remote classrooms, or a corporate officer can make a presentation from company headquarters to employees in a number of remote office locations. Most of these applications consist of one-way television with two-way audio conference facilities to permit audience interaction. The development of economical video tape cassettes and associated recording, editing, and playback mechanisms has had a significant impact on the growth of closed-circuit television systems for educational purposes. The most rapid growth has been in connection with corporate training and employee information applications. Many businesses have found the use of video tape cassettes, coupled with internal closed-circuit television systems or a self-contained playback unit with a TV monitor, to be an economical, efficient way to communicate corporate policy, introduce a new product or sales program, and even conduct specialized training courses. In many of these applications, the video-taped program material is augmented by an individual who either broadcasts over the internal closed-circuit television system, as described earlier, or is actually present in the remote location. This person answers questions and makes possible audience interaction.

Theater television. Because of nationwide and sometimes worldwide interest in certain sporting or entertainment events, one-way closed-circuit television systems are being used to permit a number of geographically dispersed audiences to view the event simultaneously. In this application, the original television signal is transmitted and distributed to theaters which are equipped with large-screen television projection systems for displaying the received image to the paying audience.

Video conferencing. Two-way closed-circuit television systems are being used to permit groups of people in geographically separated locations to interact on both a visual and audio basis. With these systems, conference rooms or studios with specially equipped and controlled television cameras and monitors are used so that people can see and hear one another and exchange graphical information. Most of the conference rooms presently in operation can comfortably accommodate four to nine active participants per location, with provision for additional people in the background. The systems are being used by a number of corporations, government organizations, and educational institutions as a substitute for, or supplement to, travel between the locations.

Video telephone. Although usually intended to be used as a visual adjunct to telephony, video telephone equipment could serve many of the roles normally associated with closed-circuit television. However, video telephone systems presently serve only very limited areas on an experimental basis. *See* VIDEO TELEPHONE.

Cable television. Community antenna television (CATV) or cable television systems are normally considered a type of closed-circuit television. As originally envisioned and implemented, these systems employ standard broadcast television signals which are received by a suitably located antenna and amplified and ditributed through coaxial cables. This type of system can be used to provide reliable television reception in a community located beyond the useful range of broadcast television transmitters, or it can be used in an apartment building to avoid having each tenant install a separate antenna.

There are more than 3000 operating CATV systems franchised to provide service within defined geographic areas. Decisions by the Federal Communications Commission (FCC) have strongly influenced the development of these systems. As a result of FCC requirements, new systems must provide a capability of 20–40 television channels and two-way capability. Those cable television systems serving more than 3500 subscribers must be capable of television program origination and must devote some of this capacity to programming of community interest. Many of the largest cable television operators are offering pay TV service, whereby subscribers have an adjunct to their home television set which permits them to receive special programs (first-run movies, sporting and cultural events) for an additional charge to their monthly service fee. Also, there is extensive ongoing experimentation to determine the feasibility of providing subscribers with customized services, including information retrieval, shop-at-home services, fire and burglar alarm monitoring, and other services which the large bandwidth and two-way nature of the new systems make possible.

Technical considerations. The range of signal formats and transmission means used in closed-circuit television is great. The signal formats are chosen to suit the particular use, and extend from high-resolution closed-circuit television applications, requiring bandwidths of 5–15 MHz, down to slow-scan television signals which require bandwidths as small as 2.5 kHz. The means of transmission range from broadband radio circuits to standard telephone message circuits.

The amount of visual information transmitted by a closed-circuit television system varies greatly, from color motion pictures for entertainment to monochrome still pictures for conveying information such as bank balances and signatures.

Pictures of moving objects require wide bandwidths and rapid scanning to provide adequate resolution and to avoid introducing a jerky characteristic into the motion. Common theater television systems, educational television systems, and video conferencing systems use the same scanning standards employed by broadcast television, with a bandwidth of about 4.2 MHz. A good-quality still picture can be transmitted over a narrowband circuit if the required transmission time can be accommodated.

The equation relating frame time T in seconds, lines per picture height N, and bandwidth B in hertz for a picture with a 4×3 aspect ratio (width to height) is $T/0.508$ N^2/B. For a 525-line picture and a bandwidth of 4.2 MHz, $T = 1/30$ sec, which is the frame period of standard broadcast television. For the same number of lines per picture and a bandwidth reduced to 2.5 kHz, $T = 56$ sec.

Transmission facilities for systems located completely within a user's location and for distances up to 50 mi (80 km) normally use coaxial cable, shielded twisted-pair cable, or a short-haul microwave radio system. In certain cases, when conditions are favorable, laser-beam transmission systems are used. As a result of developments in the production of low-loss fiber optic cables, extensive experimentation is under way on the use of this medium for short-haul applications. Long-distance and intercity transmission are commonly provided as a service by a common carrier. Terrestrial microwave radio facilities are normally used in transmitting television signals for long distances. Increasingly, domestic and international satellite systems are used for transmission over greater distances or to reach otherwise isolated communities. Satellite systems are also being used to interconnect cable television systems for program sharing and for the distribution of signals for theater television applications. *See* COAXIAL CABLE.

Technological advances in terminal equipment have broadened the potential applications for closed-circuit television systems. Among these developments are cameras that are capable of operating under very low ambient light conditions, and color cameras that are portable and simple to use. Developments in video tape cassette recording, editing, and playback mechanisms have made the production of video program material easier and less expensive. Developments in video disk technology appear to have a great many potential applications, from mass entertainment to individual instruction in a variety of special fields.

The TV signal can be transmitted as a baseband video signal, with frequencies extending from only a few hertz to whatever bandwidth is required for the signal, or it can be modulated upon a carrier. For cable television, vestigial sideband modulation is normally used, and the signals are often placed at the same frequencies as those used for standard very-high-frequency (vhf) television broadcast channels. The latter approach permits the simultaneous transmission through a system of up to 12 different television signals, and a standard home television receiver can select the desired signal. The disadvantages of modulating the signal upon a carrier are the additional equipment required to perform the modulation, the wider bandwidth required in the transmission system, and the higher electrical attenuation, at higher frequencies, in coaxial cables. [JOSEPH J. HORZEPA]

Bibliography: R. Armstrong, *Closed-Circuit TV Installation, Maintenance, and Repair*, 1978; Electronic Technician-Dealer Magazine Editors, *A Practical Guide to MATV-CCTV Design and Service*, 1974; D. H. Hamsher (ed.), *Communication System Engineering Handbook*, 1967; W. A. Rheinfelder, *CATV Circuit Engineering*, 1975; R. Veith, *Talk-Back TV: Two-Way Cable Television*, 1976; L. A. Wortman, *Closed-Circuit Television Handbook*, 3d ed., 1974.

Coaxial cable

A two-conductor transmission line with an outer metal tube or braided shield concentric with and enclosing the center conductor. The inner conductor is supported by some form of dielectric insulation, solid, expanded plastic, or semisolid. Semisolid supports are polyethylene disks (Fig. 1), helical tapes, or helically wrapped plastic strings. Beads, supporting pins, or periodically crimped plastic tubes are used in some designs.

The significant feature of the coaxial cable is that it is a shielded structure. The electromagnetic field associated with each coaxial unit is nominally confined to the space between the inner and outer conductors. Since alternating current concentrates on the inside of the outer conductor as the frequency of the current increases (skin effect), a coaxial unit is a self-shielded transmission line whose shielding improves at higher frequencies. Unshielded lines, such as the pairs of multipair cable, share the space for the electromagnetic fields. Thus, for equivalent transmission loss, pairs occupy less space than coaxials. The major use of the coaxial cable is for transmitting high-frequency broadband signals. Coaxial cables are rarely used

at or near voice frequency since the shielding properties are poor and they are more expensive than twisted pairs having the same transmission loss.

Types and uses. Coaxial units are made in three general types for different applications: flexible, semirigid, or rigid. In general, the more rigid the unit, the more predictable and stable its electrical properties. Since loss on a transmission line increases approximately as the square of the frequency, low loss in a coaxial cable is important. Also, physical and electrical irregularities, especially if they are periodic, must be kept to a minimum, or the transmission loss of the coaxial will be very high at certain frequencies.

Flexible coaxials. Flexible coaxials have a braided outer conductor, solid or expanded plastic dielectric, and a stranded or solid inner conductor. The electrical properties of flexible coaxials, particularly those due to physical irregularities, vary widely. Flexible coaxial cable is used mostly for short runs, as in interconnecting high-frequency circuits within electrical equipment.

Semirigid coaxials. This class of coaxials has thin tubular outer conductors which may be corrugated to improve bending. The dielectric can be either insulating disks or an expanded plastic insulation to lower the dielectric constant and hence the loss of the coaxial. Expanded insulation can be as much as 80% air in the form of very fine bubbles. This design finds wide usage in closed-circuit television applications. It is commonly used as the distribution cable from the transmitter to the subscriber's home. *See* CLOSED-CIRCUIT TELEVISION.

Another variation of the semirigid coaxial, insulated with polyethylene disks approximately 1 in. (1 in. = 2.54 cm) apart and enclosed inside a 10- or 12-mil-thick (1 mil = 0.0000254 m) copper tube, is widely used for transcontinental carrier transmission. One version used in Canada has a corrugated tubular outer conductor. The disk-insulated coaxial unit has very low loss and is now used in carrier systems in the United States, Japan, and Europe to transmit up to 10,800 two-way voice-frequency channels per pair of coaxial units. Figure 2 shows a 20-unit coaxial cable with 9 working coaxial pairs and 2 standby coaxials, which automatically switch in if the electronics of the regular circuits fail.

At these broad bandwidths the slightest physical irregularities introduced during manufacture of the unit or stranding of the cable can produce catastrophic variation in loss. Much worldwide development effort has been expended on developing manufacturing processes, quality control procedures, and installation techniques to make these cables suitable for signals as high as 60 MHz. These coaxial carrier systems, together with microwave radio, provide virtually all long-distance communications facilities in the continental United States.

An interesting example of a semirigid coaxial cable, with a bending radius of 4 ft (1 ft = 0.3 m), is the solid dielectric coaxial units 1.3 to 1.7 in. in diameter used for ocean cable systems. Instead of heavy external armoring for strength, these coaxials have center conductors composed of a copper tube welded around a stranded core of high-strength steel wires (Fig. 3). The center conductor provides the strength to lay and recover the cable

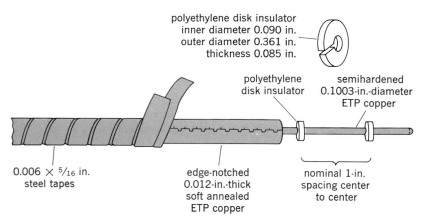

polyethylene disk insulator
inner diameter 0.090 in.
outer diameter 0.361 in.
thickness 0.085 in.

polyethylene disk insulator

semihardened 0.1003-in.-diameter ETP copper

0.006 × 5/16 in. steel tapes

edge-notched 0.012-in.-thick soft annealed ETP copper

nominal 1-in. spacing center to center

Fig. 1. An air dielectric disk-insulated coaxial.

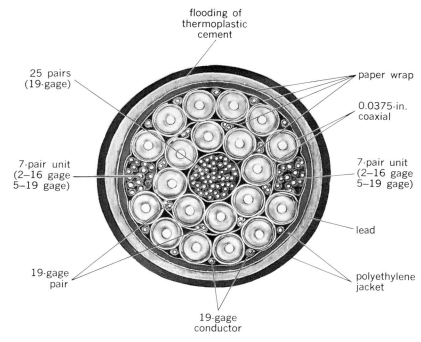

Fig. 2. Construction of coaxial transmission line.

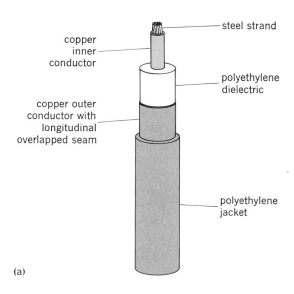

(a)

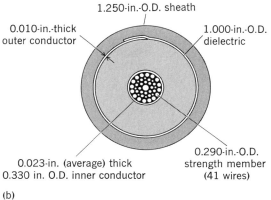

(b)

Fig. 3. Armorless ocean cable coaxial construction. (a) Vertical view; (b) cross section.

at depths up to 3000 fathoms (1 fathom = 1.83 m).

Special cable-laying machinery which grips the outer polyethylene jacket was developed and installed on a specially designed cable-laying ship to enable this unique undersea coaxial design to be used.

Rigid cables. When utmost electrical performance is required and no bending is needed to install the cable, a heavy, rigid, precise outer conductor is used. The inner conductor is either solid wire or a hollow tube which can be liquid-cooled in high-power applications. As little dielectric as possible is used to minimize losses, generally a semisolid such as pins or disks. The rigid construction allows the dielectric supports to be spaced far apart. This coaxial can have propagation velocities up to 99.7% of the speed of light. For high-voltage applications, the unit can be gas-filled to reduce corona and to increase dielectric strength.

Special coaxials. Coaxial units with periodic slots in the outer conductor have been developed since the mid-1960s to provide mobile communications with moving trains in tunnels and subways. These "leaky" coaxials transmit signals along their length while radiating a controlled amount of energy to antennas on passing trains. The Japanese have developed 3-in.-diameter slotted coaxials that are used in the United States for paging systems, vehicular tunnels, and subway communications, and for control of mine locomotives.

[JOHN R. APEN]

Coil

One or more turns of wire used to introduce inductance into an electric circuit. At power line and audio frequencies a coil has a large number of turns of insulated wire wound close together on a form made of insulating material, with a closed iron core passing through the center of the coil. This is commonly called a choke and is used to pass direct current while offering high opposition to alternating current.

At higher frequencies a coil may have a powdered iron core or no core at all. The electrical size of a coil is called inductance and is expressed in henries or millihenries. In addition to the resistance of the wire, a coil offers an opposition to alternating current, called reactance, expressed in ohms. The reactance of a coil increases with frequency. *See* INDUCTOR.

[JOHN MARKUS]

Coincidence amplifier

An electronic circuit that amplifies only that portion of a signal present when another enabling or controlling signal is simultaneously applied. The controlling signal may be such that the amplifier functions during a specified time interval, called time selection, or such that it functions between specified voltage or current levels, called amplitude selection. For example, the amplifier in a transmission gate operates on the basis of time selection where a control voltage of fixed amplitude is applied during a specific time interval. This voltage is the enabling waveform; it allows the amplifier to function normally only during the time it is applied. *See* GATE CIRCUIT.

A linear amplifier used to amplify a section of a waveform above or below specified voltage or current limits, as determined by a control signal, is an example of a coincidence amplifier that operates by amplitude selection. Limiting and clipping circuits that incorporate the function of amplification may be considered coincidence amplifiers. *See* CLIPPING CIRCUIT; LIMITER CIRCUIT.

A simple example of a coincidence amplifier that functions between two input voltage limits is shown in the illustration. This specific coincidence amplifier is sometimes referred to as a slicer or slicer amplifier. The transistor amplifier, in this case an emitter-coupled amplifier, is biased such that, until time T_1, input stage Q_1 functions as an emitter follower, but the voltage at the base of Q_1 is such that no current flows in its collector circuit until input level V_1 is reached. Between levels V_1 and V_2 both transistors are active, and an amplified output signal appears at the collector of Q_2. At time T_2, corresponding to level V_2, the input transistor becomes nonconducting and no further output is produced.

The same operation may be performed using p-n-p transistors with all bias and signal polarities reversed. The same function may also be performed using field-effect transistors or vacuum tubes as source-coupled or cathode-coupled amplifiers.

Such coincidence amplifiers, with their gains sometimes greatly magnified by positive feedback, are often used as elements in linear time-delay circuits. When such an amplifier is used and the interest is primarily in a pulse or other indication when a desired level is reached rather than in preservation of the waveform itself, the term comparator is more often applied. However, the term coincidence circuit is sometimes used interchangeably when the concept of amplification is not specifically considered. In digital logic terminology the AND gate is an example of such a coincidence or comparator circuit. *See* COMPARATOR CIRCUIT.

[GLENN M. GLASFORD]

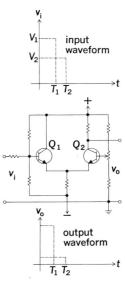

COINCIDENCE AMPLIFIER

Simple coincidence amplifier.

Color television

Transmission and reception of transient visual images in full color. The technique used for color television broadcasting in the United States is compatible with monochrome, or black-and-white, television. That is, color television signals are sufficiently similar to monochrome signals so that they produce acceptable monochrome pictures on black-and-white receivers while producing full-color pictures on color receivers. Conversely, a standard monochrome signal produces a satisfactory monochrome image on a color receiver. Compatible color television is based on a combination of the primary-color principle, conventional television technology, and special encoding techniques for combining primary-color television signals in such a way that they can be transmitted through a single broadcast channel.

Use of primary colors. Color television images are formed by combining three primary-color images. Although many sets of primary colors are theoretically possible (provided only that no combination of any two primaries in a set be capable of matching the third), the standard set which has proved to be most practical for color television consists of highly saturated red, green, and blue colors. The color kinescope, or picture tube, used to display the picture in a color television receiver can produce independent red, green, and blue images, usually in the form of tiny dots so closely intermingled that they cannot be resolved individually at the normal viewing distance. The three images are thus effectively fused by the visual system of the observer so that he sees only the combined, full-color image.

Color television cameras. In the simplest type of color television system the three primary-color images in the tricolor picture tube are controlled directly by video signals developed by three separate pickup tubes, adjusted to respond to red, green, and blue light (Fig. 1). Close registration must be maintained among the three images so that the scanning beams in the three pickup tubes are always at the same relative points in their respective image areas. In most practical cameras beam splitters are used to form three separate optical images, and a common set of scanning circuits is used to drive three deflection yokes in parallel. The scanning pattern, horizontal and vertical deflection frequencies, and deflection-synchronizing techniques for compatible color television are the same as for monochrome television. *See* TELEVISION SCANNING.

Need for encoding and decoding. Because of the great demand for television broadcast channels in the United States, it is not practical to utilize three separate channels for transmitting the red, green, and blue signals required to control a color kinescope. It is necessary, therefore, to encode the primary-color signals at the transmitter in such a way that they can be transmitted through a single television channel. The encoded color video signal contains three independent components, and is enough like a conventional monochrome signal to produce good images on monochrome receivers. Each color receiver contains decoder circuits (between the second detector and the kinescope) to retrieve from the color signal its red, green, and blue components.

Color encoder. In a compatible color encoder, the red (*R*), green (*G*), and blue (*B*) signals are first applied to a matrix, where they are cross-mixed in accordance with Eqs. (1)–(3). The signal

$$M = 0.30\,R + 0.59\,G + 0.11\,B \qquad (1)$$
$$I = 0.60\,R - 0.28\,G - 0.32\,B \qquad (2)$$
$$Q = 0.21\,R - 0.52\,G + 0.31\,B \qquad (3)$$

designated *M* conveys luminance information; it is nominally equivalent to the signal that would be generated by a monochrome camera viewing the same scene as the color camera. That is, the *M* or luminance signal controls the sensation of brightness. The *I* and *Q* signals convey chrominance information, which controls hue and saturation in the final image. The *I* and *Q* signals are reduced in bandwidth to about 1.5 and 0.5 MHz, respectively, in recognition of the fact that the human eye has limited resolving power for chrominance differences between small picture areas.

The *I* and *Q* signals are converted to modulated waves by means of a pair of balanced modulators. These modulated waves (including the necessary sidebands) are designed to fall within the 0–4.2 MHz video passband of a conventional television channel. The subcarrier inputs for the two modulators are both at the same frequency (3.579545 MHz), but are 90° apart in phase. When the modulated *I* and *Q* signals are added together, they form a two-phase-modulated wave (variable in both amplitude and phase) known as the chrominance subcarrier.

The complete color video signal is formed as the sum of (1) the chrominance subcarrier, (2) the monochrome or *M* signal, (3) conventional synchronizing pulses, and (4) a special color synchronizing signal consisting of a brief sample (or burst) of the subcarrier frequency timed (by means of burst flag pulses) to occur shortly after each horizontal sync pulse. The purpose of the burst is to establish a frequency and phase reference for the receiver circuits that demodulate the chrominance subcarrier.

Frequency interlace. The addition of a two-phase-modulated chrominance subcarrier does not prevent the monochrome component of a color signal from producing a satisfactory picture on black-and-white receivers because the subcarrier itself is frequency-interlaced. The subcarrier frequency is chosen as an odd multiple of one-half the horizontal scanning frequency, and is held in this relationship by counting circuits in the color sync generator. The waveform sketches of Fig. 2 show how the frequency-interlace technique works. The top wave (Fig. 2*a*) is a small section of a typical luminance signal for one scanning line. Next is the modulated subcarrier signal (Fig. 2*b*) to be transmitted during the same interval. Because of its harmonic relationship to the scanning frequencies, the subcarrier reverses in polarity between successive scans (1/30 sec), as indicated by the dotted line. There is usually no great change in the picture information from one frame to the next, so the luminance waveform and the modulation on the subcarrier remain essentially the same for the second scan. In Fig. 2 the complete color video signal (*c*) differs slightly from the luminance signal (*a*), but the subjective effect produced when two successive signals (*c*) are applied to a monochrome kinescope is approximately that represented by the average waveform (*d*). Persistence of vision

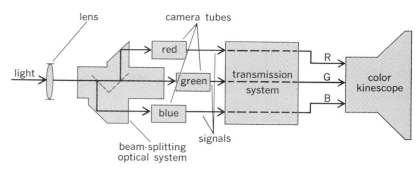

Fig. 1.   Simplified block diagram of a color television system.

tends to average the light output from several scans of each picture area, automatically canceling most of the effect of the added subcarrier. The residual dot pattern, which results from imperfect cancellation, is not particularly objectionable.

Color decoder. The basic functions performed in a color television receiver are shown in Fig. 3. The tuner, i-f amplifier, detector, sound channel, sync separator, and deflection circuits are all based on conventional designs. *See* TELEVISION RECEIVER.

The color decoder consists of the following circuits: (1) a burst separator, consisting of a gate circuit turned on by horizontal retrace pulses; the separated bursts are used to control a local subcarrier oscillator, which provides two carriers, 90° apart in phase, for demodulation of the chromi-

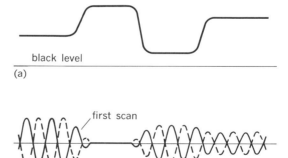

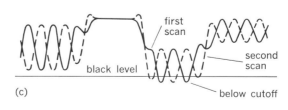

Fig. 2.   Several typical waveforms illustrating the frequency interlace technique. (*a*) Luminance signal. (*b*) Modulated subcarrier signal. (*c*) Sum of luminance and modulated subcarrier signals. (*d*) Average light output from two successive scans.

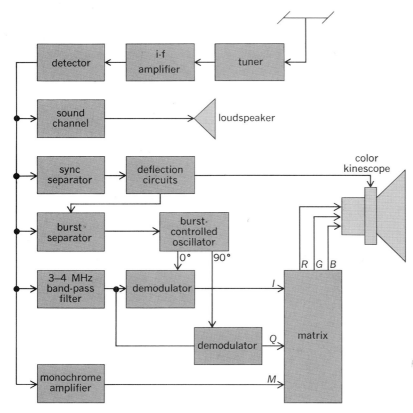

Fig. 3. Simplified block diagram of a color television receiver.

nance subcarrier; (2) the monochrome channel, which passes the luminance signal component and rejects or attenuates the subcarrier frequency; (3) the chrominance channel, which consists of a band-pass filter centered at 3.579545 MHz, plus a pair of synchronous detectors to recover two independent components from the modulated subcarrier (these components need not necessarily be the same as the original I and Q signals; (4) a matrix that cross-mixes the monochrome signal and the demodulated chrominance signals to produce red, green, and blue signals suitable for driving the corresponding guns of a color kinescope.

[JOHN W. WENTWORTH]

Color picture tube. The color picture tube is an evacuated electronic device that transforms the electrical video signal into a full-color display. Its basic operation is similar to that of a black and white picture tube (or kinescope) in which a beam of high-velocity electrons, generated in an electron gun, is magnetically deflected so as to scan line by line the inside of the faceplate of the evacuated bulb. Light is produced by electron bombardment of the luminescent layer of phosphors. The intensity of the light is controlled by the video signal applied to the electron gun, which controls the number of electrons emitted at any given instant. See CATHODE-RAY TUBE.

To produce color, the color picture tube has a means to selectively excite one or more phosphors, each of which emits light of one of the three additive primary colors: red, green, and blue. Essentially, all present-day color picture tubes used in commercial television receivers employ the sha-

dow-mask principle for selection of the primary colors.

In the shadow-mask tube, three electron guns provide three closely spaced, individually controllable electron beams which are deflected by a common magnetic deflection yoke. A shadow mask formed to the approximate spherical contour of the face of the picture tube is located about 15 mm (0.6 in.) behind the phosphor screen, which is deposited on the inner surface of the faceplate. The shadow mask is made of sheet steel and contains several hundred thousand small apertures. These apertures are associated with groups of screen elements of red, green, and blue light-emitting phosphors.

The relative position of the gun, shadow-mask apertures, and phosphor elements is such that the electron beam from one gun passes through the mask aperture and strikes only elements of one color-emitting phosphor. The mask shadows the other two phosphors. Thus, each gun can produce a picture in one of the three primary colors. The small size and large number of the phosphor elements make them blend together and appear indiscernible to the viewer's eye. By varying the relative brightness of each of the primary colors, a full-color picture or a black and white picture can be produced.

For many years, the arrangement of phosphor elements was in the form of interlaced phosphor dots as shown in Fig. 4. For each trio of red, green, and blue light-emitting phosphor dots, there was a round aperture in the shadow mask. The diameters of the mask apertures were about 0.25 mm (0.010 in.), and each of the phosphor dots was about 0.33 mm (0.013 in.) in diameter.

The dot screen has been replaced by a line screen. In this system, vertical rows of alternate red, green, and blue phosphor lines form the

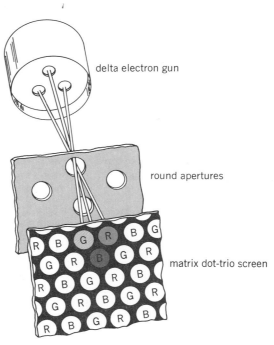

Fig. 4. Diagram of dot-trio system, with red (R), green (G), and blue (B) phosphor dots. (*RCA*)

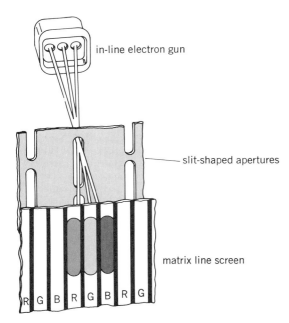

in-line electron gun

slit-shaped apertures

matrix line screen

R G B R G B R G

Fig. 5. Diagram of in-line system with red (R), green (G), and blue (B) phosphor lines. (*RCA*)

the phosphor screen is deposited and the shadow mask inserted. Tube manufacturing requires great accuracy: relative placement of phosphor lines and the shadow mask must be controlled to tolerances of a few micrometers. *See* TELEVISION.

[A. M. MORRELL]

Bibliography: R. L. Barbin and R. H. Hughes, New color picture tube system for portable TV receivers, *IEEE Trans. Broadcast Telev. Receivers*, BTR-18(3):193–200, August 1972; W. E. Burke (ed.), *Color TV Guidebook*, 1967; H. W. Coleman, *Color Television: The Business of Colorcasting*, 1968; *Electronic Engineers' Handbook*, 1975; D. G. Fink (ed.), *Television Engineering Handbook*, 1957; C. N. Herrick, *Color Television: Theory and Servicing*, 2d ed., 1977; M. S. Kiver, *Color Television Fundamentals*, 1964; R. G. Middleton, *Color TV Servicing Guide*, 1965; A. M. Morrell et al., *Color Television Picture Tubes*, 1974; A. M. Morrell, Design principles of RCA large screen 110° precision in-line color picture tube, *IEEE Trans. Consumer Electr.*, 22(1):1–7, February 1976; C. P. Oliphant and V. M. Ray, *Color TV Training Course*, 1965; H. W. Sams (ed.), *Color-TV Field Service Guide*, 2 vols., 1971–1974; H. W. Sams (ed.), *Color-TV Training Manual*, 4th ed., 1977; J. W. Wentworth, *Color Television Engineering*, 1955.

Communication cables

An assembly of from 6 to 4200 twisted wire pairs enclosed in a composite metal-plastic sheath. The major use of communications cables is in the telephone industry, to connect customers to their local switching centers (subscriber loops) and to interconnect local and long-distance switching centers (trunk plant). In the United States, more than 200,000,000 circuit miles of cable provide the path by which any one of 120,000,000 telephones can be linked to any other. The average telephone subscriber is 11,000 ft (1 ft = 0.3 m) from the local switching center; a pair of wires in an aerial or belowground cable transmits signaling and voice information to that switching center. The instrument most commonly connected to a cable is the telephone set, but teletypewriters, television cameras, and data sets all convert information into electrical signals that can be sent over cables.

Interference between circuits in multipair cables becomes more likely as larger bandwidths are transmitted. Very-high-frequency signals, such as television, high-speed data, or broadband carrier signals, are frequently transmitted over coaxial cables in which one conductor of the pair is a metal tube concentric with and enclosing the other. As higher frequencies are transmitted, the outer conductor becomes more effective in containing radiated electromagnetic energy and preventing one circuit in the cable from interfering with others. *See* COAXIAL CABLE; CROSSTALK.

However, multipair cable has several advantages when compared with coaxial: (1) a lower loss per circuit cross section, (2) mechanical ruggedness, and (3) rejection of interference from stray power currents. (This is due to the balanced mode of transmission, in which the wires of a pair are energized so that the signal voltages on them at any given instant are of equal magnitude but opposite sign with respect to ground. Outside inter-

---

screen. The shadow mask is composed of elongated apertures arranged in vertical rows, one for each trio of red, green, and blue phosphor lines on the screen. Small webs or crossties provide structural strength between the mask openings (Fig. 5). The width of the openings in the mask is about 0.15 mm (0.006 in.), and the width of each phosphor line is about 0.25 mm (0.010 in.).

The line screen is used with an in-line arrangement of the electron guns. This in-line arrangement permits ease in obtaining convergence or coincidence of the three beams at all points on the screen. Convergence is required to prevent color fringing that would result from noncoincidence of the three beams. The deflection yoke which scans the beams over the screen produces the convergence of the beam by its special field configuration. The position of the yoke must be precisely matched with the beam to obtain this condition.

To enhance contrast and light output, a black-matrix-type screen is used on most color tubes. In this type a black, light-absorbing material surrounds each phosphor element (Figs. 4 and 5). This black material, which covers about 30% of the screen area, reduces the reflection of ambient light and hence increases the contrast. The black matrix provides a "guard band" for any misalignment of the electron beam with respect to the desired phosphor element. A further improvement has been the use of a filtered or pigmented phosphor in which a red pigment is combined with the red light-emitting phosphor and a blue pigment with the blue light-emitting phosphor. These pigments reduce the reflection of ambient light and also enhance the contrast.

The bulb structure is made in two pieces (funnel and faceplate) for ease in fabrication of the tube. The mask is welded to a steel frame which is supported on leaf springs that engage tapered studs sealed in the sidewall of the faceplate. The two bulb parts are sealed together by a glass frit after

ference induces currents of equal magnitude which flow in the same direction on both conductors and cancel each other at the load.) For these reasons, the twisted, balanced pair is the choice for transmission from telegraph frequencies up through a few megahertz.

Copper is the most commonly used electrical conductor, but aluminum is widely used in England and is becoming more prevalent in the United States. Metallic conductors are commonly insulated with paper pulp or plastics, although paper strip, helically or longitudinally wrapped around the wire, which was introduced in 1890, is still used. Sheaths protect the transmission properties of the cable pairs by providing mechanical protection, electrical protection from lightning-induced currents, and shielding. In paper-insulated cables sheaths also function to keep out moisture.

Cable was first used with telegraph systems in the 1850s, in order to to communicate across large bodies of water. The invention of the telephone in 1876 placed severe demands on existing transmission lines. Telegraph signals require a bandwidth of about 100 Hz, whereas normal speech requires about 3500 Hz. Consequently, voice signals encounter greater loss and a higher probability of one circuit interfering with an adjacent circuit. Early telephone systems used open wire lines, but since the major market for the telephone was in cities, an incredible proliferation of wires (300 wires on 90-ft poles with 30 crossarms were used in New York City in 1890) made the development of more suitable cables necessary. By 1892, cables with 19-gage conductors insulated with paper tape and twisted into pairs were being produced. One hundred pairs fit into a 2½-in.-diameter pipe (1 in. = 2.54 cm). This cable was not vastly different from the type used during the following 35 years.

TYPES AND USES

Communications cables or telephone cables are divided into four general classes, as discussed below.

Local and exchange area cable. This cable is placed aerially or underground in ducts, or is directly buried. It is used to connect one switching office to another or to connect the switching office and the subscriber. It has pulp paper, plastic, or expanded plastic insulation, with either copper or aluminum conductors. Cables filled with petroleum jelly, that is, waterproofed, are rapidly supplanting unfilled cables in buried subscriber applications.

Toll cable. This cable is used to connect metropolitan areas and to bring toll circuits through cities to toll switching offices. It is generally designed to have superior transmission characteristics compared to exchange area designs. Frequently, this cable carries many voice or data channels multiplexed onto high-frequency carrier systems. Older versions still being manufactured have helically applied paper insulation. Newer versions have expanded plastic insulation. Sheaths are lead, lead-plastic composites, or special multilayer metal-plastic composites. Heavy-duty sheaths and elaborate gas-pressurization schemes are used to protect and monitor the integrity of the cable sheaths because of the large number of circuits.

Switchboard cable. This cable is intended to be used as equipment cabling in switching offices, business premises, and office buildings. Polyvinyl chloride (PVC) insulation has largely supplanted older insulations composed of cotton and acetate yarns impregnated with lacquers. These insulations are chosen primarily for fire retardance rather than for transmission efficiency, since the cables are short and are located inside buildings. PVC sheaths are common.

Special cables. Examples of special cables are coaxial cables used for high-frequency carrier systems on heavy toll routes, video pairs for baseband video transmission, stub cables used to connect apparatus such as loading coils and repeater cases to main cables, terminating cables from the central office cable vault to the equipment, and building cables used to distribute telephone service within commercial buildings. Also included in this group are composite cables having various combinations of coaxials, video pairs, and regular pairs.

Design. The features of the most common cable types are described below.

Pulp-insulated cable. About 1930 specialized papermaking machinery was developed to apply pulp directly on 60 copper conductors at one time. Combined with a lead sheath, which gave excellent hermetic protection when maintained under pressure with either nitrogen or dry air, pulp cable provided a relatively cheap, reliable transmission medium that allowed rapid expansion of telephone service in the decades prior to World War II. Because pulp insulation contains air, pulp cables are smaller than solid plastic cables are. Pulp is still the most common cable used in underground duct plants in metropolitan areas.

Plastic-insulated conductor (PIC) cables. After World War II, the development of methods to apply molten polyethylene to single conductors at speeds of 3000–4000 ft/min made plastic-insulated cables feasible and caused marked changes in the way service was brought to the customer. Since paper and pulp insulation must be dry in order to be effective, hermetically sealed terminals are needed at connection points between cables and telephone customers' drop wires. Because plastic-insulated cables are not affected by humidity, aerial terminals open to the atmosphere can be used for access to the cable pairs. This "ready-access" principle was fundamental to making the more expensive PIC cable the choice for subscriber cable starting in 1954.

The plastics used for wire insulation (first, low-density polyethylene and, later, high-density polyethylene and polypropylene) have excellent dielectric properties, low dielectric constant, low dissipation factor, and high dielectric strength. It is easy to obtain 10 colors that are distinguishable even in poor light. If two colors are combined in a twisted pair, 25 identifiable pairs can be readily obtained. This even-count color code based on 25 different pairs in a unit is universally used. Cables larger than 25 pairs are made up of groups of units bound with binders of different colors. Color coding makes each subscriber circuit readily identifiable, regardless of how many cables make up the circuit from the switching office to the customer's home.

Standard practice with PIC cable is to use 25

unique twist lengths corresponding to the different color pairs. If these pairs are arranged in a care-- fully controlled relationship with one another, crosstalk can be kept at a minimum. Properly chosen twists, positioned carefully, also reduce capacitance unbalance, important in the rejection of external disturbances. PIC cable can be designed to be satisfactory not only for voice but also for carrier frequency transmission in the several-megahertz range.

Waterproof cable. Since about 1970 more than 90% of new residential customers have been served by buried cable. Although the improved appearance is extremely desirable, and in some areas the reduction in problems caused by storm damage is also an important advantage, large-scale burial of PIC cable, even with premium-quality sheaths, has led to increasing numbers of problems caused by moisture, water, and corrosion. When water enters a PIC cable, it does not immediately short-circuit the pairs, as it would in a pulp cable. Water does increase transmission loss, but most serious is its corrosive effect on conductors. Prohibitive costs would be incurred in manufacturing cables with no insulation defects; and when these inevitable defects are connected by a water path, the central office battery provides the electromotive force to corrode the conductors. In addition, while the corrosion process is proceeding, the conductor, grounded by the water, creates an electrical unbalance which causes a noisy circuit. Filling the air spaces between the pairs with petroleum jelly during manufacture can eliminate these water problems. (Fig. 1).

Greaselike polypropylenes or low-density polyethylene grease, or mixtures of these materials with petroleum jelly, have been used as filling compounds. The requirements that the filling compound be compatible with the wire insulation, that the viscosity of the compound be suitable for manufacturing, and that the compound be neither too stiff when the cable is installed in cold weather nor too fluid for occasional aerial use required much effort for the development of suitable petroleum jelly greases.

Because the filling compound, with a dielectric constant of 2.3, replaces air (dielectric constant 1.0), the insulation diameter and cable size must be increased about 20% so that the loss will remain the same as in air-core cable. Filled cable thus requires more petroleum-based materials than its air-core equivalent, not only for insulation but also for the sheath. Expanded insulation containing as much as 50% air in the form of tiny bubbles, which reduces the dielectric constant, is

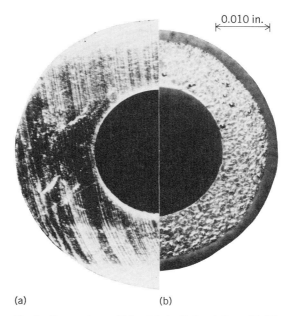

Fig. 2. Comparison of (a) solid plastic insulation with (b) dual-layer expanded insulation (0.002-in. or 50-μm solid skin over 45% expanded core) showing both the structure and the smaller diameter of the dual-layer insulation compared with the solid insulation. Transmission properties are equivalent. (*Bell Telephone Laboratories*)

used to reduce the cable size and the amount of oil-based materials which are required in waterproof cables.

A further improvement is a dual-layer insulation with a 0.002 − 0.004-in.-thick solid skin extruded over a foamed inner layer (Fig. 2). It retains the size advantages of ordinary expanded insulation, since the inner layer can have a higher percentage of bubbles, while retaining the desirable features of solid insulation. Its dielectric strength is higher than that of expanded insulation, and the color is confined to the skin, and can be made more brilliant without the danger of the pigments interfering with the expanding process. The skin also retards the absorption of oils from the filling compound, stabilizing the long-term electrical properties of the cable.

Aluminum conductor cables. The only economically practical conductor alternative to copper is aluminum. Unstable copper prices and supply, combined with possible substantial savings in using aluminum since 1 lb (0.45 kg) of aluminum replaces 2 lb of copper on a conductivity basis, have always made aluminum conductor a feasible alternative. The major problems with aluminum are rapid corrosion, when water and the central office battery are present, and larger size, since the conductivity of aluminum is only 61% of that of copper. Corrosion problems with pulp-insulated aluminum cables in the 1950s led to the use of PIC-insulated air-core aluminum cables in the 1960s. Although these cables are technically satisfactory, filling them in order to prevent all corrosion problems appears to be the most viable design. The larger aluminum conductor and the thicker insulation used for filled cables makes an aluminum-filled cable substantially larger than its

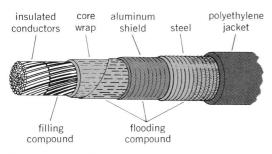

insulated conductors | core wrap | aluminum shield | steel | polyethylene jacket

filling compound | flooding compound

Fig. 1. Waterproof cable design.

copper equivalent. The use of expanded insulation recoups some of this size penalty. The British are using this design extensively, and it is beginning to be used in the United States. Because of the inherently larger conductor size, aluminum cable is not preferred for use in ducts. However, in this case, cable savings must be balanced against increased duct costs due to the smaller number of pairs per duct.

Lead, extruded over the cable core in a continuous manufacturing operation invented in the 1880s, was the common sheath material until after World War II. At that time increasing cable demand coupled with lead shortages and high prices led to the first application of polyethylene in telephone cable. Since polyethylene will not prevent diffusion of water vapor, a composite sheath called ALPETH (an acronym formed from "aluminum" and "polyethylene") was invented. From this basic idea of using metals and plastic composites to obtain a combination of protective functions, many sheath designs have evolved.

Steel and aluminum are used in these designs for mechanical and electrical protection, and the metals are generally corrugated to improve bending performance (Fig. 3). Polyethylene is utilized to protect the metals from corrosion and abrasion. Since this outer jacket may be punctured by lightning, especially in rural areas, some buried cable designs have inner polyethylene jackets to prevent such punctures from penetrating to the cable core.

Gas pressure maintenance. Whereas waterproof cable is the preferred method for reducing maintenance costs on subscriber loops, pulp or plastic cables in underground ducts, are maintained under 6–10-psi (1 psi = 6895 Pa) gas pressure by air compressors and driers in the central office. Gas pressure is primarily a preventive maintenance tool, although the pressure will keep water out of small leaks and imperfections. The gas fed from the central office is monitored by a flowmeter which indicates excessive flow if a cable leak develops. Pressure-activated switches are also used to send alarms to a central point if the pressure drops below preset levels. Pressure transducers are used to monitor the pressure changes along a cable run; they can be read remotely, and potential problems located, by analyzing the pressure changes. Elaborate remote telemetry systems combined with minicomputers allow all pressurized cable in a statewide area to be monitored centrally. The computer makes complex analyses of the transducer readings, ignoring minor fluctuations. The computer can be queried at any time as to the status of a cable, thus allowing cable repair teams to be dispatched efficiently. These techniques allow the use of plastic-insulated cable in the duct plant. When water enters pulp or paper cables through a sheath break or splice closure leak, it immediately short-circuits pairs, and the source of the problem can be quickly and easily identified. The pulp also swells and blocks the flow of water, localizing the problem.

PIC cable is not immediately affected by water entry, and water can travel many hundreds of feet from the entry point. When water damage is finally detected because of a noisy circuit or a conductor corroded open, the original sheath defect is difficult to locate, and the cable core may be damaged at several points. The inherent reliability but fairly poor maintainability of PIC cable makes it less practical than pulp, which has poorer reliability but better maintainability. Centralized pressure monitoring and analysis systems permit PIC cable, with its superior transmission qualities and, in its expanded form, smaller size, to be utilized in the duct plant. *See* TRANSMISSION LINES.

[JOHN R. APEN]

COAXIAL CABLE SYSTEMS

A coaxial cable system is a broadband transmission system that consists of the coaxial cable, frequency division electronic equipment, auxiliary amplifiers, electronic switching equipment, ac and dc power supplies, manholes, and buildings and related facilities, all properly spaced and connected to permit the transmission of signals millions of hertz in bandwidth.

The Philadelphia–New York coaxial cable, which was experimentally demonstrated in October, 1936, was the first broadband long-distance transmission system. In November, 1937, a similar coaxial cable was used to transmit television between a local broadcasting studio and the Empire State Building in New York.

Coaxial cable was introduced on a commercial basis in June, 1941, over the 195-mi distance between Minneapolis, Minn., and Stevens Point, Wis. This was the first use of the coaxial-cable carrier system known as L-1, which was designed to provide 480 two-way telephone circuits in a pair of

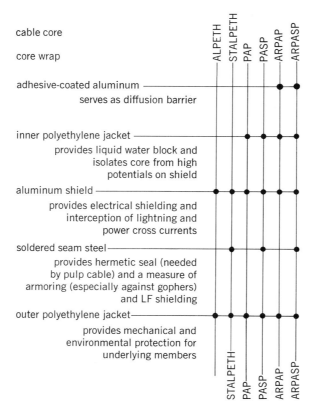

	ALPETH	STALPETH	PAP	PASP	ARPAP	ARPASP
cable core						
core wrap						
adhesive-coated aluminum serves as diffusion barrier	●	●	●	●	●	●
inner polyethylene jacket provides liquid water block and isolates core from high potentials on shield		●		●		●
aluminum shield provides electrical shielding and interception of lightning and power cross currents	●	●	●	●	●	●
soldered seam steel provides hermetic seal (needed by pulp cable) and a measure of armoring (especially against gophers) and LF shielding		●	●		●	
outer polyethylene jacket provides mechanical and environmental protection for underlying members		●	●	●	●	●

Fig. 3. Representative sheath designs. The choice of layers to be used depends on the type of cable core and the environment.

coaxials. This cable had four coaxial tubes, two for working purposes and two for protection, and its amplifiers or repeaters, many of which were not manned, were spaced at 5-mi intervals. In 1946 the size of the coaxial tubes was enlarged by about 40%, from 0.270-in. diameter to 0.375-in. diameter, thereby permitting the L-1 repeater spacing to be increased to 8 mi. In 1948 technical developments made it possible to increase the capacity of the L-1 system to 600 circuits. *See* COAXIAL CABLE.

Greater capacity. The rising demand for communication services after World War II triggered the development of a coaxial system of still greater capacity. After 6 years of work in Bell Telephone Laboratories and more than a year of field trials, such a system (coded L-3) was achieved. One pair of tubes in this system can furnish up to 1860 telephone circuits. Its amplifiers are spaced at shorter intervals, 4 rather than 8 mi (Fig. 4).

An L-3 system was placed in service in 1953 between New York and Philadelphia. Since then the use of this system has been expanded extensively. Shortly after World War II the number of tubes in coaxial cables was increased to 6, then to 8, and ultimately to 12 — a figure which remained the maximum through 1962.

L-4 system. The 1860 circuits per coaxial pair provided by L-3, and the increase in cable size from 4 to 12 tubes, provided adequate capacity to handle service requirements through the 1950s. By 1960, however, it became apparent that greater capacity was needed. To meet this demand, Bell Laboratories began the development of L-4, which was the most extensive redesign of the coaxial system since its initial development in the 1930s.

The L-4 system provides nearly twice the circuits available from L-3 — 3600 per coaxial pair compared with 1860. To do this, the L-4 system transmits a much wider frequency bandwidth, that is, much higher frequencies than were used previously. When frequency levels are raised in coaxial systems, there is a concurrent rise in transmission

Fig. 5. Sketch of a buried solid-state auxiliary amplifying station. (*American Telephone and Telegraph Co.*)

loss; therefore, the signal must be amplified more often. As a result, the L-4 repeater spacing was reduced to every 2 mi, or one-half that required for L-3.

The L-4 repeaters represented a major part of the system redesign. Solid-state plug-in-type amplifiers designed for long life are housed in apparatus cases designed for mounting in buried prefabricated concrete manholes (Fig. 5). Power for repeater operation was changed from the alternating current used in the L-3 system to direct current, which is fed to the repeaters along the coaxial center conductor from manned main stations located 150 mi apart (Fig. 6). The low power demand of the solid-state repeaters made possible the switch to direct current, which offers more efficient transmission and low noise amplification of the telephone signal — a most important factor in high-capacity systems.

The first major L-4 project was the Boston-Miami cable, which was completed in 1967.

L-5 system. The long-distance coaxial transmission system designated L-5 provides a threefold increase in capacity over that provided by L-4 — 10,800 circuits per coaxial pair as opposed to 3600.

The L-5 system was developed by Bell Laboratories in the early 1970s to meet the ever-increasing communications demand for high-quality telephone and data service. The first L-5 system went into service in January 1974 between Pittsburgh and St. Louis. Again, each successive generation of repeater operates over wider transmission bandwidths to provide greater circuit capacity. With L-5, the repeater spacing is halved again — to 1 mi between manholes — in order to counter transmission losses experienced with the high frequencies used — up to 70 MHz. L-5, like L-4 before it, employs solid-state repeaters and dc power.

Cable conversions. Concurrent with the development of L-5, a program of upgrading older

Fig. 4. Sketch of an L-3 auxiliary amplifying station. (*American Telephone and Telegraph Co.*)

Fig. 6.   Underground coaxial cable main station. (*American Telephone and Telegraph Co.*)

coaxial cable systems was begun. This program involves converting older cables, installed as L-1, L-3, or L-4, to high-capacity L-5 operation. This conversion process requires the placing of new manholes between those existing on older cables, in order to meet the 1-mi L-5 repeater spacing, and replacing the electronics with new L-5 equipment in all repeater locations and main stations. This procedure extends the life of older cable systems and obviates the need for large-scale cross-country construction programs, thereby avoiding the attendant environmental disruption.

Design characteristics. Each coaxial cable system is made up of two coaxial tubes, one for each direction of transmission. These tubes consist of a solid-copper-wire center conductor, surrounded by a coaxially positioned copper tube held in place by polyethylene disk spacers set 1 in. apart. The coaxials are called air-dielectric; that is, dry air or nitrogen under a pressure of 9 psi is maintained within the tubes to act as an insulator and to prevent the entrance of moisture. The air also aids in the rapid detection of cable sheath damage, for an air leak, with its accompanying pressure drop, triggers an alarm in the main station.

To make up a coaxial cable, up to 11 coaxial tube pairs are spirally wound around a core of individually insulated copper wires which are used for system command and control functions. The protective sheath of the cable is made up of a translucent inner polyethylene layer as a moisture barrier, which is surrounded by a solid lead shield and covered again by a dense black polyethylene jacket for protection against mechanical damage and corrosion.

When placed in service, the cable is buried 4 ft underground to guard against hazards such as accidental physical damage by construction equip-

ment and to ensure a constant temperature environment for proper operation (Fig. 7).

Lightning strikes in the vicinity of the cable are the most frequent cause of cable damage. Lightning current passing to the buried cable heats and vaporizes soil moisture, and the resulting gas expansion adjacent to the cable often dents the protective sheath, thereby crushing the coaxial tubes. To protect against lightning damage, two bare copper wires are buried along the route during construction. They are placed 2 ft above and 1½ ft to each side of the cable. These wires act to intercept and dissipate the lightning current before it reaches the cable, thereby greatly reducing its damaging potential.

Cable development. As was previously stated, 12-tube cables were the largest type in use through the early 1960s. With the advent of the L-4 system, it became apparent that a larger cable would be needed to accommodate the forecast growth in communications traffic. A 20-tube cable was developed to meet this need; it consisted of an 8-tube center ring, around which 12 additional tubes were stranded. The same machinery employed in manufacturing the 12-tube cable was used for the 20-tube cable. The resulting cable, operating with L-4 equipment, was able to carry 32,400 telephone circuits. Each L-4 coaxial pair carried 3600 circuits, nine working pairs, and one pair held in reserve for protection. In all cable systems, automatic switching equipment is available to transfer communications traffic from working pairs to the reserve pair when electronic troubles develop. This switch is instantaneous and cannot be detected by the telephone user.

When L-5 came into service in the 1970s, the 20-tube cable was further developed by the addition of two tubes in the outer ring, for a total of 10 working systems plus one protection pair. A fully loaded 22-tube cable equipped with L-5 electronics is capable of carrying 108,000 telephone circuits— 10 systems of 10,800 circuits each (Fig. 8).

With the development of the 20- and 22-tube cables, it was recognized that the 3-in. outside diameter of the new product was too large to fit into the hundreds of miles of previously built ducts that exist under the streets of urban areas in the United States. A special 2⅝-in.-diameter, 18-tube cable was developed for use in these areas to complement the larger cables used in direct-burial applications. The 18-tube urban cable equipped with L-5 electronics can carry 86,400 circuits—eight working systems plus one protection pair.

Various forms of the basic polyethylene-jacketed cable are in service. Special-duty variations include cable wrapped with single and double layers of steel armor wire, which is used when cable is placed on hazardous locations such as on a riverbed crossing or in mountainous areas in which mud slides are likely to occur. Another variation of the basic cable is wrapped with steel tape for protection against gopher gnawing, which is a serious problem in the western United States.

Fig. 8. Cross section of 22-tube coaxial cable. (*American Telephone and Telegraph Co.*)

Survivability. In the early days of telephone route construction, survivability measures were limited to protecting the outside plant from weather hazards. As the network expanded, alternative routing, that is, more than one access to a given area, provided increased service protection. Still later, as the state of the art became more advanced, and as telephone routes carried much larger numbers of circuits, more extensive survivability measures were developed.

Today's modern coaxial cable systems are completely buried 4 ft underground and are able to survive the most severe natural disasters and damage caused by humans, such as hurricanes, earthquakes, and nuclear war, short of a direct hit on the cable itself. Main interstate routes purposely avoid large metropolitan areas so that any disruptive event occurring in one city does not affect service to the rest of the country.

If an emergency such as a natural disaster were to occur, the manned underground main stations could be operated continuously in a "buttoned-up" condition for as long as 3 weeks, if necessary, to ensure continuity of service. Each of these main stations maintains a 21-day fuel supply for emergency electric power generation, as well as food, clothing, and other life-support facilities including a filtered air supply, so that the personnel required to operate the telephone system could remain underground continuously for the duration of the emergency.   [DANIEL P. MAHONEY]

Bibliography: American Telephone and Telegraph Co., *Principles of Electricity*, 1961; American Telephone and Telegraph Co., *Telecommunications Transmission Engineering*, 1974; Belden Manufacturing Co., *Electronic Cable Handbook*, 1966; *Bell Lab. Rec.*, July – August 1967; Bell Telephone Laboratories, *Transmission Systems for Communications*, 1965; G. W. A Dummer and W. T. Blackband, *Wire and R. F. Cables*, 1961; General Electric Co., *The Protection of Transmission Systems against Lightning*, 1950; F. W. Horn, ABC of the telephone cable, inside and out, in *Lee's ABC of the Telephone*, 1974; A. King and V. H. Wentworth, *Raw Materials for Electric*

Fig. 7. Tractor pays out coaxial cable into a 4-ft-deep trench on the Boston-Miami Project. (*American Telephone and Telegraph Co.*)

Cables, 1954; L4 system, *Bell Sys. Tech. J.*, April 1969; L5 system, *Bell Sys. Tech. J.*, December 1974; Rome Cable Corp., *Rome Cable Manual of Technical Information*, 1957; Simplex Wire and Cable Co., *Simplex Manual*, 1959.

Comparator circuit

An electronic circuit that produces an output voltage or current whenever two input levels simultaneously satisfy predetermined amplitude requirements. A comparator circuit may be designed to respond to continuously varying (analog) or discrete (digital) signals, and its output may be in the form of signaling pulses which occur at the comparison point or in the form of discrete direct-current levels.

Linear comparator. A linear comparator operates on continuous, or nondiscrete, waveforms. Most often one voltage, referred to as the reference voltage, is a variable dc or level-setting voltage and the other is a time-varying waveform. One common application of the comparator is in a linear time-delay circuit. Inputs consist of a sawtooth waveform of linearly increasing magnitude (ramp function) and a variable dc reference voltage. The reference voltage can be calibrated in units of time, as measured from the beginning of the sawtooth.

A clipper and a coincidence amplifier, together with a resistance-capacitance (RC) differentiating circuit, can perform the function of comparator. In Fig. 1 the series clipper, usually called a pick-off diode for this aplication, does not conduct until the input reaches level V_R. The diode input is a sawtooth as shown. Consequently, only the portion of the sawtooth above V_R appears at the output of the clipper. This output is applied to the RC differentiating network, which passes only the initial part of the rise. This short pulse is then amplified to produce the resultant output waveform. *See* CLIPPING CIRCUIT: COINCIDENCE AMPLIFIER.

The particular amplifier illustrated is a two-transistor high-gain amplifier with a relatively high input impedance and a low ouptut impedance. A sharper pulse can be obtained if the amplifier is made regenerative. It may even take the form of a multivibrator or blocking oscillator to increase the gain at the point of coincidence. *See* BLOCKING OSCILLATOR: MULTIVIBRATOR.

Regenerative comparator. Multivibrators can be used in several ways directly as comparators without need for the pick-off diodes; such comparators sense the required coincidence accurately and introduce little additional delay. A simple type is the direct-coupled bistable circuit, sometimes known as the Schmitt circuit, one form being shown in Fig. 2. This example employs enhancement mode *p*-channel field-effect transistors and can be made to function from either negative or positive-going input waveforms. The example compares a negative-going input waveform with reference voltage V_R.

Under a variety of choices of supply voltage and resistances, the circuit will be bistable, that is, either of the two transistors can be conducting for a particular voltage at input gate G_1. Until a predetermined value of reference voltage is reached, Q_1 is nonconducting, and at time T it switches from nonconducting to conducting while Q_2 simultaneously switches from conducting to nonconducting. With dc coupling as shown in Fig. 2, three outputs of differing dc levels and polarities are produced. If RC differentiating circuits are added as indicated, sharp pulses can then be obtained. When the input waveform ends, all points in the circuit return to their initial states.

Direct-coupled regenerative comparators such as the Schmitt circuit are usually bidirectional, responding to inputs approaching the reference level V_R from either the positive or the negative side. If the input starts at a value lower than V_R, the output voltage, V_1 will be at its high value until V_R is reached and then shift to its low value. Polarities of the other output signals will be correspondingly reversed. Thus at the voltage coincidence of V_i and V_R there will be generated one of two possible output states definable as logic level (1) or logic level (0) in digital terminology. Because of design limitations in practical circuits, the input voltage at which the bistable circuit changes state is slightly less or greater than V_R depending upon whether the input signal is positive-going or negative-going. This slight difference in level is referred to as the hysteresis of the circuit.

Integrated circuit comparators. High-gain dc operational amplifiers operated in the nonfeedback mode are often used to perform the comparator function, and many such amplifiers are classified as comparators because they are specifically designed to meet the needs for accurate voltage comparison applications. Such "op-amps" have two inputs, the output being inverting with respect

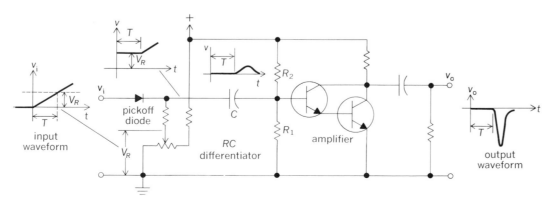

Fig. 1.   Simple comparator circuit using pick-off diode.

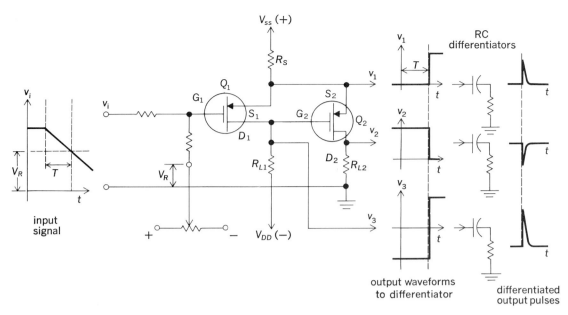

Fig. 2. Comparator using field-effect transistors in source-coupled bistable circuit.

to one and noninverting with respect to the other as shown in Fig. 3. The voltage gain (amplification) of the amplifier is so high that its output will swing through its entire dynamic range, V_{min} to V_{max}, for very small changes in input voltage. Thus, for $V_i < V_R$ the amplifier will be cut off and the output voltage will be at V_{max}, and for $V_i > V_R$ the amplifier will saturate and the output will be at V_{min}. For digital system applications, the output levels may be designed to coincide with logic level (0) and logic level (1) of the specific digital system and thus be suitable for converting a specific level in a continuously varying signal to a specific logic number assigned to the level. Arrays of such comparators connected to a common input, each designed to respond at a distinct reference voltage, and with the outputs connected to appropriate logic gates may be used to convert a range of signal levels to a specific digital code and as such form the basic building block of analog-to-digital converters. *See* AMPLIFIER; ANALOG-TO-DIGITAL CONVERTER.

The voltage gain and hence the timing precision of the operational amplifier comparator can be increased by converting it to a regenerative comparator as shown in Fig. 4. In this manner it becomes an integrated circuit form of the Schmitt circuit shown in Fig. 2.

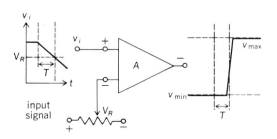

Fig. 3. Comparator circuit using integrated operational amplifier.

Digital comparator. The term "digital comparator" has historically been used when the comparator circuit is specifically designed to respond to a combination of discrete level (digital) signals, for example, when one or more such input signals simultaneously reach the reference level which causes the change of state of the output. Among other applications, such comparators perform the function of the logic gate such as the AND, OR, NOR, and NAND functions. More often, however, the term "digital comparator" is used to describe an array of logic gates designed specifically to determine whether one binary number is less than or greater than another binary number. Such digital comparators are sometimes called magnitude comparators or binary comparators.

Applications. Comparators may take many forms and can find many uses, in addition to those which have been discussed. For example, the electronically regulated dc voltage supply uses a circuit which compares the dc output voltage with a fixed reference level. The resulting difference signal controls an amplifier which in turn changes the output to the desired level. In a radio receiver the automatic gain control circuit may be thought of broadly as a comparator; it measures the short-term average of the signal at the output of the detector, compares this output with a desired bias level on the radio-frequency amplifier stages, and changes that bias to maintain a constant average level output from the detector. *See* AUTOMATIC GAIN CONTROL (AGC).   [GLENN M. GLASFORD]

Bibliography: J. A. Connelly (ed.), *Analog Integrated Circuits*, 1975; G. M. Glasford, *Electronic Circuits Engineering*, 1970; J. Millman and C. Halkias, *Electronic Fundamentals and Applications for Engineers and Scientists*, 1976; J. Millman and H. Taub, *Pulse, Digital and Switching Waveforms*, 1956; J. Millman and H. Taub, *Pulse, Digital and Switching Waveforms*, 1965; S. Prensky, *Manual of Linear Integrated Circuits*, 1974; S. Seely, *Electronic Circuits*, 1968.

COMPARATOR CIRCUIT

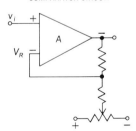

Fig. 4. Regenerative integrated circuit comparator.

Computer

A device that receives, processes, and presents information. The two basic types of computers are analog and digital. Although generally not regarded as such, the most prevalent computer is the simple mechanical analog computer, in which gears, levers, ratchets, and pawls perform mathematical operations—for example, the speedometer and the watt-hour meter (used to measure accumulated electrical usage). The general public has become much more aware of the digital computer with the rapid proliferation of the hand-held calculator and a large variety of intelligent devices, ranging from typewriters to washing machines.

Analog computer. An analog computer uses inputs that are proportional to the instantaneous value of variable quantities, combines these inputs in a predetermined way, and produces outputs that are a continuously varying function of the inputs and the processing. These outputs are then displayed or connected to another device to cause action, as in the case of a speed governor or other control device.

The electronic analog computer is increasingly used for the solution of complex dynamic problems. Electrical circuits, usually transistorized, perform the processing. Electronic amplifiers allow signals to be impressed upon cascaded circuits without significant electrical loss or attenuation through loading of prior stages, a feature absent in purely mechanical computers. Friction in a mechanical analog computer builds up and limits the complexity of the device.

Small electronic analog computers are frequently used as components in control systems. Inputs come from measuring devices which output an electrical signal (transducers). These electrical signals are presented to the analog computer, which processes them and provides a series of electronic outputs that are then displayed on a meter for observation by a human operator or connected to an electrical action device to ring a bell, flash a light, or adjust a remotely controlled valve to change the flow in a pipeline system. If the analog computer is built solely for one purpose, it is termed a special-purpose electronic analog computer.

General-purpose electronic analog computers are used by scientists and engineers for analyzing dynamic problems. A general-purpose analog computer receives its degree of flexibility through the use of removable control panels, each of which carries a series of mating plugs. Outputs from one component are routed to the input of another component by connecting an electrical conductor from one mating plug on the removable board (output) to another plug on the removable board (input). This process is called patching, and the removable panel is frequently called a patch board.

Thus, in any analog computer the key concepts involve special versus general-purpose computer designs, and the technology utilized to construct the computer itself: mechanical or electronic. In any case, an analog computer receives inputs that are instantaneous representations of variable quantities and produces output results dynamically to a graphical display device, a visual display device, or in the case of a control system a device which causes mechanical motion. *See* ANALOG COMPUTER.

Digital computer. In contrast, a digital computer uses symbolic representations of its variables. The arithmetic unit is constructed to follow the rules of one (or more) number systems. Further, the digital computer uses individual discrete states to represent the digits of the number system chosen.

Electronic versus mechanical computers. The most prevalent special-purpose mechanical digital computers have been the supermarket cash register, the office adding machine, and the desk calculator. Each of these is being widely replaced by electronic devices allowing much greater logical decision making and greatly increased speed. For example, many products now carry a bar code, the Universal Product Code (UPC); in suitably equipped supermarkets, the code is scanned by a light-sensitive device, bringing information about each product into the point-of-sale (POS) terminal that has replaced the mechanical cash register. The POS terminal then computes total charges and provides a receipt for the customer. It may also communicate with a centralized computer system that controls inventory, accounts payable, salaries and commissions, and so on. While a mechanical cash register could carry out only a small number of operations each minute, and some electromechanical devices might handle several hundred operations per second, even a small general-purpose electronic computer can carry out its computations at speeds up to a million operations per second. *See* CALCULATORS.

Stored program operation. A digital computer works with a symbolic representation of variables; consequently, it can easily store and manipulate numbers, letters, or graphical information represented by a symbolic code. Typically, a general-purpose electronic digital computer operates on numbers by using both decimal and binary number systems, and on symbolic data expressed in an alphabet. It contains both an arithmetic unit and a storage unit. As the digital computer processes its input, it proceeds through a series of discrete steps called a program. The storage unit serves to retain both the values of the variables and the program to process those variables. The arithmetic unit may operate on either variables or coded program instructions interchangeably, since both are usually retained in the storage unit in the same form. Thus, the digital computer has the capability to be adaptive because processing can be determined both by the previously prepared program and by the data values supplied as input to the computation, and by the values generated during the course of the computation. Through the use of the stored program, the digital computer achieves a degree of flexibility unequaled by any other computing or data-processing device. *See* DIGITAL COMPUTER PROGRAMMING.

Applications. Most digital computers are occupied with performing applications related to bookkeeping, accounting, engineering design, or test data reduction. On the other hand, applications previously considered esoteric are beginning to lead to industrial applications, such as the use of robots on manufacturing assembly lines. Many of the heuristic techniques employed by these robots are based on algorithms developed in such artificial-intelligence applications as chess playing and remotely controlled sensing devices. Modern chess programs are capable of defeating even excellent human players, and computer chess tournaments are held routinely at national and international computer conferences. Of lesser difficulty, perhaps, but illustrative of the kind of information that can be manipulated is the solution of algebraic equations in symbolic form, to provide general algebraic solutions instead of numeric instances of solutions. *See* ARTIFICIAL INTELLIGENCE.

Another area of major societal impact is the growing application of digital computers to word processing, including the entire concept of office automation. Given the ability of even modest computer systems to store, organize, and retrieve very large amounts of information, one may expect the very nature of business offices to change radically. Since 1950 the computer industry has grown into a multibillion-dollar business employing hundreds of thousands of people to build or maintain computers and to program or operate them to perform commercial data-processing tasks or computations related to science or engineering. *See* DATA-PROCESSING SYSTEMS; DIGITAL COMPUTER; WORD PROCESSING.   [BERNARD A. GALLER]

Bibliography: R. W. Hamming, *Computers and Society*, 1972. G. A. Silver, *The Social Impact of Computers*, 1979.

Computer-aided design and manufacturing

Application of the computer in combining design and manufacturing functions continues to gain momentum in diminishing the time between concept and finished product. Specifically, computer-aided design (CAD) refers to the use of computers to perform design calculations for determining an optimum shape and size for a variety of applications ranging from mechanical structures and tiny integrated circuits to maps of huge areas. Computer-aided manufacturing (CAM) employs computers to communicate the work instructions for automatic machinery in the handling and processing technology used to produce a workpiece. Computer-based automation is drastically changing the way things are made and profoundly changing the jobs of people who make them. Among the benefits of an integrated CAD/CAM system are increased productivity, a significant reduction in nonproductive time, improved product quality, and a payback potential obtained by lowering the cost per piece. This modern concept of manufacturing management has led to important advances in the design and production of components used in aerospace, automotive, electronics, and other industries throughout the world

CAD. This first component of the totally integrated CAD/CAM system yields an impressive number of benefits. One benefit is that the designer is equipped with vastly more efficient computer equivalents of the many drafting tools needed in the performance of the job. Computer programs can be written to generate hard copy of drawings at speeds and accuracies far beyond human skills. The computer has liberated the designer from countless, tedious drafting details. Designers are free from the repetitive task of drawing various

lines and, often, even from tasks such as calculating workpiece sizes. In fact, a CAD system not only eliminates the need for drawing dimensions on the views of a part, but also saves the monotonous labor involved in making the arrowheads.

Another time-saving feature of considerable importance is that sectional views as well as auxiliary views can be automatically conceived and drawn by computer methods. The unique system accepts commands like "erase line," "move circle," or "insert dotted or crosshatch lines." CAD systems can be programmed to generate and display a variety of symbols, characters, and points, lines, arcs, and circles—in virtually any form required for the construction of a geometric image. Once developed, part drawings can be stored in computer memory as dynamic, three-dimensional forms or as conventional, shop-type multiview projections.

Another significant benefit of CAD is that a computer-created design can be instantly recalled, either partially or totally. The graphic image can be displayed on a CRT (cathode-ray tube) screen to permit an analysis of the workability of the designer's ideas. So versatile is a computer-graphics system that mirror images of mating parts may be quickly produced and displayed. Computer-refreshed views on a CRT may be manipulated, reduced, enlarged, or viewed from different angles for possible design modifications. With this technique, a great number of design alternatives may be examined within an astonishingly short period of time. Also, when required, a CAD system can automatically generate a control tape that in some manufacturing systems can be employed to drive a numerical control metal cutting or forming machine.

When all of the elements of the design concept are in final and acceptable form, an assembly drawing can be readily generated by recalling each part from computer memory and placing the parts in their appropriate assembled graphical position. A hard copy bill of material may be automatically produced with lines ruled off, correctly spaced, and with all of the components of the design accurately recorded in the proper position.

CAM. The second component of CAD/CAM enables the utilization of processes that allow the machines to perform productive chip removal operations over a much larger percentage of time than heretofore. Using a concept which some machine tool firms call total processing, the manufacturing engineer identifies each processing requirement that interfaces with the computer data base corresponding to the original design. As an example, functional tool design data (that is, proper tool path generation ensured by selected datum plane locations on the workpiece) is integrated with program data relating to the final part design. CAM programs are written that automatically command an optimum machining sequence of processing operations, control the cutter type and size, turn coolant on or off, select appropriate feeds and speeds, and regulate a number of other machining parameters. Correlation of the design phase with the part-processing phase has significantly affected the production cycle. CAM permits a part to progress at a more rapid rate from raw material to finished product.

Programming languages. Computer programs, called software, are the principal form of communication between the programmer and the computer. Compatability of a computer-integrated design with the machine and control begins at the design stage with each aspect taking full advantage of the relative capabilities of the other. While there are many languages available for this purpose, most CAD/CAM software programs use either FORTRAN, COBOL, BASIC, PASCAL PL/1, or NUFORM.

Of all the programs, FORTRAN is currently considered by most to be the easiest programming language to work with in engineering applications. It is the program ordinarily used for solving complex numerical calculations and is a particularly useful language for solving engineering analysis problems. COBOL is mainly a business language. Its principal application is commercial data processing. Despite a somewhat limited arithmetic capability, COBOL is an effective programming language for certain kinds of engineering applications. BASIC, an acronym for Beginners All-purpose Symbolic Instruction Code, was originally developed as an educational tool to be used in teaching the use of remote-control time-sharing systems. The original language has recently undergone significant improvements and standardization. PASCAL was also originally introduced as a teaching tool for computer programming. The principal advantage of PASCAL is in its effectiveness in structured programming. It is currently being evaluated by several agencies as a possible standard. PASCAL is used extensively in universities and colleges and is gaining wide acceptance in industry and government installations. The PL/1 language in some ways resembles FORTRAN. There are also similarities in block structures to ALGOL, while the data types suggest the influence of COBOL. PL/1 is a large general-purpose language well adapted to engineering and scientific applications. NUFORM features a fixed format with numeric input. This versatile system significantly reduces the syntax and vocabulary requirements associated with an alphanumeric input. NUFORM was designed to simplify the programmer's task, thus saving time and promoting accuracy.

Vendor computer programs. Most CAD/CAM programs that have been developed by industry are proprietary and thus are not available for general use. Often the applications of such programs are limited because of their specialized and restricted nature. Commercially available preprogrammed software in the area of mechanism design has been found to be helpful in certain specific problem-solving applications. There are a number of software firms which can supply a full range of computer services. These range from a surprisingly large assortment of comparatively simple, straightforward software programs to a complete data-base manufacturing management system that, in addition to an engineering and manufacturing system, includes software for inventory control bill of material processing, material requirements planning, work in progress, and production costing modules.

Examples of software that can be purchased from software vendors are Automatic Dynamic Analysis of Mechanical Systems (ADAMS), Dynamic Response of Articulated Machinery (DRAM),

Integrated Mechanisms Program (IMP), and Kinematic Synthesis (KINSYN). Examples of other software programs currently available are NASTRAN and CSMP. These programs deal with aspects of engineering analysis. There are also two computer programs in extrusion die engineering currently available at no cost. Developed by Battelle-Columbus under the sponsorship of the U.S. Army, they are ALEXTR and EXTCAM. Both programs have been written in FORTRAN language and operate on a minicomputer. Also available are programs such as STRUDL, ANSYS, and SAP which relate to a newly developed analytical technique called finite element analysis.

Numerical control. Before the advent of numerical control (NC) the regulation of product accuracy and repeatability was directly related to the skill of the operator. The maintenance of precision is no longer considered a serious problem. The limits of accuracy on a workpiece are now controlled entirely by NC machines. The result is that the operator is no longer responsible for positioning and repeating the operation of the tool. Instead of manual controls, electric signals on NC machines precisely guide the movements of the tools and control the position of the workpiece.

NC machines may be operated by manually dialing the machine setting at a console and letting the electronic signals execute the operation. As might be expected, manual operation not only is a very slow production method but is very inefficient for large-volume production of parts. Improved versatility for NC machines is possible by using a punched paper or Mylar tape. In this method the tool and machine instructions are punched onto the tape according to an alphanumeric code. In operation, signals from the punched tape are sent to a data storage unit. As the tape unwinds past the tape reader, electric impulses are sent to the drive mechanisms which automatically control the machine functions. Information on tape can be used to establish the feed rate, spindle speed, machine table positions, traverse rates, tool selection, depth of cut, stops, and dwell intervals for selected periods of time. Unfortunately, there are some inherent problems associated with punched tape. Some of these include difficulties in tape preparation, tape breakage, limitations in adapting programs for a sufficiently wide range of workpiece conditions, and problems associated with editing and with making program changes.

The NC machine was first introduced in the 1950s. It was found that NC machines could be depended upon to operate more hours a day than was possible with traditional machine tools. Also, NC machines were more accurate. Finally, of considerable advantage was the fact that tools and workpieces could be installed and positioned automatically. As programming requirements increased in volume and complexity, however, the tape control machine proved impractical—if not impossible—for many potential jobs. Computer-assisted programming for machine tools appeared to offer a solution. In 1964 graphic computer display terminals were introduced, thus giving the system engineering staff a unique opportunity to visually verify each step in the production sequence of a workpiece.

Direct numerical control. The first industrial direct numerical control (DNC) system became operational in 1968. A DNC system, the lifeline of the CAM concept, includes both the hardware and the software required to drive one or more NC machines simultaneously while connected to a common memory in a computer. The computer may be a minicomputer, several minicomputers linked together, a minicomputer connected to a large computer, or a single large computer. Unlike an NC system, a punched tape is not used. An important advantage of the DNC system is that more than one machine can be operated at a time because the computer can be time-shared. The early DNC systems were produced out of the need to ensure optimum utilization of NC machines.

Some DNC systems consist of combinations with capabilities that encompass CRT display, part program storage, part program edit, and maintenance diagnostics. On some machines, as the program runs, the CRT provides visible part-program information, operating data, current-status messages, error messages, and diagnostic instructions. A typewriterlike keyboard facility located at the NC machine permits the operator to input, delete, or correct data. After the editing and optimizing phases have been completed, the program changes are stored in memory. The sequence of processing operations in CAM systems begins immediately after the program is recalled from storage in the computer memory. DNC programs are commanded entirely by electronic signals sent from the computer memory.

In addition to the primary function, that of controlling the manufacturing sequence of parts, DNC computers can perform a wide range of other useful data transmission functions: parts program development and verification and job scheduling. Thus, DNC systems may be applied to an almost unlimited range of product management activities.

Computer numerical control. The newest CAM innovation is called computer numerical control (CNC). This sophisticated manufacturing concept first appeared in 1970. Unlike DNC, each machine tool has its own computer. The principal advantage of CNC over DNC is that the computer software systems are less expensive. While one standard control can be used for a range of different types of machine tools, the software can be written in such a way as to adapt the control to each particular machine requirement.

Today, extremely large and versatile, totally integrated CAM machining centers are available from a number of well-established machine tool firms. Advantages cited for this trend toward complete manufacturing centers include shorter design lead time and shorter manufacturing make-ready time. In combination, these aspects result in a dramatic reduction of production throughput time. The economic pressures for computer-managed parts technology have never been more apparent. The proof is currently reflected in substantial investments in CAD/CAM research and development activities by machine tool builders both in the United States and in many other countries. The field is changing at an ever-increasing rate, and most forward-thinking parts processers are changing with it. *See* COMPUTER GRAPHICS; DIGITAL COMPUTER PROGRAMMING.

[HERBERT W. YANKEE]

Bibliography: Happy marriage of CAD and CAM, *Machine Design*, 51(2):36–42, Jan. 25, 1979; J. Harrington, *Computer Integrated Manufacturing*, 1974, reprinted 1979; G. L. Petoff, The look of modern CAM standards, *Manufact. Eng.*, 82(4):78–80, April 1979; Society of Manufacturing Engineers, *Numerical Control in Manufacturing*, 1975.

Computer graphics

This term refers to the process of pictorial communication between men and computers. The traditional (and still most widely used) means of inputting problems to computers and receiving answers has been the alphanumeric form—the letters, numbers, punctuation marks, and other symbols which can be handled by such input/output devices as punched card and tape equipment, teletypewriters, and line printers. While any problem and its solution can be reduced to the alphanumeric form by suitable coding conventions, it is often an unnatural and time-consuming way for the human to think and work. An analogous situation would be the publication of a technical textbook without illustrations of any kind, with diagrams replaced by word descriptions, graphs by tables of numbers, and so forth. Anyone who has tried to visualize an odd-shaped plot of land from the usual word form of deed description "starting at bound mark, 321.26 feet northerly, thence 195.81 feet southeasterly, 82.64 feet easterly, . . ." can appreciate the difficulty of quickly comprehending a computer result presented in this way.

This situation has begun to change with the introduction of new types of output devices such as televisionlike cathode-ray tube (CRT) displays and automatic mechanical plotting boards which operate under computer control and permit the computer to draw almost any conceivable form of pictorial output. When a computer is equipped with such devices and suitable controlling programs, the user (the person desiring computations to be performed) can request in preparing his program and input data that the results of his calculations be output graphically, that is, as charts, drawings, or appropriate pictorial representation. Use of these new output devices is showing a rapid increase, particularly in engineering applications, and to many this is what is meant by computer graphics. However, graphical output alone is only half the story—graphical input is also desired in many situations.

Operating modes. Before discussing graphical input, it is first necessary to distinguish between the two major ways of using a computer: "batch" operation and "on-line" operation. In batch operation the user has no direct access to the computer and is usually not present during the time when his job is running in the computer. A computing job must be completely prepared ahead of time, with detailed specification of input data and program, desired output, and alternate courses of action in case of certain error conditions. If an error occurs which prevents completion of the job during the run, the computer outputs partial results, terminates the job, and goes on to the next one in line. The user must then review what results he receives, make corrections in his input, and resubmit the job. This process may have to be repeated several times before all errors are corrected and a satisfactory run is achieved. Although graphical output as described in the preceding paragraph is quite feasible, the forms of graphical input which require close human-computer interaction are not usually feasible.

In on-line operation, on the other hand, the user has direct input/output access to the computer through a console or terminal of some sort during the time his job is running in the computer. In this situation a result or partial result may be immediately presented to the user for decision as to correctness or the need for further input to correct errors or extend the solution. In this on-line situation the computer waits to receive further information before proceeding with the job. The advantage of on-line operation is that the turn-around time is reduced from the hours or days typical of batch operation to a few seconds, providing the intimate human-computer relationship needed for two-way (interactive) graphical communication. It is obvious, however, that having a large computer which can perform millions of operations per second wait idly while the human user makes a decision and takes action is an inefficient use of valuable computer resources. Thus, on-line operation was at first restricted to experimental research facilities and special situations, such as air-traffic control and military systems, where rapid, joint human-computer decision-making is essential. However, two developments in computer technology have acted to increase adoption of on-line computation: small, inexpensive computers, which can be used economically in a dedicated (one-user) mode of operation; and time-sharing techniques for large computers, which permit a computer to economically service a number (perhaps 10–100) of on-line users by switching instantly to work for another user when a particular job reaches a point where human decision is required. Both of these developments have sparked the growing interest in interactive computer graphics. *See* COMPUTER; DIGITAL COMPUTER.

Graphical input. Returning now to the subject of graphical input, this may be divided into two types: noninteractive and interactive. In the first, specialized off-line equipment such as curve tracers and coordinate digitizers (human-guided or automatic) permit a sketch or drawing to be converted to the alphanumeric form on punched cards, punched tape, or magnetic tape. Such data can be used as input in batch computing but will be subject to the slow error-correction process.

The second, or interactive, form of graphical input requires on-line access to the computer. In this case hand-held input devices, such as writing styli used with electronic tablets and light pens used with on-line CRT displays, permit the human to sketch his problem description, say a bridge truss he wishes to analyze, in an on-line interactive mode in which the computer acts as a drafting assistant with unusual powers. For example, since the computer can analyze each element of a sketch as it is entered and apply precoded rules and constraints such as "all lines shall be either exactly vertical or exactly horizontal," it can instantly convert a rough freehand motion of the light pen into an accurate picture element. The computer can also establish connectivity between various picture elements so that the shape of the displayed object, say a rectangle, can be altered by simply indicating with the light pen that one of the sides should be moved. Since the computer knows which lines are connected together, it can automatically adjust the lengths of the adjoining sides as any one side is moved.

There are two key concepts in this process. One is that the computer be continuously informed of the position of the hand-held input device as it is moved over the display screen or writing surface, and the other is that a displayed picture element may be identified to the computer for action such as moving it or erasing it by simply pointing to it with the input device. The action to be taken at any particular time is made known to the computer by additional manual input actions such as pushing one of a number of button switches on a panel, or pointing with the input device to one of a number of alternate commands (called light buttons or a pick list) displayed on the viewing screen along with the picture (usually along the bottom or one edge of the screen).

Graphical output. The simplest forms of computer-controlled plotting boards or CRT displays are the point-plotting types. In these devices the computer specifies numerically the x and y coordinates of a point on the drawing surface or viewing screen, which causes an ink dot or spot of light to be made at the appropriate location. In the case of a CRT it is usual to convert the numerical values to analog voltages by means of digital-to-analog converters, and then to use these voltages to drive the deflection system of the CRT, as shown in Fig. 1. Magnetic deflection systems are usually used in preference to electrostatic deflection because of the better spot focus and deflection linearity which can be achieved. It is common to divide each deflection axis into 1024 (10 binary digits) distinct addresses, which provides over a million possible dot locations. The speed of plotting depends upon the distance that the electron beam must move between dots, and varies from 1 μsec for adjacent addresses to perhaps 30 μsec for full-screen deflection between dots. Screen sizes are usually about 10 in. square, limited primarily by CRT cost and technology.

Mechanical plotting boards, which must physically move the pen and the writing surface relative to each other, bear considerable resemblance to numerically controlled machine tools. Mechanical plotting is of course much slower than moving an electron beam in the CRT, and point-plotting rates

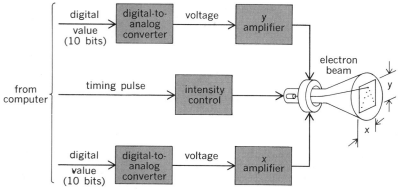

Fig. 1. Point-plotting CRT output display.

vary from a few per second up to several seconds per point, depending upon the distance to be moved between points. This slowness of plotting, coupled with the fact that a picture element once drawn cannot be erased, makes plotting boards unattractive for interactive graphics as previously described. However, the resolution of plotting can be increased by simply making the writing surface bigger (up to 6 × 8 ft in some models), permitting much more detailed drawings than can be displayed on a CRT.

The point-plotting mode of operation described above, while seemingly rudimentary, provides a complete drawing capability. Closely spaced dots plotted in a row merge into continuous appearing lines, letters and numbers can be formed from arrays of dots similar to those used in football scoreboards, and so forth. The main disadvantage of point-plotting devices is that large amounts of data are required to produce a picture (20 or more binary digits for each point) which places a burden on the computer, and the plotting speed is limited. This latter factor is most important in CRT displays, where the entire picture must be "refreshed" (continuously rewritten) 30 or more times per second in order to provide a flicker-free presentation. For example, if the CRT is refreshed 30 times per second and the plotting time is 30 μsec per point, pictures containing only up to about 1000 points can be displayed. Of course, more points can be displayed if one is willing to accept a lower refresh rate with attendant flicker. This is not a desirable situation, however, and has led to development of additional hardware to provide high-speed symbol and line (vector) drawing capabilities (Fig. 2).

Most display systems now include point, line, and symbol drawing modes. The point mode operates as described previously, but the other two modes produce an entire picture element with a single command from the computer, and at a much higher plotting rate than if it were drawn with individual points. Line generators operate either with specification of both end points or of the length of line to be plotted relative to current beam position. In either case all of the intermediate points on the line are automatically filled in by the line generator of the display system, either by rapid point plotting (incremental digital technique) or by continuously sweeping the beam (analog technique).

Of the two techniques the analog technique is considerably faster, permitting a line to be drawn clear across the screen in as little as 30 μsec. Symbol generators take many forms, the more common being based on arrays of dots, stroke techniques, or incremental tracing of the symbol shapes. The key feature of any of these is a local fixed memory which contains the detailed information about the shape of each symbol. Thus, the only input the symbol generator needs is the symbol code (6 to 8 binary digits) to tell it which symbol to plot. Current symbol generators operate from about 100 down to as little as 5 μsec per symbol, permitting as many as 4000 symbols to be displayed at a refresh rate of 30 frames per second. The capacity of a point-pointing display at the same refresh rate would be only 50–100 symbols, since 10–20 points are needed per symbol.

In addition to the basic display functions de-

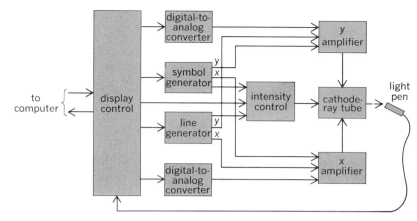

Fig. 2. General display system organization.

scribed above, some display systems have been designed with additional display functions to improve performance in special situations. Examples are generators for drawing circles or other common geometric curves with one command, hardware for scaling or rotating three-dimensional representations without requiring recomputation of the display data, and automatic insertion of perspective into three-dimensional representations. The state of the art is changing rapidly, and new ideas and capabilities are being introduced at a rapid pace.

Graphic output devices of the type described above are called cursive, that is, the computer draws narrow lines as if with a pencil, and produces line drawings. The ability of such display hardware to produce area-coverage tonal pictures is limited

Fig. 3. On-line interactive graphics console.

by the total length of line that can be drawn at a flicker-free rate. Interest in more pictorial output has led to the development of CRT computer displays borrowing television technology—raster displays, which include color capability. In these systems, the computer builds up a synthesized video signal bit by bit in a memory (this usually takes much longer than the refresh cycle), and the stored data are then used to drive the CRT at a flicker-free rate. Rather spectacular advances in computer algorithms for solving the hidden line problem and control of shading permit computer graphic output of almost photographic quality, with all of the interactive control features of the cursive displays except real-time motion of the displayed images.

Interactive input/output arrangements. The light pen, used for graphical input in conjunction with on-line CRT displays, consists of a photodetector mounted in a pen-shaped holder. When the pen is held over a displayed picture element, the pen senses the light created while the electron beam is writing that particular element and sends a signal to the computer (Fig. 3). Since the computer knows which element it is writing at any given time, the time of receipt of the signal from the pen immediately identifies to the computer which element in the picture the pen is pointing

at, providing a convenient and natural method for the human user to "discuss" a picture with the computer. Because the light pen is a passive device which can only respond to displayed information, it provides no information to the computer when held over a blank part of the screen. Thus, in order to draw with a light pen, it is necessary to perform a process known as pen tracking. One way to do this is to continuously display a "tracking cross," a cross-shaped pattern made up of dots, which is slightly larger than the field of view of the pen. As each dot is plotted, the pen reports whether or not it "saw" it, and by comparing the number of dots seen on each arm of the cross, the computer can tell whether the cross is centered under the pen. If not, it makes an appropriate positional correction before the next tracking cycle, and by this process it keeps the cross continuously aligned with the pen as the pen is moved over the viewing screen. Since the tracking cross is generated by the computer, its position (and thus that of the pen) is always known, providing the continuous input needed for drawing. The pen-tracking process is fairly effective for input drawing, but suffers from lack of precision and the use of a substantial amount of computing time to perform the tracking function.

Other devices. The effort to find a better input means than the light pen has produced a number

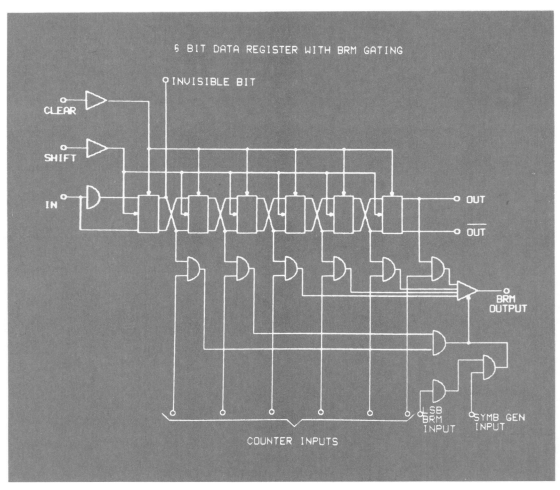

Fig. 4. Display of electronic circuit.

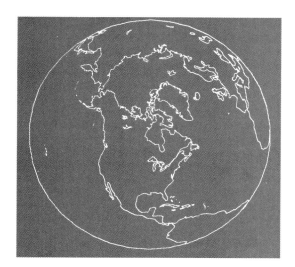

Fig. 5. Example of computer-produced map.

of new devices, the most popular of which is the electronic tablet. A number of different electronic detection principles are used in making tablets, but the end result is the same: The circuitry produces a pair of digital *x-y* coordinate values corresponding continuously to the position of a writing stylus upon the surface of the tablet. The advantage of the tablet is that no computer time is required for operation as in pen tracking (except to sample the input from the tablet), and resolution equal to or exceeding that of CRT display is achieved. Also, the tablet can be placed on the table in front of the display screen, producing a more natural writing position for the human user. In one application of tablets, numbers and block letters written on the tablet as one would on a sheet of paper are recognized by a special computer program, providing a form of alphanumeric input which some users find more convenient than typing on a keyboard. Another advantage of tablets is that in most of them the stylus is capacitively coupled to the tablet surface, permitting the curves or drawings which already exist in paper form to be placed on the tablet surface and traced as input. Tablets are considerably more costly than light pens, however.

To provide interaction between the tablet and the display, it is usual to display a cursor (a dot or tracking symbol) corresponding to the current position of the stylus on the writing surface. The user then uses the stylus to steer this cursor around the screen to point to displayed objects, draw additional picture elements, and so forth. Other devices such as joy sticks, track balls, and slewing buttons are also used to steer cursors on the screen in many displays. These are all of relatively low cost and provide satisfactory pointing capability; however, the two-axis form of control that is characteristic of these devices is too awkward to permit drawing.

Other computer graphics devices of note are CRT recorders and direct-view storage-tube (DVST) displays. CRT recorders operate like CRT displays, but are designed to record graphic images on film rather than display them for human viewing. This permits use of small, high-resolution

CRTs with up to 4000 dot addresses in each axis. Essentially, CRT recorders act like high-speed plotting boards capable of producing 2–10 pictures per second. One interesting use of CRT recorders is to produce animated movies. The DVST displays have come to the fore as a means of avoiding the refresh requirements of standard CRT displays, particularly in low-cost remote display terminals driven over low bandwidth telephone-line connections. The DVST is analogous to the familiar "Etch-a-Sketch" toy, maintaining any picture element once written until the entire screen is electronically erased.

Applications. Graphic output from computers operated in batch mode has received wide acceptance in a variety of scientific and business applications (Figs. 4 and 5). Computer-produced graphs and charts are sufficiently readable to permit direct use in reports, eliminating laborious and costly manual plotting and drafting. In applications such as highway layout and piping layouts in ships and refineries, direct pictorial output is of great value. Interactive graphics is not yet used as widely because of its dependence on the availability of on-line computing facilities, but has already been applied in such diverse fields as integrated circuit design, dress-pattern layout (with automatic scaling of patterns for different sizes), aircraft design, and study of molecular structures. A whole field, called computer-aided design, depends on interactive computer graphics for the intimate, pictorial person-computer communications required. *See* COMPUTER-AIDED DESIGN AND MANUFACTURING.

[JOHN E. WARD]

Bibliography: J. D. Foley and A. Van Dam, *Fundamentals of Interactive Computer Graphics*, 1982; S. Harrington, *Computer Graphics: A Programming Approach*, 1983; W. M. Newman and R. F. Sproull, *Principles of Interactive Computer Graphics*, 2d ed., 1979; J. E. Scott, *Introduction to Interactive Computer Graphics*, 1982.

Computer security

The preservation of computing resources against abuse or unauthorized use, especially the protection of data from accidental or deliberate damage, disclosure, or modification. Safeguards are both technical and administrative. Threats are both external (physical) and internal (logical) to the computer system.

Computing systems provide powerful tools for the storage and use of information. They are equally powerful tools for the misuse of the same information. Because of the pervasiveness of computers in modern society, the potential for abuse has triggered public awareness of the need for computer security. Computer abuse, financial fraud, invasion of individual privacy, industrial espionage, and problems with national security, are primary concerns. The right of privacy is the right of individuals or organizations to determine the degree to which information about them is collected, stored, and shared with others. Protection of privacy is an ethical and legal issue rather than a technical one, and is not unique to computing systems; however, computer security mechanisms can be used to protect information for privacy as well. Protection of data integrity is assurance that stored data will be accurate and consistent.

Computer security policies are rules of access to computing resources. Security mechanisms are the means by which policies are implemented. Discretionary access-control policies support the concept of data ownership by individual users and their right to share the data with other users. Nondiscretionary policies, used primarily by the military, rely on a built-in hierarchy of security classifications used to make access decisions. Security based on classification hierarchies is also called multilevel security. Authorization is the process of granting access rights to, and revoking the access rights of, various users of a discretionary system. Enforcement is the process of ensuring that all accesses which occur are in accordance with the current state of authorization.

Maintaining computer security is a management responsibility involving both technical and administrative problems. Management tasks include policy formulation, arrangements for physical security, cost modeling, and auditing procedures to validate the security controls and keep a record of system transactions. The primary function of management is to evaluate security threats, risks, policies, and mechanisms.

Physical and logical security. Physical and logical security are two important parts of computer security. Physical security measures, external to the hardware and software, are intended to prevent, or to facilitate recovery from, physical disasters such as fire, floods, riots, sabotage, and theft of physical resources. Logical security mechanisms, internal to the computing system, are used to protect against internal misuse of computing resources, such as the theft of computing time and unauthorized access to data.

Physical security includes the control of physical access to computing resources. In order to make decisions about who can have access to each terminal and each user account, as well as to the computer room itself, the installation must have one or more mechanisms for authenticating personal identification. Some physical access-control mechanisms are based on something a person has, such as a key for a lock or a magnetically striped card; or something a person knows, such as a password or the correct answers to an authentication dialog. Keys and cards can be lost or stolen, and passwords copied or forgotten—or sometimes even guessed. Thus, some personal identification mechanisms are based on something a person is, such as dermatoglyphics (for example, fingerprints or lip prints) or anthropometrics (for example, individual hand or head geometry). Analysis of the dynamics of pen movement and pressure during the signing of a person's signature is a promising research direction. Cryptographic digital signatures provide unforgeable and verifiable user authentication. Most of the physical security mechanisms, of course, are not limited to computer security.

The problems of logical security, on the other hand, are more specific to computing. Logical computer security has several facets, including access controls, information flow controls, inference controls, and communication controls.

Logical access controls. All modern multiuser operating systems have some access controls for the sharing of resources, especially data. The objects protected by many systems are files. A coarse granularity of protection allows access to all of a file or none of it, and governs only a few access types, or operations on the objects, such as READ, WRITE, EXECUTE, or DELETE.

The access matrix (Fig. 1) is the basis for many methods of representing discretionary authorization information. The rows of the access matrix represent subjects, or users, of the system. The columns correspond to objects, or resources, of the system (mainly files). The authorization information contained in cell (x, y) of the table is a list of access types (allowable operations) that subject x may apply to object y. Access matrix entries are created, propagated, modified, and revoked as part of the authorization process. The enforcement process uses the matrix to determine whether to allow a request by a given subject to perform a specified operation on a certain object. A row of an access matrix is a capability list, which contains the access rights of a given subject to various objects. A column of a matrix is an access control list, which contains the access rights of various subjects to a given object.

The configuration of its access matrix at any point in time defines a system's protection state. The protection state is changed as a result of an authorization operation granting or revoking access rights, creating or deleting subjects or objects. These authorization operations are of course also subject to rights in the matrix. Theoretical results show that there is no general way to prove whether a given state of protection is safe, in the sense of whether it can leak an access right to an unauthorized cell in the access matrix. In practice, however, the safety question is trivially decidable for most real protection systems.

Access controls are also used in data-base management systems, but there are additional requirements. Data-base security needs a finer granularity (separate access decisions for small units of data). Access decisions must be based on a broader range of dependencies. For example, access might be conditional on data content, as in a policy which states that users in group X may not read any record from the personnel file for which the

Objects / Subjects	File-1	File-2	•••	SORT ROUTINE	•••	File-n
SYSADMIN	--	--		OWN READ EXECUTE		--
USER 1	OWN READ	OWN READ WRITE		EXECUTE		--
SORT ROUTINE	READ	WRITE		--		--
•••						
USER 17	--	READ		--		OWN READ WRITE

Fig. 1. Access matrix.

salary value is greater than $20,000. This policy cannot be implemented by the access matrix described above. Instead, the access condition (salary must be less than or equal to $20,000) is stored in the matrix and is evaluated at access time for each record that is about to be retrieved. This approach is called predicate-based protection. Procedural monitors, called formularies, have been used to evaluate access conditions. In another approach, called query modification, the access conditions are automatically combined with data-base queries before the data is retrieved. Queries thus modified cannot request unauthorized data. *See* DATA-BASE MANAGEMENT SYSTEMS.

Information flow controls. While access controls protect access to data, they do not solve the problem of how to regulate what happens to the data after it is accessed. For example, someone authorized to read a top-secret file can make a copy of it. If access to the copy is not controlled as strictly as access to the original, the security of the information in the original can be compromised through the copy. Flow controls are concerned with labeling outputs of processes with authorization information consistent with the protection given the inputs. A multilevel protection policy called the star property expresses a flow restriction by not allowing information to flow from a given security classification to one that is lower (so that no one can later read the same information under a less strict access requirement). This property is extended in a lattice model of information flow, based on a hierarchy of security classes, which include both levels (for example, unclassified, confidential, secret, top-secret) and sets of compartments (for example, NATO, nuclear, medical, financial) representing a need-to-know policy. Output from a program is not allowed to be sent to objects with security levels lower in the lattice than the highest level which has flowed from the inputs so far during the execution of the program. Alternatively, the security level of the file to which output is to be written can be raised, allowing the writing to proceed.

Security kernels. The difficulty of verifying a piece of software as large as an operating system had led to the concept of security kernels into which all security-related functions have been concentrated. An operating system kernel is a small, certifiably secure nucleus which is separate from the rest of the operating system. Nonkernel operating system software can then be allowed to run in a less protected environment. This new structure for operating systems, though achieved at the cost of a slight reduction in performance, has increased the trustworthiness of operating systems greatly over that of earlier systems with their defect-ridden, overly complicated patchworks of software. Most kernels implement multilevel, nondiscretionary security policies for access and flow controls. The simple security condition controls access by preventing a user operating at a given security level from reading data classified at a higher level. The star property controls flow by preventing the user from writing to a higher level. *See* OPERATING SYSTEMS.

Inference controls. Statistical data bases are intended to provide statistical information (for ex-

ample, demographic data) about groups of people without allowing access to data about specific individuals. However, it is often possible to infer data about an individual, given the statistical results for a group and some amount of prior knowledge about the individual. As an example, consider this query and its result:

$$\text{COUNT}(\text{COLOR}\_\text{EYES} = \text{GREEN AND}$$
$$\text{TOWN} = \text{LYONSDALE}, \text{NY}) = 1$$

If it is known that a given individual fits this description, other details about that individual can illegally be deduced, for example, by:

$$\text{COUNT}(\text{COLOR}\_\text{EYES} = \text{GREEN AND}$$
$$\text{TOWN} = \text{LYONSDALE}, \text{NY}$$
$$\text{AND OCCUPATION} = \text{BOOKMAKER})$$

If this count is 1 also, that individual is a bookmaker. If small counts are not allowed in the responses, inference can still be made if information about the target individual is revealed in the difference between two large sets of people.

A mathematical device called a tracker provides a more general approach to inferring individual values. This approach is based on submitting statistical queries whose responses overlap. The overlapping process partitions the data base to form sets of records which act by inclusion and exclusion to isolate an individual record. Theoretical results indicate that very few statistical data bases can be made safe from all inference without becoming too restrictive for intended uses. For very large statistical data bases (for example, census data), it is effective to allow access only to randomly selected sample subsets of data which summarize the statistical characteristics of the full data base. Sometimes inference can be forestalled by introducing noise, in the form of slight random variations, into the data values.

Cryptographic controls. Cryptographic controls provide the means to encode data for storage or transmission so that even if an unauthorized person sees the data it will still be unreadable. Cryptographic controls thus protect transmitted data against electronic eavesdropping (for example, wiretapping). They also protect against the reading of data from tapes or disk packs that have been removed from their intended computers and

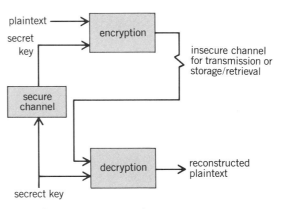

Fig. 2. Conventional encryption.

therefore from access controls. Cryptographic controls protect the secrecy of the data and can provide proof of authenticity, so that a receiver of the data can be certain who sent it and that it is unmodified. Encryption is an encoding of the data, to be stored or transmitted, into a secret message by means of a key, which is the map to the encryption method. Decryption, also by a key, is necessary to return the data to its plainly readable form, called plaintext (Fig. 2). Transposition is a method of encryption that rearranges the symbols of the message. Substitution methods replace each message symbol with another symbol. Polyalphabetic substitution alternates between two or more sets of substitution alphabets for each plaintext symbol. If substitution is made from a table of symbols, the table is the key. Typically, the key is a parameter to an algorithm which computes the substitutions. For example, in a Caesar Cipher, for each letter in the alphabet the substitute is the letter which appears x letters later in the alphabet. Here, x is the key. If x is 1, substitute B for A. If x is 2, substitute C for A, and so on. The success of the system depends not on hiding the method of encryption, but on keeping the value of the key hidden. Thus, although the messages can be transmitted over open channels (channels susceptible to eavesdropping), the keys must be sent securely (perhaps hand-carried). Computer-generated pseudorandom sequences (sequences of nearly random numbers whose generation can be repeated) provide large, variable keys that are combined arithmetically (by multiplication, and so forth) with the numeric codes for the plaintext symbols to make encryptions which are very difficult to break.

In the interest of common hardware and communication interfaces, the National Bureau of Standards has produced a standard encryption algorithm, the Data Encryption Standard (DES), which has been mass-produced in integrated circuit microchips. The DES algorithm uses a combination of substitution and transposition and has been criticized for being too easy to break.

A recently developed method, called public-key encryption, uses two keys, one for encryption and a different one for decryption (Fig. 3). The advantage is that the encryption key does not have to be transmitted in a secret way (for example, hand-carried). Anyone sending a message to party X uses X's encryption key, which is known publicly. The decryption key, however, is private to X. The two keys are related mathematically, but it is intended to be computationally too difficult to derive the decryption key from the public encryption key. *See* CRYPTOGRAPHY.

[H. REX HARTSON]

Bibliography: D. E. Denning, *Cryptography and Data Security*, 1982; L. J. Hoffman, *Modern Methods for Computer Security and Privacy*, 1977; E. B. Fernandez, R. C. Summers, and C. Wood, *Database Security and Integrity*, 1981; H. R. Hartson, Database security: Systems architecture, *Inform. Sys.* 6:1–22, 1981.

Computer storage technology

The techniques, equipment, and organization for providing the memory capability required by computers in order to store instructions and data for processing at high electronic speeds. In early computer systems, memory technology was very limited in speed and high in cost. Since the 1970s, the advent of high-speed random-access memory (RAM) chips has significantly reduced the cost of computer main memory by more than two orders of magnitude. Chips are no larger than ¼ in. (6 mm) square and contain all the essential electronics to store thousands of bits of data or instructions. Traditionally, computer storage has consisted of three or more types of memory hierarchy storage devices (for example, magnetic core memories, disks, and magnetic tape units).

MEMORY HIERARCHY

Memory hierarchy refers to the different types of memory devices and equipment configured into an operational computer system to provide the necessary attributes of storage capacity, speed, access time, and cost to make a cost-effective, practical system. The fastest-access memory in any hierarchy is the main memory in the computer. In most computers manufactured after the late 1970s, RAM chips are used because of their high speed and low cost. Magnetic core memories were the predominant main memory technology in the 1960s and early 1970s prior to the RAM chip. The secondary storage hierarchy usually consists of disks (both movable-arm and head-per-track). Significant density improvements have been achieved in disk technology so that disk capacity has doubled every 3–4 years. Between main memory and secondary memory hierarchy levels, however, there has always been what is called the memory gap. It is characterized by a wide gap in both performance and cost. The last, or bottom level (sometimes called the tertiary level) of storage hierarchy is made up of magnetic tape transports and mass storage tape systems (Fig. 1).

The memory hierarchy appears as three groupings which are characterized by performance and price (Figs. 2 and 3). Performance is broken down into two parameters: capacity and access time. (Speed or data rate is a third parameter, but it is not so much a function of the device itself as of the

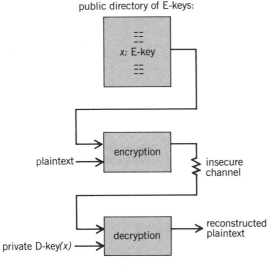

Fig. 3. Public key encryption.

overall memory design.) Capacity is the maximum on-line capacity of a single connectable unit. Access time is the time required to obtain the first byte of a randomly located set of data. The price is end user–oriented and is based on some useful arbitrary amount of memory (for example, 1000 bytes).

MEMORY ORGANIZATION

The efficient combination of memory devices from the various hierarchy levels must be integrated with the central processor and input-output equipment, making this the real challenge to successful computer design. The resulting system should operate at the speed of the fastest element, provide the bulk of its capacity at the cost of its least expensive element, and provide sufficiently short access time to retain these attributes in its application environment. Another key ingredient to a successful computer system is its operating system (that is, software) that allows the users to execute jobs on the hardware efficiently. Operating systems are available which achieve this objective to a reasonably high level of optimization.

The computer system hardware and the operating system software must work integrally as one resource. In many computer systems, the manufacturer provides what is called a virtual memory system. It gives each programmer automatic access to the total capacity of the memory hierarchy without specifically moving data up and down the hierarchy and to and from the central processing unit. During the early years of computing, each programmer had to incorporate storage allocation procedures by determining at each moment of time how information would be distributed among the different hierarchic levels of memory, whenever the totality of information was expected to exceed the size of main memory. These procedures involved dividing the program into segments which would overlay one another in main memory. The programmer was intimately familiar with both the details of the machine and the application algorithms of the program. This all changed in the 1970s when sophisticated higher-level program languages and data-base management software became well established to provide significantly greater problem-solving capability. Thus evolved manufacturer-supported operating systems with complete built-in virtual memory support capabilities, which made it possible for the user to ignore the details of memory hierarchy internal software and hardware operations. *See* DIGITAL COMPUTER PROGRAMMING; MULTIACCESS COMPUTER.

In the area of memory organization, two types of main memory augmentation have been employed in the more widely used computers to enhance the central processing unit's (CPU) performance. One is called a cache memory, which speeds up the flow of instructions and data into the central processing unit from main memory. This function is important because the main memory cycle time is typically slower than the central processing unit clocking rates. To achieve this rapid data transfer, cache memories are usually built from the faster bipolar RAM devices rather than the slower metal oxide semiconductor (MOS) RAM devices. The other type of supplementary main memory is called bulk store (also known as large core storage

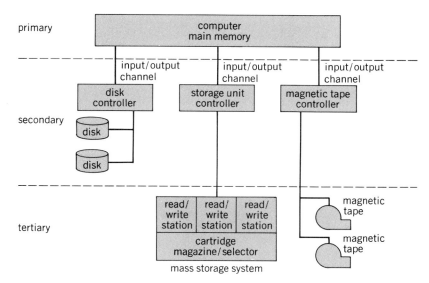

Fig. 1. Memory hierarchy levels and equipment types.

and as extended storage memory). Typically it consists of magnetic core memories or MOS RAM chips. Whatever the implementation, bulk store is intended to extend the main memory capacity for programs and data at a price per byte that is less than main memory but not as cheap as disk. Bulk store is in fact an attempt to fill the memory gap with whatever technology seems practical.

MAIN MEMORY

The rapid growth in high-density large-scale integrated (LSI) circuits has advanced to a point where only a few applications require the tens to hundreds of thousands of transistors that can be placed on a chip. One obvious exception is computer main memory, in which there is a continual demand for higher and higher capacity at lower cost.

Magnetic core memories, which dominated main memory technology in the 1960s and early 1970s, have been replaced by semiconductor RAM chip devices of ever-increasing density. This process started with the introduction of the first MOS 1K-bit RAM memory in 1971. (1K is equal to 1024 bits.) This was followed with the 4K-bit RAM chip in 1973, the 16K-bit RAM chip in 1976, the 64K-bit RAM chip in 1979, and the 256K-bit RAM chip in 1982. Figure 4 shows the progression of RAM chips, which follows the "rule of four," according to which the cost of development of a new RAM device generation can be justified only by a factor-of-four increase in capacity. *See* INTEGRATED CIRCUITS.

The 256K-bit MOS RAM has pushed photolithographic fabrication techniques to their practical limits, with feature sizes of 2 to 1 micrometer. For RAM chip densities of 256K bits or more per device, the integrated circuit industry has had to make a twofold improvement: (1) better means, such as electron-beam imaging, to achieve smaller component dimensions, and (2) a dry processing technique for selectively removing excess materials. Considerable progress has been made in the use of both x-ray and electron-beam techniques to

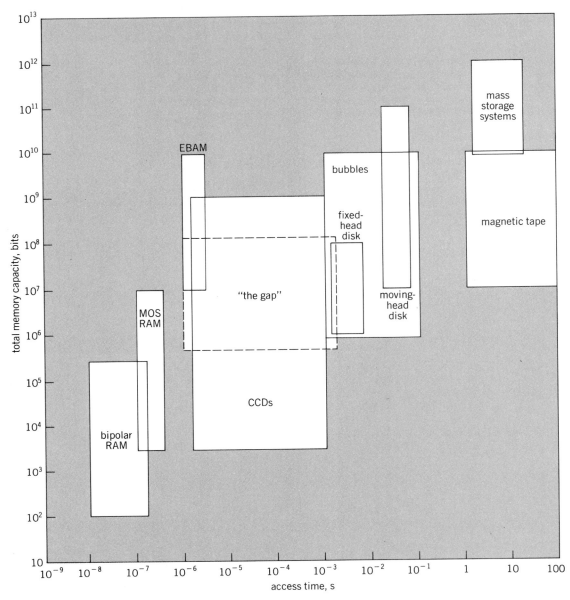

Fig. 2. Memory capacity versus access time for various storage technologies.

achieve the submicrometer-size features needed to make higher-density RAM chips. In addition, industry needs to meet the challenge of increased fabrication complexity and still achieve submicrometer dimensions in volume production.

RAM chips. RAM chips come in a wide variety of organizations and types. Computer main memories usually consist of random addressable words in which the word length is fixed to some power-of-2 bits (for example, 4 bits, 8 bits, 16 bits, 32 bits, or 64 bits). But there are exceptions such as 12-bit, 18-bit, 24-bit, 48-bit, and 60-bit word-length machines. Usually RAMs contain $NK \times 1$ (for example $64K \times 1$) bits, so the main memory design consists of a stack of chips in parallel with the number of chips corresponding to that machine's word length. There are two basic types of RAMs, static and dynamic. The differences are significant. Dynamic RAMs are those which require their con-

tents to be refreshed periodically. They require supplementary circuits to do the refreshing and to assure that conflicts do not occur between refreshing and normal read/write operations. Even with those extra circuits, dynamic RAMs still require fewer on-chip components per bit than do static RAMs (which do not require refreshing).

The fact that they require fewer components makes it possible to achieve higher densities with dynamic RAMs than are possible with static RAMs. These higher densities also lead to lower costs per bit. The production techniques for the two kinds of chips are identical. Therefore, the cost per unit quickly becomes the cost for mass-producing one chip. Since building costs per chip are about the same whether they store 4K bits or 16K bits, higher densities lead to lower costs per bit.

Static RAMs are easier to design, and compete

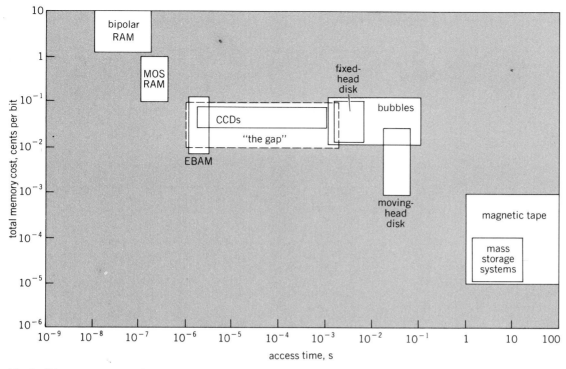

Fig. 3. Price versus access time for various storage technologies.

well in applications in which less memory is to be provided, since their higher cost then becomes less important. They are often chosen for minicomputer memory, or especially for microcomputers. Because they require more components per chip, making higher bit densities more difficult to achieve, the introduction of static RAMs of any given density follows behind that of dynamic versions.

There is another trade-off to be made with semiconductor random-access memories in addition to the choice between static and dynamic types, namely that between MOS and bipolar chips. Bipolar devices are faster, but have not yet achieved the higher densities (and hence the lower costs) of MOS. Within each basic technology, MOS and bipolar, there are several methods of constructing devices, and these variations achieve a variety of

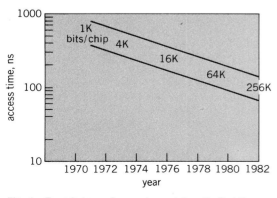

Fig. 4. Trends in performance and density (in bits per chip) of dynamic RAM chips.

memory speeds and access times, as well as power consumption and price differences. Table 1 provides a sampling of the many different technologies within the basic silicon MOS and bipolar semiconductor technologies, and their varying characteristics. *See* LOGIC CIRCUITS; TRANSISTOR.

The RAM memory cycle time is defined as that time interval required to read or write one word of data. Access time is defined as the time interval in which the data are available after the initiation of the read cycle. From the user's point of view, cycle time is also an important characteristic because it has more impact on overall computational speed of the system. A new data read/write cycle does not begin until the previous read cycle is completed. The specified timing, signal overlap, and tolerances allowed vary with each RAM main memory system.

Error checking and correction. Most computer main memories have a memory fault control consisting of a memory controller and a fault control subsystem which automatically corrects all single-bit errors and detects and reports all double-bit or three-bit errors. Each bit in a word is read from a separate RAM device. The fault control permits computer operation to continue even if a memory module is malfunctioning, and thus the computer will operate even if a memory chip is removed from the board. Failures are pinpointed to the specific chip by fault-indicating light-emitting diodes built into each array board.

ROMs, PROMs, and EPROMs. Microcomputers have evolved their own special set of semiconductor memory chips to suit their application needs. Whereas large, medium-size, and "mini" computers primarily use only RAMs that have read/write capability, microcomputers have found significant

Table 1. Characteristics of different RAM memory semiconductor technologies

Semiconductor technology*	Bipolar type of LSI semiconductor				MOS type of LSI semiconductor			
	ECL	I²L	ST²L	T²L	CMOS	HMOS	NMOS	VMOS
Static or dynamic	Static	Dynamic	Static	Static	Static	Static	Dynamic	Static
Storage capacity, bits	4K × 1	4K × 1	4K × 1	4K × 1	4K × 1	4K × 1	4K × 1	4K × 1
Access time, nanoseconds	30	120	70	50	170	70	150	55
Operating power, milliwatts per chip	1000	450	600	900	30	500	462	500
Cents per bit	0.85	0.58	1.0	0.79	0.72	0.91	0.33	0.61

*ECL = emitter coupled logic; I²L = integrated injection logic; ST²L = Schottky transistor-to-transistor logic; T²L = transistor-to-transistor logic; CMOS = complementary metal oxide semiconductor; HMOS = high-performance MOS; NMOS = N-channel MOS; VMOS = vertical MOS.

use for read-only memory (ROM) chips and programmable read-only memory (PROM) chips. Because microcomputers are typically used in dedicated applications and not in general-purpose programming installations, their data and program storage can be separately allocated between RAMs and ROMs. ROMs are cheaper and provide protection, since the contents are fixed or hardwired. During microcomputer program development, PROMs are used because they can be rewritten to modify the contents. Typically, PROMs are cleared by putting the PROM chip under an ultraviolet light to zero out its contents and then using a PROM programmer to write in the new bit pattern for the modified program. After the program has completed final test and acceptance, the final bit pattern can be put into ROMs (using a chip mask) for making production units. There are also chip devices called electrically alterable programmable read-only memories (EPROMs); they have an electrical switch to allow one to rewrite new contents into them. This means EPROMs do not have to be removed from the circuit to clear their contents in order to put in the new or modified bit pattern representing the program instruction.

MEMORY GAP TECHNOLOGIES

The memory gap noted by J. P. Eckert (a developer of the ENIAC computer) and others in the 1940s still presents a problem for the data-processing system designer. The wide gulf between fast, expensive main memory devices and slow, inexpensive secondary memories can be seen in the following memory gap performance/cost ranges: access time 1 microsecond−30 milliseconds; capacity $10^7 - 10^9$ bits; price per bit 0.1−0.001¢. The memory gap (Figs. 2 and 3) for access time is bounded on one side by a 1-μs typical computer main memory cycle time and on the other by the 30-ms typical disk-drive access time. Capacity is bounded by a reasonable main-frame memory capacity of about 1 megabyte and by the typical disk capacity of 100 megabytes. The price is then determined by these existing technologies, that is, 0.1¢ per bit for RAMs and 0.001¢ per bit for disks.

The semiconductor RAM and disk technologies are continuing to improve in performance and price, narrowing the gap. It is unlikely, however,

that the proved technologies will spread far enough to completely close it. Charge-coupled devices (CCDs) of 64K bits and 256K bits per chip have been used integrally in a few commercial computer systems as bulk store between main memory and secondary disks. Some disk replacement units have been designed by using CCDs. Electron-beam-accessed memories (EBAM) were only developed and tested in the laboratory. Bubble memory units are on the slow-access-time side of the memory gap. Bubble memories are most promising as low-capacity disk replacements and for special applications.

On the slow-access, low-cost, high-capacity side of the gap, there is especially tough competition for the new technologies from existing head-per-track disks. Any new competitor must be at least as good as these disks in terms of reliability, error rates, and maintenance costs. Also, any new subsystem must be fully developed in terms of hardware, software, and interfacing, and it must fit into existing environments, before users will buy and install it.

Charge-coupled devices. Charge-coupled devices (CCDs) first became available in 1975, but their actual use in computer-related equipment has been minimal. (CCDs, which are memory devices, should not be confused with another kind of CCD which is used as an optical sensor device.) CCDs, because of their lower circuit-component requirements, can potentially achieve higher densities than the RAMs with which they usually compete. See CHARGE-COUPLED DEVICES.

Electron-beam-accessed memories. There are no commercially available electron-beam-accessed memories (EBAM). They have been made to work only in laboratory environments. This is in spite of the fact that some of the earliest computers were equipped with cathode-ray tube (CRT) memories called Williams tubes. The CRT devices were used on the early computers produced at Manchester University, England, and on Princeton's IAS, all of which appeared shortly after World War II, and on other early machines.

Since their "demise," EBAMs have seen a great deal of improvement, if only in the laboratory. Access times for modern versions are typically 100 μs (although it takes twice that long to switch from reading to writing or vice versa), and transfer rates are typically 10 megabits per second (which is readily compatible with host-computer channel rates). Data are written onto a MOS storage chip (target) by means of an electron beam, either using a single tube or a configuration of parallel tubes where each holds one bit of a data word. The MOS target inside the tube stores the data as a positive charge on its silicon dioxide plane. Readout is accomplished by using a lower-energy electron beam to interrogate each memory cell. The read beam causes a high or low current to flow off the target, depending on whether a 1 or 0 was stored in that cell.

Magnetic-domain bubble devices. Bubbles are different from semiconductor memories in that they are magnetic devices, in which the absence or presence of a magnetic domain is the basis for a binary 1 or 0. A magnetic bubble is in reality a cylindrical magnetic domain with a polarization opposite that of the magnetic film in which it is contained. These cylinders appear under a microscope as small circles (which give them their name) with diameters from 2 to 20 micrometers. The size of a bubble is determined by the material characteristics and thickness of the magnetic film.

Bubbles are moved or circulated by establishing a magnetic field through a separate conductor mounted adjacent to the bubble chip. A large portion of the bubble chip must be given over to circuitry for generating, detecting, and annihilating the bubbles. Magnetic-domain bubble devices typically operate in an endless serial loop fashion or in an organization with bit serial minor loops feeding major loop registers. Bubble memories are particularly well suited to applications such as portable recorders because of their physical advantages (low power requirements and light weight) and speed advantage over electromechanical devices such as cassettes and floppy disks. Like their electromechanical counterparts, bubble devices also are nonvolatile; that is, they retain their contents when the power goes off. *See* MAGNETIC BUBBLE MEMORY.

SECONDARY MEMORIES

Secondary memories consist of disk units which vary in capacity from the small 250-kilobyte floppy disks (used with microcomputers), through 5-megabyte cartridge disks (used with minicomputers), to large-capacity (up to 600 megabyte) disks used with mini, medium, and large-scale computers. Disk memories are also characterized by access times in the millisecond region (versus access time of 1 μs to hundreds of nanoseconds for RAMs) and a price of 0.001¢ per bit (versus 0.1¢ per bit for RAMs). All these disk units have a movable arm which can be positioned to any track. This device minimizes the cost of the unit, because there are less read/write head electronics to be amortized over the large capacity, at the expense of longer access times to position the arm. Disk memories store information along tracks or cylinders traced in a circular fashion by a transducer which is held in a fixed position during the write or read process. The density of the data is referred to as the number of bits packed along the track in the recording process. Magnetic recording heads are constructed with one or more tracks, and disk data capacity is also increased by increasing the number of tracks per radial inch.

Disk-drive memory capacity has increased from the 100 megabytes of the IBM 3330, introduced in 1970, to 600 megabytes in 1979, and this increase was accompanied by reduction in cost and by the development of improved performance features. Rotational position-sensing capability allows explicit determination of angular position relative to an index. This releases the controller and input-output channel to do other functions during seek time and part of the rotational delay time. In addition, the IBM 3330 provides the capability to stack channel command sequences for more efficient operation. Automatic error correction and automatic error recovery are also significant improvements in disk equipment. Not only is storage capacity greater, but price is lower. For example, the IBM 3350 controller can handle, on line, 16 dual drive units. This gives a capacity of 10^{10} bytes at a price of 0.001¢ per bit.

Winchester technology. Many significant advances in the capabilities of commercially available disk memory equipment were made during the late 1970s. One of the major trends was toward larger-capacity disks from 100 megabytes to 2400 megabytes per disk drive. The majority of technology improvements to provide higher capacities, lower costs, and better reliabilities resulted from what is known as the Winchester technology. Prior to this technology, removable disk packs were established as the most advanced technology for large-capacity disks. The Winchester technology is characterized by nonremovable or sealed disk packs, which inherently provide more reliability due to less handling and make more stringent alignment tolerances practical. The extremely narrow Winchester track has been achieved by a combination of factors. The Winchester head weighs only 0.25 g and floats above the surface. The iron oxide particles on the disk surface have also been magnetically oriented so that both the track and the bits along the track are more precisely defined. Another improvement which characterizes the new technology is a lubricated surface that allows the lightweight head to rest on the disk during start and stop operations. The in-contact start/stop capability of Winchester drives has largely eliminated the "head crash" problems encountered with the earlier technologies.

As the spacing between tracks has decreased, additional demands have been made on the accuracy of the track positioning mechanism. The close spacing of IBM 3330 and Winchester tracks has led to the development of electronic sensing mechanisms. The source of the controlling information for head positioning has been moved from the disk drive to the disk pack itself. In similar fashion, the hydraulic head actuator used in early IBM 2314 technology drives gave way to the voice coil positioner. Winchester technology has carried this a step further and uses a rotary actuator that is based on the voice coil principle, but requires 40% less power, generates less heat, and offers significant advantages in terms of size, service life, and reliability. By 1980 achievable density increased to 6000 bits per inch (240 bits per millimeter) and track density to 600 tracks per inch (24 tracks per millimeter), representing an increase in area density by more than an order of magnitude in a decade. Sophisticated servomechanisms have been developed which utilize a magnetic pattern recorded on the medium itself as the feedback element, thus allowing precise head positioning referenced to the data tracks, as opposed to an outside reference.

The head, or read/write transducer, has undergone substantial refinement. Central to the advance to increase the disk packing density was the reduction in the flying height of the read/write head. Because flux spreads with distance, reduced separation between the read/write head and the magnetic surface will decrease the area occupied by an information bit. This obviously limits the number of bits that can be defined along an inch of track and increases the minimum spacing between tracks. The ideal would be direct contact between head and surface, which is the case with magnetic tape. But this is not possible with a rigid aluminum disk whose surface is traveling at rates that can exceed 100 mi per hour (45 m/s). By applying a load on the head assembly, proper spacing is maintained. All movable media memories utilize an air-bearing mechanism known as the slider or flying head to space the transducer in close proximity to the relatively moving media. The slider, when forced toward the moving surface of the medium, rests on a wedge of air formed between it and the medium by virtue of their relative motion. The thickness of the wedge can be controlled by the applied force, and the wedge in effect is an extremely stiff pneumatic spring. Head-to-medium spacings, commonly up to 250 microinches (6 mm) in 1965, were 50 μin. (1.2 mm) in 1975 and 20 μin. (0.5 mm) in 1980. *See* TRANSDUCER.

As a result of the demand for precision and material stability, sliders have evolved from hand-assembled combinations of various metals, bonded with organic adhesive, to monolithic batch-processed structures of ceramic with tailored magnetic properties, bonded with glass-welding techniques. The design process formerly required slide rule, pencil, and paper; it is now impossible without sophisticated computer simulation which optimizes dimensions, sets tolerances, establishes operating points, and determines stability. The skilled manual techniques for generating and finishing the slider surface contour have progressed to nearly automated lens-lapping procedures. Optical interferometry, first crudely applied to observe the contour of the head surface through an optical flat, is used to measure the head-to-medium spacing dynamically. From the earliest days of disk technology, clean room conditions have been required within the drive enclosure. A single smoke particle can damage the disk and destroy data. Moreover, with each reduction in the flying height, the contamination problem becomes more severe, requiring more stringent control over the disk environment.

Functionally, all disks have the advantage of immediate playback, immediate reuse, and long-term storage. But the most important quality is reliability, and thus the trend toward user-remove-able-disk packs declined after the mid-1970s. A controlled environment always provides conditions conducive to fewer problems. An example is the data-module disk storage unit, in which the magnetic read/write heads are integral with the storage medium, not separate as in conventional disk units. One version of this unit, the IBM 3340, per-manently mates the heads and disks in a special pack, providing better performance and the option of a portion of one surface being dedicated to head-per-track operation. This unit also uses a method called defect skipping when factory tests are run. Its penalty is requiring an extra revolution when writing on a track which contains a defect. In addition to this, the 3340 has an error-correction code which is capable of correcting read errors of up to 3 bits in length.

Fixed-head disks. The head-per-track disk unit can make available many data tracks on the disk without physically moving the read/write head and thus can provide faster access times. The disk storage media is nonremovable. These types of disks are called swapping disks because different user programs can be more rapidly exchanged than by using moving-arm disks. One prime application for head-per-track disks is in the communications equipment portion of data-processing systems. The communications subsystem uses these disks for message storage and data formats where fast access and nonvolatility are of prime importance. Other major application areas include minicomputer systems, ruggedized environments, and special-purpose applications. Under the competition from the memory gap technologies, head-per-track disk products are increasing their capacities from 5 megabytes to 50 megabytes; they also have higher data rates because the bit densities have increased.

Cartridge disks. Removable storage cartridges containing single-disk packs are used extensively with minicomputer systems. The cartridge disk drive unit has capacities ranging from 1 megabyte to 10 megabytes, depending on the recording density. Average access times range from 35 to 75 ms and data transfer rates from 200 to 300 kilobytes per second. Physically, these plastic-encased cartridges are either top-loaded or front-loaded units. These units, besides having a removable disk, also have a fixed disk. The price and performance features of cartridge disks compete favorably with the more expensive, larger-capacity disk pack storage equipment.

Floppy disks. Floppy disks have become the most widely used secondary memory for microcomputer systems with random-access capability. The floppy disk was originally developed by IBM in 1965 for internal use on its large 370 computers. The disk was designed to be a permanent nonvolatile memory storage device, written at the factory. However, large numbers of new and varied applications have evolved for these relatively small, compact disk systems. Floppy disks fall into two categories: IBM-compatible and non-IBM-compatible. The latter is often called OEM (original equipment manufacturer). Floppy disks derive their name from the recording media itself, which is an oxide-coated flexible disk (similar to, but more flexible than, plastic 45-rpm music records) enclosed within a protective plastic envelope. The flexible disk Mylar is 7.875 in. (200 mm) in diameter, 0.003 in. (0.076 mm) thick, and records 3200 bits per inch (125 bits per millimeter) and 48 tracks per inch (1.9 tracks per millimeter). The protective envelope contains alignment holes for spindle loading, head contact, sector/index detection, and write protect detection. The capacity of this easily transportable media is 3 megabits laid out on 77 tracks.

The flexible-disk drive subassembly consists of the metal base frame, the disk drive motor (360 rpm) that rotates the spindle through a belt drive and a stepping motor with an integral lead screw to form the head positioning actuator. The read/write head is mounted on a carriage that the lead screw moves. Head load is achieved by moving a load pad against the disk and restraining the disk between the head and the load pad to accomplish the head-to-disk compliance. The mean time between failures is about 2000 h and the mean time to repair a floppy disk system is half an hour.

Optical disks. Optical or video disks have the potential to provide very inexpensive mass-produced read/write mass memory capability. These rotating devices are used to provide read-only television storage on easily replicable disks. The pressed disk has a spiral groove at the bottom of which are submicrometer-sized depressions. These depressions are read out either through a capacity pickup in a needle-in-the-groove or through a light-beam technique. The storage capacity is equivalent to 10^{10} bits per disk. Access is serial, but some pseudorandom access method by groove selection is possible. When semiconductor lasers (that is, laser diodes) become available, data rates of up to 100 megabits per second could be obtained by paralleling or multichanneling the data from several grooves simultaneously.

MAGNETIC-TAPE UNITS

In magnetic-tape units the tape is held in intimate contact with the fixed head while in motion, allowing high-density recording. The long access times to find user data on the tape are strictly due to the fact that all intervening data have to be searched until the desired data have been found. This, of course, is not true of rotating disk memories or RAM word addressable main memories. The primary use of tape storage is for seldom-used data files and as back-up storage for disk data files.

Half-inch tapes. Half-inch (12.7-mm) tape has been the industry standard since it was first used commercially in 1953. Half-inch magnetic-tape drive transports are reel-to-reel recorders with extremely high tape speeds (up to 200 in. or 5 m per second), and fast start, stop, reverse, and rewind times (on the order of 1 ms).

Performance and data capacity have improved by several orders of magnitude. Just prior to the 1970s came the single-capstan tape drive, which improved access times to a few milliseconds. These vacuum-column tape drives have such features as automatic hub engagement/disengagement, cartridge loading, and automatic thread operation. There are two primary recording techniques, namely, nonreturn-zero-inverted (NRZI) and phase-encoded (PE), used with 800 and 1600 bytes per inch (31.5 and 63 bytes per millimeter) packing density. These typically use a 0.6-in. (15-mm) interrecord gap. In phase encoding, a logical "one" is defined as the transition from one magnetic polarity to another positioned at the center of the bit cell. "Zero" is defined as the transition in the opposite direction also at the center of the bit cell, whereas NRZI would involve only one polarity. The advantage of the phase-encoding scheme over NRZI is that there is always one or more transition per bit cell, which gives phase encoding a self-clocking capability, saving the need for an external clock. The disadvantage of phase encoding over NRZI is that at certain bit patterns the system must be able to record at twice the transition density of NRZI.

Computer-compatible magnetic-tape units are available with 6250 bytes per inch (246 bytes per millimeter) tape drives. They have nine tracks where eight bits are data and 1 bit is parity. Each track is recorded using a technique called group-coded record (GCR), which uses a 0.3-in. (7.6-mm) record gap. Every eighth byte in conjunction with the parity track is used for detection and correction of all single-bit errors. Thus, GCR offers inherently greater reliability because of its coding scheme and error-correction capability. It does, however, involve much more complex circuitry. Because of the need for media-compatible tapes with existing formats, the combination of GCR/phase-encoded drives were still in demand over the late 1970s.

The on-line capacity is strictly a function of the number of tape drives that one controller can handle (eight is typical). The price per bit is minimized over one controller with its maximum number of drives. A 2400-ft (730-m) nine-track magnetic tape at 6250 bytes per inch per track provides a capacity of approximately 10^9 bits, depending on record/block sizes. The price per bit of the tape media is 10^{-6} cent, and the on-line storage cost per bit ranges from 0.001 to 0.003¢.

Cassettes and cartridges. The most frequently used magnetic-tape memory devices for microcomputer systems are cassette and cartridge tape units. Both provide very-low-cost storage, although their access times are high and their overall throughput performance is not as great as that of floppy disks. Cassette and cartridge units both use quarter-inch-wide (6.35-mm) magnetic tape, but the packing densities used vary considerably between the two types of units. Typically a two-track cassette holds 5 megabits (over a length of 300 ft or 90 m) and has a data transfer rate of 24 kilobits per second (corresponding to a density of 800 bits per inch or 31.5 bits per millimeter); whereas four-track cartridge tape has a capacity of over 20 megabits (over a length of 300 ft) with up to 48 kilobits per second data-transfer rate (corresponding to 1600 bits per inch or 63 bits per millimeter).

The digital cassette transport was originally an outgrowth of the Philips audio cassette unit. Unfortunately, the very-low-cost audio-designed transport did not lend itself to true digital-computer-application endurance needs. There are two basic design approaches to digital cassette transports: capstan drive and reel-to-reel drive. Capstan tape drives are better for maintaining long-term and short-term tape-speed accuracy, while reel-to-reel transports have better mechanical reliability.

During the 1960s the 3M Company developed what has become known as the first true digital tape cartridge. With the cassette, a capstan must penetrate the plastic case containing the tape in order to make contact. There is no such penetration system with the cartridge because the transport capstan simply drives an elastomer belt by pressing against the rim of the belt's drive wheel.

Table 2. Comparison of mass storage system characteristics

Mass-memory storage characteristics	Mechanical selection and mounting unit for standard tape drives	Mechanical selection and accessing unit for honeycomb short-tape units	Special 2-in. (50.8-mm) video-tape drive
Maximum capacity	8×10^{12} bits	3.81×10^{12} bits	2.94×10^{12} bits
Media	0.5-in. (12.7-mm) wide magnetic tape on 10.5-in. (267-mm) reels	50-megabyte cartridges containing 770 in. (19.6 m) of 3-in. -wide (76-mm) magnetic tape	2-in. video tape, 10.5-in. (267 mm) reel, 3800 ft/reel (1158 m/reel) 4.5×10^{10} bits per reel
Maximum transfer rate	1.2×10^6 bytes/s	8.06×10^5 bytes/s	7.00×10^5 bytes/s
Access time	10 s (max 20.5 s)	15 s	15 s
Throughput	150 reels/h	100 files/h	150 files/h
Cost per bit (on-line system)	0.0000175¢/bit	0.00017¢/bit	0.000644¢/bit

This simplicity eliminates a major source of tape damage and oxide flaking.

MASS STORAGE SYSTEMS

With the gradual acceptance of virtual memory and sophisticated operating systems, a significant operational problem arose with computer systems, particularly the larger-scale installations. The sheer expense and attendant delays and errors of humans storing, mounting, and demounting tape reels at the command of the operating system began to become a problem. Mass-storage facilities are designed to alleviate this problem. A number of these mass-storage devices, of various configurations, have been installed. Their common attributes are: capacity large enough to accommodate the current data base on line; access times between those of movable head disks and tapes; and operability, without human intervention, under the strict control of the operating systems. The mass-storage facility is included within the virtual address range. All such configurations mechanically extract from a bin, mount on some sort of tape transport, and replace in a bin, following reading or writing, a reel of magnetic tape.

Mass-storage systems are hardware devices that need operating-system and data-base software in order to produce a truly integral, practical hardware/software system. Users require fast access to their files, and thus there is a definite need to cue up (stage) files from the mass-storage device onto the disks. The data-base software must function efficiently to make this happen efficiently. In general, users base their storage-device selection on the file sizes involved and the number of accesses per month. Magnetic-tape units are used for very large files accessed seldomly or infrequently. Mass-storage devices are for intermediate file sizes and access frequencies. Disk units are used for small files which are accessed often.

In practice, most users have been satisfied with tapes and disks, and have not chosen the middle ground of mass-storage systems. During the 1970s only three basic kinds were delivered, although some others were built and installed in the earlier years. In different ways, these units combine the inexpensive cost of tape as a storage medium with the operating advantages of on-line access. Table 2 compares the characteristics of the three types of delivered systems. One is a mechanical selection and mounting unit to load and unload tape reels onto and off standard magnetic-tape units; another is a mechanical selection and accessing unit to operate with special honeycomb short-tape units; the last is a 2-in. (50.8-mm) wide video-tape unit in which read/write operations are done with a transversal recording technique. The first two units (sometimes called automated tape libraries) eliminated manual tape mounting operations. The main objective of the short-tape honeycomb cartridge system is to improve access time. The shorter tape (770 in. or 19.6 m versus 2400 ft or 730 m) results in better access time. Operationally, the honeycomb cartridge tape system and the mechanical standard tape mounting units are capable of handling 100 to a peak of 300 cartridge/tape loads per hour. The cost of storage media can be a very major consideration in choosing between these mass-memory systems.

The first type of fully automated tape library uses standard magnetic tape (½ in. or 12.7 mm wide). Under computer control, this equipment automatically brings the tapes from storage, mounts them on tape drives, dismounts the tapes when the job is completed, and returns them to storage. Accessing up to 150 reels per hour, this unit can store up to 7000 standard tapes or 8000 thin-line reels in a lockable self-contained library that can service up to 32 tape drives and which can interface with up to four computers.

The honeycomb mass-storage system uses a storage component called a data cartridge. Housed in honeycomb storage compartments, these 2×4 in. (25×50 mm) plastic cartridges can each hold up to 50,000,000 bytes of information on 770 in. (19.6 m) of approximately 3-in. (76-mm) wide magnetic tape. Whenever information from a cartridge is needed by the computer, a mechanism selects the desired cartridge and transports it to one of up to eight reading stations. There the data are read out and transferred to the staging disk drives. *See* COMPUTER; DATA-PROCESSING SYSTEMS; DIGITAL COMPUTER.

[DOUGLAS J. THEIS]

Bibliography: R. Brechtlein, Comparing disc technologies, *Datamation*, 24(1):139–150, January 1978; H. L. Caswell et al., Basic technology, *IEEE (CS) Computer*, pp. 10–17, September 1978; D. Chen and J. D. Zook, An overview of optical data storage technology, *Proc. IEEE*, 63(8):1207–1229, August 1975; C. S. Chi, Advances in computer mass storage technology, *IEEE Computer Mag.*, pp. 60–74, May 1982; M. S. Cohen and H. Chang, The frontiers of magnetic bubble technology, *Proc.*

IEEE, 63(8):1196–1206, August 1975; G. C. Feth, Memories: Smaller, faster and cheaper, *IEEE Spect.*, 13(6):130–145, June 1976; D. Hodges, Microelectronic memories, *Sci. Amer.*, 237(3):130–145, September 1977; C. T. Johnson, The IBM 3850: A mass storage system with disk characteristics, *Proc. IEEE*, 63(8):1166–1170, August 1975; C. P. Lecht, *The Waves of Change*, Advanced Computer Techniques Corp., 1977; J. A. Rodriguez, An analysis of tape drive technology, *Proc. IEEE*, 63(8):1153–1159, Aug. 8, 1975; D. J. Theis, An overview of memory technologies, *Datamation*, 24(1):113–131, January 1978; M. Wildmann, Terabit memory systems: A design history, *Proc. IEEE*, 63(8):1160–1165, August 1975.

Concurrent processing

The conceptually simultaneous execution of more than one sequential program on a computer or network of computers. The individual sequential programs are called processes. Two processes are considered concurrent only if they interact in some way. This interaction may range from cooperation (the exchange of information) to competition for scarce resources (such as the processor, the memory, or a printer). Concurrent processing is achieved on a single computer, that is, on a single processor, by interleaving the execution of the individual processes.

In comparison, distributed processing requires multiple processors for its implementation. According to this definition, all distributed programs are also concurrent programs, but the converse is not true. *See* DISTRIBUTED PROCESSING.

Advantages and disadvantages. A common example of concurrent processing is a multiuser operating system, which allows more than one user process to be executed concurrently on the same processor. Many such systems do not allow the processes to communicate directly; they only coordinate the use of shared resources by the processes (for example, the static allocation of memory to the active processes or the dynamic allocation of processors to processes). This form of concurrency allows multiple users to share one processor, which is a necessity for time-sharing systems, without having the user programs complicated by codes dealing with the sharing of resources, scheduling, and so forth. This modularity is an advantage of most concurrent systems over sequential systems: different parts of a problem can be relegated to semi-independent processes that will communicate with one another only when necessary. *See* MULTIACCESS COMPUTER.

This form of concurrency has disadvantages as well. The operating system causes an overhead in the use of the processor; an operating system that executes user processes sequentially would spend very little time in allocating the processor's resources or in switching between tasks.

Distributed systems present additional advantages and disadvantages in comparison with centralized concurrent systems. Distributed systems eliminate some of the contention for the processor, because there are more processors available in the system. Furthermore, a centralized system does not lend itself to graceful degradation as does a distributed system. The failure of one processor in a distributed system need not cause the entire system to fail, if the remaining processors can take up where the failed processor left off.

Though distributed systems reduce contention for processors, they create contention for the communication network. In a centralized system, interaction can take place through shared memory protected by some synchronization mechanism; in a distributed system all interaction and communication must pass through the network. There also is a cost for the higher availability of distributed systems, especially where the processes are cooperating closely. Each processor (or its operating system) must keep track of the progress of the other processors in order to recover in the event of a failure.

Specifying concurrency. Many programming languages provide the user with the ability to write concurrent programs. Different languages allow parallelism with different degrees of concurrency.

Languages such as ADA, CSP, Concurrent PASCAL, MESA, and MODULA provide for statement-level (or procedure-level) concurrency. These languages have either a construct that causes a new process to be initiated, executing some statement or procedure in parallel with the current process, or a construct that causes a set of statements or procedures to be executed in parallel. In either case, the user tells the system what portions of the code will be executed concurrently.

In contrast, some data flow languages provide for operator-level concurrency. In such languages, the program is specified without reference to concurrency; the compiler breaks the program down into its component operations, often as small as add or multiply. When the program executes, all operators can be executed in parallel; an operator can execute as soon as it has received values for all of its input operands. Upon termination it forwards its result to those operators requiring that result as input. *See* DATA FLOW SYSTEMS; PROGRAMMING LANGUAGES.

Communication and synchronization. Interprocess communication invokes two classes of concurrent systems: synchronous and asynchronous. In a synchronous system, all processors work in lockstep; that is, they all execute the same instruction at the same time (on different data). In asynchronous systems each process (and processor) works independently of the others. Asynchronous processes need only synchronize temporarily to exchange information or to request a service. Such synchronization and communication takes place in one of two ways: by reading and writing shared memory or by exchanging messages.

In shared memory systems, processes communicate by modifying shared variables. Such variables contain values, such as the number of free tape drives or the identity of the next process to control the processor. Though shared, these variables must be protected from simultaneous access by multiple processes. The processor hardware usually prevents simultaneous writing of the same memory location (thereby preventing meaningless data from being stored), but that is not enough. Higher-level synchronization is required to prevent interference between processes. Problems arise when one process reads a shared variable, intending to change it, but a second process changes the variable before the first process performs its change. When the first process finally makes its change, the second process's update is lost. It is necessary for a process to be able to gain exclusive access to a variable for more than a read or write. The portion of a program in which a process has exclusive control over shared data is called a critical section. There are many mechanisms for obtaining such protection; one example is the monitor. Monitors protect shared variables by allowing processes to access these variables only through special procedures; only one process may execute a monitor's procedure at a time, thus preventing the destructive interference.

Message-passing systems communicate by send- and receiving messages. Sending a message is more expensive (in terms of computer time and resources) than modifying a shared variable. A message must be built out of the data provided by the sending process, and then copied to the address space of the receiving process. This may entail multiple copies if the data must be passed through intermediate processes and networks. On the other hand, message systems have no shared variables to protect from concurrent interference. There are different types of message-passing systems that depend on whether the sending process must wait until the receiving process actually receives the message, and if so, whether it must wait for a reply.

[BRENT T. HAILPERN]

Bibliography: G. R. Andrews, Synchronizing resources, *ACM Transactions on Programming Languages and Systems*, 3(4):405–430, October 1981; R. H. Kuhn and D. A. Padua (eds.), *Tutorial on Parallel Processing*, IEEE Computer Society, 1981; B. Liskov, Primitives for distributed computing, *ACM Proceedings of the 7th Symposium on Operating Systems*, pp. 33–42, 1979; P. D. Stotts, Jr., A comparative survey of concurrent programming languages, *ACM SIGPLAN Not.*, 17(9):76–87, September 1982.

Conductance

For a series circuit the real, or in-phase, part of the complex representation for the admittance Y of a circuit. This is shown in Eq. (1), where B is the

$$Y = G \pm jB \qquad (1)$$

imaginary part, called the susceptance. Since $Y = 1/Z = 1/(R + jX)$, where R is the resistance and X is the total reactance $X_L - X_C$, Eq. (2) holds, and conductance G is given by Eq. (3). This is the general expression for conductance and shows that conductance is a function involving both resistance and reactance.

$$Y = \frac{R}{R^2 + X^2} - j\frac{X}{R^2 + X^2} \qquad (2)$$

$$G = \frac{R}{R^2 + X^2} \qquad (3)$$

If reactance is negligible, then $G = R/R^2 = 1/R$, or conductance is the reciprocal of resistance. This is called simple conductance and is strictly correct only where the impedance contains no reactance. Thus in a dc circuit, conductance is the reciprocal of resistance. These reciprocal functions find application chiefly in computations of parallel circuits.

The total admittance of a number of parallel admittances, if expressed in complex quantities, is equal to the sum of the individual admittances. *See* ADMITTANCE; ALTERNATING-CURRENT CIRCUIT THEORY.

[ARTHUR A. WELCH]

Conduction (electricity)

Electrical conduction may be defined as the passage of electric charge. This can occur by a variety of processes.

In metals the electric current is carried by free electrons. These are not bound to any particular atom and can wander throughout the metal. In general, the conductivity of metals is higher than that of other materials. At very low temperatures certain metals become superconductors, possessing infinite conductivity. The free electrons are able to move through the crystal lattice without any resistance whatsoever. *See* ELECTRICAL RESISTANCE; SUPERCONDUCTIVITY.

In semiconductors (germanium, silicon, and so on) there are a limited number of free electrons and also "holes," which act as positive charges, available to carry current. The conductivity of semiconductors is much smaller than that of metals and, in contrast to most metals, increases with rising temperature. *See* HOLES IN SOLIDS; SEMICONDUCTOR.

Aqueous solutions of ionic crystals readily conduct electricity by means of the positive and negative ions present, for example, Na^+ and Cl^- in an ordinary solution of sodium chloride. Solid ionic crystals are themselves fair conductors. These crystals have sufficient lattice vacancies so that a few of the ions are able to migrate through the crystal under the influence of an applied electric field.

A strong electric field ionizes gas molecules, and thereby permits a flow of current through the gas in which the ions are the charge carriers. If sufficient ions are formed, there may be a spark. *See* ELECTRIC SPARK; ELECTRICAL CONDUCTION IN GASES.

Electric current can flow across a vacuum, for example, in a vacuum tube. The charge carriers are electrons emitted by the filament. The effective conductivity is low because of low available current densities at the normal temperatures of electron-emitting filaments. *See* ELECTRON EMISSION; ELECTRON MOTION IN VACUUM.

[JOHN W. STEWART]

Conduction band

The electronic energy band of a crystalline solid which is partially occupied by electrons. The electrons in this energy band can increase their energies by going to higher energy levels within the band when an electric field is applied to accelerate them or when the temperature of the crystal is raised. These electrons are called conduction electrons, as distinct from the electrons in filled energy bands which, as a whole, do not contribute to electrical and thermal conduction. In metallic conductors the conduction electrons correspond to the valence electrons (or a portion of the valence electrons) of the constituent atoms. In semiconductors and insulators at sufficiently low temperatures, the conduction band is empty of electrons. Conduction electrons come from thermal excitation of electrons from a lower energy band or from impurity atoms in the crystal. *See* BAND THEORY OF SOLIDS; SEMICONDUCTOR; VALENCE BAND.

[H. Y. FAN]

Consumer electronics

A variety of consumer electronic products are becoming available. The products simulate human characteristics of listening, talking, amusing, and educating. Among these microprocessor-controlled devices are held-hand language translators, programmable personal computers, educational talking toys, intelligent television receivers, and portable remote terminals that utilize the telephone system to gain access to various computer data bases.

These electronic devices are made possible through electronic microminiaturization called large-scale integration (LSI). Large-scale integration allows more features to be implemented in compact, relatively inexpensive products that were bulky and expensive only a few years ago. Circuits that had to be fabricated with large components and much wiring are now formed on smooth silicon wafers that are split up into "chips" or integrated circuits. These are then interconnected on a printed circuit board that has prefabricated silver connections for the circuits. These integrated circuits are programmable, have the capacity to store data, and perform needed calculations in a fraction of a second. Hence it is left to the imagination of the designer and the economic good sense of the marketeer to develop salable consumer electronic products. Some of the major products under development will be discussed. *See* INTEGRATED CIRCUITS.

Language translators. Portable language translators register keyboard entry of phrases in one language and speak the translated phrase in another. Each language for translation is preserved in memory modules called read-only-memory (ROM), which are basically tables storing corresponding words in both languages. One type of translator uses a speech synthesizer chip that was originally developed as an educational toy. Plug-in ROM modules are available for English, Spanish, French, German, and Russian.

Computers. Hand-held computers have been developed that fit into a briefcase whch has receptacles for acoustically coupling the computer with any telephone (Fig. 1). A variety of information can thereby be accessed while the user is out of reach of data bases. The interface protocol is compatible with most major main-frame computers. Some companies have combined the translator with a computer in the same portable terminal. One terminal has a capacity to hold up to eight language modules simultaneously. A random-access memory (RAM) is used for storing up to 500 characters. In addition, the terminal also acts as a calculator and a clock. *See* MICROCOMPUTER.

Electronic games. Two types of electronic games exist: those that measure users' dexterity and those that challenge their intelligence. Dexterity games are primarily aimed at the younger generation: they include popular games in electronic form, such as anagram, word matching, and sound matching, and sports games such as baseball and football. Intelligence games such as chess and backgammon build on the users' intelligence and ingenuity and hence appeal to a wider age group. Both types of games can be played alone against the computer or between two players.

Television receivers. Television receivers are now constructed almost entirely of integrated circuits with only the picture tube remaining as a reminder of electron tube technology. All receiver manufacturers have models that incorporate electronic tuners that lock precisely to the channel frequency. Colors have become more vivid, and displays show less color distortion. Delay line/comb filter integrated circuits that improve luminance resolution and reduce luminance-chroma crosstalk have been incorporated into the receiver. One type of delay line is based on charge-coupled-device technology.

In the National Television System Committee (NTSC) color television system, luminance and chrominance information is interlaced on a common channel. The two components have to be separated for enhanced picture quality and sharpness. The comb filter separates luminance signals by adding two composite video signals, one of which is delayed by one horizontal scan line relative to the other. Up to now this procedure involved adding a delayed and undelayed signal external to the charge-coupled-device integrated circuit. This causes delay variations that can affect the filter characteristics significantly. In the new comb filter integrated circuit, the delay and the filter functions are combined, thus ensuring that the delay be-

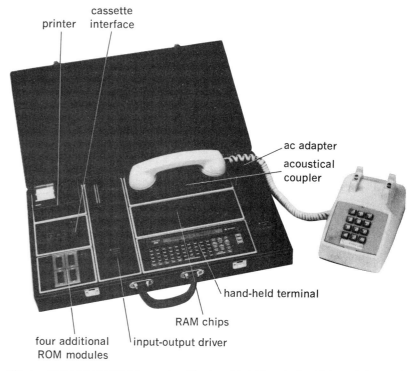

Fig. 1. Quasar hand-held computer. The hand-held terminal, which contains read-only-memory (ROM) modules for translation between languages, fits into a briefcase containing various optional components, including an acoustical telephone coupler to facilitate access to remote data bases by dialing the appropriate code.

tween the two signals is 63.55 microseconds, exactly one horizontal scan line.

For separating chrominance signals, the composite video signals must be subtracted rather than added. Subtraction is accomplished by inverting the composite signal prior to the chrominance circuit in the receiver. By separating both the luminance and chrominance, picture quality and sharpness are considerably enhanced. The outputs of the processing circuits contain components that are free from "dot-crawl" and cross-color contaminations.

In a television receiver system (Fig. 2), the comb filter is interposed between the intermediate frequency (i-f) section and the luminance and chrominance processors. In general in a color television the received signal is fed to the tuner, where it is amplified and converted to the intermediate frequency (45.75 MHz). A wide variety of tuners are still used ranging from a microprocessor-controlled synthesizer to mechanical switching. The intermediate-frequency section recovers the baseband composite video signal and also produces the sound signal. The inputs to the comb filter are composite video from the intermediate-frequency section and the regenerated chroma subcarrier from the chroma oscillator. The comb filter outputs are the combed luminance, vertical detail, and combed chrominance signals. The luminance processing circuits ensure that the display is turned off during those portions when no picture information is present. A color subcarrier and chrominance signals drive three synchronous detectors which produce three color-difference signals: R-Y, G-Y, and B-Y. The luminance signal (Y) is subtracted from these, leaving the three red (R), green (G), and blue (B) signals which are applied to the cathodes of a picture tube. After being modulated in intensity each beam passes through a shadow mask that ensures it will strike only the phosphors on the faceplate which can glow with the appropriate color. The vertical and horizontal deflection drivers scan the beams simultaneously from left to right beginning at the top center; 525 horizontal lines are scanned in two fields during one frame, and each frame takes one-thirtieth of a second to scan.

Another improvement in the television receiver that is a direct result of LSI chips is a quasistereophonic system that simulates true stereo sound by generating two spectrally distant sound channels from the monophonic audio source of the receiver. Each sound channel drives one of two separate full-range loudspeakers located symmetrically about the television screen. *See* COLOR TELEVISION; TELEVISION RECEIVER.

Home data retrieval. Decoders are being developed that will enable text and graphics to be retrieved on the home television screen. At first the decoders will be top-mounted on the set, and as LSI chips incorporate more functions, the decoders will be mounted inside the sets. Prototype decoders are already being displayed. As soon as the Federal Communications Commission establishes standards, United States television manufacturers will begin producing sets with built-in decoders.

The image, whether text or graphics, is synthetic video that is created locally. The display will show up to 24 rows, with 40 characters in each row, of constantly updated data on such diverse subjects as news, weather, sports, cooking instructions, child-adoption eligibility, and theater listings (Fig. 3).

The information is received by the television set as part of the regular broadcast but cannot be seen by the viewer unless a decoder is used. The decoder transforms information taken from the unseen (blanking) portion of the screen to the visible portion. Between each frame on the screen are about

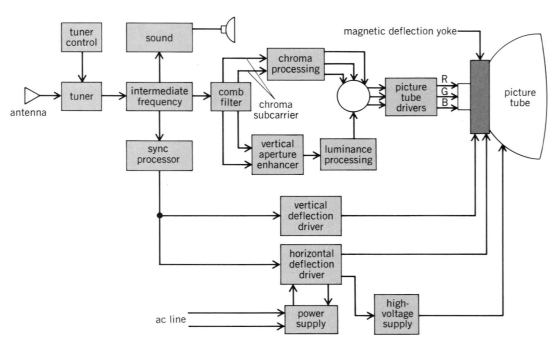

Fig. 2. Color television receiver.

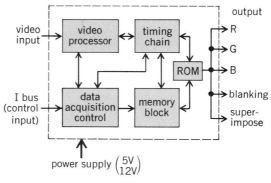

Fig. 3. Examples of teletext displays. (*a*) Real-time sports scores. (*b*) Constantly updated market indicators. (*c*) Theater ticket list. (*d*) Captioning of regular television programs for the hearing-impaired.

10 blanking lines, not used by the picture, but needed to account for the time during which the vertical retrace of the electronic beam travels from the bottom right side of the display back to the top left side for the next frame. A couple of the lines have been used in the past to carry captions for the hearing-impaired. Special different decoders are needed so that these people can see the captions. *See* VIDEOTEXT AND TELETEXT.

Set-top versus built-in decoders. Set-top decoders have the disadvantage that their outputs are fed into the receiver together with the video signal from the antenna. As a result, distortions are observed on the display because the decoding signal must pass through all other circuits in the television set, causing interference patterns. The built-in decoders, however, are connected directly to the three color drivers controlling the cathodes of the picture tube. Thus the signal bypasses most of the causes for distortion, and a clear representation of text and graphics can be observed on the screen.

Viewdata. A more sophisticated system, called viewdata, is an outgrowth of the business community. Here the television again serves as the receiver of data and graphics and displays them when the decoder is activated; but the information reaches the set through the telephone. By modulating the voice line on the telephone, digital bit streams are sent through these circuits to the decoder. As a result, different data bases can be accessed by dialing their respective codes. This practice has long been utilized by businesses with on-line time-sharing systems. Soon, with decoders and modems available at affordable prices to the consumer, these same data bases can be accessed from the home. The services provided by viewdata systems are similar to teletext services with one major difference. In teletext a remote keypad is used to access the pages as they are broadcast so that, for example, if the viewer knows that the theater guide is on page 40 the viewer presses

Fig. 4. Block diagram of one type of teletext decoder. (*From T. Suzuki et al., Television receiver design aspects for employing teletext LSI, IEEE Trans. Consumer Electr., CE-25(3):400–405, 1979*)

that number on the keypad, and after the strobe signal picks up that command it will then switch to that page on the next transmission. (Teletext data are transmitted constantly and repeated after each transmission.) This switching process takes a few seconds and limits the viewer's participation in the service.

In viewdata, on the other hand, the keypad is used as a two-way transmitter. Hence the same entry on page 40 in a viewdata system can be accessed instantly, provided the page for the category is known. Moreover, the keypad can be used to engage in a two-way conversation with the system. For instance, where the subject of child-adoption eligibility is being examined, consecutive frames might ask the viewer questions to be answered through the keypad. In the end it might even appear as if viewdata decides whether the viewer is eligible to adopt. In the future, viewdata systems dispersed around the country might even be used to play interactive video games such as national chess tournaments.

Implementation. Teletext decoders are already available for receivers in Great Britain, where this technology was established several years ago (Fig. 4). The decoder is a 160×120 mm circuit board containing four LSI chips and two 4000-bit RAMs for storing one page of data locally. When the decoder receives the remote control signal (Ibus), it extracts the teletext data signal from the video detector output and obtains red, blue, and green signals. These are synchronized with the television picture. The decoder also generates the necessary blanking signal so that the picture is invisible while teletext information is being displayed.

A number of semiconductor manufacturers are ready to supply chips for teletext and viewdata decoders as soon as international standards are established. The success of these information retrieval systems for home use will then depend only on consumers' appreciation of this new form of information dissemination and general acceptance of this valuable service.

[NICOLAS MOKHOFF]

Bibliography: Color TV, special issue of *RCA Eng.* 25(6):1–88, 1980; Consumer Text Display Systems, special issue of *IEEE Trans. Consumer Electr.*, CE-25(3):233–423, 1979; Videotext and Teletext Systems, special issue of *IEEE Trans. Consumer Electr.*, CE-26(3), 1980.

Contact potential difference

The potential difference that exists across the space between two electrically connected metals; also, the potential difference between the bulk regions of a junction of two semiconductors. In Fig. 1a consider two metals M_1 and M_2 with work functions ϕ_1 and ϕ_2, respectively. When the metals are brought in contact, the Fermi levels, that is, the levels corresponding to the most energetic electrons, must coincide, as in Fig. 1b. Consequently, if $\phi_1 > \phi_2$, metal M_1 will acquire a negative and metal M_2 a positive surface charge at the contact area. In the circuit indicated in Fig. 2 there will thus exist a potential difference between the plates such that M_1 is negative with respect to M_2; the magnitude of the contact potential is equal to $(\phi_1 - \phi_2)/e$ (of the order of a few volts or less)

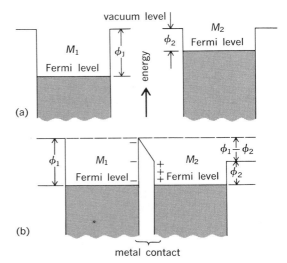

Fig. 1.   Energy-level diagram for conduction electrons of two metals. (*a*) Before contact. (*b*) After contact.

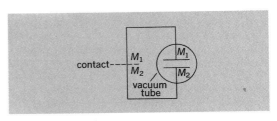

Fig. 2.   Two metallic plates in vacuum between which a contact potential difference may be measured.

where $e = 1.6 \times 10^{-19}$ coulomb. Note that the contact potential is independent of the number of different materials used in the circuit external to the plates in vacuum. The contact potential difference may be measured by placing a variable electromotive force in the external circuit, which can be balanced to make the potential difference between the plates equal to zero. Contact potentials must be taken into account in analyzing physical electronics experiments.

[A. J. DEKKER]

Continuous-wave radar

A radar in which the transmitter output is uninterrupted, in contrast to pulse radar, where the output consists of short pulses. The chief disadvantage of a continuous-wave (CW) radar is that it is difficult to use a single antenna for both reception and transmission, because both the transmitter and receiver are on all the time. This difficulty has been overcome by means of the use of isolation circuitry, which gives the receiver protection from a transmitter output up to 200 watts. *See* RADAR.

Among the advantages of CW radar is its ability to measure velocity with extreme accuracy by means of the Doppler shift in the frequency of the echo. *See* DOPPLER RADAR.

In order to measure the range of targets, some form of frequency modulation of the CW output must be used. In one very effective form of modulation, the carrier frequency of the transmitted signal is varied at a uniform rate. Range is determined by comparing the frequency of the echo with that of the transmitter, the difference being proportional to the range of the target that produced the echo. Systems in which this is done are known as FM-CW radars. *See* FREQUENCY MODULATION.

A modified form of FM-CW radar employs long, but not continuous, transmission. This might be regarded as the same as transmitting extremely long pulses on a frequency-modulated carrier. Systems of this type are referred to as pulse compression radars.

Missile guidance. The principal use of CW radar is in short-range missile guidance. Typically the missile's course is tracked from the ground while the missile is simultaneously illuminated. CW radar, in some cases the same radar, can be used both for tracking and illumination, although it is more common for pulse Doppler radar to be used for tracking.

Advances in the design of phased-array antennas have led to pulse Doppler radar becoming more attractive than CW radar even for illumination. However, at least one system allows the operator to select either CW or pulse Doppler radar for illumination.

Illumination radar is used for target acquisition. Two signals are of interest. One, the directly received illumination of the missile, is called the rear signal. The other, the signal reflected from the target, is called the front signal. The front signal is shifted in frequency as the missile closes with the target. This shift occurs because of the Doppler effect. The shift and therefore the range to the target is obtained by coherently detecting the front signal against the rear signal. In an active guidance system, detection is performed on board the missile. Semiactive missile guidance is more frequently used, and in this system the signals are relayed over a data link for processing at the ground station.

CW tracking and acquisition was used with the MPQ-34 radar for the Hawk ground-to-air missile. A late version of this system incorporates the choice of either CW or pulse Doppler radar illumination. The ability of CW radar to discriminate against clutter made it attractive in the Hawk system, which was intended to be used against low-flying hostile aircraft.

Pulsed tracking radar with CW acquisition radar is used with the Sparrow air-to-air missile system and the Tartar shipborne missile system.

Detection of hostile targets. Modulated CW radar has been used to detect hostile military vehicles and personnel.

Laser radar system. An experimental modulated CW laser radar has been built to track low-flying airborne targets from the ground. The system can determine both range and the rate at which the range is changing. An advantage of this system is its extreme precision. The divergence of the beam is less than 1 milliradian. The transmitter is a carbon dioxide laser operating at a wavelength of 10.6 μm. The infrared carrier is frequency modulated with 10-MHz excursions occurring at a frequency of 1 kHz. *See* LASER.

Automobile safety. An experimental CW radar which exploits the Doppler principle has been developed for use in automobiles. The radar can anticipate a crash when an obstacle is 10 m away in order to deploy air-cushion type passive restraints. It can sense obstacles 150 m away to govern automatic braking and headway control. The source of carrier power is an X-band Gunn-type solid-state oscillator. *See* MICROWAVE SOLID-STATE DEVICES.

Surveillance of personnel. A variant of CW radar is used to illuminate persons under surveillance by techniques employing semiconductor tracer diodes. These devices are either secreted on the subject's person without his or her knowledge or else concealed in protected objects the subject has stolen. Despite the fact that pulsed X-band sources would provide the necessary high power levels more conveniently, the requirement to reduce clutter makes it essential to use CW power.

Implementations of tracer-diode surveillance utilize a single carrier frequency and look for reflections of the third harmonic as evidence that the subject sought has been acquired by the beam. These surveillance operations have been carried out from fixed posts, vans, and low-flying aircraft.

An improved version makes use of two carrier frequencies for illumination and depends upon coherent detection of the third and fourth harmonics of the sum and difference frequencies to signify acquisition of the target. *See* ELECTRONIC LISTENING DEVICES.

[JOHN M. CARROLL]

Bibliography: H. E. Penrose and R. S. H. Boulding, *Principles and Practice of Radar*, 6th ed., 1959; *Proceedings of the IEEE International Radar Conference*, 1975.

Controlled rectifier

A three-terminal semiconductor junction device with four regions of alternating conductivity type (*p-n-p-n*), also called a thyristor. This switching device has a characteristic such that, once it conducts, the voltage in the circuit in which it con-

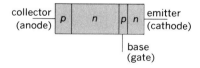

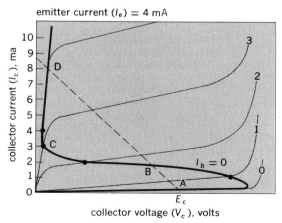

Controlled rectifier diagram and characteristics.

ducts must drop below a threshold before the controlled rectifier regains control. Such devices are useful as high-current switches and may be used to drive electromagnets and relays. *See* THYRATRON.

The principle of operation can be understood by referring to the illustration. The central junction is reverse-biased (positive collector, grounded emitter). The wide *n* region between collector and base regions prevents holes injected at the collector junction from reaching the collector-to-base barrier by diffusion. The junction between emitter and base is the emitter. When operated as a normal transistor, this device shows a rapid increase of current gain of α (equal to I_c/I_e) with collector current. This effect may be due to a field-induced increase of transport efficiency across the floating *n* region, or to increased avalanching in the high-field barrier region, or to increased injection efficiency at the two forward-biased junctions, or to a combination of these phenomena.

With a floating-base region ($I_b = 0$), the device is a two-terminal device and collector current I_c must equal emitter current I_e.

By selecting the points on the illustrated characteristics where $I_c = I_e$, the characteristic for the base current $I_b = 0$ is shown as the heavy curve with the negative-resistance characteristic. This characteristic will be found in any transistor which shows an integrated α increasing from below unity at low collector currents to above unity at high collector currents.

If this device is operated as a three-terminal device, the switching between the nonconducting and conducting states can be controlled by the base. If in the grounded-emitter case the collector is biased to $+E_c$ as shown, the device will remain at point *A* until the base is pulsed positive by at least enough current to carry the emitter current to point *B*. At this point α exceeds unity and the device will spontaneously switch to point *D* in the conducting state. To reset the unit, either the emitter must be cut off or the collector voltage must be reduced so that the load line falls below the valley point *C*. The current of point *C* is called the holding current. Either of these results can be achieved by appropriate pulses on the base. Modern terminology usually refers to the rectifier terminals as anode and cathode and to the control terminal as the gate. *See* SEMICONDUCTOR RECTIFIER.

Overall current gain α may be maintained below unity for low anode-cathode currents by designing the junction between the anode and the floating *n* region and the junction between the gate and the cathode so each has a low-current injection efficiency below 0.5. For further discussion of four-layer devices *See* TRANSISTOR.

[LLOYD P. HUNTER]

Bibliography: S. M. Sze, *Physics of Semiconductor Devices*, 1969; E. S. Yang, *Fundamentals of Semiconductor Devices*, 1978.

Corner reflector antenna

A directional antenna consisting of the combination of a reflector comprising two conducting planes forming a dihedral angle and a driven radiator or dipole which usually is in the bisecting plane. It is widely used both singly and in arrays, gives good gain in comparison with cost, and covers a relatively wide band of frequency. For a

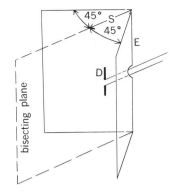

Corner reflector antenna.

discussion of directional antennas *see* ANTENNA (ELECTROMAGNETISM).

The illustration shows the general configuration for a 90° corner reflector. The distance S from the driven radiator D to edge E need not be critically chosen with respect to wavelength; for a 90° reflector D may lie between 0.25 and 0.7 wavelength. The overall dimensions of the reflector need not exceed 2 wavelengths in order to approximate the performance obtained with very large reflectors. Gain, as compared with an isotropic radiator, is about 12 dB.

The reflecting planes may be metal sheets or parallel wires separated by a small fraction of a wavelength and extending in the direction of current flow. [J. C. SCHELLENG]

Corona discharge

A type of electrical conduction that generally occurs at or near atmospheric pressure in gases. A relatively strong electric field is needed. External manifestations are the emission of light and a hissing sound. The particular characteristics of the discharge are determined by the shape of the electrodes, the polarity, the size of the gap, and the gas or gas mixture.

In some cases corona discharge may be desirable and useful, whereas in others it is harmful and attempts are made to minimize it. The effect is used for voltage division and control in direct-current nuclear particle accelerators. On the other hand, the corona discharge that surrounds a high-potential power transmission line represents a power loss and limits the maximum potential which can be used. Because the power loss due to i^2r heating decreases as the potential difference is increased, it is desirable to use the maximum possible voltage.

The shape of the electrodes has a very profound effect on the potential current characteristic. If the radius of curvature of the positive electrode is small compared to the gap between electrodes, the transition from the dark current region to the field-sustained discharge will be quite smooth. The effect here is to enable the free electrons to ionize by collision in the high field surrounding this electrode. One electron can produce an avalanche in such a field, because each ionization event releases an additional electron, which can then make further ionization. To sustain the discharge, it is necessary to collect the positive ions and to pro-

CORONA DISCHARGE

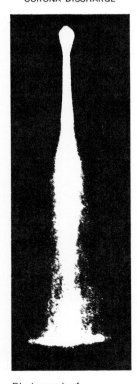

Photograph of breakdown streamers in positive-point corona crossing from positive point to cathode below. (*From L. B. Loeb, Fundamental Processes of Electrical Discharge in Gases, Wiley, 1939*)

duce the primary electrons far enough from the positive electrode to permit the avalanche to develop. The positive ions are collected at the negative electrode, and it is their low mobility that limits the current in the discharge. The primary electrons are thought to be produced by photoionization (see illustration). For a discussion of ionic mobility *see* ELECTRICAL CONDUCTION IN GASES. *See also* DARK CURRENT.

The situation at the negative electrode is quite important. The efficiency for ionization by positive ions is much less than for electrons of the same energy. Most of the ionization occurs as the result of secondary electrons released at the negative electrode by positive-ion bombardment. These electrons produce ionization as they move from the strong field at the electrode out into the weak field. This, however, leaves a positive-ion space charge, which slows down the incoming ions. This has the effect of diminishing the secondary electron yield. Because the positive ion mobility is low, there is a time lag before the high field conditions can be restored. For that reason the discharge is somewhat unstable.

From the preceding it may be seen that the electrode shape is important. The dependence on the gas mixture is difficult to evaluate. Electronegative components will tend to reduce the current at a given voltage, because heavy negative ions have a low mobility and are inefficient ionization agents. The excess electron will not be tightly bound, however, and may be released in a collision. The overall effect is to reduce the number of electrons that can start avalanches near the positive electrode. Further, if a gas with low-lying energy states is present, the free electrons can lose energy in inelastic collisions. Thus it is more difficult for the electrons to acquire enough energy to ionize. In electrostatic high-voltage generators in pressurized tanks, it is quite common to use Freon and nitrogen to take advantage of this effect to reduce corona. In a pure monatomic gas such as argon, corona occurs at relatively low values of voltage.

In the potential current characteristic, the corona region is found above the dark current region and is field-sustained. Near the upper end it goes into either a glow discharge or a brush discharge, depending on the pressure. Higher pressure favors the brush discharge. *See* BRUSH DISCHARGE; GLOW DISCHARGE.

For still higher potential difference, breakdown takes place and a continuously ionized path is formed. *See* ELECTRIC SPARK.

[GLENN H. MILLER]

Bibliography: L. B. Loeb, *Electrical Coronas*, 1965.

Counting circuit

One of several types of functional circuits used in switching or digital data-processing systems, such as dial telephone systems and digital computers. The general function of a counting circuit is to receive and count repeated current or voltage pulses, which represent information arriving from some other circuit within the same switching system or from a source external to the system. Counting circuits may be devised with a variety of switching circuit devices, such as electromagnetic relays, vacuum tubes, and transistors. *See*

SWITCHING CIRCUIT; SWITCHING SYSTEMS (COMMUNICATIONS).

The basic requirement of a counting circuit is to detect each incoming pulse and to establish within itself a unique state for each successive appearance of such a pulse. The number of pulses that have arrived can then be indicated to another circuit by an output which reflects the particular state of the counting circuit.

Obviously, the simplest counter is one that counts a single pulse. With electromagnetic relays the counting of a single pulse requires at least two relay actions, one to record the arrival of the input signal, and a second to record its disappearance. A simple embodiment of such a counter is the relay circuit of Fig. 1.

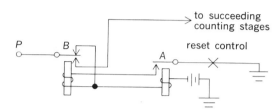

Fig. 1.   Single-pulse relay counter.

In this circuit an incoming ground pulse on lead P operates relay A. The contact of A applies ground to relay B, but as long as the input ground is present, B is short-circuited and cannot operate. When the input ground disappears, B operates in series with the winding of A, which remains operated. The count of one pulse is now completed, and the input lead is steered to a succeeding similar circuit stage. A number of such stages can, of course, be cascaded to count a train of pulses.

A frequently used counting circuit configuration is a two-pulse, or binary, counter which recycles or returns to its original state after the second pulse. A relay circuit of this kind is shown in Fig. 2.

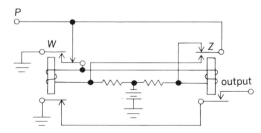

Fig. 2.   Two-pulse relay counter.

In this circuit the first ground pulse on lead P is applied to the windings of both relays W and Z through the normally closed contact on W. However, at this time only W operates, because the input ground short-circuits Z through its normally closed contact. When this first pulse disappears, W remains operated because of the ground connection through its own closed contact. This same ground now operates Z. When the second ground pulse appears on lead P, it short-circuits W through Z's closed contact, thereby restoring W to

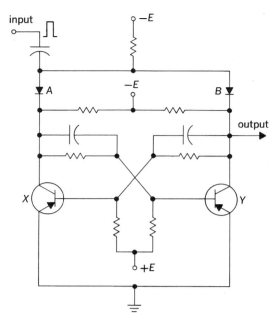

input

Fig. 3. Two-pulse transistor counter.

normal. *Z* holds operated on the input ground until this ground disappears. Thus, at the end of a second pulse, the circuit is back to normal, that is, both relays are unoperated. The output of this binary counter is a ground, which appears during every second input pulse. For this reason a binary counter is often referred to as a pulse or frequency divider.

Figure 3 shows a transistor type of binary counter. If in this circuit the normal (no count) state is designated as transistor *Y* conducting and *X* cut off, the first positive input pulse will pass through diode *A* and on to the base of *Y*, causing *Y* to cut off and *X* to conduct. The second positive input pulse will now pass through diode *B* to the base of *X*, causing *X* to cut off and *Y* to conduct. After two input pulses, therefore, the circuit is back to normal. The second pulse causes the output lead to rise from a negative voltage (*Y* cutoff) to ground (*Y* conducting). Therefore, the binary counter stage produces a positive output pulse for every two positive input pulses, thus acting as a frequency divider.

Several single-stage binary counters may be connected so that the output of one counter stage acts as the input to the next counter stage. A two-stage circuit of this type can count four pulses; a three-stage, eight pulses, and so on.

[JOHN MESZAR]

Coupled circuits

Two or more electric circuits are said to be coupled if energy can transfer electrically or magnetically from one to another. If electric charge, or current, or rate of change of current in one circuit produces electromotive force or affects the voltage between nodes in another circuit, the two circuits are coupled.

Between coupled circuits there is mutual inductance, resistance, or capacitance, or some combination of these. The concept of a mutual parameter is based on the loop method of analysis. A mutual parameter can be one that carries two or more loop currents; such a network has conductive coupling

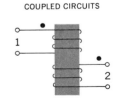

Fig. 1. Polarity of coils.

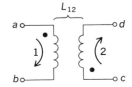

Fig. 2. Mutual inductance.

because electricity can flow from one circuit to the other. *See* NETWORK THEORY.

Also, there can be purely inductive coupling, which appears if the magnetic field produced by current in one circuit links the other circuit. A two-winding transformer is an application of inductive coupling, with energy transferred through the magnetic field only.

It is also possible to have mutual capacitance, with energy transferred through the electric field only. Examples are the mutual capacitance between grid and plate circuits of a vacuum tube, or the capacitive interference between two transmission lines, as a power line and a telephone line, that run for a considerable distance side by side.

Polarity. With inductive coupling, polarity may need to be known, particularly if two circuits are coupled in more than one way. Do two kinds of coupling produce voltages that add or subtract?

There are several ways to show the relative polarities of inductive coupling. Figure 1 shows, somewhat pictorially, two coils wound on the same core. Current flowing into the upper end of coil 1 would produce magnetic flux upward in the core, and so also would current flowing into the upper end of coil 2. For this reason the upper ends of the two coils are said to be corresponding ends.

If a wiring diagram is drawn with some such semipictorial sketch of the coils, it is not difficult to determine which are corresponding ends. However it is easier, both in representation and in interpretation, to indicate the corresponding ends symbolically. For this purpose dots are placed on a diagram at the corresponding ends of coupled coils. Such dots are shown in Fig. 1, though they are not needed in this figure. Dots are also shown in Fig. 2, where they give the only means of identifying corresponding ends of the coils shown.

Note that the bottom ends of the coils shown in Fig. 1 are also corresponding ends, and that the two lower ends might have been dotted instead of the two upper ends; it makes no difference. But if the sense of winding of either coil 1 or coil 2 in Fig. 1 were reversed, then one dot (either one) would have to be shifted correspondingly.

Voltage equations. In addition to showing corresponding ends of two coils, Fig. 2 indicates that the two coils couple to each other. This coupling is identified as L_{12}. There is voltage from *a* to *b* in Fig. 2 if current i_1 is changing through the self-inductance L_{11} of the coil in circuit 1. There is additional voltage from *a* to *b* if current i_2 is changing in circuit 2 because of the mutual inductance L_{12} between the circuits. Thus, with circuit 1 coupled to circuit 2 and to any number of other circuits, Eqs. (1) and (2) hold. Here L_{11} and L_{22},

$$v_{ab} = \left(L_{11}\frac{di_1}{dt} \pm L_{12}\frac{di_2}{dt} \pm \cdots \right) \qquad (1)$$

$$v_{cd} = \left(\pm L_{21}\frac{di_1}{dt} + L_{22}\frac{di_2}{dt} \pm \cdots \right) \qquad (2)$$

the self-inductances of the circuits, are positive numbers (inductances measured in henrys). With regard to the mutual terms, two questions of polarity must be asked. Are the mutual inductances such as L_{12} and L_{21} always positive numbers? And are the signs before the mutual terms, shown as $\pm$ in Eq. (1), actually + or are they −? Answers to these questions are not always the same, and may be different as given by different authors, but the most usual simple procedure to answer them is as follows:

1. Draw a circuit diagram, such as Fig. 2, with nominal positive directions of currents shown by arrows and with dots at corresponding ends of coupled coils.

2. Let each mutual inductance such as L_{12} be a positive number.

3. Write equations such as Eqs. (1) and (2) with the following signs: First, if both arrows enter dotted ends of a pair of coupled coils, or if both arrows enter undotted ends, use $+$ before the corresponding mutual-impedance term. Second, if one arrow enters a dotted end and the other an undotted end, use $-$ before the corresponding mutual-impedance term.

However, this simple procedure sometimes fails, for it may be impossible to dot corresponding ends of all pairs of coupled coils in a network if there are three or more coils. A more general method is the following:

1. Draw a circuit diagram with nominal positive directions of currents shown by arrows; place dots at coil ends arbitrarily, as may be convenient.

2. Determine corresponding ends of pairs of coupled coils, considering the coils two at a time. For the mutual inductance between a pair of coils with corresponding ends dotted, use a positive number, such as $L_{12} = 5$. For mutual inductance between a pair of coils with noncorresponding ends dotted, use a negative number, such as $L_{12} = -5$.

3. Write equations according to rule 3, above.

Steady-state equations. The differential Eqs. (1) and (2) are quite general. For steady-state operation at a single frequency, it is often simpler to have phasor or transform equations. Such equations can be in terms of reactances instead of inductances ($X = \omega L = 2\pi fL$ where f is frequency in cycles per second, or hertz). With the usual interpretation of phasor or transform equations, the steady-state relations corresponding to Eqs. (1) and (2) are shown in Eqs. (3) and (4). The rules in

$$V_{ab} = j\omega(L_{11}I_i \pm L_{12}I_2 \pm \cdots)$$
$$= j(X_{11}I_1 \pm X_{12}I_2 \pm \cdots) \qquad (3)$$
$$V_{cd} = j\omega(\pm L_{21}I_1 + L_{22}I_2 \pm \cdots)$$
$$= j(\pm X_{21}I_1 + X_{22}I_2 \pm \cdots) \qquad (4)$$

the foregoing paragraphs determine the choice of $+$ or $-$ in these equations; also, a mutual reactance such as X_{12} is a positive number (of ohms) if the corresponding mutual inductance such as L_{12} is a positive number (of henrys), as determined by the above rules. *See* ALTERNATING-CURRENT CIRCUIT THEORY.

Equality of mutual inductance. Because of this equality it is not uncommon, if there are only two coupled coils in a network, to use the letter M in place of both L_{12} and L_{21}. This use of M will be adopted in the following paragraphs.

Coefficient of coupling. The coefficient of coupling between two circuits is, by definition, Eq. (5).

$$k = M/\sqrt{L_{11}L_{22}} \qquad (5)$$

If the circuits are far apart or, because of orientation, have little mutual magnetic flux, they are

loosely coupled and k is a small number. Values of k for circuits with loose coupling may be in the range between 0.01 and 0.10. For closely (or tightly) coupled circuits with air-core coils, k may be around 0.5. In a transformer with a ferromagnetic core, k is very nearly 1.00.

Ideal transformers. An ideal transformer is defined as one in which primary and secondary currents are related inversely as the number of turns in the windings; this is shown in Eq. (6). Voltages across the primary and secondary windings are in direct proportion to the numbers of turns, as shown in Eq. (7).

$$I_1/I_2 = N_2/N_1 \qquad (6)$$

$$V_1/V_2 = N_1/N_2 \qquad (7)$$

An actual transformer may be almost ideal or not at all ideal, depending on how it is made, and construction in turn depends on the purpose for which it is to be used. In an ideal transformer $k = 1$; in any actual transformer k is less than 1. In an actual transformer there is magnetizing current, so Eq. (6) is not exact. Also, an actual transformer has resistance and leakage reactance, so Eq. (7) is not exact. Nevertheless, the relations of ideal transformers are closely approximated by 60-Hz power transformers, and these relations are more or less close for other transformers that have ferromagnetic cores.

Whereas coupling k is desirably as close as possible to unity in a transformer that couples power from one circuit to another, k may purposefully be considerably less than unity in a transformer used for another purpose. In an oscillator, coupling need only be sufficient to sustain oscillation. In a band-pass amplifier, coupling is determined by bandwidth requirements. *See* OSCILLATOR; RESONANCE (ALTERNATING-CURRENT CIRCUITS).

Equivalent circuits. It is often convenient to substitute into a network, in place of a pair of inductively coupled circuits, an equivalent pair of conductively coupled circuits. Circuits so substituted are equivalent if the network exterior to the coupled circuits is unaffected by the change; in many cases this requirement implies that input current and voltage and output current and voltage are unaffected by the change.

Voltages and currents at the terminals of the coupled circuits of Fig. 3 are related by Eqs. (8)

$$(R_1 + j\omega L_{11})I_1 - j\omega MI_2 = V_{ab} \qquad (8)$$

and (9). Voltages and currents at the terminals of

$$(R_2 + j\omega L_{22})I_2 - j\omega MI_1 = V_{cd} \qquad (9)$$

the network of Fig. 4 are also related by Eqs. (8) and (9), so the network of Fig. 4 is equivalent to the coupled circuits of Fig. 3. This is not immediately

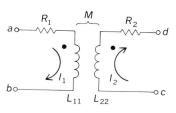

Fig. 3. A pair of coupled circuits.

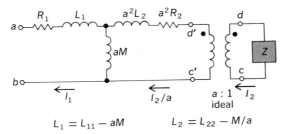

$$L_1 = L_{11} - aM \qquad L_2 = L_{22} - M/a$$

Fig. 4. A network equivalent to coupled circuits, feeding a load of impedance Z.

obvious, but it appears when loop equations for the network are simplified to a form that shows the requirement that $L_{11} = L_1 + aM$ and $L_{22} = L_2 + M/a$ and that $V_{c'd'} = aV_{cd}$, where a is the turn ratio of the ideal transformer shown, equal to N_1/N_2 in Eqs. (6) and (7).

It seems at first that this substitution, resulting in the network of Fig. 4, has produced something more complicated than the circuit of Fig. 3, but the value of substituting will now be shown.

Mathematically, a could be any number, but practically there are two particularly advantageous values for a. When the coupling between coils is loose so that k, the coefficient of coupling, is small, and if the coils have somewhere near the same number of turns, it is advantageous to let $a = 1$. With $a = 1$ the network of Fig. 4 is simplified to Fig. 5, leaving only the conductively coupled circuits with parameters R_1 and L_1, M, and L_2 and R_2, where $L_1 = L_{11} - M$, and $L_2 = L_{22} - M$. Many loosely coupled circuits, particularly in radio circuits, are conveniently represented by this equivalent network with $a = 1$.

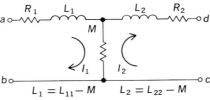

$$L_1 = L_{11} - M \qquad L_2 = L_{22} - M$$

Fig. 5. A conductive network equivalent to the coupled circuits of Fig. 3 ($a = 1$).

On the other hand, if coupling is close and especially when the two coupled coils have widely different numbers of turns (a situation that is typical of transformers), it is more convenient to let a equal the actual turn ratio of the coils. If, for example, there are 10 times as many turns in the primary winding of a transformer as there are in the secondary winding, it is well to let $a = 10$. (Letting $a = 1$ in such a transformer would result in a negative value for L_2 which, though correct for analysis, is difficult to visualize.)

With a equal to the actual transformer turn ratio, the following interpretation of Fig. 4 is usual and convenient. All causes of power loss and voltage drop have been put into the equivalent T network. Only the turn ratio, the actual transforming function, remains in the ideal transformer. It now becomes possible to consider, to study, and even to design the separate functions independently.

Transformers. The equivalent circuit of Fig. 4 is so commonly used in transformer work that a number of the quantities have been given special names. With the concept that power is supplied to one winding of the transformer, this is called the primary winding; the other, the secondary winding, provides power to a load. There may be a third, or tertiary, winding, perhaps providing power to another load at a different voltage or with a different connection, and even other windings, on the same transformer core.

However, it is only when a transformer has become part of a system that there is any meaning in designating the windings as primary and secondary, for a two-winding transformer can be used to carry power in either direction. The terms "high-voltage side" and "low-voltage side" are preferable for a transformer that is not part of a system. In the following discussion the words primary and secondary really mean nothing more than the windings of the transformer that are designated by the subscripts 1 and 2.

Speaking of a transformer with N_1 primary turns and N_2 secondary turns, let a be the turn ratio N_1/N_2. Then, referring to Fig. 4, R_1 is primary resistance; L_1 is primary leakage inductance; aM is primary exciting inductance; a^2L_2 is secondary leakage inductance referred to the primary side; and a^2R_2 is secondary resistance referred to the primary side. Usually, operation of power transformers is described in terms of these leakage inductances (or leakage reactances) and in terms of resistances, inductances, or reactances all referred to one side of the transformer or the other.

Transformation of impedance. Figure 6 is the same as Fig. 4 except that the ideal transformer with turn ratio of $a:1$ has been eliminated and the impedance of the load has been changed from Z to a^2Z. This conversion makes the network of Fig. 6 equivalent to that of Fig. 4 at the input terminals, and it suggests a concept that is quite useful in communications.

If a load with impedance Z is preceded by an ideal transformer with turn ratio a, the input impedance to the transformer is a^2Z. Thus a load of impedance Z can be made to act like a load with impedance a^2Z by using an ideal transformer with turn ratio a. A useful application is to connect a load of one impedance to an incoming line that has a different terminal impedance through an impedance-matching transformer. By this means a transformer with turn ratio a will match a load with resistance R to a source with resistance a^2R, thereby providing for maximum power transfer to the load.

The preceding paragraph assumes an ideal transformer to provide a perfect impedance match. Figure 6 shows the T network of resistances and inductances that an actual transformer

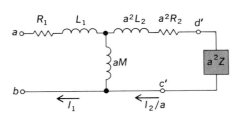

Fig. 6. Transformation of impedance from Fig. 4.

introduces into the network. In practice, the transformer used must be good enough so that these resistances and inductances are acceptable.

Core loss. Neither the equations nor the circuits of this article have taken into account the power loss in a ferromagnetic core. Both eddy-current loss and hysteresis loss may be appreciable; these iron losses are commonly great enough to affect the economics of power transformers, and even to prohibit the use of metal cores with radio-frequency current.

The relations of core loss are not linear and cannot accurately be included in linear equations, but a satisfactory approximation is used with power transformers for which frequency and applied voltage change little if at all. With these restrictions, core loss can be represented by the loss in a resistance shunted around mutual inductance aM of Fig. 4. This approximation of loss is much better than neglecting core loss entirely, and is usual in work with power transformers.

[HUGH H. SKILLING]

Bibliography: R. Boylestad, *Introductory Circuit Analysis*, 3d ed., 1977; S. Ramo, J. R. Whinnery, and T. R. Van Duzer, *Fields and Waves in Communication Electronics*, 1965; H. H. Skilling, *Electrical Engineering Circuits*, 1965.

Critical path method (CPM)

A diagrammatic network-based technique, similar to the program evaluation and review technique (PERT), that is used as an aid in the systematic management of complex projects. The technique is useful in: organizing and planning; analyzing and comprehending; problem detecting and defining; alternative action simulating; improving (replanning); time and cost estimating; budgeting and scheduling; and coordinating and controlling. It has its greatest value in complex projects which involve many interrelated events, activities, and resources (time, money, equipment, and personnel) which can be allocated or assigned in a variety of ways to achieve a desired objective. It can be used to complete a multifaceted program faster or with better utilization of resources by reassignments, trade-offs, and judiciously using more or less assets for certain of the individual activities composing the overall project.

CPM was introduced in 1957–58 by E. I. du Pont de Nemours & Company. Present applications include: research and development programs; new product introductions; facilities planning and designs; plant layouts and relocations; construction projects; equipment installations and start-ups; major maintenance programs; medical and scientific researches; weapons systems developments; and other programs in which cost reduction, progress control, and time management are important. It is best suited to large, complex, one-time or first-time projects rather than to repetitive, routine jobs.

A simplified example will show the features of the critical path method and the steps involved in constructing and using the network-based system:

Project: Design personal home-use computer.
Objective: Complete project as soon as possible.
Procedure: The CPM procedure for this project involves eight steps:

Step 1: List required events; arrange as to best guess of their sequence; assign a code letter or number to each. (See center column of Table 1.) [Events are accomplished portions or phases of the project. Activities are the resources applied to progress from one event to the next. This example has been greatly simplified for illustrative purposes. A real case could have several hundred events and activities and would be a candidate for computer analysis, for which programs are available.]

Step 2: Determine which event(s) must precede each event before it can be done and which event(s) must await completion of that event before it can be done. Construct a table to show these relationships. (See left and right columns of Table 1.)

Step 3: Draw the network, which is a diagram of the relationship of the events and their required activities that satisfies the dictates of Table 1 (see Fig. 1).

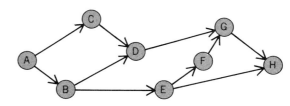

Fig. 1. Network of relationship between events (circles) and their activities (arrows).

Table 1. CPM events table

Prior event	Event	Following event
—	A = Authorization to start received	B and C
A	B = Computing circuits designed	D and E
A	C = Video circuits designed	D
B and C	D = Keyboard and cabinets designed	G
B	E = Programming completed	F and H
E	F = Operating systems completed	G
D and F	G = Testing and debugging finished	H
E and G	H = Design specifications and programs finalized	—

Table 2. Time and cost factors for activities

Activity	Expected time to complete (weeks)	Maximum possible time reduction (weeks)	Cost (per week) to reduce time
AB	5	1	$1000
AC	10	3	500
BD	12	3	1000
BE	15	2	500
CD	2	0	—
DG	18	4	500
EF	3	0	—
EH	4	1	2000
FG	2	0	—
GH	5	2	500

Table 3. Costs of shortening activities

	Reduce			Cost			
AB from	5 weeks to	4 weeks		1 week × $1000 =			$1000
BD	12	9		3	×	1000 =	3000
DG	18	14		4	×	500 =	2000
GH	5	3		2	×	500 =	1000
TOTAL	40	30		10			$7000

Step 4: List the required activities and enter the best estimate available of the time required to complete each activity. In PERT, three time estimates are entered: the most optimistic, the most likely, and the most pessimistic. This is the basic way in which PERT differs from CPM. Some users of CPM enter a second time estimate for each activity, the first being the normal time and the second being the crash (or fast at any price) time. Table 2 shows the list of activities (AB designates the activity required in going from event A to event B), the expected normal time, the maximum number of weeks that each activity can be shortened under a crash program, and the cost per week to shorten, if it can be shortened.

Step 5: Write the normal expected time to complete each activity on the diagram (Fig. 2). Note that the relative lengths of the arrows (activities) have no relationship to their time magnitudes.

Step 6: Find the critical path, that is, the *longest* route from event A to event H. The longest path is the total time that the project can be expected to take unless additional resources are added or resources from some of the other paths in which there may be slack can be reallocated to activities which lie in the critical path. In the example shown in Fig. 2, the initial critical path is A – B – D – G – H, or 5 + 12 + 18 + 5 = 40 weeks. If the critical path time works out to be less or more than the target time to complete the project, there is then a positive or negative float.

Step 7: Shorten the critical path. The stated objective for this example was to complete the project as soon as possible. Other projects might have different objectives. An examination of Table 2 and Fig. 2 suggests that the activities lying along the critical can be shortened for the costs shown in Table 3. The total expected time to complete the project can be shortened from 40 weeks to 30 weeks—a 25% improvement. A judgment can be made by those responsible for the project as to whether the 10-week time savings is worth the additional cost of $7000. In some cases, the critical path is so shortened that it is no longer the longest path, and another path becomes the critical one. Only reductions in the critical path time reduce the total project time, and activities in noncritical paths may run over or late to the limits of their floats without causing the project to be delayed. The floats are the differences between the total expected time required for each possible path in the network and the total time required for the critical path, and are measures of the spare time available.

Step 8: Construct a schedule of the project showing the earliest and latest permissible start and completion dates of each activity. To do this, the individual floats are calculated. With the slack in the critical path set at zero at the onset of the project's implementation, any slippage in the completion of any event along that path will result in a delay in the completion of the whole project, unless made up later.

The use of CPM adds another step to a project, and it requires continual updating and reanalysis as conditions change, but experience has shown that the effort can be a good investment in completing a project in less time, with less resources, with more control, and with a greater chance of on-time, within-budget completion.

[VINCENT M. ALTAMURO]

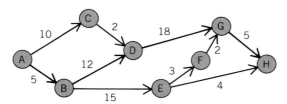

Fig. 2. Network diagram modified to show time to complete activities. The critical path is indicated by the black arrows.

Crossover network

A selective network used to divide the audio-frequency output of an amplifier into two or more bands of frequencies. The frequency separation is employed to feed two or more loudspeakers, each operating in a restricted frequency band and thereby operating more efficiently and with less distortion. Most crossover networks used in high-fidelity audio systems are designed for crossover frequencies somewhere between 400 and 2000 cycles.

These networks consist of coils and capacitors connected to provide a low-pass filter for feeding the low-frequency woofer, a high-pass filter for feeding the high-frequency tweeter, and perhaps a band-pass filter for feeding a middle-frequency-range speaker. At the crossover frequency the power outputs of the two filters are equal. The crossover network is often connected between the secondary winding of the output transformer and the voice coils of the loudspeakers. *See* LOUD-SPEAKER.

[JOHN MARKUS]

Crosstalk

A term originated for the sound produced at a receiving terminal of one voice-frequency circuit by speech signals carried by one or more other such circuits. The term is used commonly for the induction of speech signals in one circuit from a neighboring circuit. It is also used for cross modulation. *See* DISTORTION (ELECTRONIC CIRCUITS).

Crosstalk is unwanted; it impairs communication service to a degree determined by the tolerance of the receiving system or the listener, by the frequency of occurrence, and by the duration of high ratios of crosstalk to normal-speech signal intensity. If crosstalk causes loss of presumptive secrecy, the condition is intolerable.

Types. Crosstalk is often one of the most controlling factors in the design of a multichannel communication system. Several forms of crosstalk are often distinguished. Intelligible crosstalk is sufficiently understandable under pertinent circuit and room noise conditions that meaningful information can be obtained by the more sensitive observers. Nonintelligible crosstalk cannot be understood regardless of its received volume, but because of its syllabic nature is more annoying subjectively than thermal-type noise. Near-end crosstalk is that type whose energy travels in the opposite direction to the speech signal in the disturbing circuit (Fig. 1). Far-end crosstalk is that type whose energy travels in the same direction as the speech signal in the disturbing circuit (Fig. 1).

Sources. Figure 1 shows two paralleling circuits in the east-west direction between terminals E_1 and W_1 and between E_2 and W_2. Speech signals impressed at E_1 will induce crosstalk in the E_2-to-W_2 circuit. Resulting crosstalk at E_2 is called near-end and that at W_2 far-end.

Crosstalk coupling refers to the unwanted transmission path between reference points on the disturbing and disturbed circuits, and is measured either in crosstalk units (cu) or in decibels (dB). The value in cu is 10^6 times the square root of the ratio of receiving-end to sending-end power with a specified frequency of the power source. For a sin-

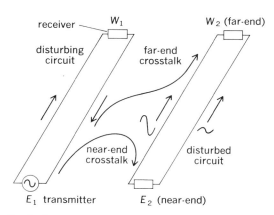

Fig. 1. Crosstalk arising in short section of two parallel circuits. Conductor closer to disturbing circuit has larger crosstalk voltage than farther conductor; voltage difference is effective at E_2 and W_2 terminals. Energy in disturbing circuit assumed to flow from east to west.

gle frequency measurement, a crosstalk (coupling) of 1000 cu means a current ratio of 0.001 for equal sending and receiving impedances. A 60-dB loss measures this same ratio. For special purposes a derived figure of 90 minus the coupling loss in dB identified by the symbol dBx is used.

The mutual impedances of closely spaced paralleling wire circuits supply coupling paths among them. Because these impedances are almost wholly reactive, coupling values in cu tend to increase linearly with frequency. The importance of crosstalk coupling and its control has been greatly increased by the development of the carrier method for multiplexing communication transmission lines; this method employs higher frequencies.

Crosstalk can also result from other than inductive relations among wire circuits. For example, with the carrier method of multiplexing, interchannel coupling can arise through imperfect functioning of modulating, amplifying, or frequency-selective apparatus at terminal or intermediate points.

Remedies. With open-wire circuits, coupling control is effected most practically by transposing (interchanging) positions of the wires of each pair in accordance with a systematic pattern seeking to oppose the coupling of one interval by that of another (Fig. 2). The design of patterns to accomplish a satisfactory result for all pair combinations is a specialty calling for a high order of skill and ingenuity. Coupling between two pairs involves the effect of another circuit or circuits called a tertiary, and there always is at least one tertiary in the form of all wires in multiple to ground.

With the carrier frequencies in general use on open-wire lines, far-end coupling is controlled by transposition; but near-end coupling is controlled by adequate frequency segregation of the paths for the two directions of transmission.

Control of coupling among insulated twisted pairs compacted within the metallic sheath of a cable is the same in principle as for open-wire lines. The transposition effect is manufactured into a cable by using different pitches of twist for closely spaced pairs. Remaining coupling can be reduced by the installation at intervals of small, adjustable coupling units. The higher dimensional stability of the cable structure permits a final re-

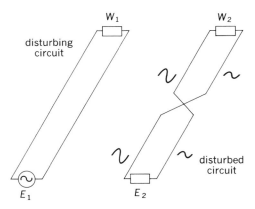

Fig. 2. Single transposition of circuit. If wires are transposed so that in every short length each is the closer conductor for an equal distance, the total crosstalk voltages for the two wires will be made nearly equal and their difference greatly reduced.

sult superior to that obtained on open-wire lines.

Peculiarly useful with telephony is the compandor, an effective device that reduces crosstalk by compressing the volume range of speech signals before they are impressed upon the line and expanding the range to its original naturalness at the other end. This process reduces the crosstalk the listener hears during intervals of silence or weak speech. During intervals of high volumes, crosstalk actually is intensified along with speech, but its presence is masked. The effect of equipping telephone circuits with compandors approximates the result of increasing coupling losses 20 dB.

[A. FISCARELLI]

Bibliography: W. C. Babcock, E. Rentrop, and C. S. Thaeler, *Crosstalk on Open-Wire Lines*, Bell Teleph. Syst. Monogr. no. 2520, 1955; A. G. Chapman, Open-wire crosstalk, *Bell Syst. Tech. J.*, 13: 19–58, 195–238, 1934; W. B. Davenport, Jr., and W. L. Root, *Introduction to Random Signals and Noise*, 1958.

Cryosar

A two-terminal semiconductor switching device which operates at very low temperatures. The thyratronlike characteristic of this device is the result of the avalanche ionization of normal semiconductor impurity centers (donors and acceptors) which, at the low temperature of operation, are no longer thermally ionized. It has a relatively high impedance because only a moderate amount of current can be drawn in the conducting state before overheating occurs, with an attendant rise in voltage. The operating temperature is the liquid-helium range. *See* SEMICONDUCTOR.

Typical characteristics show a breakdown voltage of 3 volts and a sustaining voltage of 1 volt. The nonconducting impedance is of the order of 10 megohms, and the maximum current in the conducting state is about 1 milliamp for devices of normal size. Under suitable conditions of load and voltage overdrive, it is possible to switch the device in a few millimicroseconds. It is possible to obtain the device with a breakdown voltage either equal to the sustaining voltage or much greater as the application requires.

A compound cryosar, consisting of two normal cryosars with different electrical characteristics in series, has been developed in the form of a single semiconductor wafer. This wafer contains two layers of differently doped material. The contacting electrodes are placed so that any current flowing between them must traverse both layers in series. The resulting characteristic shows a double peak before the sustaining voltage region. The valley voltage between the peaks can be made lower than the sustaining voltage so that two stable states exist at a given voltage.   [LLOYD P. HUNTER]

Bibliography: J. J. Brophy, *Semiconductor Devices*, 1964; A. L. McWhorter and R. H. Rediker, The cryosar: A new low-temperature computer component, *Proc. IRE*, 47:1207–1213, 1959; R. P. Turner, *Semiconductor Devices*, 1961.

Cryotron

A current-controlled switching device based on superconductivity for use primarily in computer circuits. The early version has been superseded by the tunneling cryotron, which consists basically of a Josephson junction. In its simplest form (see illustration) the device has two electrodes of a superconducting material (for example, lead) which are separated by an insulating film only about 10 atomic layers thick. For the electrodes to become superconducting, the device has to be cooled to a few degrees above absolute zero. The tunneling cryotron has two states, characterized by the presence or absence of an electrical resistance. They can be considered as the "on" and "off" states of the switch, respectively. Switching from on to off is accomplished by a magnetic field generated by sending a current through the control line on top of the junction. The device can switch in a few picoseconds and has a power consumption of only some microwatts. These properties make it an attractive switching device for computers, promising performance levels probably unattainable with other devices. The tunneling cryotron is still in the research stage, but various computer circuits have already been realized. *See* SUPERCONDUCTING DEVICES; SUPERCONDUCTIVITY.   [P. WOLF]

Bibliography: W. Anacker, Computing at 4 degrees Kelvin, *IEEE Spect.* 16:26–37, May 1979.

Cryptography

The various methods for writing in secret code or cipher. As society becomes increasingly dependent upon computers, the vast amounts of data communicated, processed, and stored within computer systems and networks often have to be protected, and cryptography is a means of achieving this protection. It is the only practical method for protecting information transmitted through accessible communication networks such as telephone lines, satellites, or microwave systems. Furthermore, in certain cases, it may be the most economical way to protect stored data. Cryptographic procedures can also be used for message authentication, personal identification, and digital signature verification for electronic funds transfer and credit card transactions. *See* DATA-BASE MANAGEMENT SYSTEMS; DATA COMMUNICATIONS; DIGITAL COMPUTER; ELECTRICAL COMMUNICATIONS.

Cryptographic algorithms. Cryptography must resist decoding or deciphering by unauthorized personnel; that is, messages (plaintext) transformed into cryptograms (codetext or ciphertext) have to be able to withstand intense cryptanalysis. Transformations can be done by using either code or cipher systems. Code systems rely on code books to transform the plaintext words, phrases, and sentences into codetext or code groups. To prevent cryptanalysis, there must be a great number of plaintext passages in the code book and the code group equivalents must be kept secret. To isolate users from each other, different codes must be used, making it difficult to utilize code books in electronic data-processing systems.

Cipher systems are more versatile. Messages are transformed through the use of two basic elements: a set of unchanging rules or steps called a cryptographic algorithm, and a set of variable cryptographic keys. The algorithm is composed of enciphering (**E**) and deciphering (**D**) procedures which usually are identical or simply consist of the same steps performed in reverse order, but which can be dissimilar. The keys, selected by the user, consist of a sequence of numbers or characters. An enciphering key (Ke) is used to encipher plaintext (X) into ciphertext (Y) as in Eq. (1), and a deciphering key (Kd) is used to decipher ciphertext (Y) into plaintext (X) as in Eq. (2).

$$\mathbf{E}_{Ke}(X) = Y \tag{1}$$

$$\mathbf{D}_{Kd}[\mathbf{E}_{Ke}(X)] = \mathbf{D}_{Kd}(Y) = X \tag{2}$$

Algorithms are of two types—conventional and public-key (also referred to as symmetric and asymmetric). The enciphering and deciphering keys in a conventional algorithm either may be easily computed from each other or may be identical (Ke = Kd = K, denoting $\mathbf{E}_K(X) = Y$ for encipherment and $\mathbf{D}_K(Y) = X$ for decipherment). In a public-key algorithm, one key (usually the enciphering key) is made public, and a different key (usually the deciphering key) is kept private. In such an approach it must not be possible to deduce the private key from the public key.

When an algorithm is made public, for example, as a published encryption standard, cryptographic security completely depends on protecting those cryptographic keys specified as secret.

Unbreakable ciphers. Unbreakable ciphers are possible. But the key must be randomly selected and used only once, and its length must be equal to or greater than that of the plaintext to be enciphered. Therefore such long keys, called one-time tapes, are not practical in data-processing applications.

To work well, a key must be of fixed length, relatively short, and capable of being repeatedly used without compromising security. In theory, any algorithm that uses such a finite key can be analyzed; in practice, the effort and resources necessary to break the algorithm would be unjustified.

Strong algorithms. Fortunately, to achieve effective data security, construction of an unbreakable algorithm is not necessary. However, the work factor (a measure, under a given set of assumptions, of the requirements necessary for a specific analysis or attack against a cryptographic algorithm) required to break the algorithm must be sufficiently great. Included in the set of assumptions is the type of information expected to be available for cryptanalysis. For example, this could

be ciphertext only; plaintext (not chosen) and corresponding ciphertext; chosen plaintext and corresponding ciphertext; or chosen ciphertext and corresponding recovered plaintext.

A strong cryptographic algorithm must satisfy the following conditions: (1) The algorithm's mathematical complexity prevents, for all practical purposes, solution through analytical methods. (2) The cost or time necessary to unravel the message or key is too great when mathematically less complicated methods are used, because either too many computational steps are involved (for example, in trying one key after another) or because too much storage space is required (for example, in an analysis requiring data accumulations such as dictionaries and statistical tables).

To be strong, the algorithm must satisfy the above conditions even when the analyst has the following advantages: (1) Relatively large amounts of plaintext (specified by the analyst, if so desired) and corresponding ciphertext are available. (2) Relatively large amounts of ciphertext (specified by the analyst, if so desired) and corresponding recovered plaintext are available. (3) All details of the algorithm are available to the analyst; that is, cryptographic strength cannot depend on the algorithm remaining secret. (4) Large high-speed computers are available for cryptanalysis.

In summary, even with an unlimited amount of computational power, data storage, and calendar time, the message or key in an unbreakable algorithm cannot be obtained through cryptanalysis. On the other hand, although a strong algorithm may be breakable in theory, in practice it is not.

Privacy and authentication. Anyone can encipher data in a public-key cryptographic system (Fig. 1) by using the public enciphering key, but

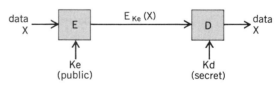

Fig. 1. Public-key cryptographic system used for privacy only.

only the authorized user can decipher the data through possession of the secret deciphering key. Since anyone can encipher data, message authentication is necessary in order to identify a message's sender.

A message authentication procedure can be devised (Fig. 2) by keeping the enciphering key secret and making the deciphering key public, pro-

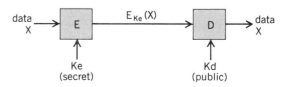

Fig. 2. Public-key cryptographic system used for message authentication only.

$E_{KBe}(D_{KAd}(X))$

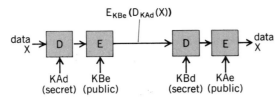

Fig. 3. Public-key cryptographic system used for both message authentication and privacy. KAe and KAd are enciphering and deciphering keys of the sender (A). KBe and KBd are enciphering and deciphering keys of the receiver (B).

vided that the enciphering key cannot be obtained from the deciphering key. This makes it impossible for nondesignated personnel to encipher messages, that is, to produce $E_{Ke}(X)$. By inserting prearranged information in all messages, such as originator identification, recipient identification, and message sequence number, the messages can be checked to determine if they are genuine. However, because the contents of the messages are available to anyone having the public deciphering key, privacy cannot be attained.

A public-key algorithm provides privacy as well as authentication (Fig. 3) if encipherment followed by decipherment, and decipherment followed by encipherment, produce the original plaintext, as in Eq. (3). A message to be authenticated is first deci-

$$D_{Kd}[E_{Ke}(X)] = E_{Ke}[D_{Kd}(X)] = X \qquad (3)$$

phered by the sender (A) with a secret deciphering key (KAd). Privacy is ensured by enciphering the result with the receiver's (B's) public enciphering key (KBe).

Effective data security with public-key algorithms demands that the correct public key be used, since otherwise the system is exposed to attack. For example, if A can be tricked into using C's instead of B's public key, C can decipher the secret communications sent from A to B and can transmit messages to A pretending to be B. Thus key secrecy and key integrity are two distinct and very important attributes of cryptographic keys. While the requirement for key secrecy is relaxed for one of the keys in a public-key algorithm, the requirement for key integrity is not.

In a conventional cryptographic system, data are effectively protected because only the sender and receiver of the message share a common secret key. Such a system automatically provides both privacy and authentication (Fig. 4).

Digital signatures. Digital signatures authenticate messages by ensuring that: the sender cannot later disavow messages; the receiver cannot forge messages or signatures; and the receiver can prove to others that the contents of a message are

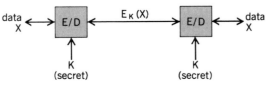

Fig. 4. Conventional cryptographic system in which message authentication and message privacy are provided simultaneously. K represents a common secret key.

genuine and that the message originated with that particular sender. The digital signature is a function of the message, a secret key or keys possessed by the sender, and sometimes data that are nonsecret or that may become nonsecret as part of the procedure (such as a secret key that is later made public).

Digital signatures are more easily obtained with public-key than with conventional algorithms. When a message is enciphered with a private key (known only to the originator), anyone deciphering the message with the public key can identify the originator. The originator cannot later deny having sent the message. Receivers cannot forge messages and signatures, since they do not possess the originator's private key.

Since enciphering and deciphering keys are identical in a conventional algorithm, digital signatures must be obtained in some other manner. One method is to use a set of keys to produce the signature. Some of the keys are made known to the receiver to permit signature verification, and the rest of the keys are retained by the originator in order to prevent forgery.

Data encryption standard. Regardless of the application, a cryptographic system must be based on a cryptographic algorithm of validated strength if it is to be acceptable. The Data Encryption Standard (DES) is such a validated conventional algorithm already in the public domain. (Since public-key algorithms are relatively recent, their strength has yet to be validated.)

During 1968–1975, IBM developed a cryptographic procedure that enciphers a 64-bit block of plaintext into a 64-bit block of ciphertext under the control of a 56-bit key. The National Bureau of Standards accepted this algorithm as a standard, and it became effective on July 15, 1977.

Conceptually, the DES can be thought of as a huge key-controlled substitution box (S-box) with a 64-bit input and output. With such an S-box, 2^{64} different transformations or functions from plaintext to ciphertext are possible. The 56-bit key used with DES thus limits the number of usable functions to 2^{56}.

A single huge S-box is impossible to construct. Therefore DES is implemented by using several smaller S-boxes (with a 6-bit input and a 4-bit output) and permuting their concatenated outputs. By repeating the substitution and permutation process several times, cryptographic strength "builds up." The DES encryption process consists of 16 iterations, called rounds. At each round a cipher function (f) is used with a 48-bit key. The function comprises the substitution and permutation. The 48-bit key, which is different for each round, is a subset of the bits in the externally supplied key.

The interaction of data, cryptographic key (K), and f is shown in Fig. 5. The externally supplied key consists of 64 bits (56 bits are used by the algorithm, and up to 8 bits may be used for parity checking). By shifting the original 56-bit key, a different subset of 48 key bits is selected for use in each round. These key bits are labeled K_1, K_2, . . . , K_{16}. To decipher, the keys are used in reverse order (K_{16} is used in round one, K_{15} in round two, and so on).

At each round (either encipherment or decipherment), the input is split into a left half (designated L) and a right half (designated R), (Fig. 5). R is

Table 1. Modulo 2 addition

A	B	A $\oplus$ B
0	0	0
0	1	1
1	0	1
1	1	0

transformed with f, and the result is combined, using modulo 2 addition (also called the EXCLUSIVE OR operation; see Table 1), with L. This approach, as discussed below, ensures that encipher and decipher operations can be designed regardless of how f is defined.

Consider the steps that occur during one round of encipherment (Fig. 6). Let the input block (X) be denoted $X = (L_0, R_0)$, where L_0 and R_0 are the left and right halves of X, respectively. Function f transforms R_0 into $f_{K_1}(R_0)$ under control of cipher key K_1. L_0 is then added (modulo 2) to $f_{K_1}(R_0)$ to obtain R_1, as in Eq. (4). The round is then completed by setting L_1 equal to R_0.

$$L_0 \oplus f_{K_1}(R_0) = R_1 \qquad (4)$$

The above steps are reversible without introducing any new parameters or requiring that f be a one-to-one function. The ciphertext contains L_1, which equals R_0, and therefore half of the original plaintext is immediately recovered (Fig. 7). The remaining half, L_0, is recovered by recreating $f_{K_1}(R_0)$ from $R_0 = L_1$ and adding it (modulo 2) to R_1, as in Eq. (5). However, to use the procedure in Fig.

$$R_1 \oplus f_{K_1}(R_0) = [L_0 \oplus f_{K_1}(R_0)] \oplus f_{K_1}(R_0) = L_0 \quad (5)$$

6 for encipherment as well as decipherment, the left and right halves of the output must be interchanged; that is, the ciphertext (Y) is defined by Eq. (6). This modified procedure easily extends to

$$Y = [L_0 \oplus f_{K_1}(R_0)], R_0 \qquad (6)$$

n rounds, where the keys used for deciphering are $K_n, K_{n-1}, \ldots, K_1$.

RSA public-key algorithm. The RSA algorithm (named for the algorithm's inventors, R. L. Rivest, A. Shamir, and L. Adleman) is based on the fact that factoring large composite numbers into their prime factors involves an overwhelming amount of computation. (A prime number is an integer that is divisible only by 1 and itself. Otherwise, the number is said to be composite. Every composite number can be factored uniquely into prime factors. For example, the composite number 999,999 is factored by the primes 3, 7, 11, 13, and 37, that is, $999,999 = 3^3 \cdot 7 \cdot 11 \cdot 13 \cdot 37$.) The RSA algorithm uses modulo arithmetic. Two integers a and b are congruent for the modulus m if their difference $(a - b)$ is divisible by the integer m. This is expressed by the symbolic statement $a \equiv b \pmod{m}$, where mod is short for modulus.

To describe the RSA algorithm, the following quantities are defined:

p and q are primes	(secret)
$r = p \cdot q$	(nonsecret)
$\phi(r) = (p-1)(q-1)$	(secret)
Kd is the private key	(secret)
Ke is the public key	(nonsecret)
X is the plaintext	(secret)
Y is the ciphertext	(nonsecret)

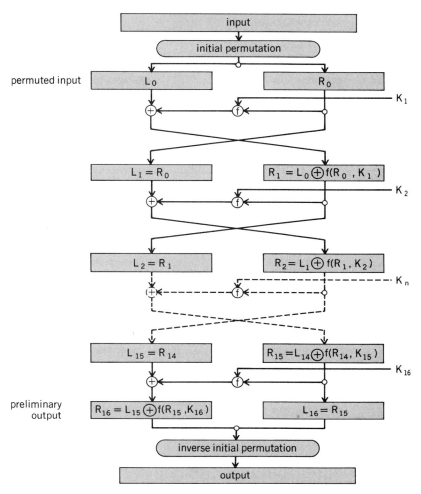

Fig. 5. Enciphering computation in the Data Encryption Standard. (*From Data Encryption Standard, FIPS Publ. no. 46, National Bureau of Standards, 1977*)

Based on an extension of Euler's theorem, Eq. (7),

$$X^{m\phi(r)+1} \equiv X \pmod{r} \qquad (7)$$

the algorithm's public and private keys (Ke and Kd) are chosen so that Eq. (8) or, equivalently, Eq.

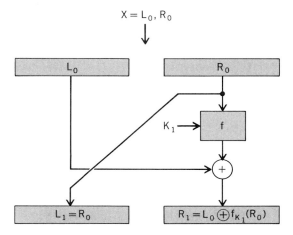

Fig. 6. Transformation of input block (L_0, R_0). (*From C. H. Meyer and S. M. Matyas, Cryptography: A New Dimension in Computer Data Security, copyright © 1980 by John Wiley and Sons; used with permission*)

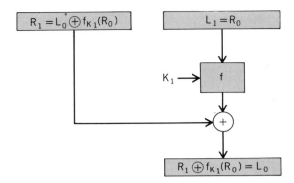

Fig. 7. Recovery of L_0. (*From C. H. Meyer and S. M. Matyas. Cryptography: A New Dimension in Computer Data Security. copyright © 1980 by John Wiley & Sons; used with permission*)

(9) is satisfied. By selecting two secret prime numbers p and q, the user can calculate $r = p \cdot q$, which is made public, and $\phi(r) = (p-1)(q-1)$, which remains secret and is used to solve Eq. (9). (Tests are available to determine with a high level of confidence if a number is prime or not.) To obtain a unique solution for the public key (Ke), a random number, or secret key (Kd), is selected that is relatively prime to $\phi(r)$. (Integers a and b are relatively prime if their greatest common divisor is 1). Ke is the multiplicative inverse of Kd, modulo $\phi(r)$, and Ke can be calculated from Kd and $\phi(r)$ by using Euclid's algorithm. Equation (7) can therefore be rewritten as Eq. (10), which holds true for any plaintext (X).

$$\mathrm{Kd \cdot Ke} = m\phi(r) + 1 \quad (8)$$

$$\mathrm{Kd \cdot Ke} \equiv 1 \, [\mathrm{mod} \, \phi(r)] \quad (9)$$

$$X^{\mathrm{Kd \cdot Ke}} \equiv X \, (\mathrm{mod} \, r) \quad (10)$$

Encipherment and decipherment can now be interpreted as in Eqs. (11) and (12). Moreover, because multiplication is a commutative operation (Ke · Kd = Kd · Ke), encipherment followed by decipherment is the same as decipherment followed by encipherment [Eq. (3)]. Thus the RSA algorithm can be used for both privacy and digital signatures.

$$\mathbf{E}_{\mathrm{Ke}}(X) = Y \equiv X^{\mathrm{Ke}} \, (\mathrm{mod} \, r) \quad (11)$$

$$\mathbf{D}_{\mathrm{Kd}}(Y) \equiv Y^{\mathrm{Kd}} \, (\mathrm{mod} \, r) \equiv$$
$$X^{\mathrm{Ke \cdot Kd}} \, (\mathrm{mod} \, r) \equiv X \, (\mathrm{mod} \, r) \quad (12)$$

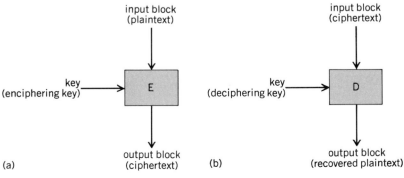

Fig. 8. Block cipher. (*a*) Enciphering. (*b*) Deciphering.

Finally, since $X^{\mathrm{Ke}} \, (\mathrm{mod} \, r) \equiv (X + mr)^{\mathrm{Ke}} \, (\mathrm{mod} \, r)$ for any integer m, $\mathbf{E}_{\mathrm{Ke}}(X) = \mathbf{E}_{\mathrm{Ke}}(X + mr)$. Thus the transformation from plaintext to ciphertext, which is many-to-one, is made one-to-one by restricting X to the set $\{0, 1, \ldots, r-1\}$.

Block ciphers. A block cipher (Fig. 8) transforms a string of input bits of fixed length (termed an input block) into a string of output bits of fixed length (termed an output block). In a strong block cipher, the enciphering and deciphering functions are such that every bit in the output block jointly depends on every bit in the input block and on every bit in the key. This property is termed intersymbol dependence.

The following example (using DES) illustrates the marked change produced in a recovered plaintext when only one bit is changed in the ciphertext or key. Hexadecimal notation (Table 2) is used. If

Table 2. Hexadecimal and binary notation

Hexadecimal digit	Binary digits
0	0000
1	0001
2	0010
3	0011
4	0100
5	0101
6	0110
7	0111
8	1000
9	1001
A	1010
B	1011
C	1100
D	1101
E	1110
F	1111

the plaintext 1000000000000001 is enciphered with a (56-bit) key 30000000000000, then the ciphertext 958E6E627A05557B is produced. The original plaintext is recovered if 958E6E627A05557B is deciphered with 30000000000000. However, if the leading 9 in the ciphertext is changed to 8 (a one-bit change) and the ciphertext 858E6E627A05557B is now deciphered with key 30000000000000, the recovered plaintext is 8D4893C2966CC211, not 1000000000000001. On the other hand, if the leading 3 in the key is changed to 1 (another one-bit change) and the ciphertext 958E6E627A05557B is now deciphered with key 10000000000000, the recovered plaintext is 6D4B945376725395. (The same effect is also observed during encipherment.)

In the most basic implementation of DES, called block encryption or electronic codebook mode (ECB), each 64-bit block of data is enciphered and deciphered separately. Every bit in a given output block depends on every bit in its respective input block and on every bit in the key, but on no other bits.

As a rule, block encryption is used to protect keys. A different method, called chained block encryption, is used to protect data. In chaining, the process of enciphering and deciphering is made dependent on other (prior) data, plaintext, ciphertext, and the like, also available at the time enciphering and deciphering takes place. Thus every bit in a given output block depends not only on every bit in its respective input block and every bit in the key, but also on any or all prior data bits, either inputted to, or produced during, the enciphering or deciphering process.

Sometimes data to be enciphered contain patterns that extend beyond the cipher's block size. These patterns in the plaintext can then result in similar patterns in the ciphertext, which would indicate to a cryptanalyst something about the nature of the plaintext. Thus, chaining is useful because it significantly reduces the presence of repetitive patterns in the ciphertext. This is because two equal input blocks encipher into unequal output blocks.

A recommended technique for block chaining, referred to as cipher block chaining (CBC), uses a ciphertext feedback (Fig. 9). Let X_1, X_2, . . . , X_n

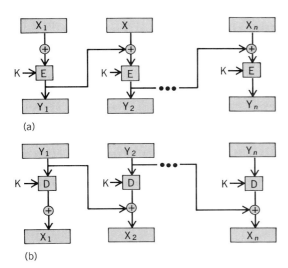

(a)

(b)

Fig. 9. Block chaining with ciphertext feedback. (a) Enciphering. (b) Deciphering. (*From C. H. Meyer and S. M. Matyas. Cryptography: A New Dimension in Computer Data Security. copyright © 1980 by John Wiley & Sons; used with permission*)

denote blocks of plaintext to be chained using key K; let Y_0 be a nonsecret quantity defined as the initializing vector; and let Y_1, Y_2, . . . , Y_n denote the blocks of ciphertext produced. The *i*th block of ciphertext (Y_i) is produced by EXCLUSIVE ORing Y_{i-1} with X_i and enciphering the result with K, as in Eq. (13), where $\oplus$ denotes the EXCLUSIVE OR operation, or modulo 2 addition. Since every bit in Y_i

$$\mathbf{E}_K(X_i \oplus Y_{i-1}) = Y_i \qquad i \geq 1 \qquad (13)$$

ation, or modulo 2 addition. Since every bit in Y_i depends on every bit in X_1 through X_i, patterns in the plaintext are not reflected in the ciphertext.

The *i*th block of plaintext (X_i) is recovered by deciphering Y_i with K and EXCLUSIVE ORing the result with Y_{i-1}, as in Eq. (14). Since the recovered

$$\mathbf{D}_K(Y_i) \oplus Y_{i-1} = X_i \qquad i \geq 1 \qquad (14)$$

plaintext X_i depends only on Y_i and Y_{i-1}, an error occurring in ciphertext Y_j affects only two blocks of recovered plaintext (X_j and X_{j+1}).

Stream ciphers. A stream cipher (Fig. 10) employs a bit-stream generator to produce a stream of binary digits (0's and 1's) called a cryptographic bit

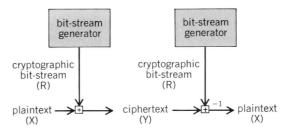

Fig. 10. Stream cipher concept. (*From C. H. Meyer and S. M. Matyas. Cryptography: A New Dimension in Computer Data Security. copyright © 1980 by John Wiley & Sons; used with permission*)

stream, which is then combined either with plaintext (via the $\boxplus$ operator) to produce ciphertext or with ciphertext (via the $\boxplus^{-1}$ operator) to recover plaintext. (Traditionally, the term key stream has been used to denote the bit stream produced by the bit-stream generator. The term cryptographic bit stream is used here to avoid possible confusion with a fixed-length cryptographic key in cases where a cryptographic algorithm is used as the bit-stream generator.)

Historically, G. S. Vernam was the first to recognize the merit of a cipher in which ciphertext (Y) was produced from plaintext (X) by combining it with a secret bit stream (R) via a simple and efficient operation. In his cipher, Vernam used an EXCLUSIVE OR operation, or modulo 2 addition (Table 1), to combine the respective bit streams. Thus encipherment and decipherment are defined by $X \oplus R = Y$ and $Y \oplus R = X$, respectively. Therefore $\boxplus = \boxplus^{-1} = \oplus$. Modulo 2 addition is the combining operation used in most stream ciphers, and for this reason it is used in the following discussion.

If the bit-stream generator were truly random, an unbreakable cipher could be obtained by EXCLUSIVE ORing the plaintext and cryptographic bit stream. The cryptographic bit stream would be used directly as the key, and would be equal in

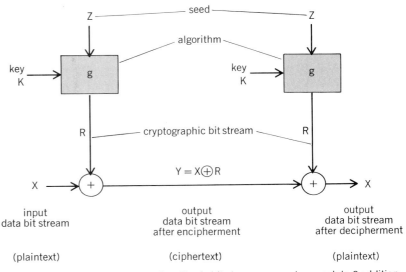

Fig. 11. Stream cipher using an algorithmic bit stream generator, modulo 2 addition, and seed. (*From C. H. Meyer and S. M. Matyas. Cryptography: A New Dimension in Computer Data Security. copyright © 1980 by John Wiley & Sons; used with permission*)

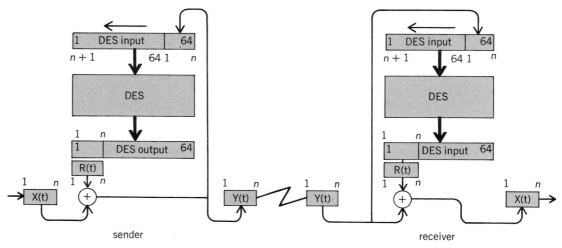

Fig. 12. Cipher feedback. (*From C. H. Meyer and S. M. Matyas. Cryptography: A New Dimension in Computer Data* Security. copyright © 1980 by John Wiley & Sons; used with permission)

length to the message. But in that case the cryptographic bit stream must be provided in advance to the communicants via some independent and secure channel. This introduces insurmountable logistic problems for heavy data traffic. Hence, for practical reasons, the bit-stream generator must be implemented as an algorithmic procedure. Then both communicants can generate the same cryptographic bit stream—provided that their algorithms are identically initialized. Figure 11 illustrates a cryptographic bit stream produced with a key-controlled algorithm.

When modulo 2 addition is used as the combining operation, each bit in the output ciphertext (recovered plaintext) is dependent only upon its corresponding bit in the input plaintext (ciphertext). This is in marked contrast to the block cipher which exhibits a much more complex relationship between bits in the plaintext (ciphertext) and bits in the ciphertext (recovered plaintext). Both approaches, however, have comparable strength.

In a stream cipher the algorithm may generate

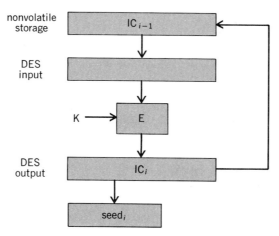

Fig. 13. Seed generation using the DES algorithm. (*From C. H. Meyer and S. M. Matyas. Cryptography: A New Dimension in Computer Data Security, copyright © 1980 by John Wiley and Sons; used with permission*)

its bit stream on a bit-by-bit basis, or in blocks of bits. This is of no real consequence. All such systems are stream ciphers, or variations thereof. Moreover, since bit streams can be generated in blocks, it is always possible for a block cipher to be used to obtain a stream cipher. However, because both the sender and receiver must produce cryptographic bit streams that are equal and secret, their keys must also be equal and secret. Therefore public keys in confirmation with a public-key algorithm cannot be used in a stream-cipher mode of operation.

For security purposes, a stream cipher must never predictably start from the same initial condition, thereby producing the same cryptographic bit stream. This can be avoided by making the cryptographic bit stream dependent on a nonsecret quantity Z (known as seed, initializing vector, or fill), which is used as an input parameter to the ciphering algorithm (Fig. 11).

In a stream cipher, Z provides cryptographic strength and establishes synchronization between communicating cryptographic devices—it assures that the same cryptographic bit streams are generated for the sender and the receiver. Initialization may be accomplished by generating Z at the sending device and transmitting it in clear (plaintext) form to the receiver.

Cipher feedback. A general approach to producing cryptographic bit streams is the automatic modification of the algorithm's input using feedback methods. In a key autokey cipher, the cryptographic bit stream generated at time $t = \tau$ is determined by the cryptographic bit stream generated at time $t < \tau$. In a ciphertext autokey cipher the cryptographic bit stream generated at time $t = \tau$ is determined by the ciphertext generated at time $t < \tau$. A particular implementation of a ciphertext autokey cipher, recommended by the National Bureau of Standards, is called cipher feedback (Fig. 12).

In cipher feedback, the leftmost n bits of the DES output are EXCLUSIVE ORed with n bits of plaintext to produce n bits of ciphertext, where n is the number of bits enciphered at one time ($1 \leq n \leq 64$). These n bits of ciphertext are fed back into the

iteration	DES input at sender								DES input at receiver							
0	0	0	0	S1	S2	S3	S4	S5	0	0	0	S1	S2	S3	S4	S5
1	0	0	S1	S2	S3	S4	S5	$Y(t_1)$	0	0	S1	S2	S3	S4	S5	$Y(t_1)$
2	0	S1	S2	S3	S4	S5	$Y(t_1)$	$Y(t_2)$	0	S1	S2	S3	S4	S5	$Y(t_1)$	$Y(t_2)$
3	S1	S2	S3	S4	S5	$Y(t_1)$	$Y(t_2)$	$Y(t_3)$	S1	S2	S3	S4	S5	$Y(t_1)$	$Y(t_2)$	$Y(t_3)$
4	S2	S3	S4	S5	$Y(t_1)$	$Y(t_2)$	$Y(t_3)$	$Y(t_4)$	S2	S3	S4	S5	$Y(t_1)$	$Y(t_2)$	$Y(t_3)$	$Y(t_4)$
5	S3	S4	S5	$Y(t_1)$	$Y(t_2)$	$Y(t_3)$	$Y(t_4)$	$Y(t_5)$	S3	S4	S5	$Y(t_1)$	$Y(t_2)$	$Y(t_3)$	$Y(t_4)$	$Y(t_5)$
6	S4	S5	$Y(t_1)$	$Y(t_2)$	$Y(t_3)$	$Y(t_4)$	$Y(t_5)$	$Y(t_6)$	S4	S5	$Y(t_1)$	$Y(t_2)$	$Y(t_3)$	$Y(t_4)$	$Y(t_5)$	$Y(t_6)$
7	S5	$Y(t_1)$	$Y(t_2)$	$Y(t_3)$	$Y(t_4)$	$Y(t_5)$	$Y(t_6)$	$Y(t_7)$	S5	$Y(t_1)$	$Y(t_2)$	$Y(t_3)$	$Y(t_4)$	$Y(t_5)$	$Y(t_6)$	$Y(t_7)$
8	$Y(t_1)$	$Y(t_2)$	$Y(t_3)$	$Y(t_4)$	$Y(t_5)$	$Y(t_6)$	$Y(t_7)$	$Y(t_8)$	$Y(t_1)$	$Y(t_2)$	$Y(t_3)$	$Y(t_4)$	$Y(t_5)$	$Y(t_6)$	$Y(t_7)$	$Y(t_8)$
9	$Y(t_2)$	$Y(t_3)$	$Y(t_4)$	$Y(t_5)$	$Y(t_6)$	$Y(t_7)$	$Y(t_8)$	$Y(t_9)$	$Y(t_2)$	$Y(t_3)$	$Y(t_4)$	$Y(t_5)$	$Y(t_6)$	$Y(t_7)$	$Y(t_8)$	$Y(t_9)$

Fig. 14. Self-synchronizing feature in cipher feedback. (From C. H. Meyer and S. M. Matyas. Cryptography: A New *Dimension in Computer Data Security. copyright © 1980 by John Wiley & Sons; used with permission)*

algorithm by first shifting the current DES input n bits to the left, and then appending the n bits of ciphertext to the right-hand side of the shifted input to thus produce a new DES input used for the next iteration of the algorithm.

A seed value, which must be the same for both sender and receiver, is used as an initial input to the DES in order to generate the cryptographic bit stream. Federal Standard 1026 (proposed) allows seed length to vary from 8 to 64 bits, but in order to ensure compatibility among users, it requires that all cipher feedback implementations be capable of using a 48-bit seed. The communicating nodes are synchronized by left-justifying the seed in the input to the DES and setting the remaining bits equal to 0.

One method to generate seed values with the DES is illustrated in Fig. 13. IC_0 (for initial condition) is a starting value supplied by the user and is placed in nonvolatile storage (where data remain permanent). IC_1 is produced by enciphering IC_0 with the (stored) cryptographic key, IC_2 is produced by enciphering IC_1, and so forth. At each iteration, IC_i replaces IC_{i-1}, and seed$_i$ is the left-most m bits ($m \leq 64$) of IC_i.

The cipher feedback approach is self-synchronizing, since any bit changes occurring in the ciphertext during transmission get shifted out of the DES input after 64 additional ciphertext bits are sent and received. If, for example, 8 bits are enciphered at one time, as shown in Fig. 12, and a bit is altered in $Y(t_1)$, changing it to $Y^*(t_1)$, then the sender's and receiver's inputs are as shown in Fig. 14, where the 40-bit seed is defined as S1, S2, . . . , S5. In this case, the blocks of ciphertext, given by $Y^*(t_1)$, $Y(t_2)$, . . . , $Y(t_8)$, can be correctly deciphered only at the receiver by chance since the DES input is incorrect in each case. After eight blocks of uncorrupted ciphertext have been received, given by $Y(t_2)$, $Y(t_3)$, . . . , $Y(t_9)$, both the sender's and receiver's cryptographic devices will have equal DES inputs again.

In general, any bit changes in an n-bit block of ciphertext can cause a change in any of the corresponding n bits of recovered plaintext and in any of the 64 bits of recovered plaintext immediately following. However, a permanent "out-of-sync" condition will result if a ciphertext bit is added or dropped, since the integrity of the block boundary is lost. To recover from such an error, the sender and receiver would have to establish the beginning and ending of blocks of bits that are enciphered at one time ($n = 8$ bits in the given example).

On the other hand, if enciphering takes place on a bit-by-bit basis ($n = 1$), then the property of self-synchronization is maintained even when bits are lost or added. This is because blocks are bits, and therefore the block boundary cannot be disturbed. *See* COMPUTER SECURITY; INFORMATION THEORY.

[CARL H. MEYER; STEPHEN M. MATYAS]

Bibliography: *Data Encryption Standard*, FIPS Publ. no. 46, National Bureau of Standards, January 1977; W. Diffie and M. Hellman, New directions in cryptography, *IEEE Trans. Inform. Theory*, IT-22:644–654, November 1976; H. Feistel, Cryptography and computer privacy, *Sci. Amer.*, 228(5):15–23, May 1973; M. Hellman, The mathematics of public-key cryptography, *Sci. Amer.*, 241(2):146–157, August 1979; C. H. Meyer and S. M. Matyas, *Cryptography: A New Dimension in Computer Data Security*, 1980; R. L. Rivest, A. Shamir, and L. Adleman, Method for obtaining digital signatures and public-key cryptosystems, *Comm. ACM*, 21(2):120–126, February 1978; C. E. Shannon, Communication theory of secrecy systems, *Bell Syst. Tech. J.*, 28:656–715, 1949.

Current measurement

The measurement of the flow of electric current. The unit of measurement, the ampere (amp), is defined in the system of electromagnetic units through its electromagnetic reaction. One abampere (10 amp) of unvarying unidirectional current (direct current) flowing through a wire and returning through an adjacent parallel wire will produce, when the two parallel wires are 1 cm apart on centers, a force between the wires of 2 dynes for each cm length so positioned. This electromagnetic reaction was first studied by André Marie Ampère in 1820. The reaction or force is proportional to the current in each wire, or to the square of the current where it returns on itself, as in the definition. The force is also inversely proportional to the square of the distance between the wires.

In the practical determination of current value the current-carrying wires are formed into circular coils. The moving coil may move up and down as in the Rayleigh balance, where the reaction is balanced by the force of gravity on known weights, or it may rotate as in the Pellat balance, being supported on a knife edge; here the force is balanced by an appropriate mass acting a known distance from the support.

Direct and low-frequency currents. The current balance is tedious to operate, and is ordinarily used only by the national laboratories of the various countries for the determination of the absolute ampere, whereby simpler (but less accurate) current-measuring devices may be calibrated.

The moving coil may rotate in jeweled bearings against the counter torque of a spiral spring. If a pointer is attached to the moving system rotating over an appropriately marked scale, calibrated in terms of the absolute current determinations of an ampere balance, the result is an indicating electrodynamometer ammeter. Such instruments are widely used as laboratory standards for both direct and alternating-current measurements and for the calibration of secondary instruments.

Direct current (dc) may also be measured by placing the movable coil in the magnetic field of a permanent magnet. This is the construction of the conventional type of dc ammeter. *See* AMMETER.

Structures incorporating a permanent magnet give a reversed deflection with reversed direction of current, and therefore are usable only on direct current. For the measurement of alternating current (ac) the instrument must take no account of current direction and, further, must give a reading of that function of the current oscillation which is to be measured. For example, only the peak value of current may be desired. An example is in the measurement of the maximum value of a lightning discharge to a transmission tower where a small magnetic link is magnetized by the peak value of the current; a test of the degree of its magnetization can be correlated to the peak current value.

In most cases it is desired to measure that value of the alternating current which is a measure of the power in a motor or of heat in a stove or lamp filament, namely, its root-mean-square (rms) value. The electrodynamometer ammeter indicates in these terms. If calibrated on direct current, it becomes a transfer instrument and may be used to calibrate other types of ac instruments in terms of true rms current value.

For ordinary switchboard or test indications of alternating current the fixed-coil, moving-iron type of instrument, with its high torque and quite adequate accuracy, is widely used. A fixed iron vane and a similar vane on a staff are magnetized by the current in the coil, the force between the vanes acting to rotate the staff against a spring as a function of the rms value of the current.

The hot-wire ammeter, indicating the mechanically amplified stretch of a fine wire heated by the current, has been wholly replaced by the thermoammeter. In this instrument a short metal strip or wire is heated by the current in question; its temperature rise is measured by a contacting thermocouple connected, in turn, to a sensitive dc meter which deflects in proportion to the temperature rise. The scale is marked in terms of the heating current. This type of instrument is insensitive to variations in frequency and ambient temperature.

When direct currents of over a fraction of 1 A are to be measured, the bulk of the current is diverted around the instrument through a shunt or bypass circuit or, simply, a shunt. Large alternating currents, on the other hand, are connected to meters rated at 5 A through current stepdown transformers, or current transformers. Both types of instrument are available in great variety and are of good accuracy.

Very low direct currents, below 10^{-10} A, can be measured by charging a capacitor with the current in question and calculating its value from the rate of voltage rise. Vacuum-tube methods are likewise of value, although they are more useful in the measurement of low alternating currents.

Extremely heavy direct currents, above 10^4 A, can be measured with a transductor, essentially an ac transformer with the iron core surrounding the dc bus. By properly proportioning the windings, the ac input, as indicated on an ac ammeter, is a measure of the direct current, and the scale can be so calibrated.

[JOHN H. MILLER/JESSE BEAMS]

Very small currents. Small direct currents in low-resistance circuits can be measured with a standard-resistance shunt and a potentiometer. In high-resistance circuits, vacuum-tube devices are now usually preferred to electrometers. Currents as low as 10^{-15} amp can be measured with shunts of 10^{11} or 10^{12} ohms, and lower values can be measured by observing the rate at which the input capacitance is charged by the flow of current. Negative feedback can be employed to reduce the voltage drop in the input loop when a shunt is used, or to reduce the effective input capacitance and so increase the current sensitivity when the rate-of-charge method is used. Small alternating currents can be measured with a shunt and an amplifier-type vacuum-tube voltmeter. Great improvement in commercial instruments for measuring very small currents has been made possible by improved solid-state electronics.

Current at high frequencies. High-frequency current is usually measured by its heating effect. Thermocouple meters are most common, and available single-range commercial instruments are accurate to 1% up to 10 MHz. Special-purpose designs have been used to about 100 MHz. Bridges incorporating thermistor elements, in which resistance varies greatly with the heating, are also widely used. One of the best methods is to measure the voltage drop across a shunt with a diode-rectifier-type vacuum-tube voltmeter. At frequencies over a few hundred megahertz, current is seldom measured directly but is usually computed from impedance and power measurements. *See* VACUUM-TUBE VOLTMETER; VOLTAGE MEASUREMENT.

[W. NORRIS TUTTLE]

Bibliography: M. Braccio, *Basic Electrical and Electronic Tests and Measurement*, 1979; W. D. Cooper, *Electronic Instrumentation and Measurement Techniques*, 2d ed., 1978; F. K. Harris, *Electrical Measurements*, 1975; F. B. Silsbee, The ampere, *Proc. IRE*, 47(5):643–649, 1959.

Cyclotron resonance experiments

The measurement of charge-to-mass ratios of electrically charged particles from the frequency of their helical motion in a magnetic field. Such experiments are particularly useful in the case of conducting crystals, such as semiconductors and metals, in which the motions of electrons and holes are strongly influenced by the periodic potential of the lattice through which they move. Under such circumstances the electrical carriers often have "effective masses" which differ greatly from the mass in free space; the effective mass is often different for motion in different directions in the crystal. Cyclotron resonance is also observed in gaseous plasma discharges and is the basis for a class of particle accelerators. *See* BAND THEORY OF SOLIDS.

The experiment is typically performed by placing the conducting medium in a uniform magnetic field H and irradiating it with electromagnetic waves of frequency ν. Selective absorption of the radiation is observed when the resonance condition $\nu = qH/2\pi m^*c$ is fulfilled, that is, when the radiation frequency equals the orbital frequency of motion of the particles of charge q and effective mass m^* (c is the velocity of light). The absorption results from the acceleration of the orbital motion by means of the electric field of the radiation. If circularly polarized radiation is used, the sign of the electric charge may be determined, a point of interest in crystals in which conduction may occur by means of either negatively charged electrons or positively charged holes. *See* HOLES IN SOLIDS.

For the resonance to be well defined, it is necessary that the mobile carriers complete at least $1/2\pi$ cycle of their cyclotron motion before being scattered from impurities or thermal vibrations of the crystal lattice. In practice, the experiment is usually performed in magnetic fields of 1000 to 100,000 oersteds (1 oersted = 79.6 amperes per meter) in order to make the cyclotron motion quite rapid ($\nu \sim 10-100$ GHz, that is, microwave and millimeter-wave ranges). Nevertheless, crystals with impurity concentrations of a few parts per million or less are required and must be observed at temperatures as low as 1 K in order to detect sharp and intense cyclotron resonances.

The resonance process manifests itself rather differently in semiconductors than in metals. Pure, very cold semiconductors have very few charge carriers; thus the microwave radiation penetrates the sample uniformly. The mobile charges are thus exposed to radiation over their entire orbits, and the resonance is a simple symmetrical absorption peak.

In metals, however, the very high density of conduction electrons present at all temperatures prevents penetration of the electromagnetic energy except for a thin surface region, the skin depth, where intense shielding currents flow. Cyclotron resonance is then observed most readily when the magnetic field is accurately parallel to the flat surface of the metal. Those conduction electrons (or holes) whose orbits pass through the skin depth without colliding with the surface receive a succession of pulsed excitations, like those produced in a particle accelerator. Under these circumstances cyclotron resonance consists of a series of resonances $n\nu = qH/2\pi m^*c$ ($n = 1,2,3 \ldots$) whose actual shapes may be quite complicated. The resonance can, however, also be observed with the magnetic field normal to the metal surface; it is in this geometry that circularly polarized exciting radiation can be applied to charge carriers even in a metal.

Cyclotron resonance is most easily understood as the response of an individual charged particle; but, in practice, the phenomenon involves excitation of large numbers of such particles. Their net response to the electromagnetic radiation may significantly affect the overall dielectric behavior of the material in which they move. Thus, a variety of new wave propagation mechanisms may be observed which are associated with the cyclotron motion, in which electromagnetic energy is carried through the solid by the spiraling carriers. These collective excitations are generally referred to as plasma waves. In general, for a fixed input frequency, the plasma waves are observed to travel through the conducting solid at magnetic fields higher than those required for cyclotron resonance. The most easily observed of these excitations is a circularly polarized wave, known as a helicon, which travels along the magnetic field lines. It has an analog in the ionospheric plasma, known as the whistler mode and frequently detected as radio interference. There is, in fact, a fairly complete correspondence between the resonances and waves observed in conducting solids and in gaseous plasmas. Cyclotron resonance is more easily observed in such low-density systems since collisions are much less frequent there than in solids. In such systems the resonance process offers a means of transferring large amounts of energy to the mobile ions, which is a necessary condition if nuclear fusion reactions are to occur.

[WALTER M. WALSH]

Bibliography: C. Kittel, *Introduction to Solid State Physics*, 4th ed., 1971; P. M. Platzman and P. A. Wolff, *Waves and Interactions in Solid State Plasmas*, 1973.

Dark current

An ambiguous term used in connection with both gaseous-discharge devices and photoelectric cells or tubes. In gaseous-conduction tubes it refers to the region of operation known as the Townsend discharge. The name is derived from the fact that photons produced in the gas do not play an important part in the production of ionization. The initial ionization arises from independent effects such as cosmic rays, radioactivity, or thermionic emission. When applied to photoelectric devices, the term applies to background current. This is current which may be present as the result of thermionic emission or other effects when there is no light incident on the photosensitive cathode. *See* ELECTRICAL CONDUCTION IN GASES: TOWNSEND DISCHARGE.   [GLENN H. MILLER]

Data-base management systems

Special data-processing systems or parts of data-processing systems which are developed to aid in the storage, manipulation, reporting, and managing of data. They may be considered as building blocks constructing a data-processing system which also acts as a mechanism for the effective control of the data.

Cost of information systems. An information (or data-processing) system consists of such hardware or physical devices as computers, storage units and input-output media controllers; also of software and the people who build, maintain, and use the system. In the early days of computing, the cost of the system was mainly in the hardware, but during the 1970s the cost of hardware dropped radically. The ratio of software-to-hardware cost now falls between 1 and 5 for most industrial and commercial operations. The cost of people has not dropped; thus the concept of "hardware is expensive and should be used sparingly" has become obsolete. Emphasis must be placed on reducing the cost of software, both in building the system and in maintaining it, that is, all parts of the so-called software life cycle.

Maintenance or modification occurs because of hardware changes, new ways of doing business, or different user orientation. Software cost can be reduced in two ways:

1. While building the system, by using general system software. This spreads effort and cost over several applications. Early examples of this were such special packages or routines that took a square root, evaluated a trigonometric function, or even sorted (ordered) a set of data according to some numeric or alphabetic key.

2. When making modifications. This is achieved by making the system modular, such that only parts of it are affected by changes in the hardware configuration or by user-induced alterations. This modularity isolates system parts from one another, for example, the square-root routine need not be changed when its calling routine is modified.

The success of early packages gradually led to the adoption of more complex ones, or system software to perform management functions. The most complex set of such packages is an operating system.

Some common functions have been identified to help with management, manipulation, and control of the data itself. Such a set of packages is generally called a data-base management system (DBMS) by its manufacturers although their specific design and implementation may differ.

Management control of data. Without data, the process of controlling any system is impossible. Data are the physical representation interpreted through work orders, experimental results or analyses, costs of operation, or tax schedules. Data can provide information to a worker on what job to perform next and what special problems to expect;

data help the engineer see how to improve a design; for the manager data can highlight operational inefficiencies; data help the bureaucrat to decide how to govern and what new laws are needed.

The computer provides business, industry, and government with the means to improve the speed and efficiency of the collection, manipulation, storage, reporting, and dissemination of data, and this has generally been advantageous. However, the advent of the data-base concept constitutes both an advantage and a potential threat. The threat has several aspects which show why control or management of data is extremely important.

Organization of data. It has been estimated that a medium-sized corporation may need to retain 10^{10} characters of current important data in high-speed automated storage. By comparison, a book of 300 pages, 40 lines per page, and 12 words (each of about 6 characters) per line, contains about 10^6 characters (including spaces between words) and thus the 10^{10}-character "data base" could be stored in 10^4 books, which is indeed a good-sized library. However, the organization of a library makes it relatively easy to find any book, but imagine a library that had no catalog, no librarians and no one putting the books back on the shelves. The cost of a library staff is similar to the cost of organizing data in a corporate data base: it is a necessary cost if data must be found and used effectively. Indeed the old systems, without well organized data, suffered from the problem that people knew data existed, but could not find it easily and became frustrated. One of the reasons for the development of data-base management systems was to solve this need.

Data become truly available when a large cross section of potential users know of their existence, when their location and format are known, and when there is a mechanism for retrieving the data. If the mechanism to locate its format is used by humans, the location is normally termed a directory, while the format is stored in a dictionary. If the locating device is embedded in the data-base system, the mechanism is termed a schema.

A schema is therefore a machine readable version of the location and format of the stored data. The difference between the schema and its "instances" or "occurrences" is shown in Figure 1. This simple schema contains information about a person: name, date of birth, and salary. The representation is: name 10 characters in length, with blanks for shorter names; date of birth as year, month, and day, with two decimal digits representing each; and salary of up to $9999.99 per month. The "model" is defined in the schema, while the data are recorded in each instance.

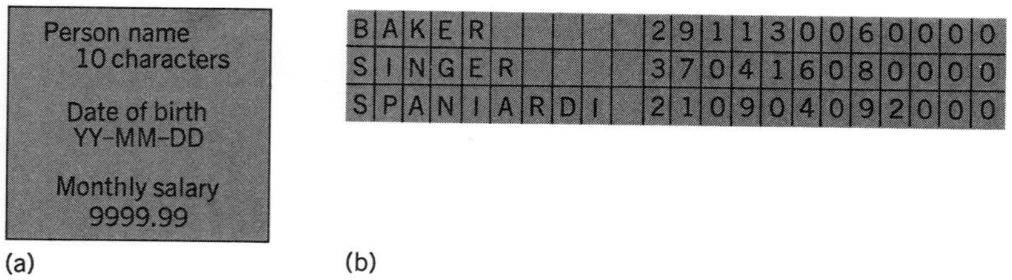

In order for the data-base management system to retrieve or store data about a person, it must consult the schema to determine its form as well as its location in machine storage. This means that a data-base management system with a "query language" interface can respond to a request like FOR PERSON, FIND MONTHLY-SALARY, WHERE PERSON-NAME = "SINGER," or FOR PERSON, FIND PERSON-NAME AND DATE-OF-BIRTH, WHERE SALARY GREATER THAN 700.00

Data security. This powerful capability of the DBMS enables a wide variety of potential users to

(a)   (b)

Fig. 1. Simple schema representing (a) a very simple record structure and (b) its stored data base of three instances.

obtain information that would otherwise be unavailable. However, this also creates some social problems, because access to the data may become too easy. *See* MULTIACCESS COMPUTER.

The concept of privacy has been defined as the right to be left alone. If anyone can obtain someone else's date of birth, that person's privacy has been violated. Thus there must be some restriction on who can obtain what data under which conditions (and maybe when). Thus a certain personnel clerk may be allowed to look at salary data and both read and update it during the working day (but not at night or during the weekend). This authority to access data must be vested in and controlled through the data-base management system; it is usually termed a security technique.

Typically, security may be enforced by insisting that users of the data-base management system identify themselves both by name or employee number and by some password. Other mechanisms may involve the insistence on use of particular terminal devices at predefined locations, allowance of only certain types of command or programs to be initiated, and "hiding" of sensitive parts of the data base except to certain highly privileged users. Thus the data may be screened from the general user through various levels of security, even to the extent of encoding them (requiring automated decoding routines to make them meaningful for the authorized user).

Integrity of data. Because the data are often stored in a device with moving parts they can be destroyed through mechanical or other failures. Data may also be lost through user carelessness or program malfunction. To protect the integrity of the data, several defensive steps may be taken. One major precaution is to take dumps, that is, copies of the data. This action is taken at regular intervals if the data are frequently changed; for example, the data may represent the quantity of various goods in a warehouse which changes every time an item is received or dispatched. In order to be able to restore a data base after it has been destroyed, it is merely necessary to get a copy of the last dump and update it. The updating makes use of an audit trail or log tape, which contains a history of the transactions affecting the data and generally also snapshots of data base segments before and after a change. Thus by remaking all changes that were on the audit trail since the dump, the data base will be correctly restored to its most current image. The audit trail also allows the user to "back-out" an error (that is, unmake changes that were made in error) by going backward through the audit trail undoing all mistaken updates by using "before images" of the data.

Improved performance. The introduction of a data-base management system changes also the system's orientation. The emphasis on programs (that is, programs using data), shifts to data (that is, data used by programs). A new measurement of efficiency determines how effectively the data are used and stored, and how readily available the data are. The data-base management system becomes the focal point from which such measurements may be taken. They allow the data administrator to determine how well the schema fits user needs, and how well the storage structuring of the

instances reflects the needs. Thus the data-base management system measurements help improve performance.

In summary, the data-base management system provides a means for management of data by making available data to a large range of authorized users, while preserving the integrity of the data and improving the overall system performance.

Architecture. The somewhat simplified architecture of a data-base management system and its environment is shown in Fig. 2. It illustrates how a data request is satisfied. For example, consider the previous request for person-name and date of birth in the PERSON data base. An ad hoc user (that is, one who can query the system without having recourse to a previously coded program) goes to an input/output device with a keyboard and/or cathode-ray-tube combination. The person types in SIGNON JONES PASSWORD 51z5QP, and then (after some response) AD HOC DBMS FOR PERSON, FIND PERSON-NAME AND DATE-OF-BIRTH WHERE SALARY GREATER THAN 700.00.

The following operations will be performed: (1) The word SIGNON is transmitted by the checking program to the scheduler program which requests the SIGNON program from its library. (2) The SIGNON program is run by the scheduler program to check that Jones has given the correct password. (3) The AD HOC program is called from the library and starts to look for more inputs. (4) The word PERSON is taken by this program as the name of the schema. (5) The scheduler program requests the data-base routine to load the PERSON schema from storage. (6) The next sentence is now checked to see whether PERSON-NAME, DATE-OF-BIRTH, and SALARY are all valid words in the PERSON schema. (7) A search of the instances is made

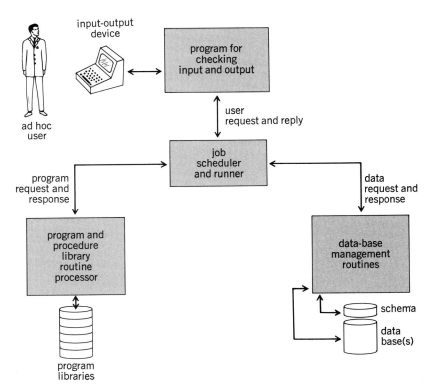

Fig. 2.   Architecture of a data-base management system.

Person

Date of birth	Name	Salary ($)
29/11/30	BAKER	600.
37/04/16	SINGER	900.
21/09/04	SPANIARDI	920.

Fig. 3. Relational data base.

to determine which have a value of more than 700.00 in the SALARY item. In the above example, two people qualify. (8) The names and dates of birth of these two people are displayed for the user.

In this example, the authority of the user (Jones) to view any element (for example, date of birth) could be checked at step 6, if necessary.

This example demonstrates that the data-base management system consists of many routines that process data according to the schema. Many other special programs are associated with the data-base management system (such as the ad hoc query processor, the utilities to dump and restore the data base, and so forth). The data-base routines, in turn, use other operating system software such as access methods, input and output routines, and so forth.

Modeling data. Data can be used to represent important facts; a particular set of facts, with their relationships, are usage-dependent. For hiring a person, the attributes of importance (that is, the model needed) may be: date of birth, phone, address, sex, skills, and salary; however, for treating a person as a patient, the important model attributes are probably: date of birth, phone, address, sex, diseases, and so forth. Thus the important attributes represent the model of the entity under investigation.

Different data-base management systems model the data in different ways. The so-called relational system represents data as tables (Fig. 3). No entry in the table contains more than one elementary value (for example, it would be impossible to have two salaries for BAKER). The hierarchical system allows more structure. In this type of system, groupings of elements are connected by parent-child relationships (Fig. 4). Here, a particular person may have one or several different educational categories (high school, bachelor's degree, and so forth), as well as several job histories (office assistant, clerk, manager, president). The network system has more structure and complexity, because it uses named relationships and more complexity. In the example given in Fig. 5, there are four groups (or record types) with four relations (or sets). Because these three types of systems are structurally different, most current commercial data-base management systems service only one

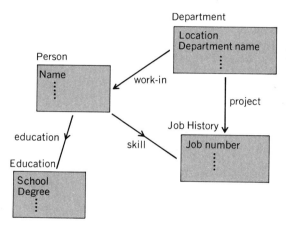

Fig. 5. Network schema.

type; future systems are expected to be able to allow the users to choose the best system to suit their special needs.

Data-base administrator. In order to utilize a data-base management system effectively, there must be one person or group within the organization that looks after the system and makes decisions about its control. Typical questions are: What is a good schema (that is, one that is effective for a mix of users with different needs)? How often should the data be dumped? How well is the system operating (that is, monitoring the time and volume of the data accesses, and so forth)? Such control parameters suggest the need for special skills, other than the programming and management skills that were prevalent in the "old" data-processing environment. The role of data-base administrator (DBA) has evolved to fill this need. Thus the DBA is a person or group that deals with the corporate data, making decisions on the use, quality, access control, integrity needs, and so forth of the entire group of data users of the organization.

Data administration. The management of the data in an organization involves both technical and administrative aspects, and implies more than a DBA function, since the DBA deals only with the technical part (for example, designing a data

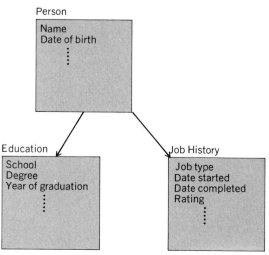

Fig. 4. Hierarchical schema.

structure that is good for the set of organizational users). The more managerial role is assigned to a higher-level function, termed a data administrator (DA). This person is usually the manager of the DBA.

The DA's position normally involves setting data policy, allocating resources in the DA group, and ensuring that the policies are carried out through control procedures enforced through the DA group. Naturally, because the data policy must mesh with the overall organizational policy, the DA acts as liaison officer with upper-level management in concerns affecting the data policy. The DA also acts as a liaison with user groups in all conflicts over data (such as ownership or sharing).

Data dictionary. Another essential function in a DBMS is the recording of proper definitions of data entities with their attributes and relations with other entities. This implies some uniform way of storing and retrieving the information, and the software mechanism to achieve this is termed a data dictionary. The data dictionary requires a person or function termed the data dictionary administrator, reporting to the DA.

The data dictionary is like a schema; it records the names of data items and their structural relations, but it also carries definitions (in the form of a humanly understandable sentence or paragraph that describes the meaning or usage). This explanation aids in reducing misunderstanding due to homonyms and synonyms. A homonym occurs when two systems or users mean different things when they reference the same name (for example, a "full-time-student" in an undergraduate school may differ from a "full-time-student" in a medical school). A synonym occurs when two people give different names to the same entity (because neither knew that the other needed to use the same material). This may be an expensive problem because data may be unnecessarily collected and stored (when it already exists elsewhere), or it may be misused (when one user calls for it with the "wrong" name).

In addition, a data dictionary allows entities other than data to be stored and manipulated, and relations may be defined between all entities. The implementor may ask: "What programs use the personnel file?" or "Which programmer wrote the accounts-receivable program?" This implies that there are entities like program, user, programmer, and so forth in the dictionary.

Design and implementation. Although commercial offerings of DBMS have existed since the early 1960s, they have not always been successful. The degree of success depends on the expectations and resources, and often the expectations have been too high and the resources too low.

Many early failures were due to poor planning, lack of corporate commitment to the concept, or misapplication. The truly integrated data base requires a good management structure, as described for the DA, and good technical support through the DBA and systems programmers. If integration needs are not carefully assessed, unnecessarily large, monolithic data bases are formed, the result being poor performance and extra expense. Equally, the DBMS should not be considered a new toy or merely a sophisticated access method. Thus good planning is needed to decide what is right for each organization. The degree of commitment of upper-level management is also critical to the success of a DBMS. Data policy must not be changed rapidly or radically; otherwise the data-base structure and procedures will be forced to change just as quickly and problems will result. Thus, the major success factor in the implementation of the information system is a good, stable organizational plan for the management of data, with proper resource allocation, and structural change to allow for the new data administrator and data-base administrator roles in the organization.

The second success factor results from everyone understanding the meaning and use of the data. This strongly implies the use of a data dictionary. The third success factor is a good data structure. There are several design aids that have been developed for specific DBMS. They allow the DBA to try various data structures for an anticipated usage, and thus they provide a means of tuning the structure for good performance. Such design aids have become the most important tools for the DBA. The fourth success factor is programming good systems to access the data. This implies the use of a professional application systems group who know and understand the use of DBMS in general, and the selected DBMS in particular.

The final success factor is providing user facilities, which might involve query language interfaces. These are available in many modern commercial DBMS, and they can be added on. Report generators (RPG) are also effective tools for the end user. The usefulness of these facilities on the education and support functions provided by the DA office so that noncomputer professionals can formulate simple queries and request reports without contacting the DP professionals. *See* DATA-PROCESSING SYSTEMS; DIGITAL COMPUTER.

[EDGAR H. SIBLEY]

Bibliography: C. J. Date, *An Introduction to Database Systems*, 1977; J. Martin, *Principles of Database Management*, 1976; E. H. Sibley (ed.), *ACM Computing Surveys: Special Issue on Database Management Systems*, vol. 8, no. 1, March 1976.

Data communications

The function of electronically conveying digitally encoded information from a source to a destination where the source and destination may be displaced in space (distance) and time. Encoded information can represent alphabetic, numeric, and graphic forms. Such information may originate as keystrokes at a terminal, as images on a page, as motion of a hand-held stylus, as a signal from some electrical or mechanical measurement device, or as output from a computer. The data communications function includes management and control of connections as well as the actual transfer of information.

Digitally encoded information typically takes the form of a series of 1s and 0s called bits (contraction for binary digit). Groups of eight bits are often referred to as bytes. Standard codes (such as ASCII and EBCDIC) have been adopted to establish a correspondence between bytes (patterns of eight bits) and alphabetic, numeric, and commonly used graphic and control characters. These codes are

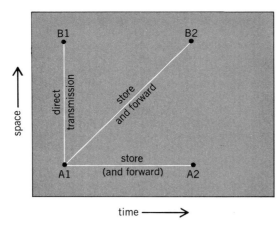

Fig. 1. Data communications involving displacements in both space (distance) and time. Points A1 and B1 are separated by distance; the movement of data from A1 to B1 involves a displacement in space. Points A1 and B2 are separated by both distance and time; transfers between A1 and B2 require data to be stored by the data communications system before delivery. Points A1 and A2 are displaced in time but not in distance; for this case, data are transmitted, stored, and then returned to the originating station.

used for the interchange of information among equipments made by different manufacturers. *See* BIT.

Data communications usually do not refer to the conveyance of voice and video, although these too may be digitally encoded and then conveyed in digital form just as data are conveyed. The increased use of computers in the design and implementation of communications systems and the efficiencies realized from the handling of digitally encoded information have markedly broadened the range of data communications, and tend to mask old distinctions between data communications and other forms of information transfer.

Data communications has evolved from a function serving simple terminal-to-terminal communications (telegraphy) to one dominated by communications with and among computers. Terminal-to-computer communications remains a sizable component of traffic, but the migration of intelligence into terminal systems (an aspect of distributed processing) is causing terminals to take on more of the communications characteristics of computers. Networks of terminals and computers are emerging to serve the needs of business, education, government, industry, medicine, recreation, and other forms of human endeavor. These systems, both publicly and privately owned and operated, manage and direct the movement of data between stations (terminals, computers, and other sources of digitally encoded information) and across networks. Where two stations are geographically separated, data communications manages the transfer of information as a displacement in space. Where there is value in delaying the transfer of information, data communications manages the transfer also as a displacement in time (Fig. 1). Access to and the movement of information have become activities of growing social and commercial importance. *See* COMPUTER; MULTIACCESS COMPUTER.

Types of applications. Data communications applications fall into three principal categories: transaction-oriented, message-oriented, and batch-oriented (Fig. 2). Transaction-oriented applications (Fig. 2a) involve the transfer of small amounts of data in the range of tens to hundreds of bytes and frequently require minimal delay in the transfer process. Transaction-oriented applications are typically bidirectional or conversational. Information is exchanged between source and destination stations. Examples are inquiry-response, information retrieval, and time-sharing, where people at terminals are communicating with large computers. For these applications, data communications serves stations displaced in space. Travel agents inquiring about seat availability, insurance agents inquiring about policy status, researchers accessing data banks, assembly-line process control, 24-hour bank tellers, merchants performing credit checks, computers monitoring life support systems, and retailers updating inventory status are examples of transaction-oriented applications.

Message-oriented applications (Fig. 2b) involve data transfers in the range of hundreds to a few thousand bytes or characters and are usually unidirectional, information flows from source to destination. Message applications frequently take advantage of time displacement (store and forward) as well as space displacement, and they exhibit data transfer options which go beyond the simple movement of information between two points. Examples of message-oriented applications are the transmittal of administrative messages within a corporate entity, the distribution of press releases, and price list updates being sent to field offices. Data entry uses, such as sales order processing, where information is collected from many sources for delivery to one or several destinations at some later time represent a large component of

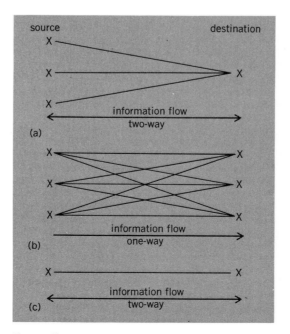

Fig. 2. Three principal categories of data communications. (a) Transaction-oriented. (b) Message-oriented. (c) Batch-oriented.

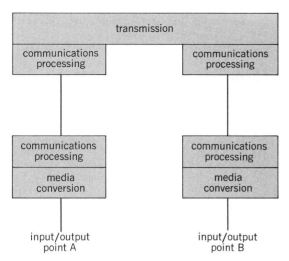

Fig. 3. Relationship of three functional components of data communications: media conversion, communications processing, and transmission.

activity. Single messages carrying multiple addresses, addresses in the form of mnemonics (easily remembered names), requirements to journal (store) messages and to confirm the delivery of messages are examples of functions being served by data communications systems. Messages may be originated at terminals or in computers.

Batch-oriented applications (Fig. 2c) involve the transfer of thousands or even millions of bytes of data and are usually point-to-point (place-to-place) and computer-to-computer. Moving work from one computer to another (remote job entry), distributing the results of some centralized data-processing function, facsimile transmission, and updating distributed copies of data bases from some central site are examples of batch-oriented applications.

Transaction, message, and batch-oriented applications exhibit different needs in terms of total bytes transferred, delay, and special features. Consequently, they impose different demands upon the functional components of data communications: media conversion, communications processing, and transmission (Fig. 3).

Media conversion. Media conversion achieves the conversion of some input, such as a keyboard keystroke, graphic material, a signal representing some measurement, and information recorded on magnetic media such as tape or disk, into a digitally encoded signal which may be transmitted. Likewise, it achieves the conversion of a digitally encoded signal into similar forms of useful output. Typical media conversion devices such as card reader/punches, paper tape reader/punches, magnetic tape systems, and printers remain in use. However, the proliferation in the late 1970s of data communications applications where people at terminals converse with computers has caused media conversion to be most often associated with the interactive terminal (2,000,000 in use in 1980). Input is typically supported by a keyboard device, while output takes the form of a hard-copy printer or screen display. The latter may be a cathode-ray-tube device (television monitor), a plasma panel, or a light-emitting diode (LED) or liquid-crystal array. Point-of-sale stations are interactive terminals

which include both keyboard entry and laser- or magnetic-actuated pattern recognition devices for inputs.

The migration of intelligence into terminals in the form of storage and processing has enabled the use of all kinds of sensory devices as inputs and outputs. Optical character recognition, spoken-word recognition, tactile panels, light pens, facsimile devices, and x-y encoders all serve as input devices. Multicolor graphic displays, voice synthesizers, microfiche generators, and robots (mechanical servos) represent less usual but increasingly important forms of output. Storage in the forms of floppy disks (2,500,000 characters) and magnetic bubble arrays (a few hundred thousand characters) are supplementing the more conventional cassette tape, and highly compact forms of storage such as video (optical memory) disks (12,500,000,000 characters) can be expected to be used as well. *See* CHARACTER RECOGNITION; COMPUTER GRAPHICS; COMPUTER STORAGE TECHNOLOGY; ELECTRONIC DISPLAY; SPEECH RECOGNITION; VOICE RESPONSE.

Communications processing. Communications processing provides the management and control in data communications systems. In the data terminal or media conversion system, it must at least manage the protocol (rules) used for intelligibly exchanging information with other stations including the administration of techniques designed to detect errors in transmission. Depending on the application, it may be necessary to perform code conversion and to multiplex or concentrate inputs from several terminals into a single output stream. It may also be desirable to augment variable data input via keyboard with fixed or automatically generated data such as date, time, operator identity, or location. The communications-processing function organizes encoded data into units suitable for transmission. Simple protocols demand that data be organized into single bytes. More complex and efficient protocols require the organization of bytes into blocks (hundreds of bytes) or bits into frames (Fig. 4). Blocks and frames typically begin with recognizable patterns of bytes and bits respectively. Still more sophisticated protocols include the organization and management of logical units of information ranging in size from 50 to about 1000 bytes called packets or messages.

For point-to-point transaction- or batch-oriented data communications applications involving only

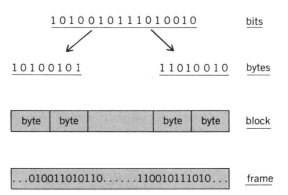

Fig. 4. Organization of data into units suitable for transmission.

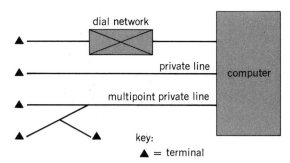

Fig. 5. Connection arrangements for terminals and computers.

space displacement, little more is required of communications processing. For network applications, communications processing uses storage and processing to assure the proper addressing and formatting of packets and messages. To support time displacement, communications processing enables information to be stored for later retrieval, scheduled for later delivery, and delivered with return confirmation.

Transmission. Transmission accomplishes the physical transfer of data. It includes the provision of transmission channels, the signal conversions necessary to use such channels, and any required switching among channels.

Terminals and computers may be connected to transmission facilities via standardized interfaces. Where dedicated wire paths of up to a few miles (1 mi = 1.6 km) can be established, so-called limited distance or baseband data sets (modems) enable the digital transmission of data at rates up to tens of thousands of bits per second (bps). At shorter distances, transmissions at more than 1,000,000 bits per second can be supported. Dedicated paths using optical fibers can support transmissions at tens of millions of bits per second. Approaches such as infrared transmission enable open-air channels to be supported. Radio and satellite transmission are used as well. *See* OPTICAL FIBERS.

Use of telephone network. Since it is generally impractical to install privately owned cables beyond single buildings or campus environments and because transmission distances of more than a few miles are commonly of interest, the conventional approach is to use the telephone network for transmission facilities. Where such facilities are analog (such as channels which support voice signals), modems are required to convert the digitally encoded information produced at the terminal or computer into modulated analog form, and vice versa. Where digital facilities are available at a user's location, no analog/digital conversion is necessary, but a device to assure synchronization of timing signals and to perform other control functions must be used. *See* ANALOG-TO-DIGITAL CONVERTER; DIGITAL-TO-ANALOG CONVERTER; MODULATION.

Channels may be provided on a dedicated (private line) or switched (dial) basis. Dedicated channels may be used for multipoint connections which serve several terminals (Fig. 5). For such multipoint configurations, a polling scheme is used to assure that the channel is used by only one ter-

minal at a time. Dedicated channels are commonly available to support rates up to 56,000 bits per second. Special arrangements can result in channels to support rates of 230,400 and 1,544,000 bits per second. Dial channels typically support up to 4800 bits per second; higher rates may be supported under special circumstances.

Asynchronous and synchronous transmission. Asynchronous and synchronous modes of transmission are used. In asynchronous transmission, bytes of data are transmitted independently of one another. Each byte begins with a start bit and ends with at least one stop bit. The time between bytes of data transmitted is variable. Asynchronous transmission is used at low speeds to support simple protocols used by low-cost terminals. Typical speeds are 75, 110, 134.5, 150, 300, 600, and 1200 bits per second full duplex (in both directions simultaneously). Synchronous transmission is more efficient than asynchronous transmission and is used to support more complex and efficient protocols. There are no start or stop bits. Data bits [for bit-synchronous protocols like X.25, derived by CLITT, an international standards body] or bytes [for byte-synchronous protocols like BSC (Binary Synchronous Communications) X3.28, defined by ANSI, an American standards body] follow one another in a precisely timed and correlated way. Typical speeds for dedicated channels are 2400, 4800, 9600, 19,200, and 56,000 bits per second full duplex. Typical dial speeds are 2400 and 4800 bits per second.

Switching. To support applications where communications among multiple stations is required, data communications systems offer switching beyond that available in the dial switching capabilities of the telephone network. Circuit switching, packet switching, and message switching are the common forms (Fig. 6). Circuit (line) switching

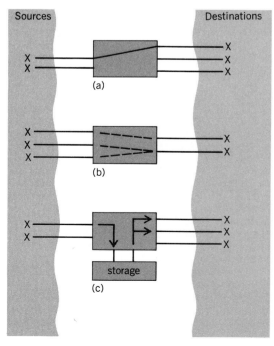

Fig. 6. Switching in data communications system. (*a*) Circuit switch. (*b*) Packet switch. (*c*) Message switch.

(Fig. 6*a*) results in the establishment of point-to-point channels, single-circuit paths where there is a direct connection between the incoming and outgoing lines to the switch, for the duration of a desired transmission. It is most suitable for batch-oriented applications.

Packet switching (Fig. 6*b*) permits a single access channel to be used for multiple concurrent transmissions. Data to be transmitted to different destinations are formed into packets to which addresses are assigned. These packets with variable addresses are output onto a transmission channel to be switched by a packet switch to their intended destinations. Likewise, packets received on a single transmission channel may have originated at different sources. In packet switching, packets are typically received, instantaneously stored, and then retransmitted over an appropriate outgoing line. There is no direct connection between the incoming and outgoing lines as in circuit switching. It is the role of a protocol handler to multiplex (mix or intersperse) packets into a single stream of data to be transmitted and to demultiplex (sort or separate) received packets into independent streams of data. Packet switching is most suitable for short transaction-oriented transmissions where multiple concurrent transmissions must be supported between a serving computer and remote terminals.

Message switching (Fig. 6*c*) is most suitable for message-oriented applications. A message switch performs communications processing as well as switching in managing the storage and distribution of messages. Whereas packets are most often created for the benefit of conveying information, including messages, messages are logical entities which represent the full informational content of a transaction and thus may be manipulated in more complex ways than packets. In message switching, a message is typically received, stored, and then retransmitted over an appropriate outgoing line at an appropriate time. As in packet switching, there is no direct connection between the incoming and outgoing lines. *See* ELECTRICAL COMMUNICATIONS; SWITCHING SYSTEMS (COMMUNICATIONS).

[MARK M. ROCHKIND]

Bibliography: D. W. Davies et al., *Computer Networks and Their Protocols*, 1979; T. Housley, *Data Communications and Teleprocessing Systems*, 1979; J. Martin, *Systems Analysis for Data Transmission*, 1972.

Data flow systems

An alternative to conventional programming languages and architectures, in which values rather than value containers are dealt with, and all processing is achieved by applying functions to values to produce new values. These systems can realize large amounts of parallelism (present in many applications) and effectively utilize very-large-scale-integration (VLSI) technology. Much progress has been made in developing these systems, and several prototype systems are operational, but several important questions remain unanswered.

Basic concepts. Data flow systems use an underlying execution model which differs substantially from the conventional one. The model deals with values, not names of value containers. There is no notion of assigning different values to an object which is held in a global, updatable memory location. A statement such as $X: = B + C$ in a data flow language is only syntactically similar to an assignment statement. The meaning of $X: = B + C$ in data flow is to compute the value $B + C$ and bind this value to the name X. Other operators can use this value by referring to the name X, and the statement has a precise mathematical meaning defining X. This definition remains constant within the scope in which the statement occurs. Languages with this property are sometimes referred to as single assignment languages. The second property of the model is that all processing is achieved by applying functions to values to produce new values. The inputs and results are clearly defined, and there are not side effects. Languages with this property are called applicative. Value-oriented, applicative languages do not impose any sequencing constraints in addition to the basic data dependencies present in the algorithm. Functions must wait for all input values to be computed, but the order in which the functions are evaluated does not affect the final results. There is no notion of a central controller which initiates one statement at a time sequentially. The model described above can be applied to languages and architectures.

Data dependence graphs. The computation specified by a program in a data flow language can be represented as a data dependence graph, in which each node is a function and each arc carries a value. Very efficient execution of a data flow program can be achieved on a stored program computer which has the properties of the data flow model. The machine language for such a computer is a dependence graph rather than the conventional sequence of instructions. There is no program counter in a data flow computer. Instead, a mechanism is provided to detect when an instruction is enabled, that is, when all required input values are present. Enabled instructions, together with input values and destination addresses (for the result), are sent to processing elements. Results are routed to destinations, which may enable other instructions. This mode of execution is called data-driven.

The illustration is an example of a machine language program for a data flow computer. Values are carried on tokens which flow on arcs of the graph. The graph in the illustration has four tokens initially. Tokens x and y carry input values; two control tokens have the value F. Iteration and conditional execution are achieved using the SELECT and DISTRIBUTE operators. SELECT routes a token to its output arc from either arc T or arc F, depending on the value of the control token on the horizontal input arc. DISTRIBUTE routes a token on its input arc to either the T or F output, depending on the value of the control token. All operators remove the input tokens used and produce a number of identical result tokens, one for each destination. In the illustration the initial output of the upper SELECT is a token with value x, since the control token has a value F. The input tokens x and F are removed, and two output tokens are produced, one for the function f and the other for the predicate p. It is a useful exercise to follow through the execution of the graph assuming that each arc is a first-in−first-out queue and can hold an unbounded

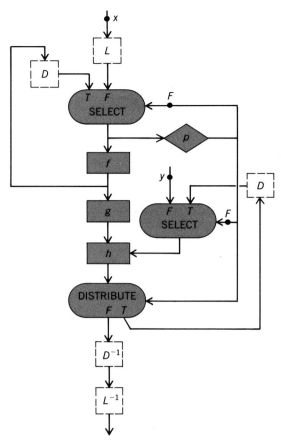

Machine language program for a data flow computer, represented as a data dependence graph.

number of tokens; and that L, L^{-1}, D, and D^{-1} are the identity operations.

Static and dynamic architectures. The execution model of a data flow computer as described above, though radically different from conventional processing, is the basis for most of the data flow machines currently being investigated. Individual differences arise because of the amount of parallelism that can be realized, the mechanisms for detecting and scheduling enabled instructions, and the handling of data structures.

In static architectures, an instruction (like h in the illustration) is represented in memory by a packet which has an operation code, and space to hold two input values and one or more destination addresses. Hardware is provided with the memory to detect the arrival of both input operands. One restriction in static data flow computers is that no arc in the graph can carry more than one token during execution. Control signals are sent from destination nodes to source nodes to indicate the consumption of previous values. Nevertheless, static machines can realize several different forms of parallelism. For example, f and p can be evaluated simultaneously. Also, g and h can be executed in parallel since they form two stages in a pipeline. However, one form of parallelism cannot be realized. Assume that f and h are simple functions which compute very fast and that g is relatively slow. If the single-token-per-arc limitation is removed, then several tokens would accumulate at the input of g and the possibility of invoking mul-

tiple, simultaneous instantiations of g would arise. Dynamic data flow architectures can realize this form of parallelism.

Dynamic data flow architectures allow multiple tokens per arc. A token carries a value and a label. The label specifies a context, an iteration number, and a destination address. Each instruction knows its successors and sets the destination field of the result token appropriately. In addition, the L, D, D^{-1}, and L^{-1} operators (ignored so far in the illustration) modify the context and iteration number. The L operator creates a new context by stacking the previous context and iteration number, and also sets the new iteration number to 1. The D operator increments the iteration number. The D^{-1} operator resets the iteration number to 1, and the L^{-1} operator restores the old context and iteration number (stacked by L). Dynamic data flow computers use an associative memory to hold tokens which are waiting for their partners to be produced. This mechanism is used to bring together tokens with identical labels. When this event occurs, the destination instruction is fetched and, together with the input values, is sent to a processing element. On completion, result tokens are produced with appropriate values. With this mechanism, simultaneous evaluations of g are possible. Since h may not be associative, successive evaluations of h must proceed in the specified order. The token-labeling mechanism guarantees this, irrespective of the order in which the simultaneous evaluations of g are completed. The token-labeling concept can be extended to handle recursion and generalized procedure calls.

Data-driven and demand-driven execution. In data-driven execution, the sources of a node N produce the input values and execution of N is triggered when all input values are produced. With the second execution rule, called demand-driven, nothing happens until a result is demanded at a primary output of a graph. The corresponding node then demands its inputs. These demands flow opposite to the arcs in the graph until the primary inputs are reached. A node executes only if its result has been demanded and its own demands satisfied. An advantage of this approach is that the computations which occur are exactly those that are required. This rule is also called lazy evaluation.

Comparison of conventional systems. Conventional languages and architectures are characterized by the existence of a sequential controller and a global addressable memory which holds objects. Languages such as FORTRAN and PL/1 allow aliasing and side effects and impose sequencing constraints not present in the original algorithm. The compile step attempts to recover the parallelism obscured by the language and to generate a data dependence graph. Depending on the parallelism to be exposed, this can be a complex step. Code is then generated for a scalar processor, a vector processor, or a multiprocessor consisting of several uniprocessors sharing storage. Each alternative is examined briefly below.

In uniprocessors, sequential decoding of instructions is necessary to place appropriate interlocks on storage and thereby guarantee the logical correctness of results. Techniques such as overlapping, pipelining, and out-of-sequence execution are used to design high-performance processors. However, because of the sequential decode, concurrency can be obtained only in a small window around the program counter. Furthermore, high-performance uniprocessors cannot effectively utilize VLSI technology.

The decoder limitation can be circumvented if a single instruction can initiate multiple operations on a data structure. This leads quite naturally to vector architectures. Portions of a program coded in instructions which have vector operands can be executed at very high speeds, limited only by hardware and memory bandwidth. However, not all applications with parallelism can be vectorized, and very sophisticated compiler analysis is needed to generate vector code automatically from sequential programs.

Conventional multiprocessing can utilize VLSI technology, but the key problem is the execution model with its global updatable memory. Since the processors execute asynchronously, the race conditions which can arise are prevented by embedding synchronization primitives in the code. Usually, the overhead for synchronization is large, and low-level parallelism cannot be realized. The code must be partitioned into relatively large blocks of computation with few synchronizing primitives. Moreover, if large amounts of parallelism are not obtained, the performance of the entire system can be critically dependent on processor-processor and processor-memory communications latency. The current state of development does not support compilation of sequential programs on a multiprocessor system. Also, multiprocessor code with embedded synchronization is extremely difficult to verify.

The complexity of the compile step for both vector and multiprocessor architectures can be reduced by extensions to sequential languages. However, this forces the programmer to consider parallelism explicitly, an added complexity.

Advantages and limitations. Data flow systems can overcome many of the disadvantages of conventional approaches. In principle, all the parallelism in the algorithm is exposed in the program and thus the programmer does not have to deal with parallelism explicitly. Since programs have mathematical properties, verification is simpler, and generation of the dependence graph from the program is a simple step. Systems of large numbers of slow-speed processors are possible, and the approach therefore exploits VLSI technology. If large amounts of parallelism are realized, then processor-memory and processor-processor communications latency is not as critical. Since there are constraints on the production and use of information, protection and security can be more naturally enforced.

Several important problems remain to be solved in data flow systems. The handling of complex data structures as values is inefficient, but there is no complete solution to this problem yet. Data flow computers tend to have long pipelines, and this causes degraded performance if the application does not have sufficient parallelism. Since the programmer does not have explicit control over memory, separate "garbage collection" mechanisms must be implemented. The space-time overheads of managing low levels of parallelism have

not been quantified. Thus, though the parallelism is exposed to the hardware, it has not been demonstrated that it can be effectively realized. The machine state is large, and without the notion of a program counter, hardware debugging and maintenance can be complex. Data flow also shares problems with conventional multiprocessor approaches: program decomposition, scheduling of parallel activities, establishment of the potential of utilizing large numbers of slow processors over a variety of applications, and system issues such as storage hierarchy management and disk seek-time limitations. *See* DIGITAL COMPUTER PROGRAMMING; PROGRAMMING LANGUAGES.

[TILAK AGERWALA]

Bibliography: T. Agerwala and Arvind (eds.), Data Flow Systems, *Computer*, special issue, 15(2):10–69, February 1982; P. C. Treleaven et al., Data-driven and demand-driven computer architecture, *Comput. Surv.*, 14(1):93–143, March 1982.

Data-processing management

Managing the data-processing function, its people, and its equipment. This activity follows the well-recognized principles of planning, control, and operation. The basic prerequisites for data-processing management are threfore the same skills which are needed to manage any other enterprise.

Data processing is a continually evolving technology, and this fact differentiates data-processing management from many other managerial environments. Furthermore, data processing is most often a service function within an organization, and not an end product. Constant awareness of this aspect is essential to ensure successful discharge of data-processing managerial responsibilities.

Organization of people. Despite the apparently overwhelming and complex presence of computing equipment and its associated technology, the major and critical component in the data-processing function is people. To select, develop, and organize people for maximum effectiveness and efficiency should be the major concern. Many technical skills—if not experience—have been in short supply, making the organizational task more pressing. Providing an effective organizational environment which offers motivation through ongoing training and a competitive reward structure is one avenue to success. The other is to match the data-processing organization to the structure, the style, and the goals of the enterprise served. To achieve this, the products, the customers, the marketplace, even the politics and problems of the enterprise, must be understood by data-processing management. In turn, data-processing management must be a recognized, participating, and accepted part of the overall management structure of the enterprise.

Products of data processing. The products of the data-processing function are the developed application systems, which should be viewed as a portfolio of corporate assets, and the information output, processed efficiently through the computing equipment and often supported by telecommunications facilities.

Application systems. Effective application systems development again depends on knowledge and understanding of the business activity, which implies full participation by and interaction with the users. Methodology and structure, that is, standards, and necessary to make systems developers—analysts and programmers—optimally productive. A formal plan, spanning 1–3 years, is needed to give direction and priority to the development tasks.

Application software for a particular need may also be available from vendors. The decision process to purchase rather than develop software should, of course, take the cost difference into consideration, but equally important is the fit of vendor software to the particular requirements. However, added to the acquisition price of the application software is the inevitable cost of conversion, education, modification, and installation.

Information output. Converting input data into meaningful and relevant output is the real function of data-processing equipment. The technology employed, as well as the internal processes of the computer, are very complex. It is here that technical knowledge is paramount, making it very likely the most elusive area to manage. A high degree of reliance on a technically competent staff, either within or outside of the organization, is unavoidable. A fundamental knowledge of the technology on the part of data-processing management and frequent interaction with the technical staff are the best insurance that the technology stays within bounds and does not become the major driving force.

Managerial control. Continuous awareness of products offered by manufacturers and vendors, their capabilities, and their cost is a useful and effective means of maintaining the level of technical knowledge needed to exercise managerial control.

Inadequate control, manifested by failure to deliver output and projects within budget, is often cited as a problem of data-processing management. It is frequently unrecognized that the data-processing function contains an appreciable creativity factor, particularly in the development area. The inability to schedule creativity contributes to an on-time performance problem. The structuring of the work effort into manageable parts and clearly established responsibilities; keeping a constant watch on progress and work effort; immediately identifying problems; and determining resolutions can overcome this.

Demands of computer technology. Computer technology requires an unequivocal and exact representation of facts and figures. Its effective use makes particular demands on technicians and managers alike. Only what has been predetermined and planned will happen. Malfunctions aside, only that which has been programmed will be produced—no more, no less. Surprises can be minimized only by attention to detail and by exploration of "what if" consequences. Recognizing this is an essential ingredient of successful data-processing management. *See* COMPUTER; DATA-PROCESSING SYSTEMS; DIGITAL COMPUTER.

[MICHAEL J. SAMEK]

Data-processing systems

Electronic, electromechanical, or mechanical machines for transforming information into suitable forms in accordance with procedures planned in advance. The term data-processing system is also applied to the scheme, or procedure, that prescribes the sequence of operations to be performed in processing the information.

Typical business applications of data-processing systems include record keeping, financial accounting and planning, processing personnel information (including payroll processing), sales analysis, inventory control, production scheduling, operations research, and market research. Data-processing systems are also used to process correspondence and other information in business offices. These systems are called word processors. *See* WORD PROCESSING.

In science and engineering, data-processing systems are used in data reduction, statistical analysis of experimental data, planning and design of engineering projects, and the display in tabular and graphic form of the results of research and development. Special-purpose computer systems, which also fall within the definition of data-processing systems, are used to monitor and control such things as chemical processes, machine tools, typesetting equipment, and power generation plants. *See* DATA REDUCTION.

Data-processing systems can be broadly classified as manual, semiautomatic, or automatic, depending on the degree of human effort required to control and execute the procedures. The trend is to reduce human effort, and automatic methods are becoming the dominant mode of processing. In a manual system the operations are performed by one or more individuals without the aid of mechanical devices. Varying degrees of mechanization can be introduced into manual systems through the use of machines such as calculators and bookkeeping machines. Semiautomatic systems use machines to a greater extent. The principal semiautomatic systems employ punched cards, and are referred to as a tabulating system or electronic accounting machine (EAM) systems. Automatic systems are usually built around electronic digital computers and termed electronic data-processing (EDP) systems. *See* DIGITAL COMPUTER.

Data-processing functions. Virtually every data-processing system, regardless of the degree of automation, consists of six basic functions: recording, transmission, manipulation, reporting, storage, and retrieval—collectively termed the data-processing cycle.

Recording. Before information can be processed, it must be recorded in some form that is meaningful to a person or a machine. The information may be written or typed on a paper document, or it may be coded as a pattern of punched holes on a card, pulses on magnetic disk, or varying-width vertical dark bars on the label of a food product. Many business machines, such as typewriters and cash registers, can be equipped to record data in a machine-readable form as a by-product of their normal function. Patterns of holes on cards or pulses on disks can be created by using a keyboard entry device. Optical scanners are being used to record the sequence of bars on the label of the food product.

At one time the principal medium used to enter information into computers was the punched card. The principal device used to record information for processing by computers was the keypunch. Subsequently keyboards linked to magnetic tapes (key to tape) or to magnetic disk (key to disk) systems became widely used. The information is usually recorded on a small reel of tape or a small disk first and then entered into the computer system later. These recording systems are being replaced in turn by key entry devices that are linked directly to the computer via communication lines. Intermediate recording, whether on punched cards, tape, or disk, is no longer required.

Transmission. Once information has been recorded, it will usually need to be transmitted to another location for processing. The distance involved may range from a few feet to thousands of miles, and the information can be conveyed by personal delivery, mail, or wire. Where fast response is desirable and economically feasible, the data can be transmitted at high speeds by means of telephone or telegraph circuits, microwave links, or satellite channels. Numerous specialized business machines, such as airline reservation and credit authorization terminals, now combine the recording and transmission functions by accepting data entered through a keyboard or read from credit cards or both and transmitting it directly to a distant computer system. *See* DATA COMMUNICATIONS.

Manipulation. This is the stage in which most of the actual processing is performed. The operations involved can range from simple to highly complex. The most common types of operations are classifying, sorting, calculating, and summarizing. More specifically, the manipulation stage frequently involves (1) arrangement of the information into a sequence that will facilitate further processing; (2) determination of the exact procedure to be followed in processing each item of information; (3) references to files containing data that must be associated with the current information; (4) arithmetic operations upon the current information or file data or both; and (5) updating (changing) of the file data to reflect the current information.

Report preparation. After information has been processed, it is usually necessary to report the results in a meaningful form. Reports are the people-oriented products of the data-processing activity.

They should be timely, complete, understandable, and in a convenient format. Checks, letters that are ready to be mailed, and lists of purchases at supermarket checkouts, as well as more routine summaries of the results of data processing, are being prepared. The routine summaries are frequently presented in graphical as well as tabular form, and include substantial amounts of text in order to make the report more understandable. Most reports are printed-paper documents, but other media, such as cathode-ray-tube displays or microfilm, are used.

Storage. Some or all of the processed information will need to be stored for future reference or retrieval. Information that is generated during one data-processing step frequently serves as a part of the input data for a later step, and must therefore be retained in a conveniently reusable form. The records of an organization may consist mainly of the contents of the storage generated by its data-processing system. Depending on the system, the information may be stored manually (as in ledgers or notebooks), electromechanically (as in punched cards or punched paper tape), or electronically (as in disk packs or magnetic tape).

Retrieval. The importance of the contents of storage and the availability of transmission facilities have resulted in many automatic data-processing systems providing immediate retrieval of stored information at work stations. Questions regarding the contents of storage can be posed at remote locations, and responses can be prepared and transmitted to these locations in a few seconds. This is called an on-line query facility. Special capabilities that permit retrieval of stored information based on a more time-consuming, complex sequence of searches through the stored information are also becoming more widely available.

Punched-card systems. The first successful punched-card data-processing equipment was developed by Herman Hollerith for the U.S. Bureau of the Census in the 1880s. By 1930, a full line of machines was available for processing data recorded in cards, and punched-card accounting sys-

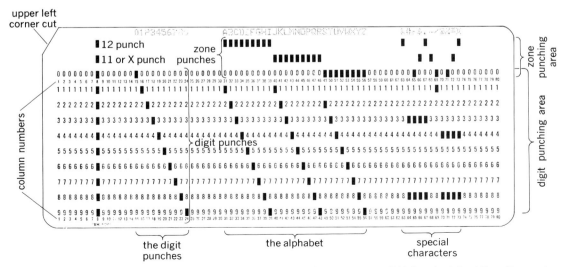

Fig. 1. An 80-column punched card showing the Hollerith coding for numeric and alphabetic information. (*From An Introduction to IBM Punched Card Data Processing, IBM Publ. no. F20-0074, November 1964*)

tems were in use around the world. Electronic data-processing systems, which offer greatly improved flexibility, speed, and economy, have been rapidly supplanting punched-card systems. The punched card itself, however, remains one of the principal means for recording data for entry into an EDP system.

The punched card is a piece of stiff paper in which data can be represented by punching holes in prescribed positions. The most commonly used type of card involves the Hollerith coding system and contains 80 vertical columns. Each column can represent a character (a digit, letter, or punctuation mark) by means of holes in 1, 2, or 3 of the 12 positions (Fig. 1). The card is 3.25 in. (82.55 mm) high, 7.375 in. (187.32 m) wide, and 0.070 in. (1.78 mm) thick. Each card normally holds the data describing a single transaction, account, or record.

Punched-card accounting systems generally contain a separate machine to perform each processing function. Though many of the machines can process hundreds of cards per minute, a fairly high degree of manual effort is required to load and unload each machine and to convey the cards from one machine to the next. The operation of most punched-card machines is controlled by plugboards. These are perforated boards whose holes (termed hubs) are manually interconnected by wires (termed patchcords) in a manner that will cause the machine to perform the desired func-

tions. This method of control is far less flexible than the internally stored programs used by most digital computers.

There are many types of punched-card machines, but a fully mechanized system can contain as few as three machines: a key punch to record the information in punched cards, sorter to arrange the cards in the proper sequence, and a tabulator to prepare and print the reports.

Electronic data-processing (EDP) systems. EDP systems (Fig. 2) take advantage of the great speed and versatility of stored-program digital computers to process large volumes of data with little or no need for human intervention. There is no fundamental difference between the computers currently used for business data processing and those used for scientific calculations; in fact, a single computer is often used for both types of applications. Scientific computers, however, tend to emphasize high computational speeds, while computers used for business data processing often place primary emphasis on fast, flexible input and output equipment.

Although large, high-performance computer systems cost hundreds of thousands or even millions of dollars, a wide variety of smaller systems is available at much lower prices for users with smaller volumes of data to be processed. These systems and others used in data communications take advantage of developments in microelectronics (Fig.

Fig. 2. Large-scale electronic data-processing system. Disk drives are at the right, and tape drives at the left. Printers are in the foreground and to the right of the tape drives. The central processing unit and storage are in the center background. (*Sperry Univac Division of the Sperry Rand Corporation*)

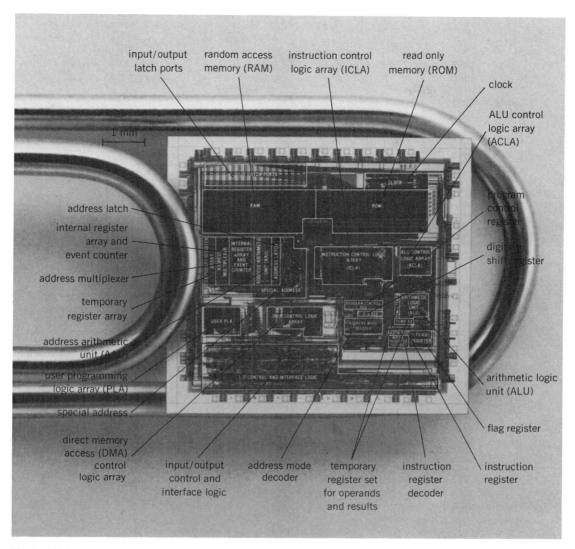

input/output
latch ports

random access
memory (RAM)

instruction control
logic array (ICLA)

read only
memory (ROM)

clock

ALU control
logic array
(ACLA)

address latch

internal register
array and
event counter

address multiplexer

temporary
register array

address arithmetic
unit (AAU)

user programming
logic array (PLA)

special address

direct memory
access (DMA)
control
logic array

input/output
control and
interface logic

address mode
decoder

temporary
register set
for operands
and results

instruction
register
decoder

instruction
register

flag register

arithmetic logic
unit (ALU)

digital
shift register

program
control
register

Fig. 3.   MAC-4 one-chip computer, developed for a variety of telecommunications applications, compared to a standard-sized paper clip. (*Bell Laboratories*)

3) to deliver surprisingly high performance, along with compact dimensions and low prices (Fig. 4).

Every computer system has four basic functional parts: input equipment to permit data and instructions to be entered into the system; a storage unit to permit data and instructions to be stored until called for; a control unit, called the central processing unit (CPU), to interpret the stored instructions and direct their execution; and output equipment to permit the processed data to be removed from the system. *See* COMPUTER STORAGE TECHNOLOGY; DIGITAL COMPUTER PROGRAMMING.

The remainder of this article discusses the input and output equipment used in EDP systems. The common types of input and output devices, their recording media, and their typical speeds are summarized in the table. Magnetic tape and disk can also be viewed as storage media that supplement the storage available in the computer itself. They have been termed auxiliary storage.

Disk pack units. Introduced in 1962, the disk pack has become the preeminent high-speed computer input/output medium. A disk pack typically consists of a stack of from 1 to 20 round stainless-steel plates mounted on a vertical spindle. Data are magnetically recorded on some or all of the surfaces of the disks. The disk pack can be conveniently mounted on a drive unit that spins it at 2400–3600 revolutions per minute. The drive unit also contains magnetic read/write heads mounted on a comb-like access mechanism that moves horizontally between the disks in order to read and record information on them.

Disk packs have transfer rates from about 200,000 to nearly 2,000,000 characters per second and capacities of from about 10,000,000 to more than 600,000,000 characters per pack. The key advantage of disk packs that accounts for their preeminence is their rapid-access capability; any record stored in a disk pack can be located and read into the computer in a fraction of a second, whereas it may take as long as several minutes to locate a particular record stored in a reel of magnetic tape. Conversely, disk packs, on the basis of cost per character stored, currently cost several times as much as magnetic tape.

DATA-PROCESSING
SYSTEMS

Fig. 4. Relatively low-cost EDP system. The system includes a keyboard, a cathode-ray-tube display, and a pair of diskette units. (*Digital Equipment Corp.*)

Input/output devices

Unit	Medium	Input or output	Typical approximate transfer rate in characters per second
Disk pack unit	Magnetic disks housed in interchangeable cartridges	Both	200,000 – 2,000,000
Diskette unit	Flexible magnetic disks	Both	Tens of thousands
Magnetic tape unit	Plastic tape with magnetizable coating, housed on reels	Both	Thousands – 1,250,000
Magnetic tape cartridge unit	Plastic tape with magnetizable coating, housed in cartridges	Both	Tens of thousands
Card reader	Punched cards	Input	20 – 2000
Card punch	Punched cards	Output	20 – 700
Paper tape reader	Perforated paper or plastic tape	Input	10 – 1000
Paper tape punch	Perforated paper or plastic tape	Output	10 – 200
Printer	Continuous paper forms	Output	10 – 20,000
Optical character reader	Paper or card documents	Input	10 – 3000
Magnetic ink character reader	Paper or card documents	Input	200 – 3000
Display unit	Cathode-ray tube or other display medium	Output*	10 – 100,000
Computer output microfilmer	Microfilm	Output	2000 – 60,000

*Input capabilities are usually provided by an associated keyboard or light pen.

One development in magnetic disk technology deserves special mention. Flexible plastic disks, called diskettes or floppy disks, are coming into widespread use for applications in which low cost and ease of use are more important than high speeds and large data-storage capacities. Most diskettes are 8 in. (20 cm) in diameter, and may be housed in thin plastic envelopes; the envelope remains stationary while the drive unit spins the disk. Each diskette holds approximately 300,000 characters, and the data-transfer rate is about 30,000 characters per second.

Magnetic tape units. Magnetic tape is an important input/output medium for EDP systems because it permits large quantities of information to be stored in a highly compact, economical, and easily erasable form. The tape is usually made of plastic with a magnetizable oxide coating on which data can be recorded in the form of magnetic spots. The tape transports must be capable of moving the tape past the read/write heads at high speeds (up to 250 in./s or 6.35 m/s) and of starting and stopping the tape movement within a few milliseconds.

The magnetic tape currently in widest use with EDP systems is ½ in. (12.7 mm) wide, is supplied in 2400-ft (732-m) reels with a diameter of 10.5 in. (26.67 mm), and is recorded with 9 parallel tracks across the tape at a density of 800, 1600, or 6250 frames per inch (31.5, 63.0, or 246.1 frames per millimeter). Each frame thus holds 9 bits (binary digits) and can represent one character of information.

Transfer rates of magnetic tape units range from a few thousand to 1,250,000 characters per second. Though impressive, the transfer rates of both tape and disk fall far short of the internal processing speeds of the computer. As a result, the overall productivity of some data-processing systems are limited by the speed at which information can be entered into and removed from the system.

As an alternative to high-performance magnetic tape units, many EDP systems include smaller, less costly tape transports in which the data are recorded on shorter lengths of narrow magnetic tape, typically 0.15 or 0.25 in. (3.81 or 6.35 mm) wide, housed in conveniently interchangeable cartridges. The most commonly used type of cartridge is an adaptation of the Philips-type cassettes that are widely used in audio recording. The cartridge tape units have comparatively slow read/write speeds and small data-storage capacities; their offsetting advantages are economy and convenience of use.

Magnetic tape is widely used to provide back-up for disk units and to permit the exchange of large volumes of information by sending reels of tape from one place to another. For example, computer programs that are purchased are often delivered on one or more reels of magnetic tape.

Card reader and punch. The fact that most early computers replaced or augmented punched-card systems gave the punched card a strong head start as a computer input/output medium, and it is still widely used. The punched card is a highly flexible medium that has many advantages for use as a source document and as a storage medium for permanent records. In EDP systems its principal drawbacks are the difficulty of correcting errors in the punched data, the fixed upper limit on the amount of data a card can hold (usually 80 or 96 characters), and the comparatively low speeds of even the fastest card readers and punches (see table).

Paper tape readers and punches. Paper tape can be punched and read by relatively simple, inexpensive equipment, making it practical to produce tape records as a direct by-product of the normal operations of many business machines, such as teletypewriters and cash registers.

Though generally called paper tape, the tape may be made of paper, plastic, metal, or laminated combinations of these materials. Data are recorded by punching round holes into the tape and read by sensing the holes either mechanically or photoelectrically. Nearly all paper tape in use at present is either 11/16, 7/8, or 1 in. (17.46, 22.22, or 25.4 mm) in width, and is recorded with 5, 7, or 8 data tracks at a density of 10 frames per inch (3.94 frames per centimeter). Each frame normally represents a single character using a code of 5–8 bits.

As with punched cards, the principal disadvantage of paper tape is the low speeds of the tape readers and punches available for use with computers. As a result, paper tape is being replaced in many applications by magnetic tape cassette units.

Printers. Electromechanical printing devices are the primary means for making information processed by computers available to people. However, display units are overtaking printers in this regard. Computer printers can be broadly classified as either serial or line printers and as either impact or nonimpact printers. A serial unit prints one character at a time in the manner of a conventional typewriter, whereas a line printer prints a full line, usually consisting of 80 to 132 characters, at the same time. An impact printer uses direct mechanical force to produce character images on the paper, whereas nonimpact printers utilize various electrical or chemical processes to form the characters.

Most high-speed computer printing is currently done by line printers using on-the-fly impact printing techniques. In these devices, multiple rapid-action print hammers press continuous paper forms against an inked ribbon and a moving type element at the precise instants when the selected characters are in the appropriate positions. The type element may be a rotating drum, a horizontally moving chain of type slugs (Fig. 5), or a horizontally oscillating typebar. Speeds of 300 to 2000 lines per minute are typical for line printers.

Matrix printers may be impact or nonimpact. The most successful are nonimpact and use thermal-chemical or electrostatic processes to create an image. Typically a 5-by-7 matrix of dots is used to print each character. The set of dots that is printed provides an image of the character (Fig. 6). A matrix printer that is widely used, especially in offices using word processing, is the ink jet printer.

A thermal matrix printer uses either heat-sensitive paper or wax-coated dark paper as the medium for printing. The head-sensitive paper undergoes a chemical change when subjected to heat, producing a visible image. Speeds of about 10 to 100 characters per second are common for thermal printers, and they are relatively inexpensive—costing from a few hundred to a few thousand dollars. However, the quality of the result is poor, and they may give off an offensive odor or accumulate wax debris.

Electrostatic printers are among the fastest and most versatile printers available. Speeds of several thousand characters per second can be achieved. The matrix printing approach accounts for their versatility. A 10-by-14 or even finer matrix can be used to obtain better printed images and a greater variety of characters.

A laser can be used to write the printed output on the light-sensitive surface of a drum. The image on this surface is then transferred to paper. Speeds of tens of thousands of characters per second can

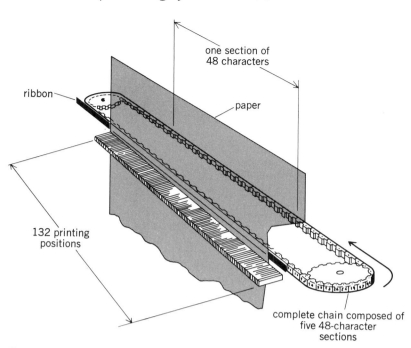

Fig. 5. Schematic representation of a chain printer. *(From H. Maisel, Computers: Programming and Applications, Science Research Associates, 1976)*

be realized by this type of printer. *See* LASER.

Character reader. Optical character recognition (OCR) has been heralded as the solution to the great computer input bottleneck because it promises to reduce, and in some cases eliminate, the manual keystroke operations which are normally required in preparing the input data for machine processing. Machines capable of reading characters printed on paper documents by typewriters, cash registers, computer printers, and many other business machines are in use in applications such as utility billing and credit card processing. Moreover, machines that can read hand-printed characters are on the market, and extensive development work is in progress on machines to recognize ordinary handwriting. The principal barriers to wider use of optical character readers have been reliability problems and comparatively high costs, and significant progress has been made in both areas.

Optical recognition techniques can also be used effectively to read hand-made pencil marks or printed bar codes. Optical mark readers and bar-code readers are considerably less costly than character readers and are widely used in applications such as test scoring, survey analysis, credit-card processing, and supermarket checkout.

Magnetic-ink character recognition (MICR), unlike OCR, requires that the characters be printed with a special magnetic ink. This requirement limits the flexibility and applicability of the MICR technique, though it permits reliable readers to be built at a lower cost. The banking field, which adopted magnetic-ink encoding as the common language for checks in 1959, remains virtually alone in preferring MICR to the more flexible optical techniques. *See* CHARACTER RECOGNITION.

Displays. Display units are widely used as "electronic blackboards" that provide rapid access to data stored in computer systems and facilitate

DATA-PROCESSING SYSTEMS

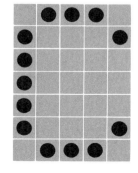

Fig. 6. Matrix of dots used to represent the letter C.

human-machine communication. The great majority of current display units use cathode-ray tubes, but other technologies such as light-emitting diodes, liquid-crystal displays, and plasma (gaseous) displays are also used. The display units may be either connected directly to the computer or located remotely and connected by way of a data communications link. The displayed information may consist of characters or lines or both on the face of a cathode-ray tube 6−22 in. (15−56 cm) in diameter. Most display units can also be used for input to the computer by means of either a keyboard or handheld light pen which the operator focuses upon particular points on the screen.

Two broad types of applications for which display units are particularly valuable are obtaining quick responses to inquiries (for example, about bank balances or airline reservations) through access to large, continually updated files of information; and providing convenient human-machine "conversations" which make it practical to construct programs and designs in step-by-step, trial-and-error fashion. Display units represent one of the most flexible and economical means for achieving communications between machines and people. Their most noteworthy weakness is their inability to produce permanent copies of the displayed information for future reference, and this can be overcome by equipping the display unit with an auxiliary printer.

Computer output microfilmers. Microfilm has two noteworthy advantages over the printed paper reports that serve as the principal output from most EDP systems: printing on microfilm can be performed at considerably higher speeds, and the microfilmed reports require far less storage space. Conversely, the acceptance of microfilm as a computer output medium has suffered because the information cannot be read by humans without the use of special microfilm readers. The advantages of microfilm clearly outweigh the drawbacks for many computer users, and equipment is now available that can record computer-generated reports on microfilm at speeds ranging from 1200 to 30,000 lines per minute.

[HERBERT MAISEL]

Bibliography: E. M. Awad, *Business Data Processing*, 4th ed., 1975; G. B. Davis, *Computers and Information Processing*, 1978; H. Maisel, *Computers: Programming and Applications*, 1976; A. Mowshowitz, *The Conquest of Will*, 1976; D. Sanders, *Computer Essentials for Business*, 1978.

Data reduction

The transformation of information, usually empirically or experimentally derived, into corrected, ordered, and simplified form.

The term data reduction generally refers to operations on either numerical or alphabetical information digitally represented, or to operations which yield digital information from empirical observations or instrument readings. In the latter case data reduction also implies conversion from analog to digital form by human reading and digital symbolization or by mechanical means. *See* ANALOG-TO-DIGITAL CONVERTER; DIGITAL COMPUTER.

Data reduction is used to prepare data in a form suitable for scientific computation, statistical analysis, and control of industrial processes and operations, and for data processing in business applications. Examples are the preparation of data obtained from test runs in missile development, wind tunnel experiments, industrial product sampling, or from readings of sensing instruments in process control.

In applications where the raw data are already digital, data reduction may consist simply of such operations as editing, scaling, coding, sorting, collating, and tabular summarization.

More typically, the data reduction process is applied to readings or measurements involving random errors. These are the indeterminate errors inherent in the process of assigning values to observational quantities. In such cases, before data may be coded and summarized as outlined above, the most probable value of a quantity must be determined. Provided the errors are normally distributed, the most probable (or central) value of a set of measurements is given by the arithmetic mean or, in the more general case, by the weighted mean.

Data reduction may also involve operations of smoothing and interpolation, because the results of observations and measurements are always given as a discrete set of numbers, while the phenomenon being studied may be continuous in nature.

In a smoothing problem a function is empirically given (for example, positions of a body as a function of time) as a collection of points (t_1,x_1), (t_2,x_2), . . ., (t_n,x_n) where the values of the variables, perhaps both independent and dependent, are inaccurate. A common procedure is to fit an nth- (commonly second-) order parabola by least squares to the data points, thus obtaining a representation that will satisfy as nearly as possible all of the given pairs, but perhaps none exactly.

In interpolation a function is known in tabular form. The problem is to determine values between the tabulated points.

Any of the above-mentioned procedures may be carried out on a digital computer or built into a process control system or procedure.

[RAYMOND J. NELSON]

Bibliography: E. Horowitz and S. Sahn, *Fundamentals of Data Structures*, 1976; D. E. Knuth, *The Art of Computer Programming*, vol. 1: *Fundamental Algorithms*, 1969; S. Runyon, Converters are finally blasting off, *Electron. Des.*, 28(8):38−50, 1978; P. P. Uhrowczik, Data dictionaries/directories, *IBM Syst. J.*, 12(4):332−350, 1973.

Decision theory

A broad spectrum of concepts and techniques which have been developed to both describe and rationalize the process of decision making, that is, making a choice among several possible alternatives. In the broadest sense, one can differentiate between prescriptive decision theory, which formulates how decisions should be made, and descriptive decision theory, which deals with how people actually make decisions. Behavioral and social scientists and psychologists are trying to discover more elaborate descriptive models of the decision process in order to provide mathematicians, economists, strategists, business managers, and others with more sophisticated prescriptive decision-making procedures.

Concepts and terminology. The need for decisions in choosing one of several possible actions or series of actions is continuously present in everyday life. It starts with the decision on whether to get up in the morning when the alarm rings or to sleep an extra 15 min before deciding what to have for breakfast, or whether to skip breakfast altogether. Depending on the importance of an individual's position, decisions on various alternatives throughout the day may govern not only his personal future and well-being but also the fate and well-being of any number of other individuals. Thus decisions, having various degrees of importance, represent a critical part of an individual's life.

However, it is not the decisions themselves which are important but rather the consequences of the decisions. These consequences or outcomes are determined by two types of independent factors: (1) the controlled factors, represented by the courses of action or alternate strategies (in decision theory terminology) available to the decision maker; and (2) uncontrolled factors, or states of nature, which interact with the strategies.

The consequences of the chosen strategies are dependent quantities commonly termed outcomes (O). These can be measured in a variety of quantitative parameters such as dollars or number of items produced, and by subjective measures of outcome or, in decision theory terminology, by expected relative values or expected utilities.

Conceptually the decision problem is straightforward: From a set of (1) alternate strategies (S_n), choose the one which when combined with the relevant (2) states of nature (N_j) will produce the (3) outcomes (O_{nm}) which best satisfy the (4) goals or objectives of the decision maker. Some type of penalty or benefit criterion is used to evaluate the value or utility of the outcome. Decision theory implies that all four factors should be known to make a completely rational and meaningful decision. At the same time decision theory tries to explain why one often makes seemingly irrational decisions because one is either unable to adequately define the four ingredients or chooses to ignore the procedure.

By taking just the relatively simple decisions which have to be made before getting to work, one can demonstrate the possible difficulties and more of the relevant terminology. In Fig. 1, assume that the decision maker has three basic decisions to make on his way to work. (1) He can get up at 7:00 A.M. (choice 1) or 7:15 A.M. (choice 2), with the associated consequence that after dressing he would depart at 7:15 or 7:30 A.M., respectively. (2) He can choose routes A or B (choices 3 and 4) with the expected outcomes that route A always takes 30 min while route B can take 15, 30, or 45 min (a, b, or c), depending on traffic conditions. (3) He can stop for breakfast, which takes 15 min or skip breakfast (choices 5 and 6).

The decision sequence is illustrated in Fig. 1 in the form of a decision tree. It shows all the possible intermediate and final outcomes which would result from all possible choices of alternatives posed by this example. The decision nodes at which decisions have to be made are marked D.

Several interesting observations can be made about the decision tree. (1) Although there are 16 possible outcomes, there are only 8 possible initial strategies since the choice a, b, c is independent and is determined by a state of nature. (2) The decision problem has been bounded by enumerating all admissible strategies, states of nature, and outcomes. This has already involved some judgment in excluding other strategies, states of nature, and outcomes on the basis that they are unlikely to occur, or because the decision maker is specifically interested only in the alternatives presented. (3) Although the alternate strategies and outcomes are readily apparent from this particular decision tree, the fourth ingredient of a decision problem, the objectives of the decision maker, has not yet been defined. The objectives which generate the criteria for evaluating the decision problem

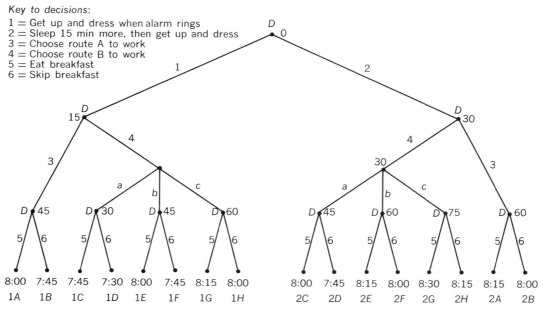

Key to decisions:
1 = Get up and dress when alarm rings
2 = Sleep 15 min more, then get up and dress
3 = Choose route A to work
4 = Choose route B to work
5 = Eat breakfast
6 = Skip breakfast

Fig. 1. Decision tree for the decision problem given in text. Alarm rings at 7:00 A.M. Numbers at nodes represent times past 7:00 A.M. Numbers at branches refer to decision made or state of nature.

represent the critical link in reaching the best decision for that particular decision maker at this particular time. In Fig. 1 it is clear that the decision on the best strategy to be followed is a function of the following objectives.

1. If the objective is to positively arrive at work at 7:45 A.M. sharp, no sooner, no later, then one can immediately eliminate all strategies which do not have 7:45 A.M. as their final outcome as inadmissible strategies. In this case the only admissible strategy would then be $(1-3-6)$ since any other strategy might result in coming to work late.

2. If, on the other hand, the decision maker decides that breakfast is a vital function and will always insist on breakfast, he will eliminate all strategies which have a branch 6 in the decision tree. If his objective is still to get to work as early as possible, he will now be faced with the decision between strategy $(1-3-5)$ and strategy $(1-4-5)$, which may under certain circumstances get him to work faster than $(1-3-5)$, that is, outcome IC. These two strategies dominate all other admissible strategies because under all circumstances they will get him to work faster than any of the other strategies. Thus when one strategy shows dominance over some other strategies, one can eliminate all but the dominant strategies from consideration. Strategies $(1-3-5)$ and $(1-4-5)$ do not dominate each other since under certain circumstances one or the other will have the more desirable outcome.

If strategy $(1-3-5)$ is chosen, this is called decision making under certainty (DMUC) since the outcome is uniquely predictable and all states of nature are known. The outcome of strategy $(1-4-5)$ depends on the traffic conditions $(a, b, $ or $c)$ or states of nature which are not under the decision maker's control. If he has no idea what the likelihood or probability of occurrence (p_m) of states a, b, or c is, he will be faced with decision making under uncertainty (DMUU) and will employ one of the techniques described in the next section. If he can predict the probability of the states of nature a, b, or c from past experience or from observations, then he can associate a probability with each of the possible outcomes. This represents decision making under risk (DMUR) with expected outcomes. The techniques used in these three types of decision making will be discussed further.

3. Except in very well-defined decision problems in which the outcomes are clearly represented by an objective parameter (such as dollars or time) there are conflicting objectives which require value judgments to determine the value or utility of each objective with respect to the others. This unescapable aspect of decision theory has led to the development of value theory normally associated with decision theory. Value theory strives to evaluate relative utilities of simple and mixed parameters which can be used to describe outcomes. For example, the utility or value of $100,000 to a multimillionaire in a gambling decision is obviously quite different from its value to a man whose total possessions amount to $100,000. Similarly, the question of the relative values of 15 min extra sleep versus no breakfast versus coming to work late requires some resolution before the objectives and evaluation criteria for the decision problem can be established.

The general concepts of decision theory in trying to formalize and rationalize the decision-making process were indicated by the above discussion. Some techniques of actually using these concepts are described below. It should be emphasized that the major contributions of decision theory are to structure the decision problem and to define the scope and importance of various components so that each part of the decision can be evaluated and discussed before synthesizing the available information into a rational decision.

Application and techniques. Decision theory in various disguises has been applied with varying degrees of success to so many diverse areas of interest that it would be impossible to provide an inclusive list of the methods and techniques used. The bibliography at the end of the article represents a broad spectrum of application techniques and current research work. The formal procedures of decision theory are described below with an indication of the relative use and value of the techniques in various areas of modern society.

The information necessary for a rational decision is usually summarized in a decision matrix, as shown in Fig. 2. The alternate pure (independent from each other) or mixed (combinations of pure strategies) strategies are listed in the first column. These strategies $(S_1$ to $S_n)$ can be generated by the decision maker, by his advisers, in group sessions, and so on, and should contain all possible and admissible courses of action which the decision maker can control. In business management these strategies could be money-borrowing strategies, sales and promotion strategies, and production and product mix strategies. In government it could be foreign policy strategies, military strategies, crime-fighting strategies, or government appointments.

Across the top all possible states of nature $(N_1$ to $N_m)$ which might affect the outcome of the various strategies are enumerated. A business manager trying to raise capital may be interested in possible states of nature such as different interest rates and inflationary and deflationary trends. A military planner may be concerned with the possibilities of war in the Middle East or Europe or both. One can easily see that the decision matrix can become quite large and the trade-off between time and money expended on a decision compared with the value of a more or less reliable decision can become a decision problem in itself.

The probabilities of occurrence of various states of nature $(p_1$ to $p_m)$ must be estimated and listed with the states of nature. Since these values will critically influence the decision, they should be as reliable as possible. In value theory one distinguishes between the subjective probability of occurrence which reflects the decision maker's opinion on what the probability is and the objective probability which is derivable mathematically or as a result of observations. Thus an optimist may consider the probability that he will get detained in traffic very unlikely and make his decision based on this assumption, while observation of traffic conditions over a period of time may indicate that 90% of the time he will be detained by traffic. Statistical theory plays an important role in defining these numbers.

Corresponding to each interaction of a strategy

States of nature		N_1	N_2	— —	N_j	— — — — — — —	N_m	Expected value
Probabilities of occurrence		p_1	p_2	— —	p_j	— — — — — —	p_m	
Alternate strategies	S_1	O_{11}	O_{12}	— —	O_{1j}	— — — — — —	O_{1m}	EV_1
	S_2	O_{21}	O_{22}	— —	O_{2j}	— — — — — —	O_{2m}	EV_2
	S_i	O_{i1}	O_{i2}	— —	O_{ij}	— — — — — —	O_{im}	EV_i
	S_n	O_{n1}	O_{n2}	— —	O_{nj}	— — — — — —	O_{nm}	EV_n

Fig. 2. Form of decision matrix. O_{ij} = outcome of jth state of nature (N_j) intersecting with ith strategy (S_i).

(S_i) and a state of nature (N_j), one determines a consequence or outcome (O_{ij}) in terms of some value or utility parameters. The outcome may sometimes be easily computed, for example, the interest cost on a business loan, but quite often must be based on insufficient data, past experience, or questionably valid assumptions, such as would be the case in trying to assess the outcome of a limited war. The generation of reliable outcome data can be expensive and time-consuming, and a probability of accuracy is often associated with each outcome. The question of how to assess the relative value or utility of mixed parameters to present a single parameter for the outcome is still a paramount question in decision and utility or value theory, that is, establishing the utility of two apples versus one apple and one orange versus two oranges and so forth. The unit used to describe such a value parameter is a "utile"; its meaning must be adapted to each situation.

The information in the decision matrix is used to generate the last column which describes the overall expected outcome, or expected value (EV), or expected utility (EU), or expected profit for each strategy. Depending upon the confidence one places in the prediction of the probabilities of the states of nature, the decision process is classified as one of three types.

Decision making under certainty (DMUC). In this case the state of nature which will interact with the strategies can be defined with sufficient certainty so that its probability is close to 1. This case reduces the decision matrix to one column showing the outcomes (profit, cost, and utility) of each strategy with the state of nature. Decision theory indicates that, assuming the alternate strategies are equal in all other respects, the one which will maximize the utility (maximize profit or minimize cost) should be selected. Since in many cases the dependent variable (outcome) is continuously related to the independent variable (strategy), one can obtain relationships which can be used to analytically find the strategy which will correspond to maximum utility. This process is normally termed optimization, which in its broadest interpretation refers to decision making under certainty.

Decision making under risk (DMUR). This implies that, although a number of possible states of nature may interact with each strategy, the probability of occurrence of each of these is known with some degree of certainty. In the previous example, for instance, observation may show that 80% of the time it takes 15 min if decision 4 (Fig. 1) is chosen, 10% of the time 30 min, and 10% of the time 45 min. The objective (observed) probabilities for outcomes a, b, and c would then be .8, .1, .1, respectively. The probabilities of occurrence of all states of nature must always add up to 1 (since one postulates that one of these states of nature will definitely interact with each strategy).

The expected value or utility for each strategy is obtained as a weighted average sum of the individual possible outcomes or, in mathematical terms,

$$(EV_i) = \sum_{j=1}^{m} p_j O_{ij}$$

Thus in the example the expected value for the branch 4 strategy would be $EV = (.8)15 + (.1)30 + (.1)45 = 19.5$ min.

When making decisions under risk, the intent is to choose the strategy which will minimize the expected value or utility. It should be emphasized again that the outcomes and expected values must be expressed by a parameter which adequately presents the objectives of the decision maker. If it does not, then a utility or value transformation is performed on the decision matrix to transform the outcomes into meaningful quantities. If minimum time to get to work is the major objective in the above example, then strategy 4 would be used since its expected value of 19.5 is lower than the expected value of 30 min for strategy 3 (in this particular case, strategy 3 involves no risk so that one is mixing DMUC with DMUR).

The condition of dominance is usually used to

eliminate all strategies which are dominated. A dominant strategy has more desirable outcomes under any of the possible states of nature than the dominated strategy.

An additional insight into the reliability of a given decision can be obtained by a sensitivity analysis on the decision matrix. This involves incremental change of one or several probabilities of the states of nature to determine what effect this would have on a decision. This is particularly appropriate for probabilities which are of questionable accuracy.

Decision making under uncertainty (DMUU). If the probabilities of the states of nature cannot be predicted, then the decision has to be made under conditions of uncertainty. For example, the probabilities of political incidents, riots, legal suits, wars, or accidental deaths cannot be accurately predicted.

Decision theory provides a number of techniques to deal with these. They are briefly outlined below without elaborating on their merits or shortcomings. Which of these will work best in a given situation is questionable, and their application depends primarily on the philosophical outlook of the decision maker.

1. Equal probability criterion: This is also known as the Laplace criterion and assumes that all states of nature considered are equally likely (that is, $p_j = 1/m$). The decision matrix is then used just as in DMUR.

2. Maxim criterion: This represents a pessimistic outlook in that it prescribes the strategy which will maximize the minimum profit; that is, in each row of the decision matrix one finds the worst possible outcome and then chooses the strategy which will give the best of the worst outcomes. Conceptually analogous is the minimax criterion which minimizes the maximum possible cost of the decision.

3. Maximax criterion: The optimist will like this one since it chooses the strategy which will maximize the maximum possible profit (the best of the best outcomes in each row).

4. Hurwicz criterion: A coefficient of optimism is defined as $\alpha = 1$ corresponding to a complete optimist, and $\alpha = 0$ corresponding to a complete pessimist (α can vary between 0 and 1 depending on the decision maker's state of optimism). The maximum possible profit for each strategy is then multiplied by α and the minimum profit by $(1 - \alpha)$ to find the best expected value.

5. Regret criterion: Also known as the Savage principle, this criterion is used to construct a regret matrix in which each outcome entry represents a regret defined as the difference between the best possible outcome and the given outcome. The matrix is then used as in DMUR with expected regret as the decision-determining quantity.

The techniques mentioned above only give an outline of the basic concepts of decision theory. The details of defining and quantifying the entries which constitute the decision matrix are determined by the area of application of the decision theory.

Historical note. Since humans have existed, they have been faced with decisions. The earliest decision theorists were concerned with gambling decisions as advisers to gamblers in the French court. The theory of games developed from these beginnings. Until fairly recently the bulk of decision theory was concerned with prescriptive decision making and was developed by economists and mathematicians for specific applications in economics, business, applied statistics, law, medicine, politics, and so on. The earliest notions of utility are attributed to Bernoulli, and these notions have developed into the modern utility and value theory concepts. Behavioral scientists are increasingly concerned with descriptive decision theory and, in particular, with the problems of measuring and describing the concepts of utility. Ordinal ranking, metric ranking, bounded interval measures, and other techniques are being developed for this purpose. Operations research has strong roots in decision theory both conceptually and in its wide use of optimization, statistical decision theory, and other decision-making techniques. *See* GAME THEORY.

Although decision theory has found extensive use in the past in various government agencies, for war games, operations research in the Defense Department, and so on, there appears to be a major interest in providing a more rational basis for government decisions at all levels. Since resource allocation is a primary function of government agencies, cost-benefit analysis as a prescriptive technique of decision making has found increasingly sophisticated application in government decisions. It too has its concepts and techniques rooted in decision theory. Modern developments in information-processing techniques, data storage and retrieval, and high-speed computers have made it possible to automate the decision process for certain applications with vast amounts of operational and research data continuously updating the decision matrix and modifying strategies in response to new information. *See* DIGITAL COMPUTER; OPTIMIZATION.

[IGOR PAUL]

Bibliography: H. Chernoff and L. E. Moses, *Elementary Decision Theory*, 1959; P. C. Fishburn, *Mathematics of Decision Theory*, 1972; R. D. Luce and H. Raiffa, *Games and Decisions*, 1957; W. T. Morris, *Decision Analysis*, 1977; M. K. Starr and M. Zeleny (eds.), *Multiple Criteria Decision Making*, 1977.

Delay line

A transmission line (as nearly dissipationless as possible) or an electric network approximation of it which, if terminated in its characteristic impedance, will reproduce at its output a waveform applied to its input terminals with little distortion but at a time delayed by an amount dependent upon the electrical length of the line.

If a transmission line is dissipationless, which will be the case if its series resistance is zero and its shunt conductance is also zero, it will have a characteristic impedance Z_0, as shown in Eq. (1),

$$Z_0 = \sqrt{L/C} \tag{1}$$

where L is the series inductance and C the shunt capacitance per unit length of the line. *See* TRANSMISSION LINES.

The velocity v of propagation of a signal along the line is shown by Eq. (2).

$$v = \frac{1}{\sqrt{LC}} \qquad (2)$$

Therefore the time required for the pulse to propagate a distance x along the line is shown by Eq. (3). Such a line, terminated in its character-

$$T_d = x\sqrt{LC} \qquad (3)$$

istic impedance, is shown in Fig. 1; and output pulses reproduce input at a delayed time T_d.

The lumped-circuit approximation of the transmission line is shown in Fig. 2. If the inductance and capacitance per section are L_1 and C_1, then the total time delay is shown by Eq. (4), where n is the number of sections.

$$T_d = n\sqrt{L_1 C_1} \qquad (4)$$

If the delay line is not terminated in its characteristic impedance, there is multiple reflection back and forth along the line. For example, if the receiving end of the line is unterminated, as in Fig. 3, but the sending end is terminated in its characteristic impedance, the receiving-end reflection coefficient is positive, and a delayed pulse appears at the output, as given by Eqs. (3) and (4). In addition, a pulse of the same polarity appears at the input terminals, delayed by twice this amount (the time required for a pulse to travel to the end and back).

If the receiving end is terminated in a short circuit as shown in Fig. 4, the receiving-end reflection coefficient is negative, and no pulse appears at the output, although an inverted pulse will appear at the input terminals with the same time delay as before.

Various applications make use of the short circuit and open circuit of delay lines, including line-controlled pulse generators. *See* PULSE GENERATOR.

Delay lines are also used for establishing a time sequence for the occurrence of events. A delay line with a total length equal to the greatest time delay required in a system may be used as a basic element. Pulses occurring at intermediate times may be obtained from taps at various points along the line. A specific application is found in the synchronizing signal generator of the television system. Also the lumped-circuit delay line is an essential element of the wide-band distributed amplifier.

When a signal is digital in nature, or consists of a series of pulses, the series of pulses may be delayed by using a shift register, which might, for example, consist of a chain of cascaded type D flip-flops. If the register has n stages, the pulse

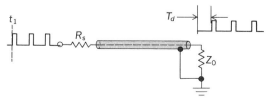

Fig. 1. Transmission line as delay line.

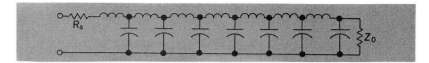

Fig. 2. Lumped-circuit delay line.

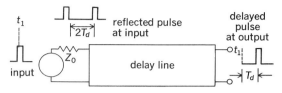

Fig. 3. Reflection due to unterminated receiving end.

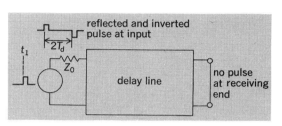

Fig. 4. Reflection due to short-circuited receiving end.

series will appear at the output delayed by a time $(n-1)T$, where T is the periodicity of the pulses. The same function can be realized by using an array of charge-coupled devices (CCDs).

[GLENN M. GLASFORD]

Bibliography: G. M. Glasford, *Fundamentals of Television Engineering*, 1955; G. M. Glasford, *Linear Analysis of Electronic Circuits*, 1965; J. Millman, *Microelectronics*, 1979; J. Millman and H. Taub, *Pulse Digital and Switching Waveforms*, 1965.

Demodulator

The stage in a radio, television, radar, or other receiver at which demodulation of the received signal takes place. Thus, in a tuned radio-frequency (rf) receiver the demodulator separates the audio-frequency (af) signal from the amplified incoming rf carrier signal. A demodulator is often called a detector. In a superheterodyne receiver the af signal is separated from the carrier signal at the second detector, because the converter or first detector merely serves to change the modulated rf carrier signal to a modulated intermediate-frequency carrier signal. In a frequency-modulation receiver the demodulator converts carrier frequency changes into corresponding audio signals. In a color television receiver the demodulator extracts the color difference signals from the incoming modulated carrier signal. *See* DETECTOR.

[JOHN MARKUS]

Detector

A device used to recover the original modulating signal from a modulated wave, also called a demodulator. In communications systems, and in some automatic control systems and data storage systems, the information to be transmitted is first impressed upon a periodic wave called the carrier; the carrier is then said to be modulated. After reception of the modulated carrier, the original modulating signal is recovered by the process of demodulation or detection. *See* MODULATION; MODULATOR.

The amplitude, frequency, or phase of a carrier may be changed in the modulation process. Therefore the process of detection and the practical circuits for accomplishing it differ in each case. Detection of an amplitude-modulated carrier is carried out by a rectifier, which produces pulsating currents, either the peak or average value of which are proportional to the amplitude of the desired modulating signal. Detection of frequency- and phase-modulated signals is usually accomplished by a discriminator or a phase-locked oscillator. The discriminator first converts the frequency- or phase-modulated carrier into an amplitude-modulated carrier; a rectifier then recovers the desired modulation. *See* AMPLITUDE MODULATOR; FREQUENCY MODULATOR; PHASE MODULATOR.

Detectors are also commonly used in various classes of measuring instruments to produce an indication of a signal or its relative magnitude. The indicating devices employed in impedance bridges or slotted lines illustrate such applications. In some cases, such as an electronic radio-frequency voltmeter, the carrier may not be modulated; the detector is then used to indicate the amplitude of the carrier.

Types. All detectors require use of nonlinear devices for recovering the desired signal from a modulated carrier. The simplest and most commonly used of these are semiconductor, or crystal, diodes. Thermionic, or vacuum, diodes were once popular but are seldom used in newly designed equipment because of their comparatively large size and the inconvenience of providing cathode heating power. The semiconductor diodes are useful from low-frequency high-power applications through microwave frequencies. The diode must be carefully selected for the application, however. Diodes which have very small internal capacitance, or fast diodes, must be used for higher radio frequencies or microwaves.

The nonlinear characteristics of transistors or multielement vacuum tubes are sometimes employed in special applications where simultaneous detection and amplification are most convenient. One such application is the regenerative or super-regenerative detector which is sometimes used as a simple high-gain narrow-band receiver.

Detection fidelity. In most applications it is important to consider the fidelity of the detection process. An ideal linear detector is one which faithfully reproduces the waveform of the amplitude variations of an amplitude-modulated carrier. A detector may fail to accomplish this function in one or more of the following ways. (1) Amplitude distortion results if the output of the detector contains frequencies not present in the modulation envelope. (2) Frequency distortion results if the various frequency components of the modulation envelope are not reproduced with the same relative amplitudes in the detector output. (3) Phase distortion results if the frequency components of

the modulation envelope are reproduced with altered phase relationships. These distortions apply particularly to amplitude-modulation detectors.

An ideal square-law detector is used in some applications. With it the amplitude of the output signal is proportional to the square of the carrier amplitude. This type of detector is required to recover the original modulating signal without distortion when single-sideband transmission is used. Any nonlinear device becomes a square-law detector if the proper combination of bias point, load resistance, and signal amplitude is used. Square-law detectors are also useful in measurement applications. *See* AMPLITUDE-MODULATION DETECTOR; FREQUENCY-MODULATION DETECTOR; PHASE-MODULATION DETECTOR.

[CHARLES L. ALLEY]

Bibliography: C. L. Alley and K. W. Atwood, *Electronic Engineering*, 3d ed., 1973; W. I. Orr, *Radio Handbook*, 21st ed., 1978.

Digital computer

Any device for performing mathematical calculations on numbers represented digitally; by extension, any device for manipulating symbols according to a detailed procedure or recipe. The class of digital computers includes microcomputers, conventional adding machines and calculators, digital controllers used for industrial processing and manufacturing operations, store-and-forward data communication equipment, and electronic data-processing systems. *See* CALCULATORS; COMPUTER.

In this article emphasis is on electronic stored-program digital computers. These machines store internally many thousands of numbers or other items of information, and control and execute complicated sequences of numerical calculations and other manipulations on this information in accordance with instructions also stored in the machine. The first section of this article discusses digital system fundamentals, reviewing the components and building blocks from which digital systems are constructed. The following section introduces the stored-program general-purpose computer in more detail and indicates the characteristics by which system performance is measured. The final section traces the history of stored-program digital computer systems and shows how the requirements of new applications and the development of new technologies have influenced system design.

DIGITAL SYSTEM FUNDAMENTALS

A digital system can be considered from many points of view. At the lowest level it is a network of wires and mechanical parts whose voltages and positions convey coded information. At another level it is a collection of logical elements, each of which embodies certain rules, but which in combination can carry out very complex functions. At a still higher level, a digital system is an arrangement of functional units or building blocks which read (input), write (output), store, and manipulate information.

Codes. Numbers are represented within a digital computer by means of circuits that distinguish various discrete electrical signals on wires inside the machine. Theoretically, a signal on a wire

Table 1. Counting from 0 to 19 by decimal and binary numbers

Decimal number	Binary number
00	00000
01	00001
02	00010
03	00011
04	00100
05	00101
06	00110
07	00111
08	01000
09	01001
10	01010
11	01011
12	01100
13	01101
14	01110
15	01111
16	10000
17	10001
18	10010
19	10011

could be made to represent any one of several different digits by means of the magnitude of the signal. (For example, a signal from 0 to 1 volt could represent the digit zero, a signal between 1 and 2 volts could represent the digit one, and so on up to a signal between 9 and 10 volts, the digit nine.) In practice, the most reliable and economical circuit elements distinguish between only two signal levels, so that a signal between 0 and 5 volts may represent the digit zero and a signal between 5 and 10 volts, the digit one. These two-valued signals make it necessary to represent numbers and symbols using a corresponding base-two or binary system. Table 1 lists the first 20 binary numbers and their decimal equivalents. For a detailed discussion of binary numbers *see* NUMBER SYSTEMS.

Data are stored and manipulated within a digital computer in units called words. The binary digits (called bits), which make up a word, may represent either a binary number or a collection of binary-coded alphanumeric characters. For example, the two-letter word "it" may be stored in a 16-bit computer word as follows, making use of the code shown in Table 2:

0100100101010100

The computer word merely contains a binary pattern of alternating 1's and 0's, and it is up to the

Table 2. American standard alphabetic code for binary representation of letters

11000001	A	11001110	N
11000010	B	01001111	O
01000011	C	11010000	P
11000100	D	01010001	Q
01000101	E	01010010	R
01000110	F	11010011	S
11000111	G	01010100	T
11001000	H	11010101	U
01001001	I	11010110	V
01001010	J	01010111	W
11001011	K	01011000	X
01001100	L	11011001	Y
11001101	M	11011010	Z

computer user to determine whether that word should be interpreted as the English word "it" or as the decimal number 18,772.

Logical circuit elements. Two kinds of logical circuits are used in the design and construction of digital computers: decision elements and memory elements. A typical decision element provides a binary output as a function of two or more binary inputs. The AND circuit, for example, has two inputs and an output which is 1 only when both inputs are 1. A memory element stores a single bit of information and is set to the 1 state or reset to the 0 state, depending on the signals on its input lines. And because such a circuit can be caused alternately to store 0's and 1's from time to time, a memory element is commonly called a flip-flop. *See* SWITCHING THEORY.

These two basic logical elements are all that are required to construct the most elaborate and complex digital arithmetic and control circuits. A simple example of such a circuit is shown in Fig. 1. Here the object is to perform a simple binary count, as shown in the table at the bottom of Fig. 1. As long as control signal C is equal to 1, the counting continues. When the control input is 0, the counter is to remain in whatever state it had last counted to. Two flip-flops are used, labeled $Q1$ and $Q2$, and will be made to count through the sequence 0,1,2,3,0,1, ... To understand the design, it is necessary to introduce one more concept, the complementary output of a flip-flop. Each flip-flop generally has two output wires, which are always of opposite polarity. When flip-flop $Q1$ is storing a 1, output $Q1$ is 1 and the complementary output (which is labeled $\overline{Q1}$ and pronounced $Q1$ bar) is 0. When the flip-flop contains a 0, the $\overline{Q1}$ output is 1 and the $Q1$ output is 0.

To analyze the circuit, note first that, when con-

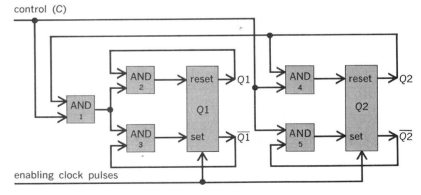

control (C)

enabling clock pulses

Binary count:		Logic equations:
Q1	Q2	$SQ1 = C \cdot Q2 \cdot \overline{Q1}$
0	0	
0	1	$RQ1 = C \cdot Q2 \cdot Q1$
1	0	
1	1	$SQ2 = C \cdot \overline{Q2}$
0	0	
0	1	$RQ2 = C \cdot Q2$
. . .	. . .	Note: $C \cdot Q2$ means C AND $Q2$

Fig. 1. Simple digital counting circuit.

trol input C is 0, the outputs of all AND gates are 0 and, because the reset and set inputs to both flip-flops are 0, the flip-flops will remain in whatever state they last reached. Now suppose that $Q1$ and $Q2$ both contain 0 and that the control input becomes 1. While flip-flop $Q2$ contains a 0, its $Q2$ output is also 0 and AND gate number 1 (labeled AND 1) is effectively turned off so that the reset and set inputs to $Q1$ are both 0. Thus flip-flop $Q1$ will remain in the 0 state. For the same reason AND gate 4 will also be turned off, and the reset input to flip-flop $Q2$ will be 0. However, from flip-flop $Q2$ complementary output $\overline{Q2}$ will be in the 1 state, and (while the control input is 1) AND gate 5 will be turned on and the set input to $Q2$ will be 1. Flip-flop $Q2$ will thus be turned on by the first clock pulse to occur after C is turned on; and from one clock pulse time to the next the two flip-flops will change from the $(0,0)$ state to the $(0,1)$ state. A careful review of the indicated circuits will show that the counter will indeed go through the count sequence as shown, as long as the control input is 1. The logic equations in Fig. 1 represent another way of describing the circuit and may be used in place of the more cumbersome diagram. *See* DIGITAL COUNTER.

Physical components. The logical elements described in the paragraphs above are the fundamental conceptual components used in virtually all digital systems. The actual physical components which were used to realize conceptual gates and flip-flops in some specific piece of equipment are dependent on the status of electronic technology at the time the equipment was designed. In the 1950s the earliest commercial computers used vacuum tubes, resistors, and capacitors as components. A flip-flop typically required a dozen or more such components in these first-generation computers. Between the late 1950s and middle 1960s, solid-state transistors and diodes replaced the vacuum tubes, and the resulting second-generation systems were considerably more reliable than their first-generation predecessors; they were also smaller and consumed less power. But the number of electronic components per conceptual logical component remained about the same—a dozen or more for a flip-flop.

Since the mid-1960s the integrated circuit (IC) has been the principal logical building block for digital systems. Early digital integrated circuits contained a single flip-flop or gate, and the use of these components permitted designers of the early third-generation systems to provide much more capability per component than was possible with the first- or second-generation technology. Since the mid-1960s integrated circuit technology has consistently improved, and by 1980 typical large-scale-integration (LSI) integrated circuits contained thousands of flip-flops and gates. *See* INTEGRATED CIRCUITS.

System building blocks. On a completely different conceptual level, a digital computer can be regarded as being composed of functional, system building blocks, containing (among other things) subassemblies of the fundamental logical components. A computer viewed at this level may be described in an oversimplified fashion by the diagram of Fig. 2. The computation and control block (often called the central processing unit, or CPU)

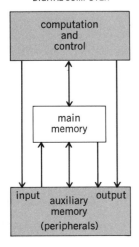

DIGITAL COMPUTER

Fig. 2. Block diagram of a digital computer.

is constructed entirely of logical elements of the kind described above. The main memory, which may store from a few thousand to several million binary digits, and the input/output and auxiliary memory devices (the so-called peripheral equipment) are specialized devices available over a range of speeds and operating characteristics.

Main memory is a building-block capable of storing data or instructions in bulk for use by the computation and control portion of the computer. The important characteristics of a memory are capacity, access time, and cost. Capacity is the amount of data that the computer can store. Access time is the maximum interval between a request to the memory for data and the moment when the memory can provide that data. Cost is measured for each bit stored. For first-generation systems, designers used a variety of technologies in realizing main memory: mercury delay lines, electrostatic storage tubes, and magnetic drums all appeared in various products. But second- and third-generation systems were almost exclusively built using magnetic core main memories. Starting in the early 1970s, the integrated circuit memory was introduced, and is now the most widely used technology. *See* COMPUTER STORAGE TECHNOLOGY.

Input/output and auxiliary memory peripherals represent the other major computer building blocks. Equipment is now and has from the beginning been available for feeding information to the computer from paper tape and punched cards, and for receiving data from the computer and printing it, or punching it on tape or cards. But in the intervening years, designers have provided additional output devices which record computer data on microfilm, or plot data on graphs, or use data to control physical devices such as valves or rheostats. They have also designed input equipment which feeds the computer data from laboratory instruments, and from devices which scan documents and "read" printed characters. Data can be transmitted to and from the computer over ordinary telephone lines, and a wide variety of devices generally called terminals, make it possible for people to send data to, or receive requested data from, a computer system located hundreds or thousands of miles away. *See* CHARACTER RECOGNITION; COMPUTER GRAPHICS; DATA COMMUNICATIONS.

The earliest auxiliary memory equipment recorded data on reels of magnetic tape. Magnetic tape units are still very widely used, for although they are slow in comparison to the operating speeds of modern computers—it typically takes 2–30 min to read all the data on a 2400-ft (732-m) reel of tape, depending on the speed of the tape unit—they make it possible to store large volumes of data at low cost by virtue of the low cost of the tape itself. The other widely used auxiliary memory devices are the magnetic disk and drum, both of which provide faster access to data than do the tape units, but at higher cost per bit of data stored, *See* DATA-PROCESSING SYSTEMS.

STORED PROGRAM COMPUTER

Components and building blocks described in the preceding paragraphs could be organized in a multitude of different ways. The first practical electronic computers, constructed during the latter part of World War II, were designed with the specific purpose of computing special mathematical functions. They did their jobs very well, but even while they were under construction, engineers and scientists had come to realize that it was possible to organize a digital computer in such a way that it was not oriented toward some particular computation, and could in fact carry out any calculation desired and defined by the user. The basic machine organization invented and constructed at that time was the stored-program computer, and it continues to be the fundamental basis for each of the hundreds of thousands of computing systems in use today. It has also become a system component, since the microcomputer is simply a stored program computer on a single integrated-circuit chip.

The concept of the stored-program computer is simple and can be described with reference to Fig. 2. Main memory contains, in addition to data and the results of intermediate computations, a set of instructions (or orders, or commands, as they are sometimes called); these specify how the computer is to operate in solving some particular problem. The computation and control section reads these instructions from the memory one by one and performs the indicated operations on the specified data. The instructions can control the reading of data from input or auxiliary memory peripherals, and (when the prescribed computations are completed) can send the result to auxiliary memory, or to output devices where it may be printed, punched, displayed, plotted, and so forth. The feature that gives this form of computer organization its great power is the ease with which instructions can be changed; the particular calculations carried out by the computer are determined entirely by a sequence of instructions stored in the computer's memory; that sequence can be altered completely by simply reading a new set of instructions into the memory through the computer input equipment.

Instructions. To understand better the nature of the stored-program computer, consider in more detail the kinds of of instructions it can carry out and the logic of the computation and control unit which interprets and implements the instructions. Because the instructions, like the data, are stored in computer words, one begins by examining how an instruction is stored in a word. As an example, assume one is looking at a small computer with words 16 bits long, and assume further that an instruction is organized as shown in Fig. 3. In this simple computer an instruction has two parts: the first 5 bits of the word specify which of the computer's repertoire of commands is to be carried out, and the last 11 bits generally specify the address of the word referred to by the command. A 5-bit command permits up to 32 different kinds of instructions in the computer, and an 11-bit address permits one to address up to 2048 different memory locations directly.

Typical instruction types for a computer of this kind are listed below.

Load. Load the number from the prescribed memory location into the arithmetic unit.

Store. Store the number from the arithmetic unit in the memory at the prescribed memory location.

Add. Add the contents of the addressed memory location to the number in the arithmetic unit, leaving the result in the arithmetic unit.

Subtract. Subtract the contents of the addressed memory location from the number in the arithmetic unit, leaving the result in the arithmetic unit.

Branch. If the number in the arithmetic unit is zero or positive, read the next instruction from the address in the next-instruction register as usual. If the number in the arithmetic unit is negative, store the address from the branch instruction itself in the next-instruction register, so that the next instruction carried out will be retrieved from the address given in the branch instruction.

Halt. Stop; carry out no further instructions until the operator presses the RUN switch on the console.

Input. Read the next character from the paper tape reader into the addressed memory location and then move the tape so a new character is ready to be read.

Output. Type out the character whose code is stored in the right-hand half of the addressed memory location.

With the exception of the branch command, the preceding instructions are easy to interpret and to understand. The load and store commands move data to and from the arithmetic unit, respectively. The add and subtract commands perform arithmetic operations, each using the number previously left in the arithmetic unit as one operand, and a number read from a designated memory location as the other. The halt command simply tells the computer to stop and requires intervention by the operator to make the computer initiate computation again. The input and output commands make possible the reading of information into the computer memory from a paper tape input device, and the printing out of the results from previous computations on an output typewriter.

To understand the branch command, consider how the computation and control unit of Fig. 4 uses the instructions in the memory. To begin with, the instructions which are to be carried out must be stored in consecutive storage locations in memory. Assume that the first of a sequence of commands is in memory location 100. Then the "next-instruction address register" in the computation and control unit (Fig. 4) contains the number 100, and the following sequence of four events takes place: (1) read, (2) readdress, (3) execute, and (4) resume.

Read. The control logic reads the next instruction to be carried out from the memory location whose address is given by the next-instruction address register. The instruction coming from memory is stored in another register called the instruction register. (In this example the next-instruction address register started out containing the number 100, and so the instruction in memory location 100 is transferred to the instruction register.)

Readdress. The control logic now adds unity to

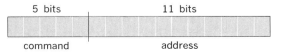

Fig. 3. Sixteen-bit instruction.

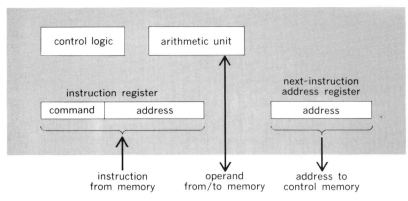

Fig. 4. Computation and control unit.

the number in the next instruction address register. (In the present example this changes the number in the next instruction address register from 100 to 101. The result is that, when the computer has interpreted and carried out the instruction from location 100, following the rules given in the third and fourth steps below, it will return to the read step above and next interpret and carry out the instruction from location 101.)

Execute. The instruction from location 100 is now in the instruction register and must be carried out. The control logic first looks at the command portion of the instruction in the left-hand 5 bits of the register and interprets or decodes it to determine what to do next. If the instruction is add, subtract, load, or output, the control logic first uses the address in the instruction register—the address associated with the command—and reads the word from that addressed location in memory; it then proceeds to load the word into the arithmetic unit, add it to or subtract it from the number in the arithmetic unit, or transfer it to the output typewriter, depending on the command. If the command is store or input, the control logic collects a number from the arithmetic unit or the paper tape reader and then transfers that number to a location in the memory whose address is given in the instruction register. If the command is halt, the control logic simply prevents all further operations, pending a signal from the operator console.

If the command is branch, the control logic begins by looking at the number in the arithmetic unit. If that number is zero or positive, the control logic goes on to the fourth step below. If the number is negative, however, the control logic causes the address in the instruction register to be transferred to the next-instruction address register before going on to the fourth step. The computer will then continue with one sequence of commands if the previous arithmetic result was positive, and with another if the result was negative. This seemingly simple operation is one of the most important features a computer possesses. It gives the computer a decision-making capability that permits it to examine some data, compute a result, and continue with one of two sequences of calculations or operations, depending only on the computed result.

Resume. As the fourth and final step in the sequence, when the command has been interpreted

and carried out properly, the control logic returns to the read step and repeats the entire series of steps.

A sequence of instructions intended to carry out some desired function is called a program; collections of such programs are called software (as distinguished from the equipment, or hardware), and the act of preparing such programs is called programming. Because a computer can perform no useful function until someone has written a program embodying that function, the programming activity is an exceedingly important one and provides a basic limitation to the facility with which the computer can be applied to new areas. *See* DIGITAL COMPUTER PROGRAMMING.

Computer characteristics. A computer installation is complex. Consequently it is difficult to describe a system or to compare the characteristics of two systems without listing their instruction types and describing their modes of operation at some length. Nevertheless, certain important descriptors are commonly used for comparison purposes and are shown in Table 3, where salient characteristics of two typical systems are shown. Definitions of these characteristics can be stated as follows.

Memory cycle time is the time required to read a word from main memory. Most modern computers have integrated-circuit memories with cycle times in the ranges shown in Table 3. Add time is the time required to perform an addition, including the time necessary to extract the addition instruction itself and the operand from memory. Main memory storage capacity is the number of words of storage available to the computation and control unit. Typically, a computer manufacturer gives the buyer some choice; the buyer can purchase enough memory to meet the needs of the expected application. This internal capacity refers to the high-speed internal storage only, and does not include disks, drums, or magnetic tape.

Word length is the number of bits in a computer word.

System cost may vary over a range of 5 or even 10 to 1 for a particular computer because of the great variety of options offered the buyer by the manufacturer—options such as memory size, special instructions for efficiency in certain calculations, and number and type of peripheral devices.

There are obviously a number of other measures which may be used to describe a computer. They include such characteristics as multiplication time, transfer rates between input/output equipment and memory, physical size, power consumption, and the availability of a variety of computing options and special features.

Table 3. Typical computer characteristics

| Characteristics | Typical systems | |
	Large	Small
Memory cycle time	$0.08\ \mu s$	$0.90\ \mu s$
Add time	$0.17\ \mu s$	$2.84\ \mu s$
Main memory storage capacity	750,000 words	128,000 words
Word length	64 bits	16 bits

EVOLUTION OF CAPABILITIES

The process by which new circuit and peripheral equipment technologies led to the development of a series of generations of computers was discussed above. But simultaneous with the changes in technology, there came changes in the structure or architecture of computers. These changes were introduced to improve the capability and efficiency of systems, as designers came to understand how computers were actually used.

Computer efficiency. One way of looking at system efficiency is indicated in Fig. 5, where the operation of a computer is shown broken down into the following four parts. (1) Operator time includes such activities as inserting cards into a card reader, loading magnetic tapes onto a tape unit, setting up controls on a computer operator's panel, and reviewing printed results. (2) Input comes to the computer from peripheral devices or from auxiliary memory. The inputs include instructions from the operator, inputs of programs to be run, and inputs of data. (3) Computation, being the principal activity, should occupy relatively much of the total time. (4) Output includes storage of intermediate and final results in auxiliary memory, and printing of results along with instructions or warnings to the computer operator.

In the first generation of computer equipment only one of these activities could be carried out at a time. Between jobs the computer was idle while an operator made ready for the next task. When the operator was ready, the program was read into the computer from some input device and the input data were then loaded. The program operated upon the data and performed necessary calculations. When the calculations were complete, the computer printed out answers, and the operator took steps to set up the next problem.

This series of operations was inefficient, and the designers of second-generation equipment removed some of the inefficiency by arranging input and output operations to be performed directly between the input/output peripherals and the computer memory without interfering with computations. As a result, second-generation computers were able to perform computations while reading in data and printing out replies, and efficiency was greatly enhanced. Figure 6 indicates schematically the organizational change between generations of computers.

First-generation equipment was most efficient while performing tedious and lengthy computations. The input/output capabilities of the second generation made them useful in applications where large volumes of data had to be handled with relatively little computation—applications such as billing, payroll, and inventory control. At the same time, the great capability and increased reliability of second-generation systems encouraged engineers to apply them to situations where the computer acts as a control element. In military aircraft, in oil refineries and chemical plants, in research laboratories, and in factories, the compu-

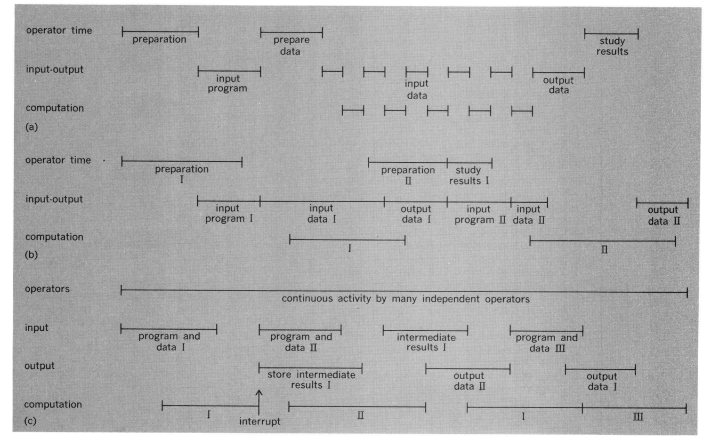

Fig. 5.   Comparison of efficiency for three generations of computers: (a) first, (b) second, and (c) third.

ter received data directly from measuring instruments, performed appropriate calculations, and as a result made adjustments in the aircraft engine thrust, the flow of raw materials in the plant or factory, or the experimental setup in the laboratory. These new application areas led to two important developments in computer design. The first was a new set of input/output equipments that could be connected to process instruments, converting instrument signals into digital quantities and back again. *See* ANALOG-TO-DIGITAL CONVERTER; DIGITAL-TO-ANALOG CONVERTER.

The second development, which evolved from the use of the computer in second-generation control applications, was the interrupt. The processes or activities under control provided data to, and required action from, the computer at random times. These random requests to the computer required rapid response, either because the process required quick control action on the part of the computer or because the process data supplied by instruments were rapidly changing and had to be stored before the data were lost. The interrupt feature, built into the computation and control unit, solved these problems by providing circuits that could stop the computer after any given instruction, enable it to carry out some special program in response to the external interrupt, and finally permit it to return to precisely the position in the program where it had originally been interrupted.

The logic to achieve these ends is quite simple. When an interrupt signal is received from an external device, the logic circuitry waits until the computer has completed the current instruction and then stores the next instruction address in a specially reserved storage location. The logic substitutes a standard predetermined address in the next instruction address register and continues. The computer of course next executes the instruction at the standard address, and the rest is up to the programmer. The programmer must have inserted a special program at this standard address, and the program must respond to the conditions that caused the interrupt (by inputting data from instrumentation or by taking previously specified control action, for example) and then must return control to the original program at exactly the same place where the program was interrupted. Though the interrupt was originally used largely in control applications, it is now employed in virtually all systems to notify the computer when transfers to and from peripheral equipment are completed.

Third-generation computer. In the mid-1960s a new set of trends in computer applications was becoming apparent, and a third generation of computer systems became available to cope with those trends. The usefulness and flexibility of the stored-program computer, together with its improved cost-performance ratio, made it apparent that the computer had the basic power to perform a great variety of small and large tasks simultaneously. For example, the speed and capability of a computer were such that it could simultaneously collect data from a test run or experiment; maintain a file of inventory records on a disk memory; answer inquiries on status of specific items in inventory, such as inquiries entered at random from a dozen different cathode-ray-tube display devices; and assemble or compile programs for

users at numerous terminals, all remote from the computer and all working independently on different problems.

A computer system which serves a number of users in this way is called a time-sharing system. To perform in such complex applications, third-generation computers required elaborations of the features found in second-generation computers (Fig. 6). Main and auxiliary memories became bigger and cheaper, more input/output equipments became available, input/output channels improved so that a larger number of simultaneous operations were possible, and interrupt structures grew more flexible. In addition, many new features appeared. Three of them, memory protection circuitry, rapid context switching, and the operating system, are worth discussing briefly.

To understand the implications of third-generation computers and the usefulness of these new features, consider the last portion of Fig. 5. Here

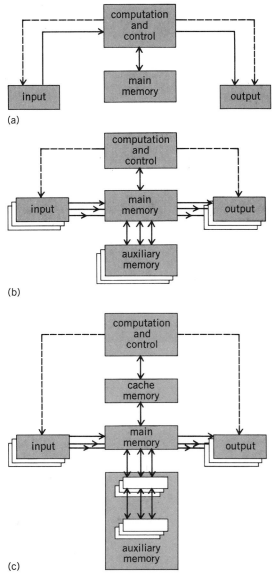

Fig. 6. Evolution of organization for four generations of computers: (*a*) first; (*b*) second and third; (*c*) fourth.

the computer is engaged in a variety of different tasks, and it switches back and forth between them as it finishes one portion of a job or as it is interrupted to perform some higher-priority job. This rapid switching from job to job led to the development of context-switching equipment. At the time of a changeover from one program to another, the programmer is able with a single instruction to interchange the instruction address register, together with the contents of various arithmetic and control registers, between the job he or she had previously been working on and the new job to be performed. Second-generation computers, without this context-switching feature, require a long sequence of commands every time a change is made from one task to another.

The memory protect feature is important for other reasons. Some programs executed in a time-sharing system such as that depicted by the third-generation portion of Fig. 6 would be new programs being run for the first time. Errors are common in such new programs, and it is the nature of the computer that such errors could have serious effects on system operation. For example, a user's program might accidentally store data in memory space reserved for supervisory or monitor programs—the programs that determine job priorities and reconcile conflicting input/output requirements. To keep such critical memory areas from being destroyed or modified by unchecked programs, designers have made it possible for the user to designate certain areas of memory as protected, and have ensured that only the monitor or supervisory programs can access these particular areas.

The operating system is the set of supervisory programs which manages the system, keeps it operating efficiently, and takes over many of the scheduling and monitoring functions which had previously been the function of the computer operator. It is typically supplied by the computer hardware manufacturer, and was first used with some second-generation systems, where it included little more than interrupt processing and error-handling routines, together with programs to control input/output operations for the user. Its size and importance have both grown with time, and it currently includes those early functions, as well as facilities for accessing compilers and utilities, scheduling jobs to maximize system throughput, managing data-communication facilities, controlling user access to protected data, maintaining records on system usage, and so forth.

Fourth-generation systems. Computer memory has become increasingly important with the passing generations, as users have found it useful to store more and more business, government, engineering, and scientific data in machine-readable form. Starting with the third generation, but increasingly with computers introduced in the early 1970s, computer-system architects have provided a hierarchy of memory devices to improve system performance and give the user access to very large memory capacities. Figure 6c shows how a very large system may make use of such a hierarchy. A so-called cache memory has been interposed between the central processing unit and the main memory, and auxiliary memory has been split into two parts.

The cache is a relatively small but very-high-speed memory which stores the data and programs currently being used by the central processing unit. When the central processing unit needs an instruction or data, it sends a main memory address to the cache memory. If the requested information is stored in the cache, it is immediately delivered to the central processing unit; if it is not in the cache, special cache memory hardware requests a block of information from main memory, and that block, which includes the data requested by the central processing unit, is delivered to the cache and displaces another block stored there. Because programs contain many branches, a typical program may repeatedly access a relatively small portion of the main memory, and so a small fast cache can be very effective in increasing the central processing unit's instruction-processing rate. (The cache's great speed makes it correspondingly expensive, and therefore not economical for use as main memory.)

A typical large fourth-generation system may make use of a similar hierarchical arrangement between main and auxiliary memory. Main memory in a large system may contain 750,000 words. But a system equipped with a virtual storage translator can give the user-programmer the illusion of working with 16,000,000 words. This is done by supplying the 750,000 words of real memory with temporary virtual addresses, and transferring data automatically from auxiliary to main storage as it is needed. For example, suppose the central processing unit needs a program which starts at address 2,500,000. The virtual storage translator may assign the virtual memory address 2,500,000 to the real address 700,000, and then transfer a block of (say) 1000 words from auxiliary memory to the main memory locations starting with 700,000. Now when the central processing unit next requests, for example, information from location 2,500,001, the translator locates it in real memory location 700,001, and sends it off to the cache.

Finally, in some very large systems there may also be an auxiliary memory hierarchy, in which a very large, cheap, but relatively slow auxiliary memory delivers data or programs as required to the primary auxiliary memory.

In all these hierarchical arrangements, the guiding principle is that the larger, slower, cheaper memory supplies data as needed to the smaller, faster, more expensive memory. The various levels of memory are invisible to the user; the hardware and, where necessary, the operating system make the various hierarchical levels deliver data and programs to the user without the necessity for any special action on his or her part.

Industry growth. The versatility of the digital computer has led to its application in a wide variety of industries and activities. The evolution of the integrated circuit has made it possible for designers to provide ever cheaper systems, and as a result the computer has become economical for use even in small organizations. Since they are so numerous, the number of digital computers in use has increased remarkably. The development of the microcomputer—the stored-program computer on an integrated-circuit chip—has made possible the introduction of the personal computer, which can supply entertainment, educational,

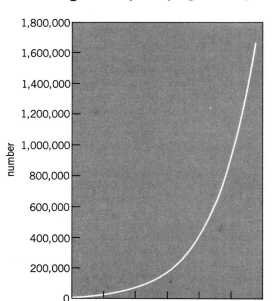

Fig. 7. Computers in use worldwide, made by United States companies.

record-keeping, and computing capabilities in the home at a relatively low cost. The net result of these developments is indicated in Fig. 7, which shows the growth in the number of computers in use each year, and forecasts how that growth will continue. *See* COMPUTER.

The influence of the digital computer is extending far beyond the realm of computing. The microcomputer is being used, and will increasingly be used, as a component in a variety of apparently noncomputing applications. Already it is used in games and toys. Increasingly it will find use in automobiles, appliances, tools, and many other artifacts. *See* MICROCOMPUTER; MICROPROCESSOR.

[MONTGOMERY PHISTER, JR.]

Bibliography: C. G. Bell and A. Newell, *Computer Structures: Readings and Examples*, 1971; M. Phister, Jr., *Data Processing Technology and Economics*, 1979; Special issue on computer architecture, *Commun. ACM*, vol. 21, no. 1, January 1978.

Digital computer programming

The art of writing instructions to control the operation of a stored-program digital computer. More generally, programming refers to the analysis and planning of a problem solution, in which the phase of instruction writing is referred to as coding.

Binary operation. The programming process will be described in terms of the most common form of digital computer. This machine operates internally entirely in binary (base 2) arithmetic and logic, and is arranged so that information is stored and accessed in units termed words. A typical word size is 32 binary digits, or bits, although word sizes range in current practice from 16 to 60 bits. *See* INFORMATION THEORY.

Decimal operation. The other broad class of computers operates logically in the decimal system (although decimal digits are fabricated within the machine by combinations of bits) and stores and accesses its information in units of individual characters such as digits or letters. This machine is referred to as a character-addressable or variable-word-length computer.

Notation. The binary, word-oriented machines (now the dominant type) call for a notation that is either pure binary, or octal (base 8) or hexadecimal (base 16), the latter two being simply conveniences for programmers while they work. The discussion here is in terms of decimal, to simplify understanding of computer principles and to avoid lengthy digression to number systems other than decimal, such as binary, but it must be kept in mind that the basic nature of the machines is binary. *See* COMPUTER.

Number operations. The purpose of a computer is to manipulate information that is stored within the machine in the form of binary numbers. These numbers can be treated as symbols and can represent alphabetic information or anything else the programmer wishes to manipulate. In the simplest case they are numbers and are to be treated arithmetically. Each word contains one number and its associated algebraic sign. The instructions which dictate the operations to be performed on such numbers (data) are also numbers and are also stored one per word in the same physical medium as the data. The stored programming concept presumes that there is no physical difference between these two types of numbers and that the instruction numbers may also be manipulated, in the proper context, as data.

Addresses. Each word in storage is assigned an address in order to refer to it, in a manner analogous to postal addresses. The addressing scheme is part of the hardware of the machine, and is wired in permanently. Word addresses range from zero to the storage size of the machine, typically from 00000 through 32,767. Any word may be loaded with any desired information. Some of this information is the data of the problem being processed; the rest is the instructions to be executed. The machine is designed basically to execute instructions in the order in which they are stored, advancing sequentially through the instruction words at addresses that increase by one.

Instruction format. Each instruction contains two basic parts: (1) a coded number that dictates what operation is to be performed, for example, ADD, MULTIPLY, and STORE; and (2) the address of the word (data) on which to perform that operation. Thus, suppose the data stored in words 12345, 12346, and 12347 are to be added together. The work is performed in a storage device called the accumulator, which is the same size as one storage word. Three operation codes are involved; namely, LOAD ACCUMULATOR, ADD, and STORE ACCUMULATOR, for which the corresponding operation codes might be 100, 123, and 234. The instructions must also reside in storage, say at addresses (called locations for instructions) 01017, 01018, 01019, and 01020. Then, the program shown in Fig. 1 applies.

All the data and instruction numbers are in storage before execution begins. The logical sequence is controlled by a counting device called the instruction counter, which is set, for this program, to 01017. Line 1, when executed, causes the contents of the first data word to be moved to the accumula-

Location	Operation	Address	Line no.
01017	100	12345	1
01018	123	12346	2
01019	123	12347	3
01020	234	12348	4

Fig. 1. A straight-line program. Line numbers are for reference with comments made in the text.

tor, destroying the information previously stored there. During this operation the instruction counter advances by one, thus completing one cycle of operation, and the process repeats. Line 2 calls for the information in word 12346 to be added to the contents of the accumulator, and line 3 does the same for the information stored at 12347. Each of these operations involves moving stored information; all such moves follow the pattern that read-in (for example, to the accumulator) is destructive, but read-out (for example, from a referenced storage word) is nondestructive. Thus, after the additions, the data words are still in storage in their original form. Line 4 is the logical inverse of line 1; the information now in the accumulator is moved to word 12348 (and still remains in the accumulator).

Each computer type has a wired-in set of operation codes. Inexpensive machines may offer as few as 16; large machines have several hundred.

Programming. The program of Fig. 1 is a straight-line program; that is, its instructions are executed strictly in sequence as stored. The stored-program computer gains its power from two additional concepts. The first concept is derived naturally in that the words of information referenced by an instruction can be other instruction words; thus, a computer can manipulate its own instructions and alter them dynamically, particularly in their address portions, during the execution of a program. The second concept concerns the ability to branch, based on the condition of the accumulator at any given moment. Basically, the machine can interrogate the accumulator and, on the basis of its contents being negative, zero, or positive, call for operating on the instruction counter so that the next executed instruction is not at the location that is one more than the last executed location, but is any other desired location. This branch of control may also be called for arbitrarily, that is, not based on the condition of the accumulator. For the available branch instructions the address portion is then the location of the next

instruction rather than that of a data word. The operation codes are of the form BRANCH, BRANCH ON MINUS, BRANCH ON ZERO, BRANCH ON NONZERO, and so forth.

The program of Fig. 1 would serve to form the sum of three words of information, but would form a poor pattern if there were 10,000 words to be added. In the latter situation there would be 10,001 instruction words, all but two of which would be monotonously similar, differing only in the regular progression of 10,000 addresses. The capabilities of the computer allow for the larger task to be looped; that is, a few instructions can be used over and over.

Flow-charting. The logic of any problem situation can be expressed graphically in a flow chart. For the problem of summing the contents of 10,000 words, the flow chart of Fig. 2 applies. Instructions are written to add the contents of one word (call its address X) to the word SUM. Prior to executing those instructions, it is arranged to set the contents of SUM to zero, and the value of address X to that of the first of the 10,000 words of data. After the contents of any word of data are added to SUM, a check is made to determine whether the word just added was the last word of data. If so, the repetitive task has been completed. If not, the address X is incremented by one, and the process continues. Programmed loops thus follow the pattern of initialization, performance of one case, a test for the last case, modification from case to case, and branch back to the performance block.

The program of Fig. 3 implements, in the same hypothetical machine code, the loop shown. Following is a list of the operation codes used:

  100 Load accumulator
  234 Store accumulator
  235 Store the address portion of
       the accumulator
  123 Add to the accumulator
  122 Subtract from the accumulator
  002 Branch if accumulator is zero
  001 Branch unconditionally

The distinct stages of the loop are marked off. The performance block is at locations 01004–01006. Line 5 calls for the word designated for the SUM to be loaded into the accumulator. The contents of one word, X, is added to it. The address of X is shown as zero, but will never be executed as such. The result of the addition is stored back at SUM (word 20001).

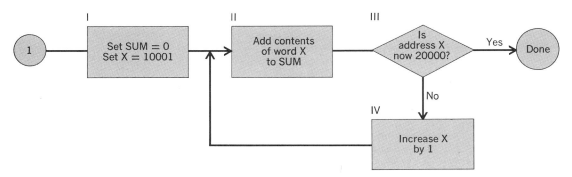

Fig. 2. Flow chart for the program of Fig. 3.

Data and constants:

Address	Value
10001–20000	data words
20001	SUM
20002	000 00000
20003	000 10001
20004	123 20000
20005	000 00001

Instructions:

Location	Operation	Address	Line no.
01000	100	20002	1
01001	234	20001	2
01002	100	20003	3
01003	235	01005	4
01004	100	20001	5
01005	123	(00000)	6
01006	234	20001	7
01007	100	01005	8
01008	122	20004	9
01009	002	01014	10
01010	100	01005	11
01011	123	20005	12
01012	234	01005	13
01013	001	01004	14
01014	Continue with the problem		

Fig. 3. A programmed loop. Line numbers are for reference with comments made in the text.

The contents of word 20001 must be set to zero; this is done by lines 1 and 2. Lines 3 and 4 ensure that x is preset to 10001. Lines 8–10 examine the instruction containing x (that is, all of the word at 01005) and effect an exit from the loop when the value of x is 20000. If the exit conditions are not met (as they will not be for 9999 traverses of the loop), the branch to 01014 indicated on line 10 is not taken and the modification steps of lines 11–13 are executed, whose function is to increase the value of x by one. The four sections ruled off in Fig. 2 correspond directly to the boxes of the flow chart for Fig. 3; line 14 performs the unconditional branch back to block II.

Programming languages. Programming for a digital computer could be done as illustrated in Fig. 3, but is far more awkward in binary and, moreover, becomes increasingly difficult as programs get longer. For these and other reasons, programmers prefer to work in languages that are at a higher level than the machine language shown. The use of mnemonic operation codes (BZE for BRANCH ON ZERO, in place of the number 002) and the replacement of all absolute machine addresses by symbols greatly speed up the work of the programmer at the modest cost of an extra computer run whose sole function is the translation back to the language used in Fig. 3. The translating program is called an assembler, and the language used is assembly language. Generally, assembly language follows the format of the machine language and is one-for-one with it; that is, for every instruction written in assembly language, one machine language instruction will be produced.

Most programming is done in the language of compilers, for example, FORTRAN or COBOL. Compiler language permits a format that fits the problems rather than the machine, and is many-to-one.

For example, a statement (instruction) in FORTRAN can be written

$$Y = B ** 2 - 4.0 * A * C$$

to correspond to the algebraic expression $b^2 - 4ac$ (an asterisk being used for multiplication, and two adjacent asterisks denoting the raising to powers). As with assemblers, a separate machine run is needed to translate from the compiler language to the language of the machine; the translating program is called a compiler. The better compilers achieve a measure of machine independence in that programs in compiler language can be compiled into other machine languages without alteration.

The looping technique illustrated in Fig. 3 is one of the two basic building blocks of the programmer. The other is the closed subroutine. Both techniques have the purpose of allowing a small number of instructions to be executed many times. The typical example of the closed subroutine is the short program for the calculation of square root. Such a program might consist of 30 instructions and be needed at eight different places within a large program. The set of 30 instructions could be inserted eight times into the main program, but this would waste the programmer's time and storage space in the machine. The 30 instructions can be stored once, and simple arrangements can be made in the main program to cause their linkage and return at each of the eight points at which the square-root operation is needed.

The concept of the subroutine leads to systematic libraries of tested subroutines that build up for each machine type. Further extensions of this idea lead to whole packaged programs furnished by the manufacturers, to proprietary programs that are offered for sale, and to higher languages designed for specific problem areas.

While the cost of nearly everything else in computing (central processors, storage devices, and input/output devices) continues to fall over the years, the cost of software (that is, computer programs) has risen. This is due in part to the greater complexity of the problems being attacked, and in part to the complexity of a computer system, which is the entity made up of the computer itself plus large control programs called operating systems. Attention has recently turned to methods and techniques, such as structured programming, for improving this situation. *See* PROGRAMMING LANGUAGES. [FRED J. GRUENBERGER]

Bibliography: T. C. Bartee, *Basic Computer Programming*, 1981; W. S. Brainerd et al., *Introduction to Computer Programming*, 1979; W. Collins, *Introduction to Computer Programming with PASCAL*, 1983; J. Reynolds, *Craft of Programming*, 1981.

Digital control

The use of digital or discrete technology to maintain conditions in operating systems as close as possible to desired values despite changes in the operating environment. Traditionally, control systems have utilized analog components, that is, controllers which generate time-continuous outputs (volts, pressure, and so forth) to manipulate process inputs and which operate on continuous signals from instrumentation measuring process variables (position, temperature, and so forth). In

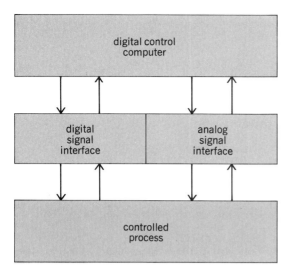

Fig. 1. Generalized digital computer control system.

the 1970s, the use of discrete or logical control elements, such as fluidic components, and the use of programmable logic controllers to automate machining, manufacturing, and production facilities became widespread. In parallel with these developments has been the accelerating use of digital computers in industrial and commercial applications areas, both for logic-level control and for replacing analog control systems. The development of inexpensive mini- and microcomputers with arithmetic and logical capability orders of magnitude beyond that obtainable with analog and discrete digital control elements has resulted in the rapid substitution of conventional control systems by digital computer–based ones. With the introduction of microcomputer-based control systems into major consumer products areas (such as automobiles and video and audio electronics), it is clear that the digital computer will be widely used to control objects ranging from small, personal appliances and games up to large, commercial manufacturing and production facilities. Hence the focus of this discussion will be on computer-based control systems. *See* MICROCOMPUTER.

Computer/process interface. The object that is controlled is usually called a device or, more inclusively, process. A characteristic of any digital control system is the need for a process interface to mate the digital computer and process, to permit them to pass information back and forth (Fig. 1).

Digital control information. Measurements of the state of the process often are obtained naturally as one of two switch states; for example, a part to be machined is in position (or not), or a temperature is above (or below) the desired temperature. Control signals sent to the process often are expressed as one of two states as well; for example, a motor is turned on (or off), or a valve is opened (or closed). Such binary information can be communicated naturally to and from the computer, where it is manipulated in binary form. For this reason the binary or digital computer/process interface usually is quite simple: a set of signal-conditioning circuits for each measured or controlled signal and a set of registers to transfer the bits of digital information in each direction. Each

register usually would contain the same number of bits as would be manipulated and stored within the digital computer.

Analog control information. Process information also must be dealt with in analog form; for example, a variable such as temperature can take on any value within its measured range or, looked at conceptually, it can be measured to any number of significant figures by a suitable instrument. Furthermore, analog variables generally change continuously in time. Digital computers are not suited to handle arbitrarily precise or continuously changing information; hence, analog process signals must be reduced to a digital representation (discretized), both in terms of magnitude and in time, to put them into a useful digital form.

The magnitude discretization problem most often is handled by transducing and scaling each measured variable to a common range, then using a single conversion device—the analog-to-digital converter (ADC)—to put the measured value into digital form. An ADC suitable for measurement and control purposes typically will convert signals in the range −10 to +10 volts direct current, yielding an output with 12 bits of accuracy in 10 to 50 μs. A multiplexer often is used to allow a number of analog inputs to be switched into a single ADC for conversion. High-level signals (on the order of volts) can be switched by a solid-state multiplexer; low-level signals (on the order of millivolts, from strain gages or thermocouples) would require mechanical relays followed by an amplifier to boost the signal to an acceptable input level for the ADC. Microcomputers are now available which contain integral analog conversion circuitry for several channels on the processor chip. *See* ANALOG-TO-DIGITAL CONVERTER.

Discretization in time requires the computer to sample the signal periodically, storing the results

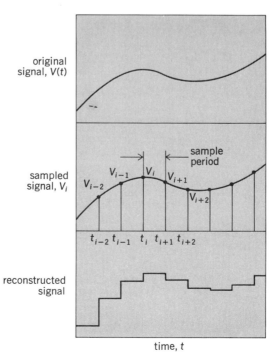

Fig. 2. Discretization in time of an analog signal.

in memory. This sequence of discrete values yields a "staircase" approximation to the original signal (Fig. 2), on which control of the process must be based. Obviously, the accuracy of the representation can be improved by sampling more often, and many digital systems simply have incorporated traditional analog control algorithms along with

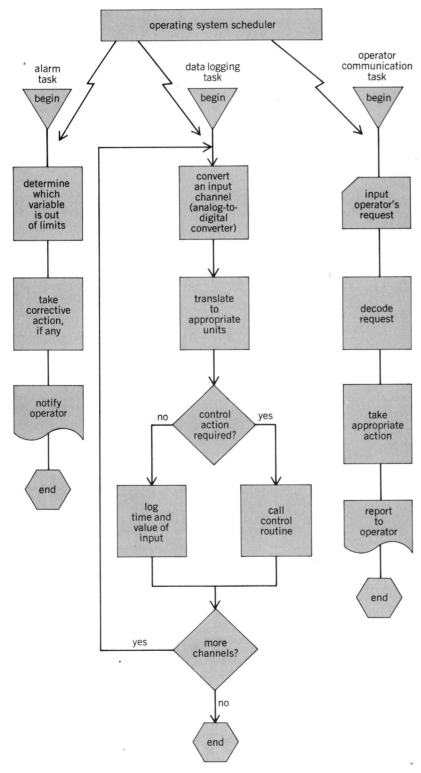

Fig. 3.   Example of a multitasked control program.

rapid sampling. However, newer control techniques make fundamental use of the discrete nature of computer input and output signals. Analog outputs from a computer most often are obtained from a digital-to-analog converter (DAC), a device which accepts a digital output from the computer, converts it to a voltage in several microseconds, and latches (holds) the value until the next output is converted. Usually a single DAC is used for each output signal. *See* DIGITAL-TO-ANALOG CONVERTER.

Real-time computing. In order to be used as the heart of a control system, a digital computer must be capable of operating in real time. Except for very simple microcomputer applications, this feature implies that the machine must be capable of handling interrupts, that is, inputs to the computer's internal control unit which, on change of state, cause the computer to stop executing some section of program code and begin executing some other section. Using its extraordinary computational abilities (on the order of a million instructions per second), the computer, which basically is a sequential or serial processor, can be made to appear to perform operations in parallel by proper design of its hardware and executive software. By attaching a very accurate oscillator (a so-called real-time clock) to the interrupt line, the computer can be programmed to keep track of the passage of real time and, consequently, to schedule process sampling and control calculations on a periodic basis.

The requirements of real-time computing also imply that the computer must respond to interrupts from the process. Thus a key process variable may be used to trigger an interrupt when it exceeds preset limits. The computer might be programmed to service such an interrupt immediately, taking whatever control action is necessary to bring the variable back within limits. The ability to initiate operations on schedule and to respond to process interrupts in a timely fashion is the very basis of real-time computing; this feature must be available in any digital control system.

Programming considerations. Much of the programming of computer control systems is done in a high-level language such as BASIC or FORTRAN; however, many microcomputer applications are carried out in machine (assembly) language. In either case the programmer must utilize program routines which access the devices in the process interface, for example, fetch the contents of a particular digital input register so as to check the status of some process digital element, or write out a digital result to a particular DAC channel. Additionally, the programmer will have the capability to schedule operations periodically or at particular times.

Computer control systems for large or complex processes may involve complicated programs with many thousands of computer instructions. Several routes have been taken to mitigate the difficulty of programming control computers. One approach is to develop a single program which utilizes data supplied by the user to specify both the actions to be performed on the individual process elements and the schedule to be followed. Such an executive program, supplied by a control system vendor, might be utilized by a variety of users; and this approach often is taken for relatively standardized operations such as machining and sequential processing (manufacturing).

Another approach is to develop a rather sophisticated operating system to supervise the execution of user programs, scheduling individual program elements for execution as specified by the user or needed by the process. The multiple program elements, called tasks when used with a multitasking operating system, or called programs with a multiprogramming system, will be scheduled individually for execution by the operating system as specified by the user (programmer) or needed by the process. A simple example is given in Fig. 3 for three tasks—a data logging task which might be scheduled every 30 s, a process alarm task which would be executed whenever a process variable exceeds limits and causes an interrupt, and an operator communication task which would be executed whenever the operator strikes a key on the console. The operating system scheduler would resolve conflicts, caused by two or three tasks needing to execute simultaneously, on the basis of user-supplied priorities; presumably the alarm task would have the highest priority here. *See* DIGITAL COMPUTER PROGRAMMING.

Control algorithms. Many applications, particularly machining, manufacturing, and batch processing, involve large or complex operating schedules. Invariably, these can be broken down into simple logical sequences, for example, in a high-speed bottling operation the bottle must be in position before the filling line is opened. Hence the control program reduces to a set of interlocked sequential operations. Some applications—in the chemical process industries, in power generation, and in aerospace areas—require traditional auto-

matic control algorithms. *See* ALGORITHMS.

Automatic control algorithms fall into two major categories: feedforward techniques, where process disturbances which would affect the controlled variable are measured and their effect canceled out or compensated by appropriate manipulation of a process input variable; and feedback techniques. In feedback control, which makes up the vast majority of applications, the controlled variable itself is measured and subtracted from a reference (set-point) variable equal to its desired value. The resulting deviation is the input to the controller, which then manipulates a process input variable through the final control element. If the controller is designed correctly, it should maintain the deviation quite small despite operator changes to the set point or despite environmental changes which cause disturbances to enter the control loop (Fig. 4*a*).

Digital implementation of traditional analog algorithms. Traditional algorithms for analog control of processes have been developed over many years. One very important example is the three-mode controller, so called because the algorithm for the controller output U operates proportionally on the deviation ϵ, the time integral of ϵ, and the time derivative of ϵ as indicated in Eq. (1), where

$$U = U_0 + K_c \left[\epsilon + \frac{1}{\tau_I} \int_0^t \epsilon \, dt + \tau_D \frac{d\epsilon}{dt} \right] \quad (1)$$

U_0 is the constant value that the controller output assumes when the controller first is turned on, and K_c, τ_I, and τ_D are the controller proportional gain, and integral and derivative time constants, respectively. These controller parameters are chosen by

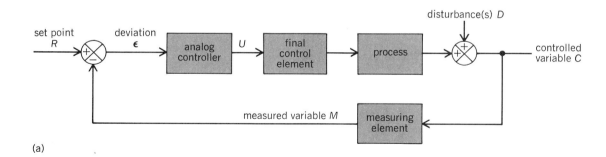

(a)

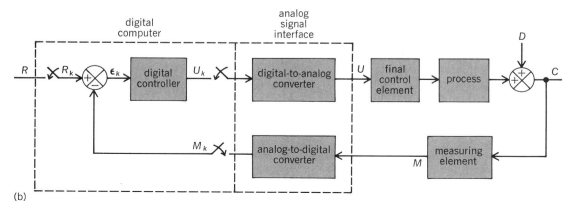

(b)

Fig. 4. Single-loop feedback control system. (*a*) Traditional analog implementation. (*b*) Digital implementation.

the system designer appropriately for each control application.

In the digital implementation of a single-loop feedback control system (Fig. 4b), the digital computer implements the controller algorithm by using process information sampled through the analog-to-digital converter and a set point supplied by the operator. The symbolic switches and the subscript on the controller input ϵ_k (obtained from R_k and M_k) in Fig. 4b are intended to indicate that the information is obtained by the computer only at discrete sampling times, T time units apart. Similarly, the controller generates an output U_k at each sampling time which it transmits through the digital-to-analog converter to the process's final control element. Implementation of the three-mode algorithm in the digital (discrete) environment is straightforward. In the "velocity form" of the proportional-integral-derivative algorithm, so called because the change in U from its previous value is calculated directly (Eq. 2), the computer need only store the

$$U_k = U_{k-1} + K_c \left[(\epsilon_k - \epsilon_{k-1}) + \frac{T}{\tau_I}\epsilon_k + \frac{\tau_D}{T}(\epsilon_k - 2\epsilon_{k-1} + \epsilon_{k-2}) \right] \quad (2)$$

controller output and two previous values of ϵ in order to calculate U at the next sampling time. In actual application, digital versions of continuous control algorithms often utilize rapid sampling (T is small) so that they respond in much the same way as analog systems do; hence the operational advantages and disadvantages of the two methods are equivalent.

Purely digital algorithms. One disadvantage of the classical feedback algorithm of Eq. (1) or (2) is that it does not function well when the dynamics of the process contain significant time delay, that is, when a change in the process input has no effect on the controlled variable for a significant period of time (known as the dead time) So-called dead-time compensation algorithms have been available in analog form; however, only with the digital computer as controller has the implementation become so inexpensive that such applications can be widely justified. Hence the use of the three-mode algorithm in conjunction with a dead-time compensator is a common digital control approach, particularly for control of processes (where dead time is an important component of many, if not most, systems).

Attempts to expand the digital control medium through development of strictly digital control algorithms is an important and continuing trend. Such algorithms typically attempt to exploit the sampled nature of process inputs and outputs, significantly decreasing the sampling requirements of the algorithm (that is, making T as large as possible) For example, in the "deadbeat" algorithm. Eq. (3), the controller attempts to follow set-point

$$U_k = U_{k-1} + \frac{1}{K(N+1)(1-e^{-T/\tau})}(\epsilon_k - e^{-T/\tau}\epsilon_{k-1}) \quad (3)$$

changes in minimum time. The use of this algorithm presupposes that the process can be modeled approximately as a first-order-plus-dead-time system, where K and τ are the process gain and time constant and NT is the dead time. Hence a

stepwise change in set point will cause the controlled variable to obtain its new, desired value in $N+1$ sampling instants, the absolute minimum.

The deadbeat algorithm has a number of disadvantages, including its sensitivity to errors in the assumed process model. These often result in undesirable oscillations (ringing) Dahlin's algorithm, Eq. (4), is one of a number of attempts to eliminate

$$U_k = U_{k-1} + \frac{1}{K[N(1-e^{-\lambda T})+1](1-e^{-T/\tau})}(\epsilon_k - e^{-T/\tau}\epsilon_{k-1}) \quad (4)$$

the undesirable features of optimal, deadbeat control. In this instance, the use of an empirical tuning parameter, λ, relaxes the requirement of minimal response time, resulting in less ringing in the system response, particularly when the process dynamics change slightly from those assumed in the first-order model.

Dahlin's and the deadbeat algorithms, and other similar algorithms which are designed on the basis of set-point response, often respond more poorly to disturbances than the three-mode algorithm. Hence there is a continuing search for the "best" algorithm—one which is simple, responds well to both set-point changes and disturbances, requires little knowledge of process dynamics and is insensitive to changes in process dynamics, tolerates process constraints and nonlinearities, can be extended to multi-input/multi-output situations, is easy to implement and program, and so forth. Such a universal digital control algorithm likely will not be found, but the search has led in a number of interesting directions, including the development of self-tuning algorithms and the application of time convolution techniques for predictive control strategies. These approaches would not be possible without the digital computer and the ability to design and implement computer control systems. Additional demands are made as well in the analysis of digital control systems: discrete algebraic (difference) equations substitute for differential equations used traditionally to describe processes and analog controllers; and discrete transform techniques (the Z transform) often replace the Laplace transform approach used in analyzing continuous systems. *See* DIGITAL COMPUTER.   [DUNCAN A. MELLICHAMP]

Bibliography: D. M. Auslander et al., Direct digital process control: Practice and algorithms for microprocessor application, *Proc. IEEE*, 66:199–208, 1978; T. J. Harrison (ed.), *Minicomputers in Industrial Control*, Instrument Society of America, 1973; D. A. Mellichamp (ed.), *CACHE Monogr. Ser. Real-Time Computing*, vols. 1–8, 1977–1979.

Digital counter

An instrument which, in its simplest form, provides an output that corresponds to the number of pulses applied to its input. Counters may be categorized into two types: the Moore machine or the Mealy machine. The simpler counter type, the Moore machine, has a single count input (also called the clock input or pulse input), while the Mealy machine has additional inputs that alter the count sequence. Digital counters take many forms, including geared mechanisms (tape counters and odometers are examples), relays (old pinball machines and old telephone switching systems), vacuum tubes (old test equipment), and solid-state semiconductor circuits (most modern electronic counters). This article will stress solid-state electronic counters.

Most digital counters operate in the binary number system, since binary is easily implemented with electronic circuitry. Binary allows any integer (whole number) to be represented as a series of binary digits, or bits, where each bit is either a 0 or 1 (off or on, low or high, and so forth). *See* NUMBER SYSTEMS.

Figure 1 shows a four-bit binary counter that can count from 0 to 15; the sixteenth count input causes the counter to return to the 0 output state and generate a carry pulse. This action of the counter to return to the 0 state with a carry output on every sixteenth pulse makes the four-bit binary counter a modulo 16 counter. The four binary-digit outputs

Q_D, Q_C, Q_B, and Q_A are said to have an 8-4-2-1 "weighting" because, if Q_D through Q_A are all ones, then the binary counter output is $1111_2 = 1 \times 2^3 + 1 \times 2^2 + 1 \times 2^1 + 1 \times 2^0 = 8 + 4 + 2 + 1 = 15_{10}$, where the subscripts indicate the base of the number system. In Fig. 1a, the counter state-flow diagram is shown. The word "state" refers to the number, the datum, in the counter; this datum is stored in four flip-flop devices, each of which is a memory bit that stores the current value of one binary digit. Each possible state is represented by the numerical output of that state inside a circle. Upon receiving a count pulse, the counter must change state by following an arrow from the present state to the next state. In Fig. 1b, a table is given showing the counter output after a given number of input pulses, assuming that the counter always starts from the 0 state. The counter output is listed in binary, octal, decimal, and hexadecimal.

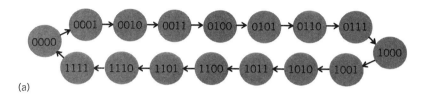

(a)

number of count pulses	binary output	octal (base 8)	decimal (base 10)	hexadecimal (base 16)
0	0000	0	0	0
1	0001	1	1	1
2	0010	2	2	2
3	0011	3	3	3
4	0100	4	4	4
5	0101	5	5	5
6	0110	6	6	6
7	0111	7	7	7
8	1000	10	8	8
9	1001	11	9	9
10	1010	12	10	A
11	1011	13	11	B
12	1100	14	12	C
13	1101	15	13	D
14	1110	16	14	E
15	1111	17	15	F
16	0000	0	0	0

(b)

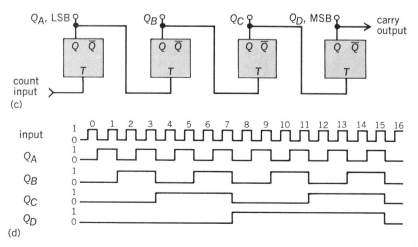

(c)

(d)

Fig. 1. Four-bit binary counter using trigger flip-flops; this is a Moore machine. (a) State-flow diagram. (b) Table of the counter output in various number systems. (c) Circuit block diagram. LSB = least significant bit; MSB = most significant bit. (d) Output waveforms.

Figure 1c shows a block diagram of the counter built with T flip-flops, and Fig. 1d shows the counter output waveforms through time, with a periodic count input. The T flip-flop is a device that has either a 0 or a 1 on its Q output at all times. When the count input T moves from the 1 state to the 0 state, the flip-flop output must change state, from a 0 to a 1 or a 1 to a 0. The carry output produces a 1-to-0 transition on every sixteenth count input, producing a divide-by-16 function. *See* MULTIVIBRATOR.

The four bits of the counter of Fig. 1 can be grouped together and used to represent a single hexadecimal digit; in Fig. 1b, each counter output state represents one hexadecimal digit. A two-digit hexadecimal counter requires two sets of four-bit binary counters, the carry output from the first set of counters driving the count input of the second set of counters.

A decimal counter built from four binary counters is shown in Fig. 2. Let four bits of data from the binary counter represent one decimal digit. The counter will work in the same way as the counter of Fig. 1, except that all the flip-flops are reset to the 0 state when the counter moves from the $1001_2 = 9$ state, instead of advancing to the $1010_2 = 10$ state. Besides the AND gate that is now used to detect the 1001 state of the counter and enable the resets, the circuit block diagram shows a new type of flip-flop. The "SR" flip-flop acts like a T flip-flop with an additional input that forces the Q output to a 1 state when the S (set) input is high and the T input has had a 1-to-0 transition applied. An R (reset) input acts as the S input does, except that the Q output goes to 0. This example decimal counter has an 8-4-2-1 weighted output that is known as binary coded decimal (B.C.D.). A seven-segment display is easily interfaced to the binary-coded-decimal counter using a binary-coded-decimal-to-seven-segment decoder/driver circuit that is widely available.

Applications. Digital counters are found in much modern electronic equipment, especially equipment that is digitally controlled or has digital numeric displays. A frequency counter, as a test instrument or a channel frequency display on a radio tuner, consists simply of a string of decade counters that count the pulses of an input signal for a known period of time, and display that count on a seven-segment display. A digital voltmeter operates by using nearly the same idea, except that the counter counts a known frequency for a period of time proportional to the input voltage. *See* ANALOG-TO-DIGITAL CONVERTER; DIGITAL VOLTMETER.

A digital watch contains numerous counter/dividers in its large-scale integration (LSI) chip, usually implemented with complementary metal oxide semiconductor (CMOS) technology. A typical watch generates a 32,768-Hz crystal oscillator signal that is divided by 2^{15} (15 T flip-flop binary counters) to produce a 1-Hz signal that is counted as seconds by using two decimal counters that reset and produce a carry at every sixtieth count. That carry output is counted, similarly, as a minutes display, generating another modulo 60 carry pulse for the hours display. The hours counters produce a carry at every twenty-fourth count to run the days counter and display; that counter generates a carry pulse every 28 to 31 days, depending on the month. *See* INTEGRATED CIRCUITS.

Digital computers may contain counters in the form of programmable interval timers that count an integral number of clock pulses of known period, and then generate an output at the end of the count to signal that the time period has expired. Most of the counters in a microprocessor consist of arithmetic logic units (ALU) that add one many-bit number to another, storing the results in a memory location. The program and data counters are examples of this kind of counter. *See* DIGITAL COMPUTER; MICROPROCESSOR.

Counter specifications. Counters have progressed from relays to light-wavelength-geometry very-large-scale-integrated circuits. There are several technologies for building individual digital counters. Single counters are available as integrated circuit chips in emitter-coupled logic (ECL), transistor-transistor logic (TTL), and CMOS. The three technologies are listed in the order of decreasing speed and decreasing power dissipation. ECL will operate to 600 MHz, TTL to 100 MHz

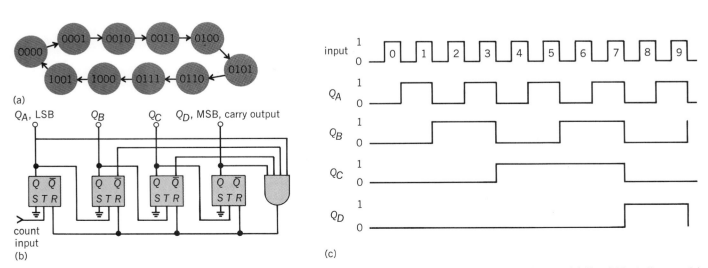

Fig. 2. Four-bit binary counter modified to be a decimal counter. (*a*) State-flow diagram. (*b*) Circuit block diagram. (*c*) Output waveforms.

(Schottky-clamped), and CMOS to 5 MHz. Standard, high-volume production n-channel metal oxide semiconductor (NMOS) LSI can implement a 1-bit binary counter in a 100×100 μm^2 (3.9×3.9 mil^2) area that will operate to 10 MHz. A gallium arsenide metal-semiconductor field-effect transistor (MESFET) master-slave JK flip-flop has been produced that operates at 610 MHz in a 390×390 μm^2 (15×15 mil^2) surface area while consuming the power of an NMOS.

[MICHAEL E. WRIGHT]

Bibliography: C. H. Roth, *Fundamentals of Logic Design*, 1975; K. Suyama et al., Design and performance of GaAs normally-off MESFET integrated circuits, *IEEE Trans. Electr. Devices*, ED-27 (6): 1092–1097, 1980.

Digital optical disk recorder

A data storage device that uses laser light to machine micrometer-sized holes (Fig 1a) on a thin tellurium alloy film, providing a permanent spatial representation of the data on a rotating disk. The digital optical disk recorder developed as an extension of the technology of the consumer video disk player first described in 1973. Gas lasers are used when maximum data rate and disk data capacity are desired, and semiconductor diode lasers are used when low cost and compactness are more important. These recorders can store between 10^{10} and 10^{11} bits of user data on a 30-cm-diameter disk, and data can be retrieved from any location in less than 1 s. Data can be recorded or played back at rates as high as 10^7 bits per second with an error rate of 1 in 10^9 bits. This article describes a gas laser recorder and optical disk which exists in prototype form.

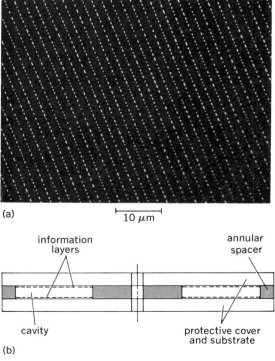

(a)

|— 10 μm —|

information layers   annular spacer

cavity   protective cover and substrate

(b)

Fig. 1. Disk used for data storage in optical recorder. (a) Detail of pits. (b) Air sandwich structure.

Disk. The sensitive material on the disk, a tellurium alloy, is deposited as a 30-nm-thick microcrystalline film on a plastic substrate. The sensitivity of the material for hole burning is between 100 and 300 millijoules/cm^2 for all visible laser light wavelengths. The film is considered suitable for use in archives requiring a record lifetime in excess of 10 years.

The disk is constructed to protect the sensitive material from dust without degrading sensitivity and to allow normal handling by the operator. The disk protective mechanism, called the Philips air sandwich (Fig. 1b), consists of two disks separated by ring spacers at the inner and outer radii of the information band. The annular cavity between the two disks is essentially a miniature clean room protecting the tellurium alloy from dust. Because the information is recorded and played back through the plastic substrate, approximately 1 mm thick, dust and fingerprints on the outer surfaces are out of focus and are essentially transparent to the system.

Drive. The recorder drive converts electrical recording signals to a spiral of holes on the active material of the disk, and retrieves the data from the disk, converting it to electrical playback signals. The driver consists primarily of a turntable and linear sled or translator to generate the spiral and optics for recording and playback.

Optical system. In the optical system, two light beams are generated, one for recording and one for playback. These beams share a common objective lens for focusing on the disk.

In the optical layout of Fig. 2, light from the laser is elliptically polarized by a half-wave plate and divided by a polarizing beam-splitting prism into a recording beam and a playback (read) beam. Usually 8 to 10% of the total laser light is in the playback beam. The intensity of the playback beam is limited by the need for nondestructive retrieval of data; the intensity of the recording beam is maximized for high-data-rate recording.

The recording beam is focused by a lens to a narrow beam waist at the modulator. The modulator is of the acoustooptic type, which is suited for hole burning because of its high extinction ratio (contrast). For digital recording, the modulator is excited by a series of high-frequency bursts in synchronism with the digital data.

The modulated recording beam is collimated by a pair of lenses and is reflected off a mirror and a second polarizing beam-splitting prism to the tracking mirror and objective lens. The tracking mirror and objective lens are mounted on the linear sled and, therefore, move with respect to the other optical elements as the read head travels from inner to outer radius.

The read beam passes through a phase grating to produce a three-beam array. The zero-order (undiffracted) beam is used for reading and focusing. The two first-order diffracted beams are used for radial tracking. The three-beam array is expanded and collimated by a pair of lenses forming a telescope, and is focused on the disk by the objective.

Electronic subsystems. The rest of the recorder consists of electronic subsystems for modulation, signal processing, error detection and correction, and computer control. An intermediate modulation

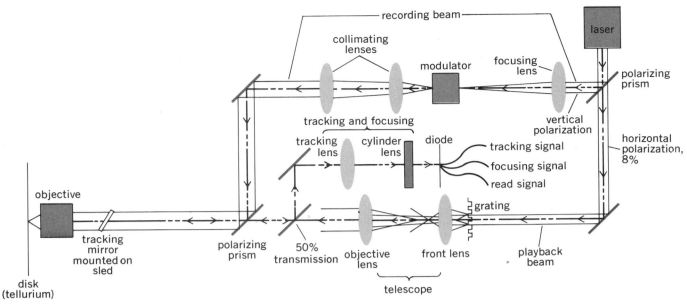

Fig. 2.   Optical layout of recorder.

of the recording signal (Miller modulation) is required before optical recording to provide for self-clocking of retrieved data and for a spectral match of the recording channel. Signal processing is required to compensate for the nonlinear recording process and to reshape the playback waveform for clock recovery.

Error correction. Error correction is accomplished by three means. First, the data are encoded for double error detection in a limited string of bits; then the encoded data are interleaved to provide immunity to long-burst errors caused by scratches. Finally, the recorded data of each block or sector are verified bit by bit by examining the playback signal which is obtained after a small delay (direct read after write, or DRAW). If the recorded data quality falls below the correcting capacity of the system, the block is rewritten.

Complete recorder. The completed recorder is shown in Fig. 3. In addition to the subsystems already described, there is a controller which allows the recorder to communicate with the PDP 11 minicomputer with random access on playback. The system records 10^{10} user bits of data per disk with record and playback rates of 5 Mbit/s. Random access to any data block (1 kilobyte) is accomplished on an average of 1 s. *See* COMPUTER STORAGE TECHNOLOGY; VIDEO DISK RECORDING.

[ROBERT MC FARLANE]

Bibliography: R. Bartolini et al., *IEEE Spectrum*, 15(8):20–28, 1978; K. Broadbent, *J. SMPTE*, 83(7):553–559, July 1974; K. Bulthius et al., *IEEE Spectrum*, 16(8):26–33, 1979; K. Compaan and P. Kramer, *Philips Tech. Rev.*, 33(7):178–180, 1973; G. Kenney et al., *IEEE Spectrum*, 16(2):33–38, 1979.

Digital-to-analog converter

A device for converting information in the form of combinations of discrete states or a signal, often representing binary number values, to information in the form of the value or magnitude of some characteristics of a signal, in relation to a standard or reference. Most often, it is a device which has electrical inputs representing a parallel binary number, and an output in the form of voltage or current.

Figure 1 shows the structure of a typical digital-to-analog converter. The essential elements, found

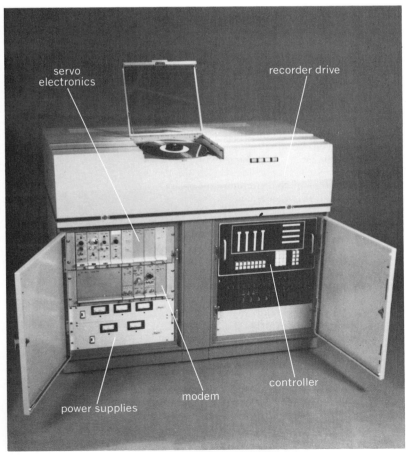

Fig. 3.   Optical recorder showing major components.

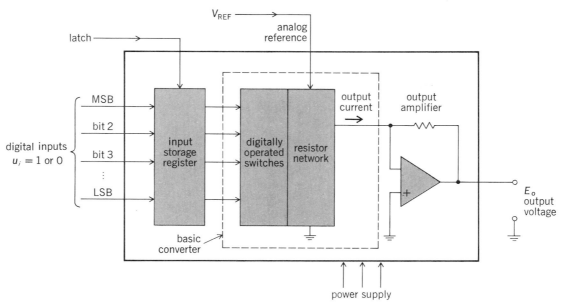

Fig. 1. Typical digital-to-analog converter.

even in the simplest devices, are enclosed within the dashed rectangle. The digital inputs, labeled u_i, $i = 1, 2, \ldots, n$, are equal to 1 or 0. The output voltage E_o is given by Eq. (1), where V_{REF} is an

$$E_o = KV_{REF} (u_1 2^{-1} + u_2 2^{-2} + u_3 2^{-3} + \cdots + u_n 2^{-n}) \tag{1}$$

analog reference voltage and K is a constant. Thus, for a 5-bit binary converter with latched input code 10110, the output is given by Eq. (2).

$$E_o = \left(\frac{16}{32} + \frac{0}{32} + \frac{4}{32} + \frac{2}{32} + \frac{0}{32}\right) KV_{REF} = \frac{11}{16} KV_{REF} \tag{2}$$

Bit 1 is the "most significant bit" (MSB), with a weight of 1/2; bit n is the "least significant bit" (LSB), with a weight of 2^{-n}. The number of bits n characterizes the resolution. *See* ANALOG-TO-DIGITAL CONVERTER; NUMBER SYSTEMS.

Uses. Digital-to-analog (D/A) converters (sometimes called DACs) are used to present the results of digital computation, storage, or transmission, typically for graphical display or for the control of devices that operate with continuously varying quantities. D/A converter circuits are also used in the design of analog-to-digital converters that employ feedback techniques, such as successive-approximation and counter-comparator types. In such applications, the D/A converter may not necessarily appear as a separately identifiable entity. such applications, the D/A converter may not necessarily appear as a separately identifiable entity.

Circuitry. The fundamental circuit of most D/A converters involves a voltage or current reference; a resistive "ladder network" that derives weighted currents or voltages, usually as discrete fractions of the reference; and a set of switches, operated by

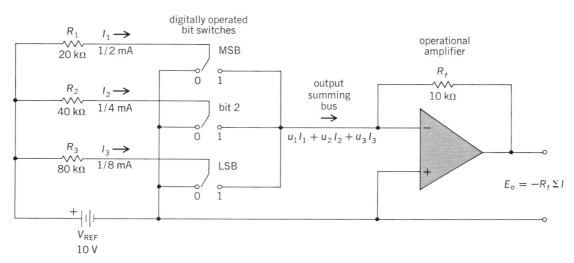

Fig. 2. Circuit of elementary 3-bit digital-to-analog converter.

Input and output of converter in Fig. 2.

Digital input code			Analog output	
u_1	u_2	u_3	$-E_0$ (volts)	$\dfrac{E_0}{V_{REF}}$
0	0	0	0	0
0	0	1	1.25	1/8
0	1	0	2.5	2/8
0	1	1	3.75	3/8
1	0	0	5.0	4/8
1	0	1	6.25	5/8
1	1	0	7.5	6/8
1	1	1	8.75	7/8

the digital input, that determine which currents or voltages will be summed to constitute the output.

An elementary 3-bit D/A converter is shown in Fig. 2. Binary-weighted currents developed in R_1, R_2, R_3 by V_{REF} are switched either directly to ground or to the output summing bus (which is held at zero volts by the operational-amplifier circuit). The sum of the currents develops an output voltage of polarity opposite to that of the reference across the feedback resistor R_f. The table shows the binary relationship between the input code and the output, both as a voltage and as a fraction of the reference.

The output of the D/A converter is proportional to the product of the digital input value and the reference. In many applications, the reference is fixed, and the output bears a fixed proportion to the digital input. In other applications, the reference, as well as the digital input, can vary; a D/A converter that is used in these applications is thus called a multiplying DAC. It is principally used for imparting a digitally controlled scale factor, or "gain," to an analog input signal applied at the reference terminal. *See* AMPLIFIER; ANALOG COMPUTER.

Construction. Except for the highest resolutions (beyond 16 bits), commercially available D/A converters are generally manufactured in the form of dual in-line-packaged integrated circuits, using bipolar, MOS, and hybrid technologies. A single chip may include just the resistor network and switches, or may also include a reference circuit, output amplifier, and one or more sets of registers (with control logic suitable for direct microprocessor interfacing).   [DANIEL H. SHEINGOLD]

Bibliography: D. J. Dooley (ed.), *Data Conversion Integrated Circuits*, 1980; *IEEE Transactions on Circuits and Systems*, special issue on analog/digital conversion, CAS-25, July 1978; D. Sheingold (ed.), *Analog-Digital Conversion Notes*, 1977; E. Zuch (ed.), *Data Acquisition and Conversion Handbook (1979)*.

Digital voltmeter

An instrument for measuring potential difference and displaying the measurement directly in digital form. Among the many devices used to provide a visual digital output are incandescent lamps with projection systems, edge-illuminated plates, neon-glow devices, and electroluminescent panels. If this readout provides five or six decimal digits, the resolution of the indicator need not be a significant factor in the accuracy of the measurement. Thus, measurement of potential difference can be made to better than 0.01% accuracy.

Digital voltmeters are used not only for their accuracy but also for their ease and speed of reading and their relative freedom from the operator error always present in reading moving-pointer voltmeters. Digital voltmeters are also widely employed in digital systems, where the voltmeter may be automatically switched to measure the voltage from many sources, and where the output is used for a visual readout and is also supplied to a printer for a permanent record. The digitally coded output often is further used to feed a digital computer either directly or through punched paper tape or cards.

A wide variety of both alternating- and direct-current instruments exists capable of measuring from microvolts to kilovolts, at rates from a measurement per few seconds to over 1000 measurements per second with an accuracy ranging from 1% to better than 0.005%.

Many different operating principles are utilized in digital voltmeters. Electronically operated precision voltage dividers or amplifiers are used to scale the input voltage to a basic range. Potentiometric and feedback methods are employed to compare voltage on this basic range to a solid-state zener-diode reference voltage. In one type, a linearly increasing voltage ramp is generated for each measurement. The time between the start of the ramp and the moment of equality between the ramp and the scaled input voltage is measured by counting pulses from a high-frequency clock oscillator. The digital count is then directly proportional to the input voltage.

In a second type, electronic switches sum the digitally weighted outputs (usually binary) from the voltage reference until they equal the scaled input voltage. These switches also control the decimal output indication through a decoding matrix. *See* NUMERICAL INDICATOR TUBE; VOLTAGE MEASUREMENT.   [MALCOLM C. HOLTJE]

Bibliography: W. D. Cooper, *Electronic Instrumentation and Measurement Techniques*, 2d ed., 1978; G. R. Cruzan and E. M. Billinghurst, For a versatile digital voltmeter: A microprocessor plus software, *Electronics*, 51(12):133–139, June 8, 1978; W. C. Kahler and H. Loeman, User-controllable filtering cuts measurement errors in D.V.M., *Electronics*, 51(15):137–141, July 20, 1978; P. Kantrowitz et al., *Electronic Measurements*, 1979; Modern D.V.M.S. simplify systems, *New Electron.*, 11(22):102, 104, 107, November 14, 1978; G. R. Parsons and S. E. Albon, Hand held D.V.M. design, *New Electron.*, 12(10):104, 108, May 15, 1979.

Diode

A two-terminal electron device exhibiting a non-linear current-voltage characteristic. Although diodes are usually classified with respect to the physical phenomena that give rise to their useful properties, in this article they are more conveniently classified according to the functions of the circuits in which they are used. This classification includes rectifier diodes, negative-resistance diodes, constant-voltage diodes, light-sensitive diodes, light-emitting diodes, and capacitor diodes.

Rectifier diodes. A circuit element is said to rectify if voltage increments of equal magnitude but opposite sign applied to the element produce unequal current increments. An ideal rectifier diode is one that conducts fully in one direction (forward) and not at all in the opposite direction (reverse). This property is approximated in junction and thermionic diodes. Processes that make use of rectifier diodes include power rectification, detection, modulation, and switching.

In power rectification, a rectifier diode is connected in series with an alternating-voltage supply and a load. The current through the load consists of unidirectional pulses, which can be converted into essentially constant direct current by means of a filter network that removes the varying component of current. In the detection of amplitude-modulated voltage, a similar process transforms an alternating voltage of varying amplitude into a direct voltage, the magnitude of which is proportional to the amplitude of the alternating input voltage. In amplitude modulation, diode rectification is used to vary the amplitude of a carrier voltage in response to a signal voltage. *See* AMPLITUDE-MODULATION DETECTOR; AMPLITUDE MODULATOR; ELECTRIC FILTER; ELECTRONIC POWER SUPPLY; RECTIFIER; SEMICONDUCTOR RECTIFIER.

Diode switching circuits include circuits in which output is obtained only in the presence or absence of one or more impressed control voltages. This is accomplished by the use of diodes that are normally maintained in the nonconducting (or conducting) state and are made conducting (or nonconducting) by the control voltages. Thus the diodes either provide a connection between the input and the output terminals or short-circuit the output terminals. Diodes can also be used to connect or disconnect two or more resistances in series or parallel between two terminals at selected values of input voltage and thus to synthesize a circuit that has a desired nonlinear current-voltage characteristic. *See* SWITCHING CIRCUIT.

Negative-resistance diodes. Negative-resistance diodes, which include tunnel and Gunn diodes, are used as the basis of pulse generators, bistable counting and storage circuits, and oscillators. *See* NEGATIVE-RESISTANCE CIRCUITS; OSCILLATOR; TUNNEL DIODE.

Constant-voltage diodes. Breakdown-diode current increases very rapidly with voltage above the breakdown voltage; that is, the voltage is nearly independent of the current. In series with resistance to limit the current to a nondestructive value, breakdown diodes can therefore be used as a means of obtaining a nearly constant reference voltage or of maintaining a constant potential difference between two circuit points, such as the emitter and the base of a transistor. Breakdown diodes (or reverse-biased ordinary junction diodes) can be used between two circuit points in order to limit alternating-voltage amplitude or to clip voltage peaks. Severe clipping of a sinusoidal voltage wave produces an approximately rectangular voltage wave. *See* LIMITER CIRCUIT.

Light-sensitive and light-emitting diodes. Light-sensitive diodes, which include phototubes, photovoltaic cells, photodiodes, and photoconductive cells, are used in the measurement of illumination, in the control of lights or other electrical devices by incident light, and in the conversion of radiant energy into electrical energy. Light-emitting diodes (LEDs) are used in the display of letters, numbers, and other symbols in calculators, watches, clocks, and other electronic units. *See* LIGHT-EMITTING DIODE; PHOTOCONDUCTIVE CELL; PHOTODIODE; PHOTOELECTRIC DEVICES; PHOTOTUBE; PHOTOVOLTAIC CELL.

Capacitor diodes. Semiconductor diodes designed to have strongly voltage-dependent shunt capacitance between the terminals are called varactors. The applications of varactors include the tuning and the frequency stabilization of radio-frequency oscillators. *See* JUNCTION DIODE; MICROWAVE SOLID-STATE DEVICES; POINT-CONTACT DIODE; SEMICONDUCTOR DIODE; SILICON DIODE; VARACTOR.　　　　　[HERBERT J. REICH]

Bibliography: C. A. Holt, *Electronic Circuits*, 1978; H. L. Krauss, H. J. Reich, and J. G. Skalnik, *Theory and Applications of Active Devices*, 1966; J. Millman, *Microelectronics*, 1979; H. J. Reich, *Functional Circuits and Oscillators*, 1961; D. Schilling and C. Belove, *Electronic Circuits*, 2d ed., 1979.

Direct-coupled amplifier

A device for amplifying signals with dc components. There are many different situations where it is necessary to amplify signals having a frequency spectrum which extends to zero frequency. Some typical examples are amplifiers in electronic differential analyzers (analog computers), certain types of feedback control systems, and some medical instruments such as the electrocardiograph. Amplifiers which have capacitor coupling between stages are not usable in these cases, because the gain at zero frequency is zero. Therefore, a special form of amplifier, called a dc or direct-coupled amplifier, is necessary. These amplifiers will also amplify ac signals. *See* AMPLIFIER.

Direct-coupled dc amplifiers. Some type of coupling circuit must be used between successive amplifier stages to prevent the relatively large supply voltage of one stage from appearing at the input of the following stage. These circuits must pass dc signals with the least possible amount of attenuation.

Transistor direct-coupled amplifiers have a number of advantages and disadvantages relative to vacuum-tube amplifiers. Interstage direct cou-

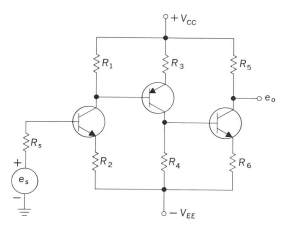

Fig. 1. Direct-coupled dc amplifier of the type which uses *npn* and *pnp* transistors.

pling is much easier with transistors because of the availability of both *pnp* and *npn* transistors and zener diodes. The circuit of Fig. 1 is a direct-coupled amplifier. The main disadvantage of this amplifier is the temperature dependence of the transistor parameters, specifically the current gain h_{FE}, the base-to-emitter voltage V_{BE}, and the leakage current I_{CBO}. Very often a zener diode is used to couple between stages, and in that case it is not necessary to use both *pnp* and *npn* transistors. The zener diode has a range of values of current through the diode for which the voltage across it is nearly constant. In effect, the zener diode acts like a battery.

It is generally recognized that the differential amplifier is the most stable dc amplifier circuit available. This is true because in this circuit the performance depends on the difference of the device parameters, and transistors can be manufactured using the planar epitaxial technique with very close matching of their parameters. Figure 2 shows a differential dc amplifier with a constant current source in the emitter circuit.

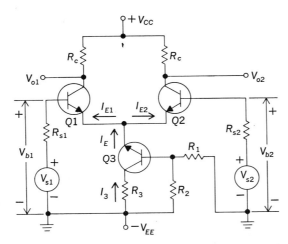

Fig. 2. Differential dc amplifier with constant-current stage in the emitter circuit.

One method of coupling in vacuum-tube dc amplifiers is by means of gas tubes, such as neon or voltage-regulator tubes. Often, in place of the gas tubes a series resistor may be used, but this reduces the gain of the amplifier. This can be avoided by returning the grid-leak resistor to a negative supply voltage instead of ground. Differential vacuum-tube dc amplifiers are possible; it is, however, impossible to manufacture vacuum tubes as closely matched as transistors, and hence differential transistor dc amplifiers now dominate the field.

Carrier dc amplifier. A method of amplifying dc (or slowly varying) signals by means of ac amplifiers is to modulate a carrier signal by the signal to be amplified, amplifying the modulated signal, and demodulating at the output. (In some applications, such as instrument servomechanisms, output in the form of a modulated carrier is required and no demodulation is necessary.) One arrangement is illustrated in block-diagram form in Fig. 3.

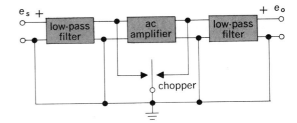

Fig. 3. Chopper amplifier.

An analysis of an actual circuit would show that in order to have an output e_o free from harmonics introduced by the modulation, the output low-pass filter cutoff frequency must be small compared to the modulation frequency. This limits the bandwidth of the input signal e_s to a small fraction of the bandwidth of signals which can be amplified by the types previously discussed. This is a disadvantage in most cases, but there are cases, such as the output voltage from a thermocouple, where the required bandwidth is small. Various types of choppers (bipolar transistors, field effect transistors, and electromechanical devices) are used as modulators. An additional disadvantage is that the chopper must be carefully designed to reduce to a minimum the hum and noise which can be introduced by an electromechanical element and the offset voltage and leakage current of a transistor. However, this amplifier provides stable amplification of the input signal, and it is used in many industrial recording instruments. *See* MODULATION; VOLTAGE AMPLIFIER.

Chopper-stabilized amplifiers. A chopper-stabilized amplifier is composed of two amplifiers: One amplifier is a carrier dc amplifier employing a chopper to modulate and demodulate a portion of the signal to be amplified; the second amplifier is a straight dc amplifier. The success of this circuit in reducing to a very low value the amount of drift appearing at the output depends upon the fact that the drift is a very-low-frequency phenomenon. A block diagram of a chopper-stabilized amplifier is illustrated in Fig. 4.

A component of the output signal e_o is due to drift in the dc amplifier, and if it is fed back through the feedback resistor to the input of the chopper-modulated amplifier (it cannot be fed

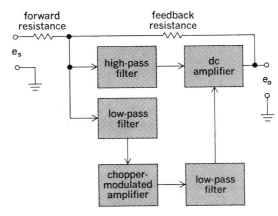

Fig. 4. Chopper-stabilized dc amplifier.

back to the dc amplifier because of the high-pass filter), the drift component will be amplified and returned to the dc amplifier. An analysis of the circuit shows that the input voltage e_s to the dc amplifier is altered by a factor equal and opposite to the equivalent drift voltage appearing at the input. The net result is that the drift can be reduced to a negligible value.

A signal appearing at the input terminals is amplified partly in the dc amplifier and partly in the chopper-modulated amplifier, because the high-pass and low-pass filters direct the frequency components of the input signal into the separate amplifiers. Thus, for dc and very-low frequency signals the gain is equal to the gain of the ac amplifier times the gain of the dc amplifier, while for higher-frequency signals the gain is that of the dc amplifier alone. However, if the gains are large, then when the feedback network is connected, the gain is equal, to a good approximation, to the feedback impedance divided by the impedance in the forward path. The gain is, therefore, essentially independent of the open-loop gain.

Chopper-stabilized amplifiers find wide application in analog computers, where they are used in integrating and summing amplifiers. In many problems solved on analog computers the time scale is reduced. This imposes strict requirements on the freedom from drift of the dc amplifiers. These requirements are met with amplifiers where a typical value for the dc amplifier gain is 100,000 and a typical value for the gain of the chopper-modulated amplifier is 1000.

[CHRISTOS C. HALKIAS]

Bibliography: C. A. Holt, *Electronic Circuits: Digital and Analog*, 1978; D. L. Schilling and C. Belove, *Electronic Circuits: Discrete and Integrated*, 1979; R. F. Shea, *Amplifier Handbook*, 1966.

Direct current

Electric current which flows in one direction only through a circuit or equipment. The associated direct voltages, in contrast to alternating voltages, are of unchanging polarity. Direct current corresponds to a drift or displacement of electric charge in one unvarying direction around the closed loop or loops of an electric circuit. Direct currents and voltages may be of constant magnitude or may vary with time.

Batteries and rotating generators produce direct voltages of nominally constant magnitude (illustration *a*). Direct voltages of time-varying magnitude are produced by rectifiers, which convert alternating voltage to pulsating direct voltage (illustration *b* and *c*). *See* RECTIFIER.

Direct current is used extensively to power adjustable-speed motor drives in industry and in transportation (illustration *d*). Very large amounts of power are used in electrochemical processes for the refining and plating of metals and for the production of numerous basic chemicals.

Direct current ordinarily is not widely distributed for general use by electric utility customers. Instead, direct-current (dc) power is obtained at the site where it is needed by the rectification of commercially available alternating current (ac) power to dc power. Solid-state rectifiers ordinarily are employed to supply dc equipment from ac

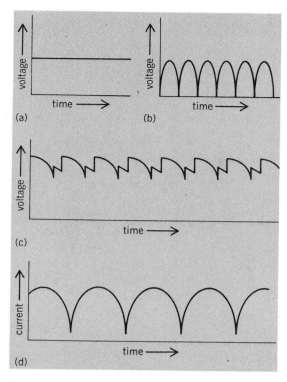

Typical direct currents and voltages. (*a*) Output from a battery; (*b*) full-wave rectified voltage; (*c*) output of the rectifier station of a high-voltage dc transmission link (*from E. W. Kimbark, Direct Current Transmission, vol. 1, Wiley–Interscience, 1971*). (*d*) Current in a rectifier-supplied dc motor (*from A. E. Fitzgerald, C. Kingsley, and A. Kusko, Electric Machinery, 3d ed., McGraw-Hill, 1971*).

supply lines. Rectifier dc supplies range from tiny devices in household electronic equipment to high-voltage dc transmission links of at least hundreds of megawatts capacity. *See* SEMICONDUCTOR RECTIFIER.

Many high-voltage dc transmission systems have been constructed throughout the world since 1954. Very large amounts of power, generated as ac and ultimately used as ac, are transmitted as dc power. Rectifiers supply the sending end of the dc link; inverters then supply the receiving-end ac power system from the link. High-voltage dc transmission often is more economical than ac transmission when extremely long distances are involved. *See* ALTERNATING CURRENT; ELECTRIC CURRENT; TRANSMISSION LINES.

[D. D. ROBB]

Direct-current circuit theory

An analysis of relationships within a dc circuit. Any combination of direct-current (dc) voltage or current sources, such as generators and batteries, in conjunction with transmission lines, resistors, inductors, capacitors, and power converters such as motors is termed a dc circuit. Historically the dc circuit was the first to be studied and analyzed mathematically. *See* CIRCUIT (ELECTRICITY).

Classification. Circuits may be identified and classified into simple series and parallel circuits. More complicated circuits may be developed as combinations of these basic circuits.

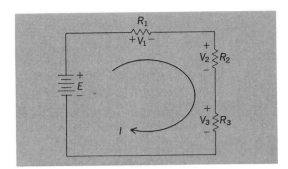

Fig. 1. Simple series circuit.

Series circuit. A series circuit is illustrated in Fig. 1. It consists of a battery of E volts and three resistors of resistances R_1, R_2, and R_3, respectively. The conventional current flows from the positive battery terminal through the external circuit and back to the negative battery terminal. It passes through each resistor in turn; therefore the resistors are in series with the battery.

Parallel circuit. The parallel circuit, shown in Fig. 2, consists of a battery paralleled by three resistors. In this case the current leaving the positive terminal of the battery splits into three components, one component flowing through each resistor, then recombining into the original current and returning to the negative terminal of the battery.

Physical laws of circuit analysis. The operation of the basic series and parallel circuits must obey certain fundamental laws of physics. These laws are referred to as Ohm's law and Kirchhoff's laws in honor of their originators.

Voltage drops. When an electric current flows through a resistor, a voltage drop appears across the resistor, the polarity being such that the volt-

age is positive at the end where the conventional current enters the resistor. This voltage drop is directly proportional to the product of the current in amperes and the resistance in ohms. This is Ohm's law, expressed mathematically in Eq. (1). Thus in Fig. 1 the drop across R_1 is V_1 and has the polarity shown. *See* OHM'S LAW.

$$V = IR \qquad (1)$$

Summation of voltages. The algebraic sum of all voltage sources (rises) and voltage drops must add up to zero around any closed path in any circuit. This is Kirchhoff's first law. In Fig. 1, the sum of the voltages about this closed circuit is as given by Eqs. (2), where the minus signs indicate a voltage

$$E - V_1 - V_2 - V_3 = 0 \qquad (2a)$$

$$E = V_1 + V_2 + V_3 \qquad (2b)$$

drop. Written in terms of current and resistance, this becomes Eq. (3). From this results the impor-

$$E = I(R_1 + R_2 + R_3) = IR_{\text{total}} \qquad (3)$$

tant conclusion that resistors in series may be added to obtain the equivalent total resistance, as shown in Eq. (4). *See* KIRCHHOFF'S LAWS OF ELECTRIC CIRCUITS.

$$R_{\text{total}} = R_1 + R_2 + R_3 + \cdots \qquad (4)$$

Summation of currents. The algebraic sum of all currents flowing into a circuit junction must equal the algebraic sum of all currents flowing out of the junction. In the circuit shown in Fig. 2, the current flowing into the junction is I amperes while that flowing out is the sum of I_1 plus I_2 plus I_3. Therefore the currents are related by Eq. (5). This is

$$I = I_1 + I_2 + I_3 \qquad (5)$$

Kirchhoff's second law. In this case the same voltage appears across each resistor. If the currents are expressed in terms of this voltage and values of the individual resistors by means of Ohm's law, Eq. (5) becomes Eq. (6). The equivalent resistance

$$I = \frac{E}{R_1} + \frac{E}{R_2} + \frac{E}{R_3} = \frac{E}{R_{\text{eq}}} \qquad (6)$$

R_{eq} that can replace the resistances in parallel can be obtained by solving Eq. (6) in the form of Eq. (7).

$$R_{\text{eq}} = \left[\frac{1}{R_1} + \frac{1}{R_2} + \frac{1}{R_3} + \cdots \right]^{-1} \qquad (7)$$

Therefore, resistances in parallel are added by computing the corresponding conductances (reciprocal of resistance) and adding to obtain the equivalent conductance. The reciprocal of the equivalent conductance is the equivalent resistance of the parallel combination. *See* CONDUCTANCE.

Sources. Sources such as batteries and generators may be connected in series or parallel. Series connections serve to increase the voltage; the net voltage is the algebraic sum of the individual source voltages.

Sources in parallel provide the practical function of increasing the net current rating over the rating of the individual sources; the net current rating is the sum of the individual current ratings.

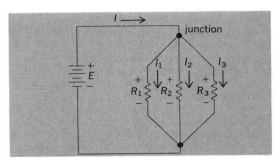

Fig. 2. Simple parallel circuit.

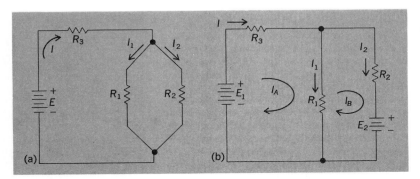

Fig. 3. Series-parallel circuits. (*a*) Single source. (*b*) Multiple sources.

Series-parallel circuits. More complicated circuits are nothing more than combinations of simple series and parallel circuits as illustrated in Fig. 3.

Single source. Circuits that contain only a single source are readily reduced to a simple series circuit. In the circuit of Fig. 3a the parallel combination of R_1 and R_2 is computed and used to replace the parallel combination. The resultant circuit is now a simple series circuit consisting of R_3 and R_{eq} and can readily be solved for the series current if the voltage is known.

Multiple sources. Circuits that contain two or more sources located in various branches cannot be reduced to a simple series circuit (Fig. 3b). The three basic laws of circuit theory still hold and may be directly applied to provide a simultaneous solution to the loop currents I_A and I_B flowing in each basic series circuit present in the overall network.

In this example the summations of voltages around the individual series circuits or loops are given by Eqs. (8) and (9), which may be solved for

$$E_1 = (R_1 + R_3)I_A - R_1 I_B \qquad (8)$$

$$-E_2 = -R_1 I_A + (R_1 + R_2)I_B \qquad (9)$$

the mathematical loop currents I_A and I_B. These loop currents can in turn be identified by reference to the circuit where I_A is identical to I, and I_B therefore, as stated by Eq. (10).

$$(I_A - I_R) = I_1 \qquad (10)$$

This method may be used to solve any complicated combination of simple circuits. Other methods are also available to the circuit analyst. *See* NETWORK THEORY.

Power. The electric power converted to heat in any resistance is equal to the product of the voltage drop across the resistance times the current through the resistance, as stated by Eq. (11a). By means of Ohm's law, Eq. (11a) may also be written as Eq. (11b). The total power dissipated in a circuit

$$P = VI \qquad (11a)$$

$$P = VI = V^2/R = I^2 R \qquad (11b)$$

is the arithmetic sum of the power dissipated in each resistance.

Circuit response. In the circuits mentioned thus far, the circuit responds in an identical manner from the moment the circuit is excited (switches closed) through any extended period of time. This is not true of circuits typified by those of Figs. 4 and 5.

For instance, when the switch of Fig. 4 is first

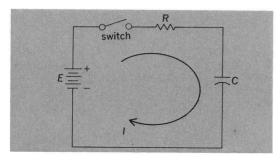

Fig. 4. A series *RC* circuit.

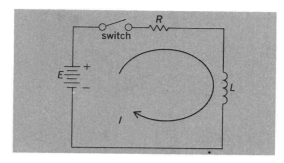

Fig. 5. A series *RL* circuit.

closed, a momentary current limited only by resistance R flows. As time passes, the capacitor, with capacitance C, charges and the voltage across it increases, eventually reaching a value equal to the applied voltage, at which time all flow of current ceases. The circuit current, given by Eq. (12), is in

$$i = (E/R)\epsilon^{-t/RC} \qquad (12)$$

amperes. The product RC is known as the time constant of the circuit. The energy W in joules stored in a capacitance at any time is given by Eq. (13). *See* TIME CONSTANT.

$$W = CE^2/2 \qquad (13)$$

For the circuit of Fig. 5, the initial current upon closing the switch is zero, since any attempt to cause a rate of change of current through the coil whose inductance is L induces a counter emf across the coil or inductor. Eventually this counter emf disappears and a steady-state current E/R flows indefinitely in the circuit.

The current at any time after closing the switch is given by Eq. (14) in amperes. The factor R/L is the

$$i = (E/R)(1 - \epsilon^{-Rt/L}) \qquad (14)$$

time constant of the circuit. The energy W stored in an inductance at any time is given by Eq. (15).

$$W = LI^2/2 \qquad (15)$$

For a complete discussion of transient phenomena *see* ELECTRIC TRANSIENT. [ROBERT L. RAMEY]

Bibliography: W. Hayt and J. Kemmerly, *Engineering Circuit Analysis*, 1978; S. L. Oppenheimer and J. P. Borchers, *Direct and Alternating Currents*, 2d ed., 1973; C. S. Siskind, *Electrical Circuits*, 2d ed., 1964.

Directional coupler

A waveguide network of four guide terminals A, B, C, and D (see Fig. 1) such that there is a complete isolation between A and C and between B and D, but no isolation between the two terminals of any other combination. The principle and usefulness of directional couplers will be illustrated below.

One common type of directional coupler is based on the interference of waves. Consider two rectangular waveguides having their narrow sides joined as shown in Fig. 2. Two small holes of the same size and shape are placed on their common narrow side spaced at a distance of one-quarter wavelength ($\lambda/4$) apart. Suppose a wave propagates into the network from terminal A. Most of the wave will reach terminal B. At either hole 1 or 2, a portion of the wave will leak into the second waveguide and divide into two equal but oppositely

DIRECTIONAL COUPLER

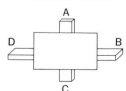

Fig. 1. A network with four waveguide terminals.

DIRECTIONAL COUPLER

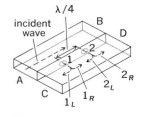

Fig. 2. A two-hole directional coupler.

directed components. Waves 1_R and 2_R will be combined in phase as they travel toward terminal D. However, wave 2_L lags behind wave 1_L by 180° in phase, and no wave will reach terminal C because of cancellation. Therefore there is a complete isolation between terminals A and C, as there is between terminals B and D by the same reasoning.

In the example cited here, there is a large difference in the coupling factors between different pairs of terminals. Thus terminals A and B (or C and D) are almost directly coupled with a coupling factor nearly equal to 1, whereas the coupling between A and D (or between B and C) can be made as small as desired. A directional coupler of this type is usually rated by the coupling factor in decibels (typically 10 or 20).

A magic tee junction is another form of directional coupler where the transmission coupling factor is 1/2. For a discussion of the properties of a magic tee and other wave guide junctions *see* WAVEGUIDE.

A directional coupler is most useful in selectively measuring either the forward (incident) or the backward (reflected) wave. In the case of a small hole coupling (large decibel rating), this objective can be achieved with a negligible attrition of the main wave.   [C. K. JEN]

Bibliography: H. A. Atwater, *Introduction to Microwave Theory*, 1962; A. J. Baden-Fuller, *Microwaves: Introduction to Microwave Theory and Techniques*, 1979; J. Helszajn, *Passive and Active Microwave Circuits*, 1978; S. Liao, *Microwave Devices and Circuits*, 1980.

Discriminator

A circuit that transforms a frequency-modulated or phase-modulated carrier into a wave that is amplitude-modulated as well as frequency-modulated. The amplitude modulation is then detected by a linear diode detector. Commonly used discriminators are the Foster-Seely discriminator and the ratio detector, which are popularly used in fm and television receivers. For these and other circuits *see* FREQUENCY-MODULATION DETECTOR.

[CHARLES L. ALLEY]

Disk recording

The process of inscribing suitably transformed acoustical or electrical signals on a flat circular plate that may be played back at a subsequent time. Virtually all modern disk recorders and reproducers are used to record or reproduce sound signals, mainly music and voice.

This article discusses monophonic, stereophonic, and quadraphonic recording and reproducing systems and their compatibility, commercial phonographs, the manufacture and specifications of disk records, distortion and noise in record reproduction, and digital recording. For related information *see* AMPLIFIER; LOUDSPEAKER; MAGNETIC RECORDING; MICROPHONE.

MONOPHONIC SYSTEM

A monophonic disk recording system consists of a disk record rotated by a turntable mechanism and a cutter for producing undulations in a groove in the disk corresponding to the sound signals. A monophonic disk reproducing system consists of a

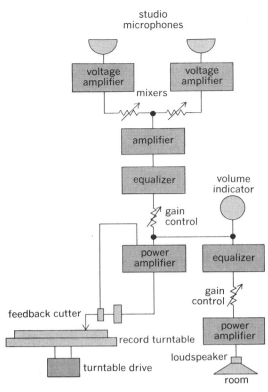

Fig. 1. Schematic arrangement of apparatus in a complete monophonic disk recording system.

pickup and mechanism for rotating the disk record by means of which the recorded undulations in the disk record are converted into electrical signals of approximately like form.

Recording system. The elements in a complete monophonic disk recording system are shown in Fig. 1. The first element is the acoustics of the studio. The output of each microphone is amplified and fed to a mixer, a device having two or more inputs and a common output. If more than one microphone is used (for example, when an orchestra accompanies a soloist, there is one microphone for the orchestra and one for the singer), the outputs of the two microphones may be adjusted for the proper balance by means of the mixers. An electronic compressor is used to reduce a large

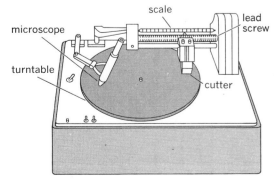

Fig. 2. Disk phonograph recorder. Microscope is used for periodic inspection of groove. (*After H. F. Olson, Acoustical Engineering, Van Nostrand, 1957*)

amplitude range to that suitable for reproduction in the home. A corrective electrical network called an equalizer provides the recording characteristics, which are described later. The attenuator, or gain control, provides a given control on the overall level fed to the power amplifier. The cutter, actuated by the amplifier, cuts a wavy path in the groove of the revolving record corresponding to the undulations in the original sound wave striking the microphone. A monitoring system consisting of a volume indicator, complementary equalizer, attenuator or gain control, power amplifier, and loudspeaker is used to control the recording operation. The volume indicator employs a logarithmic or decibel scale calibrated in volume units (VU). The volume unit is defined as 10 times the common logarithm of the power ratio p_2/p_1, where the reference power level p_1 is selected as 1 mw (0.001 watt) and p_2 is the signal power level.

Recorder. A phonograph recorder is an instrument for transforming acoustical or electrical signals into motion of approximately like form and inscribing such motion in an appropriate medium

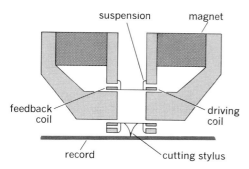

Fig. 3. Lateral feedback phonograph cutter.

by cutting or embossing. For the recording of disk phonograph records, the electrical phonograph recorder (Fig. 2) replaced the mechanical recorder in the late 1920s. The lacquer disk used in recording the master record is placed on the recording turntable. The turntable is heavy, to ensure against spurious rotational motions. A suitable mechanical filter is placed between the driving motor and the turntable so that uniform rotational motion of the turntable will be obtained. The drive system is arranged so that records of all standard speeds can be cut. In general, the recording turntable is driven with a synchronous motor to ensure constant speed of rotation. The lead screw drives the cutter in a radial direction so that a spiral groove is cut in the record. Lead screws of different pitches are used, ranging from 100 to 500 grooves per inch. In some recordings a variable pitch is used. In this procedure the spacing between the grooves is made to correspond to the amplitude—small spacing for small amplitudes and large spacing for large amplitudes. Under these conditions the maximum amount of information can be recorded on a record. The material removed in the cutting process, in the form of a fine thread, is pulled into the open end of a pipe, which is located near the cutting stylus and is connected to a vacuum system.

Cutters. The electromechanical transducers used as cutters can be of either the lateral or the vertical type. In the lateral type of disk recording, the undulations are cut in a direction parallel to the surface of the record and perpendicular to the groove. A sectional view of a lateral feedback phonograph cutter is shown in Fig. 3. The vibrating system is of the dynamic type with two coils, one the driving coil and the other the sensing (feedback) coil, wound on a common cylinder. The cutting stylus is attached to the coil cylinder. The vibrating system is designed so that it exhibits a single degree of freedom (single type of movement) over the frequency range from 30 to 16,000 hertz (Hz) with a fundamental resonant frequency at 700 Hz. The output of the sensing coil is fed to the input of the amplifier, as shown in Fig. 1. The ouput of the amplifier is fed to the driving coil in an out-of-phase relationship. With the feedback in operation, the velocity of the driving system is practically independent of frequency over the range from 30 to 16,000 Hz. The input to the amplifier is compensated to provide the desired recording characteristic.

The cutting stylus consists of a sapphire, synthetic ruby, or other hard material fashioned in the form of a pointed chisel (Fig. 4). The stylus is heated in recording and thereby imparts a smooth sidewall to the groove. This expedient results in considerable reduction in noise in reproduction. The stylus may be heated by a few turns of fine wire wound around it and operated from a low-voltage dc supply.

The original recording of disk records is made on a lacquer disk. The lacquer disk consists of a coating of an acetate plastic on two sides of an aluminum disk. The grooves are cut in the plastic.

In the vertical (hill-and-dale) type of disk recording, the undulations of the groove are cut in a direction perpendicular to the surface of the record and perpendicular to the groove. The vertical cutter is similar to the lateral cutter except that the stylus is located on the end of the cylinder of Fig. 3, and the entire system is turned 90°.

The vertical disk phonograph system is used to a limited extent in broadcasting stations, but it is not used in home disk phonograph systems.

Recording characteristics. The velocity—frequency-response characteristic of the groove in

DISK RECORDING

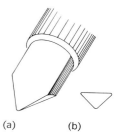

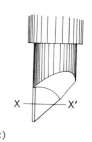

(c)

Fig. 4. Cutting stylus. (*a*) Perspective, (*b*) section, and (*c*) side views.

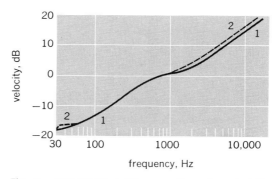

Fig. 5. Velocity-frequency recording characteristics used in commercial records. Curve 1, Record Industry Association of America standard. Curve 2, Orthacoustic standard. (*From L. C. Smeby, Recording and reproducing standards, Proc. IRE, 30(8):355–356, 1942*)

the phonograph disk record provides the velocity at the tip of the stylus of the phonograph pickup as a function of the frequency.

Electrical transcriptions (a term applied to professional disk recordings) are cut with an orthacoustic recording characteristic on records 16 in. in diameter turning at 33⅓ rpm. The orthacoustic velocity-frequency characteristic for constant voltage input to the microphone voltage amplifier is shown in Fig. 5. This is essentially a constant amplitude frequency characteristic. In reproduction of the disk record, an inverse frequency-response characteristic is used to obtain a uniform overall frequency-response characteristic.

In the recording of commercial phonograph records, some form of high-frequency compensation has always been employed. The compensation that has been used since the advent of the disk phonograph has varied over wide limits. Fortunately, in 1954 the Record Industry Association of America standardized the velocity–frequency-response characteristic of the groove in the commercial lateral disk record. The RIAA standard velocity-frequency-response characteristic is shown in Fig. 5. In the reproduction of commercial disk phonograph records, an inverse frequency-response characteristic is employed in order to obtain a uniform overall frequency-response characteristic. Commercial standard frequency records exhibiting the RIAA frequency-response characteristic are used in the development, design, and service of disk phonograph instruments. *See* FREQUENCY-RESPONSE EQUALIZATION.

Record manufacture. The processes for the mass production of disk phonograph records are depicted in Fig. 6. The original lacquer disk, termed the original, is metalized and then electroplated. The plating is separated from the lacquer and reinforced by backing with a solid metal plate. The assembly, called the master, is electroplated. This plating is separated from the master and reinforced by backing with a solid plate. The resulting assembly, the mother, is electroplated and rein-

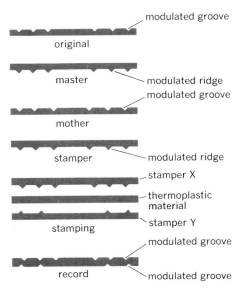

Fig. 6. Steps in mass production of disk phonograph records from lacquer originals.

(labels in figure: modulated groove — original — master — modulated ridge — modulated groove — mother — modulated ridge — stamper — stamper X — thermoplastic material — stamper Y — stamping — modulated groove — record — modulated groove)

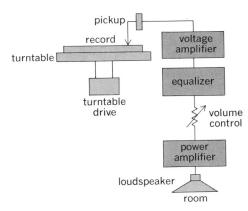

Fig. 7. Schematic arrangement of apparatus in a complete monophonic disk sound-reproducing system.

(labels in figure: pickup — record — turntable — voltage amplifier — equalizer — turntable drive — volume control — power amplifier — loudspeaker — room)

forced by a solid metal plate, forming an assembly termed the stamper. Several stampers are made from each mother. One stamper containing a sound selection to be placed on one side of the final record is mounted in the upper jaw, and another stamper containing a sound selection to be placed on the other side of the record is placed in the lower jaw of a hydraulic press equipped with means for heating and cooling the stampers. A preform, or biscuit, of thermoplastic material such as a shellac compound or vinyl compound is placed between the two stampers.

The stampers are heated, and the jaws of the press are closed to bring the two stampers against the thermoplastic material. When an impression of the stampers has been obtained in the thermoplastic material, the stampers are cooled, thus cooling and setting the plastic record. The jaws of the hydraulic press are opened, and the record is removed from the press. The modulated grooves in the record correspond to those in the original lacquer disk. The stamping procedure is repeated again and again until the desired number of records is obtained. This process constitutes the mass-production technique for the production of phonograph records.

Reproducing system. The elements of a complete monophonic disk sound-reproducing system are shown in Fig. 7. The first element is the motor-driven turntable which turns the record at a constant rotational speed. The stylus or needle of the pickup follows the wavy groove in the record and the pickup transducer generates a voltage corresponding to the undulations in the record. The output of the pickup is amplified by a voltage amplifier. The amplifier is followed by an equalizer which complements the recording characteristic of the record. Filters and tone controls are provided for further equalization of the response according to the taste of the listener. The tone controls provide means for increasing or decreasing the low- and high-frequency response. In general, the increase or decrease in response starts at 1000 Hz with a gradual increase or decrease in both high and low directions. The maximum increase or decrease in response at the extreme ends of the frequency range covered is about ±15 decibels (dB). A volume control provides means for obtaining the desired level of sound in reproduction. The volume control is followed by an amplifier which drives the loudspeaker.

Turntable and record changer. In an electrical record player and changer, the record is rotated by the turntable at the same angular speed as that used in recording. The turntable is rotated by means of an electric motor. The stylus or needle of the pickup follows the wavy spiral groove and generates a voltage corresponding to the undulations in the groove. (Pickups used in disk-record reproduction are described later.) A record player will play recordings at one or more of four rotational speeds: $16\frac{2}{3}$, $33\frac{1}{3}$, 45, and 78 rpm. A record changer may accommodate as many as eight records. A manual record player is the simplest type of disk-record reproducer, and involves placing the pickup arm on the record by hand and removing the arm at the conclusion of the record side. Such manual players include elaborate transcription types with very uniform rotational velocity and high-quality pickups.

Pickups or cartridges. A phonograph pickup is an electromechanical transducer actuated by a phonograph record and delivering energy to an electrical system, the electric current having frequency components corresponding to those of the wave in the record. The following systems are used for converting the mechanical vibrations into the corresponding electrical variations: magnetic, condenser, electronic, dynamic, ceramic, and crystal.

Modern lateral pickups are of the crystal, ceramic, magnetic, or dynamic type. A crystal phonograph pickup depends for its operation on the piezoelectric effect. The crystal used is Rochelle salt. Top and sectional views of a typical crystal pickup used for commercial phonographs are shown in Fig. 8a and b. The stylus driven by the record is coupled through an arm to the crystal and thereby produces a twist in the crystal. The open-circuit output of the crystal is proportional to the twist or displacement. The open-circuit voltage of a crystal pickup for an amplitude of 0.001 in. is about 0.5 volt. The open-circuit voltage displacement characteristic makes the frequency equalization exceedingly simple because the recording characteristic shown in Fig. 5 exhibits a practically constant displacement frequency characteristic. The electrical capacitance of the crystal is of the order of 1000–2000 picofarads. The equivalent electrical circuit is the open-circuit voltage in series with the electrical capacitance. The electrical impedance presented to the crystal pickup must be larger than the electrical impedance of the crystal in order to prevent reduction of low-frequency response. Since the electrical impedance of the crystal is large in the low-frequency range, the electrical impedance presented to the crystal must be relatively large to prevent frequency discrimination against the low-frequency range. *See* PIEZO-ELECTRICITY.

A ceramic phonograph pickup depends for its operation on the electrostrictive effect. The ceramic used is barium titanate. The ceramic pickup may be made in designs similar to that of the crystal pickup. The characteristics are similar in all essential respects save that the sensitivity is somewhat lower than that of the Rochelle salt crystal. *See* ELECTROSTRICTION.

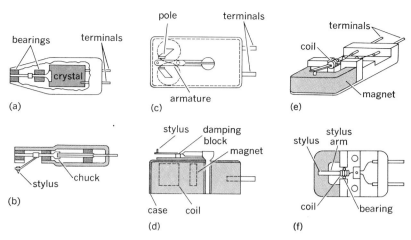

Fig. 8. Phonograph pickups. (*a*) Crystal type in cutaway top view and (*b*) sectional view. (*c*) Magnetic type in bottom view and (*d*) sectional view. (*e*) Dynamic type in perspective view and (*f*) bottom view.

A phonograph pickup which depends for its operation on the variation in magnetic flux through a stationary coil is called a magnetic pickup. A modern type is shown in Fig. 8c and d. The horizontal stylus also serves as the armature. This design makes it possible to obtain a relatively low mechanical impedance. The steady flux is supplied by a small permanent magnet. As the armature is deflected from the central position, the flux through one coil is increased and the flux through the other coil is decreased. The coils are connected in series so that the resultant voltages generated in the two coils are added. The open-circuit voltage generated in a coil is proportional to the time rate of change of magnetic flux through the coil, which in turn is

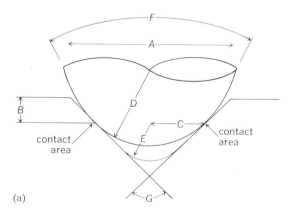

Value	Coarse groove	Fine groove	Ultrafine groove
A, in.	0.006	0.0027	0.001
B, in.	0.0008	0.0004	0.0004
C, in.	0.0019	0.0007	0.00017
D, in.	0.0027	0.001	0.00025
E, in.	0.0023	0.00027	0.00015
F	45°	45°	45°
G	90°	90°	90°

Fig. 9. Stylus in groove. (*a*) Sectional view. (*b*) Values of dimensions and angles. (*After H. F. Olson, Acoustical Engineering, Van Nostrand, 1957*)

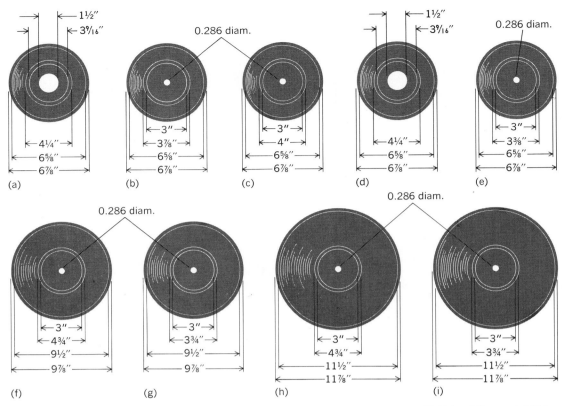

Fig. 10. Typical dimensions of the commercial disk phonograph records. (a) 7 in., 16⅔ rpm, large hole. (b) 7 in., 16⅔ rpm, small hole. (c) 7 in., 33⅓ rpm. (d) 7 in., 45 rpm. (e) 7 in., 78 rpm. (f) 10 in., 33⅓ rpm. (g) 10 in., 78 rpm. (h) 12 in., 33⅓ rpm. (i) 12 in., 78 rpm. (*After H. F. Olson, Acoustical Engineering, Van Nostrand, 1957*)

proportional to the velocity of the armature. Thus, the open-circuit voltage generated in the coil will be independent of the frequency if the velocity of the armature is independent of the frequency. In a properly designed magnetic pickup, the open-cir-

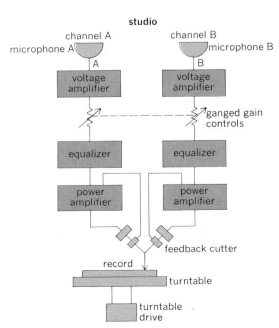

Fig. 11.   Schematic of complete stereophonic disk recording system.

cuit voltage frequency-response characteristic will correspond to the groove velocity frequency-response characteristic. For the recording characteristic of Fig. 5, frequency-compensating networks must be employed in order to obtain a uniform output frequency-response characteristic. The output of a typical magnetic pickup for an amplitude of 0.001 in. at 1000 Hz is of the order of 0.010 volt. The electrical impedance is highly inductive and therefore the electrical impedance is nearly proportional to the frequency. The electrical impedance of a typical magnetic pickup for a 0.010-volt output is 2500 ohms at 1000 Hz. The equivalent electrical circuit is the open-circuit voltage in series with the electrical impedance of the coils.

A dynamic phonograph pickup depends for its operation on the motion of a conductor in a magnetic field. A typical dynamic pickup is shown in Fig. 8e and f. The stylus arm is coupled to a coil located in a magnetic field. The open-circuit voltage developed in the coil is proportional to the rate of change of magnetic flux through the coil, and will be independent of frequency if the velocity is independent of frequency. In a properly designed dynamic pickup, the open-circuit voltage frequency-response characteristic will correspond to the groove velocity frequency-response characteristic. For the recording characteristic of Fig. 5, frequency-compensating networks must be employed in order to obtain a uniform output frequency-response characteristic. Since the vibrating system may be very small and light, it is possible to obtain a uniform velocity–frequency-response characteristic over a wide frequency range. The electrical impedance is practically an electrical resistance of about 25 ohms. The output at 1000 Hz for an amplitude of 0.001 in. is about 0.001 volt. In general, a transformer is used to step up the output voltage and electrical impedance.

In the vertical type of disk recording, the undulations of the groove are perpendicular both to the plane of the disk and to the groove. Therefore, the vibration of the stylus of the pickup is in the vertical direction. Any of the lateral pickups just described may be used by turning the transducer 90°.

Groove and stylus dimensions. A sectional view of the groove of a lateral disk record and the stylus is shown in Fig. 9a. The width of the groove at the surface of the record A, the angle of the walls of the groove G, the radius C of the stylus, the distance B below the surface of the record to the contact point of the stylus with the groove, and the width of the contact points are shown in the figure. The dimensions and angles for a coarse groove, fine groove, and ultrafine groove are given in Fig. 9b. The coarse groove is used in 78-rpm records, the fine groove in 45- and 33⅓-rpm records, and the ultrafine groove in 16⅔-rpm records.

The maximum nominal grooves per inch for the

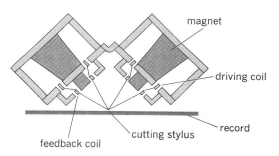

Fig. 12. Sectional view of a feedback stereophonic disk phonograph cutter.

different-sized grooves are as follows: coarse groove, 125; fine groove, 275; and ultrafine groove, 550.

The maximum amplitudes, in inches, in the frequency range 200–2000 Hz for the different-sized grooves are as follows: coarse groove, 0.004–0.005 in.; fine groove, 0.0015–0.002 in.; and ultrafine groove, 0.0007–0.001 in.

COMMERCIAL DISK RECORDS

Commercial phonograph records are made in four speeds, namely, 78 (approximately), 45, 33⅓, and 16⅔ rpm. The 78-rpm records are made in three diameters, 12, 10, and 7 in. The normal maximum playing times for full-width records are 5, 3½, and 2½ min, respectively. The 33⅓-rpm records are made in three diameters, 12, 10, and 7 in. The normal maximum playing times are 25, 17, and 8 min, respectively. The 45-rpm records are made in a diameter of 7 in. and have a normal maximum playing time of 8 min. The 16⅔-rpm records are made in a diameter of 7 in. The normal maximum playing time of the records with the large center hole is 30 min, while that for the small-hole records is 45 min for music and 60 min for speech. The overall diameter, the diameters of the first and last program groove, the label

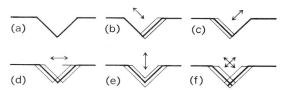

Fig. 13. Schematic views of groove undulations in stereophonic disk phonograph system. Heavy line indicates zero amplitude or unmodulated groove. Light lines indicate maximum limit of groove modulation. Arrows indicate direction of motion of recording cutter and reproducing stylus. (a) Unmodulated groove. (b) Modulation in right channel. (c) Modulation in left channel. (d) Lateral modulation, combination of b and c in phase. (e) Vertical modulation, combination of b and c out of phase. (f) Combination of equal vertical and lateral amplitudes, combination of b and c with 90° phase shift.

diameters, and the diameter of the center hole of the different records are shown in Fig. 10. It should be mentioned that the specifications of Fig. 10 and of the preceding discussion are representative and do not include all the variations.

DISTORTION AND NOISE

Distortion is any undesired change in the waveform of a signal in a sound-reproducing system. Noise is any foreign, erratic, intermittent, or statistically random signal produced within a sound-reproducing system. Distortion and noise are the two major problems that must be solved to achieve high-fidelity performance in disk sound-reproducing systems.

Distortion in reproduction. The recording and reproducing of a phonograph record constitute a complicated process, with many sources of nonlinear distortion. The record does not present an infinite mechanical impedance to the stylus. As a consequence, the vibrating system of the pickup is shunted by the effective mechanical impedance of the record at the stylus. Nonlinear distortion will be introduced if this impedance of the record is variable. Other sources of distortion are tracking error and tracing distortion.

Tracking distortion. A nonlinear distortion due

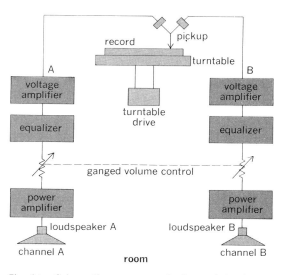

Fig. 14. Schematic arrangement of complete stereophonic disk phonograph reproducing system.

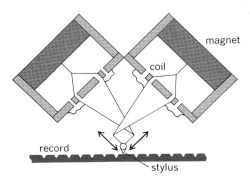

Fig. 15. Sectional view of a dynamic stereophonic disk phonograph pickup.

to a deviation in tracking is commonly termed tracking error. The angle between the vertical plane containing the vibration axis of the pickup and the vertical plane containing the tangent to the must execute symmetrical motion about the center line, which means that there should be no even harmonics. However, odd harmonics are produced.

Other distortions. If the force which the stylus presents to the record is of such magnitude that it exceeds the yield point of the record material, the record is a measure of the tracking error. If the vibration axis of the pickup passes through the tone arm pivot, the tracking error can be zero for only one point on the record. Tracking error can be reduced if the vibration axis of the pickup is set at an appropriate angle with respect to the line connecting the stylus point and the tone arm pivot, together with provisions for a suitable overhang distance between the stylus and the record axis.

Tracing distortion. A form of distortion in lateral disk-record reproduction known as tracing distortion is a function of the diameter of the stylus, the lateral velocity, and the linear groove velocity. This distortion is due to the fact that there is not a one-to-one correspondence between the shapes of the cutting and reproducing styli (Figs. 4 and 9). The cutting stylus presents a triangular shape as it cuts the groove. Thus, it is seen that the groove narrows as the cutting stylus approaches the center position because the cutting stylus is moving at an angle with respect to the motion of the record. The reproducing stylus presents a spherical surface to the groove; therefore, as the reproducing stylus moves in the groove, it will rise as the groove narrows. The frequency of the rise is twice the frequency of the modulation. The narrowing of the groove is termed the pinch effect. The two sides of the groove are symmetrical; therefore, the stylus mechanical impedance of the record will not be a constant. The result is production of nonlinear distortion. Furthermore, if the force exceeds the yield point by a considerable amount, the record may be permanently damaged.

As the needle or stylus is worn by the groove, the shape of the point changes from a spherical surface to a wedge shape. The wedge-shaped stylus introduces nonlinear distortion and a loss in the high-frequency response.

A consideration of the load and needle forces at the stylus tip shows that there is a force which is proportional to the tracking angle. This force, known as side thrust, is usually directed toward the center of the record and is applied to the inner boundary of the record groove; it is responsible for the unequal wear on the two sides of the stylus.

Another source of distortion results from the lack of correspondence between the linear groove speed in the recording and the ultimate reproduction. This type of distortion, which leads to a frequency modulation of the reproduced signal, is termed wow. Wow may be due to a nonuniform speed of the record turntable during recording or reproduction, misplacement of the center hole, or configuration distortion during the processing. In general, the major source of wow is the nonuniform speed of the reproducing turntable.

Surface noise. The record surface noise, in the absence of any signal, is one of the factors which limits the volume range and the frequency range of shellac phonograph records. The amount of surface noise for a given record is proportional to the frequency bandwidth. In order to reduce the surface noise to a tolerable value in shellac records, it is usually necessary to limit the high-frequency range in reproduction. A method of decreasing the effective surface noise consists of increasing the amplitude of the high-frequency response in recording and introducing complementary equalization, as shown in the recording characteristic of Fig. 5.

The noise of Vinylite records is extremely low and, in general, is not a problem. Adequate volume ranges can be obtained with Vinylite and similar plastics.

STEREOPHONIC SYSTEM

Two-channel disk phonograph sound reproduction was commercialized in 1958. The stereophonic disk phonograph provides the reproduction of the original sound sources in auditory perspective; that is, the spatial relations of the original sound are substantially retained in the reproduction of the recorded sound.

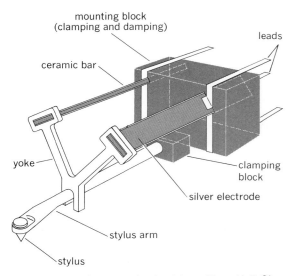

Fig. 16. Ceramic stereophonic pickup. (*From H. F. Olson, Music, Physics and Engineering, Dover, 1967*)

Recording system. The elements of a complete stereophonic disk recording system are shown in Fig. 11. There are two channels identical to the type shown in Fig. 1 except for the cutter. A two-channel disk phonograph dynamic-type feedback cutter is shown in Fig. 12. The two vibrating systems are arranged at right angles; therefore, the two channels in the groove are recorded at right angles. The modes of vibration in a plane normal to the surface of the record and normal to the groove axis are shown in Fig. 13. The motion of the cutting stylus is also shown in Fig. 13.

The same type of recorder described in the beginning of this article may be used in the recording of stereophonic records.

The recording characteristics that are employed in stereophonic disk recording are essentially the same as those that are used in monophonic disk recording.

Reproducing system. The elements of a complete stereophonic disk reproducing system are shown in Fig. 14. There are two identical channels following the pickup of the type shown in Fig. 7. A two-channel disk phonograph dynamic pickup is shown in Fig. 15. Each element of the stereophonic pickup consists of a transducer of the type employed in the single-channel lateral dynamic pickup shown in Fig. 8e and f. Reference to Figs. 13 and 15 shows that a vibration which excites one element will not excite the other. Other types of pickups have also been developed, for example, ceramic and magnetic pickups. A perspective view in schematic form depicting the elements of a ceramic pickup is shown in Fig. 16. The two ceramic elements are arranged with the vibrating planes at right angles and coupled to the stylus in such a manner that a vibration which excites one element will not excite the other. A perspective view in schematic form depicting the elements of a magnetic pickup is shown in Fig. 17. The stylus is connected to a magnet. There are two magnetic circuits with coils arranged at right angles. The vibration of the magnet leads to a variation in flux in the coils which in turn generates a voltage. A vibration

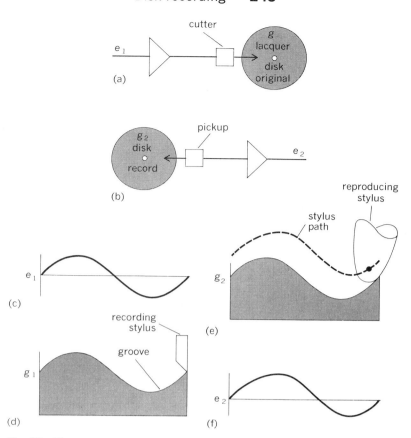

Fig. 18.   The recording and reproducing process in a stereophonic disk system. (*a*) Recording system. (*b*) Reproducing system. (*c*) Input to recording system. (*d*) Groove cut in original lacquer disk record. (*e*) Path of reproducing stylus in the groove of the disk record. (*f*) Output of reproducing system.

which generates a voltage in one set of coils will not generate a voltage in the other set. The three types of stereophonic pickups shown in Figs. 15, 16, and 17 are the most common in use.

The groove used for the stereophonic disk record is shown in Fig. 9. A stylus with a tip radius of 0.00075 in. is a standard for use in the reproduction of stereophonic disk records. A stylus with an elliptical tip with a large "radius" of 0.00075 in. and a small "radius" of 0.00025 in. in contact with the record groove has also been used. The latter, smaller radius leads to a reduction in tracing distortion.

Tracing distortion. The master stereophonic disk record is cut with a chisel-type stylus and the replica is reproduced by a ball-tipped stylus. Therefore, there is a discrepancy between the motion of the cutting stylus which becomes more pronounced at the shorter wavelengths. This type of distortion is termed tracing distortion. The recording and reproducing process in one channel of a stereophonic disk system is depicted in Fig. 18. The input signal to the recording system e_1 is a sine wave. The groove g_1 cut in the original disk record is also a sine wave. The disk record g_2 is a replica of the original disk record g_1 and is also a sine wave. However, the path of the ball-tipped reproducing stylus is not a sine wave, but distorted in a manner characteristic of a vertical recording system. The wave of the electrical output of the

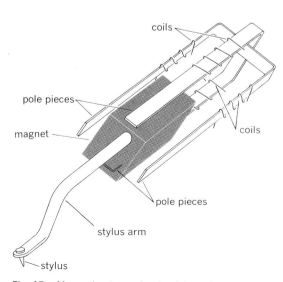

Fig. 17.   Magnetic stereophonic pickup. (*From H. F. Olson, Music, Physics and Engineering, Dover, 1967*)

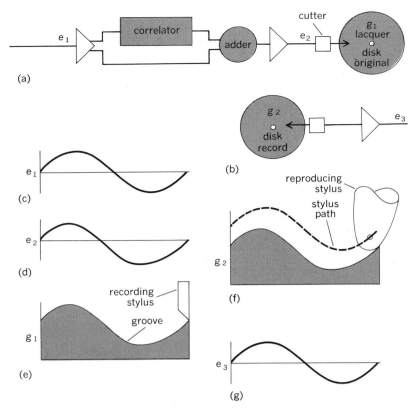

(a)

(b)

(c)

(d)

(e)

recording stylus

groove

g_1

(f)

reproducing stylus

stylus path

g_2

(g)

e_3

Fig. 19. The recording and reproducing process in a stereophonic disk system in which complementary distortion is introduced to reduce tracing distortion in reproduction. (*a*) Recording system. (*b*) Reproducing system. (*c*) Input to recording system. (*d*) Input to cutter. (*e*) Groove cut in original lacquer disk record. (*f*) Path of reproducing stylus in the groove of the disk record. (*g*) Output of reproducing system.

tem of the pickup and the plane of the record should be 15°.

Side thrust. A consideration of steady forces acting upon the stylus shows there is a force which is proportional to the horizontal tracking angle and is directed toward the center of the record. This places a permanent mechanical bias upon the pickup and may lead to distortion in stereophonic reproduction. This force can be corrected by the application of an equal and counter steady force applied to the pickup arm.

Surface noise. The signal-to-noise ratio in commercial stereophonic disk records is very large, of the order of 60 to 70 dB. Therefore, the noise in stereophonic disk records is not a problem for reproduction of sound in the home by means of disk records.

Compatibility. Monophonic and stereophonic records and phonographs are compatible as follows.

When a single-channel monophonic record is reproduced by a single-channel monophonic phonograph system, the single-channel output is reproduced by the loudspeaker.

When a single-channel monophonic record is reproduced by the two-channel stereophonic phonograph system, the single-channel output is reproduced by the two loudspeakers.

When a two-channel stereophonic record is reproduced by a single-channel monophonic phonograph reproducing system, the sound reproduced

reproducing system e_2 is unsymmetrical, which means that the major distortion component is a second harmonic.

The distortion can be reduced by the introduction of complementary distortion, as depicted in Fig. 19. The input signal to the recording system is again a sine wave e_1. However, the correlator introduces nonlinear distortion which is added to e_1. The output of the adder and the input to the cutter e_2 is not a sine wave. The groove cut in the original disk record g_1 corresponds to e_2. The disk record g_2 is a replica of the original record g_1. The groove in the record is distorted by the correlator so that the path of the reproducing stylus is a sine wave. The output of the reproducing system is a sine wave e_3. The mechanism of Fig. 19 indicates that the introduction of complementary distortion in the recording process will reduce distortion in the reproducing process. An electronic system which provides this type of distortion is termed a dynamic styli correlator. The distortion is reduced by a factor of more than 4 in actual practice.

Vertical tracking angle. A disparity between the effective vertical angle in the recording cutter cutting the modulated groove in the record and the vertical tracking angle of the pickup introduces harmonic and intermodulation in the output of the pickup. The standard vertical angle in stereophonic disk records is 15°. For this reason the angle between the axis of rotation of the vibrating sys-

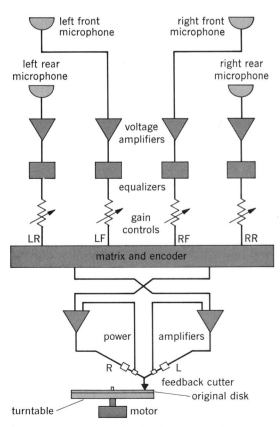

Fig. 20. Complete quadraphonic disk recording system employing matrixing and encoding for converting from four channels to two channels.

by the single loudspeaker is the sum of the two sound programs originally recorded on the two channels of the stereophonic recording system.

When the two-channel stereophonic record is reproduced by a two-channel stereophonic reproducing system, the stereophonic sound program is reproduced on the separate channels corresponding to the recording channels, and the sound from the two loudspeakers corresponds to the sound picked up by the respective microphones.

QUADRAPHONIC SYSTEM

There are two approaches to quadraphonic sound reproduction by means of the disk record: first, the matrixing and encoding from four channels, to recording on a two-channel disk, and reproducing from a two-channel disk and decoding and matrixing to four channels; and, second, the recording and reproducing of four discrete channels by means of the disk record.

Coding and matrixing system. There are many different ways of matrixing and encoding from four channels to two channels and decoding and matrixing from two channels to four channels. The generic recording and reproducing systems employing coding and matrixing in these processes are depicted in Figs. 20 and 21. The four inputs LF, LR, RF, and RR, representing left front, left rear, right front, and right rear channels, are matrixed and encoded to the left, L, and the right, R, channels of the conventional two-channel stereophonic disk recording system, as shown in Fig. 20.

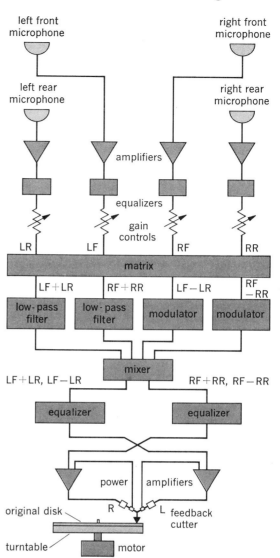

Fig. 22. Complete discrete four-channel quadraphonic disk matrixing and recording system in which two channels are recorded in the conventional two-channel manner and the other two channels are recorded by modulation means.

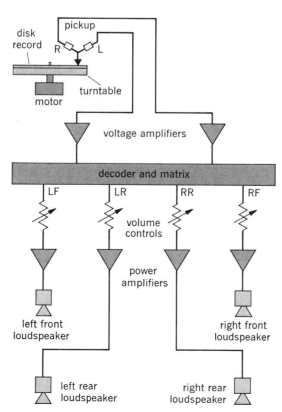

Fig. 21. Complete quadraphonic disk reproducing system employing decoding and matrixing for converting two channels to four channels.

Each of the two channels R and L contains a mixture of the four channels in such a manner that the four channels can be reconstituted in reproduction by means of the decoder and matrix which are illustrated in Fig. 21. In all of the coding and matrixing systems, the crosstalk separation that is achieved between LF, LR, RR, and RF signals is on the average 6 dB.

Discrete system. A discrete quadraphonic disk system, in which two channels are recorded in the conventional manner and the other two channels are recorded by modulation means in the region above 15,000 Hz, is depicted in Figs. 22–24. The left front, LF, plus the left rear, LR, signals are recorded in the left groove wall, and the right front, RF, plus the right rear, RR, signals are recorded in the right groove wall. These sets of signals are recorded in the conventional two-channel stereophonic mode and frequency range. The LF-minus-LR signals are recorded on a modulated carrier in the left groove wall, and the RF-minus-RR signals are recorded on a modulated carrier in the right groove wall, as shown in Fig. 22. The frequency-response characteristics are shown in Fig. 23. *See* MODULATION.

In reproduction, the RF-plus-RR signals are separated by means of a low-pass filter. RF-minus-RR signals are separated and detected by means of a high-pass filter and detector, as shown in Fig. 24. The two signals are fed to a matrix. The output is the discrete RF and RR signals. The discrete LF and LR signals are obtained from the left groove in the same way. The cross-talk separation between the four discrete signals LF, RF, LR, and RR is of the same order as that between the LF and RF signals in the conventional two-channel stereophonic system. *See* DETECTOR; ELECTRIC FILTER.

In general, the performance of the system with four discrete channels, as exemplified by auditory perspective and other spatial effects, is superior to that of the four-channel system derived from two channels by matrixing, encoding, and decoding processes.

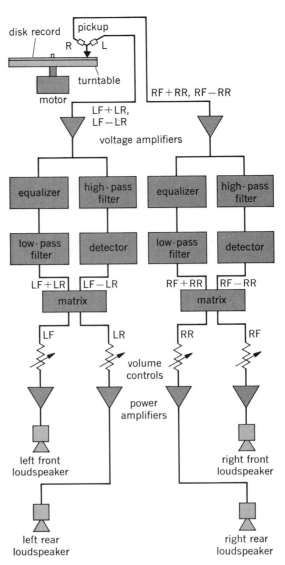

Fig. 24. Complete discrete four-channel quadraphonic disk reproducing system in which two channels are reproduced in the conventional two-channel manner and the other two channels are reproduced by demodulation means employing a detector. Matrixing provides the four discrete channels.

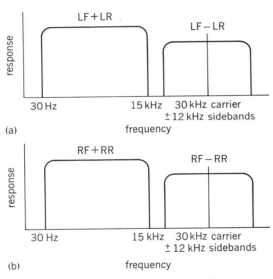

Fig. 23. The spectrum allocation depicts the frequency ranges of the four channels. (*a*) Left side of groove. (*b*) Right side of groove.

LR channels, and the R signal will be reproduced on RF and RR channels.

[HARRY F. OLSON]

DIGITAL RECORDING

Advances in solid-state devices and in recording systems make it practical to extend the benefits of digital techniques to the recording and reproduction of music on disk records. During recording, the output of the microphone amplifier is converted to a stream of digital bits. Typically the sampling rate lies between 44,000 and 55,000 samples per second with 16 bits per sample. The digitized program is recorded on magnetic tape. Special encoding schemes involving the insertion of additional bits permit error correction in playback of errors due to loss or misinterpretation of bits. Thus the digital signal may be rerecorded through many generations without deterioration of quality. Edit-

Compatibility. The quadraphonic system is compatible with monophonic and stereophonic systems. For example, if a quadraphonic record is reproduced on a one-channel monophonic system, the output will be the sum of the four signals LF, LR, RF, and RR. If a quadraphonic record is reproduced on a two-channel stereophonic system, the output on the L channel will be the sum of the LF and LR signals, and the output in the R channel will be the sum of RF and RR signals. If a monophonic record is reproduced on a quadraphonic system, all four channels LF, LR, RF, and RR will reproduce the same monophonic signal. If a stereophonic record is reproduced on a quadraphonic system, the L signal will be reproduced on LF and

ing and mixing are done with the signals in digital form. *See* ANALOG-TO-DIGITAL CONVERTER.

After editing and mixing, the digital signals are transferred from the tape to a disk rotating on the disk-recorder turntable. The recorded signal is carried in the form of microscopic indentations in a spiral track on the surface of the disk. Typically the indentations are less than 1 μm across and 0.15 μm deep (Fig. 25). The disk format and the recording techniques are those developed for video-disk recording of television signals. The master disk is used to form stampers for pressing vinyl records following the same processes used in the production of conventional phonograph records.

Players for the digital records resemble, and in some cases may be identical to, video-disk players. Depending on the record format offered by the manufacturers, the pickup in the player may be a capacitive, mechanical, or optical type. The digital output of the pickup and player circuits is error-corrected and stored temporarily in a memory. It is removed from the memory at a bit rate controlled by a precise quartz crystal, and is converted to analog audio signals to drive loudspeakers. Because the output bit rate is precisely controlled, any effects due to small fluctuations in speed of the tape recorders or of the disk turntables are removed from the analog output. *See* DIGITAL-TO-ANALOG CONVERTER.

The analog-to-digital and digital-to-analog converters are highly accurate devices, so that negligible distortion of the audio program results from digital processing. An effective dynamic range of about 90 dB is achieved in the magnetic recording stages. The bit rate is usually reduced to 14 bits per sample in the tape-to-disk transfer to yield an effective dynamic range of about 80 dB in disk playback, this being adequate for music reproduction in living rooms.

In the various prototype systems, disk diameters range from 4.5 to 12 in. (114 to 305 mm), and turntable rotation rates range from 200 to 900 rotations per minute. It is technically possible to digitally record more than 10 h of high-quality music on one side of a 12-in. disk, but this is considered imprac-

tical. Most offerings provide about 1 h of stereo per side. Digital records cannot be used on conventional phonographs and require their own players. Some record companies employ digital recording in making the master tapes, converting back to analog audio when recording the master disk. This procedure eliminates the loss of quality due to multiple generations of tape recordings, but does not remove the residual distortions and restrictions on dynamic range inherent in conventional phonograph reproduction.   [J. G. WOODWARD]

Bibliography: B. B. Bauer, D. W. Gravereaux, and A. J. Gust, A compatible stereoquadraphonic (SQ) record system, *J. Audio Eng. Soc.*, 19:638, 1971; J. B. Halter and J. G. Woodward, Vertical tracking errors in stereodisk systems, *J. Audio Eng. Soc.*, 12(1):8–14, 1964; T. Inove, N. Takahashi, and I. Owaki, A discrete four-channel disk and its reproducing system, *J. Audio Eng. Soc.*, 18(1):696, 1970; *NAB Disk Recording and Reproducing Standard*, 1963; H. F. Olson, *Modern Sound Reproduction*, 1972; H. E. Roys, *Disk Recording and Reproduction*, 1978; G. Stock, Digital recording: State of the market, *Audio*, 63(12):70–76, 1979; J. G. Woodward and E. C. Fox, Tracing distortion: Its cause and correction in stereodisk systems, *J. Audio Eng. Soc.*, 11(4):294–301, 1963.

Distortion (electronic circuits)

Any undesired change in the waveform of an electric signal passing through a circuit, including the transmission medium. In the design of any electronic circuit one important problem is to modify the input signal in the required way without producing distortion beyond an acceptable degree. Amplifier and loudspeaker systems are examples where maximum effort has been expended to produce a design for faithful amplification of speech and music input signals. There are four general types of distortion: amplitude, frequency, phase, and cross modulation.

Amplitude distortion. This is generally considered to mean distortion produced by a nonlinear relationship between the input and output amplitudes of a device. Amplitude distortion is usually introduced by a transistor or vacuum tube. The change in collector current of a transistor is nearly proportional to the change in signal voltage only over a small range of signal voltage amplitude. For further increase, the collector current change begins to depart from proportionality. *See* TRANSISTOR.

To predict the amount of amplitude distortion that may be produced in a given case, a power series representation for the characteristic of the active device is often used. The relationship between output signal current y and input signal x is expressed as

$$y = G_1 x + G_2 x^2 + G_3 x^3 + \cdots$$

The coefficients must be evaluated for each active device and the operating point used.

The harmonic distortion is expressed in terms of percentages of the fundamental component of the collector current when the input signal is a sinusoid. If $A \sin \omega t$ is substituted for x in the terms of the power series and the powers of $\sin \omega t$ are reduced by trigonometric identities to a fundamental and harmonic components, then the relative ampli-

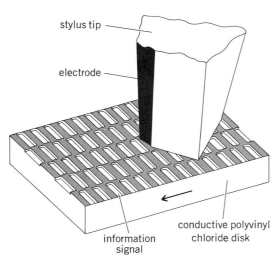

stylus tip

electrode

conductive polyvinyl chloride disk

information signal

Fig. 25. Schematic representation of high-density recorded surface of disk with capacitive-type pickup. (*Japan Victor Co.*)

tudes of each of the harmonic components can be calculated for known values of the coefficients in the power series.

In addition to distortion introduced by nonlinear transistor or tube characteristics, distortion can be introduced in several other ways, all of which occur for large amplitude signals.

Grid current. Distortion is produced when the signal applied to the input of a vacuum tube is so large that the grid becomes positive with respect to the cathode. The grid and cathode then form a diode, and current flows through the grid circuit. When this occurs, the dynamic input impedance can change appreciably and the stage amplification is adversely affected.

An additional form of distortion, called blocking, occurs in a resistance-capacitance coupled vacuum tube amplifier stage when grid current flows in the following tube. Under these conditions the coupling capacitor charges to nearly the plate voltage. When the plate voltage drops and grid conduction stops, the grid-to-cathode voltage of the following stage can become more negative than cutoff. This condition persists until the capacitor discharges (through a large resistance and therefore at a much slower rate than it charged) sufficiently to unblock the following stage. During the time that a stage is cut off, there will be no amplification. *See* VOLTAGE AMPLIFIER.

Saturation. When the input signal to a transistor or tube is large enough to drive the device to saturation, a further increase in the signal voltage produces little change in the collector or plate voltage.

Bottoming. This form of distortion is produced by a large input signal from a low-impedance source. No matter how large the input signal, the output current cannot exceed that value given by the ratio of the supply voltage divided by the load resistance. This type of distortion is called bottoming because the output voltage has fallen as low as possible.

Cutoff. This form of distortion is produced by large input signals which drive the transistor or tube to cutoff. The collector or plate voltage is then at its maximum value. If the input signal goes more negative, the collector or plate voltage cannot increase and distortion results.

Crossover distortion. An additional type of distortion is of importance in class B transistor power amplifiers. The input impedance varies inversely with emitter current and, for low values of current, may become appreciable compared with driver impedance. Under this circumstance a sinusoidal input voltage would not result in a sinusoidal output current. This is called crossover distortion.

Frequency distortion. This form of distortion is an inherent feature of all amplifiers but can be minimized by proper design. It occurs because the reactive elements and inherent reactances in the amplifier circuit do not allow the same amplification for all frequencies, and therefore some components of a signal are amplified more than others. Furthermore, in the case of audio amplifiers, the loudspeaker and enclosure characteristics affect the load presented to the amplifier in a manner which depends upon frequency. For a discussion of frequency response *see* AMPLIFIER.

The effects of frequency distortion may be considered in terms of an input signal composed of a fundamental and harmonic components, such as a square wave. If the amplifier gain is not a constant value for each frequency component, the output will not be an amplified replica of the input.

Phase distortion. Like frequency distortion, phase distortion is caused by the reactive elements in the circuit producing a phase shift that is not the same for each frequency component of the input signal. It is possible for an input signal to have frequency components with approximately constant magnitude amplification but not with a phase shift for each component which is proportional to the frequency. As a result, the output is not an amplified replica of the input. Fortunately, the ear is more tolerant of phase distortion than frequency distortion. This fact simplifies the design of a high-fidelity amplifier system.

If the gain magnitude is constant with frequency while the phase shift is proportional to the frequency, the output will be a replica of the input but delayed in time.

Cross modulation. Also called intermodulation, this effect is caused by nonlinear device characteristics. If two signals of different frequencies are applied to the input of a nonlinear transistor or vacuum tube stage, the output will contain the fundamental and harmonic components of each signal, frequency components equal to the sum and difference of the input signal frequencies, and sums and differences of the harmonics of the two input signals. Therefore, if the input is a signal composed of several frequencies, the nonlinearity will produce new frequencies not integrally related to those of the input signal. The distortion, if bad enough, is generally more noticeable than harmonic distortion.

Reduction by feedback. Distortion caused by nonlinear transistor or vacuum tube characteristics and by the frequency response of the amplifier can usually be reduced by the use of negative feedback. Amplitude distortion introduced in the last or next to last stages of a multistage amplifier can be reduced; distortion produced by the input stage cannot be reduced. Fortunately, in audio amplifiers distortion is generally produced in the last stage (the power output stage).

The use of negative feedback in a properly designed amplifier will make the amplitude more nearly constant and the phase shift more nearly proportional to frequency over a wider frequency range than in an amplifier without feedback. This extended frequency range can be made to include for all practical purposes the frequency range of music and speech signals. *See* FEEDBACK CIRCUIT.

[C. C. HALKIAS]

Bibliography: J. Millman, *Microelectronics: Digital and Analog Circuits and Systems*, 1979; J. D. Ryder, *Electronic Fundamentals and Applications*, 1975; R. F. Shea, *Amplifier Handbook*, 1966.

Distributed processing

The implementation of related information-processing functions in two or more programmable devices, connected so that data can be exchanged. It is often referred to as DDP, or distributed data processing.

Distinguishing characteristics. Distributed processing systems are distinguished from other types of computer-based systems by three characteristics. First, information-processing (application) functions are distributed. This excludes systems which consist of one information-processing computer surrounded by a network with programmable concentrators, network switches, or similar network-control devices. While such systems certainly include distributed functionality, they do not include distributed processing.

The second important characteristic is programmability in the distributed devices. Hard-wired terminals or terminal controllers, which provide management of only the terminal devices, do not fit within this definition. A central computer surrounded by terminals or terminal controllers plus terminals is not a distributed processing system.

The final important characteristic is connection for data exchange. Two independent computer systems operated by the same organization, but without any linking network, are an example of decentralized, not distributed, processing. Admittedly there is some degree of ambiguity in this definition. If the two example computer systems exchange data via reels of magnetic tape, which are created on one system and manually moved to the other for use as input, does this form a connection and transform these computers into a distributed system? The consensus is that in spite of the data transfer this is a decentralized system. Only a communications, cable, or bus connection between the computers would change this configuration into a distributed processing system.

There are situations in which components are linked together at some times and decoupled at others. For example, an organization might use personal computers or word-processing equipment mainly for local operations, in a free-standing, decentralized mode. At certain times, however, the local systems might be linked via dial-up connection to a central computer system, either to access data or services there or to send in data for processing. Whether this is considered a distributed or a decentralized system depends upon its major focus. If the local systems are selected and implemented as part of an overall system effort, but linked to the central site only occasionally to lower cost (or because only occasional connection is needed), this is best described as a distributed system. If the local systems are independently acquired, and operate mainly on local projects, but sometimes access the central system because it is convenient to do so, they are best described as decentralized systems.

Evolution. Centralized and decentralized processing were the typical modes of computer use prior to the introduction of minicomputers. Until then, the high cost of computer equipment made it most logical to concentrate processing to achieve economy of scale in both equipment and staff. In addition, data communications was a sufficiently unfamiliar technique, both to computer professionals and to the suppliers of voice-telephone-network facilities, that linking multiple computers into networks was seldom attempted.

At about the same time that minicomputers became generally available (in the late 1960s and early 1970s), the growing experience with data networks made these practical for use in more organizations. These two conditions came together to make possible the early attempts at distributed processing. Not long thereafter, the introduction of the microprocessor made it still more attractive to distribute functions rather than retaining and controlling them all centrally. *See* DATA COMMUNICATIONS; MICROPROCESSOR.

Forms of distributed processing. Although distributed systems are, in practice, extremely diverse, there are only two distributed processing structures, or ways in which to organize the functions within a distributed system. These can be used independently or can be combined.

Hierarchical systems. Hierarchical systems represent the earliest form of distribution processing. In these systems, processing functions are distributed outward from a central computer, called the host system, to intelligent terminal controllers or satellite information processors. A system of this type is shown in Figure 1. It is sometimes called a host-centered structure or a host/satellite system.

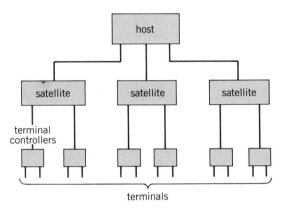

Fig. 1. Hierarchical system. Processing functions are distributed outward from the host system to satellite processors and intelligent terminal controllers.

The terms host and satellite are often equated to large system and minicomputer; however, those are not the true meanings. A host processor is typically located at some central site, where it serves as a focal point for the collection of data and often for the provision of services which cannot economically be distributed (for example, the management of large data bases, or simulations). Satellite processors are typically placed at, or near, point-of-transaction locations and are designed to serve the users at those locations. A satellite processor differs from a programmable terminal controller by providing file-management or database-management functions and also by executing application logic.

The first hierarchical systems consisted of a (usually) large host computer surrounded by mini-computer-based satellite processors, or in some cases by data-entry devices which might be either mini- or micro-based. Sometimes those structures resulted from conscious planning to install a distributed system; other times they resulted from linking independently acquired local processors to an existing central system. In the latter case, it was often discovered that the local applications

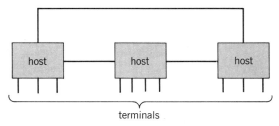

Fig. 2. Horizontal system. Logically equivalent computers are connected.

needed access to data or services available on an existing, larger central system.

Today the point-of-transaction (satellite) processors may be micro-based; for example, personal computers, data-entry devices, or word processors. They may also be minicomputers which are likely to combine two or more functions; for example, mini-based satellite processors can combine office automation functions such as word processing with data-processing functions such as transaction processing or query facilities. Satellite processors may also be general-purpose computers, and are sometimes smaller models of the same computer used as the host system.

Horizontal systems. Horizontal systems form the other major structure for distributed processing. In this structure, two or more computers which are logically equivalent are connected, as shown in Fig. 2. The term logically equivalent means that there is no connotation of hierarchy or master/slave relationship, even though the systems may not be physically equal in their functional capability or processing capacity. This form is also called a peer structure or a host/host connection.

Sometimes the goal in a horizontal system is to provide load leveling, to share the total workload among the available systems by transferring jobs and transactions (with their supporting data, if

necessary) as needed. More often, the intersystem linkage allows resources to be shared among the systems; for example, remote access may be provided to unique data bases, programs, or peripheral devices. In many of these systems, the linkage supports data transfer, so that data collected or generated at dispersed locations can be consolidated at one computer system.

Hybrid systems. Hybrid systems include both horizontal and hierarchical distribution, as shown in Fig. 3. Although this complex structure was originally very rare, it is becoming more common, especially as the merger of data processing and office automation becomes a goal for many organizations.

Advantages. Distributed processing can provide a number of advantages as compared to centralized or decentralized processing. These must, however, be considered potential advantages, as not all apply to every information system. The potential advantages are as follows: flexibility to change, because of the modularity of the distributed functions (modularity typically improves flexibility); the possibility of rapid expansion of capacity, if this can be accomplished by adding more satellite processors to the system; increased speed of response to terminal users (especially to large numbers of terminal users), because the processors are close to the users, and many processors can work in parallel; improved availability, because a distributed system is relatively invulnerable to single-point failures, and it may be possible to provide selective (rather than total) backup of only vital parts of the system; the potential for customization of functions, interfaces, reports, and so forth, to meet varying user needs; the possibility of less expensive data communications, especially if the satellite processors are close enough to the users and their terminals can be connected via local-area networks; and the fact that distributed processing mirrors the management control of most organizations, which is distributed (local managers have considerable responsibility and authority) rather than centralized.

Prospects. Many systems are still being implemented using centralized or decentralized structures. However, the increasing degree of computerization of main-line functions in many organizations accelerates the installation of more point-of-transaction satellite processors. Because most organizations require that data be transferred among all of their locations, this leads to the interconnection of the satellite processors and the formation of a distributed processing system. In addition, there is usually the need for central systems, at headquarters locations, and connections to those systems from the satellite processors. This situation applies both to traditional data-processing functions and to office automation functions as well. Over time, therefore, distributed processing can be expected to become the predominant form of computer-based information systems. *See* DIGITAL COMPUTER.

[GRAYCE M. BOOTH]

Bibliography: G. M. Booth, *The Distributed System Environment*, 1981; H. Katzen, Jr., *Distributed Information Systems*, 1979; T. Scannell and T. Henkel (eds.), Distributed data processing: Putting the pieces together, *Computerworld*, vol. 16, no. 8, Spec. Rep. pp. 1–47, Feb. 22, 1982.

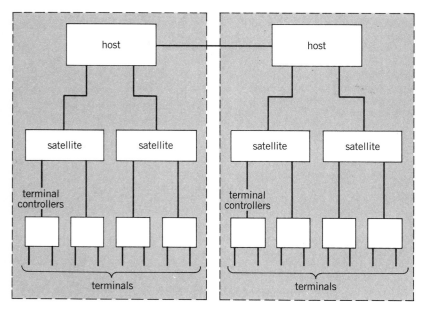

Fig. 3. Hybrid system, including both horizontal and hierarchical distribution.

Donor atom

An impurity atom in a semiconductor which can contribute or donate one or more conduction electrons to the crystal by becoming ionized and positively charged. For example, an atom of column V of the periodic table substituting for a regular atom of a germanium or silicon crystal is a donor because it has one or more valence electrons which can be detached and added to the conduction band of the crystal. Donor atoms thus tend to increase the number of conduction electrons in the semiconductor. The ionization energy of a donor atom is the energy required to dissociate the electron from the atom and put it in the conduction band of the crystal. See ACCEPTOR ATOM: SEMICONDUCTOR.   [H. Y. FAN]

Doppler radar

A radar system used to measure the relative velocity of the system and the radar target. The operation of these systems is based on the fact that the Doppler frequency shift in the target echo is proportional to the radial component of target velocity.

Airborne vehicular systems. Airborne systems are used to determine the velocity of the vehicle relative to the Earth for such purposes as navigation, bombing, and aerial mapping, or relative to another vehicle for fire control or other purposes. Ground or ship equipment is used to determine the velocity of vehicular targets for fire control, remote guidance, intercept control, traffic control, and other uses.

The Doppler frequency shift Δf is an extremely small fraction of the transmitter frequency f. It is given by the equation shown, where V is the relative speed, C is the speed of signal propagation, and γ is the angle between the velocity and the direction of propagation (Fig. 1). The only practical way to measure the frequency shift is by adding the echo signal to a reference signal derived from

$$\Delta f = \frac{2Vf}{C}\cos\gamma$$

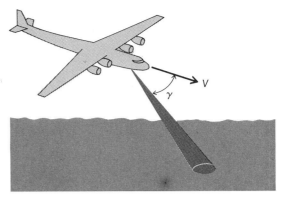

Fig. 1. Doppler frequency measurement geometry.

the transmitter and observing the difference, or beat frequency. Some means of obtaining coherent detection is required.

Practical techniques have been devised for obtaining the requisite coherence in continuous-wave (CW), pulsed, and frequency-modulated transmission systems. For general discussion of these techniques see CONTINUOUS-WAVE RADAR: RADAR. For a discussion of techniques similar to those in Doppler radar see MOVING-TARGET INDICATION.

Doppler navigation radar is a type of airborne Doppler radar system for determining aircraft velocity relative to the Earth's surface. Such a Doppler velocity sensor (Fig. 2) consists of at least the following elements: transmitter, antenna, receiver, Doppler frequency measuring device, and output signal generators or displays. It is generally used with a navigation computer.

The signal from a single beam can provide only the velocity component in the direction of that beam. Complete velocity determination requires, therefore, the use of at least three beams. Most systems use four beams for symmetry (Fig. 3).

To relate the beam directions, and hence the measured velocity to an Earth-oriented coordinate system, a vertical reference must be provided. In newer systems the antennas are fixed to the aircraft. The Doppler frequencies and the vertical

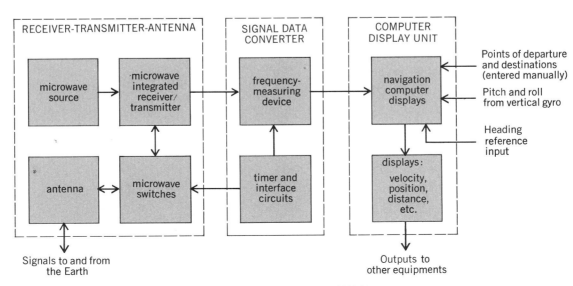

Fig. 2. Block diagram of aircraft Doppler navigation system, the AN/ASN-128.

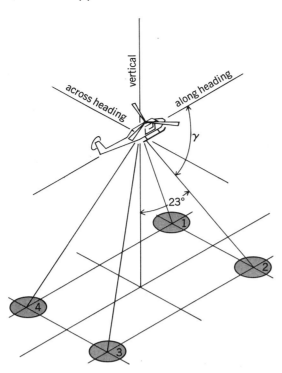

Fig. 3. Typical antenna beam arrangement.

data, such as roll and pitch, are fed to a computer. Its outputs are electrical signals or displays representing the components of velocity. The components commonly displayed are those along-heading, across-heading, and vertical.

Various types of antennas have found use in Doppler navigation radar. These include paraboloids, microwave lenses, and linear and planar arrays. Since both volume and radome cutout area should be small, various techniques are employed to enable each antenna to form (simultaneously or sequentially) more than one beam. Pencil beams of 3–5° width are used. The beams are directed 15–25° from the vertical. Larger values result in insufficient echo power over water.

Continuous-wave systems are coherent and are theoretically the most efficient. The chief difficulty is control of leakage of spurious signals from transmitter to receiver.

Pulsing enables the receiver to be rendered insensitive during transmission, thereby avoiding leakage signals. Coherence is achieved either by driving a transmitting power amplifier from a CW oscillator or by mixing at the detector a pair of pulse echoes received over different propagation paths.

A preferred technique is to employ sinusoidal frequency modulation. A sideband of the detected beat between echo and transmitter signal is used. Modulation index and rate and the sideband order are chosen such that echoes from nearby objects are rejected, while those from distant objects are accepted. Leakage noise is reduced at the expense of lowered efficiency.

An example of a system (Fig. 4) is the U.S. Army's standard airborne Doppler navigator which bears the nomenclature AN/ASN-128. It weighs 28 lb (12.7 kg), requires 89 W of power, and operates at 13.325-GHz frequency. In addition to velocities, outputs include present position, distance to go, desired track and bearing to destination, and steering signals. For helicopter use a steering hover indicator can also be provided. The system uses all-solid-state electronics and a printed grid planar array antenna. The velocity accuracy specifications are around a quarter of a percent of ground speed (and 0.2 knot or 0.1 m/s when hovering). The navigation accuracy specification is 1.3% of distance traveled when using a 1° accuracy heading reference.   [FRANCE B. BERGER]

Meteorological systems. Pulse Doppler radars are useful tools for the observation of the movements of precipitation particles. The Doppler frequency shift associated with the velocity of atmospheric targets, such as precipitation particles or artificial chaff, is always a very small fraction (10^{-6} to 10^{-8}) of the radar operating frequency. The observation and measurement of such small frequency shifts require excellent radar system frequency-stability characteristics that are not usually found in conventional radars but can be added without a drastic increase in equipment cost. Pulse Doppler radar effectively samples the backscattered signal at the radar pulse repetition rate and can therefore provide unambiguous Doppler frequency observations only in the frequency range allowed by the sampling rate. The Doppler frequency shift for the radial velocity of a given target increases with the transmitted frequency. Decreasing the radar operating frequency thus increases the effective velocity coverage for the same sampling rate. Since the need for increasing velocity coverage would otherwise require an increase of the pulse repetition rate (which reduces the radar maximum range), it is advantageous to use the longest possible wave-

(b)

(a)

(c)   (d)

Fig. 4. AN/ASN-128 Doppler navigation system. (a) Receiver-transmitter antenna. (b) Computer display unit. (c) Steering hover indicator. (d) Signal data converter. (Singer Company. Kearfott Division)

length. The choice of wavelengths is limited, however, to centimetric waves if production of narrow radar beams with antennas of reasonable size is desired.

Doppler spectrum. Since Doppler radars can observe and measure the radial velocity of targets, they are well suited for the observation of the motion of raindrops (precipitation). However, precipitation has the form of a "distributed" target; that is, there are numerous independent scatterers distributed in space. The backscattered signal that is selected at any given time after the transmitted pulse by a sampling circuit called a range gate is therefore due to the contribution of a finite scattering region that is determined by the radar beam cross-section area and half the radar pulse length in space (pulse volume). The backscattered signal sampled by the range gate is thus composed of a large number of separate scattering amplitudes, each having a Doppler frequency shift associated with the radial velocity of a particular scatterer. Therefore, a spectrum of frequency shifts (Doppler spectrum) is observed. This Doppler spectrum constitutes the basic velocity information acquired by the Doppler radar when it is used for the observation of precipitation systems.

At each range gate, a Doppler spectrum can be evaluated by processing the signal sampled at that range. The basic processing operation is a Fourier transform of the sampled signal. In order to take advantage of the radar resolution, the processing must be done at a large number of range gates, with the only limitation being that the gates must be spaced by more than one radar pulse width. The signals obtained at successive range gates can be digitized by analog-to-digital converters operating at high speed, with the digitized signals stored in a memory and then processed by a fast arithmetic system. Fast Fourier transform techniques are used for greatest computing efficiency. Although these systems are capable of processing several thousand spectra per second, they require large input memories and produce a volume of information that is difficult to handle.

Mean Doppler. The amount of information can be reduced by considering only the first moment of the Doppler spectrum (mean radial velocity, or mean Doppler) and the spectrum width. Although the mean Doppler can be evaluated directly from the Doppler spectra, it can be more easily obtained by use of mean-frequency estimators that do not involve the computation of the whole spectrum through the Fourier transform operation.

With this technique, the mean Doppler can be easily processed simultaneously at a large number of range gates with relatively simple digital circuits. The spectrum variance can be measured by a slightly different approach.

Probing motion in convective storms. It can be assumed that precipitation particles move with the air in their environment and are therefore good tracers for air motion. Observing the motion of precipitation particles inside convective storms therefore offers a unique technique for the study of kinematic processes in storms. However, the use of a single radar beam to scan the three-dimensional structure of a storm is of limited value, since only the radial velocity of precipitation particles can be observed. A more complete study of three-dimensional storm structure can be made by the simultaneous use of several scanning radars. For example, two Doppler radars installed at different locations, and operating independently with intersecting beams, will provide two different radial velocity components at the region where the beams intersect. These two components can be used to evaluate the two-dimensional velocity in the tilted plane that is common to the two radar beams and the baseline between the radars. *See* RADAR.

[ROGER LHERMITTE]

Bibliography: D. J. Barlow, Doppler radar, *Proc. IRE*, 37(4):340–355, 1949; F. B. Berger, The design of airborne Doppler velocity measuring systems, *IRE Trans.*, ANE-4(4):157–175, 1957; F. B. Berger, The nature of Doppler velocity measurement, *IRE Trans.*, ANE-4(3):103–112, 1957; H. Buell, Doppler radar systems for helicopter navigation, 36th Annual Meeting, Institute of Navigation, June 23–26, 1980 (to be published in the *Journal of the Institute of Navigation*); W. R. Fried, Principles and performance analysis of Doppler navigation systems, *IRE Trans.*, ANE-4(4):176–196, 1957: M. Kayton and W. R. Fried (eds.), *Avionics Navigation Systems*, 1969; R. Lhermitte, Meteorological Doppler radar, *Science*, 182(4109):258–262, 1973.

Earphones

A class of energy transducers capable of receiving alternating current and generating acoustic waves resembling very closely the characteristics of that current. The movement of an element (diaphragm) is accomplished by magnetic attraction, electrostatic attraction, or the piezoelectric effect (the expansion or contraction of certain crystalline substances in response to electric charges). Earphone systems include the driver element with its diaphragm and arrangements for magnetic flux, or electrostatic or direct electric charge, plus a casing, one or more acoustic cavities and ports, acoustic damping and insulation, and some arrangement for coupling the driver to the human ear. The wiring connecting the precedent amplifier to the driver, which may be incorporated into the earphone system, may in modern systems be a complicated circuit which feeds, for each of two stereophonic channels, some part of the current to each ear. The time delays and energy ratios at the two earphones (each of which may contain two drivers) can be appropriately adjusted so that the listener is given

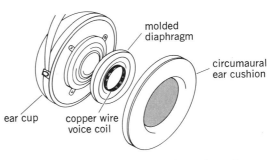

Fig. 1. Exploded view of conventional dynamic earphone. (*From P. Milton, Headphones: History and Measurement, Audio, 26(5):90–94, May 1978*)

the illusion that the sound sources are not "within the head" but are externalized appropriately in all three planes of space. The various types of earphones are described below. *See* TRANSDUCER.

Magnetic. In this type, a permanently magnetized diaphragm is moved in and out by an electromagnet energized by alternating current. Early models had two poles attracting and repelling a relatively heavy metal diaphragm with complex

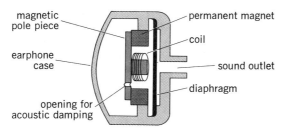

Fig. 2. Cross section of external-type hearing aid earphone. (*From W. R. Hodgson and P. H. Skinner, eds., Hearing Aid Assessment and Use in Audiologic Habilitation, Williams and Wilkins. 1977*)

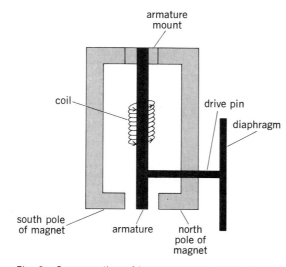

Fig. 3. Cross section of internal-type hearing aid earphone. (*From W. R. Hodgson and P. H. Skinner, eds., Hearing Aid Assessment and Use in Audiologic Assessment, Williams and Wilkins, 1977*)

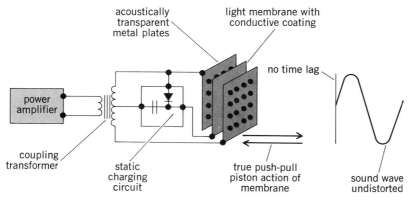

Fig. 4. Block diagram of electrostatic earphone. (*From H. Souther, An Adventure in Headphone Design: The Model ESP-6 Electrostatic Headphone, Koss Electronics Inc., Cat. no. 111, 1969*)

vibrational modes, the frequency response being limited to about 300–2500 Hz. In one development, the heavy metal diaphragm was replaced by a thin sheet of aluminum held by a ring of magnetic metal. The coil through which the signal current traveled had to move only the relatively low-weight ring, at greatly increased efficiency and with improved frequency response.

Dynamic. These earphones are actually small dynamic loudspeakers. In some, a small coil fed by the sound source is bonded to the membranous diaphragm (Fig. 1). Alternating current in the coil thus drives the diaphragm toward and away from a permanent magnet in the rear of the casing. In another configuration, the coil is relatively large and is attached to the diaphragm only at its edge; the diaphragm is a molded dome with a rolled edge and moves freely in the magnetic field. Such earphones are standard for high-fidelity communications.

Miniature magnetic. A common earphone is a small unit of which the output port fits snugly into a plastic olive in the ear canal (Fig. 2). These are widely used with small radios, with the more powerful hearing aids, by television commentators, in business transcription devices, and in many other communications situations. A coil fed by the signal current is wrapped around a pole and imparts in-and-out motion to the diaphragm proportional to the alternating current flow.

Another configuration (Fig. 3) is incorporated inside the case of most ear-level and all-in-the-ear hearing aids. A metal armature has a free end between the two poles of a magnet. When alternating current is flowing in one direction, the armature at the free end is turned into a south pole and the armature moves toward the north pole of the permanent magnet. As the current changes direction, the free end is turned into a north pole and the armature moves in the other direction. The armature moves a diaphragm, creating sound which is ported into the ear canal. *See* HEARING AID.

The diaphragm can be of any light material, not necessarily magnetic, and can be formed into an efficient shape for producing and directing an acoustic wave. Such miniature earphones are now made so efficient as to compete favorably with much larger types with respect to acceptably flat frequency response in the octave above 10 kHz and with respect to sound power levels in the ear canal with acceptably low total harmonic distortion.

Electrostatic. Efficiency is increased if the mass of the diaphragm is reduced to a minimum. A thin (2.5–12 μm) metallized plastic film can be used, on which a large constant electrostatic charge is placed by an auxiliary unit, and the motion of the diaphragm is controlled by the audio signal impressed on perforated wire mesh plates on either side (the push-pull arrangement reduces second-harmonic distortion). The assembly is mounted in a relatively large cavity and coupled to the head by a circumaural cushion (that is, one which completely surrounds the auricle). Such earphones (Fig. 4) are light and comfortable and have excellent frequency-response and transient-distortion characteristics.

Dynamic-electrostatic. The diaphragm of a dynamic-electrostatic (or orthodynamic) earphone (Fig. 5) is a permanently polarized electret of fluo-

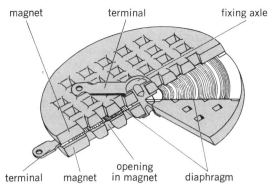

Fig. 5. Construction of orthodynamic earphone driver. *(From P. Milton, Headphones: History and Measurement, Audio, 26(5):90–94, May 1978)*

rocarbon. Consequently the need for an added source of polarization voltage (a drawback inherent to electrostatic earphones) is eliminated. In one development, certain problems of such a system are avoided by separating the functions of the diaphragm and of the electret. The diaphragm can be of the thinnest (2.5 μm) metallized polyester, stretched between two electrets in push-pull. The diaphragm is matched in acoustic impedance very closely to the surrounding air, and the audio signal can be led directly to the electrets. *See* ELECTRET TRANSDUCER.

Piezoelectric. Certain crystalline substances expand and contract when alternating voltage is applied. In some piezoelectric earphones, a crystal element is coupled mechanically to the center of a small (about 2.5 cm diameter) cone. Such earphones can be lightweight and cheap and may be acceptable for speech communication.

Substances other than quartz crystals, notably high-polymer films such as vinylidene fluoride, will expand and contract when an alternating current is applied. A thin sheet of such material can be stretched by a factor of 4, and aluminum vapor deposited over it. If a high dc voltage is now applied, and removed, the sheet will become and remain piezoelectric. Two sheets are bonded together back to back, and when alternating current is applied between the metallized faces the sheet will expand and contract. If a rectangular sheet is slightly curved and mounted rigidly top and bottom, much as a playing card might be grasped between the fingers, the movement in response to alternating current will be in and out in proportion and an acoustic wave will be created (Fig. 6). Earphones on this principle have desirable characteristics of stability, acoustic response, and cost, but require more amplification than an electret or other earphone to operate as efficiently. *See* PIEZOELECTRIC CRYSTAL; PIEZOELECTRICITY.

Real-ear response. An earphone system contains components which are in some respects as important as the driver. The configuration of the acoustic ports, for example, or the acoustic or other damping built into the system will help determine the evenness of the frequency response, while the cushions, whether circumaural or supraaural (fitting against the auricle), help determine the low-frequency real-ear response.

The effects of changes in these elements as re-

flected in acoustic differences at the eardrums of actual persons are not easy to determine. It is not difficult to couple an earphone system to a microphone by means of a standard cavity, perhaps incorporating a flat plate for the larger circumaural systems, but such a practice is useful primarily for manufacturer's quality control, and has limited relevance to the performance of the system on even the average human head. The reference equivalent threshold sound-pressure level of a certain earphone system, expressed in voltage applied as frequency changes, has to be determined by having a panel of normal-hearing listeners loudness-balance that system at each frequency against another (arbitrarily chosen) reference system.

The acoustic impedance of the several elements of the external ear, head, and torso have been measured, and artificial heads incorporating the results of such measurements are available. These devices allow fairly successful predictions as to how an earphone system will perform on an anthropometrically average person.

Realistic simulation. Any single-channel recording of a real acoustic event will, when played back to an earphone in one ear, sound "in the head." Even stereo recordings from two channels played back to earphones on two ears, while furnishing the illusion of movement, still are not externally localized by the listener. Externalization is improved if some of the signal from the left channel is time-delayed and applied also to the right earphone (the same, of course, for the right channel to the left earphone), and it is also improved if a realistic ratio is achieved between the acoustic energy density ratio from the "direct" versus the "indirect" sound sources (as from reflective walls). Stereophonic earphone systems have been built which incorporate these time delays and ratios, and which furthermore feed the signal from the right channel of an artificial-head stereophonic recording to the left earphone (and the left channel to the right earphone) through frequency filters which simulate the differential acoustic effects at various frequencies of the head and external ears in the original recording situation. Thus the ear-

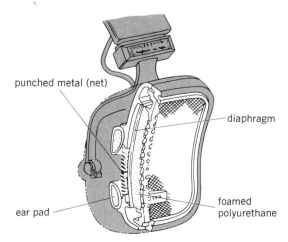

Fig. 6. Cutaway view of rear of high-polymer piezoelectric earphone. *(From P. Milton. Headphones: History and Measurement. Audio. 26(5):90–94. May 1978)*

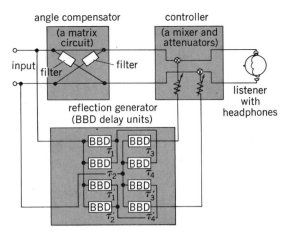

Fig. 7. Block diagram of projected sound localization earphone system for stereophonic sources. BBD-bucket brigade device. (*From N. Sakamoto, T. Gotoh, and Y. Kimura, On "out-of-head" localization in headphone listening, J. Audio Eng. Soc., 24:710–716, 1976*)

drums under earphones are presented with the exact acoustic conditions generated by a loudspeaker or other sound source in an actual room, and such earphone systems very materially advance the important psychoacoustic feature of externalization of sound and of acoustic realism generally (see Fig. 7). *See* LOUDSPEAKER; MICROPHONE.

[J. DONALD HARRIS]

Echo box

A device used to check the output power and spectrum of a radar transmitter. It consists of a low-loss, tunable, resonant cavity connected to the antenna feed line through a fixed coupling circuit so that the fraction of output power supplied to the cavity is always constant. The signal level within the cavity depends on the strength of the portion of the transmitter output spectrum lying within the cavity's narrow passband, which can be tuned to traverse the entire frequency range of interest. A microammeter connected through a crystal rectifier to a loop within the cavity permits reading the signal level. The spectrum can be measured as the cavity is tuned to different frequencies.

A single test of the performance of the entire radar system, excluding only the antenna and antenna feed, can be made with the cavity tuned to the carrier frequency. Each transmitter output pulse causes a slowly damped oscillation in the cavity which feeds a signal back through the coupling circuit to the receiver. This signal appears as an echo at the receiver output (hence the name echo box). The time required for the echo to decay to the level of the receiver noise is proportional to the logarithmic difference, or difference in decibels, between the transmitter power level and the receiver noise level. This is an excellent overall figure of merit for the transmitter and receiver performance. If a dummy load is substituted in place of the antenna to absorb the output power, the radar can be tested without actually radiating, which may be useful in some military situations. *See* RADAR. [ROBERT I. BERNSTEIN]

Eddy current

An electric current induced within the body of a conductor when that conductor either moves through a nonuniform magnetic field or is in a region where there is a change in magnetic flux. It is sometimes called Foucault current. Although eddy currents can be induced in any electrical conductor, the effect is most pronounced in solid metallic conductors. Eddy currents are utilized in induction heating and to damp out oscillations in various devices.

Causes. If a solid conductor is moving through a nonuniform magnetic field, electromotive forces (emfs) are set up that are greater in that part of the conductor that is moving through the strong part of the field than in the part moving through the weaker part of the field. Therefore, at any one time in the motion, there are many closed paths within the body of the conductor in which the net emf is not zero. There are thus induced circulatory currents that are called eddy currents (see illustration). In accordance with Lenz's law, these eddy currents circulate in such a manner as to oppose the motion of the conductor through the magnetic field. The motion is damped by the opposing force. For example, if a sheet of aluminum is dropped between the poles of an electromagnet, it does not fall freely, but is retarded by the force due to the eddy currents set up in the sheet. If an aluminum plate oscillates between the poles of the electromagnet, it will be stopped quickly when the switch is closed and the field set up. The energy of motion of the aluminum plate is converted into heat energy in the plate.

Eddy currents are also set up within the body of a material when it is in a region in which the magnetic flux is changing rapidly, as in the core of a transformer. As the alternating current changes rapidly, there is also an alternating flux that induces an emf in the secondary coil and at the same time induces emfs in the iron core. The emfs in the core cause eddy currents that are undesirable because of the heat developed in the core (which results in high energy losses) and because of an undesirable rise in temperature of the core. Another undesirable effect is the magnetic flux set up by the eddy currents. This flux is always in such a

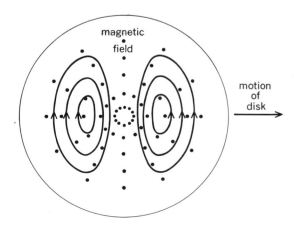

Eddy currents which are induced in a disk moving through a nonuniform magnetic field.

direction as to oppose the change that caused it, and thus it produces a demagnetizing effect in the core. The flux never reaches as high a value in the core as it would if there were no eddy currents.

Laminations. Induced emfs are always present in conductors that move in magnetic fields or are present in fields that are changing. However, it is possible to reduce the eddy currents caused by these emfs by laminating the conductor, that is, by building the conductor of many thin sheets that are insulated from each other rather than making it of a single solid piece. In an iron core the thin iron sheets are insulated by oxides on the surface or by thin coats of varnish. The laminations do not reduce the induced emfs, but if they are properly oriented to cut across the paths of the eddy currents, they confine the currents largely to single laminae, where the paths are long, making higher resistance; the resulting net emf in the possible closed path is small. Bundles of iron wires or powdered iron formed into a core by high pressure are also used to break up the current paths and reduce the eddy currents.   [KENNETH V. MANNING]

Bibliography: B. Bleaney and B. I. Bleaney, *Electricity and Magnetism*, 3d ed., 1976; E. M. Pugh and E. W. Pugh, *Principles of Electricity and Magnetism*, 2d ed., 1970.

Electret

A solid dielectric possessing persistent dielectric polarization. An electret is the analog of a magnet. Electrets are made by cooling suitable dielectrics from elevated temperatures in strong electric fields. A special class called photoelectrets is produced by the removal of light from an illuminated photoconductor in an electric field.

Electrets can be prepared from certain organic waxes and resins (for example, carnauba wax) or from ferroelectric crystals or ceramics such as barium titanate. Photoelectrets have been prepared from sulfur, cadmium and zinc sulfides, and anthracene. Electrets are metastable; their polarizations decay slowly after removal of the applied field and more rapidly with increasing temperature. Space-charge polarization is the principal mechanism involved in electret formation, except for ferroelectric substances. *See* FERROELECTRICS.

Electret transducer

A device for the conversion of acoustical or mechanical energy into electrical energy, and vice versa, which utilizes a quasi-permanently charged dielectric material (electret). Examples are certain microphones, earphones, and phonograph cartridges. In the simplest implementation, such a transducer consists of a metal backplate (first electrode) covered by a mechanically tensioned diaphragm. The diaphragm is a foil electret carrying a metal coating (second electrode) on the side facing away from the backplate. Provisions are made to maintain a shallow air gap between electret and backplate. The air gap is occupied by an electrostatic field originating from the electret charges. Upon acoustical or mechanical deflection of the diaphragm, such a device generates an electrical output signal between its two electrodes; similarly, application of an electrical signal results in diaphragm deflections. Electret devices are therefore self-biased electrostatic or condenser transducers. They thus exhibit all the advantages of this transducer class, such as wide dynamic range and flat response over a frequency range of several decades, without requiring the external bias necessary in conventional transducers of this kind. *See* ELECTRET; TRANSDUCER.

Foil electret. The basic component of all modern electret transducers is a "foil electret" consisting of a thin film of insulating material that has been electrically charged to produce an external electric field. Strongly insulating materials capable of trapping charge carriers, such as the halocarbon polymers, in particular polyfluoroethylenepropylene (Teflon), are best suited for this purpose. Before charging, the material is either metallized on one side or backed up with a metal electrode.

Charging can be achieved in a number of ways, for example, by treatment with a corona discharge or by voltage application by means of a wet-contact electrode. Another charging technique, allowing greater control than other methods, is electron injection with an electron beam having a range less than the film thickness. In all cases, carriers of one polarity are injected into the insulator and trapped. A compensation charge of equal magnitude but opposite sign flows into the back electrode, thus forming an electric double layer. *See* CORONA DISCHARGE.

In the presence of electrodes, the electret exhibits an external electric field extending from the electret to the electrodes. Such a foil electret is thus, as far as its external field is concerned, an electrostatic analog of a permanent magnet.

Charge storage on halocarbon polymers is relatively permanent. At charge densities of about 20 nC/cm^2, as required in electret transducers, the time constant of the charge decay of Teflon is about 200 years if stored under shielded conditions at room temperature and low relative humidity. This time constant drops to about 10 years at 50°C and 95% relative humidity. However, Teflon has less favorable mechanical properties because of stress relaxation, which causes a decrease in the mechanical tension of stretched films. As discussed below, this effect can be minimized in transducer applications.

Electret microphones. A simple implementation of the most widely used electret transducer, the foil-electret microphone, is shown in the cutaway drawing of Fig. 1. The diaphragm, typically 12- or 25-μm Teflon metallized on one surface, is charged to $10-20$ nC/cm^2, corresponding to an external bias of about 200 V. The nonmetallized surface of this foil electret is placed next to a backplate, leaving a shallow air layer, the thickness of which (about 20 μm) is controlled by ridges or raised points on the backplate surface. The stiffness of the air layer can be decreased (and thus the sensitivity of the microphone can be improved) by connecting the air layer to a larger cavity by means of small holes through the backplate. The backplate is either a metal disk or a metal-coated dielectric with a thermal expansion coefficient about equal to that of the diaphragm. The electrical output of the microphone is taken between the backplate, which is insulated from the outer case, and the metal side of the foil. The output is fed into a high-impedance preamplifier.

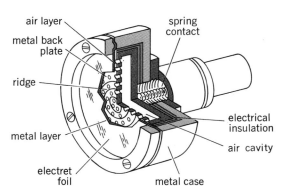

Fig. 1. A cutaway drawing of a foil-electret microphone showing the basic elements. (*From G. M. Sessler and J. E. West, Electret transducers: A review, J. Acoust. Soc. Amer., 53:1589, 1973*)

At frequencies below the resonance frequency the acoustical properties of electret microphones are largely governed by the restoring force on the diaphragm. Since the mechanical tension of the foil is generally kept at a relatively low value (about 10 N/m), the restoring force is determined by the compressibility of the air layer. Controlling the restoring force by the air layer is advantageous because changes in tension due to stress relaxation thus have only a minor effect on the sensitivity, which is largely independent of transducer area.

The problem of stress relaxation can be virtually eliminated by using a modified microphone design consisting of a metallic backplate coated with a layer of permanently charged Teflon. A metal (or metallized polyester) diaphragm is stretched over this backplate. In this approach, the excellent electrical properties of the Teflon layer and the good mechanical behavior of the metal or polyester diaphragm are used to advantage. A cross-sectional view of an electret microphone of such design is shown in Fig. 2.

Under open-circuit conditions, a displacement d of the diaphragm of an electret microphone causes a frequency-independent output voltage given by Eq. (1), where E is the (constant) electric field in

$$v = Ed = \frac{\sigma D_1 d}{\epsilon_0 (D_1 + \epsilon D_2)} \quad (1)$$

the air layer of the transducer; σ is the surface-

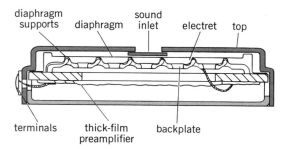

Fig. 2. Schematic cross section of a modern electret microphone for use in hearing aids. The microphone has a Teflon electret bonded to a properly shaped backplate. (*From M. C. Killion, Vibration sensitivity measurements on subminiature condenser microphones, J. Audio Eng. Soc., 23:123–127, 1975*)

charge density of the electret projected onto the surface; D_1 and D_2 are the thicknesses of electret film and air layer, respectively; ϵ_0 is the permittivity of free space; and ϵ is the relative dielectric constant of the electret material. As in conventional electrostatic transducers, the displacement is proportional to the applied pressure in a wide frequency band extending from a lower cutoff given by a pressure-equalization leak in the back cavity to an upper cutoff determined by the resonance frequency. The voltage response for constant sound pressure is therefore frequency-independent in this range.

Typical electret microphones designed for the audio-frequency range have constant sensitivities of 1–5 mV/μbar or 10–50 mV/Pa in the frequency range from 20 to 15,000 Hz. Nonlinear distortion is less than 1% for sound-pressure levels below 140 dB, and the impulse response is excellent, owing to the flat amplitude and phase characteristics. Other properties of electret microphones are their low sensitivity to vibration, owing to the small diaphragm mass, and their insensitivity to magnetic fields. Compared with conventional electrostatic transducers, electret microphones have the following advantages: they do not require a dc bias; they have three times higher capacitance per unit area, resulting in a better signal-to-noise ratio; and they are not subject to destructive arcing between foil and backplate in humid atmospheres and under conditions of water condensation.

Various electret microphones for operation at infrasonic and ultrasonic frequencies, covering the range from 0.001 Hz to 200 MHz, have been designed by properly positioning the upper and lower cutoff frequencies discussed above. Furthermore, transducers with directional characteristics such as cardioid, bidirectional, toroidal, and second-order unidirectional have been developed. *See* MICROPHONE.

Electret headphones. Another group of electret transducers of interest are electret headphones. The above microphone designs can also be utilized in this case. They produce a certain amount of nonlinear distortion due to the quadratic dependence of the force F per unit area, exerted on the diaphragm, on the applied ac voltage v, as can be seen from Eq. (2). This drawback can, however,

$$F = -\frac{1}{2}(D_1\sigma + \epsilon\epsilon_0 v)^2/\epsilon_0 (D_1 + \epsilon D_2)^2 \quad (2)$$

be minimized in a well-designed system, and high-voltage sensitivities have been achieved by making the air-gap thickness small. The distortion can be eliminated by the use of push-pull transducers and monocharge electrets.

The principle of a push-pull transducer is shown in Fig. 3a. A diaphragm consisting of two electrets, each metallized on one side and with these metal layers in contact, is sandwiched between two perforated metal electrodes, forming a symmetrical system. Application of a signal voltage $\bar{v}$ in antiphase to the two electrodes causes the electrodes to exert forces F_1 and F_2 on the diaphragm. These forces are given by Eq. (2) if $\pm\bar{v}$ is substituted for v. The net force $F = F_1 - F_2$ on the diaphragm is determined by Eq. (3), which is linear in $\bar{v}$, indi-

$$F = -2\bar{v}\sigma\epsilon D_1/(D_1 + \epsilon D_2)^2 \quad (3)$$

cating the absence of nonlinear distortion. The

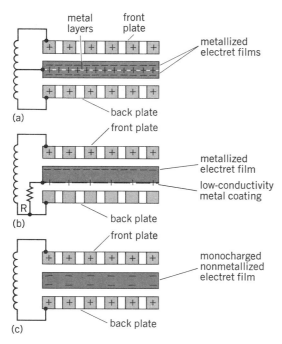

Fig. 3. Schematic cross sections of push-pull electret headphones, showing static charge distributions. (a) Conventional double-electret transducer. (b) Single-electret transducer with metallized conventionally charged electret. (c) Single-electret transducer with nonmetallized monocharged electret.

linearity, however, is maintained only if both air-layer thicknesses remain at the same value D_2. This is only possible if D_2 is relatively large, which makes such systems less sensitive than the above-described single-ended transducers.

This transducer can be simplified and improved by using the scheme shown in Fig. 3b. In this asymmetric system, a single electret coated on one side with a low-conductivity layer is used. The diaphragm holds an excess charge because of the presence of an induction charge on the front electrode. Thus, application of an ac signal causes a net force between diaphragm and plates which is linear if the resistivity of the coating (or the resistor R) prevents charge equalization during a period of lowest-frequency signal applied.

An interesting modification of the above push-pull transducers can be achieved by using non-metallized monocharge electrets, which differ from customary electrets in that they carry only a single-polarity charge and no compensation charge. It is possible to form these electrets by electron injection or by the use of wet-contact charging methods. Since a compensating charge is absent, monocharge electrets have a strong external field that is independent of electrode distance. Thus, if such electrets are used in transducer configurations (Fig. 3c), large air gaps of the order of 1 mm can be used without loss in field strength. This in turn results in highly efficient transducers that can operate at large displacement amplitudes.

Electromechanical transducers. Apart from its use in electroacoustic transducers, such as microphones and earphones, the electret principle has been applied to electromechanical transducers. Examples are phonograph cartridges, touch or key transducers, and impact transducers.

Two types of phonograph cartridges have been suggested. In one implementation, the stylus is coupled by a cavity to an electret microphone. The cavity serves as an acoustic transformer that converts the large-amplitude stylus vibrations to sound waves sensed by the microphone. Such cartridges have favorably low mass and are capable of high trackability and linearity over a wide frequency range, and yet are immune to magnetic fields. Another design consists essentially of a cantilever, to which a stylus is attached, and two electrets. The electrets, together with the cantilever, form a pair of electret transducers actuated by the vibrations of the cantilever. Advantages of these cartridges are low intermodulation distortion and a dynamic range of 100 dB. *See* DISK RECORDING.

Touch-actuated electret transducers similar in design to electret microphones have also been implemented. These transducers generate output signals on the order of 10–100 V due to direct or key operation resulting in a deflection of the diaphragm. Because of the small diaphragm deflections possible in such a device, tactile feedback is almost totally absent. However, feedback can be provided by actuating a mechanically compliant movable electrode by a mechanical spring.

Impact-sensitive electret transducers have been implemented as more or less conventional systems similar in design to the above-described microphones and as line transducers consisting essentially of a coaxial cable with polarized dielectric. In the line transducers the center conductor and the shield serve as electrodes. Mechanical excitation resulting in a deformation of the shield at any point along the length of such a cable produces an electrical output signal.

Applications. Because of their favorable properties, simplicity, and low cost, electret transducers have been used in many applications, both as research tools and in the commercial market.

Among the research applications are microphones for use in acousto-optic spectroscopy, applied to the detection of air pollution and to the study of reaction kinetics of gases and optical absorption of solids. Because of the favorable noise performance of electret microphones, the detection threshold for air pollutants has been lowered by more than an order of magnitude. Other applications of electret microphones have been reported in aeronautics and shock-tube studies, in which the low vibration sensitivity of these transducers is crucial. The wide frequency range of electret transducers, discussed above, made possible their application in infrasonic atmospheric studies and in ultrasonic investigations of liquids and solids. In addition, ultrasonic arrays of electret microphones have been used in acoustic holography. Research uses of electromechanical electret transducers have been reported in such diverse areas as vibration analysis and leak detection in space stations. In these applications, the simplicity and reliability of electret transducers are of importance. *See* PHOTOACOUSTIC SPECTROSCOPY.

Of all commercial applications of electret devices, the high-fidelity electret microphone for amateur, professional, studio, and tape recorder use is most prominent. Although the electret microphone was introduced only about 1970, tens of millions of the instruments are produced annually, accounting for about one-half of the entire output of high fidelity microphones. Other uses of electret microphones are in sound-level meters, noise dosimeters, movie cameras, and speaker phones. Miniature microphones are used in hearing aids, in telephone operators headsets, and in communication headsets, often as noise-canceling (first-order gradient) transducers. Ultrasonic electret microphones are used in burglar alarm systems and in remote TV control units. The success of the electret microphone in these applications is primarily due to its acoustic quality and low cost. It is noteworthy that, owing to their low vibration sensitivity, foil-electret microphones were the first transducers to be built directly into widely used tape recorders. In addition to electret microphones, single-backplate and push-pull electret earphones, also with monocharge electrets, have been introduced to the market. Under study for commercial application are electret phonograph cartridges, optical display systems with mechanical electret gates, electret key transducers, and underwater transducers.

[GERHARD M. SESSLER]

Bibliography: G. M. Sessler and J. E. West, Applications, in G. M. Sessler (ed.), *Electrets*, 1980.

Electric charge

A basic property of elementary particles of matter. One does not define charge but takes it as a basic experimental quantity and defines other quantities in terms of it. The early Greek philosophers were aware that rubbing amber with fur produced properties in each that were not possessed before the rubbing. For example, the amber attracted the fur after rubbing, but not before. These new properties were later said to be due to "charge." The amber was assigned a negative charge and the fur was assigned a positive charge.

According to modern atomic theory, the nucleus of an atom has a positive charge because of its protons, and in the normal atom there are enough extranuclear electrons to balance the nuclear charge so that the normal atom as a whole is neutral. Generally, when the word charge is used in electricity, it means the unbalanced charge (excess or deficiency of electrons), so that physically there are enough "nonnormal" atoms to account for the positive charge on a "positively charged body" or enough unneutralized electrons to account for the negative charge on a "negatively charged body."

The rubbing process mentioned "rubs" electrons off the fur onto the amber, thus giving the amber a surplus of electrons, and it leaves the fur with a deficiency of electrons.

In line with the previously mentioned usage, the total charge q on a body is the total unbalanced charge possessed by the body. For example, if a sphere has a negative charge of 1×10^{-10} coulomb, it has 6.24×10^8 electrons more than are needed to neutralize its atoms. The coulomb is the unit of charge in the meter-kilogram-second (mks) system of units.

The surface charge density σ on a body is the charge per unit surface area of the charged body. Generally, the charge on the surface is not uniformly distributed, so a small area ΔA which has a magnitude of charge Δq on it must be considered. Then σ at a point on the surface is defined by the equation below.

$$\sigma = \lim_{\Delta A \to 0} \frac{\Delta q}{\Delta A}$$

The subject of electrostatics concerns itself with properties of charges at rest, while circuit analysis, electromagnetism, and most of electronics concern themselves with the properties of charges in motion. *See* CAPACITANCE. [RALPH P. WINCH]

Electric current

The net transfer of electric charge per unit time. It is usually measured in amperes. The passage of electric current involves a transfer of energy. Except in the case of superconductivity, a current always heats the medium through which it passes. For a discussion of the heating effect of a current *see* JOULE'S LAW.

On the other hand, a stream of electrons or ions in a vacuum, which also may be regarded as an electric current, produces no local heating. Measurable currents range in magnitude from the nearly instantaneous 10^5 or so amperes in lightning strokes to values of the order of 10^{-16} amp, which occur in research applications.

All matter may be classified as conducting, semiconducting, or insulating, depending upon the ease with which electric current is transmitted through it. Most metals, electrolytic solutions, and highly ionized gases are conductors. Transition elements, such as silicon and germanium, are semiconductors, while most other substances are insulators.

Electric current may be direct or alternating. Direct current (dc) is necessarily unidirectional but may be either steady or varying in magnitude. By convention it is assumed to flow in the direction of motion of positive charges, opposite to the actual flow of electrons. Alternating current (ac) periodically reverses in direction.

Conduction current. This is defined as the transfer of charge by the actual motion of charged particles in a medium. In metals the current is carried by free electrons which migrate through the spaces between the atoms under the influence of an applied electric field. Although the propagation of energy is a very rapid process, the drift rate of the individual electrons in metals is only of the order of a few centimeters per second. In a superconducting metal or alloy the free electrons continue to flow in the absence of an electric field after once having been started. In electrolytic solutions and ionized gases the current is carried by both positive and negative ions. In semiconductors the carriers are the limited number of electrons which are free to move, and the "holes" which act as positive charges.

Displacement current. When alternating current traverses a condenser, there is no physical flow of charge through the dielectric (insulating material), but the effect on the rest of the circuit is as if there were a continuous flow. Energy can pass through the condenser by means of the so-called displacement current. James Clerk Maxwell introduced the concept of displacement current in order to make complete his theory of electromagnetic waves. *See* ALTERNATING CURRENT; CONDUCTION (ELECTRICITY); DIRECT CURRENT; ELECTRICAL RESISTANCE; SEMICONDUCTOR; SUPERCONDUCTIVITY.   [JOHN W. STEWART]

Electric filter

A transmission network used in electrical systems for the selective enhancement (or reduction) of specified components of an input signal. The filtering is accomplished by selectively attenuating those frequency components of the input signal which are undesired, relative to those which it is desired to enhance.

The performance of any transmission network is measured by its transfer function F, which is the ratio of output to input signal (when these are described appropriately). The general form of such a transfer function is given by Eq. (1), where s is the

$$F(s) = \frac{Hs(s^2 + \omega_0{}^2) \cdots}{(s^2 + a_1 s + a_0)(s^2 + b_1 s + b_0) \cdots} \quad (1)$$

complex frequency variable. (For the sinusoidal steady state, $s = j\omega$, where $j = \sqrt{-1}$ and ω is the angular frequency.) Each quadratic factor in the denominator determines a pair of poles; the factors in the numerator determine the zeros.

A typical filter network having two ports is shown in Fig. 1. Since the output is generally

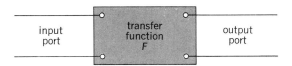

Fig. 1. Two-port filter network.

smaller than the input, it is customary to specify the amplitude information as the amount of reduction of the input signal; the loss or attenuation is defined by Eq. (2), and is measured in decibels (dB).

$$\alpha = -20 \log |F| \quad (2)$$

Whenever the output signal has an amplitude smaller than that of the input signal, $|F|$ will be less than 1, and its logarithm will be negative; thus the attenuation will have a positive value. *See* DECIBEL.

Filter types. An optical filter transmits, or passes, certain colors in the light spectrum and rejects, or stops, others. Electrical filters perform in a similar way. A band of frequencies which is rejected or greatly attenuated is called a stopband; a band which is transmitted with little attenuation is called a passband. Ideally, there should be zero attenuation in the passband and infinite attenuation in the stopband. More practically, a small amount of attenuation, say α_p, can be tolerated in the passband, and a noninfinite attenuation, say α_s, in the stopband. Two cases are illustrated in Fig. 2. The tolerable levels are shown by the horizontal dashed lines. The solid lines represent possible

attenuation characteristics approximating the tolerable levels.

The first, in Fig. 2a, is the attenuation characteristic of a low-pass filter—one that transmits low frequencies with little attenuation, but stops high frequencies. The second (Fig. 2b) represents a band-pass filter—one that transmits an intermediate band of frequencies with little attenuation, but which greatly attenuates both low and high frequencies. A filter having the inverse performance of that of a low-pass filter—that is, one that stops low frequencies but passes high ones—is called a high-pass filter. The inverse of a band-pass filter is a band-stop filter, which stops an intermediate band but passes both low and high frequencies. Finally, there are filter-type networks which pass all frequencies equally well but modify the phase of input signals; they are called all-pass networks, or phase-correcting networks.

The region between the edge of the passband and that of the stopband—with angular frequencies labeled ω_p and ω_s, respectively, in Fig. 2a—is a transition region. Similarly, the transition regions in Fig. 2b lie between ω_{p1} and ω_{s1}, and between ω_{p2} and ω_{s2}. The sharpness of the filter cutoff depends on the width of this transition region; the narrower the region, the sharper the cutoff is said to be. A particularly significant frequency is the half-power frequency. This is the frequency at which the power transmitted through the filter is half the power input. This occurs when the attenuation has risen to 3 dB from its nominal value (zero) in the passband. In Fig. 2a it is indicated by angular frequency $\omega_h = 2\pi f_h$, where h is half power (and f_h is the corresponding frequency).

The maximum attenuation in the passband can be approximated in a maximally flat manner, as in the low-pass filter in Fig. 2a, or in an oscillatory equal-ripple (Chebyshev) manner, as in Fig. 2b.

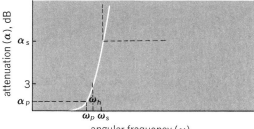

(a)

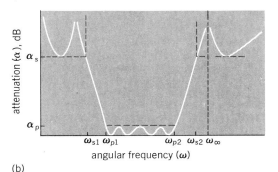

(b)

Fig. 2. Attenuation characteristics. (a) Low-pass filter; (b) band-pass filter.

Similar approximations (monotonic or oscillatory) can be made to the stopband behavior.

Applications. Electrical filters are incorporated in almost all systems of electrical communications and control. Radio and television receivers contain filters in order to pass the particular channel selected and to stop others. The telephone network incorporates millions of filters in order to pass each conversation, which is carried by a different band of frequencies, to the appropriate telephone receiver and to stop all others. Bandwidths vary from a fraction of a hertz in control systems, to 3 kilohertz (kHz) in telephone systems, to 20 kHz in AM radio systems, to 6 megahertz (MHz) in television systems. *See* AMPLITUDE-MODULATION DETECTOR; FREQUENCY-MODULATION DETECTOR; RADIO RECEIVER; TELEVISION RECEIVER.

Narrow-band band-rejection filters are often used to reduce interference from extraneous electrical noise, which may come from power lines or from industrial equipment. One example of such a null network is the twin-tee filter shown in Fig. 3. Ideally, this filter has infinite attenuation at an angular frequency $\omega = 1/RC$.

A common use of a filter is to smooth the output from a rectifier to make it more closely constant by attenuating the alternating-current component. The power supply in every radio and television receiver has such a low-pass filter. The transfer function of this power-supply filter is given in Eq. (3), under the assumption that there is negligible loading at the output port.

$$F(s) = \frac{V_2}{V_1} = \frac{1}{1 + s^2 LC} \tag{3}$$

For direct current (dc), $s = j\omega = 0$; thus the transfer function is 1, and, according to Eq. (2), there is no attenuation. The components of ac voltage present at the input are 60 Hz or its harmonics, or both. (For a full-wave rectifier, the lowest frequency present is the second harmonic.) The L and C values are chosen so as to give an attenuation of at least 40 dB at the lowest ac frequency present. The ripple, which is the ac component of the output, will then be no more than 1% of the dc value.) If the transfer function of Eq. (3) is used in Eq. (2), with $s = j\omega$, this condition leads to the result shown in inequality (4). For a full-wave rectifier, $\omega = 2\pi \times 120$. *See* ELECTRONIC POWER SUPPLY; RECTIFIER.

$$20 \log |1 - \omega^2 LC| > 40 \tag{4}$$

$$LC > \frac{10^2}{\omega^2}$$

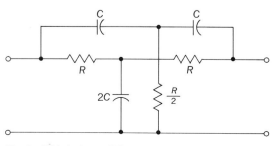

Fig. 3. *RC* twin-tee null filter.

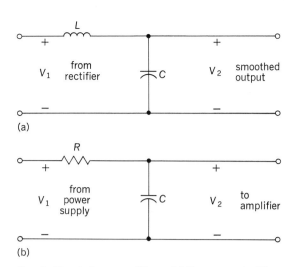

Fig. 4. Simple low-pass filters. (*a*) Power-supply filter; (*b*) decoupling filter.

Another common use of a low-pass filter is as a decoupling network when several amplifier stages share the same power supply. A low-pass decoupling filter is placed between the power supply and each amplifier stage, thereby reducing interaction between the stages (Fig. 4*b*). For this filter the transfer function is given by Eq. (5), again

$$F(s) = \frac{V_2}{V_1} = \frac{1}{1 + sCR} \tag{5}$$

under the assumption that there is negligible loading at the output, which is usually the case. With direct current, $s = j\omega = 0$, and the transfer function equals 1. Hence, according to Eq. (2), direct current passes with zero attenuation. The half-power frequency is found by substituting the magnitude of F from Eq. (5), when $s = j\omega$, into the attenuation relation, Eq. (2), and setting the attenuation equal to 3 dB. The result is $\omega_h = 1/RC$.

Design. The filters mentioned in the preceding section are all relatively simple, with few components. Their design does not require deep knowledge and advanced techniques; the performance requirements imposed on such filters are not very stringent. However, in many applications in which sharp cutoffs and high stopband attenuation is required, simple procedures are inadequate, and more complicated design procedures are necessary.

From prescribed values of passband and stopband attenuation, half-power frequency, sharpness of cutoff, and so on, a transfer function is determined such that the filter constructed so as to have this transfer function meets the specifications to the required degree of approximation. Different design approaches can lead to different locations of poles and zeros of the transfer function; therefore these locations are not uniquely determined from a given set of specifications. The higher the order of the transfer function, the more extensive the filter. Thus the order of the transfer function should be as low as possible, consistent with the requirements of the specifications.

A very common filter design utilizes reactive components (inductors and capacitors), usually in

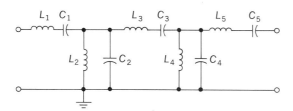

Fig. 5. *LC* band-pass ladder filter.

a structure called a ladder. An example of a band-pass ladder is shown in Fig. 5. Low-pass and high-pass filters would have a similar structure except the values of the components would be different (and possibly some of the components would be absent). An efficient design procedure for such filters is the insertion-loss design. Extensive tables based on the insertion-loss design are available which give values of inductance *L* and capacitance *C* for ladder filters of this type for denominator orders from 3 to 10.

Once a filter has been designed and constructed, it is possible that the values of the components will change. This may occur because of aging, temperature variations, and so on. An important concern for a filter designer is the sensitivity of the attenuation to changes in the component values, that is, the percentage change in the attenuation as a result of a given percentage change in component values. In the most useful design, the sensitivity is as small as possible, other factors being equal. The low sensitivity of the ladder structure makes it a desirable design.

However, the trend in all electronic equipment is toward minaturization. In fact, the trend is toward constructing as much as possible in the form of integrated circuits (tiny semiconductor chips which can be designed to perform rather complicated functions). Both because of their relatively bulky and heavy nature, and because they cannot be constructed in the form of integrated circuits, inductors are no longer popular. Designs that utilize resistors, capacitors, and active devices called active *RC* filters are increasingly being used. *See* INTEGRATED CIRCUITS.

Inductor simulation. One appealing type of design consists of the conventional *LC* filter in which the inductors have been replaced by appropriate combinations of active devices and resistors or capacitors. One such combination is called a gyrator. The physical construction of a gyrator can take many forms, some of which are more complicated than others. A particularly simple gyrator design, requiring just two operational amplifiers and four resistors, is shown in Fig. 6*a*. Whatever the design might be, a simple schematic symbol, shown in Fig. 6*b*, is used to represent the gyrator in circuit diagrams. In some ways, a gyrator is like a transformer, and the gyration ratio *r* is like the turns ratio *n* of a transformer. A transformer transforms impedances; the equivalent impedance at one of the ports is directly proportional to the impedance connected to the other port, and the proportionality constant is the square of the turns ratio. The gyrator, on the other hand, has the property of impedance inversion; the equivalent impedance at one of the ports is inversely proportional to the impedance connected to the other port, the propor-

tionality constant being the square of the gyration ratio *r*.

The connections of a capacitor and one or more gyrators which simulate an inductor are also shown in Fig. 6. Whenever an inductor is needed in a filter design, one of these combinations can be used instead.

Thus, in the filter of Fig. 5, each inductor with one terminal connected to ground can be replaced by the gyrator-capacitor combination in Fig. 6*c*. The "floating" inductors, like L_1, L_3, and L_5, which do not have one terminal connected to ground, require the more extensive arrangement with two gyrators shown in Fig. 6*d*. Although replacing the inductors in this manner seems to be a matter of converting a simple network to something more complex, this is not the case. Gyrators can be easily fabricated in the form of integrated circuits, which can be used to replace inductors in a classical design. *See* GYRATOR; TRANSFORMER.

In fact, the gyrator-capacitor combinations can be fabricated in a single integrated-circuit unit. Whenever the classical design calls for an inductor, such a simulated inductor can be used in its place.

Other active devices can also be used to eliminate the inductors in a classical *LC* design. Among these are the general converter and the frequency-dependent negative-resistance converter. Each of these devices in itself is a rather extensive network. Nevertheless, each can be constructed in the form of an integrated circuit, either as a single unit or as a partial unit to which a few capacitors are connected externally. The design procedures utilizing these devices first require the *LC* filter to be designed. Then, appropriate combinations of these devices with resistors and capacitors eliminate the inductors.

Another virtue of inductor-elimination procedures, besides providing the ability to use the integrated circuit form, and consequently reducing costs, is that the low sensitivity of the ladder structure to changes in component values is retained.

Feedback amplifier design. A variety of design procedures exist for obtaining an active *RC* filter directly, and the active device can take on one of several forms. For a given set of specifications,

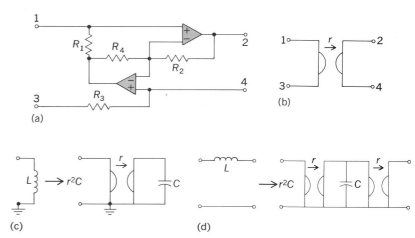

Fig. 6. Gyrators and inductor simulation. (*a*) Gyrator design; (*b*) gyrator circuit symbol; (*c*) simulation of grounded inductor; (*d*) simulation of floating inductor.

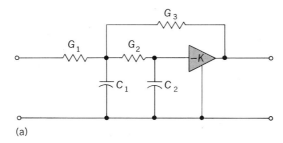

(a)

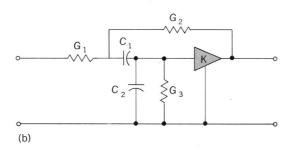

(b)

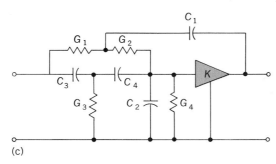

(c)

Fig. 7. Feedback amplifier design. (a) Low-pass section; (b) band-pass section; (c) attenuation-peak-producing section.

leading to a transfer function in the form of Eq. (1), two alternatives are possible. The entire transfer function can be realized as a filter all at once, or the transfer function can be written as a product of second-order functions (second-order denominator factors with appropriate factors of the numerator up to second order). Each second-order function is designed separately, and these partial designs are connected in cascade. (Other types of interconnection are also possible, requiring different ways of

breaking up the original transfer functions.) Most direct active *RC* filter design procedures use the second alternative in order to reduce the sensitivity of the filter to component-value variations.

One method of design uses an ordinary amplifier in a feedback configuration. Three such second-order filters are shown in Fig. 7. Each triangle represents an amplifier; the amplifier gain is negative in the first one and positive in the other two. The third one provides a peak of attenuation, as in the attenuation curve in Fig. 2b, at the frequency labeled ω_x. (The element values have been labeled in terms of their conductances G instead of resistances R because the design equations are thereby simplified.) Many other variations of these structures are possible. *See* FEEDBACK CIRCUIT.

The biquad. One of the most important active RC filter designs is shown in Fig. 8. It has been called the biquad because it can realize a general second-order (biquadratic) transfer function of the form shown in Eq. (6). The active devices in this

$$F = \frac{a_2 s^2 + a_1 s + a_0}{s^2 + b_1 s + b_0} \tag{6}$$

design are operational amplifiers. *See* AMPLIFIER.

For the simple low-pass characteristic—such as that obtained by the filter section in Fig. 7a—the summing amplifier at the bottom in Fig. 8 can be omitted and the output taken between the point labeled 5 and ground. Similarly, for a simple band-pass characteristic—such as that obtained by the section in Fig. 7b—the summing amplifier can again be omitted and the output taken between the point labeled 3 and ground.

The virtues of this design are that it can be easily constructed in the form of integrated circuits and that the sensitivity to variations in component values is quite small. Furthermore, the biquads can be interconnected so that higher-order filters can be obtained without interference between the sections occurring.

Special filters. Special categories of filters can be classified in accordance with either their application or the components used in their construction.

Crystal and ceramic filters. Piezoelectric crystalline materials such as quartz have the property of electrical resonance when mechanically stressed along certain axes. This property can be used to replace electrical components in a filter. The resulting crystal filters are almost always used in band-pass applications. The incoming (and outgoing) electrical signals must be coupled to the mechanical vibrations of the crystal through transducers, which are themselves electrical elements.

The frequency range of crystal filters is from 2 MHz to 15 MHz, but it can reach as high as 200 MHz. Crystal filters can be made with extremely narrow bandwidths, as low as 0.01%. They have very low loss (with Qs ranging from 10,000 up to 300,000; the higher the Q, the lower the loss). For these reasons they find application also in frequency standards. Although they are not smaller and lighter than active *RC* filters, they have the advantage of being passive and so do not generate internal noise.

Ceramic materials can also be used in place of crystals; the result is a ceramic filter. The losses are not quite as low as those of crystal filters (Qs are of the order of 2000 or 3000), but this range is

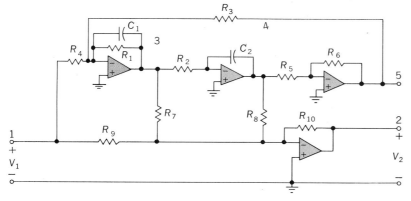

Fig. 8. The biquad, an active *RC* filter.

still superior to that of *LC* filters. Both crystal and ceramic filters compete favorably in cost both with conventional *LC* filters and with active *RC* filters. *See* PIEZOELECTRIC CRYSTAL; Q (ELECTRICITY); RESONANCE (ALTERNATING-CURRENT CIRCUITS).

Mechanical filters. These filters have some characteristics in common with crystal filters, in that the resonant elements are mechanical. But they differ from crystal filters in that the coupling mechanisms between mechanical vibrations and electrical signals (the transducers) are also mechanical. Ferrites and iron-nickel alloys whose elastic modulus is constant over a temperature range are common materials from which mechanical filters are constructed. These filters also have narrow bandwidths (as low as 0.1%) and low loss (with *Q*s ranging from 10,000 to 20,000), and their operating frequency range extends down to 3 kHz and up to 600 kHz. *See* FERRITE DEVICES.

Microwave filters. These filters operate at microwave frequencies, generally considered to lie above 1 GHz. However, they can also be used in the much lower high-frequency (hf) range of 3–30 MHz. The components are no longer lumped elements, but consist of sections of transmission lines (called stubs), strip lines, coupled lines, and waveguide structures such as irises and posts. *See* MICROWAVE TRANSMISSION LINES; WAVEGUIDE.

Distributed RC filters. These filters are made of thin-film integrated circuits or various large-area semiconductor devices. Since active devices, as well as resistors and capacitors, can be fabricated from integrated circuits, distributed *RC* filters (designated $\overline{RC}$) can be active or passive. All the components are fabricated by depositions of various thicknesses on the same substrate.

Matched filters. A filter that is matched to the characteristics of an input signal of known time dependence is called a matched filter. The term is also applied to describe an impedance match between a filter and its termination. A matched filter has the property that, when the input consists of noise in addition to the desired signal, the output signal-to-noise ratio is the maximum that can be obtained by any linear filter. Accordingly, such filters are used (in radar and similar systems) to optimize the detection of weak signals. The impulse response of a matched filter is simply related to the input signal when the spectrum of interference is uniform.

Switched filters. These filters are similar to other electrical filters except that a switch is incorporated into one of the branches. The switch is operated (turned on or off) in a periodic manner so that the switched filter is a time-varying linear network. Switched filters find particular application in modulators used in communications systems, both in frequency-division multiplexing (FDM) and in time-division multiplexing (TDM).

Optimum filters. In some applications the desired output from a filter may be specified either as a function of frequency or as a function of time, more precisely than the attenuation maximum in the passband and minimum in the stopband in Fig. 2. Then the appropriately defined error between the actual output and the desired output is a measure of the performance of the filter. The most common measure is the mean square of the error. A filter designated so as to give the least mean-square error is called an optimum filter. This

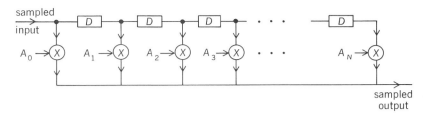

Fig. 9. Tapped delay-line digital filter.

mean-square error minimization can be achieved under specific conditions, for example, in the presence of noise interference having a specified power spectrum. Optimum filters are used in predicting the future values of the input on the basis of its past values and the mean-square error.

Digital filters. An increasing amount of signal processing is being performed in digital form. A digital signal is one which takes on only a limited number of discrete values and only at discrete and equally spaced intervals of time. Signals may already be in digital form (for example, sent between digital computers and most business machines), or a continuous signal may be converted to digital form by sampling it at uniform intervals and digitally coding its values, for ease of transmission or some other reason. This digital signal is then processed in accordance with a set of instructions in order to regain the original information. The processing is performed by a digital filter.

Since the function performed by a digital filter is the processing of digital signals, the building blocks from which a digital filter is constructed are the same as those in a digital computer—adders, multipliers, shift registers, memory units, and so on. A digital computer is a machine which processes data in accordance with a set of instructions, called the program. Therefore, not only is a digital filter a piece of equipment specially designed to perform a function but also from a conceptual standpoint, it can be viewed as the software (the program) which instructs a general-purpose digital computer. Thus a digital filter is inherently different from the other filters discussed in this article. *See* DIGITAL COMPUTER.

Digital filters can be designed by dealing with the representation of input and output signals either in the time domain or in the frequency domain. In the time domain, the filter is represented by its response in time to an impulse, called its impulse response. One form in which this representation can be implemented is shown in Fig. 9. The top row represents a delay line, with taps at uniform intervals in which each *D* represents a fixed delay related to the sampling interval. (Mathematically, $D = e^{-sT}$, where *T* is the sampling interval.) Each circle with an *x* is a multiplier, and the *A*'s are the coefficients by which the delayed signal coming from the corresponding tap is multiplied.

The sampled input signal is a sequence of numbers. (In mathematical terms, it is said to be a sequence of impulses whose weights are the values of the input at the corresponding times.) The output is again a sequence of numbers obtained by adding the delayed outputs from the taps.

In the frequency domain, it is possible to express the digital transfer function as a sum of first-

ELECTRIC FILTER

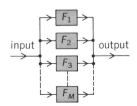

Fig. 10. Digital filter in parallel form.

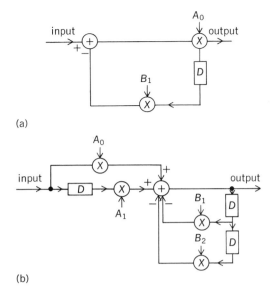

(a)

(b)

Fig. 11. Block diagrams of digital filter sections. (*a*) First order; (*b*) second order.

order and second-order terms. Each term is designed separately; the terms are then connected in parallel, as shown in Fig. 10. Designs of first-order and second-order digital filters are shown in Fig. 11. The circles with enclosed plus signs represent adders. Many other designs are also possible.

[NORMAN BALABANIAN]

Bibliography: C. Childers and A. Durling, *Digital Filtering and Signal Processing*, 1975; L. P. Huelsman, *Theory and Design of Active-RC Circuits*, 1970; D. E. Johnson, *Introduction to Filter Theory*, 1976; G. S. Moschytz, *Linear Integrated Networks: Design*, 1975; J. K. Skwirzinski, *Design Theory and Data for Electrical Filters*, 1965; C. Temes and S. K. Mitra (eds.), *Modern Filter Theory and Design*, 1973.

Electric spark

A transient form of gaseous conduction. This type of discharge is difficult to define, and no universally accepted definition exists. It can perhaps best be thought of as the transition between two more or less stable forms of gaseous conduction. For example, the transitional breakdown which occurs in the transition from a glow to an arc discharge may be thought of as a spark.

Electric sparks play an important part in many physical effects. Usually these are harmful and undesirable effects, ranging from the gradual destruction of contacts in a conventional electrical switch to the large-scale havoc resulting from lightning discharges. Sometimes, however, the spark may be very useful. Examples are its function in the ignition system of an automobile, its use as an intense short-duration illumination source in high-speed photography, and its use as a source of excitation in spectroscopy. In the second case the spark may actually perform the function of the camera shutter, because its extinction renders the camera insensitive.

Mechanisms. The spark is probably the most complicated of all forms of gaseous conduction. It is exceedingly difficult to study, because it is a

transient and because there are so many variables in the system. Some of these variables are the components of the gaseous medium, the gas pressure, the chemical form of the electrodes, the physical shape of the electrodes, the microscopic physical surface structure, the surface temperature, the electrode separation, the functional dependence of potential drop on time, and the presence or absence of external ionizing agents. One or more of these conditions may change from one spark to the next. Because of the great complexity, it will be impossible to do more than touch on some of the main features in this article.

The dependence of breakdown, or sparking, potential on pressure p and electrode separation d is considered first. It was shown experimentally by F. Paschen and theoretically by J. S. Townsend that the sparking potential is a function of the product pd and not of p or d separately (Fig. 1). Further, there is a value of pd for which the sparking potential is a minimum. Thus, if it is desired to prevent sparking between two electrodes, the region may be evacuated or raised to a high pressure. The latter method is used in accelerators of the electrostatic generator variety. Here the entire apparatus is placed in a pressurized tank.

Qualitatively, one of the aspects of a spark is that the entire path between electrodes is ionized. It is the photon emission from recombination and decay of excited states which gives rise to the light from the spark. Further, if the spark leads to a stable conduction state, the cathode must be capable of supplying the needed secondary electrons, and the conduction state produced must permit the discharge of the interelectrode capacitance at the very minimum. *See* ARC DISCHARGE; ELECTRICAL CONDUCTION IN GASES; GLOW DISCHARGE.

In a consideration of the mechanism involved in the spark, the time required for the breakdown of the gas in a gap is an important element. L. B. Loeb pointed out that this time is often less than that required for an electron to traverse the gap completely. This implies that there must be some means of ionization present other than electron impact and that the velocity of propagation of this ionizing agent or mechanism must be much greater than the electron velocity. It seems

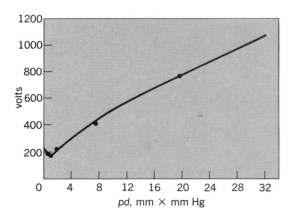

Fig. 1. Dependence of sparking potential on *pd* for a sodium cathode in hydrogen gas. (*From L. B. Loeb and J. M. Meek, The Mechanism of the Electric Spark, Stanford University Press, 1941*)

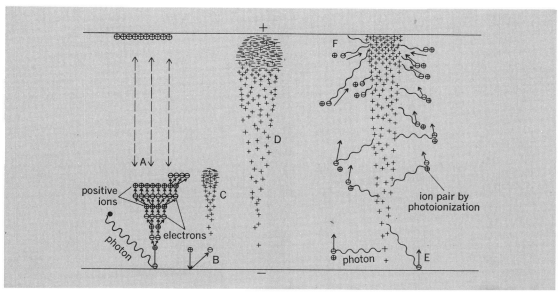

Fig. 2. Electron multiplication and avalanching during an electric spark discharge. (*From L. B. Loeb and J. M. Meek, The Mechanism of the Electric Spark, Stanford University Press, 1941*)

definitely established that this additional method must be photoionization. In the intense electric field which is necessary for the spark, the initial electron will produce a heavy avalanche of cumulative ionization. Light resulting from the decay processes will produce ionization throughout the gas and electrons at the surfaces by the photoelectric effect (Fig. 2). The electrons resulting from this will in turn produce further avalanches through the entire region, so that in a time of the order of 10^{-8} sec the entire path becomes conducting. If the pressure is approximately atmospheric, the spark will be confined to a relatively narrow region, so that the conducting path, while not straight, will be a well-defined line. If the external circuit can supply the necessary current, the spark will result in an arc discharge. At lower pressure the path becomes more diffuse, and the discharge takes on either a glow or arc characteristic.

Figure 2 shows A, the electron multiplication of electrons by the cumulative ionization of a single electron liberated from the cathode by a photon; B, a secondary electron emitted from the cathode by a positively charged ion; C, the development and structure of an avalanche, with positively charged ions behind electrons at the tip; D, the avalanche crossing the gap and spreading by diffusion; and F, an older avalanche when electrons have disappeared into the anode. A positive space-charge boss appears on the cathode at F. Ion pairs out from the trail indicate the appearance of photoelectric ion pairs in the gas produced by photons from the avalanche. E shows a photoelectron from the surface of the cathode produced by the avalanche.

Theory. Mathematically, the theory of Townsend predicts that the current in a self-sustained discharge of the glow variety will follow Eq. (1),

$$I = I_0 \frac{e^{ax}}{1 - \gamma e^{ax}} \qquad (1)$$

where I is the current with a given plate separation x, I_0 is the current when x approaches zero, and α

and γ are constants associated with the Townsend coefficients. This equation represents the case where the electrode separation is varied while the ratio of electric field to pressure is held constant. The condition for a spark is that the denominator approach zero, which may be stated as in Eq. (2).

$$\gamma = e^{-\alpha x} \qquad (2)$$

Loeb indicated that this criterion must be handled carefully. Townsend's equation really represents a steady-state situation, and it is here being used to explain a transient effect. If the processes which are involved are examined more carefully, it appears that there should be a dependence on I_0 as well.   [GLENN H. MILLER]

Bibliography: J. Beyon, *Conduction of Electricity through Gases*, 1972; J. D. Cobine, *Gaseous Conductors*, 1941; L. B. Loeb, Statistical factors in spark discharge mechanisms, *Rev. Mod. Phys.*, 20:151–160, 1948; D. Roller and D. H. C. Roller, *The Development of the Concept of Electric Charge*, 1954.

Electric transient

A temporary component of current and voltage in an electric circuit which has been disturbed. In ordinary circuit problems, a stabilized condition of the circuit is assumed and steady-state values of current and voltage are sufficient. However, it often becomes important to know what occurs during the transition period following a circuit disturbance until the steady-state condition is reached. Transients occur only in circuits containing inductance or capacitance. In general, transients accompany any change in the amount or form of energy stored in the circuit. Both direct-current (dc) transients and alternating-current (ac) transients are treated following the introduction. *See* ALTERNATING-CURRENT CIRCUIT THEORY; DIRECT-CURRENT CIRCUIT THEORY.

Introduction. The study of transient phenomena is very broad. The mathematical requirements become severe and go far beyond the borders of all known mathematics. Transient analysis often requires the use of calculating machines, models, and tests. Fourier and Laplace transforms have proven indispensable in the modern treatment of transients and these disciplines need be mastered by anyone going far in the study of transients. The analysis of lumped-parameter circuits is comparatively easy and is all that will be described here, but transients of a much more complex nature occur on distributed-parameter circuits, such as transmission lines.

An electric circuit or system under steady-state conditions of constant, or cyclic, applied voltages or currents is in a state of equilibrium. However, the circuit conditions of voltage, current, or frequency may change or be disturbed. Also, circuit elements may be switched in or out of the circuit. Any change of circuit condition or circuit elements causes a transient readjustment of voltages and currents from the initial state of equilibrium to the final state of equilibrium. In a sense the transient may be regarded as superimposed on the final steady state, so that Eq. (1) applies.

$$\begin{pmatrix} \text{Instantaneous} \\ \text{condition} \end{pmatrix} = \begin{pmatrix} \text{final} \\ \text{condition} \end{pmatrix} + \begin{pmatrix} \text{transient} \\ \text{terms} \end{pmatrix} \quad (1)$$

Furthermore, since the instantaneous condition at the first instant of disturbance (time zero) must be the initial condition, it may be described by Eq. (2).

$$\begin{pmatrix} \text{Initial} \\ \text{condition} \end{pmatrix} = \begin{pmatrix} \text{final} \\ \text{condition} \end{pmatrix} + \begin{pmatrix} \text{transient terms} \\ \text{at time-zero} \end{pmatrix} \quad (2)$$

A great deal of information may be obtained from these two "word equations" without recourse to mathematics. For example, if the weight on the end of a vertical spring is suddenly increased, its final displacement can be determined. Since the spring-weight combination is known to be an oscillating system, the amplitude of the transient oscillation follows from Eq. (2) as the difference between the initial and final displacements.

The nature of an electric transient is determined by three things: (1) the circuit or network itself — the interconnections of its elements and the circuit parameters (resistances R, inductances L, capacitances C, mutual inductances M); (2) the initial conditions of voltages, currents, charges, and flux linkages at the start of the transient; and (3) the nature of the disturbance which initiated the transient.

The circuit, or network, is usually defined by a diagram of connections showing all of the interconnections, junctions, meshes, circuit parameters, voltage and current sources and their polarities, and switches. Corresponding to the network a differential (or integral-differential) equation, or a set of such equations, may be written. These equations may also be written in terms of operational calculus, or as Laplace transforms, or in any other suitable mathematical equivalents. They are established in accordance with Kirchhoff's laws:

I. The sum of the currents i at a junction is equal to zero: $\Sigma i = 0$.

II. The sum of the voltages e around a mesh is equal to zero: $\Sigma e = 0$. *See* KIRCHHOFF'S LAW OF ELECTRIC CIRCUITS.

The voltage drops associated with a resistance, inductance, mutual inductance, and capacitance, respectively, are given by Eqs. (3). In applying

$$e_R = Ri \qquad e_L = L\frac{di}{dt}$$
$$e_1 = M_{12}\frac{di_2}{dt} \qquad e_C = \frac{1}{C}\int i\,dt \qquad (3)$$

Kirchhoff's law to a closed mesh, it is merely necessary (with due regard for signs and polarities) to equate the sum of all the voltage sources (such as generators and batteries) to the sum of all the voltage drops in the mesh, as in Eq. (4), where m refers

$$\sum_m e_m = \sum_m \sum_n \left(R_{mn}i_n + L_{mn}\frac{di_n}{dt} + \frac{1}{C_{mn}}\int i_n\,dt \right) \quad (4)$$

to any branch of the mesh in question, and n to any current causing a voltage drop in that branch (a branch may carry currents from meshes other than the mesh in question, or be mutually coupled with other branches). Equation (4) formulates the differential equation of the circuit in terms of the voltages and currents. It is sometimes convenient to make use of charges q and fluxes ϕ by the substitutions of Eqs. (5).

$$i = \frac{dq}{dt} \quad \text{or} \quad q = \int i\,dt \quad \text{and} \quad N\phi = Li \quad (5)$$

Since Eq. (4) contains an integral, it is convenient to eliminate it either by differentiating once or by substituting Eq. (5). Then Eqs. (6) follow. *See* COUPLED CIRCUITS.

$$\sum_m \frac{de_m}{dt} = \sum_m \sum_n \left(R_{mn}\frac{di_n}{dt} + L_{mn}\frac{d^2i_n}{dt^2} + \frac{i_n}{C_{mn}} \right)$$
$$\sum_m e_m = \sum_m \sum_n \left(R_{mn}\frac{dq_n}{dt} + L_{mn}\frac{d^2q_n}{dt^2} + \frac{q_n}{C_{mn}} \right) \quad (6)$$

It is customary to factor out the current from Eq. (4) and write the equation in terms of a generalized impedance operator Z_{mn} defined by Eq. (7). Since an equation of this type can be written

$$\sum_m e_m = \sum_m \sum_n \left(R_{mn} + L_{mn}\frac{d}{dt} + \frac{1}{C_{mn}}\int dt \right) i_n$$
$$= \sum_m \sum_n Z_{mn}i_n \quad (7)$$

for each of the N-meshes of the network, the totality of the differential equations for the entire network constitutes a system of simultaneous differential equations in the form of Eqs. (8).

$$\begin{aligned} e_1 &= Z_{11}i_1 + Z_{12}i_2 + \cdots + Z_{1N}i_N \\ e_2 &= Z_{21}i_1 + Z_{22}i_2 + \cdots + Z_{2N}i_N \\ &\cdots\cdots\cdots\cdots\cdots\cdots\cdots \\ e_N &= Z_{N1}i_1 + Z_{N2}i_2 + \cdots + Z_{NN}i_N \end{aligned} \quad (8)$$

In general, the process of solution of such a set of equations for any current leads to a differential equation of order $2N$; there will be $2N$ integration constants associated with it, and these integration constants must be determined from the initial conditions at the first instant of the disturbance.

Equation (9) is the solution of an ordinary differential equation with constant coefficients and is in two parts.

$$\begin{pmatrix} \text{Complete} \\ \text{solution} \end{pmatrix} =$$
$$\begin{pmatrix} \text{particular} \\ \text{integral} \end{pmatrix} + \begin{pmatrix} \text{complementary} \\ \text{solution} \end{pmatrix} \quad (9)$$

The particular integral depends on the form of the applied voltage, and represents the final steady-state solution.

The complementary solution is independent of the form of the applied voltages, but depends on the initial conditions in the circuit, and represents the transient terms. Thus Eq. (9) is the mathematical equivalent of Eq. (1).

The opening of a switch may be simulated by superimposing an equal and opposite current and thereby canceling the current through the switch. Likewise, the closing of a switch may be simulated by canceling the voltage across its terminals by an equal and opposite voltage. The effect of a cancellation current (or voltage) may be superimposed on the currents and voltages already existing in the system, as in Eq. (10).

$$\begin{pmatrix} \text{Resultant} \\ \text{voltages} \\ \text{and} \\ \text{currents} \end{pmatrix} = \begin{pmatrix} \text{those which} \\ \text{would exist if} \\ \text{switches were} \\ \text{not operated} \end{pmatrix} + \begin{pmatrix} \text{those due to} \\ \text{"cancellation"} \\ \text{voltages or} \\ \text{currents} \end{pmatrix} \quad (10)$$

In many practical cases transients are one of three types:

1. Single-energy transients, in which only one form of energy storage (either electromagnetic or electrostatic) is present; the transient exhibits simple exponential decay from the initial to the final conditions.

2. Double-energy transients, in which both forms of energy storage are present; the transient is either aperiodic or a damped sinusoid.

3. Combination of 1 and 2.

DC TRANSIENTS

Transients are initiated in dc circuits either when a switch is closed on a dc voltage, or when a switch is opened and the detached circuit is permitted to discharge its energy sources. The assumption is made that the suddenly applied dc voltage is sustained at a constant value for an indefinite period after the switching operation, and time is counted from the instant, $t=0$, at which the switch is operated. It is further usually assumed, where possible, that all circuit parameters (resistances R, inductances L, and capacitances C) are constant, so that the resulting differential equations are linear.

The energy sources in dc systems comprise dynamoelectric machines, mercury-arc rectifiers, vacuum tubes, and electric batteries. Energy W is stored in a circuit in inductances ($W=1/2Li^2$) and capacitances ($W=1/2Ce^2$). Energy is dissipated in circuit resistances ($W=Ri^2$).

Resistance and inductance in series. A lumped parameter circuit comprising a dc voltage source E, a resistance R, and an inductance L is shown in Fig. 1. When the switch S is closed, the voltage E is suddenly impressed on the series circuit. By Kirchhoff's second law the differential equation is given by Eq. (11).

$$E=Ri+L\frac{di}{dt} \quad (11)$$

Equating the right-hand side of Eq. (11) to zero,

the complementary solution is found by assuming that Eq. (12), in which I and a are constants to be

$$i=Ie^{at} \quad (12)$$

determined, applies. Upon substitution in Eq. (11). Eq. (13) results, from which follows $a=-R/L$.

$$0=RIe^{at}+aLIe^{at}=(R+aL)i \quad (13)$$

The particular integral is the final steady-state solution or $i(\infty)=E/R$. The complete solution, Eq. (14), is the sum of the complementary solution and

$$i=E/R+Ie^{-(R/L)t} \quad (14)$$

the particular integral. But this expression contains the unknown integration constant I which must be found from the initial conditions. At $t=0$, the instant at which the switch is closed, there can be no current in the circuit, since it is impossible to store energy in an inductance instantaneously. Therefore $i(0)=0$ and when this is put in Eq. (14), Eq. (15) is obtained.

$$i(0)=0=E/R+I \quad \text{or} \quad I=-E/R \quad (15)$$

The complete solution by Eq. (14) therefore is given by Eq. (16), which is in the form of Eq. (1).

$$i=\frac{E}{R}-\frac{E}{R}\epsilon^{-(R/L)t} \quad (16)$$

Here E/R is the final (steady-state) condition and $-(E/R)\epsilon^{-(R/L)t}$ is the transient term. Note also that at $t=0$ this equation gives $i(0)=0$, agreeing with Eq. (2). The graph of the transient is shown in Fig. 1, where the current starts at zero, rises exponentially, and approaches its steady-state value of E/R as t approaches infinity. *See* ELECTRICAL RESISTANCE; INDUCTANCE.

Resistance and capacitance in series. A resistance-capacitance series circuit is shown in Fig. 2. A residual charge q_0 is on the capacitor just prior to closing the switch S.

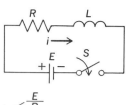
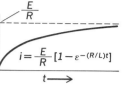

Fig. 1. Resistance and inductance in series.

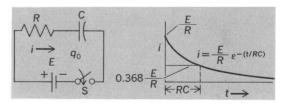

Fig. 2. Resistance and capacitance in series.

Equation (17) is the differential equation of the

$$E=Ri+\frac{1}{C}\int i\,dt \quad (17)$$

circuit. This equation may be solved as it is, or converted to a linear differential equation, Eq. (18),

$$0=R\frac{di}{dt}+\frac{i}{C} \quad (18)$$

by differentiation or restated in terms of the instantaneous charge on the capacitor by substituting $i=dq/dt$, giving Eq. (19).

$$E=R\frac{dq}{dt}+\frac{q}{C} \quad (19)$$

The solution to this equation, following precisely the steps of the previous paragraph, is given by Eq. (20), in which Q is the integration constant. But initially, at $t=0$, there was a residual charge q_0 on the capacitance; therefore the relationship given by Eq. (21) pertains, and upon the substitution of Eq.

$$q = Q\epsilon^{-(t/RC)} + CE \tag{20}$$

tially, at $t=0$, there was a residual charge q_0 on the capacitance; therefore the relationship given by Eq. (21) pertains, and upon the substitution of Eq.

$$q(0) = q_0 = Q + CE \quad \text{or} \quad Q = q_0 - CE \tag{21}$$

(21) in Eq. (20), Eq. (22) is obtained. This equation

$$q = CE - (CE - q_0)\,\epsilon^{-(t/RC)} \tag{22}$$

is in the form of Eq. (1), in which CE is the final (steady-state) condition and the other term is the transient. Initially, at $t=0$, $q=q_0$ in agreement with Eq. (2). The current is given by Eq. (23).

$$i = \frac{dq}{dt} = \frac{CE - q_0}{RC}\epsilon^{-(t/RC)} = \left(\frac{E}{R} - \frac{q_0}{RC}\right)\epsilon^{-(t/RC)} \tag{23}$$

Equations (22) and (23) have been plotted in Fig. 2 for the case $q_0 = 0$. *See* CAPACITANCE.

Resistance, inductance, and capacitance in series. This circuit is shown in Fig. 3. The capacitor is assumed to have an initial charge q_0, or an

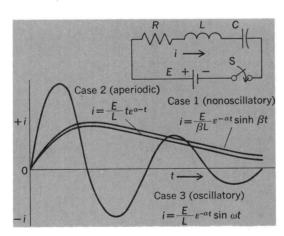

Fig. 3. Resistance, inductance, and capacitance in series for three cases.

initial voltage $V = q_0/C$. The differential equation of the circuit is given by Eq. (24).

$$E = Ri + L\frac{di}{dt} + \frac{1}{C}\int i\,dt \tag{24}$$

Differentiating once to clear the integral, Eq. (25)

$$0 = R\frac{di}{dt} + L\frac{d^2i}{dt^2} + \frac{i}{C} \tag{25}$$

is obtained. Equation (25) is a second-order linear differential equation with constant coefficients; since it is equal to zero, it will possess only a complementary solution. Equation (26) is assumed, which, upon substitution in Eq. (25), yields Eq. (27).

$$i = A\epsilon^{at} \tag{26}$$

$$0 = aRA\epsilon^{at} + a^2LA\epsilon^{at} + (A/C)\epsilon^{at} \tag{27}$$

Canceling the constant A and the exponential, there results a quadratic in a whose solution gives

the two possible values shown in Eqs. (28).

$$a_1 = \frac{-RC + \sqrt{R^2C^2 - 4LC}}{2LC}$$
$$a_2 = \frac{-RC - \sqrt{R^2C^2 - 4LC}}{2LC} \tag{28}$$

Associating the integration constant A_1 with a_1 and A_2 with a_2, the solution takes the form of Eq. (29).

$$i = A_1\epsilon^{a_1 t} + A_2\epsilon^{a_2 t} \tag{29}$$

The voltage across the capacitor, Eq. (30), is obtained from Eq. (24).

$$e_C = \frac{1}{C}\int i\,dt = E - Ri - L\frac{di}{dt}$$
$$= E - (R + a_1 L)A_1\,\epsilon^{a_1 t} - (R + a_2 L)A_2\epsilon^{a_2 t} \tag{30}$$

Initially, at $t=0$, the current must be zero because of the inductance, and the capacitor voltage is $e_C = V$. By the first of these conditions, $i(0) = 0$, in Eq. (29) it is seen that $A_2 = -A_1$. And by the second condition, $e_C(0) = V$, in Eq. (30), Eq. (31) is obtained.

$$V = E - (R + a_1 L)A_1 - (R + a_2 L)A_2$$
$$= E - (a_1 - a_2)LA_1$$
$$A_1 = \frac{E - V}{(a_1 - a_2)L} = \frac{C(E - V)}{\sqrt{R^2C^2 - 4LC}} \tag{31}$$

Then Eq. (32) is the complete solution.

$$i = \frac{C(E - V)}{\sqrt{R^2C^2 - 4LC}}\,(\epsilon^{a_1 t} - \epsilon^{a_2 t}) \tag{32}$$

There are three special cases of this solution, depending on the nature of the radical in Eq. (32).

Nonoscillatory case: $R^2C > 4L$. In this case the radical is positive, and the exponents a_1 and a_2 are real and negative, $a_1 = -\alpha + \beta$, $a_2 = -\alpha - \beta$, where $\alpha = R/2L$ and $\beta = \sqrt{R^2C^2 - 4LC}/2LC$. Then i is given by Eq. (33), which is shown in Fig. 3, where q_0 is assumed to be 0 and $V = 0$.

$$i = \frac{C(E - V)}{\sqrt{R^2C^2 - 4LC}}\epsilon^{-\alpha t}(\epsilon^{+\beta t} - \epsilon^{-\beta t})$$
$$= \frac{2C(E - V)}{\sqrt{R^2C^2 - 4LC}}\epsilon^{-\alpha t}\sinh\beta t \tag{33}$$

Aperiodic case: $R^2C = 4L$. In this case $a_1 = a_2 = -R/2L$, and the radical in the denominator of Eq. (32) is zero, as is the numerator. The indeterminant is easily evaluated from Eq. (33) upon letting $\beta \to 0$, thus giving Eq. (34). This is the aperiodic or critical case, and is illustrated in Fig. 3.

$$i = \frac{2C(E - V)}{2LC\beta}e^{-\alpha t}\sinh\beta t|_{\beta \to 0} = \frac{E - V}{L}e^{-\alpha t}t \tag{34}$$

Oscillatory case: $R^2C < 4L$. In this case the radical in Eq. (28) becomes imaginary and the exponents take the form of Eqs. (35), and Eq. (32) be-

$$a_1 = \frac{-R}{2L} + j\frac{\sqrt{4LC - R^2C^2}}{2LC} = -\alpha + j\omega$$
$$a_2 = \frac{-R}{2L} - j\frac{\sqrt{4LC - R^2C^2}}{2LC} = -\alpha - j\omega \tag{35}$$

comes Eq. (36). This is a damped oscillation and is illustrated in Fig. 3.

$$i = \frac{C(E-V)}{2j\omega LC} e^{-\alpha t}(\epsilon^{j\omega t} - \epsilon^{-j\omega t})$$

$$= \frac{E-V}{\omega L} \epsilon^{-\alpha t} \sin \omega t \qquad (36)$$

AC TRANSIENTS

Alternating-current transients differ from direct-current transients in two important respects: (1) The final condition, or steady state, is an alternating or cyclic one, and (2) the amplitudes of the transient terms depend on the point on the ac applied voltage wave at which the transient is initiated, and can therefore have many different values, or even change sign.

In the section on dc transients, solutions were carried out for RL, RC, and RLC circuits switched onto a voltage source. In the present section, the solution will be carried out in detail for the RLC circuit only; the others will be regarded as special cases of this general solution by putting $C = \infty$ and $L = 0$, respectively.

Resistance, inductance, and capacitance in series. Consider the circuit of Fig. 4 in which an RLC circuit is suddenly switched onto an ac voltage $e = E \sin(\omega t + \gamma)$ at an electrical angle γ dis-

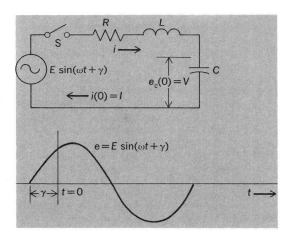

Fig. 4. *RLC* circuit with alternating current source.

placed from $\omega t = 0$. Assume that a current I is flowing in the circuit and a voltage V is across the capacitance at the instant the switch is closed on the alternator. The voltage equation and the initial conditions then are given by Eqs. (37) and (38).

$$E \sin(\omega t + \gamma) = Ri + L\frac{di}{dt} + \frac{1}{C}\int i \, dt \qquad (37)$$

$$i = I \quad \text{and} \quad e_C = V \quad \text{at} \quad t = 0 \qquad (38)$$

Differentiating Eq. (37) once to clear it of the integral gives Eq. (39). The complementary solution

$$\omega E \cos(\omega t + \gamma) = L\frac{d^2i}{dt^2} + R\frac{di}{dt} + \frac{i}{C} \qquad (39)$$

of this equation is the same as in the dc case of Eq.

(25), that is, Eq. (29), where a_1 and a_2 are given in Eq. (28).

The particular integral of Eq. (39) is its final steady-state current. This is most easily obtained as Eq. (40), the ordinary ac solution for the current, to which Eqs. (41) and (42) apply.

$$i_{ac} = \frac{E}{Z} \sin(\omega t + \gamma - \theta) \qquad (40)$$

$$Z = \sqrt{R^2 + (\omega L - 1/\omega C)^2} \qquad (41)$$

$$\tan \theta = \frac{\omega L - 1/\omega C}{R} = \frac{\omega^2 LC - 1}{\omega CR} \qquad (42)$$

The complete solution then is given by Eq. (43), in which A_1 and A_2 are integration constants and a_1 and a_2 are given in Eq. (28).

$$i = \frac{E}{Z} \sin(\omega t + \gamma - \theta) + A_1 \epsilon^{a_1 t} + A_2 \epsilon^{a_2 t} \qquad (43)$$

The capacitor voltage is given by Eq. (44).

$$e_C = \frac{1}{C}\int i \, dt = -\frac{E}{\omega CZ} \cos(\omega t + \gamma - \theta)$$

$$+ \frac{A_1}{Ca_1} \epsilon^{a_1 t} + \frac{A_2}{Ca_2} \epsilon^{a_2 t} \qquad (44)$$

Now applying the initial conditions of Eq. (38) at $t = 0$, there results from Eqs. (43) and (44), respectively, Eqs. (45) and (46).

$$i(0) = I = \frac{E}{Z} \sin(\gamma - \theta) + A_1 + A_2 \qquad (45)$$

$$e_C(0) = V = \frac{-E}{\omega CZ} \cos(\gamma - \theta) + \frac{A_1}{Ca_1} + \frac{A_2}{Ca_2} \qquad (46)$$

Solving Eqs. (45) and (46) simultaneously for A_1 and A_2 and using Eq. (28), Eqs. (47)–(49) result.

$$A_1 = \frac{1}{\sqrt{R^2C^2 - 4LC}}\left\{\frac{I}{a_2} - CV\right.$$

$$\left. - \frac{E}{Z}\left[\frac{1}{\omega}\cos(\gamma - \theta) + \frac{1}{a_2}\sin(\gamma - \theta)\right]\right\} \qquad (47)$$

$$A_2 = \frac{-1}{\sqrt{R^2C^2 - 4LC}}\left\{\frac{I}{a_1} - CV\right.$$

$$\left. - \frac{E}{Z}\left[\frac{1}{\omega}\cos(\gamma - \theta) + \frac{1}{a_1}\sin(\gamma - \theta)\right]\right\} \qquad (48)$$

$$a_1 = \frac{-R}{2L} + \sqrt{\left(\frac{R}{2L}\right)^2 - \frac{1}{LC}}$$
$$a_2 = \frac{-R}{2L} - \sqrt{\left(\frac{R}{2L}\right)^2 - \frac{1}{LC}} \qquad (49)$$

These values, together with Eqs. (41) and (42), inserted in Eqs. (43) and (44) give the solution in a form suitable for the critically damped case. However, if $R^2C < 4L$, then the radicals in Eqs. (47)–(49) become imaginary and these expressions become the complex numbers given in Eqs. (50),

$$A_1 = \tfrac{1}{2}(M + jN) \qquad A_2 = \tfrac{1}{2}(M - jN)$$
$$a_1 = -\alpha + j\omega \qquad a_2 = -\alpha - j\omega \qquad (50)$$

whereupon Eq. (43) takes the damped oscillatory form given in Eq. (51).

$$i = \frac{E}{Z} \sin(\omega t + \gamma - \theta)$$
$$+ \epsilon^{-\alpha t}(M \cos \omega t - N \sin \omega t) \qquad (51)$$

It is evident from Eq. (43) or Eq. (51) that the final steady-state value, after the transient has died out, is the alternating current of Eq. (40). It is also clear from Eqs. (47) and (48) that the amplitudes of the transient terms depend upon the angle γ on the ac applied voltage wave at $t = 0$.

Resistance and inductance in series. This may be regarded as a special case of Eq. (43), in which $C = \infty$ (the capacitance short-circuited) and $V = 0$. Consequently, the solution is given by Eq. (52).

$$i = \frac{E}{Z} \sin (\omega t + \gamma - \theta)$$
$$+ \left[I - \frac{E}{Z} \sin (\gamma - \theta) \right] \epsilon^{-(R/L)t} \quad (52)$$

Thus the transient starting from an initial value I decays exponentially to its final ac steady-state value.

There will be no transient if the switch is closed at the angle γ on the voltage wave such that Eq. (53) applies.

$$I = \frac{E}{Z} \sin (\gamma - \theta) \quad (53)$$

Resistance and capacitance in series. This may be regarded as a special case of Eq. (43), in which $L = 0$ and $I = 0$. Under these conditions the solution is given by Eq. (54) and the capacitor volt-

$$i = \frac{E}{Z} \sin (\omega t + \gamma - \theta)$$
$$- \left[\frac{V}{R} + \frac{E}{\omega CRZ} \cos (\gamma - \theta) \right] \epsilon^{-t/RC} \quad (54)$$

age is, by Eq. (44), defined by Eq. (55). Thus the

$$e_c = \frac{-E}{\omega CZ} \cos (\omega t + \gamma - \theta)$$
$$+ \left[V + \frac{E}{\omega CZ} \cos (\gamma - \theta) \right] \epsilon^{-t/RC} \quad (55)$$

transient, starting with a voltage V, decays exponentially to its final ac steady-state value. If the switch is closed at an angle γ on the voltage wave such that Eq. (56) applies, there will be no transient.

$$V = -\frac{E}{\omega CZ} \cos (\gamma - \theta) \quad (56)$$

[LOYAL V. BEWLEY]

Bibliography: R. Rudenberg, *Transient Performance of Electric Power Systems: Phenomena in Lumped Networks*, 1950, reprint 1969.

Electric uninterruptible power system

A system that provides protection against commercial power failure and variations in voltage and frequency. Uninterruptible power systems (UPS) have a wide variety of applications where unpredictable changes in commercial power will adversely affect equipment. This equipment may include computer installations, telephone exchanges, communications networks, motor and sequencing controls, electronic cash registers, hospital intensive care units, and a host of others. The uninterruptible power system may be used on-line between the commercial power and the sensitive load to provide transient free well-regulated power, or off-line and switched in only when commercial power fails.

Types of system. There are three basic types of uninterruptible power system. These are, in order of complexity, the rotary power source, the standby power source, and the solid-state uninterruptible power system.

Rotary power source. The rotary power source consists of a battery-driven dc motor that is mechanically connected to an ac generator. The battery is kept in a charged state by a battery charger that is connected to the commercial power line. In the event of a commercial power failure, the battery powers the dc motor which mechanically drives the ac generator. The sensitive load draws its power from the ac generator and operates through the outage.

Standby power source. The standby power source (Fig. 1) consists of a battery connected to a dc-to-ac static inverter. The inverter provides ac power for the sensitive load through a switch. A battery charger, once again, keeps the battery on full charge. Normally, the load operates directly from the commercial power line. In the event of commercial power failure, the switch transfers the sensitive load to the output of the inverter.

Solid-state system. The solid-state uninterruptible power system has a general configuration much like that of the standby power system with one important exception. The sensitive load operates continually from the output of the static inverter. This means that all variations on the commercial power lines are cleaned and regulated through the output of the uninterruptible power system. A commercial power line, known as a bypass, is provided around the uninterruptible power system through a switch. Should the uninterruptible power system fail at some point, the commercial power is automatically transferred to the sensitive load through the switch. This scheme is known as an on-line automatic reverse-transfer uninterruptible power system (Fig. 2).

Subsystems. An uninterruptible power system consists of four major subsystems (Fig. 3): a method to put energy into a storage system, a battery charger; an energy storage system, the battery; a system to convert the stored energy into a usable form, the static inverter; and a circuit that electrically connects the sensitive load to either the output of the uninterruptible power system or to the commercial power line, the transfer switch.

Fig. 1. Standby power source. (*Topaz Electronics*)

Battery charger. The purpose of the battery charger, also called the rectifier charger, is to deliver the power required to drive the static inverter and to charge the battery. When the commercial power line is present, the charger circuit supplies all the power used by the inverter and maintains the battery at full charge. Additionally the charger is sized to allow a safety margin for overload conditions at the output of the static inverter. Uninterruptible power systems that have a very long battery charge time require a charger that is only slightly larger than that of the inverter. In an uninterruptible power system where a short recharge time is required, the charger becomes much larger.

The battery charger is generally of the float type and exhibits very-well-regulated output voltage over a wide output current demand. This tight regulation is performed through the use of silicon controlled rectifiers for voltage regulation and silicon rectifiers to convert ac to dc. Current limiting is built into the circuit to protect the charger. The rating of the charger is determined by the requirements of the static inverter and the current necessary to recharge the battery after an uninterruptible power system has gone through a commercial power outage. *See* SEMICONDUCTOR RECTIFIER.

Batteries. The battery used on an uninterruptible power system is sized to provide power to the inverter any time that the commercial power is removed. Power may be provided for anywhere from a few seconds to an hour or more. The battery necessary for power is dependent upon the sensitive load to be powered and the desired backup time.

The batteries are arranged in a series of racks called a battery bank. Each battery is made up of individual cells containing lead alloy plates and an electrolyte. Conductor size between batteries and current to be handled by the static inverter determine the battery bank voltage. Small uninterruptible power systems use low voltages, while large uninterruptible power systems use battery banks of several hundred volts. In the smaller uninterruptible power systems the battery bank may be mounted adjacent to or inside the system.

Batteries may be any one of four different types. They are, in order of their frequency of use: lead antimony, nickel cadmium, gel electrolyte, and lead calcium. The cells within the lead antimony battery contain grids constructed of an alloy of lead and antimony and are surrounded by an electrolyte of sulphuric acid. This type of battery requires specialized charging techniques and a high level of maintenance. The nickel cadmium battery offers a high storage level at high cost. It is best used where the battery bank is subject to extremes of cold and physical space constraints. Once again, specialized charging techniques are required for this type of battery. The gel electrolyte type of battery is usually used in smaller uninterruptible power system installations where voltage and current demands will be at a relatively low level. Although generally more expensive than a liquid electrolyte lead acid battery, it is sealed and maintenance-free. In the lead calcium battery each cell uses calcium alloyed with lead and is filled with an electrolyte of sulfuric acid. This battery type has wider recharge tolerances, longer life, and substantially lower maintenance.

Fig. 2. A 250-kW on-line automatic reverse-transfer uninterruptible power system. (*Exide Electronics*)

Static inverter. The purpose of the static inverter is to change the dc voltage derived from the battery charger or battery bank to an ac voltage in order to power the sensitive load. The static inverter determines the quality of power used to drive the sensitive load. The ac output voltage must be stable and free from all interruptions. It is the most important subsystem of the uninterruptible power system. If the inverter should fail, the system is out of operation.

The most common static inverter types are, in order of complexity, the ferroresonant type, the quasi-square-wave type, the pulse-width-modulated type, and the step-wave type. The ferroresonant approach starts with a square-wave inverter system and a tuned output transformer. The output transformer performs all filtering, voltage regulation, and current limiting with magnetic regulation, allowing for the design of a very simple inverter. This technique is generally found in smaller, lower-power uninterruptible power systems. The quasi-square-wave approach uses a true electronic regulation technique. It consists of two square waves superimposed on each other to approximate the form of an ac sine wave. Regulation is achieved

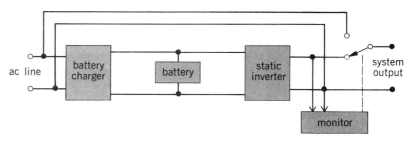

Fig. 3. Block diagram of an on-line automatic reverse-transfer uninterruptible power system, showing subsystems. (*Topaz Electronics*)

with a silicon controlled rectifier bridge and control circuit that change the relationship between the two square waves. This changes the pulse width and amplitude of the square wave, achieving regulation. An output LC filter is employed to filter and wave-shape the output sine wave. The pulse-width-modulated approach is essentially a square-wave inverter operating at high frequency. The pulse width, and not the amplitude, of the square wave is adjusted to approximate the sine wave. Once again, output filtering is employed to shape the output waveform. The stepped-wave approach is an extension of the pulse-width-modulated approach. Several pulses are provided per half cycle of the sine wave and are combined to develop an output voltage resembling a sine wave that needs very little filtering. This approach is complex and is found only in high-powered uninterruptible power system units. *See* ELECTRIC FILTER; FUNCTION GENERATOR; PULSE MODULATION; TRANSFORMER.

Transfer switch. The switch that connects the sensitive load to either the inverter or the commercial power line is called the transfer switch. The major function of the switch is to provide an alternate source of ac power to the critical load should any of the components in the uninterruptible power system fail. The position of the switch is controlled by a monitor circuit as shown in Fig. 3. Generally the switch in an uninterruptible power system is a high-speed solid-state device that can transfer the load from one ac source to another with little or no break in power. The switch may be designed for uninterrupted make-before-break or interrupted break-before-make transfers.

[JOHN SULLIVAN]

Bibliography: *AC Power Handbook of Problems and Solutions*, 1979; K. G. Brill, *Mini-Micro Syst.*, 10(7):38–45, July 1977; N. L. Conger. *Instrum. Technol.*, 20(9):57–63, September 1973; J. J. Waterman, *Specifying Eng.*, 43(2):60–64, February 1980.

Electrical communications

The science and technology by which information is collected from an originating source, transformed into electric currents or fields, transmitted over electrical networks or through space to another point, and reconverted into a form suitable for interpretation by a receiving entity.

Sources. Information to be transmitted comes from many sources. Some sources produce essentially continuously varying signals. Examples are speech, music, the output of TV cameras, and sensors of temperature, pressure, and the like. Others produce signals whose distinguishing characteristic is that they are on or off. Examples are the hand telegraph key, many operations of computers, above-or-below sensors of temperature, pressure, and so on.

Transducers. For transmission over an electrical communication channel, the original physical manifestation must be transformed into an analogous electrical signal—a speech sound in air by a telephone transmitter, variations in light intensity in a scanned picture by a television camera, temperature observations by the opening and closing of the switch of a thermostat. *See* TRANSDUCER.

Information theory. Many workers, especially the pioneers R. V. L. Hartley, N. Wiener, and C. E. Shannon, have striven for a mathematical definition of information applicable to all the various sources, in a form that could be directly related to the fundamental characteristics of an electrical communication channel. The information unit chosen is the bit—an abbreviation of binary digit—which is the specification of a choice between only two alternatives, such as "yes" or "no." Finer gradations needed to describe the output from a continuously varying signal source can be determined by a sufficient but definite number of such choices or bits. Hartley and Shannon showed that the maximum practicable capacity C which a communication channel can deliver is $C = B \log_2 (1 + S/N)$ bits/sec, where B is the bandwidth in hertz, and S/N is the power ratio of the signal to the noise at the receiving end. Bandwidth and received signal-to-noise ratio are thus the fundamental characteristics of the electrical communication channel, however implemented. *See* BANDWIDTH REQUIREMENTS; ELECTRICAL NOISE; INFORMATION THEORY; SIGNAL-TO-NOISE RATIO.

Transmission media. Media commonly used for electrical communication can be divided into two classes: (1) Those carrying electrical currents, such as a pair of wires or coaxial cable. (2) Those in which a transducer must be used to change electrical currents into a form suited to the medium. Examples are electromagnetic fields in space, such as radio, for which the antennas are the transducers; electromagnetic fields in guiding structures, such as a waveguide; pressure waves in water (sonar); light waves in space or optical fibers. The various media have characteristics that are very different; relating these characteristics to the underlying physics is a very important field of study. But with appropriate transducers, the different media are simply alternative choices for carrying electrical communication channels. *See* ANTENNA; COAXIAL CABLE; COMMUNICATION CABLES; ELECTROMAGNETIC WAVE TRANSMISSION; MICROWAVE TRANSMISSION LINES; OPTICAL COMMUNICATIONS; TRANSMISSION LINES; WAVEGUIDE.

Network configurations. Networks may be configured in several ways. Perhaps the most common is a channel between point A and point B. These may be direct connections for relatively long periods of time. They may, as in the telephone network, be switched on a call or message basis among the very large number of points available. Most point-to-point channels carry information in both directions. *See* SWITCHING SYSTEMS (COMMUNICATIONS).

Another configuration is a treelike structure in which the channel from one source is divided and subdivided to connect to many receivers. A cable television (CATV) system is an example of this type of configuration. Generally, though not always, the tree structures are one-way. *See* CLOSED-CIRCUIT TELEVISION.

The third configuration is typified by radio and TV broadcasting: a single source sends out radio waves, and any number of receivers in the area can pick them up on a one-way basis.

Multilink channels. The simplest channels are those having amplification, if any, only at the ends. One example is a telephone call to a neighbor involving only a few miles of wire between telephone instruments. Another is the radio channel from an astronaut on the Moon back to Earth. But whatever the transmission medium, the signal becomes weaker with distance, dropping nearer to the inevitable noise in the medium, thus reducing the S/N ratio, and so limiting the information-carrying capacity of the channel. It becomes desirable to place one or more amplifiers at periodic intervals along the channel to raise the signal power and hence maintain the desired S/N at the receiving end. For cable channels thousands of miles long, it would be completely impractical to pour in enough signal power at one end to have a useful S/N at the other, whereas amplifiers with quite small power output at appropriate intervals can be satisfactory. Of course, the problems of designing amplifiers such that hundreds or thousands can be operated in tandem are themselves formidable. *See* AMPLIFIER.

Frequency division multiplex. Radio broadcasting is the best-known example of many channels sharing a common medium. Each transmitter is assigned a position in the frequency spectrum and is permitted to occupy a specified bandwidth. The receivers include selective networks or filters, which, when tuned to a selected channel, accept that channel and reject the others. This is an example of sharing the common medium by dividing the available frequency spectrum among the sharing channels, a technique known as frequency division multiplex. The process of shifting the input signal to the assigned frequency slot is called modulation, and the inverse process at the receiving end is called demodulation. The same processes are applied to the sharing of a common wire or coaxial cable medium. *See* ELECTRIC FILTER.

In amplitude modulation (AM), the amplitude of a steadily applied, continuous wave (carrier) is varied in accordance with the amplitude of the input signal. Mathematical analysis of the resulting wave shows that it includes three components: a large component of the carrier frequency, an upper sideband with a frequency width equal to that of the input signal, and an identical lower sideband. Obviously, all of the desired information is included in either sideband, so it is not necessary to transmit either of the other two components.

On cable systems, where the available useful bandwidth is limited, only one sideband is sent. This requires extra complexity of equipment, which is well justified by the more efficient utilization of the cable and its amplifiers. However, in radio broadcasting, the overriding consideration is to simplify the millions of home receivers, so double sideband is sent. *See* AMPLITUDE MODULATION; SINGLE SIDEBAND.

E. L. Armstrong pioneered the use of frequency modulation (FM). In this process, the steadily applied carrier wave is varied in frequency (rather than amplitude) according to the amplitude of the input signal. The resulting output wave occupies more bandwidth than with amplitude modulation, but it can be received successfully with a poorer S/N ratio. The system is used in radio broadcasting in the FM band (88–108 MHz) and has found wide application in many radio communication systems. *See* FREQUENCY MODULATION.

In both AM and FM frequency division multiplex, the amplitude of the input signal controls the output. Hence they are well adapted to carry the continuously varying inputs like speech. However,

the on-off sources require some special treatment to adapt them to the essentially analog nature of AM or FM. *See* MODULATION.

Time division multiplex. There is another way for a multiplicity of channels to share a common medium, called time division multiplex. Each channel is assigned a short time slot during which it occupies the whole available bandwidth, with the channels occupying time slots in turn. When all channels have had their turn, a frame has been completed, and the roll call is repeated. Obviously, there has to be an identifying symbol at the beginning of each frame, so that the channel time slots can be identified.

It is not immediately evident that a continuously varying input signal can be transmitted through these repetitive, short time slots. However, H. Nyquist showed mathematically that if the amplitude of a wave is sampled for short intervals at a rate at least twice that of the highest frequency in the input, the wave can be reproduced without distortion from the samples. For speech, this "Nyquist rate" is commonly accepted to be 8 kHz, thus setting the length of the frame at 125 μsec. The short samples of the varying amplitude of the input wave could be sent directly in the allotted time slots, a technique known as pulse amplitude modulation (PAM). But consideration of the tradeoffs that exist between bandwidth and S/N in the shared medium usually leads to more complicated methods. One example is pulse code modulation (PCM). The amplitude of the sample is not sent directly. Instead, a code of an appropriate number of pulses is sent, with one code word defining one amplitude. This involves a quantization of the signal, so that it is necessary to use enough pulses in the code word to keep distortion at an acceptable level. *See* PULSE MODULATION.

Clearly, the processing of continuously varying input sources is much more complicated than in frequency-division AM or FM systems. It involves sampling, quantizing, coding, and organizing into words and frames, with framing symbols and precise timing. When only vacuum tubes were available, the economics of the resulting tradeoffs were completely unfavorable. With the advent of the transistor, and the subsequent flowering of the solid-state art into integrated circuits, the whole economic picture changed dramatically. Complicated processing could be accomplished by tiny, inexpensive chips. The extra complexity in many cases was well justified by savings in the cable system and its amplifiers.

At the same time, there was a rapidly growing demand for communication of essentially on-off signals, as between computers and machines. The time division multiplex is very well suited to these signals, with little processing. For these reasons, there is a very rapidly expanding use of time division systems. However, the choice of method to be used is very complex, depending on the signals to be transmitted, the medium to be used, and the economics of the state of the art. In the case of radio, considerations of national policy affect the choices, since the radio spectrum is a limited and valuable national resource. *See* ELECTRONICS; INTEGRATED CIRCUITS; TRANSISTOR.

For a discussion of major forms of electrical communication systems *see* RADAR; RADIO; TELEVISION.   [A. C. DICKIESON]

Bibliography: Communications, special edition of *Sci. Amer.*, September 1972; S. Haykin, *Communication Systems*, 1978; G. Kennedy, *Electronic Communication Systems*, 2d ed., 1977; A. Marcus and W. Marcus, *Elements of Radio*, 6th ed., 1973; P. Z. Peebles, *Communication System Principles*, 1976; J. R. Pierce and E. C. Posner, *Introduction to Communication Science and Systems*, 1980; R. L. Shrader, *Electronic Communication*, 4th ed., 1980; H. Taub and D. L. Schilling, *Principles of Communication Systems*, 1971.

Electrical conduction in gases

The process by means of which a net charge is transported through a gaseous medium. It encompasses a variety of effects and modes of conduction, ranging from the Townsend discharge at one extreme to the arc discharge at the other. The current in these two cases ranges from a fraction of 1 microampere in the first to thousands of amperes in the second. It covers a pressure range from less than 10^{-4} atm to greater than 1 atm. *See* ARC DISCHARGE; TOWNSEND DISCHARGE.

In general, the feature which distinguishes gaseous conduction from conduction in a solid or liquid is the active part which the medium plays in the process. Not only does the gas permit the drift of free charges from one electrode to the other, but the gas itself may be ionized to produce other charges which can interact with the electrodes to liberate additional charges. Quite apparently, the current voltage characteristic may be nonlinear and multivalued. *See* SEMICONDUCTOR.

The applications of the effects encountered in this area are of significant commercial and scientific value. A few commercial applications are thyratrons, gaseous rectifiers, ignitrons, glow tubes, and gas-filled phototubes. These tubes are used in power supplies, control circuits, pulse production, voltage regulators, and heavy-duty applications such as welders. In addition, there are gaseous conduction devices widely used in research problems. Some of these are ion sources for mass spectrometers and nuclear accelerators, ionization vacuum gages, radiation detection and measurement instruments, and thermonuclear devices for the production of power.

The discussion of this complicated process will be divided into two parts. The first will deal with the basic effects involved, including production and loss of charges within the region and the motion of charges in the gas. The second part will deal with the mechanism of conduction.

BASIC EFFECTS

To produce gaseous conduction, two conditions must obtain. First, there must be a source of free charges. Second, there must be an electric field to produce a directed motion of these charges. Considering the first of these, one finds that the free charge concentration is a result of a number of processes which produce and remove charges.

Sources of free charges. In many gaseous devices, a thermionic-emission cathode is included. The process of electron emission from a heated electrode is well known. Closely related to this as a source of electrons is field emission. Here, a strong positive field at a metallic surface lowers the barrier for electron emission. Thus, the electron current from a surface at a given temperature may be significantly increased. *See* THERMIONIC EMISSION.

Both of these effects result in electron production. Another effect, photon absorption, may result only in electrons if the absorber is a solid. However, if the photon interacts with a gas molecule or atom, ionization may result and both an electron and a positive ion be obtained. The photon may come from some external source or it may be a secondary effect of the gaseous conduction. It may have a wavelength in the visible, ultraviolet, or x-ray region. *See* PHOTOELECTRICITY.

Conduction in flames is largely a result of the thermal production of ionization. This is a specialized field which has long been of interest in chemistry and combustion studies. To produce appreciable thermal ionization, the temperature must be high, as in a flame. If the effective temperature is known, the ionization concentration may be determined from statistical mechanics. Thermal ionization is also of tremendous importance in devices for production of power by thermonuclear processes, and in ion-propulsion equipment. A special form of this is surface ionization, in which a hot surface may cause ionization of a gas atom that comes in contact with it.

Another source of ionization is particle radiation due to cosmic rays, radioactive material in the walls or in the gas, or particles produced from an external source. These particles may then produce an ionized track in the gas. An example is the ionization produced by an α-particle from a polonium source in an ionization chamber.

In most of these methods of charge production, the sources are primary. That is, the presence of other free charges is not important in the production of ionization or electrons. Other processes are secondary in origin, although they may be of prime importance. It was pointed out that photons could originate either externally or as a secondary effect. Field emission could be a secondary effect, also. Other methods of ionization, however, are generally thought of as being secondary in origin. Ionization of the gas by electron impact is such a case. Here free electrons may gain enough energy in an electric field to interact with an atomic or molecular electron to produce an ion pair.

Cumulative ionization is an extension of ionization by impact. If the original electron and its offspring gain enough energy so each may produce another electron, and if this process is repeated over and over, the result is called an avalanche, and the ionization thus produced is referred to as cumulative ionization. This is the basis for particle detection in some ionization devices.

Another secondary source is electron emission from either electron or positive ion bombardment of a surface. This should not be confused with thermionic emission resulting from heating under bombardment. *See* SECONDARY EMISSION.

Other sources which may be important are atomic collisions, sputtering, and collisions of the second kind. In the first case, an atom or heavy ion may collide with an atom to produce ionization. This is quite unlikely until an energy of many times the ionization energy is obtained. The second is somewhat analogous to secondary electron emission. Here the positive ions strike a surface and knock out atoms or groups of atoms. Some of these come off as ions. In the third case, an excited atom may interact with an atom or molecule which is chemically different and has an ionization potential lower than the excitation potential of the excited atom. The result may be the decay of the excited state with ionization of the struck molecule or atom. Symbolically, this is shown by reaction (1),

$$A^* + B \rightarrow A + B^+ + e^- \tag{1}$$

where A^* is the excited atom, B the struck atom or molecule, and e^- an electron. *See* EXCITATION POTENTIAL; IONIZATION POTENTIAL; SPUTTERING.

Free-charge removal. The net free-charge concentration is a balance between charge-production and charge-removal processes. Recombination is one such process. Here, an electron or heavy negative ion and a positive ion may recombine. The energy transition may appear as electromagnetic radiation or may be carried off by a third body, if one is present. There are a wide variety of conditions which may lead to recombination. Where the temperature and electric field are high, the recombination will occur predominantly at the walls.

The method of charge removal is important from the aspect of conduction, however. If the charges move to the appropriate electrodes under the influence of the field and there recombine, then they contribute to the current. If they simply diffuse to the walls and recombine there with ions of the opposite sign, or if they recombine in the gas volume, they may not appear as part of the external current.

Motion of the. charges. The motion of the charges within the gas will be largely influenced by the potential function. For the usual regular geometries, this could be calculated in principle if there were no charges present. However, in a gas with free charges distributed throughout, the problem is quite different. The charges modify the charge-free potential, but the potential itself determines how the charge will move. The motion of the charges further modifies the potential and so on. Although the situation can be described physically by Poisson's equation, it is generally impractical to carry out an analysis. As a practical result, the potential function must be determined by measurements which are made with probes. This requires careful procedures to obtain significant results.

Diffusion of ions. Ion diffusion is a type of random motion which is always present and is the result of thermal or agitation energy. The randomness of the motion is brought about by the many collisions with molecules and other ions. A great difference exists in the motions of electrons and heavy ions. Because of low mass, electrons are easily deflected, so they move erratically. They diffuse badly, and follow field lines only generally. Again because the mass is small, an electron can give up appreciable energy only in an inelastic collision, in which excitation or ionization takes place. Hence, electron agitation and diffusion will be much greater in a pure inert gas than in a gas having many low-energy molecular states. Conversely, heavy ions exchange energy effectively at every collision. Diffusion is much less, so that they follow the electric field lines more closely than do electrons.

MECHANISM OF CONDUCTION

The ionic mobility μ relates drift velocity v to electric field X by Eq. (2). For electrons, the mobili-

$$v = \mu X \tag{2}$$

ty is high, and a drift velocity of 10^6 cm/sec or greater may be obtained. The electronic mobility is not a true constant, but varies with field, pressure, temperature, and gas composition. For heavy ions, the mobility is much more nearly constant, but is still dependent on these quantities to some extent. Drift velocities are usually of the order of $10^3 - 10^4$ cm/sec. Thus, in a typical conduction device, an electron may move from one electrode to the other before a heavy ion is displaced appreciably.

It would appear from the foregoing that if accurate information about the important processes existed, one could predict the characteristics of the conduction process under given conditions. Unfortunately this is not the case. Generally, the situation is so complicated that the theory can yield only qualitative predictions. Accordingly, most of the information concerning the various forms of gaseous conduction is empirical. In the present description, it will be possible to mention the main features of a few of these modes.

The illustration shows a sample voltage-current characteristic for a two-electrode device with constant pressure. It is assumed that there is a constant source of ionization which could be any of the primary sources previously discussed. In region A, the current first rises and then over a limited range is relatively constant as the voltage across the electrodes is increased. The initial rise is the result of the collection of charges which were either recombining or diffusing to the walls. The nearly constant current region is the result of the collection of almost all of the charges.

In region B, further increase in voltage produces an increase in current. Here, ionization by electron impact is occurring. The situation is described by specifying that each free electron makes α additional ion pairs in traveling 1 cm in the direction of the field. The number of ion pairs produced per second in 1 cm at a distance x from the cathode (assuming parallel plate electrodes) is given by Eq. (3), where n_0 is a constant depending on the initial

$$n = n_0 e^{\alpha x} \tag{3}$$

number of electrons. This is a form of the Townsend equation, and α is the first Townsend cofficient. In the region B in the figure, the increase in current represents an increase in α. Near the end of this region, the current increases more rapidly with applied field. Here, additional effects are taking place, such as the photoelectric process and secondary emission. This situation is described by Eq. (4), where β is the second Town-

$$i = i_0 \frac{(\alpha - \beta) e^{(\alpha - \beta)x}}{\alpha - \beta e^{(\alpha - \beta)x}} \tag{4}$$

send coefficient, i_0 is the initial electron current at the cathode, and i is the anode current as a function of plate separation x; β is also a function of electric field.

At the end of the region, the slope becomes infinite, and if the external resistance is not too large, the current will jump in a discontinuous

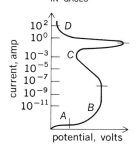

ELECTRICAL CONDUCTION IN GASES

Current-potential characteristics for a two-electrode device with constant pressure.

fashion. The transition is referred to as a spark, and the potential at which it occurs is the breakdown or sparking potential. The region B is called a Townsend discharge and is not self-sustained. Thus, if the source of primary ionization were removed, the discharge would cease. *See* BREAKDOWN POTENTIAL; ELECTRIC SPARK.

As the potential reaches the sparking potential, a transition occurs to the region C. This is the self-sustained glow discharge region. Over an extensive current range, the voltage drop remains substantially constant. During the current increase, a glow occurs at the cathode, and at the upper end of the range, the cathode is completely covered. At this point, a further current increase can be achieved only if the potential drop across the discharge is increased. This portion of the characteristic is known as the abnormal glow. Throughout this portion of the discharge characteristic curve, secondary effects are quite important. Particularly vital are the effects of cumulative ionization and secondary emission at the cathode. *See* GLOW DISCHARGE.

Further increase in current leads to another mode of discharge, the arc. This is shown as region D in the illustration. Characteristic of this mode is the low cathode potential fall and the very high current density. It is generally felt that the predominant effect in the production of the large number of electrons at the cathode necessary for the arc is thermionic emission. This is consistent with the very high temperatures known to exist either generally or locally on the cathode. Although the arc type of discharge has very great commercial value, the mechanism of its operation is not very well understood.

In addition to these general types of conduction, there are very special cases of considerable interest. Some of these are the corona discharge, radio-frequency or electrodeless discharge, hot-cathode discharge, and discharges in the presence of a magnetic field.

[GLENN H. MILLER]

Bibliography: J. Beynon, *Conduction of Electricity through Gases*, 1972; J. D. Cobine, *Gaseous Conductors*, 1941.

Electrical impedance

The total opposition that a circuit presents to an alternating current. Impedance, measured in ohms, may include resistance R, inductive reactance X_L, and capacitive reactance X_C. *See* REACTANCE.

The impedance of the series RLC circuit is given by Eq. (1).

$$Z = \sqrt{R^2 + (X_L - X_C)^2} \text{ ohms (magnitude)} \quad (1)$$

In terms of complex quantities, this impedance is given by Eq. (2). The two components of Z are

$$Z = R + j(X_L - X_C) \quad (2)$$

at right angles to each other in an impedance diagram. Therefore, impedance also has an associated angle, given by Eq. (3). The angle is called the

$$\theta = \arctan \frac{X_L - X_C}{R} \quad (3)$$

phase, or power-factor, angle of the circuit. The current lags or leads the voltage by angle θ de-

pending on whether X_L is greater or less than X_C.

Impedance may also be defined as the ratio of the rms voltage to the rms current, $Z = E/I$. This is a form of Ohm's law for ac circuits. For further discussion of impedance *see* ALTERNATING-CURRENT CIRCUIT THEORY.   [BURTIS L. ROBERTSON]

Electrical instability

A persistent condition of unwanted self-oscillation in an amplifier or other electrical circuit. Instability is usually caused by excessive positive feedback from the output to the input of an active network. If, in an audio-frequency amplifier, instability is at a low audible frequency, the output will contain a putt-putt sound, from which such instability is termed motorboating. The instability may also be at a high audible frequency, or it may be at frequencies outside the audible range. Although such oscillations may not be heard directly, they produce distortion by driving the amplifier beyond its linear range of operation. *See* AMPLIFIER; MOTORBOATING.

Instability can arise if the load on an amplifier has a critical phase or magnitude. Similarly, instability arises in a closed-loop control system if the damping is too light relative to the response time. Such instability may result in the system hunting about a control condition instead of remaining steadily at the condition. *See* NEGATIVE-RESISTANCE CIRCUITS.

In a power distribution system, if the mechanical load on synchronous motors exceeds the steady-state stability limit or if an abruptly changed mechanical load causes the synchronous machines to exceed the transient stability limit, the power system becomes unstable. It can also lose stability from a three-phase short circuit between generators and motors. High-speed circuit breakers and other protective devices guard against such instability.

[FRANK H. ROCKETT]

Electrical loading

The addition of inductance to a transmission line in order to improve its transmission characteristics over the required frequency band. Loading coils are often inserted in telephone lines, at spacings as close as 1 mi, to counteract the capacitance of the line and thus make the line impedance more closely equivalent to a pure resistance. Similar coils are used between sections of lines used for carrier transmission. *See* TRANSMISSION LINES.

With coaxial cables and waveguides, loading is placed at the end of the line to absorb all power reaching the end, thereby achieving a nonreflecting termination. The lossy sections used for this purpose may be tapered metal vanes, wedges of lossy dielectric material, or tapered sections of iron-dust core material that partly or completely fill the end of the line. *See* WAVEGUIDE.

When an antenna is too short to give resonance at the desired frequency, a coil can be inserted in series with the antenna to give the required amount of loading needed for resonance. As an example, practically all auto radios have a loading coil in the antenna circuit, because the usual whip antenna is much too short for broadcast-band frequencies. *See* ANTENNA.

[JOHN MARKUS]

Electrical measurements

The measurement of any one of the many quantities by which the behavior of electricity is characterized. The knowledge of the quantitative behavior of electricity is essential to scientific and technical progress. Electrical measurements play a major role in industry, communications, and even in such unrelated fields as medicine.

Many electrical measurements can be made with direct-indicating instruments merely by connecting the instrument properly in the circuit. Thus a volt-meter provides a pointer which moves over a scale calibrated in volts, and an ammeter in the same way presents a reading of current in amperes. Other direct-reading instruments are wattmeters, frequency meters, power-factor or phase-angle meters, and ohmmeters. Many electrical quantities are measured both as instantaneous values and as values integrated over time. Some electrical measurements must be made with various specialized devices or systems requiring adjustment or balancing to obtain the measured value. Typical of these are potentiometers and bridges in many standard and specialized forms.

Because of differences in instruments and techniques, it is convenient to divide measurements into direct-current (dc) and alternating-current (ac) classes.

DC measurements. In dc circuits the measurement of voltage and current often suffices to define the operation of the circuit. The product of the two represents power. In the commercial sale of dc electricity the measurement of energy must be made with a dc watt-hour meter. Occasional use is made of a dc ampere-hour meter in battery-charging installations.

To measure high values of current, shunts are used to bypass all but a small fraction of the current around the measuring instrument. A newer technique employs a form of saturable reactor energized by alternating current to measure large direct currents. *See* Current measurement; Voltage measurement.

AC measurements. Alternating-current circuits involve more variables and hence more measurements than dc circuits. The most common measurements are voltage, current, and power; the last requires a wattmeter, as ac power cannot always be calculated directly from voltage and current. Also measured are frequency and power factor (or phase angle) and sometimes waveform or harmonic content. Energy is measured by means of the ac induction watt-hour meter. In general, ac instruments differ in principle and design from dc instruments, although many ac instruments may be used to measure dc quantities. Direct-current instruments do not respond to ac quantities, but some may be adapted by the addition of rectifiers to convert alternating current to direct. The thermocouple is another form of convertor by which a dc instrument may be made to read ac quantities. *See* Phase-angle measurement.

If alternating voltages and currents above the normal ranges of self-contained instruments are to be measured, instrument transformers may be used to extend the ranges of those instruments. In the study of ac waveform a qualitative evaluation may be made with an oscillograph or a cathode-ray oscilloscope. Quantitative measurement of harmonic content requires the use of a harmonic analyzer. *See* Harmonic analyzer.

Accuracy of measurements. Accuracy denotes the degree of compliance of the instrument reading with the true value of the measured quantity. It is common to describe the instrument's accuracy by stating the maximum allowable error. Thus an instrument with a maximum error of 2% is often described as having an accuracy of 2%. For many applications a small panel or miniature electric instrument with a maximum error of 2–5% of full-scale calibration will suffice. More refined instruments are available with maximum errors of 1, 0.5, or 0.25%. When measurements of higher accuracy than this are required, measurement systems, such as potentiometers and bridges, must be used. Direct current and voltage can readily be measured in this way to an accuracy of 0.01%. Alternating-current measurements can be readily made to an accuracy of 0.1% or better.

Laboratory measurements. In a laboratory, emphasis is normally placed on accuracy and on completeness of facilities to deal with all types of measurements. There is relatively little limitation on the size and complexity of equipment used. If standardizing service is a function of the laboratory, the equipment must include standard cells and precision standard resistors and also suitable potentiometers, bridges, shunts, and volt boxes (voltage-dividing resistors to extend the range of voltage measurements). With these, the calibration of dc instruments can be performed with high accuracy. Extension of calibration service to ac instruments requires transfer standards (instruments having negligible difference in performance when operated on alternating current or direct current).

Field measurements. This term is used to designate all measurements made outside a laboratory as in generating stations and substations, service shops, factory testing areas, ships, and aircraft. For these uses equipment is chosen to perform only specialized services. Accuracy well below that of laboratory measurements is usually permissible. Convenience, compactness, and often portability are prime considerations in choosing equipment. Electric instruments for this kind of service are often of the panel type, sometimes in miniature sizes. Multipurpose and multirange instruments, like the volt-ohm-milliammeter, are handy for service measurements where 2–5% error is permissible.

For field measurements of alternating current and voltage at power frequencies, the hook-on volt-ammeter provides readings within 2% maximum error. As a voltmeter it is connected to the circuit with spring clips; as an ammeter it operates on the current-transformer principle. The core, which is circular and extends outside the instrument case, has a hinged link that may be opened to slip around a conductor carrying current and then closed again. Thus there is no need to break the circuit under measurement; the conductor itself becomes a one-turn primary winding in the transformer measurement system. The measurement of power can also be made with a hook-on wattmeter.

On power systems, field measurements may be desired continuously over a period of time, sometimes at unattended locations. For this purpose recording instruments may be used or the reading may be telemetered to a manned station. Any instrument that is made in indicating form can be made in recording form, but the greater power necessary to drive a marking device over a chart may call for some kind of amplification.

The operation of a power system also calls for the recording of disturbances due to lightning strokes, insulation flashover, short circuits, and other transient phenomena. Recording oscillographs used for this purpose can be triggered by any condition that deviates from normal operation, thus making a record of the disturbance.

Frequency considerations. The measurement of voltage and current is commonly made over a frequency range from a few hertz (Hz) up through 2000 megahertz (MHz). The frequency at which measurements are to be made dictates the type of equipment needed, the precautions to be taken, and the degree of accuracy which may reasonably be expected. Alternating-current instruments of the moving-iron, fixed-coil type are intended primarily for 60 Hz applications but may be used with only moderate errors up to several hundred hertz. Electrodynamic (moving-coil, fixed-coil) instruments, which are generally of greater accuracy, may also be used in this frequency range. Errors in such instruments result mainly from reactance effects, which may be minimized by special design to permit operation to several kilohertz (kHz). Rectifier-type instruments possess only small frequency errors up to several kilohertz in relation to their overall accuracy rating, which is of the order of 2–5% error. Electronic voltmeters, which are of the same general accuracy, are especially suited for use over a wide range of frequencies.

Circuit loading. All electric instruments draw some power from the circuits to which they are connected, and ac instruments generally take more power than dc instruments. This circuit loading may alter appreciably the quantity being measured. For instance, an ac voltmeter rated 150 volts and having a resistance of 3000 ohms is perfectly suitable to measure voltage on a 120-volt house lighting circuit. However, if the same voltmeter is connected to the terminals of a small power amplifier with a maximum output of 200 milliamperes (ma), the voltmeter will load the source and seriously reduce its voltage. To avoid this error the measurement should be made with a rectifier voltmeter (about 150,000 ohms resistance) or an electronic voltmeter (above 0.5 megohm resistance).

In the measurement of current, consideration must be given to the voltage drop in the ammeter. If it is appreciable in relation to the source voltage, the current is not the same as that if the ammeter were not connected in the circuit. The magnitude of this error can usually be evaluated and minimized by proper choice of instruments.

[ISAAC F. KINNARD/EDWARD C. STEVENSON]

Time dependence. When voltage and current are variable functions of time, the measurement of frequency, wavelength, and waveform is of importance, in addition to phase-angle measurements. At frequencies where the physical dimensions of the electric circuit are small compared to the wavelength of the voltage and current, the frequency is said to be low and various forms of frequency meters are employed, depending upon the range involved.

At higher frequencies it often becomes necessary to measure frequency and wavelength independently since their product is not always a constant under this condition. *See* FREQUENCY MEASUREMENT; WAVELENGTH MEASUREMENT.

The waveform of the electrical quantity being measured is of importance. Many indicating instruments are calibrated to give correct readings only for sine-wave inputs. If the waveform is nonsinusoidal, it is necessary to consider the principles of operation of the particular instrument to interpret the meter indication correctly. For example, an electronic voltmeter, which measures peak values but is calibrated in root-mean-square (rms) values based on sinusoidal wave shape, would not give correct rms values for a nonsinusoidal wave. *See* WAVEFORM DETERMINATION.

Measurement of parameters. The parameters of any electric circuit are the resistances, inductances, and capacitances along and between the conducting branches of the circuit, including any ground plane that may be near or surrounding the circuit. The measurement of these parameters may be classified according to the apparent disposition of the parameters, which is a function of frequency. For any given circuit there is some frequency below which the circuit can be treated as having lumped parameters or circuit elements; above this frequency the parameters must be considered as being distributed throughout the circuit.

Lumped parameters. The measurement of lumped parameters may be subdivided according to the measurement accuracy desired. If errors of several percent are permissible, direct-reading instruments, which indicate the value of the parameter directly on a calibrated meter scale, are available. Inductance and capacitance measurements are made at some convenient frequency. Of this class of instruments the ohmmeter, used for measuring resistance at zero frequency (direct current), is the only one in common use.

For greater accuracies bridge measurements are preferred. Direct-current measurements are made with the Wheatstone bridge with maximum errors on the order of 0.01%. Resistances of less than a few ohms can be satisfactorily measured only with a bridge, regardless of the accuracy desired. *See* RESISTANCE MEASUREMENT.

Most circuit designers prefer to measure inductance and capacitance in the particular operating frequency range under consideration, and the ac bridge is commonly used. Bridge measurements provide numerous advantages over other methods, including high accuracy and the ability to compare the unknown to a known standard. Bridges are designed to operate in various frequency ranges from direct current to several hundred kilohertz, from several kilohertz to several megahertz, from 1 or 2 to several hundred megahertz, and from several hundred to several thousand megahertz. At the higher frequencies the application of the bridge becomes more complicated, and considerable caution and planning are necessary if reliable results are to be obtained. *See* CAPACITANCE MEASUREMENT; HIGH-FREQUENCY IMPEDANCE MEASUREMENTS; INDUCTANCE MEASUREMENT.

A unique instrument known as a Q meter is

available for measuring inductance or capacitance and effective resistance at radio frequencies. *See* Q METER.   [ROBERT L. RAMEY]

Distributed parameters. Electrical systems can be completely described by their associated electric and magnetic fields, the properties of the materials involved, the physical dimensions, and the velocity of light. When the dimensions are small compared with the wavelength, however, it is more convenient to treat them as circuits composed of lumped parameters.

At low frequencies lumped inductance and capacitance can be used, although they are rigorously derived only for nonvarying currents and voltages, respectively. At high frequencies the finite propagation velocity of electromagnetic waves cannot be neglected, and the derivations break down. If only one system dimension is comparable with the wavelength of the electrical disturbance, restricted conditions permit a rigorous definition of distributed inductance and capacitance. These distributed parameters combine with the resistance of a pair of conductors and the conductance between them to define the behavior of a transmission line for plane-wave propagation and to relate the voltage between conductors at any point on the line to the voltage at any other point. *See* TRANSMISSION LINES.

The concept of distributed parameters is also useful at low frequencies when it must be recognized that a circuit component, nominally representable by a single parameter, is actually modified by the presence of residual parameters. Thus a coil has not only inductance, but capacitance and resistance as well. This capacitance is definable by low-frequency analysis, since the dimensions are small, but it cannot be localized and represented as a unique lumped parameter because the winding is not an equipotential surface.

For example, a coil mounted over a ground plane has one terminal grounded. The voltage between winding and ground increases from zero at the grounded terminal to maximum at the other. Capacitance near the grounded terminal is therefore less effective than capacitance near the other. The resultant effective terminal capacitance is, in consequence, only one-third of the total capacitance for uniformly distributed capacitance. For other conditions of grounding, the ratio of effective capacitance to total distributed capacitance will again be different.

Values of distributed parameters are inferred from the behavior of the system that they define. For transmission lines, measurements may involve observing the voltage distribution along the line under different terminal conditions. For circuit elements, impedance may be measured at different frequencies or under different conditions of adjustment of some known lumped parameter.   [DONALD B. SINCLAIR]

Bibliography: W. D. Cooper, *Electronic Instrumentation and Measurement Techniques*, 2d ed., 1978; R. F. Field and D. B. Sinclair, A method for determining the residual inductance and resistance of a variable air condenser at radio frequencies, *Proc. IRE*, 24(2), 1936; F. K. Harris, *Electrical Measurements*, 1975; P. Kantrowitz et al., *Electronic Measurements*, 1979; T. Laverghetta, *Microwave Measurements and Techniques*, 1976.

Electrical noise

Interfering and unwanted currents or voltages in an electrical device or system. Electrical noise, usually simply called noise, has an important effect on any electrical system which is used to gather, transmit, process, or present information. In such systems as telephone, radio, television, radar, radionavigation, telemetering, electronic control, or electronic computing, the desired signals carrying intelligence may be masked or distorted by noise.

Noise may originate either externally to the device in which it appears, as atmospheric static, or internally, as thermal noise in a resistor. It may result either from natural phenomena, as do both types of noise just mentioned, or from interference from man-made devices, such as nearby electric motors or generators.

Man-made interference can usually be nearly eliminated by good engineering design and proper location of equipment. Noise due to natural phenomena often cannot be reduced below certain fixed levels, and good engineering design can ensure only that the equipment will function as effectively as possible in the face of this irreducible noise. For example, a radio receiver cannot operate on received signals which are very weak (compared to some level determined by the thermal noise in the receiver), no matter how much amplification is used in the receiver, because the noise is amplified along with the signal.

Sources. Noise may conveniently be classified as either random noise or nonrandom noise. Random noise is defined as noise which is not predictable, although it may exhibit statistical regularities.

Random noise in electron-tube circuits. Thermal noise is the random voltage which appears at the terminals of a resistor or any component with internal resistance because of the random motion of thermally excited electrons in the resistor. Shot noise is caused by fluctuations of current in a thermionic vacuum tube caused by random emission of electrons from its hot cathode. Partition noise is the result of fluctuations in current to one electrode in a multielement vacuum tube caused by random division of the electron stream between two or more collecting electrodes (for example, the screen-grid and anode of a tetrode). Flicker noise is the result of low-frequency fluctuations in current in a vacuum tube apparently caused by relatively slowly changing emission conditions at various points on the cathode. Contact noise, the noise in carbon resistors and carbon microphones, for example, is caused by randomly varying fluctuations in resistance. *See* VACUUM TUBE.

Random noise in semiconductors. Random noise also originates in transistors and other semiconductor devices. There are various mechanisms at work, and the terminology is not completely standard. Thermal noise, as above, is the noise caused by the random motion of thermally excited electrons. Shot noise, or generation-recombination noise, is produced by fluctuations in free carrier densities when an electric field is applied. Excess noise or modulation noise is caused by slow fluctuations in conductivity. Noise also originates in photoconductors. Most mechanisms of random noise generation share the principle that observable gross currents and voltages are the result of many random actions at microscopic level. *See* SEMICONDUCTOR.

Radiated random noise. Any electric device which must receive electromagnetic radiation will pick up radiated random noise as well as signals. Electrical disturbances in the atmosphere cause noise which is very irregular in character, often appearing in sharp bursts. Aside from atmospheric disturbances, an antenna receives a steady background of noise which is of thermal origin, thermal radiation from the gases of the atmosphere, and thermal radiation from heavenly bodies and systems. This last is sometimes called interstellar noise. The Sun radiates noise at all times, but during sunspot activity the intensity of its noise radiation is greatly increased.

Nonrandom noise. This type of noise is usually the result of radiation from other electric equipment, unwanted coupling with other systems, or spurious oscillations within an electrical circuit. *See* CROSSTALK; ELECTRICAL INTERFERENCE.

Noise measurement. The term noise measurement is applied to a wide range of measurements of random and nonrandom noise. It usually refers to the measurement of noise power averaged over some brief interval of time (called quadratic content). This kind of measurement may be made to check the noise level against which a system must operate, or it may be made to yield information about the physical world. In radio astronomy, for example, very delicate measurements are made of the noise radiation from particular plantets, stars, or galaxies to gain information about their constitution and their relative velocity. In this application the noise-measuring device is called a radiometer.

Noise-power measurements are most conveniently made by amplifying the noise from the source in a linear amplifier and then using a quadratic detector followed by a low-pass filter and an indicating device to determine the average noise power.

indicator

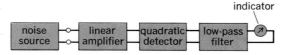

Diagram of setup for noise-power measurement.

The illustrated circuit measures noise power in the band of frequencies passed by the linear amplifier. The low-pass filter must have a narrower bandwidth than that of the linear amplifier to give an averaging of the fluctuating voltage from the detector (the low-pass filtering may occur naturally in the indicating device). In such a circuit, if the indicator responds linearly to its input voltage, its reading is proportional to the noise power. Thermocouples or thermistors are good quadratic detectors. Electron-tube and crystal-diode detectors can be made to have nearly a quadratic response over a certain range. If the amplifier is nonlinear or the detector does not follow a quadratic law, the device will still indicate noise power, but its entire response curve must be calibrated.

Noise-power measuring devices may be either direct-reading or they may make comparisons between the noise from the source being measured and that from a calibrated source. *See* ELECTRICAL NOISE GENERATOR.

Mathematical analysis of noise. The effects of nonrandom noise in an electric circuit or system can be determined mathematically in the same way that responses to signals in the circuit are determined. Random noise, being unpredictable and having only certain statistical properties, must be treated differently. Usually probabilistic methods are used.

The application of probability theory to the analysis of random noise rests on the fact that for most mechanisms of noise generation (in particular, all those listed above except perhaps atmospherics) the resulting noise waveforms have statistical properties which remain invariant with time. These statistical properties may be deduced from an analysis of the generating mechanism or determined empirically by experiment.

One characterization of a random noise waveform $v(t)$ is in terms of the probabilities that the amplitude of the noise waveform will fall between any pair of specified values at specified instants of time. For example, the probabilities at a single instant t can be expressed in terms of a probability density function $f_t(x)$ defined by notation (1). Then

$$f_t(x)\,dx = \text{probability} \atop \text{that } [x < v(t) \leqq x + dx] \tag{1}$$

the probability that $v(t)$ lies between two values a and b is given by notation (2). Most kinds of ran-

$$\int_a^b f_t(x)\,dx \tag{2}$$

dom noise, including particularly those which result from the summation of a great many microscopic effects such as thermal noise and shot noise, have what are called gaussian or normal statistics. The term gaussian noise is used for such noise. The probability density function for gaussian noise, $f_t(x)$, is given by Eq. (3), where m_t and

$$f_t(x) = \frac{1}{\sigma_t \sqrt{2\pi}} \epsilon^{-(x - m_t)^2/2\sigma_t^2} \tag{3}$$

σ_t^2 are the parameters which determine $f_t(x)$ and are called the mean value (at time t) and the variance (at time t), respectively.

A random noise is said to be stationary if all its probability relations are unchanged by a translation in time. Thus, in particular, for a stationary noise $f_{t1}(x) = f_{t2}(x)$ for any times t_1 and t_2. Hence $f_t(x)$, m_t, and σ_t^2 do not depend on time and the subscript t may be dropped. Thermal noise from a resistor at constant temperature or shot noise from a diode under fixed operating conditions are examples of stationary noise. Usually a noise voltage $v(t)$ is stationary, and in addition has the property that the probability that $v(t)$ lies between any two values a and b at any time t is nearly equal to the fraction of time the noise voltage lies between these two values if a sufficiently long observation interval is recorded. This property, more precisely defined, is called ergodicity. For a stationary ergodic random noise the mean value m used above is just the average value of the noise over a long time interval, and the variance σ^2 is the average of the square of the fluctuations about the mean.

Other quantities that are useful in characterizing random noise are the covariance function $R(\tau)$ and (for a stationary noise) the power spectral density $N(f)$. The covariance function $R(\tau)$ of a noise waveform $v(t)$ gives a measure of how closely related on the average are the present and future values of $v(t)$. For a stationary ergodic noise $R(\tau)$ is defined as the time average of a product of amplitudes, Eq. (4). The power spectral density $N(f)$

$$R(\tau) = \lim_{T \to \infty} \frac{1}{2T} \int_{-T}^{T} v(t)v(t+\tau)\, dt \qquad (4)$$

gives the distribution of average power in the noise waveform $v(t)$ as a function of frequency f; that is, $N(f)\, df$ is the average power a voltage $v(t)$ applied across a 1-ohm resistance would dissipate in the incremental frequency range f to $f + df$. $N(f)$ can be calculated as the Fourier transform of $R(\tau)$, as in Eq. (5), where $j = \sqrt{-1}$.

$$N(f) = 2 \int_{-\infty}^{\infty} R(\tau)\epsilon^{-j2\pi f\tau}\, d\tau \qquad (5)$$

If a stationary noise waveform $v(t)$ with power spectral density $N(f)$ is passed through a linear time invariant filter with a transfer function $H(f)$, where $H(f)$ is the complex gain of the filter at the frequency f, the power spectral density of the output noise is given by $N(f)|H(f)|^2$. The output noise is also stationary, and if the input noise is gaussian, so also is the output. However if gaussian noise is passed through a nonlinear device, such as a square-law detector or a limiter, its statistics do not in general remain gaussian.

In analyzing the response of an electrical system to signals-plus-noise a quantity of interest is the ratio of signal power to noise power, or briefly signal-to-noise ratio, at various points of the system. If $S(f)\, df$ is the signal power in the incremental frequency band df at f, the incremental signal-power-to-noise ratio is $S(f)/N(f)$. This ratio is unchanged by passing the signal-plus-noise through a linear filter. If $S(f)$ and $N(f)$ are each integrated over the total band of frequencies being grated over the total band of frequencies being passed, the ratio of the integrals is the total signal-power/noise-power ratio.

Thermal noise. In the narrow sense thermal noise refers to the random voltage at the terminals of a resistor caused by thermal excitation of electrons in the resistor. It can be shown from statistical mechanics that the spectral density $N(f)$ of such a noise voltage is given by Eq. (6), where k is Boltzmann's constant (1.38×10^{-23} joule/°C), R is the resistance in ohms of the resistor, T is its temperature in degrees Kelvin, and $\gamma(f)$ is given by Eq. (7), where h is Planck's constant (6.62×10^{-34} joule·sec).

$$N(f) = 4kTR\gamma(f) \qquad (6)$$

$$\gamma(f) = \frac{hf}{kT}\epsilon^{-(hf/kT)+1} \qquad (7)$$

At room temperature and even at microwave frequencies, $\gamma(f)$ is very close to unity, so the noise voltage spectral density across a resistor is usually approximated by Nyquist's formula, Eq. (8).

$$N(f) = 4kTR \qquad (8)$$

Nyquist's formula shows that the thermal noise

voltage spectral density is constant (to very high frequencies). Such noise with a uniform distribution of power with frequency is called white noise.

Nyquist's formula remains valid if the simple resistor is replaced by a two-terminal electrical network with all resistance elements at the same temperature and with R standing for the input resistance at the terminals. Generalized forms are valid when the temperature is not constant throughout the network.

In a broader sense thermal noise refers to other kinds of noise of thermal origin. Important among these is antenna noise. Nyquist's formula can be used for a two-terminal system in which the terminals are connected to an antenna in a perfectly black chamber, the whole system in thermal equilibrium. The R of Nyquist's formula is then the radiation resistance of the antenna. In practice, Nyquist's formula is often used to define the equivalent radiation temperature of the portion of sky at which the antenna is looking, in terms of the radiation resistance and the measured noise.

Shot noise. In a vacuum tube in which electrons are emitted from a heated cathode, the electron stream from cathode to anode does not have uniform density since the individual electrons are emitted randomly and with random initial velocities. Although the fluctuations in the electron stream are essentially thermal in origin, the current fluctuations, called the shot effect, are usually not classified as thermal noise. One important difference between the shot effect in tubes and thermal noise in resistors is that the former is not present until a voltage difference is applied between cathode and anode, whereas the latter is present in a quiescent circuit with no externally applied voltages. Similarly in transistors the fluctuations in the rate of flow of carrier electrons or holes is a type of noise not present in the quiescent device, and this is also often called shot effect. *See* SCHOTTKY EFFECT.

A satisfactory mathematical model for the emission of electrons from a hot cathode yields the result that the number of electrons emitted in any interval of time follows the Poisson law, Eq. (9),

$$\text{Probability } \binom{K \text{ electrons}}{\text{are emitted in } \tau \text{ sec}} = \frac{(\bar{n}\tau)^K \epsilon^{-\bar{n}\tau}}{K!} \qquad (9)$$

where $\bar{n}$ is the average number of electrons emitted per second. If a sufficiently great positive electric field is established in the cathode-anode space, all the electrons emitted at the cathode travel to the anode and the fluctuations in the current induced in the anode circuit are proportional to the fluctuations in electron emission. This happens, for example, in a diode with low cathode emission and high anode potential. In this case the power spectral density of the shot noise is given approximately at low frequencies by the Schottky formula, Eq. (10), where e is the charge on an electron and $\bar{I}$

$$S(f) = 2e\bar{I} \qquad (10)$$

is the average anode current. This formula is valid if f is small compared to the reciprocal of the transit time of an electron from cathode to anode.

In most electron tubes, for example, receiving tubes operating with negative grid bias, a negative space charge develops near the cathode. This space charge has a damping effect on the shot effect fluctuations. The low-frequency spectral density is then approximately that shown by Eq. (11), where Γ^2 depends on tube geometry, poten-

$$S(f) = 2e\bar{I}\Gamma^2 \tag{11}$$

tials, and cathode temperatures, and has a value between 0 and 1.

Noise figure. The noise figure of an amplifier is a figure of merit which measures the noisiness of the amplifier relative to the noisiness of the source driving the amplifier. It is important in radio and radar receivers, for example, that the noise figure be low if weak signals in noise are to be received. The operating noise figure F_o, which is a function of frequency, may be defined to be the ratio of the available incremental signal-to-noise power ratio of the source (over a frequency band df) to the available incremental signal-to-noise power ratio of the amplifier output (over the same frequency band df) when the amplifier is driven by the source. If the source noise is ascribed to thermal origins, F_o depends on the effective noise temperature of the source. A standard noise figure F is sometimes defined to be the operating noise figure with the source at 290 K. See ELECTRONIC-EQUIPMENT GROUNDING.   [WILLIAM L. ROOT]

Bibliography: W. B. Davenport, Jr., and W. L. Root, *Introduction to Random Signals and Noise*, 1958; M. S. Gupta (ed.), *Electrical Noise: Fundamentals and Sources*, 1977; F. N. Robinson, *Noises and Fluctuation in Electronic Devices and Circuits*, 1975; M. Schwartz, *Information, Transmission, Modulation and Noise*, 3d ed., 1980; C. A. Vergers, *Handbook of Electrical Noise: Measurement and Technology*, 1979; A. van der Ziel, *Noise in Measurement*, 1976; A. van der Ziel, *Noise: Sources, Characterization and Measurement*, 1970.

Electrical noise generator

A device which produces (usually random) electrical noise for use in electrical measurements. Electrical noise generators are commonly used in measuring the noise figure of a radio receiver or other amplifier. They are also used in other tests of the response of an electrical system to random noise, and in measurements of noise intensity. *See* ELECTRICAL NOISE.

Some standard types of noise generator are: hot-wire, diode, gas-discharge tube, and klystron. A hot-wire noise generator is commonly the filament of a lamp heated by a direct current. The filament is connected across the terminals where the noise is to be introduced, for example, the antenna terminals of a radio receiver. The noise generated by the filament is thermal, so its intensity N can be calculated from the Nyquist formula $N = 4kTR$, where T is temperature (°K), R is resistance of the filament, and k is Boltzmann's constant. A diode noise generator relies on the shot effect to produce noise. At frequencies appreciably less than the reciprocal of the transit time, the noise intensity N can be calculated from the Schottky formula $N = 2e\bar{I}$ if, as is customary in this application, the anode current is emission-limited. In this formula e is the charge of an electron and $\bar{I}$ is the average anode current. A gas-discharge noise generator is usually a fluorescent light tube enclosed in a

waveguide. The mechanism of noise production is essentially thermal; the electrons in the gas discharge acquire high random velocities, corresponding to a high equivalent noise temperature. This equivalent noise temperature varies with the gas in the tube, but does not depend very much on tube dimensions or on the discharge current. A reflex klystron, with reflector grid connected to the cavity to prevent oscillation, generates noise because of shot effect in the cathode current. *See* SCHOTTKY EFFECT.

It is convenient if the noise generated by a noise generator has a nearly constant intensity; certainly it must not fall off very much over the entire frequency range of operation. Thermal noise sources do potentially provide such a flat spectrum of noise over all radio frequencies, being frequency-limited only by capacitances and inductances inherent in the source device and its circuitry. Diodes, however, are ultimately limited to frequencies of the order of the reciprocal of the transit time. In practice, special diodes can be used up to a few hundred megahertz. For the measurement of noise figure at microwave frequencies, where most amplifiers (triodes, traveling-wave tubes, and klystrons) have a relatively high noise figure (as high as 20 dB above basic thermal noise), gas-discharge tubes are preferable to hot-wire noise sources because they produce more available noise power. Klystrons are also suitable noise generators at microwave frequencies, but they are not absolute standards and require calibration. *See* KLYSTRON; SEMICONDUCTOR.

[WILLIAM L. ROOT]

Bibliography: M. S. Gupta (ed.), *Electrical Noise: Fundamentals and Sources*, 1977; J. Markus, *Sourcebook of Electronic Circuits*, 1968; F. N. Robinson, *Noises and Fluctuations in Electronic Devices and Circuits*, 1975; C. A. Vergers, *Handbook of Electrical Noise: Measurement and Technology*, 1979; A. van der Ziel, *Noise: Sources, Characterization and Measurement*, 1970.

Electrical resistance

That property of an electrically conductive material that causes a portion of the energy of an electric current flowing in a circuit to be converted into heat. In 1774 A. Henley showed that current flowing in a wire produced heat, but it was not until 1840 that J. P. Joule determined that the rate of conversion of electrical energy into heat in a conductor, that is, power dissipation, could be expressed by the relation given in notation (1).

$$H/t \propto I^2R \tag{1}$$

The day-to-day determination of resistance by measuring the rate of heat dissipation is not practical. However, this rate of energy conversion is also VI, where V is the voltage drop across the element in question and I the current through the element, as in Eq. (2), from which the more conventional

$$H/t \propto I^2R = VI \tag{2}$$

relationship implied by Ohm's law, Eq. (3), is ap-

$$R = V/I \tag{3}$$

parent. *See* OHM'S LAW; RESISTANCE MEASUREMENT.   [CHARLES E. APPLEGATE]

Electrical shielding

A means of avoiding pickup of undesired signals or noise, suppressing radiation of undesired signals, and confining wanted signals to desired paths or regions. These shielding objectives cannot be realized without some degree of modification of the electric and magnetic fields involved, although the effects of these modifications may not seriously interfere with wanted objectives.

A change in an electric field gives rise to magnetic effects; similarly, a change in a magnetic field gives rise to electric effects. As a result, electromagnetic shielding is more commonly encountered than either electrostatic shielding or magnetostatic shielding.

A shield is a sheet, tube, screen, grid, or other object, usually of conducting material. Sometimes a shield is magnetic, or both magnetic and conducting, or even laminated. Fields may be largely confined to, or suppressed from, a specified region by shields. The ratio of the unwanted signal in a communication circuit when the source of shielding is present to the same signal when the shielding is absent is called the shield factor.

Electrostatic shielding. Virtually complete shielding from external electrostatic fields is obtained by totally enclosing the space to be shielded with a highly conducting surface, usually grounded. Conversely, no change in electrostatic conditions within the shield can appreciably influence the state of conductors, circuits, or fields external to the shield. In effect, a highly conducting shield substantially terminates electrostatic fields from either within or without.

Magnetostatic shielding. Substantially complete shielding from an external and virtually constant magnetic field is obtained by shunting the flux around the space to be shielded through a low-reluctance path. Conversely, a constant magnetic field enclosed by the shield is largely constrained from reaching the region outside the shield. Ideally, the shield should present a complete magnetic path to the flux; that is, the space to be shielded should be surrounded by a ferromagnetic material, preferably of high permeability. The shield factor may be reduced by increasing the thickness of the shield, by increasing its permeability, or by using several nested shields. Relatively thick, single-layer shields are not economical.

Electromagnetic shielding. The magnetostatic shielding such as that described above is likewise effective when the magnetic field is changing with time. A coil, for example, might be surrounded by a shield of this type. Such a shield would increase the inductance of the coil, illustrating that effective shielding necessarily modifies the fields. This kind of magnetic shielding is necessary for shielding constant or slowly varying fields.

A different yet highly effective method of realizing electromagnetic shielding is by means of an electrostatic shield. For frequencies above the upper audio range, shield factors of the order of 10^{-4} are readily obtainable, for example, between electric wave filters mounted in close proximity. Here the shield might be a closed box made of copper, about 1/16 in. thick. The main precautions are good soldered joints and absence of small cracks. A magnetic field impinging upon the copper sets up eddy currents, which oppose the penetration of the flux. The better the conductivity of the eddy-current paths, the more perfect the shielding action. In contrast to a magnetostatic shield, this type of shield reduces the inductance of a shielded coil.

In situations where space occupied by the shield is at a high premium, a laminated shield consisting of layers alternating between low-reluctance and high-conductance materials can be used.

Substantial electrostatic shielding may be obtained without appreciable magnetic shielding by making the electrostatic shield nonmagnetic and shaped so that eddy-current paths are interrupted.

Shielded wires and cables. The cable sheath, which encloses many communication lines within a telephone cable, constitutes one of the most striking examples of electromagnetic shielding. The sheath is grounded at one or more points along its length. Even with high-resistance grounds, the cable conductors are well shielded from external fields. At higher carrier frequencies, the shielding is so effective that significant conductor noise does not appreciably exceed resistance noise.

At moderately high frequencies, a coaxial line is an excellent example of a shielded wire, but at sufficiently low frequencies, the shielding effect is negligible. For this and other reasons, shielded twisted wires are preferred for transmitting television signals at video frequencies. At low frequencies, where a nonmagnetic shield is ineffective, transpositions act to attenuate noise, interference, and unwanted radiation. *See* ELECTRICAL NOISE; ELECTRONIC-EQUIPMENT GROUNDING.

[HAROLD S. BLACK]

Electroacoustics

That part of electronics concerned with the conversion of acoustical energy into electrical energy or vice versa. The means for conversion from acoustical to electrical systems and vice versa are termed electroacoustic transducers or electroacoustic systems. Electroacoustic systems are used in the telephone, phonograph, radio, sound motion picture and magnetic tape reproducers, and sound-reinforcing systems for converting acoustical waves at the input into corresponding electrical waves, and for converting electrical waves at the output into corresponding sound waves.

[HARRY F. OLSON]

Electrochromic displays

Devices that employ a reversible electrochemical reaction to cause a change in color of segments patterned to form alphanumeric characters.

Electrochromic displays (ECDs) are passive devices that only modulate ambient light, in contrast to a light-emitting diode (LED). Hence, they operate at low voltages, and have a low enough energy requirement that a watch-size display can be operated for about 1 year from a small commercial battery. In the off state, the segments are typically colorless; in the on state, they are brightly colored, for example, blue or purple. *See* LIGHT-EMITTING DIODE.

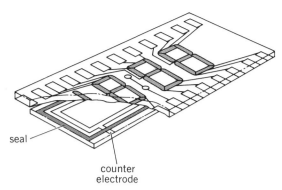

seal

counter
electrode

Structure of an electrochromic display.

The structure of an electrochromic display package is shown in the illustration. It consists of a top glass piece with the transparent electrodes, typically vacuum-deposited indium and tin oxides, on the inside surface. The lead from each segment continues to the edge of the glass. An insulating layer of silicon dioxide (SiO_2) is deposited onto the leads to prevent their coloration. A counter electrode is deposited on the bottom piece. The two pieces are held together with a solder-glass, or epoxy, seal. To operate the device, a dc potential of 1–1.5 V is applied between the segments on the top plate and the counter electrode. Coloration and bleaching times are in the 200–500-millisecond range. Electrochromic displays can be categorized by the type of reaction that occurs when the potential is applied. Viologen electrochromic displays operate in an electroplating mode; with tungsten trioxide (WO_3) displays, the electrode material itself undergoes a color change. Electrochromic displays would be most efficiently utilized in applications requiring the display of large-format and slowly changing information, such as in clocks or message boards.

Viologen displays. The electrolyte is an aqueous solution of a dipositively charged organic salt, for example, a dialkyl derivative of 4.4′-bipyridinium bromide. When the appropriate negative potential is applied to the transparent electrode, the colorless cation in the electrolyte undergoes a one-electron reduction process to produce a purple radical cation. This species is insoluble in the aqueous electrolyte, hence it remains deposited on the electrode where it is formed. Since this precipitate is stable, the system possesses an open-circuit memory. To erase the segments, an appropriate positive potential is applied to the transparent electrodes. This oxidizes the purple film back to the soluble, colorless, doubly positive cation.

WO_3 displays. In these devices, a thin, colorless film of WO_3 is vacuum-deposited onto the transparent, patterned electrodes. Electrolytes are typically solutions of acids or lithium salts, such as H_2SO_4/glycerin or $LiClO_4$/propylene carbonate. When a negative potential is applied to this electrode, the reaction below occurs. Here a cation

$$WO_3 + xM^+ + xe^- \longrightarrow M_xWO_3 \qquad M^+ = H^+, Li^+$$

from the solution and an electron from the electrode are simultaneously introduced into the WO_3 lattice. This changes the transparent WO_3 into a blue M_xWO_3, generally called a tungsten bronze. Since both states of this oxide are stable, the system has an open-circuit memory. To erase a segment, a positive potential is applied to the top electrode. This reverses the above reaction and renders the film colorless.

There are several other versions of these two concepts under development. However, fundamental to all these devices is the requirement that some chemical reaction take place reversibly, for millions of cycles. Due to various technical problems, for example, stability of the electrodes, side reactions, and the recrystallization of deposited films, the reliability necessary for commercial production of electrochromic displays has not yet been achieved. *See* ELECTRONIC DISPLAY.

[GABRIEL G. BARNA]

Bibliography: A. R. Kmetz and F. K. von Willisen (eds.), *Nonemissive Electrooptic Displays*, 1976.

Electroluminescence

A luminescence resulting from the application of an electric field to a material, usually solid. Several different types of electroluminescence can be distinguished.

Destriau effect. In 1936, G. Destriau in France observed that when suitably prepared zinc sulfide phosphor powders, activated with small additions of copper, are suspended in an insulator and an intense alternating electric field is applied with capacitorlike electrodes, light emission results. The effect is the basis for modern electroluminescent lamps. The insulator may be either liquid (oil) or solid (glass or plastic). Thin films of tin oxide or metal are used as transparent electrodes.

Electroluminescent phosphors (or electroluminors) differ from phosphors that are excitable by cathode rays or ultraviolet radiation in that small regions of excess Cu_2S cause the electrical properties to vary with location in the phosphor. This characteristic results in local intensification of the electric field, causing emission of light from a few small spots in each powder particle. A similar type of emission may be obtained from thin films of ZnS prepared by vaporization in a vacuum; such films also emit if a direct voltage is applied.

Electroluminescence makes possible very thin large-area light sources. The active layer in a lamp for operation at 120 V may be about 1 mil (25 μm) thick, whereas evaporated films 1 μm thick may be operated at 20 V. The light output L increases rapidly with increasing voltage V according to the equation below, where A and B are constants.

$$L = A \exp(-B/V^{1/2})$$

Also the output increases slightly less than linearly as the frequency of the alternating voltage is increased until a saturation output is reached at frequencies of the order of 100,000 Hz. Maximum efficiency is not obtained for the same operating conditions as maximum brightness, and the efficiency is normally low compared to that of other light sources; for example, electroluminescence, 3–5 lumens/watt; incandescent lamp, 16 lm/W; fluorescent lamp, 75 lm/W. Some phosphors exhibit a change in emission color as the operating frequency is varied. In addition to their use in lamps, electroluminescent materials may be employed in display devices and, when used in conjunction with photoconductors, in light amplifiers and in logic circuits for computers, which are, however, slow compared to semiconductor logic circuits. *See* LIGHT AMPLIFIER; LIGHT PANEL.

Injection electroluminescence. Injection electroluminescence results when a semiconductor *pn* junction or a point contact is biased in the forward direction. This type of emission was first observed from SiC in England by H. J. Round in 1907. It is the result of radiative recombination of injected minority carriers with majority carriers across the energy band gap of the material. Such emission has been observed in a large number of semiconductors, including Si, Ge, diamond, CdS, ZnS, ZnSe, ZnTe, ZnO, and many of the so-called III-V compounds, such as GaP, GaAs, InP, InAs, InSb, BP, BN and AlN. The wavelength of the emission corresponds to an energy equal, at most, to the forbidden band gap of the material, and hence in most of these materials the wavelength is in the infrared region of the spectrum. In some cases the efficiency is high, approaching one emitted photon for each carrier passing through the junction. In suitable structures the excitation intensity can become so high that stimulated rather than spontaneous emission predominates, and laser action results, with spectral narrowing and coherent emission. Direct electrical excitation is more convenient than optical excitation, but the beam divergence of injection lasers is normally much greater than that of gas lasers or the optically pumped solid type. *See* JUNCTION DIODE; LASER; SEMICONDUCTOR; SEMICONDUCTOR DIODE.

Other effects. If a *pn* junction is biased in the reverse direction, so as to produce high internal electric fields, other types of emission can occur, but with very low efficiency. The presence of very energetic ("hot") carriers can result in emission at energies greater than the band gap of the material (avalanche emission). The emission in this case may be correlated with small active regions called microplasmas. Light emission may also occur when electrodes of certain metals, such as Al or Ta, are immersed in suitable electrolytes and current is passed between them. In many cases this "galvanoluminescence" is electroluminescence generated in a thin oxide layer formed on the electrode by electrolytic action. In addition to electroluminescence proper, other interesting effects (usually termed electrophotoluminescence) occur when electric fields are applied to a phosphor which is concurrently, or has been previously, excited by other means. These effects include a decrease or increase in steady-state photoluminescence brightness when the field is applied, or a burst of afterglow emission if the field is applied after the primary photoexcitation is removed.

[CLIFFORD C. KLICK; JAMES H. SCHULMAN]

Bibliography: P. Goldberg (ed.), *Luminescence of Inorganic Solids*, 1966; H. K. Henisch, *Electroluminescence*, 1962; H. F. Ivey, *Electroluminescence and Related Effects*, 1963; H. F. Ivey, Electroluminescence and semiconductor lasers, *IEEE J. Quantum Electron.*, 2:713, 1966; S. Larach (ed.), *Photoelectronic Materials and Devices*, 1965; J. I. Pankove (ed.), *Electroluminescence* (1977); R. K. Willardson and A. C. Beer (eds.), *Semiconductors and Semimetals*, vol. 2, 1966.

Electromagnetic compatibility

The capability of electronic equipments or systems to be operated in the intended electromagnetic environment at design levels of efficiency. The electromagnetic environment existing at the input or antenna terminal of a receiver is a mixture of signals from many sources. The mixture consists of both a desired signal and many undesired signals to which the receiver may be sensitive. These undesired or interference signals may come from equipment intended to radiate for the purpose of communications or from sources not intended to radiate electromagnetic energy for the purpose of transferring information. In addition to mutual interference between systems, the possibility of extraneous energy generated within a system interfering with its own operation also exists. The subject of electromagnetic compatibility deals with reducing both the source of interference and its effect on the reception of signals.

Since the early days of wireless communication at the turn of the century, the growth in use of the spectrum has been phenomenal. In those early days spurious emissions, spurious response, and other interference were manifested as an annoyance in one form or another. During that period spectrum usage was low and corrective action was obtained by moving to higher frequencies. As new parts of the spectrum were explored, more and more uses became apparent. The usable radio spectrum has been expanded many times since the turn of the century.

Interference potential. In addition to increased frequency usage, a second situation involving increased power for transmitting equipments and improved sensitivity for receivers with the objective of improving the range of communications has caused an increase of interference potential in the environment. A transmitter not only produces energy at the frequency at which it is intended to operate, but also at other frequencies such as multiples of the desired frequency (harmonics) and other spurious emissions not necessarily related to the desired frequency. Installation of a new transmitter is then equivalent to installing many transmitters at that location. Although the power radiated at these spurious frequencies is generally well below that at the desired frequency, effects of this power can be felt at a significant range.

A similar situation exists for receivers in that they are sensitive to spurious response frequencies, as well as to the desired frequencies. Again, installing a receiver is the equivalent of installing many receivers which, although operating at reduced sensitivities, can receive signals from significant distances.

A new system in the environment then adds to spectrum congestion at not only the design frequency but at a number of other frequencies. This compounds the problem immensely.

Design consideration. The design of electronic equipment requires serious consideration of the components, circuits, equipment case (from a shielding standpoint), the system of which it is a part, and other systems which are to operate in the environment. A thorough knowledge of technical characteristics of the components and equipment, as well as how the equipment is to be used, is required to achieve a good design from the electromagnetic compatibility point of view.

Although most signals detected by a receiver would come through the antenna of the equipment, some come through the cases or through the power lines. A similar situation exists in leakage from a transmitter through cases and extraneous energy coupled to power lines. Efforts are concentrated on improving both the shielding effectiveness of cases and the filtering of power line circuits.

The design of systems components to reduce interference is also important for electromagnetic compatibility. Reduction of side-lobe levels of antennas, incorporation of linear mixing circuits to reduce intermodulation, and incorporation of transmission line filters to reduce spurious and harmonic effects are some examples of actions taken in component design to reduce interference.

In addition to the intended signals in the environment, many equipments used by industry and the public generate electromagnetic energy. This unintended radiation of man-made noise is caused by such devices as electric motors, ignition systems of internal combustion engines, diathermy machines, plastic sewing machines, welders, electric shavers, fluorescent lights, and power lines. In general, man-made noise has been increasing in urban areas because of the great density of the aforementioned devices. Reduction of this man-made noise requires improved design of these devices.

Analysis and prediction. The expanded use of the radio spectrum has placed great emphasis on analysis and prediction of potential interference situations. Because of the numerous equipments in the environment and massive number of calculations required, the use of the computer has become imperative. In general, the mathematical model accounts for the characteristics of the transmitter, its associated antenna, the propagation path, the receiving antenna, the receiver circuits, and the detection circuits. *See* INFORMATION THEORY; MODULATION.

Calculations are made considering each receiver in the environment paired with every transmitter. The many combinations require use of a simplified model for the initial computation. The purpose of this model is to sort out those cases in which interference cannot possibly occur, leaving a much smaller number of potential interference cases for more detailed computations. Such environmental computations are becoming more and more necessary and, as spectrum congestion becomes severe, increased use of these techniques will be required.   [STANLEY I. COHN]

Bibliography: Institute of Electrical and Electronic Engineers, *IEEE International Symposium on Electromagnetic Compatibility, Atlanta, GA, June 20–22, 1978*, Cat. no. 78-CH-1304-5-EMC, 1978; B. J. Keiser, *Principles of Electromagnetic Compatibility*, 1979; B. Priestly, Electromagnetic compatibility from the design engineer's point of view, *Radio Electron. Eng.*, 49(2):108–110, February 1979.

Electromagnetic wave transmission

The transmission of electrical energy by wires, the broadcasting of radio signals, and the phenomenon of visible light are all examples of the propagation of electromagnetic energy. Electromagnetic energy travels in the form of a wave. Its speed of travel is approximately 3×10^8 m/sec (186,000 mi/sec) in a vacuum and is somewhat slower than this in liquid and solid insulators. An electromagnetic wave does not penetrate far into an electrical conductor, and a wave that is incident on the surface of a good conductor is largely reflected.

Electromagnetic waves originate from accelerated electric charges. For example, a radio wave originates from the oscillatory acceleration of electrons in the transmitting antenna. The light that is produced within a laser originates when electrons fall from a higher energy level to a lower one. *See* LASER.

The waves emitted from a source are oscillatory in character and are described in terms of their frequency of oscillation. Local telephone lines (not using carrier systems) carry electromagnetic waves with frequencies of about 200–4000 Hz. Medium-wave radio uses frequencies of the order of 10^6 Hz, radar uses frequencies of the order of 10^{10} Hz, and a ruby laser emits light with a frequency of 4.32×10^{14} Hz. The method of generating an electromagnetic wave depends on the frequency used, as do the techniques of transmitting the energy to another location and of utilizing it when it has been received. *See* RADAR.

The communication of information to a distant point is generally accomplished through the use of electromagnetic energy as a carrier. A familiar example is the telephone, in which sound waves in the range of frequencies from a few hundred to a few thousand hertz are converted into corresponding electromagnetic waves, which are then guided to their destination by a pair of wires. Another familiar example is radio, in which the signals are caused to modify an identifiable characteristic, such as the amplitude or frequency, of an electromagnetic carrier wave. The electromagnetic wave, thus modified, or modulated, is radiated from an antenna and can be received over a considerable region. *See* MODULATION.

Features of electromagnetic waves. Figure 1 illustrates schematically some of the essential features of an electromagnetic wave. Shown in Fig. 1 are vectors that represent the electric field intensity E and the magnetic field intensity H at various points along a straight line taken in the direction of propagation of the wave. The electric field is in a vertical plane and the wave is said to be vertically polarized. The magnitude of the field, at a given instant, varies as a sinusoidal function of distance along the direction of propagation. The magnetic field intensity H lies in a plane normal to that of E and, at each point, is proportional in magnitude to E, as shown in Eq. (1), where H is the magnetic

$$\frac{E}{H} = \sqrt{\frac{\mu}{\epsilon}} \qquad (1)$$

field intensity in amp/m, E is the electric field in-

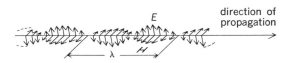

Fig. 1.   Representation of an electromagnetic wave at a particular instant of time.

tensity in volts/m, ϵ is the permittivity, or absolute dielectric constant, of the medium, and μ is the absolute permeability of the medium. For a vacuum, $\epsilon = 8.854 \times 10^{-12}$ farad/m and $\mu = 4\pi \times 10^{-7}$ henry/m; therefore for a vacuum the ratio E/H is approximately 377 ohms. This ratio is termed the wave impedance of the medium.

The E and H waves travel along a straight line, as suggested in Fig. 1. Of the two possible directions along this line, the actual direction of travel can be determined by imagining a screw with a right-hand thread placed along the axis and turned from E toward H; then the longitudinal direction of travel of the screw is the direction of propagation of the energy.

The velocity of travel of the wave is shown in Eq. (2). In a vacuum this is approximately 3×10^8 m/sec. The velocity in air is only slightly smaller.

$$v = 1/\sqrt{\mu\epsilon} \qquad (2)$$

The wavelength is the distance between two successive similar points on the wave, measured along the direction of propagation. The wavelength is denoted by λ in Fig. 1.

As the wave travels past a stationary point, the values of E and H at the point vary sinusoidally with time. The time required for one cycle of this variation is termed the period, T seconds. The number of hertz is the frequency f; and $f = 1/T$. In one cycle the wave, traveling at the velocity v, moves one wavelength along the axis of propagation. Therefore, $\lambda = vT$, or may be calculated by Eq. (3). Assuming a velocity of 3×10^8 m/sec, an

$$\lambda = v/f \qquad (3)$$

electromagnetic wave having a frequency of 60 Hz has a wavelength of 5×10^6 m, or approximately 3100 mi. At a frequency of 3 MHz (3×10^6 Hz), λ is 100 m, and at 3000 MHz, λ is 10 cm. Visible light has a frequency of the order of 5×10^{14} Hz and a wavelength of approximately 6×10^{-5} cm.

The density of energy in an electric field is $\epsilon E^2/2$ joules/m³, and that in a magnetic field is $\mu H^2/2$ joules/m³. With the aid of Eq. (1), the relationships become those shown in Eq. (4). Therefore the elec-

$$\mu H^2/2 = \mu(\sqrt{\epsilon/\mu}\ E)^2/2 = \epsilon E^2/2 \qquad (4)$$

tric and magnetic fields carry equal energies in the electromagnetic wave. The total energy density at any point is equal to ϵE^2 joules/m³. Since this is transported with a velocity equal to $1/\sqrt{\mu\epsilon}$, the rate of flow of energy per square meter normal to the direction of propagation is $\epsilon E^2 v$ or $E^2\sqrt{\epsilon/\mu}$ watts/m². In radio broadcasting a field of 50 mV/m is considered to be strong. An electromagnetic wave with this intensity has an average energy density of 2.2×10^{-14} joule/m³, and the average rate of energy flow per square meter is 6.6×10^{-6} watt/m².

Radiation from an antenna. Figure 2 illustrates the configuration of the electric and magnetic fields about a short vertical antenna in which flows a sinusoidal current of the form $i = I_{max} \sin 2\pi ft$ A. The picture applies either to an antenna in free space (in which case the illustration shows only the upper half of the fields), or to an antenna projecting above the surface of a highly conducting plane surface. In the latter case the conducting plane represents to a first approximation the surface of the Earth. The fields have symmetry about

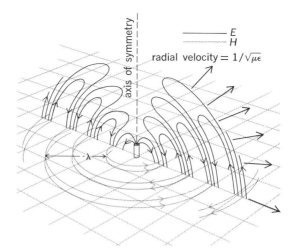

Fig. 2.   Configuration of electric and magnetic fields about a short vertical antenna.

the axis through the antenna. For pictorial simplicity only selected portions of the fields are shown in Fig. 2. The magnetic field is circular about the antenna, is perpendicular at every point to the direction of the electric field, and is proportional in intensity to the magnitude of the electric field, as indicated by Eq. (1). All parts of the wave travel radially outward from the antenna with the velocity given by Eq. (2); the wave is described as spherical, with the antenna located at the center of the wave. The wavelength of the radiation is given by Eq. (3).

If a short antenna projecting above a highly conducting plane surface carries a current of $i = I_{max} \sin 2\pi ft$ that is uniform throughout the length of the antenna, the intensity of the electric field in the radiated wave is that shown in Eq. (5), where l is

$$E_{max} = \sqrt{\frac{\mu}{\epsilon}} \frac{I_{max}}{r} \frac{l}{\lambda} \cos\ \theta\ \text{volts/m} \qquad (5)$$

the length of the antenna, r is the radial distance from the antenna, and θ is the angle measured from the horizontal. The radiated field intensity is zero directly above the antenna, is greatest along the conducting plane where θ is 0, and varies inversely with the distance from the antenna.

If the rate of flow of energy per unit area $E^2\sqrt{\epsilon/\mu}$ is integrated over an imaginary spherical surface about the antenna, the average power radiated is that shown by Eq. (6). The factor $(l/\lambda)^2$ is of particu-

$$P_{av} = \frac{2\pi}{3}\sqrt{\frac{\mu}{\epsilon}}I^2_{max}\left(\frac{l}{\lambda}\right)^2\ \text{watts} \qquad (6)$$

lar importance, for it indicates that a longer antenna is required at the longer wavelengths (lower frequencies). The radiation of appreciable energy at a very low frequency requires an impracticably long antenna.

The foregoing relations assume a uniform current throughout the length of the antenna. An approximation to this can be achieved in practice by connecting a long horizontal conductor to the top of the antenna. Where some such construction is not utilized, the current will not be uniform in the antenna and will, in fact, be zero at the tip. The results given by Eqs. (4) and (5) must be modified,

ELECTROMAGNETIC WAVE
TRANSMISSION

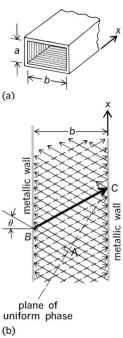

(a)

(b)

Fig. 3. A hollow metallic waveguide of rectangular cross section. (a) The guide. (b) Paths of electromagnetic energy in the simplest mode of propagation.

but the qualitative features of the radiation remain as shown in Fig. 2. For further discussion *see* ANTENNA.

Often it is desired to concentrate the radiated energy into a narrow beam. This can be done either by the addition of more antenna elements or by placing a large reflector, generally parabolic in shape, behind the antenna. The production of a narrow beam requires an antenna array, or alternatively a reflector, that is large in width and height compared with a wavelength. The very narrow and concentrated beam that can be achieved by a laser is made possible by the extremely short wavelength of the radiation as compared with the cross-sectional dimensions of the radiating system.

Propagation over the Earth. The foregoing discussion shows some of the important features of the radiation of electromagnetic energy from an antenna, but is oversimplified insofar as communication to and from positions on or near the Earth is concerned. The ground is a reasonably good, but not perfect, conductor; hence, the actual propagation over the surface of the Earth will show a more rapid decrease of field strength than that indicated by the factor of $1/r$ in Eq. (5). Irregularities and obstructions may interfere. In long-range transmission the spherical shape of the Earth is important. Inhomogeneities in the atmosphere refract the wave somewhat. For long-range transmission, the ionized region high in the atmosphere known as the Kennelly-Heaviside layer, or ionosphere, can act as a reflector. The electric field of the wave produces oscillation of the charged particles of the region, and this causes the refractive index of the layer to be smaller than that of the atmosphere below. The result is that, if the angle of incidence is not too near the normal and if the frequency of the wave is not too high, the wave may be refracted back toward the Earth. Successive reflections between ionosphere and Earth can provide communication for long distances around the periphery of the Earth.

Hollow waveguides. When an electromagnetic wave is introduced into the interior of a hollow metallic pipe of suitably large cross-sectional dimensions, the energy is guided along the interior of the pipe with comparatively little loss. The most common cross-sectional shapes are the rectangle and the circle. The cross-sectional dimensions of the tube must be greater than a certain fraction of the wavelength; otherwise the wave will not propagate in the tube. For this reason hollow waveguides are commonly used only at wavelengths of 10 cm or less (frequencies of 3000 MHz or higher).

A single wave of the type in Fig. 1 cannot propagate longitudinally inside a tubular conductor since, at some portions of the inner surface of the conducting tube, the E vector of the wave necessarily would have a component tangential to the surface. This is impossible because an electric field cannot be established along a good conductor, such as the wall of the tube. An electromagnetic wave can propagate along the interior of the tube only by reflecting back and forth between the walls of the tube. This reflection is a comparatively simple one between the plane surfaces of a rectangular tube, but is a complex reflection in tubes of other cross-sectional shapes.

A dielectric rod can also be used as a wave-

guide. Such a rod, if of insufficient cross-sectional dimensions, can contain the electromagnetic wave by the phenomenon of total reflection at the surface.

A hollow metallic waveguide of rectangular cross section is shown in Fig. 3a. The simplest mode of propagation is indicated in Fig. 3b. The entire space is filled with a plane electromagnetic wave which moves obliquely to the left in the direction shown by the solid arrows. This wave has its E vector normal to the paper and its H vector in the plane of the paper. Any plane normal to the direction of propagation is a plane of uniform phase (thus the name plane wave), and one such plane is indicated in the illustration by a broken line. The wave strikes the wall at an angle θ from the normal and is reflected at an equal angle. As the wave is reflected, the direction of its E vector reverses so as to make the tangential component of the electric field equal to zero at the conducting wall. The wave incident on the left wall thus is reflected to the right, where it is again reflected and moves to the left. By successive reflections the energy propagates longitudinally along the interior of the guide. As the wave incident upon the wall reflects and reverses the direction of its E vector, electric currents are caused to flow in the conducting wall. Since the wall is not a perfect conductor, some of the energy of the wave is transformed into heat. Consequently the amplitude of the wave diminishes exponentially as it passes down the guide; this phenomenon is termed attenuation. For an electromagnetic wave with a frequency of 3000 MHz (wavelength of 10 cm) propagating down the interior of a rectangular copper waveguide with cross-sectional dimensions of 4 cm by 8 cm, half the power is lost in a distance of approximately 150 m. Hollow waveguides are used chiefly for short-distance transmission, as from a transmitter to an antenna. *See* MICROWAVE TRANSMISSION LINES; WAVEGUIDE.

The requirements on the reflection of the wave, as outlined above, restrict the wavelength that can be propagated in a hollow guide. Consider the ray ABC in Fig. 3. The wave propagates from A to B, where it is reflected with reversal of the E vector; thereupon it propagates from B to C, where it is again reflected with another reversal of the E vector. But AC is a line of equal phase, and so the wave emerging from C must have the same phase as that at A. Thus the distance ABC must be an integral multiple of a wavelength, or $n\lambda$, where n is a positive integer. The distance ABC is $2b \cos \theta$ where b is the breadth of the guide; hence, $n\lambda = 2b \cos \theta$. The condition for propagation down the axis of the guide is that $\theta > 0$; hence $\cos \theta < 1$, and the restriction on wavelength is $\lambda < 2b/n$. The greatest ratio of wavelength to breadth of guide is obtained when $n = 1$, whence $\lambda/b < 2$. Therefore, the breadth of the waveguide must be somewhat greater than $\lambda/2$.

In the simple mode of propagation described above, the fields are independent of distance in the direction of the dimension a, and this dimension has no influence on the propagation. The net electric vector caused by the sum of the two waves is everywhere transverse to the longitudinal axis of the guide, and so the mode is described as transverse electric (TE).

If the wavelength of the radiation is small enough in comparison with the cross-sectional dimensions of the guide, more complex modes of propagation are possible, in which the wave reflects obliquely against a side wall, proceeds to the top of the guide, and reflects from there to the other side wall, then to the bottom wall, and so on. With this type of reflection it is possible to have both transverse-electric and transverse-magnetic (TM) modes. In the latter the net H vector is everywhere transverse to the axis of the guide.

When the dimensions of the guide are such that complex modes are possible, so also are the simple ones. The transmission of energy by a combination of modes introduces complications in abstracting the energy from the guide at the receiving end. Propagation in only the simplest mode is ensured by selecting the dimension b to be greater than $\lambda/2$ but not as large as λ, and also by restricting the dimension a so as to render complex modes impossible.

Waveguides of circular cross section are sometimes used. Analysis of these shows that the first TE mode is propagated if the diameter of the guide is greater than 0.586λ, and that the first TM mode is propagated if the radius is greater than 0.766λ.

Two-conductor transmission lines. Electromagnetic energy can be propagated in a simple mode along two parallel conductors. Such a waveguiding system is termed a transmission line. Three common forms are shown in Fig. 4. If the spacing between conductors is a small fraction of the wavelength of the transmitted energy, only one mode of propagation is possible. This corresponds to the wave of Fig. 1, with the direction of propagation taken longitudinally along the line. The E and H vectors are in the plane of the cross section, and the mode is termed transverse electromagnetic (TEM). The E vector must be at right angles to a highly conducting surface, and the oscillating H vector must be parallel with such a surface. With two separated conductors, there is for each geometrical arrangement of conductors one and only one cross-sectional field configuration which will satisfy the boundary conditions at the metal surface. The field configurations for coaxial and two-wire lines are shown in Fig. 4. At each point the ratio of E to H is as given by Eq. (1), and the velocity of propagation of the wave is as given by Eq. (2). Half of the propagated energy is contained in the electric field and half in the magnetic field. This mode of propagation is in contrast with the more complex modes required in a hollow metal pipe, where the conditions required at the boundaries can be satisfied only by means of reflections at the metal walls. As a result, the two-conductor transmission line does not have the upper limit on wavelength that was imposed on the hollow waveguide by the requirement of reflections; in fact, the two-conductor line operates completely normally at zero frequency (direct current).

At wavelengths that are small enough to be comparable with the cross-sectional dimensions of the line, more complex modes, involving reflections from the surfaces of the conductors, become possible. High-frequency energy can thus be propagated in several modes simultaneously. In a coaxial cable a rough criterion for the elimination of higher modes is that the wavelength should be

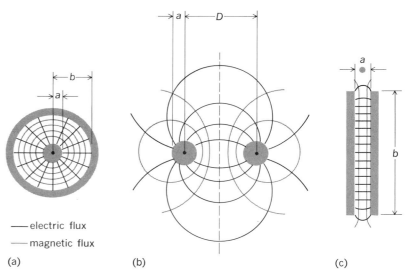

— electric flux

— magnetic flux

(a)   (b)   (c)

Fig. 4. Cross sections of common two-conductor transmission lines. (a) Coaxial cable. (b) Two-wire line. (c) Parallel-strip line.

greater than the average of the circumferences of the inner and outer conductors.

As the wave propagates along the line, it is accompanied by currents which flow longitudinally in the conductors. These currents can be regarded as satisfying the boundary condition for the tangential H field at the surface of the conductor. The conductors have a finite conductivity, and so these currents cause a transformation of electrical energy into heat. The energy lost comes from the stored energy of the wave, and so the wave, as it progresses, diminishes in amplitude. The conductors are necessarily supported by insulators which are imperfect and cause additional attenuation of the wave. In a typical open-wire telephone line operating at voice frequencies, half the energy is lost in a distance of perhaps 60 mi. The losses increase with frequency, and for a typical air-insulated coaxial line operating at 5 MHz, half the energy is lost in a distance of less than 1 mi. At a frequency of 3000 MHz ($\lambda = 10$ cm), typical distances in which half the energy is lost are, for air-insulated coaxial cable, 25 m; for flexible coaxial cable insulated with polyethylene, 10 m.

Noise. In a transmission line intended for the transmission of large amounts of power, such as the cross-country lines joining electrical generating stations to centers of population, the loss of an appreciable proportion of the power en route is a serious matter. In a communication system, however, the average rate of flow of energy is rather small, and the intrinsic value of the energy itself is not of prime importance. The important characteristic of such a system is the accurate transmission of information, and the limiting factor is noise. Noise is always present in a transmission channel. Two common causes are thermal agitation and nearby electrical discharges. In a transmission system conveying information by an electromagnetic wave, the loss of energy in transmission becomes a serious matter if the wave is attenuated to the point where it is not large enough to override the noise. Amplifiers must be inserted in the transmission system at sufficiently close intervals so that the signal never falls into the noise level, from which it could not be recovered and interpreted accurately.

Circuit analysis of transmission lines. Because the conductors of a transmission line are almost always spaced much closer together than a quarter wavelength of the electromagnetic energy that they are guiding, it is possible to analyze their performance quantitatively by circuit theory. It is then possible to deal with the voltages between the conductors and the currents flowing along the conductors, instead of with the electric and magnetic fields that exist in the insulating medium.

The waveguiding properties of the transmission line can be examined most conveniently if losses of energy are ignored. If L is defined as the inductance of the pair of conductors per unit length and C the capacitance between the conductors per unit length, field theory shows that L is μF_g and C is ϵ/F_g, where F_g is a geometrical factor that depends on the cross-sectional configuration of the conductors. For a coaxial line (Fig. 4a), $F_g = (1/2\pi)$ $\log_\epsilon (b/a)$. For a two-wire line (Fig. 4b), $F_g = (1/\pi)$ $\log_\epsilon (D/a)$. For a parallel-strip line (Fig. 4c), neglecting edge effects, $F_g = a/b$.

In the circuit analysis of a transmission line, the line can be visualized as being composed of a cascaded set of sections, each of short length Δx, as shown in Fig. 5b. The partial differential equations which describe the voltage e and the current i are shown by Eqs. (7) and (8).

$$\frac{\partial e}{\partial x} = -L\frac{\partial i}{\partial t} \qquad (7)$$

$$\frac{\partial i}{\partial x} = -C\frac{\partial e}{\partial t} \qquad (8)$$

The solution of these equations is shown by Eqs. (9) and (10), where f_1 and f_2 are any finite, single-

$$e = f_1(x - t/\sqrt{LC}) + f_2(x + t/\sqrt{LC}) \qquad (9)$$

$$i = \frac{1}{\sqrt{L/C}}[f_1(x - t/\sqrt{LC}) - f_2(x + t/\sqrt{LC})] \qquad (10)$$

valued functions of the arguments $x - t/\sqrt{LC}$ and $x + t/\sqrt{LC}$, respectively. These are interpreted physically as traveling waves, the first traveling in the positive x direction with the speed $1/\sqrt{LC}$ and the second traveling in the negative x direction at the same speed. Substitution of the values for L and C for any configuration of conductors yields the velocity $1/\sqrt{LC}$, which equals $1/\sqrt{\mu\epsilon}$.

The quantity $\sqrt{L/C}$ has the dimensions of ohms, termed the characteristic impedance Z_0 of the line, as shown by Eq. (11). Thus, Z_0 is a real quantity

$$Z_0 = \sqrt{L/C} = \sqrt{\mu/\epsilon}\, F_g \qquad (11)$$

(a resistance) and is equal to the wave impedance of the insulating medium, $\sqrt{\mu/\epsilon}$, multiplied by the geometrical factor, F_g, characteristic of the particular configuration of conductors. For the traveling waves of voltage and current, Eqs. (9) and (10), the ratio of voltage to current of the forward-traveling wave is Z_0; that of the backward-traveling wave is $-Z_0$.

In Fig. 5a, a source of electrical energy is connected at one end of a transmission line and an electrical load is connected to the other. Electromagnetic energy is propagated from the sending

end to the receiving end, and a portion of the energy is reflected back toward the sending end if the load impedance Z_R is different from the characteristic impedance Z_0 of the line. If Z_R equals Z_0, there is no reflection of energy at the load, and in Eqs. (9) and (10), the function f_2, representing leftward-traveling energy, is absent. This is the condition desired when the purpose of the line is to deliver energy from the source of the load. The sending-end impedance of the line is then equal to Z_0.

In addition to impedance matching to reduce reflections (echoes) along a transmission line, it is also necessary to minimize signal distortion, which consists of amplitude and phase (delay) distortion. If the line attenuation is frequency-dependent, then a signal consisting of a group of different-frequency components will undergo amplitude distortion due to the unequal attenuation of each component of the signal. Similarly, if the velocity of propagation along the line is frequency-dependent, then a delay in phase of each component will result in associated phase distortion of the signal.

Signal distortion can be minimized by the use of line loading, which is the addition of series impedances along the line and which is used to adjust the line parameters to obtain the so-called distortionless condition. Under distortionless operation the attenuation and velocity of propagation are independent of frequency. For a discussion of the distortionless line *see* TRANSMISSION LINES.

Instead of loading a line, one may employ equalizing circuits to compensate for the phase distortion along the line.

Short sections of transmission line are sometimes used to provide low-loss reactive impedances and resonant circuits at high frequencies. This is done by open-circuiting or short-circuiting the receiving end of the line to provide complete reflection of the incident energy. A short-circuited low-loss line provides the sending-end impedance shown by Eq. (12), measured in ohms. When l

$$Z_s = jZ_0 \tan (2\pi fl/v) \qquad (12)$$

equals $v/4f$, the line is a quarter wavelength long, the argument $(2\pi fl/v)$ of the tangent function in Eq. (12) is $\pi/2$, and Z_s approaches an infinite value. In actual practice, losses keep Z_s to a finite value. However, at high frequencies the quarter wavelength is short and the losses are small, and such a short-circuited quarter-wave section can be used successfully as a low-loss insulator. Such a section is a resonant one and can be used as a substitute for a parallel-resonant LC circuit, for example, a tank circuit for a high-frequency oscillator. At low frequencies the required quarter wavelength is so large that losses impair the performance; also, the length becomes inconveniently great. At a frequency lower than $v/4l$, the sending-end impedance of the short-circuited line is inductive, and at frequencies between $v/4l$ and $v/2l$ the impedance is capacitive. This provides the possibility of using sections of short-circuited line as reactive elements in circuits.

[WALTER C. JOHNSON]

Bibliography: C. W. Davidson, *Transmission Lines for Communications*, 1978; W. C. Johnson, *Transmission Lines and Networks*, 1950; E. C. Jordan and K. G. Balmain, *Electromagnetic Waves and Radiating Systems*, 2d ed., 1968; J. D. Kraus,

ELECTROMAGNETIC WAVE
TRANSMISSION

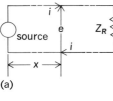

(a)

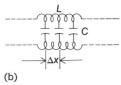

(b)

Fig. 5. Schematic representation of a transmission line. (a) Circuit diagram. (b) Visualization of L and C.

Electromagnetics, 2d ed., 1973; R. K. Moore, *Traveling-Wave Engineering*, 1960; S. Ramo, J. R. Whinnery, and T. Van Duzer, *Fields and Waves in Communication Electronics*, 1965; S. R. Seshadri, *Fundamentals of Transmission Lines and Electromagnetic Fields*, 1971; H. H. Skilling, *Electric Transmission Lines*, 1951, reprint 1979.

Electromotive force (emf)

The electromotive force, represented by the symbol ε, around a closed path in an electric field is the work per unit charge required to carry a small positive charge around the path. It may also be defined as the line integral of the electric intensity around a closed path in the field. The abbreviation emf is preferred to the full expression since emf, also called electromotance, is not really a force. The term emf is applied to sources of electric energy such as batteries, generators, and inductors in which current is changing.

The magnitude of the emf of a source is defined as the electrical energy converted inside the source to some other form of energy (exclusive of electrical energy converted irreversibly into heat), or the amount of some other form of energy converted in the source into electrical energy, when a unit charge flows around the circuit containing the source. In an electric circuit, except for the case where an electric current is flowing through resistance and thus electrical energy is changed irreversibly into heat energy, electrical energy is converted into another form of energy only when current flows against an emf. On the other hand, some other form of energy is converted into electrical energy only when current flows in the same sense as an emf.   [RALPH P. WINCH]

Bibliography: E. M. Purcell, *Electricity and Magnetism*, Berkeley Physics Course, vol. 2, 1965; R. Resnick and D. Halliday, *Physics*, 3d ed., 1978; F. W. Sears et al., *University Physics*, 5th ed., 1976.

Electron emission

The liberation of electrons from a substance into vacuum. Since all substances are built up of atoms and since all atoms contain electrons, any substance may emit electrons; usually, however, the term refers to emission of electrons from the surface of a solid.

The process of electron emission is analogous to that of ionization of a free atom, in which the latter parts with one or more electrons. The energy of the electrons in an atom is lower than that of an electron at rest in vacuum; consequently, in order to ionize an atom, energy must be supplied to the electrons in some way or other. By the same token, a substance does not emit electrons spontaneously, but only if some of the electrons have energies equal to, or larger than, that of an electron at rest in vacuum. This may be achieved by various means. If a substance is heated, the atoms begin to vibrate with larger amplitudes, and electrons may absorb sufficient energy from these vibrations to be emitted in the process known as thermionic emission. Electrons may also be liberated upon irradiation of the substance with light (photoemission). Electron emission from a substance may be induced by bombardment with charged particles such as electrons or ions in the phenomenon called secondary emission. Field emission, or cold emission, refers to the emission of electrons under influence of a strong electric field. Electrons may also be emitted from one solid into another, but this process is usually referred to as electron injection. For example, a metal may inject electrons into an insulator under certain circumstances. *See* FIELD EMISSION; PHOTOEMISSION; SECONDARY EMISSION; THERMIONIC EMISSION.

[ADRIANUS J. DEKKER]

Bibliography: L. N. Dobretsov and M. V. Gomoyunova, *Emission Electronics*, 1971.

Electron lens

An electric or magnetic field, or a combination thereof, which acts upon an electron beam in a manner analogous to that in which an optical lens acts upon a light beam. Electron lenses find application for the formation of sharply focused electron beams, as in cathode-ray tubes, and for the formation of electron images, as in infrared converter tubes, various types of television camera tubes, and electron microscopes.

Any electric or magnetic field which is symmetrical about an axis is capable of forming either a real or a virtual electron image of an object on the axis which either emits electrons or transmits electrons from another electron source. Hence, an axially symmetric electric or magnetic field is analogous to a spherical optical lens.

The lens action of an electric and magnetic field of appropriate symmetry can be derived from the fact that it is possible to define an index of refraction n for electron paths in such fields. This index depends on the field distribution and the velocity and direction of the electrons. It is given by the equivalencies shown below, Here e is the charge

$$n = \sqrt{\phi + \frac{2e\phi^2}{mc^2}} - \sqrt{\frac{e}{2m}}\, A \cos \chi$$
$$= \sqrt{\phi + 0.978 \cdot 10^{-6}\phi^2} - 0.297 A \cos \chi$$

of the electron, m its rest mass, ϕ the potential of the point in space under consideration (so normalized that the kinetic energy of the electron vanishes for $\phi = 0$), c the velocity of light, A the magnetic vector potential, and χ the angle formed by the path of the electron with the direction of the magnetic vector potential. For an axially symmetric field the magnetic vector potential is perpendicular to the plane passing through the axis of symmetry and the reference point. Its magnitude is equal to the magnetic flux through the circle about the axis through the reference point divided by the circumference of that circle. The numerical coefficients in the final expression for n apply if ϕ is measured in volts and A in gauss-centimeters.

Electron lenses differ from optical lenses both in the fact that the index of refraction is continuously variable within them and that it covers an enormous range. Furthermore, in the presence of a magnetic field, n depends both on the position of the electron in space and on its direction of motion. It is not possible to shape electron lenses arbitrarily. *See* ELECTROSTATIC LENS; MAGNETIC LENS.

[EDWARD G. RAMBERG]

Bibliography: S. Wischnitzer, *Introduction to Electron Microscopy*, 2d ed., 1970.

Electron motion in vacuum

Motion of electrons in a space freed sufficiently from matter so that collisions with other particles play a negligible role. The motion of electrons in vacuum is controlled by electric and magnetic fields whose force on the electrons is proportional to their magnitude. Electric and magnetic fields may arise from the presence of electrodes, currents, and magnets surrounding the evacuated space in which a particular electron moves, as well as from the presence of other charged particles within this space. This article deals with the nonrelativistic motion of electrons in static electric and magnetic fields, the effect of space charge on the electron paths, and motion in the time-varying fields that are encountered, for example, in the cathode-ray oscilloscope.

Static electric fields. An electron moving in a plane of symmetry of an electric field will remain indefinitely in that plane because the electrical forces acting on the electron lie within the plane. A plane of symmetry is here defined as one for which the mirror image of the potential distribution in front of the plane coincides with the potential distribution in back of the plane. Newton's second law of motion for the electron moving in the plane with rectangular coordinates x, y takes the form of Eqs. (1) and (2). Here $-e$ and m are the charge and

$$m\frac{d^2x}{dt^2} = -eE_x = e\frac{\partial\phi}{\partial x} \tag{1}$$

$$m\frac{d^2y}{dt^2} = -eE_y = e\frac{\partial\phi}{\partial y} \tag{2}$$

mass of the electron, t is time, E_x and E_y are the x and y components of the electric field, and ϕ is the electric potential, normalized as stated by Eq. (3). Newton's law as stated in Eqs. (1) and (2)

$$\frac{m}{2}\left[\left(\frac{dx}{dt}\right)^2 + \left(\frac{dy}{dt}\right)^2\right] = e\phi \tag{3}$$

implies that the speed of the electron is small enough so that its mass can be regarded as constant. For an electron having an energy of 10 kv, the increase in mass is about 1%.

Path equation. Elimination of time from Eqs. (1) and (2) leads to the path equation, Eq. (4). If poten-

$$\frac{d^2y}{dx^2} = \frac{1}{2\phi}\left[1 + \left(\frac{dy}{dx}\right)^2\right]\left[\frac{\partial\phi}{\partial y} - \frac{dy}{dx}\frac{\partial\phi}{\partial x}\right] \tag{4}$$

tial distribution $\phi(x,y)$ is known, and if position and velocity of the electron at one point within the field are also known, the electron's path can be determined by integrating Eq. (4). For simple electrode structures, the potential distribution can be determined analytically by solving Laplace's equation in the form of Eq. (5). More generally, it can be

$$\nabla^2\phi = \frac{\partial^2\phi}{\partial x^2} + \frac{\partial^2\phi}{\partial y^2} + \frac{\partial^2\phi}{\partial z^2} = 0 \tag{5}$$

found by constructing a large-scale model of the electrode structure and immersing it in an electrolytic tank so that the surface of the liquid (usually slightly acidified tap water) coincides with the plane of symmetry of interest. With potentials

proportional to the actual potentials applied to the model electrodes, an equipotential line on the surface can be found by determining the points at which a probe at the potential in question draws no current.

The path equation, Eq. (4), leads to Eq. (6) for the

$$R = \frac{2\phi}{(\partial\phi/\partial n)} \tag{6}$$

radius of curvature R of the paths. Here $-\partial\phi/\partial n$ is the component of the electric field normal to the electron path. If an equipotential plot has been prepared, this relation permits graphical plotting of an electron path (Fig. 1). The path is approximated by a series of circular arcs, the radius of curvature between successive equipotential lines being computed from the preceding relation for R.

Paraxial-ray equation. Electrostatic fields having not only a plane of symmetry but an axis of symmetry, which represents the intersection of an infinite family of planes of symmetry, have particular practical importance. Equation (4) still applies here, provided that y is identified with r, the distance from the axis, and x with z, the distance measured along the axis. The Laplace equation in the new coordinates z, r takes the form of Eq. (7).

$$\frac{\partial^2\phi}{\partial r^2} + \frac{1}{r}\frac{\partial\phi}{\partial r} + \frac{\partial^2\phi}{\partial z^2} = 0 \tag{7}$$

Equation (7) is solved quite generally by the series shown as Eq. (8), where $\Phi(z)$ is the potential on the axis of symmetry. Thus, the potential everywhere within the axially symmetric electrode structure is fully determined by the potential variation along the axis. Substitution of ϕ from Eq. (8) into Eq. (4)

$$\phi(z,r) = \Phi(z) - \frac{r^2}{4}\frac{d^2\Phi}{dz^2} + \frac{r^4}{64}\frac{d^4\Phi}{dz^4} \cdots \tag{8}$$

with retention of terms of the first order only in r

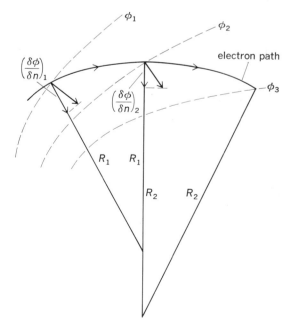

Fig. 1. Path plotting in an electrostatic field. Dashed lines ϕ_1, ϕ_2, and ϕ_3 are equipotential lines. R_1 and R_2 are radii of curvature of the electron path.

and dr/dt leads to the paraxial ray equation, Eq. (9).

$$\frac{d^2r}{dz^2} + \frac{1}{2}\frac{d\Phi}{dz}\frac{dr}{dz} + \frac{1}{4}\frac{d^2\Phi}{dz^2}r = 0 \qquad (9)$$

Equation (9) applies to electrons whose paths depart relatively little, both in slope and in distance, from the axis of the field.

Equation (9) is linear in r. Thus, if one electron path intersects the axis in two points, all electron paths passing through one of the points also pass through the other. In brief, the electric field images the one point into the other. It can be shown further that this imaging property is not limited to the axis but applies to extended areas about the axis, so that the axially symmetric electric field acts on the paths of electrons in the same manner as glass lenses act on light rays. Departures from the exact path from the paraxial equation result in image defects or aberrations similar in character to those observed for glass lenses. *See* ELECTRO-STATIC LENS.

Magnetic fields. A magnetic field exerts on an electron of velocity **v** a force **F** perpendicular to both the direction of motion and the direction of the field. In vector notation this Lorentz force is given by Eq. (10). Here, **b** is the magnetic induc-

$$\mathbf{F} = -e(\mathbf{v} \times \mathbf{b}) \qquad (10)$$

tion. The components of the Lorentz force are given by Eqs. (11)–(13). Because this force is perpen-

$$F_x = e\left[b_y\frac{dz}{dt} - b_z\frac{dy}{dt}\right] \qquad (11)$$

$$F_y = e\left[b_z\frac{dx}{dt} - b_x\frac{dz}{dt}\right] \qquad (12)$$

$$F_z = e\left[b_x\frac{dy}{dt} - b_y\frac{dx}{dt}\right] \qquad (13)$$

dicular to the direction of motion, it does no work on the electron, whose velocity consequently remains unchanged in magnitude. A uniform magnetic field parallel to the z axis is described by Eqs. (14). Newton's second law leads to a constant z

$$b_z = B \qquad b_x = b_y = 0 \qquad (14)$$

component of the velocity. The magnitude of the velocity component v_{xy} in the xy plane is similarly constant, the square of the total velocity being equal to the sum of the squares of the components. For the motion projected on the xy plane, Newton's second law thus takes the form of Eq. (15).

$$mv_{xy}^2/R = ev_{xy}B \qquad (15)$$

Here R is the radius of curvature of the projected path. R is seen to be a constant, so that the projected path is a circle with radius given by Eq. (16).

$$R = \frac{mv_{xy}}{eB} = \frac{\sin\alpha}{B}\left(\frac{2m\phi}{e}\right)^{1/2} = \frac{3.37\phi^{1/2}}{B}\sin\alpha \qquad (16)$$

Here α is the angle which the electron path makes with the field direction, and ϕ is the accelerating potential of the electrons. If B is measured in gauss, and ϕ in volts, R is in centimeters. The frequency with which the circle is traversed by the electron is given by Eq. (17). This frequency, in

$$f = v_{xy}/(2\pi R) = eB/(2\pi m) = 2.8 \times 10^6 B \qquad (17)$$

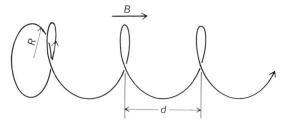

Fig. 2. Motion of an electron in a uniform magnetic field. Its path is in general a helix with pitch d, radius R, and axis parallel to the field.

$\sec^{-1}$, the cyclotron frequency, thus depends only on the magnetic field strength.

The complete motion of the electron (Fig. 2) is thus a helix about a magnetic line of force, with a pitch d in centimeters given by Eq. (18). All electrons passing through a point with equal axial ve-

$$d = v_z/f = \frac{2\pi\cos\alpha}{B}\left(\frac{2m\phi}{e}\right)^{1/2}$$

$$= 21.08\frac{\phi^{1/2}}{B}\cos\alpha \qquad (18)$$

locity components pass through a series of points separated by d on the same magnetic field line. An initially divergent electron beam is held together by a uniform magnetic field, because an electron path which intersects a particular field line can never depart from it by more than twice the radius R of the helix. Uniform magnetic fields are widely used for keeping electron beams from spreading, for example, in traveling-wave tubes and klystrons. Electron beams will also follow magnetic field lines, if these are gently curved. This property is utilized in the magnetic deflection of beams in certain television camera tubes, such as the image orthicon and the vidicon. In these tubes, a weak transverse magnetic deflection field is superposed on a strong longitudinal magnetic focusing field. *See* TELEVISION CAMERA TUBE.

Motion in nonuniform magnetic fields with axial symmetry is conveniently treated as a special case of motion in combined electric and magnetic fields.

Combined fields with axial symmetry. Motion is now expressed by Eqs. (19)–(21). Coordinates z, r,

$$m\frac{d^2z}{dt^2} = e\left[\frac{\partial\phi}{\partial z} + b_r r\frac{d\theta}{dt}\right] \qquad (19)$$

$$m\left[\frac{d^2r}{dt^2} - r\left(\frac{d\theta}{dt}\right)^2\right] = e\left[\frac{\partial\phi}{\partial r} - b_z r\frac{d\theta}{dt}\right] \qquad (20)$$

$$m\frac{1}{r}\frac{d}{dt}\left(r^2\frac{d\theta}{dt}\right) = e\left[b_z\frac{dr}{dt} - b_r\frac{dz}{dt}\right] \qquad (21)$$

and θ represent distance along the axis, perpendicular distance from the axis, and azimuthal angle about the axis. Terms b_z and b_r are the axial and radial components of the magnetic induction. From Eqs. (19)–(21), Eq. (22a), a path equation expressing the variation of the radial distance r with the axial distance z, is derived, with ϕ^* defined by Eq. (22b). Here $\phi\,D^2$ is a shorthand symbol for the last term in Eq. (22b) and C is, ex-

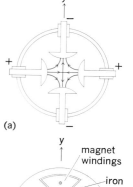

(a)

magnet
windings

iron

N   S

S   N

→x

(b)

Fig. 3. Two types of
quadrupole lens. (a)
Electrostatic quadrupole.
(b) Magnetic quadrupole.

cept for a universal multiplying constant, the angular momentum of the electron about the axis at a point where the magnetic field vanishes, given by Eq. (22c). Here A is the magnetic vector potential,

$$\frac{d^2r}{dz^2} = \frac{1}{2\phi^*}\left[1 + \left(\frac{dr}{dz}\right)^2\right]\left[\frac{\partial\phi^*}{\partial r} - \frac{dr}{dz}\frac{\partial\phi^*}{\partial z}\right] \quad (22a)$$

$$\phi^* = \phi(1 - D^2) = \phi - \left[\frac{C}{r} + \left(\frac{e}{2m}\right)^{1/2}A\right]^2 \quad (22b)$$

$$C = r^2\frac{d\theta}{dz}\phi^{1/2}\left[\left(\frac{dr}{dz}\right)^2 + r^2\left(\frac{d\theta}{dz}\right)^2 + 1\right]^{-1/2}$$

$$- \left(\frac{e}{2m}\right)^{1/2}rA \quad (22c)$$

which is numerically equal to the magnetic flux through a circle about the axis through the reference point divided by the circumference of that circle.

At the same time, the azimuth θ of the electron changes according to Eq. (23). Equations (22) can be

$$\theta = \theta_0 + \int_{z_0}^{z}\frac{D}{r(1 - D^2)^{1/2}}\left[1 + \left(\frac{dr}{dz}\right)^2\right]^{1/2}dz \quad (23)$$

solved by graphical and numerical methods useful for determining electron paths in electrostatic fields. The general paraxial equation is obtained by

substituting the expansion of Eq. (24) into Eq. (22b).

$$A = \frac{r}{2}B(z) - \frac{r^3}{16}\frac{d^2B(z)}{dz^2}\cdots \quad (24)$$

Here $B(z)$ is the magnetic induction along the axis. Substitution of Eq. (24) and that for the electrostatic potential and retention of terms of the first order in r and dr/dz only lead to Eqs. (25). With

$$\frac{d^2r}{dz^2} + \frac{1}{2\Phi}\frac{d\Phi}{dz}\frac{dr}{dz}$$
$$+ \left(\frac{1}{4\Phi}\frac{d^2\Phi}{dz^2} + \frac{eB^2}{8m\Phi} - \frac{C^2}{\Phi r^4}\right)r = 0 \quad (25a)$$

$$\theta = \theta_0 + \int_{z_0}^{z}\left[\frac{C}{r^2\Phi^{1/2}} + \left(\frac{e}{8m\Phi}\right)^{1/2}B\right]dz \quad (25b)$$

$$\frac{C}{r_0^2\Phi_0^{1/2}} = \left(\frac{d\theta}{dz}\right)_0 - \left(\frac{e}{8m\Phi_0}\right)^{1/2}B(z_0) \quad (25c)$$

B in gauss, Φ in volts, and z in centimeters, $e/8m$ equals 0.022 volt/(gauss-cm)².

Quadrupole fields. The principal use of axially symmetric electrostatic and magnetic fields is to converge electron pencils; thus, in the cathode-ray tube, electrons diverging from a point in front of the emitter, called the crossover, are converged to a small spot on the viewing screen. Similarly, in an electron-imaging device such as the electron microscope, electrons diverging from a point on an object are brought to focus at a corresponding point of the image. However, this converging effect is only secondary; the primary effect of an axially symmetric electrostatic field is to accelerate or decelerate the electrons parallel to the axis, and the primary effect of an axially symmetric magnetic field is to give electrons diverging from a point on the axis a rotation about the axis. Correspondingly, the converging action or refractive power of a conventional electrostatic lens is proportional not to the first power but to the square of the ratio of the electrostatic field to the accelerating voltage, whereas the converging action of a conventional magnetic lens is proportional to the ratio of the square of the magnetic field to the accelerating potential at low electron energies (≪0.5 MeV) and to the square of the ratio of the field to the accelerating potential at high electron energies (≥0.5 MeV). Because breakdown phenomena place practical limits on the magnitude of electrostatic fields, and because pole-piece saturation limits (static) magnetic fields, conventional electron lenses are relatively ineffective in converging high-energy particles; they can be properly characterized as weak-focusing lenses.

Quadrupole fields, on the other hand, constitute strong-focusing lenses, insofar as their converging action is directly proportional to the electrostatic or magnetic field respectively. Consequently they (particularly magnetic quadrupole fields) have assumed great importance in the development of high-energy particle accelerators. In this application they permit the narrow confinement of the particle beam and consequently greatly reduce the cost of construction. Apart from the concentration and focusing of high-energy particle beams, strong-focusing lenses have found application as projector lenses (but not as objectives) of electron

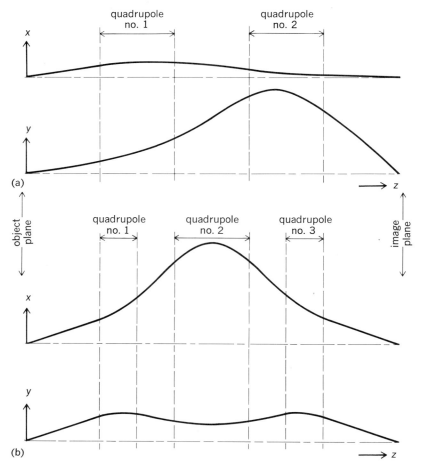

Fig. 4. Diagrammatic representation of electron paths in principal planes of image-forming symmetrical (a) doublet and (b) triplet.

microscopes. Weak quadrupole fields (stigmators) also are used to compensate residual asymmetrics in conventional electron lenses.

Figure 3 shows sections normal to the lens axis of an electrostatic and a magnetic quadrupole lens. In the first instance the electrostatic potential is symmetric with respect to the two principal (xz and yz) planes and antisymmetric with respect to the planes through the axis forming an angle of 45° with respect to the principal planes. In the second instance the magnetostatic potential is antisymmetric with respect to the principal planes and symmetric with respect to the two 45° planes. Fulfillment of Laplace's equation, Eq. (5), now demands the form of Eq. (26) for the electrostatic potential. Magnetic field B, on the other hand, may be written as Eqs. (27)–(29). In Eqs. (26)–(29),

$$\phi(x,y,z)=\frac{a}{2}(x^2-y^2)-\frac{d^2a}{dz^2}2(x^4-y^4)\cdots \quad (26)$$

$$B_x=Ay-\frac{1}{6}\frac{d^2A}{dz^2}y(3x^2+y^2)\cdots \quad (27)$$

$$B_y=Ax-\frac{1}{6}\frac{d^2A}{dz^2}x(x^2+3y^2)\cdots \quad (28)$$

$$B_z=0\quad-\frac{1}{6}\frac{d^3A}{dz^3}xy(x^2+y^2)\cdots \quad (29)$$

a and A are functions of z. Neglecting the higher-order terms, which become negligible for a long lens, one finds that the force exerted by the field on the electron is given by Eqs. (30)–(32).

Electrostatic lens		*Magnetic lens*	
$F_x=-ea\,x$	(30a)	$F_x=-e\,A\,v_z x$	(30b)
$F_y=ea\,y$	(31a)	$F_y=e\,A\,v_z y$	(31b)
$F_z=0$	(32a)	$F_z=0$	(32b)

The force is thus, to a first approximation, entirely in a plane normal to the axis. In one principal plane it is such as to produce convergence (direction of force opposite to that of the displacement from the axis), and in the other, such as to produce divergence (direction of force equal to that of the displacement). Two successive quadrupole fields of opposite polarity (a doublet) are required to form a sharp, real image of an object on the axis. This image will have different magnifications in the two principal directions. If the image is to be undistorted, a triplet of three successive quadrupole fields is required. These electron paths are illustrated in Fig. 4.

The shaping of the poles influences only terms of the fifth and higher orders in the x and y coordinates in the field expressions. For long lenses all terms except the linear terms vanish if the poles are rectangular hyperbolic cylinders. In practice, circular-cylinder pole caps are more easily realized and give practically the same results. Like axially symmetric lenses, quadrupole lens systems exhibit aberrations, although these are more complex in character. Octupole fields (Fig. 5), which produce transverse forces proportional to the third power of the displacement from the axis, opposite in polarity in the principal planes and in the 45° planes, are commonly employed to correct aberrations of quadrupole lens systems.

Effect of space charge. Space charge of either positive or negative sign can influence the paths of

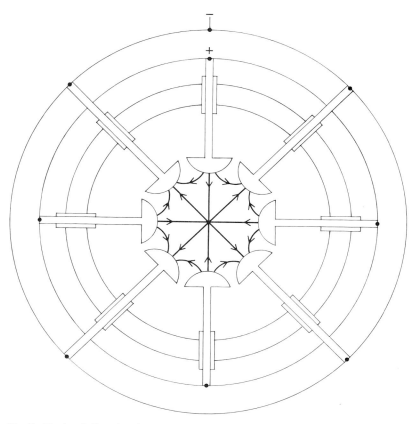

Fig. 5.   Electrostatic octupole.

electrons. Space charge of positive sign is formed by electron beams passing through an imperfectly evacuated space. The beam electrons collide with gas atoms and ionize them. The heavy ions remain in the path of the electron beam for some time and prevent it from spreading. The luminous nodular or thread beams which are produced in this way are favorite objects for demonstration. *See* SPACE CHARGE.

Electron beams in high vacuum, on the other hand, are subject only to the mutually repulsive forces between the electrons themselves. The repulsion is reduced, but never canceled, by the action of the magnetic fields that surround charges in motion; for two electrons moving with the same velocity v parallel to each other, the ratio of the magnetic attractive force to the electrostatic repulsive force is v^2/c^2, where c is the velocity of light. Hence, the magnetic force is significant only for electrons of very high energy.

The action of the remainder of the electrons in the beam upon any one electron can be approximated adequately by that of a continuous charge distribution equal to the average space-charge distribution. The behavior of the edge ray of a uniform circular beam of current I aimed at a point of convergence a distance L from the initial cross section of radius r_B may serve as an example (Fig. 6). If the variation of the potential along the axis of the beam is neglected (that is, if Φ is assumed to be constant), and the charge density ρ is regarded as uniform within any beam cross section, ρ is given by Eq. (33) and the path equation becomes Eq.

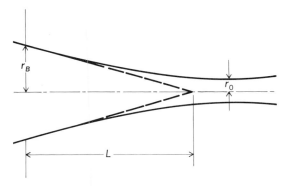

Fig. 6. Widening of electron beam by space-charge repulsion as it traverses from left to right.

(34). Here ϵ is the dielectric constant of vacuum. As the result of the repulsive force of space charge, the ray under consideration does not cross the axis, but reaches a minimum separation r_0 from the axis and diverges from this point on. Integration of the differential equation, Eq. (34), gives the radius as Eq. (35). For example, if $r_B = 1$ mm, $\Phi =$

$$\rho = \frac{I}{\pi r^2} \left(\frac{m}{2e\Phi}\right)^{1/2} \qquad (33)$$

$$\frac{d^2 r}{dz^2} = \frac{\pi\rho}{\epsilon\Phi} r = \left(\frac{m}{2e}\right)^{1/2} \frac{I}{\Phi^{3/2}} \frac{1}{\epsilon r} \qquad (34)$$

$$r_0 = r_B \exp\left[-\epsilon\left(\frac{e}{2m}\right)^{1/2} \frac{r_B^2 \Phi^{3/2}}{L^2 I}\right]$$
$$= r_B \exp\left[-3.3 \times 10^{-5} \frac{r_B^2 \Phi^{3/2}}{L^2 I}\right] \qquad (35)$$

10,000 volts, $L = 10$ cm, and $I = 0.001$ amp, then $r_0 = 0.037$ mm.

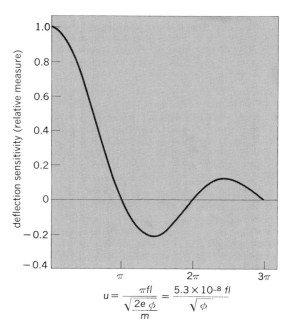

Fig. 7. Deflection sensitivity of cathode-ray oscilloscope as a function of frequency f.

Time-varying fields. In the preceding discussion, it was assumed that the electric and magnetic fields traversed by the electrons were constant in time. The total energy of the electron, or the sum of the kinetic energy and the potential energy, is then a constant. Because the potential energy is a function of position only, so is the kinetic energy. This is no longer true if the fields change appreciably in a period corresponding to the transit time of the electrons. For discussion of electron motion in time-varying fields, such as are encountered in microwave tubes, *see* KLYSTRON; MAGNETRON; TRAVELING-WAVE TUBE.

Beam deflection in the cathode-ray oscilloscope ceases to be proportional to the potential difference V applied to the deflection plates if V changes appreciably during the passage of the electron beam through the deflection field. If $V = V_0 \cos 2\pi ft$, an integration of the transverse impulse impressed on the electron passing between two parallel plates of length l and separation d expresses the deflection angle α in the form of Eq. (36). Here ϕ is the accelerating potential of the

$$\tan \alpha = \frac{V_0 l}{2\phi d} \frac{\sin u}{u} \cos(2\pi ft) \qquad (36)$$

beam, and t is the time the electron passes through the center of the deflection field. If f (the frequency of the applied voltage) is measured in sec^{-1}, l in centimeters, and ϕ in volts, then Eq. (37) is appli-

$$u = \pi fl / (2e\phi/m)^{1/2} = 5.3 \times 10^{-8} fl/\phi^{1/2} \qquad (37)$$

cable. The quantity $(\sin u)/u$ represents the ratio of the deflection sensitivity at frequency f to that at low frequencies $(f \to 0)$ (Fig. 7). Thus, for a 10-kV beam and deflection plates 1 cm in length, the response is 95.4% at 1000 MHz (10^9 sec^{-1}), 82.3% at 2000 MHz, 40.2% at 4000 MHz, and it drops to zero at about 5930 MHz. In this discussion, the deflection field is assumed to be a sharply cutoff uniform field, with the effects of fringe fields neglected.

[EDWARD G. RAMBERG]

Bibliography: P. Dahl, *Introduction to Electron and Ion Optics*, 1973; A. B. El-Kareh and J. C. El-Kareh, *Electron Beams, Lenses and Optics*, 1970; P. Grivet, *Electron Optics*, 1965; A. Septier, Strong-focusing lenses, *Advan. Electron. Electron Phys.*, 14:85–170, 1961; E. Hartung and F. H. Read, *Electrostatic Lenses*, 1976; O. E. Klemperer and M. E. Barnett, *Electron Optics*, 3d ed., 1970.

Electron optics

The branch of physics concerned with the motion of free electrons under the influence of electric and magnetic fields. The term electron optics is derived from the fact that the laws governing electron paths in such fields are formally identical with those governing light rays in media of varying refractive index. Both may be derived from Fermat's law. This law states that the actual light ray or electron path passing through two prescribed points A and B is that which makes the integral (1)

$$\int_A^B n \, ds \qquad (1)$$

carried out over it a minimum. The refractive index n is, for electrons, shown by Eq. (2). Here $-e/m$

$$n = \sqrt{\phi + \frac{2e\phi^2}{mc^2}} - \sqrt{\frac{e}{2m}} A \cos \chi \qquad (2)$$

is the specific charge of the electron, c the velocity of light, ϕ the accelerating potential of the electron, A the magnetic vector potential, and χ the angle formed by the electron path with the direction of the magnetic vector potential. Since for electrons of given kinetic energy ϕ and A are unique functions of position, the refractive index n is also a function of position and, in the presence of a magnetic field, of the direction of the electron path. A similar dependence of the refractive index on the direction of a light ray is encountered in crystal optics.

The study of electron paths, analogous to the study of light rays, is more properly called geometrical electron optics. The electron paths may be regarded as normals to electron waves, whose amplitude determines the statistical density of electrons, just as the amplitude of a light wave determines the density of light quanta, or photons. The study of the wave motion associated with electrons is called electron wave optics. It describes diffraction and interference effects between electron beams which are in every way similar to the diffraction and interference effects observed with light and x-rays.

Electron optics finds application in the formation of electron beams, as in cathode-ray tubes and television camera tubes; in the deflection of such beams by electric and magnetic fields; and in the formation of electron images, as in electron microscopes and image tubes. *See* CATHODE-RAY TUBE; ELECTRON LENS; ELECTRON MOTION IN VACUUM; ELECTROSTATIC LENS; MAGNETIC LENS.

[EDWARD G. RAMBERG]

Bibliography: P. Dahl, *Introduction to Electron and Ion Optics*, 1973; P. Grivet, *Electron Optics*, 2d ed., 1972.

Electron tube

A device in which conduction takes place by the movement of electrons or ions between electrodes through a vacuum or ionized gas within a gas-tight envelope. Electron tubes include all partially evacuated tubes whose electrical characteristics are derived from the flow of electrons through the tube. Two subclasses of electron tubes are vacuum tubes and gas-filled tubes. Vacuum tubes are evacuated to such a degree that their electrical characteristics are essentially unaffected by the presence of any residual gas or vapor. Gas-filled tubes have electrical characteristics that are substantially dependent upon the ionization of deliberately introduced gas or vapor. *See* VACUUM TUBE.

Application. The importance of electron tubes grew rapidly during the first half of the 20th century. Although semiconductor devices have largely replaced vacuum tubes in many areas of application, tubes are still essential in a number of areas. Thus, cathode-ray tubes, which are a specialized form of vacuum tube, are at present indispensable to television receivers, oscilloscopes, and radar display units; camera tubes are used in television cameras; high-power tubes are used in broadcasting transmitters; and microwave tubes find many applications in radar, telephony, space communication and control, scientific research, and high-frequency ovens. Vacuum tubes are primarily used in applications where low noise and high frequency are involved. In contrast, gas tubes are used for high-current, low-frequency applications. They may be either simple diodes, which are used primarily as rectifiers, or control-type tubes having three or more electrodes for a variety of purposes. Gas tubes are used mainly in industrial applications where high power-handling ability and efficiency overshadow their frequency limitations. *See* CATHODE-RAY TUBE; SEMICONDUCTOR.

Characteristics. The electrical characteristics of electron tubes vary greatly among the numerous types, depending upon the number and configuration of the electrodes, upon the degree of evacuation, and upon the type and capability of the electron source. The power capability ranges from milliwatts to peak values of hundreds of megawatts, and the frequency of operation ranges from zero to the order of 10^{11}Hz. The evacuated envelope may be of glass, quartz, ceramic, or metal.

History. Some of the phenomena associated with electron tubes were first noticed in the latter part of the last century. In 1883 T. A. Edison observed some peculiar effects in light bulbs which were later recognized as due to electrons. H. Hertz observed photoelectric emission in 1887. W. C. Roentgen observed x-rays in 1895. The electron itself was probably first identified by J. J. Thomson, who in 1897 also measured its ratio of charge to mass. Probably the first electron tube was a cathode-ray tube built by K. F. Braun in 1897. A. Wehnelt discovered oxide emission in 1903, and this led to the development of the vacuum diode by J. A. Fleming in 1904. With the invention of the vacuum triode by L. De Forest in 1906, the age of electronics was ushered in. The tetrode was developed in 1919 by W. Schottky and the pentode by G. Jobst and B. D. H. Tellegen in 1926. A. W. Hull developed the thyratron in 1929 and the magnetron in 1921. R. Varian and S. Varian invented the klystron in 1938, and R. Kompfner introduced the traveling-wave tube in 1946. *See* GAS TUBE; MICROWAVE TUBE; NUMERICAL INDICATOR TUBE; PHOTOTUBE.

[HERBERT J. REICH]

Bibliography: D. J. Harris and P. D. Robson, *The Physical Basis of Electronics*, 2d ed., 1974; Y. Koike, *Electron Tubes*, 1972; H. J. Reich, J. C. Skalnik, and H. L. Krauss, *Theory and Applications of Active Devices*, 1966.

Electronic countermeasures

Techniques, devices, and equipment necessary for one adversary to deny or counteract an enemy's use of radar, communications, guidance, or other radio-wave devices. Because of the growing use of optical and infrared techniques for communications, guidance, detection, and control, electronic countermeasures (ECM) are sometimes called electromagnetic, rather than electronic, countermeasures to convey more adequately the idea that countermeasures are not confined to the portion of the spectrum where electronic techniques alone are applicable but may be used throughout the electromagnetic spectrum.

Since the end of World War II, ECM has become an essential instrument in electromagnetic warfare—the worldwide struggle by major military powers for supremacy in the electromagnetic spectrum. This largely silent, secret struggle occurs continually, independent of whether its participants are at war with one another. They seek control of the spectrum because of the increasing dependence of military forces on its use for surveillance, communications, detection, measurement, guidance, and control. With mastery of the spectrum, one adversary could achieve an indispensable ingredient for conquering an enemy or discouraging a potential aggressor.

Occasionally, electromagnetic warfare takes ironic, even tragic, twists. After the U-2 reconnaissance aircraft piloted by Gary Powers was shot down over the Soviet Union in 1960, the Soviets displayed ECM equipment purportedly carried by the aircraft to monitor Soviet military electromagnetic activity. When the U.S. Navy ship *Pueblo* was seized early in 1968 by North Korean patrol boats, the *New York Times* reported, "The intelligence ship Pueblo was one unit in a vast network of electronic eavesdropping devices that the United States operates on land and sea, in the air and in space."

Similarly, Soviet trawlers loaded with electronic equipment totally inconsistent with usual needs of fishing vessels are observed in increasing numbers and with ever-growing frequency near American missile-testing areas, presumably snooping on command, tracking, and telemetry transmissions.

These occasional inadvertent public disclosures of a largely covert struggle indicate the expense and risks major nations take in eavesdropping on and monitoring a potential enemy's radar and communications by means of ECM techniques.

Categories of ECM. Traditionally, ECM equipment and techniques are categorized as active and passive, depending on whether or not they radiate their own energy. The passive category includes reconnaissance or surveillance equipment that detects and analyzes electromagnetic radiation from radar and communications transmitters in a potential enemy's aircraft, missiles, ships, satellites, and ground installations. The reconnaissance devices may be used to identify and map the location of emitters without in any way altering the nature of the signals they receive. Other types of passive ECM do seek to enhance or change the nature of the energy reflected back to enemy radars, but they do not generate their own energy.

Active ECM equipment generates energy, either in the form of noise to confuse an enemy's electromagnetic sensors or by generating false or time-delayed signals to deceive radio or radar equipment and their operators. *See* ELECTRICAL NOISE.

Surveillance systems. Reconnaissance or surveillance ECM systems are carried by Earth-orbiting satellites, aircraft, ships, unmanned (drone) aircraft, and automotive vehicles. Some are located on the ground. A few systems are small enough to be carried by a man. Reconnaissance systems are interchangeably called ferret or electronic intelligence (elint) systems.

They consist of sensitive receivers electromechanically or electronically tuned over desired portions of the spectrum in search of transmissions of interest. Bearing to an intercepted signal can be determined by direction-finding techniques. Once secured, the signals can be displayed for analysis by an operator or stored on tape or other storage media for subsequent analysis or both.

The type of information obtained by modern surveillance systems is illustrated by a standard system in use by the Strategic Air Command. It has a superheterodyne receiver and five separate radio-frequency tuning heads that span the spectrum from 1 to 18 GHz. (Some efforts have been directed towards obtaining coverage from 12 to 40 GHz.) The receiver sequentially scans through those five heads at sweep rates up to 20 Hz, selectable by an operator. The operator has a panoramic display of the five separate amplitude-versus-frequency traces projected by a multigun cathode-ray tube. When he sees a signal of interest, he can superimpose a cursor over it and cause the receiver to discontinue the sweep mode and lock onto the intercepted signal. At this point, his display automatically switches into an analytical mode, offering an enlarged time display with repetitive pulsed signals on five separate logarithmically spaced time scales. Then he can visually identify the pulse shape, its repetition rate, and pulse width. Separately, he can obtain an automatic digital indication of the signal's frequency in 100-kHz increments, pulse duration to 0.1 μsec, and pulse repetition period to 1 μsec. *See* RADIO.

Large U.S. Air Force bombers were equipped in the late 1960s with an electronically swept, tuned, radio-frequency ECM receiver that gave the electronic-warfare officer on the airplane a panoramic display of all types of airborne and ground-based electromagnetic threats to the aircraft over six frequency bands. He received an indication of carrier frequency, relative signal strength, modulation characteristics, and relative pulse-repetition frequency of detected signals. If any of these signals posed a threat, the officer could turn on one or more of the ECM jammers carried by the bomber.

Not all reconnaissance equipment is intended for spectral observation. Another form of reconnaissance equipment, carried by some U.S. Air Force aircraft, aids photointerpretation by detecting and partially identifying electromagnetic signals. Intelligence information gathered by the set is directly recorded on the film of a reconnaissance camera. The system, which covers six military frequency bands, locates an emitter in any one of eight sectors of the photograph. It helps in identifying the frequency, pulse-repetition frequency, and pulse width of the unknown intercepted radar signal.

Warning receivers. Warning-receiver systems, which came into widespread use on United States tactical and transport aircraft during the Vietnam-

Fig. 1. ECM pods carried under wings of a military aircraft. (*Lundy Electronics and Systems*)

ese war, are a more limited form of the elint system. Unlike the latter, which is intended to search for signals over a broad range of the spectrum, the warning receiver is programmed to alert a pilot when his aircraft is being illuminated by a specific radar signal above predetermined power thresholds anticipated by elint systems. When the pilot has been alerted, he can maneuver his aircraft to evade the threat or initiate counteraction with onboard ECM capability (Fig.1). The principal radar threats, which appeared in the Vietnamese war and in the Middle East wars of 1971 and 1973, were Soviet-made surface-to-air missile (SAM) radar and anti-aircraft gunfire direction radars. Although both are ground-based threats, the warning-receiver concept is equally applicable to radar threats from opposing aircraft. Similarly, warning receivers may be carried by ships or other vehicles.

Airborne warning systems usually are crystal video receivers that indicate the presence of a radar threat and offer a coarse relative bearing to it. An associated receiver and extra antennas augment this capability, automatically determining direction to the threat and giving the pilot a means of homing on the target for weapons delivery by nulling an indicator that centers the aircraft on the signal.

The warning systems also exploit a particular weakness in SAM weapons, which requires that a command radar be turned on prior to missile launch so that the missile can be guided to its airborne target after launch. The receiver can detect characteristic changes in power level of the radar, thereby warning the pilot that a missile is about to be fired. SAM commanders often counter this United States ECM technique by simply turning on the command radar frequently even when there is no impending target intercept. This can prompt aircraft to prematurely jettison fuel tanks or weapons in an effort to evade what is thought to be the imminent launch of a SAM missile.

These procedures illustrate how each advance in radar technology has a countermeasure and how each countermeasure is followed by a counter-countermeasure. Usually the countermeasure and the counter-countermeasure involve a technological advance, but frequently simple ingenuity or changes in tactics supply the new measure.

Intruding aircraft have a time advantage over defense radars because they can detect enemy radar at considerably greater distances than the hostile radar can spot them. The reason for this is that energy emitted by radar decays as a function of the square of the range from the radar to the target. *See* RADAR.

As technology progresses, the difference between the warning system and limited-capability elint receivers is narrowing with the advent of small digital processors, particularly for tactical military aircraft. The tactical aircraft needs to be able to react quickly to changes in power levels of specific radar threats, movement of transportable SAM radars, and frequency shifts by frequency agility or frequency hop radars.

Reflectors. One of the oldest passive ECM techniques, made famous by a Royal Air Force raid on Hamburg in 1943, is the use of chaff. These are metallic strips cut to lengths resonant at the

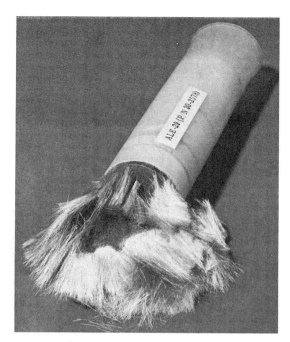

Fig. 2. Metallized glass chaff, which returns spurious radar echoes. (*Lundy Electronics and Systems*)

defense radar frequency so that they return spurious radar echoes to enemy radar (Figs. 2 and 3). Chaff can confuse an enemy by generating false targets, or noise, forcing him to take time to analyze the returns and sort real from false targets. Chaff can screen or mask aircraft or higher-speed ships so that the enemy is unable to determine their presence, or chaff can help an aircraft break track once it is alerted by its warning receiver that it is being tracked by radar.

United States bombers relied heavily on chaff

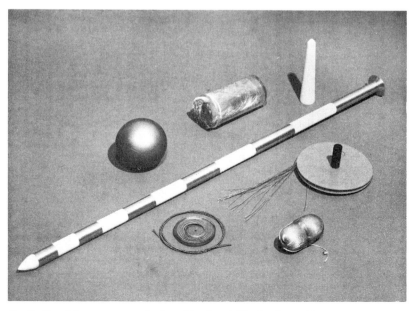

Fig. 3. Special-purpose penetration aids. (*Lundy Electronics and Systems*)

and noise jammers in penetrating North Vietnamese defenses during raids against Hanoi and Haiphong in the closing days of the war in Southeast Asia. The Israeli navy scored spectacular successes with chaff in thwarting Soviet-made radar-guided antiship missiles in the October 1973 war; of approximately 50 antiship missiles fired by Syrian and Egyptian naval ships, not one struck a target. By contrast, in the brief conflict over Bangladesh 2 years before, the Indians fired 13 Soviet-supplied Styx missiles at Pakistani ships, achieved 12 hits, and routed the Pakistani navy. Israeli success in countering the same type of weapon is attributable to the use of rocket- and mortar-fired chaff cartridges which effectively screened their maneuvering gunboats and lured the Styx missiles off course.

Chaff has evolved from the aluminum strips hand thrown from bombers during World War II to 1-mm-diameter metallized glass fibers that are automatically ejected in great quantities from aircraft by electromechanical, pyrotechnic, or pneumatic dispensers. Typically, dispensing mechanisms aboard modern military aircraft forcibly eject cardboard packages, roughly 3 in. by 5 in. by 1/2 in. (1 in. = 2.54 cm). Each package, containing thousands of strands of aluminum-coated glass fibers, is torn open at release. The contents then blossom into clouds within the aircraft's slipstream.

In Vietnam, naval pathfinding aircraft would seed air corridors with chaff through which attacking aircraft would subsequently fly. The chaff masked their specific presence from enemy radars. In addition, strategic bombers are outfitted with chaff-dispensing rockets fired ahead of the bombers to assure that an adequate screen blooms before the aircraft arrives.

By the mid-1970s, U.S. Air Force and Navy tactical fighters were being outfitted to carry large chaff pods capable of dispensing 300 lb (136 kg) of lightweight chaff contained on 6 separate drums and blown out through separate exit pipes at the rear of the pod (Fig. 4). The dispensing rate and the duration and frequency of chaff bursts can be programmed into the pod on the flight line.

To assure availability aboard an aircraft of chaff cut to proper lengths, the military services have investigated dispensing systems that cut chaff to lengths comparable to the frequencies of radars that warning receivers indicate are illuminating the aircraft. A technique known as delayed-opening chaff (DOC), by which chaff packages are deployed from aircraft or ballistic missiles approaching their targets, also is currently being developed. A DOC package is descended by parachute and then blown open by a time fuse. This might deceive an enemy into believing that there are targets far removed from the dispensing aircraft.

Other passive ECM techniques include the use of special radar-absorbing materials, such as ceramics or ferrites, which reduce reflection coefficients so that the amount of radar energy returned to the illuminating radar is reduced. A standard technique is the special shaping of bodies, specifically in missile reentry systems, that reduces the vehicle's radar cross section (the area of the vehicle that effectively returns echoes to the transmitter). The use of corner reflectors, or Luneberg lenses, which concentrate the energy they reflect back to the radar, can provide a useful ECM aid by confusing an enemy. These reflectors, fixed to drone aircraft released from bombers, produce large returns on enemy radar screens, causing the drones to appear on radar like the larger bombers that released them.

Fig. 4. Bulk-chaff dispensing pod which ejects chaff through six exit pipes.

United States aircraft carry sensitive infrared ECM receivers to protect them from heat-seeking infrared missiles. Because such missiles normally are launched toward the rear of the target so that they can home on the heat from the aircraft's jet exhaust, the receivers are located on the aircraft's tail. When the pilot is warned of impending danger from the infrared missile, infrared flares can be injected through the chaff dispenser to divert the heat-seeking missile.

U.S. Navy and Army aircraft are equipped with active infrared transmission devices designed to induce infrared missiles tracking an aircraft's engine exhaust to break target lock. These transmitters generate high-data-rate infrared pulses to jam signals derived within the missile seeker from the rotating action of the seeker's reticle in chopping incoming infrared energy.

Active ECM. The many active ECM techniques can be classified broadly either as noise or deception jamming. The former is the oldest, simplest, and most straightforward, but requires higher average power levels and is more expensive. Deception jamming is the more artful and sophisticated technique, operating on the characteristics of the pulse train generated by threat radars.

Noise jamming has a disadvantage in that it can alert the enemy to the fact that he is being jammed. Yet in tactical situations where the enemy has little reaction time and noise contributes to his confusion, this distinction has little importance.

Communications-jamming techniques include use of white-noise jamming, in which artificially produced white noise modulates the output of a transmitter. This is one of the simplest and easiest approaches, despite its excessive use of power. White-noise jamming is one of the more prevalent forms of noise jamming. In sweep-through jamming the transmitted carrier frequency is swept through a part of the spectrum at a high rate, producing pulses masking the incoming signal. This is more economical of transmitter power.

Deception-jamming techniques are predicated on the idea of operating on pulses received from the enemy so that the signal reradiated from the target deceives the enemy radar or its operators. For instance, the ECM deception set may receive an enemy radar pulse, circulate it through a delay line, amplify it, and reradiate it back toward the enemy. Because the enemy determines the position of the target by the round-trip transit time of the radar energy, his radar decision circuitry will conclude that the target is at a greater distance than it actually is because of the deceptive pulse delay inserted in that round-trip period by the active ECM set.

Similarly, the ECM deception set may operate on the radar pulse train, returning many pulses instead of one, in an effort to deceive the enemy into believing there are many targets spaced at different positions. The genuine return may be blanked out, or false signals may be generated at higher levels so that the enemy may think the genuine return is simply noise. By changing the character of the pulses in the train, the ECM set can deceive the antagonist in a number of ways.

Both deception and noise jamming have their adherents. Together the two types of active ECM system can be used effectively in defense environments characteristic of modern warfare. Newer ECM systems combine both noise and deception features.

The levels of sophistication being approached in deception systems is suggested by the following equipment desired by the U.S. Navy: a deceptive radar jammer that could produce in a hostile radar what would appear to the enemy to be an equipment malfunction; an automatic data-link (communications) jammer that inserts undetectable wrong numerals in messages sent by an enemy over data link; high-frequency jammers that disrupt enemy communications in such a way that the enemy believes the disruption is due to atmospheric disturbances; and active retroreflective chaff that amplifies and reradiates a hostile signal back in the direction from which it was received.

ECM system. Because of the complex electromagnetic environment in warfare, no single technique is satisfactory for all occasions. A late-model warplane such as the controversial F-111 may have a penetration-aids subsystem with a number of separate systems operating interdependently to give aural and visual warning of radar threats. A cryogenically cooled infrared receiving set with its own decision circuitry located in the top of the aircraft's vertical stabilizer warns the pilot of threats from heat-seeking missiles. An ECM deception set automatically generates false target returns to tracking-radar threats in several frequency bands.

A pneumatically driven chaff dispenser, working with other subsystems, selects and automatically ejects disposable packages including chaff to confuse tracking radars and flares to mislead infrared missiles. A wideband radar homing and warning set detects enemy ground and airborne radar threats, identifies and determines the detected threat location, and calculates azimuth and elevation of the threat. It monitors the effectiveness of the ECM deception set and automatically supplies signals to deploy disposables. It also provides bearing inputs to the aircraft's inertial navigation computer and drives the lead computing gunsight in an attack mode so that the aircraft can attack a threatening emitter detected by the warning set.

Trends in ECM. Future ECM systems will display greater integration among the various subsystems to tighten up the defense capability. They will probably incorporate infrared warning receivers and new active infrared jamming systems, as well as attack warning from a data link.

Active chaff would enhance the ability of chaff by investing in it some of the advantageous properties of active ECM. With the ability to reradiate signals, chaff would have longer life as a legitimate false target. It similarly would be an effective antidote to home-on-jam weapons which can seek out radiating active ECM on aircraft.

Active chaff might be a small, solid-state repeater, possibly using tunnel diode amplifiers, that picks up radar signals too weak to be detected by the enemy, perhaps delays them for fixed periods, and then reradiates them toward the hostile radar. The enemy then gets either a reinforced echo from the chaff or a false indication as to the whereabouts of the target. Disposable jammers ejected from chaff dispensers are a variation on this idea; devices radiate spurious jamming signals without first being illuminated by hostile radar.

As part of a continuing practice of updating the ECM capability of military aircraft to meet newly recognized threats, the U.S. Air Force has modified its strategic bomber force. Its B-52 bombers (Fig. 5) have been retrofitted with new jammers to counter higher-frequency radars carried by Soviet interceptors or located on the ground for control of surface-to-air missiles. Continuing attempts have been made to find suitable sensors—either infrared or pulse Doppler radar—to provide warning of missile and aircraft threats to a bomber's rear.

The need to ensure penetration capability of United States ballistic missiles is placing growing stress on techniques derived from conventional ECM. The techniques are similar to what was previously described, although active ECM techniques are less practical because of the severe environments and power limitations. Several ways of using active ECM have been investigated, however, including the possibility of installing jammers within small rocket-fired precursor decoys which would be fired ahead of a warhead. The jammers, each of which is tuned within different frequency bands, spread confusion among enemy radars.

The difficulty in securing extremely rugged microwave transmitting tubes to withstand missile environments and in generating a maximum amount of broadband power with a minimum of

Fig. 5. Strategic Air Command B-52 bomber testing countermeasures devices.

improved-efficiency noise jammers

tail warning radar

higher-frequency jammers

weight are formidable problems. A major hurdle was overcome during the mid-1960s by developing sources of kilowatt dc power having a high specific weight and high specific volume.

To use active ECM properly, an adversary must have prior knowledge about the defensive radars so that ECM jammers can be tuned. In an aircraft or on a ship, there is more likely to be space and power for devices to sense hostile signals and tune transmitters. In a missile, where space and power are at a premium, this ability is far more costly.

One possibility is for the missile to eject at an appropriate time a collection of small solid-state jammers, each of which is tuned within different frequency bands, to spread confusion among enemy radars.

In general, however, the military relies for missile penetration more heavily on passive ECM techniques, such as chaff and inflated metallized balloons, which reinforce enemy-radar returns. Other methods, such as tethered chaff, which would be ejected from a missile system on atmospheric reentry and carried along near the vehicle to create large masking radar echoes, are to be tested. Regenerative chaff, a technique by which a missile or powered decoy would sequentially eject clouds of chaff, also is slated for evaluation.

The sinking of an Israeli destroyer by Soviet-made United Arab Republic cruise missiles in 1967 revealed an ECM inadequacy in Western naval vessels that prompted the resurgence in shipboard ECM activities. A primary element of defense against a cruise missile is the time needed to direct shipboard antiaircraft weapons against it. This element of time requires that the ship command know when a missile is directed against the ship and the direction of its flight. ECM signal detection and direction-finding techniques offer one solution to these needs.

The military events of the 1960s and early 1970s have established and reaffirmed the crucial role

played in modern warfare and in strategic and tactical defense by ECM.

[BARRY MILLER]

Bibliography: Electronic warfare gains key Viet role, *Aviat. Week Space Technol.*, 88:48–62, Jan. 1, 1968; H. F. Eustace and K. R. Schoniger, *The International Countermeasures Handbook*, 1975–1976; B. Gripstad (ed.), *Electronic Warfare*, Research Institute of Swedish National Defense, Stockholm; B. Miller, Major role in electronic air war earned by radar chaff, *Aviat. Week Space Technol.*, 88:54–62, Mar. 11, 1968; A. Price, *Instruments of Darkness*, 1967; Special report on electronic warfare, *Aviat. Week Space Technol.*, 102(4):40–144, Jan. 27, 1975; System aids RF spectrum surveillance, *Aviat. Week Space Technol.*, 86:61–63, Jan. 9, 1967.

Electronic display

An electronic component used to convert electric signals into visual imagery in real time suitable for direct interpretation by a human operator. It serves as the visual interface between human and machine. The visual imagery is processed, composed, and optimized for easy interpretation and minimum reading error. The electronic display is dynamic in that it presents information within a fraction of a second from the time received and continuously holds that information, using refresh or memory techniques, until new information is received. The image is created by electronically making a pattern from a visual contrast in brightness on the display surface without the aid of mechanical or moving parts.

The use of electronic displays for presentation of graphs, symbols, alphanumerics, and video pictures is doubling every several years. The biggest growth rate is for utilitarian and industrial users. Electronic displays are replacing traditional mechanical and hardcopy (paper) means for presenting information. This change is due to the increased use of computers, microprocessors, low-cost large-scale integration (LSI) electronics, and digital mass memories. The success of the handheld calculator is directly attributable to the availability of low-cost LSI electronics and low-cost electronic numeric displays. *See* CALCULATORS; COMPUTER; INTEGRATED CIRCUITS; MICROCOMPUTER; MICROPROCESSOR.

Electronic transducers and four-digit (or more) flat-panel displays are replacing the galvanometer movement, thermometer scale, barometer movement, and other forms of scientific instrumentation. Large signs, arrival and departure announcements, and scoreboards are using electronic means to portray changing messages and data. One of the major electronic displays applications is in home color television.

The computer terminal using a cathode-ray tube is one of the most important industrial applications of electronic displays. The standard computer terminal displays 25 lines of 80 characters, for a total of 2000 characters, which correspond to a typed page. The computer terminal with a microprocessor, minicomputer, and mass memory is beginning to replace the office paper, typewriter, and file cabinet.

Cathode-ray tube. The primary applications of the cathode-ray tube (CRT) are in home entertainment television, scientific and electrical engineering oscilloscopes, radar display, and alphanumeric and graphic electronic displays. *See* OSCILLO-SCOPE; RADAR; TELEVISION.

The basic elements of the cathode-ray tube are shown in Fig. 1. The viewing screen is coated with a phosphor which emits light when struck with a beam of high-energy electrons. The electrons are emitted from the cathode at the rear of the tube in a beam that is focused electrostatically (or magnetically) to a dot or spot on the phosphor screen and positioned in horizontal and vertical coordinates by magnetic (or electrostatic) forces. The cathode grids and electron-focusing lenses are incorporated into a subassembly called the gun. The beam is accelerated toward the phosphor by a high voltage (20 kV or more) at the anode grid conductor.

The cathode-ray-tube raster is shown in Fig. 2. Imagery is created on the raster as it is traced out. The video signal is high-frequency. It is applied, after amplification, directly to the gun cathode, and controls the amplitude of the electron beam and thus the luminous output at the display surface. The deflection coils steer the beam to trace out the raster. The horizontal-scan deflection signal causes the beam to trace out the horizontal lines and then fly back for the next line. The vertical-scan deflection signal causes the beam to be stepped down the raster and then retraced to the top left corner at the beginning of each frame. Two fields are typically interlaced where the raster lines are traced in the spaces of the other. This minimizes flicker in the picture. *See* CATHODE-RAY TUBE; PICTURE TUBE; TELEVISION RECEIVER.

Flat-panel displays. Because of the depth dimension of the cathode-ray tube, there has been a concentrated effort to invent a flat-panel display. A primary motivating factor has been to achieve a flat television receiver which could be hung on a wall. The electrical phenomena most extensively developed for flat-panel displays are gas discharge (plasma), electroluminescence, light-emitting diode, cathodoluminescence, and liquid crystallinity. The cost of flat-panel displays is higher than cathode-ray tubes on a per-character basis for higher information content displays such as the 2000-character computer terminal. The flat-panel technologies are utilized extensively in portable displays and numeric displays using several hundred characters and less. *See* CATHODOLU-MINESCENCE; ELECTROLUMINESCENCE; LIGHT-EMITTING DIODE; LIQUID CRYSTALS.

Flat-panel displays are typically matrix-addressed. A row is enabled to accept display information in parallel via the column lines. The electronics commutate through the rows, serving the same purpose as the vertical deflection amplifier of the cathode-ray tube. The column data are shifted into the column drivers, and then at the proper time applied to the column lines.

The flat-panel drive is shown in Fig. 3. The thickness is approximately 1.3 cm. The row and column lines are spaced at 24 lines per centimeter. The intersection of each row line with each column line defines a pixel. (A pixel or pel is a contraction of "picture element," and is used to denote the smallest addressible element in an electronic display.)

Display categories. Electronic displays can be categorized into four classifications, as shown in the table. Each classification is defined by natural technical boundaries and cost considerations. The categorization is useful in visualizing the extent to which electronic displays are used.

Special-purpose displays. The categorization of the table emphasizes direct-view-type electronic displays which are of primary interest to industry. There are other special-purpose displays used in very sophisticated applications.

Projection display. In the projection display an image is generated on a high-brightness cathode-ray tube or similar electronic image generator, and then optically projected onto a larger screen. To illuminate screens larger than approximately 1 × 1.3 m (3 × 4 ft) and in color, multiple cathode-ray tubes or light valves are used. The light valve is any direct-view display optimized for reflecting or transmitting the image, with an independent collimated light source for projection purposes. Light valves create images to control the reflection of light to be projected onto the screen. This permits powerful light sources such as xenon lamps to be used independent of the image-generating technique. Oil-film light valves and liquid-crystal light

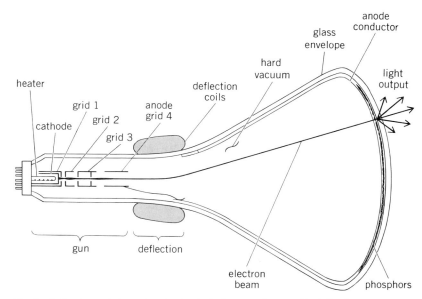

Fig. 1. Cathode-ray tube using electrostatic focus and magnetic deflection.

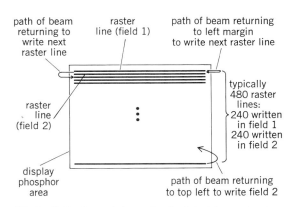

Fig. 2. Cathode-ray-tube raster for television, using interlace.

Electronic display spectrum of applications

Classification	Characteristics	Applications	Electronic technologies
Pseudoanalog	Dedicated arrangement of discrete pixels used to present analog or qualitative information	Meterlike presentations, go/no-go messages, legends and alerts, analoglike (watch) dial	Gas discharge, light-emitting diodes, liquid crystal, incandescent lamps
Alphanumeric	Dedicated alphanumeric pixel font of normally less than 480 characters; most common is 4- and 8-character numeric displays	Digital watches, calculators, digital multimeters, message terminals, games	Liquid crystal, light-emitting diodes, vacuum fluorescent, gas discharge, incandescent lamps
Vectorgraphic	Large orthogonal uniform array of pixels which are addressable at medium to high speeds; normally, monochromatic with no gray scale; may have memory; normally, over 480 characters and simple graphics	Computer terminals, TWX terminals, arrivals and departures, scheduling terminals, weather radar, air-traffic control, games	Cathode-ray tube, plasma panels, gas discharge, vacuum fluorescent, electroluminescence, other technologies in advanced development
Video	Large orthogonal array of pixels which are addressed at video rates (30 frames/s); monochromatic with gray scale or full color; standardized raster scan addressing interface, arrays of pixels approximately 512 rows by 512 columns	Entertainment television, graphic arts, earth resources, video repeater, medical electronics, aircraft flight instruments, computer terminals, command and control, games	Cathode-ray tube, other technologies in advanced development

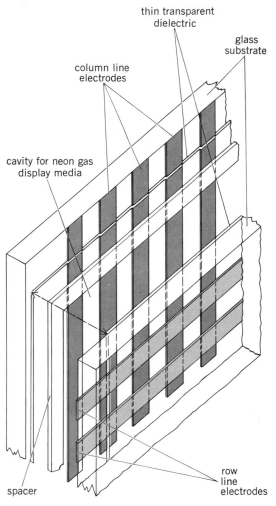

Fig. 3. Exploded section of a plasma flat-panel display.

Labels: thin transparent dielectric; glass substrate; column line electrodes; cavity for neon gas display media; spacer; row line electrodes

valves are examples of devices used in large command and control and theater-size electronic display presentations.

Three-dimensional imagery. True three-dimensional imagery can be created electronically by several techniques. One technique requires goggles using PLZT (lead zirconate titanate modified with lanthanum) electrooptical ceramic eyepieces over each eye. The eyepieces are electronically controllable shutters, with the ability to be reversibly switched from open to closed in microseconds. Two images from two television cameras placed to obtain the desired stereoscopic effects are electronically interlaced and displayed on one cathode-ray tube monitor. Each image is sequentially displayed from each camera at tv video rates. The goggles are synchronized to be opened and closed so that only the right eye sees the image from the right camera and the left eye sees the image from the left camera. The viewer sees true three-dimensional perspective while looking at the cathode-ray tube monitor through the goggles.

Helmet-mounted and heads-up displays. Helmet-mounted displays (sometimes called visually coupled displays) and heads-up displays are used in aircraft. In both of these displays the image is projected, usually from a cathode-ray tube onto a combining glass, and collimated to be in focus at infinity. The combining glass screen is designed to reflect the display imagery to the viewer, usually at selected wavelengths of light, while being sufficiently transmissive for the viewer to see the scene beyond. The primary application for the heads-up display is to present critical aircraft performance, such as speed and altitude, on a combining glass at the windscreen for pilot monitoring while permitting the pilot to look out the window for other aircraft or the runway. The primary application for the helmet-mounted display is to present, on a combining glass within the visor of the helmet of a helicopter gunner, primary information for directing firepower. The angular direction of the helmet is sensed and used to control weapons to point in the same direction in which the gunner is looking.

Color. Color can be created on a cathode-ray tube equipped with a shadow mask duplicating quite closely all colors that occur in nature. This is done in the cathode-ray tube by using three different electron guns and three phosphors in a triad of red, green, and blue on the screen at each pixel. The shadow mask is a metal screen with a hole for each pixel. It is located in the path of the beam between the deflection area and phosphor screen. Electron beams from each of three guns are constrained by each shadow mask hole to hit each respective phosphor dot. The gun, shadow-mask holes, and phosphor dots are aligned during manufacturing so that the three beams converge to pass through the single hole (or slit) in the shadow mask and then diverge as the beams emerge with sufficient separation to impact the three different phosphor dots. If all three guns are on simultaneously, the eye, upon close inspection, sees a red, a green, and a blue dot of light at each pixel. However, at a normal viewing distance, the three dots merge together in the retina of the eye, and from the laws of additive color, the pixel appears white. *See* COLOR TELEVISION.

Penetration phosphors are also used to create color on cathode-ray tube displays to eliminate the need for the shadow mask and extra guns. However, the color is limited and the brightness is low. Normally, two phosphors are placed on the screen in two layers or in microspheres of two layers. The gun and cathode-ray-tube anode are operated in two energy states to produce either a high-energy or a low-energy electron beam switchable in time. The high-energy beam penetrates the first phosphor layer and is stopped at the second layer. It then excites the second layer to produce its characteristic color. The low-energy beam is stopped by the first phosphor layer and excites it to produce its characteristic color. The two phosphor colors most often used are red and green. Intermediate-energy beams make it possible to fractionally excite both layers to get color combinations of red and green such as yellow and orange. Practical considerations limit this color approach to these four. Full color would require at least three primary colors such as red, green, and blue. Three layers of phosphors are limited in brightness due to practical considerations and have not been commercially available.

Monochromatic color is readily produced by flat-panel display technologies. Gas discharge is normally orange, owing to the neon gas. Other monochromatic colors are feasible. Light-emitting diode (LED) luminance is normally red, yellow, or green. A red or green state has been achieved which is switchable in a single diode by current control. Blue light-emitting diodes have been made in research devices. Electroluminescence is normally yellow or green, owing to the manganese or copper activator, respectively. Other monochromatic colors have been demonstrated in research devices. A wide range of color effects has been demonstrated in research devices with passive flat-panel displays such as liquid-crystal, electrophoretic, and electrochromic technologies. In flat-

panel as in cathode-ray-tube displays, full color is produced by using a three-color triad of red, green, and blue for each pixel.

Full color is very important for entertainment television displays. Most industrial electronic displays do not need or use full color. The most efficient monochrome from each display technology is normally used; for example, red for light-emitting diode, orange for gas discharge, and orange-yellow for electroluminescence. Limited color displays are sometimes used in special industrial applications such as aircraft weather radar and artificial horizons, computer-aided multilayer circuit design, earth resources studies, air-route traffic control, and medical electronic displays. In all of these applications, the display instrument is usually a cathode-ray tube using the shadow-mask color technique.

Display technique. The essence of electronic displays is based upon the ability to turn on and off individual picture elements (pixels; Fig. 4). The pixel is the smallest controllable element of the display. A typical high-information-content display will have a quarter million pixels in an orthogonal array, each under individual control by the electronics. The pixel resolution is normally just at or below the resolving power of the eye. Thus, a good-quality picture can be created from a pattern of activated pixels.

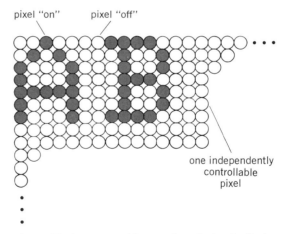

Fig. 4. Pixel array used for creating electronic display images.

The pixel concept for electronic displays has evolved from the modern flat-panel display technologies and digital electronics. It has been extended to the analog-raster-scan cathode-ray tube in the following way: The electron beam from the gun is deflected magnetically (or electrostatically), so as to sweep across the phosphor and thereby cause a line to luminesce on the face of the cathode-ray tube. In digitally modulated cathode-ray tubes, the cathode is modulated by a sine wave as the beam is swept across the face of the cathode-ray tube. Here, instead of a continuous line, a string of dots results. Each dot corresponds to a pixel. Each pixel is on when the beam density is high, and off when the beam is off. The beam is turned off between each pixel. The pixels are re-

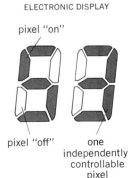

Fig. 5. Seven-bar numeric font.

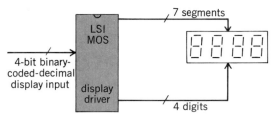

Fig. 6. Block diagram of LSI MOS electronic drive for numeric display.

freshed approximately 30 times per second on a cathode-ray tube.

Pixels are created in all the rows of the entire cathode-ray-tube raster in what is called a CRT-digital raster. This approach is commonly used in industrial applications and computer terminals, since it is easily interfaced with digital electronics. Home entertainment television uses an analog-raster-scan approach, as do nearly all video systems.

There are some applications in which a nonraster approach is used to create alphanumeric characters and vectors on cathode-ray tubes. The electron beam is deflected under control of the deflection amplifiers to stroke out each line of the image. When characters and vectors are generated this way, they are like Lissajous figures, as opposed to (digital or analog) raster characters and vectors. The Lissajous characters and vectors are best suited to large (25-in. or 63.5-cm diagonal) cathode-ray tubes and where there are numerous vectors, straight lines, and curves. Vectors and curves drawn with the raster technique have stair steps. Lissajous vectors and curves are always smooth and continuous. Lissajous techniques are yielding to the digital raster in newer designs as the cost of digital electronics improves.

Font. With flat-panel displays and cathode-ray-tube digital-raster displays, alphanumeric character fonts are created by turning on the appropriate pixels in an array. One standard size is a 5 × 7 array with two pixels between characters and two pixels between rows, as shown in Fig. 4. All the letters and numbers can be created on this common array format. Several other combinations of pixels may be used to create the letter A. The viewer soon becomes accustomed to the minor variation. Readers do not read pixels but read letters and words, and therefore the exact detail of the character pixel pattern is of a secondary consideration. In general, the more pixels available in the basic array, the more esthetically pleasing is the character, at the cost of additional electronics to control the extra pixels. The viewing distance is normally far enough so that the pixels blur together.

A very efficient and elegant array has evolved for portraying numeric characters only, called the seven-bar font (Fig. 5). Each bar is a pixel by definition. This font was initially considered crude when compared to the Leroy font and other more esthetic printer fonts. It is now universally accepted. A similar 14-bar font is sometimes used for alphanumeric characters.

Display electronic addressing. The numeric or alphanumeric display electronic drive may be performed in a single LSI metal oxide semiconductor (MOS) chip mounted in a single dual-in-line package suitable for direct assembly on a printed circuit board. All the timing, logic, memory, resistors, and drivers are contained in the single chip. A four- or seven-bit character code is serially fed into the chip for display, as shown in Fig. 6.

A smart computer terminal (Fig. 7) will incorporate a microprocessor unit and a cathode-ray-tube controller to perform master control and cathode-ray-tube housekeeping tasks. A line of video data is loaded into the shift register for serial drive of the video amplifier at the time a raster line starts. The shift register is loaded during flyback time from the character generator. The character generator is a decoder transforming the alphanumeric information in coded form such as ASCII (American National Standard Code for Information Interchange) into pixel signals for each raster line. The ASCII code is stored in the display refresh memory. The entire cathode-ray-tube frame is stored in this memory, and is continuously displayed until changed under control of the microprocessor unit. New information can come from

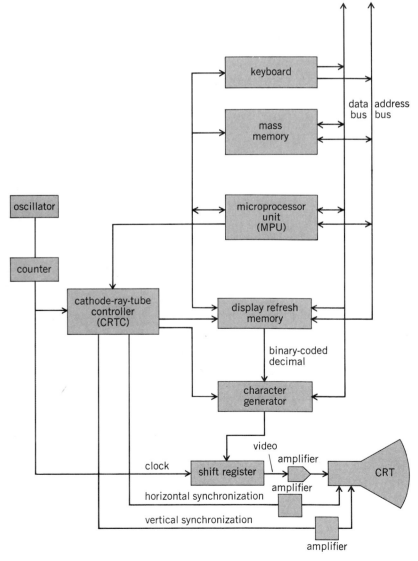

Fig. 7. Block diagram of LSI electronics for a smart computer terminal cathode-ray-tube display.

the mass memory, the keyboard, or other subsystems on the data bus and address bus, all under control of the microprocessor unit.

The cathode-ray-tube controller performs all the housekeeping tasks for proper cathode-ray-tube operation. These include the vertical- and horizontal-raster synchronization signals, the cursor control, blinking, blanking, interlace, paging, and scrolling. The cathode-ray-tube controller is a single LSI MOS chip, as are most of the other blocks shown in Fig. 7.   [LAWRENCE E. TANNAS, JR.]

Bibliography: L. E. Tannas, Jr., *Flat-Panel Displays*, 1981.

Electronic equipment grounding

The connecting of electronic equipment to an electromagnetic reference common to itself, its power source, its environment, and the environment of its users. Electronic equipment is grounded to protect users from shock, to protect the equipment from spurious currents or voltages, and especially to isolate it from the noise level that contaminates its environment.

Need for thorough grounding. Usually electronic equipment is powered from the electric service that supplies the building where the equipment is used. To provide security against hazards of electric shock or fire ignition, equipment frames and power-conductor enclosures are connected and grounded. *See* GROUNDING.

In addition, grounding (together with the shielding, isolation, compensation, and equalization) is applied to minimize the entrance of extraneous signals (noise) into the equipment. Basically, electronic equipment can be treated as a sensor or signal source, a signal circuit, and central equipment (Fig. 1). The sensor may be any of a wide variety of input devices, such as a tape reader, card reader, memory readout, medical sensor, industrial transducer, or microphone. Information from the sensor travels along the signal circuit in digital or analog form at low level. At the central equipment the signal is processed to be used for such purposes as recording or control. Reliability of the output use depends directly on the integrity of the signal received at the central equipment.

The sophistication of electronic equipment has advanced tremendously. The necessary intensity of electrical signals has been diminished generally to below 1 volt. The amount of time allocated for the transmission of one bit of information has been progressively diminished to less than 1 μsec. These advances aggravate the likelihood that spurious error signals will be of sufficient magnitude to create an error response. The effect of an error response, if not identified and corrected for, can produce tragic results in critical areas such as manned space flight. The controlling parameter can be expressed as the signal-to-noise ratio. The problem is ultimately one of ensuring acceptably high levels of signal-to-noise ratio. *See* ELECTRICAL NOISE.

The most vulnerable circuits are typically those associated with the information-gathering function, because they generally operate at low signal level and are followed by high-gain amplifiers within the data-processing system. Such an input circuit is shown in Fig. 1. The signal-sensing element may be in the next room, on a different floor of the same building, in a different building, or even in a

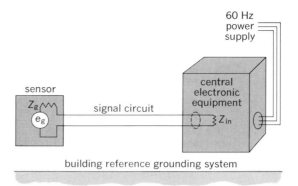

Fig. 1.   The three essential components for communicating information electrically.

different city. Techniques that are effective in establishing a high signal-to-noise ratio in this critical circuit, and techniques needed to handle the less critical circuits, are reviewed here.

Noise coupling into signal circuit. Spurious unwanted electric impulses (noise) may be mixed with true information signals in numerous ways, the most common and important being electromagnetic induction, electrostatic induction, and conductive coupling (Fig. 2).

Magnetic induction. A changing magnetic field

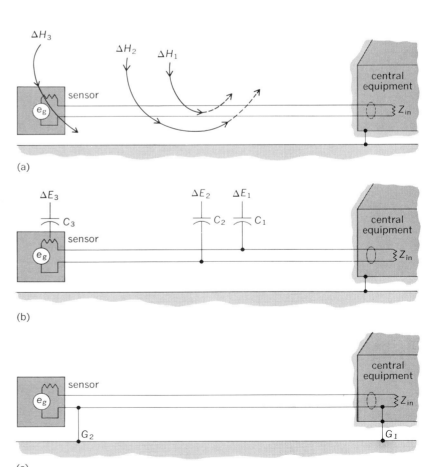

Fig. 2.   Common sources of error signals. (a) Electromagnetic coupling. (b) Electrostatic coupling. (c) Conductive coupling.

intensity in the space through which the signal circuits run (Fig. 2a) creates noise in the equipment. Such fields are present in the space surrounding any open conductor carrying substantial magnitudes of changing current. Intense fields exist around the anode leads of arc furnaces and the throat conductors of flash and spot welders. A helpful approach is to recognize that the inductive reactance of a current-carrying conductor, shown on the electrical engineer's one-line diagram as a property of the conductor, is in fact a property of the space-distributed magnetic field which surrounds the current-carrying conductor. Any electrical circuit which links a portion of such a space field will display a fractional part of the reactance voltage drop of the power conductor.

In Fig. 2a the fractional part of the changing space-distributed magnetic field ΔH_1 which threads between the two signal-circuit conductors will induce a normal-mode error signal in that circuit. The terminal sensor is often an electromagnetic device which, in the presence of a changing space-distributed magnetic field ΔH_3 in a particular direction, will display a normal-mode induced error voltage. A similar space field ΔH_2 threading between the signal circuit conductors and the nearby conducting structure which connects the central processing framework with the remote terminal structure will induce a common-mode potential which can appear on the signal circuit.

Electrostatic induction. A changing electrostatic space field (Fig. 2b) can induce a similar error voltage in the signal circuit by capacitive coupling. Electrostatic space-distributed fields will be created by unshielded, energized electrical conductors. Insulation over the conductor is not an attenuator. Intense fields can be expected in areas where ac high-potential testing is being done. Unshielded gaseous discharge lamps develop annoying space fields rich in harmonics and sharp step changes.

A changing-magnitude electrostatic field ΔE_1, if coupled to one of the signal-circuit conductors through capacitance C_1, will impart to that conductor an error voltage. To the other signal conductor will be communicated a similar error voltage produced by space field ΔE_2 through coupling capacitance C_2. Even though ΔE_1 equals ΔE_2 and C_1 equals C_2, a normal-mode error signal voltage will be created if the signal circuit is not symmetrically balanced in its coupling to the central-equipment reference ground plane. Also, the terminal sensor may contain such parts as terminals, coils, and winding surfaces that are capacitively coupled to the ambient space field ΔE_3. Unless coupling capacitance C_3 displays a symmetrical balance to the two signal conductors, a portion of the induced error voltage will be of the normal-mode type. Even an intentional grounding connection on the signal circuit at the central equipment, if not precisely balanced, will cause a fractional part of the common-mode, electrostatically induced voltage to appear as a normal-mode type.

The presence of large-magnitude, common-mode voltages on signal circuits may directly damage or even destroy electronic components and may create troublesome normal-mode error signals if only slight deviations from perfect symmetry occur on signal circuits.

Conduction between units. Normal-mode error voltages are also created by conductive coupling (Fig. 2c). For the most part this origin of noise is identical with that commonly referred to as ground loop problems. The error voltage owes its origin to the fact that the reference grounding conductor system is not at a common potential throughout. Consider that one signal-circuit conductor is grounded at the central equipment in location G_1. This would be conventional practice when using coaxial signal circuits and is sometimes used on two-conductor wire circuits. The presence of a second ground connection on the same signal conductor at a different location G_2 will likely result in severe noise in low-level signal circuits. After making the first ground connection at G_1, it will be unlikely that another spot can be found along that conductor with a zero-voltage difference to the adjacent reference ground. When a second ground connection is made along a signal conductor, it forces the voltage difference, which had previously existed here, to vanish and appear as an impedance voltage drop (IZ drop) around the loop circuit formed by the second ground connection. *See* ELECTRICAL IMPEDANCE.

By reviewing the character of the impedance elements forming this loop, it becomes quite evident that the signal-circuit conductor between grounding points G_1 and G_2 will principally account for the ground loop impedance. The remainder of the loop will be made up of such low-resistance conductors as heavy building structural members and large pipes. Thus most of the voltage difference which, prior to the second ground connection, had existed between the signal conductor and the adjacent reference ground G_2 has, after the second ground connection, become an impedance voltage drop along only one signal conductor between locations G_1 and G_2. This voltage represents a normal-mode error voltage in the signal circuit.

Conduction within one unit. At times the separation of only inches in the two ground connections can result in intolerable noise, as the following example illustrates. In a vacuum-tube amplifier (Fig. 3) the cathode and one heater lead are bonded to the chassis at point 1. One side of the filament transformer is bonded to the chassis at point 3; therefore, heater current flows through the chassis, as shown by the current flow lines. Normal to the current flow lines are equipotential lines, as shown. The grid resistor is bonded to the chassis at point 2. A noise signal e_{12} is thereby inserted into the grid circuit, resulting from the potential drop along the chassis between cathode and grid bonding points. This drop may amount to only a few microvolts, but if it enters an early stage of a high-gain amplifier, excessive noise may be noticed in the output.

To avoid such problems, all bonding connections for a given stage should be made to only one point on the chassis. Also, the flow of large currents through the chassis of any high-gain amplifier, whether vacuum-tube or solid-state, must be avoided. Where high-gain amplifiers are controlling high-power ac or dc circuits having grounded terminals, it is best to insulate the amplifier chassis from the frames of the rest

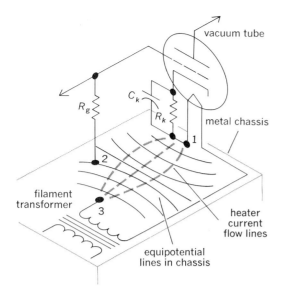

Fig. 3. Improper grounding of an individual amplifier stage to a metal chassis, so that the noise voltage is coupled into the signal circuit.

of the machinery and to provide a single ground strap from chassis to frame. This will prevent large currents from passing through the amplifier chassis. *See* CIRCUIT (ELECTRONICS).

In general, each stage of a high-gain amplifier should have its own chassis ground. Each individual amplifier should have its own ground lead to a common amplifier ground point. All machinery and other power devices should be bonded together, with a single lead to the amplifier ground point. Finally, if additional shielding is provided, such as a metal cabinet, it should have a single ground lead to the common ground point for the amplifier and the power equipment. This common ground point would then be grounded to earth by bonding to a cold-water pipe or by other methods in conformance with the local electrical codes.

Ground gradients. The medium- to high-voltage substation may display large voltage differences between the potential on the local station grounding conductors and remote earth. Information-gathering (or sensing) circuits associated with electronic equipment extending into or close to such a substation area will generally require extreme design care to avoid dangerous common-mode potentials. It is not unreasonable to expect steady-state voltage levels as much as 50 volts. During a line-to-ground fault on an out-going overhead line, the potential of the substation ground mat might be elevated to several thousand volts, relative to remote mean earth potential by voltage gradients. Common-mode voltages from such causes can be compensated by insulating transformers or equalizing transformers or by other means of bridging large-magnitude, common-mode voltages.

The ordinary commercial building may present unexpected problems relative to common-mode voltage components due to voltage gradients in the building grounding conductor system. It is widely believed that the grounding requirements as spelled out in section NEC 250–23(a) of the Na-

tional Electrical Code ensure that all electrical load current within the building will be returned to the service equipment (the point of electric service entrance) on power conductors, independent of the building structure.

An innocent-appearing exception can violate this concept. Section NEC 250–60 defines a grounding exception (rather generally used) for the frames of electric ranges, wall-mounted ovens, and counter-mounted cooking units. The code exception allows these appliance frames to be grounded by connection to the power system grounded conductor (white wire) if the electric service is a 240/120-volt, single-phase, three-wire system or is taken from a 208Y/120-volt, three-phase, four-wire system. It is unrealistic to assume that the appliance frames in question will not also be in contact with the building structure. The result is that the white load-current-carrying conductor becomes connected to the building frame through the heating appliance.

Thus a building devoted essentially to commercial activities, served with 208Y/120-volt electric power, may have a snack bar installed on the sixth floor. Contained therein may be appliances grounded to the white wire as allowed by NEC 250–60. Instead of an electrically dry building frame, it may be found that between the sixth floor and the ground floor there is distributed the same electrical voltage drop as exists on the grounded power-system conductor between the snack bar connection and the service entrance.

When the service is three-phase and four-wire, all third-harmonic currents (and their multiples) in the entire array of line-to-neutral connected loads combine in an additive fashion in the neutral conductor (white wire) to aggravate the harmonic voltage drop along the white wire. In both the three-phase and single-phase power-supply cases, the increasing use of time-modulated (SCR-controlled) current in line-to-neutral connected devices (such as fans and lighting units) makes for the presence of much step-front hash in the voltage drop along the neutral conductor, which becomes voltage gradient in the building structure if the NEC 250–60 exception is employed. *See* SEMICONDUCTOR RECTIFIER.

Efforts are being made to restrict the use of the NEC 250–60 exception to appliance circuits which originate as branch circuits at the service equipment. In the meantime it must be recognized that the ground reference potential on one floor of an office building may differ from that of another floor by as much as several volts (third harmonic) and contain substantial fast-front hash.

Section NEC 250-23 of the National Electrical Code clearly prescribes that there shall be no grounding connection to the grounded power conductor of the electrical system downstream of the service equipment for that establishment. When this rule is respected, all of the load-system power current is returned to the grounded conductor at the service via insulated conductors. There is no opportunity for electrical noise, as it exists on the grounded power conductor (the white wire) to be conductively transferred to the system grounding conductors. There are three exceptions in the NEC text which may permit that rule to be bypassed. The first one applies to an in-plant separately derived electrical system located remote from the service equipment of the establishment. It is common to apply a permanent grounding connection at the supply-machine neutral. This will create a cross bond between the grounded conductor (white wire) and the grounding conductor downstream of the main service equipment. The second one applies to the case of an electric-supply circuit run to a second independent building. The NEC requires a cross bond between the grounded and the grounding conductor at the point of entry to the second building, unless an independent grounding conductor has been included with the power conductors from the supply point in the main building. The third exception, pertaining to ranges, counter-mounted cooking units, wall-mounted ovens, and clothes dryers, can be troublesome. While the National Electrical Code says merely that the power-circuit grounded conductor (white wire) may be used as the grounding conductor, it is commonly found that the appliance frames so grounded by connection to the white wire are also in contact with building metal frame or metal piping systems. It is this "back door" connection which allows communication of the electrical noise to the establishment grounding conductor system.

Progress is being made in restricting the use of the white wire as a grounding conductor.

A few instances of a cross-bond between the grounded power conductor and the equipment grounding conductor downstream of the service equipment have been observed as a result of an in-plant emergency or standby generator. These have been associated with 208Y/120-V and 480Y/277-V solidly grounded power systems serving hospitals, telephone exchanges, police centers, and other such important establishments. When operating on an emergency or standby basis, it is proper that a grounding connection exist at the local generator as prescribed for a separately derived electric power system in section NEC 250-26 of the National Electrical Code. It is common to find an intentional permanent grounding connection at the local-generator neutral junction. Unless the neutral line of the supplied load is switched, along with the phase conductors, the result is a permanent grounding connection on the "white wire" at the generator location, contrary to the planned design pattern. Location of the in-plant generator adjacent to the service equipment would eliminate the problem, as would also inclusion of the neutral conductor in the transfer switching operation.

There will occasionally occur brief intervals of much greater than normal voltage gradients along the building grounding conductor system. An insulation failure on a power conductor may permit a large-magnitude, ground-short-circuit current to flow through the building structure toward the service equipment location for the brief interval permitted by the overcurrent protector.

Importance of low-noise environment. The noise problems presented to the electronic equipment designer will increase directly with increased ambient noise levels in the area. The quality level employed in grounding and shielding of power-system conductors will have a marked effect on the ambient noise levels.

In an occupied building much can be learned of the general ambient noise levels by a study of the

electric facility practices employed and by a general inspection of the building. Additional specific information may be obtained by test. It is important to look for unusual sources of high-energy radiation which may penetrate the building interior space. A nearby radio station antenna may create strong fields at one fixed frequency. High-power radar-scanning installations may be a source of annoying ambient fields of pulsed high frequency.

After an appraisal of the severity of the ambient electrical noise level, the problem of designing the grounding and shielding practices for the electronic equipment becomes a straightforward procedure.

Grounding and shielding practices. To establish an acceptably high signal-to-noise ratio demands that normal-mode noise be maintained at levels below a specified tolerable value.

Choice of conductor. The construction geometry of the signal conductor (Fig. 4) plays a prominent part. Of the many possible varieties, the four most commonly used are random-selected wires in a multiconductor cable, parallel-wire twin-lead, twisted-pair, and coaxial lines. Because of the superiority of the twisted-pair relative to random wires or twin-lead at modest cost increase, its use is universal when noise reduction is a factor. A half twist every inch is both typical and effective.

An encircling metal raceway accomplishes substantial additional attenuation, particularly if it consists of steel conduit. In securing high-quality rejection of capacitively coupled noise voltages, it is common practice to avoid any grounding connection on the signal conductor (coaxial lines excepted) and to incorporate a balanced impedance connection to the central-equipment reference ground terminal in the input circuits of a differential amplifier.

The coaxial line (Fig. 4d) is a theoretically superior construction but may be faced with operating conditions which deteriorate these qualities. As it is inherently an unbalanced line, the exterior shell is commonly connected to the reference ground terminal at the central equipment. The existence of any connection to adjacent ground along the coaxial line (not easy to avoid) creates the ground loop situation described in Fig. 2c with a likely inevitable insertion of normal-mode error signal (noise). Even in the absence of a second ground connection, large-magnitude varying electric fields (Fig. 2b) can create measurable values of charging current flowing along the outer shell to the grounding connection at the central equipment, and in so doing can create an *IR* voltage drop along that conductor which will appear as a normal-mode noise voltage. An enclosing metal raceway grounded to the central-equipment ground reference terminal can be used to effectively suppress this noise source. *See* IR DROP.

Conductor shielding. A multiplicity of signal circuits may be run within a common metal raceway, providing that the signal magnitude on these conductors is not great enough to create cross-coupled interference (cross talk). Of the various conductor geometries considered in Fig. 4, the random wire case in *a* would be most susceptible and would offer little hope for correction. With the conductor patterns in *b* and *c* the application of wrapped-on

ELECTRONIC
EQUIPMENT GROUNDING

(a)

(b)

(c)

(d)

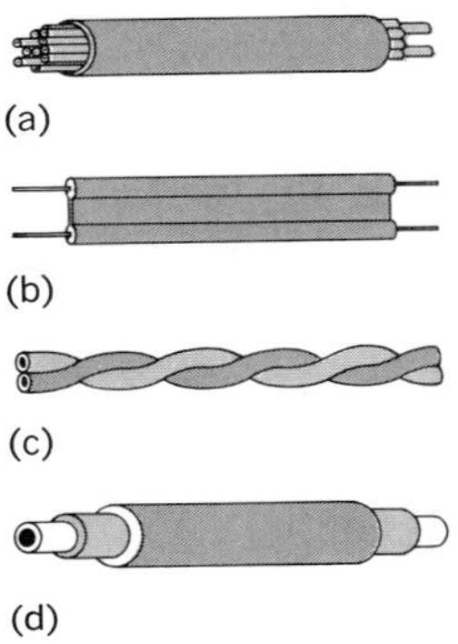

Fig. 4. Common signal-circuit geometries. (a) Two random wires. (b) Straight twin-lead. (c) Twisted pair. (d) Coaxial line.

shielding tape (or equivalent concentric conducting shell) to each signal circuit, grounded at the central-equipment grounding terminal, would only accomplish high-quality cross-talk suppression. The coaxial line is inherently free of cross-talk interference if the outer conductor is grounded only at the central equipment.

It is possible (and not unlikely) that the outer terminal sensor may itself respond to ambient magnetic field noise (Fig. 2a) or electrostatic field noise (Fig. 2b) or both, and create normal-mode, signal-circuit noise. The magnetic field problem can be resolved by the application of appropriate magnetic shielding at the sensor, the enclosure of the complete sensor within a closed shell of conductive metal, or a combination of the two. The electrostatic problem can be resolved by enclosing the sensor winding and its circuit leads within a thin-wall shielding shell maintained at zero reference ground potential. The enclosing shell can be the same one used for magnetic induction suppression. The required connection to a zero potential reference point may require a unique grounding circuit extending from the central equipment (Fig. 5).

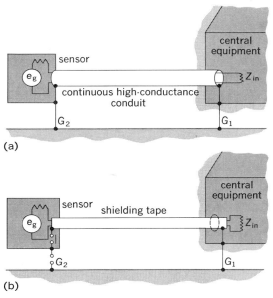

Fig. 5. Practical grounding techniques. (a) High-conductance raceway. (b) Electrostatic shielding.

Control of grounds. The remaining perplexing noise abatement problem relates to the creation of reference grounding terminals at peripheral equipment locations which appear to be at the same potential as the central-equipment grounding terminal; an example is the electrostatic suppression problem discussed in the previous paragraph. An earlier discussion (Fig. 2c) develops the fact that between a building ground-reference point G_2 and a widely separated central-equipment grounding terminal G_1 there will, more often than not, appear a difference in electrical potential. The difference voltage can create normal-mode, error-signal voltages in the several ways described.

If the remote terminal G_2 is not in the same building as is terminal G_1, or if it is located in an area where large common-mode noise voltages are expected, the use of insulating or equalizing transformers in the signal lines or the adoption of other common-mode suppression techniques should be considered.

If the remote equipment is in the same metal-frame building or in a location where common-mode voltage difference between reference ground terminals is known not to exceed perhaps 10 volts, acceptable attenuation should be possible without the introduction of isolation techniques on the signal wires themselves.

The installation of a grounding conductor, run with the signal circuits, which interconnects grounding terminal G_2 with G_1 will be helpful but not as effective as might be expected. A substantial fraction of the original noise voltage will remain in the form of an IZ drop along the grounding conductor.

High-conductance raceway. By modifying the form of this grounding conductor, a tremendous gain in attenuation can be achieved. By forming the conductive metal of the grounding conductor into a hollow tube through which the signal conductors are passed, its effectiveness is increased enormously. The resulting tubular grounding conductor, called a high-conductance raceway (Fig. 5a), interconnects grounding terminals G_1 and G_2 as did the cable grounding conductor. Voltage difference $e_{G1} - e_{G2}$ appears as an IZ drop along the raceway. The conductance (tube diameter, wall thickness, and type of metal) is selected so that only a small portion of the total IZ drop appears as a resistance drop along the metal tube. Almost the entire voltage difference between grounding terminals G_1 and G_2 is accounted for by a reactive-voltage drop IX created by an induced magnetic field encircling the tubular raceway. This magnetic field also links each signal conductor contained within the raceway and thus induces in each one identically the same IX voltage drop as in the raceway. With the IX voltage component equalized, or canceled out, only the IR drop along the high-conductance raceway appears as a voltage difference between signal conductors and adjacent ground at G_2.

It is important to note that this cancellation of the IX drop occurs only on those signal conductors run within the high-conductance raceway. If some critical signal circuits must take a different route, an independent high-conductance raceway to contain them will be needed to accomplish the desired high-magnitude attenuation.

The high-conductance raceway technique is applied extensively to interconnect the several independent housings making up the central electronics equipment to establish a nearly quiet ground reference potential throughout the central equipment. One prominent item of the central equipment is designated as the reference unit and is bonded to the building reference ground, preferably at only one point. All other equipment frames in the group would, by design intent, permit conductive connection only to the central reference structure in the form of high-conductance raceways. Power conductors and other high-noise-level conductors should be run in raceways independent of the sensitive signal circuits.

In extremely severe situations where the IR

drop in the high-conductance raceway creates intolerable noise levels, a dual concentric system may be substituted. The exterior high-conductance raceway is identical in design and function with the one just described. Within this is a second metallic tubular raceway, insulated from the former except for an intentional cross bond at the central equipment grounding terminal G_1. At the remote equipment locations, the construction consists of an outer enclosure bonded to the building ground terminal G_2. The electronic equipment chassis containing the active electronics components is insulated from the outer housing and bonded only to the internal concentric raceway. By this modification even the resistance voltage drop along the outer tubular raceway is prevented from appearing as a voltage difference between signal conductors and their enclosing raceway.

Should the remote sensor terminal be small and of simple character, with possible noise being essentially of electrostatic coupling origin, the simple circuit pattern illustrated in Fig. 5b may prove adequate. A lightweight shielding tape enclosing the signal conductors preserves a zero-potential reference plane surrounding the signal conductors. The sensor frame (or enclosing electrostatic shield) is connected to the signal-circuit shielding tape and kept insulated from the local grounding terminal G_2. The presence of high-intensity electric field noise at the outer terminal may demand a higher-conductance signal-circuit shield to avoid objectionable noise voltages of IR drop origin.

Grounding, electromagnetic shielding (and cancellation), electrostatic shielding, and insulation are intimately intertwined in effecting an acceptable solution.

Data transmission. To an ever-increasing extent the high-density channels of information handling (data transmission) are being accommodated by means other than metallic conductors. The prominent transmission avenues are microwave links and optic fibers. The problems of circuit grounding along these links are simply nonexistent. The interconnecting wire circuits between the distributed sensors and the central processing center, and between the various functional systems at the center, and the transmission to the output terminal at that location will face the same signal-to-noise ratio problem as in the past. *See* Microwave; Optical communications.

Optical isolator. In dealing with the aggravated problems associated with the large magnitudes of common-mode noise voltage encountered with circuits run to an outdoor electric substation, the optical isolator is a valuable tool. This device introduces a short optical transmission path into the electric signal circuit with a capability of withstanding several thousand volts of common-mode voltage without ill effects. *See* Optical isolator.

[R. H. Kaufmann]

Bibliography: W. W. Brown et al., System and circuit considerations for integrated industrial fiber optic data links, *IEEE Trans. Commun.*, Com-26(7):976–982, 1978; B. F. Burch, Jr., Protection of computers against transients, interruptions, and outages, *IEEE IGA Trans.*, 34C62, p. 471, 1966; Communications and microwave technology, *IEEE Spectrum*, p. 38, January 1979; R. E. Goers, Quiet-wiring zone, *IEEE IGA Trans.*, 68C27-IGA,

4:249, 1968; *Guide for Safety in Alternating Current Substation Grounding*, IEEE Stand. no. 80, 1976; Institute of Electrical and Electronics Engineers, The isolation concept for protection of wire line facilities entering electric power stations, *IEEE Trans. Power Apparatus Syst.*, PAS-95(4):1216, 1976; Institute of Electrical and Electronics Engineers, The neutralizing transformer concept for protection of wire line facilities entering electric power stations, *IEEE Trans. Power Apparatus Syst.*, PAS-96(4)1256, 1977; Reduction of electrical noise in computers and control installations in steel mills, *IEEE IGA Trans.*, 34C36, p. 667, 1966; A. Wavre, Application of integrated circuits to industrial control systems with high noise environments, *IEEE IGA Trans.*, 68C27-IGA, 4:255, 1968.

Electronic listening devices

Devices which are used to capture the sound waves of conversation originating in an ostensibly private setting in a form, usually as a magnetic tape recording, which can be used against the target by adverse interests.

There are two kinds of electronic listening devices. One takes advantage of equipment already present on the target's premises, such as a telephone, radio, phonograph, television set, public-address loudspeaker, or tape recorder, to act as a microphone, transmitter, or power supply. The other does not. In the former case, the target's equipment is said to have been compromised.

These practices are unlawful in the United States and Canada except when carried out by law enforcement officers acting under authority of a warrant. In the United Kingdom, electronic eavesdropping by private parties may contravene the laws against trespass and unauthorized use of telephone equipment or radio frequencies, but it is not unlawful in and of itself. The same is generally true in western European countries.

Compromise. Compromise of the target's own equipment takes advantage of the fact that any loudspeaker is capable of functioning just as well as a microphone, that convenient sources of dc power are available within the equipment, or that the equipment is connected to power or signal lines that can transmit the intercepted conversation to some place where recording can conveniently be accomplished.

The equipment most frequently compromised is the telephone handset. The act of compromise may be as simple as bypassing the switch hook with a Zener diode so that the instrument can transmit conversations to a wiretap, or high-impedance parallel connection off the premises, while not signaling a busy tone to someone attempting to dial in. *See* Telephone; Zener diode.

It can be as complex as the infinity transmitter, which is activated by the eavesdropper's dialing in and transmitting, before the telephone rings, an audio tone to a tuned relay concealed in the instrument.

Bugs. Eavesdropping devices that can stand alone are known commonly as "bugs." They take advantage of many developments of modern technology, such as microcircuits, miniature ceramic microphones, and miniature batteries. *See* Integrated circuits; Microphone.

The art of designing bugs has achieved its most

advanced state of development in the national intelligence services of the Great Powers. One bug was discovered to have been inserted in the heel of a diplomat's shoe.

The smallest bug available to the private citizen is probably the Hong Kong "spider." It is no larger than a common postage stamp and less than a quarter-inch (6 mm) thick. It is sold widely in the boutiques of tourist hotels in Victoria and Kowloon.

Electronically a bug is often just a two-stage frequency-modulated transmitter: an audio amplifier and a variable-frequency radio-frequency (rf) oscillator. See AMPLIFIER; OSCILLATOR; RADIO TRANSMITTER.

Bugs may operate on any frequency from 20 to 1000 MHz, but usually they snuggle up beside a powerful local FM or vhf television station.

A popular hybrid between a compromise device and a bug is the telephone drop-in. In this design, an FM transmitter is made in the form of a telephone microphone. The eavesdropper can casually unscrew the mouthpiece of his target's telephone handset and substitute the drop-in for the original microphone. The range of this device is about 75 m. It has the added advantage of drawing its dc power from the telephone company central battery.

Defenses. The telephone line analyzer is used to defeat compromise devices. Under the control of a microcomputer, the line analyzer examines in turn each of the 50 or more telephone lines used by a typical commercial firm. It can detect the tiny amount of current flow into a Zener or other compromise device. A monotonically rising audio signal is then applied to trigger any infinity transmitter. If a compromise device is discovered, a pulse of 800 V is impressed upon the line to burn it out.

Technical surveillance sweeps, as they are known to the trade, are performed alternately with external signal lines connected and disconnected and with telephones on-hook and off-hook.

Often a sniffer is used to sense the presence of a bug. A sniffer may be a simple vhf diode detector with an output meter that reveals the presence of a local carrier. More sophisticated instruments sweep the frequency bands of interest and compare the audio output with locally generated signals, usually a tape recording of simulated business activity. The newer type of sniffer affords discrimination between a genuine clandestine device and a strong local broadcasting station. See DETECTOR.

If the presence of a clandestine device is indicated, the sweep team may employ a panoramic intercept receiver. This instrument displays the amplitude of received signals on a cathode-ray-tube trace calibrated in units of frequency. The frequency of the bug can be determined by observing which of the pips on the CRT trace appears to diminish and grow in direct response to the locally generated audio signal. Once its frequency is determined, the physical location of the bug can be found by use of a tuned rf field-strength meter having a loop antenna. See ANTENNA; CATHODE-RAY TUBE.

The objective of a sweep such as the one described is to locate the device and discover its technical characteristics with a view toward feeding the device false information that may discredit or compromise the individuals responsible for planting it.

National security forces increasingly are encountering clandestine devices which are switched on and off by an agent using a control transmitter. Such a device cannot be detected by the methods described above because they presuppose that the bug is continually on the air or can be activated by locally generated audio.

To detect silent bugs, advantage is taken of the fact that all subminiature transmitters contain one or more semiconductor junctions connected to an electromagnetic radiator. Inasmuch as these junctions are nonlinear impedances, a locally generated low-power ultra-high-frequency carrier will be reradiated by the bug, and the reradiated signal will contain strong harmonic components.

Natural devices such as rusty bedsprings will reradiate second-harmonic components, but only a real semiconductor junction will reradiate third (and higher) harmonics.

A uhf sweeper is equipped with a meter that produces a negative deflection in the presence of reradiated second-harmonic energy and a positive deflection in the presence of reradiated third-harmonic energy. Such a deflection discloses the presence of a bug, albeit a silent one not detectable by sniffers or panoramic receivers. The uhf sweeper is portable; it bears a resemblance to a household appliance for vacuuming drapes. Increasing positive deflection indicates to the sweep team member that he is approaching the bug. See HARMONIC ANALYZER.

Advanced devices. Human capacity for designing electronic instruments to spy on one's neighbors appears to be boundless. Some of the more exotic include laser radar which responds to infinitesimal motion of windowpanes caused by conversation in a room, bugs which include miniature television cameras, and minicomputers which are programmed to intercept messages to a resource-sharing computer and to return answers which are spurious.

[JOHN M. CARROLL]

Bibliography: J. M. Carroll, *The Third Listener*, 1969; D. A. Pollack, *Methods of Electronic Audio Surveillance*, 2d ed., 1979.

Electronic power supply

A source of electric energy employed to furnish the tubes and semiconductor devices of an electronic circuit with the proper electric voltages and currents for their operation. The more common sources of energy are chemical batteries and alternating-current mains or lines. Batteries are useful as portable sources but are expensive and have small capacities. Alternating-current mains are not portable but are relatively inexpensive and have a large capacity. See ALTERNATING CURRENT.

One common method of classifying power supplies is by their use in electronic circuits. Most vacuum and gas tubes require a source of energy to energize their filaments or heaters; this type is known as a filament or heater power supply. Tubes require plate or anode voltages, and transistors need voltages for their collectors and emitters. These voltages are from a direct-current power supply which may provide about 5 to 40 volts for circuits involving transistors and about 50 to 400 volts for circuits using tubes. *See* GAS TUBE; SEMICONDUCTOR; TRANSISTOR; VACUUM TUBE.

If the source is one of alternating voltage, a transformer is usually used to raise or lower the voltage to the required level. The alternating current (ac) usually must be changed to a direct current (dc); this is accomplished by a rectifier. A rectifier allows current to flow mostly in one direction, and a pulsating current results. This pulsating direct current is not suitable for most purposes and must be smoothed by a power-supply filter or voltage regulator. *See* RECTIFIER.

A power-supply filter stores energy when the current is high and gives it up when the current is lower. The net result is to smooth out the variations in the current. The voltage regulator performs a similar function, but its operation is quite different. The zener diode type of voltage regulator has a voltage which is nearly constant as the current is varied over a considerable range of values. The gas-tube type of voltage regulator has a characteristic constant voltage over a large range of current values. Vacuum-tube and transistor regulators usually operate as variable resistances; the resistance decreases when the load is heavy and increases when it is lighter. Voltage regulators deliver an almost constant output voltage in spite of variation in load or in input ac supply voltage. *See* VOLTAGE REGULATOR; ZENER DIODE.

Filament or heater power supply. The filament or heater power supply is used to heat the filaments of vacuum or gas tubes so that electrons may be emitted from the filaments. If an indirectly heated cathode is used, a heater element inside the cathode must heat the cathode to a temperature at which electrons will be emitted.

Most of the present-day tubes heated by alternating current use 6.3- or 12.6-volt heaters. A step-down transformer is employed to change the 115 volts of the ac lines to these voltages. As an example, the heaters of four tubes may be connected in parallel across the secondary of a heater transformer whose primary is connected to the 115-volt ac lines. The center tap or one side of the secondary is often grounded. The tubes might take 0.5, 0.5, 1.0, and 2.0 amperes (amp), respectively, for a total of 4.0 amp. The heater transformer must have a rated capacity of at least 4.0 amp to supply these tubes. The voltage that the insulation must withstand is also often stipulated. Some tubes require other heater voltages, such as 2.5, 5, and 25 volts. Therefore, heater transformers with more than one secondary are useful.

Heater transformers are both heavy and expensive. To eliminate transformers, heaters may be connected in series across the ac line. Here each heater must take the same current, and the sum of the heater voltage ratings must be nearly equal to the ac line voltage. Other tubes use heaters made for direct connection across the ac lines.

For portable operation, batteries are usually used to heat the filaments. The tubes usually require 1–3 volts, so that one or two cells of a dry battery or one cell of a wet battery will supply sufficient voltage. Tubes having 6.3- or 12.6-volt heaters may be operated from a wet storage battery, such as an automobile battery, but the weight

of the battery decreases its portability. The filaments of the tubes are connected in parallel and one side of the battery is connected to the plate supply, so a complete circuit exists for the electrons flowing in the tube.

Direct-current power supply. The dc power supply usually supplies direct current of about 5 to 30 or 40 volts to transistor circuits, depending on the transistor type and its application. Similarly for tube circuits, the supply may deliver voltages of about 50 to 300 or 400 volts. Figure 1 shows the schematic diagram of a typical supply. The ac line

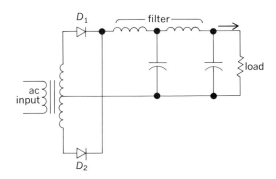

Fig. 1. Typical direct-current power supply.

energizes the primary of the transformer; the secondary is connected to the anodes of the rectifier diodes D_1 and D_2. The common diode connection is connected to the double-section L filter. This is the positive side of the filter. The negative side, usually grounded, is connected to the center tap of the transformer secondary. The load resistance is shown at the other end of the filter. A bleeder resistor is sometimes connected across the load to prevent the output voltage from increasing to dangerous values when the load is light.

The power transformer should be selected to have a sufficiently high secondary voltage to supply the desired output voltage. Allowance must be made for the voltage drop or rise caused by the type of rectifier circuit employed, and for the voltage drop in the rectifier diodes and the filter elements caused by the combined load and bleeder currents. Also, the transformer secondary current rating must be sufficient to supply the combined currents. The rectifier diodes must have voltage ratings sufficiently large for the transformer secondary voltages, and current ratings sufficiently large for the combined load and bleeder currents and for the peak currents encountered in the particular rectifier circuit used. The filter capacitors should have sufficient capacitance for smoothing the ripple of the output voltage and sufficient voltage rating. The filter inductors should have sufficient inductance for smoothing, low resistance to the current carried, and high insulation voltage.

As the line-voltage fluctuates, the output dc voltage will also fluctuate. One method of reducing this fluctuation is to regulate the voltage of the ac lines by static ac voltage regulators.

High-voltage power supply. High-voltage power supplies are employed to supply dc voltages of 1–20 kilovolts or more, usually at currents of a few milliamperes or less. Voltage-doubling and voltage-multiplying rectifier circuits are useful in such applications, which include the cathode-ray tube power supplies of television and radar. Another method of obtaining high voltages is to use a flyback power supply circuit with a half-wave rectifier and filter or with a voltage-multiplier circuit. For heavy-current high-voltage supplies, circuits such as that of Fig. 1 are almost always used. All components of the circuit must then be selected for the requirements of the high voltage and high currents. *See* Voltage-multiplier circuit.

Battery power supplies. Batteries are useful for portable applications where lower voltages and lower currents are used, and this is particularly true for transistor circuits. Probably the voltages most used are 6 volts produced by 4 dry cells in series and 9 volts produced by 6 dry cells in series. The ordinary dry cells have a voltage that gradually decreases with use until the voltage is too low for satisfatory operation of the circuit. The more expensive mercury cells last several times longer than the ordinary cells, and the voltage remains more constant with use until the end of its life when the voltage falls abruptly. Dry battery units may also be used for some tube supply requirements, although wet storage batteries would be more common. The high dc voltages required for tube circuits may be obtained by the use of vibrator or dynamotor power supplies.

Power-supply filters. Power-supply filters are usually of the low-pass *LC* variety, with the cutoff frequency of the filter made as much below the fundamental frequency of the ripple voltage as is economically feasible. Occasionally, low-pass *RC* filters will be used if the load current requirements are small. Lower cutoff frequencies generally require larger inductances and capacitances.

There are two subdivisions of these filters. The first, called an inductive-input filter, has a series inductor immediately after the rectifier. The second is the capacitive-input filter, which has a shunt capacitor immediately after the rectifier. One or more sections may be used, as shown in Fig. 2. Figure 2*a* shows a single shunt capacitor as the smoothing element. Figure 2*b* shows a two-section inductive-input filter. Figure 2*c* gives a two-section capacitive-input filter. The inductive-input filter has the disadvantage of giving a lower output voltage but the advantage of a better voltage regula-

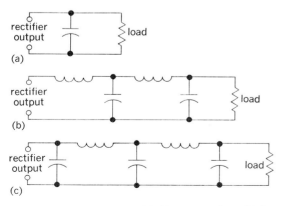

Fig. 2.   Smoothing filters. (*a*) Shunt capacitor. (*b*) Inductive-input filter. (*c*) Capacitive-input filter.

tion. The capacitive-input filter has the important disadvantage of producing a much higher peak current in the rectifiers. The inductors, sometimes called choke coils, almost always have iron cores, and the resistances of the windings are kept low for high efficiency and good voltage regulation. Ironcored inductors can be made to have an inductance that increases considerably as the load current approaches zero. These inductors, called swinging choke coils, help prevent the large rise in voltage at low load currents in capacitive-input filters. Capacitors for voltages below 500 volts are almost always electrolytic capacitors, but above that voltage they have paper or plastic insulation. *See* Capacitor; Inductor. For other types and general treatment of electronic curcuits *see* Circuit (electronics).   [donald l. waidelich]

Switching regulators. Continuing advances in semiconductor technology are increasing the uses of power supply switching regulators, in applications that formerly employed linear voltage regulators or did not incorporate regulators at all. Power supply regulators perform the function of maintaining their output constant in spite of variations of input supply or load. They are usually capable of compensating for rapid variations, thus functioning as a filter.

Power supply switching regulators control power supply output by rapidly switching an electrical power source to an energy storage component. The stored energy is extracted and applied to the load. The advantage of this approach is that high efficiencies can be achieved in spite of large variations in power supply input or load, because the power-regulating circuits are operated in a nonlinear mode. Partly as a result of their high efficiency, switching regulators can be more compact and lighter than their linear equivalents.

The control element in a switching regulator is an electronic switch, usually solid-state, capable of rapidly switching a current on and off (switching times are generally less than 10 μs). The resulting pulses of current are fed to a circuit containing an energy storage device, usually an inductor, which acts to smooth time variations in the current. The average amount of current is adjusted by changing the percentage of time that the switch is on (the duty cycle). This adjustment is made by a control circuit that maintains output voltage (current) of the power supply constant. The efficiency of a switching regulator can be very high (60–90%) because power control circuitry is not required to dissipate any power. Practical switches, of course, dissipate some power, as do feedback and control circuits. Switching rate is usually in the range of 2000 to 40,000 pulses per second. The faster the switching rate, the smaller the energy storage components need be; however, higher switching rates decrease efficiency because of the finite switching times of practical solid-state devices.

Step-down switching regulator. The possible implementations of power supply switching regulators are too numerous to be discussed exhaustively. Input to the regulator may be alternating current (ac) or direct current (dc). Only circuits utilizing dc inputs will be discussed because ac input can be considered as a special case. Figure 3 illustrates one of the most common configurations, a

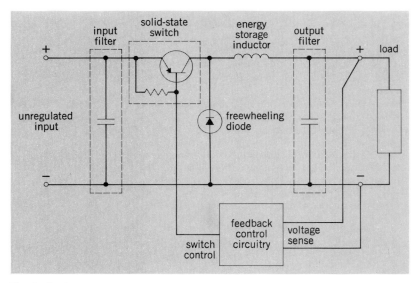

Fig. 3.   Basic step-down switching regulator.

basic step-down switching regulator capable of reducing its unregulated input voltage to a regulated output voltage. A solid-state switch controls current flow from input to inductor. Although the switch shown is a bipolar transistor, other solid-state devices such as silicon-controlled rectifiers or power field-effect transistors could be used. The freewheeling diode is actually part of the switch circuit. It allows current to continue to flow in the output circuit when the switch is not conducting input current.

The inductor stores energy when the switch is conducting and delivers it to the load when the switch is not conducting. When the switch is on, current flows from the input filter through the switch, inductor, output filter, and load, and then returns to the input filter. When the switch is off, current flows through the freewheeling diode, inductor, output filter, and load, and then returns to the freewheeling diode. Thus current flows continuously in the load. The output filter smooths the ripple in output voltage caused by the switching action. The input filter provides a low-impedance path for the input switching-frequency currents, and reduces the intensity of the radio-frequency interference conducted into the input.

The control circuit employs feedback from the power supply output to adjust the duty cycle of the switch. The output voltage of the regulator is compared with a reference voltage. The error, the difference between the output voltage and the reference, is used to adjust the duty cycle. The duty cycle of the switch may be controlled in various ways. In the constant-frequency mode, ON time is varied while repetition rate is held constant. In the constant-width mode, repetition rate is varied while ON time is held constant. In the variable frequency – variable width mode, both ON time and repetition rate are varied by holding OFF time constant and varying ON time, for example. In the gated mode, a pulse waveform with a fixed duty cycle and repetition rate is synchronously gated to the switch. The gate signal controls the average number of ON periods of the fixed waveform that are passed to the switch. The

constant-frequency mode makes the design of the filters easier, although the other modes may be simpler to implement. These various waveforms can be produced by an oscillator or multivibrator internal to the controller (driven control), or by the interaction of the controller with the entire system (regenerative control). The driven types offer greater versatility and higher performance than the regenerative types at the cost of greater complexity. Most regenerative controllers operate in the variable frequency – variable width mode, whereas driven controllers operate in all the above-mentioned modes.

Step-up switching regulator. Another common configuration is the step-up circuit of Fig. 4. This circuit is capable of providing an output voltage equal to or greater than its input voltage. In this configuration the inductor *L* is in series with the supply. When the switch conducts, current flows from the supply through the inductor, and the switch then returns to the supply. The capacitor supplies current to the load, while the diode prevents the capacitor from discharging through the switch. When the switch turns off, current flows from the supply through the inductor, diode, and parallel combination of the capacitor and load, and then returns through the supply. The voltage across the inductor adds to input voltage to charge the capacitor through the diode. The inductor can be replaced by an autotransformer for higher voltages.

Other configurations. In the configurations discussed so far, input and output voltages have had the same polarity. The circuit of Fig. 5 produces an output of opposite polarity from the input. Its operation is similar to that of the step-up circuit of Fig. 4. Many more variations are possible. For example, the input and output circuits of a switching regulator can be isolated by means of a transformer inserted between the switch and the energy-storage components; the energy-storage components may be entirely capacitors; and so forth.

Applications. The principal advantages of power supply switching regulators are efficiency, versatility, and small physical size. Their principal disadvantages are electromagnetic interference, audible noise, complexity, and cost. Electromagnetic interference is a result of rapid switching. Interference can be minimized by proper filtering and shielding techniques. Noise can be eliminated by raising the switching rate above the range of

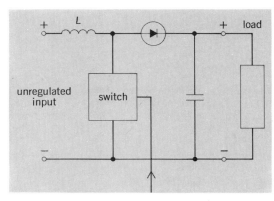

Fig. 4.   Step-up switching power supply.

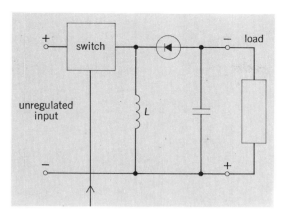

Fig. 5. Polarity-inverting switching power supply.

human audibility. Costs can be reduced through use of integrated circuits and advanced power semiconductors.

Switching regulators find applications in portable and battery-operated equipment where the efficiency, size, and weight of the power supply are important. They are used in computer equipment because of their capability of accommodating wide input voltage variations while maintaining high efficiency. Switching regulators are also used as preregulators ahead of a linear regulator to reduce the power dissipation of the linear regulator. Consumer devices that utilize power supply switching regulators include calculators and color television receivers.

[DAVID A. SOSS]

Bibliography: J. M. Carroll, *Modern Transistor Circuits*, 1959; D. J. Cramer, *Modern Electronic Circuits*, 1976; L. Dixon and R. Patel, *EDN*, 19(20):53–59, 19(21):37–40, and 19(22):76–82, 1974; C. A. Holt, *Electronic Circuits: Digital and Analog*, 1978; J. Millman, *Microelectronics*, 1979; W. E. Newell et al., in D. G. Fink and A. A. McKenzie (eds.), *Electronics Engineers Handbook*, sect. 15: Power Electronics, 1975; S. Olla, *Electronics*, 46(17):91–95, 1973; D. L. Schilling and C. Belove, *Electronic Circuits: Discrete and Integrated*, 2d ed., 1979; Texas Instruments, Inc., *Transistor Circuit Design*, 1963; J. B. S. Waugh, in L. P. Hunter (ed.), *Handbook of Semiconductor Electronics*, 3d ed., sect. 17: Power Supplies, 1970.

Electronic publishing

A technology encompassing a variety of activities which contain or convey information with a high editorial and value-added content in a form other than print. Included in the list of presently practiced electronic publishing activities are: on-line data bases, videotext, teletext, videotape cassettes, videodisks, cable TV programming (direct broadcast satellite delivery is optional), and electronic mail and messaging.

For convenience and ease of understanding, these may be classified according to the amount of interactivity the user is allowed. One-way systems deliver information in a continuous manner and allow the user only passive viewing and crude control over sequence or position. All television and sequential-access image-archiving media, like videotape, are one-way. One-way-plus systems are continuous but offer some option for sequencing and selective access, with teletext being the best example. Two-way systems access information randomly under direct user control.

Although there are few who claim profits from their activities in this area, a great deal of work is under way in developing the technologies and market positions which will ultimately fuel the economics of this new industry segment.

Electronic publishing has become a reality over the last decade partly because of progress in digitization of both text and images and partly because of the availability of new media for information or communications. The availability of new media for delivery of soft copy depends on both technology and changes in the regulatory environment. Once both areas coincide, these new media will intensify the potential for nonprint publishing. Direct-home-delivery channels like cable TV or direct broadcast satellite (DBS), and retail distribution media like videodisks, are examples of information vehicles unavailable prior to 1978. The decision by the Federal Communications Commission to allow broadcasting of data on the 650 local frequencies previously reserved for voice telecommunications (for example, pocket message transmitters for doctors) is an example of a regulatory decision which opens new horizons for the use of portable computer terminals to both originate and access electronic information.

Publishing environment. The immense growth of interest in electronic publishing has resulted primarily from technological advances affecting the areas of information distribution, display, and storage, combined with a dramatic decrease in the cost, and increase in the speed and power, of telecommunications.

In the area of communications, satellites have already begun to revolutionize commercial long-distance data transmissions. In the late 1970s and early 1980s the United States witnessed the rise of a new industry, dedicated to creation of nationwide data communication networks using packet switching technology, such as GTE's Telenet and Tymeshare's Tymenet networks. These networks (Fig. 1) are easy and inexpensive to use when compared to long-distance telephone lines. For most users, access to the network is through a local telephone call on a regular telephone. The networks are designed to allow data communications and computing devices manufactured by many different companies to connect with one another intelligently, that is, without the users having to concern themselves with the compatibility of individual devices. Due to their low cost, ease of use, and effect on the device compatibility issue, packet switching networks have created a strong and stable data telecommunications industry. This industry now seeks to encourage the growth of all kinds of electronic information services to increase its own revenues and profits. *See* DATA COMMUNICATIONS.

The technical differences between packet switching networks and long-distance telephone lines show up in the rate structures. Long-distance telephone charges are a function of the distance between parties and the connect time, while satellite-linked, packet-switching network rates are a function of the volume of data transmitted and connect time. Since this type of communica-

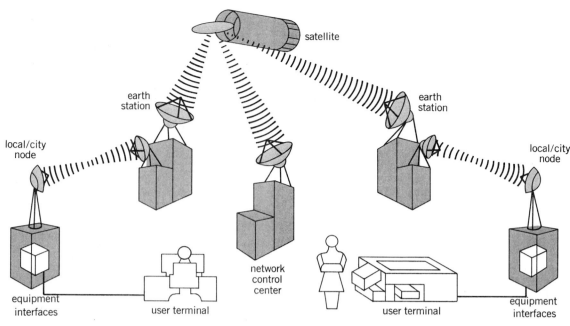

Fig. 1. Satellite-based network. (*Harris Corporation*)

tion link depends for most of its operation on satellites which are in a geostationary orbit 22,500 mi (36,000 km) above the Earth, it is not difficult to see why surface distances are relatively immaterial to the cost of telecommunications using packet switching networks. Other resources used are equivalent to those used in a local telephone call.

In this environment, electronic publishers can reach a national market of potential subscribers without the costs of logistical problems associated with the nationwide distribution of printed material (Fig. 2). Since electronic publishing is based on a pay-per-use concept and the data base of text or images is continuously available, there are dramatically different economics from those associated with traditional publishing, where manufacturing and distribution of books, magazines, or newspapers represent by far the majority of costs. The effectiveness of the distribution network for the printed artifact was also the major determinant of economic success for publishing operations. Paper has long been the single highest cost in the entire complex set of operational steps which make up the publisher's business.

Some of the main issues unresolved in the electronic publishing industry relate to the cost of human factors engineering of the users' terminal equipment. In this regard, several significant advances have been made. One such advance is the introduction of the pocket-sized IXO Telecomputing System. It has a built-in phone modem with autodialing and autologging so that users only need to enter a password to gain entry to an online data base. Its ease of use is indicated by the keyboard, which includes YES, NO, DON'T KNOW, and HELP keys. As the price of pocket-sized devices drops, they will be perceived as consumer, rather than business or luxury, items. This will have an enormous impact on the progress of market demand for electronic publishing products. Innovative information display devices, coupled

with the continued proliferation of personal computers fostered by such giants as IBM and DEC, also will have a favorable impact on the future development of the electronic information industry.

Production of electronic information products. Crucial to the economical production of new information packages is the ability to manipulate images and type in totally digital formats. This was available only experimentally and at great expense until mid-1981. Moreover, software and engineering expertise for type- and image-processing systems, most commonly referred to as pagination or image-assembly devices, are maturing at a rapid rate (Figs. 3 and 4). This is occurring just as the new generation of high-powered microprocessors of both 16- and 32- bit varieties are becoming realistic components for lower-cost systems.

Publishing organizations who currently have a need and justification for completely integrated pagination to automate production of their print products may find themselves in the most favorable position to exploit new market opportunities. Once master images have been composed for the production of printing plates, it is (or will be) a simple matter to reprocess them into the formats used for the electronic publishing market. The standard home television set or its equivalent, incorporated into a computer terminal, will probably remain the standard for delivery of electronic information until at least 1988.

Conduit versus content. Several approaches to an analysis of electronic publishing have been made which are useful in providing different perspectives on this multifaceted subject. Tony Oettinger has developed an analysis of the information business which seems especially useful. This is a grid along which the various electronic publishing offerings can be plotted. It places conduit or media-oriented offerings at one end, and content-oriented offerings at the other. The other axis places products at one end and services at the other (Fig. 5).

Electronic mail characterizes the conduit end of the spectrum. It is a person-to-person communication system, without any information content inherent in the service. Each user is assigned an electronic mailbox, the equivalent of a computer directory or queue, to which other people can send messages, data, or documents instantaneously. The message remains in the recipient's mailbox until it is retrieved. A user can establish standing distribution or mailing lists and send the same message to everyone on the list instantaneously. In more sophisticated systems, a user's mailbox may also be intelligent, and know which subjects are most important, flagging certain items for priority attention or forwarding messages on certain subjects to the user at other locations, depending on the time of day or week.

The content end of the spectrum is well characterized by on-line data bases such as the New York Times Information System. A subscriber to this electronic publisher's offering can retrieve news stories based on key words and combinations contained either in a master index or in the story contents.

Until recently, content-rich media meant only collections of alphanumeric information. Now, however, the new technologies of video image digitization, storage, manipulation, and compression are bringing pictures into the domain of electronic publishing. Although the time delay inherent in creation and manufacturing of available formats of optical videodisks makes them less than a real-time medium, they are a very effective means by which interactive video publishing can be accomplished.

On-line data bases. Most mature of the electronic publishing activities is that of on-line data base publishing. On-line data bases utilize standard magnetic media and are composed of text characters stored in ASCII. They are accessed remotely by using standard data terminals via existing public telecommunications facilities at low to moderate speed, with most hard-copy output being done on line printers at central facilities. Direct output of full-text data from electronic text-processing systems to on-line data bases is already a reality in highly automated newspaper or magazine publishing environments.

This segment of the industry has been active since 1975. There are presently more than 500 producers offering more than 1400 individual products on a pay-per-use or subscription basis. Growth has held steady at 30% per year since 1975, and this growth trend is expected to continue at a 25–40% annual growth rate for the next few years.

Bibliographic data bases accounted for only 10% of revenues in 1980, a figure which is disproportionate to the percentage of products in this category. These data bases are published predominantly by government, quasigovernment, and not-for-profit suppliers. Source data bases (or factual packages) provide better revenue than bibliographic data bases. However, they are more expensive to market than bibliographic data bases, which rely on well-trained librarians who understand and use them as professional tools. Source data bases, like Value Line or Standard and Poor's, with usage costs of over $100 per hour, require knowledgeable sales people and considerable

customer training. They also demand some pre-planning to obtain most effective results, eliminating casual usage. From a producer standpoint, factual packages also require a high degree of currency, accuracy, and powerful indexing. These are all expensive in terms of time and human judgment. However, factual data bases focus more on current data and can have less expensive on-line storage requirements than bibliographic varieties.

Inability to incorporate graphic material has been a major limitation and block to greater acceptance of on-line data bases. In 1982 the first attempt was made to use videodisk, attached locally to the access terminal, as a way of integrating image material into the on-line data base presentation. This is exciting theoretically but may

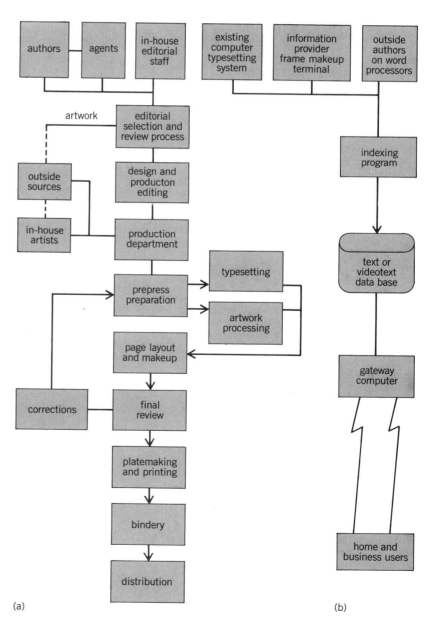

(a) (b)

Fig. 2. Comparison of work flows for (a) traditional and (b) electronic publishing for online data base or videotext.

Fig. 3.   Portion of master image of a fully composed page on Scitex's Vista system.

be difficult in practice, since each disk contains only 54,000 or 108,000 images, necessitating multiple disks for storage of a complete archive. This becomes especially inconvenient when disk volumes must be changed during a single research sequence.

A major impediment to industry growth and data base proliferation is lack of standardization in the user access protocol—the group of directives for specifying what operations are to be performed and which variable data are to be used. Nonstan-

dard command codes to perform the same operations on different data bases create a Tower of Babel situation in which it can take as much as 6 months to train a fully qualified searcher, and each new data base has its own specialized commands which must be learned.

Availability of a new generation of self-contained data base processors will make entrance to this field easier for publishers. A multiprotocol processor, and batch search techniques which use intelligent microprocessor terminals and store-and-forward to reduce communications charges, will improve the user's situation and decrease costs. *See* DATA-BASE MANAGEMENT SYSTEMS.

Videotex. Videotex is the generic-term for electronic home information delivery systems. Within this broad term there are two specific approaches, called teletext and viewdata, or videotext. Teletext is a one-way communications medium. Images, each constituting a single frame of TV data in a special, compressed format, are transmitted in a continuous sequence. Users indicate which frame they would like to see by interaction with the decoding unit in their local TV set. This "grabs" the desired frame or sequence of frames on its next cycle. Users have no way of transmitting information to the central broadcast facility. Individual teletext services have brand names like CEEFAX, the BBC's pilot program, or Teletel, the French service.

Videotext, or viewdata, is a potentially more powerful and lucrative service since it involves users' ability to directly access a central data base interactively from their local TV (Fig. 6). Users may

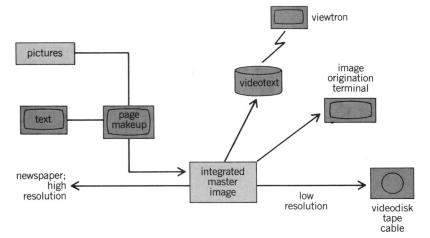

Fig. 4.   Migration of text/image masters. Images and text are placed into machine-readable form and transmitted to various output devices. (*Inter/Consult*)

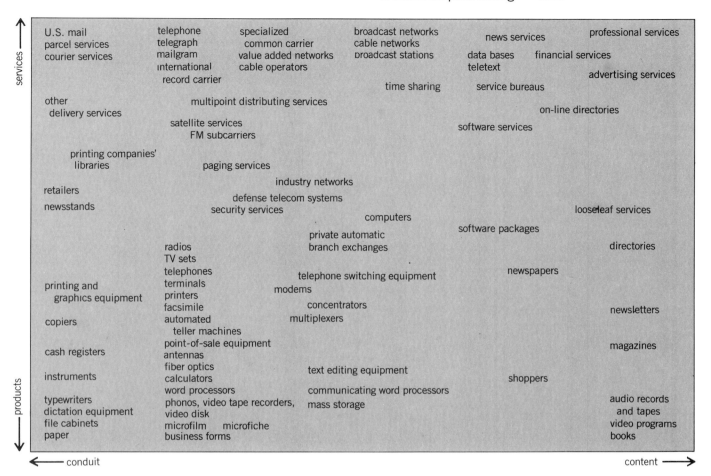

Fig. 5. The information business characterized in terms of conduit versus content as a function of available products and services.

request specific frames of information. More importantly, they can directly access what they want, meaning quicker response time and a more structured usage of the medium. Most exciting of all, users can communicate with other users via the system. They can also utilize transactional services, including banking or shopping based on information provided by the system.

The enormous potential of videotext services has not been realized as yet in any activity above the level of experiment, despite the avalanche of publicity which is heralding the arrival of a new era in home information delivery. Many fundamental problems exist in making the concept truly useful and commercially viable. Most of these relate to needed production capabilities, which require competent digital type/image manipulation, cheap hard copy, and delivery systems which need adequate bandwidth to deal with pictures or productization of present information resources into videotext data bases. Recent estimates indicate that by 1990 a total of 10% of United States homes (and a much larger percentage of businesses) will be utilizing videotext in a significant way.

Unfortunately, the strategic and regulatory environment around videotext services has become the battleground for both national telecommunica-tions policy and national technology prestige. This has overshadowed deficiencies in present product and production technology. Lack of acceptance by information providers and users has been unheard beneath the din of congressional arguments between the newspaper industry and AT&T over rights to revenues from home-accessed information bases. The economics of the new medium are wildly overestimated, with claims based on subsidized probes which rarely approach real-life market conditions.

Even for those systems which are actually operational, capital cost is high on both the publishers' and users' side. A new generation of data base and computing machinery will help the industry, as will new terminals designed for use by graphic designers and visual artists. In June 1982, AT&T introduced its new Frame Creation Terminal (FCT). This is a color-page or frame-makeup terminal which can operate either as part of an on-line system or as a stand-alone work station. It serves the needs of information providers for composing and formatting frames to be placed into the videotext data base. The FCT is equipped with a computer, color video display, keyboard, and graphics tablet. Storage of up to 600 frames on dual floppy drives is supported. *See* VIDEOTEXT AND TELETEXT.

Videodisk. Laser videodisk is the most exciting of the new visual media. It has rich potential since it employs a random-access technique. Available videodisks store single-frame television images in the form of analog FM-encoded TV signals. These are decoded and played back by using a low-power laser. Future optical disks will be memory systems rather than frame storage devices. However, because of the enormous quantities of digital data they will be capable of storing in binary form, the creation of all-digital image archives will be their prime application.

These kinds of picture data bases in digital format are impractical with present technologies because of the enormous amount of data contained in images. As an example, the number of ASCII characters in a standard office document in $8\frac{1}{2} \times 11$ in. (216×279 mm) format is about 3840 characters. This page could be transmitted over a 9600-baud phone line connection in 3.4 seconds. The same page in a digital scanned format would take 3,800,000 bits of storage, or 475,000 characters. This data would require 63 seconds of transmission time. A color picture would have four of five times the amount of information.

Laser-format videodisks contain 54,000 color video frames per side of the disk, and each frame is directly accessible by its frame number. Some player devices allow simultaneous access to both sides of the disk, making all 104,000 frames accessible simultaneously. When played at 30 frames per second, the laser disk functions much like a 30-minute video cassette, with improved video and two-channel stereo sound. However, the unique aspect of laser disks lies in their ability to display any individual frame, or sequence of frames, as a result of the interaction between a viewer and a computer linked to the laser disk player. Further sophistication may be added by use of a data-base management system in the control computer. This allows users to retrieve images based on their content, historical usage, source, or any other factor

deemed to be important, and encoded during construction of the videodisk or its retrieval index base.

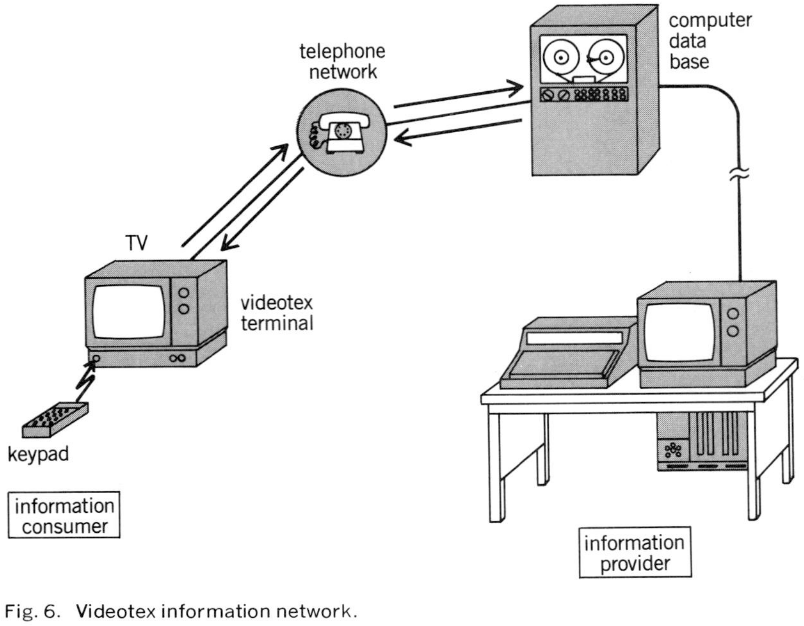

Videodisks are produced in large quantities by a stamping process, much like the manufacture of LP records. They are similar to books from a publishing perspective in that they nicely accommodate the situation where static information needs to be distributed to many different locations.

Unfortunately, the videodisk business has evolved into a state where its one solution is looking for a problem and its big problem is waiting for a solution. The laser consumer disk with its low cost and high on-line storage capability has been perceived as a suitable medium for several publishing products, including catalogs, training programs, and archive documentation, and as a source for locally stored graphics. However, all these applications accept the idea of a content-static package with a long production lead time being acceptable. An interesting prototype for industrial, research, or training applications of analog disks has been introduced by Vision Machine Research. This consists of two color-display screens controlled by a third data terminal accessing from four to ten laser disk players simultaneously. Multiple microprocessors offer data base management for search and retrieval of desired images. Full audio output for sound tracks and data communications makes this appear to be a powerful tool for a medium where access has been the problem.

Most applications where image archiving is desired need a dynamically updatable and expandable storage capability. Picture morgues for newspapers or magazines, electronic document-on-demand systems, and electronic color systems all share this requirement. However, no systems for such applications are currently available, and even when the problems of recordability, erasability, data reliability, and archival durability are solved, an enormous amount of research and development on packaging of such large data bases will be required. See VIDEO DISK RECORDING.

Case studies. CompuServe is a model for the electronic information utility of the future. Access is made by a local telephone call in over 300 United States cities presently concentrated in the northeastern seaboard, Silicon Valley (the region in California where many electronics manufacturers are based), Los Angeles, and Chicago. CompuServe offers users 301 categories of information, from African weather to used-car purchase procedures (Fig. 7).

Telecable. Telecable is a teletext operation recycling the print operation's news over a channel leased from Wisconsin CATV. Brief news stories are displayed on the screen for 25 seconds with 25-second advertising slots between stories. On important stories, viewers are encouraged to become readers by referring them to the print product for more in-depth coverage. However, poor understanding of how to use the teletext medium for advertising seems to be a major problem. For this reason, Telecable introduced an hourly 5-minute live anchored newscast for which it sells traditional 30-second advertising spots. Surveys now show the frequency of use rising, with 73% of cable subscribers presently viewing Telecable an average of 1.5 times daily, a 57% increase over the

Fig. 6. Videotex information network.

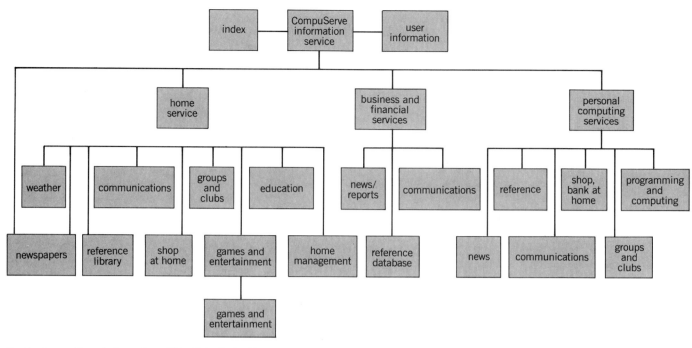

Fig. 7. CompuServe information utility. (*Inter/Consult*)

first year of operation.

Videodisk for how-to literature. One strong electronic publishing opportunity for laser disk is the traditional how-to book or magazine. Not only can the textual material and static pictures be presented, but animated sequences of a process can also be incorporated. The viewer has the option of viewing the film sequences at regular speeds or in slow motion, in both forward and reverse directions. The viewer also can stop the action at any point without additional wear on the disk. Users can follow different levels of instruction through the disk according to their level of skill or knowledge of the tools required for the task. The first commercially available interactive videodisk ever produced was an instructional package called "How To Watch Pro Football."

Applications in this mode of use range from the teaching of surgical procedures to physicians to the teaching of origami (Japanese art of paper folding) to children. Tests by the U.S. Army have found that while experts can repair a piece of complex machinery as quickly using a printed manual, novices did far better using interactive videodisks in analyzing and fixing complex problems on unfamiliar machinery or systems. Avon Products has started a program to implement videodisk-based training programs for their sales representatives.

Sears videodisk catalog. The first attempt to replace or augment full-color merchandise catalogs with videodisk was undertaken by Sears Roebuck in their Summer 1981 video edition. The catalog contained not only traditional text and graphic information frames, but TV-like action sequences and user-originated search/query capability. Action sequences included fashion modeling and operation of mechanical equipment like lawn

mowers. Production values were extremely high, with TV advertising standards prevailing throughout. There was also a heavy mix of computer-generated graphic effects.

The catalog was made available to the public for home use or at public-access facilities in Sears retail locations. Reactions were extremely favorable, although specific sales data or comparisons with traditional catalog merchandising remained proprietary in nature. Unfortunately, slow acceptance by consumers of videodisk as a component in home entertainment systems will be a gating factor in use of the disk for catalog purposes.

Electronic magazines. Nonprint formats for magazines are an appealing and seemingly successful application of electronic publishing. Magazines are presently being published in videotape format, and expansion into videodisk is imminent. Several magazines are using their print product as the basis for a cable or broadcast television program. Videotex seems very suitable for certain kinds of magazines, especially those that are how-to–oriented.

American Baby magazine has now piloted a series of programs for cable distribution in major markets, making use of their name recognition and in some cases offering video versions of articles which originally appeared in print. Classic TV ad spots are used and heavy cross promotion of the print product is done as well.

Time Incorporated launched a series of videotext news and feature magazines in 1982. *Popular Science* has started making content available through CompuServe. Some material is specially edited for the on-line format, including certain headline/capsule-oriented features. Automobile road tests and car repair articles are also offered. *Popular Science* encourages its readers to offer suggestions or pass on requests for more information via CompuServe's electronic messaging services. This is a forerunner of the kind of transactional services in responding to advertisers which will make video magazines an extremely attractive and powerful medium.

As pagination of the traditional print product matures, electronic recycling of material will give publishers with such multifunctional equipment strong incentives for content recycling in other forms.

Opportunities and obstacles. It appears that a new electronic publishing industry is at the beginning stages of its evolution. Many of the larger participants in the traditional sectors have already made strategic moves to position themselves strongly in the new sector. Activity by traditional equipment suppliers has been less enthusiastic. Most capital-intensive product development for new equipment needed to supply this sector is being undertaken by data processing suppliers.

Is electronic publishing a threat to traditional printed information distribution? As of the mid-1980s, the answer is probably no. Production of printed matter will probably experience a short-term increase as a result of the packaging and collateral materials needed to deliver, sell, or use electronic information. By the early 1990s, however, color magazine and catalog printers may begin to experience a shift in their revenue base. Certainly vendors of graphic arts equipment and supplies, especially plates, films, and inks, are expected to be strongly impacted by 1990. This may seem remote until the payback periods on a film-coating alley or a large rotogravure printing facility are taken into considerations.

Electronic publishing is the most important sector of the emerging information industry. It is on the threshold of recognition as a major industry. When seen in combination with traditional publishing, it could, by the year 2000, represent the largest single sector of the United States industrial economy. *See* CONSUMER ELECTRONICS.

[DAVID H. GOODSTEIN]

Bibliography: T. S. Dunn (ed.), *Proceedings of the 1981 Lasers in Graphics, Electronic Publishing in the 80's Conference*, Vista, CA, vols. 1 and 2, 1981; D. Goodstein, Output alternatives, *Datamation*, 26(2):122–130, February 1980; D. H. Goodstein, The outlook for optical videodisk in graphic arts applications, *TAGA Proceedings 1981*, Rochester, NY, pp. 204–221, 1981; D. H. Goodstein, Videodisk, in *Auerbach Electronic Office: Management and Technology*, 1981; *The Seybold Report on Publishing Systems*, 1971.

Electronic switch

An electronic device to which two input waveforms can be applied and which delivers at a pair of output terminals a signal that is alternately a replica of each of the input signals. The transmission gate performs the basic switching function of the electronic switch, and sometimes such a gate circuit is itself defined as an electronic switch. However, the electronic switch is usually considered to be a separate instrument to which periodically recurrent input signals, of arbitrary waveform but with synchronously related periods, are connected. The output is switched between the two waveforms at a rate that is synchronous with

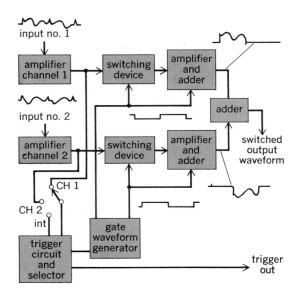

Elements of electronic switch.

the period of the input waveforms. One of the input signals often provides synchronizing information to supply periodic trigger pulses, which are then applied to the gate signal generators within the switch. *See* GATE CIRCUIT.

A frequent use of the electronic switch is to provide means for displaying two time-related signals on the screen of a cathode-ray oscilloscope without requiring two independent deflection systems within the cathode-ray tube. For this application, the time-based circuit of the display device is synchronized with the repetition rate of one of the two input signals, and the internal transmission gate of the switch alternately switches the signals to the output. At the same time a dc component, or pedestal, is added to each of the two signals so that they will appear at distinct levels on the screen of the display device.

A block diagram of an electronic switch is shown in the illustration. The two input waveforms are applied to amplifier channels. The amplified waveforms are then applied to switching devices and also to a trigger circuit, which generates trigger pulses that can be synchronized either internally or with one of the input signals. A gate waveform generator, usually a bistable multivibrator, is used to actuate a gating waveform to operate the switching devices alternately.

A controllable portion of the gating pulse may be combined with each switched channel to provide an amplitude separation of the two switched signals where separation is necessary for such applications as the cathode-ray tube display.

The trigger circuit might supply a trigger to the time-base generator of an oscilloscope, or when internal triggers are generated, it might be used to control the repetition period of the phenomena to be displayed. *See* SWITCHING THEORY; WAVE-SHAPING CIRCUITS.   [GLENN M. GLASFORD]

Electronics

The branch of science and technology relating to the conduction and control of electricity flowing through semiconducting materials or through vacuum or gases. Electronics is concerned with the study and applications of the motions of charge carriers (electrons, holes, and ions) under the influence of externally applied voltage or current, or in relation to the incidence or production of radiant energy. While electronics is properly a part of electrical engineering, the latter term is often reserved for applications involving power generation, distribution, and use at low frequencies, for example, in utility systems and industry. Since the 1960s, the dominant segment of electronics has been that known as solid-state, which involves transistors and other semiconductor devices and assemblies.

The electronics industry is concerned with the design, manufacture, and application of: transistors, diodes, integrated circuits, and other semiconductor devices; and to a much lesser extent, electron tubes. The major uses of electron tubes, not taken over by solid-state devices, are the generation of large amounts of radio-frequency power and the perception and display of images, as in television camera and picture tubes and computer displays.

These devices are used in a rapidly expanding range of applications making use of so-called digital electronics, that basic to the electronic calculator and computer. Reductions in the size and cost of powerful computers have produced, in a single large-scale integrated circuit, devices used in the automotive field to control fuel use, in "intelligent" control of household appliances, satellite communications, financial transactions, industrial instrumentation, control and automation, medical devices and implants, and the aerospace industry and numerous aspects of national defense. To a degree not envisaged as late as 1970, electronics pervades nearly every avenue of human endeavor.

History. The history of electronics divides into two periods. The first began with the invention of the electron tube; the second, occupying the latter half of the 20th century, is based on the invention of the transistor.

Electron tubes. In the first period the initial discovery, made by Thomas Edison in 1883, was that negative electric charges are emitted by a hot filament and are attracted to a nearby positively charged metal plate, when filament and plate are enclosed in a partial vacuum. When the plate is charged negatively, it repels the electrons and current ceases. Thus current flows from filament to plate in one direction only. The Edison effect was the first demonstration of the presence of freely moving negative-charge carriers, which were later named electrons. In 1897 J. A. Fleming made the first use of the unidirectional nature of the current flow in a detector for radio signals. The Fleming valve is the prototype of the electron-tube diode, and of the solid-state diode which has replaced it in nearly all applications.

The second major step in electronics technology, taken by Lee DeForest in 1906, was the introduction of a third electrode (known as a grid) between the filament and the plate of the Fleming valve. The voltage between the filament and the grid is used to control the flow of electrons, repelling them (decreasing the current) when the grid is charged negatively, while attracting them (increasing the current) when the grid is charged positively.

Since the variation in voltage applied to the grid is smaller than the resulting voltage appearing in the circuit attached to the plate, this device, known as the triode after its three electrodes, is in fact an amplifier. When the grid voltage is appropriately associated with a current or power source, the triode may also act as a current or power amplifier. Several modifications of the triode were developed by employing additional grids, but in most applications these have been replaced by solid-state devices. *See* AUDION.

Over the half century that electron tubes dominated the field, there were three disadvantages associated with their use: they required a source of power to heat the filament; they required high voltage on the plate; and the filament (or its later version, the indirectly heated cathode) eventually lost its power to emit electrons, that is, the vacuum tube had a limited life-span. *See* ELECTRON TUBE; VACUUM TUBE.

Transistors. These limitations were overcome in the transistor, invented in 1947 by John Bardeen, Walter Brattain, and William Shockley. They found that the semiconductor germanium, when in the form of highly purified crystals of essentially perfect structure, is capable of carrying simultaneously two electric currents of opposite polarity when certain specified impurities are present in extremely minute amounts. Certain impurities, for example, phosphorus or antimony, produce so-called n-type germanium, which has an excess of electrons, so that the major current is carried by these negative electrons. But a lesser positive current is also carried by "holes," that is, electron vacancies in the crystal lattice. This current of minority carriers can be controlled by externally applied voltage, thus producing an amplification effect. Other impurities, for example, boron or indium, produce p-type germanium, which has a deficiency of free electrons, and they become the carriers of the minority current, while the holes carry the majority current. When n- and p-type germanium are placed in intimate contact in the crystal lattice, by changing the impurity during crystallization, a so-called pn junction is formed. It has the property, like that of the Fleming valve, of preventing the reverse flow of the majority carriers, while allowing the minority carriers to pass. *See* JUNCTION DIODE; SEMICONDUCTOR; SEMICONDUCTOR DIODE.

The transistor (the bipolar type most widely used) consists of a single piece of germanium in which two such junctions are present, forming two closely spaced parallel planes, so that the semiconductor is present in layered n, p, and n form (npn transistor) or in p, n, and p form (pnp). The respective segments of the germanium are known as the emitter, base, and collector, which have functions roughly analogous to the filament, grid, and plate of the triode tube. When a control voltage is applied, for example, between emitter and base, it controls the flow of minority carriers reaching the collector, and amplification occurs. No filament heating power is required, the applied voltages are much smaller than in vacuum tubes, and the operating life of the transistor is extremely long. *See* TRANSISTOR.

Silicon, a much more abundant semiconductor than germanium, serves the same purpose at lower cost and with advantages in the production process, and is therefore now the preferred material for solid-state devices. Additionally, compound semiconductors, such as gallium arsenide, serve for special applications, particularly at microwave frequencies. *See* Microwave solid-state devices.

Integrated circuits. Following the invention of the transistor as a discrete device, it was found that an assemblage of diodes, transistors, and other circuit elements, with metallic interconnections, could be produced on a flat "substrate" of silicon. This is the integrated circuit (IC). Steady improvements in the methods of diffusion, ion implantation, and photodeposition used to produce integrated circuits have led to marked increases in the density of the elements. By the late 1970s, it had become possible to produce integrated circuits with tens of thousands of circuit elements on a substrate having an area of only about a quarter of a square centimeter.

By the early 1980s, certain integrated circuits, known as microprocessors, were being used by the millions in automobile control systems and in hundreds of other applications, including electronic computers, at prices permitting their use in the home. Integrated circuits have also become the universal method of storing computer information during data processing, although magnetic films are used for long-term storage. *See* Computer; Computer storage technology; Integrated circuits.

Growth of electronics industry. The extent to which high-density, large-scale, solid-state integrated circuits can permeate industry, commerce, and home life is potentially so great that electronics may well be on its way to being one of the largest industries, as measured in the value of its products. In 1922 the total annual sales of electronic components and equipment amounted to $60,000,000. By 1960, following the development of the television industry and appearance of sophisticated electronic weapons systems, the figure had risen to $10,000,000,000; by the mid-1970s, it had risen to $35,000,000,000, and this figure had doubled before the end of that decade. Sales volume continues to grow, aided by the growth of the computer industry, new forms of communication, office electronics, the application of electronics to automobile engines, and extensive uses of electronics in the home.

Applications. The manifold applications of electronic devices depend fundamentally on the physical fact that the force exerted on an electron or a hole by the voltage applied to such devices is very great in proportion to the mass of the charge carrier. Consequently the electric charge is very rapidly accelerated and can achieve its maximum velocity in a billionth of a second or less. This rapid reaction permits the device to amplify and generate currents alternating at frequencies of billions of cycles per second (gigahertz) or to respond to stimuli in less than a billionth of a second (nanosecond).

While these numbers apply primarily to low-power devices, lower-frequency semiconductor devices, such as the silicon-controlled rectifier (SCR), can control as much as 1000 MW of power at commercial frequencies. This requires the use of large arrays of silicon-controlled rectifiers. These devices are also finding use in the transfor-mation and control of power in electric locomotives and in similar heavy-current uses. These have given rise to the term power electronics. *See* Semiconductor rectifier.

Communications. The range of communications applications has greatly expanded since the 1960s. Communications satellites, which carry voice, television, and data signals throughout the world, are reaching a possible saturation point, as the number of geostationary satellites begins to exceed the number of parking orbits available.

The limits of the traffic capacity of conventional cable and microwave circuits will be expanded manyfold by the introduction of optical fiber circuits. Light waves, carrying the signal in infrared form, are produced by solid-state lasers, and injected at one end of a very small fiber of glass several kilometers long. Many such fibers may be connected end on end. At the far end of each segment another semiconductor device, the photodiode, converts the infrared signal back to its electrical form. At this point the signal may, after amplification, be fed to the conventional telephone system, or passed on to the next section of optical fiber through another laser. Optical fiber cables, each containing typically 12 fibers, are planned in the 1980s to carry the heavy telephone traffic in the corridor from Washington to Boston. A cable in this service may carry as many as 80,000 telephone conversations simultaneously, far more than can copper cable or microwave circuits. *See* Optical communications.

In still another communications application, systems have been developed to permit owners of cable-connected television sets to place a telephone call to a central data bank and request that needed information be displayed on the receiver screen. Such interactive cable systems were in limited use in the late 1970s, and rapid expansion was expected during the 1980s. *See* Data communications; Electrical communications.

Consumer electronics. Aside from such communications uses, the consumer now has available a wide variety of devices and equipment which employ the digital technology of the computer industry. The hand-held electronic calculator, the digital-display quartz-controlled wristwatch, electronic games, including chess-playing automatons, and devices capable of translating between languages and even of speaking the words, were avaliable by the late 1970s and were rapidly enlarging their scope and variety. The use of the computer as a teaching aid, even in elementary grades, has also been demonstrated. To the existing color television technology have now been added home video tape recorders (since 1976) and the home video disk system (introduced in 1979). *See* Artificial intelligence.

Industry and commerce. Impressive as this variety of communications and consumer-goods applications is, it is likely that the most rapid expansion of the electronics industry will occur in the fields of industry and commerce. The basic commodity is information in all its forms. The collection, processing, storage, retrieval, organization, display, and communication of information by electronic devices and systems, based on computer technology, have already come to dominate the affairs of all forms of business and commerce. Information-controlled processes and automation methods have been introduced to nearly all branches of manufacture, and to many extractive methods and processing of materials.   [DONALD G. FINK]

Electrostatic lens

An electrostatic field with axial or plane symmetry which acts upon beams of charged particles of uniform velocity as glass lenses act on light beams. The action of electrostatic fields with axial symmetry is analogous to that of spherical glass lenses, whereas the action of electrostatic fields with plane symmetry is analogous to that of cylindrical glass lenses. Plane symmetry as used here signifies that the electrostatic potential is constant along any normal to a family of parallel planes.

The action of an electrostatic lens on the paths of charged particles passing through it is most readily visualized with the aid of an equipotential plot of the fields in a plane of symmetry of the lens. The equipotential lines in the plot indicate the intersection with the plane of the drawing of surfaces on which the electrostatic potential is a constant. The paths of charged particles in the electrostatic field are bent toward the normals of the equipotentials as the particles are accelerated, and away from the normals as the particles are decelerated. *See* ELECTRON MOTION IN VACUUM.

Axially symmetric lenses. These lenses are generally formed at or between circular apertures and cylinders maintained at suitable potentials. A number of such lenses are shown with characteristic path plots in Fig. 1. For any of these it is possible to define focal points, principal planes, and focal lengths in the same manner as for light lenses and to determine with their aid image magnification for any object position (Fig. 2). For a thin electrostatic lens in particular, that is, a lens for which the extent of the variation in potential is small compared to its focal length, the object side focal length f_o and the image side focal length f_i are given by Eq. (1). Here Φ (z) is the potential

$$\frac{\Phi_o{}^{1/2}}{f_o} = \frac{\Phi_i{}^{1/2}}{f_i} = \frac{3}{16}(\Phi_o\Phi_i)^{1/4}\int\left(\frac{\Phi'}{\Phi}\right)^2 dz \quad (1)$$

along the axis of the lens, Φ' its derivative with respect to z (that is, the electric field along the axis), and Φ_o and Φ_i are the potential in object and image space, respectively. The integration is extended over the lens field. The quantity Φ is here normalized so that it is equal to the accelerating potential of the particle in question.

Axially symmetric lenses are commonly divided into the four classes that follow.

Simple aperture lenses. These are the lens fields formed about circular apertures in a plane metallic electrode at potential ϕ with different electrostatic fields $-\phi'_o$ and $-\phi'_i$ on the two sides. In most cases the focal length f of such a lens is given to a sufficient degree of accuracy by the Davisson-Calbick formula for an aperture, Eq. (2). Simple

$$\frac{1}{f} = \frac{\phi'_i - \phi'_o}{4\phi} \quad (2)$$

aperture lenses are encountered as parts of more complex electrostatic lens systems, as well as at the mesh openings of metal screens employed as electrostatic shields in vacuum tubes.

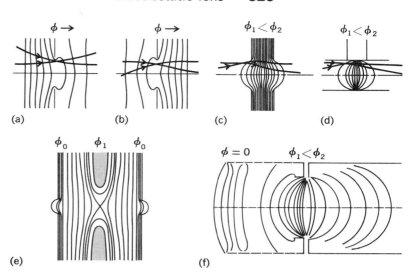

Fig. 1. Axially symmetric electrostatic lenses. (*a*) Single-aperture lens (decreasing field). (*b*) Single-aperture lens (increasing field). (*c*) Two-aperture lens. (*d*) Two-cylinder lens. (*e*) Unipotential lens. (*f*) Cathode lens (image tube). (*From E. G. Ramberg and G. A. Morton, J. Appl. Phys., vol. 10, 1939, and V. K. Zworykin et al., Electron Optics and the Electron Microscope, Wiley, 1945*)

Bipotential, or immersion, lenses. In these lenses image space and object space are field-free, but at different potentials. Typical examples are the lenses formed between apertures or cylinders at different potentials (Fig. 1c and d). If the separation d of the two apertures is large compared to their diameters and if each component aperture lens satisfies the conditions for a validity of the Davisson-Calbick formula, the focal lengths of the bipotential aperture lens are given by Eq. (3). The

$$\frac{1}{f_o} = \left(\frac{\phi_i}{\phi_o}\right)^{1/2}\frac{1}{f_i} = \frac{3}{8d}\left[1 - \left(\frac{\phi_o}{\phi_i}\right)^{1/2}\right]\left(\frac{\phi_i}{\phi_o} - 1\right) \quad (3)$$

distances of the principal planes from the plane of symmetry are given by Eqs. (4) and (5). Generally,

$$h_o = -d/2 - 4d\phi_o/[3(\phi_i - \phi_o)] \quad (4)$$

$$h_i = d/2 - 4d\phi_i/[3(\phi_i - \phi_o)] \quad (5)$$

the principal planes are displaced from the plane of symmetry toward the low-potential side, with the image-side principal plane closer to object space than the object-side principal plane.

For two cylinders of equal diameter D, whose difference in potential is small compared to their mean potential, Eq. (6) gives the focal lengths.

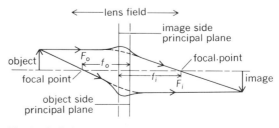

Fig. 2. Definition of principal planes, focal points, and focal lengths for axially symmetric electrostatic lens.

$$\frac{1}{f_o} = \left(\frac{\phi_i}{\phi_o}\right)^{1/2} \quad \frac{1}{f_i} = \left(\frac{\phi_i}{\phi_o}\right)^{1/4} \quad 0.66 \quad \left(\frac{\phi_i - \phi_o}{\phi_i + \phi_o}\right)^2 \frac{1}{D} \qquad (6)$$

Bipotential lenses, in particular lenses formed between two cylinders at different potentials, find wide application in beam-focusing devices such as electron guns. Like unipotential lenses, they invariably act as converging lenses.

Unipotential lenses. For this type the potentials are equal in object and image space. In their simplest form these lenses consist of three apertures of which the outer two are at a common potential ϕ_o and the central aperture is at a different, generally lower, potential ϕ_1. For such lenses with a central aperture of diameter D and the two outer apertures, of smaller diameter, separated a distance D from the plane of symmetry, the weak-lens focal length is given by Eq. (7). As ϕ_1 approaches zero,

$$\frac{1}{f} = \frac{0.2}{D}\left(\frac{\phi_o - \phi_1}{\phi_o}\right)^2 \qquad (7)$$

the quantity $1/f$ increases more rapidly than indicated by this formula; it attains a value of $0.7/D$ for $\phi_1 = 0$. Unipotential lenses operated at high potentials (for example, $\phi_o = 50$ kv, $\phi_1 = 0$) are employed as objectives and projection lenses in electrostatic electron microscopes. The electrodes are commonly made out of stainless steel and given a high polish.

Cathode lenses or immersion objectives. Here the lens field extends from the emitter surface up to field-free image space. Examples are the cathode region of an electron gun, the electron-optical system of an electrostatic image tube or image converter, and the objective of an emission electron microscope. In the electron gun the cathode lens converges the electrons emitted by the cathode to a small spot, the crossover, which is imaged by a second electron lens as the scanning spot on the cathode-ray tube screen. *See* CATHODE-RAY TUBE.

In the image tube the cathode itself – a transparent photoemissive surface on which a light picture is projected – is imaged on a fluorescent screen beyond the cathode lens. Frequently a cathode lens consists essentially of a uniform accelerating field followed by a short lens. The image magnification m is then given by Eq. (8). Here u is

$$m = v/2u \qquad (8)$$

the distance between cathode and short lens and v is the distance between short lens and image. The quantity v is given by Eq. (9), where f is the focal length of the short lens.

$$1v = 1/f - 1/(2u) \qquad (9)$$

Lenses of plane symmetry. These lenses, analogous to cylindrical glass lenses, are formed between parallel planes and at slits, replacing the circular cylinders and apertures of lenses with axial symmetry. For the simple slit in an electrode at potential ϕ separating two regions of field $-\phi'_o$ and $-\phi'_i$, the focal length is given by the Davisson-Calbick formula for a slit as shown by Eq. (10).

$$\frac{1}{f} = \frac{\phi'_i - \phi'_o}{2\phi} \qquad (10)$$

[EDWARD G. RAMBERG]

Bibliography: See ELECTRON MOTION IN VACUUM.

Electrostriction

A form of elastic deformation of a dielectric induced by an electric field; specifically, the term applies to those components of strain which are independent of reversal of the field direction. Electrostriction is a property of all dielectrics and is thus distinguished from the converse piezoelectric effect, a field-induced strain which changes sign upon field reversal and which occurs only in piezoelectric materials. *See* PIEZOELECTRICITY.

Electrostrictive strain is approximately proportional to the electric susceptibility, elastic compliance, and the square of the field strength, and is extremely small for most materials.

The electrostrictive effect in certain ceramics is employed for commercial purposes in electromechanical transducers for sonic and ultrasonic applications. *See* MICROPHONE.

[ROBERT D. WALDRON]

Embedded systems

Computer systems that cannot be programmed by the user because they are preprogrammed for a specific task and are buried within the equipment they serve. The term derives from the military, where computer systems are generally activated by the flip of a toggle switch or the push of a button. For example, an airplane pilot may wish to turn on the countermeasures equipment with the flip of a switch. There is no need for the pilot to be involved with the computer. The same holds true for a soldier who may direct a ground-to-ground missile against a target tank by the simple push of a button. In both cases, an embedded computer quickly goes to work.

The emergence of extremely small microcomputers on silicon chips about $1/4$ in. (6 mm) square has made the concept of embedded systems for the military all the more desirable. As recently as 1960, it was difficult or impossible to embed a computer the size of a room into equipment which had to fit into, say, an airplane cockpit. The microcomputer has changed all that.

The military establishment thus depends on microelectronics technology for the development of ever more sophisticated embedded systems. With very inexpensive microcomputer chips, not one but dozens of computers can be squeezed into a small piece of equipment the size of a lunch box, each preprogrammed to do a specific job. By doing only its own task, each microcomputer can devote its entire computational and analytical powers to that task, making possible extremely powerful military systems. *See* INTEGRATED CIRCUITS; MICROCOMPUTER; MICROPROCESSOR.

VHSIC program. A manifestation of the military's interest in microelectronics technology to further progress in embedded systems is the Department of Defense's Very High Speed Integrated Circuit (VHSIC) program. The military expects the VHSIC program to provide it with integrated circuit chips sufficiently advanced to enable the United States to maintain a qualitative superiority in weapons and armaments over its adversaries. The Department of Defense has concluded that electronics technology, particularly integrated circuits, will form the foundation for systems that will answer the military's needs.

The VHSIC program is a joint government–

Table 1. Prime contractors and integrated circuit chip technologies used in the VHSIC program*

Prime contractor and armed services branch	Brassboard	Chip set	Chip technology
Honeywell Inc.; Air Force	Electrooptic signal processor	Parallel programmable pipe-line controller	Bipolar
Hughes Aircraft Co.; Army	Antijam communications; bat-tlefield information and distribution system	Digital corellator; algebraic en-coder/decoder; spread-spectrum subsystem	CMOS/SOS
IBM Corp.; Navy	Acoustic-signal processor	Complex multiplier/accumu-lator	NMOS
Texas Instruments Inc.; Army	Multimode fire-and-forget mis-sile	Gate array (with and without memory); array controller and sequencer; data pro-cessor; vector address generator; vector arithmetic and logic unit; static random-access memory	Bipolar, NMOS
TRW Inc.; Navy	Electronic-warfare signal processor	Contents and window-address-able memories; register arithmetic and logic unit; address generator; micro-controller; matrix switch; 16-bit multiplier/accumu-lator; four-part memory	Bipolar, CMOS
Westinghouse Corp.; Air Force	Advanced tactical radar processor	Arithmetic unit; pipeline arithmetic unit; extended arithmetic unit; controller arithmetic unit; gate array; 64-kilobit static random-access memory; memory-management unit	CMOS

*From Department of Defense.

private industry effort, spread over four phases, 0 through III. The first three phases have been running concurrently. Phase 0 was a 1-year effort to analyze approaches to be employed in phases I and II for the development of very-high-speed digital signal–processing integrated circuits, with speeds on the order of 10^{10} Hz and line defini-tions on the order of 1 μm or less. Line definition refers to the smallest dimensions of the various structures that make up the integrated circuit chip.

Table 1 lists six major aerospace and semicon-ductor firms that are prime contractors in the VHSIC phase I program. All of them have con-tracted with the various armed services branches shown in the table. Phase I started in 1980 and is to run for 6 years. The aim of phase I is the develop-ment of a brassboard version of the VHSIC de-vices that will later be reduced to even smaller sizes on integrated circuit chips.

Phase II will provide a system demonstration of the subsystems constructed in phase I, as well as the capability for improved system performance with higher-performance design concepts. Phase III supplements phases I and II and provides new or alternate technology directions not covered in phases I and II.

In phase I, integrated circuits with on-the-chip structural geometries of 1.25 μm will be employed. Devices made under this phase would operate at a minimum clock frequency of 30 MHz and a mini-mum gate-frequency product of 10^{11} gate-Hz/cm². (The gate-frequency product is the number of logic gates — the most basic elements of a chip — times the clock frequency at which the integrated circuit on the chip operates. The cm² denotes the area of silicon on the chip required to implement this performance.)

The gate-Hz/cm² quantity, also known as throughput capacity, is a measure of the capability expected from a VHSIC device. By comparison, conventional, commercially available micropro-cessor chips have throughput capacities of 10^{10} gate-Hz/cm², or an order of magnitude slower than that expected from VHSIC integrated circuits in phase I. The military, on the other hand, is plan-ning weapons systems with throughput capacities of 10^{13} gate-Hz/cm² or more. Such throughput rates mean that the smallest definable geometries of a VHSIC integrated circuit must be on the order of 1 μm or less, or total-system weights and power dissipation levels become too prohibitive.

All three branches of the armed services have major weapons programs that will use VHSIC chips, as can be seen from Table 2. The six prime contractors are using different integrated circuit technology approaches (Table 1). If the VHSIC program goes according to the schedule, it will produce demonstration products in 1984.

Although six major corporations are listed as the prime contractors in Table 1, many other firms, universities, and laboratories are assisting in the VHSIC program, either as subcontractors or as consultants.

Ada programming language. In addition to the hardware digital-signal processors the military expects to acquire from the VHSIC program, the Department of Defense is working on the related issue of software. The agency is trying to stan-dardize on a single high-level software program-ming language known as Ada for all three branches of the service to simplify procurements, the inter-change of softward programs among the services' branches, and training. The Army has taken de-livery of an Ada Language System (ALS) and

Table 2. Major defense areas and weapons systems served by the VHSIC program*

System area	Air Force	Army	Navy
Radar	Advanced medium-range air-to-air missile; advanced on-board signal processor; airborne warning and control system (AWACS); autonomous cruise-missile guidance; multi-function radar signal processor	Multimode fire-and-forget missile	Tactical air-to-air ground radar; surveillance radar
Imaging	Autonomous cruise-missile guidance	Target-acquisition/fire-control system; multi-mode fire-and-forget missile	Digital image processor
Antisubmarine warfare			Acoustic signal processor
Electronic warfare	Advanced defensive avionic system	High-mobility electronic-warfare weapons-targeting systems	Electronic systems module signal sorter
Command, Control and Communications (C³)	Signal processor chip set for Joint Tactical Information and Distribution System	Battlefield information distribution system	Antijam modem; NATO identification system

*From IBM Federal Systems Division, Manassas, VA.

hopes to have it available to its programmers by January 1984. The Air Force has contracted for its Ada Integrated Environment (AIE) and will also begin using the language in 1984. The Navy is investigating an Ada language based on the best features of ALS and AIE. *See* PROGRAMMING LANGUAGES. [ROGER ALLAN]

Bibliography: R. Allan, Special issue on military electronics, *Electr. Des. Mag.*, 29(16):87–119, August 1981; W. H. Lee, VHSIC functions for signal processing, *1982 Southcon Professional Program*; L. W. Sumney, Special issue on war and peace, *IEEE Spectrum Mag.*, 19(10):93–94, October 1982.

Emitter follower

A circuit that utilizes a common-collector transistor which provides less than unity voltage gain but high input resistance and low output resistance.

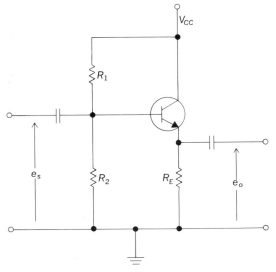

Typical emitter follower.

This circuit is used extensively to provide isolation or impedance matching between two electronic circuits.

A circuit diagram of a common-collector transistor amplifier is shown in the figure. This amplifier is similar to the cathode follower in its operations, although there are four important differences. (1) This amplifier has a voltage gain which is very close to unity (much closer to unity for typical loads than the cathode follower). (2) The voltage drop across the emitter resistance (from emitter to ground) may be either positive or negative, depending on whether an *npn* or *pnp* transistor is used. In the case of the cathode follower the drop across the cathode resistor is always positive. (3) The input resistance of the emitter follower, although high (tens or hundreds of kilohms), is low compared with that of the cathode follower. (4) The output resistance of the emitter follower is much lower (perhaps by a factor of 10) than that of the cathode follower.

In cascading transistor amplifier stages, the emitter follower is not used as an intermediate stage because its input-to-input voltage gain is in general less than that of the common-emitter amplifier. This is due to the relatively high input resistance of the emitter follower.

The most common use for the emitter follower is as a device which performs the function of impedance transformation over a wide range of frequencies with voltage gain close to unity. In addition, the emitter follower increases the power level of the signal.

A typical emitter follower circuit is shown in the illustration. The input signal is applied across the parallel combination of R_1 and R_2, which constitute the biasing arrangement for this circuit. *See* BIAS OF TRANSISTORS.

The collector in the figure is connected directly to the power supply and hence is at ground potential for ac signals. In many practical applications a protective resistance R_C is connected in series between the collector and the power supply to guard against transistor damage in the case of an accidental short circuit across R_E.

The emitter follower possesses negative feedback and results in reduced distortion. Its output is always in phase with the input, and typical values of voltage gain are about $0.95-0.99$.

[CHRISTOS C. HALKIAS]

Equivalent circuit

A representation of an actual electric circuit or device by a simplified circuit whose behavior is identical to that of the actual circuit or device over a stated range of operating conditions. These conditions depend on the elements contained in the actual circuit and may include such variables as frequency, temperature, and pressure, in addition to voltage and current. Equivalent circuits are often used by engineers to simplify circuit analysis since they show the relation between the variables more clearly than the actual circuit. *See* NETWORK THEORY.

Two types of equivalent circuits are used. One type is a simplification of an actual linear device, such as a transformer. In general, there are no restrictions on the magnitudes of applied voltages and currents in this type of equivalent circuit; therefore the equivalence holds over a wide range of operating conditions. *See* TRANSFORMER.

The second type of equivalent circuit is used to represent nonlinear devices, most commonly vacuum tubes and transistors. The actual circuit of the nonlinear device can be replaced by a fictitious generator and a simple network representing the operation of the actual device. Both the generator and network quantities depend on the parameters of the device. These parameters vary; therefore calculations made are necessarily limited to a portion of the characteristics over which the values of the parameters are reasonably constant. This type of circuit can, therefore, be used only for analysis of small signals. For large-signal analysis of nonlinear devices, it is necessary to use graphical methods on the characteristic curve of the device.

[ROBERT L. RAMEY]

Excitation

Application of energy to one portion of a system or apparatus in a manner that enables another portion to carry out a specialized function. Excitation establishes a condition essential to that function. In atomic physics, excitation means the addition of energy to an atom at ground state to produce an excited state. *See* EXCITATION POTENTIAL.

Excitation of a system or apparatus serves either to permit a transfer of energy or to control a flow of energy elsewhere in the system. Excitation energy may differ from the output energy in source, form, level, or location. Excitation produces a primary effect that is linked through an intermediate physical phenomenon to a dependent secondary effect (see illustration).

Some examples will make this dependency apparent. A dynamic loudspeaker uses an excitatory current in a field coil to generate a magnetic field; when the voice coil receives an audio signal, it generates a second magnetic field, which interacts with the first to cause motion of the voice coil and the voice cone of the speaker. A sound motion-picture projector uses an excitatory lamp to provide a source of illumination; when the photosensitive

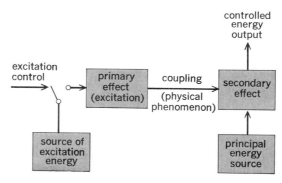

System diagram showing function of excitation.

pickup receives variations in illumination resulting from changes in the opaque area of the sound track through which the illumination must pass, it generates an audio signal output. A fluorescent lamp uses an excitatory current passing through an enclosed gas to generate ultraviolet radiation; when the phosphor coating inside the tube absorbs ultraviolet energy, it emits visible light. The human heart, to cite an example from biology, uses for excitation the decreasing electrical transmembrane potential of the pacemaker to generate an electrical impulse; when the surrounding striated muscle tissue receives this impulse, it transmits the electrical stimulus to other adjacent tissue and utilizes metabolic energy to contract, a response which produces the heartbeat.

[WILLIAM W. SNOW]

Excitation potential

The difference in potential between an excited atomic or molecular state and the ground state. The term is most generally used in connection with electron excitation, but it can be applied to excited molecular vibrational and rotational states.

A closely related term is excitation energy. If the unit of potential is taken as the volt and the unit of energy as the electron volt, then the two are numerically equal. According to the Bohr theory, there is a relationship between the wavelength of the photon associated with the transition and the excitation energies of the two states. Thus the basic equation for the emission or absorption of energy is as shown below, where h is Planck's con-

$$\frac{hc}{\lambda} = E_i - E_f$$

stant, c the velocity of light, λ the wavelength of the photon, and E_f and E_i the energies of the final and initial states, respectively. If the final state is the ground state, then the difference in energies is just the excitation energy of the initial state. If neither of the two states is the ground state, then the difference is numerically equal to the difference in excitation energies of the two states. This suggests that the excitation energy for many states may be determined spectroscopically. In fact, the careful measurement of the wavelengths associated with the transitions, along with the identification of the levels involved, permits the assignment of an excitation potential to each available energy state. This may be done by either emission or absorption spectra.

Measurements have also been made by using electron impact means. Here electrons from a hot filament are accelerated by a grid and collected on an outer electrode after passing through an intervening space where they can interact with a gas of molecular or atomic species. The potentials are so adjusted that the electrons can reach the collector only if they lose no energy after passing the grid. As the accelerating grid potential is increased, a series of sharp drops is noted in the collector current. Each of these is interpreted as representing a case where the electrons have obtained just enough energy to produce excitation, and hence no longer have the energy needed to climb the potential barrier to the collector. The evidence thus obtained is very direct, but the accuracy is not as good as that for spectroscopic data. *See* IONIZATION POTENTIAL.

[GLENN H. MILLER]

Bibliography: S. Gasiorowicz, *Structure of Matter: A Survey of Modern Physics*, 1979; F. K. Richtmyer et al., *Introduction to Modern Physics*, 6th ed., 1969; M. R. Wehr et al., *Physics of the Atom*, 3d ed., 1978.

Expert system

Methods and techniques for constructing human-machine systems with specialized problem-solving expertise. The pursuit of this area of artificial intelligence research has emphasized the knowledge that underlies human expertise and has simultaneously decreased the apparent significance of domain-independent problem-solving theory. In fact, a new set of principles, tools, and techniques have emerged that form the basis of knowledge engineering.

Expertise consists of knowledge about a particular domain, understanding of domain problems, and skill at solving some of these problems. Knowledge in any specialty is of two types, public and private. Public knowledge includes the published definitions, facts, and theories which are contained in textbooks and references in the domain of study. But expertise usually requires more than just public knowledge. Human experts generally possess private knowledge which has not found its way into the published literature. This private knowledge consists largely of rules of thumb or heuristics. Heuristics enable the human expert to make educated guesses when necessary, to recognize promising approaches to problems, and to deal effectively with erroneous or incomplete data. The elucidation and reproduction of such knowledge are the central problems of expert systems.

Importance of expert knowledge. Researchers in this field suggest several reasons for their emphasis on knowledge-based methods rather than formal representations and associated analytic methods. First, most of the difficult and interesting problems do not have tractable algorithmic solutions. This is reflected in the fact that many important tasks, such as planning, legal reasoning, medical diagnosis, geological exploration, and military situation analysis, originate in complex social or physical contexts, and generally resist precise description and rigorous analysis. Also, contemporary methods of symbolic and mathematical reasoning have limited applicability to the expert system area; that is, they do not provide the means for representing knowledge, describing problems at multiple levels of abstraction, allocating problem-solving resources, controlling cooperative processes, and integrating diverse sources of knowledge in inference. These functions depend primarily on the capacity to manipulate problem descriptions and apply relevant pieces of knowledge selectively. Current mathematics offers little help in these tasks.

The second reason for emphasizing knowledge is pragmatic: human experts achieve outstanding performance because they are knowledgeable. If computer programs embody and use this knowledge, they too attain high levels of performance. This has been proved repeatedly in the short history of expert systems. Systems have attained expert levels in several tasks: mineral prospecting, computer configuration, chemical structure elucidation, symbolic mathematics, chess, medical diagnosis and therapy, and electronics analysis.

The third motivation for focusing on knowledge is the recognition of its intrinsic value. Knowledge is a scarce resource whose refinement and reproduction creates wealth. Traditionally, the transmission of knowledge from human expert to trainee has required education and internship periods ranging from 3 to 20 years. By extracting knowledge from humans and transferring it to computable forms, the costs of knowledge reproduction and exploitation can be greatly reduced. At the same time, the process of knowledge refinement can be accelerated by making the previously private knowledge available for public test and evaluation.

In short, expert performance depends critically on expert knowledge. Because knowledge provides the key ingredient for solving important tasks, it reflects many features characteristic of a rare element: it justifies possibly expensive mining operations; it requires efficient and effective technologies for fashioning it into products; and a means of reproducing it synthetically would be "a dream come true."

Distinguishing characteristics. Expert systems differ in important ways from both conventional data processing systems and systems developed by workers in other branches of artificial intelligence. In contrast to traditional data-processing systems, artificial intelligence applications exhibit several distinguishing features, including symbolic representations, symbolic inference, and heuristic search. In fact, each of these characteristics corresponds to a well-studied core topic within artificial intelligence. Often a simple artificial intelligence task may yield to one of the formal approaches developed for these core problems. Expert systems differ from the broad class of artificial intelligence tasks in several regards. First, they perform difficult tasks at expert levels of performance. Second, they emphasize domain-specific problem-solving strategies over the more general, "weak" methods of artificial intelligence. Third, they employ self-knowledge to reason about their own inference processes and provide explanations of justifications for the conclusions they reach. As a result of these distinctions, expert systems represent an area of artificial intelligence research with specialized paradigms, tools, and system-development strategies.

Accomplishments. There have been a number of notable accomplishments by expert systems: PROSPECTOR discovered a molybdenum deposit whose ultimate value will probably exceed $100 million; R1 configures customer requests for computer systems; DENDRAL supports hundreds of international users daily in chemical structure elucidation; CADUCEUS embodies more knowledge of internal medicine than any human (approximately 80% more) and can correctly diagnose complex test cases that baffle experts; PUFF integrated knowledge of pulmonary function disease with a previously developed domain-independent expert system for diagnostic consultations and now provides expert analyses at a California medical center.

Types of systems. Most of the knowledge-engineering applications fall into a few distinct types, summarized in the table.

Generic categories of knowledge engineering applications

Category	Problem addressed
Interpretation	Inferring situation descriptions from sensor data
Prediction	Inferring likely consequences of given situations
Diagnosis	Inferring system malfunctions from observables
Design	Configuring objects under constraints
Planning	Designing actions
Monitoring	Comparing observations to plan vulnerabilities
Debugging	Prescribing remedies for malfunctions
Repair	Executing a plan to administer a prescribed remedy
Instruction	Diagnosing, debugging, and repairing students' knowledge weaknesses
Control	Interpreting, predicting, repairing, and monitoring system behaviors

Interpretation systems infer situation descriptions from observables. This category includes surveillance, speech understanding, image analysis, chemical structure elucidation, signal interpretation, and many kinds of intelligence analysis. An interpretation system explains observed data by assigning symbolic meanings to them which describe the situation or system state accounting for the data. *See* CHARACTER RECOGNITION; SPEECH RECOGNITION.

Prediction systems present the likely consequences of a given situation. This category includes weather forecasting, demographic predictions, traffic predictions, crop estimations, and military forecasting. A prediction system typically employs a parameterized dynamic model with parameter values fitted to the given situation. Consequences which can be inferred from the model form the basis of the predictions.

Diagnosis systems predict system malfunctions from observables. This category includes medical, electronic, mechanical, and software diagnosis. Diagnosis systems typically relate observed behavioral irregularities with underlying causes by using one of two techniques. One method essentially uses a table of associations between behaviors and diagnoses, and the other method combines knowledge of system design with knowledge of potential flaws in design, implementation, or components to generate candidate malfunctions consistent with observations.

Design systems develop configurations of objects that satisfy the constraints of the design problem. Such problems include circuit layout, building design, and budgeting. Design systems construct descriptions of objects in various relationships with one another and verify that these configurations conform to stated constraints. In addition, many design systems attempt to minimize an objective function that measures costs and other undersirable properties of potential designs. This view of the design problem can subsume goal-seeking behavior as well, with the objective function incorporating measures of goal attainment. *See* COMPUTER-AIDED DESIGN AND MANUFACTURING.

Planning systems design actions. These systems specialize in problems with the design of objects that perform functions. They include automatic programming, robot, project, route, communication, experiment, and military planning problems. Planning systems employ models of agent behavior to infer the effects of the planned agent activities. *See* ROBOTICS.

Monitoring systems compare observations of system behavior to features that seem crucial to successful plan outcomes. These crucial features, or vulnerabilities, correspond to potential flaws in the plan. Generally, monitoring systems identify vulnerabilities in two ways. One type of vulnerability corresponds to an assumed condition whose violation would nullify the plan's rationale. Another kind of vulnerability arises when some potential effect of the plan violates a planning constraint. These correspond to malfunctions in predicted states. Many computer-aided monitoring systems exist for nuclear power plant, air traffic, disease, regulatory, and fiscal management tasks, although no fielded expert systems yet address these problems.

Debugging systems prescribe remedies for malfunctions. These systems rely on planning, design, and prediction capabilities to create specifications or recommendations for correcting a diagnosed problem. Computer-aided debugging systems exist for computer programming in the form of intelligent knowledge base and text editors, but none of them qualify as expert systems.

Repair systems develop and execute plans to administer a remedy for some diagnosed problem. Such systems incorporate debugging, planning, and execution capabilities. Computer-aided systems occur in the domains of automotive, network, avionic, and computer maintenance, as well as others, but expert systems are just entering this field.

Instruction systems incorporate diagnosis and debugging subsystems which specifically address student behavior. These systems typically begin by constructing a hypothetical description of the student's knowledge. Then they diagnose weaknesses in the student's knowledge and identify an appropriate remedy. Finally, they plan a tutorial interaction intended to convey the remedial knowledge to the student.

An export control system adaptively governs the overall behavior of a system. To do this, the control system must repeatedly interpret the current situation, predict the future, diagnose the causes of anticipated problems, formulate a remedial plan, and monitor its execution to ensure success.

Problems addressed by control systems include air traffic control, business management, battle management, mission control, and others.

Systems components. The illustration shows an idealized representation of an expert system. No existing expert system contains all the components shown, but one or more components occur in every system. The ideal expert system contains: a language processor for problem-oriented communications between the user and the expert system; a blackboard for recording intermediate results; a knowledge base comprising facts plus heuristic planning and problem-solving rules; an interpreter that applies these rules; a scheduler to control the order of rule processing; a consistency enforcer that adjusts previous conclusions when new data or knowledge alter their bases of support; and a justifier that rationalizes and explains the system's behavior.

The user interacts with the expert system in problem-oriented languages, usually some restricted variant of English, and in some cases through means of a graphics or structure editor. The language processor mediates information exchanges between the expert system and the human user. Typically, the language processor dissects, or parses, and interprets user questions, commands, and volunteered information. Conversely, the language processor formats information generated by the system, including answers to questions, explanations and justifications for its behavior, and requests for data. Existing expert systems generally employ natural language parsers written in INTERLISP to interpret user inputs, and use less sophisticated techniques exploiting "canned text" to generate messages to the user. *See* PROGRAMMING LANGUAGES.

The blackboard is a global data base that records intermediate hypotheses and decisions

which the expert system manipulates. Every expert system uses some type of intermediate decision representation, but only a few explicitly employ a blackboard for the various types of ideal expert system decisions. The illustration shows three types of decisions recorded on the blackboard: plan, agenda, and solution elements. Plan elements describe the overall or general attack the system will use to solve the current problem, including current plans, goals, problem states, and contexts. For example, a plan may recommend processing all low-level sensor data first; formulating a small number of most promising hypotheses; refining and elaborating each of these hypotheses until the best one emerges; and, finally, focusing exclusively on that candidate until the complete solution is found. This kind of plan has been incorporated in several expert systems. The agenda elements record the potential actions awaiting execution. These generally correspond to knowledge-base rules which seem relevant to some decision placed on the blackboard previously. The solution elements represent the candidate hypotheses and decisions that the system has generated thus far, along with the dependencies, called links, that relate decisions to one another.

The scheduler maintains control of the agenda and determines which pending action should be executed next. Schedulers may utilize considerable abstract knowledge, such as "do the most profitable thing next" and "avoid redundant effort." To apply such knowledge, the scheduler needs to prioritize each agenda item according to its relationship to the plan and to other solution elements. To do this, the scheduler generally needs to estimate the effects of applying the potential rule.

The interpreter executes the chosen agenda item by applying the corresponding knowledge-base rule. Generally, the interpreter validates the relevance conditions of the rule, binds variables in these conditions to particular solution blackboard elements, and then makes rule-prescribed changes to the blackboard. Interpreters of this sort are generally written in LISP because of its facilities for manipulating and evaluating programs. However, other languages are also suitable.

The consistency enforcer attempts to maintain a consistent representation of the emerging solution. This may take the form of likelihood revisions when the solution elements represent changing hypothetical diagnoses and when some new data are introduced. Alternatively, the enforcer might implement truth maintenance procedures when the solution elements represent changing logical deductions and their truth-value relationships. Most expert systems use some kind of numerical adjustment scheme to determine the degree of belief in each potential decision. This scheme attempts to ensure that plausible conclusions are reached and inconsistent ones avoided.

The justifier explains the actions of the system to the user. In general, it answers questions about why some conclusion was reached or why some alternative was rejected. To do this, the justifier uses a few general types of question-answering plans. These typically require the justifier to trace backward along blackboard solution elements

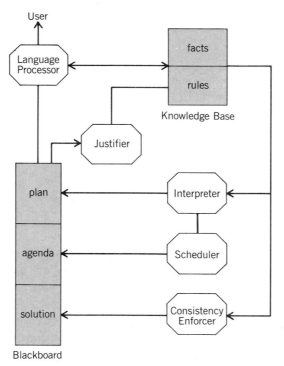

Anatomy of an ideal expert system.

from the questioned conclusion to the intermediate hypotheses or data that support it. Each step backward corresponds to the inference of one knowledge-base rule. The justifier collects these intermediate inferences and translates them to English for presentation to the user. To answer "why not" questions, the system uses a heuristic variant of this technique. Supposedly, it can identify a possible chain of rules that would reach the questioned conclusion but which did not apply because the relevance condition of some rule failed. The justifier explains the system's decision to reject a possible conclusion by claiming that such failed conditions impede all reasoning chains that can support the conclusion.

Finally, the knowledge base records rules, facts, and information about the current problem that may be useful in formulating a solution. While the rules of the knowledge base have procedural interpretations, the facts play only a passive role. *See* ARTIFICIAL INTELLIGENCE; DATA-PROCESSING SYSTEMS. [FREDERICK HAYES-ROTH]

Bibliography: E. A. Feigenbaum, Themes and case studies of knowledge engineering, in D. Michie (ed.), *Expert Systems in the Microelectronic Age*, pp. 3–25, 1979; F. Hayes-Roth, D. Waterman, and D. Lenat (eds.), *Building Expert Systems*, in print. E. Shortliffe, *Computer Based Medical Consultations: MYCIN*, 1976; R. Webster and L. Miner, Expert systems: Programming problem-solving, *Technology*, 2(1):62–73, 1982.

Facsimile

Transmission of a fixed image by wire or radio. The image or subject copy is usually in the form of a photograph, handwriting, map, or drawing. Most facsimile equipment in the United States is used for telegram pickup and delivery, but other uses are growing rapidly. Business document transmission has the largest growth. Conventional telephone handsets are sometimes used for sending documents, with an acoustic coupler providing the connection from facsimile scanner to handset; transmission is then over a regular telephone voice-quality channel. In just one weather facsimile network, more than 900 recorders throughout the United States copy weather charts 7 days a week, 24 hr a day. Almost all newspaper photographs of events outside the local city in which the newspaper is published are sent by facsimile. Complete newspapers are printed from facsimile transmissions originating in distant cities. Railroads employ high-speed facsimile (several pages per minute) for transmission of waybills. Cloud-cover photographs and other types of graphic data are recorded by facsimile from signals sent by orbiting and stationary satellites. X-ray photographs have been sent from hospitals to radiological interpretation centers. A new use that is expected to prove important is information storage and retrieval. Facsimile will provide rapid, remote access to documents stored on microfilm at central locations such as libraries. *See* RADIO.

Transmission system. The subject copy is divided into dots, or elemental areas, in the scanning and transmitting process. Each elemental area (1/100 × 1/100 in. in a typical scanning raster) is transmitted as an electrical signal that represents its blackness.

In an amplitude-modulated system, elemental

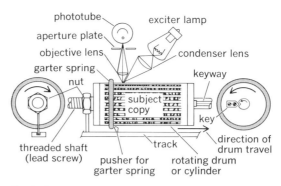

Fig. 1. Rotary facsimile scanner.

areas are transmitted at a constant rate, usually 1800–4000 per second, when transmission is over a voice-frequency circuit. Newspaper pages are transmitted at rates of up to 1,500,000 elemental areas per second over microwave circuits.

Some circuits cannot retain the true amplitude relationship of the signal in transmission. For example, high-frequency radio circuits are subject to considerable fading. In such cases amplitude modulation is unsatisfactory. Frequency-shift modulation is used so that the received copy will not be degraded by fading or level changes.

The received signals are recorded on a record sheet by any one of many processes. In a direct-recording facsimile system, an image is visible while being recorded. In photofacsimile systems using photographic film or paper, processing is necessary before the image can be viewed.

Scanning. A photosensor, such as a phototube, photodiode, or phototransistor, views the subject copy through an optical system that generally consists of an objective lens and an aperture plate (Fig. 1). The lens forms an image on an aperture plate of the subject copy. The aperture in the plate permits light from only one elemental area to pass to the photosensor. To allow the photosensor to scan the entire area of the subject copy, the copy may be placed on a rotating cylinder and the drum slowly moved the length of its shaft so as to scan a line about 1/100 in. wide during each turn. Several mechanical equivalents are used.

If the copy is to be scanned in the flat (flat-bed scanner), the beam to the photosensor is optically swept across the width of the slowly advancing copy. In an electronic scanner using the cathode-ray tube, the copy remains stationary. The elemental area to be scanned is illuminated by a spot swept across the face of the cathode-ray tube (Fig.

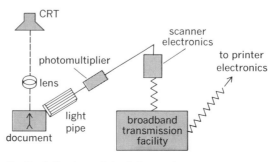

Fig. 2. Cathode-ray-tube flying-spot scanner.

2). When high brilliance and long life are essential, a phosphor-coated rotating cylinder in the tube serves as target. *See* CATHODE-RAY TUBE.

Modulation. The signal from the photosensor or photomultiplier is maximum when light from a white portion of the image passes through the aperture. This signal includes frequencies down to direct current (baseband) and thus is suitable for transmission only over channels that retain the dc level, such as direct connections. This baseband signal is preferably furnished to a modem. The modem (a contraction of modulator and demodulator) is an interface unit which translates outgoing data into forms suitable for transmission over telephone or radio channels. *See* MODULATOR; MULTIPLIER PHOTOTUBE.

The modem produces a signal suitable for transmission over regular channels with no distance limit. Amplitude, frequency-shift, or pulse-code modulation is used. For dedicated telephone channels amplitude modulation is used; for the switched public telephone network or radio channels, frequency-shift modulation is used. Where many signals are time multiplexed, for example, in digital transmission, pulse-code modulation is used.

Amplitude modulation may be performed in one of several ways. (1) The voltage produced by the photosensor under control of the copy density may modulate a carrier in a ring diode modulator. (2) Mechanical interruption of the light beam by a "chopper" may produce the carrier frequency at the photosensor. (3) The photosensor may have two cathodes which are equally illuminated by the light from the copy being scanned. When connected as one arm of a balanced bridge circuit, this photosensor directly modulates the carrier.

In a typical frequency-shift-modulated channel, 1500 Hz represents white and 2300 Hz represents black. These signals may be transmitted over an amplitude-modulated radio transmitter. This type of modulation is referred to as audio-frequency shift. Alternately, the dc picture signal may serve to shift a radio carrier frequency over a range of 800 Hz. At the receiving station, a heterodyne develops the 1500–2300 Hz audio-frequency signal, which then passes through a limiter and discriminator before energizing the recorder.

Direct recording. Direct recording is usually accomplished by one of several techniques. (1) A direct current may be passed through a paper dampened with an electrolyte (Fig. 3). Metal ions from the printing bar transfer to the paper to produce a mark at the intersection of the helix wire and the printing bar. (2) A white coating covers a conducting carbon coating on dry paper. A small stylus wire in contact with this paper passes current through the white coating at the recording point. The white coating is burned off to produce a mark by exposing the black carbon. (3) Recording may be on pressure-sensitive paper by a mechanically actuated stylus (Fig. 4). Three or more styli are carried on a belt between two pulleys. When one stylus completes the recording of a line, the next stylus starts recording the following line. (4) A cathode-ray tube forms an electrostatic image by the xerographic process (Fig. 5). The image on a selenium-coated drum is dusted with a powder that adheres to the charged areas to form the visible image. This image is trans-

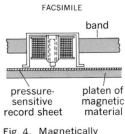

FACSIMILE

Fig. 4. Magnetically driven stylus for recording or pressure-sensitive record sheet.

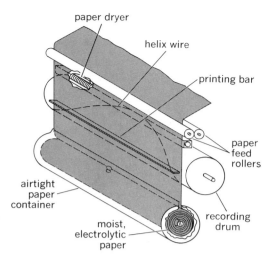

Fig. 3. Electrolytic facsimile recorder.

ferred to paper and fused to produce the output copy. (5) A fiber optics faceplate cathode-ray tube may be used to write on photosensitive recording paper as it passes over the fiber ends. Light from the fibers selectively discharges the specially coated paper which has been previously charged by passing through a corona field. The paper then passes through a toner bath where black charged particles are deposited on the recorded areas. Alternately, direct writing oscillograph paper may be substituted with image development by ultraviolet light, or dry silver paper may be used with development by heat.

Photographic recording. Most of the photographic facsimile recorders in service use a drum technique. Photographic paper or film is clamped to a cylinder which then rotates in a light-tight chamber or darkroom. A glow modulator tube (crater lamp) illuminates an aperture plate that is focused onto the drum surface. As the drum rotates, it advances one elemental area per rotation until the whole photographic surface has been exposed. Normal photographic development and fixing then produces the completed copy. Newer types of photographic recorders are completely automatic. The photographic paper used has the developer built in. The total process of recording, including loading of the photographic paper, adjusting signal level, phasing, unloading, and photographic processing, is done without attention from

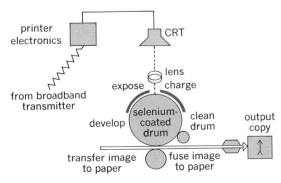

Fig. 5. Diagram of a cathode-ray-tube flying-spot xerographic recorder.

the operator. The unattended recorder delivers a finished photographic print.

Color facsimile. For transmission of color photographs, three separate facsimile transmissions are made from the original color print. Color-separation filters in the optical system of the facsimile transmitter produce black-and-white color-separation negatives at the facsimile recorder. The technique is similar to that used for color-separation negatives made by photography. These received films may be used for making matrices for dye-transfer-process color prints or for engraving printing plates for newspaper reproduction. Newsphoto services send some of their pictures in color by this method.

In another color facsimile technique the recording is directly on color film using a line sequential color system. First a line is recorded through a red filter, then a green, followed by blue. A rotating color wheel is used at the recorder between the light source and the writing point. A fiber optics circle to line converter is used to sweep the recording spot across the color film. The transmitter splits the beam from the color original into its three color components which are sensed by three separate photomultipliers. A switching arrangement sequentially selects the color to be sent. The received color print may be automatically processed in 1 min after completion of recording.

Digital facsimile. One type of digital facsimile system sends only black or white picture information; no gray scale shades are reproduced. At the facsimile transmitter the picture is sampled at a fixed rate controlled by a clock. All signals above a threshold value are sent as +6 volts for black, and all below as −6 volts for white. Because changes from black to white can occur only at discrete points along a recording line, vertical lines have a more ragged appearance than in comparable analog facsimile systems. However, noise on long circuits is less apt to degrade the received picture.

Another advantage is secrecy. For example, digital signals may be connected to military encrypting equipment for sending secret information. The output of the encrypting equipment connects to modems which process the signal for transmission. It is common to have a 2400-bits/sec speed of transmission for voice channels, but use of 4800-bits/sec and higher speed modems is growing. When the signal is sent over a distance, it is distorted by echoes, delay distortion, and other noise. Regenerative repeaters may be used to sample the signal again and eliminate these distortions unless they are severe enough to produce errors. The error rate that can be tolerated by facsimile signals is much higher than can be used for other forms of data transmission.

Instead of black or white sampling of each elemental area, gray scale shadings may be represented as the signal is digitized. Black, white, and two intermediate gray shades require two bits per elemental area. Some systems send 32 shadings which require five bits per elemental area.

Synchronization. The transmitted elemental areas are placed in the correct positions on the record sheet by properly synchronizing the scanning and recording mechanisms. In mechanically driven facsimile systems, the two mechanisms are driven at the same speed by synchronous motors.

The receiving motor must be supplied by electrical power that is synchronous with the power supplying the transmitter motor. When the transmitting and receiving stations are supplied by the same power system, synchronism is accomplished. If different power systems are involved, local power is generated at each station, and exceptionally precise control of the frequency is necessary.

If the synchronous motors are driven by local oscillators, the frequency difference between the transmitting and receiving oscillators should not exceed 0.001% when the system is transmitting letter-size message copy (scanned at 100 lines/in.); 0.0003% when transmitting 12×18 in. weather maps (scanned at 100 lines/in.); and 0.00003% when transmitting 16×22 in. newspaper copy scanned at the rate of 600 lines/in. (rate required for high definition). In facsimile systems employing cathode-ray tubes in the scanning or recording systems, the synchronism is usually controlled by a synchronizing pulse, as in television.

[KEN R. MC CONNELL]

Bibliography: A. G. Cooley, Facsimile, in K. Henney (ed.), *Radio Engineering Handbook,* 1959; D. M. Costigan, *Electronic Delivery of Documents and Graphics,* 1978.

Faraday's law of induction

A statement relating an induced electromotive force (emf) to the change in magnetic flux that produces it. For any flux change that takes place in a circuit, Faraday's law states that the magnitude of the emf ε induced in the circuit is proportional to the rate of change of flux as in expression (1).

$$\varepsilon \propto -\frac{d\Phi}{dt} \qquad (1)$$

The time rate of change of flux in this expression may refer to any kind of flux change that takes place. If the change is motion of a conductor through a field, $d\Phi/dt$ refers to the rate of cutting flux. If the change is an increase or decrease in flux linking a coil, $d\Phi/dt$ refers to the rate of such change. It may refer to a motion or to a change that involves no motion.

Faraday's law of induction may be expressed in terms of the flux density over the area of a coil. The flux Φ linking the coil is given by Eq. (2),

$$\Phi = \int B \cos \alpha \, dA \qquad (2)$$

where α is the angle between the normal to the plane of the coil and the magnetic induction B. The integral is taken over the area A enclosed by the coil. Then, for a coil of N turns, Eq. (3) holds.

$$\varepsilon = -N\frac{d\Phi}{dt} = -N \int \frac{d(B \cos \alpha)}{dt} A \qquad (3)$$

[KENNETH V. MANNING]

Fault-tolerant systems

Systems, predominantly computing and computer-based systems, which tolerate undesired changes in their internal structure or external environment. Such changes, generally referred to as faults, may occur at various times during the evolution of a system, beginning with its specification and proceeding through its utilization. Faults that occur during specification, design, implementation, or modification are called design faults; those occurring during utilization are referred to as operational faults. Design faults are due to mistakes made by humans or by automated design tools in the process of specifying, designing, implementing, or modifying a system. The source of an operational fault may be either internal or external to the system. Internal operational faults are due to physical component failures or human mistakes (when humans are integral parts of the system). External operational faults are due to physical phenomena (temperature, radiation, and so forth) or improper actions on the part of external systems and humans interacting with the system.

Fault prevention and fault tolerance. Both design faults and operational faults affect a system's ability to perform during use, typically resulting in reduced performance or failure. Faults are thus the primary concern in efforts to enhance system reliability (the probability of no system failures during a specified period of time) or, more generally, system performability (the probability distribution of system performance during a specified period of time). In order to satisfy a given reliability or performability requirement, faults can be attacked by two basic approaches: fault prevention and fault tolerance. Fault prevention is the more traditional approach; it comprises techniques which attempt to eliminate faults by avoiding their occurrence (for example, design methodologies, quality control methods, radiation shielding) or by finding and removing them prior to system utilization (for example, testing or verification methods). The use of fault tolerance techniques is based on the premise that a complex system, no matter how carefully designed and validated, is likely to contain residual design faults and to encounter unpreventable operational faults.

Faults and errors. To describe how systems tolerate faults, it is important to distinguish the concepts of fault and error. Relative to a description of desired system behavior at some specified level of abstraction, an error is a deviation from desired behavior caused by a fault. Since desired behavior may be described at different levels of abstraction (for example, the behavior of computer hardware may be described at the circuit level, the logic level, or the register-transfer level), a variety of errors can be associated with a given fault. Moreover, a fault can cause an error at one level without causing an error at another level. A fault is latent until it causes an error, and it is possible for a fault to remain latent throughout the lifetime of a system without ever causing an error. The distinction between fault and error is particularly important in the case of design faults. For example, a software design fault (commonly referred to as a bug) may remain latent for years before the software is finally executed in a manner that permits the bug to cause an execution error. The sense in which a system tolerates a fault is typically defined as some form of restriction on the errors caused by that fault. This term is usually used to refer to the prevention of errors at a level that represents the user's view of desired system behavior. Errors at this level are known as system failures and therefore a fault is tolerated if it does not cause a system failure. A system is fault-tolerant if all faults in a specified class of faults are tolerated.

Fault tolerance techniques. A variety of fault tolerance techniques are currently employed in the design of fault-tolerant systems, particularly fault-tolerant computing systems. Generally, fault tolerance techniques attempt to prevent lower-level errors (caused by faults) from propagating into system failures. By using various types of structural and informational redundancy, such techniques either mask a fault (no errors are propagated to the faulty subsystem's output) or detect a fault (via an error) and then effect a recovery process which, if successful, prevents system failure. In the case of a permanent internal fault, the recovery process usually includes some form of structural reconfiguration (for example, replacement of a faulty subsystem with a spare or use of an alternate program) which prevents the fault from causing further errors. Typically, a fault-tolerant system design will incorporate a mix of fault tolerance techniques which complement the techniques used for fault prevention. The choice of these techniques and, more generally, the specific nature of the system's hardware and software architecture is highly dependent on both the types of faults anticipated and the reliability or performability requirements imposed by the system's application environment. *See* SOFTWARE ENGINEERING.   [JOHN F. MEYER]

Bibliography: A. Avizienis, Fault-tolerant systems, *IEEE Transac. Comput.*, C-25(12):1304–1312, December 1976; T. Anderson and B. Randell (eds.), *Computing Systems Reliability*, 1979; T. Anderson and P. A. Lee, *Fault Tolerance: Principles and Practice*, 1981; D. P. Siewiorek and R. S. Swarz, *The Theory and Practice of Reliable System Design*, 1982.

Feedback circuit

A circuit that returns a portion of the output signal of an electronic circuit or control system to the input of the circuit or system. When the signal returned (the feedback signal) is at the same phase as the input signal, the feedback is called positive or regenerative. When the feedback signal is of opposite phase to that of the input signal, the feedback is negative or degenerative.

The use of negative feedback in electronic circuits and automatic control systems produces changes in the characteristics of the system which improve the performance of the system. In electronic circuits, feedback is employed either to alter the shape of the frequency-response characteristics of an amplifier circuit and thereby produce more uniform amplification over a range of frequencies, or to produce conditions for oscillation in an oscillator circuit. It is also used because it stabilizes the gain of the system against changes in temperature, component replacement, and so on. Negative feedback also reduces nonlinear distortion. In automatic control systems, feedback is used to compare the actual output of a system with a desired output, the difference being used as the input signal to a controller. These two points of view will be considered separately in the following discussion. However, the analysis of both feedback amplifiers and electromechanical control systems can be made on a frequency-response basis; from the point of view of analysis the two have much in common.

Amplifier feedback. Feedback can be introduced into an amplifier by connecting to the input a fraction of the output signal. An amplifier will, in

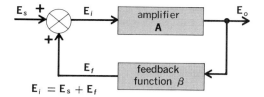

$$E_i = E_s + E_f$$

Block diagram of feedback circuit.

general, have better frequency-response characteristics when the system has feedback than when there is no feedback. The system can be designed to have a wider bandwidth and more nearly ideal frequency-response characteristics. Further, harmonic distortion due to nonlinear transistor or tube characteristics can be reduced by using feedback. *See* DISTORTION (ELECTRONIC CIRCUITS).

The use of feedback in an improperly designed system, however, can produce a system with worse characteristics. Amplifiers can become oscillators when feedback is used in an improperly designed system. For a discussion of amplifier frequency response and bandwidth *see* AMPLIFIER.

A system with feedback can be analyzed by using the block diagram representation shown in the illustration. The sinusoidal input signal is E_s and the amplifier gain is A, which is a function of frequency. When there is no feedback, $E_o = AE_s$ because $E_i = E_s$. When there is feedback, $E_i = E_s + E_f$.

Since $E_o = AE_i$, and $E_f = \beta E_o$, the overall gain of the system with feedback is then shown by the equation below.

$$\frac{E_o}{E_s} = \frac{A}{1 - A\beta}$$

This formula for the gain of an amplifier with feedback indicates the effect of feedback upon the frequency response of the amplifier without feedback. The $(1 - A\beta)$ term in the denominator is a complex number. Therefore, the magnitude and phase angle of the gain of the amplifier with feedback will differ from the gain of the amplifier without feedback. The amount of the difference depends upon the value of the A and the β terms, and no general statements can be made.

If there is a frequency for which $A\beta = 1$, the denominator of the expression for the gain will be zero while the numerator will not be zero. When this occurs, the amplifier will oscillate at approximately the frequency for which $A\beta = 1$. Furthermore, if there is a frequency for which the magnitude of $A\beta$ is greater than unity and the phase angle of $A\beta$ is 0°, 360°, or any integral multiple of 360°, the amplifier will oscillate.

Positive and negative feedback. The terms positive feedback and negative feedback are used to denote the type of feedback found in certain electronic circuits. *See* NEGATIVE-RESISTANCE CIRCUITS.

When the magnitude of the denominator of the feedback equation is greater than unity, the overall gain with this (negative) feedback will be less than the gain with no feedback, and stable operation will result.

If the magnitude of the denominator in Eq. (1) is less than unity, then the overall gain with this posi-

tive feedback will be greater than the gain without feedback. Under these circumstances the circuit may be unstable and oscillate.

It follows that both the magnitude and the phase angle θ of the product $A\beta$ are important in determining whether the feedback is positive.

The effect of feedback upon the frequency response of an amplifier can be determined from the expression $A/(1 - A\beta)$. The performance in a particular case depends upon the behavior of A and β as functions of frequency.

The usual method for analyzing the feedback amplifier to determine the possibilities of oscillation is the examination of the amplitude and phase of $A\beta$ over the frequency range from zero to a sufficiently high frequency (on the order of 100,000 Hz). The examination is often made by plotting the magnitude of $A\beta$ against the phase angle in polar coordinates.

An alternate method of analysis is to plot the logarithm of the magnitude of $A\beta$ against the logarithm of frequency and the phase angle $A\beta$ against the logarithm of frequency. The value of frequency at which the phase angle equals 180° is determined from the phase-angle-versus-log-frequency curve. If for this value of frequency the logarithm of $|A\beta|$ is greater than zero (the magnitude of $A\beta$ is greater than one), the amplifier will oscillate.

Several practical problems are found in the analysis of feedback amplifiers. One is that the values of the various circuit components, such as resistors, capacitors, and transformers, are only approximately known. For example, resistors may have a 20% tolerance and the values for transistor parameters may vary widely around a nominal value. Another problem is that all of the parameters which affect the performance, in particular stray capacitance introduced by the physical layout of components and wiring, are not included in the expression for $A\beta$. This means that, in practice, mathematical analysis is only a guide; final design is refined by experiment.

Input and output impedances. The discussion has considered voltage feedback where the output voltage, or some function of it, is fed back to the input. Current feedback, where a voltage proportional to the output current or some function of the output current is fed back, may also be used. Furthermore, an amplifier may have both voltage and current feedback.

The input and output impedances of a feedback amplifier depend upon whether the output current or voltage is sampled and upon whether this signal is fed back in series or in parallel with the input excitation. It is possible for the impedance with feedback to be greater than or less than what it was before the feedback was added. An emitter-follower is an example of a feedback amplifier where the output impedance has been reduced to a very low value and the input impedance has been increased drastically.

Oscillator feedback. An oscillator can be viewed as a feedback amplifier with an input voltage supplied from its output. Referring to the illustration, this would mean that E_s equals zero and E_i equals E_f.

From this viewpoint the condition for oscillation at a frequency f_o is (to a first and usually very good approximation) that $A\beta = 1$ at $f = f_o$. This means that the feedback must be positive.

Servomechanism feedback. The purpose of feedback in a servomechanism is to obtain better control of the system. As an example, consider a position control system which is used to position an object at a point determined by a reference signal.

The command to move the object is derived by comparing the reference signal with a signal indicating the instantaneous location of the object. The command signal is an error signal derived from the comparison of the actual and desired signals; any error signal drives the system in such a direction as to reduce the error, that is, to make the actual position agree with the desired reference position.

If feedback were not used in the position control system, a precisely calibrated control device would be needed to position the object in each position dictated by the reference. In general the required control could not be built with sufficient accuracy.

[HAROLD F. KLOCK]

Bibliography: J. D'Azzo and C. Houpis, *Linear Control Systems: Conventional and Modern*, 1975; J. Millman, *Microelectronics*, 1979; D. Schilling and C. Belove, *Electronic Circuits: Discrete and Integrated*, 1979.

Ferrite devices

Electrical devices whose principle of operation is based upon the use of ferrites, which are magnetic oxides. By common usage, ferrite devices are referred to as those using magnetically soft ferrites, with the spinel or garnet crystal structure. Since the electrical resistivity of ferrites is typically $10^6 - 10^{11}$ times that of metals, ferrite components have much lower eddy current losses and are hence used at frequencies generally above about 10 kHz.

Chemistry and crystal structure. Modern ferrite devices stemmed from the contributions on spinel ferrites made by the Japanese and Dutch scientists during World War II. The general formula for the spinel is MFe_2O_4, in which M is a divalent metal ion. In special cases, the divalent ion M can be replaced by an equal molar mixture of an univalent and trivalent ion. Thus lithium ferrite can be thought of as having the formula $Li_{0.5}Fe_{0.5}Fe_2O_4$. However, the commercially practical ferrites are those in which the divalent ion represents one or more Mg, Mn, Fe, Co, Ni, Cu, Zn, and Cd ions. The trivalent Fe ion may also be substituted by other trivalent ions such as Al. The compositions are carefully adjusted to optimize the device requirements, such as permeability, loss, ferromagnetic-resonance line width, and so forth.

The ferrimagnetic garnets were discovered in France and the United States in 1956. The general formula is $M_3Fe_5O_{12}$, in which M is a rare-earth or yttrium ion. Single-crystal garnet films form the basis of bubble domain device technology. Bulk garnets have applications in microwave devices.

With the exception of some single crystals used in recording heads and special microwave applications, all ferrites are prepared in polycrystalline form by ceramic techniques.

Applications. A summary of the applications of ferrites is given in the table. These may be divided into nonmicrowave and microwave applications. Further, the nonmicrowave applications may be divided into categories determined by the magnet-

ic properties based on the B/H behavior, that is, the variation of the magnetic induction or flux density B with magnetic field strength H, as shown in the illustration. The plot in the illustration is termed a hysteresis hoop, with the area encompassed by the loop being proportional to the power loss within the ferrite. The categories are: linear B/H, with low flux density; nonlinear B/H, with medium to high flux density; and highly nonlinear B/H, with a square or rectangular hysteresis loop.

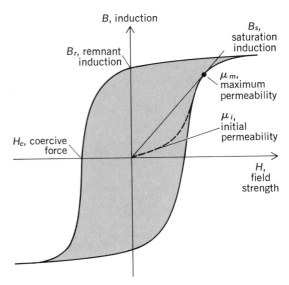

Plot of magnetic induction B as a function of magnetic field strength H.

Linear B/H devices. In the linear region, the most important devices are high-quality inductors, particularly those used in filters in frequency-division multiplex telecommunications systems and low-power wide-band and pulse transformers. Virtually all such devices are made of either manganese zinc (MnZn) ferrite or nickel zinc (NiZn) ferrite, though predominantly the former.

In the design of inductors, the so-called μQ product of a material has been found to be a useful index of the quality of the material. In this product, μ is the initial permeability and Q is equal to $\omega L/R$, where ω is the angular frequency, L the inductance, and R the effective series resistance arising from core loss. The higher the value of μQ, the better is the material, and ferrites have the highest μQ product of any commercially available magnetic material. Typical μQ values measured at 20 kHz for various materials are iron dust, 2000; Permalloy powder, 10,000; 12.5-μm-thick molybdenum Permalloy tape, 100,000; and MnZn ferrite, 250,000. Values of μQ greater than 10^6 at 20 kHz have been achieved commercially in specially prepared MnZn ferrites.

As compared with NiZn ferrites, MnZn ferrites have lower residual and hysteresis losses, higher permeabilities, and lower resistivities. A high permeability lowers the frequency of ferromagnetic resonance, which is accompanied by a large rise in losses. This factor, along with increased eddy current losses coming from lower resistivities, restricts the useful upper frequency range for

Summary of ferrite applications*

Ferrite chemistry	Device	Device function	Frequencies	Desired ferrite properties
Linear B/H, low flux density				
MnZn, NiZn	Inductor	Frequency selection network	<1MHz (MnZn)	High μ, high μQ, high stability of μ
		Filtering and resonant circuits	~1 – 100 MHz (NiZn)	with temperature and time
MnZn, NiZn	Transformer (pulse and wideband)	Voltage and current transformation Impedance matching	Up to 500 MHz	High μ, low hysteresis losses
NiZn	Antenna rod	Electromagnetic wave receival	Up to 15 MHz	High μQ, high resistivity
MnZn	Loading coil	Impedance loading	Audio	High μ, high B_s, high stability of μ with temperature, time, and dc bias
Nonlinear B/H, medium to high flux density				
MnZn, NiZn	Flyback transformer	Power converter	<100 kHz	High μ, high B_s, low hysteresis losses
MnZn	Deflection yoke	Electron-beam deflection	<100 kHz	High μ, high B_s
MnZn, NiZn	Suppression bead	Block unwanted ac signals	Up to 250 MHz	Moderately high μ, high B_s, high hysteresis losses
MnZn, NiZn	Choke coil	Separate ac from dc signals	Up to 250 MHz	Moderately high μ, high B_s, high hysteresis losses
MnZn, NiZn	Recording head	Information recording	Up to 10 MHz	High μ, high density, high μQ, high wear resistance
MnZn	Power transformer	Power converter	<60 kHz	High B_s, low hysteresis losses
Nonlinear B/H, rectangular loop				
MnMg, MnMgZn, MnCu, MnLi, etc.	Memory cores	Information storage	Pulse	High squareness, low switching coefficient, and controlled coercive force
MnMgZn, MnMgCd	Switch cores	Memory access transformer	Pulse	High squareness, controlled coercive force
MnZn	Magnetic amplifiers			
Microwave properties				
YIG†, MgMn‡, Li§, NiZn	Isolators, attenuators, circulators, switches, modulators	Impedance matching, power level control, power splitting	1 – 5 GHz (YIG), 2 – 30 GHz (YIG, MgMn, Li), 30 – 100 GHz (NiZn)	Controlled B_s, high resistivity, high Curie temperature, narrow resonance line width

*From P. I. Slick, Ferrites for non-microwave applications, in I. P. Wohlfarth (ed.), *Ferromagnetic Materials*, vol. 2, North-Holland, 1980.
†May contain Al, Gd.   ‡May contain Al, Zn.   §May contain Ti, Zn.

MnZn ferrites in high-Q inductors to about 1 MHz. Above that, NiZn ferrites are preferred. *See* ELECTRIC FILTER; INDUCTOR; Q (ELECTRICITY); TRANSFORMER.

For transformer applications the highest value of μ over the operating frequency range is desired. Values of μ in the 18,000 range (10 kHz) for MnZn ferrite are commercially available.

Nonlinear B/H devices. The largest usage of ferrite measured in terms of material weight is in the nonlinear B/H range, and is found in the form of deflecting yokes and flyback transformers for television receivers. The cores for these devices must have high saturation induction B_s along with high maximum permeability μ_m at the knee of the B/H curve to frequencies as high as 100 kHz, the effective flyback frequency used in scanning a television tube. Again, MnZn and NiZn ferrites dominate the use in these devices. *See* TELEVISION RECEIVER.

Highly nonlinear B/H devices. The largest usage of ferrite measured in terms of number of parts is in the highly nonlinear B/H range, and is used as square-loop toroidal cores for digital computer memories. The two stable states of magnetization represent the binary one and zero codes. The properties which are important here are a square hysteresis loop (with a high ratio of remanent induc-

tion to saturation induction, B_r/B_s), controlled coercivity, and fast switching time. For high packing density and fast access time, toroids with outside diameters as small as 0.5 mm have been achieved. The ferrites are primarily Mn, such as MnMg, MnCu, MnMgZn, MnCuNi, and MnLi. The last is exclusively used in small-size cores because of high Curie temperature for improved temperature stability, greater mechanical strength, and higher B_s for greater output. The future for memory ferrites, however, is dimmed by the large-scale emergence in 1977 of semiconductor memories, which have progressively taken over the rapidly growing computer market. *See* COMPUTER STORAGE TECHNOLOGY.

Microwave devices. Microwave devices make use of the nonreciprocal propagation characteristics of ferrites close to or at a gyromagnetic resonance frequency in the range of 1 – 100 GHz. The most important of such devices are isolators and circulators. Materials having a range of B_s are needed for operation at various frequencies since for resonance, B_s is less than ω/γ where γ is the gyromagnetic ratio. In the 1 – 5-GHz range, yttrium iron garnet of $B_s = 0.02 - 0.18$ tesla is used. In the 2 – 30-GHz range, MgMn, MgMnZn, MgMnAl, and Li ferrites of $B_s = 0.06 - 0.25$ T are used along with the garnets. At 30 – 100 GHz, NiZn ferrites with B_s up to 0.50 T are used. The garnets have highly desirable, small, ferromagnetic-resonance line widths, particularly in single-crystal form. In device development there is a strong trend toward realizing the conventional waveguide components in microstrip form whereby both the transmission and the gyromagnetic function are provided by a ferrimagnetic substrate, or the gyromagnetic function is provided by a ferrimagnetic insert on a nonmagnetic ferrite substrate. *See* Gy-RATOR; MICROWAVE; MICROWAVE TRANSMISSION LINES.   [GILBERT Y. CHIN]

Bibliography: C. Heck, *Magnetic Materials and Their Applications*, 1974; J. Smit and H. P. J. Wijn. *Ferrites*, 1959; E. C. Snelling, *Soft Ferrites*, 1969; E. P. Wohlfarth (ed.), *Ferromagnetic Materials*, vol. 2, 1980.

Ferroelectrics

Crystalline substances which have a permanent spontaneous electric polarization (electric dipole moment per cubic centimeter) that can be reversed by an electric field. In a sense, ferroelectrics are the electrical analog of the ferromagnets, hence the name. The spontaneous polarization is the so-called order parameter of the ferroelectric state, just as the spontaneous magnetization is the order parameter of the ferromagnetic state. The names Seignette-electrics or Rochelle-electrics,

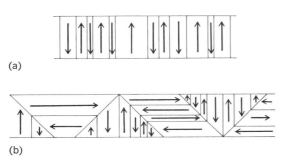

(a)

(b)

Fig. 1. Domain configurations (simplified) encountered in ferroelectric crystals. (*a*) First class. (*b*) Second class.

Fig. 2. Ferroelectric domains in BaTiO₃ photographed through a polarizing microscope. Ferroelectric domains range from macroscopic to submicroscopic size.

which are also widely used, are derived from the name of the first substance found to have this property, Seignette salt or Rochelle salt.

The reversibility of the spontaneous polarization is due to the fact that the structure of a ferroelectric crystal can be derived from a nonpolarized structure by small displacements of ions. In most ferroelectric crystals, this nonpolarized structure becomes stable if the crystal is heated above a critical temperature, the ferroelectric Curie temperature; that is, the crystal undergoes a phase transition from the polarized phase (ferroelectric phase) into an unpolarized phase (paraelectric phase). The change of the spontaneous polarization at the Curie temperature can be continuous or discontinuous. The Curie temperature of different types of ferroelectric crystals range from a few degrees absolute to a few hundred degrees absolute. As a rule, the ferroelectric phase is the low-temperature phase; however, there are crystals which are ferroelectric in a relatively narrow temperature range only, and others stay polarized up to the temperature of decomposition or melting.

Classification. From a practical standpoint ferroelectrics can be divided into two classes. In ferroelectrics of the first calss, spontaneous polarization can occur only along one crystal axis; that is, the ferroelectric axis is already a unique axis when the material is in the paraelectric phase. Typical representatives of this class are Rochelle salt, KH_2PO_4, $(NH_4)_2SO_4$, guanidine aluminum sulfate hexahydrate, glycine sulfate, colemanite, and thiourea.

In ferroelectrics of the second class, spontaneous polarization can occur along several axes that are equivalent in the paraelectric phase. The following substances, which are all cubic above the Curie point, belong to this class: $BaTiO_3$-type (or perovskite-type) ferroelectrics; $Cd_2Nb_2O_7$; $PbNb_2O_6$; certain alums, such as methyl ammonium alum; and $(NH_4)_2Cd_2(SO_4)_3$. Some of the BaTiO₃-type ferroelectrics have, below the Curie temperature, additional transition temperatures at which the spontaneous polarization switches from one crystal axis to another crystal axis. For example, $BaTiO_3$ and $KNbO_3$ polarize with decreasing temperature first along a [100] axis, then the polarization switches into a [110] axis, and finally into a [111] axis.

From a scientific standpoint, one can distinguish proper ferroelectrics and improper ferroelectrics. In proper ferroelectrics, for example, $BaTiO_3$, KH_2PO_4, and Rochelle salt, the spontaneous polarization is the order parameter. The structure change at the Curie temperature can be considered a consequence of the spontaneous polarization. The unit cell of the crystal in the ferroelectric phase contains the same number of chemical formula units as the unit cell in the paraelectric phase. In improper ferroelectrics, the spontaneous polarization can be considered a by-product of another structural phase transition. The unit cell in the ferroelectric phase is an integer multiple of the unit cell in the paraelectric phase. Examples of such systems are $Gd(MoO_4)$ and boracites. The dielectric, elastic, and electromechanical behavior of the two types of ferroelectrics differ significantly.

Ferroelectric domains. The spontaneous polarization can occur in at least two equivalent crystal

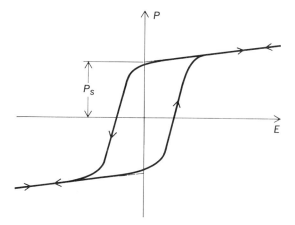

Fig. 3. Net polarization P of a ferroelectric crystal versus externally applied electric field E.

directions; thus, a ferroelectric crystal consists in general of regions of homogeneous polarization that differ only in the direction of polarization. These regions are called ferroelectric domains. Ferroelectrics of the first class consist of domains with parallel and antiparallel polarization (Fig. 1*a*), whereas ferroelectrics of the second class can assume much more complicated domain configurations (Fig. 1*b*). The region between two adjacent domains is called a domain wall. Within this wall, the spontaneous polarization changes its direction. The wall between antiparallel domains is probably only a few lattice spacings thick, whereas the wall between domains that are polarized at a right angle to each other is probably thicker. Ferroelectric domains can be observed in a number of substances by means of the polarizing microscope (Fig. 2) because of their birefringence, or double refraction. The ferroelectric domains range in size from macroscopic (millimeters) to submicroscopic.

Ferroelectric hysteresis. When an electric field is applied to a ferroelectric crystal, the domains that are favorably oriented with respect to this field grow at the expense of the others, for example, by sidewise motion of domain walls. In addition, favorably oriented domains can nucleate and grow until the whole crystal becomes one single domain. When the field is reversed, the polarization reverses through the same processes. The relation between the resulting polarization P of the whole crystal and the externally applied electric field E is given by a hysteresis loop (Fig. 3). The shape of the hysteresis loop depends strongly upon the perfection of the crystal as well as upon the rate of

change of the externally applied field E. A simple circuit that permits the observation of ferroelectric hysteresis loops by means of an oscilloscope is shown in Fig. 4. In some ferroelectrics, the polarization can be reversed within a fraction of 1 μsec.

Spontaneous polarization. The magnitude of the permanent or spontaneous polarization P_s of a domain can be obtained from the hysteresis loop by extrapolating the saturation branch to zero external field (Fig. 3). For most ferroelectrics, the values of P_s are between 10^{-7} and 10^{-4} coulomb/cm^2 (Fig. 5). In nonferroelectric dielectrics, electric fields between 10^5 and 10^8 volts/cm would be necessary in order to achieve such large polarizations.

Dielectric properties. As a rule, the dielectric constant ϵ measured along a ferroelectric axis increases in the paraelectric phase when the Curie temperature is approached. In many ferroelectrics, this increase can be approximated by the

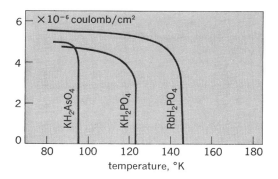

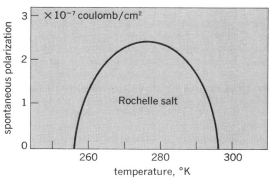

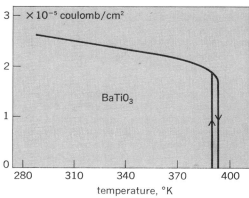

Fig. 5. Dependence upon temperature of the spontaneous polarization of some ferroelectrics.

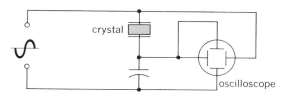

Fig. 4. Circuit for the display of ferroelectric hysteresis loops on an oscilloscope.

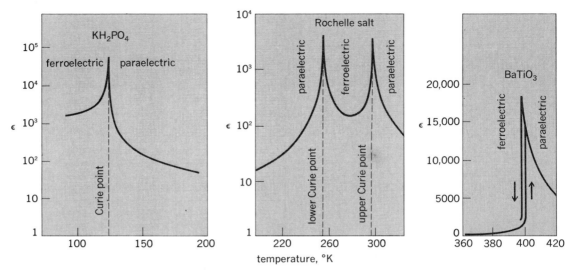

Fig. 6. Anomalous temperature dependence of relative dielectric constant of ferroelectrics at transition temperature.

Curie-Weiss law, shown in the equation below. Here T designates the temperature of the crystal, and T_0 is equal to or somewhat smaller than the

$$\epsilon = \frac{C}{T - T_0}$$

transition temperature. C is the so-called Curie constant. For $BaTiO_3$, this law holds unaltered up to frequencies of 2.4×10^{10} Hz. Dispersion sets in in the far-infrared. The dielectric constant drops when the crystal becomes spontaneously polarized (Fig. 6). In the ferroelectric phase, the dielectric constant has two components. The first component is the dielectric constant of the individual domains. It is independent of the frequency and of the electric field generally up to far-infrared frequencies. The second component is due to domain wall motions, that is, to partial reversal of the spontaneous polarization. This process can give rise to large dielectric losses, and it depends strongly upon the frequency, the electric field strength, the domain structure, and the temperature. In uniaxial ferroelectrics, the dielectric constant measured perpendicular to the ferroelectric axis generally does not show a very pronounced anomaly near the Curie temperature, and in some cases it has even the same order of magnitude and temperature dependence as for any normally behaving dielectric crystal.

Piezoelectric properties. Ferroelectrics can be divided into two groups according to their piezoelectric behavior.

The ferroelectrics in the first group are already piezoelectric in the unpolarized phase. Those piezoelectric moduli which relate stresses to polarization along the ferroelectric axis have essentially the same temperature dependence as the dielectric constant along this axis, and hence become very large near the Curie point. The spontaneous polarization gives rise to a large spontaneous piezoelectric strain which is proportional to the spontaneous polarization. In KH_2PO_4-type ferroelectrics and in Rochelle salt, for example, this strain

is a shear in the plane perpendicular to the axis of polarization. It reaches 27 min of arc in KH_2PO_4 and about 1.8 min of arc in Rochelle salt. The piezoelectric modulus decreases as the spontaneous polarization increases. But with sufficiently large stresses, it is possible to align the domains and reverse the spontaneous polarization (Fig. 7). The relation between the resulting polarization of the whole crystal and the mechanical stress is given by a hysteresis loop analogous to the loop of Fig. 3 (piezoelectric hysteresis). This effect can simulate a very large piezoelectric modulus.

The ferroelectrics in the second group are not piezoelectric when they are in the paraelectric phase. However, the spontaneous polarization lowers the symmetry so that they become piezoelectric in the polarized phase. This piezoelectric activity is often hidden because the piezoelectric effects of the various domains can cancel. However, strong piezoelectric activity of a macroscopic crystal or even of a polycrystalline sample occurs when the domains have been aligned by an electric field. The spontaneous strain is proportional to the square of the spontaneous polarization. In $BaTiO_3$, for example, the crystal (which has cubic symmetry in the unpolarized phase) expands along the axis of polarization and contracts at right angles to it. The strain is of the order of magnitude of 1%. The spontaneous polarization cannot be reversed by a mechanical stress in ferroelectrics of this group. See PIEZOELECTRIC CRYSTAL; PIEZOELECTRICITY.

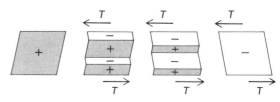

Fig. 7. Schematic representation of the reversal of the spontaneous polarization by a mechanical shear stress T in KH_2PO_4 and Rochelle salt.

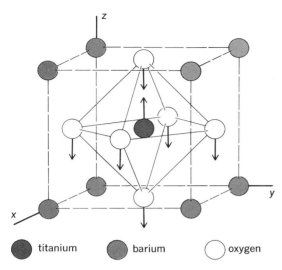

titanium   barium   oxygen

Fig. 8. Crystal lattice of BaTiO₃. Arrows indicate displacements of the ions when crystal becomes polarized.

Crystal structure. The structures of different types of ferroelectrics are entirely different, and it is not possible to establish a general rule for the occurrence of ferroelectricity. The structures of a number of ferroelectrics and the minute changes that they undergo when spontaneous polarization occurs are known in great detail from x-ray diffraction and neutron diffraction studies. In a qualitative way, the process of polarization is best understood for ferroelectrics of the $BaTiO_3$ type. Figure 8 shows schematically the structure of the unit cell of a $BaTiO_3$ crystal in the unpolarized state, and the arrows indicate the direction in which the ions are slightly displaced when the lattice becomes spontaneously polarized along the axis z. The order of magnitude of the displacements is 1% of the unit cell dimension. However, these displacements do not account quantitatively for the observed polarization, because other changes of the electronic structure occur as well.

In KH_2PO_4-type ferroelectrics, hydrogen bonds $O—H \cdots O$ play an important part in the ferro-

electric effect. Above the Curie temperature, the hydrogen ions are statistically distributed over the two possibilities $O—H \cdots O$ and $O \cdots H—O$, whereas below the Curie point, one or the other of these two possibilities is strongly favored, depending upon the sign of the spontaneous polarization.

Antiferroelectric crystals. These materials are characterized by a phase transition from a state of lower symmetry (generally low-temperature phase) to a state of higher symmetry (generally high-temperature phase). The low-symmetry state can be regarded as a slightly distorted high-symmetry state. It has no permanent electric polarization, in contrast to ferroelectric crystals. The crystal lattice can be regarded as consisting of two interpenetrating sublattices with equal but opposite electric polarization. This state is referred to as the antipolarized state.

In a certain sense, an antiferroelectric crystal is the electrical analog of an antiferromagnetic crystal. In the high-symmetry phase, the sublattices are unpolarized and indistinguishable. In general, antiferroelectric crystals have more than one axis along which the sublattices can polarize. Therefore, the low-symmetry phase consists of regions of homogeneous antipolarization which differ only in the orientation of the axis along which antipolarization has occurred. These regions are called antiferroelectric domains and can be observed by the polarizing microscope. Because these domains have no permanent electric dipole moment, an electric field generally has little influence on domain structure.

The dielectric constant of antiferroelectric crystals is generally larger than it is for nonferroelectric crystals and has an anomalous temperature dependence. It increases as the transition temperature is approached and drops when antipolarization occurs (Fig. 9). In some antiferroelectrics the phase transition is discontinuous; in others it is continuous.

The structure of antiferroelectric crystals is generally closely related to the structure of ferroelectric crystals. Some antiferroelectrics even undergo phase transitions from an antipolarized state into a spontaneously polarized, ferroelectric state; in others a sufficiently strong electric field applied

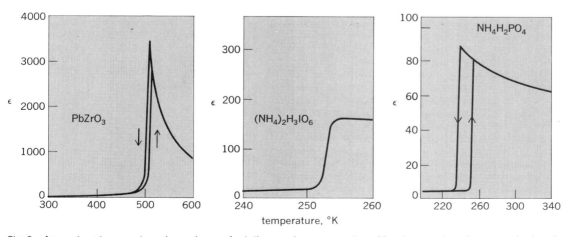

temperature, °K

Fig. 9. Anomalous temperature dependence of relative dielectric constant of antiferroelectric crystals. Note increase as transition temperature is approached and drop when antipolarization occurs.

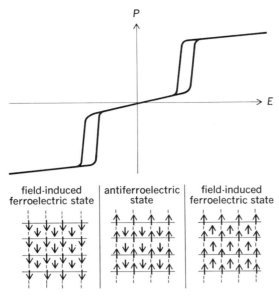

Fig. 10.   Polarization *P* of antiferroelectric PbZrO$_3$ versus externally applied electric field *E*. Strong fields "switch" the antiferroelectric crystal into a ferroelectric state, as shown here schematically.

along an antiferroelectric axis reverses the polarity of one of the sublattices so that a ferroelectric state results. The crystal reverts, however, to the antiferroelectric state when the electric field is removed. Figure 10 shows net polarization versus externally applied field for such a case.

Compounds with antiferroelectric properties are PbZrO$_3$, PbHfO$_3$, NaNbO$_3$ (isomorphous with ferroelectric BaTiO$_3$), WO$_3$ (structure related to BaTiO$_3$), NH$_4$H$_2$PO$_4$ and isomorphous NH$_4$ salts (isomorphous with ferroelectric KH$_2$PO$_4$), (NH$_4$)$_2$-H$_3$IO$_6$, Ag$_2$H$_3$IO$_6$, and certain alums.

Origin of phase transition. The ferroelectric phase transition results from an instability of one of the normal lattice vibration modes. On approaching the transition temperature, the frequency of the relevant normal mode decreases (soft mode). The restoring force of the mode displacements tends to zero. When the stability limit is reached, the displacements corresponding to the soft mode freeze in, and the ferroelectric phase results. The ferroelectric soft mode is polar (infrared-active) and of infinite wavelength. The antiferroelectric phase transition, on the other hand, emerges from a soft lattice mode with a finite wavelength equal to an integer multiple of a lattice period.

Applications. The piezoelectric effect of ferroelectrics (and certain antiferroelectrics) finds numerous applications in electromechanical transducers. The large electrooptical effect (birefringence induced by an electric field) is used in light modulators. In certain ferroelectrics (for example, BaTiO$_3$, LiNbO$_3$, KTaNbO$_3$, and LiTaO$_3$), light can induce changes of the refractive indices. These substances can be used for optical information storage and in real-time optical processors. The temperature dependence of the spontaneous polarization corresponds to a strong pyroelectric effect which can be exploited in thermal and infrared sensors.

[WERNER KANZIG]

Bibliography: R. Blinc and B. Žekš, *Soft Modes in Ferroelectrics and Antiferroelectrics*, 1974; V. M. Fridkin, *Ferroelectric Semiconductors*, 1980; V. M. Fridkin, *Photoferroelectrics*, 1979; M. E. Lines and A. M. Glass, *Principles and Applications of Ferroelectrics and Related Materials*, 1977; T. Mitsui, *An Introduction to the Physics of Ferroelectrics*, 1976; T. Mitsui et al., *Ferro- and Antiferroelectric Substances*, in K. M. Hellwege and A. M. Hellwege (eds.), *Landolt-Bornstein Series*, group 3, vol. 9, 1975.

Fidelity

The degree to which the output of a system accurately reproduces the essential characteristics of its input signal. Thus, high fidelity in a sound system means that the reproduced sound is virtually indistinguishable from that picked up by the microphones in the recording or broadcasting studio. Similarly, a television system has high fidelity when the picture seen on the screen of a receiver corresponds in essential respects to that picked up by the television camera. Fidelity is achieved by designing each part of a system to have minimum distortion, so that the waveform of the signal is unchanged as it travels through the system. *See* DISTORTION (ELECTRONIC CIRCUITS).

[JOHN MARKUS]

Field emission

The emission of electrons from a metal or semiconductor into vacuum (or a dielectric) under the influence of a strong electric field. In field emission, electrons tunnel through a potential barrier, rather than escaping over it as in thermionic or photoemission. The effect is purely quantum-mechanical, with no classical analog. It occurs because the wave function of an electron does not vanish at the classical turning point, but decays exponentially into the barrier (where the electron's total energy is less than the potential energy). Thus there is a finite probability that the electron will be found on the outside of the barrier. This probability varies as $e^{-cA^{1/2}}$, where c is a constant and A the area under the barrier. *See* PHOTOEMISSION; THERMIONIC EMISSION.

For a metal at low temperature, the process can be understood in terms of the illustration. The metal can be considered a potential box, filled with electrons to the Fermi level, which lies below the vacuum level by several electronvolts. The distance from Fermi to vacuum level is called the

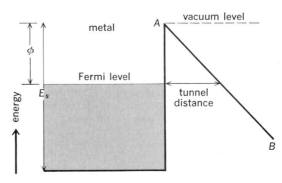

Diagram of the energy-level scheme for field emission from a metal at absolute zero temperature.

work function, ϕ. The vacuum level represents the potential energy of an electron at rest outside the metal, in the absence of an external field. In the presence of a strong field F, the potential outside the metal will be deformed along the line AB, so that a triangular barrier is formed, through which electrons can tunnel. Most of the emission will occur from the vicinity of the Fermi level where the barrier is thinnest. Since the electron distribution in the metal is not strongly temperature-dependent, field emission is only weakly temperature-dependent and would occur even at the absolute zero of temperature. The current density J is given by the Fowler-Nordheim equation below, where B is a

$$J = BF^2 e^{-6.8 \times 10^7 \phi^{3/2}/F}$$

field-independent constant of dimensions A/V² (A = amperes or state), ϕ is work function in electronvolts, and F is applied field in V/cm. The factor $\phi^{3/2}/F$ is proportional to the square root of the area under the barrier at the Fermi level. Appreciable emission requires fields of $4 - 7 \times 10^7$ V/cm, depending on ϕ.

Field emission is most easily obtained from sharply pointed metal needles whose ends have been smoothed into nearly hemispherical shape by heating. Tip radii r_t equal to or less than 100 nm can be obtained in this way; because of its small size, the emitter is generally a single crystal. If an emitter is surrounded by a hemispherical anode raised to a voltage V, then the field F is approximately $V/5r_t$ at the emitter. Thus, modest voltages suffice for emission. The electric lines of force diverge radially from the tip; since the electron trajectories initially follow the lines of force, they also diverge, and a highly magnified emission map of the emitter surface can be obtained, for instance, by making the anode a fluorescent screen. This constitutes a field-emission microscope, invented by E. W. Müller in 1936. Since work function and hence emission are affected by adsorbed layers, the field-emission microscope is very useful for studying adsorption, particularly surface diffusion. Field emitters are widely used in "ordinary" and scanning electron microscopes as high-brightness quasi-point sources of electrons, since emission occurs as if it originated from the center of the emitter cap.

Field emission can also occur from electrode asperities into insulating liquids, and thus initiates electrical breakdown. It can occur from isolated atoms or molecules in high fields, and is then called field ionization. Field ionization forms the basis of the field-ion microscope, and is a useful method of generating ions in analytical mass spectrometry. Internal field emission can occur from the valence to the conduction band of a semiconductor in a high field, and is then known as Zener breakdown. *See* ELECTRON EMISSION.

[ROBERT GOMER]

Frequency (wave motion)

The number of times which sound pressure, electrical intensity, or other quantities specifying a wave vary from their equilibrium value through a complete cycle in unit time. The most common unit of frequency is the hertz (Hz); 1 Hz is equal to 1 cycle per second. In one cycle there is a positive variation from equilibrium, a return to equilibrium, then a negative variation, and return to equilibrium. This relationship is often described in terms of the sine wave, and the frequency referred to is that of an equivalent sine-wave variation in the parameter under discussion.

Frequency is a convenient means for describing the various ranges of interest in wave motion. For example, audible sound is between approximately 20 and 20,000 Hz. Infrasonic frequencies are below approximately 20 Hz; sound having frequencies above the audible is termed ultrasonic. Electromagnetic waves vary in frequency from less than 1 Hz for commutated direct current up to 10^{23} Hz for the most energetic γ-rays that have been observed. Within this range, typical approximate frequency ranges are: AM radio in the United States, 550 to 1700 kHz: FM radio, 88 to 108 MHz; visible light, 4×10^{14} to 7.5×10^{14} Hz. *See* FREQUENCY MEASUREMENT.

[WILLIAM J. GALLOWAY]

Frequency counter

An electronic device capable of counting the number of cycles in an electrical signal during a preselected time interval. The modern high-speed electronic counter is a useful tool in the measurement of frequency when an accurate time base is available. It provides a digital counting or scaling device for registering the total number of events occurring during a given time interval.

Such electronic counting circuits operate reliably at rates of 10^7 counts/sec in many commercially available units and at rates up to approximately 5×10^8 counts/sec by use of direct input with prescaling or aperiodic frequency-dividing input stages. For measurement of a frequency within the counting range of the digital circuits, the counter may be used directly with a standard frequency oscillator as a time-base control.

Figure 1 shows a block diagram of an electronic digital counter. When f_1 is a standard frequency and $1/R$ is chosen to provide a standard time interval, the counter counts a number of cycles, or pulses, proportional to the frequency of f_2 in hertz (or other time intervals). When f_1 is not a standard frequency, the counter counts a number proportional to the ratio of f_2 to f_1. When f_1 is a standard frequency and $1/R$ is chosen to provide a time interval which is a fraction or a multiple ($1/N$ or N) of the nominal standard time interval, the counter indicator may be made direct reading for a harmonic (N) or submultiple ($1/N$) of f_2. When f_1 is a relatively low frequency, that is, its period corre-

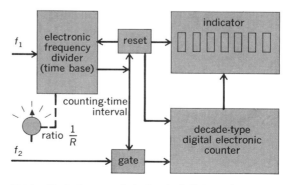

Fig. 1. Block diagram of electronic digital counter.

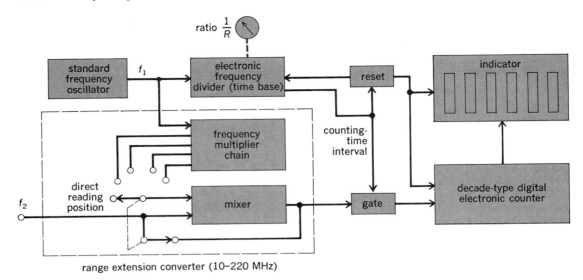

Fig. 2. Electronic digital frequency counter with heterodyne extension of useful frequency range.

Fig. 3. Electronic digital counter which has been designed to accept plug-in units for function selection and for range extension. Phase-locking transfer oscillator shown installed, at right. (*Hewlett-Packard Co.*)

Fig. 4. Electronic counter with plug-in and add-on auxiliary units for range extension. (*Hewlett-Packard Co.*)

Fig. 5. Electronic digital counter incorporating multiple functions which are selectable from panel, and using integrated circuits. (*General Radio Co.*)

sponds to the time interval between two pulses, and f_2 is a standard frequency, the counter may be used to read the period or time interval by setting $R = 1$ and arranging the gate to open on the first pulse and close on the second.

If the frequency to be measured is too high for direct counting, it is necessary to extend the range. A means commonly used is the heterodyning of the unknown signal by a standard frequency within 10 MHz of the unknown and the subsequent measuring of the beat frequency (Fig. 2). If the digital counter can count frequencies up to 10 MHz and the standard-frequency multiples are spaced by 10 MHz, continuous coverage is possible as far up in frequency as the standard frequencies are available. Typical electronic digital counters are shown in Figs. 3 and 4, with plug-in converter units for range extension. A multipurpose frequency, period, and time-interval electronic digital counter is shown in Fig. 5.

The principal requirements are for signals of adequate level, free from interference and noise. With these conditions satisfied it is usually possible to obtain a precision of reading well beyond the stability of the overall system if a sufficiently long counting time is used. An extension of the frequency-measurement range is made possible by extending the range of the converter or heterodyning system. It is also feasible to add a transfer oscillator (Fig. 6) with harmonics available up into the microwave region.

The microwave frequencies are then measured as described above (transfer-oscillator method), using the counter to measure the fundamental frequency of the transfer oscillator and with a beat detector and indicator to determine the beat note between the unknown and a harmonic of the transfer oscillator. A variation of this transfer-oscillator system provides for setting the time-base and zero-set conditions of the counter to give direct reading numbers valid for the particular harmonic of the transfer oscillator which is in use.

The electronic digital counter has been developed in commercial form to operate at counting rates up to 100 MHz or more, the cost increasing rapidly above 20–50 MHz. Extension of the frequency range covered by a given counter is possible by (1) prescaling, (2) using heterodyne systems which beat the signal against standard reference frequencies, (3) or using manual or automatic transfer oscillator systems. Transfer-oscillator and heterodyne range-extension systems operate satisfactorily into the microwave region (18 GHz or above). The limiting measurement accuracy depends on the excellence of the standard frequency source used for the time base and heterodyne converter reference. *See* ELECTRICAL MEASUREMENTS; FREQUENCY MEASUREMENT.

[FRANK D. LEWIS]

Bibliography: W. D. Cooper, *Electronic Instrumentation and Measurement Techniques,* 2d ed., 1978; P. Kantrowitz et al., *Electronic Measurements,* 1979; L. Klein and K. Gilmore, *It's Easy To Use Electronic Test Equipment,* 1962.

Frequency divider

An electronic circuit that produces an output signal at a frequency that is an integral submultiple of the frequency of the input signal.

Several information-processing and transmitting techniques require frequency division. In television, for example, it is essential to maintain a precise relationship between the horizontal-scanning frequency and the vertical-scanning frequency. Frequency division can be conveniently accomplished in two ways, digital division and division by triggering a subharmonic.

Digital division. Many circuits are available to count pulses. A bistable or flip-flop circuit produces one output pulse for every two input pulses. By cascading successive flip-flops, any desired degree of division can be obtained. Division by powers of 2 can be achieved simply by monitoring the output of the proper stage in the cascade. Division by other numbers can be achieved by gating to obtain the proper set of flip-flop conditions. Division by 10, or powers of 10, is readily obtained by using decade rings. *See* DIGITAL COUNTER; GATE CIRCUIT.

Reduction of the input to pulses can be accomplished when needed by amplifying, clipping, and differentiating the original signal. To provide an output that is a desired shape, the signal from the frequency divider can be used to synchronize a multivibrator to obtain a square wave and additional power. The square wave can be fed into a tuned circuit or filtered to obtain a sine wave, if that is needed. In many applications the input and output signals need no alteration. *See* MULTIVIBRATOR; WAVE-SHAPING CIRCUITS.

Subharmonic triggering. Any circuit which has a characteristic resonance responds to certain types of input energy by ringing, that is, by going through one or more cycles of electrical activity caused by the nature of the circuit rather than by the nature of the input. This characteristic can be used to accomplish frequency division, provided the input frequency does not vary over any extensive frequency range.

A triggered multivibrator can be used in this manner. Either an astable or a free-running type is

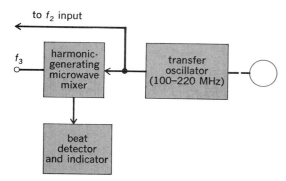

Fig. 6. Microwave transfer oscillator.

selected such that the period for one cycle of the multivibrator will allow triggering only by a pulse that is an exact integral number of pulses from the last effective trigger. Thus one cycle of the multivibrator is initiated every *n* input pulses.

Similarly, other resonant or tuned circuits can be induced to supply submultiple outputs. Because the input may go through a large number of cycles for each output cycle, greater division per stage can be realized by this method than by digital division. [WILLIAM W. SNOW]

Frequency measurement

Measurement of the frequency of a periodic quantity, defined as the number of times a cyclic phenomenon occurs per unit of time. The second is the universally used unit of time. Conversely, time may be measured by observing the number of cycles occurring at constant frequency. The ordinary pendulum and household electric clocks are examples of such time-measuring devices. In the pendulum clock the pendulum completes one cycle of mechanical motion when it returns to its original position from the opposite direction. The electric clock completes one cycle in the electrical sense when the rotor of its motor has turned 360 electrical degrees. The alternating-voltage driving force starts a cycle from zero, passes through a maximum in one direction, returns to zero, passes through a maximum in the opposite direction, and again returns to zero, thereby completing one cycle of 360 electrical degrees. A two-pole synchronous motor consisting of one pair of north and south poles will turn 360 mechanical degrees for each cycle of 360 electrical degrees.

Standards of time and frequency. The standard second is defined as the length of time required for 9,192,631,770 oscillations at the transition frequency of the cesium atom at zero magnetic field. This definition is thought to provide a constant time interval based on atomic constants. Ephemeris time is a uniform time scale based on the astronomical measurement of the time required for the Earth to orbit the Sun. The ephemeris second is 1/31,556,925.975 of the tropical year 1900 and is substantially equal to the atomic second. The time scale most widely used for navigation and daily living is based on the rotation of the Earth with respect to the Sun and is known as universal time (UT), and also as "mean solar time" or Greenwich mean time. Sidereal time is Earth rotation time measured with respect to a distant star. Universal time coordinated (UTC) is a standardized uniform time scale having an agreed offset value with respect to atomic time and being maintained in approximate agreement with observed universal time by step corrections. The agreed offset value and the step corrections are administered by the Bureau International de l'Heure (BIH), the UTC time scale being used by a majority of those countries providing radio dissemination of standard time. Time intervals on the atomic time scale may be derived from the frequencies of any of several atomic standard frequency broadcasting stations or from UTC broadcasts with a knowledge of the offset between UTC and atomic frequencies.

For calibration purposes, or for observations requiring accurate timing, a knowledge of the particular time scale being used is required. In any case calibration of a local time standard may be carried out by (1) reception of radio time signals, (2) use of a portable precision clock such as an atomic clock, or (3) astronomical measurement and computation. In most cases calibration to a given accuracy with respect to a specific time scale requires reference to a national observatory either through radio signals or a portable precision clock. In the United States the National Bureau of Standards maintains a standard frequency broadcasting service and also broadcasts standard time signals which are maintained within approximately ±100 msec of corrected UT. The time signals broadcast by the U.S. Naval Radio Service are also maintained to this accuracy. Accurate clock calibration may be obtained by carrying atomic clocks between the observatory and the clock to be calibrated or by such systems as satellite-borne radio relays using special terminal equipment.

Information on the accuracy of transmitted time and frequency broadcasts is available from the operating agencies. United States agencies are the National Bureau of Standards and the Naval Observatory. In Canada the service is provided by the Dominion Observatory, Ottawa.

Primary frequency standard. The only primary frequency standards acceptable for use in national standards laboratories for frequency reference are atomic standards of the cesium-beam type. These devices provide intrinsic calibrations, independent of other reference standards. Atomic hydrogen masers are expected eventually to be developed as primary reference frequency standards. Cesium-beam atomic clocks have superseded other types

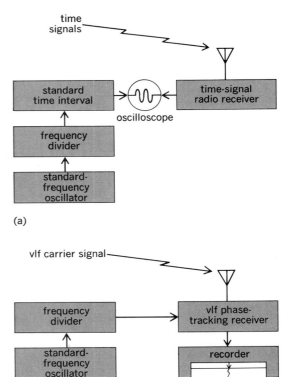

(a)

(b)

Fig. 1. Secondary frequency standard calibration methods. Block diagram of (*a*) time signal method and (*b*) phase-recording method.

of primary frequency standards as a consequence of the adoption of the atomic second as the unit of time.

Secondary standard. The frequency of a secondary standard is determined by comparison with a primary standard or by comparison with another secondary standard originally compared with a primary standard. Secondary standards include (1) quartz-crystal frequency standards, with or without clock indicators, and (2) gas-cell atomic standards such as rubidium-cell standards, and (3) ammonia masers. Rubidium-cell standards are relatively rugged and can be made to exhibit low drift rates, thus enabling relatively long intervals between calibrations. Quartz-crystal frequency standards are the least expensive initially and may provide the most satisfactory working standards if calibration means are available. *See* QUARTZ CLOCK.

Checking and calibrating a secondary standard may be accomplished by (1) use of a very-low-frequency (vlf) phase recording receiver with a vlf standard frequency signal, (2) comparison with loran-C signals on 100 kHz, or (3) use of an accurate clock indicator for comparison with radio time signals (Fig. 1). Somewhat less accuracy may be achieved using the high-frequency signals broadcast by station WWV and setting the standard to be calibrated to zero beat with the carrier frequency as received. The reduced accuracy arises from the propagation variations of the high-frequency signal.   [FRANK D. LEWIS]

Audio-frequency meters. A fairly wide variety of frequency meters has been developed and is commercially available. Broadly speaking they can be grouped into two classes, the resonant and ratiometer types. The resonant type may be further subdivided into instruments employing resonant reeds and those having electrically resonant circuits. The latter, as well as the ratiometer type, are classified as deflection-type instruments embodying moving systems with pointers and scales.

The moving systems of the conventional deflection type meters develop two opposing deflecting forces which cause the movement to come to rest when the two opposing torques are balanced. The net torque is directly dependent on the frequency being measured. Deflection-type frequency meters, commonly used in power frequency applications around 25, 50, and 60 Hz , are also available in other ranges up to about 900. Their accuracies may be in the order of 0.14 Hz in the low frequency ranges and about 4 Hz in the higher frequency ranges.

Resonant-reed-type frequency meters are available in various ranges between 10 and 1000 Hz, with special designs ranging as low as about 7 Hz and as high as about 1500 Hz. Their accuracies, which are independent of their frequency ranges, can be in the order of ±0.1% of specified frequency.

Reed-type frequency meters. This type of meter uses the principle of vibrating reeds. The reeds for such instruments, assembled into a so-called reed comb, are made of specially selected and properly tempered steel. They have bent tips which are enameled white for visibility. When the supply voltage is applied to the instrument, all of the reeds receive vibrational impulses; the effect is

visible only in the reed or reeds which are in resonance. The reed in resonance behaves somewhat like a whip, the tip swinging through a readily visible arc, the natural period of vibration depending upon length and thickness. Resonance vibration extends over a limited visible range from about 2% below to 2% above the actual frequency as shown in Fig. 2. Individual reeds can be tuned to a range of about ±0.1% of their rated frequencies.

The reeds in the comb are vibrated by an electromagnet energized from the source of which the frequency is to be measured. Depending upon de-

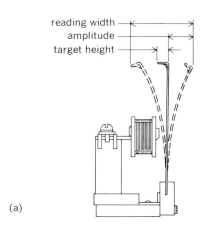

(a)

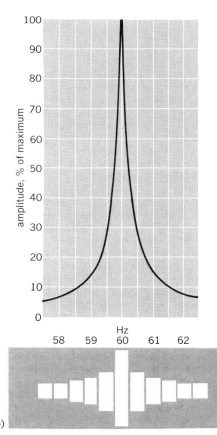

(b)

Fig. 2. Reed-type frequency meter. (*a*) Actuating system of a Frahm frequency meter. (*b*) Resonance curve for 60-Hz reed showing the amplitude of vibration for the various applied frequencies.

Fig. 3. Cutaway view of an indirect-drive, switchboard-type frequency meter. (*James G. Biddle Co.*)

sign considerations, a permanent magnet may be used in conjunction with soft iron in the magnetic circuit. The magnetic-drive system usually takes one of two forms. The direct-drive method drives the reeds magnetically by including them in the magnetic circuit as shown in Fig. 2a. In the indirect-drive method the magnetic circuit includes a soft-iron pole piece or armature attached to the base of the reed comb, the armature indirectly transmitting its mechanical vibrations to the reeds. If no permanent magnet is used in the magnetic circuit, the reeds are vibrated at a frequency twice that of the source of voltage and consequent magnetic field. Biasing the magnetic field by use of a permanent magnet causes the reeds to vibrate at the same frequency as the source voltage. Figure 3 is a cutaway view of a commercial switchboard-type frequency meter using an indirect drive.

Moving-coil meters. Figure 4 shows one design of a moving-coil resonant-type instrument. Two coils, A and B, are placed at right angles to form a moving element and are supplied through a resonant circuit. The coils are tuned to different frequencies, coil A to a frequency slightly below the lowest scale point and coil B an equal amount above the upper scale point. A fixed coil C divided into two parts carries the sum of the currents of the two circuits of coils A and B. When the frequency of the applied voltage equals the midscale frequency the currents in the two resonant circuits are practi-

cally equal and the phase lead of one coil equals the phase lag of the other with respect to the fixed-coil current. At frequencies near the low end of the scale, the moving element comes to rest with coil A approaching a position parallel to the fixed coil. As the circuit of coil A approaches resonance, the current in coil A is relatively high and in phase with the fixed-coil current and the current in coil B leads with respect to the fixed-coil current. At frequencies in the higher range, coil B approaches resonance and takes an equilibrium position parallel to the plane of the fixed coil. If the supply frequency is very low, the impedance of the circuit A and B being equal, the pointer will indicate a fictitious midscale reading. To rectify this misleading measurement, a third coil D is introduced in parallel with coil A and resonates at frequencies much less than the lowest scale point.

Figure 5 shows a pictorial diagram of another moving-coil resonant-type frequency meter based on the electrodynamometer principle and usually called a mutual-induction-type instrument. The field structure comprises a laminated magnetic core which is shaped to provide space for a field winding. This construction provides a scale length of approximately 250° and covers a range of 56–65 Hz. An alternating flux produced by the field coil induces an electromotive force in the moving coil which varies with the deflection. This emf is opposed by the voltage drop across the tuned cir-

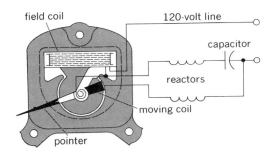

Fig. 5. Mutual-inductance-type of the frequency meter. (*General Electric Co.*)

cuit. The resultant voltage causes a current to pass through the moving coil. A torque is developed by the interaction between the moving-coil current and the field flux and deflects the coil until it reaches a position where the in-phase component of its current becomes zero, which causes the torque also to become zero. The equilibrium position changes with frequency because the drop across the tuned circuit and the induced emf both vary with frequency.

Moving-iron meters. A resonant-type frequency meter operating on the moving-iron principle is shown in Fig. 6. Two field coils are mounted opposite to each other and are connected so that their fluxes oppose. If i_1 and i_2 are moving-coil currents, the resultant field current will be $i_1 - i_2$. The moving system which lies between the field coils in comprised of an iron vane of magnetic material centered within the armature coils and rigidly attached to the shaft. Both the field coils are connected to series-resonance circuits in which reso-

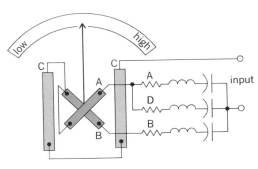

Fig. 4. Resonant-type moving-coil frequency meter.

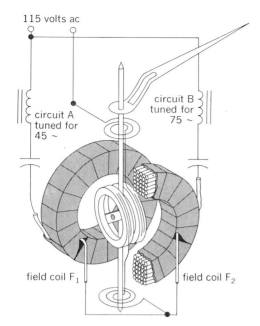

Fig. 6. Resonant-type frequency meter employing the moving iron principle. (*General Electric Co.*)

nance is produced below the operating range for one circuit and above the operating range for the other. Thus for a frequency range of 55–65 Hz the resonance may occur around 50 Hz for one circuit and 70 Hz for the other.

Deflecting torque is produced by the reaction of the armature, or moving coil, field with the in-phase component of the field coils. This torque is proportional to the product of the resultant field flux, armature flux, and the cosine of the phase angle of the two fluxes.

In addition to this deflecting torque there is a countertorque produced by the action of the field flux in the magnetic vane of the moving element. When the deflecting torque causes the iron vane on the moving system to be out of alignment with the field flux, the iron vane exerts a torque opposite in direction (to the deflecting torque) because of its tendency to realign itself with the physical

$$f = 1/2\pi\sqrt{LC}$$

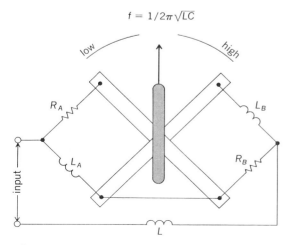

Fig. 7. Ratiometer-type Weston frequency meter which uses the moving-iron principle.

direction of the field flux. This countertorque is required to establish a position of equilibrium. This occurs when the magnitude of the countertorque is equal to the deflecting torque, which causes the pointer to take a position indicating the frequency.

Another design of a moving-iron frequency meter, but of the ratiometer type, is shown in Fig. 7. Two coils A and B are arranged at right angles to each other. Coil A is in series with a resistance R_A, and this combination is in parallel with an inductance L_A. Similarly, coil B is in series with L_B and the combination is in parallel with R_B. The complete circuit thus acts as a bridge network. When the bridge is balanced by suitable values of parameters, the currents in the two coils are equal and the moving element, a pivoted soft-iron needle, will indicate the mean-frequency position as shown in the diagram. When the frequency increases the current increases in coil A and decreases in coil B. The interaction of the fields of the two coils will cause the needle to take up a new position. Hence for each change in frequency the needle tends to align itself along the resultant magnetic fields of the two coils. In this kind of meter the effects of distorted waveforms due to higher harmonics are almost nonexistent.

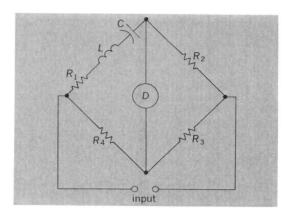

Fig. 8. Resonance-frequency bridge.

Audio-frequency bridge methods. An ac bridge is generally made up of a four-arm network, where an ac source is applied to a pair of opposite terminals and a current-detecting device is connected to the remaining two terminals. With the bridge network energized, the current in the detecting device can be made zero by adjusting suitable values of resistance, capacitance, or inductance in the four bridge arms, a process known as balancing.

Bridge networks in which the balancing action depends upon the supply frequency may be used for frequency measurement. The choice of frequency bridge is dependent on the frequency range, the available apparatus, and the ease with which the bridge can be set up and balanced. *See* BRIDGE CIRCUIT.

A simple form of resonant-frequency bridge is shown in Fig. 8. Balance is obtained by adjustment of R_4 and C, the detector D showing zero current. This will occur when $R_1/R_2 = R_4/R_3$ and $\omega L = 1/\omega C$

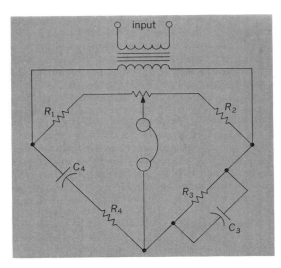

Fig. 9. Wien frequency bridge.

or $\omega^2 = 1/LC$. From this the frequency is found to be as in Eq. (1).

$$f = \tfrac{1}{2}\pi\sqrt{LC} \tag{1}$$

The most commonly employed bridge is the Wien bridge, because it has many advantages of other bridge methods. Figure 9 is the schematic diagram. R_3 and R_4 can be varied together. Condition of balance will be obtained when $R_3 = R_4$, $C_3 = C_4$, and $R_1 = 2R_2$.

The frequency of the source is written as Eq. (2).

$$f = \tfrac{1}{2}\pi R_3 C_4 \tag{2}$$

If the frequency to be measured is in the audio range, a pair of headphones may be used as a detector; for frequencies beyond the audio range an electronic voltmeter can be used.

Campbell's bridge is shown in Fig. 10. This bridge is used because of its simplicity and fairly large range of frequency. Balance is obtained by adjusting the value of M so that the detector current is zero. Under this condition the voltage induced in the secondary of the mutual inductor M is equal in magnitude and 180° out of phase with the voltage drop across the capacitance C.

Hence at balance $\omega M i = i/\omega C$ or $\omega^2 = 1/MC$, from which the frequency can be written as Eq. (3).

$$f = \tfrac{1}{2}\pi\sqrt{MC} \tag{3}$$

Direct-comparison methods. The most widely used of the simpler methods is aural comparison.

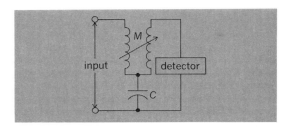

Fig. 10. Campbell's frequency bridge.

An unknown frequency can be determined by direct comparison with a known adjustable standard frequency by using the ear. Figure 11 shows a typical scheme where headphones are connected directly to the two audio sources. The unknown frequency is fed into one earpiece while the standard frequency supplied by a resistance-tuned or a beat-frequency oscillator is fed into another earpiece. A beat note between the two frequencies can be detected by the ear when the frequencies are not exactly equal.

A meter or an electric-eye indicator can take the place of the ear as a detector. An indicating ac voltmeter with a fast d'Arsonval movement may be used. Under the influence of the beat note the pointer of the voltmeter vibrates about a mean value of the output voltage. For low values of frequency difference it swings back and forth; for high values it vibrates rapidly. Such a meter or an electric-eye indicator can detect frequency differences of very low values accurately, whereas in the ear method of detection the accuracy of adjustment is limited by the skill of the operator.

Cathode-ray oscilloscopes provide a most convenient means of comparing two frequencies. The usual method consists of applying a voltage of unknown frequency to one pair of deflecting plates and a voltage of known frequency to the other. The resulting patterns, known as Lissajous figures, are

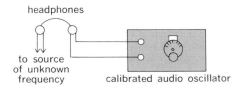

Fig. 11. Aural method of frequency measurement.

shown in Fig. 12. These patterns are a function of the difference in amplitude, frequency, phase, and the waveforms of the two applied voltages and their frequency ratio. Knowing one of the frequencies and determining the difference or the ratio of the frequencies, it is possible to calculate the second frequency. For example, the pattern with five vertical loops is produced by signals whose frequency ratio is 5:3 with the lower frequency on the horizontal plates. Any value of frequency in the audio range can be determined by this method, providing the ratio of the two frequencies does not exceed an approximate value of 10:1.

Frequency counters. For the basic measurement of audio frequencies one of the simplest methods is provided by frequency counters. In the simplest form, a mechanical counter is actuated by a polarized relay through an escapement or a linkage mechanism. The relay operates on each cycle of low-frequency voltage. The frequency values within the operating range can be determined by counting the number of cycles during a known period of time. Such counters are suitable only for very low frequencies; they are not satisfactory for frequencies even as high as 60 Hz.

For higher frequencies, electronic counters are useful. A block diagram of a commercial design is shown in Fig. 13. It shows a block diagram of an electronic counter, where the four stages consist essentially of so-called flip-flop circuits. Stages A, B, C, and D represent 1, 2, 4, and 8 counts, respectively. All the stages can be in two possible states, 1 or 2, signifying OFF and ON position of their neon lamps.

Starting with all the stages in state 1, when a pulse is sent to the first stage A, it triggers into state 2 and its neon lamp starts glowing. A second pulse sent to stage A will put it back into state 1; however, the pulse will be directed to stage B and its lamp will glow. Stage B is now in state 2, signifying a count of 2. A third pulse will put stage A again into state 2, and now both the lamps of stages A and B will glow, indicating a count of three (1 + 2). A fourth pulse triggers both the stages A and B into state 1. A pulse is sent to stage C which triggers into state 2. The lamp of stage C starts glowing, representing a count of four. The next pulse will put stage A into state 2. The lamp for stage A and C will glow, indicating a count of five (1 + 4). The total count in all cases is equal to the sum of the counts represented by the stages which glow. Thus, if three stages A, B, and C glow, seven pulses will be received, indicating seven counts. Usually four-stage counters of this type are designed so that after nine pulses are received a pulse will be sent to the next counter.

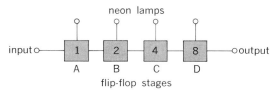

Fig. 13. Block diagram of electric counter.

In this type of counter the usual frequency range is 10 Hz to 100 kHz; however, in some cases counting rates of 10^6 Hz (1 MHz) are possible. *See* DIGITAL COUNTER; FREQUENCY COUNTER.

[EDWARD B. CURDTS]

Radio-frequency measurements. Precise measurement of frequencies above the audible range may be performed by various techniques. Basic measuring systems may consist of (1) a calibrated oscillator with some means of comparison with the unknown frequency, such as an oscilloscope or a heterodyne detector unit, (2) a digital counting or scaling device which registers the total number of events occurring during a given time interval, or (3) an electronic circuit for producing a direct current proportional to the frequency of its input signal which then may be indicated by a dc meter. In practice, each of these basic systems has an upper frequency limit for practical operation.

In order to extend the range of frequency measurement above the practical limits mentioned above, it is customary to generate a fixed standard frequency and to select a harmonic of it near the unknown frequency, after which the unknown frequency may be reduced to a lower value by subtraction of the standard-frequency harmonic in a mixer or heterodyne detector. This lower beat frequency may then be measured by application of the basic methods outlined above, which thus serve to interpolate between the known standard frequencies.

The complexity of the equipment required for a given frequency measurement is usually related to the nature of the unknown signal, the amount of manual operation of the controls required, and the precision required. If the unknown frequency is a pure sine wave of local origin or locally controllable, of adequate amplitude, and free from interfering signals, it is possible to measure its frequency by using (1) a digital counter with appropriate range-extending converter and frequency standard, (2) a heterodyne detection and interpolation system with a suitable frequency standard, (3) a calibrated oscillator and detector unit, called a heterodyne frequency meter, or (4) a calibrated absorption circuit such as a wavemeter or grid-dip meter if slightly lower precision is adequate. *See* WAVEMETER.

If the unknown frequency is subject to interference, is intermittent in nature, is too low in amplitude to override noise, or otherwise requires selective filtering and identification, the simple systems outlined above require additional equipment and manipulation to produce acceptable measurements. Additional knowledge and skill are required of the operator. Measurements to high orders of precision require longer time intervals than less precise measurements.

Microwave frequency measurements. The measurement of most electrical quantities becomes more difficult as the frequency becomes higher. Fortunately, the measurement of frequency, even in the microwave region, is not of unusual difficulty.

In the microwave range of frequencies the wavelengths involved are so short that it is easy to measure the wavelength and thus obtain the frequency by a simple calculation. Many absorption-type wavemeters are calibrated directly in frequency. The principal difference in technique at microwave frequencies is concerned with the generation of usable harmonics of fixed standard frequency oscillators or of tunable transfer oscillators in the range where measurements are required. Otherwise the use of either the digital counter or heterodyne system interpolation between standard-frequency harmonics is carried out as at lower radio frequencies.

If a microwave signal is locally controllable and of a reasonably steady, strong, and stable nature, it is possible to measure it directly by heterodyning it against a known standard frequency and measuring the resulting beat frequency. It is necessary to have a frequency standard, harmonic-generating multiplier chain, mixer, and either a heterodyne system or digital counter system for measuring the beat frequency. Added complexities requiring tunable receivers for selecting the beat frequency and tunable local oscillators for measuring the beat note are sometimes necessary if the beat frequency is not strong. A block diagram of a decade frequency standard and measuring system is shown in Fig. 14.

Transfer oscillator methods. Calibrated oscillators may be used for interpolation between standard-frequency harmonics by using their dial calibrations and conventional interpolation methods. Microwave frequencies are usually mea-

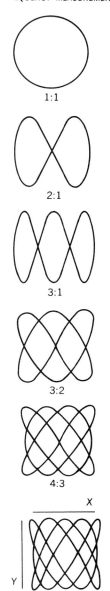

1:1

2:1

3:1

3:2

4:3

5:3

Fig. 12. Lissajous figures for various frequency ratios and phase angles.

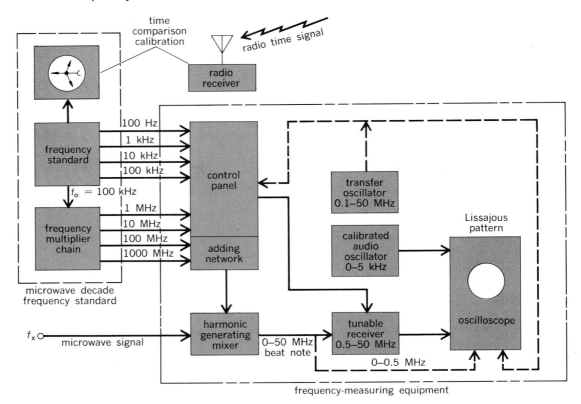

Fig. 14. Block diagram of microwave frequency standard and frequency-measuring equipment.

sured by setting the frequency of the transfer oscillator so that a harmonic of it is at zero beat with the unknown frequency. In this way, it is possible to use the tunable oscillator to transfer the signal to a lower frequency range where it is easy to measure against a frequency standard by direct means. It is usual to provide a tuning range slightly greater than 2:1 for such a transfer oscillator in order to give continuous harmonic coverage at high frequencies.

Transfer oscillators require both high stability and high harmonic output, conditions which are not mutually compatible in one circuit. The stability is generally obtained by using a low-power oscillator with high-Q tuned circuit, the harmonics by separate harmonic generators.

The development of controlled-frequency or synthesizer-type generators enables generation of accurately known stabilized frequencies and hence makes possible directly calibrated transfer oscillators or self-calibrated heterodyne frequency meters. The precision and accuracy of a frequency synthesizer depend on several factors, the limiting item ultimately being the frequency standard.

Heterodyne frequency meters. The calibrated tunable oscillators mentioned above for use as transfer oscillators may be used directly for frequency measurement by their dial calibrations alone. A beat detector and indicator are necessary, as well as a harmonic generator covering the required frequency range; a diode mixer and suitable amplifier with headphones or meter indicator may be used. Unless a synthesizer-type stabilized controlled oscillator is used, the accuracy will not be as good as that obtainable by either the direct measurement or the transfer-oscillator method.

Accuracies of 0.1% can be obtained with a well-designed portable heterodyne frequency meter. If an internal calibrating crystal oscillator is used, better accuracy may be obtained. Accuracy of the synthesizer-type approaches that of its reference frequency standard.

[FRANK D. LEWIS]

Bibliography: W. D. Cooper, *Electronic Instrumentation and Measurement Techniques*, 2d ed., 1978; E. Frank, *Electrical Measurement Analysis*, 1959, reprint 1977; F. K. Harris, *Electrical Measurements*, 1952, reprint 1975; P. Kantrowitz et al., *Electronic Measurements*, 1979; T. Laverghetta, *Microwave Measurements and Techniques*, 1976.

Frequency modulation

A special kind of angle modulation in which the instantaneous frequency of a sine-wave carrier is varied by an amount proportional to the magnitude of the modulating wave. In many so-called frequency-modulation (FM) applications, the angle modulation is neither FM nor phase modulation (PM), but one in which the angle changes in some other way in accordance with the modulating wave. *See* ANGLE MODULATION; PHASE MODULATION.

Either amplitude modulation (AM) or FM offers a solution to the important problem of how to impress the message wave to be communicated upon a high-frequency oscillation. Also, either AM or FM permits detection and faithful reproduction of the original message. However, unlike AM, FM does more than just enable communication. FM offers additional important advantages in exchange for extra bandwidth occupancy. Also, FM with negative feedback minimizes noise problems and receiver distortion. *See* MODULATION.

Instantaneous frequency. Frequency modulation is defined in terms of a generalized concept known as instantaneous frequency which is directly proportional to the time rate of change of the angle of a sine function, the argument of which is a function of time. When the argument is expressed in radians and the time in seconds, the instantaneous frequency in hertz is the time rate of change of the angle divided by 2π.

In frequency modulation the instantaneous frequency is linearly proportional to the magnitude of the modulating wave.

Principles. The illustration depicts typical waveforms of AM and FM for increasing magnitudes of a sine-wave modulating signal. Louder tones with AM mean greater changes in amplitude. Louder tones with FM mean greater changes in frequency.

In AM, as the audio volume increases, the peak power of the modulated wave increases, the average power of each sideband increases, the carrier power is unchanged, and the zero crossings (zeros) of the modulated wave are unchanged and correspond to those of the unmodulated carrier, provided the modulation is short of being complete. *See* AMPLITUDE MODULATION.

In FM, as in any kind of angle modulation, the variable zero crossings of the modulated wave carry the information of the message to be communicated. Neither peak nor average power of the modulated wave changes. Even though the total average power remains constant, its distribution with frequency in the transmitted band changes continuously and in a nonlinear manner as a function of the modulating wave.

J. R. Carson in 1922 was the first to present a mathematical analysis demonstrating the wide band of frequencies involved in FM. However, E. H. Armstrong, in his patent in 1933, was the first to have a real appreciation of the noise-reducing properties of FM. At first, there was considerable skepticism about the practical utility of FM. Today, its vast multitude of uses and widespread acceptance are ample proof that for some purposes FM is better than either AM or single-sideband modulation (SSB). Single-sideband modulation conserves bandwidth, whereas FM uses extra bandwidth to reduce noise still lower.

Advantages. In radio broadcasting, provided the frequency deviation (peak difference between instantaneous and carrier frequencies) is large and provided multipath transmission effects are small, FM is capable of high-fidelity reception combined with the advantages of reduced noise, less interference between stations, and less transmitter power to cover a given area. Constant average power and constant peak power that is only twice the average power are two factors that permit a ready realization of a simple high-efficiency transmitter, simplify problems of automatic volume control, and allow amplifiers and other devices to operate closer to their maximum power capability without the penalties normally associated with nonlinearities.

Spectrum. The spectrum of any angle-modulated wave extends above and below the carrier frequency by amounts which theoretically extend indefinitely. For example, suppose the modulating signal is a single-frequency tone, 1000 Hz, and the

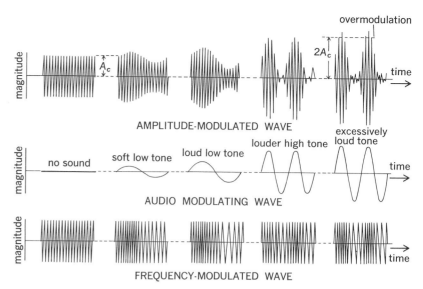

Waveforms of AM carriers and FM carriers with varying degrees of modulation. (*From H. S. Black, Modulation Theory, Van Nostrand, 1953*)

carrier frequency is 1,000,000 Hz. The lower side-frequencies in Hz will be 999,000, 998,000, 997,000, and so on, corresponding to the difference between the frequency of the carrier and the frequency of the modulating signal and its harmonics. If C and V are the carrier and modulating-signal frequencies, the lower side frequencies are $C - V$, $C - 2V$, $C - 3V$, and so on. The upper side frequencies in hertz are 1,001,000, 1,002,000, 1,003,000, and so on, corresponding to $C + V$, $C + 2V$, $C + 3V$, and so on. In addition, the carrier amplitude is reduced when the modulating tone is applied and may even become zero.

This example might suggest that the side-frequency components of an FM wave would always be symmetrical about the carrier, independent of the waveform of the modulating wave. Such is not the case. Unlike the symmetrical spectra of the upper and lower sidebands in AM, in FM unsymmetrical side-frequency spectra are a necessary consequence of an asymmetrical modulating wave.

Practically, when estimating approximate bandwidth occupancy, a rule of thumb states that angle modulation requires the band traversed by the instantaneous frequency plus the bandwidth of the modulating wave added at both top and bottom. For some purposes an even wider band may be required.

Moreover, unambiguous representation and recovery of the wanted message by angle-modulation techniques also require that the unmodulated carrier frequency comfortably exceed the sum of the frequency deviation in the down direction plus the bandwidth of the modulating wave. In other words, in FM the carrier frequency must be high compared to the maximum frequency deviation.

Noise advantage. For certain types of noise disturbance characterized by a noise spectrum that is uncorrelated and independent of frequency, the ratio of average signal power to average noise power in the output of the FM receiver will be proportional to the square of the peak-frequency deviation. Therefore under certain important conditions, the signal-to-noise ratio of an angle-modulation system improves 6 decibels for each 2:1 increase in bandwidth occupancy.

However, the noise advantage of FM cannot be increased indefinitely. As the bandwidth occupancy is continually increased to accommodate an increased frequency deviation, more noise reaches the FM detector. Presently, the assumption that the noise is less than the so-called improvement threshold is violated, whereupon the noise advantage of FM is quickly lost.

FM with negative feedback acts differently in that the improvement threshold is minimized and held constant, independent of the bandwidth occupied by the incoming FM signal.

Improvement threshold. Any system that reduces noise in exhange for extra bandwidth occupancy is characterized by a threshold effect which becomes pronounced when the reduction is large. For a wide-band FM system this threshold is a critical function of the signal-to-noise ratio at the receiver, after selection and before any nonlinear process such as amplitude limiting. When noise perturbations appreciably exceed this critical value, which is termed improvement threshold, the signal-to-noise ratio at the output of the FM receiver rapidly deteriorates. As the noise is still further increased, noise abruptly grabs control and the desired signal tends to be suppressed from the output.

FM with negative feedback. This is accompanied by a decrease in distortion originating within the FM receiver and by an increased tolerance to noise falling within the frequency band occupied by the incoming FM signal. These two important advantages are not possessed by nonfeedback receivers. Substantial benefits are realized only when the amount of negative feedback is large. In common with nonfeedback FM systems, any large reduction in noise must be paid for by a corresponding increase in the bandwidth occupied by the transmitted FM signal.

FM with negative feedback provides an efficient method for the detection and tracking of narrow-band signals in the presence of wide-band noise, especially when the receiver must operate in the presence of large and continuously varying Doppler-frequency effects. In numerous other situations where the saving of power is vital and where extra bandwidth is available, FM with negative feedback may prove attractive, for example, in communications by means of satellites.

When the FM signal is carrying a number of communication channels as in a high-grade multiplex system, nonlinear distortion must be extremely small. This extraordinary degree of linearity is readily obtained with negative feedback. An FM receiver with negative feedback is also characterized by improved stability. With enough feedback, the output becomes virtually independent of such factors as fading of the input signal, power supply voltages, or changes in receiver amplification or detector efficiency.

In a typical nonfeedback receiver, the incoming FM signal is shifted to an intermediate frequency (i-f) by a product modulator fed by a beating oscillator. The instantaneous frequency variations of the i-f signal are then detected, thereby reproducing the original audio (modulating) signal.

With feedback, a portion of the receiver output is fed back and caused to modulate the frequency of the beating oscillator. The receiver output is increased or decreased according to whether the feedback is positive or negative. The amount of feedback is measured by the decibel change in output due to feedback. Negative feedback reduces the instantaneous frequency difference between the incoming FM signal and beating oscillator, thereby reducing the frequency swing (degree of modulation) of the i-f signal.

With sufficient negative feedback, the degree of modulation of the i-f signal is reduced to near zero, thereby reducing the i-f bandwidth to little more than twice the top audio frequency. By this means, the i-f signal is separated from most of the noise existing in the wider bandwidth occupied by the incoming FM signal. Under these conditions, regardless of the bandwidth occupied by the incoming FM signal, the transmitter power need only be enough to override noise in the narrow i-f band.

Production and detection. Many schemes are possible and nearly all use spectrum translation. For the production of FM most schemes resort to spectrum multiplication.

Spectrum translation. Spectrum translation of an angle-modulated wave is accomplished by single-sideband modulation. The translated spectrum, with or without inversion, is centered about a new carrier. Otherwise its significant properties are unchanged.

Spectrum multiplication. This implies angle multiplication. By generating the xth harmonic of an angle-modulated wave, the angle is multiplied by x. If the required multiplication is too great, then after a convenient number of multiplications the resulting spectrum may be translated downward and the multiplication resumed.

Angle multiplication by x multiplies spectrum parameters by x. Typical parameters are significant band-width occupancy, carrier frequency, initial carrier phase, instantaneous frequency, instantaneous-frequency deviation, instantaneous-phase deviation, peak-frequency deviation, and peak-phase deviation.

Preemphasis and deemphasis. Widespread use is made of the fundamental principle that by linearly distorting (equalizing) the modulating wave in a predetermined manner (thereby in effect creating a new modulating wave) any given angle modulator may be converted to another angle modulator having an arbitrary response. Similarly, appropriate equalization of the filtered output of a particular angle detector permits realization of an arbitrary detector response. In radio broadcasting, these procedures are termed preemphasis and deemphasis and are used to improve signal-to-noise ratio. *See* PHASE MODULATION.

FM modulators. For the most part FM modulators fall into two classes. In Class I, a phase modulator is converted to a frequency modulator by first passing the audio signal through a circuit whose output is inversely proportional to frequency. The FM wave thus created normally goes to a spectrum multiplier to increase its frequency deviation. In Armstrong's experiments, this phase modulator consisted of a suppressed-carrier amplitude modulator. Carrier was not only supplied to this balanced modulator in the usual way but in addition was shifted in phase by 90° and then added to the modulator output. When this added carrier is sufficiently large, the resultant output approximates the output of a phase modulator. Armstrong intentionally limited the peak phase modulation resulting from this first step to a maximum range of only 30° in order to ensure reasonable linearity. Consequently, the phase had to be multiplied by several thousand by a succession of spectrum multipliers in order to produce a frequency swing of ±75,000 Hz with peak audio input. Amplitude limiting in the first multiplier attenuated any residual amplitude modulation associated with the phase-modulation process.

In Class II, a capacitance, inductance, potential of a control electrode, or other parameter is varied directly to change an oscillator frequency. Class II modulators usually require spectrum multiplication. *See* FREQUENCY MODULATOR.

FM detectors. Ideally, these would produce output proportional to changes in instantaneous frequency, would not respond to amplitude modulation, and would limit output band to that allocated to the wanted message. In a practical receiver, frequency selection (filtering) is usually followed by spectrum translation and i-f amplification.

Finally, frequency detection is usually accomplished by an amplitude limiter followed by a balanced discriminator, followed by a low-pass filter. The discriminator converts frequency variation to amplitude variation. *See* FREQUENCY-MODULATION DETECTOR. [HAROLD S. BLACK]

Bibliography: J. Betts, *Signal Processing, Modulation and Noise*, 1971; M. Schwartz, *Information, Transmission Modulation and Noise*, 2d ed., 1970.

Frequency-modulation detector

Detection or demodulation of a frequency-modulated (FM) wave. FM detectors operate in several ways. In one class of detector, known as a discriminator, the frequency modulation is first converted to amplitude modulation, which is then detected by an amplitude-modulation detector. Another type of FM detector employs a phased-locked oscillator to recover the modulation. A still different type converts the frequency modulation to pulse-rate modulation, which can be converted to the desired signal by use of an integrating circuit. Examples of these types will be discussed here. *See* AMPLITUDE-MODULATION DETECTOR.

Discriminator. An amplitude-modulation (AM) detector will demodulate an FM wave, providing

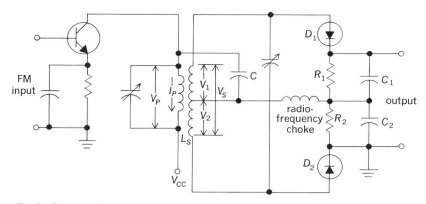

Fig. 2. Phase-shift, or Foster-Seely, discriminator circuit.

the detector is detuned so the carrier frequency is on one side of the passband instead of in the center of the passband. Frequency variations are then converted into amplitude variations or modulation, as shown in Fig. 1, and an AM detector produces the desired output signal. This detection technique has serious limitations, however. Low distortion will be realized if the amplitude variation is linearly related to the frequency variation or deviation of the FM wave. This requirement confines the frequency deviation to a linear or straight-line portion of the tuned-circuit response curve. This restriction seriously limits the permissible deviation, and thus nullifies the advantages of FM that result from high deviation ratios. However, amateur radio operators sometimes use this detuning technique to receive narrow-band FM on an AM receiver. *See* FREQUENCY MODULATION.

An improved demodulator, known as a phase-shift or Foster-Seely discriminator, is shown in Fig. 2. The transistor amplifier is not part of the discriminator, but is shown for completeness. In the circuit, voltage is induced in secondary coil L_S, and primary voltage V_P is coupled to the center tap of secondary L_S through capacitor C. The relative phase of the secondary current, and hence secondary voltage, changes rapidly with changes in input frequency because the secondary is tuned to the carrier or center frequency. A phasor diagram of primary voltage V_P, primary current I_P, and voltage V_S induced in the secondary is given in Fig. 3a. Primary current I_P lags primary voltage V_P by 90° because of the inductive reactance of the primary. Secondary voltage V_S is either in phase or 180° out of phase with the primary, depending on the winding direction, because the induced voltage is proportional to the rate of change of the primary current. *See* ALTERNATING-CURRENT CIRCUIT THEORY.

This situation is shown in Fig. 3b. The phasor sum of primary voltage V_P and voltage V_1 across coil L_1 is applied to upper diode D_1 and its load resistor R_1. Similarly, voltage V_{D2} is applied to lower diode D_2 and its load resistor R_2. Capacitors C_1 and C_2 bypass the radio frequency (rf) around the load resistors. The output voltage is the algebraic sum of the two load voltages, but because the current flows in opposite directions through these diode loads, the output voltage is actually the difference between the load voltages. Thus, if load resistances are equal, the output voltage is zero when

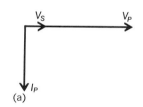

(a)

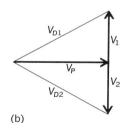

(b)

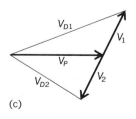

(c)

Fig. 3. Phasor diagrams in a Foster-Seely discriminator. (*a*) Input conditions at reasonance. (*b*) Output voltages at resonance. (*c*) Output voltages for input above resonance.

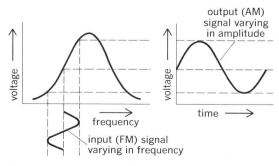

Fig. 1. Resonant circuit used to convert frequency modulation to amplitude modulation.

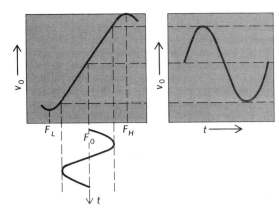

Fig. 4. Linearity and tuning characteristics of phase-shift, or Foster-Seely, discriminator.

the secondary is tuned to the input frequency.

When the input frequency rises above the resonant or center frequency, the secondary circuit is inductive, so the secondary current and the voltage across the secondary coils lag behind their resonant position, as shown in the phasor diagram of Fig. 3c. Voltage V_{D1} applied to upper diode D_1 and its load resistor becomes larger than voltage V_{D2} applied to lower diode D_2 and its load resistor. The difference between these two voltages produces an output voltage of positive polarity. Conversely, if the input frequency decreases from the resonance, the secondary becomes capacitive and the secondary current shifts phase in a leading direction. Therefore, a net output voltage having negative polarity is produced.

The output voltage is essentially proportional to the frequency deviation as long as the frequency remains in the linear portion of the response curve of the coupled circuit. As the frequency approaches the edge of the passband, the change in output voltage for a given change in input frequency becomes less than near resonance because of the reduced amplitude of primary voltage V_p. Thus, for excursions beyond those for which the tuned circuits are designed, frequency discrimination in the coupling circuit produces waveform distortion in the output.

Linearity and tuning characteristics of the Foster-Seely discriminator are shown in Fig. 4. The carrier should be tuned to center frequency F_0. Edges of the useful passband are shown at F_L and F_H. As illustrated, the output voltage is linearly related to the frequency deviation, providing the total deviation does not exceed about 70% of the passband. This degree of linearity is achieved with a double-tuned coupling circuit, as in Fig. 2, and a coefficient of inductive coupling adjusted to the critical value so that maximum linearity is obtained in the response curve. *See* RESONANCE (ALTERNATING-CURRENT CIRCUITS).

The Foster-Seely discriminator is sensitive to amplitude variations, or modulation, as well as to frequency modulation, as can be seen from the foregoing discussion. Therefore, a limiter should always precede the discriminator. The limiter removes amplitude variations from the FM signal. *See* LIMITER CIRCUIT.

A ratio detector (Fig. 5) is a discriminator operating on the same principle as the Foster-Seely discriminator but less sensitive to amplitude modulation. The coupling circuit and the addition of the primary voltage to the center tap of the secondary may be identical to the corresponding portion of a Foster-Seely discriminator. However, in the ratio detector, the two diodes are connected so that their load voltages are additive rather than subtractive. In addition, a large capacitor (perhaps $20-100$ μf) is connected across the series combination of the load resistors. Consequently, the total load voltage cannot follow short-term amplitude variations of the input signal, and the circuit is insensitive to amplitude modulation. However, the individual diode load voltages must vary in the manner described for the Foster-Seely discriminator. Thus, the ratio of the two load voltages changes with the input frequency, even though their sum is forced to remain essentially constant, and hence the name ratio detector.

The output voltage, which is obtained across either of the load resistors, is essentially a linear function of input frequency, providing the requirements placed on the Foster-Seely discriminator are met. In contrast to the Foster-Seely discriminator, the ratio detector has a dc component in the output even when the carrier is unmodulated and the circuit is properly tuned. This dc voltage is usually used for automatic gain control (AGC) when a limiter circuit is not used.

A third winding on the coupling coil, known as a

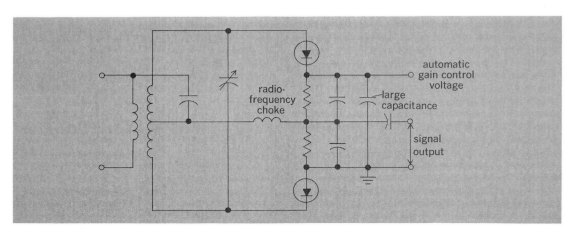

Fig. 5. Typical ratio-detector circuit.

tertiary winding, is often used to produce the voltage which is added to the secondary center tap in a discriminator instead of coupling the primary voltage through a capacitor. In fact, there are many variations of the typical discriminators shown, but they all operate on the same basic principles.

Integrating detector. A frequency-modulated wave can be converted to a pulse-rate modulated wave from which the original modulating signal can be recovered by the use of an integrator. The FM input signal is first converted to a low intermediate frequency (i-f) by mixing the input signal with a local oscillator signal in a nonlinear amplifier. *See* AMPLITUDE MODULATION.

The local oscillator frequency is nearly equal to the carrier frequency, so the difference, or intermediate frequency, is low compared with the carrier frequency. Then the frequency deviation is a high percentage of the i-f. For example, the maximum frequency deviation of a standard broadcast FM signal at 100 MHz may be 150 kHz, or 1.5 parts per thousand. Then, if the i-f is only 200 kHz, the frequency deviation of the i-f is 750 parts per thousand. Thus, the i-f amplifier is normally an untuned low-pass amplifier. The i-f amplifier is followed by a limiter which clips the signal peaks and converts the sinusoidal i-f to a nearly square wave. This square wave is differentiated to obtain pulses which trigger the bistable multivibrator. The multivibrator then produces a rectangular pulse of constant amplitude and width for each input pulse. Therefore, the pulse rate is equal to the input frequency. If the frequency deviation is such that the input frequency is minimum, the pulses are widely spaced for a low-frequency input; on the other hand, the pulses are closely spaced when the deviation produces maximum input frequency. The integrator is basically a low-pass RC filter, which produces an output voltage proportional to the average value of the input voltage and filters out the high-frequency components of the pulses. Therefore, the output voltage is proportional to the pulse rate. The simple RC integrator can be replaced by an operational-amplifier-type integrator for improved filtering, linearity, and output voltage.

Locked-oscillator detector. A third type of FM detector is the locked-oscillator. In this type, a local oscillator follows, or is locked to, the input frequency. The phase difference between local oscillator and input signal is proportional to the frequency deviation, and an output voltage is generated proportional to this phase difference. The phase-locked loop is one of the most dependable and most linear of this type detector. *See* PHASE-LOCKED LOOPS.

[CHARLES L. ALLEY]

Bibliography: C. Alley and K. Atwood, *Electronic Engineering*, 3d ed., 1973; F. M. Gardner, *Phase Locked Loop Techniques*, 1967. General Electric Co., *G. E. Transistor Manual*, 1967.

Frequency-modulation radio

Radio transmission accomplished by symmetrical variation of the carrier frequency by an analog input signal. The amount of swing from center frequency is dependent upon the peak value of the modulating voltage, as well as upon its frequency. The frequency of the modulation signal governs the rate at which the changes in carrier frequency occur. Frequency-modulation (FM) sidebands are formed during the modulation process and are separated from one another by an amount equal to the audio frequency. The amplitude of the sidebands diminishes progressively as the sidebands occur farther and farther from the center frequency, and the number of significant sidebands depends on the amplitude of the modulating signal. The deviation ratio of the carrier-frequency variation to the highest signal frequency transmitted may be any selected value, from fractional to large values. The greater the deviation ratio, the larger the bandwidth required for the transmission. Deviation ratios are standardized for various classes of service for optimum use of the radio-frequency spectrum by a large number of stations.

Because there is no amplitude change in the output of an FM transmitter, whatever the deviation ratio, this mode is an almost perfect cure for amplitude-related interference problems that plague radio-frequency reception. The FM receiver should respond only to frequency modulation; therefore, the signal is not amplified in a linear manner. In fact, the signal is amplified, clipped in a limiter, and amplified and clipped in several stages so as to remove completely any amplitude variation. The FM signal is converted to an amplitude-modulated (AM) signal in a frequency-discriminator detector circuit. The threshold for full quieting, noise-free reception occurs when the carrier-plus-noise-to-noise power ratio is about 12 dB. Another property of FM receivers is the relative freedom from interference between distant and local stations using the same channel; only the strongest signal is received, even if the wanted signal is only 3–6 dB stronger than the interfering one. This characteristic is known as the capture effect. In contrast, AM radio signals differing in strength by 35 dB result in noticeable interference. *See* FREQUENCY-MODULATION DETECTOR.

Since most FM receivers are equipped with a muting circuit (squelch) that silences the audio channel when no signal is present, an irritating hiss noise is not emitted from the loudspeaker when the frequency (or channel) being monitored is not in use. Frequency modulation is used mainly for transmissions above 25 megahertz (MHz). Typical uses are in broadcasting, television sound, mobile radio telephony, radio paging systems, space telemetry, intercity microwave relaying of all classes of public traffic including voice channels, teleprinting, facsimile, broadcast network programs, and television and computer data, and intercontinental telecommunications via satellite. Frequency modulation is used for both analog and digital communications, and phase modulation as well as frequency modulation is employed. The end result, with regard to output waveform, detection, and signal-to-noise improvement, is essentially the same whether phase-modulation (PM) or FM techniques are employed. *See* PHASE MODULATION.

FM broadcasting. The frequency band 88–108 MHz is allocated to FM broadcasting in a large part of the world by international agreement. For a channel spacing of 200 kHz, there are 100 allocable channels for transmission of an audio range of 50–15,000 Hz, with a frequency-deviation ratio of 5. This means that there are five significant sidebands above and below the carrier, the carrier is deviated a maximum of ±75 kHz, and the emitted spectrum is twice this value. Because of the relatively small signal power in the modulating frequencies above 4 kHz, the received signal-to-noise ratio is improved substantially by preemphasis of the audio signal in transmission, necessitating complementary deemphasis in the receiver to restore natural program balance. In fact, preemphasis produces a sort of hybrid form of modulation, being pure frequency modulation at the lowest audio frequencies, and gradually changing to phase modulation at the highest. *See* FREQUENCY MODULATION.

Subscription service. The large emission bandwidth of FM broadcast stations permits simultaneous transmission of other services by modulating at frequencies above the highest program frequencies. This is accomplished by frequency-division multiplexing. In one such use, one to three separate transmissions use FM subcarriers located between 25 and 75 kHz for subscription or other special services. Normal broadcast receivers will not respond to these multiplexed signals, but special receivers are employed to select at least one of them, to the exclusion of the normal broadcast program. These multiplexed special emissions are called special communication authorizations (SCA).

Stereophonic broadcasting. Most FM stations broadcast stereophonic sound, that is, auditory perspective, because two microphones are used to pick up the program. The output of each microphone is separately routed through the transmitter (by multiplexing), transmitted on the same carrier, separated in the receiver, amplified, and reproduced by two loudspeakers.

Stereophonic transmission must also provide normal reception for nonstereophonic FM receivers. To accomplish this, the main channel modulation consists of combining, by additive polarities, the sum of the outputs from the left L and right R microphones. The multiplexed complementary stereophonic channel transmits the outputs for both L and R microphones, but with their polarities reversed. The main channel $L+R$ provides a complete natural signal from ordinary FM receivers. For stereophonic reception, the multiplexed channel contains the $L-R$ signal. Because the $L+R$ and $L-R$ signals are antiphased, high-volume peaks occur in one channel when the other is at its lowest. Each channel has access to almost half the total frequency deviation when its signal peaks to its highest volume. The available power is shared between the two, and the nonstereophonic receiver is unaffected by the multiplex channel.

In this system, the $L-R$ channel is amplitude-modulated on a suppressed 38-kHz subcarrier, so that only symmetrical sidebands are transmitted. But transmission of a low-level pilot carrier at 19 kHz, from which the 38-kHz suppressed carrier frequency (2 times the 19-kHz pilot) was initially derived, also takes place.

Both the pilot frequency and the AM sidebands in turn frequency-modulate the main carrier. To be detected in the receiver, the 38-kHz subcarrier must be derived as a second harmonic of the 19-kHz pilot frequency, combined with the AM sidebands in proper phase and amplitude. An FM stereophonic system requires strict technological control throughout, especially in transmission.

A stereophonic FM station may also transmit an SCA service, which must be shifted to frequencies above those employed by the stereomultiplexed channel; a 67-kHz SCA subcarrier is commonly employed.

Quadraphonic sound. There was widespread interest in four-channel sound in the mid-1970s. The technology involved the employment of four microphones, left front, right front, left rear, and right rear, and the use of four similarly placed loudspeakers in playback was popularly nicknamed surround sound.

The primary difference between the various discrete and discrete/matrix systems proposed is the number and allocation of subcarrier signals used to scramble and unscramble the signal. Any system must work with existing monophonic and stereophonic equipment. However, there is a question as to whether such a system is worth the technical complexity. If a listener only wants surround sound, a version of it can be obtained by using the halfer connection, which feeds a rear pair of loudspeakers with a difference information available across the outputs of a conventional FM stereo receiver. This simple method provides signals from the rear from almost any program material, and the results can be highly acceptable.

FM mobile transmission. Millions of land, maritime, and aeronautical mobile FM transceivers are employed by police, firemen, public safety agencies, industrial and commercial enterprises, private citizens, and radio amateurs who desire the benefits of en-route telephony. The intensity of such usage has grown exponentially, mainly because of the availability of reliable, small, low-power-consumption solid-state equipments that are economical, and also because of the public realization of the benefits of having such communications. In fact, the increase in such usage became a major concern to managers of the radio-frequency spectrum when it became apparent that all available channels would be assigned in major city areas.

The channel width assigned for mobile radio telephony has been reduced in some bands to 15 kHz (12.5 kHz in the United Kingdom), from the wide-band frequency modulation initially employed, and state-of-the-art receivers provide IF (intermediate-frequency) selectivities of −100 dB for ±15-kHz channel spacings. New bands have been allocated for mobile radio (800 MHz), and new techniques are being investigated as an alternative to FM, such as single-sideband (SSB), spread-spectrum, stored speech, and the virtual elimination of speech by the use of data transmission in those applications where standard forms of message predominate. SSB transmission, employing linear compression and expansion (Lincompex) to combat the adverse effects of fading on SSB systems, has so far demonstrated that a minimum channel separation of about 6 kHz is possible compared with 15 kHz for FM transmission. In addition, research is being carried out to develop new automatic methods to more efficiently manage the radio spectrum. *See* SINGLE SIDEBAND.

Typically, mobile telephony is intermittent, each station transmitting only to send. Receivers are open to receive incoming calls on their channel, or the receiver is activated only when certain signaling (touch) tones are received. Simplex transmission is the the rule; that is, by means of a press-to-talk button on the handset or microphone, the transmitter is switched on to talk. Many stations can share a channel, each being informed of when the channel is clear.

Mobile transmitters have output powers of 10–30 W. Hand-held units employing powers of 1–6 W are also used. Base stations employ powers up to 250 W. Vertical polarization is used for FM radio telephony, since a mobile whip, which uses the vehicle as a ground plane, is the most practical antenna configuration. The geographical range of such stations is as much as 40 km with the most favorable terrain, but less in hilly or mountainous regions and heavily treed areas (for example, tropical forests). The use of repeaters favorably located on the top of high-rise buildings, or on the top of hills or mountains, increases the useful range for mobile-to-repeater communications to distances of 100 km. In the ideal situation, therefore, mobile-to-mobile communication to twice this distance is possible. The repeater, as its name suggests, receives the input signal on one frequency and rebroadcasts it on another frequency, with higher transmitter power or high antenna gain, or both. Amateur radio 2M repeaters employ a frequency spacing of 600 kHz between input and output frequencies, necessitating high-Q cavity duplexers so that the receiver and transmitter can employ a common high-gain antenna. The transmission signal must be attenuated by more than 100 dB, even with adequate receiver IF selectivity, since the receiver must respond to signals as small as 0.2 μV or so in the presence of a transmission signal of tens of volts.

With increasing occupancy of the mobile radio bands by many nearby users operating on equally spaced channels, problems of intermodulation are a major limitation to further expansion in some congested areas. Intermodulation occurs when two or more frequencies combine external to, but usually in, the receiver front end to produce a difference frequency that is the same as the desired frequency. Typically, communication receivers experience intermodulation problems when the combination frequencies are about 80 dB above the receiver threshold, although improved performance is possible with state-of-the-art design.

Amateur radio repeaters have become very sophisticated, with tone burst, subaudible tones which can be used for remote control of the repeater functions, linking, and so on, and push-button autopatch (automatic connection and push-button dialing into the public switched telephone system). The numbers of amateur radio repeaters (simple or complex) have grown at a rate that exceeds development of commercial repeaters, and there are one or more repeaters serving most cities and surrounding areas in the United States and Canada. *See* MOBILE RADIO.

Cellular systems. Cellular systems employ equipments operating in the new 800-MHz band. The mobile user can move from cell to cell in an urban environment, and these movements are kept track of, whether the user is monitoring, initiating, or receiving a radio-to-telephone interconnect call.

Such systems are necessarily oriented toward microprocessor control; in fact they must be if all the functions proposed are to be implemented. Cellular systems and trunking have a great deal in common.

Radio paging service. These paging transmitters, which employ high power, a 600-W centrally located transmitter, or several 60-W transmitters frequency-synchronized to within a few hertz throughout the service area, so that reception is possible employing shirt-pocket-sized receivers and inefficient antennas inside buildings and automobiles, have increased difficulties due to intermodulation. "Beeper" and voice pagers are employed. In the voice pager system, the voice of the calling telephone party is broadcast, so that the portable receiver, selectively called by a digital or dual-tone-multifrequency (DTMF) code sequence, receives the message. Technically it is possible with light-emitting diodes (LED) to digitally display the telephone number that the paged party is being asked to call. *See* LIGHT-EMITTING DIODE.

Radio relaying. Frequency modulation is used for microwave radio relaying over land, over water, and to great distances using satellites, sometimes carrying thousands of simultaneous telephone conversations or several television channels.

The advent of requirements for short-haul services, local distribution networks within cities, television relay, and a wide variety of optical communication services including high-speed computer communications, electronic mail, data transmission, and other services, where it may be cost-effective to avoid the local telephone loop, presents another application for FM radio relaying. While 11 GHz has been used to provide these office communications, an unused band centered on 18 GHz has been proposed as being well suited for such applications.

Frequency modulation is of special importance in this application because the signal amplitude remains constant throughout a substantial range of inevitable propagation path fading due to varying weather conditions. Instead, the background noise varies as the transmissions fade, but the signal level at the output of the limiter remains steady, a highly essential characteristic in a communication system.

In microwave communications, frequencies above 1 GHz and to about 18 GHz are used for terrestrial and satellite relaying. By these means, hundreds of telephone, teleprinter, television, and computer data channels of communications can be transmitted on a single microwave carrier. In large systems, as many as six or eight carriers operating in parallel over the same route, to achieve high capacity, are used.

High communications capacities are obtained by multiplexing, using large bandwidths. Two types of frequently employed carrier modulation are analog frequency-division multiplexing–frequency modulation (FDM/FM), which is more common, and digital phase-shift keying (PSK), which is growing in use. In digital communications the analog signal is sampled at some predetermined sampling rate, and the amplitude is quantized at this sample interval by an analog-to-digital (AD) converter. The main advantage of digital transmission consists in the possibility of "overcoming" noise and interference caused by human activities by means of special "coded signals" and by reception techniques which are optimum for the given conditions. In uncoded systems the signal-to-noise ratio increases linearly with increased signal bandwidth, whereas in coded systems the signal-to-noise ratio rises exponentially with increased signal bandwidth. This phenomenon can be explained as follows: Doubling of the spectrum width means doubling the number of pulses (or switch transitions) in the code combination, which, according to the rules governing combinations, squares the number of signals which can be transmitted. *See* ANALOG-TO-DIGITAL CONVERTER.

When PSK or FSK (frequency-shift keying) are used with coherent reception, high frequency and phase synchronization are essential. This is a complicating requirement for this form of modulation, particularly since the PSK signal does not normally contain a component at the carrier frequency, and thus some method of synchronization is necessary.

Radio and television broadcasting using digital communications channels is possible. Although such systems are based on the principles of speech transmission, the system parameters are different, owing to the wider band of continuous signals and the necessary increase in the number of quantizing levels. Sophisticated high-speed digital communications equipments employ not just two-level PSK (0 or 180°) but quadraphase-shift keying (QPSK) modulation (0 or 180° as well as 90 or 270°), and thus provide double the transmission capacity.

Most systems employ analog FDM/FM. For example, a normal telephone signal occupying a range of 300–3400 Hz is used to modulate a suppressed carrier, commonly between 60–108 kHz. By means of electronic filters, only a single sideband (SSB) is selected, this being the original signal translated to a higher frequency. These subcarriers are spaced at 4 kHz, so that 12 conversations can fill the 60–108-kHz range. In turn, the 12 channels are multiplexed with other groups of 12, to form a supergroup of 60 channels, similarly transposed. Master groups of 600, 980, 1200, 1800, or 2700 channels are further extensions of the principle.

Finally, the carrier is frequency-modulated by this aggregate of individual channels to occupy a frequency band that may reach 8 MHz. The carrier peak-deviation ratio is fractional for overland relaying. Typical deviation per telephone channel is 2 kHz. Larger per-channel deviations are used for troposcatter systems and in satellite communications. The Canadian Anik satellite system has a maximum capacity of 4800 two-way voice circuits or 10 television channels. The American Intelset IVA satellite was designed to have a capacity of about 10,000 two-way voice circuits or 20 television channels. Satellite communications technologists expect spacecraft capacities to exceed 100,000 voice channels by 1990.

Telegraphy. Telegraphy, including teleprinting and binary digital data transmission, is based on shifting the carrier frequency or its phase between two limiting values, one of which represents a mark signal and the other a space signal. This frequency shift (or phase shift) is a form of FM signal-ing used over a wire, cable, or radio. A common approach is to use multiplexed audio tones, each deviated a few tens of cycles (or shifted in phase between two discrete levels), to carry individual messages. A normal voice channel can carry 16–24 such telegraph channels operating at standard teleprinter speeds. High-speed data transmission requires larger bandwidths and more elaborate equipment, but is based on the same principle of frequency or phase shift.

Facsimile. Black-and-white images (line drawings and typed copy) can be transmitted by employing the principles used in FM telegraphy; one limit frequency corresponds to black, and the other to white, on the image to be transmitted. It is necessary only that the original document be scanned (for example, by a photosensitive cell), that the scan at the recorder be synchronized with that at the transmitter, and that some means be employed, for example, using an electrosensitive or electrostatic (dry) process, to make the recording paper black when a black signal is received and leave it white for a white signal. A continuous gray scale can be transmitted and recorded if, instead of just two frequencies, a continuous frequency shift is employed between some low frequency (say, 1500 Hz) and some higher frequency (say, 2700 Hz), the exact frequency at any instant being proportional to the gray level of the image. The recorder employs a frequency-amplitude transducer, the response of which is the inverse of that at the transmitter. The audio-frequency tones can be sent over telephone lines; they can modulate an FM radio transmitter, or they can modulate an SSB transmitter. *See* FACSIMILE.

Telemetry. Frequency modulation is the preferred method for transmission of information or data from a remote or inaccessible location such as a rocket vehicle in flight. Each condition to be remotely observed actuates one subchannel, which, when multiplexed with other channels reporting other status conditions, modulates the radio carrier by frequency modulation. Telemetry is not much different in principle from FM and PM telegraphy, except that it is a one-way transmission system, with each channel containing a transducer which converts sensed phenomena into electrical signals. The signals may be either digitized coded pulses or continuous (analog) changes. At the receiver, each channel is suitably instrumented to record directly in real time the information reported by the transducer on that channel; or the FM signal can be frequency-translated and recorded directly on magnetic tape for subsequent analysis. Frequency or phase modulation is very useful for digital data recording, since, for example, phase modulation is effective against varying oxide thicknesses on the magnetic tape and varying signal strength. Telemetry systems exist in all degrees of sophistication, up to hundreds of data channels.

[JOHN S. BELROSE]

Bibliography: N. H. Crowhurst, *FM Stereo Multiplexing,* 1961; *FM and Repeaters for the Radio Amateur,* ARRL, 1974; P. F. Porter, *Modulation, Noise and Spectral Analysis,* 1965; *Reference Data for Radio Engineers,* 6th ed., 1975; S. S. Sviridenko, *Fundamentals of Synchronization in Reception of Digital Signals,* 1974; C. E. Tibbs and

G. G. Johnstone, *Frequency Modulation Engineering*, 1956; T. W. Washburn et al., *Development of HF Sykwave Radar for Remote Sensing*, AGARD Conf. Proc. no. 263, 1979; W. R. Young et al., Advanced mobile phone service, *Bell Sys. Tech. J.*, 58(1):1–14, 1979.

Frequency modulator

An electronic circuit or device producing frequency modulation. The frequency modulator changes the frequency of an oscillator in accordance with the amplitude of a modulating signal. If the modulation is linear, the frequency change is proportional to the modulating voltage.

High-frequency oscillators usually employ LC tuned circuits to establish the frequency of oscillation, and this frequency can be controlled by changing the effective capacitance or inductance in accordance with the modulating signal. A simple example of this type of modulator is a capacitor microphone used as the capacitive tuning element in an oscillator. The acoustic waves that strike the diaphragm of the microphone change the capacitance, which in turn changes the oscillator frequency. The acoustic pressures are usually so small that the capacitance change and the frequency change, or frequency deviation, are linearly related to the acoustic pressures. However, this modulation method is physically cumbersome and is seldom used. Practical circuits usually employ either a reactance tube or a varactor diode to change the oscillator frequency in accordance with a modulating voltage.

Varactor modulator. The varactor diode is used commonly as a frequency modulator in modern equipment. A typical varactor modulator is shown in the illustration. In this circuit the collector volt-

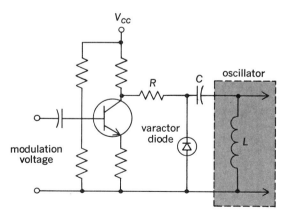

Elementary varactor-diode modulator.

age of the modulating amplifier controls the reverse bias across the diode. But since the junction capacitance of an abrupt junction diode is inversely proportional to the square root of the junction voltage, the oscillator tuning capacitance, which consists of capacitor C and the diode in series, is therefore controlled by the modulation signal. The resistance of R must be small in comparison with the reactance of either capacitor C or the diode at all modulating frequencies. Otherwise, the high modulating frequencies will be attenuated. Also,

the resistance of R must be large in comparison with either the reactance of C or the reactance of the diode at the oscillator frequency. Otherwise the Q of the tuned circuit will be degraded.

Linearity and bandwidth. Because the frequency of oscillation is not linearly related to the tuning capacitance, modulation distortion will occur in either the reactance tube or the varactor modulators. This distortion can be limited to any desired value, however, by limiting the amplitude of the modulating signal. For example, the modulation distortion of the varactor-diode modulator will be less than 4% if the peak amplitude of the modulating signal is limited to one-tenth of the average bias voltage.

Frequency-modulation systems can be easily devised for low-frequency systems by controlling the base or gate bias supply voltages to relaxation oscillators such as multivibrators or unijunction oscillators. The frequency deviation of these oscillators is a fairly linear function of the modulating voltage. Sinusoidal voltages may be obtained, if desired, from these circuits by passing the output signal through a tuned amplifier. Frequency modulation can also be obtained by phase modulation.

Voltage-controlled oscillators. Oscillators having a linear relationship between their output frequency and a controlling voltage have been developed for use in phase-locked loops. Known as voltage-controlled oscillators, they are available in integrated-circuit form either as a separate chip or as a separately usable section of an integrated phase-locked loop. Voltage-controlled oscillators have excellent linearity over very wide frequency deviations and are available for frequencies up to 60 MHz. They may have many applications as frequency modulators. *See* PHASE-LOCKED LOOPS.

There is no frequency-modulation index m_f corresponding to the 100% amplitude modulation. The index m_f is the ratio of the frequency deviation to the modulation frequency, and the optimum frequency deviation depends upon the bandwidth of the receiving system. This bandwidth must be approximately two times the sum of the frequency deviation and the modulation frequency. *See* FREQUENCY MODULATION; PHASE MODULATION.

[CHARLES L. ALLEY]

Bibliography: C. L. Alley and K. W. Atwood, *Electronic Engineering*, 3d ed., 1973.

Frequency multiplier

An electronic circuit that produces an output frequency which is an integral multiple of the input frequency. There are two basic types of frequency multipliers. The first type is a nonlinear amplifier which generates harmonics in its output current and a tuned load that resonates at one of these harmonics. The second type uses the nonlinear capacitance of a junction (semiconductor) diode to couple energy from the input circuit, which is tuned to the fundamental, to the output circuit, which is tuned to the desired harmonic.

Nonlinear amplifier. Harmonic-generating amplifiers are usually operated class C, biased so that the output current flows for only a small part of the input cycle, perhaps 90°. The output is then richer in harmonics, and higher efficiency is attained than with normal class C bias. The output circuit may be tuned to any harmonic, but the

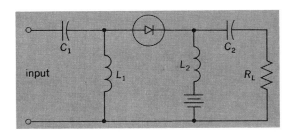

A varactor-diode multiplier.

higher harmonics are weaker and the power output may be less than the power input if the selected harmonic is higher than the third or fourth.

A highly efficient doubler can be devised by using two amplifiers, such as transistors, with their inputs driven with opposite polarity, obtained from opposite ends of a center-tapped coil, and their outputs connected in parallel.

These amplifier-type frequency multipliers are often used in crystal-controlled high-frequency transmitters. The crystal oscillator is desirable because of its excellent frequency stability, but crystals are available for only a limited frequency range. Therefore, a moderate crystal-oscillator frequency is multiplied to a high output frequency.

Nonlinear coupler. Diode-type multipliers are usually used in vhf or uhf solid-state transmitters. The frequency limit at which a transistor has useful power gain is highly dependent upon the power output capability of the transistor. Varactor-diode

multipliers are often used to provide an output frequency for the transmitter that is several times as high as the frequency limit of the power transistor or transistors in the transmitter.

Efficiency is of prime importance in this type of multiplier because the transmitter power output is the power output of the final transistor amplifier minus the power loss in the multiplier. Fortunately, the efficiency of varactor-diode multipliers may be as high as 80–90% if the circuit and diode Qs are high and the multiplication in each stage is no greater than two or three. The multipliers can be advantageously connected in cascade if higher multiplication ratios are desired. A typical varactor-diode multiplier circuit is shown in the illustration. Coil L_1 and capacitor C_1 resonate near the input frequency, while coil L_2 and capacitor C_2 resonate near the output frequency. *See* AMPLIFIER; SEMICONDUCTOR.   [CHARLES L. ALLEY]

Frequency-response equalization

The process of obtaining a desired overall frequency-response characteristic in an audio-frequency circuit by introducing corrective electrical networks of various types, termed equalizers, into the circuit.

Equalizers are used both in communications networks and in systems for the recording and reproducing of photographic film, magnetic tape, and disk phonograph records. These equalizers are in the form of electrical networks of resistance, inductance, and capacitance. The position of the frequency-response characteristics with respect to

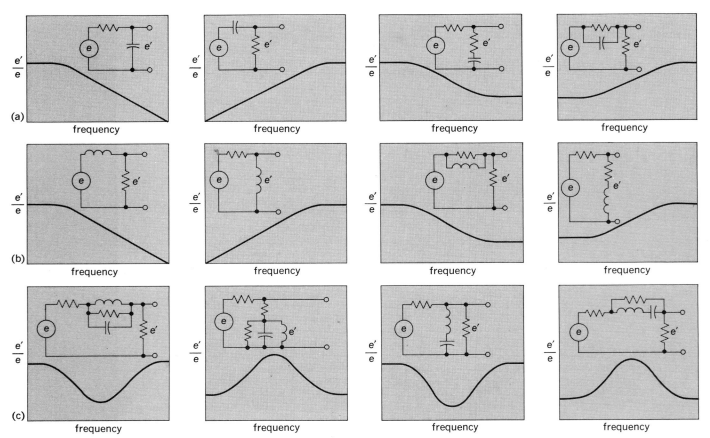

Circuit diagrams and frequency-response characteristics of equalizers employing combinations of (a) resistance and capacitance, (b) resistance and inductance, and (c) resistance, inductance, and capacitance.

frequency can be varied by a choice of values of R, L, and C. The maximum variation is 6 dB/octave. Larger variations can be obtained by cascading equalizers. Equalizers employing resistance and capacitance are shown in illustration a. Equalizers employing resistance and inductance are shown in illustration b. Equalizers employing resonant circuits consisting of resistance, inductance, and capacitance are shown in illustration c.

Referring to the figure, it will be seen that practically any type of equalization can be obtained from the systems and combinations of the systems shown. All major record companies in the United States use the RIAA (Record Industry Association of America) standard in their recordings; thus the provision of a large number of equalizing or playback settings on hi-fi amplifiers has become unnecessary. *See* AMPLIFIER; DISK RECORDING.

[HARRY F. OLSON]

Function generator

An electronic instrument which generates periodic voltage or current waveforms that duplicate various types of well-defined mathematical functions. The simplest function generator usually generates a combination of square waves, triangular waves, and sine waves (Fig. 1).

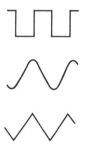

Fig. 1.   Waveforms by a simple function generator.

One electronic circuit approach to the design of a simple function generator is to begin with a bistable multivibrator or "flip-flop" controlled in time by a succession of clock pulses which generates the square wave (Fig. 2). The triangular waveform is obtained by integrating the square wave through the use of the operational amplifier integrator. The sine wave is obtained by applying the triangular wave to a shaping circuit consisting of a combination of resistors and diodes. *See* AMPLIFIER; MULTIVIBRATOR; WAVE-SHAPING CIRCUITS.

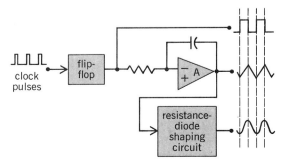

Fig. 2.   An elementary function generator.

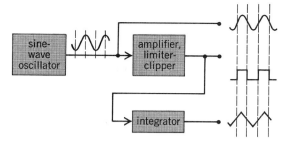

Fig. 3.   An alternate form of a function generator.

Alternatively the sine wave may be generated by a sinusoidal oscillator (Fig. 3). From this output, the square wave may be obtained by amplication, limiting, and clipping of the sine wave. Then the triangular wave may be obtained using an integrator as before. *See* CLIPPING CIRCUIT; LIMITER CIRCUIT.

A still different configuration of a function generator consists of a circuit, which simultaneously generates triangular waves and square waves, whose rate is controllable by a direct-current level (Fig 4). Such a circuit is identified as a voltage-controlled oscillator (VCO). One of the outputs can be applied to a resistor-diode shaping circuit to obtain the sine wave output. *See* FREQUENCY MODULATOR; PHASE-LOCKED LOOPS.

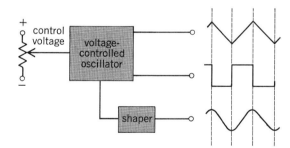

Fig. 4.   A voltage-controlled oscillator as a function generator.

More sophisticated function generators have circuits built in which will allow many other waveforms to be generated, such as parabolic waveforms and logarithmic functions, using combinations of resistor-diode combinations.

A combination of counters, programmed read-only memories (PROMS), and a digital-to-analog converter can be used as a function generator, generating almost any function desired to almost any degree of accuracy. *See* COMPUTER STORAGE TECHNOLOGY; DIGITAL-TO-ANALOG CONVERTER.

[GLENN M. GLASFORD]

Bibliography: J. A. Connelly, *Analog Integrated Circuits*, 1975; J. Millman and H. Taub, *Pulse, Digital*, and *Switching Waveforms*, 1965.

Gain

A general term used to denote an increase in signal power or voltage produced by an amplifier in transmitting a signal from one point to another. The amount of gain is usually expressed in decibels above a reference level. *See* AMPLIFIER.

Antenna gain is a measure of the effectiveness of a directional antenna as compared to a nondirectional antenna. It is usually expressed as the ratio in decibels of standard antenna input power to the power input to a directional antenna that will produce the same field strength in the desired direction. The more directional an antenna is, the higher is its gain. *See* ANTENNA (ELECTROMAGNETISM). [JOHN MARKUS]

Gallium arsenide integrated circuits

Gallium arsenide, GaAs, is a compound semiconductor whose unusual electronic properties make it attractive for applications to microwave semiconductor devices (Gunn, IMPATT, and Schottky diodes and field-effect transistors) and optoelectronic devices (light-emitting diodes and photodiodes). Significant advances in material technology and fabrication methods such as ion implantation, projection alignment, plasma etching, and ion milling techniques have made it possible to begin to fabricate gallium arsenide digital and analog integrated circuits. These circuits have provided very high switching speeds. Propagation delay per logic gate as low as 65 picoseconds has been observed on circuits fabricated by optical lithography, while electron-beam lithographic techniques have yielded circuits with delay as low as 34 ps per gate. Speed can also be exchanged for increased power dissipation. Power-delay products under 10 femtojoules (10^{-14} J)/gate have been reported for minimum-area ring oscillators. Current research is directed toward higher speed, lower power, and large-scale integration (LSI) of gallium arsenide digital circuits and analog circuit integration. *See* MICROWAVE SOLID-STATE DEVICES.

Clearly there would be applications possible at all levels of integration for circuits exhibiting the performance described above. At the small-scale integration (SSI) level (under 20 gates), high-speed frequency dividers, sample-hold amplifiers, latches, and comparators have been demonstrated in several laboratories. Frequency dividers which operate at input frequencies above 4 GHz and sample-hold circuits with acquisition times less than 300 ps have been reported. High-speed medium-scale integration (MSI) circuits (20–100 gates) such as 1 gigabit/s multiplexers, demultiplexers, and shift registers, multipliers, and 1.6-GHz variable-modulus dividers have all been demonstrated in the laboratory, and have operated at data rates consistent with the 100–200 ps/gate propagation delays expected for MSI/LSI gallium arsenide integrated circuits and the chosen circuit architecture. At the large-scale integration complexity level (approximately 1000 gates), a much wider range of potential applications becomes available, covering computational, signal processing, frequency synthesis, data conversion, and other areas. Low power-delay products are essential to the realization of LSI high-speed gallium arsenide integrated circuits because total power dissipation per chip must be less than 1–2 W if a practical, reliable circuit is to be obtained.

In addition, in order to effectively utilize high-speed, low-power integrated circuits, it is essential to build at the highest possible level of integration so that power-consuming interconnects and interconnect delays will be minimized. The optimum level of integration is that in which all input and output data are provided at a relatively low rate while the chip itself operates on those data at the maximum possible speed.

There are three major gallium arsenide digital integrated circuit research areas being actively addressed. First, the above-mentioned LSI design and fabrication area is being pursued using the Schottky diode field-effect-transistor logic (SDFL) planar approach described below (Fig. 1). Second, ultrahigh-speed circuits are being developed with the use of electron-beam lithography to provide submicrometer device dimensions. Here all emphasis has been on speed, with little effort being directed toward minimizing power dissipation. Finally, gallium arsenide integrated circuits with propagation delays less than 500 ps are being developed with emphasis on ultralow-power dissipation by use of normally-off field-effect transistor (FET) devices.

There is one further area of research being pursued in the gallium arsenide integrated circuit area, which is an outgrowth of extensive gallium arsenide microwave field-effect transistor development and application over the past decade. This is the development of high-speed analog circuits and monolithic microwave integrated circuits (MMIC). These circuits will ultimately find use in high-speed data acquisition and microwave receiver types of applications. Sample-hold amplifiers, comparators, and operational amplifiers are being developed for future integration into gigabit analog-to-digital converters. Also, integrated broadband microwave low-noise amplifiers, mixers, local oscillators, and intermediate-frequency amplifiers, which will ultimately lead to complete microwave

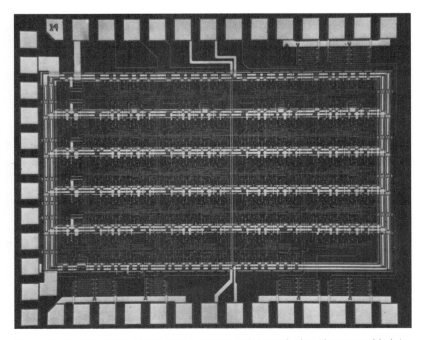

Fig. 1. Photomicrograph of 250-gate (large-scale integration) gallium arsenide integrated circuit fabricated with the planar Schottky diode field-effect-transistor logic (SDFL) technology.

receivers on a single chip, have been demonstrated. This will lead to significant cost savings and may prove to be essential for the economics of consumer satellite television receiver (cheap downlink receivers to convert 12-GHz satellite television broadcasts to the UHF range).

Gallium arsenide versus silicon. The same properties which make gallium arsenide attractive as a microwave device material also make it desirable for high-speed logic circuits. First, the electron mobility of gallium arsenide (4000 cm²/V-s at 1×10^{17} donors/cm³) is $8-10$ times greater than that of silicon. Second, the peak electron velocity (2×10^7 cm/s) is reached at an electric field of only 3.5 kV/cm in gallium arsenide. The electron velocity of silicon continues to increase as higher electric fields are applied, and slowly approaches a saturated drift velocity of 6.5×10^6 cm/s at very high fields, while the GaAs electron velocity decreases slowly at high electric fields and ultimately approaches 1 to 1.4×10^7 cm/s.

These very divergent properties between gallium arsenide and silicon are quite significant for very high-speed, low-power digital circuits. In order to minimize power dissipation, a low logic voltage swing (ΔV) is desired since dynamic power increases as ΔV^2. For high speeds, high electron velocities are needed since the gain-bandwidth product of the field-effect transistor is inversely proportional to the transit time under the gate electrode. Gallium arsenide, because of its electron-transport properties, yields its maximum peak velocity at relatively low bias voltages, thus providing the capability for achieving high speed and low power in the same device. Silicon, on the other hand, requires large electric fields to achieve high electron velocity. Therefore high-speed silicon devices dissipate much greater power per logic gate than gallium arsenide.

Finally, gallium arsenide possesses a high-resistivity (greater than 10^8 Ω cm) semiinsulating substrate which provides isolation between devices without the need for pn junction or heteroepitaxial (silicon-on-sapphire; SOS) isolation. Therefore circuit density can be greater than with comparable silicon integrated circuit approaches. Also, the insulating substrate reduces interconnect line parasitic capacitance, thereby increasing speed and decreasing dynamic power dissipation.

MESFET. The Schottky-barrier gate metal-semiconductor field-effect transistor (MESFET) is the main active device used in gallium arsenide integrated circuits to provide current gain and inversion. Figure 2 shows a modern planar, ion-implanted gallium arsenide MESFET fabricated by localized implantations of dopants into a semiinsulating gallium arsenide substrate. The source and drain regions receive a deep, low-sheet-resistance n-type implant to reduce parasitic channel and contact resistance. The channel region is implanted with a shallow (100- to 200-nm) n-type implant. Drain-source current scales directly with channel width. The metal gate electrode, which is typically 0.5 to 1μm long, forms a Schottky barrier on the 3-μm-long transistor channel, in which it is centered. This gate can be used to pinch off the channel (depletion mode) when reverse-biased. In the case where the channel receives a lighter implant dose and is pinched off at gate-source voltage

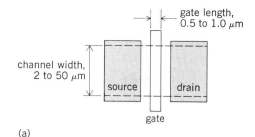

(a)

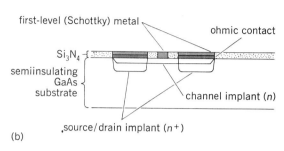

(b)

Fig. 2. Planar ion-implanted gallium arsenide MESFET. (a) Top view. (b) Cross section.

$V_{gs} = 0$, the gate enables current flow to occur as it is forward-biased (enhancement mode).

The gallium arsenide MESFET is well suited for high-speed integrated circuit applications because of its high current gain–bandwidth product ($15-20$ GHz for a 1-μm-gate-length transistor) and the relative ease of fabricating MESFET with micrometer or submicrometer gate lengths. Gate alignment tolerances for gallium arsenide MESFET are relatively less demanding than silicon metal oxide semiconductor field-effect transistors (MOSFET) of equivalent gate length, and the requirement for the extremely thin (25-nm) gate oxide is eliminated. Threshold control is achieved in the MESFET by precise control of the channel implant.

GaAs integrated circuit approaches. Three main logic gate designs, the buffered FET logic (BFL), Schottky diode FET logic (SDFL), and direct-coupled FET logic (DCFL) approaches have been employed in gallium arsenide digital integrated circuits. NOR gates using these three designs are illustrated in Fig. 3. The BFL and SDFL approaches are applicable for depletion-mode MESFET and therefore require two power supplies, V_{DD} and $-V_{SS}$. Level-shift diodes are necessary so that input and output voltage levels are compatible. In the BFL approach this level shift is achieved at the output by Schottky-barrier diodes as shown in Fig. 3a. Either NOR or NAND functions are achieved at the input by parallel or series combinations of MESFET. For practical purposes, a NOR fan-in of 4 or NAND fan-in of 2 is the maximum usable with BFL. BFL circuits have achieved very low propagation delays (34 ps/gate for 0.5-μm-gate-length MESFET), but their power dissipation (40 mW/gate) is too large for even most MSI applications. Lower-power (5.5 mW/gate) BFL gates have been designed, but with some sacrifice in speed (82 ps/gate). Two-level logic gates (NAND/NOR) can also be realized in BFL. MSI frequency dividers fabricated with two-level BFL gates have operated as high as 4.5 GHz. Other MSI circuits, such as multiplexers, have also been demonstrated.

The SDFL circuit (Fig. 3b) uses very small-area Schottky diodes at the gate input to provide the logical OR function (diodes $D_A–D_D$) and level shifting (diode D_S). Schottky diodes make very-high-speed logic elements in gallium arsenide because

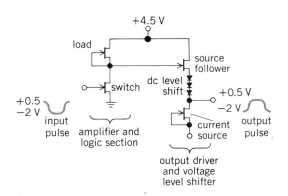

(a)

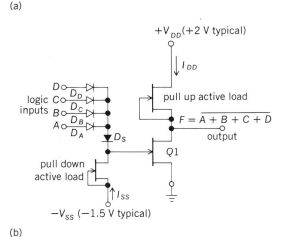

(b)

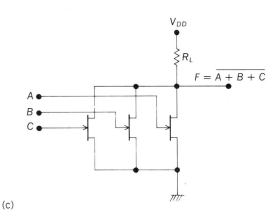

(c)

Fig. 3. Three types of NOR gate commonly utilized in gallium arsenide integrated circuits. (a) Buffered FET logic (depletion mode) (*from R. L. Van Tuyl et al., GaAs MESFET logic with 4-GHz clock rate, IEEE J. Solid-State Circuits, SC-12:485–496, 1977*). (b) Schottky diode FET logic (depletion mode); (c) direct-coupled FET logic (enhancement mode) (*from R. C. Eden et al., The prospects for ultrahigh-speed VLSI GaAs digital logic, IEEE J. Solid-State Circuits, SC-14:221–239, 1979*).

of their extremely high cutoff frequencies. MESFET Q1 provides current gain and inversion; 1-μm SDFL gates (62 ps/gate) have also demonstrated low propagaion delays, while maintaining very-low-power dissipations as well (less than 1 mW/gate). Power-delay products as low as 16 fJ have been obtained. This approach has the greatest potential for LSI; 500–1000 gate circuits were expected in 1980. Multiple-level gates have also been achieved in SDFL (OR/NAND, OR/NAND/AND) and can be effectively used to further increase circuit speed and reduce power dissipation.

The DCFL circuit (Fig. 3c) utilizes enhancement-mode field-effect transistors to achieve the NOR functions. Because the gate-source voltage V_{gs} is greater than 0 for these devices, no level shifting is required, and only one power supply (V_{DD}) is needed. Thus the logic gate requires fewer components than those required in the depletion-mode logic approaches, and will, in most cases, require less wafer area than a BFL gate. Very-small-area ring oscillators have demonstrated propagation delays below 200 ps with power-delay products of 4 fJ. Frequency dividers have also been fabricated with DCFL NOR gates, but with somewhat higher propagation delay (400–500 ps) due to fan-in and fan-out loading. *See* LOGIC CIRCUITS.

Monolithic microwave integrated circuits have all focused on the use of the MESFET as the primary active circuit element. Schottky diodes have also been used used for switching and mixers. Impedance-matching networks are often included in the chip so that inputs and outputs can directly drive a transmission line. Circuit designs have generally emphasized lumped element rather than distributed matching networks because of the smaller chip area requirements and broader bandwidth of the lumped element approach. These passive circuit elements include capacitors (interdigitated, metal-insulator-metal, or Schottky diode), spiral and simple loop inductors, and implanted or mesa resistors. Broadband, low-noise, single-stage amplifiers which provide 6.0 ± 0.5 dB gain from 8 to 18 GHz using these techniques have been reported. Feedback amplifiers providing essentially flat gain from direct current to 4 GHz, and monolitic signal generator chips have been made. Operation of a 1-W monolithic X-band amplifier chip has also been reported. Further efforts are expected to achieve complete receiver front-end integration, including low-noise amplifier, mixer, local oscillator, and intermediate-frequency amplifier in the near future. *See* INTEGRATED CIRCUITS; TRANSISTOR.

[STEPHEN I. LONG]

Bibliography: D. A. Abbott et al., *Microwave Syst. News*, 9(8):72–96, August 1979; P. Greiling et al., *Microwave Syst. News*, 10(1):48–60, January 1980; C. A. Liechti, *Microwaves*, 17(10):44–49, October 1978; S. I. Long et al., *IEEE Trans. Microwave Theory Techniques*, MTT-28(5):466–471, May 1980.

Gas tube

An electron tube that contains gas or vapor at low pressure in which an electrical discharge takes place. Gas tubes are of two general types: cold-cathode tubes, in which the phenomenon known as glow discharge serves to maintain a conducting

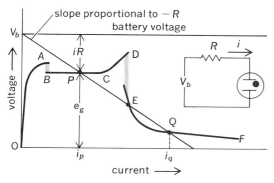

Fig. 1. Typical volt-ampere characteristics diagrammed for a cold-cathode gas tube.

path between the electrodes, and hot-cathode tubes, in which an arc discharge conducts the current. The cold-cathode glow tubes are characterized by a relatively high voltage drop and a low current, while the hot-cathode arc tubes are characterized by a low voltage drop and relatively high current.

Volt-ampere characteristics. The two types of discharges may be most easily illustrated by means of the typical volt-ampere characteristics shown in Fig. 1. The portion OA represents the prebreakdown characteristic and is a function of electrode spacing. At A, which is the breakdown voltage, the nearly nonconducting gas ionizes and becomes a good conductor, and a self-sustained discharge is established. The portion BC is known as a normal glow. This glow is characterized by a nearly constant voltage, of the order of $80-400$ volts (depending on gas and cathode material), for a range of current of the order of $1-100$ milliamperes (mA). The portion CD, having a rising volt-ampere characteristic, is an abnormal glow. The characteristic EF having a negative volt-ampere characteristic is an arc discharge. For a given battery voltage V_b and circuit resistance R, two stable points exist. At P the discharge is a glow with current i_p while at Q the discharge is a low-voltage arc having a current i_q. Both P and Q represent stable points of operation, and whether a glow-arc transition can occur depends on critical physical conditions at the cathode. The point E represents an unstable condition where operation is not possible for the indicated values of R and V_b.

The electron emission at the cathode in the glow region is largely due to positive-ion bombardment. The normal glow is characterized by a constant current density at the cathode surface. The normal glow occurs as long as current requirements of the external circuit are insufficient to cover completely the cathode surface with a glow, called the negative glow. At the point C the cathode becomes completely covered with the negative glow, and a further increase in current results in an increase in voltage, which is necessary to give the bombarding positive ions the energy required to produce the greater electron current required, giving the abnormal glow characteristic. Most of the voltage drop across the tube occurs between the cathode and the negative glow. The balance of the distance between the cathode and anode is filled with a glowing gas called a plasma, that requires only a small voltage to maintain its conducting state. *See* ELECTRICAL CONDUCTION IN GASES.

The voltage drop across an arc discharge is quite low compared with that of a glow discharge, being of the order of $10-20$ volts. The low drop of the arc is due to an efficient mechanism of electron emission from the cathode and to the presence of large numbers of positive ions that neutralize the space charge of the electrons.

Cold-cathode gas tubes. Cold-cathode gas tubes, or glow tubes, require no cathode heater power and are therefore always ready for instant service. Glow tubes are commonly constructed as two-element (diode) or three-element (triode) tubes.

Cold-cathode diode. This type of tube is usually either a glow lamp or a voltage-regulator tube.

The glow lamp has two electrodes, usually of equal size. The light from the negative glow makes the tube useful as an indicator lamp. These lamps have a small series resistance built in the base and are designed to operate directly on 120 or 240 volts. Similar tubes are also available with special bases that have no built-in resistor and are used in circuits containing their own ballast resistor.

The voltage-regulator tube operates in the normal glow region and therefore has a nearly constant anode-to-cathode voltage drop over a range of current. These tubes are used to maintain a dc potential substantially constant for adjustments of input voltage or load current. *See* VOLTAGE-REGULATOR TUBE.

Cold-cathode triode. This tube is sometimes called a grid-glow tube. Tubes of this type have a third starting electrode, which surrounds the anode (Fig. 2) and gives the device control properties. This electrode electrostatically shields the cathode from the anode potential so that the discharge cannot start until the starter is raised to a sufficiently positive voltage to break down the cathode-starter gap, after which the discharge transfers to the main anode. The anode potential need be no higher than that necessary to sustain the discharge. A typical tube of this type has a

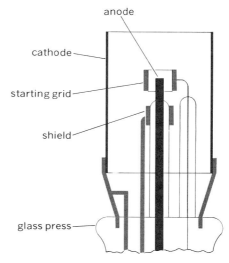

Fig. 2. The cold-cathode gas triode.

maximum anode voltage drop of 85 volts, average current of 25 mA, and a starter breakdown voltage in the range of 73–105 volts.

Hot-cathode gas tubes. These low-pressure arc-discharge devices operate with the characteristic EF of Fig. 1. Three representative types may be distinguished: the Tungar (sometimes called a Rectigon), the phanotron, and the thyratron. The first two are simple rectifier tubes while the third is a control tube having one or two grids between cathode and anode. Both phanotrons and thyratrons may be built with glass, ceramic, or metal envelopes.

Tungar tube. This tube is a low-voltage rectifier having a hot thoriated-tungsten filament, an anode spaced about 1 cm from the filament, and a gas filling of argon at a pressure of from 5 to 10 cm Hg. The name was chosen to suggest the tungsten filament and the argon gas. This tube has a low arc drop of the order of 8 volts and a current range of 1–15 A, which is well suited to battery charging service. The relatively high gas pressure permits higher than normal operating temperature of the filament by suppressing filament evaporation. Such atoms of tungsten and thorium as are evaporated are largely ionized near the cathode and returned to the filament so that a long life is obtained. However, the high pressure results in slow deionization of the gas. During inverse voltage periods (the portion of the cycle in which the anode is negative), a glow may be started at voltages of 200–300 volts with consequent danger of the glow-arc transition. The tungar tube starts at a relatively low forward voltage because of the copious supply of electrons available and the low gas density in the vicinity of the cathode where the gas temperature is relatively high. *See* Arc discharge.

Phanotron tube. The phanotron is a thermionic (hot-filament) diode rectifier tube utilizing an arc discharge in mercury vapor or an inert gas, usually xenon. Mercury is the preferred filling where a low arc drop is desired, whereas one of the inert gases is used where independence of ambient temperatures over a wide range is desired. The arc drop for the mercury tube is about 10 volts and that for a xenon tube is about 12 volts. In both cases the arc drop is about equal to the ionization potential of the filling gas or vapor. The inert-gas tubes suffer from one defect not present in mercury tubes: The gas pressure eventually is reduced to a low value and the tube becomes inoperative. The gas pressures in the hot-cathode arc tubes are of the order of 6×10^{-3} mm Hg, which for a mercury tube corresponds to a temperature of about 40°C at the coldest point. The point at which mercury condenses determines the average vapor pressure in the tube. At gas pressures of this order the Paschen breakdown curve for the electrode spacing usually found in tubes has a steeply increasing voltage relation as the pressure is decreased. The arc tubes are designed so that the product of gas pressure times separation between electrodes is such that, in the absence of electron emisson, a high breakdown voltage exists. This is the condition existing during the inverse voltage period of a rectifier when the anode is negative. Thus these low-pressure arc tubes may be used to rectify relatively high voltages. When electron emission is present, as at the cathode during a normal period for con-

duction (when the cathode is negative and the anode positive), the arc discharge is easily started. When the forward voltage increases to a value of the order of the ionization potential of the gas, or vapor filling, some electrons gain sufficient energy to ionize when they collide with the gas atoms. This results in one or more electron avalanches and breakdown of the gas is effected.

In a high-vacuum rectifier only those electrons directly acted on by electrostatic field lines from the anode can be utilized. The presence of gas in the arc tubes permits the utilization of electrons emitted in rather deep cavities in the cathode structure, so a high value of electron emission current can be obtained. This is true because electrons emitted thermionically deep in a cavity will undergo many collisions with gas atoms with resulting changes in direction so that they can reach the main discharge path. Figure 3 shows the structure of such a cavity cathode. The surfaces of the vanes, the outer side of inner cylinder, and the inner side of outer cylinder are coated with an oxide type of emitter. The emitting coating is usually a mixture of barium and strontium carbonates reduced to the oxides. The maximum current that can be drawn from the cathode is the temperature-limited, or saturation, current. If the circuit conditions require a higher current than this, the arc drop will increase, largely at the cathode, and the energy of the bombarding positive ions will be increased so as to cause the required increase in electron emission. This will result in destruction of the oxide cathode. The critical value of voltage producing this destructive effect is about 25 volts for mercury-vapor and inert-gas tubes. Such overloads should be avoided. *See* Thermionic emission.

It is sometimes desirable to use several rectifier tubes in parallel to increase the total current. This offers no difficulty with high-vacuum tubes which have a positive, or rising, volt-ampere characteris-

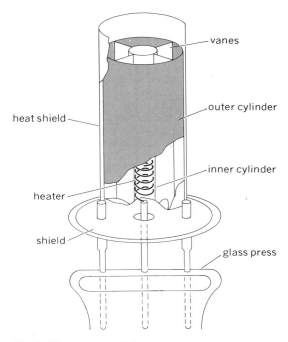

Fig. 3.   The phanotron tube.

tic. However, it is impossible to do so with many arc tubes because of their negative, or falling, volt-ampere characteristics. For these tubes sufficient ballast reactance (resistance or inductance) must be placed in each anode lead to make the combined arc-drop and reactance-voltage characteristic positive. Typical phanotrons have average current ratings from $1.25-10$ A and peak inverse voltage ratings of $10,000-22,000$ volts.

Thyratron tube. The thyratron is a hot-cathode gas tube containing one or more grid elements between the cathode and anode. It is an important current-control tube, having many applications. *See* THYRATRON.

Noise, oscillations, and surges. Hot-cathode gas tubes generate both electrical noise and oscillations that are characteristic of the tube construction and of the gas. The noise spectrum is substantially flat from frequencies of 25 Hz to 1 MHz per second or higher. One or more fundamental modes of oscillation at fairly discrete frequencies are usually present, often with many harmonics for each fundamental mode. The peak-to-peak variational component of the anode, or arc, voltage is of the order of the ionization potential of the gas or vapor present. The rms value of the noise voltage is of the order of several volts. The presence of these noise voltages in the anode potential of a gas tube during its conducting period is sometimes the source of considerable disturbance in nearby amplifiers, receivers, and other sensitive circuits either by direct electrical coupling or by electromagnetic radiation. Usually the unwanted signal can be bypassed or shielded when its source is recognized. Hot-cathode thyratrons are sometimes used as noise-source generators when placed in suitably selected and oriented magnetic fields. The coherent oscillations of many tubes can be suppressed by the use of suitably oriented magnetic fields. *See* ELECTRICAL NOISE.

When the arc current in a thyratron is increased above a critical value, any constriction of the arc path, such as the holes in the grids, becomes overloaded and serious voltage surges result. This phenomenon is caused by the complete ionization of the gas in the grid holes. When this occurs, the positive ions produced are driven out of the grid area, leaving a "vacuum" zone, that is, gas starvation results. The sudden stopping of current produces very high voltages in associated inductances.

[JAMES D. COBINE]

Bibliography: J. D. Cobine, *Gaseous Conductors*, 1958; J. D. Cobine and C. J. Gallagher, Effects of magnetic field on oscillations and noise in hot-cathode arcs, *J. Appl. Phys.*, 18(1):110–116, 1947; J. D. Cobine and C. J. Gallagher, Noise and oscillations in hot-cathode arcs, *J. Franklin Inst.*, 243: 41–54, 1947; J. D. Cobine and C. J. Gallagher, Noise in gas tubes, *Electronics*, 20(3):144, 1947; A. W. Hull and F. R. Elder, The cause of high-voltage surges in rectifier circuits, *J. Appl. Phys.*, 13(6):372–377, 1942; R. G. Kloeffler, *Industrial Electronics and Control*, 1960; L. Malter, E. O. Johnson, and W. M. Webster, Studies of externally heated hot-cathode arcs, *RCA Rev.*, 12(3):415–435, pt. 1, 1951; A. Schure (ed.), *Gas Tubes*, 1958; L. Tonks, Theory and phenomena of high current densities in low-pressure arcs, *Trans. Electrochem. Soc.*, 72:167–182, 1937.

Gate circuit

An electronic circuit having one output and one or more inputs, in which the output is a function of the inputs in a prescribed manner and in a controllable time sequence. Gate circuits may be classified as transmission gates or as switching gates, of which the logic gate is the most notable example.

Transmission gate. The transmission gate, sometimes called a sampling gate, delivers an output waveform that is a replica of a selected input during a specific time interval. The control signal, which determines the time interval, is called the gating signal or waveform; the gating generator is often a multivibrator. A simple transmission gate having one transmission signal input, G, is shown in Fig. 1. The gating signal from an external source, not shown, exists for period T. This gating signal is applied to the field-effect transistor, the gating signal allowing the transistor to amplify the signal which is applied to G only during interval T. *See* TRANSISTOR.

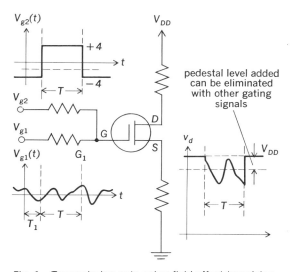

Fig. 1. Transmission gate using field-effect transistor, showing typical operating levels.

As a result of turning the device on from a completely off state, an additional level, or pedestal, appears at the output above the dashed line, as shown. Other signals could be mixed or gated in at other times by connecting the outputs of additional transistors in parallel and then applying such signals to the respective inputs of these transistors with gating waveforms applied to their G inputs. Also, gating signals applied to the parallel devices can be used to balance out the pedestal described above, if its existence is undesirable.

The function can also be performed by using the bases of emitter-coupled amplifiers for the signal and gate sources. Source-coupled field-effect transistors can function similarly.

Use of diodes in transmission gates is illustrated in Fig. 2. Here, gating waveforms of both positive and negative polarity are required from the gate generator. During the transmission interval, the diodes D_5 and D_6 are back-biased, or nonconducting. Meanwhile, dc voltages $+V$ and $-V$ ensure

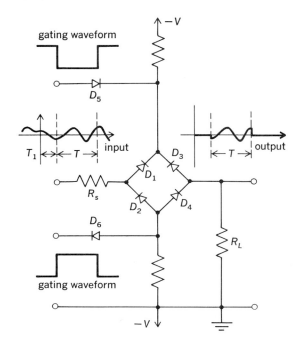

Fig. 2. Diodes used in transmission gate.

that D_1, D_2, D_3, and D_4 are forward-biased, or conducting. Thus, the signal waveform will be connected to the output through the low resistance of D_1, D_2, D_3, and D_4. During the nontransmission interval, the outputs of the gate generator are connected through D_5 and D_6, and the signal-path diodes are reverse-biased, or nonconducting. The balanced nature of the circuit is such that no pedestal is generated.

The circuit of Fig. 1 illustrates the use of a clipping circuit as a transmission gate, while Fig. 2 uses a clamping circuit as a gate. *See* CLAMPING CIRCUIT; CLIPPING CIRCUIT.

An operational amplifier with a separate strobe input which internally disconnects the input from the output sometimes functions as a transmission gate. The term analog switch is sometimes used synonymously with transmission gate. *See* AMPLIFIER.

Switching gate. In the switching, or logical, gate an output having a constant amplitude is registered during a time interval if a particular combi-

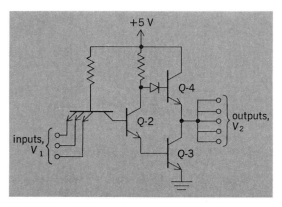

Fig. 3. Transistor-transistor logic (T²L) circuit forming a NAND gate.

nation of input signals exists. The output indicates only the existence of a particular combination of similar input signals. Examples are the OR, AND, NOR, and NAND circuits, which are basic building blocks of digital computers. Such circuits use combinations of transistors and diodes in much the same manner as transmission gates, and combinations of such functions are usually referred to as switching circuits.

The transistor-transistor logic (T²L) gate (Fig. 3) is an example. When all inputs are low, $V_1(0) = 0$, the output transistor, Q-3 is held nonconducting and the output is high, $V_2(0) = 3.5$ V. When all inputs are high, that is $V_1(1) = 3.5$ V, the output transistor is at $V_2(1) = 0$. Such a circuit is referred to as a NAND gate. *See* LOGIC CIRCUITS.

[GLENN M. GLASFORD]

Bibliography: G. M. Glasford, *Fundamentals of Television Engineering*, 1955; J. Millman and H. Taub, *Pulse, Digital, and Switching Waveforms*, 1965; H. Taub and B. Schilling, *Digital Integrated Electronics*, 1978.

GERT

A procedure for the formulation and evaluation of systems using a network approach. Problem solving with the GERT (graphical evaluation and review technique) procedure utilizes the following steps:

1. Convert a qualitative description of a system or problem to a generalized network similar to the critical path method – PERT type of network.

2. Collect the data necessary to describe the functions ascribed to the branches of a network.

3. Combine the branch functions (the network components) into an equivalent function or functions which describe the network.

4. Convert the equivalent function or functions into performance measures for studying the system or solving the problem for which the network was created. These might include either the average or variance of the time or cost to complete the network.

5. Make inferences based on the performance measures developed in step 4.

Both analytic and simulation approaches have been used to perform step 4 of the procedure. GERTE was developed to analytically evaluate network models of linear systems through an adaptation of signal flow-graph theory. For nonlinear systems, involving complex logic and queuing situations, Q-GERT was developed. In Q-GERT, a simulation of the network is performed in order to obtain statistical estimates of the performance measures of interest.

GERTE. The components of GERTE networks are directed branches and exclusive-or nodes. Two parameters are generally associated with each branch. These are the probability p that the branch is taken, given that the node from which it emanated is realized; and the time t (or cost, profit, and so on) required, if the branch is taken, to accomplish the activity which the branch represents. The time parameter can be a random variable.

For this type of network, the probability and time parameters can be combined into a single parameter, as in the equation below, where $M_t(s)$ is

$$w(s) = pM_t(s)$$

the moment-generating function of the time parameter. The calculation of the equivalent w function is shown in Fig. 1 for three basic types of networks. For general GERTE networks, an equivalent w function can be calculated by using the topology equation or Mason's rule of signal flow–graph theory.

Q-GERT. For nonlinear systems, different node and branch types are required in order to obtain realistic network models. Q-GERT is used for such systems where branches represent activities and nodes are used to model milestones, decision points, and queues. Flowing through the Q-GERT network are items referred to as transactions. Transactions can represent physical objects, information, or a combination of the two. Different types of nodes are included in Q-GERT to allow for the modeling of complex queuing situations and project management systems. For example, activities can be used to represent servers of a queuing system, and Q-GERT networks can be developed to model sequential and parallel service systems.

The symbol to represent an activity is a branch with a syntax, as shown in Fig. 2. The distribution type and parameter set number characterize the time delay involved in the activity. The activity number is a label, and the number of parallel servers specifies a limit on the number of transactions that can proceed through the branch concurrently. The probability or condition specifies when the activity is to be taken and is used in conjunction with a routing type specified for a node. Routing characteristics are the means for directing transactions across activities and hence through the network. The routing types which may be specified for any node are shown in Fig. 3.

There are seven node types included in Q-GERT. These node types are shown in Fig. 4. The BASIC node type is used to create transactions, to terminate the existence of transactions, to accumulate transactions, and to collect statistics. The release requirement associated with the node specifies the number of incoming transactions required before an output is generated from the node. The choice criterion specifies the characteristics to be associated with any transaction leaving the node after it is released. Branching from the BASIC node is done in accordance with one of the routing types shown in Fig. 3. The Q-node provides a means for storing transactions waiting for service. The SELECT node is a means for selecting from among parallel queues and parallel servers. The MATCH node provides a mechanism for matching transactions prior to their continuation through the network.

The last three node types relate to the allocation of resources to transactions. Transactions requiring resources wait in Q-nodes which precede the ALLOCATE node. If resources are available or when they become available, they are allocated to transactions at the ALLOCATE nodes. The number of units and the type of resource required are characteristics of the ALLOCATE node. Since many Q-nodes may precede an ALLOCATE node, a queue selection rule is specified for the ALLOCATE node. After a transaction has completed the activities for which the resources are required, the resources can be made available by routing the transaction through a FREE node. At the FREE node, the num-

network type	graphical representation	paths	loops	equivalent function
series	w_a → → w_b	$w_a\,w_b$	—	$w_a\,w_b$
parallel	w_a / w_b	$w_a\,,w_b$	—	$w_a + w_b$
self-loop	w_b / w_a	w_a	w_b	$\dfrac{w_a}{1 - w_b}$

Fig. 1. Calculation of the equivalent w function.

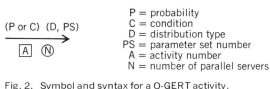

P = probability
C = condition
D = distribution type
PS = parameter set number
A = activity number
N = number of parallel servers

Fig. 2. Symbol and syntax for a Q-GERT activity.

ber of units of the particular resource to be freed is specified. Also indicated is the order in which ALLOCATE nodes should be polled in the attempt to reassign the freed resources. The last node type is the ALTER node, which allows the capacity of resources to be changed. When the capacity of a resource type is increased, a list of ALLOCATE

GERT

deterministic

probabilistic

conditional, take-first

conditional, take-all

Fig. 3. Routing types and symbols.

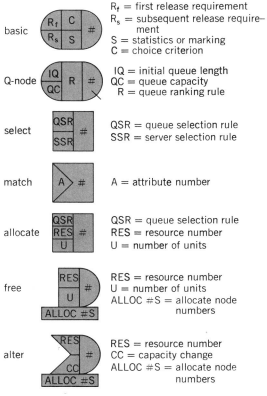

basic
R_f = first release requirement
R_s = subsequent release requirement
S = statistics or marking
C = choice criterion

Q-node
IQ = initial queue length
QC = queue capacity
R = queue ranking rule

select
QSR = queue selection rule
SSR = server selection rule

match
A = attribute number

allocate
QSR = queue selection rule
RES = resource number
U = number of units

free
RES = resource number
U = number of units
ALLOC #S = allocate node numbers

alter
RES = resource number
CC = capacity change
ALLOC #S = allocate node numbers

Fig. 4. Symbols and syntax of Q-GERT nodes.

nodes is appended to the ALTER node to specify where the newly created units are to be considered for allocation.

Applications. Many applications of GERT have been made. GERT networks have been designed, developed, and used to analyze the following situations: claims processing in an insurance company, production lines, quality control in manufacturing systems, assessment of job performance aids, burglary resistance of buildings, capacity of air terminal cargo facilities, judicial court system operation, equipment allocation in construction planning, refueling of military airlift forces, planning and control of marketing research, planning for contract negotiations, risk analysis in pipeline construction, effects of funding and administrative strategies on nuclear fusion power plant development, research and development planning, and system reliability. *See* CRITICAL PATH METHOD (CPM); DECISION THEORY; PERT; SIMULATION.

[A. ALAN B. PRITSKER]

Bibliography: S. E. Elmaghraby, *Activity Networks*, 1978; L. J. Moore and E. R. Clayton, *Introduction to Systems Analysis with GERT Modeling and Simulation*, 1976; A. A. B. Pritsker, *GERT: Graphical Evaluation and Review Technique*, RM-4973-NASA, 1966; A. A. B. Pritsker, *Modeling and Analysis Using Q-GERT Networks*, 2d ed.; G. E. Whitehouse, *Systems Analysis and Design Using Network Techniques*, 1973.

Glass switch

A glassy, solid-state device used to control the flow of electric current. Useful solid-state devices can be made from glassy as well as crystalline semiconductors. Crystals possess long-range order; that is, given the position of any particular atoms and the orientation of the neighboring atoms, the location of any other atom is known, no matter how far away from the atom under consideration. A glass is a special case of a noncrystalline class of materials, namely, amorphous solids. These do not exhibit long-range order, although they tend to have the same local structure (that is, short-range order) as the corresponding crystal. A glass is an amorphous solid that is formed by cooling rapidly from the liquid phase.

The first applications of glassy semiconductors were switches made from chalcogenide (that is, alloys containing tellurium, selenium, or sulfur) glasses. The two basic structures are known as the Ovonic Threshold Switch (OTS) and the Ovonic Memory Switch (OMS). They are active devices consisting simply of a thin film (about 1 μm thick) of glass between two metallic contacts. The device characteristics depend on the bulk properties of the semiconductor material rather than on the contacts. Consequently, the switches are symmetrical in that they respond identically to voltages and currents of both polarities.

Amorphous semiconductors, as opposed to crystalline semiconductors, can be doped or made insensitive to the effects of impurities, depending on the desired application. For switching devices, it is ordinarily preferable to use an impurity-independent structure to take advantage of lower costs and stable operation. Ovonic switches are also highly resistant to the effects of radiation.

Device characteristics. Both the OTS and OMS show a rapid and reversible transition between a highly resistive and a conductive state effected by applied electric fields. The main difference between the two devices is that, after being brought from the highly resistive state to the conducting state, the OTS returns to its highly resistive state when the current falls below a holding current value. On the other hand, the OMS remains in the conducting state until a current pulse returns it to its highly resistive state. The OMS thereby remembers the last applied switching command, and it is from this property that the device receives its name.

Intermediate resistance states are also possible for OMS devices, which can be used in applications requiring a "gray scale." In all of these devices, the transitions between states are completely reversible.

The composition of the active material determines whether the device functions as an OTS or OMS, and also affects the values of certain device parameters. The device geometry, such as thickness and cross-sectional area of the active film, also affects the numerical values of the device parameters.

The current-voltage characteristics of the OTS are shown in Fig. 1. Conduction in the highly resistive state follows Ohm's law at fields below about 10^4 V/cm. At higher fields the dynamic resistance R_{dyn} decreases monotonically with increasing voltage. Typical values are $R_{dyn} = 2 - 10$ megohms at 1 V and $R_{dyn} = 0.1 - 0.5$ megohm just prior to breakdown.

When the applied voltage exceeds a threshold voltage V_T, the OTS switches along the load line to the conducting state. The transition time τ_t of this switching process has been shown to be less than 150 picoseconds. V_T is a function of both film thickness and active material composition and can be obtained in the range $2 - 300$ V.

Current in the conducting state can be increased or decreased without significantly affecting the voltage drop across the device; the dynamic resistance is of the order of $1 - 10$ ohms. Most of the voltage falls near the two contacts, due to barriers induced prior to switching; this accounts for

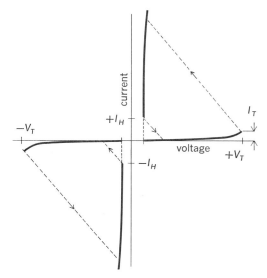

Fig. 1. OTS current-voltage characteristics.

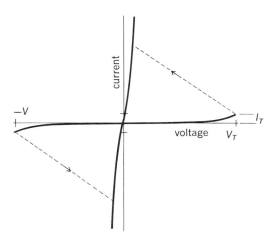

Fig. 2. OMS current-voltage characteristics.

about a 0.4-V drop. The field across the bulk is only about 1 kV/cm.

If the current is decreased below a critical value I_H, the OTS switches back to the original highly resistive state. I_H depends on circuit parameters and also can be varied; typical values are 0.1–1 mA.

The foregoing description of the static characteristics of the OTS holds for a slowly varying applied voltage. Upon application of a fast-rising pulse somewhat in excess of threshold voltage, the OTS ordinarily does not immediately switch to the conducting state but remains in the high-resistance state for a period of time τ_d, called the delay time. The magnitude of this delay is strongly dependent on the extent to which the threshold voltage is exceeded. For an applied voltage pulse slightly greater than V_T, τ_d can be several microseconds. However, it rapidly decreases with increasing voltage in excess of threshold, and essentially vanishes above a critical voltage that is proportional to film thickness. Above this point, the speed of switching is only circuit-limited, and total switching times less than 150 picoseconds have been observed.

Switching is an electronic effect, induced by the appearance of a critical electric field across a part or all of the film. The field induces a sharp transition which increases the free-carrier concentration by a factor of about 10^8. This results in a constant current density of approximately 10 kA/cm². At such current densities Joule heating effects are negligible. The high conduction occurs primarily through a central filament whose area varies proportionally with the current.

In the OMS the properties of the highly resistive state are essentially the same as for the OTS. As indicated in the current-voltage characteristics of Fig. 2, the OMS is switched to the conductive state when the threshold voltage is exceeded. It is switched back by a current pulse. The device remains in either state without power, a memory quality known as nonvolatility. Memory switching is a reversible amorphous-crystalline transition. It is preceded by a threshold switching, which provides a high density of free carriers. Under such conditions, crystallization proceeds at a greatly enhanced rate, a phenomenon known as electro-

crystallization. Only the central part of the conducting filament crystallizes, but this converts to a semimetallic phase. Erasing is accomplished by providing a high-current pulse that preferentially appears through the conducting filament, thus exciting the crystalline phase. The surrounding amorphous material remains essentially unchanged during the entire process, and consequently acts as a heat sink which provides the rapid cooling necessary to reform the glass after the termination of the current pulse.

Amorphous switch materials. Materials used in an OTS are specifically chosen to meet several important requirements. First, the film should have high electrical resistivity to avoid Joule heating effects. Second, the amorphous phase should be sufficiently stable to prevent crystallization or other structural changes. Third, the material should be chosen so that irreversible breakdown does not occur. All of these conditions can be fulfilled by the use of chalcogenide glasses. Chalcogenides are alloys one of whose major components comes from one of the elements in group VI of the periodic table, ordinarily selenium or tellurium. These atoms ordinarily form polymeric-type chains in the solid state. Their electronic structure is such that they necessarily contain large but equal densities of positively and negatively charged traps in the amorphous phase. These traps keep the film resistivity high and retard irreversible breakdown. Structural stability is maintained by using alloys with relatively large densities or cross-linking atoms, ordinarily from groups IV and V of the periodic table. *See* TRAPS IN SOLIDS.

Memory material is chosen to contain relatively weaker bonds and a smaller density of cross-linking atoms, so that structural changes are more easily attained. Crystallization involves only a small spatial diffusion of atoms. Under certain conditions, materials can be tailored such that other states intermediate between amorphous and crystalline can be attained. The lengths of the polymeric chains or the size of the crystalline grains can be varied depending on the amount and form of the energy of excitation.

Applications. Although Ovonic switches can be fabricated from bulk amorphous material, they are most conveniently produced as thin-film structures in which the active material and electrodes are vacuum-deposited or sputtered films, photolithographically defined. This economical process is compatible with transistor technology and with the methods used to produce passive components. It also lends itself to the fabrication of densely packed arrays.

Read-only memories. Integrated arrays of OMSs can be used as electronically alterable read-only memories. They fill the gap in the computer memory spectrum between permanent read-only memories and volatile random-access memories. The array is nonvolatile, so the data are as permanent as in a hard-wired memory. Readout is extremely rapid (15 nanoseconds, limited only by the circuit) and nondestructive. The memory has many advantages over conventional technology, including a higher packing density, lower programming currents (5 mA), lower programming power (5 mW), and no fatigue problems. High-temperature operation (350°C) and long lifetime (over 10^9 write/erase cycles) have been achieved. *See* COMPUTER STORAGE TECHNOLOGY.

Photographic films. Films have been developed in which the quality and amount of structural changes can be controlled by the amount of energy incident upon the film. Since these films do not contain silver, their cost does not vary with that of any precious metal. The MicrOvonic File is a completely updatable system for transferring original material onto a microfiche card. In addition, the MicrOvonic File Terminal, which accomplishes the same for electronic information, has also been developed.

Transistors. A transistor, using an OTS as the emitter, has been developed. This can be used as a threshold amplifier, as a threshold latching amplifier, or as the basis for a computer using ternary logic. In the conducting state, OTSs can be used to provide a constant current density and to inject hot electrons into crystalline-semiconductor devices. Their threshold characteristics provide the opportunity for another important control function in these applications.

Other promising application areas include ac control where use is made of the inherent symmetry of these components, and microwave generation made possible by the inherently fast switching transition. It is anticipated that the exploration of new applications for amorphous switches will accelerate as knowledge of them becomes more widespread. *See* AMORPHOUS SOLID; SEMICONDUCTOR.

[STANFORD R. OVSHINSKY; DAVID ADLER]

Bibliography: D. Adler, *Sci. Amer.*, 236(5): 36–48, May 1977; D. Adler et al., Threshold switching in chalcogenide-glass thin films, *J. Appl. Phys.* 51:3289–3309, 1980; S. R. Ovshinsky, Reversible electrical switching phenomena in disordered structures, *Phys. Rev. Lett.*, 21:1450–1453, 1968.

Glow discharge

A mode of electrical conduction in gases. Glow discharge commonly occurs under conditions of relatively low pressure and generally in the pressure range of 1–10 mm of mercury. The discharge typically gives off light, so that the region of the discharge appears to glow with considerable intensity. This glow is quite diffuse as contrasted to a higher-pressure discharge, such as a high-pressure arc. Typical currents may be of the order of tens or hundreds of milliamperes, whereas the potential drop may be of the order of 100 volts.

The most important application of the glow discharge is in the so-called voltage regulator or voltage reference tube. This device maintains a relatively constant difference of potential across itself as the current is varied over an appreciable range, and consequently is very useful in cases where a constant reference potential is required.

In terms of the potential-current characteristic, the glow discharge occurs after the potential has been increased so that the Townsend region has been passed. Thus the discharge is field-sustained. On the other hand, a continued increase in current leads first to the region referred to as the abnormal glow and beyond this to the arc discharge. The transition from the abnormal glow to the arc generally is almost discontinuous and is accompanied by a spark. For a discussion of this relationship *see* ARC DISCHARGE; ELECTRIC SPARK; ELECTRICAL CONDUCTION IN GASES; TOWNSEND DISCHARGE.

Regions of discharge. There are three main regions of interest in the glow discharge, similar to those in the arc. These are the cathode fall, the positive column, and the anode region. These will be discussed separately, but it is appropriate first to examine some of the general features of the mode (see illustration). The appearance is that of successive more or less well-defined luminous and dark regions. Starting from the cathode, there is a dark space which generally extends for a few millimeters, the Aston dark space. This is followed by a luminous region, also of limited extent, known as the cathode glow. This is succeeded by a somewhat longer dark space, designated the Crookes or Hittorf dark space. After this comes the negative glow region, the boundaries of which are rather poorly defined. Following this is the Faraday dark space, which is also more extensive and poorly defined. This changes gradually into the positive column which is luminous and of length determined by the pressure and distance between electrodes. This region may or may not contain striations, and if present they may be either stationary or moving. At the end of the positive column is a thin layer of greater luminosity, designated the anode glow. Between this and the anode is the anode dark space.

Cathode fall. The events occurring at the cathode are vital to the discharge. The current in the cathode circuit is primarily due to positive ions. However, it is necessary to produce enough electrons at the cathode to maintain the discharge. These electrons gain energy as they move in the electric field toward the anode, and produce excitation and ionization. It appears that these electrons are secondary electrons resulting from positive ion bombardment of the cathode surface. The drop in potential which occurs at the cathode depends on the kind of gas and the cathode material. Generally, this potential drop is a large fraction of the total potential drop across the discharge. The production of secondary electrons by this means is rather inefficient, which explains why the drop must be large.

Electrons starting at rest from the cathode must gain energy before they can produce excitation. This can be accomplished only by motion in the electric field, and hence there is a minimum distance which the electrons must move before they can produce excitation and consequent light emission. This explains the existence of the Aston dark space. It might be thought that the cathode glow could be explained by this also, but it is not likely that much of the light from this region is brought about by the secondary electrons. It appears that most of this light results from the positive ions that have struck the cathode and are returning to the ground state as neutral atoms. There are two facts of importance in this connection. First, the electron density is still rather low at such a short distance from the cathode. Second, the wavelengths present in the radiation indicate transitions involving states of a rather high degree of excitation. These high energy states probably could not be produced by the electrons from the cathode at this point.

The Crookes dark space is actually a region of

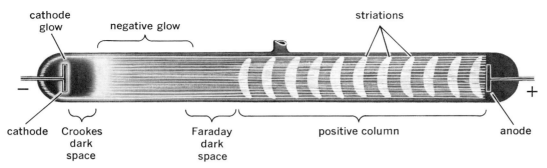

A glow discharge at approximately 0.1 mm pressure, showing successive more or less well-defined luminous and dark regions. (*From J. B. Hoag and S. A. Korff, Electron and Nuclear Physics, 3d ed., Van Nostrand, 1948*)

nearly uniform electric field. Most of the cathode drop occurs in this region, and here the positive ions gain most of their energy before striking the cathode. The electrons from the cathode gain enough energy here to produce cumulative ionization near the end of the space. In the negative glow, which follows, the potential is relatively constant. Here electrons, both from the cathode and from cumulative ionization, lose energy by inelastic collisions and produce a large amount of excitation. The boundary at the anode end of this space is poorly defined because of the broad distribution in electron energy. The slowing down of the electrons at the end of this region results in a negative space charge. Thus the electrons that move into the Faraday dark space gain energy.

Positive column. The beginning of the positive column is the result of excitation by these electrons. The situation in the positive column is the result of a balance between several processes. There is a nearly uniform potential drop which results in ionization throughout this entire region. On the other hand, there must be a loss of ions to make up for this production. This takes place primarily by diffusion to the walls, although there is also recombination. The electrons with their greater mobility diffuse to the walls, producing a slight negative potential relative to the center of the discharge. This negative potential both limits further electron diffusion and produces positive ion diffusion outward. This process is known as ambipolar diffusion. The positive column is not essential in maintaining the discharge. If the distance between electrodes is changed, with the pressure and current held constant, the extent of the positive column and the potential across the discharge change accordingly. The features of the anode and cathode regions remain unaltered under such a change up to the point where the positive column no longer exists.

A feature of this region is a succession of alternately luminous and dark regions, called striations, which usually occur when the discharge is operated at relatively high pressure; they may be stationary or moving. Their presence is related to the fact that in general the atomic species in the discharge are deexcited in times short compare to those required for them to diffuse through the positive column. Pure, inert gases do not show the effect, probably because they are excited into metastable, long-lived states.

Anode region. At the anode end of the positive column, the positive ions are repelled. This produces an increase in electric field, which causes the electrons to gain energy and excite more effectively. Thus the positive column ends in a region of increased luminosity, the anode glow.

Other aspects. There are many other aspects of the glow discharge that are interesting and important. One such phenomenon is cathode sputtering. Here the positive ions that are accelerated into the cathode knock out atoms or groups of atoms from the surface. Another aspect is that of abnormal glow. The voltage across the discharge remains nearly constant while the current is increased in the normal glow mode. This current increase is accompanied by an increase in the area of the cathode glow. When the glow has completely covered the cathode, a further current increase results in an increase in the cathode potential drop, and hence the potential drop across the discharge. This is the abnormal glow. It is characterized by more intense light emission and increased sputtering. *See* SPUTTERING.

It should be stated that many of the details of the discharge are uncertain. The processes are generally quite complicated. Reliable and accurate measurements are difficult at best, and most of the information is qualitative.   [GLENN H. MILLER]

Bibliography: J. Beynon, *Conduction of Electricity through Gases*, 1972; P. Llewellyn-Jones, *The Glow Discharge and an Introduction to Plasma Physics*, 1966; L. B. Loeb, *Basic Processes of Gaseous Electronics*, 1955.

Grain boundaries

Regions of intersection of two perfect but orientationally mismatched crystallites. In this region the atomic arrangements deviate from a perfect periodic lattice.

The rapid growth of the electronics industry during the past quarter century is due in large part to the development of processes for growing large, high-quality, single crystals of semiconductors such as germanium, silicon, and gallium arsenide. However, several recent applications of semiconductors—low-cost flat-plate solar cells and varistors (voltage-controlled resistors)—are committed, either for economic reasons or by the physics of the device operation, to the use of polycrystalline semiconductors. Since the regions between grains—the grain boundaries—largely control the electronic properties of polycrystalline materials, a considerable amount of basic research has been carried out during the past several years in order to understand the properties of these boundaries. As a result, sufficient progress has been achieved to enable the major electronic properties of these intergrain regions to be predicted with some degree of accuracy.

Atomic arrangements. Studies on single grain boundaries of germanium and silicon have shown that the most commonly observed grain boundary structure consists of a narrow (less than 5 nm) band of regularly spaced edge dislocations. Associated with this regular array are alternating regions of compression and dilation which account for the considerable elastic energy stored in the boundary structure. The boundary surface between two crystallites is sometimes curved rather than planar; this happens because of differences in the atomic arrangements at different points along the interface. This nonuniformity is thought to be caused by variations in the growth conditions as the boundary is formed.

As has been observed in metals, grain boundaries in semiconductors can act as precipitation sites for any impurities present in the material. For example, oxygen and carbon have been detected at silicon grain boundaries in amounts in excess of bulk concentrations.

Electronic structure. From the preceding discussion it is clear that rather severely perturbed electronic energy levels might be expected in the vicinity of grain boundaries. For instance, when edge dislocations are present, a significant number of grain boundary atoms find themselves in environments with less than the normal number of nearest neighbors. This leads to severe disruption of the normal crystal bonding, or sharing of valence electrons, and usually gives rise to states for electrons in the forbidden energy gap of the semiconductor. The consequences of this are shown in Fig. 1a, which illustrates a thin, electrically neutral, grain boundary region as it would exist if it could be placed between two semi-infinite, nearly adjacent crystals without any electrons being allowed to transfer from one to any other of the parts, so that the grain boundary region remains in a neutral state. The adjacent crystals are shown as being separated physically from the grain boundary region in Fig. 1a to emphasize that no charge flow to the boundary has yet occurred. Here E_V and E_C are the energies of the top of the valence band and the bottom of the conduction band, ζ is the energy separation between the highest filled (Fermi) level and the bottom of the conduction band. E_{FG} is the energy of the highest filled electron state in the grains, and E_{FB} is the corresponding quantity for the boundary region. Because of the defective structure of the grain boundary region, it contains numerous empty electron states (drawn as open circles) lying below the filled level of the crystalline regions. An n-type semiconductor which has electrons as the majority carriers is shown here, but similar phenomena occur in p-type materials. *See* SEMICONDUCTOR.

If the hypothesis is continued, and the electrons in all three parts are allowed to find a common energy level (by joining), the structure of Fig. 1b results. Electrons (shown as minus signs) flow from the crystals to fill some of the empty states of lower energy in the grain boundary. This flow results in a negative charging of the center of the grain boundary and the formation of an electrostatic repulsive barrier which stops further electron flow by raising the energy levels of all the electron states at the center of the boundary until the energy of the highest filled state is the same everywhere. Here ϕ_B is the amount of electrostatic bending of the energy bands. If some of the defect states at the center of the boundary remain unfilled, they will play an important role in the electronic behavior of the boundary.

Transport of charge. The accumulation of majority carriers at semiconductor grain boundaries has been verified in a variety of materials, including silicon (Si), germanium (Ge), gallium arsenide (GaAs), cadmium sulfide (CdS), lead sulfide (PbS), and zinc oxide (ZnO). This charging effect has a profound influence on the electrical properties of polycrystalline semiconductors. In order for current to flow across such charged boundaries, the carriers must have an energy considerably in excess of their normal thermal energy (0.025 eV at room temperature). The number of electrons having this extra energy at any time is small, and hence the conductance such a potential barrier has when current flows across it is low and strongly temperature-dependent. The temperature dependence frequently seen is an exponential function of inverse absolute temperature—much like the leakage current across a *pn* junction device. Because of the small magnitude of this grain boundary conductance, polycrystalline aggregates frequently offer conductance to current flow which is many decades lower than that of single crystals of similar purity.

The exponential temperature dependence of the

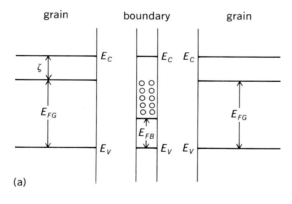

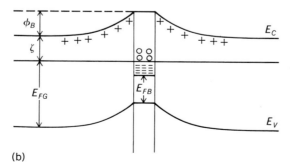

Fig. 1. Process of grain boundary charging in semiconductors. (a) Electronic energy levels of two adjacent crystallites and their intervening grain boundary region, as they would exist if electron transfer between the three regions could not take place. (b) Electronic energy levels as they actually exist, after the three regions are joined.

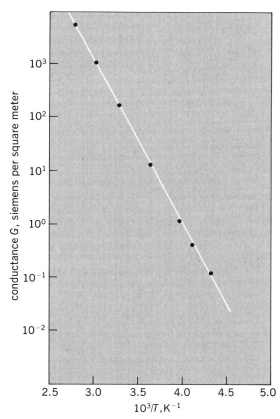

Fig. 2. Temperature dependence of the electrical conductance measured across a single silicon grain boundary.

electrical conductance G measured across a single silicon grain boundary is shown in Fig. 2. The grains in this case are doped n-type to a level of 1.3×10^{16} cm^{-3}. The data points fit a formula of the type given below (indicated by the straight line

$$G = G_0 \exp\left[-(\phi_B + \zeta)/kT\right]$$

in Fig. 2), where k is Boltzmann's constant, G_0 is a constant, T is the absolute temperature, and $(\phi_B + \zeta)$ is 0.62 eV in this case. This result shows the electrons crossing the grain boundary are thermally emitted over a potential energy barrier of total height $(\phi_B + \zeta)$. This is the energy difference between the highest filled energy state and the top of the barrier (Fig. 1b).

Because of the narrow width of the grain boundary space charge regions, the capacitance of such a structure can also be large. This effect is put to use in boundary-layer capacitors, which are polycrystalline devices made of materials like barium titanate. These are useful for high-capacitance-value applications where circuit element size is an important consideration.

Another useful property of grain boundaries — employed in devices called varistors — is their voltage-dependent resistance. As voltage is applied across a grain boundary, the presence of unfilled trapping states causes the potential barrier for current flow to remain large. However, continued application of larger and larger voltages eventually fills these states, and further bias causes a rapid

reduction in this barrier. Drastic increases in current occur above this filled trap threshold, and the grain boundary turns on. Varistors made of materials such as polycrystalline zinc oxide exhibit this effect. They are used as voltage regulators and surge protectors in a variety of applications. *See* VARISTOR.

Minority carriers are also strongly influenced by the presence of grain boundaries. An example is the photovoltaic cell, in which light-generated minority carriers must diffuse substantial distances to a collecting junction for current to be generated; here the presence of grain boundaries is quite harmful. The same grain boundary potential barrier which repels majority carriers attracts minority carriers and traps them until they recombine. This results in a short minority carrier lifetime, a property which has traditionally made polycrystalline semiconductors less than ideal for a variety of semiconductor devices such as transistors and photovoltaic cells. *See* PHOTOVOLTAIC CELL; SOLAR CELL; TRANSISTOR.

Recent research has shown that hydrogen diffused into silicon removes grain boundary defect states, thereby considerably improving the electrical properties of polycrystalline silicon; chemical methods to alter grain boundary properties in other semiconductors are also under investigation. It is believed that improved polycrystalline devices will result from these efforts.

[C. H. SEAGER]

Bibliography: H. K. Charles and A. P. Ariotedjo, *Solar Energy*, 24:329–334, 1980; L. L. Kazmerski and P. J. Ireland, *Solar Cells*, 1:178–182, 1980; G. D. Mahan, L. M. Levinson, and H. R. Phillip, *J. Appl. Phys.*, 50:2799, 1979; C. H. Seager and G. E. Pike, *Appl. Phys. Lett.*, 35:709–711, 1979.

Graph theory

A branch of mathematics that belongs partly to combinatorial analysis and partly to topology. Its applications occur (sometimes under other names) in electrical network theory, operations research, organic chemistry, theoretical physics, and statistical mechanics, and in sociological and behavioral research. Both in pure mathematical inquiry and in applications, a graph is customarily depicted as a topological configuration of points and lines, but usually is studied with combinatorial methods.

Origin of graph theory. Graph theory and topology are said to have started simultaneously in 1736 when L. Euler settled the celebrated Königsberg bridge problem. In Königsberg, there were two islands linked to each other and to the banks of the Pregel River by seven bridges. Figure 1 illustrates both this setting and its topological abstraction as

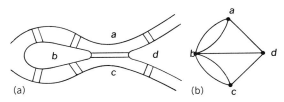

Fig. 1. Königsberg bridge problem. (a) The seven bridges of Königsberg. (b) Corresponding graph.

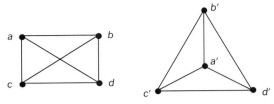

Fig. 2. Two isomorphic graphs.

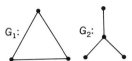

Fig. 3. Two graphs for which either a 120° rotation or a vertical reflection is an automorphism.

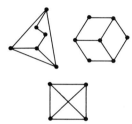

Fig. 4. Homeomorphic but nonisomorphic graphs.

a graph. The points a, b, c, d correspond to land areas, and the connecting lines to bridges. The problem is to start at one of the land areas and to cross all seven bridges without ever recrossing a bridge. Euler proved that there is no solution, and he established a rule that applies to any connected graph: such a traversal is possible if and only if at most two points are odd, that is, each is the terminus for an odd number of lines. Euler also proved that the number of odd points in a graph is always an even number. Thus, a complete traversal without recrossing any lines is possible if the number of odd points is zero or two. If zero, the complete traversal ends at the starting point.

In geometry a graph might arise as the set of vertices and edges of a convex, three-dimensional polyhedron, such as a pyramid or a prism. Euler derived an important property of all such polyhedra. Let V, E, and F be the numbers of vertices, edges, and faces of such a polyhedron. Euler proved that $V - E + F = 2$, which is now called the Euler equation. For instance, a cube has $V = 8$, $E = 12$, and $F = 6$, so that $8 - 12 + 6 = 2$. Euler's observations have been extended to a theorem about imbeddings of graphs in surfaces and to the Euler-Poincaré characteristic for cell complexes in combinatorial topology.

Definitions. A graph consists of a set of points, a set of lines, and an incidence relation that designates the end points of each line. In many applications no line starts and ends at the same point. (Such a line would be called a loop.) Also, no two lines have the same pair of end points. A graph whose lines satisfy these conditions is called simplicial. The valence of a point is the number of lines incident on it, calculated so that a loop is twice incident on its only end point. Two graphs are isomorphic if there is one-to-one correspondence from the point set and line set of one onto the point set and line set, respectively, of the other that preserves the incidence relation. The point correspondence $a \to a'$, $b \to b'$, $c \to c'$, $d \to d'$ indicates an isomorphism between the two graphs of Fig. 2.

An automorphism of a graph is an isomorphism of a graph with itself. For instance, a plane rotation of 120° would yield an automorphism of either of

the two graphs in Fig. 3. A plane reflection through a vertical axis would also yield an automorphism of either of them. The set of all automorphisms of a graph G forms the automorphism group of G. R. Frucht proved in 1938 that every finite group is the automorphism group of some graph. Two graphs are homeomorphic (Fig. 4) if, after smoothing over all points of valence 2, the resulting graphs are isomorphic. *See* GROUP THEORY.

Map coloring problems. Drawing a graph on a surface decomposes the surface into regions. One colors the regions so that no two adjacent regions have the same color, rather like a political map of the world. It is a remarkable fact that for a given surface, there is a single number of colors that will always be enough no matter how many regions occur in a decomposition of the surface. The smallest such number is called the chromatic number of that surface. It is easy to draw a plane map, as in Fig. 5, that requires four colors. In 1976 K. Appel and W. Haken settled a problem dating back to about 1850, by showing that four colors are always enough for plane maps.

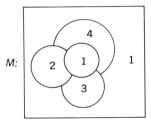

Fig. 5. Plane map requiring four colors.

Some maps on more complicated surfaces require more than four colors. For instance, Fig. 6 illustrates a map on a torus (the surface of a doughnut) that needs seven. To obtain the toroidal map from the rectangular drawing, first match the top to the bottom to get a cylindrical tube. Then match the left end of the cylinder to the right end to complete the torus. Whereas, before this matching, region 7 meets only regions 1 and 6, after the matching it also meets region 2 along FG, region 3 along GH, region 4 along AB, and region 5 along BC. In fact, after the matching, each of the seven regions borders every other region. It follows that seven colors are necessary. No map on the torus needs more than seven colors, as P. J. Heawood proved in 1890. G. Ringel and J. W. T. Youngs completed a calculation in 1968 of the chromatic numbers of all the surfaces except the plane or sphere.

Planarity. A graph is planar if it can be drawn in the plane so that none of its lines cross each other. Neither of the two graphs in Fig. 7 can be drawn in the plate. K. Kuratowski proved in 1930 that a graph is planar if and only if it contains no subgraph homeomorphic to either of those two graphs. Testing all the subgraphs might be a very tedious process, even on a fast computer. In 1974 J. Hopcroft and R. Tarjan obtained an extremely fast alternative planarity test. The time it takes a computer to perform the Hopcroft-Tarjan test is linear-

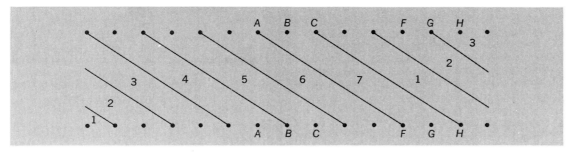

Fig. 6. Map on a torus (doughnut) that requires seven colors. To form the torus, paste opposite sides of the rectangle together.

Fig. 7. Prototypes of all nonplanar graphs.

ly proportional to the time it takes to read its point set into the computer.

There are methods to decide for any graph and any surface whether the graph can be drawn on the surface without edge crossings. The time to execute such methods is unfeasibly large for most graphs and most surfaces except the plane or the sphere. Ringel has constructed many important special drawings on higher-genus surfaces.

Variations. In a directed graph, or digraph, each line ab is directed from one end point a to the other end point b. There is at most one line from a to b. The adjacency matrix $M = (m_{ij})$ of a digraph D with points $b_1, b_2, \ldots, b_n$ has the entry $m_{ij} = 1$ if the line $b_i b_j$ occurs in D; otherwise $m_{ij} = 0$ (Fig. 8).

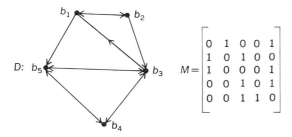

$$D: \qquad M = \begin{bmatrix} 0 & 1 & 0 & 0 & 1 \\ 1 & 0 & 1 & 0 & 0 \\ 1 & 0 & 0 & 0 & 1 \\ 0 & 0 & 1 & 0 & 1 \\ 0 & 0 & 1 & 1 & 0 \end{bmatrix}$$

Fig. 8. A digraph and its adjacency matrix.

An oriented graph is obtained from an ordinary graph by assigning a unique direction to every line. If there is one line between every pair of points and no loops, an ordinary graph is called complete. An oriented complete graph is called a tournament (Fig. 9).

Applications. A. Cayley reformulated the problem of counting the number of isomers of saturated hydrocarbons ($C_n H_{2n+2}$) in graphical language (Fig. 10). Each isomer is a tree all of whose vertices have valence 1 for hydrogen, or 4 for carbon. G.

Polya devised a general theorem in 1937 for enumeration to provide a solution to such problems. F. Harary and others have solved many related problems by applying and extending Polya's theorem. Extremely effective use of Polya's theorem occurs in theoretical physics, where G. Ford and G. Uhlenbeck solved several graphical enumeration problems arising in statistical mechanics.

Suppose that some of the points of a graph correspond to workers $x_1, \ldots, x_m$, that the rest of the points correspond to jobs $y_1, \ldots, y_m$, and the presence of a line between x_i and y_j means that worker x_i is capable of performing job y_j. The personnel assignment problem is to find m lines so that each worker x_i is matched to exactly one possible job. In the optimal assignment problem, labels on the lines tell how well a worker can do a particular job. An algorithm due to H. Kuhn and J. Munkres solves the optimal assignment problem.

If the points of a graph represent cities and the lines between them are labeled with distances, one might want to find the shortest path from one point to another. An efficient method to determine a shortest path was developed by E. Dijkstrain in 1959. K. Menger proved in 1927 that if A and B are disjoint sets of a connected graph G, then the minimum number of points whose deletion separates A from B equals the maximum number of disjoint paths between A and B. L. Ford and D. Fulkerson have generalized Menger's theorem into a method for solving network flow problems. *See* LINEAR PROGRAMMING.

According to the physical laws of G. Kirchhoff and G. Ohm, any set of voltages applied to the input nodes of an electrical network determines the voltages at all other nodes and the currents on every branch. Kirchhoff also proved the result known as the matrix-tree theorem: Let G be a connected graph with points $b_1, \ldots, b_n$, and let $A = (a_{ij})$ be the matrix such that a_{ij} is the valence of b_i and, for $i \neq j$, $a_{ij} = -1$ if b_i is adjacent to b_j or 0 otherwise. Then the cofactor of each entry a_{ij}

GRAPH THEORY

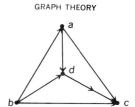

Fig. 9. A tournament.

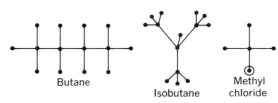

Fig. 10. The two isomers of a saturated hydrocarbon.

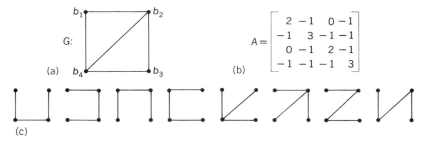

(a)

$$A = \begin{bmatrix} 2 & -1 & 0 & -1 \\ -1 & 3 & -1 & -1 \\ 0 & -1 & 2 & -1 \\ -1 & -1 & -1 & 3 \end{bmatrix}$$

(b)

(c)

Fig. 11. Matrix-tree theorem. (*a*) Graph *G*. (*b*) Corresponding matrix *A*. (*c*) Spanning trees of *G*.

equals the number of spanning trees of *G*, that is, the number of trees in *G* that includes every point of *G* (Fig. 11). *See* NETWORK THEORY.

Numerous applications of graph theory to social and behavioral science have been developed, many by Harary and his coauthors. If points represent persons and lines represent such interrelationships as communication, linking, or power, then a graph may depict various aspects of social organization. For instance, anthropologists use graphs to describe kinship, and management scientists use them to display corporate hierarchy.

Graph theory is presently in a phase of rapid growth. Two of the major branches not described here are external graph theory, founded by P. Turan and developed by P. Erdös, and hypergraph theory, developed by C. Berge. Material theory, originated by H. Whitney and expanded by W. Tutte, is closely related to graph theory. Tutte is also responsible for important results in many other areas of graph theory and combinatorial research, including connectivity, decomposition, and chromatic numbers.

[JONATHAN L. GROSS]

Bibliography: L. Beineke and R. Wilson (eds.), *Selected Topics in Graph Theory*, 1978; C. Berge, *Graphs and Hypergraphs*, 1973; N. Biggs, E. Lloyd, and R. Wilson, *Graph Theory 1736–1936*, 1976; B. Bollobas, *Extremal Graph Theory*, 1978; A. Bondy and R. Murty, *Graph Theory with Applications*, 1976; F. Harary, *Graph Theory*, 1969; F. Harary and E. Palmer, *Graphical Enumeration*, 1973; G. Ringel, *Map Color Theorem*, 1974; W. Tutte, *Introduction to the Theory of Matroids*, 1971.

Gyrator

A linear, passive, two-port electric circuit element whose transmission properties are such that it is effectively a half wavelength longer for one direction of transmission than for the other direction of transmission. Thus a gyrator is a device that causes a reversal of signal polarity for one direction of propagation but not for the other. (A two-port element has a pair of input terminals and a pair of output terminals.) This device is novel, since it violates the theorem of reciprocity. *See* RECIPROCITY PRINCIPLE.

Until the early 1950s, all known linear passive electrical networks obeyed the theorem of reciprocity. However, several different types of nonreciprocal networks are now widely applied, principally at microwave frequencies. These devices are used to control the direction of signal flow and to protect or isolate components from undesired signals. One common application of a three-port nonreciprocal network, called a circulator, is to permit connection of a transmitter and a receiver to the same antenna. This is accomplished with minimum interference and virtually no power loss of either transmitted or received signal. *See* CONTINUOUS-WAVE RADAR; NETWORK THEORY.

Reciprocity. The theorem of reciprocity can be stated in many equivalent forms, and perhaps the simplest to consider and understand is the particular form it takes when it is expressed especially for a two-port microwave network, as shown in Fig. 1*a*. Here d_1 represents the amplitude and phase of a wave incident on the network from the left, r_1 represents the wave propagating away from the network on the same side, and d_2 and r_2 represent the same quantities on the other side. In a linear network the relationship between the waves can be expressed as shown in Eqs. (1). Thus r_1 consists of

$$\begin{aligned} r_1 &= S_{11}d_1 + S_{12}d_2 \\ r_2 &= S_{21}d_1 + S_{22}d_2 \end{aligned} \tag{1}$$

a part which arises from the partial reflection of d_1 at the input port and a part contributed by the fraction of d_2 which is transmitted through the network. The reflection and transmission properties of the network are described by the coefficients S_{ij}, which are termed the scattering parameters of the network.

The network can be characterized by selectively setting one of the input waves to zero and measuring the resulting response. Thus when d_2 is zero, Eq. (2) holds. The quantity S_{21} gives the amplitude

$$S_{21} = \left(\frac{r_2}{d_1}\right) \tag{2}$$

and phase of the wave emerging from the right-hand side of the network when a unit wave, with zero phase angle, is incident on the left-hand side. The significance of the other scattering parameters can be deduced by similar arguments. If there is no reflection of the incident waves, then S_{11} and S_{22} are zero, and the network is said to be matched.

The theorem of reciprocity is stated by Eq. (3).

$$S_{12} = S_{21} \tag{3}$$

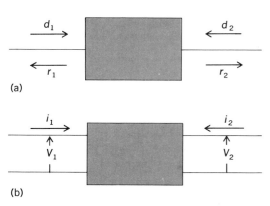

(a)

(b)

Fig. 1. Two-port microwave network. (*a*) Scattering matrix representation. (*b*) Currents and voltages.

That is, the network has the same transfer characteristics for one direction of propagation as it has for the other. Thus, if a matched network causes a particular insertion loss and phase shift for signals transmitted from left to right, signals transmitted from right to left must suffer the same loss and phase shift. An ideal gyrator is one which has the scattering matrix given in Eq. (4).

$$[S] = \begin{bmatrix} S_{11} & S_{12} \\ S_{21} & S_{22} \end{bmatrix} = \begin{bmatrix} 0 & -S_{21} \\ S_{21} & 0 \end{bmatrix} \qquad (4)$$

The terminal currents i_1, and i_2 and voltages of V_1, and V_2 of the network (Fig. 1b) are related by impedances Z_{ij} as shown in Eqs. (5). The theorem of reciprocity now requires the condition of Eq. (6).

$$\begin{aligned} V_1 &= Z_{11}i_1 + Z_{12}i_2 \\ V_2 &= Z_{21}i_1 + Z_{22}i_2 \end{aligned} \qquad (5)$$

$$Z_{12} = Z_{21} \qquad (6)$$

Theoretical gyrators. The first comprehensive treatment of nonreciprocal two-port networks was given in 1948 by B. D. H. Tellegen, who applied the word gyrator to describe such networks. Tellegen restricted his analysis to an ideal gyrator whose impedance matrix has the form given in Eq. (7). It

$$\begin{bmatrix} Z_{11} & Z_{12} \\ Z_{21} & Z_{22} \end{bmatrix} = \begin{bmatrix} 0 & -R \\ R & 0 \end{bmatrix} \qquad (7)$$

may be easily shown that for such a gyrator to be nondissipative, the Z_{ij} must be real. However, it is not necessary that the diagonal terms, Z_{11} and Z_{22}, be zero. Furthermore, this restriction on the diagonal terms is not needed in order to realize two of the most important properties of the gyrator, namely, the construction of one-way transmission systems and the phase inversion of signal polarity for one direction of transmission.

Tellegen's ideal gyrator has the additional property of impedance inversion. If the gyrator is terminated by an impedance Z_L, Eq. (8), the input im-

$$Z_L = \frac{V_2}{i_2} \qquad (8)$$

pedance Z_{in} of the gyrator for Eq. (5) is given by Eq. (9). This property of impedance inversion is

$$Z_{in} = \frac{V_1}{i_1} = \frac{R^2}{Z_L} \qquad (9)$$

not, however, unique to nonreciprocal networks. Indeed, if one makes the same requirement on a nondissipative reciprocal two-port, that $Z_{11} = Z_{22} = 0$, the input impedance Z_{in} of the terminated networks is that given by Eq. (10), where X_{12} is the

$$Z_{in} = \frac{V_1}{i_1} = \frac{X_{12}^2}{Z_L} \qquad (10)$$

transfer reactance of a nondissipative network. $Z_{12} = jX_{12}$. Such a reciprocal impedance-inverting network can be easily realized.

Practical gyrators. The theorem of reciprocity has in the past been considered so universally valid that present-day textbooks still make the statement that if the condition stated in Eq. (6) is valid, the two-port is passive, and that if the condition does not hold for a particular network, it cannot be a passive one.

Although reciprocity is as universally valid in mechanical, acoustical, or optical systems as it is in electrical ones, there are passive systems that can be constructed in each of these areas which are nonreciprocal. For example, it has been long known that a mechanical system which contains a gyroscopic coupler does not obey the theorem of reciprocity.

Perhaps the first passive nonreciprocal system was an optical one proposed by Lord Rayleigh, making use of the rotation of the plane of polarization of light when it passed through a transparent material in the presence of a magnetic field. This phenomenon is called Faraday rotation. If polarized light is propagated through a transparent medium along the direction of the magnetic field, the plane of polarization of the light is rotated through some angle θ per unit length, which is determined by the properties of the medium and the strength of the magnetic field. Faraday rotation is unusual in that it is nonreciprocal. Thus the sense

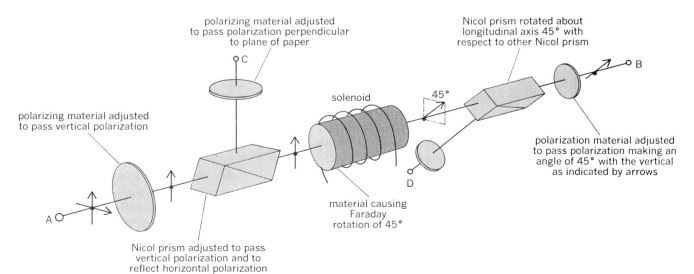

Fig. 2. Nonreciprocal optical device.

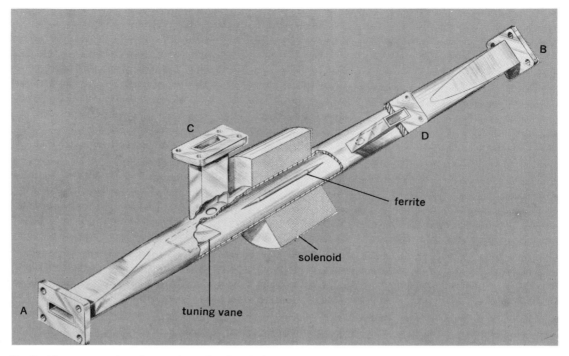

Fig. 3. Microwave analog of nonreciprocal optical device.

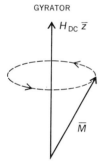

GYRATOR

Fig. 4. Precessional motion of the ferrite internal magnetization vector, $\bar{M}$, with a static magnetic field H_{dc} applied in the $\bar{z}$ direction.

(clockwise or counterclockwise) of the rotation is the same whether the light travels parallel to the applied magnetic field direction or contraparallel to it. Hence, if the plane of polarization is rotated through an angle θ in traversing the Faraday cell and the ray is reflected back through the cell toward its source. it will again be rotated through an angle θ, so that when it arrives back at the source, the plane of polarization will have been rotated through a total angle 2θ.

Lord Rayleigh's one-way system. shown in Fig. 2, consisted of two polarizing Nicol prisms oriented so that their planes of acceptance made an angle of 45° with each other. The material causing the Faraday rotation was placed between them. If the Faraday rotator is adjusted to cause a 45° rotation, light which is passed by the first crystal is passed by the second also. In the reverse direction, however, the 45° rotation added by the Faraday cell produces light rays polarized horizontally which are reflected by the Canada balsam cement in the first Nicol prism and directed toward point C in the figure. Thus light admitted to the device at point A is transmitted to point B; light admitted at point B is transmitted to point C; light admitted at point C is transmitted to point D; and light admitted at point D is transmitted to point A.

The microwave analogy of Lord Rayleigh's device was proposed by C. L. Hogan and is shown in Fig. 3. Since it circulates microwave power from waveguide A to B, from B to C, from C to D, and from D to A, it has been called a (four-port) circulator. The nonreciprocal medium used here is a ferrimagnetic material called ferrite. In such a material, infinitesimal magnetic dipole moments which arise from the electronic structure of the material act gyroscopically when a steady magnetic field is applied, as shown in Fig. 4. They precess about the applied field direction in a counterclockwise

sense, thus permitting strong coupling to the component of a microwave-frequency magnetic field which is circularly polarized in the same sense. The component with the opposite sense of polarization is weakly coupled. Thus energy exchange between magnetic dipoles and microwave field is polarization-sensitive. *See* FERRITE DEVICES.

Present-day circulators utilize the properties of electromagnetic fields in ferrite loaded microwave circuits. Consider the three-port circulator shown in Fig. 5, which comprises a circular disk resonator filled with ferrite connected to three transmission lines. When microwave energy is transmitted to the resonator along one of the transmission lines, an electromagnetic field is established in the resonator which is stationary in space, as shown in Fig. 6. The application of a dc magnetic field perpendicular to the plane of the disk resonator will

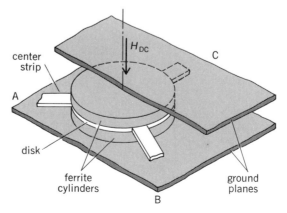

Fig. 5.   Stripline Y-junction circulator. (*From C. E. Fay and R. L. Comstock, Operation of ferrite junction circulator, IEEE Trans. Microwave Theory Tech., MTT-13:1–13, January 1965*)

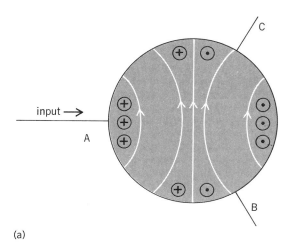

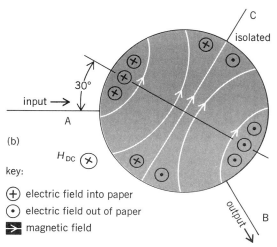

key:

$\oplus$ electric field into paper

$\odot$ electric field out of paper

▶ magnetic field

Fig. 6. Stationary electromagnetic field pattern in disk resonator of stripline circulator. (a) Junction is not magnetized, resulting in symmetric coupling to ports B and C. (b) Junction is magnetized to rotate pattern 30°, for circulation. Port C is isolated. (From C. E. Fay and R. L. Comstock, Operation of ferrite junction circulator, IEEE Trans. Microwave Theory Tech., MTT-13:1–13, January 1965)

rotate this stationary pattern through an angle dependent on the strength of the applied field. The field pattern is dipolar in nature, and hence has a region where the microwave magnetic field intensity is low. If the pattern is oriented to position this low-intensity region at one of the output transmission line ports, very little microwave power will leave the resonator via this port: it is isolated from the input. From the symmetry of the junction, this stationary pattern will advance 30° if the ports are excited sequentially. Thus energy at port A will be transmitted to port B, energy incident on port B will be transmitted to port C, and energy incident on port C will be transmitted to port A. The ideal three-port scattering matrix of this junction circulator is given in Eq. (11).

$$[S] = \begin{bmatrix} 0 & 0 & S_{13} \\ S_{21} & 0 & 0 \\ 0 & S_{32} & 0 \end{bmatrix} \qquad (11)$$

Devices based on this pattern rotation principle have been realized in a variety of transmission line geometries, including rectangular waveguide, stripline, and microstrip. High performance can be obtained over wide microwave-frequency bandwidths (with ratio of upper to lower frequency limit less than or equal to 2:1) with isolation in the order of 0.01 of the incident power and with about 95% of the power transmitted to the desired port. These devices are found in most microwave systems, where they are employed to control the flow of microwave signals. For example, their use permits a radar transmitter and a radar receiver to share the same antenna. In another use, if a microwave absorber is connected to one of the ports of a three-port circulator, a one-way transmission device results, known as an isolator.

[FRED J. ROSENBAUM]

Bibliography: H. Bosma, Junction circulators, *Advan. Microwaves*, vol. 6, 1971; C. L. Hogan, The microwave gyrator, *Bell Syst. Tech. J.*, 31(1):1–32, 1952; B. Lax and K. J. Button, *Microwave Ferrites and Ferrimagnetics*, 1962; B. D. H. Tellegen, The gyrator: A new electric network element, *Phillips Res. Rep.*, 3(2):81–101, 1948.

Gyrotron

One of a family of microwave generators, also called cyclotron resonance masers, in which cyclotron resonance coupling between microwave fields and an electron beam in vacuum is the basis of operation. This type of coupling has the advantage that both the electron beam and the associated microwave structures can have dimensions which are large compared with a wavelength. Thus, cyclotron resonance masers are potentially greatly superior to conventional microwave tubes with respect to power capability at short wavelengths.

The development of these power sources is particularly significant for magnetically confined plasma fusion experiments. Microwave heating is considered an attractive method of supplying the energy needed to bring a reactor to ignition temperature, and gyrotrons provide a potential means of producing sufficient microwave power at the very short wavelength required. Gyrotrons also have potential application in millimeter-wave radar and communications systems.

Basic characteristics. The basic cyclotron resonance condition is given by Eq. (1), where ω is the operating frequency, n is an integer, and ω_c is the cyclotron frequency or angular velocity of the electron given by Eq. (2). Here, B is the dc magnetic

$$\omega = n\omega_c \qquad (1)$$

$$\omega_c = \frac{eB}{\gamma m_0} \qquad (2)$$

field, e is the electron charge, m_o is the rest mass, and γ is the relativistic mass factor. The fundamental cyclotron resonance occurs when $n = 1$. This is the strongest and most useful interaction. The resonance condition requires that very high magnetic fields be used for high-frequency devices. For example, a frequency of 120 GHz requires a magnetic field of about 45 teslas. Generally, the very-high-frequency gyrotrons have used superconducting magnets.

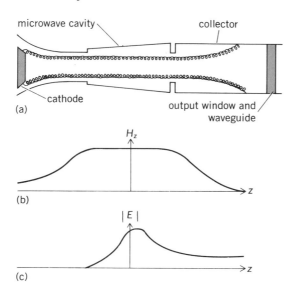

Fig. 1. Gyrotron. (a) Schematic diagram, showing elements. (b) Plot of typical dc magnetic field H_z as function of distance z along axis. (c) Plot of representative microwave electric field intensity $|E|$.

Larger values of n allow corresponding reductions in the required dc magnetic field. Practical devices have generally been limited to values of n no larger than 2 (second-harmonic operation).

The most important microwave field component in the gyrotron is the electric field tangential to the orbit of the electron. With the fundamental cyclotron resonance interaction, any spatial variation of the microwave fields is of little importance. It is this property that allows the gyrotron to use cross-section areas which are large compared with a wavelength.

Electron bunching in the gyrotron occurs by virtue of the relativistic mass effect included in Eq. (2). The transverse microwave electric field intro-

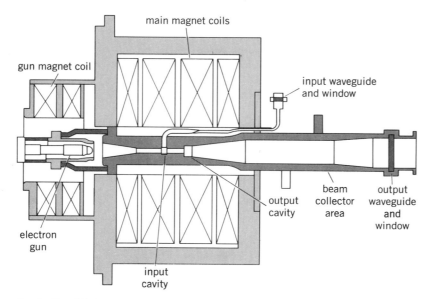

Fig. 2. Simplified cross section of pulsed gyroklystron amplifier.

duces a sinusoidal modulation of γ depending on the angular position of the electron in its orbit relative to the direction of the electric field. The modulation of γ results in a modulation of angular velocity as given by Eq. (2). As the beam drifts, this converts to angular bunching in the coordinate system centered on each electron orbit. By proper adjustment of phase conditions, the bunched beam can give up most of its energy to microwave energy.

A number of tube configurations are possible using the cyclotron resonance interaction. The simplest form, and that used for most practical gyrotrons to date, is an oscillator using a single resonant cavity. A gyroklystron amplifier employing two or more cavities is another alternative; and in a third variation a traveling-wave circuit, in analogy to a traveling-wave tube, is used.

A schematic representation of a gyrotron (single-cavity oscillator) is shown in Fig. 1, along with a

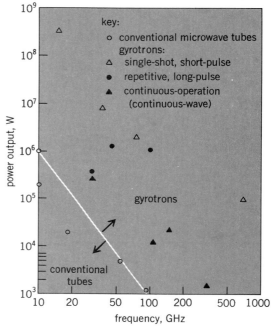

Fig. 3. Power output versus frequency of gyrotrons compared with that of conventional microwave tubes.

typical dc magnetic field profile H_z and a representative microwave electric field distribution $|E|$. The electron beam is a hollow beam with all electrons having helical motion. For efficient operation, all electrons must have a large fraction of their total energy contained in motion transverse to the device axis. A simplified cross section of a pulsed gyroklystron amplifier is shown in Fig. 2.

Capabilities. A summary of gyrotron demonstrated power output versus frequency as of 1980 is shown in Fig. 3, along with power capabilities of conventional klystrons and traveling-wave tubes. In the case of the pulsed devices, it is the peak power output that is shown on the graph. The single-shot, short-pulse results refer to devices using intense relativistic beams which are not suitable for repetitive pulsing. The gyrotron results indicate a clear capability for pro-

ducing higher power at high frequency. *See* MICRO-
WAVE TUBE; TRAVELING-WAVE TUBE.

[HOWARD R. JORY]

Bibliography: A. A. Andronov et al., *Infrared
Phys.* 18:385–394, December 1978; V. A. Flyagin
et al., *IEEE Trans.*, MTT-25:514–521, June 1977;
J. L. Hirshfield and V. L. Granatstein, *IEEE Trans.*,
MTT-25:522–527, June 1977; H. R. Jory, *Digest
of the 1977 International Electron Devices Meeting*,
IEEE, December 1977.

Hall effect

The development of a transverse electric potential
gradient in a current-carrying conductor placed in
a magnetic field when the conductor is positioned
so that the direction of the magnetic field is perpendicular to the direction of current flow. Analysis of the Hall effect yields important information
on the band structure of metals and semiconductors and on the nature of the conductivity process.
In semiconductor research, the magnitude of the
Hall effect in simple cases provides a direct estimate of the concentration of charge carriers. The
Hall effect is one of the so-called galvanomagnetic
effects.

The experimental arrangement for observing the
Hall effect is shown schematically in Fig. 1. A voltage V_H appears between the sides of the specimen
whenever the current density J_x and the magnetic
field B_z are nonvanishing. The electric field $E_H =
V_H/d = E_y$, using the coordinates of Fig. 1, is found

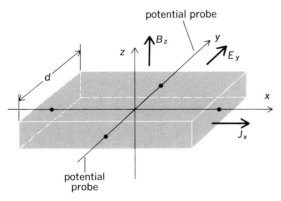

potential probe

Fig. 1. Hall effect.

to be proportional to the product of J_x and B_z. The
quantities V_H and E_H are called the Hall voltage
and Hall field, respectively.

Physical interpretation. A simple interpretation of the Hall effect may be given by the following argument. Each charge carrier in the solid is
assumed to move with a drift velocity $v_{dx} = J_x/ne$.
Here n is the number of charge carriers per unit
volume and e is the charge of each carrier. Each
charge carrier experiences a Lorentz force

$$F_{yB} = -\frac{e}{c} v_{dx} B_z \qquad (1)$$

where c is the velocity of light. Unless this force
is compensated by some other force, the charge
carriers will acquire a drift velocity in the negative
y direction, giving rise to a component of current in the y direction. The experimental arrangement, however, requires that $J_y = 0$. The necessary

compensating force is provided by the Hall field
E_y, whose magnitude and direction is such that

$$F_y = F_{yB} + F_{yE} = 0 = -\frac{e}{c} v_{dx} B_z + e E_y \qquad (2)$$

Hence

$$E_y = \frac{v_{dx} B_z}{c} = \frac{J_x B_z}{nec} \qquad (3)$$

The Hall coefficient R is defined

$$R = E_y / J_x B_z \qquad (4)$$

Thus

$$R = 1/nec \qquad (5)$$

A similar result can be obtained by solving the
Boltzmann transport equation. The Hall coefficient
of a metal provides the most direct information of
both sign and number of charge carriers, provided
the free electron model is valid. The sign of R is
the same as the sign of the charge e, and the magnitude of R is inversely proportional to n. The
number of conduction electrons per unit volume
calculated from measured Hall coefficients agrees
reasonably well with the product of valence and
number of atoms per unit volume for the mono-
valent metals. Anomalous results are often obtained for polyvalent metals, as is apparent from
the table.

In these cases a two-band model is frequently
invoked in the interpretation of the Hall effect and
other transport phenomena. Because n is almost
independent of the temperature T, R also changes
only slightly with temperature.

At high magnetic fields and low temperatures
($T < 4$ K), the Hall effect in a metal single crystal
sample generally depends sensitively on the orientations of the field and current relative to the crystal axes. This anisotropy is intimately related to
the topology of the Fermi surface; hence, such
measurements are useful in the study of the Fermi
surface of metals.

Thermal side effects. It can be shown that an
isothermal condition can be maintained in a specimen only if a thermal current can flow freely in the
transverse direction. Unless the sample is im-
mersed in a constant temperature bath, $\partial T/\partial y \neq 0$.
Under nonisothermal conditions, the finite temper-
ature difference between the sides of the sample
may give rise to spurious thermal emfs at the potential probes. The generation of a transverse
temperature gradient by a current flowing in the

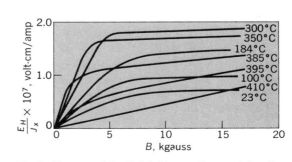

Fig. 2. Diagram of the Hall field per unit current density
in nickel as a function of magnetic induction between
room temperature and 410°C. (*After A. W. Smith, Phys.
Rev., vol. 30, no. 1, 1910*)

Number of valence electrons and $n = 1/Rec$ for several metals*

Metal	$N \times 10^{-22}$	$NZ \times 10^{-22}$	$(1/Rec) \times 10^{-22}$
Monovalent metals			
Lithium	4.6	4.6	3.7
Sodium	2.5	2.5	2.5
Potassium	1.3	1.3	1.5
Cesium	0.85	0.85	0.8
Copper	8.5	8.5	11.4
Silver	5.8	5.8	7.5
Gold	5.9	5.9	8.7
Polyvalent metals			
Beryllium	12.3	24.6	−2.5
Magnesium	4.3	8.6	6.7
Zinc	6.6	13.2	−18
Cadmium	4.6	9.2	−10.5
Aluminum	6.0	18.0	21
Indium	3.8	11.4	89
Thallium	3.5	10.5	−26
Tin	3.7	14.8	156

*N is number of atoms per cubic centimeter; Z is valence of atoms.

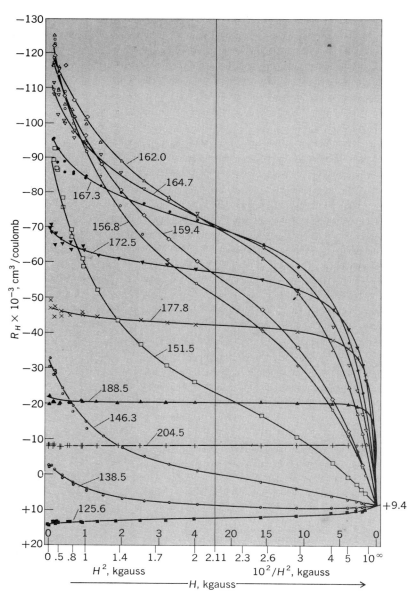

Fig. 3. Hall coefficient of p-type InSb as a function of magnetic field at various temperatures. Note change in scale on abscissa. Temperatures are in Kelvins. (*From G. Fischer, Helv. Phys. Acta, 33:472, 1960*)

presence of a transverse magnetic field is known as the Ettingshausen effect, and proper correction for this effect should be made. A convenient way of avoiding these thermal difficulties is to use an alternating current as the primary current J_x, thereby preventing the establishment of a significant temperature gradient in the sample.

Effect in ferromagnetic metals. Ferromagnetic metals exhibit anomalously large Hall effects at temperatures below the ferromagnetic Curie temperature. Moreover, the Hall voltages are not linear functions of B. Typical curves are shown in Fig. 2. The knees of the curves occur at saturation, and the shapes of the curves suggest that E_H be written in the form $E_H = R_0 H + R_1 M$. Here M is the magnetization, and R_0 and R_1 are designated the ordinary and extraordinary—or anomalous—Hall coefficients, respectively. The coefficient R_0 is not a sensitive function of temperature; R_1 changes by orders of magnitude as the temperature is increased from 4 K to the Curie temperature.

Effect in semiconductors. As in metals, the sign and magnitude of R determine the sign and number of charge carriers, provided only one type of carrier (electrons or holes) is present. At high temperatures (intrinsic region) R decreases approximately exponentially with increasing temperature, largely because of the exponential increase in carrier density with temperature. At low temperatures (extrinsic region) R is approximately constant. In the intrinsic, and in some cases also in the extrinsic region, the Hall coefficient must be calculated with the aid of a two-band model because more than one type of carrier is responsible for charge transport. The Hall coefficient, especially in a multicarrier situation, is field-dependent (Fig. 3). This field dependence provides valuable information on mobility ratios and relaxation times. *See* SEMICONDUCTOR.

Effect in ionic crystals. Alkali halides become electronic conductors only under the influence of light that will excite electrons from the valence band to the conduction band. Because the photoelectrons are few in number, they form a classical electron gas. Measurements of the Hall effect and of photoconductivity provide information on the mobility of these electrons. The scattering mechanisms and the effective masses of conduction electrons in ionic crystals are greatly influenced by the local polarization of the crystal by these charge carriers, and therefore the usual theories of mobility are not directly applicable here. For this reason, Hall effect and conductivity measurements in ionic crystals are of considerable interest. The measurements require rather special techniques and are more difficult than in metals or semiconductors. The data that are available indicate that current theories correctly describe the interaction between conduction electrons and lattice vibrations in alkali halides.   [FRANK J. BLATT]

Bibliography: N. W. Ashcroft and N. D. Mermin, *Solid State Physics,* 1976; *Effects in Semiconductors,* 1963; F. J. Blatt, *Physics of Electronic Conduction in Solids,* 1968; F. C. Brown, *Physics of Solids,* 1967; E. Fawcett, High-field galvanomagnetic properties of metals, *Advan. Phys.,* 13:139, 1964; C. Kittel, *Introduction to Solid State Physics,* 5th ed., 1976; F. Seitz and D. Turnbull (eds.), *Solid State Physics,* vol. 1, 1955, and vols. 4 and 5, 1957.

Harmonic analyzer

A device for separating and measuring the frequencies and amplitudes of the Fourier-series components of a complex periodic wave. A complete harmonic analysis would include finding the phase of each component; some analyzers do this, especially the "rolling sphere" devices, now largely of historical interest. Most electrical analyzers find component amplitude only. *See* NONSINUSOIDAL WAVEFORM.

Wattmeter measurement. At low frequencies, harmonics may be measured with a wattmeter. The voltage to be analyzed is impressed upon the potential coil, and a current of adjustable frequency is passed through the current coil. When the frequency of the current I is set approximately equal to that of a harmonic component E_h of the voltage, the meter will deflect to indicate the product $E_h I \cos \theta$, where θ is the phase angle between the voltage and current. As θ varies through 360° during the period of a beat between the two frequencies, the wattmeter will swing between the maximum positive and negative readings for which $\cos \theta = \pm 1$.

By adjusting the frequency of the current to be close to that of the harmonic, a very slow beat can be produced and an accurate maximum reading can be obtained. The instrument can be calibrated, for any given amplitude and frequency of the search current, by impressing a sine-wave voltage of known amplitude on the potential coil and adjusting it in frequency and amplitude to produce the same maximum deflection. This method is simple and accurate, even for small harmonic components, but is limited to the lower audio frequencies for which dynamometer-type instruments are suitable.

Tunable filters. Two widely used types of instrument provide measurement of component amplitude at adjustable frequencies and nonmeasurement at other frequencies. The first of these, generally referred to as a distortion and noise meter, is designed to amplify uniformly over the frequency range of interest. A narrowband rejection filter, which can be tuned to the fundamental frequency of the signal to be analyzed, is used to eliminate that component from the output, leaving a signal that comprises the residual harmonics and noise (Fig. 1). It is widely used in maintaining broadcast-station circuits in proper operating condition.

The second type of instrument is shown in Fig. 2. A tunable narrow-band rejection filter is again used, but in this application it reduces the negative feedback in a heavily degenerated amplifier to a minimum at its rejection frequency. The amplifier

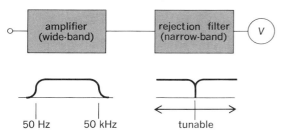

Fig. 1. Basic distortion and noise meter.

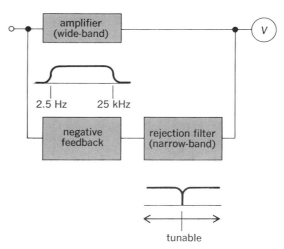

Fig. 2. Sound and vibration analyzer.

then has maximum gain and will select any signal occurring at this frequency while rejecting others. Instruments of this type are often used for analysis of noise spectra, as well as for harmonic selection and measurement. For a general analysis of this nature an important characteristic is that the passband is a constant fraction of the tuning frequency.

Heterodyne measurements. Another widely used instrument employs a highly selective fixed-tuned filter in the heterodyne system, as illustrated in Fig. 3. Because the narrow-passband filter is not

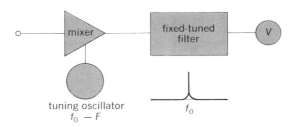

Fig. 3. Wave analyzer.

varied, it can be very precisely designed for exactly the right bandwidth, edge cutoff, and so on. The heterodyning oscillator covers a frequency range from the center frequency of the filter f_0 to a frequency $f_0 - F$, making possible measurements at any frequency from 0 to F.

Instruments of this type, generally called wave analyzers, are used for a wide variety of measurements requiring high precision and accuracy. In contrast to instruments using tunable filters, they maintain a measurement bandwidth Δf that is constant, independent of frequency. They are therefore particularly useful for measuring noise spectra, where the energy content per cycle of bandwidth is important. *See* ELECTRICAL MEASUREMENTS; FREQUENCY MEASUREMENT.

High-speed analysis. The user of the analyzers described above must search out each component in a spectrum and read an indicating meter. Time is required to allow the analyzer filter to settle as each spectral component is selected, as well

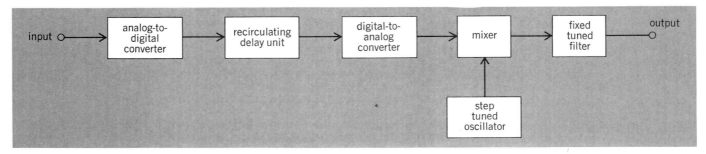

Fig. 4. Time compression analyzer.

as to read a meter, so this process can be extremely time-consuming, especially when many components are involved. Two methods can greatly reduce analysis time.

Time compression analyzer. The time compression method allows all components in a signal to be measured almost simultaneously. This type of analyzer (Fig. 4) uses a heterodyne system similar to that shown in Fig. 3 preceded by circuits that store the input signal (time sequence), compress it in time, and then repeatedly play back this replica of the original input signal as the tuning oscillator is stepped through the frequency range of interest. Compressing the time scale of the input signal is equivalent to dilating its spectrum. This means that a filter which is broader than would be required for the original signal by a factor equal to the compression ratio provides equivalent resolution on the compressed signal. Because the analyzing filter is broad, it responds rapidly and can be stepped rapidly through the frequency range. The speed of analysis is increased by a factor equal to the compression ratio.

The signal is converted to digital form at the analyzer input and stored either in a recirculating mechanical delay line or semiconductor memory prior to playback for analysis. Real-time operation over a frequency range of 10 kHz is typical in commercially available units.

Digital computer resources. The essence of harmonic analysis is that any time-varying quantity likely to be found in nature can be described and manipulated in either of two ways: as the time history of its successive values, or as the set of sinusoidal (Fourier) "components" which will add together to produce that time variation. Often a desired modification or adjustment of such a waveform (such as noise reduction or interference rejection) can be done more effectively by adjusting ("processing") the Fourier components than by shaping the time sequence of that waveform. Either the "time-domain" or the "frequency-domain" description can be converted into the other by mathematical logic; this can be done by digital computer, on a discrete-sample basis, rapidly enough for "real-time" processing in many practical cases.

The logic for such computer conversion is the discrete Fourier transform of Eqs. (1) and (2) con-

$$A_r = \sum_{k=0}^{k=n-1} X_k \exp(-2\pi jrk/n) \qquad (1)$$

$$r = -n, -n+1, \ldots, 0, 1, \ldots, n-1$$

$$j = \sqrt{-1}$$

$$X_k = \frac{1}{2n} \sum_{r=-n}^{r=n-1} A_r \exp(+2\pi jrk/n) \qquad (2)$$

$$k = 0, 1, 2, \ldots, n-1$$

necting the set of "sample" values X_k (the quantity x at successive time instants kT/n, where T is the duration of the interval in which x is analyzed) with the corresponding set of "sample" frequency components A_r (amplitude and phase of the Fourier component at frequency r/T).

Computer algorithms are available which compute the discrete Fourier transform much more rapidly than would direct substitution in the ordered sums given here. The most used of these is the fast Fourier transform, (FFT). These procedures take advantage of internal duplications in the sinusoidal weighting functions, grouping certain X_k's and A_r's together before weighting so the number of computer multiplications is drastically reduced: from n^2 to about $n \log_2 (n^2)$. Another algorithm is even faster.

[MARK G. FOSTER]

Bibliography: R. M. Bracewell, *The Fourier Transform and its Application*, 2d ed., 1978; W. T. Cochran et al., What is the fast Fourier transform?, *IEEE Trans. Audio Electroacoust.*, vol. AU-15, June 1967; W. D. Cooper, *Electronic Instrumentation and Measurement Techniques*, 2d ed., 1978; P. Kantrowitz et al., *Electronic Measurements*, 1979; L. R. Rabiner et al., The Chirp Z algorithm, *Proc. IEEE*, 60(7):809–820, 1972.

Hearing aid

An instrument used by a person who is hard of hearing to amplify sounds, particularly the sounds of speech, which he wishes to hear. Ear trumpets and speaking tubes used before about 1920 were simple and effective, but they were inconvenient. Modern electronic hearing aids are small enough to be worn comfortably and inconspicuously behind the ear, within the ear canal and concha, in the frames of spectacles, or in the clothing. A behind-the-ear aid is shown in the illustration. Essential components of a hearing aid are a microphone, an electronic amplifier, an earphone, and a plastic ear mold which serves to couple energy from the earphone to the eardrum either directly or through plastic tubes. Sometimes a vibrator is held by a spring headband behind the ear and delivers sound to the ear by bone conduction. *See* AMPLIFIER; EARPHONES; MICROPHONE.

Greater emphasis is given to high or to low tones through adjustment of tone controls in the amplifier circuit or by adjusting the acoustic circuit in

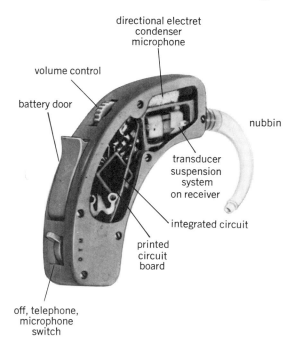

directional electret
condenser
microphone

volume control

battery door

nubbin

transducer
suspension
system
on receiver

integrated circuit

printed
circuit
board

off, telephone,
microphone
switch

Behind-the-ear hearing aid with case open to show internal components. (*Beltone Electronics Corp.*)

the ear mold and coupling tubes. Practically all hearing aids have adjustable gain controls, and in some a compression amplifier regulates gain automatically. Directional microphones are sometimes used to enhance sounds coming from the front. The range of effective amplification extends from about 200–600 Hz to about 3000–6000 Hz. The maximum output can be adjusted in some hearing aids to suit the tolerance of the user for loud sound.

Typically only one hearing aid is worn, with the microphone, amplifier, and earphone all located near the ear that receives the amplified sound. When true binaural amplification is required, two monaural aids are used, one near each of the ears. There are many intermediate arrangements in which the relative locations and the numbers of microphones, amplifiers, and earphones are chosen to accommodate the special hearing loss and lifestyle of the user.

Group hearing aids are widely used in schools for the deaf to provide amplification with greater fidelity and with less distortion and noise than would be achieved with the use of individual hearing aids. In simple wired group aids, the teacher wears a microphone, and a single amplifier drives earphones for the entire class. Each student controls the sound intensity from his or her earphones with a conveniently located gain control. With teacher and students tethered to the amplifier by wires, their mobility is limited.

Complete mobility within a classroom or outdoors is achieved with the use of a miniature battery-operated radio transmitter for the teacher, who broadcasts to students equipped with miniature radio receivers. These wireless systems are most appropriate for small children, but the electronic complexity and cost are significantly greater than they are for simple wired systems.

[ARTHUR F. NIEMOELLER]

Bibliography: H. Davis and S. R. Silverman (eds.), *Hearing and Deafness*, 4th ed., 1978.

Heterodyne principle

The basic principle underlying the operation of a superheterodyne radio, television, or other receiver, wherein two alternating currents that differ in frequency are mixed in a nonlinear device to produce two new frequencies, corresponding to the sum and the difference of the input frequencies. Only the difference frequency is commonly used in a superheterodyne receiver, where it serves as the input to the intermediate-frequency amplifier. The heterodyne principle permits conversion of a wide range of different input frequencies to a predetermined, lower intermediate-frequency value that can be amplified more efficiently. Some frequency meters also use the heterodyne principle when comparing an unknown input frequency with a calibrated frequency standard. *See* RADIO RECEIVER: SUPERHETERODYNE RECEIVER.   [JOHN MARKUS]

High-frequency impedance measurements

The electrical measurement of the complex ratio of voltage to current in a given circuit at frequencies from several hundred kilohertz (kHz) to 100,000 megahertz (MHz). This frequency range includes medium- and high-frequency bands. *See* ELECTRICAL MEASUREMENTS.

At lower frequencies impedances may be accurately measured by standard techniques for measuring resistance, capacitance, and inductance. The most precise measurements are those in which the unknown impedance is compared with a resistance, capacitance, or inductance standard of nearly equal value: at lower frequencies such standards can be very accurate. *See* CAPACITANCE MEASUREMENT; INDUCTANCE MEASUREMENT; RESISTANCE MEASUREMENT.

At higher frequencies, details of measurement-standard shape, terminal geometry, and component-interconnection wiring provide series inductance and shunt capacitance as well as stray couplings that may dominate the situation unless

Fig. 1. An rf capacitance standard fitted with a GR900 precision connector. (*General Radio Co.*)

great care is used. These undesired parameters cause particular difficulty if they have different values each time a connection is made. Highly repeatable connections may be made at high frequencies by the use of precision connectors. These connectors are coaxial to remove external effects and make a butt joint to ensure repeatable inductance and capacitance (Fig. 1). Some resistance, inductance, and capacitance standards using modified low-frequency construction are usable up into vhf range. Distributed-parameter standards consisting of coaxial transmission lines are useful to over 10 GHz. At the highest frequencies, where mechanical dimensions become comparable with the wavelength, resonant cavities and waveguides are used because they incorporate simple boundary conditions for field calculations.

For measurements over the whole high-frequency range from below 1 MHz to over 10 GHz, the measurement methods used can be classified as voltmeter-ammeter methods, resonance methods, null methods, standing-wave methods, and reflection methods.

VOLTMETER-AMMETER METHODS

The impedance of a device is defined as the ratio of the voltage across it to the current through it, and a basic method of impedance measurement is, therefore, to measure voltage and current on appropriate meters and to calculate impedance. A modification of this method is to supply a known and constant voltage or current and measure the resulting current or voltage. These methods can give a direct indication of admittance or impedance, respectively.

Vector impedance meter. This instrument not only determines the ratio between the voltage and current to give impedance magnitude but also determines the phase difference between these quantities to give the phase angle of the component under test. This principle has been applied to several instruments at low frequencies and is used to 108 MHz in the Hewlett-Packard rf vector impedance meter. Because it is difficult to perform the necessary measurement operations at high frequencies, a synchronous sampling technique is used to convert the measured voltage and current to a 5-kHz intermediate frequency. The sampled current is used to control the level of the applied test current so that the voltage measured is a direct indication of impedance magnitude and is displayed on one meter. A phase detector drives another meter which indicates phase angle. Such methods are limited to accuracy of the circuitry performing the measurement operations and of the meter movement itself. Basic accuracy of the vector impedance meter is 4% of full scale. While other methods are more accurate, the wide range and ease of use makes such instruments popular.

Potentiometer method. Another modification of the basic method defined above is to put a known impedance in series with the unknown impedance, drive them with an appropriate source, and measure the voltage across each separately. Because the same current is passed through them, the ratio of the measured voltages is the ratio of the magnitudes of the two impedances. This is the same principle as the potentiometer method of resistance measurement, long used at direct current; the advent of low-level rf voltmeters makes this

method practical to quite high frequencies. If the standard and unknown are approximately equal, the error caused by the voltmeter is small because both readings are on the same part of the scale. *See* Potentiometer (voltage meter).

RESONANCE METHODS

Resonance is a typical phenomenon at radio frequencies. It is readily observed and reproduced and defines a circuit condition for which the interrelationship among the component impedances is known. It is therefore an excellent indicator for measurement purposes. Either series-resonant or parallel-resonant circuits can be used. Series-resonant circuits are best suited for measuring low impedances, and conversely, parallel-resonant circuits for low admittances. Both methods determine the quadrature or reactive component of impedance from the change in resonant capacitance when the unknown impedance is inserted into the circuit. They differ in the method of measurement of the real, or resistive, component of impedance. *See* Resonance (alternating-current circuits).

Series-resonance methods. Two methods of obtaining the necessary data to permit solving for the unknown impedance are employed.

Resistance-variation method. A short-circuiting link is first connected across the terminals shown in Fig. 2, and the circuit is tuned to resonance as

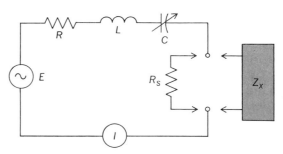

Fig. 2. Series-resonance circuit for resistance-variation and reactance-variation measurements.

indicated by a maximum current reading. This current I_1 is then given by Eq. (1), where E is the

$$I_1 = \frac{E}{R} \tag{1}$$

source voltage and R the total circuit resistance.

The short-circuiting line is then replaced by the unknown impedance Z_x and resonance reestablished. The new current I_2 is given by Eq. (2),

$$I_2 = \frac{E}{R + R_x} \tag{2}$$

where R_x is the resistive component of the unknown impedance Z_x.

Finally, the unknown impedance is replaced by a known standard resistance R_s and resonance is reestablished. The current I_3 is given by Eq. (3).

$$I_3 = \frac{E}{R + R_s} \tag{3}$$

Combining Eqs. (1), (2), and (3), one obtains for the unknown resistance Eq. (4).

$$R_x = R_s \frac{I_3(I_1 - I_2)}{I_2(I_1 - I_3)} \qquad (4)$$

The unknown reactance X_x is given by Eq. (5),

$$X_x = \frac{1}{\omega}\left(\frac{1}{C_2} - \frac{1}{C_1}\right) \qquad (5)$$

where C_1 and C_2 are the settings of the variable capacitor for resonance with the short-circuiting link and the unknown impedance in circuit, respectively, and ω is the angular frequency.

Reactance-variation method. This method differs from the resistance-variation method only in the measurement of the unknown resistive component. The circuit resistance is deduced from the capacitance change necessary to detune the circuit by a known amount.

The circuit is first tuned to resonance and the current I noted; then the circuit is detuned and the capacitance values C' and C'' for which the current becomes $0.707I$ are determined. At these settings, one on each side of resonance, the circuit reactance equals the circuit resistance R, and it can be shown that Eq. (6) holds.

$$R = \frac{1}{2\omega}\left(\frac{1}{C'} - \frac{1}{C''}\right) \qquad (6)$$

A measurement with the short-circuiting link in place yields a resistance R_1 which is equal to the total circuit resistance R. A second measurement, with the short-circuiting link replaced by the unknown impedance Z_x yields a resistance R_2 which is equal to $R + R_x$. This substitution measurement then gives Eqs. (7) and (8).

$$R_x = R_2 - R_1 \qquad (7)$$

$$X_x = \frac{1}{\omega}\left(\frac{1}{C_2} - \frac{1}{C_1}\right) \qquad (8)$$

Parallel-resonance methods. The parallel-resonance methods are duals of the series-resonance methods. The same general techniques are used except that resonance is defined in terms of maximum voltage and the circuit losses are measured by the open-circuit conductance G instead of the short-circuit resistance R (Fig. 3).

Conductance-variation method. This method is the dual of the resistance-variation method. Measurements with the terminals open-circuited, with the unknown admittance Y_x connected, and then with the unknown admittance replaced by a known conductance standard G_s give Eqs. (9) and (10),

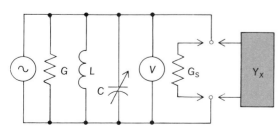

Fig. 3. Parallel-resonance circuit for conductance-variation and susceptance-variation measurements.

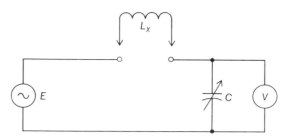

Fig. 4. Series-resonance circuit used for the resonant-rise method of measuring Q.

$$G_x = G_s \frac{V_3(V_1 - V_2)}{V_2(V_1 - V_3)} \qquad (9)$$

$$B_x = \omega(C_1 - C_2) \qquad (10)$$

where the subscripts refer to the three measurements, respectively.

Susceptance-variation method. This method is the dual of the reactance-variation method. At the capacitance settings C' and C'' for which the circuit susceptance equals the circuit conductance G, the voltage becomes $0.707V$ and Eq. (11) holds.

$$G = \frac{\omega}{2}(C'' - C') \qquad (11)$$

The same sequence of measurements as in the reactance-variation method then yields Eqs. (12) and (13).

$$G_x = G_2 - G_1 \qquad (12)$$

$$B_x = \omega(C_1 - C_2) \qquad (13)$$

The resonance methods give accurate results with fairly simple equipment but are not readily made direct-reading and are therefore not used as the basis for commercial instruments.

Resonant-rise method. A circuit that has been widely used commercially to measure the storage factor Q of coils is shown in Fig. 4. Commercially the instrument is known as a Q meter. *See* Q METER.

The resonant current I in this series-resonant circuit, as before, is given by $I = E/R$, and the voltage V across the tuning capacitor C by Eq. (14),

$$V = IX_C = IX_L \qquad (14)$$

where R is the total circuit resistance and X_C and X_L are the reactances of the capacitor C and inductor L, respectively. The voltage ratio V/E is therefore given by Eq. (15).

$$\frac{V}{E} = \frac{X_L}{R} \qquad (15)$$

If the resistance and inductance of the rest of the circuit are made negligibly small compared with the resistance and inductance of the coil L, the voltage V will be directly proportional to the storage factor Q of the coil. In many Q meters the voltmeter scale is calibrated directly in Q, and the value of the unknown inductance is determined from the calibrated capacitor setting by Eq. (16). When L is known, the effective resis-

$$L = \frac{1}{\omega^2 C} \qquad (16)$$

tance of the inductance is given by Eq. (17).

$$R_L = \frac{\omega L}{Q} \qquad (17)$$

Q of tuned circuit or cavity. Several methods of measuring Q are available.

Resonance-curve method. If frequency is varied so that the current in a tuned circuit or cavity goes through resonance, and the frequencies f' and f'' for which the current is reduced to 0.707 of its maximum (half-power points) are noted, it can be shown that Eq. (18) holds. Here the subscript 0

$$Q_0 = \frac{\omega_0}{\omega'' - \omega'} = \frac{f_0}{f'' - f'} \qquad (18)$$

refers to values at resonance. For high-Q cavities at microwave frequencies this measurement is usually performed by coupling the generator and detector to the magnetic field with small pickup coils, so oriented that direct pickup from one to the other is negligible. If the couplings are too strong, the resonance curve will be broadened by losses coupled in from the generator and detector. As they are weakened, however, this broadening will disappear and the observed curve will be that of the cavity alone.

Decrement method. The storage factor Q of a resonant device represents the ratio of the maximum energy stored in the electric or magnetic field during a cycle to the amount of energy dissipated in that cycle. It can also be measured, therefore, by observing the decay in oscillation amplitude when the exciting signal is cut off. The current in the tuned circuit then follows the law in Eq. (19),

$$I = I_0 \epsilon^{-(R/2L)t} = I_0 \epsilon^{-(\omega_0/2Q_0)t} \qquad (19)$$

where t is time, I_0 the initial resonant current, and the other symbols carry their previous connotation. The cavity Q_0 is then related to the time interval Δt during which the current decays by a factor of $\epsilon = 2.71828 \ldots$ by Eq. (20). *See* TIME CONSTANT.

$$Q_0 = \frac{\omega_0 \Delta t}{2} \qquad (20)$$

This measurement is carried out with a pulse-modulated generator, a detector having a large bandwidth compared with that of the cavity to be measured, and a cathode-ray oscilloscope. The detected signal, which measures the cavity current, is applied to the vertical plates of the oscilloscope, and the horizontal deflection is synchronized with the modulating pulse to produce a stationary pattern. This pattern may be scaled directly off the screen for rough measurements, or in a more complex setup, the pattern may be compared with the discharge curve of an RC network excited by the modulating pulse.

R_0/Q_0 of resonant cavity. At microwave frequencies, where a resonant cavity may be difficult to analyze as an equivalent LC resonant circuit, it is often necessary to measure the quantity in Eq. (21) to obtain, in conjunction with a measurement

$$\frac{R_0}{Q_0} = \sqrt{\frac{L}{C}} = \omega_0 L = \frac{1}{\omega_0 C} \qquad (21)$$

of Q_0, a value for R_0, the effective shunt resistance

of the cavity. This is most often done by the perturbation method. If C can be varied in an LC circuit, it can be shown that Eq. (22) holds.

$$\frac{R_0}{Q_0} = -\frac{2}{\omega_0^2} \frac{d\omega}{dC} \qquad (22)$$

Analysis of the field in a microwave cavity when a perturbing object is introduced can, by analogy, be used to relate the resultant change in frequency to R_0/Q_0.

NULL METHODS

A phenomenon at least equal in importance to resonance as an indicator of prescribed circuit conditions is balance. The precision with which the difference between two alternating voltages can be reduced to zero can easily reach a few parts in a million. Null methods are almost universally used for the most precise measurements.

Radio-frequency bridges. At radio frequencies the problems arising from residual parameters make use of variable air capacitors as impedance standards in substitution methods as desirable for null devices as for resonant circuits. Special circuits particularly well adapted for these frequencies have therefore been developed. Three widely used commercial instruments are discussed.

General Radio rf bridge. The balance conditions for this bridge, shown in Figs. 5 and 6, yield the unknown resistive component R_x and the unknown reactive component X_x of impedance in terms of the known bridge components R_B, C_N, C_A, and C_P, as in Eqs. (23) and (24). As indicated, both

$$R_x = \frac{R_B}{C_N} (C_{A2} - C_{A1}) \qquad (23)$$

$$X_x = \frac{1}{\omega} \left(\frac{1}{C_{P2}} - \frac{1}{C_{P1}} \right) \qquad (24)$$

the resistive and reactive components are measured in terms of capacitance differences; the subscripts 1 and 2 refer respectively to balances with the terminals, first, short-circuited, and second, connected to the unknown impedance. The dial of the capacitor C_A is calibrated in ohms, independent of frequency, and the dial of the capacitor C_P in

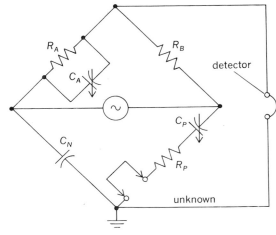

Fig. 5. Circuit schematic of General Radio radio-frequency bridge.

ohms at a frequency of 1 MHz. At other frequencies, the reading of this reactance dial must be divided by the frequency in megahertz. The instrument covers the frequency range from 400 kHz to 60 MHz; other versions extend these limits to 50 kHz and 120 MHz.

Wayne-Kerr rf bridge. This bridge uses a center-tapped transformer with a high degree of coupling between windings to develop equal and opposite voltages in the standard and unknown arms (Fig. 7). A similar transformer with one winding reversed couples the output to the detector. As shown in Fig. 7, taps are used to modify the effectiveness of the capacitance and conductance standards, C_s and G_s, respectively. In terms of the number of turns n between the center tap and the unknown admittance (n_1), the conductance stand-

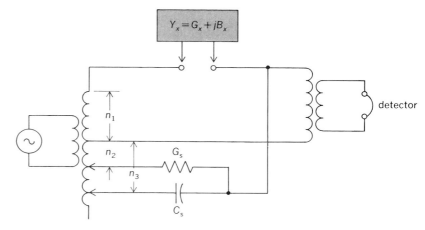

Fig. 7. Simplified inductive-ratio-arm bridge.

Fig. 6. Radio-frequency bridge. (*General Radio Co.*)

ard (n_2), and the capacitance standard (n_3), the balance conditions are given by Eqs. (25) and (26).

$$G_x = \frac{n_2}{n_1} G_s \qquad (25)$$

$$B_x = \frac{n_3}{n_1} \omega C_s \qquad (26)$$

Combinations of switched fixed-value standards and continuously adjustable standards yield scales that are calibrated directly in conductance and capacitance, independent of frequency. Similar taps on the unknown side of the center tap are used to switch admittance ranges. The instrument covers the frequency range from 15 kHz to 5 MHz; other versions extend the range from 15 kHz to 250 MHz.

Bridges of this kind are particularly well adapted to the measurement of the direct component of balanced and other three-terminal admittances because the shunt components can be thrown across the low-impedance transformer windings, by grounding of the center tap, and eliminated from the measurement. They can also be adapted to the measurement of the transfer impedance of four-terminal devices.

Hewlett-Packard RX meter. This bridge uses the same configuration of bridge arms as the General

Radio bridge but measures the unknown as an admittance in the A arm rather than an impedance in the P arm. For this inversion, the balance equations become Eqs. (27) and (28). The *RX* meter

$$G_x = \frac{C_N}{R_B}\left(\frac{1}{C_{P2}} - \frac{1}{C_{P1}}\right) \qquad (27)$$

$$B_x = \omega(C_{A1} - C_{A2}) \qquad (28)$$

differs from conventional bridges in that the bridge arms are excited by out-of-phase voltages from a transformer. At balance the junction-point voltages are equal in magnitude but opposite in phase. Null voltage is obtained between the center point of a capacitive voltage divider and ground (Fig. 8).

Microwave null devices. At frequencies so high that the distributed nature of parameters must be taken into account, the principle of null comparison can still be used to effect precise adjustment. Two commercial instruments for this frequency range are discussed here.

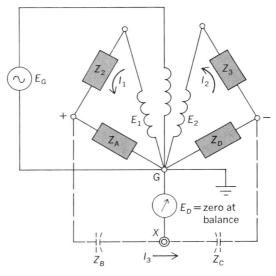

Fig. 8. Circuit schematic of Hewlett-Packard *RX* meter. Three-winding transformer couples out-of-phase voltages from generator to two halves of bridge; two capacitors Z_B and Z_C provide a center tap to feed the detector.

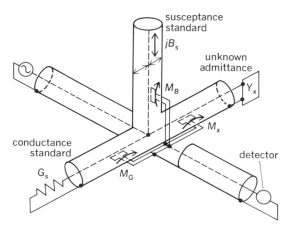

Fig. 9. General Radio admittance bridge. The three loops M_B, M_G, and M_x sample magnetic fields in susceptance, conductance, and unknown-admittance arms, respectively, and combine outputs in detector arm.

HIGH-FREQUENCY
IMPEDANCE
MEASUREMENTS

Fig. 10. Admittance meter with variable capacitor connected as susceptance standard. (*General Radio Co.*)

General Radio admittance bridge. This instrument (Figs. 9 and 10) samples the magnetic field arising from the current in each of three coaxial transmission lines through adjustable loops (M_B, M_G, and M_x) coupled to their center conductors at their junction point in the T configuration shown in the figures. All the lines are fed at this point from a common voltage, and thus the currents bear the same relationship to each other as the respective admittances seen looking into each of the lines. One of the lines is terminated in its characteristic impedance Z_0 to present a standard conductance G_s, equal to $1/Z_0$; one is terminated in an eighth-wavelength transmission line to present a standard susceptance B_s, equal to $1/jZ_0$; and the third is terminated in the unknown admittance Y_x, equal to $G_x + jB_x$. The coupling loops can be rotated by means of shafts carrying dials, the first two being calibrated in conductance and susceptance, respectively, and the third in admittance range. The loop that couples to the susceptance line is adjustable over a 180° range to indicate either positive or negative susceptances; the eighth-wavelength line is set to the proper length for the operating frequency, so that both conductance and susceptance scales are direct-reading, independent of frequency. If the unknown is connected to the instrument by a quarter-wave line, the readings of these scales become directly proportional to impedance. The instrument is direct-reading for the frequency range from 40 to 1000 MHz.

Transfer-function and immittance bridge. A modification of the admittance bridge compares the output current of a four-terminal device with admittance standards in a T configuration similar to that described above, thereby making possible the measurement of transfer admittance by a null method. Through the use of quarter- and half-wave transmission lines in the input and output circuits, it is possible to measure the ratio of output current ·to input voltage, output voltage to input current, output to input voltage, and output to input current. An interchangeable head converts the instrument to an admittance meter, making possible measurement of two-terminal admittance and impedance over the frequency range from 25 to 1000 MHz.

STANDING-WAVE METHODS

Standing-wave detectors are devices used at microwave frequencies to measure impedances in terms of electromagnetic field distributions in guided-wave systems.

It is desirable to make such measurements with devices of simple geometrical configuration so that their performance can be readily analyzed in terms of field theory. At the lower microwave frequencies, where transverse dimensions are small compared to the wavelength, the performance of coaxial lines is easy to compute, and these devices can be conductively connected to circuits characterized by two-terminal connections. At the higher microwave frequencies, where all dimensions are comparable with the wavelength and the concept of lumped, or even distributed, parameters is no longer readily usable; waveguides have simple boundary conditions that facilitate computation and visualization of field distributions. They can be coupled to other waveguides or to cavities through electric or magnetic fields rather than with conductive connections. *See* TRANSMISSION LINES; WAVEGUIDE.

Impedance can be measured with these transmission lines by analysis of the behavior of electric fields propagating along their axes. In an idealized coaxial line or waveguide with lossless, unity-dielectric-constant insulation and perfectly conducting walls, a plane wave will propagate along the axis at the speed of light without alteration so long as the cross-sectional dimensions remain uniform. The relationship between the electric and magnetic fields will remain constant, and because they are respectively proportional to voltage and current, the ratio of these fields will define a characteristic impedance for the line.

When the wave reaches the end of the transmission line, a portion determined by the impedance discontinuity at that point is reflected backwards toward the source. The incident and reflected waves then add together to form a stationary interference pattern, or standing wave, that has maxima and minima occurring alternately at intervals of a quarter wavelength. The maxima occur at those points at which the waves reinforce each other and measure the sum of the two amplitudes; the minima, conversely, measure their difference. The ratio of maximum to minimum voltage is called the voltage–standing-wave ratio (VSWR).

The amount of reflection depends upon the relation between the impedance in which the line is terminated and the characteristic impedance of the line. If the impedance is resistive and equal to the characteristic impedance, the current will flow into it just as it would into a further extension of the line itself and there is no reflection; if the transmission line is short-circuited, the reflected voltage will be equal in magnitude, but of reversed polarity, to the incident wave, because the net voltage must be zero at the termination; if the transmission line is open-circuited, the reflected voltage will again be equal in magnitude, but of like polarity, to the incident wave, so that the voltage doubles at the termination. When the terminating impedance equals the characteristic impedance, the VSWR is unity and the line is said to be matched. When the line is short-circuited or open-circuited, the VSWR would be infinite if the line were lossless. The distance from the termination to the first minimum is respectively zero and a quarter wavelength for these terminations.

It can be shown that these two quantities, the VSWR and the distance to the first minimum, uniquely define the terminating impedance. They are easy to determine experimentally, but the mathematical conversion to the conventional impedance components is somewhat involved. Graphical methods of interpretation have therefore been developed, and one known as the Smith chart is in wide use, for determining impedances from standing-wave measurements and analyzing the effects of finite lengths of connecting lines.

Slotted line. A commercial slotted line for measuring impedances at the lower microwave frequencies is shown in Fig. 11. It comprises a cylindrical coaxial line having a slot in the outer conductor into which a small capacitive probe extends to sample the electric field. As the probe slides along the line, its position can be measured to find the distance from the termination to the first voltage minimum, and the maximum and minimum voltages determined to find the VSWR. This line is suitable for measurements from about 300 to 8500 MHz.

Slotted section. A commercial slotted section for measuring impedances at the higher microwave frequencies is shown in Fig. 12. Its function is similar to that of the coaxial slotted line, but it differs in certain practical respects. Coaxial lines cover the frequency range from direct current to the frequencies at which the higher-order modes of propagation used in waveguides occur. Waveguides, on the other hand, are restricted to the relatively narrower frequency ranges over which a single, selected dominant mode is useful. Replaceable slotted sections are therefore provided so that a wide frequency range can be covered with a single carriage mechanism. These various sections cover the frequency range 2600–40,000 MHz with two models of carriage.

REFLECTION METHODS

Reflection methods are based on a separation of the incident and reflected waves in a transmission line. The individual amplitudes are measured and the ratio of these amplitudes is equal to the magnitude of the reflection coefficient $|\Gamma|$ of the terminating impedance. The VSWR can be determined from Eq. (29).

$$\text{VSWR} = \frac{1 + |\Gamma|}{1 - |\Gamma|} \qquad (29)$$

To measure the terminating impedance itself, an additional measurement of phase shift at the point of reflection is necessary. The reflection coefficient can then be represented as a vectorial quantity $\Gamma\underline{/\theta}$ related to the terminating impedance $\mathbf{Z}_L$ and the characteristic impedance $\mathbf{Z}_0$ of the transmission line by Eq. (30), from which Eq. (31) is obtained.

$$\mathbf{\Gamma} = \Gamma\underline{/\theta} = \frac{\mathbf{Z}_L - \mathbf{Z}_0}{\mathbf{Z}_L + \mathbf{Z}_0} \qquad (30)$$

$$\mathbf{Z}_L = \frac{\mathbf{\Gamma} + 1}{\mathbf{\Gamma} - 1} \mathbf{Z}_0 \qquad (31)$$

A basic component of most reflectometers is the directional coupler, which performs the actual

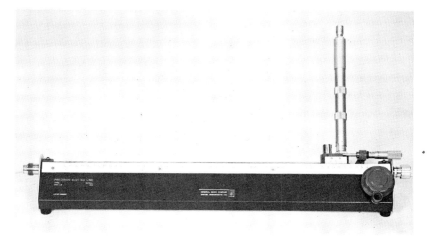

Fig. 11. A precision slotted line fitted with a precision rf connector. (*General Radio Co.*)

separation of the two waves. Two of these, coupled to the transmission line in opposite directions, yield independent measurements of the two waves from which their amplitude ratio can be computed. A convenient arrangement uses a ratio meter to display the magnitude of reflection coefficient directly. The angle θ of the reflection coefficient can be measured with an auxiliary slotted line to determine the distance from the termination to the first voltage minimum; with an auxiliary capacitance probe to indicate the vector sum of the incident and reflected waves at a known point on the line; or by comparing the phases of the outputs of the directional couplers themselves.

Similar separation of incident and reflected signals is accomplished with hybrid and hybridlike

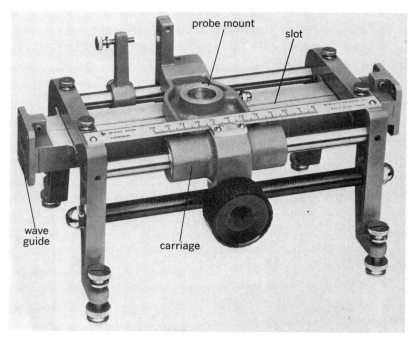

Fig. 12. Hewlett-Packard slotted section for microwave measurements. Different waveguide elements can be inserted in the carriage for frequency ranges between 3950 and 18,000 MHz. (*Hewlett-Packard Co.*)

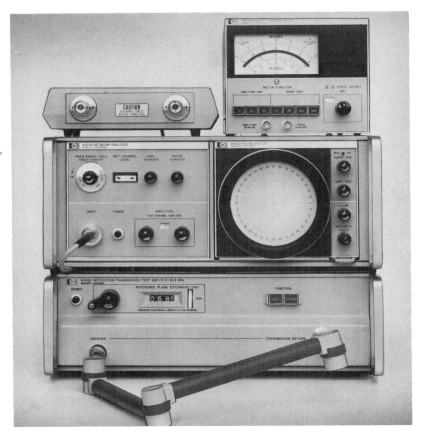

Fig. 13. Network analyzer for impedance measurements from 110 MHz to 12.4 GHz. A directional coupler is used as measuring element. (*Hewlett-Packard Co.*)

devices which utilize a "balun" (a balanced-to-unbalanced transformer).

Some commercial examples of instruments that measure impedance by a reflection method are described below.

Hewlett-Packard network analyzer. This instrument (Fig. 13) employs a directional coupler as the measuring element. The frequency of the incident and reflected signals from the coupler are

converted by a phase-coherent sampling technique to a 20-MHz intermediate frequency, where the phase-and-amplitude measurement is performed. This system and the ones described below operate on a sweep-frequency basis with readout on a meter or cathode-ray tube.

General radio reflectometer. This instrument (Fig. 14) employs a hybrid bridge to cover the 20–1500-MHz band and a directional coupler to cover the 0.5–7-GHz band. It provides amplitude measurements of the reflected signals by the use of square-law detectors.

Automatic plotter. The automatic impedance and transfer-character plotter of the Alford Manufacturing Co. employs a hybridlike device (Hybridge) as the measurement element. The phase-and-amplitude measurements of impedance are made directly at the high frequency by means of a broadband phase detector. A number of such devices cover the frequency range 0.025–7000 MHz.

Impedance plotter. Waveguide impedance plotters covering waveguide bands from 5.4 to 18.0 GHz are offered by Rantec Corp. They utilize directional couplers as measuring elements and broadband rf phase detectors to generate the phase and amplitude of the impedance.

[MARK G. FOSTER]

Bibliography: W. D. Cooper, *Electronic Instrumentation and Measurement Techniques*, 2d ed., 1978; P. Kantrowitz et al., *Electronic Measurements*, 1979.

High-polymer transducer

A transducer that uses a piezoelectric high-polymer film, consisting of polyvinylidene fluoride. Of particular interest are high-polymer loudspeakers and high-polymer stereophonic headphones. As the film is very thin and flexible, an omnidirectional high-polymer loudspeaker has been developed by making the film into a cylindrical shape. The loudspeaker has perfect omnidirectional patterns in the horizontal plane. The high-polymer headphones have wide frequency range, flat frequency response, low distortion, light weight, thin shape, and high stability in the presence of moisture and dust.

Piezoelectric high polymers. Studies on the piezoelectric properties of the high polymers have been conducted since the discovery of the piezoelectricity of wood in the 1940s. In the 1950s the piezoelectricity of biological substances, such as collagen, bone, and silk, was investigated. In the 1960s the piezoelectricity of synthetic high-polymer polypeptides was found by E. Fukada.

The piezoelectricity of polyvinylidene fluoride film (PVF_2 film) was discovered by H. Kawai in 1969. Studies to improve the piezoelectricity of polyvinylidene fluoride have been conducted mainly in Japan. The first commercial product, a high-polymer stereophonic headphone, was disclosed in 1974.

At present, the largest piezoelectric constant is found in uniaxially stretched and polarized PVF_2 films. Polyvinylidene fluoride is a fluorocarbon resin, in which every other pair of hydrogen atoms in polyethylene is replaced by a pair of fluorine atoms. Because of its high weather resistance, one of the main applications of polyvinylidene fluoride is as a coating material.

The process which gives piezoelectricity to poly-

Fig. 14. Reflectometer for voltage—standing-wave ratio measurements in coaxial systems from 20 MHz to 7 GHz. (*General Radio Co.*)

Comparison of piezoelectric materials in the case of the transverse effect*

Material	Density (ρ), g/cm³ or (kg/m³) $\times 10^{-3}$	Young's modulus (E), (dyne/cm²) $\times 10^{10}$ or (N/m²) $\times 10^{9}$	Relative dielectric constant (ϵ/ϵ_0)	Piezoelectric constants		Electro-mechanical coupling coefficient (k), %
				d, (C/N) $\times 10^{-12}$	g, (m²/C) $\times 10^{-3}$	
Usual piezoelectric materials						
Quartz	2.66	77.3	4.5	2.0	50.1	10
Rochelle salt	1.77	17.7	350	272.3	90	73
PZT† (lead zirconate titanate) ceramics	7.5	75	1700	126.7	11.1	33.4
Piezoelectric high polymers (PVF₂ film)	1.79	3.6	12	26.4	251.1	15.5

*From M. Tamura et al., Electroacoustic transducers with piezoelectric high polymer films. *J. Audio Eng. Soc.*, 23:21–26, 1975; courtesy of Journal of the Audio Engineering Society.

†PZT is the trademark for piezoelectric ceramics produced by the Clevite Corporation in the United States.

vinylidene fluoride consists of the following steps: (1) Extruded films are uniaxially stretched up to four times the original length at 60 to 100°C. (2) Aluminum is evaporated on both surfaces of the film as electrodes. (3) The films are polarized with a high direct-current electric field of 600 kV/cm (600 V for a film 10 μm thick) at 80 to 100°C for 1 hr. This procedure is very similar to that used for piezoelectric ceramics. This indicates that the piezoelectricity in polyvinylidene fluoride is due to remanent polarization after the polarization process, as in the ceramics. If a high alternating-current field is applied to the film, a hysteresis loop between the applied field and the polarization is observed similar to that of ferroelectric crystals. *See* FERROELECTRICS.

The properties of PVF₂ film and of common piezoelectric materials are compared in the table. Piezoelectric properties are compared in the case of the transverse piezoelectric effect, in which the direction of the applied electric field is perpendicular to the direction of the resultant stress. The Young's modulus of the PVF₂ film is more than one order smaller than those of the usual piezoelectric materials. This property of the film is very important for the applications to high-quality audio transducers. The d constant of the film, which is a coefficient of an induced strain by an applied field, is rather small in comparison with that of the ceramics. But this order of magnitude of the d constant is sufficient to use the material in a loudspeaker because the applied electric field between the surfaces of the thin film is about one order higher than in the case of ceramics. The g constant is a coefficient of an induced voltage by an applied force, and is important for the microphone application. The relative dielectric constant and the dimensions of a transducer determine the transducer's electrical impedance. The electromechanical coupling coefficient is defined, by analogy with an alternating-current transformer, as the ratio of the mutual elastodielectric energy density in the piezoelectric material to the square root of the product of the stored elastic and dielectric energy densities. Hence, the efficiency of the piezoelectric transducer depends on this coefficient directly. *See* PIEZOELECTRIC CRYSTAL; PIEZOELECTRICITY.

The piezoelectricity of the PVF₂ film is quite stable, even at a high temperature of 100°C for many months. Since piezoelectricity is a bulk effect, the PVF₂ film is not affected by moisture and dust. This property is one of the good features of the PVF₂ film compared with electrets, whose use in transducers depends on surface charges of dielectrics. *See* ELECTRET TRANSDUCER.

Direct radiator loudspeakers. In order to use the PVF₂ film effectively in transducers, a very simple structure has been devised by utilizing the advantage of the thin-film form. In Fig. 1a, when a sinusoidal electric field is applied to the PVF₂ film along the z-axis, the PVF₂ film vibrates in a transverse direction (along the x-axis), that is, in the stretching direction. Next, if the PVF₂ film is curved as shown in Fig. 1b, the transverse vibration is converted into a pulsating movement and generates sound waves. This is the common structure for the direct radiator loudspeakers, stereophonic headphones, and microphones using the PVF₂ films.

One of the most interesting applications of the above-mentioned structure is a perfect omnidirectional high-frequency loudspeaker as indicated in Fig. 2. To make a cylindrical shape, a perforated cylinder is placed between top and base plates. Next, a polyurethane foam backing is put on the perforated cylinder. Finally, a polarized PVF₂ film, whose thickness is typically 27 μm, is stretched over the polyurethane foam backing.

In order to cover the frequency range from 2 to 20 kHz, two coaxial cylinders are used, with one mounted above the other. The diameter of the larger unit is 120 mm; the frequency range is 2 to 5 kHz. The diameter of the smaller one is 60 mm;

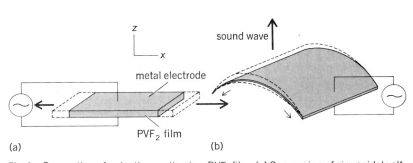

Fig. 1. Generation of pulsating motion in a PVF₂ film. (*a*) Conversion of sinusoidal voltage along z-axis into transverse motion along x-axis. (*b*) Conversion of transverse vibration into pulsating motion. (*From M. Tamura et al., Electroacoustic transducers with piezoelectric high polymer films, J. Audio Eng. Soc., 23:21–26, 1975; courtesy of Journal of the Audio Engineering Society*)

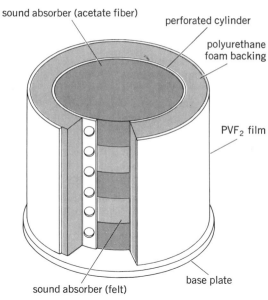

sound absorber (acetate fiber)

perforated cylinder

polyurethane foam backing

PVF$_2$ film

base plate

sound absorber (felt)

Fig. 2. Structure of the omnidirectional high-polymer direct radiator loudspeaker. (*From M. Tamura et al., Electroacoustic transducers with piezoelectric high polymer films, J. Audio Eng. Soc., 23:21–26, 1975; courtesy of Journal of the Audio Engineering Society*)

the frequency range is 5 to 20 kHz. To utilize these high-polymer loudspeakers in combination with a conventional low-frequency loudspeaker with impedance of 4 to 8 Ω, a step-up transformer is required because the impedances of the PVF$_2$ films are essentially capacitive and the loudspeaker impedances are relatively high in the operating frequency ranges. The step-up ratio of the transformer is about 1:10.

Because of the polyurethane foam backing, a sufficient mechanical damping is provided; therefore a smooth frequency response is easily obtained between 2 and 20 kHz. Perfect omnidirectional patterns have been realized up to 20 kHz. Another favorable feature of the high-polymer loudspeaker is its linearity over a wide range of input power up to 100 W. *See* LOUDSPEAKER.

Stereophonic headphones. The first practical application of the PVF$_2$ film to audio transducers was a stereophonic headphone. The structure of the high-polymer headphone is quite simple, as

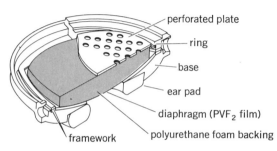

perforated plate

ring

base

ear pad

diaphragm (PVF$_2$ film)

framework

polyurethane foam backing

Fig. 3. Cross-sectional view of high-polymer headphone. (*From M. Tamura et al., Electroacoustic transducers with piezoelectric high polymer films, J. Audio Eng. Soc., 23:21–26, 1975; courtesy of Journal of the Audio Engineering Society*)

shown in Fig. 3. The PVF$_2$ film, typically 7 μm in thickness, is stretched loosely over a rectangular framework aligning the stretching axis (x-axis) to the shorter axis of the rectangle. The PVF$_2$ film is then placed in the base, and a bias force is applied through a polyurethane foam backing by a perforated plate.

In this headphone, the whole surface of the PVF$_2$ film is driven by the applied voltage, and the headphone has as good sound quality as the electrostatic type. A small input voltage, such as 3 V, is sufficient to produce a sound pressure level of 100 dB; thus, the headphones can be driven directly from the headphone terminals of a conventional amplifier in the same manner as in the moving coil type without using a step-up transformer, which must be employed in the electrostatic type.

In addition, the high-polymer headphones have the following advantages: no bias source is needed, and no irritating noises are produced even at a high input power. Also, the harmonic distortions are not greater than 1% even at an output sound pressure level of 110 dB. *See* EARPHONES.

Microphones. Another application of the PVF$_2$ film to electroacoustic transducers is in microphones. In recent years various types of electret microphones have been used because of their wide and flat frequency response, simple structure, and low cost.

An experimental high-polymer microphone has been developed which has a structure similar to that of the high-polymer headphone. The output signal can be taken from the terminals connected to both surfaces of the film. The diameter is 15 mm; the thickness of the PVF$_2$ film is 27 μm. The microphone can be simply built with only a thin PVF$_2$ film and a polyurethane foam backing. Furthermore, the accuracy of each part may be less exact that that required in the case of electrostatic microphones.

The response is similar to that of the electrostatic type. The sensitivity is −74 dB (0 dB = 1 V/μbar); the capacitance is about 700 pF. This capacitance value is about two orders greater than those of conventional electrostatic microphones. This means that the signal-to-noise ratio of the high-polymer microphone is fairly good in comparison with electrostatic microphones.

A noise-canceling microphone using a PVF$_2$ bimorph diaphragm has been proposed. The diameter is 5 mm; the thickness of the film is 32 μm. The fundamental resonance frequency is about 4 kHz.

Even with many good features, the high-polymer microphone has not yet been commercialized, because the cost of the PVF$_2$ film is not competitive with the electret film.

An experimental hydrophone has also been tested. The result indicates that the high-polymer hydrophone is a fairly promising one. *See* MICROPHONE.

Electromechanical transducers. Apart from its use in electroacoustic transducers such as described so far, piezoelectric high polymer has been tested for electromechanical transducers. Examples are phonograph cartridges and accelerometers.

Phonograph cartridges. Because of the low stiffness of PVF$_2$ film, a very-high-compliance vibrating system can be easily designed. A high resonance frequency is realized due to the light weight of the small, thin element. The PVF$_2$ film is cut into small strips 1 mm wide and 2 mm long. Two small PVF$_2$ film elements have been used in an experimental high-polymer phonograph cartridge with a structure similar to that of the ceramic cartridge. *See* DISK RECORDING.

Accelerometers. An experimental accelerometer has been developed. The PVF$_2$ film is cut out into a strip 2 mm wide and 15 mm long. The longitudinal axis of the strip is perpendicular to the stretching axis (*x*-axis). The ends of the strip are cemented together to form a ring. A small mass is attached to the top of the ring to produce a force. The output voltage can be taken from the terminals connected to the conductive layers on both surfaces of the film. The sensitivity and the frequency range are almost the same as those of conventional accelerometers.

Range of applications. Of all commercial applications of PVF$_2$ films, high-frequency loudspeakers and stereophonic headphones are most prominent. These devices were introduced only a few years ago. The range of applications of the PVF$_2$ film is not very wide yet because of the cost of this polarized product. But many applications to various transducers can be expected in the near future because the film has many good features, such as flexibility, light weight, thin shape, flat frequency response, low distortion, good sound quality, and stability against moisture and dust. *See* TRANSDUCER.

[TAKEO YAMAMOTO; MASAHIKO TAMURA]

Bibliography: J. F. Sear and R. Carpenter, *Electron. Lett.*, 11:532–533, 1975; M. Tamura et al., *J. Appl. Phys.*, 48:513–521, 1977; M. Tamura et al., *J. Audio Eng. Soc.*, 23:21–26, 1975; Y. Wada and R. Hayakawa, *Japan. J. Appl. Phys.*, 15:2041–2057, 1976.

High-temperature electronics

The technology of electronic components capable of operating at high temperatures (above 300°C). The need for electronic devices for geothermal well probes, planetary space probes, jet-engine controls, and nuclear power plant instruments is supplying the impetus for advancements in this area, since conventional silicon diodes, transistors, and integrated circuits will not function in this temperature range. The research has been carried out in three primary areas: new silicon devices, compound semiconductor devices, and integrated thermionic circuits. Some of these devices have the potential of extending the operational range of electronic circuits up to 800°C.

Silicon devices. Semiconductor devices can be divided into two categories: minority carrier devices, such as diodes and bipolar transistors, and majority carrier devices, such as the field-effect transistor (FET). Minority carrier devices made from silicon do not work well above 250°C. This is because the energy bandgap in silicon is only 1.1 eV, and at temperatures above 250°C a large number of minority carriers are generated by thermal excitation.

The density of these thermally generated carriers in silicon is approximately $10^{10}/cm^3$ at room temperature and increases by about 6 orders of magnitude at 400°C (to approximately $10^{16}/cm^3$). The large density increase in thermally generated carriers causes a large increase in reverse leakage current of *pn* junctions and makes the performance of minority carrier silicon devices, such as bipolar transistors and diodes, marginal above 250°C.

Majority carrier devices in silicon (such as FETs) do not depend on minority carriers for their operation and therefore can be made to operate at higher temperatures. Enhancement-mode metal-oxide semiconductor FET (MOSFET) devices have been made to operate at 350°C. To achieve this operating temperature it is necessary to remove the input-gate-protection bypass diodes and use dielectrically isolated (rather than junction-isolated) devices because of diode leakage. The upper limit of operating temperature for the silicon MOSFET is not known definitely. Problems that occur at high temperatures are oxide degradation in the gate and leakage at the source-drain *pn* junctions. It is generally felt that small-area devices may be eventually made to operate at 400°C. *See* JUNCTION DIODE; TRANSISTOR.

Compound semiconductor devices. The basic need for a simple high-temperature rectifier diode and moderate power devices with gain have led to high-temperature research on compound semiconductors. Many of the group III–V compound semiconductors have energy bandgaps larger than silicon. For example, the semiconductor bandgap in gallium phosphide (GaP) is 2.2 eV. The thermally generated minority carrier density in GaP at 400°C is approximately $10^{12}/cm^3$ (4 orders of magnitude less than silicon). This level is small enough that it is possible to build minority carrier devices at high temperatures by using this and other wide-bandgap semiconductor materials. Diodes and transistors have been demonstrated at 450°C in GaP material. Figure 1 shows a scanning

Fig. 1. Gallium phosphide bipolar transistor for use at temperatures up to 450°C. Center contact is the emitter, base contacts are two outside pads, and collector contact is the bottom of the chip. Chip size is $500 \times 750\ \mu$m.

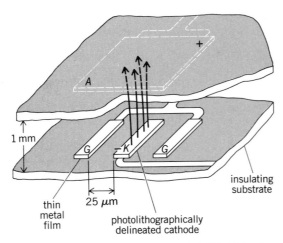

1 mm

thin
metal
film

25 μm

photolithographically
delineated cathode

insulating
substrate

Fig. 2. Integrated thermionic circuit (ITC). (*After Proceedings of the High Temperature Electronics and Instrumentation Conference, December 1981, Sandia National Laboratories Rep. SAND82-0425*)

electron photograph of a GaP bipolar transistor. The device in the photograph is a grown-junction *pnp* transistor made by liquid-phase epitaxy. The center contact is the emitter, the two outside pads are base contacts, and the collector contact is the bottom of the chip. Compound semiconductor devices of this type are estimated to be operational up to approximately 525°C, with moderate power-handling ability. Allowable current densities in these devices can be up to 100 A/cm², even at 400°C. While this current-carrying capacity is about 1 order of magnitude lower than that used in room-temperature silicon devices, it is still adequate for most high-temperature applications.

Research is also being carried out on compound semiconductor heterojunction devices. These are *pn* junctions made from dissimilar semiconductor materials such as gallium arsenide and gallium aluminum arsenide (GaAs and GaAlAs). The physics of junctions of dissimilar semiconductors can be exploited to reduce the leakage currents which plague high-temperature devices, making successful high-temperature minority carrier devices possible. *See* SEMICONDUCTOR HETEROSTRUCTURES.

Integrated thermionic circuits. A dramatic departure from semiconductor technology is the integrated thermionic circuit (ITC). This technology combines photolithographically defined subminiature thin-film metal patterns with planar vacuum-tube technology. The result is a technology that allows fabrication of active circuits with a density approaching that of present room-temperature silicon integrated circuits. Figure 2 shows a cut away view of a single ITC active device. In this structure the grids (*G*) are coplanar with the cathode (*K*) on the lower plane; the anode (*A*) is on the upper plane, and the space between the cathode and anode is evacuated. The grid metal and oxide-coated cathode are photolithographically defined on a sapphire substrate. During normal operation a resistance heater is used to heat this substrate to approximately 800°C to produce electron emission from the cathode. A positive voltage is applied to the anode to collect

these electrons, and a voltage is applied to the grid lines to set up an electric field which controls this current flow in a manner similar to that of a conventional triode vacuum tube. The resulting devices produce current densities of approximately 0.1 A/cm². Because of this limitation these devices are low-power devices, operating at power densities about 3 orders of magnitude less than room-temperature silicon devices. Thus, they are not suitable for use in power-conditioning or line-driving applications. The small size of these small-signal devices does allow the integration of analog and digital circuits with device densities of several thousand per square centimeter. ITCs have been life-tested at 500°C for several thousand hours with no detectable degradation. An upper temperature limit is not known, but it is pressured that, with suitable packaging techniques to reduce outgassing problems, device temperatures should be able to approach the cathode temperature of 800°C. These devices are also capable of surviving severe radiation environments and may be applicable to nuclear power reactor instrumentation. *See* INTEGRATED CIRCUITS; VACUUM TUBE.

Use of amorphous metal films. One of the weak links in semiconductor technology at high temperatures is the metallization used. All semiconductor metallizations are presently made from polycrystalline metals. Failures that occur with room-temperature semiconductor devices are usually attributed to metal failures due to diffusion, corrosion, or electromigration of the thin metal films used to contact and interconnect the semiconductor devices. Since these failures are due to thermally activated processes, they occur even more rapidly at elevated temperatures. In order to make reliable high-temperature semiconductor devices with reasonable lives this problem must be solved. It is known that most of these metal failures occur along grain boundaries in the polycrystalline metals. A potential solution to this problem is the use of amorphous metal films (which have no grain boundaries) in place of the polycrystalline metals. Recent experimental results have demonstrated that amorphous metal films on semiconductors are several orders of magnitude less susceptible to these failures at high temperatures, and these films can be produced by processes compatible with the semiconductor industry. *See* GRAIN BOUNDARIES.   [ROGER J. CHAFFIN]

Bibliography: *Proceedings of the High Temperature Electronics and Instrumentation Conference*, Houston, December 1981, Sandia National Laboratories Rep. SAND82-0425; *Proceedings of the Conference on High Temperature Electronics*, Tucson, March 1981, IEEE Rep. 81CH1658-4.

Holes in solids

Vacant electron energy states near the top of an energy band in a solid are called holes. A full band cannot carry electric current; a band nearly full with only a few unoccupied states near its maximum energy can carry current, but the current behaves as though the charge carriers are positively charged. The situation can be understood in terms of the definition of the effective mass: if the energy band is specified by a function $E(k)$, where k is the magnitude of the wave vector $\mathbf{k}$, the effective mass for a spherical band is given by the equation here, where $\hbar$ is Planck's constant di-

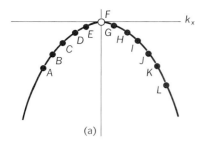

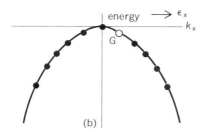

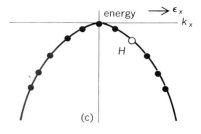

Process of hole conduction. (*a*) At time $t = 0$, energy states *A* through *L* are filled except *F*. (*b*) An electric field ϵ_x is applied in the $+x$ direction. The force on the electrons is in the $-k_x$ direction, and all electrons make transitions in the $-k_x$ direction, moving the hole to G. (*c*) Aft-er a further interval, the electrons move farther along, and the hole is now at *H*. (*After C. Kittel, Introduction to Solid State Physics, 3d ed., copyright © 1966 by John Wiley and Sons, Inc.; used with permission*)

$$m^* = \hbar^2 \left(\frac{\partial^2 E}{\partial k^2} \right)^{-1}$$

vided by 2π. Near a maximum of the band, the second derivative of the energy is negative, so the effective mass is negative. States for which the effective mass is negative are defined as hole states. Carriers in such states behave under the influence of an external electromagnetic field as though they carry positive charge. *See* BAND THEORY OF SOLIDS.

The process of conduction in such a system may be visualized in the following way. An electron moves against an applied electric field by jumping into a vacant state. This transfers the position of the vacant state, or propagates the hole, in the direction of the field, as shown in the diagram. Whether conduction occurs by electrons or holes is determined experimentally from the sign of the Hall emf. If a current is carried in the presence of a magnetic field perpendicular to the current, an emf is developed perpendicular to the current and to the field. The sign of this emf depends on the charge on the carriers. *See* HALL EFFECT.

Hole conduction is important in many semiconductors, notably germanium and silicon. The occurrence of hole conduction in semiconductors can be favored by alloying with a material of lower valence than the "host." Semiconductors in which the conduction is primarily due to holes are called *p* type. Hole conduction is also observed in some metals, including iron and chromium. In other metals, including aluminum and bismuth, both holes and electrons may be present in equilibrium. *See* SEMICONDUCTOR.

[JOSEPH CALLAWAY]

Bibliography: N. W. Ashcroft and N. D. Mermin, *Solid State Physics*, 1976; C. Kittel, *Introduction to Solid State Physics*, 5th ed., 1976; W. Shockley, *Electrons and Holes in Semiconductors*, 1950, reprint 1976.

Image orthicon

A high-sensitivity television camera tube that can be used to pick up scenes of widely varying light values. It is used singly for monochrome television or in sets of three for color television. An ordinary camera lens produces an optical image of the scene being televised on a sensitive photoelectric surface at one end of the tube. Photoelectrons released from the back of this surface are electrically focused on one side of a separate storage target. The other side of the target is scanned by an electron beam produced by an electron gun at the far end of the tube. Beam electrons are reflected back from the target in proportion to the charge at each point on the target image, and are returned to an electron multiplier surrounding the electron gun. The electron multiplier amplifies the signal many thousands of times to produce the video output signal of the camera tube. Image orthicons are used in broadcast television for both indoor and outdoor work. *See* TELEVISION CAMERA TUBE.

[JOHN MARKUS]

Image processing

A term that can mean many things, from adjusting the contrast on a television set to sophisticated nonlinear enhancement of moon images or automated detection of black lung disease from chest radiographs. This article is restricted to image enhancement systems in which the input into the system is an image and the output is also an image that has been improved in some way.

An image is any two-dimensional representation of a physical quantity. Thus, it may be a photographic negative in which the darkening of the emulsion represents brightness in an object scene. Or it can be a medical thermogram in which the color at any point is determined by the temperature at a corresponding point on a patient's skin. Or it can be a nuclear medicine image consisting of a collection of dots, with the density of dots related to the distribution of a radioactive pharmaceutical in the body. Even the restriction to two-dimensional representations is arbitrary, because many of the image-processing techniques discussed here can also be profitably applied to one-dimensional data such as spectrograms.

An image-processing system may have one of two goals: it either makes the image more pleasing or more useful to a human observer, or it actually performs some of the interpretation and recognition tasks usually performed by humans. The latter goal is often referred to as image analysis or pattern recognition.

Hardware. Of the various types of hardware that can be used for image enhancement, the digital computer is the most generally useful and perhaps also the easiest to understand conceptually. Suppose that the initial image is a photographic transparency. The first step is to convert the density levels in the image into numbers that can be entered into the computer memory. This can be accomplished with a microdensitometer in which a small, rectangular aperture is used to isolate a single picture element, or pixel. The light transmitted through this element is measured with a photodetector, digitized, and stored. Then the aperture is moved over by its width, and the density of the adjacent picture element is recorded. In this way the entire picture is converted into a matrix of numbers suitable for computer manipulation. The number of picture elements required to faithfully represent the image depends on both the overall size of the image and the size of the fine detail in it. Once the computer manipulations are complete, the processed image may be displayed on a cathode-ray tube, a mechanical printer, or any other suitable device. *See* DIGITAL COMPUTER.

Operation types. The operations that can be performed on images can be classified into point, global and local operations.

Point operations. In these operations the image value at a point in the output image depends on the value at only a single point in the input image. The most common example of this type of operation is contrast enhancement, which is very useful if the input image is underexposed or depicts a low-contrast scene. A second example is a pseudocolor display, in which each gray level in the input is encoded as a color in the output, the theory being that the eye can distinguish colors more easily than shades of gray.

Global operation. At the opposite extreme from point operations are global operations, in which the value at one picture element in the output depends on all, or almost all, of the picture elements in the input. An example is reconstruction of an image from a hologram.

Local operations. However, the most useful operations are frequently the local operations in which a relatively small neighborhood of a point on the input contributes to the value at a point on the output. To illustrate, consider the problem of edge detection. If the image were ideal, the following sequence of numbers might be generated when the aperture of the microdensitometer is scanned across the edge:

$$3, 3, 3, 3, 3, 3, 6, 6, 6, 6, 6, 6.$$

The edge is readily apparent between the sixth and seventh picture elements. Suppose, however, that noise is present, so that there is an uncertainty in the value at each element. Then the input data might look like

$$4, 3, 2, 5, 2, 3, 5, 7, 6, 4, 6, 8.$$

The position of the edge is now less apparent. One simple smoothing operation is to replace each number by its average with its two neighbors. Mathematically, this corresponds to convolution of the data with the sequence (1/3, 1/3, 1/3). The result of performing this operation on the noisy data above is

$$2,\ 3, 3, 3, 3, 3, 5, 6, 6, 5, 6, 5,$$

where all numbers have been rounded (quantized) to the nearest integer. Note that the noise has been reduced considerably, but the edge is not quite as abrupt as in the original object.

Now suppose that the problem is not noise but blur (see illustration). Here the data might look like

$$30, 30, 30, 31, 34, 41, 50, 56, 59, 60, 60, 60,$$

where now two digits are used to represent each

picture element for reasons that will become apparent shortly. To sharpen up this edge, the data can be convolved with (−2, 5, −2). In other words, the value at each picture element is multiplied by 5, and twice the sum of its neighbors is subtracted from it. This produces the following result:

$$90, 30, 28, 31, 26, 37, 56, 62, 63, 60, 60, 180.$$

It is now clear that the edge is between the sixth and seventh elements. Note also the large values at the ends. The edge detection routine has found the edges of the data.

Suppose the original blurred data had been rounded off to one digit, so that it looked like

$$3, 3, 3, 3, 3, 4, 5, 6, 6, 6, 6, 6.$$

Now application of the deblurring filter (−2, 5, −2) yields

$$9, 3, 3, 3, 1, 4, 5, 8, 6, 6, 6, 18.$$

The round-off error has been greatly magnified, and the processed image is poorer than the original. Thus, the deblurring filter (−2, 5, −2) improves sharpness but increases the noise level, whereas the smoothing filter (1/3, 1/3, 1/3) has just the opposite effect.

The action of these filters can also be described in terms of spatial frequencies. Just as an electrical signal can be decomposed by Fourier analysis into a sum of sine waves of varying amplitude, phase, and frequency, so too can a two-dimensional image be broken up into a superposition of sinusoidal bar patterns. However, the spatial frequency of one of these patterns is not uniquely specified by a single number, such as the number of bars per centimeter. The orientation of the pattern must also be given; spatial frequency is a vector.

But leaving this matter of dimensionality aside, a strong formal analogy exists between spatial and temporal frequencies. The smoothing filter described above is a low-pass filter that suppresses the higher spatial frequencies in the image. Conversely, the deblurring filter emphasizes the higher frequencies.

The practical utility of the spatial frequency concept lies in the ease with which modern computers can perform Fourier analysis. Using the so-called fast Fourier transform algorithms, a 128 × 128 image can be decomposed into its frequency components in a matter of seconds, even on a small computer. Then each frequency component can be weighted appropriately, and the output image is easily reassembled.

As powerful as these frequency-domain methods are, they still do not represent the ultimate in image-processing technology. The best results to date have been obtained with iterative nonlinear techniques in which the computer systematically searches for the "best" output image that is consistent with the input data and with any other information that is known about the object. Although spectacular results can be obtained by these nonlinear methods, they frequently require large machines and exorbitant computing times.

Many practical image-processing tasks can also be performed without the use of digital computers. For example, there is a commercial instrument that uses a closed-circuit television system to sharpen radiographs. And rather striking exam-

IMAGE PROCESSING

The processing of a blurred image. (*a*) Original object. (*b*) Blurred image of original object. (*c*) Processed image with linear deblurring filter. (*d*) Processed image using iterative nonlinear method. (*Courtesy of J. Burke and R. Hershel, University of Arizona, Optical Sciences Center*)

ples of both image enhancement and pattern recognition have been produced by holographic filtering.

Thus, image processing is an active and fruitful area of research, with the problems ranging from basis mathematics to practical hardware engineering.   [HARRISON H. BARRETT]

Bibliography: H. C. Andrews, *Computer Techniques in Image Processing*, 1970; A. Rosenfeld, *Picture Processing by Computer*, 1969.

Impedance matching

The use of electric circuits and devices to establish the condition in which the impedance of a load is equal to the internal impedance of the source. This condition of impedance match provides for the maximum transfer of power from the source to the load. In a radio transmitter, for example, it is desired to deliver maximum power from the power amplifier to the antenna. In an audio amplifier, the requirement is to deliver maximum power to the loudspeaker. *See* ELECTRICAL IMPEDANCE.

The maximum power transfer theorem of electric network theory states that at any given frequency the maximum power is transferred from the source to the load when the load impedance is equal to the conjugate of the generator impedance. Thus, if the generator is a resistance, the load must be a resistance equal to the generator resistance for maximum power to be delivered from the generator to the load. When these conditions are satisfied, the power is delivered with 50% efficiency; that is, as much power is dissipated in the internal impedance of the generator as is delivered to the load.

Impedance matching network. In general, the load impedance will not be the proper value for maximum power transfer. A network composed of inductors and capacitors may be inserted between the load and the generator to present to the generator an impedance that is the conjugate of the generator impedance. Since the matching network is composed of elements which, in the ideal case of no resistance in the inductors and perfect capacitors, do not absorb power, all of the power delivered to the matching network is delivered to the load. An example of an L-section matching network is illustrated. Matching networks of this type are used in radio-frequency circuits. The values of inductance and capacitance are chosen to satisfy the requirements of the maximum power transfer theorem. The power dissipated in the matching network is a small fraction of that delivered to the load, because the elements used are close approximations of ideal components.

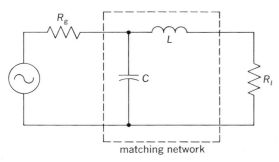

L-section impedance matching network.

Transformers. The impedance measured at the terminals of one winding of an iron-cored transformer is approximately the value of the impedance connected across the other terminals multiplied by the square of the turns ratio. Thus, if the load and generator impedances are resistances, the turns ratio can be chosen to match the load resistance to the generator resistance for maximum power transfer. If the generator and load impedances contain reactances, the transformer cannot be used for matching because it cannot change the load impedance to the conjugate of the generator impedance (the L-section matching network can). The turns ratio can be chosen, however, to deliver maximum power under the given conditions, this maximum being less than the theoretical one.

Iron-cored transformers are used for impedance matching in the audio and supersonic frequency range. The power dissipated in the core increases with frequency because of hysteresis. Above the frequency range at which iron-cored transformers can be used, the air-core transformer or transformers with powdered-iron slugs can be used effectively. However, in these cases the turns-ratio-squared impedance-transforming property is no longer true. Since the transformer is usually part of a tuned circuit, other factors influence the design of the transformer.

The impedance-transforming property of an iron-cored transformer is not always used to give maximum power transfer. For example, in the design of power-amplifier stages in audio amplifiers, the impedance presented to the transistor affects distortion. A study of a given circuit can often show that at a given output power level, usually the maximum expected, there is a value for the load resistance which will minimize a harmonic component in the harmonic distortion, such as the second or third harmonic. The transformer turns ratio is selected to present this resistance to the transistor. *See* TRANSFORMER.

Emitter follower. In electronic circuitry a signal source of large internal impedance must often be connected to a low-impedance load. If the source were connected directly to the load, attenuation of the signal would result. To reduce this attenuation, an emitter follower is connected between the source and the load. The input impedance of the emitter follower is high, more nearly matching the large source impedance, and the output impedance is low, more nearly matching the low load impedance. If the object were the delivery of maximum power to the load, it might be possible to design the emitter follower to have an output resistance equal to the load resistance, assuming that the load is a resistance. (Special audio amplifiers have been designed to use emitter followers, rather than a transformer, to connect the loudspeaker to the power amplifier.) In many cases, maximum power transfer is not the goal; the emitter follower is introduced primarily to reduce to a minimum the attenuation of the signal.

There exist a number of applications where the emitter follower is not useful as an impedance matching circuit. For example, if a very-low-impedance source must be matched to a high-impedance load, then a transistor is used in the common base configuration. *See* EMITTER FOLLOWER; VOLTAGE AMPLIFIER.

[CHRISTOS C. HALKIAS]

Bibliography: J. Millman, *Microelectronics: Digital and Analog Circuits and Systems*, 1979; R. L. Thomas, *A Practical Introduction to Impedance Matching*, 1976; M. E. Van Valkenburg, *Network Analysis*, 3d ed., 1974.

Inductance

That property of an electric circuit or of two neighboring circuits whereby an electromotive force is induced (by the process of electromagnetic induction) in one of the circuits by a change of current in either of them. The term inductance coil is sometimes used as a synonym for inductor, a device possessing the property of inductance. *See* ELECTROMOTIVE FORCE (EMF); INDUCTOR.

Self-inductance. For a given coil, the ratio of the electromotive force of induction to the rate of change of current in the coil is called the self-inductance L of the coil, given in Eq. (1), where e is

$$L = -\frac{e}{dI/dt} \qquad (1)$$

the electromotive force at any instant and dI/dt is the rate of change of the current at that instant. The negative sign indicates that the induced electromotive force is opposite in direction to the current when the current is increasing (dI/dt positive) and in the same direction as the current when the current is decreasing (dI/dt negative). The self-inductance is in henrys when the electromotive force is in volts, and the rate of change of current is in amperes per second.

An alternative definition of self-inductance is the number of flux linkages per unit current. Flux linkage is the product of the flux Φ and the number of turns in the coil N. Then Eq. (2) holds. Both

$$L = \frac{N\Phi}{I} \qquad (2)$$

sides of Eq. (2) may be multiplied by I to obtain Eq. (3), which may be differentiated with respect to t,

$$LI = N\Phi \qquad (3)$$

as in Eqs. (4). Hence the second definition is equivalent to the first.

$$L\frac{dI}{dt} = N\frac{d\Phi}{dt} = -e$$

or
$$L = -\frac{e}{dI/dt} \qquad (4)$$

Self-inductance does not affect a circuit in which the current is unchanging; however, it is of great importance when there is a changing current, since there is an induced emf during the time that the change takes place. For example, in an alternating-current circuit, the current is constantly changing and the inductance is an important factor. Also, in transient phenomena at the beginning or end of a steady unidirectional current, the self-inductance plays a part. *See* ELECTRIC TRANSIENT.

Consider a circuit of resistance R and inductance L connected in series to a constant source of potential difference V. The current in the circuit does not reach a final steady value instantly, but rises toward the final value $I = V/R$ in a manner that depends upon R and L. At every instant after the switch is closed the applied potential difference is the sum of the iR drop in potential and the back emf $L\, di/dt$, as in Eq. (5), where i is

$$V = iR + L\frac{di}{dt} \qquad (5)$$

the instantaneous value of the current. Separating the variables i and t, one obtains Eq. (6). The solution of Eq. (6) is given in Eq. (7).

$$\frac{di}{(V/R) - i} = \frac{R}{L}\,dt \qquad (6)$$

$$i = \frac{V}{R}\left(1 - e^{-(R/L)t}\right) \qquad (7)$$

The current rises exponentially to a final steady value V/R. The rate of growth is rapid at first, then less and less rapid as the current approaches the final value.

The power p supplied to the circuit at every instant during the rise of current is given by Eq. (8).

$$p = iV = i^2R + Li\,di/dt \qquad (8)$$

The first term i^2R is the power that goes into heating the circuit. The second term $Li\,di/dt$ is the power that goes into building up the magnetic field in the inductor. The total energy W used in building up the magnetic field is given by Eq. (9). The

$$W = \int_0^t p\,dt = \int_0^t Li\frac{di}{dt}\,dt = \int_0^I Li\,di = \tfrac{1}{2}LI^2 \qquad (9)$$

energy used in building up the magnetic field remains as energy of the magnetic field. When the switch is opened, the magnetic field collapses and the energy of the field is returned to the circuit, resulting in an induced emf. The arc that is often seen when a switch is opened is a result of this emf, and the energy to maintain the arc is supplied by the decreasing magnetic field.

Mutual inductance. The mutual inductance M of two neighboring circuits A and B is defined as the ratio of the emf induced in one circuit $\mathscr{E}$ to the rate of change of current in the other circuit, as in Eq. (10).

$$M = -\frac{\mathscr{E}_B}{(dI/dt)_A} \qquad (10)$$

The mks unit of mutual inductance is the henry, the same as the unit of self-inductance. The same value is obtained for a pair of coils, regardless of which coil is taken as the starting point.

The mutual inductance of two circuits may also be expressed as the ratio of the flux linkages produced in circuit B by the current in circuit A to the current in circuit A. If Φ_{BA} is the flux threading B as a result of the current in circuit A, Eqs. (11) hold. Integration leads to Eq. (12).

$$\mathscr{E}_B = -N_B\frac{d\Phi_{BA}}{dt} = -M\frac{dI_A}{dt}$$

or
$$N_B\,d\Phi_{BA} = M\,dI_A \qquad (11)$$

$$M = \frac{N_B\Phi_{BA}}{I_A} \qquad (12)$$

See INDUCTANCE MEASUREMENT.

[KENNETH V. MANNING]

Inductance bridge

A device for comparing inductances. The inductance bridge is a special case of an alternating-current impedance bridge. Just as the Wheatstone bridge is used to compare resistances, the impedance bridge is used to compare impedances which may contain inductance, capacitance, and resistance.

General impedance bridge. A general impedance bridge is shown in Fig. 1. Four impedances Z_a, Z_b, Z_c, and Z_d are connected into a square array. A source of ac voltage v is applied across one diagonal of the square, and a null detector D is connected across the other diagonal. The bridge is in balance when the voltage across D is zero. At balance the ac current through D is also zero, which means that Eqs. (1) hold. For the voltage

$$I_a = I_b \qquad I_c = I_d \qquad (1)$$

across D to be zero, the instantaneous voltage drop across Z_b must equal that across Z_d; that is, the two instantaneous sine-wave voltages must have equal amplitudes and be in phase with each other.

Equating the magnitudes, $Z_a I_a = Z_c I_c$, and by similar reasoning, $Z_b I_a = Z_d I_c$. Dividing the first of these equations by the second gives Eq. (2), one of the two equations of balance of the bridge.

$$Z_a Z_b = Z_c / Z_d \qquad (2)$$

The voltage across Z_a leads I_a by the power-factor angle ϕ_a, and the voltage across Z_c leads I_c by the power-factor angle ϕ_c. If I_a leads I_c by the angle ϕ, then $\phi + \phi_a$ must equal ϕ_c if the two voltage drops are to be in phase. Similarly, $\phi + \phi_b = \phi_d$. By eliminating ϕ from these two equations the second equation of balance for the impedance bridge, Eq. (3), is obtained.

$$\phi_a + \phi_d = \phi_b + \phi_c \qquad (3)$$

Several important properties can be recognized by considering the second equation of balance. If Z_a and Z_b are resistors, with ϕ_a and ϕ_b both equal to zero, then for balance $\phi_d = \phi_c$. This means that

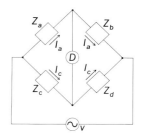

Fig. 1. General impedance bridge.

Z_c and Z_d must both be inductive or both capacitive for balance. If ϕ_a and ϕ_d are both equal to zero, the second equation for balance becomes $\phi_b + \phi_c = 0$. This means that Z_b is inductive and Z_c is capacitive or vice versa.

General inductance bridge. The inductance bridge of Fig. 2 has resistors R_a and R_b as ratio arms and compares an unknown Z_c to a standard consisting of R_d and L_d. If the standard L_d is variable, it and R_d are varied to reduce the detector voltage to zero. This balances the bridge, and the equations of balance become Eq. (4).

$$L_c / L_d = R_c / R_d = R_a / R_b \qquad (4)$$

Sometimes a substitution method is preferred. In this case the balance is obtained, as above, with any good quality inductance for L_d. The unknown is then replaced by a standard L_s in series with R_s. The bridge is balanced a second time by varying L_s and R_s. When balance is obtained, L_s equals L_c and R_s equals R_d.

If the standard is not adjustable, it becomes necessary to vary one of the ratio arms R_a or R_b or both, as well as R_d, in order to obtain balance. Obviously the substitution method cannot be applied where only fixed standards are available.

Theoretically the condition for balance is independent of frequency. In practice, the capacitance between turns and between layers of wire in the two coils will be different, and at high frequencies the bridge can be balanced at only one frequency at a time. Under such conditions, harmonics in the supply voltage will make absolute balance impossible. When harmonics are present, the bridge should be balanced by reducing the fundamental component in the detector voltage to zero. This condition does not necessarily correspond to minimum detector voltage. A technician can sometimes achieve a balance by ear at audio frequencies. An oscilloscope provides a better means of seeing when the fundamental is reduced to zero.

Anderson bridge. A suitable variable capacitor may often be available when an appropriate inductometer is not. The Anderson bridge, shown in Fig. 3, can be used to measure the inductance L_b in terms of a standard capacitance C. The equations of balance of this bridge are Eqs. (5) and (6).

$$L_b = \left(R_d + r + \frac{r R_d}{R_c} \right) R_a C \qquad (5)$$

$$R_b + r_b = \frac{R_a R_d}{R_c} \qquad (6)$$

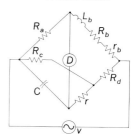

Fig. 3. Anderson bridge.

The bridge is usually balanced by varying r_b and C. The second equation indicates that for some choice of R_a, R_d, and R_c a negative r_b might be required. Consequently, if a balance does not seem possible by varying C and r_b, R_a or R_d should be increased, or R_c should be reduced. This will increase $R_a R_d / R_c$, which is needed because a balance cannot be obtained unless this quantity is at least as large as R_b.

Carey-Foster bridge. This useful bridge for determining mutual inductance is shown in Fig. 4. It is theoretically independent of frequency. When the bridge is balanced, Eqs. (7) and (8) hold.

$$M = R_a (R_M + R) C \qquad (7)$$

$$L = (R_c + R_a)(R_M + R) C \qquad (8)$$

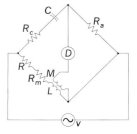

Fig. 4. Carey-Foster bridge.

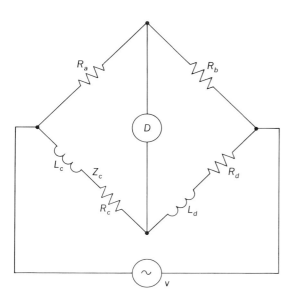

Fig. 2. General inductance bridge.

If these equations are to be used directly, the resistance R_M of the coil must be measured by some other circuit arrangement. If a known large resistance R is connected in series with the coil, it may be possible to neglect R_M in comparison with R, and the equations become Eqs. (9) and (10).

$$M = R_a RC \qquad (9)$$

$$L = (R_c + R_a)RC \qquad (10)$$

Resistances R_a and R_c are varied to obtain balance. An obvious advantage of this bridge is that all balancing operations can be performed by varying resistances. The capacitance C may be a constant standard.

Any discussion of impedance-bridge operation should include mention of Wagner ground precautions. Each portion of a bridge has a capacitance to ground as well as to all other portions. In high-impedance circuits and at high frequencies, the effects of these capacitances are not negligible, and may make balance impossible if the detector is not at ground potential. To produce this ground, a divider, which may include reactive elements, can be placed across the source with a point near its middle connected to ground. The divider is adjusted until the detector is brought to ground potential at the same time the bridge is balanced. This method of obtaining a ground at the detector is called the Wagner ground connection. *See* HIGH-FREQUENCY IMPEDANCE MEASUREMENTS; INDUCTANCE MEASUREMENT.

[HARRY SOHON/EDWARD C. STEVENSON]
Bibliography: M. Braccio, *Basic Electrical and Electronic Tests and Measurement*, 1977; W. D. Cooper, *Electronic Instrumentation and Measurement Techniques*, 2d ed., 1978; P. Kantrowitz et al., *Electronic Measurements*, 1979; M. B. Stout, *Basic Electrical Measurements*, 2d ed., 1960.

Inductance measurement

The determination of an electromagnetic parameter of an electric circuit. The electric current in a circuit produces a magnetic field which is considered to consist of lines of magnetic flux that link the circuit. Whenever the magnetic field linking a circuit changes, a voltage is induced in the circuit. The faster the change in the field, the larger is the induced voltage. When there is no ferromagnetic material present, the magnetic field is proportional to the current i, and the induced voltage v is proportional to the rate of change of current, as in Eq. (1). The proportionality factor L is, by definition,

$$v = L\, di/dt \qquad (1)$$

the self-inductance of the circuit. If v is measured in volts and if the rate of change of current is in amperes per second, the inductance is in henries.

The direction of the induced voltage is specified by Lenz's law, which states that the current that would be produced by the induced voltage would be in a direction that opposes the change in the magnetic field. This is a consequence of the fact that energy is stored in a magnetic field and, as it collapses, it feeds out power. To do this, the induced voltage must be in the direction of the current. Conversely, as a current builds up creating a magnetic field, the induced voltage must be oppos-

ite in direction to the current so that power will be delivered to the field. *See* INDUCTANCE.

If two circuits are so close together that the magnetic field of one will link the other, a changing current in one will cause a voltage proportional to the rate of change of this current to be induced in the other. The proportionality factor in this case is called the mutual inductance, and is the same whichever circuit has the changing current. *See* COUPLED CIRCUITS.

If ferromagnetic material is present, saturation and hysteresis effects may be evident for some values of current. For such cases several definitions may be introduced, or the concept may be discarded entirely. An effective inductance is defined as the ratio of the effective induced voltage to the effective rate of change of current, and is a function of a maximum current. When the current changes about a certain average value and the magnitude of the change is small compared to the average, this definition specifies an incremental inductance. This inductance is a function of the average current and the magnitude of the change.

Inductance standards. Coils constructed so that their dimensions and consequently their inductance remain constant over long periods of time are used as inductance standards. If the dimensions are known, the inductance can be computed from Eq. (2), where N is the number of turns

$$L = N^2 P \qquad (2)$$

of wire comprising the inductor and P is a proportionality coefficient. Such an inductance is called a primary standard. Standards are usually maintained at constant temperature to keep the coil size and therefore the inductance from changing in value.

When the inductance cannot be computed precisely, it may be measured by comparing it with a primary standard, and it would then be called a secondary standard.

The effect of inductance is manifested only when the current in a circuit is changing with time. Any means of measuring inductance must employ changing current. Usually the current is an alternating current of frequency f, given by Eq. (3), with a maximum instantaneous value of I_m.

$$i = I_m \sin 2\pi f t \qquad (3)$$

The voltage induced by the changing magnetic field is given by Eq. (4), where L is the self-inductance in henries.

$$v_x = 2\pi f L I_m \cos 2\pi f t = V_x \cos 2\pi f t \qquad (4)$$

The ratio of V_x to I_m is called the reactance of the circuit and is represented by X in Eq. (5).

$$X = 2\pi f L = V_x / I_m \qquad (5)$$

Impedance measurement. This is the determination of the total effect of a circuit element. The wire of which an inductor is wound has resistance, and there will be capacitance between turns. Thus every inductor exhibits resistance and capacitance as well as inductance. Also, every resistor exhibits some inductance and some capacitance, and every capacitor has some inductance and some resistance. Even if the capacitance of an inductor can be neglected, a measurement to determine the inductance is usually an impedance measurement giving the resistance as well as the inductance. *See* INDUCTANCE BRIDGE.

If the circuit element has only inductance L with resistance R in series, the total voltage drop across the element is given by Eq. (6), where $Z = \sqrt{R^2 + X^2}$

$$v = RI_m \sin 2\pi ft + XI_m \cos 2\pi ft$$

$$= ZI_m \sin (2\pi ft + \phi) \qquad (6)$$

is the impedance of the circuit and ϕ is the phase angle or the power-factor angle of the circuit given by the relation $\tan \phi = X/R$.

It is impossible to construct an inductor with an ohmic resistance of zero; however, by careful design it is possible to minimize the resistance for a given amount of inductance. The factor of merit of any inductor is called the Q of the inductor and is defined as the ratio of the reactance to the effective resistance at any specified operating frequency, as in Eq. (7). The effective resistance R_{eff}

$$Q = \frac{2\pi fL}{R_{eff}} \qquad (7)$$

represents the combined ac resistance of the wire plus eddy-current losses in the wire at the specified operating frequency. The Q of a well-designed radio-frequency inductor will be on the order of several hundred. *See* Q METER.

If the Q of an inductor is high enough that the effective resistance may be neglected, then Eq. (8) holds.

$$L = \frac{V_x}{2\pi fI_m} \qquad (8)$$

Measurement of distributed inductance of cables, lines, and waveguides is accomplished indirectly. For very low frequencies the cable currents are distributed uniformly throughout the conductors and the inductance per unit length is a maximum. At higher frequencies the skin effect causes the current to concentrate along the conductor surfaces. This concentration does not affect the magnetic field between the conductors, but it does tend to eliminate the field within each conductor itself. This reduction in magnetic field brings about a reduction in inductance. Consequently, as the frequency is increased the inductance decreases, approaching a fixed value.

The capacitance per unit length of a cable or open-wire transmission line is independent of frequency and can be measured at low frequencies. If direct measurements of attenuation and velocity of propagation are made, the inductance per unit length of transmission line may be computed at any frequency of operation. *See* ELECTRICAL MEASUREMENTS.

[HARRY SOHON/EDWARD C. STEVENSON]

Bibliography: M. Braccio, *Basic Electrical and Electronic Tests and Measurement*, 1979; W. D. Cooper, *Electronic Instrumentation and Measurement Techniques*, 2d ed., 1978; P. Kantrowitz et al., *Electronic Measurements*, 1979; M. B. Stout, *Basic Electrical Measurements*, 2d ed., 1960.

Inductor

A device for introducing inductance into a circuit. The term covers devices with a wide range of uses, sizes, and types, including components for electric-wave filters, tuned circuits, electrical measuring circuits, and energy storage devices.

Inductors are classified as fixed, adjustable, and variable. All are made either with or without magnetic cores. Inductors without magnetic cores are called air-core coils, although the actual core material may be a ceramic, a plastic, or some other nonmagnetic material. Inductors with magnetic cores are called iron-core coils. A wide variety of magnetic materials are used, and some of these contain very little iron. Magnetic cores for inductors for low-frequency, or high-energy storage, use are most commonly made from laminations of silicon steel. Some iron-core inductors with cores of compressed powdered iron, powdered permalloy, or ferrite are more suitable for higher-frequency applications.

Fixed inductors. In fixed inductors coils are wound so that the turns remain fixed in position with respect to each other. If an iron core is used, any air gap it has is also fixed and the position of the core remains unchanged within the coil.

A toroidal coil is a fixed inductor wound uniformly around a toroidal form (see illustration). Because of the closed magnetic circuit, such an inductor has practically no leakage flux and is little affected by the presence of stray magnetic fields. High-accuracy standard inductors are commonly made in this form. Powdered cores are used to increase the Q of the coil and reduce the size required for a specified inductance. Ceramic-core toroidal coils supported in cork are used as standard inductors of high stability and accuracy.

Adjustable inductors. These either have taps for changing the number of turns desired, or consist of several fixed inductors which may be switched into various series or parallel combinations.

Variable inductors. Such inductors are constructed so that the effective inductance can be changed. Means for doing this include (1) changing the permeability of a magnetic core; (2) moving the magnetic core, or part of it, with respect to the coil or the remainder of the core; and (3) moving one or more coils of the inductor with respect to one or more of the other coils, thereby changing mutual inductance. *See* ALTERNATING-CURRENT CIRCUIT THEORY; INDUCTANCE.

[BURTIS L. ROBERTSON; WILSON S. PRITCHETT]

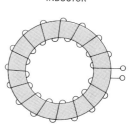

INDUCTOR

Toroidal coil.

Information theory

A branch of communication theory devoted to problems in coding. A unique feature of information theory is its use of a numerical measure of the amount of information gained when the contents of a message are learned. Information theory relies heavily on the mathematical science of probability. For this reason the term information theory is often applied loosely to other probabilistic studies in communication theory, such as signal detection, random noise, and prediction. *See* ELECTRICAL COMMUNICATIONS.

Information theory provides criteria for comparing different communication systems. The need for comparisons became evident during the 1940s. A large variety of systems had been invented to make better use of existing wires and radio spectrum allocations. In the 1920s the problem of comparing telegraph systems attracted H. Nyquist and R. V. L. Hartley, who provided some of the philosophy behind information theory. In 1948 C. E. Shannon published a precise general theory which formed the basis for subsequent work in information theory. *See* MODULATION.

In information theory, communication systems are compared on the basis of signaling rate. Finding an appropriate definition of signaling rate was itself a problem, which Shannon solved by the use of his measure of information, to be explained later. Of special interest are optimal systems which, for a given set of communication facilities, attain a maximum signaling rate. Optimal systems provide communication-systems designers with useful absolute bounds on obtainable signaling rates. Although optimal systems often use complicated and expensive encoding equipment, they provide insight into the design of fast practical systems.

Communication systems. In designing a one-way communication system from the standpoint of information theory, three parts are considered beyond the control of the system designer: (1) the source, which generates messages at the transmitting end of the system, (2) the destination, which ultimately receives the messages, and (3) the channel, consisting of a transmission medium or device for conveying signals from the source to the destination. Constraints beyond the mere physical properties of the transmission medium influence the designer. For example, in designing a radio system only a given portion of the radio-frequency spectrum may be available to him. His transmitter power may also be limited. If the system is just one link in a larger system which plans to use regenerative repeaters, the designer may be restricted to pulse-transmission schemes. All such conditions are considered part of the description of the channel. The source does not usually produce messages in a form acceptable as input by the channel. The transmitting end of the system contains another device, called an encoder, which prepares the source's messages for input to the channel. Similarly the receiving end of the system will contain a decoder to convert the output of the channel into a form recognizable by the destination. The encoder and decoder are the parts to be designed. In radio systems this design is essentially the choice of a modulator and a detector.

Discrete and continuous cases. A source is called discrete if its messages are sequences of elements (letters) taken from an enumerable set of possibilities (alphabet). Thus sources producing integer data or written English are discrete. Sources which are not discrete are called continuous, for example, speech and music sources. Likewise, channels are classified as discrete or continuous according to the kinds of signals they transmit. Most transmission media (such as transmission lines and radio paths) can provide continuous channels; however, constraints (such as a restriction to use pulse techniques) on the use of these media may convert them into discrete channels.

The treatment of continuous cases is sometimes simplified by noting that a signal of finite bandwidth can be encoded into a discrete sequence of numbers. If the power spectrum of a signal $s(t)$ is confined to the band O to W hertz (cycles per second) then Eq. (1) applies. Equation (1) reconstructs

$$s(t) = \sum_{n=-\infty}^{\infty} s\left(\frac{n}{2W}\right) \frac{\sin 2\pi W\left(t - \frac{n}{2W}\right)}{2\pi W\left(t - \frac{n}{2W}\right)} \quad (1)$$

$s(t)$ exactly from its sample values (Nyquist samples), at discrete instants $(2W)^{-1}$ sec apart. Thus, a continuous channel which transmits such signals resembles a discrete channel which transmits Nyquist samples drawn from a large finite set of signal levels and at the rate of $2W$ samples per second.

Noiseless and noisy cases. The output of a channel need not agree with its input. For example, a channel might, for secrecy purposes, contain a cryptographic device to scramble the message. Still, if the output of the channel can be computed knowing just the input message, then the channel is called noiseless. If, however, random agents make the output unpredictable even when the input is known, then the channel is called noisy. *See* PRIVACY SYSTEMS (SCRAMBLING).

Encoding and decoding. Many encoders first break the message into a sequence of elementary blocks; next they substitute for each block a representative code, or signal, suitable for input to the channel. Such encoders are called block encoders. For example, telegraph and teletype systems both use block encoders in which the blocks are individual letters. Entire words form the blocks of some commercial cablegram systems. The operation of a block encoder may be described completely by a function or table showing, for each possible block, the code that represents it.

It is generally impossible for a decoder to reconstruct with certainty a message received via a noisy channel. Suitable encoding, however, may make the noise tolerable. For illustration, consider a channel that transmits pulses of two kinds. It is customary to let binary digits 0 and 1 denote the two kinds of pulse. Suppose the source has only the four letters A, B, C, D. One might simply encode each single-letter block into a pair of binary digits (code I of table). In that case the decoder would make a mistake every time noise produced an error. If code II is used, the decoder can at least recognize that a received triple of digits must contain errors if it is one of the triples 001, 010, 100, or 111 not listed in the code II column. Because an error in any one of the three pulses of code II always produces a triple that is not listed, code II provides single-error detection. Similarly code III provides double-error detection, because errors in a single pulse or pair of pulses always produce a quintuple that is not listed.

Three possible binary codes for four-letter alphabet

Letter	Code I	Code II	Code III
A	0 0	0 0 0	0 0 0 0 0
B	0 1	0 1 1	0 0 1 1 1
C	1 0	1 0 1	1 1 0 0 1
D	1 1	1 1 0	1 1 1 1 0

As an alternative, code III may provide single-error correction, an idea due to R. W. Hamming. In this usage, the decoder picks a letter for which code III agrees with the received quintuple in as many places as possible. If only a single digit is in error, this rule chooses the correct letter.

Even when the channel is noiseless, a variety of encoding schemes exists and there is a problem of picking a good one. Of all encodings of English letters into dots and dashes, the Continental Morse encoding is nearly the fastest possible one. It achieves its speed by associating short codes with the most common letters. A noiseless binary channel (capable of transmitting two kinds of pulse 0, 1, of the same duration) provides the following example. Suppose one had to encode English text for this channel. A simple encoding might just use 27 different five-digit codes to represent word space (denoted by #), A, B, . . . , Z; say # 00000, A 00001, B 00010, C 00011, . . . , Z 11011. The word #CAB would then be encoded into 00000000110000100010. A similar encoding is used in teletype transmission; however, it places a third kind of pulse at the beginning of each code to help the decoder stay in synchronism with the encoder. The five-digit encoding can be improved by assigning four-digit codes 0000, 0001, 0010, 0011, 0100 to the five most common letters #, E, T, A, O. There are 22 quintuples of binary digits which do not begin with any of the five four-digit codes; these may be assigned as codes to the 22 remaining letters. About half the letters of English text are #, E, T, A, or O; therefore the new encoding uses an average of only 4.5 digits per letter of message. *See* TELETYPEWRITER.

More generally, if an alphabet is encoded in single-letter blocks, using $L(i)$ digits for the ith letter, the average number of digits used per letter is shown in Eq. (2), where $p(i)$ is the probability of the ith letter. An optimal encoding scheme will minimize L. However, the encoded messages must be decipherable, and this condition puts constraints on the $L(i)$. B. McMillan has shown that the code lengths of decipherable encodings must satisfy the relationships shown in inequality (3). The real

$$L = p(1)L(1) + p(2)L(2) \\ + p(3)L(3) + \cdots \quad (2)$$

$$2^{-L(1)} + 2^{-L(2)} + 2^{-L(3)} + \cdots \leq 1 \quad (3)$$

numbers $L(1), L(2), \ldots$, which minimize L subject to inequality (3) are $L(i) = -\log_2 p(i)$ and the corresponding minimum L is shown in Eq. (4),

$$H = -\sum_i p(i) \log_2 p(i) \quad (4)$$

which provides a value of H equal to a number of digits per letter.

The $L(i)$ must be integers and $-\log_2 p(i)$ generally are not integers; for this reason there may be no encoding which provides $L = H$. However, Shannon showed that it is always possible to assign codes to letters in such a way that $L \leq H + 1$. A procedure for constructing an encoding which actually minimizes L has been given by D. A. Huffman. For (27-letter) English text $H = 4.08$ digits per letter, as compared with the actual minimum 4.12 digits per letter obtained by Huffman's procedure.

By encoding in blocks of more than one letter, the average number of digits used per letter may be reduced further. If messages are constructed by picking letters independently with the probabilities $p(1), p(2), \ldots$, then H is found to be the minimum of the average numbers of digits per letter used to encode these messages using longer blocks.

Information content of message. The information contained in a message unit is defined in terms of the average number of digits required to encode it. Accordingly the information associated with a single letter produced by a discrete source is defined to be the number H. Some other properties of H help to justify using it to measure information. If one of the $p(i)$ equals unity, only one letter appears in the messages. Then nothing new is learned by seeing a letter and, indeed, $H = 0$. Second, of all possible ways of assigning probabilities $p(i)$ to an N-letter alphabet, the one which maximizes H is $p(1) = p(2) = \cdots = 1/N$. This situation is the one in which the unknown letter seems most uncertain; therefore it does seem correct that learning such a letter provides the most information. The corresponding maximum value of H is $\log_2 N$. This result seems reasonable by the following argument. When two independent letters are learned, the information obtained should be $2H = 2 \log_2 N$. However, such pairs of letters may be considered to be the letters of a larger alphabet of N^2 equally likely pairs. The information associated with one of these new letters is $\log_2 N^2 = 2 \log_2 N$. Although H given by Eq. (4) is dimensionless, it is given units called bits (a contraction of binary digits). Occasionally the information is expressed in digits of other kinds (such as ternary or decimal). Then bases other than 2 are used for the logarithm in Eq. (4).

The majority of message sources do not merely pick successive letters independently. For example in English, H is the most likely letter to follow T but is otherwise not common. The source is imagined to be a random process in which the letter probabilities change, depending on what the past of the message has been. Statistical correlations between different parts of the message may be exploited by encoding longer blocks. The average number of digits per letter may thereby be reduced below the single-letter information H given by Eq. (4). For example, by encoding English words instead of single letters, 2.1 digits/letter suffice. Encoding longer and longer blocks, the number of digits needed per letter approaches a limiting minimum value. This limit is called the entropy of the source and is interpreted as the rate, in bits per letter, at which the source generates information. If the source produces letters at some fixed average rate, n letters/sec, the entropy may also be converted into a rate in bits per second by multiplying by n. The entropy may be computed from tables giving the probabilities of blocks of N letters (N-grams). If in Eq. (4) the summation index i is extended over all N-grams, then the number H represents the information in N consecutive letters. As $N \to \infty$, H/N approaches the entropy of the source. The entropy of English has been estimated by Shannon to be about 1 bit/letter. However, an encoder might have to encode 100-grams in order to achieve a reduction to near 1 digit/letter. Comparing English with a source which produces 27 equally likely letters independently (and hence has entropy $\log_2 27 = 4.8$ bits/letter), this result is often restated: English is 80% redundant. Other common sources are also very redundant. Facsimile, for example, can be speeded by a factor of 10 by means of practical encoding techniques.

Capacity. The notion of entropy is more widely applicable than might appear from the discussion of the binary channel. Any discrete noiseless channel may be given a number C, which is called the capacity. C is defined as the maximum rate (bits per second) of all sources that may be connected directly to the channel. Shannon proved that any given source (which perhaps cannot be connected directly to the channel) of entropy H bits/letter, can be encoded for the channel and run at rates arbitrarily close to C/H letters/sec.

By using repetition, error-correcting codes, or similar techniques, the reliability of transmission over a noisy channel can be increased at the expense of slowing down the source. It might be expected that the source rate must be slowed to 0 bits/sec as the transmission is required to be increasingly error-free. On the contrary, Shannon proved that even a noisy channel has a capacity C. Suppose that errors in at most a fraction ϵ of the letters of the message can be tolerated ($\epsilon > 0$). Suppose also that a given source, of entropy H bits/letter, must be operated at the rate of at least $(C/H) - \delta$ letters/sec ($\delta > 0$). No matter how small ϵ and δ are chosen, an encoder can be found which satisfies these requirements.

For example, the symmetric binary channel has binary input and output letters; noise changes a fraction p of the 0s to 1 and a fraction p of the 1s to 0 and treats successive digits independently. The capacity of this channel is shown by Eq. (5), where

$$C = m\{1 + p \log_2 p + (1-p) \log_2 (1-p)\} \quad (5)$$

m is the number of digits per second which the channel transmits.

A famous formula is shown by Eq. (6), which

$$C = W \log_2 \left(1 + \frac{S}{N}\right) \quad (6)$$

gives the capacity C of a band-limited continuous channel. The channel consists of a frequency band W Hz wide, which contains a Gaussian noise of power N. The noise has a flat spectrum over the band and is added to the signal by the channel. The channel also contains a restriction that the average signal power may not exceed S.

Equation (6) illustrates an exchange relationship between bandwidth W and signal-to-noise ratio S/N. By suitable encoding, a signaling system can use a smaller bandwidth, provided that the signal power is also raised enough to keep C fixed. See BANDWIDTH REQUIREMENTS (COMMUNICATIONS).

Typical capacity values are 20,000 bits/sec for a telephone speech circuit and 50,000,000 bits/sec for a broadcast television circuit. Speech and TV are very redundant and would use channels of much lower capacity if the necessary encodings were inexpensive. For example, the vocoder can send speech, only slightly distorted, over a 2000-bits/sec channel. Successive lines or frames in television tend to look alike. This resemblance sug-

gests a high redundancy; however, to exploit it the encoder may have to encode in very long blocks.

Not all of the waste in channel capacity can be attributed to source redundancies. Even with an irredundant source, such as a source producing random digits, some channel capacity will be wasted. The simplest encoding schemes provide reliable transmission only at a rate equal to the capacity of a channel with roughly 8 dB smaller signal power (the 8-dB figure is merely typical and really depends on the reliability requirements). Again, more efficient encoding to combat noise generally requires larger-sized blocks. This is to be expected. The signal is separated from the noise on the basis of differences between the signal's statistical properties and those of noise. The block size must be large enough to supply the decoder with enough data to draw statistically significant conclusions. See ELECTRICAL NOISE.

Algebraic codes. Practical codes must use simple encoding and decoding equipment. Error-correcting codes for binary channels have been designed to use small digital logic circuits. These are called algebraic codes, linear codes, or group codes because they are constructed by algebraic techniques involving linear vector spaces or groups.

For example, each of the binary codes I, II, and III in the table contains four code words which may be regarded as vectors $C = (c_1, c_2, \ldots, c_n)$ of binary digits c_i. Define the sum $C + C'$ of two vectors to be the vector $(c_1 + c'_1, \ldots, c_n + c'_n)$ in which coordinates of C and C' are added modulo 2. Codes I, II, and III each have the property that the vector sum of any two code words is also a code word. Because of that, these codes are linear vector spaces and groups under vector addition. Their code words also belong to the n-dimensional space consisting of all 2^n vectors of n binary coordinates. Codes II and III, with $n = 3$ and 5, do not contain all 2^n vectors; they are only two-dimensional linear subspaces of the larger space. Consequently, in Codes II and III, the coordinates c_i must satisfy certain linear homogeneous equations. Code II satisfies $c_1 + c_2 + c_3 = 0$. Code III satisfies $c_3 + c_4 = 0$, $c_2 + c_3 + c_5 = 0$, $c_1 + c_2 = 0$, and other equations linearly dependent on these three. The sums in such equations are performed modulo 2; for this reason the equations are called parity check equations. In general, any r linearly independent parity check equations in $c_1, \ldots, c_n$ determine a linear subspace of dimension $k = n - r$. The 2^k vectors in this subspace are the code words of a linear code.

One may transform the r parity checks into a form which simplifies the encoding. This transformation consists of solving the original parity check equations for some r of the coordinates c_i as expressions in which only the remaining $n - r$ coordinates appear as independent variables. For example, the three parity check equations given for Code III are already in solved form with c_1, c_4, c_5 expressed in terms of c_2 and c_3. The $k = n - r$ independent variables are called message digits because the 2^k values of these coordinates may be used to represent the letters of the message alphabet. The r dependent coordinates, called check digits, are then easily computed by circuits which perform additions modulo 2.

At the receiver the decoder can also do additions modulo 2 to test if the received digits still satisfy the parity check equations. The set of parity check equations that fail is called the syndrome because it contains the data that the decoder needs to diagnose the errors. The syndrome depends only on the error locations, not on which code word was sent. In general, a code can be used to correct e errors if each pair of distinct code words differ in at least $2e + 1$ of the n coordinates. For a linear code that is equivalent to requiring the smallest number d of "ones" among the coordinates of any code word (excepting the zero word $(0,0, \ldots ,0)$) to be $2e + 1$ or more. Under these conditions each pattern of $0,1, \ldots ,e-1$, or e errors produces a distinct syndrome; then the decoder can compute the error locations from the syndrome. This computation may offer some difficulty. But at least it involves only r binary variables, representing the syndrome, instead of all n coordinates.

Hamming codes. The r parity check equations may be written concisely as binary matrix equation (7). Here C^T is a column vector, the transpose

$$HC^T = 0 \tag{7}$$

of $(c_1, \ldots ,c_n)$. H is the so-called parity check matrix, having n columns and r rows. A Hamming single-error correcting code is obtained when the columns of H are all $n = 2^r - 1$ distinct columns of r binary digits, excluding the column of all zeros. If a single error occurs, say in coordinate c_i, then the decoder uses the syndrome to identify c_i as the unique coordinate that appears in just those parity check equations that fail.

Shift register codes. A linear shift register sequence is a periodic infinite binary sequence $\ldots , c_0,c_1,c_2, \ldots$ satisfying a recurrence equation expressing c_j as a modulo 2 sum of some of the b earlier digits $c_{j-b}, \ldots ,c_{j-1}$. A recurrence with two terms would be an equation $c_j = c_{j-a} + c_{j-b}$, with a equal to some integer $1,2, \ldots$, or $b - 1$. The digits of a shift register sequence can be computed, one at a time, by very simple equipment. It consists of a feedback loop, containing a shift register to store $c_{j-b}, \ldots ,c_{j-1}$ and a logic circuit performing modulo 2 additions. This equipment may be used to implement a linear code. First, message digits $c_1, \ldots ,c_b$ are stored in the register and transmitted. Thereafter the equipment computes and transmits successively the $n - b$ check digits c_j obtained from the recurrence equation with $j = b + 1, \ldots ,n$. By choosing a suitable recurrence equation, one can make the period of the shift register sequence as large as $2^b - 1$. Then, with n equal to the period $2^b - 1$, the code consists of the zero code word $(0,0, \ldots ,0)$ and $2^b - 1$ other code words which differ from each other only by cyclic permutations of their coordinates. These latter words all contain $d = 2^{b-1}$ "ones" and so the code can correct $e = 2^{b-2} - 1$ errors. *See* SWITCHING CIRCUIT.

Intermediate codes. The Hamming codes and maximal period shift register codes are opposite extremes, correcting either one or many errors and having code words consisting either mostly of message digits or mostly of check digits. Many intermediate codes have been invented. One of them, due to R. C. Bose, D. K. Ray-Chaudhuri, and A. Hocquenghem, requires $n + 1$ to be a power of 2; say $n + 1 = 2^q$. It then uses at most qe check digits to correct e errors.

Perfect codes. Although each pattern of $0,1, \ldots ,e$ errors produces a distinct syndrome, there may be extra syndromes which occur only after more than e errors. In order to keep the number of check digits small, extra syndromes must be avoided. A code is called perfect if all 2^r syndromes can result from patterns of $0,1, \ldots ,e-1$, or e errors. Hamming codes are all perfect. M. J. E. Golay found another perfect binary code having $n = 23$, $r = 11$ check digits, and correcting $e = 13$ errors.

Orthogonal parity codes. Orthogonal parity codes are codes with especially simple decoding circuits which take a kind of majority vote. Suppose the parity check equations can be used to derive $2e + 1$ linear equations in which one digit, say c_1, is expressed with each of the remaining digits $c_2, \ldots ,c_n$ appearing in at most one equation. If at most e errors occur, then the received digits satisfy a majority of the $2e + 1$ equations if and only if c_1 was received correctly. For example, the recurrence $c_j = c_{j-2} + c_{j-3}$ generates a maximal period shift register code with $n = 7$, $r = 4$, $e = 1$. Set $j = 1,3$, and 4 in the recurrence equation. One obtains three of the parity check equations, $c_1 = c_5 + c_6$, $c_1 = c_3 + c_7$, and $c_1 = c_2 + c_4$, after using the fact that the shift register sequence has period 7. These three equations are already in the form required for decoding c_1 by majority vote. Similar equations, obtained by permuting $c_1, \ldots ,c_7$ cyclically, apply for $c_2, \ldots ,c_7$. Then the decoder can be organized so that most of the equipment used to decode c_1 can be used again in decoding $c_2, \ldots ,c_7$.

[EDGAR N. GILBERT]

Bibliography: R. B. Ash, *Information Theory*, 1965; E. R. Berlekamp, *Algebraic Coding Theory*, 1968; C. Cherry, *On Human Communications: A Review, a Survey, and a Criticism*, 3d ed., 1980; S. Guiasu, *Information Theory With New Applications*, 1977; R. W. Hamming, *Coding and Information Theory*, 1980; S. Lin, *An Introduction to Error-Correcting Codes*, 1970; M. Schwartz, *Information Transmission, Modulation, and Noise*, 2d ed., 1970; C. E. Shannon and W. Weaver, *The Mathematical Theory of Communication*, 1949.

Infrared imaging devices

Devices that convert an invisible infrared image into a visible image. Infrared radiation is usually considered to span the wavelengths from about 0.8 or 0.9 micrometer to several hundred micrometers; however, most infrared imaging devices are designed to operate within broad wavelength regions of atmospheric transparency, that is, the atmospheric windows. At sea level, for horizontal paths of a few kilometers' length, these are approximately at $8-14$ μm, $3-5$ μm, $2-2.5$ μm, $1.5-1.9$ μm, and wavelengths shorter than 1.4 μm. The radiation available for imaging may be emitted from objects in the scene of interest (usually at the longer wavelengths called thermal radiation) or reflected. Reflected radiation may be dominated by sunlight or may be from controlled sources such as lasers used specifically as illuminators for the imaging device. The latter systems are called active, while those relying largely on emitted radiation are called passive. Active optical imaging systems were developed to achieve a nighttime aerial photographic capability, and work during World War II pushed such systems into the near-infrared spectral region. Development of passive infrared imaging systems came after the war, but only the advent of lasers allowed creation of active infrared imagers at wavelengths much longer than those of the photographic region. Striking advances have been made in active infrared systems which utilize the coherence available from lasers, and hybrid active-passive systems have been studied intensively. *See* LASER.

Although developed largely for military purposes, infrared imaging devices have been valuable in industrial, commercial, and scientific applications. These range from nondestructive testing and quality control to earth resources surveys, pollution monitoring, and energy conservation. Infrared images from aerial platforms are used to accomplish "heat surveys," locating points of excessive heat loss. An example is shown in Fig. 1a. As discussed below, calibration allows association of photographic tones in this figure with values of apparent (that is, equivalent blackbody) temperatures. Dark areas in the figure are "colder" than light ones.

Scanning systems. Infrared imaging devices may be realized by electrooptical or optomechanical scanning systems. All have an optical head for receiving the infrared radiation and a display for the visible image. These are connected by electronics for the passage of electrical signals from the detector element(s) to the display input. Signal processing may be incorporated in the electronics to selectively enhance or reduce features in the produced visible image. For example, in Fig. 1b a "level-slice" technique presents in white all areas (mainly rooftops) with apparent temperatures between −7.9 and −8.9°C. (The ambient air temperature was −5°C.) Black regions in the figure correspond to apparent temperatures below or above the narrow "sliced" temperature range of the white regions.

Optomechanical methods such as rotating prisms or oscillating mirrors may be used to sample or scan the spatial distribution of infrared radiation in either the object or image plane. Electrooptical imaging devices may use an electron beam (for example, vidicons) or charge transport in solids (for example, charge-coupled devices, or CCDs) to scan the infrared image formed by the optics of the device. This image-plane scanning places more stringent requirements upon the optics for image quality off-axis than does use of mechanically moved optical elements placed before the entrance aperture of the system. Intensive development of pyroelectric vidicons, detector arrays, and infrared charge-coupled devices (IRCCDs) has taken place, reflecting the critical role played by the detector element in all infrared systems. The spectral, spatial, and temporal responses of detectors are the major factors in determining the wavelength regions, the spatial resolution, and the frequency response (that is, the time constant) of imaging devices. *See* CHARGE-COUPLED DEVICES.

Detector arrays. Optomechanical scanning methods often stress the response time of the detector-electronics package. As a result, multiple detectors or detector arrays are sometimes incorporated in the focal planes, resulting in partially electronically scanned optomechanical systems. The technology for use of a linear array of detector elements (often lead selenide and indium antimonide detectors for the $3-5\text{-}\mu\text{m}$ region, and mercury-doped germanium and mercury cadmium telluride for the $8-14\text{-}\mu\text{m}$ window) is well developed, and the use of a two-dimensional array or matrix of detectors has been studied. Optomechanical imagers incorporating such arrays allow the use of time delay and integration (TDI) of the signals to improve the resulting signal-to-noise ratios.

Solid-state components such as charge-coupled devices afford the opportunity for implementation of signal processing directly at the focal plane. Two approaches have been undertaken to attain the focal-plane array technology of infrared charge-coupled devices. In one, the development of a hybrid device, an infrared detector matrix of any suitable photodetector material, for example, indium antimonide, mercury cadmium telluride, and lead tin telluride, is mated with a conventional silicon charge-coupled device. Thus two solid-state wafers or "chips" are integrated to obtain an infrared charge-coupled device. In the other, the goal is a monolithic chip, one incorporating the photodetection, charge generation, and charge transfer in a structure made of essentially one material. Candidate materials include impurity-doped silicon, indium antimonide, and mercury cadmium telluride. The hybrid device technology can be implemented more readily than that needed for monolithic infrared charge-coupled devices. The development of infrared charge-coupled devices with the number of detecting elements in a sufficiently closely packed array required for high-performance infrared imaging devices involves additional difficulties.

Scanning motion. Some optomechanical imagers produce a two-dimensional scan entirely by

Fig. 1. Thermal imagery in the wavelength range 10.4–12.5 μm obtained during flights over Ypsilanti, MI, at 2400 hours, Nov. 23, 1975, by the Airborne Multispectral Scanner operated by the Environmental Research Institute of Michigan. (a) Calibrated thermal imagery. (b) Signal-processed thermal imagery of same scene. (*From F. Tanis, R. Sampson, and T. Wagner, Thermal imagery for evaluation of construction and insulation conditions of small buildings, Environmental Research Institute of Michigan, ERIM Rep. 116600-12-F, July 1976*)

movement of components of the device itself; others utilize the motion of a platform such as an aircraft or satellite. The first kind of system includes the forward looking infrared (FLIR) or framing imagers which usually scan in television-like raster patterns and display, synchronously if done in real time, a visible image corresponding to the spatial distribution of infrared radiation. These visible image outputs have been named thermographs and thermograms. Commercially available imaging devices of this type have used horizontally and vertically rotating germanium prisms, mirrors oscillated in orthogonal directions, two mirrors and a six-sided rotating prism, and other schemes to produce images at relatively slow rates from 16 per second to less than a quarter of a frame per second. Higher-performance systems have been produced for military purposes. The second class of imaging systems includes those often called line scanners or thermal mappers. One such system, the 12-channel Airborne Multispectral Scanner operated by the Environmental Research Institute of Michigan (ERIM; Fig. 2), includes two thermal radiation channels, at 8.2–9.3 μm and 10.4–12.5 μm, whose magnetic-tape recorder output was processed to produce the thermal imagery in Fig. 1.

Characterization of output. The instantaneous field of view (IFOV) or resolution element of imaging systems is always geometrically filled by the radiating source, so that the output of the device is a response to changes in amount of radiation from field of view to field of view. These changes are best characterized in terms of radiance L, the radiant flux per unit area per unit solid angle, usually in a selected spectral band of wavelengths, $\Delta\lambda$. Even in the infrared regions, the radiance variation may be ascribed to changes in reflectance, incident radiation, emissivity, or temperature. By restriction to wavelengths dominated by emission, the so-called thermal wavelengths longer than 3.5 μm, the radiance change can be described by the equation below, where T is the absolute tempera-

$$\Delta L = \frac{\partial L}{\partial T}\Delta T + \frac{\partial L}{\partial \epsilon}\Delta\epsilon$$

ture and ϵ is the emissivity. Contributions due to $\Delta\epsilon$ are usually treated as changes in an equivalent blackbody temperature by setting $\epsilon = 1$ and $\Delta\epsilon = 0$. Then T represents an equivalent blackbody temperature, and the radiance variation can be ascribed entirely to a value of ΔT. That value of ΔT corresponding to a radiance difference which will just produce a signal-to-noise ratio of 1 is called the noise equivalent temperature difference (NETD). One can also characterize the performance of an imaging system by a noise equivalent emissivity difference or even a noise equivalent reflectivity difference. The use of noise equivalent temperature difference as a figure of merit for thermal imagers is obviously more appropriate. For the higher-performance forward-looking infrared systems, a useful figure of merit is the minimum resolvable temperature difference (MRTD), a figure of merit which includes the characteristics of the display and of the observer as well.

Display. The visible image which is the output of infrared imaging devices may be displayed in the same manner as a conventional television picture by means of a cathode-ray tube (CRT). Cathode-ray-tube technology has been developed to a level

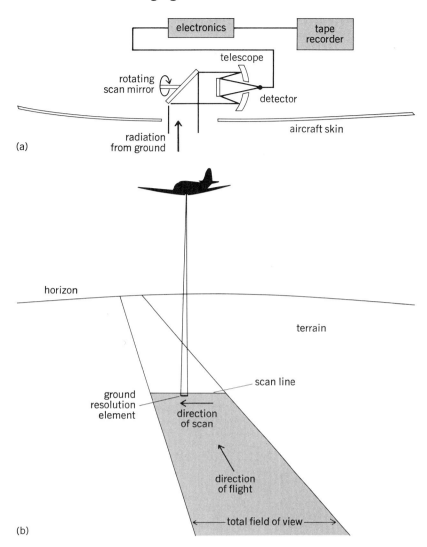

Fig. 2. Airborne Multispectral Scanner. (a) Schematic diagram of equipment. (b) Scanning operation, utilizing motion of aircraft. (*From F. Tanis, R. Sampson, and T. Wagner, Thermal imagery for evaluation of construction and insulation conditions of small buildings, Environmental Research Institute of Michigan, ERIM Rep. 116600-12-F, July 1976*)

that is suitable, and research has been undertaken toward creation of satisfactory flat panel displays using liquid crystal elements, light-emitting diodes, or plasma panels. Systems not requiring a real-time image display may utilize analog or digital data storage or transmission systems, which then are used to produce permanent visual records such as photographs. High-resolution "hard copy" images can be produced by sophisticated systems using electron-beam or laser recording on film. Complex signal-processing techniques are easily introduced before the final image recording is made. *See* CATHODE-RAY TUBE; LIGHT-EMITTING DIODE; LIQUID CRYSTALS.

[GEORGE J. ZISSIS]

Bibliography: J. M. Lloyd, *Thermal Imaging Systems*, 1975; *Proceedings of the IEEE: Special Issue on IR Technology for Remote Sensing*, vol. 63, pp. 1–208, January 1975; F. Tanis, R. Sampson, and T. Wagner, *Thermal Imagery for Evaluation of Construction and Insulation Conditions of Small*

Buildings, Environmental Research Institute of Michigan, ERIM Rep. 116600-12-F, July 1976; W. Wolfe and G. Zissis (eds.), *The Infrared Handbook*, Infrared Information and Analysis Center, Environmental Research Institute of Michigan, 1978.

Integrated circuits

Miniature electronic circuits produced within and upon a single semiconductor crystal, usually silicon. Integrated circuits range in complexity from simple logic circuits and amplifiers, about ¹⁄₂₀ in. (1.3 mm) square, to large-scale integrated circuits up to about ⅓ in. (8 mm) square. They contain tens of thousands of transistors and other components which provide computer memory circuits and complex logic subsystems such as microcomputer central processor units.

Since the mid-1960s, integrated circuits have become the primary components of most electronic systems. Their low cost, high reliability, and speed have been essential in furthering the wide use of digital computers. Microcomputers have spread the use of computer technology to instruments, business machines, automobiles, and other equipment. Other common uses of large-scale integrated circuits are in pocket calculators and electronic watches. For analog signal processing, integrated subsystems such as FM stereo demodulators and switched-capacitor filters are made. *See* CALCULATORS: DIGITAL COMPUTER.

Integrated circuits consist of the combination of active electronic devices such as transistors and diodes with passive components such as resistors and capacitors within and upon a single semiconductor crystal. The construction of these elements within the semiconductor is achieved through the introduction of electrically active impurities into well-defined regions of the semiconductor. The fabrication of integrated circuits thus involves such processes as vapor-phase deposition of semiconductors and insulators, oxidation, solid-state diffusion, ion implantation, vacuum deposition, and sputtering.

Generally, integrated circuits are not straightforward replacements of electronic circuits assembled from discrete components. They represent an extension of the technology by which silicon planar transistors are made. Because of this, transistors or modifications of transistor structures are the primary devices of integrated circuits. Methods of fabricating good-quality resistors and capacitors have been devised, but the third major type of passive component, inductors, must be simulated with complex circuitry or added to the integrated circuit as discrete components. *See* TRANSISTOR.

Simple logic circuits were the easiest to adapt to these design changes. The first of these, such as inverters and gates, were produced in the early 1960s primarily for miniaturization of missile guidance computers and other aerospace systems. Analog circuits, called linear integrated circuits, did not become commercially practical until several years later because of their heavy dependence on passive components such as resistors and capacitors. The first good-quality operational amplifiers for analog computers and instruments were produced in 1966. *See* AMPLIFIER; ANALOG COMPUTER; LOGIC CIRCUIT.

TYPES OF CIRCUITS

Integrated circuits can be classified into two groups on the basis of the type of transistors which they employ: bipolar integrated circuits, in which

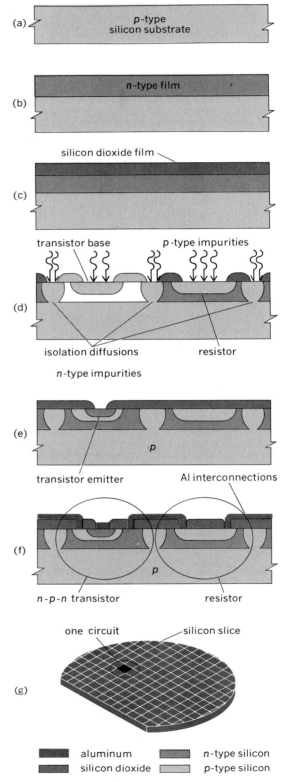

Fig. 1.   Steps *a–f* in fabrication of bipolar inverter circuit; *g* shows numerous circuits incorporated on a silicon slice.

the principal element is the bipolar junction transistor; and metal oxide semiconductor (MOS) integrated circuits, in which the principal element is the MOS transistor. Both depend upon the construction of a desired pattern of electrically active impurities within the semiconductor body, and upon the formation of an interconnection pattern of metal films on the surface of the semiconductor.

Bipolar circuits are generally used where highest logic speed is desired, and MOS for largest-scale integration or lowest power dissipation. Linear circuits are mostly bipolar, but MOS devices are used extensively in switched-capacitor filters.

Bipolar integrated circuits. The principal steps involved in the fabrication of a bipolar inverter circuit, representative of the simplest logic function, are illustrated schematically in Fig. 1. An inverter requires only a transistor and resistor, shown in cross section. Complete digital integrated circuits generally contain tens to hundreds of inverters and gates interconnected as counters, arithmetic units, and other building blocks. As indicated by Fig. 1g, hundreds of such circuits may be fabricated on a single slice of silicon crystal. This feature of planar technology—simultaneous production of many circuits—is responsible for the economic advantages and wide use of integrated circuits.

The starting material is a slice of single-crystal silicon, more or less circular, up to 5 in. (125 mm) in diameter, and a fraction of an inch (1 in. = 25 mm) thick. Typically, this material is doped with p-type impurities. A film of semiconductor, less than 1/1000 in. (25 μm) thick, is then grown upon this substrate in a vapor-phase reaction of a silicon-containing compound. The conditions of this reaction are such that the film maintains the single-crystal nature of the substrate. Such films are called epitaxial (Greek for "arranged upon"). By incorporating n-type impurities into the gas from which the film is grown, the resulting epitaxial film is made n-type. See SEMICONDUCTOR.

Next, the silicon slice is placed into an oxygen atmosphere at high temperatures (1200°C). The silicon and oxygen react, forming a cohesive silicon dioxide film upon the surface of the slice that is relatively impervious to the electrically active impurities.

To form the particular semiconductor regions required in the fabrication of electronic devices, however, p- and n-type impurities must be introduced into certain regions of the semiconductor. In the planar technology, this is done by opening windows in the protective oxide layer by photoengraving techniques, and then exposing the slice to a gas containing the appropriate doping impurity. In the case of an integrated circuit, the isolation regions—p-type regions which, together with the p-type substrate, surround the separate pockets of the n-type film—are formed first by the diffusion of a p-type impurity. This is followed by a shorter exposure to p-type impurities during which the base region of the transistors and the resistors are formed. See JUNCTION TRANSISTOR.

Next, the slice is again covered with oxide, smaller windows are cut over the transistor base regions, and n-type impurities are permitted to diffuse in these regions to form the emitters of the transistors. As the lateral and vertical dimensions

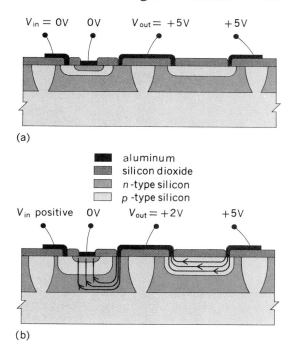

Fig. 2. Operation of bipolar inverter circuit. (a) Input voltage V_{in} is zero. (b) Positive input voltage applied. Arrows indicate direction of current flow.

of these devices become smaller, control of the number and position of these impurities becomes more important. This has resulted in the increased use of ion implantation to introduce these impurities into the silicon. During ion implantation, high-energy (20–200 keV) impurity ions bombard the silicon surface and penetrate the solid.

Openings are cut in the oxide layer at all places where contact to the silicon is desired. Then a metal, aluminum for example, is deposited over the entire slice by vacuum evaporation or sputtering, and finally the undesired aluminum is removed by photoengraving, leaving behind aluminum stripes which interconnect the transistor and the resistor, as indicated in the lowest part of Fig. 1. See ION IMPLANTATION.

The preparation of the thousands of inverter circuits on the silicon slice is now complete. The slice

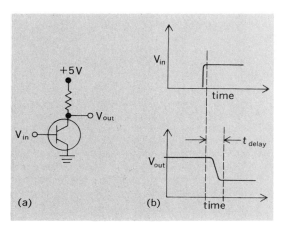

Fig. 3. Characteristics of the inverter circuit of Fig. 2. (a) Circuit symbol. (b) Switching waveforms.

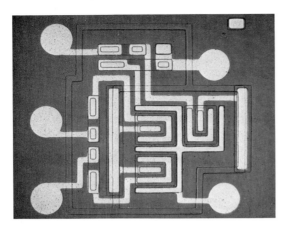

Fig. 4. Photomicrograph of early bipolar logic gate circuit. (*Fairchild Semiconductor*)

The way an integrated inverter circuit operates is illustrated in Fig. 2. The input voltage V_{in} is applied to the base of the transistor. When V_{in} is zero or negative with respect to the emitter, no current flows. As a result, no voltage drop exists across the resistor, and the output voltage V_{out} will be the same as the externally applied biasing voltage, $+5$ volts in this example. When a positive input voltage is applied, the transistor becomes conducting. Current now flows through the transistor, hence through the resistor: as a result, the output voltage becomes more negative. Thus, the change in input voltage appears inverted at the output.

The circuit symbol and the changes in input and output voltages during the switching process just described are illustrated in Fig. 3. Note that the change in the output voltage occurs slightly later than the change in the input voltage. This time difference, called propagation delay, is an important characteristic of all integrated circuits. Much effort has been spent on reducing it, and values less than one-billionth of a second have been achieved.

Most simple digital circuits can be fabricated much as the inverter circuit described above. As an example, a photomicrograph of an early logic gate circuit is shown in Fig. 4. This circuit is one of the earliest digital integrated circuits, introduced commercially in 1961. For comparison, a digital integrated circuit introduced in the early 1970s is shown in Fig. 5. This circuit, although hardly larger than the former, contains more than 100 times as much circuitry. Further increases in density continue to occur as shown in Fig. 6 by

is cut apart, much as a pane of glass is cut, and the individual circuits are tested and packaged.

Both the transistor and the resistor are formed within a separate pocket of *n*-type semiconductor surrounded on all sides by *p*-type regions. When the inverter circuit is operated, a reverse bias develops between the *n*-type pocket and its surroundings. The depletion region separating the *n*- and *p*-type regions has a very high resistance; consequently, the individual transistors and resistors are electrically isolated from each other, even though both of them have been formed within the same semiconductor crystal.

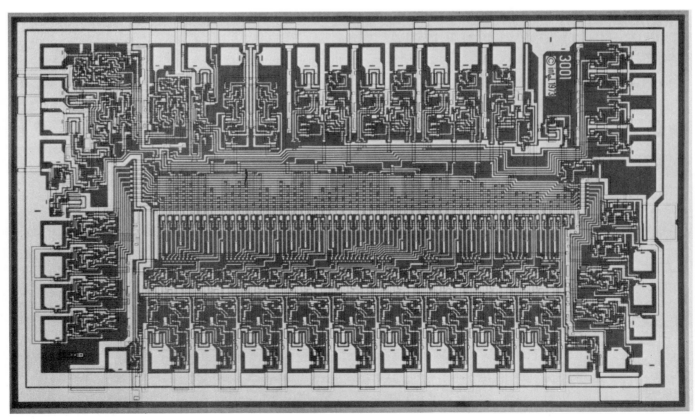

Fig. 5. Photomicrograph of bipolar logic circuit forming section of microcomputer central processor unit. (*Intel*)

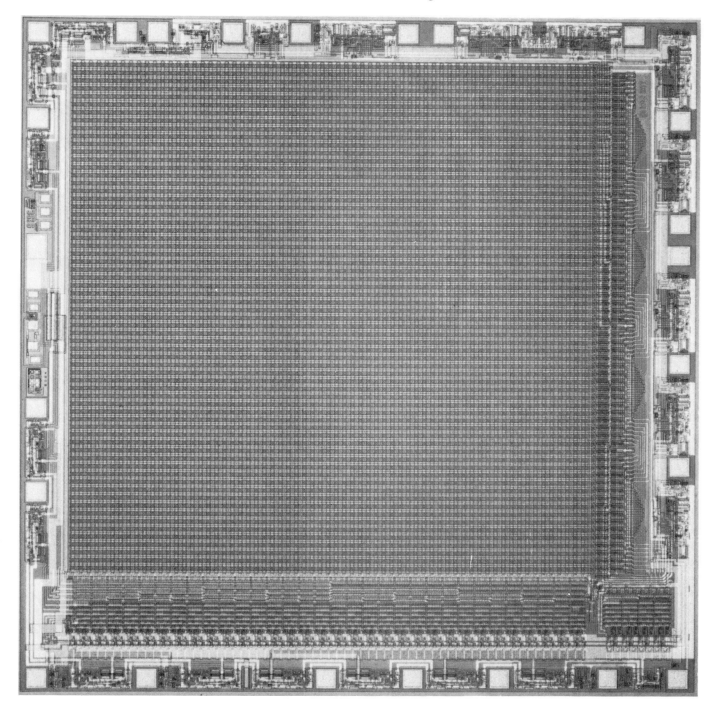

Fig. 6. Photomicrograph of a bipolar 16,384-bit programmable read-only memory (ROM). (*Intel*)

the 16,384-bit programmable read-only memory (ROM), which was introduced in 1979.

Figures 4, 5, and 6 indicate the tendency toward increasing complexity in development of digital integrated circuits. This tendency is dictated by the economics of integrated circuit manufacturing. Because of the nature of this manufacturing process, all circuits on a slice are fabricated together. Consequently, the more circuitry accommodated on a slice, the cheaper the circuitry becomes. Because testing and packaging costs depend on the number of chips, it is desirable, in order to keep costs down, to crowd more circuitry onto a given chip rather than to increase the number of chips on a wafer. Circuits such as the one in Fig. 4 are the building blocks of digital computers. Their development and that of large computers were closely interrelated. *See* DIGITAL COMPUTER.

When subsystem complexity is too great for production in a small area of crystal, the subsystem is "sliced" into building blocks. The regular structure of memory circuits allows memory blocks to contain several times as much logic as other circuits. For instance, a 1024-bit random-access memory (RAM) has 1024 flip-flop circuits for data storage, plus address decoders. The random structure of circuits such as Fig. 5 requires more interconnection area, and thus contains fewer logic elements. This is a 2-bit slice of a microcomputer central processor unit (CPU). Eight such slices are simply parallel-connected to form a 16-bit CPU, for instance.

Linear circuits. Integrated circuits based on amplifiers are called linear because amplifiers usually exhibit a linearly proportional response to input signal variations. However, the category includes memory sense amplifiers, combinations of analog and digital processing functions, and other circuits with nonlinear characteristics. Some digital and analog combinations include analog-to-digital converters, timing controls, and modems (data communications modulator-demodulator units).

Linear circuits cannot yet match the power and frequency range of discrete transistors. However, they meet most other technical requirements. One long-standing drawback was the lack of inductors for tuning and filtering. That was overcome by the use of resistor-capacitor networks and additional circuitry. For low-frequency circuits the resistor in these networks is being replaced by the switched capacitor. At the higher frequencies, an oscillator-based circuit called the phase-locked loop provides a general-purpose replacement for inductors in applications such as radio transmission demodulation. *See* PHASE-LOCKED LOOP.

At first, the development of linear circuits was slow because of the difficulty of integrating passive components and also because of undesirable interactions between the semiconductor substrate and the operating components. Thus, much greater ingenuity was required to design and use the early linear circuits.

In addition, manufacturing economics favors digital circuits. A computer can be built by repetitious use of simple inverters and gates, while analog signal processing requires a variety of specialized linear circuits.

Semiconductor devices. In the continuing effort to increase the complexity and speed of digital circuits, and the performance characteristics and versatility of linear circuits, a significant role has been played by the discovery and development of new types of active and passive semiconductor devices which are suitable for use in integrated circuits. Among these devices is the *pnp* transistor which, when used in conjunction with the standard *npn* transistors described earlier, lends added flexibility to the design of integrated circuits.

Two types of *pnp* transistor structures, both compatible with standard integrated circuit technology, are shown in Fig. 7. Figure 7a shows the so-called lateral *pnp*, which is a *pnp* transistor formed between two closely spaced *p*-type diffused regions. Since these can be formed using the same diffusion step by which the base of the *npn* transistor and the resistor are formed, this structure can be fabricated simultaneously with the rest of the circuit. The same is true of the substrate *pnp* shown in Fig. 6*b*, except here the *p*-type substrate is employed as the collector.

Other possibilities involve means of producing integrated circuits with resistors of high resistance

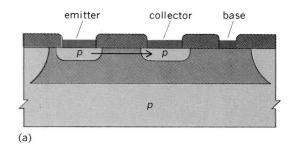

(a)

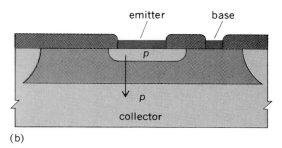

(b)

Fig. 7. Types of *pnp* transistor structures. (a) Lateral *pnp* transistor. (b) Substrate *pnp* transistor. Arrows indicate direction of useful transistor action.

values. By using the standard resistor process, as described in connection with Fig. 1, high resistances require long resistors, hence large chips are required in order to accommodate these long resistors. An alternate approach is to use the *n*-region between the diffused *p*-region and the substrate as the resistor. This semiconductor region has a significantly higher resistance than that formed by the *p*-type diffusion. However, due to the reverse bias between the *n*-region and the two *p*-regions enclosing it from top and bottom, such resistors have highly nonlinear current-voltage characteristics. An alternate scheme involves the use of a very thin film of high-resistivity metal deposited by evaporation or sputtering on top of the insulating silicon dioxide layer. Such films can display resistivities as much as a hundred times higher than that of the diffused *p*-region, and since their deposition is performed after the entire semiconductor structure is complete, they can be used in conjunction with any type of integrated circuits.

Several alternatives to *pn*-junction isolation are in use. These involve etching around the transistor regions, so that oxide or air isolates the components. Called dielectric isolation, the technique was initially used in military circuits to stop radiation-induced currents from flowing through and destroying the circuits. It is now used commercially to reduce circuit capacitance, thus speeding up operation, and to reduce the silicon area required for each transistor. It is not yet widely used, however, because it complicates the basic process.

MOS integrated circuits. The other major class of integrated circuits is called MOS because its principal device is a metal oxide semiconductor field-effect transistor (MOSFET). It is more suitable for large-scale integration (LSI) than bipolar circuits because MOS transistors are self-isolating and can have an average size of less than a millionth of a square inch (6×10^{-4} mm²). This has made it practical to use over 100,000 transistors per circuit. Because of this high density capability, MOS transistors are used extensively for high density RAMs (Fig. 8), ROMs, and microprocessors.

Several major types of MOS device fabrication technologies have been developed since the mid-1960s. They are: metal-gate p-channel MOS, which uses aluminum for electrodes and interconnections (Fig. 9); silicon-gate p-channel MOS, employing polycrystalline silicon for gate electrodes and the first interconnection layer; n-channel MOS, which is usually silicon gate; and complementary MOS, which employs both p-channel and n-channel devices. In 1980 the silicon gate n-channel and the complementary MOS were the dominant technologies.

The basic principles of MOS technology can be illustrated with the metal-gate p-channel MOSFET in Fig. 9. It consists of two p-type diffused regions (such as can be obtained by the process used to form the base region of the npn transistors described earlier), separated by an n-type region which is under control of an electrode, called the gate. When there is no voltage applied to the gate, the two p-type regions, called source and drain, are electrically insulated from each other by the n-type region which surrounds them. When, however, a negative voltage is applied to the gate, the electric field induces a very thin p-type region at the surface of the silicon. This thin p-type region now connects the source and the drain, permitting the passage of current between them. The path is called a channel.

Both conceptually and structurally the MOS transistor is a much simpler device than the bipolar transistor. In fact, its principle of operation has been known since the late 1930s, and the research effort that led to the discovery of the bipolar transistor was originally aimed at developing the MOS transistor. What kept this simple device from commercial utilization until 1964 is the fact that it depends on the properties of the semiconductor surface for its operation, while the bipolar transistor depends principally on the bulk properties of the semiconductor crystal. Hence MOS transistors became practical only when understanding and control of the properties of the oxidized silicon surface had been perfected to a very great degree.

Figure 10 shows an MOS inverter circuit. One of the valuable features of MOS devices in integrated circuit applications is that with some minor modification, such as increasing the distance between the two p-type regions, they can also be employed as resistors with adequately high resistance values. Thus, in an MOS inverter circuit both the transistor and the resistor are realized using an MOS device of suitable geometry, as shown in Fig. 9. In this example a separate voltage, $V_{resistor}$, is applied to the gate of the resistor to set its resistance at the desired value.

The operation of an MOS inverter circuit is very similar to the operation of the bipolar inverter circuit discussed earlier. The input voltage V_{in} is applied to the gate of the transistor. When it is zero, no conduction can take place between the source and drain of the transistor, and therefore no current flows in the resistor. As a result, no voltage drop will exist across the resistor and the output voltage V_{out} will be the same as the externally applied biasing voltage, −10 volts in this example.

When a negative input voltage is applied, a field-induced conduction path is created between source and drain. Current now flows through the transistor and hence the resistor, and as a result the output voltage will become more positive. Thus, the change in input voltage appears inverted at the output.

A comparison of the bipolar inverter circuit of Figs. 1 and 2 with its MOS counterpart in Fig. 10

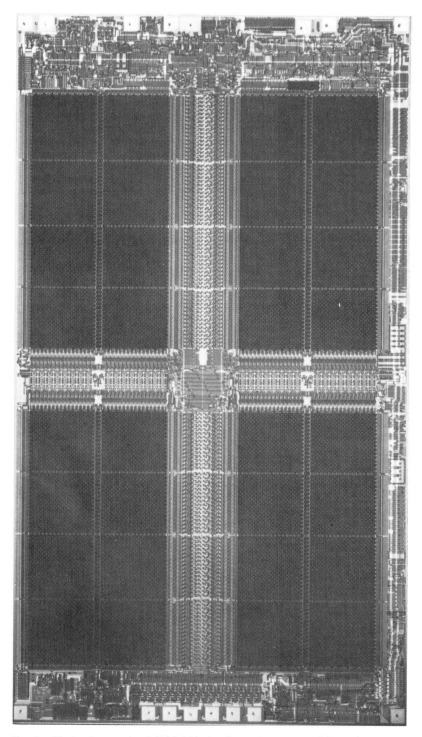

Fig. 8. Photomicrograph of MOS LSI circuit used as 65,536-bit random-access memory (RAM). (*Intel*)

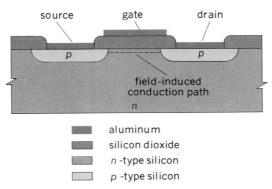

Fig. 9. Diagram of *p*-channel MOS field-effect transistor.

reveals most of the principal advantages of the latter. Because all biased junctions are always reverse-biased, there is no need for additional means of isolation between the elements. In addition, no emitters are required in MOS integrated circuits. As a result, the fabrication process is simpler. Resistor and transistor are connected through a common *p*-type diffused region in the MOS case, but an added aluminum metallization stripe is needed in the bipolar case. Thus, interconnection is generally much simpler in an MOS circuit. This becomes an especially significant advantage in connection with complex circuits.

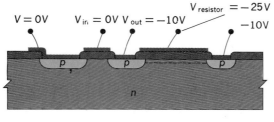

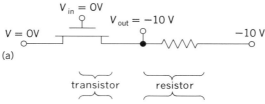

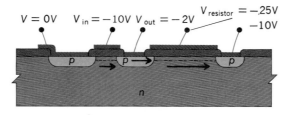

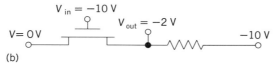

Fig. 10. Operation of a MOS inverter circuit. (*a*) Input voltage V_{in} is zero. (*b*) Input voltage applied. Arrows indicate direction of current flow.

The n-channel MOS. An *n*-channel MOSFET has *n*-type source and drain diffusions in a *p*-type region. A positive voltage on the gate electrode converts the *p*-type region to a conducting *n*-type channel. Because *n*-channel properties were more difficult to control than *p*-channel, this form of MOS did not become commercially viable until about 1972.

The *n*-channel MOS is faster and more compatible with bipolar logic than *p*-channel MOS. It is the preferred MOS for LSI logic and memory, which are often interconnected with bipolar logic.

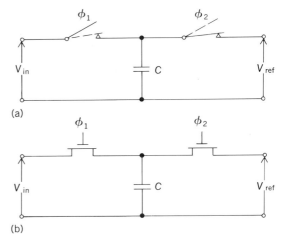

Fig. 11. Switched-capacitor circuit. (*a*) Schematic of an RC section. (*b*) Implementation of the switches with MOS transistors with clocks applied to their gates.

The *n*-channel MOS circuits can operate at speeds approaching bipolar circuits. Moreover, switching thresholds and other operating voltages can be made the same as bipolar voltages, eliminating the special interfaces and power supplies which are required by *p*-channel MOS.

Complementary MOS (CMOS). The basic CMOS logic switch is a pair of *p*-channel and *n*-channel transistors. A voltage applied to both gate electrodes makes only one of the two conductive. Thus, significant current flows through the device only during the switching operations, reducing average power dissipation to the range of microwatts per circuit. Also, like *n*-channel MOS, CMOS is compatible with the voltage levels used in bipolar logic circuits.

CMOS is preferred for equipment with limited power supply or cooling problems, such as battery-powered portable equipment and aerospace systems. In such equipment, its advantages often offset the added cost of fabricating the complementary transistors.

Silicon-gate MOS. The newer MOS devices generally have silicon rather than metal gates. Silicon-gate technology has greatly accelerated LSI development by simplifying the MOS process, allowing smaller MOSFETs to be made, and serving as a first layer of interconnections. The 65,536-bit RAM in Fig. 8 is a silicon-gate *n*-channel MOS circuit.

Polycrystalline silicon is deposited on the oxidized silicon wafer prior to diffusion, then etched

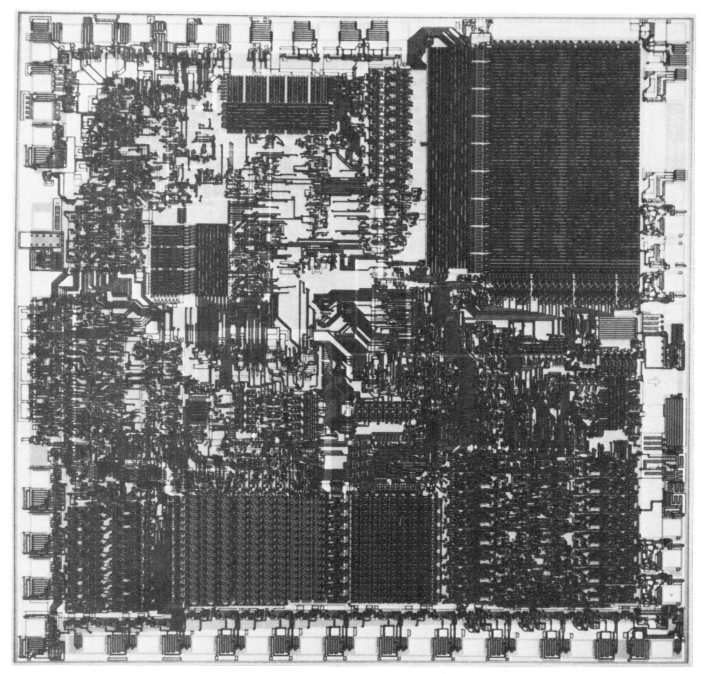

Fig. 12. A 16-bit microprocessor circuit. (*Intel*)

to form diffusion windows, electrodes, and interconnections. The result is automatic alignment of source, drain, and channel with the gate, making very small transistor geometries practical. An aluminum interconnection layer is finally deposited on the oxidized silicon-gate layer.

MOS sampled-data devices. MOS structures are also being used extensively in signal-processing applications that are based on sampled-data techniques. Two classes of devices, namely, charge-coupled devices (CCD) and switched-capacitor networks, play the major role in these applications.

In CCDs the stored charge at the semiconductor surface can also be made to propagate along the surface via potential wells created by a series of these MOS structures. The storage cell in the RAM circuit of Fig. 8 can be viewed as using a single CCD element for each bit. *See* CHARGE-COUPLED DEVICES.

The capacitance C from the MOS structure can be integrated with MOS transistors, which are used as switches, to form a switched-capacitor circuit (Fig. 11). One of the switches ϕ_1 is closed when the other switch ϕ_2 is open, and vice versa. The circuit is equivalent to a resistor with a value of R-T/C, where T is the sampling interval. The advantage of the technique is the high quality and the high value of resistors that can be put on an integrated circuit. These resistors combined with capacitors and operational amplifiers (all made with MOS technology) can be used to make active filters. Since many stages of these filters can be integrated into one chip, precise filters are now possible with integrated circuit technology. These filters are having a tremendous impact on low-frequency filters (frequencies less than 100 kHz) and will be used extensively in telecommunication equipment. *See* COMPUTER STORAGE TECHNOLOGY; SWITCHED CAPACITOR.

INTEGRATED OPTICAL DEVICES

Semiconductors have long been used as light sensors. Advances in LSI have enabled large arrays of sensors, such as solid-state television cameras, to be made commercially. MOS LSI is now preferred for sensor arrays because it permits large amounts of control logic to be fabricated in the same circuit as the sensors.

In MOS optical arrays, the sensors are MOS devices or CCD elements. The sensor portion of the circuit views the scene through a transparent window. Light variations in the scene viewed cause variations in charge, producing signals transferable in shift-register fashion to processing stages such as amplifiers. These arrays have numerous applications, from measuring and sorting objects on production lines to optical character readers, which automatically translate written information to digital computer input codes.

MICROCOMPUTERS

The LSI development having the most profound effect on electronic equipment economics and design in general is the microcomputer. The complexity and capability of these circuits, introduced in 1971, have expanded greatly. Initially these integrated circuits were applied to dedicated functions with the tailoring of that function being done with the ROM circuits on the chip. However, there are now many different types of

microprocessor circuits ranging from the simple microcontroller types of functions to very complex computers on a chip such as the 16-bit microprocessor shown in Fig. 12.

This rapid development has been a result of both the technology which allows more circuits to be placed on a chip and circuit innovation. The technology has increased the number of transistors on a chip from a few thousand in 1971 to over 100,000 in 1980. The circuit innovations include both digital techniques and the combination of analog and digital processing functions on the same chip. Both analog-to-digital and digital-to-analog data conversion circuits have been combined with the digital processor into one integrated circuit.

As the technology enables the development of more and more complex circuits, the problem becomes not one of how to build these complex circuits but of what to build. The application of these circuits is rapidly becoming limited by the effort in programming them, rather than by their inherent capability. The emphasis in the future will be to apply circuit design features to reduce the rapidly increasing complexity of programming.

Although these LSI circuits are being used in calculators, automobiles, instruments, appliance controls and many other applications, realization of their potential is just beginning. As the power of the computer is captured in the relatively inexpensive integrated circuit, the role of integrated circuits will continue to expand rapidly. *See* CIRCUIT (ELECTRONICS); MICROCOMPUTER; MICROPROCESSOR; PRINTED CIRCUIT.

[RON BURGHARD; NEIL BERGLUND; YOUSSEF EL-MANSY]

FABRICATION

Integrated-circuit fabrication begins with a thin, polished slice of high-purity, single-crystal semiconductor (usually silicon) and employs a combination of physical and chemical processes to create

Fig. 13. Vertical laminar-flow clean room for integrated-circuit fabrication.

the integrated-circuit structures described above. Junctions are formed in the silicon slice by the processes of thermal diffusion or high-energy ion implantation. Electrical isolation between devices on the integrated circuit is achieved with insulating layers grown by thermal oxidation or deposited by chemical deposition. Conductor layers to provide the necessary electrical connections on the integrated circuit are obtained by a variety of deposition techniques. Precision lithographic processes are used throughout the fabrication sequence to define the geometric features required.

Requirements. The integrated-circuit fabrication process is quite sensitive to both particulate and impurity contamination. Airborne particulates must be minimized during the fabrication sequence, since even small (1-μm) particles on the wafer surface can cause defects. A particulate-free fabrication ambient is normally achieved by the use of vertical laminar-flow clean rooms or benches (Fig. 13). Lint-free garments are worn to minimize operator-borne particulates. To minimize impurity contamination effects, the chemicals, solvents, and metals which are used must be of the highest possible purity (electronic grade). Yellow light is necessary in the clean room because of the ultraviolet-sensitive photolithographic processes employed.

The precision and cleanliness requirements of integrated-circuit processing necessitate high discipline throughout the process sequence. This is achieved by extensive operator training, in-process tests and inspection with continual feedback, and a high degree of equipment calibration and control. The physical environment and operator attitude in an integrated-circuit fabrication facility are important factors for successful operation.

Processes. The basic relationship between the major processes in integrated-circuit fabrication is shown schematically in Fig. 14. Film formation is normally followed by impurity doping or lithography. Lithography is generally followed by etching, which in turn is followed by impurity doping or film formation. Impurity doping is normally followed by film formation or lithography. A complete integrated-circuit process sequence requires many cycles through the flow diagram in Fig. 14. For example, metal gate CMOS requires seven cycles through lithography. The complete flow time for a CMOS process is approximately 2 to 6 weeks, depending on process complexity.

Film formation. Film formation employs thermal oxidation to produce silicon dioxide (SiO_2) films, chemical vapor deposition to produce silicon, silicon dioxide, or silicon nitride (Si_3N_4) films, or vacuum evaporation/sputtering to produce metal films. Thermal oxidation is performed in quartz-walled furnace tubes (Fig. 15a) in an ambient of oxygen or steam, at temperatures ranging from 800 to 1200°C.

Chemical vapor deposition (CVD) is a gas-phase process where a film deposit is obtained by combining the appropriate gases in a reactant chamber at elevated temperatures. A typical CVD reaction (silox process) is given below. Figure 15b shows a

$$SiH_4 + O_2 \xrightarrow{(400-500°C)} SiO_2 \downarrow + 2H_2 \uparrow$$
$$\text{(Deposits}$$
$$\text{as film)}$$

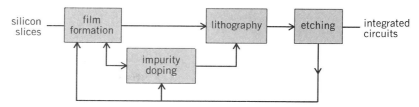

Fig. 14. Integrated-circuit fabrication sequence in schematic form. Fabrication normally proceeds in the direction of the arrows.

cold-walled, atmospheric-pressure CVD system where the silicon slice is heated by rf energy. Figure 15c shows a low-pressure CVD system where the slices and process gases are heated in a partially evacuated furnace tube. This low-pressure process produces very uniform film thicknesses.

Evaporation or sputtering of metal coatings is performed in a vacuum, with metal transport being produced either by heat (evaporation) or bombarding ions (sputtering). The vacuum evaporator in Fig. 15d uses fixturing with planetary motion during evaporation. This achieves uniform metal thickness over surface topology on the silicon slice.

Impurity doping. Impurity doping is performed either by thermal diffusion in high-temperature furnaces (Fig. 15a) or by ion implantation. Boron and phosphorus are dopants which can be introduced by thermal diffusion at temperatures from 800 to 1200°C. Ion implantation is performed in a specialized accelerator designed to impinge an intense, uniform beam of a particular ion onto the silicon slice (Fig. 15e). Ion implantation is more controllable than diffusion and is performed at lower temperatures. Ion implantation is used when greater precision of dopant is required or when a reduced temperature cycle is advantageous.

Lithography. Lithography is necessary to define the small geometries required in integrated circuits. In lithography the silicon slice is coated uniformly with a thin film of photosensitive material called resist. If the lithography is to be performed optically, the integrated-circuit pattern to be transferred to the resist is first created on a glass plate or "mask." This pattern can then be transferred to the resist by a number of optical techniques. These techniques range from direct contact printing using a collimated source of ultraviolet light (Fig. 15f), to optical projection of a single integrated-circuit pattern with associated reduction (for example, 10:1, 1:1) and precise x-y motion of the silicon slice (direct step-on-wafer; Fig. 15g). Electron-beam direct patterning can be performed, without a mask, by using a controllable electron beam and an electron-sensitive resist. Lithography has also been achieved with x-rays, by their projection through a special mask in close proximity to the slice. Direct step-on-wafer photolithography, the most advanced of the optical lithographic techniques, is capable of defining 1 μm geometries. Electron-beam and x-ray lithography have demonstrated the capability to define features substantially smaller than 1 μm.

Etching. Etching is necessary to transfer the resist pattern achieved through lithography to the underlying surface. Traditionally, integrated-circuit fabrication has employed wet chemical pro-

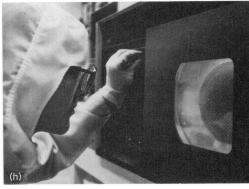

Fig. 15.   Modern integrated-circuit fabrication equipment. (*a*) Diffusion furnace. (*b*) Atmospheric-pressure chemical vapor deposition (CVD). (*c*) Low-pressure CVD. (*d*) Vacuum evaporator. (*e*) Ion implanter. (*f*) Contact mask aligner. (*g*) Direct step-on-wafer machine. (*h*) Plasma etcher.

cesses to etch lines and features. In general, these processes are inadequate to etch lines or features less than $3-4$ μm in size. Dry plasma etching, reactive ion etching, and ion milling are advanced techniques being developed to realize smaller geometries. Plasma etching (Fig. 15*h*), the most advanced of these new techniques, operates by creating active ion species in an rf plasma. The vertical etch rate in this process can be adjusted to be much greater than the lateral etch rate. This enables the etching of fine lines and features without loss of definition, even in films approaching 1 μm in thickness.

Changes in technology. Integrated-circuit fabrication technology changes rapidly due to the steadily increasing chip and silicon wafer sizes and the steadily decreasing size of the individual circuit elements on the integrated circuit. Because of these rapid changes a fabrication facility will remain state-of-the-art in capability for no more than $3-5$ years without major equipment and process changes.   [BOB L. GREGORY]

Bibliography: A. B. Glaser and G. E. Subak-Sharpe, *Integrated Circuit Engineering*, 1977; A. S. Grove, *Physics and Technology of Semiconductor Devices*, 1967; S. M. Sze, *Physics of Semiconductor Devices*, 1969; R. M. Warner, Jr., and J. N. Fordemwalt (eds.), *Integrated Circuits: Design Principles and Fabrication*, 1965.

Integrated optics

The study of optical devices, singly and in combinations, that are based on light transmission in waveguides, that is, structures that confine the propagating light to a region with one very small dimension (or sometimes two), of the order of the wavelength of the light. The principal motivation for these studies is to enable the combination of individual devices thus miniaturized, through waveguides or other means, into a functional optical system mounted on a small substrate. The resulting system is called an integrated optical circuit (IOC) by analogy with the semiconductor type of integrated circuit. An integrated optical circuit could, for example, include a laser, switches, polarizers, modulators, detectors, and so forth. An important use envisioned for integrated optical circuits is in connection with optical communications through the medium of glass fibers, which are themselves waveguides. Integrated optical circuits could be used in such a system as optical transmitters, repeaters, and receivers. *See* OPTICAL COMMUNICATIONS.

The advantages of having an optical system in the form of an integrated optical circuit rather than a conventional series of components on an optical bench, include, apart from miniaturization, the reduced sensitivity to ambient-temperature gradients and changes, to airborne acoustical effects, and to mechanical vibrations of the separately mounted parts. As in the case of semiconductor integrated circuits, the integrated optical circuit might be made of essentially one material, modified for the different components by incorporating suitable substituents or dopants, or of many different materials. The latter option, called hybrid, has the advantage that each component could be optimized, for example, by using a gallium arsenide or neodymium yttrium-aluminum-garnet (YAG) laser, a lithium niobate modulator or switch, a silicon detector, and so forth. In the former case, the integrated optical circuit is called monolithic and is expected to have the advantage of ease of processing, similar to the situation for monolithic semiconductor integrated circuits. The most promising material for monolithic integrated optical circuits is gallium arsenide, since with suitable substituents this material may be made into a laser, switch, modulator, detector, and so forth. *See* INTEGRATED CIRCUITS.

Guided waves. The simplest type of waveguide is a three-layer or sandwich structure with the index of refraction largest in the middle layer, designated "film" in Fig. 1. The top layer or "cover" is very frequently air, and it will be assumed that is the case in what follows. In a guided wave the light is not distributed uniformly across the guide, but in a pattern that depends on the indices of refraction of all three layers and the height of the film. The pattern may be a simple one, such as one of those shown in Fig. 1a, or higher numbers in that sequence, or it may be a superposition of such patterns. In the simplest possible guided waves, called modes of the guide, a pattern with m zeroes, such as one of those in Fig. 1a, is swept down the guide with a speed characteristic for that particular pattern. The mode may also be described by a characteristic wave vector k_m for the mth mode, directed along the path of the light, that is, along the guide.

A pattern such as one of those in Fig. 1a represents a standing wave, which is equivalent to two progressing waves moving in opposite directions. To form the guided wave, one superimposes on these the appropriate uniform speed along the guide. The guided wave is thus equivalent to two rays bouncing back and forth with a characteristic angle θ_m, as indicated in Fig. 1b. From this point of view it is readily seen that the larger index of the middle layer keeps the wave confined because there is total internal reflection of the two rays every time they strike the surfaces. *See* WAVEGUIDE.

Materials and fabrication. Waveguides have been made of many different materials with many different techniques. The film has been made of sputtered glass, sputtered oxides of tantalum or zinc, epitaxial gallium arsenide, ion-bombarded gallium arsenide, epitaxial garnets, sputtered and epitaxial lithium niobate, nitrobenzene liquid, nematic liquid crystals, and a number of other organics and polymers.

Most of the materials with desirable electrooptic or nonlinear properties for active waveguides or waveguide devices are single crystals, high-quality thin films of which are difficult to obtain. A good way to obtain waveguiding in such a case is to create a thin layer of higher refractive index at the top surface of a suitable single crystal by diffusion or ion exchange. With lithium niobate, for example, satisfactory guides may be made by heating in vacuum to diffuse out lithium oxide, or by diffusing in various metallic impurities, such as titanium. Although this type of treatment yields a region in which the refractive index varies with distance in a thin layer, rather than being constant as in the middle layer of the sandwich guide, guided waves can be obtained provided the refractive index decreases with distance below the surface. The modes are rather similar to those shown in Fig. 1.

Waveguides that confine light in two dimensions, rather than one, consist of a channel or strip of higher refractive index than its surroundings. A channel guide can be made by ion implantation or diffusion through a mask into a thin film, a raised strip guide by masking the desired strip region and removing the surrounding film by sputtering or etching.

Coupling of external light beams. An external light beam may be coupled into a waveguide by focusing the light on the end of the guide, but this is an inefficient process because the height of the film is much less than the width of the usual laser beam. It is clear, however, that light cannot be coupled in by illuminating the top of the guide

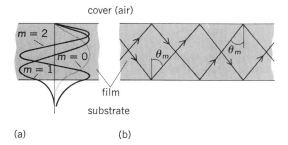

Fig. 1. Optical waveguide. (*a*) Light amplitude variation across waveguide for first few guided modes. Each mode is assigned a number *m* equal to number of times the light amplitude goes to zero inside film. (*b*) Ray picture of guided waves. (*From E. Conwell, Integrated optics, Physics Today, 29(5):48–56, 1976.*)

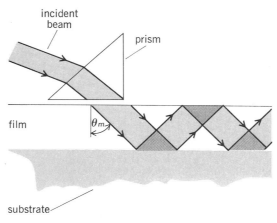

Fig. 2. Prism coupler. (*From E. Conwell, Integrated optics, Phys. Today, 29(5):48–56, 1976*)

from any angle because, as noted earlier, the guided light is kept inside by total internal reflection. (A light ray must be able to travel the same path in either direction.) It can, however, be coupled in from the top if it first goes through a prism made of higher-refractive-index material (Fig. 2). If the prism is close enough to the surface of the guide, within a wavelength of light, light may leak through the air gap into the guide. The exact condition to be satisfied is that the component of the propagation vector in the prism parallel to the guide be equal to the characteristic wave vector k_m of some guide mode. This is achieved by varying the angle of the incident beam. Prism coupling is capable of over 90% efficiency, but does not appear to lend itself well to integration. A coupler that is generally less efficient but more suitable for integration is the grating coupler (Fig. 3). The grating is usually made of a thin layer of photoresist that has been exposed to an optical interference pattern and then developed. Its action can be described by saying that, as a periodic structure (with period d), it exchanges momentum with the light wave, making it possible to obtain a match with some wave vector k_m. Output coupling is provided by the same type of couplers.

Modulators. Modulators are used to impress information on the guided light wave. Many integrated optics modulators are based on the same principles as many bulk modulators, notably the change in indices of refraction caused by an elec-

tric field in materials such as lithium niobate. The main difference is that the light is confined in the small area of the waveguide, providing the advantage of a much smaller power requirement. Another type of modulator, which does not have a bulk analog, is based on the principle of the directional coupler used in microwave systems. This acts by controlling electrically the switching of a guided wave from one channel waveguide to a parallel and identical one, spaced within about a wavelength of light from the first. *See* OPTICAL MODULATORS.

Lasers. The gallium arsenide (GaAs) diode laser is already an integrated optics device in a sense, since its efficiency is due partly to waveguiding arising from refractive index differences between regions of different impurity content. More deliberate progress has been made toward integrated-optics types of lasers by using waveguides with a periodic variation in their properties. This is most easily achieved as a periodic variation in width, which may be created by the technique described for the grating coupler. The periodic variation results in a strong reflection of waves with wave vector equal to π/d, d being the spatial period. Thus, by using for the active medium a periodic waveguide with d chosen so that the laser frequency is strongly reflected, it is possible to dispense with end mirrors. The feedback provided by the end mirrors, in other words, is replaced by distributed feedback. Alternatively, periodic structures of this kind can be used at the ends of the active medium to act as mirrors. *See* LASER.

[ESTHER M. CONWELL]

Bibliography: E. M. Conwell and R. D. Burnham, Materials for integrated optics: GaAs, in *Annu. Rev. Mater.*, 8:135–179, 1978; Special issue on integrated optics, *IEEE Trans. Microwave Theory Tech.*, MTT-23(1):1–180, 1975; T. Tamir (ed.), *Integrated Optics*, 1975; P. K. Tien, Integrated optics and new wave phenomena in optical waveguides, *Rev. Mod Phys.*, 49(2):361–420, 1977.

Intermediate-frequency amplifier

A tuned amplifier employed in the amplification of the signals produced by the mixer in a radio receiver. Because the carrier frequency of the modulated signal from the mixer is essentially constant, the resonant frequency of the amplifier is fixed.

The proper design of the intermediate-frequency (i-f) amplifier is essential for good selectivity and reproduction of the original transmitted signal. If the amplifier is tuned too sharply, the high-frequency components of the modulating signal will be lost. To avoid this, stagger-tuning of the individual stages may be used. In a stagger-tuned amplifier the resonant frequency of each stage is slightly different from the carrier frequency, with the result that the gain is essentially constant over the bandwidth of the modulated signal. The gain decreases rapidly at frequencies outside this band.

The standard i-f frequency for broadcast radio receivers is 455 kHz; other frequencies used depend upon the application, such as television receivers or radar receivers. *See* AMPLIFIER.

[HAROLD F. KLOCK]

Ion implantation

A process of introducing impurities into the near-surface region of solids by directing a beam of energetic ions at the solid (Fig. 1). When ions of suffi-

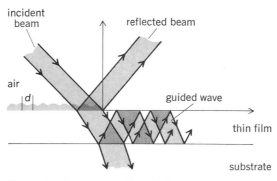

Fig. 3. Grating coupler. (*From E. Conwell, Integrated optics, Phys. Today, 29(5):48–56, 1976*)

cient energy are directed toward a crystal surface, they will penetrate the surface and slow to rest within the solid. Ion implantation differs from the normal thermal equilibrium means of introducing atoms into a solid, which is usually carried out either by adding the elements during growth of a solid from a liquid or vapor phase, or by diffusing the atoms into the solid at elevated temperatures. The advantages of the implantation process are: precise control of the type of impurity to be introduced, the amount of impurity introduced, and the impurity distribution in depth. In addition, since any atoms can be added to any solid, mixtures can be formed which would not normally be found in nature, that is, systems which are not in thermal equilibrium. Most of the work in developing and understanding the ion implantation processes has taken place since the middle 1960s. A major area of application of ion implantation has been to semiconductor device fabrication, where the process is a standard technique. Additional applications have been developed in metals, where unexpectedly beneficial effects on surface-sensitive properties have been found, and in insulators, and research in ion implantation has been undertaken for the purpose of developing scientific understanding and technological applications.

IMPLANTATION PROCESS

Any atom can be implanted into a solid. An ion implantation system is shown schematically in Fig. 2. First the atom must be ionized, which is the process of changing the number of electrons associ-

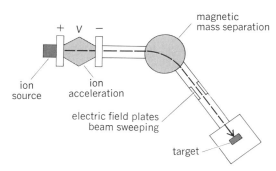

Fig. 2. Ion implantation system.

ated with an atom and thus leaving it with a net positive or negative charge. This is done in an ion source, usually by a plasma discharge. As the ions are formed, they are continuously extracted from the plasma and accelerated through a voltage difference V. This gives each ion an energy $E = qV$, where q is the charge of the ion. Typical acceleration voltages are 10,000 to 500,000 V. The ion beam is then passed through a transverse magnetic field so that the different mass ions of energy E will be deflected by different angles, and in this way ions of particular energy and mass are selected. Varying electric fields are often used to sweep the selected ion beam laterally so that an area can be uniformly implanted with the ions. The region between the ion source and the solid target is kept under vacuum since the ions would travel only very short distances in air. The total number of ions incident on the target is determined by mea-

suring the current to the sample during ion implantation. By integrating this current to obtain the total ion charge and using the known charge per ion, the number of ions implanted in the target can be precisely controlled.

As the energetic ions penetrate a solid, they encounter electrons and atoms with which they have collisions. In this manner they transfer their energy to the electrons and atoms in the solid until they finally come to rest. In the process of transferring energy to electrons and atoms, radiation damage is produced in the target. For example, in a single crystal target some atoms will be given enough energy to be knocked off normal lattice sites, and the ion thus leaves a track of damage as it comes to rest. Each of the collision events usually results in a loss of only a small fraction of the ion energy, and in many cases the collisions can be treated as statistically independent events. In this case, the final profile of the ions after they come to rest will be very close to a Gaussian distribution in depth. The concentration of implanted atoms n as a function of depth x can then be described by the equation shown, where N is the total number of

$$n = \frac{N}{\sqrt{2\pi}\Delta R_p} \exp\left[-\left(\frac{x - R_p}{\sqrt{2}\Delta R_p}\right)^2\right]$$

atoms implanted per unit surface area (Fig. 3). In this description there are two important parameters which describe the final distribution of the ions which have been implanted into the solid: R_p is called the projected range of the ion, and ΔR_p is referred to as the spread in this projected range. For a given ion energy, ion mass and target material R_p and ΔR_p have unique values, which can be predicted theoretically. For example, boron ions implanted into silicon with an energy of 100,000 eV would have a projected range of $R_p = 290$ nm and a range spread of $\Delta R_p = 71$ nm. This description of the composition in the implanted region is important in understanding the changes which result in an ion-implanted solid.

APPLICATIONS

The ion implantation process changes the chemical composition of the near-surface region of a solid and introduces radiation damage into this region. These changes provide the possibility of modifying an extremely wide range of near-surface physical and chemical properties of solids.

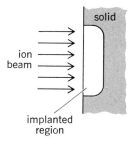

Fig. 1. Ion implantation into a solid.

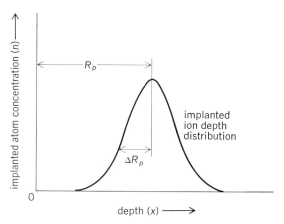

Fig. 3. Implanted ion depth profile.

Semiconductors. Ion implantation is most widely used in the controlled doping of semiconductors for the microelectronics industry. Here ion implantation is used instead of diffusion to dope semiconductors chemically, for the fabrication of such things as resistors, diodes, and MOS transistors. The advantage of implantation over diffusion is its ability to control very precisely the number of impurities that are introduced, and to form depth distributions of the impurities which may not be easily obtainable by diffusion. A complication introduced by implantation is the introduction of radiation damage, but this has been overcome for most applications by heating the semiconductor material to sufficiently high temperatures. The compatibility with the planar technology used in forming integrated circuits has allowed rapid acceptance of ion implantation in semiconductor device manufacturing. Examples of use include the formation of more accurate and higher-value resistors within smaller areas, and more precise control of the gate voltages of MOS transistors. The selected use of ion implantation to form better devices in terms of their type of operation, speed, reliability, and precision has been an important contribution to the steady advance of microelectronics technology. *See* SEMICONDUCTOR DIODE; INTEGRATED CIRCUITS; SEMICONDUCTOR; TRANSISTOR.

Insulators. Changes in properties of insulators can also be accomplished by ion implantation. One potential application involves a class of materials known as magnetic garnets. Films of these materials are used for storing and moving reversed magnetic domains called magnetic bubbles and can be used for memories in much the same way that certain forms of large silicon integrated circuits are used. The implantation process allows better control of the nature and paths of motion of magnetic bubbles. Implantation in optical materials involving both insulators and semiconductors may find important applications as the technology of integrated optics is developed. Here the processing and handling of light signals is envisioned in much the same way that electrical signals are stored and processed in integrated circuits. *See* COMPUTER STORAGE TECHNOLOGY; INTEGRATED OPTICS.

Metals. Most applications to metals require the introduction of many more ions than for semiconductors, so that the final impurity-atom concentrations may be as high as 1 to 10% of the target, in contrast to semiconductor applications where impurity concentrations in the parts per million are usually desired.   [SAMUEL T. PICRAUX]

Surface-sensitive mechanical properties. The friction, wear rates, fatigue life, and cavitation erosion of various metals have been found to be beneficially affected by an ion-implanted alloyed layer to an unexpectedly high degree. It would be anticipated that any improvement in properties would be short-lived because of the limited thickness of the alloyed layer. However, this is not found to be the case, as exemplified by the improvements observed in the wear resistance of steel implanted by various species, such as titanium. The depth of the wear track far exceeds the range of implantation and, in the case of implanted nitrogen in steel, nitrogen is detected at the base of tracks 100 times deeper than the initial ion range.

This suggests that deformation-induced transport of the implanted species may be responsible for the persistence of the beneficial effects. However, such a mechanism is not likely to be available for all implanted species in all host materials and, therefore, the persistence of the effects may be very specific. For example, the implantation of boron has no effect on the wear resistance of steel, while the implantation of titanium imparts considerable improvement.

Oxidation and corrosion resistance. Ion implantation is exploited both for potential practical applications to enhance resistance and as a research tool for increasing the understanding of oxidation and corrosion processes. High-temperature oxidation resistance in metallic alloys requires the formation of a stable, adherent oxide film which has a low ionic conductivity (and, preferably, a low electronic conductivity also) and, therefore, a very low growth rate. Such properties are generally obtained by either bulk alloying, for example, of aluminum into copper alloys or chromium into steels, by diffusing the appropriate species into the surface layer at high temperatures, or by coating the part. Ion implantation of a species which enhances the growth of such a film has the following advantages over these conventional techniques. (1) The properties of the alloy are not affected as they are by bulk alloying. (2) Very little of the alloying species is required, thus minimizing the use of expensive or strategic materials. (3) There are no solid solubility limitations to the choice of solute, thereby permitting implantation of species which enhance the oxidation resistance but are normally insoluble. (4) The process is a low temperature one and so is not limited by diffusion rates, segregation of solutes, or annealing of the part. (5) The resultant surface alloy is an integral part of the material and will not suffer from decohesion as do many coatings. (6) The process does not alter the dimension of the part and so may be applied to the finished product.

These advantages of ion implantation also hold for improving aqueous corrosion in which the corrosion rate is determined, not simply by the oxide film, but also by the potential at its surface. Species may be implanted into a base metal, therefore, either to alter the corrosion potential or to provide a passive film. Noble metals such as platinum have been investigated for the former process. For the latter process, tantalum, which is insoluble in iron, can be implanted into the surface, and has been found to be even more effective than chromium in passivating iron and reducing its corrosion rate.

Other properties. Other areas in which ion implantation is being used or is being considered for use include the simulation of neutron radiation damage, the formation of catalysts, the synthesis of high-critical-temperature (T_c) superconductors, and research on basic metallurgical parameters. In the first application, short exposures of a material to an ion beam can produce damage similar to that produced by many years' exposure to neutrons in a nuclear reactor and, thereby permitting accelerated testing of the response of materials to the nuclear environment. For catalytic action, the impetus for using ion implantation is the potential cost saving of precious metals such as platinum. The efficiency of many chemical reactions such as the electrolysis of water or the oxidation of carbon monoxide is enhanced by various metals, particularly those of the platinum group, platinum being the most effective. Because of its costs and limited availability, there is every incentive to use as little as possible. Laboratory tests indicate that implantation of platinum ions, which uses much smaller amounts of the metal than do coating techniques, provides a surface electrocatalytic activity approaching that of platinum itself. For superconductivity, ion implantation is being used to synthesize metallic compounds, for example, Nb_3Si, which do not exist in equilibrium but are theoretically predicted to have high superconducting transistion temperatures. As a research tool, ion implantation makes it possible to study such factors as lattice site location of impurities, defect structures, diffusion characteristics, and solute interactions in alloys which are otherwise impossible to prepare. *See* SUPERCONDUCTIVITY.

Prospects. The number and type of engineering components treated by ion implantation are currently very limited. Components include: nitrogen-implanted press tools and mill rolls, to reduce their adhesive wear; nitrogen-implanted dies and molds used in the injection molding of plastic, to reduce their abrasive wear; chromium-implanted ball bearings, to reduce their corrosion, and boron-implanted beryllium bearings, to increase their hardness and reduce wear. Nevertheless, the rapid increase in the understanding of the ion implantation process and its advantages and limitations, coupled with a greater availability of implantation facilities, is being accompanied by its increased exploitation as a competitive industrial process.

[CAROLYN M. PREECE]

Bibliography: G. Dearnaley et al., *Ion Implantation*, 1973; J. K. Hirvonen, *Treatise on Materials Science and Technology*, vol. 18: *Ion Implantation*, 1980; J. W. Mayer, L. Eriksson, and J. A. Davies, *Ion Implantation in Semiconductors*, 1970; F. F. Morehead and B. L. Crowder, Ion implantation, *Sci. Amer.*, 228(4):64–71, 1973; S. T. Picraux, E. P. EerNisse, and F. L. Vook, *Applications of Ion Beams to Metals*, 1974; C. M. Preece and J. K. Hirvonen, *Ion Implantation Metallurgy*, 1980.

Ionization

The process by which an electron is removed from an atom, molecule, or ion. This process is of basic importance to electrical conduction in gases and liquids. In the simplest case, ionization may be thought of as a transition between an initial state consisting of a neutral atom and a final state consisting of a positive ion and a free electron. In more complicated cases, a molecule may be converted to a heavy positive ion and a heavy negative ion which are separated.

Ionization may be accomplished by various means. For example, a free electron may collide with a bound atomic electron. If sufficient energy can be exchanged, the atomic electron may be liberated and both electrons separated from the residual positive ion. The incident particle could as well be a positive ion. In this case the reaction may be considerably more complicated, but may again result in a free electron. Another case of considerable importance is the photoelectric effect. Here a photon interacts with a bound electron. If the photon has sufficient energy, the electron may be removed from the atom. The photon is annihilated in the process. Other methods of ionization include thermal processes, chemical reactions, collisions of the second kind, and collisions with neutral molecules or atoms. *See* ELECTRICAL CONDUCTION IN GASES.

[GLENN H. MILLER]

Bibliography: H. D. Beckey, *Field Ionization and Field Desorption Mass Spectrometry*, 1980; E. Nasser, *Fundamentals of Gaseous Ionization and Plasma Electronics*, 1971; H. Rusotti, *The Study of Ionic Equilibria: An Introduction*, 1978; A. H. Von Engel, *Ionized Gases*, 2d ed., 1965.

Ionization potential

The potential difference through which a bound electron must be raised to free it from the atom or molecule to which it is attached. In particular, the ionization potential is the difference in potential between the initial state, in which the electron is bound, and the final state, in which it is at rest at infinity.

The concept of ionization potential is closely associated with the Bohr theory of the atom. Although the simple theory is applicable only to hydrogenlike atoms, the picture furnished by it conveys the idea quite well. In this theory, the allowed energy levels for the electron are given by the equation below, where E_n is the energy of the state

$$E_n = -k/n^2 \quad n = 1, 2, 3, \ldots$$

described by n. The constant k is about 13.6 ev for atomic hydrogen. The energy approaches zero as n becomes infinite. Thus zero energy is associated with the free electron. On the other hand, the most tightly bound case is given by setting n equal to unity. By the definition given above, the ionization potential for the most tightly bound, or ground, state is then 13.6 ev. The ionization potential for any excited state is obtained by evaluating E_n for the particular value of n associated with that state.

The ionization potential for the removal of an electron from a neutral atom other than hydrogen is more correctly designated as the first ionization potential. The potential associated with the removal of a second electron from a singly ionized atom or molecule is then the second ionization potential, and so on.

Ionization potentials may be measured in a number of ways. The most accurate measurement is obtained from spectroscopic methods. The transitions between energy states are accompanied by the emission or absorption of radiation. The wavelength of this radiation is a measure of the energy difference. The particular transitions that have a common final energy state are called a series. The series limit represents the transition from the free electron state to the particular state common to the series. The energy associated with the series limit transition is the ionization energy.

Another method of measuring ionization potentials is by electron impact. Here the minimum energy needed for a free electron to ionize in a collision is determined. The accuracy of this type of measurement cannot approach that of the spectroscopic method.

[GLENN H. MILLER]

Bibliography: S. Gasiorowicz, *Structure of Mat-*

ter: A Survey of Modern Physics, 1979; F. K. Richtmyer et al., *Introduction to Modern Physics*, 6th ed., 1969; M. R. Wehr et al., *Physics of the Atom*, 3d ed., 1978.

IR drop

That component of the potential drop across a passive element (one which is not a seat of electromotive force) in an electric circuit caused by resistance of the element. This potential drop, by definition, is the product of the resistance R of the element and the current I flowing through it. The *IR* drop across a resistor is the difference of potential between the two ends of the resistor.

In a simple direct-current circuit containing a battery and a number of resistors, the sum of all the *IR* drops around the circuit (including that of the internal resistance of the battery itself) is equal to the electromotive force of the battery. This is an important circuit theorem useful in the analytic solution of electrical networks. *See* ELECTRICAL RESISTANCE; KIRCHHOFF'S LAWS OF ELECTRIC CIRCUITS; NETWORK THEORY.

[JOHN W. STEWART]

Josephson effect

The passage of paired electrons (Cooper pairs) through a weak connection (Josephson junction) between superconductors, as in the tunnel passage of paired electrons through a thin dielectric layer separating two superconductors.

Nature of the effect. Quantum-mechanical tunneling of Cooper pairs through a thin insulating barrier (on the order of a few nanometers thick) between two superconductors was theoretically predicted by Brian D. Josephson in 1962. Josephson calculated the currents that could be expected to flow during such superconductive tunneling, and found that a current of paired electrons (supercurrent) would flow in addition to the usual current that results from the tunneling of single electrons (single or unpaired electrons are present in a superconductor along with bound pairs). Josephson specifically predicted that if the current did not exceed a limiting value (the critical current), there would be no voltage drop across the tunnel barrier. This zero-voltage current flow resulting from the tunneling of Cooper pairs is known as the dc Josephson effect. Josephson also predicted that if a constant nonzero voltage V were maintained across the tunnel barrier, an alternating supercurrent would flow through the barrier in addition to the dc current produced by the tunneling of unpaired electrons. The frequency ν of the ac supercurrent is given by Eq. (1), where e is the magni

$$\nu = 2eV/h \qquad (1)$$

tude of the charge of an electron and h is Planck's constant. The oscillating current of Cooper pairs that flows when a steady voltage is maintained across a tunnel barrier is known as the ac Josephson effect. Josephson further predicted that if an alternating voltage at frequency f were superimposed on the steady voltage applied across the barrier, the ac supercurrent would be frequency-modulated and could have a dc component whenever ν was an integral multiple of f. Depending upon the amplitude and phase of the ac voltage, the dc current-voltage characteris

tic would display zero-resistance parts (constant-voltage steps) at voltages V given by Eq. (2), where

$$V = nhf/2e \qquad (2)$$

n is any integer. Finally, Josephson predicted that effects similar to the above would also occur for two superconducting metals separated by a thin layer of nonsuperconducting (normal) metal. In 1963 the existence of the dc Josephson effect was experimentally confirmed by P. W. Anderson and J. M. Rowell, and the existence of the ac Josephson effect was experimentally confirmed by S. Shapiro. *See* TUNNELING IN SOLIDS.

Theory of the effect. The superconducting state has been described as a manifestation of quantum mechanics on a macroscopic scale, and the Josephson effect is best explained in terms of phase, a basic concept in the mathematics of quantum mechanics and wave motion. For example, two sine waves of the same wavelength λ are said to have the same phase if their maxima coincide, and to have a phase difference equal to $2\pi\delta/\lambda$ if their maxima are displaced by a distance δ. An appreciation of the importance that phase can have in physical systems can be gained by considering the radiation from excited atoms in a ruby rod. For a given transition, the atoms emit radiation of the same wavelength; if the atoms also emit the radiation in phase, the result is the ruby laser.

According to the Bardeen-Cooper-Schrieffer (BCS) theory of superconductivity, an electron can be attracted by the deformation of the metal lattice produced by another electron, and thereby be indirectly attracted to the other electron. This indirect attraction tends to unite pairs of electrons having equal and opposite momentum and antiparallel spins into the bound pairs known as Cooper pairs. In the quantum-mechanical description of a superconductor, all Cooper pairs in the superconductor have the same wavelength and phase. It is this phase coherence that is responsible for the remarkable properties of the superconducting state. The common phase of the Cooper pairs in a superconductor is referred to simply as the phase of the superconductor. *See* PHASE (PERIODIC PHENOMENA).

The phases of two isolated superconductors are totally unrelated, while two superconductors in perfect contact have the same phase. If the superconductors are weakly connected (as they are when separated by a sufficiently thin tunnel barrier), the phases can be different but not independent. If ϕ is the difference in phase between superconductors on opposite sides of a tunnel barrier, the results of Josephson's calculation of the total current I through the junction can be written as Eq. (3), where I_0 is the current due to single elec

$$I = I_0 + I_1 \sin\phi \qquad (3)$$

tron tunneling, and $I_1 \sin\phi$ is the current due to pair tunneling. The time dependence of the phase is given by Eq. (4). In general, the currents I, I_0,

$$\partial\phi/\partial t = 2\pi(2eV/h) \qquad (4)$$

and I_1 are all functions of the voltage across the junction. For $V = 0$, I_0 is zero and ϕ is constant. The value of I_1 depends on the properties of the tunnel

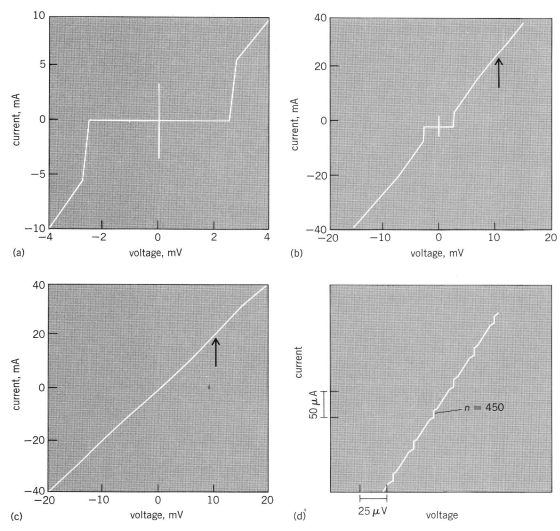

Fig. 1.   DC current-voltage characteristics of lead–lead oxide–lead Josephson tunnel junction at 1.2 K. (*a*) Without microwave power. (*b*) Same characteristic with reduced scale. (*c*) 11-GHz microwave power applied. (*d*) Expanded portion of *c*; arrow indicates a constant-voltage step near 10.2 mV corresponding to $n = 450$ in Eq. (2). This voltage is also indicated by arrows in *b* and *c*. (*From T. F. Finnegan, A. Denenstein, and D. N. Langenberg, AC-Josephson-effect determination of e/h: A standard of electrochemical potential based on macroscopic quantum phase coherence in superconductors, Phys. Rev. B,4:1487–1522, 1971)*

barrier, and the zero-voltage supercurrent is a sinusoidal function of the phase difference between the two superconductors. However, it is not the phase difference that is under the control of the experimenter, but the current through the junction, and the phase difference adjusts to accommodate the current. The maximum value $\sin \phi$ can assume is 1, and so the zero-voltage value of I_1 is the critical current of the junction.

Integration of Eq. (4) shows the phase changes linearly in time for a constant voltage V maintained across the barrier, and the current through the barrier is given by Eq. (5), where ϕ_0 is a constant.

$$I = I_0 + I_1 \sin\left[2\pi(2eV/h)t + \phi_0\right] \qquad (5)$$

The supercurrent is seen to be an ac current with frequency $2eV/h$. The supercurrent time-averages to zero, so the dc current through the barrier is just the single-electron tunneling current I_0.

If the voltage across the junction is $V + v \cos 2\pi ft$, Eqs. (3) and (4) give Eq. (6) for the current.

$$I = I_0 + I_1 \sin\left[2\pi(2eV/h)t + \phi_0 + (2ev/hf)\sin 2\pi ft\right] \qquad (6)$$

The expression for the supercurrent is a conventional expression in frequency-modulation theory and can be rewritten as expression (7), where J_n is

$$I_1 \sum_{n=-\infty}^{n=\infty} (-1)^n J_n(2ev/hf) \sin\left[2\pi(2eV/h)t - 2\pi nft + \phi_0\right] \qquad (7)$$

an integer-order Bessel function of the first kind. This expression time-averages to zero except when $V = nhf/2e$, in which case the supercurrent has a dc component given by $(-1)^n I_1 J_n(2ev/hf) \sin \phi_0$. As for the zero-voltage dc supercurrent, the phase difference ϕ_0 adjusts to accommodate changes in current at this value of V, and the dc current-voltage characteristic exhibits a constant voltage step. The dc characteristic of a Josephson tunnel junction with and without a microwave-frequency ac voltage is shown in Fig. (1). The straightening of the current-voltage characteristic in the presence

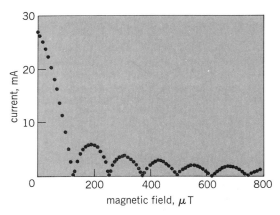

Fig. 2. Magnetic field dependence of the critical current of a Josephson tunnel junction. Data are for a tin–tin oxide–tin junction at 1.2 K, with the magnetic field in the plane of the barrier. (*From D. N. Langenberg, D. J. Scalapino, and B. N. Taylor, The Josephson effects, Sci. Amer., 214(5):30–39, May 1966*)

of microwave power displayed in Fig. 1c is due to the phenomenon of photon-assisted tunneling, which is essentially identical to classical rectification for the junction and frequency in question.

Josephson pointed out that the magnitude of the maximum zero-voltage supercurrent would be reduced by a magnetic field. In fact, the magnetic field dependence of the magnitude of the critical current is one of the more striking features of the Josephson effect. Circulating supercurrents flow through the tunnel barrier to screen an applied magnetic field from the interior of the Josephson junction just as if the tunnel barrier itself were weakly superconducting. The screening effect produces a spatial variation of the transport current, and the critical current goes through a series of maxima and minima as the field is increased. Figure 2 shows the variation of the critical current with magnetic field for a tunnel junction whose length and width are small in comparison with the characteristic screening length of the junction (the Josephson penetration depth, λ_J). The mathematical function which describes the magnetic field dependence of the critical current for this case is the same function as that which describes the diffraction pattern produced when light passes through a single narrow slit.

Josephson junctions. The weak connections between superconductors through which the Josephson effects are realized are known as Josephson junctions. Historically, superconductor-insulator-superconductor tunnel junctions have been used to study the Josephson effect, primarily because these are physical situations for which detailed calculations can be made. However, the Josephson effect is not necessarily a tunneling phenomenon, and the Josephson effect is indeed observed in other types of junctions, such as the superconductor – normal metal – superconductor junction. A particularly useful Josephson junction, the point contact, is formed by bringing a sharply pointed superconductor into contact with a blunt superconductor. The critical current of a point contact can be adjusted by changing the pressure of the contact. The low capacitance of the de-

vice makes it well suited for high-frequency applications. Thin-film microbridges form another group of Josephson junctions. The simplest microbridge is a short narrow constriction (length and width on the order of a few micrometers or smaller) in a superconducting film known as the Anderson-Dayem bridge. If the microbridge region is also thinner than the rest of the superconducting film, the resulting variable-thickness microbridge has better performance in most device applications. If a narrow strip of superconducting film is overcoated along a few micrometers of its length with a normal metal, superconductivity is weakened beneath the normal metal, and the resulting microbridge is known as a proximity-effect or Notarys-Mercereau microbridge. Among the many other types of Josephson junctions are the superconductor-semiconductor-superconductor and other artificial-barrier tunnel junctions, superconductor – oxide – normal metal – superconductor junctions, and the so-called SLUG junction, which consists of a drop of lead-tin solder solidified around a niobium wire. Some different types of Josephson junctions are illustrated in Fig. 3.

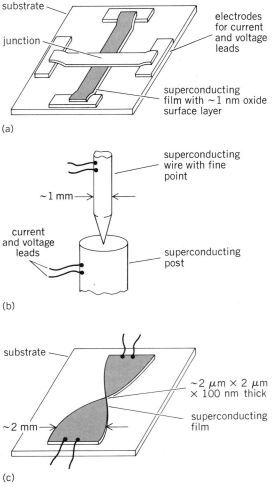

Fig. 3. Some types of Josephson junctions. (*a*) Thin-film tunnel junction. (*b*) Point contact. (*c*) Thin-film weak link. (*From D. N. Langenberg, AC Josephson tunneling-experiment, in E. Burstein and S. Lundqvist, eds., Tunneling Phenomena in Solids, Plenum, 1969*)

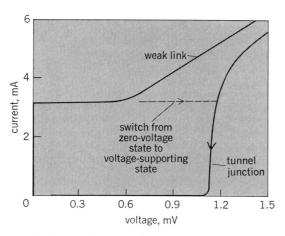

Fig. 4. DC current-voltage characteristics for a weak link and a tunnel junction. (From D. N. Langenberg, AC Josephson tunneling-experiment, in E. Burstein and S. Lundqvist, eds., Tunneling Phenomena in Solids, Plenum, 1969)

The dc current-voltage characteristics of different types of Josephson junctions may differ, but all show a zero-voltage supercurrent, and constant-voltage steps can be induced in the dc characteristics at voltages given by Eq. (2) when an ac voltage is applied. The dc characteristics of a microbridge and a tunnel junction are compared in Fig. 4.

Applications. The United States legal volt, V_{NBS}, is now defined by Eq. (1) through the assigned value given by Eq. (8), and it is maintained at the National Bureau of Standards to an accuracy of within a few parts in 10^8 using the ac Josephson effect;

$$2e/h = 483593420 \text{ MHz/V}_{NBS} \qquad (8)$$

the standards of voltage of most other nations as well as the international volt are similarly defined and maintained. This developed as a natural consequence of extremely precise measurements of $2e/h$ via the Josephson effect, and the recognition that a Josephson junction is a precise frequency-to-voltage converter and that atomic frequency standards are inherently more stable than electrochemical voltage standards.

The Josephson effect permits measurement of absolute temperature: a voltage drop across a resistor in parallel with a Josephson junction causes the junction to emit radiation at the frequency given by Eq. (1), but voltage fluctuations resulting from thermal noise produce frequency fluctuations which depend on absolute temperature. The temperature scale below 1 K is maintained at the National Bureau of Standards via this noise thermometry in conjunction with nuclear-orientation thermometry.

Josephson junctions, and instruments incorporating Josephson junctions, are used in other applications for metrology at dc and microwave frequencies, frequency metrology, magnetometry, detection and amplification of electromagnetic signals, and other superconducting electronics such as high-speed analog-to-digital converters and computers. A Josephson junction, like a vacuum tube or a transistor, is capable of switching signals from one circuit to another; a Josephson tunnel junction is capable of switching states in as little as 6 ps and is the fastest switch known. Josephson

junction circuits are capable of storing information. Finally, because a Josephson junction is a superconducting device, its power dissipation is extremely small, so that Josephson junction circuits can be packed together as tightly as fabrication techniques will permit. All the basic circuit elements required for a Josephson junction computer have been developed. It has been predicted that the first computer to be made will fill a cube 5 cm on a side and will have a cycle time of 2 nanoseconds, at least 10 times faster than an equivalent high-speed semiconductor-based computer. *See* SUPERCONDUCTING DEVICES; SUPERCONDUCTIVITY.

[LOUIS B. HOLDEMAN]

Bibliography: E. Burstein and S. Lundqvist, *Tunneling Phenomena in Solids*, 1969; D. N. Langenberg, D. J. Scalapino, and B. N. Taylor, The Josephson effects, *Sci. Am.*, 214(5):30–39, May 1966; J. Matisoo, The Superconducting computer, *Sci. Am.*, 242(5):50–65, May 1980; L. Solymar, *Superconductive Tunnelling and Applications*, 1972.

Joule's law

A quantitative relationship between the quantity of heat produced in a conductor and an electric current flowing through it. As experimentally determined and announced by J. P. Joule, the law states that when a current of voltaic electricity is propagated along a metallic conductor, the heat evolved in a given time is proportional to the resistance of the conductor multiplied by the square of the electric intensity. Today the law would be stated as $H = RI^2$, where H is rate of evolution of heat in watts, the unit of heat being the joule; R is resistance in ohms; and I is current in amperes. This statement is more general than the one sometimes given that specifies that R be independent of I. Also, it is now known that the application of the law is not limited to metallic conductors.

Although Joule's discovery of the law was based on experimental work, it can be deduced rather easily for the special case of steady conditions of current and temperature. As a current flows through a conductor, one would expect the observed heat output to be accompanied by a loss in potential energy of the moving charges that constitute the current. This loss would result in a descending potential gradient along the conductor in the direction of the current flow, as usually defined. If E is the total potential drop, this loss, by definition, is equal to E in joules for every coulomb of charge that traverses the conductor. The loss conceivably might appear as heat, as a change in the internal energy of the conductor, as work done on the environment, or as some combination of these. The second is ruled out, however, because the temperature is constant and no physical or chemical change in a conductor as a result of current flow has ever been detected. The third is ruled out by hypothesis, leaving only the generation of heat. Therefore, $H = EI$ in joules per second, or watts. By definition, $R = E/I$, a ratio which has positive varying values. Elimination of E between these two equations gives the equation below,

$$H = RI^2$$

which is Joule's law as stated above. If I changes

to a new steady value I', R to R', and H and H', then $H' = R'I'^2$ as before. The simplest case occurs where R is independent of I. If the current is varying, the resulting variations in temperature and internal energy undoubtedly exist and, strictly speaking, should be allowed for in the theory. Yet, in all but the most exceptional cases, any correction would be negligible.

This phenomenon is irreversible in the sense that a reversal of the current will not reverse the outflow of heat, a feature of paramount importance in many problems in physics and engineering. Thus the heat evolved by an alternating current is found by taking the time average of both sides of the equation. Incidentally, the changes in internal energy, if they were included in the theory, would average out. Hence the equation continues to have a similar form, $\overline{H} = \overline{RI^2}$, for ac applications. *See* OHM'S LAW.

[LLEWELLYN G. HOXTON/JOHN W. STEWART]

Junction diode

A semiconductor rectifying device in which the barrier between two regions of opposite conductivity type produces the rectification (Fig. 1). Junction diodes are used in computers, radio and television, brushless generators, battery chargers, and electrochemical processes requiring high direct current and low voltage. Lower-power units are usually called semiconductor diodes, and the higher-power units are usually called semiconductor rectifiers. For a discussion of conductivity types, carriers, and impurities *see* SEMICONDUCTOR.

Junction diodes are classified by the method of preparation of the junction, the semiconductor material, and the general category of use of the finished device. By far the great majority of modern junction diodes use silicon as the basic semiconductor material. Germanium material was used in the first decade of semiconductor diode technology, but has given way to the all-pervasive silicon technology, which allows wider temperature limits of operation and produces stable characteristics more easily. Other materials are the group III–V compounds, the most common being gallium arsenide, which is used where its relatively large band-gap energy is needed. A partial list of silicon types includes the diffused silicon switching diode, alloyed silicon voltage reference diode, epitaxial planar silicon photodiode, and diffused silicon rectifier. Other types include the ion-implanted varactor diode and the gallium arsenide light-emitting diode.

In silicon units nearly all categories of diodes are made by self-masked diffusion, as shown in Fig. 2a. Exceptions are diodes where special control of the doping profile is necessary. In such cases, a variety of doping techniques may be used, including ion implantation, alloying with variable recrystallization rate, silicon transmutation by neutron absorption, and variable-impurity epitaxial growth. The mesa structure shown in Fig. 2b is used for some varactor and switching diodes if close control of capacitance and voltage breakdown is required. *See* DETECTOR; ELECTRONIC SWITCH; SEMICONDUCTOR DIODE; RECTIFIER.

Fabrication methods. The alloy and mesa techniques are largely historical, but were important in the development of junction diodes. The alloy junction section (Fig. 1) is produced by placing a pill of doped alloying material on the clean flat surface of a properly oriented semiconductor wafer and heating it until the molten alloy dissolves a portion of the semiconductor immediately beneath it. Upon cooling, the dissolved semiconductor, now containing the doping impurity, recrystallizes upon the surface of the undissolved semiconductor, reproducing its crystal structure and creating a *pn* junction at the position marking the limit of the solution of the original wafer. If such a junction is held at the peak temperature of its alloying cycle for sufficient time to allow diffusion of the alloy impurity beyond the limit of the dissolved semiconductor into the solid semiconductor, the junction produced is called alloy-diffused.

The planar diffused junction section (Fig. 2a) is produced in silicon by first polishing the top surface of a large silicon wafer and then oxidizing the surface by heating the wafer at about 1000°C in the presence of wet oxygen. After about 0.5 μm of oxide has grown on the surface, the wafer is cooled, and an array of holes is opened in the oxide by high-precision etching geometrically controlled by a photoresist technique. A very heavily doped thin oxide layer is chemically deposited in the holes

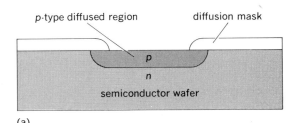

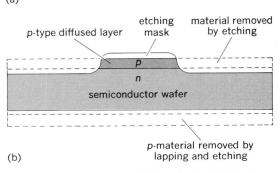

Fig. 2. High-speed diffused silicon diodes. (*a*) Mesaless structure. (*b*) Mesa structure.

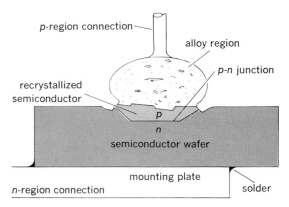

Fig. 1. Section of a bonded or fused junction diode.

opened in the oxide. This predeposition step is followed by a drive-in diffusion at a higher temperature, causing the deposited impurity to penetrate the substrate, thereby forming diffused *pn* junctions beneath each hole. Subsequently the individual junctions are separated out of the large wafer by scribing and breaking and are encapsulated as individual diodes. Such planar diffused diodes have relatively high breakdown voltages and low leakage currents. The ends of the junction are automatically protected by the oxide mask so that such diodes show long-term stability. This protection by the oxide is often referred to as passivation.

Planar diodes and planar transistors are used in integrated circuits. The diodes in integrated circuits usually consist of the emitter junction or collector junction of a transistor structure rather than being fabricated as a separate diode. Most discrete diodes are power rectifiers, voltage regulators, varactors, or light-emitting diodes. *See* Integrated circuits; Varactor; Voltage regulator.

The mesa structure (Fig. 2*b*) is produced by diffusing the entire surface of the large wafer and then delineating the individual diode areas by a photoresist-controlled etch that removes the entire diffused area except the island or mesa at each diode site.

Still another method of doping control used in modern diodes is through epitaxially deposited material. In this process the polished wafer is subjected at an elevated temperature to a vapor containing a compound of the semiconductor together with a compound containing the appropriate doping element. These compounds decompose upon contact with the surface of the wafer and cause the semiconductor to grow a layer of doped material on its surface. Under proper conditions of cleanliness and growth rate, the underlying crystal structure is propagated into the growing layer, which is then said to be epitaxial in character. In this way either localized areas or entire surfaces of either conductivity type may be produced. In diode fabrication it is typical to use the epitaxially grown material as a lightly doped layer over the original substrate material of the same conductivity type. The junction is then formed in the lightly doped layer by masked diffusion of the opposite-conductivity-type material. By this means the thickness of the web of lightly doped material immediately beneath the diffusion can be controlled to give both a desired reverse breakdown voltage and a relatively constant capacitance. Forward-bias recovery time can be controlled in a trade-off with reverse breakdown voltage in such a structure.

A method of doping control used when special doping concentration profiles are needed, or when localized doping must be accomplished without self-masking oxide, is ion implantation. At present the largest use of this technique in *pn* junction fabrication is to replace the chemical predeposition step in the planar diffusion process. Here ion implantation gives a much more precise control of the sheet resistivity of the diffusion, and it can be accomplished without opening holes in the protective oxide. Crystal damage is automatically healed during the subsequent drive-in diffusion. *See* Ion implantation.

Junction rectification. Rectification occurs in a semiconductor wherever there is a relatively abrupt change of conductivity type. In any semiconductor the product of the concentrations of the majority and minority current carriers is a temperature-dependent equilibrium constant. The conductivity is proportional to the majority carrier concentration and inversely proportional to the minority-carrier concentration. When a *pn* junction is reverse-biased (*p*-region negative with respect to the *n*-region), the majority carriers are blocked completely by the barrier, and only the minority carriers can flow under the barrier. This minority carrier current is the sum of the individual currents from the *n*- and *p*-regions, and each component is inversely proportional to the conductivity of its region. In addition, there is a thermal regeneration current of minority carriers generated in the depletion region of the reverse-biased junction. In silicon the regeneration current dominates and is about 10^{-3} A/m^2 at room temperature.

When a *pn* junction is forward-biased (*p*-region positive with respect to the *n*-region), the majority hole and electron distributions can flow into the opposite region because the bias has markedly lowered the barrier. Since electrons flowing into a *p*-region or holes flowing into an *n*-region represent a great increase in minority-carrier concentration, the thermodynamic equilibrium of the holes and electrons is disturbed, and the product of their concentrations increases as the junction is approached. The resistivity of both the *n*- and *p*-type regions is considerably lowered by these excess minority carriers, and the forward current is greater than the current through a geometrically equivalent bar of material containing no *pn* junction.

The electrons in an *n*-type semiconductor are given up to the conduction process by donor impurity atoms which remain as fixed, positively charged centers. Similarly, the holes of a *p*-region are created by the capture of electrons by acceptor impurity atoms which remain as fixed, negatively charged centers. In both cases the space charge of the ionized impurity centers is neutralized by the space charge of the majority carriers.

At a *pn* junction the barrier that keeps majority carriers away consists of a dipole layer of charged impurity centers, positive on the *n*-type side and negative on the *p*-type side. When a reverse bias is applied, the barrier height increases and requires more charge in the dipole layer to produce the required step in voltage. To add to the charge, the layer must widen, because ionized impurities are in fixed positions in the crystal. As the layer widens, the capacitance of the junction decreases since the plates of the capacitor are farther apart. Therefore, a *pn* junction acts as a variable capacitance as well as a variable resistance.

Optical properties. When light of sufficient energy is absorbed by a semiconductor, excess minority carriers are created. In a *pn* junction device these excess carriers will increase the reverse-bias leakage current by a large factor if they are within diffusion distance of the junction. If the junction is open-circuited, a forward voltage will develop to oppose the diffusion of the excess carriers generated by the light absorption. This photovoltaic response is the basis of the operation of solar cells and most photographic exposure meters. *See* Photovoltaic cell; Photovoltaic effect; Solar cell.

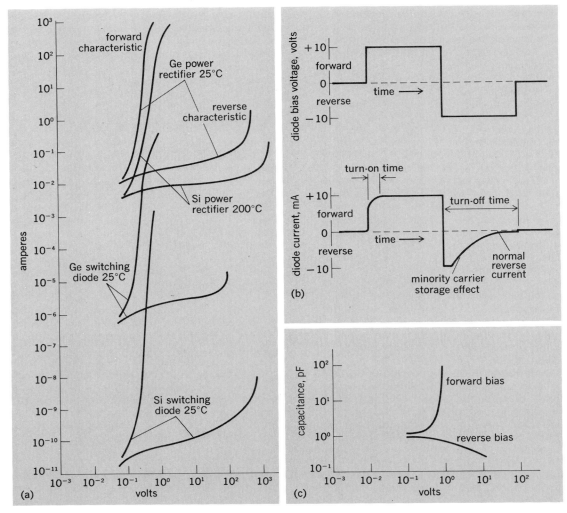

Fig. 3. Junction diode characteristics. (*a*) Rectification. (*b*) Switching. (*c*) Silicon switching diode capacitance.

The inverse of the above effect also exists. When a *pn* junction in a so-called direct-gap semiconductor is forward-biased, the electrically injected excess minority carriers recombine to generate light. This is the basis of light-emitting diodes and injection lasers. Typical direct-gap semiconductors (suitable for light-emitting diodes) are compounds between III and V group elements of the periodic table such as gallium arsenide. *See* LASER.

Characteristics. Figure 3 shows typical rectification, switching, and capacitance characteristics of a junction diode. Rectification characteristics (Fig. 3*a*) show that silicon units provide much lower reverse leakage currents and higher voltage breakdowns and can operate up to 200°C.

For switching purposes, turn-on and turn-off times are most important (Fig. 3*b*). The turn-on time of a diode is governed by its junction capacitance and is usually short. The turn-off time, usually the critical characteristic, is governed by the time required to remove all of the excess minority carriers injected into the *n*- and *p*-regions while the diode was in the forward-bias state. This is called the minority carrier storage effect, and it is of the order of a few microseconds for good switching

diodes. Silicon diodes are usually somewhat superior to germanium units in this respect. The limits of operation of present junction diodes are about 2500 V reverse-standoff voltage and 1500 A forward current in power rectifiers; about 1.0 nanosecond reverse recovery time and 100 picoseconds rise time for fast-switching diodes; a minimum reverse leakage current in a small signal diode is about 0.01 nA.

For further discussion of the properties of *pn* junctions *see* JUNCTION TRANSISTOR; TRANSISTOR. *See also* POINT-CONTACT DIODE.

[LLOYD P. HUNTER]

Bibliography: E. S. Yang, *Fundamentals of Semiconductor Devices*, 1978.

Junction transistor

A transistor in which emitter and collector barriers are formed by *pn* junctions between semiconductor regions of opposite conductivity type. These junctions are separated by a distance considerably less than a minority-carrier diffusion length, so that minority carriers injected at the emitter junction will not recombine before reaching the collector barrier and therefore be effective in modulating the collector-barrier impedance. Junction transistors are widely used both as discrete devices and in integrated circuits. The discrete devices are found in the high-power and high-frequency applications. Junction transistors range in power rating from a few milliwatts (mW) to about 300 W, in characteristic frequency from 0.5 to 2000 MHz, and in gain from 10 to 50 dB. Silicon is the most widely used semiconductor material, although germanium is still used for some applications. Junction transistors are applicable to any electronic amplification, detection, or switching problem not requiring operation above 200°C, 700 V, or 2000 MHz. Not all these limits can be achieved in one device, however. Junction transistors are classified by the number and order of their regions of different conductivity type, by the method of fabricating and structure, and sometimes by the principle of operation. Most modern transistors are fabricated by the silicon self-masked planar double-diffusion technique. The alloy technique and the grown-junction technique are primarily of historical importance. For a general description and definition of terms used here and a description of the mechanism of operation *see* TRANSISTOR.

Alloy-junction transistors. Also called fused-junction transistors, these are made in the *pnp* and *npn* forms. The emitter and collector regions are formed by recrystallization of semiconductor material from a solution of semiconductor material dissolved in some suitably doped metal alloy. The major metal of the alloy serves as the solvent for the semiconductor, while the minor element serves as a source for doping impurity in order to render the recrystallized material opposite in conductivity type to the original wafer.

Alloy junctions are abrupt and allow for bidirectional operation. They usually show a low series resistance, and were therefore used in high-power transistors.

Figure 1 compares several transistor profiles which show how the impurity content varies through the structure. In these profiles C_p is the concentration of the *p*-type impurity; C_n is the concentration of the *n*-type impurity. The net impurity content determines the conductivity type and magnitude. The profile of the alloy transistor shows that there are abrupt changes of impurity concentration at emitter and collector junctions and that the conductivities of emitter and collector regions are therefore high compared to those of the base region. Such a structure shows good emitter-injection efficiency but only moderate collector-voltage rating and relatively high collector capacitance. *See* SEMICONDUCTOR.

Grown-junction transistors. These are made in the *pnp* and *npn* forms, as well as in more complicated forms. There are several variations of the grown-junction technique. The simplest consists of successively adding different types of impurities to the melt from which the semiconductor crystal is being grown.

A semiconductor crystal is usually grown by dipping the end of a seed crystal into molten semiconductor and by arranging the thermal gradients so that new semiconductor solidifies on the end of the seed as it is slowly withdrawn. The solid-liquid interface is roughly a plane perpendicular to the axis of withdrawal. A *pnp* structure can be grown by starting with a *p*-type melt; by adding, at one point in the crystal growth, enough *n*-type impurity to give a slight excess over the *p*-type impurity originally present; and, after growth has continued for a few micrometers, by adding an excess of *p*-type impurity to the melt. The last-grown region will be the emitter region, and the original *p*-type crystal will be the collector region. The impurity profile of such a structure is shown in Fig. 1*b*.

The high-conductivity emitter region gives a good injection efficiency, and the junction between the base and collector regions is gradual enough so that the unit will show a relatively low collector

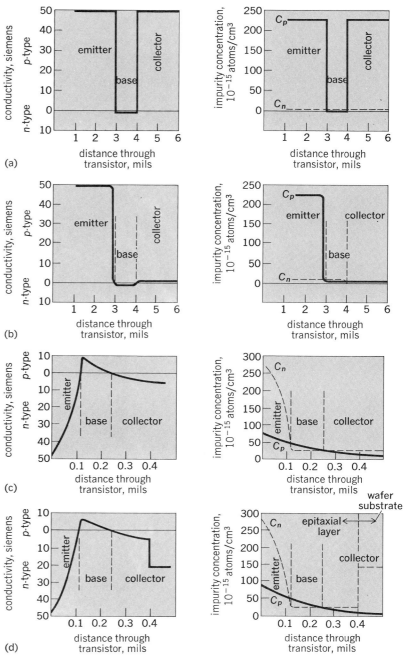

Fig. 1. Conductivity and impurity profiles of typical junction transistors; 1 mil = 25.4 μm. (*a*) *pnp* alloy-junction type. (*b*) *pnp* grown-junction type. (*c*) *npn* double-diffused-junction type. (*d*) *npn* epitaxial double-diffused-junction type.

capacitance and a high breakdown voltage. The one disadvantage of this method is that both the collector and base regions show relatively high series resistances.

Planar diffused epitaxial transistors. The structure of this transistor is shown in section in Fig. 2, and the doping profile through the emitter, base, and collector is shown in Fig. 1d. In this structure both collector and emitter junctions are formed by diffusion of impurities from the top surface, as shown in Fig. 2. Using silicon, the structure is formed by growing a diffusion mask of native oxide (silicon dioxide) on the previously polished wafer. A hole is opened in the oxide by a photoresist etch technique (Fig. 2a) to define the area of the collector buried layer. For a p-type substrate a heavy concentration (n^+) of n-type impurity such as phosphorus is diffused into the substrate through the opening in the masking oxide. The oxide is etched away, and an epitaxial layer of lightly doped (n-) silicon is grown over the entire wafer by vapor decomposition at a temperature low enough to prevent significant diffusion of the n^+ material out of the buried layer (Fig. 2b). A new oxide layer is grown on the surface of the epitaxial layer, and an opening is etched in it to define the limits of the p-type base diffusion (Fig. 2c). (This automatically controls the collector junction geometry and capacitance.) The masking oxide is again stripped and regrown for the masking of the n^+ diffusion used to form the emitter and collector contact region (Fig. 2d). Next the emitter mask is removed, and an impervious layer of oxide is formed over the surface of the crystal. A layer of glass is bonded to the crystal by means of the oxide layer. The glass must match the expansion coefficient of silicon fairly well, and the oxide must be sufficiently impervious to the glass at the bonding temperature to prevent the diffusion of impurities from the glass into the silicon transistor structure. Finally, holes are etched in the glass-oxide structure so that electrical contact can be made to the various regions of the transistor (Fig. 2e).

In modern technology the above-described base and emitter diffusions are carried out in two steps: a predeposition step, in which a very thin layer of heavily doped oxide is chemically deposited over the open surface of the silicon in the hole opened in the masking oxide; and a drive-in diffusion step, in which the deposited dopant is diffused into the silicon at a higher temperature than that used for the predeposition. This typically controls the sheet resistance of the final diffusions to about ±10% of the design value. The forward current-transfer ratio of the transistor is determined by the ratio of these sheet resistances through the medium of the injection efficiency of minority carriers.

The chemical predeposition step is being replaced by ion implantation directly through the oxide. The masking is accomplished by placing a layer of photoresist on top of the oxide. This eliminates the oxide etching step and allows an accurately metered deposition by controlling the time of bombardment and the current of the ion beam. This modification of the process promises to keep the emitter and base region sheet resistivities within about ±1% of the design value. *See* ION IMPLANTATION.

In this transistor, formation of the base region

by diffusion from the emitter side produces a steep doping gradient and thereby a strong electric field in the base region. In the typical alloy-junction transistor (uniform base doping) the minority-carrier transport across the base is achieved by a relatively slow diffusion process. In this diffused base type (sometimes called a drift transistor) the base region shows a high conductivity gradient, decreasing from the emitter to the collector (Fig. 1d). This conductivity gradient means that the majority-carrier concentration is much greater near the emitter than near the collector. In order to cancel the natural diffusion of majority carriers from the high- to the low-concentration region, an electric field must exist of such a polarity as to tend to drive majority carriers back toward the emitter. This polarity of field then tends to drive minority carriers from the emitter to the collector; when normal bias is applied to the device, excess inject-

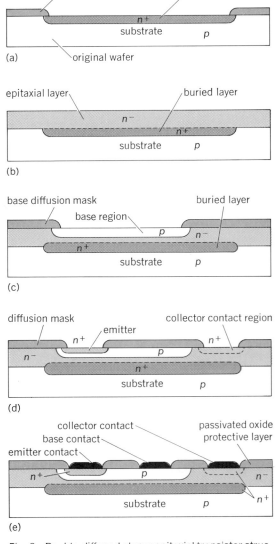

Fig. 2. Double-diffused planar epitaxial transistor structure and method of fabrication. (a) Buried layer. (b) Epitaxial layer. (c) Collector junction formation. (d) Emitter junction. (e) Contact stripe placement.

ed minority carriers will be accelerated across the base by this field.

The buried layer of n^+ doped material has very low resistance and acts as a shorting bar between the area immediately beneath the emitter and the area immediately beneath the collector contact stripe, thus maintaining a low collector series resistance even if the n^- material of the epitaxial region is of quite high resistivity. The collector breakdown voltage may be maintained at a reasonably high value and the collector capacitance at a low value by controlling the thickness of the n^- material between the base and the buried layer and by keeping the doping level of the n^- material quite low.

Mesa transistors. These transistors minimize the collector capacitance by limiting the collector junction area. This area limitation is achieved by etching away the surrounding material so that the entire transistor structure stands up from the body of the wafer like a small mesa. This structure gave the highest frequency response for many years. It is now replaced by the planar type of structure.

Power transistors. These are used in the output stage of an electronic circuit both as switches and as amplifiers. Depending on the load, a high voltage rating, a high current rating, or a high power rating may be required. With any of these, heat dissipation within the device is a serious limitation. In order to obtain high current capability in a power transistor, a large emitter junction area is required. The base-region recombination current produces a lateral voltage drop between the center of the emitter area and the center of the base contact area in a planar device. This voltage tends to bias the center of the emitter area to off and concentrate the injection current at the periphery of the emitter. Modern silicon power transitors minimize this effect by using emitter junctions with a large perimeter-to-area ratio, usually in the form of a multipronged fork with base contacts interdigitated between the tines. This preserves high forward current-transfer ratio to large values of emitter current. *See* CONTROLLED RECTIFIER.

Unijunction transistor. This device is really a diode in which the resistance of a portion of the base region is modulated by minority carriers injected by forward-biasing its single junction. Its structure typically consists of a lightly doped base region with ohmic contacts at opposite ends. The single junction is formed over a narrow range near the center of the base region by a shallow diffusion of heavily doped material of the opposite conductivity type. If a bias current is set up from end to end in the base, the potential at the junction can be set at a desired reverse bias relative to ground. If a signal is applied to the junction electrode, the device will turn on when the signal exceeds the original reverse-bias potential of the base at that point. Once forward-biased, the junction injects sufficient minority carriers into the base to short the region beneath the junction to the ground end of the base, and the device remains conducting until reset, either by the base bias or by the emitter signal. These devices show a typical negative resistance characteristic and are used for timing, control, and sensing circuits.

Summary. Silicon planar passivated transistors show a wide range of performance with character-istic frequencies up to 2000 MHz, voltage ratings of 12–700 volts and power dissipation ratings of 100 mW–300 watts. The highest-frequency devices range up to 4000 MHz. Silicon planar technology is used in fabricating integrated circuit chips. The general form of the transistor structure displayed in Fig. 2 is used in integrated circuits. Such a structure is used for diodes as well as transistors since, for example, it is necessary only to connect the base and collector contacts to use the collector junction as a diode. *See* INTEGRATED CIRCUITS.   [LLOYD P. HUNTER]

Bibliography: L. P. Hunter, *Handbook of Semiconductor Electronics*, 3d ed., 1970; E. S. Yang, *Fundamentals of Semiconductor Devices*, 1978.

Kelvin bridge

A specialized version of the Wheatstone bridge network designed to eliminate, or greatly reduce, the effect of lead and contact resistance and thus permit accurate measurement of low resistance. The circuit shown in the figure accomplishes this by effectively placing relatively high-resistance-ratio arms in series with the potential leads and contacts of the low-resistance standards and the unknown resistance. In this circuit R_A and R_B are the main ratio resistors, R_a and R_b the auxiliary ratio, R_x the unknown, R_s the standard, and R_y a heavy copper yoke of low resistance connected between the unknown and standard resistors.

By applying a delta-wye transformation to the network consisting of R_a, R_b, and R_y, the equivalent Wheatstone bridge network shown in the illustration is obtained, where Eqs. (1) hold. By an anal-

$$R'_s = \frac{R_y R_a}{R_y + R_a + R_b}$$

$$R'_x = \frac{R_y R_b}{R_y + R_a + R_b} \tag{1}$$

$$R'_G = \frac{R_a R_b}{R_y + R_a + R_b}$$

ysis similar to that for the Wheatstone bridge, it can be shown that for a balanced bridge Eq. (2) holds. If Eq. (3) is valid, the second term of Eq. (2) is zero, the measurement is independent of R_y, and Eq. (4) is obtained.

$$R_x = \frac{R_B}{R_A} R_s + R_y \left(\frac{R_b}{R_a + R_b + R_y}\right)\left(\frac{R_B}{R_A} - \frac{R_b}{R_a}\right) \tag{2}$$

$$\frac{R_B}{R_A} = \frac{R_b}{R_a} \tag{3}$$

$$R_x = \frac{R_B}{R_A} R_s \tag{4}$$

As with the Wheatstone bridge, the Kelvin bridge for routine engineering measurements is constructed using both adjustable ratio arms and adjustable standards. However, the ratio is usually continuously adjustable over a short span, and the standard is adjustable in appropriate steps to cover the required range. *See* WHEATSTONE BRIDGE.

Sensitivity. The Kelvin bridge sensitivity can be calculated similarly to the Wheatstone bridge. The open-circuit, unbalance voltage appearing at the

KELVIN BRIDGE

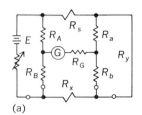

(a)

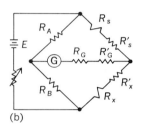

(b)

Kelvin bridge.
(a) Actual circuit.
(b) Equivalent Wheatstone bridge circuit.

detector terminals may be expressed, to a close degree of approximation, as in Eq. (5).

$$e = E \frac{r}{(r+1)^2} \left[\frac{\Delta R_x}{R_x + R_y \left(\dfrac{r}{r+1} \right)} \right] \qquad (5)$$

The unbalance detector current for a closed detector circuit may be expressed as in Eq. (6).

$$I_G = \frac{E \left(\dfrac{\Delta R_x}{R_x} \right)}{\dfrac{R_G}{r/(r+1)^2} + R_A + R_B + R_a + R_b} \qquad (6)$$

The Kelvin bridge requires a power supply capable of delivering relatively large currents during the time a measurement is being made. The total voltage applied to the bridge is usually limited by the power dissipation capabilities of the standard and unknown resistors.

Errors. Kelvin bridge resistance-measurement errors are caused by the same factors as for the Wheatstone bridge. However, additional sources of error, as implied by the second term of Eq. (2), must be evaluated since these factors will seldom be reduced to zero. For minimum error the yoke resistance should be made as low as possible by physically placing the commonly connected current terminals of the unknown and standard as close together as possible and connecting with a low-resistance lead.

The ratio resistors each include not only the resistance of the resistors but also that of the interconnecting wiring and external leads and the contact resistance of the potential circuit contacts. The external leads are most likely to cause errors, and they should therefore be of the same resistance so that Eq. (3) will be fulfilled as nearly as possible. In addition, they should be relatively short, since the addition of a large resistance (long leads) will introduce an error in the calibrated ratio R_B/R_A. For precise measurements, trimmer adjustments are required in the ratio-arm circuits and provision is made to connect the bridge resistors into two different Wheatstone bridge configurations. By successively balancing first the Kelvin network and then each of the Wheatstone networks, these additive errors are virtually eliminated. *See* BRIDGE CIRCUIT; RESISTANCE MEASUREMENT.   [CHARLES E. APPLEGATE]

Kirchhoff's laws of electric circuits

Fundamental natural laws dealing with the relation of currents at a junction and the voltages around a loop. These laws are commonly used in the analysis and solution of networks. They may be used directly to solve circuit problems, and they form the basis for network theorems used with more complex networks.

In the solution of circuit problems, it is necessary to identify the specific physical principles involved in the problem and, on the basis of them, to write equations expressing the relations among the unknowns. Physically, the analysis of networks is based on Ohm's law giving the branch equations, Kirchhoff's voltage law giving the loop voltage equations, and Kirchhoff's current law giving the node current equations. Mathematically, a

network may be solved when it is possible to set up a number of independent equations equal to the number of unknowns. *See* CIRCUIT (ELECTRICITY); NETWORK THEORY.

When writing the independent equations, current directions and voltage polarities may be chosen arbitrarily. If the equations are written with due regard for these arbitrary choices, the algebraic signs of current and voltage will take care of themselves.

Kirchhoff's voltage law. One way of stating Kirchhoff's voltage law is: "At each instant of time, the algebraic sum of the voltage rise is equal to the algebraic sum of the voltage drops, both being taken in the same direction around the closed loop."

The application of this law may be illustrated with the circuit in Fig. 1. First, it is desirable to

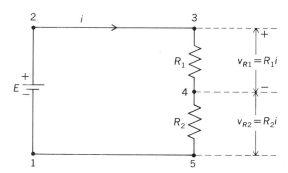

Fig. 1. Simple loop to show Kirchhoff's voltage law.

consider the significance of a voltage rise and a voltage drop, in relation to the current arrow. The following definitions are illustrated by Fig. 1.

A voltage rise is encountered if, in going from 1 to 2 in the direction of the current arrow, the polarity is from minus to plus. Thus, E is a voltage rise from 1 to 2.

A voltage drop is encountered if, in going from 3 to 4 in the direction of the current arrow, the polarity is from plus to minus. Thus, $v_{R1} = R_1 i$ is a voltage drop from 3 to 4. The application of Kirchhoff's voltage law gives the loop voltage, Eq. (1).

$$E = v_{R1} + v_{R2} = R_1 i + R_2 i \qquad (1)$$

In the network of Fig. 2 the voltage sources have the same frequency. The positive senses for the branch currents I_R, I_L, and I_C are chosen arbitrarily, as are the loop currents I_1 and I_2. The voltage equations for loops 1 and 2 can be written using instantaneous branch currents, instantaneous loop currents, phasor branch currents, or phasor loop currents.

The loop voltage equations are obtained by applying Kirchhoff's voltage law to each loop as follows.

By using instantaneous branch currents, Eqs. (2)

$$e_{g1} = R i_R + L \frac{di_L}{dt} \qquad (2)$$

and (3) may be obtained. By using instantaneous

$$e_{g2} = \frac{1}{C} \int i_C \, dt + L \frac{di_L}{dt} \qquad (3)$$

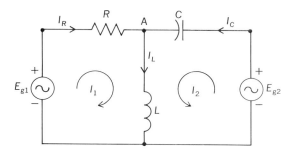

Fig. 2. Two-loop network demonstrating the application of Kirchhoff's voltage law.

loop currents, Eqs. (4) and (5) are obtained. Equations (6) and (7) are obtained by using phasor

$$e_{g1} = Ri_1 + L\frac{d(i_1 + i_2)}{dt} \qquad (4)$$

$$e_{g2} = \frac{1}{C}\int i_2\,dt + L\frac{d(i_2 + i_1)}{dt} \qquad (5)$$

tions (6) and (7) are obtained by using phasor

$$\mathbf{E}_{g1} = R\mathbf{I}_R + j\omega L\mathbf{I}_L \qquad (6)$$

$$\mathbf{E}_{g2} = -j\frac{1}{\omega C}\mathbf{I}_C + j\omega L\mathbf{I}_L \qquad (7)$$

branch currents. By using phasor loop currents, Eqs. (8) and (9) may be obtained.

$$\mathbf{E}_{g1} = R\mathbf{I}_1 + j\omega L(\mathbf{I}_1 + \mathbf{I}_2) \qquad (8)$$

$$\mathbf{E}_{g2} = -j\frac{1}{\omega C}\mathbf{I}_2 + j\omega L(\mathbf{I}_2 + \mathbf{I}_1) \qquad (9)$$

Kirchhoff's current law. Kirchhoff's current law may be expressed as follows: "At any given instant, the sum of the instantaneous values of all the currents flowing toward a point is equal to the sum of the instantaneous values of all the currents flowing away from the point."

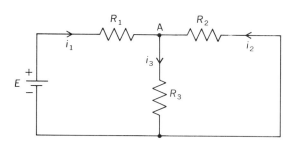

Fig. 3. Circuit demonstrating Kirchhoff's current law.

The application of this law may be illustrated with the circuit in Fig. 3. At node A in the circuit in Fig. 3, the current is given by Eq. (10).

$$i_1 + i_2 = i_3 \qquad (10)$$

The current equations at node A in Fig. 2 can be written by using instantaneous branch currents or phasor branch currents.

By using instantaneous branch currents, Eq. (11) is obtained.

$$i_R + i_C = i_L \qquad (11)$$

By using phasor branch currents, Eq. (12) is obtained.

$$\mathbf{I}_R + \mathbf{I}_C = \mathbf{I}_L \qquad (12)$$

See DIRECT-CURRENT CIRCUIT THEORY.

[K. Y. TANG/ROBERT T. WEIL]

Klystron

An evacuated electron-beam tube in which an initial velocity modulation imparted to electrons in the beam results subsequently in density modulation of the beam. A klystron is used either as an amplifier in the microwave region or as an oscillator. For use as an amplifier, a klystron receives microwave energy at an input cavity through which the electron beam passes. The microwave energy modulates the velocities of electrons in the beam, which then enters a drift space. Here the faster electrons overtake the slower to form bunches. In this manner, the uniform current density of the initial beam is converted to an alternating current. The bunched beam with its significant component of alternating current then passes through

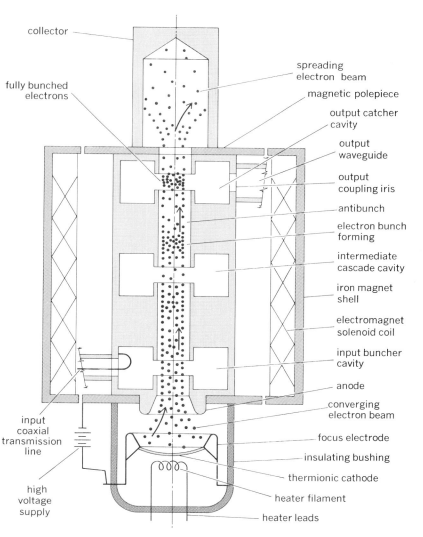

Fig. 1. Cross section of cascade klystron amplifier. (*Varian Associates*)

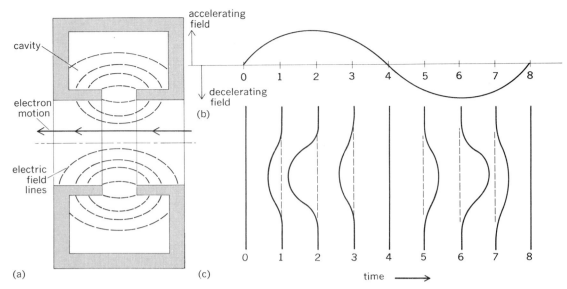

Fig. 2. Cavity concentrates electric field between the reentrant noses. As field varies sinusoidally with time, an electron crossing in the gap between the noses experiences an electric field whose strength and direction depend on the instantaneous phase of the field. (*a*) Map of instantaneous field. (*b*) Cycle variation of field. (*c*) Profile of field on the axis at various times in the cycle. (*Varian Associates*)

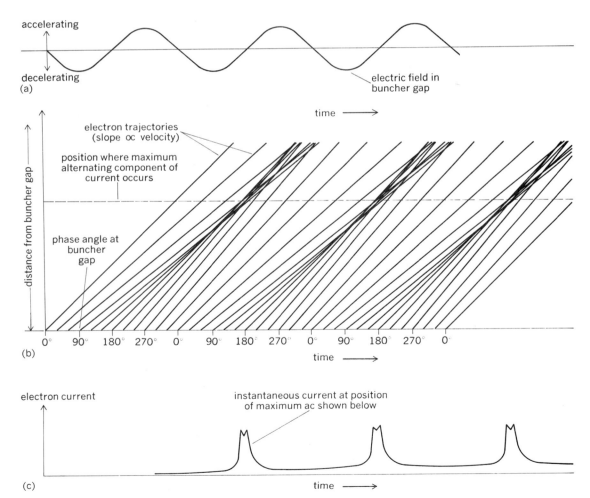

Fig. 3. Bunching the electron stream. (*a*) As the electric field varies periodically, electrons traversing the cavity are speeded up or slowed down. (*b*) Distance-versus-time lines graph the gradual formation of bunches. (*c*) Current passing a fixed point becomes periodic, that is, it becomes alternating. (*Varian Associates*)

an output cavity to which the beam transfers its ac energy. *See* MICROWAVE.

Klystron amplifier. In a typical klystron (Fig. 1), a stream of electrons from a concave thermionic cathode is focused into a smaller cylindrical beam by the converging electrostatic fields between the anode, cathode, and focusing electrode. The beam passes through a hole in the anode and enters a magnetic field parallel to the beam axis. The magnetic field holds the beam together, overcoming the electrostatic repulsion between electrons which would otherwise make the beam spread out rapidly. The electron beam goes through the cavities of the klystron in sequence, emerges from the magnetic field, spreads out, and is stopped in a hollow collector where the remaining kinetic energy of the electrons is dissipated as heat. *See* ELECTRON MOTION IN VACUUM.

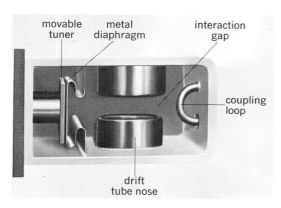

Fig. 5. Detail of input cavity for four-cavity klystron amplifier of Fig. 4. (*Varian Associates*)

The signal wave to be amplified is introduced into the first, or buncher, cavity through a coaxial transmission line. This hollow metal cavity is a resonant circuit, analogous to the familiar inductance-capacitance combination, with the electric field largely concentrated in the reentrant noses so that the highest voltage occurs between them. The inductance may be considered as a single-turn conductor formed by the outer metal walls. In Fig. 1 the current in the center conductor of the input transmission line flows through a loop inside the cavity and back to the outer conductor. The magnetic flux generated in the loop links through the cavity inductance, as in a transformer. At the resonant frequency of the cavity, the voltage across the reentrant section through which the electron beam passes is built up by the cavity configuration to 10–100 times the voltage in the input line. *See* CAVITY RESONATOR.

Figure 2 shows the pattern of the electric field in the cavity and how it varies cyclically with time. As electrons pass through the gap in which the cavity field is concentrated (Fig. 2*a*), they are accelerated or decelerated (Fig. 2*b*), depending on the instantaneous direction of the field.

Figure 3 illustrates graphically the effect of these velocity changes. Each slanted line represents the flight of an electron as a function of time measured in electrical degrees as the electron travels from the buncher gap. The slope of a line is

thus the velocity of that electron. The velocities leaving the buncher vary sinusoidally with time, as determined by the instantaneous field. The horizontal broken line represents a fixed point beyond the buncher. The flow of electrons past this point is given by the time sequence in which the electron paths cross the broken line. Figure 3 shows how the electrons have gathered into bunches. The rate of current flow is now periodic with time so that the current has an alternating component.

When the bunched beam passes through a second cavity, its space charge induces in the walls of this cavity an alternating current of opposite sign to the electron current. The cavity is tuned to the input frequency so that it has a high resonant impedance; the induced current flowing through this impedance generates voltage in the cavity.

In Fig. 1 the second cavity is not coupled to any outside circuits. Voltage built up here by the beam current produces further velocity modulation in the beam. The resulting alternating current component is about 10 times greater than the initial current. More of these uncoupled cascade cavities can be added for increased amplification.

The final output cavity is coupled into a transmission line (a waveguide in Fig. 1) which carries off the generated power to its useful destination. Because the cavity is tuned to resonance, its reactance is canceled. The induced current flowing through its pure resistive impedance generates inphase voltage in the direction opposing the current flow. Thus the field in the gap is at its maximum decelerating value at the time a bunch of electrons

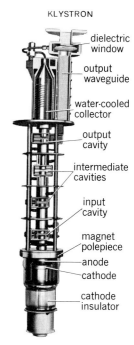

Fig. 4. Cutaway view of four-cavity amplifier. (*Varian Associates*)

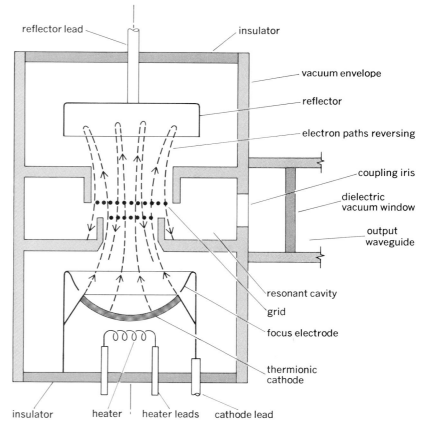

Fig. 6. Schematic cross section of reflex oscillator. (*Varian Associates*)

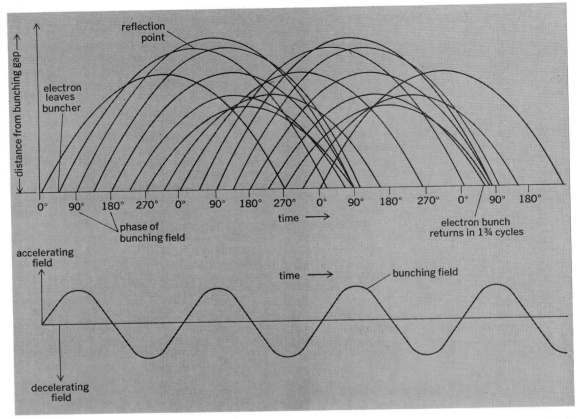

Fig. 7. Trajectories show bunching in a reflex oscillator. Electrons are turned back by a retarding field, faster ones going farther and taking longer. The bunched beam returns through the cavity. (*Varian Associates*)

passes. Most of the electrons therefore are slowed down, and there is a net transfer of kinetic energy of the electrons into electromagnetic energy in the cavity. Klystron amplifiers are used in transmitters for radar and one-way radio communication, for driving particle accelerators, and for dielectric heating. The useful range of frequencies is from 400 MHz to 40 GHz. Power levels range from a few watts up to 400 kw of continuous power or 20 Mw for short pulses. Amplification is about 10 dB for a two-cavity tube. With more cavities, gains up to 60 dB are practical.

Figure 4 shows the construction of a four-cavity amplifier rated at 2 Mw pulsed output at 2.8 GHz. It operates in a solenoid magnet, as in Fig. 1. The cavities are tuned to the operating frequency by moving one flexible inner wall of the box-shaped cavity, changing its volume and its effective inductance. In Fig. 5 the details of the input cavity are enlarged.

Reflex oscillator. Klystrons may be operated as oscillators by feeding some of the output back into the input circuit. More widely used is the reflex oscillator in which the electron beam itself provides the feedback. Figure 6 illustrates the operation. The beam is focused through a cavity, as in the amplifier. No magnetic field is needed to keep the beam focused because the total travel distance is short and the amount of natural spreading is tolerable. The cavity usually has grids with open mesh through which the electrons can penetrate. The purpose of the grids is to concentrate the electric field in a short space so that the field can inter-

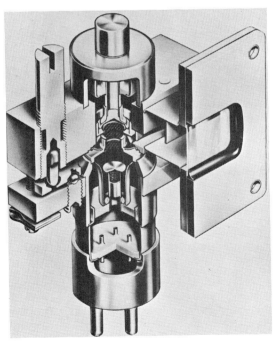

Fig. 8. Cutaway of reflex oscillator shows (at left) the screw that deforms the flexible bottom wall of the cavity to change its resonant frequency. The beam passes through honeycomb grids in the cavity. The cathode leads come out at the bottom socket, and the reflector lead comes out at the top. The output waveguide has a mica vacuum window. (*Varian Associates*)

act with a slow, low-voltage electron beam.

In the cavity the beam is velocity-modulated as in the amplifier. Leaving the cavity, the beam enters a region of dc electric field opposing its motion, produced by a reflector electrode operating at a potential negative with respect to the cathode. The electrons do not have enough energy to reach the electrode, but are reflected in space and return to pass through the cavity again. The points of reflection are determined by electron velocities, the faster electrons going farther against the field and hence taking longer to get back than the slower ones.

A trajectory plot for the reflex oscillator is shown in Fig. 7. In a uniform retarding field the space-versus-time curves are parabolas. As in Fig. 3, velocity modulation produces bunches of electrons. If the voltages are adjusted so that the average time to return is $n + 3/4$ cycles ($n =$ integer), the bunches cross the cavity when the alternating field is maximum in the decelerating direction. This transfers beam energy to the cavity.

Because the reflex klystron has only one cavity, it is easy to tune its frequency with a single adjustment. Power output is from 10 mw to a few watts. Reflex oscillators are used as signal sources from 3 to 200 GHz. They are also used as the transmitter tubes in line-of-sight radio relay systems and in low-power radars.

A cutaway of a typical reflex klystron is shown in Fig. 8. This tube is tuned by deforming the upper cavity wall, which varies the spacing between the grids and hence the effective capacitance of the cavity resonator.

[RICHARD B. NELSON]

Bibliography: L. Ginzton, The klystron, *Sci. Amer.*, 190(3):84–88, March 1954; S. Liao, *Microwave Devices and Circuits*, 1980; J. R. Pierce and W. G. Shepherd, Reflex oscillators, *Bell Syst. Tech. J.*, 26:460–681, 1947; R. F. Soohoo, *Microwave Electronics*, 1971.

Langmuir-Child law

A law governing space-charge-limited flow of electron current between two plane parallel electrodes in vacuum when the emission velocities of the electrons can be neglected. It is often called the three-halves power law, and is expressed by the formula shown below.

$$j(\text{A/cm}^2) = \frac{\epsilon}{9\pi}\left(\frac{2e}{m}\right)^{1/2}\frac{V^{3/2}}{d^2}$$

$$= 2.33 \times 10^{-6}\frac{V(\text{volts})^{3/2}}{d(\text{cm})^2}$$

Here ϵ is the dielectric constant of vacuum, $-e$ the charge of the electron, m its mass, V the potential difference between the two electrodes, d their separation, and j the current density at the collector electrode, or anode. The potential difference V is the applied voltage reduced by the difference in work function of the collector and emitter. The Langmuir-Child law applies, to a close approximation, to other electrode geometries as well. Thus for coaxial cylinders with the inner cylinder the cathode, it leads to a deviation from the true value of the current density of 13% at most. *See* SPACE CHARGE.

[EDWARD G. RAMBERG]

Laser

A device that uses the principle of amplification of electromagnetic waves by stimulated emission of radiation and operates in the infrared, visible, or ultraviolet region. The term laser is an acronym for light amplification by stimulated emission of radiation, or a light amplifier. However, just as an electronic amplifier can be made into an oscillator by feeding appropriately phased output back into the input, so the laser light amplifier can be made into a laser oscillator, which is really a light source. Laser oscillators are so much more common than laser amplifiers that the unmodified word "laser" has come to mean the oscillator, while the modifier "amplifier" is generally used when the oscillator is not intended. *See* MASER.

The process of stimulated emission can be described as follows: When atoms, ions, or molecules absorb energy, they can emit light spontaneously (as in an incandescent lamp) or they can be stimulated to emit by a light wave. This stimulated emission is the opposite of (stimulated) absorption, where unexcited matter is stimulated into an excited state by a light wave. If a collection of atoms is prepared (pumped) so that more are initially excited than unexcited, then an incident light wave will stimulate more emission than absorption, and there is net amplification of the incident light beam. This is the way the laser amplifier works.

A laser amplifier can be made into a laser oscillator by arranging suitable mirrors on either end of the amplifier. These are called the resonator. Thus the essential parts of a laser oscillator are an amplifying medium, a source of pump power, and a resonator. Radiation that is directed straight along the axis bounces back and forth between the mirrors and can remain in the resonator long enough to build up a strong oscillation. (Waves oriented in other directions soon pass off the edge of the mirrors and are lost before they are much amplified.) Radiation may be coupled out as shown in Fig. 1 by making one mirror partially transparent so that part of the amplified light can emerge through it. The output wave, like most of the waves being amplified between the mirrors, travels along the axis and is thus very nearly a plane wave.

Comparison with other sources. In contrast to lasers, all conventional light sources are basically hot bodies which radiate by spontaneous emission. The electrons in the tungsten filament of an incandescent lamp are agitated by, and acquire excitation from, the high temperature of the filament.

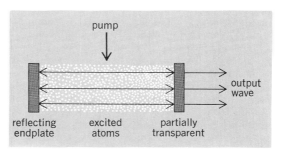

Fig. 1. Structure of a parallel-plate laser.

Once excited, they emit light in all directions and revert to a lower energy state. Similarly, in a gas lamp, the electron current excites the atoms to high-energy quantum states, and they soon give up this excitation energy by radiating it as light. In all the above, spontaneous emission from each excited electron or atom takes place independently of emission from the others. Thus the overall wave produced by a conventional light source is a jumble of waves from the numerous individual atoms. The phase of the wave emitted by one atom has no relation to the phase emitted by any other atom, so that the overall phase of the light fluctuates randomly from moment to moment and place to place. The lack of correlation is called incoherence.

Hot bodies emit more or less equally in all directions radiation whose wavelength distribution is dictated by the Planck blackbody radiation curve. For example, the surface of the Sun radiates like a blackbody at a temperature of about 6000 K, and emits a total of 7 kW/cm², spread out over all wavelengths and directions. Light from gas lamps can be more monochromatic (wavelengths radiated are restricted by the quantized energies allowed in atoms), but radiation still occurs in all directions. In contrast to this, an ideal plane wave would have the same phase all across any wavefront, and the time fluctuations would be highly predictable (coherent). The output of the parallel-plate laser described above is very nearly such a plane wave and is therefore highly directional. This arises because in the laser oscillator atoms are stimulated to emit in phase with the stimulating wave, rather than independently, and the wave that builds up between the mirrors matches very closely the mirrors' surfaces. The output is powerful because atoms can be stimulated to emit much faster than they would spontaneously. It is highly monochromatic largely because stimulated emission is a resonance process that occurs most rapidly at the center of the range of wavelengths that would be emitted spontaneously. Since atoms are stimulated to emit in phase with the existing wave, the phase is preserved over many cycles, resulting in the high degree of time coherence of laser radiation.

The various types of lasers are discussed below, classified according to their pumping (or excitation) scheme. The function of the pumping system is to maintain more atoms in the upper than in the lower state, thereby assuring that stimulated emission (gain transitions) will exceed (stimulated) absorption (loss transitions). This so-called population inversion ensures net gain (amplification greater than unity).

Optically pumped lasers. One way to achieve population inversion is by concentrating light (for example, from a flash lamp or the Sun) onto the amplifying medium. Alternatively, lasers may be used to optically pump other lasers. For example, powerful continuous-wave (cw) ion lasers can pump liquid dyes to lase, yielding watts of tunable visible and near-visible radiation. Molecular lasers, like carbon dioxide (CO_2), can pump gases, like deuterium cyanide (DCN), to lase powerfully in the far infrared, operating out to several hundred micrometers wavelength. Optical pumping can be employed to pump gases at very high pressures (for example, 42 atm or 4.3 MPa) to ob-

tain tunability where other excitation methods would be difficult if not impossible to implement.

Many lasers are three-level lasers; that is, ground-state atoms are excited by absorption of light to a broad upper-energy state, from which they quickly relax to the emitting state. Laser action occurs as they are stimulated to emit radiation and so return to the original ground state. Crystalline, glass, liquid, and gaseous systems have been found suitable, but many possible materials remain to be explored. Solid three-level lasers usually make use of ions of a rare-earth element, such as neodymium, or of a transition metal, such as chromium, dispersed in a transparent crystal or glass. For example, ruby, which is crystalline aluminum oxide containing a fraction of a percent of trivalent chromium ions in place of aluminum ions, has been used for lasers to produce red light with wavelengths of 693 to 705 nm. The chromium ions have broad absorption bands in which pumping radiations can be absorbed. Thus broad-band white light can be used to excite the atoms.

Many rare-earth ion lasers use a fourth level above the ground state. This level serves as a terminal level for the laser transition, and is kept empty by rapid nonradiative relaxation to the ground state. This means that, relative to three-level systems, a population inversion is easier to maintain, and therefore such materials require relatively low pumping light intensity for laser action. Neodymium ions can provide laser action in many host materials, producing outputs in the infrared, around 1 μm. In glass, which can be made in large sizes, neodymium ions can generate high-energy pulses or very high peak powers (Fig. 2). Lasers using neodymium ions in crystals such as yttrium aluminum garnet (YAG) can provide continuous output powers up to a kilowatt. The output of either type (in glass or in crystals) can be converted to visible light, near 500 nm, by a harmonic generator crystal, as discussed below. Optically pumped solid-state lasers provide relatively high peak output powers. Tens of kilowatts can easily be obtained in a pulse lasting 100 μs. Much higher peak powers can be obtained by special techniques described below.

The structure of an optically pumped solid-

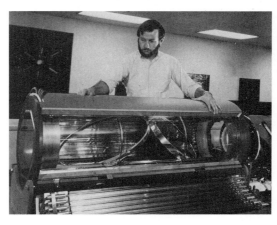

Fig. 2.   High-power laser amplifier stage, with a 30-cm aperture, using liquid-cooled slabs of neodymium glass. (*Lawrence Livermore Laboratory*)

state laser can be as simple as a rod of the light-amplifying material with parallel ends polished and coated to reflect light. Pumping radiation can enter either through the transparent sides or the ends. Other structures can be used. The mirror ends can be spherical rather than plane, with the common focal point of the two mirrors lying halfway between them. Still other structures make use of internal reflection of light rays that strike the surface of a crystal at a high angle.

Liquid lasers have structures generally like those of optically pumped solid-state lasers, except that the liquid is generally contained in a transparent cell. Some liquid lasers make use of rare-earth ions in suitable dissolved molecules, while others make use of organic dye solutions. The dyes can lase over a wide range of wavelengths, depending upon the composition and concentration of the dye or solvent. Thus tunability is obtained throughout the visible, and out to a wavelength of about 0.9 μm. Fine adjustment of the output wavelength can be provided by using a diffraction grating or other dispersive element in place of one of the laser mirrors. The grating acts as a good mirror for only one wavelength, which depends on the angle at which it is set. With further refinements, liquid-dye lasers can be made extremely monochromatic as well as broadly tunable. They may be pumped by various lasers to generate either short, intense pulses or continuous output. Dyes may also be incorporated into solid media such as plastics or gelatin to provide tunable laser action. Then the tuning may be controlled by a regular corrugation in the refractive index of the host medium, which acts as a distributed Bragg reflector. The reflection from any one layer is small, but when the successive alternations of refractive index provide reflections in phase, the effect is that of a strong, sharply tuned reflection.

Certain color centers in the alkali halides can be optically pumped to make efficient, tunable lasers for the region 0.8 to 3.3 μm, thus taking over just where the organic dyes fail. Both pulsed (10 kW) and continuous (watts) operation have been achieved.

In several infrared regions, tunable laser action can be obtained by using an infrared gas laser to pump a semiconductor crystal in a magnetic field, giving amplification by stimulating spin-flip Raman scattering from the electrons in the semiconductor. Tuning is achieved by varying the magnetic field.

Gas-discharge lasers. Another large class of lasers makes use of nonequilibrium processes in a gas discharge. At moderately low pressures (on the order of 1 torr or 10^2 Pa) and fairly high currents, the population of energy levels is far from an equilibrium distribution. Some levels are populated especially rapidly by the fast electrons in the discharge. Other levels empty particularly slowly and so accumulate large numbers of excited atoms. Thus laser action can occur at many wavelengths in any of a large number of gases under suitable discharge conditions. For some gases, a continuous discharge, with the use of either direct or radio-frequency current, gives continuous laser action. Output powers of continuous gas lasers range from less than 1 μW up to about 100 W in the visible region. Wavelengths generated span the ultra-

violet and visible regions and extend out beyond 700 μm in the infrared. They thus provide the first intense sources of radiation in much of the far-infrared region of the spectrum.

The earliest, and still most widely used, gas-discharge laser utilizes a mixture of helium and neon. Various infrared and visible wavelengths can be generated, but most commonly they produce red light at a 632.8-nm wavelength, with power outputs of a few milliwatts or less, although they can be as high as about a watt. Helium-neon lasers can be small and inexpensive. Argon and krypton ion gas−discharge lasers provide a number of visible and near-ultraviolet wavelengths with continuous powers typically about 1 to 10 W, but ranging up to more than 100 W. Unfortunately, efficiencies are low.

Many molecular gases, such as hydrogen cyanide, carbon monoxide, and carbon dioxide, can provide infrared laser action. Carbon dioxide lasers can be operated at a number of wavelengths near 10 μm on various vibration-rotation spectral lines of the molecule. They can be relatively efficient, up to about 30%, and have been made large enough to give continuous power outputs exceeding tens of thousands of watts.

Many gas-discharge lasers, for example, helium-neon, produce only very small optical gain; thus losses must be kept low. Consequently, mirrors with very high reflectivity must be used. Diffraction losses can be kept low by using curved mirrors. One common arrangement, which combines relatively low diffraction losses with good mode selection, uses a flat mirror at one end and a spherically concave mirror at the other. The spacing between mirrors is made equal to the radius of curvature of the spherical mirror. On the other hand, in some of the higher-power lasers even the plane-parallel mirror structure does not give sufficient discrimination against those undesired modes of oscillation which cause the beam to be excessively divergent. It is then helpful to use "unstable" resonators, with at least one of the mirrors convex toward the other.

A smooth, small-bore dielectric tube can guide a light wave in its interior with little loss. Thus a light wave can be amplified by a long, narrow medium without spreading or diffraction. A gas discharge in a hollow optical waveguide can be run at high pressure and benefits from cooling by the nearby walls, so that relatively high-power outputs (watts) can be obtained from a small volume. Waveguide structures are also useful when a medium is pumped optically by another laser, whose narrow beam can be confined within the bore of the tube. For example, pumping of various molecules, such as methyl fluoride, by a carbon dioxide laser has been used to generate coherent light in the very-far-infrared (submillimeter wavelength) region.

Pulsed gas lasers. Pulsed gas discharges permit a further departure from equilibrium. Thus pulsed laser action can be obtained in some additional gases which could not be made to lase continuously. In some of them the length of the laser pulse is limited when the lower state is filled by stimulated transitions from the upper level, and so introduces absorption at the laser wavelength. An example of such a self-terminating laser is the nitrogen laser, which gives pulses of several nanoseconds duration with peak powers from tens of kilowatts up to a few megawatts at a wavelength of 337 nm in the ultraviolet. They are easy to construct and are much used for pumping tunable dye lasers throughout the wavelength range from about 350 to 1000 nm. Very powerful laser radiation in the vacuum ultraviolet region, between 100 and 200 nm, can be obtained from short-pulse discharges in hydrogen, and in rare gases such as xenon at high pressure. (High pressure leads to higher power.) When the gas pressure is too high to permit an electric discharge, excitation may be provided by an intense burst of fast electrons from a small accelerator, the so-called E beam.

Some gases, notably carbon dioxide, which can provide continuous laser action, also can be used to generate intense pulses of microsecond duration. For this purpose, gas pressures of about 1 atm (10^5 Pa) are used, and the electrical discharge takes place across the diameter of the laser column, hence the name transverse-electrical-atmospheric (TEA) laser.

Chemical lasers. It is also possible to obtain laser action from the energy released in some fast chemical reactions. Atoms or molecules produced during the reaction are often in excited states. Under special circumstances there may be enough atoms or molecules excited to some particular state for amplification to occur by stimulated emission. Usually the reacting gases are mixed and then ignited by ultraviolet light or fast electrons. Both continuous infrared output and pulses up to several thousand joules of energy have been obtained in reactions which produce excited hydrogen fluoride molecules.

Pulsed laser action in the ultraviolet (193 to 353 nm) has been obtained from excimer states of rare gas monohalides (for example, KrF, XeF, KrCl, XeCl, XeBr). Such molecules have ground states which are unstable, thereby making a population inversion easy to achieve. Although these lasers require an electrical source for initial gas reaction, laser pumping is dependent on chemical reactions.

Photodissociation lasers. Intense pulses of ultraviolet light can dissociate molecules in such a way as to leave one constituent in an excited state capable of sustaining laser action. The most notable examples are iodine compounds, which have given peak 1.3-μm pulse powers above 10^9 W from the excited iodide atoms.

Nuclear lasers. Laser action in several gases has also been excited by the fast-moving ions produced in nuclear fusion. These fusion products excite and ionize the gas atoms, and make it possible to convert directly from nuclear to optical energy.

Gas-dynamic lasers. When a hot molecular gas is allowed to expand suddenly through a nozzle, it cools quickly, but different excited states lose energy at different rates. It can happen that, just after cooling, some particular upper state has more molecules than some lower one, so that amplification by stimulated emission can occur. Very high continuous power outputs have been generated from carbon dioxide in large gas-dynamic lasers.

Semiconductor lasers. Another method for providing excitation of lasers can be used with certain semiconducting materials. Laser action takes place when free electrons in the conduction band are stimulated to recombine with holes in the valence band. In recombining, the electrons give up energy corresponding nearly to the band gap. This energy is radiated as a light quantum. Suitable materials are the direct-gap semiconductors, such as gallium arsenide. In them, recombination occurs directly without the emission or absorption of a quantum of lattice vibrations. A flat junction between p-type and n-type material may be used. When a current is passed through this junction in the forward direction, a large number of holes and electrons are brought together. This is called recombination, and is accompanied by emission of radiation. A light wave passing along the plane of the junction can be amplified by stimulating such recombination of electrons and holes. The ends of the semiconducting crystal provide the mirrors to complete the laser structure. In indirect-gap semiconductors, such as germanium and silicon, only a small amplification by stimulated emmision is possible because of their requirement for interaction with the lattice vibrations.

Semiconductor lasers can be very small, less than 1 mm in any direction. They can have efficiencies higher than 50% (electricity to light). Power densities are high, but the thinness of the active layer tends to limit the total output power. Even so, maximum continuous powers are comparable with those of other moderate-size lasers. Since semiconductor lasers are so small, they can be assembled into compact arrays of many units, so as to generate higher peak powers. An alternative excitation method, bombardment of the semiconductor by a high-voltage beam of electrons, may provide laser action in larger crystal volumes, but it is likely to cause damage to the crystal. *See* SEMICONDUCTOR.

Free-electron lasers. Free-electron lasers are of interest because of their potential for efficiently producing very high-power radiation, tunable from the millimeter to the x-ray region. The principle of operation, so far demonstrated only at 3.4 μm and at very low-power levels, involves passage of electrons through a spatially varying magnetic field which causes the electron beam to "wiggle" and hence to radiate. The large Doppler upshift due to relativistic electron velocities can be adjusted, resulting in tunable emission at optical frequencies.

High-power, short-pulse lasers. The output power of pulsed lasers can be greatly increased, with correspondingly shorter pulse durations, by the Q-switch technique. In this method, the optical path between one mirror and the amplification medium is blocked by a shutter. The medium is then excited beyond the degree ordinarily needed, but the shutter prevents laser action. At this time the shutter is abruptly opened and the stored energy is released in a giant pulse (1–100 MW peak power, lasting 1–30 ns for optically pumped solid-state lasers). Still higher peak powers can be obtained by passing this output through a traveling-wave laser amplifier (without mirrors). Peak powers in excess of 100 MW have been obtained in this way.

Still shorter, and higher-power, pulses can be generated by mode-locking techniques. A typical laser without mode locking usually oscillates simultaneously and independently at several closely spaced wavelengths. These modes of oscillation can be synchronized so that the peaks of their waves occur simultaneously at some instant. The result is a very short, intense pulse which quickly ends as the waves of different frequency get out of step. Mode-locked lasers have generated pulses shorter than 1 ps. Since such brief pulses tend to produce somewhat less damage to materials, they can be amplified to very high peak intensities. Power outputs of picosecond pulses as high as 10–100 MW have been obtained, limited as in the Q-switch case by material damage. *See* OPTICAL PULSES.

For the highest peak power, the output may be further intensified by additional stages of laser amplification. The beam diameter is increased by some optical arrangement, such as a telescope, so as to expand (dilute) the beam and thereby prevent damage to the laser material and optics. Sometimes the amplifying medium is divided into flat slabs separated by cooling liquid (Fig. 2). Here the open faces of the light-amplifying slabs present a large area to receive pumping light and liquid cooling.

Development of very large multistage lasers has been undertaken for research on thermonuclear fusion. In a particularly large one, at Lawrence Livermore Laboratory, a single neodymium-glass laser oscillator is designed to drive 20 amplifier chains of glass rods and disks, each delivering a pulse of more than 10^{12} W. Focusing all these pulses onto the surface of a small pellet of heavy hydrogen is designed to heat and compress the pellet by ablation, until it is so hot that the hydrogen nuclei fuse to produce helium and release large amounts of nuclear energy. Ultimately this type of controlled laser fusion may become an important source of thermal, electrical, and chemical energy.

The technology of scaling up lasers to higher and higher powers has been so successful that the limitations are often set by material damage thresholds of the laser medium or associated optics. At the high-power densities attainable by focusing laser beams, the electric fields of the light can be large. Thus when the light intensity is 10^{12} W/cm^2, the corresponding electric field is 10^7 V/cm. To such large fields, many transparent materials have a nonlinear dielectric response. This nonlinearity can be large enough to permit the generation of optical harmonics. It is possible, with careful design and good nonlinear materials, to obtain substantially complete conversion of a laser's output to the second harmonic, at twice the frequency, even for continuous lasers near 1-W output power. Nonlinear dielectrics can be used as mixers to give the sum of difference of two laser frequencies. They also permit the construction of optical parametric oscillators which, when pumped by a laser, can generate coherent light tunable over a wide range of wavelengths.

Applications. The variety of technological uses for lasers has increased steadily since their appearance in 1960. Among the noticeable applications are those that utilize high-speed controllability of the tiny focal spot of a laser beam. For example, high-speed automatic scanners identify library cards, ski passes, and supermarket purchases and perform a variety of functions known

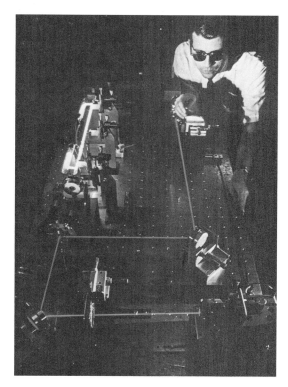

Fig. 3. Experimental arrangement to impress pulse-code modulation on a laser beam for optical communication. (*Bell Laboratories*)

as optical processing. Other uses for the laser beam's programmable control include information storage and retrieval (including three-dimensional holography and video disk reading), laser printing, micromachining, and automated cutting. Further applications involving high power include weaponry, laser welding, laser surgery (self-cauterizing), laser fusion, and materials processing. Optical communications utilize the laser's high frequency, which makes possible high information capacity (Fig. 3.) Except for space applications, laser light communication is primarily done through glass fibers. Low-loss optical fibers are far more compact and economical than copper wires. Bright laser beams are used by the construction industry to align straight excavations and for surveying. Some of the more specialized laser applications include laser gyros, laser velocity sensing (laser infrared radar or lidar), optical testing, metrology (including the distance to the Moon), laser spectroscopy (including pollution monitoring), and lasers to pump other lasers, and nonlinear processes are used to produce beams of coherent light at other wavelengths from millimeters to x-rays, and for exploration of ultrafast phenomena (by using picosecond laser pulses). *See* CHARACTER RECOGNITION; CONTINUOUS-WAVE RADAR; INTEGRATED OPTICS; OPTICAL COMMUNICATIONS; PHOTOACOUSTIC SPECTROSCOPY; VIDEO DISK RECORDING.

[STEPHEN F. JACOBS; ARTHUR L. SCHAWLOW]

Bibliography: S. F. Jacobs et al. (eds.), *Free Electron Generators of Coherent Radiation*, 1980; S. F. Jacobs et al., *Laser Photochemistry, Tunable*

Lasers, and Other Topics, 1976; B. A. Lengyel, *Lasers*, 1969; L. Marton and C. Marton, *Methods of Experimental Physics*, vol. 15, 1979; D. C. O'Shea, W. R. Callen, and W. T. Rhodes, *An Introduction to Lasers and Their Applications*, 1977; A. E. Siegman, *Introduction to Lasers and Masers*, 1971; D. C. Sinclair and W. E. Bell, *Gas Laser Technology*, 1969; A. Yariv, *Introduction to Optical Electronics*, 1976.

Light amplifier

In the broadest sense, a device which produces an enhanced light output when actuated by incident light. A simple photocell relay – light source combination would satisfy this definition. To make the term more meaningful, common usage has introduced two restrictions: (1) a light amplifier must be a device which, when actuated by a light image, reproduces a similar image of enhanced brightness; and (2) the device must be capable of operating at very low light levels without introducing spurious brightness variations (noise) into the reproduced image. The term is used synonymously with image intensifier. The light amplifier increases the brightness of an image which is below the visual threshold to a level where it can be readily seen with the unaided eye. It is, of course, impossible to see under conditions of complete darkness. Indeed, there is a fundamental lower limit of illumination under which an image of a given quality can be recognized. This limitation prevails because of the corpuscular nature of light.

Photons arriving through a lens, or other optical system, onto an image area are random in time. If the number of photons per unit area arriving during the time allotted for image formation (for example, the period of the persistence of vision) is too low, the statistical fluctuation will be greater than the variation in number due to true differences in image brightness. Under these circumstances, image recognition is impossible.

Image-intensifier tubes. Intensifier tubes may consist of a semitransparent photocathode which emits electrons with a density distribution proportional to the distribution of light intensity incident on it. Thus, when a light image falls on one side of the cathode, an electron-current image is emitted from the other side. An electron optical system, which acts on electrons in much the same way as does a glass lens system on light, focuses the electron image onto an intensifier element. This electron optical system may be purely electrostatic or may utilize magnetic focusing. The intensified electron image from the other side of the intensifier element is again focused, by a second electron lens, onto a second intensifier element (if additional amplification is required) or onto a fluorescent viewing screen, where the electron energy is converted to visible light.

Proximity tubes are a class of intensifiers in which the photocathode and the fluorescent viewing screen are parallel and separated by only a short distance. No focusing is required for a proximity tube, since the electrons generated at the photocathode are accelerated along the tube axis by the high electric fields.

The spectral response of the image-intensifier tube depends on the type of photocathode employed, and the tube may serve as a wavelength converter as well as an intensifier. *See* ELECTRON OPTICS.

Intensifier elements. Two types of intensifier elements are ordinarily employed. One consists of a transparent support (either a thin film or a fiber optics plate) with a phosphor screen on the side on which the electron image is incident, and a photocathode on the other. An electron striking the phosphor produces a flash of light which causes a release of 50 or more electrons from the photoemitter directly opposite the point of impact. Thus, the intensifier element increases the electron-image current density by a factor of 50. Two such intensifier elements in cascade give a gain of 2500. The illustration shows diagrammatically a three-stage electrostatic image-intensifier tube.

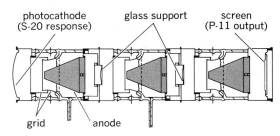

Three-stage electrostatically focused image tube.

A second type of high-gain intensifier element is a thin secondary-emission current amplifier called a microchannel plate, which is placed between the photocathode and screen. It consists of a parallel bundle of small, hollow glass cylinders, where the inside walls of the cylinders are coated with a secondary emitting material. Electrons emitted from the photocathode strike the inside walls of the cylinders, causing secondary electron generation. These secondary electrons in turn continue to cascade down the inside walls of the cylinders, resulting in a high total current gain.

Television camera tube. The intensifier image device can be combined with television pickup tubes to make a television camera whose sensitivity is very close to the threshold determined by the photon "shot noise" from the scene being televised. A SIT (silicon intensifier target) camera tube has a wide dynamic range and operates at very low light levels. In the SIT tube, the photoelectrons accelerated from the photocathode are focused onto a thin silicon target consisting of an array of *pn* junctions. The high-energy photoelectrons generate multiple hole-electron pairs in the silicon target, resulting in gain. The signal is read out by scanning the reverse side of the silicon target with an electron beam, similar to the method used in vidicon camera tubes. In other types of EBS (electron-bombarded silicon) tubes, the silicon *pn* junction target is replaced with scanned photodiode arrays or with CCD (charge-coupled device) imagers. In these cases, the electronic readout of the silicon targets is accomplished without the use of a scanned electron beam, and this results in even lower light-level performance. *See* TELEVISION CAMERA TUBE.

Solid-state image intensifiers. A great deal of exploratory work has been done on solid-state light amplifiers. The form which has been extensively investigated consists of a photoconductive film in contact with an electroluminescent screen. Current flows through the photoconductor at illuminated areas, causing light to be generated in the electroluminescent layer. This type of intensifier gives considerable image enhancement at intermediate light levels, but fails at very low light levels because of extreme time lag. *See* ELECTROLUMINESCENCE.

Another type of solid-state light amplifier which works at intermediate light levels is the liquid-crystal light valve. The light valve consists of an electrically biased multilayer sandwich which has a liquid-crystal layer on the front face and a photoconductive layer on the rear face. In this system, the incident light is focused on the photoconductive side, while a bright, polarized light is projected onto the liquid-crystal side. The incident light from the photoconductive side causes an increase in voltage drop across the liquid crystal. This increased voltage causes a rotation in the orientation of the liquid-crystal molecules, which in turn modulates the polarization of the reflected projection light. Only the modulated reflected light is allowed to be viewed.

Applications. In addition to their application for night vision, light amplifiers have been useful in many fields of science, such as astronomy, nuclear physics, and microbiology.

[DEAN R. COLLINS]

Bibliography: D. R. Collins et al., Development of a CCD for ultraviolet imaging using a CCD photocathode combination, *Proceedings of the Symposium on Charge Coupled Device Technology for Scientific Imaging Applications*, Jet Propulsion Laboratory, pp. 163–174, 1975; A. D. Jacobson et al., A new color-TV projector, *SID77 Digest*, Society for Information Displays, pp. 106–107, 1977; *RCA Electro-Optics Handbook*, 1974; Symposium of Photoelectronic Image Devices, *Advances in Electronics and Electron Physics*, 1960.

Light-emitting diode

A rectifying semiconductor device which converts electric energy into electromagnetic radiation. The wavelength of the emitted radiation ranges from the green to the near infrared, that is, from about 550 to over 1300 nm. Blue light-emitting diodes (LEDs) have also been reported, but they are not available commercially.

Fabrication methods. Most commercial light-emitting diodes, both visible and infrared, are fabricated from group III–V compounds. These compounds contain elements such as gallium, indium, and aluminum of group III and arsenic and phosphorus of group V of the periodic table. With the addition of the proper impurities, III–V compounds can be made *p*- or *n*-type, to form *p-n* junctions. They also possess the proper band gap to produce radiation of the required wavelength and are efficient in the conversion of electric energy into radiation. The fabrication of light-emitting diodes begins with the preparation of single-crystal substrates usually made of gallium arsenide (GaAs), 250–350 μm thick. Both *p*- and *n*-type layers are formed over this substrate by depositing layers of semiconductor material from a vapor or from a melt.

The most commonly used light-emitting diode is the red light-emitting diode, made of gallium arsenide-phosphide on gallium arsenide substrates (Fig. 1*a*). An *n*-type layer is grown over the substrate by vapor-phase deposition followed by a diffusion step to form the *p-n* junction. Ohmic contacts are made by evaporating metallic layers to both *n*- and *p*-type materials. The arrows indicate the emitted light at the *p-n* junction. The light generated at any point is uniformly distributed in all directions. Only a small fraction of the light striking the top surface of the diode can escape, however, due to the large difference in the refractive indices between semiconductor and air. Most of the light is internally reflected and absorbed by the substrate. Hence a typical red light-emitting diode has only a few percent external quantum efficiency, that is, only a few percent of the electric energy results in useful light. More efficient and therefore brighter light-emitting diodes can be fabricated on a gallium phosphide substrate, which is transparent to the emitted radiation and permits the light to escape upon reflection from the back contact. Although more expensive than light-emitting diodes made of gallium arsenide, these high-efficiency light sources compete favorably in brightness with miniature incandescent lamps.

Extracting the infrared radiation into an optical fiber is even more challenging since the fiber has a narrow (10–15°) acceptance angle. It is also necessary to shorten the response time of the light-

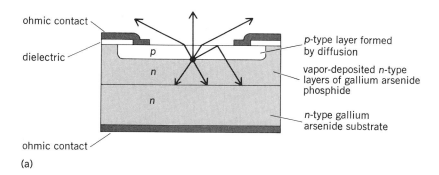

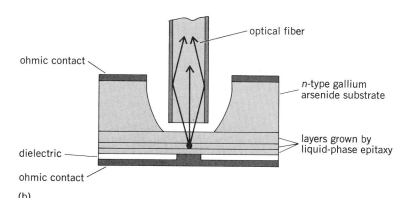

Fig. 1. Diagrams of light-emitting diodes. (*a*) Red-emitting gallium arsenide-phosphide LED. (*b*) Infrared, etched-well LED designed for coupling radiation into an optical fiber.

emitting diode for high data-rate transmission. This is accomplished by introducing several changes into the light-emitting diode design (Fig. 1 b). First, the light-emitting diode structure is turned around compared with that of the red diode (Fig. 1a), and the radiation is extracted through an etched well on the side of the substrate. A dielectric layer separates most of the ohmic contact on the other side, limiting the active area of the junction to the size of the optical fiber. Finally, one or several additional layers are deposited (4 in Fig. 1b) to increase quantum efficiency and response time. The optical power from the light-emitting diode is limited by the heat dissipation of the semiconductor chip, which is usually mounted on a gold heat sink. Although some light-emitting diodes of this type are commercially available, development is being directed toward longer-wavelength devices (peak emission at 1.3 μm compared with 0.85 μm for the above light-emitting diodes) using layers containing four elements, gallium, indium, arsenic, and phosphorus, grown on indium phosphide substrates. These light-emitting diodes can send higher data rates over longer distances in the optical fibers.

Applications. Visible light-emitting diodes are used as solid-state indicator lights and as light sources for numeric and alphanumeric displays. Infrared light-emitting diodes are used in optoisolators and in optical fiber transmission in order to obtain the highest possible efficiency.

Indicators and displays. The advantages of light-emitting diodes as light sources are their small size, ruggedness, low operating temperature, long life, and compatibility with silicon integrated circuits. They are widely used as status indicators in instruments, cameras, appliances, dashboards, computer terminals, and so forth, and as nighttime illuminators for instrument panels and telephone dials. Visible light-emitting diodes are commercially available in red, orange, yellow, and green. Blue-light emitters have been made in the laboratory on wide band-gap materials such as silicon carbide, gallium nitride, and zinc sulfide. These devices have different structures than the p-n junctions described above, their fabrication is more expensive, and their brightness is lower by approximately two orders of magnitude than that of other light-emitting diodes, and hence they are not employed commercially.

Some of the most commonly used light-emitting diode structures are shown in Fig. 2. The metal-flanged, single-lead design (Fig. 2a) is very rugged and easy to insert; the lead-frame package (Fig. 2b) can easily incorporate built-in voltage regulators so that the light-emitting diodes can be operated over a range of input voltages such as 3–15 V. Some packages have provisions to focus or redistribute the light, such as the lead frame with a built-in reflector (Fig. 2c). Light-emitting diodes can also be employed to light up a segment of a large numeric display, used for example on alarm clocks; or a small numeric display with seven light-emitting diodes can be formed on a single substrate, as commonly used on watches and hand-held calculators.

Optoisolators. Infrared light-emitting diodes with solid-state photodetectors provide an optical interface in electric circuits. In the simplest optical interface, the optoisolator, a light-emitting diode and a photodetector are optically coupled, but electrically isolated, in a small package. This device can be used, for example, at the interface between two different circuits, such as the switching equipment in a telephone central office and the connecting loop circuit which carries the signals to the telephone sets. The electric signal from the central office is converted to radiation by the light-emitting diode, which in turn is converted back into an electric signal by the photodetector before it enters the loop circuit. This type of interface is traditionally provided by electromechanical relays or isolation transformers. The electrical isolation resulting from the optical path protects the central office from electromagnetic interferences such as lightning which hits telephone wires or surge currents from electromechanical relays. Light-emitting diodes are ideal for this application because they are small, rugged, efficient, reliable, and can be modulated to carry high-frequency signals.

In a typical optoisolator structure (Fig. 3), the light-emitting diode and the phototransistor are

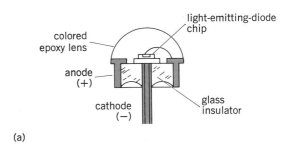

(a)

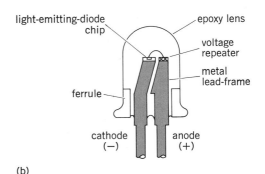

(b)

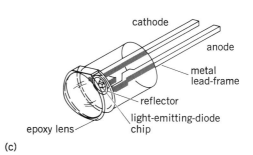

(c)

Fig. 2. Light-emitting diode lamps. (a) Metal-flanged single-lead header. (b) Lead-frame package with built-in resistor. (c) Lead-frame package with built-in reflector.

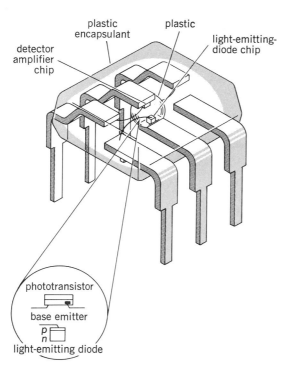

Fig. 3. Typical optoisolator in plastic-encapsulated dual in-line package.

mounted on separate metal lead frames, and the two components are coupled optically through a transparent plastic encapsulant. This plastic is also the source of electrical isolation, typically on the order of 2500 V. Final encapsulation is completed with a black, opaque overmold which also provides mechanical stability. Optoisolators are compatible with silicon integrated circuits in size, reliability, and performance parameters, giving them a prominent role in modern solid-state circuits. See OPTICAL ISOLATOR.

Optical fiber transmission. Another rapidly evolving application of infrared light-emitting diodes is in optical fiber transmission. The optical signal is fed into a thin (50–100 μm in diameter) optical fiber and transmitted over distances ranging from several hundred meters to over 10 km. At the other end of the fiber, a photodetector converts the optical signals back to electric signals similar to those in optoisolators. This fiber, which replaces coaxial cables, is smaller in volume, less expensive, and immune to electromagnetic interference. It also transmits higher data rates and provides longer repeater spacings than metal conductors. Since doped silica fibers exhibit both low loss and minimum material dispersion near 1.3 μm, light-emitting diode sources at this wavelength can achieve a repeater spacing of 10 km at a data rate of 250 megabits per second. This will provide a major market for high-performance light-emitting diodes and support the development of long-wavelength light-emitting diodes. See OPTICAL COMMUNICATIONS.

For further discussion of the properties of *p-n* junctions and light generation in solid-state devices see ELECTROLUMINESCENCE; JUNCTION DIODE; JUNCTION TRANSISTOR; LASER.   [A. A. BERGH]

Bibliography: A. A. Bergh and P. J. Dean, *Light Emitting Diodes*, 1976; C. H. Gooch, *Injection Electroluminescent Devices*, 1973; S. E. Miller and A. G. Chynoweth, *Optical Fiber Telecommunications*, 1979.

Light panel

A surface-area light source that employs the principle of electroluminescence to produce light. Light panels are composed of two sheets of electrically conductive material, one a thin conducting backing and the other a transparent conductive film, placed on opposite sides of a plastic or ceramic sheet impregnated with a phosphor, such as zinc sulfide, and small amounts of compounds of copper or manganese. When an alternating voltage is applied to the conductive sheets, an electric field is applied to the phosphor. Each time the electric field changes, it dislodges electrons from the edges of the phosphor crystals. As these electrons fall back to their normal atomic state, they affect the atoms of the slight "impurities" of copper or manganese, and radiation of the wavelength of light is emitted. See ELECTROLUMINESCENCE.

In contrast to incandescent, vapor-discharge, and fluorescent lamps, which are essentially point or line sources of light, the electroluminescent light panel is essentially a surface source of light. Complete freedom of size and shape is a fascinating aspect of luminescent cells (see illustration).

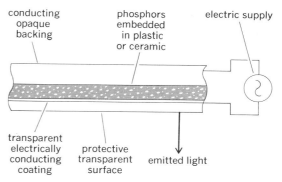

Simplified diagram of an electroluminescent cell; the sketch is not drawn to scale.

Brightness of the panel depends upon the voltage applied to the phosphor layer and upon the electrical frequency. In general, higher voltage and higher frequency both result in a brighter panel. Blue, green, red, or yellow light can be produced by the choice of phosphors, and the proper blend of these colors produces white light. Color can be varied for a particular phosphor by changing the frequency of the applied voltage. Increasing the frequency shifts the color toward the blue end of the spectrum.

The efficiency of these light panels is only a fraction of that of the most efficient fluorescent lamps. Theoretical limits indicate, however, that the efficiency can be further improved, probably to exceed that for fluorescent lamps. Because panel lights employ no filaments and no evacuated or gas-filled bulbs, replacement of units is virtually eliminated. Glareless uniform distribution of light from large-area sources is possible without shades, louvers, or other control devices.

[WARREN B. BOAST]

Limiter circuit

An electronic circuit used to prevent the amplitude of an electrical waveform from exceeding a specified level while preserving the shape of the waveform at amplitudes less than the specified level. The limiting action takes place by effectively shunting a normal load resistance with a much lower resistance at and above the specified limiting level.

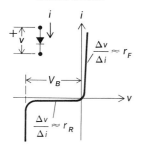

Fig. 1. Voltage-current characteristic of junction diode shows instantaneous diode current i as a function of instantaneous voltage v.

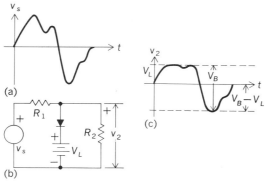

Fig. 2. In combined forward-bias and reverse-breakdown limiting, (a) input voltage applied to (b) limiting circuit produces (c) output voltage.

Diode limiters. Limiting action is usually accomplished by use of the highly nonlinear or "switching" voltage-current characteristic of an electronic device. A semiconductor diode used as a clamp is often the essential element in a limiter circuit. The action of such a diode may be understood with reference to the voltage-current characteristic of a typical semiconductor diode shown in Fig. 1. In the first quadrant, the diode is forward-biased, conduction depends on forward resistance r_F, and the characteristic can be approximated by a linear resistance of a very low value, ranging from a few ohms to several hundred ohms. In the third quadrant, the diode is reverse-biased. Until the breakdown point is reached, this portion of the characteristic can be approximated by a very large resistance r_R, ranging up to several megohms. At the breakdown point a sharp change occurs. For the semiconductor diode, this point is known as the Zener or avalanche breakdown region and can range from a few volts upward. After the breakdown, current becomes essentially independent of voltage. Either the sharp transition between reverse- and forward-bias regions or the avalanche transition may be used as a limit. *See* CLAMPING CIRCUIT; SEMICONDUCTOR DIODE.

Use of both limits. Use of both limits simultaneously is illustrated in Fig. 2, where input voltage v_s and output voltage v_2 are shown as functions of time t. The positive amplitude of the output waveform is limited at reference voltage V_L; negative amplitude is limited at a level $V_L - V_B$. Sharp limiting in the forward-bias direction is achieved when

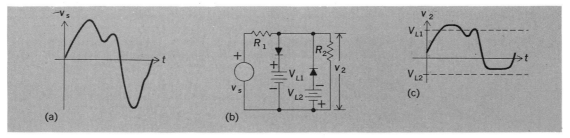

Fig. 3. For bidirectional limiting (a) input voltage can be limited in both directions by (b) circuit. (c) Bias voltage V_{L1} sets positive limit level and independent voltage V_{L2} sets negative level to produce output voltage.

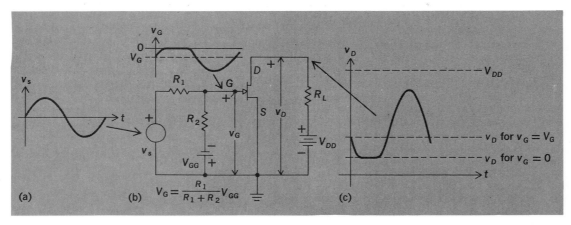

Fig. 4. In a gate-current limiter using a junction FET, (a) input voltage v_s is limited in (b) the gate circuit to produce gate voltage v_G. This voltage is then amplified and inverted in the drain circuit to produce (c) output voltage v_D. Gate bias voltage V_{GG} and the divider composed of the resistances R_1 and R_2 determine the quiescent level V_G which corresponds to $v_s = 0$.

forward resistance r_F is small compared to reverse-bias resistance r_R and when r_F is small compared to series limit resistance R_1. Any shunting load resistance R_2 is assumed large compared to R_1. A vacuum diode may be used as the limiter at its transition between reverse- and forward-bias regions, but it has no well-defined breakdown level comparable to the avalanche region of a junction diode. Either positive or negative limiting at the reverse–forward-bias transition can be obtained with semiconductor or vacuum diodes by choice of

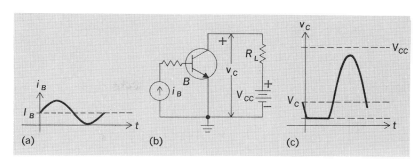

Fig. 6. When (a) the input voltage is applied to (b) the circuit with an *npn* transistor, the collector current limiting produces (c) the output voltage.

polarities, voltage, and diode. If bidirectional limiting is desired, the circuit of Fig. 3 can be used with two diodes and two separate bias voltages.

Triode limiters. Bipolar or field-effect transistors can also be used as limiters, in either the input or output circuits. For example, in the circuit of Fig. 4, if the input waveform becomes sufficiently positive, the gate-source circuit becomes a forward-biased diode, and limiting takes place as it does in the diode circuit. The drain current is limited at a value corresponding to the positive gate drive if R_L is sufficiently low that output limiting does not take place simultaneously. *See* TRANSISTOR.

Limiting by saturation. Saturation characteristics of the ouput circuit of a field-effect transistor or bipolar transistor can also produce limiting action. Figure 5 illustrates this action for an *n*-channel junction field-effect transistor. The characteristics shown are those of drain current i_D as a function of drain voltage v_D for successive values of input, or gate voltage V_G through the normal range. The input voltage bias level and the load resistance are so chosen that limiting takes place in the output circuit when drain current saturation level is reached. The corresponding output voltage waveform is plotted.

The bipolar transistor functions in a similar manner, with the output characteristics that are similar to those of Fig. 5 being collector current as a function of collector voltage, for a range of values of input or base current (rather than input voltage) because the bipolar transistor is a low-input-resistance device. Such a circuit is shown in Fig. 6 for an *npn* transistor with the base-emitter circuit forward-biased.

All polarities can be reversed by using *p*-channel field-effect transistors or *pnp* transistors in place of the transistors of Figs. 5 and 6, respectively.

The term limiter is often defined in sufficiently broad terms to include the related operation of clipping, which also results in deletion of a portion of a waveform. Some authors even use the terms interchangeably. Limiters and clippers are often used together, rather than bidirectional limiters alone, where both negative and positive peaks of a waveform are to be removed. *See* CLIPPING CIRCUIT.

[GLENN M. GLASFORD]

Bibliography: G. M. Glasford, *Fundamentals of Television Engineering,* 1955; J. Millman, *Microelectronics,* 1979; H. Taub and D. Schilling, *Digital Integrated Electronics,* 1977.

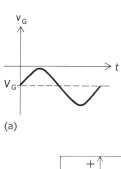

(a)

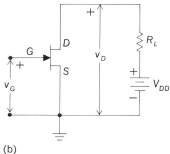

(b)

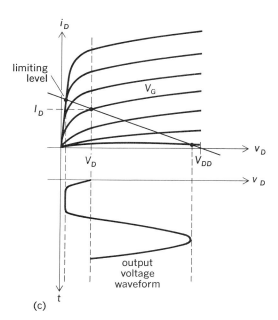

(c)

Fig. 5. Output saturation limiting using *n*-channel junction field-effect transistor limits (a) input voltage in (b) circuit because of saturation of drain current, as shown on (c) the $v_D - i_D$ characteristic curves, to produce output voltage.

Linear programming

A mathematical subject whose central theme is finding the point where a linear function defined on a convex polyhedron assumes its maximum or minimum value. Although contributions to this question existed earlier, the subject was essentially created in 1947, when G. B. Dantzig defined its scope and proposed the first, and still most widely used, method for the practical solution of linear programming problems. The largest class of applications of linear programming occurs in business planning and industrial engineering. Its basic concepts are so fundamental, however, that linear programming has been used in almost all parts of science and social science where mathematics has made any penetration. Furthermore, even in cases where the domain is not a polyhedron, or the function to be maximized or minimized is not linear, the methods used are frequently adaptions of linear programming. *See* NONLINEAR PROGRAMMING; OPERATIONS RESEARCH.

General theory. A typical linear programming problem is to maximize expression (1), where

$$c_1 x_1 + \cdots + c_n x_n \qquad (1)$$

$x_1, \ldots, x_n$ satisfy the conditions shown by notation (2). The a_{ij}, c_j, and b_i are constants; $x_1, \ldots, x_n$ are variables.

$$
\begin{aligned}
x_j &\geqq 0 \\
a_{11} x_1 + \cdots + a_{1n} x_n &\leqq b_1 \\
&\cdots \cdots \cdots \\
a_{m1} x_1 + \cdots + \alpha_{mn} x_n &\leqq b_m
\end{aligned}
\qquad (2)
$$

Sometimes the problem may be to minimize rather than to maximize; sometimes some of the variables may not be required to be nonnegative; sometimes some of the inequalities $a_{i1} x_1 + \cdots + a_{in} x_n \leqq b_i$ may be reversed, or be equalities.

The linear inequalities, which the variables satisfy, correspond algebraically to the fact that the variable point $x = (x_1, \ldots, x_n)$ lies in a convex polyhedron. Hence, the principal mathematical bases for linear programming are the theory of linear equalities, a part of algebra, and the theory of convex polyhedra, a part of geometry. The most important theoretical foundations are as follows.

If the function $c_1 x_1 + \cdots + c_n x_n$ does not get arbitrarily large for points $x = (x_1, \ldots, x_n)$ on the convex polyhedron, then a maximum is attained at a vertex of the polyhedron.

If there is a maximum in the stated problem, then there is a minimum in the dual problem of minimizing $b_1 y_1 + \cdots + b_m y_m$, where notation (3)

$$
\begin{aligned}
y_i &\geqq 0 \\
a_{11} y_1 + \cdots + a_{m1} y_m &\geqq c_1 \\
&\cdots \cdots \cdots \cdots \\
a_{1n} y_1 + \cdots + a_{mn} y_m &\geqq c_n
\end{aligned}
\qquad (3)
$$

applies. Further, the maximum and minimum are the same number. This duality theorem is essentially equivalent to the minimax theorem. Both the duality theorem and the minimax theorem are algebraic paraphrases of the geometric fact that if a point is not contained in a convex polyhedron then a hyperplane can be found which separates the point from the polyhedron. The duality theorem is also closely related to the concept of Lagrange multipliers. *See* CALCULUS OF VARIATIONS; GAME THEORY.

In an important class of linear programming problems, there is imposed an additional requirement that some or all of the variables must be whole numbers. If the numbers $b_1, \ldots, b_n$ are whole numbers, and if every square submatrix of (a_{ij}) has determinant 0, 1, or -1, then the integrality requirement will be automatically satisfied. In any case, the faces of the convex hull of the integral points inside a polyhedron can be found by a finite process in which, at each stage, a new inequality is determined by taking a nonnegative linear combination of previous inequalities and replacing each coefficient by the largest integer not exceeding it.

Methods of calculation. The most popular method for solving linear programs, that is, the finding of the $x = (x_1, \ldots, x_n)$ which maximizes $c_1 x_1 + \cdots + c_n x_n$, is the simplex method, developed by Dantzig in 1947. The method is geometrically a process of moving from vertex to neighboring vertex on the convex polyhedron, each move attaining a higher value of $c_1 x_1 + \cdots + c_n x_n$, until the vertex yielding the greatest value is reached. Algebraically, the calculations are similar to elimination processes for solving systems of algebraic equations. The computer programs used try to take advantage of the fact that almost all the matrices (a_{ij}) arising in practice have very few nonzero entries.

The popularity of the simplex method rests on the empirical fact that the number of moves from vertex to vertex is a small multiple (about $2-4$) of the number of inequalities in most of the thousands of problems handled, although it has never been proved mathematically that this behavior is characteristic of "average" problems.

There have also been developed methods of calculation specifically tailored for special classes of problems. One such class (network flows) is exemplified by the transportation problem described below. Another important class is where most of the columns of (a_{ij}) are not known explicitly in advance, but consist of all columns which satisfy some particular set of rules. The idea used here (known in different contexts as the column-generation or the decomposition principle) is that the new vertex prescribed by the simplex method can be found if a solution is achieved for the problem of maximizing a linear function defined on the set of columns satisfying the particular rules.

The technology of solving integer linear programming problems is not stabilized. Methods vary from testing integral points near the optimum fractional point to systematic methods for generating relevant faces of the convex hull of the integral points satisfying the given inequalities.

Applications. Let a_{ij} denote the number of units of nutrient j present in one unit of food i. Let c_j be the minimum amount of nutrient j needed for satisfactory health, b_i be the unit price of food i. Let the variables y_i denote respectively the number of units of food i to be bought. Then the dual problem is the so-called diet problem, to maintain adequate nutrition at least cost. It has been used in planning feeding programs for several varieties of livestock and poultry. With a different interpretation of the symbols, the same format describes the problem of combining raw materials in a chemical process to produce required end products as cheaply as possible.

In another situation, let c_{ij} be the cost of shipping one unit of a given product from warehouse i to customer j for all m warehouses i and all n customers j pertaining to a business. Let a_i be the amount available at warehouse i, and b_j the amount required by customer j. The transportation problem is to ship the required amounts to the customers at least cost. Thus, if x_{ij} is the amount to be shipped from i to j, then minimize notation (4),

$$c_{11}x_{11} + \cdots + c_{mn}x_{mn} \qquad (4)$$

where $x_{i1} + \cdots + x_{in} \leqq a_i$ $(i = 1, \ldots, m)$ and $x_{1j} + \cdots + x_{mj} = b_j (j = 1, \ldots, n)$.

The foregoing are representative of the business applications of linear programming, several hundred of which have now been reported. The best-known applications outside business have been to economic theory and to combinatorial analysis. In both instances, the duality theorem has been used to illuminate and generalize previous results.

[ALAN J. HOFFMAN]

Bibliography: S. I. Gass, *Linear Programming*, 4th ed., 1975.

Linearity

The relationship that exists between two quantities when a change in one of them produces a directly proportional change in the other. Thus, an amplifier has good linearity when doubling of its input signal strength always doubles the output signal strength. A transistor displays good linearity when doubling of the instantaneous base voltage serves to double the instantaneous value of collector current.

In television, uniform linearity of scanning is essential for uniform spacing of picture elements in each horizontal line and uniform vertical spacings between scanning lines. Poor linearity causes the picture to be compressed or stretched in some area, so that circles of test patterns become egg-shaped or even more greatly distorted. *See* SAWTOOTH-WAVE GENERATOR. [JOHN MARKUS]

Liquid crystals

A state of matter that mixes the properties of both the liquid and solid states. Liquid crystals may be described as condensed fluid states with spontaneous anisotropy. They are categorized in two ways: thermotropic liquid crystals, prepared by heating the substance, and lyotropic liquid crystals, prepared by mixing two or more components, one of which is rather polar in character (for example, water). Thermotropic liquid crystals are divided, according to structural characteristics, into two classes, nematic and smectic. Nematics are further subdivided into ordinary and cholesterics.

When the solid which forms a liquid crystal is heated it undergoes transformation into a turbid system that is both fluid and birefringent. The consistency of the fluid varies with different compounds from a paste to a free-flowing liquid. When the turbid system is heated, it is converted into an isotropic liquid (optical properties are the same regardless of the direction of the measurement). These changes in phases can be represented schematically as follows:

Solid $\underset{\text{cool}}{\overset{\text{heat}}{\rightleftarrows}}$ liquid crystal $\underset{\text{cool}}{\overset{\text{heat}}{\rightleftarrows}}$ liquid

On cooling the system, the process reverses and

goes from isotropic liquid to liquid crystal and finally to the solid.

Lyotropic liquid crystals often have an amphiphilic component, a compound with a polar head attached to a long hydrophobic tail. Sodium stearate and lecithin (a phospholipid) are typical examples of amphiphiles. Starting with a solid amphiphile and adding water, the lamellar structure (molecular packing in layers) is formed. By stepwise addition of water, the molecular packing may take on a cubic structure, then hexagonal, then micellar, followed by true solution. The process is reversed by withdrawing water. Thousands of compounds will form liquid crystals on heating, and still more will do so if two or more components are mixed. A few representative compounds are listed in the table.

Classification and structure. Conventionally, matter is considered to have only three states: solid, liquid, and gas. In the solid state, the molecules or atoms show small vibrations about rigidly fixed lattice positions, but they cannot rotate. A liquid will take the shape of its container and will bound itself at the top with its own free surface. The liquid state is characterized by relatively unhindered rotation and no long-range order. The space in a system constituting a gas is sparsely occupied. The molecules are free to occupy the entire volume of their container.

Liquid crystals are a state of matter that combines a kind of long-range order (in the sense of a solid) with the ability to form droplets and to pour (in the sense of waterlike liquids). They also exhibit properties of their own such as the ability to form monocrystals with the application of a normal magnetic or electric field; an optical activity of a magnitude without parallel in either solids or liq-

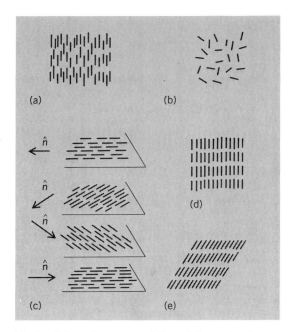

Fig. 1. Schematic representation of the molecular arrangement in the (*a*) ordinary nematic liquid crystal; (*b*) isotropic liquid; (*c*) cholesteric-nematic liquid crystal; (*d*) smectic A liquid crystal; and (*e*) smectic C liquid crystal. (*From G. H. Brown and J. J. Wolken, Liquid Crystals and Biological Structures, Academic Press, 1979*)

Name, formula, and liquid crystalline range of selected compounds

THERMOTROPIC LIQUID CRYSTALS

1. Nematic liquid crystals

$H_3C-O-\bigcirc-\underset{\underset{H}{|}}{C}=N-\bigcirc-C_4H_9-n$ 21–47°C

p-Methoxybenzylidene-*p'*-butylaniline (MBBA)

$H_{11}C_5-\bigcirc-\bigcirc-CN$ 24–35°C

4-Cyano-4'-*n*-pentyl-biphenyl

$H_{11}C_5-\bigcirc-\bigcirc-\bigcirc-CN$ 131–240°C

4-Cyano-4''-*n*-pentyl-*p*-terphenyl

$H_3C-O-\bigcirc-N=\overset{\overset{O}{\uparrow}}{N}-\bigcirc-O-CH_3$ 117–137°C

p-Azoxyanisole (PAA)

2. Cholesteric esters

145–179°C

Cholesteryl nonanoate

3. Noncholesteryl, chiral-type compound

$H_3C-O-\bigcirc-\underset{\underset{H}{|}}{C}=N-\bigcirc-\underset{\underset{H}{|}}{C}=\underset{\underset{H}{|}}{C}-\underset{\underset{O}{\|}}{C}-O-CH_2-\underset{\underset{H}{\overset{\overset{CH_3}{|}}{|}}}{C}-C_2H_5$

76–125°C

(−)-2-Methylbutyl-*p*-(*p'*-methoxy-benzylideneamino) cinnamate

uids; and a temperature sensitivity which results in a color change in certain liquid crystals. Thermotropic liquid crystals are either nematic or smectic.

Nematic structures. Nematic liquid crystals are subdivided into the ordinary nematic and the cholesteric-nematic. The molecules in the ordinary nematic structure maintain a parallel or nearly parallel arrangement to each other along the long molecular axes (Fig. 1*a*). They are mobile in three directions and can rotate about one axis. This structure is one-dimensional.

When the nematic structure is heated, it is generally transformed into the isotropic liquid (Fig 1*b*). The nematic structure is the highest-temperature mesophase in thermotropic liquid crystals. The energy required to deform a nematic liquid crystal is so small that even the slightest perturbation caused by a dust particle can distort the structure considerably.

In the cholesteric-nematic structure (Fig. 1*c*), the direction of the long axis of the molecule in a given layer is slightly displaced from the direction of the molecular axes of the molecules in an adjacent layer. If a twist is applied to this molecular packing, a helical structure is formed. The helix has a pitch which is temperature-sensitive. The helical structure serves as a diffraction grating for visible light. Chiral compounds show the cholesteric-nematic structure (twisted nematic), for example, the cholesteric esters.

Smectic structures. The term smectic covers all thermotropic liquid crystals that are not nematics. At least seven smectic structures have been described. Indications are that two more can be added, making a total of nine from smectic A(S_A) to smectic I (S_I). The alphabetic subscripts only indicate the order in which the smectic structures were first recognized and identified. In most smectic structures, molecules are arranged in strata. The molecules (except in smectic D) are arranged in layers with their long axes parallel to each other. They can move in two directions in the plane and can rotate about one axis. Those within layers can be in neat rows or randomly distributed.

Smectic liquid crystals may have structured or

Name, formula, and liquid crystalline range of selected compounds (cont.)

THERMOTROPIC LIQUID CRYSTALS (cont.)

4. Smectic A

Ethyl *p*(*p*′-phenylbenzalamino) benzoate   121– 131°C

5. Smectic B

Ethyl *p*-ethoxybenzal-*p*′-aminocinnamate   77– 116°C

6. Smectic C

n—$H_{17}C_8$—O—⟨◯⟩—COOH   108– 147°C

p-n-Octyloxybenzoic acid

LYOTROPIC LIQUID CRYSTALLINE COMPOUNDS

1. Sodium stearate

$CH_3(CH_2)_{16}COO^- Na^+$

2. α-Lecithin

CH_2—O—CO—$(CH_2)_{16}$—CH_3

CH—O—C—O—$(CH_2)_{16}$—CH_3

CH_2—O—P—O—CH_2—CH_2—$N^+(CH_3)_3$

(Stearic acid derivative)

---

unstructured strata. Structured smectic liquid crystals have long-range order in the arrangement of molecules in layers to form a regular two-dimensional lattice. The most common of the structured liquid crystals is smectic B. Molecular layers are in well-defined order, and the arrangement of the molecules within the strata is also well ordered. The long axes of the molecules lie perpendicular to the plane of the layers. In the smectic A (Fig. 1*d*) structure, molecules are also packed in strata, but the molecules in a stratum are randomly arranged. The long axes of the molecules in the smectic A structure lie perpendicular to the plane of the layers. Molecular packing in a smectic C (Fig. 1*e*) is the same as that in smectic A, except the molecules in the stratum are tilted at an angle to the plane of the stratum.

There is also a unique kind of liquid crystal known as the smectic D which is isotropic, but nevertheless shows three-dimensional order in the molecular packing of the structure.

Applications. Liquid crystals have many applications. They are used as displays in digital wrist-watches, calculators, panel meters, and industrial products. They can be used to record, store, and display images which can be projected onto a large screen. They also have potential use as television displays.

Displays. The two features which make them more desirable for displays than other materials are lower power consumption and the clarity of display in the presence of bright light. The power requirements are often so low that a digital display on a wristwatch requires about the same power as does the mechanism which runs the watch. The two modes most widely used in liquid crystal displays are dynamic-scattering and field-effect.

In displays, the liquid crystal cell design usually begins with a thin film of a room-temperature liquid crystal sandwiched between two transparent electrodes (glass coated with a metal or metal oxide film). The thickness of the liquid crystal film is $6-25$ μm and is controlled by a spacer which is chemically inert. The cell is hermetically sealed to eliminate oxygen and moisture, both of which may chemically attack the liquid crystalline material.

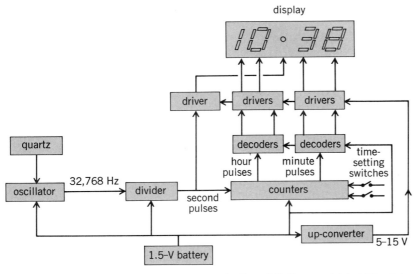

display

Fig. 2. Block diagram of a liquid crystal watch. (*From F. D. Saeva, ed., Liquid Crystals: The Fourth State of Matter, Marcel Dekker, 1979*)

32,768 Hz. Each of the time pulses is decoded to give outputs that are needed for seven-segment displays. All electronic watches operate on the same principles, regardless of the display technique (for example liquid crystals and light-emitting diodes). New display techniques, such as multiplexing, will reduce the number of electrical leads to the segments in digital and bar-graph displays.

The electronics for the wristwatch display is complex, and the dividers, driver circuits, decoders, and counters contain about 1500 transistors. All of these transistors can be collected on a silicon area of approximately 0.25 by 0.40 cm. The display for field-effect requires low voltages (1.5 to 5.0 V), and little power.

The optics of the liquid crystal display are shown in Fig. 3. The cell with the nematic liquid crystal is placed between two crossed polarizers. Polarized light entering the cell follows the twist of the nematic liquid crystal, is rotated 90°, and as such can allow passage of the light through the second polarizer (Fig. 3*a*). Application of an electric field changes the molecular alignment in the liquid crystal such that the polarization is not altered in the cell and no light is transmitted. If a mirror is used behind the second polarizer, the display will appear black (Fig. 3*b*) when voltage is applied. If, in addition, one of the electrodes is shaped in the pattern of segments of digits, then a numeric display will appear when the voltage is on. By changing the direction of the polarizers, the digit can be made to appear white on a black background. *See* ELECTRONIC DISPLAY.

Nondestructive testing. Cholesteric-nematic liquid crystals undergo a change in color with a small change in temperature, a property that can be used for nondestructive testing.

Thermometers. Desk thermometers are available which use cholesteric-nematic liquid crystals. Observation of skin temperature changes following blockage of the sympathetic nervous system enables the physician to determine if neurological and vascular pathways are open. Monitoring skin temperature over extended areas provides a more detailed and readily interpreted indication of circulatory patterns than point measurements with thermocouples and thermistors.

[GLENN H. BROWN]

Bibliography: G. H. Brown and J. J. Wolken, *Liquid Crystals and Biological Structures*, 1979; J. D. Margerum and L. J. Miller, Electro-optical applications of liquid crystals, *J. Coll. Int. Sci.* 58: 559–580, 1977; F. D. Saeva (ed), *Liquid Crystals: The Fourth State of Matter*, 1979; A. Skoulios, Amphiphiles: Organization et diagrammes de phases, *Ann. Phys.*, 3:421–450, 1978.

In dynamic scattering, if no electric field is applied, the cell is transparent. However, on addition of electric field to the liquid crystal, the cell becomes opaque. The field-effect display utilizes twisted nematic liquid crystals. The cell is prepared by rubbing the glass surface directionally or by chemically treating the surface and by adding a chiral compound to nematic liquid crystals. Digital displays are made by photoetching a seven-segment pattern onto one or both of the indium–tin oxide–coated glass plates. Reflection displays have one of the plates coated with a reflecting layer. Transmissive displays have etched surfaces on both plates.

The field-effect display is more widely used in watch and pocket-calculator displays. In a liquid crystal watch that displays hours and minutes (Fig. 2), the quartz crystal accurately controls the oscillating circuit. The oscillator frequency is typically

Load line

A line drawn on the output characteristic curves of a vacuum tube or transistor, used to determine the operating range and the quiescent point (the operating point for zero signal input voltage or current). In the case of vacuum tubes, this graphical construction is necessary because the mathematical function relating plate current to plate voltage and grid voltage is not known analytically. A set of plate current curves with a load line drawn is illustrated. *See* VACUUM TUBE.

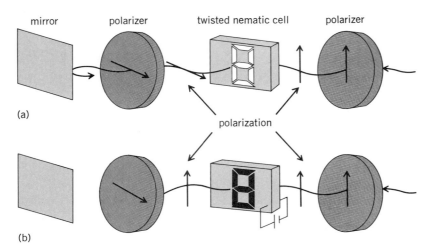

Fig. 3. Optics of field-effect display device when a twisted nematic cell is utilized. (*a*) Optics without electric field applied to liquid crystal cell. (*b*) Optics with electric field applied.

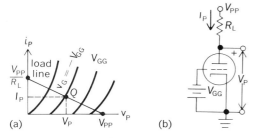

Graphical construction for load line. (*a*) Load line drawn on plate current characteristics. (*b*) Tube with load resistor.

If the tube has fixed bias, as shown in the circuit diagram in the illustration, the procedure for drawing the load line is simple. The load line is drawn to intersect the plate-voltage axis at a value equal to V_{PP}, the plate supply voltage, and the vertical axis at a value of plate current which is equal to V_{PP}/R_L, where R_L is the load resistance. The quiescent plate current I_p and plate voltage V_p are read at the point Q, where the curve for the value of grid bias intersects the load line.

If the tube is self-biased with a cathode resistor, R_K, then the load line is drawn to intersect the plate-voltage axis at V_{PP} and the vertical axis at $V_p/(R_L + R_k)$. A bias line is constructed corresponding to the equation $v_c = -i_p R_k$ (assuming that $V_{GG} = 0$ and a grid-leak resistor is used). The point of intersection of the bias line and load line gives the quiescent point on the tube characteristics.

Load lines similarly constructed may also be used with transistor circuits and other nonlinear devices to determine the quiescent point. *See* TRANSISTOR.   [CHRISTOS C. HALKIAS]

Logic circuits

The basic building blocks used to realize consumer and industrial products that incorporate digital electronics. Such products include digital computers, video games, voice synthesizers, pocket calculators, and robot controls.

The change that has enabled widespread economical use of digital logic once found only in very expensive, room-sized computers has been a dramatic evolution in device technology. Logic circuits which comprise several basic electronic devices (typically transistors, resistors, and diodes) were once designed with each device as a separate physical entity. Now, very-large-scale integration of devices offers up to several hundred thousand equivalent basic devices on one small piece of silicon, typically rectangular with maximum dimensions of a few tenths of an inch (1 in. = 25 mm) per side. *See* INTEGRATED CIRCUITS.

This dramatic reduction in size has been accompanied by a number of effects. The cost and power consumption per logic device have been greatly reduced. The modestly priced digital watch that can run for over a year on a tiny battery exemplifies these effects.

Logic circuit operation. Logic circuits process information encoded as voltage or current levels. The adjective "digital" derives from the fact that symbols are encoded as one of a limited set of specific values. While it would be feasible to en-

code information using many distinct voltage levels (for example, each of 10 voltage levels might correspond to a distinct decimal digit value), there are several reasons for using binary of two-valued signals to represent information: the complexity of the sending and receiving circuit is reduced to a simple on-off switch type of operation corresponding to the two-valued signals; the speed of operation is far greater since it is not necessary to wait until a changing signal has had time to converge to a final value before interpreting its binary value; and the on-off operation makes the circuitry very tolerant of both changing characteristics of devices due to aging or temperature-humidity environment and electromagnetically induced noise added to the voltage levels.

Since signals are interpreted to be one of only two values (denoted as 0 and 1), a group of four such signals must be used to represent a decimal digit, as an example. A very commonly used convention in computers and terminals is to use a group of eight binary signals to represent alphabetic, numeric, and in general any keyboard-derived symbol. The eight signals enable encoding any of 256 character or symbol values.

More generally, the binary digits, or bits, are used in a number system with digit weights that change by multiples of 2 rather than multiples of 10 as in the decimal system. In this system, successive digits or bits have weights of 1, 2, 4, 8, ... in contrast to the decimal system where successive digits have weights of 1, 10, 100, ... *See* NUMBER SYSTEMS.

Types of logic functions. All logic circuits may be described in terms of three fundamental elements, shown graphically in Fig. 1. The NOT element has one input and one output; as the name

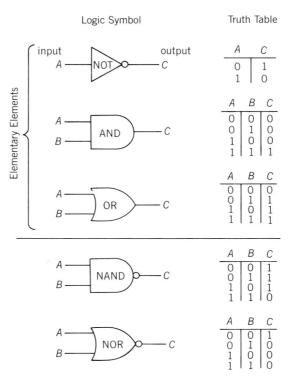

Fig. 1. Logic elements.

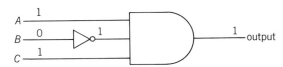

Fig. 2. Logic circuit whose output is 1 only when input signals *A*, *B*, and *C* are 1, 0, and 1, respectively.

suggests, the output generated is the opposite of the input in binary. In other words, a 0 input value causes a 1 to appear at the output; a 1 input results in a 0 output. Two NOT elements in series, one output driving the input to the other, simply reproduce a copy of the input signal.

The AND element has an arbitrary number of inputs and a single output. (Electrical characteristics of devices used or physical packaging of devices result in some practical limit to the number of inputs allowed.) As the name suggests, the output becomes 1 if, and only if, all of the inputs are 1; otherwise the output is 0. The AND together with the NOT circuit therefore enables searching for a particular combination of binary signals. If, for example, a signal is needed that goes to 1 only when signals *A*, *B*, and *C* are 1, 0, and 1, respectively, then the circuit indicated in Fig. 2 will generate the desired output.

The third element is the OR function. As with the AND, an arbitrary number of inputs may exist and one output is generated. The OR output is 1 if one or more inputs are 1.

The operations of AND and OR have some analogies to the arithmetic operations of multiplication and addition, respectively. The collection of mathematical rules and properties of these operations is called Boolean algebra.

While the NOT, AND, and OR functions have been designed as individual circuits in many circuit families, by far the most common functions realized as individual circuits are the NAND and NOR circuits of Fig. 1. A NAND may be described as equivalent to an AND element driving a NOT element. Similarly, a NOR is equivalent to an OR element driving a NOT element. (The reason for this strong bias favoring inverting outputs is that the transistor, and the vacuum tube which preceded it, are by nature inverters or NOT-type circuit devices when used as signal amplifiers.) An interesting property of the NAND or NOR circuits is that all logic functions may be accessed using either type of circuit. A NOT element, for example, is realized as a one-input NAND. An AND element is realized as NAND followed by a NOT element. An OR element is realized by applying a NOT to each input individually and then applying the resulting outputs to the NAND as inputs.

Combinational and sequential logic. As the names of the logic elements described suggest, logic circuits respond to combinations of input signals. In an arithmetic adder, for example, a network of logic elements is interconnected to generate the sum digit as an output by monitoring combinations of input digits; the network generates a 1 for each output only in response to those combinations in the addition table that call for it.

Logic networks which are interconnected so that the current set of output signals is responsive

only to the current set of input signals are appropriately termed combinational logic.

An important further capability for processing information is memory, or the ability to store information. In digital systems, the memory function has been provided for by a variety of technologies, including magnetic cores, stored charge, magnetic bubbles, and magnetic tape. The logic circuits themselves must provide a memory function if information is to be manipulated at the speeds the logic is capable of. The logic elements defined above may be interconnected to provide this memory by the use of feedback. *See* COMPUTER STORAGE TECHNOLOGY.

The circuit indicated in Fig. 3 illustrates a basic and perhaps the most commonly used form of memory circuit. Normally both inputs *A* and *B* to the cross-coupled NAND gates are at the voltage corresponding to logic 1; if a 0 is applied to signal *A*, outputs *C* and *D* assume the values 0 and 1, respectively, based on the definition of NAND operation; the outputs retain these values after input *A* is returned to the quiescent value of 1. Conversely, if *B* is set to 0 while *A* is 1, then the outputs *C* and *D* become 1 and 0, respectively, and again retain these values after *B* is returned to a value of 1. In the jargon of logic circuit design, this circuit is variously called a flip-flop, toggle, or trigger. Like the light switch on the wall, the circuit retains one of two stable states after the force initiating change to that state is removed. This circuit is therefore well matched to the storage of binary or two-valued signals. Logic circuit networks that include feedback paths to retain information are termed sequential logic networks, since outputs are in part dependent on the prior input signals applied and in particular on the sequence in which the signals were applied.

Logic circuit embodiment. Several alternatives exist for the digital designer to create a digital system.

Ready-made catalog-order devices can be combined as building blocks. In this case, manufacturers have attempted to provide a repertoire of logic networks that will find common usage and generality. Some examples of the types of circuits found in digital circuit manuals or catalogs are: arrays of individual logic gates (typically four to six) with outputs and inputs available as external connections to the device; arrays of memory elements (flip-flops) used to hold a collection of binary signals or digits; arithmetic adders capable of adding four corresponding binary digits from each

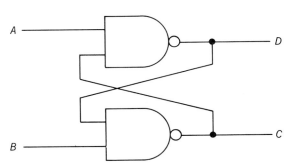

Fig. 3. Memory element.

of two binary numbers and of being cascaded to produce an adder for any multiple of four digits; and multipliers capable of forming the product of two 16-bit numbers in a single step (that is, a combinational multiplier).

The significance of these examples is that the number of gates or logic elements realized on individual devices ranges from a few to several thousand. Clearly a design that can use the devices with larger gate counts is likely to be more compact and more economical.

A second and increasingly common realization of logic gates is custom-designed devices. Two factors which make this practical are computer-aided-design (CAD) tools and libraries of subnetworks that can be used as building blocks. Custom design of a logic network implies the generation of highly complicated artwork patterns, called masks, for use in photographic-development-like steps used to produce the integrated circuit networks. The availability of a library of subnetworks for which this artwork has already been generated reduces the additional artwork required to interconnection patterns among these library elements to set up the desired logic network. *See* COMPUTER-AIDED DESIGN AND MANUFACTURING.

A third option that is rapidly growing in usage is gate-array devices. In this case, the device vendor manufactures devices comprising a two-dimensional array of logic cells. Each cell is equivalent to one or a few logic gates. The final layers of metallization that determine the exact function of each cell and interconnect the cells to form a specific network are deferred until the customer orders such a device. This procedure uses the advantage of mass production for the majority of processing steps necessary to manufacture a device, including most of the necessary artwork. Since the interconnecting metallization layers are a relatively small and simple part of the total device fabrication, the customization cost and time can be reduced significantly by this method relative to a totally custom fabrication. The sacrifice in this approach is that the packing density of the two-dimensional array will tend to be less than a custom-designed layout, since routing channels must allow for reasonably general interconnection patterns.

A fourth option in realizing logic functions is programmable logic arrays, or PLAs. The manufactured part has the potential for realizing any of a large number of different sets of logic functions; the particular functions produced are determined by actually blowing microscopic fuses built into the device, in effect removing unwanted connections. In one device, each of 16 device-input signals is an input to each of 48 AND gates; the 48 AND outputs are input signals to 8 OR gates that generate the device outputs. Selective removal of inputs to the AND gates, and between the AND and OR gates, enables implementation of a wide variety of functions.

A fifth embodiment of logic network functions involves the use of table look-up. A logic network has several binary input signals and one or more binary outputs. The transformation of input signal combinations to output signal values need not be realized by AND-OR-type logic elements at all;

instead, the collection of input signals can be grouped arbitrarily as address digits to a memory device. In response to any particular combination of inputs the memory device location that is addressed becomes the output. For example, a logic network function involving 10 binary inputs and 8 binary output signals could be realized with a single memory device that holds 1024 memory cells, with 8 bits stored in each cell. (Ten bits addresses $2^{10} = 1024$ locations.)

The form of memory device used for table look-up is usually read-only, that is, memory contents are defined as part of the manufacturing process by means of the last layer of metallization, or one of a number of other means is used to fix the memory cell contents so that it is not necessary to reload the memory after each turn-off of power.

Typically, this approach may lead to slower operation than the use of logic circuits, but in some cases can lead to very significant economies.

The last form of logic network embodiment that will be discussed is the microcomputer. A microcomputer is a single device that includes a read-only memory to hold a program, a processor capable of reading and executing that program, and a small read-write memory for scratch working space. Just as a memory device provides an efficient realization of a combinational network, the equivalent of a highly complex sequential network can be had with a microcomputer, as evidenced by electronic games. Two important advantages of this approach are that little or no custom-device fabrication is required, and the programmability permits utilization of complicated and modifiable equivalent networks. *See* MICROCOMPUTER; MICROPROCESSOR.

As with table look-up, the principal disadvantage of this approach is that the speed of operation may be slower than if an actual network of high-speed logic circuits were used.

Technology of logic circuits. There are basically two logic circuit families in widespread use: bipolar and metal-oxide-semiconductor (MOS).

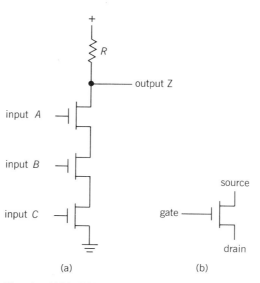

Fig. 4. MOS NAND gate. (*a*) Circuit configuration. (*b*) MOS transistor (*n* or *p*), the basic circuit device.

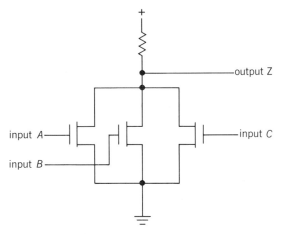

Fig. 5. MOS NOR gate.

The basic MOS device is formed by using a silicon substrate which has been doped in such a way as to greatly increase the hole/electron relative density, forming a p-type substrate. In addition to the substrate, two diffused n-type regions are formed. A layer of metal, insulated from the substrate by a deposit of oxide, is situated between the two n regions. This metal connection is known as the gate, and the two n regions are called the source and drain; since the device is symmetrical in every respect, source and drain are functionally interchangeable. When a sufficient voltage exists between the metal gate and the substrate, electrons are conducted between the drain and source connections, essentially shorting the two together. Otherwise, a high resistance exists between the two. Thus, the gate input is analogous to the control lever of a switch. In like manner, an n-type doping of the substrate, along with a p-type source and drain, may be used to implement the switch. This will result in hole migration between source and drain when the gate is turned on. *See* TRANSISTOR.

The simplest MOS logic structure is the transmission gate (also known as pass transistor and steering logic); only one transistor is used to im-

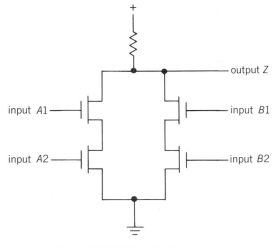

Fig. 6. MOS A-O-I (AND-OR-INVERT) gate.

plement the function, where source and gate leads represent the two input variables, and the drain becomes the output value. The transistor is turned on when the gate voltage is above the threshold voltage that assures sufficient current flow through the channel; in other words, it acts as a short between source and drain. This is analogous to a mechanical switch in the "on" position. When the gate is below the threshold potential the switch is opened, thus performing an AND-like function on the inputs.

The NAND logic function is also easily derived with MOS circuits. Figure 4 shows a sample NAND circuit in which the transistors $A, B,$ and C are configured in series and must therefore all be in the "on" state for a conduction path from output Z to ground to be established. If A, B and C are all above threshold, output Z is shorted to ground or is forced to logical zero. On the other hand, if any of $A, B,$ or C are off, an open circuit essentially exists between Z and ground. In this case, output node Z exhibits much less resistance to the power source ($R << R_a + R_b + R_c$) and therefore will be at the voltage potential of the power source.

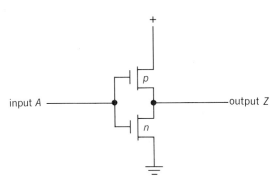

Fig. 7. CMOS inverter.

This NAND circuit can be reconfigured into the NOR circuit of Fig. 5. Here, the transistors $A, B,$ and C are in parallel instead of in series, and it is easily seen that when any of the inputs are turned on the output will be grounded. Otherwise, Z will be pulled up to the power supply voltage.

Another useful MOS logic structure is the AND-OR-INVERT, which performs the logic functions exactly as stated in the name. The A-O-I is shown in Fig. 6 to be a combination of the NAND and NOR configurations. Output Z is low only if the function ($A1$ AND $A2$) OR ($B1$ AND $B2$) are high.

As described earlier, either an n-channel substrate (NMOS) or a p-channel substrate (PMOS) may be used to implement the logic structures. Additionally, both types may be used concurrently in the logic circuit to fabricate a more sophisticated device. Figure 7 shows an inverter built with both NMOS and PMOS devices. When the input is low, the NMOS transistor is cut off and the PMOS transistor conducts, shorting the output to the power source. On the other hand, when the input is high, the PMOS gate is shut off and the NMOS gate shorts the output to ground, thus performing the inverter function. The advantage of such a complementary configuration, known as

CMOS, is primarily one of power dissipation. Owing to the fact that only one complementary transistor conducts at any time (other than the short overlap during switching), there is always a very high resistance from the power source to ground, and therefore very little current flows through the pair. The primary disadvantage of CMOS over the common MOS circuit is the increased fabrication complexity due to the n and p substrates occupying the same chip, and the secondary disadvantage is the increased area needed to build the CMOS structure because of the use of complementary pairs.

The charge carriers in an MOS transistor are either free electrons or holes. For this reason, MOS is known as a unipolar device. In contrast, devices which utilize both free electrons and hole migration are made. These are known as bipolar devices, such as the junction transistor. Although general comparisons are hard to make, these devices typically exhibit much less input resistance and demand a more sophisticated fabrication technique than do MOS devices, but offer the advantage of accentuated high-frequency response. In terms of logic realization, the bipolar family encompasses many common circuit types. Perhaps the best-known and most widely used implementation of logic switches is the bipolar transistor-transistor logic, or TTL. Shown schematically in Fig. 8, the basic TTL NAND gate is formed by a multiemitter transistor (turned on only if every input is high) followed by an output transistor that acts as a pullup/buffer. Thus, the first transistor performs an AND operation on the inputs and the second transistor completes the NAND by performing an inversion. *See* JUNCTION TRANSISTOR.

TTL transistors are operated in the saturation mode; in other words, the transistors are driven hard to either the cutoff or the saturation limits. This overdriving introduces a time delay that does not exist if the transistors are operated in the nonsaturated mode. Such nonsaturating logic, while inherently faster, is more susceptible to noise since it is biased in the linear region.

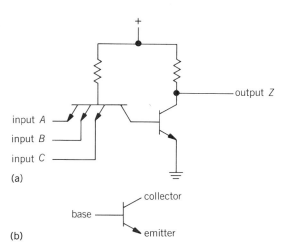

Fig. 8. TTL (transistor-transistor-logic) NAND gate. (*a*) Circuit configuration. (*b*) Bipolar junction transistor (*npn*), the basic circuit device.

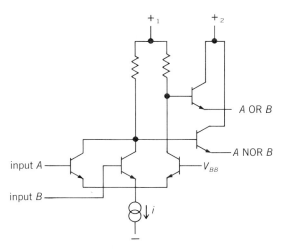

Fig. 9. ECL (emitter-coupled-logic) OR/NOR gate.

Current-mode logic, or CML, is a popular form of nonsaturation logic and most often takes the configuration of emitter-coupled logic, or ECL. The basic ECL gate is shown in Fig. 9 to be composed of current-steering transistors that perform an OR operation on the inputs. Typically, the gate output is amplified by an emitter-follower transistor, and both the true and complement signals can be made available with no added delay as outputs.

[R. R. SHIVELY; W. V. ROBINSON]

Loudspeaker

A device that converts electrical signal energy into acoustical energy, which it radiates into a bounded space, such as a room, or into outdoor space. The essential characteristic of the signal energy must be preserved in the energy-conversion process. The shorter term speaker is also used. The term unit, as in horn or driver unit, is applied to devices comprising a motor and diaphragm but lacking the acoustic element such as a horn to complete the speaker. The term speaker system is normally reserved for a plurality of speaker units and the associated accessories, horns, cabinets, and electrical networks required for an integrated system.

Requirements for speech and music. The two most important variables in a sound wave are pressure and the frequency of the pressure alternations, because the auditory response of the listener is most closely related to these variables. The approximate auditory limits of the ear are shown in Fig. 1. These limits cover an enormous range. Frequencies differ by a factor of 10^3, pressures by a factor of 10^6, and intensities (power) by a factor of 10^{12}. Fortunately, commonly reproduced sounds such as music and speech cover the more limited ranges in Fig. 1. The need for reproducing even these ranges is further modified by psychoacoustic factors and noise.

Tests have shown that frequency ranges of 80–11,000 Hz for music and 150–8000 Hz for speech, if uniform throughout the listening area, give high-quality reproduction in an otherwise distortionless system. The pressure range requirements are set primarily by listener preference. The listener is seldom interested in hearing a wanted sound in the

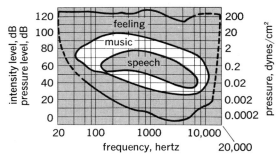

Fig. 1. Approximate auditory limits of the ear. The colored areas shown indicate the intensity and pressure levels that are found in speech and music. The standard reference levels are 10^{-16} watt/cm² intensity (plane wave) and 0.0002 dyne/cm² pressure.

presence of a competing unwanted sound (by definition, noise) of the same pressure. Noise, therefore, commonly limits the minimum pressure of interest, particularly in music reproduction.

Alternatively, the requirements of the speaker may be considered from the viewpoint of the complete-system designer, who normally desires a speaker that does not noticeably narrow the frequency bandwidth, lower the signal-to-noise ratio, or increase the distortion of the system. Probably the most important requirement is that the speaker not introduce noises that are clearly spurious and unrelated to the signal, such as rattles and buzzes. Some physical performance characteristics can be adequately measured and specified; others cannot, however, and a thorough listening test is an essential part of any speaker-testing program. The final judge of the value and adequacy of the speaker (and system) is the ear of the ultimate user.

Physical performance characteristics. Because of the importance of pressure and frequency in hearing, the loudspeaker output in terms of these variables is also important.

Pressure response. Measured at a specified frequency, the pressure response is the sound pressure in dynes/cm² at a designated location with respect to the speaker, per volt input. The pressure is commonly measured under actual or simulated outdoor free-field conditions in the absence of unintended reflecting or diffracting surfaces.

Pressure response–frequency characteristic. Commonly called simply the frequency response, this is the pressure response obtained as the test-signal frequency is varied slowly over the audible frequency range (20–20,000 Hz). The response characteristic should be free of abrupt changes, which are accompanied by transient and sometimes nonlinear distortion. A smooth response curve also simplifies electrical equalization or compensation.

Directional characteristics. These are determined by measuring the pressure response at a sufficient number of points in the intended listening region and at a sufficient number of frequencies to predict the frequency response at any listener's position with the desired precision. The most common practice is to obtain frequency-response characteristics at a few points equidistant from the speaker but subtending different angles with the principal axis of the speaker. On-axis

curves are usually published because the pressure is normally a maximum, particularly at high frequencies when focusing is marked. In typical environments the response characteristic at 30° is much more useful for predicting the average response throughout the listening region. In use, the speaker is positioned and oriented to give the most favorable average pressure response–frequency characteristic in the listening region.

Distortion. Although distortion may be more broadly defined, loudspeaker distortion is commonly limited to nonlinear, transient and frequency-modulation distortion.

Nonlinear distortion arises from lack of proportionality between the electrical input and acoustical output signal with a sustained (steady-state) input signal. With a sinusoidal input, integral multiples (harmonics) and rarely submultiples (subharmonics) of the fundamental frequency are generated. If two or more sinusoidal signals are applied, the nonlinearity gives rise to interaction between the two (intermodulation distortion), resulting in the generation of harmonics and sum and difference frequencies of the fundamentals and harmonics.

Transient distortion arises from the inability of the speaker to generate an acoustic wave which exactly follows sudden changes in the wave shape of the electrical signal. Sharp peaks and valleys in the pressure-frequency response curve are to be avoided, because they indicate the presence of inadequately damped resonances which give rise to this distortion. These poorly damped resonances color the reproduction and, when present at high frequencies, can be detected as a ringing sound.

Frequency modulation of a signal is the alteration, or modulation, of its apparent frequency by a second, or modulating, signal. It occurs when the radiating diaphragm is vibrated toward and away from the listener by the modulating signal. The velocity component with respect to the listener produces a Doppler shift.

Electrical speaker impedance. This is the complex ratio of the applied sinusoidal voltage across the signal terminals to the resulting current. The normal or loaded impedance Z_N, the free or unloaded impedance Z_f, and the blocked impedance Z_b correspond respectively to operation with the normal acoustic load, in a vacuum, and with the motor and diaphragm blocked or immobilized.

Force factor. The force factor M of the motor is (1) the complex ratio of the force required to block or immobilize the motor to the signal input current to the motor, and (2) the complex ratio of the resulting open-circuit voltage to the velocity of the mechanical system. In a moving-coil speaker the first quantity equals Bl, where B is the average radial flux density through the conductor and l is the conductor length. The second quantity equals $-Bl$. When consistent units are used, these quantities have the same magnitude but opposite signs indicating an antireciprocal relationship, that is, $M_{21} = -M_{12}$.

The normal impedance of a speaker is given by Eq. (1), where Z_m is the total effective mechanical

$$Z_N = Z_b + \frac{M^2}{Z_m} \qquad (1)$$

impedance of the moving system, including diaphragm and air load. In a moving-coil speaker $M^2 = B^2l^2$. In a two-electrode electrostatic speaker $M^2 = -(E_0/\omega d_0)^2$, where E_0 is the dc bias voltage, d_0 the average biased electrode separation, and $\omega = 2\pi$ times frequency. The second term on the right side of Eq. (1) arises from motion of the moving system and is called the loaded or normal motional impedance.

Efficiency. The speaker efficiency is the ratio of the useful acoustic-energy output to the electrical signal-energy input. The system, or available power, efficiency is the ratio of the useful acoustic-energy output to the energy the signal system will supply to a designated resistance.

The acoustic output may be obtained by determining the intensity (see below) of the sound wave at a number of equidistant points from the speaker, under free-field (outdoor) conditions, and summing or integrating the intensity in all directions. The distance should be several times the maximum dimension of the speaker and a sufficient number of points must be used to permit accurate interpolation and integration. Because sound-intensity meters are rarely available, the pressure is measured. Under the above test condition, but not in a normal listening room, the intensity I in the direction of propagation is given by Eq. (2), where p is

$$I = p^2/\rho c = 2.42 \times 10^{-9} p^2 \text{ watt/cm}^2 \qquad (2)$$

the pressure in dynes/cm²; ρc is the characteristic impedance (here resistance) of air, ρ is the density of dry air (0.0012 g/ml), and c is the velocity of propagation of the sound wave (34,280 cm/sec), both at 20°C and 760 mm Hg pressure. The electrical input power is the rms input current squared times the resistive part of the electrical normal impedance.

Speaker power rating (input). Unless otherwise stated, this is the maximum output power rating of a speaker's intended associated amplifier when the signal is speech and music and the amplifier is operated in its linear (undistorted) rated region. Under these conditions the average power over 15-sec intervals will be 1–2%, and the maximum power at any one frequency over a 2-sec interval will be 10–15% of the rated value. As an example, a 10-watt speaker is intended to work with a normally operated 10-watt amplifier with speech and music input. It is not expected to operate continuously with 10 watts input at all frequencies in its transmitted frequency band without distortion or failure.

Speaker types. Speakers are commonly classified by terms which describe their three important functional parts: the motor, diaphragm, and acoustic radiation-controlling element. The motor converts electrical energy into mechanical energy and couples the electrical signal source, commonly an amplifier, to the diaphragm. Common motor types are moving-conductor (principally moving-coil or dynamic, rarely ribbon conductor) and electrostatic. Moving-conductor speakers are those in which the mechanical forces result from magnetic reactions between the field of the current in a moving conductor and a steady magnetic field. In a moving-coil speaker the conductor is in the form of a coil conductively connected to a source of electrical energy and mechanically attached to the diaphragm. An electrostatic speaker is

one in which the mechanical displacements are produced by the action of electrostatic fields. Other motor types, which include the magnetic-armature, pneumatic, piezoelectric, and ionic types, are discussed in the references listed in the bibliography.

Speakers are also classified by their radiation-controlling elements into direct-radiator (hornless) and horn types. The direct-radiator types commonly employ baffles or enclosures.

The diaphragm is the element which, vibrated by the motor, causes the air to vibrate; hence it couples the air load (radiation impedance) to the motor.

Radiation impedance. The added impedance to motion of a diaphragm or sound-radiating surface, arising from its contact with air, is called the radiation impedance. Radiation resistance is a component arising from energy radiation. Radiation reactance is the reactive component. A major problem to the speaker designer and one of importance to the user is that of obtaining efficient energy transfer or coupling from a relatively high motor-and-diaphragm mechanical impedance to a low air-radiation impedance.

The average radiation impedance per unit area of a flat piston is shown in Fig. 2. When the length of the radiated wave λ exceeds the perimeter of the diaphragm (when $\lambda > 2\pi R$ at low frequencies), the radiation reactance is mass reactance and proportional to frequency, and the radiation resistance per unit area is proportional to the frequency squared. The acoustic power P_a radiated by a pis-

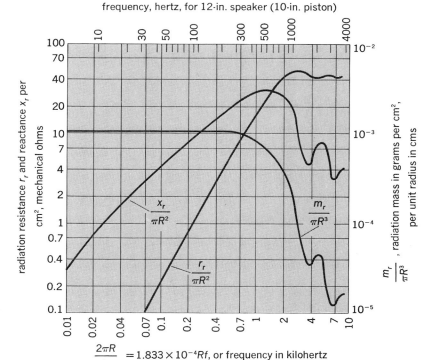

Fig. 2. Average radiation resistance, reactance, and mass per square centimeter of a flat, rigid piston vibrating in an infinite, rigid, nonabsorbing baffle. Piston radiates into a solid angle $\Omega = 2\pi$ steradians (hemisphere). $\lambda =$ length of radiated wave in centimeters. 1 in. = 2.5 cm. (*From K. Henney, ed., Radio Engineering Handbook, 5th ed., McGraw-Hill, 1959*)

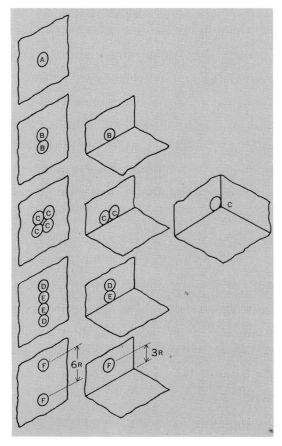

Fig. 3. The effect of adding pistons and reflecting planes on radiation impedance. All pistons that are marked with the same letter see the same radiation impedance. (*From K. Henney, ed., Radio Engineering Handbook, 5th ed., McGraw-Hill, 1959*)

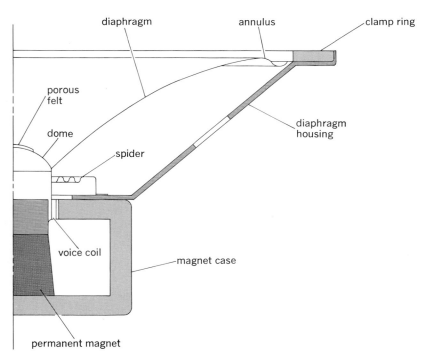

Fig. 4. Sectional view of small moving-coil permanent magnet speaker.

ton is given by Eq. (3), where r_r is the total radia-

$$P_2 = r_r v^2 \times 10^{-7} \text{ watt} \qquad (3)$$

tion resistance (obtained by multiplying the per unit area value from Fig. 2 by the area) and v the rms piston velocity, cm/sec.

If two or more identical speakers are mounted close together in a common plane and supplied with signal energy from the same source, the combined diaphragms will have substantially the same radiation impedance at low frequencies as a single diaphragm of the combined area. Differences occur as the wavelength diminishes (as frequency rises) and approaches half the diaphragm perimeter. By mounting a speaker adjacent to one or two large, rigid, nonabsorbing surfaces, such as a floor or floor-wall combination, the radiation resistance may be made that of one speaker of a two- or four-speaker combination, as shown in Fig. 3. Each surface, by complete wave reflection, adds an image (by analogy with an optical mirror) which the diaphragm cannot distinguish from a true speaker or pair of speakers. An alternate way of considering this effect is to note that the solid angle into which the speaker radiates is 2π, π, and $\pi/2$ steradians, or half, quarter, or eighth spheres, respectively. A further advantage of locating a speaker in the corner is the narrower angle over which uniform high-frequency response is required.

Direct-radiator speakers. These are the most widely used speakers. Of these, substantially all of the general-purpose and special-purpose low-frequency (woofer) speakers are of the moving-coil type. Electrostatic speakers are sometimes used as high-frequency speakers (tweeters) in broadband systems and infrequently as general-purpose speakers.

Moving-coil speakers. Widely used because they can cover a broad frequency range (5–8 octaves), these speakers are compact, reliable, and low in cost (Fig. 4). Their available power or system efficiency is low, typically 0.5–4%.

Their relatively uniform efficiency up to 1000 Hz or more is attained by placing the fundamental mechanical resonant frequency near the lowest frequency of interest. This is the frequency at which the diaphragm and air mass are resonated by the diaphragm suspension and air stiffness, if present. At frequencies that are higher than this by a factor of from 2 (1 octave) to 10 or so, the normal impedance is substantially the blocked impedance, which in a moving-coil speaker is largely resistive. A constant-voltage and constant-internal-resistance signal source will give a fairly uniform current and hence, force, in this range. The effective moving mass is nearly constant. The impedance is therefore proportional to frequency and the diaphragm velocity varies inversely with frequency. Consequently, the radiated power is independent of frequency.

Electrostatic speakers. In these speakers the mechanical forces are produced by the action of electrostatic fields. Structurally they are capacitors in which one electrode is free to move and serve as a diaphragm. The movable electrode is commonly a thin plastic sheet with a conductive surface, which may be a metallic foil or evaporated

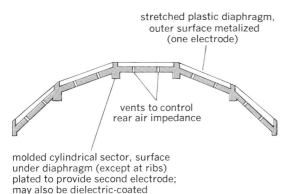

Fig. 5. Simplified cross-sectional view of a two-electrode electrostatic speaker of the type that is used to obtain an approximately cylindrical wave front at high frequencies. (*From K. Henney, ed., Radio Engineering Handbook, 5th ed., McGraw-Hill, 1959*)

film. Two fixed electrodes, one on each side of the movable electrode, are used in push-pull types when the diaphragm amplitude may be relatively large. A single fixed electrode is used in single-ended types (Fig. 5). These are commonly used as high-frequency speakers (tweeters). A high dc bias potential, of the order of kilovolts, is applied between the fixed and movable electrodes to increase sensitivity and reduce distortion. The ac potential is superimposed on the dc, resulting in an alternating force on the diaphragm.

Some advantages of the electrostatic speaker are (1) the diaphragm can be very light, (2) the force is distributed over the diaphragm, (3) the transient and phase distortion may be kept low by proper design, and (4) the cost may be low.

Some of the disadvantages are (1) the permissible diaphragm amplitude is normally low, leading to large-area diaphragms or limiting the lower cutoff frequency, (2) the normal impedance is substantially that of a capacitor and is difficult to match properly, except in a limited frequency range, and (3) because of the high potentials, dielectric problems are involved.

Radiation-controlling structures. These are rigid, nonvibrating, normally nonabsorbing structures which control the divergence of the sound wave from one or both sides of the diaphragm and prevent or control interaction between the front and back radiation. Their principal function is to increase the sound pressure in the listening region, both by increasing the energy-conversion efficiency and by directing the emitted wave in the desired direction. Vibration of the diaphragm produces slight alternating increases and decreases in the air pressure above and below atmospheric pressure. Increases occur on one side as decreases occur on the other. At low frequencies the front pressure wave can reach the back of the diaphragm (and the back pressure wave can reach the front of the diaphragm) before the diaphragm has moved appreciably. Hence almost complete pressure neutralization, or destructive interference, occurs unless a rigid surface, or baffle, is interposed as an extension of the diaphragm support.

The term baffle is usually applied to a relatively flat structure of substantial area. Three types are

shown in Fig. 6. Either the speaker is mounted eccentrically, or the baffle shape is made irregular so that serious destructive interference does not occur in a narrow frequency range. Destructive interference may be prominent in outdoor environments but is usually obscured indoors by reflected waves.

Enclosures. These are radiation-controlling structures of the hornless type. Enclosures may be classified as total, in which the rear radiation is completely absorbed by the closed rear enclosure, and vented, or reflex, in which the rear radiation is shifted in phase, modified in amplitude, and combined with the front radiation to increase the total radiation in a frequency band of approximately an octave just above the lowest transmitted frequency. At still lower frequencies the total radiation is substantially reduced by pressure neutralization or destructive interference.

Total enclosures are used primarily when the volume is smaller than desirable for a reflex type. Because the air stiffness limits the minimum diaphragm–motor system stiffness, the mass may be raised to reduce the resonant frequency. In the reflex type, higher efficiency is obtained in the lowest octave by using a larger cabinet with less air stiffness and by making use of the reflex, or rear, acoustic by network phase shift and radiation. A simplified sketch of a reflex enclosure is shown in Fig. 7. The cabinet air stiffness couples the diaphragm to the air mass in the vent, or port region. The total enclosure has no port or opening.

Horns. Structurally a horn is a rigid, nonabsorbing, tapered duct or passage. It may be straight, bent, or folded with concentric sound passages. The small end, to which the horn or driver unit (speaker) is attached, is called the throat and the large radiating end is called the mouth.

Functionally it is both a tapered acoustical transmission line used to match the relatively high impedance looking back into the diaphragm to the relatively low radiation impedance seen at the mouth, and a device which by virtue of its mouth size and shape controls directional properties.

V. Salmon has discovered a family of horns with interesting properties to which many common horns belong. In these Salmon or hyperbolic exponential horns, the cross-sectional area A is related

Fig. 6. Irregular baffle shapes used outdoors to broaden frequency band of destructive interference between speaker front and back waves at listener's position.

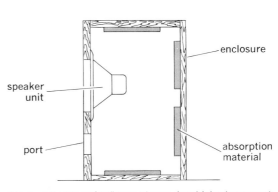

Fig. 7. One type of reflex enclosure in which a large port area is placed near the diaphragm to obtain maximum aid from the port radiation. Phase shift of backside radiation is obtained by choice of enclosure volume and port mass. (*From K. Henney, ed., Radio Engineering Handbook, 5th ed., McGraw-Hill, 1959*)

to the axial distance by Eq. (4), where A_t is the

$$A = A_t \left[\cosh\left(\frac{x}{x_0}\right) + T \sinh\left(\frac{x}{x_0}\right) \right]^2 \quad (4)$$

throat area, x_0 is a constant fixing the axial scale of length, T is a constant determining a member of the general family, and the cosh and sinh are the hyperbolic exponential cosine and sine functions, respectively. The longitudinal sections of these horns for various values of T are shown in Fig. 8 for straight-axis circular horns.

The performance of a horn depends principally on the throat impedance and its dependence on frequency. Although waves reflected from the mouth introduce variations with frequency into the throat impedance, it has been found that the average impedance is close to that of a horn with no reflected wave. The throat impedance in mechanical ohms of catenoidal horns with rigid, nonabsorbent walls and negligible reflected waves is given by Eq. (5), where ρ, c, and T have been defined

$$Z_t = R_t + jx_t$$
$$= A_t \rho c \frac{[1 - (f_c/f)^2]^{1/2} - j(Tf_c/f)}{1 - (1 - T^2)(f_c/f)^2} \quad (5)$$

previously, f is the frequency, and f_c is the cutoff frequency, given by $f_c = c/2x_0$ and $j = \sqrt{-1}$. Thus the reference axial length x_0 is of fundamental importance in determining the behavior of the impedance. In Fig. 9 the behavior of R_t for various values of T is shown.

Speaker systems. Two or more identical speakers may be used to improve low-frequency efficiency and increase permissible input and output powers or, by proper relative orientation, improve the directional properties. A source of sound which is very small compared to the length of the radiated wave (point source) is nondirectional. The source becomes increasingly directional as the wave length is diminished; that is, the frequency is increased. Noncircular sources tend to be most directional in the plane of the maximum dimension. Diaphragm and horn mouth shapes are therefore used to control directivity. Acoustic lenses, analogous to optic lenses, are also useful. Several substantially identical speakers may be used in combination to influence the directivity pattern.

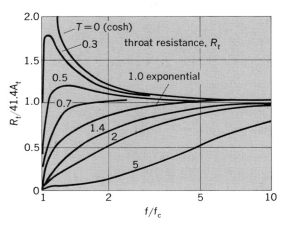

Fig. 9. Frequency dependence of throat resistance for horns of various values of T. Below f_c the resistance is zero. The factor 41.4 is the value of ρc in cgs units at room temperature. (*From K. Henney, ed., Radio Engineering Handbook, 5th ed., McGraw-Hill, 1959*)

More commonly, the two or more speakers cover complementary frequency ranges. The more important advantages are (1) improved frequency response, because each type of unit covers a moderate range, (2) higher system efficiency, for the same reason, (3) improved directivity characteristic, because the diaphragm (or horn mouth) for the highest-frequency speaker may be made relatively small, (4) improved transient response, because many of the artifices used to obtain extended frequency ranges in single units make the transient response worse, particularly at high frequencies, (5) reduced intermodulation, because large amplitudes are confined to speakers reproducing low frequencies, and (6) reduced frequency modulation.

Coaxial units comprise two, or less commonly three, speaker units mounted on substantially the same axis in an integrated mechanical assembly. Appropriate wave filters or dividing networks control the amplitude and phase of the electrical signal reaching each unit.

The term coaxial speaker should be reserved for a coaxial unit with its required acoustic radiation controlling structure.

Speaker placement. In single-channel, or monophonic, systems speaker placement is largely controlled by the desirability of uniform response in the listening area. A corner location in a room is desirable, improving the low-frequency efficiency and reducing the angle the high-frequency radiation must cover. In stereophonic systems multiple speakers and signal channels are used to reproduce the spatial effects present in the original sound environment. In these, speaker placement is more critical, and the region for satisfactory reproduction is relatively limited. The simpler conventional systems employ two channels and two speakers or speaker systems. These speakers are located symmetrically in the room, preferably along the shorter wall. In typical small listening rooms the speakers are 6–10 ft apart. Properly connected, identical speakers will give identical pressure and phase along the plane of symmetry between the speakers which coincides with a median vertical plane of the room. Optimum listener

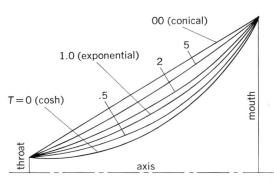

Fig. 8. Longitudinal section of straight-axis circular cross-sectional horns of hyperbolic exponential or catenoidal family. (*From K. Henney, ed., Radio Engineering Handbook, 5th ed., McGraw-Hill, 1959*)

locations are in the plane of symmetry approximately 1.5–2.5 times farther away from the plane of the speakers than the speakers are from each other. Transverse locations are limited by the need of having the more remote speaker less than 5 ft farther from the listener to prevent the precedence effect, in which the earlier signal takes substantial control of the ear if its arrival precedes the latter by approximately 5–35 msec, corresponding to path-length differences of 5.6–39 ft.

[HUGH S. KNOWLES]

Bibliography: L. L. Beranek, *Acoustics*, 1954; M. Colloms, *High-Performance Loudspeakers*, 2d ed., 1979; M. L. Gayford, *Electroacoustics*, 1971; H. S. Knowles, Loud-speakers and room acoustics, in K. Henney (ed.), *Radio Engineering Handbook*, 5th ed., 1959; H. F. Olson, *Music, Physics and Engineering*, 1967.

Magnetic bubble memory

A memory storage technology which uses localized magnetized regions to store information. Magnetic bubble memory (MBM) is the integrated-circuit analogy to rotating magnetic memories such as disks and tapes.

Principles of operation. Magnetic bubble memories employ materials that are easily magnetized in one direction but are hard to magnetize in the orthogonal direction. The most commonly used materials are magnetic garnets which are deposited on a nonmagnetic substrate to form the basis of a bubble memory chip. A thin film of one of these materials, with the easy direction perpendicular to the surface, will allow only two natural directions of magnetization — up or down in the easy direction. When no external magnetic field is applied, the magnetic garnet film forms sepentine patterns of upward and downward magnetization. When a magnetic field is applied by sandwiching the film between two permanent magnets, cylindrical-shaped magnetic domains are formed in place of the serpentine structure. These cylinders are called magnetic bubbles and have a magnetization pointing in the opposite direction of the surrounding area. In effect, magnetic bubbles are magnetic islands in a magnetic sea of opposite magnetic polarity.

Storage locations. To make a memory of these magnetic bubbles, several functions must be performed. A bit of storage is represented as the presence or absence of a bubble at a given storage location. There are several methods of defining the storage locations, but the most common way is to use the chevron patterns shown in Fig. 1. These chevrons are made of a soft ferromagnetic material such as permalloy and are deposited on the surface of the magnetic garnet. There can be one bubble for each chevron.

Loop organization. The chevrons also serve as the path for bubble movement when the information is accessed. To organize the bubble memory bits, the chevron patterns are deposited to form loops along which the bubbles can travel. The simplest organization is one single, large loop. The drawback is that the bubbles must travel the entire length of the loop before the information can be retrieved, and this is time-consuming. Most bubble memory chips now use a more sophisticated architecture called the major/minor loop (Fig. 2). In

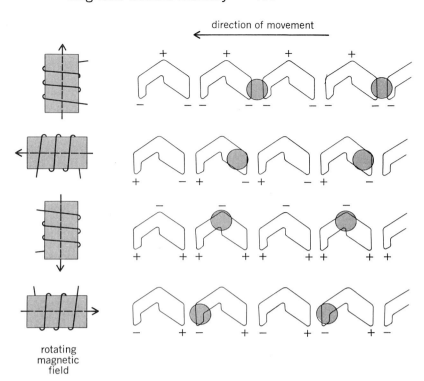

rotating magnetic field

Fig. 1. Movement of magnetic bubbles under pattern of conductive chevrons.

direction of movement

these multiple storage loops, the length and the time taken to traverse a minor loop is much smaller than for one major, large loop. The information is retrieved by waiting until the desired bubbles are at the top of the minor loops. At that point one bubble from each minor loop is transferred to the major loop. This "bubble train" traverses the major loop, during which time the information can be

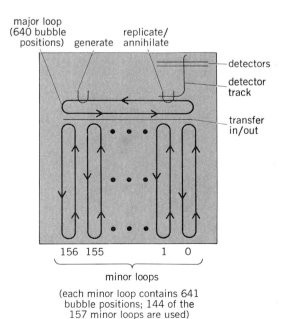

major loop (640 bubble positions)   generate   replicate/annihilate   detectors   detector track   transfer in/out

156 155   1   0

minor loops

(each minor loop contains 641 bubble positions; 144 of the 157 minor loops are used)

Fig. 2. Diagram of the major/minor loop architecture of a 100,000-bit magnetic bubble memory chip. With multiple storage loops, the time to traverse a minor loop is much reduced as compared with one major, large loop.

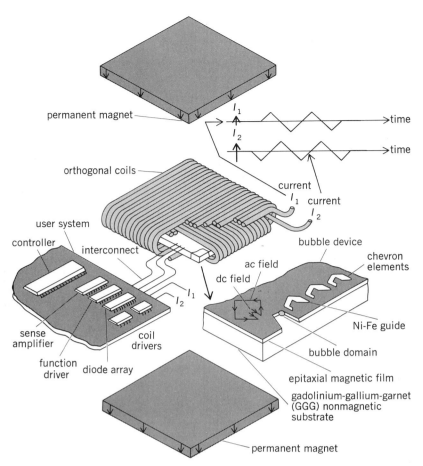

permanent magnet

I_1

→time

I_2

→time

orthogonal coils

current I_1

current I_2

user system

controller

interconnect

bubble device

ac field

chevron elements

dc field

sense amplifier

coil drivers

Ni-Fe guide

function driver

diode array

bubble domain

epitaxial magnetic film

gadolinium-gallium-garnet (GGG) nonmagnetic substrate

permanent magnet

Fig. 3. Principal components and operation that are utilized in magnetic bubble memory.

read or new information can be substituted. When the bubble train has traveled once around the major loop, all the bubbles are transferred back into the minor loops. The commensurate lengths of the major and minor loops are such that the bubble train retains its original position in the minor loops.

Information retrieval. To move the bubbles along the loops, a rotating magnetic field must be generated. For each rotation of the magnetic field each bubble moves synchronously from one chevron to the next. In other words, the magnetic bubbles move within the material, but no physical motion takes place. This is the opposite of what takes place in other magnetic memories such as disks and tapes. Disks and tapes also store information as small magnetized regions. However, to retrieve this information, disks and tapes must physically move the magnetized regions past a read/write head. The "motion" of bubble memories is accomplished by using two orthogonal coils which are wound around the bubble memory chip.

Figure 3 shows the main components and operation of magnetic bubble memories, including the memory chip, the magnetic garnet material where bubbles are formed, the permalloy patterns which guide and control bubble movements, the two coils that power bubble movement, and the two permanent magnets that retain the information when power is removed giving the memory the property of nonvolatility. Various types of interfacing cir-

cuitry which are needed to make a complete bubble memory system are also shown.

Technology status. The fundamentals of bubble memories were discovered in the late 1960s. The bubble technology advanced very rapidly thereafter, with the state of the art for 1980 being 1,000,000 bits on a chip which is about 15 mm square. This chip, when enclosed inside two coils and two permanent magnets, is about a 1 in. square (25 mm square) package.

Bubble memory manufacturing is similar to semiconductor production. The materials are different, but the manufacturing steps are nearly identical. For this reason, major semiconductor manufacturers, as well as computer companies in the United States and other countries, have invested in the magnetic bubble memory technology.

The average time to retrieve the information depends on the length of the minor loops and speed of bubble memory shifting. The shift rate is typically 100 kHz and this is also the rate of information transfer, that is 100 kilobits per second. The 1,000,000-bit chips have minor loops lengths of about 2000 bits, and this gives an average information access time of approximately 10 milliseconds.

Compared to competing technologies, magnetic bubble memories have many desirable features and few disadvantages. Compared with both disks and floppy disks, magnetic bubble memories have lower access time, lower entry price, smaller physical size, and simpler interfacing, but have the disadvantage of lacking media removability. However, removable bubble memory cartridges have been introduced. Magnetic bubble memories also have the disadvantages of higher bit price and lower transfer rate compared with disks (other than floppy disks). On the other hand, compared with semiconductor memories, magnetic bubble memories have nonvolatility, lower bit price and more bits per chip, but have the disadvantages of higher access time, complex interfacing, and lower transfer rate. The advantages for bubble memories are strongest when a relatively small amount of secondary storage is needed—a few million bits or less. Application examples are terminals, desk-top computers, computer-controlled machinery, test equipment, and similar microcomputer-based systems. *See* COMPUTER STORAGE TECHNOLOGY; INTEGRATED CIRCUITS. [J. EGIL JULIUSSEN]

Bibliography: H. Chang, Major activity in MBM technology, *Comput. Des.*, 18(11):117–125, November 1979; G. Cox, Trio of dense bubble memories has large supporting cast, *Electronics*, 52(23): 123–128, Nov. 8, 1979; J. E. Juliussen, The competitive status of bubble memories, *Proceedings of Electro '79 Conference*, New York, Apr. 26, 1979.

Magnetic lens

A magnetic field with axial symmetry capable of converging beams of charged particles of uniform velocity and of forming images of objects placed in the path of such beams. Magnetic lenses are employed as condensers, objectives, and projection lenses in magnetic electron microscopes, as final focusing lenses in the electron guns of cathode-ray tubes, and for the selection of groups of charged particles of specific velocity in velocity spectrographs.

As illustrated, magnetic lenses may be formed

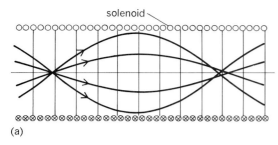

(a)

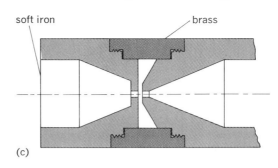

soft iron   brass

(c)

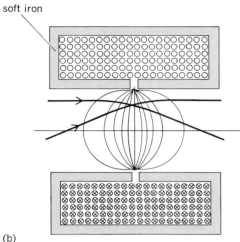

soft iron

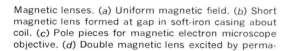

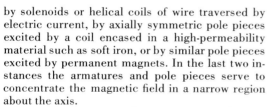

(b)

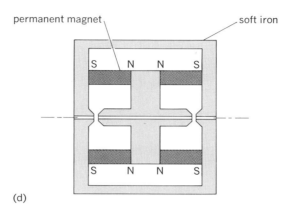

permanent magnet   soft iron

(d)

Magnetic lenses. (*a*) Uniform magnetic field. (*b*) Short magnetic lens formed at gap in soft-iron casing about coil. (*c*) Pole pieces for magnetic electron microscope objective. (*d*) Double magnetic lens excited by perma- nent magnets. (*From E. G. Ramberg and G. A. Morton, J. Appl. Phys., vol. 10, 1939, and J. Hillier and E. G. Ramberg, J. Appl. Phys., vol. 18, 1947*)

by solenoids or helical coils of wire traversed by electric current, by axially symmetric pole pieces excited by a coil encased in a high-permeability material such as soft iron, or by similar pole pieces excited by permanent magnets. In the last two instances the armatures and pole pieces serve to concentrate the magnetic field in a narrow region about the axis.

Magnetic lenses are always converging lenses. Their action differs from that of electrostatic lenses and glass lenses in that they produce a rotation of the image in addition to the focusing action. For the simple uniform magnetic field within a long solenoid the image rotation is exactly 180°. Thus a uniform magnetic field forms an erect real image of an object on its axis. This image has unity magnification and is at a distance from the object equal to $(8\pi^2 m\phi/eB^2)^{1/2} = 21.08\phi^{1/2}/B$, where m is the mass of the particles, e their charge, ϕ the accelerating potential of the particles, and B the magnetic induction on an axis of symmetry of the field. The numerical coefficient 21.08 applies for electrons, with ϕ in volts and B in gauss.

For short magnetic lenses, or lens fields which are short compared to the focal length, both the magnification and image position depend on the position of the object. The focal length f is given by Eq. (1). The integration is carried out over the ex-

$$\frac{1}{f} = \frac{e}{8m\phi}\int B^2\, dz = \frac{0.022}{\phi}\int B^2 \cdot dz \quad \text{cm}^{-1} \quad (1)$$

tent of the lens field along the axis of symmetry

(the z axis). The numerical coefficient 0.022 again applies for electrons. At the same time the field produces an image rotation through an angle θ, as given in Eq. (2). Thus the magnetic lens field

$$\theta = \left(\frac{e}{8m\phi}\right)^{1/2}\int B\, dz = \frac{0.147}{\phi^{1/2}}\int B\, dz \quad (2)$$

formed by two identical coils in tandem, traversed by oppositely directed currents, is rotation free. As a specific example of a short lens, a single circular loop of wire of radius r traversed by current I produces a lens with focal length $f = 96.8r\phi/I^2$ cm. Here ϕ is in volts and I in amperes. The image rotation for this lens is $\theta = 0.185I/\phi^{1/2}$ radian. *See* ELECTRON MOTION IN VACUUM. [E. G. RAMBERG]

Bibliography: P. Dahl, *Introduction to Electron and Ion Optics*, 1973; A. B. El-Kareh and J. C. El-Kareh, *Electron Beams, Lenses and Optics*, 2 vols., 1970; P. Grivet, *Electron Optics*, 2d ed., 1972; O. E. Klemperer and M. E. Barnett, *Electron Optics*, 3d ed., 1970.

Magnetic recording

The technique of storing electric signals as a magnetic pattern on a moving magnetic surface. In magnetic recording a time-varying electric current produces a corresponding time-varying magnetic field in a recording head. This field then lays down a spatially varying pattern of magnetization in a sheet or strip of magnetic material moving past the head. When this recorded pattern moves past a playback head, a time-varying electric output is generated in the head corresponding to the original

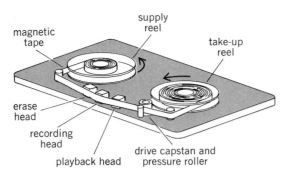

Fig. 1. Common form of magnetic tape recorder.

electric signal. A common form of magnetic recorder is diagrammed in Fig. 1. Magnetic tape from the supply reel moves past an erase head which removes any previously recorded pattern and leaves the tape in a demagnetized condition. An electric signal fed to the recording head lays down the desired new pattern on the tape. This recorded pattern induces an electric output as it moves over the playback head. Since the magnetization pattern remains in the tape until erased or otherwise altered, the tape may be rewound onto the supply reel and replayed as often as desired, with the erase and recording heads inactivated. The motor-driven capstan and pinch roller pull the tape over the heads at a uniform, controlled speed during recording and playback.

Magnetic recorders have become indispensable in innumerable areas of modern society. Audio recorders are sold by the millions as consumer products in reel-to-reel, cartridge, and cassette formats. Tape recorders are central in professional sound-recording studios and in radio and television broadcasting. They serve as mass-storage memories in computer systems and are used as instrumentation recorders for collecting data in research, industry, and military applications. They are employed in satellites and spacecraft for delayed transmission of data to Earth stations.

PRINCIPLES OF OPERATION

Regardless of the nature of the application and of the mechanical configuration of the recorder, all magnetic recorders rely on the same basic magnetic behavior for their operation.

Recording medium. The principles of magnetic recording were first demonstrated in 1898 using steel wire as the recording medium. However, the principles did not find practical application until the 1930s and 1940s. The recorders developed then used steel ribbon or steel wire as the medium on which the magnetic patterns were laid down. These media were superseded by paper tapes carrying a coating of iron oxide on one surface. This was a major improvement, but the relatively rough surface of the paper caused nonuniformity in the coating, giving rise to a high noise level in the reproduced electric signal. Paper tapes were soon supplanted by thin plastic tapes, which are the medium now in universal use in tape recorders. Depending on the application, the thickenss of the plastic substrate usually is between 0.0005 and 0.0015 in. (12.5 and 37.5 μm).

The magnetic coating consists of very tiny parti-cles (Fig. 2) dispersed in a thermosetting resin that is spread evenly on the surface of the plastic as a viscous liquid. The tape, in a web 12 to 24 in. (30 to 60 cm) wide, with its freshly applied coating, passes through an oven for curing. The web is then slit into individual tapes of various widths as required. Depending on the application, the coating thickness usually lies between 0.00005 and 0.0002 in. (1.2 and 5 μm). The most frequently used materials for the magnetically active particles are iron oxide (Fe_2O_3) and chromium dioxide (CrO_2). The coatings used in magnetic disk files for computer memories employ similar materials.

Each of the particles in a tape coating is a tiny permanent magnet embedded in the resin so it cannot move. However, the direction of the particle's magnetization can be reversed by an external magnetic field of sufficient strength. The bulk magnetic characteristics measured for a tape coating are the sum of the states of all of the particles as an aggregate. Thus, when a tape is demagnetized it means that the contributions of some particles exactly cancel the contributions of other particles to yield zero field outside the coating. When an external magnetizing field is applied and gradually increased in strength, more and more particles will switch the direction of their magnetization to conform to the direction of the applied field. The tape is magnetically saturated when all of the particles have been switched. When the external field is removed, some of the particles will reverse their magnetization due to the influence of the fields of neighboring particles, and the tape will be left in a remanent state of magnetization that is somewhat weaker than the saturated state. If an external field in the opposite direction is applied, increased,

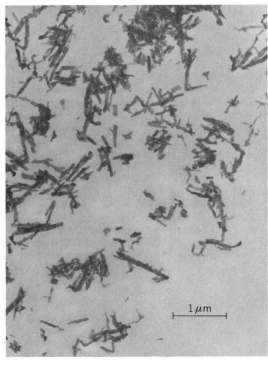

Fig. 2. Photomicrograph of particles of the type used in magnetic recording media. In an actual coating the particles would be much more densely packed.

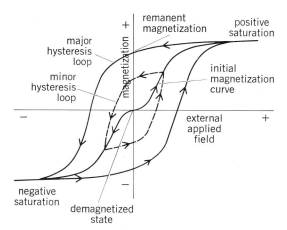

Fig. 3. Magnetization characteristics of recording tape.

and removed, saturation and remanance of the tape occur in the opposite polarity. The bulk result yields the familiar form of a magnetic major hysteresis loop shown in Fig. 3.

Magnetic heads. While magnetic heads may differ widely in geometrical details, they are basically of the form sketched in Fig. 4, whether they are recording or playback heads. Frequently, the same head is used for both functions. As indicated in the figure, a coil of wire is wound on a core of material that very readily carries magnetic flux (that is, a high-permeability material). Iron and some of its alloys are such materials. Certain ceramiclike ferrites are others. A nonmagnetic gap is formed in the core at some point. When an electric current passes through the coil, magnetic field lines associated with the current follow paths around the high-permeability core. At the gap, some of the lines spread outside the core to form a fringing field. The recording medium passes through the fringing field above the gap and is magnetized there during recordings.

The width of the track on the recording medium is determined by the thickness of the core at the gap. Several cores, each with its own coil, may be mounted side by side in a single structure to form a multitrack head.

Recording process. During recording, the fringing field is the external applied field mentioned in

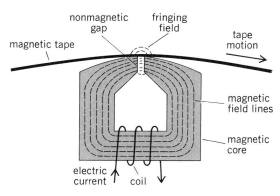

Fig. 4. Schematic representation of a magnetic recording head.

connection with Fig. 3. Starting with erased (demagnetized) tape, if the strength of the fringing field is gradually increased, the tape magnetization will increase while following the initial magnetization curve depicted in Fig. 3. As a segment of tape passes away from the recording gap, the strength of the fringing field falls to zero and the tape is left with a remanent magnetization. If a smaller value of current is fed to the coil, the remanent magnetization will have a lesser value, as indicated by the remanence for a minor hystersis loop in Fig. 3. The remanence always corresponds to the strongest magnetizing field experienced by an element of tape in passing through the fringing field.

Playback process. The structure represented in Fig. 4 can function as a playback head. In this case no current is fed to the coil and no fringing field is produced at the gap. Rather, the "magnets" previously recorded on the tape are moved over the head. These generate their own external magnetic field lines, many of which enter the core of the head. Some of these pass directly across the gap in the core and reenter the tape to complete their loop. These lines have no role in the playback process. Other lines follow the longer route around the core and pass through the coil. This time-varying playback flux threading through the coil generates a voltage in the coil that is amplified to produce the output signal of the recorder. *See* FARADAY'S LAW OF INDUCTION.

High-frequency bias. The recording process described above produces in the tape a remanent magnetization having a one-to-one correspondence with the current in the coil. However, the process is highly nonlinear because of the sharp curvature in the initial magnetization curve near the origin. This introduces distortion that is unacceptable for some applications, including audio recording. The process can be linearized and the distortion reduced to very low values by the use of a high-frequency bias during recording. The frequency of the bias must be several times higher than the highest frequency to be recorded. For example, if the highest signal frequency of interest in audio recording is 15 kHz, the bias frequency should be in the 75–100-kHz range. The signal and bias currents are simply added together in the recording head, with the bias current being the larger of the two and being of sufficient strength to cause saturation of the tape. As an element of tape passes through the fringing field, the bias swings the element through several trips around the major hysteresis loop before the element moves beyond the gap and the fringing field falls to zero. This "anhysteretic" magnetization process leaves the tape magnetized in direct proportion to the strength of the signal current, and the high-frequency bias itself is recorded only weakly, if at all, and does not pass through the playback circuits. This magnetization process is quite complicated and, while several theoretical models offer insights and explanations, analysis complete in all details is still elusive. Nevertheless, high-frequency-bias recording has been routinely and successfully used in audio recorders for several decades.

Digital recording. Magnetic recording is naturally adapted to the use of digital signals consisting of binary states represented by ones and zeros. The two digital states are made to correspond to the two polarities of magnetic saturation of the recording medium. Since only saturation is involved, there is no concern with linearity of recording and high-frequency bias is not needed, nor is erasure of the recording medium. While nearly all digital recording systems depend on switching between the two saturation states, several data-encoding schemes have been devised to use the magnetic medium more efficiently by detecting the time at which each magnetism reversal occurs instead of, or in addition to, the direction of reversal for identifying ones and zeros.

While the most widespread use of digital recording is in computers, it is finding increasing use in instrumentation recorders, satellite and telecommunication recorders, machine-tool controllers, electronic games, and audio and video recorders. *See* ANALOG-TO-DIGITAL CONVERTER.

FM recording. In frequency-modulation (FM) recording the information to be stored in the magnetic tape is carried as frequency modulation of a high-frequency carrier. The carrier is recorded on and played back from the tape, being demodulated after playback to retrieve the original information. Since only changes in the frequency of the carrier must be detected, linearity in recording is not important, and high-frequency bias is not needed. However, the tape should be erased prior to recording a new signal. FM recording is in widespread use in instrumentation recorders and in video recorders. *See* FREQUENCY MODULATION.

Noise and wavelength limitations. The information-carrying capacity of every communications or recording system is limited by the background noise inherent in the system and by the range of frequencies it can handle. In magnetic recording, the basic noise limitation in a well-designed system arises from nonuniformities in the recording medium. Inevitable statistical variations in the size and dispersion of the particles are reflected in small fluctuations of the magnetic properties in the coating, as well as in imperfect smoothness of its surface. When manufacturing tapes to be used in high-quality systems, great pains are taken in dispersing the particles and in polishing the coating surface.

The range of signal frequencies that can be handled in a magnetic recorder is limited by the abrupt falling off of the record-playback response at high frequencies. The fundamental factors responsible for the falloff are geometrical in nature and are more correctly described as short-wavelength losses. When a signal consisting of only a single frequency is recorded, the distance occupied along the track by one cycle of the signal is its wavelength. The wavelength is equal to the speed of the tape moving past the head divided by the frequency. Therefore, shorter wavelengths correspond to higher frequencies at any head-to-tape speed. Some of these short-wavelength losses are instrinsic in the recording process, arising from self-demagnetization of the recorded signals and from a too gradual falling off of the fringing field at the trailing edge of the gap. Both effects cause a smearing out of the recorded pattern, with smearing being more severe at shorter wavelengths.

Two further losses occur in playback. One is a separation loss due to imperfect contact between the tape coating and the surface of the playback head. This loss amounts to 55 dB (about 1/600) for a separation of one wavelength. The other loss is called gap loss and is related to the length of the head gap in the direction of tape motion. Gap losses are increasingly serious as the recorded wavelengths are made short enough to be comparable to the gap length. The output falls to zero when the wavelength and gap length are equal, since the head now receives equal fluxes from the two opposite magnetization polarities of each recorded cycle on the tape.

The principal motivation for extending system response to ever-shorter wavelengths is to increase the density of information on the recording medium. Greater densities mean that less tape is needed for recording a given amount of information. All magnetic recording systems can benefit by extending the response to shorter wavelengths, since the short-wavelength response is usually the limiting factor in performance. The benefit may be used, as it becomes available through improved tapes and heads, to extend the response to higher frequencies, to increase the signal-to-noise ratio, to improve the system stability, to reduce the tape speed, or to achieve some optimal combination of these, depending on the requirements of the system.

TYPES OF RECORDING SYSTEM

Some of the most common and most important recording systems are described below.

Audio recorders. The earliest significant use of magnetic recording was in the audio-frequency domain, and today audio tape recorders greatly outnumber other types. Machines for professional use in recording studios require the highest possible linearity, signal-to-noise ratio, and speed constancy. They operate at standard tape speeds of 15 in./s (38.1 cm/s) and 30 in./s (76.2 cm/s), since short-wavelength losses are reduced at these speeds. By contrast, the cassette recorders sold in the mass consumer market have a tape speed of only $1^7/8$ in./s (4.8 cm/s). Certain other recorders requiring only low-grade voice quality, such as dictation machines and recorders for long-term monitoring purposes, operate at even slower tape speeds. Stereo-8 cartridge players operate at $3^3/4$ in./s (9.5 cm/s). Reel-to-reel consumer machines operate at selectable speeds of $3^3/4$ and $7^1/2$ in./s (9.5 and 19.1 cm/s).

Of the numerous attempts to eliminate the inconvenience of threading tape through its path from the supply reel to the take-up reel in a recorder, only two have received widespread acceptance. They are the endless-loop cartridge and the coplanar-hub cassette.

Cartridge. In the cartridge, the tape is wound on a single hub. Tape is drawn out of the wound stack at the hub, pulled past a playback head by a capstan, and returned to the outside of the stack, with the two ends of the length of tape being spliced together to form an endless loop (Fig. 5). Since adjacent layers of the tape in the stack are continuously sliding past one another, a low-friction coating is applied to the side of the tape opposite the magnetic coating during manufacture of the tape to permit smooth motion of the tape without application of excessive pulling forces. The hub, with its wound stack, is contained in a plastic case that

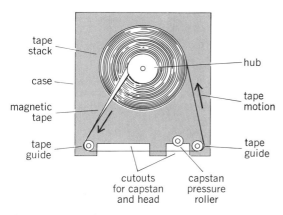

tape stack

case

magnetic tape

hub

tape motion

tape guide

tape guide

cutouts for capstan and head

capstan pressure roller

Fig. 5.   Diagram of a Stereo-8 audiotape cartridge.

also has posts and rollers to guide the tape past the head and capstan on the player. Cutouts in one edge of the case allow a short span of tape to engage the head and capstan when the cartridge is inserted in a slot in the player.

A common use of endless-loop cartridges is for commercial messages in radio stations. Endless-loop cartridges have received popular acceptance for music reproduction in the home and in automobiles. In this application eight tracks are recorded side by side along the $\frac{1}{4}$-in.-wide (0.67-cm) tape, and the package is designated Stereo-8. The tracks are used in pairs for stereophonic sound reproduction. Thus, four passes of the loop occur before a recorded program repeats itself. Usually a player carries a sensor to detect the splice in the loop, and the playback head is automatically advanced to the next pair of tracks at that point. Any desired pair of tracks may be selected manually at any time. Stereo-8 players usually do not have provision for recording and are used with prerecorded cartridges.

Cassette. In an audio cassette the tape is wound in a stack on a hub, with one end of the tape being permanently attached to the hub. The other end of the tape is permanently attached to an identical hub in the same plane as the first and symmetrically located in the flat plastic case. The tape path between the hubs is determined by guide posts in the case, and the tape travels from one hub to the other. Cutouts in one edge of the case allow a span of tape to engage the capstan and heads in the player when the cassette is inserted, (Fig. 6) The tape is 0.15 in. (0.38 cm) wide and carries four

tracks. Depending on the player, the tracks may be used for monophonic reproduction or for stereophonic reproduction. One pair of tracks is used for one direction of tape motion and the other pair for the opposite direction, the direction reversal usually being achieved by removing the cassette and reinserting it with the other side up. Most cassette players have a provision for recording as well as playback.

Reel-to-reel recorders. Reel-to-reel recorders were the earliest form of audio-tape machine, and they still are the form employed when the highest-possible quality of reproduced sound is desired. The high quality in reel-to-reel machines is due partly to the higher tape speeds available, partly to wider tracks on the tape, and partly to the more precise construction of the elements in the tape transport mechanism. The consumer products use tape that is $\frac{1}{4}$ in. (0.64 cm) wide. Monophonic machines use the entire width of the tape as a single track. Much more common are stereophonic machines that have two tracks. Some quadraphonic machines are available with four tracks.

There is much greater diversity in the equipment used professionally in recording studios. The simplest machine in this category records two tracks on tape having a width of $\frac{1}{4}$ in. (0.64 cm). Next in the hierarchy are machines that record two, four, or eight tracks on $\frac{1}{2}$-in. (1.27-cm) tape. Other machines record up to 16 tracks on 1-in. (2.54-cm) tape and, finally, the most complex machines can record 24 or 32 tracks on 2-in. (5.08-cm) tape. The large number of tracks are desired in studio recording sessions to permit an exclusive track assignment for an individual instrument or performer in a musical ensemble.

Sound quality. Professional-grade equipment running at a tape speed of 30 in./s (76.2 cm/s) can yield a signal-to-noise ratio between 60 and 70 dB in the original recording. However, in each subsequent generation of rerecording, the tape copy suffers some reduction in signal-to-noise ratio, along with an increase in distortion. Since several generations are normally involved in the production of a working master tape, the sound quality of this tape may be marginal. An improvement of 10 dB or so in the signal-to-noise ratio is obtained by employing electronic noise-reduction systems. Various approaches are used in these systems but, in essence, relatively quiet signals are raised in level during recording and reduced in a complementary way in playback so that even low-level signals are kept above the background noise generated by the tape.

Consumer recorders operating at lower tape speeds cannot provide such low noise and distortion levels. In the case of prerecorded music on cassettes which are played at only $1\frac{7}{8}$ in./s (4.76 cm/s) and are several generations removed from the original studio recording, signal-to-noise ratios between 45 and 50 dB are to be expected. However, improvements in tapes and the incorporation of electronic noise reduction in prerecorded cassette tapes and in some of the more expensive players can keep the ratio close to or somewhat above the upper value of this range.

Digital recorders. The problems of distortion and noise have prompted the development of digital magnetic tape recorders for audio use. In digital

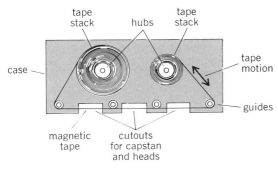

tape stack

hubs

tape stack

case

tape motion

guides

magnetic tape

cutouts for capstan and heads

Fig. 6.   Diagram of a coplanar-hub audiotape cassette.

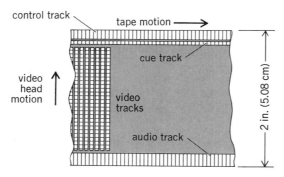

Fig. 7. Track format of a Quadruplex video recorder.

recording, the output of the microphone amplifiers in the studio is immediately converted from its analog form to digital, and the signals are recorded in digital codes on the tracks on the tape. The signals remain in digital form throughout all subsequent mixing, editing, and rerecording procedures and are not converted to analog form until the final playback from digital tape or digital disk for reproduction from loudspeakers. The advantages of digital recording are that signal-to-noise ratios between 80 and 90 dB are readily attainable and that there is no degradation in quality during rerecording, regardless of the number of generations.

Video recorders. The purpose of video tape recorders is to record and play back television signals. The television signal frequency-modulates a carrier which is recorded on the magnetic tape, with frequencies in excess of 15 MHz being involved. High head-to-tape speeds are required for recording these high frequencies. High head-to-tape speeds are obtained by mounting the heads on a rapidly rotating wheel that sweeps the heads more or less transversely across the tape from edge to edge. Two types of recorders that function in this manner have come into use, namely, the Quadruplex and the helical scan (sometimes called slant-track).

Quadruplex recorders. In Quadruplex recorders, four equally spaced heads are mounted on the rim of a headwheel having a diameter of 2 in. (5.08 cm). In the United States, Canada, Japan, and a few other countries that adopted the National Television System Committee (NTSC) standards, the headwheel rotates at 14,400 revolutions/min to give a peripheral head-to-tape speed of 1556 in./s (3952 cm/s). In countries following other television standards the headwheel speed is somewhat different. The tape is 2 in. (5.08 cm) wide. It is pulled by a capstan and moves from a supply reel to a

take-up reel at a uniform speed of 15 in./s (38.1 cm/s) or 7½ in./s (19.0 cm/s). As it approaches the headwheel, the tape is formed into a circular, concave surface by a vacuum guide. This allows the heads to maintain contact with the tape throughout their scan across the tape. The combination of rapid scan across the tape and slow forward motion of the tape results in a series of parallel video tracks making a nearly 90° angle with the direction of tape motion (Fig. 7). Downstream from the headwheel a straight guide restores the tape to travel in a plane before reaching fixed heads that record audio signals in a conventional way on tracks along the edges of the tape. Quadruplex machines are used only in broadcasting and other television-related enterprises.

Helical-scan recorders. While helical-scan recorders have been available for a number of years, only in the late 1970s did they become sufficiently advanced to meet the stringent requirements of broadcast television. Recorders are available in several formats, but all are designed around the same underlying principle, which will be discussed with reference to the type of helical-scan recorder preferred for professional use in broadcast studios. In this recorder the tape is 1 in. (2.54 cm) wide and moves from a supply reel to a take-up reel at a speed of 9.6 in./s (24.4 cm/s), being pulled through its path by a capstan and pressure roller. In one portion of its path the tape makes a nearly 360° helical wrap around the outside of a hollow circular cylinder, with the magnetic coating against the surface of the cylinder. The recording heads are carried on a drum rotating inside, and concentric with, the cylinder. The heads protrude through a slot in the cylinder wall to contact the coating of the tape. The drum rotation is 1800 revolutions/min, and the peripheral head-to-tape speed is 1000 in./s (25.4 m/s). The head-to-tape speed and the helix combine to lay down a series of parallel video tracks across the tape, making an angle of 2°34′ with the edges of the tape (Fig. 8). After leaving the cylinder, the tape moves past a series of stationary heads that lay down audio and control tracks along the edges of the tape.

Other types of helical-scan recorders are in widespread use for educational and entertainment purposes. Most of these use ½-in.-wide (1.27-cm) tape. Nearly all require the tape to be stored in a cassette that is basically similar to, though larger than, the audio cassette described above. The video tape players incorporate ingenious mechanical devices to extract a length of tape from the cassette and wrap it in the helix on the cylinder before forward motion of the tape commences. *See* TELEVISION.

Recorders for computers. From the inception of electronic computers, magnetic tape recorders have been an indispensable element in the hierarchy of computer memories. During the 1960s and 1970s, magnetic disk files assumed an equally important function in the hierarchy. All recording associated with computers employs digital signals. Since the loss of even a single bit in a digital number can have near-catastrophic results, much effort goes into assuring precision and reliability in the manufacture and fabrication of tapes, disks, heads, and mechanical structures involved in the recorders.

Computer tape recorders. Computer tape record-

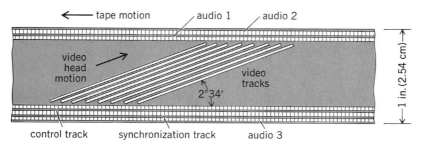

Fig. 8. Track format for a television-broadcast helical-scan (slant-track) video recorder.

ers usually are of the reel-to-reel type, although cassettes are employed in certain applications. The computer tape recorders differ from the reel-to-reel recorders described previously in that tape motion in both directions is required, together with very short start-up intervals. Computer recorders have two counterrotating capstans, one on each side of the record/playback head. Solenoid-operated pressure rollers engage the tape with one or the other capstan to determine the direction of motion. A few meters of tape from the supply and take-up reels are stored loosely in bins between the capstans and the reels, so that when a "start" command is received the activated capstan must accelerate only the miniscule mass of a short length of tape. The high-inertia reel can be accelerated much more slowly by a separate motor to eventually make up the supply of tape removed from the bin by the capstan. With this arrangement, a reversal from full speed in one direction to stop, and to full speed in the opposite direction, can be accomplished in milliseconds.

The most common computer-tape width is $1/2$ in. (1.27 cm). Multiple tracks, typically nine, are used with stacked heads to record the bits of a digital word in parallel in a line across the tape, with the tape moving at 120 in./s (305 cm/s). To increase reliability, the packing density is kept low, seldom exceeding 6250 bits/in. (2460 bits/cm) along each track.

Disk file. In a disk file the most common recording medium is a very smooth aluminum disk, 14 in. (35.6 cm) in diameter, having both of its surfaces coated with magnetic oxide. Data are recorded in concentric tracks on both surfaces with track densities of up to 600 tracks per inch (236 tracks per centimeter) along the radius. The disk rotation rate is 3600 revolutions/min. Up to 10 disks are contained in a module, called a disk pack, that may be mounted on the spindle of a disk drive unit. The disks in the pack are clamped one above the other, with space between disks to allow entry of two record/playback heads, one for an upper and one for a lower surface of adjacent disks. The tiny ferrite heads are carried in a highly polished ceramic plate called a slider, having a carefully contoured surface facing the surface of the disk. The high-speed motion of the disk surface carries air along with it, and the contour of the slider gives rise to an aerodynamic flotation providing a constant, controlled separation between disk and head. The separation is required to prevent rapid wear of disks and heads, since the disk files run continuously whenever their computer is in operation. Aerodynamic separations between 0.00001 and 0.0001 in. (0.25 and 2.5 μm) are used in various disk systems.

The head assemblies are mounted on an actuator that moves them in and out radially, so any track on any surface may be selected for use upon command from the computer. The average time to change from one track to any other is measured in milliseconds. Modern disk packs have a capacity of as much as 5×10^9 bits, and the packs are removable and interchangeable on the drives. Several drives are usually used in each computer installation to give even greater storage capacity.

Floppy disk. The floppy disk was introduced as a computer storage medium in the 1970s. In these disks the substrate is a flexible plastic sheet with a magnetic oxide coating. Since the mechanical precision achievable with these disks is much less than that customary with the aluminum substrates, the packing density must be greatly reduced. However, the relatively low cost of flexible disks makes them attractive for certain applications. *See* COMPUTER STORAGE TECHNOLOGY.

Instrumentation recorders. The purpose of an instrumentation recorder is the storing of electric signals so they may be reproduced at a later time for detailed analysis. Examples for which instrumentation recorders are used are geophysical exploration; medical diagnosis; radar surveillance; measurement of explosions and supersonic shock waves; measurement of mechanical stress and vibration of structures; testing of jet, rocket, and internal combustion engines; and flight testing of aircraft and missiles. Any environmental, physical, chemical, or physiological parameter that can be converted to an electric signal by a suitable transducer is a candidate for instrumentation recording.

The design of instrumentation recorders is as varied as the applications in which they are employed. All of the types of recorder discussed in this article, as well as several not mentioned, have been used in instrumentation work. Depending on the nature of the signal and the type of recorder, the signals enter the recording medium in any of the following forms: direct analog, digital, pulse-code modulation, frequency modulation, and multiplexed frequency modulation.

Tape duplication. The ability to copy a recorded program from one tape onto another is an important feature of magnetic recording. In principle all that is required for the copying process is two recorders, one for playback and a second one for recording the output of the first. When multiple copies are desired, this simple procedure is inefficient and time-consuming. It is preferable to feed the playback output to several recorders and to record several copies simultaneously. The playback unit is the "master" and the recording units are "slaves." Simple duplicators are available to churches, educational institutions, and other enterprises that may rather frequently wish to make only a few copies of material recorded on a cassette.

In commercial duplication large numbers of copies are required, and the duplicating procedures are modified in the interests of manufacturing efficiency. There may be as many as 20 slave units, and the tape is moved at faster-than-normal speed to reduce the copying time. The same arrangement is used for duplicating audio tapes for cassettes, Stereo-8, and reel-to-reel products.

In a large operation the program to be copied is recorded on a working-master tape, $1/2$ in. (1.27 cm) wide, at its normal speed. This tape is unwound from its reel and fed into a loose stack in a storage bin. The ends of the tape are spliced to form an endless loop that runs continuously through the master playback machine, generating the signals that are fed to the bank of slave recorders. The supply of tape in each slave is in a large reel carrying up to 7200 ft (2194 m) of tape, which is sufficient to make a number of copies before reloading. The master tape is run at 120 in./s (304.8 cm/s), which is 64 times the normal cassette speed, 32 times the Stereo-8 speed, and either 32 times or 16 times the reel-to-reel speeds. When the copying is finished, the tape is transferred from the slave to an editing machine where leaders are spliced on, and the tape is cut at the end of each program and wound onto hubs or reels for the final product.

In this high-speed duplication process all signal frequencies from the master tape are raised in the same ratio as the tape speeds. Thus, in the 64-times case a 15 kHz signal becomes 960 kHz. Also, the high-frequency bias in the slave recorder now ought to be in the neighborhood of 6.4 MHz. The duplicator must be provided with heads and electronics capable of operating at these high frequencies.

Video cassette tapes also are duplicated by the master/slave procedure. Some of the video-frequency components that must be copied lie in the 5–10-MHz range. Since no head materials are known that perform efficiently above 20 MHz, it is not possible to greatly reduce the copying time by high-speed duplication as in audio duplication. High-speed contact-printing techniques have been demonstrated as alternatives to the master/slave approach, but are not yet widely used.

[J. G. WOODWARD]

Bibliography: S. W. Athey, *Magnetic Tape Recording*, National Aeronautics and Space Administration, NASA SP-5038, 1966; G. L. Davies, *Magnetic Tape Instrumentation*, 1961; A. S. Hoagland, *Digital Magnetic Recording*, 1963; C. B. Mee, *The Physics of Magnetic Recording*, 1964; J. F. Robinson, *Videotape Recording*, 1975.

Magnetic thin films

Sheets of magnetic material with thicknesses of a few micrometers or less, used in the electronics industry. Magnetic films can be single-crystalline, polycrystalline, or amorphous in the arrangement of their atoms. Applications include magnetic bubble technology, magnetoresist sensors, thin-film heads, and recording media. *See* MAGNETIC BUBBLE MEMORY.

Fabrication. Thin films are usually directly deposited on a substrate. The techniques of deposition vary from high-temperature liquid-phase epitaxy such as that used in the fabrication of garnet films, through electroplating, to vapor deposition and sputtering. Permalloy, for example, can be both electroplated and sputtered or vapor-deposited.

Magnetic order. Both ferro- and ferrimagnetic films are used. The ferromagnetic films are usually transition-metal-based alloys. For example, permalloy is a nickel-iron alloy. The ferrimagnetic films, such as garnets or the amorphous films, contain transition metals such as iron or cobalt and rare earths. The ferrimagnetic properties are advantageous in bubble applications where a low overall magnetic moment can be achieved without a significant change in the Curie temperature. They are also useful for magnetooptic applications.

Magnetic anisotropy. In an isotropic bulk magnetic sample the overall magnetization does not prefer a particular direction. However, the same material in the form of a thin film has a lower energy when the magnetization lies in the plane of the film. This shape-induced anisotropy is useful in applications such as permalloy drive or sensor patterns used in bubble technology. However, there are other situations where it is advantageous to have perpendicular rather than in-plane anisotropy. For example, magnetic bubble and magnetooptic memory materials have perpendicular anisotropy. This is obtained by using anisotropic crystals, by growing films in a magnetic field or directional deposition, and by annealing in a magnetic field or applying an anisotropic stress.

Domain structure. Depending upon the nature of anisotropy (perpendicular to planar), a variety of domain patterns can be obtained. For example, when the perpendicular anisotropy is large relative to shape-induced anisotropy, stripe domains are observed. These form cylindrical domains if an external magnetic field is applied perpendicular to the plane of the films. Such cylindrical domains are called magnetic bubbles. The domain structure can be observed in an optical microscope with polarized light using either the Faraday or Kerr effect. It can also be observed using ferrofluids. Domains and domain wall structure are also observable by electron microscope techniques.

Coercivity. The coercivity of thin film varies with applications. For most applications a low coercive force is desired. Coercivity is generally associated with inhomogeneities such as stress fields, compositional fluctuations, defects, and surface roughness.   [PRAVEEN CHAUDHARI]

Bibliography: P. Chaudhari and D. Turnbull, Structure and properties of metallic glasses, *Science*, 199:11–21, 1978; I. S. Jacobs, Magnetic materials and applications: Quarter century overview, *J. Appl. Phys.*, 50:7294–7306, 1979.

Magnetron

The oldest of a family of crossed-field microwave electron tubes wherein electrons, generated from a heated cathode, move under the combined force of a radial electric field and an axial magnetic field. By its structure a magnetron causes moving electrons to interact synchronously with traveling-wave components of a microwave standing-wave pattern in such a manner that electron potential

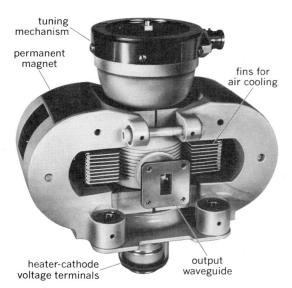

Fig. 1. Coaxial cavity magnetron with horseshoe-shaped magnets. (*From J. W. Gewartowski and H. A. Watson, Principles of Electron Tubes, Van Nostrand, 1965*)

energy is converted to microwave energy with high efficiency. Magnetrons have been used since the 1940s as pulsed microwave radiation sources for radar tracking. Because of their compactness and the high efficiency with which they can emit short bursts of megawatt peak output power, they have proved excellent for installation in aircraft as well as in ground radar stations. In continuous operation, a magnetron can produce a kilowatt of microwave power which is appropriate for rapid microwave cooking.

The magnetron is an oscillator; it generates radio-frequency energy, usually over limited portions of the microwave frequency range of 1–40 gigahertz (GHz). It can produce microsecond pulses of peak power as high as 10 Mw but with low duty-cycle ratios (ratio of pulse length to pulse repetition period). Mechanical tuning of the magnetron microwave circuit can change the center frequency of the output radiation by about ±5%.

Configuration. Raw materials of a magnetron are chiefly copper, iron, and ceramic welded and brazed together to form a vacuum tube. A slow-wave microwave circuit, made of copper to reduce ohmic losses, serves to provide a traveling wave that can interact with the electrons. Iron pole pieces guide the magnetic flux from externally attached horseshoe-shaped magnets (Fig. 1). Pole pieces shape the kilogauss magnetic field in the electron-microwave interaction region to be perpendicular to the electric field and to constrain the electrons to the interaction region. Ceramic (or sometimes glass) serves as a standoff insulator for the tens of kilovolts applied between cathode and anode. Ceramic is also used as the microwave window to couple the power generated inside the magnetron through the vacuum enclosure to externally attached waveguides. The breakdown limit of this window may be the chief factor in setting the peak power output obtainable from a magnetron.

The magnetron is a device of essentially cylindrical symmetry (Fig. 2). On the central axis is a hollow cylindrical cathode. The outer surface of the cathode carries electron-emitting materials, primarily barium and strontium oxides in a nickel matrix. Such a matrix is capable of emitting electrons when current flows through the heater inside the cathode cylinder. *See* VACUUM TUBE.

At a radius somewhat larger than the outer radius of the cathode is a concentric cylindrical anode. The anode serves two functions: (1) to collect electrons emitted by the cathode and (2) to store and guide microwave energy. The anode consists of a series of quarter-wavelength cavity resonators symmetrically arranged around the cathode. *See* CAVITY RESONATOR.

A radial dc electric field (perpendicular to the cathode) is applied between cathode and anode. This electric field and the axial magnetic field (parallel and coaxial with the cathode) introduced by pole pieces at either end of the cathode, as described above, provide the required crossed-field configuration.

Microwave generation. Provided that the magnitudes of the electric field and the magnetic field are properly adjusted, an electron emerging from the heated cathode proceeds to orbit the cylindrical cathode, moving on the average in a direction perpendicular to both applied fields (Fig. 3). The

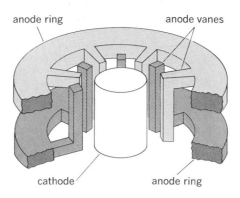

Fig. 2. An interdigital-vane anode circuit and cathode, indicating the basic cylindrical geometry. (*From G. D. Sims and I. M. Stephenson, Microwave Tubes and Semiconductor Devices, Blackie and Son, London, 1963*)

totality of electrons emitted by the cathode form a swarm or hub of negative charge rotating about the cathode axis. The hub thickness extends from the cathode surface outward to a radius intermediate between cathode and anode radii. Motion of the rotating swarm of electrons past the surface of the concentric-anode microwave circuit induces a noise current in the copper circuit. This noise shock excites the resonators so that microwave fields build up at the resonant frequency. The velocity of the electrons, as determined by the applied voltage and magnetic field, is made to be close to or synchronous with the slow-wave phase velocity characteristic of the microwave circuit. Consequently, as the electron swarm rotates, it concentrates into bunches that deliver microwave energy to the resonators. Thus, as in most oscillators, the process begins with noise; then electromagnetic fields build up at the resonant frequency. These fields react, in turn, upon the electrons, remove some of their potential energy, and cause them to approach the circuit more closely, where-

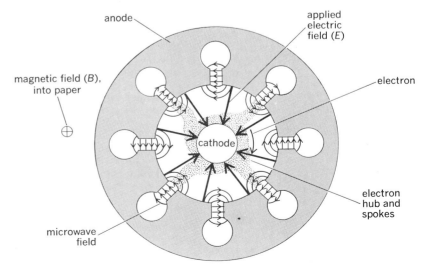

Fig. 3. Electron motion in magnetron interaction region under action of applied electric and magnetic fields and generated microwave field. (*After C. H. Dix and W. H. Aldous, Microwave Valves, Iliffe Books, London, 1966*)

upon the process intensifies. Buildup of the microwave field is limited by internal ohmic losses in the circuit, external loading of the output coupling, and the electron energy available for conversion to microwave energy. *See* TRAVELING-WAVE TUBE.

Electron bunching. Bunching of electrons plays a prominent role in the generation of microwave power in the magnetron, as it does in most microwave tubes. The applied radial electric field E and the axial field B constrain the average motion of the hub electrons to a direction perpendicular to both fields and hence circumferential about the cathode. (There is cycloidal motion superimposed on the average, as well.) The average circumferential electron velocity, called drift velocity, has a magnitude $u = E/B$. The microwave circuit is designed to support a traveling wave whose phase velocity v is equal to u. If $u = v$, the electron hub rotation is said to be synchronous with the rotating traveling wave. Under these conditions, one can consider the interaction of electrons with microwave fields from the anode in a frame of reference which rotates at the synchronous velocity.

The quarter-wave cavity resonators, which support the rotating microwaves and surround the electron hub, present at the hub a rotating microwave field configuration consisting of electric fields having strong circumferential components. These components are 180° out of phase (oppositely directed) for neighboring cavities, as shown in Fig. 3. This set of alternating fields can be considered as composed of two sets of uniformly directed fields (each internally inphase), one set 180° out of phase with the other and the two interlaced. (Such a microwave field configuration is called a π mode, the two sets of fields being phased π radians from each other.) In the frame of reference rotating in synchronism with these two sets of microwave fields, the circumferential components of the microwave electric fields can be regarded as steady or static fields crossed with the applied axial magnetic field.

Those hub electrons rotating in phase and under the influence of one of the sets of microwave fields, because of the crossed-field situation, are constrained to move radially toward the anode (average electron motion is always perpendicular to both electric and magnetic crossed-field directions). In doing so they contribute to the strength of the microwave field, converting their potential energy in the applied electric field to microwave energy. The process continues until the electrons finally fall through the entire hub-to-anode potential and are collected on the anode. As a consequence, the distribution of electrons in the interaction region between cathode and anode looks like a hub with spokes, all rotating together at drift velocity E/B (Fig. 3). In this sense the electrons are bunched. Throughout the interaction the electrons continue to rotate at close to synchronous velocity, remaining in phase with the wave and continuing to convert dc potential energy into microwave energy until striking the anode. This staying-in-synchronism results in high energy conversion efficiency (values may be as high as 65–75%) of the magnetron in contrast to those microwave tubes that depend on differential velocities of electrons for bunching. It also means that the anode must be rugged enough to absorb the full kinetic

energy of the electrons ($mu^2/2$, where m is the electron mass). *See* KLYSTRON.

Those electrons in phase with the remaining set of microwave fields (the set 180° out of phase with the first) will be constrained to move radially back through the hub toward the cathode. This motion results in an increase of electron potential energy at the expense of the microwave field. However, the loss in microwave energy is small because such electrons move out of high microwave field regions toward weaker ones and so move to minimize their interaction. On striking the cathode these electrons heat up the cathode and reduce the required heater power. The phenomenon of electron bunching, wherein certain in-phase electrons give up their potential energy to the microwave field while other out-of-phase electrons are forced out of the microwave field region to prevent their extracting microwave energy, is further abetted by the shape of the microwave fields near the anode. The field shape is such as to cause electrons approaching the anode to become even more closely in phase so as to deliver maximum power to the circuit.

Microwave circuit. The only aspect of the magnetron that is not symmetric is the output coupler. Being an oscillator and generating its own signal starting from noise-induced fields, the magnetron must maintain enough energy in its resonant circuit to keep the generation process going. On the other hand, only the microwave energy coupled

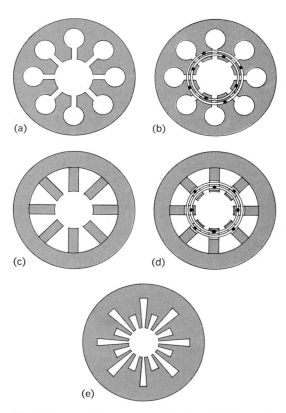

Fig. 4. Magnetron slow-wave circuits. (*a*) Unstrapped cavity. (*b*) Strapped cavity. (*c*) Unstrapped vane block. (*d*) Strapped vane block. (*e*) Rising-sun block. (*From C. H. Dix and W. H. Aldous, Microwave Valves, Iliffe Books, London, 1966*)

out of the magnetron is useful to the radar operator. Thus some compromise must determine the fraction of the generated power which is coupled out through the output coupling slot, the ceramic window, and output waveguide.

The π-mode electromagnetic field configuration is desirable for interaction with the electrons as described above, and the output coupling is generally designed for such a resonant condition. However, the microwave circuit can sustain other resonant modes which can interact with the electrons and which are generally less efficient. Some of these undesirable modes can so orient themselves as to present a null in the microwave field pattern at the output. All the energy converted from interaction with electrons is then stored inside the magnetron in such modes, and none is usefully coupled out. The fields in the interaction region can become high enough to block operation in the π mode. Such undesirable modes can be close in frequency to the π mode, and unless special circuit design is incorporated, they are difficult to discriminate against. Much of the design of magnetron microwave structures has to do with maximizing efficiency of π-mode interaction while inhibiting electron interaction with other undesirable modes.

A variety of shapes have been used in the design of the quarter-wavelength cavities whose open-end, capacitative section provides the electric fields for interaction with the electrons. There are vane-type cavities and hole-and-slot versions (Fig. 4). These may be strapped (alternate cavities tied together with a copper strap to promote π-mode operation), or alternate cavities can be elongated (rising-sun configuration) to separate the undesirable modes from the π mode. Such circuits are generally tuned in frequency by inserting copper pins into the induction sections of the quarter-wave cavities. Voltage tuning has been achieved in certain low-power magnetrons.

Coaxial cavity magnetron. A more recent magnetron design, the coaxial cavity magnetron (Fig. 5), achieves mode separation, high efficiency, stability, and ease of mechanical tuning by coupling a coaxial high Q cavity to a normal set of quarter-wavelength vane cavities. *See* CAVITY RESONATOR; Q (ELECTRICITY).

The high Q cavity is resonant in the TE_{011} mode, a mode in which current flow on the inner conductor is circumferential and phase is not a function of circumferential position. To couple microwave energy from the slow-wave circuit into the coaxial cavity, slots are cut into the backs of alternate quarter-wavelength cavities. If the rotation of the electron hub induces current flow on the vanes, so that a π-mode resonant buildup of microwave fields commences there, the slot coupling geometry is correct to also stimulate the buildup of the TE_{011} mode in the outer coaxial cavity. Because alternate vane cavities are in phase in π-mode operation, the current coupled into the high Q cavity through slots cut only in alternate cavities will also be in phase. This is the condition required to induce the TE_{011} mode in the coaxial cavity. The output frequency can be tuned merely by altering the axial length of the outer coaxial cavity (a tuning piston is employed). Most stored microwave energy is in the coaxial cavity, and its high Q ensures

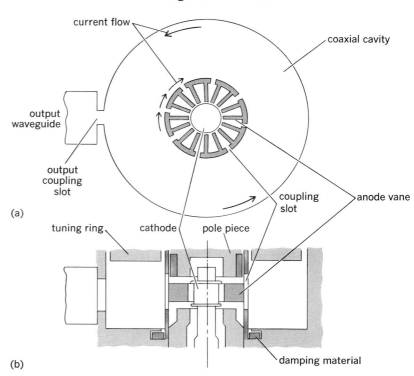

Fig. 5. Coaxial cavity magnetron. (a) End view. (b) Cross-sectional view. (*After J. W. Gewartowski and H. A. Watson, Principles of Electron Tubes, Van Nostrand, 1965*)

efficient storage. The coaxial cavity magnetron has replaced many earlier designs for airborne radar.

Some characteristics typical for the coaxial cavity magnetron are as follows.

Frequency (tunable)	15.5 – 17.5 GHz
Peak output power	125 kw
Peak applied voltage	17.5 kv
Peak anode current	19 amp
Pulse duration	3 μsec
Duty cycle	0.001
Magnetic field	0.74 weber/in.^{2}
Overall efficiency	43%

Other crossed-field tubes. The magnetron is a member of the crossed-field family of devices. Other members are platinotrons, amplitrons, forward-wave crossed-field amplifiers, non-reentrant-beam crossed-field amplifiers, and carcinotrons. Some of these devices are oscillators and others amplifiers (Fig. 6). Some interact with backward slow waves, others with forward waves. Some employ an electron gun, others an emitting sole. The magnetron oscillator itself has both a reentrant electron beam and a reentrant microwave circuit (the electron flow is continuous in the circumferential direction, orbiting the axial cathode, while the quarter-wavelength cavities are symmetrically disposed around the cathode with no beginning or end. *See* BACKWARD-WAVE TUBE.

Power generation in an oscillator begins with noise, and the resonant nature of the circuit produces the high electric fields required to further react with the electrons. Amplifiers, on the other hand, have non-reentrant circuits with a definite input and output. The input signal provides the initial microwave field to begin interaction between

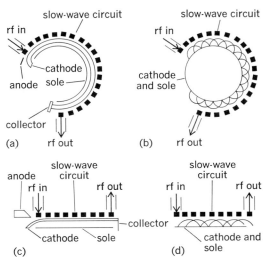

Fig. 6. Magnetron amplifiers. (a) Circular type with beam injection. (b) Circular type with continuously emitting sole. (c) Linear type with beam injection. (d) Linear type with continuously emitting sole.

electrons and circuit. The electron beam may be reentrant or not, depending upon the geometry of the circuit, that is, whether the cavities are arranged with cylindrical or linear symmetry.

The magnetron and related crossed-field amplifiers and oscillators are rather noisy devices and operate best in the high-output-power region. As amplifiers, they have generally low gain, but their output power is high. Compactness and high efficiency are the most attractive features of crossed-field electron tubes. Conversion of electron potential energy to microwave energy is characteristic of all crossed-field devices (M-type carcinotrons) in contrast to conversion of kinetic energy in the case of traveling-wave tubes (O-type devices). *See* TRAVELING-WAVE TUBE.

[R. J. COLLIER]

Bibliography: C. H. Dix and W. H. Aldous, *Microwave Valves*, 1966; J. W. Gewartowski and H. A. Watson, *Principles of Electron Tubes*, 1965; S. Liao, *Microwave Devices and Circuits*, 1980; E. Okress (ed.), *Crossed-Field Microwave Devices*, vols. 1 and 2, 1961; R. F. Soohoo, *Microwave Electronics*, 1971.

Maser

A device for coherent amplification or generation of electromagnetic waves by use of excitation energy in resonant atomic or molecular systems. The word is an acronym for *m*icrowave *a*mplification by *s*timulated *e*mission of *r*adiation. The device uses an unstable ensemble of atomic or molecular particles which may be stimulated by an electromagnetic wave to radiate excess energy at the same frequency and phase as a stimulating wave, thus providing coherent amplification. Masers, however, are not limited to the microwave region; this type of amplification has been extended to include a frequency range from audio to infrared or optical frequencies. Maser-type amplifiers and oscillators are also sometimes referred to as molecular, or quantum-mechanical, since they involve processes on a molecular scale, and since

some types cannot be adequately described by classical mechanics, but show characteristic quantum-mechanical phenomena.

Maser amplifiers can have exceptionally low noise, and come close to effectively amplifying a single quantum of radiation in the microwave region; that is, they approach the limits, set by the uncertainty principle, on the precision with which phase and energy of a wave may be amplified. Their inherently low noise makes maser oscillators that use very narrow atomic or molecular resonances extremely monochromatic, providing a basis for frequency standards. Since atoms or molecules may have resonances and effective amplification over a wide frequency range and to very short wavelengths, masers are useful as coherent amplifiers of millimeter, infrared, optical, and perhaps also ultraviolet wavelengths, where older types of circuit elements are not effective.

Because of their low noise and high sensitivity, maser amplifiers are particularly useful for reception and detection of very weak signals in radio astronomy, microwave radiometry, long-distance radar, and long-distance microwave communications. They also provide research tools for very sensitive amplification or detection of electromagnetic radiation.

Thermodynamic equilibrium of an ensemble of particles—such as atoms, molecules, electrons, or nuclei—which have discrete energy levels and which may radiate electromagnetic energy, requires that the number n_1 of particles in a lower level 1 be related to the number n_2 in an upper level 2 by the Boltzmann distribution condition $n_1/n_2 = e^{(E_1 - E_2)kT}$, where E_1 and E_2 are the respective energies of the two levels; k is Boltzmann's constant; and T is the absolute temperature, a positive number. Thermodynamic equilibrium also requires that phases of oscillation of the particles, or relative phases of quantum-mechanical wave functions for the various states, be random. A violation of either condition can result in instabilities which may release electromagnetic radiation. The frequency ν of radiation released is characteristically given by $h\nu = E_2 - E_1$, where h is Planck's constant.

The particles may be stimulated by an electromagnetic wave to make transitions from the lower to the upper level, absorbing energy from the wave, or from the upper to the lower level, imparting energy to the wave, and thereby increasing the wave amplitude coherently. Stimulated transitions from the upper to the lower state and those from the lower to the upper state are equally probable. For equilibrium at any positive temperature, the Boltzmann distribution requires that n_1 be greater than n_2. Therefore, there is a net absorption of energy from the wave, because particles which absorb are more numerous than those which emit. If the condition $n_2 > n_1$ occurs, the system may be said to have a negative absolute temperature, because the Boltzmann condition is fitted only by a negative value of T. If there are not too many counterbalancing losses from other sources, this condition allows a net amplification, because particles which emit energy are more numerous than those which absorb.

Gas masers. An amplifier where $n_2 > n_1$ is the beam-type maser (Fig. 1). Operated in 1954, this

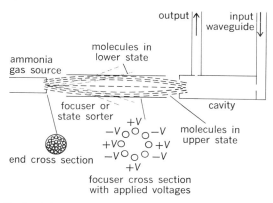

Fig. 1. Schematic of beam-type maser.

was the first type of maser to be suggested. Ammonia gas issues from a small orifice into a vacuum system to form a molecular beam. Molecules in the lower of the two states are deflected away from the axis of the state sorter or focuser by inhomogeneous electric fields which act on their dipole moments. Those molecules in the upper state are deflected toward the axis and sent into the microwave-resonant cavity. If losses in the cavity walls and coupling holes are sufficiently small, or if the number of molecules is sufficiently large, amplification or oscillation will occur. This maser is particularly useful as a frequency or time standard because of the relative sharpness and invariance of resonances of the ammonia beam.

The condition for oscillations to occur is given by Eq. (1), where h is Planck's constant, V the cavi-

$$n_2 - n_1 \gtreqqless hV \Delta\nu/8\pi \mu^2 Q \qquad (1)$$

ty volume, $\Delta\nu$ the width at half maximum of resonant response of the molecules, μ the molecular dipole moment (matrix element), and Q the quality factor of the loaded cavity. The maximum power output is approximately $h\nu$ multiplied by the rate at which molecules enter the cavity, and is very small. A wave impinging on the cavity will be amplified on reflection if it occurs near the resonant response and if $n_2 - n_1 \gtreqqless hV \Delta\nu/8\pi\mu^2 Q_0$, where Q_0 is the quality factor of the unloaded cavity.

Other masers using gaseous molecules have been proposed which involve production of a nonequilibrium distribution by excitation of the gas by means of externally applied radiation of shorter wavelength. Normally, such a system requires molecules with at least three energy levels, two used for amplifying a wave and a third higher level to which molecules are excited from the lowest level, as indicated in Fig. 2. In decaying from the higher level, molecules return, at least in part, to fill up the intermediate level and to satisfy the condition $n_2 > n_1$. If it is light radiation which excites molecules to higher levels, the system is said to use optical pumping.

Solid-state masers. Solid-state masers usually involve the electrons of paramagnetic atoms or molecules in a static or slowly varying magnetic field. In the simplest case, the two-level solid-state maser, only one electron on each molecule is affected. The energy of the electron is quantized into two levels, according to whether the magnetic

moment, associated with the electron spin, is parallel or antiparallel to the magnetic field. At thermal equilibrium there are more magnetic moments parallel than antiparallel to the field, corresponding to $n_1 > n_2$. This situation may be reversed, so that $n_2 > n_1$, by interchanging the two populations n_1 and n_2. The interchange is accomplished by rapid variation of the frequency of an intense electromagnetic field through resonance, by application of a pulse of resonant electromagnetic radiation, or in principle by sudden reversal of the magnetic field.

Electron-spin moments are more weakly coupled to the electromagnetic field than are molecular electric-dipole moments (by a factor of about 10^4). A much larger preponderance in the upper state is required than for the maser of Fig. 1. If requirements for amplification are met, however, electron-spin moments give correspondingly greater power output. Furthermore, their resonant frequencies are easily tunable by variation of the magnetic field, because their energies involve interaction between electronic magnetic moments and the field. In the simplest cases, the resonant frequency in megacycles is approximately 2.8 times the magnetic field strength in oersteds. Electron-spin resonances in paramagnetic materials allow amplification over broader bandwidths (one to a few hundred megahertz) than do gas systems. Favorable conditions are usually obtainable only at very low temperatures, such as occur in a liquid-helium cryostat; hence, cryogenic problems are often involved and materials used are normally solids rather than liquids. The two-level solid-state maser is most easily operated in pulses, between which the populations of the two levels are readjusted.

The popular three-level solid-state maser also uses paramagnetic material containing electronic magnetic moments in a magnetic field. It has many of the characteristics of the two-level solid-state maser, but can be operated continuously with much more convenience. In Fig. 2, the spacing between the three levels is shown to correspond to microwave frequencies ν_1, ν_2, and ν_3. Usually a few milliwatts of microwave power at frequency ν_3, called the pumping frequency, are sufficient and amplification occurs at a lower frequency with a maximum power output of a few microwatts. Under the simplest assumptions, the number of systems in levels 1 and 3 is equal and the number in level 2 is greater if $\nu_1 < \nu_2$, giving amplification at frequency ν_1. If $\nu_1 > \nu_2$, simple assumptions predict amplification for frequency ν_2.

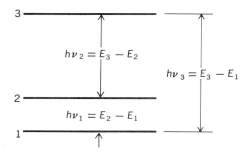

Fig. 2. Energy levels for a three-level maser.

For three suitable levels to occur in a paramagnetic material, each paramagnetic center must involve the magnetic moment of more than one electron, and must interact with a surrounding array of atoms which are not cubically arranged. The energy levels and frequencies still respond to an externally applied magnetic field. They are, however, no longer simply related to it but may vary widely in accordance with fields internal to the crystalline material. This allows responses at high frequencies with relatively low applied fields.

Three-level solid-state maser amplifiers have been made which have a noise temperature less than about 5 K (noise figure $\leqq$ 1.02). Although a wide variety of paramagnetic materials may be used, synthetic ruby, containing paramagnetic chromic (Cr^{3+}) ions, is favored. It has provided amplification both in resonant cavities and in traveling-wave structures.

Optical and infrared masers. Optical and infrared masers utilize a variety of principles for excitation of the nonequilibrium distribution. All involve multimode cavities, usually consisting of two optically flat plates between which the radiation is reflected many times through the excited medium. Such masers are particularly valuable as oscillators, producing coherent light that is extremely monochromatic and can be focused either to a very intense small spot or radiated in a remarkably parallel beam. *See* LASER.

Maser oscillators in the megahertz and audio range have been proposed; they use nuclear moments in an applied magnetic field or, with pumping at a higher frequency, in internal crystalline fields. The magnetic moments of protons in liquid water have provided a successful maser of this type. Small impurities of a paramagnetic ion furnish the higher energy level needed and transfer their excitation to the protons. The proton resonances in a magnetic field must be extremely narrow. Such a maser may be used as a very monochromatic oscillator with frequency proportional to the magnetic field strength. Hence it serves as a very precise magnetometer.

Deviation from thermodynamic equilibrium of the second type, involving phase coherence, also allows maser-type amplification and is present in many masers. In the beam-type maser oscillator, molecules decay toward the lower state. They continue to amplify after the probability of their being found in the lower state is greater than that of being found in the upper, because they oscillate coherently and in such a phase that they transfer energy to the electromagnetic field.

Raman-type masers rely on a type of phase coherence. Molecules with two levels, separated by energy $h\nu_1$, may be strongly driven by an electromagnetic field of frequency ν_2. If the majority of systems is in the lower state, and if $\nu_2 > \nu_1$, the Raman effect can allow amplification at frequency $\nu_2 - \nu_1$. This requires an intense driving field or a very strong coupling of the systems to the field, such as occurs in ferromagnetic electron resonances. In ferromagnetic materials, large numbers of electrons act in unison, thus providing coupling to an oscillating field which is strong enough to make Raman effects prominent. This can also be discussed in terms of classical theory if nonlinearities are allowed for, and is closely related to fer-

romagnetic amplifiers and to other parametric amplifiers. *See* PARAMETRIC AMPLIFIER.

Circuits. Maser circuits characteristically involve atoms or molecules which provide resonant reactances, positive or negative resistances, and coupling between two or more frequencies or circuit components. They also use certain elements of spectroscopic systems and a wide variety of components found in other radio-frequency and microwave devices.

In the simplest cases the molecular resonances behave like a resonant LC circuit with a positive or negative series resistor, or like a large number of such circuits in parallel and tuned over a distribution of frequencies.

The more classical circuit elements normally involved in masers supply the following functions:

1. Means of ensuring sufficiently strong interaction of an electromagnetic wave with material which amplifies the wave by stimulated emission.

2. Input and output coupling for the wave.

3. Auxiliary circuits which take appropriate advantage of maser characteristics in an overall system.

4. Where electromagnetic excitation is used, circuits which supply energy to the material and produce an unstable state which can radiate.

5. Magnetic field or other components for controlling frequencies of resonance of the material. The schematic of a three-level solid-state maser amplifier, shown in Fig. 3, illustrates each of these functional parts.

The resonant cavity, fulfilling function 1, must have sufficiently low internal losses and produce a sufficiently intense oscillating field (in this case a magnetic field) in the region of the amplifying material. If the cavity is uniformly filled with this material, the condition that amplification be obtainable is given by Eq. (2), where Q_0 is the unload-

$$Q_0 \geqq -Q_M = \frac{h\Delta\nu V}{8\pi\mu^2(n_1 - n_2)} \qquad (2)$$

ed quality factor of the cavity, h is Planck's constant, $\Delta\nu$ is the width of the molecular or atomic resonance, V is the cavity volume, n_1 is the

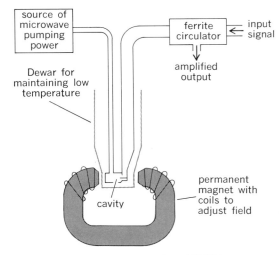

Fig. 3. Maser amplifier system for 9000 MHz.

number of particles in the upper state, n_2 is the number in the lower state, and μ is the effective dipole moment (matrix element) of the atoms or molecules.

A maser of this type is very similar to any other amplifier with positive feedback. There is an effective negative resistance with which may be associated the negative quantity Q_M, giving a fractional gain per cycle in energy stored of $2\pi/|Q_M|$. As the losses, characterized by the loaded Q of the cavity, are decreased, the amplifier gain increases until it becomes unstable and oscillates when $1/Q + 1/Q_M \leqq 0$. Here $1/Q = 1/Q_0 + 1/Q_E$, where Q_E applies to the external coupling. The power gain is $G = [(2Q/Q_E) - 1]^2$. The bandwidth B decreases with gain in such a way that the bandwidth-voltage gain product is nearly constant under typical conditions, as in Eq. (3), where ν is

$$(\sqrt{G} + 1)\, B = 2\nu/Q_E \qquad (3)$$

the frequency for maximum gain. Thus, for given characteristics of the maser material specified by Q_M, the gain and bandwidth can be adjusted over certain limits by variation of the cavity losses, or Q_0, and the coupling, or Q_E.

The noise temperature T for the amplifier alone (Fig. 3) is given by Eq. (4), where k is Boltzmann's

$$T = \frac{h\nu}{k \ln\left[\dfrac{n_2}{n_1} + \dfrac{n_1 - n_2}{n_1}\dfrac{Q_M}{Q_0}\right]} \qquad (4)$$

constant. Minimum noise temperatures require a Q_0 appreciably larger than Q_M and a small ratio n_2/n_1. A noise temperature of $h\nu/k$, or near 1K for microwave frequencies, is the minimum needed, since this allows effective amplification of approximately one quantum, the limit set by quantum mechanics.

The input and output use the same coupling hole in Fig. 3, and require a directional coupler, or preferably a circulator, for their separation. Separate input and output coupling holes may be used, but this tends to decrease the gain-bandwidth product and to increase noise.

To take full advantage of the low noise in the maser amplifier, the input and output circuits must be prevented from radiating excess noise into the amplifier. For example, attenuation in parts of the input wave guide and in the circulators, which are not at low temperature, results in noise radiation into the amplifier. If there is 0.1 dB loss in some part which is at room temperature, noise radiation into the amplifier corresponding to about 7K will result. Since most solid-state maser amplifiers operate at low temperatures, some components of the input or output circuits may conveniently be cooled to minimize their noise radiation. If input and output coupling holes are used, the output wave may be passed through a cooled isolator to avoid noise radiation from warm components of the output circuit.

If energy is supplied to the amplifying material by an electromagnetic drive, the circuits used for this purpose must provide a sufficiently strong controllable drive (usually constant in amplitude and frequency). If the interaction between the driving or pumping radiation and the material is strong, a relatively small amount of power may be needed. In Fig. 3 the cavity must be resonant at both the pumping and the amplifying frequencies. In some cases the orientation of fields at both frequencies must also be carefully controlled. Usually excess pumping power is available so that careful design for coupling it to the amplifying material is not essential.

Masers using a resonant cavity may also be operated as superregenerative amplifiers, for which various components such as the driving field, the cavity losses, the couplings, or the "static" magnetic field may be modulated.

Traveling-wave masers normally use slow-wave structures, which need only be effective over the range of response of the maser material. Amplification of a reflected wave may be controlled in the usual ways by matching, by attenuators or isolators, or by arrangement of the maser material itself to absorb a reflected wave.

The overloading of a maser amplifier is qualitatively like that of any other amplifier, the overload occurring when the molecular energy becomes exhausted. Recovery times vary between microseconds and seconds, depending especially on the maser material used; *TR* (transmit-receive) devices are sometimes necessary. *See* RADAR.

Maser oscillators used as frequency standards require especially stable cavities, in which the field is strongly coupled to the excited molecules. The cavities should be decoupled from external circuits, as by an attenuator or isolator.

Amplification or generation of electromagnetic waves at infrared or optical frequencies by maser techniques requires circuits which superficially appear quite different from those at lower frequencies but which serve the same functions. The resonant circuit is usually provided by two surfaces between which the radiation is reflected. The short wavelengths in this region imply that all man-made circuit elements are large compared with a wavelength. Partially transparent surfaces, multilayers of dielectrics, lenses, and other normal infrared and optical components fill the role of circuit elements.   [CHARLES H. TOWNES; J. P. GORDON]

Bibliography: T. K. Ishii, *Maser and Laser Engineering*, 1980; A. E. Siegman, *Introduction to Lasers and Masers*, 1971; H. G. Unger, *Introduction to Quantum Electronics*, 1970; A. Yariv, *Quantum Electronics*, 1975.

Metallic-disk rectifier

A rectifier that consists of one or more disks of metal in contact with coatings or layers of a semiconductor material. Alternating current is changed into pulsating direct current by the rectifying action that occurs at the junction interface between a metal disk and its mating semiconductor layer. The most common examples are the selenium and copper-oxide rectifiers. In a selenium rectifier a thin layer of selenium is deposited on one side of an aluminum plate and a conductive metal coating is sprayed or otherwise deposited on the selenium; electrons flow much more readily in the direction from the metallic coating to the selenium. In a copper-oxide rectifier rectification occurs at the junction between a copper disk and a coating of cuprous oxide. For a more detailed discussion *see* SEMICONDUCTOR RECTIFIER.   [JOHN MARKUS]

Microcomputer

A digital computer whose central processing unit (CPU) is a microprocessor. Generally, microcomputers handle data in the form of 4-bit, 8-bit, or 16-bit words. The difference between microcomputers and minicomputers has become increasingly blurred, as contemporary microcomputers provide features and performance once offered only by minicomputers. They may even resemble each other physically.

Microcomputers are usually divided into three categories: single-board (or single-card) microcomputers; personal (home or business) microcomputers; and microcomputer development systems. *See* MICROCOMPUTER DEVELOPMENT SYSTEM.

Both single-board and personal microcomputers have the same basic elements. The CPU is a microprocessor. Commonly used microprocessors include 8-bit devices such as the Zilog Z80, the Intel 8080, the MOS Technology 6502, and the Motorola 6809; and 16-bit devices such as the Motorola 68000 and Intel 8086. They have random-access memory (RAM) devices as well as read-only memory (ROM) devices. RAM is for storage of data used during execution of a program, while ROM devices are for data frequently needed by the microprocessor (such as a high-level language interpreter or bootstrapping instructions, which reset the microprocessor whenever power is switched on). *See* MICROPROCESSOR.

Single-board microcomputer. A single-board microcomputer (Fig. 1) usually has a numeric keypad for data input. Output must be added through a printing or video display terminal. Single-board microcomputers also normally have space for the addition of other semiconductor devices, such as extra memory or peripheral devices, which can be added as needed by the user. Single-board microcomputers are often used for teaching or design purposes, or as the starting point for custom microcomputer designs or systems.

Personal microcomputer. Personal microcomputers include such units as the Radio Shack TRS-80 (Fig. 2*a*), the Apple II (Fig. 2*b*) and III, and the IBM Personal Computer. These systems include a full alphanumeric keyboard for data input; video display terminals for output; high-level language capabilities, such as BASIC; provision for data storage on magnetic media, such as diskettes; and the ability to be used in conjunction with peripherals, such as printers and modems.

Elements of a typical personal microcomputer include the disk drive, modem, and printer. Disk drives are used to store and retrieve data on magnetic disks. Typical sizes used with microcomputers include 5¼- and 8-in. (133- and 203-mm) disks. A modem is an acoustic device which converts data from a microcomputer into audio tones for transmission over telephone lines and then converts audio tones back into digital form for processing by computer. A printer is an electromechanical device similar to an electric typewriter for making a permanent record of the output of a microcomputer. Some printers can reproduce artwork and symbols generated by a microcomputer; these are known as graphics printers. *See* COMPUTER STORAGE TECHNOLOGY; DATA-PROCESSING SYSTEMS.

A recent trend in microcomputer development is expanded graphics capabilities, including the use of color. This makes it possible to generate special visual effects, graphs, art, and so forth, on a microcomputer. Another area of interest has been sound and voice synthesis. This has led to the generation of music by microcomputer and

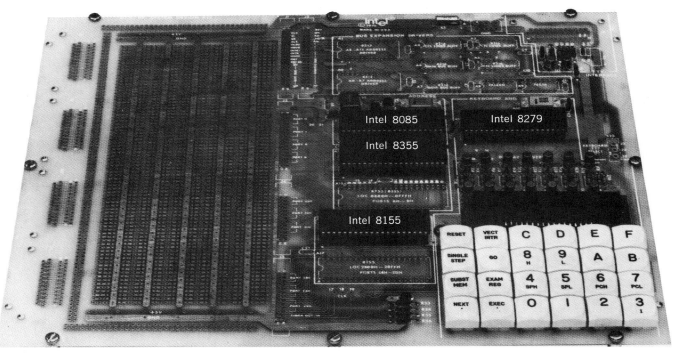

Fig. 1.   SDK-85 one-card microcomputer. (*Intel Corp.*)

Fig. 2. Personal microcomputers. (*a*) TRS-80 (*Radio Shack*). (*b*) Apple II (*Apple Computer, Inc.*).

speech output to users of microcomputers. *See* COMPUTER GRAPHICS; VOICE RESPONSE.

The most common high-level language used by microcomputers is BASIC, with manufacturers of microcomputers using implementations which take advantage of the special features offered by their hardware. Second in popularity is PASCAL. Other languages, such as FORTRAN and COBOL, are used, although these tend to be much less powerful versions of the languages than those implemented on larger computers. *See* PROGRAMMING LANGUAGES.

A trend to portable microcomputer systems is also underway. In 1980, both Radio Shack and Sharp offered pocket-sized microcomputers powered by batteries. These units were very limited in their capabilities, however. Another significant entry was the Osborne I microcomputer, a system using a Z80 microprocessor, two disk drives, and numerous software packages in a case small enough to fit under an airplane seat.

A significant problem facing users and programmers of microcomputers is the incompatibility of programs written for various microcomputer systems. For example, a program written for an Apple computer will likely not run on an IBM Personal Computer, and vice versa. Efforts to develop standards for the microcomputer industry and to enhance the portability of programs and operating systems are ongoing. *See* DIGITAL COMPUTER; SOFTWARE ENGINEERING.

[HARRY L. HELMS]

Bibliography: L. E. Frenzel, Jr., *Crash Course in Microcomputers*, 1981; A. Osborne, *An Introduction to Microcomputers*, vol. 1, 1977.

Microcomputer development system

A complete microcomputer system used to test and develop both the hardware and software of other microcomputer-based systems, from initial development through debugging of final prototypes.

A typical microcomputer development system (MDS) includes assembler facilities, a text editor, debugging facilities, and hardware emulation capabilities. Often an MDS will also include a test board for peripheral interfaces and other hardware, a floppy disk storage system, a video dis-

play terminal, a keyboard or keypad input device, and a provision for using a printing peripheral. A few microcomputer development systems include provisions for programmable read-only memory (PROM) devices.

The assembler permits instructions for the microcomputer to be written as mnemonics, which are then translated by the assembler into hexadecimal or binary machine language. The machine language program, known as an object file, is loaded into the microcomputer for execution. The program written in mnemonics is known as the source code, and is generated by the text editor. *See* DIGITAL COMPUTER PROGRAMMING.

A key component is the emulator, a combination of hardware and software which permits the MDS to emulate programs written for other microcomputer systems. Many emulators can execute source code programs for different microprocessors. A few emulators can also execute high-level languages such as BASIC, FORTRAN, and PASCAL. *See* PROGRAMMING LANGUAGES.

An MDS normally includes several software development and debugging tools. One common debugging facility is the trace function. This allows program execution to be halted at desired points so that central processing unit (CPU) registers or memory contents may be examined.

If true emulator and debugging facilities are not included in an MDS, a cross assembler will usually be provided. This is an assembler which accepts source code for one microprocessor type and compiles it by using a system for another microprocessor. However, it is not possible to run or debug programs by using a cross assembler.

Typical MDS random-access memory (RAM) capacity is 64K bytes or greater. A floppy disk–based mass storage system is now part of almost all available microcomputer development systems. *See* COMPUTER STORAGE TECHNOLOGY.

Most microcomputer development systems support the development of just one microprocessor or family of microprocessors. However, both GenRad and Tektronix offer microcomputer development systems which support several different microprocessors and high-level languages. *See* MICROCOMPUTER; MICROPROCESSOR.

[HARRY L. HELMS]

Microphone

An electroacoustic device containing a transducer which is actuated by sound waves and delivers essentially equivalent electric waves.

Modern conventional microphones may be classified as pressure, gradient, combination pressure-gradient, and wave types. A pressure microphone is one in which the electrical response is caused by variations in pressure in the actuating sound wave. In a gradient microphone the electrical response corresponds to some function of the pressure difference between two points in a sound wave. A wave microphone is one in which the response depends upon sound wave interaction.

This article will discuss transducers used in microphones, some of the common types of microphones, the characteristics of microphones, and the calibration of microphones.

MICROPHONE TRANSDUCERS

A transducer is a device which, when actuated by power from one system, will supply power to one or more other systems. For the conversion of the acoustical variations into the corresponding electrical variations, the transducers most commonly used in microphones are dynamic, magnetic, electrostatic, piezoelectric, electrostrictive, or carbon. *See* TRANSDUCER.

Dynamic transducer. This consists of a moving conductor located in a magnetic field (Fig. 1). The motion of the conductor leads to the induction of an electromotive force (emf) in the conductor, the magnitude of the emf being proportional to the velocity of the conductor. The conductor may be in the form of a coil, a straight wire, or a ribbon. The electrical impedance of the dynamic transducer is relatively low, ranging from 0.1 to 60 ohms in practical structures. The electrical impedance is practically all resistive and is therefore independent of frequency. When the dynamic transducer is connected to a load, it may be considered to consist of the open-circuit voltage, the impedance of the transducer, and the impedance of the load, all connected in series. The dynamic transducer is reversible; that is, it can be actuated by electrical energy and deliver corresponding sound energy.

MICROPHONE

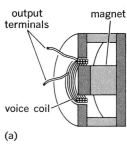

(a)

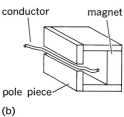

(b)

Fig. 1. Dynamic transducers. (a) Perspective sectional view of moving coil transducer. (b) Perspective view of straight-line conductor.

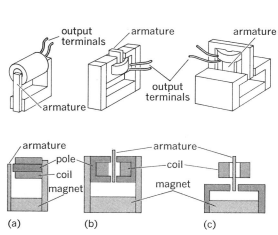

Fig. 2. Magnetic transducers, perspective and sectional views. (a) Single pole and armature and (b, c) multiple pole and balanced armature types.

Magnetic transducer. This consists of a magnetic field including a variable reluctance path and a coil surrounding all or a part of this path. The variation in reluctance leads to a variation in the magnetic flux through the coil and a corresponding induced emf. There are an enormous number of arrangements and configurations possible using this principle. The two most common, namely, the single pole and armature and the multiple pole and armature types, are shown in Fig. 2. The electrical impedance of this type is governed by the size of the wire and the coil, and may vary from a few ohms to thousands of ohms. The impedance is proportional to the frequency, because the inductance of the coil is the predominant electrical element. When the magnetic transducer is connected to a load, it may be considered to consist of the open-circuit voltage, the impedance of the transducer, and the impedance of the load, all connected in series. The magnetic transducer is reversible.

Electrostatic transducer. The electrostatic or condenser transducer consists of a fixed electrode and a movable electrode. The electrodes are charged electrostatically in opposite polarity. Motion of the movable electrode leads to the production of a voltage which corresponds to the amplitudes of the motion of the electrode (Fig. 3). The

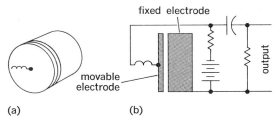

Fig. 3. Electrostatic transducer. (a) Perspective view. (b) Sectional view and electrical circuit diagram.

impedance of the electrostatic transducer is due to the capacitance between the two electrodes. When the electrostatic transducer is connected to a load, it may be considered to consist of the open-circuit voltage, the capacitance of the transducer, and the impedance of the load, all connected in series. The electrostatic transducer is reversible.

Piezoelectric transducer. The piezoelectric or crystal transducer consists of a crystal element having piezoelectric properties. A deformation of the crystal leads to the generation of a voltage which corresponds to the amplitude of the deformation (Fig. 4). The impedance of this type is due to the electrical capacitance of the crystal. When the crystal transducer is connected to a load, it may be considered to consist of the open-circuit voltage, the capacitance of the crystal, and the electrical impedance of the load, all connected in series. The crystal transducer is reversible. *See* PIEZOELECTRICITY.

Electrostrictive transducer. The electrostrictive or ceramic transducer consists of a ceramic element having electrostrictive properties. A deformation of the ceramic leads to the generation of a voltage corresponding to the amplitude of the

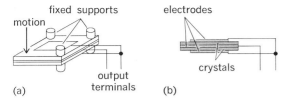

(a)

(b)

Fig. 4. Crystal or piezoelectric transducer. (a) Perspective view. (b) Sectional view.

deformation (Fig. 5). The impedance of this type is due to the capacitance of the ceramic. When the electrostrictive transducer is connected to a load, it may be considered to consist of the open-circuit voltage, the capacitance of the ceramic, and the electrical impedance of the load, all connected in series. The ceramic transducer is reversible. *See* ELECTROSTRICTION.

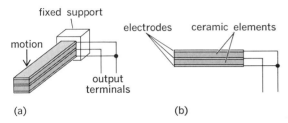

(a)

(b)

Fig. 5. Ceramic or electrostrictive transducer. (a) Perspective view. (b) Sectional view.

Carbon transducer. This consists of carbon granules in contact with a fixed electrode and a movable electrode (Fig. 6). Motion of the movable electrode varies the resistance of the granules. If the transducer is connected in series with an external steady voltage and a resistor, a voltage corresponding to the amplitude of the motion of the movable electrode will be developed across the resistor. The impedance of this transducer is an almost pure electrical resistance, which is governed by the dimensions of the carbon granule aggregate. In general, the resistance is of the order of 100 ohms. The carbon transducer is not reversible.

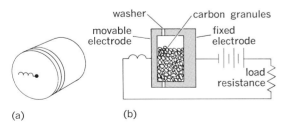

(a)

(b)

Fig. 6. Carbon transducer. (a) Perspective view. (b) Sectional view and electrical circuit diagram.

MICROPHONE TYPES

Microphones may be classed in many different ways, for example, by the type of electrical response, by the type of transducer, or by the direc-

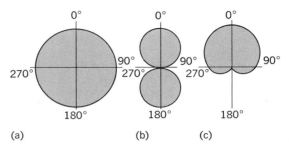

tivity pattern. In the following discussion, microphones will be classified primarily according to the type of response.

Pressure microphone. In this microphone the electrical response is caused by variations in sound pressure. Pressure microphones are inherently nondirectional (omnidirectional), because pressure is a scalar and not a vector quantity. Three of the most common directivity patterns of microphones are shown in Fig. 7.

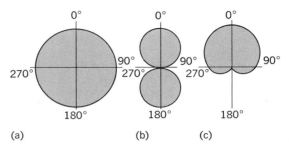

(a)   (b)   (c)

Fig. 7. Polar directivity patterns. (a) Nondirectional (omnidirectional). (b) Bidirectional. (c) Unidirectional.

Dynamic types. The three principal types of dynamic microphones are the moving coil, the moving conductor, and the ribbon-type pressure microphone.

A moving coil microphone consists of a voice coil located in a magnetic field and coupled to a diaphragm acted upon by sound waves (Fig. 8). To

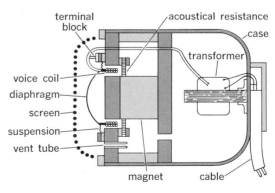

Fig. 8. Sectional view of moving coil microphone.

obtain constant output for constant sound pressure on the diaphragm, the velocity of the coil must be independent of the sound wave frequency. To accomplish this objective, the system must be acoustical-resistance-controlled. The resonant frequency of the system is usually set at 700 Hz. The acoustical resistance is made larger in magnitude than the acoustical reactance due to the mass of the diaphragm and coil and the compliance of the suspension by placing an acoustical resistance behind the voice coil. This acoustical resistance may be a slit or a silk cloth or piece of felt.

The materials most commonly used for the diaphragms of pressure microphones are aluminum alloys, paper, and bakelite, Mylar, and other plas-

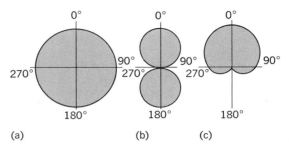

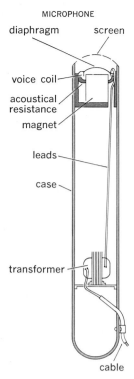

MICROPHONE

diaphragm — screen

voice coil

acoustical resistance

magnet

leads

case

transformer

cable

Fig. 9. Sectional view of unobtrusive moving coil microphone.

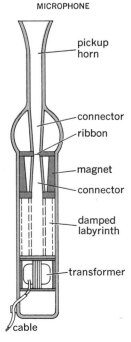

MICROPHONE

pickup horn

connector

ribbon

magnet

connector

damped labyrinth

transformer

cable

Fig. 10. Sectional view of ribbon-type pressure microphone.

tics. In order to obtain a minimum density-resistivity product, aluminum is almost universally used for the voice coil. The aluminum may be in the form of wound ribbon or wound wire.

The impedance of the voice coil may vary from 1 to 30 ohms. In general, a transformer is included to step up the impedance to any of the standard values used for transmission over a line, namely, 30, 150, and 250 ohms, or to a high impedance (25,000 ohms) for direct feed to the grid of a vacuum tube.

For many applications an unobtrusive microphone is desirable. In the unobtrusive design, a small dynamic unit is located in one end of a long slender tube (Fig. 9). The long tube provides sufficient volume to ensure adequate low-frequency response.

A small microphone supported by means of a slender band around the user's neck in the form of a pendant is known as a lavalier microphone. The miniature dynamic unit is housed in a case similar to that in Fig. 9, except that the diameter is smaller and the length is shorter. The principal purpose of the lavalier microphone is to allow a person to move freely without introducing any appreciable change in output, as would be the case if a stationary microphone were used. It also frees the speaker's hands.

A moving conductor microphone consists of a straight-line conductor located in a magnetic field and coupled to a V-shaped diaphragm acted upon by sound waves. The action of this microphone is similar to that of the moving coil microphone.

A ribbon-type pressure microphone consists of a metallic ribbon located in a magnetic field and terminated in an acoustical resistance (Fig. 10). To obtain constant output for constant sound pressure, the velocity of the ribbon must be independent of the frequency. This objective is accomplished by making the acoustical resistance, which terminates the ribbon, the controlling element. The acoustical resistance is in the form of a folded damped pipe. In order to improve the high-frequency response, a small horn is used at the pickup point to couple the ribbon to the sound field. A transformer is used to raise the electrical impedance of the ribbon to a value that is suitable for transmission over a line.

Magnetic type. A magnetic microphone consists of a diaphragm acted upon by sound waves and connected to an armature which varies the reluctance in a magnetic field surrounded by a coil (Fig. 11). To obtain constant output for constant sound pressure on the diaphragm, the velocity of the armature must be independent of the frequency. Therefore, the system must be acoustical-resistance-controlled. This can be accomplished by placing an acoustical resistance behind the diaphragm. In order to obtain stability of the magnetic armature system, considerable stiffness must be introduced into the system.

Electrostatic (condenser) type. An electrostatic or condenser microphone consists of a fixed plate and a movable plate or diaphragm exposed to the actuating sound waves (Fig. 12). The impinging sound wave produces a motion of the diaphragm and a corresponding variation in capacitance. In Fig. 12a a polarizing voltage is applied to the diaphragm and backplate. Under these conditions the electrical output of the diaphragm corresponds to

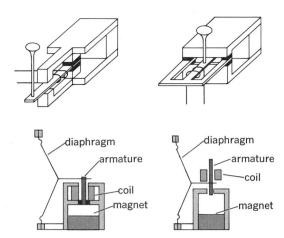

diaphragm

armature

coil

magnet

diaphragm

armature

coil

magnet

Fig. 11. Two magnetic microphones of the balanced-armature type, perspective and sectional views. (*From H. F. Olson, Acoustical Engineering, Van Nostrand, 1957*)

the motion of the diaphragm. The output is amplified by a vacuum tube or transistor amplifier. In Fig. 12b the capacitance of the diaphragm and backplate is part of the resonant circuit of a vacuum tube or transistor oscillator. Motion of the diaphragm produces a variation in the frequency of the oscillator. The output of the oscillator is detected by means of a frequency modulation (FM) detector. The resultant electrical output of the detector corresponds to the motion of the diaphragm.

The capacitance of the condenser microphone is very small, 50–300 picofarads (pf). The electrical impedance is relatively very large at low frequencies. Therefore, the polarizing resistor in Fig. 12a must be of the order of tens of megohms in order to maintain adequate low-frequency response. In view of the low electrical capacitance it is almost imperative that the amplifier in Fig. 12a or the os-

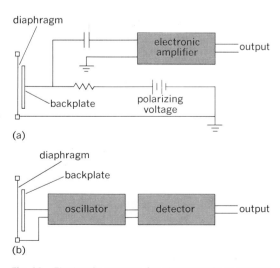

diaphragm

electronic amplifier — output

backplate

polarizing voltage

(a)

diaphragm

backplate

oscillator   detector — output

(b)

Fig. 12. Electronic systems for electrostatic or condenser transducers, which consist of fixed plate and movable diaphragm. (a) Polarized electrostatic or condenser transducer and electronic amplifier. (b) Electrostatic or condenser transducer as an element of an oscillator and a frequency modulation detector.

cillator in Fig. 12*b* be located next to the transducer to obtain a relatively small fixed capacitance, because any additional fixed capacitance reduces the sensitivity of the systems shown in Fig. 12.

Crystal type. This consists of a crystal having piezoelectric properties acted upon by sound waves either directly or through a diaphragm connected to the crystal. A diaphragm-type crystal microphone is shown in Fig. 13. To obtain constant voltage with respect to frequency for constant sound pressure, the amplitude of the motion of the crystal must be independent of frequency. Therefore, the diaphragm must be stiffness-controlled. This is accomplished by placing the resonant frequency of the diaphragm and the crystal assembly at or near the upper limit of the response range.

The piezoelectric element may be Rochelle salt or ammonium dihydrogen phosphate (ADP). In general, two crystals are cemented together so as

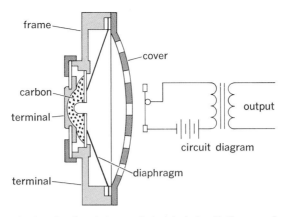

Fig. 14.   Sectional view and electrical circuit diagram of carbon microphone, widely used in telephones.

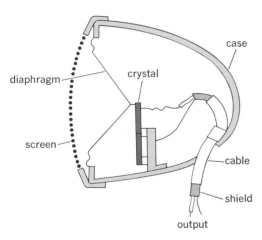

Fig. 13.   Sectional view of a crystal microphone.

to form a "bimorph" structure and thereby increase the ratio of stress to amplitude. In this way the output is increased.

The capacitance of the crystal unit is of the order of 1000–2000 pf. This makes it possible to transmit over several feet of low-capacitance cable without appreciable attenuation.

Ceramic type. This consists of a ceramic, usually barium titanate, having electrostrictive properties. The ceramic is acted upon by sound waves transmitted through a diaphragm connected to the crystal. The general arrangement of the elements is similar to that of the crystal microphone shown in Fig. 13 except that a ceramic transducer is substituted for the crystal transducer. The performance and characteristics of the ceramic microphone are similar to those of the crystal microphone.

Carbon type. This type consists of a diaphragm connected to a movable electrode which together with a stationary electrode forms a cup containing carbon granules (Fig. 14). A polarizing current is passed through the electrodes and the carbon granules. Motion of the diaphragm, caused by incident sound waves, varies the resistance of the carbon element. To obtain constant current varia-

tion in the electrical circuit of Fig. 14 for constant sound pressure upon the diaphragm, the amplitude of the diaphragm must be independent of the frequency. Therefore, the system must be stiffness-controlled. This can be accomplished by placing the resonant frequency of the diaphragm and carbon element assembly at or near the upper limit of the response range.

The electrical impedance of the carbon element is a resistance of the order of 100–200 ohms. A transformer is usually used with the carbon microphone, as shown in Fig. 14.

The carbon microphone is universally employed in all telephone applications. It is also used for communication systems in which high sensitivity is a requirement. The second harmonic distortion is relatively high. Double-button microphones, that is, microphones having two cups of carbon granules, have been used in a push-pull arrangement to reduce nonlinear distortion.

Contact type. This can be attached directly to

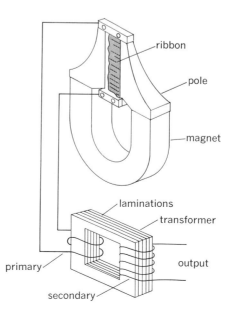

Fig. 15.   Elements of a velocity microphone.

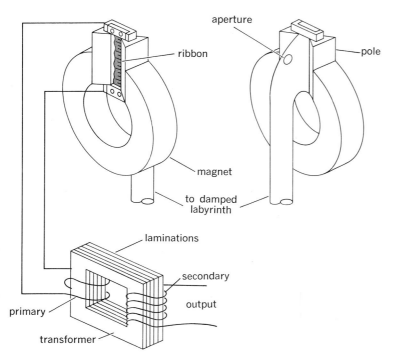

Fig. 16.   Elements of a single-unit ribbon-type unidirectional microphone.

the ribbon. If the effective distance between front and back is small compared to a wavelength, the magnitude of the pressure is the same on both sides of the ribbon. However, there is a phase shift between the two sides and hence a difference in pressure. The pressure difference is proportional to the frequency and also to the cosine of the angle of the incident sound with respect to an axis normal to the plane of the ribbon. In order to obtain constant voltage output, the velocity must be independent of frequency. Since the actuating pressure is proportional to frequency, the vibrating system must be mass-controlled to obtain a constant relationship between the sound pressure in free space and the velocity of the ribbon. This can be accomplished by placing the resonant frequency of the ribbon below the response range of the microphone. The resonant frequency is usually about 12 Hz in practical microphones.

One of the fundamental advantages of the ribbon-type velocity microphone is that the ribbon serves as both the diaphragm and conductor. For this reason, it is possible to obtain greater sensitivity than in a dynamic system, in which the diaphragm and voice coil have separate functions.

The directivity pattern of a velocity microphone is the bidirectional cosine characteristic in Fig. 7.

Unidirectional types. These microphones have a substantially unidirectional pattern over their response range. Unidirectional microphones may be constructed by combining a bidirectional microphone and a nondirectional microphone or by combining a single-element microphone with an appropriate acoustical delay system.

The directivity pattern of a unidirectional microphone is given by Eq. (1), where e is the voltage

$$e = a + b \cos \theta \tag{1}$$

output of the microphone, a the voltage output of the pressure element, b the voltage output of the velocity element, and θ the angle between the normal to the plane of the ribbon and the incident sound wave.

If $a = b$ the directivity pattern is the cardioid pattern shown in Fig. 7.

The elements of a single-unit unidirectional microphone are shown in Fig. 16. The transducer is of the dynamic ribbon type. The back of the ribbon is coupled to an acoustical resistance in the form of a damped pipe which is folded to fit in the lower part of the case. The addition of the aperture in the portion of the pipe covering the back of the ribbon

the body or sounding board of a musical instrument to provide solo pickup from the instrument. Contact microphones may employ magnetic, crystal, or ceramic transducers. The transducer is coupled directly to the instrument.

Gradient microphone. A pressure-gradient microphone is one in which the electrical response corresponds to some function of the difference in pressure between two points in space. In general, when the distance between the two points is small compared to the wavelength, the pressure gradient corresponds to the particle velocity in the sound wave. A velocity microphone is one in which this condition holds.

Velocity type. This consists of a metallic ribbon which is located in a magnetic field and is freely accessible to sound waves on both sides. A schematic diagram of the elements of a velocity microphone is shown in Fig. 15. The ribbon is actuated by the difference in sound pressure between the two sides. The motion of the ribbon in the magnetic field induces a voltage between the two ends of

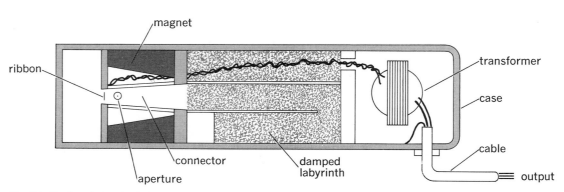

Fig. 17.   Sectional view depicting elements of unidirectional microphone employing a ribbon dynamic transducer.

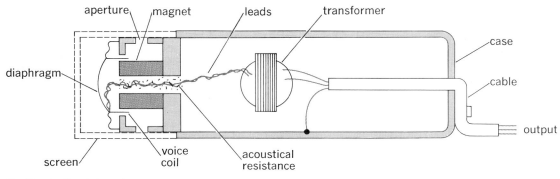

Fig. 18. Sectional view of elements of unidirectional microphone employing voice coil dynamic transducer.

introduces an acoustical element in the form of an acoustic inertance which shifts the phase so that the response from the back is very low. The directional pattern under these conditions is a cardioid, as shown in Fig. 7. This characteristic is useful in picking up sounds from the front and discriminating against sounds originating from the rear.

Other directional patterns may be obtained from the microphone shown in Fig. 7. For example, if the aperture is closed, a pressure-responsive microphone with a nondirectional characteristic is obtained. If the aperture is very large, a velocity microphone with bidirectional characteristics is obtained.

The sectional view of Fig. 17 depicts another version of the single-unit unidirectional microphone employing a ribbon-type dynamic transducer coupled to an acoustical resistance in the form of a damped folded pipe. The directivity pattern is the cardioid of Fig. 7. The maximum sensitivity of the ribbon unidirectional microphone of Fig. 17 is along the axis of the cylinder.

A unidirectional microphone employing a voice coil dynamic transducer is illustrated by the sectional view shown in Fig. 18. The back of the diaphragm is coupled to the acoustic inertance of the ports in the side of the case and to an acoustical resistance and acoustical capacitance of the air volume of the case. The phase shift introduced by the acoustical network consisting of the acoustic inertance, acoustical resistance, and acoustical capacitance is the same as the distance from the ports to the front of the diaphragm. As a result, for sound arriving from the back, the forces on the front and back of the diaphragm are almost equal

in phase and amplitude, and the response is low. The phase shift is at a maximum for sound arriving in a forward direction, with a resultant maximum sensitivity. The directivity pattern of the dynamic-unidirectional microphone of Fig. 18 is the cardioid of Fig. 7.

A unidirectional microphone employing a condenser or electrostatic transducer is depicted by the sectional view of Fig. 19. An acoustical resistance is located behind the diaphragm to provide resistance control of the vibrating system. The back of the diaphragm is coupled to the acoustic inertance of the ports in the side of the case and an acoustical resistance and acoustical capacitance of the air volume in the case. The acoustical network is similar to that of the dynamic-unidirectional microphone. The directional pattern is the cardioid of Fig. 7. The maximum sensitivity is along the cylindrical axis of the microphone. The electronic systems which may be used with this microphone are the same as those for the condenser microphone of Fig. 12.

The unidirectional microphone with the cardioid characteristic of Fig. 7 is almost universally employed in sound motion picture recording, television sound stage pickup, and sound reinforcing systems. The unidirectional microphone is useful in any application in which the desired sound originates in front of the microphone and the undesired sounds originate in the rear; this extends its use to almost all sound pickup applications.

Higher-order gradient types. The velocity microphone described in the preceding section is a first-order gradient microphone. Higher-order gradient microphones may be obtained by carrying the

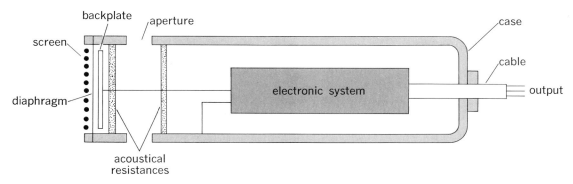

Fig. 19. Sectional view of unidirectional microphone employing condenser or electrostatic transducer.

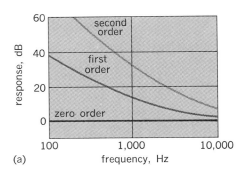

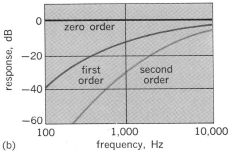

Fig. 20. Frequency response characteristics of (a) zero-, first-, and second-order gradient microphones at a distance of 3/4 in. from a sound source, and (b) of the same microphones for a plane wave. The microphones are compensated so that the responses of all three are the same and independent of the frequency when operating at a distance of 3/4 in. from a sound source.

differential process to higher orders. The directivity patterns are cosine functions in which the power of the cosine is the order of the gradient.

The response of a gradient microphone is a function of both the distance from the sound source and the frequency. The frequency response characteristics of a zero-order gradient or pressure microphone, a first-order or velocity microphone, and a second-order gradient microphone for a distance of 3/4 in. are shown in Fig. 20a. Figure 20b depicts the response of these microphones at a large distance from the sound source when the microphones are compensated to yield uniform response at 3/4 in. This shows the discrimination of

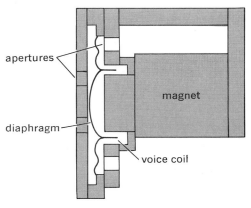

Fig. 21. Sectional view of a moving-coil close-talking first-order gradient microphone. Such microphones are useful when there is much background noise.

the gradient microphone against sounds originating at a distance. There is also an additional 3- and 5-dB attenuation by the first- and second-order gradient microphones, respectively, due to their directivity. These considerations are for the use of the gradient microphone as a close-talking microphone, that is, 3/4 in. from the mouth. Thus close-talking gradient microphones are useful for picking up sound under very noisy conditions. A close-talking dynamic-type first-order gradient microphone is shown in Fig. 21. A close-talking second-order gradient microphone employing two first-order gradient microphones and connected in opposition is shown in Fig. 22. The close-talking gradient microphones in Figs. 21 and 22 are for the pickup of speech. Therefore, the frequency range can be limited as compared to that of high-quality applications.

A second-order gradient microphone for use in wide frequency range high-quality applications is depicted in Fig. 23. The microphone consists of two first-order gradient unidirectional microphones of the type shown in Fig. 17 connected in opposition and combined with suitable frequency compensating networks. For this application the first-order gradient microphones are designed to be more directional than the cardioid characteristic of Fig. 7. Under these conditions the directivity pattern of the microphone of Fig. 23 is given by Eq. (2), where e_0 is the sensitivity constant of the mi-

$$e_0 = e_0 (1 + \cos \theta) \cos \theta \qquad (2)$$

crophone, and θ is the angle between the cylindri-

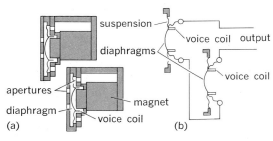

Fig. 22. Second-order gradient microphone. (a) Sectional. (b) Electrical connections of the moving coils.

cal axis of the microphone and the direction of the incident sound wave.

The applications for the second-order gradient uniaxial microphone are for the pickup of sound over large distances or under acoustically difficult conditions in which a high degree of directivity is desired, for example, in sound motion pictures and television. The size and weight of the microphone are such that it may be mounted on a conventional boom. The response of the second-order gradient microphone to random sounds is one-eighth that of a nondirectional microphone. The increased directional efficiency makes it possible to use a pickup distance of about three times that of a nondirectional microphone and 1.8 times that of a unidirectional microphone with a cardioid directional pattern.

Wave microphone. This consists of a system in which the directivity depends upon some type of

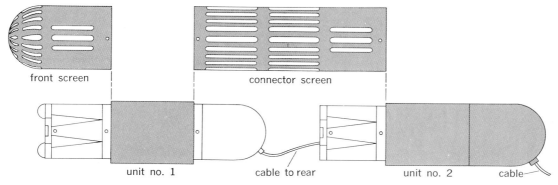

Fig. 23. Second-order microphone consisting of two first-order microphones and front and connector screens.

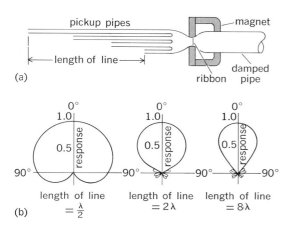

Fig. 24. Line microphone. (a) Sectional view, showing open-ended pipes. (b) Directivity patterns. The maximum voltage response is arbitrarily chosen as unity.

wave interference. The most common wave microphones are line, reflector, and lens types. Line and reflector microphones are highly directional in the speech-frequency range and may therefore be used for long-distance pickup of speech under conditions of high ambient noise and excessive reverberation. High-sensitivity transducers are employed in line and reflector microphones to provide an adequate signal-to-noise ratio.

Line type. This consists of a number of small tubes or pipes with the open ends, as pickup points, equally spaced along a line and the other ends connected at a common junction to a transducer. In the system shown in Fig. 24, the transducer is a ribbon element terminated in an acoustical resistance. Under these conditions, the output of the pipes can be added vectorially. In the case of a uniform line, the directivity characteristic, that is, the ratio R_θ of the response for an angle θ to the response for $\theta = 0$, is given by Eq. (3), where l

$$R_\theta = \frac{\sin \frac{\pi}{\lambda}(l - l \cos \theta)}{\frac{\pi}{\lambda}(l - l \cos \theta)} \tag{3}$$

is the length of line, λ the wavelength, and θ the angle between the direction of the incident sound and the axis of the line. The directional patterns for a simple line for line lengths of $\lambda/2$, 2λ, and 8λ are shown in Fig. 24.

Reflector type. This consists of a surface which reflects the pencils of the impinging sound in phase to a common point, termed the focus, at which a microphone is located (Fig. 25). The directional characteristics of a parabolic reflector 3 ft in diameter are shown in the figure.

Hot-wire microphone. This consists of a fine wire heated by the passage of an electric current. The cooling due to the motion of air past the wire causes a change in electrical resistance of the wire. In a sound wave the particle velocity cools the wire. There are also minor cooling effects produced by the sound pressure. The change in resistance due to the passage of a sound wave may be used to detect its presence. However, the frequency of the electrical output is twice the frequency of the sound wave because the wire is cooled equally by both positive and negative particle velocities. The use of a direct-current airstream for polarization appears to be impractical. Therefore, this microphone cannot be used for the reproduction of sound.

PERFORMANCE CHARACTERISTICS

The performance of a microphone is determined by the following principal factors: the open-circuit voltage response frequency characteristic, the electrical impedance characteristic, the directional characteristic, the nonlinear distortion characteristic, and the noise characteristic. The important

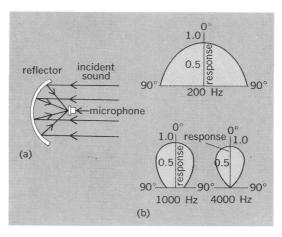

Fig. 25. Parabolic reflector-type microphone. (a) Schematic view. (b) Directivity patterns.

Characteristics of microphones

Type	Directivity pattern	Transducer	Internal voltage	Internal impedance	Operating impedance*
Pressure	Nondirectional	Carbon	$e \propto x$	Resistive	L, M, H
		Magnetic	$e \propto \dot{x}$	Inductive	L, M, H
		Dynamic	$e \propto \dot{x}$	Resistive	L, M, H
		Ribbon	$e \propto \dot{x}$	Resistive	L, M, H
		Condenser	$e \propto x$	Capacitive	H
		Crystal	$e \propto x$	Capacitive	H
		Ceramic	$e \propto x$	Capacitive	H
Velocity, first-order gradient	Bidirectional, given by $\cos\theta$	Ribbon	$e \propto \dot{x}$	Resistive	L, M, H
		Condenser	$e \propto x$	Resistive	H
First-order gradient	Unidirectional, given by $(1 + \cos\theta)$	Ribbon	$e \propto \dot{x}$	Resistive	L, M, H
		Dynamic	$e \propto \dot{x}$	Resistive	L, M, H
		Condenser	$e \propto x$	Capacitive	H
Second-order gradient	Unidirectional, given by $(1 + \cos\theta)\cos\theta$	Ribbon	$e \propto \dot{x}$	Resistive	L, M, H
Antinoise, first-order gradient	Bidirectional, given by $\cos\theta$	Carbon	$e \propto x$	Resistive	L, M, H
		Dynamic	$e \propto \dot{x}$	Resistive	L, M, H
Antinoise, second-order gradient	Bidirectional, given by $\cos^2\theta$	Dynamic	$e \propto \dot{x}$	Resistive	L, M, H
Line	Unidirectional, highly directional	Ribbon	$e \propto \dot{x}$	Resistive	L, M, H
		Dynamic	$e \propto \dot{x}$	Resistive	L, M, H
Reflector	Unidirectional, highly directional	Dynamic	$e \propto \dot{x}$	Resistive	L, M, H

*Operating impedance is that obtainable with a transformer. L = low impedance, <100 ohms; M = medium impedance, >100 ohms and ≤1000 ohms; and H = high impedance, > 1000 ohms.

characteristics of microphones are summarized in the table.

Open-circuit response. This is the open-circuit voltage output of a microphone as a function of sound frequency when the microphone is placed in a plane wave sound field of constant pressure. It is expressed in decibels.

Electrical impedance. This characteristic is the complex ratio of the voltage applied across the electrical terminals to the resulting current as a function of frequency.

Directional characteristic. This is the open-circuit voltage output with respect to some reference axis of the microphone. The directional characteristics may be depicted as either polar characteristics at different frequencies or frequency-response characteristics at different angles.

Nonlinear distortion. This characteristic is the ratio of the total rms harmonic distortion output to the rms of the fundamental output. It may be expressed as a function of the frequency at different sound pressures or as a function of the sound pressure at different frequencies.

Noise characteristic. This is the open-circuit voltage generated in a given frequency band in the absence of an acoustical or electrical input.

The noise spectrum level at a given frequency is the open-circuit voltage in a one-cycle bandwidth centered on that frequency in terms of the rms sound pressure which would generate the same voltage.

Other characteristics. There are other characteristics such as the acoustical impedance characteristic and the transient response characteristic which are also important but more difficult to measure and specify. In some cases the type of cable used also affects the performance.

CALIBRATION

The performance characteristics of microphones have been outlined in the preceding section. The means for determining the performance characteristics will be described in the following sections.

Response frequency characteristic. The pressure response frequency characteristic of a microphone is the ratio e/p as a function of the frequency where e is the open-circuit voltage, in volts, generated by the microphone and p is the sound pressure, in microbars, which actuates the microphone. The ratio e/p is usually expressed with respect to some arbitrary reference level. The pressure on the diaphragm may be generated by a pistonphone, thermophone, or an electrostatic actuator.

A schematic arrangement of a pistonphone for use in calibrating a pressure-type microphone is shown in Fig. 26a. The small piston is driven by a crank. The pistonphone method is useful for calibrating microphones in the low-frequency range. The upper limit is governed by the permissible speed of the mechanical system, which is approximately 200 Hz.

The thermophone consists of one or more strips of thin gold leaf mounted upon terminal blocks, as shown in Fig. 26b. The thermophone strip carries a steady current upon which the sinusoidal current

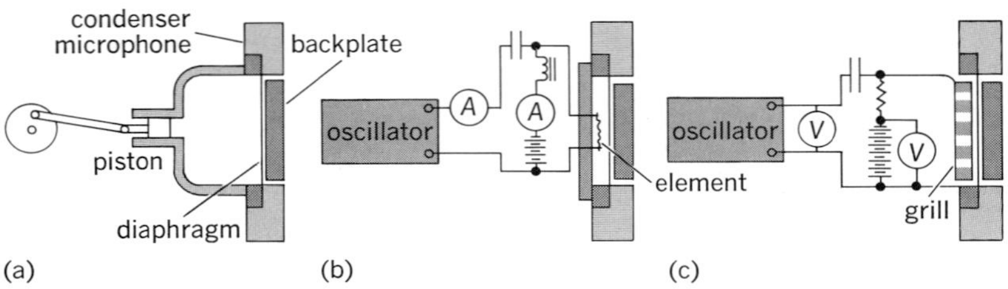

Fig. 26. Apparatus for obtaining the pressure-frequency of a condenser-type microphone. (a) Pistonphone. (b) Thermophone. (c) Electrostatic actuator. The pistonphone and thermophone may be used for other types of pressure microphones.

is superimposed. The periodic heating of the gold leaf produces a variation in sound pressure in the chamber which acts upon the diaphragm.

The electrostatic actuator consists of an auxiliary electrode in the form of a grill mounted in front of the microphone diaphragm, as shown in Fig. 26c. A large polarizing voltage is applied between the grill and microphone diaphragm. The sinusoidal voltage is applied effectively in series with the polarizing voltage and thereby produces an alternating force upon the microphone of the frequency of the alternating voltage.

The acoustical reciprocity procedure has replaced the pistonphone, thermophone, and electrostatic actuator for most calibration procedures. The reciprocity procedures are depicted in Fig. 27.

In Fig. 27a an alternating current is fed to the loudspeaker S_2. A sound pressure p_1 having an acoustical capacitance C_A is produced in the volume. Let the open-circuit voltage, in statvolts of S_1 used as a microphone be designated as e_s. In Fig. 27b the same alternating current is fed to the loudspeaker S_2, as in Fig. 27a. Let the open-circuit voltage in statvolts of the microphone M be designated as e_M. In Fig. 27c, S_1 is used as a loudspeaker. A current of i statampères is fed to S_1. The output voltage of the microphone M is designated as e'_M.

The response K_M of the microphone M, in statvolts per microbar, is given by Eq. (4), where e_s, e_M,

$$K_M = \sqrt{\frac{j\omega C_A e_M e'_M}{e_s i}} \qquad (4)$$

e'_M, and i are determined from the experiments depicted in Fig. 27 and $\omega = 2\pi f$, f being the frequency in hertz. The acoustical capacitance C_A is obtained from the dimensions of the cavity. The units are as follows: the voltages in statvolts, the currents in statampères, and the acoustical capacitance in (centimeters)⁵ per dyne. Since all transducers in Fig. 27 are assumed to be electrostatic, it is convenient to use the centimeter-gram-second

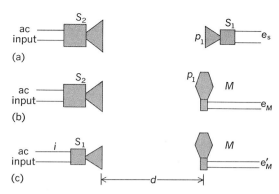

Fig. 28. The three experiments of the reciprocity procedure for obtaining the free-field calibration of a microphone. (a) The open-circuit voltage e_s of the reversible microphone loudspeaker S_1 when used as a microphone and actuated by a sound pressure p_1. (b) The open-circuit voltage e_M of the microphone M to be calibrated, when actuated by a sound pressure p_1. (c) The open-circuit voltage e'_M of the microphone M to be calibrated, when actuated by a sound pressure produced by the reversible microphone loudspeaker S_1 used as a loudspeaker with a current input i and a spatial separation d.

(cgs) electrostatic system. However, Eq. (4) applies to other transducers and other unit systems, provided the proper stipulations are satisfied.

The free field calibration by means of the reciprocity procedure is shown in Fig. 28. The transducer S_2 is a dynamic loudspeaker. S_1 is a dynamic microphone-loudspeaker. The microphone to be calibrated is designated as M.

The response K_M of the microphone M, in abvolts per microbar, is given by Eq. (5), where e_s, e_M,

$$K_M = \sqrt{\frac{2d\lambda e_M e'_M}{r_A i e_s}} \qquad (5)$$

e'_M, and i are obtained from the experiments in Fig. 28. The units are as follows: voltages in abvolts, currents in abampères, distance d in centimeters, wavelength λ in centimeters, and acoustical resistance $r_A = 41.5$.

Electrical impedance characteristics. The electrical impedance frequency characteristic of a microphone is the electrical impedance at the output terminals as a function of the frequency. Any convenient method for measuring electrical impedance may be used for determining the electrical impedance frequency characteristic.

Directional characteristic. The directional characteristic of a microphone is an expression of the variation of behavior of the microphone with respect to direction. A polar diagram showing the output variation of the microphone with direction is usually employed. These polar diagrams are obtained at specific frequencies and are determined from the response obtained at different angles with respect to some reference axis of the microphone.

Nonlinear distortion characteristic. The nonlinear distortion characteristic of a microphone is a plot of the total nonlinear distortion in percent of the fundamental with respect to the frequency. The nonlinear distortion components at the output of the microphone are measured for a pure tone sound input to the microphone.

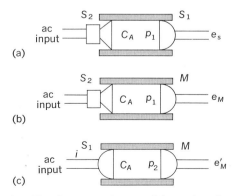

Fig. 27. The three experiments of the reciprocity procedure for obtaining the pressure calibration of a microphone. (a) The open-circuit voltage e_s of the reversible microphone loudspeaker S_1 when used as a microphone and actuated by a sound pressure p_1. (b) The open-circuit voltage e_M of the microphone M to be calibrated, when actuated by a sound pressure p_1. (c) The open-circuit voltage e_M' of the microphone M to be calibrated, when actuated by a sound pressure p_2 produced by the reversible microphone loudspeaker S_1 used as a loudspeaker with a current input i and a volume coupling C_A.

Noise characteristic. The noise generated by a microphone is measured by determining the electrical output in the absence of a sound input to the microphone.   [HARRY F. OLSON]

Bibliography: M. L. Grayford, *Electroacoustics*, 1971; H. F. Olson, *Modern Sound Reproduction*, 1971, reprint 1978; H. F. Olson, *Music, Physics and Engineering*, 1967; G. Porges, *Applied Acoustics*, 1979.

Microprocessor

A central processor unit (CPU) on a single integrated circuit (IC) chip. The place of the microprocessor in the development of computer technology is shown in the time chart in Fig. 1. The physical size of a typical microprocessor wafer is around 5 mm × 5 mm, but may vary from one microprocessor to another. Figure 2 is a magnified view of a typical microprocessor. Based on its word size (bits), the commercially available microprocessors can be classified as 4-bit, 8-bit, 16-bit, or bit-sliced. The physical size and the number of pins of a microprocessor chip are proportional to its word size. The longer the word size, the bigger the chip, and the greater the number of pins required. The first and shortest word-size microprocessor was the Intel 4004, manufactured by Intel Corporation in 1970. Since then, the microprocessor has grown continually bigger and more powerful. Table 1 shows the years in which the various microprocessors became available and the number of pins of each of the microprocessor chips.

Architecture and instruction set. Each microprocessor has its own architecture, which encompasses the general layout of its major components, the principal features of these components, and the manner in which they are interconnected. How-

Fig. 2. A typical microprocessor (Intel 8085). (*Intel Corp.*)

Table 1. Types of microprocessors

Microprocessor	Year available	Number of pins
4-bit	1970	24
8-bit	1973	40
16-bit	1977	40, 48, 64
bit-sliced	1975	40, 48

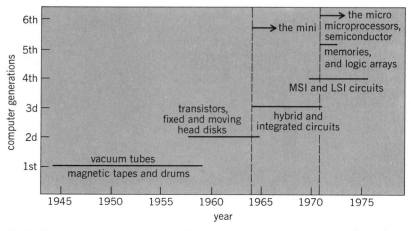

Fig. 1. Time chart of computer technology.

ever, the major components of all microprocessors are the same: (1) the clock; (2) control unit, comprising the program counter (PC), instruction register (IR), processor status word (PSW), and stack pointer (SP); (3) control memory; (4) bus control; (5) working register; (6) arithmetic/logic unit (ALU); (7) internal memory or stack. Their general layout is shown in Fig. 3. They are also called the hardware of a microprocessor. Besides the hardware, a microprocessor has its instruction set or software. According to the type of operations that the instructions perform, an instruction set can be subdivided into group operations for arithmetic, data transfer, branching, logic, and input/output (I/O). A set of logically related instructions stored in memory is referred to as a program. The microprocessor "reads" each instruction from memory in a logically determinate sequence, and uses it to initiate processing actions.

Fundamentals of operation. Even though the architectures and instruction sets of different microprocessors are different, their fundamentals of operation are the same. For example, the pin configuration and the block diagram of the architecture of Intel 8085, one of the commonly used microprocessors, are shown in Fig. 4. The 8085 contains a register array with both dedicated and general-purpose registers: (1) a 16-bit program counter (PC); (2) 16-bit stack pointer (SP); (3) six 8-bit general-purpose registers arranged in parts: BC, DE, HL; (4) temporary register pair: WZ; (5) serial I/O register pair; (6) interrupt control register.

The 16-bit program counter fetches instructions from any one of 2^{16} or 65,536 possible memory locations. When the RESET IN pin of 8085 is made logic 0, the program counter is reset to zero; when

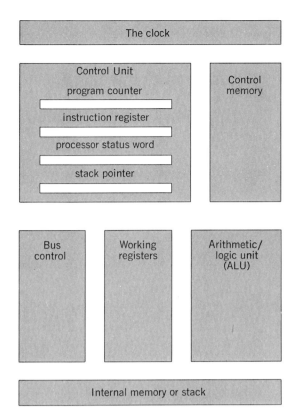

Fig. 3. Components of a microprocessor.

the RESET IN pin is returned to logic 1, the control unit transfers the contents of the PC to the address latch, providing the address of the first instruction to be executed. Thus, program execution in the 8085 begins with the instruction in memory location zero.

8085 instructions are 1 to 3 bytes in length. The first byte always contains the operation code (OP code). During the instruction fetch, the first byte is transferred from the memory by way of the external data bus through the data bus buffer latch into the instruction register. The PC is automatically incremented so that it contains the address of the next instruction if the instruction contains only 1 byte, or the address of the next byte of the present instruction if the instruction consists of 2 or 3 bytes.

In the case of a multibyte instruction, the timing and control section provides additional operations to read in the additional bytes. The timing and control section uses the instruction decoder output and external control signals to generate signals for the state and cycle timing and for the control of external devices. After all the bytes of an instruction have been fetched into the microprocessor, the instruction is executed. Execution may require transfer of data between the microprocessor and memory or an I/O device. For these transfers, the memory or I/O device address placed in the address latch comes from the instruction which was fetched or from one of the register pairs used as a data pointer: HL, BC, or DE.

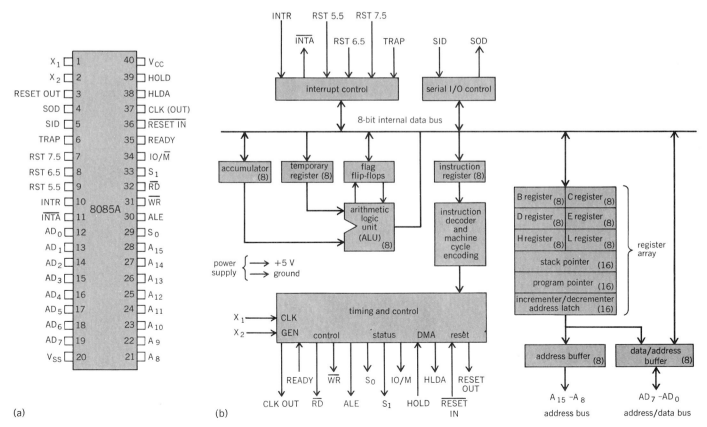

(a)

(b)

Fig. 4. Intel 8085 microprocessor. (a) Pin configuration. (b) Block diagram. (*Intel Corp.*)

The six general-purpose registers in the register array can be used as single 8-bit registers or as 16-bit register pairs. The temporary register pair, WZ, is not program-addressable and is only used by the control unit for the internal execution of instructions. For example, to address an external register for a data transfer, WZ is used to hold temporarily the address of an instruction read into the microprocessor until the address is transferred to the address and address/data latch.

The 16-bit stack pointer, SP, always points to the top of the stack allocated in external memory. The stack, as previously indicated, primarily supports interrupt and subroutine programming.

The 8085's arithmetic/logic unit performs arithmetic and logic operations on data. The operands for these operations are stored in two registers associated with the ALU: the 8-bit accumulator and the 8-bit temporary register. The accumulator is loaded from the internal bus and can transfer data to the internal bus. Thus, it serves as both a destination and source register for data. The temporary register stores one of the operands during a binary operation. For example, if the contents of register B are to be added to the contents of the accumulator and the result left in the accumulator, the temporary register holds a copy of the contents of register B while the arithmetic operation is taking place.

Associated with the ALU is the 5-bit flag register, F, which indicates conditions associated with the results of arithmetic or logic operations. The flags indicate zero, a carry-out of the high-order bit, the sign (most significant bit), parity, and auxiliary carry (carry-out of the fourth bit). *See* DIGITAL COMPUTER PROGRAMMING.

Bit-sliced microprocessors. Bit-sliced microprocessors achieve a high performance level which the single-chip microprocessors are unable to provide. The major logic of a central processor is partitioned into a set of large-scale-integration (LSI) devices as opposed to being placed on a single chip. The chip set is used as the basic building block to construct a microprogrammed central processor which can be configured in various ways. Unlike the single-chip microprocessor, several identical processor elements can be wired in parallel to achieve a desirable word length. The microprogrammed architecture allows the addition of new instructions without modifying the hardware wiring, and consequently provides a flexibility that is very desirable.

Digital system design. A list of some available microprocessors is given in Table 2. The main dif-

Table 2. Some available microprocessors

Manufacturer	Device	Pins	Bit slice	Data word size	Memory capacity, bits*	Microprogrammed
Burroughs	Mini-D	16	No	8	256	Yes
Fairchild	F-8	40	No	4/8	64K	No
Semiconductor	PPS-25	16/18/24/40	No	4	12K	No
General Instrument	CP-1600	40	No	16	64K	Yes
Intel	4004	16/24	No	4	4K	No
	4040	24	No	4	4K	No
	8008	18	No	8	16K	No
	8080	40	No	8	64K	No
	8048/8748	48	No	8	64K	Yes
	8085	40	No	8	64K	No
	8086	40	No	8/16	64K	No
	8088	40	No	8/16	256	No
	8089	40	No	8/16	256/64K	No
	3000	28	Yes	2-bit slice	64K	Yes
Intersil	6100	40	No	12	4K	No
Monolithic Memories	5701 – 6701	40	Yes	4-bit slice	64K	No
Mostek	5065	40	No	8	32K	No
Motorola	M6800	40	No	16	64K	No
	M68000	64	No	16	16,000K	No
	M10800	48	Yes	4-bit slice	64K	Yes
National	IMP4	24	Yes	4	64K	Yes
	IMP8	24	Yes	8	64K	Yes
	IMP16	24	Yes	16	64K	Yes
	PACE	40	No	8/16	64K	Yes
RCA	COSMAC	28/40	No	8	64K	No
Raytheon	RP16	48	Yes	4-bit slice	64K	No
Rockwell	PPS-4	42	No	4	12K	No
	PPS-8	42	No	8	12K	No
Signetics	2650	40	No	8	32K	No
Texas Instruments	TMS 1000	28/40	No	4	8K	No
	SBP 0400	40	Yes	4-bit slice	64K	Yes
	TMS 9900	64	No	16	32K	No
Toshiba Transistor Works	TLCS 12	16/24/26/42	No	4-bit slice	4K	Yes
Zilog	Z80	40	No	8	64K	No
	Z8000	48	No	16	8000K	No

*1K = 1024 bits.

ference between the microprocessor digital system design and the hard-wired logic digital system design is that the former uses the microprocessor to replace hard-wired logic by storing program sequences in the read-only memory (ROM) rather than implementing these sequences with gates, flip-flops, counters, and so on. After the design is completed, any modifications or changes may be made by simply changing the program in the ROM. The microprocessor digital system design is now widely used because of the following advantages:

1. Manufacturing costs of industrial products can be significantly reduced.

2. Products can get to the market faster, providing a company with the opportunity to increase product sales and market share.

3. Product capability is enhanced, allowing manufacturers to provide customers with better products, which can frequently command a higher price in the marketplace.

4. Development costs and time are reduced.

5. Product reliability is increased, while both service and warranty costs are reduced. *See* SEMICONDUCTOR MEMORIES.

Applications. The list of possible application areas includes: industrial sequence controllers; machine tool controllers; point-of-sale terminals; intelligent terminals; instrument processors; traffic light controllers; weather data collection systems; and process controllers. *See* DIGITAL COMPUTER; INTEGRATED CIRCUITS; MICROCOMPUTER.

[SAMUEL C. LEE]

Microwave

An electromagnetic wave which has a wavelength in the centimeter range. Microwaves occupy a region in the electromagnetic spectrum which is bounded by radio waves on the side of longer wavelengths and by infrared waves on the side of shorter wavelengths (Fig. 1). There are no sharp boundaries between these regions except by arbitrary definition. In the microwave region, a further delineation is sometimes made with such names as decimeter, centimeter, or millimeter waves.

The historical development of microwaves is only a particular phase of the gradual evolution of the concept and application of electromagnetic waves in general. The foundation for the entire field was laid by James Clerk Maxwell in 1864, when he formulated a set of equations governing electromagnetic phenomena which became known as Maxwell's equations. A solution of these equations led Maxwell to predict the existence of electromagnetic waves when none were known to exist at that time. The next decisive development came when Heinrich Hertz demonstrated in 1888 an experimental proof for the existence of electromagnetic waves and verified substantially all aspects of Maxwell's predictions. Once the validity

of the basic theory was established, further generalizations and applications became a matter of technical development.

In his classical experiments Hertz used damped electromagnetic waves in the decimeter or meter range. It might seem that this demonstration was the starting point for modern microwave techniques. Actually, the detailed development of microwaves lagged many years behind that of radio waves and got its start only in the 1930s when continuous-wave microwave generators were invented and hollow waveguides and cavity resonators were introduced. The drastic change occurred not because of the selection of any particular portion of the electromagnetic spectrum but in the exercise of a new approach which was particularly amenable to the microwave range. This new approach resulted from an application of the solutions of Maxwell's equations for apparatus dimensions which are comparable to the microwave wavelength.

Frequency and velocity. Although an electromagnetic wave is commonly referred to in terms of its wavelength, a more fundamental characteristic of a wave lies in its frequency. As in any type of wave motion, these two quantities are related to a third quantity, velocity of propagation, by the simple relation $f\lambda = v$, where $f =$ frequency, $\lambda =$ wavelength, and $v =$ velocity of propagation.

It is usually simpler and more definitive to refer to a wave by its frequency than by its wavelength. The only exception is in the case of wave propagation in free space (in vacuum or, approximately, in air) where a wave propagates with the velocity of light (approximately 3×10^{10} cm/s); the wavelength is then a perfectly definitive quantity. According to usual practice, when a wavelength is mentioned without qualification, it means the free-space wavelength. Thus, for a microwave wavelength of 3 cm, the corresponding frequency is 10^{10} Hz.

Generation. Microwaves can be generated as direct radiation from electrical sparks across gaps by applying a high electric potential. The spark gap can also be a part of a very-high-frequency oscillating circuit which radiates electromagnetic waves. Microwaves can also be derived from the thermal radiation of warm bodies. But all these sources are unsatisfactory because of the lack of purity of the wave and the low power of the radiation. In contrast to these, all modern microwave generators are electronic devices which produce continuous-wave (CW) oscillations of a single tunable frequency. Some important microwave generators are known as klystrons, magnetrons, and traveling-wave oscillators. Their power outputs range from microwatts to thousands of kilowatts, depending upon the type and design of the generator and the operating frequency. *See* KLYSTRON; MAGNETRON; MICROWAVE SOLID-STATE DEVICES; MICROWAVE TUBE; TRAVELING-WAVE TUBE.

Circuit elements. Any particular grouping of physical elements which are arranged or connected together to produce certain desired effects on the behavior of microwaves is known as a microwave circuit. It should be noted that a microwave circuit or any of its elements is not "closed" in the sense of a low-frequency electric circuit. Figure 2

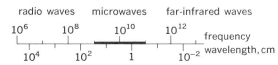

Fig. 1. A portion of the electromagnetic spectrum.

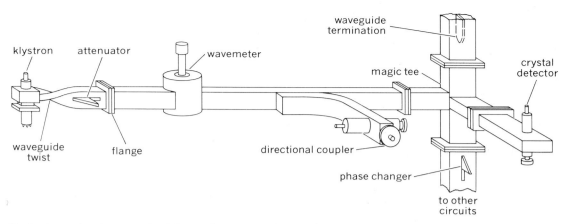

Fig. 2. A sample microwave circuit illustrating the use of some microwave components.

shows a microwave circuit in which some circuit elements to be described later are illustrated.

Microwave waveguides. A waveguide is a circuit element which constrains or guides the propagation of microwaves along a path defined by the physical construction of the guiding element. It can be, for instance, a coaxial cable having an outer conductor of annular cross section coaxially placed with respect to an inner conductor. By prevailing usage, however, a microwave waveguide usually means a hollow metallic tube which can confine and guide the propagation of microwaves.

When an electromagnetic wave propagates in a hollow waveguide, the electric and magnetic intensities (denoted by E and H, respectively) must satisfy both Maxwell's equations and certain boundary conditions. The result is that only certain specific patterns for the distribution of E and H (taken together) can exist in the waveguide. Each unique pattern of field distribution is called a mode. There are two types of mode possible in a hollow waveguide—one of them being the transverse electric (TE) mode and the other the transverse magnetic (TM) mode. *See* WAVEGUIDE.

One of the most important characteristics of a TE or TM wave is that it has a cutoff wavelength for each mode of transmission. If the free-space wavelength is longer than the cutoff value, that particular mode cannot exist in the waveguide. For any given waveguide, the mode that has the longest cutoff wavelength is known as the dominant mode. Figure 3 illustrates the dominant mode TE_{01} for a rectangular waveguide.

Rectangular waveguides having cross sections uniform through their entire length are by far the commonest in use. Waveguide sections may be straight, bent, or twisted. Several waveguides may form a network such as a T junction, a magic tee, or a directional coupler. *See* DIRECTIONAL COUPLER.

Attenuators. A microwave attenuator is a device that causes the field intensity of a wave to decrease by absorbing a part of the incident power. This objective is usually accomplished by inserting a piece of lossy material in the waveguide along the direction of electric intensity. The lossy material can be simply a sheet of material containing powdered carbon whose electrical conductivity is in the proper range to cause the desired attenua-

tion. For an adjustable amount of attenuation, it is usually arranged that the absorbing sheet can be mechanically moved by varying degrees in or out of the path of wave propagation.

Phase changers. A phase changer is a device that causes the field intensity to shift its phase without attenuating its amplitude. This can be done by inserting a thin slab of low-loss dielectric material parallel to the direction of electric intensity or, for small phase shifts, a metal pin in the form of a screw can be used to penetrate variable amounts into the waveguide. Still another way of changing the phase of a wave is to vary the path length of a reflected wave by moving a short-circuiting plunger (a metal block fitted snugly in the waveguide to reflect totally all incident waves).

Detectors. A microwave detector is a device that can demonstrate the presence of a microwave by a

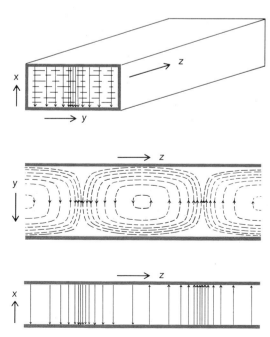

Fig. 3. Instantaneous field pattern for the TE_{01} wave in a rectangular waveguide. The wave propagates in the z direction. Solid lines indicate the electric intensity E, and dashed lines the magnetic intensity H.

specific effect that the wave produces. One of the most effective means of detection makes use of a nonlinear device which converts the microwave field intensity into either a direct current or a low-frequency alternating current. A silicon crystal making a pinpoint contact with a tungsten wire is perhaps the most commonly used detector for microwaves. This semiconducting material has the nonlinear characteristic that the resistance for a potential of one polarity is drastically different from that for the opposite polarity. Hence, an alternating potential produces unequal currents in the positive and negative portions of the cycle, giving an average current which can be indicated by a direct-current meter. A typical crystal detector gives a current of the order of 1 milliampere (ma) for a microwave power input of 1 milliwatt (mw). One important reason why a crystal detector is particularly adaptable to the detection of high-frequency waves, including microwaves, is the exceedingly small capacity in the contact area between the tungsten point and the crystal. The contrary case of a larger capacity would cause the high-frequency currents to bypass the detector, which would then become ineffective.

A crystal detector for microwaves is usually placed in a special waveguide mount which consists of a wire post positioned along the direction of strongest electric intensity in the waveguide. *See* DETECTOR.

Another type of microwave detector in common use is called a bolometer. It is a device whose resistance changes sensitively with temperature, the latter being a function of the absorbed microwave power. The bolometer can be used as one arm of a resistance bridge circuit in which any change in resistance causes an unbalance in potential or current, which can be measured by a suitable meter.

Probes. Either of the aforementioned microwave detectors can be used in a waveguide where the detector is designed to absorb almost all the power that impinges on it. In this case the detector is said to form a matched termination for the waveguide. At other times, however, the object of the operation is to probe into the character of the field intensity at any point with a minimum disturbance of the field distribution. Under these conditions a detector is used in conjunction with a small pin which is inserted into the field by the least amount that makes detection possible. Such a detector is called a probe. It is used primarily for measuring the standing-wave ratio, which gives a measure of the interference pattern formed by the incident and reflected waves.

Wavemeters. A microwave wavemeter is usually made of a tunable resonant cavity which measures the free-space wavelength (or frequency) of a wave by the position of tuning at resonance. The resonance frequency relative to the tuning position can be calculated accurately from the known dimensions of the cavity or calibrated against a known frequency standard. *See* CAVITY RESONATOR.

Transmission. A microwave transmitter is similar in all its principal aspects to an ordinary radio transmitter. It consists of a microwave generator, a power amplifier (if necessary), a circuit containing all necessary elements, means of modulation to impart some form of information or program to the waves, and an antenna network to send the waves

out into space. Microwave generators and various circuit elements have been mentioned in the preceding sections. Since World War II, klystron amplifiers have come into practical use at high powers, and traveling-wave amplifiers at low powers.

Modulation processes. A microwave can be transmitted as a continuous wave without any form of modulation. As a means of communication, a CW transmission is useful only in coded messages such as dots and dashes when the transmission is keyed on and off. To enrich the information content of a transmission, methods involving amplitude modulation (AM), frequency modulation (FM), or phase modulation (PM) can be used. In amplitude modulation the amplitude of the wave is varied at some rate (audio or video) while the frequency is kept constant. In frequency modulation, the frequency of the wave is varied at some rate while the amplitude is kept constant. Phase modulation is so similar to frequency modulation in many respects that a distinction between the two will not be drawn here. Generally speaking, the net result of each type of modulation is to form two sidebands (each sideband containing instantaneously more than one side frequency for frequency modulation) that are symmetrically located around the central carrier frequency, which is provided by the generator. Thus, the whole transmission occupies a channel width twice as wide as each sideband. *See* AMPLITUDE MODULATION; FREQUENCY MODULATION; PHASE MODULATION.

Pulse operation of microwave transmitters is a special modulation process which is important enough to deserve particular mention. Transmission by pulsed waves is an efficient method of sending an extraordinarily large peak power at the expense of a small average power. For example, if a microwave generator is operated at the rate of 500 pulses per second and each pulse lasts only for 1 μsec, then it can deliver a peak power of 200 kw at the expense of only 100 watts. But even apart from the power considerations, pulse transmission is the most effective way of securing echo signals by reflection, as in radar. *See* PULSE MODULATION.

Microwave antennas. A transmitting antenna is a circuit element which transfers the wave excitation from the generator into waves propagating in free space. There are two major considerations in the design of an antenna system: sensitivity and directivity. Sensitivity of a transmitting antenna is measured in terms of the ratio of the power delivered into space to the available excitation power and naturally should be made as high as possible. Directivity refers to the pattern of the field intensity distribution in all directions at a fixed radial distance from the antenna. Although the degree of directivity depends upon the specific requirements of a given situation, highly directive antennas are necessary in most microwave applications. Figure 4 illustrates a simple microwave antenna using a parabolic reflector. *See* ANTENNA; MICROWAVE TRANSMISSION LINES.

Propagation in space. The most important fact about electromagnetic wave propagation in space is that a wave tends to travel along a straight line in the absence of obstructions. In the application of microwaves, straight line, or line of sight, propagation in the atmosphere is often implicitly assumed.

MICROWAVE

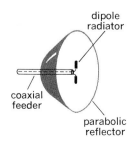

Fig. 4. A microwave antenna consisting of parabolic reflector, coaxial feeder, and dipole radiator.

However, if the transmitted beam either wholly or partially hits the Earth's surface or any physical obstacle, it will be reflected, with a possible loss of energy, depending on the shape, size, and the reflective properties of the obstructing medium. In fact, it would be even more correct to say that the wave is in general scattered by various objects with the particular phenomenon variously described as reflection, refraction, or diffraction. For an explanation and discussion of these phenomena *see* MICROWAVE OPTICS.

A similar situation arises when a microwave has a "sky" component which will be little affected by the atmosphere in the troposphere and stratosphere regions but will be more strongly influenced by the ionosphere. Unlike radio waves of lower frequencies, which are often reflected by the ionosphere, a microwave has too high a frequency to be reflected by this medium at angles near vertical incidence. However, for microwaves slanted toward the horizon, reflection by the ionosphere becomes possible.

Although the attenuation of a microwave by the Earth's atmosphere is usually small, it may be selectively large for certain bands of microwaves. For instance, the absorption by water vapor is selectively high for waves in the K-band (wavelength around 1 cm) and that by oxygen is high for waves in the so-called K/2-band (wavelength around 1/2 cm). These considerations are important in the choice of wave bands for the purpose of microwave communication.

Reception. The microwave transmitting antennas mentioned earlier can be used for reception, where the direction of wave propagation is in the reverse order. Since the received wave is often appreciably weaker than the transmitted wave, reliable reception depends upon the sensitivity of the receiver in picking up weak signals in the midst of noises. The receiver sensitivity is specified by the minimum detectable voltage at the input end of the receiver. A typically good receiver has a sensitivity of about 1 μv. In most cases such a receiver makes use of a superheterodyne scheme which involves a local microwave oscillator beating with the incoming wave to form a so-called intermediate-frequency (i-f) signal for amplification and further detection. A crystal is generally used for the mixer, while automatic frequency control (AFC) is often used to keep the i-f constant in frequency. The bandwidth for the amplifier should be wide enough to pass both sidebands of the signal, but not much wider than necessary in order to minimize the noise.

Radar. Radar is the abbreviated name for radio detection and ranging. This apparatus is used to detect the presence and the range of a reflecting object by transmitting a pulsed microwave and receiving the reflected wave of the same pulse. Also, the direction of the reflecting object can be determined from the orientation of the directive antenna at maximum reception intensity. Initial development of radar began secretly around 1932, but active use of radar first occurred around 1940 during World War II. *See* RADAR.

There are many types of oscilloscope presentations used for various purposes in radar. There is in particular a special presentation known as plan-position indicator (PPI) in which the echo strength is made to intensify a spot on the oscilloscope screen while the position of the spot corresponds to specific orientations of the scanning antenna. When used in an airborne radar, PPI presentation gives a panoramic view of the surface structures of the ground.

Advantages of microwaves. From the point of view of radio communication, a great advantage of microwaves lies in the immense spaciousness of useful frequencies. For instance, the frequency difference between the S-band (wavelength around 10 cm) and K-band (wavelength around 1 cm) is roughly 20,000 MHz. This frequency "space" is about 100 times the combined frequency range of present-day radio broadcasting, communication, and television. There is a good prospect that this immense range can be increased 5–10 times by further improvement of power generation at the high-frequency end of the microwave range. Another advantage of microwaves is related to the high directivity and resolving power of microwave wavelengths. Narrow microwave beams can be readily formed by antennas of physically convenient sizes. The superiority of microwaves over the waves of much longer wavelengths, with respect to radiation directivity, is made evident by the very large antenna structures necessary for beam transmission at longer wavelengths. Moreover, the resolving power, as measured by the ability of a wave to differentiate one reflecting object from another, is larger for microwaves than for longer waves.

The reader may wonder at this point: If the advantages of microwaves are attributed to properties of very short waves, why is it that shorter waves like infrared radiation would not be even superior for various applications? The answer is that infrared radiation is being used in many applications where its advantages are well proved. However, there are some major differences between these regions of radiation. Unlike microwaves, infrared radiation is usually nonmonochromatic in frequency, noncoherent in phase, and not easily subject to manipulations like frequency modulation, amplification, and electronic control. Also, infrared radiation probably cannot compare favorably with microwaves with regard to power concentration at a given frequency and range of propagation in the atmosphere. These factors tend to make microwaves a more useful tool for communication purposes.

Applications of microwaves. The most important practical application of microwaves is radar. In fact, microwave techniques have been developed mainly under the incentive of radar work. Next to radar the use of microwaves as carrier waves in relay links for multichannel transmission of telephone, telegraph, and television has already become practically important.

In postwar years, application of microwaves has flourished in many scientific and basic research fields. Microwave spectroscopy has become an established science chiefly because of the availability of microwave instruments and techniques. The knowledge of the structure of numerous molecules and crystals has been greatly increased by this new branch of spectroscopy. To mention a celebrated case, the fine structure of the hydrogen atom known as the Lamb shift was discovered by the use of microwaves. Another important advance came with the development of the so-called atomic clocks, which use microwave resonance interactions with either cesium atoms or ammonia molecules. The introduction of the ammonia maser led to the development of the solid-state maser, which is virtually a noiseless amplifier. One of the most important applications of the maser is in the field of radio astronomy, which is in its own right a fertile ground for microwave application. For instance, the hyperfine line of the hydrogen atom at 1420 MHz has been observed in stellar radiation with refined microwave techniques. In the field of nuclear physics instrumentation, microwave electronics has been applied in a most significant manner to high-energy linear accelerators and similar devices. *See* MASER.

[C. K. JEN]

Bibliography: A. J. Baden-Fuller, *Microwaves: An Introduction to Microwave Theory and Techniques,* 1979; K. C. Gupta, *Microwaves,* 1980; J. Helszajn, *Passive and Active Microwave Circuits,* 1978; T. Laverghetta, *Microwave Measurements and Techniques,* 1976; S. Liao, *Microwave Devices and Circuits,* 1980; R. F. Soohoo, *Microwave Electronics,* 1971; H. E. Thomas, *Handbook of Microwave Techniques and Equipment,* 1972.

Microwave optics

The study of those properties of microwaves which are analogous to the properties of light waves in optics. The fact that microwaves and light waves are both electromagnetic waves, the major difference being that of frequency, already suggests that their properties should be alike in many respects. But the reason microwaves behave more like light waves than, for instance, very low-frequency waves for electrical power (50 or 60 Hz) is primarily that the microwave wavelengths are usually comparable to or smaller than the ordinary physical dimensions of objects interacting with the waves.

In his classical experiments to verify Maxwell's theory, H. Hertz first demonstrated the optical properties of damped decimeter waves, such as rectilinear propagation, reflection, refraction, and polarization. It is virtually taken for granted that microwaves inherently possess all these properties, and the language of geometrical or physical optics is freely used wherever allowed by the situation.

Rectilinear propagation. As is the case with light, a beam of microwaves propagates along a straight line in a perfectly homogeneous infinite medium. This phenomenon follows directly from a general solution of the wave equation in which the direction of a wave normal does not change in a homogeneous medium. In the use of radar, a microwave beam is justifiably presumed to travel in a straight line before and after reflection by an object. For microwave communication in cases when two distant stations are not along the line of sight, the waves would be blocked by the Earth's surface. The difficulty is remedied by the use of microwave relay links so that straight-line propagation is maintained in each section. *See* MICROWAVE TRANSMISSION LINES.

Reflection and refraction. Consider a plane boundary between two semi-infinite media having different physical properties (Fig. 1). If a plane-po-

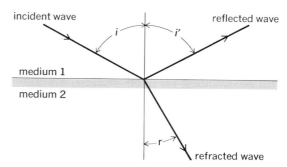

Fig. 1. Reflection and refraction of microwaves at a plane boundary between two insulating media.

larized microwave is incident from medium 1, the boundary conditions generally require the presence of a reflected wave back to medium 1 and a transmitted (refracted) wave into medium 2. In the case of two insulating media, the familiar relations in optics shown in Eqs. (1) and (2) hold, where the

$$i = i' \qquad \text{for reflection} \qquad (1)$$

$$\frac{\sin i}{\sin r} = \sqrt{\frac{\epsilon_2}{\epsilon_1}} = N \quad \text{for refraction} \qquad (2)$$

angles i, i', and r are as indicated in Fig. 1, ϵ_1 and ϵ_2 are the dielectric constants of media 1 and 2, respectively, and N is the index of refraction of medium 2 relative to medium 1. The reflected and refracted intensities depend upon whether the incident electric intensity is polarized in the plane of incidence or perpendicular to it. In any case, the well-known Fresnel equations of optics can all be applied to this case.

With some modification the laws of reflection and refraction can be applied to the propagation of microwaves inside a dielectric-filled metallic waveguide. The usual case is that of a vertical incidence to the plane boundary between two dielectric media perpendicular to the lengthwise direction of the waveguide. The reflection coefficient can be obtained by measuring the standing-wave ratio in the waveguide. Another interesting application is associated with the microwave analog of total internal reflection in optics. It may be seen from Eq. (2) that if $\epsilon_1 > \epsilon_2$ (that is, if the wave is incident from a denser medium), there is a total internal reflection for the wave when $i > \sin^{-1} \sqrt{\epsilon_2/\epsilon_1}$. This means that a properly designed dielectric rod (without metal walls) can serve as a waveguide by totally reflecting the elementary plane waves. Still another case of interest is that of a microwave lens. By using either a natural dielectric of a certain shape or an artificial dielectric consisting of an array of thin metal plates of a certain design, a microwave lens can be constructed which has the required index of refraction. Such lenses have been used as microwave antennas. *See* ANTENNA; WAVEGUIDE.

The reflection of a microwave by a conducting plane has all the characteristics of the reflection of a light wave by a metallic mirror. The amplitude of the reflected wave is practically identical to that of the incident wave, with the angle of reflection equal to the angle of incidence. The wave in the conducting medium does not go much beyond a

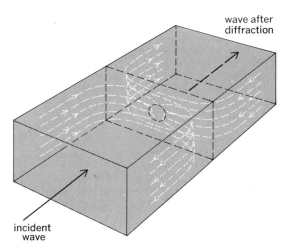

wave after
diffraction

incident
wave

Fig. 2. Diffraction of microwaves through a small aperture between two waveguides. Only magnetic lines of force are shown for the field pattern.

"skin depth" and is of little practical consequence. Examples of reflection of microwaves by conductors are found in parabolic reflectors used as antennas and in targets for radar beams.

Diffraction. In an analogous manner to light, a microwave undergoes diffraction when it encounters an obstacle or an opening which is comparable to or somewhat smaller than its wavelength. Diffraction problems have been much studied but the results are generally too complicated for a simple description. One case of considerable importance, however, may be cited as an illustration. Let two waveguides be coupled through a small hole in a metallic partition as shown in Fig. 2. The radius of the hole is assumed to be much smaller than $\lambda/2\pi$, where λ is the wavelength. A wave in one waveguide will leak through the hole by diffraction into the other waveguide. While an exact calculation is difficult, a satisfactory treatment can be worked out by regarding the diffraction effect as being equivalent to the presence of an electric and magnetic dipole placed at the position of the hole. The radiation fields of the dipoles supply the necessary wave coupling between the two waveguides.

Polarization. The polarization of an electromagnetic wave is specified by the direction of the electric and magnetic intensities. For simplicity, consider a plane wave propagating in the z direction. The electric and magnetic intensities are always mutually perpendicular to each other in the xy plane. It is, consequently, only necessary to consider the polarization of the electric intensity alone. If the electric intensity is polarized along one direction, say the x axis, then the wave is said to be plane polarized. If there are components E_x and E_y equal in amplitude but differing in phase by $90°$, the wave is circularly polarized (right-handed for E_y lagging E_x and left-handed for E_y leading E_x). Lastly, if the wave is neither plane nor circularly polarized, it must be elliptically polarized.

The preceding essentially optical description holds true for microwaves in free space or wherever there is a TEM wave. Since a hollow waveguide does not support a TEM wave, the description of the polarization of the wave is much more complicated. However, the general notions expressed here are still valid and useful. For instance, the electric intensity of the dominant mode (TE_{01}) in a rectangular waveguide is plane polarized. This situation is also approximately true for the dominant mode (TE_{11}) in a circular waveguide. A circular waveguide is particularly useful in transforming plane-polarized electric intensity into circularly polarized electric intensity by a technique equivalent to the use of a quarter-wave plate in optics. For this purpose, it suffices to use a thin slab of dielectric material, such as mica, to introduce a 90° phase shift for one of the two equal components of the electric intensity. For information on TEM waves *see* MICROWAVE.

Faraday effect for microwaves. In optics, the Faraday effect is the rotation of the plane of polarization of a light beam which propagates in a dense transparent medium placed in a magnetic field along the direction of propagation. In microwaves, a similar phenomenon has been discovered and has led to many interesting applications.

Consider, for example, a circular waveguide which contains a slender rod of ferrite (a magnetic material of very low conductivity), as shown in Fig. 3. Place a steady magnetic field along the axial direction. If a wave with vertical electric polarization is incident from the left, then in passing through the ferrite zone its plane of polarization is rotated by an angle θ as indicated in the figure. This Faraday effect can be explained by the action of precessing elementary magnets in the magnetized ferrite upon the phase of the propagating wave. The initial plane-polarized electric intensity is equivalent to two oppositely rotating circularly polarized components. Only one of the components is affected (principally in the phase factor) by the precessing magnets because the latter can only precess in a unique direction corresponding to a given direction of the magnetic field. In the output, the combination of two circularly polarized waves with a relative phase shift is equal to a plane-polarized wave having its polarization rotated by an angle with respect to the initial plane.

If the output wave is sent back to pass by the ferrite from right to left, the plane of polarization of the backward wave rotates another θ degrees in the same direction as the initial rotation. In other words, the polarization of the new output wave on the left is rotated from the initial plane by 2θ, instead of zero degrees as might be expected by the principle of reciprocity. Thus this system constitutes a nonreciprocal circuit element and is sometimes called a gyrator to signify the gyrating mo-

cylindrical wave guide

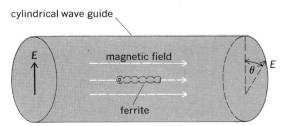

E

magnetic field

ferrite

θ

E

Fig. 3. Rotation of the plane of polarization of a microwave by a ferrite rod in a longitudinal magnetic field. E represents electric intensity.

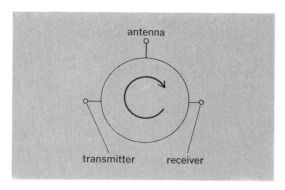

Fig. 4. A circulator formed by one or more gyrator elements for nonreciprocal application.

tion of the elementary magnets. *See* GYRATOR; RECIPROCITY PRINCIPLE.

One of the most important applications of a gyrator is found in unidirectional transmission. Suppose a plane-polarized wave is introduced from a rectangular to a circular waveguide containing a gyrator and finally to another rectangular waveguide oriented to suit the polarization of the output wave. If the angle of rotation, which defines the output polarization, is adjusted to 45°, any wave from the output which is fed back to the input will be at right angles with the input intensity and hence will not be accepted by the rectangular waveguide. Another important application of gyrators is known as a circulator, a simple example of which is given in Fig. 4. By means of one or more gyrators, a circulator such as the one depicted will allow a wave from the transmitter to go to the antenna but not to the receiver and will let a wave from the antenna go to the receiver but not to the transmitter. *See* MICROWAVE TRANSMISSION LINES. [C. K. JEN]

Bibliography: A. J. Baden-Fuller, *Microwaves: An Introduction to Microwave Theory and Techniques*, 1979; K. C. Gupta, *Microwaves*, 1980; L. Lewin, *Theory of Waveguides*, 1975.

Microwave reflectometer

A form of directional coupler that is used for measuring the power flowing in both directions in a waveguide. A pair of single-detector couplers appropriately positioned on opposite sides of the waveguide can be used for this purpose, with one detector positioned to monitor transmitted power and the other detector positioned to measure power reflected back from a discontinuity in the line. Each coupler receives a constant small fraction of the energy flowing in one direction in the waveguide, the energy being extracted from the waveguide through two small holes drilled 1/4 wavelength apart along the length of the guide. *See* HIGH-FREQUENCY IMPEDANCE MEASUREMENTS.

[JOHN MARKUS]

Microwave solid-state devices

Semiconductor devices for the generation, amplification, detection, and control of electromagnetic energy in the frequency range of about 1–100 GHz. Devices that generate or amplify microwave energy are referred to as active microwave de-

vices. The most important active solid-state microwave devices are microwave transistors, transferred electron (Gunn) devices, avalanche diodes, and tunnel diodes. Devices that detect or control microwave energy are referred to as passive microwave devices. The most important passive solid-state microwave devices include PIN diodes, varactor diodes, and point-contact and Schottky-barrier detector diodes. *See* MICROWAVE; MICROWAVE TUBE; SEMICONDUCTOR DIODE; TRANSISTOR.

ACTIVE DEVICES

At the lower microwave frequencies, silicon bipolar transistors and gallium arsenide (GaAs) field-effect transistors (FETs) are the most widely used active solid-state devices. At the higher microwave frequencies, the usefulness of transistors is severely limited by high-frequency effects, and two-terminal negative resistance devices such as transferred-electron devices, avalanche diodes, and tunnel diodes predominate. In a negative resistance, the current and voltage are 180° out of phase. A current I flowing through a negative resistance $-R$ produces a voltage rise $V = -IR$, and a power $P = I^2R$ will be generated by the power supply associated with the negative resistance. In microwave applications, the negative-resistance device is usually mounted in a circuit placed at the end of a transmission line, and a ferrite circulator

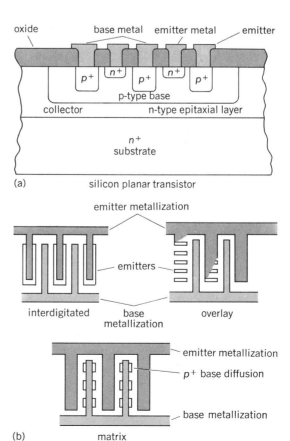

Fig. 1. Planar silicon bipolar-power microwave transistor. (*a*) Cross section of transistor. (*b*) Surface geometries. (*From H. Sobol and F. Sterzer, Microwave power sources, IEEE Spectrum, 9(4)20–33, 1972*)

is used to separate the input power from the amplified output power. *See* FERRITE DEVICES.

Silicon bipolar transistors. Most modern silicon bipolar microwave transistors are epitaxial diffused *npn* structures. Fig. 1*a* shows a cross-sectional view of a modern microwave transistor, a double-diffused epitaxial device. An epitaxial *n* layer is deposited on a heavily doped n^+ substrate. A *p* base is diffused into the epitaxial layer and many heavily doped *n* emitters are diffused into the base with p^+ plugs used to reduce the base spreading resistance. Silicon oxide layers are used as diffusion masks and for insulation. Contacts are made through openings in the oxide. Figure 1*b* shows three surface geometries that are widely used in microwave power transistors. The width of the emitter in Fig. 1*b* may be as small as 1 μm.

The basic principles of operation of bipolar microwave transistors are the same as those of low-frequency bipolar transistors. However, because of the higher frequencies involved, the requirements on dimensions, process control, packaging, heat sinking, and radio-frequency (rf) circuitry are much more severe for microwave transistors than for low-frequency transistors. The table lists the

Typical CW performance of microwave silicon power transistors (28-V bias)

Frequency, GHz	Power, W	Gain, dB	Efficiency, %
1	35	10	60
2	20	7	50
4	5	6	35

continuous-wave (CW) performance of typical silicon microwave power transistors. The pulsed power outputs that can be obtained are significantly higher than the CW power outputs. For example, peak power outputs of 250 W can be obtained from single transistors operating at 1 GHz. The noise figure of silicon bipolar transistors specifically designed for low-noise operation ranges from about 1.2 dB at 1 GHz to about 3.5 dB at 4 GHz.

GaAs field-effect transistors. Microwave GaAs FETs are a type of microwave transistor that provides improved performance over silicon microwave transistors. Low-noise, low-power, and medium-power GaAs FETs are commercially available. Figure 2 shows the cross section of a typical GaAs FET. An expitaxial *n* layer of GaAs is deposited on a semi-insulating GaAs

substrate. Ohmic source and drain contacts and a Schottky-barrier gate contact are formed on the *n* layer. In a typical power microwave FET, the *n* layer is 0.3 μm thick, the source-drain spacing is about 2 μm, and the gate length ranges from about 0.5 μm to 1 μm. In operation, the gate is biased negative with respect to the source, while the drain is biased positive. The microwave input signal is applied between gate and source, and modulates the depth of the depletion layer that is formed underneath the gate. Since the electrons flowing from source to drain must move around the depletion layer, the source-drain current flow is modulated by the microwave input signal. The amplified signal is taken from a load resistor in the source-drain circuit. GaAs rather than silicon is used in the construction of FETs designed for operation at microwave frequencies, because the mobility of electrons in GaAs is several times as high as in silicon. This higher mobility is particularly important at microwave frequencies, since it leads to lower series resistance and larger device dimensions for devices designed for operation at a given frequency.

The best noise figures obtained from GaAs FETs range from about 0.5 dB at 4 GHz to about 1.5 dB at 12 GHz and 6 dB at 30 GHz. CW power outputs range from 20 W at 4 GHz to 4 W at 12 GHz and 0.5 W at 20 GHz. Conversion efficiencies as high as 40% at 10 GHz and 70% at 4 GHz have been obtained with devices operating class B (dc gate bias just sufficient to shut off direct current flow between source and drain).

Transferred-electron (Gunn) devices. In transferred-electron devices (TEDs), negative resistance is achieved by taking advantage of the negative differential mobility ($\mu = dv/dE$; $v =$ velocity of charge carriers, $E =$ electric field) of electrons in certain *n*-type III − V compounds, particularly GaAs. The basic theoretical concepts underlying the operation of TEDs were developed during the early 1960s, and the first experimental TED was described in 1963 by J. B. Gunn.

Figure 3 shows graphs of electron drift velocity versus electric field for silicon and GaAs. The drift velocity for silicon exhibits a "normal" behavior, increasing monotonically with increasing electric field. In GaAs, on the other hand, the drift velocity decreases with increasing field above about 3 kV/cm. The negative differential mobility of GaAs is caused by the transfer of electrons from high- to low-mobility energy bands, hence the name "transferred electron devices." *See* SEMICONDUCTOR.

Structurally, TEDs are particularly simple semiconductor devices. They consist merely of a bar of transferred electron material of length *L* with ohmic cathode and anode contacts. The anode is biased positive with respect to the cathode at a magnitude larger than the threshold voltage $V_{TH} = E_{TH} \times L$ (E_{TH} is the field at which maximum electron drift velocity occurs). Therefore, part of the device is biased into the negative mobility region. Excess electron space charge introduced at the cathode moves through the device under the influence of the applied field. The charge grows exponentially as it traverses the negative mobility region, since in a negative mobility region charges of like polarity attract one another and any accumulation of space charge grows with time at the rate

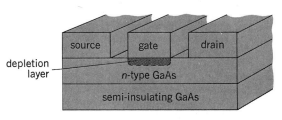

Fig. 2. Cross section of GaAs Schottky-barrier FET.

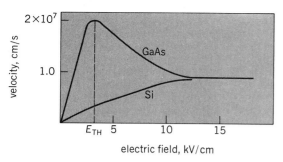

Fig. 3. Curves of electron drift velocity versus electric field for silicon and gallium arsenide. (*From H. Sobol and F. Sterzer, Microwave power sources, IEEE Spectrum, 9(4):20–33, 1972*)

exp $t/|\tau_r|$ (τ_r = dielectric relaxation time = $\epsilon/en\mu$; ϵ = dielectric constant, e = electronic charge, n = carrier concentration). If the space-charge growth in the device is not too large ($nL \gtrsim 7.6 \times 10^{11}/\text{cm}^2$ for GaAs), then the space-charge distribution in the device will be stable with time and the device will exhibit a stable negative resistance over a wide range of frequencies, and can be used as the active element in a negative-resistance amplifier. For GaAs, the center frequency of the negative-resistance region is approximately given by Eq. (1).

$$f(\text{GHz}) \approx \frac{100}{L(\mu m)} \tag{1}$$

If the space-charge growth in a TED is sufficiently large (nL greater than between 1 and a few times 10^{12} cm^2 for typical GaAs TEDs), the electric field distribution becomes a function of time. Several different time-varying field distributions (modes) useful for oscillator applications are then possible. For example, in the transit time mode, the domain of charges is nucleated at the cathode of the TED. The domain grows exponentially while moving toward the anode until the voltage across it is so large that the field in the parts of the TED outside the domain falls below threshold. The fully formed or "mature" domain disappears when it is collected at the anode. As it is being collected, the electric field throughout the TED rises above threshold, a new domain is nucleated at the cathode, and the cycle repeats itself. Short current pulses are generated while this process is going on. The spacing of the current pulses is approximately the transit time of the domains ($\sim L/v$). Another mode is the limited space-charge accumulation (LSA) mode. In the LSA mode, the formation of traveling domains is suppressed by an rf voltage whose amplitude is large enough to drive the TED below threshold during every rf cycle. The frequency of the rf voltage must also be high enough so that there is not enough time for a domain to form during the part of the rf cycle when the device voltage is above threshold. Frequency of oscillation in the LSA mode is independent of the transit time of the charge carriers and is determined solely by the circuit around the TED. At any given frequency, a TED operating in the LSA mode can be made much thicker than a TED operating in the domain mode.

The CW power outputs that can be obtained from TEDs range from 1 to 2 W at 10 GHz to about

60 mW at 90 GHz. The latter results were obtained with TEDs fabricated from indium phosphide (InP). Conversion efficiencies of commercially available devices are typically only a few percent, but in the laboratory, efficiencies in the 10–20% range have been obtained. The noise figure of transferred-electron amplifiers ranges typically from 10 to 20 dB. Several kilowatts of peak power can be obtained at the lower microwave frequencies from TEDs operating in the LSA mode. Unfortunately, high-power pulsed TEDs require thick n layers that are difficult to heat-sink. The duty cycle of such thick devices is usually limited to a small fraction of 1%.

Avalanche diodes. Avalanche diodes are junction devices that produce a negative resistance by appropriately combining impact avalanche breakdown and charge-carrier transit time effects. Avalanche breakdown in semiconductors occurs if the electric field is high enough for the charge carriers to acquire sufficient energy from the field to create electron-hole pairs by impact ionization. For silicon and gallium arsenide, the semiconductors most commonly used to fabricate avalanche diodes, the threshold field for breakdown is on the order of a few hundred thousand volts per centimeter. Transit time effects occur if the time the charge carriers spend traversing the diode becomes an appreciable fraction of an rf period. If the current injected at the cathode of a thick diode is $I_0 \cos \omega t$, then the current flowing through the external circuit of the diode is given by Eq. (2),

$$I_T = I_0 \frac{\sin (\omega\tau/2)}{\omega\tau/2} \cos [\omega\tau - \omega\tau/2] \tag{2}$$

where ω is the angular frequency and τ is the time it takes the charge carriers to move from the cathode to the anode of the diode.

The two most important classes of avalanche diodes are IMPATT diodes (impact avalanche and transit time diodes) and TRAPATT diodes (trapped-plasma avalanche-triggered transit diodes). In IMPATT diodes, best operation is usually obtained when the transit angle of the carriers traversing the diode is of the order of 180°. In TRAPATT diodes, the transit angle is usually much smaller than 180°.

The basic operation of IMPATT diodes can be most easily understood by reference to the first proposed avalanche diode, the Read diode. The theory of this device was presented in 1958 by W. T. Read, and the first experimental diode was described in 1965. A model of the Read diode and the dc electric field distribution when a large reverse bias is applied across the diode are shown in Fig. 4. It is assumed for purposes of the illustration that the diode has been placed in an rf circuit or cavity and that the total field across the diode is the sum of the dc and ac fields. The diode bias is chosen so that the field in the avalanche region is high enough to cause breakdown during the positive half of the rf voltage cycle, but is below the critical field required for breakdown during the negative part of the voltage cycle. As a result, the hole current generation in the avalanche region grows during the positive half of the rf voltage cycle and dies down during the negative half. The hole current therefore reaches its maximum value one-quarter of a cycle after the voltage reaches its maximum;

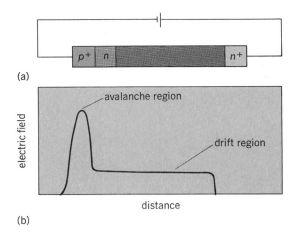

(a)

(b)

Fig. 4. Two characteristics of a Read-type IMPATT diode. (a) Doping profile. (b) Electric field distribution. (*From H. Sobol and F. Sterzer, Microwave power sources, IEEE Spectrum, 9(4)20–33, 1972*)

that is, the hole current lags the rf voltage by 90°. During the entire rf cycle, the electric field in the drift region is kept high enough to cause the injected holes to travel with the constant limiting drift velocity. The length of the drift region is chosen to make $\omega\tau = \pi$, so that from the transit time equation, Eq. (2), I_T lags I by 90°, and the phase difference between the applied voltage and I_T is 180°; that is, the diode acts as a negative resistance.

Most practical IMPATT diodes differ from the Read diode in that avalanching is not confined to a narrow, well-defined region, but occurs over a significant portion or over the entire depletion region. These types of diodes behave similarly to Read diodes and also exhibit negative resistance when the transit time of the charge carriers becomes a significant fraction of an rf period. The semiconductor materials favored for microwave IMPATT diodes are silicon and gallium arsenide.

TRAPATT diodes are typically p^+nn^+ or n^+pp^+ silicon diodes. In operation, a diode is placed in a circuit containing high-Q resonators and is back-biased into avalanche. As the diode breaks down, a highly conducting electron-hole plasma quickly fills the entire n region, and the voltage across the diode drops to a low value. The plasma is then extracted from the diode by the low residual electric field, thus causing a large current flow even though the voltage is low. Once extraction of the plasma is completed, the current becomes very small and the voltage rises to a high value. Eventually, the reactive energy stored in the resonant circuits raises the voltage above breakdown, and the cycle repeats.

The CW power outputs that can be obtained from IMPATT diodes range from several watts at 10 GHz to 2 W at 40 GHz and 50 mW at 250 GHz. Peak powers as high as 50 W can be obtained at 10 GHz. The best efficiencies obtained to date are on the order of 35%. TRAPATT diodes are particularly well suited for generating high peak powers. Peak power outputs of TRAPATT diodes range from several hundred watts at 1 GHz to tens of watts in the 10–15-GHz range. The best pulsed efficiencies range from about 60% at 1 GHz to about 30% at 10 GHz.

Tunnel diodes. Tunnel diodes were first described by L. Esaki in 1958. They are heavily doped *pn* junctions that exhibit a negative differential resistance over part of their current-voltage characteristics. The negative resistance is caused by processes (quantum-mechanical tunneling) that are so fast that there are no transit time effects even at the highest microwave frequencies. Tunnel diodes can be fabricated from a host of different semiconductor materials, but only diodes from germanium, gallium arsenide, and gallium antimonide are used at microwave frequencies.

Tunnel diodes are used in oscillators, low-noise amplifiers, and detectors at microwave frequencies. However, the power output of tunnel diode oscillators and amplifiers is limited to a few milliwatts at most, and for this reason tunnel diodes are being replaced by some of the newer solid-state microwave devices, particularly by GaAs field-effect transistors. *See* TUNNEL DIODE.

PASSIVE DEVICES

The oldest passive microwave device is the point-contact diode, a diode that was widely used in World War II radars. More modern passive microwave solid-state devices include Schottky-barrier diodes, PIN diodes, and varactor diodes.

Point-contact and Schottky-barrier diodes. Point-contact and Schottky-barrier diodes are majority-carrier rectifiers that are used as the nonlinear element in passive microwave-frequency down-converters and in rectifiers. The rectification characteristics of these diodes are similar to those of *pn* junction diodes, but unlike *pn* junction diodes they do not exhibit any minority-carrier charge storage capacitance. This lack of charge-storage capacitance makes point-contact and Schottky-barrier diodes particularly attractive for down-converter applications, since variations in capacitance are in general undesirable in down-conversion. In point-contact diodes, a pointed tungsten wire is forced down on a semiconductor wafer (usually silicon). In Schottky-barrier diodes, the rectifying junction is formed at the interface of a deposited metal layer and a semiconductor crystal, usually silicon or gallium arsenide. The metal-semiconductor interface is planar, and as a result the contact potential and the current distribution are nearly uniform over the entire junction. This is in contrast to point-contact diodes, where the interface between the sharp tungsten point and the semiconductor wafer is nearly hemispherical, and both contact potential and current distribution are therefore in general nonuniform. Because of their greater uniformity, Schottky-barrier diodes have lower series resistance, lower noise figures, greater dynamic range, and greater resistance to pulse burnout than point-contact diodes. With GaAs Schottky-barrier diodes, conversion losses as low as 2.5 dB can be obtained when down-converting a 12 GHz signal to 70 MHz. *See* POINT-CONTACT DIODE.

PIN diodes. PIN diodes are junction diodes in which the *p* and *n* regions are separated by a relatively thick high-resistivity intrinsic layer. These diodes are widely used in microwave switches, electronically variable microwave attenuators, and microwave limiters. When a PIN diode is back-biased, the intrinsic region of the diode is swept free of holes and electrons, and the series resistance of the diode is very large. Under forward bias, on the other hand, the *p* and *n* regions inject holes and electrons into the intrinsic region, forming a highly conductive plasma, and the series resistance becomes very small. Because of the high resistivity and thickness of the intrinsic layer, the junction capacitance of a PIN diode is not a function of bias voltage at microwave frequencies, and the value of capacitance for a given junction area is much smaller in a PIN diode than in a conventional *pn* junction diode. The power-handling capability of PIN diode arrays can be as high as 1 MW peak and 1 kW average. *See* JUNCTION DIODE.

Varactor diodes. A varactor (variable-capacitance) diode is a *pn* junction diode that is specifically designed for applications that make use of the voltage dependence of the diode depletion-layer capacitance. The voltage-dependent junction capacitance C_j of an ideal varactor diode is given by Eq. (3), where K is a constant, ϕ is the contact

$$C_j = K(\phi - V_j)^{-n} \qquad (3)$$

potential, V_j is the voltage across the junction, and n is a coefficient whose value depends on the doping profile of the junction ($n = 1/3$ for a graded junction, $n = 1/2$ for an abrupt junction, $n = 1$ for a hyperabrupt junction).

Microwave varactor diodes are usually fabricated from either silicon or gallium arsenide. They are widely used as tuning elements in voltage-controlled filters and oscillators, in harmonic generators, and in parametric amplifiers and frequency converters. An important figure of merit of varactor diodes is their cutoff frequency f_c, which is defined in Eq. (4), where R_S is the series resistance

$$f_c = \frac{1}{2\pi R_S C_j} \qquad (4)$$

of the diode. GaAs varactor diodes with cutoff frequencies (measured at breakdown) of many hundreds of gigahertz are commercially available. *See* VARACTOR.   [FRED STERZER]

OPTICALLY CONTROLLED DEVICES

Optically controlled microwave solid-state devices utilize light to modulate or otherwise control the microwave energy output. The basic idea of such "optical control" is to use light to directly regulate the internal dynamics of a microwave source by the optical generation of charge carriers within that source. This is in marked contrast to the conventional approach whereby control is achieved indirectly through the interaction of some external device (for example, a varactor or PIN diode) with the microwave energy produced by the source.

Principle of optical control. When a semiconductor is illuminated with light having a photon energy slightly greater than the band gap, the incident photons are absorbed to a considerable depth in the material. The absorption of this light results in the generation of charge carriers (electron-hole pairs) within the semiconductor. Since the operation of a solid-state device depends on the dynamics of its internal carriers, the presence of such optically generated carriers alters the behavior of the device. For example, in an avalanche diode oscillator the rate of carrier buildup during an oscillation cycle depends on the number of carriers

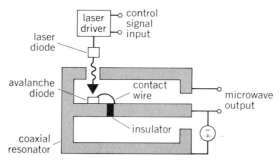

Fig. 5. Basic configuration of an optically controlled microwave oscillator.

present at the beginning of the cycle. The additional carriers present when the device is illuminated with light will therefore result in a change in the carrier buildup rate which can produce changes in the frequency and amplitude of the oscillation.

Device configuration. A simple configuration for an optically controlled microwave source is shown in Fig. 5. A silicon avalanche diode is mounted in a coaxial microwave resonator which has a small opening in the outer wall near the diode. The diode is illuminated through the opening by a controllable light source, such as a miniature semiconductor laser. The structure of the avalanche diode, shown in Fig. 6, is essentially the same as that of conventional diodes, except that the top metal contact has been made small to allow light to reach the silicon surface. The depth to which this light is absorbed in the silicon depends on its wavelength. Hence, in order to produce optically generated carriers in the active *n* layer of the diode, the wavelength emitted by the light source must be tailored to the diode structure. For the dimensions of typical microwave devices (micrometers), this requirement is well met by the gallium-arsenide infrared lasers and light-emitting diodes recently developed for optical communication systems.

Optically modulated oscillators. Optical control is of particular interest for oscillators like the TRAPATT diode, for which high peak power levels severely hamper conventional control schemes. Amplitude modulation depths of 90% and frequency modulation of several percent have been demonstrated with optically controlled TRAPATTs operating at about 1 GHz with microwave output powers of nearly 100 W. Modulation at such high microwave power levels has been possible with optical power levels as low as a few milliwatts. The relative amounts of frequency and amplitude control can be adjusted by tuning the microwave circuit. Thus, it is possible to produce predominantly frequency or amplitude modulation in accordance with what is desired for a particular application. Similar results have also been obtained with IMPATT diodes operating at 10 GHz. For the IMPATT, however, frequency modulation has so far been limited to several tenths of 1%.

The capability for very-high-speed modulation of these devices has been demonstrated by illuminating the device with rapidly varying optical signals. Both frequency and amplitude shifts have been produced in nanosecond time intervals corresponding to the sharp rise and fall of the controlling

MICROWAVE
SOLID-STATE DEVICES

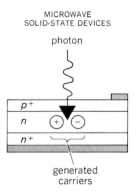

Fig. 6. Optical carrier generation in an avalanche diode.

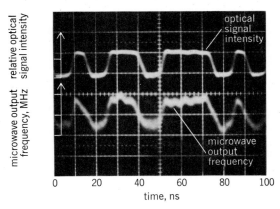

Fig. 7. Oscilloscope display showing frequency modulation with an optically modulated TRAPATT diode oscillator. *(From R. A. Kiehl and R. E. Hibray, IEEE Proc., 66: 708–709, June 1978)*

optical signal itself. Figure 7 illustrates high-speed digital frequency modulation with an optically modulated TRAPATT diode. The top oscilloscope trace in this figure is proportional to the intensity of the optical control signal, and the bottom trace is proportional to the microwave oscillation frequency.

Optically induced phase control. Phase control of microwave oscillators by optical signals has also been achieved. In this case, carriers generated by a modulated optical signal are used to periodically disturb the normal operation of the oscillator. As with any oscillator, such a periodic disturbance causes the oscillation to lock in phase with the disrupting signal, provided that the signal frequency is close enough to the oscillation frequency or a subharmonic thereof. Such optically induced locking of transistor oscillators has been achieved at frequencies as high as 2 GHz with optical signals modulated at frequencies of about 100 MHz.

Applications. Optically modulated microwave devices are particularly attractive for application in high-speed data links and short-pulse radars. For such systems, the high-speed capabilities of optical techniques promise the realization of higher data rates and narrower microwave pulse widths than are possible with conventional schemes. Furthermore, the electrical isolation that exists between the microwave source and the control device (the light source) eliminates unwanted coupling of microwave energy into the modulator circuitry, a troublesome problem in conventional high-speed systems.

Optical phase control is attractive for active phased-array radars where it is necessary to control the phase of a great number of microwave sources spaced some distance apart. The use of a modulated optical signal rather than a microwave signal to achieve phase control allows the control signal distribution to be done with tiny optical fibers instead of coaxial or waveguide lines. This gives optical control an important advantage in lightweight airborne systems.

[RICHARD A. KIEHL]

Bibliography: B. G. Bosch and R. W. H. Engelman, *Gum-Effect Electronics*, 1975; R. A. Kiehl, *IEEE Trans. Electron Devices*, ED-25:703–710, June 1978; R. A. Kiehl and E. P. EerNisse, in *Digest of 1977 IEEE International Electron Devices Meeting*, pp. 103–106, December 1977; C. M. Lee, G. I. Haddad, and R. J. Lomax, A comparison between n^+-p-p^+ and p^+-n^+-n^+ silicon IMPATT diodes, *IEEE Trans. Electron Devices*, ED-21(2): 137–141, 1974; C. P. Snapp, Bipolars quietly dominate, *Microwave Sys. News*, pp. 45–67, November 1979; H. Sobol and F. Sterzer, Microwave power sources, *IEEE Spectrum*, 9(4):20–23, 1972; F. Sterzer, GaAs field effect transistors, *Microwave J.*, p. 73, November 1978; H. A. Watson, *Microwave Semiconductor Devices and Their Circuit Applications*, 1969; H. W. Yen and M. K. Barnoski, *Appl. Phys. Lett.*, 32:182–184, February 1978.

Microwave transmission lines

Structures used for transmission of electromagnetic energy at microwave frequencies from one point to another. A transmission line may be defined more precisely as a system of material boundaries forming a continuous path from one place to another and capable of directing the transmission of electromagnetic energy along this path. At microwave frequencies, a wide variety of metallic and dielectric structures are used, the choice depending upon the specific application and frequency range. The wavelengths at microwave frequencies are small, ranging from a few millimeters to under 1 m. It is therefore typical of microwave transmission lines that even when they are physically rather short, their length measured in electrical wavelengths ranges from an appreciable fraction of a wavelength to many wavelengths. Partly for this reason, if microwave transmission lines are not carefully designed, substantial losses of energy by radiation, reflection, and attenuation may be encountered.

Circuit elements used in electrical networks at ordinary radio frequencies, such as coils and capacitors, become so small and inefficient as to be impractical for applications at microwave frequencies. Circuits and networks at microwave frequencies are therefore usually designed with lengths of transmission line serving as circuit elements. These microwave circuit elements can be highly efficient. Using transmission lines, it is possible to construct networks for microwave frequencies equivalent to networks constructed with coils and capacitors that function at lower frequencies.

Microwave structures. The structures in most widespread use as microwave transmission lines are coaxial lines and hollow-pipe waveguides. Other structures, including striplines, open-wire lines, and dielectric rods, are used in special applications.

For a detailed discussion of the properties of waveguides *see* WAVEGUIDE.

Coaxial transmission lines. Coaxial lines are widely used at lower frequencies, and are satisfactory for many applications at microwave frequencies. In a coaxial line (Fig. 1) the electromagnetic waves are transmitted through the dielectric medium bounded by two conducting, coaxial cylinders. Because of skin effect, the currents in the conductors are concentrated in the surfaces of the conductors bounding the dielectric medium. To permit transmission of energy in only a single mode of

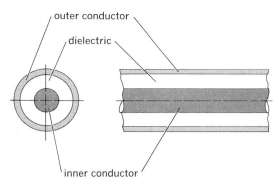

Fig. 1. Coaxial transmission line.

propagation, it is necessary to restrict the mean circumference to less than 1 wavelength. At higher frequencies and shorter wavelengths, the maximum permissible dimensions are small and the losses relatively high. For these reasons, the hollow-pipe waveguide and other types of transmission lines are increasingly preferred at the higher frequencies.

In a coaxial line, the inner conductor may be physically supported by a continuous solid dielectric, as in a flexible cable, or if the dielectric losses are excessively large, they can be reduced by supporting the center conductor by spaced dielectric beads. The center conductor can also be supported by sections of shorted transmission line 1/4 wavelength long, called stub supports. Because of their resonant properties, these stubs do not interfere with the propagation of electromagnetic waves along the line.

Hollow-pipe waveguides. Waveguides are extensively used as microwave transmission lines. Electromagnetic waves are transmitted through the interior of hollow metal pipes, and electric currents flow on the inner surfaces. In contrast to coaxial lines, hollow-pipe waveguides are characterized by cutoff frequencies; that is, for a waveguide of given dimensions, the operating frequency must be higher than a critical cutoff frequency for energy to propagate. Waveguides therefore become impractically large at lower frequencies, and the dimensions are reasonably small only in the microwave range.

Although a metal pipe of any cross section will function as a waveguide, a rectangular cross section is most extensively used. Waveguides of circular and ridged cross section are also in widespread use (Fig. 2). The principal advantages of waveguides as compared to coaxial lines are their lower attenuation and structural simplicity.

Microwave striplines. Dielectric rods and tubes can also be used as microwave transmission lines, as can single-conductor lines, which are sometimes coated with dielectric. These structures can have very low attenuation. The electromagnetic fields, however, extend outward from these transmission lines, and serious problems are presented by supports, bends, and shields, which inevitably perturb the fields.

Striplines (Fig. 3), which consist of a metal strip supported above a metal plane or between two metal planes, have somewhat higher attenuation than hollow-pipe waveguides. They are, however,

frequently used where structures of unusual compactness and physical simplicity are desired.

Traveling and standing waves. A microwave transmission line performs its primary function of transmitting electromagnetic energy by guiding electromagnetic waves along the line. Normally, the energy is carried along the line in a single mode of transmission. For any transmission line, an infinite number of modes can be excited or established on the line, each characterized by a unique configuration of electric and magnetic fields. The dimensions of the line are usually chosen, however, so that at the operating frequency the line is capable of efficient energy transmission in only a single mode. When the fields of the other modes are excited, they decay rapidly from the point of excitation, and energy is not transmitted any appreciable distance along the line in these modes. If the dimensions of the transmission line are such that energy can be efficiently transmitted in several modes, the structure by which energy is delivered onto the transmission line is usually designed so that only a single one of the propagating modes is excited. *See* TRANSMISSION LINES.

An electromagnetic wave traveling along a transmission line is weakened or attenuated with distance because energy is dissipated in resistive losses in the conductors and dielectric losses in the dielectric media. The magnitude of this attenuation is one of the important characteristics of a transmission line. For an efficient transmission line, the attenuation is low.

Another important characteristic of a transmission line is its phase velocity, computed by multiplying the wavelength of the traveling wave on the line by the operating frequency. Because the wavelength on the transmission line may be greater or less than the wavelength in free space, the phase velocity on the line may be either greater or less than the velocity of light in free space.

A third important characteristic of a transmission line is its characteristic impedance, defined as the input impedance to a uniform line of infinite length. A unique numerical value of impedance cannot be specified for many microwave transmission lines, because no unique definitions of voltage between conductors or total current in conductors exist. However, the ratio of the characteristic impedances of two different transmission lines is often more easily computed, and the knowledge of this ratio is sufficient for many purposes.

The attenuation along a transmission line varies with changes in the operating frequency. For some types of transmission line, the characteristic impedance and phase velocity also change with frequency; for others they are nearly independent of frequency.

MICROWAVE
TRANSMISSION LINES

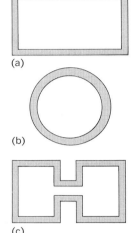

(a)

(b)

(c)

Fig. 2. Typical waveguide cross sections. (*a*) Rectangular. (*b*) Circular. (*c*) Ridged.

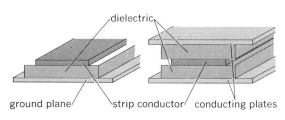

Fig. 3. Microwave striplines.

Waves can travel along a transmission line in either direction. The presence of two waves on a transmission line, traveling in opposite directions, gives rise to standing waves on the line as shown in Fig. 4. If a wave is launched on a transmission line, it continues to travel along the line as long as the line is uniform in cross section. If the wave encounters a discontinuity, or abrupt change in impedance, part of the wave is usually reflected from the discontinuity back toward the source of the wave. A standing-wave pattern then exists between the source and the discontinuity. If the terminal or load impedance of a transmission line is equal to the characteristic impedance of the line, all of the energy in the wave traveling toward the line is absorbed in the load impedance, and the line is said to be matched. If the load impedance is not equal to the characteristic impedance, part of the traveling wave is reflected. The reflection coefficient is the ratio of the intensities of the reflected and incident waves. When the load impedance is incapable of dissipating energy, as in a short-circuit, an open-circuit, or a pure reactance load, all of the incident wave is reflected. The standing-wave pattern on the line is then of maximum amplitude, and the reflection coefficient is unity. *See* STANDING-WAVE DETECTOR.

A discontinuity may exist on a transmission line at some point between the source and the load impedance, causing a partial reflection of the traveling wave back toward the source. If the load impedance is matched to the transmission line, there are no standing waves between the load and the discontinuity, but standing waves exist between the discontinuity and the source.

Impedance matching. A matched transmission line is one on which only a single wave traveling from source to load exists; that is, there is no standing wave. For most applications this is an optimum condition, as it minimizes the undesired losses of energy in the line itself and maximizes the energy delivered to the load impedance. Frequently, however, the load impedance does not match the characteristic impedance of the transmission line, and various matching networks and devices must then be employed. One such device is a discontinuity with a reflection coefficient equal in magnitude to the reflection coefficient of the mismatched load impedance. If this discontinuity is placed at the proper distance from the load impedance, the reflected waves from the load and from the discontinuity, which are equal in amplitude, will be opposite in phase and will cancel each other. The line is then matched between the source and the discontinuity. Another impedance-matching device consists of a section of transmission line 1/4 wavelength long connecting the load impedance to the main transmission line. If the characteristic impedance of this quarter-wave section is the geometric mean of the load impedance and the transmission-line impedance, the quarter-wave section matches the load to the transmission line. *See* IMPEDANCE MATCHING.

One of the most useful impedance-matching devices is the taper. An abrupt transition from one transmission line to another generally produces a mismatch and introduces standing waves. If, however, the dimensions are tapered sufficiently gradually from one line to the other, the reflection is minimized or largely eliminated. Shorter tapers can also be used successfully. A taper that is an integral number of half-wavelengths long introduces the minimum reflection.

Transmission-line structures, such as bends, corners, and junctions, introduce reflections unless they are carefully designed. Transitions from one type of transmission line to another must also be carefully designed if they are to be matched.

Lines as circuit elements. Lengths of transmission line are used as circuit elements in microwave networks. A short length of short-circuited transmission line is equivalent to an inductance. The input impedance of a short-circuited line 1/4 wavelength long is very high and is electrically equivalent to a parallel resonant circuit. A short length of open-circuited line presents a capacitive input impedance. If the open-circuited line is 1/4 wavelength long, the input impedance is very small, equivalent to a series resonant circuit.

Graphical solutions. Various graphical charts have been developed as an aid to computation for problems involving transmission lines. Perhaps the most widely used is the Smith chart, developed by P. H. Smith. With this chart, if the impedance at any point on a transmission line is known, relative to its characteristic impedance, the impedance at any other point on the line can be computed. Other problems, such as those involving impedance discontinuities and design of matching elements, can also be solved with the aid of these charts. *See* MICROWAVE.

[THEODORE MORENO]

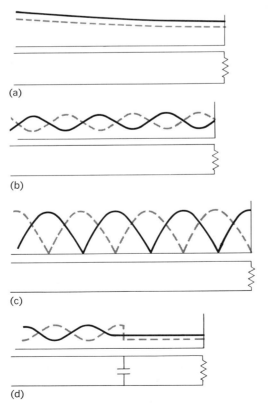

Fig. 4. Standing waves on transmission lines. (*a*) Impedance matched to load impedance. (*b*) Small mismatch. (*c*) Large mismatch. (*d*) Matched to load with discontinuity. Solid lines, voltage; dashed lines, current.

Bibliography: A. J. Baden-Fuller, *Microwaves: An Introduction to Microwave Theory and Techniques*, 1979; R. G. Brown et al., *Lines, Waves and Antennas: The Transmission of Electric Energy*, 2d ed., 1973; K. C. Gupta, *Microwaves*, 1980; L. Lewin, *Theory of Waveguides*, 1975; S. Liao, *Microwave Devices and Circuits*, 1980.

Microwave tube

A high-vacuum tube designed for operation in the frequency region from approximately 3000 to 300,000 MHz. Two considerations distinguish a microwave tube from vacuum tubes used at lower frequencies: the dimensions of the tube structure in relation to the wavelength of the signal that it generates or amplifies, and the time during which the electrons interact with the microwave field.

Effect of tube geometry. In a vacuum tube the active region is where the electrons travel through the evacuated space from the cathode, through the grid, to the plate (Fig. 1). The circuit in which the tube operates extends from this active region along the internal tube structure, through the enclosing vacuum-tight envelope, and onto the portions of the circuit external to the tube. In the microwave region wavelengths are in the order of centimeters; resonant circuits are in the forms of transmission lines that extend a quarter of a wavelength from the active region of the microwave tube. With such short circuit dimensions the internal tube structure constitutes an appreciable portion of the circuit.

For these reasons a microwave tube is made to form part of the resonant circuit. Leads from electrodes to external connections are short, and electrodes are parts of surfaces extending through the envelope directly to the external circuit that is often a coaxial transmission line or cavity. Design of the tube and of the circuit in which it is to operate thus become closely related. *See* CAVITY RESONATOR; TRANSMISSION LINES.

Effect of transit time. At frequencies well below the microwave region, the time during which any one electron travels through the active region within a tube is so short compared to the period of the signal as to be negligible. At microwaves the period of the signal is in the range of 0.001–1 nanosecond. If transit time is comparable to the period of the signal, an electron experiences an inappreciable net change in energy. Even if transit time is reduced to half the signal period, an electron that is in transit when the signal reverses polarity experiences little net change in energy (Fig. 2). Only if transit time is less than a quarter of the signal period do significant numbers of electrons exchange appreciable energy with the signal field.

Transit time is reduced in several ways in microwave tubes. Electrodes are closely spaced and made planar in configuration. Spacings of 0.025 mm are practical, but to be effective require that electrodes be closely parallel to each other. High interelectrode voltages also decrease transit time by their acceleration of electrons; however, the voltage stresses that glass vacuum seals can withstand place a practical limit on the voltage. *See* VACUUM TUBE.

Alternative designs. Tubes designed by the foregoing principles are effective for wavelengths from a few meters to a few centimeters. At longer wavelengths lumped-constant circuits are effective and tubes can be designed for optimum internal characteristics. At shorter wavelengths different principles are necessary. To obtain greater exchange of energy between the electron beam and the electromagnetic field several alternative designs have proved practical.

Instead of collecting the electron beam at a plate formed by the opposite side of the resonant circuit, the beam is allowed to pass into a field-free region before reacting further with an external circuit. The electron cloud can be deflected by a strong static magnetic field so as to revolve and thereby react several times with the signal field before reaching the plate. *See* KLYSTRON; MAGNETRON.

Instead of producing the field in one or several resonant circuits, the field can be supported by a distributed structure along which it moves at a velocity comparable to the velocity of electrons in the beam. The electron beam is then directed close to this structure so that beam and field interact over an extended interval of time. *See* BACKWARD-WAVE TUBE; TRAVELING-WAVE TUBE.

Such structures as these greatly extend the region over which useful gains, low signal-noise ratios, and significant powers can be produced, although the design of any one microwave tube can usually be optimized for only one of these characteristics at a time. Even so, dimensions and tolerances limit the wavelengths for which such tubes can be manufactured. For operation at shorter

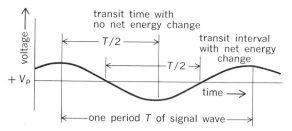

Fig. 2. Diagram of transit time. For efficient tube operation, the transit time during which an electron passes through the signal field needs to be short compared to the period of the signal.

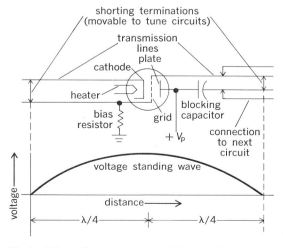

Fig. 1. Schematic of microwave triode in a circuit of two transmission lines. Usually the tube has circular symmetry and fits directly into coaxial lines.

wavelengths entirely different techniques are necessary, such as those using quantum behavior within molecules or the intermodulation of signals within a nonlinear device. *See* MASER; MICROWAVE SOLID-STATE DEVICES; MICROWAVE TRANSMISSION LINES; PARAMETRIC AMPLIFIER.

[FRANK H. ROCKETT]

Bibliography: R. E. Collin, *Foundations for Microwave Engineering*, 1966; S. Liao, *Microwave Devices and Circuits*, 1980; R. F. Soohoo, *Microwave Electronics*, 1971.

Mixer

A device having two or more signal inputs, usually adjustable, and one common output. It is used in audio amplifiers to combine the outputs of individual microphones and other audio-signal sources linearly and in desired proportions to produce one audio output signal. In television, a mixer serves similarly to combine the outputs of two or more television cameras or other video signal sources. The mixer stage in a superheterodyne receiver combines the incoming modulated rf signal with the signal of a local rf oscillator to produce a modulated i-f signal. Crystal diodes are widely used as mixers in radar and other microwave equipment. *See* MICROPHONE; RADIO RECEIVER.

[JOHN MARKUS]

Mobile radio

Radio communication in which at least one of the transmitters or receivers is installed in a vehicle and can be operated while in motion. Mobile radio is the short form of the term "land mobile radio service."

Early development. The first practical use of radio transmission to a moving automobile for dispatch service is credited to the Detroit Police Department in 1922. The Detroit and other early police radio installations provided only one-way communication to the vehicles, and employed amplitude modulation first in the broadcast band and later in the medium-frequency bands. By the middle to late 1930s, the concept of using radio to direct the movement of people in cars had advanced to the point that the Federal Communications Commission (FCC) set aside some frequencies in the vhf band (30–40 MHz)—thereby eliminating interference from sky-wave propagation (though not "skip" interference) and reducing the level of ignition noise interference that had previously hampered operations at lower frequencies. *See* AMPLITUDE-MODULATION RADIO.

Narrow-band frequency modulation (±15 kHz) was proposed not only as a remedy against noise but also as a means for providing more precise limits on area coverage by virtue of the capture effect, in which a strong fm signal in an fm receiver suppresses a weaker signal on the same frequency. The first statewide fm system was installed by the Connecticut State Police just before World War II. At first the modulation deviation was ±15 kHz for maximum capture effect and minimal noise impact, but in the mid-1950s as the radio spectrum became increasingly crowded, ±5 kHz became the standard. *See* FREQUENCY-MODULATION RADIO.

Spectrum allocation and growth. At the conclusion of World War II, the Radio Technical Planning Board laid out a comprehensive spectrum allocation plan that set aside some 43 MHz for commercial use in the frequency bands of 25–50, 152–162, and 450–470 MHz; frequencies for nonmilitary Federal government use were also designated at 162–174 and 406–420 MHz.

Growth in the use of mobile radio since 1945 has exceeded such planning, no matter how ample it might have seemed at the time. There are over 3,000,000 mobile units in operation. To accommodate this growth, channel-spacing assignments have been halved and, in the 150- and 450-MHz bands, halved again. However, in spite of these changes, in some cities business users are forced to share a single channel with as many as 200–300 mobile units.

Most of the growth has taken place in what are termed dispatch systems, where one station talks to many mobile units, and vice versa. But in the early 1950s, the Bell System started to install a nationwide network of base stations whereby specially designed mobile units can be used in the car to provide normal personal and business two-way public telephone service. Provided the car is within range of a base station, it can, in effect, carry a person's extension telephone whereby calls can be made to any other telephone, not just in the United States but worldwide. Similarly, any public telephone can be used to contact a particular mobile telephone—again, provided it is within range of a telephone company base station. By 1975 mobile telephone service had grown to about 75,000 mobile units and was expected to expand considerably using new frequencies and new technology.

The pressure built until the FCC had to find some additional space—first, by sharing the lower seven uhf TV channels in certain cities, and second, by reassigning the top 10 uhf TV channels, plus other spectrum, covering the band 806–947 MHz (excepting the industrial, scientific, and medical or ISM band, 902–928 MHz), to the land mobile services.

Spectrum engineering. In an environment where the growth of demand for spectrum has long exceeded supply, spectrum engineering has had to play a major role. The development engineers on the manufacturers' staffs have continually had to find new ways to get more "talk per square mile" without more spectrum space. This pressure has led to many interesting advances. Narrow-band filters have been developed with a broad response band that will not aggravate impulse noise and with rejection capabilities of over 100 decibels (dB). High-stability oscillators have been developed to keep both the transmitter and receiver frequencies from wandering and thereby creating either interference or susceptibility to adjacent channels. These provide protection from the off-frequency effect of ignition noise; quartz filters can keep frequencies from varying more than ±2 parts per million (ppm) at temperatures from −40 to +70°C. Spurious transmissions have been reduced to the point where all unintended radiations are more than 80 dB down from the carrier. In an effort to keep the combination of two or more signals from producing undesired interference, transmitter and receiver intermodulation has been reduced, while the receiver has been made more sensitive. Cavity filters, circulators, and antenna design and positioning have been perfected in order to provide better means to share the best antenna locations without creating additional spurious signals via intermodulation. A combination of preemphasis, clipping, and deemphasis has been developed to prevent overmodulation without destructive distortion. The use of tone-coded squelch allows many users to share a channel without having to listen to all conversations on the channel. This is a system in which 19 audio tones below 300 Hz are used to code various transmitters operating on the same frequency in the same area so that each will open the squelch circuit only of its own mobiles. *See* CLIPPING CIRCUIT; DISTORTION (ELECTRONIC CIRCUITS); ELECTRIC FILTER; FREQUENCY MODULATION; OSCILLATOR.

There are many other steps that have been taken beyond those cited, but there is no more important activity than that of evaluating proposed systems in light of the physical and electromagnetic environment in which they will operate before the plan is firmed up. The FCC operates a Regional Spectrum Center in the Chicago area to assist users and manufacturers in finding the best frequency and antenna location for proposed systems; their technical performance files of the actual status of systems in operation are of great value in getting more effective "talk per square mile" within the allocated bands. *See* ELECTROMAGNETIC COMPATIBILITY.

Uses. While the growth of land mobile communications since 1945 has been almost explosive, it is nowhere near saturation. First, excluding citizens band, approximately 20% of the trucks and 8% of all vehicles were equipped in 1975; and of the total units in service, only about 300,000 could be classified as personal or portable. Second, the original motivation for the use of mobile or personal radio has become increasingly significant: urban society is much more complex and emergencies have tended to become more serious; yet there is also a national trend to deurbanization, thereby increasing the need for radio in the delivery of governmental and commercial services. Third, the United States economy is built on the use of capital equipment to produce gains in efficiency and cost, and mobile radios increase the return on investment in police cars, trucks, and delivery vans, and gain better utilization of the driver's time. Finally, an increasing number of people, in every type of professional activity, must do business while traveling.

The list of users includes—but is not limited to—police, fire, state highway maintenance, electric utility maintenance, telephone company maintenance, gas and petroleum pipeline maintenance, outside construction, doctors, realtors, forestry, commercial agriculture, long-haul trucking, railroads, short-haul delivery, commercial urban services (plumber, cleaner, florist, and so forth), taxis, news services, airline loading coordination, river barges, emergency aid service, emergency medical services, commercial buses, and school buses.

Technological trends. Few land mobile systems consist simply of a base station and a fleet of mobile units. As systems have become larger in number of vehicles and areas to be covered, and as telephone service and dispatch service have had to be provided with even faster response and more

privacy, land mobile system design has come to involve all the tools of modern electronics.

Large public-safety systems are designed to incorporate data processing (with its associated memory) to provide assistance to dispatchers by automatically keeping the log on assigned cars, assigned areas, and unassigned cars. Area information has to be manually fed into the system, but experiments have been undertaken in several systems to ultimately find a means by which a car's latest location can automatically be entered into the computer that is used to assist the dispatcher. That information can visually be displayed on a map, thereby providing an additional tactical tool.

Radio spectrum time can be used more efficiently by arranging for push-buttons, either at the dashboard control unit or on a personal two-way radio, to transmit digitally coded signals for the more routine identification and status replies. The same technique can be used in the reverse direction with the output, such as an address, being read on a small printer. Since these messages are in digital code, the information can be tied in directly with all other local or nationwide data sources (over land lines). A number of police departments provide direct computer access, not only to the radio dispatchers at the control center but also to the officers in the field. Computer terminals, consisting of a typewriterlike keyboard and read-out facility, have been installed in police cars. Washington, DC has undertaken experiments with miniature computer terminals as an adjunct to personal radio sets, so that even an officer on foot patrol may have direct access to a computer for such information as stolen cars, guns and other property, missing and wanted persons, warrants, and so on. Inquiries are transmitted in less than two seconds, and the read-out appears in a matter of seconds, even though as many as three computer systems may be queried in sequence—local, state, and the National Crime Information Center. Digital communications are much more efficient than voice communications in conserving air time. The use of digital messages provides a simple form of privacy and security against scanning receivers sold to the general public, ostensibly for "entertainment" in monitoring local activities. Whenever security is required, voice signals, as well as digital codes, can be scrambled quite readily with microelectronic devices. *See* PRIVACY SYSTEMS (SCRAMBLING).

The band at 900 MHz has provided a previously unused spectrum, enabling two new techniques to be worked into the frequency assignment plan. Each is designed to provide for more effective spectrum use—one targeted primarily on dispatch systems and the other on mobile telephone systems. The former uses the principle of improving access to a free channel by automatic frequency selection circuitry in both the mobile unit and base station transmitter, which may simultaneously operate on one or more channels out of, say, four. The radio link, in effect, becomes a trunk with many access lines, and the availability of a free channel is increased significantly over that which would prevail if the same channels were preassigned among a given fleet or fleets of cars and trucks. The multiplying factor can be quite high though using only a few trunked channels, if the messages are short as in dispatch service.

The latter approach divides a given area (say, a city) into many relatively small zones. The combination of transmitter power and antenna height is adjusted so that frequencies may be reused in alternate (as opposed to adjacent) zones. Such a system is well suited to mobile telephone service where the messages may be long. The most challenging technical problem is that of making the base station/frequency change for a conversation in process at the time the vehicle moves from one zone to the next (the hand-off problem).

Technical progress has also been made in providing higher power transmitters in hand-held units with reasonable battery life and in increased reliability, for mobile radio becomes essential in emergencies and must always be ready to operate.
[RICHARD P. GIFFORD]

Modulation

The process or result of the process whereby a message is changed into information-bearing signals that not only unambiguously represent the message but also are suitable for propagation over the transmitting medium to the receiver.

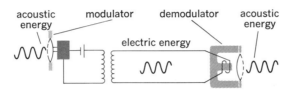

Fig. 1. Diagram of ordinary telephony sequence. (*From H. S. Black, Modulation Theory, Van Nostrand, 1953*)

The vehicle for the propagation of electric signals from one region in space to another is always an electromagnetic field, and when this field changes with time it takes the form of a wave.

At the distant end the receiver is waiting to be informed; this is accomplished by the arrival of the propagated wave, which, it is important to note, must change in a way the receiver cannot predict. Here, too, modulation is the process whereby in response to the received wave either the original message or information pertaining to the original message is made available in the form desired and is delivered when and where it is wanted. The terms demodulation and detection are often used to denote the recovery of the wanted message from a modulated signal.

Modulation is fundamental to communication. No matter how, when, or where communication takes place, modulation is encountered. Many kinds of modulation are possible, but a characteristic common to all is change. *See* ELECTRICAL COMMUNICATIONS.

Modulation implies bandwidth occupancy. For any signal to change in a way that cannot be predicted implies necessarily that the signal occupy a nonzero band of frequencies. For example, the spoken word occupies a band from a few hundred to several thousand hertz.

Ordinary telephony (Fig. 1) is a good example of these modulation concepts. Longitudinal sound waves generated by the spoken word constitute the information-bearing signals to be communicated. The telephone transmitter, acting as a modulator, changes this acoustic energy into electric energy suitable for high-speed propagation to a distant point. At the receiving end, demodulation in the telephone receiver changes the electric signals back to pressure waves in the air.

Defined broadly, modulation is the process or result of the process whereby some parameter of one wave is varied in accordance with another. As is customary in the treatment of modulation, the word "wave" is used as a generic term intended to include such concepts as signal, voltage, current, pressure, displacement, and the like, whether these are constant or changing.

This broad definition of modulation, which may be illustrated by a familiar example of amplitude modulation, implies three fundamental concepts: modulating wave, carrier, and modulated wave (Fig. 2). A modulating wave changes some parameter of the wave to be modulated; a carrier is a wave suitable for modulation by the modulating wave; a modulated wave has some parameter changed in accordance with the modulating wave.

Amplitude modulation (AM). In amplitude modulation the amplitude of a carrier is the parameter subject to change by the modulating wave.

In a more restrictive sense, AM is defined to mean modulation in which the amplitude factor of a sine-wave carrier is linearly proportional to the modulating wave. Analysis shows that the modulated wave depicted in Fig. 2 is composed of the

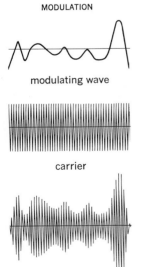

Fig. 2. Typical waveforms in amplitude modulation of a sine-wave carrier by a voice-frequency–modulating wave. (*From H. S. Black, Modulation Theory, Van Nostrand, 1953*)

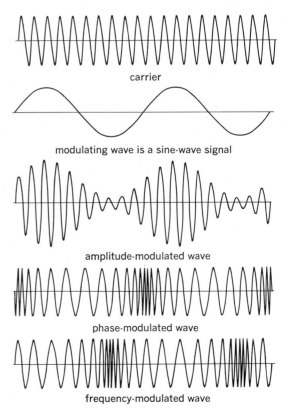

Fig. 3. Amplitude, phase, and frequency modulation of a sine-wave carrier by a sine-wave signal. (*From H. S. Black, Modulation Theory, Van Nostrand, 1953*)

transmitted carrier, which conveys no information (apart from its amplitude, frequency, and phase), plus the familiar upper and lower sidebands, which convey identical and therefore mutually redundant information. Thus, AM doubles bandwidth occupancy. *See* AMPLITUDE MODULATION; SIDEBAND.

Assuming adequate knowledge of the carrier, either sideband alone unambiguously defines the message. Accordingly, carrier-suppressed single-sideband transmission (SSB) conserves both power and frequency space and, when combined with frequency-division multiplexing, is widely used for long-distance telephony.

Angle modulation. Instead of conserving bandwidth, bandwidth occupancy may be intentionally increased in exchange for improved performance. Angle modulation is one of the simplest ways of sacrificing frequency space for reduced noise. In angle modulation the angle (entire argument) of a sine-wave carrier is the parameter changed by the modulating wave. Frequency and phase modulation are particular forms of angle modulation. Often the term frequency modulation is used to connote angle modulation. *See* ANGLE MODULATION.

A fundamental concept in angle modulation is that of instantaneous frequency. It is proportional to the time rate of change of the angle of a sine function, the argument of which is a function of time. When the argument is expressed in radians and time in seconds, the equation shown here may be formulated.

$$\text{Instantaneous frequency} = \frac{1}{2\pi}\frac{d}{dt}\ \text{(angle)}$$

Frequency modulation (FM). FM is angle modulation in which the instantaneous frequency of a sine-wave carrier is caused to depart from the carrier frequency by an amount proportional to the instantaneous value of the modulating wave (Fig. 3). *See* FREQUENCY MODULATION.

Phase modulation (PM). PM is angle modulation in which the angle of a sine-wave carrier is caused to depart from the carrier angle by an amount pro-

Typical kinds of modulation, with principal advantages, disadvantages, and uses

Kind	Advantages	Disadvantages	Uses
Amplitude modulation			
Double-sideband plus carrier (AM)	Simplifies receiver; preserves waveform of wanted message	Doubles bandwidth occupancy; requires extra signal power	Radio broadcasting, TV broadcasting, telephony, telegraphy, telemetering
Single-sideband, suppressed carrier (SSB)	Saves bandwidth occupancy; conserves signal power	Unable to handle relatively low frequencies; adds inherent delay; waveform of wanted message is not preserved	Long-distance telephony and telegraphy over land and submarine cables
Vestigial sideband (VSB)	Avoids disadvantages of single sideband	Modest increase in bandwidth occupancy as compared with single sideband	TV broadcasting
Day's system (special phase discrimination multiplexing applicable to pairs of channels)	Conserves bandwidth and signal power	Sensitive to transmission impairments	To multiplex the two so-called color components in color TV broadcasting
Phase-discrimination multiplexing (applicable to many channels)	Conserves bandwidth and signal power	Sensitive to transmission impairments	Minor
Angle modulation			
Narrow-band	Constant signal power	Extra bandwidth occupancy	Telecommunications, particularly broad-band carrier and TV over microwave radio relay systems
Wide-band	Reduces noise in exchange for extra bandwidth occupancy; constant signal power; channel-grabbing property	Extravagant of bandwidth occupancy; sensitive to some forms of transmission impairment; signal power must be adequate to override wide-band noise	Telecommunications generally, including such fields as telegraphy, telephony, radio broadcasting, telemetering, mobile communications for military and peacetime services, navigational aids, maritime beacons, meteorological aids
Wide-band with negative feedback receiver	Reduces noise in exchange for extra bandwidth occupancy; constant signal power; stabilizes receiver and improves its linearity; signal power need only override noise in a narrow band directly above and below the carrier	Extravagant of bandwidth occupancy; sensitive to some forms of transmission impairment	Negligible; suitable where the noise advantage of wide-band angle modulation is important and signal power must be conserved

(continued)

Typical kinds of modulation, with principal advantages, disadvantages, and uses (cont.)

Kind	Advantages	Disadvantages	Uses
Pulse modulation			
Amplitude (PAM)	Permits multiplexing channels by time division	Sensitive to some forms of transmission impairment	Radio, radar, telegraphy, telephony, telemetering, time-multiplexed sampled-data systems, computers, switching systems
Duration (PDM)	Permits multiplexing channels by time division; reduces noise in exchange for extra bandwidth occupancy; constant signal power	Extra bandwidth occupancy; pulses vary in position	Microwave radio relay systems, telemetering
Position (PPM)	Permits multiplexing channels by time division; reduces noise in exchange for extra bandwidth occupancy; constant signal power; saves signal power as compared with PDM	Extra bandwidth occupancy; pulses vary in duration	Microwave radio relay systems, telemetering
Code (PCM)	Permits multiplexing channels by time division; in exchange for extra bandwidth occupancy, tolerates considerable noise and serious transmission impairments, and may be repeatered again and again without significant distortion; constant signal power	Extra bandwidth occupancy; transmits digital instead of analog signals, thereby introducing quantization noise if the receiver delivers analog signals	Multiplex telephony and telegraphy, TV, data processing, combined transmission and switching systems, telemetering
Multiple modulation	Many, depending upon circumstances; often accomplishes what cannot be done in one step	Extra steps usually add extra complexity	A feature of all but the simplest transmitters and receivers

portional to the instantaneous value of the modulating wave. The instantaneous frequency deviates from its unmodulated value by an amount proportional to the time derivative of the modulating wave (Fig. 3). *See* PHASE MODULATION.

FM and PM are similar in the sense that any attempt to shift frequency or phase is accomplished by a change in the other. The terms FM and PM simply indicate which parameters of the complete argument (angle) of the sine function are being modulated.

Pulse modulation. In pulse modulation the carrier may be a train of regularly recurrent pulses. Modulation may control the amplitude, duration, position, or mere presence of the pulses so as to represent the message to the communicated. These forms of pulse modulation are commonly called, respectively, pulse-amplitude modulation, pulse-duration modulation, pulse-position modulation, and pulse-code modulation. Ease of multiplexing channels by time division is one of the important economic advantages of all forms of pulse modulation. All but pulse-amplitude modulation can exchange extra bandwidth occupancy for noise reducion. Pulse-code modulation is a digital system transmitting "on" or "off" pulses and thereby offering major transmission advantages not possible with an analog system. *See* PULSE MODULATION.

Multiple modulation. Practical applications of modulation often utilize what is commonly referred to as multiple modulation, a succession of modulating processes in which the modulated wave from one process becomes the modulating wave for the next. A typical example is a system in which position-modulated pulses are used to modulate the amplitude of a sine-wave carrier. Such a system is abbreviated PPM-AM (pulse-position modulation – amplitude modulation) inasmuch as it is customary to list the processes in the order in which the message to be conveyed encounters them. A subcarrier is any carrier used in an intermediate step of multiple modulation.

Engineering applications. The principal application and major fields of use of these various kinds of modulation are indicated by the technical comparisons and typical examples in the table.

[HAROLD S. BLACK]

Bibliography: J. Betts, *Signal Processing, Modulation and Noise*, 1971; F. R. Connor, *Modulation*, 1973; Howard W. Sams Engineering Staff, *Reference Data for Radio Engineers*, 6th ed., 1975; M. Schwartz, *Information, Transmission, Modulation and Noise*, 1970.

Modulator

Any device or circuit by means of which a desired signal is impressed upon a higher-frequency periodic wave known as a carrier. The process is called modulation. The modulator may vary the amplitude, frequency, or phase of the carrier. *See* MODULATION.

An amplitude modulator varies the amplitude of the carrier in accordance with the modulating signal. The envelope of the carrier then has the same waveform as the modulating signal if the modulation is distortionless. There are many ways to accomplish amplitude modulation, but in all cases a nonlinear element or device must be employed. The modulating signal controls the characteristics of the nonlinear device and thereby controls the amplitude of the carrier. Class B or class C amplifiers are frequently used as the nonlinear element so that power gain may be obtained in the modulated amplifier. However, other devices, such as a diode or a nonlinear resistor, may be used as the modulating element. See AMPLIFIER; AMPLITUDE MODULATION; AMPLITUDE MODULATOR; AMPLITUDE-MODULATION DETECTOR.

The frequency modulator changes the frequency of the carrier in accordance with the modulating signal while the amplitude remains constant. The frequency modulator usually changes the effective capacitance or inductance in the frequency-determining LC circuit of the oscillator. However, other techniques can be used. For example, a multivibrator can be used to generate carrier frequencies up to a few megahertz, and the multivibrator frequency can be modulated by controlling the base, gate, or grid bias supply voltage. See FREQUENCY MODULATION; FREQUENCY MODULATOR; FREQUENCY-MODULATION RADIO.

The phase modulator varies the phase of the carrier in accordance with the modulating signal. Phase modulation is very closely related to frequency modulation. Phase modulation will produce the characteristics of frequency modulation if the frequency characteristics of the modulating signal are altered. See PHASE MODULATION; PHASE MODULATOR; PULSE MODULATION; PULSE MODULATOR.

[CHARLES L. ALLEY]

Monopulse radar

A radar that obtains a complete measurement of the target's angular position from a single echo pulse. Together with the range measurement performed with the same pulse, the target position in three dimensions is determined completely. Usually a train of echo pulses is then employed to make a large number of repeated measurements and produce a refined estimate, but this is not intrinsically necessary.

The antenna receiving characteristics are especially pertinent to monopulse performance. It is usually convenient, but not necessary, to use the same antenna to illuminate the target. The monopulse operation is implemented by means of two pairs of feed points appropriately located at the antenna. The two feeds for determining azimuth information are located in the same horizontal plane on either side of the beam axis. The main lobe belonging to one is directed slightly to the left of the beam axis, and the main lobe belonging to the other is tilted slightly to the right. A target located exactly on the beam axis produces the same signal in both feeds, but a target located to one side of the beam axis produces a stronger signal in one feed than the other. The difference between the signals received by the two feeds indicates the azimuth separation between the beam axis and the line of sight to the target. The sum of the signals

received by the two feeds indicates the gross signal strength and is used as a normalizing factor. See ANTENNA.

A similar arrangement is used to measure elevation angle, with the two feeds located in the same vertical plane above and below the beam axis. There is nothing to prevent simultaneous measurement of azimuth and elevation.

The same antenna can also be used for transmission by connecting the four feed points in parallel. The radiation pattern thereby produced is almost the same as the pattern produced by a single feed, except that the beam width is slightly smaller, the gain slightly higher, and the lobe pattern is symmetrically positioned with respect to the beam axis instead of being tilted.

The sum and difference signals can be formed either at the carrier frequency or after conversion to a lower frequency. It is of some advantage to obtain the sum and difference results immediately at the carrier frequency to minimize the possibility of subsequent circuits introducing errors. The operations can be performed conveniently at microwave frequencies by the use of waveguide hybrid mixers. Only the sum and difference signals, in both azimuth and elevation, would then be conveyed beyond this point in the receiver.

An important advantage of a monopulse tracking radar over one employing conical scan is that the instantaneous angular measurements are not subject to errors caused by target scintillation. In a conical-scan system the angular information is derived from the phase and amplitude of the modulation envelope on the echo pulse train. The conical nutation frequency usually lies in the range between 20 and 40 Hz, and the scintillation spectrum of most targets possesses nonnegligible components within this band. The scintillation produces random modulation at the nutation frequency which cannot subsequently be differentiated from the signal, leading to tracking errors. A monopulse tracking radar is not susceptible to this difficulty because each pulse provides an angular measurement without regard to the rest of the pulse train. There is no opportunity for radar cross-section fluctuations to affect the measurement.

An additional advantage of monopulse tracking as compared to conical scan is that no mechanical action is required in monopulse. In a precise and highly maneuverable tracking radar it is helpful not to have to accommodate a whirling scanner. For discussion of conical scan see RADAR.

[ROBERT I. BERNSTEIN]

Bibliography: D. K. Barton (ed.), *Radar: Monopulse Radars*, vol. 1, 1974; M. I. Skolnik, *Introduction to Radar Systems*, 2d ed., 1980.

Motorboating

A form of oscillation that occurs at a very low audio frequency in a system, circuit, or component. It is caused by excessive amount of audio feedback at low frequencies. This oscillation is a succession of pulses; when these occur in a circuit that is feeding a loudspeaker, the pulses result in putt-putt sounds resembling those made by a motorboat.

Feedback through a common power supply is one cause of motorboating and can be suppressed by using resistance-capacitance decoupling filters between the power supply and the plate circuit of each tube. Another solution is to cause the amplification to fall off sharply immediately below the useful range of frequencies, so that the motorboating frequencies are inherently suppressed. *See* AMPLIFIER.   [JOHN MARKUS]

Moving-target indication

A method of presenting pulse-radar echoes in a manner that discriminates in favor of moving targets and suppresses stationary objects. Moving-target indication (MTI) is almost a necessity when moving targets are being sought over a region from which the ground clutter echoes are very strong. The most common presentation of the output of a radar with MTI is a plan-position indicator (PPI) display. The moving targets appear as bright echoes, while ground clutter is suppressed.

Principles of operation. MTI is based upon the use of the Doppler effect; that is, the carrier frequency of the echo from a target moving toward or away from the radar shifts by an amount proportional to the product of radial velocity and transmitted frequency (see illustration). A stable oscillator in the receiver is synchronized with the transmitter, providing a continuous reference of the transmitter frequency and phase. Echoes are heterodyned with the reference oscillator. Stationary objects supply echoes having a carrier phase shift which is constant from pulse to pulse, because the time for the signal to travel to a station-

ary object and back is always the same. Therefore the heterodyner output for stationary echoes does not change in phase from pulse to pulse. However, phase shift in echoes from moving targets changes from pulse to pulse because of the change in signal propagation time. *See* DOPPLER RADAR.

To utilize the lack of pulse-to-pulse phase difference to discriminate against clutter echoes, the heterodyner output is fed to a delay line which stores the signal for a period of time exactly equal to the period between pulses. Then the output of the delay line is subtracted from the freshly produced output of the heterodyner. If the two outputs are identical, as will be the case for clutter, the difference is zero. But this difference is not zero for moving targets because of their pulse-to-pulse phase shift. The difference between consecutive echoes is presented as the MTI output. The delay line usually employed utilizes acoustic propagation within fused quartz.

A variety of internal arrangements are in use for assuring that the reference oscillator is synchronized with the transmitter output. The reference oscillator may be at the carrier frequency but usually operates at an intermediate frequency, because better stability can be obtained in a low-frequency oscillator. This arrangement requires that the local oscillator used for conversion from the carrier frequency to the intermediate frequency be extremely stable, and the term stalo (stable local oscillator) is used to denote circuits that satisfy this requirement. The reference oscillator is called the coho (coherent oscillator) to denote that it remains coherent (fixed phase relation) with the transmitter output.

Measures of merit. The figure of merit of an MTI system, clutter suppression factor, is defined as the ratio of rms clutter amplitude before MTI to rms clutter fluctuation after MTI. It is the amount by which the clutter is suppressed. It depends in turn upon several other factors, some of which are system parameters and thus controllable, and some of which are functions of terrain and weather.

Another measure of merit for MTI radar is the signal-to-clutter ratio (SCR). However, it has been found that increasing the SCR simultaneously reduces the signal-to-noise ratio (SNR), a measure of merit of the sensitivity of the radar. Targets in a clutter-free environment are less detectable with MTI than without it because of reduced SNR. Scientists have determined that a parallel bank of Doppler filters can provide a simultaneous improvement in both SCR and SNR.

Use of charge-transfer devices. Charge-transfer devices (CTD) can be utilized to implement MTI with excellent capability for canceling clutter. A CTD is a capacitorlike semiconductor device of the bucket-brigade variety. As used in MTI, samples of the returned radar signal are stored and processed as discrete quantities of electrical charge. This implementation has become known as discrete-signal moving-target indication (DSMTI). Metal oxide semiconductor field-effect transistors (MOSFET) are used as CTDs to provide the delay that permits subtracting a radar return delayed by one pulse-repetition interval (PRI) from the currently incoming radar return.

Adaptive MTI. Radar systems designers have long appreciated that the best target range and

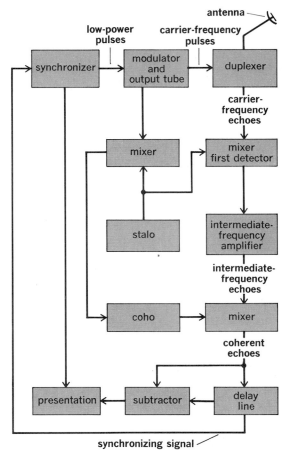

Diagram of radar system with moving-target indication.

velocity resolution are obtained by using recurrent pulses whose duration is short compared with the PRI. However, this practice gives rise to ambiguities in determining the range and velocity of the target. In fact, it causes the radar to be blind to targets moving at certain velocities when MTI is in use. A way to avoid blind velocities is to implement a random staggering of the radar's pulse-repetition frequency (PRF). Fast changes in the PRF combined with filtering systems and integration methods that adapt automatically to these changes permit ambiguities in determining the range and velocity of the target to be resolved in a single look at the target. Clutter filters switched in synchronism with the PRF also reduce undesirable clutter sidebands even when the PRF is switched at a high rate, and the irregular pulse spacing alleviates the problem of blind velocities. The clutter filter can be a notch-type filter that has its rejection band centered about the carrier frequency. *See* ELECTRIC FILTER.

This adaptive MTI (AMTI) has proved to be useful in achieving acceptable detection of targets in the presence of natural precipitation static and chaff dispersed by the enemy as a defensive measure against radar. *See* ELECTRONIC COUNTER-MEASURES; RADAR.   [JOHN M. CARROLL]

Bibliography: A. Andrews, *ABC's of Radar*, 1961; *Proceedings of the IEEE International Radar Conference*, 1975.

Multiaccess computer

A computer system in which computational and data resources are made available simultaneously to a number of users. Users access the system through terminal devices, normally on an interactive or conversational basis. A multiaccess computer system may consist of only a single central processor connected directly to a number of terminals (that is, a star configuration), or it may consist of a number of processing systems which are distributed and interconnected with each other as well as with the user terminals.

The primary purpose of multiaccess computer systems is to share resources. The resources being shared may be simply the data-processing capabilities of the central processor, or they may be the programs and the data bases they utilize. The earliest examples of the first mode of sharing are the general-purpose, time-sharing, computational services. Examples of the latter mode are airlines reservation systems in which it is essential that all ticket agents have immediate access to current information. Both of these classes of systems are still popular, with the proportion and importance of distributed application systems continuing to increase in areas such as corporate management and operations. *See* DATA-BASE MANAGEMENT SYSTEMS.

There are a number of economic factors supporting the growth of multiaccess systems. Primary among these are improvements in hardware performance and economics, with the latter being the principal factor. Digital-computer hardware continues to exhibit rapidly decreasing costs while its operational capabilities increase. At the same time most of the other cost factors involved with corporate operations such as labor, travel, and communications are increasing, and the cost of

time delays in operations such as inventory management and confirming orders, as well as the cost of investment funding, is becoming significant.

System components. The major hardware components of a multiaccess computer system are terminals or data entry/display devices, communication lines to interconnect the terminals to the central processors, a central processor, and on-line mass storage.

Terminals may be quite simple, providing only the capabilities for entering or displaying data. Also utilized in multiaccess systems are terminals having an appreciable amount of "local intelligence" to support simple operations like editing of the displayed text without requiring the involvement of the central processor. Some terminals provide even more extensive support, such as the local storage of small amounts of text. It is also possible to assemble clusters of terminals to share local logic and storage.

The interconnecting communication lines can be provided by utilizing the common-user telephone system or by obtaining leased, private lines from the telephone company or a specialized carrier. Another important source of data circuits is a value-added network specializing in providing data transmission services. *See* DATA COMMUNICATIONS.

The communications interface for the processor may be provided by an integral hardware component or by a separate device known as a communications controller or front-end (communications) processor. The detailed operations involved in the control of a single communications line are not very great; however, servicing a large number of such lines may present an appreciable load to the central processor; therefore, separate hardware for communications interfacing is almost essential in a multiaccess computer system.

A central processor suitable for use in a multiaccess system must include the capability to support a large central memory as well as the communications interface mentioned above and the on-line mass storage discussed below.

The desirable characteristics of a mass storage device are: the ability to quickly locate any desired data; the ability to transfer data at a very high speed; the provision of economical storage for a large quantity of data; and a high degree of reliability.

The most common form of mass storage is rotating magnetic disks. Originally, magnetic drums were utilized for this purpose; however, the price, performance, and capacity of disks have resulted in their replacing drums. *See* COMPUTER STORAGE TECHNOLOGY.

System operating requirements. A multiaccess system must include the following functional capabilities: (1) multiline communications capabilities that will support simultaneous conversations with a reasonably large number of remote terminals; (2) concurrent execution of a number of programs with the ability to quickly switch from executing the program of one user to executing that of another; (3) ability to quickly locate and make available data stored on the mass storage devices while at the same time protecting such data from unauthorized access.

The ability of a system to support a number of

simultaneous sessions with remote users is an extension of the capability commonly known as multiprogramming. In order to provide such service, certain hardware and software features should be available in the central processor. Primary among these is the ability to quickly switch from executing one program to another while protecting all programs from interference with one another. These capabilities are normally provided by including a large central memory in the processor, by providing hardware features that support rapid program switching, and by providing high-speed transfers between the central memory and the mass storage unit on which the programs and the data they utilize are stored.

Memory sharing is essential to the efficient operation of a multiaccess system. It permits a number of programs to be simultaneously resident in the central memory so that switching execution between programs involves only changing the contents of the control registers. The programs that are resident in central memory are protected from interfering with one another by a number of techniques. In earlier multiprogramming systems, this was accomplished by assigning contiguous memory space to each program and then checking every access to memory to ensure that a program was accessing only locations in its assigned space. The drawback to this type of memory allocation is that the entire program had to be loaded into memory whenever any portion of it was to be executed. A popular memory management technique is the utilization of paging. The program is broken into a number of fixed-size increments called pages. Similarly, central memory is divided into segments of the same size called page frames. (Typical sizes for pages and page frames are 512 to 4096 bytes.) Under the concept known as demand paging, only those pages that are currently required by the program are loaded into central memory. Page frames may be assigned to a given program in a random checkerboard fashion. Hardware capabilities are provided to automatically manage the assignment of page frames as well as to make them appear to the program as one continuous address space. At the same time, the hardware provides memory access protection.

Software capabilities. The control software component of most interest to an interactive user is the command interpreter. It is only one portion of a larger control program known as the operating system; however, in multiaccess systems it differs greatly from the command interpreter in a system providing batch multiprogramming service on a noninteractive, non-terminal-oriented basis. This routine interacts directly with users, accepting requests for service and translating them into the internal form required by the remainder of the operating system, as well as controlling all interaction with the system.

The operating system must also have the capability for controlling multiprogramming; that is, the concurrent execution of a number of user programs quickly switching from one to another during their execution as well as controlling memory sharing. The capability to page the memory as outlined above can be utilized to provide users with the impression that each has available a memory space much larger than is actually assigned. Such a system is said to provide a virtual memory environment. Similarly, the ability of the operating system to quickly change context from one executing program to another will result in users' receiving the impression that each has an individual processor. This is especially true when considering the large difference between the response and thinking time of a human compared to the computer's processing time. For the interactive users, such a capability results in the impression of having a private virtual machine.

The mass storage device mentioned above as hardware required for a multiaccess system must be supported by an efficient file management system. The latter is responsible for maintaining current information as to the physical location of the data stored on the mass storage device as well as providing a capability for quickly locating those data and controlling their transfer to the central memory for utilization. In addition, the file management system must provide protection of data from unauthorized access. *See* DATA-PROCESSING SYSTEMS; DIGITAL COMPUTER.

[PHILIP H. ENSLOW, JR.]

Bibliography: D. Davies et al., *Computer Networks and Their Protocols*, 1979; D. Doll, *Data Communications*, 1978; J. Martin, *Introduction to Teleprocessing*, 1972.

Multimeter

A common term for a volt-ohm-milliammeter, also called an analyzer or circuit analyzer. A much less common usage applies to a self-contained test instrument containing two or more single- or multi-scale indicating instruments for measuring simultaneously two or more electrical quantities. *See* VOLT-OHM-MILLIAMMETER.

[ISAAC F. KINNARD/EDWARD C. STEVENSON]

Multiplexing

The means or result of the means whereby a number of independent messages, each unambiguously represented by information-bearing signals, may be sent and received over a common transmitting medium. Viewed broadly, a multiplex system is any part of a communication system in which two or more distinct signal channels are combined. Multiplexing is generic to most present-day systems of telecommunications.

Theoretically, many types of multiplexing are possible. Typical examples are time division, frequency division, and phase discrimination. In time division, messages are multiplexed by sequentially allocating different time intervals for the transmission of each message; in frequency division, different frequency bands are allocated to different message channels; and in phase discrimination there need be no separation of channels in either frequency or time, that is, each message channel may utilize the entire available frequency spectrum all of the time. *See* AMPLITUDE MODULATION; FREQUENCY MODULATION.　[HAROLD S. BLACK]

Multiprocessing

The situation in which a computer system includes two or more processing units under the control of a single copy of the operating-system software. This term has sometimes been used to describe the situation in which two processors are connected via bus or cable, with each operating under its own copy of the operating system (sometimes identical copies, sometimes different). The latter configuration is more correctly termed a multicomputer system, in which each of the computers may be either a uniprocessor or multiprocessor configuration.

A typical multiprocessor system is shown in the illustration. In some cases these systems make use of entirely shared main memory (as in the system in the illustration); that is, all processors can access all areas of memory. In other cases there are some shared-memory areas and some areas which are private to each processor. When the latter approach is used, the shared memory includes the operating system and other common software (language compilers, for example) as well as communication regions through which the processors can exchange data.

Advantages. Originally all computer systems were single-processor configurations. The use of multiprocessing began for two reasons: to increase processing capacity and to improve availability.

The use of a multiprocessor configuration to increase capacity is an alternative to building equivalently powerful uniprocessors. During the 1960s and 1970s several computer manufacturers chose the multiprocessor approach while others concentrated on higher-capacity uniprocessor systems. One manufacturer emphasized loosely coupled multicomputer configurations of uniprocessors; for example, the attached processor (AP) in which one computer performed input/output data management for another computer which carried out the major processing functions. In the 1980s the trend appears to be to the more general use of multiprocessing, partly because the overall requirement for computing capacity is increasing more rapidly than the computer industry can provide through increased capacity in single processors. In addition, availability—the second reason for multiprocessing—is becoming increasingly important.

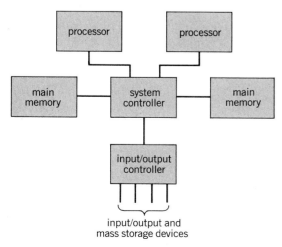

Typical multiprocessor system.

Improved availability has usually resulted from the use of multiprocessing, because the failure of one processor need not cause the system to fail. The ability to continue operation (although with lower capacity) even when one or more components fail is referred to as graceful degradation. Until the mid-1970s, the graceful degradation possible through the use of multiprocessing tended to be an underrated advantage. As long as batch processing dominated computer use, availability was considered somewhat less important than throughput. However, with the accelerated move to on-line computing during the 1970s, higher availability than was possible with a uniprocessor system became essential.

Types of systems. Many of the pre-1980s multiprocessor systems employed an independent-processor technique. Each processor operated essentially independently, with a central set of dispatching routines (part of the operating system) used to determine which task each processor would carry out next. For example, processor no. 1 might begin execution of an application program and continue that execution until an input/output operation was required. While that operation was being carried out, processor no. 1 might work on another application. On completion of the input/output for the first application, processor no. 2 or no. 3 might pick up that processing and continue it until the next interruption. Depending upon the specific hardware design, all processors might be equally able to carry out all instructions, or a master (or control) processor might be designated to handle specific operations such as the management of input/output interrupts.

There are some multiprocessor-system designs which use different approaches. One example is the Illiac system, built as a one-of-a-kind computer for the National Aeronautics and Space Agency. The Illiac, which has now been retired, included 64 processors, all of which executed exactly the same instructions in parallel at the same time. For many common business-oriented applications, this type of parallel processing is of no value. However, there are problems, such as calculations related to nuclear reactions, in which parallel processing makes an enormous difference in the ability to obtain solutions in a practical amount of processing time. [The replacement of Illiac by a Cray-1, which is essentially a uniprocessing system (although the main processor is supported by peripheral processors which handle the tasks associated with input/output and similar functions), is an illustration of the advances which have taken place in computer technology.]

In some cases multiprocessing is used mainly, or entirely, because of its ability to improve system availability. One system has taken advantage of the requirement for increased availability by introducing a multiprocessor (two processors) system which also includes the dualing of other components, as well as dual copies of the software. That system is specifically designed for high uptime transaction processing.

A slightly different form of multiprocessor system was subsequently introduced to compete in the same market. Unlike the previous configuration, which uses an independent-processor approach, with fallback to a uniprocessor system in case of failure, this system uses two processors which operate in parallel on the same instructions. If either fails, the system continues to operate with the same level of performance as before. Only the continued improvements which have been made in computer-hardware price/performance have made it commercially feasible to offer this level of redundancy at a competitive price.

A dyadic-processor configuration includes two processors which operate under control of the same copy of the operating system and share access to all system resources. Fallback to a uniprocessor configuration is possible if either processor fails, although throughput is reduced if this occurs. This is an example of the independent-processor approach, but so far is limited to two processors (hence the term dyadic).

Prospects. Multiprocessing will continue to be an important capability in large-scale computer systems, and in some smaller systems as well, largely because of availability concerns. On-line systems often need very high availability, and as applications come to support the main-line operational functions of many organizations this need will intensify. When uniprocessor systems are employed in high up-time situations, a complete backup system may be required. Airline reservation systems are a good example of this situation. Typically, they run on large uniprocessor configurations, with a second (approximately) equivalent configuration available for backup. If the main computer fails, a switchover to the backup computer takes place. Often manual intervention is involved, and switchover takes more time than if a multiprocessor system with the ability to gracefully degrade in case of processor failure were used. In the future, therefore, multiprocessing will generally be used for on-line systems, even if uniprocessor systems of sufficient capacity to handle the workload are available.

Very large scientific computer systems, in contrast, will probably continue to be predominantly uniprocessor configurations. Multiprocessing does involve additional overhead, because of the hardware or software required to coordinate the actions of the processors. In scientific systems, the minimization of overhead to increase raw processing capacity is very important. Scientific systems, therefore, will typically be of the same type as today's with a very powerful uniprocessor (or perhaps a parallel-processing array) used for major calculations, while subsidiary processors handle input/output and other minor functions.

[GRAYCE M. BOOTH]

Bibliography: An application-oriented multiprocessing system, *IBM Sys. J.*, special issue, vol. 6, no. 2, 1967; E. L. Glaser, J. F. Couleur, and G. A. Oliver, System design of a computer for time sharing applications, in *Proceedings of the AIFPS Fall Joint Computer Conference*, Part I, pp. 197–201.

Multivibrator

A form of relaxation oscillator consisting normally of two or more active devices, such as transistors, interconnected by electric networks. In a multivibrator a portion of the output voltage or current of each active device is applied to the input of the other with such magnitude and polarity as to maintain the devices alternately conducting over controllable periods of time. The transition time of each device from one state to the other is extremely short. As a consequence the voltage waveform from the output of each of the devices is essentially rectangular in form.

Multivibrators are classified by the manner in which the reversal-of-state action of each device is initiated and by the method of control of the time interval in each state, whether from external sources or from the decay of voltage across a capacitor in circuits containing RC time constants within the multivibrator itself.

Symmetrical bistable multivibrator. In bistable multivibrators either of the two devices may remain conducting, with the other nonconducting, until the application of an external pulse. Such a multivibrator is said to have two stable states.

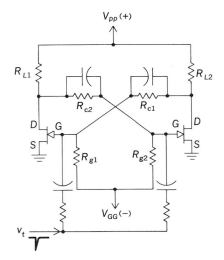
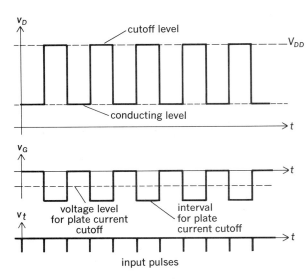

Fig. 1. Bistable multivibrator with triggering, grid, and plate waveforms shown for one tube.

JFET circuits. The original form of bistable multivibrator made use of vacuum tubes and was known as the Eccles-Jordan circuit, after its inventors. It was sometimes called a flip-flop or binary circuit because of the two alternating output voltage levels. The junction field-effect transistor (JFET) circuit (Fig. 1) is a modern version of the Eccles-Jordan circuit. Its resistance networks between positive and negative supply voltages are such that, with no current flowing to the drain of the first JFET, the voltage at the gate of the second is slightly negative, zero, or limited to, at most, a slightly positive value. The resultant current in the drain circuit of the second JFET causes a voltage drop across the drain load resistor; this drop in turn lowers the voltage at the gate of the first JFET to a sufficiently negative value to continue to reduce the drain current to zero. This condition of the first device OFF and the second ON will be maintained as long as the circuit remains undisturbed.

If a fast negative pulse is then applied to the

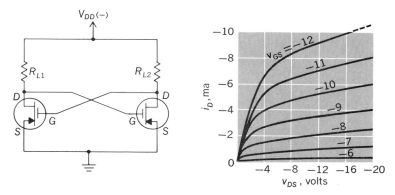

Fig. 3. Direct-coupled, bistable multivibrator which uses *p*-channel enhancement-mode, field-effect transistors, resulting in considerable simplification.

base of the transistor corresponds to the gate, the emitter to the source, and the collector to the drain. Although waveforms are of the same polarity and the action is roughly similar to that of the JFET circuit, there are important differences. The effective resistance of the base-emitter circuit, when it is forward-biased and being used to control collector current, is much lower than the input gate resistance of the JFET when used to control drain current (a few thousand ohms compared to a few megohms). This fact must be taken into account when the divider networks are designed. If *pnp* transistors are used, all voltage polarities and current directions are reversed.

IGFET circuits. Insulated-gate field-effect transistors (IGFETs) may be used effectively in multivibrators. Use of enhancement-mode insulated-gate field-effect transistors permits direct cross-coupling with considerable simplification (Fig. 3).

Unsymmetrical bistable circuits. Bistable action can be obtained in the emitter- or source-coupled circuit with one of the set of cross-coupling elements removed (Fig. 4). In this case, regenerative feedback necessary for bistable action is obtained by the one remaining common coupling element, leaving one emitter or gate free for triggering action. Biases can be adjusted such that device 1 is ON, forcing device 2 to be OFF. In this

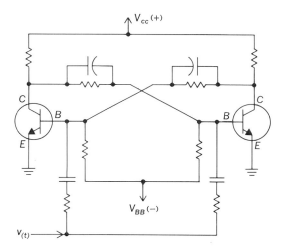

Fig. 2. Bistable multivibrator using *npn* bipolar transistors. Action is similar to vacuum-tube circuit.

gate of the ON transistor, its drain current decreases and its drain voltage rises. A fraction of this rise is applied to the gate of the OFF transistor, causing some drain current to flow. The resultant drop in drain voltage, transferred to the gate of the ON transistor, causes a further rise at its drain. The action is thus one of positive feedback, with nearly instantaneous transfer of conduction from one device to the other. There is one such reversal each time a pulse is applied to the gate of the ON transistor. Normally pulses are applied to both transistors simultaneously so that whichever device is ON will be turned off by the action. The capacitances between the gate of one transistor and the drain of the other play no role other than to improve the high-frequency response of the voltage divider network by compensating for the input capacitances of the transistors and thereby improving the speed of transition.

Bipolar transistor circuits. A bipolar transistor counterpart of the JFET bistable multivibrator with *npn* bipolar transistors is shown in Fig. 2. The

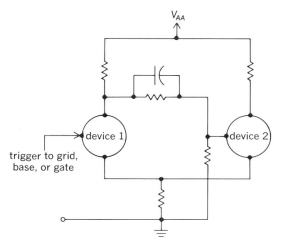

Fig. 4. Unsymmetrical bistable multivibrator.

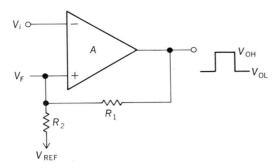

Fig. 5. Operational amplifier comparator used as Schmitt trigger circuit.

case, a pulse can be applied to the free input in such a direction as to reverse the states. Alternatively, device 1 may initially be OFF with device 2 ON. Then an opposite polarity pulse is required to reverse states. Such an unsymmetrical bistable circuit, historically referred to as the Schmitt trigger circuit, finds widespread use in many applications.

A Schmitt trigger circuit often employs a high-gain operational amplifier of the type normally used as a comparator in a positive-feedback or regenerative mode (Fig. 5). When $V_i > V_F$, the output is in its low state V_{OL}, with V_F in turn determined by V_{OL}, V_{REF}, R_1, and R_2. When $V_i < V_F$, the output will be in its high state V_{OH}, and V_F will be at a correspondingly higher level. Thus if the input V_i switches between two levels, the output will switch between its low and high states. These

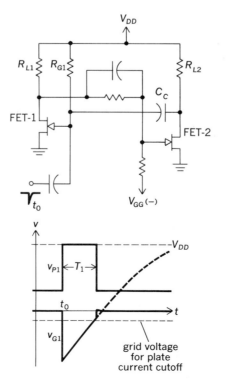

Fig. 6. Basic monostable multivibrator using vacuum triodes. Almost identical characteristics are obtained with n-channel junction, field-effect transistors.

switching levels are different depending upon whether V_i is increasing or decreasing. The difference in $V_i = V_F$ for the two levels is known as the hysteresis of the circuit. When the input V_i is between the low and high V_F levels, the output can be in either its high or low state depending upon the previous turn-on history; hence the term bistable circuit.

Monostable multivibrator. A monostable multivibrator has only one stable state. If one of the normally active devices is in the conducting state, it will remain so until an external pulse is applied to make it nonconducting. The second device is thus made conducting and remains so for a duration dependent upon RC time constants within the circuit itself.

A typical monostable multivibrator is shown in Fig. 6. The input of field-effect transistor (FET) 1 is

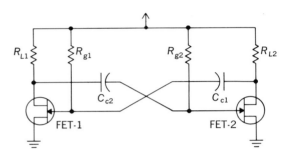

Fig. 7. Astable or free-running multivibrator.

capacitance-coupled to the output of FET 2. In the absence of external pulses, FET 1 is conducting, with its gate at zero potential, or limited to a slight positive value by saturation. The resultant drain current limits the drain voltage to a value that makes the gate of FET 2 sufficiently negative to keep FET 2 cut off. If a negative pulse is applied to the gate of FET 1, this FET is cut off and FET 2 conducts. The circuit action is similar to that of the bistable circuits except that the voltage drop at the drain of FET 2 is transferred to the gate of FET 1 through the capacitance C_c. This change, or transition, cannot be maintained indefinitely because the current flowing through C_c and R_{G1} causes a decrease in voltage with time across C_c and rise in voltage at the gate of FET 1, as shown. The initial drop is of the same magnitude as that at the drain of FET 2 at the time the trigger is applied. The ensuing rise is exponential in form and (if R_{G1} is much greater than R_{L2}) is given by Eq. (1). When

$$V_{G1} = (V_{GG} - V_{DD}) \exp\left[-t/R_{G1}C_c\right] + V_{DD} \quad (1)$$

the rising voltage reaches the level V_{c1}, drain current again flows in FET 1. Positive feedback quickly causes FET 1 to become fully conducting and limited by gate current saturation. The duration of the nonconducting interval for FET 1 time (T_1 in Fig. 6) is found by solving Eq. (1) for time $t = T_1$ and for $V_{G1} = V_{c1}$. The result is Eq. (2). The circuit will

$$T_1 = -R_{G1}C_c \ln \frac{V_{c1} - V_{DD}}{V_{GG} - V_{DD}} \quad (2)$$

remain in this state until another initiating trigger is applied.

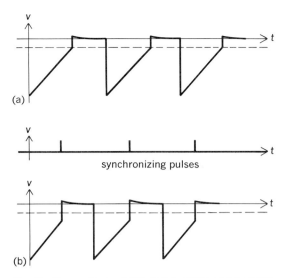

(a)

v

synchronizing pulses

v

(b)

Fig. 8. Comparison of the free-running and synchronized multivibrator waveforms. (a) Free-running waveform. (b) Synchronized waveform.

Astable multivibrator. The astable multivibrator has capacitance coupling between both of the active devices and therefore has no permanently stable state. Each of the two devices functions in a manner similar to that of the capacitance-coupled half of the monostable multivibrator, as shown in Fig. 7. It will therefore generate a periodic rectangular waveform at the output with a period equal to the sum of the OFF periods of the two devices. The duration of each of the two periods is governed by an equation of the form of Eq. (2), with the appropriate values for each of the two parts of the circuit used. The transistor astable multivibrator similarly functions as the combination of two transistor monostable sections coupled together.

Astable multivibrators, although normally free-running, can be synchronized with input pulses recurrent at a rate slightly faster than the natural recurrence rate of the device itself. This is illustrated in Fig. 8, which shows the relation between the internal waveform and the applied synchronizing pulses. If the synchronizing pulses are of sufficient amplitude, they will bring the internal wave-

form to the conduction level at an earlier than normal time and will thereby determine the recurrence rate.

Triggering of multivibrators. The period of the bistable, the monostable, and the synchronized astable multivibrator is controlled by pulses (triggers) from an external source. These triggers may be applied to the circuit in various fashions. The initiating trigger should be sufficiently wide for the circuit to respond (as limited by its high-frequency response) before the pulse is over, but not so wide as to interfere with normal action of the multivibrator once the transition has taken place. The trigger should be coupled to the multivibrator through a small capacitance so that loading by the trigger source is negligible. Usually a faster transition can be achieved if triggers turn off a normally ON device. Triggers are usually applied to the appropriate input, but they may also be applied

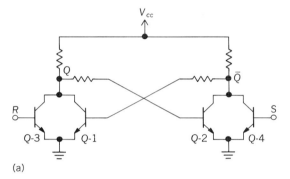

(a)

Q_n	$\bar{Q}_n$	R	S	Q_{n+1}	$\bar{Q}_{n+1}$
1	0	0	0	1	0
0	1	0	0	0	1
1	0	1	0	0	1
0	1	0	1	1	0
0	1	1	0	0	1
1	0	0	1	1	0
1	0	1	1	?	?
0	1	1	1	?	?

(b)

Fig. 10. Cross-coupled RTL gates as R-S flip-flop. (a) Circuit. (b) Truth table.

to the outputs and reach the inputs through the coupling networks. In some cases any coupling between trigger source and multivibrator is objectionable, and an isolating amplifier is used with its plate and the plate of the multivibrator connected together. Auxiliary diodes are frequently used to provide trigger isolation and to improve triggering stability. *See* TRIGGER CIRCUIT.

Also, particularly where bistable circuits are used in computer logic systems, it is necessary to reset initial states so that the desired device is ON before a subsequent set of triggers are applied. Such a multivibrator provides extra available terminals for triggering purposes, as illustrated in Fig. 9. *See* DIGITAL COMPUTER.

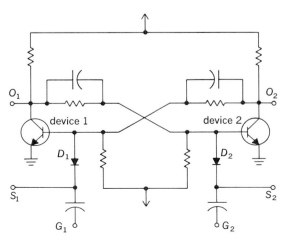

Fig. 9. Multivibrator triggering techniques.

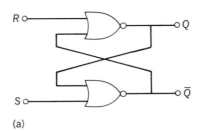

(a)

R	S	Q_{n+1}	$\bar{Q}_{n+1}$
0	0	Q_n	$\bar{Q}_n$
1	0	0	1
0	1	1	0
1	1	not used	

(b)

Fig. 11. NOR gate R-S flip-flop. (a) Symbolic representation. (b) Abridged truth table.

Logic gate multivibrators. Multivibrators may be formed by using two cross-coupled logic gates, with the unused input terminals used for triggering purposes. Such circuits are usually referred to as flip-flops. For example, the circuit of Fig. 10, which is like that of Fig. 2, may be thought of as the cross connection of two, two-input RTL (resistor-transistor-logic) gates. The two extra inputs are available for set-reset triggering functions. The inputs and outputs are at standard logic levels, (0) and (1). The output levels after inputs at R and S are removed depend upon the input gate combinations. This is illustrated by the accompanying truth table, where Q_n and $\bar{Q}_n$ are the output states before the input gate signals are applied, and Q_{n+1} and $\bar{Q}_{n+1}$ are the outputs after signals at R and S are removed. The outputs are independent of the prior states unless R and S are both zero, in which case they remain unchanged. Also if both inputs are logic (1) the output is indeterminate, since if they both go to (0) at the same time the outputs may return to either bistable condition. This combination is generally indicated as "not used." The RTL flip-flop is representative of two cross-coupled NOR gates shown symbolically in Fig. 11 with an abridged

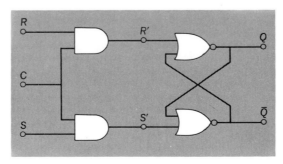

Fig. 12. Clocked R-S flip-flop.

truth table leaving out the unnecessary initial state columns. The circuit function illustrated by these two circuits is generally referred to as an R-S flip-flop. Such flip-flops identified by the same truth table can be constructed by using cross-coupled NAND gates with inverting stages added at the inputs.

If the R-S flip-flop inputs are preceded by AND gates [Fig. 12, where R' or S' can be (1) only if both inputs to the AND gate are (1)] with a precisely timed "clock" pulse applied during the existence of the R and S functions, the output state of the flip-flop is set at the time that the clock pulse is applied. This is called a clocked R-S flip-flop and has the same truth table as the unclocked version.

If a clocked R-S flip-flop uses a three-input NAND gate at the inputs with a feedback path consisting of a single-gate noninverting delay (Fig.

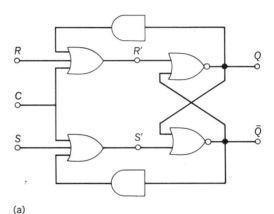

(a)

R	S	Q_{n+1}	$\bar{Q}_{n+1}$
0	0	Q_n	$\bar{Q}_n$
1	0	0	1
0	1	1	0
1	1	$\bar{Q}_n$	Q_n

(b)

Fig. 13. Pulse-triggered R-S flip-flop. (a) Circuit. (b) Truth table.

13), the inputs are "steered" in such a way that the not-used state is removed as indicated by the accompanying truth table. This works only if the clock pulse is very narrow so that it is removed before the feedback action is completed. Such a circuit is often referred to as a pulse-triggered R-S flip-flop.

Additional circuits which make the operation of the flip-flop independent of pulse width result in a circuit known as a J-K flip-flop, shown symbolically in Fig. 14 with a truth table which identifies J and K inputs in a standard manner.

Sometimes problems exist with J-K flip-flops for slow rising pulses at inputs J and K. Additional circuits are sometimes used whereby one flip-flop

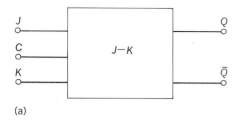

(a)

J	K	Q_{n+1}	$\bar{Q}_{n+1}$
0	0	Q_n	$\bar{Q}_n$
0	1	0	1
1	0	1	0
1	1	$\bar{Q}_n$	Q_n

(b)

Fig. 14. J-K flip-flop. (a) Symbolic representation. (b) Truth table with standard output designation.

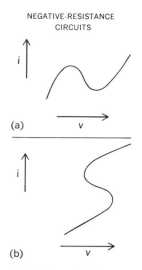

NEGATIVE-RESISTANCE CIRCUITS

(a)

(b)

Fig. 1. Examples of (a) voltage-stable and (b) current-stable current-voltage characteristic curves.

called a master drives another flip-flop called a slave with feedback connections such that relative independence of clock pulse width or rise and fall times is achieved. Such a circuit is called a master-slave J-K flip-flop. The truth table is that of the ordinary J-K flip-flop. *See* LOGIC CIRCUITS; TRANSISTOR.

[GLENN M. GLASFORD]

Bibliography: W. H. Eccles and F. W. Jordan, A trigger relay utilizing three-electrode thermionic vacuum tubes, *Radio Rev.*, no. 1, p. 143, 1919; J. Millman, *Microelectronics*, 1979; H. Taub and D. Schilling, *Digital Integrate Electronics*, 1977.

Negative-resistance circuits

Circuits or devices whose static (dc) current-voltage characteristic at one or more ports has a range in which the slope is negative. Figure 1 shows two typical negative-resistance characteristics in which a positive increment of voltage is accompanied by a negative increment of current. Strictly, the term also includes circuits whose static current-voltage characteristics do not have negative-slope ranges, but in which the input impedance or admittance at one or more ports has a negative real component. The negative resistance may arise from inherent properties of a device or from the proper choice of electrical circuit configuration and operating conditions. Only circuits having static characteristics of the general form of Fig. 1 are treated in this article. Negative-resistance circuits and the circuits based upon them find many applications in counting or information storage, in the generation of pulses or alternating voltages, in instrumentation, in the control of current and power, and in physical realization of idealized (lossless) or hypothetical circuits.

Basic circuits. A port having a characteristic of the form of Fig. 1a is said to be voltage-stable (voltage-controllable) because at any value of voltage there is only one value of current, but over a

range of current there are three values of voltage, between which transition may take place spontaneously or under excitation. Similarly, a port having a characteristic of the form of Fig. 1b is said to be current-stable (current-controllable) because for each value of current there is only one value of voltage, but over a range of voltage there are three values of current, between which transition may take place. In general, negative-resistance circuits may have both voltage-stable and current-stable ports. Because the negative-slope portion of the current-voltage characteristic, or the entire characteristic, may be located anywhere with respect to the origin of coordinates, the origin is not shown in Fig. 1.

Figure 2 shows one type of negative-resistance circuit. The indicated direction of input current Δi resulting from a positive increment of input voltage Δv is that which would obtain if the input resistance of the port were positive (dissipative). If the voltage amplification A of the amplifier is positive and greater than unity, the output voltage $A\Delta V$ is of the same sign as the input voltage and greater in magnitude. The current Δi_r through the resistance R is then in the direction shown. If this current exceeds the current Δi_i into the amplifier input, Δi must be opposite in direction to that shown. In other words, the positive increment of impressed voltage produces a negative increment of current, and the slope of the characteristic relating current i with voltage v is negative. Because the range of input voltage over which the amplifier has an amplification greater than unity is limited, the negative-slope portion of the characteristic is also limited with respect to voltage, and the characteristic is the voltage-stable form shown in Fig. 1a. A similar analysis shows that a voltage-stable port can be formed across the output terminals of the amplifier and that current-stable ports can be formed in place of resistor R or by opening the connection between the lower input and output terminals. Voltage-stable and current-stable ports can in general also be formed within the amplifier itself. *See* AMPLIFIER.

A more general form of the circuit of Fig. 2 is shown in Fig. 3a. As in the circuit of Fig. 2, feedback must be positive (regenerative), and the open-loop amplification of the circuit must be positive and greater than unity. *See* FEEDBACK CIRCUIT.

Another generalized form of negative-resistance circuit is shown in Fig. 3b, in which positive feedback from output to input of the amplifier is pro-

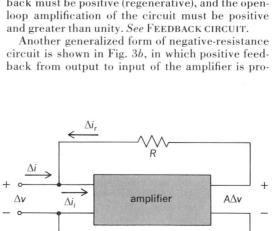

Fig. 2. Example of a circuit that has a voltage-stable current-voltage characteristic.

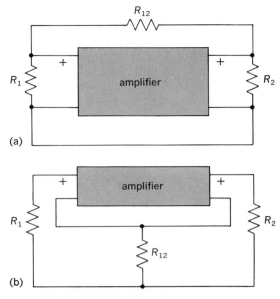

(a)

(b)

Fig. 3.   Generalized negative-resistance circuits. (a) Generalization of circuit in Fig. 2 (b) Circuit in which positive feedback is produced by the coupling resistance R_{12}.

duced by the coupling resistance R_{12}. In this circuit, voltage-stable ports can be formed in place of R_{12} or across the input or output terminals of the amplifier, and current-stable ports can be formed in place of R_1 or R_2 or in series with the lower input or output terminal of the amplifier.

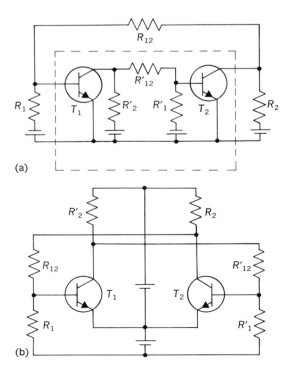

(a)

(b)

Fig. 4.   Eccles-Jordan negative-resistance circuit (a) drawn to show amplifier in Fig. 3 (a), and (b) drawn in symmetrical form.

The two essentials of a negative-resistance circuit are amplification and positive feedback; any circuit modification that tends to increase the open-loop amplification of the amplifier and feedback network is favorable to the production of negative resistance. In order that negative resistance can be obtained under static conditions, it is necessary that the amplifier be capable of amplifying increments of direct voltage.

Typical circuits. All negative-resistance circuits are of the general form of Fig. 3a or b or variants thereof. They differ with respect to details of the amplifier. Although only transistor circuits are discussed in the remainder of this article, any type of dc amplifier may be used. In the most commonly used circuit, the Eccles-Jordan circuit, the amplifier consists of two resistance-coupled common-emitter stages. The amplifier is used in the circuit of Fig. 3a, as shown in Fig. 4a. The com-

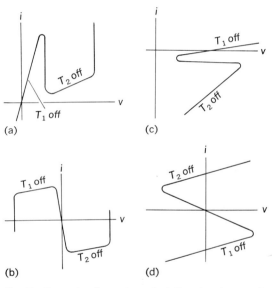

(a)

(b)

(c)

(d)

Fig. 5.   Current-voltage characteristics at various ports of the circuit of Fig. 4: (a) at port formed in place of R_2 or R'_2; (b) between the two collectors; (c) at port formed in place of either R_{12} or R'_{12}; (d) between the two emitters when additional equal resistances are inserted between each emitter and the voltage supply.

plete circuit is usually drawn in the symmetrical form of Fig. 4b. Because of the symmetry, ports formed across or in place of R'_1, R'_{12}, and R'_2 have the same forms of current-voltage characteristics as those formed across or in place of R'_1, R'_{12}, and R_2, respectively. This fact justifies for this circuit the statement that negative-resistance ports can in general be formed within the amplifiers of Fig. 3.

Typical current-voltage characteristics measured in the circuit of Fig. 4 are shown in Fig. 5. Characteristic a is that of a port formed in place of R_2 or R'_2. Characteristic b is obtained between the two collectors, and the current-stable characteristic c is observed at a port formed in place of either R_{12} or R'_{12}. Symmetrical current-stable characteristic d is obtained between the two emitters when additional equal resistances are inserted between each emitter and the voltage supply.

Figure 6 shows a form of the circuit of Fig. 3a in which the amplifier consists of a common-base stage directly coupled to a common-collector stage. Voltage-stable characteristic b is observed at a port formed in place of R_{12}; current-stable characteristic c is observed at a port formed in place of R_1. A current-stable characteristic is also observed between the two emitters.

Bistable circuits. Any negative-resistance circuit or device can be converted into a bistable circuit simply by connecting a resistance R of proper magnitude in series with a properly chosen voltage V_s across any port, as in Fig. 7a. That such a circuit can be multistable is shown by the diagrams of Fig. 7. In these diagrams the straight line of slope $-1/R$ is the locus of all corresponding values of current through resistance R and of resulting voltage v across the port, determined by the relation $v = V_s - iR$. In Fig. 7b the negative-resistance characteristic is that of a voltage-stable port of the internal negative-resistance circuit. It is the locus of all possible combinations of voltage across the port and current into the port. In the complete circuit, because the current through the resistance is also the current into the port, intersections of the resistance line with the negative-resistance characteristic give possible equilibrium values of current and voltage across the port.

If the slope of the resistance line in Fig. 7b is less than the maximum slope of the negative-slope portion of the characteristic, as it will be if resistance R exceeds the magnitude of the negative resistance, there are three intersections. Analysis of the circuit with inherent capacitances taken into consideration shows that point 3 is always unstable. If the current i and the voltage v initially have values corresponding to point 3, any small circuit disturbance causes the voltage to decrease or increase until it reaches the value at point 1 or that

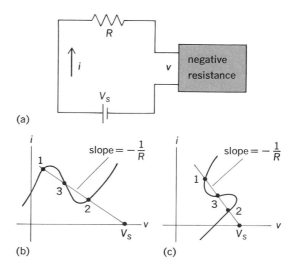

Fig. 7. Bistable circuit. (a) Circuit diagram. (b, c) Current-voltage characteristics.

at point 2, both of which correspond to stable conditions of equilibrium. Similarly, three intersections may be obtained with a current-stable characteristic if the resistance R is less than the magnitude of the negative resistance, as in Fig. 7c. Transition from one stable state to the other can be initiated by changing (1) the magnitude of R and therefore the slope of the resistance line, (2) the value of V_s and thus the voltage intercept of the resistance line, or (3) one or more biasing voltages in the amplifier and hence the position or shape of the current-voltage characteristic. It follows that the basic negative-resistance circuits of Fig. 3 are in themselves bistable circuits if resistances and supply voltages are properly chosen. Reference to Fig. 5, or to similar characteristic curves for other two-transistor negative-resistance circuits, shows that one transistor is off and the other conducts in each stable state.

The usual means of initiating transition from one stable state to the other in bistable circuits is the application of short triggering pulses of voltage to one or more electrodes of the amplifier transistors. The fact that a change from one stable state to the other is an indication that a triggering pulse has occurred makes possible the use of bistable circuits in pulse counting and in the storage of information in digital computers. Bistable circuits have also found numerous applications in electronic instrumentation.

Astable circuits. An astable circuit (relaxation oscillator for the generation of nonsinusoidal voltages and currents) can be formed by shunting any voltage-stable port of a negative-resistance circuit by inductance in series with resistance, provided that this resistance is smaller than the magnitude of the negative resistance. In general, such a circuit oscillates over a wide range of inductance and with values of resistance ranging from zero to thousands of ohms. In a similar manner an astable circuit can be formed by shunting any current-stable port by capacitance in parallel with resistance, provided that this resistance is larger than the negative resistance. Relaxation oscillation occurs over wide ranges of capacitance and resistance.

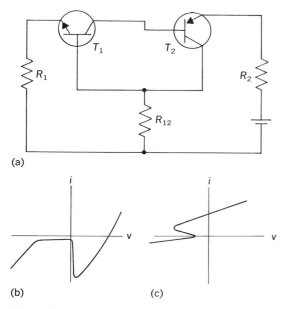

Fig. 6. Negative-resistance circuit based on circuit of Fig. 3b. (a) Circuit diagram. (b) Characteristic at port formed in place of R_{12}. (c) Characteristic at port formed in place of R_1.

The frequency of oscillation of these oscillators is varied by changing the time constant L/R or RC, or one or more operating voltages. The output waveform in general consists of two or more exponential sections or of exponential and sinusoidal sections. A simple oscillator of this type is formed by shunting R_1 or R_2 in the circuit of Fig. 6a by a capacitor, or inserting an inductor in series with, or in place of, R_{12}.

Astable circuits may be looked upon as modifications of bistable circuits. Addition of one or more capacitors or inductors to a bistable circuit converts the two stable states of equilibrium into two quasi-stable states, in each of which the circuit remains for periods determined by the time constants. Transition from one state to the other is followed by charging or discharging of the capacitors or inductors, and thus by gradual changes of electrode voltages. When the voltage of a control electrode reaches a critical value, transition to the previous state of equilibrium occurs. This transition is again followed by discharging or charging of the capacitors or inductors until transition occurs once more, and the cycle repeats.

One of the most useful relaxation oscillators is the multivibrator, which is formed from the Eccles-Jordan circuit of Fig. 4 by replacing the resistors R_{12} and R'_{12} by capacitors. Thus capacitors shunt the current-stable ports between the collector of each transistor and the base of the other. If the circuit is made symmetrical with respect to resistances, capacitances, and voltages, symmetrical voltage waves are produced between the collectors and between the bases. The frequency of oscillation is adjusted by means of the resistors or capacitors or by means of the base supply voltage.

Monostable circuits. The operating voltages of the transistors of an astable circuit can be adjusted so that only one state of equilibrium is quasi-stable while the other is stable. Under standby conditions the circuit remains in the stable state. Application of a triggering pulse to the circuit causes transition to the quasi-stable state, followed by charging or discharging of one or more capacitors or inductors, and thus by a gradual change of electrode voltages. At a critical value of an electrode voltage the circuit rapidly returns to the stable state. The output of such a circuit is a pulse of length determined by the RC or L/R time constants and by one or more supply voltages. The output voltage can be made to approximate a rectangular pulse by proper choice of circuit resistances, capacitances, and supply voltages.

Sine-wave oscillators. A sine-wave oscillator can be formed from a negative-resistance circuit or device by shunting any voltage-stable port by a resonator consisting of inductance and capacitance in parallel, or any current-stable port by a resonator consisting of inductance and capacitance in series. Circuit parameters and voltages must be such that the resistance line corresponding to the dc resistance between the terminals of the resonator intersects the current-voltage characteristic only in the negative-resistance range. A particularly useful oscillator of this type is formed by connecting a parallel resonator between the collectors of the Eccles-Jordan negative-resistance circuit of Fig. 4, as shown in Fig. 8. *See* Os-CILLATOR.

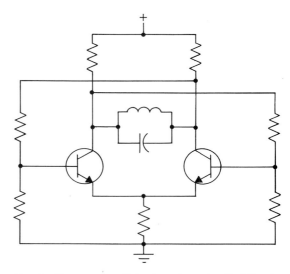

Fig. 8.   Sine-wave oscillator based on circuit of Fig. 4.

Advantages of negative-resistance sine-wave oscillators include relatively small dependence of oscillation frequency upon supply voltages, and the simplicity of the circuits in comparison with feedback oscillators. The symmetry of the current-voltage characteristic used in the oscillator of Fig. 8 and the monotonically decreasing slope of the characteristic with increase of voltage magnitude (Fig. 5b) make possible low amplitude of oscillation and associated low harmonic content of the voltage across the resonator.

Improved amplitude limitation and stabilization can be achieved by any of a number of circuit modifications. One means of limiting the amplitude of oscillation is by rectifying the resonator voltage and using the rectified voltage to reduce the forward bias of the transistor bases, and hence the amplification of the transistors. This in turn reduces the slope of the negative-resistance portion of the current-voltage characteristic as the amplitude of oscillation increases.

Another method of amplitude limitation is using in series with the inductor a resistor that has a positive temperature coefficient of resistance, or in parallel with the resonator a resistor that has a negative temperature coefficient of resistance. Amplitude limitation occurs because heating of the resistor by the current through it causes the resonator losses to increase with amplitude of oscillation. A third method of amplitude limitation is the use of two oppositely connected diodes in parallel with the resonator. Decrease of the effective resistance of the diodes with increase of voltage across them causes the circuit losses to increase rapidly with increase of amplitude.

Simultaneous oscillation at two or more frequencies can be produced by a modification of the circuit of Fig. 8 in which parallel resonators of different frequencies are connected in series between the collectors. Each resonator must incorporate its own temperature-sensitive resistor or pair of diodes. *See* CIRCUIT (ELECTRONICS); NETWORK THEORY.

Negative-resistance devices. Two examples of current-stable negative-resistance devices are the *pnpn* transistor and the unijunction transistor; two examples of voltage-stable devices are the tunnel (Esaki) diode and the Gunn diode. An advantage of negative-resistance devices over circuits that incorporate two transistors to achieve negative resistance is the simplicity of bistable, monostable, and oscillator circuits in which they are used. Tunnel and Gunn diodes have the additional very important advantage that their rapid response makes them ideal for use in amplification and power generation at very high microwave frequencies (as high as 100 GHz) and in high-speed switching circuits.

The *pnpn* transistor, which consists of four alternate layers of *p*-type and *n*-type semiconductor, is equivalent to the two directly coupled transistors of Fig. 6a, the inner *n*-type layer serving both as the collector of T_1 and the base of T_2, and the inner *p*-type layer as the collector of T_2 and the base of T_1. As the *pnpn* transistor is usually used, R_1 is zero. The supply voltage V_s and the load resistance R_2 are then directly between the two outer elements (layers), which correspond to the two emitters in the circuit of Fig. 6a. Typical current-voltage characteristics observed between these elements at two values of current i_G into one of the inner elements, called the gate, are shown in Fig. 9, together with the resistance line corresponding to the load resistance R_2 and supply voltage v_s. Transition from a very low value of current at point 1 to a high value at point 2 can be initiated by a change of gate current from i_{G1} to i_{G2}. Because the voltage across the device is small in the high-current state, this device affords an efficient means of controlling the current of motors and lights. Another important application is in the conversion from alternating to direct current (rectification) and from direct current to alternating current (inversion). Units designed for the control or rectification of large currents are called (silicon) controlled rectifiers. *See* CONTROLLED RECTIFIER; SEMICONDUCTOR RECTIFIER.

The unijunction transistor owes its negative resistance to increase of conductivity of a semiconductor region as the result of increase in the number of injected minority carriers with increase of current. The negative resistance of the tunnel diode is produced by tunneling of electrons across

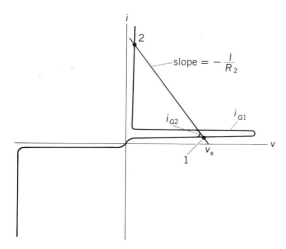

Fig. 9. Current-voltage characteristic of *pnpn* transistor at two values of gate current, i_{G1} and i_{G2}.

a junction that has a high potential barrier. Small capacitance and the rapidity with which tunneling takes place account for the excellent high-frequency response of the tunnel diode. In microwave negative-resistance amplifiers, the negative resistance of a diode is used to compensate for the positive resistance of other circuit elements and thus increase the amplification. *See* MICROWAVE SOLID-STATE DEVICES; TRANSISTOR; TUNNEL DIODE.   [HERBERT J. REICH]

Bibliography: J. J. Brophy, *Basic Electronics for Scientists*, 3d ed., 1977; H. J. Reich, *Functional Circuits and Oscillators*, 1961; H. J. Reich, H. L. Krauss, and J. G. Skalnik, *Theory and Applications of Active Devices*, 1966.

Neper

A unit of attenuation used in transmission-line theory. On a uniform transmission line having waves traveling in only one direction, the magnitudes of voltage E and of current I decrease with distance x traveled, as given by Eq. (1), where E_0,

$$\frac{E}{E_0} = \frac{I}{I_0} = \epsilon^{-\alpha x} \qquad (1)$$

I_0, and α are constants. The attenuation in nepers between the points where E_0 and I_0 are measured and where E and I are measured is given by Eq. (2),

$$\alpha x = \ln \frac{E_0}{E} = \ln \frac{I_0}{I} \qquad (2)$$

in which ln denotes the natural (or Napierian) logarithm.

The word neper originated from a misspelling of the proper name Napier. One neper equals 8.686 dB, the decibel being the practical unit of attenuation. *See* TRANSMISSION LINES.

[EDWARD W. KIMBARK]

Network theory

The systematizing and generalizing of the relations among the elements of an electrical network. To be precise, certain terms will be defined. *See* ALTERNATING-CURRENT CIRCUIT THEORY.

Elements. The elements of a network model are resistance, inductance, and capacitance (the passive elements) and sources of energy (the active elements), which may be either independent sources or controlled, that is, dependent, sources. An independent-voltage source produces voltage between its terminals that is independent of all currents and voltages, although it may be a function of time, as an alternating source would be; an independent-current source carries current that is independent of all voltages or currents but may be a function of time. *See* CAPACITANCE; ELECTRICAL RESISTANCE; ELECTROMOTIVE FORCE (EMF); INDUCTANCE.

A number of definitions, together with theorems that relate them, are taken from the mathematical subject of topology, which deals with certain aspects of form. Two or more elements are joined at a node (Fig. 1). If three or more elements are connected together at a node, that node is called a junction. (The term major node may be used instead of junction; topological terminology varies among authors.) An element extends from one node to another. A branch of a network extends from one junction to another and may consist of

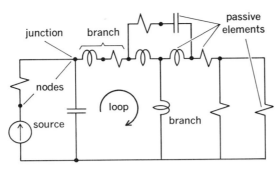

junction   branch
nodes
source   loop
branch
passive elements

Fig. 1. Parts of a network. (*From H. H. Skilling, Electrical Engineering Circuits, 2d ed., copyright © 1965 by John Wiley and Sons, Inc.; used with permission*)

one element or several elements in series. A loop, or circuit, is a single closed path for current. A mesh, or window, is a loop with no interior branch.

Figure 1 shows a network with 13 elements, 12 passive and 1 active. Nodes are indicated by dots; of the 9 nodes, 6 are junctions. There are 10 branches of which 5 come together in a single node or junction at the bottom of the figure.

Equation (1) is an extension of Ohm's law, where

$$V = IZ \tag{1}$$

I is the current in an element, Z is the impedance of that element, and V is the voltage or potential difference between the nodes that terminate that element; this applies to every passive element of an electrical network. [Equation (1) may be a phasor equation for steady alternating current, in which case impedance is a function of frequency and $V(j\omega) = I(j\omega)Z(j\omega)$, or it may more generally be the transform of the differential equation, $V(s) = I(s)Z(s)$.] The number of elements (active or passive) in a network may be designated E. *See* ELECTRICAL IMPEDANCE; OHM'S LAW.

At every node of an electrical network the sum of currents entering that node is zero. Equation (2)

$$\Sigma I = 0 \tag{2}$$

expresses Kirchhoff's current law. An equation of this form can be written for each node, but in a fully connected network one of these equations can be derived from the others; hence the number of independent node equations is one less than the number of nodes. The number of independent node equations, called N, equals in a fully connected network the number of nodes minus one. *See* KIRCHHOFF'S LAWS OF ELECTRIC CIRCUITS.

Around every loop of an electrical network the sum of the voltages across the elements is zero. Equation (3) expresses Kirchhoff's voltage law. A

$$\Sigma V = 0 \tag{3}$$

network such as that of Fig. 1 can have many possible loops and hence many equations of this form, but only a limited number are independent. If L is the number of independent loops, topology gives Eq. (4), from which L can be computed, E and N

$$E = N + L \tag{4}$$

having been counted. Thus in Fig. 1 there are 13 elements and 9 nodes, hence 8 independent nodes,

so that there are $13 - 8$ or 5 independent loops; that is, $L = 5$.

If a network is planar (flat) and fully connected, the number of independent loops is equal to the number of meshes, or windows. Figure 1 shows such a network, and the number of meshes is obviously 5, so again $L = 5$.

Branch equations. There are E elements in a network. Suppose that all impedances are known. Across each element there is a voltage, and through each element there is a current; hence there are $2E$ voltages and currents to be known. One equation is provided by each element; either the element is a source for which voltage or current is given, or it is an impedance for which there is a relation given by Ohm's law in the form of Eq. (1). Hence there are E equations from the elements. From Eqs. (2) and (3) the nodes and loops provide $N + L$ connection equations, and by Eq. (4) $N + L = E$. Thus there are $2E$ equations relating voltages and currents.

In an actual solution for current and voltage in a network, it is probably desirable to reduce the number of equations by combining elements that are in series. This can reduce the number of elements to the number of branches and the number of nodes to the number of junctions; the network is then described by $2B$ branch equations.

These branch equations are easy to write but tedious to solve, unless an electronic computer is used. Two modifications have been devised, however, that eliminate a great deal of the labor and reduce the number of equations from $2B$ to either L, the number of independent loops, or N, the number of independent nodes, as will be described in following paragraphs.

Linearity. Although branch equations can be written for networks of either linear or nonlinear elements, the solution is tremendously more difficult for nonlinear networks. A linear network is one that gives rise to linear systems of equations, which are subject to special methods of solution, and with which the principle of superposition can be applied, permitting the use of loop or node equations. In a linear system the values of resistance, inductance, and capacitance are constant with respect to voltage and current, and a controlled source produces a voltage or a current that is proportional to another voltage or current. *See* SUPERPOSITION THEOREM (ELECTRIC NETWORKS).

Fortunately, many electrical networks are linear or are nearly enough linear to be so considered, at least in the useful range of operation or in a piecewise linear fashion. Examples of branch, loop, and node equations are given below for linear systems.

Examples. Using Ohm's law six times, the branch equations for the network of Fig. 2 are shown in Eqs. (5).

$$V_{BC} = Z_{BC}I_{BC} \qquad V_{CD} = Z_{CD}I_{CD}$$
$$V_{BD} = Z_{BD}I_{BD} \qquad V_{DA} = Z_{DA}I_{DA} \tag{5}$$
$$V_{CA} = Z_{CA}I_{CA} \qquad V_{BA} = Z_{BA}I_{BA} + E$$

The source electromotive forces and branch impedances are assumed to be known, and all currents and branch voltages are to be found. There are, then, 6 equations with 12 unknowns, the unknowns being voltage and current of each branch. (Although a source is indicated in only one of these branches, the method can be applied in the same

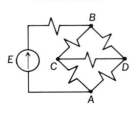

Fig. 2. A network of six branches. (*From H. H. Skilling, Electrical Engineering Circuits, 2d ed., copyright © 1965 by John Wiley and Sons, Inc.; used with permission*)

way if there are sources in any or all of the branches.) Clearly, 6 more equations are needed.

The 6 needed equations may be called connection equations, and they are found from Kirchhoff's laws. The current law gives Eqs. (6), called node

$$I_{BC} + I_{BD} + I_{BA} = 0 \qquad (6a)$$

$$I_{CB} + I_{CD} + I_{CA} = 0 \qquad (6b)$$

$$I_{AC} + I_{AD} + I_{AB} = 0 \qquad (6c)$$

equations or junction equations. Equation (6a) shows the sum of the currents flowing out of junction B to be zero. Equations (6b) and (6c) make similar statements with regard to junctions C and A.

Three more equations may be obtained from Kirchhoff's voltage law. The sum of the voltages around loop BCD must be zero, Eq. (7a); the sum of the voltages around loop CAD must be zero, Eq. (7b); and the sum of the voltages around loop ABC must be zero, Eq. (7c).

$$V_{BC} + V_{CD} + V_{DB} = 0 \qquad (7a)$$

$$V_{CA} + V_{AD} + V_{DC} = 0 \qquad (7b)$$

$$V_{AB} + V_{BC} + V_{CA} = 0 \qquad (7c)$$

There are now 12 equations. Recognizing that $I_{BC} = -I_{CB}$, and so on with other currents and voltages, there are still 12 unknowns, for the 6 connection equations of Eqs. (6) and (7) have added no new unknowns. The 12 equations can be solved simultaneously for the 12 unknowns. The actual solution is not particularly interesting and will not be pursued.

It would seem to be possible to write a fourth junction equation at junction D: $I_{BD} + I_{CD} + I_{AD} = 0$. This proposed equation contains no new information, however, for it could have been derived from the other three equations. It results from adding Eqs. (6a), (6b), and (6c) and canceling equal and opposite quantities. Although it is a true equation, it is not an independent equation. The solution for 12 unknowns requires the use of 12 independent equations.

A somewhat similar observation can be made about the independence of loop equations. The three loops for which equations are written are not the only possible ones. For example, there is also loop $ACBD$, but this loop will not give another independent equation, nor will any of the other possible loops.

Loop equations. The ingenuity of the loop method lies in the selection of currents to be determined. It is necessary to find only as many different currents as there are independent loops, instead of finding as many different currents as there are branches.

Thus in the network of Fig. 3 a current which can be called I_1 flows around loop 1. (This is the current in the source E_1 and in the impedance Z_a.) The convention generally adopted is to take the reference direction for all loop currents as clockwise, and this will be taken as the reference direction of I_1. The reference direction of the current called I_2 that flows around loop 2 is also clockwise. This is the current in the source E_2 and in impedance Z_c. Loop currents I_1 and I_2 both flow in Z_b. The reference direction of one is downward and that of the other is upward. Thus the total current downward in Z_b is $I_1 - I_2$. When I_1 and I_2 are

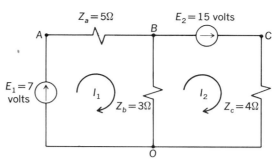

Fig. 3. A network of two loops. (*From H. H. Skilling, Electrical Engineering Circuits, 2d ed., copyright © 1965 by John Wiley and Sons, Inc.; used with permission*)

known, every current in the network is determined.

The first step in solving for these currents is to write network equations in terms of I_1 and I_2. Kirchhoff's voltage law is used to write Eq. (8), which says that the sum of all voltages about loop 1 is zero. Rearranged, Eq. (8) becomes Eq. (9). This

$$E_1 - Z_a I_1 - Z_b (I_1 - I_2) = 0 \qquad (8)$$

$$(Z_a + Z_b) I_1 - Z_b I_2 = E_1 \qquad (9)$$

equation is valid when there is current in both loops of the network.

Similarly, Eq. (10) is written for loop 2 of the

$$(Z_b + Z_c) I_2 - Z_b I_1 = E_2 \qquad (10)$$

network. This is done from Fig. 3 without further explanation. The first term is the voltage in loop 2 when the only current is I_2, and the second term is the voltage in loop 2 caused by I_1. Since E_2 has the same reference direction as I_2, it is positive.

Equations (9) and (10) are a pair of equations with only two unknowns, I_1 and I_2. They can be solved simultaneously by Eqs. (11a) and (11b). By

$$(Z_a + Z_b) I_1 - Z_b I_2 = E_1 \qquad (11a)$$

$$-Z_b I_1 + (Z_b + Z_c) I_2 = E_2 \qquad (11b)$$

way of illustration let the electromotive forces and impedances be given numerical values and solve for the currents as follows. It is assumed that the equations are linear (see above), which implies that the resistances are constant, and that the electromotive forces are independent of current and voltage.

Examples. If the impedances shown in Fig. 3 are pure resistances and with the values shown, Eqs. (12) and (13) may be written. The electromotive

$$8I_1 - 3I_2 = 7 \qquad (12)$$

$$-3I_1 + 7I_2 = 15 \qquad (13)$$

forces may be either dc values or phasors of alternating voltages (the solution is the same). These are linear equations that can be solved by any convenient means. Using determinants and Cramer's rule, they are solved in Eqs. (14) and (15), where D,

$$I_1 = \frac{\begin{vmatrix} 7 & -3 \\ 15 & 7 \end{vmatrix}}{\begin{vmatrix} 8 & -3 \\ -3 & 7 \end{vmatrix}} = \frac{49 + 45}{56 - 9} = \frac{94}{47} = 2 \qquad (14)$$

$$I_2 = \frac{\begin{vmatrix} 8 & 7 \\ -3 & 15 \end{vmatrix}}{D} = \frac{120 + 21}{D} = \frac{141}{47} = 3 \quad (15)$$

the denominator of I_2, is the same as the denominator of I_1. The results are two loop currents in amperes.

Now all unknown quantities in the network may be easily found. Current in the central branch is $I_{BO} = I_1 - I_2 = -1$ amp. The negative sign indicates that 1 amp is flowing upward. If the bottom node (node O) is taken to be the reference node at an assumed zero potential, then the potential at node B is -3 volts. The potential at node A is $-3 + 5 \cdot 2 = 7$ volts, which is also the electromotive force of the source E_1. The potential at node C is $-3 + 15 = 12$ volts, which can also be found (across Z_c) as $4 \cdot 3 = 12$ volts. The most convenient way to specify all the voltages of a network is to give the potentials at the various nodes with reference to some one node that is arbitrarily assumed to be at zero potential.

Standard notations. It is customary to use a standard system of symbols for writing loop equations. Equations (11) are specific examples of the general form shown in Eqs. (16). Equations (16) are

$$Z_{11}I_1 + Z_{12}I_2 + Z_{13}I_3 + \cdots + Z_{1L}I_L = V_1 \quad (16a)$$

$$Z_{21}I_1 + Z_{22}I_2 + Z_{23}I_3 + \cdots + Z_{2L}I_L = V_2 \quad (16b)$$

$$Z_{31}I_1 + Z_{32}I_2 + Z_{33}I_3 + \cdots + Z_{3L}I_L = V_3 \quad (16c)$$

$$\cdots\cdots\cdots\cdots\cdots\cdots\cdots\cdots\cdots\cdots\cdots$$

$$Z_{L1}I_1 + Z_{L2}I_2 + Z_{L3}I_3 + \cdots + Z_{LL}I_L = V_L \quad (16d)$$

a set of L simultaneous linear equations, applying to the L loops of a network; the network may be any network and L may be any number. In writing the equations, the following conventions are used. Each loop current is numbered, as I_1, I_2, and so on. V_1 is a voltage in loop 1 that is not taken into account by the terms of the left-hand side of the equation. It may be, as it was in Eq. (9), an independent electromotive force. It may be the sum of several voltages, and must include all voltages that appear in loop 1 when all the other loops are open. Note that its nominal positive direction is taken to be that of I_1.

The total impedance about loop 1 is Z_{11}. In Eq. (11), which applies to Fig. 3, Z_{11} is $Z_a + Z_b$. It might include many more elements if the network were more complicated. Z_{11} is called the self-impedance of loop 1. It could be measured by means of a bridge or other impedance-measuring instrument connected in place of the source in loop 1, all other loops of the network being opened during the measurement. Each loop has its self-impedance: Z_{22}, Z_{33}, and so on.

Certain branches are common to two loops. Thus Z_b in Fig. 3 is in both loop 1 and loop 2. In such a case, current in one loop produces voltage in another loop, and there is said to be mutual impedance. By definition, the mutual impedance is the ratio of such a voltage in one loop to the current in another loop that produces it. That is, if current in loop 2 is I_2 and mutual impedance with loop 1 is Z_{12}, then the resulting voltage in loop 1 is $Z_{12}I_2$. For example, Eq. (11) shows that the voltage produced in loop 1 of Fig. 3 by the current in loop 2 is $-Z_b I_2$. By definition, then, $Z_{12} = -Z_b = -3$. The negative sign results from the fact that in the common element the reference direction of I_2 is opposite to the reference direction of I_1. If positive I_1 produces a positive voltage, positive I_2 in the same element produces a negative voltage. In network computations the reference directions are commonly assumed in such a way that the mutual impedances (such as Z_{12}, Z_{23}, and so on) are negative quantities. However, it is not wrong to direct the arrows that indicate the nominal positive direction of current in such a way that mutual impedances are positive.

In Eq. (16b) the first term contains Z_{21}. Comparison with Eq. (11) shows that, for the network of Fig. 3, $Z_{21} = -Z_b$. Z_{21} is therefore equal to Z_{12}. When two loops contain resistors or coils or capacitors in a common branch, current in loop 1 will produce the same voltage in loop 2 that equal current in loop 2 would cause in loop 1; therefore $Z_{21} = Z_{12}$. The general form for this relation is given in Eq. (17). Although less evident, this relation

$$Z_{pq} = Z_{qp} \quad (17)$$

is still true if two circuits are coupled by a magnetic field (as in a transformer); whatever the turn ratio, the mutual inductance L_{21} equals the mutual inductance L_{12}. If the coupling between the loops is by means of an electric field through mutual capacitance with no conductive connection, Eq. (17) is again valid. Indeed, it fails only for circuit elements that are not bilaterally symmetrical. (For example, a transistor or a vacuum tube is not bilateral.)

The order of the subscripts attached to Z has the following significance. The first subscript is the number of the equation in the array of Eqs. (16) and therefore agrees with the subscript attached to V in that equation. The second subscript is the number of the term in the equation and therefore agrees with the subscript attached to I in that term. Thus $Z_{pq}I_q$ is a voltage in circuit p produced by a current in circuit q.

Node equations. Loop equations are written starting with the concept of loop currents. This makes it unnecessary to give any attention to Kirchhoff's current law, for loop currents necessarily add to zero at every node, and Kirchhoff's current law is automatically satisfied. The loop-current concept therefore reduces the number of equations that must be solved simultaneously from the $2B$ equations of the branch method to L, the number of independent loops, which is usually about one-fourth as many.

In the node-equation method, the simplifying concept is the idea of measuring voltage from all the nodes of the network to one particular node that is called the reference node, or the datum node. This makes it unnecessary to give any attention to Kirchhoff's voltage law. It is only necessary to satisfy Kirchhoff's current law at each node, for the voltage law is automatically satisfied. Thus the number of simultaneous equations is reduced to the number of independent nodes N, a number much smaller than $2B$ and comparable with L.

Whether the node method or the loop method is the more convenient depends on the network. Some networks have fewer loops than nodes, and some have fewer nodes than loops. Other factors also affect the relative convenience, as will be seen.

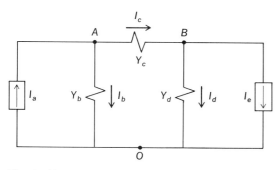

Fig. 4. Network with two independent nodes; a third one is the reference node O. *(From H. H. Skilling, Electrical Engineering Circuits, 2d ed., copyright © 1965 by John Wiley and Sons, Inc.; used with permission)*

Figure 4 shows a network with two independent nodes; that is, it has three nodes, one of which is the reference node O, and the others are marked A and B. Kirchhoff's current law is used for node A, Eq. (18), and node B, Eq. (19). Assume that the

$$I_a = I_b + I_c \qquad (18)$$

$$I_e = -I_d + I_c \qquad (19)$$

potential at node O is zero; if the potential at node A is called V_A, then $I_b = Y_b V_A$. Also $I_d = Y_d V_B$ and $I_c = Y_c(V_A - V_B)$. Equations (20) and (21) may now be solved for I_a and I_e.

$$I_a = Y_b V_A + Y_c(V_A - V_B)$$
$$= (Y_b + Y_c)V_A - Y_c V_B \qquad (20)$$

$$I_e = -Y_d V_B + Y_c(V_A - V_B)$$
$$= Y_c V_A - (Y_c + Y_d)V_B \qquad (21)$$

It will be found with practice that it is easy to write the final form of Eqs. (20) and (21) directly. To write Eq. (20), note that the independent value I_a of incoming current is set equal to the current that would flow from node A if all other nodes were at zero potential, that is, $(Y_b + Y_c)V_A$, from which is subtracted the current that would flow to node A if it were at zero potential while all other nodes were at their actual potentials, that is, $Y_c V_B$. To write the first term, assume that all nodes except A are short-circuited to node O; to write the second term, assume that node A alone is short-circuited to node O.

To write Eq. (21) at node B, the independent incoming current $-I_c$ is equated to current that would flow from node B if node A were at zero potential, that is, $(Y_c + Y_d)V_B$, less the current that would flow to node B if it alone were at zero potential, in this case $Y_c V_A$; the result, with all signs changed, is Eq. (21).

Standard notations. There is a standard form for writing node equations similar to the standard form for loop equations. For a network of N independent nodes, Eqs. (22) may be written. Y_{AA} is called the

$$Y_{AA}V_A + Y_{AB}V_B + Y_{AC}V_C + \cdots + Y_{AN}V_N = I_A$$
$$Y_{BA}V_A + Y_{BB}V_B + Y_{BC}V_C + \cdots + Y_{BN}V_N = I_B$$
$$Y_{CA}V_A + Y_{CB}V_B + Y_{CC}V_C + \cdots + Y_{CN}V_N = I_C$$
$$\cdots\cdots\cdots\cdots\cdots\cdots\cdots\cdots\cdots \qquad (22)$$
$$Y_{NA}V_A + Y_{NB}V_B + Y_{NC}V_C + \cdots + Y_{NN}V_N = I_N$$

self-admittance at node A, and in the example it is equal to $(Y_b + Y_c)$. Note that Y_{AA} is the sum of all admittances attached to node A. Y_{BB}, Y_{CC}, $\cdots$ are self-admittances at the other nodes.

Y_{AB} is the mutual admittance between nodes A and B. In the example $Y_{AB} = -Y_c$; Y_{BA} also equals $-Y_c$. Both Y_{AB} and Y_{BA} are the sum of all admittances connected directly between nodes A and B but written with a negative sign.

I_A is another current flowing toward node A. In the example it is the source current designated I_a in the diagram. Similarly, I_B is current toward node B, and in the example given it is the source current $-I_c$.

It will be noted that every term in the node Eqs. (22) is a current, as every term in the loop Eqs. (16) is a voltage.

Examples. As a numerical example, the following values are given to the impedances and the source currents of Fig. 4, and then voltage at the two independent nodes A and B and current in the three impedances are obtained. Note that the source currents are taken to be real numbers, implying either that they are direct current, or (if ac sources) that they are in phase with each other; the impedances are taken to be real, and this implies pure resistance. In a practical problem one voltage and all impedances might well be complex, and this would complicate the arithmetic but not change the method of solution. Given that $I_a = 2$ amp, $I_e = 5$ amp, $Y_b = 1/4$ mho, $Y_c = 1$ mho, and $Y_d = 1/2$ mho, node equations (22) are written as Eqs. (23) and (24).

$$(1/4 + 1)V_A - 1V_B = 2 \qquad (23)$$

$$-1V_A + (1 + 1/2)V_B = -5 \qquad (24)$$

These equations may be solved for V_A and V_B using Cramer's rule; they are shown as Eqs. (25) and (26).

$$V_A = \frac{\begin{vmatrix} 2 & -1 \\ -5 & 3/2 \end{vmatrix}}{\begin{vmatrix} 5/4 & -1 \\ -1 & 3/2 \end{vmatrix}} = \frac{-2}{7/8} = -\frac{16}{7} \text{ volts} \qquad (25)$$

$$V_B = \frac{\begin{vmatrix} 5/4 & 2 \\ -1 & -5 \end{vmatrix}}{7/8} = \frac{-17/4}{7/8} = -\frac{34}{7} \text{ volts} \qquad (26)$$

Currents I_b, I_c, and I_d are given in Eqs. (27)–(29).

$$I_b = V_A Y_b = -\frac{16}{7}\frac{1}{4} = -\frac{4}{7} \text{ amp} \qquad (27)$$

$$I_c = (V_A - V_B)Y_c = \left(-\frac{16}{7} + \frac{34}{7}\right)1 = \frac{18}{7} \text{ amp} \qquad (28)$$

$$I_d = V_B Y_d = -\frac{34}{7}\frac{1}{2} = -\frac{17}{7} \text{ amp} \qquad (29)$$

Equations (30) and (31) confirm these results. Note

$$I_a = I_b + I_c = -4/7 + 18/7 = 2 \text{ amp} \qquad (30)$$

$$I_e = I_c - I_d = 18/7 + 17/7 = 5 \text{ amp} \qquad (31)$$

that currents I_b and I_d both turn out to be upward because of the large value of the source current I_e, and that therefore the two nodes A and B are both at negative potential compared with node O.

Thévenin's theorem. It is often convenient before applying network theory to simplify a problem

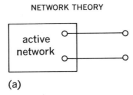

(a)

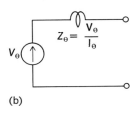

(b)

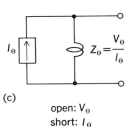

(c)

open: V_θ
short: I_θ

Fig. 5. Networks
(a) active, (b) Thévenin
equivalent, and (c) its
Norton equivalent. (*From
H. H. Skilling, Electrical
Engineering Circuits, 2d
ed., copyright © 1965 by
John Wiley and Sons, Inc.;
used with permission*)

by means of Thévenin's theorem. This theorem and its dual, Norton's theorem, may be expressed in many ways, but the following is among the more useful: Open-circuit voltage V_θ and short-circuit current I_θ are measured (or computed) at a pair of terminals of an active linear network. The active network is equivalent at these terminals to either an independent voltage source V_θ in series with an impedance $Z_\theta = V_\theta/I_\theta$, or alternatively to an independent-current source I_θ in parallel with the same Z_θ (Fig. 5). The former alternative is Thévenin's theorem, the latter is Norton's. A proof is not given; however, Eqs. (32)–(37) illustrate the application. *See* THÉVENIN'S THEOREM (ELECTRIC NETWORKS).

For the example of node equations Fig. 4 is used with the data given above to find currents and voltages in the network, using Thévenin's theorem.

Thévenin's theorem is applied twice to Fig. 4 to obtain the circuit of Fig. 6. Voltage V_1 and impedance Z_1 are equivalent to current I_a and admittance Y_b; similarly V_2 and Z_2 are equivalent to I_e and Y_d, as follows.

In Eqs. (32) the open-circuit voltage is computed at each pair of terminals with Z_c open.

$$V_1 = \frac{I_a}{Y_b} = \frac{2}{1/4} = 8 \qquad (32a)$$

$$V_2 = \frac{I_e}{Y_d} = \frac{5}{1/2} = 10 \qquad (32b)$$

Short-circuit current is computed with each of nodes A and B short-circuited to node O in Eqs. (33).

$$I_1 = 2 \qquad (33a)$$

$$I_2 = 5 \qquad (33b)$$

The impedances, Eqs. (34), may be obtained from Eqs. (32) and (33).

$$Z_1 = \frac{V_1}{I_1} = \frac{8}{2} = 4 \qquad (34a)$$

$$Z_2 = \frac{V_2}{I_2} = \frac{10}{5} = 2 \qquad (34b)$$

The current I_c may be obtained from Fig. 6, Eq. (35).

$$I_c = \frac{V_1 + V_2}{Z_1 + Z_c + Z_2} = \frac{8+10}{4+1+2} = \frac{18}{7} \text{ amp} \qquad (35)$$

Voltages V_A and V_B at nodes A and B are given by Eqs. (36) and (37). These answers are, of course,

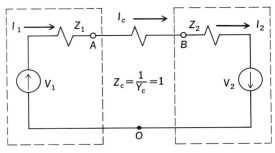

Fig. 6. Thévenin equivalent of Fig. 4.

the same as those obtained by node equations (25) and (26).

$$V_A = V_1 - Z_1 I_c = 8 - 4(18/7) = -16/7 \text{ volts} \qquad (36)$$

$$V_B = -(V_2 - Z_2 I_c)$$
$$= -[10 - 2(18/7)] = -34/7 \text{ volts} \qquad (37)$$

Many other network theorems are available, but perhaps none are as useful as Thévenin's and Norton's.

[HUGH HILDRETH SKILLING]

Bibliography: V. K. Aatre, *Network Theory and Filter Design*, 1980; H. H. Skilling, *Electric Networks*, 1974; G. Temes and J. Lapatra, *Introduction to Circuit Synthesis*, 1977; T. N. Trick, *Introduction to Circuit Analysis*, 1977; D. Tuttle, *Circuits*, 1977; M. E. Van Valkenburg, *Network Analysis*, 3d ed., 1974.

Neutron transmutation doping

The introduction of certain impurity atoms into semiconducting materials by nuclear transmutation as a consequence of thermal neutron absorption in a nuclear reactor. Silicon is a particularly attractive candidate for transmutation doping because it has only one isotope which transmutes to a new element, phosphorus, a standard dopant used in Si in the fabrication of electronic devices. The use of neutron transmutation doping (NTD) has proved to be a very effective method for obtaining a precisely controlled and very uniform distribution of P in Si. It has been shown that the performance and yield of high-power Si rectifiers and thyristors, and of avalanche detectors, can be significantly improved when NTD Si rather than conventionally doped material is used in these devices. In each case higher operating voltages are possible because of the extremely uniform impurity distribution in the NTD Si. Commercial producers of Si in the United States and other countries are now marketing NTD Si. *See* MICROWAVE SOLID-STATE DEVICES; SEMICONDUCTOR RECTIFIER.

Neutron transmutation doping process. Normal isotopic Si contains 3.05% of ^{30}Si, which transmutes to ^{31}P after thermal neutron absorption, with a radioactive half-life of only 2.6 hr. The ^{28}Si and ^{29}Si isotopes also transmute to ^{29}Si and ^{30}Si, respectively, but these processes do not directly alter the electrical properties of the material. The range of a thermal energy neutron in Si is 2 m, so that it is possible to obtain a desired concentration of ^{31}P in large-volume ingots or large-area devices. There are no significant problems associated with induced radioactivity for total ^{31}P concentrations up to about 10^{16} cm^{-3}. Above this concentration, second thermal neutron absorptions by the ^{31}P induce an additional radioactivity from ^{32}P, which transmutes to ^{31}S with a half-life of decay of 14.3 days and requires an extended radioactivity cooling period. The rate of introduction of ^{31}P into Si by the transmutation process is such that doping concentrations of interest to the semiconductor industry are readily attainable in many nuclear reactors. NTD Si has the disadvantage that the lattice damage introduced by the neutron irradiation must be annealed before the desired electrical properties can be recovered. However, it appears that in most applications the diffusion schedules used in device fabrication are adequate to anneal lattice damage.

Advantages. NTD Si provides four clearly identifiable advantages over Si doped by more traditional chemical techniques. These are areal and spatial uniformity of dopant distribution; precise control of doping level; elimination of dopant segregation at grain boundaries in polycrystalline Si; and superior control of heavy-atom contaminants. The first of these items has proved to be of particular importance in device applications because there is no known method of crystal growth that provides a uniform distribution of any chemical impurity dopant in Si. It has been demonstrated that the distribution of any such impurities takes the form of swirls or striations which serve to degrade the electrical properties of the *pn* junctions required in most devices. NTD Si completely eliminates this problem.

Use in solar cells. Significant improvements in the performance of many types of Si electronic devices may be attainable through the use of NTD Si but many applications are still in the research stage. One interesting use is in solar cell development. The presence of inhomogeneities in the dopant concentration is considered by a number of device theorists to be one of the primary reasons that solar cell efficiencies (presently about 15%) closer to the theoretically predicted values (about 20%) have not been attained. Since these inhomogeneities do not occur in NTD Si, it can serve as a useful research tool to investigate at least one aspect of the efficiency problem. It is true that a small increase in efficiency may not be of particular significance under standard sunlight conditions. However, it is likely that concentration up to 100 suns may be employed in many terrestrial solar cell applications, and even a small improvement in efficiency will then be significant. *See* SOLAR CELL.

Doping of thin layers. The technique of chemical vapor deposition (CVD) is widely used to deposit thin layers of semiconducting materials on metal, semiconductor, and insulator substrates in order to fabricate a wide variety of semiconductor devices. Typical industrial terminology for such devices includes CVD, epitaxial (epi), bipolar, metal oxide semiconductor (MOS), and silicon on sapphire (SOS). The electrical properties of CVD (polycrystalline) or epitaxial (single crystal) Si layers on substrate materials have been investigated, but attempts to incorporate a precisely controlled and very uniform dopant distribution have not been very successful to date. It is believed that the dopant migrates to the grain boundaries during CVD, or that free-carrier trapping occurs at the grain boundaries. It has also been demonstrated that the same type of impurity swirls or striations are introduced in epitaxial growth as occur in conventional single-crystal growth techniques.

Introduction of P by transmutation doping into thin polycrystalline or single-crystal Si, either freestanding or as deposited on a substrate, offers the obvious advantage of a uniform dopant distribution regardless of the sample or polycrystalline grain size. It has been demonstrated that there is no aggregation of dopant at the grain boundaries in NTD polycrystalline Si as a consequence of annealing to remove lattice damage, and macroscopic and mi-

croscopic resistance measurements on NTD Si epitaxial layers as deposited on Si substrates indicate that any fluctuations in dopant distribution are virtually nonexistent. Of course, the choice of substrate material must be compatible with any reactor irradiation requirements as regards integrity and induced radioactivity, but reasonably pure metal, semiconductor, or insulator materials are available as suitable substrates for NTD thin-film Si device applications. One obvious advantage of a uniform dopant distribution is that many individual units can be fabricated on a single chip, and uniformity of response from each of the individual units is ensured.

The use of NTD Si is currently under extensive investigation at a number of laboratories, and it is anticipated that many additional applications will be discovered. *See* INTEGRATED CIRCUITS.

[J. W. CLELAND; R. F. WOOD]

Bibliography: J. W. Cleland, K. Lark-Horovitz, and J. C. Pigg, *Phys. Rev.*, 78:814–815, 1950; W. E. Haas and M. S. Schnoller, *J. Electron. Mater.*, 5:57–58, 1976; *IEEE Transactions on Electron Devices*: *Special Issue on High-Power Semiconductor Devices*, vol. ED-23, no. 8, 1976; R. T. Young, J. W. Cleland, and R. F. Wood, *Proceedings of the 12th IEEE Photovoltaic Specialists Conference*, IEEE Conf. Rec. no. 76CH 1142 9ED, 1976.

Noise filter (radio)

A filter used in communications receivers to reduce noise. Usually it is an auxiliary low-pass filter which can be switched in or out of the audio system. The noise filter may also be equipped with a switch to vary the effective receiving bandwidth to meet the existing conditions of interfering noise. The tone control of a radio or record player can act as a noise filter, as when high-frequencies are cut down to reduce record noise. A bandpass filter may also be used, if the noise has a band-limited spectrum. *See* ELECTRIC FILTER.

[WILBER R. LEPAGE]

Nonlinear programming

An area of applied mathematics concerned with finding the values of the variables which give the smallest or largest value of a specified function in the class of all variables satisfying prescribed conditions. The function which is to be optimized is called the objective function, and the functions defining the prescribed conditions are referred to as the constraint functions or constraints. This general problem is called the nonlinear programming problem. The study of the theoretical and computational aspects of the nonlinear programming problem is called nonlinear programming, mathematical programming, or optimization theory. When there are no constraints, the nonlinear programming problem is said to be unconstrained; otherwise the problem is said to be constrained. If the objective function and the constraint functions are linear, the nonlinear programming problem is said to be a linear programming problem. When the objective function is quadratic and the constraints are linear, the nonlinear programming program is said to be a quadratic programming problem. *See* LINEAR PROGRAMMING.

General theory. The general nonlinear programming problem can be stated as notation (1), where

$$\text{minimize } f(x)$$
$$\begin{array}{ll} \text{subject to } g_i(x) \geq 0, & i = 1, \ldots, m \\ h_j(x) = 0, & j = 1, \ldots, p \end{array} \qquad (1)$$

$x = (x_1, \ldots, x_n)$ are the variables of the problem, f is the objective function, g_i are the inequality constraints, and h_j are the equality constraints. By changing f to $-f$, a maximization problem is transformed to a minimization problem. Moreover, an inequality constraint in the form $g_i(x) \geq 0$ is equivalent to $-g_i(x) \leq 0$. Consequently the format of problem (1) can handle both types of inequality constraints and both minimization and maximization problems.

The basic theory deals with conditions which a solution of the nonlinear programming problem (1) must satisfy. In the case when the objective function and the constraint functions are differentiable, the most important necessary conditions are the Karush-Kuhn-Tucker conditions. Consider the lagrangian function (2) whose added variables

$$L(x,u,w) = f(x) - \sum_{i=1}^{m} u_i g_i(x) + \sum_{j=1}^{p} w_j h_j(x) \qquad (2)$$

$u = (u_1, \ldots, u_m)$ and $w = (w_1, \ldots, w_p)$ are called Lagrange multipliers. The Karush-Kuhn-Tucker theory says that if x solves problem (1), then there exists u and w so that the triple (x,u,w) satisfies system (3) of equations and inequalities. The nota-

$$\nabla f(x) - \sum_{i=1}^{m} u_i \nabla g_i(x) + \sum_{j=1}^{p} w_j \nabla h_j(x) = 0$$
$$\begin{array}{ll} g_i(x) \geq 0, & i = 1, \ldots, m \\ h_j(x) = 0, & j = 1, \ldots, p \\ u_i g_i(x) = 0, & i = 1, \ldots, m \\ u_i \geq 0, & i = 1, \ldots, m \end{array} \qquad (3)$$

tion $\nabla f(x)$ is used to denote the gradient of f at x or the vector of partial derivatives at x with respect to the n independent variables. In order for this theory to be valid, an additional assumption called the constraint qualification must be made. Numerous constraint qualifications have been formulated; a useful but somewhat restrictive constraint qualification requires the gradients of the constraints which are actively involved in the solution to form a linearly independent set.

While the Karush-Kuhn-Tucker conditions (3) are only necessary conditions—meaning that, in general they are not sufficient to ensure that x will solve problem (1)—in the case that f and the g_i are convex and the h_j are affine, these necessary conditions are also sufficient conditions. The Lagrange multipliers are also called dual variables. In many problems in business, economics, and engineering, the dual variables have a useful interpretation in terms of sensitivity of the objective function to a particular constraint.

The branch of nonlinear programming which does not require the functions in problem (1) to be differentiable is called nondifferentiable programming or nondifferentiable optimization. Although this area has not received as much attention as the differentiable case, it has become the subject of considerable research activity and has increased in importance.

Scope of application. Many of the quantitative problems in business, economics, and engineering design can be expressed as nonlinear programming problems. General computational methods have been designed and implemented on large digital computers. Many problems which were considered large in 1970 are now considered small and can be solved efficiently. While considerable progress has been made in both the design and algorithms and computer technology, there are still many practical large nonlinear programs which cannot be solved today. Current research is aimed at taking advantage of the structure of the particular problem in question. One important example of structure is sparseness. Sparseness would occur if the nonlinear programming problem had many variables, but the objective function and each constraint function involved only relatively few variables.

Computational methods. Computational methods for the unconstrained optimization problem are well understood. For the unconstrained optimization problem, the Karush-Kuhn-Tucker conditions specialize to $\nabla f(x) = 0$. The preferred class of algorithms for this problem is the class of quasi-Newton methods which approximate the solution by the iterates given by Eq. (4), where B_k is an $n \times$

$$x^{k+1} = x^k - \alpha^k B_k^{-1} \nabla f(x_k) \qquad (4)$$

n matrix and the superscript -1 denotes the process of matrix inversion. The scalar α_k in Eq. (4) is chosen to approximately solve the one-dimensional optimization problem in α: minimize $f[x^k - \alpha B_k^{-1} \nabla f(x^k)]$. The special case of the quasi-Newton method, Eq. (4), which arises by choosing B_k equal to the identity matrix is the gradient method, while the special case which arises by choosing $B_k = \nabla^2 f(x^k)$ (the hessian matrix of f at x^k, that is, the matrix of second-order partial derivatives of f at x^k) is Newton's method. The gradient method is so slow that it should generally not be used, and Newton's method requires so much work per iteration that it is of questionable value. The class of quasi-Newton methods called secant methods requires that $B_{k+1}^{-1} y = s$ in Eq. (4), where $s = x^{k+1} - x^k$ and $y = \nabla f(x^{k+1}) - \nabla f(x^k)$. This class of methods seems to be the best currently available, and the preferred formula for B_{k+1}^{-1} is the Broyden-Fletcher-Goldfarb-Shanno (BFGS) secant update formula, Eq. (5), where the superscript T denotes matrix transposition.

$$B_{k+1}^{-1} = B_k^{-1} - [s y^T B_k^{-1} + (B_k^{-1} y - s) s^T]/$$
$$s^T y + s s^T (y^T B_k^{-1} y)/(s^T y)^2 \qquad (5)$$

The BFGS secant method is substantially faster than the gradient method, but not as fast as Newton's method. It requires more work than the gradient method, and less than Newton's method per iteration. Overall it is the preferred algorithm. The choice of α_k in Eq. (4) is important and the subject of current research.

There is lack of agreement of computational methods for attacking the constrained optimization problem. However, a reasonable approach to extending the successful secant methods from unconstrained optimization to constrained optimization appears to be the method of successive quadratic programming. Specifically the solution x is approximated by the iterates x^k which satisfy Eq. (6), where $\triangle x$ solves the quadratic programming problem of notations (7). The BFGS secant update for the matrix B_k is given by Eq. (8), where $s =$

$$x^{k+1} = x^k + \alpha_k \triangle x \tag{6}$$

minimize $q(\triangle x) = \nabla f(x^k)^T \triangle x + \tfrac{1}{2}\triangle x^T B_k \triangle x$

subject to $\nabla g_i(x^k)\triangle x + g_i(x^k) \geq 0, \quad i = 1, \ldots, m$ (7)

$$\nabla h_j(x^k)\triangle x + h_j(x^k) = 0, \quad j = 1, \ldots, p$$

$$B_{k+1} = B_k + yy^T/s^Ty - B_kss^TB_k/s^TB_ks \tag{8}$$

$x^{k+1} - x^k$ and $y = \nabla_x L(x^{k+1}, u^{k+1}, w^{k+1}) - \nabla_x L(x^k, u^{k+1}, w^{k+1})$, with the multipliers u^{k+1}, and w^{k+1} obtained from the solution of the quadratic programming problem (7). The notation $\nabla_x L$ denotes the vector of partial derivatives of L taken with respect to the x variables. The choice of the scalar α_k in Eq. (6) is the subject of current research. It is known that in the case of unconstrained optimization and constrained optimization the choice $\alpha_k = 1$ for large k is optimal.

There is considerable activity in the general area of algorithms for constrained optimization problems, and the development of effective and efficient algorithms for this general problem can be expected. However, there is still a considerable amount of work to be done in this area. *See* OPTIMIZATION.

[RICHARD A. TAPIA]

Bibliography: M. Avriel, *Nonlinear Programming*, 1976; J. E. Dennis and J. J. Moré, Quasi-Newton methods, motivation and theory, *SIAM Rev.*, 19:46–89, 1977; O. Mangasarian, R. Meyer, and S. Robinson (eds.), *Nonlinear Programming 3*, 1978.

Nonsinusoidal waveform

The representation of a wave that does not vary in a sinusoidal manner. Electric circuits containing nonlinear elements, such as electron tubes, iron-core magnetic devices, and transistors, commonly produce nonsinusoidal currents and voltages. When these are repetitive functions of time, they are called nonsinusoidal electric waves. Oscillograms, tabulated data, and sometimes mathematical functions for segments of such waves are often used to describe the excursions throughout one cycle. A cycle corresponds to 2π electrical radians and covers the time interval T sec in which the wave repeats itself.

These electric waves can be represented by a constant term, the average or dc component, plus a series of harmonic terms in which the frequencies of the harmonics are integral multiples of the fundamental frequency. The fundamental frequency f_1, if it does exist, has the time span $T = 1/f_1$ sec for its cycle. The second-harmonic frequency f_2 then will have two of its cycles within T sec, and so on.

Fourier series representation. The series of terms stated above is known as a Fourier series

and can be expressed in the form of Eq. (1), where

$$
\begin{aligned}
y(t) &= B_0 + A_1 \sin \omega t + A_2 \sin 2\omega t + \cdots \\
&\quad + A_n \sin n\omega t + B_1 \cos \omega t + B_2 \cos 2\omega t \\
&\quad \quad \quad + \cdots + B_n \cos n\omega t \\
&= B_0 + C_1 \sin (\omega t + \phi_1) + \cdots \\
&\quad \quad \quad + C_n \sin (n\omega t + \phi_n) \\
&= \sum_{n=0}^{\infty} C_n \sin (n\omega t + \phi_n)
\end{aligned}
\tag{1}
$$

$y(t)$, plotted over a cycle of the fundamental, gives the shape of the nonsinusoidal wave. The terms on the right-hand side show the Fourier series representation of the wave where Eqs. (2) and (3) apply.

$$C_n = \sqrt{A_n{}^2 + B_n{}^2} \tag{2}$$

$$\phi_n = \arctan \frac{B_n}{A_n} \tag{3}$$

Here A_0 is identically zero, and $B_0 = C_0$ in Eq. (1). The radian frequency of the fundamental is $\omega = 2\pi f_1$, and n is either zero or an integer. C_1 is the amplitude of the fundamental ($n = 1$), and succeeding C_ns are the amplitudes of the respective harmonics having frequencies corresponding to $n = 2$, 3, 4, and so on, with respect to the fundamental. The phase angle of the fundamental with respect to a chosen time reference axis is ϕ_1, and the succeeding ϕ_ns are the phase angles of the respective harmonics.

The equation for $y(t)$ shows all its separate components, which, in general, include an infinite number of terms. In order to represent a given nonsinusoidal wave by a Fourier series, it is necessary to evaluate each term, that is, B_0, and all A_ns and all B_ns. In practical problems the first several terms usually yield an approximate result sufficiently accurate for portrayal of the actual wave. The degree of accuracy desired in representing faithfully the actual wave determines the number of terms that must be used in any computation.

The constant term B_0 is found by computing the average amplitude of the actual wave over one cycle. Assuming any reference time $t = 0$ on the wave, B_0 is given by Eq. (4), where the angle ωt is

$$B_0 = \frac{1}{T} \int_0^T y(t)\, dt = \frac{1}{2\pi} \int_0^{2\pi} y(\omega t)\, d\omega t$$

$$= \frac{1}{2\pi} \int_0^{2\pi} y(\theta)\, d\theta \tag{4}$$

replaced by θ. Since B_0 is a constant, or dc, term, it merely raises or lowers the entire wave and does not affect its shape.

The coefficients of the sine series are obtained by multiplying the wave $y(\theta)$ by $\sin n\theta$, integrating this product over a full cycle of the fundamental, and dividing the result by π. Thus, A_n is given by Eq. (5).

$$A_n = \frac{1}{\pi} \int_0^{2\pi} y(\theta) \sin n\theta\, d\theta \tag{5}$$

The coefficients of the cosine terms are obtained in like manner, except that $\cos n\theta$ replaces $\sin n\theta$. Thus, B_n is given by Eq. (6).

$$B_n = \frac{1}{\pi} \int_0^{2\pi} y(\theta) \cos n\theta\, d\theta \tag{6}$$

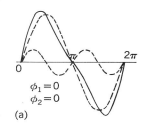

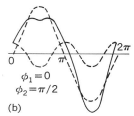

Fig. 1. Addition of a fundamental and a second harmonic. (a) $\phi_1 = 0$, $\phi_2 = 0$. (b) $\phi_1 = 0$, $\phi_2 = +\pi/2$ radians.

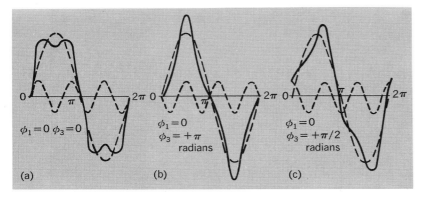

Fig. 2. Addition of a fundamental and a third harmonic. (a) $\phi_1 = 0$, $\phi_3 = 0$. (b) $\phi_1 = 0$, $\phi_3 = +\pi$ radians. (c) $\phi_1 = 0$, $\phi_3 = +\pi/2$ radians.

If mathematical expressions describe $y(\theta)$, Eqs. (4), (5), and (6) give the coefficients of the series directly through analytical methods. If oscillograms or tabulated data describe $y(\theta)$, then graphical or tabular forms of integration are used.

Effect of even harmonics. Figure 1 shows waves composed of a fundamental and a second harmonic only. In Fig. 1a, both ϕ_1 and ϕ_2 are zero. In Fig. 1b, ϕ_1 is zero and ϕ_2 is $+\pi/2$ radians with respect to one cycle of the second harmonic. In the example given in Fig. 1b the negative part of the overall wave is completely unlike the positive portion. Also, in general, these two sections will have different time intervals. Even harmonics give unsymmetrical waves.

Effect of odd harmonics. Figure 2 shows waves composed of a fundamental and a third harmonic. In Fig. 2a, both ϕ_1 and ϕ_3 are zero. In Fig. 2b, ϕ_1 is zero and ϕ_3 is $+\pi$ radians with respect to one cycle of the third harmonic. In Fig. 2c, ϕ_1 is zero and ϕ_3 is $+\pi/2$ radians on the third-harmonic time scale. In these diagrams the negative and positive parts of the overall waves are alike and both embrace π radians of the fundamental. Odd harmonics lead to symmetrical waves.

Symmetry. To determine symmetry of nonsine waves, the constant term B_0 is first removed. This means moving the wave down or up by the value of B_0. After this, if the wave from π to 2π is rotated about the horizontal axis and moved forward π radians, and then coincides exactly with the section from 0 to π, the total wave is said to have half-wave symmetry. Often the wave is said merely to be symmetrical. This means that in such cases $y(\theta + \pi) = -y(\theta)$. If, in turn, each half of the wave is symmetrical about the vertical axes of $\pi/2$, or $3\pi/2$, the wave is said to have quarter-wave symmetry as well. Half-wave and quarter-wave symmetry do not necessarily accompany each other.

All three waves of Fig. 2 have half-wave symmetry. The first two have quarter-wave symmetry also, but that of Fig. 2c does not. Waves having only a fundamental and odd harmonics show half-wave symmetry. Conversely, half-wave symmetry indicates that only the fundamental, if it exists, and odd harmonics are present in the total wave. Half-wave symmetry permits Eqs. (5) and (6) to be integrated over the interval π, with the result multiplied by two. Quarter-wave symmetry permits integration over one-quarter cycle of the fundamental, with the result multiplied by four.

Half-wave symmetry means that the fundamental and all odd harmonics may pass through their zero values at times quite distinct from each other. With quarter-wave symmetry the fundamental and the odd harmonics all pass through their zero values at the same time; therefore all phase angles ϕ_n are either zero or 180°.

The wave of Fig. 1a has no symmetry of the kind discussed above. Waves containing only the fundamental and even harmonics, or even harmonics alone, are unsymmetrical. Although half-wave symmetry is absent, quarter-wave symmetry may exist. A wave which is the same from π to 2π as it is from 0 to π, that is, $y(\theta) = y(\theta + \pi)$, is unsymmetrical. Only even harmonics are present, and the wave has no fundamental component. The output from a full-wave rectifier, for example, contains only the average term B_0 and even harmonics.

Waves that do not meet any of the special conditions noted above can be expected to contain both even and odd harmonics, and probably both sines and cosines also. Any doubt arising on the harmonic content of a wave is resolved by assuming all components of the Fourier series to be present. Analysis will show exactly those that actually exist.

Even and odd functions. The time origin of a wave can be chosen arbitrarily. If the reference axis $t = 0$ is such that the wave to the left is merely the wave to the right rotated about this axis, then $y(-\theta) = y(\theta)$, which is said to be an even function. Only cosine terms will be present in the Fourier series for the wave. On the other hand, if the wave to the left is the wave to the right rotated about the $t = 0$ axis, and then rotated about the horizontal axis, $y(-\theta) = -y(\theta)$, which is said to be an odd function. Only sine terms will appear in the Fourier series. Neither case precludes the possibility of the presence of both even and odd harmonics.

The rms value of nonsinusoidal wave. A nonsinusoidal wave has an rms value obtained through the following steps:

1. Combine all terms of the same frequency so as to have a single $A_1, A_2, \ldots, A_n; B_1, B_2, \ldots, B_n$; or a single $C_1, C_2, \ldots, C_n$. Terms such as $\sin(n\omega t \pm \alpha)$ and $\cos(n\omega t \pm \beta)$ each have sine and cosine components which can be separated out by trigonometric expansion.

2. Form the series $y(\theta)$ as in Eq. (1).

3. The rms value of the wave is then given by Eq. (7).

$$y_{\text{rms}} = \sqrt{B_0^2 + \frac{A_1^2}{2} + \frac{A_2^2}{2} + \cdots + \frac{A_n^2}{2} + \frac{B_1^2}{2} + \frac{B_2^2}{2} + \cdots + \frac{B_n^2}{2}} = \sqrt{B_0^2 + \frac{C_1^2}{2} + \frac{C_2^2}{2} + \cdots + \frac{C_n^2}{2}} \qquad (7)$$

If y_{rms} represents a voltage or a current, this value is shown by an electrodynamometer or iron-vane voltmeter or ammeter. The rms of the wave is merely the square root of the sum of the squares of the rms values of all of the frequency components.

Power. An indicating wattmeter with a nonsinusoidal voltage impressed on its potential circuit and a nonsinusoidal current in its current coils indicates the average power taken by the circuit. Designating peak values of the component voltages and currents by E_ns and I_ns in place of C_ns results in Eq. (8). Each coefficient is simply the

$$\text{Average power} = \frac{1}{2\pi} \int_0^{2\pi} ei \, d\theta$$

$$= E_0 I_0 + \frac{E_1 I_1}{2} \cos \theta_1 + \cdots + \frac{E_n I_n}{2} \cos \theta_n \quad (8)$$

product of rms voltage and current. No cross-product terms involving different frequencies result from the integration. That is, no power can be contributed by a voltage of one frequency and a current of another frequency.

Power factor. The apparent power taken by a circuit carrying nonsinusoidal voltage and current is the product of the rms values of these quantities. Power factor for such a case is defined only by the ratio of the average power to the apparent power. Thus, the power factor is given by Eq. (9). Power

$$\text{Power factor (pf)} = \frac{\text{watts average power}}{\text{rms volts} \times \text{rms amperes}} \quad (9)$$

factor is hence the ratio of instrument readings as stated. All circuits have a power factor, but pf = $\cos \theta$ only for a sine wave voltage and current of the same frequency. There is no average or representative phase angle for a circuit carrying nonsinusoidal waves.

Example of nonsinusoidal waves. Assume a series circuit to have 8 ohms resistance and 15.91 millihenries inductance and that the impressed voltage is given by Eq. (10). The problem is to cal-

$$e = 100 \sin 377t + 80 \sin 1131t \quad (10)$$

culate the rms voltage and current, the average power, and the power factor. The voltage has a fundamental component of 60 cycles ($f_1 = \omega_1/2\pi = 377/2\pi$) and a third harmonic of 180 cycles ($f_3 = \omega_3/2\pi = 1131/2\pi$).

At 60 cycles:

$X_{L_1} = 377 \times 0.01591$
$\quad = 6.0$ ohms inductive reactance
$Z_1 = \sqrt{8^2 + 6^2} = 10$ ohms impedance
$I_1 = 100/10 = 10$ A max fundamental current
$\theta_1 = \arctan(6/8) = 36.87°$

At 180 cycles:

$X_{L_3} = 3 \times 6 = 18$ ohms inductive reactance
$Z_3 = \sqrt{8^2 + 18^2} = 19.70$ ohms impedance
$I_3 = 80/19.7$
$\quad = 4.06$ A max third-harmonic current
$\theta_3 = \arctan(18/8)$
$\quad = 66.06° \ (= 22.02°$ on fundamental scale)

The equation for the current is

$$i = 10 \sin(377t - 36.87°)$$
$$+ 4.06 \sin(1131t - 22.02°)$$

$$E_{rms} = \sqrt{\frac{100^2}{2} + \frac{80^2}{2}} = 90.06 \text{ volts}$$

$$I_{rms} = \sqrt{\frac{10^2}{2} + \frac{4.06^2}{2}} = 7.63 \text{ amperes}$$

Apparent power = $90.06 \times 7.63 = 687$ volt-amperes
Average power = $I^2 R = 7.63^2 \times 8 = 466$ watts

$$\text{Power factor} = \frac{466}{682} = 0.678$$

See ALTERNATING-CURRENT CIRCUIT THEORY.

[BURTIS L. ROBERTSON; W. S. PRITCHETT]
Bibliography: C. S. Siskind, *Electric Circuits, Direct and Alternating Current*, 2d ed., 1965; D. D. Weiner and J. F. Spina, *The Sinusoidal Analysis and Modeling of Weakly Nonlinear Circuits: With Application to Nonlinear Interference Effects*, 1980.

Number systems

Integral numbers may be represented as linear combinations of powers of any convenient and arbitrarily chosen base. The choice of the base is not always made on a rational basis, the number systems have been based on 5, 6, 10, and 60. More recently, systems based on 2 and 8 have proved quite useful in computer applications. The duodecimal number system, in which numbers are represented as linear combinations of powers of 12, has certain advantages because 12 has the factors 1, 2, 3, 4, 6, and 12.

Decimal system. Every positive integer is uniquely a polynomial in 10 with coefficients, called digits, taken from 0, 1, . . . , 9. The fact that

$$205714 = 4 + 1 \cdot 10 + 7 \cdot 10^2 + 5 \cdot 10^3$$
$$+ 0 \cdot 10^4 + 2 \cdot 10^5$$

is nearly always lost in present-day teaching, and in the hurried application of ordinary arithmetic. In fact, numbers are likely to be thought of as merely an orderly arrangement of decimal digits.

The decimal method of representing numbers comes from India and Arabia and is only a few centuries old in Europe. The base, 10, is due to the biological fact that man has that many articulate fingers and thumbs. The positional significance, including the meaning and usefulness of zero, is of oriental origin.

The operations of addition and multiplication consist of the corresponding operations with polynomials, together with rules that serve to keep the results inside the system so that they can be used in future operations. In the case of addition of two numbers, use is made of either the familiar "carry" rule or the addition table, while for multiplication, use is made of the multiplication table to help represent the product of two digits as a two-digit number. These apparently nonalgebraic operations are so dominant that the basic polynomial structure of the numbers is obscured. Thus, the multiplication of polynomials is done by a more intelligent method than that used for numbers. For example, the multiplication of 2057 by 3416 can be carried out as follows:

```
        2 0 5 7
        3 4 1 6
      _____
      6,1 7,3 3,4 2
        8,5 3,3 7
      _____
      7 0 2 6 7 1 2
```

Commas separate those pairs of digits that arise from sums of products of pairs of digits taken one each from the original numbers, and having equal significance. Thus 53, the fourth most significant contribution, is given by

$$53 = 2 \cdot 6 + 0 \cdot 1 + 4 \cdot 5 + 3 \cdot 7$$

This process can be carried out either from right to left or from left to right, and in the latter case, may be terminated when half done if the least significant half on the product is not needed.

In connection with the design and use of automatic computers, in which numbers of a limited size only may be added and multiplied at one time, precautions against overflow in addition, and approximation by rounding in multiplication further complicate the execution of ordinary arithmetic. This creates a system that, strictly speaking, fails to satisfy the axioms of arithmetic. This causes serious difficulties in some problems involving millions of additions and multiplications.

Subtraction introduces negative numbers that may be handled by introducing a special digit called a sign digit with its own rules of combination, or by introducing complementation in which the digits of a number are subtracted from 9, except for the last nonzero digit which is subtracted from 10. Thus to subtract 20570 from 34162, 20570 may be complemented, and 34162 added to it to obtain the desired difference, 13592, as follows:

```
. . . 99979430
. . . 00034162
      _____
. . . 00013592
```

Numbers that begin with a run of nines are considered negative in this system. Of course care must be taken to guard against overflow in which a very large positive number might be confused with a very small negative one.

Division is a process that can be carried out only rarely with absolute exactness in the decimal system, the process usually being nonterminating. This introduces the notion of infinite decimal expansions and the more or less theoretical operations with such numbers. In practice, truncation and rounding are used as in $2/3 = .66667$, with consequent errors and departure from the axioms of arithmetic. In this case a quantity like ab/c is not unique but may depend upon the order in which the indicated operations are performed. For complicated and extensive problems involving only the four rational operations of arithmetic, an adequate analysis of the errors involved may be very costly indeed.

Automatic calculation in this simulated real number system may be facilitated by the use of a normalizing coding device called "floating arithmetic." In this system a positive real number is expressed as a truncated decimal between .1 and 1 times the appropriate power of 10. Thus the number π on a 10-digit decimal machine could be coded 3141592751. In interpreting this "word," the machine separates the last two digits, 51, and subtracts 50 to get the exponent (possibly negative) of the power of 10 by which the mantissa .31415927 would have to be multiplied to obtain π correct to eight decimals. Rules for multiplying and adding in this system are easily formulated. They involve inspection, comparison, and manipulation of the exponents, followed by appropriate shifting right or left of the mantissas, followed next by ordinary decimal arithmetic on the mantissas, and finally a normalization and reassembly of the answer as a "floating word." The system has the advantage of greater control over numbers of widely varying orders of magnitude. The disadvantages include slower operation, often by a factor of 5 or more, and occasional unpredictable loss of information.

Besides the operation of complementation, there are other nonarithmetic operations with decimal numbers, for example, comparison. Two numbers may be compared for size by a simple inspection of their corresponding digits, beginning from the left and stopping at the first case of inequality. This simple but important property is worth mentioning because comparison is almost impossible in certain other systems. An unusual use of decimal digits is the so-called middle-of-the-square method of generating random numbers. By this method the next 10-digit random number is obtained from the preceding one by squaring the latter and selecting from the square the central 10 digits.

There are many interesting properties of the digits of integer numbers. The simpler ones depend on the theory of congruences. The most familiar fact of this sort is the statement that a number is even if, and only if, its last digit is even. A similar statement is true with respect to divisibility by 5. If a number is diminished by the sum of its digits, the result is a multiple of 9. This fact is the basis for the scheme for checking arithmetic by "casting out nines," at one time known to every school boy. Elevens may be cast out in like manner if the digits are added with alternating signs. Thus, $34162 - (2 - 6 + 1 - 4 + 3) = 34166$, is a multiple of 11. Similarily, grouping the digits by threes, $34535599 - (599 - 535 + 34) = 34535501$ is a multiple of $1001 = 7 \cdot 11 \cdot 13$. This fact is sometimes used to check desk calculator computations by casting out 1001s. It is also used to decide quickly whether a given number is divisible by 7, 11, or 13. The number 24535599 is not divisible by 7, 11, or 13 since $599 - 535 + 24 = 108$ is not.

Squares of integers have digital properties. For example, the final digit of a square is either 0, 1, 4, 5, 6, or 9, never 2, 3, 7, or 8. There are only 22 combinations of two digits in which a square can end, and so on. Such facts are sometimes used in finding the factors of a given number by expressing it as a difference of two squares. The rapid recognition of nonsquares is also helpful in many other diophantine problems.

The representation of real numbers requires infinite, that is, unending, decimals. If the digits of such a decimal ultimately become periodic, the decimal is the ratio P/Q of two integers, and conversely. The length of the period is a complicated function of Q, depending on the prime factors of numbers of the form $10^n - 1$. If, and only if, Q is of the form 2^a5^b, the decimal expansion of P/Q terminates. In such cases P/Q has in reality two expansions. Thus $7/5 = 1.4000. . . = 1.3999. . . .$

The great majority of decimals do not become periodic, or in other words, almost all real numbers are irrational. The class of irrational algebraic numbers, such as the square root of 2, that are roots of polynomials with integer cofficients, is almost completely obscured by other real numbers in their decimal representation. There are only a few statements that can be made about the digits of such numbers other than the obvious one of nonperiodicity. For example, if k consecutive zeros occur, then they cannot occur "too soon" for an infinity of k. On the other hand, almost all real numbers have perfectly normal decimal expansions in the sense that each digit occurs, on the average, one-tenth of the time, each ordered pair one-hundredth of the time, and so on. Whether π, e, or $\sqrt{2}$ are normal is not known. The totality of all known examples of normal numbers is countable.

Almost everything that has been said so far about the decimal system applies with equal force and very little modification to a general system based on an integer $b > 1$ instead of 10. The fact that people "know" all the powers of 10 but not the powers of 7 or 12 is purely psychological and based on tradition. Beyond the fact that 10 is even, there is little to recommend it as a base. The Babylonians used 60, a large but useful base that is still in vogue for measurement of time and angles. The mathematician J. d'Alembert and many others after him urged the adoption of $b = 12$ with its six divisors. The advent of electronic computers has made a good case for $b = 8$ or some other power of 2. Probably base 8 is used by humans more than any base except 10. For $b > 10$, new characters are needed to represent the extra digits. Although there is no agreement as to which characters to adopt, the modern tendency is to use roman letters because they are easily available on the typewriter. The adoption of a second system brings up the question of translating or converting numbers from one system to the other. Methods for doing this are explained in following sections on binary and octal systems.

Binary system. In the binary system every positive integer is the sum of distinct powers of 2 in just one way. Thus $434 = 2^8 + 2^7 + 2^5 + 2^4 + 2^1$, and this is expressed by writing 110110010. The digits corresponding to 2^0, 2^2, 2^3, and 2^6 are zero, since these powers do not occur in 434. The first dozen integers are written as follows:

1	1	4	100	7	111	10	1010
2	10	5	101	8	1000	11	1011
3	11	6	110	9	1001	12	1100

The great advantage of the binary system lies in the fact that there are only two kinds of binary digits, or "bits," namely 0 and 1. This not only gives a simplified arithmetic but provides a language in which to treat two-valued functions or bistable systems. Among its disadvantages is the fact that the binary system requires nearly three times as many digits to represent a given number as does the familiar decimal system.

Digital computers invariably use the binary system. The so-called decimal computers code the decimal digits into binary form, while the purely binary machines use full binary arithmetic.

The physical representation of binary numbers, or information, is possible in many forms. A row of lights, some on and some off, may be interpreted as a binary number. A set of condensers, some charged and some not, a set of high and low voltages, or a set of magnets with fluxes in one direction or another are electronic examples of media for the processing and retention of data in the binary system. The fact that there are only two states to recognize accounts for the great reliability of such computing systems.

The conversion of decimal, or base 10, integers into the binary system can be done in two ways. First, one may subtract from the given integer the highest power of 2 not exceeding this number and record a 1 in the binary position corresponding to this power of 2. The remainder of this subtraction, if not zero, now replaces the original number and the process is repeated until a zero remainder is obtained.

Alternatively, one may divide the given number by 2 and record the remainder, 0 or 1, as the final binary digit. The quotient in this division now replaces the given number and the process is repeated and continued until a quotient of zero is reached. The two methods are illustrated in the case of converting 434 to the binary system:

434		434	
−256		217	0
178	1	108	10
−128		54	010
50	11	27	0010
−32		13	10010
18	1101	6	110010
−16		3	0110010
2	11011	1	10110010
−2		0	110110010
0	110110010		

Both processes have obvious inverses for going from the binary system to the decimal system. In the first case, the indicated powers of 2 are simply added together, and in the second case, a sequence of doubling operations is used, followed by the addition of 0 or 1 as specified by the given binary number. For numbers between 0 and 1 similar procedures are available. Either the subtraction of powers of 2 (negative powers) can be continued or the given number can be doubled, followed by subtraction of whichever of the numbers 0 or 1 will make the remainder lie between 0 and 1, and the operation continued with the remainder as before. The reader may wish to test his understanding by verifying that 43.4294 has the binary representation

$$101011.01101101111011010. . .$$

Arithmetic in the binary system is remarkably simple. For addition, only $1 + 1 = 10$ is needed. while the multiplication table reduces to $1 \cdot 1 = 1$.

Examples of addition and multiplication are

$$
\begin{array}{ll}
110101 & (53) \\
11001 & (25) \\
\hline
1001110 & (78)
\end{array}
\qquad
\begin{array}{ll}
1101 & (13) \\
1011 & (11) \\
\hline
1101 \\
1101 \\
1101 \\
\hline
10001111 & (143)
\end{array}
$$

Such simple operations are readily performed electronically with extreme rapidity and reliability.

The binary system is useful not only to represent numbers but also to record and process information. In fact the unit of information is a binary digit. For example, given a set S of objects and a property P, it is possible to record which objects have the property P, and which do not, by assigning a binary position to each object of S and recording there a 1 or 0 according as the property P is, or is not, possessed by the corresponding object. Thus if the objects are the first odd numbers and P is the property of primality, the binary number

$$N = .0111011011010011010\ldots$$

is equivalent to the list of odd primes

$$3, 5, 7, 11, 13, 17, 19, 23, 29, 31, 37, \ldots$$

A binary computer, with its ability to extract and examine a given binary digit, can use this compact method of storing information. The operation $N + N$ replaces N, which shifts the digits one place to the left and produces overflow if, and only if, the corresponding number is a prime, can be used in general to select the successive members of S having a property P. Other combinatorial processes involving several coded binary numbers can be used to advantage with a binary computer. For example, one can make a search for those objects of S that have a set of specified properties P_1, $P_2, \ldots$.

The binary system is implicit in a number of different arithmetical operations and games. The so-called Russian peasant method of multiplying by doubling and halving is a case in point. To multiply 323 by 146, form two columns of figures (in the decimal system)

$$
\begin{array}{rr}
146 & \cancel{323} \\
73 & 646 \\
36 & \cancel{1292} \\
18 & \cancel{2584} \\
9 & 5168 \\
4 & \cancel{10336} \\
2 & \cancel{20672} \\
1 & 41344 \\
\hline
& 47158
\end{array}
$$

Each term of the first column is the integer part of half the preceding term. Each term of the second column opposite an even number in the first column is struck out. The sum of the remaining numbers gives the desired product 47158. The method works because, in forming the first column, one is, in effect, converting 146 to the binary system.

Another operation in which binary representation is effective is that of raising a given base B to a high integer power. Suppose that

$$n = b_k b_{k-1} \cdots b_2 b_1 b_0$$

is the binary representation of the integer n. To compute B^n most efficiently, form recursively the numbers w_i, defined by

$$
\begin{aligned}
w_0 &= B^{b_k} = B \\
w_1 &= B^{b_{k-1}} (w_0)^2 \\
&\cdots\cdots\cdots\cdots \\
w_i &= B^{b_{k-i}} (w_{i-1})^2
\end{aligned}
$$

Then $w_k = B^n$. In fact

$$w_k = B^{b_0} (w_{k-1})^2 = B^{b_0 + 2[b_1 + 2(b_2 + \cdots)]}$$

so that the exponent is

$$b_0 + 2b_1 + 2^2 b_2 + \cdots + 2^k b_k = n$$

Octal system. To write a number in the octal system, once it has been expressed in the binary system, one merely groups the binary digits by threes, beginning at the binary point and working to the left and right. Thus the decimal number 43.4294 gives

$$(101)(011).(011)(011)(011)(110)(110)(10.)$$

or simply 53.333664, where the last digit should perhaps be 5. On the other hand, decimal to octal conversion can be accomplished directly by either of the two methods that correspond in an obvious way to those given for decimal to binary conversion. Thus, by subtracting appropriate multiples of powers of 8, beginning with the largest possible power,

$$5280 = 1 \cdot 8^4 + 2 \cdot 8^3 + 2 \cdot 8^2 + 4 \cdot 8$$

so that in the octal system there are 12240 feet in a mile. Alternatively, one may divide 5280 by 8, getting 0 as remainder and 660 as quotient. Dividing 660 by 8, 4 and 82 are obtained. Dividing 82 by 8 gives 2 and 10. Dividing 10 by 8 gives 2 and 1. This gives the digits in reverse order.

Octal to decimal conversion may be effected by the use of a convenient table of powers of 8, a sample of which follows:

n	8^n	n	8^n
0	1	-1	.125000
1	8	-2	.015625
2	64	-3	.001953
3	512	-4	.000244
4	4096	-5	.000031
5	32768	-6	.000004

The octal system with its eight digits $0, 1, \ldots, 7$ affords a convenient way of condensing the lengthier display of the binary system. Arithmetic in the octal system resembles the familiar decimal arithmetic. The addition and multiplication tables are as shown:

Addition

	0	1	2	3	4	5	6	7
0	0	1	2	3	4	5	6	7
1	1	2	3	4	5	6	7	10
2	2	3	4	5	6	7	10	11
3	3	4	5	6	7	10	11	12
4	4	5	6	7	10	11	12	13
5	5	6	7	10	11	12	13	14
6	6	7	10	11	12	13	14	15
7	7	10	11	12	13	14	15	16

Multiplication

	0	1	2	3	4	5	6	7
0	0	0	0	0	0	0	0	0
1	0	1	2	3	4	5	6	7
2	0	2	4	6	10	12	14	16
3	0	3	6	11	14	17	22	25
4	0	4	10	14	20	24	30	34
5	0	5	12	17	24	31	36	43
6	0	6	14	22	30	36	44	52
7	0	7	16	25	34	43	52	61

Examples of addition and multiplication in the octal system are

$$
\begin{array}{r}
4375 \\
3704 \\
\hline
10301
\end{array}
\qquad
\begin{array}{r}
5734 \\
16 \\
\hline
43450 \\
5734 \\
\hline
123010
\end{array}
$$

Octal arithmetic can be checked by "casting out sevens" (instead of nines) by adding the digits. Thus for the addition problem above,

$$
\begin{aligned}
4375 &\equiv 4+3+7+5=23 \equiv 2+3=5 \quad &(\bmod\ 7) \\
3704 &\equiv 3+7+0+4=16 \equiv 1+6 \equiv 0 \quad &(\bmod\ 7) \\
10301 &\equiv 1+3+1=5 \quad &(\bmod\ 7)
\end{aligned}
$$

Checking by casting out nines involves taking the octal digits with alternating signs. Thus,

$$
\begin{aligned}
4375 &\equiv 5-7+3-4=-3 \equiv 6 \quad &(\bmod\ 9) \\
3704 &\equiv 4-0+7-3=\ 8 \quad &(\bmod\ 9) \\
10301 &\equiv 1+3+1=5 \equiv 6+8 \quad &(\bmod\ 9)
\end{aligned}
$$

The octal system requires only 10% more digits than the decimal system to represent the same amount of information. Some computing systems use base 16, in which case binary information is handled in sets of four bits. This system is more compact than the decimal system, 100 hexodecimals being equivalent to 120 decimals, but it requires a multiplication table with nearly three times as many entries.

Computing systems of binary type have subroutines for the conversion of any kind of decimal information into binary information during input, and vice versa during output, so that a facility in octal arithmetic is needed only rarely during checking and testing of a new problem. *See* DIGITAL COMPUTER; NUMERICAL ANALYSIS.

[DERRICK H. LEHMER]

Bibliography: R. H. Bruck, *Survey of Binary Systems*, 3d ed., 1971; T. Danzig, *Number, the Language of Science*, 4th ed., 1967; C. Reid, *From Zero to Infinity*, 2d ed., 1960; H. Schmid, *Decimal Computation*, 1974.

Numerical analysis

The development and analysis of computational methods (and ultimately of program packages) for the minimization and the approximation of functions, and for the approximate solution of equations, such as linear or nonlinear (systems of) equations and differential or integral equations. Originally part of every mathematician's work, the subject is now often taught in computer science departments because of the tremendous impact which computers have had on its development. Research focuses mainly on the numerical solution of (nonlinear) partial differential equations and the minimization of functions. *See* COMPUTER.

Numerical analysis is needed because answers provided by mathematical analysis are usually symbolic and not numeric; they are often given implicitly only, as the solution of some equation, or they are given by some limit process. A further complication is provided by the rounding error which usually contaminates every step in a calculation (because of the fixed finite number of digits carried).

Even in the absence of rounding error, few numerical answers can be obtained exactly. Among these are (1) the value of a piecewise rational function at a point and (2) the solution of a (solvable) linear system of equations, both of which can be produced in a finite number of arithmetic steps. Approximate answers to all other problems are obtained by solving the first few in a sequence of such finitely solvable problems. A typical example is provided by Newton's method: A solution c to a nonlinear equation $f(c) = 0$ is found as the limit $c = \lim_{n \to \infty} x_n$, with x_{n+1} a solution to the linear equation $f(x_n) + f'(x_n)(x_{n+1} - x_n) = 0$, that is, $x_{n+1} = x_n - f(x_n)/f'(x_n)$, $n = 0, 1, 2, \ldots$. Of course, only the first few terms in this sequence $x_0, x_1, x_2, \ldots$ can ever be calculated, and thus one must face the question of when to break off such a solution process and how to gauge the accuracy of the current approximation. The difficulty in the mathematical treatment of these questions is exemplified by the fact that the limit of a sequence is completely independent of its first few terms.

In the presence of rounding error, an otherwise satisfactory computational process may become useless, because of the amplification of rounding errors. A computational process is called stable to the extent that its results are not spoiled by rounding errors. The extended calculations involving millions of arithmetic steps now possible on computers have made the stability of a computational process a prime consideration.

Interpolation and approximation. Polynomial interpolation provides a polynomial p of degree n or less which uniquely matches given function values $f(x_0), \ldots, f(x_n)$ at corresponding distinct points $x_0, \ldots, x_n$. The interpolating polynomial p is used in place of f, for example in evaluation, integration, differentiation, and zero finding. Accuracy of the interpolating polynomial depends strongly on the placement of the interpolation points, and usually degrades drastically as one moves away from the interval containing these points (that is, in case of extrapolation).

When many interpolation points (more than 5 or 10) are to be used, it is often much more efficient to use instead a piecewise polynomial interpolant or spline. Suppose the interpolation points above are ordered, $x_0 < x_1 < \ldots < x_n$. Then the cubic spline interpolant to the above data, for example, consists of cubic polynomial pieces, with the ith piece defining the interpolant on the interval $[x_{i-1}, x_i]$ and so matched with its neighboring piece or pieces that the resulting function not only matches the given function values (hence is continuous) but also has a continuous first and second derivative.

Interpolation is but one way to determine an approximant. In full generality, approximation involves several choices: (1) a set P of possible approximants, (2) a criterion for selecting from P a particular approximant, and (3) a way to measure the approximation error, that is, the difference between the function f to be approximated and the approximant p, in order to judge the quality of approximation. Much studied examples for P are the polynomials of degree n or less, piecewise polynomials of a given degree with prescribed breakpoints, and rational functions of given numerator and denominator degrees. The distance between f and p is usually measured by a norm, such as the L_2 norm $(\int |f(x) - p(x)|^2 dx)^{1/2}$ of the uniform norm $\sup_x |f(x) - p(x)|$. Once choices 1 and 3 are made, one often settles 2 by asking for a best approximation to f from P, that is, for an element of P whose distance from f is as small as possible. Questions of existence, uniqueness, characterization, and numerical construction of such best approximants have been studied extensively for various choices of P and the distance measure. If P is linear, that is, if P consists of all linear combinations

$$\sum_{i=1}^{n} a_i p_i$$

of certain fixed functions $p_1, \ldots, p_n$, then determination of a best approximation in the L_2 norm is particularly easy, since it involves nothing more than the solution of n simultaneous linear equations.

Solution of linear systems. Solving a linear system of equations is probably the most frequently confronted computational task. It is handled either by a direct method, that is, a method which obtains the exact answer in a finite number of steps, or by an iterative method, or by a judicious combination of both. Analysis of the effectiveness of possible methods has led to a workable basis for selecting the one which best fits a particular situation.

Direct methods. Cramer's rule is a well-known direct method for solving a system of n linear equations in n unknowns, but it is much less efficient than the method of choice, elimination. In this procedure the first unknown is eliminated from each equation but the first by subtracting from that equation an appropriate multiple of the first equation. The resulting system of $n - 1$ equations in the remaining $n - 1$ unknowns is similarly reduced, and the process is repeated until one equation in one unknown remains. The solution for the entire system is then found by back-substitution, that is, by solving for that one unknown in that last equation, then returning to the next-to-last equation which at the next-to-last step of the elimination involved the final unknown (now known) and one other, and solving for that second-to-last unknown, and so on.

This process may break down for two reasons: (1) when it comes time to eliminate the kth unknown, its coefficient in the kth equation may be zero, and hence the equation cannot be used to eliminate the kth unknown from equations $k + 1$, $\ldots, n$; and (2) the process may be very unstable. Both difficulties can be overcome by pivoting, in which one selects, at the beginning of the kth step, a suitable equation from among equations $k, \ldots, n$, interchanges it with the kth equation, and then

proceeds as before. In this way the first difficulty may be avoided provided that the system has one and only one solution. Further, with an appropriate pivoting strategy, the second difficulty may be avoided provided that the linear system is stable. Explicitly, it can be shown that, with the appropriate pivoting strategy, the solution computed in the presence of rounding errors is the exact solution of a linear system whose coefficients usually differ by not much more than roundoff from the given ones. The computed solution is therefore close to the exact solution provided that such small changes in the given system do not change its solution by much. A rough but common measure of the stability of a linear system is the condition of its coefficient matrix. This number is computed as the product of the norm of the matrix and of its inverse. The reciprocal of the condition therefore provides an indication of how close the matrix is to being noninvertible or singular.

Iterative methods. The direct methods described above require a number of operations which increases with the cube of the number of unknowns. Some types of problems arise wherein the matrix of coefficients is sparse, but the unknowns may number several thousand; for these, direct methods are prohibitive in computer time required. One frequent source of such problems is the finite difference treatment of partial differential equations (discussed below). A significant literature of iterative methods exploiting the special properties of such equations is available. For certain restricted classes of difference equations, the error in an initial iterate can be guaranteed to be reduced by a fixed factor, using a number of computations that is proportional to $n \log n$, where n is the number of unknowns. Since direct methods require work proportional to n^3, it is not surprising that as n becomes large, iterative methods are studied rather closely as practical alternatives.

The most straightforward iterative procedure is the method of substitution, sometimes called the method of simultaneous displacements. If the equations for $i = 1, \ldots, n$ are as shown in Eq. (1),

$$\sum_{j=1}^{n} a_{ij} x_j = b_i \qquad (1)$$

then the rth iterate is computed from the $r - 1$st by solving the trivial equations for $x_i^{(r)}$ shown in Eq. (2) for $i = 1, \ldots, n$, where the elements $x_i^{(0)}$

$$\sum_{j \neq i} a_{ij} x_j^{(r-1)} + a_{ii} x_i^{(r)} = b_i \qquad (2)$$

are chosen arbitrarily. If for $i = 1, \ldots, n$, the inequality

$$\sum_{j \neq i} |a_{ij}| \leq |a_{ii}|$$

holds for some i, and the matrix is irreducible, then $x_i^{(r)} \xrightarrow{r} x_i$ is the solution. For a matrix to be irreducible, the underlying simultaneous system must not have any subset of unknowns which can be solved for independently of the others. For practical problems for which convergence occurs, analysis shows the expected number of iterations required to guarantee a fixed error reduction to be proportional to the number of unknowns. Thus the total work is proportional to n^2.

The foregoing procedure may be improved several ways. The Gauss-Seidel method, sometimes called the method of successive displacements, represents the same idea but uses the latest available values. Equation (3) is solved for $i = 1, \ldots,$

$$\sum_{j<i} a_{ij}x_j^{(r)} + a_{ii}x_i^{(r)} + \sum_{j>i} a_{ij}x_j^{(r-1)} = b_i \qquad (3)$$

n. The Gauss-Seidel method converges for the conditions given above for the substitution method and is readily shown to converge more rapidly.

Further improvements in this idea lead to the method of successive overrelaxation. This can be thought of as calculating the correction associated with the Gauss-Seidel method and overcorrecting by a factor ω. Equation (4) is first solved for y.

$$\sum_{j<i} a_{ij}x_j^{(r)} + a_{ii}y + \sum_{j>i} a_{ij}x_j^{(r-1)} = b_i \qquad (4)$$

Then $x_i^{(r)} = x_i^{(r-1)} + \omega(y - x_i^{(r-1)})$. Clearly, choosing $\omega = 1$ yields the Gauss-Seidel method. For problems of interest arising from elliptic difference equations, there exists an optimum ω which guarantees a fixed error reduction in a number of iterations proportional to $n^{1/2}$, and thus in total work proportional to $n^{3/2}$.

A number of other iterative techniques for systems with sparse matrices have been studied. Primarily they depend upon approximating the given matrix with one such that the resulting equations can be solved directly with an amount of work proportional to n. For a quite large class of finite difference equations of interest, the computing work to guarantee a fixed error reduction is proportional to $n^{5/4}$. The work requirement proportional only to $n \log n$ quoted earlier applies to a moderately restricted subset.

Overdetermined linear systems. Often an over-determined linear system has to be solved. This happens, for example, if one wishes to fit the model

$$p(x) = \sum_{j=1}^{n} a_j p_j$$

to observations $(x_i, y_i)_{i=1}^{m}$ with $n < m$. Here one would like to determine the coefficient vector $\mathbf{a} = (a_1, \ldots, a_n)^T$ so that $p(x_i) = y_i$, $i = 1, \ldots, m$. In matrix notation, one wants $A\mathbf{a} = \mathbf{y}$, with A the m-by-n matrix $[p_j(x_i)]$. If $n < m$, one cannot expect a solution, and it is then quite common to determine $\mathbf{a}$ instead by least squares, that is, so as to minimize the "distance" $(\mathbf{y} - A\mathbf{a})^T(\mathbf{y} - A\mathbf{a})$ between the vectors $\mathbf{y}$ and $A\mathbf{a}$. This leads to the so-called normal equations $A^TA\mathbf{a} = A^T\mathbf{y}$ for the coefficient vector $\mathbf{a}$. But unless the "basis functions" $p_1, \ldots, p_n$ are chosen very carefully, the condition of the matrix A^TA may be very bad, making the elimination process outlined above overly sensitive to rounding errors. It is much better to make use of a so-called orthogonal decomposition for A.

Assume first that A has full rank (which is the same thing as assuming that the only linear combination p of the functions $p_1, \ldots, p_n$ which vanishes at all the points $x_1, \ldots, x_m$ is the trivial one, the one with all coefficients zero). Then A has a QR decomposition, that is, $A = QR$, with Q an orthogonal matrix (that is, $Q^T = Q^{-1}$), and R an m-by-n matrix whose first n rows contain an invertible upper triangular matrix R_1, while its remaining m-n rows are identically zero. Then $(\mathbf{y} - A\mathbf{a})^T \cdot$

$(\mathbf{y} - A\mathbf{a}) = (Q^T\mathbf{y} - R\mathbf{a})^T(Q^T\mathbf{y} - R\mathbf{a})$ and, since the last m-n entries of $R\mathbf{a}$ are zero, this is minimized when the first n entries of $R\mathbf{a}$ agree with those of $Q^T\mathbf{y}$, that is, $R_1\mathbf{a} = [(Q^T\mathbf{y})(i)]_1^n$. Since R_1 is upper triangular, this system is easily solved by back-substitution, as outlined above. The QR decomposition for A can be obtained stably with the aid of Householder transformations, that is, matrices of the simple form $H = I - (2/\mathbf{u}^T\mathbf{u})\mathbf{u}\mathbf{u}^T$, which are easily seen to be orthogonal and even self-inverse, that is, $H^{-1} = H$. In the first step of the process, A is premultiplied by a Householder matrix H_1 with $\mathbf{u}$ so chosen that the first column of H_1A has zeros in rows $2, \ldots, m$. In the next step, one premultiplies H_1A by H_2 with $\mathbf{u}$ so chosen that H_2H_1A retains its zeros in column 1 and has also zeros in column 2 in rows $3, \ldots, m$. After $n - 1$ such steps, the matrix $R: = H_{n-1} \ldots H_1A$ is reached with zeros below its main diagonal, and so $A = QR$ with $Q: = H_1 \ldots H_{n-1}$.

The situation is somewhat more complicated when A fails to have full rank or when its rank cannot be easily determined. In that case, one may want to make use of a singular value decomposition for A, which means that one writes A as the product USV, where both U and V are orthogonal matrices and $S = (s_{ij})$ is an m-by-n matrix that may be loosely termed "diagonal," that is, $s_{ij} = 0$ for $i \neq j$. Calculation of such a decomposition is more expensive than that of a QR decomposition, but the singular value decomposition provides much more information about A. For example, the diagonal elements of S, the so-called singular values of A, give precise information about how close A is to a matrix of given rank, and hence make it possible to gauge the effect of errors in the entries of A on the rank of A.

Differential equations. Classical methods yield practical results only for a moderately restricted class of ordinary differential equations, a somewhat more restricted class of systems of ordinary differential equations, and a very small number of partial differential equations. The power of numerical methods is enormous here, for in quite broad classes of practical problems relatively straightforward procedures are guaranteed to yield numerical results, whose quality is predictable.

Ordinary differential equations. The simplest system is the initial value problem in a single unknown, $y' = f(x,y)$, and $y(a) = \eta$, where y' means dy/dx, and f is continuous in x and satisfies a Lipschitz condition in y; that is, there exists a constant K such that for all x and y of interest, $|f(x,y) - f(x,z)| \leq K|y - z|$. The problem is well posed and has a unique solution.

The Euler method is as follows: $y_0 = \eta$, Eq. (5)

$$y_{i+1} = y_i + hf(x_i, y_i) \qquad (5)$$

holds, and $i = 0, 1, 2, \ldots, (b - a)/h$. Here h is a small positive constant, and $x_i = a + ih$. Analysis shows that as $h \to 0$, there exists a constant C such that $|y_k - y(x_k)| \leq Ch$, where $y(x_k)$ is the value of the unique solution at x_k, and $a \leq x_k \leq b$. This almost trivial formulation of a numerical procedure thus guarantees an approximation to the exact solution to the problem that is arbitrarily good if h is sufficiently small, and it is certainly easy to implement. A trivial extension of this idea is given by the method of Heun, Eq. (6).

$$y_{i+1} = y_i + \frac{1}{2}h[f(x_i,y_i) + f(x_i + h, y_i + hf(x_i,y_i))] \quad (6)$$

The method of Heun similarly is guaranteed to approximate the desired solution arbitrarily well since there exists another constant C_1 such that the $\{y_i\}$ satisfy relation (7). This is clearly asymptotically better than the method of Euler. It is

$$|y_k - y(x_k)| \leq C_1 h^2 \quad (7)$$

totically better than the method of Euler. It is readily found to be practically superior for most problems. Further improvement of this type is offered by the classical Runge-Kutta method, Eq. (8), where $\phi(x,y,h) = \frac{1}{6}[k_1 + 2k_2 + 2k_3 + k_4]$,

$$y_{i+1} = y_i + h\phi(x_i,y_i,h) \quad (8)$$

and $k_1 = f(x,y)$, $k_2 = f(x + h/2, y + hk_1/2)$, $k_3 = f(x + h/2, y + hk_2/2)$, and $k_4 = f(x + h, y + hk_3)$. For this method there exists a constant C_2 such that $|y_k - y(x_k)| \leq C_2 h^4$.

The foregoing methods are called single-step since only y_i is involved in the computation of y_{i+1}. The single-step methods yielding the better results typically require several evaluations of the function f per step. By contrast, multistep methods typically achieve high exponents on h in the error bounds without more than one evaluation of f per step. Multistep methods require use of $y_{i-\alpha}$ or $f(x_{i-\alpha}, y_{i-\alpha})$ or both for $\alpha = 0, 1, \ldots, j$ to compute y_{i+1}. Typical is the Adams-Bashforth method for $j = 5$, Eq. (9).

$$y_{i+1} = y_i + \frac{h}{1440}[4277 f(x_i,y_i) - 7923 f(x_{i-1},y_{i-1})$$
$$+ 9982 f(x_{i-2},y_{i-2}) - 7298 f(x_{i-3},y_{i-3})$$
$$+ 2277 f(x_{i-4},y_{i-4}) - 475 f(x_{i-5},y_{i-5})] \quad (9)$$

Analysis shows the solution of this Adams-Bashforth procedure to satisfy relation (10) for

$$|y_k - y(x_k)| \leq C_3 h^6 \quad (10)$$

some constant C_3. A large number of valuable multistep methods have been studied. If f is nontrivial to evaluate, the multistep methods are less work to compute than the single-step methods for comparable accuracy. The chief difficulty of multistep methods is that they require j starting values and, therefore, cannot be used from the outset in a computation.

Partial differential equations. Methods used for partial differential equations differ significantly, depending on the type of equation. Typically, parabolic equations are considered for which the prototype is the heat flow equation, Eq. (11), with $u(x,0)$

$$\frac{\partial^2 u}{\partial x^2} = \frac{\partial u}{\partial t} \quad (11)$$

given on $x\epsilon[0,1]$, say, and $u(0,t)$ and $u(1,t)$ given for $t > 0$. A typical finite difference scheme is shown in Eq. (12), where $i - 1, \ldots, 1/h - 1$, $w_{0,n}$

$$(w_{i,n})_{x\bar{x}} = \frac{w_{i,n+1} - w_{in}}{k} \quad (12)$$

$u(0,t_n)$, and $w_{1/h,n} = u(1,t_n)$, with $(w_i)_x = (w_{i+1} - w_i)/h$, $(w_i)_{\bar{x}} = (w_{i-1})_x$, and $w_{i,n}$ is the function defined at $x_i = ih$, $t_n = nk$. Analysis shows that as $h,k \to 0$, the solution $w_{i,n}$ satisfies $|w_{i,n} - u(x_i,t_n)| < C(h^2 + k)$

for some constant C if $k/h^2 \leq 1/2$, but for k/h^2 somewhat larger than 1/2, $w_{i,n}$ bears no relation at all to $u(x_i,t_n)$.

The restriction $k/h^2 \leq 1/2$ can be removed by using the implicit difference equation, Eq. (13), but

$$(w_{i,n+1})_{x\bar{x}} = \frac{w_{i,n+1} - w_{in}}{k} \quad (13)$$

now simultaneous equations must be solved for $w_{i,n+1}$ each step. The inequality, $|w_{i,n} - u(x_i,t_n)| < C(h^2 + k)$, still holds for some constant C. An improved implicit formulation is the Crank-Nicolson equation, Eq. (14).

$$\frac{1}{2}(w_{i,n} + w_{i,n+1})_{x\bar{x}} = \frac{w_{i,n+1} - w_{i,n}}{k} \quad (14)$$

As $h,k \to 0$, solutions satisfy relation (15) for

$$|w_{i,n} - u(x_i,t_n)| < C(h^2 + k^2) \quad (15)$$

some constant C; again h/k is unrestricted. Such techniques can readily extend to several space variables and to much more general equations. The work estimates given above for iterative solution of simultaneous equations, such as Eq. (14), apply for two-space variables.

Work using variational techniques to approximate the solution of parabolic and elliptic equations has been most fruitful for broad classes of nonlinear problems. The technique reduces partial differential equations to systems of ordinary differential equations. Analysis shows that solutions obtained approximate the desired solution, as within a constant multiple of the best that can be achieved within the subspace of the basis functions used. Practical utilization suggests this to be the direction for most likely future developments in the treatment of partial differential systems.

[CARL DE BOOR]

Bibliography: P. G. Ciarlet, *The Finite Element Method for Elliptic Problems*, 1978; S. Conte and C. de Boor, *Elementary Numerical Analysis*, 3d ed., 1980; J. M. Ortega and W. C. Rheinboldt, *Iterative Solution of Nonlinear Equations in Several Variables*, 1970; R. D. Richtmyer and K. W. Morton, *Difference Methods for Initial-Value Problems*, 2d ed., 1967; J. H. Wilkinson, *The Algebraic Eigenvalue Problem*, 1965.

Numerical indicator tube

Any electron tube capable of visually displaying numerical figures. Some varieties also display alphabetical characters and commonly used symbols. In many electronic circuits and equipments it is desirable to indicate numbers or characters. Such a tube can be one of a set which displays, for instance, the magnitude and polarity of voltage in digital form. This is similar to electrical clocks which display the time in numerical form.

A simple and extensively used form is a cold-cathode gas tube in which there is a series of cathodes which light up because of the cathode glow that surrounds any cathode in a gas discharge (see illustration). A common anode is used. The cathodes are shaped to correspond to the different numerical digits, zero through nine. These cathodes may also be shaped to form alphabetical characters or other symbols, such as (+) and (−). The desired cathode is select-

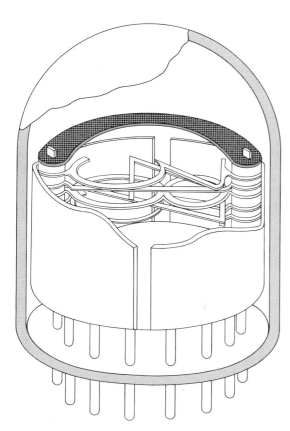

Cold-cathode gas-type numerical indicator tube.

ed by any suitable switching scheme. The cathodes are made by photoetching the desired shape from a suitable metal. They are insulated and stacked one above the other, so that any particular number can be read when the corresponding cathode is energized. *See* ELECTRICAL CONDUCTION IN GASES; GAS TUBE.

A cold-cathode gas tube is shown in the illustration. The metal cup surrounding the cathodes is used as the common anode and also provides a dark background to minimize reflections from the tube base and to increase the viewing contrast. This type of indicator tube is bright enough to be used in daylight, although there are sometimes disturbing reflections on the glass envelope. These reflections can be overcome by placing a polaroid screen in front of the tubes.   [ROGER W. WOLFE]

Octal number system

A system in which numbers are represented as linear combinations of powers of 8. Below is shown the relationship between the octal system and the decimal system.

Decimal notation	Octal notation
$12 = 8^1 + 4$	$= 14$
$31 = 3 \times 8^1 + 7$	$= 37$
$123 = 1 \times 8^2 + 7 \times 8^1 + 3$	$= 173$

The octal system has come into importance because of its value in computer applications. *See* NUMBER SYSTEMS.   [MARVIN YELLES]

Ohmmeter

A small, portable instrument using a microammeter and associated circuitry to measure resistance by the voltmeter-ammeter method. Additional circuits are usually included to measure alternating and direct-current volts and amperes, and the instrument is called a volt-ohm-milliammeter, or multimeter (Fig. 1). *See* RESISTANCE MEASUREMENT.

A typical resistance-measuring circuit is shown in Fig. 2a. Figure 2b shows a simplified schematic diagram of the ×100 range of the circuit. In operation, the instrument is first adjusted for full-scale deflection of the meter with the measuring leads shorted. When the unknown resistance is added to the circuit, the current through the meter decreases according to the equation below.

$$R_x = \left[\frac{I_{RO}}{I_{RX}} - 1\right]\left[\frac{R_M R_s}{R_M + R_s} + R_p\right]$$

The various resistances are identified in Fig. 2b. I_{RO} is the current required for full-scale deflection of the meter when R_x is 0, and I_{RX} is the meter current with the unknown resistor R_x in the circuit. The meter deflects according to the value of I_{RX}.

Since R_x is a function if $1/I_{RX}$, the scale, calibrated in resistance units, is a reciprocal function. Full-scale deflection of the meter always corresponds to zero ohms, and no deflection indicates an open circuit.

The scale relation between these two points is arbitrary and is governed by the choice of one other scale point, such as the resistance for half-scale current. The scale of the instrument in Fig. 1 is based on a half-scale resistance of about 12 ohms. In common with any measuring equipment having an error stated in terms of full-scale

Fig. 1.   Volt-ohm-milliammeter. (*Simpson Electric Co.*)

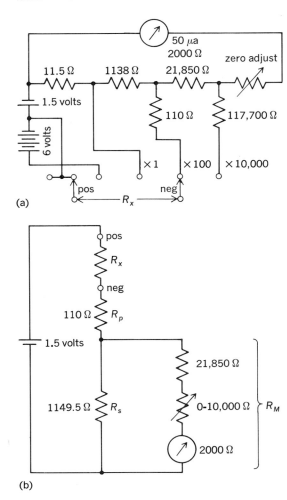

Fig. 2. Ohmmeter circuits. (*a*) Typical basic circuit. (*b*) Simplified circuit of the ×100 range.

deflection, the absolute error (error of the reading) is large for small deflections.

Voltage and current ranges for this equipment are obtained in the usual fashion with resistance multipliers for extended voltage ranges, current shunts for multiple current ranges, and a copper-oxide rectifier for alternating-current operation. A vacuum-tube voltmeter (VTVM) usually incorporates a resistance-measuring circuit for convenience. *See* AMMETER; SAMPLING VOLTMETER; VACUUM-TUBE VOLTMETER; VOLTMETER.

[CHARLES E. APPLEGATE]

Ohm's law

The direct current flowing in an electrical circuit is directly proportional to the voltage applied to the circuit. The constant of proportionality R, called the electrical resistance, is given by the equation below, in which V is the applied voltage and I is the current.

$$V = RI$$

This relationship was first described by Georg Simon Ohm in 1827 and was based on his experiments with metallic conductors. Since that time numerous deviations from this simple, linear relationship have been discovered. *See* ELECTRICAL RESISTANCE; RESISTANCE MEASUREMENT; SEMICONDUCTOR.

[CHARLES E. APPLEGATE]

Open circuit

A condition in an electric circuit in which there is no path for current between two points: examples are a broken wire and a switch in the OPEN, or OFF, position. *See* CIRCUIT (ELECTRICITY).

Open-circuit voltage is the potential difference between two points in a circuit when a branch (current path) between the points is open circuited. Open-circuit voltage is measured by a voltmeter which has a very high resistance (theoretically infinite), such as a vacuum-tube voltmeter.

[CLARENCE F. GOODHEART]

Operating system

A set of programs to control and coordinate the operation of hardware and software in a computer system. Operating systems (OS) typically consist of two sets of programs. Control programs direct the execution of user application programs and support programs. They also supervise the location, storage, and retrieval of data (including input and output to various peripherals) and allocate the resources of the computer system to the different tasks entered into the system. Processing programs include applications programs; utility programs; system diagnostics; and language translators, such as compilers and interpreters.

There are five general types of operating systems:

1. Real-time. Processing functions are executed within narrow time limits, as in updating a data bank or reading and evaluating input from a sensor or transducer.

2. Serial or batch. Tasks or jobs are run one at a time according to a schedule determined by the operating system.

3. Multiprogramming. Two or more jobs may be run concurrently in the same system.

4. Time sharing. Multiple users have simultaneous access to the system. A user is not aware that others have access to the system; the impression is that of exclusive use of the system. *See* MULTIACCESS COMPUTER.

5. Multiprocessing. Two or more processing units are coupled together to execute a single job. *See* MULTIPROCESSING.

Control programs. The control program has an executive function and a system resource allocation and management function. The executive function arranges the computer system environment necessary to run a job. One task of the executive function is to schedule the jobs. Jobs may be scheduled sequentially, but many operating systems permit scheduling on the basis of assigned priorities. The executive function directs communications between the computer system and human operator, and communications between the processor and various input and output peripherals.

The system resource allocation and management function is sometimes referred to as the supervisor. A key task of the supervisor is to allocate, monitor, and control the computer system's main memory space. The supervisor controls and synchronizes the tasks necessary to execute the instructions in a program as well. The supervisor also manages data within the system. For example, it allocates space on disks for data storage. It is responsible for constructing and developing data structures, and it maintains directories of data files within the system.

Modern supervisors are increasingly sensitive to the problem of data security, particularly unauthorized access to data files. The most common security device is the password, a code which must be correctly entered into a computer system before a user can have access to a computer system's data files. The supervisor may also record all users of a system along with the data files accessed and the amount of time each user spent using the system. *See* COMPUTER SECURITY.

Processing programs. Processing programs include applications developed for use with a computer system, and translators which convert high-level languages such as COBOL, FORTRAN, and BASIC into the assembly language used by the processor. *See* DIGITAL COMPUTER PROGRAMMING; PROGRAMMING LANGUAGES.

Several software development and diagnostic support aids are also part of the processing programs. A common feature is the presence of several debugging aids to assist in the proper development and use of applications programs. Commonly, these aids will indicate why a program failed to execute properly. These aids give an output to the user, known as a diagnostic. Most systems also include several utilities, which permit the user to accomplish such tasks as transferring data from one storage medium to another or duplicating storage disks.

Time-sharing applications. Much effort in recent years has been directed toward the development of operating systems for time sharing and multiprogramming tasks. Many commercial companies have been organized which sell access time to a large computer system on a time-sharing basis. The customers have a terminal at their place of business and are linked to the computer by telephone lines. Customers are billed according to the amount of time they use the computer. Such systems require a supervisor which restricts access to customers and adequately protects against unauthorized or fraudulent use. Large commercial users of computers need multiprogramming capability, which allows jobs to be run according to priority. Some multiprogramming systems permit programs to be run during off-peak hours. This allows a computer system to be utilized around the clock, giving more efficient use of computing resources.

Microcomputer systems. The growth of microcomputers has resulted in the development of operating systems specifically designed for them. Perhaps the most widely used is CP/M (control program for microprocessors), an operating system developed by Digital Research Corporation. CP/M is used with 8-bit microprocessor microcomputers. Another system is Unix, developed by Bell Labs for use with the C programming language. Unix has been adopted for several 16-bit microprocessor microcomputers.

Work has been done on the development of multiprocessing operating systems for microcomputers. In these systems, for example, a 16-bit microprocessor could be used for such tasks as actual computation and an 8-bit microprocessor could be used for such tasks as input and output. The essential task of these operating systems is to coordinate the activities of the two microprocessors. This task is often complicated by the fact that various microprocessors operate at different speeds. However, interest in multiprocessing operating systems for microcomputers can be expected to spread since multiple microprocessors can perform certain tasks more effectively than single microprocessors. *See* DIGITAL COMPUTER; MICROCOMPUTER; MICROPROCESSOR.

[HARRY L. HELMS]

Bibliography: G. M. Wiederhold, *Database Design*, 2d ed., 1982.

Operations research

The application of scientific methods and techniques to decision-making problems. A decision-making problem occurs where there are two or more alternative courses of action, each of which leads to a different and sometimes unknown end result (Fig. 1). Operations research is also used to maximize the utility of limited resources. The objective is to select the best alternative, that is, the one leading to the best result. Often, however, it is not simply a matter of searching a table, as there may literally be an infinity of outcomes. More intelligent means are needed to seek out the prime result.

To put these definitions into perspective, the following analogy might be used. In mathematics, when solving a set of simultaneous linear equations, one states that if there are seven unknowns, there must be seven equations. If they are independent and consistent and if it exists, a unique solution to the problem is found. In operations research there are figuratively "seven unknowns and four equations." There may exist a solution space with many feasible solutions which satisfy the equations. Operations research is concerned with establishing the best solution. To do so, some measure of merit, some objective function, must be prescribed.

In the current lexicon there are several terms

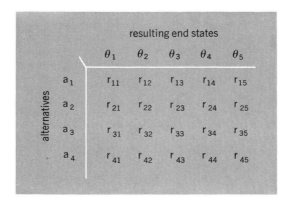

Fig. 1. End states for alternatives in a decision-making process.

associated with the subject matter of this program: operations research, management science, systems analysis, operations analysis, and so forth. While there are subtle differences and distinctions, the terms can be considered nearly synonymous.

The field can be divided into two general areas with regard to methods. These are those that can be termed mathematical programming and those associated with stochastic processes. While computers are heavily used to solve problems, the term programming should not be considered in that sense, but rather in the general sense of organizing and planning. Also, the tools of probability and statistics are used to a considerable extent in working with stochastic processes. These areas will be explored in greater detail in a later section. With regard to areas of applications, there are very few fields where the methods of operations research have not been tried and proved successful. Following is a brief history of the field, and then the general approach to solving problems.

HISTORY

While almost every art and every science can reach back into antiquity for its roots, operations research can reach back only a half-century to find its beginnings. During World War I, F. W. Lancaster developed models of combat superiority and victory based on relative and effective firepower. Thomas Edison studied antisubmarine warfare, and in 1915 F. Harris derived the first economic order quantity (EOQ) equation for inventory. Starting in 1905 and continuing into the 1920s, A. Erlang studied the flow of calls into a switchboard and formed the basis of what is now known as queueing theory.

Empiricists. The formal beginning of operations research was in England during World War II, where the term was and still is operational research. Early work concerned air and coastal defense—the coordination of fighter aircraft, antiaircraft guns, barrage balloons, and radar. Typical of the research groups formed was "Blackett's Circus." This interdisciplinary group consisted of three physiologists, two math physicists, an astrophysicist, an army officer, one surveyor, one general physicist, and two mathematicians. The basic mode of operation was to observe the problem area, and then call on the expertise of the various disciplines to apply methods from other sciences to solve the particular problem. In retrospect, this was an era of "applied common sense," and yet it was novel and highly effective.

Pragmatists. Following World War II, operations research continued to exist mainly in the military area. The operations research groups formed during the war stayed together, and a number of civilian-staffed organizations were established— RAND (1946), ORO (1948), and WSEG (1948). The real impetus to this era and to the whole field was the work done by George Dantzig and colleagues at RAND on Project Scope, undertaken for the U.S. Air Force in 1948. In attacking the problem of assigning limited resources to almost limitless demands, they developed the techniques of linear programming. Perhaps no other method is more closely associated with the field. Its use quickly spread from the military to the industrial area, and a new dimension was added to operations research. No longer was it simply "observe, analyze, and try." For the first time the field became "scientific." It could now "optimize" the solutions to problems. *See* LINEAR PROGRAMMING.

It was during this "dynasty" that formal courses in operations research were first offered. It was also during this time that operations research suffered its first lapse. Linear programming soon was looked to as the cure for too many of industry's ills. Unfortunately, industry's problems were not all linear, and "straightening them out" to fit resulted in many aborted projects and reports that were simply shelved instead of implemented. The early 1950s saw a growth in the number of industrial operations research groups; by the end of the decade, many had disappeared.

Theorists. As if in reaction to the failures of the third dynasty, toward the end of the 1950s a number of highly skilled scientists emerged who made some substantial contributions to the field. Operations research came of age and matured, and its adherents now strove for respectability. In fact, the movement was so far advanced toward developing a sound theoretical base that a new problem arose—the practicality gap.

METHODOLOGY

Operations research today is a maturing science rather than an art—but it has outrun many of the decision makers it purports to assist. The success of operations research, where there has been success, has been the result of the following six simply stated rules:

1. Formulate the problem.
2. Construct a model of the system.
3. Select a solution technique.
4. Obtain a solution to the problem.
5. Establish controls over the system.
6. Implement the solution (Fig. 2).

The first statement of the problem is usually vague, inaccurate, and sometimes not a statement of the problem at all. Rather, it may be a cataloging of observable effects. It is necessary to identify the decision maker, the alternatives, goals, and constraints, and the parameters of the system. At times the goals may be many and conflicting; for example: Our goal is to market a high-quality product for the lowest cost yielding the maximum profit while maintaining or increasing our share of the market through diversifications and acquisitions; yielding a high dividend to our stockholders while maintaining high worker morale through extensive benefits, without the Justice Department suing us for constraint of trade, competition, price fixing, or just being too big.

More properly, a statement of the problem contains four basic elements that, if correctly identified and articulated, greatly eases the model formulation. These elements can be combined in the following general form: "Given (the system description), the problem is to optimize (the objective function), by choice of the (decision variable), subject to a set of (constraints and restrictions)."

In modeling the system, one usually relies on mathematics, although graphical and analog models are also useful. It is important, however, that the model suggest the solution technique, and not the other way around. Forcing a model to fit a

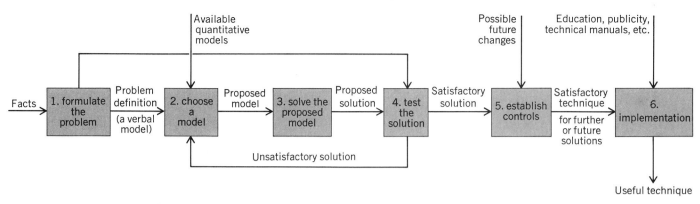

Fig. 2. Operations research approach—the six basic rules of success.

preferred technique led to some of the bad operations research of the third dynasty.

With the first solution obtained, it is often evident that the model and the problem statement must be modified, and the sequence of problem-model-technique-solution-problem may have to be repeated several times. The controls are established by performing sensitivity analysis on the parameters. This also indicates the areas in which the data-collecting effort should be made.

Implementation is perhaps of least interest to the theorists, but in reality it is the most important step. If direct action is not taken to implement the solution, the whole effort may end as a dust-collecting report on a shelf. Given the natural inclination to resist change, it is necessary to win the support of the people who will use the new system. To do this, several ploys may be used. Make a member of the using group also a member of the research team. (This provides liaison and access to needed data.) Educate the users about what the "black box" does—not to the extent of making them experts, but to alleviate any fear of the unknown. Perhaps the major limitation to successful use of operations research lies in this phase.

MATHEMATICAL PROGRAMMING

Probably the one technique most associated with operations research is linear programming. The basic problem that can be modeled by linear programming is the use of limited resources to meet demands for the output of these resources. This type of problem is found mainly in production systems, but definitely is not limited to this area. Since this method is so basic to the operations research approach, its use will be illustrated.

Linear program model. Consider a company that produces two main products—X and Y. For every unit of X it sells, it gets a $10 contribution to profit (selling price minus direct, variable costs), and for Y it gets a $15 contribution. How many should they sell of each? Obviously there must be some limitations—on demand and on productive capacity. Suppose they must, by contract, sell 50 of X and 10 of Y, while at the other end the maximum sales are 120 of X and 90 of Y. Unfortunately they have only 40 total hours of productive capacity per period, and it takes 0.25 h to make an X and 0.4 h to make a Y. What is their optimal strategy? First, the problem must be formally stated: "Given a production system making two products, the problem is to maximize the contribution to profit, by the choice of how many of each product to make, subject to limits on demand (upper and lower) and available production hours."

If X equals the number of first products made and Y the number of the second products made, the system can be modeled as follows:

Maximize
Profit contribution $= 10X + 15Y$
Subject to
$$X \geq 50$$
$$X \leq 120$$
$$Y \geq 10$$
$$Y \leq 90$$
$$.25X + .40Y \leq 40$$

The first two constraints are the lower and upper bounds of demand on X, the next two are for Y, and the last constraint refers to the productive capacity.

The problem is illustrated in Fig. 3. Any point in the feasible region of solution and the edges will satisfy all the demand and capacity constraints. To pick the best, the objective—"maximize contribution to profit"—is used. Several isoprofit lines, that is, lines where the profit is a constant value, have been added to Fig. 3. Note that the $600 line is below the feasible region, the $1800 line is above it, and the $1200 line runs through it. If other lines were formed between $1200 and $1800, one could graphically find the maximum-profit-level line that

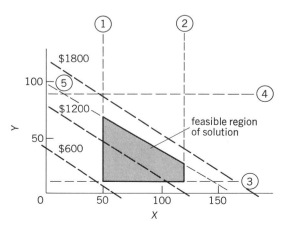

Fig. 3. Graphical presentation of production problem; circled numbers refer to the constraints.

still had one point in the feasible region. This would occur at the intersection of the second and fifth constraints. At this point, the solution is:

Profit contribution $= \$1575$
$$X = 120$$
$$Y = 25$$

It is fairly easy to verify that this is the optimal solution, that all the constraints are satisfied, and that no larger profit can be found.

While this simple example did not pose much of a computational problem, actual cases do, in that there may be thousands of variables and thousands of constraints. Problems of scheduling the product mix from a refinery where a blend of different crudes can be used (each with different costs and properties) are in this category. Efficient algorithms and computer codes have been developed to solve these problems.

Other models. There are a number of other linear programming models that relate to specific types of problems. However, advantage is taken of the special structure of the model to develop special and more efficient methods of solution.

1. Transportation problem. Given a set of sources (as, for example, factories) and a set of demands (such as regional warehouses), the problem is to minimize the cost of transporting product to destination, by choice of how much will be supplied from each source to each destination, subject to limits on demand and capacity.

2. Assignment problem. Given a set of tasks to be performed and a set of workers (machines), the problem is to maximize the overall efficiency by choice of assignment of task to worker, subject to the constraint that all tasks must be done and a worker can do at most one task.

Network models. There is another set of linear programming models that fall under the general category of network models:

The shortest-path problem determines the shortest path from one node to another. (The reverse of this is the critical path method used in project management.)

The max flow problem determines the capacity of a network such as a pipeline or highway system.

The min cost problem is a variation of the transportation model.

Integer linear programming. There is a class of problems that have the linear programming form, but in which the variables are limited to be integer values. These represent the most difficult set to solve, especially where the variables take on values of 0 or 1. (This problem occurs, for instance, in project selection—the project is either funded, 1, or not funded, 0.) While a number of algorithms have been developed, there is no assurance that a given problem will be solved in a reasonable amount of computer time. These algorithms fall into two general categories—the cutting plane method, which adds constraints that cut off noninteger solution points, and the branch and bound method, which examines a tree network of solution points.

Nonlinear programming. Another category of problems arise where either the objective function or one or more of the constraints, or both, are nonlinear in form. Again, a series of methods have been developed that have varying degrees of success, depending on the problem structure.

Geometric programming. One of the new techniques developed relates to a certain class of nonlinear models that use the arithmetic-geometric mean inequality relationship between sums and products of positive numbers. Such models result from modeling engineering design problems, and at times can be solved almost by inspection. Because of the ease of solution, a considerable effort has been made to identify the various problems that can be structured as a geometric programming model.

Dynamic programming. This technique is not as structured as linear programming, and more properly should be referred to as a solution philosophy rather than a solution technique. Actually, predecessors of this philosophy have been known for some time under the general classification of calculus of variation methods. Dynamic programming is based on the principle of optimality as expressed by Richard Bellman: "An optimal policy has the property that whatever the initial state and initial decision are, the remaining decision must constitute an optimal policy with regard to the state resulting from the first decision." *See* OPTIMIZATION.

The operative result of this principle is to start at the "end" of the problem—the last stage of the decision-making process—and "chain back" to the beginning of the problem, making decisions at each stage that are optimal from that point to the end. To illustrate this process, the fly-away-kit problem (also known as the knapsack problem) will be briefly described: Given a set of components, each with a unit weight and volume, the problem is to maximize the value of units carried to another location, by choice of the number of each to be taken, subject to limitations on volume and weight that can be carried.

In this problem each component will represent a stage at which a decision is to be made (the number to be included in the kit), and the amount of volume and weight left are defined as state variables. The end of the problem, then, is where there is only one more component to consider (Fig. 4a).

Usually at this stage the solution is almost trivial. As many of the last components (X_1) are selected as can be within the limits of available volume (V_1) and weight (W_1). In practice, all possible values of these two state variables are solved for, since it is not known how much will be left when the last stage is reached. To this solution is now

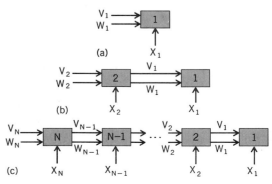

Fig. 4. Dynamic programming problems. (*a*) Final decision stage. (*b*) Two-stage problem. (*c*) "N"-stage problem.

"chained" the problem of how to select the second-last component (Fig. 4b).

Here one considers the problem that, for each level of volume and weight, one must choose between taking one or more of component 2 or passing it on to the last stage. Combinations of components 1 and 2 must now be looked at to find the best mix, considering the available resources. In like manner, one "chains back" to the beginning of the problem (Fig. 4c).

The measures of value can be in any terms. If the components are simple cargo, they could have a monetary value. If they are spare parts for some system, a reliability measure could be used.

STOCHASTIC PROCESSES

A large class of operations research methods and applications deals with stochastic processes. These processes can be defined as where one or more of the variables take on values according to some, perhaps unknown, probability distribution. These are referred to as random variables, and it takes only one to make the process stochastic.

In contrast to the mathematical programming methods and applications, there are not many optimization techniques. The techniques used tend to be more diagnostic than prognostic; that is, they can be used to describe the "health" of a system, but not necessarily how to "cure" it. This capacity is still very valuable.

Queueing theory. Probably the most studied stochastic process is queueing. A queue or waiting line develops whenever some customer seeks some service that has limited capacity. This occurs in banks, post offices, doctors' offices, supermarkets, airline check-in counters, and so on. But queues can also exist in computer centers, repair garages, planes waiting to land at a busy airport, or at a traffic light.

Queueing theory is the prime example of what can be said about the state of the system but not how to improve it. Fortunately, the improvements can be made by increasing the resources available. For example, another teller opens up a window in a bank, or another check-out stand is opened in the supermarket. Other possible changes are more subtle. For example, the "eight items or less" express lane in a supermarket minimizes the frustration of a customer in line behind another with two full shopping carts. Some post offices and banks have gone to a single queue where the person at the head of the line goes to the next available window.

While it is not possible to generalize the results of queueing analysis, it is possible to provide some general measures. One is the load or traffic factor. This is the ratio of arrival rate to combined service rate, considering the number of service centers in operation. It is possible to plot queueing statistics against this factor as shown in Fig. 5.

When the load factor (ρ) is low, there is excess serving capacity. There may be occasional lines, but not often. As the load factor rises, a fairly linear rise is obtained in any queueing statistics until approximately the point where $\rho = 0.75$. After this there is a sharp and continuous rise in the lengths of line and waits. The system is becoming saturated. If $\rho = 1.00$, the best that can be said is, "the larger the line, the larger the wait."

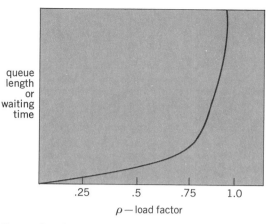

Fig. 5. Plot for general queues relating load factor to serving capacity.

The mathematics of the probability distributions of arrival rates and service times often define closed-form solutions, that is, being able to directly solve for an answer. In these cases another technique has proved very useful.

Simulation. Simulation is defined as the essence of reality without reality itself. With stochastic processes, values are simulated for the random variables from their known or assumed probability distributions, a simpler model is solved, and the process is repeated a sufficient number of times to obtain statistical confidence in the results.

As a simple example of this, assume that someone makes the statement that average female student at a university is of a certain height. To verify this assumption the actual heights of all female students could be measured, but this may be a long process. As an alternative, a hopefully random sample may be taken, the heights measured, the average determined, and an inference made as to the whole female population.

Many stochastic processes do not lend themselves to such a direct approach. Instead an assumption is made as to what the underlying probability distributions of the random variables are, and these are sampled. This sampling is done with random numbers. True random numbers are difficult to obtain, so pseudorandom numbers are generated by some mathematical relationship. This apparent paradox is justified by the fact that the numbers appear to be random and pass most tests for randomness. As an example, a queueing system could be modeled by having two wheels of fortune—one with numbers in random order from 1 to 15, and the other from 1 to 20. The first could represent the number of arrivals into the system per 10-min period, and the other the number served during the same period. In this manner a system with a load factor of 0.75 could be simulated. Many simulation computer languages have been developed to ease the work of analyzing queueing systems.

Markov processes. This class of stochastic processes is characterized by a matrix of transition probabilities. These measure the probability of moving from one state to another. Such methods have been used to determine the reliability of a system, movement of a stock in the stock market, aging of doubtful accounts in credit analysis, and brand switching. This last application will be illustrated.

A company is considering marketing a new product. A preliminary market survey indicates that if a customer uses the product, there is a 60% probability that he or she will continue to do so. Likewise, the advertising campaign is such that 70% of those using all other products will be tempted to switch. The question is, what market share will the product be expected to finally obtain? The probability matrix is shown in Fig. 6.

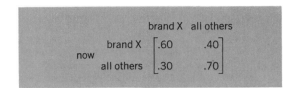

		brand X	all others
now	brand X	.60	.40
	all others	.30	.70

Fig. 6. Probability matrix for brand-switching model.

Initially the market share of brand X is 0%, but after one period this will rise to 30%; that is, the advertising campaign will induce 30% to try brand X. In the next period, only 60% of these 30% will continue to purchase brand X, but 30% of the other 70% will switch, for a total market share of 39%.

Period by period, there will be transitions from state to state, but soon the variations will dampen out and a steady state will be achieved. This can be found for this problem by matrix value, and the final market share for brand X is 42.8%.

Decision trees. While not originated by the field of operations research and not strictly in its domain, decision trees are an important tool in the analysis of some stochastic processes. More properly, they are a part of what may be considered statistical decision making. Their use can be illustrated with an example from the oil industry.

An oil company is developing an oil field and has the option to lease the mineral rights on an adjacent block of land for $100,000. An exploratory well can be drilled at a cost of $250,000. If the well

is considered moderate or good, it will cost an additional $50,000 to complete before production can start.

For simplicity, one can assume that the well, if good, has either moderate or good production. Also, the present worth of the net returns on all the future production is $1,000,000 and $3,500,000 for these two states. The problem facing the oil company is what they should do.

To analyze this problem, some subjective probabilities must be estimated. Assume the probability that the well is dry is 70%, that the flow is moderate is 20%, and that it is good is 10%. With these percentages, a decision tree can be constructed (Fig. 7).

By a process of "folding back" the values and probabilities, the expected value at each decision point is seen to be represented by a square in Fig. 7. Actually, while there are three sequential decisions — lease, drill, and complete the well — once the decision to lease is made, the sequence is fixed unless the well is discovered to be dry. The expected value of this decision can be calculated as follows:

$$\begin{aligned} E\,[\text{lease}-\text{drill}] = &-100,000-250,000+0.70\,(0)\\ &+0.20\,(-50,000+1,000,000)\\ &+0.10\,(-50,000+3,500,000)\\ =&\ \$185,000 \end{aligned}$$

Note that this expected value is slightly positive. In reality one of three things will happen — the well is dry, and the loss is $350,000; the well is moderate, with a net gain of $600,000; or the well is good, with a net gain of $3,100,000. This expected value can be interpreted as follows: if a large number of wells are drilled, 70% will be dry at a loss of $350,000 each, 20% will be moderate, and 10% will be good. The average gain will be $185,000.

SCOPE OF APPLICATION

There are numerous areas where operations research has been applied. The following list is not intended to be all-inclusive, but is mainly to illustrate the scope of applications: optimal depreciation strategies; communication network design; computer network design; simulation of computer time-sharing systems; water resource project selection; demand forecasting; bidding models for offshore oil leases; production planning; assembly line balancing; job shop scheduling; optimal location of offshore drilling platforms; optimal allocation of crude oil using input-output models; classroom size mix to meet student demand; optimizing waste treatment plants; risk analysis in capital budgeting; electric utility fuel management; public utility rate determination; location of ambulances; optimal staffing of medical facilities; feedlot optimization; minimizing waste in the steel industry; optimal design of natural-gas pipelines; economic inventory levels; random jury selection; optimal marketing-price strategies; project management with CPM/PERT/GERT; air-traffic-control simulations; optimal strategies in sports; system availability/reliability/maintainability; optimal testing plans for reliability; optimal space trajectories.

It can be seen from this list that there are few facets of society that have not had an application for operations research. *See* DECISION THEORY; GERT; PERT.

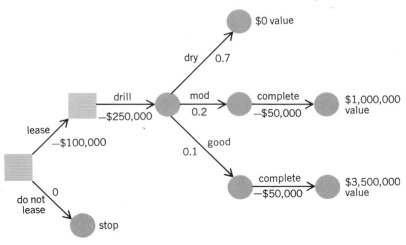

Fig. 7. Decision tree for an oil well problem.

[WILLIAM G. LESSO]

Bibliography: R. E. Bellman and S. E. Dreyfus, *Applied Dynamic Programming*, 1962; V. N. Bhat, *Elements of Applied Stochastic Processes*, 1969; F. S. Hillier and G. Lieberman, *Introduction to Operations Research*, 3d ed., 1980.

Optical bistability

A phenomenon in which an optical device can exist in either of two stable states. Optical bistability is an expanding field of research because of its potential application to all-optical logic and because of the interesting physical phenomena it encompasses. A bistable optical device can function as a variety of logic devices. It can have two stable output states labeled 0 and 1 for the same input, as shown in Fig. 1a. Thus it can serve as an optical memory element. Under slightly modified operating conditions, the same device can exhibit the optical transistor characteristic of Fig. 1b. For input intensities close to I_{gain}, small variations in the input light are amplified, in much the same way that a vacuum tube triode or transistor amplifies electrical signals. The characteristic of Fig. 1b can also be used as a discriminator; inputs above I_{gain} are transmitted with far less attenuation than those below. Finally, there is limiting action above I_{gain}: large changes in the input hardly change the output. There is hope that bistable devices will revolutionize optical processing, switching, and computing. Since the first observation of

optical bistability in a passive, unexcited medium in 1974, bistability has been observed in many different materials, including tiny semiconductor etalons. Current research is focused on optimizing these devices and developing smaller devices of better materials which operate faster, at higher temperatures, and with less power.

Optical bistability has also attracted the attention of physicists interested in fundamental phenomena. In fact, a bistable device often consists of a nonlinear medium within an optical resonator; it is thus similar to a laser except that the medium is unexcited (except by the coherent light incident on the resonator). Such a device constitutes a simple example of a strongly coupled system of matter and radiation. Counterparts of many of the phenomena studied in lasers such as fluctuations, regenerative pulsations, and optical turbulence can be observed under better-controlled conditions in passive bistable systems. *See* LASER.

All-optical systems. The transmission of information as signals impressed on light beams traveling through optical fibers is replacing electrical transmission over wires. The low cost and inertness of the basic materials of fibers and the small size and low loss of the finished fibers are important factors in this evolution. Furthermore, for very fast transmission systems, for example, for transmitting a multiplexed composite of many slow signals, optical pulses are best. This is because it is far easier to generate and transmit picosecond optical pulses than electrical pulses. With optical pulses and optical transmission, the missing component of an all-optical signal processing system is an optical logic element in which one light beam or pulse controls another. Because of the high frequencies of optical electromagnetic radiation, such all-optical systems have the potential for subpicosecond switching and room-temperature operation. Although any information processing and transmitting system is likely to have electrical parts, especially for powering the lasers and interfacing to humans, the capability of subpicosecond switching appears unique to the optical part of such a system. *See* INTEGRATED OPTICS; OPTICAL COMMUNICATIONS.

Bistable optical devices. Recently, bistable optical devices have been constructed which have many of the desirable properties of an all-optical logic element. A device is said to be bistable if it has two output states for the same value of the input over some range of input values. Thus a device having the transmission curve of Fig. 1a is said to be bistable between I_{down} and I_{up}. This device is clearly nonlinear; that is, I_{out} is not just a multiplicative constant times I_{in}. In fact, if I_{in} is between I_{down} and I_{up}, knowing I_{in} does not reveal I_{out}. To accomplish this behavior, an all-optical bistable device requires feedback. Even though a nonlinear medium is essential, the nonlinearity alone only means that I_{out} versus I_{in} is not a straight line. The feedback is what permits the nonlinear transmission to be multivalued, that is, bistable.

Fabry-Perot interferometer. The most nearly practical devices so far are tiny semiconductor etalons, that is, tiny Fabry-Perot interferometers consisting of a gallium arsenide (GaAs) or indium antimonide (InSb) crystal with flat parallel faces sometimes coated with dielectrics to increase the reflectivity to about 90% (Fig. 2). In these etalons

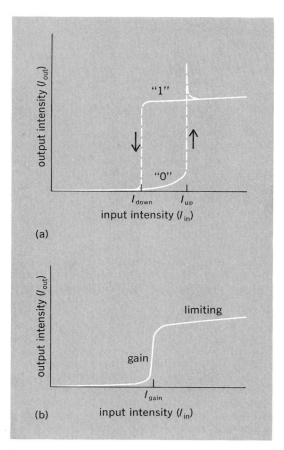

Fig. 1. Transmission of a typical bistable optical device, under conditions of (a) bistability (memory) and (b) high ac gain (optical transistor, discriminator, or limiter).

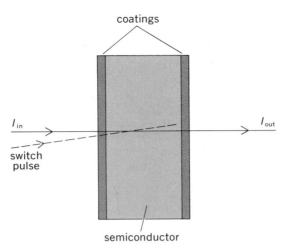

coatings

I_{in} → → I_{out}

switch
pulse

semiconductor

Fig. 2. Bistable optical etalon.

the Fabry-Perot cavity provides the optical feedback, and the nonlinear index of refraction n_2 is the intensity-dependent parameter. The bistable operation of such a Fabry-Perot cavity containing a medium with a nonlinear refraction can be pictured as follows. In the off or low transmission state, the laser is detuned from one of the approximately equally spaced transmission peaks of the etalon and most of I_{in} is reflected (Fig. 3a). The refractive index is approximately n_0, its value for weak light intensity. In the on state the index is approximately $n_0 + n_2 I_c$, where the intensity inside the cavity is I_c. This change in index shifts the etalon peak to near coincidence with the laser frequency, permitting a large transmission and a large I_c. Clearly there must be a consistency between the index and the laser frequency; each affects the other through the feedback. As the input is increased from low values, the frequency ν_{FP} of the Fabry-Perot peak begins to shift when $n_2 I_c$ becomes significant. But this shift increases I_c, which further increases $n_2 I_c$, and so on. This positive feedback continues until, at I_{up}, the effect runs

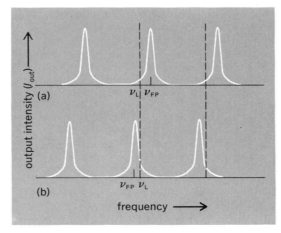

Fig. 3. Transmission function of a nonlinear Fabry-Perot interferometer: (a) device off, index $n \cong n_0$, and (b) device on, $n \cong n_0 + n_2 I_c$. The laser frequency ν_L is fixed, but the cavity peak frequency n_{FP} changes with light intensity via n_2.

away, the device turns on, and the transmission reaches a value on the negative feedback side of the etalon peak consistent with $n = n_0 + n_2 I_c$ (Fig. 3b). Once the device is turned on, I_c is larger than I_{in} because of the storage property of the cavity. Therefore I_{in} can now be lowered to a value below I_{up}, and the large I_c will keep the device on. Thus the hysteresis of the bistability loop arises from the fact that, for the same input intensity, the intracavity intensity and index contribution are small in the off or detuned state and large in the on or in-tune state.

Clearly, a device with the transmission characteristic of Fig. 1 can serve as an optical memory. If I_{in} is maintained between I_{down} and I_{up}, the value of I_{out} reveals the state of the device. Light pulses can be used to switch the device on and off, just as for an electrical flip-flop. A simple analysis of a nonlinear Fabry-Perot cavity and experiments show that by changing the initial detuning this device can perform a whole host of optical operations. Bistable optical devices have been operated as an all-optical differential or ac amplifier (also called an optical transistor and a transphasor), limiter, discriminator, gate, oscillator, and pulse shaper.

Properties. The GaAs device was only 5 μm thick, and the laser beam diameter on the etalon was only 10 μm—about one-tenth the diameter of a human hair. The device was switched on by a 10-ps pulse in a time shorter than the 200-ps detector response time, and a time of a few picoseconds is believed possible. The GaAs device operates at a convenient wavelength (0.83 μm) and is constructed of a material used for electronics and diode lasers, facilitating interfacing and integration. But all the properties of the demonstrated device are not ideal. The switch-down time is about 40 nanoseconds, presumably arising from the lifetime of the free carriers and excitons created by the intense light in the on state. The minimum input power is about 100 mW, and the minimum switch-on energy about 1 nanojoule, far more than a practical device can have. And the highest operating temperature is 120 K, because the free exciton resonance used to produce n_2 decreases, and the undesirable band-tail absorption increases, with increased temperature. However, measurements of the intrinsic properties of pure GaAs reveal that the switching times and energy should be greatly reduced in optimized etalons.

The InSb etalons are longer (a few hundred micrometers), operate at longer wavelengths ($\approx$ 5 μm), and hence have larger transverse dimension limitations than GaAs, have demonstrated switching times of a few hundred nanoseconds, operate at least up to 77 K, and have functioned with as little as 8-mW input power. And, just as for GaAs, it is unlikely that the InSb devices are optimized.

The results of these first experiments seeking practical devices point to obvious areas currently under intense research. The physics of semiconductors is being challenged to identify giant nonlinearities at convenient wavelengths and at higher temperatures. In addition to the GaAs free exciton and InSb below-edge band-to-band saturation mechanisms already observed, electron-hole plasma and biexciton mechanisms have been proposed. Other bistable configurations are being sought to better utilize the nonlinearities and minimize the power required: thin etalons, nonlinear interfaces, self-focusing devices, and guided-wave structures.

Fundamental studies. In addition to work on the application of optical bistability to practical devices, a comparable effort is being directed toward the properties of these devices, whose operation depends on the nonlinear coupling between electronic material and light. Figure 1 shows a discontinuity in the transmission, reminiscent of a first-order phase transition. The input light maintains the system in equilibrium which is far from thermal. Fluctuations and transient behavior of such systems are of great interest. A sluggishness in response, called critical slowing down, has already been observed for I_{in} close to I_{up}. Regenerative pulsations have also been seen in which competing mechanisms cause the device to switch on and off repeatedly, forming an all-optical oscillator.

Hybrid devices. Many of these fundamental studies have been conducted with hybrid (mixed optical and electronic) bistable devices in which the intensity dependence of the intracavity index results from applying a voltage proportional to the transmitted intensity across an intracavity modulator. In fact, the cavity is not necessary, since electrical feedback is present. Placing the modulator between crossed polarizers provides the required nonlinearity. Integrated hybrids are being considered for practical devices, but the ultimate in shortening the detector-to-modulator wire length is to place the detector inside the cavity. The distinction between hybrid and intrinsic then fades. The best device may have a voltage across the semiconductor to increase its speed and sensitivity. Arrays of hybrid devices have been used to study image processing using bistable elements; eventually bistable arrays could be used for parallel computing, for example, for propagation problems with transverse effects.

Optical turbulence. Hybrid devices have been considered to be completely analogous to intrinsic devices. Recently it has been predicted that an intrinsic ring-cavity device subjected to a steady input will exhibit periodic oscillations and turbulence or chaos if the medium response time is short compared with the cavity round-trip time. The hybrid analogy is to delay the feedback by a time longer than the detector-feedback-modulator response time. Optical chaos, observed in such a hybrid, has been used to study the evolution, from a stable output, through periodic oscillations, to chaos as the input is increased. The optical turbulence studies interface to phenomena in many other disciplines, such as mathematics, genetics, and hydrodynamics, emphasizing the basic similarity of these seemingly diverse phenomena, in agreement with some recent mathematical findings.

With such important potential applications, with such significant possibilities for improvements, and with so many opportunities for experiments of fundamental interest, the field of optical bistability is likely to continue its explosive growth.

[HYATT M. GIBBS]

Bibliography: H. M. Gibbs, S. L. McCall, and T. N. C. Venkatesan, *Opt. Eng.*, 19:463–468, 1980; H. M. Gibbs et al., *Phys. Rev. Lett.*, 46:474–477, 1981; S. D. Smith and D. A. B. Miller, *New Sci.*, 85:554–556, Feb. 21, 1980; P. W. Smith and W. J. Tomlinson, *IEEE Spectrum*, 18(6):26–33, 1981.

Optical communications

The transmission of speech, data, pictures, or other information by light. An information-carrying light wave signal originates in a transmitter, passes through an optical channel, and enters a receiver which reconstructs the original information. The ensemble of these three items constitutes an optical communication system. A block diagram of the basic optical communication system is shown in the illustration.

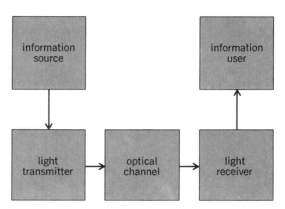

Block diagram of a simplified optical communication system.

The study of such systems, their components, and their application is generally understood to constitute the field of optical communications. Optical communication is one of the most advanced forms of communication by electromagnetic waves. Light waves are generally understood to occupy the part of the electromagnetic spectrum with wavelengths in the range $0.2–100 \ \mu m$. The advantages sought in using light waves instead of electromagnetic waves with longer wavelengths are threefold:

1. Because of its short wavelength, light can be focused into narrower beams or confined in smaller waveguides than radiation of longer wavelengths.

2. The information-carrying capacity is potentially greater than that of longer-wavelength radiation.

3. The highest transparency for electromagnetic radiation achieved in any solid material is that of silica glass in the wavelength range $1–2 \ \mu m$. This transparency is orders of magnitude higher than that of any other solid material in any other part of the spectrum.

Much effort has been made toward realizing these advantages ever since the introduction of the telephone demonstrated the importance of worldwide communication networks. But progress was held back by the lack of a suitable light source until the 1960s. Until then, the only available sources generated light from a multitude of independent atomic radiators. Such light cannot be focused effectively to form a narrow beam or to propagate along the axis of a waveguide. In 1960 the demonstration of the first laser removed this shortcoming and made feasible in the optical wavelength region all the communication techniques formerly limited to microwaves and other electromagnetic waves of longer wavelengths. *See* LASER.

Another technical achievement which had a deciding impact on the effort in the field of optical communications occurred in 1970 with the fabrication of a fiber waveguide which demonstrated the potential transparency of silica glass in the light spectrum. A third influence was the increasing need for systems suited for the transmission of digital signals which permit the integration of speech, data, pictures, or any other information in a universal communications format. The physical dimensions and characteristics of fiber lightguides and of semiconductor lasers are uniquely matched with modern digital circuits, and with the associated processing and transmission techniques. The most urgent need for this advanced communication capability arose in metropolitan communication networks where optical fiber systems began to appear in the late 1970s, supplementing other digital communication systems of a more conventional kind. *See* DATA COMMUNICATIONS.

Although optical fiber systems presently appear as the most significant outgrowth of intense exploration in the field of optical communications, they were not the only or first optical systems considered or utilized. Optical communication systems can be categorized according to the optical channel used, which may be a beam in free space, in the atmosphere, or in a pipe rather than a continuously guided wave in a waveguide of glass. Optical systems can also be categorized according to the source used, the characteristics of the information transmitted, or the choice of the wavelength. Accordingly, one speaks of laser or light-emitting diode (LED) systems, digital or analog optical systems, or systems operating in the visible or infrared part of the light spectrum.

Free-space optical communications. A channel of this kind exists, for example, between orbiting satellites. For satellite-to-satellite communications, the region between the information source and information use is ideal, essentially a vacuum, which does not distort or attenuate the light beam. For this application, the laser is the best optical source because of its ability to radiate spatially coherent light. Lasers particularly well suited to this application are the carbon dioxide infrared gas laser and the neodymium yttrium-aluminum-garnet (Nd:YAG) solid-state laser (near infrared or green).

The beam can be launched at the transmitter and picked up at the receiver by telescopes with apertures limited only by weight and by the precision of the means used to aim at each other. Aiming is a real problem since light beams less than 1 second of arc wide may be required—a beam width attainable with a 4-in.-diameter (10-cm) telescope at visible wavelengths. When the need for communications between satellites arises, the laser system must be compared with millimeter-wave systems in which larger, less precise antennas (15-cm diameter for a 1° beam) replace telescopes, and less precision is needed to aim the antennas, but in which a higher-power transmitter would be required.

Atmospheric optical communications. For satellite-to-Earth communications and other communication through air, the Earth's atmosphere strongly influences the light transmission between information source and destination. In the visible-wavelength band, $0.4-0.7$ μm, and in bands near 2.3, 3.8, and 10 μm, transmission losses are low for clear weather conditions. However, minute temperature differences of the air along the light path cause beam broadening, so that even in clear weather, laser-beam widths no less than about 20 seconds of arc are achievable for a typical overland path several kilometers long. Scintillations of the received beam occur at rates below 1000 Hz, and systems can be designed to tolerate them. Rain, fog, or snow cause much more serious transmission changes. Beam broadening to as much as 80 seconds of arc (caused by snow) and attenuation of the laser-beam power to less than 1/1000 the clear weather value can occur in a 3-km path. Thus, in systems requiring transmission through the Earth's atmosphere, the weather introduces considerable transmission variation. This leads to the occasional need for orders-of-magnitude more transmitted light power than during clear weather conditions.

While the high incidence of attenuation on the Earth-space path is caused by clouds, it is fog and snow, rather than rain, that produce most of the attenuation on a terrestrial communications path. In neither case would the transmission reliability be considered satisfactory for most communication purposes. On the other hand, the probability of outage of an atmospheric propagation system can be arbitrarily low if the transmission path is short. For example, optical communication links confined to rooms or buildings for the remote control of television sets or video recorders have become quite popular. In this case, a well-collimated light beam would degrade operation rather than improve it because such a beam would demand an unreasonable pointing accuracy by the transmitter. As a result, these links utilize the more conventional light-emitting diode as a source rather than a laser. The wavelength of operation is typically in the near infrared to avoid any perception by the human eye.

In a shielded atmosphere such as a room, a building, or a pipe, the effects of precipitation can be avoided so that attenuation is quite low. Optical communication systems have, therefore, been studied which guide light waves along a specific path inside a pipe with the help of lenses and mirrors. In this case, many laser beams can be sent along spatially separable paths through the same pipe if the refocusing elements are not more than a few tens of meters apart. Even in a confined atmosphere, however, thermal effects must be extremely well controlled to prevent degradation of the laser beams, and continuous automatic control of the directing elements may be necessary. Since many separate laser beams can be maintained, the information-carrying capacity is great indeed, probably well over a million telephone conversations. On the other hand, the installation and maintenance costs are high, and the demand for such a communication highway may never be sufficient to justify the expense of its development.

Optical fiber communication. Early applications of glass fibers include the transmission of images through short fiber bundles in the medical practice of endoscopy for the purpose of inspecting internal organs. These early fibers, however, were not nearly transparent enough to carry light for typical communication distances, nor could they deliver high-speed information with sufficient fidelity for error-free reconstruction in the receiver. For fibers to become useful light waveguides or "light guides" for communication applications, transparency and the control of signal distortion had to be improved by orders of magnitude. Two processes were successful in achieving the transparency objective: (1) Painstaking purification of conventional multicomponent glasses, notably sodium borosilicates, before drawing them into fiber from the molten phase has resulted in light guides which attenuate a light wave at 850 nm wavelength by not more than 60–70% in 1 km. (2) Glass rods made by a special process and consisting almost entirely of silica can be pulled into fiber at temperatures approaching 2000° C, and produce light guides that attenuate light at 850 nm by 40–50% in 1 km and light at 1550 nm by as little as 5% in 1 km. Since the reconstruction of digital messages is possible even after attenuation in the waveguide by three to four orders of magnitude, the transmission distance between transmitter and receiver can surpass 10 km at 850 nm and be several times that distance near 1550 nm.

Maintaining signal fidelity over such distances required modifications of the principal guidance concept employed in early fibers. These early fibers were made from two different glasses such that the core glass exhibited a slightly higher index of refraction than the outer cladding glass. Light rays that strike the core-cladding interface at a grazing angle are reflected into the core and thus confined there. Depending on the angle of incidence, therefore, the rays follow different zigzag paths inside the core. Rays that bounce back and forth more often take longer to travel the length of the fiber than rays that propagate more directly along the axis. The difference in propagation time results in a distortion of the message, a broadening and interleaving of adjacent pulses in digital systems, for example. Clearly, the difference in propagation time increases with the distance traveled so that this effect places a limit on the transmission distance for a given pulse rate. Fibers of the kind described above are, therefore, used for shorter communication links in which their application is motivated by virtues other than transparency and lack of signal distortion. These advantages may be the freedom from environmental electrical noise and the small size of fibers in the case of optical data links used between automobile instrument panels, computer frames, and other control apparatus which must function in an environment congested by electromagnetic radiation.

Two modifications of the "step-index" light guide described above have led to essentially distortion-free transmission of rapid information flow over many kilometers:

1. It is possible by various fabrication means to produce an index of refraction that decreases radially from the core center outward until it matches the cladding index at the core-cladding interface. Light rays entering the core of the "graded-index" fiber at a small angle with respect to the axis follow undulating paths. The speed of light along the undulating paths increases in the regions of lower refractive index so that the travel time along these paths is nearly equal to that along the straight axial path.

2. The core size and the index step at the core-cladding interface can be reduced to the point at which only essentially axial propagation is possible. In this condition, only one mode of propagation exists and determines the travel time. The core diameters of such "single-mode" fibers must not exceed six to eight wavelengths for stable transmission. This small dimension requires extreme precision of alignment in splices and in connections of the fiber to the light source and the detector. See INTEGRATED OPTICS; WAVEGUIDE.

The fidelity of information transmission in the single-mode fiber is limited only by the fact that the travel time of light energy in silica varies as a function of the wavelength. This variation has a very small value near the wavelength of 1300 nm in practical single-mode fibers and increases at shorter or longer wavelengths. Since practical optical sources, including most semiconductor lasers, radiate at several wavelengths within a narrow spectral region, operation at 1300 nm is desirable to guarantee fidelity of the message over long distances. Single-mode fiber cables are likely to find an application in optical communication systems in which extremely long transmission distances between the transmitter and the receiver are important. Examples are high-capacity communication systems carrying speech and data information between cities or across oceans in undersea installation. If the transmission distance is so long that the information would become unintelligibly weak after enhanced transmission through a fiber, intermediate repeaters can be installed which amplify or regenerate the information before it is sent on. This method has served well in conventional communication systems and has been applied successfully to many fiber systems now in service.

The latter systems utilize graded-index fibers which are easier to splice and to connect to sources and detectors since the core diameter measures 50 μm or more. These early systems use a wavelength of operation in the region between 800 and 900 nm. This wavelength region does not coincide with the spectral region in which silica fibers offer the highest transparency or fidelity of transmission. The choice was simply dictated by available semiconductor lasers and detectors.

Optical transmitters. The preferred sources for use in optical communication systems are semiconductor diodes which generate light in forward-biased operation. The simplest device is the light-emitting diode which radiates in all directions from a fluorescent area located in the diode junction. Since fibers accept only light entering the core at a small angle with respect to the axis, only a small part of the light radiated from a light-emitting diode can be captured and transmitted by the fiber. See LIGHT-EMITTING DIODE.

In the case of the semiconductor laser, two ends of the junction plane are furnished with mirror surfaces which form an optical resonator and enhance the light bouncing back and forth between the mirrors by stimulated emission. As a result, the light emitted through the partially transparent end mirrors is well collimated within a narrow cone of angles. A large fraction of the light can be captured and transmitted by optical fibers. In practical semiconductor lasers, the semitransparent mirrors are simply formed by careful cleaving of two sides of the junction. Electric current flow is confined to a narrow strip between the mirrors, and various techniques of composition change during the growth of the semiconductor crystal are used to eliminate nonradiative processes that do not participate in the stimulated emission process.

The most efficient lasers are fabricated by growing crystal layers of varying composition on a substrate crystal. The first devices of this kind were based on the gallium-aluminum-arsenic system. Light generated by these sources is confined to wavelengths shorter than 900 nm. When it was discovered that the transparency of silica fibers in the vicinity of 1300 nm is several times that at 900 nm, a new material system based on indium, gallium, arsenic, and phosphorus was found which can produce semiconductor sources radiating in the wavelength range between 1100 and 1600 nm.

Both light-emitting and laser diodes can be modulated by varying the forward diode current. Typically, the message is a digital sequence of coded electrical pulses which are simply used to turn the source diode on and off. The light injected into the optical channel is a faithful replica of the information sequence. See OPTICAL MODULATORS.

Optical receivers. The simplest photodetectors are designed to capture light in a reverse-biased junction where it generates electron-hole pairs. The electrical carriers are swept out by the internal electrical field and induce a photocurrent in the external circuit. The minimum light level required for correct reconstruction of the message is limited by noise currents superimposed on the signal photocurrent in subsequent amplifier stages. For example, a photodiode receiver must collect in excess of 100 nW of signal light in order to reconstruct a digital message of 10^8 pulses per second with sufficient fidelity. See PHOTODIODE.

Greater receiver sensitivity may be achieved by multiplying the photo-generated carriers in the diode junction. This can be done by virtue of an internal electrical field that is sufficiently high to cause avalanche multiplication of free carriers. Devices of this kind are made from silicon and have been operated successfully in optical communication systems. Unfortunately, silicon is sensitive to light only at a wavelength shorter than 1000 nm. Photodiodes which operate in the important spectral region between 1300 and 1600 nm have been fabricated in the indium-gallium-arsenic materials system. Avalanche gain has also been demonstrated in these diodes, but so far is noisier and less stable than in silicon avalanche photodiodes. See OPTICAL DETECTORS.   [DETLEF C. GLOGE]

Bibliography: D. Gloge, *Optical Fiber Technology*, IEEE Reprint Ser., 1976; S. E. Miller and A. G. Chynoweth, *Optical Fiber Telecommunications*, 1979; S. E. Miller, E. A. J. Marcatili, and T. Li, Research toward Optical-Fiber Transmission Systems, *Proc. IEEE*, 61(12):1703–1751, 1973.

Optical detectors

Devices that generate a signal when light is incident upon them. The signal may be one observed visually in reflected or transmitted light, or it may be electrical.

The eyes of animals and the leaves of plants, for example, perform this function. Light, as well as other forms of electromagnetic energy, can also be detected by observing the temperature change which it produces in a body absorbing it. Furthermore, it can be detected by the color changes its absorption produces in certain materials. However, the need for high sensitivity or fast response or both usually leads, in optical technology, to detectors based on other mechanisms.

One of the most widely used optical detectors is photographic film. This method of detection is based on a photochemical process in particles embedded in the film. This fairly sensitive technique is more effective than others in detecting or recording, or both, the large amount of information usually contained in an image or a complicated optical spectrum. Disadvantages are that it is insensitive to wavelengths substantially longer than those in the optical range, and that the film requires development after exposure.

In many cases, fast detector response is needed to detect rapid changes in incident light; furthermore, quite often an electrical output signal is desired. In such cases, the optimum detector is one based on either external or internal photoemission.

In an externally photoemitting device, the light incident upon a surface causes an emission of electrons from that surface into a vacuum. These electrons can be amplified in number by using electric and magnetic fields to cause them to impinge on other surfaces, thereby producing secondary electrons. Detectors of this type can be fabricated in such a way as to detect images. These detectors are limited to wavelengths in the visible region or shorter, but they are capable of a sensitivity and speed of response higher than any other type. See IMAGE ORTHICON; PHOTOMULTIPLIER; PHOTOTUBE.

In an internally photoemitting device, incident light produces free charge carriers within a body and is detected through the effect of these charge carriers on the electrical impedance of the body. In its simplest and most widely used form, the body is a photoconductor whose impedance is high in the absence of incident light. Such detectors are less sensitive and slower than external photoemitters, but they require no vacuum envelope and operate at low voltages. They also have the advantage that some of them are sensitive to all wavelengths shorter than about 40 μm, far into the infrared. In general, however, operation at the longer wavelengths requires that the device be cooled. Detectors of this type can also be built in such a way as to detect images. This, however, is generally done with an electron beam as part of the read-out circuit and does require a vacuum envelope. See PHOTOCONDUCTIVE CELL; PHOTOCONDUCTIVITY.

A development among internal photoemitting detectors is the reverse-biased semiconductor diode, which takes advantage of a large electric field at the reverse-biased semiconductor pn junction to give a response faster than that of the devices simply based on photoconductivity in bulk samples. The device, when made of germanium, is sensitive to all wavelengths shorter than 1.6 μm. Its advantages are further increased by operating it at voltages which cause avalanche processes to follow each absorption of a photon. This gives rise to internal gain, and thereby increases the figure of merit of the device if, as is often the case, other sources of noise are dominant. It is difficult to produce avalanche diodes so that the properties of successive diodes are identical. This problem has been solved for the nonavalanching devices, however. As a result it has been possible to fabricate arrays which have potential for use in vidicons as image detectors. *See* OPTICAL COMMUNICATIONS; OPTICAL MODULATORS; PHOTODIODE; SEMICONDUCTOR DIODE.

[J. K. GALT]

Bibliography: L. K. Anderson and B. J. McMurty, High speed photodetectors, *Appl. Opt.*, 5:1573–1587, October 1966; H. Melchior, M. B. Fisher, and F. R. Arams, Photodetectors for optical communication systems, *Proc. IEEE*, 58:1466–1486, 1970; G. H. S. Rokos, Optical detection using photodiodes, *Opto-Electronics*, 5:351–366, 1973; G. E. Stillman and C. M. Wolfe. Avalanche photodiodes, in R. W. Willardson and A. C. Beer (eds.), *Semiconductors and Semimetals*, vol. 12, 1977; L. R. Tomasetta et al., High sensitivity optical receivers for 1.0–1.4 μm fiber-optic systems, *IEEE J. Quantum Electron.*, QE-14:800–304, 1978.

Optical isolator

A very small four-terminal electronic circuit element that includes in an integral package a light emitter, a light detector, and, in some devices, solid-state electronic circuits. The emitting and detecting devices are so positioned that the majority of the emission from the emitter is optically coupled to the light-sensitive area of the detector. The device is also known as an optoisolator, optical-coupled isolator, and optocoupler. The device is housed in an integral opaque package so that the only optical emission impinging on the detector is that produced by the emitter. This configuration of components can perform as a solid-state electronic transformer or relay, since an electronic input signal causes an electronic output signal without any electrical connection between the input and the output terminals.

Optical isolators are particularly useful to electronic circuit and systems designers because two circuits often have a large voltage difference between them and yet it is necessary to transfer a small signal between them without changing the basic voltage level of either.

Before optical isolators were developed, this circuit function was performed by such devices as isolation transformers or signal relays. The optical isolator provides a number of advantages over such devices in that it is much smaller, much faster, has no contact bounce (unlike a relay), has no inductance (unlike a transformer), and provides a very high voltage isolation between the input and output circuits.

In order to discuss applications of the optical isolator, it is necessary to identify the different types of optical isolators that can be produced through the use of different light emitter-detector combinations.

Optical emitters. The optical emitter most commonly selected for use in optical isolators is the gallium arsenide light-emitting diode (LED), which emits in the infrared or near-infrared regions of the spectrum. The typical wavelength of the emission is 850 nm. The LED is extremely small (about the size of a transistor chip) and provides extremely fast infrared light pulses. Photodetectors fabricated from silicon are particularly sensitive to the wavelength of light emitted by the gallium arsenide LED, so that the transfer of a signal from the input to the output of the optical isolator is particularly efficient.

The other input devices commonly used are gas-discharge and incandescent lamps. These emitters tend to be somewhat larger and slower than LEDs, but are used when their particular electronic characteristics are desired, and in conjunction with detectors that have a spectral peak in the visible portion of the spectrum.

Optical detectors. The light detectors that are used in the construction of optical isolators include light-dependent resistors (such as photocells), light-sensitive devices that generate a voltage without any electrical input (such as photovoltaic devices), light-sensitive devices that switch from one state to another (such as photothyristors), and light-sensitive devices that modify a voltage or current (such as phototransistors, photodiodes, and photodetector-amplifier combinations).

Having such a wide variety of emitter and detector combinations available means that circuitry and system designers can use them over a wide range of applications. For example, the combination of a photocell and an incandescent lamp produces a device with entirely different characteristics from those of the combination of an LED and a phototransistor. Each combination is described below and its primary application identified.

Incandescent lamp–photocell optical isolators. In these devices the resistance of the photocell varies from about 10 megohms to 1000 megohms, as the input light is varied from off to maximum brightness. If the input changes instantaneously, the output resistance change occurs in 20 ms (typical). If the input voltage changes gradually, the resistance of the photocell varies somewhat proportionally to the input voltage. As a result of this variable-resistance characteristic, the device whose schematic is shown in Fig. 1 can be used as a remote-controlled rheostat where there is no electrical connection between the control wire and the circuit being controlled. In applications where signals are required for two separate circuits, without an electrical connection between any of the three circuits, devices are used with two photocells built into the case of the optical isolators.

Neon lamp–photocell combinations. These devices have output characteristics similar to the

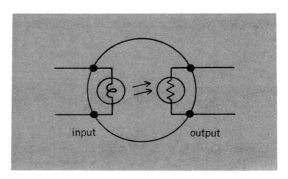

Fig. 1. Schematic diagram of incandescent lamp–photocell optical isolator.

incandescent lamp–photocell devices. However, the input characteristics vary dramatically, because the neon lamp does not emit light (ignite) until the input voltage reaches about 70 V. Also, the current after ignition is very small. These devices are employed where large input voltage swings are typical and where long life is required. A schematic diagram is shown in Fig. 2.

LED-silicon detector combinations. All of the optical isolators that employ silicon photodetectors have LED emitters. The most common photodetector is the phototransistor. A schematic diagram of a type of LED-phototransistor optical isolator is shown in Fig. 3. In such devices the light from the LED causes the phototransistor current to vary as a function of the amount of light impinging on the photosensitive area of the phototransistor. The devices can isolate circuits that differ by as much as 5000 V. The devices are very fast, with the output current changing in some devices in as little as 10–20 ns after the occurrence of an input pulse. This fast response time means that these devices are particularly useful in computer circuit applications where such fast pulses are common. Not only do these devices provide high voltage isolation between circuits, but because the phototransistor has built-in gain, the detected signal is amplified. These devices are also immune to certain types of noise and are extremely small, the package being just larger than a typical transistor can. The devices can be used in most circuits that require an isolation transformer, and provide the additional advantages of higher speed and higher voltage ratings.

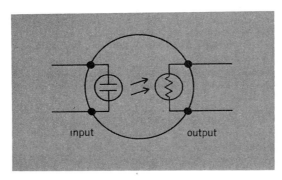

Fig. 2. Schematic diagram of neon lamp–photocell optical isolator.

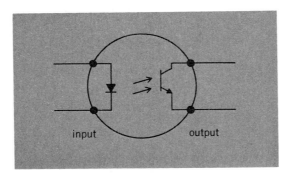

Fig. 3. Schematic diagram of LED-phototransistor optical isolator.

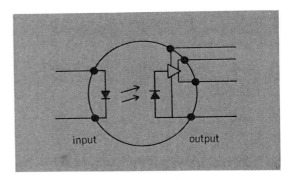

Fig. 4. Schematic diagram of LED–photodiode–integrated-circuit optical isolator.

Such devices also can perform as a solid-state relay (no mechanical wear, so no contact bounce), as a high-speed chopper, and as a pulse amplifier.

A similar type of optical isolator that produces a higher output amplitude for the same input employs a photo-Darlington sensor and output device. The photo-Darlington is in reality two phototransistors that are integrally connected. Although it does have increased amplification, it is a much lower speed device.

LED-photodetector-amplifier. In some optical isolators, integrated circuits are also included in the package to provide specific output characteristics. For example, one optical isolator includes an integrated circuit amplifier so that the optical isolator can perform like a broadband pulse transformer that is compatible with diode-transistor-logic (DTL) computer circuitry as well as provide a frequency response to zero frequency. A schematic diagram of an LED–photodiode–integrated-circuit optical isolator is shown in Fig. 4. *See* CIRCUIT (ELECTRONICS); OPTICAL DETECTORS; SEMICONDUCTOR DIODE. [ROBERT D. COMPTON]

Optical modulators

Devices that serve to vary some property of a light beam. The direction of the beam may be scanned as in an optical deflector, or the phase or frequency of an optical wave may be modulated. Most often, however, the intensity of the light is modulated.

Rotating or oscillating mirrors and mechanical shutters can be used at relatively low frequencies (less than 10^5 Hz). However, these devices have too much inertia to operate at much higher frequencies. At higher frequencies it is necessary to take advantage of the motions of the low-mass electrons and atoms in liquids or solids. These motions are controlled by modulating the applied electric fields, magnetic fields, or acoustic waves in phenomena known as the electrooptic, magnetooptic, or acoustooptic effect, respectively.

For the most part, it will be assumed that the light to be modulated is nearly monochromatic— either a beam from a laser or a narrow-band incoherent source.

Electrooptic effect. The quadratic or Kerr electrooptic effect is present in all substances and refers to a change in refractive index, Δn, proportional to the square of the applied electric field E. The liquids nitrobenzene and carbon disulfide and the solid strontium titanate exhibit a large Kerr effect.

Much larger index changes can be realized in single crystals that exhibit the linear or Pockels electrooptic effect. In this case, Δn is directly proportional to E. The effect is present only in noncentrosymmetric single crystals, and the induced index change depends upon the orientations of E and the polarization of the light beam. Well-known linear electrooptic materials include potassium dihydrogen phosphate (KDP) and its deuterated isomorph (DKDP or KD*P), lithium niobate (LiNbO$_3$) and lithium tantalate (LiTaO$_3$), and semiconductors such as gallium arsenide (GaAs) and cadmium telluride (CdTe). The last two are useful in the infrared ($1-10$ μm), while the others are used in the near-ultraviolet and visible regions ($0.3-3$ μm). *See* FERROELECTRICS; SEMICONDUCTOR.

The phase increment Φ of an optical wave of wavelength λ that passes through a length L of material with refractive index n is given by Eq. (1).

$$\Phi = \frac{2\pi}{\lambda} nL \qquad (1)$$

Thus phase modulation can be achieved by varying n electrooptically. Since the optical frequency of the wave is the time derivative of Φ, the frequency is also shifted by a time-varying Φ, yielding optical frequency modulation. *See* MODULATION.

The refractive index change is given in terms of an electrooptic coefficient r by Eq. (2), where typi-

$$n(E) = n(0) - \tfrac{1}{2}n^3rE \qquad (2)$$

cal values for n and r are 2×10^{-11} and 3×10^{-11} m/V, respectively.

Electrooptic intensity modulation. Intensity modulation can be achieved by interfering two phase-modulated waves as shown in Fig. 1. The electrooptically induced index change is different for light polarized parallel (*p*) and perpendicular (*s*) to the modulating field; that is, $r_p \neq r_s$. The phase difference for the two polarizations is the retardation Γ given by Eq. (3), which is proportional

$$\Gamma(V) = \Phi_p - \Phi_s = \frac{2\pi}{\lambda}(n_p - n_s)L \qquad (3)$$

to the applied voltage V. If it is assumed that $n_p(0) = n_s(0)$ for $V = 0$, then $\Gamma(0) = 0$. Further assume that $\Gamma(V') = \pi$. Then incident light polarized at 45° to the field may be resolved into two equal, in-phase components parallel and perpendicular to the field. For $V = 0$, they will still be in phase at the output end of the electrooptic crystal and will recombine to give a polarization at +45° which will not pass through the output polarizer in Fig. 1. For $V = V'$, however, the two components will be out of phase and will recombine to give a polarization angle of −45° which will pass through the output polarizer. The switching voltage V' is called the half-wave voltage and is given by Eq. (4), with d

$$V' = (\tau d/n^3r_cL) \qquad (4)$$

the thickness of the crystal and r_c an effective electrooptic coefficient. A typical value for n^3r_c is 2×10^{-10} m/V.

Electrooptic devices can operate at speeds up to several GHz.

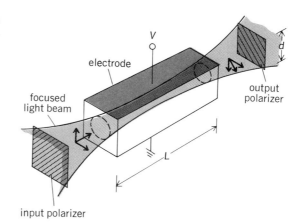

Fig. 1. Electrooptic intensity modulator.

Acoustooptic modulation and deflection. All transparent substances exhibit an acoustooptic effect—a change in n proportional to strain. Typical materials with a substantial effect are water, glass, arsenic selenide (As$_2$Se$_3$), and crystalline tellurium dioxide (TeO$_2$).

An acoustic wave of frequency F has a wavelength given by Eq. (5), where C is the acoustic

$$\Lambda = C/F \qquad (5)$$

velocity, which is typically 3×10^3 m/s. Such a wave produces a spatially periodic strain and corresponding optical phase grating with period Λ that diffracts an optical beam as shown in Fig. 2. A short grating ($Q < 1$) produces many diffraction orders (Raman-Nath regime), whereas a long grating ($Q > 10$) produces only one diffracted beam (Bragg regime). The quantity Q is defined by Eq. (6).

$$Q = (2\pi L\lambda)/n\Lambda^2 \qquad (6)$$

If one detects the Bragg-diffracted beam (diffraction order $+1$), its intensity can be switched on and off by turning the acoustic power on and off.

The Bragg angle θ through which the incident beam is diffracted is given by Eq. (7). Thus the

$$\sin \tfrac{1}{2}\theta = \lambda/2\Lambda \qquad (7)$$

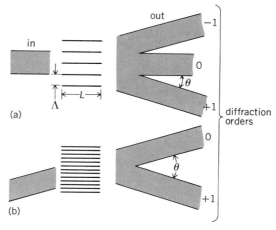

Fig. 2. Acoustooptic modulator-deflector. (*a*) Raman-Nath regime. (*b*) Bragg regime.

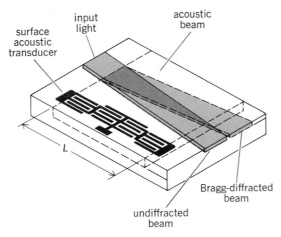

Fig. 3. Waveguide acoustooptic modulator-deflector.

beam angle can be scanned by varying F.

Acoustooptic devices operate satisfactorily at speeds up to several hundred megahertz.

Optical waveguide devices. The efficiency of the modulators described above can be put in terms of the electrical modulating power required per unit bandwidth to achieve, say, 70% intensity modulation or one radian of phase modulation. For very good bulk electrooptic or acoustooptic modulators, approximately 1 to 10 mW is required for a 1-MHz bandwidth; that is, a figure of merit of 1 to 10 mW/MHz. This figure of merit can be improved by employing a dielectric waveguide to confine the light to the same small volume occupied by the electric or acoustic field, without the diffraction that is characteristic of a focused optical beam. This optical waveguide geometry is also compatible with optical fibers and lends itself to integrated optical circuits. *See* INTEGRATED OPTICS; OPTICAL COMMUNICATIONS.

An optical wave can be guided by a region of high refractive index surrounded by regions of lower index. A planar guide confines light to a plane, but permits diffraction within the plane; a strip or channel guide confines light in both transverse dimensions without any diffraction.

Planar guides have been formed in semiconductor materials such as $Al_xGa_{1-x}As$ by growing epitaxial crystal layers with differing values of x.

Since the refractive index decreases as x increases, a thin GaAs layer surrounded by layers of $Al_{0.3}Ga_{0.7}As$ will guide light. If one of the outer layers is p-type and the other two layers are n-type, a reverse bias may be applied to the pn-junction to produce a large electric field in the GaAs layer. Very efficient electrooptic modulators requiring about 0.1 mW/MHz at $\lambda = 1\ \mu m$ have been realized in such junctions. Further improvement has been realized by loading the planar guide with a rib or ridge structure to provide lateral guiding within the plane. Ridge devices 1 mm long require about 5 V to switch light on or off, with a figure of merit of $10\ \mu W/MHz$.

Both planar and strip guides have been formed in LiNbO$_3$ by diffusing titanium metal into the surface in photolithographically defined patterns. Planar guides with surface acoustic wave transducers (Fig. 3) have been employed in waveguide acoustooptic devices. Strip guides (Fig. 4) have been used to make electrooptic phase modulators 3 cm long requiring 1 V to produce π radians of phase shift at $\lambda = 0.63\ \mu m$; the figure of merit is 2 $\mu W/MHz$. In order to make an intensity modulator or a 2×2 switch, a directional coupler waveguide pattern is formed, and electrodes are applied as shown in Fig. 5. With no voltage applied and L the

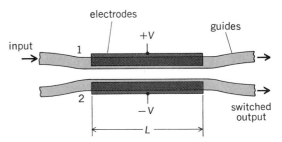

Fig. 5. Directional coupler waveguide modulator switch.

proper length, light entering guide 1 will be totally coupled to and exit from guide 2; with the proper applied voltage, the phase match between the two guides will be destroyed, and no light will be coupled to guide 2, but will all exit from guide 1. Switches and modulators operating at speeds up to 5 GHz have been realized in this way. *See* SURFACE ACOUSTIC-WAVE DEVICES.

[IVAN P. KAMINOW]

Bibliography: R. C. Alferness, Guided wave devices for optical communication, *IEEE J. Quart. Elec.*, 17(6), June 1981; I. P. Kaminow, *An Introduction to Electrooptic Devices*, 1974; J. F. Nye, *Physical Properties of Crystals*, 1960; T. Tamir, *Integrated Optics*, 1975.

Optimization

In its most general meaning, the efforts and processes of making a decision, a design, or a system as perfect, effective, or functional as possible. Formal optimization theory encompasses the specific methodology, techniques, and procedures used to decide on the one specific solution in a defined set of possible alternatives that will best satisfy a selected criterion. Because of this decision-making function the term optimization is

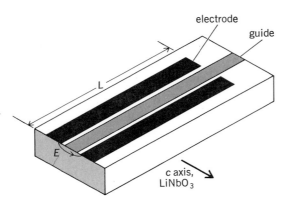

Fig. 4. Titanium-diffused LiNbO$_3$ strip waveguide with coplanar electrodes for electrooptic phase modulation.

often used in conjunction with procedures which more appropriately belong in the more general domain of decision theory. Strictly speaking, formal optimization techniques can be applied only to a certain class of decision problems known as decision making under certainty.

CONCEPTS AND TERMINOLOGY

Conceptually, the formulation and solution of an optimization problem involves the establishment of an evaluation criterion based on the objectives of the optimization problem, followed by determination of the optimum values of the controllable or independent parameters that will best satisfy the evaluation criterion. The latter is accomplished either objectively or by analytical manipulation of the so-called criterion function, which relates the effects of the independent parameters on the dependent evaluation criterion parameter. In most optimization problems there are a number of conflicting criteria and a compromise must be reached by a trade-off process which makes relative value judgements among the conflicting criteria. Additional practical considerations encountered in most optimization problems include so-called functional and regional constraints on the parameters. The former represents physical or functional interrelationships which exist among the independent parameters (that is, if one is changed it causes some changes in the others); the latter limits the range over which the independent parameters can be varied.

Application. Before proceeding with a detailed description of the steps involved in formulating and solving optimization problems in general, a simple problem is briefly discussed below to demonstrate the overall concepts and terminology.

A common optimization problem is solved in tuning a bad-reception TV channel. Because the tuning knob position for the best picture is usually different from the position giving the best sound quality, it is impossible to have both optimum sound and picture simultaneously. In this case the position of the knob becomes the independent parameter (it can be controlled by the person performing the optimization) whose value determines the resulting value of the criteria, which in this case are picture clarity and sound quality. The relative importance of the two conflicting criteria in regard to his objectives must be judged by the optimizer in order for him to arrive at an optimum setting of the tuning knob. If, for example, the viewer is interested in reading some printed information displayed on the screen he will optimize the picture resolution at the expense of sound quality, while sound quality may be the primary criterion in watching a concert on TV. For general viewing a setting which represents a trade-off between the two criteria might be chosen. This setting would be suboptimum with respect to picture resolution and sound quality, but would be optimum with respect to the combined criteria of sound and picture. In this example picture resolution and sound quality have been considered as the conflicting criteria. Both of these can have rather broad interpretation and for a more formalized optimization it is necessary to define some measurable, quantitative parameters which would serve to describe the criteria. In that case, it is possible

experimentally to obtain two criterion functions which would show the behavior of the picture criterion and the sound criterion as a function of tuning knob setting, as shown schematically in Fig. 1a and b. Regional constraints on the range of possible knob positions might be imposed by the fact that an unbearable noise develops at knob settings below a certain value or that the picture blacks out completely at knob settings above a certain value.

After a value judgment with respect to the relative importance of the two criteria has been made, an overall criterion function can be established. Figure 1c shows two overall criterion functions:

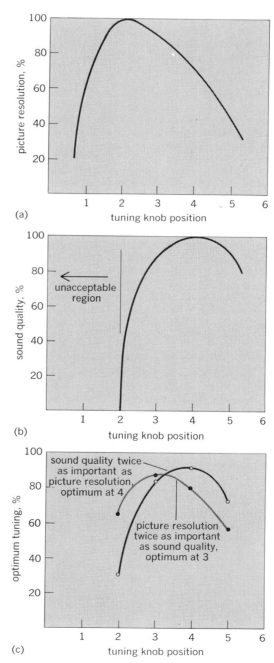

Fig. 1. Effects of tuning knob positions on the quality of the television picture. (a) Picture quality. (b) Sound quality. (c) Two overall criterion functions.

one with units of picture resolution twice as important as units of sound quality, and the other with the relative importance of the criteria reversed. It is easily observed that the optimum setting of the knob (at which the value of the overall criterion is at a maximum) is different for the two cases.

Other parameters. It should be emphasized that optimization techniques only work for a specific system or configuration which has been described to a point where all criteria and parameters are defined within the system and are isolated and independent from other parameters outside the boundaries of the defined system. Formal optimization is not a substitute for creativity in that it depends on a clear definition of the system to be optimized, and even though an optimum solution for a given system configuration is obtained, this does not guarantee that a better solution is not available.

This can best be demonstrated by expanding the example given above. After the optimum tuning knob setting has been determined, someone may come along and change the direction of the TV antenna (a parameter which was not included in the original problem), thus producing perfect picture and sound, or someone may suggest that the viewer watch the program on a portable TV which has perfect reception of the particular channel in question. Both of these solutions are obviously more optimal than the original one but would not be identified by formal optimization techniques. This very vital deficiency of optimization theory is frequently overlooked. Realistically speaking, formal optimization theory should be used only to find optimum parameters in well-defined, preferably analytically describable systems with unique, quantitative-dependent and -independent parameters, criterion function, and functional and regional constraints. When several system configurations are possible, each should be optimized and then compared with the others.

OPTIMIZATION PROBLEM FORMULATION

To further demonstrate formal optimization procedures and solution techniques, a step-by-step description of the procedures follows and is applied to an example.

Definition. This is a critcal part of problem formulation and consists of:

(1) A description of the system configuration to be optimized, including definition of system boundaries to an extent that system parameters become isolated and independent of external parameters.

(2) Definition of a single, preferably quantitative, parameter which will serve as the overall evaluation criterion for the specific optimization problem. This is the dependent parameter (it depends upon the choice of the optimum solution) which measures how well the solution satisfies the desired objectives of the problem and this is the parameter which will be maximized or minimized to satisfy the objectives. As mentioned earlier, the process of finding a single evaluation parameter can become quite complex because normally specification of the optimum solution is complicated by conflicting criteria.

A compromise or trade-off between conflicting criteria must be reached by making value deci-

sions based on value judgment of the relative importance of each criterion. The relative value of each criterion with respect to the others depends on the objectives of the optimization problem. Thus, in a complicated optimization problem all criteria relevant to the objectives must be enumerated, their relative importance established, and this information combined into a single parameter that will serve as the optimization criterion.

(3) Definition of controlled or independent parameters that will have an effect on the criterion. These are the parameter or parameters whose values determine the value of the criterion parameter and these should include all controllable parameters which influence the criterion and are within the boundaries of the defined problem.

Criterion function. The most critical aspect of formulating a formal optimization problem is the establishment of a satisfactory criterion function that describes the behavior of the evaluation criterion as a function of the independent parameters. The criterion function, often also referred to as the pay-off or objective function, usually takes the form of a penalty or cost function (which attempts are made to minimize) or a merit, benefit or profit function (which attempts are made to maximize by choosing the optimum values of the independent variables). In order to apply formal solution methods, the criterion function should be expressed graphically or analytically. In symbolic form the criterion function can be expressed as: criterion parameter = function of independent parameters $x_1, x_2, \ldots x_n$. This may be represented by Eq. (1).

$$C = f(x_1, x_2, \ldots x_n) \tag{1}$$

At this point it should be helpful to introduce another example to illustrate the steps described so far. The problem of optimizing the shape of a vegetable garden so as to maximize the available planting area which can be enclosed by a given length of fence will be considered.

To formalize the problem, a system configuration must be chosen. Several obvious possibilities come to mind, such as triangular, rectangular, elliptical, or even polyhedral shapes, or randomly shaped, closed curves. Consideration will be limited to the two shapes shown in Fig. 2, namely, a rectangular configuration and a circular configuration. It should be noted that, based on some objective judgments, some possible configurations which may give a better solution than the chosen shapes have been eliminated.

The implied criterion is the maximization of the area of the garden, which emerges as the principal criterion even though secondary considerations such as esthetics and compatibility with the surroundings may have already influenced the choice of the configurations actually being optimized.

The independent parameters in the two suboptimization problems which have developed become the length a and width w of the rectangular plot and the radius r of the circular plot. The respective criterion functions for the three choices are given by Eqs. (2).

$$A = aw \tag{2a}$$

$$A = \pi r^2 \tag{2b}$$

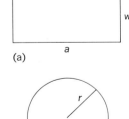

OPTIMIZATION

(a)

(b)

Fig. 2. Possible shapes for garden plots. (a) Rectangular. (b) Circular.

To continue the example it is necessary to return to the optimization steps.

Functional constraints. Physical principles of operation which govern the relationship among the various independent parameters of the problem represent functional constraints. Symbolically, this becomes functional constraints = functions of independent parameters $x_1, x_2, \ldots x_n$. This is represented by Eqs. (3). The number of independ-

$$fc_1 = f_1 (x_1, x_2, \ldots x_n) \qquad (3a)$$

$$fc_2 = f_1 (x_1, x_2, \ldots x_n) \qquad (3b)$$

$$fc_m = f_m (x_1, x_2, \ldots x_n) \qquad (3c)$$

ent functional constraints must always be less than the number of independent parameters; otherwise all independent parameters can be eliminated from the criterion function, resulting in a single solution without a choice for optimization.

In the example the respective functional constraints relating the problem parameters become Eqs. (4).

$$2a + 2w = L \text{ (rectangle)} \qquad (4a)$$

$$2\pi r = L \text{ (circle)} \qquad (4b)$$

Since the circle only has one independent parameter, r, the single functional constraint eliminates that parameter when combined with the criterion function; for example, $A = \pi r^2$ and $r = L/2\pi$ (from the constraint) so that Eq. (5) is obtained.

$$A = \frac{\pi L^2}{(2\pi)} = \frac{L^2}{4\pi} \qquad (5)$$

This result indicates that, regardless of the size of the radius chosen, the relationship between the area and the total length of the fence is unique and given by Eq. (6).

$$A = L^2/4\pi \qquad (6)$$

In general, when an optimization problem includes functional constraints, these can be used to eliminate the number of independent parameters that can be varied (as will be indicated further in discussing solution methods).

Regional constraints. Most optimization problems involve practical limits on the range over which each parameter or function of the parameters can be varied. Symbolically these regional constraints are expressed as the inequalities in Eqs. (7).

$$l_1 \leq r_1 (x_1, x_2, \ldots x_n) \leq l_2 \qquad (7a)$$

$$l_3 \leq r_2 (x_1, x_2, \ldots x_n) \leq l_4 \qquad (7b)$$

$$l_{2p-1} \leq r_p (x_1, x_2, \ldots x_n) \leq l_{2p} \qquad (7c)$$

There can be any number of regional constraints to limit the analytical region of independent parameters over which one can search for the optimum. For the respective configurations in the above example, it is possible to stipulate the regional constraints given in Eqs. (8).

$$L \leq 100 \text{ ft} \qquad (8a)$$

$$a \leq 50 \text{ ft} \qquad w \leq 50 \text{ ft} \qquad (8b)$$

Solution of optimization problems. The reliability and the sophistication of the solutions of optimization problems increase as the problem becomes better defined. In the early stages of optimization of a complex problem, the considerations may be primarily objective judgments based on the optimizer's judgments of the relevant parameters. As alternative subsystem and system configurations evolve through a process of conceptualization, analysis and evaluation, and elimination of undesirable system configurations, a clearer configuration of parts of the problem can be defined, together with the relevant criteria, parameters, and constraints. Although these initial stages are often also referred to as optimization, the techniques of searching for alternate strategies and determining the single best choice are more generally referred to as decision theory. Procedures for defining competing systems or alternate strategies and formulating the problem in order to apply formal optimization techniques are also common to operations research, systems analysis, and systems design.

The most general formal optimization problem is formulated in terms of the three types of equations discussed above: the criterion function, the functional constraints, and the regional constraints [Eqs. (1), (3), and (7), respectively]. The techniques used for solving the optimization problem depend on the complexity of the optimization equations. Detailed treatment of the various techniques for solving Eqs. (1), (3), and (7) represent the major part of the literature on optimization. Only a very cursory description of the most popular techniques is presented here since the mathematics is beyond the scope of this discussion.

Graphical analysis. Simple optimization problems with one or two independent variables can be solved by graphically representing the objective function as a function of the independent variables (over the constrained parameter range) and then picking the appropriate maximum or minimum. Figure 1a demonstrates this technique for one independent parameter where picture resolution represents the evaluation criterion to be maximized and tuning knob position represents the independent parameter. For two independent parameters the profit or penalty parameter can be graphically represented as contour lines with the two independent parameters as axes. This is shown in Fig. 3a for the example of the rectangular garden plot of Fig. 2a. Each contour line indicates the combination of length and width of the plot which will result in the value of the profit variable (area in this example) shown on the contour lines. These profit function contours can be shown over the full permissible range of the independent parameters. To find the solution, however, the functional constraint must also be considered. The functional constraint $2a + 2w = 100$ is shown as the straight line in Fig. 3a. Only combinations of length and width lying on this line are permissible. One can easily see that the maximum area is obtained when the functional constraint line intersects the highest profit function contour. In this case it occurs at $a = w = 25$ ft, for which the area is 625 ft² (1 ft² = 0.09 m²). Thus the optimum solution for the rectangular plot is a square.

In many cases the functional constraints can be substituted directly into the criterion function to reduce the number of independent parameters. In the example (Fig. 2), Eq. (4a) can be solved for

a and substituted into Eq. (2a) to obtain Eq. (9),

$$A = 50w - w^2 \qquad (9)$$

which has only one independent parameter, w. Solutions to Eq. (9) plotted in Fig. 3b again give 625 ft² as the optimum solution.

Although in the solutions illustrated, the evaluation parameter has a simple maximum, in more general cases the criterion parameter may be complicated, as shown in Fig. 4, with relative maxima and minima, inflection points, and so on.

Graphical methods result in the most descrip-

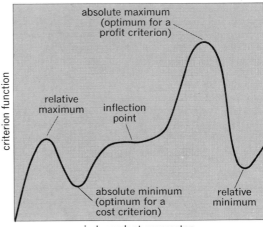

Fig. 4. Maximum and minimum values of a criterion function. Concepts of local and absolute minima and maxima can be extended to n-dimensional criterion functions.

tive solutions. However, they are usually restricted to problems with at most two independent parameters, because one can only visualize in three dimensions (one for the evaluation criterion and two for the independent parameters). But graphical solutions serve as a good conceptual starting point for the more general optimization problem in which one is looking for a maximum or minimum of a function in n-dimensional space. Analytical and numerical techniques must be used for these purposes.

Analytical approach. The classical analytical methods involve reduction of the number of independent parameters by substituting the functional constraints into the objective function, differentiating the resulting objective function with respect to each of the n-independent variables, and setting these first derivatives equal to zero. This results in a set of n simultaneous equations in n unknowns which are then solved by various mathematical techniques (for example, Jacobian or Newton-Raphson methods). The classical Lagrange multiplier technique is a variation of these differential methods of finding an extremum for a function in n-dimensional space.

For the simple example (Fig. 2), equating the first derivative of Eq. (9) to zero gives Eqs. (10) and (11). It should be noted that although the opti-

$$\frac{dA}{dw} = 50 - 2w = 0 \qquad (10a)$$

$$w = 25 \text{ ft} \qquad (10b)$$

$$A = aw = 625 \text{ ft}^2 \qquad (11)$$

mum solution for a rectangular plot gives 625 ft², this is only a suboptimum in the example, since a circular plot would give an area of $A = L^2/4 = 796$ ft², using 100 ft of fence.

Some optimization problems involve optimization of a whole function rather than a single operating point (for example, optimum lunar landing trajectories). The calculus of variations is used in that case to maximize or minimize the objective function.

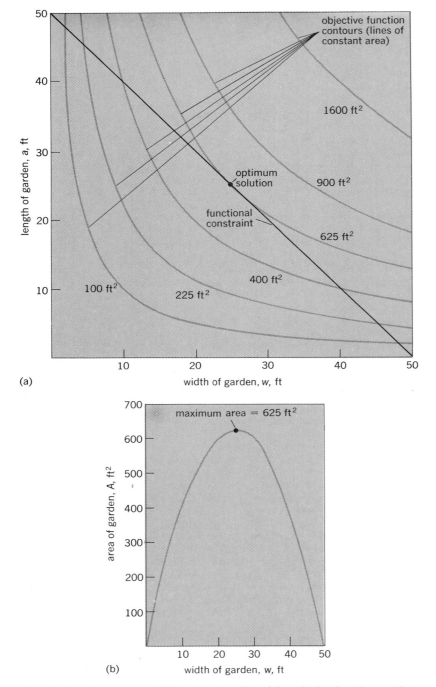

(a)

(b)

Fig. 3. Graphical analysis. (a) Graphical solution of the criterion function and functional constraint for the rectangular garden plot of Fig. 2a. (b) Graphical solution with single independent variable. 1 ft = 0.3 m. 1 ft² = 0.09 m².

Recursive methods. Among the recursive, step-by-step techniques for finding an extremum, the best known are linear and dynamic programming and random search methods.

Linear programming can be applied only to problems having linear objective functions and linear functional constraints. For a number of practical problems, such as production scheduling, product mix decisions, capacity allocation, and dispatching strategies, which can be formulated in terms of linear equations, linear programming has become an important optimization technique.

The more sophisticated dynamic programming technique transforms sequential *n*-dimensional optimization problems into a sequence of single-stage problems having only one variable each. It can be used on both linear and nonlinear problems.

Both of these techniques are programmed on digital computers and are routinely used to solve problems with 30–40 independent variables and 50–70 constraints.

Random search methods. A number of random search techniques have been developed to arbitrarily sample the *n*-dimensional space to find an extremum of the function. The Monte Carlo method is one of these. The more sophisticated random search techniques use secondary information, such as local gradients, to narrow the random search region.

With the development of computers the application of optimization techniques (particularly linear and dynamic programming and random search techniques) to multivariable optimization problems has prospered, and these techniques are routinely used in business management, resource allocation, military and economic planning, engineering design and manufacture, and other areas. *See* DECISION THEORY; DIGITAL COMPUTER.

[IGOR PAUL]

Bibliography: W. Conley, *Computer Optimization Techniques*, 1980; J. R. Dixon, *Design Engineering: Inventiveness, Analysis, and Decision Making*, 1966; J. H. Faupel and F. E. Fisher, *Engineering Design*, 2d ed., 1980; N. L. Svensson, *Introduction to Engineering Design*, 1977; J. P. Vidosic, *Elements of Design Engineering*, 1969.

Oscillator

An electronic circuit that converts energy from a direct-current source into a periodically varying electrical output. If the output voltage is a sine-wave function of time, the generator is called a sinusoidal, or harmonic, oscillator. Only sinusoidal oscillators are discussed in this article. If the output waveform contains abrupt changes in voltage, such as occur in a pulse or square wave, the device is called a relaxation oscillator. *See* RELAXATION OSCILLATOR; WAVE-SHAPING CIRCUITS.

Basic principles. The fundamental laws governing sinusoidal oscillators are the same for all oscillator circuits. These basic concepts are illustrated in Fig. 1. The amplifier provides an output voltage v_o as a consequence of an external input signal voltage v_s. The voltage v_o is applied to a circuit called a feedback network whose output is v_f. If the feedback voltage v_f were made identically equal to the input voltage v_s, and if the external input were disconnected and the feedback voltage connected to the amplifier input terminals 1 and 2,

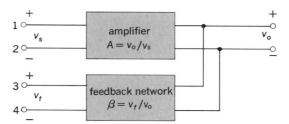

Fig. 1. An amplifier and feedback network which is not yet connected to form a closed loop.

the amplifier would continue to provide the same output voltage v_o as before. This requires that the instantaneous values of v_f and v_s be exactly equal at all times. Since no restriction was made on the waveform, it need not be sinusoidal.

If the entire circuit operates linearly and the amplifier or feedback network or both contain reactive elements, the only periodic wave that will preserve its form is the sinusoidal waveform, and such a circuit will be a sinusoidal oscillator. For sinusoidal oscillators the condition where v_s equals v_f requires that amplitude, phase, and frequency of v_s and v_f be identical. The phase shift introduced in a signal while being transmitted through a reactive network is invariably a function of the frequency, and there is usually only one frequency at which v_f and v_s are in phase. Therefore, a sinusoidal oscillator operates at the frequency for which the total phase shift of the amplifier and feedback network is precisely zero (for an integral multiple of 2π). The frequency of a sinusoidal oscillator, provided the circuit oscillates at all, is therefore determined by the condition that the loop phase shift is zero.

Another condition, which must clearly be met if the oscillator is to function, is that the magnitude of v_s and v_f must be identical. If the amplifier has a voltage amplification, or gain, A, then v_o equals Av_s. The fraction of the voltage v_o applied to the input through the feedback network is called the feedback factor β. Therefore, Eqs. (1) are obtained. If v_f is to equal v_s, then βA must equal 1. βA is called the loop gain.

$$v_f = \beta v_o \qquad v_f = \beta A v_s \qquad (1)$$

An oscillator will not function if, at the oscillator frequency, the magnitude of the product of the gain of the amplifier and the feedback factor of the feedback network is less than unity. The condition of unity loop gain is called the Barkhausen criterion.

Referring again to Fig. 1, if βA at the oscillator frequency is precisely unity and the feedback voltage is connected to the input terminals, the circuit will operate with the external generator removed. If βA is less than unity, the removal of the external generator will immediately result in a cessation of oscillations. If βA is greater than unity, 1 volt appearing initially at the input terminals will, after a trip around the loop, appear at the input as a voltage larger than 1 volt. After another trip around the loop this larger voltage will be still larger, and so on. It seems, then, that if βA is larger than unity, the amplitude of the oscillations will continue to increase without limit. Of course, such increases in the amplitude can continue only as long as it is not limited by nonlinearity of operation in the active devices associated with the amplifier. Such a nonlinearity becomes more marked as the amplitude of oscillations increases. This onset of nonlinearity to limit the amplitude of oscillations is an essential feature of the operation of all practical oscillators because all oscillators operate with βA greater than one. The condition that βA equal 1 imposes a single and precise condition of operation which is not practical in electronic design. Even if the circuit were initially designed to satisfy this condition, it could not be maintained because circuit components (especially active devices) change characteristics (drift) with age, temperature, and voltage. Therefore, if the oscillator is left to itself, in a short time βA will become either less than or larger than unity. An oscillator in which the loop gain is exactly unity is an abstraction that is completely unrealizable in practice. A practical oscillator always has a βA somewhat larger than unity (say 5%) to ensure that, with incidental variations in transistor and circuit parameters, βA does not fall below unity.

Phase-shift oscillator. The phase-shift oscillator, Fig. 2, exemplifies the principles set forth above. A field-effect transistor (FET) amplifier of conventional design is followed by three cascaded arrangements of a capacitor C and a resistor R, the output of the last RC combination being returned to the gate. The phase of the signal is shifted 180° by the amplifier, and the network of resistors and capacitors shifts the phase by an additional amount. At some frequency the phase shift introduced by the RC network is precisely 180°, and the total phase shift around the circuit is exactly zero. At this particular frequency the circuit will oscillate, provided that the magnitude of the amplification is sufficiently large.

The frequency at which the phase shift for the RC network is 180° is given by Eq. (2). At this frequency of oscillation β equals $-(1/29)$. For βA to be greater than unity, A must be at least 29. The oscillator then cannot be made to work with an FET for which $g_m R_d < 29$, where g_m is the transductance of the FET.

$$f = 1/(2\pi RC\sqrt{6}) \tag{2}$$

The phase-shift oscillator is particularly suited to frequencies from several hertz to several

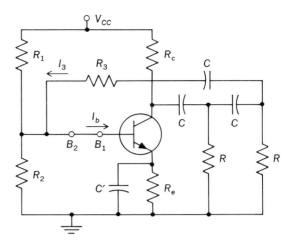

Fig. 3. A transistor phase-shift oscillator.

hundred kilohertz and so includes the range of audio frequencies. At frequencies in the range of megahertz it has no marked advantage over circuits employing a tuned LC network. In fact, at the higher frequencies, the impedance of the phase-shifting network may become quite small, and the loading of the amplifier by the phase-shifting network may become serious. On the other hand, if R and C are made large but still well within the range of commercially available values, frequencies of 1–2 Hz are easily attained. Inductors suitable for use in LC tuned oscillators for this frequency range are often impractical.

The frequency of the oscillator may be varied by changing the value of any of the impedance elements in the phase-shifting network. For variations of frequency over a large range the three capacitors should be varied simultaneously. Such a variation will keep the input impedance to the phase-shifting network constant and keep constant also the magnitude of β and βA. Hence the amplitude of oscillation will not be affected as the frequency is adjusted.

If a bipolar transistor were used as the active element in Fig. 2, the output R of the feedback network would be shunted by the relatively low input impedance of the transistor. Hence the circuit of Fig. 3 is used, where the feedback signal is the current I_3 applied in shunt with the base current I_b. The resistor $R_3 = R - R_i$, where R_i is the input resistance of the transistor. This choice makes the three RC sections of the phase-shifting network alike and simplifies the calculations. It is assumed that the biasing resistors R_1, R_2, and R_e have no effect on the signal operation, and these are neglected in the following analysis.

The Barkhausen condition that the phase of I_3/I_b must equal zero leads to Eq. (3), the expression for

$$f = \frac{1}{2\pi RC}\frac{1}{\sqrt{6+4k}} \tag{3}$$

the frequency of oscillation, where $k \equiv R_c/R$. The requirement that the magnitude of I_3/I_b must exceed unity in order for oscillations to start leads to the inequality in Eq. (4), where h_{fe} is the small-

$$h_{fe} > 4k + 23 + \frac{29}{k} \tag{4}$$

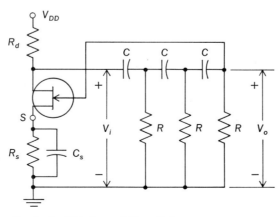

Fig. 2. An FET phase-shift oscillator.

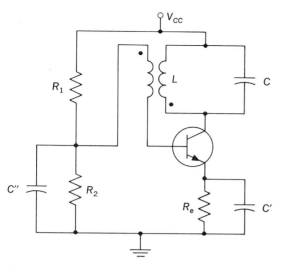

Fig. 6. A tuned-collector oscillator.

signal common-emitter short-circuit current gain. The value of k which gives the minimum h_{fe} turns out to be 2.7, and for this optimum value of R_c/R, it is found that $h_{fe} = 44.5$. A transistor with a small-signal common-emitter short-circuit current gain less than 44.5 cannot be used in this phase-shift oscillator.

A general form of oscillator circuit. Many radio-frequency oscillator circuits fall into the general form shown in Fig. 4. The active device may be a bipolar transistor, an FET, or an operational amplifier. In the analysis that follows it is assumed that the active device has very high input resistance which does not load down Z_1. If it is assumed that the impedances Z are pure reactances X (either inductive or capacitive), then, from the Barkhausen criterion, the circuit will oscillate at the resonant frequency of the series combination of X_1, X_2 and X_3. Also, the loop gain is given by Eq. (5).

$$A_v = \frac{+X_3}{X_2} \qquad (5)$$

Since β must be positive and at least unity in magnitude, then X_1 and X_2 must have the same sign. In other words, they must be the same kind of reactance, either both inductive or both capacitive. Then $X_3 = -(X_1 + X_2)$ must be inductive if X_1 and X_2 are capacitive, or vice versa.

If X_1 and X_2 are capacitors and X_3 is an inductor, the circuit is called a Colpitts oscillator. If X_1 and X_2 are inductors and X_3 is a capacitor, the circuit is called a Hartley oscillator. In this latter case, there may be mutual coupling between X_1 and X_2. If X_1 and X_2 are tuned circuits and X_3 represents the gate-to-drain interelectrode capacitance, the circuit is called a tuned-drain, tuned-gate oscillator. Both gate and drain circuits must be tuned to the inductive side of resonance.

A practical form of a Hartley oscillator is shown in Fig. 5. The drain voltage V_{DD} is applied through the inductor L, whose reactance is high compared with X_2. The capacitor C has a low reactance at the frequency of oscillation. However, at zero frequency it acts as an open circuit. Without this capacitor the supply voltage would be short-circuited by L

in series with L_2. The parallel combination of C_g and R_g acts to supply the bias. The circuit operates in class C, and the gate current charges up C_g to provide the X_2 bias voltage.

The Hartley oscillator may be modified by substituting L_1 and L_2 with a transformer having separate primary and feedback windings. A bipolar version of such a circuit is shown in Fig. 6, which is known as the tuned-collector oscillator. In this design, the quiescent bias is determined by R_1, R_2, and R_e. If R_1 were omitted, then initially the transistor currents would be zero, g_m would be zero, and the circuit would not oscillate. With R_1 in place, the transistor is biased in its active region, oscillations build up, and the dynamic self-bias is obtained from the R_2C'' combination due to the flow of base current. This action results in class C operation.

A transistor Colpitts oscillator is indicated in Fig. 7. Qualitatively, this circuit operates in the manner described above. However, the detailed analysis of a transistor oscillator circuit is much more difficult than that of an FET circuit, for two fundamental reasons. First, the low input impedance of the transistor shunts Z_1 in Fig. 4 and hence complicates the expressions for the loop gain given above. Second, if the oscillation frequency is beyond the audio range, the simple low-frequency h-parameter model is no longer valid and the more complicated high-frequency hybrid-π model must be used.

Very-high-frequency (vhf) oscillators. These operate in the range of from a few to several hundred megahertz. The basic configuration of these oscillators is similar to that indicated in Fig. 4. However, usually the impedances in the circuit are not lumped elements but are rather distributed (a parallel-wire transmission line or a coaxial cable). These elements are adjusted so that they appear as pure reactances. Sometimes a tuning element called a butterfly is used with a vhf oscillator. This element is similar to a variable air capacitor except that the stator plates have holes cut in them in the shape of the wings of a butterfly. As the rotor is turned the inductance (the magnetic energy storage) as well as the capacitance (the

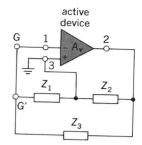

Fig. 4. The basic configuration for many resonant-circuit oscillators.

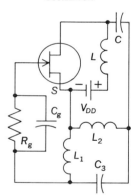

Fig. 5. FET Hartley oscillator.

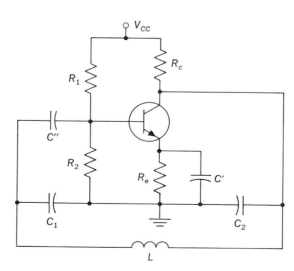

Fig. 7. A transistor Colpitts oscillator.

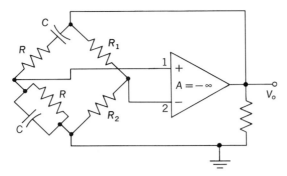

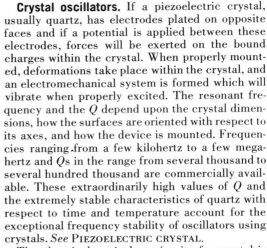

Fig. 8. A Wien-bridge oscillator.

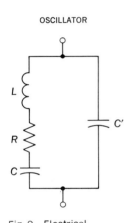

OSCILLATOR

Fig. 9. Electrical equivalent circuit of a piezoelectric crystal.

electrostatic energy storage) is varied. Hence, tuning over a wide frequency range is possible.

High-frequency transistors, both bipolar and FET types, are available in the 100-MHz to 5-GHz range. In particular, the gallium arsenide (GaAs) and metal-semiconductor field-effect transistor (MESFET) is capable of oscillating beyond 15 GHz.

Bridge oscillators. In a bridge circuit the output is in phase with the input at the balance frequency ω_0. Hence, this circuit may be used as the feedback network for an oscillator, provided that the phase shift through the amplifier is zero. This condition is satisfied in the Wien-bridge oscillator shown in Fig. 8. The frequency of oscillation is precisely the null frequency of the balanced bridge, namely, $f_0 = 1/(2\pi RC)$.

The output of a balanced bridge is zero when ω equals ω_0; therefore β and βA are both 0 at this frequency. To satisfy the Barkhausen condition ($\beta A = 1$), the bridge must be unbalanced, but in such a way that the phase shift remains zero. This is accomplished by making the ratio $R_2/(R_1 + R_2)$ smaller than 1/3. In Fig. 8, the coupling capacitors are made large enough so that they introduce no appreciable phase shifts even at the lowest frequencies of operation. Maximum frequency of oscillation is limited by the slew rate of the amplifier.

Continuous variation of frequency is accomplished by varying simultaneously the two capacitors C (ganged variable air capacitors). Changes in frequency range are accomplished by simultaneously switching in different values for the two identical resistors R.

If in Fig. 8 the resistor R_2 is replaced by a sensistor, the amplitude is stabilized against variations due to the aging of transistors and circuit components. The regulation mechanism introduced by the sensistor automatically changes β to keep βA more nearly constant whenever the value of A should change, as when amplifier loading changes. The resistance of a sensistor increases with temperature, and the temperature is in turn determined by the root-mean-square value of the current which passes through it. An alternate method makes use of the nonlinear characteristic of diodes so that the gain is reduced when the output amplitude exceeds a given value.

Other types of bridge networks, such as the twin-T and bridge-T, may be used as feedback elements to form an oscillator. The general principles enunciated above are applicable to these bridge-type oscillators, although the practical details are different.

Crystal oscillators. If a piezoelectric crystal, usually quartz, has electrodes plated on opposite faces and if a potential is applied between these electrodes, forces will be exerted on the bound charges within the crystal. When properly mounted, deformations take place within the crystal, and an electromechanical system is formed which will vibrate when properly excited. The resonant frequency and the Q depend upon the crystal dimensions, how the surfaces are oriented with respect to its axes, and how the device is mounted. Frequencies ranging from a few kilohertz to a few megahertz and Qs in the range from several thousand to several hundred thousand are commercially available. These extraordinarily high values of Q and the extremely stable characteristics of quartz with respect to time and temperature account for the exceptional frequency stability of oscillators using crystals. *See* PIEZOELECTRIC CRYSTAL.

The electrical equivalent circuit of a crystal is indicated in Fig. 9. The inductor L, capacitor C, and resistor R are the analogs of the mass, the compliance (the reciprocal of the spring constant), and the viscous damping factor of the mechanical system. Typical values for a 90-kHz crystal are in L of 137 henrys (H), a C of 0.0235 picofarads (pF), and an R of 15 kilohms, corresponding to a Q of 5500. The dimensions of such a crystal are $30 \times 4 \times 1.5$ mm. Since C' represents the electrostatic capacitance between electrodes with the crystal as a dielectric, its magnitude ($\cong 3.5$ pF) is much larger than C.

If one neglects the resistance R, the impedance of the crystal is a reactance jX whose dependence upon frequency is given by Eq. (6), where $\omega_s^2 =$

$$jX = -\frac{j}{\omega C'}\frac{(\omega^2 - \omega_s^2)}{(\omega^2 - \omega_p^2)} \qquad (6)$$

$1/LC$ is the series resonant frequency (the zero-impedance frequency) and $\omega_p^2 = (1/L)(1/C + 1/C')$ is the parallel resonant frequency (the infinite-impedance frequency). Since C' is much greater than C, then $\omega_p \cong \omega_s$. For the crystal whose parameters are specified above, the parallel frequency is only 0.3 of 1% higher than the series frequency. For $\omega_s < \omega < \omega_p$ the reactance of the crystal is inductive; outside this frequency range it is capacitive, as indicated in Fig. 10.

A variety of crystal oscillator circuits is possible. If in the basic configuration of Fig. 4, a crystal is

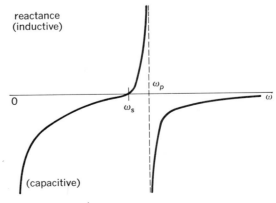

Fig. 10. The reactance function of a crystal (whose resistance has been neglected).

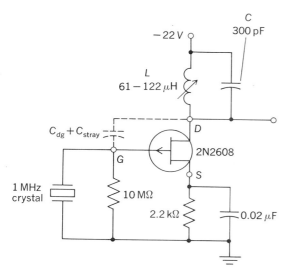

Fig. 11. A 1-MHz FET crystal oscillator. (*Siliconix Co.*)

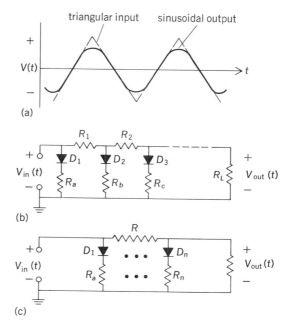

Fig. 13. Conversion of a symmetrical triangular wave-form into a sinusoid by the use of a diode-resistor clipping network. (*a*) Input and output waveforms. (*b*) Lumped diode-resistor chain. (*c*) Distributed diode-resistor chain. Sections shown for positive half-cycle clipping. (*From A. S. Grebene, Monolithic waveform generation, IEEE Spect., 9 (4):30–40, April 1972*)

used for Z_1, a tuned LC combination for Z_2, and the capacitance C_{dg} between drain and gate for Z_3, the resulting circuit is as indicated in Fig. 11. From the theory given above, the crystal reactance, as well as that of the LC network, must be inductive. In order for the loop gain to be greater than unity, X_1 cannot be too small. Hence the circuit will oscillate at a frequency which lies between ω_s and ω_p but close to the parallel-resonance value. Since $\omega_p \approx \omega_s$, the oscillator frequency is essentially determined by the crystal and not by the rest of the circuit.

Integrated oscillators. The unavailability of an integratable inductance necessitates a fundamental change in the approach of oscillator design in integrated circuits. In general, an integrated sinusoidal oscillator is obtained by using a relaxation oscillator to generate a periodic waveform which is then processed through a wave-shaping network. For example, a triangular waveform is readily available in the RC-coupled multivibrator shown in Fig. 12. The triangular waveform may be converted into a sine wave by rounding off its peaks with the aid of a diode-resistor network (Fig. 13).

By using eight or more diodes, a symmetrical triangular wave input can be converted into a sinusoidal output with less than 0.5% harmonic distortion. Furthermore, one may use the distributed nature of the *pn* junction in an integrated diffused resistor structure to effect the conversion.

An alternative approach to realize a harmonic oscillator makes use of the digital signal synthesis technique. By combining a digital clock, a programmable read-only memory (ROM) and control circuitry, nearly ideal sinusoidal waveforms are obtainable.

Heterodyne oscillator. In the heterodyne, or beat-frequency, oscillator circuit the voltage from one radio-frequency oscillator is mixed with the output from a similar device tuned to a slightly different frequency. The difference frequency, or beat note, may be varied over the audio or video range by means of a tuning capacitor.

Microwave oscillators. Special tubes and semiconductor devices are used for generating waveforms whose frequency range lies between a few hundred and several tens of thousands of megahertz. *See* MICROWAVE SOLID-STATE DEVICES; MICROWAVE TUBE. [EDWARD S. YANG]

Bibliography: A. B. Grebene, Monolithic waveform generation, *IEEE Spect.*, 9(4):34–40, April 1972; J. Millman, *Microelectronics*, 1979.

Oscilloscope

An electronic instrument which produces a luminous plot on a fluorescent screen showing the relationship of two or more variables. In most cases it is an orthogonal (x,y) plot with the horizontal axis being a linear function of time. The vertical axis is normally a linear function of voltage at the signal input terminal of the instrument. Because trans-

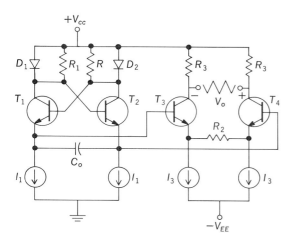

Fig. 12. Oscillator circuit for generating a symmetrical triangular waveform.

ducers of many types are available to convert almost any physical phenomenon into a corresponding electric voltage, the oscilloscope is a versatile tool in all forms of physical investigation.

The primary advantage of a cathode-ray oscilloscope over other forms of plotting devices is its speed of response. Commercially available instruments in the general-purpose category can display frequencies as high as 100 MHz while special high-speed oscilloscopes can respond as high as 2000 MHz. The horizontal linear time axis of one general-purpose oscilloscope may be varied in 25 calibrated ranges from a slow speed of 10 cm in 50 sec to a high of 10 cm in 0.2 μsec. The same oscilloscope can record on photographic film a single trace at a rate of 250 cm/μsec.

In its normal form the cathode-ray oscilloscope is made up of five basic elements (Fig. 1):

1. The cathode-ray tube and associated controls in focus, intensity or brightness, and astigmatism.

2. The vertical or signal amplifier (plug-in unit in Fig. 1) and its associated devices such as input terminal, attenuators, position control, and ac or dc amplifier operation selector.

3. Horizontal-axis time-base circuits, frequently called the sweep generator. Included in this group are the sweep-base control, trigger or synchronizing circuits, and usually a circuit for turning on the cathode-ray tube beam only when the sweep is going in the left-to-right direction on the screen.

4. Auxiliary facilities, such as amplitude or time calibrators and repetition rate generators.

5. Power supplies furnishing the correct operating voltages for the above circuits.

Cathode-ray tube. The central component in a cathode-ray oscilloscope is the cathode-ray tube (CRT) which, in its simplest form, consists of an evacuated glass container with a fluorescent screen at one end and a focused electron gun and deflection system at the other.

Either magnetic or electric fields may be used for focusing and deflection. Oscilloscopes almost always use electric fields for both functions. Electrostatic focusing is used largely because of its convenience and simplicity. Electrostatic deflection is almost universally used for oscilloscopes because it is capable of superior high-frequency response. The only magnetically deflected oscilloscopes are those using TV picture tubes for application where 100-kHz bandpass or less is adequate, but a large image is needed.

The CRT designer is faced with four primary technical objectives: (1) high deflection sensitivity; (2) a bright image for ease of observation and photography; (3) small spot size relative to the image area; and (4) accurate deflection geometry. All of these are related in various ways so that each tube is the result of compromises which seem least undesirable to its designer. *See* CATHODE-RAY TUBE.

Signal amplifier. The signal is usually applied to the vertical axis of the oscilloscope; thus this amplifier is frequently called the vertical amplifier. Signals commonly observed by means of the cathode-ray oscilloscope vary in amplitude from microvolts to kilovolts and in frequency from direct current to many megahertz. The deflection sensitivity of common types of cathode-ray tubes lies in the range of 0.01−0.2 cm/volt, with 4−10

cm of deflection available. Thus, for many purposes, it is necessary to amplify the signal in order to get sufficient deflection on the tube for accurate observation or measurement. The prime requirement of this amplifier is that it produce an amplified replica of the signal applied to its input with a minimum of distortion or variation in wave shape. This requires an amplifier with adequate frequency and phase response, in addition to a very linear transfer characteristic.

Most oscilloscopes use dc amplifiers, that is, amplifiers whose low-frequency response extends to zero. This feature is valuable because slowly changing phenomena can be accurately observed, and the relation of a waveform to essential dc reference levels can be observed. *See* AMPLIFIER; DIRECT-COUPLED AMPLIFIER; VOLTAGE AMPLIFIER.

Signal delay networks. In order to view widely spaced pulses, especially those whose spacing is not uniform, it is necessary to trigger the time sweep directly from the pulse being observed. Since the sweep takes a small but finite time to begin it is necessary to delay the signal for a slightly longer time so it will appear away from the extreme edge of the screen. This is accomplished by inserting a delay network (adjusted by control knobs as shown in Fig. 1) in the signal channel after the trigger takeoff point.

If this delay network is to transmit the signal without waveform distortion, it must be carefully designed and constructed. Three types are used:

1. Coaxial cable, which has little signal distortion but is bulky—200 ft would be necessary for 0.25 μsec delay.

2. Continuously wound delay lines, usually on a ferrite core. These are compact and can be made in convenient impedances but have rather high attenuation at frequencies of 15 MHz or more. Coupling networks on both ends are usually needed to eliminate reflectors from the capacitances of the input and output circuits.

3. Distributed constant networks. This type is used in commercial oscilloscopes with bandwidths as high as 50 MHz. They are usually of the M-derived type with an adjustable capacitance in each section. With push-pull amplifiers, a balanced network is used, having two sets of inductances with the capacitors connected between points of equal delay. This type of network may consist of as many as 90 sections. When properly adjusted, the aberrations on a square pulse are less than 1% of the pulse amplitude.

Attenuators and gain controls. To obtain a suitable image size on the cathode-ray tube, a convenient means of varying the amplifier gain or sensitivity is necessary. Two methods are used in oscilloscopes. First, a compensated step attenuator is placed in front of the amplifier. There are usually two or three steps per decade with a total attenuation such that signals of several hundred volts can be observed. So that the attenuator will have the same attenuation for all frequencies, it usually consists of resistance and capacitance dividers in parallel. When both dividers have the same ratio, the attenuation is independent of frequency. If only the resistive section were used, the stray shunt capacitance would increase the attenuation with frequency. These strays become a part of the

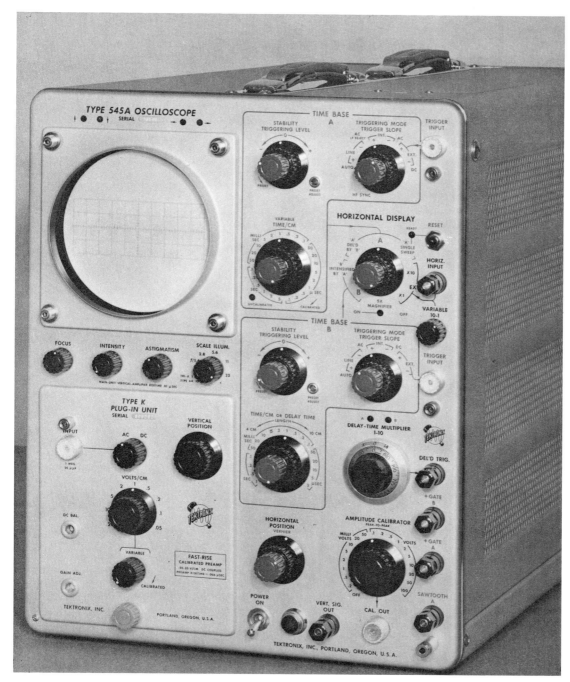

Fig. 1. Typical commercial oscilloscope with precision sweep delay and plug-in preamplifiers. (*Tektronix, Inc.*)

capacitance divider and are rendered harmless. The second gain control is usually an uncalibrated, variable one having a range of 3 to 1 or less. This fills in between the steps of the fixed attenuators. It is usually a low-impedance control so that high-frequency compensation is not necessary.

Differential input amplifiers. Most oscilloscope signal amplifiers have one terminal, usually called the ground terminal, connected to the chassis of the instrument. This is normally satisfactory because most waveforms being observed also have a common or ground reference. Many times, however, it is necessary to observe the waveforms

between two points, neither of which is grounded. For this purpose balanced, or differential, amplifiers are needed.

The output of this type of amplifier is proportional to the algebraic difference of the signals applied to its two input terminals. Thus, signals common to both terminals are cancelled, but potential differences between the terminals are amplified. In high-sensitivity differential amplifiers the amplification of the desired signal may be as much as 10,000 times the common mode signal. This property often makes it possible to observe a small signal in the presence of large interfering signals.

A differential amplifier can be inserted in place of the vertical amplifier.

Distortion in an oscilloscope. For purposes of this discussion, distortion is said to occur if the waveform on the screen is not a replica of the input waveform except, of course, for a change in scale. Among the causes for distortion are the following:

1. The cathode-ray tube may have different deflection factors in various portions of the screen or may not have orthogonal axes.

2. The amplifier may not have a linear relationship between input and output amplitudes.

3. The frequency response of the oscilloscope may be inadequate at either or both ends of the frequency spectrum of the signal being observed.

4. All components of the signal may not arrive at the deflection plates at the same time.

The remedy for the first cause is a correctly designed and built cathode-ray tube. In the second case, the designer should first use tubes with the most linear transfer characteristics, operated at optimum grid bias, screen, and plate voltages. Further reduction in amplitude distortion can then be obtained by using balanced or push-pull circuits and negative feedback when possible.

If the high-frequency response is inadequate, the slope of the steeply rising and falling portions of the waveform will be decreased. When an infinitely steep wave front, or step, is applied to a circuit having a finite bandpass, it will rise or fall according to the approximate formula, Eq. (1),

$$T_r = \frac{0.35}{f} \qquad (1)$$

where T_r is rise time in microseconds from 10 to 90% amplitude, and f is frequency in megahertz on the bandpass curve where response falls to 70% of the midfrequency amplitude.

Thus, a zero-rise-time step displayed on a 1-MHz bandpass oscilloscope will appear to have a rise time of 0.35 μsec. The bandpass necessary to observe a signal with the desired accuracy is obtained from a formula relating the resultant rise time T_r for a signal of finite rise time T_{r1} going through an oscilloscope of known rise time T_{r2}. This is written as Eq. (2). Thus, a pulse with a 1-

$$T_r = \sqrt{T_{r1}{}^2 + T_{r2}{}^2} \qquad (2)$$

μsec rise in passing through a 1-MHz oscilloscope would appear to have a rise time of approximately 1.05 μsec.

Inadequate low-frequency response causes a slope in the horizontal portions of the waveform. If only one RC coupling is involved, an oscilloscope with a 16-cycle, 70% response will cause a 10% slope in a 0.001-sec flat portion of the waveform.

The distortion caused by various frequency components arriving at different times is often called phase distortion. This causes overshoots or spurious damped oscillation following sudden steep portions of a waveform. The analytical problem of determining the effect of phase distortion on a given waveform is quite complicated, but the oscilloscope designer has a simple and direct way to observe it in practice. He observes the response to a clean, sharp amplitude step whose rise time is short compared with that of the oscilloscope under test and makes adjustments to the high-frequency compensating circuits until the waveform on the cathode-ray tube has the desired precision.

Horizontal sweep and synchronization. The most useful oscilloscope presentation is that having a linear horizontal time axis accurately synchronized with the signal being observed. Prior to 1946, most oscilloscopes, except special-purpose pulse monitors and synchroscopes, used a sweep generator of the recurrent type, which is one that continues to operate in the absence of synchronizing signals. The circuit is adjusted so that its natural frequency is slightly lower than an integral fraction of the signal frequency. When the signal frequency is injected into the circuit, the sweep is caused to terminate at a fixed point on the signal waveform, immediately returns to its initial value, and starts a new sweep. For closely spaced signals, such as sine waves, square waves, and the like, this method is simple and satisfactory. For observation of pulses or other widely spaced waveforms, it cannot be used. Consider the case of a 1-μsec pulse at a repetition rate of 1000 pulses/sec. The sweep would operate at the pulse repetition rate and be almost 1000 μsec long. A 1-μsec pulse would be hardly visible on such a sweep and certainly no detail could be observed.

Triggered sweep is the solution to this problem. In this circuit the sweep is inoperative except when started by a trigger signal. When the sweep is completed, the circuit returns to its original state and awaits another trigger. With this circuit there is no necessary relationship between the repetition rate and sweep speeds so that sweep-speed controls may be varied at will without affecting the synchronization.

In modern oscilloscopes, the sweep generator usually consists of the following elements: (1) trigger selector and amplifier; (2) sweep waveform generator; (3) sweep amplifier; and (4) CRT unblanking circuits.

Trigger selector and amplifier. If the oscilloscope is to provide a stable image, each sweep must start at the same point on the signal waveform. The information needed by the sweep generator to accomplish this may come from several sources: first and most useful, from the signal waveform itself; second, from a separate waveform which has an accurate time relationship with the signal, for example, the synchronizing signals in a television system; and third, from the power-line frequency, because many waveforms such as those found in power supplies are accurately related to it. A selector switch is frequently provided to enable convenient choice of these sources (trigger-mode adjusting knobs in Fig. 1). These trigger signals may be too small to activate the sweep generator; thus an amplifier is usually provided. This amplifier is frequently of the regenerative type whose output is always a rectangular wave of fixed amplitude regardless of the shape of the input signal. A single steep portion of this wave provides a sharp definite starting signal for the sweep generator. The sweep amplifier frequently provides facilities for causing the sweep generator to start at any selected portion of the waveform.

Sweep waveform generator. To provide a linear time axis, it is necessary to apply a sawtooth voltage waveform to the CRT horizontal plates. This waveform is one which starts at a fixed voltage, rises or falls at a linear rate to a second fixed voltage, and then returns to the first reference to repeat the cycle. The sweep waveform is generated by the charge and discharge of a capacitor. The linear portion used on the sweep is obtained by charging or discharging the capacitor with a constant current. When the linear portion is completed, the capacitor is brought back to its original voltage as rapidly as possible. In order to make sure that the linear portion starts from a stable reference on each sweep, auxiliary circuits are frequently used which prohibit the trigger signals from reacting the sweep generator until the timing capacitor has had adequate time to stabilize at its initial reference voltage. This circuit is usually called the hold off. *See* SWEEP GENERATOR.

Sweep amplifiers. Two functions are usually provided by the sweep amplifier. First, it amplifies the sweep generator waveform to that required to deflect the CRT beam; and second, it provides balanced signals to the deflection plates of the CRT. Balanced signals imply that as one plate goes in a negative direction, the other goes positive an equal voltage. This is essential for accurate image geometry.

In order to accomplish its function, a sweep amplifier must have a very linear transfer characteristic, high gain stability, and a frequency response adequate to amplify the linear portion of the sweep waveform uniformity from the slowest to the fastest sweep rates.

CRT unblanking circuits. The CRT beam is normally turned on only during the linear portion of the sweep waveform. This is especially necessary in pulse observation because the space between pulses may be several hundred times the sweep length. If the beam were not turned off between sweeps, it would rest in a bright spot at the edge of the screen, causing much extraneous light and probably damaging the screen.

Time-interval measurement. There are several methods of measuring time intervals.

Calibrated sweeps. Practically all laboratory oscilloscopes have calibrated horizontal sweeps so that time interval measurements may be read directly from a graticule over the cathode-ray tube screen. Probable error of this method ranges from 1 to 10%, depending on the precision of the oscilloscope and adequacy of maintenance and calibration. This method combines convenience with an ability to cover a wide range of time intervals. One popular oscilloscope has 25 ranges accurate to within 3% covering a range of 5 sec/cm to 0.02 μsec/cm.

Precision sweep delay. For greater accuracy, some form of expanded scale is needed. This may be accomplished by an accurately calibrated sweep-delay circuit. To use this method, the oscilloscope sweep is set to a speed so that adequate resolution of the waveform is available. The sweep-delay dial (usually having 1000 divisions) is then turned so that the start of the waveform being measured is under the center mark of the screen. The dial reading is noted and the dial turned so that the end of the waveform is now under the center mark. The time interval is thus the difference of the two dial readings. This method is capable of accuracy within 1% or better.

Time marker. Another method which is used, principally in connection with radar range measurement and television, uses a synchronized precision marker generator. This is usually an accurate sine-wave oscillator started and stopped by the sweep generator. The sine waves are put through shaping circuits which produce a sharp pulse or pips at the same point on each cycle. These pips are usually applied to the cathode-ray tube so as to brighten the trace once during each cycle of the oscillator. These bright pips are referred to as time markers. The advantage of this method is that its accuracy is dependent on the calibration of a sine-wave generator rather than the usually less stable sweep generator and cathode-ray tube circuits. If the time-mark pips are sharp and the waveform being measured has steep rises and falls, the reading accuracy of this system is excellent, because a small horizontal displacement of a pip will move it a large vertical distance on the waveform. Time-mark generators are not widely used on general-purpose oscilloscopes for several reasons. First, many ranges would be needed, each with a different time between marks. This means a large number of components to be switched. Second, on the faster sweep ranges the pips must be very short. On a 0.02 μsec/cm sweep, the pip should be 0.005 μsec or less. This requires a bandwidth for the circuits carrying the pip of about 200 MHz.

Dual-trace oscilloscopes. Frequently it is desirable to compare two waveforms on the face of a single cathode-ray tube. For this purpose several methods are used.

Dual-trace amplifier and single-gun CRT. This method uses a circuit that is frequently termed an electronic switch. The oscilloscope amplifier is switched electronically between the two signals under observation. If this switching is made to take place in the interval between sweeps when the beam is blanked out, the presentation on the screen is indistinguishable from a dual-beam oscilloscope. This method of presentation is simpler and less expensive than a true dual-beam oscilloscope and also provides more accurate time comparisons between the two signals because time-base errors are common to both waveforms.

Split-beam CRT. This oscilloscope has been popular in England for many years. It uses a single electron beam with a splitter plate in front of the final aperture. The beam is divided into two parts, each receiving separate vertical deflection. Both parts of the beam, however, pass through a single pair of horizontal deflection plates.

Dual-gun CRT. This method is similar to the split-beam method but has the additional flexibility of separate brightness, focus control, and separate balanced vertical-deflection plates. Dual-trace amplifiers can be provided in both vertical systems so that four traces are available, if desired.

The most versatile arrangement is that with two separate oscilloscopes presenting their waveforms on a common screen for convenient comparison. A 0–30 MHz instrument with versatile sweep facilities is available. One sweep is available to provide a precision delay for the other sweep. Thus, an entire waveform and a highly magnified portion of it can be shown simultaneously on the same screen.

High-speed oscilloscopes. As technology progresses, the need for a higher degree of time resolution becomes increasingly important. When signal components reach into the hundreds or thousands of megacycles, conventional amplification becomes impractical. Great care must be taken to avoid mismatches in the signal channel because the resulting reflections would distort the signals

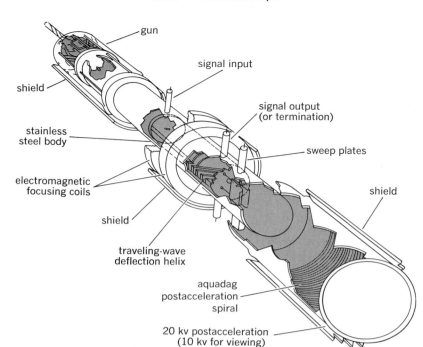

Fig. 2. Cutaway drawing of a commercial traveling-wave deflection cathode-ray tube. (*Edgerton, Germeshausen, and Grier, Inc.*)

nal to be observed is repetitive. This involves an amplitude-sampling technique using a short gating pulse, less than 1 mμsec. Samples are taken at a slightly later time at each recurrence of the signal. These samples are amplified and lengthened in time and displayed on a much slower sweep. The time resolution of this type of instrument is limited by the length of the gating pulse and the accuracy with which successive pulses may be positioned along the signal. It also has relatively high sensitivity, limited by random noise. The limitation of this type is its inability to observe single transients, and the slowness of the presentation for low repetitive rates. For example, a 60-cycle signal would need 2 sec to produce a trace made up of 120 samples. The performance specifications of a commercial sampling oscilloscope are (1) rise time, 0.4 mμsec; (2) vertical sensitivity, 0.4 cm/mv; and (3) maximum apparent sweep rate, 0.04 mμsec/cm.

Electrical measurements. In addition to providing a visual indication of waveform, the oscilloscope provides an accurate method of electrical measurement.

Measurement of frequency. If the oscilloscope has a calibrated horizontal axis or time base, frequency may be measured by reading the time necessary for one complete cycle and inverting the result. The accuracy is essentially that of the time-base calibration, whose errors are usually between 1 and 10%. For greater accuracy, the usual method is to substitute a variable calibrated sinusoidal oscillator for the time base and adjust it until a stationary pattern is obtained.

Measurement of phase difference. A common method of measuring the relative phase of two signals of the same frequency is to apply them to the two axes of the oscilloscope and compute the answer by means of the method shown in Fig. 3. The sine of the phase angle θ between the two signals is given directly by measuring the relative heights A and B as indicated in Fig. 3 and by Eq. (3). Any suitable scale, such as inches, may be used.

$$\sin \theta = \pm B/A \qquad (3)$$

A more convenient method is to use a dual-beam oscilloscope having a common calibrated time base. The phase angle can be read directly from the pattern on the screen.

Measurement of voltage. Complementing their calibrated sweeps, laboratory oscilloscopes have directly calibrated amplitude scales. Thus the voltage difference in dc or peak-to-peak volts is easily read from the calibrated graticule. This method is made possible by the development of highly stable amplifiers.

To aid in maintaining the accuracy of calibration, many oscilloscopes have built-in precision reference voltage sources. These are usually square-wave generators whose limits are set by precision dc voltages and voltage dividers.

An oscilloscope always measures in peak-to-peak volts. If the form factor is known for any particular waveform, the peak-to-peak may be converted into rms, average, or any other voltage system.

Photography of oscilloscope patterns. Oscilloscope patterns are frequently photographed to preserve a repetitive image for future study and comparison, and to record for study single brief

being observed. Very fast sweeps must be provided so that rise times of millimicroseconds or less may be measured. The problem of taking a trigger from the signal being observed without distorting or seriously attenuating it is very difficult. To make matters even more difficult, it is frequently necessary to have sufficient light output from the CRT so that single traces may be photographed with the spot traveling at speeds approaching the velocity of light. This requires very high accelerating voltages, which reduce the deflection sensitivity. Since the deflection sensitivity increases as the scan angle decreases, most millimicrosecond (mμsec) oscilloscopes have small deflection areas with a correspondingly small spot so that resolution is maintained. In these small display areas, the usual measure of sensitivity, volts per centimeter, becomes meaningless. A more significant measure is to use volts per trace width. This is termed sensibility.

The problem of transit time of the electron beam through the deflection plates is solved in most cases by some form of a traveling-wave system. Here the deflection plates are broken up into a number of segments and arrangements made so that the signal travels from one segment to the next at the same speed as the electrons in the beam. Thus any electron is deflected by the same signal component throughout its entire time in the deflection system.

Figure 2 shows a cutaway drawing of a commercial traveling-wave CRT which has the following performance: (1) vertical (TW) sensibility 0.03 volt/trace width; (2) maximum vertical scan 0.4 in.; (3) vertical frequency response of approximately 2000 MHz; (4) horizontal sensibility 0.2 volt/trace width; (5) horizontal scan 0.6 in.; and (6) writing speed 10^{11} trace widths/sec.

A different approach is possible when the sig-

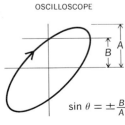

Fig. 3. Measuring phase between two signals by means of Lissajous figure.

OSCILLOSCOPE

$$\sin \theta = \pm \frac{B}{A}$$

transient waveforms which could otherwise not be studied in detail. The first case is relatively easy because an exposure of sufficient length can be made so that wide-aperture lenses and high-intensity oscilloscopes are not needed. A typical exposure might be 1 sec at $f/8$.

For photography of all but the faster single transients cameras using the Polaroid-Land system are popular. A finished positive paper print is available about 1 min after the exposure is made, and if the picture is poor, additional exposures can be made until a good print is obtained.

In order to photograph single transients at speeds of 25 cm/μsec or more, conventional 35-mm miniature cameras with wide-aperture lenses ($f/1.4$ or $f/2$) and high-speed films are often used.

High contrast, rather than a good tonal range, is wanted, so extended development in a high-energy developer is usual to obtain maximum film speed. Exposure is less critical than in normal photography for the same reason. [HOWARD VOLLUM]

Bibliography: D. Bapton, *Modern Oscilloscope Handbook*, 1979; J. Douglas-Young, *Practical Oscilloscope Handbook*, 1979; E. M. Noll, *Oscilloscope Applications and Experiments*, 1979; S. Prentiss, *Oscilloscopes*, 1980; R. Van Erk, *Oscilloscopes*: *Functional Operation and Measuring Examples*, 1978.

Parallel circuit

An electric circuit in which the elements, branches (elements in series), or components are connected between two points with one of the two ends of each component connected to each point. The illustration shows a simple parallel circuit. In more complicated electric networks one or more branches of the network may be made up of various combinations of series or series-parallel elements. *See* CIRCUIT (ELECTRICITY).

In a parallel circuit the potential difference (voltage) across each component is the same. However, the current through each branch of the parallel circuit may be different. For example, the lights and outlets in a house are connected in parallel so that each load will have the same voltage (120 volts), but each load may draw a different current (0.50 amp in a 60-watt lamp and 10 amp in a toaster).

For a discussion of parallel circuits *see* ALTERNATING-CURRENT CIRCUIT THEORY; DIRECT-CURRENT CIRCUIT THEORY.

[CLARENCE F. GOODHEART]

Parametric amplifier

A highly sensitive low-noise amplifier for ultrahigh-frequency and microwave radio signals, utilizing as the active element an inductor or capacitor whose reactance is varied periodically at another microwave or ultrahigh frequency. A varactor diode is most commonly used as the variable reactor. The varactor is a semiconductor *pn* junction diode, and its junction capacitance is varied by the application of a steady signal from a local microwave oscillator, called the pump. Amplification of weak signal waves occurs through a nonlinear modulation or signal-mixing process which produces additional signal waves at other frequencies. This process may provide negative-resistance amplification for the applied signal wave and in-

creased power in one or more of the new frequencies which are generated. *See* VARACTOR.

The parametric amplifier operates through a periodic variation of a circuit parameter, in this case, the capacitance of the varactor. Similar amplification is possible by using a periodically variable inductor which may utilize a saturable magnetic material such as a ferrite. The inductive form of this device has not found application because it requires much larger amounts of pumping power to vary the inductance significantly at microwave frequencies.

Effects. Parametric effects are also found with variable resistive elements such as the varistor or rectifying diode. These elements do not commonly provide amplification of signals but are widely used as frequency converters in radio receivers to translate signals to high frequencies to lower or intermediate frequencies for more convenient signal amplification. *See* VARISTOR.

The varactor diode is a simple *pn* junction semiconductor device designed to maximize the available reactance variation, to minimize signal losses in electrical resistance, and to require low microwave power from the pump. The junction is biased with a low dc voltage in the nonconducting direction and the applied voltage of the microwave pump wave causes the total bias to vary periodically at the microwave pumping frequency. The capacitance of a reverse-biased semiconductor diode varies with voltage, decreasing as the bias voltage is increased. Common diodes with an abrupt junction between the *p* and *n* regions have a capacitance variation according to the equation shown below, where C is the capacitance in farads, k is a

$$C = \frac{k}{\sqrt{V_i + V_b}}$$

constant depending upon the area of the junction and the doping concentrations of impurities in the two semiconductor regions, V_i is an internal or "built-in" voltage (a few tenths of a volt in most diodes), and V_b is the applied bias voltage (in the nonconducting direction). When used in a parametric amplifier, V_b includes a steady dc component plus a microwave ac value, varying periodically with time at a very high rate.

There are several possible circuit arrangements for obtaining useful parametric amplification. The two most common are the up-converter and the negative-resistance amplifier. In both types, the pump frequency is normally much higher than the input-signal frequency. In the up-converter, a new signal wave is generated at a higher power than the input wave. In the negative-resistance device, negative resistance is obtained for the input-signal frequency, causing an enhancement of signal power at the same frequency. *See* NEGATIVE-RESISTANCE CIRCUITS.

When a varactor is energized with two simultaneous voltage waves of different frequencies, such as a pump wave at a frequency f_p and a signal wave at a frequency f_s, the current induced may have a complex waveform which contains other frequencies. Mathematical analysis shows that the first-order and most significant of these new frequencies are the sum ($f_p + f_s$) and the difference ($f_p - f_s$). Depending upon the varactor characteristics and the strength of the applied waves, still other fre-

PARALLEL CIRCUIT

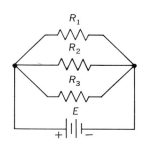

Schematic of a parallel circuit.

quencies can be produced which can have frequencies such as $nf_p \pm mf_s$, where m and n are integers of any value, including zero. Normally only the sum or the difference frequency is utilized.

Analysis also shows that, if the varactor and its surrounding network are so tuned that significant power is generated by the varactor and dissipated at the new frequency $f_p - f_s$ and not at $f_p + f_s$, the varactor will present a negative-resistive input impedance to an input wave at the frequency f_s. Instead of being absorbed by the amplifier network, the input wave will be reflected back down the input transmission line with an increased power. The wave at the frequency $f_p - f_s$ is not utilized outside the amplifier and is called, therefore, the idler frequency. This is the mechanism of the negative-resistance form of the parametric amplifier, the most common type. A functional block diagram of this type of amplifier is shown in the illustration.

Alternatively, if the network is so designed that the sum frequency $f_p + f_s$ is generated and transmitted out by a second transmission line, and if significant energy is not generated or dissipated at the difference frequency $f_p - f_s$, the power output at $f_p + f_s$ can be greater than the power input at f_s by a factor which may approach the ratio of these two frequencies $(f_p + f_s)/f_s$. This is the mechanism of the up-converter form of the parametric amplifier, so named because the signal wave is converted to a new higher frequency band.

Advantages. The most important advantage of the parametric amplifier is its low level of noise generation. In most other amplifiers, current flows through the active device by discretely charged electrons passing from one electrode to another. Random fluctuations in the rate of such flow normally occur, providing a finite background level of electrical noise (called shot noise) in the output which may override very weak signals, destroying their usefulness. Similar weak fluctuation noise is generated by resistive elements at normal temperatures in amplifier networks (called thermal noise). In the parametric amplifier, relatively very few electrons pass from one electrode to another in the varactor, reducing the shot-noise effects. Resistance and thermal noise generation effects are also minimized by proper diode and network design.

The parametric amplifier finds its greatest use as the first stage at the input of microwave receivers where the utmost sensitivity is required. Its noise performance has been exceeded only by the maser. Maser amplifiers are normally operated under extreme refrigeration using liquid helium at about 4° above absolute zero. The parametric amplifier does not require such refrigeration but in some cases cooling to very low temperatures has been used to give improved noise performance that is only slightly poorer than the maser. *See* MASER.

[M. E. HINES]

Bibliography: J. C. Decroly et al., *Parametric Amplifiers*, 1973.

Parasitic oscillation

An undesired oscillation which may occur in any type of circuit, such as an audio-, video-, or radiofrequency amplifier, oscillator, modulator, or pulse waveform generating circuit. For example, it often happens that with no apparent input signal to an amplifier an output voltage of considerable magnitude is obtained. The amplifier may be oscillating because some part of the output is inadvertently being fed back into the input. This feedback may result from the output impedance of the power supply. If feedback does occur through the power supply impedance, the oscillations can usually be stopped by the use of appropriately placed decoupling networks. Such a filter is obtained by placing a resistor in series with the plate load and bypassing the resistor to ground with a large capacitor. *See* AMPLIFIER; FEEDBACK CIRCUIT; OSCILLATOR.

Feedback may also occur through the interelectrode capacitance from grid to plate of a tube, through lead inductances, stray wiring, and other paths, which are often difficult to determine exactly. Parasitic oscillations are particularly prevalent in circuits where physically large tubes are used, in circuits where tubes or transistors are operated in parallel or push-pull, and in power stages. The frequency of oscillation may be in the audio range but is usually much higher. Often it is so high (hundreds of megahertz) that its presence cannot be detected with an oscilloscope. A low-wattage neon bulb insulated from ground may be used as an indicator. When the lamp is brought near that portion of the circuit which is oscillating, it will glow.

Parasitic oscillations represent a waste of power, a distortion of the desired waveform, or a complete malfunctioning of the circuit. Hence these oscillations must be eliminated. This can usually be accomplished by a change in circuit parameters, a rearrangement of wiring, some additional bypassing or shielding, a change of tube or transistor, the use of an rf inductor in the plate circuit, the use of rf chokes in series with filament lead, and other methods. A small resistance (50–1000 ohms) placed in series with a grid and as close to the grid terminal as possible is often effective in reducing high-frequency oscillations.

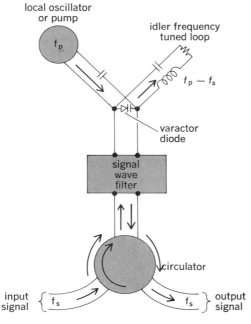

local oscillator or pump

f_p

idler frequency tuned loop

$f_p - f_s$

varactor diode

signal wave filter

circulator

input signal $\{ f_s$

output signal $f_s \}$

Negative-resistance-type parametric amplifier.

Even if the circuit is not oscillating, an output voltage may be present in a vacuum-tube amplifier in the form of hum from the use of ac heated filaments. Some hum may also appear from pickup resulting from stray magnetic fields of the power transformer or from the fields produced by the heater current in the connecting leads. It is also possible to pick up rf signals radiated through space. Spurious output voltages caused by vibrations of the electrodes because of mechanical or acoustical jarring of the tube are called microphonics. The undesired waveforms discussed in this paragraph should not be confused with true parasitic oscillations. *See* ELECTRICAL NOISE; ELECTRICAL SHIELDING; ELECTRONIC EQUIPMENT GROUNDING.

[JACOB MILLMAN]

Passive radar

A technique for surveillance, mapping, navigation, and guidance that employs the reception of microwave-frequency energy radiated by warm bodies, or reflected from other sources. It is similar in principle to infrared systems used for the same purposes. *See* MICROWAVE.

Passive radar has some similarity to both passive infrared systems and active (radiating) radar systems. Like infrared systems, passive radar emits no energy of its own, and therefore does not give away its position or existence. Because they employ no transmitter, passive radar systems are smaller, lighter, and less complex than active radar. However, an infrared system is smaller, lighter, and less complex than passive radar. The antenna of a passive radar system is comparable to that of an active radar at the same operating frequency; it is larger than that of an infrared system, which operates at higher frequencies. Target resolution is inferior to that of active radar and infrared systems. However, the ability of a passive radar system to discriminate between different targets and backgrounds can be better than that of either active radar or infrared.

Every object at a temperature above absolute zero radiates electromagnetic energy. Most of it is in the infrared region, but small amounts are emitted throughout the spectrum. The amount of radiation is determined by the absolute temperature and emissivity of the radiating body; emissivity is defined as the ratio of the object's radiation to that of an equivalent blackbody at the same absolute temperature.

In addition, objects reflect microwave energy from the sky and other sources. The total radiation received at a receiving antenna is, therefore, the sum of the radiation emitted directly as a result of the object's temperature and the energy reflected into the antenna's beam from the sky and other sources. The amount of reflected energy depends on the reflectivity of the body; this provides another way of discriminating between two objects at the same temperature.

The ability of a passive radar to discriminate between different objects depends primarily on (1) the apparent temperature differential between the objects (this takes into account emissivity and reflectivity), (2) the grazing angle between the antenna beam and the object, (3) antenna polarization, (4) antenna beam width, and (5) the minimum detectable signal level of the receiver. The grazing angle, or angle of incidence, is a factor when viewing smooth objects, such as bodies of water. Polarization is most pronounced at small grazing angles. The effect of polarization varies with the object, providing an additional method of discriminating between objects at the same temperature. *See* RADAR.

[PHILIP J. KLASS]

PERT

An acronym for program evaluation and review technique; a planning, scheduling, and control procedure based upon the use of time-oriented networks which reflect the interrelationships and dependencies among the project tasks (activities). The major objectives of PERT are to give management improved ability to develop a project plan and to properly allocate resources within overall program time and cost limitations; and to control the time and cost performance of the project, and to replan when significant departures from budget occur.

Background. In 1958 the U.S. Navy Special Projects Office, concerned with performance trends on large military development programs, introduced PERT on the Polaris weapons system. Since that time the use of PERT has spread widely throughout the United States and the rest of the industrialized countries. At about the same time that the Navy was developing PERT, the DuPont Company, concerned with the increasing cost and time required to bring new products from research to production, and to overhaul existing plants, introduced a similar technique called the critical path method (CPM). *See* CRITICAL PATH METHOD (CPM).

Requirements. The basic requirements of PERT, in its time or schedule form of application, are:

1. All individual tasks required to complete a given program must be visualized in a clear enough manner to be put down in a network composed of events and activities. An event denotes a specified program accomplishment at a particular instant in time; in effect, it represents a state of the project system. An activity represents the time and resources that are necessary to progress from one event to the next. Emphasis is placed on defining events and activities with sufficient precision so that there is no difficulty in monitoring actual accomplishment as the program proceeds. Figure 1 shows a typical operating-level PERT network from the electronics industry. Events are shown by squares, and activities are designated by arrows leading from predecessor to successor events.

2. Events and activities must be sequenced on the network under a logical set of ground rules. The activity sequencing is not arbitrary, but rather it is based on technological constraints; a foundation must be dug before the concrete can be poured. The network logic is merely the requirement that an event is said to occur when all predecessor activities are completed, and only then can the successor activities begin. The initial event, without predecessors, is self-actuated when the project begins, and the occurrence of the final event (without successor activities) denotes completion of the project. This logic requires that all activities in a network must be completed before

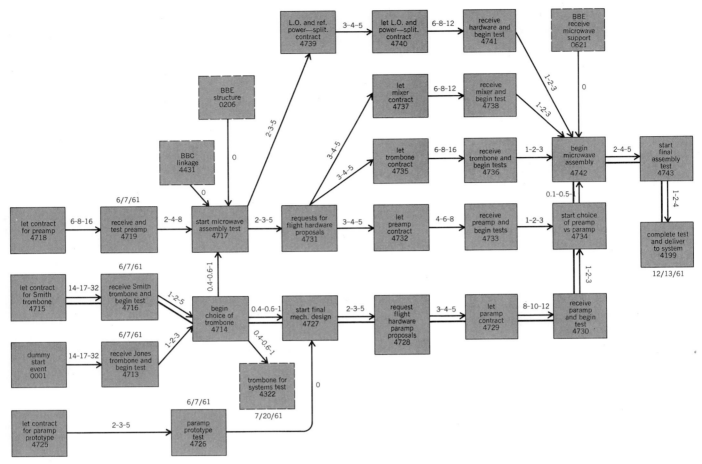

Fig. 1. Typical PERT network of an electronic module development project. Arrows and lines denote the criti- cal path. (*Applied Physics Laboratory, Johns Hopkins University*)

the project is complete, and no "looping" of activities in the network is allowed. Another technique, called GERT, relaxes these logic constraints. *See* GERT

3. Time estimates can be made for each activity of the network on a three-way basis (the three numbers shown along the arrows in Fig. 1). Optimistic (minimum), most likely (modal), and pessimistic (maximum) performance time figures are estimated by the person or persons most familiar with the activity involved. The three-time estimates are used as a measure of uncertainty of the eventual activity duration; they represent the approach used in PERT to express the probabilistic nature of many of the tasks in development-oriented and nonrepetitive programs. It is important to note, however, that for the purposes of critical path computation and reporting, the three-time estimates are reduced to a single expected time T_E, and it is used in the same way that CPM employs a single (deterministic) time estimate of activity duration time.

4. Finally, critical path and slack times are computed. The critical path is that sequence of activities and events on the network that will require the greatest expected time to accomplish. Slack time is the difference between the earliest time that an activity may start (or finish) and its latest allowable start (or finish) time, as required to complete

the project on schedule. Thus, for any event, it is a measure of the spare time that exists within the total network plan. If total expected activity time along the critical path is greater than the time available to complete the project, the program is said to have negative slack time. This figure is a measure of how much acceleration is required to meet the scheduled program completion date.

5. The difference between the pessimistic (b) and optimistic (a) activity performance times is used to compute the standard deviation (σ) of the hypothetical distribution of activity performance times ($\hat{\sigma} = (b - a)/6$). The PERT procedure employs these expected times and standard deviations (σ^2 is called variance) to compute the probability that an event will be on schedule, that is, will occur on or before its scheduled occurrence time. The procedure merely adds the expected time (T_E) and variances (σ^2) of the activities on the critical path to get the mean and variance of the hypothetical distribution of project duration times. [See columns headed EXP TIME and EXP VAR in Fig. 2, and the total values of 28.6 and 1.9, respectively, for the last activity (4004-743 to 4004-199) on the critical path.] The normal distribution is then used to approximate the probability of meeting the project schedule, as the area under the normal distribution curve to the left of (earlier than) the project scheduled completion date.

RUN 1	ENDING EVENT

BY PATHS OF CRITICALITY

CHART AJ   LR SN 9 ELECTRONIC MODULE (ILLUSTRATIVE NETWORK)   SYSTEM W 034   DATE 06-07-61

EVENT PREDECESSOR	SUCCESSOR	NOMENCLATURE	DEP	DATE EXPECTED	ALLOWED	DATE SCHD/ACT	PROB	SLACK	EXP TIME	EXP VAR
4004-715	4004-716	REV DATE (SMITH TROMBONE RECD-BEG TEST)	98		05-26-61	A06-07-61		-1.6	+	
4004-716	4004-714	SMITH TROMBONE TESTED	0146	06-23-61	06-12-61			-1.6	+ 2.3	.4
4004-714	4004-727	TROMBONE CHOSEN-BEGIN MECH DESIGN	0146	06-28-61	06-16-61			-1.6	+ 3.0	.5
4004-727	4004-728	RFP PARAMP FLIGHT HARDWARE	0146	07-20-61	07-08-61			-1.6	+ 6.1	.7
4004-728	4004-729	PARAMP CONTRACT LET	0146	08-17-61	08-05-61			-1.6	+10.1	.8
4004-729	4004-730	PARAMP RECEIVED		10-26-61	10-14-61			-1.6	+20.1	1.3
4004-730	4004-734	PARAMP TESTED	0146	11-09-61	10-28-61			-1.6	+22.1	1.4
4004-734	4004-742	CHOICE BETWEEN PREAMP-PARAMP	0146	11-13-61	11-01-61			-1.6	+22.6	1.4
4004-742	4004-743	COMPL MICROWAVE ASSY	0146	12-09-61	11-28-61			-1.6	+26.5	1.6
4004-743	4004-199	COMPL FINAL TEST MICWAVE ASSY-DELIVERED	0146	12-25-61	12-13-61	12-13-61	.12	-1.6	+28.6	1.9
4000-001	4004-713	REV DATE (JONES TROMBONE RECD-BEG TEST)	99		05-29-61	A06-07-61		-1.3	+	
4004-713	4004-714	JONES TROMBONE TESTED	0146	06-21-61	06-12-61			-1.3	+ 2.0	.1
4004-714	4004-717	TROMBONE CHOSEN-BEGIN MICWAVE ASSY TEST	0146	06-28-61	07-02-61			+ .5	+ 3.0	.5
4004-717	4004-731	RFP FOR FLIGHT HDW-MIXER-TROMB-PREAMP	0146	07-20-61	07-24-61			+ .5	+ 6.1	.7
4004-717	4004-739	COMPL MICWAVE ASSY TEST-RFP LOC OSCIL	0146	07-20-61	07-24-61			+ .5	+ 6.1	.7
4004-731	4004-735	TROMBONE CONTRACT LET	0146	08-17-61	08-21-61			+ .5	+10.1	.8
4004-731	4004-737	MIXER CONTRACT LET	0146	08-17-61	08-21-61			+ .5	+10.1	.8
4004-739	4004-740	CONTRACT LET FOR LOC OSCIL AND PWR SPLT	0146	08-17-61	08-21-61			+ .5	+10.1	.8
4004-735	4004-736	TROMBONE RECEIVED		10-14-61	10-18-61			+ .5	+18.5	1.8
4004-737	4004-738	MIXER RECEIVED		10-14-61	10-18-61			+ .5	+18.5	1.8
4004-740	4004-741	LOC OSC-PWR SPLITTER RECEIVED		10-14-61	10-18-61			+ .5	+18.5	1.8
4004-736	4004-742	TROMBONE TESTED	0146	10-28-61	11-01-61			+ .5	+20.5	1.9
4004-738	4004-742	MIXER TESTED	0146	10-28-61	11-01-61			+ .5	+20.5	1.9
4004-741	4004-742	LOC OSC-PWR SPLITTER TESTED	0146	10-28-61	11-01-61			+ .5	+20.5	1.9

Fig. 2. Typical PERT computer output. First three paths of Fig. 1 are shown here. (*From J. J. Moder and C. R. Phillips, Project Management with CPM and PERT, 2d ed., Van Nostrand–Reinhold, 1970*)

A computer-prepared analysis of the illustrative network contains data on the critical path (first group of activities in the table) and slack times for the other, shorter network paths. Note that the events (points in time) are labeled in the network, but the computer output is by activities (identified by event numbers) which also have descriptive labels. Under the column heading PROB(ability), note the figure 0.12 for the final activity on the critical path (4743-4199). This analysis indicates that the expected completion time of 12/25/61 results in a low probability of meeting the scheduled time of 12/13/61. This computer output (slack order report) is the most important of a number of outputs provided by most PERT computerized systems. Other reports may give greater or lesser details for different levels of management; they may deal with estimated and actual costs by activities (system called PERT COST); and so forth.

In the actual utilization of PERT, review and action by responsible managers is required, generally on a biweekly basis, concentrating on important critical path activities. Where necessary, effective means of shortening critical path time must be found by applying new resources or additional funds, often obtained from those activities that can afford them because of their slack condition. Alternatively, sequencing of activities along the critical path may be compromised to reduce overall duration. A final alternative may be a change in the scope of the work along the critical path to meet a given program schedule. Utilization of PERT requires constant updating and reanalysis, since the outlook for completion of activities in a complex program is constantly changing. Systematized methods of handling this aspect have been developed.

Advantage. A major advantage of PERT is the kind of planning required to create an initial network. Network development and critical path analysis reveal interdependencies and problem areas before the program begins that are often not obvious or well defined by conventional planning methods. Another advantage, especially where there is a significant amount of uncertainty, is the three-way estimate. If there is a minimum of uncertainty, the single-time approach may be used while retaining the advantages of network analysis.

In summary, it should be stated that while the developments of PERT and CPM were independent, they are both based on the same network logic to represent the project plan. PERT emphasizes the time performance of a project, including a probabilistic treatment of the uncertainty in the activity performance times and scheduled completion dates, while CPM treats time deterministically and addresses the problem of minimizing total (direct plus indirect) project cost as a function of scheduled project duration. The acronym PERT is now used as a generic term for network-based project management schemes that have evolved over the years. Today these schemes are hybrids of both PERT and CPM, but they are most often referred to as PERT. Finally, mention should be made of GERT, which denotes a generalization of the PERT/CPM network logic to complex situations where branching at events and closed loops of activities are required to adequately portray a complex project plan in the form of a network.

[JOSEPH J. MODER]

Bibliography: D. G. Malcolm et al., Applications of a technique for R and D program evaluation, *Operations Res.*, 7(5):646–669, 1959; J. J. Moder and S. E. Elmaghraby, *Handbook of Operations Research, Models and Applications*, vol. 2, 1978; J. J. Moder and C. R. Phillips, *Project Management with CPM and PERT*, 2d ed., 1970.

Phase

The fractional part of a period through which the time variable of a periodic quantity (alternating electric current, vibration) has moved, as measured at any point in time from an arbitrary time origin. In the case of a sinusoidally varying quantity, the time origin is usually assumed to be the last point at which the quantity passed through a zero position from a negative to a positive direction. It is customary to choose the origin so that the fractional part of the period is less than unity.

In comparing the phase relationships at a given instant between two time-varying quantities, the

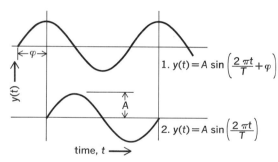

1. $y(t) = A \sin\left(\dfrac{2\pi t}{T} + \varphi\right)$

2. $y(t) = A \sin\left(\dfrac{2\pi t}{T}\right)$

An illustration of the meaning of phase for a sinusoidal wave. The difference in phase between waves 1 and 2 is φ and is called the phase angle. For each wave, A is the amplitude and T is the period.

phase of one is usually assumed to be zero, and the phase of the other is described, with respect to the first, as the fractional part of a period through which the second quantity must vary to achieve a zero of its own (see illustration). In this case, the fractional part of the period is usually expressed in terms of angular measure, with one period being equal to 360° or 2π radians. Thus two sine waves of a given frequency are said to be 90°, or $\pi/2$, out of phase when the second must be displaced in time, with respect to the first, by 1/4 period in order for it to achieve a zero value. *See* PHASE-ANGLE MEASUREMENT. [WILLIAM J. GALLOWAY]

Phase-angle measurement

The determination of the relative times at which alternating currents and voltages in a circuit take on zero values. If two voltages v_1 and v_2 are zero at the same instant, they are in phase, with zero phase difference (or out of phase with 180° difference). If one voltage v_1 passes through zero 1/8 cycle before a second voltage v_2, it leads by 360°/8 or 45° (Fig. 1). The common phase meter, a commercial device for determining the angle between current and voltage, can be used when its

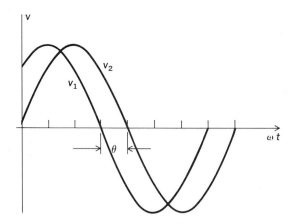

Fig. 1. Phase angle between two voltages.

presence will not disturb the circuits under measurement. When phase angles to be measured are in high-impedance or low-power circuits, this device is unsatisfactory and other measurement methods must be used.

Three-voltmeter method. This method can be used when the voltages involve a common point. Figure 2a shows three terminals a, b, and c. If the voltages v_{ab}, v_{bc}, and v_{ca} are measured by a high-impedance voltmeter (one voltmeter is sufficient), the magnitudes can be plotted to give a triangle, Fig. 2b. The angle θ between v_{ab} and v_{bc} can be determined from the law of cosines in trigonometry, as shown by Eq. (1).

$$v_{ca}^2 = v_{ab}^2 + v_{bc}^2 + 2v_{ab}v_{bc}\cos\theta \qquad (1)$$

Electronic phase-angle meter. This instrument gives the angle $(\pi - \theta)$ of Fig. 2 directly. One such instrument converts the two voltage waves to square waves by repeated amplification and limiting. The zero crossings of the square waves are identical to the zero crossings of the original voltage waves. The two square waves are applied to the input of a circuit that will pass current only

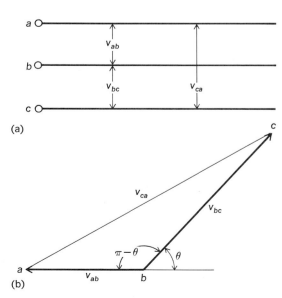

Fig. 2. Voltages employed in three-voltmeter method. (a) Circuit diagram. (b) Vector diagram.

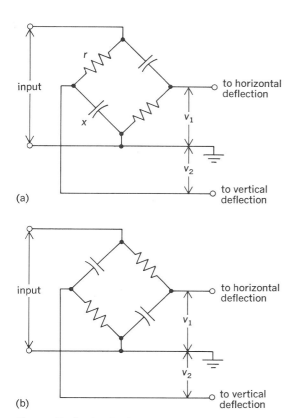

(a)

(b)

Fig. 3. Basic circuits for measuring the phase angle with an oscilloscope. (*a*) Clockwise circular sweep generated. (*b*) Counterclockwise circular sweep generated.

when both square waves are positive. In this case the greater the lag of one voltage, the smaller the overlap of the positive portions and the lower the average current. The current in this case is proportional to θ of Fig. 2.

This circuit has the theoretical limitation that each input voltage must be greater than a critical minimum value. In practice the critical value is determined by the noise on the amplifier input. If the voltage is too low, this noise causes a random zero-crossing shift and the results would be subject to this uncertainty.

A precision phase-angle meter for high-frequency voltages uses a variable delay line, and its operation is based on the fact that the difference of two voltages of constant amplitude is a minimum when the two are in phase. One of the two voltages to be compared is connected to both inputs of a variable-delay line, which is then adjusted to give a minimum output. The two voltages to be compared are then connected to the two terminals and the delay line is readjusted to give a minimum output. The change in the delay-line setting gives the time delay of one voltage relative to the other. When the frequency is known, the time delay can be computed as angle of lag. If Δt is the change in the delay-line setting and the frequency is f hertz, the phase angle is given by $2\pi f \,\Delta t$ radians or $360f \,\Delta t$ degrees.

Oscilloscope methods. Phase-angle measurements by oscilloscopes are popular in the laboratory when quick approximate results are required. If one voltage is connected to the vertical amplifier and the other to the horizontal amplifier, a Lissajous figure is obtained.

If the two voltages are of the same frequency, as is the case when phase angles are measured, the basic figure is an ellipse. A straight line with a positive slope implies that the two waves are in phase. If one leads the other, the cathode-ray beam starts back in one direction before it reaches a maximum in the other and the ray traces an ellipse.

If the amplifiers are adjusted so that the horizontal amplitude is equal to the vertical amplitude, the slope of the straight line for in-phase signals is 45°. The ellipse widens to a circle when the phase angle is 90°. Intermediate values are indicated by Eq. (2), where ϕ is phase angle sought, b is width of

$$\tan \frac{\phi}{2} = \frac{b}{a} \qquad (2)$$

the ellipse, and a is its length. *See* OSCILLOSCOPE.

Another method of utilizing an oscilloscope, developed by H. Sohon, may be illustrated by the following considerations. If a signal is applied across the input of the bridge circuit shown in Fig. 3a, the output voltages v_1 and v_2 will be 90° apart, and if r equals x (equals $1/2\pi fC$), they will be equal in amplitude. Now if v_1 and v_2 are applied to the vertical and horizontal amplifiers of an oscilloscope, the resulting trace will be a circle. If r and x are interchanged as in Fig. 3b, the trace will again be a circle, but the spot will trace the circle in the opposite direction.

If these two circuits are combined so that the sums of the outputs are applied, the spot cannot go around the circles in opposite directions at the same time, and it will trace a straight line instead. As the phase of one voltage is advanced, say by the angle $\Delta\theta$, the straight line will rotate on the screen through the angle $2\Delta\theta$. A scale can be marked on the screen and either end of the straight line used for reference. The presence of harmonics and slight errors in the $r = x$ relationship or in the equality of the input voltages will cause the line to open into a narrow ellipse. The slope of the major axis of the ellipse in this case is used for the slope of the straight line. Figure 4 shows a working circuit. It is built so that $r = x$ at the operating frequency, and R is several times r so that the voltages are added without appreciable loading of the bridge circuits.

Electronic switch. An electronic switch can be used with an oscilloscope as a phase-angle meter. The switch permits the oscilloscope to display first one wave and then the other. If the linear sweep is at the same frequency as the waves being

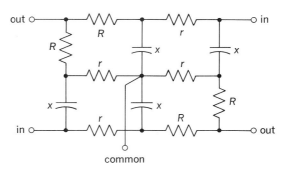

Fig. 4. Diagram showing an actual circuit for measuring phase angle with an oscilloscope.

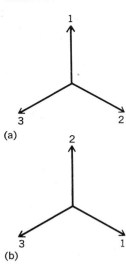

Fig. 5. Phase sequence.
(a) Sequence 1–2–3.
(b) Sequence 1–3–2.

compared, the two waves will be superimposed on the screen. The phase difference and the period can be measured on the screen in inches, the ratio giving the phase angle as a fraction of 360°.

Phase-order indicators. These devices are used to indicate which phase voltage of a polyphase circuit leads or lags another. If the voltage vectors of a three-phase generator are as indicated in Fig. 5a, the voltage from neutral to line 2 reaches a maximum 1/3 cycle (or period) after the voltage of line 1 and 1/3 cycle before the voltage of line 3. The voltage of phase 2 lags that of phase 1 and leads that of phase 3. The phase order or phase sequence is then said to be 1–2–3.

Relative motion between the armature conductors and the magnetic field induces the voltages in an alternator. Therefore, if the alternator were rotated in the opposite direction, the order in which the phase voltages reached maximum would be reversed. The phase sequence would be 1–3–2 as shown in Fig. 5b.

A miniature three-phase motor designed to rotate clockwise when connected to a three-phase system possessing a phase sequence 1–2–3 is used as a phase-sequence indicator. Counterclockwise rotation would indicate a 1–3–2 phase sequence.

A common type of phase-order indicator consists of an inductance and two lamps connected in Y to the three-phase lines as in Fig. 6a. If it is assumed that the inductive reactance is very high, the common connection on the Y is at a voltage nearly equal to the midpoint of line 2–3 in Fig. 6b and c. The voltage across the reactor is v_{n1}, and the current lags by 90° and lies on n2 in Fig. 6b and on n3 in Fig. 6c. This current divides, part going through each lamp. The result is to increase the current in lamp 2 when the phase sequence is 1–2–3, and to increase the current in lamp 3 when the phase sequence is 1–3–2.

Phase-relation indicators. These devices are used to indicate the instant when two generators or

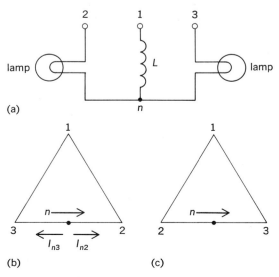

Fig. 6. Phase-sequence indicator. (a) Circuit diagram. (b) Vector diagram for sequence 1–2–3. (c) Vector diagram for sequence 1–3–2.

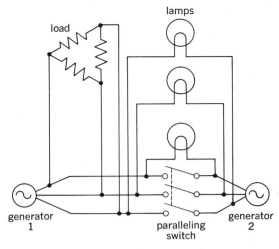

Fig. 7. Paralleling a second generator with a first generator which is already under load.

sources of alternating voltage are in phase with one another. If two voltages reach maximum at the same time, they are in phase. When two sources are to be connected in parallel, they should have the same voltage, frequency, and phase. A voltmeter and a tachometer can be used to indicate when the voltages and frequencies are nearly equal. The phase relation between the two sources is shown by means of phasing lamps or by means of a synchroscope or synchronizer.

Phasing lamps placed across the open switch used to parallel two generators will often suffice to indicate an in-phase condition (Fig. 7). Depending upon the relative phase of the two machines, the lamp voltage varies from the sum to the difference of the machine voltages. As the two frequencies approach one another, the lamps flicker, changing from full bright to dim at a decreasing rate. If the two frequencies are equal, the lamp will maintain a fixed brilliance. Usually the oncoming machine is set with a slightly higher frequency so that it will take up some load rather than be an additional load on the system. As the lamps slowly go through the dim phase, the switch is closed, connecting the machine to the system.

Synchronizer or synchroscope. This is a variation of the Tuma phase meter. The current in the fixed coil is supplied by one machine; the current in the movable crossed coils is supplied by the other machine. If the two machines are in synchronism, their frequencies are equal and the crossed coils will take a position depending upon the relative phase angle. If the frequency of one machine is slightly higher, the phase will continue to vary and the crossed coils will rotate in a direction determined by whether the speed is too low or too high. Generally the incoming machine is given a slightly higher speed and is connected to the line when the synchroscope pointer drifts past the zero mark. *See* ELECTRICAL MEASUREMENTS.

[HARRY SOHON/EDWARD C. STEVENSON]

Bibliography: M. Braccio, *Basic Electrical and Electronic Tests and Measurement,* 1979; W. D. Cooper, *Electronic Instrumentation and Measurement Techniques,* 1978; P. Kantrowitz, *Electronic Measurements,* 1979.

Phase inverter

A circuit having the primary function of changing the phase of a signal by 180°. The phase inverter is most commonly employed as the input stage for a push-pull amplifier. Therefore, the phase inverter must supply two voltages of equal magnitude and 180° phase difference. A variety of circuits are available for the phase inversion. The circuit used in any given case depends upon such factors as the overall gain of the phase inverter and push-pull amplifier, the possibility that the input to the push-pull amplifier may require power, space requirements, and cost. *See* PUSH-PULL AMPLIFIER.

Overall fidelity of a phase inverter and push-pull amplifier can be adversely affected by improper design of the phase inverter. The principal design requirement is that frequency response of one input channel to the push-pull amplifier be identical to the frequency response of the other channel. In this respect popular phase-inversion circuits are capable of providing precisely 180° phase difference only after careful selection of components. Some phase-inverter circuits can perform inversion at only one frequency; at other frequencies distortion is introduced because of unequal frequency response characteristics.

Transformer inverter. The simplest form of phase-inverter circuit is a transformer with a center-tapped secondary (Fig. 1). Careful design of the transformer assures that the secondary voltages are equal. The transformer forms a good in-

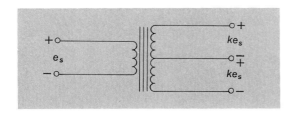

Fig. 1. Transformer, simplest form of phase-inverter circuit. The turns ratio is represented by k.

verter when the inverter must supply power to the input of the push-pull amplifier. The turns ratio can be adjusted for maximum power transfer. *See* TRANSFORMER.

The transformer inverter has several disadvantages. It usually costs more, occupies more space, and weighs more than a transistor or vacuum-tube circuit. Furthermore, some means must be found to compensate for the frequency response of the transformer, which may not be as uniform as that which can be obtained from solid-state or vacuum-tube circuits.

Paraphase amplifiers. An amplifier that provides two equal output signals 180° out of phase is called a paraphase amplifier. If coupling capacitors can be omitted, the simplest paraphase amplifier is illustrated in Fig. 2. Approximately the same current flows through R_L and R_E, and therefore if R_L and R_E are equal, the ac output voltages from the collector and from the emitter are equal in magnitude and 180° out of phase. The gain of the circuit is less than unity, which is one factor that limits its

applicability. A second important factor is that the addition of coupling capacitors and biasing resistors, necessary when the circuit is coupled to the push-pull stage, causes the phase inversion to be other than 180° over the frequency range of expected operation. One of the most stable and important paraphase amplifiers is the emitter-coupled phase inverter (Fig. 3). If the emitter resistance R_E is large compared to the impedance seen looking into the emitter of each transistor, current i_1 will equal i_2. Under this condition the voltage at one collector is exactly the negative of that at the other collector, and push-pull operation is achieved. The collector-to-collector gain is equal to the gain which would be provided by a single-transistor grounded-emitter amplifier with collector load R_L.

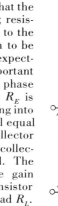

Fig. 2. Single-transistor inverter.

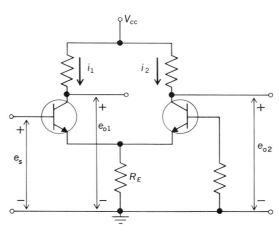

Fig. 3. Emitter-coupled phase inverter.

If the phase-inverter circuit is to produce two voltages 180° out of phase, the equivalent circuits governing the behavior of the two output voltages must be identical. The mid-frequency gain of each must be identical and the phase-shift functions must be identical. The phase-shift requirements are often compromised in the interests of simplicity of the final circuit and freedom from critical adjustments of key circuit parameters. *See* PHASE; PHASE-ANGLE MEASUREMENT; SEMICONDUCTOR.

[HAROLD F. KLOCK]

Bibliography: C. A. Holt, *Electronic Circuits, Digital and Analog*, 1978; D. L. Schilling and C. Belove, *Electronic Circuits, Discrete and Integrated*, 2d ed., 1979.

Phase-locked loops

Electronic circuits for locking an oscillator in phase with an arbitrary input signal. A phase-locked loop (PLL) is used in two fundamentally different ways: (1) as a demodulator, where it is employed to follow (and demodulate) frequency or phase modulation, and (2) to track a carrier or synchronizing signal which may vary in frequency with time. When operating as a demodulator, the PLL may be thought of as a matched filter operating as a coherent detector. When used to track a carrier, it may be thought of as a narrow-band filter for removing noise from the signal and regenerat-

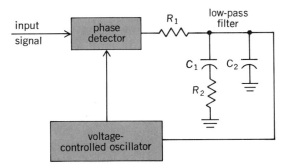

Fig. 1. Phase-locked loop.

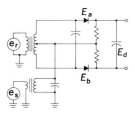

PHASE-LOCKED LOOPS

Fig. 2. Diode phase detector.

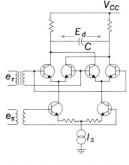

PHASE-LOCKED LOOPS

Fig. 3. Double-balanced phase detector.

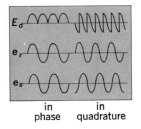

PHASE-LOCKED LOOPS

in phase   in quadrature

Fig. 4. Waveforms in double-balanced phase detector.

ing a clean replica of the signal. *See* DEMODULATOR; ELECTRIC FILTER.

Basic operation. The basic components of a phase-locked loop are shown in Fig. 1. The input signal is a sine or square wave of arbitrary frequency. The voltage-controlled oscillator (VCO) output signal is a sine or square wave of the same frequency as the input, but the phase angle between the two is arbitrary. The output of the phase detector consists of a direct-current (dc) term, and components of the input frequency and its harmonics. The low-pass filter removes all alternating-current (ac) components, leaving the dc component, the magnitude of which is a function of the phase angle between the VCO signal and the input signal. If the frequency of the input signal changes, a change in phase angle between these signals will produce a change in the dc control voltage in such a manner as to vary the frequency of the VCO to track the frequency of the input signal.

Two qualities of the loop specify its performance. One is the hold-in range, while the other is the pull-in range. The hold-in range is the maximum change in input frequency for which the loop will remain locked. It is governed by the dc gain of the loop. As the input frequency is changed, the change in phase of the two signals to the phase detector will produce a dc control voltage that will change the frequency of the VCO. As the input frequency is further changed, the phase angle will continue to increase until it reaches 0 or 180°, when the loop will unlock. If an amplifier is added to the loop, a greater control voltage will be generated which will decrease the phase error in the phase detector, and hence further detuning can occur before unlocking takes place.

The pull-in range is that range of frequencies that the loop will lock to if it is initially unlocked. Suppose the loop is unlocked and the VCO is running at frequency f_1. If the input signal f_2 is applied, but out of the pull-in range, a beat note $f_1 - f_2$ will appear at the output of the phase detector. The filter components will govern the amplitude of this beat note at the input to the VCO. If the frequency difference is reduced, the frequency of the beat note will decrease and the amplitude at the VCO input will increase. At some point, the amplitude will drive the VCO far enough over in frequency to match the input frequency and locking will occur. The lower the roll-off frequency of the loop filter, above which its attenuation begins to increase, the less will **be the** pull-in range.

Phase-locked loops are generally designed to have narrower pull-in range than hold-in range. This is the advantage of a PLL over more conventional types of filters. The pull-in range is analogous to filter bandwidth and may be made as narrow as desired by suitable choice of the resistance-capacitance (RC) filter, while the center frequency may be any desired value.

Voltage-controlled oscillators. Voltage-controlled oscillators may take many forms, two common types being the reactance modulator and the varactor modulator. These circuits are commonly used in narrow-band PLLs where narrow hold-in and pull-in ranges are desired. *See* VARACTOR.

Another common form of voltage-controlled oscillator utilizes a multivibrator type of circuit where the timing capacitor charging current may be varied by a dc control voltage. This circuit relies on the fact that the time required to charge the timing capacitance is inversely proportioned to the frequency of oscillation, hence the control voltage will vary the frequency of the oscillator. *See* MULTIVIBRATOR.

Reactance- and varactor-type modulators are used with crystal-controlled oscillators where tuning ranges of $0.25-0.5\%$ are desired, or with LC oscillators where tuning ranges of 5% can be achieved. Multivibrator types of circuits are used where wide tuning ranges of up to 1000 to 1 have been achieved. Conversely, phase noise and stability are best with crystal-controlled oscillator and least with multivibrator-type circuits.

Phase detectors. Two commonly used phase detectors are the diode phase detector and the double-balanced phase detector. The diode phase detector has been used historically because of lower cost and better performance at high frequencies (above 50 MHz). With the advent of monolith integrated circuits, the double-balanced phase detector has become useful at lower frequencies.

In the diode detector (Fig. 2) the voltage applied to the two peak rectifiers are the sum and difference of the two input signals e_r and e_s. Rectified output voltages are equal to the amplitudes of each of the sums E_a and E_b. The phase detector output voltage E_d is equal to the difference of the two rectified voltages, $E_a - E_b$. Considering that e_r and e_s are sinusoidal waves of the same frequency but varying phase angle, the dc output E_d will be zero when the two input signals are in quadrature (90°), a maximum positive value when in phase, and a maximum negative value when 180° out of phase.

The double-balanced phase detector (Fig. 3) works well at low frequencies. The upper four transistors operate as a double-pole double-throw switch driven by the reference signal e_r. The lower transistors operate as an amplifier, their collector currents being proportional to the input signal e_s. Figure 4 shows the output waveform for different phase angles between the two input signals. The output signal shown is that which would be obtained if the filter capacitor C were removed.

The loop filter determines the dynamic characteristics of the loop: its bandwidth and response time. This filter is generally composed of a lag-lead network (Fig. 1). The transfer function of the VCO has the characteristic of an integrator: a step change in the dc control voltage will produce a ramp change in output phase, increasing without bound. The low-pass filter characteristic begins

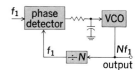

Fig. 5. Frequency multiplier.

attenuating frequencies above its break point determined by R_1C_1; at a higher frequency determined by R_2C_1, no further attentuation occurs. This is important because the phase shift associated with the VCO is 90°, and that associated with the roll off due to R_1C_1 is also approaching 90°. With a phase shift of 180° the loop will oscillate and become unstable. The leading phase shift associated with R_2C_1 reduces the phase shift around the loop and enhances stability.

Uses of phase-locked loops. The most widespread use of phase-locked loops is undoubtedly in television receivers. Synchronization of the horizontal oscillator to the transmitted sync pulses is universally accomplished with a PLL; here it is desirable to have a stable, noise-free reference source to generate the scanning line since the eye is intolerent of any jitter (or phase noise). The color reference oscillator is often synchronized with a phase-locked loop; here it is necessary to maintain less than 5° of phase error between the transmitted color reference signal and the locally generated reference in the receiver. In both of these applications, the VCO signal is the useful output, and the design of the loop filter is such as to produce a narrow-band characteristic so that noise is reduced and the filter "remembers" the proper phase during short periods of noise and loss of signal.

Phase-locked loops are also used as frequency demodulators. Since the VCO is locked to the incoming carrier frequency, and the control voltage is proportional to the VCO frequency, as long as the loop remains locked, the control voltage will be a replica of the modulating signal. Phase-locked loops may also be used as frequency multipliers. The VCO is operated at N times the input frequency f_1 and divided by N in a digital divider (Fig. 5). If a wide-range VCO is used, the circuit will operate over a wide range of input frequencies. The loop filter must remove the components of carrier appearing at the output of the phase detector or phase modulation of the VCO will occur.

Phase-locked loops have been applied to stereo decoders made on silicon monolithic integrated circuits. This design technique eliminates the coils used in previous decoder designs. The circuit consists of a 76-kHz VCO, which, after being divided by 4, is locked to the transmitted 19-kHz pilot carrier. The PLL therefore acts as a carrier regenerator (or narrow-band filter). The 76 kHz is also divided by 2 to obtain 38 kHz that is applied to a double-balanced demodulator, which decodes the stereo signal into left and right channels.

High-performance amplitude demodulators may be built using phase-lock techniques. If, instead of using a diode detector, a double-balanced demodulator is used, the lower port of which is driven with the signal to be demodulated, and the upper port is driven with a noise-free replica of the carrier, a very linear (low distortion) demodulator results. The phase-locked loop (Fig. 6) consisting of phase detector 1 and the VCO locks to the carrier frequency: the loop filter should be narrow enough that modulation is not present in the PLL. Since the first phase detector is driven in quadrature with the input signal, very little amplitude information is demodulated in the loop. The VCO signal is also phase-shifted 90° and applied to phase detector 2, which is also driven with the signal to be demodulated. Since the carrier and reference are in phase in this detector, the output will be proportional to the amplitude of the incoming carrier. Demodulators using this technique have been built with linearity better than 0.5% distortion for 90% modulated signals. *See* AMPLITUDE-MODULATION DETECTOR.

[THOMAS B. MILLS]

Bibliography: F. M. Gardner, *Phaselock Techniques*, 2d ed., 1979.

Phase modulation

A special kind of angle modulation in which the linearly increasing angle of a sine-wave carrier has added to it a phase angle that is proportional to the instantaneous value of the modulating wave (message to be communicated). Phase modulation (PM) is a scheme for impressing the message to be communicated upon a high-frequency sine-wave carrier. There is a direct proportionality between the message to be communicated and the phase variations imparted to the modulated wave propagated to the receiver. For basic concepts, technical terms, and supplementary information *see* MODULATION. *See also* ANGLE MODULATION.

Advantages and applications. Like other forms of angle modulation, PM reduces noise at the cost of extra bandwidth occupancy, transmits constant average signal power, transmits constant peak power equal to twice the average signal power, and has a channel-grabbing property whereby, if two signals reach the PM detector, the larger signal is accepted to the near exclusion of the smaller.

Important applications include certain types of telegraph, telemetering, and data-processing systems. PM is used in certain microwave radio relay systems, some of which carry telephone conversations and television programs simultaneously. PM techniques are used in many types of measuring and control systems.

The sharp limitation of range (channel grabbing) is especially important for some services. Typical examples using PM include mobile and fixed radio systems for police, airway, and military applications.

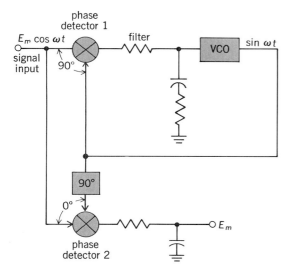

Fig. 6. Amplitude-modulation demodulator.

Noise response. Noise appearing in the output of an angle-modulation detector depends upon the kind of angle modulation, other factors being equal. When the noise disturbance has the characteristics of resistance noise, the average noise power in the output of a PM detector is uniformly distributed with respect to frequency. Under the same conditions, the distribution of root-mean-square noise currents in the output of a frequency-modulation (FM) detector is a distribution increasing linearly with frequency. Other kinds of angle modulation are characterized by other kinds of noise spectra. Normally, the kind of angle modulation used would be that giving the best signal-to-noise ratio. *See* ELECTRICAL NOISE; FREQUENCY MODULATION.

Fundamental properties. Instantaneous phase variations imply and are necessarily accompanied by uniquely related instantaneous frequency variations, and conversely. Also, given one, it is possible to reproduce the other.

For example, in PM the instantaneous phase variations imparted to the modulated wave are directly proportional to the modulating wave. The resulting variations in instantaneous frequency are, however, directly proportional to the time derivative of the modulating wave.

Similarly, in FM the instantaneous frequency of the modulated wave is linearly proportional to the modulating wave. However, the resulting variations in instantaneous phase are directly proportional to the time integral of the modulating wave.

Actually, PM and FM are not essentially different. A circuit whose output is inversely proportional to frequency (zero frequency excepted) preceding a phase modulator converts PM to FM, and following an FM detector, converts frequency to phase detection. Similarly, a circuit whose output is directly proportional to frequency (zero frequency excepted) preceding a frequency modulator converts FM to PM, and following a PM detector, converts phase to frequency detection.

Angle modulation manifests itself by the zeros of the angle-modulated wave. These zeros are the exact instants of time that the angle-modulated wave passes through zero. Theoretically, given the zeros, it is possible to determine for all values of time the instantaneous frequency deviations from the fixed frequency of the unmodulated carrier, and also the instantaneous phase deviations corresponding to the instantaneous frequency deviations. In other words, the zeros, which are nothing more than a distribution of points along the time axis, unambiguously identify the original message.

When detecting an angle-modulated wave perturbed by noise, nonsignificant information must be ignored, because only by this means can the full noise advantage of angle modulation be realized. The limiter in a conventional PM detector ignores nonsignificant information by completely destroying the waveform of the received wave, leaving only the zeros. *See* PHASE-MODULATION DETECTOR; PHASE MODULATOR.

[HAROLD S. BLACK]

Bibliography: J. Betts, *Signal Processing, Modulation and Noise*, 1971; F. R. Connor, *Modulation*, 1973; M. Schwartz, *Information, Transmission, Modulation and Noise*, 2d ed., 1970.

Phase-modulation detector

A device which recovers or detects the modulating signal from a phase-modulated carrier. Any frequency-modulation (FM) detector with minor modifications will detect phase-modulated waves. *See* FREQUENCY-MODULATION DETECTOR; PHASE MODULATION.

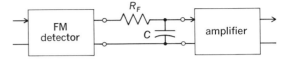

An *RC* filter which is used to convert a frequency-modulation detector into a phase-modulation detector.

The only difference between FM and phase modulation (PM) is the manner in which the modulation index varies with the modulating frequency. The modulation index is independent of the modulating frequency in PM but is inversely proportional to the modulating frequency in FM. Therefore an FM detector, when used to detect a phase-modulated wave, produces an output voltage which is proportional to the modulating frequency, assuming the original modulating signal to be of constant amplitude. Consequently, a low-pass filter with a single reactive element, such as an *RC* (resistance-capacitance) filter, is needed in the output of the FM detector which is used to detect a phase-modulated wave.

A simple *RC* filter which might be used to convert an FM detector into a PM detector is shown in the figure. The *RC* time constant of the filter should equal the reciprocal of the lowest modulating frequency component in radians per second. However, the resistance *R* involved in the time constant is the total resistance in parallel with the capacitance *C*, not just the resistance R_f. Commercial FM broadcasting ordinarily utilizes the characteristics of FM for modulating frequencies below about 2100 Hz and the characteristics of PM for frequencies above 2100 Hz for the purpose of maintaining improved signal-to-noise ratio at high modulating frequencies, as compared with pure FM. This technique is usually known as preemphasis. Then the FM detector must include a filter such as the *RC* filter shown in the figure, but the crossover frequency must be approximately 2100Hz instead of the lowest modulating frequency. This filter is usually known as a deemphasis filter and has a time constant of approximately 75 μsec ($RC = 1/6.28 \times 2100$). This deemphasis filter must be added to an FM detector to convert it into a PM detector. [CHARLES L. ALLEY]

Phase modulator

An electronic circuit that causes the phase angle of the modulated wave to vary (with respect to the unmodulated carrier) in accordance with the modulating signal. Since frequency is the rate of change of phase, a phase modulator will produce the characteristics of frequency modulation (FM) if the frequency characteristics of the modulating signal are so altered that the modulating voltage is inversely proportional to frequency.

Commercial FM transmitters normally employ a phase modulator because a crystal-controlled oscillator can then be used to meet the strict carrier-frequency control requirements of the Federal Communications Commission. The chief disadvantage of phase modulators is that they generally produce insufficient frequency-deviation ratios, or modulation index, for satisfactory noise suppression. Frequency multiplication can be used, however, to increase the modulation index to the desired value, since the frequency deviation is multiplied along with the carrier frequency. Many different types of phase modulators have been devised. A few of the typical and more commonly used modulators are described in this article. *See* FREQUENCY MODULATION; PHASE MODULATION; PHASE-MODULATION DETECTOR.

Types. A simple modulator is shown in Fig. 1. In this circuit the modulating voltage changes the capacitance of the varactor diode. The phase shift depends upon the relative magnitudes of the capacitive reactance of the varactor diode and the load resistance R. Therefore the phase shift varies with the modulating voltage and phase modulation (PM) is accomplished. However, the phase shift is not linearly related to the modulating voltage if the PM exceeds a few degrees, because the phase shift is not linearly related to the capacitance and the capacitance of the varactor diode is not linearly related to the modulating voltage. *See* VARACTOR.

Figure 2 illustrates the principle of operation of a phase modulator which provides greater phase shift for a given distortion than the simple phase modulator of Fig. 1. The block diagram of Fig. 2*a* shows that the oscillator signal is passed through two channels, 1 and 2. In channel 1 the phase of the signal is shifted 90°, and in channel 2 the signal is amplitude-modulated. The outputs of channel 1 and channel 2 are then recombined and passed through a limiter to remove the residual amplitude modulation (AM). The phasor diagram shows how PM is produced when the amplitude-modulated carrier voltage V_2 is added to the 90° phase-shifted carrier voltage V_1. The phasor sum V_0 is the hypotenuse of the phasor triangle and varies in phase with respect to the channel 1 voltage as the channel 2 voltage varies in amplitude. The phasor sum V_0 also varies somewhat in amplitude, but a limiter can be used to remove this AM, as previously mentioned and shown in Fig. 2. This phase modulator

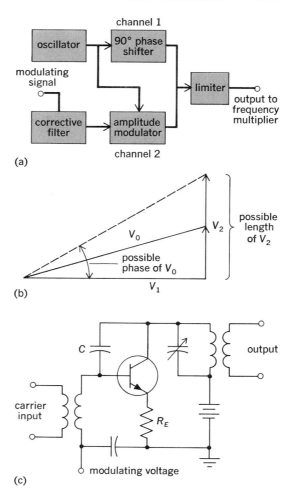

(a)

(b)

(c)

Fig. 2. A method for the production of phase modulation. (*a*) Block diagram. (*b*) Phasor diagram showing the combination of voltages. (*c*) A circuit which performs the necessary functions, except corrective filtering and limiting, in a single transistor circuit.

is linear, or distortionless, providing first, that the AM is distortionless and second, that the change in phase is proportional to the change in amplitude of the amplitude-modulated signal from channel 2.

But elementary trigonometry and Fig. 2*b* show that the phase of V_0 is the angle whose tangent is V_2/V_1. Therefore linear modulation is attained only for small phase angles where the tangent of the angle is approximately equal to the angle itself. This requirement limits the total phase variation to about 20° if the distortion is limited to 5%. This limit would permit only 10° phase deviation on each side of the unmodulated, or carrier, position.

A simple circuit employing a single transistor can perform the basic functions of a phase modulator, as shown in Fig. 2*c*. The transistor is a base-modulated amplifier which has low gain because of the unbypassed resistor R_E in the emitter circuit. The current through the small capacitance C leads the input voltage by approximately 90° and adds to the amplitude-modulated collector current in the collector load. The small voltage gain, perhaps less than 1, is required in the transistor amplifier for two reasons. First, the current through capacitor C

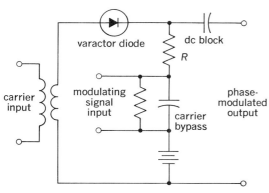

Fig. 1. A simple phase modulator.

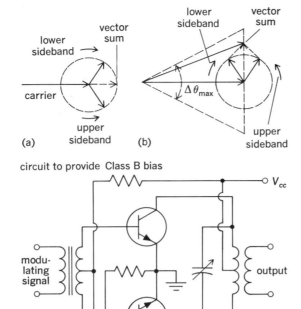

circuit to provide Class B bias

(c)

Fig. 3. Illustration of the principle of phase modulation using a balanced modulator. (*a*) Amplitude modulation. (*b*) Phase modulation. (*c*) A balanced modulator.

is then essentially proportional to the carrier input, which is essentially constant, as assumed in the phasor diagram of Fig. 2*b*. Second, the capacitance C can cause the amplifier to be unstable unless the voltage gain is low.

The phase deviation can be doubled for a given distortion level if the base-modulated amplifier of Fig. 2 is replaced by a balanced modulator. The reason for this improvement is that the carrier is suppressed in the balanced modulator output and only the sideband frequencies are present in channel 2 output to add to channel 1 output. Thus the carrier is not present to produce the unmodulated phase offset shown by the solid line labeled V_0 in Fig. 2*b*. The improvement gained by suppressing the AM carrier can be visualized if the modulating signal is a single frequency so there are only two side frequencies. Each of these side frequencies can be viewed as a rotating phasor which adds vectorially to the carrier phasor. The upper side frequency is higher than the carrier. Therefore when the carrier phasor is used as a reference, the upper side frequency appears to rotate counterclockwise. Similarly, the lower side-frequency phasor appears to rotate clockwise because its frequency is lower than the carrier frequency. The vector sum of these three phasors produces AM as shown in Fig. 3*a*. However, PM is produced when the side frequencies are shifted 90° in phase with respect to the carrier (or vice versa) as shown in Fig. 3*b*. Then the phase is seen to deviate symmetrically on either side of the phase-shifted carrier and the peak-to-peak deviation can be at least twice as great for a given distortion as when the phase de-

viates only in one direction. The peak-phase deviation may be about 30° if 8% harmonic distortion is tolerable. A simple balanced modulator circuit is shown in Fig. 3*c*. *See* AMPLITUDE MODULATION.

When the balanced modulator is used in the block diagram of Fig. 2*a* and the frequency multiplier and a power amplifier follows the limiter, a complete FM transmitter results and is known as the Armstrong system of phase (or frequency) modulation.

Phasitron. In this system of phase modulation the phase shift of the carrier is obtained by means of a special electron tube, the phasitron. The principal difficulty with normal phase modulators is a result of the relatively small amount of phase shift that can be produced without introducing nonlinearities and the resultant large amount of frequency multiplication which is usually needed. This difficulty is alleviated to a large degree by means of the phasitron in which a phase swing of as much as ±200° can be obtained with low distortion.

The principle of operation of the phasitron tube can be explained with the aid of Fig. 4. Suppose that an electron stream is produced in the form of wheel spokes from a central cathode in which the

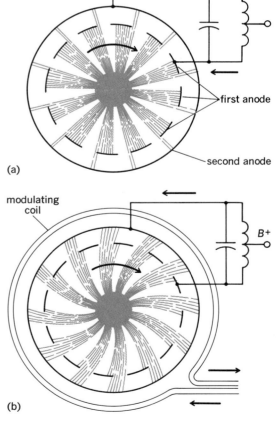

Fig. 4. Diagram illustrating principle of operation of phasitron tube. (*a*) Rotating electron beams originate from a central cathode (shown in gray). (*b*) Bending of the electron beams by an axial magnetic field, supplied by the modulating coil, advances or retards arrival of electrons to the second anode. (*From R. Adler, A new system of frequency modulation, Proc. IRE, 35(1), 1947*)

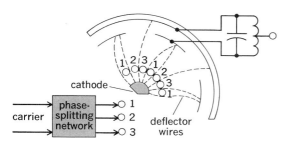

Fig. 5. Arrangement of deflector wires in vicinity of cathode of phasitron which permits generation of rotating electron beams. Three-phase potentials are applied to the three sets of wires. (*From R. Adler, A new system of frequency modulation, Proc. IRE, 35(1), 1947*)

electrons move outward along radial lines and rotate with uniform velocity. Figure 4*a* shows that the electron beams alternatively will be intercepted by the first anode or permitted to pass to the second anode. The tuned circuit connected between the two anodes is therefore excited at a constant frequency equal to the speed of rotation of the electron spokes by the segmented first anode. If a magnetic field is applied parallel to the axis of the cathode, the electron beams are deflected as shown in Fig. 4*b* and arrive at the second anode either earlier or later than normal, depending upon the direction of the magnetic field. The advance caused by the magnetic field corresponds to the width of one segment of the first anode, so that the second-anode current is advanced in phase by 180°. The amount of phase shift that can be obtained depends upon the magnetic field produced by the modulating signal.

In the phasitron the rotating electron spokes are produced by surrounding the cathode by a squirrel cage of conductors to which three-phase voltages derived from a crystal oscillator are applied. A possible arrangement of such wires is shown in Fig. 5. If all the wires numbered 1 are connected together, and all number 2, and all number 3, and the three sets are excited from a three-phase voltage source, a rotating electron stream can be produced. This can be seen by imagining that a set of wires numbered 1 are positive, whereas 2 and 3 are negative; this causes most of the electron current to be emitted in the direction of the wire numbered 1, thus causing the spoke to form. The field along the squirrel cage rotates in a manner

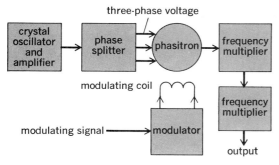

Fig. 6. Block diagram showing method of producing frequency modulation by using phasitron tube.

similar to that in the synchronous motor.

Figure 6 shows a block diagram of a frequency-modulated transmitter using a phasitron tube. A crystal oscillator operating at some low radio frequency (rf) supplies a voltage to the network consisting of inductances and capacitors from which three voltages 120° apart are derived. The introduction of this three-phase voltage to the squirrel-cage elements of the phasitron creates the rotating electron beam. The rf output from the phasitron is introduced into subsequent frequency-multiplier stages. The phase of the output carrier is shifted by means of the signal derived from a modulator amplifier as indicated.

[CHARLES L. ALLEY]

Bibliography: C. L. Alley and K. W. Atwood, *Electronic Engineering*, 3d ed., 1973; D. J. Comer, *Modern Electronic Circuit Design*, 1976; S. Seely, *Electronic Circuits*, 1968.

Photoacoustic spectroscopy

A technique for measuring small absorption coefficients in gaseous and condensed media, involving the sensing of optical absorption by detection of sound. It is frequently called optoacoustic spectroscopy. Although the technique dates back to 1880 when A. G. Bell used chopped sunlight as the source of radiation, it remained dormant for many years, primarily because of the lack of suitable powerful sources of tunable radiation. However, the usefulness of optoacoustic detection for spectroscopic applications was recognized early in its development, and pollution monitoring instruments (called spectrophones) dedicated for detection of specific gaseous constituents have been used intermittently since Bell's work.

Methods of measuring absorption. During the transmission of optical radiation through a sample (gas, liquid, or solid), the absorption of radiation by the sample can be measured by at least three techniques. The first one is the straight-forward detection technique which requires a measurement of the optical radiation level with and without the sample in the optical path. The transmitted power P_{out} and the incident power P_{in} are related through Eq. (1), where α is the absorption coefficient and l

$$P_{out} = P_{in}e^{-\alpha l} \qquad (1)$$

is the length of the absorber. With this technique, the minimum measurable αl is of the order of 10^{-4} unless special precautions have been taken to stabilize the source of radiation.

The second of the techniques is the derivative absorption technique where the frequency of the input radiation is modulated at a low radio frequency or audio frequency, ω_m. The transmitted radiation then contains a time-varying component at ω_m, if the optical path contains absorption which has a frequency-dependent structure. (For structureless absorption, modulated absorption spectroscopy does not provide a signal that can characterize the amount of absorption.) For situations where the absorption has well-defined structure, the modulation absorption spectroscopy can be used to measure αl as small as about 10^{-8} for sufficiently high input powers. The ability to measure the small absorption effects is independent of the input and output power levels for the straight-forward measurement technique as long as the noise

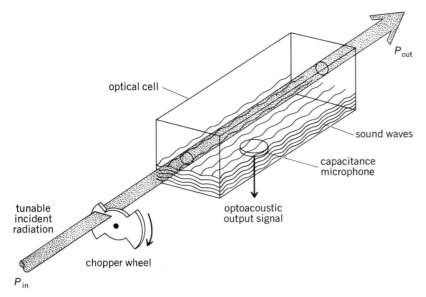

Fig. 1. Optoacoustic cell for gaseous spectroscopy.

contributed by the detector is not a factor in determining the signal-to-noise ratio. For the derivative absorption technique, the smallest αl that can be measured varies as $(P_{in})^{-1}$ until the shot noise of the detector begins to be appreciable.

The third technique, optoacoustic detection, is a calorimetric method where no direct detection of optical radiation is carried out but, instead, a measurement is made of the energy, with power P_{abs}, absorbed by the medium from the incident radiation, Eq. (2). Thus the optoacoustic signal, V_{oa}, is

$$P_{abs} = P_{in}(1 - e^{-\alpha l}) \qquad (2)$$

given by Eq. (3), where K is the a constant describ-

$$V_{oa} = K \{P_{in}(1 - e^{-\alpha l})\}$$
$$\approx KP_{in}\alpha l \qquad (\text{for } \alpha l \ll 1) \quad (3)$$

ing the conversion factor for transforming the absorbed energy into an electrical signal using an appropriate transducer. It has been tacitly assumed that the absorbed energy is lost by nonradiative means rather than by reradiation. The optoacoustic detection scheme implies that the absorbed energy will be converted into acoustic energy for eventual detection.

From Eq. (3), the optoacoustic signal is proportional to the incident power and the absorption-length product αl. Thus, for given sources of noise from the detection transducers, the signal-to-noise ratio improves as the incident energy is increased. Put differently, the smallest amount of absorption that can be measured using the optoacoustic technique varies as $(P_{in})^{-1}$ with no limitation on the level to which P_{in} can be increased for detecting small absorptions. Values of αl as small as 10^{-10} can be measured in the gas phase. The techniques generally used for gases and those used for condensed-phase optoacoustic spectroscopy differ in detail somewhat.

Gases. If optical radiation, is amplitude-modulated at an audio frequency, the absorption of such radiation by a gaseous medium that has been confined in a cell with appropriate optical windows for the entrance and exit of the radiation, and nonradiative relaxation of the medium, will cause a periodic variation in the temperature of the column of the irradiated gas (Fig. 1). Such a periodic rise and fall in temperature gives rise to a corresponding periodic variation in the gas pressure at the audio frequency. The audio-frequency pressure fluctuations (that is, sound) are efficiently detected using a sensitive gas-phase microphone. The intrinsic noise limitation to the optoacoustic detection scheme arises from the Brownian motion of gas atoms/molecules, and Kreuzer showed that the minimum detectable absorbed power is $P_{min} \approx 3.6 \times 10^{-11}$ W for a 11.9-cm-long cell. Substituting P_{min} for P_{abs} in Eq. (2), and noting, as in Eq. (3), that $(1 - e^{-\alpha l}) \cong \alpha l$ for $\alpha l \ll 1$, it follows that α_{min} varies as (P_{min}/P_{in}) as indicated above. The usefulness of the optoacoustic detection for measurement of small absorption coefficients became evident with the development of a variety of tunable high-power laser sources which could take advantage of the $(P_{in})^{-1}$ dependence. Using a spin-flip Raman laser tunable in the 5.0- to 5.8-μm range, with a power output of approximately 0.1 W, C. K. N. Patel and R. J. Kerl were able to detect α_{min} of approximately 10^{-10} cm^{-1} for a cell length of 10 cm. These studies used a miniature optoacoustic cell (Fig. 2) with a total gas volume of approximately 3 cm^3. The absorber used was nitric oxide diluted in nitrogen. It is estimated that for a signal-to-noise ratio of approximately 1, and a time constant of 1 s, it is possible to detect a nitric oxide (NO) concentration of approximately 10^7 molecules cm^{-3}, corresponding to a volumetric mixing ratio of approximately $1:10^{12}$ at atmospheric pressure. *See* LASER.

The capability of measuring extremely small absorption coefficients and correspondingly small concentrations of the absorption gases has many applications, including high-resolution spectroscopy of isotopically substituted gases, excited states

Fig. 2. Sensitive optoacoustic gaseous spectroscopy cell.

of molecules and forbidden transitions, and pollution detection. In the last application, both continuously tunable lasers, such as the spin-flip Raman laser and dye lasers, and step tunable infrared lasers, such as the carbon dioxide (CO_2) and carbon monoxide (CO) lasers, have been used as sources of high power radiation. The pollution measurements have demonstrated that the optoacoustic spectroscopy technique in conjunction with tunable lasers can be routinely used for on-line real-time in-place detection of undesirable gaseous constituents at sub-parts-per-billion levels. Specific examples include the measurement of nitric oxide on the ground and in the stratosphere (where nitric oxide plays an important role as a catalytic agent in the stratospheric ozone balance) and measurements of hydrocyanic acid (HCN) in the catalytic reduction of $CO + N_2 + H_2 + \ldots$ over platinum catalysts. These studies point toward expanding use in the future of optoacoustic spectroscopy in pollution detection. *See* LASER.

Condensed-phase spectroscopy. A straightforward application of the gas-phase optoacoustic spectroscopy technique to the study of condensed phase (liquid or solid) spectra involves enclosing the condensed phase material within the gas-phase optoacoustic cell (Fig. 3). The "photoacoustic" signals generated in the sample due to the absorption of optical radiation are communicated to the gas-phase microphone via coupling through the gas filling the chamber. The inefficiency of such a system is high because of the very poor acoustical match (coupling efficiency approximately 10^{-5}) between the condensed-phase sample and the gas. In reality, because of the large acoustical mismatch, the detection scheme is really "photothermal" rather than "photoacoustic," and this scheme provides a capability of measuring fractional absorption at a level of approximately 10^{-4} when a continuous-wave laser power of approximately 10 W is used. A more severe drawback of the scheme, however, lies in the difficulty of interpretation of the data because of the intimate dependence of the observed optoacoustic signal from the microphone on the chopping frequency, absorption depth, and heat diffusion depth. However, in spite of its shortcomings, the gas-phase microphone technique for condensed-phase optoacoustic spectroscopy has found application.

A very sensitive calorimetric spectroscopic

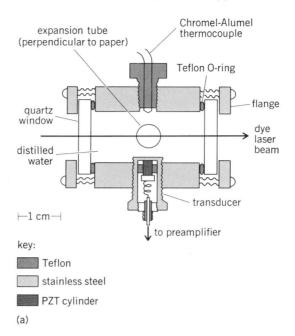

key:

■ Teflon

▢ stainless steel

■ PZT cylinder

(a)

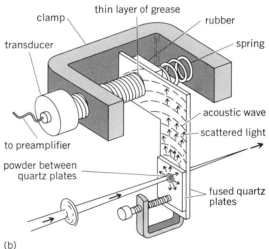

(b)

Fig. 4. Arrangement for pulsed-laser (*a*) immersed and (*b*) contacted piezoelectric transducer optoacoustic spectroscopy.

technique has been developed for the study of weak absorption in liquids and solids. This technique uses a pulsed tunable laser for excitation and a submerged piezoelectric transducer, in the case of a liquid, or a contacted piezoelectric transducer, in the case of a solid, for the detection of the ultrasonic signal generated due to the absorption of the radiation and its subsequent conversion into a transient ultrasonic signal (Fig. 4). The major distinction between the above condensed-phase "photoacoustic" spectroscopy technique and the pulsed-source, submerged or contacted piezoelectric transducer technique, is the high coupling efficiency of approximately 0.2 for the ultrasonic signal in the liquid to the submerged transducer, or an efficiency of approximately 0.9 for coupling the ultrasonic wave in a solid to a bonded transducer. Because of this high efficiency, the pulsed-laser, submerged or bonded optoacoustic spectros-

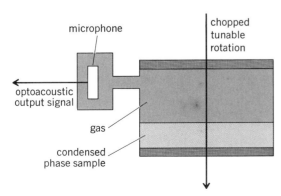

Fig. 3. Arrangement for condensed-phase photoacoustic spectroscopy.

copy technique has been shown to be useful for measuring fractional absorptions as small as 10^{-7} when using a laser source with pulse energy of approximately 1 mJ, pulse duration of approximately 1 μs, and a pulse repetition frequency of 10 Hz. There is room for improvement by increasing the laser pulse energy. There is a possibility of the electrostriction effect giving rise to an unwanted background signal, but this signal is not dependent on the light wavelength, and can be minimized by proper choice of experimental parameters.

The pulsed-laser, submerged or bonded piezoelectric transducer technique has a further advantage that time-gating the ultrasonic signal output can be utilized for the rejection of spurious signals since the sound velocity in condensed media is known and hence the exact arrival time of the real optoacoustic pulse can be calculated. This technique has been used for measurement of very weak overtone spectra of a variety of organic liquids, optical absorption coefficients of water and heavy water in the visible, two-photon absorption spectra of liquids, Raman gain spectra in liquids, absorption of thin liquid films, spectra of solids and powders, and weak overtone spectra of condensed gases at low temperatures. Because of the capability of measuring very small fractional absorptions, the technique is clearly applicable to the area of monitoring water pollution, impurity detection in thin semiconductor wafers, transmission studies of ultra-pure glasses (used in optical fibers for optical communications), and so forth. Further, even though in all of the present studies use is made of only the optical radiation, there is no reason to restrict "optoacoustic" spectroscopy to the optical region. By using pulsed x-ray sources, such as the synchrotron light source or pulsed electron beams, the principle described above for a pulsed-light-source, submerged or bonded piezoelectric transducer, gated-detection technique can be extended to x-ray acoustic spectroscopy and electron-loss acoustic spectroscopy. These extensions are likely to have major impact on materials and semiconductor fabrication technology.

[C. K. N. PATEL]

Bibliography: A. G. Bell, On the production and reproduction of sound by light, *Proc. Amer. Ass. Adv. Sci.*, 29:115–129, 1880; L. B. Kreuzer, The physics of signal generation and detection, in Y.-M. Pao (ed.), *Optoacoustic Spectroscopy and Detection*, 1977; C. K. N. Patel, Spectroscopic measurements of stratosphere using tunable infrared lasers, *Opt. Quantum Electr.*, 8:145–154, 1976; C. K. N. Patel and R. J. Kerl, A new optoacoustic cell with improved performance, *Appl. Phys. Lett.*, 30:578–579, 1977; C. K. N. Patel and A. C. Tam, Pulsed optoacoustic spectroscopy of condensed matter, *Rev. Mod. Phys.*, 53:517–550, 1981; M. B. Robin and W. R. Harshbarger, The opto-acoustic effect: Revival of an old technique for molecular spectroscopy, *Acc. Chem. Res.*, 6:329, 1973; M. Rosencwaig, Photoacoustic spectroscopy of solids, *Optics Comm.*, 1:305, 1973.

Photoconductive cell

A device for detecting electromagnetic radiation (photons) by variation of the electrical conductivity of a substance (a photoconductor) upon absorption of the radiation by this substance. During opera-

tion the cell is connected in series with an electrical source and current-sensitive meter, or in series with an electrical source and resistor. Current in the cell, as indicated by the meter, is a measure of the photon intensity, as is the voltage drop across the series resistor.

Photoconductive cells are made from a variety of semiconducting materials in the single-crystal or polycrystalline form. There are elemental types such as germanium, silicon, and tellurium; binaries such as lead sulfide, cadmium sulfide, and indium arsenide; ternaries such as mercury cadmium telluride, lead tin telluride, indium arsenide antimonide, and other combinations of elements. The semiconductor detectors respond to photons that exceed a material-related threshold energy. The bolometer is another important detector and consists of a blackened material having a temperature-sensitive conductivity. This device detects photons over a broad spectrum, utilizing the heating effect of photons. The cells are prepared by growth of the semiconducting materials as nearly pure single crystals, or as polycrystalline films deposited chemically or by evaporation on suitable substrates.

Cadmium sulfide cells are extensively used in the visible spectrum for street lighting control and camera exposure meters. Lead sulfide and mercury cadmium telluride, sensitive to infrared radiation, are used for night vision. Infrared photoconductive cells are becoming more important for detecting energy loss from buildings and for early detection of breast cancer.

The choice of cell type depends on the application requirements, which include operating temperature, wavelength to be detected, and response time. *See* PHOTOCONDUCTIVITY; PHOTOELECTRIC DEVICES.

[SEBASTIAN R. BORRELLO]

Bibliography: S. Larch (ed.), *Photoelectronic Materials and Devices*, 1965; W. L. Wolfe and G. J. Zissis (eds.), *The Infrared Handbook*, Environmental Research Institute of Michigan, 1978.

Photoconductivity

The increase in electrical conductivity caused by the excitation of additional free charge carriers by light of sufficiently high energy in semiconductors and insulators. Effectively a radiation-controlled electrical resistance, a photoconductor can be used for a variety of light- and particle-detection applications, as well as a light-controlled switch. Other major applications in which photoconductivity plays a central role are television cameras (vidicons), normal silver halide emulsion photography, and the very large field of electrophotographic reproduction. The phenomena related to photoconductivity have also played a large part in the understanding of electronic behavior and crystalline imperfections in a variety of different materials. *See* OPTICAL DETECTORS; OPTICAL MODULATORS; TELEVISION CAMERA TUBE.

Since the electrical conductivity σ of a material is given by the product of the carrier density n, its charge q, and its mobility μ [Eq. (1)], an increase in

$$\sigma = nq\mu \tag{1}$$

the conductivity $\Delta\sigma$ can be formally due to either an increase in n, Δn, or an increase in μ, $\Delta\mu$. Although cases are found in which both types of effects are observable, photoconductivity ($\Delta\sigma$) in single-crystal materials is due primarily to Δn, with only small effects at low temperatures due to $\Delta\mu$ if photoexcitation decreases the density of charged impurities that scatter charge carriers. In polycrystalline materials, on the other hand, where transport may be limited by potential barriers between the crystalline grains, an increase in mobility $\Delta\mu$ due to photoexcitation effects on these intergrain barriers may dominate the photoconductivity.

The increase in carrier density Δn can be conveniently related to the photoexcitation density f (excitations per unit volume per second) by the simple relation (2), where τ is the lifetime of the

$$\Delta n = f\tau \qquad (2)$$

photoexcited carrier, that is, the length of time that this carrier stays free and able to contribute to the conductivity before it loses energy and returns to its initial state via recombination with another carrier of opposite type (that is, electrons with holes, or holes with electrons). In Eq. (2) the photoexcitation term f includes all the processes of optical absorption (excitation across the bandgap of the material, excitation from or to imperfection states in the material, generation of excitons that are thermally dissociated to form the free carriers), and the lifetime τ includes all the processes of recombination (free electron with free hole, free electron with trapped hole, free hole with trapped electron). An understanding of the detailed processes of photoconductivity therefore requires a comprehensive understanding of the variation of optical absorption with photon energy, and of the dependence of recombination on imperfection density, capture cross section, photoexcitation intensity, and temperature. *See* BAND THEORY OF SOLIDS; HOLES IN SOLIDS.

Photosensitivity. Although all insulators and semiconductors may be said to be photoconductive, that is, they show some increase in electrical conductivity when illuminated by light of sufficiently high energy to create free carriers, only a few materials show a large enough change, that is, show a large enough photosensitivity, to be practically useful in applications of photoconductors. There are several ways that the magnitude of the photosensitivity can be defined, depending on the application in mind.

Lifetime-mobility product. Comparison of Eqs. (1) and (2) shows that the basic measure of material photosensitivity is given by the $\tau\mu$ product in the common case where $\Delta\sigma$ results primarily from Δn. The mobility does vary from material to material, but in most practical photoconductors μ has values between 10^2 and 10^4 cm²/V-s at room temperature, and of course the choice of a particular material is usually dominated by its desirable optical absorption characteristics. The free carrier lifetime τ, on the other hand, can take on a wide range of values from 10^{-9} to 10^{-2} s, depending on the particular density and properties of imperfections present in the material.

Detectivity. One of the major applications for photoconductors has been in the detection of small signals in the infrared portion of the spectrum, where the principal objective is to be able to detect the smallest signal possible with the detecting sys-

tem. Since in this case the photoconductivity is usually much smaller than the dark conductivity, an ac technique is used in which the light signal is chopped and ac amplification stages are used. The limit to detectability is reached when the light-generated signal is comparable to the electrical noise in the photoconductor. Thus the photosensitivity in this particular case is often defined as a detectivity, which is a normalized radiation power required to give a signal equal to the noise.

Gain. A third device-oriented definition of photosensitivity is that of photoconductivity gain. The gain is defined as the number of charge carriers that circulate through the circuit involving the photoconductor for each charge carrier generated by the light. The time required for a charge carrier to pass through the photoconductor from one electrode to the other, called the transit time, t_r, is given by Eq. (3), where L is the distance between elec-

$$t_r = L^2/\mu V \qquad (3)$$

trodes and V is the applied voltage. The gain is given then by τ/t_r for each type of possible charge carrier, giving Eq. (4) for the total gain if both elec-

$$\text{Gain} = (\tau_e\mu_e + \tau_h\mu_h)V/L^2 \qquad (4)$$

trons and holes contribute. Gains of hundreds or thousands can be readily achieved if the lifetimes are sufficiently long. Gains greater than unity require electrical contacts to the photoconductor that are able to replenish charge carriers that pass out of the opposite contact in order to maintain charge neutrality; such contacts are called ohmic contacts. If nonohmic contacts are used, so that charge carriers cannot be replenished, the maximum gain is simply unity, since only the initially created charge carrier contributes to the current flow. Historically, unity-gain currents of this latter type have been called primary photocurrents, whereas high-gain currents described by Eq. (4) have been called secondary photocurrents.

Spectral response. The variation of photoconductivity with photon energy is called the spectral resonse of the photoconductor. The spectral response of 10 typical photoconductors is shown in the illustration. These curves typically show a fairly well-defined maximum at a photon energy close to that of the bandgap of the material, that is, the minimum energy required to excite an electron from a bond in the material into a higher-lying conduction band where it is free to contribute to the conductivity. This energy ranges from 3.7 eV, in the ultraviolet, for zinc sulfide (ZnS) to 0.2 eV, in the infrared, for cooled lead selenide (PbSe). Photoconductivity associated with excitation across the bandgap of the material is called intrinsic photoconductivity. For photon energies smaller than the bandgap, the light is not strongly absorbed by the material, and the photoconductivity decreases. For photon energies larger than the bandgap, the optical absorption is large, and absorption takes place close to the surface of the material; since the surface has in general more imperfections than the bulk, the carrier lifetime at the surface is generally smaller, and hence the photoconductivity decreases. If the bulk of the material contains a sufficiently high density of imperfections contributing localized levels within the bandgap of the material, it is often possible to detect photoconductivity corre-

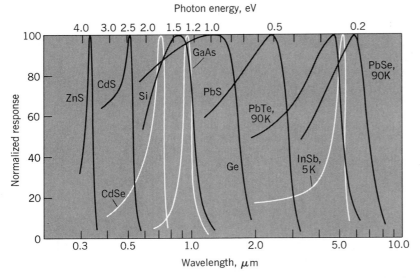

Spectral response of photoconductivity for 10 common photoconducting materials (*From R. H. Bube, Photoconductivity of Solids, John Wiley and Sons, Inc., 1960; reprint, Krieger, 1978*)

sponding to optical excitation from an occupied imperfection level to the conduction band or to an unoccupied imperfection level from the valence band of the material. This photoconductivity occurs for photon energies smaller than the bandgap, and is called extrinsic photoconductivity.

Speed of response. A third major characteristic of a photoconductor of practical concern is the rate at which the conductivity changes with changes in photoexcitation intensity. If a steady photoexcitation is turned off at some time, for example, the length of time required for the current to decrease to $1/e$ of its initial value is called the decay time of photoconductivity, t_d. The magnitude of the decay time is determined by the lifetime τ and by the density of carriers trapped in imperfections as a result of the previous photoexcitation, which must now also be released in order to return to the thermal equilibrium situation. If the photoexcitation intensity is high, or the density of imperfections is small, the decay time t_d approaches the lifetime τ as a minimum limiting value. For low light intensities or high imperfection densities, where the density of trapped carriers is much larger than the density of free carriers, the decay of photoconductivity is controlled not by the free carrier recombination rate but by the rate of thermal freeing of trapped carriers, and can be many orders of magnitude larger than the lifetime.

Device forms. Photoconductive detectors are made as single-crystal or polycrystalline film homogeneous materials, or as *pn* or *npn* semiconductor junctions. The behavior of a *pn* junction is similar to that of a homogeneous material with nonohmic contacts; the maximum gain is unity. The behavior of an *npn* junction is similar to that of a homogeneous material with ohmic contacts; a photoexcited hole remaining in the *p*-type base causes injection of electrons which traverse the device until the hole diffuses out of the base. Thus the gain for an *npn* junction can be much larger than unity. The advantage of the junctions is that

they can be made with standard silicon semiconductor technology. *See* JUNCTION DIODE.

Polycrystalline film photodetectors can be made from a variety of methods involving vacuum evaporation, powder sintering, and chemical solution deposition. The photoconductive behavior can be dominated by quite different effects in different materials systems. For example, the photoconductivity in cadmium sulfide (CdS) films deposited by spray pyrolysis is usually controlled by the modulation of intergrain barriers by photoexcitation, so that $\Delta\mu \gg \Delta n$. On the other hand, standard infrared detecting films of lead sulfide (PbS), deposited from chemical solution, exhibit a photoconductivity for which $\Delta n \gg \Delta\mu$, even though the effects of intergrain barriers are clearly measurable in the mobility.

The television camera vidicon and electrophotography are two applications of photoconductivity in which the device form is dictated by the specific nature of the information processing system involved. In both cases an electrical charge is deposited on one side of a high-resistivity photoconducting material; subsequent illumination of the material increases the conductivity locally and allows the charge to leak off through the material. The local absences of charge are then detected and used to produce or reproduce the original light pattern. The material involved must have special characteristics: it must have a high enough dark resistivity so that the deposited charge does not leak off by ordinary dark conduction, and a high enough photosensitivity so that the charge will leak off as quickly as desired. In the vidicon the charge is deposited by scanning by an electron beam; absence of charge is detected by current flow in a subsequent scanning by the beam. In electrophotography the charge is deposited by a corona discharge; absence or presence of charge is fixed by a subsequent printing process.

[RICHARD H. BUBE]

Bibliography: R. H. Bube, Photoconductivity of semiconductors, in H. Eyring, D. Henderson, and W. Jost (eds.), *Physical Chemistry*, vol. 10: *Solid State*, pp. 515–578, 1970; R. H. Bube, *Photoconductivity of Solids*, 1960, reprint 1978; J. Mort and D. M. Pai (eds.), *Photoconductivity and Related Phenomena*, 1976; A. Rose, *Concepts in Photoconductivity and Allied Problems*, 1963.

Photodiode

A semiconductor two-terminal component with electrical characteristics that are light-sensitive. All semiconductor diodes are light-sensitive to some degree, unless enclosed in opaque packages, but only those designed specifically to enhance the light sensitivity are called photodiodes.

Most photodiodes consist of semiconductor *pn* junctions housed in a container designed to collect and focus the ambient light close to the junction. They are normally biased in the reverse, or blocking, direction; the current therefore is quite small in the dark. When they are illuminated, the current is proportional to the amount of light falling on the photodiode. For a discussion of the properties of *pn* junctions *see* JUNCTION DIODE.

Photodiodes are used both to detect the presence of light and to measure light intensity. *See* PHOTOELECTRIC DEVICES.   [W. R. SITTNER]

Photoelectric devices

Devices which give an electrical signal in response to visible, infrared, or ultraviolet radiation. They are often used in systems which sense objects or encoded data by a change in transmitted or reflected light. Photoelectric devices which generate a voltage can be used as solar cells to produce useful electric power. The operation of photoelectric devices is based on any of the several photoelectric effects in which the absorption of light quanta liberates electrons in or from the absorbing material. *See* PHOTOVOLTAIC CELL; SOLAR CELL.

Photoconductive devices are photoelectric devices which utilize the photoinduced change in electrical conductivity to provide an electrical signal. Thin-film devices made from cadmium sulfide, cadmium selenide, lead sulfide, or similar materials have been utilized in this application. Single-crystal semiconductors such as indium antimonide or doped germanium are used as photoelectric devices for the infrared spectrum. The operation of these devices requires the application of an external voltage or current bias of relatively low magnitude. *See* PHOTOCONDUCTIVE CELL.

Photoemissive systems have also been used in photoelectric applications. These vacuum-tube devices utilize the photoemission of electrons from a photocathode and collection at an anode. Photoemissive devices require the use of a relatively large bias voltage. *See* PHOTOEMISSION.

Many photoelectric systems now utilize silicon photodiodes or phototransistors. These devices utilize the photovoltaic effect, which generates a voltage due to the photoabsorption of light quanta near a *pn* junction. Modern solid-state integrated-circuit fabrication techniques can be used to create arrays of photodiodes which can be used to read printed information. *See* PHOTODIODE; PHOTOTRANSISTOR.

Photoelectric devices can be used in systems which read coded or printed information on data cards and packages. Similar systems are used to sense and control the movement of objects in perimeter guard systems which sense an intruder by the interruption of a light beam.

Most visible and ultraviolet photoelectric devices operate at room temperature. Photoelectric devices for use in the infrared must be cooled with the longer-wavelength response devices requiring the most cooling. *See* PHOTOELECTRICITY.

[RICHARD A. CHAPMAN]

Photoelectricity

The process by which electromagnetic radiation incident on a solid, liquid, or gas liberates electrical charge, which is detectable in an electric field. The process is strictly quantum in nature.

Historically, this phenomenon was first explained by Albert Einstein, who invoked the quantum character of electromagnetic radiation, namely the photon, with characteristic energy $h\nu$, where h is Planck's constant and ν the frequency. The liberation of electrons from matter is governed by Einstein's photoelectric equation $E = h\nu - \phi$. where E is the kinetic energy of the emitted electron and ϕ represents the binding energy of the electron.

Early photoelectric experiments involved the photoemission of electrons from the surface of metals into a vacuum, or the liberation of electrons and positively charged ions in a gas by the process of photoionization. Values of ϕ for these processes typically are a few electronvolts, and as such these experiments were limited primarily to the visible and ultraviolet regions of the spectrum. The common photomultiplier tube is an example of a photoemissive device.

The quantum nature of electromagnetic radiation also manifests itself in the liberation of electrons and positive holes within the interior of a solid, giving rise to photoconductive and photovoltaic effects. These phenomena are readily observable in the class of solids known as semiconductors. The binding energy ϕ of electrons and holes in semiconductors can be as little as a few millielectronvolts, and as such the region of the spectrum involved transcends both the visible and the infrared, out into the far infrared. Devices based upon these effects are in widespread use in the fields of thermal imaging, solar energy conversion, security monitoring, data readout, and product control. *See* PHOTOCONDUCTIVITY; PHOTOELECTRIC DEVICES; PHOTOEMISSION; PHOTOVOLTAIC EFFECT. [MICHAEL A. KINCH]

Photoemission

The ejection of electrons from a solid (or less commonly, a liquid) by incident electromagnetic radiation. Photoemission is also called the external photoelectric effect. The visible and ultraviolet regions of the electromagnetic spectrum are most often involved, although the infrared and x-ray regions are also of interest. For important practical applications of photoemission *see* PHOTOTUBE; TELEVISION CAMERA TUBE.

The salient experimental features of photoemission are the following: (1) There is no detectable time lag between irradiation of an emitter and the ejection of photoelectrons. (2) At a given frequency the number of photoelectrons ejected per second is proportional to the intensity of the incident radiation. (3) The photoelectrons have kinetic energies ranging from zero up to a well-defined maximum, which is proportional to the frequency of the incident radiation and independent of the intensity.

Einstein photoelectric law. These characteristics cannot be explained by J. C. Maxwell's theory of electromagnetic waves. In 1905 Albert Einstein made the clarifying assumption that the radiation had characteristics like those of particles when it delivered energy to electrons in the emitter. In Einstein's approach the light beam behaves like a stream of photons, each of energy $h\nu$, where h is Planck's constant, and ν is the frequency of the photon (Fig. 1). The energy required to eject an

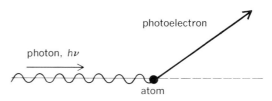

Fig. 1. External photoelectric effect.

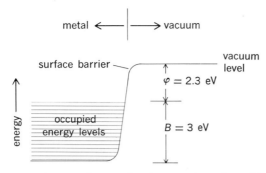

metal ⟵ | ⟶ vacuum

Fig. 2. Energy diagram for electrons in sodium. The photoelectric threshold energy is φ; in a metal φ is equal to the electronic work function. The band of energy levels occupied by almost free electrons has a width B.

electron from the emitter has a well-defined minimum value φ called the photoelectric threshold energy. When a photon interacts with an electron, the latter absorbs the entire photon energy.

For $h\nu$ values below the threshold, photoelectrons are not ejected. Even though the electrons absorb photon energy, they do not receive enough to surmount the potential barrier at the surface, which normally holds the electrons in the solid. The threshold energy φ is associated with a threshold frequency φ/h and a threshold wavelength ch/φ, where c is the velocity of light. For photon energies above φ, the kinetic energies of photoelectrons range from zero up to a maximum value, $E = h\nu - \varphi$. This is the Einstein photoelectric law, and E is commonly termed the Einstein maximum energy. Careful photoelectric experiments by R. A. Millikan in 1916 fixed h in Einstein's law with considerable precision and furthered its identification with the constant which M. Planck had used in his theory of blackbody radiation. For a discussion of the surface potential barrier *see* SCHOTTKY EFFECT.

Metals. The Einstein law is based only on the photon hypothesis and on the conservation of energy. It does not take into account momentum, which must also be conserved. The incident photon has a momentum $h\nu/c$ which is negligible compared to the change in momentum of the electron when it gains the energy $h\nu$. Thus, it is not possible for a free electron to absorb the entire energy of a photon. In order for this to happen the electron must be bound to another body, which takes up the recoil momentum.

Figure 2 shows an energy diagram of the electrons in the metal sodium. There is a potential barrier at the surface, which the electrons must surmount before they can escape. The most easily ejected electrons must acquire 2.3 electronvolts (eV) of additional energy from photons in order to do this. This 2.3 eV is the electronic work function, which for a metal is equal to the photoelectric threshold energy. Inside the metal the electrons occupy a band of energies about 3 eV wide. These electrons are said to be quasi-free. This means that they behave in many ways like a gas of free, noninteracting electrons; nevertheless they move in the periodic potential due to the positive sodium ions, and in this sense they are bound.

In this situation two types of photoemission are theoretically possible, the surface effect and the

volume effect. In the surface effect, recoil momentum is communicated to the crystal because the electron is coupled to the barrier at the surface during photon absorption. In the volume effect the electron is coupled to the internal periodic potential.

Experimental determination of the relative importance of surface and volume photoeffects in metals is difficult. Experiments by H. Mayer and his collaborators indicate that for potassium the volume effect is predominant for photon energies from the threshold value at 2.1 eV to at least 4 eV. Thus the photoelectric emission increases as the thickness of a potassium film increases. (This would not be true for the surface effect.) Photoelectrons can escape from depths greater than 10^{-6} cm when excited by light in the threshold region.

Thus far the photoelectric threshold has been treated as a sharply defined quantity. This is precisely true for metals only at temperatures near absolute zero. At higher temperatures the upper edge of the band of occupied electron energy states in Fig. 2 is no longer sharp. It becomes diffuse because of thermal agitation. Electrons may be then emitted for photon energies less than the threshold value φ. At ordinary room temperatures, for example, measurable photoemission appears for photon energies as much as 0.2 eV below the threshold. R. Fowler has developed a convenient graphical technique, known as a Fowler plot, for determining the absolute-zero threshold from data taken at higher temperatures on the spectral dependence of the photoelectric yield, which is the number of photoelectrons ejected per incident photon. L. A. DuBridge has developed a similar technique using either the temperature dependence of the photoelectric yield or the distribution of photoelectrons in energy at fixed frequency. These treatments show that the photoelectric yield is approximately proportional to the quantity $(h\nu - \varphi)^2$ when the photon energy $h\nu$ is within about 1 eV of the threshold energy φ. Figure 3 shows a graph of the spectral dependence of pho-

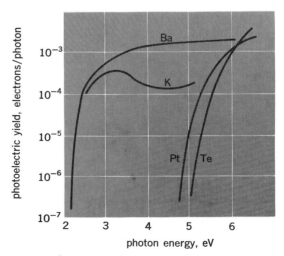

Fig. 3. Spectral distribution of the photoelectric yield from typical samples of barium, Ba; potassium, K; platinum, Pt; and tellurium, Te. Platinum and tellurium have practically the same electronic work function. Note the higher threshold and more steeply rising curve for tellurium, which is a typical elemental semiconductor.

toelectric yield for some typical emitters. Figure 4 shows typical energy distributions. Photoelectric yields from metals are of the order of 10^{-3} electron per incident photon when $h\nu - \varphi$ is 1 eV. Photoelectric threshold energies range from 2 eV for cesium to values such as 5 eV for platinum. They vary for different types of crystal faces on the same crystal, and they are exceedingly sensitive to small traces of adsorbed gases.

Semiconductors. The photoelectric behavior of semiconductors, such as germanium or tellurium, differs from that of metals. As shown in Fig. 5, the electrons in a semiconducting emitter completely occupy a closed band of energies, which lies just below a so-called forbidden energy band. The electrons behave quite differently from those in metals. As a result, the photoelectric threshold energy φ is larger than the electronic work function W. Thus, a semiconductor exhibits a higher photoelectric threshold energy than a metal having the same work function. An example of this is shown for the metal platinum and the semiconductor tellurium in Fig. 3. Both this particular platinum sample (Pt) and the tellurium (Te) have the same electronic work function, about 4.8 eV. The photoelectric threshold of the platinum is equal to the work function, whereas that for the tellurium is clearly higher. Spectral and energy distributions are shown in Figs. 3 and 4. Clean surfaces of silicon, germanium, and certain semiconducting chemical compounds have been made by cleaving single crystals in ultra-high vacuum. Both the surface photoelectric effect and the volume effect have been measured. From the measurements of the volume effect, valuable information on the detailed nature of the electron energy bands has been deduced from structure that occurs in photoelectron energy distributions. *See* BAND THEORY OF SOLIDS.

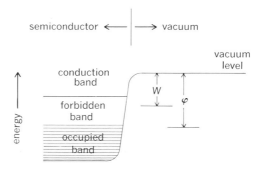

Fig. 5. Energy diagram for electrons in a semiconductor. Occupied band is filled with bound electrons that behave differently from the electrons in metals. As a result, the electronic work function W is smaller than the photoelectric threshold energy φ.

A particularly interesting and important kind of photoemitter is typified by cesium antimonide, Cs_3Sb. This material is a semiconductor having a forbidden energy band about 1.5 eV wide. The photoelectric threshold energy is only slightly higher than this. Electrons excited from the occupied energy band by incident photons cannot assume energies lying in the forbidden band. They must remain in the conduction band shown in Fig. 5. Thus even the slowest ones must retain energies only slightly less than that required for escape. The probability of photoemission is higher than for metals (or for semiconductors that have threshold energies greater than twice the width of the forbidden energy band). Cs_3Sb is sensitive over much of the visible range and can give very high yields, in excess of 0.2 electron per incident photon. It is widely used in practical phototubes. Related compounds can be made with enhanced photoelectric response in the red or ultraviolet regions of the spectrum.

Alkali halides. Three basically different kinds of photoemission are possible for alkali halides: intrinsic, extrinsic, and exciton-induced photoemission.

Intrinsic photoemission. This is characteristic of the ideally pure and perfect crystal. It is thus analogous to the emission already described for metals and semiconductors. It appears only for photon energies higher than the intrinsic threshold. For example, potassium iodide, KI, is an alkali halide having this intrinsic threshold in the far ultraviolet near 7 eV. Apparently the width of the forbidden electron energy band in KI is only about 1 eV less than this. For the same reason that was mentioned for the semiconductor Cs_3Sb, the photoelectric yields are high, in excess of 0.1 electron per incident photon, as shown by section C of the curve in Fig. 6.

Extrinsic photoemission. A second kind of emission occurs when a KI crystal contains imperfections in the form of negative ion vacancies (lattice sites from which negative iodine ions are missing). These vacancies can be filled by electrons. Color centers, which absorb visible light, are formed. They may reach concentrations as high as 10^{20} per cm^3. External photoelectrons may be ejected directly from these centers by photons. It is termed an extrinsic process since the light is absorbed by a crystal defect; it is also called direct ionization.

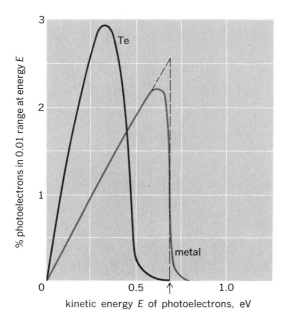

Fig. 4. Energy distributions of photoelectrons from tellurium and a metal having the same work function. Solid line for the metal shows results for room temperature, and dashed line for absolute zero. Arrow marks Einstein maximum energy; photon energy is 5.42 eV.

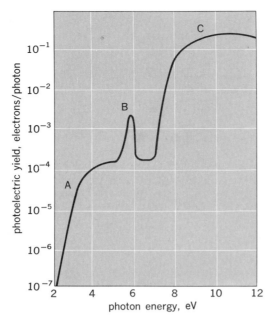

Fig. 6. Spectral distribution of photoelectric yield from potassium iodide containing color centers. Region *A* of the curve is due to direct ejection of photoelectrons from color centers; the peak *B* is due to exciton-induced emission; *C* is due to intrinsic emission.

The threshold energy for this process is about 2.5 eV. The yields can reach values of the order of 10^{-4} electron per incident photon, as shown in section A of the curve in Fig. 6. The exact value of the yield depends on the concentration of color centers. Most of the incident radiation is lost because it is not intercepted by the centers, which present a limited cross-section to the incident photons.

Exciton-induced photoemission. When color centers are present, another photoelectric process takes place in two stages. Potassium iodide has a sharp optical absorption band, peaking at a photon energy of 5.6 eV. This is the first fundamental or intrinsic optical absorption band. Energy absorbed in this peak does not release free electric charges in the crystal. Rather, it leads to a kind of nonconducting excited state called an exciton state. The exciton can transfer enough energy to color centers to eject photoelectrons from the crystal. This two-stage process is termed exciton-induced photoemission. It appears in the peak B on the curve in Fig. 6. It is more efficient than direct ejection of photoelectrons from color centers. The entire crystal is capable of the primary photon absorption, and the energy can be transferred rather efficiently to color centers. Thus the process avoids much of the loss in incident energy that arises from the limited cross section of color centers when they absorb photons directly.

Other compounds. Other ionic crystals, such as barium oxide, behave much like the alkali halides. Direct ejection of photoelectrons from chemical impurities and from energy levels or defects localized at the crystal surface can be important. In addition to these extrinsic processes, exciton-induced emission and intrinsic photoemission both occur.

Compounds such as zinc sulfide behave some-

what like germanium but have higher intrinsic threshold energies, of the order of 7 eV. The photoelectric yields are comparatively low, as for germanium. Extrinsic processes, such as direct ejection of electrons from chemical impurities (or defects), are sometimes detectable but are usually weak.

Certain complex photoemitters are made by letting cesium react with silver oxide to form cesium oxide and silver. They are valuable because they have threshold energies below 1 eV, and thus they are sensitive in the infrared. The photoelectrons appear to be directly ejected either from cesium adsorbed on the oxide surface or from discrete energy levels in the cesium oxide. The yields are about 10^{-3} electron per incident photon. Intrinsic emission from cesium oxide (with yields above 0.01) occurs for photon energies above the intrinsic threshold at about 4 eV.   [L. APKER]

Bibliography: L. Azaroff and J. J. Brophy, *Electronic Processes in Materials*, 1963; M. Cardona and L. Ley (eds.), *Photoemission in Solids*, 2 vols., 1978, 1979; A. Sommer, *Photoemissive Materials*, 1968, reprint 1980; A. Van der Ziel, *Solid State Physical Electronics*, 3d ed., 1976.

Photoferroelectric imaging

The process of storing an image in a ferroelectric material by utilizing either the intrinsic or extrinsic photosensitivity in conjunction with the ferroelectric properties of the material. Specifically, photoferroelectric (PFE) imaging is a process of storing photographic images or other optical information in transparent lead lanthanum zirconate titanate (PLZT) ceramics. The PFE imaging device consists simply of a thin flat plate (0.2–0.3 mm thick) of optically polished PLZT ceramic with transparent conductive indium–tin oxide (ITO) electrodes sputter-deposited on the two major surfaces. The

Fig. 1. Photographic image with high resolution and gray scale stored in a PFE imaging device.

image to be stored is exposed onto one of the ITO electroded surfaces by using near-ultraviolet illumination in the intrinsic photosensitivity region (corresponding to a bandgap energy of approximately 3.35 eV) of the PLZT. Simultaneously, a voltage pulse is applied across the electrodes to switch the ferroelectric polarization from one stable remanent state to another. Images are stored both as spatial distributions of light-scattering centers in the bulk of the PLZT and as surface deformation strains which form a relief pattern of the image on the exposed surface. Both the light scattering and surface strains are related to spatial distributions of ferroelectric domain orientations introduced during the image-storage process. These spatial distributions correspond to brightness variations in the image to which the PLZT is exposed.

The stored image may be viewed directly or it may be projected onto a screen by using either transmitted or reflected light. For projection, the image contrast is usually improved by using collimated light and a Schlieren optical system.

Either total or spatially selective erasure of stored images is accomplished by uniformly illuminating the area to be erased with near-ultraviolet light and simultaneously applying a voltage pulse to the electrodes to switch the ferroelectric polarization back to its initial remanent state, that is, the polarization state prior to image storage.

Important potential applications of PFE imaging are temporary image storage and display. Various types of image processing, including image contrast enhancement, are also offered by the capability of switching from a positive to a negative stored image in discrete steps. Nonvolatile photographic images with gray scale ranges extending from an optical density of about 0.15 to more than 2.0 and with resolutions as high as 40 line pairs per millimeter are routinely stored in PLZT ceramic plates by using the intrinsic PFE effect (Fig. 1).

Intrinsic PFE effect. The intrinsic PFE effect is characterized by a photoinduced change in the ferroelectric coercive field E_c produced by irradiating the PLZT surface with bandgap or higher-energy light. The near-ultraviolet irradiation photoexcites carriers into the conduction state. These carriers then diffuse (with no field applied) or drift under the influence of an applied field to deep trapping sites beyond the near-ultraviolet absorption depth. Retrapped carriers establish a space charge field E_{sc} which aids the domain switching process and effectively decreases the coercive field E_c, thus altering the domain pattern in regions where the applied image is bright. Photoexcited carriers which remain in the conduction state contribute to a steady-state photovoltaic current which is driven by the anomalous bulk photovoltaic effect. The photovoltaic current assists in domain nucleation, which also aids the domain switching process. The most important mechanism contributing to the PFE imaging capabilities of PLZT is the photoinduced space charge field E_{sc}.

Photosensitivity enhancement. The exposure energy W_{ex} (product of near-ultraviolet light intensity I and image storage time t) required to store high-quality images in unmodified PLZT is from 100 to 500 mJ/cm². This relatively high value of W_{ex} severely limits the scope of practical applications of PFE imaging devices. Significant reduc-

Fig. 2. Resolution chart stored in a PLZT plate. The lower half of the plate was implanted with 2×10^{16} —200-keV protons (H⁺ ions)/cm²; the upper half was unmodified. The image was exposed on both upper and lower halves of the plate at approximately 20 mJ/cm².

tions of the W_{ex} required for high-quality image storage have been obtained by implanting hydrogen, helium, or argon ions into the image-storage surface of PLZT plates.

A vivid illustration of the image-storage sensitivity enhancement achieved by proton implantation is presented in Fig. 2. Only the lower half of the ceramic plate was implanted with protons. Both halves of the plate were exposed identically during image storage. A high-quality image is stored in the implanted region of the plate at exposure levels insufficient to store a detectable image in the unimplanted region.

An indication of the relative photosensitivities of unimplanted and hydrogen-, helium-, and argon-implanted PLZT is shown in Fig. 3, in which the effective coercive field E_c is plotted as a function of near-ultraviolet light intensity I. The effective coercive field E_c was measured from the ferroelectric polarization P versus applied electric field E hysteresis loop at $P = 0$. Each measurement involved switching P through the complete loop while exposing the ion-implanted surface to spatially uniform near-ultraviolet light. The effective coercive field E_c is obtained by dividing the applied voltage at $P = 0$ by the PLZT plate thickness. The values of E_c plotted on the $I = 10^{-3}$ mW/cm² line actually correspond to $I = 0$. A deviation from the value of E_c at $I = 0$ is taken as the initial response to the near-ultraviolet illumination. The curves indicate that the onset of photosensitivity for unimplanted PLZT is about 10 mW/cm²; for hydrogen-implanted PLZT, about 1 mW/cm²; for helium-implanted PLZT, about 100 μW/cm²; and for argon-implanted PLZT, about 10 μW/cm². Image-storage thresholds have been determined from

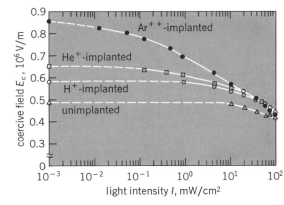

Fig. 3. Photosensitivity of PLZT in terms of reduction of coercive field E_c produced by near-ultraviolet irradiation while ferroelectric polarization P is switched through a complete P versus ferroelectric field E hysteresis loop.

other measurements to be: unimplanted PLZT, about 85 mJ/cm²; hydrogen-implanted PLZT, about 10 mJ/cm²; helium-implanted PLZT, about 3 mJ/cm²; and argon-implanted PLZT, about 500 μJ/cm². Ion implantation can thus be used to increase the photosensitivity by up to three orders of magnitude, which yields sensitivities approaching those of fine-grained holographic film. *See* ELECTRONIC DISPLAY; FERROELECTRICS; PHOTOCONDUCTIVITY.

[CECIL E. LAND]

Bibliography: C. E. Land, Optical information storage and spatial light modulation in PLZT ceramics, *Opt. Eng.*, 17:317–326, 1978; C. E. Land and P. S. Peercy, Photoferroelectric effects in PLZT ceramics, *Ferroelectrics*, 22:677–679, 1978; C. E. Land, P. D. Thacher, and G. H. Haertling, in R. Wolfe (ed.), *Applied Solid State Science*, vol. 4: *Electrooptic Ceramics*, pp. 137–233, 1974; P. S. Peercy and C. E. Land, A model for ion implantation induced improvements of photoferroelectric imaging in lead lanthanum zirconate titanate ceramics, *Appl. Phys. Lett.*, 37:815–818, 1980.

Photomultiplier

A very sensitive vacuum-tube detector of light or radiant flux containing a photocathode which converts the light to photoelectrons; one or more secondary-electron-emitting electrodes or dynodes which amplify the number of photoelectrons; and an output electrode or anode which collects the secondary electrons and provides the electrical output signal. Because of the very large amplification provided by the secondary-emission mechanism, and the very short time variation associated with the passage of the electrons within the device, the photomultiplier is applied to the detection and measurement of very low light levels, especially if very high speed of response is required.

The first photomultipliers were developed during the late 1930s; applications were generally in astronomy and spectrometry. Development of photomultipliers was stimulated beginning in the late 1940s by their application to scintillation counting. Variations in electron-optical design and the introduction of a variety of photocathode, dynode, and envelope materials led to the use of photomultipliers in a wide variety of applications. Active design programs continue despite the competition

from various solid-state detectors. Photomultiplier tubes are available in sizes ranging from ½ in. (13 mm) in diameter to 5 in. (127 mm), with photocathodes useful for detecting radiation having wavelengths from 110 nm, the deep ultraviolet (using a lithium fluoride window), to 1100 nm, the infrared (with special photocathodes), and with amplification factors ranging from 10^3 to 10^9.

Operation and design. Figure 1 is a schematic of a typical photomultiplier and illustrates its operation. Light incident on a semitransparent photocathode located inside an evacuated envelope causes photoelectron emission from the opposite side of the photocathode. (Some photomultipliers are designed with the photocathode mounted inside the vacuum envelope to provide for photoemission from the same side of the photocathode on which the light is incident.) The efficiency of the photoemission process is called the quantum efficiency, and is the ratio of emitted photoelectrons to incident photons (light particles). Photoelectrons are directed by an accelerating electric field to the first dynode, where from 3 to 30 secondary electrons are emitted for each incident electron, depending upon the dynode material and the applied voltage. These secondaries are directed to the second dynode, where the process is repeated and so on until the multiplied electrons from the last dynode are collected by the anode.

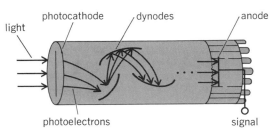

Fig. 1. Schematic of a photomultiplier. (*From P. W. Engstrom, Photomultipliers—then and now, RCA Eng., 24(1): 18–26, June–July 1978*)

A typical photomultiplier may have 10 stages of secondary emission and may be operated with an overall applied voltage of 2000 V. In most photomultipliers the focusing of the electron streams is done by electrostatic fields shaped by the design of the electrodes. Some special photomultipliers designed for very high speed utilize crossed electrostatic and magnetic fields which direct the electrons in approximate cycloidal paths between electrodes. Because the transit times between stages are nearly the same, anode pulse rise times as short as 150 picoseconds are achieved.

Another special photomultiplier design is based on the use of microchannel plates. A single-channel multiplier is schematically shown in Fig. 2. The channel is coated on the inside with a resistive secondary-emitting layer. Gain is achieved by multiple electron impacts on the inner surface as the electrons are directed down the channel by an applied voltage over the length of the channel. The gain of the channel multiplier depends upon the ratio of its length l to its diameter d; a typical ratio is 50:1. A microchannel plate is formed by combining a large number of the channels in parallel with

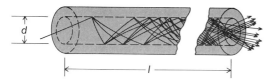

Fig. 2. Continuous-channel multiplier structure. (*From RCA Photomultiplier Manual, Tech. Ser. PT-61, 1970*)

spacings on the order of 40 μm. A very high-speed photomultiplier utilizes a microchannel plate mounted in a closely spaced parallel arrangement between a photocathode and an anode. Time resolutions of less than 100 ps are achieved.

Dynode materials. Common dynode materials used in photomultiplier tubes are cesium antimonide (Cs_3Sb), magnesium oxide, and beryllium oxide. The surface of the metal-oxide dynodes is modified by activation with an alkali metal such as cesium which lowers the surface potential barrier, permitting a more efficient escape of the secondary electrons into the vacuum.

Materials characterized as negative-electron affinity (NEA) have been developed and utilized in photomultipliers for both dynodes and photocathodes. A typical NEA secondary-emission material is gallium phosphide (GaP) whose surface has been treated with cesium (GaP:Cs). Figure 3 shows the energy band model of an NEA material. The surface barrier is reduced by the electropositive cesium layer, and band-bending occurs so that the conduction band may actually lie above the vacuum level as indicated.

When a primary electron impacts an NEA material, secondary electrons are created within the material, and even low-energy electrons escape to the vacuum. The result is a much higher secondary-emission ratio than that achieved in other secondary emitters, especially at higher primary energies when the primary electrons penetrate more deeply into the material. Secondary electrons within an emitter lose energy before reaching the surface, so that in ordinary materials the secondary emission actually decreases with increase in voltage above an energy of perhaps 500 eV. For example, Cs_3Sb reaches a maximum secondary emission of about 8:1 at 500 V. At the same primary voltage GaP:Cs has a secondary emission of 25:1, and secondary emission increases linearly with voltage to a ratio of at least 130:1. *See* SECONDARY EMISSION; SEMICONDUCTOR.

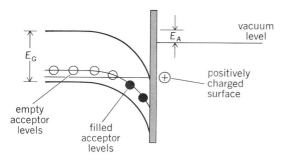

Fig. 3. Semiconductor energy-band model showing negative electron affinity, E_A. Band-gap energy $= E_G$. (*From RCA Photomultiplier Manual, Tech. Ser. PT-61, 1970*)

Photocathode materials. The first photomultipliers developed utilized a photocathode of silver oxide activated with cesium (Ag-O-Cs). Although the quantum efficiency of this photocathode is quite low (less than 1% at the wavelength of its maximum response), it is still used because of its near-infrared response out to 1.1 μm.

Most photomultipliers today have photocathodes classed as alkali-antimonides: Cs_3Sb, K_2CsSb, Na_2KSb, $Na_2KSb:Cs$, Rb-Cs-Sb. These photocathodes have good quantum efficiency in the visible-wavelength range—in some cases exceeding 20%. The NEA type of photocathodes such as GaAs:Cs and InGaAs:Cs have even higher quantum efficiency and response through the visible and into the near infrared. However, they are difficult to fabricate and are more readily damaged by excessive photocurrents. *See* PHOTOTUBE.

Detection limits. Because there is very little noise associated with the amplification process in a photomultiplier, detection limits are primarily determined by the quantum efficiency of the photocathode, by the thermionic or dark emission of electrons from the photocathode, and by the bandwidth of the observation. A typical photomultiplier limited by dark emission is capable of detecting 10^{-12} lumen—the equivalent of the flux on 1 cm^2 from a 1-candlepower lamp at a distance of 6 mi (10 km). By selecting a photomultiplier having a photocathode with low dark emission, restricting the effective photocathode area magnetically, and cooling the tube to reduce dark emission further, a background count as low as one or two electrons emitted from the photocathode per second can be achieved. Against this background, with a photocathode having 20% quantum efficiency, the incidence of a flux of only 10 photons per second can be detected in an observation time of perhaps 10 s. In this technique, referred to as photon counting, each electron originating at the photocathode results in an output pulse which is counted. An evaluation must be made by statistically comparing the cases of source-on and source-off. Such techniques were applied successfully in 1969 in the laser-ranging experiments using a retroreflector placed on the Moon by the *Apollo 11* astronauts. *See* THERMIONIC EMISSION.

Applications. Ever since their invention, photomultipliers have been found useful in low-level photometry and spectrometry. The many applications of photomultipliers include high-speed inspection of small objects such as fruits, seeds, toys, and other industrial products; pollution monitoring; laser ranging; and process control with transmitted or reflected light to detect flaws in various solid, liquid, or gaseous manufacturing operations. *See* LASER.

Scintillation counting. The most important applications of photomultipliers are related to scintillation counting. In a scintillation counter, gamma rays produce light flashes in a material such as NaI:Tl. A photomultiplier is optically coupled to the scintillator and provides a count of the flashes and a measure of their magnitude. The magnitude of the scintillation flash is proportional to the energy of the gamma ray, thus enabling the identification of particular isotopes. A whole science of tracer chemistry has developed using this technique and is applied to agriculture, medicine, and industrial problems.

Gamma-ray camera. The gamma-ray camera developed in the late 1950s has proved to be a very valuable medical tool. In this application, a radioactive isotope combined in a suitable compound is injected into or fed to a patient. Certain compounds and elements concentrate preferentially in particular organs, glands, or tumors: for example ^{131}I concentrates in the thyroid gland. Gamma rays which are then emitted are detected by a large scintillator, and an array of photomultiplier tubes provides spatial data on the gamma-ray source. The technique is widely used in diagnostic medicine, especially in locating tumors.

Computerized tomography. The CT or CAT (computerized axial tomographic) scanner is a similar medical instrument developed during the 1970s. This instrument uses a pencil or fan-beam of x-rays that rotates around the patient. Several hundred photomultiplier-scintillator combinations surround the patient and record density from many different positions. These data are analyzed by a computer, thus providing a cross-section density map of the patient. The CT scanner is particularly useful because of the two-dimensional data provided, although it does not supply the functional information of the gamma-ray camera.

Positron camera. A similar development is the positron camera. This device uses radioisotopes, such as ^{11}C, ^{13}N, and ^{15}O, that have short half-lives and emit positrons. When a disintegration occurs, the positron is annihilated by an electron, and the result is a pair of oppositely directed gamma rays detected in coincidence with photomultiplier-scintillator detectors on opposite sides of the patient. Functional information is provided and analyzed in a manner similar to that of a CT scanner.

[RALPH W. ENGSTROM]

Bibliography: Detectors: Photo/optical, *IEEE Trans. Nucl. Sci.*, vol. NS-26, no. 1, pt. 1, pp. 338–421, 1979; *RCA Photomultiplier Handbook*, Tech. Ser. PMT-62, 1980; H. Rougeot and C. Baud, Negative electron affinity photoemitters, in L. Marton (ed.), *Advances in Electronics and Electron Physics*, vol. 48, pp. 1–36, 1979.

Phototransistor

A semiconductor device with electrical characteristics that are light-sensitive. Phototransistors differ from photodiodes in that the primary photoelectric current is multiplied internally in the device, thus increasing the sensitivity to light. For a discussion of this property *see* TRANSISTOR.

Some types of phototransistors are supplied with a third, or base, lead. This lead enables the phototransistor to be used as a switching, or bistable, device. The application of a small amount of light causes the device to switch from a low current to a high current condition. *See* PHOTOELECTRIC DEVICES. [W. R. SITTNER]

Phototube

An electron tube containing a photocathode from which electrons are emitted when it is exposed to light or other electromagnetic radiation. An elementary vacuum phototube consists of a photocathode, an anode or electron collector, and an evacuated envelope through which radiation is transmitted to the photocathode. A gas phototube contains, in addition, an inert gas which may be ionized by electron current from the photocathode. For a description of a phototube in which the electron current is amplified by means of a secondary-emission electron multiplier *see* PHOTOMULTIPLIER.

Phototubes serve as sensing elements in the detection and measurement of light and ultraviolet or infrared radiation. Phototubes also convert variations in intensity of incident radiation into corresponding variations in electron output current, as in light-controlled relay circuits, and in the conversion of sound modulation of photographic film into audio-frequency currents, as in the sound tracks on motion-picture film.

The fundamental characteristics of a phototube are its spectral sensitivity characteristic, or output current, expressed as a function of the wavelength of incident radiation at constant anode voltage, and its anode-current characteristics, which show the dependence of anode current on applied voltage and radiant flux input. Gas phototubes do not differ from vacuum phototubes with regard to spectral sensitivity characteristics, which are described below; however, the anode-current characteristics of gas and vacuum phototubes are essentially different.

Principles of operation. The anode current of a vacuum phototube is directly proportional to the intensity of radiation incident on its photocathode. The anode is normally connected to a positive potential of at least 20 volts relative to the photocathode in order that the anode current be limited by photoelectric emission rather than by space charge or electron emission velocity. *See* ELECTRON MOTION IN VACUUM.

In a gas phototube the photoelectric current is amplified by partial ionization of a gas contained in the tube at low pressure. An inert gas such as neon or argon is used because photocathodes react chemically with other gases. At low anode voltages the anode current of a gas phototube is emission-limited like that of a vacuum phototube. At anode potentials greater than 25 volts, electrons emitted from the photocathode acquire sufficient energy to ionize some of the gas atoms. The total current is then the sum of the free-electron current, the positive-ion current, and the current due to secondary electrons which are produced by ion bombardment of the photocathode. The ratio of anode current, at an anode voltage sufficient to cause ionization, to the emission-limited current measured at a lower voltage is the gas amplification factor of a gas phototube. This factor ranges between 3 and 10 in commercial gas phototubes. *See* ELECTRICAL CONDUCTION IN GASES.

Gas amplification increases with anode voltage and with the intensity of incident radiation. Because of their nonlinear characteristics gas phototubes seldom are used in photometric applications. Gas amplification increases with anode voltage up to breakdown voltage, which is that voltage at which the ion current becomes self-sustaining. A self-sustained glow discharge in a gas phototube causes sputtering of the photocathode and rapidly impairs its sensitivity.

The response of a gas phototube to rapid changes in light intensity decreases with increasing modulation frequency, particularly at frequencies above 10–15 kHz. Factors which govern the speed of response of a gas phototube to modulated radiation input are ion transit time between anode and cathode, and delayed secondary emission produced at the photocathode by gas atoms which are excited to metastable energy states. The speed of response of a vacuum phototube is limited only by electron transit time across the interelectrode space.

Sensitivity to incident light. The photocathode of a vacuum, gas, or multiplier phototube is selectively sensitive; that is, it emits electrons photoelectrically only when it is exposed to radiation having wavelengths in specific regions of the spectrum. Cathode radiant sensitivity is the photoelectric current emitted per unit of incident monochromatic radiant power. The spectral sensitivity characteristic of a phototube exhibits cathode radiant sensitivity as a function of the wavelength of radiation incident upon the window of the phototube.

Spectral sensitivity characteristics are shown in the figure and are designated by standard symbols S-1, S-3, and so on. The sharp cutoff of sensitivity on the short-wavelength side of the curves is determined primarily by the transmission characteristic of the glass envelope or window of the phototube. The long-wavelength limit of radiant sensitivity is the threshold wavelength of the photocathode.

The sensitivity of a phototube is more easily measured as cathode luminous sensitivity, the photoelectric current emitted per unit of incident luminous flux from a specified source of light. It is expressed in microamperes per lumen. The source commonly used is a tungsten-filament lamp operated at a color temperature of 2870 K. Cathode-radiant and cathode-luminous sensitivities are measured with a collimated beam of radiation perpendicular to the window of the tube. Summarized in the table are radiant sensitivity at the wavelength of maximum response, luminous sensitivity, and other data relative to the spectral sensitivity characteristics in the illustration.

The quantum efficiency of a photocathode is the number of electrons emitted per incident photon of a given wavelength. Because the energy per photon of wavelength λ is hc/λ, where h is Planck's constant and c is the velocity of light in vacuum,

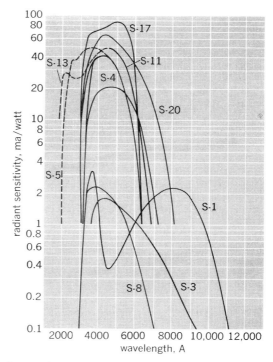

Curves of the average spectral sensitivity characteristics of some typical phototubes.

the quantum efficiency at a wavelength λ is simply the radiant sensitivity of the photocathode multiplied by the factor hc/λ in appropriate units. Typical quantum efficiencies are listed in the table. *See* PHOTOEMISSION.

Photocathode material. Photocathodes of practical importance contain one or more of the alkali metals lithium, sodium, potassium, rubidium, or cesium in complex combination with other metals or with oxides of certain metals. Because of their high reflectivity and conductivity, the pure alkali metals have lower quantum efficiencies than do the more complex photocathodes, which invariably are semiconductors.

Average cathode characteristics

Spectral sensitivity characteristic*	Cathode material	Wavelength of maximum response, A	Peak radiant sensitivity, ma/watt	Peak cathode quantum efficiency, %	Luminous sensitivity, μamp/lumen†	Remarks
S-1	Cs_2O, Ag	8000	2.2	0.3	25	
S-3	Rb_2O, Ag	4200	1.8	0.5	6.5	
S-4	Cs_3Sb	4000	40	12.4	40	
S-5	Cs_3Sb	3400	49	17.8	40	Ultraviolet transmitting window
S-8	Cs_3Bi	3650	2.3	0.8	3	
S-10	Bi, Ag, O, Cs	4500	20.3	5.6	40	Semitransparent
S-11	Cs_3Sb	4400	48	13.5	60	Semitransparent
S-13	Cs_3Sb	4400	47	13.2	60	Semitransparent; ultraviolet transmitting window
S-17	Cs_3Sb	4900	85	21.4	125	Semitransparent, on reflecting substrate
S-20	(NaKCs)Sb	4200	64	18.8	150	Semitransparent

*These characteristics, shown in the figure, refer to typical phototubes rather than to photocathodes.
†Light source is a tungsten-filament lamp operated at a color temperature of 2870 K.

Practical photocathodes may be classified broadly under two prototypes: the cesium oxide–silver cathode and the cesium antimonide cathode. The cesium oxide–silver cathode is obtained by permitting cesium to react with a thin layer of silver oxide. The resultant cathode layer is cesium oxide containing silver, possibly oxides more complex than Cs_2O, and a critical excess of cesium. Phototubes which contain this photocathode have the S-1 spectral sensitivity characteristic. The rubidium oxide–silver cathode is produced in a similar manner.

The cesium antimonide photocathode is obtained by exposing a thin layer of antimony to cesium vapor at elevated temperatures. The cathode surface is an intermetallic compound, cesium antimonide, containing an excess of cesium. This cathode has maximum sensitivity in the blue and ultraviolet regions of the spectrum and a threshold wavelength at about 6500 A. The cesium-bismuth cathode is similar to the cesium antimonide cathode in composition and in method of preparation.

The bismuth-silver-oxygen-cesium cathode is formed by cesium activation of an oxidized layer of silver and bismuth. The S-10 characteristic associated with this type of photocathode is an effective combination of the blue response of the cesium-bismuth cathode and the red response of the cesium oxide–silver cathode. This broad characteristic, which extends over most of the visible spectrum, is a desirable feature for phototubes used in photometry and colorimetry.

The sodium-potassium-cesium-antimony, or tri-alkali, photocathode is produced by exposing a thin antimony layer to vapors of the alkali metals. This cathode has a higher peak radiant sensitivity than do any of the photocathodes mentioned above. Its spectral sensitivity characteristic extends from ultraviolet to infrared wavelengths. The tri-alkali photocathode is therefore an excellent panchromatic detector of visible and near-visible radiation.

Photocathode construction. A photocathode may be either an opaque layer of the emissive material on a metal electrode or a semitransparent layer on glass. A semitransparent layer is deposited directly on the window or envelope of the phototube. A portion of the layer overlaps a high-conductivity layer of aluminum or other metal which provides electrical contact to the photocathode. Radiation transmitted through the window is incident upon one side of the layer while electrons are emitted photoelectrically from the opposite, or vacuum, surface of the layer. This type of photocathode is commonly used in multiplier phototubes having S-1, S-10, S-11, S-13, or S-20 characteristics.

A semitransparent cathode may also be formed on an opaque, highly reflecting metal surface. Radiation transmitted through the cathode layer is reflected into and partially absorbed by a relatively thin layer from which photoelectrically excited electrons can readily escape. The high radiant sensitivity obtainable in this manner is illustrated by the S-17 spectral sensitivity characteristic.

Dark current. Dark current is the current measured at the terminals of a phototube when it is shielded from all radiation capable of causing photoelectric emission from its photocathode. Two common causes of dark current are electrical conductivity across or through the insulation supporting the electrodes or tube terminals, and thermionic emission from the photocathode. Electrical conductivity can be reduced to very small values by placing the cathode and anode terminals at opposite ends of the tube envelope. Dark current due to thermionic emission is proportional to the area of the photocathode and increases almost exponentially with cathode temperature. At 25°C the thermionic emission from the cesium oxide–silver photocathode is of the order of 10^{-11} amp/cm^2, whereas that of the cesium antimonide photocathode is about 1×10^{-15} amp/cm^2. At low light levels such that the photoelectric current and dark current are of similar magnitudes, discrimination between the two is readily achieved by modulating the input flux and measuring only the ac component of anode current.

Types of service. Phototubes are used in general in relay operation, detection of intensity-modulated radiation, and photometry. Response of a phototube to a change in light level causes a relay to close in safety and warning devices, and in counting or sorting equipment. Since the change in photocurrent may be considerably less than 1 microampere (μamp) in a vacuum phototube, a gas phototube is frequently used to trigger a thyratron in relay applications. Typical applications involving the detection of intensity-modulated light are reproduction of sound-on-film and facsimile. Gas phototubes are commonly used in sound reproduction because of their inherent gas amplification and acceptable audio-frequency response characteristic. Vacuum phototubes are used in facsimile service for which high-frequency response is required.

Vacuum phototubes are used whenever linear response to radiant flux input is required as, for example, in photometry and in colorimetry. In photometric applications a spectral sensitivity characteristic similar to that of the human eye is frequently required and may be approximated by means of special optical filters used in combination with phototubes which are sensitive over the visible spectrum. In spectrophotometry two vacuum phototubes may be used, one having S-5 and the other S-1 response, in order to provide sensitivity at wavelengths from 2500 to 11,000 A. In applications such as colorimetry and the measurement of density of color film, narrow-band filters are used with appropriate vacuum phototubes to establish tri-stimulus values independently of the particular sensitivity characteristics of the tubes. These sensitivity characteristics differ from tube to tube as a result of small uncontrollable variations in the composition of the photocathode that is being used.

Vacuum and gas phototubes are capable of providing reliable service over thousands of hours of operation at moderate temperatures and radiation levels. Temperatures above 75–100°C and average cathode-current densities greater than about 30 μamp/cm^2 cause a gradual decrease in sensitivity. The life of a phototube is limited largely by slow effusion of gaseous contaminants which cannot be baked out of the phototube after the photocathode is formed. [JAMES L. WEAVER]

Bibliography: Detectors: Photo/optical, *IEEE Trans. Nucl. Sci.*, vol. NS-26, no. 1, pt. 1, pp. 338–421, 1979; A. Schure (ed.), *Phototubes*, 1959.

Photovoltaic cell

A device that detects or measures electromagnetic radiation by generating a current or a voltage, or both, upon absorption of radiant energy. Specially designed photovoltaic cells are used in solar batteries, photographic exposure meters, and sensitive detectors of infrared radiation. An important advantage of the photovoltaic cell in these particular applications is that no separate bias supply is needed—the device generates a signal (voltage or current) simply by the absorption of radiation.

Most photovoltaic cells consist of a semiconductor pn junction or Schottky barrier in which electron-hole pairs produced by absorbed radiation are separated by the internal electric field in the junction to generate a current, a voltage, or both at the device terminals. The influence of the incident radiation on the current-voltage characteristics of a photovoltaic cell is to shift the current-voltage characteristic downward by the magnitude of the photogenerated current as shown in the illustration. Under open-circuit conditions (current $I = 0$)

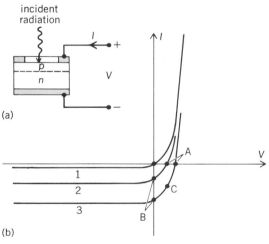

The pn junction photovoltaic cell. (a) Cross section. (b) Current-voltage characteristics. Curve 1 is for no incident radiation, and curves 2 and 3 for increasing incident radiation.

the terminal voltage increases with increasing light intensity (points A), and under short-circuit conditions (voltage $V = 0$) the magnitude of the current increases with increasing light intensity (points B). When the current is negative and the voltage is positive (point C, for example), the photovoltaic cell delivers power to the external circuit. In this case, if the source of radiation is the Sun, the photovoltaic cell is referred to as a solar battery. When a photovoltaic cell is used as a photographic exposure meter, it produces a current proportional to the light intensity (points B), which is indicated by a low-impedance galvanometer or microammeter. For use as sensitive detectors of infrared radiation, specially designed photovoltaic cells can be operated with either low-impedance (current) or high-impedance (voltage) amplifiers, although the lowest noise and highest sensitivity are achieved in the current or short-circuit mode. *See* JUNCTION DIODE; PHOTODIODE; PHOTOELECTRIC DEVICES; PHOTOVOLTAIC EFFECT; SEMICONDUCTOR; SOLAR CELL.   [GREGORY E. STILLMAN]

Bibliography: H. J. Hovel, *Solar Cells*, vol. 11 of R. K. Willardson and A. C. Beer (eds.), *Semiconductors and Semimetals*, 1975; D. Long, Photovoltaic and photoconductive infrared detectors, in R. J. Keyes (ed.), *Optical and Infrared Detectors*, pp. 101–145, 1977; W. F. Wolfe and G. J. Zissis (eds.), *The Infrared Handbook*, 1978.

Photovoltaic effect

A term most commonly used to mean the production of a voltage in a nonhomogenous semiconductor, such as silicon, by the absorption of light or other electromagnetic radiation. In its simplest form, the photovoltaic effect occurs in the common photovoltaic cell, used, for example, in solar batteries and exposure meters. The photovoltaic cell consists of an np junction between two different semiconductors, an n-type material in which conduction is due to electrons, and a p-type material in which conduction is due to positive holes. When light is absorbed near such a junction, new mobile electrons and holes are released, as in photoconduction. An additional feature of a photovoltaic cell, however, is that there is an electric field in the junction region between the two semiconductor types. The released charge moves in this field. This current flows in an external circuit without the need for a battery as required in photoconduction. If the external circuit is broken, an "open-circuit photovoltage" appears at the break.

In certain rather complex electrolytic systems, illumination of the electrodes may give rise to a voltage classed as photovoltaic. *See* PHOTOCONDUCTIVITY; PHOTOVOLTAIC CELL; SEMICONDUCTOR; SOLAR CELL.

[L. APKER]

Bibliography: L. Azaroff and J. J. Brophy, *Electronic Processes in Materials*, 1963; A. Van der Ziel, *Solid State Physical Electronics*, 3d ed., 1976.

Picture tube

A cathode-ray tube used as a television picture tube. It might also be referred to as a television picture reproducer. Modern television picture tubes usually have large glass envelopes (Fig. 1) on the inner face of which a light-emitting layer of luminescent materials is deposited. A modulated stream of high-velocity electrons is made to scan this layer of luminescent materials in a series of horizontal lines so that it recreates the picture elements (light and dark areas).

The number of electrons in the stream at any instant of time is varied by electrical impulses corresponding to the signal sent out by the television transmitter. These electrical impulses (picture information) were originally generated by a studio television camera. The home television receiver picks up the signal, and after suitable amplification and detection the picture information is supplied to the picture tube so that it recreates the original picture. *See* TELEVISION.

Construction. For detailed discussion of cathode-ray tube construction *see* CATHODE-RAY TUBE.

This article discusses construction specific to kinescopes.

Deflection angle. This is the maximum angle through which the electron beam must be

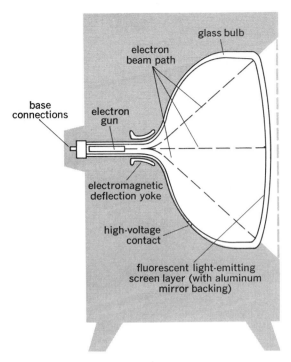

Fig. 1. Black-and-white kinescope.

deflected to scan the picture area. Most tubes in use today have deflection angles of 90 or 114°.

Focus. Picture tubes are defined by the method used to focus the stream of electrons to a small spot. There are two general methods, magnetic focus, using a strong external electromagnet; and electrostatic focus, accomplished by applying voltage differences to the electron-gun electrodes. Electrostatic focus has become universal because of its lower cost.

Electromagnetic deflection. The electron beam is deflected electromagnetically to cause it to scan the picture area. This deflection is accomplished by a deflecting yoke, made up of two pairs of shaped coils which fit around the small neck of the picture tube (Fig. 1). When pulsating electric currents of proper wave shape and phase are supplied to these coils, they generate magnetic fields which cause the electron beam to bend as it passes through them. By changing the magnitude and direction of the magnetic field, the electron beam can be made to arrive at any point on the face of the picture tube.

Glass envelope. The glass envelope is made of a special composition to minimize optical defects and to provide electrical insulation for high voltages. It also provides protection against x-radiation and has a light-absorption characteristic which improves the contrast of the picture when it is viewed in brightly illuminated locations. The structural design of the glass bulb is made to withstand 3–6 times the force of atmospheric pressure to provide a safety margin over normal atmospheric pressure. However, care must be taken in handling these evacuated glass bulbs, or a dangerous implosion may result.

Aluminized screen. The luminescent screen is made of a thin layer of phosphors 0.001 in. thick (3 mg/cm²). The phosphor materials are primarily zinc cadmium sulfide (emits yellow light) and zinc sulfide (emits blue light). By careful proportioning and mixing of these two phosphors the resultant emanation is a blue-white light which most manufactures have adjusted to a color temperature of between 9000 and 12,000 K.

The luminescent screen is aluminized by vacuum evaporation from a small molten aluminum pellet. The thin layer of aluminum, approximately 2000 A, is deposited on a smooth plastic film placed on top of the luminescent screen. The plastic film is subsequently volatilized and removed in the high-temperature processing of the tube. In the operation of the completed tube, the high-velocity electron beam penetrates the aluminum film, and its energy is transferred primarily to the luminescent screen. Only a small percentage (10–20%) of the electron-beam energy is converted into useful light energy, but even this amount is sufficient to produce brightness of several hundred footlamberts in the picture highlights. The reflection of light by the aluminum mirror increases the picture brightness and improves picture contrast by preventing stray light from illuminating the back side of the luminescent screen.

Wall coating. Graphite coatings are placed on the inside walls of the bulb to provide electrically conducting surfaces between the screen and the electron gun and to provide an unipotential field through which the electron beam may travel without being disturbed by stray electrostatic fields. Graphite is a sufficiently poor conductor to minimize eddy-current power absorption from the electromagnetic fields generated by the external deflecting yoke.

Electron gun. The electron gun produces a stream of high-velocity electrons which are focused to a small spot at the screen. In currently used commercial tubes, the spot produced by the electron beam at the screen is approximately 0.025–0.125 in. in diameter, depending on beam current required (usually 50–1000 microamperes). The beam current is controlled by an electrical signal supplied to the cathode with respect to grid no. 1, called cathode drive, or to grid no. 1 with respect to the cathode, called grid drive. Conventional tubes require 50–125 volts of signal to modulate the beam from zero beam current (picture dark areas) to maximum beam current (picture highlights).

Transistorized portable battery-operated TV receivers use picture tubes requiring less signal voltage, much less power for deflection of the electron beam, and smaller amounts of power to heat the cathode from which the electrons are emitted.

Color kinescope. A color television picture tube is similar to the black-and-white picture tube but differs in two ways. (1) The light-emitting screen is made up of small elemental areas, each capable of emitting light in one of the three primary colors (red, green, or blue) laid in interlaced arrays. (2) Means are provided for selectively exciting any one or more of these phosphor primaries.

Color may be controlled by any of three methods: (1) sensing, which is an accurate electron beam scanning of desired phosphor areas, with a feedback signal from the screen through circuitry which maintains accurate positioning of the beam(s); (2) a switching device near, and registered with, the screen, which deflects the beam to the proper phosphor; or (3) a screen structure which provides phosphor excitation according to incident beam direction, generally used with three angularly spaced beams. The third device is termed a shadow-mask tube.

Compared with black-and-white tubes all color picture tubes are somewhat limited in light output because their phosphors are chosen for excellence of primary colors rather than high luminous efficiency. Some color systems have additional output limitations; for these reasons, color tubes are usually operated at higher currents and voltages than their black-and-white counterparts. Although each color-tube system has certain advantages, only the shadow-mask type has been successfully developed for high production. Its advantages are its ability to produce colors as pure as its phosphor primaries, its excellent contrast, the feasibility of using it in a fine barely perceptible dot screen structure, and the fact that the cur-

rent from three beams is available for simultaneous presentation. Commercially available shadow-mask tubes are manufactured in a number of different picture sizes, such as 25 in. picture diagonal, 22 in., 18 in., 15 in., and so on. Also, the tube lengths vary depending upon the deflection angle used, which is commonly 90° or 110°. The tubes use a triple-beam gun with electrostatic focus and magnetic means for keeping the three electron beams converged to a point; and a curved aperture mask registered with and close to the phosphor dot screen. The dots are in hexagonal arrays, deposited on the spherical faceplate. Magnetic scanning and synchronized dynamic convergence are used (Fig. 2). *See* COLOR TELEVISION.

[C. PRICE SMITH]

Bibliography: E. G. Bylander, *Electronic Displays*, 1979; A. M. Dhake, *Television Engineering*, 1980; A. M. Morrell et al., *Color Television Picture Tubes*, 1974; S. Sherr, *Electronic Displays*, 1979.

Piezoelectric crystal

A crystalline substance which exhibits the piezoelectric effect. This "pressure electricity" was first positively identified by the Curies in 1880, when they discovered that some crystals produced electric charges on parts of their surface when the crystals were compressed in particular directions, the charge disappearing when the pressure was removed. It was later discovered that these crystals become strained when subjected to electric fields; the piezoelectric deformation is directly proportional to the field and it reverses in sign as the sign of the field is reversed. These basic properties of piezoelectric crystals are used in electromechanical transducers, such as ultrasonic generators, microphones, phonograph pickups, and electromechanical resonators, such as frequency-controlling quartz crystals. *See* PIEZOELECTRICITY.

Piezoelectric materials. The principal piezoelectric materials used commercially are crystalline quartz and rochelle salt, although the latter is being superseded by other materials, such as barium titanate. Quartz has the important qualities of being a completely oxidized compound (silicon dioxide), and is almost insoluble in water. Therefore, it is chemically stable against changes occurring with time. It also has low internal losses when used as a vibrator. Rochelle salt has a large piezoelectric effect, and is thus useful in acoustical and vibrational devices where sensitivity is necessary, but it decomposes at high temperatures (55°C) and requires protection against moisture. Barium titanate provides lower sensitivity, but greater immunity to temperature and humidity effects. Other crystals that have been used for piezoelectric devices include tourmaline, ammonium dihydrogen phosphate (ADP), and ethylenediamine tartrate (EDT). *See* BARIUM TITANATE; ROCHELLE SALT.

The quartz crystal resonator is the most important class of piezoelectric device. Its principal application is in the fields of frequency control and electric-wave filters. It is also used in transducers, especially where heat or moisture are factors.

Characteristics and manufacture. The electrical properties of quartz crystals as circuit elements, including their temperature coefficient of frequency, motional inductance and capacitance, series resistance, and electrode or shunt capaci-

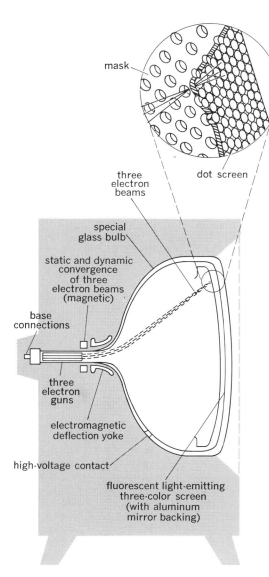

mask

three electron beams

dot screen

special glass bulb

static and dynamic convergence of three electron beams (magnetic)

base connections

three electron guns

electromagnetic deflection yoke

high-voltage contact

fluorescent light-emitting three-color screen (with aluminum mirror backing)

Fig. 2. Color kinescope.

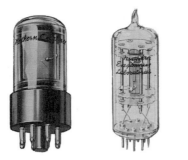

Fig. 1.   Three typical vacuum-mounted quartz resonators. (*Northern Engineering Laboratories*)

tance, are largely determined by the dimensions and angles of rotation of the resonator surfaces with respect to the crystal axes. In commercial practice, raw quartz crystals are oriented by means of x-ray goniometers, the required angles of rotation are measured on the mounting jig, and the required blanks (unfinished slabs) cut from the mother crystal by diamond-faced saws. The blanks are then ground to the required frequency using lapping techniques with graduated sizes of abrasives to obtain a smooth finish. Some high-quality crystals are given optically polished surfaces. It is common practice to etch the surface of the crystal with fluorine compounds to remove microscopic surface irregularities.

Electrodes may be applied directly to the surface of the crystal, or they may be mounted externally in close proximity to the quartz element. Crystals with electrodes on their surfaces are frequently mounted by means of wires, which may also provide the connections to the electrodes. Containers for crystals may be hermetically sealed for protection from atmospheric effects, and some crystals are mounted in evacuated envelopes to improve Q and reduce aging drift. Examples of vacuum-mounted quartz crystals are shown in Fig. 1. Vacuum-mounted crystals are not capable of handling much power (usually less than 1 or 2 mW maximum dissipation). Therefore, sealed crystals for power oscillators are usually mounted in an inert gas, such as nitrogen or helium.

The crystal-resonator cut chosen for a given application usually is dictated by the frequency at which the crystal must operate and the temperature range over which the crystal must work. A plot of typical frequency versus temperature

curves for several different crystal cuts is shown in Fig. 2. A chart of frequency ranges covered by widely used crystal cuts is shown in Fig. 3.

Applications. Quartz crystal resonators are used for stabilizing the frequency of oscillators. The degree of stabilization depends on several factors, the principal ones being the Q of the resonator, the type of crystal cut used and temperature range of operation, the type of circuit used, and the amount of power dissipated in the resonator. Thermostatic ovens are often used to enhance oscillator stability. The application of quartz crystals for oscillator stabilization has made possible the modern radio and television broadcasting industry and mobile radio communications with aircraft and ground vehicles. *See* Oscillator.

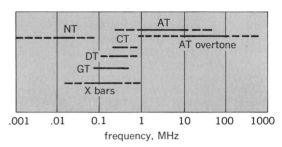

Fig. 3.   Frequency range of various crystal cuts.

Quartz crystal resonators are also used in electric-wave, or frequency-separation, filters. Many thousands of such crystals are used in telephone systems for carrier-frequency separation, and in radio communication equipment for selecting a desired signal frequency band while rejecting undesired frequencies. *See* Electric filter.

Transducers using piezoelectric elements are used for converting vibrations into electrical signals in such applications as crystal microphones, phonograph pickups, vibration pickups, and dynamic pressure-sensing elements. The inverse piezoelectric effect is used for converting electrical signals into mechanical vibrations. Thus, piezoelectric transducers are used in such applications as underwater sound ranging equipment (sonar, asdic), and in ultrasonic cleaning devices, which use a liquid medium for washing small to medium-sized objects.   [FRANK D. LEWIS]

Bibliography: W. G. Cady, *Piezoelectricity*, 1964; J. Van Randeraat and R. E. Setterington, *Piezoelectric Ceramics*, 2d ed., 1974.

Piezoelectricity

Electricity, or electric polarity, resulting from the application of mechanical pressure on a dielectric crystal. The application of a mechanical stress produces in certain dielectric (electrically nonconducting) crystals an electric polarization (electric dipole moment per cubic meter) which is proportional to this stress. If the crystal is isolated, this polarization manifests itself as a voltage across the crystal, and if the crystal is short-circuited, a flow of charge can be observed during loading. Conversely, application of a voltage between certain faces of the crystal produces a mechanical distortion of the material. This reciprocal relationship is referred to as the piezoelectric effect. The phe-

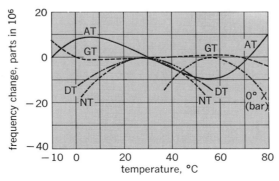

Fig. 2.   Crystal frequency as a function of temperature for several different crystal cuts.

nomenon of generation of a voltage under mechanical stress is referred to as the direct piezoelectric effect, and the mechanical strain produced in the crystal under electric stress is called the converse piezoelectric effect.

Piezoelectric materials are used extensively in transducers for converting a mechanical strain into an electrical signal. Such devices include microphones, phonograph pickups, vibration-sensing elements, and the like. The converse effect, in which a mechanical output is derived from an electrical signal input, is also widely used in such devices as sonic and ultrasonic transducers, headphones, loudspeakers, and cutting heads for disk recording. Both the direct and converse effects are employed in devices in which the mechanical resonance frequency of the crystal is of importance. Such devices include electric wave filters and frequency-control elements in electronic oscillator circuits. See DISK RECORDING; MICROPHONE; PIEZOELECTRIC CRYSTAL.

Necessary condition. The necessary condition for the piezoelectric effect is the absence of a center of symmetry in the crystal structure. Of the 32 crystal classes, 21 lack a center of symmetry, and with the exception of one class, all of these are piezoelectric. In the crystal class of lowest symmetry, any type of stress generates an electric polarization, whereas in crystals of higher symmetry, only particular types of stress can produce a piezoelectric polarization. For a given crystal, the axis of polarization depends upon the type of the stress. There is no crystal class in which the piezoelectric polarization is confined to a single axis. In several crystal classes, however, it is confined to a plane. Hydrostatic pressure produces a piezoelectric polarization in the crystals of those 10 classes that show pyroelectricity in addition to piezoelectricity. The pyroelectric axis is then the axis of polarization.

The converse piezoelectric effect is a thermodynamic consequence of the direct piezoelectric effect. When a polarization P is induced in a piezoelectric crystal by an externally applied electric field E, the crystal suffers a small strain S which is proportional to the polarization P. In crystals with a normal dielectric behavior, the polarization P is proportional to the electric field E, and hence the strain is proportional to this field E. Superposed upon the piezoelectric strain S is a much smaller strain which is proportional to P^2 (or E^2). This strain is called the electrostrictive strain. It is present in any dielectric. See ELECTROSTRICTION.

Matrix formulation. The relation of the six components T_j of the stress tensor (three compressional components and three shear components) to the three components P_i of the polarization vector can be described by a scheme (matrix) of 18 piezoelectric moduli d_{ij}. The same scheme (d_{ij}) also relates the three components E_i of the electric field to the six components S_j of the strain:

The direct effect is obtained by reading this scheme in rows, as in Eq. (1). The converse effect is obtained by reading it in columns, as in Eq. (2).

$$P_i = -\sum_{j=1}^{6} d_{ij}T_j \qquad i = 1, 2, 3 \qquad (1)$$

$$S_j = \sum_{i=1}^{3} d_{ij}E_i \qquad j = 1, 2, \ldots, 6 \qquad (2)$$

An analogous matrix (e_{ij}) relates the strain to the polarization and the electric field to the stress, as in Eqs. (3).

$$P_i = \sum_{j=1}^{6} e_{ij}S_j \qquad i = 1, 2, 3$$

$$T_j = -\sum_{i=1}^{3} e_{ij}E_j \qquad j = 1, 2, \ldots, 6 \qquad (3)$$

The matrices (d_{ij}) and (e_{ij}) are not independent, but are related by expressions involving the elasticity tensor $c_{jh}{}^E$ (for constant electric field E), as in Eq. (4).

$$e_{mh} = \sum_{j=1}^{6} d_{mj}c_{jh}{}^E \qquad \begin{array}{l} m = 1, 2, 3 \\ h = 1, 2, \ldots, 6 \end{array} \qquad (4)$$

Alternative formulations can be made by introducing the dielectric displacement D or visualizing the simultaneous action of electrical and mechanical stresses.

The number of independent matrix elements d_{ij} or e_{ij} depends upon the symmetry elements of the crystal. For the lowest symmetry, all 18 matrix elements are independent, whereas piezoelectric classes of higher symmetry can have as few as one independent element in the matrix (d_{ij}). The matrix takes its simplest form if the natural symmetry axes of the crystal are chosen for the coordinate system.

Electromechanical coupling. The direct piezoelectric effect makes a crystal a generator, and the converse effect makes it a motor. Consequently, a piezoelectric crystal has many properties in common with a motor-generator. For example, the electrical properties, such as the dielectric constant, depend upon the mechanical load; conversely, the mechanical properties, such as the elastic constants, depend upon the electric boundary conditions. The electromechanical coupling factor k can be defined as follows. Suppose electrodes are attached to a piezoelectric crystal and connected to a battery. Then the ratio of the energy stored in mechanical form to the electrical energy delivered by the battery is equal to k^2. In general, k ranges from below 1 to about 30%. In quartz, for example, the coupling is roughly 10%. In ferroelectric crystals, k can approach unity in certain circumstances. See FERROELECTRICS.

In quartz, a stress of 1 newton/m applied along the diad axis produces a polarization of about 2×10^{-12} coulomb/m^2 along the same axis. Conversely, an electric field of 10^4 volts/m produces a strain of about 2×10^{-8}. In ferroelectric crystals, such as rochelle salt and KH_2PO_4, and in certain antiferroelectrics, such as $NH_4H_2PO_4$ (ADP), these effects can be several orders of magnitude larger.

Molecular theory. Quantitative theories based on the detailed crystal structure are very involved. Qualitatively, however, the piezoelectric effect is readily understood for simple crystal structures.

		Compression			Shear		
		S_1 T_1	S_2 T_2	S_3 T_3	S_4 T_4	S_5 T_5	S_6 T_6
E_1	P_1	d_{11}	d_{12}	d_{13}	d_{14}	d_{15}	d_{16}
E_2	P_2	d_{21}	d_{22}	d_{23}	d_{24}	d_{25}	d_{26}
E_3	P_3	d_{31}	d_{32}	d_{33}	d_{34}	d_{35}	d_{36}

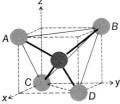

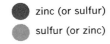

zinc (or sulfur)

sulfur (or zinc)

Fig. 1. Tetrahedral structure of zincblende, ZnS. Only part of unit cell is shown. Size of circles has no relation to size of ions.

Figure 1 illustrates this for a particular cubic crystal, zincblende (ZnS). Every Zn ion is positively charged and is located in the center of a regular tetrahedron *ABCD*, the corners of which are the centers of sulfur ions, which are negatively charged. When this system is subjected to a shear stress in the *xy* plane, the edge *AB*, for example, is elongated, and the edge *CD* of the tetrahedron becomes shorter. Consequently, these edges are no longer equivalent, and the Zn ion will be displaced along the *z* axis, thus giving rise to an electric dipole moment. The dipole moments arising from different octahedrons sum up because they all have the same orientation with respect to the axes *x*, *y*, and *z*.

Another simple type of piezoelectric structure is encountered in barium titanate, $BaTiO_3$, as shown in Fig. 2. The positive Ti ions are surrounded by an almost regular octahedron of negative

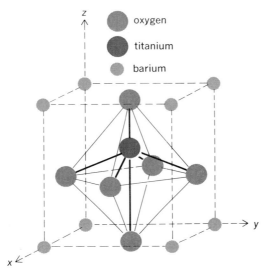

oxygen

titanium

barium

Fig. 2. Unit cell of tetragonal barium titanate, $BaTiO_3$. Deviation from cubic symmetry is exaggerated. Size of the circles has no relation to the size of the ions.

oxygen ions. The Ti ions are not in the center of the octahedron, but somewhat displaced along the *z* axis. This structure already has a dipole moment or spontaneous polarization in the absence of externally applied stresses. It is clear from Fig. 2 that the Ti ion is pushed more off center when the crystal is mechanically compressed in the *xy* plane or elongated along *z*. The additional polarization associated with this deformation is the piezoelectric polarization. *See* BARIUM TITANATE.

Piezoelectric ceramics. Barium titanate and a few related compounds have the remarkable property that, by means of a sufficiently strong electric field, the direction of the spontaneous polarization can be switched to any one of the *x*, *y*, or *z* axes. This makes it possible to produce polycrystalline samples (ceramics) which are piezoelectric. The electromechanical coupling factors of such ceramics can reach about 50%.

Piezoelectric resonator. The piezoelectric strains that can be induced by a static electric field

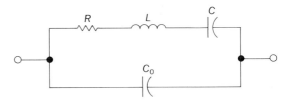

Fig. 3. Network equivalent to a piezoelectric resonator near and at a resonance frequency.

are very small, except in certain ferroelectrics. Larger strains can be obtained when a piezoelectric crystal is driven by an alternating voltage, the frequency of which is equal to a mechanical resonance frequency of the crystal. The vibrating crystal reacts back on the circuit through the direct piezoelectric effect. In the range of a mechanical resonance, this reaction is equivalent to the response of the network shown in Fig. 3, provided that the series resonance frequency of the network is equal to a mechanical resonance frequency of the crystal, as in Eq. (5). An important difference

$$f_R = 1/(2\pi\sqrt{LC}) \tag{5}$$

between the network of Fig. 3 and the piezoelectric resonator is that the latter has many discrete modes of vibration, whereas the network has only one resonance frequency.

Network elements. The elements L, C, and C_0 of the equivalent network can be calculated from the physical constants of the crystal. Consider, for example, the simple resonator shown in Fig. 4. A rectangular crystal bar with the dimensions $l_1 \gg l_2 \gg l_3$ is excited to compressional lengthwise vibrations. The *xy* faces have adherent electrodes, and the bar is oriented with respect to the natural crystal axes so that an electric field E_3 along *z* causes a strain S_1 along the bar according to the equation $S_{1(\text{piezoel})} = d_{31}E_3$. A mechanical stress T_1 along the bar causes a strain $S_{1(\text{mech})} = s_{11}{}^E T_1$, where $s_{11}{}^E$ is the elastic compliance measured at constant electric field E_3. The resonance frequency for the fundamental lengthwise compressional mode is then given by Eq. (6), where ρ is

$$f_R = 1/(2l_1\sqrt{\rho s_{11}{}^E}) \text{ Hz} \tag{6}$$

the density of the crystal. The parallel capacitance C_0 is the static capacitance of the crystal, as in Eq. (7). Here ϵ is the relative dielectric constant along

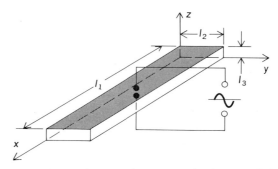

Fig. 4. Simple piezoelectric resonator. A voltage applied to the electrodes shortens or lengthens the bar, thus exciting longitudinal vibrations.

$$C_0 = 8.85\epsilon l_1 l_2/l_3 \text{ picofarad} \qquad (7)$$

z. For C and L, the analysis yields Eqs. (8) and (9).

$$C = \frac{70.8 d_{31}{}^2 \int_1 \int_2}{\pi^2 S_{11} E \int_3} \text{ picofarad} \qquad (8)$$

$$L = \frac{\rho (S_{11}{}^E)^2 \int_1 \int_3}{8 d_{31}{}^2 \int_2} \text{ henry} \qquad (9)$$

(All physical constants are in mks units.) For the nth overtone, C_0 and L remain the same, whereas C must be divided by n^2. The losses (damping) represented by the resistance R in Fig. 3 arise, for example, from ultrasonic radiation, friction in the crystal mount, internal friction in the crystal originating in various imperfections, and dielectric relaxation.

At the mechanical resonance frequency f_R, the alternating current is maximum and is determined by R. At the antiresonant frequency, given by Eq. (10), the current is minimum. The difference $\Delta f =$

$$f_A = \sqrt{(C_0 + C)/LCC_0} \qquad (10)$$

$f_A - f_R$ increases with increasing electromechanical coupling according to relation (11).

$$\Delta f \approx 4k^2/\pi^2 \qquad (11)$$

The reactance depends upon frequency, as shown in Fig. 5. For a typical piezoelectric crystal such as quartz, resonating at about 10^5 Hz, the orders of magnitude given by relations (12) are typical for the elements of the equivalent network.

$$\begin{aligned} L &\approx 10^2 \text{ henry} \\ C &\approx 0.02 \text{ picofarad} \\ C_0 &\approx 5 \text{ picofarad} \end{aligned} \qquad (12)$$

The damping resistance R varies from about 10^2 to 10^4 ohms; that is, the Q factors, given by Eq. (13),

$$Q = \frac{1}{R} \sqrt{\frac{L}{C}} \qquad (13)$$

are in the range between 10^6 and 10^4, and the resonances are very sharp. These characteristics cannot be achieved with conventional coils and condensors as circuit elements.

Vibration modes. With piezoelectric resonators of various types, the range from audio frequencies

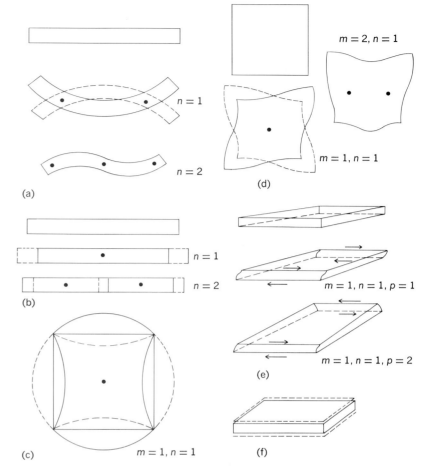

Fig. 6. Diagrammatic representation of examples of vibration modes of bars and plates. (a) Flexural vibrations of a bar. (b) Longitudinal vibrations of a bar. (c) Longitudinal vibration of a plate. (d) Face shear vibrations of a plate. (e) Thickness shear vibrations of a plate. (f) Thickness vibration of a plate.

to many megahertz can be covered. The vibration modes frequently used are, in order of increasing frequency: (1) flexural vibrations of bars and plates, (2) longitudinal vibrations of bars and plates, (3) face shear vibrations of plates, and (4) thickness shear vibrations and compressional vibrations of plates. Figure 6 illustrates some of these modes. The excitation of particular vibration modes can be achieved by proper orientation of the resonator with respect to the natural crystal axes, by proper positioning of the electrodes, and by proper mounting. A simple example is illustrated by Fig. 7. A bar is oriented so that an electric field along x causes an expansion or contraction along y. The electrodes are split and cross-connected so that the bar flexes in the yz plane when a voltage is applied. The fundamental flexure mode is easily excited with this arrangement; however, excitation of higher even-numbered flexural modes is also possible. Interesting resonators are possible with piezoelectric ceramics ($BaTiO_3$ type) because different parts of the resonator can be polarized in different directions.

Common applications. The sharp resonance curve of a piezoelectric resonator makes it useful in the stabilization of the frequency of radio oscillators. Quartz crystals are used almost exclusively

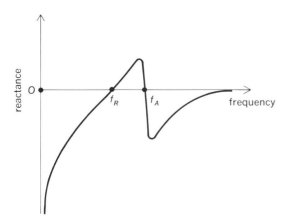

Fig. 5. Diagram showing reactance versus frequency for a piezoelectric resonator.

in this application. The main advantages of quartz are high Q factor, stability with respect to aging, and the possibility of orienting the resonator with respect to the natural crystal axes so that the tem-

perature coefficient of the resonance frequency vanishes near the operating temperature. Figure 8 illustrates the orientation of commonly used cuts.

In vacuum-tube oscillators, the crystal generally is part of the feedback circuit. In the circuit proposed by G. W. Pierce, the conditions for oscillation are not satisfied unless the crystal reactance is positive. Hence, the oscillation frequency is between the resonant and antiresonant frequency of the crystal (Fig. 5). Circuits of this type hold the frequency within a few parts per million. Much greater stability can be achieved with the bridge circuit of L. A. Meacham. Here the oscillation conditions are fulfilled by zero phase shift in the feedback circuit, that is, at the exact series resonance frequency of the crystal. Long-term frequency stability of about one part in 10^8 and short-term stability of one part in 10^9 can be achieved with such oscillators; for an example of this *see* QUARTZ CLOCK. For detailed information on the Pierce and Meacham circuits *see* OSCILLATOR.

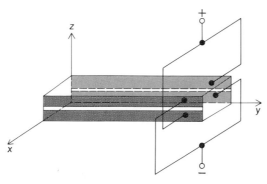

Fig. 7. Excitation of flexure mode by split electrodes.

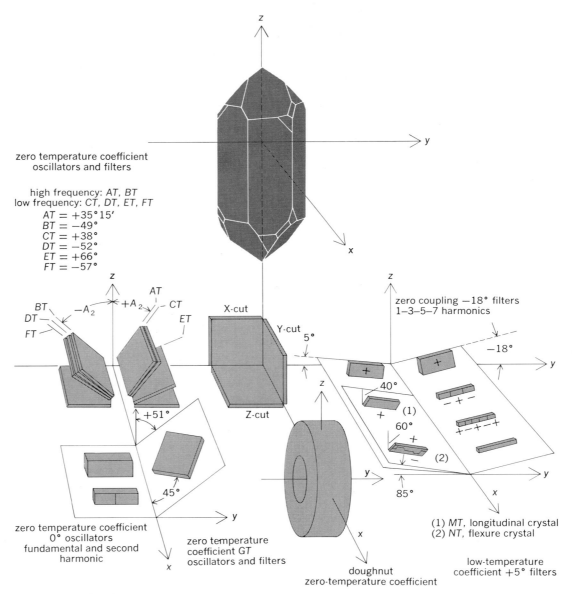

Fig. 8. Orientation with respect to the natural crystal axes of some of the more commonly used special cuts of quartz. (*From W. P. Mason, Piezoelectric Crystals and Their Application to Ultrasonics, Van Nostrand, 1950*)

Selective band-pass filters with low losses can be built by using piezoelectric resonators as circuit elements. With a simple network consisting of resonating crystals only, a passband of twice the difference between resonant and antiresonant frequency can be obtained. For quartz resonators, this passband is about 0.8%. At relatively low operating frequencies, this band is too narrow, and combinations of crystal resonators with coils and condensors are generally used. A synthetic piezoelectric crystal which is often substituted for quartz in this application is ethylene diamine tartrate.

Piezoelectric crystals provide the most convenient means for generation and detection of vibrations in gases, liquids, and solids at frequencies above 10^4 Hz. Quartz, ammonium dihydrogen phosphate, rochelle salt, and barium titanate are frequently used in sonic and ultrasonic transducers. The mechanical impedances of liquids and solids are generally close enough to the mechanical impedance of the piezoelectric crystal so that efficient energy transfer is possible. The intensity of ultrasonic radiation that can be achieved is mainly limited by the mechanical strength of the piezoelectric crystal. The maximum ultrasonic intensity theoretically obtainable in water by means of quartz or ammonium dihydrogen phosphate is of the order 2000 watts/cm^2 and 200 watts/cm^2, respectively. For gases, the mechanical impedance match is so poor that the corresponding values are about 4000 times smaller. However, the mechanical impedance match can be greatly improved by using piezoelectric devices consisting of two differently oriented crystal cuts cemented together in such a way that a voltage applied to the electrodes causes the elements to deform in opposite directions, and a twisting or bending action results. Assembles of this type (bimorphs) with $BaTiO_3$ ceramics or rochelle salt are widely used in such devices as microphones, earphones, and phonograph pickup cartridges.

Ultrasonic waves at microwave frequencies up to 2.4×10^{10} Hz have been generated by means of the piezoelectric effect. The arrangement is shown in Fig. 9. The end surface of a piezoelectric crystal rod is exposed to a strong microwave electric field in a resonant reentrant cavity. The ultrasonic waves travel through the crystal rod in a guided

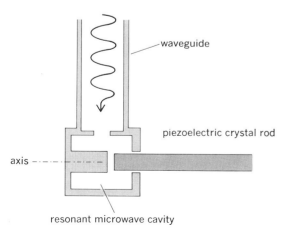

Fig. 9. Diagram showing experimental arrangement for generation of ultrasound at microwave frequencies by means of a piezoelectric crystal.

wave mode. The attenuation is low only at very low temperatures.   [H. GRANICHER]

Bibliography: D. A. Berlincourt, D. R. Curran, and H. Jaffe, Piezoelectric and piezomagnetic materials and their function in transducers, in W. P. Mason (ed.), *Physical Acoustics*, vol. 1, pt. A, 1964; W. G. Cady, *Piezoelectricity*, 1964; W. P. Mason, *Crystal Physics of Interaction Processes*, 1966; J. F. Nye, *Physical Properties of Crystals: Their Representation by Tensors and Matrices*, 1957.

Point-contact diode

A semiconductor rectifier using the barrier formed between a specially prepared semiconductor surface and a metal point to produce the rectifying action. The contact is usually maintained by mechanical pressure, but in some instances it may be welded or bonded. The rectifying action implies that the resistance of the contact is significantly greater for one direction of applied voltage (reverse direction) than for the other (forward direction). *See* SEMICONDUCTOR RECTIFIER.

Point-contact diodes have been widely used in radio and television, and most notably in computers, microwave detectors, and ultrahigh-frequency mixers. Today their use is very small, having been usurped by junction diodes and Schottky barrier diodes.

Whenever a metal-semiconductor contact is made, an electrical barrier generally exists between the two. Only specially prepared ohmic contacts show no barrier. This barrier impedes the flow of majority carriers. For a definition of majority carriers and a discussion of conductivity type *see* SEMICONDUCTOR.

In an *n*-type semiconductor the majority electrons are immobilized by the barrier, and a bias, which renders the semiconductor positive with respect to the metal, repels the positive holes (electron vacancies) in the metal. Very little current flows under this condition and the resistance is high. If the *n*-type semiconductor is negative with respect to the metal, most of the electrons are still immobilized by the barrier. However, holes can now enter from the metal and a relatively large current flows, that is, low resistance is present. For a *p*-type semiconductor the barrier impedes holes, and the bias polarities are reversed for the high- and low-resistance conditions.

The small physical size of the point contact forces a high current density in the neighborhood of the point. The current distribution in the semiconductor gives rise to a spreading resistance in series with the barrier. The forward current is limited by this resistance while the reverse current is limited by the barrier. The spreading resistance steadily decreases with increasing forward current because of heating and the injection of minority carriers. *See* TRANSISTOR.

In the reverse direction there is a breakdown phenomenon at relatively high voltages due either to heating or avalanching of the minority carriers passing through the high field barrier region.

[LLOYD P. HUNTER]

Point-contact transistor

A transistor in which the emitter and collector consist of metal point contacts closely spaced on the surface of a block of semiconductor. The usual

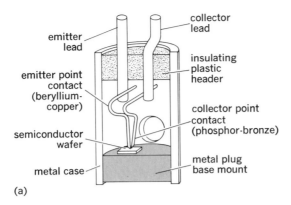

(a)

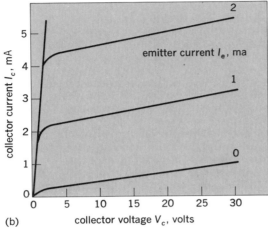

(b)

Point-contact transistor. (a) Cutaway view. (b) A graph of the characteristics.

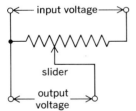

Fig. 1. Schematic of potentiometer.

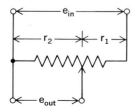

Fig. 2. Schematic of voltage divider.

compares to 1 microampere (μA) for a junction transistor under the same condition.

Point-contact transistors can be made with frequency ranges up to 100 MHz and power ratings of 200 milliwatts (mW). They can be used quite conveniently for oscillators and flip-flops, because their $\alpha > 1.0$ causes them to show a negative resistance characteristic when the base is used as an input. They have not achieved widespread acceptance because of the variability of their characteristics and because of their relatively high I_{co}. Today point-contact transistors are almost unknown in practical use.

[LLOYD P. HUNTER]

Bibliography: L. P. Hunter (ed.), *Handbook of Semiconductor Electronics*, 3d ed., 1970; W. Schockley, *Electrons and Holes in Semiconductors*, 1950, reprint 1976.

Potentiometer (variable resistor)

A variable-resistance device with three terminals used in electric circuits. As shown schematically in Fig. 1, the three terminals are the two ends of a resistor (or series combination of resistors) and a movable connection, which allows adjustment of the resistance between this movable connection and either end connection. The movable connection often consists of a sliding contact which moves along the actual resistor element. The size or rating or a potentiometer is specified by giving its total resistance in ohms and the permissible losses in watts. By using only the movable and one fixed connection, a potentiometer may be used as a rheostat. *See* RESISTOR.

The term potentiometer is also applied to a precision instrument used to measure or compare electrical voltages. For this device which depends on the same type of resistor arrangement *see* POTENTIOMETER (VOLTAGE METER).

Use. A potentiometer is used to adjust and control the electric potential difference (voltage) applied to some device or part of a circuit. The output voltage may be varied from zero to the value of the input voltage. Examples of its use are as a field-current control on an electric generator and as a volume control on a radio. Since the resistance between the input terminals is fixed (assuming the load device takes little current), potentiometers with precision resistors are used as range selector switches on vacuum-tube voltmeters and other precision electronic measuring equipment. Other uses are to divide a voltage into two parts, to compare two voltages, and to divide the total resistance between two parts of a circuit. The ratio between output and input voltage, as shown in Fig. 2, is given by the equation below.

$$\frac{e_{\text{out}}}{e_{\text{in}}} = \frac{r_2}{r_1 + r_2}$$

Construction. A potentiometer may be linear or nonlinear. In the linear case the resistor is uniform, and the voltage distribution along the resistor is the same for any fixed fraction of its total length. Therefore, the output voltage (Fig. 1) is proportional to the slider position. In a nonlinear potentiometer the resistance per unit length varies, and the output voltage varies as some function (such as the logarithm, the square, or the sine) of the slider position. Nonlinear potentiometers are

configuration is with both points on the same surface and about 2 mils apart (see illustration), although good devices have also been made with the points on opposite sides of a thin wafer of semiconductor.

This type of transistor was the first transistorlike device invented. The most common type uses *n*-type semiconductor material, beryllium-copper emitter-point material, and phosphor-bronze collector-point material. In fabrication the surface of the semiconductor is carefully lapped and etched. The sharp points are mechanically assembled with some spring pressure against the surface. The collector point is electrically pulsed in the reverse-bias direction with sufficient voltage and total energy to cause electrical breakdown. The point of contact is heated nearly to the melting point of the semiconductor. The pulse duration is a millisecond or less. The result of this electric forming procedure is to increase the current multiplication factor α^* of the collector point from something much less than unity to the order of 10.0. The injection efficiency γ of the point emitter is about 0.3 and the transfer efficiency is about 1.0 so that the overall current gain α of the device is 3.0. *See* SEMICONDUCTOR; TRANSISTOR.

The electrical forming process, besides increasing α^*, also increases the collector barrier leakage current I_{co} so that, at a collector voltage of 10 volts, a typical point-contact device will draw 1 milliampere (mA) in the absence of emitter current. This

often called tapered potentiometers. In some cases the current-carrying capacity of various parts of the potentiometer may be different.

A slide-wire potentiometer employs a movable sliding connection on a length of resistance wire.

A wire-wound potentiometer is similar to a slide-wire one, except that the resistance wire is wound on a form and contact is made by a slider which moves along an edge from turn to turn. The form may be straight or bent into a part of a circle, in which case the slider is mounted on an arm which is rotated by a knob. The form may be made of a ceramic material for heat resistance or a good grade of stiff paper or plastic (known as a card).

A carbon potentiometer uses a thin layer of carbon or graphite in place of a resistance wire.

A button-type potentiometer uses fixed contact points which are touched by the slider. Fixed resistors are then connected between the buttons.

Multiturn potentiometers. Sold under various trade names, these have the resistance material, usually coiled resistance wire, placed in the form of a cylindrical helix. A slider is moved down the helix by a lead-screw arrangement. This arrangement permits a long length of potentiometer in a small volume and gives more accurate adjustment than is possible with a single-turn potentiometer.

Trimmer potentiometer. This is a potentiometer used to provide a small percentage adjustment and is often used with a coarse control.

[CLARENCE F. GOODHEART]

Potentiometer (voltage meter)

A device for the measurement of an electromotive force (emf) by comparison with a known potential difference. The known potential difference is established by the flow of a definite current through a

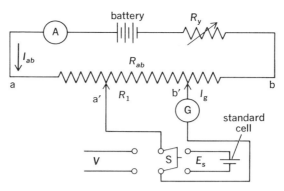

Fig. 1. Elementary dc potentiometer circuit. (*From I. F. Kinnard, Applied Electrical Measurements, copyright © 1956 by John Wiley and Sons, Inc.; used with permission*)

known resistance, using a standard cell as a reference. The principal potentiometer circuits are (1) the constant-current dc potentiometer (historically known as Poggendorff's first method); (2) the Brooks deflectional dc potentiometer (a variant of the basic constant-current potentiometer); (3) the constant-resistance dc potentiometer (known as Poggendorff's second method); (4) the Drysdale ac potentiometer; and (5) the Tinsley-Gall ac potentiometer. *See* ELECTRICAL MEASUREMENTS.

Constant-current dc potentiometer. This is widely used in the standardization of dc measuring instruments; its basic circuit is illustrated in Fig. 1.

A battery causes a current I_{ab} to flow through a calibrated resistor or slidewire R_{ab}. With the switch S connected to the standard cell and sliders a′ and b′ set for an appropriate value of resistance R_1, the slidewire current I_{ab} is adjusted until the

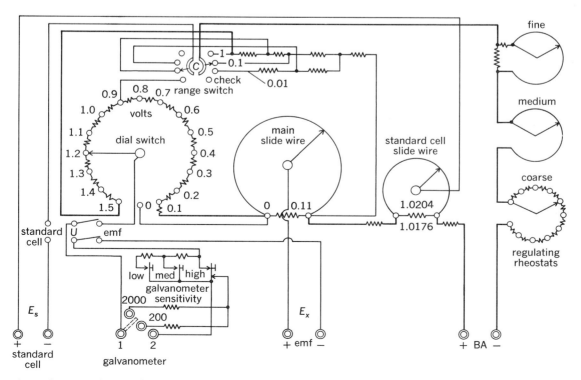

Fig. 2. Complete dc potentiometer circuit. (*Leeds and Northrup Co.*)

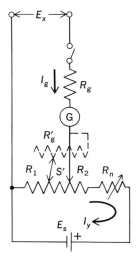

Fig. 3. Elementary Brooks deflectional potentiometer. (*From I.F. Kinnard, Applied Electrical Measurements, copyright © 1956 by John Wiley and Sons, Inc.; used with permission*)

galvanometer current I_g is equal to zero. The slide-wire voltage $R_1 I_{ab}$ is now equal to the standard cell emf E_s.

The unknown voltage V is then substituted for the standard cell and the sliders a′ and b′ are again adjusted for a null galvanometer reading, with slide-wire current I_{ab} unchanged. Let the new value of slidewire resistance be R_2. The unknown voltage is given by Eq. (1). Thus, this potentiom-

$$V = E_s R_2 / R_1 \qquad (1)$$

eter compares potential differences in terms of known resistances, which are usually calibrated in terms of volts and millivolts.

A complete potentiometer circuit is shown in Fig. 2. The resistance of the dial switch, the main slidewire, and the standard slidewire represent R_{ab}. Transfer from the standardizing to the measuring circuit is accomplished by the switch U. With the switch in the left position the standard cell, with voltage E_s, and the galvanometer are connected across the resistance composed of nine coils and part of the slidewire. Balance is obtained by adjustment of the regulating rheostats. The switch is then thrown to the right, connecting the unknown voltage E_x and the galvanometer across the dial switch and main slidewire, both of which are adjusted to obtain balance. A range switch

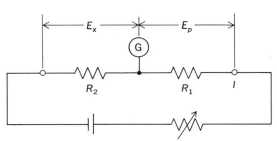

Fig. 4. Elementary constant-resistance potentiometer circuit. (*From I. F. Kinnard, Applied Electrical Measurements, copyright © 1956 by John Wiley and Sons, Inc.; used with permission*)

reduces the slidewire current to 1/10 or 1/100 of its full value when required. The measurement range of the constant-current potentiometer is 0–2.0 volts. Multipliers of 1, 0.1 and 0.01 are available. This device has undergone intensive refinement, and errors can be reduced to the order of 0.025% of the reading. Potentiometer measurement of current is accomplished by passing current through a standardized resistor of appropriate value and measuring the potential difference across this resistor.

Brooks deflectional dc potentiometer. This potentiometer eliminates the time-consuming operation of obtaining an exact balance of the slidewire. This is accomplished by circuitry and a galvanometer calibration by which the galvanometer reading may be added to or subtracted from an approximate dial setting. As illustrated in Fig. 3, E_x is the unknown voltage, E_s is the terminal voltage of a storage cell of negligible resistance, and R_g is the resistance of galvanometer plus its series resistor. Analysis of the circuit shows that if the

quantity $R_g + R_1(R_2 + R_n)/R_1 + R_2 + R_n$ is kept constant for all positions of the sliders, the galvanometer can be calibrated directly in volts. This is achieved by the addition of the auxiliary resistor R'_g in series with the galvanometer and slider. As the slider is moved, an increment of resistance ΔR_g is added to or subtracted from the above resistance to keep the ratio constant. The total unknown emf is the algebraic sum of the slider setting and galvanometer reading. The instrument is otherwise similar to the basic constant-current potentiometer. Self-contained deflectional potentiometers are available in ranges of 0–1.5 volts, which may be extended to 300 volts, and 0–0.75 amp, which may be extended to 150 amp with special multipliers. Maximum errors of 0.05% self-contained, plus 0.04% for multipliers, if used, are obtained.

Constant-resistance dc potentiometer. This instrument is especially adapted to the measurement of very low potentials. As illustrated in Fig. 4, the unknown potential E_x is applied to the known constant resistance R_2, and the current I is adjusted for null reading of the galvanometer. In series with R_2 is another constant and known resistance R_1. The potential E_p across this resistance is measured by any null-type potentiometer. Then Eq. (2) holds. Thus, the range of the null-type po-

$$E_x = \frac{E_p R_2}{R_1} \qquad (2)$$

tentiometer may be extended downward manyfold by making R_1 large and R_2 small. Reliable measurement in the microvolt range is practicable if care is taken to avoid parasitic emfs in R_2. The null potentiometer may be replaced by a dc voltmeter, but measurement accuracy is thereby diminished.

Drysdale ac polar potentiometer. This is shown in principle in Fig. 5. An ordinary slidewire is supplied with current by a phase-shifting transformer. The amount of current is measured by an ammeter which must be an accurate ac to dc transfer instrument. The potentiometer current and voltage drop along the slidewire are brought into phase with the unknown voltage by adjustment of the transformer

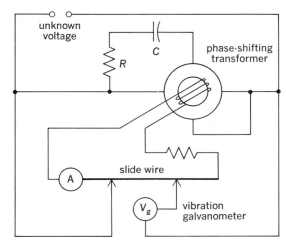

Fig. 5. Drysdale ac potentiometer circuit. (*From I. F. Kinnard, Applied Electrical Measurements, copyright © 1956 by John Wiley and Sons, Inc.; used with permission*)

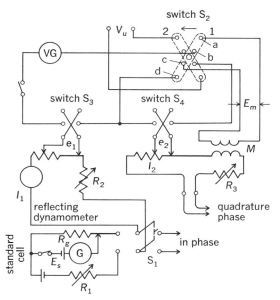

Fig. 6. Tinsley-Gall ac potentiometer circuit. *(From I. F. Kinnard, Applied Electrical Measurements, copyright © 1956 by John Wiley and Sons, Inc.; used with permission)*

rotor. The unknown voltage is then measured by observation of the slidewire setting for a null indication of the vibration galvanometer. The potentiometer is calibrated on direct current by use of a dc potentiometer. The equivalent value of ac, as measured by ammeter A, is maintained for ac measurement. Maximum errors of 0.1–0.2% are obtained depending upon the accuracy of the ammeter and sensitivity of the galvanometer.

Tinsley-Gall ac polar potentiometer. Shown in Fig. 6, this potentiometer measures the unknown emf by measurement of the in-phase and quadrature components in reference to a standard current. Two potentiometers are used, the currents in which are numerically equal but 90° displaced in phase. The in-phase potentiometer is first standardized by closing switch S_1 to the left and adjusting R_1 for null reading of galvanometer G. Then Eq. (3) holds. R_g is preadjusted so that a definite

$$E_s = I_1 R_g \qquad (3)$$

current, usually 50 ma, flows through the slidewire. The reading of the reflecting dynamometer is then observed. Alternating current from the in-phase source is then applied by closure of switch S_1 to the right, and rheostat R_2 is adjusted to reproduce the 50-ma reading of the reflecting dynamometer. The quadrature potentiometer is now standardized by closure of S_2 is position 1, together with closure of contacts ab and cd. This series connects the quadrature slidewire through the vibration galvanometer and the in-phase slidewire to the secondary of mutual inductor M in which an emf, given by Eq. (4), is generated. Reversing

$$E_m = j\omega M I_2 \qquad (4)$$

switches S_3 and S_4 are connected so that the voltage drops e_1 and e_2 are opposed. The in-phase slidewire is set to a predetermined value of induced voltage E_m as determined by the frequency and coefficient of mutual induction. The rheostat R_3 is

adjusted for null indication of G, and 50 ma now flow through the quadrature slidewire. The transfer switch S_2 is changed to position 2, cutting out the mutual inductor and series connecting the slidewire switches S_2 and S_3, as well as the vibration galvanometer to the unknown voltage V_u. The apparatus is now ready for measurement of the quadrature components of the unknown voltage. All four quadrants may be explored by use of reversing switches S_3 and S_4. This potentiometer, for best results, requires a very stable ac power supply consisting of two single-phase alternators with adjustable stators. It is used especially for such specialized laboratory procedures as instrument-transformer ratio, phase-angle determination, and ac cable testing. For other methods of measuring voltage see VOLTAGE MEASUREMENT; VOLTMETER.

[ALMON J. CORSON]

Bibliography: M. Braccio, *Basic Electrical and Electronic Tests and Measurements*, 1979; E. Frank, *Electrical Measurement and Analysis*, 1959, reprint 1977.

Power amplifier

The final stage in multistage amplifiers, such as audio amplifiers and radio transmitters, designed to deliver appreciable power to the load.

Power amplifiers may be called upon to supply power ranging from a few watts in an audio amplifier to many thousands of watts in a radio transmitter. In audio amplifiers the load is usually the dynamic impedance presented to the amplifier by a loudspeaker, and the problem is to maximize the power delivered to the load over a wide range of frequencies. The power amplifier in a radio transmitter operates over a relatively narrow band of frequencies with the load essentially a constant impedance.

The mode of operation of power amplifiers is denoted by class A, AB, B, and C. Class C operation is limited to radio frequencies with a tuned load. The other classes may be used for audio and high-frequency operation. For discussion of the modes of operation see AMPLIFIER.

Class A. Class A operation is used when the amount of power transferred to the load is relatively small, say, less than 10 watts. The amount of harmonic distortion introduced into the load voltage can be kept small by operating transistors or tubes within the nearly linear region of their characteristics. Class A operation has relatively little use because the conversion efficiency (the efficiency of a power amplifier) is low. The maximum possible efficiency is 50%. However, for the usual operating conditions the efficiency is on the order of 10%. If the power amplifier were required to deliver 10 watts with 10% efficiency, the tube or transistor would have to be capable of dissipating an average power of 100 watts. Furthermore, the power supply must be capable of supplying the power dissipated as heat plus the useful power delivered to the load. This poses an unnecessary burden upon the power supply. Other classes of operation have a higher conversion efficiency and are therefore used for higher-power applications.

Class AB. An improvement in the conversion efficiency can be had by using class AB operation. However, while a class A amplifier can be operated single-ended (one output transistor or tube), a class AB amplifier must be operated push-pull. In class AB operation the transistor or tube current does not flow for the complete cycle of the input voltage. In a single-ended circuit this would introduce excessive distortion. See PUSH-PULL AMPLIFIER.

Class B. This class is often used for the power amplifier in an audio amplifier. The amplifier in this class must be a push-pull circuit. Theoretically, with ideal transistors or tubes, the class B amplifier can have a conversion efficiency of 78.5%; practically, the efficiency is on the order of 50%, an appreciable improvement over that of class A operation.

The load is usually transformer-coupled to the two transistors or tubes operating in push-pull. For maximum power transfer the dynamic load impedance presented to the amplifier is determined by the amount of harmonic distortion that can be tolerated.

Use of more sophisticated circuitry than that considered in an elementary presentation of a push-pull amplifier operating in class B can produce nearly distortionless power amplification. This is of prime importance in the final amplifier stages of a high-fidelity audio amplifier.

Vacuum-tube power amplifiers can operate in class B_2 with an appreciable amount of grid current flowing for a small portion of the cycle of an input sinusoidal signal. This imposes additional requirements upon the driving stage of the amplifier. If the equivalent circuit of the driver has too large an equivalent output impedance, the flow of grid current through this impedance will cause a distortion of the grid waveform. This class of operation is usually encountered in driver circuits for class B amplifiers operating in the radio-frequency region where high-Q circuits counteract effects of grid current. Audio operation is usually restricted to class B_1 operation, because the usual form of phase-inverter circuitry has a large output impedance. In radio-frequency operation the transformer phase inverter can be used with tuned circuitry and air-core or powdered-iron slug coils, because the operation is essentially at one frequency.

Class C. Because the collector or plate current flows for less than one-half cycle of the input sinusoidal signal, this class of operation is restricted to radio-frequency operation where a tuned load is employed. The load is usually the input impedance of an antenna or of an antenna matching network. The load voltage will be nearly sinusoidal, even though the current in the tube flows in pulses, because of the relatively sharp tuning of the load. This phenomenon allows the amplification of large amounts of power at conversion efficiencies as high as 80%. This is extremely important for applications requiring delivery of large amounts of power to the load.

The driving source must usually be called upon to deliver power to the base circuit or grid circuit of the power amplifier, in many cases as much as 10% of the power delivered to the load. This requirement is not excessive. A class B power amplifier can be used in the grid-driving circuit to obtain an efficient combination of driver and final amplifiers.

[HAROLD F. KLOCK]

Bibliography: C. A. Holt, *Electronic Circuits,* *Digital and Analog,* 1978; J. Millman, *Micro-Electronics,* 1979; D. L. Schilling and C. Belove, *Electronic Circuits, Discrete and Integrated,* 2d ed., 1979.

Preamplifier

A voltage amplifier suitable for operation with a low-level input signal. It is intended to be connected to another amplifier with a higher input level. Preamplifiers are necessary when an audio amplifier is to be used with low-output transducers such as magnetic phonograph pickups. A preamplifier may incorporate frequency-correcting networks to compensate for the frequency characteristics of a given input transducer and to make the frequency response of the preamplifier-amplifier combination uniform. See VOLTAGE AMPLIFIER.

The design of preamplifiers is critical because the input signal level is low and the amplification is high. The hum introduced by tubes and the noise voltages from resistors and vacuum tubes must be closely controlled. Furthermore, the preamplifier must be shielded from external magnetic fields that would otherwise induce stray voltages in the circuit. For a discussion of these undesirable conditions see AMPLIFIER.

[HAROLD F. KLOCK]

Printed circuit

A generic term applied to circuits fabricated by any of several graphic art processes. Printed circuits greatly simplify mass production and increase equipment reliability. Their most important contribution, however, is the tremendous reduction achieved in size and weight of electronic devices. Printed circuits are used in practically all types of electronic equipment, radio and television sets, telephone systems units, electrical wiring behind automobile dashboards, guided-missile and airborne electronic equipment, computers, and industrial control equipment.

Advances in electronic equipment since World War II could not have been achieved without the sophisticated manufacturing processes made possible by graphic technology. Fabrication of very lightweight, extremely small equipment containing such devices as transistors, diodes, and integrated circuits depends upon the high precision attainable only through applied graphics. The rapid adoption of graphic art processes by industry is a demonstration of the effectiveness of those processes in achieving significant cost reduction and equipment miniaturization. See JUNCTION DIODE; SEMICONDUCTOR DIODE; TRANSISTOR.

Printed circuits are of interest to industry because:

1. They are the common denominator for almost all approaches to mechanized fabrication of electronic equipment.

2. Their use has greatly reduced the labor required for the wiring of an electronic circuit. This reduction is especially significant for small units used in computer and guided-missile equipment.

3. They can be manufactured uniformly because graphic art processes provide repeatable results.

4. Their uniformity improves the product through simplification of quality control.

5. Printed circuitry has helped to minimize unskilled labor and to greatly reduce one major cause

of unreliability in electronic equipment (individually soldered connections) by permitting the use of dip-soldering rather than hand-soldering processes. (In dip soldering, connections between the electronic component and the conductor are exposed to molten solder and joined in one precisely controlled operation.)

6. Precision capabilities of graphic art processes (particularly in integrated circuits) have resulted in significant size and weight reductions, which are especially important in electronic equipment for such applications as medical instrumentation and space exploration.

TECHNOLOGY

Printed circuit technology may be divided into three basic parts: engineering, photography, and manufacturing.

Engineering. The configuration in which circuit elements are located and the routing of conductor paths establish the precise circuit pattern. Using this pattern, an artwork master normally several times larger than the final size of the printed circuit is prepared. Scale depends on the type of circuit that is to be fabricated. Masters are prepared by applying matte-finished black or red tapes to a clear, stable-based material or by the cut-and-strip method, in which a photographically opaque coating on a stable base is cut and stripped away as necessary. The automation of the design for optimum location of circuit elements and of the preparation of masters using computer programs for conductor routing and the use of automated photoplotting equipment (Fig. 1) have become essential as a result of increased circuit complexity and the need for greater accuracy.

Photography. Enlarged masters are photographically reduced to the required size within the allowable tolerance, thereby reducing any dimensional inaccuracies which have occurred in the master. When a number of circuits are to be fabricated simultaneously, precision step-and-repeat equipment provides a composite master, with each circuit precisely positioned relative to key indexing locations, and all identical.

A typical camera used for thin-film reductions, glass plate tooling, and first- and second-stage reductions of integrated circuit patterns requires that the three important vertical planes, the copyboard, lens, and plane of the film, maintain alignment within 5 seconds of arc; therefore, rigid construction is essential. Precision adjustment of the location of the copyboard relative to the lens must be provided so that reduction ratios can be very accurately controlled. Precision measuring equipment, such as optical vernier and comparator systems, permits the positioning and measurement of an image on the film or glass plate within $\pm$ 0.0001 in. (± 2.5 μm).

Manufacturing. The circuit pattern masters are used to fabricate the screens and masks for the application of photoresistive materials in the actual formation of the required patterns on the finished parts. The masters are also used in the preparation of numerous types of tooling, for example, drill templates, tapes for operation of numerical-tape-controlled drilling equipment (especially common where high precision is required), routing templates and dicing fixtures for trimming printed circuits to final configuration, and laminating and holding fixtures. Numerous processes (described later), including etching, screening, plating, laminating, vacuum deposition, diffusion, and application of protective coatings, are used in combination to produce various types of printed circuits. Completed printed circuits are inspected visually and dimensionally, using such techniques as microsectioning and infrared photospectrometer measurements in determining thicknesses of critical materials; in addition, they may be x-rayed and electrically tested to assure conformance to requirements.

APPLICATIONS

Printed circuits can be divided, in terms of application, into (1) printed wiring, (2) thick and thin films, (3) hybrid circuits, and (4) integrated circuits.

Printed wiring. Printed wiring is undoubtedly the most common type of printed circuit. The printed wiring board is a copper-clad dielectric material with conductors etched on one or both sides (Fig. 2). Single-sided boards are commonly used in such commercial applications as automotive equipment, radio and television sets, and, occasionally, in military electronic ground-support equipment in which space and weight are not critical.

Two-sided boards may use plated-through holes or eyelets to provide electrical continuity between the sides. They are used in those applications in which the maximum number of interconnections (conductors) in a given area are required for minimum cost. Typically such applications include airborne, shipboard, and a large percentage of missile and spacecraft equipment.

Multilayer boards. Multilayer printed wiring boards capitalize on the reduced size of miniaturized and microelectronic parts and accommodate the increasing complexity of communication satellites which has caused increased density of circuitry. The surface area required for mounting of these subminiature parts and integrated circuit packages has decreased significantly while the number of interconnections has increased manyfold. Formerly, a part that occupied a given mounting area had only two or three leads; today a part that occupies the same area may have 20 or more leads. In a multilayer board, conductors are located on several insulated internal layers. Common

Fig. 1. In response to input commands from punched paper tape, this automatic photographic plotter accurately plots circuit patterns directly on a photographic glass plate, completing a typical 4 × 6 in. (10 × 15 cm) circuit pattern in 20–45 min. (*TRW Inc.*)

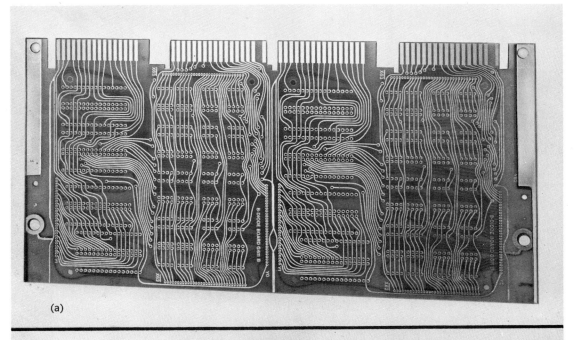

(a)

(b)

Fig. 2. Both sides of a printed circuit board after the etching and plating processes. (a) Component side. (b) Side that components are soldered to. (*Digital Equipment Corp., Maynard, Mass.*)

multilayer circuits consist of a number of two-sided etched copper foil boards, separated by an insulating layer (the prepreg) and laminated together under controlled temperature, pressure, and time. Feed-through and connection holes are drilled, and the cylindrical surface is plated by a chemical deposition process, then electroplated to form the interconnections between circuit layers or for the mounting of electronic parts (Fig. 3a). The monolithic (or plated-up) approach to multilayer printed wiring boards (Fig. 3b) is also used, but to a much lesser degree due to higher costs and more difficult production processes.

Responding to the demand for greater miniaturization by reducing the circuit board size, conductor line widths as narrow as 0.02 in. (0.5 mm) and minimum spacing between conductors of 0.015 in. (0.4 mm) with up to eight layers (16 conductor surfaces) are common in space applications. The fabrication of a multilayer printed circuit is a precision manufacturing process requiring tight control of tolerances for both the conductor pattern and the indexing of each layer of the circuit board. Thus, only dimensionally stable glass-cloth-based plastic laminates are used. Due to their much lower cost, phenolic paper laminates are used as the base material on single- or two-sided printed circuits for consumer products and on some commercial equipment. The use of higher radio frequencies (1-GHz range) or more rapid response digital circuitry has required other than the 1-oz (28-g) or 2-oz (57-g) copper on an epoxy glass laminate, standard printed circuit board material. In the gigahertz range, polytrifluoromonochlorethlene (Teflon) glass is used as the laminate base for the copper foil. At still higher frequencies (above 3 GHz) a ceramic base material (substrate) such as alumina is used, and the conductors or circuit elements such as inductance and transmission lines are fabricated by a vacuum deposition process. In the 10-GHz range, synthetic sapphire is sometimes used on account of its superior dielectric properties, although high cost has limited its use. Polyimide glass is sometimes used to replace epoxy glass as the laminate base, especially when very narrow line widths are used. The polyimide material is stronger mechanically, but, even more important, it can withstand the high temperatures used for soldering components to the circuit board. This characteristic is especially useful when conductor widths are at a minimum 0.02 in. (0.5 mm).

Heat dissipation. Heat-dissipating sinks for heat-generating components may be provided by leaving most of the copper cladding intact on the surface on which the electronic parts are mounted; conductors are then routed on the opposite side or on internal layers of multilayer board. A metal sheet coated with a thermally conductive dielectric, such as hard-anodize thermal conductive aluminum oxide, with the circuit pattern bonded on the dielectric coating, may be used where large quantities of heat must be dissipated.

Protective finish. The copper pattern on both single- and double-sided boards and exposed layers of multilayer boards are normally given a protective finish, commonly either gold or tin-lead alloy to improve solderability. Tin-lead is used more often, when the assembly is to be dip-soldered.

Flexible printed wiring. Flexible printed wiring (Fig. 4) is another form which reduces assembly line time significantly when used to interconnect subassemblies. Several layers of etched conductors may be sandwiched between thin, flexible, dielectric materials to form wiring between units in consoles and cabinets of ground-support equipment. This method of interconnection has the added feature of rigidly fixing the position of each conductor, thus controlling the electromagnetic and electrostatic coupling between circuits, and is especially advantageous at high frequencies.

Connections. Methods most often used to attach parts to printed wiring boards are: soldering leads in terminal holds to surrounding pads which are part of the conductive pattern and lap soldering or thermocompression bonding of flat leads to pads on the surface of the board (Fig. 5).

Thick-film circuits. Thick-film circuits (Fig. 6) consist of such passive elements as resistors, capacitors, and inductors deposited on wafers or substrates of such dielectric materials as ceramic, glass, porcelain-coated metal, and the like. They are used commercially for mass fabrication of passive networks for use in linear microcircuits and large-signal digital modules. A large-scale electronic computer may contain millions of these circuits. Thick-film design and manufacture are usually based on film thickness of approximately 0.001 in. (25 μm).

Each type of circuit element requires: (1) a master pattern defining the configuration of the specific element or portion of the element; this pattern is

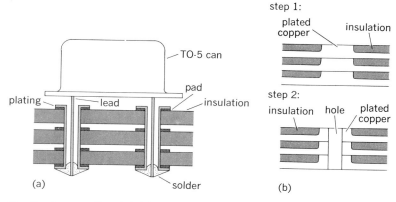

Fig. 3. Cross-sectional detail of multilayer printed wiring board. (*a*) TO-5 can is attached by its leads to a plated-through multilayer circuit. (*b*) Two steps in the fabrication of a hole in a plated monolithic multilayer circuit. (*TRW Inc.*)

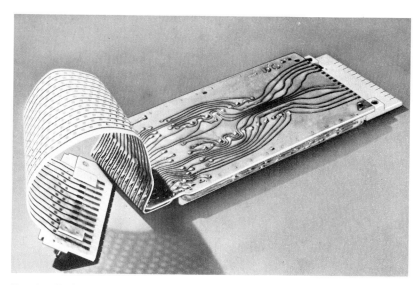

Fig. 4. Flexible printed wiring, used to interconnect subassemblies. (*Technical Liaison Co., Inglewood, Calif., and Flexible Circuits Inc., Warrington, Pa.*)

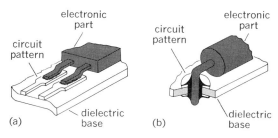

Fig. 5. Methods for attaching parts to printed wiring boards. (*a*) Swaged or welded connections. (*b*) Soldered connections. (*TRW Inc.*)

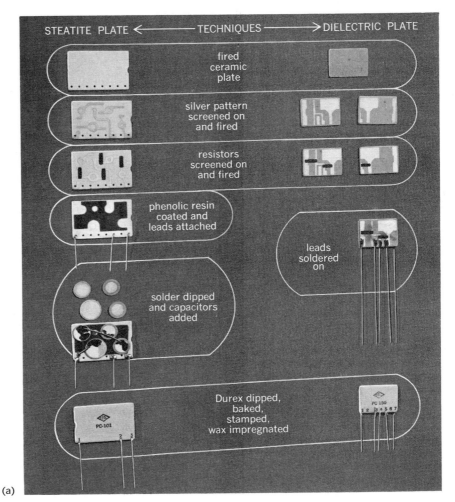

(a)

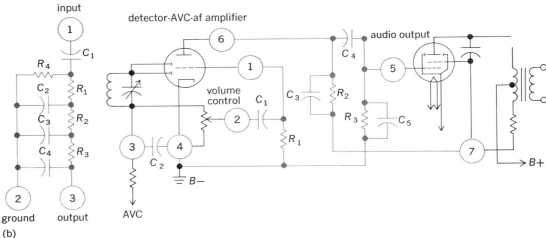

(b)

Fig. 6. A printed circuit on ceramic-base materials. (a) Circuits in left column screened on a steatite ceramic with barium titanate ceramic capacitors soldered to base. Circuits on right column screened on a high-electric-constant ceramic. The base plate is used as the dielectric for the capacitor formed by screening a conductive area on opposite sides of the plate. (b) Schematics of circuits in a (gray). Circuit on right in representative situation. AVC = automatic volume control; af = audio frequency. (*Centralab Div., Globe Union, Inc.*)

commonly 5–25 times actual size; (2) a precision photographic reduction within ±0.005 in. (±125 μm) of actual size and within ±0.002 in. (±50 μm) of the other patterns in a set; (3) a photolithographically prepared stainless steel stencil of the pattern; (4) deposition of the appropriate ink through the stencil; (5) air-drying of the film; and (6) firing at an elevated temperature.

Resistor networks. Typically, a printed resistor network consists of a pattern using resistive inks and a conductor. The inks are high-viscosity mixtures of precious metals or precious metal oxides and a vitreous binder suspended in an organic vehicle. After being printed in the designed pattern on the dielectric, these inks are fired at a temperature of approximately 750°C, and a conductor pattern of silver or gold inks is then printed in registration and fired at temperatures ranging from 850 to 1000°C. Printed resistances range from 10 to 10^6 ohms and are determined by varying the type of ink used, the aspect ratio (the ratio of length to width) of the rectangular pattern, and, to some extent, the thickness of the film.

Through careful process control and minimized master and photographic tolerances, resistive values within ± 5% can be produced. When tighter tolerances are required, the resistive value may be adjusted by abrading the surface or one edge of the pattern. Thus the resistive element is formed with a value below that required and then adjusted upward to fall within design tolerance limits. The stability of the resistance depends primarily upon the type of ink used. In general, resins or glass binders cured at higher temperatures provide the more stable resistances; therefore, an epoxy or silicone binder provides a more stable resistance than that obtained with a phenolic binder. Stability can be improved by aging at high temperatures.

Carbon inks fired at temperatures between 150 and 600°F (65 and 315°C) are frequently used in the manufacture of printed resistors in such consumer products as radios or television sets which the tolerance requirements and operating environment are not too severe. _See_ RESISTOR.

Capacitances. Printed capacitive elements are formed by first depositing a base electrode of the same material as that of the conductor patterns, then depositing a dielectric film (barium titanate, titanium dioxide, and borosilicate glass frit being among those commonly used) fired at approximately 775°C, and finally, depositing the counter electrode and connection to the conductor pattern. Capacitors fabricated in this manner generally range in value from 10 to 10^8 picofarads, depending on the area used (3000–8000 picofarads/cm²/mil of dielectric thickness or 75,000–200,000 pF/cm²/μm) and the type of dielectric material. Capacitances can be adjusted after fabrication by: (1) providing several small parallel trimming capacitances integral with the basic capacitance; (2) fabricating the capacitor on the high side of the desired value; (3) severing the trimming capacitances as required to reduce the total value to within ±1% of the nominal capacitance. _See_ CAPACITANCE.

Printed capacitors are also fabricated by screening the electrodes on opposite sides of a ceramic wafer with a high dielectric constant, such as one of the titanates. The dielectric constant of the titanates, however, varies widely with temperature; thus these capacitors are temperature-sensitive and are limited to circuits that can accommodate wide circuit tolerances.

Inductors. Inductors in the form of printed spiral patterns of conductive inks are sometimes used. Inductances ranging from 1 to 900 microhenries are attainable. Use of ferrite coupling increases these values fivefold; however, the increased surface area required for printed inductors at fre-

quencies lower than 10 megahertz (MHz) generally makes the use of other types of inductors desirable. _See_ INDUCTOR.

Protective coatings. Resistive and capacitive elements are commonly coated with glass or non-hygroscopic resins to protect the elements from environmental effects and from other processing operations, such as solder coating or electroplating, which may be applied to conductor and inductor patterns in order to improve conductivity and solderability.

Thin-film circuits. The deposition of thin films was the first application of printed circuit technology to microelectronics. Like thick-film circuits, thin-film circuits consist primarily of passive elements; however, such active elements as field-effect transistors and diode devices are becoming available. The most important operational difference between thick- and thin-film circuits is that thin films have greater precision and stability. Some important characteristics of thin-film circuits are: (1) Films with a uniform thickness in the range from 5×10^{-6} to 5×10^{-3} mm can be vacuum-deposited and controlled by measuring the resistance across a test pattern during deposition to ensure that final thicknesses are within design limits; (2) patterns formed during deposition or by selective etching afterward are much more precisely controlled than those which are stenciled, as in thick-film circuits; and (3) more stable thin-film resistive materials can be used than those in thick-film circuits. Because of this precision and stability, thin-film circuits are frequently used in such aerospace and industrial electronics as differential amplifiers, ladder networks, and dc-to-ac signal converters.

Fig. 7. Using a coordinatograph, designers prepare artwork for a thin-film circuit pattern. (_TRW Inc._)

Master artwork for the circuit pattern is prepared at 25 times actual size using cut-and-strip material. The pattern is cut on a coordinatograph (Fig. 7) which is dimensionally accurate within ±0.0015 in. (±40 μm). Temperature, humidity, and dust control of the area in which the masters are prepared and photoprocessed is essential.

Resistor networks. Resistor networks (Fig. 8) are generally formed in one of two ways: The resistor pattern is formed on a substrate by vacuum-depositing resistive material on the substrate through

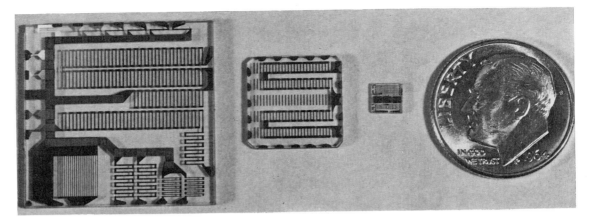

Fig. 8.   Three thin-film resistor networks. (*TRW Inc.*)

a metal mask (Fig. 9), and the conductor pattern (frequently gold over nichrome) is added in a subsequent vacuum deposition cycle; or patterns are photoetched directly on a substrate on which the resistive and conductive materials have been previously deposited. The latter technique has proved to be more practical for complex networks; however, use of standard metal masks in automatic processing equipment is less expensive. The most commonly used resistive material is nichrome (80% nickel, 20% chromium alloy), which is used to form resistive elements ranging from 100 to 75,000 ohms—current state of the art permits a maximum of approximately 6 to 8 megohms total per square inch (0.9 to 1.2 MΩ/cm^2). Resistances

are established and adjusted in basically the same manner as thick films by varying material, aspect ratio of the pattern, and film thickness; however, trimming is normally accomplished by providing in parallel several small resistive elements which may be disconnected from the primary resistive element by etching or diamond scribing as necessary to increase the total resistance. This trimming technique can be used to reduce the +5% as-fabricated tolerance to within ±1%. Because of the high stability and durability of nichrome, resistor networks of this type have been utilized in aircraft servomechanisms since World War II.

A second type of resistive material is a ceramic-metallic film commonly called cermet, which is

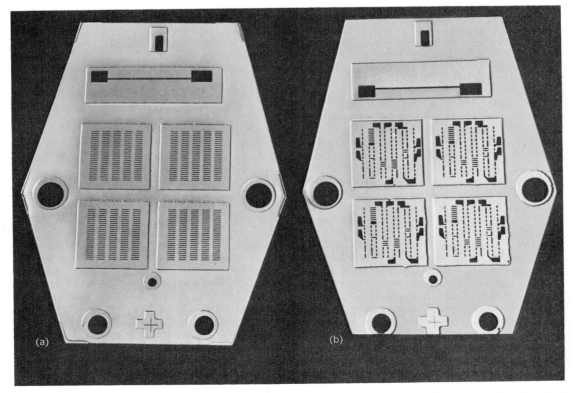

Fig. 9.   Metal masks for vacuum deposition of thin-film resistors. (*a*) Pattern for resistive elements. (*b*) Pattern for interconnection circuits. Conductor pattern is added in subsequent cycle. (*TRW Inc.*)

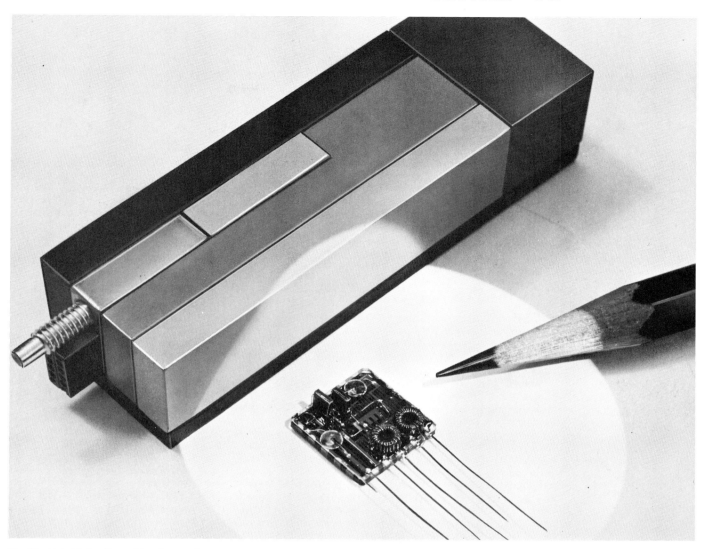

Fig. 10. Hybrid circuit combines lumped-constant components and printed-circuit techniques. (*TRW Inc.*)

formed by the simultaneous evaporation of chromium and silicone monoxide (or similar materials). Cermet has a much higher resistivity than nichrome and therefore requires less space for equivalent resistances. Cermet resistors are not as stable as those of nichrome. For these reasons, their use is generally limited to integrated circuits in which space is critical and subsequent processing of the circuit improves stability.

Capacitances. Thin-film capacitors are vacuum-deposited using masks similar to those shown in Fig. 9. The method of fabricating and adjusting values of thin-film capacitors is basically identical to that used for thick-film capacitors (by deposition of a base electrode, dielectric, and counterelectrode, with provision for trimming capacitances, and so on). Some dielectric materials that have been used include silicon monoxide, silicon dioxide, magnesium fluoride, zinc sulfide, and tantalum pentoxide. Typical capacitances range from 0.025 to 0.032 microfarad per square inch (3.9 to 5.0 nF/cm^2) with voltage ratings of $25-50$ volts per micrometer of dielectric thickness.

Post-deposition treatments. To accelerate stabi-

lization by oxidation of the surfaces, complete circuits are frequently coated or encapsulated in resins following attachment of leads to provide both mechanical support and environmental protection.

Hybrid circuits. Thick thin-film hybrid circuits (Fig. 10) consist of one or more substrates upon which the passive elements of the circuit (resistor networks and inductances) have been deposited and to which discrete components (such as diodes, transistors, integrated circuits, capacitors, and inductors) have been attached by split-tip series resistance welding or by thermocompression bonding, or by soldering (thick-film hybrids only). After components and external leads are attached, the circuit is normally encapsulated in a resin compound. Hybrid circuits allow advantageous use of thick- and thin-film fabrication techniques for such applications as complex switching circuits and differential amplifiers in the form of functional electronic building blocks.

At microwave frequencies (above 1 GHz) the spatial configuration of the circuit elements and the dielectric constant and dielectric losses of the base or substrate material become important; thus, hybrid circuits used at these frequencies are constructed on alumina (aluminum dioxide) ceramic or synthetic sapphire as the substrate. Conductors, inductances, and transmission lines are formed by vacuum deposition, etching, and electroplating using a photographic process very similar to that for the etched circuit board. Other circuit elements and active devices are then soldered to the conductor pattern.

Integrated circuits. An integrated circuit is a semiconductor device with both active and passive elements diffused into a silicon wafer to form a functional circuit. Because of the tremendous size reduction and long life, many applications are found in aerospace equipment besides consumer and industrial uses. Many of the fabrication processes for integrated circuits such as photomasking, selective etching, and photoreduction are similar to those used in the manufacture of printed circuits. *See* INTEGRATED CIRCUITS.

MANUFACTURING PROCESSES

Commonly used printed-circuit manufacturing processes may be divided roughly into three main groups listed in the order of greatest acceptance: (1) material removal, (2) film deposition, and (3) mold and die.

Material removal. Of the material-removal processes, photoetching and stencil etching are probably the most widely used techniques. They are used primarily in the fabrication of printed wiring.

Photoetching. In photoetching, the etchant-resist (conductor) pattern is formed photographically. The copper foil is coated with a photosensitive emulsion. A photographic negative of the circuit pattern is superimposed on the emulsion and exposed to ultraviolet light. This method is similar to that for the production of a photographic positive. During exposure the photoemulsion hardens. The plate is then placed in alcohol to dissolve the unexposed emulsion from the copper foil. The exposed areas are left covered by the hardened emulsion, which serves to protect the copper foil during the subsequent etching process. The uncovered copper is next dissolved in an acid or ferric chloride etchant bath. Finally, the hardened emulsion is dissolved from the exposed areas, leaving the copper conductor pattern.

An alternate method, which is as commonly used, exposes the emulsion by using a positive of the circuit pattern. The unexposed emulsion is removed; the final plating (gold or tin-lead) for the board is electroplated on the exposed circuit pattern; the emulsion is removed; and the board is etched using the gold or tin-lead plating for the etchant resist.

Stencil etching. In stencil etching, the protective film that forms the circuit pattern is applied by a printing process such as silk screening. The protective film, usually an enamel, is dried, and the exposed copper is etched as previously described. This method is not as precise as the photoetching technique; however, it is a more rapid and less expensive method of applying the etchant resist.

Film deposition. Stencil screening and electroplating are the processes most often used for deposition of materials to form resistances or capacitances. Photoelectrostatic techniques have also been developed, primarily for the fabrication of conductive elements.

Screening. Commonly used screens consist of a photosensitive film over a finely woven stainless steel screen. The stencil is usually formed by a photographic process similar to that used for the photoetching process mentioned above. Films are applied in the same manner as in silk screening.

Vacuum deposition. For vacuum deposition, films are deposited on the substrate by evaporating material (ultimately the film) in a vacuum of 10^{-5} torr (10^{-3} Pa) or lower (Fig. 11). At these pressures, the mean free path of the evaporated molecules or atoms exceeds the chamber dimensions sufficiently so that they travel in an essentially straight line between the source and the substrate. Source temperatures high enough to reach vapor pressures of 10^{-1} to 10^{-3} torr (10 to 10^{-1} Pa) for the source material are usually sufficient; such temperatures are in the 1100–1800 K range. Metal masks may be used within the vacuum to deposit specific patterns, or the entire surface of the substrates may be covered.

Plating. A plastic laminate, such as paper-base phenolic or epoxy-glass, is first coated with a material that conducts electricity. This may be done by depositing a thin (0.0001-in. or 2.5 μm) silver coating on the surface of the laminate in much the same way that mirrors are silvered. The silver film is then coated with a plating resist, usually an enamel, by stenciling, leaving exposed the area that will form the circuit pattern. The plating process is similar to that for the plating of decorative metals. After sufficient thickness (usually 0.001–0.005 in. or 25–125 μm) of copper is deposited, the plating resist is removed by a solvent. The exposed silver film is removed by acid etching, leaving the much thicker copper plating to form the circuit pattern on the plastic sheet. The bond strength between the conductor pattern and the base material resulting from this process is, however, relatively poor.

For electronic circuits that may be subjected to high humidity and continuous application of voltages (such that conditions for silver migration exist), the chemical reduction of copper instead of

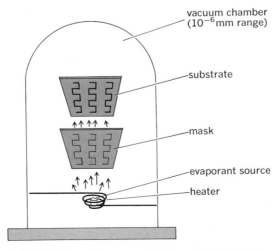

Fig. 11.   Schematic representation of basic thin-film vacuum deposition. (*TRW Inc.*)

vacuum chamber (10^{-6} mm range)

substrate

mask

evaporant source

heater

Fig. 12. Mold and die process. Closeup of machine embossing a copper foil circuit pattern onto a phenolic-paper laminate. The copper foil, feeding into the machine from left to right, has a temperature-sensitive adhesive on the undersurface. During the embossing operation a hot die cuts the circuit pattern, embeds the edges of the conductors into the phenolic, and temperature-cures the adhesive so that it forms a bond with the base plate. (*A. W. Franklin*)

silver is used to form the plating electrode. All the subsequent processes are as described above. This approach is commonly used in the formation of plated-through holes in double-sided and multilayer printed wiring boards for the purpose of connecting one layer to another.

Mold and die processes. Of the mold and die processes, embossing and stamping are the most popular (Fig. 12). In general, these two processes are used for the fabrication of conductors or inductances. Usually copper foil is embossed on a phenolic laminate base plate, or powdered silver is stamped on a plastic sheet. Because of the relatively high cost of tooling and the limitations in circuit fabrication, these processes have not gained wide use. [L. K. LEE]

Bibliography: C. J. Baer and J. R. Ottaway, *Electrical and Electronics Drawings*, 4th ed., 1980; J. S. Cook, *Printed Circuit Design and Drafting*, 1967; C. Coombs, Jr., *Printed Circuit Handbook*, 2d ed., 1979; W. S. DeForest, *Photoresist: Materials and Processes*, 1975; P. Eisler, *Technology of Printed Circuits*, 1959; R. A. Geshner and G. R. Messner (eds.), *Multilayer Printed Circuit Boards: Technical Manual*, 1966; Institute of Electrical and Electronics Engineers, *Transactions on Components, Hybrids and Manufacturing Technology* (journal); P. Lund, *Generation of Precision Artwork for Printed Circuit Boards*, 1978; T. D. Schlabach and D. K. Rider, *Printed and Integrated Circuitry: Its Materials and Processes*, 1963.

Privacy systems (scrambling)

Devices and methods for ensuring the privacy of overseas telephone conversations. Such privacy can be accomplished in a variety of ways; two are particularly widely used at circuit terminals.

The first method employs equipment for inverting speech to make overseas telephone conversations unintelligible to the casual listener. At the originating end, devices change the speech-con-

veying signal, making high frequencies low and low frequencies high, for example. At the distant end, synchronized equipment restores the inverted speech to its original form for delivery to the listener.

A more complicated method employs filters to separate the transmitted speech signal into several narrow-frequency bands. Each of these is then treated differently in a prearranged manner, the relative positions of individual portions of the signal being interchanged in the frequency spectrum by inversion and transposition. At the distant end, associated equipment puts the full signal back together in proper shape so that the listener may receive intelligible speech. This method, particularly if such changes are automatically varied every few seconds, makes possible a transmission system that is very difficult to decipher. *See* ELECTRIC FILTER.

[CHARLES C. DUNCAN; RICHARD B. NICHOLS]

Programming languages

Notations with which people can communicate algorithms to computers and to one another. To accomplish this, programmers specify a sequence of operations to be applied to data objects. *See* ALGORITHMS.

Computers process a rather low-level machine language, which has simple instructions (load, add, and so forth) and data (for example, words or partial words including those containing instructions). Example (1) shows machine code that assigns the absolute value of $B - C$ to A.

Location	Machine code	Meaning	
040003	100020040001	load B into register A1	
040004	250320040002	subtract C from register A1	
040005	741020040007	GOTO 040007 if A1 positive	(1)
040006	110020000000	negate A1	
040007	010020040000	store contents of A1 into A	

Writing in machine code places a tremendous burden on the programmers. They must write in an unusual number base (for example, eight), and transfer data from memory to registers in order to perform operations. They must know the machine's operation codes for the operations. [These appear in the first two digits of the machine code in example (1). Thus, load is 10; branch is 74; store is 01.] Finally, they must know the addresses of the operands (both registers and data words) and instructions (to be able to skip some operations with branches). In example (1) the last five digits give the address of one operand. Thus A is at 40000, B at 40001, C at 40002, and the target of the branch instruction at location 40005 is the instruction at 40007.

The introduction of assembler languages eased communication between programmers and machines by permitting symbolic rather than numeric or positional references for operation codes (opcodes) and addresses. Additional data and instructions could therefore be added to a program without changes to existing instructions. Thus, in example (2), which shows assembler code that assigns the absolute value of $B - C$ to A, instructions could be added between the branch positive, BP, and its destination.

	Op-			
Label	code	Operands	Comments	
	L	A1, B		
	AN	A1, C	· A1 := B − C	(2)
	BP	A1, NEXT		
	N	A1	· A1 := − A1	
NEXT	S	A1, A	· A := |B − C|	

Unfortunately, assembler language does not go very far in improving programmers' ability to communicate algorithms. A programmer must still be concerned with expressing algorithms in a form suitable to particular machines. For instance, if an expression requires more registers than are available, the programmer must create temporary variables to hold partial results of expressions. High-level languages allow programmers to communicate more easily with one another by providing structures that hide the architecture of the machine. Expressions in high-level languages can be written almost as they are in algebra so that the programmer need not worry about which register contains a particular computation or which registers are available. Higher-level control structures like the IF statement make it easier to see the overall structure of a program than do low-level branch instructions. High-level language code for assigning the absolute value of $B - C$ to A is shown in example (3). *See* DIGITAL COMPUTERS; DIGITAL COMPUTER PROGRAMMING.

$$
\begin{aligned}
&A := B - C; \\
&\text{IF } A < 0 \\
&\quad \text{THEN} \\
&\quad\quad A := -A
\end{aligned}
\qquad (3)
$$

STRUCTURE OF PROGRAMMING LANGUAGES

The components of a program and their relationships are shown in Fig. 1. The basic executable unit of a programming language is a program. Programs contain declarations of data and procedures, and a sequence of statements to be executed.

Objects, variables, and identifiers. Programming languages manipulate objects or values like integers or files. Objects are referred to by their names, called identifiers. Constant objects, such as the number 2 (one of whose names is the literal "2"), do not change. Variable objects may contain either other objects or the names of other objects. The same identifier may denote different objects at different places in the program or at different times during execution. Also, a single object may have several names, or aliases. Variable and constant identifiers are introduced by data declarations. Declarations (4) introduce the constant KilometersPerMile (whose value is 0.6) and the variables Miles, Hours, and MilesPerHour. Notice that in this example the object 0.6 has two names.

$$
\begin{aligned}
&\text{CONST} \\
&\quad \text{KilometersPerMile} = 0.6; \\
&\text{VAR} \\
&\quad \text{Miles, Hours, MilesPerHour;}
\end{aligned}
\qquad (4)
$$

Expressions. Expressions combine objects and operators to yield new objects. They may be written in one of several forms. Infix operators appear between their operands, as in expression (5); prefix

$$
B * B - 4 * A * C \qquad (5)
$$

operators precede their operands, as in expression (6); and postfix operators follow their operands, as

$$
(- (* B B) (* (* 4 A) C)) \qquad (6)
$$

in expression (7). Most languages adopt infix notation for binary operators and prefix notation for

$$
((B B *) ((4 A *) C *) -) \qquad (7)
$$

unary operators. However, LISP uses only prefix operators and FORTH uses only postfix operators.

Two ideas govern the order in which infix operators are applied to their operands in expressions: precedence and associativity. Operators of higher precedence are evaluated before those of lower precedence. A typical operator precedence hierarchy is shown in (8). Operators with equal precedence are usually left associative, that is, they are evaluated from left to right.

$$
\begin{array}{ll}
\text{Highest precedence} & * \text{ (multiply)}/\text{(divide)} \\
\quad | & + \text{(add)} - \text{(subtract)} \\
\quad | & < <= = <> > >= \\
\quad | & \quad \text{(relationals)} \qquad (8) \\
\quad | & \text{NOT} \\
\quad \downarrow & \text{AND} \\
\text{Lowest precedence} & \text{OR}
\end{array}
$$

Not all languages using infix notation have the same operator precedence or associativity. PASCAL has fewer levels in its precedence hierarchy: NOT has the highest precedence, AND has the same precedence as *, and OR has the same precedence as +. APL has no precedence among its operators, which are right associative. Thus $5 - 4 + 3$ has the value -2 in APL.

Statements. Important types of statements include assignment statements, compound statements, conditional statements, repetitive statements, and GOTO statements.

Assignment statements. Assignment statements assign values to variables. In an assignment of form (9), the value in the object denoted by B is

$$
A := B \qquad (9)
$$

assigned to the object denoted by A. The order of evaluation of the two sides of an assignment statement differs among languages—some insist on left-to-right evaluation, while others leave the

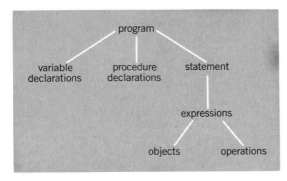

Fig. 1. Components of a high-level language.

order of evaluation undefined. Some languages permit the same value to be assigned to several variables in a single statement; for example, the value of C can be assigned to both A and B by using statements (10). Other languages, like C

$$\begin{aligned}&A, B := C &&\{PL/1\}\\&A := B := C &&\{ALGOL\ 60\}\end{aligned} \tag{10}$$

(which uses the symbol = for assignment), treat assignment just like any other operator in an expression. Thus statement (11) assigns C + 2 to B and C + 5 to A.

$$A = (B = C + 2) + 3 \tag{11}$$

Compound statements. A compound statement (12) groups a sequence of statements into a single

$$\text{BEGIN stmt1; stmt2; . . . stmtn END} \tag{12}$$

statement. Execution of statements within a sequence of statements is in the order in which the statements appear in the sequence.

Conditional statements. The IF statement (13) allows a programmer to indicate that a statement

$$\text{IF expression THEN statement} \tag{13}$$

is to be executed if a stated condition is true. If the expression is true, the statement following the THEN is executed followed by the statement after the IF statement. If the expression is false, the statement is skipped. Many languages permit execution of one of two alternate statements with the form of IF statement (14). If the expression is true,

$$\text{IF expression THEN statement1 ELSE statement2} \tag{14}$$

statement 1 is executed; otherwise statement 2 is executed. In either case, execution continues with the statement after the IF.

Repetitive statements. One of the advantages offered by computers is performing the same operations repetitively. Two statements in programming languages support this kind of execution: FOR and WHILE. The FOR statement (15) is generally used when a programmer knows how many times a statement is to be executed. The statement is executed a number of times given by (16), where expression1 and expression2 are the first and second expressions in statement (15). For example, statement (17) will write "hello" three times (2 −

$$\begin{aligned}&\text{FOR variable-identifier} := \\&\text{expression TO expression DO statement}\end{aligned} \tag{15}$$

$$\text{expression2} - \text{expression1} + 1 \tag{16}$$

$$\begin{aligned}&\text{FOR i} := 0 \text{ TO } 2 \text{ DO}\\&\quad\text{write ("hello")}\end{aligned} \tag{17}$$

0 + 1 = 3). If the first expression is greater than the second, the statement is not executed. While the statement is executed, the variable identifier (called the control variable) takes on successive values from the list (18).

$$< \text{expression1, expression1} + 1, . . ., \text{expression2} > \tag{18}$$

Usually the control variable may not be assigned

to within the loop, and its value is usually considered to be undefined at the end of execution of the FOR statement. The two expressions are evaluated once at the start of execution of the FOR statement rather than each time the statement is executed. Thus, changing the values of the variables in these expressions has no effect on the number of times the statement is executed, as is illustrated in example (19). The statement is executed five times, even though the values of First and Last are changed during execution. The results printed by this statement are shown in (20).

$$\begin{aligned}&\text{First} := 1;\\&\text{Last} := 5;\\&\text{FOR Index} := \text{First TO Last DO}\\&\quad\text{BEGIN}\\&\quad\quad\text{First} := \text{First} + 1;\\&\quad\quad\text{Last} := \text{Last} - 1;\\&\quad\quad\text{write (Index, First, Last)}\\&\quad\text{END}\end{aligned} \tag{19}$$

Index	First	Last	
1	2	4	
2	3	3	
3	4	2	(20)
4	5	1	
5	6	0	

The WHILE statement (21) is used when continued execution depends on a condition that is true

$$\text{WHILE expresion DO statement} \tag{21}$$

for an unknown number of repetitions. The statement is executed as long as the expression is true. When the expression becomes false, execution continues with the statement after the WHILE. The code fragment (22) can be used to approximate the square root of N by successive approximation.

$$\begin{aligned}&\text{WHILE abs } (X - N/X) > 0.1 \text{ DO}\\&\quad X := 0.5 * (X + N/X)\end{aligned} \tag{22}$$

While the two approximations of the square root, X and N/X, differ by more than 0.1 (that is, the absolute value of their difference is greater than 0.1), new approximations are computed. After some unknown number of iterations, the difference between the two approximations is small enough and execution of the WHILE statement ceases.

GOTO statements. The lowest-level control statement in most languages is the GOTO statement (23). This statement provides for the direct transfer of control to a statement labeled with the identifier that is the argument of the GOTO statement.

$$\text{GOTO identifier} \tag{23}$$

Any of the previous conditional or repetitive statements can be simulated with GOTO statements. For example, the previous WHILE statement would appear as (24).

$$\begin{aligned}&\text{Test: IF abs}(X - N/X) > 0.1 \text{ THEN GOTO Done;}\\&\quad\quad X := 0.5 * (X + N/X);\\&\quad\text{GOTO Test;}\\&\text{Done: . . .}\end{aligned} \tag{24}$$

Many programming languages have eliminated the GOTO statement because it widens the gap between the textual structure and execution-time behavior of a program's source text. Control can transfer to a label from almost any GOTO statement in a program, and GOTO statements can be either forward or backward jumps. In addition, GOTO statements can affect storage allocation in many languages.

Procedures. As problems grow more complex, they must be divided into subproblems that can be solved independently. This approach to problem solving is supported by procedure declarations, which group sequences of statements together and associate identifiers with the statements. A procedure identifier abstracts the details of the computation carried out by the statements of the procedure by encapsulating them. Procedures also help reduce the size of a program by permitting similar code to be written only once.

Procedure declarations, whose form is shown in (25), look just like programs, except they start with the word PROCEDURE instead of "program." In

$$
\begin{array}{ll}
\text{PROCEDURE ProcedureName;} & \\
\quad \{\text{data declarations}\} & \\
\quad \{\text{procedure declarations}\} & \\
\text{BEGIN } \{\text{ProcedureName}\} & (25) \\
\quad \{\text{sequence of statements}\} & \\
\text{END } \{\text{ProcedureName}\} &
\end{array}
$$

some languages each procedure has its own local data and procedure declarations. Since execution begins with the first statement of a program, the sequence of statements of a procedure can be executed only if they are explicitly invoked by a procedure-call statement containing the name of the procedure.

When the name of a procedure appears in a sequence of statements the procedure is called; that is, execution of the current sequence is suspended and the statement sequence associated with the procedure name in its declaration is executed. When execution of this sequence is completed, execution resumes in the previous sequence right after the procedure statement.

The designers of ALGOL 60 defined the effect of a procedure statement in terms of textual substitution. Their definition has come to be known as the copy rule. Essentially, the copy rule states that each procedure statement can be replaced by a compound statement containing the text of the procedure.

A procedure declaration is called recursive if it contains a procedure-call statement that invokes the procedure being defined. The WHILE statement that approximates the square root of N could be replaced by program fragment (26). If the

$$
\begin{array}{ll}
\text{PROCEDURE Approximate;} & \\
\quad \text{BEGIN } \{\text{Approximate}\} & \\
\quad \text{IF abs}(X - N/X) > 0.1 & \\
\qquad \text{THEN} & \\
\qquad\quad \text{BEGIN} & \\
\qquad\qquad X := 0.5 * (X + N/X); & (26) \\
\qquad\qquad \text{Approximate} & \\
\qquad\quad \text{END} & \\
\quad \text{END; } \{\text{Approximates}\} & \\
\text{BEGIN} & \\
\quad \ldots; & \\
\quad \text{Approximate;} & \\
\quad \ldots & \\
\text{END} &
\end{array}
$$

approximation was not close enough, a new estimate would be assigned to X, execution of the current sequence of statements would be suspended, and control would be transferred to the first statement of a new copy of Approximate. When the approximation is close, the current copy of the procedure terminates and control is returned to the statement following the last procedure invocation. Since there are no more statements in this copy of the procedure, it also returns to a previous copy, which returns to a previous copy, and so forth. Eventually control returns to the statement after the first call of Approximate (indicated by ellipses). As this example demonstrates, recursion is not always the most efficient technique to use in problem solving. However, it can be a valuable tool in divide-and-conquer approaches to problem solving. With the exception of FORTRAN and COBOL, most programming languages support this feature.

DATA TYPES

Data types are sets of values and the operations that can be performed on them. Data types offer programmers two principal advantages: abstraction and authentication. Data types hide representational details from programmers so they do not have to be concerned whether characters are represented by ASCII or EBCDIC codes, or whether they are left-aligned and blank-filled or right-aligned and zero-filled in machine words. The concept of data type also provides useful redundancy. The appearance of a variable as an operand in an expression implies a particular type for the variable (for example, $A + B$ implies that A and B are either integer or real) that may be checked against the type of its current value.

Scalar types. Scalars are indivisible units of data; they are treated as single complete entities. Most languages contain several scalar types. The most common are integer, real, boolean, and character.

Integer. Integer constants are positive and negative whole numbers and zero. Since machines have finite word lengths, only a subset of the integers are provided, as in (27), where MAXINT is a con-

$$
\{-\text{MAXINT}, \ldots, -2, -1, 0, 1, 2, \\
\ldots, \text{MAXINT}\} \quad (27)
$$

stant identifier corresponding to the largest integer value that can be represented on a particular machine. The usual arithmetic and relational operators are defined for integer values.

Boolean. The boolean type consists of the constant values false and true, and the operators NOT, AND, OR. NOT is a unary prefix operator, taking a boolean argument and returning a boolean result. AND and OR are binary infix operators that each take two boolean values and produce a single boolean result. *See* BOOLEAN ALGEBRA.

Character. The character constants are the uppercase and lowercase alphabetic characters, digits, and other punctuation symbols. Character constants are usually indicated by enclosing them in single quotation marks (for example, 'a', '7', or '?'). To represent the character constant, which is itself a single quote, a special notation is used (for example, doubling the internal quote '''' or using a special escape character '\'' in the C

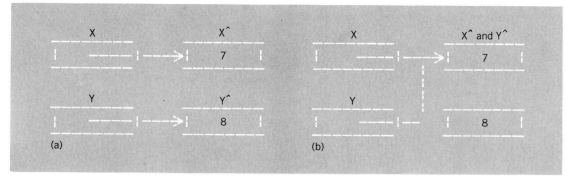

Fig. 2. Effect of the assignment Y:=X. (*a*) Before Y:=X. (*b*) After Y:=X.

language). Almost all languages allow character comparison; however, the result of comparisons may differ across languages. In some languages, expression (28) is true, while in others it is false. However, languages generally define the relational operators on character values so that expressions (29) are valid.

$$'a' < 'A' \qquad (28)$$

$$\begin{array}{l} 'A' < 'B' < 'C' \ldots < 'Y' < 'Z' \\ 'a' < 'b' < 'c' \ldots < 'y' < 'z' \\ '0' < '1' < '2' \ldots < '8' < '9' \end{array} \qquad (29)$$

Real. Real numbers may be represented either as decimal fractions (for example, 1234.56) or in scientific notation (for example, 1.23456E3). The operators are the same as for type integer. Care must be exercised in the use of the equality and inequality operators, $=$ and $<>$, on real values because the finite-word length restrictions on real machines can lead to problems of round-off and truncation so that, for example, expression (30) can occur.

$$1.0 <> (1.0/3.0 * 3.0) \qquad (30)$$

Enumerated types. New types can be defined by listing (enumerating) the constants of the type. The only operators that may be applied to operands with enumerated types are the relational operators. The enumerated type (31) defines a new type

$$\begin{array}{l} \text{(Jan, Feb, Mar, Apr, May, June,} \\ \text{July, Aug, Sept, Oct, Nov, Dec)} \end{array} \qquad (31)$$

whose constants are the identifiers Jan, Feb, and so forth. The order of appearance of the identifiers in the type declaration defines their relative values (that is, the constant identifier with the smallest value appears first). Thus, relational operators can be applied to those objects, as in expression (32).

$$\text{Jan} < \text{July} \qquad (32)$$

Constant identifiers declared in an enumerated type often replace mappings of integers to concepts in programs. For example, programmers who declare the enumerated type (31) and use its constants in their programs would not have to remember if 1 or 0 was used to represent January. **References.** While variables usually hold values of objects, they may also hold the name of some

object. Variables that can hold such values are called references or pointers. Reference constants are object names and a special value "nil" (which points to no variable). The operations defined for reference values are equality and dereferencing. Dereferencing is a unary operation that maps a pointer to the object it references. Dereferencing an object containing a reference (for example, by appending $\wedge$ to the object as in X$\wedge$) yields the object referenced. An example of dereferencing is statement (33), in which X contains

$$\text{write}(X\wedge, Y) \{\text{prints 2   2}\} \qquad (33)$$

a reference to Y and Y contains the value 2. The result of dereferencing nil is undefined.

Languages also have operations that create reference values. For example, statement (34)

$$X := \text{NEW (integer)} \qquad (34)$$

creates an object of type integer (with no initial value) and makes X point at the object.

Since reference variables hold object names, an assignment between two reference variables X and Y results in X and Y pointing at the same object (Fig. 2). Thus $X = Y$ and $X\wedge = Y\wedge$. If X and Y are both dereferenced before the assignment (Fig. 3), the objects, X$\wedge$ and Y$\wedge$, will contain the same values, but could still be distinct objects. Thus $X <> Y$, but $X\wedge = Y\wedge$.

Aggregate types. Collections of related values can be represented by aggregate types. Aggregate types can be characterized by the types of the objects that are their elements and the manner in which the elements are accessed. Elements of a collection can have homogeneous or heterogeneous types, and can be selected by their position in a collection or by a name associated with them.

Arrays. An array is a mapping from a domain specified by a sequence of values to a range whose members are the values of the elements of the array. In statement (35), the array A is declared to have the domain $\{-3, -2, \ldots, 4, 5\}$.

$$\text{VAR A: ARRAY} [-3 .. 5] \qquad (35)$$

In ALGOL, PL/1, PASCAL, and ADA, all the elements of an array must have the same type, but this restriction does not apply in SNOBOL, LISP, and APL.

Very few operations are defined for array objects. The access operation is called subscription

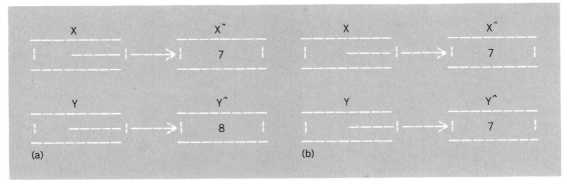

Fig. 3. Effect of the assignment Y^ := X^. (a) Before Y^ := X^. (b) After Y^ := X^.

and is usually indicated by enclosing an expression indicating the position of the desired element in brackets or parentheses. Thus, if A is an array, expression (36) selects its I + Jth element. Arrays

$$A(I+J) \text{ or } A[I+J] \qquad (36)$$

often have multiple dimensions, each of which has its own range of values as subscripts. A chess board could be represented as an 8×8 array, as in statement (37).

$$\text{VAR B: ARRAY} [1 .. 8, 1 .. 8] \qquad (37)$$

The diagonal elements of the board B are given by (38). Some languages (for example, ALGOL 68

$$\begin{array}{l} B[1, 1] \ B[2, 2] \ B[3, 3] \ B[4, 4] \\ B[5, 5] \ B[6, 6] \ B[7, 7] \ B[8, 8] \end{array} \qquad (38)$$

and ADA) permit parts of arrays to be referenced by using trimming or slicing operations. Thus, expression (39) selects the first two rows of the board.

$$B[1:2, 1:8] \qquad (39)$$

Most languages permit an array object to be assigned to an array variable; the values of the object are copied to the corresponding locations in the variables. PL/1 permits the assignment of a single value to every element in an array. The assignment (40) is equivalent to (41).

$$A := 0 \qquad (40)$$

$$A[1] := 0; A[2] := 0; . . . \qquad (41)$$

APL and ADA have array literals; the assignment (42) is equivalent to (43).

$$\text{Odds} := (1, 3, 5, 7, 9) \qquad (42)$$

$$\text{Odds}[1] := 1; \text{Odds}[2] := 3; . . .; \text{Odds}[5] := 9 \qquad (43)$$

APL has the richest collection of array operators, having extended many of its operators which take scalar operands to work on array operands as well.

Records. Records have components of different types (called fields), which are explicitly accessed by identifiers (called selectors). For instance, in PL/1, X. Name is a qualified name that selects record component Name from record X. If X is a

record object of the type given in (44), the names

```
RECORD
    Name: ARRAY [1 .. 20];
    Address: RECORD
        StNumber;
        StName: ARRAY [1 .. 10];
        City: ARRAY [1 .. 10];
        State: ARRAY [1 .. 2];      (44)
        ZipCode
    END;
    Age;
    Salary
END
```

of the components are: X. Name, X. Address, X. Address. StNumber, X. Address. StName, X. Address. City, X. Address. State, X. Address. ZipCode, X. Age, and X. Salary. Some components of the record are scalar variables, some are arrays, and one (X. Address) is a record. The term record comes from COBOL and business data processing.

BINDING ATTRIBUTES TO VARIABLES

Both constant and variable objects have values and attributes. Among the most important attributes of an object are its type, scope, and extent.

Scope. In most programming languages, the same identifier can name distinct objects in different parts of a program. The scope of an identifier is that part of a program in which the identifier has the same meaning. Generally that section of program is called a scope unit and corresponds to a procedure or compound statement. The set of objects that can be referenced in a scope unit is called the environment.

Identifiers are either declared locally (that is, in the immediately enclosing scope unit) or nonlocally (that is, in some other scope unit). Locally declared identifiers are like bound variables in formulas—their meaning does not change no matter where the procedure or compound statement containing their declaration is placed in a program, and is not available outside the scope unit. Nonlocally declared identifiers are like free variables—their meaning can depend on the placement of the text in which they are referenced or the execution sequence of the program.

In the program (45), the scope units are the program P and the procedures Q and R. The identifier I names distinct locally declared variables in P and R, and J names another variable in Q. The scope of each local identifier is marked to the right of the program.

```
PROGRAM P;          P's I    Q's J    R's I
  VAR I;                 |
  PROCEDURE Q;           |
    VAR J;               |       |
    BEGIN {Q}            |       |
      J := 0;            |       |
      I := I + 1         |       |
    END; {Q}             |       |
  PROCEDURE R;           |       |
    VAR I;                                |
    BEGIN {R}                             |
      I := 1;                             |     (45)
      Q;                                  |
      write(I)                            |
    END; {R}                              |
  BEGIN {P}              |
    read(I);             |
    IF I > 0             |
      THEN R             |
      ELSE Q;            |
    write(I)             |
  END {P}                |
```

With the exception of the reference to I in Q, all identifier references are to locally declared variables.

There are three ways to resolve references to nonlocal variables:

1. Consider the use of nonlocal variables to be illegal. This scope rule is associated with both older languages like FORTRAN and modern languages like EUCLID.

2. Bind the reference to a declaration for the identifier occurring in the innermost lexically enclosing scope (for example, I is declared in P and P encloses Q). This method is called a static scope rule because it is based on the program text which does not change during execution. Among the languages using this scope rule are ALGOL, PL/1, PASCAL, C, and ADA.

3. Bind the reference to a declaration in the procedure which has been most recently called, but has not completed execution yet. [In example (45), if the value read into I were greater than 0, the reference to I in Q would be bound to the declaration for I in R. Otherwise the reference would be bound to the declaration of I in P.] This association method is called a dynamic scope rule because nonlocal references can be bound to different variables depending on the sequence of statements executed. SNOBOL, LISP, and APL have dynamic scope rules.

Data type. Data types may be associated with objects statically via declarations or dynamically via assignment statements. In statically typed languages, the type associated with an object remains the same throughout the scope of the object. Declaration (46) binds the type INTEGER to I and

$$
\begin{aligned}
&\text{VAR} \\
&\quad \text{I: INTEGER;} \\
&\quad \text{Ch: CHARACTER;}
\end{aligned} \qquad (46)
$$

CHARACTER to Ch. In dynamically typed languages, the type of an object is usually determined by the type of its value. Assigning a new value to a vari-

able can change the type of the variable. Thus, in example (47), X initially has type integer and subsequently has type character.

$$
\begin{aligned}
&X := 5; \\
&\text{write}(X - 3); \qquad \{\text{print: } 2\} \\
&X := \text{'h'}; \\
&\text{write}(X) \qquad \{\text{print: h}\}
\end{aligned} \qquad (47)
$$

ALGOL, PL/1, PASCAL, C, and ADA are statically typed languages; SNOBOL, LISP, and APL are dynamically typed.

In statically typed languages the type of a variable may not be known when it is declared, or the variable may need to be bound to objects of several different types during the course of the program. Type unions permit variables to be bound to objects with different types. Two kinds of type unions exist: free and discriminated. In a free union, the programmer is permitted to treat a variable as being of any of the types in the union. In a discriminated union, another field, called the tag field, identifies the type of the object. Unions are often represented as records in which enumerated types are used to select alternatives. In example (48), a record type is defined that contains a single field whose name is either Ival or Cval. The number of the constant identifiers in the CASE portion of the record determines the number of alternate fields.

```
RECORD
    CASE (Int, Chr) OF
        Int: (Ival: INTEGER);         (48)
        Chr: (Cval: CHARACTER)
END
```

In example (49), X is treated as though its type were a free union of the types INTEGER and CHARACTER.

```
IF Y = 0
    THEN X.Ival := 12
    ELSE X.CVAL := 'c';              (49)
X.Ival := X.Ival + 1
```

All the assignment statements in example (49) are legal. However, the final assignment adds one to the numeric representation for the character 'c' if the statement following the ELSE is executed.

The free union in example (49) could be converted to a discriminated union by adding a field named Tag whose value is either Int or Chr, depending on the value contained in the second field, as in (50).

```
RECORD
    Tag: (Int, Chr);
    CASE (Int, Chr) OF
        Int: (Ival: INTEGER);        (50)
        Chr: (Cval: CHARACTER)
END
```

Extent. The extent or lifetime of a variable is the period of program execution during which the storage is allocated. Static storage is created at the beginning of the program execution and not deallocated until program termination. Local storage for variables is created on entry to the block or procedure containing the variable's declaration and destroyed on exit. Dynamic storage is created and destroyed by execution of special statements in the programming language.

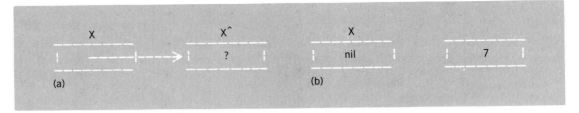

Fig. 4. Execution sequence resulting in garbage. (*a*) After new (X). (*b*) After X := nil.

Consider procedure (51). In a language like

```
PROCEDURE P;
  VAR X;
  BEGIN                          (51)
    {sequence of statements}
  END
```

FORTRAN, the same storage location can be used for X no matter how many times P is invoked. The value of X is preserved between calls to P. However, in languages like ALGOL, PL/1, PASCAL, C, and ADA, local storage for X is allocated on each entry to the procedure and freed on each exit, so that the value of X is lost between executions of P. Local storage allocation permits two procedures to share storage locations if the procedures do not call one another. In example (52), the arrays A and B can share the same storage locations.

```
PROGRAM Prog;
  PROCEDURE P;
    VAR A: ARRAY [1 . . 100];
    BEGIN {P}
      {sequence of statements not containing
        a call of Q}
    END; {P}
  PROCEDURE Q;
    VAR B: ARRAY [1 . . 75];          (52)
    BEGIN {Q}
      {sequence of statements not
        containing a call of P}
    END; {Q}
  BEGIN
    P;
    Q
  END
```

Storage can be created dynamically through the execution of a statement like (53).

$$Ptr := new ( ) \qquad (53)$$

Such storage is independent of the structure of the source code of the program and remains allocated until a corresponding statement (54) is executed.

$$dispose(Ptr) \qquad (54)$$

Care must be taken that all dynamically allocated storage can be referenced through a pointer or the storage becomes garbage. The sequence of statements (55) results in the execution sequence

$$X := new(INTEGER); X^{\wedge} := 7; X := nil \quad (55)$$

shown in Fig. 4, in which the storage formerly referenced by X becomes garbage because there is no way to reference it. Dangling references are created when storage referenced by several pointers is freed. In the sequence of statements (56), Y

$$new(X); X^{\wedge} := 7; Y := X; dispose(X) \qquad (56)$$

contains a dangling reference to the storage formerly referenced by X. X may also contain a dangling reference if its contents are not set to nil by dispose, as in Fig. 5.

PARAMETERS

Parameters communicate information between the calling environment and the environment of the called procedure. Although this can be accomplished by referencing nonlocal variables, communicating with parameters permits a procedure to work on different variables each time it is called. Formal parameters are identifiers that are locally declared in a procedure definition; actual parameters are expressions in a procedure

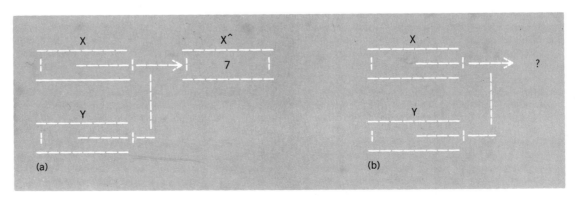

Fig. 5. Execution sequence resulting in dangling reference. (*a*) After Y := X. (*b*) After dispose (X).

call statement. In the program fragment (57), which shows two versions of a procedure named Max, X and Y are formal parameters and A, B, $A + 2$, and $B - 3$ are actual parameters. The non-

```
PROCEDURE Max (X, Y);        PROCEDURE Max;
  BEGIN {Max}                  BEGIN {Max}
    IF X > Y                     IF X > Y
      THEN                         THEN
        write(X)                     write(X)
      ELSE                         ELSE
        write(Y)                     write(Y)
  END; {Max}                   END; {Max}
BEGIN                        BEGIN                    (57)
  . . .                        . . .
  Max(A, B);                   X := A;
  Max(A + 2, B − 3);           Y := B;
  . . .                        Max;
END                           X := A + 2;
                              Y := B − 3;
                              Max;
                              . . .
                            END

  Parameter version           Nonlocal version
```

local version of Max is slightly longer than the parameter version because the calling environment must initialize the nonlocal references before Max is called. This requires the calling environment to know the implementation details of Max, thereby losing some of the advantages of abstraction gained by making Max a procedure.

The correspondence between the formal and actual parameters is positional—the first actual parameter is bound to the first formal parameter, and so forth. Several different kinds of parameter binding occur in programming languages: reference, value, result, and name. The key issues in binding are when the binding takes place (usually when the procedure is called, but possibly each time a formal parameter is referenced in the procedure body), and what is bound (the value or address of the actual, or perhaps a procedure that computes the address of the actual).

Reference binding. When an actual parameter is bound by reference to a formal parameter, an alias is set up between the name of the actual parameter and the name of the formal parameter. The word alias is used because both names identify the same object. The object identified by the actual is evaluated at the point of call. Each time a value is assigned to a formal reference parameter, the value of the corresponding actual parameter is also altered. Thus reference parameters can be used to supply input values to procedures and to return values computed by the procedure.

Value binding. Many arguments to procedures are used as inputs only; no value is returned through these parameters. Input parameters are bound by value to formal parameters. In value binding, the value of the actual parameter at the point of call is assigned to a new object which is bound to the formal parameter, but no alias is established between the formal and the actual parameters. Thus, any assignment to a formal value parameter changes the value of the formal but not that of the actual parameter.

Result binding. A procedure called to compute an output value can use result binding to assign the value to a result parameter. At the point of call, the actual parameter is bound by reference to its corresponding formal parameter. However, during execution of a procedure, assignments to a result parameter alter a local copy of the variable, but not the actual parameter. When the procedure completes execution, the result is transmitted by assigning the value of the formal parameter to the actual parameter.

Name binding. In name binding, actual parameters are passed unevaluated to formal parameters and are evaluated only when necessary. Actually, a procedure that computes the object represented by the actual parameter (called a thunk) is bound to the formal parameter. Each time the formal parameter is referenced, its corresponding thunk is called to deliver the actual object. This method of parameter transmission occurs only in ALGOL 60.

Examples of different bindings. Program (58) illustrates the differences in parameter binding mechanisms.

```
PROGRAM Bindings;
  VAR
    A: ARRAY [1 . . 2];
    I;
  PROCEDURE P (X, Y);
    BEGIN {P}
      Y := I + 1;
      X := Y + 3;
      A[1] := A[1] + 20        (58)
    END; {P}
  BEGIN
    A[1] := 1;
    A[2] := 1;
    I := 1;
    P(A[I], I);
    write (I, A[1], A[2])
  END
```

The results printed for the different binding mechanisms are summarized in table (59).

Bindings	I	A[1]	A[2]	
Value	1	21	1	
Reference	2	25	1	(59)
Result	2	5	1	
Name	2	21	5	

In value binding, only the assignment to A[1] causes a change in the variables in the calling environment. X is aliased to A[1] and Y is aliased to I in reference binding. Thus the change to Y is reflected in I, and the changes to both X and A[1] affect the value of the latter. While the result mechanism yields the same bindings as the reference mechanism, the values of the result parameters ($Y = 2$ and $X = 5$) are not assigned to the actuals until P finishes execution. Thus the result of adding 20 to the value of A[1] is lost. As long as no local identifiers of P have the same name as any of the actual name parameters, name binding can be simulated by rewriting P's text with the text of the actuals replacing that of the formals, as in (60).

$$
\begin{aligned}
&I := I + 1; \\
&A[I] := Y + 3; \qquad (60) \\
&A[1] := A[1] + 20
\end{aligned}
$$

Since I is incremented from 1 to 2 by the first

assignment statement, A[2] rather than A[1] is altered by the second assignment.

FUNCTIONS

Function declarations are similar to procedure declarations. But unlike a procedure, which communicates its results by assigning values to its parameters, a function communicates a distinguished value known as the result of the function. This result of the function is often indicated by assigning a value to the name of the function. The declaration of the function Max2 is given by (61).

$$\begin{aligned}
&\text{FUNCTION Max2 (X, Y);}\\
&\quad\{<\text{Max2} := \text{if X} > \text{Y then X else Y}>\}\\
&\quad\text{BEGIN \{Max2\}}\\
&\quad\quad\text{IF X} > \text{Y}\\
&\quad\quad\quad\text{THEN}\qquad\qquad\qquad\qquad\qquad(61)\\
&\quad\quad\quad\quad\text{Max2} := \text{X}\\
&\quad\quad\quad\text{ELSE}\\
&\quad\quad\quad\quad\text{MAX2} := \text{Y}\\
&\quad\text{END; \{Max2\}}
\end{aligned}$$

Function calls look like procedure calls; the name of the function is followed by a list of actual parameters. However, there is an important difference between function and procedure calls. Since functions deliver values, function calls are expressions which may appear either within larger expressions or alone wherever an expression could appear in a program. Procedure calls, which do not deliver values, are statements.

To find the maximum of three values, Max3 could be declared, as in (62). The expression in the

$$\begin{aligned}
&\text{FUNCTION Max3 (X, Y, Z);}\\
&\quad\{<\text{Max3}> := <\text{the largest value in [X, Y, Z]}>\}\\
&\quad\text{BEGIN \{Max3\}}\qquad\qquad\qquad\qquad(62)\\
&\quad\quad\text{Max3} := \text{Max2(X, Max2(Y, Z))}\\
&\quad\text{END; \{Max3\}}
\end{aligned}$$

assignment statement would be evaluated by first calling Max(Y, Z) to obtain the maximum of Y and Z, and then calling Max2 again with X and the result of the earlier call.

Side effects. When a statement in a procedure or function alters the value of a nonlocal variable or reference or name parameter, the procedure is said to have a side effect. It may not be quite accurate to call such assignments in procedures side effects, because procedures must include such statements in order to communicate their results. Therefore, they are really main effects of procedures. Side effects in functions are less desirable because they make the results of expression evaluation unpredictable. While languages specify the order in which operators are applied, they do not always define the order in which operands are fetched in expressions. For example, if Identity returned the value of its actual parameter, but increased the value of its reference parameter and the nonlocal variable N1, as in (63), then expression (64) might have any of the values in (65).

$$\begin{aligned}
&\text{FUNCTION Identity (Parm);}\\
&\quad\text{BEGIN \{Identity\}}\\
&\quad\quad\text{Identity} := \text{Parm;}\qquad\qquad(63)\\
&\quad\quad\text{Parm} := \text{Parm} + 1;\\
&\quad\quad\text{N1} := \text{N1} + 3\\
&\quad\text{END; \{Identity\}}
\end{aligned}$$

$$\text{N1} + \text{Identity(In)} + \text{In}\qquad(64)$$

$$\begin{array}{ll}
\text{N1} + \text{In} + \text{In} + 1 & \{\text{if the operands are fetched}\\
& \quad\text{left to right}\\
& \quad\text{T1} := \text{N1;}\\
& \quad\text{T2} := \text{Identity(In);}\\
& \quad\text{T3} := \text{In;}\\
& \quad\text{Result} := \text{T1} + \text{T2} + \text{T3}\}\\
\text{N1} + 3 + \text{In} + \text{In} & \{\text{if the operands are fetched}\\
& \quad\text{right to left}\\
& \quad\text{T1} := \text{In;}\\
& \quad\text{T2} := \text{Identity(In);}\\
& \quad\text{T3} := \text{N1;}\\
& \quad\text{Result} := \text{T3} + \text{T2} + \text{T1}\}\\
\text{N1} + \text{In} + \text{In} & \{\text{if both operands are}\quad(65)\\
& \quad\text{fetched before the call}\\
& \quad\text{T1} := \text{N1;}\\
& \quad\text{T2} := \text{In;}\\
& \quad\text{T3} := \text{Identity(In);}\\
& \quad\text{Result} := \text{T1} + \text{T3} + \text{T2}\}\\
\text{N1} + 3 + \text{In} + \text{In} + 1 & \{\text{if both operands are}\\
& \quad\text{fetched after the call}\\
& \quad\text{T1} := \text{Identity(In);}\\
& \quad\text{T2} := \text{N1;}\\
& \quad\text{T3} := \text{In;}\\
& \quad\text{Result} := \text{T2} + \text{T1} + \text{T3}\}
\end{array}$$

Thus, it is good programming practice to avoid side effects in functions and to use procedures instead when side effects are desired.

Procedure and function parameters. When procedures and functions are passed as parameters, they are bound like name parameters. Actual procedure parameters are passed unevaluated to procedures and evaluated each time they appear in procedure call statements.

Procedure parameters may contain references to nonlocal identifiers that must be bound to variables. Deep binding associates a procedure parameter's nonlocal identifiers with variables in the environment in which the procedure was created. Shallow binding associates these identifiers with variables in the environment when the procedure is invoked. In example (66), P contains a nonlocal reference to A, and Q is a function that returns a procedure that is subsequently invoked in R.

$$\begin{aligned}
&\text{PROGRAM Main;}\\
&\quad\text{VAR A, B;}\\
&\quad\text{PROCEDURE P;}\\
&\quad\quad\text{BEGIN \{P\}}\\
&\quad\quad\quad\text{A} := \text{A} + 10;\\
&\quad\quad\quad\text{write(A)}\\
&\quad\quad\text{END; \{P\}}\\
&\quad\text{FUNCTION Q(Proc);}\\
&\quad\quad\text{VAR A;}\\
&\quad\quad\text{BEGIN \{Q\}}\\
&\quad\quad\quad\text{A} := 2;\\
&\quad\quad\quad\text{Q} := \text{Proc}\\
&\quad\quad\text{END; \{Q\}}\qquad\qquad\qquad(66)\\
&\quad\text{PROCEDURE R(Proc);}\\
&\quad\quad\text{VAR A;}\\
&\quad\quad\text{BEGIN \{R\}}\\
&\quad\quad\quad\text{A} := 3;\\
&\quad\quad\quad\text{Proc; \{really invokes P\}}\\
&\quad\quad\quad\text{write(A)}\\
&\quad\quad\text{END; \{R\}}
\end{aligned}$$

```
BEGIN {Main}
   A := 1;
   B := Q(P);
   B ();
   write(A)
END {Main}
```

In ALGOL-like languages, nonlocal references are associated with the variables in the environment in which the procedure is declared. This deep binding associates the references to A in P with the A declared in Main. Thus the output of the program is: 11   3   11.

In LISP, either deep or shallow binding can be specified by applying one of the functions FUNCTION and QUOTE to the value returned by Q, as in (67).

```
FUNCTION Q(Proc);          FUNCTION Q(Proc);
   VAR A;                     VAR A;
   BEGIN {Q}                  BEGIN {Q}
      A := 2;                    A := 2;        (67)
      Q := FUNCTION              Q := QUOTE
        (Proc)                     (Proc)
   END; {Q}                   END; {Q}
   Deep binding               Shallow binding
```

The use of FUNCTION in Q binds P's A to the variable declared in Q, the most recently declared version of A since LISP has a dynamic scope rule. Thus the program output is: 12   3   1.

The use of QUOTE in Q delays the binding of P's A until P is invoked as Proc in R. At this time, P's A is bound to the most recent declaration of A, which occurs in R. Thus the program output is: 13   13   1.

While the ability to pass procedure and function parameters may seem too exotic to include in a programming language, it does facilitate construction of generic software. The alternative to passing a function parameter to a procedure is to embed the source for the function in the procedure. Thus the procedure can only operate with one specific function rather than with any function passed to it.

APPLICATIVE AND OBJECT-ORIENTED LANGUAGES

Most programming languages are imperative languages, because their programs largely consist of a series of commands to assign values to objects. ALGOL, PL/1, FORTRAN, ADA, COBOL, and PASCAL are imperative languages. Two other types of languages exist: applicative and object-oriented. No commonly used language is purely imperative, applicative, or object-oriented, so that every language has some characteristics of each type of language.

Applicative languages. A purely applicative language would be one in which there are no statements, only expressions without side effects. The name applicative refers to the repeated application of functions to the results of other functions to specify an algorithm. LISP, APL, and VAL are three languages in which programs are largely applicative.

LISP has a well-defined applicative subset known as pure LISP. Programs in LISP consist of a sequence of expressions known as S-expressions. If functions with side effects are excluded from the these expressions, then the language becomes pure LISP. Pure LISP excludes such functions as SET, which assigns a value to a variable, and DEF (or DEFUN), which declares that a variable is a procedure with a certain body.

A program to find the absolute value of B − C would be written in pure LISP as in (68). COND is

$$(\text{COND} \, ((> (- \text{B C}) \, 0) \, (- \text{B C})) \atop (\text{t} \, (- \text{C B}))) \qquad (68)$$

a function that takes a list of ordered pairs of expressions and returns the value of the second member of a pair whose first member evaluates true. The expression pairs in this example are given by (69). LISP uses prefix notation for operators, so the first member of the first pair is given by (70). The first member of the second pair, "t," or true, ensures that COND always delivers some result.

$$((> (- \text{B C}) \, 0) \, (- \text{B C})) \text{ and } (\text{t} \, (- \text{C B})) \qquad (69)$$

$$\text{B} - \text{C} > 0 \qquad (70)$$

The main advantage of applicative over imperative programming is mathematical elegance. An applicative description of an algorithm has the form of a mathematical function definition, and so all the accumulated power of mathematical analysis can be used to understand it. This is especially useful in proving programs correct.

Object-oriented languages. Programs written in object-oriented languages consist of sequences of commands directed at objects. An object receiving a command invokes an internal procedure to respond to the command. Assignment statements are replaced by commands to objects requesting that they change their values.

SMALLTALK, ACTORS, and LISP are three languages that encourage object-oriented programming. As with applicative languages, there is a object-oriented subset of LISP, called FLAVORS. Since the only way to do something in an object-oriented language is to direct a command at an object, object-oriented programming means passing messages to objects. For instance, the way to add two integers in SMALLTALK is to pass a message to the first integer requesting that it add itself to the second integer.

A program to compute the absolute value of B − C might be written in object-oriented form as in (71). This program is read: send a message to

```
(send INTEGER "new" "A")
(send A "assign" B)
(send A "−" C)            (71)
(send A "absolute")
```

the integer object asking it to make a new integer called A, send A an assign message with message body B, send A a subtract message with message body C, and finally send A a message requesting that it take the absolute value of itself.

One advantage of object-oriented programming is the encapsulation of responses to commands within each object. For instance, when adding two numbers the programmer need not be concerned with whether the numbers are integer or real. The number receiving the "add-yourself-to" message is responsible for knowing how to add itself to any reasonable other number, and for complaining if it cannot. This effect can also be achieved in imperative languages such as ADA by using encapsulated data types.

A second advantage of object-oriented programming is the power of mixing types of objects. If the programmer defined a type of object which could respond to messages to add itself to things, and another type of object which could respond to messages to print itself, then by mixing types a new kind of object would be created which could do both. In LISP FLAVORS the "vanilla flavor" is the set of messages to which all objects respond; more complex objects are created by mixing other flavors in with vanilla.

DEFINING NEW DATA TYPES

Procedure and function declarations define new operations that operate on existing objects. Type declarations add new objects and operations to a programming language. Designing programs with user-defined types allows solutions to be stated in problem-oriented rather than machine-oriented terms. ALGOL 68- and PASCAL-type declarations allow programmers to define the representation of new objects of the data types, and ALGOL 68 permits the definition of operations. For example, declaration (72) defines a new type named Rational.

$$\begin{aligned} \text{TYPE Rational} = \text{RECORD} \\ \text{Numerator,} \\ \text{Denominator:} \\ \text{INTEGER} \\ \text{END} \end{aligned} \quad (72)$$

Once this definition is made, PASCAL programmers can declare variables and parameters with type Rational, but cannot add rational numbers without manipulating implementation directly (73). Dealing with the fields of the representation negates many advantages of abstraction.

```
VAR Op1, Op2, Result: Rational;
BEGIN
    . . .
    {Result: = Op1 − Op2}
    Result. Denominator: =
        Op1. Denominator * Op2. Denominator;      (73)
    Result. Numerator: =
        Op1. Numerator * Op2. Denominator
        + Op1. Denominator * Op2. Numerator
END
```

Procedures and functions can be defined in PASCAL and operators in ALGOL 68 to encapsulate these sequences of operations, hiding the details of how functions are computed. However, these definitions are independent of the type definitions, that is, programmers can still directly manipulate the representation of objects. SIMULA 67 introduced the class concept to bind the definition of the representation and operations together, but still did not prohibit programmers from accessing the representation directly. By limiting access to objects with user-defined types to procedures defining operations, programmers can deal with data abstractly (that is, without caring how the data are represented). This enables programmers to change the representation of objects (for example, to improve the efficiency of their programs by changing the representation of rationals from a pair of integers to a single real) without changing any part of the program except the type definition and the procedures and functions implementing the operations. CLU, EUCLID, and ADA all have data abstraction functions that limit access to objects' representations. In example (74), written in ADA, the type Rational and the addition operator for rational numbers is made available through a package specification. However, the appearance of the word PRIVATE in the definition of Rational prevents programmers from accessing the fields of the record used to represent a rational number.

```
PACKAGE Rat IS
    TYPE Rational IS PRIVATE;
    FUNCTION "+" (Op1, Op2: Rational)
        RETURN Rational;
    . . .
PRIVATE
    TYPE Rational IS RECORD                    (74)
        Numerator, Denominator: INTEGER;
    END RECORD;
END; − Package Rat
PACKAGE BODY Rat IS
    . . .
    FUNCTION "+" (Op1, Op2: Rational)
        RETURN Rational IS
    . . .
    END "+";
END Rat;
```

PARTICULAR LANGUAGES

The more widely used languages include FORTRAN, ALGOL, COBOL, BASIC, PL/1, APL, SNOBOL, LISP, PASCAL, C, and ADA.

FORTRAN. FORTRAN was designed in the mid-1950s by a group at IBM headed by John Backus for efficient scientific applications. To achieve this goal, the data and control components of the language are very simple, providing programmers with low-level hardware operations.

FORTRAN control structures are particularly primitive; neither compound statements nor unbound repetitive statements are part of the language. Two conditional statements (arithmetic and logical IF's) and a bounded iteration statement (DO) provide the only relief from GOTO's and labels. An arithmetic IF, statement (75), provides a three-way branch depending on whether its expression is negative, zero, or positive. A logical IF, statement (76), permits execution of a single state-

$$\text{IF (expression) MinusLabel, ZeroLabel,} \\ \text{PlusLabel} \quad (75)$$

$$\text{IF (expression) statement} \quad (76)$$

ment, which cannot be a DO or an IF, if its expression is true. The absence of nested statements causes heavy use of GOTO's and labels, as shown in example (77).

```
    IF (X GE 0) GOTO 1    IF X > = 0
    GOTO 2                THEN
1   A = B − C                 A : = B − C
    GOTO 3                ELSE             (77)
2   A = C − B                 A : = C − B
3   CONTINUE

FORTRAN version    ALGOL version
```

A revision of the FORTRAN language, FOR-

TRAN-77, adopted many of ALGOL's control constructs.

Scaler data types include logical, integer, real, double-precision real, and complex. Arrays with homogeneous elements provide the only aggregate types. Types are associated with variables statically through declarations.

Procedures may not be nested in FORTRAN, so references are either global or local. The global variables are partitioned into named common blocks, which must be mentioned in a procedure if the procedure contains nonlocal references. Storage for both global and local variables is statically allocated before program execution, so local variables retain values between invocations of the procedures in which they are declared.

FORTRAN has both procedures and functions, but neither can be recursive. Parameter transmission is by reference, and procedures can be transmitted. Functions may only return scalar results.

ALGOL 60. ALGOL was designed by an international committee in the late 1950s and early 1960s as a language for scientific computation. The elegance of ALGOL's structure has made it the basis for several subsequent language design efforts, most notably ALGOL 68 and PASCAL, and its control structures have even been adopted by PL/1 and later versions of FORTRAN. ALGOL was also the first language whose syntax was defined with a formal grammar (called Backus-Naur Form or BNF after two of the principal designers of ALGOL). The concise definition of the language has made it an object of study in the computer science community for many years.

ALGOL control statements are very much like statements (12)–(15). However, all repetitive statements are variants of the FOR statement, and compound statements are units of scope that may contain declarations of their own local variables.

ALGOL has integer, real, and boolean scalar types; aggregate types are restricted to arrays with homogeneous elements. ALGOL has static rules for both scope- and data-type association. Unless they are declared as OWN, variables in ALGOL programs are locally allocated, coming into existence in the block or procedure in which they are declared and being deleted when that program unit finishes execution. The lack of user-defined data types and dynamic storage allocation limits the applications of ALGOL.

Procedures and functions can be recursive and accept parameters which have the standard data types as well as labels and procedures. Parameter binding can be either by name (the default) or by value.

COBOL. COBOL was designed in 1959 and 1960 as a language to support business applications on computers. As was the case with FORTRAN, efficiency was a primary concern. Because of its intended application area, COBOL has relatively poor features for specifying computations (such as expressions, functions, and parameters), but strong features for data description and input/output.

Two of the most noticeable features of COBOL programs are the English-like syntax and the division of programs into four parts. Either English-language syntax or algebraic notation can be used in expressions, as in (78).

$$\text{ADD B, C, GIVING A} \atop \text{COMPUTE A} = \text{B} + \text{C} \qquad (78)$$

The identification division provides documentation, the procedure division describes algorithms, the data division contains descriptions of the files and working storage used in the procedure division, and the environment division collects machine-dependent information.

Statements in procedure divisions are grouped into labeled paragraphs. COBOL contains the usual IF and GOTO statements, but has unusual repetitive and procedure statements. The tasks of these last two statements are accomplished by the PERFORM statement (79) that executes a sequence

$$\text{PERFORM label l THRU label N} \qquad (79)$$

of paragraphs before control is returned to the statement following the PERFORM. Bounded and unbounded repetition are also specified using PERFORMs, as in (80).

$$\text{PERFORM label l THRU label N expression TIMES} \atop \text{PERFORM label l THRU label N UNTIL condition} \qquad (80)$$

Numbers and character strings are the basic data types of COBOL. Data structuring is provided with arrays and records. Level numbers on declarations describe the hierarchical relationship of the data (with the exception of some special level numbers). Types are specified by picture clauses that specify how many characters are needed to represent the values to be assigned to the variables.

Procedures with parameters and local data are optional COBOL features. No functions are provided and procedures may not be recursive. Unless a version of COBOL contains these optional features, all variables in the data division are global and statically allocated. Like FORTRAN, the values of all variables (even those declared in local data divisions) are retained between calls.

BASIC. T. E. Kurtz and J. G. Kemeny developed BASIC at Dartmouth College in the early 1960s. The goal of their design was to provide nonscience majors with a simple, interactive language.

BASIC supports real and string data as well as one- or two-dimensional arrays of one of these primitive types. Dimension statements permit string lengths and array subscripts to be specified. Variable names are restricted to single letters, followed by an optional digit if the variable is numeric or $ if the variable is a string. Thus, types are statically associated with variables.

Statements in BASIC programs are numbered, and control passes through them sequentially unless altered by a GOTO or conditional branch statement, such as statement (81). Repeated execution can be specified by a FOR statement.

$$\text{IF expression THEN statement-number} \qquad (81)$$

There are no procedure declarations in BASIC; any sequence of statements can be considered to be a procedure. Procedure calls are indicated by the appearance of a GOSUB statement (82) that

$$\text{GOSUB statement-number} \qquad (82)$$

transfers control to the indicated statement number. The procedure is executed until a RETURN statement is encountered. The lack of procedure declarations means that the notions of parameters and scope rules are foreign to BASIC.

PL/1. PL/1 was designed by a committee organized by the IBM Corporation in the mid-1960s. The language was designed to meet the needs of a broad range of applications: scientific, business applications, and systems. PL/1 combines and extends many of the features of FORTRAN (parameter binding and formated input/output), ALGOL (scope rules, variable extents, recursive procedures, and control structures), and COBOL (record structures and picture data types). By providing redundant features from different languages, the designers have made the language difficult to learn and encouraged users to stay in one of the FORTRAN, COBOL, or ALGOL subsets of PL/1.

The control structures of PL/1 are like those of ALGOL, except that repetitive statements are introduced by DO instead of FOR.

Numeric types in PL/1 permit the programmer to specify four attributes: mode (real or complex), scale (fixed or floating point), base (decimal or binary), and precision (the number of digits). Since most programmers do not want to provide such detail in declarations (and it may hinder transportability to machines with different word lengths), PL/1 compilers provide many defaults. (There are actually more attributes describing the variable's lifetime and initial value.) PL/1 also provides bit and character strings with lengths that are either fixed or varying up to a maximum, as well as a reference type called POINTER. Both arrays and COBOL records called structures are provided.

PL/1 has static scope rules and type association. The extent of a variable may be either static (ALGOL's own), automatic (ALGOL's local), based, or controlled. The last two extent attributes provide dynamic storage allocation using the predefine procedures ALLOCATE and FREE. The based attribute can be used to provide free unions because pointer variables may reference data of any type. Example (83) illustrates this facility. The

$$
\begin{array}{ll}
\text{DECLARE Int FIXED BASED;} & \\
\text{DECLARE Chr CHARACTER(4) BASED;} & \\
\text{DECLARE P POINTER;} & (83) \\
\text{ALLOCATE Int SET P;} & \\
\text{P} - > \text{Int} = 64; & \\
\text{PUT LIST(P} - >\text{Chr});
\end{array}
$$

declaration of based variables does not cause storage to be reserved, but rather defines a template. The ALLOCATE statement reserves enough storage for an integer object and binds a reference to the object to P. The assignment statement dereferences P and treats the resulting object as an integer in order to assign the value 64 to the object. The PUT (or write) statement dereferences P and treats the resulting object as a character. Controlled storage must also be explicitly allocated and freed by the programmer, but unlike based storage only the last object allocated may be referenced. Thus, the language provides programmer-defined stack operations.

PL/1 may have recursive procedures and functions. Parameter transmission is by reference as in FORTRAN, but labels, procedures, and functions can also be transmitted as in ALGOL. PL/1 also provides primitives for parallel execution of procedures and for handling interrupts.

APL. APL was designed by Kenneth Iverson furing the mid-1950s to the mid-1960s. It is an interactive language whose operators accept and produce arrays with homogeneous elements of type number or character. The only control construct is a GOTO statement, but this is inconsequential since many APL programs have a strong applicative flavor. Although they often consist of an imperative sequence of assignments, APL programs generally operate on array objects to produce new array objects. Thus repetitive constructs are not needed as frequently as they are in ALGOL-like languages.

There is really no such thing as an APL program because functions are defined independently and invoked either interactively by the programmer or by another function. A function definition for IN, which finds the positions at which one word appears in another, is shown in (84). IN would require at least one loop if it were written in an ALGOL-like language.

$$
\begin{array}{ll}
& \triangledown \quad Z \leftarrow A \text{ IN } B; J \\
[1] & J \leftarrow (A[1]=B)/\iota\rho B \\
[2] & J \leftarrow (J \leq 1+(\rho B)-\rho A)/J \qquad (84) \\
[3] & Z \leftarrow (B[J^\circ + {}^{-}1 + \iota\rho A]^\wedge = A)/J \\
& \triangledown
\end{array}
$$

APL functions consist of sequentially numbered statements, which may have control transferred to them via GOTO statements. A and B are formal parameters (only two are allowed but they can be arrays), J is a local variable, and Z is a special identifier that is assigned the result of any function. Parameters are passed by value and nonlocal references are resolved dynamically. Types are associated with variables and storage is allocated for variables dynamically. For example, if IN is invoked with the sequence of commands (85), then X and Y are dynamically allocated and assigned values of type vector of characters.

$$
\begin{array}{ll}
& X <- \text{'THE'} \\
& Y <- \text{'THE MEN THEN} \\
& \qquad \text{WENT HOME'} \qquad (85) \\
& X \text{ IN } Y \\
1 & 9
\end{array}
$$

The assignment statements in function (84) have the following effects. $A[1] = B$ returns a vector whose length is the same as B's length, and whose values are either 1 where $A[1] = B[k]$ for $1 < = k < = $ Length (B) or 0 otherwise. The rho operator (ρ) gives the length of B, and the iota operator (ι) generates a vector of length B whose values are $1, 2, \ldots, \rho B$. The first vector is used to compress the second by selecting those values from the second vector for which the corresponding value in the first vector is one. Thus J is assigned a vector with values (1 9 17). The second statement checks that a word with the same length as A could really start at these positions (without running past the end of the vector). The final statement subscripts B with a matrix of positions of words that could match A to obtain a matrix of characters, as in (86). The rows of this matrix are compared to the vector A, yielding a matrix (87)

$$B \begin{vmatrix} 1 & 2 & 3 \\ 9 & 10 & 11 \\ 17 & 18 & 19 \end{vmatrix} = \begin{vmatrix} THE \\ THE \\ T\ H \end{vmatrix} \tag{86}$$

$$\begin{vmatrix} 1 & 1 & 1 \\ 1 & 1 & 1 \\ 1 & 0 & 0 \end{vmatrix} \tag{87}$$

containing 1's where the values of the corresponding elements are equal and 0's otherwise. The elements of the rows are "anded" to form a vector of 0's and 1's (1 1 0) that is used to compress J.

SNOBOL. During the 1960s, SNOBOL was developed by Ralph Griswold at Bell Laboratories as a string processing language. It went through several forms, only one of which is still in use: SNOBOL 4. SNOBOL has no real control constructs. Instead, every statement either "succeeds" or "fails" and may optionally GOTO a new statement depending on success or failure. SNOBOL statements provide extremely powerful operators for string manipulation. For instance, SNOBOL statement (88) will assign the value a to X and the

$$\text{"This is a, this is b" "a" . X ARB "b" . Y} \tag{88}$$

value b to Y. The operators used here are the blank before a which means pattern match; the "." which means assign, and "arb" which matches an arbitrary string.

Data types are dynamically declared and converted in SNOBOL, so that the same variable can at one instance contain an integer and the next instant contain a procedure. User-defined types permit the use of records. An interesting aggregate type is the "table," which is an array addressed by content. Variables are dynamically scoped.

SNOBOL has both procedures and functions, which may be recursive. Parameter transmission is by value, and procedures can be transmitted. Functions may return any type, including type NAME which can be used as the target of an assignment statement.

SNOBOL is used for string processing applications, such as editors, compilers, and data bases. It executes rather slowly, so once a program has been coded and debugged in SNOBOL it is often recoded in a more efficient language, such as PASCAL.

LISP. LISP was invented by John McCarthy while at MIT in the late 1950s. It is a very widely used language in the artificial intelligence community, which has never agreed on what version of LISP should be standard. The two most widely used forms are INTERLISP and MACLISP. *See* ARTIFICIAL INTELLIGENCE.

The basic type in LISP is the list, usually denoted by a sequence of items in parentheses. LISP programs also take the form of lists, thus LISP programs can manipulate themselves or other LISP codes. The only syntax in LISP is the matching of parentheses, thus LISP is easy to extend by adding additional functions.

Procedures and functions are recursive in LISP,

and parameters are transmitted by value. The power of LISP comes from the representation of everything by lists, which leads to very powerful environments of tools and extensions to LISP. LISP is a good language to write experimental programs, since its flexibility is useful for dealing with the unexpected. However, it executes slowly and uses a lot of memory.

PASCAL. PASCAL was developed by Niklaus Wirth in 1969 to obtain a language which would be suitable to teach programming as a systematic activity with constructs that could be implemented reliably and efficiently. The control structures of PASCAL are similar to statements (12)–(15) and (21). However, a CASE statement (whose design had been proposed in a language called ALGOL X designed by Wirth and C. A. R. Hoare) was added as an additional selection statement. This statement permits execution of one alternative among several statements based on the value of a selection expression. The alternative selections are labeled and selection is performed by name rather than by position to minimize the chance for errors resulting from rearranged or omitted statements. Statement (89) is a sample CASE statement. If this

$$\begin{array}{l} \text{CASE expression OF} \\ \quad \text{Reader: statement1;} \\ \quad \text{Disk, Tape: statement2;} \\ \quad \text{Printer: statement3} \\ \text{END} \end{array} \tag{89}$$

expression has the value Disk, statement2 is executed and control skips to the statement following the CASE statement.

PASCAL scalar types include integer, real, character, and boolean. PL/1's POINTER type is improved by adding information about the type of data referenced. Thus the PASCAL types "reference to type1" and "reference to type2" are not compatible, so that assignments cannot take place between variables of one of these types and values of the other. Pascal extends ALGOL 60's data-structuring facilities by adding sets, records, files, and user-defined types. Type unions can be obtained through the use of variant records.

Like ALGOL, PASCAL has a mandatory declaration of variables, static scope rules, and static association of types with variables. Except for those accessed through other reference variables, variables are locally allocated on procedure entry and deleted on procedure exit. Variables accessed via reference variables are dynamically allocated and freed by using NEW and DISPOSE, respectively.

Both procedures and functions can be recursive in PASCAL. Variables of any type can be bound by value (the default) or by reference to formal parameters. Procedures and functions can also be passed as parameters. Functions are limited to returning scalar or reference values as their results.

C. C was designed and implemented in 1972 by Dennis Ritchie to implement the Unix operating system. Similar to PASCAL in many ways, C has enumerated types, records (called structures), the case statement, and reference variables. The main difference between programming in C and other languages is the extensive use of reference variables, which are defined to be equivalent to arrays. For instance, program (90) declares an integer array named array and a pointer to array named

int array [10], *pointer = array;
array (3) = 1; (90)
*(pointer + 3) = 0;

pointer. The first assignment statement sets the fourth element of the array to 1, and the second assignment sets it to zero. The dereferencing operator in C is represented by *. Reference variables allow the programmer to specify algorithms which are particularly efficient to execute on the underlying machine, at the expense of writing code which is difficult to read.

C is usually used within the Unix operating system which is entirely written in C. Unix extends C to have storage allocation, multitasking and interrupt handling, and very flexible input/output facilities. *See* OPERATING SYSTEM.

ADA. ADA was designed by a group from Cii-Honeywell Bull headed by Jean Ichbiah in the mid-1970s. This group won a competition sponsored by the Department of Defense to design a new language to support development of embedded systems. ADA was designed using PASCAL as a base, to meet the reliability and efficiency requirements imposed by these applications.

Design decisions to include static scope rules and static association of types and variables were made in the name of reliability. ADA's most important contribution is to extend PASCAL's user-defined type to include packages; these permit users to define new operations and encapsulate the representations of their new types. While reliability required that types be statically associated with variables in ADA, efficiency required that controlled escapes from these associations be possible by specifying a target type for an expression. In example (91), a ten-element slice [similar to expression (39)] of B is assigned to A.

TYPE Seq IS ARRAY (Integer RANGE <>) OF Integer;
SUBTYPE Decile IS Seq(1 .. 10);
A: Decile; (91)
B: ARRAY (1 .. 100) OF Integer;
A := Decile(B(11 .. 20));

Decile coerces the expression B(11 .. 20) to the same type as A, making the assignment possible.

Like PL/1, ADA contains an exception-handling code so that code designed to handle exceptional situations can be separated from that for normal processing. For example, if a compound statement contained division operations and none of the divisors should be zero, the ADA structure (92)

BEGIN
. . . statements containing
divisions
EXCEPTION
WHEN Numeric__Error=>
. . . statements to inform user of (92)
error situation
WHEN OTHERS =>
. . . all other errors besides
numeric errors
END;

might be used. If a division by zero occurs in the body of the compound statement, control is transferred to the statements in the exception handler labeled Numeric__Error. After these statements

are executed, control returns to the caller. All other errors raised in the compound statement are handled by the OTHERS exception handler. If no exception handlers are specified, exceptions are propagated to the caller. Exceptions can also be declared and raised by users to handle the exceptional conditions of their applications (for example, stack overflow or singular matrix).

Efficiency considerations make it possible for programmers to describe interfaces to programs written in other languages and to access machine-level components (for example, instructions and particular memory locations).

Because many embedded systems are real-time applications, ADA provides tasks that can be concurrently executed, but communicate in very controlled ways. The ADA fragment (93) shows a task

TASK BODY Buffer IS
. . . local declarations
BEGIN
LOOP
SELECT
WHEN QLength < Max =>
ACCEPT Put(N: IN
Integer) DO
. . . statement list
END;
. . . statement list (93)
OR
WHEN QLength > 0 =>
ACCEPT Get(Result: OUT
Integer) DO
. . . statement list
END;
. . . statement list
END SELECT;
END LOOP;
END Buffer;

named Buffer with two entry points, Put and Get, defined by ACCEPT statements. Other concurrently executing tasks call Buffer through one of its two entry points, for example, Buffer Put(12). Each time the SELECT statement is executed, all the expressions guarding the operations (that is, those expressions after the WHEN's) are evaluated. From those entries whose expressions evaluate to true, one is arbitrarily chosen for execution. The task whose call was chosen and the Buffer task rendezvous until the statements between ACCEPT and END are executed. Then each task is free to proceed with its own execution.

[J. D. GANNON; M. D. WEISER]

Bibliography: C. Ghezzi and M. Jazayeri, *Programming Language Concepts*, 1982; E. Horowitz, *Fundamentals of Programming Languages*, 1983; H. Ledgard and M. Marcotty, *The Programming Language Landscape*, 1981; E. I. Organick, A. I. Forsythe, and R. P. Plummer, *Programming Language Structures*, 1978; T. W. Pratt, *Programming Languages: Design and Implementation*, 1975.

Pulse generator

An electronic circuit capable of producing a waveform that rises abruptly, maintains a relatively flat top for an extremely short interval, and then rapidly falls to zero. A relaxation oscillator, such as a multivibrator, may be adjusted to generate a rectangular waveform having an extremely short duration, and as such it is referred to as a pulse generator. However, there is a class of circuits whose exclusive function is generating short-duration, rectangular waveforms. These circuits are usually specifically identified as pulse generators. An example of such a pulse generator is the triggered blocking oscillator, which is a single relaxation oscillator having transformer-coupled feedback from output to input. *See* BLOCKING OSCILLATOR; MULTIVIBRATOR; RELAXATION OSCILLATOR.

Pulse generators sometimes include, but are usually distinguished from, trigger circuits. Trigger circuits, by means of *RC* differentiating, gated *RLC* peaking circuits, or blocking oscillators, generate a short-duration, fast-rising waveform for initiating or triggering an event or a series of events in other circuits, such as monostable or bistable multivibrators. In the pulse generator, the pulse duration and shape are of equal importance to the rise and fall times. In this sense the blocking oscillator is a circuit which can be made to perform well in both respects. *See* TRIGGER CIRCUIT.

Digitally controlled pulse generator. The term pulse generator is often applied not only to an electronic circuit generating prescribed pulse sequences but to an electronic instrument designed to generate sequences of pulses with variable delays, pulse widths, and pulse train combinations, programmable in a predetermined manner, often microprocessor-controlled.

For example, in the circuit shown in Fig. 1, the basic pulse generator is a bistable *R-S* or *J-K* flip-flop under control of a sequence of clock pulses with a digital logic block to determine the appropriate combinations of set *(S)*, reset *(R)*, and clock pulses. An output pulse can be programmed to start with the next clock pulse following $R = (1)$ and stop with the clock pulse following $S = (1)$. Thus, the pulse width which can be programmed is nT_c, where T_c is the clock pulse interval and n is any number equal to or greater than unity.

Pulse-forming networks. A network, formed in such a way as to simulate the delay characteristics of a lossless transmission line, and appropriate switching elements to control the duration of a pulse form the basis for a variety of types of pulse generators. *See* DELAY LINE.

A lossless transmission line has a characteristic impedance, given by Eq. (1), where *L* is the series

$$R_0 = \sqrt{\frac{L}{C}} \qquad (1)$$

inductance and *C* the shunt capacitance per unit length. Such a line may be approximated by a network consisting of a number *n* of cascaded *LC* elements. A pulse applied to the input of such a

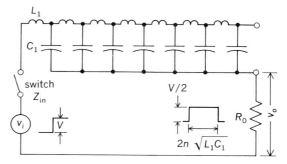

Fig. 2. Principle of line-controlled pulse generator.

line reaching the output offers a time delay T_d, given by Eq. (2).

$$T_d = n\sqrt{LC} \qquad (2)$$

An idealized circuit showing how such a network is used in a pulse generator is shown in Fig. 2. A source is connected by a switch to a simulated, unterminated transmission line in series with a resistance R_0 equal to the characteristic impedance of the line. As the time voltage is applied, and until all the capacitors in the line become fully charged, the input impedance of the line is equal to R_0, and the current I is given by Eq. (3). The step

$$I = \frac{V_i}{2R_0} \qquad (3)$$

function progresses along the line, charging each *C* in succession. When it reaches the end, it is reflected back with no change in phase and returns to the source in a time *T*, in Eq. (4).

$$T = 2n\sqrt{LC} \qquad (4)$$

At this time the line is fully charged, the impedance becomes infinite, and current ceases to flow. The pulse appearing across R_0 is suddenly terminated as shown. For discussion of switching circuits suitable for supplying the line-charging current *see* CLAMPING CIRCUIT; GATE CIRCUIT.

Similar results can be obtained from the use of a current generator as a source and a short-circuited line as the controlling circuit element.

Various forms of delay-line-controlled pulse generators can be found, and some are capable of generating pulses containing considerable amounts of power for such applications as modulators in radar transmitters. For an example of this type pulse generators *see* BLOCKING OSCILLATOR; WAVE-SHAPING CIRCUITS.

[GLENN M. GLASFORD]

Bibliography: D. A. Bell, *Solid State Pulse Circuits*, 1976; R. L. Castellucis, *Pulse and Logic Circuits*, 1976; L. Strauss, *Wave Generation and Shaping*, 2d ed., 1970; R. J. Tocci, *Fundamentals of Pulse and Digital Circuits*, 2d ed., 1977; H. C. Veatch, *Pulse and Switching Circuit Action*, 1971.

Pulse modulation

A system of modulation in which the amplitude, duration, position, or mere presence of discrete pulses may be so controlled as to represent the message to be communicated. These several forms of pulse modulation are commonly called, respec-

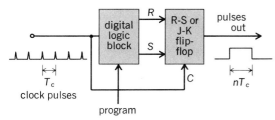

Fig. 1. Digitally controlled pulse generator.

tively, pulse-amplitude modulation (PAM), pulse-duration modulation (PDM), pulse-position modulation (PPM), and pulse-code modulation (PCM). For basic concepts, technical terms, and supplementary information *see* AMPLITUDE MODULATION; FREQUENCY MODULATION; MODULATION; PHASE MODULATION.

Of all the different forms of pulse modulation, PCM is the most outstanding. This radically new form of pulse modulation represents a major contribution to the communications art. With PCM, transmission circuits are simplified; overall transmission losses are avoided; crosstalk, interference, and distortion are virtually eliminated; signals may be repeated again and again without accumulating significant distortion; patterns of on-or-off-pulses constitute the only type of signals that are propagated, regardless of the type of message to be communicated; in the interest of reliability, no on-or-off pulse can be lost (wrongly identified); and in the interests of efficiency and economy, little time is wasted.

PRINCIPLES OF OPERATION

All forms of pulse modulation transmit message information intermittently rather than continuously. Therefore, unless the message information to be transmitted happens to be a time sequence of discrete values, it must be reduced to this form before transmission. Sampling (a process for obtaining a sequence of instantaneous values of a wave) accomplishes this by unambiguously representing a continuously varying wave by a series of distinct values (samples). Each message is momentarily sampled at regular intervals at a rate that is in excess of twice the highest message frequency to be communicated. As will be explained presently, in a PAM, PDM or PPM system (Fig. 1) a single pulse is used to specify the value of each sample.

Ease of multiplexing channels by time division is

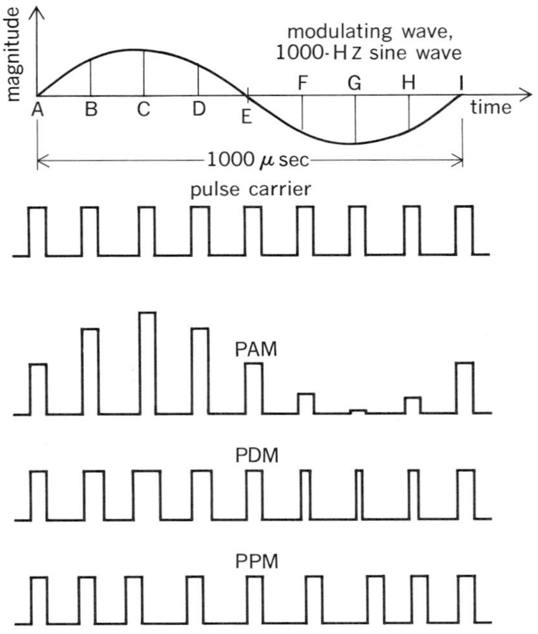

Fig. 1. Examples of pulse modulation. (*From H. S. Black, Modulation Theory, Van Nostrand, 1953*)

one of the important economic advantages of all forms of pulse modulation. Circuits for accomplishing this are simple and low in cost, and because this low cost is shared by a number of channels, the cost per channel is even lower. When pulse modulation is applied to long-distance communications, these time-division techniques also permit substantial simplification at way stations and branching points. Because each information-bearing pulse keeps its individuality in journeying from the first transmitter to the last receiver, it is comparatively easy to drop and add message channels at various intermediate points along the way.

Basic concepts. With the preceding description as background, the fundamental aspects of pulse modulation which underlie so much of present-day communication may now be presented in more detail. All forms of pulse modulation may be defined in terms of pulse carrier, and all involve sampling; in addition, PCM implies quantization and coding.

Pulse carrier. This is a carrier (Fig. 1) consisting of a series of regularly recurrent pulses. In general, the power associated with each pulse differs essentially from zero only during a limited interval of time, which is the pulse width.

Sampling. Sampling is a process of extracting successive portions of predetermined duration taken at regular intervals from a continuously varying magnitude-time wave. By sampling at a fast enough rate, namely, in excess of twice the highest significant frequency composing the sampled wave, the samples will unambiguously define the wave. And, conversely, given the samples, the wave can be reconstructed in all its detail.

For example, suppose the highest significant frequency in a voice wave is less than 4000 Hz. Then all the information necessary for its distortionless reconstruction is given by short samples of the voice wave taken at regular intervals at the rate of 8000 samples per second, that is, by samples taken every 125 microseconds (μsec). This complete process, including recovery of the voice wave, is illustrated step by step in Fig. 2.

A voice wave passes through a low-pass filter (Fig. 2a) which cuts out all frequencies above less than one-half the sampling frequency. After filtering, the wave is designated V and depicted in Fig. 2b.

The unit sampling function, designated U, is shown in Fig. 2c. This will be used presently to sample the voice. Mathematically (Fig. 2f), U is equal to dc component k plus components at the sampling frequency f_c and its harmonics. The interval between pulses is $1/f_c$, and k is the ratio of pulse duration t_0 to the interval between pulses T.

Next (Fig. 2d), U is multiplied by V. Because U is either unity or zero, the product UV is a mathematical process for sampling the voice. The result is a series of positive and negative pulses. When U is unity, the product is V. At all other times, the product is zero.

Physically, UV is an array of amplitude-modulated pulses. Consequently, an attenuated replica of V is obtained merely by passing UV through a low-pass filter. This may be demonstrated by performing the indicated multiplication UV. Amplification restores the reconstructed wave to its original value.

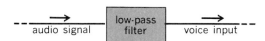

(a) source of voice input

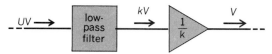

(e) passing *UV* through a low-pass filter and amplifier to obtain *V*

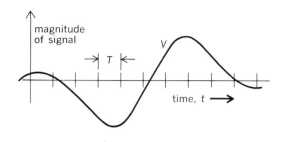

(b) typical voice input

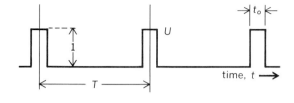

(f) enlarged diagram of *U*, unit sampling function

(c) diagram of *U*

$$U = k + 2k \sum_{m=1}^{\infty} A_m \cos mCt$$

$$k = \frac{t_0}{T} = \text{duty cycle}$$

$$f_c = \frac{C}{2\pi}$$

$$\frac{1}{f_c} = T$$

$$A_m = \frac{\sin mk\pi}{mk\pi}$$

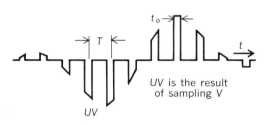

(d) diagram of *UV*

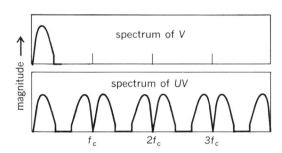

(g) spectrum analysis of *V* and *UV*

Fig. 2. Properties of wave samples. (*a*) Source of voice input. (*b*) Typical voice input. (*c*) Diagram of *U*. (*d*) Diagram of *UV*. (*e*) Passing *UV* through a low-pass filter and amplifier to obtain *V*. (*f*) Enlarged diagram of *U*, unit sampling function. (*g*) Spectrum analysis of *V* and *UV*. (*From H. S. Black, Modulation Theory, Van Nostrand, 1953*)

A spectrum analysis of *V*, and also *UV*, is depicted by Fig. 2*g*. The top diagram is the spectrum of *V*. The spectrum of *UV* is the spectrum of *V*, small but exact, plus upper and lower side bands about f_c and about harmonics of f_c. This illustrates, in terms of the familiar concepts of amplitude modulation, that passing *UV* through a low-pass filter gives an attenuated replica of the sampled wave.

Pulse-amplitude modulation (PAM). Pulse-amplitude modulation is amplitude modulation of a pulse carrier. The modulated wave (Fig. 1) is linearly proportional to equally spaced samples of the modulating wave. Another illustration is given in Fig. 2*d*.

Chief interest in PAM lies in its application to time-division multiplexing. Ordinarily, the band-width occupancy of PAM appreciably exceeds the theoretical minimum; that is, it appreciably exceeds the sum of the individual message bands. Yet, like other forms of amplitude modulation, PAM is not helped by wider bands; unlike FM and unlike other forms of pulse modulation, PAM cannot trade extra bandwidth for noise reduction.

Pulse-duration modulation (PDM). Pulse-duration modulation is modulation of a pulse carrier wherein the value of each instantaneous sample of a modulating wave produces a pulse of proportional duration (Fig. 3) by varying the leading, trailing, or both edges of the pulse. PDM is also termed pulse-length modulation or pulse-width modulation.

In constrast to PAM, PDM, which was invented

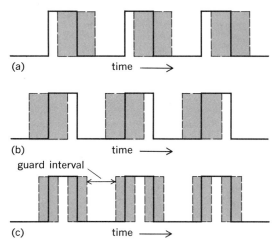

Fig. 3. Types of pulse-duration modulation. (a) Trailing edge modulated, leading edge fixed. (b) Leading edge modulated, trailing edge fixed. (c) Both edges modulated. Solid lines indicate duration of unmodulated pulses, shaded areas limits of maximum modulation. (*From H. S. Black, Modulation Theory, Van Nostrand, 1953*)

by R. A. Heising in 1924, is able to trade extra bandwidth for noise reduction. This noise advantage of PDM over PAM makes multiplexing by PDM even easier than multiplexing by PAM inasmuch as certain tolerances for controlling interchannel interference may be eased by an amount corresponding to the noise advantage. However, in order to achieve this important advantage, the instantaneous values of interference must not be permitted to exceed the so-called improvement threshold often enough to be disturbing.

Only the position of the modulated edge or edges conveys information, and the part of each PDM pulse that conveys no information represents wasted pulse power. When this wasted power is subtracted from PDM the result is PPM, which was invented by R. D. Kell in 1934. This power saving is the theoretical advantage of PPM over PDM.

Pulse-position modulation (PPM). Pulse-position modulation is modulation of a pulse carrier wherein the value of each instantaneous sample of a modulating wave varies the position in time of a pulse relative to its unmodulated time of occurrence.

Figure 4 illustrates the sinusoidal modulation of the channel 2 pulse. Successive diagrams indicate the change in the relative position in time of the channel 2 pulses from sample to sample for nine successive samples. When a particular channel is idle, that channel pulse recurs every 125 μsec. When a channel is busy, its pulse comes earlier or later depending upon the polarity of the sample. The exact displacement of the pulse from its unmodulated position is proportional to the magnitude of the sample to be communicated. All channel pulses are of constant magnitude and constant duration.

Channel pulses, one for each channel, are transmitted in turn and are preceded by a synchronizing pulse called a marker. This array of marker-plus-channel pulses repeats itself every 125 μsec and is called a frame. In Fig. 4 the synchronizing pulse is identified by its longer time duration. Its func-

tion is to control the timing of the receiver with high accuracy.

In practical applications, even though PPM is more efficient than PDM, both are highly inefficient when used for certain purposes, for example, when used for multiplexing ordinary telephone channels. Consequently, communication engineers have shown considerable interest in PCM, which not only is more efficient but also possesses many other very important advantages.

Pulse-code modulation (PCM). Pulse-code modulation, invented by H. A. Reeves in 1939, is a method of transmitting continuously varying message waves in which (1) the message wave is sampled, (2) the value of each sample is replaced by the closest one of a finite set of permitted values, and (3) these permitted values are then each unambiguously represented by some one of the possible patterns of N on-or-off pulses. These three operations are known as sampling, quantizing, and coding, respectively.

Sampling. Modern PCM is based upon the recognition that quantized samples may approximate an exact specification of continuously varying wave as closely as desired. Assume for purposes of illustration that a 7-pulse code is to be used. This is a binary system whose capacity is 2^7 permitted or quantized values. Each sample, after being coded, would be represented by some one of the 128 possible patterns of 7 on-or-off pulses.

Quantization. Quantization changes a continuously varying signal to a stepped signal. Graphically quantization means that the straight line representing the relationship between input and output has been replaced by a stair-step function. Clearly, by using small enough steps, errors may be cut to an arbitrary minimum (Fig. 5).

When speech samples are quantized, the steps are tapered as shown in Fig. 5b. By tapering most of the steps logarithmically, nearly uniform percentage precision is obtained over most of the

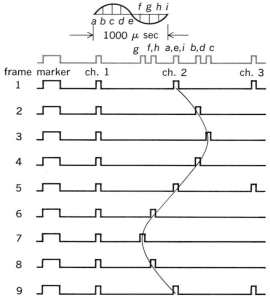

Fig. 4. Pulse-position modulation, modulation of channel 2 by a sine wave. (*From H. S. Black, Modulation Theory, Van Nostrand, 1953*)

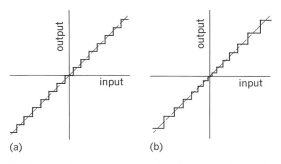

Fig. 5. Relation between input and quantized output. Quantization (*a*) uniform, (*b*) tapered. (*From H. S. Black, Modulation Theory, Van Nostrand, 1953*)

range, and far fewer steps are required. To reproduce telephonic speech to a high degree of fidelity, 128 logarithmic steps will suffice.

Coding and multiplexing. In PCM these 128 values are represented by a 7-pulse code. Each voice sample is represented with adequate accuracy by some one of the possible patterns of 7 on-or-off pulses (Fig. 6). Channels are multiplexed by time division.

Similarly, other forms of communication may also be represented by on-or-off pulses. In practice, a single PCM system may process many kinds of communications and may provide many channels of each kind.

Ordinarily, PCM systems are called upon to transmit large numbers of on-or-off pulses per second. For example, for each telephone conversation, PCM transmits 56,000 on-or-off pulses per second in each direction of transmission. This means that each one-way speech channel provided by PCM makes available a phenomenally fast one-way data channel, fast enough to operate simultaneously more than 1000 conventional teletypewriters. In general, the number of on-or-off pulses per second per message will depend upon the kind of message. A one-way television signal might, for example, be represented by 70,000,000 on-or-off pulses per second.

Repeatering. By having to make only on-or-off distinctions, PCM is able to deliver a high-quality signal even when noise and interference are so bad that it is barely possible to recognize the pulses. At each regenerative repeater, just so long as incoming pulses are correctly identified, a new pulse generator may be caused to generate new pulses of correct magnitude, waveform, and timing.

APPLICATIONS OF PCM

PCM systems are widely used in many ways and for many purposes by all branches of the military, Comsat, Intelsat, NASA, and private and public communications companies in virtually every country. Moreover, PCM systems transmitting secret information are used to protect many military and public services.

In 1979, Communications Satellite Corporation (COMSAT) installed a new type of high-speed (60,000,000 digital pulses per second) communications system on 250 seagoing ships, thereby linking each ship to every ship-to-shore receiver in the world via COMSAT's existing geostationary three-satellite system which utilizes PCM to transmit speech, teletypewriter, and facsimile messages. This means that anyone with access to a telephone can talk via satellite to someone out to sea at all times.

The Bell System uses PCM in the loop plant connecting customer premises and central offices, in the exchange plant between central offices, on microwave radio systems, on coaxial cable systems, and on optical fibers. Synchronized timing is required in every branch of each system in order to achieve superior performance and maximum economy.

Originally each one-way PCM channel used a seven-pulse code and transmitted 56 kilobits per second (kbps). An eight-pulse code has been developed (the extra pulse being used for both information and control) and is used to transmit 64

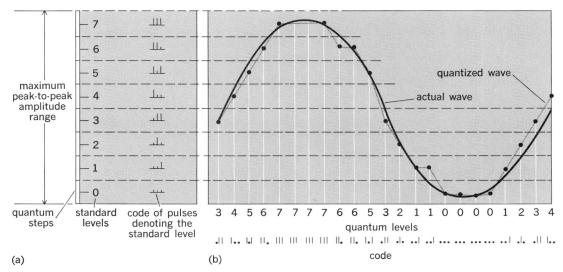

Fig. 6. Pulse-code modulation. (*a*) Quantizing of amplitude level and level designation by a code of pulses; 3-pulse code in this example permits $2^3 = 8$ levels. (*b*) Representation of a wave by a succession of coded pulse groups. (*From F. E. Terman, Electronic and Radio Engineering, McGraw-Hill, 4th ed., 1955*)

kbps. While 64 kbps occupies four times the bandwidth occupied by a one-way analog voice channel, pulse transmission provides lower cost and better performance, because it is only necessary to recognize the presence or absence of each pulse in order to reconstruct the original voice wave at the receiving end.

Radio set AN/TRC. AN/TRC (Army-Navy transportable radio transmitting and receiving communications equipment) was the first American-built microwave two-way multichannel PCM radio relay system to be used throughout the world. With many repeaters in tandem, its eight voice channels were outstanding for their high quality and remarkable stability. AN/TRC was the forerunner of microwave radio relay systems and was the first implementation of PCM.

Telephone PCM systems. The economic importance of using PCM for voice circuits on existing voice pairs in congested metropolitan ducts was recognized as early as 1945. In 1962, after 6 years of exploratory development, the T1 system went into operation. Its low-cost channel banks provide 24 one-way voice circuits which are unambiguously represented by patterns of 1,500,000 on-or-off pulses per second propagated over a twisted pair using regenerative repeaters spaced at 6000 ft (1.8 km). Millions of these systems are now in operation.

With the growing web of T1 systems in each part of an extended metropolitan area, the payoff for intercity and express-route digital lines to interconnect these T1 webs becomes attractive. This calls for the interleaving of many nonsynchronous T1 bitstreams to form a single high-bit-rate system. Bell Telephone Laboratories' engineers developed the technique of "pulse stuffing" in the 1960s to make this kind of interleaving possible.

In 1978 digital systems began to be used in customer loops, enabling 80 single-party lines in a rural district to be served over a digital line using T1 repeaters. Subsequently, a system serving 96 customers was introduced for more general application in the loop plant. Other extensions in loop plant services made possible by pulse modulation include amplification of voice signals as well as ringing and dialing pulses.

The Bell System developed nine pulse modulation systems, all of which provide the important advantage of pulse modulation and regenerative repeatering. These systems are: T1B4, T2, 1A, T1C, TCAS, DR18, DUV, WRT, and WT4. Some of the systems are designed for operation over relatively short distances and others over distances as long as 4000 mi (6400 km). They are employed on all the transmission media in current use: wire pair, coaxial cable, microwave radio, optical fibers and satellites.

Between 1983 and 1990 these nine pulse modulation systems are expected to be augmented by long-life, highly reliable optical fiber systems which will provide nationwide digital transmission. Existing microwave radio relay systems which carry 66% of the Bell System's long-distance messages are also expected to be augmented by high-capacity optical fiber transmission systems.

The L5 coaxial carrier transmission system is a long-haul, high-quality analog system designed to carry high-quality two-way telephone conversations. L5 systems carry 33% of the Bell System's long-distance messages, and similarly are to be replaced by high-capacity optical fiber transmission systems. *See* COMMUNICATION CABLES.

Synchronization. For any digital system to deliver correct information, no bit can be lost. Moreover, the connection of a transmitting channel to its proper receiving channel requires that the transmitting and receiving terminals be synchronized. This means that the timing operations at the receiver, except for the time lost in propagating and repeatering, must follow closely the timing operations at the transmitting end. To accomplish this amounts to getting a local clock to keep the same time as a distant standard clock, except that the local clock is slow corresponding to the time required to transmit the signals. Once this objective is accomplished, every in-band frequency going into the network comes out unchanged at the distant end.

The most promising solution to this difficult problem is to provide one master oscillator for the entire network of the many different kinds of systems operating throughout the United States and then to provide and control a plurality of slave oscillators from the one master oscillator. To put this idea into practice requires three identical master standard oscillators. If one standard malfunctions, its reading will not match the readings of the other two. Were there only two standards, it would be impossible to determine which was faulty when the two readings failed to match. Each of the three standard reference frequencies is measured every 2.5 min to an accuracy of 1 part in 10^{12}, and the three frequencies are simultaneously compared with each other. Should one be wrong, the integrity of the reference is guaranteed by unique fast-acting automatic switching circuits. The three standards, put into service in 1973, are based on cesium-beam atomic clocks and are buried deep underground at Hillsboro, MO.

The Hillsboro facility transmits a reference frequency of 20.48 MHz accurate to 1 part in 10^{11} to each slave oscillator, which in turn is required to deliver 20.48 MHz accurate to 1 part in 10^{10}. Three slave oscillators are required for the same reason that three master oscillators are used. The heart of a slave oscillator is a 39A digitally controlled crystal oscillator accurate to almost 1 part in 10^{11}. Of course, the incoming 20.48 MHz might be completely lost due to transmission difficulties or suffer short-time impairments due to excessive noise, switching transients, and so forth. The 39A oscillator is loosely coupled to the incoming reference signal in order to automatically correct any slow change in oscillator frequency due to aging, assuming that the reference signal is unimpaired. When the reference signal is impaired, the 39A oscillator automatically disconnects itself from the line and can operate continuously for weeks, delivering 20.48 MHz accurate to 1 part in 10^{10}.

Electronic switching systems. Electronic switching systems (ESS) handle both local and toll switching. No. 1 ESS, which was developed for general use, went into commercial service on May 30, 1965, and was the first office switching system to use a stored memory and logic programs. It was followed by No. 2 ESS for use in the suburbs and No. 3 ESS for use in rural areas. Installation of No. 4 ESS, with about three times the capacity of its predecessors, began in 1976.

ESS can complete a direct-dialed long-distance call in 12 sec. For ESS to accomplish this, a call destined for a point a thousand miles away may have to travel four times that distance over alternative routes if by chance direct lines should be busy. This is possible because every ESS is like a computer: billions of bits of information are compactly stored in the memory, and the semiconductor integrated circuitry allows almost instant access, retrieval, and logic programs. See SWITCHING SYSTEMS (COMMUNICATIONS).

Optical-fiber transmission systems. A typical all-digital optical waveguide is an optical fiber consisting of a central core of only 0.002 of an inch (50 μm) in diameter with a surrounding cladding with an outer diameter of 0.005 of an inch (125 μm). The attenuation of the 5-mil fiber is essentially constant over a frequency band wide enough to transmit 300,000,000 on-or-off light pulses per second. The fiber loss is also essentially constant over the ambient temperature ranges encountered in practice (−40 to 212°F; −40 to 100°C). There exists no significant crosstalk between optical waveguides even when bundles of one hundred are so bunched that even waveguide is in contact with others and when as many as half may transmit signals in opposite directions. Many simpler and highly useful optical waveguide systems do not possess or need most of these advantages.

Low-loss, low-dispersion, high-silica-content fibers are produced by a process called chemical vapor deposition (CVD) which controls fiber manufacture with great precision. In this technique silica and other glass-forming oxides are deposited at high temperatures on the inner wall of a fused silica tube. Once the CVD process is completed, the tube is collapsed and drawn at a high temperature into a fiber of the desired diameter. The fiber is drawn from specially prepared glass rod (preform) at typical speeds of 3 ft/sec (1 m/sec) with less than 1% deviation in diameter. This is accomplished by a monitoring system that measures the fiber 1000 times a second, automatically adjusting drawing speed to keep the diameter constant. An improved process for the production of these optical glass fibers, called modified chemical vapor deposition (MCVD), is a major contribution to fiber optics communications and is used by most of the major companies in the field.

An optical-fiber transmission system requires a reliable long-life light source. Semiconductor lasers are expected to operate unfailingly over a period in excess of 11 years. Light-emitting diodes (LEDs) are less-expensive sources of light but, unfortunately, provide less optical power coupled into the fiber. Both types of source can be directly modulated, which means that they can be turned on and off simply by turning the drive current from a semiconductor driver on and off. See LIGHT-EMITTING DIODE.

Light detection at the receiving end requires evaluation of the intensity of the incoming light pulses. This can be done with a photodiode which converts light pulses into weak electric pulses which are amplified by an amplifier of appropriate bandwidth. The photodiode may be of the *pin* type or the more sensitive, more expensive avalanche type. See PHOTODIODE.

There is a very large number of possible applications of optical-fiber transmission systems. Many important on-premise applications exist where fibers run from only a few feet to several hundred feet, so that the system requirements are not stringent. In the exchange plant hundreds of individual calls could be woven together into single-pulse streams at rates up to about 1,000,000 pulses per second for transmission from one central office to another through optical-fiber cables in underground conduits and, in heavily populated areas, the offices are so close together that repeaters may not be needed. Eventually, optical-fiber transmission systems, because of their freedom from interference, immunity to ground-loop problems, flexibility of system growth, large information capacity, and potential, will span the continent, adding significantly to the digitization of the network. See OPTICAL COMMUNICATIONS.

Nationwide digital transmission. Nationwide digital transmission means the use of digital transmission over all of the transmission and switching facilities connecting every residential and business telephone in the United States. With great expansion since the late 1940s, there exists an extensive hierarchy of switching offices and interconnecting paths over which calls utilizing digital transmission are routed automatically and efficiently. The use of digital technology for electronic switching as well as transmission is increasing at a rapid rate in the telephone network. For example, either waveguide or optical fiber systems are expected to provide digital transmission over distances of upward of 10,000 km, thereby supplying the long-haul digital link between the rapidly expanding digital exchange network and the time division digital No. 4 ESS toll switching system. Digital switching is inherently a four-wire operation.

Interwoven with, and an important part of, the nationwide digital transmission network are special services (such as teletypewriter, facsmile, data transmission, microwave radio broadcasting, and television programs) requested by particular customers or stations.

To expedite nationwide digital transmission, the Bell System will provide a 611-mi (983-km) light-wave transmission system that will serve the United States Northeast Corridor, carrying up to 80,000 digitalized simultaneous telephone conversations through optical fibers. Ultimately two-way propagation over the half-inch-thick (13-mm) light-wave cable will transmit digitalized data and visual communications as well as conversations. This system will also connect to 23 supercapacity all-digital No. 5 ESS electronic switching systems, each ESS office handling up to 850,000 calls per hour.

Global digital transmission. The availability of long-wavelength devices, single-mode filters, integrated optical repeaters, and single-mode optical fiber transmission systems will make it practical to plan to apply fiber optics to submarine cable systems transmitting 10^9 digital pulses per second. Optical fiber systems are expected to be used in all applications where existing wire, waveguide, and microwave systems are employed.

Other uses involving fiber optics. The application of fiber optics to existing types of copiers promises a 40% price reduction, accompanied by smaller and more reliable machines requiring less maintenance.

Integrated circuits designed for use in a high-speed optical fiber scanner are being developed by the U.S. Postal Services and are to be used on a system of electronic mail transmission.

During 1980 Bell Telephone Laboratories fabricated a high-speed integrated circuit called the digital speech processor (DSP) which is to make a wide range of new telephone services possible. This DSP chip contains over 45,000 transistors in an area smaller than 1 cm², and utilizes digital transmission to perform a million additions and multiplications in a second. Western Electric Company manufactures chips with 150,000 components. Smaller chips containing over 10,000,000 million components are envisioned.

Digital filters. The use of digitalized techniques has been applied to the numerous components of many different types of systems. Digital filters are a good example. Many signals are no longer represented by undulatory currents, but by sequences of numbers in digital form that represent periodic snapshots or samples of the original, continuously varying quantity. The tasks once performed by conventional filters composed of capacitors, inductors, and resistors are now accomplished by numerical operations on these samples. Each sample is represented by a number, and numerical operations on these numbers are carried out by miniaturized high-speed computers built from even smaller integrated solid-state circuits. These extremely small digital filters bear no relation to the bulky ladder and lattice structures of the 1920s and 1930s. *See* DATA COMMUNICATIONS; DIGITAL COMPUTER; ELECTRIC FILTER; ELECTRICAL COMMUNICATIONS; INTEGRATED CIRCUITS.

[HAROLD S. BLACK]

Bibliography: T. A. Abelle, D. A. Alsberg, and P. T. Hutchison, A high-capacity digital communication system using TE$_{01}$ transmission circular waveguide, *IEEE Trans. Microwave Theory Tech.*, MTT-23(4):326–333, 1975; M. Barnoski (ed.), *Fundamentals of Optical Communications*, 1976; *Bell Syst. Tech J.*, 43(5):1831–2605, special issue, September 1964; H. S. Black, *Modulation Theory*, 1953; J. H. Bobsin and L. F. Forman, the T2 digital line: Extending the digital network, *Bell Lab. Rec.*, 51(8):239–243, 1973; R. G. Buus, D. L. Rechtenbaugh, and R. B. Whipp, T-carrier administration system speeds service restoration, *Bell Lab. Rec.*, 53(5):216–255, 1975; H. Feistel, W. A. Notz, and J. Lynn Smith, Some cryptographic techniques for machine-to-machine data communications, *Proc. IEEE*, 63(11):1547–1554, 1975; J. F. Graczyk, E. T. Mackey, and W. J. Maybach, T1C carrier: The T1 doubler, *Bell Lab. Rec.*, 53(6):256–263, 1975; J. C. Hancock, *An Introduction to the Principles of Communication Theory*, 1961; L5 coaxial cable transmission system, *Bell Syst. Tech. J.*, 53(10):1897–2269, special issue, 1974; E. T. Mackey, W. J. Maybach, and S. B. Pfeiffer, Mixing data and voice on the T1 line, *Bell Lab. Rec.*, 53(2):136–142, 1975; L. Marton (ed.), *Advances in Electronics (and Electron Physics)*, 2d ed., vol. 3, 1951; S. E. Miller, E. A. J. Marcatili, and T. Li, Research toward optical-fiber transmission systems, *Proc. IEEE*, 61(12):1703–1751, 1973; New digital radio transmission system, *Bell Lab. Rec.*, 52(11):362, 1974; J. F. Oberst, Keeping Bell System frequencies on the beam, *Bell Lab.

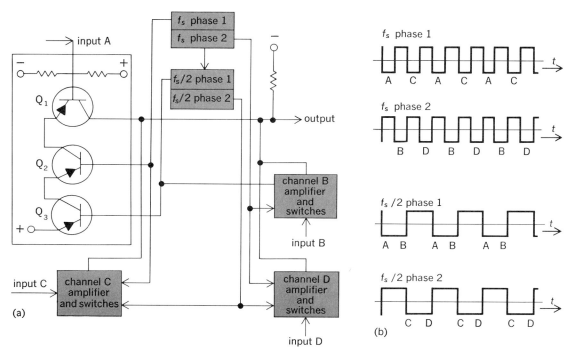

Fig. 2. Electronic distributor for a four-channel time-division multiplex transmitter. (a) Transistor amplifier Q₁ together with switches Q₂ and Q₃, which permit input A to be connected to the output line. The switches are operated by square waves of frequency f_s and $f_s/2$. (b) Waveforms needed to operate the switches.

pulse-modulation schemes and is usually used in the initial stages of other pulse-modulation methods. The amplitude of the input signal is sampled periodically at a fixed frequency, producing a sequence of pulses whose amplitude varies in accordance with the modulating signal. In a common practical system the trains of pulses representing several input signals are interlaced to form a time-division multiplex system.

Figure 1a is a block diagram of the essential parts of a four-channel time-division multiplex transmitter that uses PAM. Figure 1b depicts some of the representative waveforms. The principal elements of this system are the distributor, or commutator, which can be either electromechanical or electronic; a source of square waves at the sampling frequency f_s; and a source of square waves of one-half the sampling frequency $f_s/2$ obtained from a bistable circuit arranged to provide both phases of the square wave. The function of each is described below.

Figure 2 shows a simplified schematic diagram of an electronic distributor which can be employed in a four-channel time-division multiplex PAM system. For the purpose of explanation, only the switching operation, which occurs to connect the input channel A to the output line, will be discussed. When a positive voltage is applied to its emitter, *pnp* transistor Q₁ becomes a normal amplifier and the signals appearing at its base are amplified and appear at the output. However, the positive voltage is applied to the emitter of Q₁ only during the periods when tandem switches Q₂ and Q₃ are both made conducting. The sampling-frequency square wave f_s of phase one alternately applies negative and positive voltages to the base of the transistor switch Q₂, causing it to open

(conduct) and close, respectively. The $f_s/2$ square wave of phase one controls the transistor switch of Q₂ in a similar manner. An examination of the waveforms in Fig. 2b shows that there is only one interval during a distributor cycle when phase one of both square waves is negative. At any other interval, phase one of either one or the other of the two square-wave frequencies is positive, and the input A becomes isolated from the distributor output. Inputs B, C, and D are connected to the output line in a similar manner. Their associated switches are connected to the proper phases of the square-wave sources as indicated schematically in Fig. 2a.

In a practical PAM system, it is necessary to provide guard time between pulses to reduce interchannel crosstalk. Figure 1a shows the distributor output pulses passing through a synchronous gate operating at twice the sampling frequency. This gate, operating on the same principle as those used in the distributor proper, selects the center portion of each pulse in the train. The resultant output is a series of amplitude-modulated pulses having only one-half of their original width with blank time between each pulse. The waveform in the output line with signals due to channel A alone is shown in Fig. 1b.

Pulse-duration scheme. Pulse-duration modulation can be produced by converting a train of pulses from a PAM system by means of a circuit such as that indicated in the block diagram in Fig. 3a. Parts a and b show that if the signals generated in a PAM system are added to constant-amplitude sampling-frequency pulses, a train of pulses of varying amplitude, but with a finite minimum value, is produced at the output of the adder. The resulting pulses can be differentiated and clipped, resulting in a series of reverse sawtooth

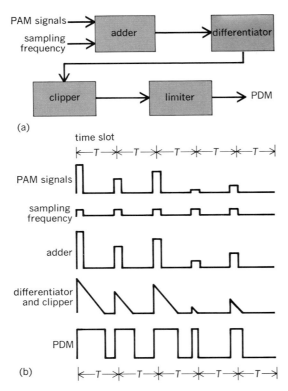

(a)

(b)

Fig. 3. Conversion of PAM signals to PDM. (a) Block diagram. (b) Representative waveforms.

shapes in which the width of each sawtooth becomes proportional to the amplitude of the particular pulse. When passed through a limiter, the desired sequence of pulses, having constant amplitude and varying in width in accordance with the signal, is generated. By properly adjusting the differentiating and clipping circuits, the pulse corresponding to zero amplitude produces a pulse of minimum width, used for reference.

Guard time between pulses is required in a practical PDM system, just as in the PAM. Such circuits should be designed so that maximum-amplitude samples, when converted to pulse width, do not occupy 100% of the available time slot. In addition, the differentiating circuit must be completely discharged during the guard time to prevent interaction between adjacent pulses.

Pulse-position scheme. PPM, sometimes called pulse-time modulation, can be derived from PDM. Referring to waveform E in Fig. 3, the position of the trailing edge of each pulse varies with the amplitude of the input signal. Figure 4 shows one method of converting PDM to PPM, together with representative waveforms. The variable-width pulses are first differentiated and then rectified to permit only the negative peaks to trigger a monostable multivibrator. In the latter, each negative trigger pulse causes a pulse of constant amplitude and width to be generated, whose time position is proportional to the amplitude of the modulating signal (input).

Pulse-code scheme. PCM is a form of pulse communication in which the signal is first sampled, as in PAM; the magnitude of the sample is then replaced by the nearest value selected from a finite set; finally, the permitted values are represented by a simple code pattern of ON or OFF pulses. The operations mentioned are called sampling, quantizing, and coding, respectively.

The sampling techniques needed for PCM can be similar to those employed in PAM. Although electronic quantizing and coding often can become highly complex, a relatively simple scheme using eight linear quantizing steps and a beam coder tube will be described as an example.

The quantizing of the pulses generated by PAM can be explained with the aid of the diagrams shown in Fig. 5. The voltage wave to be sampled and quantized is shown in Fig. 5a, where the magnitude of the voltage levels is divided into eight increments. The actual amplitudes A and B in Fig. 5a are transmitted as standard amplitudes 2 and 3, respectively. This quantizing process introduces a certain amount of error which can be shown to be insignificant if a significantly large number of standard amplitudes or more complicated quantizing procedures are employed.

To quantize a signal, the amplitude sample is first matched with a series of amplitude standards and then converted to the one with the nearest standard value. One method of achieving this is to

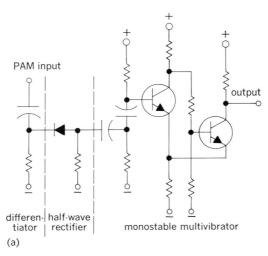

(a)

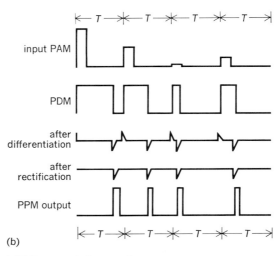

(b)

Fig. 4. Conversion of PAM to PPM. (a) Components required. (b) Representative waveforms.

utilize a multiplicity of regenerative devices, each of which changes between its OFF and ON state within a very small input voltage range. By proper biasing, these devices can be set to operate at any given input voltage level. When the inputs and outputs of eight of these level selectors work in parallel, it is possible to convert input signals of varying amplitudes to outputs of eight discrete amplitudes.

The quantized levels can be represented in this case by the ON or OFF values of three pulses, as shown in Fig. 5b. Thus the magnitude A is represented by two pulses of zero amplitude and the third pulse of unit amplitude. As indicated in the diagram, the three code pulses can merge into a continuous waveshape.

Figure 6 shows a diagram of a system that permits PAM signals to be converted to PCM. In the case illustrated, this is accomplished with the aid of the parallel quantizing circuits mentioned above and a special beam code tube. The latter consists of a cathode-ray tube with conventional gun and electrostatic deflection plates, a special aperture plate placed perpendicularly to the electron beam and an anode. The aperture plate (Fig. 6b) has horizontal slots in eight vertical rows in accordance with the three-unit pulse code shown in Fig. 5b.

The quantized pulses are applied to the vertical deflection plates of the beam coder tube and the sawtooth sweep derived from the sampling frequency is applied to the horizontal plates. The electron beam is swept at a fixed vertical height once during each sampling period, causing the anode current to appear in the form of a pulse whose shape is determined by the opening in the

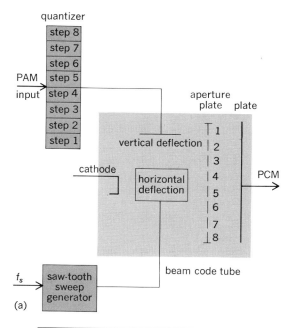

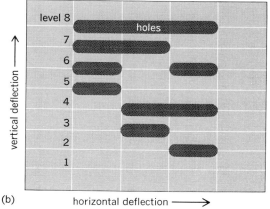

(b)   horizontal deflection ⟶

Fig. 6. An elementary pulse-code modulation system. (a) Block diagram of system that permits conversion of PAM to PCM with aid of beam code tube. (b) Arrangement of holes in aperture plate of beam code tube.

aperture plate. In this manner, the quantized signal is converted directly into the PCM signal.

[EDWARD L. GINZTON/CHARLES L. ALLEY]

Bibliography: J. Betts, *Signal Processing Modulation and Noise,* 1971; F. R. Connor, *Modulation,* 1973; M. Schwartz, *Information, Transmission, Modulation and Noise,* 1970.

Pulse transformers

Iron-cored devices which are used in the transmission and shaping of low-power pulses whose widths range from a fraction of a microsecond to about 25 μsec. Among the extensive applications of pulse transformers are the following: (1) to couple between the stages of pulse amplifiers; (2) to invert the polarity of a pulse; (3) to change the amplitude and impedance level of a pulse; (4) to differentiate a pulse; (5) to effect "dc isolation" between a source and a load; (6) to act as coupling element in certain pulse-generating circuits.

In many cases the functions listed above can be

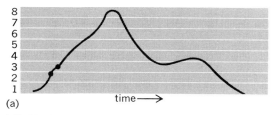

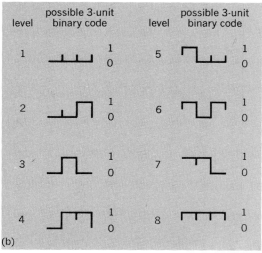

Fig. 5. Representation of signals by an eight-level code. (a) A signal wave with subdivision of its amplitude into eight levels. (b) A possible three-pulse binary code to represent the eight levels.

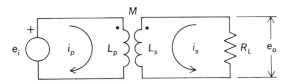

Schematic diagram of a pulse transformer.

accomplished as well or better with transistor circuits. However, the pulse transformer, being a passive element, has none of the instability associated with active circuits.

The figure gives a schematic diagram of a pulse transformer. The primary inductance is L_p, the secondary inductance is L_s, and the mutual inductance is M. In a more accurate description of the pulse transformer it is necessary to take into account the primary and secondary resistances, all capacitances, the core loss, and the nonlinearity of the magnetic circuit. The coefficient of coupling K between primary and secondary is defined by $K = M/\sqrt{L_p L_s}$. An ideal transformer is one for which $K = 1$ and L_p is infinite. In this case, the output e_o is an exact replica of the input e_i. For an ideal pulse transformer this relationship is expressed by the equation below, where N_p is the primary number of turns and N_s is the secondary number of turns. See TRANSFORMER.

$$n = \frac{e_o}{e_i} = \frac{i_p}{i_s} = \sqrt{\frac{L_s}{L_p}} = \frac{N_s}{N_p}$$

A pulse transformer behaves as a reasonable approximation to a perfect transformer when used in connection with the fast waveforms it is intended to handle. The core of a pulse transformer is usually molded from a magnetic ceramic such as sintered manganese-zinc ferrite. The maximum permeability of this material is not very great, but its resistivity is at least 10,000,000 times that of Hipersil or Permalloy. This high resistivity means that the skin effect due to eddy currents is very small, and an effective permeability of the order of 1000 is attained. The windings of the pulse transformer are placed on a circular nylon or paper bobbin, which is then inserted in the core.

[CHRISTOS C. HALKIAS]

Bibliography: J. Millman and H. Taub, *Pulse, Digital and Switching Waveforms*, 1965; L. Strauss, *Wave Generation and Shaping*, 2d ed., 1970.

Push-pull amplifier

A two-transistor or two-tube amplifier circuit often used as the power-output stage of a multistage amplifier. The power-output stage in an audio amplifier is normally expected to furnish from 5 to 50 watts or more. Use of one active device in this stage is not feasible, because the transistor or tube would have to operate class A with a low conversion efficiency, on the order of 10%. Use of two or more active devices in parallel does not improve efficiency. However, with two active devices operating in push-pull, it is possible to supply the required amount of power with conversion efficiency on the order of 50%. This higher efficiency means that the amplifier does not require as much power

from the power supply, because less power is dissipated as heat by the power amplifier active devices. See POWER AMPLIFIER.

Operation. A simplified circuit for a push-pull amplifier is shown in the illustration. The input signals must be of equal magnitude but 180° out of phase. The collector current in the transistor with the positive signal is increasing while the transistor current in the other transistor is decreasing. This relation between the collector currents gave rise to the name push-pull. Because of their phase relationship, the two currents flowing in the two halves of the primary winding of the transformer produce an output similar to that of a single source connected to the transformer.

Although the active devices are indicated as junction transistors in the illustration, other devices such as triodes, beam power tubes, or field-effect transistors (FET) can also be used in this push-pull arrangement. The discussion to follow applies equally well regardless of the particular power device employed.

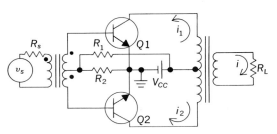

Two transistors in a push-pull arrangement.

A push-pull amplifier may be operated class A, class AB, or class B. The greatest conversion efficiency occurs when the operation is class B. Therefore, if large amounts of power are required, the amplifier is designed for class B operation. An additional merit of class B operation is that the quiescent current is zero. Another advantage of push-pull operation is that the average currents of the two transistors or tubes flow in opposite directions through the primary winding of the output transformer, resulting in no average magnetization of the core. This allows the use of smaller cores with savings in size and weight.

If the two transistors have identical characteristics and the transformer is considered ideal, the even harmonic components of distortion are absent in the output signal. A high-power class B push-pull amplifier can be designed with little harmonic distortion. See AMPLIFIER; DISTORTION (ELECTRONIC CIRCUITS).

The circuit shown in the illustration indicates the basic push-pull amplifier. The interest in high-fidelity audio amplifiers has led to the development of more complicated circuits.

Driver stages. The input circuit shown in the illustration employs a transformer to produce the necessary phase inversion. Because of the limited frequency response of the transformer and also the size of transformer required, a transformer is rarely used. Instead, a transistor phase inverter is employed. Because some phase-inverter circuits have a gain considerably greater than unity, the overall gain of the amplifier can be increased more than it could be if a transformer with a step-up turns ratio were used. *See* PHASE INVERTER.

A push-pull circuit using transistors having complementary symmetry has been used which requires neither an output nor an input transformer. [HAROLD F. KLOCK]

Bibliography: C. A. Holt, *Electronic Circuits, Digital and Analog,* 1978; J. Millman, *Microelectronics,* 1979; D. L. Schilling and C. Belove, *Electronic Circuits, Discrete and Integrated,* 2d ed., 1979.

Q (electricity)

Often called the quality factor of a circuit, *Q* is defined in various ways, depending upon the particular application. In the simple *RL* and *RC* series circuits, *Q* is the ratio of reactance to resistance, as in Eqs. (1), where X_L is the inductive reactance,

$$Q = X_L/R \quad Q = X_C/R \quad \text{(a numerical value)} \quad (1)$$

X_C is the capacitive reactance, and R is the resistance. An important application lies in the dissipation factor or loss angle when the constants of a coil or capacitor are measured by means of the alternating-current bridge.

Q has greater practical significance with respect to the resonant circuit, and a basic definition is given by Eq. (2), where Q_0 means evaluation at resonance.

$$Q_0 = 2\pi \frac{\text{max stored energy per cycle}}{\text{energy lost per cycle}} \quad (2)$$

onance. For certain circuits, such as cavity resonators, this is the only meaning *Q* can have.

For the *RLC* series resonant circuit with resonant frequency f_0, Eq. (3) holds, where *R* is the total

$$Q_0 = 2\pi f_0 L/R = 1/2\pi f_0 CR \quad (3)$$

circuit resistance, *L* is the inductance, and *C* is the capacitance. Q_0 is the *Q* of the coil if it contains practically the total resistance *R*. The greater the value of Q_0, the sharper will be the resonance peak.

The practical case of a coil of high Q_0 in parallel with a capacitor also leads to $Q_0 = 2\pi f_0 L/R$. *R* is the total series resistance of the loop, although the capacitor branch usually has negligible resistance.

In terms of the resonance curve, Eq. (4) holds,

$$Q_0 = f_0/(f_2 - f_1) \quad (4)$$

where f_0 is the frequency at resonance, and f_1 and f_2 are the frequencies at the half-power points. *See* RESONANCE (ALTERNATING-CURRENT CIRCUITS).

[BURTIS L. ROBERTSON]

Q meter

A direct-reading instrument widely used for measuring the *Q* of an electric circuit at radio frequencies. Originally designed to measure the *Q* of coils, the Q meter has been developed into a flexible, general-purpose instrument for determining many other quantities such as (1) the distributed capacity, effective inductance, and self-resonant frequency of coils; (2) the capacitance, *Q* or power factor, and self-resonant frequency of capacitors; (3) the effective resistance, inductance or capacitance, and the *Q* of resistors; (4) characteristics of intermediate- and radio-frequency transformers; and (5) the dielectric constant, dissipation factor,

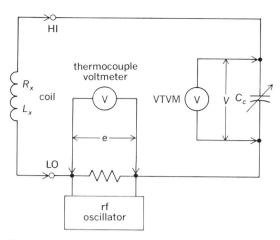

Simplified measurement circuit of a Q meter.

and power factor of insulating materials. *See* Q (ELECTRICITY).

The illustration shows in simplified form the measurement circuit of a Q meter. The coil L_x being measured is connected into the circuit by means of the external terminals HI and LO. The coil is brought into resonance by tuning the calibrated capacitor C_c. A controlled and measured input voltage *e* is introduced into the circuit by means of an rf oscillator. A thermocouple voltmeter measures the input voltage *e*, and a vacuum-tube voltmeter (VTVM) measures the voltage *V* across the calibrated capacitor. With the circuit tuned to resonance, Eqs. (1) and (2) hold, where R_x

$$\omega L_x = 1/\omega C_c \quad (1)$$

$$\frac{V}{e} = \frac{(R_x^2 + \omega L_x^2)^{1/2}}{[R_x^2 + (\omega L_x - 1/\omega C_c)^2]^{1/2}}$$
$$= (1 + \omega^2 L_x^2/R_x^2)^{1/2} \quad (2)$$

is the resistance of the coil. Since $Q = \omega L/R$, Eq. (3) is valid.

$$\frac{V}{e} = (1 + Q^2)^{1/2} \quad (3)$$

When *Q* is large, the equation may be simplified to this basic equation of the Q meter, Eq. (4). The

$$Q = \frac{V}{e} \quad (4)$$

thermocouple voltmeter and the VTVM are calibrated in such a way that the product of their readings gives the *Q* of the coil directly.

The many quantities enumerated above are determined by inserting suitable circuit elements in series with the coil or in parallel with the capacitor and measuring the effect on the circuit *Q*. Simple calculations based on the preceding equations are required to determine the desired quantity from the measured values of *Q*. *See* ELECTRICAL MEASUREMENTS; HIGH-FREQUENCY IMPEDANCE MEASUREMENTS.

[ISAAC F. KINNARD/EDWARD C. STEVENSON]

Bibliography: M. Braccio, *Basic Electrical and Electronic Tests and Measurements,* 1979; W. D. Cooper, *Electronic Instrumentation and Measurement Techniques,* 1978; T. Kantrowitz et al., *Electronic Measurements,* 1979.

Quantum electronics

A loosely defined field concerned with the interaction of radiation and matter, particularly those interactions involving quantum energy levels and resonance phenomena, and especially those involving lasers and masers. Quantum electronics encompasses useful devices such as lasers and masers and their practical applications; related phenomena and techniques, such as nonlinear optics and light modulation and detection; and related scientific problems and applications, such as quantum noise processes, laser spectroscopy, picosecond spectroscopy, and laser-induced optical breakdown.

In one sense any electronic device, even one as thoroughly classical in nature as a vacuum tube, may be considered a quantum electronic device, since quantum theory is presently accepted to be the basic theory underlying all physical devices. In practice, however, quantum electronics is usually understood to refer to only those devices such as lasers and atomic clocks in which stimulated transitions between discrete quantum energy levels are important, together with related devices and physical phenomena which are excited or explored using lasers. Other devices such as transistors or superconducting devices which may be equally quantum-mechanical in nature are not usually included in the domain of quantum electronics.

Stimulated emission and amplification. Quantum electronics thus centers on stimulated-emission devices, primarily lasers and masers. Atoms, molecules, and other small isolated quantum systems typically have discrete, well-resolved quantum-mechanical energy levels. In a collection of a large number of identical such atoms or molecules, one can speak of the number of atoms, or the population, residing in each such energy level. Under normal thermal equilibrium conditions, lower energy levels always have larger populations. One may apply electromagnetic radiation (radio waves or light waves, as appropriate) to such a collection of atoms at a frequency (or wavelength) corresponding through Planck's law to the quantum-mechanical transition frequency between any two levels. This radiation will then be absorbed by the atoms, and some of the atoms will be correspondingly lifted from the lower energy level to the upper. Measurement of the strength of this absorption versus the probing frequency or wavelength provides a powerful way of studying the quantum energy levels, and provides the basic approach of spectroscopy or of resonance physics.

Under suitable conditions it is also possible, in a wide variety of atomic systems, to create a condition of population inversion, in which more atoms are temporarily placed in some upper energy level than in a lower energy level. Population inversion is created by "pumping" the atomic system, by using a wide variety of techniques. Application of a signal at the appropriate transition frequency then produces coherent amplification rather than absorption of the applied signal, with the amplification energy being supplied by a net flow of atoms from the upper to the lower energy level through stimulated transitions. The addition of electromagnetic feedback to the atomic system by means of lumped electrical circuitry, microwave circuitry,

or optical mirrors, as appropriate to the frequency range involved, can convert this amplification into coherent oscillation at the atomic transition frequency. While this type of stimulated emission and amplification has found useful but limited application at ordinary radio-wave and microwave frequencies (primarily in ultrastable atomic clocks and very-low-noise microwave maser amplifiers), its overwhelming importance is at optical frequencies, where it makes possible for the first time the coherent amplification and oscillation of optical signals in laser devices.

Nonlinear optical phenomena. Soon after the invention of the laser, it was discovered that a wide assortment of interesting and useful nonlinear optical phenomena could be produced using coherent beams from lasers because of their unprecedentedly high power and coherence. By using nonlinear optical phenomena such as harmonic generation, subharmonic generation, parametric amplification and oscillation, as well as stimulated Raman and Brillouin scattering, it becomes possible to produce a wide range of new infrared, optical, and ultraviolet wavelengths, some of them widely tunable, in addition to obtaining important basic physical information about the nonlinear materials involved. Although the elementary description of these nonlinear optical devices is thoroughly classical in nature, they have become a major portion of the field of quantum electronics.

Applications of lasers. Lasers have found an enormous variety of important practical applications in engineering, technology, and medicine, ranging from highway surveying and supermarket checkout counters to automobile production lines and retinal surgery. As these applications have become established as routine techniques, they have generally moved out of the domain of quantum electronics considered as a scientific discipline. At the same time, the laser has been applied as the primary tool for many fundamental as well as exotic measurement techniques in science, including particularly high-resolution spectroscopy, such as saturated-absorption, tunable-laser, and picosecond spectroscopy. Many of these more complex techniques have been retained within the domain of quantum electronics, along with certain exotic engineering applications such as laser isotope separation and the study of laser interactions with plasmas, particularly for laser-induced fusion. *See* LASER; MASER; OPTICAL DETECTORS; OPTICAL MODULATORS.

[A. E. SIEGMAN]

Bibliography: F. T. Arecchi and E. O. Schulz-DuBois (eds.), *Laser Handbook*, 1972; A. E. Siegman, *An Introduction to Masers and Lasers*, 1971; A. Yariv, *Quantum Electronics*, 1975.

Quartz clock

A clock that uses the piezoelectric property of a quartz crystal. When a quartz crystal vibrates, a difference of electric potential is produced between two of its faces. The crystal has a natural frequency of vibration, depending on its size and shape, and if it is introduced into an oscillating electric circuit having nearly the same frequency as the crystal, two effects take place simultaneously: The crystal is caused to vibrate at its natural frequency, and the frequency of the entire circuit becomes the same as the natural frequency of the crystal.

In the clock the alternating current from the oscillating circuit is amplified, and the frequency is subdivided in steps such as from 100 kHz to 1 kHz. This setup finally drives a synchronous motor and gear train to display time by hands on a clock face.

The natural frequency of a quartz crystal is nearly constant if precautions are taken in cutting and polishing it, and if it is maintained at nearly constant temperature and pressure. After a crystal has been placed in operation, the frequency changes slowly as a result of physical changes. When allowance is made for such changes, the best crystals run for a year with accumulated errors of less than 0.1 sec. See OSCILLATOR.

[GERALD M. CLEMENCE]

Radar

An acronym for radio detection and ranging, the original and still principal application of radar. The name is applied to both the technique and the equipment used.

Radar is a sensor; its purpose is to provide estimates of certain characteristics of its surroundings of interest to a user, most commonly the presence, position, and motion of such objects as aircraft, ships, or other vehicles in its vicinity. In other uses, radars provide information about the Earth's surface (or that of other astronomical bodies) or about meteorological conditions. To provide the user with a full range of sensor capability, radars are often used in combinations or with other elements of more complete systems.

Radar operates by transmitting electromagnetic energy into the surroundings and detecting energy reflected by objects. If a narrow beam of this energy is transmitted by the directive antenna, the direction from which reflections come and hence the bearing of the object may be estimated. The distance to the reflecting object is estimated by measuring the period between the transmission of the radar pulse and reception of the echo. In most radar applications this period will be very short since electromagnetic energy travels with the velocity of light.

Many different kinds of radar have been developed for a wide range of purposes, but they all use electromagnetic radiation (radio waves) to detect and measure certain characteristics of objects (or targets) in their vicinity.

Historical development. The fact that radio waves produce echoes was known before 1920. During the 1930s researchers in the United States, England, France, and Germany pointed out that ships and airplanes produced radio echoes which could be used to deduce their locations. In the mid-1930s the British Air Ministry authorized the installation of some 20 radar stations on the east and southeast coasts of England to provide surveillance of these air approaches. As World War II began, a continuous radar watch was maintained over the principal air and sea approaches to Britain. The radar network was so effective in locating German bombers and directing fighters against them that it is generally credited with making it possible for the severely outnumbered Royal Air Force to defeat the Luftwaffe in the Battle of Britain. When the Germans resorted to night bombing to reduce the losses suffered in daylight encounters, airborne radar aboard British fighters enabled them to train their guns on the enemy in the dark with devastating results.

During the war, American and British scientists cooperated to develop radars for such diverse applications as surveillance of large regions and early warning of approaching ships, aircraft, and missiles; fire control for automatically directing gunfire against air or surface targets; directing gunfire at enemy aircraft from aboard radar-equipped interceptor aircraft; radar bombsights; and detection of submarines from aircraft.

Many nonmilitary applications resulted from military developments; today civil and military applications are numerous. Development has turned to improvements in radar performance resulting from applications of the computer, integrated circuits in the apparatus, and more precise design to meet the specific needs of the user. See COMPUTER; INTEGRATED CIRCUITS.

Kinds of radar. Radar has so many valuable applications that the physical nature of radars varies greatly. Several radars are available for use on small boats as a safety and navigation aid, some so small as to be carried by an operator, others a little larger in which the transmitter, antenna, and receiver are all contained in a mast-mounted weatherproof unit which feeds a small inboard display. Another familiar small radar, also seen in a hand-held form, is that used by police to measure the speed of automobiles.

Perhaps the largest radars are those covering acres of land, long arrays of antennas all operating together to monitor the flight of space vehicles or astronomical bodies. Other very large radars are designed to monitor flight activity at substantial distances around the world. These are large mainly because they must use longer-than-usual radio wavelengths associated with ionospheric containment of the signal for over-the-horizon operations; these longer wavelengths require very large antennas to form the narrow beams useful in the measurement of target positions.

More common in size are those radars seen at airports, those with rotating antennas 10–30 ft (3–9 m) wide, equipment housings the size of a small shed, and operator consoles the size of a small desk. Radars intended for mobile use, particularly airborne radars, are quite compact, usually very modular in construction and very well adapted to fit into the vehicle.

Radars intended principally to determine the presence and position of reflecting targets in a region around the radar are called search radars. Other radars examine further the targets detected: examples are height finders with antennas that scan vertically in the direction of an assigned target, and tracking radars that are aimed continuously at an assigned target to obtain great accuracy in estimating target motion. In some modern radars, these search and track functions are combined, usually with some computer control. Surveillance radar connotes operation of this sort, somewhat more than just search alone. There are also very complex and versatile radars with considerable computer control, with which many functions are performed and which are therefore called multifunction radars. Very accurate tracking radars intended for use at missile test sites or similar test ranges are called instrumentation radars. Radars designed to detect clouds and precipitation are called meteorological or weather radars.

Some radars have separate transmit and receive antennas sometimes located kilometers apart. These are called bistatic radars, the more conventional single-antenna radar being monostatic. Some useful systems have no transmitter at all and are equipped to measure, for radarlike purposes, signals from the targets themselves. Such systems are often called passive radars, but the terms radiometers or signal intercept systems are generally more appropriate. Some genuine radars may, of course, occasionally be operated in passive modes. *See* PASSIVE RADAR.

The terms primary and secondary are used to describe, respectively, radars in which the signal received is reflected by the target and radars in which the transmission causes a transponder (transmitter-responder) carried aboard the target to transmit a signal back to the radar. The Identification Friend or Foe (IFF) system in wide military and civil use is a secondary radar.

Fundamentals of operation. It is convenient to consider radars composed of four principal parts: the transmitter, antenna, receiver, and display. Each is discussed more fully below.

The transmitter provides the rf signal in sufficient strength (power) for the radar sensitivity desired and sends it to the antenna, which causes the signal to be radiated into space in a desired direction. The signal propagates (radiates) in space, and some of it is intercepted by reflecting bodies. These reflections, in part at least, are radiated back to the antenna. The antenna collects them

and routes all such received signals to the receiver, where they are amplified and detected. The presence of an echo of the transmitted signal in the received signal reveals the presence of a target. The echo is indicated by a sudden rise in the output of the detector, which produces a voltage (video) proportional to the sum of the rf signals being received and the rf noise inherent in the receiver itself. The time between the transmission and the receipt of the echo discloses the range to the target through Eq. (1), where R is the range to

$$R = \tfrac{1}{2}c\Delta T \qquad (1)$$

the target, c is the speed of electromagnetic propagation (3×10^8 m/s), and ΔT is the time between the transmission and the receipt of the echo. The direction or bearing of the target is disclosed by the direction the antenna is pointing when an echo is received.

Figure 1 shows a block diagram of a typical radar, one using pulsed transmissions to perform a search function. The transmitter shown is of the master oscillator, power amplifier (MOPA) type, discussed more completely below. The principal conversion of electrical power to transmitted signal power comes in the final power amplifier stage under the control of the modulator forming the radar pulse. In some simpler radars, the amplifier used may itself be an oscillator, and hence its modulator (or pulser) constitutes the waveform generator, and there are no successive stages of amplification. *See* AMPLIFIER; MODULATOR; OSCILLATOR; RADIO TRANSMITTER.

The duplexer permits the same antenna to be used on both transmit and receive, and is equipped with protective devices to block the very strong transmit signal from going to the sensitive receiver and damaging it. The antenna forms a beam, usually quite directive, and, in the search example, rotates throughout the region to be searched. *See* ANTENNA.

The radar reflections are among the signals received by the antenna in the period between transmissions. Most search radars have a pulse repetition frequency (prf), antenna beamwidth, and rotation rate such that several pulses are transmitted (perhaps 20 to 40) while the antenna scans past a target. This allows a buildup of the echo being received. Some radars are equipped with low-noise rf preamplifiers to improve sensitivity. Almost all radars use some form of mixing of the received signal with a local oscillator signal to produce an intermediate frequency (i-f) signal, commonly at 30 or 60 MHz, more conveniently amplified and processed. The local oscillator signal, offset from the transmitted rf by precisely the i-f, can be supplied by the waveform generator as shown. The mixer is sometimes called the first detector. The i-f signal is then fed to the detector (sometimes called the second detector) which produces a video signal, a voltage proportional to the strength of the i-f at its input. This video may be amplified and used in the cathode-ray tube (CRT) of the radar display to cause a bright spot to be formed on the face of the tube when an echo is received, in a manner indicating the range and bearing of the target. *See* CATHODE-RAY TUBE; DETECTOR; ELECTRONIC DISPLAY; MIXER; RADIO RECEIVER.

key:
- - - - rf signals
——— i-f signals
——— video, bias voltages, timing, and control

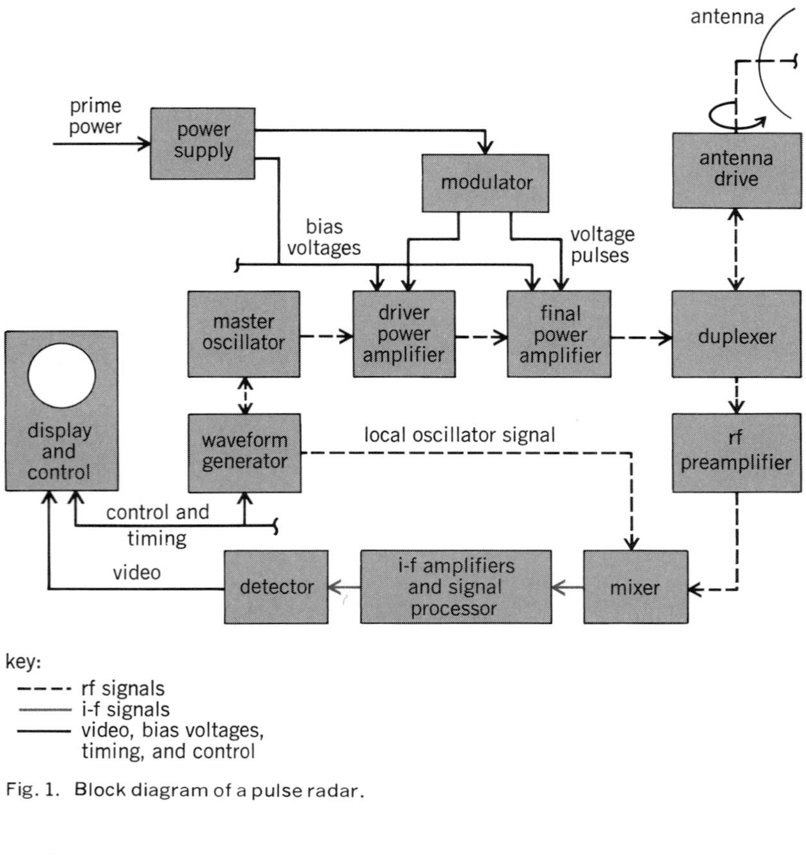

Fig. 1. Block diagram of a pulse radar.

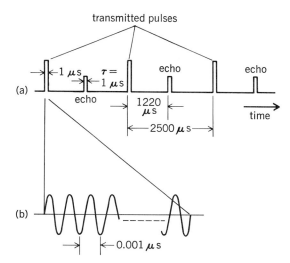

Fig. 2. Waveform of a simple pulse radar. (a) Portion of transmitted pulse train and echoes (exaggerated). Echo corresponds to a target at 100 nautical miles (183 km). Pulse period corresponds to a pulse repetition frequency of 400 pulses per second. Example pulse width is 1 μs. (b) Single transmitted pulse showing carrier cycles for an L-band (1000-MHz) radar.

Figure 2 shows a reasonable waveform of an elementary search radar. Duty factors (the ratio of pulse width τ, or transmit "on" time, to total time) in such radars are typically 0.1 to 1.0%. With modern signal processing, much more complicated waveforms are often used, and with the advent of transistors and other solid-state amplifiers, radar systems have been developed with duty factors of 10 to 50%.

Radar range equation. The equations which relate radar characteristics and performance can be derived from consideration of the echo power received, S_r. This signal results from a transmitter of peak power P_t sending its signal into space through an antenna of directive gain G_t. The spatial power density decreases with range as though an effective transmitted signal of power $P_t G_t$ were radiated isotropically, with the power density (in, say, watts per square meter) decreasing as the spherical surface $4\pi R^2$ (R being the range in meters) increases. An object of a radar cross section σ (in square meters) in the beam intercepts the incident power and reradiates it, again as if isotropically with $4\pi R^2$ decrease in power density. The reflection is intercepted by the receiving antenna of an effective aperture A_e (given, again, in square meters). At the antenna terminals, then, this received signal strength is given by Eq. (2), where the factors L ($0<L<1$) account for various

$$S_r = \frac{P_t G_t L_t}{4\pi R^2} \cdot \sigma \cdot \frac{1}{4\pi R^2} \cdot A_e L_r$$
$$= \frac{P_t G_t L_t \sigma A_e L_r}{(4\pi)^2 R^4} \qquad (2)$$

losses in signal power in the transmitting and receiving processes. Also, the ranges from transmitter to target and from target to receiver are treated identically in this monostatic radar case.

Successful detection requires that S_r meet some strength criterion, often that it exceed the level of signal produced by the receiver alone by a wide margin (a ratio, W, of 20:1, say). The receiver's noise production can be stated as kTB_n, where k is Boltzmann's constant (1.38×10^{-23} joule per kelvin), T is an effective noise temperature (Kelvin scale) representing the type of receiver and other conditions, and B_n is the noise bandwidth of the receiver. The effective temperature shown here is not just ambient temperature; it represents thermal conditions and the receiver's noise factor, a measure of its noise producing characteristics. The noise bandwidth for many receivers is proportional to the i-f passband, $B_{i\text{-}f}$. This passband is generally designed to accommodate the transmitted pulse; ideally, this means setting $B_{i\text{-}f}$ equal to $1/\tau$, but in practice $B_{i\text{-}f}$ is only approximated by that value. *See* ELECTRICAL NOISE.

Detection occurs, then, when S_r meets this criterion of Eq. (3). Solving Eq. (3) for R produces a

$$S_r = \frac{P_t G_t L_t \sigma A_e L_r}{(4\pi)^2 R^4} = WkTB_n \qquad (3)$$

familiar form of the radar range equation Eq. (4).

$$R = \left(\frac{P_t G_t L_t \sigma A_e L_r}{(4\pi)^2 WkTB_n}\right)^{1/4} \qquad (4)$$

Since antenna performance is represented in this expression by both gain and aperture, it is sometimes convenient to use Eq. (5), relating the two to the carrier wavelength λ giving Eq. (6).

$$G = \frac{4\pi A_e}{\lambda^2} \qquad (5)$$

$$R = \left(\frac{P_t G^2 \lambda^2 \sigma L_t L_r}{(4\pi)^3 WkTB_n}\right)^{1/4} \qquad (6)$$

The formulation to this point assumes that detection is based on a single pulse. If many pulses are "integrated" (as in the phosphor of a display), the factors n for the number of pulses so integrated and $E(n)$ for the efficiency of integration can be included to give Eq. (7).

$$R = \left(\frac{P_t G^2 nE(n)\lambda^2 \sigma L_t L_r}{(4\pi)^3 WkTB_n}\right)^{1/4} \qquad (7)$$

It is not always receiver noise with which the echo competes for detection. Other useful forms of this equation can be derived by substituting, for the receiver noise power kTB_n considered here, expressions for signals received from jammers or strong clutter reflections.

Radar carrier-frequency bands. Radar carrier frequencies are broadly identified by a nomenclature that originated in wartime secrecy and has since been found very convenient and widely accepted. The spectrum is divided into bands, the frequencies and wavelengths of which are given in the table.

Propagation. The above presentation of the radar range equation includes no account of the medium; propagation in free space is assumed. Actually, the Earth's surroundings have many noteworthy properties. The charged layers of the ionosphere present a highly refractive shell at radio frequencies well below the microwave frequencies of most radars. Consequently, over-the-horizon radars have been built in the 10-MHz area to

Radar carrier-frequency bands

Band designation	Nominal frequency range	Representative wavelength
HF	3–30 MHz	30 m at 10 MHz
VHF	30–300 MHz	3 m at 100 MHz
UHF	300–1000 MHz	1 m at 300 MHz
L	1000–2000 MHz	30 cm at 1000 MHz
S	2000–4000 MHz	10 cm at 3000 MHz
C	4000–8000 MHz	5 cm at 6000 MHz
X	8000–12,000 MHz	3 cm at 10,000 MHz
K_u	12.0–18 GHz	2 cm at 15 GHz
K	18–27 GHz	1.5 cm at 20 GHz
K_a	27–40 GHz	1 cm at 30 GHz
mm	40–300 GHz	0.3 cm at 100 GHz

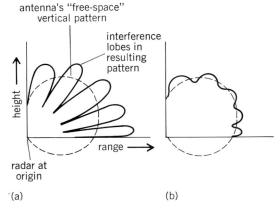

Fig. 3. Effect of a reflecting earth surface on the radiation pattern of an antenna. (a) Reflection nearly perfect. (b) Reflection much reduced.

exploit this skip path. The Earth's surface represents such a distinct medium interface that microwave reflections take place. A multipath phenomenon results, causing the lobed interference structure seen in field strength graphs of microwave antennas (Fig. 3). The lobes in the plane normal to the reflecting surface are caused by the successive phase relationships in the direct and reflected signals due to the path lengths involved. When terrain is not so reflective, the lobe structure is less pronounced. A complete account of the reflection must include the nature (conductivity) of the surface, the diffusion (due to surface roughness), and the divergence of the reflection due to the spherical shape of the Earth.

The atmosphere also acts as a lens since its dielectric constant decreases with increasing altitude. Consequently, microwave rays are bent downward slightly and the radar horizon is somewhat beyond the visual. Under certain conditions defined as standard, the effect is such that if one were to draw the Earth with a radius 4/3 its actual value, the refracted rays would be bent back to straight lines. A simple expression, Eq. (8), results

$$R_{\tan} = \sqrt{2h_1} + \sqrt{2h_2} \qquad (8)$$

for the tangent line distances, relating radar and target heights in feet (1 ft = 0.3048 m), h_1 and h_2, to the tangent range $R_{\tan}$ (Fig. 4) in statute miles (1 mi = 1.609 km).

Frequently, an atmospheric condition involving a pronounced departure from the smoothly varying dielectric constant with altitude occurs, and a superrefractive or ducting condition results with low-altitude targets detected well beyond 4/3-Earth predictions. The variation in such conditions over time of day, season, geographic location, and local weather makes reliable exploitation of the phenomenon difficult.

Attenuation of radar signals, also not accounted for in the above radar equations, is also encountered in the Earth's atmosphere. Attenuation is due to both molecular absorption by resonant excitation of uncondensed gases and scattering by particles such as dust or water droplets in fog and clouds. The attenuation due to water vapor and to oxygen molecules is shown in Fig. 5 and is seen to be negligible below the L-band and tolerable through the X-band. The resonance peaks for these molecules cause much more concern in choosing frequencies in the millimeter region for special-purpose radars.

Target characteristics. In the radar equations, the target is represented by a single term, its radar cross section σ. The discussion here assumes that the target is a vehicle such as an aircraft, a truck, or a spacecraft, rather than coastlines or terrain features which are less frequently targets of interest.

The amount of radar signal reflected by such targets is clearly a random variable, not a constant. Such targets are composed of many scatterers or reflecting surfaces, all of which contribute to the whole reflection. The individual reflections add coherently (that is, with both constructive and destructive phase interference, depending on the many path lengths involved) in composing the total reflection. Consequently, over successive observations, the reflections will differ in magnitude if even the slightest repositioning of the scatterers occurs, or if a different carrier frequency is employed.

Such targets have been categorized by the degree to which their composition encourages independent values of reflected signal strength. Targets are described as Swerling models I through IV after P. Swerling. It is still convenient, however, to use single average values of cross section in radar computations to represent even these fluctuating targets. This is made possible by adjusting the W factor of the range equation so that the same probabilities of detection and false alarm are main-

h_1 = height of radar above Earth surface

h_2 = height of target

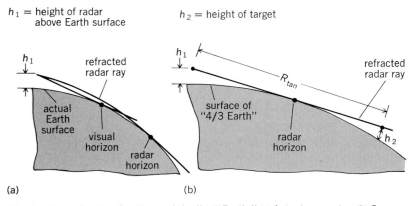

Fig. 4. Atmospheric refraction and the "4/3 Earth." (a) Actual geometry. (b) Geometry with 4/3 Earth.

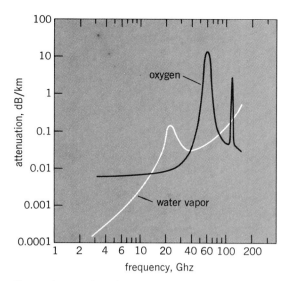

Fig. 5. Attenuation of electromagnetic energy by atmospheric gases at 1 atm (101.325 kPa) pressure. Absorption due to water vapor is for an atmosphere containing 1% water vapor molecules (7.5 g water/m³). (*From M. I. Skolnik, Introduction to Radar Systems, 2d ed., McGraw-Hill, 1980*)

tained regardless of the fluctuations. Figure 6 shows such adjustments for the single-pulse case. Similar curves for the multiple-pulse case should be used to complete the adjustment.

Noise. The detectability of a radar echo is determined by its strength relative to various competing signals also causing a detector output. Among the sources of such signals is the receiver itself. The receiver produces rf and i-f signals as a result of thermal conditions (molecular agitation in the components) and of the mixing process (unintended constituents of the local oscillator signal, for example).

A measure of the receiver's noise production is called its noise factor. In general terms, the noise factor is the ratio of noise power at the output of a network to the noise power that would have resulted from the network simply acting on the noise power at the input (that power being kT_0B_n, $T_0 = 290$ K as a standard ambient). The added noise of the network itself causes a degradation in the signal-to-noise ratio from input to output. It can be seen, then, that a sensitive radar receiver might to good advantage use a first stage of low noise factor but of considerable gain, so that subsequent stages are provided with a relatively strong signal (and noise) at their inputs, their own noise production further degrading the now-amplified signal-to-noise ratio very little.

Noise factor when expressed in decibels is more commonly called noise figure. Such conditions as network terminations and test-signal bandwidth must be better specified in actual measurement procedures than is possible here.

Other noise sources can be external to the radar. They include the Sun (in many radars, a sun strobe can be used as a bearing alignment check), other cosmic sources, electromagnetic interference as from urban areas, and deliberately energized sources intended to reduce the radar's sensitivity (jamming). *See* ELECTRONIC COUNTERMEASURES.

The minimum test-signal strength that can be fed into a receiver and reliably detected is called the radar's minimum discernible signal (MDS). Typical values are in the region of 90 to 120 dB below a milliwatt.

Clutter. While some radars are designed to examine land features and others weather phenomena, generally such radar returns are considered a nuisance, hindering the detection of targets of greater interest (such as aircraft) and producing false alarms. It is possible to state elementary formulas for the power of such returns and to use them in the radar range equation, as was done for receiver noise [Eq. (3)], but in general the simple expressions fail to account sufficiently for the statistical distribution of clutter backscatter encountered in practice.

In such approximations, backscatter statistics associated with different types of clutter are used. Surface clutter includes scattering from the sea surface (for various wave or sea-state conditions) and from land surface (for various terrain types). Volume clutter includes fog, clouds, rain, snow, and chaff (metallic threads deliberately sown in a region as a radar countermeasure). Birds and insects have been described in such computations, as have such phenomena as clear-air turbulence and thermal inversions, the returns from which are often called angels by radar users.

In the radar bands, clutter backscatter generally increases with increasing frequencies—in a given amount of rain, the clutter power will be worse for an X-band radar than for an L-band one. Radars using frequencies below L-band are called all-weather radars. Radars at S-band and higher may use high resolution (narrow beams, very short pulses) in part to reduce the amount of clutter competing with a target at any one time.

Not only the amount of backscatter but also the spread in frequency of the return, due to Doppler

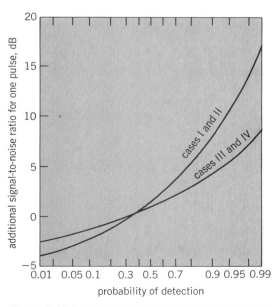

Fig. 6. Additional signal-to-noise ratio required for detecting fluctuating targets, for the single-pulse case ($n = 1$). (*From M. I. Skolnik, Introduction to Radar Systems, 2d ed., McGraw-Hill, 1980*)

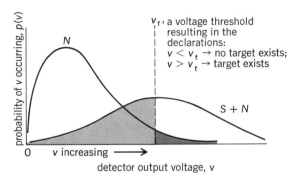

areas represent error probabilities:

▨ probability of missing a genuine target, $1 - P_D$

▨ probability of false alarm, P_{FA}

Fig. 7. Detection based on the output of a detector in the presence of noise.

shift resulting from turbulence among the scatterers, are of concern in radar design, since some signal-processing techniques intended to suppress returns from stationary clutter are made less effective by this spectral spread.

Detection process. Detecting a target in many radars means that an operator correctly judges that a bright spot on a display discloses the existence of a target and not just a spurious impulse of noise. To improve judgment, careful receiver and display adjustments are made so that the background of noise on the display is only faintly visible; this minimizes the probability that a target will be declared to exist when in fact none does (a noise-induced false alarm), and allows recognition of even slight intensifications of the display due to weaker echoes. Turning these adjustments down too far (reducing the false alarm probability to nearly zero) unfortunately increases the probability of missing a genuine target. Proper detection balances these two probabilities, the probability of detection, P_D, and the probability of false alarm, P_{FA}.

The graph of Fig. 7 represents the possible range of voltage values at the detector output for two cases: N, only noise exists; and $S + N$, a desired signal exists with the noise. The figure shows that a bright spot on a display fed properly by this voltage is possible but not likely when only noise is present. If a threshold in voltage, v_t, is imagined, the P_{FA} is given by the proportion of the area under the N curve to the right of v_t to the whole area under the N curve. The figure also shows that for such a threshold it is always possible that a true target will be missed because the output $S + N$ can indeed be less than threshold value. This probability of miss ($1 - P_D$) is similarly given by the area under the $S + N$ curve to the left of v_t. In Fig. 7 the threshold is shown for a P_{FA} of, say, 0.1 and a P_D of approximately 0.5. In practice, adjustments are made so that P_{FA} is in the 10^{-4} to 10^{-8} region.

In operator (or manual) detection, v_t depends on operator judgment of display brightness. In some modern radars, the detector output voltage is treated very precisely by computers and v_t is a specific part of the programming. Such operation is called automatic detection. Usually the design must be somewhat more conservative (less sensitive for a given false alarm rate than a good operator making manual detections), but the advantage of automatic detection is that very consistent performance and the handling of greater target densities are achieved.

Since receiver noise is not always the competing signal, an experienced operator learns to recognize clutter regions and tempers the declaring of targets; similarly an automatic detector must recognize abrupt changes of background statistics and adjust v_t to maintain a constant false alarm rate. Such techniques have been developed and are called CFAR (constant false alarm rate) techniques.

The W factor [the signal-to-noise ratio considered needed in Eq. (3) to assure reliable detection] can be seen in Fig. 7 to be related to the average values of the N and $S + N$ curves, the latter accruing from a given target size σ at a given range R. The error probabilities described above express the reliability of the detection.

Tracking. Many processes are used in radar systems to associate successive detections of a target, thereby providing an estimate of its motion or track. In a radar in which the antenna rotates regularly, a sequence of detections, in each of which the range and bearing of the target are estimated,

Fig. 8. AN/FPS-16 tracking radar, a very accurate monopulse radar, operating in the C-band, installed at more than 50 sites around the world. (*RCA Corp.*)

may be manually plotted with a grease pencil or other implement on either the display face or a plotting board. The individual estimates of location are called contacts or plots, the sequence a track. In modern systems a computer can be programmed to accept contacts from an operator-activated marking device built into the display, or from an automatic detector, and to create an estimate of the target's present position and velocity, using various track-filtering algorithms to diminish the impact of measurement errors in the individual contacts, that is, to smooth the track. Such a computer operation is called automatic tracking. Some computers are programmed to accept contacts from more than one search radar with common fields of view and to combine them into single track estimates; such a process is called integrated automatic tracking. Whenever the radar supplying the contacts is of rotating search type, the tracking is called a track-while-scan (TWS) process.

Radars have been built that are dedicated to tracking just one target but to do so very accurately (Fig. 8). In these radars a dish antenna is steered to the direction of a known target. Such a designation may have come from a search radar, and the tracking radar may have to perform a small local search in order to acquire the target. After acquisition, a continuous stream of contacts results and is used to keep the antenna steered to the target. The dish antenna produces a beam narrow in both horizontal and vertical angles (a pencil beam). If this beam is scanned in a very small cone around the target's presumed direction, the amplitude and phasing of the resulting modulations of the returns (Fig. 9) can be detected and used to drive the motors that control the position of the antenna.

In con-scan trackers, usually only the feedhorn in front of the dish is caused to nutate; the whole dish and feed assembly is repositioned by the director. The final estimate of the target's angle is made by sensing the director's position; the inertia due to its mass acts as a smoothing filter in the track estimate.

Another technique used in such tracking radars is monopulse angle measurement, in which squinted simultaneous beams (or an equivalent) are generated by a cluster of feedhorns (usually in horizontal and vertical pairs). Then, with a single radar transmission, the error signals due to the target's not being precisely on axis can be generated from the proportion of signals in each beam (separate receivers are required) and used to reposition the antenna. Such radars are used to control the pointing of guns or to provide accurate data needed in missile guidance; they are also used in flight-test facilities as instrumentation radars to provide a measurement basis for experiments.

Another kind of tracking is that performed by radars equipped with phased-array antennas in which beams may be positioned in angle under computer control and at pulse-to-pulse rates. Such electronic beam steering permits sampling rates adapted to the track dynamics being experienced on each of hundreds of tracks at a time. This process is called sampled-data tracking.

Composition of radar systems. It is convenient to discuss radar systems in terms of four principal subsystems, namely the transmitter, the

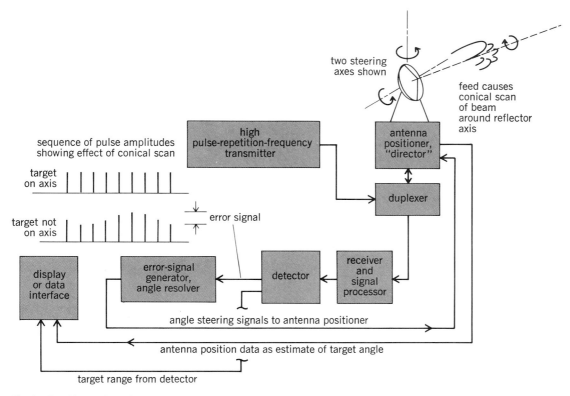

Fig. 9. Tracking radar using con-scan technique.

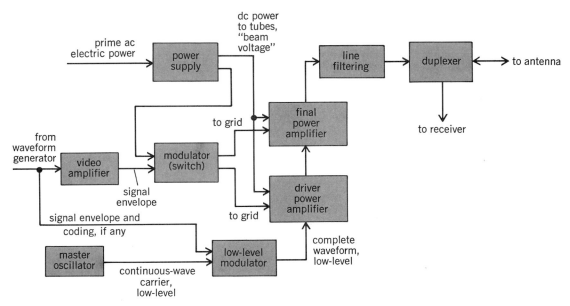

Fig. 10.   Transmitter of MOPA (master oscillator, power amplifier) type.

antenna, the receiver and signal processor, and the display and control apparatus.

Transmitter. The transmitter converts electrical power (as from a motor-generator set or a utility system) to electromagnetic power at the carrier frequency and in the waveform desired. Figure 10 shows a block diagram of a transmitter of the MOPA type. The waveform may actually be generated digitally in the receiver, then sent to the transmitter which in low-level equipment transfers the waveform to the output of an rf oscillator (a frequency synthesizer in frequency-agile radars). This signal is sent to the amplifier chain. Typical amplifiers are traveling-wave tubes (TWTs) in which the applied electrical power is converted to kinetic energy in an electron beam and the beam made to interact with the signal's electromagnetic fields, amplifying them. Figure 10 shows modulation by a grid to turn the beam on and off, not always the case. *See* TRAVELING-WAVE TUBE.

Other useful amplifiers are klystrons, in which the rf fields in resonant cavities interact similarly with the electron beam; still others are crossed-field amplifiers (CFAs), in which clouds of electrons, pulled from a cathode surface by the electric field applied by the power supply, are formed into periodically dense structures by the rf signal and similarly interact with that signal to amplify it. *See* KLYSTRON; MICROWAVE TUBE.

The power supply includes rectification and filtering to convert alternating-current prime power to very well regulated direct-current power with absolutely no ripple in the bias voltages supplied the amplifiers. The modulator must supply a very faithful replica of the pulse to the modulating grid of the amplifiers. Some amplifiers require that the bias voltage itself be modulated with the signal envelope, there being no lower-power-capable modulation grid; in this case the final stage of the modulator must handle the entire pulse energy from the power supply. Modulators generally include a switch tube of a vacuum or gas-filled sort (thyratrons, for example), but since the late 1970s more stages have been built with solid-state (semiconductor) devices. *See* ELECTRIC FILTER; GAS TUBE; MICROWAVE SOLID-STATE DEVICES; VACUUM TUBE.

Line filtering is included to eliminate undesired spectral components from the output signal. Included as well is the receiver-protecting duplexer, an assembly permitting the single antenna to be used for both transmitting, receiving, and routing the signals correctly. New designs of power supplies and modulators use modular solid-state circuits far more than the vacuum-tube circuitry of older radars. Designers stress fail-soft characteristics by these modular and fault-tolerant approaches. Amplifiers are sometimes designed with integral power supplies to achieve greater total reliability as well.

Much simpler radar transmitters are also in wide use, those in which an oscillator (such as a magnetron) is used rather than an amplifier chain. In such a case (Fig. 11), the modulator supplies a cathode-to-anode voltage pulse which causes an rf

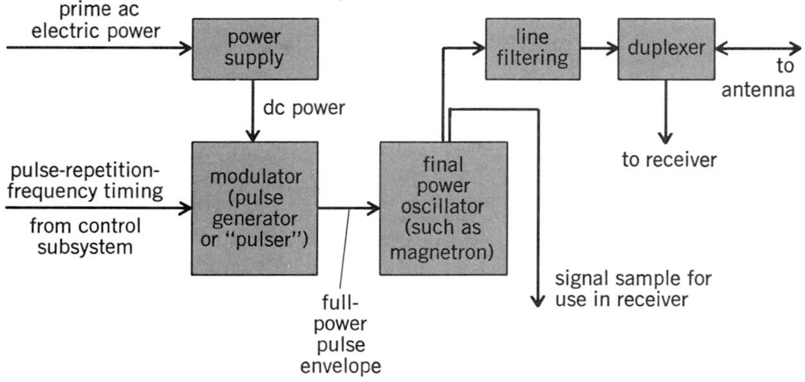

Fig. 11.   Radar transmitter of the final-power-oscillator type.

signal to be generated. *See* MAGNETRON.

The transmitter subsystem must also include many safety features because of the lethal level of voltages and dangerous radiation present. Transmitters also require very involved cooling systems, since the efficiencies in the energy conversion processes are usually in the 15 to 40% region and much damaging heat is generated.

Antenna. In a typical antenna the signal from the transmitter is carried by waveguide (usually rectangular metal piping) through a rotary joint and to a feedhorn at the focus of a parabolic reflector. Such an antenna is pictured in Fig. 12; the operation is shown in Fig. 13. *See* WAVEGUIDE.

The optical properties of antennas are such that the beam formed has a width inversely proportional to the dimension of the illuminated aperture—the wider (in units of wavelengths) the antenna, the narrower the beam in that direction. The beam shown in Fig. 13 is called a fan beam and is suitable for a two-dimensional (2D) search radar making target position estimates in just range and azimuth angle. Another parabolic antenna, the height finder (Fig. 14), forms a horizontal fan beam which can be nodded in the vertical to ascertain the elevation angle of targets to which the antenna has been directed in azimuth. Newer search radars are three-dimensional (3D), measuring azimuth, elevation, and range as they rotate in azimuth. Figure 15 illustrates two such antennas, one in which a stack of beams (requiring separate receivers) is used to measure elevation, and one in which a single pencil beam is caused to scan rapidly in elevation while the antenna rotates in azimuth only one beam width.

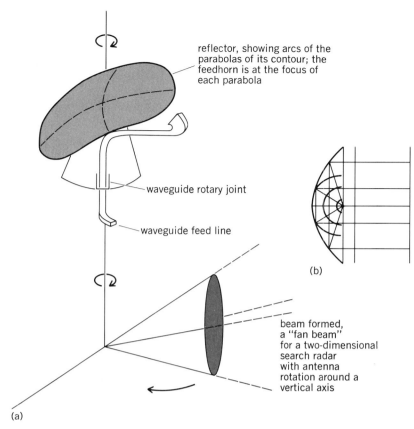

reflector, showing arcs of the parabolas of its contour; the feedhorn is at the focus of each parabola

waveguide rotary joint

waveguide feed line

(b)

beam formed, a "fan beam" for a two-dimensional search radar with antenna rotation around a vertical axis

(a)

Fig. 13. Antenna forming a vertical fan beam. (*a*) Antenna configuration. (*b*) Principle of parabolic reflection, showing spherical wavefronts converted to planar fronts.

Fig. 12. AN/SPS-49 search radar antenna, which forms a vertical fan beam for the two-dimensional L-band radar widely used in the U.S. Navy. (*Raytheon Co.*)

Many tracking radars use dish antennas with some sort of multiple-beam feed complex to permit very accurate angle measurement. In Fig. 16 are diagrammed two squinted-beam monopulse feed approaches (illustrated in one plane of measurement only), one with the feedhorns at the focus and the other of a Cassegrain type, wherein a hyperbolic subreflector is used and the feedhorn assembly can be located in the area of the parabolic reflector. When this optical scheme is combined with certain polarization rotation techniques in the two reflecting surfaces, the subreflector appears reflective to the signals from the feed but transparent after reflection from the parabola. Consequently the amount of aperture blockage (with detrimental effects in accuracy) is greatly reduced.

Several radars have been developed that use phased-array antennas, in which the transmitter power is divided among many radiating elements (possibly thousands) and in which the phase of each element is controlled by a computer. In such a fixed-aperture antenna, the beam can be steered in only microseconds from one position to another many beam widths away. Consequently a wide variety of dwell routines can be implemented that are not at all constrained by mechanical rotation. Phased-array antennas are quite complex and expensive; they are used, therefore, only where great demand for very versatile operation exists.

In all antennas, the microwave optics involved results in formation of side lobes in the radiation

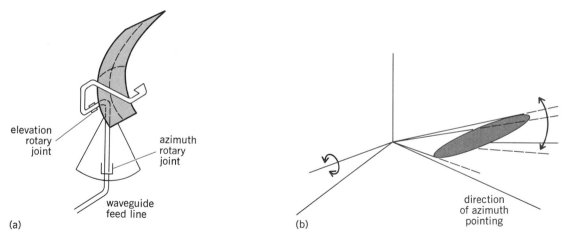

Fig. 14.   Antenna of a height-finder radar. (a) Antenna configuration. (b) Horizontal fan beam formed.

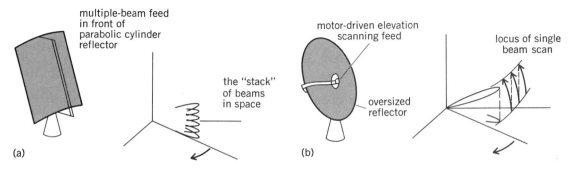

Fig. 15.   Two three-dimensional (3D) radar antennas. (a) Multiple simultaneous beams from a stacked-beam antenna. (b) Single-beam scanner.

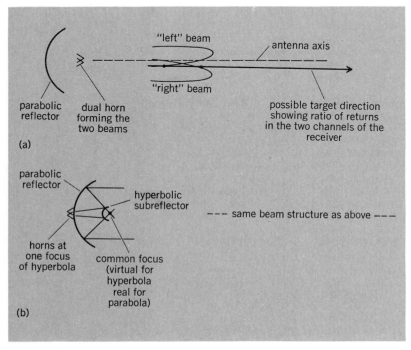

Fig. 16.   Monopulse antennas for tracking. (a) Single reflector. (b) Cassegrain feed.

pattern or gain function. The shape of the pattern is determined by the excitation function at the antenna. In Fig. 17 are sketched some common excitation functions and the resulting patterns. High side lobes are very undesirable in radar because they permit unwanted signals to enter the receiver. Simple antennas usually have modest side-lobe levels. More severely weighted excitation functions can be generated for more desirable (low side-lobe level) patterns, but only at the expense of complicating the feed. *See* MICROWAVE OPTICS.

Since radar angular resolution is related to antenna beam width which is inversely related to the size of the antenna, it has occurred to radar designers to build very large antennas where possible. One way to "build" a large antenna is to move (as in an aircraft) a smaller antenna over a great distance and, in the signal processing, to "remember" all the characteristics of the signals received. With modern digital processing, it is then possible to reconstruct the radar scene as though the antenna had been as large as the path traversed. A radar using such a technique is called a synthetic-aperture radar (SAR). SARs are being used increasingly in the Earth-monitoring fields.

Receiver and signal processor. The signals collected by the antenna are amplified in some radars (those with rf preamplifiers), then converted to the i-f for further amplification and processing. There may be several parallel channels of receiver necessary, depending on the type of radar (for the several beams in a stacked-beam radar or for the angle error channels in a monopulse tracker, for example), each requiring all the processing described here.

Many radars require a long pulse for the requisite energy on each transmission (product of pulse width and peak power, which can be limited in many transmitters), yet require the range resolution (the ability to discern two closely spaced targets) of a short pulse. The effect in range resolution of a very short pulse can be achieved by modulating the long pulse with either frequency or phase modulation so that it has the same bandwidth as the very short pulse desired. Then, on receive, the waveform generator supplies a replica of the modulation on the transmitted pulse to the receiver so that a matched-filter action takes place in circuitry called the pulse compressor. The result is that the high energy for detection sensitivity and the resolution effects of a short pulse have both been obtained (Fig. 18).

In many radar applications it is desired to determine if returns are Doppler-shifted, that is, if certain returns are coming from moving targets rather than from stationary clutter. Even the simplest radars can be equipped with moving target indication (MTI), wherein successive echoes are examined for a progressive phase change (indicative of slight round-trip path change between pulses), the Doppler shift in frequency being too slight to filter out on a single pulse. All MTIs require, then, a phase reference in the receiver. This reference can range from reflections from clutter (assumed fixed) in the vicinity of targets and simply a part of the total return, to a local oscillator signal in the receiver, carefully phase-locked to the transmitted signal with each transmission. In low–pulse repetition frequency radars, the returns may be compared on an each-succeeding-pair basis in an appropriate canceller. In high–pulse repetition frequency pulse-Doppler radars, the process results in actually measuring Doppler shift (hence the radial component of target velocity) and not just in estimating that such a shift exists, as in the simpler cancellation schemes. *See* DOPPLER RADAR; MOVING-TARGET INDICATION.

The pulse compression and Doppler processing schemes are regarded as coherent signal processing insofar as both the phase and amplitude of the signals are involved. Further stages of the whole signal processor may be noncoherent; that is, a signal detector may now act on the amplitude of the signal resulting from the coherent processing, producing video signals with no carrier to convey phase relationships. These video signals (detector output voltages described above) may be displayed for operator observation and detection judgment; in more sophisticated radars they may be processed further in an automatic detector with constant-false-alarm-rate features in the noncoherent units of the signal processor. In such units, various comparisons among the sum and difference (angle error) channels (carried in parallel to this point) may be made to refine the position estimate, or other comparisons may be made to reduce the

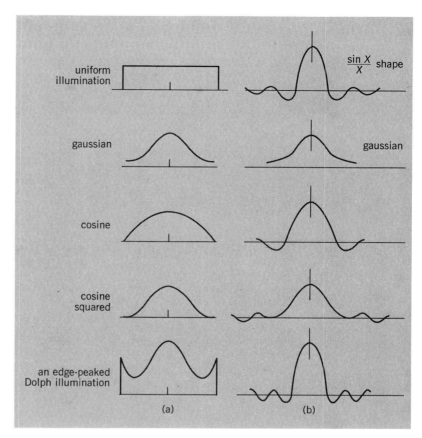

Fig. 17. Antenna excitation functions and patterns. (a) Excitation functions. (b) Resulting patterns.

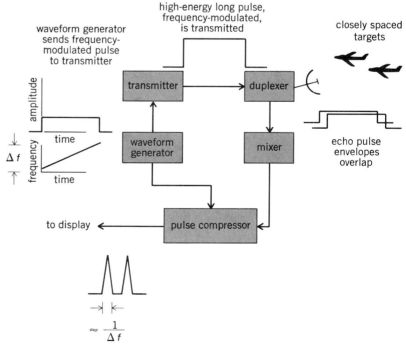

Fig. 18. Pulse compression. The pulse compressor, using knowledge of the modulation in the transmitted pulse, acts as a matched filter producing maximum amplitude narrow responses, thus resolving the closely spaced targets.

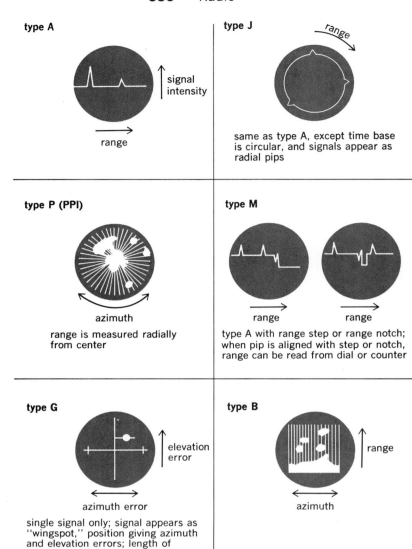

type A

signal
intensity

range

type J

range

same as type A, except time base
is circular, and signals appear as
radial pips

type P (PPI)

azimuth

range is measured radially
from center

type M

range   range

type A with range step or range notch;
when pip is aligned with step or notch,
range can be read from dial or counter

type G

elevation
error

azimuth error

single signal only; signal appears as
"wingspot," position giving azimuth
and elevation errors; length of
wings inversely proportional to range

type B

range

azimuth

Fig. 19. Cathode-ray-tube display formats.

false alarm effects of antenna side lobes or interference signals.

The object of all coherent and noncoherent signal processing is to increase the probability that correct estimates and accurate measurements are made at the user's end of the radar and that effects of undesired signals seen by the antenna are greatly reduced.

In newer radars, signal-processing equipment consists of a cabinet (or several) of printed circuit boards with tiny integrated circuit units on them, and extensive back-plane wiring to interconnect properly the many circuit boards. Both coherent and noncoherent processing can be done with digital computerlike equipment to a large degree.

Display and control. In simple radars, the operator may observe the output of an elementary video-producing detector on one of several formats on a cathode-ray-tube display. The most common format is the PPI (plan position indication), in which a radial trace (the path of the cathode-ray-tube electron beam exciting the phosphor) is synchronized to the pulse repetition frequency, resulting in radar range being displayed radially from the center, and in which this trace is moved around the display, linked electronically to the antenna rotation. The detector output is applied to the intensity grid of the cathode-ray tube so that a bright spot results when a strong video voltage is produced by the detector. The display circuitry may also superimpose upon the radar video certain calibration pulses (range marks and bearing lines, for example), maps of the local terrain, or other symbols of importance in the entire system of which the radar is a part.

The PPI format and several other useful formats for video display are shown in Fig. 19.

In a radar equipped with more automatic treatment of the video from the detector, the operator may play a more managerial role, and a suitable display would show numerous symbols and alphanumeric notation about the targets and traffic activities rather than merely displaying the radar video itself. Such information would permit the operator either to instruct the radar (as in establishing boundaries for special modes of operation) or to alert the rest of the overall system of which the radar is a part to any unusual situation observed. In such a radar, the specific data on individual targets may well be passed from the report-formatting computers of the radar to other user computers or displays without any particular judgment or action by the operator. *See* CONTINUOUS-WAVE RADAR; MONOPULSE RADAR.

[ROBERT T. HILL]

Bibliography: D. K. Barton, *Radar System Analysis*, 1964; F. E. Nathanson, *Radar Design Principles*, 1969; M. I. Skolnik, *Introduction to Radar Systems*, 2d ed., 1980; M. I. Skolnik (ed.), *Radar Handbook*, 1970.

Radio

Communication between two or more points, employing electromagnetic waves as the transmission medium.

Technical history. In 1840 two separate events began the long history of radio. Joseph Henry, an American, first produced high-frequency oscillations, at roughly the same time, Samuel F. B. Morse was demonstrating the telegraph in Washington, D.C. Needless to say, the public was more impressed with the telegraph than with Henry's experiments, but both steps were of equal importance to the development of radio.

In 1873 a British scientist, James Clerk Maxwell, first tied together the theories of Henry and Morse to explain the theory of propagation of energy from wires. He also showed, at least mathematically, that these generated waves would travel at the speed of light.

Heinrich Hertz was finally able to demonstrate practically Maxwell's theory. He did so in 1888, by passing a rapidly alternating current through a wire, creating what were to become known as Hertzian waves.

It was left to Guglielmo Marconi to make practical application of Hertz's work. He experimented with an antenna and ground setup from 1895 on, and when in 1898 he broadcast the first paid radiogram from the Isle of Wight, radio was born.

Hundreds and thousands of other inventions and discoveries were to follow, but these men had

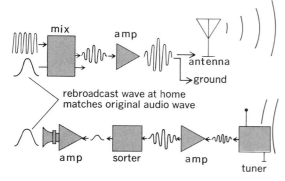

Fig. 1. Transmission of audio information by radio.

formed the basis of radio. The same formula is in use today. Figure 1 shows the audio pulse and the carrier pulse being blended into a modulated carrier wave, which is then amplified and fed into the antenna. The receiving antenna and tuner catch the weak signal, amplify it, sort the audio pulse from the carrier, and play a now reamplified audio pulse through the speaker at home.

The first significant radio applications employed frequencies of about 500 kilohertz for ship communications. Their value was demonstrated dramatically in 1909 and 1912, when assistance was obtained by the sinking passenger ships *Republic* and *Titanic*, and hundreds of lives were saved.

Roughly 10 years later, the first real overseas radio complex was established, upon discovery that frequencies of 3000–30,000 kHz traveled over great distances by reflection from the ionosphere. Today, although cable and ionosphere reflection signals are still used to a great extent, satellite relay has almost completely taken over transmission of broadcast-station quality. In later years, many other services, such as television, FM broadcasting, radar, and microwave relaying were developed as the unique properties of much higher frequencies were discovered and exploited.

Methods of information transmission. Radio waves transmitted continuously, with each cycle an exact duplicate of all others, indicate only that a carrier is present. The message must cause changes in the carrier which can be detected at a distant receiver. The method used for the transmission of the information is determined by the nature of the information which is to be transmitted as well as by the purpose of the communication system (Fig. 2).

Code telegraphy. The carrier is keyed on and off to form dots and dashes. The technique, often used in ship-to-shore and amateur communications, has been largely superseded in many other point-to-point services by more efficient methods.

Frequency-shift transmission. The carrier frequency is shifted a fixed amount to correspond with telegraphic dots and dashes or with combinations of pulse signals identified with the characters on a typewriter. This technique is widely used in handling the large volume of public message traffic on long circuits, principally by the use of teletypewriters. *See* TELETYPEWRITER.

Amplitude modulation. The amplitude of the carrier is made to fluctuate, to conform to the

fluctuations of a sound wave. This technique is used in AM broadcasting, television picture transmission, and many other services. *See* AMPLITUDE MODULATION; AMPLITUDE-MODULATION DETECTOR; AMPLITUDE-MODULATION RADIO; AMPLITUDE MODULATOR.

Frequency modulation. The frequency of the carrier is made to fluctuate around an average axis, to correspond to the fluctuations of the modulating wave. This technique is used in FM broadcasting, television sound transmission, and microwave relaying. *See* FREQUENCY MODULATION; FREQUENCY-MODULATION DETECTOR; FREQUENCY-MODULATION RADIO; FREQUENCY MODULATOR.

Pulse transmission. The carrier is transmitted in short pulses, which change in repetition rate, width, or amplitude, or in complex groups of pulses which vary from group to succeeding group in accordance with the message information. These forms of pulse transmission are identified as pulse-code, pulse-time, pulse-position, pulse-amplitude, pulse-width, or pulse-frequency modulation. Such techniques are complex and are employed principally in microwave relay systems. *See* PULSE MODULATION; PULSE MODULATOR.

Radar. The carrier is normally transmitted as short pulses in a narrow beam, similar to that of a searchlight. When a wave pulse strikes an object, such as an aircraft, energy is reflected back to the station, which measures the round-trip time and converts it to distance. A radar can display varying reflections in a maplike presentation on a cathode-ray tube. *See* RADAR.

Uses of radio. The first practical application of radio, in the 1900s between ships and shore sta-

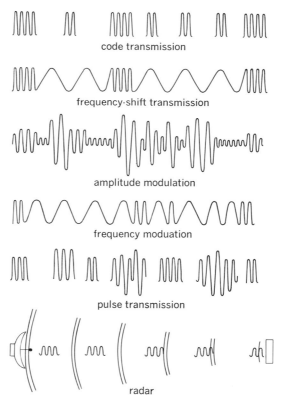

Fig. 2. Methods for transmission of information.

tions, was followed quickly by overseas communication and communication between other widely separated fixed points. Subsequently, the applications have become widely diversified. Some of the important uses are listed below.

1. Public safety: marine and aviation communications, police and fire protection, forestry conservation, and highway traffic control.

2. Industrial: power utilities, pipelines, relay services, news systems, agriculture, and petroleum processing.

3. Land transportation: railroads, motor carriers, taxicabs, and automobile emergency needs.

4. Broadcasting: television, FM broadcasting, AM broadcasting, and shortwave broadcasting.

5. Military and space: radar, communications, navigation, telemetering, missile tracking, satellite photographic surveys, and missile guidance.

6. Fixed point-to-point: long-distance message and picture transmission.

7. Relaying: television, sound, picture, and public message relaying over long distances. Most recently, and soon to be a major part of the relaying function of radio, communications satellites have handled a great deal of overseas communications. These satellites supply not one but many services, including several thousand narrow channels for telephone, telegraph, and teletype, with wide channels for television and other video transmission. With the exception of military satellites, all United States communications satellites are privately owned, by Federal decree, by the Communications Satellite Corp. Partial holders in this corporation are AT&T, IT&T, RCA., GE, GT&E, Lockheed, and Hughes Aircraft, among other large industries. *See* TELEVISION.

8. Telemetering: remote indication of water levels in reservoirs and rivers, performance of experimental aircraft, missiles, and satellites.

9. Weather reporting: early warning and location of hurricanes and other storms, trends of weather for industrial and public information.

Frequency separation. Hundreds of thousands of radio transmitters exist, each requiring a carrier at some radio frequency. To prevent interference, different carrier frequencies are used for stations whose service areas overlap and receivers are built to select only the carrier signal of the desired station. Resonant electric circuits in the receiver are adjusted, or tuned, to accept one frequency and reject others.

Each station operates within a specific radio-frequency (rf) channel. All other stations within a geographical area are excluded from using this channel. Each channel must be wide enough to accommodate the message information, provide tolerance for small carrier frequency drift, perhaps provide a guard band, and allow for imperfect receiver selectivity capabilities. The minimum usable channel widths (or bandwidths) vary from service to service, depending upon the amount of information a channel must accommodate. In television it is 6000 kHz because of the large amount of essential picture information. In FM broadcasting it is 200 kHz and in AM broadcasting 10 kHz. The great demand for authorizations requires efficient channel utilization. In the mobile transportation service in the United States, the Federal Communications Commission was compelled to reduce

channel widths 50% when technical developments made it feasible.

Federal Communications Commission. All nations have a sovereign right to use freely any or all parts of the radio spectrum. But a growing list of international agreements and treaties divides the spectrum and specifies sharing among nations for their mutual benefit and protection.

Each nation designates its own regulatory agency. Functioning within the international agreements, it issues authorizations; assigns frequencies; polices operations; creates technical standards, rules, and practices; and safeguards and protects the public interest.

In the United States all nongovernmental radio communications are regulated by the Federal Communications Commission (FCC), according to the provision of the Communications Act of 1934, as amended. Creation of a radio station or service requires authorization by the FCC. Upon completion of an authorized facility, a license to operate is issued. Radio stations are inspected regularly by engineers attached to FCC field offices. Stations must comply with the terms of their authorization regarding carrier-frequency tolerance, power limitations, permissible communications, calls signals, and control by properly licensed personnel. *See* AMATEUR RADIO.

[JOHN D. SINGLETON]

Bibliography: American Radio Relay League, *Radio Amateur's Handbook*, 1979; G. J. Augebauer, *Electronics for Modern Communication*, 1974; E. Barnouw, *A History of Broadcasting in the United States*, 3 vols., 1966, 1968, 1970; Howard W. Sams Engineering Staff, *Reference Data for Radio Engineers*, 6th ed., 1975; R. L. Shrader, *Electronic Communication*, 4th ed., 1980.

Radio-frequency amplifier

A tuned amplifier that amplifies the high-frequency signals commonly used in radio communications. The frequency at which maximum gain occurs in a radio-frequency (rf) amplifier is made variable by changing either the capacitance or the inductance of the tuned circuit. A typical application is the amplification of the signal received from an antenna before it is mixed with a local oscillator signal in the first detector of a radio receiver. The amplifier that follows the first detector is a special type of rf amplifier known as an intermediate-frequency (i-f) amplifier. *See* AMPLIFIER; INTERMEDIATE-FREQUENCY AMPLIFIER.

An rf amplifier is distinguished by its ability to tune over the desired range of input frequencies. The shunt capacitance, which adversely affects the gain of a resistance-capacitance coupled amplifier, becomes a part of the tuning capacitance in the rf amplifier, thus permitting high gain at radio frequencies. The power gain of an rf amplifier is always limited at high radio frequencies, however, for reasons which will be discussed.

Basic circuit. Two typical rf amplifier circuits are shown in Fig. 1. The conventional bipolar transistor amplifier of Fig. 1a uses tapped coils in the tuned circuits to provide optimum gain-bandwidth characteristics consistent with the desirable value of tuning capacitance. Inductive coupling provides the desired impedance transformation in the input and output circuits. The tuning capacitors are

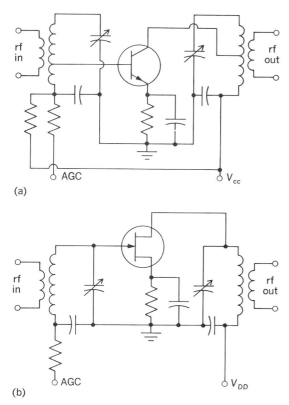

(a)

(b)

Fig. 1. Typical rf amplifiers. Circuits with (a) bipolar transistor and (b) field-effect transistor.

usually ganged so as to rotate on a single shaft, providing tuning by a single control knob. Sometimes varactor diodes are used to tune the circuits, in which case the tuning control is a potentiometer that controls the diode voltage. Automatic gain control (AGC) is frequently used on the rf amplifier, as shown. AGC voltage controls the bias and hence the transconductance of the amplifier. In the field-effect transistor (FET), circuit (Fig. 1b) tapped coils are not required because of the very high input and output resistances of the FET. Characteristics of the FET are similar to those of a pentode tube except for the lower drain voltage and higher capcitance between the input and the output. Thus the FET circuit is similar to a pentode circuit except that it is much simpler because it has no heater, screen grid, or suppressor grid equivalents. *See* AUTOMATIC GAIN CONTROL (AGC); SEMICONDUCTOR; TRANSISTOR.

High-frequency limit. The power gain of a bipolar transistor is strongly dependent upon the forward current transfer ratio β, or h_{fe}, as well as on the transconductance g_m. However, the current transfer ratio (or current amplification factor) β decreases as the frequency increases because of the base region diffusion capacitance resulting from the finite time required for the charge carriers to traverse the base region. A graph of β as a function of frequency for a typical transistor is given in Fig. 2. The current transfer ratio β remains essentially constant at the low-frequency value β_0 (200 for the transistor of Fig. 2) below the frequency known as the beta cutoff frequency f_β, at which $\beta = 0.707\,\beta_0$. As the frequency increases above f_β,

β decreases and is essentially inversely proportional to frequency. The frequency at which β decreases to 1 is known as the transitional frequency f_T, which is 300 MHz for the transistor of Fig. 2.

Because of the linear relationship between β and the frequency, $\beta_0 f_\beta = f_T$ and $\beta f = f_T$ for frequencies above f_β. Therefore, β can be determined for any desired frequency, provided β_0, or h_{fe}, and f_T are given by the manufacturer. For example, if $f_T = 300$ MHz and $\beta_0 = 200$ for the transistor of Fig. 2, then $f_\beta = 300$ MHz/200 = 1.5 MHz. Also, at 30 MHz the value of $\beta = f_T/f = 300$ MHz/30 MHz = 10. Therefore, the available current gain of a common emitter transistor is approximately equal to the ratio f_T/f, where f is the desired operating frequency, provided f is above f_β. Transistors are available with very high values of f_T, in the neighborhood of 2.5×10^9 Hz. The transistor type must be carefully chosen for high-frequency rf amplifiers. The parameter f_{fb} or f_α which is approximately equal to f_T, is sometimes given for transistors. Field-effect transistors and vacuum tubes have similar frequency limitations imposed by the carrier transit time between source and drain or cathode and plate. Special amplifying devices such as traveling-wave tubes or parametric amplifiers are used as rf amplifiers in the microwave region, or for frequencies above 10^9 Hz. *See* MICROWAVE TUBE.

Capacitive feedback. Careful design is required in rf amplifiers to avoid oscillation or instability. The inherent instability results from the capacitive coupling between output and input circuits; this coupling permits feedback in the amplifier, which can be regenerative when inductive loads or tuned circuits are used in the amplifier output. The coupling capacitance is primarily the collector junction capacitance in a bipolar transistor, the gate-drain junction capacitance in an FET, or the grid-plate capacitance in a vacuum tube. The tetrode and pentode tubes were developed primarily to reduce this capacitance and were popular as rf amplifiers until transistors with good high-frequency characteristics were developed. However, pentode tubes generate more noise than triodes or transistors and therefore are not well suited for amplifying very weak signals.

Transistor or triode amplifiers can be stabilized by either of two methods. The first method, known as neutralization, cancels the effect of the capacitive feedback in the device by providing equal feedback current of opposite polarity from the amplifier output to the input. This feedback current is usually obtained by connecting a capacitor,

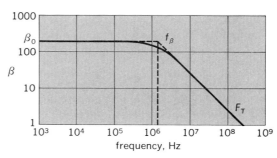

Fig. 2. Variation of current transfer ratio as a function of frequency for typical transistor.

known as a neutralizing capacitor, between the input and an output-circuit point that has opposite polarity to the collector (drain or plate). This opposite polarity voltage is available either at the secondary of a coupling transformer or opposite the collector end of a tapped coil in the output circuit. The neutralizing capacitance is adjusted until the neutralizing current is equal to the feedback current through the amplifier. Some transistors require a resistor in series with the neutralizing capacitor to provide the 180° phase difference.

The second method is to limit the voltage gain of the amplifier stage to a value at which the feedback current will be insufficient to produce oscillations or other undesirable characteristics. This method requires more stages to produce a total specified amplifier gain but eliminates the need to neutralize each stage. The criterion for stability is given by notation (1), where R'_i and R'_0 are the ef-

$$R'_i R'_0 > 2/\omega_0 C_c g_m \qquad (1)$$

fective shunt resistances in the input and output circuits respectively, ω_0 is the resonant frequency, C_c is the coupling capacitance, and g_m the transconductance. See FEEDBACK CIRCUIT.

Cascode circuit. A technique commonly used to provide stable low-noise amplification at radio frequencies is illustrated in Fig. 3. This amplifier is known as a cascode amplifier. Two transistors are

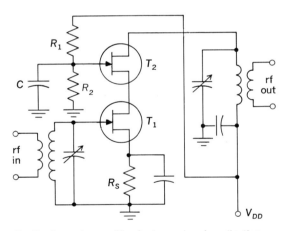

Fig. 3. Cascode amplifier for low-noise rf application.

connected in series so that the source input resistance of the second transistor T_2 is the load for input transistor T_1. Because the resistance looking into the source terminal is approximately $1/g_m$ and the voltage gain of the FET is approximately $g_m R_L$, the voltage gain across transistor T_1 is only unity. Since the gate of transistor T_2 is maintained at signal ground potential by capacitor C and the drain signal voltage of transistor T_1 is applied between the source and gate of transistor T_2, the tuned load in the drain circuit of transistor T_2 therefore provides the normally large voltage gain for this transistor.

The primary advantage of the cascode connection results from the fact that the gate of transistor T_2 is maintained at signal ground potential. Therefore, the current which flows through the gate-drain capacitance of T_2 flows through capacitor C

to ground and not into the signal input of the amplifier. Thus the effectively grounded gate of transistor T_2 essentially eliminates the capacitance between the drain of T_2 and the gate of T_1, thereby avoiding feedback. Capacitance still exists between the drain of transistor T_1 and its input gate, but because the voltage gain across this transistor is only unity, the amplifier is unconditionally stable. The proper drain-source voltage V_{DS} for transistor T_1 is obtained from Eq. (2). Resistor R_S in the

$$V_{DS} = V_{DD} R_2 / R_1 + R_2 \qquad (2)$$

source of transistor T_1 provides the desired gate-source bias voltage V_{GS} for both transistors.

Bipolar transistors or triode tubes can be used in the cascode connection with slight bias circuit modifications as compared with Fig. 3. The cascode connection eliminates another problem, which results from capacitance between the input and output of an amplifier. This capacitance has an effective value, when viewed from the input of the amplifier, which is proportional to the voltage gain of the amplifier. This effect is known as the Miller effect and causes detuning of the resonant input circuit whenever the gain of the amplifier is changed, as with AGC.

[CHARLES L. ALLEY]

Bibliography: C. L. Alley and K. W. Atwood, *Electronic Engineering*, 3d ed., 1973; D. J. Cramer, *Modern Electronic Circuits*, 1976; C. A. Holt, *Electronic Circuits: Digital and Analog*, 1978.

Radio receiver

That part of a radio communication system which abstracts the desired information from the radio-frequency (rf) energy collected by the antenna. All radio receivers must perform three basic functions: selectivity, amplification, and detection. For basic discussion of radio principles see RADIO.

Selectivity. Many radio signals transmitted at the same time are available at the antenna. From these many signals the receiver must select the single one desired. This is done by tuning the receiver to the frequency of the desired carrier. The tuning circuit contains a combination of inductances L and capacitances C, one or more of which are variable. The frequency f selected is determined by the relation below, where f is in hertz (Hz), L is in henries, and C is in farads.

$$f = 1/2\pi\sqrt{LC}$$

Tuning the receiver, therefore, consists in changing the inductance or capacitance, usually the latter. When so tuned the inductance-capacitance circuit accepts the desired frequency and rejects other frequencies. By using several such circuits in series, a high degree of selectivity may be obtained.

Amplification. Because the incoming signal may be weak and because a certain minimum energy is required to operate the loudspeaker, the headphones, or the television picture tube, considerable amplification must take place between the input of the receiver and its output. This is usually called the gain of the receiver. It may amount to 10,000,000 times in voltage or 140 decibels (dB). See AMPLIFIER.

If the detector, which abstracts the desired communication from the high-frequency amplified signals, requires 1 volt to perform its function properly and if the input to the receiver is 1 microvolt, a total amplification of 1,000,000 times is required prior to detection. If the loudspeaker requires 10 volts, another voltage amplification of 10 is necessary between the detector and the loudspeaker.

Figure 1 shows the gain between stages of a superheterodyne receiver. The voltage gain of any stage or group of stages can be obtained by taking the ratio of the ordinates of the proper curves. The gain in decibels is found by subtracting the values of the proper curves. For example, the voltage gain at 1000 kHz of the second intermediate-frequency (i-f) amplifier is the total gain to the second detector divided by the total gain to the second i-f grid, or 940,000/17,000, or 55. The dB gain is 119 − 84 = 35.

Detection. The energy collected by the antenna and presented to the input of the receiver is in the form of rf waves which act as a carrier for the information to be transmitted. The purpose of the detector in a receiver is to remove the desired communication from this carrier and to convert it into a form that will actuate the output device, such as a loudspeaker. *See* AMPLITUDE-MODULATION DETECTOR; DETECTOR; FREQUENCY-MODULATION DETECTOR.

Types of receivers. Two general types of receiver are in use today, the tuned-radio-frequency (TRF) and the superheterodyne. Both of these can be used for amplitude-modulated (AM) signals. Frequency-modulation (FM) receivers are almost always superheterodyne. *See* AMPLITUDE MODULATION; AMPLITUDE-MODULATION RADIO; FRE-QUENCY MODULATION; FREQUENCY-MODULATION RADIO.

TRF receiver. In a TRF receiver all amplification up to the detector takes place at the frequency of the incoming signal. This usually requires several stages of tuned amplification. Each stage is tuned to the same frequency and the tuning elements are ganged together for convenience when tuning. For a discussion of tuned amplifiers *see* AMPLIFIER.

TRF receivers are especially applicable to the low-frequency and very-low-frequency bands from about 10,000 to 300,000 Hz. Figure 2 shows a block diagram of a TRF receiver compared to a superheterodyne receiver.

Superheterodyne receiver. With the increased use of higher frequencies for broadcasting and communications, a high degree of selectivity became necessary. The TRF receiver was inadequate in this respect. This selectivity was obtained by circuit techniques involving frequency changing. The superheterodyne circuit changes the frequency by heterodyning, or beating, two frequencies together to get a third. When two signals of small difference in frequency are superimposed on a nonlinear device, the output consists of energy at two major combination frequencies which are the sum and difference of the original frequencies. The sum of the two beating frequencies is usually eliminated by tuned circuits following the heterodyne process. The difference frequency, referred to as the intermediate frequency, is passed on for further amplification and, what is more important, through stages of high selectivity because selectivity at lower frequencies is more easily obtained than at high frequencies.

The new i-f signal, derived from the modulated carrier frequency and an unmodulated local oscillator frequency (Fig. 2), is modulated to the same degree as the original carrier. There is very little distortion in this process, which takes place in the mixer.

The difference frequency (i-f) must be high enough so that little response is obtained to the so-called image frequency, which is an incoming (but undesired) rf signal whose difference from the

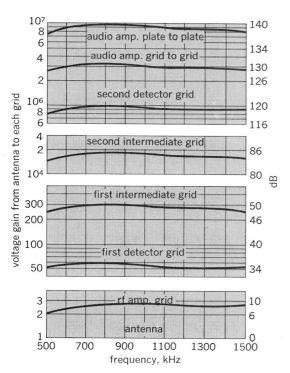

Fig. 1. Voltage gain in successive stages of superheterodyne receiver. (*From K. Henney, ed., Radio Engineering Handbook, 5th ed., McGraw-Hill, 1959*)

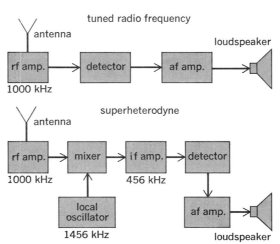

Fig. 2. Block diagrams showing comparison of TRF and superheterodyne receivers.

local beating oscillator is the same as the desired signal. The image signal differs from the desired signal by twice the i-f frequency. An example of this is a desired signal of 1000 kHz and a local oscillator of 1500 kHz beating to obtain an intermediate frequency of 500 kHz. The image frequency of 2000 kHz, if allowed by the selectivity of the TRF amplifiers ahead of the heterodyne, will cause interference by developing the same difference of 500 kHz when beating with the 1500-kHz local oscillator.

The tuning of the local oscillator in home-entertainment receivers is ganged to the tuning of the TRF amplifiers and thus does not present any more difficulty in manipulation than does a TRF receiver.

Where the ultimate in operational characteristics is desirable, it may be necessary to heterodyne the incoming signal more than once. Commercial, radiotelegraph, traffic-handling systems employ at least dual and even triple superheterodyne receivers. The first intermediate frequency is made rather high for good image suppression. The second intermediate frequency is lowered to a value where other, spurious signals are not too obtrusive. A third very low intermediate frequency, for very high selectivity, is then utilized for final detection or demodulation to the original intelligence.

The actual change of frequency in a superheterodyne receiver is performed by a frequency converter, often called a mixer, heterodyne modulator, or first detector. This device may be a tube, transistor, or other nonlinear device. Two inputs are applied to the mixer: the incoming signal and the output of the local oscillator. The frequency of this oscillator differs from that of the incoming signal by the intermediate frequency.

The mixer may be very simple, with both signals applied to the same grid of a tube, or it may be more complex. It may contain means for generating the local frequency, or it may have additional internal elements to which the local oscillator output is applied. In any case, the mixing process takes place because of the ability which each tube element has to modulate the electron stream from cathode to anode.

The greater the number of electrodes in a common electron stream, the greater will be the noise developed in the output. Therefore multielement mixers are used only in those circuits which have considerable signal amplification ahead of the mixer. This reduces the relative noise contribution of the mixer. The use of a separate local oscillator allows better frequency stabilization. Good frequency stabilization is especially required when the following i-f amplifier has a high selectivity (narrow bandwidth).

Local oscillators used in superheterodyne receivers require careful design. Assuming that the transmitted signal is kept within narrow limits of frequency tolerance, the local oscillator must keep the resultant intermediate frequency at the center of the pass band of the amplifier. This is important to reduce distortion and operator attention.

The narrower the pass band of the amplifier—that is, the greater its selectivity—the more important it is that the local oscillator does not vary in frequency. This is difficult in receivers which must be tuned by an operator so that one of several signals may be selected at will. In dual-detection superheterodynes the second oscillator may be accurately controlled in frequency by a piezoelectric quartz crystal, such as that used to maintain radio transmitters on their assigned frequencies.

Regenerative receiver. A very simple and effective form of receiver, often employed in the early days of radio communication, utilizes the phenomenon of regeneration to improve signal strength. In this system, some of the received energy is fed back into the input after amplification. If the feedback has the proper phase, the energy fed back adds to the incoming signal and produces a greater output than if no feedback were employed. Although the amplification is high in such a system, the selectivity is not enhanced, so regenerative receivers are seldom employed now.

FM receivers. Receivers for FM systems differ in several respects from those used in AM systems. In an AM system the desired communication is impressed on a high-frequency carrier by varying the magnitude of the carrier in accordance with the magnitude of the signal to be transmitted. The detector in a receiver for this system produces an audible signal corresponding to these amplitude variations.

In an FM system the carrier is modulated by the desired message by varying the frequency of the carrier instead of varying its amplitude. The receiver for such a system must (Fig. 3) have some means of producing a varying voltage amplitude to

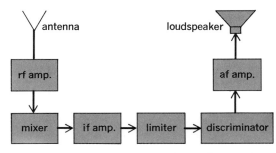

Fig. 3.   Block diagram of FM superheterodyne receiver.

correspond to the varying frequency. In other words, the FM signal must be converted to an AM signal. *See* FREQUENCY MODULATION.

Because the actual incoming frequency varies, the bandwidth to be passed by the receiver circuits, must be wide, and the greater the frequency variation for a given input voltage variation, the greater will be the advantage of the FM system compared to an AM system from the standpoint of eliminating noise.

In an FM receiver a limiter is employed to eliminate all amplitude variations of the carrier and to deliver to the final detector a signal which is free of noise. Several types of FM detectors, usually called discriminators, are employed. *See* FREQUENCY-MODULATION DETECTOR.

Single-sideband (SSB) receivers. This type of communication system is advantageous compared to FM and AM in spectrum conservation and power gain. *See* AMPLITUDE MODULATION; SINGLE SIDEBAND.

There are two forms of SSB receivers. One, commonly referred to as single-sideband (SSB), is for the reception of but one sideband of intelligence on a reduced carrier. The other, independent-sideband receiver (ISB), is used for the reception of two channels of intelligence, one on an upper and one on a lower sideband. To effect this type of reception without undue distortion, the carrier is exalted or reinforced, and in some instances a local carrier synchronized by the incoming reduced carrier is employed for demodulation, sometimes referred to as product detection.

To separate the sidebands and carrier, carefully designed filters are employed. Their characteristics are very important to reduce crosstalk between the sidebands and the sidebands and carrier channel.

ISB enables a great deal of intelligence to be received over a single rf carrier. It is possible to receive four telephone conversations in a 12-kHz total bandwidth, or a single telephone conversation, facsimile transmission, and many telegraph subcarriers at one and the same time.

Diversity reception. Signals received at a given point vary in strength (fade) because they are reflected down to the antenna from the ionosphere, which is unstable with time. Signals at two locations spaced a wavelength or more apart do not fade the same amount at the same time. They may be strong at one site and weak at the other. The correlation decreases with the number of wavelengths, 10 wavelengths being a practical limit.

Diversity reception takes advantage of this phenomenon. Antennas are located at several sites and the detected outputs derived from receivers connected to these antennas are applied, through switching devices, to a common output. Automatic gating or selection means, controlled by the strongest or best signal, assure it to be the major or only contributor to the final output. Thus, the output will have the best signal-to-noise ratio for any combination of operating conditions. The final output of the diversity receiver is derived from the voltage induced on one antenna only.

Receiver antennas. The main objective in the design of a receiving antenna is extraction of maximum power with least noise and interference from unwanted signals.

A high signal-to-noise ratio is desirable. This may not always be feasible; for example, in the case of receivers used in homes for entertainment purposes, a properly designed antenna would be an inconvenience not required in areas served by high field strengths. In the field of commercial communications, however, a properly designed antenna will reduce the requirement (and expense) of a receiver of low noise factor. A good antenna of narrow directivity will extract a large amount of power from the wanted radio wave.

The form that antennas take is determined by many factors, according to the particular requirements of the receiving instrument with which they are to be associated, the frequency coverage applicable, the service, the availability of land, and so forth. The field of radio antenna engineering is large, requiring specialists to resolve each design uniquely. *See* ANTENNA.

[WALTER LYONS]

Bibliography: American Radio Relay League, *Radio Amateur's Handbook*, 1979; G. W. Bartlett (ed.), *NAB Engineering Handbook*, 6th ed., 1975; J. J. Carr, *The Complete Handbook of Radio Receivers*, 1980; Howard W. Sams Engineering Staff, *Reference Data for Radio Engineers*, 6th ed., 1975; J. Markus, *Modern Electronic Circuits Reference Manual*, 1980.

Radio transmitter

A generator of high-frequency electric current whose characteristics of amplitude, frequency, or phase angle may be altered, or modulated, in accordance with the intelligence to be transmitted. A radio transmitter consists of several distinct major components to accomplish the objectives of a particular design for a particular requirement.

The power a transmitter delivers to the antenna may vary from a fraction of a watt to 1,000,000 watts. Lower powers are used mainly for portable or mobile services, while higher powers are required for broadcasting over large areas and in point-to-point communications.

Transmitters may be classified by the type of modulation used. Amplitude-modulation (AM) transmitters are employed for broadcast purposes at medium frequencies. Frequency modulation (FM) and phase modulation require much larger bandwidths in broadcast service, and are used mainly at very high frequencies for broadcast purposes. Frequency and phase modulation provide greater signal-to-noise ratio than amplitude modulation for the same antenna input power. There is also an advantage in operating at very high frequencies, where noise is considerably less than at the lower or medium frequency band. Single-side-band (SSB) or independent-sideband (ISB) transmitters are used for the transmission of single or independent adjacent sidebands. The carrier is suppressed or reduced to an amount which is negligible in comparison to the total power of the transmitter. The modulation of the transmitter in this mode is both amplitude and angular. The principal application of single-sideband transmission is for point-to-point long-distance telephone and telegraph circuits. A particular type of SSB transmission called compatible single-sideband may be utilized for broadcast program transmission since it can be received by the usual AM broadcast receiver. *See* AMPLITUDE MODULATION; FREQUENCY MODULATION; PHASE MODULATION; SINGLE SIDEBAND.

Amplitude-modulation transmitters. AM transmitters (Fig. 1) have two principal design types, either low-level modulated or high-level modulated. The low-level modulated transmitter is modulated at its low power stages, requiring little modulation power. The high-level modulated transmitter, which usually accomplishes modulation at the anode of the output power amplifier, requires modulation power to be equal to about 50% of the carrier's power.

In order to amplify faithfully and reproduce the modulation at low levels, the power amplifiers must be linear. These are usually class B linear amplifiers, which are much less efficient than the class C amplifiers employed in high-level modulated carrier amplifiers. Because it operates over a large band of frequencies, the high-power modulator must have linear power amplification to

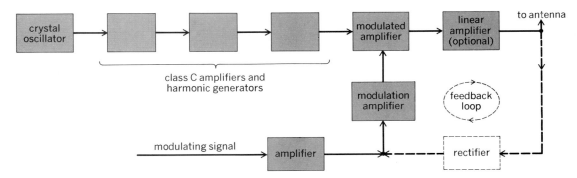

Fig. 1. AM radiotelephone transmitter. (*From F. E. Terman, Electronic and Radio Engineering, 4th ed., McGraw-Hill, 1955*)

achieve low distortion. *See* AMPLIFIER.

Typical equipment. A low-power radio-frequency (rf) oscillator, whose frequency is very accurately controlled (since it determines the final carrier frequency), is the exciter for the transmitter. The exciter is followed by several stages of power amplification, which are required to drive the final power output stage. In low-level modulated transmitters, all the amplifiers following the modulated stage are tuned to the same frequency. Those ahead of the modulated stage may be used to double or triple the frequency of the exciter. In a high-level modulated transmitter the power amplifier stages are seldom tuned to the same frequency, except for the input and output tank circuits of the power amplifier which feeds the antenna.

The modulator of a high-level transmitter derives its input from a microphone or other source of audio signal and amplifies the signal, with low distortion of the order of 1%, to a level which is usually half that of the carrier power.

Most high-power modulators utilize push-pull in either class A or linear class B to reduce distortion by balancing even harmonics and to balance out some hum and noise components. Negative feedback is also used for this purpose. *See* AMPLITUDE MODULATOR.

Antennas. Antennas used with transmitters transform the power generated to an electromagnetic field. They are designed to have a high ratio of radiation to total resistance, which determines the efficiency. Other factors in the design are a sufficient bandwidth to accommodate the frequency band transmitted, directivity, and the restriction of the solid angle of radiation. With certain types of antennas, it is possible to suppress radiation of harmonic frequencies which may cause

interference for other services. Usually, harmonic traps are used to couple the antenna to the transmitter. *See* ANTENNA.

Protection. This important design objective in transmitters assures continuity of operation and protection of personnel. The power supply and high-voltage system design are probably the major problems for any new transmitter type. Safety interlocks, circuit breakers, and warning lights are employed extensively. Automatic discharge of capacitors for the protection of personnel is important. For the protection of equipment, the temperature of the cooling water or air draft must be maintained low enough for efficient operation or the equipment is shut down automatically.

Monitoring. The operation of a transmitter must be monitored at all times to keep the operating personnel informed of the condition of transmission. Meters, in important parts of the circuit, oscilloscopes, and aural monitoring are all used.

Frequency-modulation transmitters. The transmitting frequency in an FM transmitter is varied above and below the median by an amount according to the amplitude of the modulating signal, and at a rate determined by the modulating signal. The amplitude of the transmitted radio frequency is constant; therefore the entire FM system is arranged to be insensitive to amplitude disturbances.

FM transmitters are basically similar to AM transmitters, except that the AM modulation amplifying system is dispensed with and the exciter must be a variable frequency source. One method of modulating the frequency at the exciter utilizes a reactance in the frequency-determining section of the oscillator in the exciter (Fig. 2). The reactance value is changed electronically or electric-

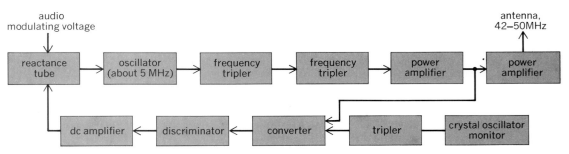

Fig. 2. FM reactance tube transmitter. (*From F. E. Terman, Radio Engineers' Handbook, McGraw-Hill, 1943*)

ally in accordance with the low-frequency modulating signal. The remainder of the transmitter, as in AM transmitters, is made up of frequency doubling and tripling stages of power amplification and a power output stage.

Carrier stabilization. For FM transmitters, this is a more difficult problem than for AM transmitters. Many schemes have been used, including those used for carrier automatic frequency control in receivers. Another means of frequency stabilization uses heterodyning a high-frequency, crystal-generated wave with a lower-frequency, well-stabilized, tank-circuit oscillator, modulated by push-pull reactance tubes. *See* AUTOMATIC FREQUENCY CONTROL (AFC); FREQUENCY MODULATOR.

Protection. Protection of FM transmitters is similar to that of AM transmitters, except that there are fewer components involved and the system is therefore not as extensive.

Monitoring. FM transmitter monitoring differs from that in the AM transmitter, because the means of detecting frequency modulation are different from the means of detecting amplitude modulation. A frequency discriminator tuned to the relevant frequency must be used to reconstitute the modulation at the transmitter monitoring position.

Single sideband/independent sideband. The use of SSB/ISB transmission for communications has become very popular because of several peculiar system characteristics. Its ability to transmit intelligence occupying only half the bandwidth, as compared to AM and FM modes, and its adaptability for modulation by many low-frequency subcarriers are principal advantages. The decreased band occupancy and reduced carrier effectively concentrate all the intelligence in half the bandwidth, greatly increasing the effective power in voice or broadcast. Decreased bandwidth required for transmission doubles the number of services which can use any portion of the frequency spectrum.

Without SSB/ISB means for transmission, radiotelegraph and radiotelephone circuits are difficult to operate on closely spaced frequency assignments. Maintaining precision and stability at high radio frequencies is much more exacting than at the subcarrier modulation frequencies in the low audio spectrum.

Typical equipment arrangement for SSB transmission comprises an SSB generator (exciter), which makes the SSB or ISB signal (using a system of balanced modulators usually) and filters to separate out the carrier for reinsertion in the desired amount, select the desired sideband, and reject the undesired sideband. The output, at low levels of several watts, is then amplified without frequency changing in linear class B amplifiers to the final power output for coupling to the antenna. This method is commonly referred to as linear SSB transmission. A principal design objective is the maximizing of power output in order to minimize spurious radiation.

A more complex transmission system employs an AM modulated transmitter of normal design, an exciter or adaptor which splits the SSB signal into its two components of amplitude and phase modulation, the normal transmitter stages (which may

employ frequency multiplication and class C power amplifiers for amplifying the phase component), and a rectifier for rectifying the AM component for insertion into the modulator as an ordinary AM signal. These two components are then recombined, after phase and amplitude equalization, at the highest power level. This method is called envelope elimination and restoration single sideband. The major design objective here is the reduction of spurious radiation, which is independent of the power output.

A so-called compatible single-sideband system, utilizing the techniques of the last paragraph, modifies the amplitude envelope by using product detection for the AM component and full instead of reduced carrier. Compatibility is effected because the amplitude envelope of this SSB wave is similar to that of a normal AM wave and can be received by the usual AM receiver.

[WALTER LYONS]

Bibliography: American Radio Relay League, *Radio Amateur's Handbook*, 1979; G. W. Bartlett (ed.), *NAB Engineering Handbook*, 6th ed., 1975; J. J. Carr, *The Complete Handbook of Radio Transmitters*, 1980; J. E. Cunningham, *The Complete Broadcast Antenna Handbook: Design, Installation, Operation and Maintenance*, 1977; Howard W. Sams Engineering Staff, *Reference Data for Radio Engineers*, 6th ed., 1975; E. M. Noll, *Radio Transmitter Principles and Projects*, 1973; E. L. Safford, Jr., *Guide to Radio-TV Broadcast Engineering Practices*, 1971.

Radome

A strong, but electrically transparent, thin shell used to house a radar antenna, or a space-communications antenna of similar structure. The shell must be large enough not to interfere with the scanning motion of the antenna. In airborne radar the radome prevents the antenna from upsetting the aerodynamic characteristics of the airplane or missile and protects the antenna against aerodynamic forces. Shipboard radars frequently require radomes to protect them against wind and water damage and blast pressures from nearby guns. Large land-based radars are usually shielded by radomes, especially in severe climatic conditions.

None of the materials available for radomes possesses a dielectric constant equal to that of the atmosphere. The resulting impedance mismatch causes reflections at the inner and outer faces of the shell. In particular, if the shell thickness is a substantial fraction of the carrier wavelength, the reflections from the inner and outer faces may reinforce, causing a standing wave between the radome and the antenna. This affects the antenna impedance in a variable manner as it scans and may change the load on the magnetron sufficiently to pull its frequency out of the passband of the receiver. The standing wave may also distort the antenna pattern, producing undesirable side lobes and changing the orientation of the main beam. Also, the reflected power is not transmitted; thus effective power output is reduced. *See* ELECTROMAGNETIC WAVE TRANSMISSION.

Several means are employed to prevent reflections. If the shell is very thin compared to the carrier wavelength, the reflections from the inner and outer faces are almost a half cycle out of phase and cancel each other. This condition can be obtained easily at frequencies below and including L-band or in the uncommon situations in which a very thin, weak radome can be employed. Alternatively, if the shell thickness is an integral number of half wavelengths, the reflections from the inner and outer faces cancel each other; this arrangement is frequently employed. Numerous multilayer and sandwich-type radomes have been developed in the reasonably successful attempt to combine the properties needed for cancellation of reflections with mechanical strength, stiffness, and lightness.

An additional requirement is that the radome material not cause so much loss as the subtract substantial power from the waves passing through it. A number of available materials satisfy this requirement and also possess chemical and mechanical properties for ease of forming and production. Most of these are organic high polymers, such as resins, rubbers, and fibrous material. Fine glass yarn is also employed. Since the usual production run is small and specialized, radomes are produced either by drawing large flat sheets to the desired shape, in the case of single-layer radomes, or by low-pressure molding and bonding in the case of multilayer and sandwich types. *See* RADAR.

[ROBERT I. BERNSTEIN]

Reactance

The opposition that inductance and capacitance offer to alternating current through the effect of frequency. Reactance alters the magnitude of current and also changes the circuit phase angle.

Inductive reactance X_L equals $2\pi fL$, where f is the frequency in hertz and L is the self-inductance in henrys. The voltage E across an inductance reaches its peak 90° before the current I reaches its peak, and $I = E/X_L$ amperes. Capacitive reactance X_C equals $1/(2\pi fC)$, where C is the capacitance in farads. The voltage E across a capacitance reaches its peak 90° after the current reaches its peak, and $I = E/X_C$ amperes.

Reactances are components of impedance which, in general, includes resistance R and reactance. Impedance is given by Eq. (1) for the series

$$Z = \sqrt{R^2 + (X_L - X_C)^2} \text{ ohms} \qquad (1)$$

RLC circuit. In terms of complex quantities, Eq. (2) holds. Both reactances have magnitude and

$$Z = R + jX_L - jX_C = R + j(X_L - X_C) \quad \text{ohms} \qquad (2)$$

angle: $+j$ means $+90°$ for X_L, and $-j$ means $-90°$ for X_C, the angles by which the voltages across them lead, or lag, the current. The resulting phase angle between voltage and current is given by Eq. (3), and current lags, or leads, the voltage de-

$$\theta = \arctan\left[(X_L - X_C)/R\right] \qquad (3)$$

pending upon whether $X_L - X_C$ is positive or negative. *See* ALTERNATING-CURRENT CIRCUIT THEORY.

[BURTIS L. ROBERTSON]

Real-time systems

Computer systems in which the computer is required to perform its tasks within the time restraints of some process or simultaneously with

the system it is assisting. Usually the computer must operate faster than the system assisted in order to be ready to intervene appropriately.

Types of systems. Real-time computer systems and applications span a number of different types.

Real-time control and real-time process control. In these applications the computer is required to process systems data (inputs) from sensors for the purpose of monitoring and computing system control parameters (outputs) required for the correct operation of a system or process. The type of monitoring and control functions provided by the computer for subsystem units ranges over a wide variety of tasks, such as turn-on and turn-off signals to switches; feedback signals to controllers (such as motors, servos, and potentiometers) to provide adjustments or corrections; steering signals; alarms; monitoring, evaluation, supervision, and management calculations; error detection, and out-of-tolerance and critical parameter detection operations; and processing of displays and outputs.

Real-time assistance. Here the computer is required to do its work fast enough to keep up with a person interacting with it (usually at a computer terminal device of some sort, for example, a screen and keyboard). These are people-amplifier-type real-time computer systems. The computer supports the person or persons interacting with it and provides access, retrieval, and storage functions, usually through some sort of data-base management system, as well as data processing and computational power. System access allows the individual to intervene (control, adjust, supply parameters, direct, and so forth) in the system's operation. The real-time computer also often provides monitoring or display information, or both. *See* DATA-BASE MANAGEMENT SYSTEM; MULTIACCESS COMPUTER.

Real-time robotics. In this case the computer is a part of a robotic or self-contained machine. Often the computer is embedded in the machine, which then becomes a smart machine. If the smart machine also has access to, or has embedded within it, artificial intelligence functions (for example, a knowledge base and knowledge processing in an expert system fashion), it becomes an intelligent machine. *See* ARTIFICAL INTELLIGENCE; EMBEDDED SYSTEMS; EXPERT SYSTEMS; ROBOTICS.

Evolution. Real-time computer systems have been evolving constantly since the interrupt function, which allowed the computer to be synchronized with the external world, was invented. There are five primary paths along which real-time systems continue to advance:

First-generation real-time control systems comprise process control (for example, an oil refinery); guidance and control (for example, antiballistic missile and intercontinental ballistic missile systems); numerical control (for example, factory machine operations); dedicated (mini) computer systems; and store-and-forward message switching. *See* DATA COMMUNICATIONS.

Second-generation real-time computer systems comprise time-shared and multiprocessor computing; interactive computing; smart and intelligent terminals; and computer networks (distributed computers, distributed smart machines, and distributed intelligent machines). *See* DISTRIBUTED PROCESSING; MULTIPROCESSING.

Third-generation real-time assistance systems comprise operating systems; CAD (computer-aided design), CAM (computer-aided manufacturing), CAI (computer-assisted instruction), MIS (management information systems), and DSS (decision support systems); personal computers; word processing and work stations; and artificial intelligence expert systems. *See* COMPUTER-AIDED DESIGN AND MANUFACTURING; MICROCOMPUTER; OPERATING SYSTEM; WORD PROCESSING.

Fourth-generation real-time machines comprise smart machines with embedded computors; intelligent machines; and robots (dumb, smart, and intelligent).

Fifth-generation real-time integrated systems comprise the factory-of-the-future (totally automated factories); just-in-time (JIT) systems (in the factory and in distribution); computer utilities and knowledge utilities; and knowledge inference processing systems.

Artificial intelligence and experts systems. A new technology, artificial intelligence (AI), is rapidly advancing beyond the research stage into practical use by scientists, management, and many other areas of business and society. One branch of artificial intelligence is directed toward the development of real-time expert systems. Thus, widespread use of artificial intelligence, knowledge bases, expert systems, and people-amplifiers is expected to develop, and decision making will be amplified with real-time and intelligent computer systems.

Real-time simulation. The real-time computer can serve as a tool allowing simulation of models of the real world. By coupling this tool with artificial-intelligence knowledge-base expertise, scientific experimentation becomes possible without cumbersome laboratory equipment and procedures. Scientists can play serious real-time experimental and mathematical games with the object of their research without first needing to learn sophisticated laboratory techniques or mathematics. The detailed mathematical and discipline-oriented experimental skills and procedures are embedded within the computerized model. Such robot simulators allow the scientist to concentrate upon the investigation, rather than the scientist becoming buried within the relevant mathematics and discipline crafts. However, the scientist must first learn his or her field plus the computer simulation/modeling language. Thus, robot simulator assistants increasingly do for the scientist what the calculator does for the average person: they remove the need to perform bulky, precise, and rote skill functions, allowing the researcher to get more quickly and easily to the core of scientific investigation.

With robot simulators the researcher can ask questions of a computer modeling system, and have simulated experiments performed that are otherwise nearly impossible or too time-consuming and costly. For example, the researcher can ask "What happens if . . . ?" After performing the simulated experiment, the computer gives an answer, while the experimenter views the progress of the computerized experiment and intervenes when desired. Then the experimenter can ask "What happens if something else is done instead?" to arrive at a different comparable answer from the robot simulator. Through such real-time interactive simulations the researcher becomes directly involved in the experiment as a surrogate participant.

Future robot simulators are expected to take the form of advanced hand-held calculators with voice dialog and artificial intelligence capabilities, and to contain specialized expert knowledge making them capable of general decision making. Such smart robot simulators will be used as people-amplifiers or electronic assistants by managers, programmers, doctors, politicians, voters, and others. *See* COMPUTER; DIGITAL COMPUTER; SIMULATION.

[EARL C. JOSEPH]

Bibliography: Articles in the journal *Computer* (IEEE Computer Society) and the *Computing Surveys* (Association for Computing Machinery).

Reciprocity principle

In the scientific sense, a theory that expresses various reciprocal relations for the behavior of some physical systems. Reciprocity applies to a physical system whose input and output can be interchanged without altering the response of the system to a given excitation. Optical, acoustical, electrical, and mechanical devices that operate equally well in either direction are reciprocal systems, whereas unidirectional devices violate reciprocity. The theory of reciprocity facilitates the evaluation of the performance of a physical system. If a system must operate equally well in two directions, there is no need to consider any nonreciprocal components when designing it.

Examples of reciprocal systems. Some systems that obey the reciprocity principle are any electrical network composed of resistances, inductances, capacitances, and ideal transformers; systems of antennas, with restrictions given according to Eq. (2); mechanical gear systems; and light sources, lenses, and reflectors.

Devices that violate the theory of reciprocity are transistors, vacuum tubes, gyrators, and gyroscopic couplers. Any system that contains the above devices as components must also violate the reciprocity theory. The gyrator differs from the transistor and vacuum tube in that it is linear and passive, as opposed to the active and nonlinear character of the other two devices.

Rayleigh's theorem of reciprocity. Reciprocity is concisely expressed by a theorem originally proposed by Lord Rayleigh for acoustic systems and later generalized by J. R. Carson to include electromagnetic systems. Both mathematical expressions of the theory of reciprocity are closely related to the mathematical theorem known as Green's theorem. The acoustical reciprocity theorem of Lord Rayleigh is as follows: In an acoustic system consisting of a fluid medium having boundary surfaces $s_1, s_2, \ldots, s_k$ and subject to no impressed body forces, surface integral (1) holds. Here p_1 and

$$\int_s (p_1 v_{2n} - p_2 v_{1n}) ds = 0 \qquad (1)$$

p_2 are the pressure fields produced respectively by the components of the fluid velocities v_{1n} and v_{2n}

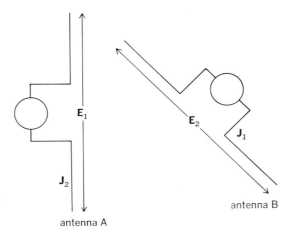

Fig. 1. Antenna system.

normal to the boundary surfaces $s_1, s_2, \ldots, s_k$. The integral is evaluated over all boundary surfaces.

For a region containing only one simple source H. L. F. Helmholtz has shown that the theorem can be expressed as follows: A simple source at A produces the same sound pressure at B as would have been produced at A had the source been located at B. In other words, the response of a human ear at B due to a vibrating tuning fork at A is the same as the response of the ear at A due to the same tuning fork when located at B. The human ear, the tuning fork, and the intervening acoustical media constitute a physical system that obeys the theory of reciprocity.

Electromagnetic systems. The generalization of Lord Rayleigh's theorem to electromagnetic systems can be mathematically expressed by volume integral (2), where E_1 and H_1 are the electric

$$\int_v \nabla \cdot (E_1 \times H_2 - E_2 \times H_1)\, dv = 0 \qquad (2)$$

and magnetic field vectors describing a state due to one electromagnetic sound and E_2 and H_2 describe another state due to a second source. The above relation is valid as long as the medium is isotropic and the field vectors are finite and continuous, and vary according to a linear law (thus excluding ferromagnetic materials, electronic space charges, and ionized gas phenomena).

By means of Maxwell's equations, relation (2) can be expressed in another form when restricted to systems of conduction current only where J_1 and J_2 are the conduction current densities in an electromagnetic system due to the action of the external electric fields E_1 and E_2, respectively.

Equation (3) is readily applied to antennas and

$$\int_v (E_1 \cdot J_2 - E_2 \cdot J_1)\, dv = 0 \qquad (3)$$

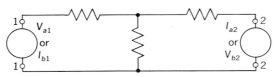

Fig. 2. Two-mesh network.

radiation. If, in Fig. 1, J_1 is the resulting current density in antenna B due to an electric field E_1 established by antenna A, and J_2 is the current density in antenna A due to electric field E_2 established by antenna B, then $J_1 = J_2$, provided $E_1 = E_2$. The two emfs need not be applied at the same instant of time. The integral in Eq. (3) over all space reduces to an integral over the two antennas since J_1 and J_2 are zero elsewhere. From this particular application of the reciprocity theorem it is seen that the transmitting and receiving patterns of an antenna are the same.

Equation (3), when evaluated over an N-mesh electrical network, reduces to Eq. (4), where a and

$$\sum_{j=1}^{N} V_{aj} i_{bj} = \sum_{j=1}^{N} V_{bj} i_{aj} \qquad (4)$$

b are two different states of the network and the j subscript denotes in which of the N meshes the voltage and current are measured. For the two-mesh network in Fig. 2, Eq. (4) gives Eq. (5). Expressed in words: If an emf source of magnitude V

$$V_{a1} i_{b1} = V_{b2} i_{a2} \qquad (5)$$

pressed in words: If an emf source of magnitude V and zero internal impedance, when applied to terminals $1-1$, produces a current I at terminals $2-2$, then the same current I will be measured at terminals $1-1$ when the emf V is applied to terminals $2-2$. This statement, that is, Eq. (5), is probably the most familiar form of the theorem of reciprocity.

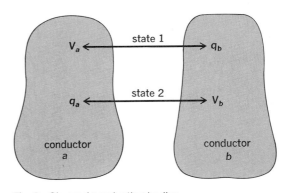

Fig. 3. Charged conducting bodies.

Electrostatic systems. The statement of reciprocity for electrostatics is given by Eq. (6), where V_1 and V_2 are the electric potentials produced at some arbitrary point due to the volume charge distributions ρ_1 and ρ_2, respectively. When the integral expression in Eq. (6) is applied to the

$$\int_v \rho_1 V_2\, dv = \int_v \rho_2 V_1\, dv \qquad (6)$$

electrostatic system of two charged conductors in Fig. 3, it becomes Eq. (7). Here V_a is the poten-

$$V_a q_a = V_b q_b \qquad (7)$$

tial on conductor a due to charge q_b on conductor b; the remaining quantities are similarly defined. In other words, if a charge q_b on conductor b raises the potential of conductor a to V, then the same charge on conductor a raises the potential of conductor b to V.

Electrical networks. A somewhat different approach to reciprocity is the so-called black box, or two-terminal pair, method illustrated in Fig. 4. The box might contain a mechanical, acoustical, optical, or electrical system. The applied excitation or cause is E and the response or effect is E'. The ratio of E/E' (or E'/E) is the transfer function G for the system within the black box. Using the subscript notation of G_{12} when E is impressed at terminals $1-1$ and E' is measured at terminals $2-2$, then G_{21} represents a response measured at $1-1$ for an excitation at $2-2$. Mathematically the general behavior of the box to excitations at both sets of terminals can be expressed by Eqs. (8), as long as the response bears a linear relation to the excitation.

$$E_1 = G_{11}E'_1 + G_{12}E'_2$$
$$E_2 = G_{21}E'_1 + G_{22}E'_2 \qquad (8)$$

If, in addition to its linear characteristic, the system satisfies Eq. (9), the principle of reciprocity is

$$G_{12} = G_{21} \qquad (9)$$

obeyed, and the device will operate equally in either direction. Whenever $G_{12} \neq G_{21}$, the system violates the theory of reciprocity, with the result that the response in one direction is different from that obtained in the other direction. *See* NETWORK THEORY.   [HUGH S. LANDES]

Bibliography: B. Bleaney and B. I. Bleaney, *Electricity and Magnetism*, 3d ed., 1976; D. E. Gray (ed.), *American Institute of Physics Handbook*, 3d ed., 1972; Howard W. Sams Engineering Staff, *Reference Data for Radio Engineers*, 6th ed., 1975; J. D. Kraus, *Antennas*, 1950; J. D. Kraus and K. R. Carver, *Electromagnetics*, 2d ed., 1973; J. A. Stratton, *Electromagnetic Theory*, 1941.

Rectifier

A nonlinear circuit component that allows more current to flow in one direction than in the other. An ideal rectifier is one that allows current to flow in one (forward) direction unimpeded but allows no current to flow in the other (reverse) direction. Thus, ideal rectification might be thought of as a switching action, with the switch closed for current in one direction and open for current in the other direction. Rectifiers are used primarily for the conversion of alternating current (ac) to direct current (dc). *See* ELECTRONIC POWER SUPPLY.

A variety of rectifier elements are in use. The vacuum-tube rectifier can efficiently provide moderate power. Its resistance to current flow in the reverse direction is essentially infinite because the tube does not conduct when its plate is negative with respect to its cathode. In the forward direction, its resistance is small and almost constant. Gas tubes, used primarily for higher power requirements, also have a high resistance in the reverse direction. The semiconductor rectifier has the advantage of not requiring a filament or heater supply. This type of rectifier has approximately constant forward and reverse resistances, with the forward resistance being much smaller. Mechanical rectifiers can also be used. The most common is the vibrator, but other devices are also used. *See* GAS TUBE; SEMICONDUCTOR RECTIFIER.

A rectifying element can be illustrated by assuming a device having a forward resistance R_1 and a reverse resistance R_2, which is much greater than R_1. A sinusoidal alternating voltage $E_m \sin 2\pi ft$ is applied to the rectifier, where E_m is the maximum value of the applied voltage, f is the frequency of the voltage wave, and t is time. The magnitude of the current in the forward direction is $(E_m/R_1)\sin 2\pi ft$. This current flows from t equals 0 to $1/(2f)$, or for one-half of the cycle of the alternating voltage wave. The forward current, averaged over one cycle, is $E_m/(\pi R_1)$. The reverse current has the magnitude $(E_m/R_2)\sin 2\pi ft$ and flows from t equals $1/(2f)$ to $1/f$, or for the other half-cycle. The reverse current, averaged over one cycle, is $E_m/(\pi R_2)$. The net forward average current is $E_m(R_2 - R_1)/(\pi R_1 R_2)$.

If reverse resistance R_2 is extremely large compared to R_1, the average current approaches $E_m/(\pi R_1)$. If the average current is subtracted from the current flowing in the rectifier, an alternating current results. This ripple current flowing through a load produces a ripple voltage which is often undesirable. Filter and regulator circuits are used to reduce it to as low a value as is required. *See* ELECTRIC FILTER; VOLTAGE REGULATOR.

Half-wave rectifier circuit. A half-wave rectifier circuit is shown in Fig. 1. The rectifier is a diode, which allows current to flow in the forward direction from A to B but allows practically no current to flow in the reverse direction from B to A. The ac input is applied to the primary of the transformer; secondary voltage e supplies the rectifier and load resistor R_L.

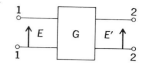

RECIPROCITY PRINCIPLE

Fig. 4. Four-terminal, black-box network.

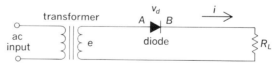

Fig. 1. Half-wave diode rectifier.

The rectifying action of the diode is shown in Fig. 2, in which the current i of the rectifier is plotted against the voltage e_d across the diode. The applied sinusoidal voltage from the transformer secondary is shown under the voltage axis; the resulting current i flowing through the diode is shown at the right to be half-sine loops. Averaging the value of these half-sine loops gives the direct current flowing; the ripple current is the variation of load current about the average value.

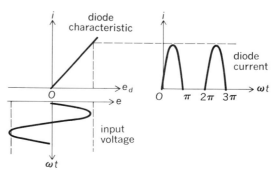

Fig. 2. Rectifying action of half-wave diode rectifier.

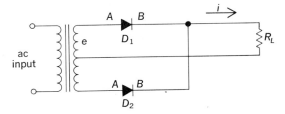

Fig. 3. Full-wave diode rectifier.

Full-wave rectifier circuit. A full-wave rectifier circuit is shown in Fig. 3. This circuit uses two separate diodes, D_1 and D_2. The center tap of the transformer is connected through the load resistance R_L to the B sides of both diodes. During one-half cycle of the ac input the A terminal of diode D_1 is positive with respect to the B terminal, and D_1 conducts current from A to B. This current passes through the load resistor R_L in the direction shown by i. During this time diode D_2 is not conducting because terminal A is negative with respect to B. When the ac potential goes through zero, A of diode D_1 becomes negative and the diode stops conducting. The potential on A of D_2 then becomes positive, and D_2 starts to conduct. The resulting current wave shape is shown in Fig. 4.

The effect of using two diodes instead of one is to produce a more continuous flow of direct current through load R_L because the first diode conducts for the positive half-cycle and the second diode conducts for the negative half-cycle, as shown in Fig. 4. Comparison of Figs. 2 and 4 indicates that a full-wave circuit will produce a more nearly uniform output than will the half-wave circuit.

Polyphase rectifier circuits. When high dc power is required by an electronic circuit, a polyphase rectifier circuit may be used. It is also desirable when expensive filters must be used. This is particularly true of power supplies for the final

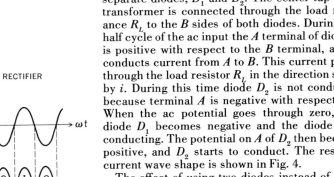

RECTIFIER

Fig. 4. Applied voltage and output current of full-wave rectifier.

radio-frequency and audio-frequency stages of large radio and television transmitters. The rectifier employed in polyphase circuits generally is a gas tube that has a low voltage drop in the forward direction and thus has a high efficiency. Semiconductor diodes are also used. The number of phases used in these circuits is most often 3, but 2, 4, 6, and 12 phases are used occasionally.

The simplest polyphase circuit is the three-phase half-wave circuit of Fig. 5. The primaries of the transformers are connected in delta to the three-phase ac line, and the secondaries are connected in wye with the common connection going to one end of the load resistor. The other end of the load resistor is attached to the B terminals of the three rectifier diodes required in the circuit. The A terminals are connected to the separate ends of the three transformer secondaries.

Operation of the circuit is such that diode D_1, connected to the first secondary, conducts for 120° of the ac cycle. As soon as the voltage on secondary 2 equals that of secondary 1, diode D_2 starts to conduct and diode D_1 stops conducting. Secondary voltages e_{o1}, e_{o2}, and e_{o3} are shown in Fig. 6, and

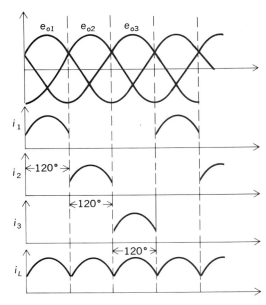

Fig. 6. Transformer voltages, diode currents, and load current in a three-phase half-wave rectifier.

diode currents are indicated as i_1, i_2, and i_3. The resulting load current i_L is also shown in Fig. 6. This current is much closer to a true direct current than is the current for the single-phase circuits of Figs. 1 and 3. Ripple voltage is much lower, and less elaborate filter circuits are needed to smooth the output wave. The diodes in Fig. 5 could be replaced by gas rectifier tubes or ignitrons if higher load currents were required.

Another common polyphase rectifier circuit is the three-phase full-wave or six-phase half-wave circuit of Fig. 7. In this circuit the tubes conduct for 60° instead of 120°, as in the circuit of Fig. 5. The ripple voltage for the full-wave circuit is much smaller.

Many other polyphase rectifier circuits are possible.

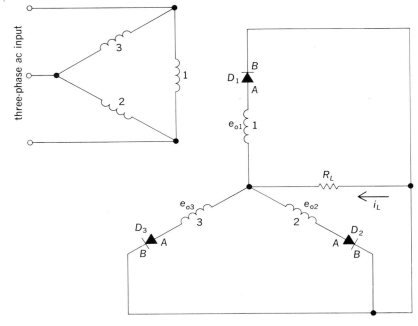
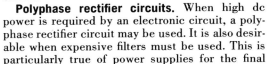

Fig. 5. Three-phase half-wave rectifier.

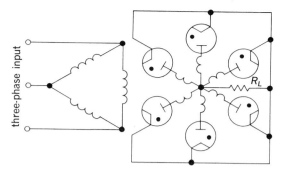

Fig. 7. Three-phase full-wave or six-phase half-wave rectifier, as example of polyphase rectifier circuit.

Bridge rectifier circuits. Bridge rectifier circuits are useful in both single-phase and polyphase applications in which a transformer must be used whose secondary has no center tap or in which dc voltages approximately equal to the total secondary voltage of the transformer must be obtained. Another use of the bridge circuit is in ac rectifier-type meters. The bridge circuit is shown in Fig. 8. Four separate half-wave rectifier diodes are used. When the left-hand side of the transformer secondary is positive, current flows through diode D_1, the load resistor R_L, and diode D_3. When the secondary voltage reverses, the current flows through diodes D_2 and D_4 passing through R_L in the same direction as during the first half cycle.

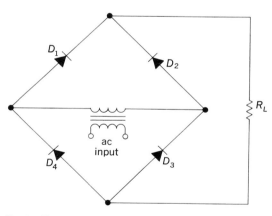

Fig. 8. Single-phase full-wave bridge circuit.

Parallel rectifiers. If greater current is desired, two or more rectifiers can be connected in parallel. In such an arrangement small resistors or inductors are put in series with the rectifiers before they are connected in parallel.

Controlled rectifiers. Controlling the current delivered by a rectifier can be accomplished by varying the primary voltage of the power transformer or by changing a resistance in series with the load resistor. The first technique has the disadvantage of being expensive; the second leads to poor efficiency. A more convenient and less expensive method is to control the angle at which the rectifier tube starts to conduct. Special gas tubes that accomplish this control are thyratrons and ignitrons. Silicon-controlled rectifiers (SCR) may

also be used for this purpose. Thyratrons are hot-cathode gas tubes with a large grid structure that prevents the arc from being ignited until the correct voltage is applied to the grid. An ignitron is a cold-cathode pool-type tube with an igniter grid actuated by an electrical pulse. The igniter of the ignitron requires a substantial amount of power, which is usually supplied by an auxiliary thyratron in the control circuit. The SCR (silicon-controlled rectifier) is a silicon semiconductor device which conducts when the trigger gate electrode is raised to the triggering potential. *See* CONTROLLED RECTIFIER; SEMICONDUCTOR RECTIFIER.

One circuit for an SCR is shown in Fig. 9, with control of the triggering point of the SCR possible

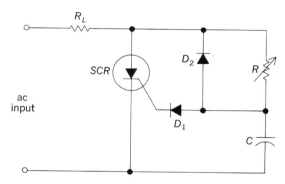

Fig. 9. One circuit for a silicon-controlled rectifier with phase control triggering network.

over a full half-cycle of 180 electrical degrees. When the upper terminal of the ac input is positive, capacitor C will charge to the triggering voltage of the SCR in a time determined by the RC time constant and the voltage across the SCR. Current then flows through load resistor R_L until the ac input voltage starts to reverse. When the lower terminal of the ac input is positive, capacitor C charges negatively through the diode D_2, and the cycle of conduction repeats. Diode D_1 prevents the negative peak voltage from appearing on the gate of the SCR. By varying the resistor R, the triggering point can be varied from zero degrees when the SCR will conduct the maximum current and to 180° when the SCR conducts zero current. Hence a continuous control from zero to maximum current is achieved. Other control circuits can also be used.

Inverse voltage. The inverse voltage of a rectifier is the voltage that the rectifier must withstand when it is not conducting or when it is conducting slightly in the reverse direction. As an example, in the full-wave rectifier circuit of Fig. 3, when diode D_2 is conducting, diode D_1 has impressed upon it the total secondary voltage of the power transformer minus the voltage drop in D_2. For a well-designed power supply the maximum value of the inverse voltage should not exceed the rated value of the rectifier specified by the manufacturer.

Current ratings. Another important rating for a rectifier is the average current through it. The average rectifier current of a half-wave single-phase rectifier is the same as the average load current. For a full-wave single-phase rectifier the average rectifier current is one-half the average load current. The maximum value of instantaneous current through the rectifier should not exceed the peak current rating of the rectifier. This is particularly true when capacitor-input filters are employed, because these filters generally produce high peak-to-average current ratios in the rectifier. For information on electronic circuits in general *see* CIRCUIT (ELECTRONICS).   [DONALD L. WAIDELICH]

   Bibliography: I. Gottlieb, *Principles and Applications of Inverters and Converters*, 1977; A. H. Lytel, *Solid-State Power Supplies and Converters*, 1965; G. Scoles, *Handbook of Rectifier Circuits*, 1980.

Regeneration

The process of feeding back a portion of the output signal of an amplifier to its input in such a way that the input signal is reinforced. The result is greatly increased amplification. The feedback must be positive; that is, the two signals must be in phase, and it must be limited in magnitude to prevent the circuit from going into oscillation. *See* FEEDBACK CIRCUIT.

   In storage devices for computers, regeneration involves the restoration of deteriorating electrostatic, magnetic, or other conditions to their original state. This is particularly essential in charge-storage cathode-ray tubes to overcome natural decay effects, as well as loss of charge by reading out the information stored. *See* STORAGE TUBE.

   In the nuclear power field, regeneration involves the purification of contaminated nuclear fuel for reuse.

                            [JOHN MARKUS]

Regulation

The process of maintaining a quantity or condition essentially constant despite variations in such factors as line voltage and load. In an industrial process-control system, the speed, temperature, voltage, or position of a critical element can be kept constant by measuring the condition being regulated and feeding back into the system a signal representing the difference between the actual and the desired quantities. For example, if the temperature of a mixture of chemicals is too low, a sensing element feeds back to the controller a signal that results in the application of more heat.

   The term regulation is also used in the opposite sense to indicate the difference between the maximum and minimum voltages at the terminals of a tube, transformer, generator, or other device over the range of normal operating conditions. When used in this manner, regulation R in percent is computed from the equation below, where E_{max}

$$R = \frac{E_{max} - E_{min}}{0.5(E_{max} + E_{min})}\,100$$

and E_{min} are, respectively, maximum and minimum voltages (or other property whose regulation is being computed) in the region of interest.

                          [FRANK H. ROCKETT]

Relaxation oscillator

An electronic circuit which has two stable states, resulting in two distinct output levels, and which switches between the two states at a rate deter-

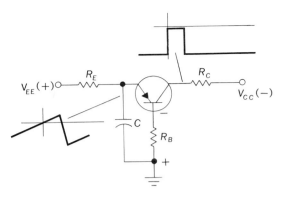

Fig. 1. Point-contact transistor relaxation oscillator.

mined by the rate of rise or decay of voltage across the storage element in an RC or RL circuit. The output waveform is usually nonsinusoidal and may be approximately a square wave, a sawtooth wave, or a series of short repeating pulses. *See* WAVE-SHAPING CIRCUITS.

   One of the most widely used forms of relaxation oscillator is the astable multivibrator, which generates a rectangular or square wave. In this circuit two devices are connected so that they are alternately ON and OFF. Connected together by a positive feedback path, they are driven rapidly from one state to the other. *See* MULTIVIBRATOR.

   There is also a class of circuits in which a single device has two stable conditions, either ON-OFF or ON with two distinct states or levels. Switching between the two states usually involves an RC time constant. The blocking oscillator is representative of such a circuit, which also includes a positive feedback path. *See* BLOCKING OSCILLATOR.

   There are other relaxation oscillators using single transistors or transistorlike devices with a current gain greater than unity, such as the point-contact transistor, which can be arranged as a relaxation oscillator as shown in Fig. 1. During the ON period a large collector current flows through R_B and R_C, making the base slightly negative, and the capacitance C charges toward a negative potential through the emitter circuit. It finally reach-

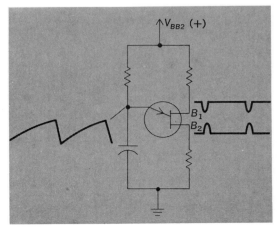

Fig. 2. Unijunction transistor as relaxation oscillator.

es a point at which the current gain drops to less than unity. The voltage drop across R_B suddenly decreases, and the emitter circuit becomes reverse biased. The emitter will remain reverse-biased until C can again charge to the conduction level through R_E. The circuit is suitable for the generation of short pulses, such as those needed for trigger circuits.

As shown in Fig. 2, a unijunction transistor, sometimes called a double-base diode, may be used as a simple relaxation-oscillator trigger generator. Negative-resistance devices such as tunnel diodes can be used as a relaxation oscillator. *See* NEGATIVE-RESISTANCE CIRCUITS; TUNNEL DIODE.

[GLENN M. GLASFORD]

Bibliography: M. Ghausi, *Electronic Circuits*, 1971; J. Millman and H. Taub, *Pulse, Digital, and Switching Waveforms*, 1965.

Reluctance

A property of a magnetic circuit analogous to resistance in an electric circuit.

Every line of magnetic flux is a closed path. Whenever the flux is largely confined to a well-defined closed path, there is a magnetic circuit. That part of the flux that departs from the path is called flux leakage.

For any closed path of length l in a magnetic field H, the line integral of $H \cos \alpha\, dl$ around the path is the magnetomotive force (mmf) of the path, as in Eq. (1), where α is the angle between H and

$$\text{mmf} = \oint H \cos \alpha\, dl \qquad (1)$$

the path. If the path encloses N conductors, each with current I, Eq. (2) holds.

$$\text{mmf} = \oint H \cos \alpha\, dl = NI \qquad (2)$$

Consider the closely wound toroid shown in the figure. For this arrangement of currents, the magnetic field is almost entirely within the toroidal coil, and there the flux density or magnetic induction B is given by Eq. (3), where l is the mean circumference of the toroid and μ is the permeability.

$$B = \mu \frac{NI}{l} \qquad (3)$$

The flux Φ within the toroid of cross-sectional area A is given by either form of Eqs. (4), which is similar in form to the equation for the electric circuit,

$$\Phi = BA = \frac{\mu A}{l} NI$$

$$\Phi = \frac{NI}{l/\mu A} = \frac{\text{mmf}}{l/\mu A} = \frac{\text{mmf}}{\mathscr{R}} \qquad (4)$$

although nothing actually flows in the magnetic circuit. The factor $l/\mu A$ is called the reluctance $\mathscr{R}$ of the magnetic circuit. The reluctance is not constant because the permeability μ varies with changing flux density. From the defining equation for reluctance, it is seen that when the mmf is in ampere-turns and the flux is in webers, the unit of reluctance is the ampere-turn/weber.

Reluctances in series. For the simple toroid, all parts of the magnetic circuit have the same μ and the same A. More complicated circuits may include parts that differ in permeability, in cross section, or in both. Suppose a small gap were cut in the core of the toroid. The flux would fringe out at the gap, but as a rough approximation, the area of the gap may be considered the same as that of the core.

The magnetic path then has two parts, the core of length l_1 and reluctance $l_1/\mu_1 A$, and the air gap of length l_2 and reluctance $l_2/\mu_2 A$. Since the same flux is in both core and gap, this is considered a series circuit and Eq. (5) holds. Since the

$$\mathscr{R} = \mathscr{R}_1 + \mathscr{R}_2 = \frac{l_1}{\mu_1 A} + \frac{l_2}{\mu_2 A} \qquad (5)$$

relative permeability of the ferromagnetic core is several hundred or even several thousand times that of air, the reluctance of the short gap may be much greater than that of the much longer core. For any combination of paths in series, $\mathscr{R} = \Sigma l/\mu A$. Then Eq. (6) holds.

$$\Phi = \frac{\text{mmf}}{\Sigma \mathscr{R}} = \frac{\text{mmf}}{\Sigma l/\mu A} \qquad (6)$$

Reluctances in parallel. If the flux divides in part of the circuit, there is a parallel magnetic circuit and the reluctance of the circuit has the same relation to the reluctances of the parts as has the analogous electric resistance. For the parallel circuit Eq. (7) is valid.

$$\frac{1}{\mathscr{R}} = \frac{1}{\mathscr{R}_1} + \frac{1}{\mathscr{R}_2} + \cdots \qquad (7)$$

[KENNETH V. MANNING]

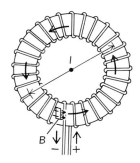

RELUCTANCE

A toroidal coil.

Resistance measurement

The quantitative determination of that property of an electrically conductive material, component, or circuit called electrical resistance. The unit of measurement is called the ohm. Measurements for engineering applications usually range from 0.1 microhm (10^{-7} ohm) to 1000 teraohms (10^{15} ohms). Measurements often required for materials development, basic research, or experiments under extreme environmental conditions extend this span to include zero ohms (superconductors) and 10^{18} ohms. *See* ELECTRICAL RESISTANCE; OHM'S LAW; SUPERCONDUCTIVITY.

Resistance may be measured by using either direct or alternating current (dc or ac). If dc is used, the true resistance of the conductor is measured. When the measurement is made with ac, the result is usually called the effective resistance. Physical factors which must be considered when measuring resistance include (1) the ambient temperature and self-heating of the conductor, (2) the connecting lead and contact resistance when measuring low-resistance conductors, (3) parallel leakage paths around a high-resistance conductor, (4) current distribution within the conductor, (5) thermally, electrolytically, and other spuriously generated voltages within the resistance being measured or in the measuring circuit, and (6) the capacitance and inductance associated with the conductor.

Except for determinations involving extreme values, practically all methods for measuring resistance are based on Ohm's law, and countless variations of electrical networks have been devised for specific resistance-measurement requirements. Ohm's law, first proven experimentally in 1826, states that, for dc circuits, the difference of potential E existing between two terminals of a

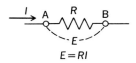

Fig. 1.   Ohm's law.

conductor is directly proportional to the current I flowing in the conductor, or $E = RI$. The constant of proportionality R is called the resistance of the conductor (Fig. 1). This relationship is valid for most dc circuits, although certain exceptions are found, as in gaseous conduction, thermionic conduction, and semiconductors, which exhibit non-ohmic characteristics. Conductors of this type must be measured under completely specified physical conditions pertinent to the phenomena, such as gas pressure, emission temperature, applied voltage, and polarity. For a discussion of resistance in a dc circuit *See* DIRECT-CURRENT CIRCUIT THEORY.

Ohm's law is also valid for ac circuits if vector quantities are used for all parameters, that is, $\mathbf{E} = \mathbf{ZI}$ where $\mathbf{Z}$ is the complex impedance $R_e + jX_e$, R_e is the effective resistance, and X_e is the effective reactance. *See* ALTERNATING-CURRENT CIRCUIT THEORY.

Measurement of low resistance. The measurement of low resistance values (10^{-7} to 10 ohms) or the accurate measurement of intermediate resistance values (1 to 10^4 ohms) requires special techniques to eliminate the resistance of the test leads and their associated contact resistance from the measurement. For example, if the 0.1-ohm resistor of Fig. 2a is measured with an ohmmeter and two test leads of no. 18 copper wire each 5 ft long, an error of about 65% would result. Since the test leads and their contact connection resistances are in series with the resistor to be measured, their resistance is also included in the measurement.

The technique usually employed to eliminate this effect is to provide the unknown resistance with four leads, as in Fig. 2b. Two of these are current connections, which supply current to the resistance; the other two are potential connections physically located between the current connections. Special measuring circuits requiring little or no current in the potential circuit are used. These techniques allow measurement of only that resistance between the potential contacts. *See* KELVIN BRIDGE.

Measurement of high resistance. The measurement of high resistances (10^5 to 10^{18} ohms) requires consideration of spurious leakage paths, which may exist in parallel with the resistance terminals,

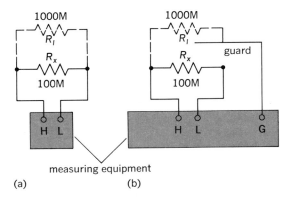

Fig. 3.   Measurement of high resistance (10^5 to 10^{18} ohms). (a) Electrical leakage resistance R_l included in measurement. (b) R_l eliminated by guard.

as shown in Fig. 3. Electrical leakage is usually the result of inadequate, moisture-sensitive, dirty, or deteriorated insulation between the terminals. Errors of several thousand percent may result unless suitable precautions are taken. Under extreme conditions, such erratic readings may be obtained that a measurement is impossible.

The effect of leakage can usually be eliminated by suitably guarding one terminal or portion of the resistance to be measured. Physically a guard consists of a low-resistance conductor, electrically insulated from the guarded terminal and located to intercept the leakage current, as shown in Fig. 3b. Measuring circuits designed to take advantage of this technique maintain the guard and the guarded terminals at equal or nearly equal potential to prevent or minimize current flow between the guard and guarded electrodes. The guard circuit of the measuring device is also designed to conduct the leakage current around the main measuring circuit so that it has little or no effect on the measurement.

Measurement of extreme resistance. Conventional techniques for measuring resistance involving the direct application of Ohm's law are not suitable for measuring extreme resistance values below 10^{-7} ohm and above 10^{15} ohms. Resistance standards are not available, the accuracy or sensitivity of meters and null detectors is inadequate, and for some applications electrical connections cannot be made to the sample without altering its properties or the experimental conditions. For these situations, electrical effects related to resistance are measured, and the resistance is then calculated from theoretical or experimentally determined relationships. The rate of decay of current or charge in inductive or capacitive circuits, determined by the circuit time constants, and the relaxation time of the eddy currents induced in a sample by a changing magnetic field both provide satisfactory measurement techniques but at a considerable inconvenience of complex experimental conditions. *See* EDDY CURRENT; TIME CONSTANT.

Measurement of thin-film resistance. Thin-film resistance is measured in units of ohms per square which is commonly called the square resistance of the film. The concept of square resistance is particularly useful in the evaluation or design of semiconductor materials; deposited, sputtered, or silk-screened components and interconnections; and

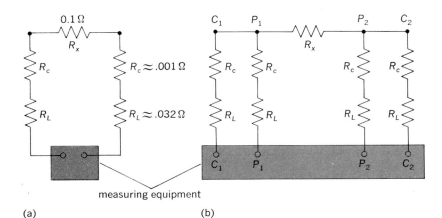

Fig. 2.   Measurement of low resistance. (a) Contact resistance R_c and lead resistance R_L included in measurement of unknown resistance. (b) R_c and R_L eliminated.

printed circuit wiring. For a conductive film having a uniform resistivity throughout its volume and any arbitrary thickness, the square resistance of a sample can be determined by measuring a strip of the film of uniform width. Knife-edge Kelvin potential connections are placed across the strip and spaced the same distance apart as the width of the strip. Kelvin current connections are made outside of the potential contacts and completely across the strip to ensure uniform current density in the sample. Resistance is then measured with either a Kelvin bridge or by the voltmeter-ammeter method. The sample width and the connection configuration determine that the measurement is made on a part of the sample having an arbitrary but unit length and width—a square. That the absolute size of the square does not influence the measurement is illustrated in Fig. 4. Each small block represents a unit square of film having a resistance of 1 ohm/square. Four unit squares connected in the series-parallel arrangement shown also measures 1 ohm, but is a square two units long and wide. The square resistance of physically small samples, for example, the bulk material used for fabrication of semiconductor devices, is measured by using very closely spaced point probes— typically 0.05 in. center to center—for the current and potential connections. Probe geometry and sample uniformity must be carefully evaluated for their effect on uniformity of current density in the sample. *See* INTEGRATED CIRCUITS; PRINTED CIRCUIT.

Measurement of inductive resistance. For dc measurement of inductive resistance, the time constant of the circuit should be considered. The time constant t in seconds is equal to the inductance in henries divided by the resistance in ohms. Inductive resistance is found in relay coils, transformers, chokes, and similar components.

Upon closing the test potential circuit, the current through this type of resistance requires approximately $5t$ to reach 99% of its steady-state value. During this time transient conditions exist, and the sensitive detectors of some circuits may be damaged if the detector is connected into the circuit prior to $4t$ or $5t$. Upon opening the supply circuit, the high voltage generated by the collapsing magnetic field of the inductance can cause serious damage to components by voltage breakdown or high surge currents.

Alternating-current resistance. The ac or effective resistance of a conductor differs from the dc value by an amount that is a complex function of (1) the self-inductance and capacitance of the conductor, (2) the effective mutual inductance and capacitance of nearby conductors and shields, and (3) skin effect. These factors are dependent upon the test frequency and waveform. It is therefore essential that ac resistance measurements be made under carefully and completely specified mounting, connection, and operating conditions.

The simplest equivalent circuit of a conductor is shown in Fig. 5, where R is the true resistance, L_R is the true inductance, and C_R is the self-capacitance of the conductor. The impedance Z of this circuit at a radian frequency ω is given by Eq. (1).

$$Z = \frac{R + j\omega[L(1-\omega^2 CL) - CR^2]}{(1-\omega^2 CL)^2 + \omega^2 C^2 R^2} \qquad (1)$$

If both L and C are small, as in a resistor, the approximate effective resistance R_e (neglecting skin effect) is given by Eq. (2).

$$R_e \approx R[1 + \omega^2 C(2L - CR^2)] \qquad (2)$$

If L is large and C is small, as in an inductance coil, Eq. (3) holds.

$$R_e \approx R[1 + 2\omega^2 CL] \qquad (3)$$

If C is large and L is small, as in a capacitor, Eq. (4) is valid.

$$R_e \approx R[1 - \omega^2 C^2 R^2] \qquad (4)$$

Complication of this simple analysis with the actual physical conditions of mutual capacitance, conductance, and inductance to the earth or to a shield requires a complex solution.

Resistance measurement methods. Commonly used methods of measuring resistance may be classified as either deflection methods or comparison methods. As implied by the names, deflection methods utilize the deflection of ammeters or voltmeters, which may be calibrated in terms of resistance under specific operating conditions, while comparison methods are based upon the use of a calibrated resistor, which can be compared to an unknown resistor. The fundamental deflection method is known as the voltmeter-ammeter method. Basic comparison methods are by potential drop, Wheatstone bridge, and Kelvin bridge.

Voltmeter-ammeter method. This method of measuring resistance is illustrated in Fig. 6. Simultaneous readings of the voltmeter and ammeter are

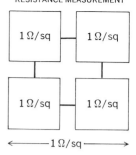

Fig. 4. Square resistance.

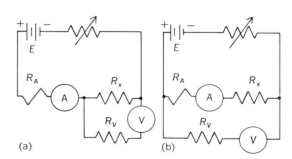

Fig. 6. Voltmeter-ammeter method which is used to measure resistance. (*a*) Least error when R_x is low. (*b*) Least error when R_x is high.

taken, and the unknown resistance R_x is calculated from Ohm's law, as shown in Eq. (5).

$$R_x = V/A \qquad (5)$$

For the voltmeter connection shown in Fig. 6*a*, the true resistance R_x is slightly larger than the calculated value, because the ammeter measures the sum of the currents in R_x and the voltmeter. For the connection of Fig. 6*b*, R_x will be slightly

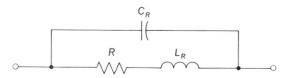

Fig. 5. Simple equivalent circuit of a conductor.

smaller than that calculated, since the voltmeter reading is larger than the potential drop across R_x. For the most accurate measurements, a correction must be applied as required by the connection. The circuit in Fig. 6a causes the least error when R_x is low; the circuit in Fig. 6b causes the least error when R_x is high.

While the voltmeter-ammeter method serves to illustrate a basic principle, the need for simultaneously reading two meters and then calculating the resistance makes the method inconvenient to use, except where resistance must be measured while a circuit is maintained in operation. A practical and widely used simplification based on this method is used in the ohmmeter. *See* OHMMETER.

Comparison by potential drop. This method, accomplished with the circuit of Fig. 7, is a logical

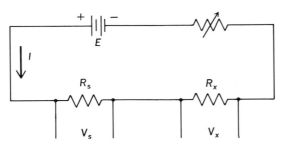

Fig. 7. Resistance comparison by potential drop.

development from the voltmeter-ammeter method. Only one meter is used for the measurement. The unknown resistance R_x is connected in series with a standard resistance R_s, which may be either fixed or adjustable. Since the same current flows in both resistors, by Ohm's law Eqs. (6) hold.

$$I = V_s/R_s = V_x/R_x$$
$$R_x = R_s V_x/V_s \qquad (6)$$

If the standard resistance is fixed, R_x may be determined from the ratio of the potential difference readings multiplied by the value of R_s. If the standard is adjustable, it may be set to that value for which V_x/V_s equals unity; R_x then equals R_s and may be read directly from the R_s setting. The potential difference may be read with either a voltmeter or a potentiometer. Several readings should be made to ensure that the current did not change between successive voltage measurements. This method is seldom used commercially, but it is capable of high accuracy for comparing like-value resistors, using a potentiometer for the potential difference measurement. *See* WHEATSTONE BRIDGE.

Time-constant method for very low resistance. The rate of decay of current in a circuit containing an inductance L and a resistance R (Fig. 8) is an exponential function determined by the time constant of the circuit L/R. When the switch is opened, the instantaneous circulating current i will decay because of the collapse of the magnetic field of the inductor L. At any time t after removing the current source, the current i is given by the relation in Eq. (7), from which Eq. (8) can be written.

$$i = i_0 \exp(-Rt/L) \qquad (7)$$

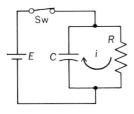

Fig. 9. Circuit for very-high-resistance measurement.

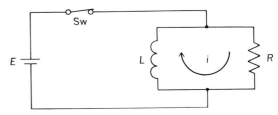

Fig. 8. Circuit for very-low-resistance measurement.

$$R = (L/t)\ln(i_0/i_t) \qquad (8)$$

ten. In one application the unknown resistance is placed in series with a superconducting coil which has zero resistance. A current is introduced into the circuit, the source is removed, and instantaneous current is measured as a function of time. Contact resistance of less than 1 nanohm (10^{-9} ohm) has been measured with this method. Current is calculated from the resulting magnetic field of the inductor, which is measured with a Hall effect sensor. *See* HALL EFFECT.

Time constant method for very high resistance. The rate of decay of charge on a capacitor C shunted with a resistance R (Fig. 9) is an exponential function determined by the time constant of the circuit RC. The "zero loss" capacitor is connected with the unknown resistance R, and the circuit charged to an initial potential E by closing the switch. When the switch is opened, the charge on the capacitor CE will be dissipated slowly by a current i flowing through R, according to the relation in Eq. (9), from which Eq. (10) can be written.

$$i = (e_0/R)\exp(-t/RC) \qquad (9)$$
$$R = (t/C)\ln(e_0/e) \qquad (10)$$

The charge remaining on C at time t is measured with an electrometer voltmeter. Special attention must be given to circuit shielding and guarding; to mechanical stability of the circuit wiring and components, as they affect the circuit capacitance; and to the selection or construction of the highest-quality capacitor and electrometer. *See* ELECTRICAL SHIELDING.

Eddy current method. The resistivity of a material can be measured without making electrical contact with the sample by inducing eddy currents in the sample and measuring their effects on coupled circuits. The simplified schematic of a typical circuit is shown in Fig. 10. The secondary coil S is

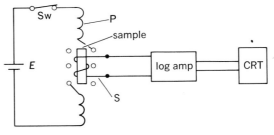

Fig. 10. Typical circuit for resistance measurement by eddy current decay.

tightly coupled to the sample and both placed in the magnetic field of primary coil P, which is energized with a constant dc current while switch Sw is closed. When Sw is opened, the collapse of the primary field will induce a voltage in the sample proportional to the rate of change of flux, and eddy currents will flow in the sample material. The apparatus and sample shape are chosen so that the relaxation time or time constant of the sample current is much longer than that of the primary flux. After the primary flux has fallen to zero, the sample eddy currents induce a measurable voltage in the secondary coil. This voltage will decrease exponentially, according to the relation in Eq. (11).

$$V_t = V_0 \exp(-t/T) \qquad (11)$$

Here T has a value which is directly proportional to the sample permeability and equipment design factors and inversely proportional to sample resistivity. Resistivity can be calculated from the relation in Eq. (12), where K is a constant which

$$\rho = (K/t)\ln(V_0/V_t) \qquad (12)$$

includes the permeability of the sample. Measurements on magnetic samples must therefore be corrected for the effect of permeability to determine resistivity. This procedure measures the dc or very-low-frequency resistivity. By measuring the change of the Q of an inductive circuit with and without a sample in the inductor field, high-frequency resistivity can be measured. These values will increase with increasing frequency because of skin effect in the sample. *See* COUPLING; Q (ELECTRICITY); Q METER.

Resistance standards. High-quality standard resistors are used in comparison methods of resistance measurement. The unknown resistance is measured by comparing it to the accurately known value of the standard resistor. These standards are specially constructed and treated to achieve (1) constancy over long periods of time, (2) a low temperature coefficient, (3) a low internal thermal voltage, and (4) stability under varying humidity conditions. When intended for use in ac circuits, an additional requirement is to reduce the self-inductance, the distributed capacitance, and the skin effect of the resistor to the lowest practical values

so that the standard will have a minimum frequency coefficient.

Because of their high degree of permanence, resistance standards are used as the basis for comparison to determine the limits of error, the precision, and the stability of other types of resistive components. Since the procedures of precision calibration are time-consuming and more exacting than required for the usual engineering determination, resistance standards are often kept and used in the standardizing laboratory to calibrate working instruments, such as bridges, meters, and less accurate working standards.

Primary resistance standards. These are usually supplied as four-terminal, wire-wound or folded-strip, manganin resistors in decade values of resistance covering the range of about 10^{-4} to 10^4 ohms. They are designed for maximum stability as their most important quality, and little attention is given to the ac characteristics of these units. They are usually adjusted to within ±0.01% of their nominal value, and a certificate furnished with each unit shows its deviation from nominal in parts per million at the time of measurement under specified conditions of measurement, such as ambient temperature and power dissipation. After several annual or semiannual certifications have been obtained from either the manufacturer or the National Bureau of Standards, an invaluable record of the average drift and random instability will have been accumulated to serve as a guide for estimating the dependable accuracy of the standard with a high degree of certainty.

The highest-quality resistance standards ever produced are hermetically sealed, wire-wound, 1-ohm resistors designed by J. L. Thomas of the National Bureau of Standards (Fig. 11a). A group of these standards is used to maintain the reference ohm at the Bureau of Standards, and some of these units have changed less than 1 part per million in several years.

Secondary resistance standards. By virtue of more difficult construction problems or less accurate comparison methods, secondary standards cannot be guaranteed to as high a degree of accuracy and stability as primary standards. They may be broadly classified as those having values below 10^{-4} ohm and those above 10^4 ohms, although the

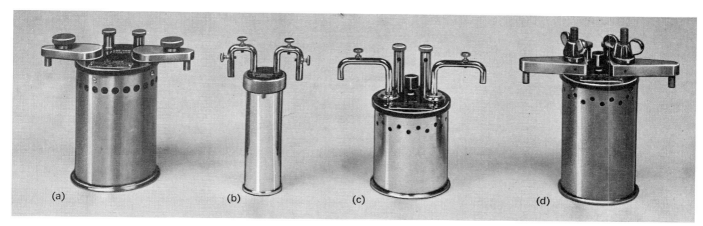

Fig. 11.   Primary resistance standards. (*a*) 1.0-ohm Thomas type. (*b*) 1.0-ohm Rosa type. (*c*) 0.1-ohm Reich- sanstalt type. (*d*) 0.001-ohm Reichsanstalt type. (*Leeds and Northrup Co.*)

Fig. 12. Secondary resistance standard 0.0004 ohm. (*Leeds and Northrup Co.*)

intermediate values may also be constructed with less care than required for primary standards and so be classed as secondary standards.

The low-resistance secondary standards are always of the four-terminal type and are usually made with multiple straight strips of manganin brazed into bus-bar-type current connectors (Fig. 12). This construction is necessary to obtain low resistance values and at the same time provide sufficient cooling surface to dissipate the internally generated heat.

Secondary standards for the higher resistance range (10^4 to 10^7 ohms) are of the wire-wound or woven type. Because of the fine wire required to construct these resistors, they should be hermetically sealed for maximum long-term stability. In addition, the winding supports and the external

cases should be treated with a nonwettable material to reduce surface leakage effects under conditions of high humidity.

For secondary resistance standards above 10^7 ohms, the carbon film, borocarbon film, or metallic-film units are the best available despite their shortcomings of high temperature coefficients, voltage sensitivity, and relative instability. These units must also be hermetically sealed in glass or ceramic tubes, whose surfaces have been treated. Great care must be used in the handling and storing of these units since the surface of the case must not become contaminated with fingerprints, condensed oil, or chemical vapors if these units are to retain their maximum stability.

Resistance decade. This assembly of resistors has a suitable circuit switching device to insert or remove one or more resistors of the circuit and change the total circuit resistance in unit or decade amounts. Although for many applications a resistance decade is used as an adjustable resistance standard, it is fundamentally a standard of resistance change ΔR.

Figure 13 illustrates various switching methods and resistor configurations which have been developed. Regardless of configuration, all resistance decades have a residual and finite "zero" resistance that includes the internal wiring and switch contact resistance. They also have an instability due to the variation in contact resistance. For these reasons, the limits of error are usually stated

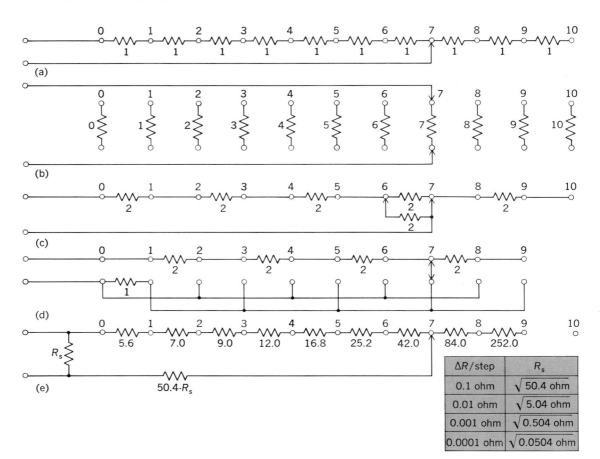

Fig. 13. Resistance decade circuits *a–e.*

in two factors. The first is a function of the error in the resistors themselves; the second is a function of the residual factors. For low resistance values, the residual term may determine the error, while for high resistance values, it usually can be neglected.

Applications. There are two major classifications of the applications of resistance measurement: those in which resistance is measured because the resistance value must be known, and those in which the measured resistance is associated with another physical quantity of interest. The first group would include such measurements as component checking during resistor manufacturing, evaluation of material properties (such as resistivity) for design data, or the testing of equipment for analytical purposes (such as generator winding resistance).

Examples falling in the second area are temperature measurement with resistance thermometers, insulation resistance for equipment condition check, fault location in cables, strain gages, circuit continuity checks, and many others. *See* RESISTOR.

[CHARLES E. APPLEGATE]

Bibliography: M. Braccio, *Basic Electrical and Electronic Tests and Measurements*, 1979; W. D. Cooper, *Electronic Instrumentation and Measurement Techniques*, 1978; F. K. Harris, *Electrical Measurements*, 1952, reprint 1975; P. Kantrowitz et al., *Electronic Measurements*, 1979.

Resistor

A component of an electric circuit that produces heat while offering opposition, or resistance, to the flow of electric current. All conductors exhibit resistance in varying degrees; however, the term resistor is generally used only to describe a device specifically used to introduce resistance into an electric circuit. The unit of resistance measurement is the ohm. Resistors are described by stating their total resistance in ohms along with their safe power-dissipating ability in watts. A more detailed description would specify the residual inductance and stray shunt capacitance of the resistor. *See* ELECTRICAL RESISTANCE.

Classification by use. Resistors may be classified according to the general field of engineering in which they are used.

Power resistors. Such resistors range in size from about 5 watts to many kilowatts and may be cooled by air convection, air blast, or water. The smaller sizes, up to several hundred watts, are used in both the power and electronics fields of engineering.

Instrument resistors. Direct-current (dc) ammeters employ resistors as meter shunts to bypass the major portion of the current around the low-current elements. These high-accuracy, four-terminal resistors are commonly designed to provide a voltage drop of 50 millivolts when a stated current passes through the resistor.

Voltmeters of both the dc and the ac types employ scale-multiplying resistors designed for accuracy and stability. The arc-over voltage rating of these resistors is of importance in the case of high-voltage voltmeters.

Resistors for electronic circuits. By far the greatest number of resistors manufactured are intended for use in the electronics field. The majority of these resistors are intended for use in frequency-selective circuits involving potentials up to several hundred volts but currents seldom over 10–100 milliamperes. Their power-dissipating ability is small, as is their physical size.

Classification by construction. Resistors are also classified according to their construction, which may be composition, film-type, or wire-wound. Further classification may be made according to whether the resistor has a fixed or adjustable resistance. Adjustable resistors may be further classified as adjustable-slide, rheostat, and potentiometer types.

Composition resistor. This resistor is by far the most widely manufactured type because of its low cost, reliability, and small size. Basically it is a mixture of resistive materials, usually carbon, and a suitable binder molded into a cylinder. Copper wire leads are attached to the ends of the cylinder, and the entire resistor is molded into a plastic or ceramic jacket. The overall length of the jacketed resistor excluding the leads is $\frac{1}{3}-1\frac{1}{3}$ in. (8.5–33.9 mm) for resistors varying in power rating from $\frac{1}{4}$ to 2 watts. After manufacture the resistors are automatically zoned according to their individual resistance values, which are indicated on the jacket of the resistor by means of a color code.

Composition resistors are commonly used in the range from several ohms to 10–20 megohms, and are available with tolerances of 20, 10, or 5%. Higher values of resistance are available but are not normally used in communication equipment or most electronic instruments. For very high resistance values, above about 100 megohms, special jacketing is often required to prevent the leakage resistance over the surface of the jacket from altering the overall resistance of the resistor.

All resistors possess a finite shunt capacitance across their terminals. This capacitance is a function of the geometry and physical size of the resistor and is essentially independent of the value of the resistor. The result is that at higher frequencies each resistor is effectively paralleled by a capacitive reactance which decreases in magnitude with increasing frequency. At approximately 100 kHz the overall impedance of a 1-megohm resistor begins to decrease with an increase of frequency. A 1000-ohm resistor will not display this effect until a frequency of about 100 MHz is reached.

The wattage rating of resistors is normally based upon the amount of thermal drift in resistance that can be tolerated. If greater thermal stability is required, the designer should use resistors with a higher power rating. Composition resistors are easily damaged permanently by overheating. Because of this, care must be exercised when soldering a resistor into a circuit.

Film-type resistor. This resistor is rapidly replacing the composition resistor in applications in which greater stability of resistance with voltage, temperature, and humidity is demanded. The design of the film resistor further lends itself to the controlled manufacture of precision resistors of any desired value. Basically this resistor consists of a conducting film of carbon, metal, or metal oxide deposited upon a ceramic cylinder. The value of the resistance is controlled by controlling the thickness and length of the film. The length of the film is often controlled by cutting a spiral groove around the resistor, the groove passing through the film to the insulating cylinder. This spiral groove increases the effective length of the resistor and thereby determines its ohmic value. By accurately controlling the pitch of the spiral the manufacturer can make a resistor of any value and can maintain close manufacturing tolerances.

The film resistor is often finished by coating it with an insulating varnish. Often a plastic sleeve is slipped over the resistor to provide mechanical protection. The spiral-cut resistor displays a small inductive effect at the higher frequencies.

Wire-wound resistors. Wire remains the most stable form of resistance material available; therefore all high-precision instruments rely upon wire-wound resistors. Wire-wound resistors are available in resistance ratings from a fraction of an ohm to several hundred thousand ohms, at power ratings from less than 1 watt to several thousand watts, and at tolerances from 10 to 0.1%. Because mechanical manufacturing problems limit the smallest wire size that can be used, these resistors are usually limited to values below about 100 kilohms. Both inductive and noninductive types of resistor are manufactured.

The inductive design is the common construction and consists of a spiral winding of wire about a cylindrical ceramic form. After winding, the entire resistor is covered with a vitreous material. The spiral winding introduces a considerable amount of inductance into the circuit, which may become objectionable at the higher audio frequencies and all radio frequencies.

The noninductive design includes several winding methods. One of the simplest and most satisfactory is to reduce the cross-sectional area of the coil by winding the wire around a thin, flat card.

Adjustable resistors. The deposited-film and wire-wound resistors lend themselves to the design of adjustable resistors or rheostats and potentiometers. Adjustable-slider power resistors are constructed in the same manner as any wire-wound resistor on a cylindrical form except that when the vitreous outer coating is applied an uncovered strip is provided. The resistance wire is exposed along this strip and a suitable slider contact can be used to adjust the overall resistance, or the slider can be used as the tap on a potentiometer. *See* POTENTIOMETER (VARIABLE RESISTOR).

Where continuous adjustment of the resistor is intended, a ring-shaped form is generally used. For power resistors the ring is wound with resistance wire. For compact 1/2- and 1-watt resistors, the ring is coated on one surface with a resistance film. Each type possesses all the advantages and disadvantages described above under fixed-value resistors of its type. In addition, adjustable resistors have the problem of maintaining a good, noise-free, electrical contact at the wiper, which is mounted on a shaft concentric with the ring.

For discussion of nonlinear resistors *see* VARISTOR.   [ROBERT L. RAMEY]

Bibliography: D. G. Fink and D. Christiansen (eds.), *Electronic Engineers' Handbook*, 2d ed., 1982; C. A. Harper, *Handbook of Components for Electronics*, 1977; Howard W. Sams Engineering Staff, *Reference Data for Radio Engineers*, 6th ed., 1975.

Resonance (alternating-current circuits)

A condition in a circuit characterized by relatively unimpeded oscillation of energy from a potential to a kinetic form. In an electrical network there is oscillation between the potential energy of charge on capacitance and the kinetic energy of current in inductance. This is analogous to the mechanical resonance seen in a pendulum.

Three kinds of resonant frequency in circuits are officially defined. Phase resonance is the frequency at which the phase angle between sinusoidal current entering a circuit and sinusoidal voltage applied to the terminals of the circuit is zero. Amplitude resonance is the frequency at which a given sinusoidal excitation (voltage or current) produces the maximum oscillation of electric charge in the resonant circuit. Natural resonance is the natural frequency of oscillation of the resonant circuit in the absence of any forcing excitation. These three frequencies are so nearly equal in low-loss circuits that they do not often have to be distinguished.

Phase resonance is perhaps the most useful in many practical situations, as well as being slightly simpler mathematically. The following discussion considers phase resonance in passive, linear, two-terminal networks.

Resonance can appear in two-terminal networks of any degree of complication, but the three circuits shown in Fig. 1 are simple and typical. The first illustrates series resonance and the second,

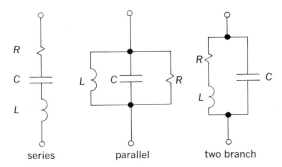

Fig. 1. Resonant circuits.

parallel resonance; the third is a series-parallel resonant circuit of two branches (sometimes referred to as antiresonance). Series resonance is highly practical for providing low impedance at the resonant frequency. Parallel resonance is the dual of series resonance, but it is not practical because it assumes an inductive element with no resistance. The third example, however, shows an eminently practical means of providing the typical characteristic of parallel resonance, which is high impedance at the resonant frequency.

Use. Resonance is of great importance in communications, permitting certain frequencies to be passed and others to be rejected. Thus a pair of telephone wires can carry many messages at the same time, each modulating a different carrier frequency, and each being separated from the others at the receiving end of the line by an appropriate arrangement of resonant filters. A radio or television receiver uses much the same principle to accept a desired signal and to reject all the undesired signals that arrive concurrently at its antenna; tuning a receiver means adjusting a circuit to be resonant at a desired frequency.

Many frequency-sensitive circuits are not truly resonant, and oscillations of a certain frequency can be produced or enhanced by networks that do not involve inductance. It is difficult and expensive to provide inductance with integrated circuits, but frequency selection can be provided by the use of capacitance and resistance, a large amount of amplification being obtained from the semiconductor material employed.

Series resonance. Figure 2 shows a phasor diagram of the voltage, resulting from a given current (steady alternating current) in a series-resonant circuit, such as shown in Fig. 1. The component voltages across the three circuit elements add to give the total applied voltage V, as shown for a frequency slightly above resonance, for the resonant frequency, and for a frequency below resonance. It is of course possible in low-loss (high Q) circuits for the voltage across the capacitance and the voltage across the inductance each to be many times greater than the applied voltage.

Analytically, the impedance of the series-resonant circuit is given by Eq. (1). The resonant frequency f_0 is the frequency at which Z is purely real (phase resonance), so $\omega_0 L = 1/\omega_0 C$, or $2\pi f_0 L = 1/2\pi f_0 C$, from which Eq. (2) obtains.

$$Z = R + j\omega L + \frac{1}{j\omega C} = R + j\left(\omega L - \frac{1}{\omega C}\right) \quad (1)$$

A more convenient notation is expressed by Eq. (3).

$$f_0 = \frac{1}{2\pi\sqrt{LC}} \quad (2)$$

$$Z = R_0\left(\frac{R}{R_0} + jQ_0\delta\frac{2+\delta}{1+\delta}\right) \quad (3)$$

In Eqs. (1)–(3):

Z = impedance at the terminals of the series-resonant circuit
R, L, C = the three circuit parameters
R_0 = resistance (effective) at resonant frequency
Q_0 = $\omega_0 L/R_0$
δ = $(\omega - \omega_0)/\omega_0$
ω = $2\pi f$, where f is frequency (hertz or cps)
ω_0 = $2\pi f_0$, where f_0 is resonant frequency

Equation (3) is true for all series-resonant circuits, but interest is mainly in circuits for which Q_0, the quality factor at the resonant frequency, is high (20 or more) and for which δ, the fractional detuning, is low (perhaps less than 0.1). Assuming high Q_0 and low δ, which means a low-loss circuit and a frequency near resonance, Eq. (4) is very nearly the relative admittance of the series-resonant circuit.

$$\frac{Y}{Y_0} = \frac{Z_0}{Z} = \frac{1}{1 + j2Q_0\delta} \quad (4)$$

Universal resonance curve. The magnitude and the real and imaginary components of Eq. (4) are usefully plotted in the universal resonance curve of Fig. 3. Since Y/Y_0 is plotted as a function of $Q_0\delta$, this curve can be applied to all series-resonant circuits. (If $Q_0 = 20$, the error in Y barely exceeds 1%

of Y_0 for any δ, and is less for small δ.)

Moreover, because of the duality of the network, the curve can also be applied to any parallel-resonant circuit (Fig. 1) provided Q_0 is now interpreted as $Q_0 = R_0/\omega_0 L$. When used for a parallel-resonant circuit, the curve of Fig. 3 gives not Y/Y_0 but the relative input impedance Z/Z_0.

Finally, the universal resonance curve of Fig. 3 can also be applied (with the same slight approximations) to the two-branch resonant circuit of Fig. 1. For this purpose the curve shows Z/Z_0 (as for the

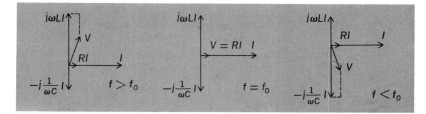

Fig. 2. Phasor diagrams at frequencies near resonance ($Q = 5$).

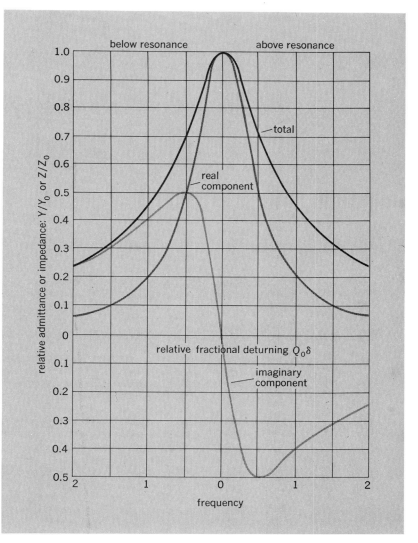

Fig. 3. Universal resonance curve. (*From H. H. Skilling, Electrical Engineering Circuits, 2d ed., copyright © 1965 by John Wiley and Sons, Inc.; used with permission*)

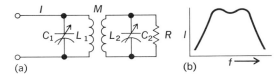

Fig. 4. (a) Resonance in a double-tuned network. (b) Current in R as function of frequency.

three-branch parallel-resonant circuit), but the value of Q to be used is $Q_0 = \omega_0 L/R_0$, exactly as with the series-resonant circuit. Note that Z_0 for this circuit is given by Eq. (5) (instead of being

$$Z_0 = (\omega_0 L)Q_0 = R_0 Q_0{}^2 \qquad (5)$$

equal to R_0 as it is in the other two circuits of Fig. 1).

Multiple resonance. If two or more coupled circuits are resonant at slightly different frequencies, many valuable characteristics can be obtained. Figure 4 shows a double-tuned network and a typical curve of current in R, the load, as a function of frequency. *See* ALTERNATING-CURRENT CIRCUIT THEORY; NETWORK THEORY.

[HUGH HILDRETH SKILLING]

Bibliography: D. Bell, *Fundamentals of Electric Circuits*, 2d ed., 1981; J. R. Duff and M. Kauffman, *Alternating Current Fundamentals*, 1980; Institute of Electrical and Electronics Engineers, *IEEE Standard Dictionary of Electrical and Electronics Terms*, 2d ed., 1977.

Response

A quantitative expression of the manner in which a microphone, amplifier, loudspeaker, or other component or system performs its intended function. A linear response means that the output signal is exactly proportional to the input signal for the entire range of frequencies over which the device is intended to operate. A logarithmic response means that the output signal is a logarithmic function of the input signal. The response of a device is often presented as a curve on a graph, indicating deviation over the frequency range from the response at some selected frequency, such as 1000 Hz. An example is the frequency-response curve of an amplifier. *See* AMPLIFIER; CHARACTERISTIC CURVE. [JOHN MARKUS]

Ripple voltage

The total voltage across the load resistor of a rectifier minus the average voltage across the same resistor. The ripple can be expressed as a Fourier series. The fundamental frequency of the ripple voltage of single-phase half-wave circuits is the same as that of the alternating-current (ac) input. For a single-phase full-wave circuit the fundamental frequency is twice that of the ac input voltage, while for a three-phase half-wave circuit the fundamental frequency is three times that of the ac supply. To reduce the ripple voltage, a low-pass filter is usually placed between the rectifier and the load. The filter is more efficient in reducing the ripple voltage if the fundamental frequency of the ripple voltage is high. *See* ELECTRIC FILTER; RECTIFIER.

[DONALD L. WAIDELICH]

Robotics

The study of problems associated with the design, application, and control and sensory systems of robots. Historically, the term robot has been used loosely, and has been applied to almost any feedback-controlled mechanical system. While the exact usage of the term is a matter of preference, the computer-controlled mechanical arm used in industrial applications probably represents a reasonable middle ground for definition by example (Fig. 1). Most concerns of practitioners of robotics are (or can be) involved in such devices, and so much of the work in robotics relates directly to these devices.

Robotics is also a broadly interpreted term. Most workers in the field would agree that it covers research and engineering activities relating to the design and construction of robots, but persons engaged in planning robot manufacturing or in studying the economic impact of robots might also consider themselves to be engaged in robotics. Much of this breadth of usage arises from the fact that robotics is a highly interdisciplinary field. Mechanical engineering, computer science, artificial intelligence, biomechanics, control theory, cybernetics, and electrical engineering are only a few of the many fields which have direct application to robotics.

The emergence of robotics as a separate discipline has been given impetus by the developing complexity of robot systems, which has emphasized the interdependence of the design of the mechanical, electronic, and computational aspects of robots. Previously it was possible for mechanical engineers to view a robot as just another numerically controlled machine tool, and for computer scientists to regard it as just another peripheral device. Although this viewpoint is possible for many simpler forms of robots, it is no longer a workable approach for the current complex, hierarchically controlled, sensory-interactive robots. In typical robotics research laboratories, specialists in many different areas work cooperatively on an integrated robot system.

Mechanical design. Almost all robots produce some sort of mechanical motion. In most cases this serves the purpose of manipulation or locomotion. For example, robot arms manipulate tools and parts to perform jobs such as welding, painting, and assembly; and robot carts are used to transport materials. The mechanical design of robots is thus of great importance in robotics. Areas of concern in the mechanics of robots include degrees of freedom of movement, size and shape of the operating space, stiffness and strength of the structure, lifting capacity, velocity, and acceleration under load. In addition, good mechanical design is a factor in other performance measures, such as accuracy and repeatability of positioning, and freedom from oscillation and vibration.

Some robots have very simple mechanical designs, involving only a few degrees of freedom of movement. However, the design of robot manipulators can also be quite complex. In a typical industrial robot arm, six degrees of freedom of movement (exclusive of gripper closure) are required to enable the gripper to approach an object from any orientation. This is usually accomplished with three arm joints, which can position the wrist at any x, y, z coordinate in the working volume. The complex three-axis wrist joint can then orient the gripper mounted on it by three independent motions (roll, pitch, and yaw) around the x, y, z location of the wrist (Fig. 2). In effect, the wrist represents the origin of a three-axis coordinate system fixed to the gripper. Moving the first three joints of the arm translates this origin to any point in a three-axis coordinate system fixed to the working volume; motion of the final three joints (in the wrist) orients the gripper coordinate system in rotation about its origin at the wrist point.

Both sliding joints and rotational joints may be included in the robot's articulation (joint structure). Many robots use only rotational joints, as does the human arm, but only limited actions can be produced by using sliding joints alone (Fig. 1). Robot mobility usually involves adaptations of traditional devices such as wheels and treads, but walking robots, usually in some stable configuration such as a six-legged hexapod, have also been developed. The mechanical design problems of articulated legs are similar to those of robot arms.

The problem of powering the robot's joints is made more difficult by the complex mechanical articulation. One approach is to place a prime mover (electrical, hydraulic, or pneumatic) at the joint itself. Power for these movers can be brought with relative ease through the joints and members of the arm. However, the weight and bulk of such motors and their associated gearing place constraints on the performance and mechanical design of the arm, particularly of joints in the wrist. A second approach is to place the prime movers in the immobile base of the robot and to transmit motion to the joints through mechanical linkages such as shafts, belts and cables, or gearing. This overcomes many of the problems associated with the first approach, but introduces a new set of problems in designing intricate, backlash-free mechanical linkages which can transmit power effectively through the complex articulations of the arm in all of its positions. No single approach has clearly dominated the field.

The purpose of the elaborate mechanical arm is to position an end effector (frequently a gripper) where it can perform some useful function. End effectors may be highly specialized for particular applications, or may be simple, general-purpose pincers. Some robots change their own end effectors to suit the job at hand from a selection of special-purpose attachments. The development of more elaborate and dexterous general-purpose end effectors, including hands with humanlike fingers, is an area of intense study. However, the mechanical and control problems associated with such effectors are exceedingly complex.

Control systems. A robot control system is the apparatus (usually electronic) which directs the activities of the mechanical parts. This may consist of only a sequencing device and a set of mechanical stops, so that the mechanism moves in a repetitive pattern between selected positions. However, more sophisticated systems employ servo-controlled positioning of the joints, and a measure of the actual joint position is obtained from a transducer, such as an optical shaft encoder, and is compared with the position specified

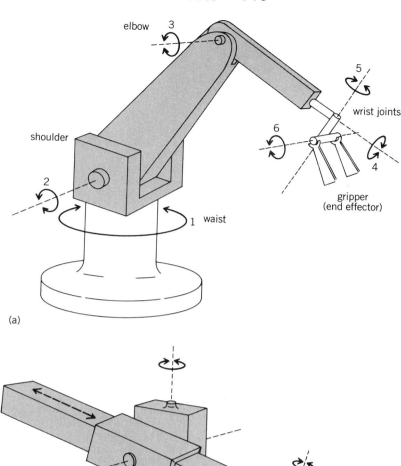

(a)

(b)

Fig. 1. Six-degree-of-freedom robot arms. (a) Arm with all rotational joints (numbered 1 to 6). (b) Arm with one sliding joint.

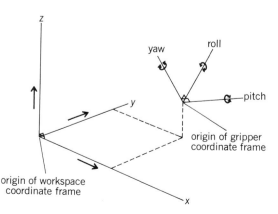

Fig. 2. Coordinate relations for a six-degree-of-freedom robot.

for the joint. If the desired position and the measured position differ, the circuitry applies a correcting drive signal to the joint motor (Fig. 3). Such servo systems may be digital or analog, and in addition to position, they may control joint velocity. Servo devices allow the robot to be moved through any selected sequence of positions on command, without the necessity of presetting mechanical stops. If the appropriate sequence of commands is generated, servo-controlled joints may be driven through continuous, smoothly varying paths. The sequence of positions defining the robot's trajectory may be preprogrammed by numerically specifying positions in the robot's coordinate frame, or they may be "taught" to the system by moving it to the desired point and recording the output of the joint-position transducers. Such teaching methods allow a human to direct a complex action once, and to have the robot repeat it indefinitely thereafter.

More sophisticated systems generate the robot's trajectory automatically by computer. Such computations may be based on mathematical descriptions of work objects or tasks contained in the computer's data base. The computer may also generate trajectories for the robot which are not fixed but vary with the state of the external world as reported by the robot's sensory system. This sensory-interactive type of control permits the robot to act appropriately in relation to conditions, rather than relying on assumptions about the world. For example, without expensive fixturing and timing of the work, the actual location of a part may vary from one instance to the next; in such cases, without sensory-interactive control, the robot could proceed blindly through a set of actions at preprogrammed, but incorrect, positions.

The most advanced robot control systems make use of hierarchical control. Each level of a hierarchy of control stages accepts, from its superior level, a statement of a goal to be achieved. The complexity of this goal will depend on the position of the level in the hierarchy; the lowest level is the joint-position servo, where the goal is simply the next position commanded for that joint. Higher levels attempt to achieve their current goal by issuing sequences of commands (subgoals) to their subordinate levels. In selecting these subgoals, each level takes into account its own goal, its sensory input describing the state of the external world, and the status of its subordinate level. Thus, each level of the hierarchy acts as a servo control on the actions of the next lower level, giving it commands to reduce the difference between the current state of events and that defined by its own goal (Fig. 4). These systems are sensory-interactive and constitute a task-decomposition hierarchy similar to that of many functions in the human nervous system. They allow the robot to be instructed with very general commands at the highest level.

Sensory systems. The purpose of a robot sensory system is to gather specific information needed by the control system and, in more advanced systems, to maintain an internal, predictive model of the environment. The joint-position transducers used in feedback control are in fact a minimal sensory system, but other sensors are usually included to gather data about the external environment. Visual, proximity, tactile, acoustic, and force or torque senses are all used.

Tactile (touch) sensors may be mounted in the robot gripper to detect contact with objects. These may take the form of simple switches, or they may be analog transducers indicating degree and direction of pressure. Arrays of transducers may be used to give a sense of patterned pressure, which enables the robot to discriminate types and orientations of objects. Force and torque sensors, frequently mounted in the robot's wrist, are used to sense the degree and direction of resistance encountered by the gripper. These resistive forces may be due to the weight of the object being manipulated, or to contact with other objects or surfaces. Such sensors are used to adjust gripping pressure, to avoid applying destructive forces, and to guide proper mating of surfaces and parts. In combination, these senses allow a robot to feel the proper fit of work parts much as a human worker does.

The most commonly used means of sensing objects at a distance is some form of visual sense. Usually this is done by computer analysis of an image from a television camera. Two important approaches are ambient light systems and structured light systems. Ambient light systems rely on normal sources of scene illumination, while structured light systems provide special patterns of illumination whose shape and orientation are known to the sensory system. The advantage of structured light is that the special illumination patterns may be chosen to simplify and speed up the processing required to interpret the image. Speed is important in robot vision because visual information is used by the control system to correct the robot's movements in real time, a process called visual servoing.

Whatever the system of illumination used, the techniques of computer image processing are similar. At low levels, they include algorithms to do thresholding, and to find lines, edges, corners, and connected regions. Procedures for determining depth in images rely fundamentally on triangulation, but in structured light systems triangulation is between the camera and the light projector, whereas in other systems corresponding points in two images from different viewing positions are

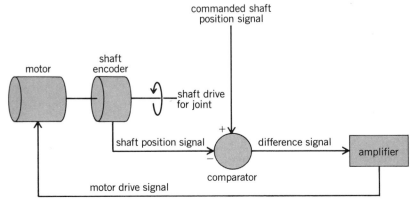

Fig. 3. Servo control for a single robot joint.

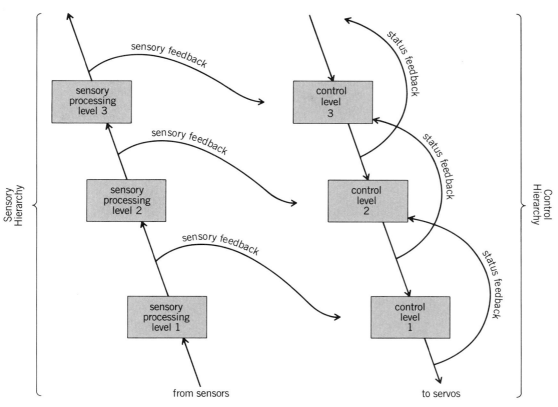

Fig. 4. Complex, hierarchical, sensory-interactive robot control system.

triangulated. These low-level vision processes are sufficient to perform many fundamental visual servoing operations. More complex systems include subsequent image-understanding operations which allow the robot to identify objects and to determine their orientation; this requires the robot to have some sort of previous knowledge about the kinds of objects in its environment. Such a knowledge base is used, together with the sensory data from vision (perhaps combined with other senses), to generate a world model describing the state of the environment.

The robot's world model can be continually updated on the basis of new sensory input, and also on the basis of expectations about how the view will be transformed by movements in progress. The world model in turn aids in the interpretation of incoming sensory data. The most advanced robot sensory systems are multimodal (based on several senses) and are also hierarchically structured to generate successive levels of description, at increasing degrees of complexity and decreasing rates, suitable for use by the successive levels of a hierarchical control system (Fig. 4).

Emerging areas. As robots become more complex, they quickly exhaust the capacity of traditional computers. Computer science in robotics is focusing on the design of parallel multicomputer systems, and ultimately may evolve special-purpose hardware for robot "brains." The growth of robotics technology is also producing new disciplines concerned with the utilization, economics, and societal impact of robots. As robotics matures, these concerns can be expected to assume a major role in the field. *See* ARTIFICIAL INTELLIGENCE; COMPUTER-AIDED DESIGN AND MANUFACTURING; DIGITAL CONTROL.   [ERNEST KENT]

Bibliography: J. S. Albus, A. J. Barbera, and R. N. Nagel, Theory and practice of hierarchical control, *23d IEEE Computer Society International Conference*, pp. 18–39, September 1981; W. B. Gevarter, *An Overview of Artificial Intelligence and Robotics*, vol. 2: *Robotics*, NBSIR 82-2479, U.S. Department of Commerce, March 1982; R. P. Paul, *Robot Manipulators: Mathematics, Programming, and Control*, 1981.

Rochelle salt

The sodium potassium salt of the *d*-tartaric acid $NaKC_4H_4O_6 \cdot 4H_2O$, also called Seignette salt, the first crystalline solid discovered to possess the properties of ferroelectricity. Such crystals are grown easily and have been widely used, for example, in microphones and phonograph pickup cartridges, because of their large piezoelectric effect. *See* PIEZOELECTRICITY.

The crystal structure has orthorhombic symmetry (point group 222) above $+24°C$ and below $-18°C$. Between these Curie temperatures the crystal is ferroelectric, having a spontaneous polarization along the **a** axis and a spontaneous shear deformation y_z in the (100) plane. This reduces the symmetry to monoclinic. In general, the crystal consists of ferroelectric domains of opposite polarization direction. By applying an electric field along the **a** axis or a shear stress Y_z the spontaneous polarization can be aligned and reversed (hysteresis). Its highest value of 2.5×10^{-3} coulomb/m² is reached at about +3°C. The dielectric constant shows peak values of several thousand at the Curie points. It drops, according to the Curie-Weiss law, on both sides of the ferroelectric temperature range. *See* FERROELECTRICS.

For technical applications the anomalously large effects are impaired by the narrow temperature range of ferroelectric behavior and by the limited stability of the crystals with respect to temperature and humidity.   [H. GRANICHER]

Bibliography: W. G. Cady, *Piezoelectricity*, 1964; M. E. Lines and A. M. Glass, *Principles and Applications of Ferroelectrics and Related Materials*, 1977; T. Mitsui, *An Introduction to the Physics of Ferroelectrics*, 1976.

Sampling voltmeter

A special class of voltmeter that uses sampling techniques and is particularly effective in measuring high-frequency signals or signals mixed with noise. High-frequency sampling instruments can operate on signals with frequencies as high as 12 gigahertz (GHz) and amplitudes as small as 1 millivolt (mv). Measurements are generally of average-absolute or root-mean-square voltage with accuracies of 5–10%. Sampling voltmeters, as conventional voltmeters, may use scale and pointer meters, graphic recorders, cathode-ray tubes, or digital-type indicators for readout of measured quantities. *See* VOLTAGE MEASUREMENT.

Sampling technique. The sampling technique detects the instantaneous value of an input signal at prescribed times by means of an electronic switch connecting the signal to a memory capacitor. Waveforms of the input signal and the sampled signal that appears across the memory capacitor are shown in simplified form in Fig. 1. At the points in time indicated by the heavy dots on the input signal voltage waveform, the switch is closed for a very short interval and the memory capacitor is thereby briefly connected to the signal. The connection generates a sample; that is, the capaci-

tor charges to a voltage proportional to the input voltage at the instant of sampling. After a sample the capacitor voltage remains constant until the next sample is taken. The sampled voltage from this type of sample-and-hold circuit is an approximate replica of the signal waveform assembled bit by bit and therefore at a much lower frequency. The input signal frequency F_I, the output sampled voltage frequency F_O, and the frequency of sampling F_S are related by $F_I - NF_S = F_O$, where N is a whole number closest to the number of cycles of input signal between samples (two in the waveforms above). By proper adjustment of frequencies and by filtering, the sampled signal waveform may be made to be an extremely exact waveform replica of the input signal. Since the reproduced waveform is on a different time scale than the original, the type of sampling described above is sometimes called equivalent-time sampling. Equivalent-time sampling is analogous to the apparent slowing of motion of rotating machinery when viewed with stroboscopic light. In that case the eye is briefly exposed to the high-frequency rotating object by the flash of light that illuminates the object approximately every N rotations. The after-image characteristic of the eye serves to hold the brief image or sample until the next sample is produced. If the flashing rate is made exactly equal to a subharmonic of the rotation rate ($NF_S = F_I$), motion of the object appears to be stopped. In the electrical case this corresponds to sampling the input signal at the same point in its cycle, with every sample producing a constant (dc) output sampled voltage. If NF_S is made greater than F_I, the rotating object appears to move backward and an electrical sampled voltage waveform will also appear reversed in time. *See* CAPACITOR.

For instrumentation purposes the principal advantage of equivalent-time sampling is that the low-frequency sampled output voltage is a replica of the high-frequency input and it can be measured by means that, because of economic or technologi-

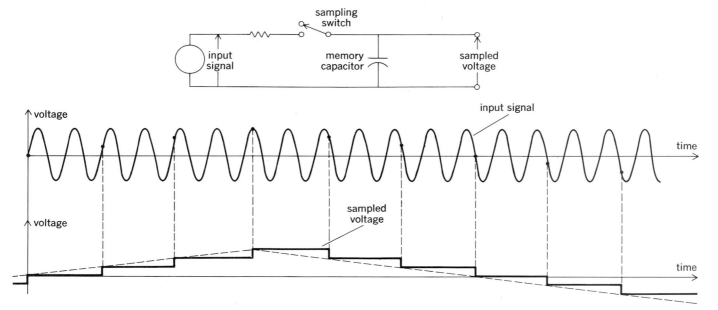

Fig. 1. Simplified sampling circuit and waveforms.

cal limitations, cannot be extended in frequency response to directly measure the input signal.

High-frequency sampling voltmeter. High-frequency voltmeters consist of the sample-and-hold circuits described in principle above, conventional voltmeter circuits, and synchronizing circuitry to adjust the sampling frequency to a value that maintains the sampled signal frequency within the proper range for the voltmeter circuits. The vector voltmeter described below synchronizes the **sampling** frequency by sweeping it until the sampled **output** is at the proper frequency. The sampling rate is then maintained at the correct value by means of a phase-lock servomechanism or feedback system that compares the output signal phase to the phase of a reference signal and uses the phase error to maintain the proper sampling rate. *See* VOLTMETER.

The sampling gate must be extremely fast to measure high-frequency signals. It must connect the input signal and the memory capacitor for a time interval that is short compared to the minimum period of the input signal. Highly specialized techniques are used to generate gate durations as short as 30 picoseconds (30×10^{-12} sec). Light, which can travel to the Moon in about 1.5 sec, moves only 0.9 cm in 30 picoseconds.

Vector voltmeter. The vector-type voltmeter is a two-channel high-frequency sampling voltmeter that measures phase as well as voltage of two input signals of the same frequency. The two signals are sampled at the same instant of time to produce corresponding sampled waveforms that have the same phase relationship as do the input signals. Since both input signals must have the same frequency to make phase measurements meaningful, only one phase-lock synchronizing circuit is needed.

Random-sampling voltmeter. This class of sampling voltmeter takes advantage of random sampling to simplify or eliminate the synchronizing portions of the instrument. If samples of an input signal are taken at random times instead of at a controlled constant rate, the result is loss of waveform information in the output. However, the amplitude statistics are preserved; that is, the average, peak, root-mean-square, and other values of a random-sampled signal are the same as those of the input signal. The effect is as though the individual samples of a regularly sampled signal were stored and then replayed in random time order. The randomness in time of sampling does not have to be exact to preserve amplitude statistics. Simple variations such as low-frequency modulation of the sampling rate are satisfactory in most cases. Random sampling requires no information on input or sampled signal frequency, and synchronizing circuits are therefore not required.

Noise reduction techniques. Conventional sampling techniques are limited to operation only on periodic input signals. Techniques that recognize this periodicity and thereby reduce the effects of noise upon a signal may be employed in voltage measurement. Instruments that use signal enhancement techniques generally operate at frequencies of less than 1 MHz and have relatively long gate durations. The sampling circuits may operate with sampling gate alternately connected to the input signal for one-half cycle and then disconnected for one-half cycle as shown in Fig. 2. During

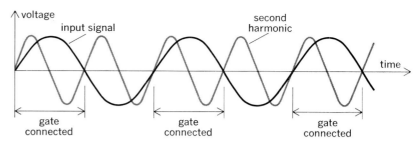

Fig. 2. Waveforms and timing for synchronous rectifier.

the connected intervals the input signal is integrated or, more properly, averaged. When gate and signal are related as shown, operation is similar to that of a half-wave rectifier circuit, and a large average results. Odd harmonics of the signal produce an average whose amplitude diminishes inversely as the harmonic number. Even harmonics such as the second harmonic shown have an average value of zero during the gate duration and therefore ideally produce no response. Most noise signals, including white noise, which has a symmetrical gaussian amplitude distribution, produce an average near zero for long averaging times. Signals that are not of the same basic frequency as the gating frequency also tend to produce small average values. The synchronous rectifier thus responds to a signal of the proper frequency and its odd harmonics, but rejects even harmonics and most noise. This narrow-band property improves small-signal capabilities of synchronous-rectifier instruments. A synchronous-rectifier type voltmeter may measure signals in the microvolt amplitude range with 3–10% accuracies even when these signals are accompanied by noise with an amplitude up to 10 times the signal amplitude. Extensive synchronizing circuits such as those described for high-frequency sampling voltmeters are used to capture and lock the sampling rate to the signal frequency.

A variety of sampling systems are available for signal enhancement; some systems with other than half-wave gating are capable of full waveform recovery.

[JAMES K. SKILLING]

Bibliography: A. B. Carlson, *Communication Systems*, 1975; B. P. Lathi, *Signals, Systems and Controls*, 1974; M. Schwartz, *Information Transmission Modulation and Noise*, 2d ed., 1970.

Saturation

The condition in which, after a sufficient increase in a causal force, further increase in the force produces no additional increase in the resultant effect. Many natural phenomena display saturation. For example, after a magnetizing force becomes sufficiently strong, a further increase in the force produces no additional magnetization in a magnetic circuit; all of the magnetic domains have been aligned, and the magnetic material is saturated.

After a sponge has absorbed all the liquid it can hold, it is saturated. In thermionic vacuum tubes thermal saturation is reached when further increase in cathode temperature produces no (or negligible) increase in cathode current; anode saturation is reached when further increase in plate voltage produces substantially no increase in anode current. *See* DISTORTION (ELECTRONIC CIRCUITS); VACUUM TUBE.

In an *npn* transistor when the collector voltage becomes more negative than the base voltage, the base-to-collector diode becomes forward-biased; the collector then emits and the transistor is saturated. Current flow is controlled by a stored charge of carriers in the base region. *See* SEMICONDUCTOR; TRANSISTOR.

In colorimetry the purer a color is, the higher its saturation. Radiation from a color of low saturation contains frequencies throughout much of the visible spectrum.

In an induced nuclear reaction, saturation exists when the decay rate of a given radionuclide is equal to its rate of production. In an ionization chamber, saturation exists when the applied voltage is high enough to collect all the ions formed by radiation but not high enough to produce ionization by collision. When addition of the dissolved species produces no further increase in the concentration of a solution at a certain temperature and pressure, the solution is said to be saturated.

[FRANK H. ROCKETT]

Saturation current

A term having a variety of specific applications but generally meaning the maximum current which can be obtained under certain conditions.

In a simple two-element vacuum tube, it refers to either the space-charge-limited current on one hand or the temperature-limited current on the other. In the first case, further increase in filament temperature produces no significant increase in anode current, whereas in the latter a further increase in voltage produces only a relatively small increase in current. *See* VACUUM TUBE.

In a gaseous-discharge device, the saturation current is the maximum current which can be obtained for a given mode of discharge. Attempts to increase the current result in a different type of discharge. Such a case would be the transition from a glow discharge to an arc discharge. *See* ELECTRICAL CONDUCTION IN GASES.

A third case is that of a semiconductor. Here again the saturation current is that maximum current which just precedes a change in conduction mode. *See* SEMICONDUCTOR.

[GLENN H. MILLER]

Sawtooth-wave generator

A device which generates a current or voltage waveform whose magnitude is a continuously increasing function of time for a fixed interval, and then repeats the sequence periodically. The most widely used sawtooth waveform is ideally a linear function of time during the forward or rising interval and appears as shown in Fig. 1a, with the total period T made up of the active forward interval T_f, a retrace interval T_r, and an inactive interval T_i. Often it is desirable to make the two intervals T_r and T_i as small as possible, with T_i often being zero.

Mathematically the sawtooth wave may be expressed in terms of a Fourier series of harmonically related components with the fundamental having a period equal to the total period T of the sawtooth. If a sufficient number of harmonics hav-

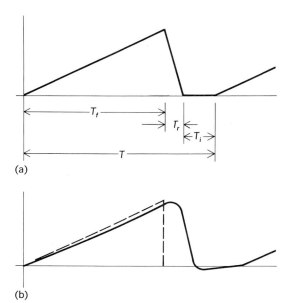

(a)

(b)

Fig. 1. Sawtooth wave. (*a*) An ideal linear sawtooth. (*b*) Approximate sawtooth generated by actual circuits.

ing the proper amplitude and phase relationships is included, the mathematical representation will be an accurate approximation of the waveform.

Electronic circuits can generate only an approximation of the idealized waveform which, as a result, tends to have deviations which are often of the type shown in Fig. 1b. Generally the delayed start (with respect to the ideal) and the extended retrace time are caused by high-frequency deficiencies (inadequate generation or transmission of the higher-order harmonics), whereas the inability to maintain a constant slope for large values of T_f is a low-frequency deficiency in the generation or transmission of components near the fundamental frequency.

Sawtooth waveforms are used as time-base elements or sweep generators and in time-delay and time-measuring equipment. *See* SWEEP GENERATOR.

Sawtooth voltage generation. The complexity of electronic circuitry required to generate linear sawtooth voltage waves depends upon the accuracy to which such generation is specified. An approximate linear sawtooth may be generated by a dc voltage source, a series RC circuit, and a switch which has a small but nonzero resistance when closed. This elementary sawtooth generator is shown schematically in Fig. 2.

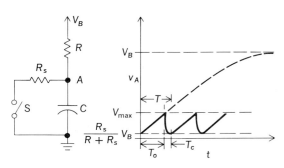

Fig. 2. Elements of sawtooth sweep circuit.

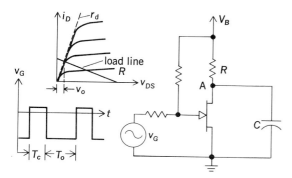

Fig. 3. Pulsed triode sawtooth generator.

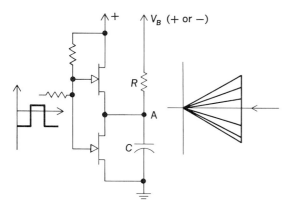

Fig. 4. Sweep generator using bidirectional clamp.

If the switch S in the illustration is suddenly opened, the voltage at point A will rise from an initial value of $V_B R_s/(R + R_s)$ toward V_B according to exponential equation as shown by the dashed curve in Fig. 2. Now if the switch is closed after a time T_o, the rise will be interrupted at a value V_{max} obtained from the solution for v_A in Eq. (1) for $t =$

$$v_A = \frac{R_s}{R + R_s} V_B + V_B \left(1 - \frac{R_s}{R + R_s}\right)(1 - \epsilon^{-(t/RC)}) \quad (1)$$

T_o. While the switch is closed during the period T_c the potential will fall in accordance with Eq. (2).

$$v_A = \frac{R_s}{R + R_s} V_B + \left(V_{max} - \frac{R_s}{R + R_s} V_B\right) \epsilon^{-(t/R_sC)} \quad (2)$$

If R_s is very small, the waveform will essentially have recovered to its initial value in a short retrace interval T_r less than T_o. This complete cycle will repeat itself for alternate opening and closing of the switch. If the time T_c during which the switch is closed is less than the time T_r required for complete recovery, the minimum value for the waveform will be higher than $V_B R_s/(R + R_s)$. The maximum and minimum excursions of the waveform can then be found by simultaneous solution of Eqs. (1) and (2) for the appropriate periods, T_o and T_c.

A periodic sawtooth waveform can be generated by using an astable relaxation oscillator as a switch in the above circuit. When used as switches in sawtooth generators, relaxation oscillators may be synchronized with external pulses to maintain an accurately controlled period. *See* RELAXATION OSCILLATOR.

If the switch of Fig. 2 is the switch in a synchro-

nous or keyed clamp each interval of the total period of the waveform may be controlled directly from an external source of pulses. A simple triode unidirectional clamp using an n-channel junction field-effect transistor (FET) is shown in Fig. 3. During the time T_c that the switch is closed, the gate is held at a slightly positive value by limiting, and the drain resistance r_d is low. During the open time, the gate voltage is sufficiently negative that no drain current will flow and the switch is open.

The use of a p-channel FET together with a negative value for V_B and negative control pulses would produce a negative-going sawtooth. *See* CLAMPING CIRCUIT.

Either a positive-going or a negative-going sawtooth may be generated by making V_B either positive or negative and replacing the clamp with a bidirectional clamp as shown in Fig. 4. If the potential V_B is variable at a rate which is slow compared to the periodicity of the waveform, a succession of sawtooth waveforms of varying amplitude will be generated. One particular application of such a circuit would be to generate one of the components of the rotating radial sweep if V_B were to be made to vary sinusoidally with the desired angular modulation.

Improvement of linearity. A linearly increasing voltage v_c, given by Eq. (3), will appear across the

$$v_c = (I_c/C)t \quad (3)$$

terminals of a capacitor C if a constant current I_c is flowing, since Eq. (4) applies. Therefore, improvement of linearity of the simple RC sawtooth genera-

$$v_c = (1/C) \int I_c\, dt \quad (4)$$

tors basically depends upon keeping the current through the capacitor more constant. For a specified sawtooth amplitude this may be done by increasing the supply voltage V_B. This is a practical solution only within narrow limits. Another method in effect replaces R with an active device, such as the output circuit of a transistor, which has a relatively low absolute resistance but an extremely high ac or incremental resistance over the limits of the desired amplitude range. In the circuit of Fig. 5, when point A is not clamped to the ground and the base-emitter bias voltage, as determined

SAWTOOTH-WAVE GENERATOR

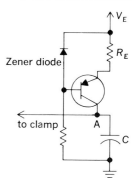

Zener diode

to clamp

Fig. 5. Constant-current sawtooth generator.

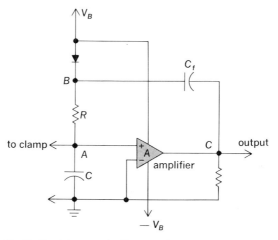

Fig. 6. Bootstrap sawtooth generator.

by the diode operating in the Zener breakdown region, is such that the transistor is in its normal range, the collector current will be relatively independent of collector voltage.

The circuit of Fig. 6 often referred to as the bootstrap sawtooth generator. When the clamp is closed, point A is at ground potential, point B is at V_B minus the small diode voltage drop, and point C is approximately at the potential of A if the amplifier is direct-coupled and noninverting with near unity gain and has high input and low output impedance. In other words, the amplifier is an emitter follower. When the clamp is opened, capacitor C starts to charge toward V_B through R and the diode resistance. Point C follows closely since the amplifier gain is close to unity. If capacitor C_f is very large compared to capacitor C, point B will rise the same amount. This causes the diode to stop conducting. Capacitor C will continue to

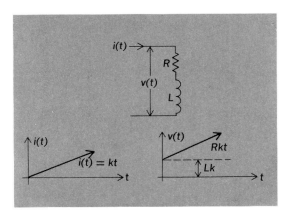

Fig. 9. Linear current in inductive circuit.

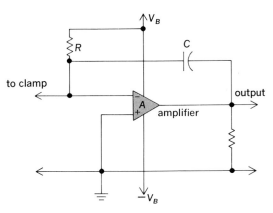

Fig. 7. Basic Miller integrator.

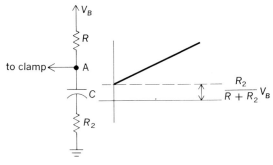

Fig. 10. Trapezoid voltage generator.

charge through R and C_f. If sufficiently large, C_f functions as a constant-voltage source, and to the extent that it does, and if the gain of the amplifier is near but less than unity, the charging current will be nearly constant and a nearly linear sawtooth of a magnitude approaching the supply voltage can be generated. To a good approximation, the voltage at point C can be expressed by Eq. (5),

$$v_C(t) =$$

$$AV_B \frac{C_f}{C + C_f(1-A)}(1 - \epsilon^{-[C + C_f(1-A)/RCC_f]t}) \quad (5)$$

where A is the voltage gain of the amplifier. If C_f is much larger than C, this is an exponential charging curve with an effective supply voltage of $(A/1-A)$ V_B. If A is near unity, this represents a great increase in effective supply voltage and a corresponding increase in linearity for a required amplitude. It can be shown that with an amplifier gain slightly greater than unity, a condition for an exactly linearly rising voltage can be established.

A circuit in which an integrating amplifier is used in addition to the clamp and RC time constant is sometimes referred to as the Miller integrating circuit (Fig. 7). If the input impedance of the amplifier is high, the output impedance low, and the gain high, the output approximates the integral of the suddenly impressed constant-amplitude supply voltage V_B. The approximate equation of the output waveform is given by Eq. (6), where A

$$v_o = -AV_B(1 - \epsilon^{-(t/ARC)}) \quad (6)$$

is the magnitude of the gain of the amplifier. The

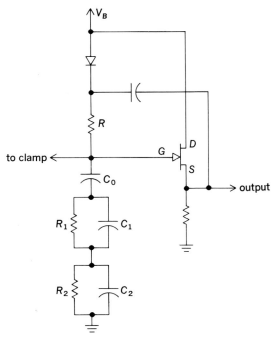

Fig. 8. Hyperbolic sweep generator.

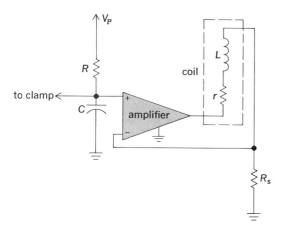

Fig. 11. Current feedback sweep generator.

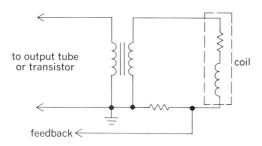

Fig. 12. Transformer-coupled deflection coil.

result will be a negative-going sawtooth, the first part of an exponential charging toward the effective supply voltage $-AV_B$. Thus the linearity is increased by the same amount that a charging voltage V_B multiplied by a factor A would increase it.

Hyperbolic and other waveforms. A sawtooth waveform other than linear may be generated by using RC circuits more complicated in form than the simple ones which have been described. For example, the bootstrap generator shown in Fig. 8 generates an approximate hyperbola.

Sawtooth current generation. Often a sawtooth current waveform must be applied to a circuit hav-

ing an inductive component, such as the deflection coil of a magnetically deflected cathode-ray device. If the coil can be represented by an inductance and resistance in series as shown in Fig. 9, the voltage appearing across the terminals of the coil will be given by Eq. (7), and if the current $i(t)$ is

$$v = Ri(t) + L\frac{di(t)}{dt} \qquad (7)$$

specified as a linear sawtooth, $i(t) = kt$, then the voltage will be given by Eq. (8). This voltage is a

$$v = Rkt + Lk \qquad (8)$$

step added to a linear sawtooth as shown. Such a voltage waveform, required for a linear sawtooth of current, may be generated by using any of the previous circuits with an additional circuit element R_2 as shown in Fig. 10. Initially, since the voltage across C cannot change instantaneously when the clamp is opened, the total voltage V_B will divide between R and R_2, making the potential at point A suddenly rise to $V_B R_2/(R + R_2)$.

A more common method of generating a current waveform is to generate a voltage waveform of the same form and apply it to a negative feedback amplifier with a large amount of current feedback, which forces the current in the coil to be approximately the form of the generated voltage. Such a system is shown in block form in Fig. 11. Such current feedback makes the output impedance of the amplifier high and thus approximates a current source which is a replica of the applied voltage.

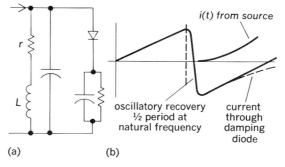

Fig. 14. Deflection system with diode damping. (*a*) Circuit diagram. (*b*) Typical waveform.

If dc levels do not have to be preserved, the deflection coils may be transformer-coupled to the output device as shown in Fig. 12.

High-frequency limitations. Any actual deflection coil can be represented as a series inductance and resistance only at relatively low frequencies. It also has distributed capacitance, which can be represented crudely by a shunt capacitance as in Fig. 13. This capacitance accounts principally for the delay in the start of the sweep and for the minimum retrace time T_r necessary for recovery. The best conditions occur when the oscillatory circuit is critically damped by the addition of the shunt resistance R. Typical sweep waveforms for critical damping and departures from it are shown.

The horizontal deflection circuit of the television system represents a special case where the circuit

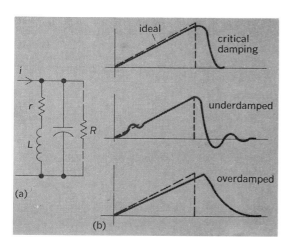

Fig. 13. Effect of shunt capacitance on deflection system. (*a*) Circuit diagram. (*b*) Typical waveforms.

is underdamped, or allowed to be oscillatory, for half a period with the beginning of the sweep waveform controlled by special diode circuits as shown in Fig. 14. For other types of waves *see* WAVE-SHAPING CIRCUITS. [GLENN M. GLASFORD]

Bibliography: G. M. Glasford, *Fundamentals of Television Engineering,* 1955; J. Millman, *Microelectronics,* 1979; J. Millman and H. Taub, *Pulse, Digital, and Switching Waveforms,* 1965; L. Strauss, *Waveform Generation and Shaping,* 2d ed., 1970.

Scaling circuit

An electronic circuit that produces one output pulse for a specific number n of input pulses. Such a circuit is referred to as a scale-of-n or an n-counter circuit.

A somewhat specialized form of counter, otherwise known as a frequency divider, employs an astable relaxation oscillator, such as a multivibrator or blocking oscillator. The natural period τ_1 of one state is made slightly longer than the period between n input pulses of an equally spaced pulse train, and the period τ_2 of the other state is shorter than the period between each pulse. If the input pulses to be counted serve as synchronizing pulses superimposed on the input waveform, they cause the relaxation oscillator to recycle in the manner shown by the input waveform of Fig. 1, instead of the natural period shown by the extended dotted lines. A bistable circuit can serve the same purpose, with the circuit switching on the $n - 1$ pulse and recycling on the nth pulse. *See* BLOCKING OSCILLATOR; MULTIVIBRATOR.

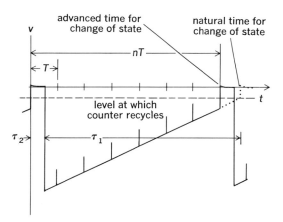

Fig. 1. Controlling waveform in astable scaling circuit. For this particular example, $n = 7$.

A class of circuits with wider use functions even when the input pulses have random spacing. The basic unit of such a scale-of-n circuit is often the bistable multivibrator, which is itself a scale-of-2, or binary, circuit. If the square wave from one output terminal of the multivibrator is differentiated and the resultant pulses of one polarity used to trigger a second such circuit, the result is a count of 4. Thus, a cascaded group of n circuits produces a count of 2^n. The waveforms of three cascaded stages are shown in Fig. 2. Feedback may be used in cascaded counter circuits to achieve a count of

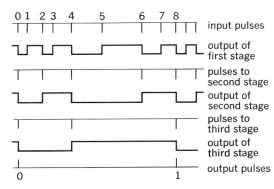

Fig. 2. Basic waveforms of three cascaded stages in random input scale-of-8 circuit.

any number less than 2^n but other than a multiple of 2.

Other forms of counters are possible, some using a different arrangement of scale-of-2 circuits known as the ring counter, others using special tubes such as beam-switching tubes, and still others using the switching properties of square-loop magnetic cores. *See* COUNTING CIRCUIT.

Scaling circuits are used for direct counting of a series of events and for basic measurements of time and of frequency. The scale-of-2 circuit is the building block in digital computers that operate internally with the binary number system. *See* DIGITAL COMPUTER; FREQUENCY COUNTER.

[GLENN M. GLASFORD]

Bibliography: G. M. Glasford, *Electronic Circuits Engineering,* 1970; J. Millman and H. Taub, *Pulse, Digital, and Switching Waveforms,* 1965.

Schottky effect

The enhancement of the thermionic emission of a conductor resulting from an electric field at the conductor surface. Since the thermionic emission current is given by the Richardson formula, an increase in the current at a given temperature implies a reduction in the work function ϕ of the emitter. *See* THERMIONIC EMISSION.

With reference to Fig. 1, let the vacuum level represent the energy of an electron at rest in free

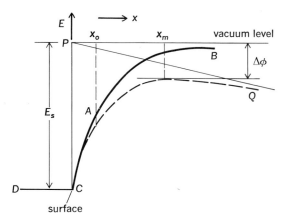

Fig. 1. Surface potential barrier and Schottky effect; the lowering of the work function $\Delta\phi$ is highly exaggerated.

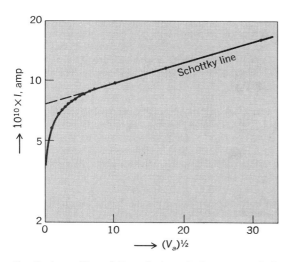

Fig. 2. Logarithm of thermionic emission current *I* of tungsten as function of square root of anode voltage V_a. (After W. B. Nottingham, Phys. Rev., 58:927–928, 1940)

space and let *CD* be the energy of a conduction electron at rest in a metal. If an electron approaches the metal surface from infinity, its potential energy *V* relative to the vacuum level is given by the well-known image potential $V(x) = -e^2/4x$, where *x* is the distance from the surface, and $e = 1.6 \times 10^{-19}$ coulomb is the electron's charge. The image potential is valid only for $x > x_0$, where x_0 is of the order of the distance between neighboring atoms in the metal; that is, x_0 is a few angstroms. In the absence of an applied field, *CAB* then represents the potential energy of an electron as a function of *x*. *AB* corresponds to the image potential; the exact shape of the curve between *C* and *A* is uncertain.

Suppose now a constant field *F* is applied externally between the surface of the emitting cathode and an anode; this produces a potential energy of an electron of $-eFx$ (line *PQ* in Fig. 1), and hence the total potential energy of an electron for $x > x_0$ is given by Eq. (1), indicated by the dashed line

$$V(x) = -(e^2/4x) - eFx \qquad (1)$$

CAQ in Fig. 1. This function has a maximum value given by Eq. (2).

$$x = x_m = (1/2)(e/F)^{1/2} \qquad (2)$$

The maximum lies below the vacuum level by an amount $\Delta\phi$, given in Eq. (3), which represents the

$$\Delta\phi = V(x_m) = -e(eF)^{1/2} \qquad (3)$$

reduction in the work function of the metal. For $F = 1000$ volt/cm, $x_m \simeq 10^{-5}$ cm and $\Delta\phi \simeq 10^{-2}$ ev; the actual change in the work function is thus small. If a field is present, the work function ϕ in the Richardson formula should be replaced by $(\phi - \Delta\phi)$. Hence, the current increases by the factor given by notation (4).

$$\exp\left[(e/kT)(eF)^{1/2}\right] \qquad (4)$$

According to this interpretation, a plot of the logarithm of the current versus the square root of the anode voltage should yield a straight line. An example is given in Fig. 2 for tungsten; the deviation from the straight line for low anode voltages is

due to space-charge effects. *See* SPACE CHARGE.

The straight portion of the line (the Schottky line) confirms the interpretation; the true saturation current for zero field is obtained by extrapolation of the Schottky line as indicated. Detailed studies have shown extremely small periodic deviations with reference to the Schottky line; these deviations are interpreted on the basis of the wave-mechanical theory describing the motion of electrons across the image potential barrier shown in Fig. 1. [ADRIANUS J. DEKKER]

Secondary emission

The emission of electrons from the surface of a solid into vacuum caused by bombardment with charged particles, in particular with electrons. The mechanism of secondary emission under ion bombardment is quite different from that under electron bombardment; the discussion here is limited to the latter case because it is in this sense that the term secondary emission is generally used. *See* ELECTRON EMISSION.

The bombarding electrons and the emitted electrons are referred to, respectively, as primaries and secondaries. Secondary emission has important practical applications because the secondary yield, that is, the number of secondaries emitted per incident primary, may exceed unity. Thus, secondary emitters are used in electron multipliers, especially in photomultipliers, and in other electronic devices such as television pickup tubes, storage tubes for electronic computers, and so on.

Secondary yield. The most thoroughly investigated property of secondary emission is the yield as a function of the energy of the primaries. The yield may be measured by means of the circuit shown schematically in Fig. 1. A beam of primary electrons strikes a target with an energy determined by the potential difference between the target and the cathode. The primary beam passes

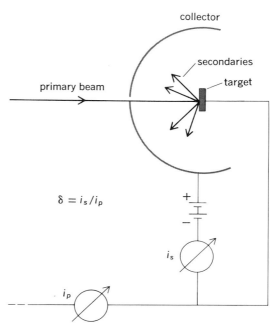

Fig. 1. Schematic circuit for measuring secondary yield; i_p and i_s represent the primary and secondary currents.

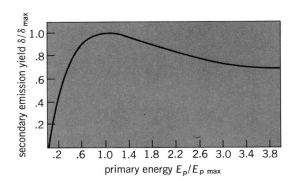

Fig. 2. Theoretical curve of secondary emission yield as a function of primary energy in normalized coordinates.

through a hole in the collector, which has been made positive with respect to the target. The secondaries emitted by the target then flow to the collector, and the yield is obtained as the ratio of the secondary current i_s to the primary current i_p.

Mechanism of the process. The emission of secondary electrons can be described as the result of three processes: (1) excitation of electrons in the solid into high-energy states by the impact of high-energy primary electrons, (2) transport of these secondary electrons to the solid-vacuum interface, and (3) escape of the electrons over the surface barrier into the vacuum. The efficiency of each of these three processes, and hence the magnitude of the secondary emission yield δ, varies greatly for different materials.

Taking into account the material characteristics that, in addition to the value of the primary energy E_p determine the yield δ, one arrives at the equation below for the dependence of δ upon E_p,

$$\delta = \frac{B_1 B_2 E_p}{\epsilon R}(1 - e^{-R/L})$$

where ϵ is the energy required to produce a secondary electron; R is the range of primary electrons; B_1 is the coefficient, taking into account that only a fraction of the excited electrons diffuse toward the surface; L is the mean free path of the secondary electron; and B_2 is the probability that an electron reaching the solid-vacuum interface can escape over the surface barrier. Here ϵ and R are associated with process 1 above, B_1 and L with process 2, and B_2 with process 3. Without giving a detailed derivation of the equation above, it is qualitatively plausible that δ increases with in-

creasing B_1, B_2, E_p, and L and decreases with increasing ϵ and R.

On the basis of the equation, a universal curve can be derived (Fig. 2) in which δ/δ_{max} is plotted versus $E_p/E_{p max}$, where δ_{max} is the maximum yield and $E_{p max}$ is the corresponding primary energy. Whereas curves for the absolute values of δ versus E_p vary over a wide range for different materials, experiments have generally confirmed the validity of the curve in Fig. 2. The peak in the curve can be interpreted as follows: With increasing E_p, the number of secondaries produced within the solid increases, but at the same time the primaries penetrate to a greater depth in the material. Because of energy loss processes, the "escape depth" of the secondaries has a finite value which is determined by some of the parameters entering the equation above. Thus the peak of the curve represents the point beyond which the number of secondaries produced at a depth greater than the escape depth exceeds the number of additional secondaries produced due to the higher primary energy. Because the escape depth is predominantly determined by L and B_2 in the equation, the variations in these two parameters are the main reasons why the δ values for different materials vary over such a wide range.

Experimental yield curves. In a discussion of measured δ versus E_p curves, it is useful to consider metals and semiconductors (or insulators) separately. In metals the secondaries lose their energy rapidly by electron-electron scattering. As a result, L and B_2 in the equation are small, and the escape depth is of the order of at most nanometers. Hence δ_{max}, and consequently $E_{p max}$, have low values, typically well below 2.

For semiconductors and insulators, the situation is more complicated and is best understood in terms of energy-band models. Referring to Fig. 3, the highest δ values that can be obtained depend on the relative position of the top of the valence band (where the secondary electrons originate), the bottom of the conduction band, and the vacuum level. Three typical band models are shown in Fig. 3. The model shown in Fig. 3a is characterized by a small ratio of band-gap energy E_G to electron affinity E_A. In the model shown in Fig. 3b the E_G to E_A ratio is large. In the model shown in Fig. 3c the bands are bent downward to such an extent that the vacuum level lies below the bottom of the conduction band in the bulk. A material with this characteristic is said to have negative effective electron affinity. This concept is also of great importance in photoelectric emission. The differences in secondary emission yields associated with each of the three band models can be qualitatively summarized as follows. *See* PHOTOEMISSION.

$E_G \ll E_A$ *model.* Secondary electrons excited from the valence band to levels above the vacuum level tend to lose their energy by exciting additional electrons from the valence band into the conduction band and thus to arrive at the solid-vacuum interface with insufficient energy to overcome the surface barrier. In other words, the escape depth is very small, and the maximum δ values are below 2, similar to those of metals. Examples of this model are germanium and silicon.

$E_G \gg E_A$ *model.* Whereas secondary electrons

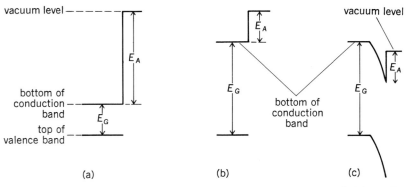

Fig. 3. Typical energy band models for semiconductors (or insulators). (a) $E_G \ll E_A$. (b) $E_G \gg E_A$ (c) Negative electron affinity.

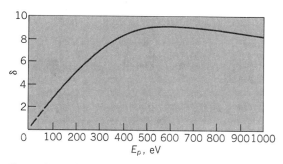

Fig. 4. Secondary emission yield versus primary energy for MgO.

excited from the valence band gradually lose energy by phonon-phonon scattering, an appreciable number of secondaries reach the solid-vacuum interface with sufficient energy to overcome the surface barrier. In other words, the escape depth is larger than in the case where $E_G \ll E_A$, of the order of tens of nanometers, and maximum δ values in the 8–15 range are typically obtained. Most of the materials used in practical devices fall into this category. Examples are MgO (see Fig. 4), BeO, Cs_3Sb (cesium antimonide), and KCl.

Negative effective electron affinity. Here the vacuum level is below the bottom of the conduction band. This case differs drastically from that shown in Fig. 3b because electrons that have dropped to the bottom of the conduction band as a result of phonon-phonon scattering still have enough energy to escape into the vacuum. Because the lifetime of electrons in the bottom of the conduction band is orders of magnitude longer than in states above this level, the escape depth

of the secondaries is orders of magnitude greater than in the case represented in Fig. 3b. The most important material in this category is cesium-activated gallium phosphide, GaP(Cs). Figure 5 shows the δ versus E_p curve for GaP(Cs) by comparison with MgO. Because of the much greater escape depth, δ values exceeding 100 are readily obtained. The curve for GaP(Cs) still follows quite closely the universal curve (Fig. 2), but the $E_{p_{max}}$ value is now in the 5–10-kV region compared with several hundred volts for materials represented in Fig. 3a or b.

Materials of the GaP(Cs) type represent a major breakthrough in the use of secondary emission for practical devices. In photomultipliers, for example, GaP(Cs) is superior to the conventional materials represented in Fig. 3b for a number of reasons, the greatest advantage being the improved signal-to-noise ratio. Negative effective electron affinity materials are not in universal use because of the more complex activation procedure and the associated higher cost. *See* BAND THEORY OF SOLIDS; SEMICONDUCTOR.

Dependence on angle of primary beam. The secondary yield for a given primary energy increases as the angle θ between the primary beam and the normal to the surface increases; the secondaries are then produced closer to the surface and consequently have a larger escape probability. At the same time, the energy for which the yield reaches its maximum value increases with increasing θ.

Secondary energies. A typical energy distribution of secondary electrons emitted by a silver target bombarded with primaries of 160-eV energy is given in Fig. 6. Note that most of the secondaries

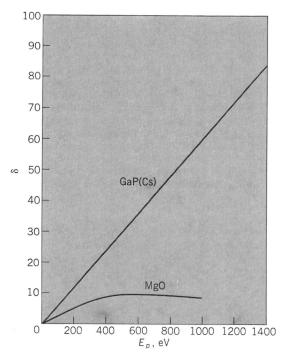

Fig. 5. Secondary emission yield versus primary energy for GaP(Cs). The curve for MgO is shown for comparison.

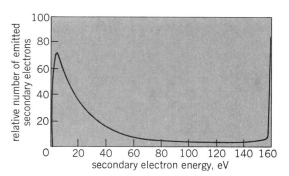

Fig. 6. Relative number of secondary electrons as a function of secondary energy for $E_p = 160$ V.

have relatively low energies. A small fraction of the emitted electrons have the same energy as the incident primaries and are called reflected primaries.

[ALFRED H. SOMMER]

Bibliography: R. L. Bell, *Negative Electron Affinity Devices*, 1973; A. J. Dekker, Secondary electron emission, in *Solid State Physics*, vol. 6, p. 251, 1958; R. U. Martinelli and D. G. Fisher, Application of semiconductors with negative electron affinity surfaces to electron emission devices, *Proc. IEEE*, 62:1339–1360, 1974; R. E. Simon and B. F. Williams, Secondary-electron emission, *IEEE Trans. Nucl. Sci.*, NS-15:167–170, 1968.

Selectivity

The ability of a radio receiver to separate a desired signal frequency from other signal frequencies, some of which may differ only slightly from the desired value. Selectivity is achieved by using tuned circuits that are sharply peaked and by increasing the number of tuned circuits. With a sharply peaked circuit, the output voltage falls off rapidly for frequencies increasingly lower or higher than that to which the circuit is tuned. *See* Q (ELECTRICITY); RADIO RECEIVER; RESONANCE (ALTERNATING-CURRENT CIRCUITS).   [JOHN MARKUS]

Semiconductor

A solid crystalline material whose electrical conductivity is intermediate between that of a metal and an insulator. Semiconductors exhibit conduction properties that may be temperature-dependent, permitting their use as thermistors (temperature-dependent resistors), or voltage-dependent, as in varistors. By making suitable contacts to a semiconductor or by making the material suitably inhomogeneous, electrical rectification and amplification can be obtained. Semiconductor devices, rectifiers, and transistors have replaced vacuum tubes almost completely in low-power electronics, making it possible to save volume and power consumption by orders of magnitude. In the form of integrated circuits, they are vital for complicated systems. The optical properties of a semiconductor are important for the understanding and the application of the material. Photodiodes, photoconductive detectors of radiation, injection lasers, light-emitting diodes, solar-energy conversion cells, and so forth are examples of the wide variety of optoelectronic devices. *See* CRYOSAR; INTEGRATED CIRCUITS; LASER; LIGHT-EMITTING DIODE; PHOTODIODE; PHOTOELECTRIC DEVICES; SEMICONDUCTOR DIODE; SEMICONDUCTOR RECTIFIER; TRANSISTOR; VARISTOR.

CONDUCTION IN SEMICONDUCTORS

The electrical conductivity of semiconductors ranges from about 10^3 to 10^{-9} ohm^{-1} cm^{-1}, as compared with a maximum conductivity of 10^7 for good conductors and a minimum conductivity of 10^{-17} ohm^{-1} cm^{-1} for good insulators.

The electric current is usually due only to the motion of electrons, although under some conditions, such as very high temperatures, the motion of ions may be important. The basic distinction between conduction in metals and in semiconductors is made by considering the energy bands occupied by the conduction electrons.

A crystalline solid consists of a large number of atoms brought together into a regular array called a crystal lattice. The electrons of an atom can each have certain energies, so-called energy levels, as predicted by quantum theory. Because the atoms of the crystal are in close proximity, the electron orbits around different atoms overlap to some extent, and the electrons interact with each other; consequently the sharp, well-separated energy levels of the individual electrons actually spread out into energy bands. Each energy band is a quasi-continuous group of closely spaced energy levels. *See* BAND THEORY OF SOLIDS.

At absolute zero temperature, the electrons occupy the lowest possible energy levels, with the restriction that at most two electrons may be in the same energy level. In semiconductors and insulators, there are just enough electrons to fill completely a number of energy bands, leaving the rest of the energy bands empty. The highest filled energy band is called the valence band. The next higher band, which is empty at absolute zero temperature, is called the conduction band. The conduction band is separated from the valence band by an energy gap which is an important characteristic of the semiconductor. In metals, the highest energy band that is occupied by the electrons is only partially filled. This condition exists either because the number of electrons is not just right to fill an integral number of energy bands or because the highest occupied energy band overlaps the next higher band without an intervening energy gap. The electrons in a partially filled band may acquire a small amount of energy from an applied electric field by going to the higher levels in the same band. The electrons are accelerated in a direction opposite to the field and thereby constitute an electric current. In semiconductors and insulators, the electrons are found only in completely filled bands, at low temperatures. In order to increase the energy of the electrons, it is necessary to raise electrons from the valence band to the conduction band across the energy gap. The electric fields normally encountered are not large enough to accomplish this with appreciable probability. At sufficiently high temperatures, depending on the magnitude of the energy gap, a significant number of valence electrons gain enough energy thermally to be raised to the conduction band. These electrons in an unfilled band can easily participate in conduction. Furthermore, there is now a corresponding number of vacancies in the electron population of the valence band. These vacancies, or holes as they are called, have the effect of carriers of positive charge, by means of which the valence band makes a contribution to the conduction of the crystal. *See* HOLES IN SOLIDS.

The type of charge carrier, electron or hole, that is in largest concentration in a material is sometimes called the majority carrier and the type in smallest concentration the minority carrier. The majority carriers are primarily responsible for the conduction properties of the material. Although the minority carriers play a minor role in electrical conductivity, they can be important in rectification and transistor actions in a semiconductor.

Electron distribution. The probability f for an energy level E to be occupied by an electron is given by the Fermi-Dirac distribution function, Eq. (1),

$$f = \left[1 + \exp\!\left(\frac{E - W}{kT}\right) \right]^{-1} \qquad (1)$$

where k is the Boltzmann constant and T is the absolute temperature. The parameter W is the Fermi energy level; an energy level at W has a probability of 1/2 to be occupied by an electron. The Fermi level is determined by the distribution of energy levels and the total number of electrons.

In a semiconductor, the number of conduction electrons is normally small compared with the number of energy levels in the conduction band,

and the probability for any energy level to be occupied is small. Under such a condition, the concentration of conduction electrons is given by Eq. (2),

$$N_n = \frac{2}{h^3} (2\pi m_n kT)^{3/2} \exp\left[\frac{(W - E_c)}{kT}\right] \quad (2)$$

where h is Planck's constant, E_c is the lowest energy of the conduction band, and m_n is called the effective mass of conduction electrons. The effective mass is used in place of the actual mass to correct the coefficient in the equation and to bring the results in line with experimental observations. This correction is necessary because the theory leading to these equations is based upon electrons moving in a field free space, which is not the exact picture. The electrostatic Coulomb potential throughout the crystal is varying in a periodic manner, the variation being due to the electric fields around the atomic centers. The concentration of holes in the valence band is given by Eq. (3),

$$N_p = \frac{2}{h^3} (2\pi m_p kT)^{3/2} \exp\left[\frac{(E_v - W)}{kT}\right] \quad (3)$$

where m_p is the effective mass of a hole and E_v is the highest energy of the valence band.

Mobility of carriers. The velocity acquired by charge carriers per unit strength of applied electric field is called the mobility of the carriers. The velocity in question is the so-called drift velocity in the direction of the force exerted on the carriers by the applied field. It is added to the random thermal velocity. In semiconductors the carrier mobility normally ranges from 10^2 to 10^5 cm²/(s) (volt). A material's conductivity is the product of the charge, the mobility, and the carrier concentration.

Electrons in a perfectly periodic potential field can be accelerated freely. Impurities, physical defects in the structure, and thermal vibrations of the atoms disturb the periodicity of the potential field in the crystal, thereby scattering the moving carriers. It is the resistance produced by this scattering that limits the carriers to only a drift velocity under the steady force of an applied field.

Intrinsic semiconductors. A semiconductor in which the concentration of charge carriers is characteristic of the material itself rather than of the content of impurities and structural defects of the crystal is called an intrinsic semiconductor. Electrons in the conduction band and holes in the valence band are created by thermal excitation of electrons from the valence to the conduction band. Thus an intrinsic semiconductor has equal concentrations of electrons and holes. The intrinsic carrier concentration N_i is determined by Eq. (4),

$$N_i = \frac{2}{h^3} (2\pi kT)^{3/2} (m_n m_p)^{3/4} \exp\left(-\frac{E_g}{2kT}\right) \quad (4)$$

where E_g is the energy gap. The carrier concentration, and hence the conductivity, is very sensitive to temperature and depends strongly on the energy gap. The energy gap ranges from a fraction of 1 ev to several electron volts. A material must have a large energy gap to be an insulator.

Impurity semiconductors. Typical semiconductor crystals such as germanium and silicon are formed by an ordered bonding of the individual atoms to form the crystal structure. The bonding is attributed to the valence electrons which pair up with valence electrons of adjacent atoms to form so-called shared pair or covalent bonds. These materials are all of the quadrivalent type; that is, each atom contains four valence electrons, all of which are used in forming the crystal bonds.

Atoms having a valence of +3 or +5 can be added to a pure or intrinsic semiconductor material with the result that the +3 atoms will give rise to an unsatisfied **bond** with one of the valence electrons of the **semicon**ductor atoms, and +5 atoms will result in an **extra** or free electron that is not required in the bond structure. Electrically, the +3 impurities add holes and the +5 impurities add electrons. They are called acceptor and donor impurities, respectively. Typical valence +3 impurities used are boron, aluminum, indium, and gallium. Valence +5 impurities used are arsenic, antimony, and phosphorus.

Semiconductor material "doped" or "poisoned" by valence +3 acceptor impurities is termed p-type, whereas material doped by valence +5 donor material is termed n-type. The names are derived from the fact that the holes introduced are considered to carry positive charges and the electrons negative charges. The number of electrons in the energy bands of the crystal is increased by the presence of donor impurities and decreased by the presence of acceptor impurities. Let N be the concentration of electrons in the conduction band and let P be the hole concentration in the valence band. For a given semiconductor, the relation $NP = N_i^2$ holds, independent of the presence of impurities. The effect of donor impurities tends to make N larger than P, since the extra electrons given by the donors will be found in the conduction band even in the absence of any holes in the valence band. Acceptor impurities have the opposite effect, making P larger than N. See Acceptor atom; Donor atom.

At sufficiently high temperatures, the intrinsic carrier concentration becomes so large that the effect of a fixed amount of impurity atoms in the crystal is comparatively small and the semiconductor becomes intrinsic. When the carrier concentration is predominantly determined by the impurity content, the conduction of the material is said to be extrinsic. There may be a range of temperature within which the impurity atoms in the material are practically all ionized; that is, they supply a maximum number of carriers. Within this temperature range, the so-called exhaustion range, the carrier concentration remains nearly constant. At sufficiently low temperatures, the electrons or holes that are supplied by the impurities become bound to the impurity atoms. The concentration of conduction carriers will then decrease rapidly with decreasing temperature, according to either $\exp(-E_i/kT)$ or $\exp(-E_i/2kT)$, where E_i is the ionization energy of the dominant impurity.

Physical defects in the crystal structure may have similar effects as donor or acceptor impurities. They can also give rise to extrinsic conductivity.

An isoelectronic impurity, that is, an atom which has the same number of valence electrons as the host atom, does not bind individual carriers as strongly as a donor or an acceptor impurity. However, an isoelectronic impurity may show an appreciable binding for electron hole pairs, excitons, and thereby have important effects on the properties. An example is nitrogen substituting for phosphorus in gallium phosphide; the impurity affects the luminescence of the material.

Hall effect. Whether a given sample of semiconductor material is n- or p-type can be determined by observing the Hall effect. If an electric current is caused to flow through a sample of semiconductor material and a magnetic field is applied in a direction perpendicular to the current, the charge carriers are crowded to one side of the sample, giving rise to an electric field perpendicular to both the current and the magnetic field. This development of a transverse electric field is known as the Hall effect. The field is directed in one or the opposite direction depending on the sign of the charge of the carrier. *See* HALL EFFECT.

The magnitude of the Hall effect gives an estimate of the carrier concentration. The ratio of the transverse electric field strength to the product of the current and the magnetic field strength is called the Hall coefficient, and its magnitude is inversely proportional to the carrier concentration. The coefficient of proportionality involves a factor which depends on the energy distribution of the carriers and the way in which the carriers are scattered in their motion. However, the value of this factor normally does not differ from unity by more than a factor of two. The situation is more complicated when more than one type of carrier is important for the conduction. The Hall coefficient then depends on the concentrations of the various types of carriers and their relative mobilities.

The product of the Hall coefficient and the conductivity is proportional to the mobility of the carriers when one type of carrier is dominant. The proportionality involves the same factor which is contained in the relationship between the Hall coefficient and the carrier concentration. The value obtained by taking this factor to be unity is referred to as the Hall mobility.

MATERIALS AND THEIR PREPARATION

The group of chemical elements which are semiconductors includes germanium, silicon, gray (crystalline) tin, selenium, tellurium, and boron.

Elemental semiconductors. Germanium, silicon, and gray tin belong to group IV of the periodic table and have crystal structures similar to that of diamond. Germanium and silicon are two of the best-known semiconductors. They are used extensively in devices such as rectifiers and transistors. Gray tin is a form of tin which is stable below 13°C. White tin, which is stable at higher temperatures, is metallic. Gray tin has a small energy gap and a rather large intrinsic conductivity, about 5×10^3 ohm^{-1} cm^{-1} at room temperature. The n-type and p-type gray tins can be obtained by adding aluminum and antimony, respectively.

Selenium and tellurium both have a similar structure, consisting of spiral chains located at the corners and centers of hexagons. The structure gives rise to anisotropy of the properties of single crystals; for example, the electrical resistivity of tellurium along the direction of the chains is about one-half the resistivity perpendicular to this direction. Selenium has been widely used in the manufacture of rectifiers and photocells.

Semiconducting compounds. A large number of compounds are known to be semiconductors. Copper(I) oxide (Cu_2O) and mercury(II) indium telluride ($HgIn_2Te_4$) are examples of binary and ternary compounds. The series zinc sulfide (ZnS), zinc selenide (ZnSe), zinc telluride (ZnTe), and the series zinc selenide (ZnSe), cadmium selenide (CdSe), and mercury(II) selenide (HgSe) are examples of binary compounds consisting of a given element in combinations with various elements of another column in the periodic table. The series magnesium antimonide (Mg_2Sb_2), magnesium telluride (MgTe), and magnesium iodide (MgI_2) is an example of compounds formed by a given element with elements of various other columns in the periodic table.

A group of semiconducting compounds of the simple type AB consists of elements from columns symmetrically placed with respect to column IV of the periodic table. Indium antimonide (InSb), cadmium telluride (CdTe), and silver iodide (AgI) are examples of III-V, II-IV, and I-VI compounds, respectively. The various III-V compounds are being studied extensively, and many practical applications have been found for these materials. Some of these compounds have the highest carrier mobilities known for semiconductors. The compounds have zincblende crystal structure which is geometrically similar to the diamond structure possessed by the elemental semiconductors, germanium and silicon, of column IV, except that the four nearest neighbors of each atom are atoms of the other kind. The II-VI compounds, zinc sulfide (ZnS) and cadmium sulfide (CdS), are used in photoconductive devices. Zinc sulfide is also used as a luminescent material. *See* PHOTOCONDUCTIVITY.

Binary compounds of the group lead sulfide (PbS), lead selenide (PbSe), and lead telluride (PbTe) are sensitive in photoconductivity and are used as detectors of infrared radiation. The compounds, bismuth telluride (Bi_2Te_3) and bismuth selenide (Bi_2Se_3), consisting of heavy atoms, are found to be good materials for thermocouples used for refrigeration or for conversion of heat to electrical energy.

The metal oxides usually have large energy gaps. Thus pure oxides are usually insulators of high resistivity. However, it may be possible to introduce into some of the oxides impurities of low ionization energies and thus obtain relatively good extrinsic conduction. Copper(I) oxide (Cu_2O) was one of the first semiconductors used for rectifiers and photocells: extrinsic p-type conduction is obtained by producing an excess of oxygen over the stoichiometric composition, that is, the 2-to-1 ratio of copper atoms to oxygen atoms. A number of oxide semiconductors can be obtained by replacing some of the normal metal atoms with metal atoms of one more or less valency. The method is called controlled valence. An example of such a semiconductor is nickel oxide containing lithium.

Some compounds with rare-earth or transition-metal ions in their composition, such as EuTe and NiS_2, are semiconductors with magnetic properties. Another interesting type of semiconductor is characterized by layered structures. The interaction within a layer is significantly stronger than that between layers. A number of semiconductors of this type are known, such as PbI_2, GaSe, and various transition-metal dichalcogenides such as $SnSe_2$ and MoS_2.

Preparation of materials. The properties of semiconductors are extremely sensitive to the presence of impurities. It is therefore desirable to start with the purest available materials and to introduce a controlled amount of the desired impurity. The zone refining method is often used for further purification of obtainable materials. The floating zone technique can be used, if feasible, to prevent any contamination of molten material by contact with crucible.

For basic studies as well as for many practical applications, it is desirable to use single crystals. Various methods are used for growing crystals of different materials. For many semiconductors, including germanium, silicon, and the III-V compounds, the Czochralski method is commonly used. The method of condensation from the vapor phase is used to grow crystals of a number of semiconductors, for instance, selenium and zinc sulfide. For materials of high melting points, such as various metal oxides, the flame fusion or Vernonil method may be used.

The introduction of impurities, or doping, can be accomplished by simply adding the desired quantity to the melt from which the crystal is grown. Normally, the impurity has a small segregation coefficient, which is the ratio of equilibrium concentrations in the solid and the liquid phases of the material. In order to obtain a desired impurity content in the crystal, the amount added to the melt must give an appropriately larger concentration in the liquid. When the amount to be added is very small, a preliminary ingot is often made with a larger content of the doping agent; a small slice of the ingot is then used to dope the next melt accurately. Impurities which have large diffusion constants in the material can be introduced directly by holding the solid material at an elevated temperature while this material is in contact with the doping agent in the solid or the vapor phase.

A doping technique, ion implantation, has been developed and used extensively. The impurity is introduced into a layer of semiconductor by causing a controlled dose of highly accelerated impurity ions to impinge on the semiconductor. *See* ION IMPLANTATION.

A growing subject of scientific and technological interest is amorphous semiconductors. In an amorphous substance the atomic arrangement has some short-range but no long-range order. The representative amorphous semiconductors are selenium, germanium, and silicon in their amorphous states, and arsenic and germanium chalcogenides, including such ternary systems as Ge-As-Te and Si-As-Te. Some amorphous semiconductors can be prepared by a suitable quenching procedure from the melt. Amorphous films can be obtained by vapor deposition.

RECTIFICATION IN SEMICONDUCTORS

In semiconductors, narrow layers can be produced which have abnormally high resistances. The resistance of such a layer is nonohmic; it may depend on the direction of current, thus giving rise to rectification. Rectification can also be obtained by putting a thin layer of semiconductor or insulator material between two conductors of different material.

Barrier layer. A narrow region in a semiconductor which has an abnormally high resistance is called a barrier layer. A barrier may exist at the contact of the semiconductor with another material, at a crystal boundary in the semiconductor, or at a free surface of the semiconductor. In the bulk of a semiconductor, even in a single crystal, barriers may be found as the result of a nonuniform distribution of impurities. The thickness of a barrier layer is small, usually 10^{-3} to 10^{-5} cm.

A barrier is usually associated with the existence of a space charge. In an intrinsic semiconductor, a region is electrically neutral if the concentration n of conduction electrons is equal to the concentration p of holes. Any deviation in the balance gives a space charge equal to $e(p - n)$, where e is the charge on an electron. In an extrinsic semiconductor, ionized donor atoms give a positive space charge and ionized acceptor atoms give a negative space charge. Let N_d and N_a be the concentrations of ionized donors and acceptors, respectively. The space charge is equal to $e(p - n + N_d - N_a)$.

A space charge is associated with a variation of potential. A drop in potential $-\Delta V$ increases the potential energy of an electron by $e\Delta V$; consequently every electronic energy level in the semiconductor is shifted by this amount. With a variation of potential, the electron concentration varies proportionately to $\exp(eV/kT)$ and the hole concentration varies as $\exp(-eV/kT)$. A space charge is obtained if the carriers, mainly the majority carriers, fail to balance the charge of the ionized impurities.

A conduction electron in a region where the potential is higher by ΔV must have an excess energy of $e\Delta V$ in order for it to have the minimum energy on reaching the low potential region. Electrons with less energy cannot pass over to the low potential region. Thus a potential variation presents a barrier to the flow of electrons from high to low potential regions. It also presents a barrier to the flow of holes from low to high potential regions.

Surface barrier. A thin layer of space charge and a resulting variation of potential may be produced at the surface of a semiconductor by the presence of surface states. Electrons in the surface states are bound to the vicinity of the surface, and the energy levels of surface states may lie within the energy gap. Surface states may arise from the adsorption of foreign atoms. Even a clean surface may introduce states which do not exist in

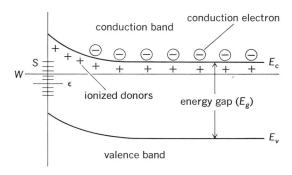

Fig. 1. Energy diagram of a surface barrier as employed in an *n*-type semiconductor.

the bulk material, simply by virtue of being the boundary of the crystal.

The surface is electrically neutral when the surface states are filled with electrons up to a certain energy level ϵ in the energy gap E_g, which is the energy difference between the bottom of the conduction band E_c and the top of the valence band E_v. If the Fermi level W in the bulk semiconductor lies higher in the energy gap, more surface states would be filled, giving the surface a negative charge. As a result the potential drops near the surface and the energy bands are raised for n-type material (Fig. 1). With the rise of the conduction band, the electron concentration is reduced and a positive space charge due to ionized donors is obtained. The amount of positive space charge is equal to the negative surface charge given by the electrons in the surface states between ϵ and the Fermi level.

Contact barrier. The difference between the potential energy E_0 of an electron outside a material and the Fermi level in the material is called the work function of the material. Figure 2 shows the energy diagram for a metal and a semiconductor, the work functions of which differ by eV. Upon connecting the two bodies electrically, charge is transferred between them so that the potential of the semiconductor is raised relative to that of the metal; that is, the electron energy levels in the semiconductor are lowered. Equilibrium is established when the Fermi level is the same in the two bodies. In this case, the metal is charged negatively and the semiconductor is charged positively. The negative charge on the metal is concentrated close to the surface, as is expected in good conductors. The positive charge on the semiconductor is divided between the increase of space charge in an extension of the barrier and the depopulation of

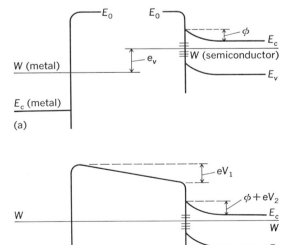

Fig. 2. Energy diagram for a metal (left) and an n-type semiconductor (right). E_0 is the potential energy of an electron outside the material, E_c is the energy at the bottom of the conduction band, and E_v is the energy at the top of the valence band. (a) Semiconductor and metal isolated. (b) Semiconductor and metal in electrical contact, $eV_1 + eV_2 = eV$.

some of the surface states. The charging of the semiconductor is brought about by a change of eV_2 in the barrier height ϕ. The sum of eV_2 and the potential energy variation eV_1 in the space between the two bodies is equal to the original difference eV between the work functions.

With decreasing separation between the two bodies, the division of eV will be in favor of eV_2. However, if there is a very large density of states, a small eV_2 gives a large surface charge on the semiconductor due to the depopulation of surface states.

It is possible that eV_2 is limited to a small value even at the smallest separation, of the order of an interatomic distance in solids. In such cases, the barrier height remains nearly equal to the value ϕ of the free surface, irrespective of the body in contact. This situation has been found in germanium and silicon rectifiers. Before the explanation was given by J. Bardeen, who postulated the existence of surface states, it had been assumed that the height of a contact barrier was equal to the difference of the work functions.

The understanding and the application of metal-semiconductor contacts have been extended to various kinds of contacts, such as that between different semiconductors, heterojunctions, and metal oxide semiconductor (MOS) junctions.

Single-carrier theory. The phenomenon of rectification at a crystal barrier can be described according to the role played by the carriers. Where the conduction property of the rectifying barrier is determined primarily by the majority carriers, the single-carrier theory is employed. Such cases are likely to be found in semiconductors with large energy gaps, for instance, oxide semiconductors. Figure 3 shows the energy diagrams of metal-semiconductor contact rectifiers under conditions of equilibrium. The potential variation in the semiconductor is such as to reduce the majority carrier concentration near the contact. If the energy bands were to fall in the case of an n-type semiconductor or to rise in the case of a p-type semiconductor, the majority carrier concentration would be enhanced near the contact, and the contact would not present a large and rectifying resistance. It is clear that in the cases shown in Fig. 3, the minority carrier concentration increases near the contact. However, if the energy gap is large, the minority carrier concentration is normally very small, and the role of minority carriers may be still negligible even if the concentration is increased.

Under equilibrium conditions, the number of carriers passing from one body to the other is balanced by the number of carriers crossing the contact in the opposite direction, and there is no net current. The carriers crossing the contact in either direction must have sufficient energies to pass over the peak of the barrier. The situations under applied voltages are shown in Fig. 4 for the case of an n-type semiconductor. When the semiconductor is made positive, its energy bands are depressed and the height of the potential barrier is increased, as shown in Fig. 4a. Fewer electrons in the semiconductor will be able to cross over into the metal, whereas the flow of electrons across the contact from the metal side remains unchanged. Consequently, there is a net flow of electrons from the metal to the semiconductor. The flow of

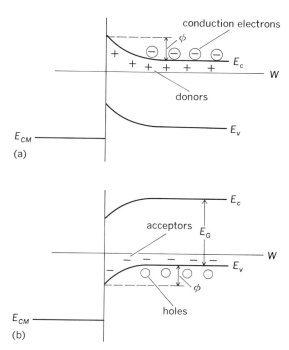

Fig. 3. Energy diagrams of a rectifying contact between a metal and a semiconductor: (a) n-type semiconductor, (b) p-type semiconductor.

electrons from the metal side is the maximum net flow obtainable. With increasing voltage, the current saturates and the resistance becomes very high. Figure 4b shows the situation when the semiconductor is negative under the applied voltage. The energy bands in the semiconductor are raised. The flow of electrons from the semiconductor to the metal is increased, since electrons of lower energy are able to go over the peak of the barrier. The result is a net flow of electrons from the semiconductor to the metal. There is no limit to the flow in this case. In fact, the electron current increases faster than the applied voltage because there are increasingly more electrons at lower energies. The resistance decreases, therefore, with increasing voltage. The direction of current for which the resistance is low is called the forward direction, while the opposite direction is called the reverse or blocking direction. A general expression for the current can be written in the form of Eq. (5), where

$$j = enC\left(\exp\frac{-\phi}{kT}\right)\left[\exp\left(\frac{eV}{kT}\right) - 1\right] \qquad (5)$$

j is the current density, n is the carrier concentration in the bulk of the semiconductor, ϕ is the barrier height, and V is the applied voltage taken as positive in the forward direction. The factor C depends on the theory appropriate for the particular case.

Diffusion theory. When there is a variation of carrier concentration, a motion of the carriers is produced by diffusion in addition to the drift determined by the mobility and the electric field. The transport of carriers by diffusion is proportional to the carrier concentration gradient and the diffusion constant. The diffusion constant is related to the mobility, and both are determined by the

scattering suffered by moving carriers. The average distance traveled by a carrier in its random thermal motion between collisions is called the mean free path. If barrier thickness is large compared to mean free path of carriers, motion of carriers in the barrier can be treated as drift and diffusion. This viewpoint is the basis of the diffusion theory of rectification. According to this theory, the factor C in Eq. (5) depends on the mobility and the electric field in the barrier.

Diode theory. When the barrier thickness is comparable to or smaller than the mean free path of the carriers, then the carriers cross the barrier without being scattered, much as in a vacuum tube diode.

According to this theory, the factor C in the rectifier equation is $v/4$, where v is the average thermal velocity of the carriers.

Tunneling theory. Instead of surmounting a potential barrier, carriers have a probability of penetrating through the barrier. The effect, called tunneling, becomes dominant if the barrier thickness is sufficiently small. This effect is important in many applications. *See* TUNNELING IN SOLIDS.

Two-carrier theory. Often the conduction through a rectifying barrier depends on both electron and hole carriers. An important case is the pn junction between p- and n-sections of a semiconductor material. Also, in metal-semiconductor rectifiers, the barrier presents an obstacle for the flow of majority carriers but not for the flow of minority carriers, and the latter may become equally or more important.

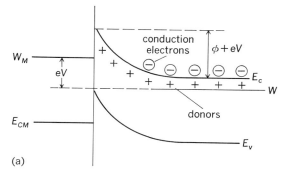

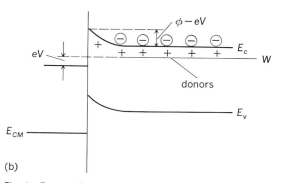

Fig. 4. Energy diagrams of a rectifying contact between a metal and an n-type semiconductor under an applied voltage V. (a) Positive semiconductor. There is a net flow of electrons from metal to semiconductor. (b) Negative semiconductor. There is a net flow of electrons from semiconductor to metal.

Rectification at pn junctions. A *pn* junction is the boundary between a *p*-type region and an *n*-type region of a semiconductor. When the impurity content varies, there is a variation of electron and hole concentrations. A variation of carrier concentrations is related to a shift of the energy bands relative to the constant Fermi level. This is brought about by a variation of the electrostatic potential which requires the existence of a space charge. If the impurity content changes greatly within a short distance, a large space charge is obtained within a narrow region. Such is the situation existing in a rectifying *pn* junction.

When a voltage is applied to make the *n*-region negative relative to the *p*-region, electrons flow from the *n*-region, where they are abundant, into the *p*-region. At the same time, holes flow from the *p*-region, where holes are abundant, into the *n*-region. The resistance is therefore relatively low. The direction of current in this case is forward. Clearly, the resistance will be high for current in the reverse direction.

With a current in the forward direction, electrons in the *n*-region and holes in the *p*-region flow toward the junction and there must be continuous hole-electron recombination in the neighborhood of the junction. The minority carrier concentration in each region is increased near the junction because of the influx of the carriers from the other region. This phenomenon is known as carrier-injection. When there is a current in the reverse direction, there must be a continuous generation of holes and electrons in the neighborhood of the junction, from which electrons flow out into the *n*-region and holes flow out into the *p*-region. Thus current through a *pn* junction is controlled by the hole-electron recombination or generation in the vicinity of the junction.

The transistor consists of two closely spaced *pn* junctions in a semiconductor with an order *pnp* or *npn*.

Contact rectification. If the height of a rectifying contact barrier is high, only a very small fraction of majority carriers can pass over the barrier. The fraction may be so small as to be comparable with the concentration of the minority carriers, provided the energy gap is not too large. The current due to the minority carriers becomes appreciable if the barrier height above the Fermi level approaches the energy difference between the Fermi level and the top of the valence band (Fig. 3).

The concentration of minority carriers is higher at the contact than in the interior of the semiconductor. With a sufficiently high barrier, it is possible to obtain at the contact a minority carrier concentration higher than that of the majority carriers. The small region where this condition occurs is called the inversion layer.

As in the case of a *pn* junction, a forward current produces injection of minority carriers. With the presence of an inversion layer, the injection can be so strong as to increase appreciably the conductivity in the vicinity of the contact. Ordinarily, contact rectifiers consist of a semiconductor in contact with a metal whisker. For large forward currents, the barrier resistance is small, and the resistance of the rectifier is determined by the spreading resistance of the semiconductor for a contact of small area. By increasing the conductivity in the vicinity of the contact where the spreading resistance is concentrated, carrier-injection may reduce considerably the forward resistance of the rectifier.

Surface electronics. The surface of a semiconductor plays an important role technologically, for example, in field-effect transistors and charge-coupled devices. Also, it presents an interesting case of two-dimensional systems where the electric field in the surface layer is strong enough to produce a potential wall which is narrower than the wavelengths of charge carriers. In such a case, the electronic energy levels are grouped into subbands, each of which corresponds to a quantized motion normal to the surface, with a continuum for motion parallel to the surface. Consequently, various properties cannot be trivially deduced from those of the bulk semiconductor. *See* CHARGE-COUPLED DEVICES; SURFACE PHYSICS.

[H. Y. FAN]

Bibliography: B. L. Crowder (ed.), *Ion Implantation in Semiconductors and Other Materials*, 1973; N. B. Hannay and U. Colombo (eds.), *Electronic Materials*, 1973; *Nuovo Cimento*, vol. 38B, no. 2, 1977; Proceedings of the 8th (1976 International) Conferences on Solid State Devices, *Jap. J. Appl. Phys.*, suppl. 16–1, 1977; F. Seiz and D. Turnbull (eds.), *Solid State Physics*, vol. 1, 1955; W. Shockley, *Electrons and Holes in Semiconductors*, 1950; *Surf. Sci.*, vol. 73, 1978; S. Sze, *Physics of Semiconductor Devices*, 1969; R. K. Willardson and A. C. Beer (eds.), *Semiconductors and Semimetals*, vols. 1–10, 1966–1975.

Semiconductor diode

A two-terminal electronic device that utilizes the properties of the semiconductor from which it is constructed. In a semiconductor diode without a *pn* junction, the bulk properties of the semiconductor itself are used to make a device whose characteristics may be sensitive to light, temperature, or electric field. In a diode with a *pn* junction, the properties of the *pn* junction are used. The most important property of a *pn* junction is that, under ordinary conditions, it will allow electric current to flow in only one direction. Under the proper circumstances, however, a *pn* junction may also be used as a voltage-variable capacitance, a switch, a light source, a voltage regulator, or a means to convert light into electrical power. *See* SEMICONDUCTOR.

Silicon and germanium are the semiconductors most often used in diodes. However, other materials may be used for special purposes: cadmium sulfide and cadmium selenide in photoconductors; gallium phosphide, gallium arsenide–phosphide, and silicon carbide in light-emitting diodes; and gallium arsenide in microwave generators.

Diodes without pn junctions. The band structure of each semiconductor contains a forbidden energy gap. That is, there is a range of energies, ΔE, in which there are no quantum states that electrons may occupy. When the semiconductor is in thermal equilibrium, its electrons are distributed among the allowed quantum states according to Fermi statistics: The free electrons tend to reside near the minimum in the conduction band, and the unoccupied states in the valence band, called free holes, tend to be concentrated near the maximum in the valence band. Figure 1 shows a schematic of the band structure in gallium arsenide, GaAs. According to Fermi statistics, the holes are located mainly at $E=0$, $p=0$; and the electrons are concentrated near $E=E_1$, $p=0$.

The conductivity of a semiconductor is proportional to the number of electrical carriers (electrons and holes) it contains. In a temperature-compensating diode, or thermistor, the number of carriers changes with temperature. For example, in an intrinsic diode the number of carriers and the conductivity are proportional to $e^{-\Delta E/kT}$. As temperature T increases, the conductivity increases. This effect may be used to cancel out other conductivity changes in an electrical circuit so that the net change is zero. *See* THERMISTOR.

In a photoconductor the semiconductor is packaged so that it may be exposed to light. Light photons whose energies are greater than ΔE can excite electrons from the valence band to the conduction band, increasing the number of electrical carriers in the semiconductor. Thus the conductivity of a semiconductor is a measure of the light intensity striking the semiconductor. *See* PHOTOCONDUCTIVITY.

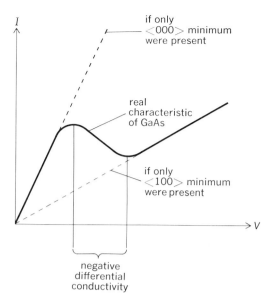

Fig. 2. Current-voltage characteristic bulk of GaAs.

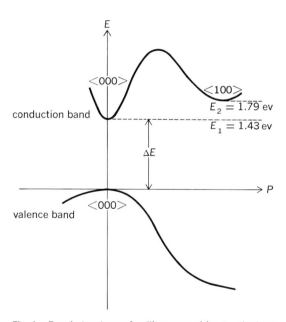

Fig. 1. Band structure of gallium arsenide; $E=$ electron energy, $P=$ momentum.

In some semiconductors the conduction band has more than one minimum. An example is gallium arsenide, as shown in Fig. 1. Here the $<000>$ minimum is located 1.43 ev above the valence band, and the $<100>$ minimum is located 0.36 ev above the $<000>$ minimum. When an electric field is applied to GaAs, the electrons start out in the $<000>$ minimum and the current-voltage relationship initially follows Ohm's law, as shown in Fig. 2. In this and other figures, I represents current and V represents voltage. When the field approaches 3200 volts/cm, a significant number of electrons are scattered into the $<100>$ minimum. Since carriers in the $<100>$ minimum

have lower mobilities than those in $<000>$, the conductivity must decrease. This results in a region of negative differential conductivity, and a device operated in this region is unstable. The current pulsates at microwave frequencies, and the device, a Gunn diode, may be used as a microwave power source. *See* MICROWAVE SOLID-STATE DEVICES.

Diodes with pn junctions. A rectifying junction is formed whenever two materials of different conductivity types are brought into contact. Most commonly, the two materials are an *n*-type and a *p*-type semiconductor, and the device is called a junction diode. However, rectifying action also occurs at a boundary between a metal and a semiconductor of either type. If the metal contacts a large area of semiconductor, the device is known as a Schottky barrier diode; if the contact is a metal point, a point-contact diode is formed. *See* POINT-CONTACT DIODE; SCHOTTKY EFFECT.

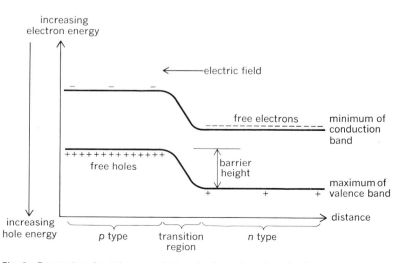

Fig. 3. Energy bands at the potential barrier for *pn* junction of a diode.

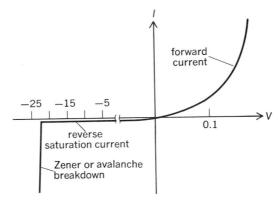

Fig. 4. Current-voltage characteristic of a *pn* junction.

The contact potential between the two materials in a diode creates a potential barrier which tends to keep electrons on the *n* side of the junction and holes on the *p* side. This barrier is shown in the energy band diagram of Fig. 3. When the *p* side is made positive with respect to the *n* side by an applied field, the barrier height is lowered and the diode is forward biased. Majority electrons from the *n* side may flow easily to the *p* side, and majority holes from the *p* side may flow easily to the *n* side. When the *p* side is made negative, the barrier height is increased and the diode is reverse-biased. Then, only a small leakage current flows: Minority electrons from the *p* side flow into the *n* side, and minority holes from the *n* side flow into the *p* side. The current-voltage characteristic of a typical diode is shown in Fig. 4. Rectifying diodes can be made in a variety of sizes, and much practical use can be made of the fact that such a diode allows current to flow in essentially one direction only. At one extreme, single devices may be used to handle thousands of watts of power in changing alternating current to direct current. At the other extreme, a small diode may be used to detect an amplitude-modulated (AM) radio signal that has only microwatts of power.

The potential barrier at a *pn* junction consists of an electric dipole made up of charged impurity atoms, positively charged on the *n* side and negatively charged on the *p* side. Because the ends of the dipole are separated by a small distance, the junction acts like a capacitor. Moreover, the capacitance is voltage-variable since the barrier height is voltage-dependent. A diode which is specifically designed to utilize this capacitance characteristic is called a varactor.

When light photons with energies greater than ΔE strike a semiconductor near a *pn* junction, the potential barrier will sweep the conduction electrons that were generated by the light into the *n* side of the diode, and the newly created holes into the *p* side. If the diode, now called a solar cell, is connected to an external circuit, the diode will supply electrical power to the circuit as long as light strikes the *pn* junction. If, on the other hand, the diode is reverse-biased, the magnitude of the reverse saturation current can be used to measure light intensity. In this configuration the diode is called a photodiode. *See* SOLAR CELL.

When a junction diode is forward-biased, the number of electrons on the *p* side and the number of holes on the *n* side are increased far above their equilibrium values. This makes the recombination of electrons with holes highly probable, and the recombination energy may be released as photons of light.

In some materials, notably GaP for red or green light and GaAs for infrared light, the conversion of electrical energy directly to light can be made quite efficient. And in materials like GaAs, where the valence band maximum and the conduction band minimum are at the same value of crystal momentum, laser action may be achieved.

An Esaki tunnel diode is formed when both the *n* and *p* sides of a junction diode are very heavily doped. In this case, as shown in Fig. 5, an electron may move from the *n* side to the *p* side under forward bias in either of two ways: It may climb over the barrier, as in an ordinary diode; or it may quantum-mechanically tunnel through the barrier, since there are some available states on the *p* side that are at the same energy as states on the *n* side. Tunneling is highly probable at low and medium reverse voltages and at low forward voltages. As shown in Fig. 6*a*, a tunnel diode has an unstable negative-resistance region. The diode may be operated either at point 1 or point 2, but may not be operated in between. This feature may be used to advantage in logic circuits, where the diode is made to switch from one point to the other. *See* TUNNEL DIODE.

A backward diode is similar to a tunnel diode except that it has no forward tunnel current (Fig. 6*b*). It is useful as a low-voltage rectifier.

Although simple diode theory predicts a constant reverse saturation current no matter how high the reverse voltage, all real diodes eventually break down either by tunneling or by avalanche (Fig. 4).

Avalanching occurs when the electrons or the holes gather so much energy that they can create new hole-electron pairs by collision. This process is cumulative and can result in a rapid buildup of current carriers. A diode that has a sharp breakdown provides a virtually constant voltage over a

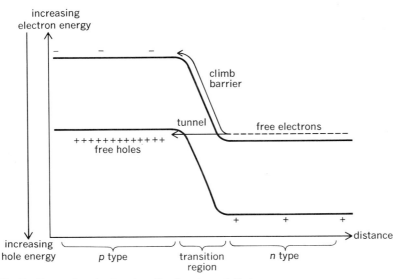

Fig. 5. Energy bands at *pn* junction in a tunnel diode.

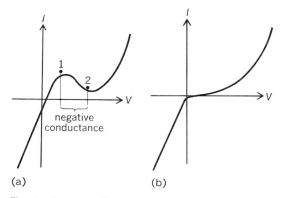

Fig. 6. Current-voltage characteristic of (a) a tunnel diode and (b) a backward diode

wide range of current. Popularly known as a Zener diode, this device may be used to provide a reference voltage or to regulate the voltage at some point in a circuit.

[STEPHEN NYGREN]

Bibliography: J. Millman, *Microelectronics*, 1976; E. S. Yang, *Fundamentals of Semiconductor Devices*, 1978.

Semiconductor heterostructures

Structures consisting of two different semiconductor materials in junction contact, with unique electrical or electrooptical characteristics. A heterojunction is a junction in a single crystal between two dissimilar semiconductors. The most important differences between the two semiconductors are generally in the energy gap and the refractive index. In semiconductor heterostructures, differences in energy gap permit spatial confinement of injected electrons and holes, while the differences in refractive index can be used to form optical waveguides. Semiconductor heterostructures have been used for diode lasers, light-emitting diodes, optical detector diodes, and solar cells. In fact, heterostructures must be used to obtain continuous operation of diode lasers at room temperature. Heterostructures also exhibit other interesting properties such as the quantization of confined carrier motion in ultrathin heterostructures and enhanced carrier mobility in modulation-doped heterostructures. Structures of current interest utilize III−V and IV−VI compounds having similar crystal structures and closely matched lattice constants. *See* BAND THEORY OF SOLIDS; LASER; LIGHT-EMITTING DIODE; OPTICAL DETECTORS; SOLAR CELL.

Carrier and optical field confinement. The most intensively studied and thoroughly documented materials for heterostructures are GaAs and $Al_xGa_{1-x}As$. Several other III−V and IV−VI systems also are used for semiconductor heterostructures. The variation of the energy gap E_g and the refractive index $\bar{n}$ with AlAs mole fraction x are shown in Fig. 1. The lattice constant a_0 is also noted in Fig. 1 to emphasize that a_0 for GaAs and AlAs differs, but by an amount less than 0.14%. A close lattice match is necessary in heterostructures in order to obtain high-quality crystal layers by epitaxial growth and thereby to prevent excessive carrier recombination at the hetero-

junction interface. *See* CRYSTAL STRUCTURE.

A $GaAs-Al_{0.3}Ga_{0.7}As$ double heterostructure is illustrated in Fig. 2. The energy gap E_g versus x given in Fig. 1a shows that E_g for $Al_{0.3}Ga_{0.7}As$ is 0.37 eV greater than for GaAs. When a positive potential is connected to the p-type $Al_{0.3}Ga_{0.7}As$, electrons are injected into the GaAs layer from the wider-energy-gap n-type $Al_{0.3}Ga_{0.7}As$, and

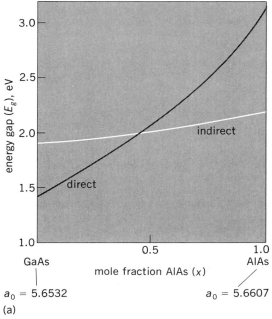

(a)

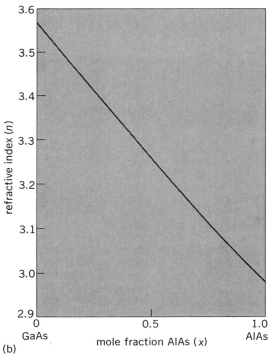

(b)

Fig. 1. Compositional dependence in $Al_xGa_{1-x}As$ of (a) the energy gap and (b) the refractive index at a photon energy of 1.38 eV. In both cases, temperature = 297 K. (*From C. A. Burrus, H. C. Casey, Jr., and T. Li, in S. E. Millet and A. G. Chynoweth, eds., Optical Fiber Telecommunications, Academic Press, 1979*)

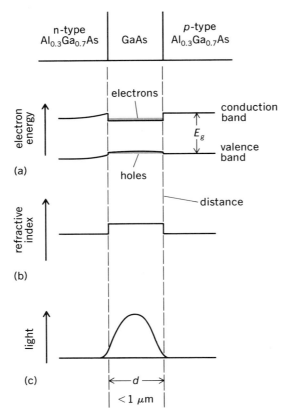

n-type Al$_{0.3}$Ga$_{0.7}$As   GaAs   p-type Al$_{0.3}$Ga$_{0.7}$As

electrons

conduction band

E_g

valence band

(a)

holes

distance

(b)

(c)

d

$< 1\ \mu$m

Fig. 2. Schematic representation of a Al$_{0.3}$Ga$_{0.7}$As-GaAs-Al$_{0.3}$Ga$_{0.7}$As double heterostructure laser showing (a) energy band diagram at high forward bias, (b) refractive index profile, and (c) optical field distribution. (From H. C. Casey, Jr., and M. B. Panish, Heterostructure Lasers, Part A, Fundamental Principles, Academic Press, 1978)

holes are injected into the GaAs from the wider-energy-gap p-type Al$_{0.3}$Ga$_{0.7}$As. The wider energy gap Al$_{0.3}$Ga$_{0.7}$As layers also confine the injected electrons and holes to the GaAs layer, where they can recombine radiatively. This confinement of the injected carriers is illustrated in Fig. 2a.

Figure 1b shows that the refractive index for the Al$_{0.3}$Ga$_{0.7}$As layers will be less than for the GaAs layer. This refractive index step is shown in Fig. 2b. Solution of the reduced wave equation shows that the optical field will be confined to the larger-refractive-index GaAs layer. The resulting optical field distribution is shown in Fig. 2c. When this heterostructure is formed as a rectangular bar with parallel reflecting surfaces, it can become a diode laser at current densities near 2×10^3 A/cm^2 at a forward voltage of 1.6 V.

Quantum well effects. When the narrow energy gap layer in heterostructures becomes a few tens of nanometers or less in thickness, new effects that are associated with the quantization of confined carriers are observed. These ultrathin heterostructures are referred to as superlattices or quantum well structures, and they consist of alternating layers of GaAs and Al$_x$Ga$_{1-x}$As (Fig. 3a). These structures are generally prepared by molecular-beam epitaxy. Each layer has a thickness in the range of 5 to 40 nm. The energy band diagram for

the quantum well structure is shown in Fig. 3b. The abrupt steps in the energy gaps form potential wells in the conduction and valence bands.

In the GaAs layers, the motion of the carriers is restricted in the direction perpendicular to the heterojunction interfaces, while they are free to move in the other two directions. The carriers can therefore be considered as a two-dimensional gas. The Schrödinger wave equation shows that the carriers moving in the confining direction can have only discrete bound states. As the thickness of the quantum wells gets large, a continuum of states then results. The discrete states in the undoped GaAs quantum wells are illustrated in Fig. 3b and are accurately predicted by the well-known quantum-mechanical problem of the particle-in-a-box.

The presence of the discrete quantum states may readily be observed by low-temperature optical absorption measurements. The observation wavelength is from 0.6 to 0.9 μm. These measurements are generally made at liquid helium temperatures with 20 or more GaAs quantum wells separated by Al$_x$Ga$_{1-x}$As layers. The absorption of photons takes place in the GaAs quantum wells by the transition of electrons in quantized states of the valence band to the quantized states of the conduction band. Sharp peaks in the absorption spectrum occur at the energies that separate the quantized states in the valence and conduction bands, and the photon energies of these absorption peaks vary with the thickness of the GaAs layer as predicted.

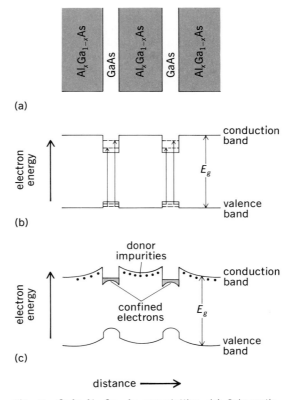

Fig. 3. GaAs-Al$_{0.3}$Ga$_{0.7}$As superlattice. (a) Schematic representation. (b) Quantum wells, with the discrete conduction and valence band states for undoped GaAs. (c) Quantum wells with modulation doping.

Modulation doping. Another property of semiconductor heterostructures is illustrated by a modulation doping technique that spatially separates conduction electrons and their parent donor impurity atoms. As illustrated in Fig. 3c, the donor impurities are incorporated in the wider energy gap $Al_xGa_{1-x}As$ layer but not in the GaAs layer. The GaAs conduction band edge is lower in energy than the donor states in the $Al_xGa_{1-x}As$, and therefore electrons from the donor impurities will move into the GaAs layers.

The useful feature of modulation doping is that the mobile carriers are the electrons in the GaAs layers and that these mobile carriers are spatially separated from their parent donor impurities in the $Al_xGa_{1-x}As$ layers. Since the carrier mobility in semiconductors is decreased by the presence of ionized and neutral impurities, the carrier mobility in the modulation-doped GaAs is larger than for a GaAs layer doped with impurities to give the same free electron concentration. Higher carrier mobilities should permit preparation of devices that operate at higher frequencies than are possible with doped layers.

Heteroepitaxy. In order to prepare optical sources and detectors that operate at a desired wavelength or to achieve high-mobility semiconductors, often it is necessary to use quaternary solid solutions. To obtain high-quality layers, the epitaxial layers must be grown on binary substrates that have the same lattice constant as the solid solutions. An example of considerable interest is the growth of $Ga_xIn_{1-x}P_yAs_{1-y}$ layers lattice-matched to InP. In this case, the heterojunction is used to obtain a similar lattice constant for dissimilar semiconductors at a desired bandgap. The heteroepitaxial growth techniques are chemical-vapor deposition, liquid-phase epitaxy, and molecular-beam epitaxy.   [H. C. CASEY, JR.]

Chemistry. The III–V and II–VI compounds and alloys commonly used in heterostructures are usually formed either by direct synthesis from the elements or by vapor-phase decomposition reactions using compounds containing hydrogen, halogen, or organic radicals. Substrates used for the growth of heterostructures are usually gotten from single-crystal boules synthesized directly from the elements. Most heterostructure materials can be doped either n or p type. However, in some cases one conductivity type can be difficult to obtain due to a "self-compensation" mechanism in which the formation of one type of vacancy is energetically favorable and compensates the opposite doping type. For example, the p-type beryllium (Be) doping of AlN is compensated by the formation of n-type nitrogen vacancies. In general, variations in stoichiometry will greatly affect electrical and optical properties. Also, the properties of heterostructures greatly depend on the lattice constant match and chemical abruptness of the interface. A lattice mismatch of greater than 0.5% usually results in an interface with a significant density of both electron and hole traps. The chemical rule that determines the allowable alloy compositions is that the sum of the atom fraction of the group III (or group II) elements must equal the sum of the atom fractions of group V (or group VI) elements in the crystal, for example, $Ga_{1-x}Al_xAs$, $0 \leq x \leq 1$.

Fabrication. Semiconductor heterostructures are usually fabricated as single-crystal structures using thin-film epitaxial crystal growth techniques such as liquid-phase epitaxy (LPE), chemical vapor deposition (CVD), and molecular-beam epitaxy (MBE).

Liquid-phase epitaxy. The LPE method has been the most widely used technique for fabricating heterostructures. The principal advantage is the ease of fabrication of high-purity heterostructures with good electrical, optical, and interface properties as a result of impurity segregation from the growing layer into the melt during growth. The main disadvantage is the poor control of layer thickness and interface morphology and abruptness. Epitaxial growth by the LPE method occurs when a melt or solvent becomes supersaturated, usually by supercooling, with respect to a solid phase in presence of a single-crystal substrate. For the growth of III–V materials, the melt usually has a large concentration of one of the group III components in epitaxial layer. For example, epitaxial growth of $Ga_{1-x}Al_xAs$ for $0 \leq x \leq 1$ can be obtained from melts containing about 90% Ga and 10% Al plus As. A "typical" schedule for the LPE growth of a $Ga_{1-x}Al_xAs$ layer on GaAs is shown in Fig. 4. The

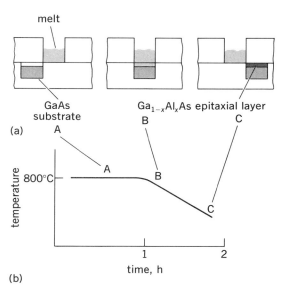

Fig. 4. Liquid-phase epitaxy method for GaAs-Ga_{1-x}-Al_xAs heterostructures. (a) Simplified diagrams of successive stages of the process. (b) Variation of temperature during the process. (*From J. M. Woodall, III-V compounds and alloys: An update, Science, 208:908–915, 1980*)

important features are a "bottomless" bin to contain the melt and a fixture with a slightly recessed substrate which can be moved with respect to the melt bin. The first step is to thermally equilibrate the melt prior to contact with the substrate (A in Fig. 4). This means that the melt is either at solid-liquid equilibrium or slightly undersaturated. Next, the melt is cooled, usually at constant rate, while the substrate is in contact with it (B of Fig. 4). The desired layer thickness is obtained by programming either a change in temperature or elapsed time and then removing the substrate from the melt (C of Fig. 4). The growth of heterostructures can also be accomplished by using multimelt fixtures, moving the substrate between the various melts during cooling. The term LPE originally referred to growth from supercooled melts. The technique now includes growth from melts supersaturated by other techniques such as electroepitaxy, in which the solid-liquid interface region becomes supersaturated due to the flow of electric current across the interface and isothermal melt mixing techniques in which solid-liquid equilibrium melts of different compositions are mixed together and become supersaturated.

Chemical vapor deposition. The CVD method refers to the formation of thin solid films as the result of thermochemical vapor-phase reactions. When the films are epitaxial, the method is sometimes called vapor phase epitaxy (VPE). For III–V materials two different chemistries have been widely studied: the group III and group V halogen compounds and group V hydrogen compounds; and the group III metal-organic compounds and V-hydrogen compounds such as $Ga(CH_3)_3$ and AsH_3. The halogen transport reactions are of the "hot" to "cold" type in which the III-halogen is produced in a high-temperature zone by the reaction of the III element with HCl. The III-halogen then diffuses to the low-temperature zone, where it combines with the V species to form an epitaxial layer of III–V material. Vapor-phase epitaxy using metal-organic compounds (MOCVD) occurs when the organic radical R of the III-R compound is "cracked" or pyrolyzed away at a "hot wall" or hot substrate in the presence of the V-H_3 compound. A schematic illustration of the MOCVD technique applied to GaAs is shown in Fig. 5. Both the halogen and metal-organic chemistries have been successfully used to fabricate heterostructures. It is thought that the MOCVD method may have long-term advantages such as better purity, composition, and thickness control and

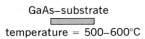

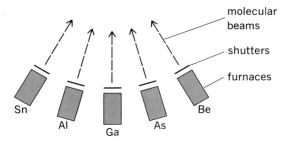

Fig. 6. Schematic illustration of a system configuration for growth of doped $Ga_{1-x}Al_xAs$ by molecular-beam epitaxy. (*From P. E. Luscher, Crystal growth by molecular beam epitaxy, Solid State Technol., 20(12):43–52, 1977*)

the ability to produce a wide range of III–V compounds and alloys.

Molecular-beam epitaxy. The MBE technique is relatively new compared with LPE and CVD. For heterostructure fabrication, it is capable of a layer thickness and interface abruptness control on a monoatomic scale. This represents almost two orders of magnitude improvement in structural resolution in the growth direction compared to LPE and CVD. The MBE technique is similar to evaporation techniques used to deposit thin metal films. The key features of the MBE method are the use of an ultrahigh-vacuum environment and epitaxial growth by the reaction of multiple molecular beams of differing flux and chemistry with a heated single-crystal substrate. A schematic illustration of MBE applied to doped $Ga_{1-x}Al_xAs$ is shown in Fig. 6. Each furnace heats a crucible which is charged with one of the constituent elements or compounds of the desired film. Tin (Sn) functions as an *n*-type dopant, and beryllium (Be) functions as a *p*-type dopant. The furnace temperature is chosen so that the vapor pressure is sufficient to produce the desired beam flux at the substrate surface. The furnaces are arranged so that the flux from each is maximum at the substrate position. The quality and composition of the epitaxial film are determined by the quality of the ultrahigh-vacuum system, the substrate temperature, and the furnace temperature. The shutters interposed between each furnace and the substrate allow the beams to be modulated. This feature, coupled with typical growth rates of a few tenths of a nanometer per second, facilitates the fabrication of very thin and very abrupt heterostructures. *See* SEMICONDUCTOR.

[JERRY M. WOODALL]

Bibliography: H. C. Casey, Jr., and M. B. Panish, *Heterostructures Lasers*, Part A: *Fundamental Principles*, and Part B: *Materials and Operating Characteristics*, 1978; R. Dingle, in H. J. Queiser (ed.), *Festkörperprobleme*, Advances in Solid State Physics, vol. 15, pp. 21–48, 1975; R. Dingle et al., Electron mobilities in modulation-doped semiconductor heterojunction superlattices, *Appl. Phys. Lett.*, 33:665–667, Oct. 1, 1978; J. W. Matthews (ed.), *Epitaxial Growth*, Materials Science and Technology Series, 1975; A. G. Milnes

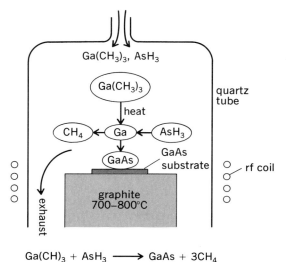

Fig. 5. Simplified diagram of the metal-organic method for chemical vapor deposition.

and D. L. Feucht, *Heterojunctions and Metal-Semiconductor Junctions*, 1972; M. B. Panish, Molecular beam epitaxy, *Science*, 208:916–922, 1980; J. M. Woodall, III-V compounds and alloys: An update, *Science*, 208:908–915, 1980.

Semiconductor memories

Devices for storing digital information that are fabricated by using integrated circuit technology. Semiconductor memories are widely used to store programs and data in almost every digital system. Initially developed as a replacement for magnetic core memories, which were used as the main computer storage memory, semiconductor memories started to appear in the early 1970s and have almost totally replaced core memories as the main computer memory elements.

Many different types of semiconductor memories are used in computer systems to perform various functions—bulk data storage, program storage, temporary storage, and cache (or intermediate) storage. Almost all of the memories are a form of random-access memory (RAM), where any storage location can be accessed in the same amount of time. However, there are many different types of RAMs, the most frequently used of which is the writable and readable memory that is simply referred to as a RAM.

Read/write RAMs. Although the RAM acronym indicates the random-access capability, it is a misnomer since almost all semiconductor memories except for a few specialty types can be randomly accessed. A more appropriate name for the memory would be a read/write RAM to indicate that data can be written into the memory as well as be read out of it.

Dynamic and static types. Even within this one class of memory, there are finer subdivisions. Basically there are two different types of read/write RAMs—dynamic and static. The RAM type refers to the structure of the actual storage circuit used to hold each data bit (the cell structure) within the memory chip. A dynamic memory uses a storage cell based on a transistor and capacitor combination, in which the digital information is represented by a charge stored on each of the capacitors in the memory array. The memory gets the name "dynamic" from the fact that the capacitors are imperfect and will lose their charge unless the charge is repeatedly replenished (refreshed) on a regular basis (typically every 2 ms). If refreshed, the information will remain until intentionally changed or the power to the memory is shut off. Static memories, in contrast, do not use a charge-storage technique; instead, they use either four or six transistors to form a flip-flop for each cell in the array. Once data are loaded into the flip-flop storage elements, the flip-flop will indefinitely remain in that state until the information is intentionally changed or the power to the memory circuit is shut off.

Capacity. In the early 1970s, dynamic memories with a capacity (density) of 1024 bits per chip were introduced. Improvements in semiconductor processing and circuit design have made practical an increase in density, first to 4096 bits on a chip, then to 16,384 bits, and in 1980 to 65,536 bits (often rounded off to 64 kbits). Samples of dynamic RAMs with 262,144 bits (known as the 256k dynamic RAM) were tested by some firms in 1982, and limited production began in 1983. Projecting these developments to the next level would indicate that a 1-megabit (1,048,676 bits) chip will be sampled by 1985—an increase of three orders of magnitude in about 15 years. Static memories require more complex structures for each cell, and thus cannot pack as many cells on a chip. They have typically lagged behind dynamic memories in density by a factor of four. For example, at the time the 1024-bit dynamic memory was introduced, static memories had densities of 256 bits per chip. In 1980, when the 64-kbit dynamic RAM was introduced, static densities were just reaching 16,384 bits on a chip. And, in 1983, as samples of the 256k dynamic RAM appeared, 64-kbit static RAMs were in limited production.

Nonvolatile memories. There are many other forms of semiconductor memories in use—mask-programmable read-only memories (ROMs), fuse-programmable read-only memories (PROMs), ultraviolet-erasable programmable read-only memories (UV EPROMs), electrically alterable read-only memories EAROMs), electrically erasable programmable read-only memories (EEPROMs), and nonvolatile static RAMs (NV RAMs). All of these memory types are also randomly accessible, but their main distinguishing feature is that once information has been loaded into the storage cells, the information stays there even if the power is shut off.

ROM. The first of these memory types, the ROM, is programmed by the memory manufacturer during the actual device fabrication. Here, though, there are two types of ROMs; one is called last-mask or contact-mask programmable, and the other is often called a ground-up design. In the last-mask type of ROM, the final mask used in the fabrication process determines the connections to the internal transistors. The connections, in turn, determine the data pattern that will be read out when the cell is accessed. The ground-up type of ROM is designed from the bottom up—all fabrication masks used in the multiple mask process are custom-generated.

Last-mask ROMs are theoretically less area-efficient (the final chip size is larger than for a ground-up design), and thus tend to cost a little bit more since chip area is directly related to price. However, they offer a rapid turnaround time (the time it takes for a user to obtain finished and programmed devices from the manufacturer from the time the code pattern was provided to the manufacturer by the user), since all circuits can be premanufactured up to the next-to-last mask step. Typically, ground-up designs offer lower cost and slightly higher performance, since they are more area-efficient. However, ROMs require that the user make a commitment to a specific code pattern or place an order for many more devices than are actually required.

Fuse PROM. As an alternative to the mask-programmable memories, semiconductor manufacturers developed all the other programmable memory types to permit the users to program the memories themselves and order as few as they want, even a single device. The first of the user-programmable devices was the fuse PROM. Offered in standard sizes ranging from a few hundred bits to over 64,000 bits, the fuse-programmable memories are one-time programmable memories — once the information is programmed in, it cannot be altered. There are basically two types of fuse-programmable memory types; one type uses microscopic fuse links that are blown open to define a logic one or zero for each cell in the memory array; the other type of fuse programming causes metal to short out base-emitter transistor junctions to program the ones or zeros into the memory.

Reprogrammable memories. The birth of the microprocessor in the early 1970s brought with it new memory types that offered a feature never before available — reusability. Information stored in the memory could be erased — in the case of the UV EPROM, by an ultraviolet light, and in the case of the EAROM, EEPROM, or NV RAM, by an electrical signal — and then the circuit could be reprogrammed with new information that could be retained indefinitely. All of these memory types are starting to approach the ideal memory element for the computer, an element that combines the flexibility of the RAM with the permanence of the ROM when power is removed. At the present time there are still some limiting factors in the various memory types that must be overcome before that ideal goal can be attained. *See* MICRO-PROCESSOR.

The ideal memory circuit should allow unlimited read and write operations without any unusual voltage levels or extra circuitry. At present, UV EPROMs require an external ultraviolet lamp to erase the stored information and a programming voltage of more than four times the read-mode supply voltage. Even the EAROMs and EEPROMs require higher-than-supply programming voltages and, in many cases, some external support logic. However, some of the newer EAROMs and EEPROMs have been able to place circuitry on the same chip to boost the standard 5-V supply voltage up to the level necessary to program or erase the cell, as well as provide the other necessary support functions. Coming closest to the ideal specification is the NV RAM, a memory that combines a static RAM with a nonvolatile memory array, so that for every stored bit there are two memory cells, one of which is volatile and the other nonvolatile. During normal system operation, the NV RAM uses the volatile memory array, but when it receives a special store signal, information held in the RAM area is transferred into the nonvolatile section. Thus the RAM section provides unlimited read and write operations, while the nonvolatile section provides back-up when power is removed.

However, the EAROM, EEPROM, and NV RAM will suffer one common failing that keeps them from reaching the ideal — they wear out. The electrical process used to store information in the nonvolatile array causes a steady deterioration in the ability of the memory to retain data for a guaranteed time period. Currently available capabilities range from about 10,000 to over 1,000,000 write cycles, but many times that number are needed for general-purpose use.

Semiconductor technology. The semiconductor technology used to fabricate all the different

Summary of memory types and technologies

Memory type	Maximum capacity commercially available 1983–1984*	Relative speed†	Programmability	Technology
Dynamic RAM	256k	Fast/medium	Read/write, volatile	NMOS
Static RAM	16k	Fastest	Read/write, volatile	Bipolar ECL
	4k	Fast	Read/write, volatile	Bipolar TTL
	64k	Fast/medium	Read/write, volatile	CMOS
	16k	Fastest/fast	Read/write, volatile	NMOS
ROM	16k	Fastest	Factory mask	Bipolar
	256k	Medium/fast	Factory mask	NMOS
	1Mbit‡	Medium/slow	Factory mask	CMOS
PROM	4k	Fastest	Fuse, one-time	Bipolar ECL
	64k	Fast	Fuse, one-time	Bipolar TTL
	16k	Medium	Fuse, one-time	CMOS
UV EPROM	256k	Medium	Ultraviolet erasable, reprogrammable	NMOS floating gate
	64k	Medium	Ultraviolet erasable, reprogrammable	CMOS floating gate
EAROM	8k	Medium/slow	Electrically reprogrammable	MNOS *p*-channel
	64k	Medium	Electrically reprogrammable	MNOS *n*-channel
EEPROM	64k	Medium	Electrically erasable, reprogrammable	NMOS floating gate
	32k	Medium	Electrically erasable, reprogrammable	CMOS floating gate
NV RAM	4k	Medium	Electrically erasable, reprogrammable	NMOS floating gate

*1k = 1024 bits.   ‡1Mbit = 1,048,676 bits.
†Fastest < 40 ns; 41 ns < fast < 200 ns; 201 ns < medium < 450 ns; 451 ns < slow < 800 ns; slowest > 801 ns.

memory types spans the entire range of available commercial processes, as summarized in the table. The dynamic memory, originally introduced in a p-channel MOS process (PMOS), has been upgraded through the use of a higher-performance n-channel process (NMOS). Typical dynamic RAMs offer access time ranging from 100 ns, for the very fast versions, up to about 300 ns, for the slow units. NMOS static RAMs range from some extremely fast devices at 25 ns all the way up to about 400 ns for the very slow models. NMOS technology is also used by most of the other memory types, although most of the EAROMs use a PMOS-based process along with a special nitride step during fabrication to form a metal-nitride MOS system (MNOS).

To save power, many memory products are fabricated by using a complementary MOS structure (CMOS) that combines both p- and n-channel devices on the same chip. For static memory types, in many cases this technology can cut power dissipation by several orders of magnitude. For the highest-performance semiconductor memories, bipolar technology still offers the fastest access times — with some units accessing as fast as 3 ns — at the expense of power. Many of the static RAMs and PROMs are fabricated by using either transistor-transistor logic (TTL) or emitter-coupled logic (ECL) to get the highest performance. *See* COMPUTER STORAGE TECHNOLOGY; INTEGRATED CIRCUITS; LOGIC CIRCUITS; MAGNETIC BUBBLE MEMORY.   [DAVE BURSKY]

Semiconductor rectifier

An electrical component which conducts current preferentially in one direction and inhibits the flow of current in the other direction by utilizing the properties of a semiconductor material such as silicon. A major use of the semiconductor rectifier is as a component in electrical equipment designed to convert electrical power from alternating current (ac) to direct current (dc). Such total equipment, consisting of not only the semiconductor components but also circuit breakers, bus work, fuses, sharing reactors, control circuits, and so forth, is often referred to as a semiconductor rectifier equipment. *See* CONTROLLED RECTIFIER; RECTIFIER; SEMICONDUCTOR; SEMICONDUCTOR DIODE.

Early rectifier components, utilizing the semiconductor properties of such polycrystalline materials as selenium and copper oxide, were called metallic rectifiers. Selenium rectifiers were introduced commercially around 1930 in Germany. In the early 1950s, following the invention of the transistor at Bell Laboratories, monocrystalline semiconductor rectifiers found commercial application — first germanium, then silicon in about 1954. Since then the silicon semiconductor rectifier diode has completely taken over the field in applications ranging from small electronic power supplies for radio and television to very large power-rectifier installations in the electrochemical and aluminum industries, which convert many millions of amperes from ac to dc for plating, refining, and reduction of aluminum oxide ore (bauxite) to pure aluminum. In doing so, the silicon rectifier has also displaced the mercury-arc rectifier.

In 1957 the first commercial silicon-controlled rectifier (SCR) was developed and introduced by General Electric in the United States. This electrical semiconductor component has, besides the anode and cathode of the rectifier diode, a third control electrode, or gate, which allows initiating conduction through the device by means of a small trigger pulse signal from a suitable control circuit. Functionally the SCR is a semiconductor version of a thyratron (gas) tube. By the late 1960s the SCR had become the dominant power-control component. It displaced the technologies of the thyratron tube, the controlled mercury-arc rectifier (ignitron), and by the 1970s the rotating motor-generator set type of power conversion device. Its application ranges from adjustable speed control of fractional horsepower motors, to lighting and heating controls, to control of the largest dc motors. *See* THYRATRON.

Silicon rectifier diodes. Rectifier diodes are usually distinguished from other types of semiconductor diodes by the nature of their application as power-rectifying devices. While this function can take place at power conversion frequencies of from 50 Hz to 50 kHz, other diodes are used in information-processing areas such as computers, and are called computer diodes and tunnel, or Esaki, diodes. Zener diodes are employed as voltage-regulating or voltage reference devices and may be used at signal or power levels. Still another class of diodes is used in microwave-frequency applications, extending beyond the X and Ku bands, to 20 GHz. These diodes are known as Impatt, PIN, and step-recovery diodes. They are used in both signal and power applications in microwave transmitters and receivers to generate, control, and switch microwave frequencies. They are also used in high-speed instrumentation equipment, such as oscilloscopes and counters. In the rectifier diode, the important properties are current- and voltage-handling capability, recovery time (a measure of speed), efficiency, and favorable heat transfer characteristics. The information-processing and microwave-frequency type of diodes are usually very small and of lower voltage capability, and are optimized for operation at frequencies very much greater than the power-frequency (50 Hz − 50 kHz) range of operation associated with rectifier diodes. *See* MICROWAVE SOLID-STATE DEVICES; SILICON DIODE; TUNNEL DIODE; ZENER DIODE.

The electrical heart of the rectifier diode is the junction between a p-type region and an n-type region of conductivity within a monocrystalline slice of silicon semiconductor material called the pellet or the chip. The processing required to form such a pn junction is accomplished in commercially available silicon devices by either a diffusion or an alloying process. For conductivity types and doping of semiconductors *see* SEMICONDUCTOR.

In the diffusion process, a chemically very pure slice of silicon is exposed to the hot vapor of a dopant material. Three-valence elements, such as gallium or boron, accept electrons and form a p-type region; 5-valence elements, such as phosphorus, donate electrons and form an n-type region of conductivity. If, for example, an n-type slice of silicon is exposed to an acceptor material, a p-type region will be formed at the surfaces. If subsequently one of these regions is removed, a pn junction will remain.

In the alloy process, a similar result is accom-

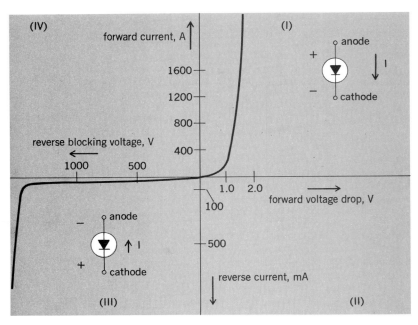

Fig. 1. Volt-ampere characteristic of a typical 250-A silicon rectifier diode.

plished by placing a die of suitable dopant material in physical contact with the silicon slice under the proper conditions.

Once the *pn* junction is formed, it exhibits its rectifying characteristic by allowing current to flow in the conventional sense from anode (*p*-

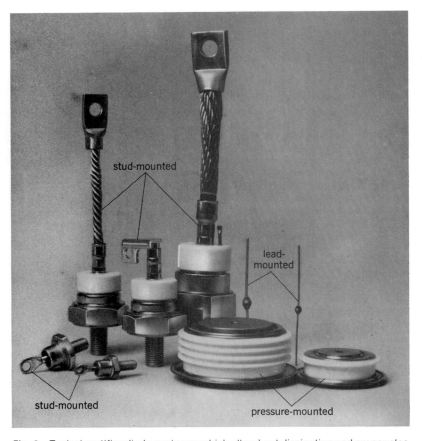

Fig. 2. Typical rectifier diode packages which allow heat dissipation and proper electrical contact. (*General Electric Co.*)

type) to cathode (*n*-type) under a very small forward bias voltage; when the voltage polarity is reversed, the *pn* junction blocks the flow of current and only a very small reverse blocking or leakage current will flow under high-reverse-voltage bias.

Figure 1 shows the volt-ampere characteristic of a typical 250-A silicon rectifier diode. Quadrant I shows the conducting characteristic (high current, low forward voltage drop); quadrant III shows the reverse blocking characteristic (high blocking voltage, low reverse blocking current). Note the difference in scale factor. The respective voltage polarities are shown on the circuit symbol of the diode in each quadrant. An ideal diode would have zero forward drop and zero reverse blocking current. Practical silicon power diodes have forward voltage drops of the order of 1 V at rated forward current, and reverse blocking current of the order of milliamperes for large diodes at their maximum rated operating temperature (usually about 150–200°C). With decreasing temperature, reverse blocking current likewise decreases, but the forward voltage drop increases, displaying the negative temperature coefficient at normal rated current densities characteristic of all silicon *pn* semiconductor junction devices.

The maximum value of forward current that a silicon diode can carry is limited by the power dissipation (the characteristic 1-V forward voltage drop times the current) capability of the diode housing and any external heat-sink system. In some types of construction it is limited by the internal joints between the silicon and the other parts necessary to mount the brittle silicon pellet (or chip) to the housing (usually copper).

The typical thickness of the rectifier diode pellet is of the order of 0.35 mm, with a diameter of up to 100 mm for some of the larger devices rated in thousands of amperes. The pellet must be mounted in a suitable protective housing which allows the heat to be dissipated to the external cooling ambient and which allows for proper electrical contact. Usually, housing or packages are lead-mounted for currents up to a few amperes, stud-mounted and flat-base-mounted for currents up to several hundred amperes, and external-pressure-mounted for the largest devices (Fig. 2). The lead-mounted package dissipates the losses directly into the ambient air, whereas all other packages are mounted on heat sinks suitable for either air or water cooling.

Figure 3 shows the external and internal views of a typical silicon pellet subassembly enclosed by a hermetically sealed package. An SCR assembly is shown; the diode assembly is similar but does not have the small auxiliary gate and cathode leads.

Single-pellet silicon diodes are commercially available in current ratings from under 1 A to the order of 5000 A. Diodes may be connected in parallel for greater current capability when suitable means are employed to provide for current sharing between individual diodes.

The maximum value of reverse voltage blocking capability of a silicon diode is limited by the inherent semiconductor properties of the blocking *pn* junction, by the surface characteristics of the silicon pellet, particularly where the blocking junction meets the surface, and to a lesser degree by the housing. If the reverse blocking voltage is

fect of a metal-to-silicon barrier. The Schottky diode, sometimes called Schottky barrier diode, overcomes a major limitation of the *pn* junction diode; being a majority-carrier device, it has both higher forward conductivity (lower forward voltage drop) and faster switching speeds than its minority-carrier *pn* junction counterpart. However, other factors confine its use to low-voltage power

(a)

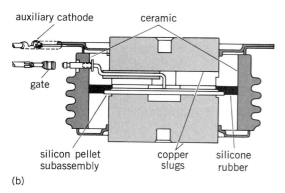

(b)

Fig. 3. High-current SCR in a pressure-mounted package. (*a*) External view. (*b*) Cross section. (*General Electric Co.*)

limited by the inherent or bulk semiconductor properties of the pellet, the volt-ampere characteristic will show a certain voltage beyond which the current increases very rapidly (Fig. 1, quadrant III). This is referred to as avalanche breakdown, usually a nondestructive phenomenon if the resulting power dissipation is limited to a safe value. If, on the other hand, the voltage capability of a silicon diode is limited by the surface properties of the pellet, the resulting breakdown following an excessive voltage stress is usually destructive.

Voltage ratings of single-pellet silicon rectifier diodes commercially available cover a range from 50 to over 5000 V. Whenever a single diode has insufficient voltage capability to meet the requirements of an application, individual diodes can be connected in series for greater voltage capability. In addition to connecting individual diodes in series, diodes can be obtained prepackaged in series strings for high-voltage applications. There are commercially available single packages containing multiple diodes with voltage ratings in excess of 100 kV and current ratings in the range of 200 mA to 2 A. Lower-voltage, multiple-diode packages can be purchased with voltage ratings in the 10–20-kV range and current ratings up to 10 A.

Schottky silicon rectifier diodes. Unlike the *pn* junction of a regular silicon rectifier diode, the Schottky diode makes use of the rectification ef-

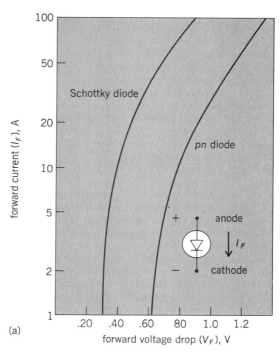

(a)

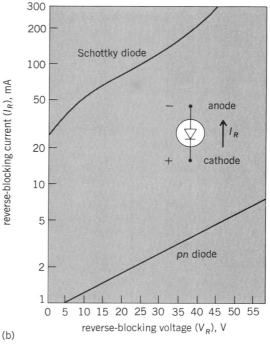

(b)

Fig. 4. Comparison of electrical characteristics of a Schottky diode with a *pn* diode of similar ratings. (*a*) Forward characteristics. (*b*) Typical reverse characteristics. Schottky diode is operated at 45 V, 30 A, with junction at 125°C in *a*, 150°C in *b*. The *pn* diode is operated at 50 V, 30 A, with junction at 200°C.

supply applications—chiefly its limited reverse blocking voltage, typically 45 V, with a maximum of 80 V presently available (1981). Secondary shortcomings are high reverse blocking current and restricted temperature of operation, with commercial devices providing a maximum of 175°C, compared with 200°C for *pn* junction diodes. Figure 4 compares the forward and reverse electrical characteristics of the two types of devices.

Integrated circuits used in computer and instrument systems commonly require voltages less than 15 V and frequently as low as 5 V. Thus, the advantage of low forward voltage drop and faster switching favors the Schottky rectifier diode, due to its inherent efficiency. This is particularly true for the high-frequency-switching regulator supply applications where power at switching frequencies of 20–50 kHz has to be rectified. The higher reverse blocking current of the Schottky diode is more than compensated for by the Schottky's superior switching speeds. However, its cooling is more critical because of its high reverse blocking current temperature coefficient and lower maximum operating temperature. Thermal runaway and consequent destruction is more likely to occur with the Schottky diode than its *pn* junction counterpart. This occurs when a diode's power dissipation reaches a level that exceeds the ability of the cooling system to maintain the diode's temperature in equilibrium. It can be thought of as analogous to that of a sailboat which, when tipped beyond a certain angle, can no longer maintain equilibrium and will capsize. Sailboats with ballasted keels have increased stability over unballasted boats. Likewise, *pn* junctions have increased thermal stability over Schottky diodes.

Rectifier circuits. The greatest usage of rectifier diodes is in the conversion of electrical power from ac to dc by means of employing the diodes in a suitable electric circuit. Figure 5 shows some basic rectifier circuits.

In Fig. 5*a* a single diode supplies power to the load only on the positive (or on half of the available) cycles of the ac supply. The load current is unidirectional and has an average value; hence, the supply current has been rectified.

Figure 5*b* shows a popular circuit for rectification of both of the available half-cycles of the supply. In both cases the waveform of the output current departs from a pure dc current which could be supplied by a battery. A measure for the quality of the output of a rectifier circuit relative to the direct current that a battery could supply is called the ripple voltage. The output ripple voltage of the full-wave circuit is less than that of the half-wave circuit. When the load requirements call for less ripple, polyphase rectifier circuits are used in high-power applications. In lower-power applications from a few watts to a few tens of kilowatts, such as are used in computers and television receivers, it is common either to use a switching regulator to generate high-frequency ac power (10–50 kHz) from the high-ripple rectified 50–60-Hz ac power supply commonly available from the local utility, or to add a relatively bulky and expensive filter to the output of the 50–60-Hz rectifier circuit. The advantage of the switching regulator approach stems from the ease of filtering high-frequency ripple resulting from rectifying high-frequency ac as opposed to filtering low-frequency ripple.

Silicon-controlled rectifier. The SCR is a triode reverse-blocking thyristor. Whereas diodes use two alternate layers of *pn*-type semiconductor material and transistors use three such layers, thyristor devices utilize four layers forming three or more junctions within a slice of silicon semiconductor material. Thyristor devices exhibit regenerative, or latching-type, switching action in one or two quadrants of their volt-ampere characteristic. They can be switched into the ON state (conducting condition) but must usually be restored to their OFF state (voltage-blocking condition) by circuit action. Figure 6 shows a family of thyristor devices.

The most widely used of all thyristor devices for power control is the SCR. For specialized ac-switching power control, such as in lamp dimmers and heating controls, the bidirectional triode thyristor, popularly called the triac, has also come into widespread usage. Other types of thyristors are discussed below.

Figure 7 shows a typical arrangement of alternate *p* and *n* layers in an SCR structure. The thickness is exaggerated in proportion for clarity of illustration. With positive voltage on its cathode with respect to its anode, the SCR blocks the flow of reverse current in a manner similar to that of a conventional silicon rectifier diode. When the voltage is reversed, the SCR blocks forward current flow until a low-power trigger signal is applied between the gate terminal and the cathode, whereupon the SCR switches into a highly conductive state with a voltage drop of approximately 1 V between anode and cathode similar to that of the rectifier diode. Once in conduction, the SCR continues to conduct even after the gate signal is removed, provided the anode (or load) current remains above the holding current level, typically

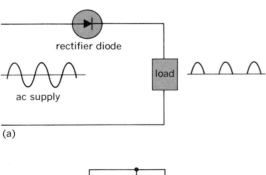

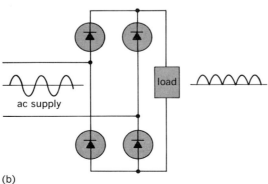

Fig. 5. Some basic rectifier circuits. (*a*) Half-wave rectifier circuit. (*b*) Full-wave bridge rectifier circuit.

in the order of milliamperes. If anode current momentarily drops below the holding current level or if the anode voltage is momentarily reversed, the SCR reverts to its blocking state and the gate terminal regains control. Typical SCRs turn on in about $1-5\ \mu$s and require $10-100\ \mu$s of momentary reverse voltage on the anode to regain their forward-blocking ability. The details of the ON and OFF switching characteristics of SCRs vary with different types made for varying applications.

Anode voltage applied to the SCR significantly in excess of the voltage rating of the SCR can trigger the device into conduction even in the absence of a gate signal. Excess reverse voltage, however, can permanently damage the SCR much as in the case of the silicon rectifier diode. SCRs, like the silicon diode and all power semiconductors, have a failure mechanism called thermal fatigue. Thermal fatigue failure is due to the thermal stresses induced during repetitive temperature changes occurring in the normal operation of the device. These stresses are inherent in all devices undergoing substantial temperature changes which contain dissimilar metals. When power semiconductors, such as rectifier diodes and SCRs, are properly applied to take into account their thermal fatigue limitations, they can be expected to perform their function faultlessly for the life of the equipment in which they are used.

Current ratings of SCRs range from under 1 to 5000 A. Blocking voltage capability of commercially available devices extends to 4400 V for the higher-power types, with voltages up to 6 kV having been demonstrated in the laboratory.

Like most semiconductor devices, SCRs are dependent on temperature in some of their characteristics. Usual operating junction temperatures are 125°C, and some devices are available up to 150°C.

The mounting considerations for SCRs are similar to those for diodes. Small devices are lead-mounted, and above 2 and 4 A SCRs are generally mounted to radiating fins or some type of heat sink for adequate cooling of the semiconductor junction. Figures 8 and 9 show two typical SCRs: a small 2-A plastic-encapsulated SCR for use in consumer applications, and a 1000-A 1400-V SCR water-cooled assembly used in glass-melting applications.

Processing techniques used in SCRs are similar to and extensions of the processing used for silicon diodes. In addition to alloy and diffusion processing technology, epitaxial processing is sometimes used. In small devices, similar to the one shown in Fig. 8, a planar structure such as that developed for signal transistors and monolithic integrated circuits is used. Higher-power SCR structures are of a mesa type of construction, with the edges of the pellet often shaped in a manner to reduce the surface field across the blocking junction for higher voltage-blocking capability.

SCR applications fall into two general categories. In one category the devices are used from an ac supply, much as the silicon rectifier diode is used. However, unlike the rectifier diode, which conducts load current as soon as the anode voltage assumes a positive value, the SCR will not conduct load current until it is triggered into conduction. If, when applied in rectifier circuits of the type shown in Fig. 5, conduction through the SCR is de-

Type of thyristor	Common circuit symbol	Electrical conduction characteristics
reverse-blocking diode thyristor	anode / cathode	
bidirectional diode thyristor (SIDAC)	A_2 / A_1	
reverse-blocking triode thyristor (SCR)	anode / gate / cathode	
asymmetrical triode thyristor (ASCR)	anode / gate / cathode	
reverse-conducting triode thyristor (RCT)	anode / gate / cathode	
commutable reverse-blocking triode thyristor (GTO)	anode / gate / cathode	
bidirectional triode thyristor (TRIAC)	A_2 / gate / A_1	

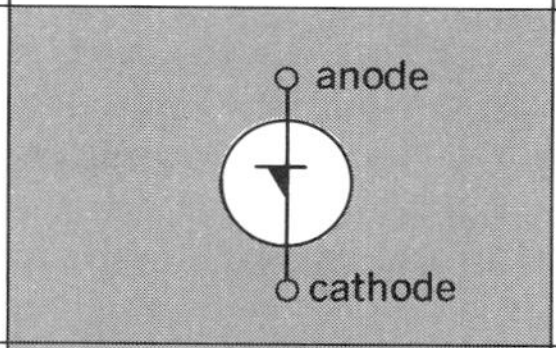

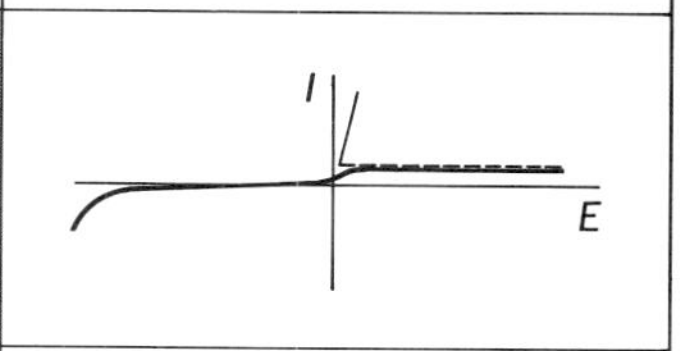

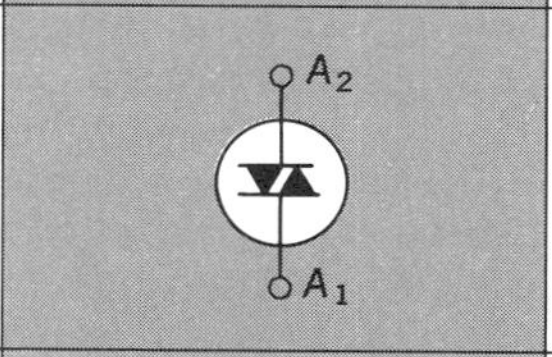

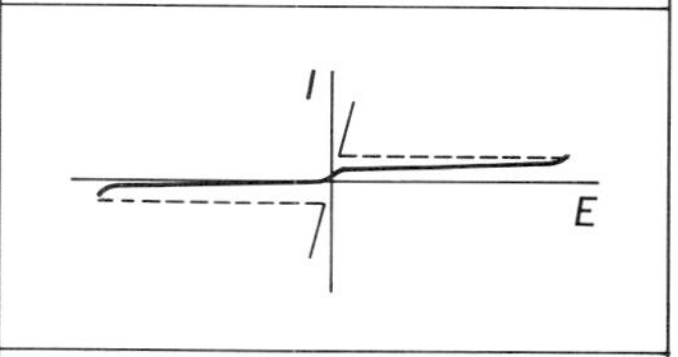

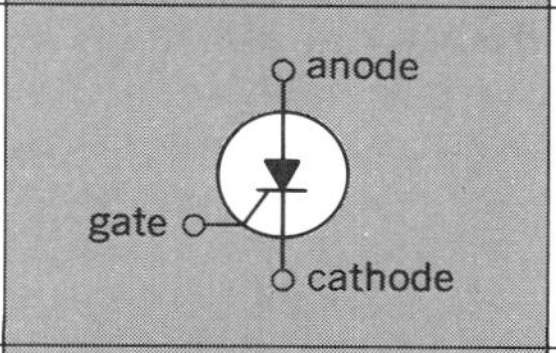

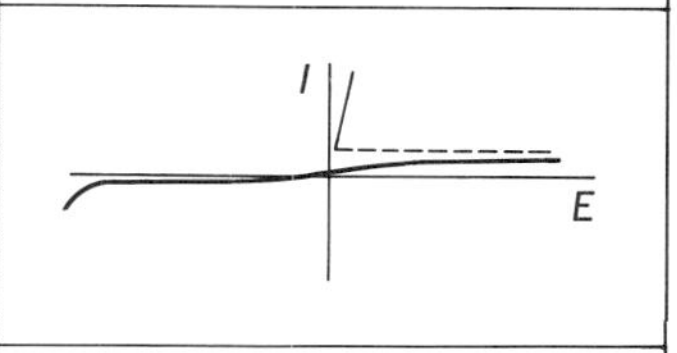

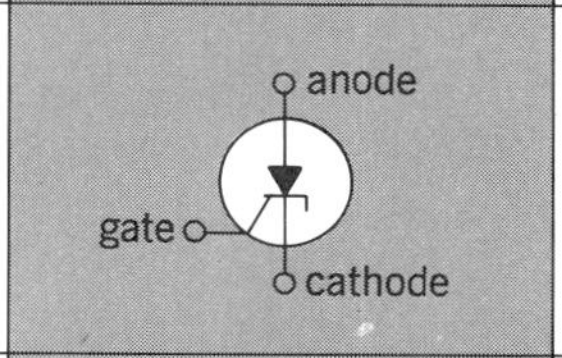

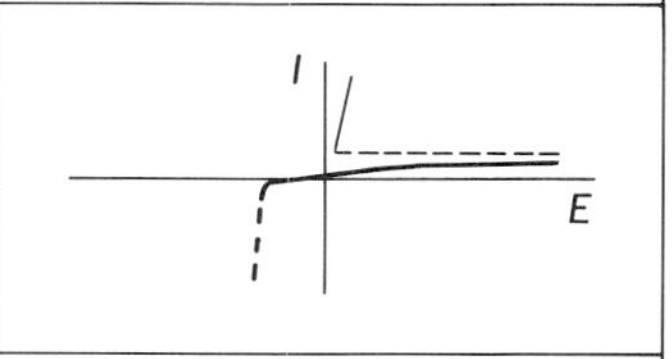

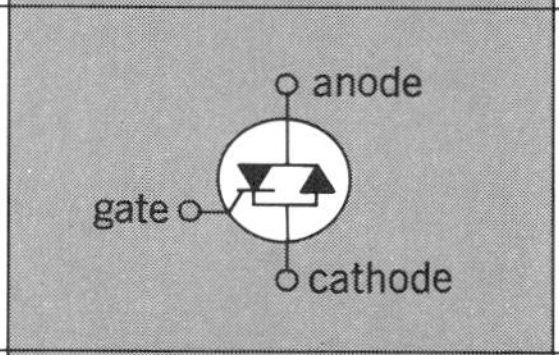

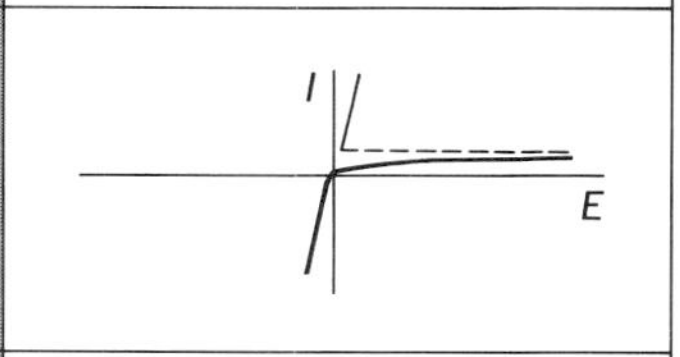

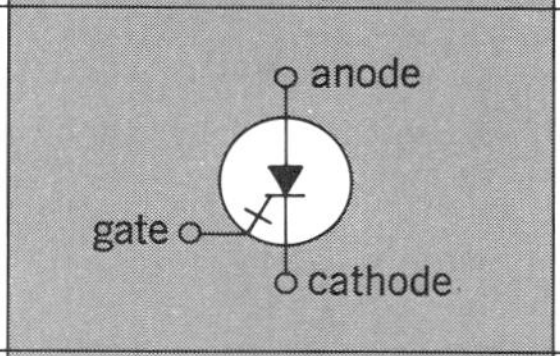

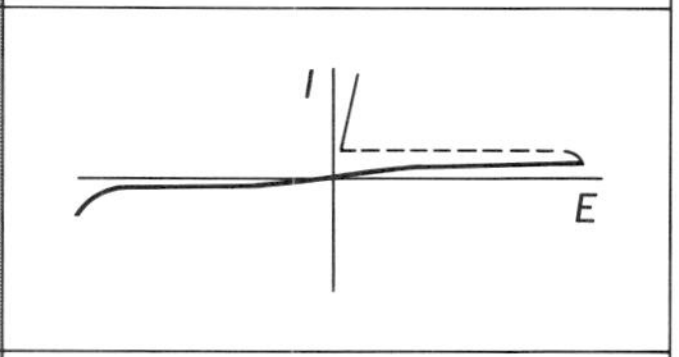

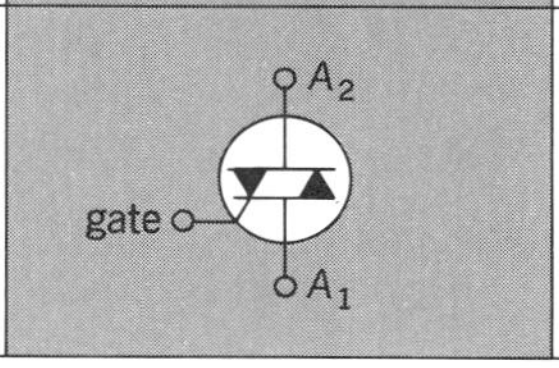

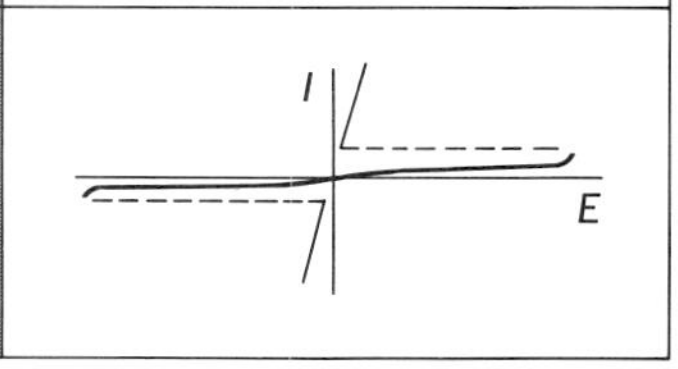

Fig. 6. Family of thyristor devices.

layed from the point of the natural zero crossing of the forward anode-to-cathode voltage, the power delivered to the load can be varied. This mode of control is referred to as ac phase control. It is used extensively in ac to variable-voltage dc output types of applications, such as the adjustable-speed dc motor drives in the circuits similar to those shown in Fig. 5a and b, in which the diodes have been replaced by SCRs and suitable control circuitry is provided to drive the gates. The circuit of Fig. 10 is the parallel-inverse, or ac-switch, circuit which can supply variable voltage to ac loads. It is used extensively in lighting and heating control (see Fig. 9).

Turnoff of the SCR in ac circuits is accomplished by the reversal in voltage polarity of the ac supply

SEMICONDUCTOR RECTIFIER

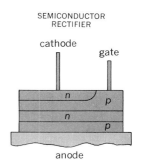

Fig. 7. Diagrammatic view of typical SCR structure.

Fig. 8. Small plastic-
encapsulated SCR.
(*General Electric Co.*)

line. Since the SCR, like the diode, blocks the flow of reverse current, there is no flow of load current during the half-cycle of applied line voltage which places reverse bias across the SCR.

The other basic category of application for SCRs from a circuit point of view is operation from a dc supply. The major distinguishing feature of dc operation from ac operation is that there is no reversal of supply voltage polarity for turnoff of the SCR, which would allow the gate electrode to regain control of the device. Therefore, auxiliary circuit means must be employed to effect turnoff of the SCR. One common way to accomplish this is to switch a previously charged capacitor across the load-carrying SCR in such a manner that the voltage on the capacitor reverse-biases the SCR sufficiently to reduce the load current through the SCR to zero, and then to allow the device a short time (about $10-50~\mu$s) before reapplying forward blocking voltage to the anode of the SCR.

ASCRs, RCTs, and GTOs. These devices are all in the thyristor family and are mainly used in place of SCRs in power circuits requiring operation from

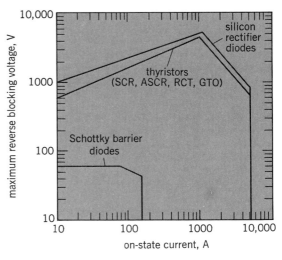

Fig. 11. Rating comparison of SCRs, rectifier diodes, and Schottky barrier diodes commercially available with single silicon discrete devices. The horizontal scale gives rms current for SCRs, average current for rectifier diodes.

a dc source. The ASCR, asymmetrical silicon-controlled rectifier, and RCT, reverse-conducting thyristor, have the advantage of faster turnoff time than the SCR, that is, $5-25~\mu$s, and thus require a less costly auxiliary circuit to effect turnoff. The RCT has an added circuit advantage, as it has a built-in reverse rectifier diode in parallel with the device. (The RCT is the integrated equivalent of a discrete ASCR in parallel with a discrete rectifier diode.) Along with faster turnoff times, the ASCR and RCT devices have lower forward-voltage drops for comparable forward-blocking voltage ratings and silicon area, thus increasing the device's efficiency.

The GTO, gate turnoff device, is also a thyristor. Like the SCR, it is a symmetrical reverse-blocking triode thyristor (unlike the ASCR and RCT, which cannot block reverse voltages), but it has the added advantage of being able to turn off current when a negative signal is applied to the gate. (Voltage and current characteristics are given in Fig. 6.) Thus the GTO does not require an auxiliary circuit to commutate it off as do the SCR, ASCR, and RCT devices. The added complexity of gate turnoff makes the GTO higher priced than similarly rated SCRs.

The GTO, ASCR, and RCT all came into commercial usage during the latter half of the 1970s, primarily at voltage levels greater than 800 V and currents exceeding 25 A. It is expected that they will soon be manufactured in all current and voltage ratings that are presently available in SCRs above the lower values stated. It is projected that the greatest utilization will be of voltage ratings exceeding 1400 V and currents exceeding 100 A. Figure 11 shows a rating comparison of rectifier diodes, thyristors, and Schottky barrier diodes now on the market.

It is necessary to operate thyristors from a dc supply in order to achieve power conversion from the dc (battery or rectified ac line) supply to a load requiring an alternating supply (dc to ac inversion) or to a load requiring a variable-voltage dc

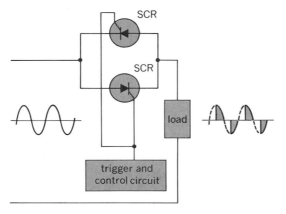

Fig. 9. High-current water-cooled SCR ac switch for glass-melting applications. (*General Electric Co.*)

Fig. 10. Phase-controlled SCR ac switch circuit.

supply (dc to dc conversion). Since the rate of switching the thyristors in dc circuits can be varied by the control circuit, a thyristor inverter circuit can supply an ac load with a variable frequency. An important application of this mode of operation is for adjustable speed operation of ac synchronous and induction motors in industrial processing.

A battery source can be converted to a variable-voltage dc source for a dc motor by "chopping" the dc source voltage either at a variable rate at constant pulse width (frequency power modulation) or by operating the chopper circuit at a constant frequency and varying the pulse width (pulse-width power modulation). An important application of chopper circuits is in battery-operated vehicles, such as forklift trucks and mining locomotives. During the mid-1980s it is expected that automotive vehicles will be added to this list. *See* CHOPPER; MULTIVIBRATOR.

[F. B. GOLDEN]

Bibliography: S. K. Ghandhi, *Semiconductor Power Devices*: *Physics of Operation and Fabrication Technology*, 1977; F. B. Golden and D. R. Grafham, *SCR Manual*, 6th ed., 1979; *Power Semiconductor D.A.T.A. Book*, 13th ed., vol. 25, book 23, 1980; J. Schaefer, *Rectifier Circuits*: *Theory and Design*, 1968.

Sensitivity

The ability of the output of a device or system to respond to an input stimulus. Mathematically, sensitivity is expressed as the ratio of the response or change induced in the output to a stimulus or change in the input. If the sensitivity varies with the level of the input signal, then sensitivity is usually expressed in terms of the derivative of the output with respect to the input at a specified input level.

The reciprocal of sensitivity is called the scale factor or figure of merit, and it represents the conversion factor by which the output indicator or scale reading must be multiplied to obtain the magnitude of the input. Occasionally the scale factor is called the sensitivity through loose usage.

Sensitivity is closely related to noise. Quite often the limiting factor in increasing the sensitivity of devices or systems is the inherent noise level. For instance, the noise level for an electric galvanometer or a gravitational weight balance arises from the random or Brownian movement of the air molecules surrounding the apparatus. In electrical circuits, the noise level arises from the random movement of electrons in resistors. *See* ELECTRICAL NOISE.

Examples of sensitivity are visual contrast sensitivity, which is the ability of the eye to distinguish between the luminances of adjacent areas; galvanometer sensitivity, expressed as the ratio of the scale deflection in millimeters per microampere input; and radio receiver sensitivity, which is actually expressed in terms of reciprocal sensitivity in the form of antenna voltage in microvolts necessary to cause a specified output.

The sensitivity of a radio receiver is a measure of its ability to reproduce weak broadcast signals with satisfactory output volume. The sensitivity of a television camera tube determines its ability to deliver a usable picture signal under poor lighting conditions. [JOHN MARKUS]

Series circuit

An electric circuit in which the principal circuit elements have their terminals joined in sequence so that a common current flows through all the elements.

The circuit may consist of any number of passive and active elements, such as resistors, inductors, capacitors, electron tubes, and transistors.

The algebraic sum of the voltage drops across each of the circuit elements of the series circuit must equal the algebraic sum of the applied voltages. This rule is known as Kirchhoff's second law and is of fundamental importance in electric circuit theory. *See* KIRCHHOFF'S LAWS OF ELECTRIC CIRCUITS.

When time-varying voltages and currents are involved, it is necessary to employ differential or integral equations to express the summation of voltages about a series circuit. If the voltages and currents vary sinusoidally with time, functions of a complex variable are used in place of the calculus. *See* ALTERNATING-CURRENT CIRCUIT THEORY; CIRCUIT (ELECTRICITY); DIRECT-CURRENT CIRCUIT THEORY. [ROBERT LEE RAMEY]

Shunting

The act of connecting one device to the terminals of another so that the current is divided between the two devices in proportion to their respective admittances. Shunting is widely used in ammeters, galvanometers, and other current-measuring instruments to bypass part of the current around the instrument so as to change the measuring range. Resistors are frequently shunted across tuned circuits to broaden the tuning characteristics.

Shunting is equivalent to connecting in parallel. Shunting one resistor with another gives a lower resistance for the combination, whereas shunting one capacitor with another gives a total capacitance equal to the sum of the individual values. *See* ALTERNATING-CURRENT CIRCUIT THEORY; DIRECT-CURRENT CIRCUIT THEORY.

[JOHN MARKUS]

Sideband

The frequency band located either above or below the carrier frequency within which fall the frequency components of the wave produced by the process of modulation. Apart from the carrier, all components of an amplitude-modulated sinusoidal carrier, when taken together, form a pair of sidebands extending on either side of the carrier frequency in mirror symmetry and containing all the frequency components of the modulating wave. The sidebands above and below the carrier frequency are called upper sideband and lower sideband, respectively. *See* AMPLITUDE MODULATION; CARRIER; MODULATION; SINGLE SIDEBAND.

[HAROLD S. BLACK]

Signal generator

A piece of electronic test equipment that delivers a sinusoidal output of accurately calibrated frequency. The frequency may be anywhere from audio to microwave, depending upon the intended use of the instrument. The frequency and the amplitude are adjustable over a wide range. The oscillator must have excellent frequency stability, and its amplitude must remain constant over the tuning range.

The Wien-bridge oscillator is commonly used for frequencies up to about 200 kHz. For a radio-frequency signal generator up to about 200 MHz, a resonant circuit oscillator is used (such as a tuned-plate tuned-grid, Hartley, or Colpitts). Beyond this range vhf and microwave oscillators are used.

Many signal generators contain circuitry that allows the output to be either amplitude- or frequency-modulated. The most common forms of amplitude modulation are sinusoidal, square-wave, or pulse. The frequency is either kept constant, is sinusoidal-modulated, or is swept linearly across a band of frequencies. For example, for testing broadcast receivers, it is important to sweep the generator frequency over a range of ±10 kHz at a low rate, say 60 times a second. *See* OSCILLATOR; PULSE GENERATOR; WAVE-SHAPING CIRCUITS. [JACOB MILLMAN]

Bibliography: C. F. Coombs, *Basic Electronic Instrument Handbook*, 1972; W. D. Cooper, *Electronic Instrumentation and Measurement Techniques*, 2d ed., 1978; J. Markus, *Sourcebook of Electronic Circuits: Modern Electronic Circuits Manual*, 1980.

Signal-to-noise ratio

The ratio, at some location, of some measure of the desired signal to the same measure of the total noise, abbreviated S/N. This is a primary consideration in the design of any communication system, and is a measure of the efficiency of the system. Distortion of the signal by noise causes errors. The objective of engineering is the maintenance of error-free communications over the smallest possible S/N value.

In radio communications, it is common to rate a receiver in noise factor (NF). The value of the factor is stated in decibel units of the noise contribution of the receiver circuitry over that of a theoretically perfect receiving device. Good engineering design can maintain a figure of approximately 2.5 dB NF in the vhf band. *See* ELECTRICAL NOISE.

[WALTER LYONS]

Silicon diode

A small silicon rectifier of either the point-contact, bonded-contact, or junction type. It is distinguished from a silicon rectifier by size only, the latter term being used for units of relatively large power-handling capacity. Point-contact types originally found application in microwave detectors and mixers. The bonded and microjunction types have shown a marginal life and stability. Almost all modern diodes are of the diffused planar type. Silicon diodes, contrasted with germanium diodes, are capable of operating at higher temperatures and therefore at higher power levels. Operation at temperatures as high as 200°C and voltages up to 1000 volts is possible. *See* JUNCTION DIODE; POINT-CONTACT DIODE; SEMICONDUCTOR; SEMICONDUCTOR DIODE.

[LLOYD P. HUNTER]

Simulation

The development and use of computer models for the study of actual or postulated dynamic systems. This definition, based on the current meaning of the word to the technological community, requires some interpretation.

The essential characteristic of simulation is the use of models for study and experimentation rather than the actual system modeled. In practice, it has come to mean the use of computer models because modern electronic computers are so much superior for most kinds of simulation that computer modeling dominates the field. "Systems," as used in the definition, refers to an interrelated set of elements, components, or subsystems. "Dynamic systems" are specified because the study of static systems seldom justifies the sophistication inherent in computer simulation. *See* ANALOG COMPUTER; COMPUTER; DIGITAL COMPUTER.

"Postulated" systems as well as "actual" ones are included in the definition because of the importance of simulation for testing hypotheses, as well as designs of systems not yet in existence. The "development" as well as "use" of models is included because, in the empirical approach to system simulation, a simplified simulation of a hypothesized model is used to check educated guesses, and thus to develop a more sophisticated and more realistic simulation of the simuland. The simuland is that which is simulated, whether real or postulated.

The impracticality of developing intercontinental ballistic missiles and spacecraft by actual flight testing gave simulation its big impetus just after World War II. Since then the equipment and the techniques have been adopted by workers in other fields where the simuland does not exist or is intractable to experimental manipulation, or where experiments with the actual system would entail high cost, danger to the system or the experimenter, or both. Among these nonaerospace systems which have been simulated, in addition to those considered later, are chemical and other industrial processes; structural dynamics; physiological and biological systems; automobile, ship, and submarine dynamics; social, ecological, political, and economic systems; electrical, electronic, optical, and acoustic systems; and learning, thinking, and problem-solving systems.

Simulations may be classified according to (1) the kind of computer used (analog, digital, hybrid); (2) the nature of the simuland (spacecraft, chemical plant, economic system); (3) the signal flow in the simuland (continuous, discrete or mixed); or (4) the temporal relation of events in the simuland (faster-than-real-time, real-time, slower-than-real-time, with real-time being clock or simuland time). Thus, for example, one may have a continuous real-time analog spacecraft simulation, or a sampled-data (discrete) faster-than-real-time digital economic system simulation, though such complete classification is seldom spelled out.

Mathematical modeling. Mathematical modeling is a recognized and valuable adjunct, and usually a precursor, of computer simulation. In practice the mathematical equations describing the interrelation of components of the simuland are written in a form suitable for the computer. In true simulation the computer is programmed to retain an identifiable correspondence between computer functions and the dynamics of the simuland. If mathematical manipulations or computer characteristics obscure this relationship, the computer may be processing information relative to the simuland, but it cannot properly be said to be simulating it.

Mathematical modeling does not necessarily precede simulation, however; sometimes the simuland is not well enough understood to permit rigorous mathematical description. In such cases it is often possible to postulate a functional relationship of the elements of the simuland without specifying mathematically what that relationship is. This is the building-block approach, for which analog computers are particularly well suited. Parameters related to the function of the blocks can be adjusted intuitively, systematically, or according to some established technique for system identification until some functional criteria are met. Thus the mathematical model can be developed as the result of, rather than as a requirement for, simulation.

Analog simulation. Simulation, as defined here, has been developed since World War II. Analog computers, in which signals are continuous and are processed in parallel, were originally the most popular for simulation. Their modular design made it natural to retain the simulation-simuland correspondence, and their parallel operation gave them the speed required for real-time operation. The result was unsurpassed human-machine rapport.

Digital simulation. However, block-oriented digital simulation languages (such as MIDAS, CSMP, and their successors) were developed which allow a pseudo–simulation-simuland correspondence, and digital computer speeds have increased to a degree that allows real-time simulation of all but very fast or very complex systems. Inadequate input/output facilities degrade human-machine interaction in all but the most sophisticated digital systems which embody expensive input/output equipment; nevertheless, digital computers are better than analog for simulating certain kinds of systems, particularly those requiring high precision or extremely wide dynamic ranges. *See* DATA-PROCESSING SYSTEMS; DIGITAL COMPUTER PROGRAMMING.

Hybrid simulation. Hybrid simulation, in which both continuous and discrete signals are processed, both in parallel and serially, is the result of a desire to combine the speed and human-machine rapport of the analog computer with the precision, logic capability, and memory capacity of the digital computer. After years during which the weak points of each kind of computer seemed to combine more readily than their strengths, hybrid simulation now makes possible the simulation of a new array of systems which require combinations of computer characteristics unavailable in either all-analog or all-digital computers.

In the past, most hybrid simulation systems were engineered by connecting a general-purpose digital computer to one or more general-purpose analog computers through interface equipment consisting largely of digital-to-analog and analog-to-digital signal converters (Fig. 1). However, the practice of hybridizing analog computers by the addition of a complement of digital logic and the development of specialized digital hardware was followed by the development of powerful hybrid systems specifically designed for the simulation of large complicated simulands. The relatively high cost of such systems is more than justified for

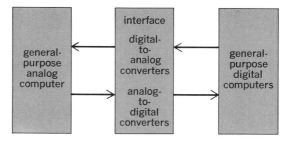

Fig. 1.   Hybrid simulation system.

some types of problems, primarily by the speed derived from the parallel operation of the analog elements, particularly the integrators, which are so prominent in any computer simulation program. Such speed pays dividends when the simulation involves high frequencies or when many iterations of an experiment are required—for instance, when it is necessary to get a statistically significant number of simulation runs of a stochastic process. In such systems the digital processer can select and program analog elements and the interface equipment, and check out the simulation. *See* ANALOG-TO-DIGITAL CONVERTER; DIGITAL-TO-ANALOG CONVERTER.

As the digital computer is capable of determining which parts of the model are simulated in the central processor and which in the remote terminal, the operator need not know what parts of the simuland are being simulated digitally and what parts by analog components.

Trends. With the advent of still more powerful digital computers, indications were that large computation centers would be designed around extremely fast time-shared digital central processors, accessed through many different special-purpose remote terminals (Fig. 2), and this has indeed been the case in many instances. Terminals for business data-processing and purely mathematical computation by such systems can consist of digital alphanumeric input/output devices, sophisticated successors to the teletype-

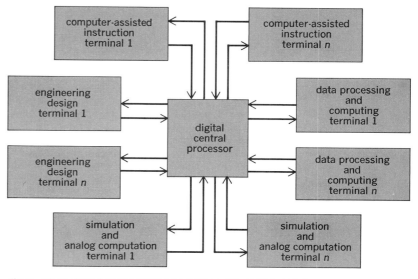

Fig. 2.   Time-shared digital central processor with multiple inputs.

writer. Terminals for engineering design, however, usually have graphic input and output, and use programs to facilitate interactive human-machine discourse. Either of the above kinds of terminals may be used for computer-assisted instruction, but still more sophisticated special-purpose terminals, designed specifically for educational purposes such as those used with the PLATO system, are coming into use. *See* COMPUTER GRAPHICS; MULTI-ACCESS COMPUTER.

Some special-purpose remote terminals which are particularly well suited for simulation are functionally similar to a small- to medium-sized hybrid computer. These "intelligent terminals" contain elements to facilitate human-machine interaction and have a limited capability for independent computation under local control. Typically, however, these terminals are used for communication with a central processor.

As the computing capacity of the remote terminals increases, they take on more of the character of satellite computers, and the overall system that of a computer network. The effect of such networks on simulation will be to provide access to more powerful simulation programs and to make more powerful equipment more widely and economically available. It should also discourage the practice of individuals who wish to simulate building their own model; ready-made models of many kinds can be accessed through the network.

The foregoing is a trend toward bigness; however, there is a countertrend. Started by the minicomputers and now getting a boost of unpredictable importance by the microcomputers, the current trend is such that it may not be long before all users can have as much computing power as they need in their own facility—if not at their own desk—so that the need to go to a network will be diminished if not eliminated. The exception may be the need for access to a large data bank, which becomes a communication, not a simulation problem. *See* MICROCOMPUTER.

Applications. Some simulationists have claimed that any system which can be adequately described in a natural language can be modeled on a computer, and thus simulated. Admittedly this seems to be an extremely broad statement, but it becomes credible if the meaning of "adequately" in this context is examined. A simulation study does not require that the simuland be completely described (an impractical if not impossible task in the case of many complicated systems). The simuland is adequately modeled when those factors—the parameters and the variables and their interrelationships which will significantly influence the results of a simulation experiment—are modeled and described with an accuracy commensurate with the accuracy required of the results.

The above is a qualitative description of adequacy as it applies to computer modeling for simulation. But computer modeling is completely quantitative; everything—parameters, variables, relationships, and even external influences—must be described by numbers. To do this, the modeler must be thoroughly familiar with the simuland and, having obtained the data required to assign the necessary numbers, must be able to program a computer to relate these factors. The modeler must also know the accuracy required of the re-

sults. Therefore most simulations of complicated systems are done by teams—an expert in the field of the simuland (economics, for example), a mathematician, a computer programmer, and an analyst working together to produce results useful to a decision maker (a business executive or a politician, for example).

Because simulation is a methodology for improving insight relative to complex issues in general, rather than in just certain fields, it has been used to investigate problems of all kinds. Only a few will be mentioned here.

Mental models of an economy have always been a tool of economic research. Three applications of simulation in economics are: (1) forecasting the effects on the economy (employment, production, consumption expenditure, inflation, balance of payments) of various policy changes (changes in government expenditure, tax rates, interest-rate ceilings, or the treatment of depreciation); (2) examining the behavior of the individual units in the economy (such as households, business firms, laborers); and (3) improving statistical tools used in estimating relationships among economic variables.

Management cannot be divorced from economics, and neither can be studied without regard to the social impacts involved. This is reflected in the computer simulation techniques that have been adopted by corporations in the advanced industrial countries. But with the increasing availability of low-cost microcomputers, simulation is also a tool for the management of small businesses. Their needs and problems are very similar to those of larger firms.

Traffic and transportation systems as well as business organizations require management, and effective management requires an understanding of the system of concern and, if possible, a means of studying the impact of alternative methods of controlling that system. Simulation fulfills these requirements.

Energy, however, has become the subject of the most intense simulation activity. Hundreds of models to support simulation studies of the generation, distribution, and use of electrical energy, triggered by the oil embargo of 1973, have been developed. The study of electrical energy systems has not been the only use of simulation to study energy-related problems. Another important application has been to study tradeoffs among the many existing and proposed sources of energy.

Computer simulation has become an indispensible tool in the study of complex biological processes ranging from intracellular chemical reactions, through the behavior of various organ systems, to the evolution of entire ecosystems. For example, computer simulation is uniquely suited to the study of plant disease epidemics, complex biological phenomena in which growth and interaction of the pathogen and host are affected by environmental factors; and the crop manager, when intervening by means such as chemical treatment, can become a third interactive component. The simulation of ecosystems via numerical models is essential in studying the way in which these ecosystems operate normally, how resistant this operation is to the short- or long-term effects of perturbation, and how such effects can be predicted; all these questions are important in current ecological research as the pressures of human economic growth threaten to impair or even destroy the function of many natural ecosystems.

[JOHN H. MC LEOD, JR.]

Bibliography: H. S. D. Cole et al., *Models of Doom*, 1973; J. W. Forrester, *World Dynamics*, 1971; G. Fromm et al., *Federally Supported Mathematical Models: Survey and Analysis—Final Report*, 1974; P. W. House and J. McLeod, *Large-Scale Models for Policy Evaluation*, 1977; J. McLeod (ed.), *Simulation in the Service of Society*, monthly; J. McLeod (ed.), *Simulation: The Dynamic Modeling of Ideas and Systems with Computers*, 1968; D. L. Meadows et al., *Dynamics of Growth in a Finite World*, 1974; D. L. Meadows et al., *The Limits to Growth*, 1974; M. Mesarovic and E. Pestel, *Mankind at the Turning Point*, 1974; *Proceedings of the Annual Summer Computer Simulation Conference*; *Proceedings of the International Scientific Forum on an Acceptable World Energy Future*, Cambridge, MA, 1979; *Proceedings of the National Computer Conference*; *Proceedings of the Pittsburgh Modeling and Simulation Conference*; *Proceedings of the Simulation Symposium*; *Proceedings of the Winter Simulation Conference*; M. Shubic and G. Brewer, *Systems Simulation and Gaming as an Approach to Understanding Organizations*, 1971; *Simulation Councils Proceedings Series*, semiannual; Society for Computer Simulation, *Simulation*, monthly.

Single sideband

An electronic signal-processing technique in which a spectrum of intelligence is translated from a zero reference frequency to a higher frequency without a change of frequency relationships within the translated spectrum. Single-sideband (SSB) signals have no appreciable carrier. After translation only the single-sideband energy remains. This form of intelligence transmission requires amplification of the SSB signal prior to transmission and occupies only the spectrum bandwidth of that intelligence. These advantages allow SSB to be selected for communication of voice and digital data wherever there is a premium on having a minimum of transmitted power and available frequency spectrum.

Amplitude-modulated (AM) signals have identical upper and lower sidebands symmetrically located on each side of the translation frequency, which is often called the carrier. The SSB spectrum differs from the AM spectrum in having little or no carrier and only one sideband. (A double-sideband signal is one in which only the carrier is suppressed, leaving both sidebands. This technique is not commonly used.) *See* AMPLITUDE MODULATION.

Generation. In the SSB signal-processing action, the intelligence spectrum to be translated is applied to the signal input port of a balanced modulator. A higher-frequency sinusoidal signal, often called a carrier, is applied to the other input port of this circuit. Its function is to translate the zero reference spectrum to the carrier frequency and to produce the upper and lower sidebands, which are symmetrically located on each side of the carrier. The carrier frequency power is suppressed to a negligible value by the balanced operation of the

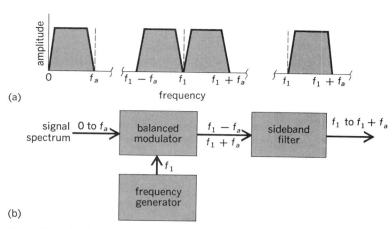

(a)

(b)

Generation of a single-sideband signal. (*a*) Spectrum. (*b*) Block diagram.

modulator and does not appear at the output. Generally, the balanced modulator operates at an intermediate frequency which is lower than the frequency of transmission. Following the balanced modulator is a sideband filter which is designed to remove the unwanted sideband signal power and to allow only the desired intelligence spectrum to pass. A block diagram and drawings of the signal characteristics at each major circuit interface are shown in the illustration. *See* AMPLITUDE MODULATOR; MODULATOR.

The outputs of several sideband filters having adjacent frequency passbands may be added to produce a succession of intelligence channel groups, each with differing information content. These may be translated to higher frequency and amplified for transmission; amplified and transmitted directly; or used as an input to a frequency, phase, or pulse code modulation system. *See* FREQUENCY MODULATION; MULTIPLEXING; PHASE MODULATION; PULSE MODULATION.

The SSB signal-handling processes are carefully designed to avoid distortion of the frequency relationships within the sideband spectrum. Amplifier circuits and components are chosen to provide a highly linear signal-handling capability. Filters are designed for equal amplitude response and uniform time delay of the frequency components in the signal passband.

Reception. Reception of SSB signals is essentially a reverse process from signal generation. The signal energy derived from the antenna is selected for a desired frequency range by radio-frequency filtering, is amplified to a desired level, and is translated down to an intermediate frequency. The signal-processing circuitry required to derive the desired intelligence channel spectrum from the intermediate frequency requires a demodulator which is similar to the modulator of the SSB generator.

Frequency stability requirements. Frequency stability of a high order is required of the signal generators that provide the translation frequencies in both SSB generation and reception. In the precise recovery of the intelligence spectrum there is a direct dependence upon total system frequency accuracy in returning the received intelligence sideband to zero reference frequency. Voice communication intelligibility suffers little with a system frequency error of up to 50 Hz, although digital data may be disrupted severely with only a few hertz error. Many SSB systems use a frequency synthesizer which makes all translation frequencies dependent on a single piezoelectric frequency generator which can be designed for frequency accuracy on the order of 1 part in 10^9 under severe environmental conditions. Atomic standards for frequency generation can provide reference frequency accuracy on the order of 1 part in 10^{12}.

The Doppler shift, caused by motion of high-velocity vehicles carrying SSB equipment, for example, supersonic aircraft or satellites, necessitates transmitting some reference frequency power (suppressed carrier) to provide at the receiver a source for frequency or phase locking of the frequency standard. *See* AUTOMATIC FREQUENCY CONTROL (AFC).

Communications. There are many advantages in the use of SSB techniques for communication systems. The two primary advantages are the reduction of transmission bandwidth and transmission power. The bandwidth required is not greater than the intelligence bandwidth and is one-half that used by amplitude modulation. The output power required to give equal energy in the intelligence bandwidth is one-sixth that of amplitude modulation.

Propagation of radio energy via ionospheric refraction provides the possibility for multiple paths of differing path length which can cause a selective cancellation of frequency components at regular frequency spacings. This produces in amplitude modulation a severe distortion of the intelligence because of the critically dependent carrier-to-sideband amplitude and phase relationships. SSB is much less affected under these conditions. SSB has proved to be a highly satisfactory communication signaling method where signal-to-noise ratio levels of under 10:1 are present, and SSB can produce intelligible voice communication at less than unity signal-to-noise ratio. SSB techniques are finding wide use in wire line, coaxial cable, long-range high-frequency communications, microwave multiplex, and ground-to-air – air-to-ground voice communication and are being used in experimental satellite-to-Earth channels.

History. The discovery of SSB can be credited to John R. Carson in 1915. He found from the mathematical equations that there was a preservation of the premodulation spectrum in the remaining sideband after the removal of the lower sideband and the carrier of an AM signal. He experimentally proved this by using the filtering capability of a frequency-selective antenna. His patent No. 1,449,382, filed in 1915 and granted in 1923, recognized this. Transoceanic experiments were also in progress by 1923, and were soon followed by commercial radio telephone channels. Wider utility followed developments in stable frequency generators and improved filter technology and amplifier components.   [DAVID M. HODGIN]

Bibliography: American Radio Relay League, *Radio Amateur's Handbook*, revised periodically; American Radio Relay League, *Single Sideband for the Radio Amateur*, revised periodically; G. W. Bartlett (ed.), *NAB Engineering Handbook*, 6th ed., 1975; F. R. Connor, *Modulation*, 1973.

Slide rule

A mechanical analog computing aid which is used extensively for multiplication and division and to a lesser degree for looking up functions. In its most common form a slide rule consists of a body formed from two parallel members rigidly fastened together, a slide which can be moved left or right between the body members, and a transparent indicator which carries a hairline and can be moved left or right over the face of the body and the slide. Scales are provided on the body and the slide as shown in the illustration.

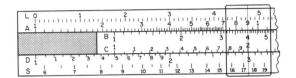

Left of slide rule, showing multiplication of 1.5 by 2.

The C and D scales, used for multiplication and division, are graduated from 1 on the left to 10 on the right, with intermediate numbers distributed logarithmically. Multiplication on a slide rule is based upon the fact that the product of two numbers can be obtained by adding the logarithms of the two numbers and then taking the antilogarithm. To perform multiplication, the index on the slider is aligned with the graduation on the body that represents the multiplicand, and the indicator hairline is placed over the position on the slider that represents the multiplier. The product appears under the hairline on the body. The illustration shows the positions of the body, slider, and indicator for performing the multiplication of 1.5 by 2.

Division, which is accomplished by subtraction of the logarithm of the divisor from that of the dividend, is performed in reverse sequence to multiplication. Thus, in the illustration the indicator would be set over the dividend 3 as read on the D scale, and then the slider would be moved until the divisor 2 as read on the C scale also appeared under the hairline. The quotient 1.5 then appears on the D scale opposite the index on the slider.

Numbers for which graduations do not exist must be estimated by eye, and a rough mental calculation must be performed to determine the position of the decimal point in the result obtained.

Many slide rules also carry scales from which sines, cosines, tangents, natural logarithms, logarithms to the base 10, squares, and cubes can be read. The S scale on the body of the rule illustrated is calibrated from approximately 6 to 90°, with the angles placed opposite the corresponding sines as read on the D scale. Thus the sine of 17.5° is 0.3. Logarithms to the base 10 can be read using the L and D scales.

The most common rule has a 10-in. scale, but larger and smaller straight rules and circular rules are used. Circular rules offer a convenience in multiplication. For example, multiplication of 2 by 6 using the left-hand C index is not possible with a straight rule. Instead, one must reverse ends and use the right-hand C index. With a circular rule this problem does not arise, because the scale is continuous.

In addition to the usual type of slide rule, many special rules have been devised to mechanize particular computations. Examples include slide rules for carrying out Ohm's law calculations and for finding the reactance of a given inductance or capacitance at a prescribed frequency. These may be thought of as mechanized nomographs.

[WILLIAM W. SEIFERT]

Soft fails

Noise bursts in microelectronic circuits, caused by cosmic-ray particles, that result in spontaneous changes in the information stored in computer memories. These changes are called soft fails. This sensitivity to cosmic rays is one of the unanticipated results of the ever-decreasing size of the components of integrated microelectronic circuits, and it presents new considerations in the development of very-large-scale integrated circuits. The problem is not necessarily catastrophic, since modern computers are usually made to continue to

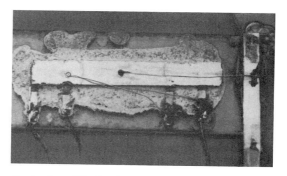

Fig. 1. One of the first integrated circuits, made in 1958. (*Texas Instruments*)

work properly, despite errors, by internal correction of electronic mistakes. However, the creation of soft fails adds an additional load to any internal corrections scheme.

Microelectronics. The transistor, invented in 1948, was initially utilized as a separate component with which to build electronic circuits by us-

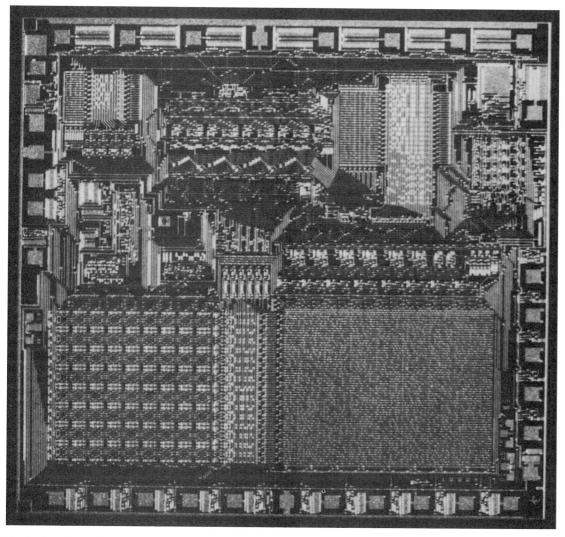

Fig. 2. Complete microcomputer built on one chip a few millimeters across. (*Texas Instruments*)

ing the same methods as had been developed for vacuum tube circuits. There was one obvious difference between vacuum tubes and transistors, however: transistors are inherently much smaller. This led to the realization that one can build integrated circuits, that is, electronic devices, with transistors, resistors, capacitors, and so forth, all made at the same time on one chip of silicon by using photolithographic techniques.

Since the first integrated circuit in 1958 (Fig. 1), the size of components in integrated circuits has decreased at a remarkable rate—almost a factor of 2 for every year. Now integrated circuits are available which have hundreds of thousands of components built on a single chip only a few millimeters across (Fig. 2). A typical state-of-the-art example is a 64 K (1K = 1024 bits) dynamic random-access computer memory (d-RAM) where binary information may be stored by charging (or not charging) a small capacitor. This capacitor stores approximately 1,500,000 electrons when charged, and is made of a thin film of conducting metal with an area of about 100 μm^2 on a thin insulator on a single-crystal silicon substrate. It is the very small amount of stored charge, along with the design of the capacitor, which makes such devices susceptible to the loss of bits of information by ionizing particles penetrating the underlying silicon. *See* INTEGRATED CIRCUITS; TRANSISTOR.

Ionization-induced soft fails. The spontaneous flipping of a bit stored in a computer memory is referred to as a soft fail, as distinguished from a hard fail in which a circuit component is permanently damaged and must be replaced. Integrated-circuit storage memories are normally extraordinarily reliable. The usual industry reliability unit is in failures per million hours per chip with nominal reliability being one fail per million hours. That means that if a chip stores 64,000 bits of binary information, the mean time to fail for each bit is 7,500,000 years!

In 1978 a new and unexpected source of soft fails became apparent to the microelectronics in-

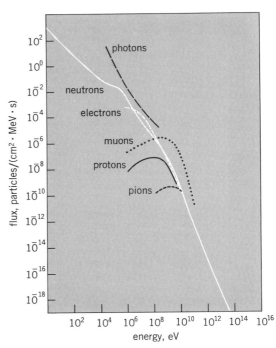

Fig. 4. Flux of cosmic-ray particles at sea level and geomagnetic latitude 45°N. These curves are average values, and large fluctuations exist, attributed to magnetic latitude, time of day, season, solar cycle, angle of incidence, and so forth. (*From J. Ziegler and W. A. Lanford, Effects of cosmic rays on computer memories. Science, 206:776–788, copyright 1979 by the American Association for the Advancement of Science*)

dustry. Memory circuits which had been designed based on seemingly reasonable extrapolations of previous devices had measured soft-fail rates far above expectation. The source of these unexpected upsets was traced to α-particles (helium nuclei) being emitted by naturally occurring radionuclides which had been inadvertently introduced as part of a ceramic support for the memory. The dramatic effect of α-particles on electronic memories is shown by the experimental array in Fig. 3. The charge-coupled device (CCD) memory matrix was filled with 1's, and then an α-particle from a radioactive source hit the memory at an angle of 20°. An array of hits spontaneously flipped to zeros in the pattern shown. This α-particle problem rapidly led to a study of the sources of the α-emitting contaminations and to the introduction of low background techniques in the semiconductor industry. *See* CHARGE-COUPLED DEVICES.

The discovery that α-particles from naturally occurring radionuclides could cause soft fails of computer memories led to the consideration of possible effects of another well-known source of naturally occurring radiation, cosmic rays. The average fluxes of various types of cosmic-ray particles at sea level are shown in Fig. 4. Investigations have demonstrated that sea-level cosmic rays have important effects which limit the reliability of memory devices.

The mechanism which causes these soft fails is familiar to all nuclear scientists because it is the basis for one of the most common types of radiation detectors, the solid-state ionization detector.

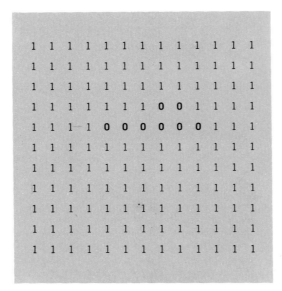

Fig. 3. Effect of α-particle ionization on a CCD memory matrix. (*Institute of Electrical and Electronics Engineers*)

While the detailed geometry, voltages, and so forth, vary considerably with memory circuit design, the underlying principle used to store information in integrated-circuit computer memories is always the same. Binary information is stored by the presence of charge on some element of the integrated circuit. In the case of d-RAM memories, the charge is on a capacitor.

In the case of charged-coupled devices, the charge is stored in a potential well at the interface between crystal silicon and a covering insulator such as silicon dioxide (SiO_2). This charge then migrates in a "racetrack" and stores information by the sequence of charges stored in this racetrack. In all such devices, the stored charges have to be periodically refreshed because of the slow but continual leakage of charge through the semiconducting silicon substrate.

The presence of charge results in an electric field in the silicon so that any electrons or holes created in this biased region are rapidly collected at the "anodes" and "cathodes," respectively. These are just the conditions needed to create a solid-state ionization chamber used as a particle detector. In the detector case, the charge pulse of electron-hole pairs created by ionizing radiation passing through the biased region of the silicon is used as a signal to measure the amount of ionization created by the particle. In the integrated-circuit memory case, the charge pulse of electron-hole pairs created by ionizing radiation decreases the stored charge (for example, on the capacitor in a d-RAM) and may result in the loss of the bit of information represented by this stored charge.

Ionization density thresholds. As the size of microelectronic circuits has decreased, the amount of charge used to store a bit of information has also decreased. For example, a 64K d-RAM may have a stored charge of order 1,500,000 electrons, and a 64K CCD has a stored charge an order of magnitude smaller. Table 1 lists some typical dimensions and charges for such devices. If this charge is decreased by an amount called $Q_{critical}$, the bit will be misread and a soft fail occurs. $Q_{critical}$ is of order 0.2 times the stored charge for most devices. While $Q_{critical}$ decreases with the size of the device, it also becomes progressively more difficult to deposit charge in the smaller active volume as dimensions decrease. For example, suppose $Q_{critical}$ is 500,000 electrons, and the active volume of the device has a mean diameter of order 10 μm. An ionizing particle must deposit charge within the sensitive volume at a rate of at least 500,000 electron-hole pairs per 10 μm = 50,000 electron-hole pairs per micrometer. Such a device would be insensitive to the ionization wakes of pro-

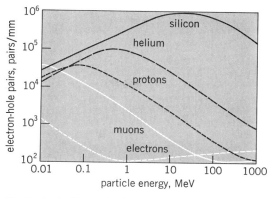

Fig. 5. Ionization wake density of electron-hole pairs in silicon following the passage of various charged particles. (*From J. F. Ziegler and W. A. Lanford, Effects of cosmic rays on computer memories, Science, 206:776–788, copyright 1979 by the American Association for the Advancement of Science*)

tons, muons, or electrons, because the maximum ionization density of any charge-one particle is 37,000 electron-hole pairs per micrometer. On the other hand, α-particles (which have charge number $Z = 2$) have a maximum ionization density of 100,000 electron-hole pairs per micrometer and, hence, could cause soft fails in such a device. The ionization wake density of electron-hole pairs in silicon following the passage of various charged particles is shown in Fig. 5.

While there are many more detailed considerations, the concept of thresholds in ionization density is central to understanding when soft fails may occur. When, as a consequence of minimization, devices were made with $Q_{critical}$ below the threshold of maximum α-particle ionization density, the α-particle soft-fail problem became important. Most early devices had $Q_{critical}$ so large that even α-particles could not deposit enough charge in the sensitive column to cause a soft fail. Such devices may, however, be susceptible to heavy-ion cosmic rays, which, while not present at sea level, are present above the Earth's atmosphere. Indeed, such heavy-ion-induced upsets were discovered in satellite computers.

Soft fail rates for typical devices. Calculations for some model devices have been carried out by assuming the characterizations given in Table 1. The results for the estimated fail rates for the mechanisms considered are shown in Table 2. For the 64k d-RAM, the single most important cause of soft fails is the production of α-particles by the interaction of cosmic-ray neutrons with silicon nuclei. The 64k CCD has such a low $Q_{critical}$ that it is sensitive to the primary ionization wake of cosmic-ray muons.

As can be seen in Table 2, both these model devices have fail rates larger than the traditional reliability standard of 1 fail per million hours with the 64k CCD memory orders of magnitude above this rate. These results are representative in the sense that, as $Q_{critical}$ falls below about 100,000 electrons for a device with dimension about 10 μm, it becomes sensitive to the ionization wake of $Z = 1$ particles and, consequently, has a rather large soft-fail rate. Because muons are so penetrating, it

Table 1. Device parameters for model computer memories

Parameter	64K d-RAM	64K CCD
Active area, μm²	100	200
Stored charge, e^-	1,500,000	180,000
$Q_{critical}$, e^-	300,000	36,000
Mean collection diameter, μm	12	16
Bits per chip	65,536	65,536

Table 2. Cosmic-ray-induced soft fails of model computer memories chip fails per million hours

Mechanism	64K d-RAM	64K CCD
Electron ionization wake	0	0
Proton ionization wake	0	140
Muon ionization wake	0	330
Si (silicon) recoils from electron scattering	0	<1
Si recoils from proton scattering	<1	<1
Si recoils from neutron scattering	1	100
Si recoils from muon scattering	<1	3
Proton + Si → α-particles	<1	<1
Neutron + Si → α-particles	6	22
Muon capture → nuclear disintegration	<1	7
TOTAL	~7	~600

is impractical to avoid this problem by shielding, and if high reliability is needed, either the device design has to be changed or error-correcting codes must be used to detect and correct soft fails as they occur.

There is another area where integrated-circuit technology is employed and where cosmic rays are causing serious concern. As indicated above, the conditions in a single memory cell are very similar to those in solid-state ionization detectors. This fact has led to the development of large-area imaging detectors consisting of arrays of CCD memory cells. Such CCD cameras have the potential of becoming important low-level light detectors for use in astronomy. However, the importance of cosmic-ray-induced "background" events in CCD cameras has been demonstrated.

In summary, it has become clear that cosmic-ray-induced soft fails present a new unanticipated problem to the future minimization of microelectronic circuits. However, the microelectronics industry has faced several seemingly more difficult problems in the past which it has successfully solved, and it is anticipated that suitable solutions will be found to the problem of cosmic-ray soft fails.

[WILLIAM LANFORD; JAMES ZIEGLER]

Bibliography: J. F. Ziegler and W. A. Lanford, *IEEE Elec. Dev. Trans.*, 1980; J. F. Ziegler and W. A. Lanford, *IEEE 1980 International Solid State Circuits Conference Proc.*, pp. 70–80, 1980; J. F. Ziegler and W. A. Lanford, *Science*, 206:776–788, 1979.

Software engineering

The process of engineering software, that is, not only the programs that run on a computer, but also the requirements definition, functional specification, design description, program implementation, and test methods that lead to this code. Engineering implies the systematic application of scientific and technological knowledge through the medium of sound engineering principles to reach a practical goal. This goal becomes practical through the process of trading off performance, reliability, cost, and other characteristics of the resulting product in the light of the funds and time available for development. In short, software engineering is based on fundamental engineering principles and guided by economic considerations.

MOTIVATION

In the early years of computer development, as computer generations succeeded one another, the hardware performance-price ratio kept climbing at an average rate of about 25% a year; it was still advancing rapidly in the early 1980s. Programming productivity, on the other hand, while hard to measure, gained at a much slower rate—something like 7–9% per year. The continuing hardware improvement created a series of computer systems on which it was economically feasible to run ever larger programs, but the slower rate of improvement in programming led to progressively larger programming organizations requiring more levels of management. Some software systems approached a million instructions and absorbed 5000 worker-years of development time. By the mid-1960s the fact that program development almost defied effective management control was becoming evident.

Programs of this era were not only extremely large, but frequently lacked clarity as well. They were typically all in one piece, that is, not modularized. The sequence of program execution from one instruction to the next in line would jump instead to an instruction that was pages away in the program listing. A reader could track one or two of these jumps or branches, but scores of them made the program logic hard to grasp.

Since the product being developed was something of a mystery, it was nearly impossible for managers to find out enough about it to supervise the work of the programmers, to coordinate them with other groups, or to estimate how much time such poorly defined tasks would take or how many programmers were needed. It was under these circumstances that the NATO Software Engineering Conference (1968) pointed out the direction in which solutions to the "software crisis" should be sought.

Several aspects of software engineering may be distinguished. From one vantage point it appears as a process taking place over time, a process that is divided into perhaps a half dozen principal stages. Each stage can be examined more closely, and the various methods employed can be identified. Finally, from a broader point of view, the software development process must be managed.

DEVELOPMENT STAGES

A programmer tends to program, that is, to write a series of instructions in a programming language, and to give less thought to other stages of what is really, considered as a whole, a design process. On the other hand, in terms of software engineering, the problem-solving steps that have been learned in the other branches of engineering may be applied also here.

In general, these steps include formulating the problem, searching for and developing solutions, evaluating the solutions in terms of the goals, and refining and verifying the solution selected. In each branch of engineering these general steps have been adapted to the needs of the branch.

In software engineering there is now considerable agreement on the six stages shown in Fig. 1. As work proceeds through this series of stages, the findings of later stages flow back to influence earlier stages.

Requirements definition. In the first stage, requirements definition, the basic question is what exactly the user expects the software to do. This problem-oriented definition provides the guide to later development stages and a checklist upon which the using and the development organizations can achieve agreement on what must be done. It provides a base against which the resulting software can eventually be compared for acceptance by the user.

If the using organization is unfamiliar with what computing systems can do and the development organization is inexperienced in the user's application area, defining the requirements becomes a learning process for both. Even if both organizations have some experience, the definition of requirements will still involve learning. In addition, this learning continues into later stages as the problem becomes better understood. Consequently the requirements definition must be revised periodically to match what has been learned.

Functional specification. At the functional specification stage, the user requirements are converted into a specification of the functions to be performed by the proposed system. While the requirements definition may have set values for some performance objectives that were closely related to the problem circumstances, the functional specification carries the goal-setting process further. Under constraints set by the current capabilities of computer technology and the user's resources, the specification sets target numbers for performance, reliability, compatibility, and other product objectives.

This specification should serve at least three purposes. It should fulfill the needs expressed by the requirements definition and at the same time characterize the system to be built. It should lead into the design stage, and it can do this more effectively if the specification mirrors the structure of the problem. Finally, its clear expression of the functions and objectives should establish the criteria against which the completed system will be functionally tested.

Design. Software design is the process of going from what a system is to do to how it is to do it. In this stage the designer transforms the functional specification into a form from which the programmer can more easily code the problem represented by the specification into a list of programming instructions. While small, simple problems can sometimes be coded directly by a programmer, experience shows that most real problems cannot be coded directly and that an intervening design step is needed.

Design includes the task of partitioning (or decomposing) the entire problem area into subsystems, routines, subroutines, modules, and so forth, which individually become easier to understand than the whole. Together these elements may be thought of as a hierarchy, like an organization chart, or alternatively, a flow diagram, like a strip road map. In either case, a key task of the design process is determining the manner in which these

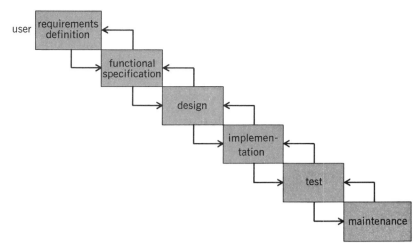

Fig. 1. Engineering development life cycle.

elements relate to one another—their interconnections. Developing the best possible arrangement—best in terms of the constraints imposed by the product specified and development resources—makes design an engineering activity.

The nature of design is suggested by Fig. 2. Input data go in, and undergo some processing; and output data come out (Fig. 2a). The data have some kind of structure, suggested by Fig. 2b. During the process of detailed design, this diagram is expanded by using a variety of methods described later.

Implementation. In its essentials the implementation stage transforms the output documents of the design stage into program instructions. These instructions are converted by a software support program called a compiler or an interpreter into machine code, that is, the binary digits that constitute the internal language of a computer.

In practice, however, if the program is of more than trivial size, the probability of making errors in programming is high. One of the most effective ways of finding these errors is the program review, conducted before the program is compiled. In a formal review the full weight of the organization is put behind the effort. The errors found are formally recorded and analyzed to provide corrective feedback to the entire programming organization. Unfortunately the reliance on the formal organization may discourage the willing cooperation of the psychologically less secure programmers.

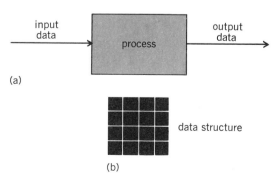

(a)

data structure

(b)

Fig. 2. Data considered from two points of view: (a) their processing and (b) their structure.

To overcome this drawback, the structured walk-through was conceived. The programmer arranges the meeting and invites his or her peers. Management is not included, and no formal record of the errors found is filed. The term walk-through is derived from the practice of assigning sample values to the data and walking these numbers through the program steps one by one. This less threatening atmosphere is believed to make it easier for a programmer to cooperate in finding errors and later correcting them. The point of view of the structured walk-through is that programming is known to be difficult and the peer group is there to help.

Errors can occur in the earlier stages, too. In fact, studies of the development process indicate that errors tend to be more frequent in the early stages than in the later ones, and more expensive to correct in later stages than in earlier ones. Thus it is important to review the results at each stage and to eliminate errors as early as possible.

Testing. After the program is compiled, it is tested, that is, a set of values is assigned to the inputs, the program is run, and the program outputs are compared to the correct outputs. Since the test is being done by computer, a large number of input sets may be run. Moreover, the task of comparing the test output with the known correct output can also be performed by computer. When this match fails, the circumstances are reported back to the programmer to find and correct the errors.

While the above outlined testing task is simple in concept, it becomes difficult in execution. The reasons are that each input may have a large number of values, combinations of these numbers become larger still, and the number of different paths through a processing network of more than trivial size is enormous. Even at computer speeds, the test of all combinations and paths might run up into years of test time. So it becomes the task of test personnel to devise test procedures that exercise the program adequately in a reasonable time and at affordable cost.

Maintenance. The term maintenance, adopted from hardware terminology, has a different meaning in software. Software maintenance takes place in response to finding a previously unknown error, or to a change in the environment in which the system functions, or to an opportunity to improve performance. The "repair" does not return the software to its original state, as hardware repair does. It advances the system to a new and sometimes poorly understood state. One result is often the introduction of further errors, leading to the need for further maintenance as soon as they are discovered, and so on.

Software has been considered, perhaps unwisely, to be easily changeable. It followed, then, that it could be adapted to new circumstances simply by changing it, unlike hardware, which has to be redesigned and manufactured as a new model. Experience has demonstrated, however, that unstructured software is rather difficult to change successfully. As a result of the ensuing errors and complications, such software tends to deteriorate over time as changes are implemented.

METHODS

The stages of the life cycle provide a framework for spreading a number of activities over the time

dimension, and consequently limiting the number which must be dealt with at any one time. Still, in each stage of a large system under development there remains too much complexity to deal with all at once. Psychological research has revealed that the human mind can only cope with about five to nine entities at a time. Above this limit people lose track of some of the factors and make errors.

Although the physical computer is a very complex machine, the software interfaces not only to this complex machine, but also to the great complexity of all the rest of the natural and human-made world, at least potentially. The development of large software systems is probably the most complex analytical task that the human race has ever encountered.

Thus one of the tasks of software engineering is to acquire from computer scientists or to develop for itself methods for partitioning a problem area into pieces small enough to be dealt with by the limited human brain. Concomitant with this task is that of finding satisfactory ways to represent this process of analysis and synthesis, both to bolster the capability of the analyst and to communicate the results of the work to others.

Structured design. At the beginning of the design stage, the designer knows that there are inputs and outputs, joined by a process, as diagrammed in Fig. 2. The methods of design are used to break out further detail. One way is by means of the flow of data, as shown in Fig. 3. The transfor-

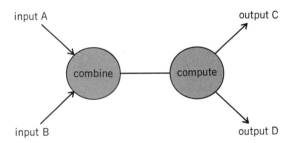

Fig. 3.   Data flow diagram. Data inputs are transformed by the operations specified in each circle.

mation of the data, symbolized by the circles, may be further refined in a series of steps into a greater number of individually simpler operations. Another way is to show the flow or movement of program control from one process or decision point to another, as the flow chart does (Fig. 4).

Thus the problem, as defined by the requirements and specification, is decomposed into smaller pieces and these, in turn, into still smaller pieces until a minimal size, called the software module, is reached.

If the decomposition process begins at the top and works downward, it is called top-down design. If it begins with other machines with which the program is to interface, it sorts out first the functions needed to manage these interfaces, then builds a structure up from there, and is called bottom-up design. The several levels of program modules form a hierarchy.

This hierarchy may be thought of as a series of

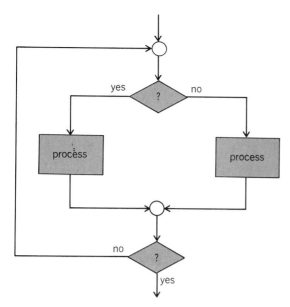

Fig. 4. Flow chart.

levels of abstraction (Fig. 5). Each level is composed of a number of modules (Fig. 6). The levels-of-abstraction concept is comparable to the working levels of an organization chart. Modules on the bottom level handle details, such as physical files of data. Modules on higher levels deal with broader matters. Only such information is abstracted from a lower-level module as is necessary for a higher-level module to perform its duties. The rest is "hidden," easing the comprehension task of programmers of higher-level modules. Similarly, all the programmer of a lower-level module need know about higher-level modules is that they may call this lower-level module in some prescribed form.

Levels and modules compose a structure. However, the fact of "structure" is not inherent in either the problem or the ultimate program. At the beginning, the problem looks like a "big buzzing confusion." At the other end of the development process, the program is a large linear list which

may be complicated by branch or jump instructions that can move the sequence of execution up or down the list at any time, sometimes pages away (Fig. 7). It may have no more perceptible structure than a long string tangled by a kitten. Modularization has not been an obvious approach, and in fact most programs before the 1970s were not modularized.

The business of design is to introduce order. The purpose of modularization is to reduce the number of entities with which the designer and later the programmer have to work at a given time. Also, a program with structure is easier to maintain and modify.

In a modular program there are two sets of entities. One set consists of the function or functions that the module is to perform internally. The other comprises the interrelationships it is to have with other modules. Complexity is reduced if the module is "cohesive," that is, performs only one function or a few closely related functions. Complexity is also reduced if the "coupling" between modules is limited and clearly defined. Under these circumstances a designer has only a few functions and a few relationships to keep in mind at one time.

Software engineers have conceived these and other ways of helping themselves think through the complexities of software design. In contrast to the earlier methods, modern practices emphasize structure, both in methodology and the result, and have come to be called structured design.

Design data representation. One achievement of other branches of engineering has been to draw systems so as to convey the intent of the designer to the manufacturer and others. Software designers, too, need methods of documenting their design data for its many users. The prime use is to convey the design to the programmers. However, the design data representation also serves as a source for test engineers, software maintenance programmers, and documentation specialists.

Software design took over the flow chart (Fig. 4) from other branches of engineering and still uses it widely. However, it has drawbacks. For one thing flow charts are not naturally structured. Hence several other forms of design data repre-

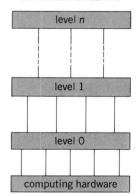

Fig. 5. Levels of abstraction. Building on the hardware, each level retreats further from detail.

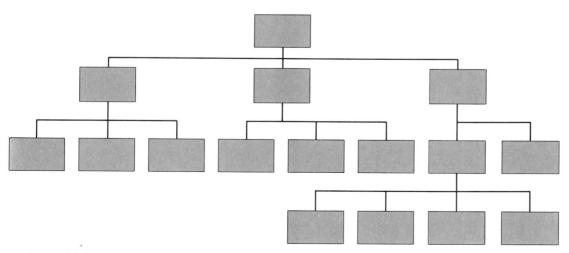

Fig. 6. Module hierarchy. Software modules are arranged in a structure, with defined relationships between the modules.

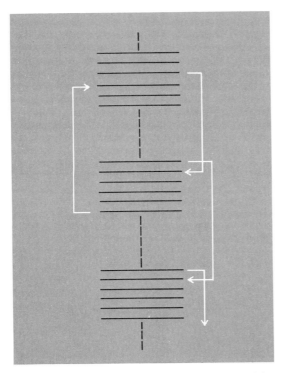

Fig. 7. Part of a sequential list of instructions containing branches or jumps that move the sequence of execution up or down the list.

sentation have been developed and gained adherents, and new methods are still being developed.

In a broad sense these representation techniques can be divided into three classes: diagrammatic, decision tables, and pseudocode. The diagrammatic techniques, of which the flow chart is one, represent module functions and relationships in some kind of visual form. Some examples are data flow diagrams, structure charts, HIPO (hierarchy plus input-process-output) charts, Nassi-Shneiderman charts, SDAT (structured design analysis technique), and data-structure charts (Warnier-Orr and Jackson method).

CONDITIONS				
newcomer to field	Y	N		
want more information on topic			Y	N
ACTIONS				
read this article carefully	X			
scan article quickly		X		
read referenced encyclopedia article			X	
no action				X

Fig. 8. Decision table. The user lists conditions at the top which have binary answers, yes or no. At the bottom are the actions that result from each state of the conditions.

The decision table is a tabular method of sorting out the logic of a problem (Fig. 8). It has the advantage over the diagram methods of considerably compressing the amount of space needed for analysis.

Pseudocode is a combination of logical structures, the flow of control, and a language (Fig. 9). The logical structures are such constructs as:

Do first operation *then* second operation
If condition *then* action *else* other action
While condition *do* action

Mathematically, all logical operations can be reduced to just these three. However, other logical operations exist, and it is often shorter and more convenient to make use of them.

The flow of control proceeds from one line of pseudocode to the next. Digressions to distant parts of the list of operations are not inherent in this method.

The language is normally English, or the user's natural language. Hence pseudocode is not a high-level programming language, since such languages are, by definition, compilable. (A compiler is a computer program that recognizes only the words permitted by the high-level programming language with which it works; it is not set up to recognize all the words of a natural language.) On the other hand, if the logical constructs available in a particular programming language are used in a pseudocode analysis and the blanks in the logical operations are filled in with terms belonging to that language (and, hence, recognizable by its compiler), then one has, in effect, moved to the implementation stage and programmed the design.

Through the use of pseudocode it is possible to start with a very broad version of the initial problem, since there is no limit to the concepts that may be employed. As the problem is gradually worked through, the pseudocode is refined in a stepwise manner to solve in more detail smaller pieces of the big problem. Ultimately a level of refinement is reached at which the conversion to programming language is deemed to be feasible without problems.

Structured programming. Contemporary programming languages incorporate the basic control structures (logic operations) in their instruction sets. In consequence, the final refinement of the design stage is simply converted to the corresponding logic structure of the language and is then compilable. However, the earlier programming languages, such as Fortran, were not based on these structures, and as a result the control structures from the design stage have no direct counterparts in the programming language. To overcome this deficiency, versions of these languages have been developed which do contain the control structures. However, these versions have to be first processed by a preprocessor (a computer operation) into the standard language, which is then compiled in the normal manner.

In a structured program there is ideally one entry point to the module and one exit point. Branch or Go-To instructions are used infrequently, if at all. The code listing is indented to signify its structure and enhance its readability. In general, each module of the design turns into only a page or two of code listing, short enough to be readily compre-

```
Begin software engineering study

    If newcomer to this field then

        Read this article carefully

    Else

        Scan article quickly

    Endif

    If want more information on a topic then

        Read referenced encyclopedia article

    Endif

    If want pursue field further then

        Consult one of bibliography items

        Do until your interest is satisfied

            Read bibliography items

            If item is irrevelant then

                Proceed to next item

            Endif

        Enddo

    Endif

    If you have understood this pseudocode then

        Congratulations! you have the idea of
        pseudocode as a design technique

    Endif

End software engineering study
```

Fig. 9. Example of pseudocode, a software design data representation technique.

hensible by persons other than the programmer. Understandability is further enhanced by preceding the actual code with a module description in natural language. The pseudocode design may also be included. Particular lines are further explained by comments. Such a program is said to be machine-documented. In this form it is more readily updated and accessible to later users.

SOFTWARE MANAGEMENT

The methods of software engineering find application in business, industry, and government. The problems are often very large. The resulting projects employ many people and substantial resources over long time periods. Personnel, money, and time are the factors with which management works, so that software development becomes subject to the methods of management: planning, organizing, scheduling, directing, monitoring, controlling, and so on. But because software engineering is relatively new, experienced managers lack working experience in it, and software engineers have not had time to grow into management. Even when some of them do, the pace of oncoming technology may outmode them.

If management could confine itself to management and engineers to engineering, the problems would be fewer and less severe. In software, how-

ever, there is reason to believe that management decisions, that is, decisions properly within the management sphere, can influence design matters, and vice versa.

It is well established, for example, that large software projects consume more development resources per unit of accomplishment than smaller projects, largely because of the increased burden of training, coordination, and communication. Mealy's "law" states: There is an incremental person who, when added to a project, consumes more energy (resources) than he or she makes available. Thus, beyond a certain point, adding resources (people) slows progress, in addition to increasing the cost.

One solution to this problem is to break large projects into small components. But management cannot make independent small projects simply by ordering them. If the small projects are all aspects of one large problem area, then there are relationships between them. In that case, coordinating relationships between projects takes just as much time as coordinating the same relationships within one large project; the technical relationships are there in both cases.

Of course, a large system can be partitioned into subsystems and eventually modules and the work divided, but this task—maximizing cohesiveness and minimizing coupling—is highly technical. The extent to which a system can be decomposed into independent systems is an engineering question, not a matter of managerial prerogative.

To take another example, the assignment of personnel is a common management responsibility. At the beginning of a project a manager might wish to assign to it right away the number of people that the project is planned to average over its lifetime, to get it off to a good start. Unfortunately, only a few highly skilled people can be usefully employed in the first design steps. More than these few just get in the way. One experienced software manager recommends sending the extra ones off to pertinent training programs. As the few sort out the problems and begin to structure the design, they create new elements to which more talent can be applied. This process repeats itself, and the staff gradually expands. It appears that the proper timing of additional personnel is largely dependent on the rate of technical progress. Again, management has the authority, but should be guided in its exercise by technical considerations.

According to Frederick P. Brooks, Jr., adding worker-power to a software project that is running behind schedule only slows it down further. In other words, under the circumstances that Brooks described, the managerial prerogatives of setting schedules and monitoring them appear to be hollow rights. There seem to be factors at work, technical in nature, more powerful than adding more people can offset.

EFFECTIVENESS

The very success of computer systems led to the software predicament: software became too costly, too error-prone, too complex, too hard to maintain, too people-heavy. In fact, these problems threaten to delay the spread of computers in new fields.

It is now clear that the methods of software engineering, while not easy to apply, do help overcome the software predicament. Formal studies of the before-and-after type show improvements in the 25–75% range in cost, error reduction, and productivity. Executives with before-and-after experience have reached similar conclusions. See DIGITAL COMPUTER; DIGITAL COMPUTER PROGRAMMING.

SOFTWARE PORTABILITY

Portability refers to the ease with which a program can be moved from one computer environment to another. If there were just one programming language, translated to just one type of computer by one compiler, then all application programs would be portable. However, with the many different types of computers and application domains and the hundreds of programming languages and compilers available, a particular program is generally not transportable between different types of computers, domains, or languages.

It follows, then, that a large number of programmers are needed, and this number would have to increase as rapidly as the number of different installations. The discrepancy between the rate at which hardware can be manufactured (very fast) and the rate at which programmers can be found and trained (very slow) is called the software crisis. One way to ameliorate this crisis is to improve the portability of software. Portability could be enhanced if the computer community were willing to settle on a limited number of languages, standardize them, and adhere to the standards. In practice, most implementations of languages have deviated to some degree from the language standard, thus restricting portability. In the past some manufacturers may have tacitly accepted incompatibility as a means to lock up a captive market, but this attitude seems to be declining as the software crisis become more evident.

Another approach is to accept at least some variety in computers and languages as a given, and develop exchange methods. Several companies translate a number of languages into a common intermediate code, variously called *p*-code or *q*-code, and then translate that into the native assembly language of each of many processors.

SOFTWARE PIRACY

Piracy refers to the process of copying commercial software without the permission of the originator. It is the dark side of portability—without portability piracy could not exist. Piracy is found among amateurs, who copy programs for friends without direct pecuniary gain, and professionals, who copy programs in quantity for sale at low prices to amateurs.

The amateurs seem to have need for more software, little money, and the feeling that just a few copies among friends will not make any difference. However, surveys reveal that thousands of amateurs are pirating software for their friends. All together they do make a difference; in fact, some suppliers have abandoned the amateur marketplace. In any case others' work is being appropriated by amateurs without recompense.

The professional pirates have a need to make money. They are not likely to be deterred by the fact that moralists consider copying to be unethical; it would have to be illegal to have an effect on them. There are three approaches in law to protecting intellectual effort: patents, copyrights, and trade secrets.

Algorithms—and thus commercial software—are unpatentable. Source programs can be copyrighted, but object code cannot, because it is not intended to be read by a human being, thus making it legal (perhaps) to copy the 0 and 1 code of a read-only memory.

Several courts have ruled in different ways on this question and it remains unclear. As for the trade secrets approach, a for-sale program can be kept secret only to the extent that it can be coded to prevent copying. Code protection is being used, leading to an escalating contest between the code writers and the code breakers. Some feel that Congress should take up the Supreme Court's invitation, repeated several times, to draft a law suited to the peculiarities of software. Others hope that the courts will eventually sort out the inconsistencies.

[WARE MYERS]

Bibliography: F. P. Brooks, Jr., *The Mythical Man-Month: Essays on Software Engineering*, 1975; P. Freeman and A. I. Wasserman, *Tutorial: Software Design Techniques*, 1980; R. W. Jensen and C. C. Tonies, *Software Engineering*, 1979; C. Jones, Alternatives to programming, in C. Jones (ed.), *Tutorial: Programming Productivity: Issues for the Eighties*, IEEE Computer Society, pp. 377–381, 1981; R. H. Stern and J. L. Squires, Can we stop software theft?, *IEEE Micro*, 2(1) 12–25, February 1982; R. C. Tausworthe, *Standardized Development of Computer Software*, 1977; E. Yourdon and L. L. Constantine, *Structured Design: Fundamentals of a Discipline of Computer Program and System Design*, 1979.

Solar cell

A semiconductor electrical junction device which absorbs and converts the radiant energy of sunlight directly and efficiently into electrical energy. Solar cells may be used individually as light detectors, for example in cameras, or connected in series and parallel to obtain the required values of current and voltage for electric power generation.

Most solar cells are made from single-crystal silicon and have been very expensive for generating electricity, but have found application in space satellites and remote areas where low-cost conventional power sources have been unavailable. Research has emphasized lowering solar cell cost by improving performance and by reducing materials and manufacturing costs. One approach is to use optical concentrators such as mirrors or Fresnel lenses to focus the sunlight onto smaller-area solar cells. Other approaches replace the high-cost single-crystal silicon with thin films of amorphous or polycrystalline silicon, gallium arsenide, cadmium sulfide, or other compounds.

Solar radiation. The intensity and quality of sunlight are dramatically different outside the Earth's atmosphere from that on the surface of the Earth, as shown in Fig. 1. The number of photons at each energy is reduced upon entering the Earth's atmosphere due to reflection, to scattering, or to absorption by water vapor and other gases. Thus, while the solar energy at normal incidence outside the Earth's atmosphere is 1.36 kW/m^2 (the solar constant), on the surface of the Earth at noon-

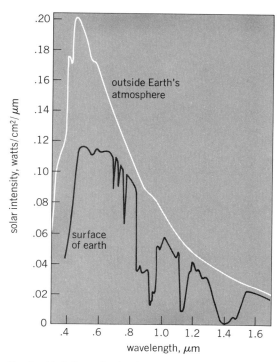

Fig. 1. Variation of solar intensity with wavelength of photons for sunlight outside the Earth's atmosphere and for a typical spectrum on the surface of the Earth.

time on a clear day the intensity is about 1 kW/m².

On clear days the direct radiation is about 10 times greater than the diffuse radiation, but on overcast days the sunshine is entirely diffuse. The mean annual solar energy falling on the Earth's surface varies greatly from one location to another. The sunniest regions of the globe receive about 2500 kWh/m² per year of total sunshine on a horizontal surface. The Earth receives about 10^{18} kWh of solar energy each year. The worldwide annual energy consumption is about 80×10^{12} kWh, so that from a purely technical viewpoint, the world energy consumption corresponds to the sunlight received on about 0.008% of the surface of the Earth.

Principles of operation. The conversion of sunlight into electrical energy in a solar cell involves three major processes: absorption of the sunlight in the semiconductor material; generation and separation of free positive and negative charges to different regions of the solar cell, creating a voltage in the solar cell; and transfer of these separated charges through electrical terminals to the outside application in the form of electric current.

In the first step the absorption of sunlight by a solar cell depends on the intensity and quality of the sunlight, the amount of light reflected from the front surface of the solar cell, the semiconductor bandgap energy which is the minimum light (photon) energy the material absorbs, and the layer thickness. Some materials such as silicon require tens of micrometers' thickness to absorb most of the sunlight, while others such as gallium arsenide, cadmium telluride, and copper sulfide require only a few micrometers.

When light is absorbed in the semiconductor, a negatively charged electron and positively charged hole are created. The heart of the solar cell is the electrical junction which separates these electrons and holes from one another after they are created by the light. An electrical junction may be formed by the contact of: a metal to a semiconductor (this junction is called a Schottky barrier); a liquid to a semiconductor to form a photoelectrochemical cell; or two semiconductor regions (called a *pn* junction).

The fundamental principles of the electrical junction can be illustrated with the silicon *pn* junction. Pure silicon to which a trace amount of a fifth-column element such as phosphorus has been added is an *n*-type semiconductor, where electric current is carried by free electrons. Each phosphorus atom contributes one free electron, leaving behind the phosphorus atom bound to the crystal structure with a unit positive charge. Similarly, pure silicon to which a trace amount of a column-three element such as boron has been added is a *p*-type semiconductor, where the electric current is carried by free holes. Each boron atom contributes one hole, leaving behind the boron atom with a unit negative charge. The interface between the *p*- and *n*-type silicon is called the *pn* junction. The fixed charges at the interface due to the bound boron and phosphorus atoms create a permanent dipole charge layer with a high electric field. When photons of light energy from the Sun produce electron-hole pairs near the junction, the build-in electric field forces the holes to the *p* side and the electrons to the *n* side (Fig. 2). This displacement of free charges results in a voltage difference between the two regions of the crystal, the *p* region being plus and the *n* region minus. When a load is connected at the terminals, an electron current flows in the direction shown by the arrow, and useful electrical power is available at the load. *See* SEMICONDUCTOR; SEMICONDUCTOR DIODE.

Characteristics. The electrical characteristics of a typical silicon *pn*-junction solar cell are shown in Fig. 3. Figure 3*a* shows open-circuit voltage and

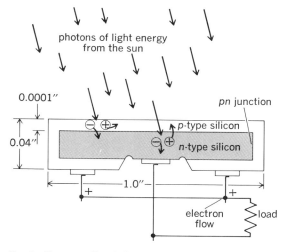

Fig. 2. Cross-sectional view of a silicon *pn* junction solar cell, illustrating the creation of electron pairs by photons of light energy from the Sun. 0.0001″ = 2.5 μm; 0.04″ = 1 mm; 1.0″ = 25 mm.

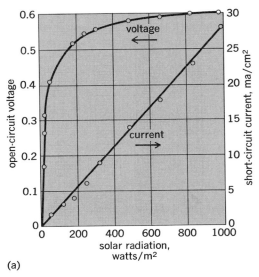

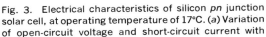

(a)

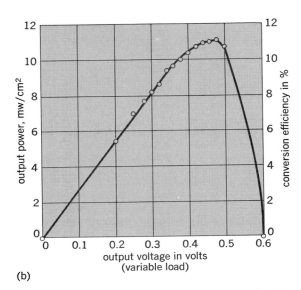

(b)

Fig. 3. Electrical characteristics of silicon *pn* junction solar cell, at operating temperature of 17°C. (*a*) Variation of open-circuit voltage and short-circuit current with light intensity. (*b*) Variation in power output as load is varied from short to open circuit.

short-circuit current as a function of light intensity from total darkness to full sunlight (1000 W/m²). The short-circuit current is directly proportional to light intensity and amounts to 28 mA/cm² at full sunlight. The open-circuit voltage rises sharply under weak light and saturates at about 0.6 V for radiation between 200 and 1000 W/m². The variation in power output from the solar cell irradiated by full sunlight as its load is varied from short circuit to open circuit is shown in Fig 3*b*. The maximum power output is about 11 mW/cm² at an output voltage of 0.45 V.

Under these operating conditions the overall conversion efficiency from solar to electrical energy is 11%. The output power as well as the output current is of course proportional to the irradiated

Fig. 4. Silicon solar cells assembled into panels and arrays to obtain higher voltage and power output. Peak power of modules is at 60°C and 15.8 V. (*U.S. Department of Energy*)

Fig. 5.   A 3-kW solar cell–powered communication link in southern California. (*Spectrolab, Inc., Sylmar, CA*)

surface area, whereas the output voltage can be increased by connecting cells in series just as in an ordinary chemical storage battery. Experimental samples of silicon solar cells have been produced which operate at efficiencies up to 18%, but commercial cell efficiency is around 10–12% under normal operating conditions.

Using optical concentration to intensify the light incident on the solar cell, efficiencies above 20% have been achieved with silicon cells and above 25% with gallium arsenide cells. The concept of splitting the solar spectrum and illuminating two optimized solar cells of different bandgaps has been used to achieve efficiencies above 28%, with expected efficiencies of 35%. Thin-film solar cells have achieved between 4 and 9% efficiency and are expected in low-cost arrays to be above 10%.

[KIM W. MITCHELL]

Arrays. Individual silicon solar cells are limited in size to about 40 cm² of surface area. At a 15% conversion efficiency, such a cell can deliver about 0.6 W at 0.5 V when in full sunlight. To obtain higher power and higher voltage, a number of cells must be assembled in panels or arrays (Fig. 4).

Fig. 6.   A 25-kW photovoltaic array which is operating in an agricultural irrigation experiment located at Mead, NE. (*U.S. Department of Energy and MIT Lincoln Laboratory*)

Cells may be connected in series to multiply their output voltage and in parallel to multiply their output current. Cells operated in series must be closely matched in short-circuit current since the overall performance of a solar cell array is limited by the cells having the lowest current.

Applications. Although the photovoltaic effect was discovered by A.C. Becquerel in 1839, practical solar cells made of silicon crystals were not developed until 1955. Beginning with *Vanguard 1*, launched in 1958, silicon solar cell arrays have become the almost exclusive power source for satellites. *Skylab*, launched in 1973, had a 20-kW solar cell array, the most powerful to be used in space so far (early 1980).

Terrestrial applications of solar cells have increased rapidly since 1970. Solar cell arrays have been used primarily to power small remote electrical loads that would otherwise be impractical or uneconomical to power by conventional means such as storage batteries or motor-generator sets. In 1979 a total of approximately 1 MW of solar cell arrays were sold worldwide to power such equipment as remote radio repeaters, navigational aids, consumer products, railroad signals, cathodic protection devices, and water pumps. Figure 5 shows a remote solar cell–powered communication link installed in 1976. Since most of the aforementioned uses require power to the load at times even when the Sun is not shining, electrical storage batteries are typically used in conjunction with solar cell arrays to provide reliable, continuous power availability.

Although extended terrestrial uses await cheaper solar arrays, a number of experiments have been undertaken to explore the use of solar cell arrays in larger agricultural, residential, commercial, and industrial applications (Fig. 6). When powering loads which require ac voltage, a static inverter is used to convert the dc voltage from the solar cell array into usable ac power.

Future prospects. The growing worldwide demand for energy, its increasing cost, and the depletion of nonrenewable energy reserves make solar cell power systems an attractive alternative for supplying electricity for a wide range of uses. Despite significant progress in this technology, the cost of solar cells must be significantly reduced before they can economically supply a substantial amount of electricity. Research has been undertaken to develop new approaches lowering the cost of solar cell materials and manufacturing processes.   [DONALD G. SCHUELER]

Bibliography: J. A. Merrigan, *Sunlight to Electricity: Prospects for Solar Energy Conversion by Photovoltaics*, 1975; W. Palz, *Solar Electricity: An Economic Approach to Solar Energy*, 1978; D. L. Pulfrey, *Photovoltaic Power Generation*, 1978.

Solion

A dynamic electronic circuit element that supplements vacuum tubes and transistors. The term solion applies to a class of devices that use ions in solution instead of electrons as the charge carriers. A solion consists of two or more electrodes sealed in an electrolyte. At the electrode surfaces, conduction changes from ionic to electronic by means of electrochemical reactions. At the anode, ions lose electrons (or are oxidized) and at the cathode,

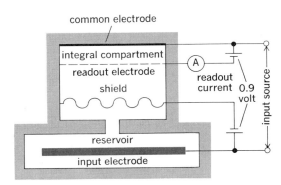

Fig. 1.   Electrical readout integrator.

they gain electrons (or are reduced). The reactions are reversible, the electrodes are not affected by the reactions, and the electrolyte contains both the oxidized and the reduced species of the ions. This is called a redox system.

Solions are low-frequency devices but where applicable they offer power reduction, circuit simplification, and increased reliability and ruggedness. In some cases, a single solion may replace a complete circuit assembly. The basic solion units are the diode, the integrator, the pressure transducer, and the electroosmotic driver. More complicated solions can be designed to perform various mathematical or process-control functions.

The solion diode uses platinum electrodes in an aqueous solution that may be iodine and potassium iodide. Forward-to-back current ratios of 500:1 are easily obtainable and are maintained at voltages below 0.9 volt. The diode exhibits a storage charge effect (hysteresis) at very low frequencies. The maximum voltage that may be applied is 0.9 volt. Its chief use is as a low-frequency switch.

The solion electrical readout integrator uses the same electrochemical system as the diode (Fig. 1). Current flowing between the input and the common electrode oxidizes iodide to iodine in the inte-

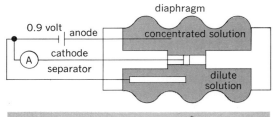

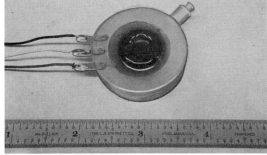

Fig. 2.   Solion linear detector.

gral compartment. The readout current through meter A is determined by the iodine concentration in the integral compartment and is proportional to the integral of the input current. The integrator is reversible and linear, and may be temperature-compensated externally. It is used as an integrator, a linear time-scale generator, an amplifier, and an adjustable constant-current element.

The solion pressure transducer (or linear detector) measures fluid flow through an orifice separating two electrolyte chambers (Fig. 2). A detecting cathode is located in the orifice. Pressure causes iodine to flow past the cathode where it is reduced to iodide. This results in a current increase in the cathode circuit. Pressure changes as small as 5 dynes/cm^2 can be detected. The transducer is used as a pressure-change detector, a vibration pickup, a hydrophone, and an accelerometer.

The solion electroosmotic driver, or micropump, converts voltage into fluid pressure. It uses a different electrochemical system, depolarizing electrodes, and operates through an effect known as the streaming potential; 10 volts can produce as much as 40 in. of water pressure. It is used as a pressure generator, an amplifier (with solion pressure transducer), and a part of mathematical function devices. [DONALD B. CAMERON]

Space charge

The net electric charge within a given volume. If both positive and negative charges are present, the space charge represents the excess of the total positive charge diffused through the volume in question over the total negative charge. Since electric field lines end on electric charge, the space-charge density ρ may also be defined in terms of the divergence of the electric field **E** or the Laplacian of the electric potential V by Eq. (1)

$$-\frac{4\pi\rho}{\epsilon} = -\text{div } \mathbf{E} = \nabla^2 V = \frac{\partial^2 V}{\partial x^2} + \frac{\partial^2 V}{\partial y^2} + \frac{\partial^2 V}{\partial z^2} \quad (1)$$

(Poisson's equation). Here ϵ is the dielectric constant of the medium and x, y, and z are rectangular coordinates defining the position of a point in space. If, under the influence of an applied field, the charge carriers acquire a drift velocity v, the space charge becomes j/v, where j is the current density. For current carried by both positive and negative carriers, such as positive ions and electrons, the space charge density is given by Eq. (2).

$$\rho = j_+/v_+ - j_-/v_- \quad (2)$$

Here the subscripts $+$ and $-$ indicate the current density and drift velocity for the positive and negative carriers, respectively. Thus a relatively small current of slow-moving positive ions can neutralize the space charge of a much larger current of high-velocity electrons. [EDWARD G. RAMBERG]

Spark gap

The region between two electrodes in which a disruptive electrical spark may take place. The gap should be taken to mean the electrodes as well as the intervening space. Such devices may have many uses. The ignition system in a gasoline engine furnishes a very important example. Another important case is the use of a spark gap as a protective device in electrical equipment. Here, surges

in potential may be made to break down such a gap so that expensive equipment will not be damaged. The dependence of spark-gap operation on such factors as pressure, length, electrode characteristics, and time dependence of potential is quite complicated. *See* BREAKDOWN POTENTIAL; ELECTRIC SPARK.

[GLENN H. MILLER]

Spectrum analyzer

A device which sweeps over a portion of the radio-frequency spectrum, responds to signals whose frequencies lie within the swept band, and displays them in relative magnitude and frequency on a cathode-ray-tube screen.

In essence, it is a superheterodyne receiver having a local oscillator whose frequency is varied cyclically, usually at the power-line frequency. The block diagram of a typical spectrum analyzer is shown in the illustration.

Signals whose frequencies lie within a range equal to the bandwidth of broadband i-f (intermediate-frequency) amplifier 1, can be heterodyned by local oscillator 1 and converted to within the passband of that i-f amplifier. The resulting spectrum is scanned by local oscillator 2 and, in a second conversion, swept back and forth across narrow-band i-f amplifier 2. This amplifier is tuned to a frequency outside the passband of i-f amplifier 1. Whenever a signal lying within the spectrum is swept across the passband of i-f amplifier 2, a burst of energy passes through that amplifier to the detector and produces a pulse that is applied to the vertical-deflection plates of a cathode-ray tube. Since the heterodyne process maintains a linear relation between input and output, the magnitude of the pulse is proportional to the strength of the input signal. The horizontal deflection of the cathode-ray spot is obtained from the same source as the sweep signal applied to local oscillator 2, thereby providing a display corresponding to frequency. *See* RADIO RECEIVER.

Spectrum analyzers are used specifically to study the spectra of pulsed transmitters, such as radar, to make sure that they are operating properly, without spurious emissions. Spectrum analyzers are often built into test equipment and military identification receivers. The output display shown in the illustration can be used to monitor radio traffic in the frequency band being covered; transmitter activity which might be overlooked by a standard receiver groping about in its tuning

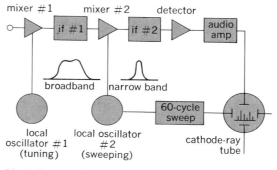

Block diagram of typical spectrum analyzer.

range will show up on the panorama display with its carrier frequency identified. Naturally the modulation on such a carrier cannot be "read" if it is other than the most primitive kind of keying. As an aid in making accurate frequency measurements, spectrum analyzers are used as comparison devices to indicate when coincidence occurs between a known and an unknown signal or to locate an unknown frequency with respect to a "picket-fence" spectrum generated from a standard frequency. *See* FREQUENCY MEASUREMENT.

[MARK G. FOSTER]

Speech recognition

The process of analyzing an acoustic speech signal to identify the linguistic message that was intended, so that a machine can correctly respond to spoken commands. Procedures must be incorporated

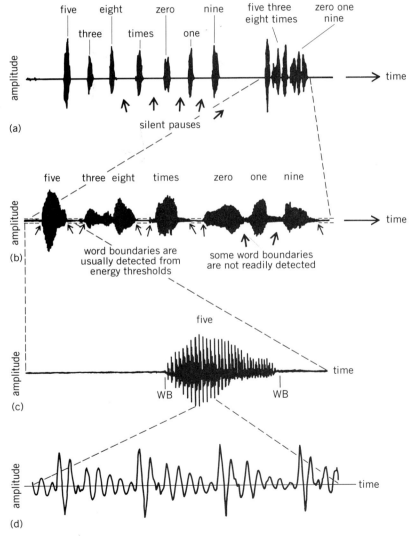

(a)

(b)

(c)

(d)

Fig. 1. A series of progressively more detailed views of an acoustic waveform, showing how a recognizer can detect utterance boundaries and identify words and messages. (a) Words isolated from each other by silent pauses, followed by a continuously spoken version of a calculator command. (b) Waveform of the continuous-speech portion of a, showing how some, but not all, word boundaries can be detected from energy thresholds. (c) Further expansion of the time scale of part of the waveform of b, so that the constantly changing sound structure within the word "five" is evident, and word boundaries (WB) are illustrated. (d) Local resonant structure of the diphthongal transition from /a/ to /I/ in the "five" of diagram c.

for distinguishing linguistic contrasts from irrelevant acoustic changes due to speaker variabilities and environmental conditions.

Motivation. Speaking commands to a computer is the ultimate in natural human-to-machine communications. Natural, spontaneous interactions are possible, with little or no user training, permitting direct access by the physically handicapped or those who cannot type well. The user can communicate in the dark or blinding light, around obstacles, and while walking about in an area, without direct contact with a computer console or need for large data-input devices. Telephones and radios can become low-cost computer input terminals, and there is interest in voice control of toys, appliances, wristwatch calculators, and other consumer products.

Speech allows rapid data entry and frees the user's hands and eyes for other tasks. Experiments show that complex tasks can be accomplished in half the time if speech is allowed as one mode of interaction. Even when one of the communicators is replaced by a restricted machine that can recognize only single words or phrases, preceded and followed by pauses, the speed and accuracy of complex data entry are still better for speech than for keyboard or graphical pen.

Speech recognizers have been advocated, sold, and used for various hands-busy applications such as: quality control and inspection; package sorting for automatic conveyor systems; control of machine tools; entry of topographical data in mapmaking facilities; and cockpit controls. Voice-actuated wheelchairs, hospital room environmental controls, and recognizer-resynthesizer translators for pathological speech are of public interest. Telephone banking, voice authorization of credit card transactions, and catalog ordering are other applications, which introduce difficulties related to telephone bandwidth, noise, and distortions.

Difficulty. Fluent conversation with a machine is difficult to achieve because of intrinsic variabilities and complexities of speech, as well as limitations in current practical capabilities. Recognition difficulty increases with vocabulary size, confusability of words, reduced frequency bandwidth, noise, frequency distortions, the population of speakers that must be understood, and the form of speech to be processed. To limit the problem, speech recognizers have been primarily confined to small vocabularies of words with large differences in sound structure, and isolated words, with pauses before and after each word. As illustrated in Fig. 1, word boundaries are then relatively easy to find, as transitions into and out of silences. Contextual influences of surrounding words on the pronunciation of each word are minimized by pausing. In contrast, the continuous flow of natural speech can make it difficult to tell when one word ends and the next word begins. Coarticulatory degenerations from ideal dictionary pronunciations (for example, "dija" for "did you," or "lissum" for "list some") can be avoided with strict constraints (formats) on allowable word sequences, and can be practically eliminated by pausing between words, but at some loss in naturalness and speed of data entry.

Isolated word recognizers are thus the easiest to develop, followed in turn by recognizers of digit strings and strictly formatted word sequences.

Sometimes the identification of every spoken word is not necessary, and detecting key words in context is enough to determine the topic of a conversation. Such a word-spotting system usually deals only with the important words, which are stressed and well articulated, easing the recognition, despite arbitrary contexts. Full understanding of the total sentence meaning and intended machine response involves cooperative use of many "knowledge sources" like acoustics, phoenetics, prosodics, restricted syntax, semantics, and task-dictated discourse constraints. The term speech-understanding system is frequently used to describe such a sentence-understanding device. An even more ambitious system goal would involve accurately identifying all the words in any possible utterance, yielding what is variously called a phonetic typewriter, an automatic transcription facility, a speech-to-text machine, or a task-independent continuous-speech recognizer.

Accuracy. The difficulty of recognition is closely correlated with required recognition accuracy, defined as the percentage of all received utterances that are correctly identified (that is, for which the machine response is as intended). Substitutions of the wrong response are usually considered more serious than machine rejections of the utterance as being too similar to other possible utterances to be reliably categorized. Studies are needed regarding necessary accuracy in various practical situations, although most field tests suggest that 4% error rates are not acceptable to serious users. One of the most important areas in speech recognition technology is the human factors aspect, concerned with determining what specific tasks warrant voice input, what accuracies are needed, and how to balance the trade-offs between versatility and cost, criticality of errors and technical feasibility, and exploitation of task constraints versus flexible enhancement for use on new tasks.

Operation of speech recognizers. Speech recognition requires the transformation of the continuous-speech signal into discrete representations which may be assigned proper meanings, and which, when comprehended, may be used to effect responsive behavior. Only some of the information in the speech signal is related to selecting the correct machine response, so that a critical task is to extract all, and only, those parts that convey the message. Recognizers primarily differ on how they reduce the data to message-distinguishing features, and how they classify utterances from that reduced data.

Figure 2 illustrates one general structure for how a machine acquires knowledge of expected pronunciations and compares input data with those expectations. Since most devices must be trained to the specific pronunciation of each talker, a precompiling stage involves the user speaking sample pronunciations for each allowable utterance (word, phrase, or sentence), while identifying each with its typewritten form or a vocabulary item number. Later, when an unknown utterance is spoken, it is compared with all the lexicon of expected pronunciations to find which training sample it most closely resembles. For complex utterances, other linguistic and situational information can be used to help guide and confirm hypothesized word sequences. Figures 1, 3, 4, and 5 detail some of the processes needed in the system components diagrammed in Fig. 2.

Word boundary detection. Figure 1 illustrates the initial process of detecting word boundaries, for word sequences with or without deliberate pauses between words. Silent pauses (Fig. 1a) permit easy detection of word onsets at transitions from no signal to high energy (above some threshold amount), and word offsets as energy dips below the threshold. In continuous speech, pauses normally occur only at boundaries between sentences or clauses, or at thoughtful hesitations just before

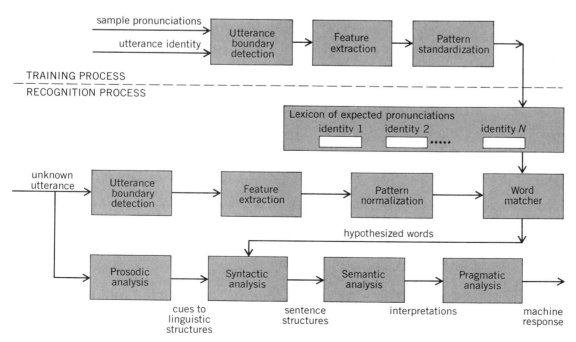

Fig. 2. One typical structure for a trainable speech recognizer that could recognize words, phrases, or sentences.

important words. Even the rapidly spoken calculator command shown in Figure 1b shows more prominent energy dips, and very brief pauses, between words than is usual in conversationally flowing speech. Word boundary confusions can occur, however, when short silences (usually less than 100 ms in duration) during stop consonants (/p, t, k/) may look like intended pauses. Thus, a word such as "transportation" could appear to divide into three "words": "transp," "ort," and "a-tion." Recognizers looking for pauses between words must thus measure the duration of the pause, to distinguish short consonantal silences from longer deliberate pauses.

Feature extraction. Once a word or other utterance unit has been delimited, its identity can be determined from the details of its pronunciation. Figures 1c and d show that the sound structure in a word is constantly changing, with frequent alternations (and transitions) between vowellike damped resonant structures and weak or noiselike consonantal sounds such as /p, t, k, f, s/. Not all aspects of the pronunciation of a word will be consistent from time to time, from one talker to another, or from one context to another. Hence, important information-carrying features are usually sought in the acoustic data to detect contrasts of linguistic significance and to detect segments such as vowels and consonants.

Figure 3 shows a few typical acoustic parameters used in recognizers. From the acoustic waveform, one can extract local peak values, such as at 1 in Fig. 3a, as an amplitude measure. Also, the sum of the squares of all waveform values over a time window can provide a measure of the energy

in that window of speech data. The number of times the signal passes through a value of zero (zero crossings) can be used as a cue to vowellike smooth waveforms (with few crossings per unit time) versus noiselike fricative segments (with many crossings per unit time). The time between the prominent peaks at onsets of pitch cycles can be used to determine the pitch period T_0, or its inverse, the rate of vibration of the vocal cords, called fundamental frequency or F_0. Resonant frequencies are evident from the number of peaks per pitch period. For example, the third pitch period in Fig. 3a shows seven local peaks, indicating a ringing resonance of about seven times the fundamental frequency. This resonance is the first formant or vocal tract resonance of the /a/-like vowel, and is one of the best cues to the vowel identity.

Figure 3b shows the superimposed results of two methods for analyzing the frequency content of the short sample of speech in Figure 3a. The computer can achieve rapid determination of the total jagged Fourier frequency spectrum, with its peaks at harmonics of the fundamental frequency, using a fast Fourier transform. However, to extract the exact positions of major spectral peaks at the formants of the speech, an advanced method called linear predictive coding (LPC) can yield the smoothed LPC spectrum shown passing through the middles of the vertical jumps of the FFT spectrum. The peaks in such a spectrum then indicate the basic resonant frequencies of the speaker's vocal tract, and can be tracked versus time, as in Fig. 3c, to indicate the nature of the vowel being articulated. Another common spectral analyzer is a filter bank, with narrow bandpass filters spaced across the frequency spectrum, to monitor the amount of energy in each frequency range versus time.

Pattern standardization and normalization. Figure 4 shows why a recognizer needs pattern standardization and pattern normalization components. The word "five" spoken on two successive occasions in Fig. 4a has different amplitudes (A1 and A2). A recognizer must neglect (normalize for) such amplitude variations that have nothing to do with intended linguistic contrasts. Also, as a comparison of Fig. 4b and c illustrates, the timing of speech events may not be the same for two repetitions of the same utterance. With identical time scales, the /w/ in 4b lines up with the timing of the /ʌ/ in c. A recognizer must realign the data, so that the proper portions of an unknown utterance are aligned with corresponding portions of the templates. This normally requires nonuniform time normalization, as suggested by the different phonemic durations in Fig. 4. Dynamic programming is a method for trying all reasonable alignments, and picking the one that yields the closest match to each template. Another possible pattern normalization might be speaker normalization, such as moving one speaker's formants in a pattern such as Fig. 3c up or down the frequency axis to match those of a "standard" speaker, for whom the machine has been trained. Channel normalization is another need; if training data were collected under broadband conditions, but unknown utterances come over the band-limited telephone, the sharp cutoff due to telephone filtering may have to be introduced into the training data before spectral comparisons. During training, the speaker's data

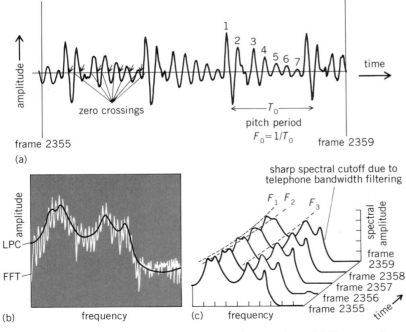

Fig. 3. Typical acoustic parameters used in speech recognizers. (a) Time waveform of Fig. 1c, showing parameters that can be extracted. (b) Frequency spectrum of the waveform of a, with the detailed frequency spectrum derived from a fast Fourier transform (FFT) being smoothed by linear predictive coding (LPC) to yield a smooth spectrum from which formants can be found as the spectral peaks. (c) Smoothed LPC spectra for five successive short time segments (frames, each 6.4 ms long), with formants F_1, F_2, and F_3 tracked as the spectral peaks.

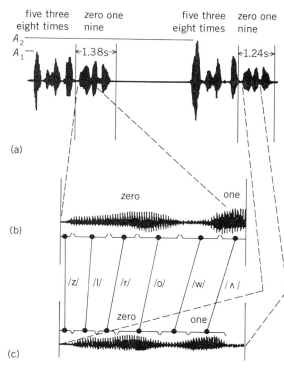

Fig. 4. Two successive utterances of the same phrase, showing need for amplitude and time normalization processes in speech recognizers. (*a*) Utterances in succession, showing difference in amplitudes A_1 and A_2. (*b, c*) Expansion of portions of the utterances, showing how rate of speaking can cause misalignment of data, so that time normalization is needed.

for amplitudes, timing, and spectral content can be set to standard values, or calibrated, so that unknowns can later be warped (normalized) to match those standard values.

Word matching. From such feature extraction, standardization, and normalization processes, an array of feature values versus time can be obtained. During training, this array can be stored as a template of expected pronunciation; during recognition, a new array can be compared with all stored arrays, to see which word is closest to it. This involves a distance measure that weights all the features and time slices in some manner, and accumulates a total difference in structure from one array to another.

Ambiguities in possible wording result from errors and uncertainties in detecting the expected sound structure of an utterance. Some "robust" segments like vowels and strong consonants may be evident, but a variety of words could usually correspond with a detected sound structure. To prevent catastrophic errors in word identifications, recognizers will often give a complete list of possible words, in decreasing order of agreement with the data (that is, increasing distance from the input pattern). For each time period of the unknown utterance, the word matcher of Fig. 2 then hypothesizes many words that are to varying degrees close to the sound structure of the incoming speech in that region. As shown in Fig. 5, the first word might look like "five," based on the correspondence between the detected vowel AY and the expected vowel AY, and the matching of initial and final consonants "f" and "v" also. The first word is also similar to "nine" and "times" in a small vocabulary of calculator terms. Later in the utterance, words such as "zero," "seven," "clear," and "one" might compete as candidates for the same region of speech. One reason for such potential confusions is that a practical recognizer may mislabel segments, fail to detect certain segments, or insert segments that are not actually in the utterance. Figure 5 shows only the more robust detected segments, such as vowels and some consonants, around which the remainder of the sound structure must be carefully (but less reliably) matched.

Higher-level linguistic components. A challenge for speech recognizers is how to keep down the

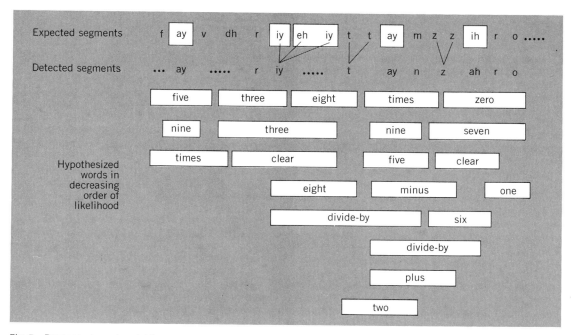

Fig. 5. Process of word matching.

combinatorial explosion of alternative word sequences to consider. That is the primary purpose for higher-level linguistic components such as prosodics, syntax, semantics, and pragmatics. Prosodic information such as intonation can help distinguish questions from commands, and can divide utterances into phrases and rule out word sequences with incorrect syllabic prominences or stress patterns. A syntactic rule may be written into the computer to disallow consideration of ungrammatical sequences like two operators in a row, so that "plus divide," "multiply clear plus," and so forth, are ruled out from hypothesized word sequences. A collection of such word-sequence constraints forms a grammar. Most practical recognizers use a simple grammar type called a finite-state grammar that determines the allowable next words based on the immediately preceding word. A semantic constraint might disallow meaningless but grammatical sequences such as "zero divide-by zero." A pragmatic constraint might eliminate unlikely or impractical sequences such as the useless initial "zero" in "zero one nine" of Fig. 1. In a chessplaying task, to take another example, syntactic rules limit word sequences to those for acceptable statements of movements of pieces, such as "pawn to queen four"; semantic rules then would rule out illegal or meaningless moves; pragmatic rules would discount the likelihood of ridiculous or counterproductive moves, such as moves that were just made in the immediately previous step. Some sophisticated recognizers have included all these sources of knowledge to help correctly identify an utterance. Isolated word recognizers usually use finite-state grammars, if any higher-level linguistics is used at all.

It is also possible to restructure the recognizer components shown in successive steps in Fig. 2. One promising structure avoids the propagation of errors through successive stages of a recognizer by having all the components intercommunicate directly through a "blackboard" or central control component that, for example, might allow syntax or semantics to affect feature-extraction or word-matching processes, or vice versa.

Prospects. Speech recognition is advancing vigorously, and each component in Fig. 2 is being improved upon, as are the total system structures. As recognition techniques advance, and the markets and users demand more, there will be a growing concern with how to systematically select, design, and effectively use speech recognizers. Procedures for thoroughly evaluating the performances of each system component and the total system will be needed, and standard tests and common data bases will allow side-by-side comparisons of devices. Market research and human factors studies will further clarify the wide range of uses for such devices that permit natural voice control of machines. *See* VOICE RESPONSE.

[WAYNE A. LEA]

Bibliography: J. L. Flanagan, Synthesis and recognition of speech: Teaching computers to listen, *Bell Lab. Rec.*, pp. 146–151, May–June 1981; W. A. Lea, *Computer Recognition of Speech*, 1981; W. A. Lea, *Selecting, Designing, and Using Speech Recognizers*, 1981; W. A. Lea, *Trends in Speech Recognition*, 1980; L. Rabiner and R. Shafer, *Digital Processing of Speech Signals*, 1979; S. E. Levinson and M. Y. Liberman, Speech recognition by computers, *Sci. Amer.*, 244(4): 64–76, April 1981.

Sputtering

The process by which atoms or groups of atoms are ejected from a metal surface as the result of heavy-ion impact. It generally takes place at the cathode of a self-maintained gaseous discharge, indicating that the important agent is the positive ion. Although sputtering is useful for certain processes, such as the generation of a clean surface, it is usually harmful. In the case of an oxide-coated thermionic cathode, sputtering by positive-ion bombardment may destroy the surface completely.

In general, there will be a threshold energy for sputtering. This depends on both the surface material and the bombarding ion. It has been found empirically that the mass m sputtered per unit time is given by the equation below. Here k and V_0

$$m = k \, (V_c - V_0)$$

are constants, while V_c is the potential through which the ion has fallen, and is therefore a measure of the energy of the ion. Further, it has been shown by A. Güntherschulze that the sputtered mass decreases as the pressure increases. The sputtering process becomes more efficient as the mass of the ion is increased. The ejected atoms may come off either as neutral particles or as ions.

The explanation of sputtering is not clear. It is thought that local heating may play a prominent part in the process.

[GLENN H. MILLER]

Bibliography: R. Behrisch et al. (eds.), *Ion Surface Interaction, Sputtering and Related Phenomena*, 1974.

SQUID

A device which, in its original form, consists of two Josephson tunnel junctions connected in parallel on a superconducting loop (Fig. 1). The term is an acronym for superconducting quantum interference device. A small applied current I flows through the junctions as a supercurrent, without developing a voltage, by means of Cooper pairs

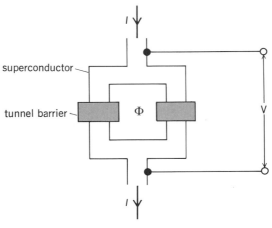

Fig. 1. Direct-current SQUID with enclosed magnetic flux Φ.

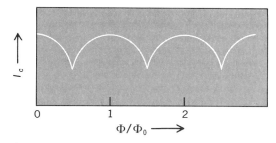

Fig. 2. Graph of maximum supercurrent I_c versus Φ/Φ_0 for a dc SQUID.

tunneling through the barriers. However, when the applied current exceeds a certain critical value I_c, a voltage V is generated. As shown in Fig. 2, the value of I_c is an oscillatory function of the magnetic flux Φ threading the loop, with a period of one flux quantum, $\Phi_0 = h/2e \approx 2.07 \times 10^{-15}$ weber. The oscillations arise from the interference of the two waves describing the Cooper pairs at the two junctions, in a way that is closely analogous to the interference between two coherent electromagnetic waves. Thus the SQUID is often also called an interferometer. *See* JOSEPHSON EFFECT; SUPERCONDUCTIVITY.

The SQUID has important device applications. When each Josephson tunnel junction is shunted with an external resistance to eliminate hysteresis on the current-voltage characteristic and the SQUID is biased with a constant current greater than the critical value I_c, the voltage across the SQUID is also an oscillatory function of Φ. If one measures the change in voltage produced by the application of a flux equivalent to a small fraction of one flux quantum, one has a very sensitive magnetometer. Since the device in this mode operates with a constant bias current, it is usually referred to as the dc SQUID. Another important potential application is as a logic element or memory cell in high-speed computers. When an unshunted (hysteretic) SQUID is appropriately current-biased, the application of a flux pulse switches it from the zero-voltage to the nonzero-voltage state, a function that can be used to perform logic; three-junction SQUIDs are also used for this purpose. The SQUID can be used as a dissipation-free memory cell to store a 1 or 0 as a clockwise or anti-clockwise circulating persistent supercurrent. *See* COMPUTER STORAGE TECHNOLOGY; LOGIC CIRCUITS.

The rf SQUID consists of a single junction interrupting a superconducting loop. It can be operated as a magnetometer by coupling it to the inductor of an LC-tank circuit excited at its resonant frequency by a rf current. The rf voltage across the tank circuit oscillates as a function of the magnetic flux in the loop, again with a period Φ. The rf SQUID is in fact misnamed, since no interference takes place. *See* SUPERCONDUCTING DEVICES.

[JOHN CLARKE]

Bibliography: R. C. Jaklevic et al., Quantum interference effects in Josephson tunneling, *Phys. Rev. Lett.*, 12:159–160, 1964; B. D. Josephson, Possible new effects in superconductive tunneling, *Phys. Lett.*, 1:251–253, 1962.

Standing-wave detector

An electric indicating instrument used for detecting standing waves along a transmission line or in a waveguide and measuring the resulting standing-wave ratio. It can also be used to measure the wavelength and hence the frequency of an electromagnetic wave in a line. The detecting device is usually a bolometer, thermocouple, or crystal, connected to an indicating meter directly or through an amplifier. The detecting device is moved along the line while observing the meter indication; the positions along the line at which maximum and minimum readings are obtained correspond to the nodes and antinodes of the standing wave that is produced by transmitted and reflected waves of equal frequency moving in opposite directions. The reflected wave is generated at a discontinuity in the transmission line or waveguide. *See* WAVELENGTH MEASUREMENT; WAVEMETER.

[JOHN MARKUS]

Static-induction transistor

A type of field-effect transistor whose drain current is controlled by an electrostatic potential barrier. The static induction transistor (SIT) is under development for use at high current and voltage. The name was introduced in 1972 by J. Nishizawa, T. Terasaki, and J. Shibata. The SIT is similar in structure to a short-channel vertical-junction field-effect transistor (JFET), which has received intermittent attention since it was introduced in 1950 by Nishizawa, and in a somewhat different form by W. Shockley in 1952. Unlike the conventional JFET, the current-voltage characteristics do not saturate. They are very similar in form to those of a vacuum triode and, for this reason, SITs are sometimes referred to as "triodelike" FETs. Present devices can handle 20 A and 500 V at 50 MHz, or 1000 W at 1 GHz. Developemets are under way on devices to deliver 100 W at 3–4 GHz, and 10 W at 9–10 GHz. The SIT is currently used as a low-distortion audio power amplifier and as a fast high-current switch in switched dc power supplies. Developments will provide devices for power amplifiers and drivers in microwave applications. The devices may eventually replace magnetrons in microwave ovens.

Device physics. Figure 1 shows a cross section of the SIT. The applied voltages are such that the space charge region around the p-type gate regions extends throughout the n^- region. Under these

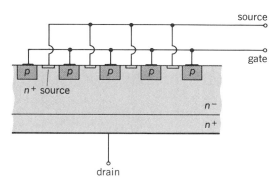

Fig. 1. Cross section of SIT showing physical structure.

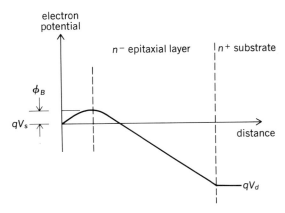

Fig. 2. One-dimensional potential distribution in SIT. V_s = source voltage; V_d = drain voltage; q = electron charge.

conditions, the electrostatic potential along a line extending from source to drain midway between the p-type gates has the shape shown schematically in Fig. 2. The SIT is a majority carrier device. The drain current is carried by electrons injected from the n^+ source into the space charge region, where they move principally by drift to the n^+ drain region. As shown in Fig. 2, these electrons must surmount a potential barrier ϕ_B in order to be injected into the space charge region. The number of source electrons that have sufficient energy to cross the barrier is an exponential function of the barrier height, which in turn is a function of the drain voltage, gate voltage, and device geometry. As a result, the drain current varies exponentially with changes in either gate or drain voltage. A detailed analysis of the operation must include other effects such as the saturation velocity of the electrons and space charge limits on the current, but the exponential nature of the current-voltage characteristics remains the same. A typical set of characteristics is shown in Fig. 3.

The temperature dependence of the drain cur-

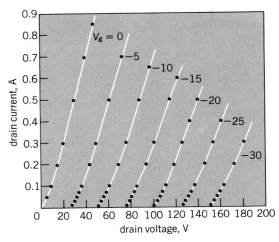

Fig. 3. Typical current voltage characteristics of SIT. Numbers next to curves give gate voltage V_g in volts. (From J. I. Nishizawa and K. Yamamoto, High-frequency high-power static induction transistor, IEEE Trans. Electron Devices, ED-25(3):314–323, 1978)

rent is found to be positive at low currents but negative at high currents, a fact that eliminates the problem of thermal runaway in high-power applications.

A related device is the field terminated diode (FTD), sometimes referred to as the field controlled thyristor (FCT). It differs from the SIT in that the n^+ substrate is replaced by a heavily doped p^+ substrate. The general features of the rest of the structure remain the same. Inclusion of the n-p^+ junction in the structure gives it a reverse-voltage-blocking capability that the SIT does not have. In the forward direction, FTD devices have the capability of blocking more than 1000 V with an applied gate bias of about 30 V, and simultaneously exhibiting a low forward voltage drop in the "on" state. The gate structure allows the forward current to be turned off in less than 1 μs.

Applications. SITs find application where high voltage and high current are simultaneously required. They are now commercially used as audio amplifiers with minimum breakdown voltages of 40 V between the source and gate, and 200 V between the gate and drain. In the low-to-medium frequency range, they are finding application as the switch in switched dc power supplies. For this application, devices that can handle 500 V and 20 A and be switched on or off in less than 1 μs are used. Use of the SIT should allow the switching frequency of the supply to be raised, thereby reducing the size of the inductance and capacitance required, and hence the cost of the supply. At microwave frequencies, the combination of high voltage at high current is the SIT's most attractive feature. Bipolar transistors can operate at high current but low voltages, while other devices, such as traveling-wave tubes, operate at high voltage but low current. It is the possibility of an inexpensive high-voltage high-current device that will operate with 60–100 V at up to 6 GHz and above that motivates the present developments. For applications at frequencies above 6 GHz, the GaAs field-effect transistor is a more likely candidate at this time. One very large commercial use of the SIT may be as a power source for microwave ovens, where it would represent a significant cost advantage over the magnetrons presently used. The requirement of 700 W at 2.45 GHz appears to be within reach of the SIT. See MAGNETRON; MICROWAVE SOLID-STATE DEVICES; TRAVELING-WAVE TUBE.

Fabrication. The fabrication of the SIT begins with a heavily doped n^+ silicon substrate. The lightly doped high-resistivity n^- layer is epitaxially grown on the n^+ substrate. The thickness of this layer determines the maximum gate-to-drain operating voltage and is about 40 μm for a 500-V device. The p gate regions are formed by diffusion, as are the n^+ source regions. The fabrication is completed by depositing and etching a layer of metal to provide electrical contact to the source and gate regions. Contact to the drain is made via the back side of the n^+ substrate when the chip is bonded into the final package. The p-type gate regions must be close enough together that the space between them can be totally depleted of movable carriers when the normal gate and drain voltages are applied. Center-to-center gate spacing in some devices is as low as 5 μm, which means that high-resolution fine-line lithography and other recently

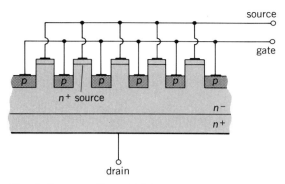

Fig. 4. Cross section of high-frequency SIT showing physical separation of source and gate.

developed processing methods must be used in fabrication. These methods include shallow ion implants, self-aligned diffusion processes, and x-ray lithography. In fact, it is the development of these methods that led to the renewed interest in the device.

The high capacitance between the gate and source is one problem of the structure which must be held to a minimum for high-frequency devices. Much of the present development effort is devoted to this problem. The input capacitance is most directly reduced by increasing the physical separation between the p-type gates and the n^+ source regions. This is usually done by placing the gate in the bottom of a narrow recessed area as shown in Fig. 4. The fabrication of such a structure with center-to-center gate spacing of a few micrometers requires very sophisticated processing methods. *See* TRANSISTOR.   [ROBERT J. HUBER]

Bibliography: J. I. Morenza and D. Esteve, *Solid State Electron.*, 21:739–746, 1978; J. I. Nishizawa, T. Terasaki, and J. Shibata, *IEEE Trans. Electron Devices*, ED-22(4):185–197, April 1975; J. I. Nishizawa and K. Yamamoto, *IEEE Trans. Electron Devices*, ED-25(3):314–322, March 1978.

Storage tube

An electron tube into which information can be introduced and then extracted at a later time; also called a memory tube. *See* ELECTRON TUBE.

Operation. The process of introducing information into a storage tube is known as writing, and that of extracting useful information is called reading. The deliberate removal of information from the storage surface is called erasing. In some tubes, reading automatically effects erasing; in others, a separate operation is required. A charge-storage tube is one in which information is retained on a surface in the form of electric charges.

The characteristics of a storage tube are largely governed by the nature of the storage surface. This surface is usually a deposit or sheet of insulating or semiconducting material. The point-by-point potential of this surface is varied, in a controlled way, by the dielectric charging processes associated with electron bombardment of an insulator. The low conductivity of the storage-surface material ensures that the charge (or potential) pattern will not be dissipated before the reading operation is initiated and completed.

In the reading operation, the potential pattern established by writing controls: (1) the percentage of the primary beam that is able to pass through the openings of a fine-mesh screen on which the storage surface may be deposited, or (2) the percentage of the primary beam which impinges on a phosphor screen, or (3) the magnitude of the secondary-electron current which results when the primary beam strikes the storage surface. In any case, the current collected by the output electrode is modulated in accordance with the stored charge pattern.

Storage tubes utilize electron guns to address the storage surface and effect the operations of reading, writing, and erasing. The electron gun will produce a fine pencil beam of electrons which may be deflected, as in a cathode-ray tube, to any desired location on the storage surface, or it will produce a flood beam which uniformly and simultaneously floods the entire storage surface with electrons. Because the use of the pencil-beam electron gun is related to its use in the cathode-ray tube, these tubes are frequently termed cathode-ray charge-storage tubes. *See* CATHODE-RAY TUBE.

Storage tubes may be classified according to the general nature of the input and output signals associated with them. Thus there are visual-electrical, electrical-electrical, and electrical-visual storage tubes. Visual-electrical storage tubes are known as camera tubes and are described elsewhere. *See* TELEVISION CAMERA TUBE.

Electrical storage tubes. Several different types of electrical-electrical storage tubes have been of commercial importance. These include the Radechon, recording storage tube, Graphechon, and silicon target storage tube. Each of these cathode-ray charge-storage tubes requires an electrical input signal and provides an electrical output signal. They have found use as devices for computer storage, signal time delay, signal comparison service, systems for the conversion of signal time bases, and scan conversion.

The widespread availability of low-cost solid-state memory devices and integrated circuits which will perform many of the same functions as electrical storage tubes has led to a general decline in the use of these tubes in new equipment designs.

Direct-view storage tubes. This group consists of cathode-ray charge-storage tubes which require an electrical input and provide a visual output. Among those of commercial importance are the display tube and the phosphor storage tube.

Display storage tube. The display storage tube employs a pencil beam of electrons, scanned across an insulating material coated on one side of a fine-mesh screen, to effect the writing process. This writing beam establishes a dielectric charge pattern on the insulating surface corresponding to the written information.

Erasing and reading processes in the display storage tube are accomplished through the use of a continuous low-velocity flood beam having a cross-sectional area equal to that of the fine-mesh-screen storage surface. The flood beam is switched between the erasing and reading modes by a change of a few volts in the beam-accelerating voltage. The flood beam in the reading mode passes through the fine-mesh storage screen, modulated point by point by the written charge pattern, and strikes a phosphor screen, producing a visible display of the pattern. Because of the action of the storage screen, these tubes are frequently termed transmission-control storage tubes.

Display storage tubes provide a bright display, including halftones, and have controlled persistence. They are used in radar displays and in oscillography. Growth in the use of display storage tubes has been restricted by the increasing availability of solid-state devices which will perform many similar storage functions when used in conjunction with conventional cathode-ray tubes.

Phosphor storage tube. The phosphor storage tube is closely akin to the display storage tube. An important difference in construction is that the phosphor screen, which is an insulator, itself serves as the storage surface, eliminating the need for a separate fine-mesh screen.

The phosphor storage tube operates in a bistable mode, that is, without halftones. Its availability in sizes up to a screen diagonal of 19 in. (48 cm) has led to its widespread use in computer graphics displays. *See* COMPUTER GRAPHICS; ELECTRONIC DISPLAY. [NORMAN W. PATRICK]

Bibliography: C. Curtin, Recent advances in direct view storage tubes, in Society for Information Display, *1977 SID Symposium*: *Digest of Technical Papers*, pp. 132–133, 1977; B. Kazan and M. Knoll, *Electronic Image Storage*, 1968.

Superconducting computers

High-performance computers whose circuits employ superconductivity and the Josephson effect to reduce cycle time.

Computer performance. A stored-program digital computer generally consists of a central processing unit (CPU) memory, and input/output (I/O) system. All problems that have to be solved on the computer are reduced to a series of elementary instructions (stored in memory) that govern the various operations to be performed on the data, such as add, multiply, and compare. The instructions are interpreted and the execution is controlled by the CPU, containing the arithmetic-logical unit (ALU) which performs the operations on the data; the CPU also has registers which store intermediate results. Ultimately the operating performance of the computer is governed by how the machine is organized to deal with the various problems and how fast the circuits used in the computer operate. Presently, high-performance, synchronous digital computers have operating cycle times in the range of 12 to 60 nanoseconds, and it normally takes from 1 to 10 cycles to execute an elementary instruction. High-performance computers are usually measured in terms of how many million instructions per second (MIPS) they can execute. This figure typically runs from a few to over 100 MIPS, depending on the machine, its technology, its organization (architecture), and the problem set (specific or general-purpose). *See* DIGITAL COMPUTER.

Josephson technology and superconductivity are being explored as ways of reducing the cycle time further; initially the goal is to reduce the cycle time below 4 ns, and ultimately it is thought that a subnanosecond cycle time will be possible.

The reasons that a change from the well-established silicon technology may be needed to enter the subnanosecond region have to do with the nature of the limitations on the computer's cycle time.

Limitations on cycle time. There are principally two components to the cycle time of a computer: the circuit delay and the package delay. The circuit delay, reflecting the time required to perform the logical operations, is a strong function of the chosen technology (CMOS, NMOS, bipolar, or Josephson), the power dissipated per circuit, and the level of integration (100 circuits per chip, 1000 circuits per chip, and so forth). Usually, for a given technology, higher circuit speed can be obtained by increasing dissipation or reducing circuit dimensions. The reduction in device size, paced by available lithographic tools, is also very desirable from the point of view of reliability and cost. However, there are two principal limitations to available circuit speed, both of which are related to power requirements. *See* INTEGRATED CIRCUITS.

The first is associated with the wasted power: the power dissipated in the form of heat. This has to be removed by a cooling medium, and the level of heat removal from circuit chips is limited on the one hand by the cost and complexity of the heat removal structures, and on the other hand by the allowable rise in temperature that the circuits can sustain while still operating correctly. Heat removal structures usually also limit the volumetric efficiency of packing circuit chips.

The second limitation to the power comes from the circuit engineers' ability to provide regulated power to the circuit chips. A large fraction of the chip input/output connections is devoted to power input—to carry the current, but even more importantly to reduce the power voltage swings caused by varying current drain on the chip as the circuits change their binary-encoded state under

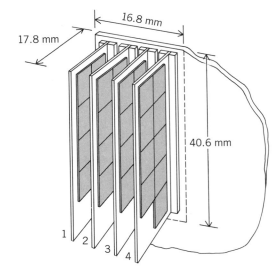

Fig. 1. Card-on-board package for high-density packing of chips. Typical dimensions expected for a small computer are indicated. (*After F. F. Tsui, JSP: A research signal processor in Josephson technology, IBM J. Res. Develop., 24:243–252, 1980*)

Fig. 2. Scanning electron micrograph of high-speed pluggable connections: array of platinum micropins, line of right-angle solder joints from mid-left to top right, and larger guidepin in foreground. (*From M. B. Ketchen et al., A Josephson technology system level experiment, IEEE Electr. Device Lett., EDL-2(10):262–265, 1981*)

as being the basis of a very promising computer technology. Initially, the very high switching speed and low power dissipation of this device were the main attractions, but it soon developed that many other aspects of the physics and engineering were very favorably inclined toward superconducting computer technology. Principal among these was the essentially lossless-transmission-line characteristics of all superconducting circuit interconnection lines. Intercircuit and interchip signals could be transmitted on matched lossless lines, allowing communication in a single pass with no reflections and with a speed of approximately one-third that of light. The characteristic wave impedances of such lines suitable from both fabrication and circuit criteria are sufficiently low (for example, 10 ohms) that interline crosstalk and disturbs are very small and, for all practical purposes, negligible. The interconnection lines have very small cross section; they are nearly lossless, and so there is no concern with respect to series resistance losses which significantly lower the projected and actual limits on the conductor cross-sectional area that semiconductor technology can employ. Furthermore, the very serious limit of electromigration is also absent in superconducting lines. *See* JOSEPHSON EFFECT.

It soon became apparent that circuits with such

program control. This inductive variation of chip voltage (sometimes called the Δi noise problem) becomes an increasingly difficult problem to solve as the circuit chip performance is improved. Of the two principal power limitations to circuit performance, this one appears the most demanding and severe.

The package delay component of the cycle time depends on the way the integrated circuit chips are arranged. A high-performance supercomputer usually consists of a large number of high-speed logic and memory chips that are required to be able to communicate with one another within a single cycle time. Signals propagate at typically one-third to three-fourths of the velocity of light in high-performance machines, and yet the total path length can easily become the principal limit to how small a cycle time can be realized. The chips must be packed closer and closer together — not only in a two-dimensional arrangement but most effectively in a three-dimensional arrangement. Put another way, the ultimate goal of high-performance computer packaging schemes is to achieve the highest possible density of circuits in a given volume, and this not only means high circuit density on each chip, but also high-density packing of chips. It is necessary to approach as closely as possible the ideal of a three-dimensional arrangement, and a good approximation is the card-on-board package (Fig. 1), where chips are attached to both sides of the card. The cards in turn are plugged closely together into the board.

Advantages of Josephson technology. The Josephson effect, predicted in 1962 and observed very soon thereafter, was quickly identified

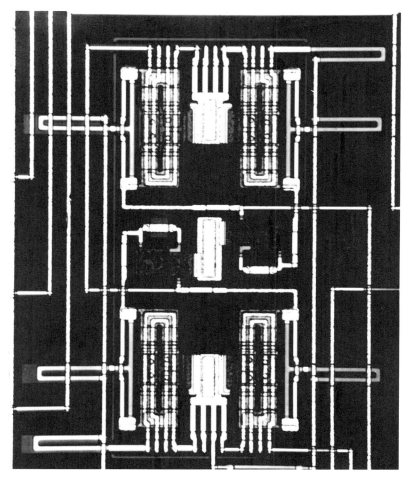

Fig. 3. Josephson chip circuit consisting of six Josephson devices configured into two logical AND gates. (*IBM*)

low power dissipation would not, for a very long time in the development of this technology, be concerned about fundamental cooling limits; the power removal problem was no longer a concern. This has enabled particularly attractive and novel techniques to be applied to power supply and regulation concerns faced by the circuit designer; active power supply and regulation circuits can be incorporated into the on-chip design, thereby removing the Δi problem.

These advantages naturally accrue from the basic physics of the superconducting state and the Josephson effect. The energy gap that characterizes these devices has shrunk from the hundreds of meV of the semiconductor electronic devices to just a few meV with the superconducting Josephson devices.

The next significant advantage that Josephson technology realized was in the techniques and technologies that are available to the circuit package engineer. With little or no cooling constraint, simple attachment of high-performance chips to the package is feasible; there are no heat removal structures required, and the package can be fabricated in a high-density three-dimensional card-on-board arrangement such as that shown in Fig. 1, where the total power dissipation is estimated at only 400 mW. Josephson technology also makes possible the high-density superconducting interconnection lines and the ability, for the first time, to make a monolithic package in the sense that the package components are all made from the same substrate material: silicon. This has significant importance in that it allows the engineer to work with little concern for differential thermal expansion in the package design. The use of silicon, not for its electrical properties but for its crystal perfection and machining properties, as well as mechanical strength and thermal properties, opens up a new door to micromachine fabrication techniques that can be applied to this novel package.

As package dimensions are shrunk and chip densities increase, it becomes increasingly difficult to provide an adequate number of very small, high-speed, pluggable connectors for each card. The techniques presently being explored consist of the use of very small solder joints to take signals around corners, and of platinum micropins that plug into extremely small sockets each containing a captured mercury drop, 200 μm in diameter. The pins are only 80 μm in diameter and 200 μm long, and they are arranged in a square array of a thousand or more pins with 300 μm between centers (Fig. 2).

Finally, there is a significant anticipation that one of the advantages in Josephson technology will be the low-temperature operation. Most of the failure-inducing mechanisms that plague present room-temperature computers will be frozen out and will not affect an operating computer at 4.2 K; thus, computer operation at this temperature holds the promise of high reliability.

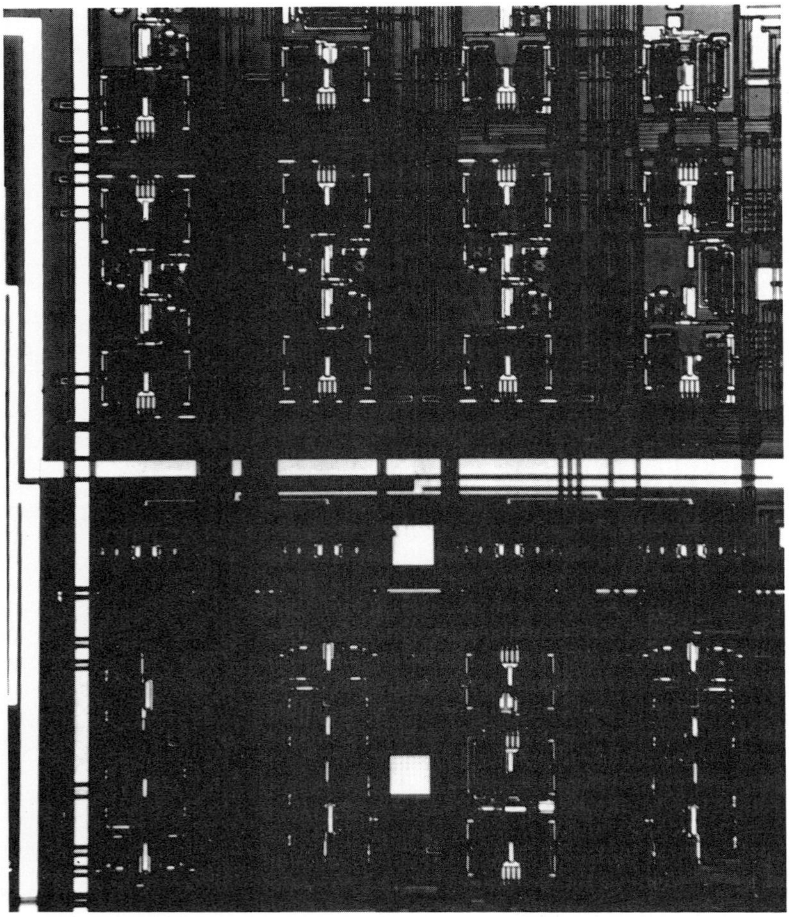

Disadvantages of Josephson technology. The significant optimism generated by the attractions of Josephson technology and superconductivity is tempered by the fact that a lot of detailed engineering and materials work remains to be done before a superconducting computer becomes a reality. Josephson devices are not simple extensions of transistor circuits. They involve a much more significant break with the past than there was in going from the triode vacuum tube to the transistor. The change involves much more than reducing power supply voltage and reducing impedances. Current and magnetic flux become of prime importance, rather than potentials and charge. Duality has been helpful only at providing physicists and engineers with hints of what must be done.

The biggest challenge is perhaps the sheer magnitude of the tasks involved in introducing a fundamentally new technology: chip technology, circuit design, package technology, design techniques, testing, and so forth. Good control over the prime circuit parameters has to be shown on a scale that allows superconducting computers to be contemplated. In this respect the control of the device threshold current presents a significant challenge. The superconducting-pair tunneling current depends exponentially on the barrier thickness to such an extent that average oxide thickness control to an accuracy of better than 0.1 nm is required to control the tunneling current to better than 10 or 20%. Also, the methods needed to ensure sufficient device isolation and power gain for reliable logic-circuit and memory design are still evolving.

There are obvious difficulties associated with the use of liquid helium. Cooling the computer from 300 to 4.2 K will undoubtedly produce stres-

Fig. 4. Portion of Josephson logic chip showing regular array of logic gates. (*IBM*)

ses. These have to be understood and taken care of; they affect the choice of materials both for the devices and for the package. The choice of more familiar but relatively ductile lead-alloy Josephson devices has to be weighed against the more robust refractory-based devices, such as niobium. More is known with respect to lead-alloy processes and device design, and this technology also provides a faster device since the device capacitance represented by lead-oxide (PbO) is approximately one-third that of niobium oxide (Nb_2O_5). However, sufficient progress has been made with the refractory-based devices that they have begun to challenge the position held in the 1970s by lead alloy, particularly with respect to low capacitance and low device leakage in the nonzero voltage state.

Perhaps the most obvious disadvantage of a superconducting computer, that of the liquid helium environment, is in fact not a particularly significant one. Admittedly, small liquid helium refrigerators are inefficient at converting kilowatts of electrical energy to only a few watts of cooling power at 4.2 K; however, this is not the bottleneck in the conventional cooling problem. That bottleneck, which is in the transistor technology of getting the heat away from the circuit without affecting correct operation and impacting packaging efficiency, has been broken with the direct cooling of Josephson chips in liquid helium.

Progress and prospects. Logic and memory test vehicles have received very active attention at a number of locations. Figure 3 shows an example of Josephson chip technology: six Josephson devices configured into two logical AND gates. This circuit was designed and fabricated by using a 2.5-μm design rule. Figure 4 is a portion of a Josephson logic chip with a regular array of logic gates, such as those shown in Fig. 3, connected together by wires running in wiring channels configured in a rectangular array. Across the center of the figure runs a utility line in the form of the on-chip power regulation and distribution system.

In 1981 some results associated with a cross-section model of a prototype machine were announced. This experiment, which attempted to explore all of the fundamental aspects of the packaging and circuit technologies, was designed to reveal potential problems associated with the fabrication of this card-on-board design. A representative path through the package was exercised with a minimum cycle time of 3.7 ns.

There is still much to be done before a superconducting supercomputer will be involved in calculations on weather forecasting, hydrodynamics, and so forth, but the direction of development appears to have been set. Fundamental considerations involving information processing and thermal noise argue that ultimately 4.2 K operation will be used for computers. *See* SUPERCONDUCTING DEVICES.

[DENNIS J. HERRELL]

Bibliography: W. Anacker, Computing at 4 degrees Kelvin, *IEEE Spectrum*, 16(5):26–37, May 1979; *IBM J. Res. Develop.*, vol. 24, no. 2, March 1980 (issue devoted entirely to Josephson computer technology); M. B. Ketchen et al., A Josephson technology system level experiment, *IEEE Electr. Device Lett.*, EDL-2(10):262–265, October 1981; J. Matisoo, The superconducting computer. *Sci. Amer.*, 242(5):50–65, May 1980.

Superconducting devices

Devices that perform functions in the superconducting state that would be difficult or impossible at room temperature, or that contain components which perform such functions. The superconducting state involves a loss of electrical resistance and occurs in many metals and alloys at temperatures near absolute zero. One of the major categories of superconducting devices is small-scale, electronic devices used in measuring instruments and computers, most of which involve Josephson tunneling. *See* SUPERCONDUCTIVITY.

The unique properties of superconductors have led to the development of tiny measuring and computing devices that are superior in performance to their nonsuperconducting counterparts. Most of the devices operate at or below 4.2 K, the temperature of liquid helium boiling under standard atmospheric pressure, and involve Josephson tunneling. The initial development of these devices took place in the second half of the 1960s. Fabrication of the devices has increasingly involved photolithography and electron-beam lithography, and superconducting junctions and circuits are now at the forefront of electronic ultraminiaturization. This article describes the four major areas of interest: SQUID magnetometers, computers, physical measurement standards, and detectors of high-frequency electromagnetic radiation.

SQUID magnetometers. There are two types of superconducting quantum interference device (SQUID) for detecting changes in magnetic flux: the dc SQUID and the rf SQUID. The dc SQUID, which operates with a dc bias current, consists of two Josephson junctions incorporated into a superconducting loop (Fig. 1a). The maximum dc supercurrent, known as the critical current, and the current-voltage (*I-V*) characteristic of the SQUID,

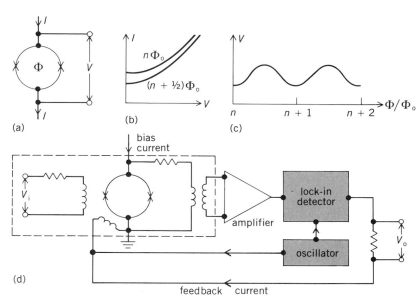

Fig. 1. The dc SQUID. (*a*) Schematic diagram. The symbol * denotes a Josephson junction. (*b*) Current-voltage (*I-V*) characteristic with applied flux $\Phi = n\Phi_0$ and $\Phi = (n + \frac{1}{2})\Phi_0$, where n is an integer. (*c*) Graph of voltage V versus Φ/Φ_0. (*d*) Block diagram of device in feedback loop with input circuit for measuring voltage. The dashed line encloses cryogenic components.

shown in Fig. 1*b*, oscillate as a function of the magnetic flux Φ threading the ring, with a period of one flux quantum, $\Phi_0 = h/2e \approx 2.07 \times 10^{-15}$ weber, where h is Planck's constant and e is the magnitude of the charge of the electron. Thus, when the SQUID is biased with a constant current, the voltage is periodic in the flux (Fig. 1*c*). The SQUID is almost invariably operated in the flux-locked loop shown in Fig. 1*d*. A change in the applied flux gives rise to a corresponding current in the coil that produces an equal and opposite flux in the SQUID. The SQUID is thus the null detector in a feedback circuit, and the output voltage V_0 is linearly proportional to the applied flux. Modern dc SQUIDs are fabricated from thin films, and usually involve Josephson tunnel junctions, resistively shunted to eliminate hysteresis in the *I-V* characteristics. For a number of years, the flux sensitivity remained at 10^{-5} to $10^{-4}\ \Phi_0 \mathrm{Hz}^{-1/2}$, but it has increased by several orders of magnitude, and approaches the limit set by the uncertainty principle. It appears that the dc SQUID is an ideal quantum-limited amplifier, although much work remains to be accomplished to exploit this property fully. *See* JOSEPHSON EFFECT.

The rf SQUID (Fig. 2*a*) consists of a single Josephson junction incorporated into a superconducting loop and operates with a rf bias. The SQUID is coupled to the inductor of an *LC*-resonant circuit excited at its resonant frequency, typically 30 MHz, although frequencies as high as 10 GHz have been used successfully. The rf voltage across the tank circuit versus the rf current is shown in Fig. 2*b*. When the amplitude of the rf current is properly adjusted, the amplitude of the rf voltage across the tank circuit oscillates as a function of applied flux (Fig. 2*c*), with a period Φ_0. The rf SQUID is also usually operated in a feedback

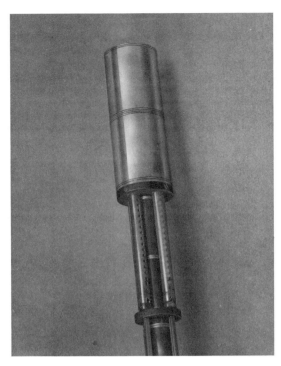

Fig. 3.   Second-derivative magnetic field gradiometer used for medical measurements. (*S. H. E. Corpo*)

mode (Fig. 2*d*). Most rf SQUIDs are made from machined niobium components with a point contact junction. Although they have been used more widely than dc SQUIDs, their sensitivity is now considerably lower, and it seems unlikely that their performance will be able to match that of the dc SQUID.

Most SQUIDs are used in conjunction with input circuits. Figure 1*d* shows a voltmeter, the sensitivity of which is usually limited by Johnson noise in the resistor to typically $10^{-15} - 10^{-12}$ V Hz$^{-1/2}$. Figure 2*d* shows a magnetic field gradiometer, in which two balanced superconducting pick-up loops are connected in series with a superconducting coil coupled to the SQUID. If a uniform magnetic field is applied, no flux is linked to the SQUID, while the application of a field gradient $\partial H_z / \partial z$ induces a flux in the SQUID proportional to the difference of the flux in the two loops. Sensitivities of 10^{-13} T m^{-1} Hz$^{-1/2}$ are typical. A gradiometer can be adapted to measure the paramagnetic susceptibility of tiny samples: one inserts the sample into one of the loops in the presence of a constant magnetic field, and measures the resulting change in flux. A system with a single pick-up loop is a magnetometer, with a typical sensitivity of 10^{-14} T Hz$^{-1/2}$; however, substantially higher sensitivities have been achieved. *See* VOLTMETER.

SQUIDs have been widely used in low-temperature research, but are now important tools for noncryogenic applications as well. One example is the measurement of magnetic signals produced by the human heart and brain: a second-order gradiometer ($\partial^2 H_z / \partial z^2$) (Fig. 3) gives sufficient rejection of noise from external magnetic sources that the measurements do not require a shielded environment. SQUID magnetometers are of growing

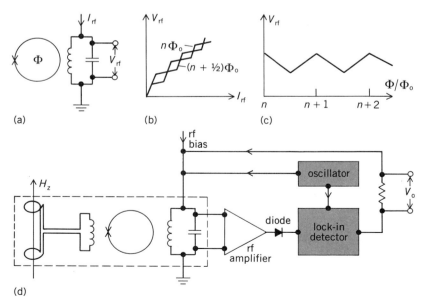

(a)   (b)   (c)   (d)

Fig. 2.   The radio-frequency (rf) SQUID. (*a*) Schematic diagram of device coupled to *LC*-resonant circuit. (*b*) Graph of rf voltage V_{rf} versus rf current I_{rf} with applied flux $\Phi = n\Phi_0$ and $\Phi = (n + \frac{1}{2})\Phi_0$. (*c*) Graph of V_{rf} versus Φ/Φ_0. (*d*) Block diagram of device in feedback loop with input circuit for measuring magnetic field gradients. The dashed line encloses cryogenic components.

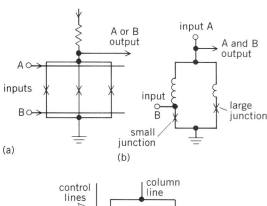

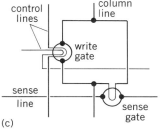

Fig. 4. Schematic diagrams of Josephson junction computer elements. (*a*) Three-junction interferometer OR gate. (*b*) Asymmetric two-junction interferometer AND gate. (*c*) High-speed memory cell.

importance in geophysics, for example, in the measurement of rock susceptibility, and in magnetotellurics. A further potential application is as the transducer for gravitational wave antennas. *See* SQUID.

Computer elements. Typical high-speed digital computers, using large-scale integration of semiconductor devices, have cycle times of 30 to 50 ns. To achieve the goal of further reducing this time to 1 ns, both the device switching time and the time required to transmit a signal between different parts of the computer must be less than 1 ns. Because the signal propagation speed is typically 10^8 ms^{-1}, the second requirement implies that the largest dimension of the computer must be no greater than 0.1 m. Although semiconducting devices with switching times substantially less than 1 ns are certainly available, they dissipate considerable amounts of power. Because of the difficulty of extracting this power from a small volume, it does not appear feasible to reduce the physical size of a main-frame computer sufficiently to achieve a cycle time of 1 ns. On the other hand, preliminary studies of Josephson junction computer elements in the mid-1960s established that these devices had the combined requirements of very high switching speed and very low dissipation. A substantial research program has subsequently been maintained. *See* INTEGRATED CIRCUITS.

The circuits consist of many layers of metals and insulators, each one patterned by a photolithographic mask, and deposited over a superconducting ground plane on a silicon wafer. The line widths are currently 2.5 or 5 μm. The Josephson tunnel junctions have hysteretic *I-V* characteristics, and are fabricated from lead alloys; their reliability is excellent.

The two essential components of a computer are logic circuits and memory cells. The logical opera-

tions necessary for a central processor involve two basic functions: OR and AND. One type of OR circuit involves a two- or three-junction interferometer or SQUID; the three-junction version is shown schematically in Fig. 4*a*, and as fabricated in Fig. 5*a*. The interferometer is biased with a current below its initial critical current. A current pulse in either of the control lines, which in practice are films overlaying the interferometer, couples a magnetic flux into the device that lowers the critical current and induces a transition into the nonzero voltage state. An output from the gate thus represents A OR B. The fastest measured switching is 6 ps, while the propagation time along the transmission line is 7 ps, giving an overall gate time of 13 ps. Figure 4*b* shows an AND gate, which contains two junctions of different areas and hence different critical currents, asymmetrically arranged on the loop. Because of the asymmetries, simultaneous signals are required on the two in-

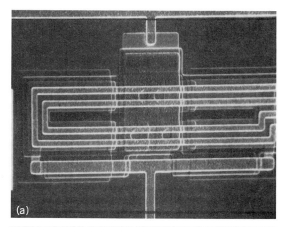

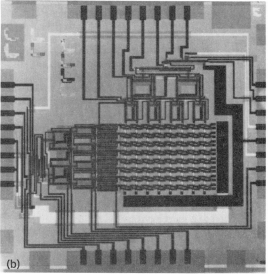

Fig. 5. Photographs of Josephson junction computer elements. (*a*) Three-junction interferometer OR gate with overlaying control lines (*from L. M. Geppert et al., Damped three-junction interferometers for latching logic, IEEE Trans. Magnet., MAG-15:412–415, 1979*). (*b*) 64-bit Josephson random-access memory on 6.35 × 6.35 mm² chip (*from W. H. Henkels and Hans H. Zappe, Experimental 64-bit decoded Josephson NDRO random access memory, IEEE J. Solid State Circuits, SC-13:591–600, 1978*).

puts to induce a transition, producing an output voltage A AND B. *See* LOGIC CIRCUITS.

There are two types of cryogenic memory, a very fast cache memory that is coupled directly to the logic circuits, and a larger but slower main memory. One design for the cache memory (Fig. 4*c*) consists of an interferometer incorporated into a superconducting loop. A "1," represented by a persistent supercurrent, or a "0," represented by zero supercurrent, can be written into the memory by the application of appropriate current pulses on the column line and the control lines. The contents of the cell can be read nondestructively by applying pulses to the column line and the sense line. The cell is designed for operation in an array, in which individual cells can be addressed; Fig. 5*b* is a photograph of an array with 64 cells used for testing purposes. The main memory loop is more compact, and consists of two-junction interferometers that are read destructively. Both types of memory dissipate power during writing or reading, but not during storage. *See* COMPUTER STORAGE TECHNOLOGY.

All of the necessary components for a computer have been constructed and successfully tested. Unless there is a major breakthrough in semiconductor technology or a new computer technology arises, it seems likely that the next generation of ultrahigh-speed computers will be based on the Josephson junction. However, such computers are unlikely to be available before 1990. *See* DIGITAL COMPUTER.

Standards. The Josephson effects have an established role in standards laboratories. The most important applications are the measurement of the fundamental constant ratio e/h, and maintaining the standard volt. When a Josephson junction is irradiated with microwaves of frequency f, constant-voltage steps are induced on the I-V characteristic at voltages $nhf/2e$, where n is an integer. Precise measurements of the voltages at which the steps are induced by a known frequency have led to the most accurate determination available of e/h. Furthermore, the standard volt at the United States National Bureau of Standards and at a number of other national laboratories is now maintained (but not defined) by these voltage steps, which may be compared periodically with the conventional standard cells. Since frequency can be measured very accurately, it is a relatively simple procedure for widely spaced laboratories to compare their voltage standard. *See* ELECTRICAL MEASUREMENTS.

Josephson devices have been used in two ways as noise thermometers for temperatures down to the millikelvin range. In the first, a junction shunted with an external resistance is biased with a stable current. The Johnson noise voltage generated by the resistance, which is proportional to the absolute temperature T, induces fluctuations in the frequency of the Josephson radiation emitted by the junction. By measuring the bandwidth of these fluctuations, one can determine T. In the second method, a SQUID voltmeter (Fig. 1*d*) is used to measure the Johnson noise voltage generated by a resistor.

Other applications of SQUIDs include the comparisons of static voltages or currents to high precision, and the measurement of rf power levels.

Josephson junction mixers can be used to synthesize frequencies up to 1 THz or more from microwave sources. Fixed-point thermometers that rely on the reproducibility of the superconducting transition temperature of a series of metals are available for the temperature range 0.015 K (tungsten) to 7.2 K (lead).

Electromagnetic radiation detectors. There has been extensive investigation of Josephson junctions, notably point contacts, as sensitive detectors of microwave and millimeter radiation. Various modes of operation have been used, including square law detectors, mixers, and parametric amplifiers. These devices have met with varying degrees of success, but none is clearly superior to other, nonsuperconducting devices. However, a superconducting device that does not involve Josephson tunneling has been developed, the superconductor-insulator-superconductor (SIS) quasiparticle mixer. The device consists of a small-area tunnel junction, fabricated from lead alloys, with the I-V characteristic shown in Fig. 6.

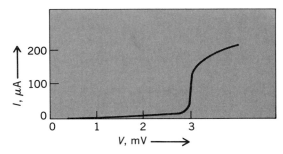

Fig. 6.   Current-voltage (I-V) characteristic of superconductor-insulator-superconductor (SIS) quasiparticle junction.

The mixer is operated near the sharp onset in the current, where the characteristics are highly nonlinear. The nonlinearity is used to mix the signal frequency f_S with the local oscillator frequency f_{LO} to produce an intermediate frequency $f_{IF} = |f_S - f_{LO}|$ that is coupled out of the junction into a low-noise preamplifier. At 36 GHz the detector has achieved the quantum limit, that is, it can detect single photons, and exhibits conversion gain. It is expected that photon-noise-limited performance can be extended to frequencies in excess of 100 GHz. Such receivers are likely to have a major impact on radio astronomy, and could be of great importance in such applications as space communications.

[JOHN CLARKE]

Bibliography: J. Matisoo, The superconducting computer, *Sci. Amer.*, 242(5):50–65, May 1980; Proceedings of the 1980 Applied Superconductivity Conference, *IEEE Trans. Magn.*, vol. 17, no. 1, January 1981; B. B. Schwartz and S. Foner (eds.), *Superconductor Applications: SQUIDs and Machines*, 1977; Special issue of Josephson computer technology, *IBM J. Res. Develop.*, 24(2):107–252, 1980; Special issue on superconducting device, *IEEE Trans. Elec. Devices*, ED-27(10):1855–2042, 1980.

Superconductivity

A phenomenon occurring at very low temperatures in many electrical conductors, in which the electrons responsible for conduction undergo a collective transition to an ordered state with many unique and remarkable properties. These include the vanishing of resistance to the flow of electric current, the appearance of a large diamagnetism and other unusual magnetic effects, substantial alteration of many thermal properties, and the occurrence of quantum effects otherwise observable only at the atomic and subatomic level.

Superconductivity was discovered by H. Kamerlingh Onnes in Leiden in 1911, while studying the variation with temperature of the electrical resistance of mercury within a few degrees of absolute zero. He observed that the resistance dropped sharply to an unmeasurably small value at a temperature of 4.2 Kamerlingh Onnes's original data are shown in Fig. 1. The temperature at which the transition occurs is called the transition or critical temperature, T_c. The vanishingly small resistance (very high conductivity) below T_c suggested the name given the phenomenon.

In 1933 W. Meissner and R. Ochsenfeld discovered that a metal cooled into the superconducting state in a not-too-large magnetic field expels the field from its interior. This discovery demonstrated that superconductivity involves more than simply very high or infinite electrical conductivity, remarkable as that alone is.

Superconductivity remained a much studied but puzzling phenomenon for nearly half a century after its discovery. A great deal of experimental in-formation was amassed on its occurrence and its properties, and several useful phenomenological theories were developed. Then, in 1957, J. Bardeen, L. N. Cooper, and J. R. Schrieffer reported the first successful microscopic theory of superconductivity. It describes how and why the electrons in a conductor may form an ordered superconducting state, and makes predictions about many properties of superconductors which are in good agreement with experimental information.

Since the appearance of the Bardeen-Cooper-Schrieffer (BCS) theory, theoretical and experimental understanding of superconductivity has continued to expand. New superconducting materials and effects have continued to be discovered. Practical applications of the phenomenon are becoming common, ranging from powerful electromagnets and machinery to ultrasensitive electronic instruments and computer elements. *See* SUPERCONDUCTING DEVICES.

BASIC EXPERIMENTAL PROPERTIES

Soon after its discovery in mercury, superconductivity was found to occur also in such common metals as lead and tin. Initially, the number of known superconductors was quite small. This was so in part because experiments were then confined to temperatures above about 1 K, the minimum temperature readily available using liquid helium (^{4}He) as a refrigerant, so that only superconductors with transition temperatures above 1 K could be discovered. It was therefore thought that superconductivity might be a relatively rare phenomenon. Scientists now know that it is not, but in fact occurs quite generally. Advances in the technology for achieving low temperatures have pushed the minimum available temperature down to about 0.001 K, and progress in the preparation of materials has greatly expanded the number and variety of materials which have been tested for superconductivity. Some 26 of the metallic elements are known to be superconductors in their normal forms, and another 10 become superconducting under pressure or when prepared in the form of highly disordered thin films (Fig. 2). The number of known superconducting compounds and alloys runs into the thousands. Superconductivity is thus a rather common characteristic of metallic conductors, so much so that its absence is often more unusual and striking than its presence.

Despite the existence of a successful microscopic theory of superconductivity, there are no completely reliable rules for predicting whether a metal will be a superconductor. Certain trends and correlations are apparent among the known superconductors, however—some with obvious bases in the theory—and these provide empirical guidelines in the search for new superconductors. Superconductors with relatively high transition temperatures tend to be rather poor conductors in the normal state. For many years, no superconductors were known among the noble metals, the alkali metals, and the alkaline earth metals. However, cesium, beryllium, and barium have been found to be superconducting under high pressure or in disordered films, and there is some evidence that at least one of the noble metals may be superconducting at extremely low temperatures.

The ordered superconducting state appears to

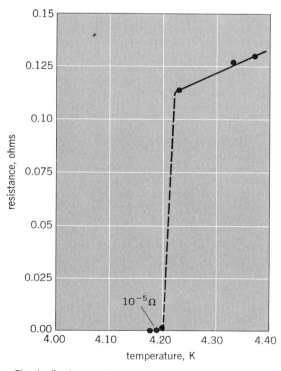

Fig. 1. Resistance in ohms of a specimen of mercury versus absolute temperature. (*From H. Kamerlingh Onnes, Akad. van Wetenschappen, Amsterdam, 14: 113, 818, 1911*)

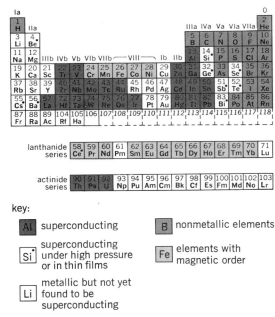

lanthanide series | 58 Ce* | 59 Pr | 60 Nd* | 61 Pm | 62 Sm | 63 Eu | 64 Gd | 65 Tb | 66 Dy | 67 Ho | 68 Er | 69 Tm | 70 Yb* | 71 Lu

actinide series | 90 Th | 91 Pa | 92 U | 93 Np | 94 Pu | 95 Am | 96 Cm | 97 Bk | 98 Cf | 99 Es | 100 Fm | 101 Md | 102 No | 103 Lr

key:

[Al] superconducting

[Si*] superconducting under high pressure or in thin films

[Li] metallic but not yet found to be superconducting

[B] nonmetallic elements

[Fe] elements with magnetic order

Fig. 2. Superconducting elements in the periodic table. (*From N. W. Ashcroft and N. D. Mermin, Solid State Physics, Holt, Rinehart and Winston, 1976*)

be incompatible with any long-range-ordered magnetic state: None of the ferromagnetic or antiferromagnetic metals are also superconducting. (Cerium displays antiferromagnetic order but is not superconducting in the phase which exists at ordinary pressures; it is superconducting but not magnetically ordered in a different phase which occurs at pressures above about 5 GPa.) This distaste of superconductors for magnetism extends to the effects of impurities. The presence of nonmagnetic impurities in a superconductor usually has very little effect on the superconductivity, but the presence of impurity atoms which have localized magnetic moments can markedly depress the transition temperature even in concentrations as low as a few parts per million.

Some semiconductors with very high densities of charge carriers are superconducting, and others such as silicon and germanium have high-pressure metallic phases which are superconducting. Many elements which are not themselves superconducting form compounds which are. Examples are CuS and a polymeric form of Sn. Although nearly all the classes of crystal structure are represented among superconductors, certain structures appear to be especially conducive to superconductivity. An example is the so-called A-15 structure shared by a series of intermetallic compounds based on niobium, for example, Nb_3Sn with $T_c = 18.1$ K. High values of transition temperature occur most frequently in elements, compounds, or alloys having three, five, or seven valence electrons per atom.

The highest transition temperature observed is 23 K for a specially prepared alloy of niobium, aluminum, and germanium. There is no widely accepted theoretical proof that superconductivity is necessarily restricted to low temperatures. Encouraged by this and motivated by visions of the immense practical and fundamental implications of a "high-temperature" superconductor, many investigators have searched for materials with higher transition temperatures.

Electrical resistance. It is, of course, not possible to establish that the dc (zero-frequency) electrical resistance of a superconductor is identically zero, but a rather stringent upper limit on the resistance can be established by inducing an electrical current in a superconducting loop or coil and observing whether it dies away in time. The decay time of such a current in a nonsuperconducting coil at low temperatures is on the order of 1 sec or less. Induced currents have been observed to persist in superconducting loops for several years. Very precise measurements of the magnetic field produced by a persistent current, using nuclear magnetic resonance over shorter periods of time, have established that the supercurrent decay time is at least 100,000 years. This implies that the resistance in the superconducting state is at least 10^{12} times less than in the normal state.

It can be shown theoretically that the persistent current state is not really absolutely stable, but only metastable. In superconducting materials such as those used to build superconducting magnets, finite persistent current decay times are often observed, due to processes which cause irreversible redistribution of magnetic flux in the material. But under many conditions the lifetime of the metastable persistent current state is so long that it is not unreasonable to say that the lifetime is infinite, and the electrical resistance is zero.

Magnetic properties. The existence of the Meissner-Ochsenfeld effect, the exclusion of a magnetic field from the interior of a superconductor, is direct evidence that the superconducting state is not simply one of infinite electrical conductivity. If this were so, a superconductor cooled in a magnetic field through its transition temperature would trap the field in its interior. If the external source of the field were subsequently removed, persistent eddy currents would be induced in the superconductor which would preserve the interior field even in the absence of the external source. Instead, the Meissner-Ochsenfeld effect implies that the superconducting state is a true thermodynamic equilibrium state, a new phase which has lower free energy than the normal state at temperatures below the transition temperature and which somehow requires the absence of magnetic flux.

The exclusion of magnetic flux by a superconductor costs some magnetic energy. So long as this cost is less than the condensation energy gained by going from the normal to the superconducting phase, the superconductor will remain completely superconducting in an applied magnetic field. If the applied field becomes too large, the cost in magnetic energy will outweigh the gain in condensation energy, and the superconductor will become partially or totally normal. The manner in which this occurs depends on the geometry and the material of the superconductor. Consider first the simplest geometry, a very long cylinder with field applied parallel to its axis. Two distinct types of behavior may then occur, depending on the type of superconductor.

Type I superconductors. Below a "critical field" H_c which increases as the temperature decreases below T_c, the magnetic flux is excluded from a type

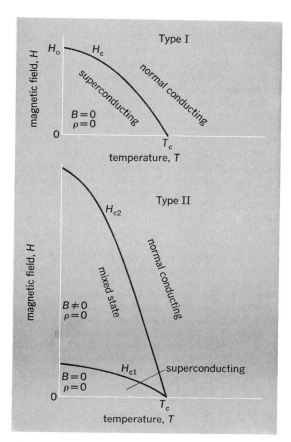

Fig. 3. The superconductive state in the magnetic-field temperature plane.

typically 10^{-7} m. In a sufficiently pure and defect-free type II superconductor, the fluxoids tend to arrange themselves in a regular lattice. This vortex state of the superconductor is known as the mixed state. It exists for applied fields between H_{c1} and H_{c2}. At H_{c2}, the superconductor becomes normal, and the field penetrates completely. (Actually, a superconducting surface sheath may persist up to an even higher critical field H_{c3}, which is approximately 1.5 H_{c2}.)

In contrast to critical fields in type I superconductors, which tend to be less than 1000 oersteds (1 Oe = 79.6 A/m), H_{c2} for type II superconductors may be several hundred thousand oersteds or more. (The maximum known H_{c2} is about 600,000 Oe). Since a zero-resistance supercurrent can flow in the mixed state in the superconducting regions surrounding the fluxoids, a type II superconductor can carry a lossless current even in the presence of a very large magnetic field. Such superconductors are therefore of considerable practical importance in high-field magnets.

A type II superconductor in the mixed state is not necessarily completely lossless, however. The presence of an electric current creates a force on the fluxoids. They therefore tend to move. Moving magnetic flux creates voltages by electromagnetic

I superconductor, which is said to be perfectly diamagnetic. If the applied field is increased above H_c, the entire superconductor reverts to the normal state and the field penetrates completely. The curve of H_c versus temperature T in Fig. 3 is thus a phase boundary in the magnetic field-temperature plane separating a region where the superconducting phase is thermodynamically stable from the region where the normal phase is stable. The curve of H_c versus T for any type I superconductor is approximately parabolic: To within a few percent, $H_c = H_0[1 - (T/T_c)^2]$, where H_0 is the value of H_c at absolute zero. Values of T_c and H_0 for some typical superconductors are given in the table. All of the known elemental superconductors except niobium are of type I.

Type II superconductors. For a type II superconductor, there are two critical fields, the lower critical field H_{c1} and the upper critical field H_{c2}. In applied fields less than H_{c1}, the superconductor completely excludes the field, just as a type I superconductor does below H_c. At fields just above H_{c1}, however, flux begins to penetrate the superconductor in microscopic filaments called fluxoids or vortices. Each fluxoid consists of a normal core in which the magnetic field is large, surrounded by a superconducting region in which flows a vortex of persistent supercurrent which maintains the field in the core. The total magnetic flux in each fluxoid is exactly equal to a fundamental quantum of magnetic flux, $\phi_\phi = 2.07 \times 10^{-7}$ gauss cm^2 = 2.07×10^{-15} Wb. The diameter of the fluxoid is

Values of T_c and H_0 for the superconducting elements*

Element	Phase	T_c (K)	H_0 (oersteds)†
Al		1.196	99
Cd		0.56	30
Ga		1.091	51
Hf		0.09	—
Hg	α(rhomb)	4.15	411
	β	3.95	339
In		3.40	293
Ir		0.14	19
La	α(hcp)	4.9	798
	β(fcc)	6.06	1096
Mo		0.92	98
Nb		9.26	1980†
Os		0.655	65
Pa		1.4	—
Pb		7.19	803
Re		1.698	198
Ru		0.49	66
Sn		3.72	305
Ta		4.48	830
Tc		7.77	1410
Th		1.368	162
Ti		0.39	100
Tl		2.39	171
U	α	0.68	—
	γ	1.80	—
V		5.30	1020
W		0.012	1
Zn		0.875	53
Zr		0.65	47

*From N. W. Ashcroft and N. D. Mermin, *Solid State Physics*, Holt, Rinehart, and Winston, 1976.
†At $T = 0$ K. 1 Oe = 76.9 A/m.
‡For Nb, a type II superconductor, the zero-temperature critical field quoted is obtained from an equal-area construction: The low-field ($H < H_{c1}$) magnetization is extrapolated linearly to a field H_c chosen to give an enclosed area equal to the area under the actual magnetization curve.

induction, and the presence of nonzero voltages together with the current implies power dissipation. This loss mechanism can often be suppressed by introducing defects into the crystal structure of the superconductor which tend to pin down the fluxoids and prevent them from moving.

Penetration depth. The way in which a superconductor excludes from its anterior an applied magnetic field smaller than H_c (type I) or H_{c1} (type II) is by establishing a persistent supercurrent on its surface which exactly cancels the applied field inside the superconductor. This surface current flows in a very thin layer of thickness λ, which is called the penetration depth. The external field also actually penetrates the superconductor within the penetration depth. Lambda depends on the material and on the temperature, the latter variation being given approximately by $\lambda = \lambda_0[1 - (T/T_c)^4]^{-1}$ (λ_0 is the penetration depth at zero temperature for the particular material, and is typically of order 5×10^{-8} m).

Intermediate state. Another kind of magnetic effect occurs in type I superconductors for all but the simplest geometries. The exclusion of magnetic flux by the superconductor distorts the field in its vicinity. As a result, the magnetic field may reach H_c at some points on the surface of the superconductor while remaining below H_c elsewhere. The superconductor near the points of highest field will tend to go normal. The magnetic flux then begins to penetrate the superconductor in nonsuperconducting lamellae separated by regions where superconductivity remains. If the applied field is further increased, the fraction of the specimen occupied by the normal lamellae increases, while the fraction occupied by the intervening superconducting regions decreases. With increasing applied field, this continues until the specimen becomes completely normal and the field penetrates everywhere. A type I superconductor with such alternating superconducting and normal lamellae is in the intermediate state.

There are superficial similarities between the intermediate state in a type I superconductor and the mixed state in a type II superconductor, but the two states are really quite different. In the intermediate state, the lamellae are of macroscopic size, sometimes an appreciable fraction of a millimeter. This size is determined by two energies, a magnetic energy which increases as the size of the lamellae increases, and a positive surface energy associated with the interface between the superconducting and normal regions, rather like the surface energy associated with the surface tension at a liquid-gas interface. This surface energy increases with the supernormal interface area and hence with the number of lamellae, that is, it decreases as the size of the lamellae increases. The magnetic energy favors small lamellae, and the surface energy favors large lamellae. The equilibrium lamellar structure is determined by a compromise between the two and is rather sensitive to the geometry and degree of perfection of the specimen. In type II superconductors, on the other hand, the supernormal interface energy turns out to be negative. It is therefore energetically favorable for the regions of flux penetration to be as small as possible. Their minimum size is limited only by a fundamental quantum constraint on the

nature of the superconducting state itself. It is this which sets the level of magnetic flux contained within a fluxoid at the flux quantum mentioned above and determines the microscopic scale of the fluxoid structure in the mixed state.

Critical current. The existence of the critical field leads to another property of superconductors which is of some practical importance. In his Nobel prize lecture, Kamerlingh Onnes referred to the possibility of constructing powerful electromagnets which would consume no electrical power, using superconductors. However, a supercurrent flowing in a superconducting wire will itself create a magnetic field, and this field will drive the superconductor normal at some critical value of the current (the Silsbee critical current). Unfortunately, the critical currents which accompany typical critical fields for type I superconducting wires of macroscopic size are so small that Kamerlingh Onnes's dream remained unrealized until the exploitation of type II superconductors nearly 40 years later.

Thermal properties. The appearance of the superconducting state is accompanied by quite drastic changes in both the thermodynamic equilibrium and thermal transport properties of a superconductor.

Heat capacity. Figure 4 shows the heat capacity of an aluminum specimen in both the normal and superconducting states. In the normal state (produced at temperatures below the transition temperature by applying a magnetic field greater than the critical field), the heat capacity is determined primarily by the normal electrons (with a small contribution from the thermal vibrations of the crystal lattice) and is nearly proportional to the temperature. In zero applied magnetic field, there appears a discontinuity in the heat capacity at the transition temperature. At temperatures just below the transition temperature, the heat capacity is larger than in the normal state. It decreases more rapidly with decreasing temperature, however,

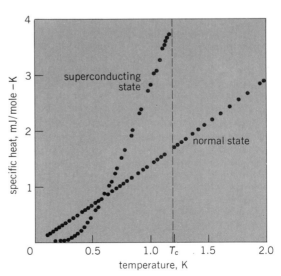

Fig. 4. Low-temperature specific heat of normal and superconducting aluminum. (*From N. E. Phillips, Heat Capacity of Aluminum between 0.1 K and 4 K, Phys. Rev., 114:676–685, 1959*)

and at temperatures well below the transition temperature varies exponentially as $e^{-\Delta/kT}$, where Δ is a constant and k is Boltzmann's constant. Such an exponential temperature dependence is a hallmark of a system with a gap Δ in the spectrum of allowed energy states. Heat capacity measurements provided the first indications of such a gap in superconductors, and one of the key features of the macroscopic BCS theory is its prediction of just such a gap.

Thermal conductivity. Ordinarily a large electrical conductivity is accompanied by a large thermal conductivity, as in the case of copper, used in electrical wiring and cooking pans. However, the thermal conductivity of a pure superconductor is less in the superconducting state than in the normal state, and at very low temperatures approaches zero. This property is applied in "heat switches" for use at low temperatures, in which the thermal contact between two bodies connected by a superconducting wire can be switched on and off simply by application of a magnetic field which switches the superconductivity on and off. Crudely speaking, the explanation for the association of infinite electrical conductivity with vanishing thermal conductivity is that the transport of heat requires the transport of disorder (entropy). The superconducting state is one of perfect order (zero entropy), and so there is no disorder to transport and therefore no thermal conductivity.

Thermoelectric properties. A combined thermal and electrical effect of interest and practical importance is the Peltier effect, which is the basis of operation of thermocouples used for temperature measurement. If the two junction regions of a loop made of two different metals are maintained at different temperatures, an electrical potential gradient is produced which will drive a current around the loop. This effect vanishes in the superconducting state. (That, at least, is the conventional view. Claims have been made for observation of some sort of thermoelectric effects in superconductors.)

High frequency electromagnetic properties. The electrical and magnetic behavior of superconductors at high frequencies differs from the zero frequency behavior described above. In the radio-frequency (up to about 10^8 Hz) and microwave-frequency (from 10^8 to about 10^{11} Hz) regions of the electromagnetic spectrum, it is found that superconductors do not have zero resistance to the flow of current. The resistance and the accompanying electrical energy loss are still much smaller than in the normal state, but they are not zero, and they increase with increasing frequency. On the other hand, in the optical region of the spectrum (about 10^{15} Hz), the electromagnetic response in the superconducting state is indistinguishable from that in the normal state. This may be confirmed simply by looking at a superconductor as it transforms from the normal state to the superconducting state; there is no change in its appearance. Clearly something interesting happens somewhere between 10^{11} and 10^{15} Hz. It is found that in the region of 10^{11} to 10^{12} Hz, depending on the material, the absorption of electromagnetic radiation by a superconductor rises quite sharply from a small value to the value characteristic of the normal state. This behavior provides another clear indica-

tion of the presence of a gap in the electronic energy spectrum of a superconductor, rather like the gaps which occur in semiconductors. (In semiconductors, however, gaps tend to be on the order of 1 eV, whereas the gaps in superconductors are typically a thousand times smaller.) The sharp rise in electromagnetic absorption occurs at the frequency for which the energy of a single photon (equal to Planck's constant times the frequency) becomes just sufficient to produce an excitation of some sort (consisting, in fact, of two "electrons") out of the superconducting state across the gap. *See* SEMICONDUCTOR.

Isotope effect. The availability of substantial quantities of separated isotopes of the elements after World War II made it possible to test whether the superconducting state depended in some way on atomic mass. The result provided a crucial key to the development of the BCS microscopic theory of superconductivity. It was found that the superconducting transition temperature for a given element was proportional to $M^{-\frac{1}{2}}$, where M is the isotopic mass. The vibration frequency of a mass M on a spring is proportional to $M^{-\frac{1}{2}}$, and the same relation holds for the characteristic vibrational frequencies of the atoms in a crystal lattice. Thus, the existence of the isotope effect indicated that, although superconductivity is an electronic phenomenon, it nevertheless depends in an important way on the vibrations of the crystal lattice in which the electrons move. Fortunately, not until after the development of the BCS theory was it discovered that the situation is more complicated than it had appeared. For some superconductors, the exponent of M is not $-\frac{1}{2}$, but near zero, and for at least one it is positive.

Absence of effects. While most of the electronic properties of a superconductor are profoundly affected by the transition to the superconducting state, many properties are changed very little if at all. These include the mechanical and elastic properties, tensile strength, sound velocity, and density, among others.

THEORY

The principal theories which have been constructed to explain the basic experimental properties of superconductors are discussed in this section. Those which preceded the BCS microscopic theory are no less useful for being phenomenological and incomplete. The original BCS theory was based on an idealized model, but nevertheless has been broadly successful in explaining the properties of real superconductors. It has been extended and elaborated to cover ever more complex and realistic situations.

Two-fluid model. C. J. Gorter and H. B. G. Casimir introduced in 1934 a phenomenological theory of superconductivity based on the assumption that in the superconducting state there are two components of the conduction electron "fluid" (hence the name given this theory, the "two-fluid model"). One, called the superfluid component, is an ordered condensed state with zero entropy, hence is incapable of transporting heat. It does not interact with the background crystal lattice, its imperfections, or the other conduction electron component and exhibits no resistance to flow. The other component, the normal component, is composed of electrons which behave exactly as they do in the normal state. It is further assumed that the superconducting transition is a reversible thermodynamic phase transition between two thermodynamically stable phases, the normal state and the superconducting state, similar to the transition between the liquid and vapor phases of any substance. The validity of this assumption is strongly supported by the existence of the Meissner-Ochsenfeld effect and by other experimental evidence. This assumption permits the application of all the powerful and general machinery of the theory of equilibrium thermodynamics. The results tie together the observed thermodynamic properties of superconductors in a very satisfying way.

Thermodynamic relations. From a thermodynamic point of view, the superconducting phase appears below the transition temperature because the free energy of the superconducting phase becomes less than the free energy of the normal phase for all temperatures below T_c. The exclusion of magnetic flux by the superconductor in an applied field H increases the free energy per unit volume of superconductor by $\mu_0 H^2/2$ (in SI units); it costs energy to push the flux lines out of the superconducting region. When this increase in free energy becomes equal to the decrease in free energy associated with the normal-to-superconducting transition, it no longer pays the superconductor to remain superconducting, and the superconductor goes normal. Hence, the superconducting condensation energy must equal $\mu_0 H_c^2/2$. This can be verified experimentally by comparing the results of critical field measurements with direct measurements of the heat capacity in the normal and superconducting states.

The zero-field heat capacity in Fig. 4 shows a finite discontinuity at the transition temperature, not an infinite singularity. This means that there is no latent heat at the transition, that is, the transition is "second-order," and therefore the entropy is the same in the superconducting and normal states at the transition temperature. At all but the lowest temperatures, the electronic heat capacity in the superconducting state is nearly proportional to T^3. In the normal state, it is proportional to T, $C_n = \gamma T$, where γ is a constant. All of these facts may be combined to yield the results that the heat capacity in the superconducting state $C_s = 3\gamma T^3/T_c^2$ and that the entropy in the superconducting state is less than that in the normal state for all temperatures below the transition temperature, that is, the superconducting phase is more ordered than the normal state. Further, consideration of the free energy yields the results $\gamma T_c^2 = 2\mu_0 H_0^2$, where H_0 is the critical field at $T = 0$ K (in SI units), and $H_c = (\gamma/2\mu_0)^{\frac{1}{2}} T_c [1 - (T/T_c)^2]$. There is thus a direct thermodynamic connection between the T^3 heat capacity and the parabolic critical field curve; one implies the other.

More general thermodynamic relations exist between the superconducting parameters H_c and T_c and the normal state heat capacity constant γ, which do not depend on any assumption about the temperature dependence of the heat capacity in the superconducting state or the form of the critical field curve. The good agreement of these relations with experimental data is the basis of the conviction that the superconducting state is indeed a thermodynamic equilibrium phase.

Another result of the two-fluid model is that the normal fluid forms a fraction $x = (T/T_c)^4$ of the total electron fluid. The superfluid fraction $1 - x = 1 - (T/T_c)^4$ therefore rises rapidly from 0 to 1 as T falls below T_c.

London equation. In order to understand the electromagnetic behavior of superconductors within the two-fluid model, it is necessary to postulate something about how the superfluid responds to electric and magnetic fields. This was done by the brothers F. London and H. London in 1935. It is natural to suppose that any electric field in a superconductor will give rise to a force on the superfluid electrons and hence will accelerate the superfluid. Since the supercurrent J_s is proportional to the superfluid velocity, this implies a rate of change with time of J_s which is proportional to the electric field. This is simply equivalent to an assumption that the superfluid has infinite conductivity. Straightforward combination of this assumption with the classical Maxwell equations relating the electric and magnetic fields yields a relation between the supercurrent and the magnetic field which does not lead to the Meissner-Ochsenfeld effect. (This indicates that the Meissner effect is not simply a consequence of infinite conductivity.) The Londons therefore postulated a modification of this relation, now called the London equation, which in one simple form is $J_s = -K\mathbf{A}$. Here $\mathbf{A}$ is the so-called vector potential from which the magnetic field can be derived, and K is a constant which contains the density of superfluid electrons. This equation does lead to the Meissner effect.

With the addition of the London equation to describe the electrodynamics, the two-fluid model predicts magnetic-field penetration depths and finite conductivity at nonzero frequencies in reasonable accord with experiment. The source of the latter is qualitatively apparent: At zero frequency all of the current will be carried by the superfluid, which will "short out" the normal fluid. At nonzero frequencies, however, the electromagnetic fields induced in the surface of the superconductor will tend to drive the normal fluid as well as the superfluid, and this will cause some dissipation.

Nonlocal theory. In the course of experiments on the electromagnetic response of superconductors at very high frequencies, A. B. Pippard in 1952 observed significant discrepancies between his results and the predictions of London electrodynamics. He traced these to a failure of the London equation $J_s = -K\mathbf{A}$. This equation is a "local" one, relating as it does the supercurrent at a given point in the superconductor to the vector potential at that point. Pippard found that the correct relation is nonlocal; the supercurrent at a point actually depends on the vector potential throughout a region around the point of size ξ. The parameter ξ is called the coherence length and may be as large as 10^{-6} m. Nonlocality turns out to be a very fundamental feature of the superconducting state, and ξ is a fundamental length of great importance in a variety of superconducting phenomena. For example, the distinction between type I and type II superconductors depends entirely on the relative sizes of the penetration depth and the coherence length: If λ is greater than ξ, the surface energy is negative and the superconductor is type II. The coherence length can be decreased from its value in a pure superconductor by adding impurities which scatter electrons, and this can change a superconductor from type I to type II.

Ginzburg-Landau theory. V. L. Ginzburg and L. D. Landau proposed in 1950 a highly innovative phenomenological theory of superconductivity which now bears their names. Their objective was to understand the situation at supernormal interfaces like those which occur in the intermediate state of type I superconductors. Since the superconducting state varies from something to nothing at such an interface, they needed a theory which describes a spatial variation of the superconducting state. They began by supposing that the "strength" of the superconducting state can be described by an "order parameter" ψ which may be spatially varying. In the normal state, ψ is zero. Then they assumed that sufficiently near T_c, where superconductivity is weak and ψ is small, the free energy of a superconductor can be expressed as a sum of a series of terms in increasing powers of $|\psi|^2$; $|\psi|^2$ is interpreted as proportional to the superfluid density. The absolute value is used here because ψ is allowed to be a complex number with an amplitude ψ_0 and a phase $\varphi, \psi = \psi_0 e^{i\varphi}$. This is a crucial if initially inexplicable feature of the order parameter. Thus ψ has something of the character of a quantum-mechanical wave function. This similarity is reinforced by the addition to the free-energy sum of a term involving the spatial gradient of ψ and the magnetic vector potential. This term has the same form as the standard representation of the kinetic energy in the Schrödinger wave equation of quantum mechanics, and here represents the kinetic energy of the superfluid electrons. The equilibrium form of ψ is then assumed to be that which minimizes the free energy. This leads to a rather complicated nonlinear differential equation for ψ. Together with an electrodynamic equation which relates the supercurrent $\mathbf{J}_s$ to ψ, this equation forms a pair of simultaneous differential equations called the Ginzburg-Landau equations.

Obtaining solutions of the Ginzburg-Landau equations is in general rather difficult. Nevertheless, they have been a remarkably powerful tool in superconductivity. They are intrinsically nonlocal because of the presence of the spatial gradient of ψ in the original free-energy equation. This has the consequence that a local disturbance (such as a sharp boundary between a normal and a superconducting metal) causes ψ to vary on the scale of a characteristic length which can be identified with the coherence length ξ. The other characteristic length in superconductivity, the penetration depth λ, also appears in a natural way. In 1957 A. A. Abrikosov reported a solution of the Ginzburg-Landau equations for the case $\lambda > \xi$ which constituted the first theoretical explanation of the mixed state in type II superconductors. Although the Ginzburg-Landau equations have many limitations, they continue to provide the basis for most understanding of the spatially varying superconducting state.

Microscopic (BCS) theory. The key to the basic interaction between electrons which gives rise to superconductivity was provided by the isotope effect. It is an interaction mediated by the back-ground crystal lattice and can crudely be pictured as follows: An electron tends to create a slight distortion of the elastic lattice as it moves, because of the Coulomb attraction between the negatively charged electron and the positively charged lattice. If the distortion persists for a brief time (the lattice may ring like a struck bell), a second passing electron will see the distortion and be affected by it. Under certain circumstances, this can give rise to a weak indirect attractive interaction between the two electrons which may more than compensate their Coulomb repulsion.

H. Frölich recognized in 1950 that such an interaction might be responsible for superconductivity. However, all initial attempts to develop a theory based on the interaction failed. The average energy per electron associated with the superconducting condensation is tiny compared with typical electron kinetic or Coulomb interaction energies. It was natural to try to exploit this by trying to arrive at the quantum-mechanical description of electrons in the superconducting state by small pertubations of the description in the normal state. In retrospect, it is clear why this approach failed. The superconducting quantum wave function is qualitatively different from any normal state wave function.

The first forward step was taken by Cooper in 1956, when he showed that two electrons with an attractive interaction can bind together to form a "bound pair" (often called a Cooper pair) if they are in the presence of a high-density fluid of other electrons, no matter how weak the interaction is. The two partners of a Cooper pair have opposite momenta and spin angular momenta. Then, in 1957, Bardeen, Cooper, and Schrieffer showed how to construct a wave function in which all of the electrons (at least, all of the important ones) are paired. Once this wave function is adjusted to minimize the free energy, it can be used as the basis for a complete microscopic theory of superconductivity.

With the discovery of the BCS superconducting-ground-state wave function, the fundamental reason for the remarkable properties of the superconducting state became clear. An analogy is useful in understanding it. Consider an enormous ballroom packed with dancers, shoulder to shoulder. Suppose each dancer is vigorously doing his or her own individual dance. The dancers will collide with each other and with any other objects which may be scattered about the dance floor. If there is some pressure on the whole group to move toward one side of the ballroom, dancing all the while, the collective motion will be random and chaotic, and a lot of energy will be lost in collisions. This represents the electrons in a normal metal, colliding with each other and with irregularities or impurities in the crystal lattice. If an electric field is applied and a current induced, collisions will dissipate energy and cause a finite conductivity.

Now suppose the dancers are paired in couples, each pair dancing together. This represents Cooper pairs, but now an important feature of the pairs enters. The pairing interaction is so weak that the two members of a pair are separated on the average by a distance which turns out to be just the coherence length. The average distance between two paired electrons not involved in the same pair, however, is about a hundred times smaller. The partners comprising each couple are not dancing cheek to cheek, but are separated by a hundred other dancers. Consequently, if every couple is going to dance together, it is clear that everybody must dance together. The result is a single coherent motion, with order extending all the way across the ballroom. The superconducting state is something like that. A localized pertubation which might deflect a single electron in the normal state, and thus give rise to some resistance, cannot do so in the superconducting state without affecting all the electrons participating in the superconducting ground state at once. That is not impossible, but extremely unlikely, so that a collective drift of the coherent superconducting electrons, corresponding to a current, will be dissipationless.

The successes of the BCS theory and its subsequent elaborations are manifold. One of its key features is the prediction of an energy gap. Excitations called quasiparticles (which are something like normal electrons) can be created out of the superconducting ground state by breaking up pairs, but only at the expense of a minimum energy of Δ per excitation; Δ is called the gap parameter. The original BCS theory predicted that Δ is related to T_c by $\Delta = 1.76\, kT_c$ at $T = 0$ for all superconductors. This turns out to be nearly true, and where deviations occur they are understood in terms of modifications of the BCS theory. The manifestations of the energy gap in the low-temperature heat capacity and in electromagnetic absorption provide strong confirmation of the theory. The theory accounts for all the thermodynamic properties of superconductors, including such details as deviations from parabolicity of the critical field curve. The theory is intrinsically nonlocal. The Ginzburg-Landau theory and the Pippard nonlocal electrodynamics can be derived from it. The Ginzburg-Landau order parameter ψ can be associated with the BCS ground-state wave function, and the coherence length with the size or range of a Cooper pair.

FURTHER EXPERIMENTAL PROPERTIES

Some experiments which illustrate some special features of the superconducting wave function are developed in the modern theory of superconductivity discussed in this section.

Flux quantization. Consider the complex order parameter or wave function ψ for a superconductor in the form of a hollow cylinder. If ψ is to be a well-defined object, its phase φ must change by an integral multiple of 2π along any closed path which lies entirely within the superconductor, including any path which surrounds the hole in the cylinder. There is a fundamental relation between the spatial gradient of ψ, hence of its phase, and the magnetic vector potential $\mathbf{A}$. A consequence of this relation and the constraint on the phase change around a closed path is that the hole in the cylinder cannot contain an arbitrary amount of magnetic flux, but only integral multiples of a fundamental quantum of flux, $\Phi_0 = h/2e = 2.07 \times 10^{-15}$ Wb in SI units, or $\Phi_0 = hc/2e = 2.07 \times 10^{-7}$ gauss cm^2 in cgs units; h is Planck's constant, c is the velocity of light, and e is the electron charge. This is the reason for the quantization of flux in fluxoids in type II superconductors. There the hole in the cylinder is

the normal core of the fluxoid. Flux quantization has been observed in macroscopic hollow cylinders. The factor of 2 in the flux quantum is related to the two electrons of a Cooper pair, so that observation of the expected magnitude of the flux quantum may be interpreted as experimental evidence for electron pairing in the superconducting state.

Quasiparticle tunneling. The phenomenon of tunneling is a direct consequence of the wave nature of material particles and was recognized very early in the development of quantum or wave mechanics. A particle, such as an electron, can pass into a region which classically would be forbidden to it and, if the region is not infinitely thick, can pass (tunnel) through it. Consider, for example, two metals separated by an insulating barrier which prevents electrons from passing between the metals. The resulting structure, called a junction, will be nonconducting. However, if the barrier is thin enough (10^{-9} to 10^{-8} m), electrons can tunnel from one metal to the other and this "tunnel junction" will become conducting. In 1960 I. Giaever discovered that if one or both of the metals in such a junction are superconducting, the dependence of the current on the voltage across the junction becomes highly nonlinear. The lower curve in Fig. 5 shows an example for a junction composed of two thin films of tin separated by a tin oxide barrier. The junction passes very little current until the voltage reaches a value of about 1.2 mV, where it rises sharply. This behavior is a direct consequence of the existence of the superconducting energy gap. The voltage V_g at which the current rises is related to the gap parameter by the simple relation $eV_g = 2\Delta$. Giaever's discovery made it possible to measure the gap parameter with nothing much more than an ammeter and a voltmeter. Previous determinations of the gap had come from much more difficult very-low-temperature heat capacity experiments or far-infrared

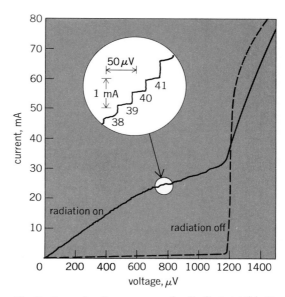

Fig. 5. Current-voltage curves of a Sn-Sn tunnel junction at 1.2 K displaying radiation-induced current steps. (*From W. H. Parker et al., Determination of e/h, Using Macroscopic Quantum Phase Coherence in Superconductors, I, Phys. Rev., 177, 639–664, 1969*)

absorption experiments. The tunnel junction has become the single most valuable tool for the study of the superconducting state. For example, the current-voltage characteristic at voltages above the gap voltage contains small structures which can be unraveled to yield the complete energy dependence of the electron-lattice interaction function responsible for the superconductivity.

Pair tunneling. In 1962 B. D. Josephson made a remarkable theoretical discovery: Not only can quasiparticles ("normal electrons") tunnel through an insulating barrier between two superconductors, but so can Cooper pairs. This implies a coupling between the superconducting wave functions of the two superconductors which leads to the existence of a tunnel supercurrent which depends on the difference between the phases of the two wave functions. Josephson's predictions were soon verified experimentally. They include the following: It is possible for a lossless supercurrent to pass through the insulating barrier; this is called the dc Josephson effect. If a voltage V is applied between the two superconductors, an oscillating supercurrent with frequency $2\,eV/h$ exists in the junction; this is the ac Josephson effect. The factor of 2 again comes from the pairing. The upper curve in Fig. 5 shows one way to detect the ac supercurrent. A microwave field of frequency ν is applied to the junction. This field frequency modulates the ac supercurrent, producing zero-frequency increments in the supercurrent when the frequency of the ac supercurrent is equal to an integer multiple of the microwave frequency. These appear as steps in the current-voltage characteristic of the junction at voltage intervals of $h\nu/2e$. This effect provides by far the most accurate way to measure the fundamental physical constant h/e, as well as the basis for the United States national standard of voltage.

A Josephson tunnel junction can be thought of as a quantum phase meter. The Josephson effects have been used to confirm experimentally that the superconducting order parameter has all the quantum-phase coherence characteristics of an atomic wave function, but on a scale 10^{10} times larger. The fundamental connection between the quantum phase and the magnetic vector potential makes the Josephson effects incredibly sensitive to magnetic fields. Josephson junctions have been used to detect fields 10^{10} times smaller than the Earth's magnetic field and the tiny magnetic fields produced by neural currents in the human brain. *See* JOSEPHSON EFFECT; TUNNELING IN SOLIDS.

Research on superconductivity has continued actively on a variety of fronts. New superconducting materials have been sought, with higher transition temperatures and more favorable properties for practical applications. There have been theoretical conjectures that much higher transition temperatures, perhaps higher than room temperature, might be achieved with exotic materials in which conduction occurs in one- or two-dimensional molecular structures, or through the exploitation of some electron-electron attractive interaction mechanism which is stronger than the electron-lattice interaction which acts in all known superconductors. These possibilities have been pursued. Superconductivity may have a cosmic role, for the interiors of neutron stars and of the planet Jupiter may be superconducting. There has been much interest in fluctuations in the properties of superconductors near the superconducting phase transition, and in the behavior of superconductors forced into states far from thermal equilibrium. An enormous range of practical applications have been investigated, from superconducting electric power transmission lines to superconducting computers. *See* SUPERCONDUCTING COMPUTERS.

[D. N. LANGENBERG]

Bibliography: A. C. Rose-Innes and E. H. Roderick, *Introduction to Superconductivity*, 2d ed., 1977; D. R. Tilley and J. Tilley, *Superfluidity and Superconductivity*, 1975; M. Tinkham, *Introduction to Superconductivity*, 1975, reprint 1980.

Superheterodyne receiver

A receiver that uses the heterodyne principle to convert the incoming modulated radio-frequency signal to a predetermined lower carrier frequency, the intermediate-frequency (i-f) value. This is done by using a local oscillator tuned simultaneously with the input stage of the receiver, so that the oscillator frequency always differs from that of the incoming carrier by the i-f value.

With a fixed and favorably chosen i-f value, the i-f amplifier can efficiently provide the major portion of the amplification and selectivity required by the receiver. After amplification, the i-f signal is demodulated by the second detector to obtain the desired audio output signal. A similar circuit arrangement is used in television and radar receivers to obtain the desired video output signal. For amplitude-modulated (AM) radio broadcast receivers the i-f value is usually 455 kHz. For frequency-modulated (FM) radio receivers the standard i-f value is 10.7 MHz. For modern television receivers the usual i-f values are about 45 MHz for video and 4.5 MHz for sound.

In a double superheterodyne receiver, the incoming carrier signal is changed to one i-f value for preliminary amplification and filtering, then changed to the final i-f value by beating with another oscillator frequency in a second mixer. This gives higher gain without instability, improves adjacent-channel selectivity, and provides greater suppression of undesired signal frequencies. *See* HETERODYNE PRINCIPLE; RADIO RECEIVER.

[JOHN MARKUS]

Superposition theorem (electric networks)

Essentially, that it is permissible, if there are two or more sources of electromotive force in a linear electrical network, to compute at any element of the network the response of voltage or of current that results from one source alone, and then the response resulting from another source alone, and so on for all sources, and finally to compute the total response to all sources acting together by adding these individual responses.

Thus, if a load of constant resistance is supplied with electrical energy from a linear network containing two batteries, two generators, or one battery and one generator, it would be correct to find the current that would be supplied to the load by one source (the other being reduced to zero), then to find the current that would be supplied to the

being reduced to zero), and finally to add the two currents so computed to find the total current that would be produced in the load by the two sources acting simultaneously. *See* NETWORK THEORY.

Note that while any one source is being considered all other independent sources are reduced to zero. An independent-voltage source is reduced to zero by making the voltage between its terminals equal zero (short circuit), or an independent-current source is reduced to zero by making current through it equal zero (open circuit).

By means of the principle of superposition, effects are added instead of causes. This principle seems so intuitively valid that there is far greater danger of applying superposition where it is incorrect than of failing to apply it where it is correct. It must be recognized that for superposition to be correct the relation between cause and effect must be linear.

In an electrical network, linearity implies that the parameters of resistance, inductance, and capacitance must be constant (with respect to current or voltage) and that the voltage or current of sources must be independent of, or directly proportional to, other voltages and currents (or their derivatives) in the network. Sources or parameters may vary as functions of some other independent variable such as time, however; an example of this is a source of sinusoidal or other time-varying voltage.

The constant resistance of a metal wire is an example of a linear relation. There is no linear relation between voltage and current in many crystalline materials, however, including those classed as diodes or transistors. In such a nonlinear conductor, the resistance R is not independent of the current i. The inductance of a coil is not independent of current if the coil has an iron core; this phenomenon of nonlinearity is termed saturation. The capacitance of certain capacitors is not independent of applied voltage, making the current-voltage relation nonlinear. *See* SATURATION.

Linearity requires a proportionality of cause and effect. Thus the relation of current to voltage may be linear, but the relation of current to power ($p = i^2R$) is nonlinear. If a load of constant resistance (as considered above) is supplied from two sources, it would be wrong to attempt to find the total power to the load by adding the two values of power supplied by each of the two sources acting individually.

Mathematical definition. Linearity is defined mathematically as follows: If a disturbance $x_1(t)$ gives a response $y_1(t)$ and a disturbance $x_2(t)$ gives a response $y_2(t)$, then if a disturbance $Ax_1(t) + Bx_2(t)$, in which A and B are constants, gives a response $Ay_1(t) + By_2(t)$, the disturbed system is linear.

Linearity implies homogeneity and additivity. If $x_1(t)$ gives $y_1(t)$ and $x_2(t)$ gives $y_2(t)$, then if $Ax_1(t)$ gives $Ay_1(t)$, the relation is said to be homogeneous. If $x_1(t) + x_2(t)$ gives $y_1(t) + y_2(t)$, the relation is said to be additive.

For example, let current and voltage be related by Eq. (1).

$$v = Ri + \frac{d(Li)}{dt} \qquad (1)$$

With a disturbance i_1, the response is given by Eq. (2).

$$v_1 = R_1 i_1 + \frac{d(L_1 i_1)}{dt} \qquad (2)$$

With a different disturbance i_2, the response is given by Eq. (3).

$$v_2 = R_2 i_2 + \frac{d(L_2 i_2)}{dt} \qquad (3)$$

If the disturbance is $Ai_1 + Bi_2$, the response is given by Eq. (4).

$$v_3 = R_3(Ai_1 + Bi_2) + \frac{d[L_3(Ai_1 + Bi_2)]}{dt}$$

$$= A\left[R_3 i_1 + \frac{d(L_3 i_1)}{dt}\right] + B\left[R_3 i_2 + \frac{d(L_3 i_2)}{dt}\right] \qquad (4)$$

If R and L are constant, so that $R_1 = R_2 = R_3$ and $L_1 = L_2 = L_3$, then $v_3 = Av_1 + Bv_2$, and hence Eq. (1) is linear. But if R or L changes with the value of current carried, Eq. (1) is not linear.

It is hardly necessary to show $p = i^2R$ is not a linear relation between current and power. Current i_1 gives $p_1 = i_1^2 R$ and i_2 gives $p_2 = i_2^2 R$, but a current of $i_1 + i_2$ gives $p_3 = (i_1 + i_2)^2 R$, which is not $(i_1^2 + i_2^2)R$.

Uses. The principle of superposition is possibly the most useful relation in physical theory. In electrical networks it is used in deriving both loop and node equations from branch equations and in many network theorems.

In numerical solutions, superposition provides a means of finding currents and voltages resulting from multiple sources in a network. This is helpful if the method of solution used does not take into account more than one source. Also, it is helpful if two sources do not add to give a convenient function, as, for example, in finding current produced by an alternating-current generator and a direct-current generator in the same network.

Perhaps most important, the concept of fundamental and harmonic frequencies each producing its own response is dependent on the principle of superposition. Thus the principle of superposition underlies network analysis by means of Fourier series, or by Fourier transformation or Laplace transformation. [HUGH H. SKILLING]

Bibliography: H. H. Skilling, *Electric Networks,* 1974.

Suppression

The elimination of some undesired component of a signal. In automobile radio installations, suppression techniques are essential to prevent ignition interference from reaching the radio circuits. Radio receivers themselves often contain special noise-suppression circuits and devices; this is particularly true for communication receivers that operate in crowded and noisy shortwave bands.

In many radar installations, suppression circuits and techniques are used to reduce clutter caused by the ground and by fixed objects close to the antenna. One radar suppression technique involves reducing the receiver gain suddenly after each high-power pulse is transmitted, then gradually and automatically restoring normal gain so that nearby echo signals are amplified much less than the desired distant echo signals. *See* ELECTRIC FILTER; ELECTRICAL NOISE; SUPPRESSOR.

[JOHN MARKUS]

Suppressor

A device used to reduce or eliminate noise or other signals that interfere with the operation of a communication system. The term may be applied to a noise filter in a radio receiver, but it is more frequently used to describe a device applied at the noise source, such as a resistor used in series with spark plugs of a gasoline engine, or a capacitor across the terminals of a commutator motor or other sparking device that acts as a noise generator. The term suppressor may also be applied to a filter used in power leads of an electronic device to eliminate unwanted signals of noise. *See* ELECTRIC FILTER; ELECTRICAL INTERFERENCE; ELECTRICAL NOISE; NOISE FILTER (RADIO); SUPPRESSION.

[WILBUR R. LE PAGE]

Surface acoustic-wave devices

Devices which employ surface acoustic waves (SAW) in the analog processing of electronic signals with frequencies in the range $10^7 - 10^9$ hertz (Hz).

Background. Surface acoustic waves which contain both compressional and shear components in phase quadrature, propagating nondispersively along and bound to solid surfaces, were discovered by Lord Rayleigh in the 1880s. As an example, earthquakes furnish sources for propagating these waves on the Earth's surface. It is of importance for electronic applications that if the solid is a piezoelectric material, the surface acoustic energy is complemented by a small amount of electric energy. This electric energy provides the physical mechanism for the coupling between conventional electromagnetic signals and propagating SAW. The coupling is attained by means of interdigital transducers (IDT). SAW devices have led to a versatile microminiature technology for analog signal processing in the frequency range $10^7 - 10^9$ Hz. Notable devices include bandpass filters, resonators, oscillators, pulse compression filters, and fast Fourier transform processors. Application areas include the color television consumer market, radar, sonar, communication systems, and nondestructive testing. *See* PIEZOELECTRIC CRYSTAL; PIEZOELECTRICITY.

Transduction. The basic arrangement is shown in Fig. 1. A piezoelectric substrate, often crystalline quartz, has a polished upper surface on which two transducers, denoted T, with terminals A and A', and B and B', are deposited. The left-hand input transducer is connected, via fine bonded leads,

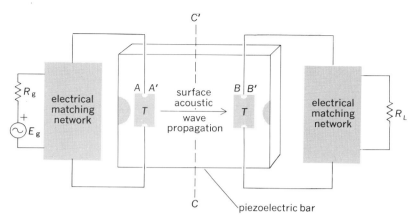

Fig. 1. Basic arrangement of surface acoustic-wave bandpass filter.

to the electric source through an electrical matching network. The right-hand output transducer drives the load R_L, usually 50 ohms, through another electrical matching network. R_g and E_g represent the resistance and voltage of the generator. Because these transducers are bidirectional, they lead to devices with at least 6 decibels (dB) loss even in the passband. The unwanted acoustic waves are absorbed by terminations at the ends of the piezoelectric substrate. At the symmetry plane C-C' a metal baffle serves to isolate electromagnetically the two transducers.

The transducers originally demonstrated by R. M. White and F. W. Voltmer in 1966 consist of a set of metal interdigital electrodes, each a few hundred nanometers thick, fed from two busbars (Fig. 2*a*). For this transducer arrangement the period p of the interdigital electrode structure is constant and equals one surface acoustic wavelength λ_0 at the center of frequency f_0 of the response. The width of the metal electrodes is typically $p/4$, being 100 micrometers (μm) at 10^7 Hz and 1 μm at 10^9 Hz. The electrode overlap distance w is also constant and defines the acoustic beam width, which is typically 40 wavelengths. *See* TRANSDUCER.

The 100-μm electrodes are readily fabricated by using techniques standard to the semiconductor integrated circuit industry of metallization: photoresist, masking, and chemical etching. The 1-μm electrodes require more sophisticated processing techniques. These include conformable optical

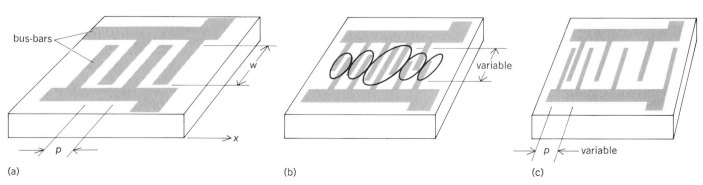

(a)　　　　　　　　　　(b)　　　　　　　　　　(c)

Fig. 2. Transducers. (*a*) Interdigital construction. (*b*) Apodization technique. (*c*) Graded periodicity technique.

masks (optical masks made on a flexible glass substrate so that they can be pulled down under vacuum into intimate contact with the piezoelectric substrate for accurate IDT fabrication) and x-ray lithography, both developed by H. I. Smith, coupled with sputter etching by radio-frequency and ion-beam methods for even finer resolutions. Important piezoelectric substrates are ST-cut, X-propagating quartz for temperature stability; Y-cut, Z-propagating lithium niobate, because of its high piezoelectric coupling; and bismuth silicon oxide, which has found application in color television filters. *See* INTEGRATED CIRCUITS.

The acoustic response at frequency f for this simplest SAW structure can be calculated approximately be regarding each N-period IDT as an end-fire array antenna. For the pair of IDTs the acoustic response is proportional to $(\sin x/x)^2$, where $x = N\pi(f-f_0)/f_0$, giving a bandpass filter characteristic. The electrical matching networks are normally arranged to minimize filter loss without overriding the acoustic response. The necessary network analysis was originally developed by W. R. Smith and coworkers. They show that the optimum number of periods, N, is inversely proportional to the piezoelectric coupling as is the filter bandpass width. *See* ANTENNA.

Delay T is an important parameter in SAW signal processing devices. Here, T is proportional to the center-to-center path separation of the IDT pair and is typically 3 microseconds per centimeter (μs/cm) of piezoelectric substrate length.

Bandpass filters. A disadvantage of constant-p, constant-w transducers is that the minimum out-of-band rejection is only 26 dB. However, the filter designer can control the characteristics of the filter by varying the overlap between adjacent IDT electrodes, a technique termed apodization (Fig. 2b), or by grading the periodicity (Fig. 2c). In contrast to conventional LC (inductive) filters, SAW filters are members of a class of nonminimum phase networks having the property that linear phase response can be achieved independently of the amplitude response. Further, SAW filters are transversal filters in which the signal is repetitively delayed and added to itself, as in antenna arrays and digital filters. This analogy has been identified by R. H. Tancrell. He has applied the optimization procedures of digital filters in the synthesis of SAW filters with equiripple bandpass response. Another basic design procedure, due to C. S. Hartmann, is to compute the inverse Fourier transform of the prescribed filter response, giving the impulse response, which is the spatial image desired of the electrodes in the IDT. Due to finite piezoelectric substrate lengths, the infinitely long time-duration impulse response is not realizable. This has necessitated the use of weighting functions to multiply and truncate the impulse response.

A vital factor in SAW filters is amplitude weighting, that is, alteration of the amplitude-frequency response, in order to improve the filter's selectivity. A popular method of achieving this, due to ease of design and relative insensitivity to fabrication errors, has been apodization of the input IDT. However, an apodized IDT generates a spatially nonuniform SAW. This necessitates an output IDT of constant electrode overlap and few electrodes to reproduce faithfully the apodized IDT response.

Undesirable consequences are high loss and poor selectivity for the filter.

The solution of Tancrell and H. Engan embodies a multistrip coupler (MSC) developed by F. G. Marshall and E. G. S. Paige. The MSC has the property of converting the spatially nonuniform SAW generated by one apodized IDT into a spatially uniform SAW received by an identical apodized IDT. The apodized transducers incorporate special electrodes to reduce SAW impedance mismatch between the unloaded and metallized substrate surface. An alternative approach, conceived by Hartmann, obtains amplitude weighting by selective withdrawal of electrodes from identical and constant-overlap IDTs arranged in nondispersive combination. In this arrangement the electrodes to the right of both IDTs are spaced by a large amount (that is, low frequencies are excited), whereas those to the left of both IDTs are spaced by a small amount (that is, high frequencies are excited; Fig. 2c). Thus, both low- and high-frequency components of the excitation travel exactly the same distance before being received, so that the nondispersive characteristics of the SAW are preserved.

Normal and projected performance of SAW bandpass filters are shown in the table. Sidelobe rejection specifies the magnitude of the loss of sidelobes, that is, unwanted but naturally occurring responses close to but outside the bandpass, with respect to magnitude of the loss at the center of the bandpass. Ultimate rejection specifies the greatest magnitude of the loss at frequencies far removed from the bandpass region with respect to that at the center of the bandpass. Linear phase deviation expresses the undesired deviation in the response of the filter from a linear phase change with frequency. The majority of the data in the table are not obtainable simultaneously.

The minimum loss of 6 dB assumes a three-IDT configuration in which the outer IDTs are connected electrically in parallel. This removes half the 6-dB bidirectionality loss. The remaining 3 dB is composed of resistive losses in the IDTs and matching networks; propagation, beam steering, and diffraction losses on the piezoelectric; and losses associated with apodization. The projected minimum insertion loss is dependent on the development of better unidirectional IDTs.

The minimum transition bandwidth data given in the table are an important indicator of the maximum selectivity inherent in SAW filters. The tran-

Surface acoustic-wave bandpass filter capabilities

Parameter	Current	Projected
Center frequency, f_0 (Hz)	$10^7 - 10^9$	$10^6 - 2 \times 10^9$
3-dB bandwidth (Hz)	$5 \times 10^4 - 0.4\,f_0$	$2 \times 10^4 - 0.8\,f_0$
Minimum loss (dB)	6	1.5
Minimum shape factor (ratio of bandpass width at 3 dB and 40 dB)	1.2	1.2
Minimum transition width from bandpass to bandstop (Hz)	5×10^4	2×10^4
Sidelobe rejection (dB)	55	65
Ultimate rejection (dB)	65	80
Amplitude of bandpass ripple (dB)	0.5	0.2
Linear phase deviation (degrees)	5	2

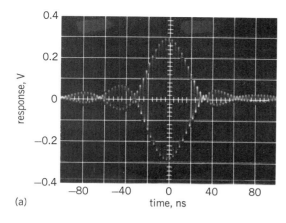

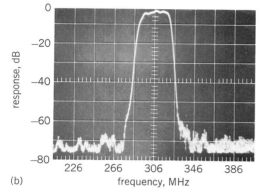

Fig. 3. Surface acoustic-wave bandpass filter performance, showing oscillograms of (a) impulse response and (b) bandpass response. (*Texas Instruments*)

sition bandwidth must exceed the inverse of the impulse response length of the IDT. Hence, a minimum transition bandwidth is equally applicable to narrow- and wide-band filters. This implies that low shape factors cannot be achieved for filters with extremely narrow bandpass widths, and this has led to the development of the SAW resonator. Figure 3 shows both the impulse response in the time domain and the bandpass response of an SAW filter made by Texas Instruments, having 3.06×10^8 Hz center frequency and a sidelobe rejection of 60 dB.

The smallest SAW filters are 0.02 in.^{3} (0.3 cm^3) in volume. Unfortunately, high performance as indicated in the table demands complex SAW structures, thus increasing both substrate size and cost. This dilemma presents a considerable challenge for applications in the consumer market. *See* ELECTRIC FILTER.

Resonators. The SAW resonator, due independently to E. J. Staples and K. M. Lakin, comprises two thin metallic strip arrays, similar to those of the MSC, each normal to the SAW beams with an IDT interposed symmetrically for coupling the electromagnetic energy in and out. The metallic strips and the gaps between are arranged to be each $\lambda_0/4$ wide. Cumulative reflection is obtained from each metallic strip array due to the slight SAW impedance discontinuity between each metallic strip and each gap. For Y-cut, Z-propagating lithium niobate, which has an impedance discontinuity of 1.2%, 200 metallic strips are required

in the array for a reflection coefficient of 98%. Both resonant and antiresonant behavior can be obtained, depending on whether the spacing between the arrays is an integral number of SAW half-wavelengths or there is an additional $\lambda_0/4$. The maximum Q and operating frequency achieved with SAW resonators is 10^4 and 4×10^8 Hz, respectively, although a Q of 10^5 seems realizable. Features of the SAW resonator not shared by its bulk-wave counterpart are the ability to control the resonator coupling precisely by the IDT, and a rugged structure. Applications are envisaged as multipole narrow-band filters and in single superheterodyne radio-frequency receivers.

Oscillators. Two classes of oscillators are in common use, the quartz crystal oscillator and the LC oscillator, including the resonant cavity type. The quartz crystal oscillator is highly stable ($Q > 10^4$) but suffers from a number of disadvantages, including mechanical fragility, low fundamental frequency operation ($<3 \times 10^7$ Hz), and limited frequency modulation (FM) capability (500 parts per million, or ppm). In contrast, the LC oscillator is much less stable ($Q \ll 10^4$) but has superior FM performance. The SAW oscillator, developed by M. F. Lewis, has demonstrated an intermediate stability and modulation capability, with practical advantages over both. The basic SAW oscillator comprises a SAW quartz delay line with a transistor amplifier as the feedback element arranged to give unity gain with an integral number of 2π phase shifts around the loop.

The frequency of operation is determined by the IDT pattern and not by a dimension of the quartz crystal. Hence, the crystal is rugged and can be firmly bonded to the encapsulating package. This good thermal contact allows output powers up to 1 watt. Performance of the SAW oscillator is strongly determined by the interaction between the amplifier characteristics and the delay line Q. Typical figures are: frequency deviation up to 2%; short-term stability of <1 in 10^9 for 1 sec; medium-term stability of 100 ppm for a temperature excursion of $\pm40°C$; and single-sideband FM noise of -138 dB/Hz, 10 kHz away from carrier. Limits to the long-term stability are not resolved. A level of ±1 ppm per month is achievable following a burn-in period of a month. *See* OSCILLATOR.

Pulse compression filters. The goal in pulse compression filters is to output an energy pulse of narrow time duration, with minimal spurious response, when the input of the filter is fed with a coded wave train having a substantially longer time duration. The ratio of the time duration of the input wave train to that of the output pulse is defined as the processing gain, which is equivalent to the improvement in signal-to-noise ratio. Pulse compression filters find application in high-resolution radar systems where the transmitter radiates energy pulses of duration ΔT which are linearly frequency-modulated over Δf. The pulse compression filter is employed in the receiver and has a processing gain of $\Delta T \times \Delta f$. SAW versions of pulse compression filters have found significant engineering application in upgrading the resolution capabilities of airborne radars. Processing gains up to 5000 and spurious responses below 40 dB have been achieved.

Two distinct techniques have been developed

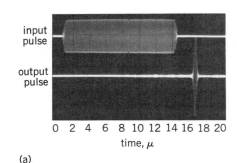

(a)

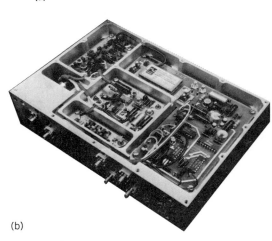

(b)

Fig. 4. SAW pulse compression filter for high-resolution radar. (*a*) Oscillograms showing input rectangular pulse centered at 75 MHz and the output compressed pulse, whose sidelobes are below 35 dB. (*b*) Module deploying filter complete with all peripheral electronics. (*Results courtesy of MESL, Scotland*)

for SAW pulse compression filters. The first uses the IDT of Fig. 2c on quartz to yield a quadratic phase-frequency response, or equivalently, a time delay which is a linear function of frequency. Apodization is used to ensure low spurious responses. The acoustic energy is received on the IDT of Fig. 2a arranged to have N small so that the waveform is undistorted. The second technique, developed by R. C. Williamson, involves passing the SAW under an array of suitably angled reflective slots. The spacing of these slots determines the frequency selectivity, and the depth of these slots determines the amplitude weighting. The first array of reflective slots are matched by a second group to reform the SAW beam, parallel to but displaced from the incident SAW beam. Thereby, the required pulse compression characteristics are obtained. Dimensionally the slots must have a depth in the region of one-thousandth of an acoustic wavelength, be approximately one-half wavelength wide, and be spaced by one wavelength between centers. Slot depths are typically 3 nanometers at 10^9 Hz, necessitating controlled ion-beam etching of the lithium niobate substrate. Results for a high-resolution radar module deploying the first technique for realizing an amplitude-weighted SAW pulse compression filter with a processing gain of 80 are shown in Fig. 4.

[J. H. COLLINS]

Bibliography: *IEEE Trans. Microwave Theory Tech.*, Special Issue on Microwave Acoustics, MTT-21(4), April 1973; *1974 Ultrasonics Symposium Proceedings*, IEEE Cat. no. 74; W. R. Smith et al., Analysis of interdigital surface wave transducers by use of an equivalent circuit model, *IEEE Trans. Microwave Theory Tech.*, MTT-17(11):856–864, November 1969; R. H. Tancrell, Analytic design of surface wave band pass filters, *IEEE Trans. Sonics Ultrasonics*, SU-21(1):12–22, January 1974.

Surface physics

The study of the structure and dynamics of atoms and their associated electron clouds in the vicinity of a surface, usually at the boundary between a solid and a low-density gas. Thus, surface physics may be regarded as a branch of solid-state physics which deals with those regions of large and rapid variations of atomic and electron density that occur in the vicinity of an interface between the two "bulk" components of a two-phase system. In conventional usage, surface physics is distinguished from interface physics by the restriction of the scope of the former to interfaces between a solid (or liquid) and a low-density gas, often at ultra-high-vacuum pressures of $p = 10^{-10}$ torr (1.33×10^{-8} N/m² or 10^{-13} atm).

More specifically, surface physics is concerned with two separate but complementary areas of investigation into the properties of such solid-"vacuum" interfaces. Ultimately, interest centers on the specification and theoretical prediction of surface composition and structure (that is, the masses, charges, and positions of surface species), of the dynamics of surface atoms (such as surface diffusion and vibrational motion), and of the energetics and dynamics of electrons in the vicinity of a surface (such as electron density profiles and localized electronic surface states). As a practical matter, however, the nature and dynamics of surface species must be determined experimentally by scattering and emission measurements involving particles or electromagentic fields (or both) external to the surface itself. Thus, a second major interest in surface physics is the study of the interaction of external entities (that is, atoms, ions, electrons, and electromagnetic fields) with solids at their vacuum interfaces. It is this aspect of surface physics which most clearly distinguishes it from conventional solid-state physics, because quite different scattering and emission experiments are utilized to examine surface as opposed to bulk properties of a given sample.

Physical principles of measurements. Since the mid-1960s, surface physics has enjoyed a renaissance by virtue of the development of a host of techniques for characterizing the solid-vacuum interface. While they might appear complicated in detail, all of these techniques are based on one of two simple physical mechanisms for achieving surface sensitivity. The first, which is the basis for field-emission and field-ionization microscopy, is the achievement of surface sensitivity by utilizing electron tunneling through the potential-energy barrier at a surface. This concept reached its apex of application in direct determinations of the energies of individual electronic orbitals of adsorbed complexes via the measurement of the energy distributions either of emitted electrons or of Auger electrons emitted in the process of neutralizing a slow (energy $E \sim 10$ eV) external ion.

The second mechanism for achieving surface sensitivity is the examination of the elastic scattering or emission of particles which interact strongly with the constituents of matter, for example, "low-energy" ($E \leq 10^3$ eV) electrons, thermal atoms and molecules, or "slow" (300 eV $\leq E \leq 10^3$ eV) ions. Since such entities lose appreciable ($\Delta E \sim 10$ eV) energy in distances of the order of tenths of nanometers (nm), typical electron analyzers with resolutions of tenths of an electron volt are readily capable of identifying scattering and emission processes which occur in the upper few atomic layers of a solid. This second mechanism is responsible for the surface sensitivity of photoemission, Auger-electron, electron-characteristic-loss, low-energy-electron-diffraction (LEED), and ion-scattering spectroscopy techniques. The strong particle-solid interaction criterion which renders these measurements surface-sensitive is precisely the opposite of that used in selecting bulk solid-state spectroscopies. In this case, weak particle-solid interactions (that is, penetrating radiations) are desired in order to sample the bulk of the speciman via, for example, x-rays, thermal neutrons, or fast ($E \gtrsim 10^4$ eV) electrons. *See* PHOTOEMISSION.

Surface preparation. An atomically flat surface, labeled by $M(hkl)$, may be visualized as being obtained by cutting an otherwise ideal, single-crystal solid M along a lattice plane specified by the Miller indices (hkl), and removing all atoms whose centers lie on one side of this plane. On such a surface the formation of a "selvedge" layer can also be envisaged. Such a layer might be created, for example, by the adsorption of atoms from a contiguous gas phase. It is characterized by the fact that its atomic geometry differs from that of the periodic bulk "substrate." From the perspective of atomic structure, this selvedge layer constitutes the "surface" of a solid. In principle, its thickness is a thermodynamic variable determined from the equations of state of the solid and of the contiguous gas phase. In practice, almost all solid surfaces are far from equilibrium, containing extensive regions (micrometers thick) of surface material damaged by sample processing and handling.

Another reason for the renaissance in surface physics is the capacity to generate in a vacuum chamber special surfaces which approximate the ideal of being atomically flat. These surfaces are prepared by cycles of fast-ion bombardment, thermal outgassing, and thermal annealing for bulk samples (for example, platelets with sizes of the order of 1 cm $\times$ 1 cm $\times$ 1 mm) or field evaporation of etched tips for field-ion microscopes. In this fashion, reasonable facsimiles of uncontaminated, atomically flat solid-vacuum interfaces of many simple metals with fcc (face-centered cubic) or bcc (body-centered cubic) structure as well as zincblende and wurtzite semiconductors have been prepared and subsequently characterized by various spectroscopic techniques. Such characterizations must be carried out in an ultra-high vacuum ($p \leq 10^{-8}$ N/m²), however, so that the surface composition and structure are not altered by gas adsorption during the course of the measurements.

Experimental apparatus. Presuming that atomically flat surfaces can be prepared, the bulk of modern experimental surface physics is devoted to the determination of the chemical composition, atomic geometry, and electronic structure of such surfaces. Since different measurements are required to assess each of these three aspects of a surface, the typical surface-characterization instrument consists of equipment for carrying out a combination of several measurements in a single ultra-high-vacuum chamber. Two types of sample geometry are common. Platelet samples are studied using scattering and emission experiments. A typical modern apparatus, such as that shown in Fig. 1, contains an electron gun, an ion gun, an electron energy analyzer, a source of ultraviolet or x-ray electromagnetic radiation, and a sample holder permitting precise control of both its orientation and temperature. Occasionally other features (such as a mass spectrometer) also are incorporated for special purposes. For specific applications in which less than a complete characterization of the surface is required, commercial instruments designed to embody only one or two measurements often are available. Such limited-capability instruments commonly are utilized to determine the chemical composition of surfaces by, for example, ion scattering, secondary ion mass spectrometry, x-ray photoemission, or Auger-electron emission. Obviously, the utility of such instruments is not limited to atomically flat or even crystalline surfaces, so that they find widespread applications in metallurgy and polymer science.

The second common sample geometry is an etched tip, about a hundred nanometers in radius. Such specimens are studied by field emission and ionization experiments, which provide a direct magnified image of the surface structure in contrast to the statistical description of platelet surfaces afforded by instruments like that shown in Fig. 1.

Data acquisition, analysis, and theory. Given the ability to perform surface-sensitive spectroscopic measurements, questions naturally arise concerning analysis of the raw spectra to extract parameters characterizing the structure of a given surface and the synthesis of such data to form a coherent picture of the behavior of electrons and atomic species at surfaces. Thus, surface physics may be divided into three types of activity: the acquisition of surface-sensitive spectroscopic data, the analysis of these data using physical models of the appropriate scattering or emission spectroscopy, and the construction of broad theoretical models of surface structure and properties to be tested via critical comparison of their predictions with the results of such data analyses.

Ground- and excited-state properties. Theoretical models have been proposed for the description of two distinct types of surface properties. The stability of surface structures is examined by calculations of ground-state properties, such as surface energies or effective potential-energy diagrams for adsorbed species. These quantities are difficult to measure experimentally, although they are the most direct manifestations of the intrinsic behavior of an undisturbed surface. The interactions of external projectiles or fields with a solid create excited states of the electrons or atoms within the solid. Consequently, the associated scattering and emission spectra indicate the nature and energies of these excited states (called excitations) rather than of the ground state. Two kinds of excitations occur. Electronic excitations

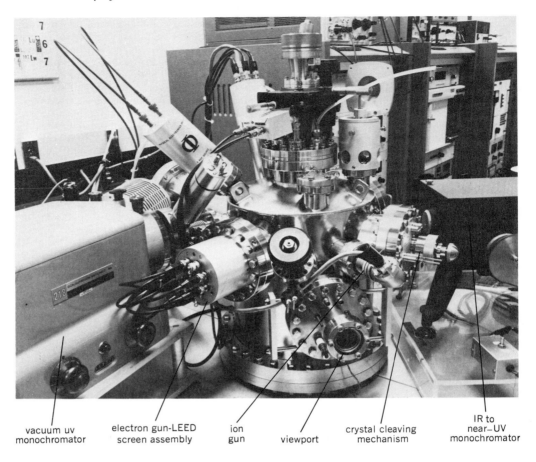

vacuum uv
monochromator

electron gun-LEED
screen assembly

ion
gun

viewport

crystal cleaving
mechanism

IR to
near—UV
monochromator

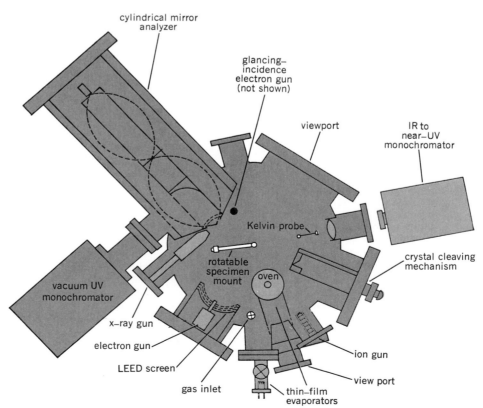

cylindrical mirror
analyzer

glancing—
incidence
electron gun
(not shown)

viewport

IR to
near—UV
monochromator

Kelvin probe

rotatable
specimen
mount

oven

crystal cleaving
mechanism

vacuum UV
monochromator

x—ray gun

electron gun

ion gun

LEED screen

view port

gas inlet

thin—film
evaporators

Fig. 1. Photograph and schematic diagram of modern
multiple-technique ultra-high-vacuum surface characterization instrument for study of insulator surfaces.
(L. J. Brillson, Xerox Corporation)

are generated when a disturbing force causes the electrons in the solid to alter their quantum states, whereas atomistic excitations are associated with the vibration or diffusion of atomic species (such as adsorbed atoms or molecules). It is important to distinguish between ground-state properties, electronic excitations, and atomistic excitations because rather different models are used to describe each of these three types of phenomena.

Quantum theory of surfaces. The structure of the theory of the properties of solid surfaces does not differ in any fundamental way from that of the quantum theory of bulk solids. Specifically, the conventional quantum theory of interacting electron systems is thought to be applicable, although technical refinements are required because of the loss of translational symmetry and the presence of large electron density gradients normal to the surface.

Macroscopic models. It is premature to speak of an embracing theory of surface phenomena. Rather, a diverse array of specific models have been proposed for the description of various properties. In the case of macroscopic models, the presence of a surface is treated as a boundary condition on an otherwise continuum theory of bulk behavior. Such models have found widespread use in semiconductor and insulator physics because the penetration depth of electrostatic fields associated with surface charges usually is large ($\lambda_e \sim 10^4 - 10^5$ nm) relative to the spatial extend of the charges themselves ($d \lesssim 1$ nm). To describe the atomistic and electronic properties associated with the upper few atomic layers at a surface, however, one must make use of a description of surfaces at the atomistic or electronic level. *See* SEMICONDUCTOR.

Microscopic models. Four major classes of microscopic models of surface properties have been explored. The simplest of these consists of models in which consideration of the electronic motion is suppressed entirely, and the solid is visualized as composed of atomic species interacting via two-body forces. While such models may suffice to describe the vibrational motion of atoms near a surface, they are inadequate to describe ground-state properties such as adsorbate potential-energy curves, although they have been utilized for such calculations. The next more sophisticated models are empirical quantum chemical models (such as "tight-binding" or "pseudopotential" models in solid-state terminology), in which electronic motions are considered explicitly but electron-electron interactions are not. Such models have proved useful in solid-state physics, although their value for surface physics is limited because the large charge rearrangements (relative to the bulk) which occur at surfaces require an accurate, self-consistent treatment of electron-electron and electron-ion interactions. The simplest model in which these interactions are treated explicitly is the "jellium" model of metals, in which the positive charge associated with the ion cores immersed in the sea of conduction electrons is replaced by a uniform positive "background" charge terminating along a plane. This model permits the most accurate (but still approximate) treatment of electron-electron interactions at the expense of losing the effects of atomic lattice structure because of the uniform-positive-background hypothesis. Elec-

tronic computers have permitted, however, the construction of semi-empirical models in which both the electron-electron and electron-ion interactions can be treated in a self-consistent, if approximate, fashion. Such models have been applied to examine ground-state electronic charge densities and localized electronic surface states at the surfaces of simple (that is, *s-p* bonded) and transition metals, as well as homopolar semiconductors. The major tests of their adequacy arise from comparisons of their predictions with measured work functions, photoemission spectra, and characteristic electron-loss spectra. The only new result (relative to comparable bulk analyses) emanating from these computations is the recognition that since the local electronic density of states in the upper few layers depends on the geometrical structure of these layers, comparison of the calculated local density of states with observed valence-electron emission spectra provides a qualitative means of assessing the adequacy of proposed models for the atomic geometry of surface species.

Theoretical models for data analysis. Another distinct but important group of theoretical models in surface physics consists of those utilized to analyze observed scattering and emission spectra in order to extract therefrom quantitative assessments of the atomic and electronic structure of surfaces. These models differ substantially from their bulk counterparts because of the necessity of strong particle-solid interactions to achieve surface-sensitive spectroscopies. Consequently, the fundamental assumption underlying the linear-response theory of bulk solid-state spectroscopies — that is, the appropriate particle-solid interaction is weak and hence can be treated by low-order (usually first) perturbation theory — is invalid. This fact results in collision theories of surface-sensitive particle-solid scattering exhibiting a considerably more complicated analytical structure in order to accommodate the strong elastic as well as inelastic scattering of the particle by the various constituents of the solid.

Applications to LEED. While the above considerations are quite general, the special case in which they have been developed in most detail is the coherent scattering (that is, diffraction) of low-energy electrons from the surfaces of crystalline solids. This is an important case because elastic low-energy electron diffraction (ELEED) is the analog of x-ray diffraction for surfaces — that is, it is the vehicle for the achievement of a quantitative surface crystallography. Since 1968 quite complete quantum-field-theory models of the ELEED process have been developed, tested, and reduced to computational algorithms suitable for the routine analysis of ELEED intensity data. From such analyses the surface atomic geometries of the low-index faces of a host of simple metals and polar semiconductors have been determined, as have the geometries of a few simple overlayer structures on the low-index faces of fcc metals. Similar quantitative analyses of inelastic low-energy electron diffraction (ILEED) intensities have yielded the energy-momentum relations of collective surface electronic excitations (surface plasmons). Therefore, adequate quantum field theories of both the elastic and inelastic diffraction of electrons have been constructed and applied to the quantitative characterization of the structure of the low-index faces of crystalline solids via the analysis of LEED intensities. Development of analogous theories of photoelectric emission has been undertaken.

Data acquisition. It is the development of a host of novel surface-sensitive spectroscopic techniques, however, which has provided the foundation for the renaissance in surface physics. Having recognized that low-energy electrons, thermal atoms, and slow ions all constitute surface-sensitive incoming or exit entities in a particle-solid collision experiment, one can envisage a wide variety of surface spectroscopies based on these plus quanta of electromagnetic radiation (photons) as possible incident or detected species. Most of these possibilities actually have been realized in some form. The selection of which technique to use in a particular application depends both upon what one wishes to learn about a surface and upon the relative convenience and destructiveness of the various measurements.

Typically, one wishes to determine the composition of a surface region, and often to ascertain its atomic geometry and electronic structure as well. For planar surfaces of crystalline solids, measurement of the rms surface-atom displacements is an ancillary to that of the atomic geometry because the values of these displacements emerge from an examination of the temperature dependence of the same ELEED intensities whose analysis yields the atomic structure of the surface. In the measurement of any of these quantities, however, important issues are the spatial and depth resolution of the possible techniques. Typically the depth resolution is determined by the particle-solid force law of the incident and exit particles, higher resolution being associated with stronger inelastic collision

processes. The lateral spatial resolution depends on the ability to focus the incident beam. It is of the order of 1 cm^2 for photon beams, 10^{-8} to 10^{-12} cm^2 for electron beams, and 10^{-8} cm^2 for ion beams. Thus, scanning microscopies are both feasible and common with electron and ion beams but not with photon beams. Depth resolution is a single monolayer for thermal-atom and slow-ion scattering, and a few monolayers for slow-electron scattering. It can become 1000 atomic layers or more, however, for fast (MeV) ions and (10 keV) electrons.

Surface composition. The elemental composition of surfaces is specified by measuring the masses or atomic numbers, or both, of resident species. Their masses may be ascertained either by the elastic backscattering of slow incident ions (ion scattering spectrometry, or ISS) or by using such ions to erode the surface, detecting the ejected surface species in a mass spectrometer (secondary ion mass spectrometry, or SIMS).

The atomic numbers of surface species are determined by measuring the energy of tightly bound "core" electrons. A schematic diagram illustrating the nature and labeling of the various physical processes which can be utilized to accomplish this task is shown in Fig. 2. An electron, photon, or chemical species incident on a surface excites a low-energy core electron. The binding energy of this electron commonly is determined directly by measuring the energy loss of the incident electron (characteristic loss spectroscopy, or CLS), the energy of the core electron ejected by an incident x-ray photon (x-ray photoelectron spectroscopy, or XPS, sometimes referred to as electron spectroscopy for chemical analysis, or ESCA), or the threshold energy of an incident particle necessary to generate a threshold in the secondary x-ray yield (soft-x-ray appearance potential spectroscopy, or SXAPS). Alternatively, the binding energy of the core electron may be ascertained by "secondary" processes in which an initially empty core state (generated by a direct process) is filled by an electron in a higher energy state. If the filling process is radiative recombination, then the energy of the emitted x-ray yields the binding energy (soft x-ray emission spectroscopy). If this process is radiationless, however, then the energy of the electron excited by the Auger process indicates the binding energy of the initially empty core state (Auger electron spectroscopy, or AES).

Essentially all of these techniques operate on the dictionary premise: that is, calibration spectra are obtained on surfaces of independently known composition, with elemental analysis on unknown samples being performed by comparison of their spectra with the reference calibration spectra. Consequently, although the detailed interpretation of observed line shapes has eluded surface physicists, the use of these spectroscopies for elemental analysis has proved both practical and eminently useful. Difficulties in interpretation have precluded the use of these techniques for quantitative chemical analysis (for example, the determination of whether C and O are adsorbed on aluminum as CO, CO_2, or C on Al_2O_3, and so on). Progress has been made, however, in developing this aspect of the core-electron spectroscopies.

Surface atomic geometry. The atomic geometry of planar surfaces of crystalline solids usually is

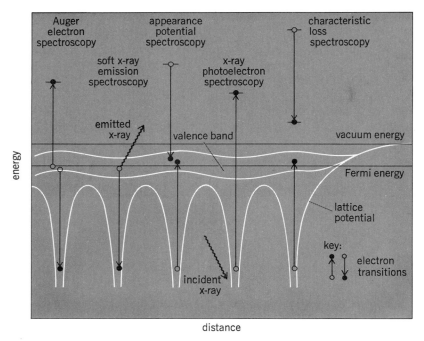

Fig. 2. Schematic diagram of the core-electron transitions utilized to ascertain the atomic number of surface species. (*From R. L. Park, Inner-shell spectroscopy, Phys. Today, 28(4):52–59, April 1975*)

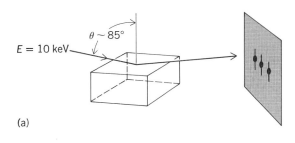

(a)

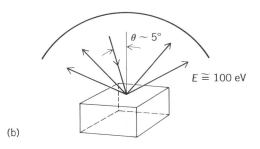

(b)

Fig. 3. Schematic diagram of the two electron diffraction techniques used to determine the geometry of single-crystal surfaces. (a) Reflection high-energy electron diffraction (RHEED). (b) Elastic low-energy electron diffraction (ELEED). (*From C. B. Duke, Determination of the structure and properties of solid surfaces by electron diffraction and emission, in I. Prigogine and S. A. Rice, eds., Adv. Chem. Phys., 27(1):1–209, Wiley-Interscience, 1974*)

obtained by electron diffraction, although in certain simple cases slow-ion backscattering or valence-electron photoemission spectroscopy also may be employed. Two experimental configurations commonly are used, as indicated in Fig. 3. The reflection high-energy electron diffraction (RHEED) configuration embodies glancing incidence electrons at keV energies. It yields only the space-group symmetry of the surface, and is quite sensitive to surface topography. The ELEED experiment consists of measuring the backscattering intensities of electrons in the energy range 50 eV $\leq E \leq$ 500 eV. The configuration of diffracted beams reveals the space group symmetry of the surface structure, whereas analysis of their intensities permits determination of their atomic geometry. In the case of tip sample geometries, field-ion microscopy permits the direct imaging of atoms on the tip surface.

Surface atomic motion. The vibrational motion of surface species may be examined either by analysis of the temperature dependence of ELEED intensities or by direct observation of small ($\Delta E \sim 0.01$ eV) electron energy losses caused by the excitation of a normal mode of vibration. The first approach provides the rms vibrational amplitudes of surface species, whereas the second yields the frequencies of localized "surface" normal modes of vibration. Nonstandard equipment is required in both cases: an ultra-high-vacuum goniometer embodying precise temperature control in the former, and a high resolution ($\Delta E \leq 0.01$ eV) electron spectrometer in the latter.

Surface electronic structure. The electronic structure of a solid-vacuum interface is studied by measuring the emission of valence electrons (induced by external fields, electrons, ions, or pho-

tons) or the inelastic scattering of an incident electron. A special situation arises when an emitted Auger electron is a valence electron. In this case, the initially empty core state is highly localized in space. Consequently, the emission line shape is a measure of the local electronic structure in the vicinity of this core state. The shifts in energy of core-level photoemission and Auger transitions (called chemical shifts) caused by the nearby electronic charge densities also yield an indication of the local electronic structure around a particular kind of surface atom.

In contrast to these emission processes involving localized core electrons, the photoemission, field-emission, ion-neutralization, and characteristic-loss spectroscopies of valence electrons provide measures of their average behavior in the vicinity of a surface. Indeed, for precisely this reason such spectra from clean surfaces are quite difficult to interpret because the distinction between "bulk" and "surface" features is often vague. Thus, their major use has occurred in the arena of chemisorption, in which case the changes in spectra upon adsorption can be monitored, and qualitative features of the electronic structure of the chemisorbed complexes inferred therefrom. Since quantitative theoretical models of these emission processes are not available, however, ambiguities in interpretation are common. Although the advent of variable-energy synchrotron radiation sources and measurements of the angular dependence of the emission intensities have reduced the probability of such ambiguities, a proper theoretical analysis of these processes is needed to convert the observed spectra into quantitative indicators of surface structure. The only case for which such a model has been constructed is that of ILEED. The availability of this theoretical model permits the analysis of ILEED intensities to extract the energy-momentum relationship of surface plasma oscillations; this is the only quantitative application developed for valence-electron surface spectroscopy.

[C. B. DUKE]

Bibliography: J. M. Blakely (ed.), *Surface Physics of Materials*, vols. 1 and 2, 1975; H. L. Davis (ed.), Surface physics, *Phys. Today*, special issue, 28(4):23–71, April 1975; H. L. Davis (ed.), Vacuum: A special report, *Phys. Today*, 25(8):23–58, August 1972; F. Garcia-Moliner and F. Flores, *Introduction to the Theory of Solid Surfaces*, 1979; F. O. Goodman (ed.), *Dynamic Aspects of Surface Physics: Proceedings of the International School of Physics 'Enrico Fermi,'* course 58, 1975; T. S. Jayadevaiah and R. Vanslow (eds.), Surface science: Recent progress and perspectives, *Crit. Rev. Solid State Sci.*, vol. 4, issues 2 and 3, 1974; H. Kumagai and T. Toya (eds.), Proceedings of the 2d International Conference on Solid Surfaces, *Japan. J. Appl. Phys.*, suppl. 2, pt. 2, 1974; S. R. Morrison, *The Chemical Physics of Surfaces*, 1977; I. Prigogine and S. A. Rice (eds.), Aspects of the study of surfaces, *Adv. Chem. Phys.*, vol. 27, 1974; Proceedings of the Annual Surface Science Symposia of the American Vacuum Society, *J. Vacuum Sci. Tech.*, January/February issues; M. Prutton, *Surface Physics*, 1975; G. A. Somorjai, Surface science, *Science*, 201:489–497, 1978; R. Vanselow and S. Y. Tong (eds.), *Chemistry and Physics of Solid Surfaces*, 1977.

Surging

A sudden and momentary change of voltage or current in a circuit. It can be due to a sudden change in the applied input signal, a sudden change in the load placed on the circuit, or to the action of a relay, switch, or other device that changes operating conditions within the circuit. The resulting surges or transients are often called pulses or impulses when they have only one polarity. An oscillatory surge includes both positive and negative polarity values. Surging in electric circuits corresponds to overshooting. Cathode-ray oscilloscopes are frequently used to obtain visual patterns of the transient voltages due to surging. *See* ELECTRIC TRANSIENT. [JOHN MARKUS]

Susceptance

The imaginary part of the complex representation for the admittance Y, defined by Eq. (1), of a cir-

$$Y = G \pm jB \qquad (1)$$

cuit, where G is the real part, called the conductance, and B is the susceptance. Since $Y = 1/Z = 1/(R + jX)$ where X is the total reactance, $X_L - X_C$, and R is the resistance, then Eq. (2) holds and sus-

$$Y = \frac{R}{R^2 + X^2} - j\frac{X}{R^2 + X^2} \qquad (2)$$

ceptance $B = X/(R^2 + X^2)$. This is the general expression for susceptance which shows that susceptance is a function involving both resistance and reactance.

If resistance is negligible, then $B = X/X^2 = 1/X$, or the reciprocal of the reactance. This is called simple susceptance and is correct only where the impedance contains no resistance. For simple susceptances Eqs. (3) hold.

Inductive susceptance $B_L = 1/jX_L = -jB_L$ mhos

Capacitive susceptance $B_C = 1/-jX_C = jB_C$ mhos (3)

These functions find application chiefly in computation of parallel circuits. *See* ADMITTANCE; ALTERNATING-CURRENT CIRCUIT THEORY.

[BURTIS L. ROBERTSON]

Sweep generator

An electronic circuit that generates a voltage or current, usually recurrent, as a prescribed function of time. The resulting waveform is used as a time base to be applied to the deflection system of an electron-beam device, such as a cathode-ray tube. Sweep generators are classified as linear, circular, rotating radial, or hyperbolic.

Linear sweep generation. A linear sweep generator provides a current or voltage that is a linear function of time. The waveform is usually recurrent at uniform periods of time (Fig. 1). The

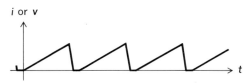

i or **v**

Fig. 1. Sawtooth wave.

deflection of the electron beam in a cathode-ray tube of the type normally used in the cathode-ray oscilloscope may be expressed by Eq. (1), where k is a constant which is inversely proportional to the beam-accelerating potential, and v_d is the potential applied between the deflection plates. Thus, if the

$$d = kv_d \qquad (1)$$

sweep waveform applied to the plates has the form $v_d = k_2 t$ at repeating intervals, the deflection of the beam between the plates will be a linear function of time, recurrent at the same rate. *See* SAWTOOTH-WAVE GENERATOR.

If the deflection system in a cathode-ray device is electromagnetic, the deflection of the beam is proportional to the transverse magnetic flux density in the deflection region. In this case, the deflection is approximately proportional to the current in the deflection coils, external to the tube, which produce the magnetic field. The sweep waveform is a waveform of current rather than voltage. *See* OSCILLOSCOPE.

The well-known raster scan of the television system is produced if simultaneously linear recurrent sweep waveforms are applied to the horizontal and vertical deflection systems of the cathode-ray device, and the vertical period is many times longer than the horizontal period (Fig. 2). Thus a number

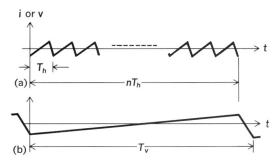

i or *v*

Fig. 2. Waveforms producing the raster scan. (*a*) Horizontal sweep. (*b*) Vertical sweep.

of equally spaced, nearly horizontal scans are produced, the number being the ratio of the horizontal sweep period to the vertical sweep period. In the cathode-ray oscilloscope, a desired time-varying function may be visually presented by applying a linear voltage sweep waveform to the horizontal deflection system and the function to be plotted to the vertical system. Where the raster scan is used, as in the television receiver, the information to be plotted usually appears on the viewing screen as intensity variations produced by modulation of the current in the cathode-ray tube beam. *See* TELEVISION SCANNING.

Circular sweep generation. A circular sweep is generated by applying a constant-frequency sinusoid to the horizontal component of a deflection system and another of the same frequency but shifted in phase by 90° to the vertical component. If the horizontal component of deflection d_x is $A \cos \omega t$, and the vertical component d_y is $A \sin \omega t$, the total deflection d is the vector sum $\sqrt{d_x^2 + d_y^2}$, which equals A. The deflection path is circular at a constant radius A. The angular position θ, with respect to the horizontal, is given by $\tan \theta = \tan \omega t$. The angular rate of change of the deflected beam is constant.

A spiral scan is generated if, in addition to the sinusoidal x and y components, a linear sawtooth waveform ($v = kt$) is used to modulate equally both the horizontal and voltage deflection waveform (Fig. 3). Thus the deflection is given by Eq. (2),

$$d = \sqrt{A^2 k^2 t^2 \cos^2 \omega t + A^2 k^2 t^2 \sin^2 \omega t} \qquad (2)$$

which reduces to Eq. (3). If the active period of the

$$d = Akt \qquad (3)$$

sawtooth is n times the period of the sinusoid, the spiral will have n revolutions for each period of the sawtooth. Circular and spiral sweeps are useful in special forms of oscilloscopes where a long time base is desirable.

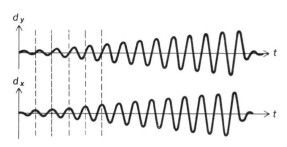

Fig. 3. Waveforms producing spiral scan.

Rotating radial sweep generation. A rotating radial sweep may be generated by applying a linear sawtooth of current to the deflection system of a magnetically deflected cathode-ray device and rotating the deflection coil producing the magnetic field at a constant angular velocity. Such a sweep may also be generated by using a fixed position deflection system having separate x and y components of deflection. The same combination of linear and sinusoidal modulation as that of the spiral sweep may be used, except that the period of the linear modulation component to the x-y deflection system must be short compared to the period of the angular deflection. As indicated in Fig. 4, such a combination of sweeps produces one complete radial line for a negligible change in angle. The

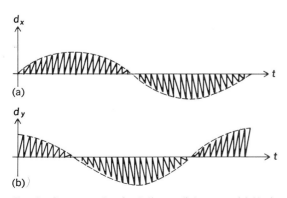

Fig. 4. Components of rotating radial sweep. (a) Horizontal sweep. (b) Vertical sweep.

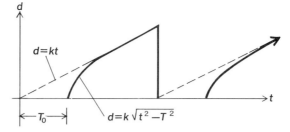

Fig. 5. Hyperbolic sweep.

rotating radial sweep is most widely used in radar display systems, where the angle of rotation is made synchronous with the scan angle of a radar antenna. The linear radial sweep is proportional to radar range. Thus intensity modulation on a display cathode-ray tube having such sweeps will present a true position of individual radar returns. *See* RADAR.

Hyperbolic sweep generation. There are applications, particularly in airborne radar systems, where hyperbolic sawtooth sweeps are preferred to linear sawtooth sweeps. Such a hyperbolic sweep, shown in Fig. 5, may be given by Eq. (4).

$$d = k \sqrt{t^2 - T_0^2} \qquad (4)$$

Such a sweep is asymptotic to a linear sweep for large values of t, but is delayed with respect to the beginning of it by a time, T_0. Such a hyperbolic sweep used in a rotating radial sweep system provides for true ground range radar mapping in an airborne system. Here the time T_0 is made equal to h/c, where h is the height of the aircraft, c is the velocity of propagation of electromagnetic energy. The factor k is a constant which represents the scale factor of the display.

Hyperbolic sweeps may be generated as a modification of the type of circuitry used in the generation of sawtooth sweep waveforms.

[GLENN M. GLASFORD]

Bibliography: J. Millman, *Microelectronics*, 1979; J. Millman and H. Taub, *Pulse, Digital and Switching Waveforms*, 1965; L. Strauss, *Wave Generation and Shaping*, 2d ed., 1970.

Switched capacitor

A module consisting of a capacitor with two metal oxide semiconductor (MOS) switches connected as shown in Fig. 1a. These elements in the module are easily realized as an integrated circuit on a silicon chip by using MOS technology. The switched capacitor module is approximately equivalent to a resistor, as shown in Fig. 1b. The fact that resistors are relatively difficult to implement gives the switched capacitor a great advantage in integrated-circuit applications requiring resistors. Some of the advantages are that the cost is significantly reduced, the chip area needed is reduced, and precision is increased. Although the switched capacitor can be used for any analog circuit realization such as analog-to-digital or digital-to-analog converters, the most notable application has been to voice-frequency filtering. *See* ANALOG-TO-DIGITAL CONVERTER; DIGITAL-TO-ANALOG CONVERTER.

Although a switch has long been used as an element in circuits and systems, it was not until

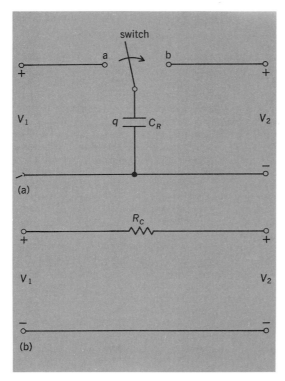

Fig. 1. Switched capacitor. (a) Basic circuit. (b) Equivalent resistive circuit.

the late 1970s that its potential and practicality in integrated-circuit design was realized.

Integrator circuits. A conventional RC integrator circuit is shown in Fig. 2a. The output voltage is given by Eq. (1), where $1/s$ indicates the opera-

$$V_{out} = \frac{1}{R_1 C_2} \frac{1}{s} V_{in} \qquad (1)$$

tion of integration (s is the differential operator), showing that integrator performance depends on R_1. In MOS integrated circuits, the value of R_1 cannot be controlled better than 20% by using standard fabrication techniques. In addition, considerable chip space is required to realize resistors in the megohm range. In contrast, the switched capacitor realizations shown in Fig 2b and c depend on the ratio of capacitors which can be controlled with great accuracy. For example, the output voltage of the differential integrator shown in Fig. 2c is given by Eq. (2), where f_c is the clock fre-

$$V_{out} = \frac{C_1}{C_2} \frac{f_c}{s} (V_2 - V_1) \qquad (2)$$

quency. With C_1 and C_2 in the range of 1 picofarad and f_c at 100 kHz, this circuit has an equivalent resistance of 10 MΩ, and the gain of the circuit is about 10^4. The silicon chip area required to implement the capacitors is about 0.01 mm². If a resistor is used in place of the switched capacitors, an area at least 100 times larger would be required.

Equivalent resistance. Returning to the switched-capacitor circuit of Fig. 1a, the operation may be visualized as follows. With the switch in

position a, the capacitor C_R is charged to the voltage V_1. The switch is then thrown to position b, and the capacitor discharged at voltage V_2. The amount of charge transferred is then $q = C(V_2 - V_1)$. If the switch is thrown back and forth at a clock frequency f_c, the average current will be $C(V_2 - V_1) f_c$. The size of an equivalent resistor to give the same value of current is given by Eq. (3). From this equation, it is seen that with $C = 1$ pF and $f_c = 100$ kHz, the value of 10 MΩ used previously is obtained.

$$R_C = \frac{1}{C f_c} \qquad (3)$$

The accuracy of the equivalence between the switched capacitor and the resistor depends on the relative size of the clock frequency f_c and the frequencies in the signal being processed. If the switching frequency is much larger than the signal frequencies of interest, the equivalence is excel-

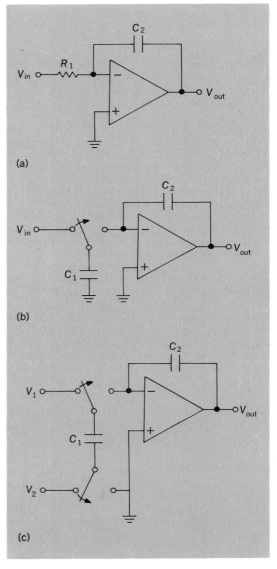

Fig. 2. Integrator circuits. (a) Conventional RC integrator circuit. (b) Single-input switched-capacitor integrator. (c) Switched-capacitor differential integrator circuit.

lent and the time sampling of the signal can be ignored in a first-order analysis, such that the switched capacitor is a direct replacement for a conventional resistor. If this is not the case, then sampled-data techniques in terms of a z-transform variable must be used for accuracy.

Analog operations. The switch that has been used in describing the switched capacitor is actually realized by an MOS transistor to which a pulse of voltage at the clock frequency is applied to produce the off and on conditions of the switch. This periodically operating switch is used for a number of analog operations, such as addition, subtraction, inversion, and integration. These operations are essential in the construction of analog filters, as well as in other applications of switched capacitors. These operations may be explained in terms of the circuits of Fig. 2. In Fig. 2a and b the analog operation of integration is accomplished. In addition, these circuits are of the inverting type, meaning that a sign reversal is accomplished in addition to integration. The sign reversal of a voltage can be accomplished directly by using switches and a capacitor, as seen in Fig. 2c. Assume that V_1 is grounded or $V_1 = 0$. The operation of the switches is such that the voltage applied to the MOS operational amplifier is the negative of V_2. With the switch operating from left to right, V_2 with respect to ground is reversed. With V_1 not grounded, the circuit of Fig. 2c is a differential integrator, meaning that the output voltage is a function of the voltage difference, $V_2 - V_1$. In conventional active-filter design, these analog operations are accomplished by means of additional stages incorporating operational amplifiers. In switched-capacitor design, these analog operations are implemented with switches. *See* AMPLIFIER; ANALOG COMPUTER; TRANSISTOR.

Filter design. Although there are many strategies for filter design, the discussion will be restricted to the case of filters based on the passive *LC* ladder with resistive terminations at both ends. Extensive tables are available giving element values to achieve various forms of frequency response, such as Butterworth, Chebyshev, and Cauer (elliptic). All tables are given in terms of a normalized termination of 1 Ω, and a normalized frequency of $\omega_o = 1$ rad/s and for the low-pass case. It is standard procedure to make use of frequency transformations to realize high-pass, band-pass, band-elimination, and similar kinds of responses, and to use frequency and magnitude scaling to give practical element values. The passive ladder structure with double terminations is chosen because it has low sensitivity of changes in transmission with changes in element sizes.

Starting with the low-sensitivity, low-pass ladder structure, a frequency transformation is first accomplished. From these steps, a structural simulation is then carried out by replacing the actual filter by its signal flow graph representation. The flow graph is chosen so that most of the operations required are integration. The elements in the flow graph are then simulated by circuits like that shown in Fig. 2c.

An example of filter design is shown in Fig. 3. The ladder network shown in Fig. 3a is known as the low-pass prototype. In the usual case, $R_1 = R_2 = 1$ Ω. The elements C_1 and L_2 are determined

from tables, depending on the form of frequency response required. Going from Fig. 3a to b accomplishes a low-pass to band-pass transformation in which all element values are determined from those in Fig. 3a, and the specification of the center frequency and bandwidth of the band-pass case. In Fig. 3c the filter of Fig. 3b is represented by its flow graph, in which the lines and arcs with arrows indicate the structure of the circuit of Fig. 3b, and the associated symbols represent the impedance or admittance. To this structural simulation of the ladder filter, an element simulation is next applied. In particular, all elements of a form such as $1/RC_A s$ are realized by using the integrator of Fig. 2c with differences of voltages accomplished by the switched capacitors. The final result, that shown in

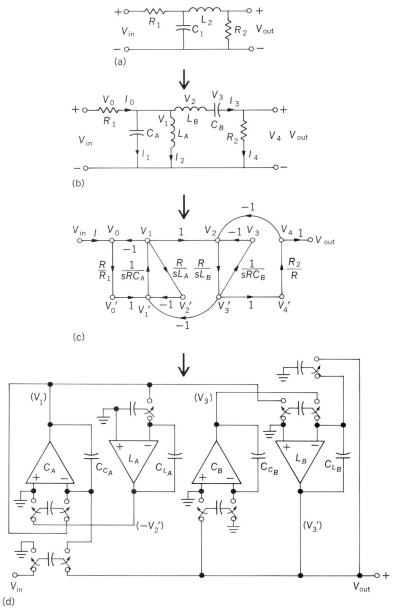

Fig. 3. Steps in the realization of a switched-capacitor filter. (a) Low-pass prototype filter. (b) Corresponding band-pass filter. (c) Signal flow graph representation of the circuit of b. (d) Final switched-capacitor band-pass filter.

Fig. 3*d*, is then implemented as an integrated circuit containing only switched capacitors, ordinary capacitors, and operational amplifiers. The chip area required to realize a filter of modest order might be 100 mils (2.5 mm) on each side. *See* ELECTRIC FILTER; INTEGRATED CIRCUITS.

[M. E. VAN VALKENBURG]
Bibliography: G. M. Jacobs et al., *Electronics*, 52(4):105–112, Feb. 15, 1979; G. M. Jacobs et al., *IEEE Trans. CAS*, 25:1014–1020, December 1978.

Switching circuit

A constituent electric circuit of switching or digital data-processing systems. Well-known examples of such systems are digital computers, dial telephone systems, automatic accounting and inventory systems. In these and other switching systems the component circuit units receive, store, and manipulate information in coded (digital) form to accomplish the specified objectives of the system. *See* SWITCHING SYSTEMS (COMMUNICATIONS); SWITCHING THEORY.

Physically, switching circuits consist of conducting paths interconnecting discrete-valued electrical devices. The most generally used switching circuit devices are two-valued or binary, such as switches and relays in which manual or electromagnetic actuation opens and closes electric contacts; vacuum and gas-filled electronic tubes, semiconductor rectifiers and transistors, which do or do not conduct current; and magnetic structures, which can be saturated in either one of two directions.

The electrical conditions controlling these switching circuit devices are also generally two-valued or binary, such as open versus closed path, full voltage versus no voltage, large current versus small current, and high resistance versus low resistance. Such two-valued electrical conditions, as applied to the input of a switching circuit, represent either (1) a combination of events or situations which exist or do not exist; (2) a sequence of events or situations which occur in a certain order; or (3) both combinations and sequences of events or situations. The switching circuit responds to such inputs by delivering at its output, also in two-valued terms, new information which is functionally related to the input information.

The two fundamental characteristics of switching circuits are logic and memory. A switching circuit embodies such logical relationships as output X is to exist only if inputs A and B occur simultaneously; and output Y is to exist if either input A or input B occurs. The factor of memory, in turn, enables a switching circuit to hold or retain a given state after the condition that produced the state has passed.

Basic combinational circuits. A combinational switching circuit is one in which a particular set of input conditions always establishes the same output, irrespective of the past history of the circuit. An example of a simple combinational circuit is the problem of controlling the entrance-hall light of a residence by three up-down wall switches located in three different rooms; that is, any one of the three wall switches must be able to turn the hall light either on or off. Analysis of this problem shows that the circuit must meet the following simple requirements. If any one or all three wall

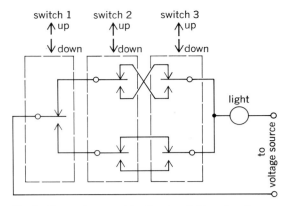

Fig. 1.   Elementary combinational switching circuit.

switches are down, the hall lamp must light; if any one or all three switches are up, the lamp must be dark. An obvious (but not the most efficient) circuit meeting these requirements is shown in Fig. 1.

In this problem the circuit inputs are, of course, the manual switch settings, and the circuit output is the control of the light.

In electronic switching circuits, so-called gates are used to perform logical functions equivalent to these series-parallel networks of switch contacts. In this sense, an electronic gate is an elementary combinational circuit. Gates do not function by physical rearrangement of interconnecting paths, as do switch or relay contacts. Instead, they function by control of voltage or current levels at their output.

The most commonly encountered gates are the AND, and the OR gates. The AND gate produces an output only if all its inputs are concurrently present; an OR gate produces an output if any one or any combination of its inputs is present. Figure 2 shows both an AND gate and an OR gate, using rectifier or diode elements.

In the AND gate the rectifiers are so oriented that current from a positive voltage source E passes through the relatively large resistance R and then through the low forward resistance of any one of the rectifiers to ground in the circuits controlling the gate. Thus, in the inactive state of the gate the

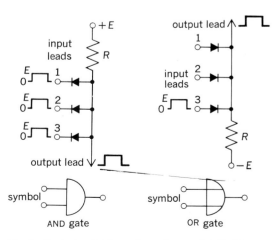

Fig. 2.   Typical switching gates using crystal diodes.

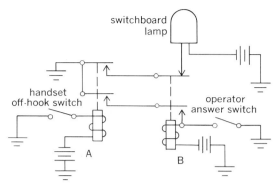

Fig. 3. Elementary sequential relay switching circuit.

output lead is at or near ground potential. If all three input leads of this gate concurrently receive a positive voltage pulse of magnitude E, the rectifiers approach open circuit, and the output lead will be raised from near ground to a positive potential for the duration of the input pulse. In other words, input leads 1 AND 2 AND 3 must all receive the positive pulse to obtain the positive output voltage.

In the OR gate the rectifiers are reversed so that current flows from ground in the input circuits through the low forward resistance of any rectifier and then through the relatively large resistance R to the negative voltage source $-E$. Thus, in the inactive state of the gate the output lead is at or near ground potential. If, however, a relatively high positive voltage pulse is applied to input leads 1 OR 2 OR 3, the remaining two diodes are cut off and the output is raised to a positive potential for the duration of the input pulse.

Gates may, of course, be constructed with other electronic devices, such as tubes, transistors, and magnetic cores.

Basic sequential circuits. A sequential switching circuit is one whose output depends not only upon the present state of its input, but also on what its input conditions have been in the past. Sequential circuits, therefore, require memory elements.

By way of illustration, consider the following simple sequential circuit problem. When a telephone customer lifts his handset, a lamp is to light in front of a switchboard operator. When an opera-

tor answers, the light should go out to avoid other operators also answering. After the operator has satisfied the customer's request for a connection, she withdraws. The light, however, should not relight now, even though the conditions existing at this time are seemingly identical with those at the start; that is, the customer has his handset lifted and no operator is on the line. A sequential relay circuit meeting these simple requirements is shown in Fig. 3. In this circuit, when the handset is lifted, the handset off-hook switch connects a ground input to relay A which operates and lights the switchboard lamp. When the operator answers, another ground input operates relay B, and this relay puts out the light. A holding circuit on relay B keeps relay B operated until the handset off-hook switch is again opened and relay A is deenergized. Relay B "remembers" that the operator has answered and prevents the relighting of the lamp

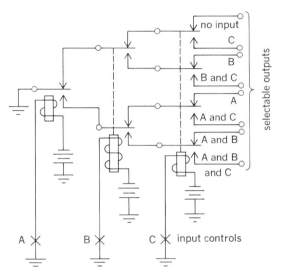

Fig. 5. Simple relay selecting circuit.

when the operator withdraws. It is, therefore, the memory element of the circuit.

A typical electronic memory element used in sequential circuits is a simple circuit called a flip-flop. A flip-flop consists of two amplifiers connected so that the output of one amplifier is the input of the other. A voltage pulse will set the flip-flop into one of two states, and that state remains until another voltage pulse resets, or returns the flip-flop to its original state. It can therefore be used to remember that an event has taken place.

Figure 4 is an *npn* transistor flip-flop. When set, transistor A is conducting and transistor B is cut off. When reset, transistor B is conducting and transistor A is cut off. A positive output voltage with respect to ground may be obtained from either transistor to indicate the condition of the flip-flop.

Relays, flip-flop, and similar memory elements provide static, or fixed, memory; they hold the stored information indefinitely, or until they are told to "forget." In contrast, a delay-line provides transient memory. A delay line has the property

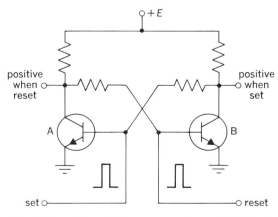

Fig. 4. Transistor switching memory element (flip-flop).

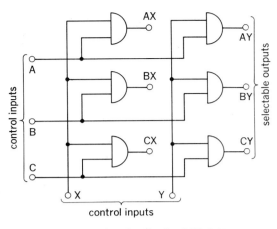

Fig. 6. Matrix selecting circuit using AND gates.

that an electrical signal applied to its input is delayed on its way to the output.

Functional switching circuits. Even in large and complex switching systems the majority of circuit requirements can be met by a relatively small number of types of circuits, each of which performs one or a limited number of somewhat distinct functions. These functional circuits are the basic building blocks of a switching system.

Selecting circuits. A selecting circuit receives the identity (called the address) of a particular item and selects that item from among a number of similar ones. The selectable items are often represented by terminals or leads. Selection usually involves marking the specified terminal or lead by

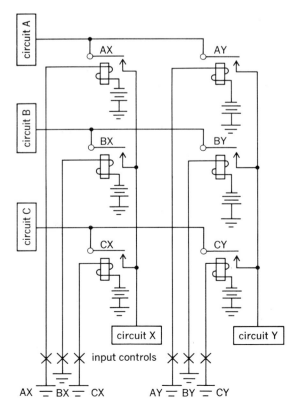

Fig. 7. Simple relay connecting circuit.

applying to it some electrical condition, such as a voltage or current pulse, or a steady-state dc signal. By means of this electrical condition, the selected circuit is alerted, seized, or controlled.

Figure 5 shows a simple relay selecting circuit. This circuit uses three relays to select one of eight outputs according to the combinations in which the three relays are operated or not operated. The input is ground or no ground on control leads A, B, C, in various combinations (the address). The output of the circuit is ground appearing on the single selected output lead.

An electronic selecting circuit using AND gates is the matrix type shown in Fig. 6. In this type of circuit an input signal appears on one of the horizontal input leads and concurrently on one of the vertical input leads. The selected output is at the crosspoint of these two leads.

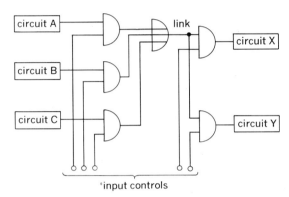

Fig. 8. Connecting circuit using AND and OR gates.

Connecting circuits. A switching system is an aggregate of functional circuit units, some of which must sometimes be directly coupled to each other to interchange information. Such a system needs, therefore, connecting circuits which establish the circuit associations dictated by the momentary needs of the system.

Figure 7 shows a simple relay circuit illustrating the principle of connectors. Any one of the three circuits A, B, or C, can be connected over a single lead with either circuit X or Y by operating the relay whose designation corresponds to the desired circuit association. These relays are operated by an external control circuit that determines which circuit association is needed and insures that only one relay is operated at a time in any row or any column. The connector relays may, of course, carry more than one interconnecting lead, and the number of interconnectable circuits could be fewer or many more.

Figure 8 shows a simple electronic connecting circuit using AND and OR gates. In this arrangement a communication path is provided over a single link from any one of the three functional circuits A, B, C, to either the X or Y circuit by an external control circuit activating the appropriate pair of AND gates. To provide a multilead link, or to provide for other simultaneous interconnections, additional AND gates would, of course, be required. The OR gate maintains separation of the inputs at the common junction point.

Lockout circuits. In switching systems, situations often arise where several similar circuit units are ready at the same instant to request collaboration with another type of functional circuit. Mutual interference among the requesting circuits is prevented by the lockout circuit. In response to concurrent inputs from a number of external circuits, a lockout circuit provides an output indication corresponding to one, and only one, of these circuits at any time.

Figure 9 illustrates a basic relay lockout circuit. The external circuits signify their requests to be allowed to proceed by grounding their respective control leads designated C. The output of the lockout circuit is ground appearing on a single lead designated B, associated with the particular external circuit whose request has been granted. The characteristics that enable this circuit to perform its function are (1) the output ground goes through a contact network chained from left to right; this ground can appear only on the output lead of the lowest numbered operated relay which represents the winning external circuit; (2) the voltage source or battery on which the relays operate, in turn goes through another contact network chained in such a manner that once any relay operates, from then on only higher numbered relays are permitted to operate; (3) an operated relay stays operated on battery through its own closed contact, until the external circuit removes the control ground as an indication that it has been satisfied.

Figure 10 shows a typical electronic lockout circuit using cold-cathode gas-filled tubes. The external circuits furnish positive potential on the input leads to the control gaps of the tubes as indications of service requests. The operation of the circuit is based on the dynamic negative-resistance characteristics of gas tubes. If such tubes are provided with a common impedance in their conduction paths (the cathode impedance in this circuit), simultaneous input signals will result in the ionization of only one tube. Once the control gap of a tube is ionized, conduction current starts flowing in its main gap and this current through the common impedance instantaneously reduces the voltage across all the other tubes below the value needed to ionize them. This reduced voltage is, however, adequate to keep the single ionized tube in the conducting state until its conduction path is opened. The identity of the particular ionized tube is derived from the anode resistance individual to each tube; the output lead whose potential has been lowered by this resistance represents the circuit whose request has been granted.

Lockout circuits are sometimes referred to as hunting or finding circuits. Irrespective of name, the problem in all applications of lockout circuits is that of concurrently competing circuits, among which one has to be picked for some action.

Translating circuits. Switching systems process information in coded form; the information they manipulate is generally in the form of numbers.

Numerical codes are many and varied, each with its own characteristics and more or less distinct advantages for different switching circuit situations. Therefore, one of the common functional circuits in switching systems is the translating circuit, which translates information received in one code into the same information expressed in another code. These translating circuits are combinational circuits; a given input signal combination representing a code to be translated always produces the same output signals, which represent the desired code.

Figure 11 shows an elementary relay translating circuit. In this circuit the input code is biquinary (ground on one of the input leads 1–5 being the quinary or five-valued part, and ground on lead A or B being the binary or two-valued part). The output of the circuit in turn is decimal; in response to a biquinary input, a ground appears on one of the 10 output leads.

Figure 12 is an example of a magnetic-core translating circuit that translates from binary code (1, 2, 4) to a one out of eight code (0, 1, 2, . . . , 6, 7). The circuit has three flip-flops which are set or reset (not set) according to the binary input code combination. The translating elements are eight magnetic cores, each with five windings, and are represented in Fig. 12 by a heavy vertical line. Each short, slanting line segment represents a separate winding on a core. These short slanting lines also symbolize a mirror action; an input current pulse coming from a flip-flop sets a core if it is

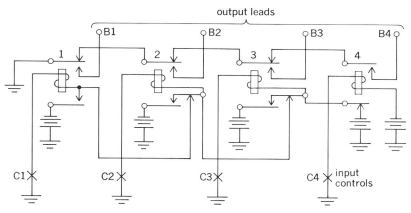

Fig. 9. Basic relay lockout circuit.

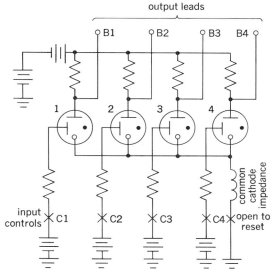

Fig. 10. Lockout circuit using cold-cathode gas tubes.

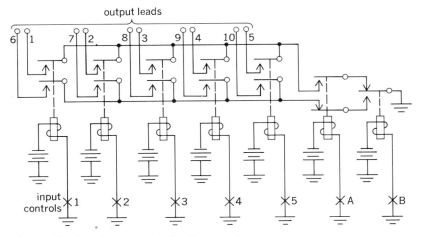

Fig. 11. Simple code translating circuit using relays.

reflected upward by the mirror and prevents setting or resets the core if reflected downward. Once set, the subsequent resetting of a core induces a current which flows upward in the vertical line (in a direction opposite to the resetting current) and is

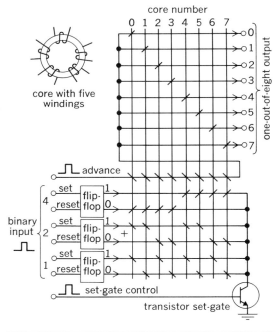

Output digit desired	Flip-flop 4		Flip-flop 2		Flip-flop 1	
	Set	Reset	Set	Reset	Set	Reset
0		x		x		x
1		x		x	x	
2		x	x			x
3		x	x		x	
4	x			x		x
5	x			x	x	
6	x		x			x
7	x		x		x	

Fig. 12. Code register and translating circuit using magnetic cores, each of which has five windings.

reflected to the left or to the right by each mirror symbol.

With this explanation of the symbolism, the circuit works as follows. The input is binary; that is, it consists of a positive voltage pulse to each of the three flip-flops either on its set or on its reset input lead, according to the table. (Note that by adding the numerical designations to those flip-flops which are set in a particular combination, the value of the output digit is determined.)

While the flip-flops are being set, their output current is prevented from flowing into the core windings by the transistor set-gate which is normally nonconducting. Shortly after the binary input combination is recorded in the flip-flops, this set-gate is pulsed for a moment into its conducting state. During this moment, output current will flow from each flip-flop either in its ONE output lead (if the flip-flop has been set) or in its ZERO output lead (if the flip-flop has been reset). As Fig. 12 shows, the output current of flip-flop 4 is always used to set the cores; that is, the current in the ZERO output lead of this flip-flop is used to magnetize the first four cores in the set direction, or the current in its ONE output lead is used to magnetize the last four cores in the set direction. In contrast, the output currents from flip-flops 2 and 1 are always used to magnetize the cores in the opposite or reset direction. Initially all cores are in the reset condition, and cores that receive both set and reset currents simultaneously will not change this initial condition. An analysis of Fig. 12 will therefore show that, for any desired digit, one and only one of the eight cores will be set by the flip-flops in combination. For instance, if output 3 is desired, the current from flip-flop 4 tends to set cores 0, 1, 2, and 3, but cores 0, 1, and 2 are prevented from being set by the output current from either or both flip-flops 2 and 1. When the translated code is needed, the current pulse on the advance lead resets the single previously set core, and consequently an induced output current pulse appears on the appropriate output lead. (The rectifiers in the input and output portions of the circuit prevent unwanted reverse current.)

Register circuits. Information received by a switching system is not always used immediately. It must be stored in register circuits for future use.

In a register circuit the coded information to be stored is applied as input, is retained by memory elements of the circuit and, when needed, the registered information is taken as output in the same code or in a different code. Figure 12 embodies a register function as well as a translating function. Register circuits are devised with a great variety of memory elements, and have capacities to store from a few to millions of information bits.

A frequently encountered form of register circuit is the shift register. This type of register has the ability to shift its stored digital information internally to positions representing higher or lower numerical values in the code employed. For example, in decimal code registration a digit may be shifted from the units to the tens position. An obvious use of such registers is in digital computers when, for example, partial multiplication products have to be lined up for addition.

Counting circuits. One of the most frequently

encountered circuits in switching systems is the counting circuit whose function, in general, is to detect and count repeated current or voltage pulses which represent incoming information. *See* COUNTING CIRCUIT; LOGIC CIRCUITS.

[JOHN MESZAR]

Bibliography: H. J. Beuscher et al., *Electronic Switching and Circuits*, 1971; F. H. Edwards, *The Principles of Switching Circuits*, 1973; A. Friedman and P. R. Menon, *Theory and Design of Switching Circuits*, 1975; M. P. Mitchell, *Switching Circuits for Engineers*, 3d ed., 1975; H. C. Torng, *Switching Circuits: Theory and Logic Design*, 1972.

Switching systems (communications)

The assemblies of switching and control devices provided so that any station in a communications system may be connected as desired with any other station. A telecommunications network consists of transmission systems, switching systems, and stations. Transmission systems carry messages from an originating station to one or more distant stations. They are engineered and installed in sufficient quantities to provide a quality of service commensurate with the cost and expected benefits. To enable the transmission facilities to be shared, stations are connected to and reached through switching system nodes that are part of most telecommunications networks. Switching systems act under built-in control to direct messages toward their ultimate destination or address. Most switching systems, known as central or end offices, are used to serve stations. A smaller number of systems serve as tandem (intermediate) switching offices for large urban areas or toll (long-distance) offices for interurban switching. These end and intermediate office functions are sometimes combined in the same switching system.

There are many types of telecommunication services. The principal ones are voice, data (record), picture (still), and video (motion pictures). For each service there is a different balance between the relative investment in transmission, switching, and station (terminal) facilities. When several of the services are offered in the same network and the network is growing, it is difficult to make general conclusions concerning the economic balance between switching and transmission. Also, as new technologies are introduced into any portion of the network, new economic balances among the elements may materialize. This article deals primarily with systems for the switching of voice. Some data services use the voice network. Since separate networks for data services are available, a brief discussion will be given of switching exclusively for voice and data services.

Public telecommunications have grown over a long period, principally for the telephone and teletypewriter (Telex), and large investments have been made. As new technology becomes available, it is usually first used for growth of the network. When a detailed cost analysis shows that the cost of continued use of old equipment exceeds the cost of providing new equipment, new technology replaces older equipment. Since telecommunications is a mature discipline in industrialized nations, and particularly in the United States, the processes of growth and replacement are much in evidence. At the same time new technology is appearing and being adapted for replacement at an increasing rate.

Switching system fundamentals. Telecommunications switching systems generally perform three basic functions: they transmit signals over the connection or over separate channels to convey the identity of the called (and sometimes the calling) address (for example, the telephone number). and alert the called station; they establish connections through a switching network for conversational use during the entire call; they process the signal information to control and supervise the establishment and disconnection of the switching network connection.

In some data or message switching when real-time communication is not needed, the switching network is replaced by a temporary memory for data storage. This type of switching is known as store-and-forward switching.

Signaling and control. The control of switching systems is accomplished remotely by a specific form of data communications known as signaling. Switching systems are connected with one another by telecommunication channels known as trunks. They are connected with the served stations or terminals by lines. Originally, most signals were sent by direct- or alternating-current pulses over the lines and trunks. Starting in 1976, a new form of signaling was introduced between toll offices in the United States that was intended to be used at end offices. It is known as common channel interoffice signaling (CCIS). As its name implies, a network of separate data communication paths is used for transmitting all signaling information between offices.

In some switching systems the signals for a call directly control the switching devices through which the transmission path is established. This direct control was the earliest and is still the most prevalent form of automatic switching around the world. The step-by-step system described below is the foremost example of direct control. It employs a switch (Fig. 1) in which metallic wipers are moved up and around by electromagnetic actions to contact one out of 100 sets of fixed terminals.

For most modern switching systems the signals for identifying or addressing the called station are received by a central control that processes calls on a time-shared basis. Central controls receive and interpret signals, select and establish communication paths, and prepare signals for transmission. These signals include addresses for use at succeeding nodes or for alerting (ringing) the called station.

Until the introduction of electronics into switching, the central controls employed complex relay logic circuits known in many systems as markers. Most electronic controls are now designed to process calls not only by complex logic, but also by the use of logic tables or a program of instructions stored in bulk electronic memory. The tabular technique is known as action translator (AT). The latter arrangement is now the most accepted and is known as stored program control (SPC). Either type of control may be distributed among the switching devices rather than residing centrally. *See* COMPUTER STORAGE TECHNOLOGY.

SWITCHING SYSTEMS (COMMUNICATIONS)

Fig. 1. Step-by-step switch.

Fig. 2. Toll regions and signal transfer points.

Common channel signaling became practical as a result of processor control. To reduce the number of data channels between all switching nodes, a signaling network with separate switching of signaling data is introduced. These signal switching nodes are fully interconnected and duplicated and are known as signal transfer points (STPs). In the United States there are 10 toll regions, each with its own pair of STPs (Fig. 2.). Each SPC toll switching system connects to the two STPs in its region.

Numbering plan. In an automatic telephone system, a numbering plan provides for uniquely identifying every main telephone station so that calls may be directed to it. The American system is based upon decimal digits. (When letters are on the dial, the telephone system actually recognizes the numerals associated with the letters.)

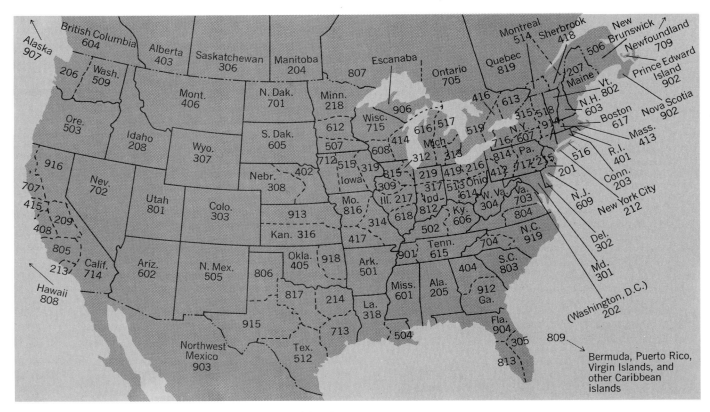

Fig. 3. Numbering plan areas with codes.

A telecommunications central office customarily has the nominal maximum capacity to serve 10,000 main stations, using the number series 0000–9999. When there are more than 10,000 main stations, more than one central office is provided, sometimes in more than one building or wire center. Each office is given a separate three-digit designation or code. The minimum requirement is for an adequate number of digits or characters in each number to address each main station in the dialing area. When a call reaches the called office, the called telephone station is determined from the last four numerals (the main station code).

A seven-digit numbering plan has adequate capacity for only a small portion of the telephones in North America. Hence, a geographical area, such as a state or a Canadian province, is selected as a numbering plan area (NPA), within which there are no duplications of seven-digit numbers. The more populous states, which have large numbers of central offices, are divided into two or more numbering plan areas.

Each numbering plan area is given a three-digit NPA code, the middle digit of which is a 1 or 0. Examples are 803 for the state of South Carolina and 415 for the portion of California that includes San Francisco (Fig. 3). In dialing the number of a subscriber outside the local or home numbering plan area, the area code is dialed ahead of the seven digit number. For example, a subscriber or operator in Asbury Park, NJ, wishing to dial 421-9000 in San Francisco would first dial area code 415 followed by 421-9000. From any other numbering plan area, the dialing would be identical to reach the 421-9000 number, except from within the 415 area, where only the seven digits of the telephone number need be dialed. With this plan the equipment uses, first, the NPA code to determine which area is desired; second, the central office code to select the office in that area; and third, the main telephone number to determine the particular telephone being called. Sometimes to aid routing or to speed completion of a call a "1" is dialed before a number.

In addition to central office and area codes, code numbers for special services, such as 411 for directory assistance and 911 for emergency calls, are used.

Switching connectives. Space and time division are the two basic techniques used in establishing connections. When an individual conductor path is established through a switch for the duration of a call, the system is known as space division. When the transmitted speech signals are sampled and the samples multiplexed in time so that high-speed electronic devices may be used simultaneously by several calls, the switch is known as time division. Space-division switching has been employed since the early manual switchboards in which the connectives were cords, plugs, and jacks.

Today most switching is automatic. Operators are required only for ancillary functions that cannot yet be economically automated. Such systems use the most modern techniques. In the United States, cord, plug, and jack switchboards have almost disappeared, having been replaced by cordless consoles in which calls are distributed

Fig. 4. Traffic service position.

automatically to operators who are provided only with keys and lamps to permit them to serve calls. Figure 4 shows a cordless traffic service position (TSP) used for the completion of person-to-person, credit card, and some coin toll calls. The positions may be located many miles from the SPC system through which the connection is established or processed.

Most switching systems now in service are space division and employ some form of electromechanical switch. These switching systems, unless controlled electronically (for example, SPC) are referred to by the name of the electromechanical device (for example, a step-by-step, panel, crossbar, EMD, codebar, XY, or rotary system). These devices are connected together in successive stages to concentrate, distribute, and select idle paths as required to establish the required connections. The simplest arrangement is the one in which, as each digit is received, the call progresses from one stage to the next by making a selection and finding an idle link at each stage until the selection process is completed.

In direct control systems such as the step-by-step, XY, and EMD systems, the switches are arranged in stages that are accessed progressively. While very few new offices of this type of system are now being installed, systems in service are serving about 50% of the world's telephones (with about 25% in the United States). The switches are of three main types following the basic connective functions, but with the names line finders (for concentration), selectors, and connectors (for distribution and selection). The line finder acts on its own to reach a line requesting service. The selector acts under control of the dial (one-out-of-ten selections) and then seeks (hunts for) an idle path to the next stage (also usually a one-out-of-ten selection). Each selector and connector switch also has a third capability to home or restore the switch to normal. Signaling and control relays are generally mounted with each switch (shown above the mechanism in Fig. 1). The connector is arranged to make two selections under direct control of dial, thereby selecting one of 100 lines for completing the connection.

Figure 5 shows such a switching system with an arrangement of switches to respond to a seven-digit number such as 595-5465. All available terminals in such a system may not always be needed. To avoid the expense of completely equipping unneeded selector stages, these selectors may be arranged to absorb specific received digits without any connection being made through the switch. By so doing, the seven-digit numbering plan may be preserved without providing selectors for the codes not in use.

Some electromechanical switching devices are not amenable to direct dial control since they are not based on the decimal system. Furthermore, with the introduction of push-button (TOUCH-TONE)* dialing, the switches are not fast enough to keep up with the pulsing of a digit at a time. As a result, a number of systems have been developed that use indirect control. With indirect control the line finder connection is extended through another stage of concentration to a signaling circuit known as a register-sender. The progressive control of the switches then takes place after the dialed digits are partially or completely received (registered). The register sends (pulses) the same or substituted digits forward to operate switches. Different types of pulsing may be used. The routing plan is thereby divorced from the numbering plan. To secure more economical trunking, a call may then be routed through a tandem office.

The register-sender may be used to provide a number of other features in addition to tandem routing, such as different routings for certain classes of lines (such as coin calls). The step-by-step system with registers is used in some large English cities, where they are known as directors. The register arrangement also lends itself to the provision for the automatic recording of call billing information known as automatic ticketing or automatic message accounting (AMA). This permits subscribers to dial their toll calls without the necessity of employing operators to record billing information. The provision of this feature also requires identifying the calling line number. Automatic number identification (ANI) equipment is often used, but operator intervention may be required when provision is not made in some of the older switches or to handle multiparty-line customers.

ELECTROMECHANICAL SWITCHES

The step-by-step switch is described above. Other switches and systems in general use around the world are described below.

XY system. Made in Europe and the United States, the XY system also employs a 100-point, two-motion switch operated by electromagnets. The switch is flat, and the first or X motion is in a horizontal plane and to the right. The second or Y motion is also horizontal, but at right angles to the first, and carries the brushes into the bank assembly. The system is directly controlled by dial pulses; the switches are operated one after another in the progressive manner. The flat construction permits the switches to be stacked one above another on frames, thereby facilitating multiple wiring of the banks.

Systems using rotary switches. The EMD system uses a 110-point single-motion rotary switch or uniselector. Arcs of terminals are rapidly passed in response to dial pulses, and individual terminal links are tested between digits. The number of links per group may be varied. This system is widely used in West Germany, Asia, Africa, and the Middle East.

In addition to the EMD system, the following two systems are the principal ones still used extensively in Sweden, other parts of Europe, and South and Central America. The switches are larger than 100 points, and their wipers are moved by clutching a constantly turning motor drive shaft. They employ indirect control.

Rotary system. The switches are of the unidirectional rotary type. Line finders may be of either 100- or 200-point capacity. Group selectors are usually 300-point, while final selectors are 200-point, although in one of the systems 100-point rotary switches are used throughout. The switch carries 10 sets of brushes (wipers), only one of which is tripped as part of the control and selection process.

AGF system. This power-driven switch is of the flat type which is driven first in a rotary direction to one of 25 positions and then in a radial direction to one of 20 sets of terminals. Thus it has access to 500 sets of terminals.

Systems using coordinate switches. In systems with a common or central system control, the connective elements or crosspoints are usually assembled in matrices. These matrices can be made up of individual metallic or nonmetallic crosspoints in a coordinate array such as sealed contact relays or $PnPn$ diodes (where P stands for heavy positive doping and n stands for negative doping) of an integrated circuit, or with common row and column operating elements to select a crosspoint as in the crossbar switch. About 35% of the nonelectronic systems of the world, as well as many electronic systems, use some form of co-

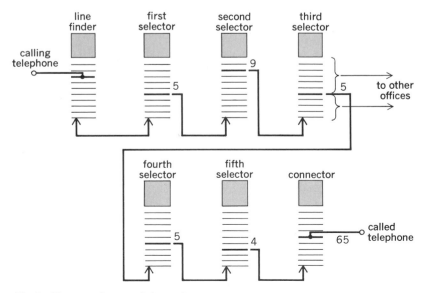

Fig. 5. Diagram of seven-digit step-by-step system.

ordinate switch. More than one crosspoint of a coordinate array may be operated simultaneously, so that, unlike the other electromechanical switches, they are used by several calls at the same time. This is also made possible by use of the common control.

Crossbar system. A crossbar switch (Fig. 6) consists of 100 or 200 contact sets, or crosspoints. Each crosspoint may have from three to six pairs of contacts. The individual crosspoints are operated by interposing a flexible select finger, moved by the rotation of a horizontal bar, between the contact set and the armature of a vertical-hold magnet. The horizontal bar with a butterfly-shaped armature is located between two contact sets and is rotated through a small arc either upward or downward by energizing either of two electromagnets. Once the horizontal bar and fingers move into position, the vertical-hold magnet is energized to close the crosspoint contacts. The flexible finger is held interposed between the operated vertical armature and contacts when the horizontal magnet is deenergized. Contact sets associated with other vertical magnets may then be actuated. The matrices are formed by wiring or multiplying the contact pairs horizontally and vertically.

The size of the matrices is generally limited due to the relatively high cost of the crosspoints. Typically sizes are 8×8, 10×10, 20×10, 16×16, and 14×25. To obtain greater access the matrices are linked in stages; for two-stage systems this is known as a primary-secondary linking arrangement (Fig. 7). Each input to the first stage of 20×10 switches can then be connected to each of the 200 outputs of the second stage of 20×10 switches. While access to a particular output may be blocked by a link being busy on another call, networks of successive frames of two-stage arrays may provide very low probability of blocking for connections between any two terminals.

While there are many crossbar switching systems in use, only the two most prominent in the United States, the No. 5 and the No. 4A crossbar systems, are described below. Variants of the No. 5 crossbar system are manufactured and used in Canada, Japan, Greece, and elsewhere.

Connections in a crossbar system are controlled by a relay marker. The time required to set up a connection is short, and consequently a small number of markers and other associated common control equipment is sufficient to handle most offered calls.

Most crossbar systems differ from progressive electromechanical switching systems. When a call is originated in a crossbar office, the location of the calling line in the switches is noted. The location of the called line or outgoing trunk to another office is marked. The marker then selects an idle talking channel through the crossbar switches to interconnect the marked points and causes all contacts in this channel to be closed simultaneously. The channel is held busy for the duration of the call.

Also, unlike other electromechanical systems, the marker of crossbar systems is arranged to look at alternate routes to the called office in case all trunks of the first-choice route are busy. The crossbar circuits are designed so that the marker

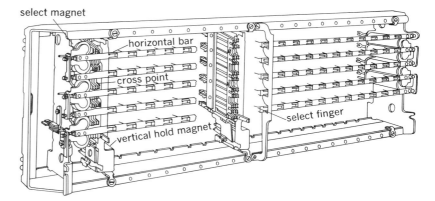

Fig. 6. Crossbar switch.

can detect certain trouble conditions in the office and make a second trial using other circuits to complete the calls. A record of the trouble and its probable location is made for subsequent analysis by maintenance people.

No. 5 crossbar switching system. This system has been the most popular form of crossbar equipment for local central office use. Although local switching is its primary use, it has been adapted for toll, tandem, private line, private branch exchange (PBX), Telex, and video switching.

The switching network of this system comprises two primary-secondary arrangements: first, the line link (LL) frames on which the telephone lines appear and, second, the trunk link (TL) frames on which the trunks appear. A switching entity may grow to a maximum of 60 LL and 30 TL frames. Each LL frame is interconnected with every TL frame by a network of links called junctors. Each LL frame has a basic capacity for 290 telephone lines and may be supplemented in 50-line increments to a maximum of 590 lines. The size used in a particular office depends upon the calling rate and holding time of the assigned lines.

Figure 8 is a block diagram of a No. 5 crossbar office. When a call is originated, the dial-tone marker causes the calling telephone to be connected through the LL and TL frames to an idle originating register. The register then places dial tone on the line as an indication for the customer to begin dialing. When the complete called number is dialed, a completing marker is chosen to estab-

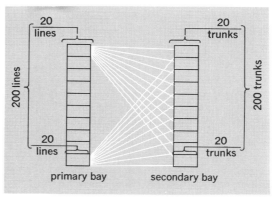

Fig. 7. Primary-secondary link arrangement.

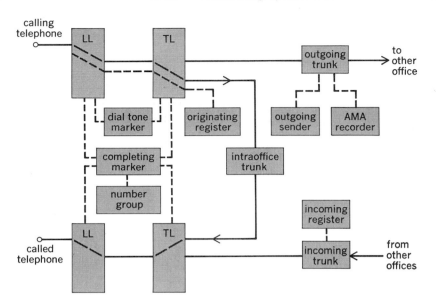

calling
telephone

called
telephone

to
other
office

from
other
offices

Fig. 8. Block diagram of the No. 5 crossbar system.

lish a connection. The completing marker examines the first three or six digits (area plus central office codes) to determine if the call is to be completed within the office, to another local office, or to a toll office. If the call is destined for the same office, the called number, together with the LL frame location of the calling telephone, is transferred into the completing marker. The completing marker consults a number group to find the LL frame location of the called number. An idle intraoffice trunk and channels through the TL and LL frames are chosen to interconnect these two locations. Crosspoints are closed and the called telephone is rung. The connection to the originating register is released in the process.

If the call is to a telephone in another switching entity (local or toll), the completing marker selects an outgoing trunk to the called office. The called number is transferred through the completing marker into the outgoing sender. The outgoing sender pulses forward the called number to the terminating point and releases.

The No. 5 crossbar system is able to interconnect with all types of switching systems. It is arranged to send out and receive different types of signals—multifrequency pulses between No. 5 crossbar offices, dial pulses to and from offices using step-by-step equipment, and other forms of pulsing to operate with electromechanical switches that operate only on an indirect basis.

A call originating in some other office for a number in the crossbar office reaches the office over an incoming trunk to which an incoming register is temporarily connected. The called number is pulsed from the originating office into the register. The register associates itself with a completing marker which, with the help of the number group, selects and closes the channels through the TL and LL frames to the called telephone. By using the number group, the line terminations need not appear on the LL in numerical order.

The system uses Automatic Message Accounting (AMA) to make call records for billing purposes. Originally AMA consisted of perforating

the data on a 3-in.-wide (7.62-cm) paper tape. More recent developments use minicomputers located in the office or reached over data links to record the AMA data on magnetic tape. On calls to the local area where a message-unit basis of charging applies, frequently only the calling telephone and the number of message units are recorded. For toll calls, the calling and called numbers, answering and disconnect times (from which the length of conversation is computed), and other data are recorded. Automatic data processors at accounting centers convert the recorded information into a form for printing the bill statement.

Centralized automatic message accounting (CAMA) has been applied to the offices used for tandem or toll switching functions. The CAMA equipment may be similar to the local office AMA equipment. The calling station is identified automatically in the local office, and this number is pulsed over the trunk to the tandem office. Where local offices are not so arranged, an operator is momentarily connected to ask the calling party for his or her number and to key it into the register.

The No. 5 crossbar system has been extended to provide switching services such as TOUCH-TONE, Centrex, private line, and call distribution to large groups of operators.

No. 4A toll crossbar switching system. This system is commonly used in the completion of toll calls between distant cities. Figure 9 shows a typical arrangement of switching systems with a No. 4A toll crossbar office in each of two cities. It also shows at the originating point two methods of placing calls from the calling telephone: that in which the call is routed from a local office directly to the nearest toll office, and that in which a call is routed from the local office through a traffic service position system (TSPS) when a customer dials "0" before the called number or makes a long-distance call from a public telephone.

The 4A crossbar system also has two main primary-secondary switching frames. Incoming trunks from originating points appear on the incoming link frame, and the outgoing trunks leave from the outgoing link frames. Using an initial set of 10×20 crossbar switches, a particular entity may have a maximum of 40 frames of each type, serving as many as 8000 incoming trunk terminations and 12,000 outgoing trunk terminations. The number of terminations may be increased. There is thus a full flexibility for any incoming trunk to be connected to any outgoing trunk.

A call arriving at the No. 4A crossbar office in city A appears at the incoming trunk and is connected to an incoming sender, into which the called telephone number (seven or ten digits) is pulsed. The electronic stored program control (SPC) processor determines the routing of the call from this number and connects to an idle marker, passing the routing information to it. The marker closes through the channel to interconnect the incoming trunk with the outgoing trunk, which in Fig. 9 is another 4A crossbar office in city B.

The signaling for the call may be sent to city B either over the trunk by pulsing seven digits from the sender A to the sender B or, since each office includes an SPC processor, it may be connected to the CCIS signaling network. In the latter case the identity of the selected outgoing trunk, the called number, and perhaps other information

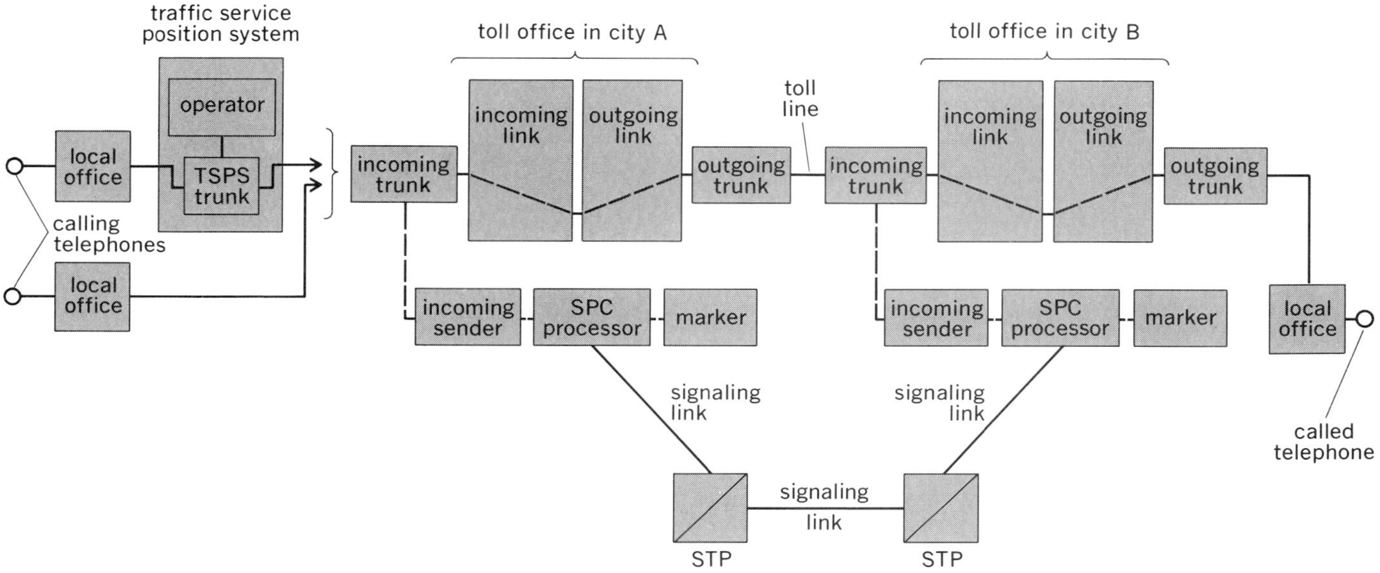

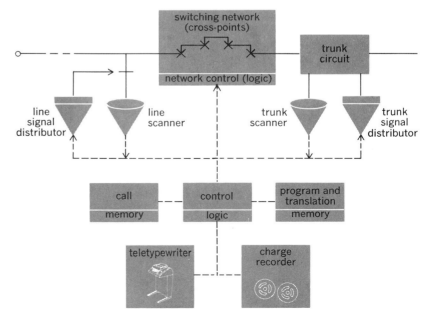

Fig. 9. Diagram of trunking in the No. 4A toll crossbar office.

about the call will be sent at high speed (4800 bits per second) over a link to a regional signal transfer point serving city A. The STP will transfer the information to a signal transfer point serving city B. This STP will then contact the 4A/SPC office at which the trunk terminates. The SPC in the office at B determines the routing and completes the connection to the local office, which in turn sets up the connection to the called telephone. The CCIS process includes a check that the transmission path selected for the call is suitable.

An important feature of crossbar systems is their ability to pick an alternate route if the first-choice route between cities A and B has no idle toll lines. For example, a route from city A to city C to city B may be selected if no direct trunk to city B is available. Several such alternate routes may be looked at, and a call may be routed through several switching points before reaching the terminating toll center. The use of this plan ensures good service and economical trunking since, if one route is either busy momentarily or out of service due to an equipment cable failure, the network is engineered to ensure that there is a good possibility that a toll line in some other route is available.

The toll switching system in cities B and C need not be of the No. 4A crossbar type. The No. 4A equipment is capable of sending pulses of the required type to operate the switches in the various switching centers. This equipment can also delete digits or add digits to the called telephone number as required to operate the various switches.

As contrasted with most electromechanical switching systems, all No. 4A and some installations of the No. 5 crossbar systems are arranged for four-wire switching. In these systems, to facilitate connection with carrier transmission systems and to ensure good transmission quality, the voice is carried over one pair of wires in one direction, and over another pair of wires in the opposite direction.

ELECTRONIC SWITCHING

The invention of the transistor spurred the introduction of electronics and semiconductor technology into switching system design. Many experiments were conducted in the decade 1954–1964. The first commercially produced SPC electronic switching system was placed in service in 1963 in the United States. Since that time over 2500 local and 250 toll stored-program-control switching systems have been placed in service in the United States, serving more than 40,000,000 lines (as of 1980). Stored-program control has become the principal type of control for all types of new switching systems throughout the world, including toll, private branch, data, and Telex systems. About

Fig. 10. Block diagram of space-division SPC system.

1200 small systems were earlier placed in service in Canada, Great Britain, and France using action translator (AT) electronic logic and memory controls.

No. 1A electronic switching system. The high speed of electronics enables the systems to be designed so that all calls in progress are processed and supervised by the same control equipment. Figure 10 is a block diagram of the No. 1A electronic switching system (No. 1A ESS) widely used in the United States for large local and small toll offices. This system has a capacity of 240,000 local calls per hour and may serve as many as 125,000 lines. The broken lines represent high-speed buses that connect the stored-program-control processor to the peripheral circuits that access lines, trunks, and the switching network control.

The SPC comprises a general-purpose assemblage of semiconductor circuits that are structured to interpret the instructions used in programs for the processing of the calls and for the maintenance of the system. These instructions are stored in a memory subsystem as coded programs that are read in sequences that determine actions to take.

Portions of the call-processing functions are used repetitively for each call. The programs are relatively fixed and remain in the systems memory while the call information is stored for relatively short periods. The ease of changing the program and other stored information provides a new degree of flexibility in the telecommunication services and administrative features the system may be designed to offer.

The No. 1A ESS is provided with integrated circuit (IC) memory for call data, translation, and program storage. Integrated circuit chips of 64,000 bits are used in a random access storage. A typical program may require 500,000 words of 32 bits each. All control circuits are duplicated to ensure service continuity should a component fail or to enable the system to grow while in service. Memory redundancy is also provided. Copies of the call processing as well as additional, less frequently used maintenance, administrative, and operations (MAO) programs are stored on magnetic disks. The periphery includes data links to centralized MAO facilities and for CCIS, as well as local maintenance consoles and tape drive units.

Space-division switching networks. Initially, most SPC systems employed space-division networks. As indicated earlier, many unique matrix arrays of devices, both metallic and nonmetallic, have been developed and used in commercial electronic switching systems. These arrays are used in the same topology of primary-secondary link arrangements and line and trunk link frames as in crossbar systems.

For some electronically controlled switching systems, metallic contact are used in the space-division network. Many of these contacts are magnetic reeds sealed in glass with an inert gas. They latch magnetically when activated by a short pulse of one polarity and release with a pulse of opposite polarity. When the contacts are made of hard magnetic material, such as those employed in the No. 1A ESS, they are known as remreeds. Other systems use miniature crossbar or similar switches (Fig. 11). In all of these space-division networks, the switching device is slow in comparison to the speed of the electronic controls. Buffering, shown as the network control, is generally required between the SPC or AT and the network.

Electronics in the form of semiconductor or integrated circuit devices have been increasingly applied to the switching network function of switching systems. Semiconductor crosspoints or gates have been developed for the two types of electronic switching networks. Generally their use has been confined to smaller systems or PBXs (less than 2000 terminals) or in combination with metallic crosspoints for larger systems.

Semiconductor crosspoints may be used in network configurations in the same manner as reed switches. They are designed to be bistable and held actuated over the established speech path. Therefore, these devices combine both transmission and control characteristics. Each crosspoint may use a pair of bistable devices as the two conductors of the connection, or a single device with one wire and a ground return. Some modern semiconductor devices can tolerate the same high voltages as metallic crosspoints.

Time-division switching networks. Time-division switching is practical only with high-speed electronic techniques. It is used in switching networks as well as in the control portions of systems. For time division, transmitted signals are sampled at a rate at least twice the highest frequency to be transmitted. Typically for voice this is 8000 times per second. Within a switching system the samples may be pulses of varying amplitude that are analogus to the electrical signals representing the voice. This is known as pulse amplitude modulation (PAM).

More robust forms of pulse modulation use dig-

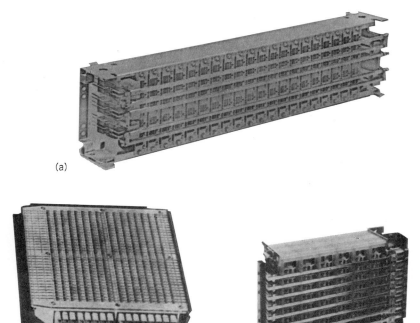

(a)

(b)

(c)

Fig. 11.   Small coordinate switches. (a) Minibar, introduced in Canada in 1968. (b) Metabar, introduced in France in 1968. (c) Minicrossbar, introduced in Japan in 1969.

ital or on-off signals that can be readily encoded from the amplitude pulse, sent over a transmission medium, and periodically reformed to eliminate most of the vagaries of transmission and switching. These types of digital sample transmission are known by the form of coding employed; the most popular is referred to as pulse-code modulation (PCM). In one type of PCM, each sample is coded into one of approximately 256 amplitudes and represented by eight binary (on/off) pulses. The eight pulses representing each speech channel may be placed in sequence or time-multiplexed into groups, typically of 24 or 32 channels, so that the line and repeaters may be used more efficiently. *See* PULSE MODULATION.

Time-multiplexed coded-voice signals reaching a switching system are switched by using two techniques. Assuming signals arrive at the switching network on different multiplexed lines, they first need to be synchronized with respect to multiplex channel indentification. This usually requires some form of time delay or buffering. This establishes uniform channel periods or timeslots. Within the switch there may be more time slots than in the lines delivering the signals.

One necessary switching network function uses further time buffering by placing successive digitized channel samples in a memory in one order and removing them as indicated by the switching selection requirements of each call. This function is known as time slot interchange (TSI).

The other switching network function provides for interchanging channels between time-multiplexed lines. Generally this function is carried out by using a high-speed space-division network of one or more stages. The space-division network control acts to change the actuated crosspoints between time slots so that the successive channels of input lines may be switched to corresponding channels in the same or other lines. Space-division stages operating at time slot rates are known as time-multiplexed switches (TMS).

Time-division switching networks are based upon the use of successive TSI, called T (for time), and TMS, called S (for space) stages. Typical systems are said to employ T-S-T or S-T-S type networks. Lines or trunks not reaching the switching system by time-multiplexed facilities must first be multiplexed with similar inputs as part of the time-division switching-network function.

Since the coded signals represent only one direction of transmission, the switching-network function in a circuit switch is duplicated by reciprocity to provide for the other direction of transmission. Also, since the elements of the network are active and may be used for hundreds of simultaneous calls, redundancy of the network and its controls, similar to that used in the call-processing portions of the system, is usually part of the system design.

Analog (PAM) systems. The first commercial electronic switching system (1963) used PAM time-division techniques since digital modulation was not then economical. The input signals on lines and trunks are sampled periodically. Each active input is associated with the desired output for a specific time slot. This type of time-division network (Fig. 12) utilizes semiconductor gate elements which, during each time slot, connect to-

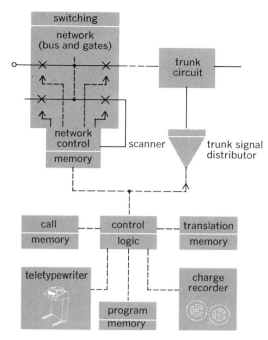

Fig. 12. Block diagram of PAM time-division SPC system.

gether on a common bus the terminals associated with a particular call. These associations for each call are read from a cyclically arranged memory which, together with the gates and bus, constitutes the network control. The number of simultaneous conversation paths which may be established through these networks is limited by the speed of the semiconductor devices and the associated circuitry that must accurately transfer the representation of the electrical amplitude of the sample with sufficient power. The sampling rate must be high enough to permit the faithful reproduction of the signal when passed through a low-pass filter. Generally, networks of the time-division type have been applied where there are less than 500 time slots.

Digital (PCM) systems. Time-division switching is a natural adjunct to digital time-division transmission where the coding is performed for purposes of multiplexing. PCM transmission was initially economical on interoffice trunk routes from 10 to 50 mi (16 to 80 km). This made time-division digital switching initially attractive where such facilities were found, namely, for tandem and toll applications.

The No. 4 ESS SPC time-division digital switching system was developed for this application and is the largest switching system developed. It has a capacity of 100,000 trunks, and its stored-program control is capable of switching 550,000 calls per hour.

The switching network for the No. 4 ESS consists of both TSI and TMS stages: the latter is reconfigured at the rate of 1,024,000 times per second (128 time slots for each frame of 8000 samples per second). Figure 13 illustrates how a sample progresses through both the memory (TSI) and four-space-division time-multiplexed switch stages for one direction of a typical time-slot connection, with the network control memory being

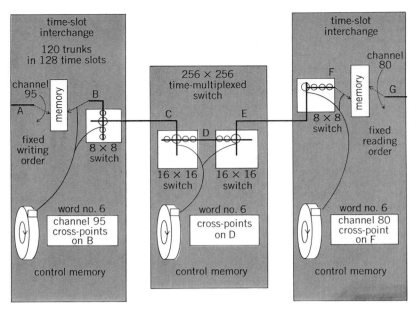

Fig. 13. Diagram of digital time-division switching network showing a typical connection. The network has 128 time slots throughout, and the sample shown is in time slot number 6 throughout the network.

read out cyclically at a rate of 8000 times per second.

As shown in the block diagram (Fig. 14), the No. 4 ESS serves three types of transmission channels: analog metallic, analog multiplex carrier, and digital multiplex carriers. The digital time-division switching network routes digital signals from incoming trunks to the desired outgoing trunks. The audio signals on the analog

channels are converted by the LT-1 connectors or D4 channel banks into pulse-code modulation digital samples. The digital interface frame (DIF) processes the digital signals into the format required for the switching network and removes the signaling information. For analog and digital channels, signaling information is detected by the equivalent of trunk circuits in the digital interface frame except where CCIS is now being used. Echo suppression on a digital basis is inserted ahead of the switching network.

All operations are directed and supervised by the SPC processor and are aided in routine tasks by the signal processors built into the DIFs. The signal processors provide the scanner and distributor functions for a portion of the trunks and, in turn, pass the significant information content in a more compact form to the central control. The use of peripheral or distributed processors enables the central processor to devote its attention to the more critical decisions in the processing of calls, thereby providing for greater call attempt capacity. System units synchronize their operations under the control of a system clock.

As the cost of digital integrated circuits is decreasing, particularly for converting an individual line or trunk from analog-to-digital (A/D) and digital-to-analog (D/A) transmission, local digital time-division switching is becoming feasible. In the meantime, some administrations, such as the French E10 (see table) started with reed or solid-state crosspoint concentrators ahead of the A/D circuit. As shown in the table, others are proceeding to develop systems with individual line circuits that convert high ringing and analog speech levels into signals more compatible with the low levels required for a local time-division switch. Canadian

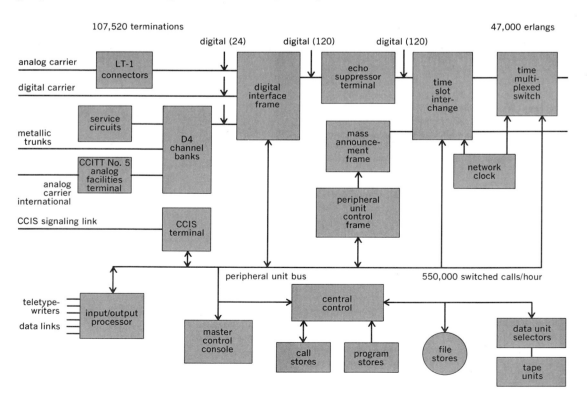

Fig. 14. Block diagram of No. 4 electronic switching system.

manufacturers and independent telephone manu-facturers in the United States led the way in developing and placing 648 of these systems in service by the end of 1980. Most are small with a capacity of less than 6000 lines. A new genera-tion of systems is being designed, and systems with more than 150,000 line capacity can be ex-pected in the early 1980s.

Service features. New service and administrative features which are made more economic by the use of stored-program techniques and bulk memories are among the most important aspects of electronic switching systems. Two types of data are stored in the memories of electronic switching systems.

One type is the data associated with the progress of the call, such as the dialed address of the called line. Another type, known as the translation data, contains infrequently changing information, such as the type of service subscribed to by the calling line and the information required for routing calls to called numbers. These translation data, like the program, are stored in a memory which is easily read but protected to avoid accidental erasure. This information may be readily changed, however, to meet service needs. The flexibility of a stored program also aids in the administration and maintenance of the service so that system faults may be located quickly.

The availability of large memories and the ease of changing the program residing in them has led to the development and deployment in most SPC systems of many new and optional services such as abbreviated dialing, call waiting, call forwarding, and three-way calling. The wide introduction of CCIS into the network broadens the range of SPC service offerings. These services will take advantage of the rapid exchange of information about calls between the originating, intermediate, and terminating offices before connections are established. As a result, calls may be directed within the entire network to suit the individual users and subscribers whose needs are likely to be nationwide rather than residing in a single central office or PBX.

One of the most impressive results of employing electronics in switching is the space savings. Even through the systems installed use mostly discrete semiconductor components, the space savings are as high as 60% when compared with electro-mechanical switching. Additional space savings can be expected as more integrated circuits are applied. *See* INTEGRATED CIRCUITS.

Types of electronic switching systems described so far provide two-way communication circuit paths (or their equivalent for voice) in real time. The systems of this type are known as circuit or line switching systems. They are used for voice and data service where instant two-way message exchange is required. Another type of system (Fig. 15), known as a message or store-and-forward switching system, is used to process and deliver messages or data, which may be stored temporarily in a memory awaiting the availability of outgoing trunk or line facilities. Depending upon the form of messages and the engineered delay time, message switching systems may provide several different capacities and access times. Sometimes short messages are stored at stations and transmitted to the

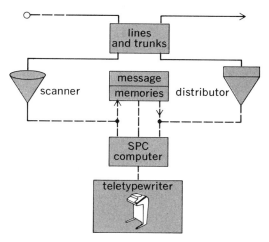

Fig. 15. Block diagram of a store-and-forward switching system.

switch in segments as determined (poled) by the switch. One such system in which the message is divided into uniform segments with an address placed on each segment is known as packet switching. The switch intermixes the packets from all sources and may route individual packets of the same message over different routes. Most message switching systems employ stored-program control supplied by commercial digital computers. *See* DATA COMMUNICATIONS.

Other electronic systems. Since electronic switching systems may use technology that is not unique to switching, the variety of systems being explored and produced have grown from the days when switching was performed only with the employment of specialized technology. The table shows some of the different combinations of technology that are being used in the major countries producing switching systems. As indicated, most systems are of the SPC type. New technology has generally appeared first in small PBX switching systems, in advance of the general trend in central office switching.

The major differences between space-division systems are the types of crosspoints used. In addition to reed switches, the next most successful devices have been smaller crossbar or crossbar-like (codebar) switches (Fig. 11). Unique among the local digital time-division multiplex (TDM) systems are those that require no space-division concentration in converting from analog to digital transmission at the line circuit.

To provide large capacity, particularly for toll system applications, additional control processors are used, usually by providing several identical processors that share the load. These are called multiprocessor systems. Also, microprocessors are being associated with the periphery of systems to distribute some of the stored-program-control functions. For smaller systems, some basic control functions are built into logic circuits rather than being stored as programs in the processor memory. These are action translator systems with call sequences determined by the use of memory on a look-up-table basis. *See* MICRO-PROCESSOR.

With local digital TDM systems, switching con-

Typical electronic circuit switching systems in service*

Country	Application	Switching network†	Control‡	Identification
United States	PBX	Analog/TDM	SPC	Dimension
	PBX	PCM/TDM	SPC	CBX
	PBX	*PnPn*-PWM	SPC	CBX 200
	PBX	Δ/TDM	AT	D 1201
	Local/toll	MH reed	SPC + SC	No. 1A ESS
	Local/toll	EH reed/PCM/TDM	SPC	DCO
	Toll	Crossbar	SPC	No. 4A/ETS
	Toll/tandem	PCM/TDM	SPC/DP	No. 4 ESS
Canada	PBX	PCM/TDM	SPC	SL 1
	Local	Crosspoint	AT	C1-EAX
	Local	Minicrossbar	SPC	SP 1
	Local/toll	PCM/TDM	SPC/DP	DMS 100/200
France	PBX	PCM/TDM	SPC	TLC 10
	PBX	*PnPn*	SPC	3750/1750
	Local	EH reed/*PnPn*/PCM/TDM	AT + SC	E 10
	Local	Minicrossbar	SPC	11 F
	Toll	PCM/TDM	SPC	MT 20
West Germany	PBX	Analog/TDM	SPC	6030
	Local	MH reed	SPC + SC	EWS-A
	Local	OC relay	AT/SPC	ESK 10,000 E
Belgium	Local	EH reed	SPC	10C
	Toll	PCM/TDM	SPC/DP	1220
Japan	Local/toll	Minicrossbar	SPC	D 10
	Local/toll	*PnPn*/PCM/TDM	SPC	NEAX 61
East Germany	Local	MH reed/PCM§	SPC	ENSAD
Netherlands	Local	EH reed	SPC + SC	PRX 205
Sweden	Local/toll	Crossbar	SPC	ARE 11/13
	Local/toll	EH reed/PCM/TDM	SPC/DP-SC	AXE 10
	Toll	Codebar	SPC/MP	AKE 13
Great Britian	Local	EH reed	AT	TXE 2/4

*At the end of 1979.
†EH = Electronically held. MH = Magnetically held. OC = Open contact. PCM = Pulse-code modulation. *PnPn* = Pairs of positive and negative semiconductor gates. PWM = Pulse-width modulation. TDM = Time-division multiplex. Δ = Delta modulation.
‡AT = Action translator system. DP = Distributed processor configuration. MP = Multi-processor configuration. SC = Separate service computer. SPC = Stored program control system.
§ = PCM digital signal through MH reed switches.

centration modules, including distributed SPC control, are designed to be located either in or remote from the central office. Using digital transmission facilities the connections to these modules appear at the host central office as trunks. If the remote module also provides for some operation independent of the central office, it is known as a remote switching module or system.

Increasingly, maintenance and administrative functions are being centralized in computers that serve several central offices. Communication with these centers is two-way, with some information for special calls being stored at the central location.

For the smaller systems, line concentrations, remote switches, and PBXs, integrated circuit crosspoints are used in space-division arrays and for time-division bus connections. The time-division systems have increasingly employed digital encoded samples. Digital samples may also pass through space-division networks. *See* TELEPHONE.

[AMOS E. JOEL, JR.]

Bibliography: *Bell. Syst. Tech. J.*, special issue on No. 1 ESS, September 1964, special issue on TSPS, December 1970, special issue on No. 1A processor, February 1977, special issue on No. 4 ESS, September 1977, special issue on CCIS, February 1978; *Computer*, special issue on circuit switching, June 1979; B. T. Fought and C. J. Fink, Electronic translator system for toll switching, *IEEE Transactions on Communication Technology*, pp. 168–175, June 1970; *IEEE Transactions on Communications*, special issue on telecommunications, September 1977; A. E. Joel, Jr. (ed.), *Electronic Switching Systems of the World, Proceedings of the IEEE*, special issue on telecommunications, September 1977; *Record of Colloque International de Communication, Paris*, 1979; *Record of International Switching Symposium, Kyoto*, 1976.

Switching theory

The theory of circuits made up of ideal digital devices. Included are the theory of circuits and networks for telephone switching, digital computing, digital control, and data processing.

Switching theory generally is concerned with circuits made of devices or elements that can be in two or more discrete conditions or states. Examples of such devices are switches or relay contacts, which can be opened or closed, rectifying diodes, which can be either forward- or back-biased, switching tubes or transistors, which can be saturated or cut off, and magnetic cores, which can be magnetized to saturation in either of two directions. Switching theory establishes an ideal representation of the digital circuit, examines the properties of the representation, then interprets these as properties of the circuit. Switching theory is not concerned with the physical phenomena of action or stability in a particular condition or with the details of transition from one state to another. It takes these as established and proceeds to examine more or less complex combinations of digital devices whose properties are assumed to be ideal.

The bulk of switching theory is concerned with circuits made of binary (two-valued) devices, since these are most common. Switching theory can be based in part on mathematical logic.

A switching circuit whose outputs are determined only by the concurrent inputs is called a combinational circuit (or logic circuit). A circuit in which outputs at one time may be affected by inputs at a previous time is called a sequential circuit.

Combinational circuits. A rule by which the outputs of a combinational circuit can be determined from its inputs is called a switching function. Since the variables are discrete, a switching function may be expressed in tabular form as a truth table, or may be indicated by a diagram or geometric pattern. If the function and variables are binary, the symbols 1 and 0 are commonly used to represent the two values. The function may then be represented by a Boolean algebraic expression. The two values of a switching function can represent closed and open circuits, as for switches or relay contacts, or high and low or plus and minus voltages, as in electronic circuits.

The simplest combinational switching functions are the NOT function, the AND function, and the OR function. The NOT function is designated by the prime in Boolean algebra; $Y = X'$ means that Y is closed (high, plus) when X is open (low, minus), and vice versa. The AND function is designated by the Boolean product: $Z = X \cdot Y$ means that Z is closed (high, plus) only if both X and Y are closed. The OR function is designated by the Boolean sum: $Z = X + Y$ means that Z is closed if either X or Y or both are closed. All other combinational switching functions can be made by combining these elementary building blocks.

For example, Fig. 1 shows a switching circuit with three switches, or contacts, X, Y, and Z', each of which can be either open or closed. These can be thought of as input variables. The circuit as a whole will be open or closed depending upon the individual positions of X, Y, and Z'. Its condition can be designated by W, an output variable. Let 0 represent the open condition, and 1 the closed condition. The table in Fig. 1 represents the switching function of the circuit. The Boolean expression for this function is $W = Z'(X + Y)$. To interpret this expression the rules of simple Boolean algebra must be used:

$$
\begin{array}{lll}
0 + 0 = 0 & 0 \cdot 0 = 0 & 0' = 1 \\
0 + 1 = 1 & 0 \cdot 1 = 0 & 1' = 0 \\
1 + 0 = 1 & 1 \cdot 0 = 0 & \\
1 + 1 = 1 & 1 \cdot 1 = 1 &
\end{array}
$$

Switching theory establishes a number of methods for analysis and synthesis of combinational circuits. A significant problem is minimization, that is, given a switching function, to synthesize the simplest circuit which will realize it. A problem of some theoretical difficulty is that of realizability, that is, given a statement of specifications, to de-

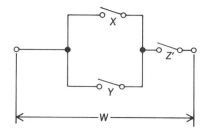

X	Y	Z'	W
0	0	1	0
0	0	0	0
0	1	1	1
0	1	0	0
1	0	1	1
1	0	0	0
1	1	1	1
1	1	0	0

Fig. 1. Combinational circuit. $W = Z'(X + Y)$. X and Y are normally open contacts. Z' is a normally closed contact.

termine whether a switching circuit exists which satisfies them.

Analysis of a series-parallel combination of switches or relay contacts can be carried out by a direct application of Boolean algebra. Variables or terms corresponding to contacts, or combinations in parallel, are added, and those in series are multiplied. The values are interpreted according to the rules of Boolean algebra. Similar methods can be applied to combinational circuits which employ diode rectifiers, vacuum tubes, or transistors. Circuits that are not series-parallel can be dealt with by an extension of the Boolean method, by the use of matrices with discrete-valued elements, or by a number of special methods.

A switching function can be simply synthesized as a series-parallel combination of contacts by giving Boolean symbols circuit interpretations explained previously. Electronic logic circuits can be synthesized in a similar fashion. This approach will lead to a method for embodying any switching function expressed in Boolean terms. The Boolean expression of a function given in tabular or diagrammatic form is easily obtained.

Synthesizing the minimal circuit, or minimization, is more difficult since for every switching function there are many possible circuits. Where the number of variables is small, the minimization problem can often be reduced to one that has already been solved. Tables of minimal or nearly minimal solutions for relay circuits and vacuum-tube circuits are available for circuits with one output and as many as four inputs. Harvard chart methods and Karnaugh map methods utilize geometrical relationships to explore systematically functions with one output and as many as six inputs.

As the number of variables increases, the possible number of functions rapidly becomes large. For example, there are more than 10^{19} different functions of six binary variables. No completely general and practical design methods have been discovered. However, a growing array of special methods for synthesis and minimization is available.

Sequential circuits. Since the outputs of sequential circuits depend on past, as well as present, inputs, they must contain means for remembering or storing the effect of past inputs, such as locking relays, flip-flops, delay lines, or magnetic cores. A device with two stable states can remember one binary digit, or bit. The amount of memory in a circuit can be measured either in bits or in internal states. An internal state of a circuit is a particular configuration of its internal memory devices. The number of internal states is equal to 2^n, where n represents its number of bits. Binary counters and shift registers are examples of sequential circuits.

It is possible to represent a sequential circuit as a combinational circuit with feedback. Thus, the combinational circuit of Fig. 2 becomes a sequential circuit with two bits of memory if two of its outputs are connected to two of its inputs. Any such closed loop must contain gain and some delay; sometimes additional delay is inserted.

If the combinational circuit and the delays in Fig. 2 are completely specified, the internal description of the circuit is known and its behavior can be analyzed. If the switching function of the combinational circuit is such that $m_1 = M_1$ and $m_2 = M_2$ for a given set of inputs, no change can occur as a result of the action of the memory loops and the circuit is stable; otherwise, it is unstable. If it is unstable, the inputs must cause a transition to a new state, which in turn may be stable or unstable. If no stable state is reached, the circuit is said to buzz. If the state to which a circuit may pass depends on which of two or more memory loops acts first, the circuit is said to have a race condition, and its performance may be ambiguous. This difficulty does not occur in circuits in which changes are caused or timed by repetitive clock pulses. Such circuits are called synchronous. Cir-

cuits which make transitions at the natural internal rate are known as asynchronous, and these asynchronous circuits must be designed with greater care.

To proceed from external circuit requirements to an internal description of a sequential circuit requires art and skill, as well as knowledge of switching theory. *See* DATA-PROCESSING SYSTEMS; DIGITAL COMPUTER; LOGIC CIRCUITS; SWITCHING CIRCUIT; SWITCHING SYSTEMS (COMMUNICATIONS).

[WILLARD D. LEWIS]

Bibliography: N. U. Biswas, *Introduction to Logic and Switching Theory*, 1975; F. J. Hill and G. R. Peterson, *Introduction to Switching Theory and Logical Design*, 1981; Z. Kohavi, *Switching and Finite Automata Theory*, 2d ed., 1974; S. C. Lee, *Modern Switching Theory and Digital Design*, 1978; R. Miller, *Switching Theory*, 1965, reprint 1979.

Synchronization

The process of maintaining one operation in step with another. The commonest example is the electric clock, whose motor rotates at some integral multiple or submultiple of the speed of the alternator in the power station. In television, synchronization is essential in order that the electron beams of receiver picture tubes are at exactly the same spot on the screen at each instant as is the beam in the television camera tube at the transmitter. Synchronism in television is achieved by transmitting a synchronizing pulse at the end of each scanning line, to make all receivers move simultaneously to the start of the next line. A similar vertical synchronizing pulse is transmitted when the camera beam reaches the bottom of the picture, to make all beams go back to the top for the start of the next field. *See* OSCILLOSCOPE; TELEVISION.

[JOHN MARKUS]

Synchroscope

An instrument used for indicating whether two alternating-current (ac) generators or other ac voltage sources are synchronized in time phase with each other. In one type, for example, the position of a continuously rotatable pointer indicates the instantaneous phase difference between the two sources at each instant; the speed of rotation of the pointer corresponds to the frequency difference between the sources, while the direction of rotation indicates which source is higher in frequency. In more modern synchroscopes, a cathode-ray tube serves as the indicating means.

The term synchroscope is also applied to a special type of cathode-ray oscilloscope designed for observing extremely short pulses, using fast sweeps synchronized with the signal to be observed. *See* OSCILLOSCOPE.

[JOHN MARKUS]

Tank circuit

An inductor and capacitor in parallel. The term is quite often used to denote the parallel resonant circuit in the output stage of a radio transmitter, but it has been applied to any parallel resonant circuit. In many cases the inductance in the tank circuit is one winding of a two-winding, air-core transformer. The secondary is connected to some

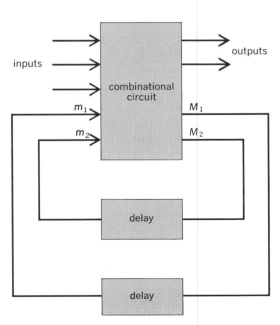

Fig. 2.   Sequential circuit with two memory loops.

load, such as an antenna. Power is delivered from the source to the load through the tank circuit, with an effort usually made to adjust the parameters for maximum power transfer. *See* IMPEDANCE MATCHING; RESONANCE (ALTERNATING-CURRENT CIRCUITS); TRANSFORMER.

Since the tank circuit is a parallel resonant circuit, the parameters can be chosen so that at a desired frequency the voltage across the tank circuit will be a maximum. In radio transmitters this is done by varying the capacitance, but in other situations, such as oscillators, the inductance may be varied by means of a tuning slug in the coil. *See* OSCILLATOR.

Tank circuits have an important role as a plate load in class C amplifiers, in which the plate current flows for only a small fraction of a cycle. If the damping in the circuit is small and the circuit is excited at its resonant frequency, the plate-current impulses produce a sustained sinusoidal voltage across the tank. If a load is transformer-coupled to the tank, the voltage across the load is sinusoidal. *See* AMPLIFIER.   [HAROLD F. KLOCK]

Telephone

The term telephone formerly referred to the telephone receiver, the instrument originally invented by Alexander Graham Bell; it is now commonly applied to the telephone set, which includes a transmitter and an electric network in addition to the receiver.

The transmitter and receiver are housed together in a handset. A cord connects the electrical components of the handset to the network in the telephone set.

Transmitter. The transmitter is a transducer which converts acoustical energy into electric energy. In most transmitters an electric current is modulated by the variations in contact resistances of carbon granules. Sound waves impinge on the diaphragm of the transmitter, causing the carbon granules to move closer together, making more contacts and decreasing resistance, or to move further apart, making fewer contacts and higher resistance.

The transmitter has a frequency-response range from 250 to 5000 hertz (Hz). The frequency response rises uniformly to a broad maximum in the region of 2500 Hz. Because of these characteristics and those of the telephone receiver, the speech heard by the listener resembles closely that of direct mouth-to-ear speech as heard by a listener a few feet from the person speaking.

The transmitter is of the direct-action granular-carbon type shown in Fig. 1. It consists essentially of a diaphragm, back cavity, and carbon chamber. The diaphragm is rigidly clamped at its periphery to obtain a high output in the upper frequency range. This achieves a quality of transmission that approximates the orthotelephonic, mouth-to-ear transmission through the air objective. The sound pressure on the diaphragm varies the pressure of the dome-shaped electrode on the carbon. Changes in pressure on the carbon granules vary the resistance of the granules, causing the current to change in magnitude in proportion to the sound. The carbon granules are specially treated to reduce the effect of aging on their resistance. The carbon chamber is designed to keep the mechani-

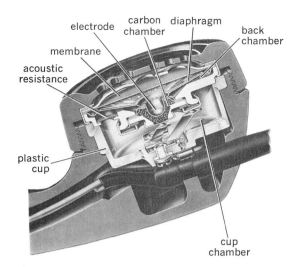

Fig. 1. Direct-action granular-carbon transmitter.

cal impedance of the carbon to a minimum, regardless of the position of the transmitter unit, to gain a high modulating efficiency.

The desired frequency response is obtained by coupling the diaphragm to a doubly resonant system, composed of a cavity within the unit behind the diaphragm and a chamber between the unit and a plastic cup. The two cavities are connected by holes covered with a woven fabric. The size of the hole and resistance of the material to the flow of air are carefully controlled to balance the acoustical impedance and therefore prevent large irregularities in the transmitter response.

The transmitter assembly is located in a specially designed handset also housing the receiver.

Receiver. The receiver transducer operates on the relatively low power used in the telephone circuit to convert electric energy into acoustical en-

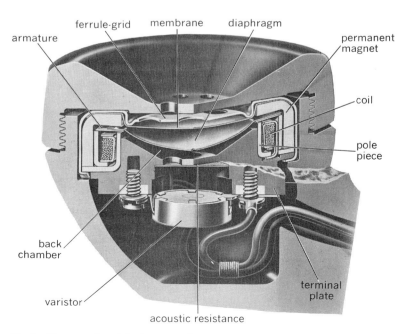

Fig. 2. Ring armature telephone receiver.

ergy. Unlike a loudspeaker, the telephone receiver is designed for close coupling to the ear. The telephone receiver has an approximate impedance of 150 ohms at a frequency of 1000 Hz. The relationship of the acoustic and electrical elements produces a desired response-frequency characteristic.

There are two types of receiver units, the bipolar receiver and the ring armature receiver, which is shown in Fig. 2. The bipolar receiver is used with operator's head telephone sets. The ring armature receiver is used with the telephone set in which high efficiency is particularly important.

The advantages of the ring armature receiver are its low acoustic impedance and high available power response over a wide frequency range (350–3500 Hz). These advantages are achieved by the piston action of a thin, nonmagnetic, lightweight, dome-shaped diaphragm, which is attached to a ring-shaped armature of magnetic material driven at its periphery by a ring-shaped magnetic coil associated with a ring-shaped permanent magnet. The diaphragm contains a small hole that introduces a low-frequency cutoff. This is desirable to reduce interference picked up from electric power circuits.

The diaphragm, magnets, and coil are encased in a ferrule grid attached to a molded terminal plate. A membrane between the ferrule grid and diaphragm protects the diaphragm from dirt and mechanical damage. Because of its mechanical impedance, the membrane acts as one of the controls over the diaphragm to help achieve the desired frequency response.

The acoustic chamber between the membrane and diaphragm is connected to a chamber molded in the terminal plate behind the diaphragm, called the back chamber. These chambers are connected by passageways having acoustic mass and resistance. The back chamber exhausts into the handset through acoustic fabric. All these controls are designed to extend the frequency range of the receiver and reduce undesirable diaphragm resonance. A click-reducing varistor is mounted on the back of the receiver.

Electric network. The electric network serves three basic functions: (1) to couple the receiver to the transmitter circuit, (2) to balance the impedances within the telephone set to those of the circuit and reduce sidetone, and (3) to provide for transmission equalization, that is, the same general loudness and quality of speech over any circuit between the telephone set and the central office.

Sidetone balance. Sidetone is the sound of the speaker's voice reaching his own ear through the electrical path to his own receiver. Sidetone tends to make the talker unconsciously reduce the level of his voice and is therefore objectionable. To keep sidetone at a minimum, any voltage developed in the local transmitter is divided in windings A and B (Fig. 3) so that the voltages induced in winding C are opposing. Also, the voltage across the network resistance arising from the current flowing in winding B opposes the resultant voltages induced in C. The overall effect is that the current in the receiver, as a result of voltages developed in the transmitter, is small. However, the key to good sidetone balance is to balance effectively the impedances

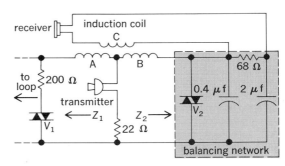

Fig. 3. Transmission circuit of a telephone set illustrating method of achieving sidetone balance.

Z_1 and Z_2 both in magnitude and phase. Z_1 varies because it is influenced by the telephone circuit to the central office. The essential elements in the impedance matching are the two silicon carbide varistors V_1 and V_2. The dc and ac resistances of these varistors vary with the voltages applied to them, which are in turn dependent on the direct current in the loop and in the telephone set. This current is a function of the length and impedance of the circuit.

Transmission equalization. The transmission equalization provided by the circuit assures that the overall transmission performance of the set is kept within reasonable limits as the loop length varies. Equalization is provided by reducing the transmitting and receiving gain on short loops. This is accomplished by a varistor V_1 across the line, which limits the current in the transmitter on short loops, and by a varistor V_2, which shunts the receiver on short loops. As the loop resistance increases and the voltages across V_1 and V_2 decrease, the shunting effect of the varistor decreases. Thus the transmitting and receiving levels are reasonably constant for loops of all lengths.

Accessory devices. In addition to the basic components, the telephone set has a switch hook to connect it to the line and operate other associated features, a dial which produces dc pulses to actuate the switching gear in the central office, and a ringer with an adjustable volume control, used to signal that a call is incoming. These are all enclosed in a common housing.

Fig. 4. Service technician making adjustment on Picturephone installation. (*Bell Telephone Laboratories*)

A variety of optional features are provided for telephone sets. Transistor amplifiers are available to amplify speech at the transmitter to aid weak voices, to amplify received speech, to overcome ambient room noise or hearing difficulties, and to provide adequate transmission at reduced currents on extremely long loops. Telephone sets also are available with dial lights. Other available sets provide up to 30 keys for holding or switching calls, and some permit local communication on the customer's premises. A limited number of Picturephone installations have come into commercial use. These installations permit the transmission of the user's image as well as his voice (Fig. 4). *See* VIDEO TELEPHONE.

[ROWLAND F. DAVIS]

Telephotography

The transmission of photographs over electrical communication channels. The channels are generally those provided by common carrier communication companies. Individual users furnish telephotograph or facsimile machines to suit their requirements. In the United States such services range from simple arrangements connecting two sets of machines within a city to large networks involving many cities and types of communication facilities. Coordination must thus be effected between the communication companies and the manufacturers of equipment used by customers for their stations to ensure compatibility of equipment and optimum system performance. Unattended reception is made possible by starting and synchronizing receiving machines with pulses or tones generated at the transmitting location.

The transmission of news photographs and weather maps is one example of a use with contrasting requirements involving different types of machines and networks. Normally, news pictures require half-tone reproductions with a number of shades of gray. At the receiving locations photographic reproducing devices are used. In contrast, weather maps and other black-and-white nonshaded material can readily be reproduced by direct recording processes. Networks used for transmitting news pictures are designed for transmitting and receiving at most points because of the unpredictable nature of the location of news events, while the weather network has only a few transmitting points at key locations with many receiving-only locations.

For coordination of operations and for making special announcements, both types of networks must handle speech. Therefore, loudspeakers and telephones are provided at all locations.

Most of the equipment in use today operates at speeds which will transmit an $8\frac{1}{2} \times 11$ in. (21.6 × 27.9 cm) page in about 6 min. There has been considerable interest in much higher speeds of transmission, which would require wider bandwidths, and in equipment that will produce sharper definition so that the resulting photographic film may be used for engraving plates suitable for printing of high quality. A number of trials have been run over special intercity facilities in which definitions up to 1000 lines/in. at high speeds have been tested. This service may require bandwidths ranging from 25 kHz to 1 MHz. The usual transmission equipment employs a double sideband or vestigial single-sideband amplitude-modulated carrier for transmission. Representative carrier frequencies are 1920, 2000, and 2400 Hz; thus, voice bandwidth circuits can be used.

Photographs, with various shades of gray, require the transmission of considerably more information than simple black-and-white material. Consequently, circuits used to transmit photographic information must be specially engineered and maintained. For example, to obtain sharp edges on picture material, it has been found important to reduce transmission distortion to the extent that the elemental areas of the scanning process are not shortened or elongated by more than one-half of their size. This means that the high-grade intercity networks must be equalized from about 1000 to about 2600 Hz to within about 300 microseconds of envelope delay distortion. Amplitude-frequency equalization is also important and must be maintained within reasonable limits over the transmitted band.

Special treatment is given facilities used for the permanent networks. Abrupt changes in level, which could cause sharp changes of shading in pictures, are minimized. Reflection, or echo currents, which could cause multiple images, is reduced. Random and impulse noise, which could cause interfering patterns, are controlled. The linearity of amplifying equipment is improved to maintain a wide gray scale.

The transmission of nonshaded black-and-white material requires the same general techniques, but broader limits are permissible with respect to level variations and certain types of noise. This is particularly true when direct-recording processes are used or where high quality is not a paramount consideration. *See* FACSIMILE. [CHARLES C. DUNCAN]

Teletypewriter

An electromechanical device, also called a teleprinter, for transmitting and receiving messages over a telegraph circuit. A sending and receiving teletypewriter performs two functions: The keyboard transmitter generates coded electrical signals for transmission over a telegraph circuit; and the typing unit converts such signals into a printed message. A page teletypewriter (Fig. 1) prints a message in page form, usually on a continuous roll of paper $8\frac{1}{2}$ in. (21.6 cm) wide.

Fig. 1. Page teletypewriter.

Baudot code. The Baudot code, widely used in printing telegraph devices, consists of five code pulses, any one of which may be either marking or spacing. In single-current signals, used for operating most teletypewriters, a marking pulse is an interval of time during which current flows through the circuit, and a spacing pulse is an interval during which no current flows. In polar signals, frequently used over long lines, a marking pulse is an interval during which negative current flows, and a spacing pulse is an interval during which positive current flows.

When the Baudot code is used to operate teletypewriters, the five code pulses are preceded by a start pulse, which is always a spacing pulse, and are followed by a rest, or stop, pulse, which is always a marking pulse (Fig. 2).

There are 32 possible combinations of the five code pulses. One combination is assigned to each of the 26 letters of the alphabet. The blank combination, in which all five code pulses are spacing signals, is not normally used. The remaining five combinations are used for the following functions: (1) figures shift, which causes the typing unit to shift to a position to print digits or punctuation marks; (2) letters shift, which shifts the typing unit into position to print letters; (3) carriage return, which causes the printing carriage to return from the right margin to the left margin at the end of a printed line; (4) line feed, which feeds the paper up one line; and (5) space, which causes the typing unit to space between words (Fig. 3).

Because only 26 code combinations are available for printing in each of the two shift positions, both capital and lowercase letters cannot be used. The uppercase position is reserved for the 10 digits, punctuation marks, and commonly used symbols. One uppercase combination is used for operating an audible (bell) signal.

Operation. An electric motor provides power for operating the keyboard and typing unit. The motor is geared to driving members of clutches on the keyboard shaft and on the receiving shaft of the typing unit. When a key lever is depressed, five code bars are positioned to the right or left (marking or spacing position) in a pattern corresponding to the code combination for the character or function represented by the depressed key lever. A universal bar, operated when any key lever is depressed, actuates a clutch-release mechanism, which causes the keyboard clutch to engage and allows a cam sleeve assembly on the keyboard transmitting shaft to rotate. During this rotation, six cams operate contacts in sequence to generate the start-stop signals. The contacts close to gener-

LC	UC	marking pulses	LC	UC	marking pulses
A		1 2	Q	1	1 2 3 5
B	?	1 4 5	R	4	2 4
C	:	2 3 4	S	'	1 3
D	$	1 4	T	5	5
E	3	1	U	7	1 2 3
F		1 3 4	V	;	2 3 4 5
G	&	2 4 5	W	2	1 2 5
H	#	3 5	X	/	1 3 4 5
I	8	2 3	Y	"	1 3 5
J	Bell	1 2 4	Z	6	1 5
K	(	1 2 3 4	Letters		1 2 3 4 5
L	)	2 5	Figures		1 2 4 5
M	.	3 4 5	Car. Ret.		4
N	,	3 4	line feed		2
O	9	4 5	Space		3
P	0	2 3 5	Blank		

Fig. 3. Page teletypewriter code assignments. Uppercase arrangement shown is one of many versions.

ate a marking pulse and are prevented from closing to generate a spacing pulse. One of these transmitting cams generates the start and rest pulses. The other five cams generate the code pulses corresponding to the selected character. At the end of the revolution of the cam sleeve, the transmitting clutch is disengaged, and the cam sleeve remains at rest until a key lever is again depressed.

The start-stop telegraph signals are received by an electromagnet on the typing unit. When current flows through the magnet coils (marking pulse), the armature is attracted to the pole piece of the magnet. When no current flows through the coil (spacing pulse), a spring pulls the armature away from the pole piece. When a start pulse is received, the armature is pulled away from the pole piece. This releases the receiving selector clutch, which then drives a cam sleeve assembly on the receiving shaft. As this assembly rotates, five selector cams sequentially position mechanical members on the typing unit either to the marking or spacing position, depending on the position of the armature at the time each selection is made. After the five code selections have been set up, a sixth cam on the cam sleeve trips another clutch which drives the mechanism for printing or performing the selected function. The character printed or the function performed is determined by the code combination set up in the selector mechanism.

Shortly after the fifth code selection is set up, the receiving cam sleeve assembly returns to its rest position and the receiving selector clutch is disengaged. The cam assembly stops rotating and remains at rest until the next start pulse is received. The receiving shaft rotates faster than the transmitting shaft, the most common speed ratios being 8:7 and 13:12. The receiving shaft makes one revolution and returns to its home position while the rest pulse is being received. Therefore, the receiving shaft always comes to rest briefly at the end of a revolution. This ensures that the receiving teletypewriter always begins each opera-

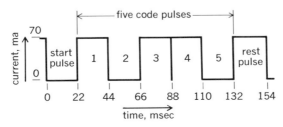

Fig. 2. Five-unit start-stop telegraph code, seven-unit code pattern. This combination is for letter F.

tion in synchronism with the transmitting unit. This start-stop synchronization prevents accumulation of any minor speed differences between the sending and receiving machines.

Only a small portion of each pulse length is required to set up a selection. The remaining length of each pulse provides an operating margin to ensure correct printing even when the received signals are distorted in transmission. A device called a range scale allows the instants of selection to be oriented with respect to the received signals so that the selections will occur in the middle of each code pulse, even when the signals are badly distorted.

ASCII Code. On June 17, 1963, the American Standards Association (now the American National Standards Institute) adopted a new American Standard Code for Information Interchange (ASCII). The new code consists of seven code pulses, or bits, instead of five as in the Baudot code. There are thus 2^7, or 128, discrete permutations of the seven bits in the code. Sixty-four of these permutations were assigned to printing characters, including the capital letters of the alphabet, the ten digits, punctuation marks, and special symbols. Thirty-five permutations were assigned to control characters, and the remainder were unassigned.

Monocase teletypewriters which use the 1963 version of ASCII are used in many communications systems, particularly in computer-oriented teletypewriter switching systems. These units operate on the same basic principles as conventional five-unit code teletypewriters, but the new code eliminates the need for letters and figures shift functions, there now being a discrete bit permutation assigned to each character.

Teletypewriters designed for use with this code use an added bit, or intelligence pulse, as an even vertical parity check bit; that is, if the ASCII character generated contains an odd number of marking pulses, the eighth pulse is made a marking pulse. Conversely, if the ASCII character contains an even number of marking pulses, the eighth pulse is made a spacing pulse. This feature permits errors to be detected by means of auxiliary equipment designed to detect parity failures or by a computer in computer switching systems. In 1968 there were no teletypewriters available which would detect receipt of a character containing an odd number of marking pulses and print an error symbol to indicate parity check failure, but such units are being developed. Auxiliary devices that perform this function are available.

The monocase ASCII teletypewriters all use a unit-length start pulse and a two-unit-length rest pulse. The seven intelligence pulses in ASCII plus the vertical parity check pulse give an 11-unit transmission pattern; that is, each character transmitted consists of the equivalent of 11-unit-length pulses.

On July 7, 1967, the U.S.A. Standards Institute (now the American National Standards Institute) approved a revised USA Standard Code for Information Interchange, X3.4-1967. In the revised code table (Fig. 4), marking pulses are shown as unit bits and spacing pulses as zero bits. In the new version of the code, both a lowercase and an uppercase of the alphabet have been assigned. In addition, five special symbols not previously included in the code have been added. The first two vertical columns of the code table are assigned to control characters, used for such purposes as communications controls, device controls, format effectors, and information separators. The bit permutations in the remaining six columns are assigned to graphic, or printing, characters, except for the delete character, which is neither a graphic nor a control.

Dual-case teletypewriters using the revised ASCII are being placed in service, particularly in computer-oriented systems for communications and data processing. Most dual-case ASCII teletypewriters are capable of operating at speeds of 150 wpm and higher, as compared to a maxi-

COLUMN →				0	1	2	3	4	5	6	7
b_7 →				0	0	0	0	1	1	1	1
b_6 →				0	0	1	1	0	0	1	1
b_5 →				0	1	0	1	0	1	0	1
ROW ↓	b_4 ↓	b_3 ↓	b_2 ↓	b_1 ↓							
0	0	0	0	0	NUL	DLE	SP	0	@	P	\	p
1	0	0	0	1	SOH	DC1	!	1	A	Q	a	q
2	0	0	1	0	STX	DC2	"	2	B	R	b	r
3	0	0	1	1	ETX	DC3	#	3	C	S	c	s
4	0	1	0	0	EOT	DC4	$	4	D	T	d	t
5	0	1	0	1	ENQ	NAK	%	5	E	U	e	u
6	0	1	1	0	ACK	SYN	&	6	F	V	f	v
7	0	1	1	1	BEL	ETB	′	7	G	W	g	w
8	1	0	0	0	BS	CAN	(	8	H	X	h	x
9	1	0	0	1	HT	EM	)	9	I	Y	i	y
10	1	0	1	0	LF	SUB	*	:	J	Z	j	z
11	1	0	1	1	VT	ESC	+	;	K	[	k	{
12	1	1	0	0	FF	FS	,	<	L	\	l	|
13	1	1	0	1	CR	GS	-	=	M	]	m	}
14	1	1	1	0	SO	RS	.	>	N	^	n	~
15	1	1	1	1	SI	US	/	?	O	_	o	DEL

Key:
NUL = null
SOH = start of heading
STX = start of text
ETX = end of text
EOT = end of transmission
ENQ = enquiry
ACK = acknowledge
BEL = bell (audible or attention signal)
BS = backspace
DC1 = device control 1
DC2 = device control 2
DC3 = device control 3
DC4 = device control 4 (stop)
NAK = negative acknowledge
SYN = synchronous idle
ETB = end of transmission block
CAN = cancel
HT = horizontal tabulation
LF = line feed
VT = vertical tabulation
FF = form feed
CR = carriage return
SO = shift out
SI = shift in
DLE = data link escape
EM = end of medium
SUB = substitute
ESC = escape
FS = file separator
GS = group separator
RS = record separator
US = unit separator
SP = space
DEL = delete

Fig. 4. The Revised American National Standard Code for Information Interchange.

mum of 100 wpm for most monocase ASCII machines. The principles of operation of the typing units on the newer sets are basically the same as those of the five-level Baudot code and monocase ASCII printers; however, many of the new keyboards utilize electronic distributors to serialize the signals generated by the keyboard instead of employing cam-operated contacts. This both simplifies the mechanical structure of the keyboard and also improves the quality of the transmitted signals.

Many dual-case ASCII teletypewriters incorporate new features such as backspacing, half-line feed, reverse line feed, and reverse half-line feed. These features permit printing of mathematical equations, chemical charts, diagrams, and graphs from signals received from either a computer or another teletypewriter.

[FRED W. SMITH]

Bibliography: American National Standards Institute, *American National Standard Code for Information Interchange*, X3.4-1977; D. G. Fink (ed.), *Electronic Engineers' Handbook*, 1975; Howard W. Sams Engineering Staff, *Reference Data for Radio Engineers*, 1975.

Television

The electrical transmission and reception of transient visual images. Like motion pictures, television consists of a series of successive images, which are registered on the brain as a continuous picture because of the persistence of vision. Each visual image impressed on the eye persists for a fraction of a second. In television in the United States, 30 complete pictures are transmitted each second, which with the use of interlaced scanning is fast enough to avoid evident flicker.

At the television transmitter, minute portions of a scene are sampled individually for brightness (and color for color television), and the information for each portion is transmitted consecutively. At the receiver, each portion is reproduced in its proper position and with correct brightness (and color) to reproduce the original scene.

The scene is focused on a photoelectric screen of a camera tube. Each portion of the screen is changed by the photoelectrons to a degree depending upon the brightness of the particular portion. The screen is scanned by an electron beam just as a reader scans a page of printed type, character by character, line by line. When so scanned, an electric current flows with an instantaneous magnitude proportional to the brightness of the portion scanned. *See* TELEVISION CAMERA; TELEVISION CAMERA TUBE.

Variations in the current are transmitted to the receiver, where the process is reversed. An electron beam in the picture tube is varied in intensity (modulated) by the incoming signals as it scans the picture-tube screen in synchronism with the scanning at the transmitter. The photoelectric surface of the picture tube produces light in proportion to the intensity of the electron beam which strikes it. In this way the minute portions of the original scene are re-created in their proper positions, brightness, and (for color transmission) color values. The elements of a television system are shown in Fig. 1. *See* TELEVISION RECEIVER; TELEVISION TRANSMITTER.

Scanning. In the United States an individual picture (frame) is considered to be made up of 525 lines, each line containing several hundred picture elements. All these lines are scanned and the light values are sent to the receiver so that each second 30 pictures are received. These figures vary from nation to nation. The picture is blanked out at the end of each line while the scanning beam is directed to the next line. During these short intervals, synchronizing signals are transmitted to keep the scanning process at the receiver in step with that at the transmitter.

To take full advantage of the persistence of vision, each frame is scanned twice, alternate lines being scanned in turn. This technique is called interlaced scanning.

Since 525 horizontal lines are scanned in $\frac{1}{30}$ s, the horizontal scanning rate is 15,750 times per second. Since two vertical fields are scanned in $\frac{1}{30}$ s, the vertical scanning rate is 60 times per second. *See* TELEVISION SCANNING.

Bandwidth. The bandwidth required for any information transmission system is a function of the number of bits of information, or the detail, to be transmitted per second. For a television picture, the greatest detail would be required if the picture consisted of a checkerboard pattern of the smallest squares the system must handle to provide acceptable resolution. The standard of 525 lines sets the vertical detail, and the standard aspect ratio (picture width to height) of 4:3 requires 700 horizontal picture elements for equal horizontal and vertical resolution, or 350 sets of alternate black-and-white squares. The picture is reproduced 30 times a second, for a total of $525 \times 350 \times 30$ or 5,512,500 complete cycles per second. Less detail than this is actually transmit-

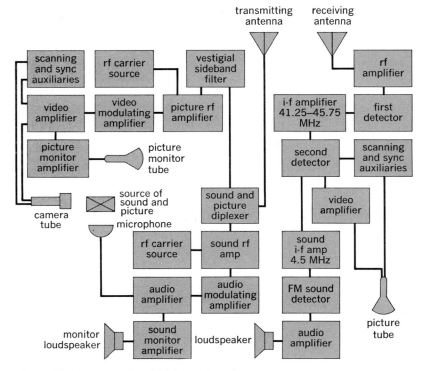

Fig. 1. The elements of a television system. (*From K. Henney, ed., Radio Engineering Handbook, 5th ed., McGraw-Hill, 1959*)

ted and received; the highest video frequency actually transmitted is 4.2 MHz.

Frequency. The band of frequencies assigned to a television station for the transmission of synchronized picture and sound signals is called a television channel. In the United States a television channel is 6 MHz wide, with the visual carrier frequency 1.25 MHz above the lower edge of the band and the aural carrier 0.25 MHz below the upper edge of the band.

Television channels in the United States are identified by numbers, starting with channel 2. The frequency assigned to channel 1 was later reassigned to other uses. The table shows that these channels are in three frequency bands. Channels 2–6 occupy the region from 54 to 88 MHz, channels 7–13 are from 174 to 216 MHz, and channels 14–83 are from 470 to 890 MHz. The first two groups of channels fall in the very-high-frequency (vhf) band; the channels in the last group are in the ultra-high-frequency (uhf) band.

Sound transmission. In the United States the sound portion of the program is transmitted by

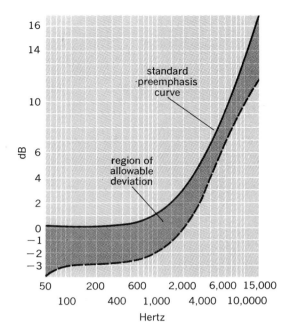

Fig. 2. Standard preemphasis curve.

frequency modulation at a carrier frequency 4.5 MHz above the picture carrier. Maximum frequency deviation (bandwidth) of the sound signals is 25 kHz. *See* FREQUENCY MODULATION.

The normal frequency response is altered in the transmitter to emphasize the higher audio frequencies with respect to the lower frequencies. Called preemphasis, this is accomplished by a circuit that causes the audio response to increase with frequency. A corresponding circuit is used in the receiver to produce an equal and opposite decrease of response to higher audio frequencies. By so doing, noise produced in the receiver is attenuated without the overall system audio-frequency response being affected. The Federal Communications Commission (FCC) requires that the response of the aural transmitting system must not exceed the limits shown in Fig. 2.

The harmonic distortion of the audio-frequency signals must not exceed the following rms values when the harmonics are measured out to 30 kHz:

Frequency range, Hz	% Distortion
50–100	3.5
100–7500	2.5
7500–15,000	3.0

Picture transmission. The visual signals are transmitted at a carrier frequency 1.25 MHz above the lower limit of the channel, using amplitude modulation and vestigial sideband transmission. The upper sideband is fully transmitted, but the lower sideband is attenuated beginning at 0.5 MHz below the carrier. Attenuation is virtually complete at 1.25 MHz below the carrier. This method of transmission reduces the required bandwidth of the channel and allows more channels to use the available space in the radio spectrum. Figure 3 is an output characteristic of a transmitter for a channel 2 station, showing how the 6-MHz band is used for picture and sound transmission. *See* AMPLITUDE MODULATION.

Television channels in the United States

Channel no.	Frequency band, MHz	Channel no.	Frequency band, MHz
2	54–60	43	644–650
3	60–66	44	650–656
4	66–72	45	656–662
5	76–82	46	662–668
6	82–88	47	668–674
7	174–180	48	674–680
8	180–186	49	680–686
9	186–192	50	686–692
10	192–198	51	692–698
11	198–204	52	698–704
12	204–210	53	704–710
13	210–216	54	710–716
14	470–476	55	716–722
15	476–482	56	722–728
16	482–488	57	728–734
17	488–494	58	734–740
18	494–500	59	740–746
19	500–506	60	746–752
20	506–512	61	752–758
21	512–518	62	758–764
22	518–524	63	764–770
23	524–530	64	770–776
24	530–536	65	776–782
25	536–542	66	782–788
26	542–548	67	788–794
27	548–554	68	794–800
28	554–560	69	800–806
29	560–566	70	806–812
30	566–572	71	812–818
31	572–578	72	818–824
32	578–584	73	824–830
33	584–590	74	830–836
34	590–596	75	836–842
35	596–602	76	842–848
36	602–608	77	848–854
37	608–614	78	854–860
38	614–620	79	860–866
39	620–626	80	866–872
40	626–632	81	872–878
41	632–638	82	878–884
42	638–644	83	884–890

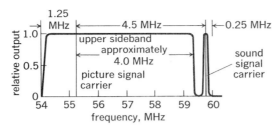

Fig. 3. Output characteristic of a television transmitter. (*From K. Henney, ed., Radio Engineering Handbook, 5th ed., McGraw-Hill, 1959*)

Negative modulation for picture transmission is used in the United States to minimize the effects of noise during synchronizing signal reception. In negative modulation an increase in brightness causes a decrease in transmitted power. Some foreign systems use positive modulation.

Ghost image. Radio waves from a transmitter to a receiver normally follow a straight path. However, it is possible for such waves to follow a long or short path to the receiver. Reflections from large objects, such as mountains or buildings, follow a long path. The reflected waves in this instance arrive later than the direct waves, so that a second picture is reproduced from a fraction of an inch to several inches to the right, depending on the length of the indirect path. The second picture is called a ghost image. There may be several such ghosts when reflecting objects are in an area. Short-path ghosts may actually be picked up by the antenna lead in wires or the actual input of the receiver and thus produce what is known as a leading ghost.

Snow (noise). All electronic circuitry produces some voltages or currents that are not related to the signal. Such voltages or currents are referred to as noise. Generally speaking, noise has its most deleterious effect on reception quality when it occurs in the input circuitry of the receiver, since this is usually where the basic signal-to-noise ratio is established. The generation of noise in this area is principally caused by thermal agitation in resistive input components. When the incoming signal is too weak to override this noise, a visual effect is displayed on the picture tube that is commonly referred to as snow. Although there are other types of noise that produce different visual effects, snow is the most common visual effect.

Scrambling of television signals. If the synchronizing signals are caused to vary at random but predetermined rates, the signal is unusable in conventional receivers. Special receivers may contain unscrambling devices which, by a specially transmitted signal or a built-in keyer, cause the receiver to synchronize and produce normal pictures. Scrambling devices were developed for pay television, generally called pay TV (PTV), to prevent reception by anyone but a subscriber. *See* CLOSED-CIRCUIT TELEVISION.

Recording television programs. Television programs are recorded for rebroadcast at a later time and for many other reasons. The principal method is video tape recording (VTR) in which television sound and pictures are recorded on magnetic tape. Electrical signals constituting the sound and picture pass through magnetic recording heads. As the tape is drawn across these heads, the signal currents produce magnetization of finely powdered iron in the tape emulsion. The signals are recovered when the magnetized tape is drawn across a reproducing head and the magnetic fields intercept a pickup coil. There are two basic types of video tape recording. One type is known as quadruplex, most commonly used in television broadcasting. The second type of video tape recording is known as helical scan (commonly called helical) and was principally used for closed-circuit applications.

VTR has the great advantage that it can be used immediately without further processing, and that the tape can be erased by demagnetization and reused. It is of unique value for prerecording programs, reruns, special events coverage, test programs, and delayed program transmission for different network station time zones.

Quadruplex. Quadruplex tape consists of a coated plastic base 2 in. (50.8 mm) wide. A 12½-in. (317.5-mm) reel provides 64 min of program time. Prior to recording, the video signal frequency modulates a carrier and produces a deviation of 4.25–6.8 MHz in low-band monochrome recording; for high-band color the deviation is 7.06 to 10.00 MHz. A vestigial sideband signal is recorded on transverse tracks 0.010 in. (0.254 mm) wide by means of four magnetic heads equally displaced on the circumference of a 2-in. (50.8-mm) wheel rotating 14,400 times per second. About 18.4 scanning lines are recorded on each transverse track. One picture frame (525 lines) comprises 32 transverse tracks or ½ in. (12.7 mm) of tape length. The accompanying sound and control signals are recorded on separate longitudinal tracks by means of separate recording heads. In reproduction the FM signal is recovered from the tape and demodulated.

Helical scan. Helical recorders use tape widths of ½, ¾, 1, and 2 in. (12.7, 17.7, 25.4, and 50.8 mm). A 1-in. (25.4-mm) C-type format, a high-performance nonsegmented helical-scan recorder, is widely used for broadcast applications. A 10½-in. (266.7-mm) reel of 1-in. (25.4-mm) tape will record 96 min, and a 9-in. (228.6-mm) reel will record 64 min. Tape speed is 24.4 cm/s, and writing speed is 25.59 m/s.

C-format VTRs contain three audio channels which allow for easy stereo recording, with the third audio channel usually assigned to record time code. These machines utilize two rotary video heads. One head records and plays the video information, and the other records and plays the vertical interval. This configuration allows for continuous recording without any dropout time and is not subject to color banding.

Grades of television service. The FCC has established by definition two grades of television service for the United States. Grade A service provides relatively high freedom from interference from other television stations and also good freedom from artificial and receiver noise. It specifies that picture quality acceptable to the average observer is expected to be available at least 90% of the time at the best 70% of all receiver locations at the outer geographical limits of this service. Grade B service recognizes that service is provided but may be more vulnerable to interference and noise. It specifies that service equal to that of grade A is available, but to only 50% of all receiver locations at the limiting distance.

Teletext and videotext. Teletext is the general term that describes the transmission of alphanumeric characters as part of the television signal. This telecommunications service can deliver textual information and simple graphics at low cost to a wide range of users. The television receiver acts as a display terminal which converts digital information into text and graphics on the television screen. This information is inserted in digital form onto unused lines which are hidden on top of the screen, commonly known as the vertical interval portion of the television picture. This form of telecommunications can deliver information such as weather, program guides, store prices, and stock market reports.

The teletext system originates with a data base that is able to compose pages, store them until needed, code them into digital pulses, and insert this data stream into the vertical interval of the normal program information. A decoder retrieves this information at the television receiver by detecting the digital pulses, removes them from the vertical interval, and composes the digital information into pages of text which are then stored in a minicomputer after being decoded into letters, numbers, and graphics. After the information has been decoded, it is fed into a memory where it is retained until the viewer recalls the various pages

desired by pressing a special button on the remote-control key pad located in the viewer's home. After viewing the index, another combination of selector buttons allows the selection to appear on the screen in detail.

Videotext is the generic term applied to those services that provide transmission using an interactive system, generally via telephone, between the television receiver and a computer. *See* DATA COMMUNICATIONS; MULTIACCESS COMPUTER.

For other aspects of television *see* COLOR TELEVISION; VIDEO TELEPHONE; VIDEOTEXT AND TELETEXT.

[STEVE DE SATNICK]

Bibliography: A. Abramson, *Electronic Motion Pictures*, 1955, reprint 1974; G. W. Bartlett (ed.), *NAB Engineering Handbook*, 6th ed., 1975; B. Grob, *Basic Television*, 4th ed., 1975; B. Hartman, *Fundamentals of Television: Theory and Service*, 1975; K. Henney (ed.), *Radio Engineering Handbook*, 5th ed., 1959; M. S. Kiver and M. Kaufman, *Television Simplified*, 7th ed., 1973.

Television camera

An electrooptical system used to pick up and convert a visual image or scene into an electrical signal called video. The video may be transmitted by cable or wireless means to a suitable receiver or monitor some distance from the actual scene.

A television camera may fall within one of sever-

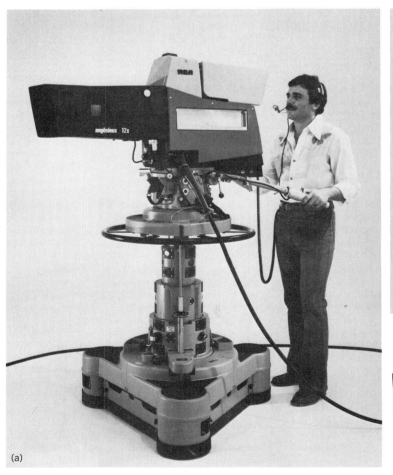

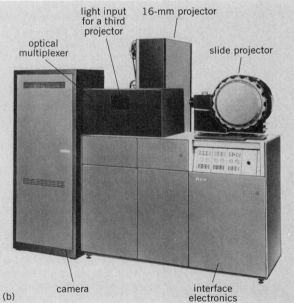

Fig. 1. Television cameras. (*a*) Studio camera. (*b*) Telecine camera. (*c*) Portable camera. (*RCA Corp.*)

Fig. 2. Modern camera control unit for an automatic camera. (*RCA Corp.*)

al categories: studio, portable, or telecine (Fig. 1). It may also be one of several highly specialized cameras used for remote viewing of inaccessible places, such as the ocean bottom or inside nuclear power reactors. The camera may be capable of producing color or monochrome (black and white) pictures. Modern cameras are entirely solid-state, except for the actual pickup tube. Even the pickup tube may be replaced by solid-state semiconductors called charge-coupled devices.

A television camera may be one-piece, with all components contained in one assembly, or it may consist of a head and a camera control unit (CCU) connected by a multiple-conductor cable, and be called a camera chain. The head unit comprises the optical system with lens, the picture pickup tubes, and a minimum of electronics necessary to generate and amplify the minute signals from the pickup tubes. The CCU (Fig. 2) may contain electronics and controls which allow a skilled operator to adjust brightness (luminance), color (balance, saturation, and hue), and certain correction circuits (registration, gamma, and aperture) which improve the picture. A specially calibrated oscilloscope, called a waveform monitor, is generally provided to the CCU operator so that accurate voltage levels may be set. Modern cameras provide for automatic as well as manual setup of some or all of the above adjustments. Some cameras have a built-in microprocessor which enables complete setup of all parameters by simply pushing one button. Triax cameras are those which utilize a small-diameter triaxial cable (three concentric conductors) to connect the camera head to the CCU, instead of the heavier, multiple-conductor cable known as TV-81, which has 81 conductors. Triax can be used because all of the normal signals are multiplexed onto the three wires. *See* MICROPROCESSOR.

Essential elements. Every camera shares certain essential elements: an optical system, one or more picture pickup devices, preamplifiers, scan-

ning circuits, blanking and synchronizing circuits, and video processing circuits and control circuits. Color cameras also include some kind of color-encoding circuit.

Optical system. The optical system consists minimally of a fixed-focal-length lens placed directly in front of a pickup device. Provision must be made to focus the image on the focal plane of the pickup device. Usually the pickup device, complete with its mounting and scanning components, is adjusted with respect to the lens, while the lens is set at the distant extreme of its focusing range. This establishes the correct back-focus so that a sharp image can be obtained by adjusting only the lens-focusing ring as camera-to-subject distances vary. More elaborate systems provide for insertion of various filters into the optical path, and usually replace the fixed-focal-length lens with a zoom type which allows smooth, continuous transition from wide-angle to telephoto focal lengths while the image remains focused. Color cameras with multiple pickup devices must split the incoming light into suitable primary colors. This is usually done by means of a relay lens and dichroic mirror system placed between the objective lens and the pickup devices (Fig. 3a). Dichroic mirrors reflect light of one color, while passing all others; that is, a red dichroic mirror reflects red and passes all other colors. An improvement on the mirror system utilizes prisms with dichroic materials coated onto the glass (Fig. 3b). The advantages of the prism system include ruggedness and, most importantly, the elimination of air-to-glass surfaces except at the entry and exit points. This reduces image deterioration due to misalignment of the mirrors and dust accumulation on the surfaces.

Pickup device. The picture pickup device used in most cameras is a photosensitive vacuum tube. The tube is oriented so that the image from the lens is focused on a light-sensitive target within the tube. The target is scanned by an electron beam, which results in an output voltage from the tube that varies in proportion to the amount of light striking each point on the target. Monochrome cameras and less-sophisticated color cameras utilize vidicon pickup tubes. High-quality broadcast cameras may use Plumbicons, Leddicons, Saticons, or other proprietary tubes. Monochrome cameras invariably have only one pickup tube; color cameras may have one, two, or three pickup tubes. Those color cameras which have only one or two pickup tubes employ a color-stripe filter in front of the photosensitive target effectively dividing the target into red, blue, or green areas. As the electron beam scans the target, it sequentially represents the red, blue, or green light levels. Electronic circuits within the camera separate the three colors. Such systems are not used in the highest-quality cameras because of the inherent loss of resolution and sensitivity which results from the stripe filter and the smaller effective area of the pickup tube. Modern broadcast color cameras employ three tubes, one for each of three primary colors used to derive the full-color spectrum. Red, green, and blue are used in additive color systems, while subtractive color systems have been devised which use white (full-spectrum luminance), red, and blue. Most color cameras use the additive scheme. Each pickup tube is scanned in synchronism so that separate red, blue, and green representations of the scene being picked up are always being generated. *See* TELEVISION CAMERA TUBE.

A few cameras have appeared which utilize charge-coupled device (CCD) arrays in lieu of pickup tubes. These solid-state devices essentially consist of a large number of photodiodes aligned in a matrix so that each diode's output voltage can be related to a particular point in the picture which is focused on the array (Fig. 4). The amount of light falling on each diode determines its output level. CCD pickups are very resistant to shock, have precise image geometry, and should have nearly indefinite life. They also offer freedom from microphonic noise and image burn-in commonly found in pickup tubes. Development work on CCD arrays is continuing, and they will find much wider application in the future, perhaps supplanting pickup tubes entirely. *See* CHARGE-COUPLED DEVICES.

Preamplifier. The pickup device's output is fed to a preamplifier, which helps to maintain a high signal-to-noise ratio and also provides a level sufficient to operate an electronic viewfinder at the camera. *See* PREAMPLIFIER.

Scanning circuits. The electron beam which scans the pickup tube's photosensitive target is caused to sweep through the action of magnetic fields impressed on the tube by means of deflection coils. These coils are assembled into an integrated yoke assembly. Yokes must be made with great precision and are often computer-matched so that all pickup tubes in a multiple-tube camera scan with precisely the same geometry. Horizontal and vertical drive circuits cause the deflection yoke to deflect the electron beam so that it scans the photosensitive target in the pickup tube according to a definite pattern, called a raster. These drive circuits are synchronized by pulses from a synchronizing generator, which also provides horizontal and vertical blanking pulses to suppress the pickup tube output during the retrace interval. If not suppressed, there would be objectionable lines through the camera's picture output. CCD cameras, on the other hand, do not require complicated deflection or blanking circuits. By their very nature, they are scanned by using computer-type addressing techniques. Pulse counters, divider chains, and memory provide the correct sequential readout of the CCD array so that a picture is recovered from the individual photodiode outputs. *See* TELEVISION SCANNING.

Blanking and synchronizing and video processing and control circuits. Before the camera output can be viewed on a conventional video monitor, horizontal and vertical synchronizing and blanking pulses must be added. In two-piece cameras this is ordinarily done in the CCU. Other functions which are necessary to obtain high-quality pictures include gamma correction, aperture correction, registration, and color balance. Gamma correction is required because the pickup tubes do not respond linearly to increasing light levels. It allows the CCU operator to capture detail in the dark areas of high-contrast scenes, essentially by "stretching" the video levels in those areas. Aperture correction provides several benefits mainly related to an even overall response to scenes with more or less detail. It also helps to improve the signal-to-noise ratio of

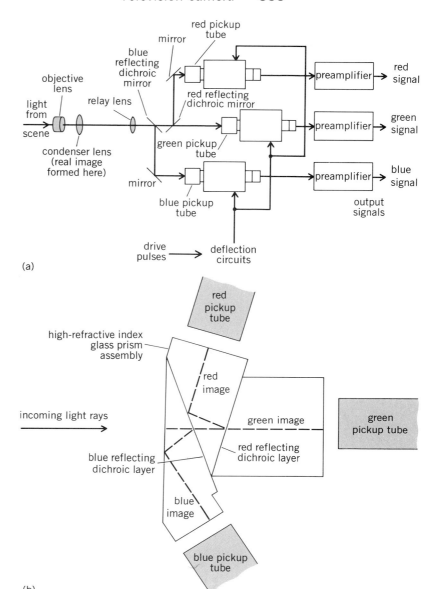

Fig. 3. Color television camera optical systems. (*a*) Dichroic mirror system (*From G. W. Bartlett, ed., NAB Engineering Handbook, 6th ed., National Association of Broadcasters, 1975*) (*b*) Prism optical beam splitter.

the camera's output video. Registration must be adjusted on multiple-tube cameras to ensure that the separate red, blue, and green images are precisely aligned on one another. Color balance must be properly set on color cameras and must be consistent from dark scenes to bright scenes, or there will be an objectionable "tint" to the camera output. This is referred to as black balance and white balance. The former is set up with all light blocked from the camera's lens, while the latter is set up by using a "white card" under the actual lighting conditions of the scene. Color balance is frequently an automatic push-button feature on modern cameras.

Color cameras also have a color encoder which combines the three primary colors into a composite signal that must conform to one of several standards used in various countries. In the additive

Fig. 4. Charge-coupled device (CCD) array. (*Fairchild Camera and Instrument*)

color NTSC (National Television Systems Committee) system used in the United States, the red (R), blue (B), and green (G) signals are matrixed by a combining network and amplified to provide three new signals, M, I, and Q. The M signal contains luminance information, while I and Q contain the color, or chrominance, information from the televised screen. Component M consists of 30% red plus 59% green plus 11% blue. The I signal consists of 60% red minus 28% green minus 32% blue. Component Q is made up of 21% red minus 52% green plus 31% blue (Fig. 5). It is possible to subtract voltage levels by inverting the phase and summing, and this is exactly what is done in the color encoder. I and Q then modulate a 3.58-MHz carrier in a two-phase balanced modulation system, and the resulting amplitude-modulated sidebands, 90° out of phase, are added to the M signal along with the synchronizing and blanking signals already mentioned. A short reference burst of 3.58-MHz carrier is also added before the start of each television line. The resultant signal, known as composite NTSC video, contains all necessary information to recreate the original color scene on a color monitor or receiver. The M signal is used by black and white monitors or receivers, which are insensitive to I and Q, to create an image in shades of gray. Color monitors and receivers have special circuits to process the I and Q signals back into the primary colors. In a monochrome scene the

red, green, and blue signals are equal, and therefore I and Q become zero and a black and white picture is produced on both monochrome and color television monitors. *See* COLOR TELEVISION.

Typical configurations. Live studio cameras (Fig. 1a) are equipped with several ancillary systems to enhance their operation. An electronic viewfinder, actually a small television monitor, shows the camera operator what the camera is seeing, making it possible to frame and focus the picture. The tally system consists of one or more red lights which illuminate when the camera's picture is "on the line" so that production and on-camera personnel know which camera is active. Generally an intercom system is built into the camera so that the director can communicate with the camera operator. The camera itself may be mounted upon a tripod, but more often it is on a dolly and pedestal, which allows the camera to be moved around on the studio floor and raised or lowered as desired. A pan head permits the camera to be rotated to the left or right and furnishes the actual mounting plate for the camera. The lens zoom and focus controls are mounted on a panning handle convenient to the operator. A common accessory on studio cameras is the videoprompter. This is a television monitor on which the program script may be displayed and read by the on-camera personnel.

Telecine cameras (Fig. 1b) are used in conjunction with film or slide projectors to televise motion pictures and still images. They generally employ vidicon tubes, one for monochrome or three for color, and many of the usual controls are automatic so as to require less operator attention. The film projectors used in television may use either a constant-rate pulldown or 3-2 intermittent mechanism to translate the motion picture's 24 frames per second to the 30 frame-per-second television frame rate. The constant-rate pulldown has a shutter which is closed before the film is rapidly pulled down. The shutter opens 120 times per second so that the camera sees each film frame in a 5:4 ratio. The 3-2 pulldown moves the film intermittently, so that one film frame is held in the film gate for two television fields (2/60 s) and the next film frame is held in the film gate for three television fields (3/60 s). Hence two film frames are displayed in 5/60 s, and 24 film frames are shown in 1 s (30 TV frames).

Generally one telecine camera serves several projectors through the use of an optical multiplexer. The optical multiplexer may use either station-

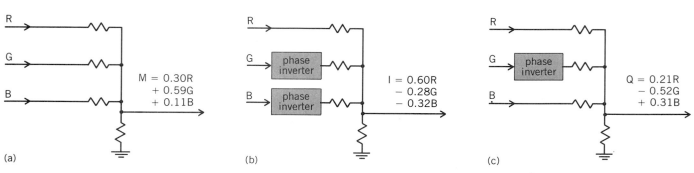

Fig. 5. Resistor matrices showing derivation of (a) M, (b) I, and (c) Q signals. (*From G. W. Bartlett, ed., NAB Engineering Handbook, 6th ed., National Association of Broadcasters, 1975*)

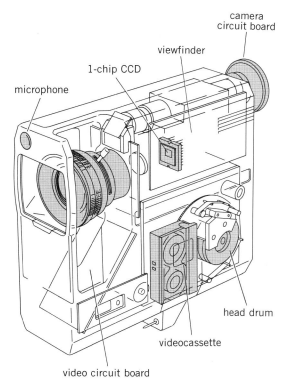

microphone
1-chip CCD
viewfinder
camera circuit board
head drum
videocassette
video circuit board

Fig. 6. Prototype portable charge-coupled device (CCD) camera with built-in videotape recorder. (*Sony Corporation*)

ary, half-silvered mirrors or movable, front-silvered mirrors to direct the image from each projector into the camera. Movable mirrors offer superior light-transmission efficiency and are preferable in color systems due to the light losses inherent in the pickup tube dichroic mirror splitter.

Portable cameras (Fig. 1c) usually combine all of the basic elements into one package and may be used for a multitude of purposes. They have found their way into electronic news gathering (ENG) for broadcast television, and into electronic field production (EFP), where they can be used for production of broadcast programs, commercials, and educational programs. The units are lightweight, often have built-in microphones for sound pickup, and can be handled by one person. Used in conjunction with modern battery-powered portable videocassette recorders, an entire video and audio pickup system can be carried and operated by one or two people. Low-cost color cameras have become popular with home video enthusiasts and may someday supplant film-based home movie cameras (Fig. 6). *See* TELEVISION; TELEVISION RECEIVER; TELEVISION TRANSMITTER.

[EARL F. ARBUCKLE, III]

Bibliography: B. Astle et al., Automatic set-up system for a broadcast color camera, *121st Technical Conference of the Society of Motion Picture and Television Engineers*, Oct. 21–26, 1979; G. W. Bartlett (ed.), *NAB Engineering Handbook*, 6th ed., 1975; CCD cameras: Coupled with the future, *Broadcast Manage./Eng.*, 16(9):47–50, September 1980; M. Hermann and J. Wolber, The P-squared CCD 500B linear imager, *Electron. Components Appl.*, 1(3):183–190, May 1979.

Television camera tube

An electron tube having a light-sensitive receptor that converts an optical image into an electrical television video signal. The tube is used in a television camera to generate a train of electrical pulses representing the light intensities present in an optical image focused on the tube. Each point of this image is interrogated in its proper turn by the tube, and an electrical impulse corresponding to the amount of light at that point of the optical image is generated by the tube. This signal represents the video or picture portion of a television signal. Television camera tubes are designed for broadcast television to pick up live programs, indoors or outdoors, as well as to reproduce motion pictures and other filmed material. *See* TELEVISION CAMERA.

The tubes are also used extensively in closed-circuit cameras for surveillance, and in training studios, schools, video tape recorder cameras, and military special-purpose cameras. Special versions are designed to work with intensifier tubes which increase the effective sensitivity so that the cameras can operate at very low light levels. These are used for nighttime surveillance work, television astronomy, and viewing low-intensity x-ray fluoroscope images in medical x-rays and in baggage inspection units in airports. In general, three tubes are used in color television cameras. A class of tubes has built-in stripe color filters which allow a single tube to develop a complete color picture, although with somewhat reduced fidelity compared to the multitube cameras. Although the television camera tube is sensitive primarily to visible light, special tubes are sensitive to radiant energy in the infrared and the ultraviolet.

Charge-coupled devices are a new generation of solid-state electronic image sensors which can produce a television signal from an optical image. These can operate independently or can be incorporated into an intensifier-type vacuum tube to achieve enhanced sensitivity. Figure 1 shows contemporary television camera tubes. *See* CHARGE-COUPLED DEVICES.

Image orthicon. The image orthicon made broadcast television practical. It was used for more than 20 years as the primary studio and field camera tube for black and white and color television programming because of its high sensitivity and its ability to handle a wide range of scene contrast and to operate at very low light levels. It is one of the most complicated camera tubes. It is an outgrowth of the earlier multiplier orthicon and image iconoscope, and was made possible by the invention of the "two-sided" storage target. The image orthicon is divided into an image section, a scanning section, and a multiplier section (Fig. 2), within a single vacuum envelope.

Image section. A light image is focused on the photoemissive layer, which is a continuous film inside the tube faceplate. Electrons absorb the energy and leave the surface in numbers proportional to the intensity of the illumination at each point. These photoelectrons flow in essentially parallel streams through the image section. The magnetic field focuses each to a sharp focus at the target plane. *See* PHOTOEMISSION.

The two-sided target consists of a fine wire mesh screen placed several thousandths of an inch from

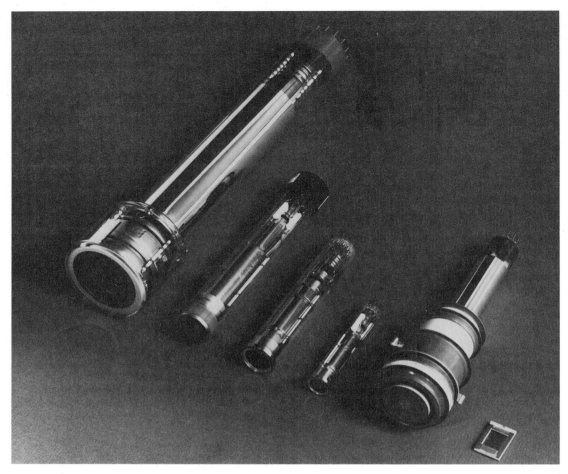

Fig. 1. Typical television tubes in modern use. From left to right: image orthicon, lead oxide vidicon, industrial-type vidicon, miniature vidicon, silicon intensifier vidicon, and charge-coupled device. (*RCA Corp.*)

a glass membrane less than 0.0002 in. (5.0 μm) thick. Most of the photoelectrons pass through the target mesh and hit the front surface of the target glass. Each photoelectron knocks several additional electrons from the target glass surface, producing a positive charge at the impact point. The secondary electrons are collected by the target mesh, which is held at a slightly more positive voltage. *See* SECONDARY EMISSION.

Scanning section. The positive charge pattern is stored on the front side of the target glass. A beam of low-velocity electrons generated by an electron gun is made to scan the rear surface of the glass by varying magnetic fields produced by the deflecting coils. As the beam moves across the glass, it deposits electrons wherever positive charges are built up on the image side. The glass resistance is controlled so that charges can move from one face to the other before the scanning beam returns to the same spot; yet the glass is of high enough resistance to inhibit lateral movement of the charges. When the scanning beam has deposited enough electrons at each point to neutralize the charge on the glass and reduce it to the voltage of the electron gun cathode, the remaining electrons return toward the electron gun. When the beam scans an uncharged (dark) area, the full beam is returned. When the beam scans a highly charged (bright) area, most of the beam is deposited and little returns. The variations in the return beam current constitute the television picture information, at low intensity. The return beam is amplified about 1000 times in the electron multiplier section of the tube and then is taken out at the anode of the multiplier as a video signal current. *See* CATHODE-RAY TUBE.

Multiplier section. The electron multiplier is of unique construction, although it operates like the multiplier used in a photomultiplier. It consists of a flat first-dynode structure and a series of pinwheel multipliers. When the return beam strikes the first dynode, a shower of secondary electrons cascades through the pinwheels, which

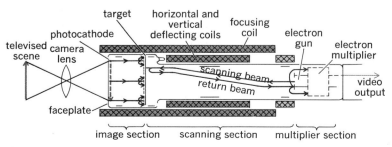

Fig. 2. The image orthicon and its associated deflecting and focusing coils. (*From D. G. Fink, ed., Television Engineering Handbook, McGraw-Hill, 1957*)

are maintained at progressively higher voltages, where repeated secondary emission multiplies their number. The final group of electrons is collected by the anode and forms the video signal current. *See* PHOTOMULTIPLIER.

Image isocon. The image isocon is a further development of the image orthicon. The excess primary electrons in the scanning beam of the image isocon are returned from the target in two components: the scattered electrons and those that are reflected (Fig. 3). The image isocon works on the principle of separating out these two components of the return scanning beam and utilizing the scattered electron component, which has the highest signal compared to the random noise in the beam current. The separation section (Fig. 4) directs only the scattered beam into the electron multiplier. This improves the signal-to-noise ratio of the output signal and allows the camera utilizing the image isocon to operate at very low light levels in such fields as astronomy and intensification of x-ray fluoroscopic images.

Photoconductive tubes. These types have a photoconductor as the light-sensitive portion. A photoconductor is a material that absorbs light and transfers the energy of the photons of light to electrons in the material. This frees some of the electrons and allows them to move through the material, and thereby changes the electrical conductivity of the material where the light is absorbed. The electron tube is designed to detect this change in electrical conductivity and develop a television signal. *See* PHOTOCONDUCTIVITY.

The name vidicon was applied to the first photoconductive camera tube developed by RCA. It is loosely applied to all photoconductive camera tubes, although some manufacturers adopt their own brand names to identify the manufacturer or the type of photoconductive material used.

The vidicon tube is a small tube that was first developed as a closed-circuit or industrial surveillance television camera tube. The development of

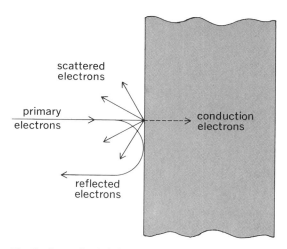

Fig. 3. Formation of the two components in the return beam of the image isocon: the scattered and the reflected electrons.

new photoconductors has improved its performance to the point where it is now utilized in one form or another in most television cameras. Its small size and simplicity of operation make it well suited for use in systems to be operated by relatively unskilled people.

The vidicon is a simply constructed storage type of camera tube (Fig. 5). The signal output is developed directly from the target of the tube and is generated by a low-velocity scanning beam from an electron gun. The target generally consists of a transparent signal electrode deposited on the faceplate of the tube and a thin layer of photoconductive material, which is deposited over the electrode. The photoconductive layer serves two purposes. It is the light-sensitive element, and it forms the storage surface for the electrical charge pattern that corresponds to the light image falling on the signal electrode.

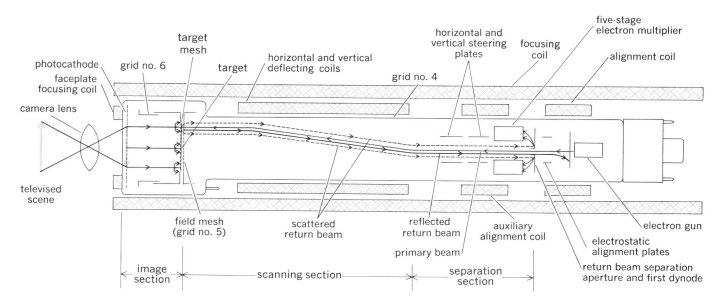

Fig. 4. Image isocon and 175 associated deflectron and focusing components. The separation section isolates the scattered and reflected return beam components. (*RCA Corp.*)

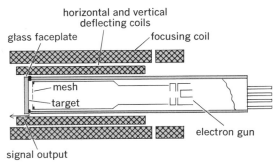

horizontal and vertical
deflecting coils

glass faceplate

focusing coil

mesh

target

signal output

electron gun

Fig. 5. Cross section of a vidicon tube and its associated deflection and focusing coils.

The photoconductor has a fairly high resistance when in the dark. Light falling on the material excites additional electrons into a conducting state, lowering the resistance of the photoconductive material at the point of illumination. A positive voltage is applied to one side of the photoconductive layer by means of the signal electrode. On the other side, the scanning beam deposits sufficient electrons at low velocity to establish a zero voltage. In the interval between successive scans of a particular spot, the light lowers the resistance in relation to its intensity. Current then flows through the surface at this point, and the back surface builds up a positive voltage until the beam returns to scan the point. The signal output current is generated when the beam deposits electrons on these positively charged areas. An equal number of electrons flow out of the signal electrode and through a load resistor, developing a signal voltage that is fed directly to a low-noise video signal amplifier.

A fine-mesh screen stretched across the tube near the target causes the electron scanning beam to decelerate uniformly at all points and approach the target in a perpendicular manner. The beam is brought to a sharp focus on the target by the longitudinal magnetic field of the focusing coil and the proper voltage for the focusing electrode. The beam scans the target under the influence of the varying magnetic fields of the deflecting coils.

Photoconductor properties determine to a large extent the performance of the different types of vidicon tubes. The first and still most widely used photoconductor is porous antimony trisulfide. The latest photoconductors are the lead oxide, selenium-arsenic-tellurium, cadmium selenide, zinc-cadmium telluride, and silicon diode arrays. All of

these either improve the sensitivity or the speed of response (ability to capture motion without "smearing") or both. The first two are barrier-layer types that operate like large-area reversed-bias junctions. Figure 6 illustrates the configuration of the barrier-layer or reversed-bias junction photoconductors; n refers to good electron conductivity, i refers to good electron and hole conductivity, and p refers to good hole conductivity. A positive voltage is applied to the n side through the signal plate, and a negative voltage is applied to the p side by the scanning beam electrons. This reverse-biases the junctions between the i and p and the i and n sections. This produces low dark current in the absence of light, a very desirable characteristic. Light is absorbed in the bulk of the material, where it produces charge carriers that provide a positive charge image on the side opposite the faceplate. The silicon diode tube is also a reversed-biased junction type, but consists of an array of

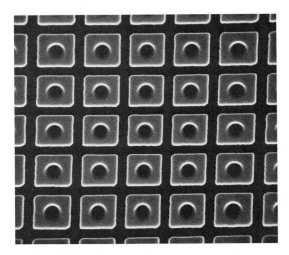

Fig. 7. Small section of the scanned side of a silicon diode target. Diodes are circles, and beam landing pads are squares. (*RCA*)

hundreds of thousands of individual diodes on a wafer of silicon (Fig. 7). The photocarriers are generated in the silicon wafer, but are collected and stored on the diode cells (Fig. 8). *See* SEMICONDUCTOR.

Most color television cameras now utilize the lead oxide or selenium-arsenic-tellurium tubes. Industrial and scientific-industrial cameras utilize the other types.

Silicon intensifier. The silicon intensifier camera tube utilizes a silicon diode target, but bombards it with a focused image of high-velocity electrons. These electrons are emitted by a photoemitter on the inside of the window on the front of the image section (Fig. 9). A fiber-optics window is utilized so that the emitting surface can be curved to produce good uniformity of focus of the high-energy electrons on the silicon diode target. Each high-energy electron can free thousands of electron carriers in the silicon wafer (compared to one carrier per photon of light on a silicon diode vidi-

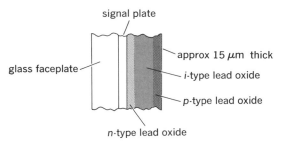

signal plate

glass faceplate

approx 15 μm thick

i-type lead oxide

p-type lead oxide

n-type lead oxide

Fig. 6. Cross-sectional view of lead oxide barrier-layer type of photoconductor.

con). This high amplification allows the camera to operate at light levels below that of the dark-adapted eye. With such a camera tube, it is possible to "see" the individual photons of light that compose a low-level optical image. The silicon intensifier tube is utilized for nighttime surveillance and other extremely low-light-level television uses in industrial, scientific, and military applications. It can be operated over a very wide range of light levels by varying the image section voltage. This changes the amplification over a range of more than 1000 to 1.

Solid-state imagers. These are solid-state devices in which the optical image is projected onto a large-scale integrated-circuit device which detects

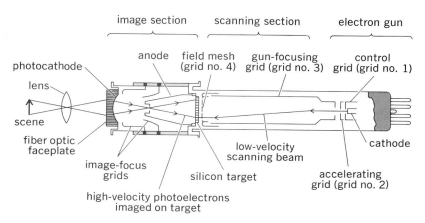

Fig. 9. Silicon intensifier vidicon. (*RCA Corp.*)

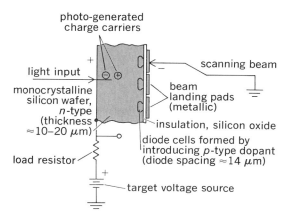

Fig. 8. Cross section of silicon diode target.

the light image and develops a television picture signal. Typical of these is the charge-coupled-device imager. The term charge-coupled device (CCD) refers to the action of the device which detects, stores, and then reads out an accumulated electrical charge representing the light on each portion of the image. The CCD transfers the individual charges to the output in the proper television scanning sequence to constitute a television video signal. The device detects light by absorbing it in a photoconductive substrate, such as silicon. The charge carriers generated by the light are accumulated in isolated wells on the surface of the silicon that are formed by voltages applied to an array of electrodes on top of an oxide insulator formed on the surface of the silicon. These wells are actually small MOS (metal-oxide-semiconductor) capacitors (Fig. 10a). Charges are transferred through the structure by varying the voltages on the metal electrodes. For example, if electrodes A and B are made more negative and C is made positive, the charges will move laterally from point 1 to point 2 (Fig. 10b). *See* INTEGRATED CIRCUITS.

A practical CCD imager employing these principles consists of a structure that forms several hundred thousand individual wells or pixels, and transfers the charges accumulated in these pixel wells out to an output amplifier in the proper sequence.

An example is shown in Fig. 11. Here the charges are accumulated by light exposure for the

time it takes to complete a single television picture, or approximately 1/60 s. Then all of the charges are rapidly transferred line by line upward into the storage register. In the storage area all charges are then moved upward one scan line at the end of each television line interval. The upper line of charges is moved into the horizontal readout register. Then the pixel charges are moved to the left through the horizontal readout register to the

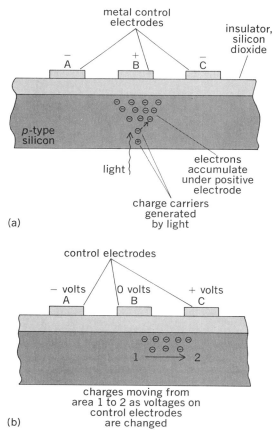

(a)

(b)

Fig. 10. Charge-coupled device. (*a*) Accumulation of an electron charge in a pixel element. (*b*) Movement of accumulated charge through the silicon by changing the voltages on the electrodes A, B, anc C.

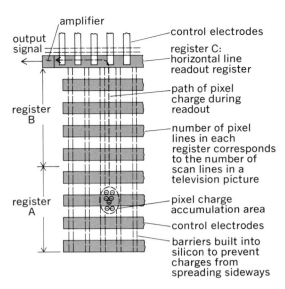

amplifier

output
signal

control electrodes

register C:
horizontal line
readout register

path of pixel
charge during
readout

register
B

number of pixel
lines in each
register corresponds
to the number of
scan lines in a
television picture

register
A

pixel charge
accumulation area

control electrodes

barriers built into
silicon to prevent
charges from
spreading sideways

Fig. 11. One type of charge-coupled-device imager. Register A accumulates the pixel charges produced by photoconductivity generated by the light image. The B register stores the lines of pixel charges and transfers each line in turn into register C. Register C reads out the charges laterally as shown into the amplifier.

output amplifier. During this readout period of the lines in the storage area, a new group of image charges are being accumulated in the imaging area, and the complete sequence is repeated every 1/60 s.

Investigation has been undertaken of many versions of solid-state imagers employing different light-sensitive materials and charge storage and readout methods.

A CCD imager has been incorporated in an image intensifier tube. In this device the charge carriers are generated by a focused image of high-energy electrons in the CCD imager in the same manner as in the silicon intensifier tube. This greatly enhances the sensitivity of the device and allows it to operate at very low light levels. *See* LIGHT AMPLIFIER.

[ROBERT G. NEUHAUSER]

Bibliography: D. G. Fink (ed.), *Television Engineering Handbook*, 1957; R. Kingslake (ed.), *Applied Optics and Optical Engineering*, vol. 2, 1965; R. G. Neuhauser, The silicon target vidicon, *J. SMPTE*, 86(6):414–418, June 1977; G. A. Robinson, The silicon intensifier target tube, *J. SMPTE*, 86(6):408–414, June 1977; H. V. Soule, *Electro-optical Photography at Low Illumination Levels*, 1968; P. K. Weimer et al., Multielement self-scanned sensors, *IEEE Spectrum*, 6(3):52–65, 1969; V. K. Zworykin and G. A. Morton, *Television*, 2d ed., 1954.

Television receiver

The equipment used to receive the transmitted modulated radio-frequency signals and produce synchronized visual images and sound for entertainment or educational purposes. The radio-frequency portion operates on the superheterodyne principle. *See* AMPLITUDE MODULATION; FREQUENCY MODULATION; RADIO RECEIVER.

The first television receivers to be mass-produced were monochrome; that is, they provided pictures in black and white only. Later, color receivers, which produce pictures in full color as well as black-and-white, became available. For basic discussion of a television system *see* TELEVISION; VIDEO TELEPHONE.

Monochrome receivers. Figure 1 shows a block diagram of a conventional monochrome television receiver, the major sections of which are discussed in the following paragraphs.

Antenna and transmission line. Since all broadcast television transmissions in the United States are horizontally polarized, the most basic type of television-receiving antenna is the horizontally mounted half-wave dipole. Because the stations serving a given area may operate on widely different frequencies, however, the dipole dimensions must be a compromise that permits reasonable performance on all the desired channels. More complex antennas combine several dipole elements of various lengths, and passive reflectors may be used to achieve some degree of horizontal directivity. Highly directive antennas are frequently mounted on remotely controlled rotators so that they can be pointed in the direction providing the best reception of the desired signal. The most common types of transmission line between the antenna and receiver are 300-ohm "twin-lead," employing polyethylene as a dielectric spacer between two uniformly spaced, unshielded wires. Also 75-ohm coaxial cable is used. *See* ANTENNA; TRANSMISSION LINES; YAGI-UDA ANTENNA.

Tuner. The tuner of a television receiver selects the desired channel and converts the frequencies received to lower frequencies within the passband of the intermediate-frequency amplifier. For very-high-frequency (vhf) reception the tuner generally has 12 discrete positions, corresponding to channels 2–13. For ultra-high-frequency (uhf) reception, continuous tuning is employed. Nearly all vhf tuners employ a radio-frequency (rf) amplifier, a mixer, and local-oscillator circuits arranged as shown in Fig. 1. In uhf tuners the rf amplifier is sometimes omitted because of the difficulty of obtaining low-noise amplification at uhf. The rf amplifier may be cascode, tetrode, or pentode. In general, the cascode circuit provides superior results.

The mixer and local-oscillator circuits may employ separate tube envelopes or may be combined in the same glass envelope. The received signal and the local oscillator signal are applied to the mixer. Difference frequencies, representing the picture and sound carriers, are produced and remain essentially constant as the rf amplifier, mixer, and oscillator circuits are tuned to the different channels. Known as intermediate frequencies (41.25 MHz for sound and 45.75 MHz for picture), they are available for further amplification. The correct oscillator frequency is approximately set at the time of channel selection. A fine adjustment is provided to permit more accurate tuning.

Such performance characteristics as noise factor, gain, bandwidth, and oscillator radiation must be optimized in the design of the tuner.

Intermediate-frequency amplifier. The output from the tuner is applied to the intermediate-frequency (i-f) amplifier. Several stages of

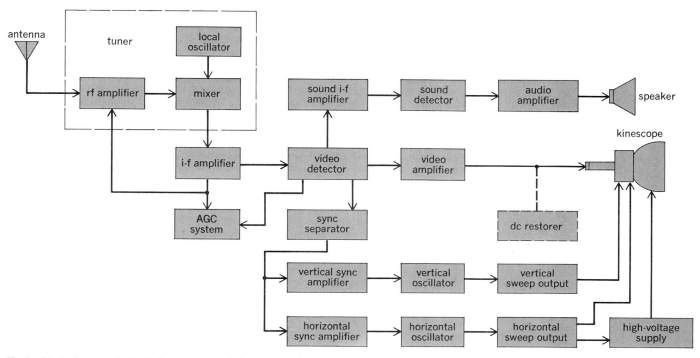

Fig. 1. Block diagram of a typical monochrome television receiver.

amplification are required to obtain the desired output signal level and selectivity.

The gain of this amplifier is essentially constant from 43 to 45 MHz. Above 45 MHz the response decreases such that at 45.75 MHz, the picture carrier frequency, it is 50%. This slope is required to compensate for the vestigial sideband transmitted signal.

Below 43 MHz the response decreases until at 41.25 MHz, the sound carrier frequency, it is 5–10% of the flat response. This minimizes cross modulation between picture and sound carriers. Fixed tuned trap circuits are used to produce sharp cutoffs at the lower and upper limits of the i-f passband. Sufficient selectivity is provided to minimize interference from signals originating in adjacent television channels.

Separation of video and audio. The output of the i-f amplifier consists of two modulated rf signals. One of these, which is amplitude-modulated, provides a varying signal corresponding to the black-and-white portions of the picture, a blanking signal to render the return trace invisible on the picture tube, horizontal sync pulses to initiate the retrace of the beam at the end of each line, and vertical sync pulses to initiate the retrace of the beam at the end of each picture field. The other signal is frequency-modulated and contains the transmitted sound information.

These two rf signals are applied to a diode, either a tube or crystal, which produces a rectified output that follows the instantaneous peak value of the amplitude-modulated picture carrier. The polarity of this output depends upon the design of the video amplifier and method of picture-tube drive. Usually, maximum picture carrier (sync-pulse modulation) produces a negative output voltage.

Coincidentally a 4.5-MHz signal results from the heterodyne beat of the picture and sound carriers. This signal contains the frequency-modulated sound information, which can be further amplified and detected in the sound channel. This is known as the intercarrier sound (ICS) system.

Following the detector is a video amplifier, which consists of one or two stages depending upon the overall receiver requirements. An output level of about 100 volts is ordinarily sufficient to assure full drive of the picture tube over its modulation range. Single-stage pentode amplifiers are generally adequate, driving the picture-tube cathode with sync positive. A 4.5 MHz trap is included in the video amplifier circuit to prevent the appearance of the intercarrier sound signal on the picture tube.

For sound reproduction, the intercarrier sound signal formed at the detector is passed through a 4.5-MHz i-f amplifier, through some form of amplitude limiting, and to an FM detector, which converts the sound carrier modulation to an audio-frequency signal. This signal is then passed through an audio amplifier to a loudspeaker and converted to acoustic output.

Automatic gain control. Since television receivers, like radio receivers, may be subjected to widely varying incoming signal strengths, some form of automatic gain control (AGC) is necessary. Circuits for this function provide a nearly constant carrier signal level to the video detector by changing the bias on the rf and i-f amplifier tubes as the strength of the incoming signal varies. A simple form may rectify the peak amplitude of the detected video signal in an *RC* load circuit. Under noisy conditions, however, its performance is not adequate. Improved forms employ an AGC amplifier and some type of gating circuit. Time gating voltage is derived from the horizontal sweep circuit during the beam scanning return time. In this way the plate current of the AGC amplifier tube is time-gated, and most noise pulses on the grid of the tube do not affect AGC action. *See* AUTOMATIC GAIN CONTROL (AGC).

Sync separator circuits. Picture synchronizing information is obtained from the video signal by means of sync separation circuits. In addition, these circuits must separate this information from noise and interference during the reception of weak signals, particularly if impulse noise is present. In general, sync separation circuits perform the following functions: (1) separation, by means of amplitude clipping, of the sync information from the picture information; (2) separation of the desired horizontal and vertical timing information by means of frequency selection; and (3) rejection of noise signals that are higher in amplitude than sync pulses by amplitude limiting or gating (noise suicide) circuits.

Sweep systems. Two independent sweep systems are employed in the vertical and horizontal sweep circuits. Each employs a timing generator, generally of the oscillatory type, controlled by the synchronizing information obtained from the sync separators. The oscillators are followed by drive and waveform-shaping circuits. These are followed by power amplifier stages capable of providing the currents required by the deflection coils of the yoke for picture-tube beam deflection. Substantially different techniques are required for vertical and horizontal scanning.

Vertical deflection. Generally the vertical oscillator is of the blocking type operating at approximately 60 Hz. Its frequency is accurately controlled by a signal obtained from the sync separator. The output waveform of the sync separator consists of a train of pulses representing the horizontal and vertical synchronizing pulses. When these are passed through a low-pass filter or integrating circuit, a sawtooth-shaped voltage wave representing vertical sync is obtained. This is applied to the grid of the vertical oscillator. A frequency control in the vertical oscillator circuit is so adjusted that its free-running frequency is slightly lower than the synchronizing signal frequency. For good interlace it is necessary that no horizontal frequency components be included in the vertical synchronizing voltage.

The vertical output stage is generally operated as a class A amplifier. The yoke is transformer-coupled to the plate of the output tube to match the yoke impedance to the output tube impedance. Since the yoke impedance is partly resistive and partly inductive, the voltage waveform across it is the sum of a sawtooth and a rectangular pulse. The current through the yoke has essentially a sawtooth waveform, but each sawtooth has a symmetrical S shape to take care of picture-tube faceplate geometry and result in a linear scan. *See* POWER AMPLIFIER.

Horizontal deflection. A more complex system is required for horizontal scanning. There are several basic reasons for this: (1) Horizontal sync pulses are of much shorter duration than are vertical sync pulses; (2) some form of automatic frequency control (AFC) of the horizontal oscillator is required to average the incoming horizontal sync information and retain accurate phase; and (3) considerably greater power output is required to generate the deflecting yoke fields as well as the high voltage, of 10–20 kV, for the picture tube. *See* AUTOMATIC FREQUENCY CONTROL (AFC).

The horizontal oscillator is generally of the blocking type. The frequency of oscillation is determined both by a time-constant control and by a bias voltage derived from an AFC circuit. The AFC circuit may be a phase comparator, in which the pulses from the sync separator are compared to the oscillator output signal. The output of the comparator is a voltage proportional to the phase departure of the two signals.

The desired current waveform in the horizontal windings of the deflection yoke is a line-frequency sawtooth, possibly modified by the addition of a small amount of S curvature to compensate for picture-tube face geometry. Energy from the horizontal drive, or output tube, is normally supplied to the yoke (through the horizontal output transformer) only during approximately the last half of each sawtooth period. At the conclusion of the sawtooth period, the horizontal driver is cut off, and the energy stored in the form of current through the yoke causes an oscillation in the self-resonant circuit consisting of the yoke, horizontal output transformer, and the associated capacitances. This oscillation is permitted to continue for only one half-cycle, during which time the current through the yoke reverses in polarity and attains a negative value almost equal to the original positive value. The self-resonant frequency of the horizontal output circuit must be high enough to permit the full current reversal to be accomplished within the horizontal blanking interval. The oscillation is stopped after the first half-cycle by the action of a damper tube (normally a diode), which controls the release of the energy stored in the yoke in such a way that the current follows the desired sawtooth waveform. In approximately the middle of the sawtooth period, the damper tube becomes nonconductive, and the horizontal driver tube takes over the task of supplying the energy required for the next cycle.

High-voltage supply. Since the impedance of the yoke at horizontal scan frequency is primarily inductive, the voltage across the horizontal deflection windings is essentially constant during active scan. During the retrace period, however, the high rate of current change causes the generation of a high-voltage pulse having a shape similar to that of a half sine wave and a duration equal to the retrace period. It is common practice to employ a stepup winding on the horizontal output transformer to raise this so-called kickback pulse up to a still higher voltage level, commonly about 18 kv, and to pass it through a simple rectifier and filter to serve as the high-voltage supply for the kinescope.

Picture tubes. The display device for a television receiver is a cathode-ray tube, consisting of an evacuated bulb containing an electron gun and a phosphor screen, which emits light when excited by an electron beam. The intensity of the electron beam is controlled by the video signal, which is applied either to the grid or the cathode of the electron gun. Ths position of the electron beam is controlled by electromagnetic fields produced by the deflection yoke placed around the neck of the tube. *See* CATHODE-RAY TUBE.

Controls. Certain controls are available to the

user for adjustment of the receiver. These are the audio volume, channel selector, fine tuning, brightness, contrast, horizontal hold, and vertical hold controls. Other controls, normally mounted on the rear of the chassis or under a removable panel, include height, width, and linearity controls.

The ON-OFF switch for the receiver is frequently mounted on the same shaft as the audio volume control, which controls the gain of the audio channel. The channel selector adjusts the tuner's selective circuits for optimum performance at the desired channel, and fine tuning is a vernier control for the frequency of the local oscillator. Brightness is usually a manual adjustment of the bias on the electron gun in the picture tube. The contrast control adjusts the level of the video signal, by some such means as a variable resistor in the cathode circuit of one of the video amplifier stages.

The horizontal and vertical hold controls adjust the free-running frequencies of the horizontal and vertical oscillators to achieve the most reliable synchronization with the incoming signal. In both cases, the controls may actually consist of variable resistors in the grid circuits of the respective blocking oscillators.

Vertical linearity is generally controlled by a variable resistance in the grid circuit of the vertical output stage, and picture height may be controlled by a variable resistor in the plate circuit of the vertical blocking oscillator. The width control may be a variable resistor in the screen grid circuit of the horizontal output tube, and horizontal linearity may be controlled by a variable inductor placed between the damper tube plate and the source of plate voltage.

Color receivers. Television receivers designed to produce images in full color are necessarily more complex than those designed to produce monochrome images only, because additional information must be handled to produce color. In monochrome systems, the video signal controls only the luminance of the various areas of the image. In color systems, it is necessary to control both the luminance and chrominance of the picture elements.

The chrominance of a color refers to those attributes which cause it to differ from a neutral (white or gray) color of the same luminance. While chrominance can be expressed in a great variety of ways, it is always necessary to employ at least two variables to express the full range of chrominance that can be perceived by the human eye. In qualitative terms, chrominance may be regarded as those properties of a color that control the psychological sensations of hue and saturation. For color television purposes, chrominance is most frequently expressed quantitatively in terms of the amounts of two hypothetical, zero-luminance primary colors (usually designated I and Q), which must be added to or subtracted from a neutral color of a given luminance to produce the color in question.

As a practical matter, color television receivers produce full-color images as additive combinations of red, green, and blue primary-color images, and it is necessary to process the luminance and chrominance information contained in a color signal in such a way as to make it usable by a practical reproducing device.

Nature of color signal. Color television broad-

casts in the United States employ signal specifications that are fully compatible with those used for monochrome, making it possible for color programs to be received on monochrome receivers and monochrome programs to be received on color receivers. (Color pictures are produced, of course, only when color programs are viewed through color receivers—in all other cases, the images are in black and white only.) Compatibility is achieved by encoding the color information at the transmitting end of a color television system in such a way that the transmitted signal consists essentially of a normal monochrome signal (conveying luminance information) supplemented by an additional modulated wave conveying chrominance information. Figure 2 shows the major components of a color television signal. Although it is added directly to the monochrome signal component before transmission, the color subcarrier signal does not cause objectionable interference, because of the use of the frequency interlace technique. Because the chrominance information involves two variables, the modulated subcarrier signal varies in both amplitude and phase, and it is necessary to employ sychronous detectors to recover the two variables. A phase reference for the special local oscillator, which provides the synchronized carriers in each color receiver, is transmitted in the form of so-called color synchronizing bursts. These are short samples of unmodulated subcarrier transmitted during the horizontal blanking periods after the horizontal sync pulses. *See* COLOR TELEVISION.

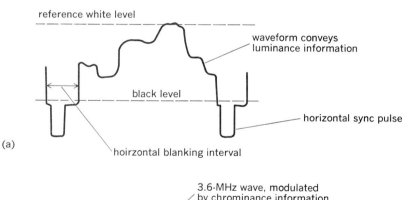

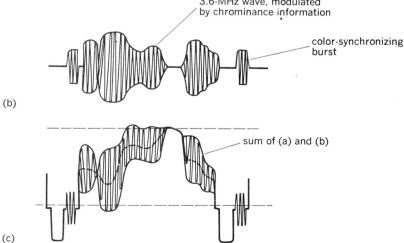

Fig. 2. Waveform sketches of major components of color television signal. (*a*) Normal monochrome signal. (*b*) Color subcarrier signal. (*c*) Complete color signal.

Overall color receiver. A simplified block diagram for a color television receiver is shown in Fig. 3. Since many of the circuits in a color receiver are the same in principle as the corresponding circuits in a monochrome receiver, it is unnecessary to redescribe them in detail. It is important to recognize, however, that all circuits handling the complete color signal must be designed for high performance standards. Because the chrominance information is received in the form of sidebands occupying the upper portion of the video spectrum (centered on approximately 3.6 MHz), it is necessary that the antenna, tuner, i-f amplifier, and video detector be designed to handle the full 4-MHz bandwidth provided in the broadcast transmission standards if degradation of the color information is to be avoided. Because the color subcarrier signal is simply added to the normal monochrome signal before transmission, it is necessary that all stages handling the complete signal be linear, so as to avoid intermodulation or distortion of the various signal components. The deflection circuits for a color receiver are similar in principle to those used in monochrome receivers, although the output stages are normally designed for a higher power level because of the greater deflection requirements for color kinescopes.

Color decoding circuits. Special decoding circuits are necessary in a color receiver to process the luminance and chrominance information in a color signal so that it can be used for the control of a practical color kinescope utilizing red, green, and blue primary colors. The major features of the most common approach to color decoding circuits are shown within the dotted lines in Fig. 3.

The video amplifier shown at the bottom handles the monochrome portion of the signal and is designed to provide attenuation in the vicinity of 3.6 MHz to block the passage of chrominance information. The chrominance information is recovered from the modulated subcarrier signal through a band-pass filter (centered at 3.6 MHz) and a pair of synchronous demodulators, in which the modulated wave is heterodyned against fixed carriers of two different phases but of the same frequency. In the most rigorous type of color decoding circuit, the chrominance components recovered from the modulated subcarrier signal are the same I and Q originally used to produce the modulated wave, but it is possible to use almost any two phase positions (not necessarily 90° apart) to recover any two independent combinations of the original I and Q signals. The bandwidths of the signals produced by the demodulators are normally adjusted somewhere between 0.5 and 1.5 MHz, and delay compensation may be required to keep all three signal components in time coincidence. The matrix circuit is essentially a linear cross-mixing network for combining the M, I, and Q signals in the proper proportions to produce red, green, and blue signals. If signals other than I and Q are produced by the chrominance demodulators, it is necessary only to design the matrix circuit with slightly different mixing constants.

The synchronous carriers required for the demodulation of the chrominance information are provided by a subcarrier regenerator, which is usually a burst-controlled oscillator operating at the subcarrier frequency. Control information for the subcarrier regenerator is obtained from a burst separator, which is a gate circuit turned on only during the horizontal blanking periods by pulses derived from the horizontal deflection system. The separated bursts are compared with the output of the local subcarrier oscillator in a phase detector. If an error exists, a correction voltage is developed, which may be applied through a reactance tube to restore the subcarrier oscillator to the proper frequency and phase. For good noise immunity, a time constant is normally provided so that control

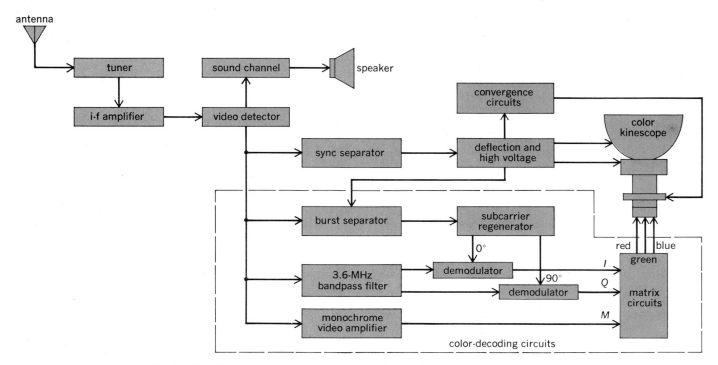

Fig. 3. Simplified block diagram of a color television receiver.

information is averaged over at least several line periods.

Color kinescope and convergence circuits. The great majority of color television receivers employ the shadow-mask color kinescope, in which color images are produced in the form of closely intermingled red, green, and blue dots. The primary-color phosphor dots are excited by three separate electron beams, which are prevented from striking dots of the wrong color by the shadowing effect of an aperture mask located about 1/2 in. (1.25 cm) behind the special phosphor screen. The three beams in such a kinescope are all deflected simultaneously by the fields produced by a single deflection yoke placed in the conventional position around the neck of the tube. It is necessary, however, to provide an auxiliary deflection system to maintain convergence of the three beams in all areas of the viewing screen, so that the primary-color images are properly registered. A special convergence yoke is commonly placed around the neck of the kinescope, just ahead of the electron guns; this yoke has separate coils for electromagnetic control of the positions of the three beams. Appropriate waveforms for the converence yoke coils are derived from the basic deflection waveforms. In general, each coil requires a different combination of sawtooth and parabolic waveforms at the horizontal and vertical scanning frequencies.

In other types of color kinescopes, not yet developed commercially, the need for convergence circuits may be eliminated by the use of a single-gun approach. Such alternative kinescopes may have other special requirements, such as precise position control of the single beam or special gating for the video signal.

Controls. In addition to the same controls required for monochrome receivers, color receivers normally have controls for convergence, hue, and saturation. The convergence controls, considered servicing adjustments only, adjust the relative amplitudes and phases of the signal components that are added together to form the proper waveforms for the convergence yoke. The hue control usually adjusts the phase of the burst-controlled oscillator and alters all the colors in the image in a systematic manner comparable to the effect achieved when a color circle diagram is rotated in one direction or the other. The proper setting for the hue control is normally determined by observing skin tones on actors and actresses. The saturation control, frequently labeled chroma or simply color, adjusts the gain of the chrominance circuits relative to the monochrome channel and controls the saturation or vividness of the reproduced colors. When this control is set too low, the colors are all pale or pastel, and when it is reduced to zero, the picture is seen in black and white only.

[CHESTER M. SINNETT/JOHN W. WENTWORTH]

Bibliography: B. Grob, *Basic Television*, 4th ed., 1975; B. Hartman, *Fundamentals of Television: Theory and Service*, 1975; M. S. Kiver, *Television Simplified*, 7th ed., 1973.

Television scanning

The process of scrutinizing the brightness of each element of detail contained in the image of a scene to be transmitted by television. In monochrome television, the process is instrumental in converting the brightness of each individual element so

scrutinized into a unique voltage-time response suitable for transmission. In color television, the brightness variations of each scene are first separated by red, green, and blue filters, after which the conversion from brightness to voltage on a time basis occurs separately for each of the three colors. Scanning also takes place in the receiver in exact synchronism with the camera-tube scanning, and synchronizing signals are transmitted for that purpose. *See* TELEVISION; TELEVISION CAMERA TUBE; TELEVISION RECEIVER.

Interlaced scanning. An image is analyzed by scanning it according to a fine structure of parallel, nearly horizontal lines called a scanning raster. The complete raster is rectangular in shape. Scanning may be done conventionally by starting at the upper left-hand corner along line 1 and moving toward the right at constant speed. At the end of line 1 a quick return is made to the left-hand side to start the scanning of line 2, again moving toward the right. When all lines have been scanned in this way, from top to bottom, the process is repeated by returning quickly to the upper left-hand corner to line 1. If all the lines are scanned in sequence, the process is called sequential scanning.

A variation of this kind of scanning, called interlaced scanning, is used to conserve bandwidth in the transmission system without introducing intolerable flicker. Flicker is a function of the frequency of repetition of coverage of the raster. With interlaced scanning, alternate (odd-numbered) lines are scanned first, and the remaining (even-numbered) lines are scanned next. The entire raster area is covered or scanned twice. Therefore, the picture repetition rate for interlaced scanning is twice that of sequential scanning, which is at the same velocity along each line, and a corresponding reduction in the sensation of flicker is obtained. This is called double interlacing and is standard in all broadcast television systems. The entire raster is covered 30 times per second; therefore, the picture area is scanned 60 times a second.

Lines and frequency of scanning. The United States standards require that 15,750 lines be scanned per second. With a vertical scanning rate of 30 times a second, there are 525 lines allocated to each frame and 262.5 lines to each field. Each line is 63.49 microseconds (μsec) in duration. A finite period of time is required to return the scanning beam of electrons in the camera or picture tube to the left edge of the scene for the next line. This period, blanking or retrace time, requires 16–18% of the total line time, or 10.16–11.43 μsec. Similarly, 7.5–8% of the vertical field period, or 1250–1333 μsec, is required for the scanning beam to return to the top of the picture. This is the equivalent of about 21 scanning lines. Blanking circuits prevent the transmission of brightness variations during both the horizontal and vertical retrace intervals.

The scanning process is shown in the illustration. Line 1 begins at the top center of the image. The beam proceeds for half a line to the right edge. Retrace to the left occurs, and line 3 follows. At the right end of line 3, 1½ lines have been scanned. Line 5 and successive odd-numbered lines are scanned, for a total of 241.5 lines, ending at the lower right corner. Twenty-one lines elapse during the vertical retrace interval while the scanning beam is moved to the upper left corner, placing it

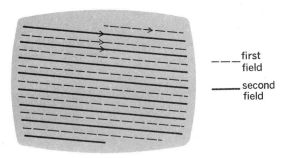

Scanning sequence for interlaced scanning. Spacing between lines is greatly exaggerated. (*From F. E. Terman, Electronic and Radio Engineering, McGraw-Hill, 1955*)

in position to start scanning line 2. This completes one full field. Then, 241.5 successive even-numbered lines are scanned, ending at the middle bottom of the scene. Again 21 lines elapse for vertical retrace, during which time the scanning beam is returned to the top center of the scene. This completes the second field and a full frame. The sequence then repeats.

Resolution. The resolution of a television system is a measure of its ability to reproduce fine detail. It is measured in terms of the number of lines appearing in the reproduced image of a test pattern at the output of a system. The diameter of the electron scanning beam in the camera tube limits the resolution. The smaller the spot, the higher is the resolution. With the 525-line system, the bandwidth f limits resolution according to the equation $f = 0.0125n$, where n is the number of lines and f is in megahertz. The resolution is 320 lines when the bandwidth is 4.0 MHz. A camera tube may have a resolution of 800 lines, but bandwidth limitations imposed by the television channel reduce that figure to about 300.

Color coding. A color television camera has separate tubes for red, green, and blue, each of which scans in the same manner as for monochrome. The three camera signals are combined in a colorplexer into a single signal containing both luminance and chrominance information. *See* COLOR TELEVISION.

[DONALD G. FINK]

Television transmitter

An electronic device that converts audio and video signals into modulated radio-frequency (rf) energy which can be radiated from an antenna and received on a television receiver. The term can also refer to the entire television transmitting plant, consisting of the transmitter proper, associated visual and aural input and monitoring equipment, transmission line, the antenna with its tower or other support structure, and the building in which the equipment is housed.

A television transmitter is really two separate transmitters integrated into a common cabinet (Fig. 1). Video information is transmitted via a visual transmitter, while audio information is transmitted via an aural transmitter. Because video and audio have different characteristics, the two transmitters differ in terms of bandwidth, modulation technique, and output power level. Nevertheless, a common transmitting antenna is generally used,

and the two transmitters feed this antenna via an rf diplexer or combiner.

Television stations are licensed to operate on a particular channel, but since it takes a very wide bandwidth to transmit a television picture, these channels are allocated over a broad range of frequencies. Channels 2 through 6 are low-band very-high-frequency (VHF) channels, while channels 7 through 13 are high-band VHF channels. Channels 14 through 83 are ultra-high-frequency (UHF) channels. Each channel is 6 megahertz wide. Because of the wide range of frequencies, television transmitters are designed to work in only one of the foregoing groups, and employ specific circuits which are most efficient for the channels involved. Nevertheless, every television transmitter, regardless of operating frequency, transmits a standard television signal in conformity with the regulations of the country in which it is operated.

Signal characteristics. In the United States the Federal Communications Commission (FCC) specifies the standard television signal in *FCC Rules and Regulations*, vol. 3, pt. 73. This specification enables receiver manufacturers to market receivers that are compatible with all television transmitters.

The FCC requires that a visual transmitter produce an amplitude-modulated (AM) carrier with an upper sideband extending to 4.2 MHz above the carrier and a lower sideband extending to only 0.75 MHz below the carrier. The lower sideband is restricted to this narrow bandwidth in order to conserve valuable frequency spectrum. Since both sidebands contain the same information, only one is required to transmit a picture. This is known as vestigial sideband transmission (Fig. 2). *See* AMPLITUDE MODULATION; SINGLE SIDEBAND.

FCC rules provide that the aural transmitter be frequency-modulated (FM) and that its carrier frequency must be 4.5 MHz above the visual carrier. This standard spacing between carriers allows use of a simplified sound receiver in most television sets. In the aural transmitter 100% modulation is

Fig. 1. Modern television transmitter. (*RCA Corp.*)

defined as equal to 25 kilohertz deviation, and the transmitter must be capable of faithfully passing audio-modulating frequencies from at least 50 Hz to 15 kHz. *See* FREQUENCY MODULATION.

Transmitter power. Television broadcast stations are limited to a specific effective radiated power (ERP) by the FCC. ERP is defined as the transmitter output power multiplied by transmitting antenna gain and an efficiency factor (less than 100%) due to the losses in the transmission line components between the transmitter and antenna.

To provide a consistent signal strength at the receiver, the FCC allows UHF stations to operate with more power (5 MW maximum visual ERP) than high-band VHF stations, and high-band VHF stations may operate with greater power (316 kW

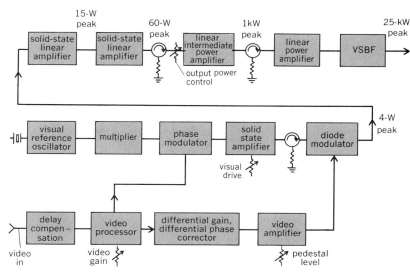

Fig. 3. Visual transmitter with modulation and vestigial-sideband filter (VSBF) at final frequency. (*From G. W. Bartlett, ed., NAB Engineering Handbook, 6th ed., National Association of Broadcasters, 1975*)

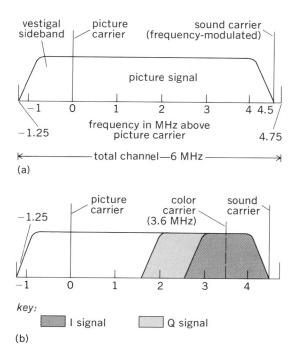

Fig. 2. Diagram of television channel showing portions occupied by color and monochrome signal components. (*a*) Monochrome television channel. (*b*) Color television channel. (*From G. W. Bartlett, ed., NAB Engineering Handbook, 6th ed., National Association of Broadcasters, 1975*)

The lowest-power transmitters are generally configured as television translators, which are used to relay the signal of a high-power primary station into areas where geography or other factors prevent viewers from being able to receive the primary station. The translator accepts the input on the primary channel and shifts its output to another channel so that one does not interfere with the other.

Transmitting antennas. Antenna gain is a function of design and the number of sections employed. Gain usually increases with physical size, due to increased radiating area, and may be increased further by stacking identical elements vertically; VHF stations usually operate with antenna gains of less than 5, while UHF stations commonly use gains of up to 50 in order to generate the

maximum) than low-band VHF stations (100 kW maximum). The maximum power that any station may utilize is reduced proportionally if its antenna height exceeds 1000 ft (305 m) above average terrain. Furthermore, due to its narrower bandwidth and other factors, the aural FM signal tends to carry better than the visual signal, so it is restricted to between 10 and 20% of the visual power by the FCC.

Because transmitting antennas usually have a power gain greater than unity, television transmitters need only provide a fraction of the ERP. Modern transmitters are rated at 10 W to 30 kW for the low-band VHF, and 10 W to 75 kW for high-band VHF; UHF transmitters are manufactured with outputs ranging from 100 W to 165 kW.

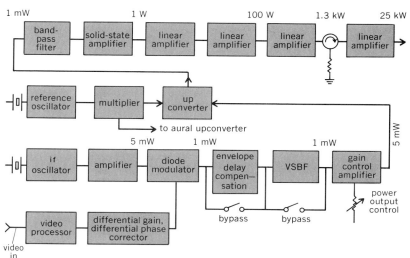

Fig. 4. Visual transmitter with modulation and vestigial-sideband filter (VSBF) at intermediate frequency. (*From G. W. Bartlett, ed., NAB Engineering Handbook, 6th ed., National Association of Broadcasters, 1975*)

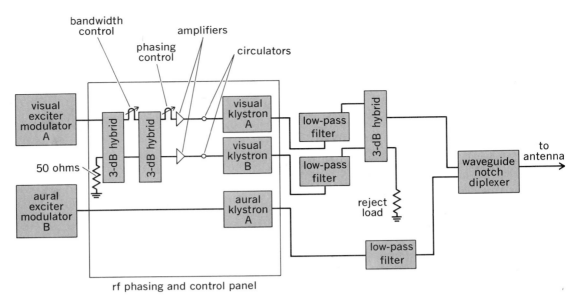

Fig. 5. Typical UHF transmitter block diagram.

much greater effective radiated power allowed by the FCC.

The horizontal radiation pattern of most television transmitting antennas is circular, providing equal radiated signal strength to all points of the compass. Higher-gain antennas achieve greater power in the direction of the horizon by reducing the power radiated at vertical angles above and below the horizon. Since this could result in weaker signals at some receivers close to the transmitter, beam tilt and null fill are often used to lower the angle of maximum radiated power.

Until about 1978, all television stations employed horizontally polarized antennas so that their signals would suffer less interference from impulse noise sources, such as automobile ignition systems, which tend to be vertically polarized. Some broadcasters have begun utilizing circularly polarized antennas in the belief that television sets with "rabbit-ear" receiving antennas will obtain a better signal.

Because television signals travel in a "line of sight," transmitting antennas are usually placed as high as possible above ground with respect to the surrounding service area. Such locations minimize signal blockage or ghosting due to tall buildings and hills. It is also desirable to locate all of the transmitting antennas serving a given locality in the same place. This allows viewers to orient their receiving antennas in one direction for the best reception from all of the stations. In Los Angeles, for example, advantage is taken of a 5700-ft (1740-m) summit, Mount Wilson, to locate most of the transmitting antennas on individual 200- to 500-ft (60- to 150-m) towers. In New York City all of the transmitting antennas are mounted on a common 365-ft (111-m) mast atop the 1366-ft (416-m) World Trade Center. In areas with neither mountains nor tall buildings, extremely tall towers have been erected, some of which exceed 2000 ft (600 m). *See* ANTENNA.

Transmitter designs. There are two broad classes of VHF visual television transmitter design philosophy. The classical approach modulates the carrier at a moderate power level, amplifies the carrier to rated output power by means of linear amplifiers, and then filters this high-power carrier to obtain the required vestigial-sideband signal Fig. 3). The more contemporary approach, used by nearly all transmitter manufacturers, employs modulation at a very low power level of an intermediate-frequency (if) signal. The required vestigial-sideband filtering is imposed on this low-level signal, generally by means of a highly stable surface acoustic-wave (SAW) filter, whereupon the signal is upconverted to the carrier frequency and amplified by linear amplifiers to rated output power (Fig. 4). *See* AMPLIFIER; SURFACE ACOUSTIC-WAVE DEVICES.

Generally, UHF transmitters employ very large tubes, called klystrons, to produce the large amounts of rf carrier power required. Figure 5 shows a block diagram of a typical UHF transmitter. *See* KLYSTRON; MICROWAVE TUBE.

Every transmitter contains a visual and aural exciter. This element determines the operating frequency of the visual and aural carriers and must be extremely stable, since the FCC requires that

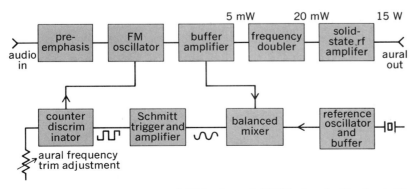

Fig. 6. Typical aural exciter. (*From G. W. Bartlett, ed., NAB Engineering Handbook, 6th ed., National Association of Broadcasters, 1975*)

they be kept within 1 kHz of the assigned frequencies. The aural exciter produces the required frequency modulation by varying the frequency of the rf carrier oscillator at an audio rate (Fig. 6).

Regardless of the visual modulation approach taken, certain parameters must be kept within FCC tolerances. Nonlinearity of the rf power amplifier stages must be compensated by suitable circuits. Certain time delays, within the transmitter, called envelope delays, must be the complement of those found in the home television receiver, so that colors in the picture are properly superimposed. Flat frequency response over adequate bandwidth is necessary to ensure that the picture has good detail and accurate color rendition. Excessively wide response is not permitted by the FCC, however, so that a low-pass filter which attenuates video above 4.75 MHz by at least 20 dB is commonly inserted in the video input circuit. *See* ELECTRIC FILTER.

Transmission of color requires that the color subcarrier (3.58 MHz above the visual carrier) not be affected by changes in the luminance level of the picture. A change in color saturation (chrominance level) brought about by a change in luminance level is termed differential gain. A change in color hue (subcarrier phase) brought about by a change in luminance level is termed differential phase. Both parameters are to be minimized. *See* COLOR TELEVISION.

The radiated rf carrier is also strictly regulated by the FCC. Harmonics, or multiples, of the carrier frequency must be attentuated at least 60 dB below the peak power level. Harmonic filters are placed on the output of both the visual and aural transmitters, ahead of any aural/visual diplexers, to ensure compliance.

In those transmitters which employ high-level vestigial-sideband filtering, a device known as a filterplexer often combines the functions of sideband filtering and aural/visual diplexing.

Modern transmitters are essentially all solid-state, with the exception of the final power amplifier stages. One manufacturer has employed semiconductor amplifiers to the 1600-W level. Beyond this point, large metal and ceramic vacuum tubes are used to generate rated output in both aural and visual transmitter sections. *See* TELEVISION; TELEVISION CAMERA; TELEVISION CAMERA TUBE; TELEVISION RECEIVER; TELEVISION SCANNING.

[EARL F. ARBUCKLE, III]

Bibliography: G. W. Bartlett (ed.), *NAB Engineering Handbook,* 6th ed., 1975; A. H. Bott, *SAW Filter Application in a TV Transmitter,* 1977; Progress report: Circularly polarized antennas for TV, *Broadcast Manage./Eng.,* 14(10):48–57, October 1978.

Thermionic emission

The emission of electrons into vacuum by a heated electronic conductor. In its broadest meaning, thermionic emission includes the emission of ions, but since this process is quite different from that normally understood by the term, it will not be discussed here. Thermionic emitters are used as cathodes in electron tubes and hence are of great technical and scientific importance. Although in principle all conductors are thermionic emitters, only a few materials satisfy the requirements set

by practical applications. Of the metals, tungsten is an important practical thermionic emitter; in most electron tubes, however, the oxide-coated cathode is used to great advantage. For a detailed discussion of practical thermionic emitters *see* VACUUM TUBE. *See also* ELECTRON EMISSION.

Richardson equation. The thermionic emission of a material may be measured by using the material as the cathode in a vacuum tube and collecting the emitted electrons on a positive anode. If the anode is sufficiently positive relative to the cathode, space charge (a concentration of electrons near the cathode) can be avoided and all electons emitted can be collected; the saturation thermionic current is then measured. Actually, the emission current increases slightly with increasing field strength at the cathode, and in order to obtain the true saturation current one should extrapolate to zero applied field. *See* SCHOTTKY EFFECT.

The emission current density J increases rapidly with increasing temperature; this is illustrated by the following approximate values for tungsten:

T (K)	1000	2000	2500	3000
J (amperes/cm^2)	10^{-15}	10^{-3}	0.3	15

The temperature dependence of J is given by Eq. (1), the Richardson (or Dushman-Richardson)

$$J = AT^2 e^{-(\phi/kT)} \qquad (1)$$

equation. Here A is a constant, k is Boltzmann's constant ($= 1.38 \times 10^{-23}$ joule/degree), and ϕ is the work function of the emitter. The work function has the dimensions of energy and is a few electronvolts for thermionic emitters.

The temperature dependence of J is essentially determined by the exponential factor, since its temperature dependence predominates strongly over that of the factor T^2. Both A and ϕ may be obtained experimentally by plotting the logarithm of J/T^2 versus $1/T$, as illustrated for tungsten in Fig. 1. The Richardson formula can be derived for

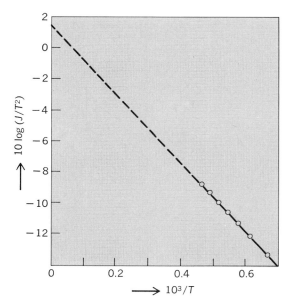

Fig. 1. Richardson plot for tungsten, an important thermionic emitter. (*After G. Herrmann and S. Wagener, The Oxide-Coated Cathode, vol. 2, Chapman and Hall, 1951*)

metals and semiconductors on the basis of relatively simple physical models.

Metals. According to quantum theory the electrons in a free atom occupy a set of discrete energy levels. When atoms are brought together to form a solid, these energy levels broaden into energy bands; the broadening is a result of the perturbing fields produced by neighboring atoms on the electrons and is most pronounced for the outer or valence electrons. In a metal, the perturbing influence on the valence electrons is so strong that they can no longer be associated with particular atoms but must be considered as moving freely throughout the crystal. These so-called free, or conduction, electrons are responsible for the high electrical and thermal conductivity of metals and also for the thermionic emission. *See* BAND THEORY OF SOLIDS.

The free electrons may be assumed to move in an approximately constant potential as indicated in Fig. 2. The bottom of the box corresponds to the energy of a conduction electron at rest in the metal; the "vacuum" level represents the energy of an electron at rest in free space. According to quantum mechanics, the electrons in this model can assume only particular states of motion which correspond to a set of very closely spaced energy levels. The probability for a given state to be occupied depends on the energy E of the state and on the absolute temperature T in accordance with Eq. (2), the so-called Fermi-Dirac distribution function.

$$F(E) = \frac{1}{1 + \exp\left[(E - E_F)/kT\right]} \qquad (2)$$

The quantity E_F is called the Fermi energy; it is determined by the number of electrons per unit volume in the metal and is of the order of a few electronvolts. Since kT at room temperature ($T = 300$ K) is only about 0.025 eV, $E_F \gg kT$ for all temperatures below the melting point of metals. Note that for $T = 0$, $F(E) = 1$ for $E < E_F$, and $F(E) = 0$ for $E > E_F$. Hence, at absolute zero all energy levels up to E_F are occupied by electrons, whereas those above E_F are empty. For temperatures different from zero, some electrons have energies larger than E_F and the thermionic emission is due to those electrons in the "tail" of the Fermi distribution for which the energy lies above the

vacuum level in Fig. 2. Note that when $E = E_F$, $F(E) = 0.5$; that is, the Fermi energy corresponds to those states for which the probability of being occupied is equal to 0.5.

When these ideas are put in a quantitative form, one arrives at the Richardson equation with the specific value $A = 120$ amperes/cm² (if one takes into account reflection of electrons against the surface potential barrier, the theoretical value of A is < 120 amperes/cm²).

Experiments by M. N. Nichols in 1940 and by G. F. Smith in 1954 on single crystals of tungsten have shown that experimental values for A and ϕ depend on the crystallographic plane from which the emission is measured; values for A (in amperes/cm²) and ϕ (in electronvolts) for two crystallographic directions are given in the table. For

Experimental values for single crystals of tungsten

Direction	Nichols		Smith	
	A	ϕ	A	ϕ
(111)	35	4.39	52	4.38
(100)	117	4.56	105	4.52

polycrystalline metals, the experimental values for A and ϕ are thus average values for the particular specimen.

Semiconductors. For semiconductors, the thermionic emission is also due to the escape of electrons which have energies above the vacuum level. The theory leads to the Richardson formula, as it does for metals. The work function measures again the difference between the Fermi level of the semiconductor and the vacuum level. *See* SEMICONDUCTOR.   [A. J. DEKKER]

Thermionic tube

An electron tube that relies upon thermally emitted electrons from a heated cathode for tube current.

Thermionic emission of electrons means emission by heat. In practical form an electrode, called the cathode because it forms the negative electrode of the tube, is heated until it emits electrons. The cathode may be either a directly heated filament or an indirectly heated surface. With a filamentary cathode, heating current is passed through the wire, which either emits electrons directly or is covered with a material that readily emits electons. Some typical filament structures are shown in Fig. 1. Filaments of tungsten or thoriated tungsten are commonly used in high-power transmitting tubes where their ruggedness and ability to withstand high voltages are essential. Oxide-coated filaments are used in a few small high-voltage rectifier tubes.

Indirectly heated cathodes have a filament, commonly called the heater, located within the cathode electrode to bring the surface of the cathode to emitting temperature. Some common forms are shown in Fig. 2. They are usually coated with barium-strontium oxide, on the periphery in receiving tubes and on the end in kinescopes. Because the emitting surface carries no heating current, there is no voltage drop along the surface. Hence such cathodes are usually known as equipo-

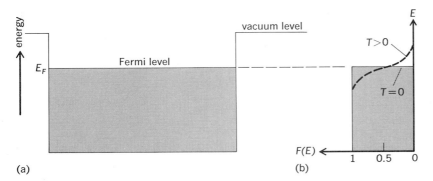

Fig. 2. Free electrons are assumed to move in approximately constant potential. (a) Occupation of electron states between the bottom of the conduction band and the Fermi level of a metal is indicated for $T = 0$ by the shaded area. (b) Fermi distribution function is represented schematically for $T = 0$ and for $T > 0$.

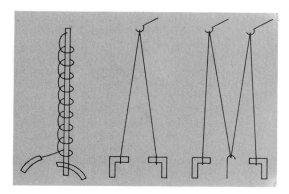

Fig. 1. Typical filament structures.

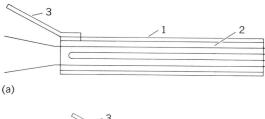

(a)

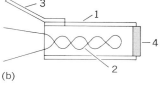

(b)

Fig. 2. Cathodes. (a) Receiving-tube cathode: 1, cathode sleeve, oxide-coated on exterior; 2, folded heater, insulated with refractory oxide; 3, cathode tab, for electrical connection. (b) Kinescope cathode: 1, cathode sleeve; 2, heater, insulated with refractory oxide; 3, cathode tab, for electrical connection; 4, emitting "button," oxide-coated on right surface.

tential cathodes. The high emission capability, the equipotential surface, and the favorable geometry of these cathodes make possible the close-spaced tube structures that lead to the high transconductances required in modern applications. Hence, oxide-coated equipotential cathodes are used in almost all receiving and medium-power transmitting tubes. They are also used in some high-power pulsed transmitting tubes, where the remarkable ability of the oxide cathode to emit very high current densities (tens of amperes per square centimeter, for microsecond periods at low repetition rates) is exploited. The majority of all vacuum tubes are thermionic tubes. It is possible to make so-called cold-cathode tubes, but they tend to be unstable in vacuum and find their main application in gas tubes, not vacuum tubes. *See* ELECTRON TUBE; GAS TUBE; THERMIONIC EMISSION; VACUUM TUBE.   [LEON S. NERGAARD]

Thévenin's theorem (electric networks)

A theorem from electric circuit theory. It is also known as the Helmholtz or Helmholtz-Thévenin theorem, since H. Helmholtz stated it in an earlier form prior to M. L. Thévenin. Closely related is the Norton theorem, which will also be discussed.

Laplace transform notation will be used.

Thévenin's theorem states that at a pair of terminals a network composed of lumped, linear circuit elements may, for purposes of analysis of external circuit or terminal behavior, be replaced by a voltage source $V(s)$ in series with a single impedance $Z(s)$. The source $V(s)$ is the Laplace transform of the voltage across the pair of terminals when they are open-circuited; $Z(s)$ is the transform impedance at the two terminals with all independent sources set to zero (Fig. 1). The Thévenin equivalent may also be found experimentally.

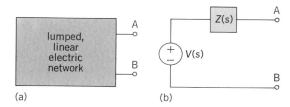

Fig. 1. Network and its Thevenin equivalent. (a) Original network. (b) Thévenin equivalent circuit.

Norton's theorem states that a second equivalent network consists of a current source $I(s)$ in parallel with an impedance $Z(s)$. The impedance $Z(s)$ is identical with the Thévenin impedance, and $I(s)$ is the Laplace transform of the current between the two terminals when they are short-circuited (Fig. 2).

Thévenin's and Norton's equivalent networks are related by the equation $V(s) = Z(s) \cdot I(s)$. This may be seen by comparing Figs. 1b and 2b. In Fig. 1b, if terminals A and B are short-circuited, a current $I(s) = V(s)/Z(s)$ will flow; this is also true in Fig. 2b. Similarly the open-circuit voltage in Fig. 2b is $V(s) = Z(s) \cdot I(s)$. *See* ALTERNATING-CURRENT CIRCUIT THEORY.

These theorems are useful for the study of the behavior of a load connected to a (possibly complex) system that is supplying electric power to that load. The system may be a power distribution system, such as in a home or office, in which case the load may be lights or appliances. The system may be an electronic amplifier, in which case the load may be a loudspeaker. However, the theorem is of no value in studying the internal system behavior, because the behavior of the equivalent network is very different from that of the original.

Examples. Two examples will be used to show how Thévenin and Norton equivalent networks may be calculated from the original network and

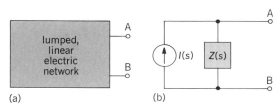

Fig. 2. Network and its Norton equivalent. (a) Original network. (b) Norton equivalent circuit.

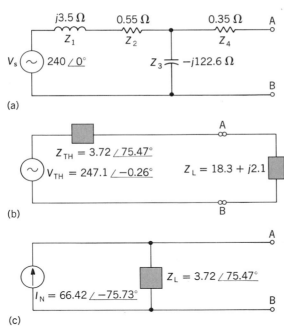

(a)

(b)

(c)

Fig. 3. Power distribution circuit and its Thévenin and Norton equivalents. (a) Original circuit. (b) Thévenin equivalent circuit with its load Z_L connected. (c) Norton equivalent circuit.

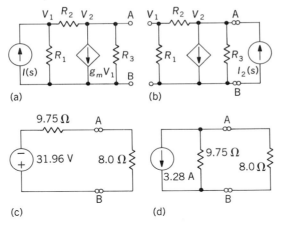

(a)　(b)

(c)　(d)

Fig. 4. Amplifier and its Thévenin and Norton equivalents. Numerical values are given in the text. (a) Original circuit. (b) Circuit constructed to find Thévenin impedance. (c) Thévenin equivalent circuit. (d) Norton equivalent circuit.

then used for some typical calculations.

Power distribution circuit. Suppose a simplified power distribution circuit contains the elements shown in Fig. 3a. Sinusoidal steady-state operation is assumed. In this circuit the voltage across A, B is given by Eq. (1), and with the source set to zero, the impedance at A, B is given by Eq. (2). Thus the

$$V_{AB} = \frac{Z_3}{Z_1 + Z_2 + Z_3} \cdot V_s$$
$$= \frac{-j122.6}{0.55 + j3.5 - j122.6} \cdot 240\underline{/0°} \qquad (1)$$
$$= 247.1\underline{/-0.26°} \; V$$

$$Z_{TH} = Z_4 + \frac{Z_3(Z_1 + Z_2)}{Z_1 + Z_2 + Z_3}$$
$$= 0.35 + \frac{(-j122.6)(0.55 + j3.5)}{0.55 + j3.5 - j122.6} \qquad (2)$$
$$= 3.72\underline{/75.47°} \text{ ohms}$$

Thévenin equivalent is given in Fig. 3b, and the Norton equivalent in Fig. 3c.

When a load $Z_L = 18.3 + j2.1 = 18.42\underline{/6.55°}$ is connected, at A, B, the current through the load is given by Eq. (3), and the power delivered to the

$$I_L = \frac{V_{TH}}{Z_{TH} + Z_L} = \frac{247.1\underline{/-0.26°}}{3.72\underline{/75.47°} + 18.42\underline{/6.55°}} \qquad (3)$$
$$= 12.32\underline{/-16.77°}$$

load is $(12.32)^2(18.3) = 2.776$ kW. Other loads are handled in a similar fashion.

Amplifier. As a second example, suppose that the circuit of Fig. 4a is a simplified model of an electronic amplifier, and that a load (loudspeaker) is to be connected at A, B. The circuit is driven by a current source $I(s)$. For analysis, the voltages $V_1(s)$ and $V_2(s)$ are the transforms of the voltages on their respective nodes, and become the dependent variables in the analysis. The dependent or controlled source $(g_m V_1)$ models the amplification. Two Kirchhoff current law equations, (4) and (5),

$$I(s) = V_1\left(\frac{1}{R_1} + \frac{1}{R_2}\right) - V_2\left(\frac{1}{R_2}\right) \qquad (4)$$

$$-g_m V_1 = -V_1\left(\frac{1}{R_2}\right) + V_2\left(\frac{1}{R_2} + \frac{1}{R_3}\right) \qquad (5)$$

may be used to find the Thévenin voltage, which is also V_2. Solution of this pair of equations gives Eq. (6). *See* KIRCHHOFF'S LAWS OF ELECTRIC CIRCUITS.

$$V_2(s) = \frac{(-g_m + 1/R_2) \cdot I(s)}{\dfrac{1}{R_1 R_2} + \dfrac{1}{R_1 R_3} + \dfrac{1}{R_2 R_3} + \dfrac{g_m}{R_2}} \qquad (6)$$

To find the Thévenin impedance, $I(s)$ must be set to zero, which leaves an infinite impedance in the branch, and an auxiliary current source $I_2(s)$ must be added between A and B (Fig. 4b). A new set of equations, (7) and (8), is written, and solved

$$0 = V_1\left(\frac{1}{R_1} + \frac{1}{R_2}\right) - V_2\left(\frac{1}{R_2}\right) \qquad (7)$$

$$-g_m V_1 + I_2(s) = -V_1\left(\frac{1}{R_2}\right) + V_2\left(\frac{1}{R_2} + \frac{1}{R_3}\right) \qquad (8)$$

for the ratio $V_2(s)/I_2(s)$, which is the desired Thévenin impedance; the controlled source must not be set to zero.

Solution of Eqs. (7) and (8) gives, after simplification, Eq. (9).

$$Z_{TH}(s) = \frac{V_2(s)}{I_2(s)} = \frac{R_3(R_1 + R_2)}{R_1 + R_2 + R_3 + g_m R_1 R_3} \qquad (9)$$

To make a numerical example in a purely resistive circuit, let $R_1 = 2.0$ kΩ, $R_2 = 8.2$ kΩ, $R_3 = 400$ Ω, $g_m = 510 \times 10^{-3}$ S (siemens), and $I = 4.0$ mA. Substitution of these into the equations gives a Thévenin voltage $V_{TH} = -31.96$ V, and a Thévenin

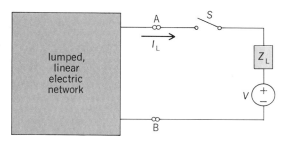

Fig. 5. Circuit constructed to demonstrate Thévenin's theorem.

impedance $Z_{TH} = 9.75\ \Omega$. This is shown in Fig. 4c, and the corresponding Norton equivalent is shown in Fig. 4d, where $I_N = -31.96/9.75 = -3.28$ A. In both cases the negative signs lead to a polarity reversal, which is reflected in Fig. 4c and d.

Suppose an 8.0-ohm speaker is connected at A, B. A current of $31.96/(8.0 + 9.75) = 1.80$ A will flow, giving a power of $(1.80)^2(8.0) = 25.95$ W.

Proof. To prove this theorem, consider a general network with two accessible terminals, as in Figs. 1a and 2a, to which an auxiliary voltage source V and an impedance $Z_L(s)$ have been added (Fig. 5). Let this source be such as to cause $I_L = 0$ when switch S is closed. By superposition, Eq. (10) is

$$I_L = 0 = \frac{V_{TH}}{Z_{TH} + Z_L} - \frac{V}{Z_{TH} + Z_L} \tag{10}$$

valid. This shows that $V_{TH} = V$, and that the current that flows when V is removed is given by Eq. (11).

$$I_L = \frac{V_{TH}}{Z_{TH} + Z_L} \tag{11}$$

See NETWORK THEORY; SUPERPOSITION THEOREM (ELECTRIC NETWORKS).   [EDWIN C. JONES, JR.]

Bibliography: M. E. Van Valkenburg, *Network Analysis*, 3d ed., 1974.

Thyratron

A hot-cathode gas-filled tube with one or more grids placed between the cathode and anode to provide control characteristics (Fig. 1). Thyratrons operate in the gas arc region using mercury vapor where temperature control is possible, or an inert gas such as argon or xenon. In some tubes both mercury and an inert gas are used to take advantage of the desirable features of each. For applications requiring great accuracy in control, fast deionization, and very high peak currents of short duration (as in radar pulse-modulators), hydrogen gas is used. *See* ELECTRICAL CONDUCTION IN GASES; GAS TUBE.

Control action. The control action of a thyratron is quite different from that of a vacuum electron tube. In the high-vacuum tube the anode current is modulated by the action of small varying voltages applied to the grid. In the thyratron the grid serves to prevent the flow of current until its potential is made less negative than a critical value. When this occurs, electrons from the cathode are accelerated toward the anode and quickly produce an arc plasma by ionization by collision. Once the arc is established the grid has no further

effect on the arc current as long as the anode is sufficiently positive relative to the cathode to supply the arc voltage. When the anode potential is made zero or negative, the arc is extinguished and the grid regains control after a period of time known as the deionization time or recovery time. This is the time necessary for the ions of the residual arc plasma to be neutralized to a sufficiently low density to permit the grid to regain control. The deionization time is affected by both applied grid voltage and the grid resistance and is about 1 msec for most tubes, although some special tubes have deionization times as low as 10 μsec. During the conducting period the grid acts essentially like a probe in a plasma. A negative grid in a plasma collects a positive ion current from the plasma, while a positive grid collects an electron current. By varying the grid potential relative to the cathode, a grid current–grid voltage characteristic can be plotted. The positive-ion space charge that forms about a negative probe has a thickness that can be calculated from the space-charge equation using the mass of the positive ions. At quite small arc currents and with grids having small holes for the arc path, this positive-ion space-charge sheath can become sufficiently thick so as to close off the conducting path and stop the current—a mode of operation that is seldom employed. Since the function of the grid in a thyratron is to permit a relatively small negative voltage to shield the region about the cathode from the positive field produced by the anode potential, it is evident that the grid must be made increasingly negative as the anode potential is increased.

Control characteristics. The relation between the voltage necessary to prevent conduction and the anode potential is called the control characteristic. Figure 2 shows a typical control characteristic for a negative-grid thyratron (triode), such as Fig. 1. The shaded area represents the range that

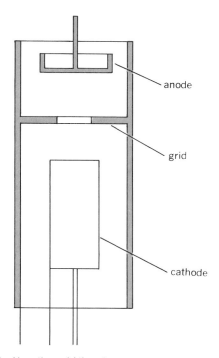

Fig. 1. Negative-grid thyratron.

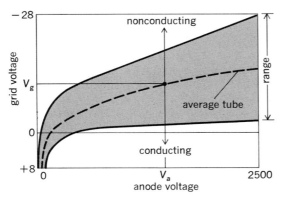

Fig. 2. Control characteristic of negative-grid thyratron.

may be expected for a group of tubes. The dotted curve represents the characteristic of an average tube. For a given value of anode voltage V_a the tube is nonconducting for grid voltages more negative than the characteristic V_g and conducting for a more positive grid voltage. The control characteristic of mercury tubes is also affected by the condensed mercury temperature, so for this type of tube the characteristics are usually given for several typical temperatures in the recommended operating range.

The construction of a shield-grid thyratron (tetrode) is shown in Fig. 3. The grid control characteristics of this tube are similar to those shown in Fig. 2, and are usually presented as a family with the shield-grid potential as a parameter. For a sufficiently negative voltage on the shield grid the tube is given positive-grid characteristics; that is, it is necessary to place a positive potential on the control grid before the tube will conduct. An important advantage of the shield-grid tube for many applications is that the control grid requires only a

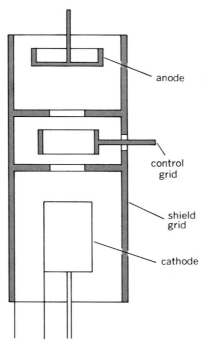

Fig. 3. Shield-grid thyratron.

small current to initiate conduction. The control grid is also so placed within the shield grid that heat from the cathode and the anode cannot reach it. Thus even if active material from the cathode reaches the control grid, its low grid-current property is maintained for a long period, since it never gets hot enough for thermionic emission.

Ionization time. The ionization time of a thyratron is the time required for the rated starting voltage on the grid to establish an arc in the tube. The ionization time for most thyratrons is from 0.5 to 10 μsec, depending largely on the gas and tube construction.

Thyratron ratings. Three current ratings are given for thyratrons: (1) the average current, set by the ability of the tube to dissipate heat; (2) the peak current, the largest instantaneous current that the tube is designed to carry—the peak current averaged over a period known as the integration time must not exceed the average current; and (3) the surge current (or fault), the greatest current that may be passed by the tube under abnormal conditions such as an arc-back or a load short circuit. The life of the tube will be shortened each time this current is attained. Circuits should have sufficient impedance and be fused so that this current is never exceeded.

[JAMES D. COBINE]

Time constant

The time required for a physical quantity to change its initial (zero-time) magnitude by the factor $(1 - 1/\epsilon)$ when the physical quantity is varying as a function of time, $f(t)$, according to the decreasing exponential function as in Eq. (1), or the in-

$$f(t) = \epsilon^{-kt} \qquad (1)$$

reasing exponential function as in Eq. (2). See Figs. 1 and 2.

$$f(t) = 1 - \epsilon^{-kt} \qquad (2)$$

The numeric ϵ has the value 2.71828. Therefore, the change in magnitude of $(1 - 1/\epsilon)$ has the fractional value 0.632121. Thus, after a time lapse of one time constant, starting at zero time, the magnitude of the physical quantity will have changed 63.2%.

When time t is zero, Eq. (1) has the magnitude 1, and when time t is $1/k$, the magnitude is ϵ^{-1}, or $1/\epsilon$. The corresponding change in magnitude is $(1 - 1/\epsilon)$. The specific time required to accomplish this change is shown in Eq. (3), where T is called

$$t = 1/k = T \qquad (3)$$

the time constant and is usually expressed in seconds. The same results are obtained for Eq. (2).

The initial rate of change of both the increasing and decreasing functions is equal to the maximum amplitude of the function divided by the time constant. Figures 1 and 2 are universal in that the plotted function is of unit height and the time scale is given in terms of time constants. To use these curves for a specific problem, the values in the ordinate axis are multiplied by the maximum amplitude of the quantity occurring in the problem, and the values in the abscissa axis are multiplied by the numerical value of the corresponding time constant.

The concept of time constant is useful when

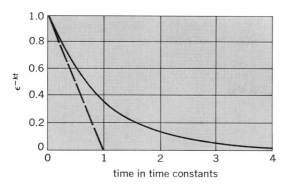

Fig. 1. Universal time-constant curve indicated for the decreasing function.

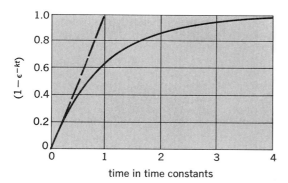

Fig. 2. Universal time-constant curve indicated for the increasing function.

evaluating the presence of transient phenomena. The relative amplitude of a transient after a lapsed time of a certain number of time constants is readily computed:

Lapsed time, time constants	Transient completed, %
1	63.2
2	86.5
3	95.0
4	98.2
5	99.3
10	99.996

Usually a transient can be considered as being over after a period of 4–5 time constants.

For electric circuits, the coefficient k and thus the time constant T is determined from the parameters of the circuit. For a circuit containing resistance R and capacitance C, the time constant T is the product RC. When the circuit consists of inductance L and resistance R, the time constant is L/R. See ELECTRIC TRANSIENT.

The concept of time constant can be applied to the transient envelope of an ac signal; however, it is more common to describe the change in amplitude in terms of logarithmic decrement.

[ROBERT L. RAMEY]

Townsend discharge

A particular part of the voltage-current characteristic curve for a gaseous discharge device named for J. S. Townsend, who studied it about 1900. It is that part for low current where the discharge can-

not be maintained by the field alone. Thus, if the agents producing the initial ionization were removed, conduction would cease.

In the lower end of this region, conduction is accomplished only by charges produced by external agents. As the electric field is increased, secondary ionization and more efficient collection of the primary ionization cause an increase in the current. After further increase in the field, the end of the Townsend region is reached. Any additional increase in the field causes a transition into a region where the discharge may be maintained by the field alone, whether it be glow, brush, or arc. See DARK CURRENT; ELECTRICAL CONDUCTION IN GASES; GLOW DISCHARGE.

[GLENN H. MILLER]

Transducer

Any device or element which converts an input signal into an output signal of a different form. An example is the microphone, which converts vibrations caused by an impinging sound wave into an electrical signal. This electrical signal can be measured to determine the magnitude of the sound wave; it can be recorded (through the use of another transducer); or it can be used to control some instrument. Although the most common transducers are designed to transform periodic signals (such as sound waves or alternating-current electrical signals), the word also applies to devices which convert static signals from one form to another. An example of this is the barometer, which produces a signal proportional to the atmospheric pressure. The input signal is the atmospheric pressure. The output signal can either be a mechanical displacement (a dial reading or a liquid level) or it can be a direct-current electrical signal. A different type of transducer is the photoelectric cell, which produces an electrical signal in response to incident light. The most widely used class of transducers is the electromechanical transducer, which converts an electrical signal into a mechanical signal (a vibration or a displacement) or vice versa. Aside from the microphone mentioned above, this class includes phonograph pickups, loudspeakers, automobile horns, doorbells, and underwater transducers. See MICROPHONE.

[M. A. BREAZEALE]

Bibliography: D. E. Gray (ed.), *Amer. Inst. Phys. Handb.*, 3d ed., 1972; J. D. Lenk, *Handbook of Controls and Instrumentation*, 1980; W. P. Mason (ed.), *Phys. Acoust.*, vol. 1, pts. A and B, 1964; P. H. Sydenham, *Transducers in Measurement and Control*, 2d ed., 1980.

Transformer

An electrical component used to transfer electric energy from one alternating-current (ac) circuit to another by magnetic coupling. Essentially, it consists of two or more multiturn coils of wire placed in close proximity to cause the magnetic field of one to link the other. In general, the transformer accomplishes one or more of the following between two circuits: (1) a difference in voltage magnitude, (2) a difference in current magnitude, (3) a difference in phase angle, (4) a difference in impedance level, and (5) a difference in voltage insulation level, either between the two circuits or to ground.

Transformers are used to meet a wide range of

requirements. Pole-type distribution transformers supply relatively small amounts of power to residences. Power transformers are used at generating stations to step up the generated voltage to high levels for transmission. The transmission voltages are then stepped down by transformers at the substations for local distribution. Instrument transformers are used to measure voltage and currents accurately. Audio- and video-frequency transformers must function over a broad band of frequencies. Radio-frequency transformers transfer energy in narrow frequency bands from one circuit to another.

Transformers are often classified according to the frequency for which they are designed. Power transformers are for power-frequency circuits, audio transformers for audio-frequency circuits, and so forth. Of course, many of the basic principles of operation apply to all.

POWER TRANSFORMERS

A power transformer consists of two or more multiturn coils wound on a laminated iron core. At least one of these coils serves as the primary winding.

Principle of operation. When the primary of a power transformer is connected to an alternating voltage, it produces an alternating flux in the core. The flux generates a primary electromotive force, which is essentially equal and opposite to the voltage supplied to it. It also generates a voltage in the other coil or coils, one of which is called a secondary. This voltage generated in the secondary will

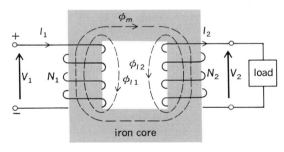

TRANSFORMER

Fig. 1. Basic transformer.

supply alternating current to a circuit connected to the terminals of the secondary winding. A current in the secondary winding requires an additional current in the primary. The primary current is essentially self-regulated to meet the power (or volt-ampere) demand of the load connected to the secondary terminals. Thus in normal operation, energy (or volt-amperes) can be transferred from the primary to the secondary electromagnetically.

Figure 1 shows a transformer with a primary of N_1 turns and a secondary of N_2 turns. A primary voltage V_1 causes a current I_1 to flow through the coil. Since all quantities shown are alternating, the arrows indicate only instantaneous polarities.

The magnetic flux ϕ set up by the primary consists of two components. One part passes completely around the magnetic circuit defined by the iron core, thus linking the secondary coil. This is the mutual flux ϕ_m. The second part is a smaller

component of flux that links only the primary coil. This is the primary leakage flux ϕ_{l1}. If the secondary circuit is completed through a load, a secondary current I_2 flows and in turn creates a secondary leakage flux ϕ_{l2}. These leakage fluxes contribute to the impedance of the transformer. If the leakage flux is small, the coupling between primary and secondary is said to be close. The use of an iron core decreases the leakage flux by providing a low-reluctance path for the flux. *See* COUPLED CIRCUITS.

In a power transformer the voltage drops due to winding resistance and leakage are small; therefore V_1 and V_2 are essentially in phase (or 180° out of phase, depending on the choice of polarity). Since the no-load current is small, I_1 and I_2 are essentially in phase (or 180° out of phase). Therefore, Eq. (1) applies, and the voltage ratio is expressed

$$V_1 I_1 \simeq V_2 I_2 \tag{1}$$

by Eq. (2), in which a is the transformation ratio.

$$\frac{V_1}{V_2} \simeq a \tag{2}$$

Substituting Eq. (2) into Eq. (1) demonstrates that the current ratio is inversely proportional to the transformation ratio, as in Eq. (3). A transform-

$$\frac{I_1}{I_2} \simeq \frac{1}{a} \tag{3}$$

er therefore may be used to step up or down a voltage from a level V_1 to a level V_2 according to the transformation ratio a. Simultaneously the current will be transformed inversely proportional to a.

Equation (1) may be rewritten in the form of Eq. (4).

$$I_1^2 \frac{V_1}{I_1} \simeq I_2^2 \frac{V_2}{I_2} \tag{4}$$

Since V_2/I_2 is the impedance Z_2 of the load on the secondary and V_1/I_1 is the impedance Z_1 of the load as measured on the primary, Eq. (5) applies.

$$I_1^2 Z_1 \simeq I_2^2 Z_2 \tag{5}$$

Equation (5) may be rewritten in the form of Eq. (6).

$$\frac{Z_1}{Z_2} \simeq \left(\frac{I_2}{I_1}\right)^2 \simeq a^2 \tag{6}$$

The transformer is thus capable of transforming circuit impedance levels according to the square of the transformation ratio; this property is used in telephone, radio, television, and audio systems.

The transmission of power from primary coil to secondary coil is via the magnetic flux. The flux is proportional to the ampere turns in either coil. Since the power in each coil is nearly the same, Eqs. (7) and (8) are obtained. The transformation

$$N_1 I_1 \simeq N_2 I_2 \tag{7}$$

$$\frac{N_1}{N_2} \simeq \frac{I_2}{I_1} \simeq a \tag{8}$$

ratio is therefore approximately equal to the turns ratio.

Construction. Transformer cores are made of special alloy steels rolled to approximately 0.014 in. thick. These thin sheets, or laminations, are

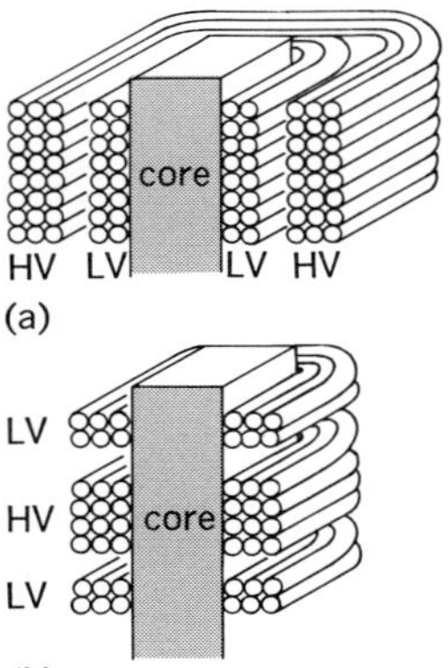

Fig. 2. Winding arrangements. (a) Concentric. (b) Interleaved.

HV LV LV HV
(a)

LV
HV core
LV
(b)

stacked to form the transformer core, each sheet being insulated from the others to reduce unwanted eddy-current loss. The steel is heat-treated to obtain low hysteresis loss, low exciting current, and low sound level. *See* EDDY CURRENT.

Copper conductors are used almost universally. Conductor wires are round in smaller transformers, and rectangular in larger ones.

The conductors are insulated with special paper or cotton covering, with enamel, or with a combination of both. Large outdoor transformers are immersed in oils to obtain good electrical insulation within small spacings and to provide a cooling medium. When lightweight or nonflammable materials are important, transformers may be made with compressed gases as the insulating and cooling medium. Increasing the pressure raises the dielectric strength. The gas is pumped through the transformer and through a gas-to-air heat exchanger for cooling.

The low-voltage (LV) winding is usually in the form of a cylinder next to the core. The high-voltage (HV) winding, also cylindrical, surrounds the LV windings as in Fig. 2a. These windings are often described as concentric windings. The number of turns N may be obtained from Eq. (9), where

$$E = \frac{fBAN}{22,500} \qquad (9)$$

E is the rms voltage, f is the frequency in cps, B is the maximum flux density in kilolines/in.², and A is the cross-sectional area of the iron core in square inches.

Some manufacturers use a winding arrangement having coils adjacent to each other along the core leg as in Fig. 2b. The coils are wound in the form of a disk, with a group of disks for the LV winding stacked alternately with a group of disks for the HV windings. This construction is referred to as interleaved windings.

The core sheets are stacked sheet by sheet to form the desired cross-sectional area. The closed magnetic circuit typically has joints between adjacent sheets, but cores of moderate cross section may be made with a long continuous sheet which has been coiled up to give the required cross section. Passages may be provided between groups of sheets for circulation of the cooling oil.

For single-phase transformers (Fig. 3), the HV and LV coils may be on one leg of a core, with the return path in one, two, or more other legs. The total area of the return legs is equal to that of the main leg. An alternative construction has two legs,

Fig. 4. Three-phase core and coils, rated at 50,000 kVA, 115,000 volts.

each with half of the primary windings and half of the secondary windings.

Figure 4 shows a typical three-phase transformer core with coils. A typical three-phase core has three legs, with the HV and LV windings for one phase on each leg. The yokes of the core connect between the two outer legs and the middle leg on top and bottom. This core-type construction is shown in Fig. 5a. The iron in another construction that is sometimes used (shell type) is as shown in Fig. 5b. Either concentric windings or interleaved windings may be used with either core.

The core and coils are placed in a steel tank with openings for the electrical connections to the windings, and for the cooling equipment.

Cooling. Small transformers are self-cooled. Radiation, conduction, and convection from the tank or from radiating surfaces remove the heat generated by the power losses of the transformer. On larger units, fans are sometimes added to the radiating surfaces. A transformer may have one rating with a basic method of cooling and a higher rating with supplemental cooling. Pumps may be added to give further cooling. An oil-to-air heat exchanger with finned tubes is used on the very large units. This equipment has a pump for circulating oil and fans for forcing the air against the heat exchanger. Water cooling may be used with cooling coils or with an oil-to-water heat exchanger having an oil pump.

Characteristics. The service conditions for a particular transformer are considered by the designer in choosing materials and the arrangement of parts.

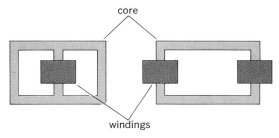

Fig. 3. Location of windings in single-phase cores.

TRANSFORMER

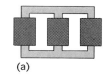

(a)

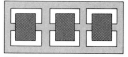

(b)

Fig. 5. Typical three-phase cores showing location of windings. (a) Core type. (b) Shell type.

The final design then may be measured by test with respect to a number of characteristics.

No-load loss. The sum of the hysteresis and eddy loss in the iron core is the no-load loss.

Exciting current. The exciting current is that supplied to the transformer at no load when operating at rated voltage. This current energizes the core and supplies the no-load loss. Owing to the characteristic shape of the *B-H* curve of iron, the current is not a true sine wave, but has higher frequency harmonics. In a typical power transformer the exciting current is so small (usually less than 1%) that I_2 is approximately $(N_1/N_2)I_1$. In this sense the ampere-turns in the two windings are said to balance.

Load loss. This is the sum of the copper loss, due to the resistance of the windings (I^2R loss), plus the eddy-current loss in the winding, plus the stray loss (loss due to flux in metallic parts of the transformer adjacent to the windings, the flux resulting from current in the windings).

Total loss and efficiency. The total loss in a transformer is the sum of the no-load and full-load losses. Representative values for a 20,000-kVa, three-phase, 115-kV power transformer are no-load loss, 42 kW; load loss, 85 kW; and total loss, 127 kW. Equation (10) expresses the efficiency of a

$$\text{Efficiency} = \frac{\text{output in kW}}{\text{input in kW}} = \frac{\text{output}}{\text{output} + \text{losses}} \quad (10)$$

transformer. For this transformer the efficiency is 20,000/20,127, or 99.37%.

Voltage ratio. This is the ratio of voltage on one winding to the voltage on another winding at no load. It is the same as the turns ratio.

Impedance. Consider a transformer having equal turns in the primary and secondary windings. If one side is connected to a generator and the other side to a typical power system load, the voltage measured on the load side will be less than that on the generator side, by the amount of the impedance drop through the transformer. *See* ELECTRICAL IMPEDANCE.

Impedance is measured by connecting the secondary terminals together (short-circuited) and applying sufficient voltage to the primary terminals to cause rated current to flow in the primary winding. The transformer impedance in ohms equals the primary voltage divided by the primary current. Impedance is usually referred to the transformer kVA and kV base and given as **percent** impedance, as in Eq. (11). Percent reactance is usually close in value to percent impedance, since the percent resistance, given by Eq. (12), is small.

$$\% \text{ impedance} = \frac{1}{10} \frac{\text{kVA}}{(\text{kV})^2} \times \text{ohms} \quad (11)$$

$$\% \text{ resistance} = \frac{\text{load loss in kVA}}{\text{kVA rating}} \times 100 \quad (12)$$

Typical values for a 20,000-kVa, three-phase, 115-kV self-cooled power transformer are resistance, 0.4% and impedance, 7.5%.

Regulation. Regulation is the change in output (secondary) voltage that occurs when the load is reduced from rated value to zero, with the primary impressed terminal voltage maintained constant. This is usually expressed as a percent of rated output voltage at full load (E_{FL}), as in Eq. (13),

$$\% \text{ regulation} = \frac{E_{NL} - E_{FL}}{E_{FL}} \times 100 \quad (13)$$

where E_{NL} is the output voltage at no load. When a transformer supplies a capacitive load, the power factor may cause a higher full-load voltage than no-load voltage.

Cooling. Temperature tests (heat run tests) are made by operating the transformer with total losses until the temperatures are constant. In the United States the standard winding rise is 55°C over a 30°C air ambient.

Insulation. Sufficient insulation strength must be built into a transformer so that it can withstand normal operation at its rated voltage and system voltage transients due to lightning and switching surges.

Audio sound. The iron core lengthens and shortens because of magnetostriction during each voltage cycle, giving rise to a hum having a frequency twice that of the voltage. This and other frequencies may cause mechanical vibrations in different parts of the transformer due to resonance.

Taps. The application of a transformer to a power system involves a correct choice of turns ratio for average operating conditions, and the selection of proper taps to obtain improved voltage levels when average conditions do not prevail.

Tap changers are frequently used in the HV winding to give plus or minus two $2\frac{1}{2}$% taps (5% above and 5% below rated voltage). These taps may be changed only when the transformer is de-energized, that is, when the service is interrupted.

A special motor-driven tap changer is used to permit tap changing when the transformer is energized and carrying full load. One of its simpler forms is shown in Fig. 6. The transformer taps are brought to a tap changer having two sets of fingers A and B. Initially these are on the same tap. When a change is required, a contactor C opens, and A moves to the next lower tap. C now closes. Next D opens and B moves down to the same tap as A. The current, which initially divided half-and-half through A and B, has changed, first to be all in B, then partly in A and partly in B, then all in A, and finally half-and-half in A and B. E is a center-tapped reactor, which limits the current when A and B are not on a common tap.

This equipment is essential where a constant voltage is required under changing loads. It is frequently applied with a tap range of plus or minus

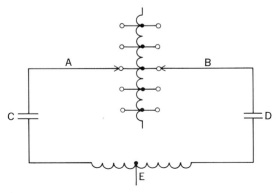

Fig. 6. Typical circuit for tap changing under load.

10% of rated voltage. It may be made to operate automatically, maintaining a specified voltage at a predetermined point remote from the transformer.

Tap changing under load equipment is used on power transformers supplying residential loads, where variations in voltage would adversely affect the use of lights and appliances. It is also used for chemical and industrial processes, such as on pot lines for the manufacture of aluminum.

Parallel operation. Two transformers may be operated in parallel (primaries connected to the same source and secondaries connected to the same load) if their turns ratios and per unit impedances are essentially equal. A slight difference in turns ratio would cause a relatively large out-of-phase circulating current between the two units and result in power losses and possible overheating.

Phase transformation. Polyphase power may be changed from 3-phase to 6-phase, 3-phase to 12-phase, and so forth, by means of transformers. This is of value in the power supply to rectifiers, where the greater number of phases results in a smoother dc voltage wave. *See* ALTERNATING-CURRENT CIRCUIT THEORY.

Overloads. Transformers have a capacity for loading above their rating. Such factors as low ambient temperature and type of load carried may be used to increase the continuous load possible on a given transformer. In emergencies it is possible to increase the load further for short times with a calculable loss of transformer life. Such a load would permit, for instance, a 50% overload for 2 hr following full load.

[J. R. SUTHERLAND]

AUDIO AND RADIO-FREQUENCY TRANSFORMERS

Audio or video (broad-band) transformers are used to transfer complex signals containing energy at a large number of frequencies from one circuit to another. Radio-frequency (rf) and intermediate-frequency (i-f) transformers are used to transfer energy in narrow frequency bands from one circuit to another. Audio and video transformers are required to respond uniformly to signal voltages over a frequency range three to five or more decades wide (for example, from 10 to 100,000 Hz), and consequently must be designed so that very nearly all of the magnetic flux threading through one coil also passes through the other. These units are designed to have a coupling coefficient k, given in Eq. (14), nearly equal to one. Here L_p and L_s are the

$$k = M/\sqrt{L_p L_s} \qquad (14)$$

primary and secondary inductances, respectively, and M is the mutual inductance (Fig. 7). The high coupling coefficient is obtained by the use of interleaved windings and a high-permeability iron core, which concentrates the flux. Typical values of k for highest quality video transformers may be as high as 0.9998; that for power transformers need not be greater than 0.98.

The rf and i-f transformers are built from individual inductors whose magnetic fields are loosely coupled together, $k < 0.30$; each inductor is resonated with a capacitor to make efficient energy transfer possible near the resonant frequency. *See* RESONANCE (ALTERNATING-CURRENT CIRCUITS).

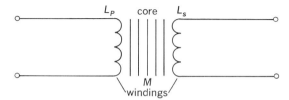

Fig. 7. Schematic of a transformer with symbols.

The audio and video transformers. Audio and video transformers have two resonances (caused by existing stray capacitances) just as many tuned transformers do. One resonance point is near the low-signal-frequency limit; the other is near the high limit. As the coefficient of coupling in a transformer is reduced appreciably below unity by removal of core material and separation of the windings, tuning capacitors are added to provide efficient transfer of energy. The two resonant frequencies combine to one when the coupling is reduced to the value known as critical coupling, then stay relatively fixed as the coupling is further reduced.

All transformers are devices for transferring energy from one circuit to another. The energy transferred is absorbed either in the circuits themselves or in an external load circuit. For this reason, proper termination is essential for achieving optimum behavior in circuits containing transformers.

Audio and video transformers have a minimum operating frequency at which the open-circuit reactance of the primary is approximately twice its effective loaded impedance. As with wide-band RC amplifiers, gain may be traded for bandwidth with transformer-coupled amplifiers. The reduction of the terminating resistance across the secondary of the transformer reduces the minimum operating frequency f_1 and, in the presence of output capacitance, raises the maximum frequency f_2. The approximate values of the minimum and maximum frequencies and the resonant frequency f_r are given by Eqs. (15), where $L_{ss} = L_s - M^2/L_p = L_s(1 - k^2)$.

$$f_1 = R_c (N_p/N_s)^2/\pi L_p$$
$$f_2 = 1/2\pi R_c C_s \qquad (15)$$
$$f_r = 1/2\pi\sqrt{L_{ss} C_s}$$

This L_{ss} is the secondary inductance with the primary short-circuited; C_s is the output capacitance, both external and internal, on the transformer; and R_c is the load resistance (Fig. 8). The resonant frequency f_r should be larger than f_2 for best performance. *See* AMPLIFIER.

A transformer used to activate terminating circuitry is called an output transformer; one to activate an input circuit is an input transformer; and others are called interstage transformers.

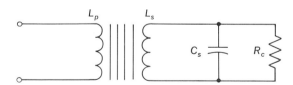

Fig. 8. Circuit of loaded transformer.

Distortion. The distortion introduced into the amplified signal by a transformer is caused primarily by its hysteresis loss. This loss may be minimized by proper loading on the secondary. The load component of current then is large compared to the magnetizing current. In addition, a resistive load keeps the amplification uniform as a function of frequency, and keeps the phase distortion to a minimum.

The magnetic core in an audio or video transformer is subject to two kinds of saturation, that due to applied direct current in the windings, and that due to excessively large signal currents. The direct current in the windings may make the hysteresis loop of the iron core nonsymmetrical, necessitating the use of a larger core having a built-in air gap. Both large signal amplitudes and low frequencies can cause signal saturation to occur in the core. The structure of audio and rf transformers is shown in Fig. 9.

The rf and i-f transformers. These use two or more inductors, loosely coupled together, to limit the band of operating frequencies. Efficient transfer of energy is obtained by resonating one or more of the inductors. By using higher than critical coupling, a wider bandwidth than that from the individual tuned circuits is obtained, while the attenuation of side frequencies is as rapid as with the individual circuits isolated from one another.

The tuning of the primary, the secondary, or

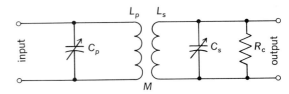

Fig. 10. Tuned rf transformer.

both may be accomplished either by the variation of the tuning capacitor or by an adjustable magnetic or conducting slug that varies the inductance of the inductor (Fig. 10).

The operating impedance of a tuned circuit of an rf transformer is a function of its Q and its tuning capacitance. In general, high-power circuits require a high capacitance for energy storage, and therefore have low values of impedance. In any application, the impedance level must be kept sufficiently small to prevent instability and oscillation. See RADIO-FREQUENCY AMPLIFIER.

[KEATS A. PULLEN]

Bibliography: American Institute of Physics, *The Transformer,* 1976; L. Anderson, *Electric Machines and Transformers,* 1980; R. Feinberg, *Modern Power Transformer Practice,* 1979; C. McLyman, *Transformer and Inductor Design Handbook,* 1978; G. McPherson, *An Introduction to Electrical Machines and Transformers,* 1981; A. J. Pansini, *Basic Electrical Power Transformers,* 1976; D. Richardson, *Rotating Electric Machinery and Transformer Technology,* 1978.

Transistor

An active component of an electronic circuit which may be used as an amplifier, detector, or switch. A transistor consists of a small block of semiconducting material to which at least three electrical contacts are made. Transistors are of two general types, bipolar and field-effect. The bipolar type involves excess minority current carrier injection. The field-effect type involves only majority current carriers. Historically the bipolar type was developed before the field-effect type. Today both are widely used. The unmodified term transistor usually refers to the bipolar type.

In a bipolar transistor, at least one contact is ohmic (nonrectifying), and at least one contact is rectifying. Usually there are two closely spaced rectifying contacts and one ohmic contact. For a discussion of rectifying contacts *See* JUNCTION DIODE; POINT-CONTACT DIODE; SEMICONDUCTOR RECTIFIER.

The operation of a simple transistor consists of the control of the current flowing in the high-resistance direction through one rectifying contact (called the collector) by the current flowing in the low-resistance direction in the other rectifying contact (called the emitter). The third contact, which is ohmic, is called the base contact.

These contacts usually consist of two or more regions. The regions in which the actual rectification processes take place are called the emitter barrier and collector barrier. The region between these two barriers is called the base region, or simply the base. The regions outside of these barriers are called the emitter and collector regions.

Transistors are used in radio receivers, in electronic computers, in electronic instrumentation

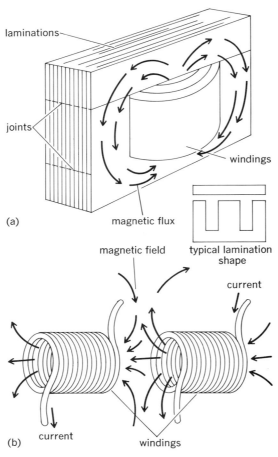

Fig. 9. Audio and rf transformers. (*a*) Iron-core audio transformer. (*b*) Air-core rf transformer.

and control equipment, and in almost any electronic circuit where vacuum tubes are useful and the required voltages are not too high. Transistors have the advantages over their vacuum-tube counterparts of being much smaller, consuming less power, and having no filament to burn out. They are at a disadvantage in that they do not yet operate at as high voltages as some vacuum tubes and their action is degraded at high temperatures.

Classification of transistors. Transistors are classified chiefly by four criteria: (1) by the type and number of structural regions of the semiconductor crystal; (2) by the technology used in fabrication; (3) by the semiconductor material used; and (4) by the intended use of the device. A typical designation following this scheme would be *npn* double-diffused silicon switching transistor. It is not necessary to include all of the above criteria in a single designation nor to rigidly follow this order.

A modern transistor type is the *npn* double-diffused silicon planar passivated transistor (Fig. 1). The term double-diffused refers to the fabrication technique in which the base region is formed by diffusion through a mask into the body of the silicon wafer which forms the collector region.

In turn, the emitter region is formed by diffusion through a second mask into the previously formed base region. The term planar refers to the fact that all three electrical connections are found on a single surface of the device. The term passivated means that the surface to which all junctions return is protected by a layer of naturally grown silicon oxide which, together with an overcoating of glass or other inert material, passivates the surface, electrically minimizing leakage currents. The double-diffusion process allows very close control of narrow base widths. The base diffusion provides a resistivity gradient in the base region which has an associated electric field. In this field charge transport is by drift. Such transistors have been called drift transistors to distinguish them from most other transistors in which the charge transport is by a diffusion process. Silicon planar transistors have power ratings in the 100 mW to 50 W range with characteristic frequencies between 50 and 2000 MHz, usually of the *npn* type. The designation *npn* stands for the conductivity type of the emitter, base, and collector regions, respectively. The *n* stands for negative since the charge on an electron is negative and electrons carry most of the current in a region of *n*-type conductivity. In a region of *p*-type conductivity most of the current is carried by electron vacancies, called holes, which behave as if they were positively charged. For a discussion of conductivity type *see* SEMICONDUCTOR.

A historically important type was the *pnp* alloy-junction germanium transistor. This type was very widely used in the first decade of the solid-state electronics era. The term alloy-junction in this transistor designation refers to the fabrication method. The emitter and collector regions were produced by recrystallization from an alloy of some suitable metal doped with a *p*-type impurity. The alloy had previously been fused in contact with the opposite surfaces of the original *n*-type semiconductor body and had dissolved some of the semiconductor material. Fused-junction is equivalent terminology. This type of transistor was made in power ratings from 50 mW to 200 W, and in frequency ranges up to about 20 MHz.

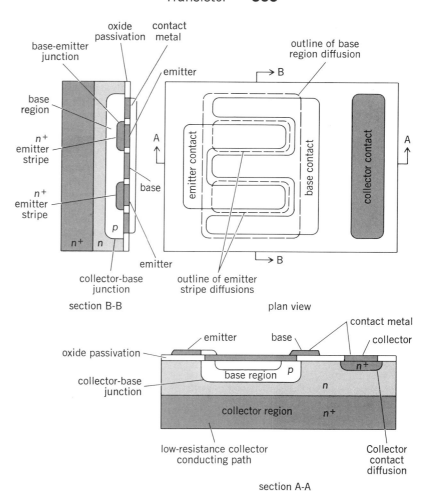

Fig. 1. Plan and sections of a planar *npn* double-diffused silicon transistor.

For further classification of transistors *see* JUNCTION TRANSISTOR; POINT-CONTACT TRANSISTOR. See also below for descriptions of other special devices.

Transistor action. To explain transistor action in more detail, some of the basic properties of a semiconductor material are first presented. An *n*-type semiconductor contains electrons, and a *p*-type semiconductor contains holes. These are called the majority carriers of the two types. Actually there are always present a small number of holes in an *n*-type semiconductor and a small number of electrons in a *p*-type semiconductor. These are called the minority carriers of the two types. At a given temperature with a given material the product of the densities of the majority and minority carriers is a constant. This means that if there is present a very high density of majority carriers (low-resistivity material), there will be a correspondingly low density of minority carriers.

The emitter current controls the collector current in a simple transistor. To understand this, first consider the magnitude of the collector current in the absence of emitter current. In normal operation the collector barrier is biased in the high-resistance (reverse) direction. Under this condition of bias the majority carriers are stopped by the barrier, and only the minority carriers are free to flow. If the collector barrier is a silicon *pn* junction, the minority-carrier diffusion current is negligible and the reverse-bias leakage current will consist of thermally generated carriers and be in the nanoampere range. If emitter current is present, the portion consisting of carriers entering the base will continue across the collector barrier and thus control the collector current.

Injection. The emitter controls the density of minority carriers by injecting extra minority carriers into the base region when the emitter is biased in the low-resistance (forward) direction. This is the fundamental process of simple transistor action. Whenever a rectifying barrier is forward-biased, extra minority carriers are added to the semiconductor near the barrier. Since the source of these minority carriers is the majority-carrier density on the other side of the barrier, it is clear that the largest part of the forward current will be carried by those carriers which come from the largest majority density. A *pn* junction will have a high injection efficiency for electrons if the *n* region has a much larger density of carriers (lower resistivity) than the *p* region. Therefore, in a *npn* transistor the emitter *n* region should have a low resistivity compared to the *p*-type base region. The phenomenon of minority-carrier injection is observed also in rectifying metal-semiconductor contacts, and such contacts may be used as emitters as well as *pn* junctions.

Current gain. The current gain α of a simple transistor may be expressed as the product of three factors: the fraction γ of the emitter current carried by the injected carriers, and fraction β of the injected carriers which arrive at the collector barrier, and the current multiplication factor α^* of the collector. For a double-diffused transistor, typical values of these factors are $\gamma = 0.985$, $\beta = 0.999$, and $\alpha^* = 1.000$, giving $\alpha = 0.984$. From this it can be seen that most of the current which flows into the emitter flows right on through the base region and out the collector, while only a small fraction (here 0.016) flows out the base connection.

For a fixed value of emitter current I_e there is a fixed value of collector current αI_e added to the collector-barrier leakage current I_{co}, giving a total collector current, $I_c = I_{co} + \alpha I_e$. This means that the slope of the dc characteristics should be the same as the slope of the collector-barrier leakage current curve for $I_e = 0$. The typical characteristics in Fig. 2 illustrate this. The slope of the collector leakage curve is very low since the collector voltage does not influence the relatively fixed number of minority carriers carrying the current. For a discussion of transistor characteristics *see* TRANSISTOR CONNECTION.

High-frequency effects. These originate in three distinct properties of transistors: the transit time of injected carriers across the base region, the charging time of the collector- or emitter-barrier capacitance through the base-region and collector-region resistances in series, and the time required to build up the proper density of injected carriers in the base region (called storage-capacity effect). In alloy-junction transistors with a base region of uniform resistivity, the transport of injected carriers across the base is usually the limiting factor. Of course, base transit time alone introduces only a phase shift between the emitter and collector signals, but this time also gives a chance for inject-

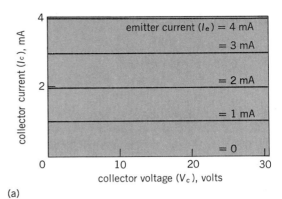

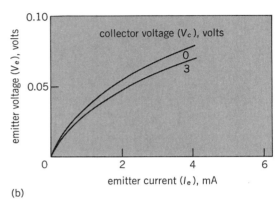

Fig. 2. Typical transistor dc characteristics. (*a*) Collector characteristics. (*b*) Emitter characteristics.

ed carriers, bunched by the emitter signal, to diffuse apart and therefore degrade the signal (Fig. 3).

In double-diffused (drift) transistors the base transit time is usually negligible compared to the charging time of the collector or emitter capacitance, and in some units the storage capacity (often called diffusion capacity) seems to be an appreciable limitation.

Storage capacity also shows up in another way in transistors used as switches. Here it introduces a time delay both in turning on and in turning off the transistor. The turn-off delay is usually longer than the turn-on delay, because the density of injected carriers in the base region has had time to build up to large values during the time the transistor was on, and therefore takes a long time to subside to the level where the transistor can turn off. These delays are only slightly related to the actual time of rise or fall of the collector level, which is determined primarily by collector-capacitance—base-resistance time constant.

A fabrication technique, called epitaxial growth, is used to minimize the storage capacity effects in high-speed transistors. In this process a transistor structure is formed entirely in a very thin skin of good semiconductor material grown upon the surface of a wafer of heavily doped material. The heavily doped material has very low lifetime for excess carriers and, therefore, a very low storage effect, as well as a low series resistance. The collector junction of such a transistor is close to this low-lifetime material but is formed in the high-quality, epitaxially grown skin so that its properties are not degraded by the heavily doped material. Such transistors are called epitaxial transistors.

Close control of the injection ratio γ, defined above, is afforded by the fabrication technique of ion implantation. In this technique a beam of ions composed of the desired dopant material is accelerated to a specific kinetic energy and caused to strike the surface of the region to be doped. The ions penetrate the surface and remain embedded in the semiconductor material. By controlling the ion-beam current and the time of bombardment, a very accurate control of the total number of dopant ions in the region is achieved. After heating the semiconductor to diffusion temperature, the ions move on into the material, creating the emitter and base regions of the double-diffused structure. These regions now have precisely controlled doping and hence show a γ-factor within $\pm 1\%$ of the design value. *See* ION IMPLANTATION.

Transistor noise. Noise is quite low if a low source impedance is used. With source impedances of about 1000 ohms, a good junction transistor will have a noise factor of about 4 dB. The noise factor is independent of the connection but rises with source impedances above 10,000 ohms and with frequencies below 1000 Hz.

Temperature effects. These are most marked in connection with the collector-barrier leakage current with no emitter current flowing I_{co}. This current increases exponentially with temperature and leads to a phenomenon called thermal runaway. If a transistor is operated at a given ambient temperature and a given initial power dissipation, this power will soon raise the temperature of the collector barrier, which then draws more current and in turn increases the dissipation. The process is cumulative, and precautions must be taken to stabilize against it. Current gain increases slightly with increasing temperature in most *npn* transistors, but this is a small effect unless the current gain is unusually close to unity.

Power switching. There are several transistor structures which are used for power switching and make use of current gains greater than unity to achieve a thyraton-like characteristic. These devices are often called four-layer devices since they usually contain four regions of alternating *n*- and *p*-type semiconductor material. Connections are made to the end regions and to one of the interior regions. The end regions are oppositely biased so that the center junction is reverse-biased. The connection to the interior region is then the control and is usually called the gate. When the gate is biased to cause injection of excess carriers across the junction between it and the nearest end connection, the device is triggered on and a saturation current is drawn between the two end connections normally called anode and cathode. Such devices are normally classified as rectifiers but in reality are a form of transistor. *See* CONTROLLED RECTIFIER; SEMICONDUCTOR RECTIFIER.

Field-effect transistor (FET). There are two major types of field-effect transistors, the junction-gate FET (JFET) and the insulated-gate FET (IGFET). The IGFET is commonly called MOSFET or MOS transistor. The acronym MOS stands for metal-oxide-semiconductor which describes, in order, the structure of the device from the gate toward the channel. The JFET was developed first, since it involved no technology beyond that of the planar bipolar silicon transistor. The development of the MOSFET was delayed while the technology was extended to stable control of silicon surface potential. The MOSFET is very widely applied in large-scale integration, particularly in implementing large random-access high-speed memories for computers. *See* COMPUTER STORAGE TECHNOLOGY.

JFET. Figure 4*a* shows a section of a JFET. The channel consists of relatively low-conductivity semiconductor material sandwiched between two regions of high-conductivity material of opposite type. When these junctions are reverse-biased, the junction depletion regions encroach upon the channel and finally, at a high reverse bias, pinch it off entirely. The thickness of the channel, and hence its conductivity, is controlled by the voltage on the two gates. This device is therefore normally on and may be switched off. It is called a depletion-mode FET. In practice this FET has an input impedance several orders of magnitude greater than that of a silicon bipolar transistor. JFETs are made in both *n*-channel and *p*-channel types. They are used in amplifiers, oscillators, mixers, and switches. The general performance limits are about 500 MHz, 1 W, 100 V, and 100 mA (saturation drain current). They also find application in integrated circuits employing bipolar transistors since their technology is compatible. *See* INTEGRATED CIRCUITS.

MOSFET. Figure 4*b* shows a section of a MOSFET. Here the source and drain regions consist of *n* diffusion in a *p*-type substrate. The gate is a metal film evaporated on a thin SiO insulator spanning the separation between the source and drain. With

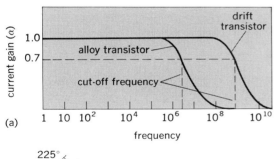

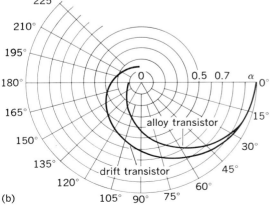

Fig. 3. Typical transistor frequency characteristics. (*a*) Frequency dependence of α. (*b*) Phase of collector current I_c versus emitter current I_e.

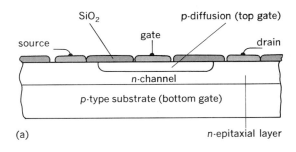

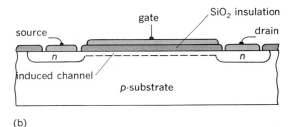

Fig. 4. Field effect transistors. (a) Junction-gate FET (JFET). (b) Insulated-gate FET (IGFET or MOSFET).

no voltage on the gate, the source and drain are insulated from each other by their surrounding junctions. When a positive voltage is applied to the gate, electrons are induced to move to the surface of the p-type substrate immediately beneath the gate, producing a thin surface of induced n-type material which now forms a channel connecting the source and drain. Such a surface layer is called an inversion layer since it is of opposite conductivity type to the substrate. The number of induced electrons is directly proportional to the gate voltage, so that the conductivity of the channel increases with gate voltage. This device is called an n-channel enhancement-mode MOSFET. It is normally off at zero gate voltage.

Because of the quality of the silicon dioxide gate insulator, the input impedance of a MOSFET is several orders of magnitude greater than that of a JFET. Typical MOSFET dc characteristics are shown in Fig. 5. The low-drain voltage channel resistance is inversely proportional to $(V_{gs} - V_{th})$, where V_{gs} is the gate-source voltage and V_{th} is the

threshold voltage, and the saturation drain current is proportional to $(V_{gs} - V_{th})^2$.

MOSFET devices are fabricated in both p-channel and n-channel types, as well as for both depletion (normally on) and enhancement (normally off) modes of operation. In a MOSFET the mode of operation is determined by a threshold voltage of the gate at which the device changes from off to on, or vice versa. In modern technology this threshold voltage can be set for a wide range of values by the use of ion implantation through the gate oxide.

MOSFET discrete devices are used for ultra-high-input impedance amplifiers such as electrometers where the input leakage current is less than 10^{-14} A. Dual-gate depletion types can be used as mixers up to 1000 MHz, and power-switching types (the VMOS discussed below) are good to 25 W, 2 A, or 100 V. Most integrated circuits using MOSFETs are called CMOS integrated circuits, where the C stands for complementary. These circuits use n-channel and p-channel types together to achieve digital logic. Typical propagation delay times through small-scale integrated building-block circuits such as three-input NAND or NOR gates is about 20 nanoseconds for a 20-picofarad load. At a 10-MHz clock rate, the power dissipation for such a gate is about 10 mW. For large-scale integration, a typical 16 kilobit random-access memory has an access time of 200 ns, an active power of 500 mW, and a standby power of 20 mW.

VMOS and SOS. There are a number of variations of the MOS technology. Two of particular interest are VMOS (V for vertical) and SOS (silicon on sapphire). The VMOS device is fabricated by etching a notch down through a planar double-diffused structure similar to that of an npn bipolar transistor. The surface of the notch is first oxidized and then covered with the gate metallization. The source contact bridges the n^+-p junction near the surface, and the drain connection corresponds to the collector contact of the bipolar structure. The channel length is now determined by the thickness of the p region. This allows controlled short channels and gives both high current and high voltage capability.

The SOS device is fabricated in a very small silicon body grown epitaxially on a sapphire substrate. An experimental MOS/SOS 1000-bit memory has shown a standby power of only 1 μW.

Double-base diode. Also called a unijunction transistor, this consists of a single rectifying contact situated approximately midway along a semiconductor bar which carries two ohmic contacts at its ends. If a steady bias is applied between the ends of the bar, a negative-resistance diode characteristic is observed between the rectifying contact and one end of the bar. This device is used primarily for switching.

Transistor manufacture. The manufacture of transistors has required a whole new field of exacting technology. Good semiconductor material requires the maintenance of chemical purities far beyond the spectroscopic range. A purity of 1 part in 10^8 is not unusual. Most devices must be made from oriented single crystals of semiconductor material which can have only very low densities of structural defects.

Physical tolerances of the high-frequency tran-

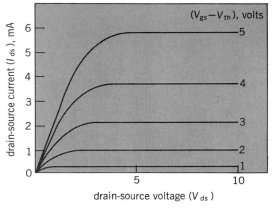

Fig. 5. Typical MOSFET drain characteristics.

sistor structures are microscopic; the separation of emitter and collector junctions must be of the order of a few micrometers in these units.

To solve these problems new techniques have appeared. Purity is achieved by melting a small zone of a bar, or ingot, and gradually passing this molten zone from one end of the bar to the other. Impurities in the material remain in the liquid phase and are carried along with the molten zone, leaving high-purity material behind.

Tolerances are achieved by a collection of new techniques, such as epitaxial growth, solid-state diffusion, ion implantation, and the photolithographic delineation of diffusion masks.

[LLOYD P. HUNTER]

Bibliography: J. Millman, *Micro-Electronics*, 1979; E. S. Yang, *Fundamentals of Semiconductor Devices*, 1978.

Transistor connection

The method of connecting a transistor into a circuit. Bipolar transistor connections and field-effect transistor (FET) connections will be discussed.

Bipolar transistor connection. The common-emitter, common-base, and common-collector

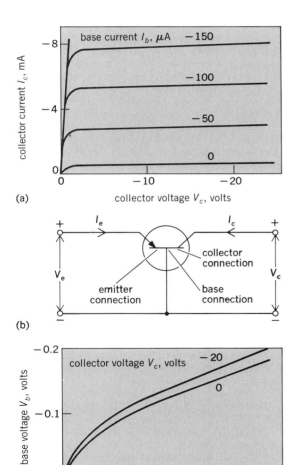

(a)

(b)

(c)

Fig. 1. Common-emitter connection of an alloy-junction transistor. (a) Collector characteristics. (b) Schematic diagram. (c) Base characteristics.

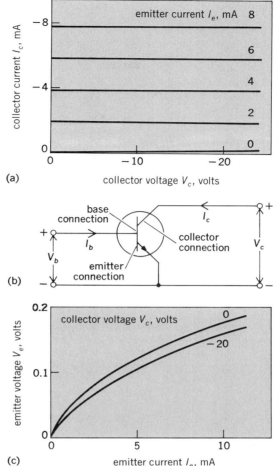

(a)

(b)

(c)

Fig. 2. Common-base connection of an alloy-junction transistor. (a) Collector characteristics. (b) Schematic diagram. (c) Emitter characteristics.

connections are the most frequently used connections of bipolar transistors, and of these the common-emitter connection is by far the most popular for transistors with current gain $\alpha < 1.0$. To compare these connections, the following should be examined: small-signal current gain, voltage gain, input resistance with shorted output, and output resistance with open-circuited input. Shorted output means an ac short; the dc bias voltage is still present. Open-circuited input means a constant dc current bias. These quantities are easily measured and are useful in calculating the performance of a transistor in a circuit connection.

Common-emitter connection. This connection (Fig. 1) has a base-to-emitter input and a collector-to-emitter output connection. The slope of the illustrated I_c versus V_c characteristics gives a commonly used conductance h_{cc}, which is the reciprocal of the output resistance with open-circuited ($I_b =$ const) input.

The current gain h_{cb} (equal to $\partial I_c/\partial I_b$, with $V_c =$ const) is related to current gain α by $h_{cb} = \alpha/(1-\alpha)$. The input resistance h_{bb}, with a short-circuited ($V_c =$ const) output, is given by the slope of the V_b versus I_b characteristics. Finally the voltage gain $1/h_{bc}$ (equal to $\partial V_c/\partial V_b$, with $I_b =$ const) can be obtained from the separation of these curves. Typical values of these small signal parameters are $h_{bb} = 1800$ ohms; $1/h_{bc} = 1600$; $h_{cb} = 50$; and $1/h_{cc} = 0.05 \times 10^6$ ohms.

Common-base connection. This has an emitter-to-base input and a collector-to-base output connection. As with the common-emitter connection, the output resistance $1/h_{cc}$ is found from the slope h_{cc} of the I_c versus V_c characteristics. The current gain h_{ce} equals $\partial I_c/\partial I_e$ ($h_{ce} = -\alpha$ because of the assigned polarity of the currents; Fig. 2). The input resistance h_{ee} is found from the slope of the V_e versus I_e characteristics. The voltage gain $1/h_{ec}$ equals $\partial V_e/\partial V_c$. Typical values of the common-base parameters are $h_{ee} = 36$ ohms; $1/h_{ec} = 1500$; $h_{ce} = -0.98$; and $1/h_{cc} = 2.5 \times 10^6$ ohms.

Common-collector connection. This connection, shown in Fig. 3, has a base-to-collector input and an emitter-to-collector output connection. Again the output resistance $1/h_{ee}$ is found from the slope h_{ee} of the I_e versus V_e characteristics. The current gain h_{eb} equals $\partial I_e/\partial I_b$. The input resistance h_{bb} is found from the slope of the V_b versus I_b

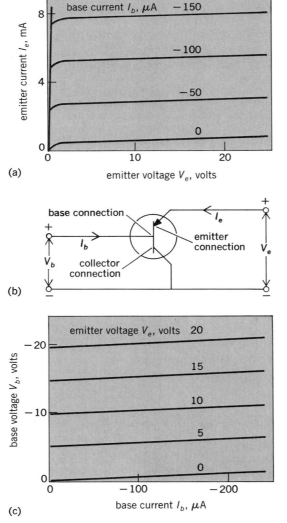

Fig. 3. Common-collector connection of an alloy-junction transistor. (*a*) Emitter characteristics. (*b*) Schematic diagram. (*c*) Base characteristics.

characteristics. The voltage gain $1/h_{bc}$ equals $\partial V_c/\partial V_b$. Typical values of the common-collector parameters are $h_{bb} = 1800$ ohms; $1/h_{be} = 1$; $h_{eb} = -50$; and $1/h_{ee} = 0.05 \times 10^6$ ohms.

Selection of bipolar transistor connections. From the foregoing definitions, it is possible to compare these connections for use in amplifiers. The power gain of the common-emitter connection (current gain times voltage gain) is the highest. This is the reason that this connection is used most frequently. The common-base connection shows the lowest input resistance and the highest output resistance. Its most useful characteristic is its linearity, which gives low distortion when driven by a current source. Accordingly, it is often used with a driver transformer. The common-collector connection shows the highest input resistance and the lowest output resistance. It is somewhat analogous to a cathode follower.

Switching circuits or pulse circuits can be monostable, bistable, or astable. The input characteristic must have a negative-resistance region to be useful in such circuits. Such a characteristic may be achieved by a single transistor if it has a current gain $\alpha > 1.0$. Four-region transistors have such an α and show a negative-resistance region in the base I_b versus V_b characteristic directly, or in the emitter I_e versus V_e characteristic if there is a sufficiently high resistance in the base circuit. It is customary to use more than one transistor to achieve a negative-resistance characteristic, with the exception of the controlled rectifier device used in power switching. The operation of the circuit is monostable if the load line intersects this characteristic in only one point on one of the positive-resistance branches of the curve, bistable if the load line intersects the curve in two such points, and astable if it intersects it in only one point and that in the negative-resistance region. The bistable circuit finds wide application in counters and computers. If transistors are used which have $\alpha < 1.0$, two transistors are required for each bistable switching circuit.

Complementary symmetry is the use of both *pnp* and *npn* transistors together to take full advantage of their opposite bias and signal polarities. For example, an emitter-follower circuit with base input and grounded collector provides a positive drive to a load for a negative base signal in the case of a *pnp* transistor and for a positive base signal in the case of an *npn* transistor. If the two are connected in parallel, they give a positive drive to a load with either polarity of input signal.

FET connections. By far the largest use of field-effect transistors (FET) today is in the large-scale integration of computer memory and logic circuits. In particular the *n*-channel MOSFET technology (NMOS) prevails. The basic circuit most used in this technology is the inverter circuit in Fig. 4*a*. In this circuit a depletion-mode MOSFET is used to load an enhancement-mode MOSFET switching device. The switching device is designated Q and is shown at the bottom of Fig. 4*a*. The load device is designated L and is shown at the top of Fig. 4*a*. The grounded arrows indicate that the substrates of both devices are grounded. The two states of the inverter are given in Fig. 4*b* and are shown in the loaded drain characteristic diagram in Fig. 4*c*.

The nature of an inverter circuit is that if the

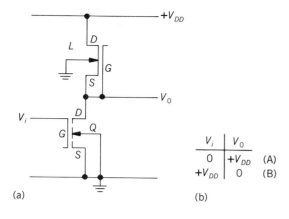

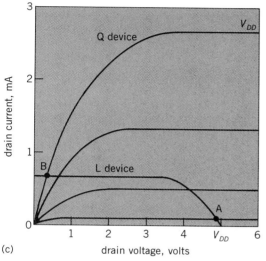

Fig. 4. Typical *n*-channel MOSFET inverter. (*a*) Circuit; S = source, D = drain, G = gate; Q = enhancement-mode switching device; L = depletion-mode load device. (*b*) Truth table for the circuit. (*c*) Loaded drain characteristics.

point of the circuit.

For a switching inverter there are two quiescent points. One, where the Q device is not conducting, is designated A in Fig. 4*c* and is called the off-state. The second, where the Q device is conducting, is designated B in Fig. 4*c* and is called the on-state. There is a single load line while there are several curves in the family of the switching characteristics. The reason for this is that the gate of the load device is connected to its source and cannot change its voltage relative to the source, whereas the gate of the switching device can take on any value of input (gate-source) voltage. In the circuit shown, however, the Q device gate voltage moves between the limits of zero and $+V_{DD}$. Some intermediate gate voltage curves are shown as a reminder that there are a multiplicity of states of the inverter between the off-state and the on-state and that considerable power may be dissipated during the switching process. The off-state (A) has negligible standby power drain. The on-state (B) dissipates typically about 0.1 mW. The switching time ratio (pull-up time to pull-down time) is about 4 to 1, and the total switching delay time of a pair of inverters is of the order of 20 nanoseconds.

In small-scale integrated circuit chip components, it is customary to use complementary MOSFET devices (CMOS). In such circuits both *n*-channel and *p*-channel devices are used together, one as the load of the other. The use of complementary devices this way greatly reduces standby power to about 10 nanowatts. *See* COMPUTER STORAGE TECHNOLOGY; INTEGRATED CIRCUITS; LOGIC CIRCUITS; SEMICONDUCTOR; TRANSISTOR.

[LLOYD P. HUNTER]

Bibliography: L. P. Hunter (ed.), *Handbook of Semiconductor Electronics*, 3d ed., 1970; J. Millman, *Microelectronics*, 1979.

Transmission lines

A system of conductors suitable for conducting electric power or signals between two or more termini. For example, commercial-frequency electric power transmission lines connect electric generating plants, substations, and their loads. Telephone transmission lines interconnect telephone subscribers and telephone exchanges. Radio-frequency transmission lines transmit high-frequency electric signals between antennas and transmitters or receivers. In this article the theory of transmission lines is considered.

Although only a short cord is needed to connect an electric lamp to a wall outlet, the cord is, properly speaking, a transmission line. However, in the electrical industry the term transmission line is applied only when both voltage and current at one line terminus may differ appreciably from those at another terminus. Transmission lines are described either as electrically short if the difference between terminal conditions is attributable simply to the effects of conductor series resistance and inductance, or to the effects of a shunt leakage resistance and capacitance, or to both; or as electrically long when the properties of the line result from traveling-wave phenomena.

Depending on the configuration and number of conductors and the electric and magnetic fields about the conductors, transmission lines are described as open-wire transmission lines, coaxial

input voltage goes up the output voltage goes down, and vice versa. Figure 4*b* illustrates this. Considering the circuit of the inverter (Fig. 4*a*), it can be seen that Q and L are in series between the supply voltage V_{DD} and ground. The load device is always conducting because it is a depletion-mode device and because its gate is permanently connected to its source. The switching device may be either conducting or nonconducting, depending on the input signal V_i on its gate terminal. When V_i is positive, electrons are collected in the channel of Q and it is conducting. When conducting, the channel resistance of Q is very much lower than that of L. This means that the output voltage V_o is held just above ground. When V_i is nearly zero, the switching device is not conducting and the conducting channel of L holds V_o just below the positive supply voltage V_{DD}. The circuit thus fulfills the criterion for inverter action. This behavior is illustrated in Fig. 4*c*. Here the drain characteristic of the load device is drawn as a nonlinear load line on the drain characteristic curves of the switching device. This load line is marked L device. The intersection of the load line with the operating characteristic of the Q device determines the quiescent

transmission lines, cables, or waveguide transmission lines.

Open-wire transmission lines. Open-wire lines may comprise a single wire with an earth (ground) return or two or more conductors. The conductors are supported at more or less evenly spaced points along the line by insulators, with the spacing between conductors maintained as nearly uniform as feasible, except in special-purpose tapered transmission lines, discussed later in this section.

Open-wire construction is used for communication or power transmission whenever practical and permitted, as in open country and where not prohibited by ordinances.

Open-wire lines are economical to construct and maintain and have relatively low losses at low and medium frequencies. Difficulties arise from electromagnetic radiation losses at very high frequencies and from inductive interference, or crosstalk, resulting from the electric and magnetic field coupling between adjacent lines accompanying the characteristic field configuration (Fig. 1).

Coaxial transmission lines. A coaxial transmission line comprises a conducting cylindrical shell, solid tape, or braided conductor surrounding an isolated, concentric, inner conductor which is solid, stranded, or (in certain video cables and delay cables) helically wound on a plastic or ferrite core. The inner conductor is supported by ceramic or plastic beads or washers in air- or gas-dielectric lines, or by a solid polyethylene or polystyrene dielectric.

The purpose of this construction is to have the shell prevent radiation losses and interference from external sources. The electric and magnetic fields shown in Fig. 1b are nominally confined to the space inside the outer conductor. Some external fields exist, but may be reduced by a second outer sheath.

Coaxial lines are widely used in radio, radar, television, and similar applications.

Sheathed cables. Also termed shielded cables, these comprise two or more conductors surrounded by a conducting cylindrical sheath, commonly supported by a continuous solid dielectric. The sheath provides both shielding and mechanical protection.

Coaxial lines, sheathed cables, or shielded cables are often termed simply cables. For cable assemblies of coaxial lines and other circuits *see* COAXIAL CABLE.

Traveling waves. When electric power is applied at a terminus of a transmission line, electromagnetic waves are launched and guided along the line. The steady-state and transient electrical properties of transmission lines result from the superposition of such waves, termed direct waves, and the reflected waves which may appear at line discontinuities or at load terminals.

Principal mode. When the electric and magnetic field vectors are perpendicular to one another and transverse to the direction of the transmission line, this condition is called the principal mode or the transverse electromagnetic (TEM) mode. The principal-mode electric- and magnetic-field configurations about the conductors are essentially those of Fig. 1. Modes other than the principal mode may exist at any frequency for which conductor spacing exceeds one-half of the wavelength of an electromagnetic wave in the medium separating the conductors. Such high-frequency modes are called waveguide transmission modes. *See* WAVEGUIDE.

In a uniform (nontapered) transmission line, the voltage or current applied at a sending terminal determines the shape of the initial voltage or current wave. In a line with negligible losses the transmitted shape remains unchanged. When losses are present, the shape, unless sinusoidal, is altered, because the phase velocity and attenuation vary with frequency.

If a wave shape is sinusoidal, the voltage and current decay exponentially as a wave progresses. The voltage or current, at a distance x from the sending end, is decreased in magnitude by a factor of $\epsilon^{-\alpha x}$, where ϵ is the Napierian base (2.718), and α is called the attenuation constant. The voltage or current at that point lags behind the voltage or current at the sending end by the phase angle βx, where β is called the phase constant.

The attenuation constant α and the phase constant β depend on the distributed parameters of the transmission line, which are (1) resistance per unit length r, the series resistance of a unit length of both going and returning conductors; (2) conductance per unit length g, the leakage conductance of the insulators, conductance due to dielectric losses, or both; (3) inductance per unit length l, determined as flux linkages per unit length of a

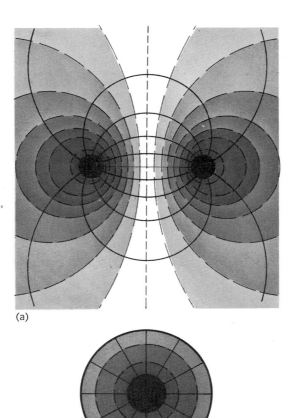

(a)

(b)

Fig. 1. Electric (solid lines) and magnetic (dashed lines) fields about two-conductor (a) open-wire and (b) coaxial transmission lines in a plane normal to the conductors, for continuous and low-frequency currents.

line of infinite extent carrying a constant direct current; and (4) capacitance per unit length c, determined from charge per unit length of a line of infinite extent with constant voltage applied.

The values of α and β may be found from complex equation (1), where j is the notation for the

$$\alpha + j\beta = \sqrt{(r + j2\pi fl)(g + j2\pi fc)} \quad (1)$$

imaginary number $\sqrt{-1}$, and f is the frequency of the alternating voltage and current. The complex quantity $\alpha + j\beta$ is often called the propagation constant γ. Since $r + 2\pi fl$ is the impedance z per unit length of line, and $g + 2\pi fc$ is the admittance y per unit length of line, the equation for the propagation constant is often written in the form of Eq. (2). The velocity at which a point of constant phase

$$\gamma = \sqrt{zy} \quad (2)$$

is propagated is called the phase velocity v, and is equal to $2\pi f/\beta$. For negligible losses in the line (when r and g are approximately zero) the phase velocity is $1/\sqrt{lc}$, which is also the velocity of electromagnetic waves in the medium surrounding the transmission-line conductors.

The distributed inductance and resistance of the lines may be modified from their dc values because of skin effect in the conductors. This effect, which increases with frequency and conductor size, is usually, but not always, negligible at power frequencies.

Characteristic impedance. The ratio of the voltage to the current in either the forward or the reflected wave is the complex quantity Z_0, called the characteristic impedance.

When line losses are relatively low, that is, when relationships (3) apply, the characteristic imped-

$$r \ll 2\pi fl \quad (3)$$
$$g \ll 2\pi fc$$

ance is given by Eq. (4), and is a quantity nearly

$$Z_0 = \sqrt{l/c} \quad (4)$$

independent of frequency (but not exactly so since both l and c may be somewhat frequency-dependent). The magnitude of Z_0 is used widely, at high frequencies, to identify a type of transmission line such as 50-ohm line, 200-ohm line, and the popular 300-ohm antenna lead-in line used with television antennas. *See* ELECTRICAL IMPEDANCE.

Distortionless line. Transmission lines used for communications purposes should be as free as possible of signal waveshape distortion. Two types of distortion occur. One is a form of amplitude distortion due to line attenuation, which varies with the signal frequency. The other, delay distortion, occurs when the component frequencies of a signal arrive at the receiving end at different instants of time. This occurs because the velocity of propagation along the line is a function of the frequency.

Theoretically, a distortionless line can be devised if the line parameters are adjusted so that $r/g = l/c$. In practice this is approached by employing loading circuits. Under these conditions the propagation constant is given by Eq. (5).

$$\gamma = \alpha + j\beta = \sqrt{r/g}(g + j2\pi fc) \quad (5)$$

The attenuation constant α is $\sqrt{rg}$, which is in-

dependent of frequency f. Therefore, there will be no frequency distortion.

The phase constant β is $2\pi f \sqrt{lc}$ which depends upon frequency. The velocity of propagation along any transmission line is $2\pi f/\beta$, and for the distortionless line this becomes $1/\sqrt{lc}$. Thus the velocity of propagation is independent of frequency, and there will be no delay distortion.

Transmission-line equations. The principal-mode properties of the transmission-line equations are described by Eqs. (6) and (7), in which e and i

$$\frac{\partial e}{\partial x} = -\left(ri + l\frac{\partial i}{\partial t}\right) \quad (6)$$

$$\frac{\partial i}{\partial x} = -\left(ge + c\frac{\partial e}{\partial t}\right) \quad (7)$$

are instantaneous values of voltage and current, respectively, x is distance from the sending terminals, and t is time.

For steady-state sinusoidal conditions, the solutions of these equations are given by Eqs. (8) and (9) for voltage E and current I at a distance x from

$$E = E_s \cosh \gamma x - I_s Z_0 \sinh \gamma x \quad (8)$$

$$I = I_s \cosh \gamma x - \frac{E_s}{Z_0} \sinh \gamma x \quad (9)$$

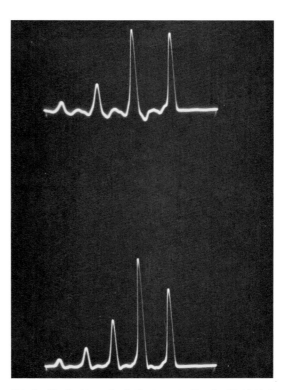

Fig. 2. Typical transient phenomena in a transmission line. These are oscillographic recordings of voltage as a function of time at the sending end of a 300-m transmission line with the receiving end open-circuited. Time increases from right to left; the first (right-hand) pulse is delivered by a generator, equivalent to an open circuit, so that a new forward wave results from each reflected wave arriving at the sending end. At the end of each 2-μsec interval, an echo arrives from the receiving end. In the upper trace, minor discontinuities in the line at intermediate points result in intermediate echos. Intermediate discontinuities are minimized in lower trace.

the sending end in terms of voltage E_s and current I_s at the sending end. In Eqs. (8) and (9) $Z_0 = \sqrt{(r + j2\pi fc)/(g + j2\pi fc)}$. All values of current and voltage in these and the following equations are complex.

In terms of receiving-end voltage E_r and current I_r, these solutions are given by Eqs. (10) and (11), where x is now the distance from the receiving end.

$$E = E_r \cosh \gamma x + I_r Z_0 \sinh \gamma x \qquad (10)$$

$$I = I_r \cosh \gamma x + \frac{E_r}{Z_0} \sinh \gamma x \qquad (11)$$

Reflection coefficient. If the load at the receiving end has an impedance Z_r, the ratio of reflected voltage to direct voltage, known as the reflection coefficient ρ, is given by Eq. (12).

$$\rho = \frac{Z_r - Z_0}{Z_r + Z_0} \qquad (12)$$

When the load impedance is equal to Z_0, the reflection coefficient is zero. Under this condition the line is said to be matched.

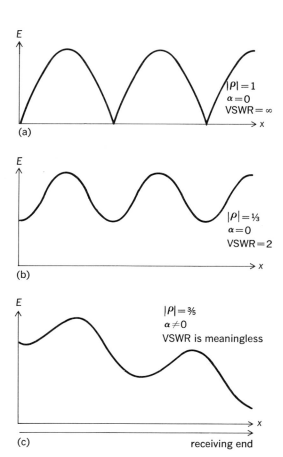

(a)

(b)

(c)   receiving end

Fig. 3. Voltage distribution under sinusoidal steady-state conditions on a section of transmission line, illustrating three standing-wave conditions: (a) line with negligible losses, reflection coefficient of unity, (b) line with negligible losses, reflection coefficient of one-third, and (c) line with finite losses, reflection coefficient of three-fifths. Position of the voltage wave in each case is dependent on angle of phasor value of reflection coefficient. In each case a current maximum (not shown) appears at a voltage minimum in the wavelength.

Pulse transients. The transient solutions of Eqs. (6) and (7) are dependent on the particular problem involved. Typical physical phenomena with pulse transients are shown in Fig. 2. The characteristic time delay in transmission is often advantageously employed in radar and other pulse-signal systems. For examples *see* DELAY LINE.

Standing waves. The superposition of direct and reflected waves under sinusoidal conditions in an unmatched line results in standing waves (Fig. 3).

Voltage standing-wave ratio. When losses are negligible, successive maxima are approximately equal; under this condition a quantity, the voltage standing-wave ratio, abbreviated VSWR, is defined by Eq. (13).

$$\text{VSWR} = \frac{V_{max}}{V_{min}} \qquad (13)$$

Power standing-wave ratio. This quantity, abbreviated PSWR, is equal to $(\text{VSWR})^2$. Measurements of voltage magnitude and distribution on a line of known characteristic impedance Z_0 can be used to determine the magnitude and phase angle of an unknown impedance connected at its receiving end. Lines adapted for such impedance measurements, known as standing-wave lines, are widely used.

Transmission-line circuit elements. The impedance Z_s at the sending end of a loss-free section of transmission line that has a length d, in terms of its receiving-end impedance Z_r, is given by Eq. (14).

$$Z_s = \frac{Z_r \cos \beta d + j Z_0 \sin \beta d}{\cos \beta d + j(Z_r/Z_0) \sin \beta d} \qquad (14)$$

This equation describes the property of a length of line which transforms an impedance Z_r to a new impedance Z_s. In the simple cases, in which Z_r is a short circuit or open circuit, Z_s is a reactance. Various lengths of line may be used to replace more conventional capacitors or inductors. These properties are widely applied at high frequencies, where suitable values of βx require only physically short lengths of line.

Tapered transmission lines. Transmission lines with progressively increasing or decreasing spacing are used as impedance transformers at very high frequencies and as pulse transformers for pulses of millimicrosecond duration. Although tapers designed to produce exponential-varying parameters, as in the exponential line, are most common, a number of other tapers are useful. *See* MICROWAVE TRANSMISSION LINES.

[EVERARD M. WILLIAMS]

Bibliography: American Radio Relay League, *ARRL Antenna Book,* revised periodically; L. N. Dworsky, *Modern Transmission Line Theory and Applications,* 1979; D. G. Fink and H. W. Beaty (eds.), *Standard Handbook for Electrical Engineers,* 11th ed., 1978.

Traveling-wave tube

A microwave electronic tube in which a beam of electrons interacts continuously with a wave that travels along a circuit, the interaction extending over an appreciable distance measured in wavelengths. Traveling-wave tubes are normally used in amplifiers with exceedingly wide bandwidths. Typ-

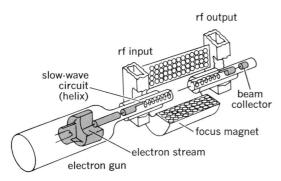

Fig. 1. Traveling-wave tube.

ical bandwidths are 10–100% of the center frequency, with gains of 20–60 dB. Such traveling-wave amplifiers serve at the inputs to sensitive radars or communication receivers. High-power traveling-wave amplifiers operated as the final stages of radars or scatter communication transmitters deliver pulsed powers exceeding a megawatt. For use in satellite transmitters, lightweight traveling-wave amplifiers develop 35 watts continuously at 45% overall efficiency.

In a traveling-wave tube amplifier (Fig. 1), a thermionic cathode produces the electron beam. An electron gun initially focuses the beam, and additional focusing means retain the electron stream as a beam throughout the length of the tube until the beam strikes the collector electrode. Microwave energy enters the tube near the electron gun and propagates along a slow-wave circuit. The tube delivers amplified microwave energy into a matched load connected near the collector end. The slow-wave circuit serves to propagate the microwave energy along the tube at approximately the same velocity as that of the electron beam. Interaction between beam and wave is continuous along the tube with contributions adding in phase.

In principle, operation in a traveling-wave tube is similar to that in a klystron: Velocity modulation of the electrons results in current modulation of the beam because of the nonzero transit time. However, unlike a klystron, the traveling-wave tube provides for the process to take place continuously along the slow-wave circuit. Thus, at one point along the circuit, the microwave field accelerates the electrons axially. As these electrons proceed along the tube, they bunch with those ahead. At the same point along the circuit, but a half-cycle later in time, the microwave field decelerates electrons; these electrons bunch with those behind. Because average velocity of the electrons is made to be slightly faster than the velocity of the energy on the slow-wave circuit, the electron bunches that form drift into a decelerating microwave field, thereby delivering energy to the field. In this manner the original energy in the electron beam is converted into microwave energy and delivered to the slow-wave circuit. Because of the continuous distributed interaction, the circuit wave grows exponentially as it travels along the tube.

Among the particular design features of a traveling-wave tube are the electron gun, means for focusing the beam, the slow-wave circuit, and the collector.

Electron gun. The usual electron gun for a traveling-wave tube is a cylindrical cathode with a plane face followed by a focusing electrode and an anode (Fig. 2a). The face of the cathode is the active emitting surface. The focusing electrode (or beam-forming electrode) is biased close to cathode potential, producing a repelling force on the electrons to counteract the beam-spreading space-charge forces within the beam. Thus, the beam leaves the gun with electron trajectories straight and parallel. Because of this beam configuration, current density is limited to that available from the cathode. For an oxide-coated cathode this density is a fraction of an ampere per square centimeter of oxide-coated cathode. To make the perveance of such a gun much above 1×10^{-6} is difficult; perveance is defined by Eq. (1).

$$\text{Perveance} = \frac{\text{dc current}}{(\text{dc beam voltage})^{3/2}} \quad (1)$$

If higher current densities are required than those obtainable directly from the cathode surface, converging guns are designed (Fig. 2b). Spherical electrodes produce radial electron flow to concentrate the beam. For perveances above 1×10^{-6}, the anode aperture becomes so large that the focusing field for simple electrode geometries is distorted. More controlled shapes of focus electrode and anode (Fig. 2c) result in perveances as high as 5×10^{-6}. Some high-perveance guns are constructed with spherical control grids close to the spherical cathode; this construction permits pulsing of the electron beam from a low-voltage source.

For higher-perveance guns in power tubes or in backward-wave tubes, hollow electron beams may be desirable. Gun design for a hollow beam is similar to that for a solid beam, except that focusing electrodes are placed inside the beam as well as outside it (Fig. 3a). Convergent-flow guns for hollow beams, although more difficult to design than those for solid beams, can be made using a similar structure (Fig. 3b).

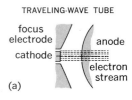

(a)

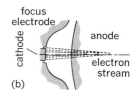

(b)

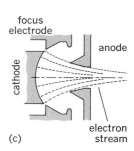

(c)

Fig. 2. Electron guns for traveling-wave tubes. (a) Pierce gun. (b) Pierce gun with converging flow. (c) High-perveance gun.

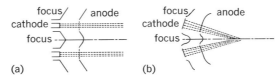

Fig. 3. Hollow-beam electron guns. (a) Parallel flow. (b) Converging flow.

For low-noise guns, additional noise-reducing electrodes are introduced to modify the space-charge waves produced by the statistical fluctuation of current and velocity of the electrons as they emerge from the cathode. Without such noise-reducing means, the noise figure of a traveling-wave tube amplifier is in the range of 15–30 dB; with low-noise guns that have a series of electrodes each at different dc potentials, noise figures as low as 3.5 dB are possible. The dc potential profile along the beam is adjusted for optimum noise reduction by means of these electrodes (Fig. 4).

Focusing methods for the beam. To hold the beam in focus throughout the interaction region of the tube, beam focusing is normally required to overcome the space-charge spreading forces of the

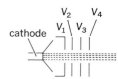

Fig. 4. Low-noise disk-type gun with graded acceleration.

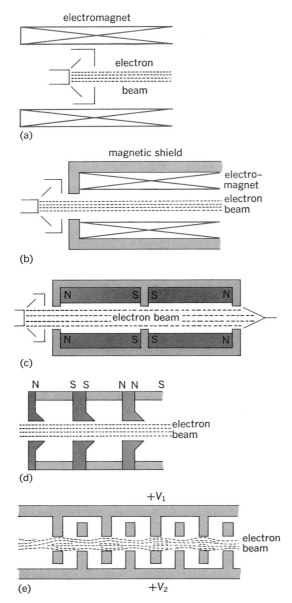

electromagnet

electron

beam

(a)

magnetic shield

electro-
magnet

electron
beam

(b)

N   S S   N

electron beam

N   S S   N

(c)

N   S S   N N   S

electron
beam

(d)

$+V_1$

electron
beam

(e) $+V_2$

Fig. 5. Beam-confining methods. (a) Confined flow. (b) Brillouin flow. (c) Single-reversal focusing. (d) Periodic magnetic lens focusing. (e) Electrostatic lens focusing.

relatively high-density beams in these tubes. The simplest focusing method is a longitudinal fixed magnetic field along the length of the tube (Fig. 5a). This method is called confined flow. In confined flow, any tendency of an electron to move radially outward is converted by the magnetic field to a tight spiraling motion. The required magnetic field is typically from a few hundred to a few thousand gauss. Although such a field strength can be obtained from permanent magnets or electromagnets, the structure may be undesirably heavy. However, this focusing scheme is usually used for low-noise tubes, the magnetic field in the critical gun region often being higher than that in the interaction region.

A similar method of focusing is called space-charge balanced flow. For this method, the magnetic field in the gun region is lower than the field

strength in the interaction region. At the transition between the two values of magnetic field strength the beam is given a rotation with a direction to produce an inward force as the beam proceeds along the axial magnetic field in the interaction region. This inward force counterbalances the outward forces from space charge and from the centrifugal force set up by the rotation. A special type of space-charge balanced flow is called Brillouin flow; there is no magnetic field at the cathode (Fig. 5b).

In some tubes with permanent-magnet focusing, two bar magnets with like poles adjacent produce a reversal of the magnetic field (Fig. 5c). Such single-reversal focusing results in considerable saving in magnet weight.

Magnetic or electrostatic lenses for focusing have been developed for many traveling-wave tubes, with considerable saving in weight. A system of magnetic lenses requires a periodic magnetic field, which is usually produced by permanent magnets (Fig. 5d). A system of electrostatic lenses is attractive because of its further reduction in weight, but it is limited in application, being easy to apply only with slow-wave circuits that have natural separations for the application of different dc potentials, such as folded-line circuits (Fig. 5e). Periodic magnetic focusing has so far proved the more practical and is widely used.

Slow-wave circuit. In order that the electromagnetic signal wave will travel along the tube at a velocity approximately equal to the beam velocity, the signal is guided by a slow-wave circuit. Beam velocity is typically 2–10% of the velocity of a free-space electromagnetic wave. The signal must be slowed down to this velocity at all wavelengths within the bandwidth of the tube. The electric field produced by the slow-wave circuit at the beam for a given power on the circuit is also of interest. This relation, termed the interaction impedance, is given by Eq. (2) in which β, which is another impor-

$$K = \frac{(\text{ac electric field at beam})^2}{2\,(\text{average power flow on circuit})\beta^2} \quad (2)$$

tant characteristic, is the phase constant.

A helix is one of the simplest and best slow-wave circuits; it is used in most traveling-wave tubes (Fig. 1). The electromagnetic signal wave travels along the wire at about the free-space velocity. Consequently the signal phase velocity along the tube axis is given by Eq. (3). As this ap-

$$v_p \approx \text{free-space velocity} \left(\frac{\text{pitch}}{\text{circumference}}\right) \quad (3)$$

proximation shows, since phase velocity is independent of wavelength, wide-band amplifiers are possible. A helix also has good interaction impedance.

Helix tubes have delivered up to 250 watts of continuous output power at frequencies as high as 18 gigahertz (GHz). For higher power outputs, coupled-cavity slow-wave circuits are generally used (Fig. 6). This circuit has a narrower bandwidth than does a helix, 10% of center frequency being typical.

Other design features of the circuit are the coupling in and out and the providing of attenuation so that reflected waves do not cause the tube to oscil-

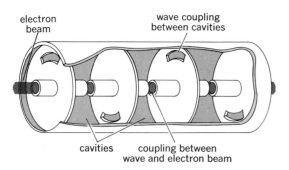

Fig. 6. Coupled-cavity slow-wave circuit for a high-power traveling-wave tube.

late. The couplings must be matched over the desired operating band to eliminate reflections. This requirement necessitates careful design. Attenuation is provided relatively easily in low-power tubes by lossy material sprayed onto the circuit or onto its dielectric supports near the center of the tube. In high-power coupled-cavity circuits, loss-impregnated ceramics are placed in selected cavities. When the lossy cavities are separated by sections of lossless circuit with 20 dB of gain, stability is good with only negligible degradation of efficiency.

Beam collector. The collector electrode and the slow-wave circuit are often connected to the same dc potential for simplicity. The interaction between electrons and signal wave has, however, produced a spread in electron velocities. If the collector potential is made low enough to collect the slowest-moving electron, overall efficiency is improved. The technique is used extensively in satellite applications.

For the backward-wave class of traveling-wave tubes *see* BACKWARD-WAVE TUBE. For other microwave tubes *see* KLYSTRON; MAGNETRON; MICROWAVE TUBE. [JAMES W. GEWARTOWSKI]

Bibliography: C. H. Dix and W. H. Aldous, *Microwave Valves*, 1966; J. W. Gewartowski and H. A. Watson, *Principles of Electron Tubes*, 1965; Y. Kano, *Electron Tubes*, 1972.

Trigger circuit

An electronic circuit that generates or modifies an existing waveform to produce a pulse of short time duration with a fast-rising leading edge. This waveform, or trigger, is normally used to initiate a change of state of some relaxation device, such as a multivibrator. The most important characteristic of the waveform generated by a trigger circuit is usually the fast leading edge. The exact shape of the falling portion of the waveform often is of secondary importance, although it is important that the total duration time is not too great. A pulse generator such as a blocking oscillator may also be used and identified as a trigger circuit if it generates sufficiently short pulses. *See* BLOCKING OSCILLATOR; PULSE GENERATOR.

Peaking (differentiating) circuits. These circuits, which accent the higher-frequency components of a pulse waveform, cause sharp leading and trailing edges and are therefore used as trigger circuits. The simplest form of peaking circuits are the simple *RC* and *RL* networks shown in Fig. 1. If

a steep wavefront of amplitude V is applied to either of these circuits, the output will be a sudden rise followed by an exponential decay according to the equation $v_o = V\epsilon^{-kt}$, where $k = 1/RC$ or R/L.

These circuits are often called differentiating circuits because the outputs are rough approximations of the derivative of the input waveforms, if the *RC* or *R/L* time constant is sufficiently small.

If a pulse is applied to the differentiating circuits, the resultant waveform shown in Fig. 2 may be used as a trigger. It is sometimes necessary, however, to remove by limiting or clipping the undesired portion of the waveform to prevent circuits from responding to it.

The *RL* circuit of Fig. 1 cannot be considered in its simplest form when extremely fast rise times are required because of the distributed capacitance and small series resistance associated with

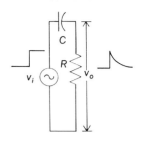

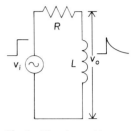

Fig. 1. Simple peaking circuit.

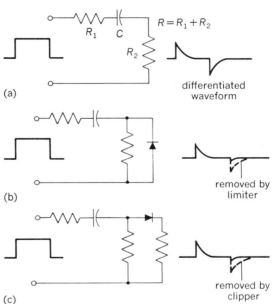

Fig. 2. Differentiated pulses. (*a*) Basic circuit. (*b*) Limiting unwanted portion. (*c*) Clipping unwanted portion.

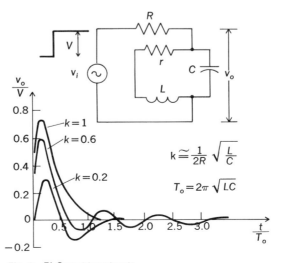

$$k \cong \frac{1}{2R}\sqrt{\frac{L}{C}}$$

$$T_o = 2\pi\sqrt{LC}$$

Fig. 3. *RLC* peaking circuit.

the inductance. A more accurate representation of the circuit is that in Fig. 3. The response is limited as shown for a fixed value of L and C. The value for $k = 1$ is referred to as critical damping. A value of k slightly less than unity provides a pulse that is a suitable trigger for many applications.

Ringing circuits. A circuit of the form shown in Fig. 3 that is highly underdamped, or oscillatory $(k \gg 1)$, and is supplied with a step or pulse input is often referred to as a ringing circuit. When used in the output of a field-effect or bipolar transistor as in Fig. 4, this circuit can be used as a trigger circuit. When the input pulse is applied, current in the output circuit is immediately cut off. Since the current in L cannot change instantaneously, it flows in the LC circuit in an oscillatory manner, gradually decaying because of the resistance in the circuit. However, if the diode is in the

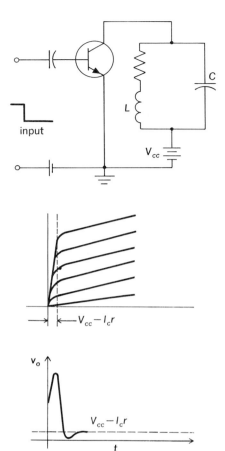

Fig. 5. Ringing circuit with transistor saturation damping.

circuit, the circuit will be highly overdamped for the negative portion of the oscillatory waveform, and the oscillations will be damped out as shown.

If a transistor is used as the current source and operated near saturation, damping will take place when the transistor goes into saturation as shown in Fig. 5. The diode is not required. For other waveforms *see* WAVE-SHAPING CIRCUITS.

[GLENN M. GLASFORD]

Bibliography: J. Millman and H. Taub, *Pulse, Digital and Switching Waveforms*, 1965; S. Seely, *Electronic Circuits*, 1968; H. Taub and D. Schilling, *Digital Integrated Electronics*, 1977.

Tuner

A device containing circuits that can be tuned to the carrier frequency of a desired transmitter in the frequency range for which that tuner is designed. The tuner serves to select the desired carrier frequency while rejecting the carrier frequencies of all other stations that may be on the air at that time. A tuner for a television receiver ordinarily contains only the radio-frequency amplifier, local oscillator, and mixer stages. A radio tuner usually contains, in addition, the i-f amplifier and second-detector stages so that it can feed an audio signal directly into a separate audio amplifier and loudspeaker system. *See* RADIO RECEIVER; TELEVISION RECEIVER; TUNING.

[JOHN MARKUS]

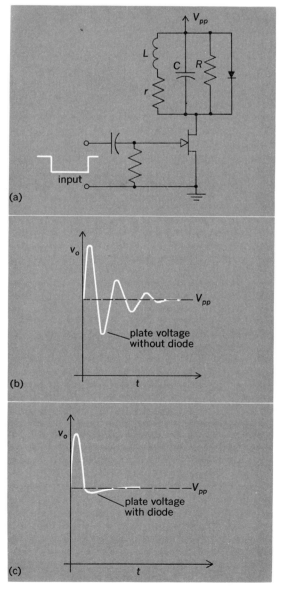

Fig. 4. Ringing circuit as trigger source. (a) Circuit diagram. (b) Plate-voltage waveform without diode limiter. (c) Plate-voltage waveform with diode limiter.

Tuning

The process of adjusting the inductance or capacitance, or both, in a tuned circuit — for example in a radio, television, or radar receiver or transmitter — so as to obtain optimum performance at a selected frequency. The tuning procedure that is carried out during manufacture or servicing of equipment generally involves the adjustment of screwdriver-type controls located inside or at the rear of the equipment. These are adjusted in such a way that the main tuning control on the front of the panel will serve to adjust all tuning circuits in the equipment simultaneously when a change of frequency is desired. *See* RESONANCE (ALTERNATING-CURRENT CIRCUITS). For an application of tuning *see* RADIO RECEIVER; TELEVISION RECEIVER.

[JOHN MARKUS]

Tunnel diode

A two-terminal semiconductor junction device (also called the Esaki diode) which does not show rectification in the usual sense, but exhibits a negative resistance region at very low voltage in the forward-bias characteristic and a short circuit in the negative-bias direction.

The short-circuit condition exists because both the *p* and *n* regions of the device are doped with such high concentrations of the appropriate impurities that the normal barrier is rendered sufficiently thin to allow the free passage of current at zero and all negative-bias conditions. The forward-bias characteristic (illustration *a*) shows a maximum and a minimum in the current with a negative-resistance region between. Band-potential diagrams show the internal electronic situation existing at the current minimum (illustration *b*), the current maximum (illustration *c*), and zero bias (illustration *d*). The top of the shaded regions of the band-potential diagrams shows the level to which electrons fill the available energy levels in the valence and conduction bands of the materials forming the *pn* junction. The bottom of the conduction band is designated E_c and the top of the valence band E_v. No electrons can penetrate the forbidden energy gap between E_c and E_v except in the barrier region, where it is thin enough to allow electron transit by tunneling.

The observed characteristic may be accounted for as follows. In illustration *d* the electron level is the same on both sides of the junction (zero bias). No net current flows because there is no difference in electronic energy across the junction. As forward bias is applied, tunneling current will flow since now the electrons in the *n*-type material on the left will rise to a level above those on the right (illustration *c*). As long as these electrons are still below the top of the valence band E_v on the right, current will increase. When the top of the elevated electron distribution exceeds the level of E_v, current will begin to decrease and the diode has entered the negative-resistance region. In illustration *c* the top of the electron distribution on the left is even with E_v on the right, and any further increase in bias will reduce the number of electrons available for the tunneling current. Therefore, illustration *c* corresponds to the current maximum. As bias continues to increase, the point is reached where the bottom of the conduction band E_c on the left is even with E_v on the right, and the entire electron distribution, being above E_v, is removed from the tunneling process. At this point, only normal forward-bias diffusion current flows. This current is composed of energetic electrons diffusing over the top of the barrier while remaining in the conduction band. Illustration *b* shows the point at which tunneling is no longer possible and thus corresponds to the current minimum at the end of the negative-resistance region. For further discussion of the properties of junction diodes *see* JUNCTION DIODE; SEMICONDUCTOR; ZENER DIODE.

[LLOYD P. HUNTER]

Bibliography: E. S. Yang, *Fundamentals of Semiconductor Devices*, 1979.

Tunneling in solids

A quantum-mechanical process which permits electrons to penetrate from one side to the other through an extremely thin potential barrier to electron flow. The barrier would be a forbidden region if the electron were treated as a classical particle. A two-terminal electronic device in which such a barrier exists and primarily governs the transport characteristic (current-voltage curve) is called a tunnel junction.

During the infancy of the quantum theory, L. de Broglie introduced the fundamental hypothesis that matter may be endowed with a dualistic nature — particles such as electrons, α-particles, and so on, may also have the characteristics of waves. This hypothesis found expression in the definite form now known as the Schrödinger wave equation, whereby an electron or an α-particle is represented by a solution to this equation. The nature of such solutions implies an ability to penetrate classically forbidden regions of negative kinetic energy and a probability of tunneling from one classically allowed region to another. The concept of tunnel-

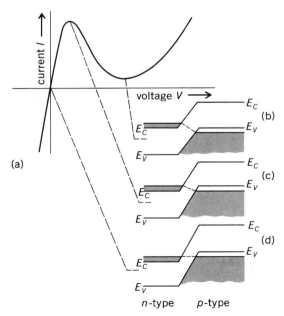

Tunnel diode characteristic. (*a*) Forward-bias voltage-current plot. (*b*) Band-potential diagram for the current minimum. (*c*) Band-potential diagram for the current maximum. (*d*) Band-potential diagram for zero bias.

ing, indeed, arises from this quantum-mechanical result. The subsequent experimental manifestations of this concept, such as high-field electron emission from cold metals, α-decay, and so on, in the 1920s, can be regarded as one of the early triumphs of the quantum theory. *See* FIELD EMISSION.

In the 1930s, attempts were made to understand the mechanism of electrical transport in resistive contacts between metals and rectifying metal-semiconductor contacts in terms of electron tunneling in solids. In the latter case, since a proposed theoretical model did not properly represent the actual situation, the theory predicted the wrong direction of rectification. In many cases, however, conclusive experimental evidence of tunneling was lacking, primarily because of the rudimentary stage of material science.

Tunnel diode. The invention of the transistor in 1947 spurred the progress of semiconductor technology. By the 1950s, materials technology for semiconductors such as Ge and Si was sufficiently advanced to permit the construction of well-defined semiconductor structures. The tunnel diode (also called the Esaki diode) was discovered in 1957 by L. Esaki. This discovery demonstrated the first convincing evidence of electron tunneling in solids, a phenomenon which had been clouded by questions for decades. This device is a version of the semiconductor *p-n* junction diode which is made of a *p*-type semiconductor, containing mobile positive charges called holes (which correspond to the vacant electron sites), and an *n*-type semiconductor, containing mobile electrons (the electron has a negative charge). Esaki succeeded in making the densities of holes and electrons in the respective regions extremely high by doping a large amount of the appropriate impurities with an abrupt transition from one region to the other. Now, in semiconductors, the conduction band for mobile electrons is separated from the valence band for mobile holes by an energy gap, which corresponds to a forbidden region. Therefore, a narrow transition layer from *n*-type to *p*-type, 5 to 15 nm thick, consisting of the forbidden region of the energy gap, provided a tunneling barrier. Since the tunnel diode exhibits a negative incremental resistance with a rapid response, it is capable of serving as an active element for amplification, oscillation, and switching in electronic circuits at high frequencies. The discovery of the diode, however, is probably more significant from the scientific aspect because it has opened up a new field of research—tunneling in solids. *See* BAND THEORY OF SOLIDS; CIRCUIT (ELECTRONICS); HOLES IN SOLIDS; JUNCTION DIODE; NEGATIVE-RESISTANCE CIRCUITS; SEMICONDUCTOR; SEMICONDUCTOR DIODE; TUNNEL DIODE.

Esaki and colleagues have explored negative resistance phenomena in semiconductors which can be observed in novel tunnel structures. One obvious question is: What would happen if two tunnel barriers are placed close together, or if a periodic barrier structure—a series of equally spaced potential barriers—is made in solids? It has been known that there is a phenomenon called the resonant transmission. Historically, resonant transmission was first demonstrated in the scattering of electrons by atoms of noble gases and is known as the Ramsauer effect. In the above-men-

tioned tunnel structures, it is clear that the resonant tunneling should be observed. In preparing double tunnel barriers and periodic structures with a combination of semiconductors, the resonant tunneling was experimentally demonstrated and negative resistance effects were observed. *See* SEMICONDUCTOR HETEROSTRUCTURES.

Tunnel junctions between metals. As discussed above, tunneling had been considered to be a possible electron transport mechanism between metal electrodes separated by either a narrow vacuum or a thin insulating film usually made of metal oxides. In 1960, I. Giaever demonstrated for the first time that, if one or both of the metals were in a superconducting state, the current-voltage curve in such metal tunnel junctions revealed many details of that state. At the time of Giaever's work, the first satisfactory microscopic theory of superconductivity had just been developed by J. Bardeen, L. N. Cooper, and J. R. Schrieffer (BCS theory). Giaever's technique was sensitive enough to measure the most important feature of the BCS theory—the energy gap which forms when the electrons condense into correlated, bound pairs (called Cooper pairs).

The tunneling phenomenon has been exploited in many fields. For example, small-area tunnel junctions are used for mixing and synthesis of frequencies ranging from dc to the infrared region of the spectrum. This leads to absolute frequency measurement in the infrared and subsequently provides the most accurate determination of the speed of light. *See* PARAMETRIC AMPLIFIER.

To study nonequilibrium superconducting properties, two tunnel junctions, one on top of the other sharing the middle electrode, are used. One junction seems to inject quasi-particles, while the other detects their effects on the important parameters. Tunnel junctions are also used as a spectroscopic tool to study the phonon and plasmon spectra of the metals and the vibrational spectra of complex organic molecules introduced inside the insulating barriers (tunneling spectroscopy). *See* SUPERCONDUCTIVITY.

Josephson effects. Giaever's work opened the door to more detailed experimental investigations—it pioneered a new spectroscopy of high accuracy to study the superconducting state. In 1962, B. Josephson made a penetrating theoretical analysis of tunneling between two superconductors by treating the two superconductors and the coupling process as a single system, which would be valid if the insulating oxide were sufficiently thin, say, 20 A (2 nm). His theory predicted, in addition to the Giaever current, the existence of a supercurrent, arising from tunneling of the bound electron pairs.

This led to two startling conclusions: the dc and ac Josephson effects. The dc effect implies that a supercurrent may flow even if no voltage is applied to the junction. The ac effect implies that, at finite voltage V, there is an alternating component of the supercurrent which oscillates at a frequency of 483.6 MHz per microvolt of voltage across the junction, and is typically in the microwave range. The dc Josephson effect was soon identified among existing experimental results, while the direct observation of the ac effect eluded experimentalists for a few years. The effects are indeed quantum phenomena on a macroscopic scale. Extraordinary sensitivity of the supercurrents to applied electric and magnetic fields has led to the development of a rich variety of devices with application in wide areas of science and technology. Superconducting quantum interference devices (SQUIDs) are made of one or more Josephson junctions connected to form one or more closed superconducting loops. Owing to their unprecedented sensitivity, SQUIDs are the main building blocks of many sensitive instruments such as magnetometers, power meters, voltmeters, gradiometers, and low-temperature thermometers. These are finding wide-range application in the fields of solid-state physics, medicine, mineral exploration, oceanography, geophysics, and electronics. Josephson junction and SQUIDs are used as switches for digital applications. They are the basic elements found in the picosecond-resolution sampling oscilloscope, as well as memory and logic circuits featuring high switching speed and ultralow power dissipation, in the order of 1 μW. In the communication field, they are used in analog applications, such as high-frequency local oscillators, detectors, mixers, and parametric amplifiers. Furthermore, the ac Josephson effect is now used to define the volt in terms of frequency in standards laboratories, eliminating the antiquated standard cell. *See* JOSEPHSON EFFECT; SQUID; SUPERCONDUCTING DEVICES.   [LEO ESAKI]

Bibliography: L. Esaki, Long journey into tunneling, *Science*, 183:1149–1155, 1974; I. Giaever, Electron tunneling and superconductivity, *Science*, 183:1253–1258, 1974; B. D. Josephson, The discovery of tunneling supercurrents, *Science*, 184: 527–530, 1974.

Vacuum tube

An electron tube evacuated to such a degree that its electrical characteristics are essentially unaffected by any residual gas or vapor.

Classification. According to function, vacuum tubes are classified as receiving tubes, transmitting tubes, phototubes, and cathode-ray tubes, microwave tubes, and storage tubes. Structurally, they are classified according to number of electrodes as diodes, triodes, tetrodes, pentodes, and so on, as discussed in later sections of this article.

Prior to the advent of semiconductor devices, hundreds of types of vacuum tubes were developed, many differing only in minor respects. With a few exceptions, most of these have been replaced by semiconductor devices or are used only as replacements in older equipment designed for vacuum tubes. Important exceptions include cathode-ray tubes, microwave tubes, high-power transmitting tubes, and x-ray tubes. In spite of the obsolescence of many types of vacuum tubes, they are still of considerable interest because the principles that underlie their operation and design have wide application in the field of electronics. *See* SEMICONDUCTOR.

Receiving tubes are low-voltage and low-power tubes designed for use in radio receivers, computers, and sensitive control and measuring equipment. Phototubes are two-electrode electron tubes in which the current is controlled by light flux incident upon one of the electrodes. They are used in sound-film equipment, in light-controlled switches, and in many industrial and scientific applications. Cathode-ray tubes function by virtue of an electron beam that can be focused to a small cross section on a screen and varied in position and intensity by means of electrical signals. They include oscilloscope tubes, television picture tubes, and camera tubes. Oscilloscope tubes make possible the visual examination of electrical signals. Picture tubes transform television signals into pictures upon a luminescent screen. Conversely, camera tubes convert light images into electrical output or, as image converters or intensifiers, produce a visible picture from incident infrared light or from x-rays.

Microwave tubes are designed for operation at frequencies of the order of 3 gigahertz (GHz = 10^9 hertz) and above. They are used in such diverse applications as radar equipment, space communication and control channels, scientific research, and radio-frequency ovens. Storage tubes, which are electron tubes designed so that information can be introduced and then extracted after a required storage interval, were developed for use in computers. *See* BACKWARD-WAVE TUBE; CATHODE-RAY TUBE; KLYSTRON; MAGNETRON; MICROWAVE TUBE; PHOTOTUBE; PICTURE TUBE; STORAGE TUBE; TELEVISION CAMERA TUBE; TRAVELING-WAVE TUBE.

Basic operation. Vacuum tubes depend upon two basic physical phenomena for their operation. The first is the emission of electrons by certain elements and compounds when the energy of the surface atoms is raised by the addition of heat (thermionic emission), by incident light photons (photoelectric emission), by kinetic energy of bombarding particles (secondary emission), or by potential energy (field emission). The second phenomenon is the control of the movement of electrons within an evacuated enclosure by forces exerted upon them by electric and magnetic fields. *See* ELECTRON EMISSION; ELECTRON MOTION IN VACUUM.

Vacuum tubes consist of an electrode capable of electron emission and one or more electrodes for collecting the emitted electrons and for establishing variable electric fields in order to control the movement of the electrons between the emitting electrode and the collecting electrode or electrodes. Where magnetic fields are required, they are produced by permanent magnets or by electromagnets, usually external to the evacuated space of the tube. This article deals almost entirely with thermionic vacuum tubes, in which the primary source of electrons is thermionic emission, and in which control is by electric fields. The emitting electrode is called the cathode, or filament when it is filamentary in form, and the collecting electrodes are called anodes. The principal anode is usually called the plate. Control electrodes are termed grids.

Thermionic vacuum tubes are all basically unilateral circuit elements; that is, the net flow of electrons can normally take place only from cathode to anode. In other words, conventional current, which is opposite in direction to the movement of electrons, is observed only from a positive electrode to the cathode.

Structure and fabrication. The mechanical structure of vacuum tubes is exemplified by the glass pentode receiving tube of Fig. 1. It consists of an assemblage of electrodes, made of nickel in low-power tubes, mounted on a base through which

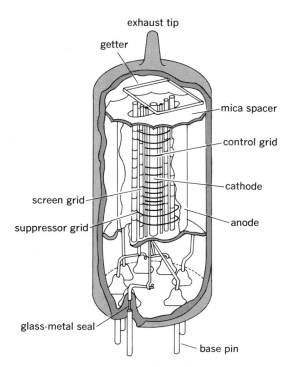

Fig. 1. A miniature glass receiving tube.

electrical connections are brought out, and enclosed in an evacuated glass or metal envelope.

After the electrode assembly has been mounted within the envelope, the tube is partially evacuated by a mechanical pump, capable of reducing the gas pressure to 10^{-3} mm of mercury, or one-millionth of atmospheric pressure (0.1 Pa). During this process, the envelope and electrodes must be heated by gas flames, or by high-frequency currents induced in metal parts by electromagnetic fields, in order to remove absorbed or adsorbed gases. The tube is then sealed off by closing the gas passage between envelope and pumping system.

When the tube has been sealed off, the vacuum must be improved; that is, the pressure must be further reduced. This is accomplished by flashing (evaporating) a "getter" material contained in a small nickel container previously mounted within the envelope. The getter vapor condenses upon the inner surface of the envelope in the form of a thin film. When gas molecules come into contact with this film, they combine with it to form stable compounds, with the result that the vacuum becomes progressively better with time until the pressure reaches a value of about one-billionth of atmospheric pressure (10^{-4} Pa). While the getter is being flashed, the emitter material of the cathode is heated by passing current through the heater winding or through the cathode itself. Gas released by the emitter material when it is heated is captured by the getter. Barium is the most extensively used getter material, although magnesium, calcium, sodium, and phosphorus have been used.

Cathode materials. Thermionic emission is negligible at room temperature. The cathode must therefore be heated either directly by an electric current through it or indirectly by an electrical heater within the cathode structure. A satisfactory

thermionic emitter must produce the required electron emission with as little heating power as possible, must have long life both in operation and in storage, must not be adversely affected by small amounts of residual gas within the envelope, and must have high resistance to mechanical shock. In addition, it should be cheap and readily fabricated.

The thermionic emission current density from an emitter surface is given by Eq. (1), in which A

$$J = AT^2 \epsilon^{q_e w/kT} \qquad (1)$$

and k are physical constants, T is absolute temperature, q_e is the magnitude of charge of an electron, and w is the work function of the emitting material, which is defined as the energy required to remove an electron through the surface of the material. The work function depends upon the emitting material and the condition of its surface. A low work function ensures large electron emission at temperatures well below the vaporization temperature of the emitter and a high ratio of emission current to heating power. Commonly used emitters are of three types: pure metals, thin films of one metal upon the surface of another metal, and coatings of rare-earth oxides upon metals. *See* THERMIONIC EMISSION.

Pure-metal cathodes are usually made of tungsten or tantalum and are filamentary in form. Although tungsten and tantalum have values of work function high in comparison with values of thin-film or oxide emitters, and must therefore be operated at relatively high temperature, these metals have the advantage of high resistance to the deleterious effects of residual gases. This feature is particularly important in tubes operated at high voltages, where bombardment of the cathode surface by ionized gas molecules accelerated in the high electric fields can cause severe damage or destruction of the cathode. For this reason, pure-metal cathodes are used in high-power tubes. Tantalum has a slightly lower work function than tungsten, but is less rugged. It is more readily worked than tungsten in the manufacture of cathodes.

A thorium film of atomic thickness, formed upon the surface of tungsten, lowers the work function considerably below the value for pure tungsten and thus makes possible emitter operation at much lower temperatures. Thoriated tungsten emitters have the serious disadvantage, however, that the thorium film may be partially or completely removed as the result of bombardment by ionized residual gas molecules or operation at too high a temperature. Formation of the film involves a number of steps and therefore complicates manufacture of tubes having this type of cathode.

Oxide-coated emitters usually consist of a mixture of barium and strontium oxides upon a nickel base. Although the manufacture of this type of emitter also involves a number of steps that must be carefully controlled, the work function of oxide-coated emitters is less than half that of thoriated tungsten emitters, and they are therefore capable of providing high emission at relatively low values of heating power and at temperatures well below that at which vaporization may damage the emitting surface. Oxide-coated emitters are capable of long life when they are properly operated. However, like thoriated tungsten emitters, they are susceptible to damage or destruction from bombardment by ionized gas molecules or excessive temperature. For this reason, they are not used in high-voltage transmitting tubes.

An instructive comparison of the efficiencies of emitters is afforded by Fig. 2, in which emission current density and emission in milliamperes per watt of heating power are shown as functions of operating temperature and of heating power.

Electrode configurations. Because characteristics and functions of vacuum tubes are determined by the number and configurations of the electrodes that the tubes incorporate, one of the most useful classifications of vacuum tubes is by number of electrodes.

The diode. The simplest tube is the diode; it has only two electrodes: the cathode, which emits electrons, and the anode or plate, which collects the electrons. In most diodes, electrodes have the form of concentric cylinders, although plane structures are also used. Utility of the diode rests mainly in the fact that conduction between plate and cathode occurs only when the plate is positive with respect to the cathode; when the plate is negative, electrons emitted by the cathode are prevented from reaching the plate.

The current-voltage characteristic of a vacuum diode with an indirectly heated cathode, which is a unipotential surface, closely approximates the relation given in Eq. (2), in which i is the current, A is

$$i = Av^{3/2} \qquad (2)$$

a constant that depends upon electrode size and configuration and upon cathode emission capability, and v is the voltage (positive) of the plate relative to the cathode. In diodes with filamentary cathodes, in which heating current flows directly through the cathode, the voltage drop along the filament causes the current-voltage characteristic to follow more nearly a 5/2-power law. The unidirectional-conduction (rectification) property of the vacuum diode is used in the conversion of alternating current into direct current in power supplies. This property is also useful in the detection of modulated voltages and in other circuit functions. *See* DEMODULATOR; DETECTOR; DIODE; LIMITER CIRCUIT; RECTIFIER.

The triode. A three-electrode tube, called a triode, is formed by the addition of a grid between the cathode and the plate of a diode. Electrons, in moving from the cathode to the plate, can pass through openings in the grid. The principal value of this third electrode is that large plate current and plate-circuit power can be controlled by small variations of grid voltage and with the expenditure of little power in the grid circuit. A commonly used triode electrode configuration is shown in basic form in Fig. 3. Although the control electrode in vacuum tubes is usually a helix (Fig. 3) rather than a grid, the term grid is used for any electrode having one or more apertures through which electrons or ions can pass.

The grid of a triode functions by changing the electric field near the surface of the cathode and thus the number of electrons that leave the cathode and arrive at the plate. If the grid voltage relative to the cathode is zero, a large fraction of the electric field produced by the positive plate voltage terminates on the surface of the cathode and accel-

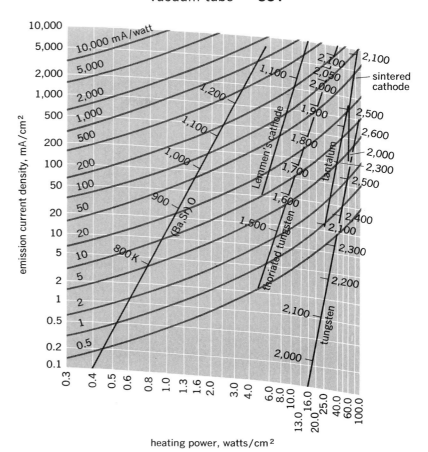

erates emitted electrons toward the plate. A few of these electrons strike the grid and thus cause a small grid current, but most of them reach the plate. If the grid is positive relative to the cathode, the accelerating field at the cathode surface is increased. This causes an increase in the number of electrons reaching the plate. However, the positive potential of the grid causes more electrons to strike the grid. The resulting increase of grid current is undesirable in many applications of the triode, especially because it causes power to be drawn from the source of grid voltage.

On the other hand, if the grid is negative relative to the cathode, the accelerating field at the cathode is reduced and fewer electrons reach the plate. Because electrons are repelled by the grid if it is at a negative potential, the grid current is negligible when the grid voltage is negative by more than one-half to three-quarters of a volt. This fact accounts for the ability of the grid to control plate current and plate-circuit power without the expenditure of appreciable power from the source of controlling voltage. If the grid voltage is more negative than a value called the cutoff grid voltage, essentially all electrons, including those that leave the cathode with a finite initial velocity of emission, are returned to the cathode, and the plate current is zero.

Typical curves relating the electrode voltages and currents of a triode are shown in Figs. 4 and 5. The curves relating plate current i_P with plate voltage v_P at constant grid voltage v_G, which are called

Fig. 2. Curves showing the relative emission efficiencies of various emitters.

VACUUM TUBE

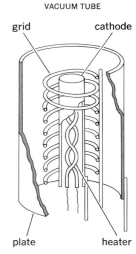

Fig. 3. Basic electrode structure of a cylindrical triode.

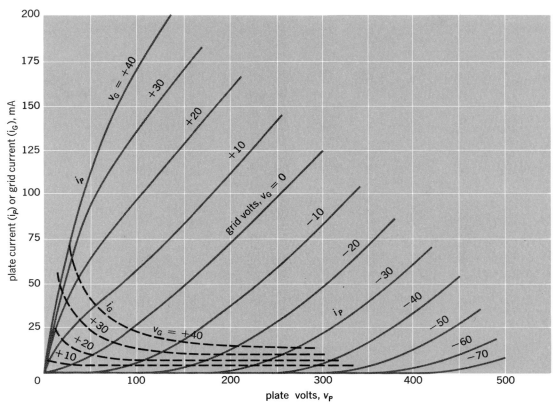

Fig. 4. Typical plate characteristics and grid-plate transfer characteristics of a triode.

plate characteristic curves, or simply plate characteristics, are shown by the solid lines in Fig. 4. The dashed lines are curves of grid current i_G as a function of plate voltage v_P at constant grid voltage v_G, and are called the plate-grid transfer characteristics. The solid curves of Fig. 5 relate plate current i_P to grid voltage v_G at constant plate voltage v_P; they are called the grid-plate transfer characteristics. The dashed curves, showing changes of grid current i_G with grid voltage v_G at constant plate voltage v_P, are called grid characteristics. Figure 5 shows clearly that the plate current can be varied over a wide range without grid conduction if the grid voltage is maintained negative. This range is used in all low-power voltage, current, and power amplifiers. The positive-grid-voltage range is used in some types of power amplifiers. *See* AMPLIFIER; OSCILLATOR.

The tetrode. In the use of vacuum triodes in very-high-frequency amplifiers, undesirable oscillation may result from the feedback of plate-circuit power to the grid circuit through the capacitance between plate and grid. Although this difficulty can be avoided by making circuit modifications, a much more desirable solution is the reduction of the grid-plate capacitance. This can be accomplished by placing between grid and plate a fourth electrode, which functions as an electrostatic shield. The name of this electrode, screen grid, is descriptive both of its function and of its physical structure. The first grid, which is used to control the plate current, is called the control grid. Such a four-electrode tube is called a tetrode. A typical low-power voltage-amplifier tetrode is shown in Fig. 6. In this tube, the capacitance between the

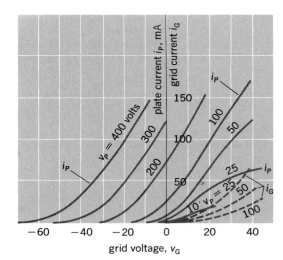

Fig. 5. Typical grid-plate transfer characteristics and grid characteristics of a triode.

control grid and the plate is reduced to a low value by the use of a screen grid that encloses the plate almost completely and thus effectively shields the control grid from the plate field.

To permit most of the electrons to pass through to the plate, the screen grid of a tetrode must be operated at a positive potential. If the plate voltage falls below the screen-grid voltage, secondary electrons emitted from the plate as the result of bombardment of the plate surface by electrons from the cathode are attracted to the screen grid. Because the movement of electrons away from the

VACUUM TUBE

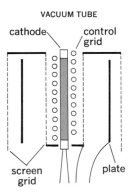

Fig. 6. Typical electrode structure of a low-power tetrode.

plate constitutes a negative component of plate current, the plate current falls markedly as the plate voltage is reduced below the screen-grid voltage. This is apparent in Fig. 7, in which the value of the screen-grid voltage is indicated by the vertical line at 90 volts. The rise in plate current as the plate voltage is reduced below about 70 volts is caused by the decrease of secondary emission from the plate as the reduction of accelerating voltage reduces the velocity with which the primary electrons strike the plate.

The pentode. The negative-slope region of the tetrode plate characteristics, which is undesirable in most applications, can be eliminated by the addition of a third grid located between the screen grid and the plate. If the potential of this grid relative to the cathode is zero or negative, the field between it and the plate is of the proper polarity to return to the plate secondary electrons emitted from the plate. The name of this grid, suppressor grid, is indicative of its function in suppressing secondary-electron plate current. The structure of a typical type of five-electrode tube, called a pentode, is shown in Figs. 1 and 8.

To minimize interception of electrons by the grids, the wires of the three grids are usually aligned. The presence of the suppressor grid reduces the capacitance between plate and control grid to an even lower value than that of the tetrode. Typical plate characteristics of a low-power voltage-amplifier pentode are shown in Fig. 9. The absence of the negative-slope region present in the tetrode at low plate voltages is apparent. It is of interest that the pentode plate characteristic curves resemble the collector characteristic curves of junction and field-effect transistors. *See* JUNCTION TRANSISTOR.

The beam tetrode. Use of a suppressor grid is not the only way to prevent the undesirable effects of secondary emission from the plate. In a type of tube called the beam power tetrode, suppression is accomplished by the use of beam-forming electrodes that focus the electron stream from the cathode into dense beams in the vicinity of the plate. The high density of electrons produces a retarding field between screen grid and plate that returns to the plate secondary electrons emitted from it. Because of the relatively low electron density in tubes designed for voltage amplification, this method of suppressing secondary-electron current is limited to tubes designed for power amplification, as the term beam power tetrode suggests. The electrode structure of a beam power tetrode is illustrated in Fig. 10, and a typical family of plate characteristics for such a tube is shown in Fig. 11.

Other multielectrode tubes. Special tubes containing four, five, or six grids were developed for special applications. The additional grids may serve as a means of impressing additional control voltages, or they may serve as shields between other electrodes. If two control voltages of different frequencies are impressed upon different control grids, the nonlinear properties of the tube cause the generation of new frequencies equal to the sums and the differences of the impressed frequencies (and their multiples). This process, called frequency conversion, is essential to the operation of certain types of radio receivers. Tubes used for

this purpose are called mixer tubes or mixers. In another type of multigrid tube, called a frequency converter, some of the electrodes are used in conjunction with an oscillator circuit to generate a high-frequency alternating voltage while the remaining electrodes serve as a mixer or as shields. The nonlinear properties of the tube result in the generation of new signals of frequencies equal to the sum of and the difference between the oscilator frequency and the frequency of an impressed signal. According to the number of electrodes that they incorporate, mixers and frequency converters are called hexodes, heptodes, and octodes. *See* MIXER; SUPERHETERODYNE RECEIVER.

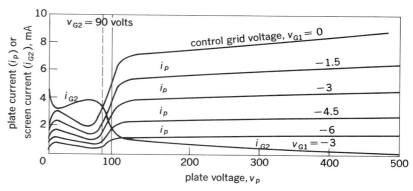

Fig. 7. Typical plate characteristics of a low-power tetrode.

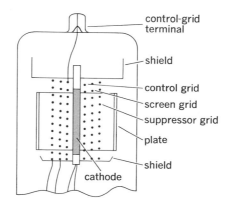

Fig. 8. Electrode structure of low-power pentode.

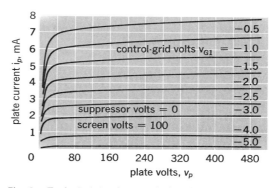

Fig. 9. Typical plate characteristics shown graphically for a low-power pentode.

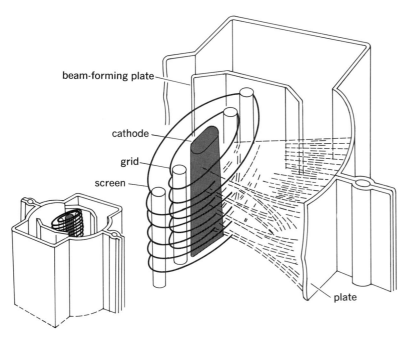

Fig. 10. Electrode structure of a beam power tetrode.

In another class of vacuum tubes, two or more complete electrode assemblies are incorporated within a single envelope, sometimes with a common cathode. Useful combinations include two diodes, two triodes, a diode and a triode or pentode, and a triode and a pentode. The circuit symbols for the more commonly used types of vacuum tubes are shown in Fig. 12.

Tube factors. Electrical behavior of vacuum tubes in specific circuits can be predicted from a knowledge of the circuit parameters, the tube interelectrode capacitances, and several important tube factors, including the amplification factor, the plate resistance, and the transconductance. The amplification factor μ, which is a measure of the voltage amplification of which the tube is capable, is defined as the negative ratio of an infinitesimal change of plate voltage v_p to the corresponding

increment of control grid voltage v_G necessary to maintain the plate current constant. Mathematically the amplification factor is expressed by Eq. (3). The amplification factor at any combination of

$$\mu = -\frac{\partial v_P}{\partial v_G}\bigg|_{i_P=\text{constant}} \qquad (3)$$

electrode voltages and currents is the negative of the slope of a curve of plate voltage as a function of control-grid voltage at constant plate current at the point corresponding to the selected values of current and voltage. This factor is a measure of the effectiveness of the control grid, relative to that of the plate, in controlling the plate current.

The amplification factor is relatively independent of the electrode voltages except in the vicinity of plate-current cutoff (zero plate current), where its value decreases. Its value depends upon electrode geometry and increases with increase of grid-wire diameter, decrease of grid-wire spacing,

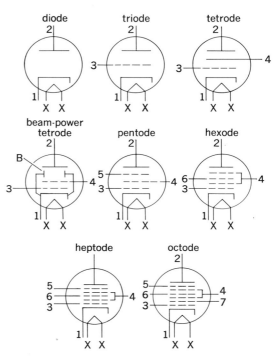

Fig. 12. Circuit symbols for common tubes.

and increase of grid-plate spacing. Geometrically similar tubes of different size have the same amplification factor. Typical magnitudes of triode amplification factor range from 10 to 100. Values for voltage-amplifier pentodes are higher.

The dynamic ac plate resistance r_p is defined as the quotient of an incremental change in plate voltage v_p by the corresponding increment of plate current i_p when the control grid voltage and all other electrode voltages are maintained constant. Usually the term dynamic plate resistance is simplified to plate resistance r_p, because the static plate resistance (quotient of the direct plate voltage by the direct plate current) is ordinarily of no interest.

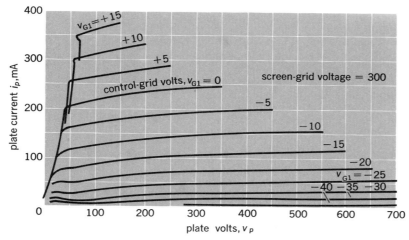

Fig. 11. Plate characteristics of a beam power tetrode.

Mathematically, the plate resistance is expressed by Eq. (4). At any combination of values

$$r_P = \frac{\partial v_P}{\partial i_P}\bigg|_{v_G = \text{constant}} \qquad (4)$$

of electrode voltages, the plate resistance is the slope of the corresponding plate characteristic curve at these values. It is essentially the resistance offered to an alternating plate current of small magnitude between the plate and the cathode when all electrode voltages except that of the plate are constant. The characteristic curves of Figs. 4, 7, 9, and 11 show that the plate resistance varies greatly with electrode voltages and is higher in tetrodes and pentodes than in triodes.

The transconductance (mutual conductance) g_m is defined as the quotient of an incremental change in plate current by the corresponding increment of grid voltage that causes it, when the plate voltage and all other electrode voltages are maintained constant. Mathematically, the transconductance is defined by Eq. (5). Thus transconductance of the

$$g_m = \frac{\partial i_P}{\partial v_G}\bigg|_{v_P = \text{constant}} \qquad (5)$$

triode whose grid-plate characteristics are shown in Fig. 5 is the slope of the characteristic at the chosen values of plate voltage and grid voltage. The term transconductance is also generalized to apply to the relation between the current to any electrode and the voltage of any other electrode, the term then being preceded by an adjective denoting the electrodes, such as suppressor-plate transconductance.

Grid-plate transconductance can be shown to be equal to the amplification factor divided by the plate resistance. Because the amplification factor varies little with electrode voltages, whereas plate resistance depends greatly upon electrode voltages, the grid-plate transconductance is also greatly dependent upon electrode voltages. This is also shown by the triode grid-plate characteristics of Fig. 5 and by the transfer characteristics of tetrodes and pentodes.

The voltage amplification or gain of a triode increases with the amplification factor and with the ratio of the external-plate-circuit resistance to the plate resistance. That of pentodes is approximately equal to the product of the transconductance and the external plate-circuit resistance. Transconductance is also important in many other applications of vacuum tubes, including power amplification, oscillation, and control.

Interelectrode capacitance. The performance of vacuum tubes in various types of circuits depends greatly upon, and may be limited by, capacitances that are invariably present between electrodes. The most important capacitances are the capacitance C_{gk} between control grid and cathode, the capacitance C_{pk} between plate and cathode, and the capacitance C_{gp} between control grid and plate. The effective input and output capacitances C_i and C_o of a vacuum tube used in an amplifier are not merely C_{gk} and C_{pk}, but include a component that is caused by the grid-plate capacitance in conjunction with the tube amplification. The value of this component of the input capacitance may approximate the product AC_{gp}, where A is the voltage amplification of the tube, if the magnitude of amplification is high. For this reason small grid-plate capacitance is essential in many applications of vacuum tubes. Shielding provided by the screen and suppressor grids of a pentode reduces the capacitance between the control grid and the plate to a low value and thus results in a low value of effective input capacitance, even though the voltage gain of pentodes may be high.

Frequency and gain limitations. The maximum frequency at which vacuum tubes can be used effectively as amplifiers or oscillators is limited by interelectrode capacitances, by inductances of the leads connecting the electrodes with the external circuits, and by transit time of the electrons between the electrodes.

At high frequencies, the input capacitance of the tube tends to short-circuit the control grid to the cathode; this reduces the alternating grid-to-cathode voltage and, consequently, the alternating output voltage. Undesirable resonances may also occur between tube capacitances and the lead or circuit inductances. It can be shown that circuit modifications to increase the voltage gain in general also reduce the bandwidth over which the gain is essentially constant and give a gain-bandwidth product that is, therefore, approximately constant. The gain-bandwidth figure of merit for a vacuum tube is the ratio $g_m/(C_i + C_o)$, which is the ratio of the transconductance to the sum of the effective input and output capacitances. It follows that pentodes, which have low input and output capacitances and relatively high transconductance, are particularly suitable for use in amplifiers having high gain over a wide frequency band. The gain-bandwidth figure of merit $g_m/(C_i + C_o)$ is also indicative of the maximum frequency of oscillation obtainable when a tube is used in an oscillator circuit and of the maximum attainable switching speed when a tube is used in a bistable circuit. *See* NEGATIVE-RESISTANCE CIRCUITS.

Lead inductance is undesirable at high frequencies because it introduces high reactance between the electrode and the external circuit. This reactance reduces the alternating voltage of the electrode and shifts the phase of the electrode voltage. In the cathode lead, it provides a feedback path between input and output circuits that may cause considerable reduction in gain and, in conjuction with tube and circuit capacitances, may lead to undesired oscillation.

Transit time of electrons between electrodes of a vacuum tube causes the alternating component of plate current to be out of phase with the alternating control-grid voltage that produces it. Consequently, the low-frequency transconductance g_m is replaced by an admittance y_m having an inductive susceptive component. Of greater importance is a reduction of the magnitude of y_m and of the conductive component g_m of y_m to a value below the low-frequency value of g_m. The reduction is the result of differences of transit time of individual electrons caused by differences of acceleration or deceleration of the electrons in the varying grid-cathode electric field. The reduction of magnitude of y_m with increase of frequency reduces the gain in amplifiers and the maximum attainable frequency of oscillation in oscillators.

Although the gain of amplifiers can be made as

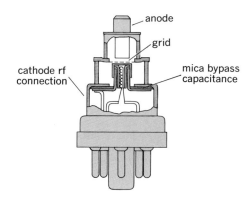

Fig. 13. Structure of a disk-seal tube.

high as desired, in practice useful gain is limited by noise generated by undesired fluctuations of voltage and current within the amplifying device and associated circuits. In vacuum tubes, noise results from random emission of electrons from the cathode and from the random partition of electron flow to two or more electrodes in multielectrode tubes such as the tetrode and the pentode. Circuit noise is mainly the result of the random motion of electrons among the thermally agitated molecules of the conductors. The magnitude of the noise generated in a circuit element having resistance is proportional to the square root of the product of the resistance, the absolute temperature, and the frequency bandwidth over which the noise is measured. The noise voltage generated in a tube is inversely proportional to the square root of the transconductance and is greater in pentodes than in triodes.

Lighthouse tube. At frequencies in the microwave range, the open-wire leads and lumped capacitances, inductances, and resistances that are used at low frequencies must be replaced by enclosed transmission lines and enclosed cavity resonators. In such circuits, the undesirable effects of electrode capacitances and lead inductances can be greatly reduced by the use of electrode and lead configurations that allow them to form portions of the conducting surfaces for lines and resonators. A structure that makes this possible is the disk-seal (lighthouse) tube, one form of which is shown in Fig. 13. In this tube, the grid lead and other electrode leads have the form of disks. The manner in which a disk-seal tube can be incorporated into an amplifier or oscillator with completely enclosed cavity resonators is shown in Fig. 14.

The disk-seal structure also allows the use of small spacing between the electrodes and thus greatly increases the frequency above which transit-time effects limit the performance of amplifiers and oscillators. At low frequencies, the disk-seal structure does not afford significant advantages over the conventional electrode structure because the required size of the cavity resonators increases inversely with decrease of wavelength and becomes prohibitive at frequencies below a few hundred megahertz.

Field-emission tubes. It is possible to obtain cold emission from a sharp point in a vacuum when the voltage gradient is large enough to pull electrons out of the metal without heating it. This usually requires a point with a radius of curvature of the order of one-millionth of an inch. With such a point it is possible to produce electric fields of the order of millions of volts per inch in the immediate vicinity of the point.

Field-emission tubes must be much more highly evacuated than ordinary vacuum tubes; otherwise the emitting point will become contaminated and fail to emit satisfactorily. Tremendous current densities can be obtained by this means, although the total currents are of the order of fractions of an ampere. Field emission has been used in high-voltage rectifier and x-ray tubes.

[HERBERT J. REICH]

Bibliography: D. G. Fink and D. Christiansen (eds.), *Electronics Engineer's Handbook*, 2d ed., 1982; Y. Kano, *Electron Tubes*, 1972; RCA Receiving Tube Manual, RC-30, 1978; H. J. Reich, J. G. Skalnik, and H. L. Krauss, *Theory and Applications of Active Devices*, 1966; G. F. Tyne, *Saga of the Vacuum Tube* 1977.

Vacuum-tube voltmeter

Any of several types of instrument in which vacuum tubes, acting as amplifiers or rectifiers, are used in circuits for the measurement of ac or dc voltage. The various types of ac vacuum-tube voltmeters all derive from the voltage to be measured a direct current to operate an indicating meter. The vacuum-tube voltmeter is the principal instrument for voltage measurements at high frequencies, as the frequency limit is determined essentially by the rectifier characteristics. It is also the usual test instrument in a wide variety of applications where little power can be taken from the source. *See* VOLTMETER.

Types. Six principal types of vacuum-tube voltmeters are described briefly here. The most important type, the diode rectifier–amplifier meter, is discussed in detail in a later section.

Plate-circuit rectification meter. In this meter the signal voltage is applied to the control grid of a tube, and rectification takes place because of the form of the grid voltage–plate current characteristic. The increase in average plate current operates the indicating meter. This was the first type used but is now rarely seen because the calibration depends greatly on the tube characteristics and the frequency range is limited in comparison with diode voltmeters. A modification, called a reflex vacuum-tube voltmeter, uses the rectified voltage in the plate circuit to increase the negative grid

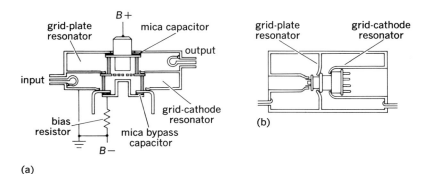

Fig. 14. Use of disk-seal tube with (a) cavity and (b) coaxial resonators.

bias. This permits much larger voltages to be handled without the grid drawing current and, at the higher voltage ranges, makes the calibration more linear and less dependent on tube characteristics.

Grid-rectification meter. The grid and cathode of a tube act as a diode rectifier, and the rectified grid voltage, amplified by the tube, operates a meter in the plate circuit.

Diode rectifier–amplifier meter. This is probably the most widely used meter. It has separate tubes for the rectification and dc amplification functions, permitting an optimum design for each. The dc amplifier, which may have more than one tube, usually employs inverse feedback for the stabilization of gain. The diode rectifier can be specially designed for high-frequency performance and is usually mounted in a small probe which can be put at the point in the circuit where the voltage is to be measured. *See* DIRECT-COUPLED AMPLIFIER.

Direct-current vacuum-tube voltmeter. This is essentially the amplifying and indicating portions of the diode rectifier–amplifier meter, which are usually designed so that the diode rectifier can be disconnected for dc measurements.

Amplifier-type meter. This meter has an ac amplifier, with gain stabilized by inverse feedback, preceding the rectifier.

Slide-back voltmeter. This meter employs a sharp-cutoff tube with adjustable negative grid bias. The bias at which plate current just commences is observed with and without the unknown voltage, and the difference in bias, read on a dc voltmeter, is taken as a measure of the applied ac voltage.

Diode rectifier–amplifier meter. A simplified circuit diagram of a typical diode rectifier–amplifier instrument is given in Fig. 1. The components to the left comprise a parallel-type diode rectifier circuit, in which the diode D_1 charges the capacitor C_1 to a voltage approximately equal to the peak value of the applied ac voltage, the rectified voltage appearing also across the diode D_1. The resistor R_1 permits C_1 to discharge when the voltage is removed. To avoid low-frequency error the product $\omega R_1 C_1$ should be about 100 or greater at the lowest frequency to be measured. The resistor R_2 and the capacitor C_2 filter out the ac component and the dc is applied to tube T_1 of a balanced two-tube dc amplifier. A second diode D_2 in the same envelope with D_1 is connected to the grid of the second amplifier tube T_2 to balance zero drift from warmup and heater-voltage variations. Plate-voltage fluctuations are also balanced out by the amplifier, so that stable operation down to 1.5 volts full scale can be obtained without employing a regulated power supply. The cathode-circuit resistance is high enough to stabilize the gain of the amplifier. The constants of the rectifier circuit are chosen so that the diode operates on the exponential-cutoff part of its characteristic. This varies only slightly from tube to tube, so that a new calibration is not required when the diode is replaced.

High-frequency performance. For diode rectifier–amplifier meters this factor depends on the geometry of the diode and of the probe in which it is mounted. The plate-to-cathode spacing determines the electron transit-time error, which is important at high frequencies and low voltages.

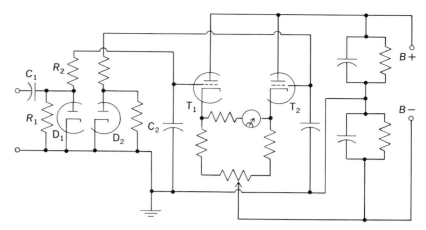

Fig. 1. Circuit diagram of a diode rectifier–amplifier meter.

The series-resonant frequency of the input loop of the rectifier determines the resonance error, which is independent of voltage. The two effects combine to give a high-frequency error depending both on the frequency and on the scale range of the instrument. A set of error curves for a commercial instrument is given in Fig. 2.

Input impedance. At low frequencies the input resistance of a diode-rectifier meter is determined by the diode current and is effectively about one-fifth the value of the discharge resistance (R_1 in Fig. 1), or of the order of 10 megohms. Dielectric losses are a controlling factor at high frequencies and give a value of roughly 50–100 kilohms at 100 MHz in high-quality instruments. This is in parallel with a capacitance of a few micromicrofarads.

Direct-current vacuum-tube voltmeters have an input resistance determined by the grid current of the input tube. For ordinary receiving tubes this is in the range 10^{-7} to 10^{-9} ampere, and the usual dc vacuum-tube voltmeter may have an input resistance of 10–100 megohms. Direct-current vacuum-tube voltmeters with special low-grid-current input tubes are called electrometer voltmeters, or vacuum-tube electrometers, and may have grid currents down to 10^{-15} ampere or lower. Such voltmeters may be used with input resistances as high as 10^{11} or 10^{12} ohms.

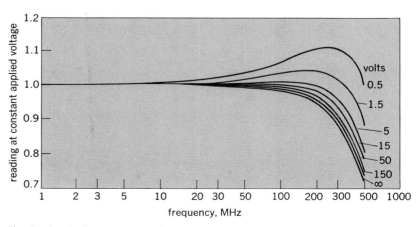

Fig. 2. A set of error curves of a commercial diode rectifier–amplifier voltmeter. (*General Radio Co.*)

Low-voltage limits. Diode rectifier—amplifier meters are limited in sensitivity by the square-law response of the rectifier at low voltages and by zero instability of the direct-current amplifier from contact potential variations in the diode and the first amplifier.

The maximum practicable sensitivity of a dc vacuum-tube amplifier with careful design is about 30 mV full scale, and electrometer voltmeters of this sensitivity are commercially available. A rectified voltage of this value requires an ac input of about 0.1 V, and the smallest detectable ac voltage would be 5–10 mV. However, this limit is hardly feasible in practice, and full-scale sensitivities lower than about 0.5 V on either ac or dc ranges are seldom seen in general-purpose instruments.

In amplifier-type ac meters, however, the amplifier is ahead of the rectifier, and measurements can be made down to about 1 mV over a 5-MHz bandwidth with full accuracy. Highly selective amplifiers, not usually classed as vacuum-tube voltmeters, are necessary to make ac measurements down to the microvolt level.

Waveform error. At voltages above about 5 V, diode rectifier voltmeters respond essentially to the peak value of the applied voltage. The calibration assumes sinusoidal waveform. On other waveforms the meter reads 0.707 times the peak value, which may deviate by as much as the percentage of harmonics present from the rms value or from the value of the fundamental component. Below about 0.3 V the response of diode rectifier meters is essentially square-law, and the waveform error is negligible. Amplifier-type voltmeters provide ample power to operate the rectifier, so that it can be designed for linear response, giving reduced waveform error. *See* CURRENT MEASUREMENT; DIGITAL VOLTMETER; SAMPLING VOLTMETER; VOLTAGE MEASUREMENT.

[W. N. TUTTLE]

Valence band

The highest electronic energy band in a semiconductor or insulator which can be filled with electrons. The electrons in the valence band correspond to the valence electrons of the constituent atoms. In a semiconductor or insulator, at sufficiently low temperatures, the valence band is completely filled and the conduction band is empty of electrons. Some of the high energy levels in the valence band may become vacant as a result of thermal excitation of electrons to higher energy bands or as a result of the presence of impurities. When some electrons are missing, the remaining ones may be redistributed among the energy levels within the valence band under an applied electric field, giving rise to an electric current. The net effect of the valence band is then equivalent to that of a few particles which are equal in number and similar in motion to the missing electrons but each of which carries a positive electronic charge. These "particles" are referred to as holes. *See* BAND THEORY OF SOLIDS; CONDUCTION BAND; HOLES IN SOLIDS; SEMICONDUCTOR.   [H. Y. FAN]

Varactor

A semiconductor diode designed to maximize the variation of its capacitance with applied reverse bias voltage. Such diodes are also often called vari-

able capacitance diodes. Today they are almost exclusively junction diodes. The free-charge depletion region of any *pn* junction in a semiconductor widens with the application of reverse bias. Since the surfaces of this free-charge depletion region represent the effective plates of the capacitance of the junction, the application of increasing reverse bias voltage will cause a decrease in the capacitance. The design and fabrication of varactors then reduces to the art of doping the two sides of the junction in such a way that the desired voltage variation of depletion region width (and hence capacitance) is obtained.

It is easily shown that, with an abrupt junction in which the doping level on the *n*-side and the *p*-side is uniform right up to the plane of the junction, the capacitance varies as $V^{-1/2}$, where V represents the reverse bias voltage. For junctions with suitable doping profiles, the power of the voltage dependence of the capacitance can be designed in the range from $-1/3$ to -1.0.

Since varactor diodes are primarily used in parametric amplifiers and subharmonic generators, it has become customary to rate varactors by a so-called cutoff frequency. This is defined as the frequency at which the Q of the diode, under the condition of zero dc bias, drops to unity. This frequency may be expressed as Eq. (1), where C_0 is

$$f_{\text{cutoff}} = \frac{1}{2\pi C_0 R_S} \tag{1}$$

the zero bias capacitance and R_S is the series resistance of the diode structure. Above this frequency the diode is predominantly resistive and will not oscillate. For an abrupt junction with the *n*-region doped lightly enough and made thin enough so that the free-charge depletion region just fills the *n*-region at the breakdown voltage, the cutoff frequency can be given in terms of the conductivity σ_n of the *n*-region and the diode capacitance at breakdown C_B, as in Eq. (2), where ϵ is the dielectric

$$f_{\text{cutoff}} = \frac{\sigma_n C_B}{2\pi\epsilon(C_0 - C_B)} \tag{2}$$

constant of the semiconductor material. From this it is seen that there is a trade-off between cutoff frequency and breakdown voltage. *See* JUNCTION DIODE; Q (ELECTRICITY); SEMICONDUCTOR; SEMICONDUCTOR DIODE.   [LLOYD P. HUNTER]

Bibliography: L. P. Hunter (ed.), *Handbook of Semiconductor Electronics*, 3d ed., 1970; J. Millman, *Microelectronics*, 1979; E. S. Yang, *Fundamentals of Semiconductor Devices*, 1978.

Varistor

Any variable resistor whose resistance depends upon voltage, current, or polarity. All semiconducting diodes are varistors. A varistor is generally made to have a symmetrical current-voltage relationship, but it can be nonsymmetrical. For many varistors, current varies as the *n*th power of voltage ($I \propto V^n$), where n is a number generally in the range 2 to 4.

Semiconductor rectifiers, of either the *pn*-junction or Schottky-barrier (hot-carrier) types, are popularly utilized for varistors. A single rectifier has a nonsymmetrical characteristic. A symmetri-

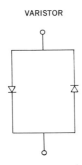

VARISTOR

Fig. 1. Symmetrical rectifier varistor.

cal varistor is made, utilizing two rectifiers connected in parallel with opposing polarity (Fig. 1).

Copper-oxide rectifiers are formed by oxidizing copper disks or washers at 1000°C to form cuprous oxide. This oxide is a semiconductor, and the contact between the oxide layer and the copper base is a rectifying contact. After further treatment, an electrode, usually of vacuum-deposited silver, is placed on top of the copper oxide to form a completed varistor.

In forming the selenium rectifier, a thin layer of elemental selenium is applied to one side of a base plate, usually made of nickel, nickel-plated iron, or aluminum. The unit is then heat-treated in order to change the selenium from the amorphous form to a crystalline form that is semiconducting. Then an electrode of cadmium or cadmium-bearing alloy is sprayed on the surface of the selenium. The contact between the selenium and cadmium is rectifying, while the contact between the base plate and selenium is ohmic.

The equivalent circuit of a varistor is shown in Fig. 2.

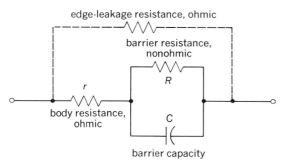

Fig. 2. Equivalent circuit of a varistor.

Silicon rectifiers are mostly fabricated by solid-state diffusion techniques similar to those utilized for transistors and integrated circuits. A thin plate of n-type silicon is exposed to a gaseous atmosphere containing a dilute p-type impurity at a high temperature (about 1200°C). Some atoms of the p-type impurity diffuse into the silicon surface by displacing silicon atoms, forming a pn junction within about 0.001 in. (0.0254 mm) of the surface. Metal electrodes are applied to p- and n-type regions by plating or evaporation. *See* JUNCTION DIODE; SEMICONDUCTOR RECTIFIER.

[I. A. LESK]

Video amplifier

A low-pass amplifier having a bandwidth in the range from 2 to 100 MHz. Typical applications are in television receivers, cathode-ray-tube computer terminals, and pulse amplifiers. The function of a video amplifier is to amplify a signal containing high-frequency components without introducing distortion.

In a single stage of an RC-coupled amplifier the high-frequency half-power limit is determined essentially by the load resistance, the internal transistor capacitances, and the shunt capacitance in the circuit. To extend the bandwidth of an RC-coupled amplifier, it is necessary to overcome the effects of these capacitances. In the past, this was

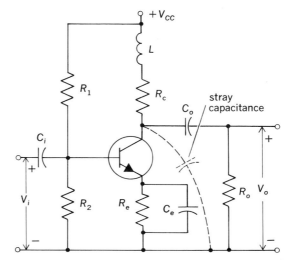

Shunt-compensated amplifier.

often done by adding an inductance L, as shown in the illustration, employing shunt peaking. The principal design requirement is to choose a value for the inductance that will extend the frequency response without introducing an undesirable hump in the gain characteristic for frequencies near the upper half-power frequency.

Modern video amplifiers use specially designed integrated circuits. With one chip and an external resistor to control the voltage gain, it is possible to make a video amplifier with a bandwidth between 50 and 100 MHz having voltage gains ranging from 20 to 500. The use of integrated video amplifiers minimizes cost and space. These amplifiers also eliminate the need for the shunt-peaking inductor shown in the illustration. *See* AMPLIFIER; INTEGRATED CIRCUITS. [HAROLD F. KLOCK]

Bibliography: C. A. Holt, *Electronic Circuits, Digital and Analog*, 1978; J. Millman, *Micro-Electronics*, 1979.

Video disk recording

A disk system used to reproduce television pictures and sound. The reproduction system consists of three parts: disk, disk player, and television receiver. The disk player rotates the disk, whose information density is several hundred times that of conventional audio long-playing records, at high speed and reads the recorded signal by a sensor. The signal can be seen on the picture tube or heard from the loudspeaker of the television receiver. The recording of television signals is not yet possible in the home; the system operates only for reproduction. Although the video tape recorder (VTR) can record and reproduce the television signal at home, the tape is expensive. A video disk can be mass-produced and is relatively inexpensive. Further, the degradation of recorded signal with elapsed time is far less than that of magnetic tapes. *See* TELEVISION RECEIVER.

The various types of video disk systems, which have been investigated since the early 1960s, have been reduced to three major types, whose developers are RCA, Philips-MCA, and the Victor Company of Japan. Unfortunately, these systems are technically quite different from each other and have no compatibility. For example, a video disk of

Comparison of video disk reproduction systems

Characteristics	VHD	VLP		SV-Disk
Disk diameter (mm)	260	301.6		300
Disk groove	Without	Without		With
Sensor	Capacitive sensor (sapphire stylus)	Optical sensor (laser)		Capacitive sensor (diamond stylus)
Reproduction time	60 min × 2	30 min × 2	60 min × 2	60 min × 2
Special effects (stop, slow, and quick motion); random access	Possible	Possible	Impossible	Impossible
Number of audio channels	2	2		1
Disk life	>1 hour even in stop motion	−		500 plays
Stylus life	>2000 hours	−		300 hours

the RCA type cannot play on the Philips disk player. This problem is one of the great barriers to the popularization of video disks. The table compares the main features of these systems.

SV-Disk. In the recording of the Selectavision Video Disk, input electrical signals are transformed into the corresponding modulated groove in an original disk record. Because the bandwidth of video signals is so wide that the high-frequency components of the signal cannot be recorded directly, a time-stretched signal, 25 times as slow as the real-time signal, is used as the input signal of the disk recorder. The cutting of the original disk is carried out in a similar manner as in the case of an audio disk. The processes for mass production of this type of video disk are also the same as those of conventional long-playing records. *See* DISK RECORDING.

In reproduction, the disk is rotated on a disk player at a speed 25 times as fast as that at the recording stage. The diamond stylus follows the groove, and an electrostatic sensor picks up the recorded signal.

This type of disk is a simple technical extension of conventional audio disk recording technology, and existing production facilities are largely available. Therefore the commercial price of disk and

disk player is expected to be the lowest among the three types. However, slow motion, stop motion, quick motion, and random access of images in reproduction are impossible because the stylus must constantly move along the groove. Further, this type is not suitable for stereophonic sound reproduction and cannot provide bilingual services because the number of audio channels is limited to one.

VHD. The Video High Density Disk System is a grooveless, capacitive pickup system. It can provide random access and special effects such as still (stop), slow, and quick motion in reproduction, and has two audio channels.

The recording system (Fig. 1) employs a smooth, flat glass disk coated with ordinary photosensitive material. Minute laser beams irradiate the disk while it rotates at a speed of 900 revolutions per minute. The source of beams is moved radially at a constant speed. A laser beam is first split in two, one half for modulation of video and audio signals, the other for modulation of tracking signals. Fine pits are recorded spirally on the glass disk (Fig. 2). A metallic master disk is produced from the original glass disk. Mother disks, which have indented pits, are made from the metallic master. Then stamper disks are produced from

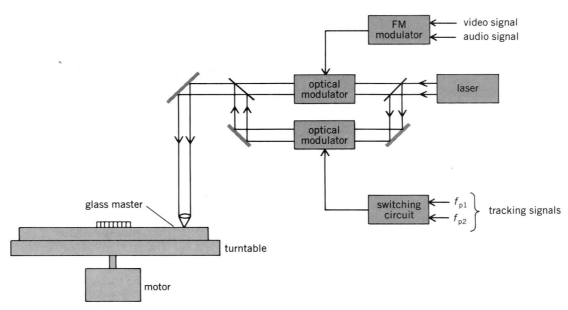

Fig. 1. Recording system of VHD. (*Victor Company of Japan*)

Fig. 2. Microscopic photograph of pit pattern in VHD. (*Victor Company of Japan*)

each mother. Each stamper, with its extended pits, produces approximately the same number of final disks as the typical audio stamper. Mass production of VHD disks is similar to that of the SV type. *See* LASER.

In reproduction, a five-sided stylus, which is far broader than that of a stylus-in-groove system, is used for signal pickup (Fig. 3). This broader construction of stylus and electrode, which runs the height of the stylus, makes long life of the stylus possible. The electrode on the stylus simultaneously picks up the video-audio and tracking signals electronically as capacitance variations between the disk surface and the electrode on the stylus. In Fig. 3, the schematic view of the pit structure is shown. The two tracking signals are indicated as f_{p1} and f_{p2}. The cantilever arm which holds the stylus is servocontrolled to track the virtual grooves on the disk and to correct for the time base error (jitter) of the rotating disk by means of an electrotracking system.

There are no actual mechanical grooves on the disk's surface to guide the stylus; instead, the sty-

lus merely slides along the surface and is guided electronically to pick up the recorded signals. This feature enables the pickup arm to move freely over the entire surface of the disk, and permits special effects such as still, slow, and quick motion, quick search, and random access to be achieved. Fig. 4 shows the loci of the stylus for various special effects in reproduction. The third pilot signal f_{p3}, which corresponds to the vertical blanking interval of a television signal, is inserted four times per revolution. The transition to the next virtual groove is carried out during the interval where f_{p3} is detected. The picked-up signal which contains video

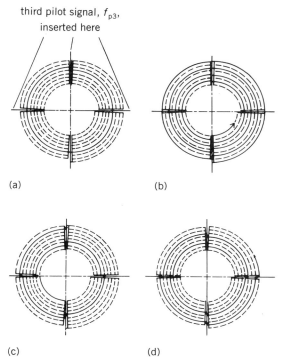

Fig. 4. Locus of stylus for various special effects in VHD reproduction: (*a*) still-picture reproduction, (*b*) fast reproduction (double speed), (*c*) fast reproduction (quintuple speed), (*d*) fast backward reproduction (treble speed). (*Victor Company of Japan*)

and audio information is applied to the television receiver after electronic processing.

The VHD system is fully equal to the VLP system in performance for home use and is expected to be cheaper than the VLP and more expensive than the SV-Disk system.

VLP. Quite different from the above two systems, the Video Long Play System is essentially an optical system using lasers for both recording and reproduction. This system is grooveless, has no stylus, has special features such as still, slow, and quick motion and random access, and provides two audio channels. It is thus quite equal in performance to the VHD system. The optical technology used in VLP is highly advanced and promising, and is expected to become widely available in various aspects of future optical communications. But the VLP disk player is the most expensive of the three systems. *See* OPTICAL COMMUNICATIONS.

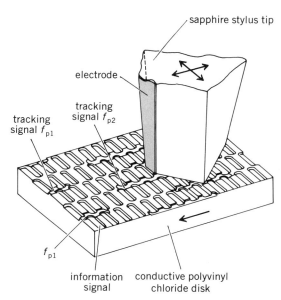

Fig. 3. Signal pickup mechanism in VHD. (*Victor Company of Japan*)

Fig. 5. Microscopic photograph of pit pattern in VLP. (*Pioneer Co. Ltd., Tokyo*)

The signal-processing scheme for video and audio signals prior to recording is quite similar to that of the VHD system except that no special tracking signals are used in this case. The microscopic photograph of the pits pattern thus obtained is shown in Fig. 5. The depth of pits is equal to a quarter wavelength of helium-neon laser light, that is, 0.15 μm. The width of the pits is 0.8 μm and the separation width of tracks is 1.6 μm. The length of each pit along a track represents the information of recorded radio-frequency signals containing one video and two audio channels.

Replication processes are similar to but slightly more complex than those of the VHD and SV-Disk systems. Fig. 6 shows the optical system of the VLP player. The complete pickup unit can move backward and forward on a carriage or rails underneath the record disk to follow the track. The player has excellent servomechanisms. The light beam from the laser is focused automatically at the record by the objectives which are controlled by a focusing servo. Other automatic control systems act on the pivoting mirror, thus keeping the spot centered on the track. A prism ensures that light reflected by the record falls on the detector. The detected electrical signal is separated into three parts by filters and FM-demodulated into a video and two audio signals. The production of special effects is very similar to that of the VHD case.

Reproduction of audio signals. Pulse-code modulated (PCM) sound has very superior quality

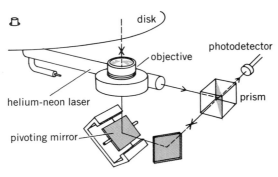

Fig. 6. Optical system of the VLP player. (*Philips*)

to any existing high-fidelity audio systems, but it requires a very wide frequency range. Video disks have enough bandwidth to record PCM audio signals. Both VHD and VLP systems provide a PCM decoder as an option to the disk player, and super-high-fidelity reproduction of audio signals can be heard at home. *See* PULSE MODULATION.

[H. DATE]

Bibliography: K. Compaan and P, Kramer, The Philips VLP system, *Philips Tech. Rev.*, 33(3): 178–189, 1973; E. Sigel et al., *Video Discs: The Technology, The Applications and the Future,* 1980; Special issue on RCA's Selecta Vision, *RCA Rev.*, vol. 39, no. 1, March 1978; Special issue on video long play systems, *Appl. Opt.*, 17:1993–2000, 1978; Victor Company of Japan, *New Technology, Video Audio High-Density Disc System, VHD/AHD,* 1980.

Video telephone

A communications instrument used for the simultaneous exchange of visual images and associated speech. A complete video telephone communication system involves three basic elements: (1) terminal equipment (camera, display, microphone, and speakerphone) to transform both the aural and visual inputs into electrical signals, and vice versa; (2) transmission facilities to carry the electrical signals over large distances; and (3) switching systems to allow a choice of terminals to be interconnected. The equipment that is used as a visual adjunct to telephony may also be arranged to serve many of the applications associated with closed-circuit television and surveillance. *See* CLOSED-CIRCUIT TELEVISION.

In 1927 H. E. Ives transmitted images one way with simultaneous two-way voice by wire from Washington, DC, to New York State and by radio from Whippany, NJ, to New York City. The television standards in those experiments were chosen to reproduce a recognizable human face. The picture had 50 lines, was sequentially scanned, and had a frame rate of 18 per second. The transmitted signal required a bandwidth of 20 kHz. In 1930 Ives demonstrated two-way video telephone over a wire path in New York City. From 1936 to 1940 a public video telephone service was provided on a local and intercity basis by the German post office. Calls could be set up by appointment between any two users in Berlin, Leipzig, Nuremberg, and Munich. This service provided a picture with 180 lines repeated 25 times a second, and required 270 kHz of bandwidth. A similar system was inaugurated by the Soviets in 1961, serving users in eight cities, including Moscow, Kiev, and Leningrad. Their system operated with the European format of 625 lines per picture and 25 pictures per second. More than 6 MHz of bandwidth was required.

Since the 1960s, most major communication agencies throughout the world have been exploring the subjective, technical, and economic feasibility of a commercial video telephone service. In 1961 an intercity appointment-type service was introduced by the Bell System to users in New York City, Washington, and Chicago. The equipment for this service provided a 275-line picture with a 30-Hz frame rate and 2:1 interlace, and required a bandwidth of about 500 kHz. The station sets were early versions of the type used by the Bell System in its later offering of commercial video telephone service. Similar service was provided within the environs of several industrial plants in the United States and in Europe.

Picturephone. This first limited commercial switched video telephone service was offered in the early 1970s by the Bell System. It provided a picture of 267 lines, 30-Hz frame rate, 2:1 interlace, and 1-MHz bandwidth. It was viewed as an extension of telephone service with the visual adjunct, and a video telephone customer was to have the same interconnection capabilities as a telephone-only subscriber. Basic telephone switching equipment was used to provide the audio connection, but the design was extended to include the ability to switch broadband video signals.

Ordinary telephone wires utilizing specially designed repeaters were used to transmit the video signals. Video required two pairs of wires, one for each direction of transmission, while both directions of the audio were transmitted on another pair of wires. For short distances between subscribers and a central office, or between central offices, baseband analog transmission was used. For longer distances, the plan was to use digital encoding of the video signal. A differential pulse-code modulation scheme was used and operated at a rate of 6.312 megabits per second. This rate was compatible with the evolving digital hierarchy of the telephone plant, and would allow the mixing of video telephone and other signals on long-distance transmission facilities.

This early service offering (in Chicago and Pittsburg) provided valuable insight into customer needs, but customer acceptance did not meet the early expectation. Later more successful operation was achieved in limited markets, for law enforcement agencies and conference room intercommunication.

Cost reduction approaches. Expense is a very important factor in the acceptance of a video telephone service, and the wide bandwidth required to transmit a video signal is a major contributor to the system cost. Continuing efforts to reduce costs center on schemes (1) to reduce the bandwidth required to handle the signal by sacrificing resolution or motion rendition in acceptable ways, and (2) to reduce the bandwidth required to handle the signal by taking advantage of the redundancy present in a video signal.

Examples of the first type of approach include an experimental system demonstrated in 1956 which presented a new 60-line snapshot picture every 2 s, and one described in the 1960s which used a 310-line picture renewed every 40 s. Both required only about 2000 Hz of bandwidth and could utilize conventional telephone lines. For a period in the early 1970s RCA marketed a similar slow-scan system called Videovoice which provided a 525-line still picture every 60 s over conventional telephone lines. It was not a commercial success, indicating a need for even more technological development to reduce cost and improve performance.

Experiments have been carried out with dot and line interlace and a slow frame rate (four per second) in order to reduce the bandwidth required for a 516-line picture. Still other methods have been explored to reduce the transmission requirements of video telephone signals by taking advantage of the inherent redundancy in typical video signals. By the early 1970s, the bit rate required to transmit a Picturephone signal had been reduced to only 1.5 megabits per second. The methods for doing this center on locally storing the picture and comparing adjacent frames, transmitting only enough to update elements of the picture that have changed from frame to frame, with logic based on recognizing correlations from element to element and line to scanning line. Research has been undertaken to electronically track object motion in the scene and predict changes in adjacent frames so that only the errors in such predictions need be transmitted.

Video conferences. Group video conferences have been introduced by the British Post Office, Nippon T&T, the Australian Post Office, the Bell System, and a number of private companies. The British Post Office experimental conference service is called Confravision. It is centered on the use of specially equipped group conference rooms in major cities in Great Britain, Sweden, and the Netherlands. The equipment operates on the European 625-line broadcast television standard.

The Bell System experimental conferencing service is called Picturephone Meeting Service and involves specially designed rooms which utilize voice-operated cameras to select the part of the local room to be televised. The design of the rooms is based on the use of 525-line cameras and monitors and includes interface equipment to switch the signals between the several facilities. Transmission facilities for video conferencing include the domestic satellite communication relay network, which has made privately owned networks possible.

[C. C. CUTLER]

Bibliography: L. E. Flory, E. G. Romberg, and V. K. Zworykin, *Television in Science and Industry*, 1958; H. E. Ives, Two-way television, *Bell Labs Rec.*, 8:399–404, 1930; N. Mockhoff, The global video conference, *IEEE Spectrum*, 17(9): 45–47, 1980; The Picturephone system, *Bell Syst. Tech. J.*, vol. 50, no. 2, 1971.

Videotext and teletext

Computer communications services that take advantage of standard television receivers and use them as display terminals to retrieve information from data bases. A "black box" is placed between the television set and the signal carrier. The signal carrier may be a telephone network, a data communications network, a television cable, or a broadcast signal. The black box receives data from the signal carrier. It formats that data into a signal suitable for display on the television set. The user interacts with the black box to determine what information is actually displayed on the screen of the television set.

Both videotext and teletext use the same display technology. The difference between them is that videotext is a two-way, or interactive, service, which allows the user to interact with the service in selecting information to be displayed, while teletext is a broadcast service. In the teletext service, preprogrammed sequences of frames of data are broadcast cyclically. The user interacts with the black box to "grab" the frames of relevant information. The textual and graphic information in teletext is inserted into the "blanking period" in television signals. While select frames of information can be viewed in the teletext service, the user cannot interact with the data base.

In videotext, the black box is actually a limited computer terminal which uses the television screen for display. Both videotext and teletext provide the user with a numeric keypad for interacting with the black box. Some videotext terminals also provide, as an option, a full alphanumeric keyboard for interacting with videotext services.

In both videotext and teletext the user will normally make selections from a series of menus to select the desired information. In videotext the user may then continue to interact with the service selected.

Display alternatives. Alphamosaic, alphageometric, and alphaphotographic are three different display technologies that may be used with videotext and teletext services.

Alphamosaic, the first to be introduced, is the kind of display technology used in many video games. The screen is divided into a series of rectangular mosaic elements. The picture to be displayed is transmitted as a series of rasters, line by line, similar to the way that a basic television signal is transmitted. Both alphanumeric characters and pictures are built up from these elements, a line at a time.

Alphageometric technology is more closely related to conventional computer graphics. The alphageometric format is the basis for the Canadian Telidon technology, which has heavily influenced the American Telephone and Telegraph videotext standard for North America. This technology stores and transmits commands to draw graphic objects on the display screen. For example, the command to draw an arc will consist of a series of instructions giving the location of the center of the circle, the radius of the circle, the starting point of the arc (in degrees), and the end point of the arc. Each object to be drawn on the screen is transmitted as a series of commands. *See* COMPUTER GRAPHICS.

The alphageometric display technology gives higher-resolution images and requires lower bandwidths for data transmission than alphamosaic. However, the black box which acts as a computer terminal is more complex. It translates graphics commands into the actual display on the television screen. The alphageometric display black box does more processing than the alphamosaic black box.

Alphaphotographic display technology provides the highest resolution of the three alternatives, making it possible to incorporate full-color photographs into the data. The display quality is similar to a normal, good-quality television frame. Conceptually, there is little difference between alphaphotographic and alphamosaic. However, in alphaphotographic the size of the displayed mosaic element is reduced to the minimum displayable picture element on the television tube. In practice, alphamosaic videotext systems are designed to operate at much lower bandwidths, while alphaphotographic systems require relatively high bandwidths to the videotext terminal.

The data which make up frames of information are different for each of the three display technologies. Some work has been done to allow conver-

sion from one format to another, but one cannot generally retrieve information stored in one format and display it on a terminal equipped for a different format.

Implementations. The United Kingdom led the world with the first operational teletext services (Ceefax and Oracle) and videotext service (Prestel). Both of these use alphamosaic format displays. The German Bildschirmtext is a compatible alphamosaic format service. The French Teletel is another alphamosaic format videotext service, but it uses the French Antiope technology, which is a different approach from Prestel in the way it does character coding. Many other European countries have begun operating videotext and teletext services based on alphamosaic technology that is compatible with either Antiope or Prestel.

One of the first major North American videotext trial systems was Bell-Canada's VISTA service, and the Trans Canada Telephone System (TCTS) trial Intelligent Network (iNET) continues to provide access across Canada to alphageometric videotext data bases. There are numerous other trial systems in Canada. All employ the Canadian Department of Communication's Telidon protocol with an alphageometric format.

Numerous technical trials, market tests, and full commercial services are now available or starting up in the United States. Some use alphageometric formats, while others use alphamosaic. Three major but incompatible trial systems in the United States include Knight-Ridder's and American Telephone and Telegraph's Viewtron in Florida, Cox Cable's Indax in San Diego, and Times-Mirror's service in southern California.

Most North American trial systems follow American Telephone and Telegraph Standard 709E, which is compatible with the Canadian Telidon standard. There is not yet a single internationally agreed-upon standard; however, there is considerable effort among international standards committees to reach agreement. Alphaphotographic formats are not being used in any current trial system.

Services. A variety of information has been made available on videotext and teletext, with varying degrees of acceptance. A few examples follow.

Data-base retrieval forms a major class of videotext and teletext services. Examples include restaurant guides, hotel information, stock market information, and travel information. These are suitable for both the broadcast mode of teletext and the interactive mode of videotext.

Typically, a user might start with a menu listing categories for hotels or restaurants. Upon selecting one of the categories, the user is given a screen with more information, such as a list of four-star hotels or a list of Chinese restaurants. The user can continue a tree-structured search until the desired information is obtained.

Another major class of services is interactive, and therefore is suitable only for videotext. Examples include calculation services, electronic mail, teleshopping, and financial services. Calculation services prompt the user for inputs to a program which does the calculation and displays the result. Electronic mail generally requires a full keyboard to input the text of a message to be sent from one user to another. Teleshopping displays items a user might want to purchase, and accepts orders, like a catalog store, when the user chooses to purchase an item. A variety of financial services, such as transferring funds from one account to another, are possible.

Because of its inherent one-way broadcast nature, teletext service can be offered only on a subscription basis. However, videotext services can be charged on the basis of individual requests. This offers a real potential for stimulation of the information marketplace. *See* CONSUMER ELECTRONICS; DATA-BASE MANAGEMENT SYSTEMS; ELECTRONIC PUBLISHING; MULTIACCESS COMPUTER; TELEVISION RECEIVER.

[RONALD P. UHLIG]

Bibliography: H. B. Thomas and M. Tyler, Videotext and Teletext: Computing and communication for the mass market, *Proceedings of the Fifth International Conference on Computer Communications*, p. 490, 1980; J. Tydeman et al., *Teletext and Videotext in the United States: Market Potential, Technology, Public Policy Issues*, 1982; *Videotext '81 Proceedings* and *'82 Proceedings*, On-Line Publications Ltd.

Voice response

The process of generating an acoustic speech signal that communicates an intended message so that a machine can respond to a request for information by talking to the human user. Communication of speech from the machine to the human requires abilities to convert machine information into a desired word sequence, with the pronunciations of all desired words carefully sequenced and inflected to produce the natural flow of continuous speech waves.

Talking machines permit natural interactions that can be valuable for announcing warnings, reporting machine status, or otherwise informing the computer user, especially when the user cannot view displays, due to concurrent visual tasks, visual handicap, or remote telephone links. The many advantages of voice interactions with machines and the expanding uses of computers in offices, factories, schools, and homes have encouraged the development, sales, and use of voice response systems as voice warning devices, cockpit advisory systems, automated telephone-directory assistance systems, time-of-day services, bank-by-phone facilities, talking clocks and appliances, toys that speak, and automatic readers of printed texts for the blind. Low-cost synthesizers on circuit boards or in microminiaturized integrated circuit chips are rapidly giving machines the ability to speak.

Reproduction of stored speech. For simple tasks involving small vocabularies, machine production of speech is not difficult to achieve. Human speech can be stored and reproduced later on demand, using techniques similar to those in familiar devices like audio tape recorders, phonograph records, and other analog or digital magnetic media. Simple machine control of the position of the pickup device on an audio recorder can determine the sequence of words or phrases. The earliest commercial voice response devices were simple machine-accessed analog recordings on tape loops and magnetic drums.

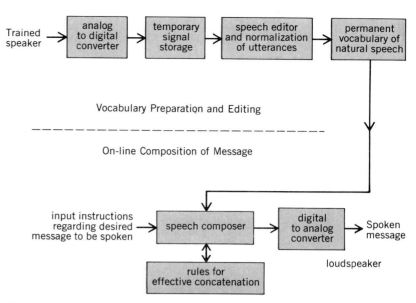

Fig. 1. Voice response system based upon concatenation of words spoken by a human.

Word concatenation systems. It is only a small conceptual step from such recorders to the simplest form of modern voice response system, called a word concatenation system (Fig. 1). Acting like "automated tape recorders," word concatenation systems can retrieve previously spoken versions of words or phrases and carefully concatenate them without pauses to approximate normally spoken word sequences.

The processes involved in word concatenation begin off-line, when a trained speaker (for example, a radio announcer) in a quiet booth speaks words or phrases from a selected vocabulary, and this high-quality speech is digitized and placed in a temporary signal store. The speech is edited (automatically or with human help) to adjust sound levels, to locate the beginning and ending of each word, to assess the naturalness and clarity of the pronunciation, and to check the inflection of the voice. If necessary, words may be spoken again until good examples are obtained. Some systems store two versions of certain words: one with flat intonation for nonfinal positions in word sequences, and another with the characteristic falling intonation of terminal parts of declarations and commands. The result is a pronunciation dictionary, or permanent store of human utterances of desired words or phrases, as illustrated on the left in Fig. 2. It is not uncommon for some vocabulary items to be long phrases like "voice response system" or "at the tone the time will be," spoken as a unit.

Voice response from the machine is accomplished later by input instructions dictating to the "synthesis program" or utterance composer that (for the example shown at the top of Fig. 2) the first word to be spoken will be some vocabulary item I_{68}, then comes item I_{82}, then I_5, then I_9, and then I_{47}. The composer produces $I_{68}I_{82}I_5I_9I_{47}$, heard as "The voice response system concatenates human speech," as a juxtaposing of the stored signals for each of the vocabulary items. Such juxtaposing

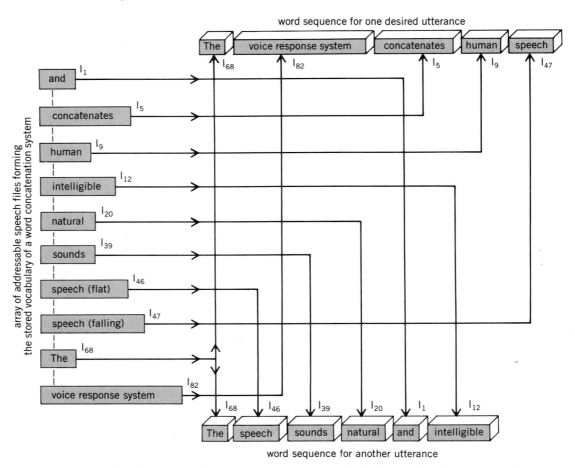

word sequence for one desired utterance

array of addressable speech files forming the stored vocabulary of a word concatenation system

word sequence for another utterance

Fig. 2.   Waveforms of words or phrases from the dictionary are connected end-to-end to achieve connected voice output from a word concatenation system.

will usually be guided by internal rules that select nonterminal versus terminal versions of words, like I_{46} versus I_{47} in Figure 2. Rules may also correct for anomalies or bad transitions at junctures between words. The concatenated word sequence then is converted into a continuous signal that drives a loudspeaker. Figure 2 shows another utterance being generated in the same fashion, but with a different word sequence.

There are problems with such a simple head-to-tail composition of word sequences. The speech may sound choppy, with unnatural intonations and timing, and with noticeable breaks at word boundaries. Also, the storage of a high-quality digitized waveform may require as much as 64,000 bits per second of speech, so that large computer disk packs can store perhaps at most 1200 seconds of speech, and microcomputer memories might be able to store only hundreds, or perhaps only a few tens, of words. Low-cost concatenation systems are thus restricted to small vocabularies.

Parametrized voice response systems. Figure 3 illustrates a spectrum of alternative solutions to the storage problem for voice response. There is a general tradeoff between storage requirements and the complexity of coding and decoding of the speech. Storage requirements can be reduced substantially by first extracting informative parameters from the original human speech and later reconstructing speech from such stored parame-

ters. Parameterized voice response systems typically use signal-processing methods like delta modulation, adaptive differential pulse-code modulation, spectral filtering through a bank of bandpass filters (channel encoding), linear predictive coding (LPC), or tracking of important speech features like natural resonant frequencies (formants) of the speaker's vocal tract and fundamental frequency (pitch) of the voice. Such parameterized approaches assume that not all the information in the original signal is important to conveying the message. Informative parameters totaling about 16,000 bits/s to as low as 600 bits/s can be stored as simplified instructions for controlling reconstitution of the waveform.

Figure 4 illustrates a parameterized voice response system based on extensive evidence that the resonant peaks (formants) of the voice are important cues to the identities of vowels and other vowellike sounds. Words or phrases spoken by a human are analyzed in off-line mode to extract time variations of the frequencies (F_1, F_2, F_3) and amplitudes (A_1, A_2, A_3) of the formants, as well as the variations of the rate of vibration (fundamental frequency F_0, or pitch) and amplitude A_0 of the vocal cords versus time. The system notices "unvoiced" periods when the vocal cords are not vibrating, so that the signal is an aperiodic noise of amplitude A_N. All these parameters may require less than 1000 bits/s for vocabulary storage.

Later, upon input of a command specifying the desired message, the formant synthesizer reconstructs the overall acoustic character of the speech by using the stored parameters to control signal generation processes in the random number generator (for unvoiced speech), the pitch pulse or glottal wave generator (for controlling fundamental frequency), and the bank of resonating digital filters (controlled by formant amplitudes and frequencies). For fricative sounds, the formant resonances are replaced by broad spectral peaks (of frequency F_P and bandwidth B_P) and antiresonances (frequency F_2 and bandwidth B_2). A separate path for the nasal tract also has its own resonances (frequency F_{NP} and bandwidth B_{NP}) and antiresonances (frequency F_{NZ} and bandwidth B_{NZ}). The combination of these processes produces a signal resembling the resonating or noiselike character of each portion of the original speech.

One advantage of the parameterized voice response system is the possibility of altering the parameters to have them differ from the original speech. For example, formant trajectories, or prosodic characteristics like pitch, durations of certain parameter configurations, and intensity of the speech, can be computer-controlled to achieve smooth transitions at word boundaries. Words spoken with flat pitch contours can be analyzed and resynthesized with the terminal intonation fall of a command or the terminal rise of a yes-no question. Such parameterized synthetic speech can thus be valuable in testing the perceptual consequences of various acoustic parameter patterns.

Phonemic synthesizers. Another type of speech generation system listed in Fig. 3 is a phonemic synthesizer. Human speech is not the direct basis for stored representation of words or phrases in this type of system. Rather, each word is abstractly represented as a sequence of expected vowels and consonants (phonemes, or phones if more detailed articulatory contrasts are included). Speech is composed or synthesized by juxtaposing the expected phonemic sequence for each word with the sequences for preceding and following words. Each phonemic unit is used as a set of instructions for the form of acoustic data to be produced at the appropriate times during the utterance. For example, the phonemic sequence /faɪr/ could dictate that a formant synthesizer will first produce a period of weak unvoiced noise like an /f/, followed by an /a/-like vowel period of voiced, resonant sound, with formant frequencies placed appropriately for the phoneme /a/, followed by an /ɪ/-like vowel period of voiced, resonant sound, with formant frequencies of an /ɪ/, followed by an /r/-like resonant period. Transitions into and out of each expected target configuration of formant data will be necessary, and must be derived either by computer rules or by human judgment. Similarly, the durations and intensities of each phoneme must be specified, as must the pitch of the voice during each voiced period. Such a synthesizer system thus requires extensive rule-governed modeling of the steady-state sound structures of phonemes, and predictions of the forms of transitions. Also, prosodic features must be specified by rules. Commercially available phonemic synthesizers have typically not provided comprehensive sets of such rules, so that uses of phonemic synthesizers have normally required extensive linguistics expertise or outstanding patience in trying out alternative phonemic sequences, until the resulting speech sounds somewhat natural and intelligible.

Text-to-speech synthesizers. Problems with the extensive hand-tailoring required for phonemic synthesis have prompted the development of complex rule sets for predicting the needed phonemic states directly from the input message and dictionary pronunciations. Such "synthesis-by-rule" permits virtually unlimited-vocabulary voice response. By providing automatic means to take a specification of any English text at the input and to generate a natural and intelligible acoustic speech signal at the output, a text-to-speech synthesizer represents the most ambitious form of voice response system listed in Fig. 3. Without the need for intervening human sequence selection such as a phonemic synthesizer needs, and with less storage requirement than other synthesizers, the text-to-speech synthesizer is probably the primary choice for large-vocabulary speech generation.

Figure 5 illustrates a typical structure for a text-to-speech synthesizer suitable for use in automatically reading books to the blind. An optical character recognizer can read typewritten characters and produce a computer-readable specification of the spellings of words in the text. Next, words must be identified, and extensive text-to-sound rules must be used to provide the basic phonetic and prosodic instructions for controlling a speech synthesizer. Each common word is stored in a pronouncing dictionary, with its phonetic form, word stress pattern, and rudimentary syntax information such as whether it is a verb or noun. Unusual words like surnames can be generated by letter-to-sound rules. The parser examines the spelled input to determine the structural role of each word in the sentence, so that word ambiguities like the noun "*con*vict" versus the verb "con*vict*", or "average" as an adjective versus "average" as a verb, can be

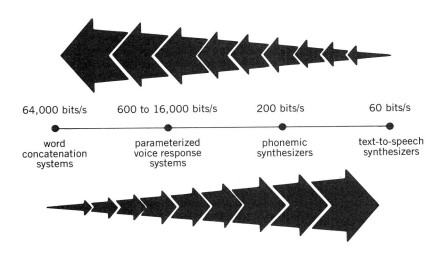

Fig. 3. Spectrum of alternative types of voice response systems.

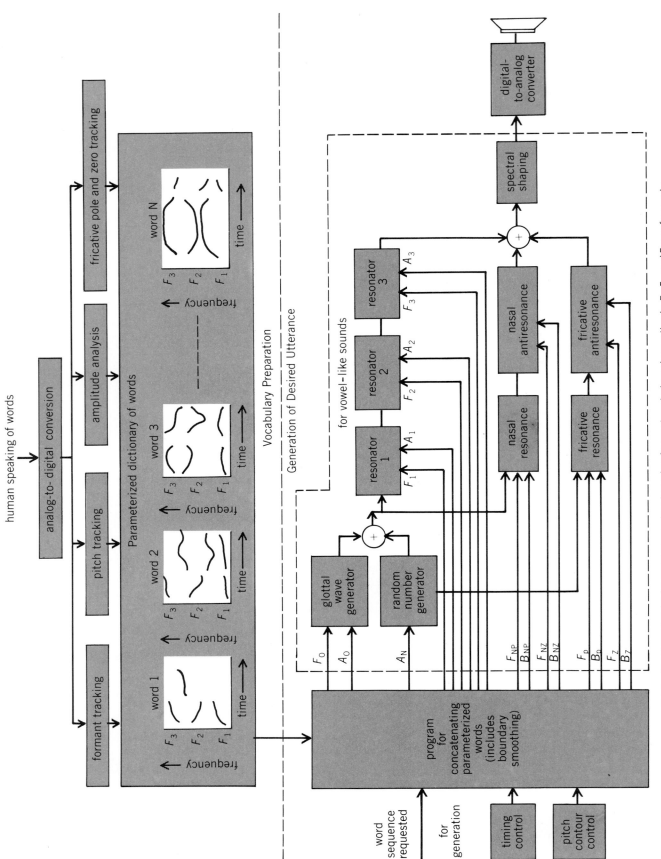

Fig. 4. Parameterized voice response system based on formant analysis and resynthesis. F_1, F_2, and F_3 are formants.

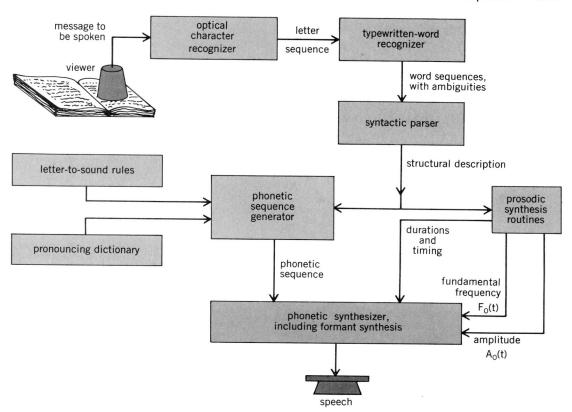

Fig. 5. Text-to-speech system for generating complex sentences with large vocabularies.

resolved. Also, homographs like "lead" (a metal) versus "lead" (to guide) can be distinguished by studying the syntactic structures in which they appear. Parsing can also control the prosody of the synthesized speech, so that sound intensity, phonemic and syllabic durations, and pitch contours throughout utterances are determined by the phonemic sequence, the lexical stress pattern, the syntactic bracketing into phrases, the placement in sentence intonation and rhythm patterns, and so forth. Indeed, rule-governed control of prosody is one of the primary aspects of text-to-speech synthesis, since prosodies are the primary acoustic correlates of linguistic structures, and prosodies are critical to the achievement of natural flow, emphasis, and inflection in synthesized speech. Text-to-speech synthesizers must also adjust word pronunciations to take into account effects of neighboring words on expected forms of word-initial or word-final phonemes. Extensive research and rule testing are needed before text-to-speech synthesizers will be able to produce arbitrary English sentences without flaw in phonetic or prosodic details.

Assessment of systems. Parameterized voice response systems and text-to-speech synthesizers are expected to dominate most applications, since one provides high-quality speech in a reliable manner while the other permits unlimited vocabulary and fully automatic production of arbitrary English utterances. Critical to the assessment of any speech generator are the intelligibility and the quality (or naturalness) of the generated speech. Methods are needed for accurate measurement of

the intelligibility of continuous speech and the subjective assessment of adequacies of machine-quality speech. In some applications, such as for cockpit warnings, the machine-generated speech should sound robotic or distinctive to attract attention and be heard above noises, while in other situations, such as with bank-by-phone or consumer products, naturalness and good quality will be essential.

Since the cost of storage is decreasing rapidly, larger-vocabulary concatenation systems (with or without parameterization) are likely to be increasingly used, but the flexibility in sound construction and the unrestricted vocabulary and discourses permitted by text-to-speech systems will promote further work on them as well. As techniques advance, and the markets and uses expand, there will be a growing interest in methods for systematically evaluating the performances of all aspects of voice response systems. Standard tests for intelligibility and quality assessment will allow side-by-side comparisons of devices. Further research in phonetics, prosodics, syntax, and application of synthesized speech will contribute improved voice response capabilities, and those capabilities will provide informative assessment of abilities to model processes of human speech production. *See* SPEECH RECOGNITION.

[WAYNE A. LEA]

Bibliography: J. L. Flanagan, *Speech Analysis, Synthesis and Perception*, 2d ed., 1972; J. L. Flanagan and L. R. Rabiner (eds.), *Speech Synthesis*, 1973; Special issue on man-machine communication by voice, *Proc. IEEE*, vol. 64, no. 4, April 1976.

Volt-ohm-milliammeter

A self-contained test instrument for measuring a wide range of voltages (both ac and dc), resistances, and currents (usually dc only), particularly in radio, TV, and electronic servicing.

All readings are obtained on a single multiscale indicating instrument. The selection of the correct ranges and function is accomplished either by means of one or two rotary switches, by multiple pin jacks, or by a combination of both.

Instrument sensitivity for dc-voltage measurements is usually 20,000 ohms per volt, although the figure ranges from 1000 to 100,000; for ac-voltage measurements the sensitivity ranges from 1000 to 20,000 ohms per volt. Accuracy is usually about 3% for dc, and 5% for ac measurements.

The terms multimeter and analyzer (or circuit analyzer) are quite commonly used as synonyms for volt-ohm-milliammeter. *See* AMMETER; OHM-METER; VOLTMETER.   [ISAAC F. KINNARD]

Voltage amplifier

An electronic amplifier that produces, with minimum distortion, an output voltage greater in magnitude than the input voltage. For a general discussion of amplifiers *see* AMPLIFIER.

Voltage amplifiers are built to amplify signals with frequency components in the range from zero hertz to thousands of megahertz. To cover this frequency range, several different types of amplifiers are required. An amplifier which cannot amplify voltage signals with zero frequency components (dc) is called *RC*-coupled. If zero frequency components (dc) are amplified, the amplifier is known as dc or direct-coupled. In the

range from a few hertz to 30 kHz, amplifiers are classified as audio. In the frequency range up to 10 MHz for use in television, pulse circuits, and electronic instruments, amplifiers are known as video. For the radio-frequency range from a few hundred kilohertz to hundreds of megahertz, tuned amplifiers are normally used. At ultrahigh frequencies, amplifiers designed around special tubes are employed. *See* KLYSTRON; MAGNETRON; TRAVELING-WAVE TUBE.

RC-coupled amplifier. The *RC*-coupled amplifier is employed in the frequency range from a few hertz to about 10 MHz. However, operation at the higher frequencies in this range is obtained only by employing frequency-compensation networks. The term *RC*-coupled amplifier generally refers to any set of amplifier stages that are coupled by a resistance-capacitance network. Figure 1*a* illustrates a grounded-source circuit. The coupling capacitor C_b is required to block the dc voltage at the drain of the first field-effect transistor (FET) from appearing at the gate of the following FET. This capacitor limits the lower frequency for which the amplifier will have a usable gain because the reactance $1/2\pi f C_b$ of the capacitor is inversely proportional to frequency f. Therefore, as the frequency decreases, the amplified signal voltage at the drain of the FET will be attenuated by the voltage divider formed by this reactance and the gate resistor.

A similar condition occurs when two transistor stages are *RC*-coupled as in Fig. 1*b*. Here, two common-emitter stages are coupled by means of capacitor C_b, and the same problem of reduced gain at lower frequencies exists.

The high-frequency limit of usable amplification is determined by the shunt capacitance and the effective drain load resistance of a FET, or output load resistance of a transistor. The shunt capacitance is composed of the input capacitance of the following FET or transistor, the drain-to-source capacitance of the FET or output capacitance of the transistor in the stage, and the stray wiring capacitance. *See* VIDEO AMPLIFIER.

The input capacitance to a FET in a grounded-source amplifier stage is composed of the capacitance formed from the gate-to-source and from the drain-to-gate capacitance multiplied by 1 minus the gain of the stage. Similarly, the input capacitance of a transistor in the grounded-emitter configuration is the sum of the emitter-junction capacitance and the collector-junction capacitance multiplied by 1 minus the gain of the stage. The gain is negative in the midband region, permitting the input capacitance to be large, on the order of 30−50 picofarads (pf).

If the upper and lower half-power (or cutoff) frequencies are, respectively, high and low, compared with the frequency spectrum of the signal to be amplified, then the gain is essentially uniform with frequency.

Midfrequency gain. In the case of the FET amplifier the midfrequency gain is shown by Eq. (1),

$$A_v = -g_m R_{\text{eq}} \qquad (1)$$

where g_m is the transconductance of the FET, the minus sign indicates a 180° phase reversal between the input gate signal and the output signal, and the equivalent load resistance R_{eq} is the parallel com-

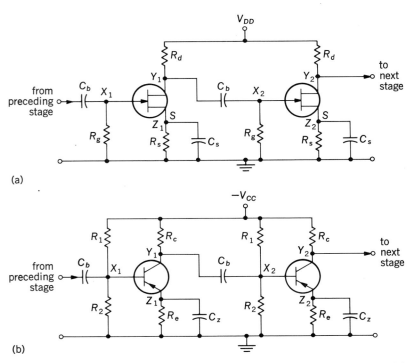

Fig. 1. A cascade of (*a*) common-source (CS) depletion-type, or JFET, stages; (*b*) common-emitter (CE) transistor stages.

bination of the dynamic drain resistance r_d of the FET, the gate resistance R_g, and the load or coupling resistance R_d.

For the transistor amplifier the computation of gain is more involved, because of the relatively lower input and output impedances of the transistors. The midfrequency voltage gain is shown by Eq. (2). The input resistance of this stage is given by Eq. (3). R_L, the total load resistance seen by this

$$A_v = A_i \frac{R_L}{R_{in}} \qquad (2)$$

$$R_{in} = \frac{\Delta^h R_L + h_{ie}}{h_{oe} R_L + 1} \qquad (3)$$

transistor, includes the parallel combination of the collector resistor R_c, the bias network R_B where $R_B = R_1 R_2/(R_1 + R_2)$, and the input resistance R_{in} of the next stage. The h_{ie} is the input impedance and h_{oe} the output admittance of the transistor; h_{re} is the voltage-feedback ratio and h_{fe} is the forward-current ratio, and the quantity Δ^h is $h_{ie}h_{oe} - h_{re}h_{fe}$. Because of the dependence of the gain on the R_{in} of the next stage, it is necessary to compute the gain per stage in a transistor amplifier by starting with the last stage and working backward toward the input. The current gain is given by Eq. (4).

$$-A_i = \frac{h_{fe}}{1 + h_{oe} R_L} \qquad (4)$$

High-frequency gain. The upper half-power frequency is that frequency for which the reactance of the effective shunt capacitance is equal to the equivalent load resistance.

For the FET amplifier the equivalent load resistance R_{eq} is the parallel combination of r_d, R_d, and R_g, the gate resistance in Fig. 1a. The upper half-power frequency is shown by Eq. (5).

$$f_2 = \frac{1}{2\pi R_{eq} C_{shunt}} \qquad (5)$$

For the transistor amplifier the equivalent resistance is the parallel combination of R_c, R_B, the input resistance of the next stage R_{in}, and the output resistance of the transistor R_o. The upper half-power frequency is formed from the same equation as given for the FET, provided there is a long chain of stages.

Low-frequency gain. The lower half-power frequency is that frequency for which the reactance of the coupling capacitor C_b equals the effective resistance R appearing in parallel with the coupling capacitor. For the FET amplifier at low frequencies, the effective resistance is the dynamic drain resistance r_d and the coupling resistance R_d in parallel, and this combination in series with the gate resistor R_g of the next stage. Therefore, the lower half-power frequency is shown by Eq. (6).

$$f_1 = \frac{1}{2\pi R C} \qquad (6)$$

In the case of the transistor amplifier, the resistors appearing across C_b are the output resistance R_o, in parallel with the collector resistor R_c, and this combination in series with the parallel combination of the bias resistor R_B and the input resis-

tance R_{in} of the next stage. The same expression for f_1 is applicable.

Passband. At the half-power frequencies the gain is equal to the midband gain divided by the square root of 2. For either the FET or transistor *RC*-coupled amplifier the gain at any low frequency f can be expressed in terms of the midfrequency gain, as shown in Eq. (7). At high frequencies,

$$A_{low} = \frac{A_{mid}}{1 - j(f_1/f)} \qquad (7)$$

gain is shown by Eq. (8). This simple relationship

$$A_{high} = \frac{A_{mid}}{1 + j(f/f_2)} \qquad (8)$$

permits drawing a universal frequency-response curve A, Fig. 2, applicable to any single stage of either a FET or transistor amplifier. The graph is made universal by plotting relative gain as a function of the ratio of the signal frequency to the half-power-point frequencies. The slope of the curve approaches -6 decibels per octave for a single stage of amplification. One octave is a change in frequency by a factor of 2. The frequency range $f_2 - f_1$ is called the bandwidth B of the stage.

The gain of a multistage amplifier is equal to the product of the gain of each of the individual stages. Curve B of Fig. 2 is for three identical stages. For this specific case the slope of the curve is -18 dB per octave, which is the sum of the -6 dB per octave slope inherent in each of the three stages. When stages are cascaded, the half-power (70.7%, or -3 dB) points are moved in toward the midfrequencies. Therefore, the cascading of stages decreases the frequency passband between the half-power points.

For reasons of economy, amplifiers used for entertainment purposes and similar applications sometimes make use of the frequency regions beyond the flat midfrequency region. Amplifiers employed in calibrated instruments, such as electronic voltmeters and cathode-ray oscilloscopes, are limited to the midfrequency range.

Interstage transfer function. One important distinction arises between the FET and the transistor amplifier. The FET is inherently a voltage amplifier, and a voltage appearing across the gate resistor R_g in Fig. 1a also appears from gate to source in the next FET and is amplified. The situation is

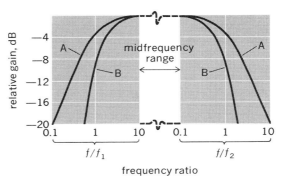

Fig. 2. Universal frequency-response curve for *RC*-coupled amplifiers. Curve A is for any single stage of amplification. Curve B is for three identical stages.

considerably different with transistors. A signal current leaving the first transistor collector follows two paths, each of which shunts a portion of this current to ground. This reduces the current gain and must be accounted for in computing the overall amplifier current gain and hence the voltage gain.

The interstage transfer function is given by the ratio of the current flowing into the input resistance R_{in} of the next stage, divided by the current from the preceding transistor. In terms of the parameters of the circuit, the transfer function is shown by Eq. (9), where R_B is as given before. For

$$a = \frac{R_c R_B}{R_c R_B + R_{in}(R_c + R_B)} \tag{9}$$

well-designed, broadband amplifiers, transfer function a is in the neighborhood of 0.7.

The overall voltage gain of a multistage FET amplifier is shown by Eq. (10) where A_{v1}, A_{v2}, A_{v3},

$$A_{total} = A_{v1} \cdot A_{v2} \cdot A_{v3} \cdots \tag{10}$$

is the voltage gain of each stage. For a transistor amplifier the overall voltage gain is shown in Eqs. (2) and (11). A_1, A_2, . . . are the current gains per

$$A_i = a_1 \cdot A_1 \cdot a_2 \cdot A_2 \cdot a_3 \cdot A_3 \cdots \tag{11}$$

stage, and a_1, a_2, . . . represent the interstage transfer functions.

Tuned amplifier. A tuned amplifier amplifies voltage signals in a selected narrow frequency band and suppresses signals outside the desired band. A tuned amplifier is thus a bandpass amplifier with the additional constraint that its bandwidth be a small fraction of the center frequency. Such amplifiers are commonly used as radio frequency (rf) or intermediate frequency (i-f) amplifiers for superheterodyne radio receivers. In the case of rf amplifiers it is required that the center frequency be tunable or variable and that, in addition, over the tuning range the overall selectivity should remain constant. Constant selectivity means that the bandwidth should remain constant. In order to obtain the necessary selectivity and high voltage gain, LC tank or resonant circuits are

used as loads with transistor or FET amplifiers as shown in Fig. 3.

The important characteristics of a tuned amplifier are the voltage gain at resonance, the variation of voltage gain with frequency near the resonant frequency, and, if the center frequency is to be varied, the way in which the voltage gain changes.

Single-tuned amplifier. The circuits of Fig. 3 show single-tuned amplifiers. Resonance occurs when the capacitive and inductive reactances are equal, that is, when $\omega_0 L = 1/\omega_0 C$, where $\omega_0 = 2\pi f_0$, or Eq. (12) is satisfied.

$$f_0 = \frac{1}{2\pi\sqrt{LC}} \tag{12}$$

The quality of the inductor in any tuned circuit can be described in terms of the quality factor Q as given in Eq. (13), where L is the inductance in

$$Q = \frac{\omega L}{R} \tag{13}$$

henries, R is the series resistance of the coil in ohms, and ω equals 2π times the frequency at which Q is evaluated.

The impedance at resonance of the tank circuit is given by Eq. (14).

$$Z = \frac{L}{RC} = \omega_0 LQ \tag{14}$$

The maximum voltage gain at resonance for the FET amplifier of Fig. 3b is given by Eq. (15), where

$$A_v = -g_m Z \tag{15}$$

g_m is the transconductance of the FET. A similar expression holds for the transistor-tuned amplifier.

The bandwidth of the parallel-tuned (or single-tuned) amplifier is given by Eq. (16).

$$B = \frac{f_0}{Q} \tag{16}$$

Double-tuned amplifier. Double-tuned amplifiers are used where the pass band is centered on a fixed frequency, as is the case for i-f radio amplifiers. The two circuits may be coupled inductively or capacitively or a combination of both. Usually one circuit is known as the primary circuit and the other as the secondary circuit, as is the case with a transformer-coupled amplifier. The transformer primary and secondary are both tuned with suitable capacitors. Often the transformer has either air or powdered metal cores. The core slug is about as long as the length of the coil form for the windings. Tuning in this case may be accomplished by moving the core slug partially in and out of the coil. *See* TRANSFORMER.

For the double-tuned transformer-coupled FET amplifier, the gain at the resonant frequency f_0 is given by Eq. (17). The subscripts indicate primary

$$A = g_m k \frac{\omega_0 \sqrt{L_p L_s}}{k^2 + 1/Q_p Q_s} \tag{17}$$

and secondary inductances and Qs. The transconductance of the FET is g_m, $\omega_0 = 2\pi f_0$, and k is the coefficient of coupling between the coils.

For critical coupling ($k = 1/\sqrt{Q_p Q_s}$), nearly identical primary L_p and secondary L_s coils ($Q_p = Q_s =$

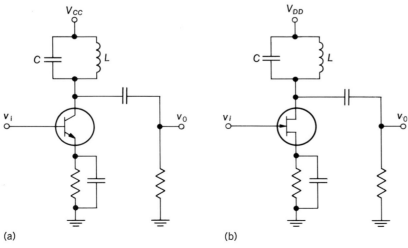

(a)   (b)

Fig. 3. Single-tuned amplifier: (a) transistor; (b) FET.

Q), and large Qs, the voltage gain at resonance is a maximum and is shown by Eq. (18). The associated

$$A = g_m \omega_0 L Q / 2 \qquad (18)$$

bandwidth between half-power points is given by Eq. (19), when the coupling is critical and B and f_0 are in the same units.

$$B = \frac{\sqrt{2} f_0}{Q} \qquad (19)$$

When transistors are employed, the possibility of obtaining considerable voltage gain from the tuned transformer, as is possible with FETs, becomes more difficult because of the relatively low input and output impedances of transistors. The tuned transformer functions more as an impedance-matching device. Taps or auxiliary windings are often provided on rf and i-f transformers for use with transistors to provide means for impedance matching.

Source and emitter-follower stages. If the drain of the FET is connected directly to the power-supply source and the output is taken from the source, the resulting circuit has special properties which are useful. Such a circuit is called a source-follower. Similarly, if collector is connected directly to the power supply, the circuit is called an emitter follower and the same overall results are obtained. Simple versions are illustrated in Fig. 4. Voltage gain of these circuits is less than unity, typical values ranging from 0.90 to 0.99. Input and output voltages are in phase. The principal properties of interest are the relatively high input impedance and the low output impedance. The amplifier is often used as an impedance-matching device. In such application an emitter follower can be inserted between a source and a load with only slight decrease in voltage gain. Current and power gain may be quite high if desired.

The input capacitance of a source-follower stage is low, because gain is nearly equal to unity and is positive. Therefore, the input capacitance is essentially the gate-to-drain capacitance, which is of the order of 2 pf. Thus the source follower is ideally suited for use in the interstage coupling of a broadband amplifier.

The midfrequency gain of a source-follower circuit is given by Eq. (20). The symbol μ represents

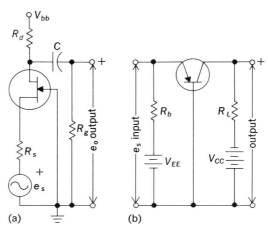

Fig. 5. Amplifiers in which the gate or base is grounded. (a) Grounded-gate amplifier. (b) Grounded-base amplifier. V_{bb} is the drain supply voltage, V_{EE} the emitter supply voltage, and V_{cc} the collector supply voltage.

$$A_v = \frac{\mu R_s}{r_d + (\mu + 1) R_s} \qquad (20)$$

the amplification factor, r_d is the dynamic drain resistance, and R_s is the source load resistor. The output impedance that the source follower presents to its output circuit is of interest and is given by Eq. (21), where g_m is the transconductance. By

$$Z_{\text{out}} = \frac{R_s}{1 + g_m R_s + R_s/r_d} \qquad (21)$$

proper selection of FET and source load resistor R_s, a wide range of output impedances may be obtained.

If transistors are used at midfrequency, the voltage gain is approximated by Eq. (22).

$$A_v = \frac{v_o}{v_s} = -\frac{(1 + h_{fe}) R_e}{h_{ie} + (1 + h_{fe}) R_e} \qquad (22)$$

The input resistance is shown by Eq. (23). The

$$R_i = R_1 \| R_2 \| [h_{ie} + (1 + h_{fe}) R_e] \qquad (23)$$

two vertical lines in Eq. (23) mean the parallel combination of the corresponding resistors. R_1 and R_2 represent the bias resistors of the transistor input.

The output admittance of the emitter follower is given by Eq. (24), where zero source resistance is assumed.

$$Y_{\text{out}} = \frac{1}{R_e} + h_{oe} + \frac{1 + h_{fe}}{h_{ie}} \qquad (24)$$

Grounded-gate (base) amplifier. If the gate (or base) is grounded and the input signal is applied to the source (or emitter) with the output taken from the drain (or collector), the circuit is called a grounded-gate (or grounded-base) amplifier (Fig. 5). This circuit features a low input impedance and a high output impedance. Because of the low input impedance, the amplifier will draw power from the source and must, therefore, be used with a low-impedance source. It has found its chief application in rf amplifiers operating at very high frequen-

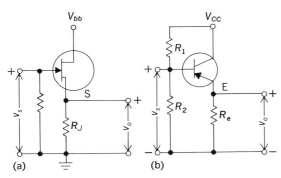

Fig. 4. Amplifiers in which the FET or transistor is connected directly to the power source. (a) Source-follower amplifier. (b) Grounded-collector amplifier. The drain supply voltage is V_{bb} and the collector supply voltage is V_{cc}.

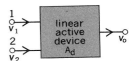

Fig. 6. The output is a linear function of v_1 and v_2. For an ideal differential amplifier, $v_0 = A_d(v_1 - v_2)$.

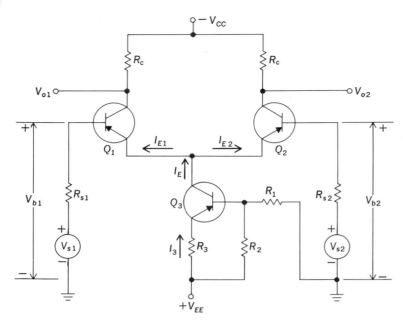

Fig. 7. Emitter-coupled difference amplifier, with a constant-current stage in the emitter circuit. Nominally, $R_{s1} = R_{s2}$.

cies, such as the tuner in a television receiver; at these frequencies it gives a high signal-to-noise ratio with low input-output capacitance. It does so at the expense of rf selectivity. This excludes it from use in a narrow-band receiver application.

Grounded-gate amplifier. Voltage gain of the grounded-gate amplifier is given by Eq. (25). R_s is the internal resistance of the signal source, μ and

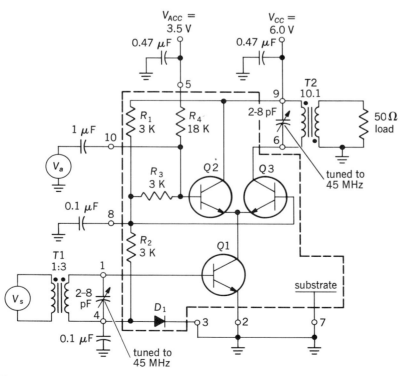

Fig. 8. A practical 45-MHz tuned amplifier (with $V_a = 0$), or an rf modulator if $V_a \neq 0$. (*Motorola Semiconductor, Inc.*)

$$A = \frac{(\mu + 1) Z_L}{r_d + (\mu + 1) R_s + Z_L} \quad (25)$$

r_d are the amplification factor and dynamic drain resistance, respectively, of the FET, and Z_L is the load impedance. For the midfrequency gain in an *RC*-coupled amplifier Z_L is the parallel combination of the coupling resistor R_d and the gate resistor of the next stage R_g.

The input impedance of the stage is shown by Eq. (26). The stage can be used to match low-impedance circuits to high-impedance circuits.

$$Z_{\text{in}} = \frac{r_d + Z_L}{\mu + 1} \quad (26)$$

Grounded-base amplifier. Voltage gain of the grounded-base amplifier is given approximately by Eq. (27), where h_{fe} and h_{ie} are the short-circuit cur-

$$A = h_{fe} \frac{R_L}{h_{ie}} = \frac{v_o}{v_s} \quad (27)$$

rent gain and input resistance for the common-emitter connection.

The input and output resistances are approximated by Eqs. (28) and (29).

$$R_i = \frac{h_{ie}}{1 + h_{fe}} \quad (28)$$

$$R_o = R_L \quad (29)$$

Direct-coupled amplifier. When the coupling between successive stages of an amplifier is such that direct current may flow from the output of one stage to the input of the following stage, the amplifier is said to be direct-coupled. *See* DIRECT-COUPLED AMPLIFIER.

Difference amplifiers. The function of a difference, or differential, amplifier is, in general, to amplify the difference between two signals. The need for differential amplifiers arises in many physical measurements, in medical electronics, in analog computers, and in direct-coupled amplifier applications.

Figure 6 represents a linear active device with two input signals v_1, v_2, and one output signal v_o, each measured with respect to ground. In an ideal differential amplifier the output signal v_o should be given by Eq. (30), where A_d is the gain of the dif-

$$v_o = A_d (v_1 - v_2) \quad (30)$$

ferential amplifier. Thus it is seen that any signal, such as noise, which is common to both inputs will have no effect on the output voltage. However, a practical differential amplifier cannot be described by Eq. (30) because, in general, the output depends not only upon the difference signal v_d of the two signals, but also upon the average level, called the common-mode signal v_c. The values of v_d and v_c are given by Eqs. (31) and (32). It can be seen that if,

$$v_d \equiv v_1 - v_2 \quad (31)$$

$$v_c \equiv \tfrac{1}{2} (v_1 + v_2) \quad (32)$$

for example, one signal is $+50$ μv and the second is -50 μv, the output will not be exactly the same as if $v_1 = 1050$ μv and $v_2 = 950$ μv, even though the difference $v_d = 100$ μv is the same in the two cases.

In general, the output of the difference amplifier is given by Eq. (33), where ρ is a quantity

$$v_o = A_d v_d \left(1 + \frac{1}{\rho}\frac{v_c}{v_d}\right) \qquad (33)$$

called the common-mode rejection ratio. An emitter-coupled difference amplifier is shown in Fig. 7. This circuit can also be used as a single-signal amplifier by grounding one base and taking the output at either v_{o1} or v_{o2}. The noise voltage still is present at each base and cancels in the output, making possible low-noise amplification of the signal v_{s1}.

Integrated amplifier. Integrated circuit amplifiers are dc amplifiers commonly referred to as operational amplifiers. The basic building block of an integrated tuned amplifier is the emitter-coupled difference amplifier of Fig. 7. Such a circuit is shown in Fig. 8. The block enclosed by a dashed line is an integrated circuit; all other components are discrete elements added externally. The input signal is applied through the tuned transformer $T1$ to the base of $Q1$. The load R_L is applied across the tuned transformer $T2$ in the collector circuit of $Q3$. The amplification is performed by the transistors $Q1$ and $Q3$, whereas the magnitude of the gain is controlled by $Q2$. The combination $Q1 - Q3$ acts as a common-emitter common-base pair, known as cascode combination. *See* INTEGRATED CIRCUITS.

[CHRISTOS C. HALKIAS]

Bibliography: C. A. Holt, *Electronic Circuits*, 1978; J. Millman and C. C. Halkias, *Integrated Electronics*, 1971.

Voltage measurement

Determination of the difference in electrostatic potential between two points. In the circuit shown in Fig. 1, the voltage rise through the battery is V, and is measured between points a and c. The voltage drop caused by resistance R_1 is $V_1 = I R_1$, and is measured between the points a and b. Similarly, the voltage drop across R_2 is $V_2 = I R_2$, and is measured between points b and c. The sum of the voltage drops around any complete circuit path back to the starting point equals zero. Thus, in Fig. 1, $V_{R_1} + V_{R_2} - V = 0$.

The measurement of voltage or potential difference between two points is accomplished by observing some reproducible effect caused by the voltage. Many different effects have been utilized in the construction of voltmeters, depending upon whether the voltage produces a direct-current (dc) or an alternating-current (ac) flow and upon the power available to operate the voltmeter. *See* VOLTMETER.

Direct current. In dc circuits the flow of current is unidirectional and the voltmeter must respond only to the magnitude of the potential difference. For voltages greater than 1000 volts, electrostatic voltmeters are often employed. These depend upon electrostatic forces to deflect a pointer against a spring restoring force. In the $1-1000$-volt range, voltage is commonly measured by using a large resistance in series with a milliammeter. The current which flows, $I = V/R$, is directly proportional to the applied voltage. Milliammeters utilize the force exerted on a conductor carrying current in a magnetic field to deflect a pointer. Thus in most voltmeters the measurement of voltage is accomplished ultimately by measuring the

force exerted against a calibrated spring. Measurement accuracy is from 0.1 – 3.0%.

To measure voltages below 1 volt, particularly in the microvolt range, these direct deflection instruments often require more power to operate than is available from the circuit under measurement. Extremely sensitive galvanometers have been devised to solve this problem, although use of an electronic amplifier between the circuit under measurement and a deflection-type instrument is more common. The modern amplifier-type voltmeter, called the vacuum-tube voltmeter (VTVM), is more likely to employ solid-state transistors or integrated circuits than the older vacuum tube. The name VTVM, however, has remained. *See* VACUUM-TUBE VOLTMETER.

In the volt range, the voltage to be measured is impressed directly upon a dc amplifier, while in the millivolt or microvolt range, a chopper-type amplifier is used. The chopper is either a mechanical switch, a photocell turned on and off by light, or a diode bridge switched by an ac voltage. It converts the low-level dc voltage to an ac voltage for amplification without drift and then reconverts this amplified signal back to direct current to drive a conventional moving-pointer voltmeter. In this manner only minute amounts of power are required from the circuit under measurement.

Potentiometric measurement. When greater accuracy than can be obtained with a moving-pointer voltmeter is required, a potentiometric method is often employed (Fig. 2). A precisely cali-

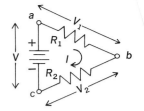

Fig. 1. Voltage measurement around closed loop.

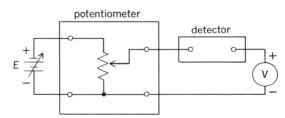

Fig. 2. Potentiometric measurement.

brated potentiometer or resistive voltage divider is used to attenuate battery voltage E to the same magnitude as voltage V to be measured. When they are equal, no current flows through the detector. A sensitive galvanometer or amplifier-type voltmeter is used as a detector to sense this null condition. A typical potentiometer would be a five-, six-, or seven-dial decade divider with the attenuation ratio indicated directly in decimal digits. In use, two measurements are normally required. First, a standard cell whose voltage is precisely known is measured by adjusting the potentiometer for a null indication on the detector. Second, a similar measurement is then made with unknown voltage V. The ratio of the unknown voltage to the standard cell voltage is then given by the ratio of the two potentiometer settings. The battery voltage E and the linearity of the detector are thus eliminated from the voltage determination. The accuracy of the measurement depends only on the linearity of the decade voltage divider, the sensitivity of the detector, and the stability of battery voltage E during the time of measurement. With care,

Common ac voltage uses with frequency range

Use	Approximate frequency range
Primary power	60 Hz
Telephone and phonograph	20 Hz – 20 kHz
Radio broadcasting (AM)	0.5 – 1.5 MHz
Television	50 – 250 MHz
Radar	100 MHz – 1 GHz
Space communication	10 GHz

accuracy on the order of a few parts per million can be obtained. *See* POTENTIOMETER (VOLTAGE METER).

In practice, the potentiometer dials are usually set to the standard cell voltage, and battery voltage E is adjusted to obtain the first null. The potentiometer is then direct-reading in volts. While this simplifies the measurement considerably, it is still a slow measurement process. Using this same basic technique, digital voltmeters are available that can automatically make many measurements per second to better than 0.01% accuracy. *See* DIGITAL VOLTMETER.

Alternating current. In ac circuits the flow of current reverses periodically. This reversal can occur at rates (frequencies) from far less than once per second to over 10^{10} times per second. The table lists some common ac voltage uses and their approximate frequency ranges. Many different voltage-measurement techniques are used over this wide range, depending upon the frequency as well as on whether the voltage is in the kilovolt, volt, or microvolt range.

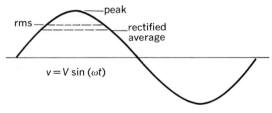

Fig. 3. Sinusoidal voltage variation.

An ac voltage not only reverses direction each cycle but also changes its instantaneous value within a cycle in a repetitive manner. The most common periodic variation is the sinusoidal waveform (Fig. 3). Ac voltage waveforms have several measurable attributes. Ac voltmeters commonly measure the peak value, the average rectified value, or the root-mean-square (rms) value of the waveform.

In the low-frequency range, where sufficient power is available, moving-pointer voltmeters that respond to the rms value are usually employed. If a semiconductor rectifier is utilized directly with a dc voltmeter, an ac voltmeter responding to the average value of the rectified voltage results, which is useful through the audio-frequency range. One of the most widely used average-responding ac voltmeters is the operational rectifier. It requires minute signal power, oper-

ates from 1 mvolt to 1000 volts and from 10 Hz to 10 MHz, and typically provides 2% accuracy. Peak voltage is measured by using a vacuum-tube diode to charge a capacitor to the peak value, which is then measured by a dc VTVM. Voltmeters of this type are employed up to 1 GHz.

Sampling measurement. To extend voltage measurement about 10 GHz, a sampling voltmeter has been used. This technique employs a diode bridge which is switched on for a few nanoseconds to sample the voltage at a moment in time. This sampled voltage is stored in a capacitor for a few microseconds and then replaced with a new sample. Since the sampling rate and high frequency being sampled are unrelated in frequency, this sample-and-hold procedure creates a totally different but lower-frequency voltage waveform of a random nature. When measured over many cycles, however, this random voltage has the same peak, average, and rms value as the original voltage and can thus be measured by a conventional ac voltmeter. *See* SAMPLING VOLTMETER.

The ac-dc transfer. For high-accuracy ac-voltage measurement, a thermocouple or dynamometer which responds identically to ac and dc voltage is used as a transfer device. The transfer is accomplished by adjusting a dc voltage to provide an identical (or counterbalancing) effect to that of the applied ac voltage. Once equality has been established, the dc voltage is compared by a potentiometric technique against a standard cell. With care, transfer accuracy of about 10 ppm can be obtained.

Special amplifier-type voltmeters employing operational rectifiers are also capable of high accuracy. Rugged instruments can be constructed which utilize high loop gain and which rival or surpass the best thermocouple or dynamometer transfer devices. *See* CURRENT MEASUREMENT; ELECTRICAL MEASUREMENTS.

[MALCOLM C. HOLTJE]

Bibliography: M. Braccio, *Electrical and Electronic Tests and Measurements*, 1979; W. D. Cooper, *Electronic Instrumentation and Measurement Techniques*, 2d ed., 1978; J. Cunningham, *Understanding and Using the VOM and EVM*, 1975; P. Kantrowitz et al., *Electronic Measurements*, 1979.

Voltage-multiplier circuit

A rectifier circuit capable of supplying a dc output voltage that is two or more times the peak value of the ac input voltage. Such circuits are especially useful for high-voltage, low-current supplies. These supplies are usually lighter in weight,

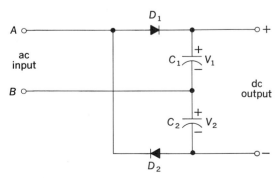

Fig. 1. Full-wave voltage doubler.

smaller in size, and less expensive than the more usual half-wave and full-wave rectifier supplies. They require either no power transformer or a much smaller transformer, but they have the disadvantage of requiring more rectifiers and capacitors. If, for safety or other reasons, input and output must be isolated, a transformer may be required at the input even though the voltage-multiplier circuit does not need it for operation. Common configurations are the half-wave and full-wave voltage-doubling circuits, but tripling and higher orders of multiplication are used.

A full-wave voltage-doubling rectifier circuit is shown in Fig.1. When the ac input voltage is positive at terminal A, diode D_1 conducts, producing voltage V_1 across capacitor C_1. On the other half cycle, diode D_2 conducts, producing the voltage V_2 across capacitor C_2. Both V_1 and V_2 will have a magnitude approaching the maximum value of the ac input voltage, so the output voltage approaches twice the maximum value of the ac input voltage. Because the output voltage receives one pulse every half cycle, ripple voltage is similar to that of a single-phase full-wave rectifying circuit. *See* RECTIFIER.

A half-wave voltage-doubling rectifying circuit is shown in Fig. 2. When top terminal A of the trans-

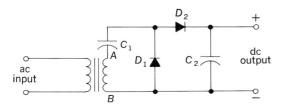

Fig. 2. Half-wave voltage doubler.

former secondary is positive, current flows through diode D_1, charging capacitor C_1 with its lower terminal positive in potential. When bottom terminal B of the transformer secondary is positive, current flows through diode D_2, charging capacitor C_2 to nearly twice the maximum value of the secondary voltage. Only one pulse per cycle is received by capacitor C_2, so the ripple voltage is similar to that of a single-phase half-wave rectifying circuit. The full-wave doubler has the advantage over a half-wave doubler of better voltage regulation and smaller ripple voltage; the half-wave doubler has the advantage of having a common ground.

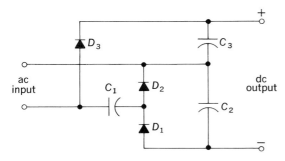

Fig. 3. Voltage-tripler circuit, a voltage doubler in series with a half-wave rectifier.

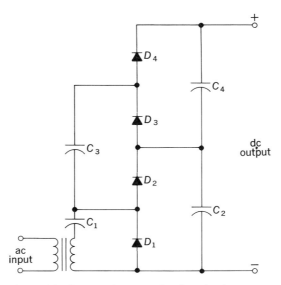

Fig. 4. A half-wave voltage-quadrupling circuit.

A voltage-tripling rectifier circuit is shown in Fig. 3, and a half-wave voltage-quadrupling rectifier circuit is shown in Fig. 4. Comparison of Figs. 2 and 4 shows that a half-wave doubler can be transformed into a half-wave quadrupler by adding two capacitors and two diodes. Theoretically, this may be continued indefinitely to achieve as high a multiplication as is desired. In practice, each stage added contributes slightly less to the output voltage than the preceding stage, so that a point is finally reached where an additional stage does not contribute enough to make it worthwhile. *See* ELECTRONIC POWER SUPPLY.

[DONALD L. WAIDELICH]

Bibliography: A. H. Lytel, *Solid-State Power Supplies and Converters*, 1965; G. Scales, *Handbook of Rectifier Circuits*, 1980.

Voltage regulator

A device or circuit that maintains a load voltage nearly constant over a range of variations of input voltage and load current. Voltage regulators are used wherever the unregulated voltage would vary more than can be tolerated by the electrical equipment using that voltage. Alternating-current distribution feeders use voltage regulators to keep the voltage supplied to the user within a prescribed range. Electronic equipment often has voltage regulators in dc power supplies.

ELECTRONIC VOLTAGE REGULATORS

In much electronic equipment the dc power supply voltage must remain constant in spite of input ac line voltage variations and output load variations. To perform this function, two types of voltage-regulator circuits are used in electronic power supplies. The first uses a nearly constant voltage device, such as a zener diode or a gas tube, as the voltage regulator; the second uses the variable-resistance properties of a transistor or a vacuum tube. Because of the rapid response of these devices, either type regulator serves not only to compensate for irregular changes in input or output but also to counteract ripple voltage. Thus an electronic regulator makes possible the use of

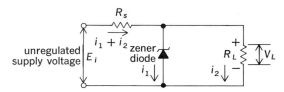

Fig. 1.   Zener-diode voltage regulator.

VOLTAGE REGULATOR

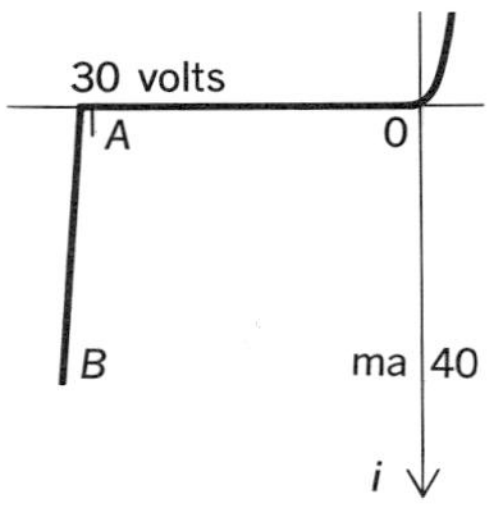

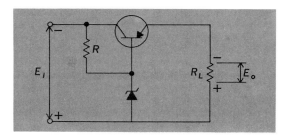

Fig. 2.   Typical zener-diode characteristic.

simpler filters. The voltage regulator circuit may also augment the filtering action of the power-supply filters, and it may also raise or lower the internal impedance of the power supply as seen from the load. The trade off for the regulator action is a considerable power loss in the regulator circuit. Most of the power is lost either in a regulator resistor or in the regulator diodes or tubes. *See* RECTIFIER; RIPPLE VOLTAGE.

Diode regulator circuit. A zener-diode regulator circuit is shown in Fig. 1. The zener diode is characterized by an almost constant voltage over its specified range of current, as shown in Fig. 2. The constant voltage depends upon the particular diode employed and may be from about 2 to 200 volts.

The circuit of Fig. 1 depends on two conditions: (1) Input voltage E_i is always the sum of voltage drop V_s across series resistor R_s and output voltage V_L which is maintained constant by the zener diode; and (2) the current through series resistor R_s is always the sum of load current i_2 and diode current i_1.

Because the voltage drop across the series resistor is constant for a given input voltage, the current through the resistor must also be constant. Therefore a change in load current must be accompanied by an equal, but opposite, change in current through the diode.

If the input voltage varies, V_s must also vary if V_L is to remain constant. This variation in V_s must be produced by a change in diode current if load current i_2 is to remain constant. The substantially constant voltage drop for a wide range of current, characteristic of a zener diode in the region A to B of Fig. 2, provides the required change in i_2 to maintain V_L nearly constant over wide changes of input voltage E_i and of load current i_2.

Transistor regulator circuits. A single transistor regulator circuit is the shunt regulator using a transistor and a zener diode as shown in Fig. 3. Resistor R is chosen so that the zener diode has a current approximately in the middle of its operating range. If voltage E_o across the load increases, the

base of the transistor rises in potential, increasing the collector current and, in consequence, the voltage across R_s. This action tends to return the load voltage to where it was initially. In this manner the circuit compensates for load voltage variations.

If input voltage E_i increases, again the base of the transistor rises in potential, so that, as before, the voltage across R_s increases. This increase largely overcomes the increase in the input voltage, so that the output voltage increases very little. Thus the circuit also compensates for input voltage variations.

The regulator is not perfect; a small change in output voltage does occur in the presence of a much larger change in input voltage. This change in output voltage can be reduced even more by adding one or more transistor amplifier stages.

A series transistor regulator circuit is shown in Fig. 4. Resistor R is chosen so that the zener diode

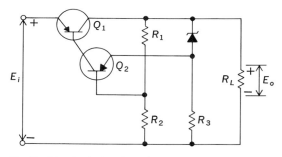

Fig. 4.   Series transistor regulator.

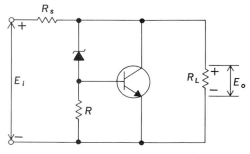

Fig. 5.   A basic electronic series regulator circuit.

is approximately in the middle of its operating range. The transistor must be capable of carrying all of the load current and withstanding a voltage equal to the difference of the input and output voltages. An increase in the output voltage E_o increases the voltage from emitter to base, in effect increasing the voltage across the transistor. Voltage E_o is thus reduced, approaching its original value. Conversely, if input voltage E_i increases, the emitter-to-base voltage increases and the transistor voltage rises to overcome most of the increase. Again, an amplifier would make the circuit more responsive to very small changes.

A simplified circuit incorporating an amplifier is shown in Fig. 5. Transistor Q_1 is the series regulating device, while the transistor Q_2 is the amplifier. If output voltage E_o increases for some reason, the emitter-to-base voltage on Q_2 increases. This

Fig. 3.   Transistor shunt regulator.

change in turn increases the base current and the potential drop across Q_1 in such a manner as to oppose the original increase in E_o. The net effect is to keep the output voltage nearly constant and to compensate for load variations. In a similar fashion, if the unregulated supply potential increases, voltage E_o across the output tends to increase. The sequence of events enumerated above again occurs, tending to increase the voltage across Q_1 and again keeping the load voltage practically constant. If the amplifier Q_2 had sufficient gain, the voltage regulation would be nearly perfect. If the current taken by the load is too high for one transistor Q_1, two or more may be connected in parallel.

An actual voltage regulator circuit using these principles is shown in Fig. 6. The voltage of the 12-volt reference diode is compared to a portion of the output voltage through a 2N1481 transistor. The difference voltage is amplified by another 2N1481 transistor and is then applied to the 2N1489 transistor, which acts as the series regulating device. To improve the response to fast changes such as those caused by pulses, square waves, high-frequency sine waves, and so forth, a capacitor of about 1 μfarad can be connected across the 12,-000-ohm resistor. Such a regulator is used in an electronic power supply, as shown in Fig. 7, to provide constant output voltage to load circuits and to produce a low impedance so that several load circuits cannot interact on each other through their common power supply. *See* ELECTRONIC POWER SUPPLY.   [DONALD L. WAIDELICH]

POWER-SYSTEM VOLTAGE REGULATORS

Voltage regulators are used on distribution feeders to maintain voltage constant, irrespective of changes in either load current or supply voltage. Voltage variations must be minimized for the efficient operation of industrial equipment and for the satisfactory functioning of domestic appliances, television in particular. Voltage is controlled at the system generators, but this alone is inadequate because each generator supplies many feeders of diverse impedance and load characteristics. Regulators are applied either in substations to control voltage on a bus or individual feeder or on the line to reregulate the outlying portions of the system. These regulators are variable autotransformers with the primary connected across the line. The secondary, in which an adjustable voltage is induced, is connected in series with the line to boost or buck the voltage.

A control and drive provide automatic operation. A voltage-regulating relay senses output voltage. When the voltage is either above or below the band of acceptable voltage maintained by the regulator, this relay causes the motor to operate and change the regulator position to raise or lower the voltage as required to bring it back within the control band. It is desirable to maintain constant voltage at the average load center out on the feeder rather than at the regulator terminals. Hence the control circuit includes a line-drop compensator with resistance and reactance elements that can be adjusted to represent line impedance. These impedances carry current proportional to circuit load, thereby simulating the voltage drop between the regulator and the load center, and modify the volt-

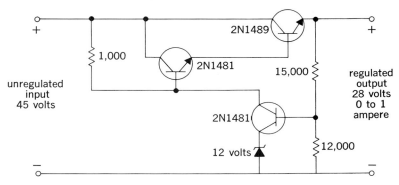

Fig. 6. An electronically regulated power supply. All resistances in ohms.

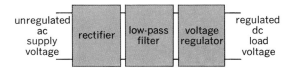

Fig. 7. If a voltage regulator is used in an electronic power supply, the rectifier circuit may be required to handle higher voltage than without the regulator, but the filter may be simpler than without the regulator.

age sensed by the voltage-regulating relay.

Two principal types of feeder regulators are used: the step regulator, which provides increments or steps of voltage change, and the induction regulator, which provides continuous voltage adjustment.

Step voltage regulator. The transformation ratio of the autotransformer in this regulator is adjusted by a voltage selector switch which changes the secondary winding tap connected to the line. Figure 8 shows the most commonly used circuit.

Switching is performed without interrupting load current by means of the two switch fingers in the selector switch. When the switch moves from the full-tap position shown in Fig. 8, one finger contacts the next tap before the other finger leaves the first tap; this constitutes a tap-bridging position. Switching reactors limit the current circulating between the bridged taps. Most regulators are designed to operate continuously in these bridging

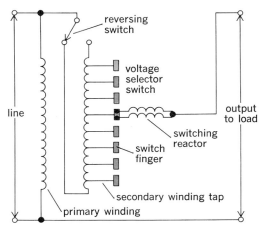

Fig. 8. Step voltage regulator circuit.

positions (as well as the full-tap positions) to provide voltages midway between the voltages of adjacent taps. Thus the voltage step between adjacent positions is half-tap voltage, and the number of winding taps required is half the number of operating positions in the boosting range. The automatic reversing switch changes the polarity of the sec-

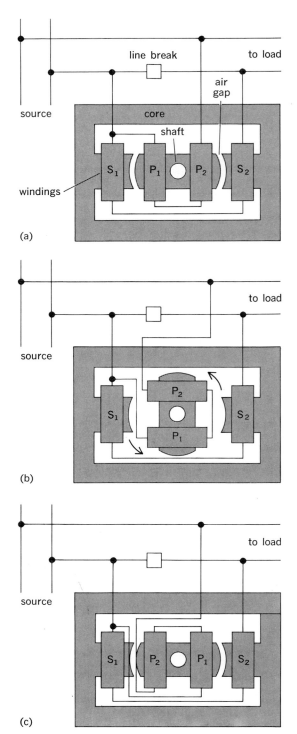

(a)

(b)

(c)

Fig. 9. Induction voltage regulator. (*a*) Full boost position. (*b*) Neutral position. (*c*) Full buck position of rotor. *P* coils are rotor primary coils, *S* coils are stationary secondary coils. (*From B. G. A. Skrotski, ed., Electric Transmission and Distribution, McGraw-Hill, 1954*)

ondary relative to the primary, thus providing a bucking range equal to the boosting range. Single- and three-phase designs are available with a range of ±10% in 16 or 32 steps. Other design variations may employ single-finger switching and operate continuously only in full-tap positions.

Load tap-changing transformers are often used to provide both regulation and transformation from one voltage level to another. They are similar to step voltage regulators, except that they have separate high- and low-voltage windings.

Induction voltage regulator. This is similar in structure to a wound-rotor induction motor with rotor restrained so that it moves only to adjust voltage. The primary winding on the rotor is magnetically coupled with the series secondary winding on the stator.

The principle of a two-pole single-phase regulator is illustrated in Fig. 9. Secondary voltage is continuously adjusted from full buck to full boost by changing the relative angular position of these windings through 180 electrical degrees.

Single-phase regulators require an additional permanently short-circuited rotor winding that is in space quadrature to the primary winding. Without this winding, the reactance of the regulator to the line current flowing in the secondary would be excessive in the neutral region between buck and boost. Three-phase induction regulators, if built on a three-phase core, do not require a short-circuited winding. Such induction regulators inherently introduce phase shift between primary and secondary voltages and are no longer supplied for feeder regulation.

Other regulators. Other types of regulators are also used outside the United States. One construction is a transformer structure with moving coils to change coupling; another has contacts moving over the exposed conductors to provide a large number of small, discrete steps.

Line voltage may be increased by drawing leading-power-factor current through the line reactance. Static capacitors, shunt-connected in fixed or automatically switched banks, are often applied near the loads to raise voltage. The increase in voltage is not limited solely to the vicinity of the capacitor. They also help compensate for the usual system condition of lagging power factor. The application of series capacitors on fluctuating loads is increasing.

Infrequently, synchronous machines called synchronous condensers, overexcited to draw leading current, are connected to lines. Their use has decreased with the availability of low-cost reliable static capacitors.

Conditions may exist where the inherent static capacity of the circuit is excessive. The line voltage may rise on cable systems or lightly loaded lines; shunt reactors may be used to neutralize the capacity of such systems.

[DONNELL D. MAC CARTHY]

Bibliography: J. J. Brophy, *Basic Electronics for Scientists*, 3d ed., 1977; B. W. Gaddis, *Troubleshooting Solid-State Electronic Power Supplies*, 1972; A. H. Lytel, *Solid-State Power Supplies and Converters*, 1965; J. Millman, *Microelectronics*, 1979; A. I. Pressman, *Switching and Linear Power Supply, Power Converter Design*, 1977; J. D. Spencer and D. E. Pippenger, *The Voltage Regulator Handbook for Design Engineers*, 1977.

Voltage-regulator tube

A gaseous glow-discharge tube operating in the normal-glow region and therefore possessing an almost constant anode-to-cathode voltage drop, as shown in Fig. 1. The purpose of these tubes is to provide a substantially constant dc voltage from a fluctuating dc source. *See* ELECTRICAL CONDUCTION IN GASES; GAS TUBE.

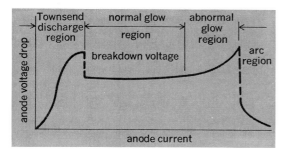

Fig. 1. The gas discharge volt-ampere characteristics of a typical VR tube.

Typical construction of a voltage-regulator (VR) tube is shown in Fig. 2a. The cathode is a large cylindrical electrode and the anode is a small wire, mounted concentric with the cathode. The gap between the starting probe and the anode, combined with adjustment of the filling-gas pressure, permits the tube to be made with a starting, or breakdown, voltage only slightly higher than the glow voltage.

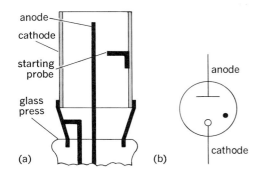

Fig. 2. (a) Construction of a typical VR tube. (b) The common symbol for a VR tube.

The volt-ampere characteristics of a VR tube (Fig. 1) show a slight increase in anode voltage drop for increasing current over the normal-glow region. For a typical tube, this voltage rise is about 5 volts. For use of this tube in voltage-regulator circuits *see* VOLTAGE REGULATOR.

Voltage-regulator tubes are available in several different values of regulated voltage, including 75, 90, 105, and 150 volts. These various voltages are attained by changing the nature of the cathode surface used and its treatment, and also the filling gas. The regulating range is usually for currents of 5–50 milliamperes (mA), but commercial tubes have been built for currents of 500 mA by using large cathodes. Voltage-reference tubes are small voltage-regulator tubes designed to maintain close voltage limits over a narrow range of current (4–5 mA). [JAMES D. COBINE]

Voltmeter

An instrument for the measurement in volts of potential difference between two points. Derivatives of the voltmeter are the microvoltmeter, millivoltmeter, and kilovoltmeter, for measurement of voltages with a measurement span of 1,000,000,000:1. Voltmeters are connected between points of a circuit, between which the potential difference is to be measured. The accuracy rating is usually stated in terms of the full-scale reading. *See* VOLTAGE MEASUREMENT.

Classifications. Voltmeters are classified in respect to the kind of voltage to be measured and the kind of mechanism used, as follows: (1) for measurement of direct voltage, (*a*) the moving-coil permanent-magnet voltmeter, (*b*) the fixed-coil moving-magnet voltmeter, and (*c*) the electrostatic voltmeter; (2) for measurement of alternating voltage, (*a*) the fixed-coil moving-magnet voltmeter, (*b*) the fixed-coil moving-iron voltmeter, (*c*) the thermovoltmeter consisting of a thermal converter and dc millivoltmeter, and (*d*) rectifier and vacuum-tube voltmeters.

Voltmeters, in general, are secondary instruments consisting of voltage-to-current transducers and mechanisms which respond to milliamperes or microamperes. In the simpler forms, the transducer consists of a resistor of constant value, which may have taps for several voltage ranges. Since by Ohm's law the current is proportional to the potential difference, calibration of the combination using a primary standard, such as the potentiometer and standard cell, is valid. In the more specialized forms, the transducer consists of a thermal converter (ac to dc), rectifier, or amplifier. The electrostatic instrument is the only voltmeter which measures potential difference directly.

Moving-coil voltmeter. This voltmeter consists of a permanent magnet, moving coils pivoted in jewel bearings, control springs, and a pointer, as shown in Fig. 1. The current I is passed through the moving coil of N turns via the two control springs which apply a restraining torque T_r equal to $K_s\theta$, where K_s is the spring constant and θ is the angle of deflection. The deflecting torque T_e is equal to the product of the magnetic moment of

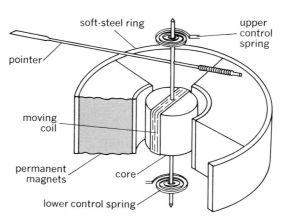

Fig. 1. Direct-current permanent-magnet, moving-coil voltmeter. (*General Electric Co.*)

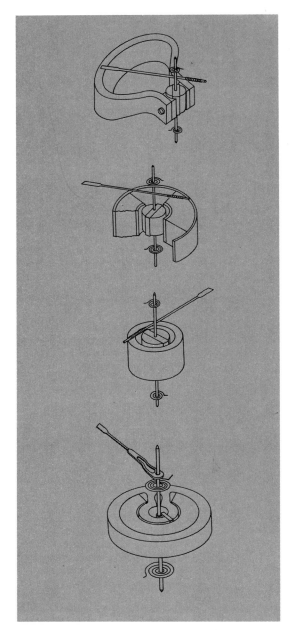

Fig. 2. Four forms of dc moving-coil mechanisms. (*From I. F. Kinnard, Applied Electrical Measurements, copyright © 1956 by John Wiley and Sons, Inc.; reprinted by permission*)

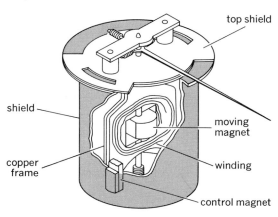

Fig. 3. Moving-magnet dc voltmeter. (*General Electric Co.*)

the coil and the field intensity B, as in Eq. (1),

$$T_e = NIwlB \qquad (1)$$

where w and l are coil width and length, respectively. The deflection θ resulting from the current I is given by Eq. (2). Since the magnetic field is usu-

$$\theta = BNIwl/K_s \quad \text{radians} \qquad (2)$$

ally uniform, the voltmeter scale is usually calibrated in uniform intervals.

The moving system must be damped to eliminate unwanted oscillations of the pointer. This is accomplished electromagnetically by a damping shell, usually aluminum, on which the moving coil is wound. As the coil moves, the shell moves in the magnetic field. A voltage and current are induced in the shell in the direction of the winding. This current reacts with the magnetic field to produce a braking action proportional to the angular velocity of the moving system. The amount of damping is controlled by the cross section and conductivity of the shell. It is customary to damp voltmeters so that a small but definite overshoot will result if full voltage is suddenly applied. In some instances, the shell is replaced by an equivalent coil and in others the instrument may be damped by the action of the driving coil itself. The latter condition is achievable only when the loop circuit resistance is low.

Permanent-magnet moving-coil instruments are made in several forms; four common ones are illustrated in Fig. 2.

The first three figures illustrate the evolution of the magnet structure, beginning with the expanded U-shaped magnet constructed of chrome or tungsten steels, and extending to the internal cylindrical magnet made with Alnico material of high-energy content. These mechanisms have scale lengths of about 90°. The fourth shows an annular magnet and offset coil with a scale length of 240°.

Voltmeters and their various derivatives are available in ranges of 10 mv to 750 volts, with resistances of 100–10,000 ohms/volt. Maximum errors range from 0.1% of full scale for secondary standards to 2.0% of full scale for panel-type instruments.

Moving-magnet voltmeter. This voltmeter, illustrated in Fig. 3, has the magnet attached to the pivoted shaft which carries the pointer. The magnet is surrounded by the field coil and aligned by a control magnet which supplies the restoring torque. A fixed copper frame damps the movement electromagnetically, and the magnetic shield minimizes the effect of external fields, as well as providing a path for the internal fluxes. As current is applied to the field coil, the coil and restoring-magnet fields produce a resultant field. The rotor aligns itself with this resultant field and indicates the magnitude of the current and thus the potential difference. The construction shown is used in commercial 2½- and 3½-in.-diameter (6.2- and 8.7-cm) panel instruments of 2% accuracy rating (maximum error 2%).

Electrostatic voltmeter. The electrostatic voltmeter action is based on the force of attraction or repulsion between two charged conductors, such as the plates of a variable air capacitor. The moving plate, when charged, tends to move so as to increase the capacitance between it and the fixed plate. If this capacitance is C farads, the voltage is

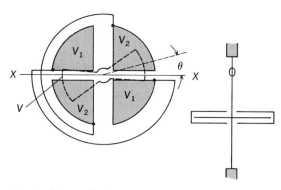

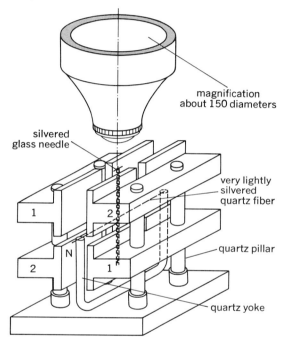

Fig. 4. Schematic of quadrant electrometer. (*From I. F. Kinnard, Applied Electrical Measurements, copyright © 1956 by John Wiley and Sons, Inc.; reprinted by permission*)

Fig. 5. Lindemann electrometer. (*From F. A. Laws, Electrical Measurements, 2d ed., McGraw-Hill, 1938*)

sion fiber at the center of the system. Opposite quadrants are electrically connected and potentials V_1 and V_2 are applied. In the heterostatic method, the vane is independently energized to potential V_3. The capacitance effect causes the vane to turn out of one pair of quadrants and into the other. This movement actuates the indicating needle and the deflection of the needle is directly proportional to the voltage difference $(V_1 - V_2)$ between the two sets of quadrants.

In the idiostatic method, the vane is connected to quadrant 1. This method has the advantage of dispensing with the auxiliary voltage, and the deflection is proportional to the square of the voltage difference.

The Lindemann electrometer, Fig. 5, is a variant of the quadrant electrometer, designed for portability and insensitivity to changes in position. The quadrants 1 and 2 are two sets of plates about 6 mm apart and mounted on insulating quartz pillars. A taut silvered quartz suspension fixes the center of rotation of the moving system. When voltage is applied to the needle, the needle rotates toward the oppositely charged plates. This movement is observed through a microscope. This electrometer has a low capacitance of about 1 picofarad, and currents of the order of 10^{-15} ampere can be observed with it.

Electrodynamic ac voltmeter. This voltmeter has two coils connected in series through a resistor to the source, the potential of which is to be measured, as shown in Fig. 6. The fixed coil provides the field in which the moving coil, supported by a pivoted shaft, operates. The deflection of the moving system is restrained by a control spring. When the circuit is closed, the current which flows through the coils produces a deflecting torque which moves the indicating needle. This current I produces a deflecting torque T_e defined by Eq. (6),

$$T_e = K_1 I^2 dM/d\theta \qquad (6)$$

where M is that portion of the total field flux which links the moving system, K_1 is a constant of proportionality, and θ is the angle between moving coil and fixed coil. Since the control spring produces a restoring torque T_r at deflection α, Eq. (7)

E volts and the spacing is s meters, the energy W in joules of the capacitor is expressed by Eq. (3). If

$$W = \tfrac{1}{2}CE^2 \qquad (3)$$

the upper plate is moved vertically a distance ds while the voltage is held constant, an energy change dW takes place, numerically equal to the work done in moving the plate. The resultant force F is given by Eq. (4). Thus, the force acting on the

$$F = dW/ds = (E^2/2) \, dC/ds \qquad (4)$$

upper plate is proportional to E^2 times the space rate of capacitance change. For a rotatable system the corresponding torque is given by Eq. (5).

$$T = (E^2/2) \, dC/d\theta \quad \text{radians} \qquad (5)$$

The quadrant electrometer, Fig. 4, is a useful embodiment of this principle. The four quadrants compose the fixed capacitor plates and surround the movable vane suspended by a conducting tor-

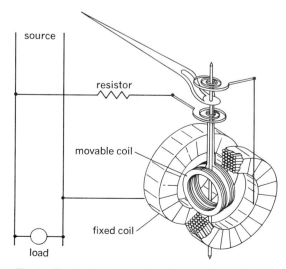

Fig. 6. Electrodynamic mechanism used as voltmeter. (*General Electric Co.*)

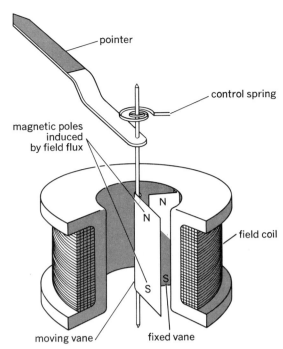

Fig. 7. The moving-iron radial-vane ac repulsion voltmeter. (*General Electric Co.*)

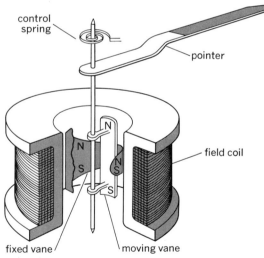

Fig. 8. The concentric-vane ac repulsion voltmeter. (*General Electric Co.*)

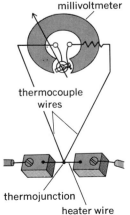

Fig. 9. Elementary thermovoltmeter circuit. (*General Electric Co.*)

$$T_r = K_S \alpha \qquad (7)$$

holds, where K_S is the constant of the spring. The moving system will come to rest at an angle where deflecting and restoring torques are equal, and the indication is given by Eq. (8). The scale distributions of electrodynamic voltmeters are in general

$$\alpha = \left(\frac{K_1}{K_S}\right)\left(\frac{I^2 \, dM}{d\theta}\right) \qquad (8)$$

contracted at low voltages and expanded at higher voltages. This is due in part to the necessary square-law response and in part to designing for scale legibility over the upper half of the scale. The

inherent accuracy of the electrodynamic voltmeter is excellent, and this construction is used in secondary standards of 0.1% error rating as well as in portable voltmeters of 0.25 or 0.50 % error rating. Voltage ranges of 10–750 volts are available in commercial instruments. Refinements in design include magnetic shielding, damping of the moving system, and compensation for temperature and frequency errors.

Moving-iron voltmeter. There are many forms of this voltmeter. Figure 7 is representative. Here, the field coil is connected to the line through a series resistor (not shown). Inside the field coil are two rectangular vanes, one fixed and the other attached to the shaft which carries the pointer. When a current traverses the coil, the vanes are similarly magnetized, and the moving vane is repelled from the fixed vane, deflecting the pointer in a clockwise direction. The deflecting torque T_e is given by Eq. (9), where I is the current traversing

$$T_e = K \frac{I^2 \, dL}{d\theta} \qquad (9)$$

the coil, L is the inductance of the field coil, and θ the angle between the vanes. Since the coil inductance is a maximum when the vanes are at maximum separation, the moving vane will, unless restrained, move 180° away from the fixed vane. The control spring exerts a restraining torque defined by Eq. (10) or the deflection given by Eq. (11).

$$T_r = K_S \alpha \qquad (10)$$

$$\alpha = \frac{K}{K_S} \frac{V^2 \, dL}{d\theta} \qquad (11)$$

The moving-iron instrument, like the electrodynamic voltmeter, when properly calibrated indicates true rms volts. The scale length and distribution can be controlled over wide limits by suitable configuration of the fixed iron. Figure 7 illustrates an arrangement for scale lengths of 90–100°. Figure 8 illustrates a circumferential fixed-vane design used in long-scale (250°) switchboard voltmeters. Moving-iron voltmeters with magnetic damping are available in commercial 2½- and 3½-in. (6.2- and 8.7-cm) panel instruments (2% error), 4- and 6-in. (10- and 15-cm) switchboard instruments (1% error), and portable instruments of ¼ or ½% error. Scale ranges vary from 0–10 to 0–750 volts.

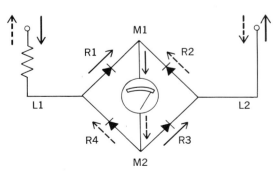

Fig. 10. Operating principle of the bridge-type rectifier voltmeter. The solid arrows show direction of current during the positive half-cycle of ac input; the dashed arrows show direction of current during the negative half-cycle. (*General Electric Co.*)

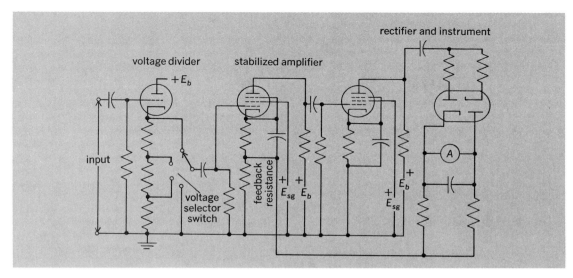

Fig. 11. Circuit for amplifier-rectifier electronic voltmeter. (*From I. F. Kinnard, Applied Electrical Measurements, copyright © 1956 by John Wiley and Sons, Inc; reprinted by permission*)

Instruments of 1% accuracy and higher are magnetically shielded. Power consumption varies from 1.5 to 10 watts, depending on size, range, and the specific design.

Thermovoltmeter. This unit utilizes a thermojunction and dc millivoltmeter (Fig. 9). A current from the voltage source is passed through a resistor (not shown) and a fine vacuum-enclosed platinum heater wire. A thermocouple, attached to the midpoint of the heater, generates millivolts proportional to the temperature.

The thermovoltmeter is a true rms instrument, in which the squaring, integration, and averaging are accomplished thermally rather than electromagnetically. The inductance of the heater wire is negligible, and the thermovoltmeter is especially suitable for measurement at high frequencies. Its sensitivity, while greater than that of electromagnetic instruments, is limited by the current necessary to heat the wire, usually several milliamps.

Rectifier voltmeter. This voltmeter consists of a dc milliammeter calibrated in volts and connected to the voltage source through a rectifier bridge (Fig. 10). The individual rectifiers are connected so that the current through the milliammeter is always in the same direction. This meter measures the average value of the current. While not an rms instrument, the rectifier voltmeter may be calibrated in rms volts for sine waves and will measure sine-wave voltages accurately. Harmonics in the voltage wave cause appreciable errors, but this circuit has the advantage of low power consumption. Commercial forms of rectifier voltmeters having resistance of 1000 ohms/volt and ratings of 10–750 volts are available.

Electronic ac voltmeter. This voltmeter consists of a dc milliammeter calibrated in volts and connected to an amplifier-rectifier circuit, of which Fig. 11 is typical. The input voltage is connected to the grid of the first triode. A voltage-range selector switch is connected to the tapped cathode resistor. The two stages of amplification are resistance-coupled to cover a wide frequency range. The amplifier output is applied to a rectifier section consisting of two diodes, milliammeter, and feedback resistor. *See* DIGITAL VOLTMETER; SAMPLING VOLTMETER; VACUUM-TUBE VOLTMETER.

By use of adequate feedback, errors due to changes in power-supply voltage and tube characteristics are minimized. Electronic voltmeters usually respond to average voltage but are calibrated in rms values for sine waves. The addition of the amplifier greatly increases the sensitivity and range of measurement, 0.001–300 volts at frequencies of 20 Hz to 2 MHz and with input impedances of several megohms.   [ALMON J. CORSON]

Bibliography: M. Braccio, *Electrical and Electronic Tests and Measurements*, 1979; W. D. Cooper, *Electronic Instrumentation and Measurement Techniques*, 2d ed., 1978; J. Cunningham, *Understanding and Using the VOM and EVM*, 1973; P. Kantrowitz et al., *Electronic Measurements*, 1979.

Volume control systems

Electronic systems that regulate the signal amplification or limit the output of circuits. Examples are volume compressors, limiters, and expanders. A volume compressor is an electronic system that reduces the amplification of an amplifier when the signal being amplified is large and increases the amplification when the signal is small. Compressors are used to reduce the volume range in sound motion picture and phonograph recording, sound broadcasting, public address and sound reinforcing systems, and so forth.

A volume limiter is an electronic system in which the relationship between the input and output signals is constant up to a certain level, but beyond this point the output remains constant regardless of the input. The limiter is useful for protection against sudden overloads, as for example, in the audio input to a radio or television transmitter.

A volume expander is an electronic system that increases the amplification of an amplifier when the signal is large and decreases the amplification when the signal is small. In reproduction, a volume expander is used to counteract or complement the

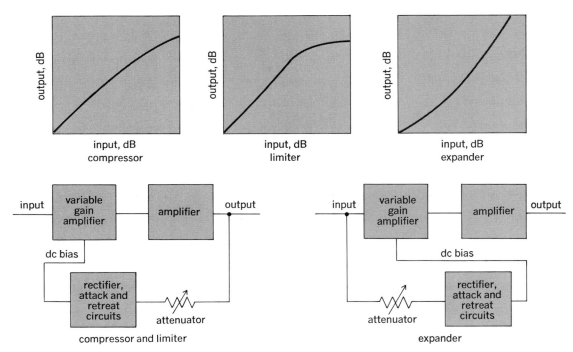

Schematic block diagrams and input-output characteristics for electronic compressors, limiters, and expanders.

effect of the compressor in recording or the transmission of an audio signal.

Volume compressors, limiters, and expanders are electronic amplifiers in which the amplification varies as a function of the average level of the signal. Compressor, limiter, and expander systems and their input-output characteristics are shown schematically in the illustration.

In all the systems, the input signal to the control circuits is rectified and applied to a capacitor-resistance network. The direct-current (dc) voltage across the capacitor is used to vary the amplification of a vacuum tube or transistor amplifier operating in a variable gain mode. In the case of a vacuum tube amplifier, the dc voltage is used to vary the bias on the control grid and as a consequence the amplification of a push-pull amplifier with variable transconductance. In the case of the transistor amplifier, the dc voltage is used to vary the bias of a field effect transistor which results in a change in the gain of the amplification of the transistor.

In the compressor characteristic shown in the illustration, there is a gradual reduction in gain with the increase of the input. This characteristic may be varied over very wide limits. A reduction in the volume range in radio and phonograph sound reproduction makes it possible to reproduce a wide range of orchestral music in the home without excessive top levels. The compressor may also be used to increase the signal-to-noise ratio in an audio signal. Use of a compressor improves the intelligibility of speech and enhances music reproduction when the ambient noise is high, for example, in sound motion picture reproduction and sound reinforcing applications.

The main application for the limiter characteristic shown in the illustration is to protect equipment against overload. Below the overload point, the limiter does not alter the signal.

In the expander characteristic shown in the illustration, there is gradual increase in gain with increase of input. The combination of a compressor and expander system, in which the two characteristics are complementary, may be used to reproduce a wide amplitude range by means of a much smaller recorded or transmitted amplitude range.

The attack time for a gain reduction of 10 dB in compressors and limiters is of the order of a millisecond. The retreat to normal is of the order of a second. *See* AMPLIFIER; AUTOMATIC GAIN CONTROL (AGC).   [HARRY F. OLSON]

Bibliography: H. F. Olson, *Modern Sound Reproduction*, 1972, reprint 1978.

Wave-shaping circuits

Electronic circuits used to create or modify a specified time-varying electrical quantity, usually voltage or current, using combinations of electronic devices, such as vacuum tubes or transistors, and circuit elements including resistors, capacitors, and inductors.

One such waveform is the square wave, in which a quantity such as voltage alternately assumes two discrete values during repeating periods of time. Where each period is composed of two equal intervals, the wave can be obtained by amplifying a sinusoidal time-varying voltage and removing all but the section near the zero-voltage axis. Also, square waves of either equal or unequal intervals, sometimes referred to as rectangular waves, can be generated by multivibrators or various other forms of vacuum-tube or transistor switching or gating circuits. *See* CLIPPING CIRCUIT; LIMITER CIRCUIT; MULTIVIBRATOR; RELAXATION OSCILLATOR.

Other wave shapes of particular interest in electronics are the linearly increasing function of time or ramp function (which if recurring at equally spaced periods is usually called the linear sawtooth waveform), the hyperbolic waveform, and the rectified sine wave. Such recurrent waveforms are shaped by the combination of electronic switches or gating circuits, resistance-capacitance time constants, and linear feedback amplifiers. *See* CLAMPING CIRCUIT; COINCIDENCE AMPLIFIER; ELECTRONIC SWITCH; ELECTRONIC DISPLAY; PULSE GENERATOR; RECTIFIER; SWEEP GENERATOR; TRIGGER CIRCUIT.

[GLENN M. GLASFORD]

Waveform

The pictorial representation of the form or shape of a wave, obtained by plotting the amplitude of the wave with respect to time. There are an infinite number of possible waveforms. Some of the more common electrical waveforms are shown in the illustration. These diagrams are plots of voltage against time. It is equally possible to show current waveforms.

It is possible to represent any periodic waveform mathematically as a Fourier series of sine and cosine terms at harmonic frequencies. Any nonsinusoidal wave is composed of a constant or dc term, plus a series of harmonic terms in which the frequencies are integral multiples of the fundamental frequency. The Fourier series for each waveform is given beside each figure as functions of time t, where E_m is the maximum value of the wave and T is the period. *See* NONSINUSOIDAL WAVEFORM.

Sine waves are obtained from sine-wave generators and LC, RC, and beat-frequency oscillators. *See* OSCILLATOR.

Square waves are obtained from square-wave generator circuits, such as multivibrators and clippers. *See* MULTIVIBRATOR.

Sawtooth waves are obtained from gas-tube relaxation oscillators and thyratron, transistor, and vacuum-tube sweep circuits. *See* RELAXATION OSCILLATOR; SAWTOOTH-WAVE GENERATOR; SWEEP GENERATOR.

Triangular waves are obtained from integrated square waves.

The output wave shape of a half-wave rectifier with resistance load is as illustrated. *See* RECTIFIER.

The output wave shape of a full-wave rectifier with resistance load is as illustrated.

[DONALD L. WAIDELICH]

Waveform determination

The defining of a curve, or waveform, that represents the variation of the magnitude of a quantity with time. Waveform is determined either with oscillographs that display and record the waveform directly, or with wave analyzers that indicate the numerical values of amplitude, frequency, and sometimes phase angle of the harmonic components of a complex wave.

The measurement and control of waveform is of real concern to the electrical industry's engineers, because the transformers, motors, lighting circuits, and other equipment are designed to operate at maximum efficiency when the voltage waveform is a sine wave with a predetermined crest voltage. Departures from the specified waveform cause losses in efficiency. In radio communication systems the carrier waveform is sinusoidal, and deviations from the sine wave introduce noise that interferes with the intelligence being transmitted.

Waveforms are of two basic types: (1) periodic or continuous waves, and (2) aperiodic or transient waves. For periodic waves *see* NONSINUSOIDAL WAVEFORM; WAVEFORM. For aperiodic waves *see* ELECTRIC TRANSIENT.

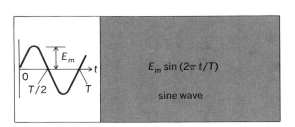

$$E_m \sin (2\pi t/T)$$

sine wave

$$\frac{4E_m}{\pi} \sum_{}^{\infty} \frac{1}{n} \sin (2\pi nt/T)$$

$$n = 1, 3, 5, \ldots$$

square wave

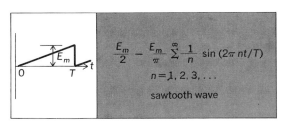

$$\frac{E_m}{2} - \frac{E_m}{\pi} \sum_{}^{\infty} \frac{1}{n} \sin (2\pi nt/T)$$

$$n = 1, 2, 3, \ldots$$

sawtooth wave

$$\frac{8E_m}{\pi^2} \sum_{}^{\infty} (-1)^{(n-1)/2} \frac{1}{n^2} \sin (2\pi nt/T)$$

$$n = 1, 3, 5, \ldots$$

triangular wave

$$\frac{E_m}{\pi} \left[1 + \frac{\pi}{2} \sin (2\pi t/T) - 2 \sum_{}^{\infty} \frac{1}{4n^2 - 1} \cos (4\pi nt/T) \right]$$

$$n = 1, 2, 3, \ldots$$

half-wave rectified sine wave

$$\frac{2E_m}{\pi} \left[1 - 2 \sum_{}^{\infty} \frac{1}{4n^2 - 1} \cos (2\pi nt/T) \right]$$

$$n = 1, 2, 3, \ldots$$

full-wave rectified sine wave

Common waveforms.

Two general classes of measurement devices are used to determine waveforms. The harmonic analyzer is widely used and is the most accurate for determining the waveform of a continuous wave. Any continuous wave can be defined by a Fourier series of sine waves, including a wave at fundamental frequency and waves at harmonics of this fundamental frequency. The harmonic analyzer indicates numerical values of the amplitude, frequency, and phase of the fundamental and each of the harmonics of the waveform under study. When required, the complete waveform may be constructed by graphically superimposing the component waves. The second type of measurement device, the oscillograph, is used to determine transient waveforms. The oscillograph may also be used to determine the waveform of continuous waves but with considerably less accuracy than that provided by the harmonic analyzer. *See* ELECTRICAL MEASUREMENTS; HARMONIC ANALYZER; OSCILLOSCOPE.

[ISAAC F. KINNARD/EDWARD C. STEVENSON]

Waveguide

A device which constrains or guides the propagation of electromagnetic waves along a path defined by the physical construction of the guide. In a broad sense, devices such as a pair of parallel wires and a coaxial cable can certainly be called waveguides. When used in a more restricted sense, however, the term waveguide usually means a metallic tube which can confine and guide the propagation of electromagnetic waves in the hollow space along the lengthwise direction of the tube. For reasons which will become clear in the following discussion, hollow waveguides of convenient sizes are best adapted to the transmission of microwaves.

The concept that hollow waveguides can transmit electromagnetic waves may seem strange to people who lean heavily on experiences with low-frequency waves. It will not appear so strange, however, if one thinks in terms of an analogy with sound waves going through pipes (for example, pipe organs). Because a sound wave can transmit through a pipe only when its wavelength is comparable to or smaller than the size of the pipe, one would expect that a similar requirement should hold true for electromagnetic waves. Indeed, if the frequency of an electromagnetic wave is high enough that the wavelength is comparable to or smaller than the waveguide dimension, then wave transmission through the hollow waveguide becomes possible.

Although hollow waveguides and coaxial cables are the commonest in application, some other types of waveguides are also occasionally used. A single conductor (called a G-string) is sometimes used as a waveguide. Another waveguide takes the form of a flat conducting strip having a certain spacing from a ground plane, known as a microwave strip. Still another example is found in a dielectric rod.

Maxwell's equations. The most basic approach to the understanding and analysis of the behavior of electromagnetic waves in any waveguide is obtained by the application of Maxwell's equations to a given physical situation. These are a set of partial differential equations relating the quantities electric intensity **E**, electric induction **D**, magnetic intensity **H**, and magnetic induction **B**. Each quantity is regarded as a vector, having a direction as well as a magnitude which is a function of space coordinates and time. In the ordinary case of a homogeneous isotropic medium, electric induction is proportional to electric intensity. The constant of proportionality is known as the dielectric constant ϵ. Similarly, magnetic induction is related to magnetic intensity by a constant known as the permeability μ.

A general solution of Maxwell's equations leads to a wave equation which points definitely to the possible existence of electromagnetic waves in the medium. Thus, for a dielectric medium of infinite extent, all solutions of the wave equation are equally admissible. One common characteristic for all these waves is that there is a velocity of propagation which is completely determined by the dielectric and permeability constants of the medium.

Any particular solution for a realizable guided wave must obey certain boundary conditions imposed by the physical situation. The walls of a hollow waveguide are almost always made of a highly conducting metal like copper, brass, or aluminum. The electrical conductivity of such materials, while always finite, is so high that it can be considered to be infinite for the present considerations. It is well known that no electric intensity can exist inside a perfect conductor. On the basis of this fact and the application of Maxwell's equations, the boundary conditions for the electric and magnetic intensities at the interface between a perfect conductor and a dielectric (usually air) in a hollow waveguide turn out to be that both the tangential component of **E** and the normal component of **H** are zero at the boundary.

Transmission modes. Consider a hollow waveguide with a given cross section which is uniform throughout its entire length. As a result of the application of these boundary conditions to the wave equation, it can be shown that only certain unique patterns for the distribution of **E** and **H** (taken together) can exist in the waveguide. Each unique pattern of the field distribution is called a mode. There are two types of mode possible in a hollow waveguide. One type is called the transverse electric (TE) mode, in which **E** has only a component transverse (that is, perpendicular) to the direction of propagation, whereas the magnetic intensity **H** has both transversal and longitudinal components. The other type is called the transverse magnetic (TM) mode, in which the magnetic intensity has only a transverse component and the electric intensity has both components. Each type (TE or TM) of mode has an infinite number of submodes which have the common characteristics of the type to which they belong, but differ among themselves in the details of field distribution. Since it is known that the transverse electric and magnetic (TEM) mode is not possible in a hollow waveguide, any arbitrary electromagnetic wave inside such a waveguide can be considered as a linear superposition of all possible modes of both the TE and TM types.

Rectangular waveguides. The type of waveguide with a rectangular cross section is not only the commonest in use but also the simplest in theoretical analysis. It is used here as a concrete example to illustrate various common properties of a waveguide. Consider a rectangular waveguide as shown in Fig. 1. The wave propagates along the z axis. The simplest and also the most commonly used mode is called the TE_{01} wave; its electric and magnetic intensities can be described by the expressions satisfying the boundary conditions of Eqs. (1), where $A=$ arbitrary constant depend-

$$E_x = A \sin\left(\frac{\pi}{b}y\right)\sin\omega\left(t-\frac{z}{v}\right)$$

$$H_y = \frac{1}{\mu v}A \sin\left(\frac{\pi}{b}y\right)\sin\omega\left(t-\frac{z}{v}\right) \qquad (1)$$

$$H_z = -\frac{\pi}{\mu\omega b}A\cos\left(\frac{\pi}{b}y\right)\cos\omega\left(t-\frac{z}{v}\right)$$

$$H_x = E_y = E_z = 0$$

ing upon the strength of wave excitation, $\omega = 2\pi\times$ frequency, $t=$ time, $v=$ velocity of propagation of the wave, and $\mu=$ permeability of the dielectric filling the waveguide. It is particularly important to note that the velocity of propagation v in a waveguide is different from that in an infinite space filled with the same dielectric material v_0. These two quantities are related for the TE_{01} wave by the Eq. (2), in which $f=$ frequency.

$$v = \frac{v_0}{\sqrt{1-(v_0/2fb)^2}} \qquad (2)$$

By applying the simple formula $v=f\lambda$, it is seen that the wavelength in the waveguide is similarly related to the corresponding quantity (for the TE_{01} wave) in an infinite dielectric (λ_0) as in Eq. (3).

$$\lambda = \frac{\lambda_0}{\sqrt{1-(\lambda_0/2b)^2}} \qquad (3)$$

The expressions in Eq. (1) reveal the wave nature of the field quantities through the sinusoidal functions of $\omega(t-z/v)$, which are a description of wave motion with velocity v along the z axis. On any cross section (constant z), these sinusoidal functions depend simply on time. Hence, each of the field quantities oscillates at the common frequency f and with an amplitude which varies with the field point (y in this particular case).

Phase velocity. Equation (2) represents the phase velocity of the guided wave because the quantity v is contained in $\omega(t-z/v)$, which is called the phase of wave propagation. If $\lambda_0 < 2b$, the phase velocity in a waveguide is larger than that in an open space filled with the same dielectric material. Correspondingly, the wavelength in a waveguide is longer than that in an infinite dielectric medium, as indicated by Eq. (3). All waves with $\lambda_0 < 2b$ belong to the transmission region of the waveguide because only the waves in this region are allowed to pass through. When $\lambda_0 = 2b$, both v and λ are infinite, and when $\lambda_0 > 2b$, they are imaginary. It can be shown that waves are not allowed to propagate in a waveguide when $\lambda_0 \geq 2b$. The critical value of $\lambda_c = 2b$ is known as the cutoff, or critical, wavelength.

Generalization to other TE waves. The preceding discussion can be carried through in a parallel fashion for the general case of any TE wave designated as TE_{nm}, where n and m are integers. Each TE_{nm} wave has its characteristic field distribution,

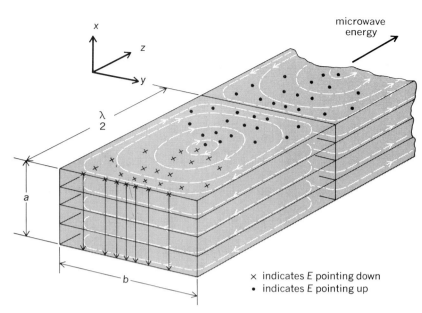

Fig. 1. Instantaneous field pattern for the TE_{01} wave in a rectangular waveguide. Solid lines indicate the electric intensity **E**, and broken lines the magnetic intensity **H**. (*From MIT Radar School Staff, Principles of Radar, 1952*)

× indicates E pointing down
• indicates E pointing up

velocity, and wavelength. The expression for the wavelength in the waveguide, for example, is given by Eq. (4). The corresponding cutoff wavelength is

$$\lambda = \frac{\lambda_0}{\sqrt{1-\left(\frac{n\lambda_0}{2a}\right)^2 - \left(\frac{m\lambda_0}{2b}\right)^2}} \qquad (4)$$

given by Eq. (5), which reduces to $\lambda_c = 2b$ for the

$$\lambda_e = \frac{2}{\sqrt{\left(\frac{n}{a}\right)^2 + \left(\frac{m}{b}\right)^2}} \qquad (5)$$

special case of TE_{01}, where $n = 0$ and $m = 1$. A TE_{00} wave ($n = 0$ and $m = 0$) would have an infinite cutoff wavelength which is characteristic of a principal mode, if such were possible. Actually, the solution of the field equation shows that the principal mode cannot exist in a hollow metallic waveguide. Equation (5) further shows that any other values of n and m in TE_{nm} would lead to a cutoff wavelength shorter than that of TE_{01}, which is called the dominant mode for this reason.

TM waves. The TM waves can be treated in almost exactly the same manner as the TE waves. Although the field distributions in the two cases are completely different, the wavelengths (similarly, the velocities of propagation) in the guide for both cases obey Eqs. (4) and (5), except that neither n nor m can become zero for a TM_{nm} wave.

Band designation for rectangular waveguides

Band	Dimension, cm²	λ_0, cm
S	7.62×2.54	$8.9 - 10.5$
C	3.48×1.58	$3.7 - 5.1$
X	2.54×1.27	$3.0 - 3.5$
K	1.06×0.43	$1.2 - 1.5$

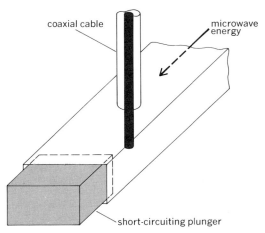

Fig. 2.   Coupling between waveguide and coaxial cable.

WAVEGUIDE

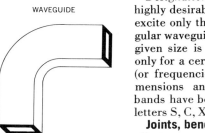

Fig. 3. **H**-plane 90° bend for a rectangular waveguide.

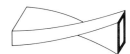

Fig. 4.   A 90° twist for a rectangular waveguide.

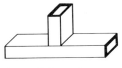

Fig. 5.   A T junction.

Designation of rectangular waveguides. It is highly desirable and customary for practical use to excite only the dominant mode (TE_{01}) in a rectangular waveguide. This means that a waveguide of a given size is useful for this one-mode operation only for a certain range of free-space wavelengths (or frequencies). Four waveguides of certain dimensions and their corresponding wavelength bands have been conventionally designated by the letters S, C, X, and K, as shown in the table.

Joints, bends, and junctions. To achieve certain effects, waveguides are frequently joined together, bent or twisted, or formed into networks known as junctions.

Joints. Waveguides are often joined together under various conditions. Identical waveguide terminals may be connected from end to end for extension or for inserting circuit elements. This can be done by providing a flange at each end of the waveguide and then butting the two flat surfaces of the flanges together so that the waveguide ends form a contact joint smoothly. Sometimes, to alleviate the difficulty of making good contact, a choke joint is operated. This is based on the principle that an artificially created current nodal line at the joint will make the physical contact of the waveguide ends unnecessary.

Another type of waveguide joint is required when a coaxial cable is to be joined to a hollow waveguide. Here, a transformation of the wave is to take place between the coaxial mode and, for instance, the TE_{01} mode of a rectangular waveguide. In this case the coaxial cable is usually led into the blocked end (or into a tunable short-circuiting plunger) of the regular waveguide, and the center conductor is extended to touch the opposite face so that the extended wire will act as an antenna for the excitation of the wave in the rectangular guide (Fig. 2).

Bends and twists. Waveguide bends are often used to change the direction of the waveguide by a desired angle. There are two types of 90° bend —one in the **H** plane and the other in the **E** plane. A 90° bend in the **H** plane is shown in Fig. 3. Twists are used to change the plane of polarization by a desired angle while maintaining the direction of the waveguide. A 90° twist is shown in Fig. 4.

Junctions. The term junction is used to denote a network of waveguides joined in a specified manner to give certain desired properties to the whole network. Common examples are a T junction, a directional coupler, and a magic-tee junction.

A T junction is a network with three waveguide terminals arranged in the form of the letter T. In rectangular waveguides, there are two ways of arranging a symmetrical T junction: Either all three broad sides are in one plane or two broad sides are in one plane and the third in a perpendicular plane; the latter arrangement is shown in Fig. 5. In any case, when a microwave is incident to one waveguide terminal, the incoming power is divided equally between the remaining two waveguides, and the two branch waves go along their separate directions.

A directional coupler is a network of four waveguide terminals. When suitably arranged, it has the property that for each terminal there is another with which it does not interact in any way. A special directional coupler known as the magic tee is described here. For a discussion of the two-hole interference type of directional coupler *see* DIRECTIONAL COUPLER.

A magic-tee junction is shown in Fig. 6. This network has a plane of symmetry as indicated. Consider only the dominant mode TE_{01} for all the waveguides. The mode in waveguide A is always symmetrical to the plane of symmetry, while the mode in waveguide C is always antisymmetrical. The symmetrical mode in A can excite symmetric modes in B and D but not in C. The antisymmetric mode in C can excite antisymmetric modes in B and D but not in A. A superposition of simultaneous excitations in A and C of equal strength must necessarily cause the intensity in either B or D to vanish because, if it is exactly in phase for one of them, it must be exactly out of phase for the other. Hence, in addition to the isolation between A and C, there is an isolation between B and D. These rather intricate properties are used in microwave circuits in a variety of ways, including the direct measurement of both incident and reflected intensities.

Waveguide shapes. Next to the rectangular type, metallic waveguides with circular cross sections are most frequently used. Because of its circular cylindrical symmetry, this type of waveguide can sustain waves of all kinds of field polarization (linear, circular, and elliptic) in many ways which are not possible with the rectangular type. However, the general notions of transmission modes, propagation velocities, cutoff wavelengths, and so on are just as applicable to circular waveguides as they are to rectangular waveguides, except that the mathematics used for circular shapes is somewhat more involved. Waveguides of elliptical cross section are almost never used on purpose, but they may represent results of deformation from the original circular waveguides. Detailed descriptions of wave behavior in elliptical waveguides are available in the literature.

Attenuation of hollow waveguides. In the preceding discussion, there has been an implicit assumption that the attenuation of a wave by the waveguide is negligible. Although this assumption is quite justifiable for interpretive purposes, the effect of attenuation should be considered for all quantitative evaluations. As a wave passes through

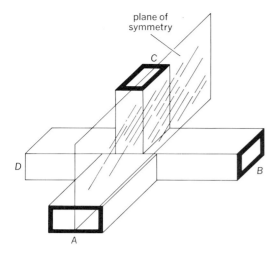

plane of
symmetry

C

D

B

A

Fig. 6. A magic-tee junction.

a waveguide (say in the z direction), its power becomes reduced exponentially according to Eq. (6),

$$P_z = P_0 e^{-\alpha z} \qquad (6)$$

where P_0 is the power at $z = 0$, P_z is the power at a distance z, and α is called the attenuation constant of the waveguide. The quantity α is a measure of the lossy character of the waveguide in absorbing the wave energy because of the currents set up on the inner walls of the waveguide and because of dielectric losses, if any. The same quantity is a function of the electrical conductivity of the waveguide walls, geometric dimensions, mode of propagation, and frequency of the wave. Other things being equal, α becomes smaller with a larger conductivity or a greater volume-to-surface ratio. For practical reasons, hollow waveguides are usually made of copper or brass, but on rare occasions they are also made of silver (or coated inside with silver) for higher conductivity.

The dominant mode (TE_{01}) of a rectangular copper waveguide in the X-band has an attenuation constant of the order of 0.1 dB/m. The overall attenuation may be regarded as small for a short guide length, but it would become undesirably large as the length increased. Consequently, rectangular waveguides are not considered suitable for long-distance microwave transmission.

In contradistinction to the rectangular waveguide, a circular waveguide operated at the TE_{01} mode has the unusual characteristic that its attenuation constant decreases indefinitely with increasing frequency. It is thus possible to choose a frequency sufficiently high to keep the attenuation constant satisfactorily low, even with small waveguide dimensions. For example, at a frequency corresponding to a free-space wavelength of 6 mm, the attenuation constant for the TE_{01} mode in a circular waveguide can be almost 100 times smaller than that in a corresponding rectangular waveguide. However, for the realization of long-distance transmission with circular waveguide, it is necessary to prevent the degradation of the low-attenuation property resulting from interconversion between the degenerate modes TE_{01} and TM_{11}, the interconversion being caused by manufacturing and laying (chiefly bending imperfec-

tions). Some preventive measures have been devised and demonstrated.

Coaxial cable. A coaxial cable consists of a hollow cylindrical conductor coaxially placed with respect to an inner cylindrical conductor. It is very extensively used as a waveguide, though seldom so called, for very high as well as very low frequencies. The adaptability of a coaxial cable to a wide range of frequencies (including zero) is due to the existence of a principal mode, which is in fact a TEM wave. The analysis of the wave propagation in a coaxial cable is simple and is analogous to that of two parallel wires or plane plates. The velocity of propagation in such systems is the same as in infinite space filled with the same dielectric material. In common with a hollow waveguide, coaxial cable has the advantage that the outside conductor acts as a shield against external electrical interference. However, it does not have the ability possessed by hollow waveguides of filtering low frequencies, when such filtering is desired. Also, for very high frequencies, the attenuation of a coaxial cable is apt to be higher than that of a comparable hollow waveguide because of higher losses due to either the dielectric medium or the supports. *See* TRANSMISSION LINES.

Other special waveguides. A single conductor, often thinly coated with a dielectric substance, can act as a waveguide under the nickname of G-string. A wave, usually of the TM mode, is guided along the surface of the conductor. The transmission characteristics are quite favorable in the frequency range of 80 to 300 MHz.

A special waveguide adaptable to microwave circuit wiring has been developed under the name of a microwave strip. It consists of a flat conducting strip separated from the ground plane by a dielectric layer. Its chief advantage is convenience of fabrication by printed circuit techniques.

Dielectric rods not involving any conductor can guide very-high-frequency (vhf) waves quite successfully. The propagating wave is partly inside the dielectric and partly outside. Such dielectric waveguides can be used for short-distance transmission. *See* ELECTROMAGNETIC WAVE TRANSMISSION; MICROWAVE; MICROWAVE TRANSMISSION LINES.   [C. K. JEN]

Wavelength measurement

The wavelength of an oscillating electromagnetic wave depends upon the frequency of the oscillation and the velocity of propagation in the medium or in the transmission system in which the wave is propagating. The wavelength in a transmission system is obtained by measuring the distance between successive wavefronts of equal phase. This measurement is most conveniently carried out in a standing-wave field in which interference occurs between forward-propagating and reverse-propagating, or reflected, waves. A distance equal to one-half wavelength exists between successive minima or maxima in the standing-wave pattern. *See* WAVEMETER.

The velocity of electric waves in free space is 299,792.5 km/sec, or approximately 3×10^8 m/sec. The presence of any dielectric material (such as air) or any magnetic matter with permeability greater than unity will cause the wave to travel at a lower velocity. Also, the presence of various types

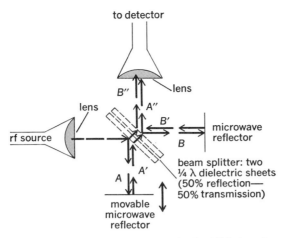

Fig. 1. Wavelength measurement by the Michelson interferometer used at millimeter wavelengths.

of transmission lines and waveguides will affect the velocity. In general the group velocity, or velocity of propagation of the wavefront of a suddenly applied signal, is less than that in free space, but it is possible for the phase velocity to be greater than the value for free space. In particular, the phase velocity in a waveguide operated with waves near the cutoff frequency is greater than the phase velocity in free space.

Wavelength by frequency measurement. In order to specify the wavelength of an electromagnetic wave, it is essential to know the medium or device through which the wave is propagating. However, unless otherwise specified, it is general practice to quote the wavelength of an oscillatory electric wave as that in air, or free space. When this convention is used, all that is necessary is a frequency measurement in order to apply the relation $\lambda_0 = c/f$, where c is the velocity of light, λ_0 is the free-space wavelength, and f is frequency.

In view of the direct relationship between frequency and wavelength, the earliest method used for specifying the tuning point of a given radio wave was the wavelength of the wave. Early experimenters found that standing waves existed in space wherever reflections occurred. In such a measurement, the distance between two maxima, or two minima, of a standing-wave field is measured with a meter stick or measuring tape, the location of the field maxima being observed by moving a suitable detector unit to various points in the field.

Interferometer methods. The standing-wave method of measurement is somewhat similar to the interferometer measurements used in optics. With shorter-wavelength radio waves it is possible to apply optically derived interferometer techniques directly to the measurement of wavelength.

An example of an interferometer used in the millimeter wavelength range is shown in Fig. 1. Essentially, a microwave beam is directed at a beam splitter, which splits the beam into two parts, A and B, by partial reflection. The A beam is reflected to a movable reflector and reflected again as A'. The beam splitter transmits part of this as A''. The transmitted part B of the original beam is reflected by a fixed microwave reflector as B'. This is partially reflected by the beam splitter as B''. The beams A'' and B'' combine to form standing waves, which are then detected. Movement of the movable reflector causes the position of the standing wave to move, which causes the detected signal to pass through successive points of maximum and minimum amplitude. The distance between points of successive maxima or minima is one-half wavelength. This distance may be determined from the motion of the movable reflector. *See* LASER.

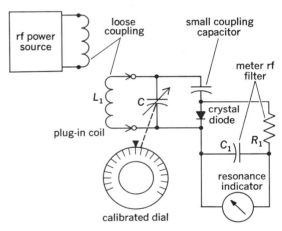

Fig. 3. Schematic diagram of inductance-capacitance type of absorption wavemeter (for frequencies between approximately 50 kHz and 1000 MHz).

Lecher wires. A more convenient, if less precise, measurement method for the determination of wavelength is the Lecher wire wavemeter (Fig. 2). With this simple device, wavelength is measured by sliding the short circuit along the line from a first to a second point of equal amplitude of effect, as indicated by absorption of signal being detected by an external detector, the variation in input power to oscillator or amplifier being checked, or by some similar method, such as the dc grid current of a negative-grid triode oscillator. The distance between two successive absorption maxima is one-half wavelength ($\lambda/2$); thus, by this length measurement, the wavelength is measured directly. *See* TRANSMISSION LINES.

Tuned circuits. At lower frequencies, inductance-capacitance resonant circuits may serve similar functions as absorption devices. With

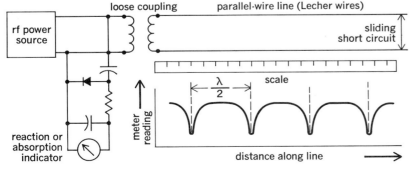

Fig. 2. Lecher-wire (resonant-line) wavemeter.

Fig. 4. Grid-dip oscillator types of wavemeters. Note plug-in coils used with calibrated capacitors in each instrument. (a) For range from 2.2 to 400 MHz (*McGraw Edison Co.*). (b) For range from 300 to 1000 MHz (*Boonton Electronics Corp.*).

calibrated values of inductance L and capacitance C, it is possible to provide a scale calibrated in wavelength or frequency. For low-loss circuits the resonant frequency is $1/(2\pi\sqrt{LC})$. Figure 3 shows a schematic diagram of a simple absorption-type wavemeter. Wavemeters of this type, often constructed with the principal inductance L_1 as a plug-in coil, are used for frequency or wavelength measurement up to frequencies of approximately 1000 MHz.

A variation of the absorption wavemeter, known as a grid-dip meter or grid-dip oscillator, provides in one instrument an absorption wavemeter and a calibrated oscillator, which may be used to determine the resonant frequency (or wavelength) of a passive network, such as an antenna system or an LC tuned circuit, by observing the dip in grid current of the oscillator as the oscillator frequency is tuned to the network resonant frequency. Figure 4 shows typical grid-dip oscillators, covering frequencies from 2.2 to 1000 MHz. *See* RESONANCE (ALTERNATING-CURRENT CIRCUITS).

Microwave wavemeters make use of resonant coaxial-line sections or cavities as tuned elements. The two general types of microwave wavemeters are the absorption, or reaction, type and the transmission type. Wavemeters of low or medium selectivity are frequently used as coarse measuring devices to establish the general range of frequency of operation of a system before applying more refined and complex methods for accurate frequency checking. *See* CAVITY RESONATOR.

A modern resonant-cavity microwave waveme-

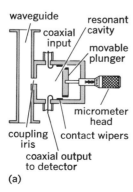

Fig. 5. Resonant-cavity wavemeter. (a) Typical construction providing either coaxial or waveguide inputs. (b) Commercial model. (*Hewlett-Packard Co.*)

ter is shown in Fig. 5. The dimensions of the cavity determine the resonant frequency of the wavemeter. A signal is fed in from either a coaxial line or waveguide, and energy is fed out to a suitable detector by a second coaxial line. The cavity is tuned by means of a micrometer-driven plunger, which may be calibrated in terms of wavelength. *See* COAXIAL CABLE. [FRANK D. LEWIS]

Bibliography: E. Hecht and A. Zajac, *Optics,* 1974; F. A. Jenkins and H. E. White, *Fundamentals of Optics,* 4th ed., 1976; A. L. Lance, *Introduction to Microwave Theory and Measurements,* 1964; Optical Society of America, *Handbook of Optics,* 1980; Wentworth Institute, *Laboratory Manual for Microwave Measurements,* 1971.

Wavemeter

A device for measuring the geometrical spacing between successive surfaces of equal phase along an electromagnetic wave. To avoid instrument calibration problems due to the dependence of the phase velocity upon the particular transmission system under measurement, it is common procedure to calibrate wavemeters in terms of free-space wavelength, which is the ratio of the velocity of light (299,792,458 m/sec) divided by the frequency of the signal in hertz.

At frequencies up to about 100 MHz, wavemeters consist basically of a tuned LC circuit and a suitable resonance indicator not noticeably different from that originally used by H. Hertz. The choice of a detector depends upon the power level of the signal and the desired accuracy. For power levels greater than several watts and for moderate accuracy, a miniature lamp bulb in series with the inductor L serves as a resonance indicator by glowing brightly when the induced current is a maximum. For low power levels or for higher accuracy the loading effect of the bulb cannot be tolerated. Therefore a suitable electronic voltmeter is used to measure the capacitor voltage, which is a maximum at resonance. The capacitor is variable and calibrated in units of wavelength or frequency.

At higher frequencies it is necessary to employ well-defined transmission systems such as open-wire or coaxial lines and waveguides. Any open-circuited or short-circuited section of transmission system can be adjusted in physical size to cause it to resonate at a given wavelength. From a construction standpoint short-circuited sections are preferred and may take the form of Lecher wires, coaxial wavemeters, and cavity resonators. Suitable electronic standing-wave detectors are employed to indicate when the wavemeter is tuned to resonance. *See* CAVITY RESONATOR; STANDING-WAVE DETECTOR; WAVELENGTH MEASUREMENT. [ROBERT L. RAMEY]

Wheatstone bridge

A device used to measure the electrical resistance of an unknown resistor by comparing it with a known standard resistance. This method was first described by S. H. Christie in 1833, only 7 years after Georg S. Ohm discovered the relationship between voltage and current. Since 1843 when Sir Charles Wheatstone called attention to Christie's work, Wheatstone's name has been associated with this network. *See* RESISTANCE MEASUREMENT.

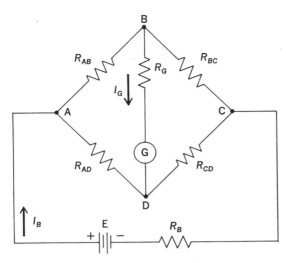

Fig. 1. Wheatstone bridge circuit.

The Wheatstone bridge network consists of four resistors R_{AB}, R_{BC}, R_{CD}, and R_{AD} interconnected as shown in Fig. 1 to form the bridge. A detector G, having an internal resistance R_G, is connected between the B and D bridge points; and a power supply, having an open-circuit voltage E and internal resistance R_B, is connected between the A and C bridge points. *See* BRIDGE CIRCUIT.

Application of Ohm's and Kirchhoff's laws to this network results in the equation

$$I_G = \frac{I_B(R_{BC}R_{AD} - R_{AB}R_{CD})}{R_G(R_{AB} + R_{BC} + R_{CD} + R_{AD}) + (R_{BC} + R_{CD})(R_{AB} + R_{AD})} \quad (1)$$

for the detector current. In this expression,

$$I_B \approx \frac{E}{R_B + \frac{(R_{AB} + R_{BC})(R_{AD} + R_{CD})}{R_{AB} + R_{BC} + R_{CD} + R_{AD}}} \quad (2)$$

It is apparent, from Eq. (1), that if the network is adjusted so that

$$R_{BC}R_{AD} - R_{AB}R_{CD} = 0 \quad (3)$$

the detector current will be zero and this adjustment will be independent of the supply voltage, the supply resistance, and the detector resistance. Thus, when the bridge is balanced,

$$R_{BC}R_{AD} = R_{AB}R_{CD} \quad (4)$$

and, if it is assumed that the unknown resistance is the one in the CD arm of the bridge, then

$$R_{CD} = \left(\frac{R_{BC}}{R_{AB}}\right) \times R_{AD} \quad (5)$$

Three methods of adjustment to achieve this condition are possible when the circuit is used as a ratio arm bridge: (1) use of a fixed ratio R_{BC}/R_{AB} and a continuously adjustable standard R_{AD}, (2) use of a continuously adjustable ratio and a fixed standard, and (3) a combination of the foregoing with the ratio usually adjustable in discrete steps of decade values. The first method provides a linear calibration of unknown versus standard resistance but is limited in resistance range to the ad-

justable range of the standard. The second method provides a wide range, since the ratio is easily adjustable from zero to infinity, but results in a nonlinear scale, highly expanded for low resistances and greatly compressed for high resistances. The third method, using a ratio adjustable in several decade steps and a resistance standard of three to five decades, provides a wide range and linear calibration and is the most practical of the combinations for general use with reasonable accuracy.

If the circuit is considered as a product arm bridge, it is seen that conductance G_{CD} of unknown resistance

$$R_{CD} = \frac{R_{BC}R_{AD}}{R_{AB}} \quad (6)$$

is measured directly in terms of the adjustable standard R_{AB} since

$$G_{CD} = \frac{1}{R_{CD}} = \frac{R_{AB}}{R_{BC}R_{AD}} \quad (7)$$

Sensitivity. The sensitivity of the bridge assembly (battery, bridge, and detector) is of interest for two purposes: (1) to determine the required detector sensitivity for a given deviation in the unknown resistance, or (2) to determine the change in resistance which can be measured using a detector of a stated sensitivity. The precision of balance is affected by the detector sensitivity, the detector resistance, the ohmic value of the bridge resistors, the bridge supply voltage, and the bridge supply resistance, which except for special cases can be neglected.

If the detector circuit BD (Fig. 1) is opened, the open-circuit voltage between points B and D is

$$e = E_{BC} - E_{CD}$$

or

$$e = E\left(\frac{R_{BC}}{R_{AB} + R_{BC}} - \frac{R_{CD} - \Delta R_{CD}}{R_{CD} - \Delta R_{CD} + R_{AD}}\right) \quad (8)$$

where ΔR_{CD} is a small, incremental change in the unknown resistance R_{CD}. To a close approximation, this open-circuit voltage due to a small bridge unbalance can be expressed in terms of the bridge ratio, the fractional change in the unknown resistance, and the applied voltage as

$$e = E\frac{r}{(r+1)^2}\left(\frac{\Delta R_{CD}}{R_{CD}}\right) \quad (9)$$

where $r = R_{BC}/R_{AB}$.

Sensitivity can also be expressed in terms of the unbalance voltage per volt applied to the bridge

$$e' = \frac{e}{E} = \frac{r}{(r+1)^2}\left(\frac{\Delta R_{CD}}{R_{CD}}\right) \quad (10)$$

If the detector circuit is closed, a current flows through the detector. Neglecting the battery resistance, this current can be calculated from Thévenin's theorem and Fig. 2 as

$$I_G = \frac{e}{R_G + \frac{R_{AB}R_{BC}}{R_{AB} + R_{BC}} + \frac{R_{AD}R_{CD}}{R_{AD} + R_{CD}}} \quad (11)$$

or

$$I_G = \frac{E\left(\frac{\Delta R_{CD}}{R_{CD}}\right)}{\frac{R_G}{\frac{r}{(r+1)^2}} + R_{AB} + R_{BC} + R_{CD} + R_{AD}} \quad (12)$$

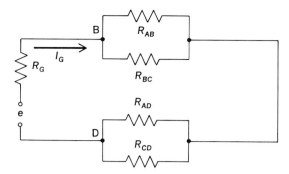

Fig. 2. Equivalent circuit of unbalanced bridge.

and in terms of unbalance current per volt applied to the bridge

$$I_{G'} = \frac{I_G}{E} = \frac{\left(\dfrac{\Delta R_{CD}}{R_{CD}}\right)}{\dfrac{R_G + R_{AB} + R_{BC} + R_{CD} + R_{AD}}{\dfrac{r}{(r+1)^2}}} \quad (13)$$

The unbalance current can thus be expressed in terms of (1) the fractional change in unknown resistance, (2) an "effective" detector resistance which depends upon the bridge ratio in use, (3) the sum of all of the bridge resistors, and (4) the applied voltage.

Accuracy. The errors in a Wheatstone bridge measurement are caused by (1) the value of unknown resistance and the conditions of measurement, (2) the ability to balance the bridge to the required precision, (3) the available bridge sensitivity, (4) the errors of the comparison resistors, ratios, or both, and (5) an accumulation of small errors resulting from practical circuit and construction problems.

Once the allowable errors of measurement have been determined, the bridge components and the detector should be selected so that the allowable limit of error, the ability to physically adjust the bridge, and the detector sensitivity are in the proportions $1:\frac{1}{2}:\frac{1}{4}$. For an allowable error of $\pm 0.1\%$, the equipment should be adjustable to at least $\pm 0.05\%$, and the detector should be sufficiently sensitive to detect at least $\pm 0.025\%$ deviation in the measurement.

If the ratio resistors have been adjusted to their individual limits of error, the error in the ratio will probably be larger than the error of each resistor. For this reason, bridge ratio arms are often adjusted for a specified error in ratio, maintaining only a nominal resistance value.

[CHARLES E. APPLEGATE]

Word processing

Word processing (WP) is a term commonly used to describe a system for accomplishing office work through people, procedures, and technologically advanced equipment. Originally limited to improved production of written communications generated from dictation in an office, WP has expanded to encompass all business communications generated by an organization, including the entire spectrum of office work. Although still commonly used, the term, which in the United States dates back to the early 1960s, fails to reflect this comprehensive expansion into the storage, retrieval, manipulation, and distribution of information and, most recently, a growing interdependence with electronic data processing. As a result, new descriptive terms such as information processing, administrative systems, and office systems are being used.

WP represents a further stage in modern society's application of automation, reaching beyond manufacturing and production lines into the office. "Automatic typewriters"—the generic predecessors of multifunction information processors—were the starting point. Now, a wide range of advanced equipment is available to electronically record keystrokes in electronic memories and on a variety of magnetic storage media—tape, card, cassette, or diskette—and electronically assist in correcting, editing, revising, and manipulating text material (Fig. 1). Retyping is limited to changes and corrections; perfect copy is automatically produced by pressing a button. Once stored, words, sentences, and paragraphs can be automatically rearranged, reassembled, and retyped repeatedly, and customized versions produced. Distribution of the finished work has also been affected. The effect in the office has become analogous to automation in the factory: higher production, lower costs, higher quality, and more efficient use of people.

More than equipment helps achieve these goals. A division of labor and specialization, particularly among secretaries, is generally evident in the use of support personnel. In addition, revised and updated procedures are required to take full advantage of the capabilities of increasingly sophisticated equipment, from initial dictation and keyboarding through the final stages of distribution and transmission. New management structures are developing to integrate this advanced equipment with both procedures and personnel.

WP's basic goal is to provide an economic and efficient solution to the paperwork explosion characterizing the 1960s and 1970s and to streamline the flow of information and knowledge in an orga-

Fig. 1. The IBM 6/450 Information Processor includes diskette storage, mag card reader/recorder, multilingual keyboard, and functional display. The print station at right contains a high-speed ink jet printer, automatic paper and envelope feeder, and stackers. (IBM)

nization. WP modernizes the office, which is becoming the dominant work environment for the American labor force. (The U.S. Department of Labor estimates that by 1985 white-collar workers will outnumber blue-collar workers by a ratio of 3 to 2.) Rising costs, increasing amounts of information, and growing numbers of office workers are major factors in the spread of WP, especially in business organizations and government units heavily involved in paperwork.

HISTORY AND BACKGROUND

In the late 1950s, Ulrich Steinhilper, of the International Business Machines Corporation, originated the term *textverarbeitung*, in Germany, to convey the notion of processing words systematically in a manner analogous to the systematic way electronic data processing handles numbers. In the United States, the term (translated as "word processing" rather than "text processing") was introduced as a marketing technique, first to sell dictation equipment, then in 1967 as an element of a sales campaign for the IBM Magnetic Tape/Selectric Typewriter.

The new automatic typewriters (introduced in 1964) increased secretarial production by recording typed material on a magnetic tape cartridge and limiting retyping to changes and corrections. In addition, form letters were customized by using stock codes which made it possible to insert personalized information. Not only was the secretary freed from the necessity of completely retyping material, but the final copy was automatically turned out in perfect form at four to five times the speed of an average typist.

Initially, organizations either bought or rented the new typewriters, but did not utilize them fully since only one-third of the traditional secretary's time is devoted to typing. In the late 1960s, as a response to emerging user requirements, a systems approach was advanced, concentrating typing and correspondence in centers where the machines were operated all day. The system was called word processing.

In the early 1970s, the concept was further refined. Adjustments were made for the kind of typing work (production or custom) and for the various ways of arranging work groups. As electronically stored memory capability increased through technological advances, the information processor emerged, comprising keyboard, video display, high-speed printer, and extensive storage capacity for keyboarded material.

The next evolutionary step during the mid-1970s involved the two-thirds of secretarial time spent on nontyping work. WP reached into daily record processing—file updating, list preparation, selection, scheduling, formatting, and printout of stored material—as information recorded on diskettes was called upon as needed. Word processors extracted selectively and rearranged stored information on demand. In 1976 communication features were added: word processors were directly linked so that copy could be printed out at the receiving end, thereby bypassing mail routing. In addition, the laser was used to speed printing.

APPLICATION

WP's objectives in any situation are specifically: increased productivity of managers and professionals; more responsive secretarial support for principals (managers and other administrators who originate written material); more efficient output of paper work; reduced costs; improved communications; and a better end product.

The application of WP varies in its specific characteristics as the office work in an organization varies. An integrated office system is designed in terms of how text matter is generated (longhand, shorthand, or machine dictation), length of documents, volume, and variety. A crucial variable is the type of work, whether it is oriented to production or custom environments. Production work is routine, predictable, and unvaried, whereas custom work is complex, unpredictable, constantly changing, and flexible. Typically, a tendency toward either production or custom work will represent most of an organization's information handling.

WP's application can be viewed in terms of the three basic elements of equipment, people, and procedures. Cumulative experience has demonstrated that neglect of any one of these elements defeats the purposes of WP.

Equipment. Generally speaking, the office is the last place where organizations have applied new technology to improve production and lower costs. This is widely cited as a major factor in low office productivity. While industrial productivity jumped by almost 90% in the past decade, productivity increased less than 5% in the office, the most labor-intensive segment of the American economy. Capital investment reflects this gap: the capital investment supporting the typical manufacturing worker is reported to be eight to ten times the investment supporting the typical office worker. In recent years, growth in sales of WP equipment reflects the increased interest in improving office productivity as labor costs have mounted steadily. Studies by the Dartnell Institute of Business Research, for example, report that more than half of the average cost of a business letter is attributed to the costs of originator's and secretary's time.

WP equipment aims at improving productivity in these two fundamental phases of office work: input and output. Input is provided by the principal who originates the ideas, words, and information for transmittal by word in written or spoken form. Output involves the production of written material and its distribution.

Input. Input originates by hand-held microphone, by built-in microphone, and by telephone activation from an outside location. Dictation equipment ranges from battery-powered portable units to centralized systems serving a large number of originators. Cartridges or tape cassettes have a capacity ranging from 6 min to more than 2 hr to record material for playback and transcription. Department of Labor statistics show that dictating to a machine is two to three time faster than dictating to a secretary and six times faster than writing longhand. Yet, many offices still commonly use face-to-face dictation and longhand to originate material. The reluctance of originators to use dictation equipment is a significant barrier to full achievement of WP's productivity potential. Overall, dictation equipment can cut by more than 25% the time needed to get material out compared with face-to-face dictation or writing in longhand.

Output. WP equipment, which is manufactured by many companies, has had its greatest impact on output, the preparation and distribution of the entire range of typed and printed communication in an organization. Electronic storage capability makes possible the storage of keyboarded material for correction and change, without time-consuming retyping. Word spacing, margins, and formatting are done automatically. Using the capability to compare, information processors select, qualify, rearrange, or place stored material in sequence, then automatically turn out printed material. The increasingly popular feature of keyboard displays enables the user to make corrections on a video screen without the use of any paper until a final draft emerges.

Equipment varies according to the amount of material which is stored magnetically and the speed with which it is automatically played out. With a printer (either as a component of an information processor or operating separately), the speed of printing, for example, can reach 1800 characters per second compared with 15.5 charac-

ters per second provided by a magnetic card typewriter. With optical character recognition (OCR) equipment, higher-function devices can store and edit typewritten material for high-speed output by specialized printers.

Whether producing individual copies, producing multiple print sets, or distributing material electronically, WP is characterized by increasing speed of production. Versatility, as well, is exemplified by an office information distributor, sometimes referred to as an "intelligent" copier, which prints with a laser and receives and transmits documents electronically over ordinary telephone lines. It also links WP with data processing, using customized formats to print typewriterlike originals of computer-based information.

Copiers routinely handle a variety of documents, drawings, photographs, and texts of thick books, and can reproduce color. Copier output has reached a speed of a page per half second, with costs reduced to pennies per page. Facsimile machines can operate as remote output printers of information transmitted electronically. In addition,

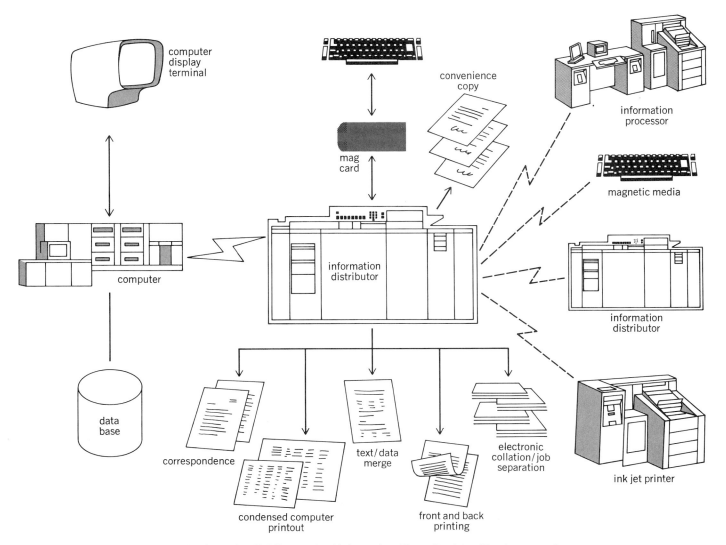

Fig. 2. The multipurpose IBM 6670 Information Distributor represents a new step toward the office of the future through its ability to access and process a wide variety of information. The unit prints with a laser, receives and transmits documents electronically, processes text and data, and can also make convenience copies. *(IBM)*

WP equipment anywhere can be linked directly for electronic communication and can be linked to central computers for access to their data bases. Meanwhile, WP has begun to go beyond paper into micrographics, by which information is stored on microfilm and microfiche for automatic retrieval.

People. Secretaries are directly affected by WP as their work is divided, specialization introduced, and their job status raised. A noteworthy result is the opening up of managerial opportunities in supervising WP. For the organization, these changes in secretarial assignments result in time, labor, and money savings, as well as increased opportunities for the individual.

A common way of reorganizing an office to achieve full WP potential involves the division of secretaries into correspondence or administrative specialists. This constitutes a move away from the traditional system in which the secretary reports directly to one or more principals for whom the secretary performs a full range of duties from typing to making travel plans. In the shift from such an arrangement, an adjustment period has been necessary for both principals and secretaries. While the particular forms of specialization vary considerably in organizations, the development of WP management is universally required. Not only does equipment change; so do work relationships, duties, secretarial responsibilites, and chains of command.

Secretarial specialization in correspondence and administration is only one of many possibilities. In a consolidated system, administrative secretaries can specialize even further by concentrating on specific office duties, such as answering the phone, scheduling, or making travel plans. In such an arrangement, all secretaries report to secretarial managers rather than to the principals they support. Typically, correspondence secretaries work in WP centers concentrated in one or more office locations. In a further variation, a widely dispersed secretarial force with their WP equipment distributed near end-user locations is coordinated by a computer to manage, control, and handle work flow. This arrangement eliminates the need for a WP center. As new equipment and more integrated applications become available, even more variations will develop.

Procedures. WP required procedures to determine the way office work is most efficiently produced by people interfacing with equipment. The step-by-step movement of information from input through output must be handled in uniform ways. This encompasses formatting, labeling, storing, and retrieving, as well as manipulation of material that is generated. Since procedures must be tailor-made for specific situations, no particular procedure is standard, but the need for these procedures is universal.

TRENDS

"Office of the future," a term being used with increasing frequency, reflects the broad implications and potentials of WP. It takes into account the full range of functions in the office and of technology currently available, as well as that anticipated in the near future. It points up the disappearance of the boundary between WP and data processing. The office of the future will link computers, information processors, work groups in correspondence and administration, varieties of printers, copiers, and telephone equipment (Fig. 2).

A key to further expansion will be the opportunities for low-cost electronic document distribution via broadband satellite communication, which will provide inexpensive transmission of the electrically coded information generated by WP.

Meanwhile, rapidly improving technology continues to narrow the gap between originators and equipment, and increase the flexibility and capacity of WP equipment. Managers already can use a WP video display to review calendars and schedules, to file references and correspondence, and to issue instructions by using keyboard on their own desk. Tomorrow, letters may be dictated directly to a voice-activated typing device, while intelligent copiers produce hard copy output directly from information processors and computers. In the projected mixture of technology and systems in the office, both WP and data processing may well blend into information processing.

[EDWARD W. GORE, JR.]

Work function (electronics)

A quantity with the dimensions of energy which determines the thermionic emission of a solid at a given temperature. The thermionic electron current density J emitted by the surface of a hot conductor at a temperature T is given by the Richardson formula, $J = AT^2 e^{-\phi/kT}$, where A is a constant, k is Boltzmann's constant ($= 1.38 \times 10^{-23}$ joule per degree Celsius) and ϕ is the work function; the last may be determined from a plot of log (J/T^2) versus $1/T$. For metals, ϕ may also be determined by measuring the photoemission as a function of the frequency of the incident electromagnetic radiation; ϕ is then equal to the minimum (threshold) frequency for which electon emission is observed times Planck's constant h ($= 6.63 \times 10^{-34}$ joule sec). The work function of a solid is usually expressed in electronvolts (1 eV is the energy gained by an electron as it passes through a potential difference of 1 volt, and is equal to 1.60×10^{-19} joule). A list of average values of work functions (in electronvolts) for metals is given in the table.

The work function of metals varies from one crystal plane to another and also varies slightly with temperature (approximately 10^{-4} eV/degree). For a metal, the work function has a simple interpretation. At absolute zero, the energy of the most energetic electrons in a metal is referred to as the Fermi energy; the work function of a metal is then equal to the energy required to raise an electron with the Fermi energy to the energy level corresponding to an electron at rest in vacuum. The

Average values of work functions for metals, in electronvolts

Metal	Value	Metal	Value	Metal	Value
Al	4.20	Cs	1.93	Na	2.28
Ag	4.46	Cu	4.45	Ni	4.96
Au	4.89	Fe	4.44	Pd	4.98
Ba	2.51	K	2.22	Pt	5.36
Cd	4.10	Li	2.48	Ta	4.13
Co	4.41	Mg	3.67	W	4.54
Cr	4.60	Mo	4.24	Zn	4.29

work function of a semiconductor or an insulator has the same interpretation, but in these materials the Fermi level is in general not occupied by electrons and thus has a more abstract meaning. *See* FIELD EMISSION; PHOTOEMISSION; THERMIONIC EMISSION. [ADRIANUS J. DEKKER]

Y-delta transformations

Electrically equivalent networks with three terminals, one being connected internally by a Y configuration and the other being connected internally by a Δ configuration.

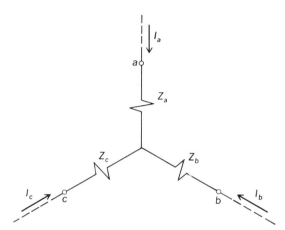

Fig. 1. Three-terminal star, or Y.

Before considering these networks, consider the simpler two-terminal network. If the path consists of a network, however complicated, of passive, linear elements, the input impedance is the ratio of the transform or phasor of input voltage to the transform or phasor of entering current. It is usually not difficult to find the input impedance of any passive, linear, two-terminal network, commonly by series and parallel combination of impedances. The input impedance will in general be a function of frequency. At one particular frequency it is possible to determine a single element to have the same impedance, and this element can be said to be equivalent to the given network at that frequency, in the sense that the

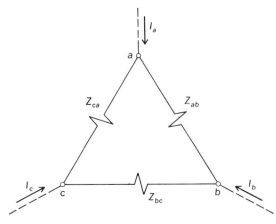

Fig. 2. Three-terminal mesh, or Δ.

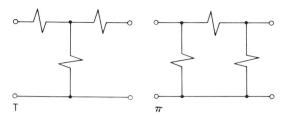

Fig. 3. Three-terminal star and mesh drawn as T and π.

terminal relations of voltage and current will be the same (at that frequency) for either. In certain particular networks the equivalence is valid at all frequencies.

If a network connects three terminals with one another, it is called a three-terminal network. The simplest configurations of three-terminal networks are the Y or star (Fig. 1) and the Δ or mesh (Fig. 2). [The star of three elements may also be called a T, and the mesh of three elements a π (Fig. 3)].

If a given three-terminal network, passive and linear, has a Y configuration, it is possible to determine an equivalent Δ network that could be substituted for the Y without changing the relations of voltage and current at the network terminals, or elsewhere external to the network. Similarly, if a Δ network is given, an equivalent Y network can be found. Impedances of equivalent networks are usually functions of frequency, and realization of these impedances by use of physically possible elements is usually limited to a single frequency.

Derivation. If terminal-to-terminal voltages V_{ab}, V_{bc}, and V_{ca} are taken to be equal in the Y and Δ of Figs. 1 and 2, then the terminal currents I_a, I_b, and I_c are equal in the two networks if and only if the impedance relations in Eqs. (1)–(3) are satisfied.

$$Z_a = \frac{Z_{ab} Z_{ca}}{Z_{ab} + Z_{bc} + Z_{ca}} \tag{1}$$

$$Z_b = \frac{Z_{bc} Z_{ab}}{Z_{ab} + Z_{bc} + Z_{ca}} \tag{2}$$

$$Z_c = \frac{Z_{ca} Z_{bc}}{Z_{ab} + Z_{bc} + Z_{ca}} \tag{3}$$

These are derived by expressing the three currents of each configuration in terms of the three voltages, equating, and solving simultaneously. Only one of these three equations is needed, for the subscripts a, b, and c are arbitrary (Fig. 4).

The same equivalence can be expressed in admittances instead of impedances by Eq. (4). Again,

$$Y_a = \frac{Y_{ab} Y_{bc} + Y_{bc} Y_{ca} + Y_{ca} Y_{ab}}{Y_{bc}} \tag{4}$$

only one equation is needed, for the subscripts are arbitrary.

To change Y to Δ, a simultaneous solution of Eqs. (1), (2), and (3) gives a set of relationships, Eqs. (5)–(7), to be used if the impedances or admit-

$$Z_{ab} = \frac{Z_a Z_b + Z_b Z_c + Z_c Z_a}{Z_c} \tag{5}$$

tances of a Y are known and those of an equivalent Δ are wanted.

For admittances, the equivalence is expressed by Eq. (6).

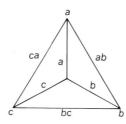

Fig. 4.   Diagram to show the pattern of subscripts used in equations.

$$Y_{ab} = \frac{Y_a Y_b}{Y_a + Y_b + Y_c} \qquad (6)$$

Example. The following example, though rather artificial, shows several interesting aspects of equivalence. Given the Δ (Fig. 5), with two sides purely resistive and the other side purely inductive, to find an equivalent Y. Given values are 20 ohms in each of two sides, and $L = 5$ henrys or $Z = j\omega 5$ ohms in the third side. From Eq. (1), relation (7) can be written. Similarly, from Eq. (2), relation (8) can be written. In this example $Z_c = Z_b$.

$$Z_a = \frac{(20)(20)}{20 + 20 + j\omega 5} = \frac{400}{40 + j\omega 5} \text{ ohms} \qquad (7)$$

$$Z_b = \frac{(20)(j\omega 5)}{20 + 20 + j\omega 5} = \frac{j\omega 100}{40 + j\omega 5} \text{ ohms} \qquad (8)$$

The impedances of the equivalent Y can be computed at any particular frequency; for example, at 60 cycles per second, or 60 Hz, $\omega = 377$. At low frequency, Z_a approaches 10 ohms of resistance, but for all finite frequencies it is somewhat less than 10 ohms and is more or less capacitive. Z_b and Z_c are equal because of symmetry in the original Δ network; at low frequency they have low inductive values, approaching zero, whereas at high frequencies each approaches 20 ohms of resistance.

At any particular frequency, each of the Y impedances can be physically realized as resistance in series with either a capacitance or an induc-

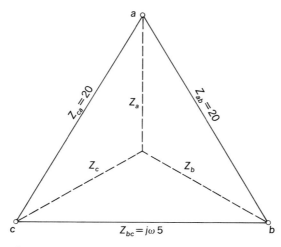

Fig. 5.   Example of a Y that is equivalent to a given Δ.

tance, but physical realization is not simple over any range of frequencies.

Difficulty with physical realization, however, does not invalidate the mathematical equivalence, and the Y network that has here been determined can be substituted for the given Δ network for all analytical or computational purposes. *See* ALTERNATING-CURRENT CIRCUIT THEORY; NETWORK THEORY.   [HUGH H. SKILLING]

Bibliography: H. H. Skilling, *Electric Networks*, 1974; D. Tuttle, *Circuits*, 1977.

Yagi-Uda antenna

A combination of a single driven antenna and a closely coupled parasitic element which may function either as a reflector as a result of inductive reactance or as a director as a result of capacitive reactance, depending on both the length and spacing of the parasitic element; also called the Yagi antenna (see illustration). Such structures are not only feasible but have a rather important place in antenna practice and concept, particularly in very-high-frequency and ultra-high-frequency ranges. Since wave propagation velocity from one element to the next is somewhat less than the velocity characteristic of free space, the power flow of the

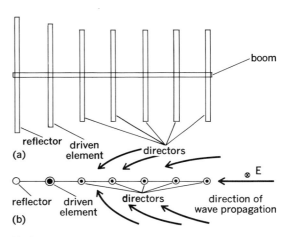

Yagi-Uda antenna. (a) View from above. (b) Side view; arrow shows energy flow close to parasitic elements.

Poynting vector tends to fall toward the array in the receiving case (illustration b).

Wide-band characteristics can be obtained by adding reactance, using thick elements, or using proper coupling to transmission line. The gain over a half-wave dipole of a Yagi-Uda antenna is shown in the table.

An array antenna of this kind was invented by H. Yagi and S. Uda at Tohoku Imperial University (now called Tohoku University) in Japan in 1926

Gain of Yagi-Uda antenna over a half-wave dipole

Antenna type	Gain, dB
Two-element Yagi-Uda antenna	$3 \sim 4.5$
Three-element Yagi-Uda antenna	$6 \sim 8$
Four-element Yagi-Uda antenna	$7 \sim 10$
Five-element Yagi-Uda antenna	$9 \sim 11$

and was reported first in Japanese in 1926 and 1927 by Uda. In 1928 it was described in English, by Yagi. *See* ANTENNA.

[KUNITAKA ARIMURA]

Zener diode

A two-terminal semiconductor junction device with a very sharp voltage breakdown in the reverse-bias region. This device is used principally in voltage regulator circuits to provide a voltage reference. It is named after C. Zener, who first proposed electronic tunneling as the mechanism of electrical breakdown in insulators. *See* SEMICONDUCTOR; VOLTAGE REGULATOR.

Most semiconductor diode applications make use of the normal *pn*-junction rectification characteristic for the functions of rectification, switching, or mixing. In the Zener diode, the normal rectifying characteristic of the junction is of no interest. The electrical breakdown in the normally blocking polarity of the junction is the critical characteristic. This breakdown must be sharp enough and sufficiently temperature-insensitive to provide a good reference voltage for voltage regulator circuits to be of use. For silicon diodes with breakdown voltages of 6 V or less, the Zener mechanism of breakdown is operative and provides excellent voltage reference diodes. Above about 6 V, the breakdown mechanism involves avalanching instead of tunneling, and the breakdown characteristic begins to show more temperature sensitivity and becomes less sharp. Up to about 15 V, the characteristic is still reasonably suitable for voltage reference use, and all such diodes today are called Zener diodes regardless of the actual mechanism of breakdown. *See* JUNCTION DIODE; TUNNELING IN SOLIDS.

[LLOYD P. HUNTER]

Bibliography: L. P. Hunter, *Handbook of Semiconductor Electronics*, 3d ed., 1970; J. Millman, *Micro-Electronics*, 1979.

CONTRIBUTORS

CONTRIBUTORS

A

Adler, Dr. David. *Department of Electrical Engineering, Massachusetts Institute of Technology.* GLASS SWITCH—coauthored.

Agerwala, Dr. Tilak. *Manager, Architecture and System Design, IBM T. J. Watson Research Center, Yorktown Heights, New York.* DATA FLOW SYSTEMS.

Allan, Roger. *Senior Editor, Electronic Design Magazine, Rochelle Park, New Jersey.* EMBEDDED SYSTEMS.

Allen, Edward W. *Formerly, Chief Engineer, Federal Communications Commission.* BANDWIDTH REQUIREMENTS (COMMUNICATIONS).

Alley, Prof. Charles L. *Department of Electrical Engineering, University of Utah.* AMPLITUDE MODULATOR; DETECTOR; DISCRIMINATOR; FREQUENCY MULTIPLIER; RADIO-FREQUENCY AMPLIFIER; other articles.

Apen, John R. *Bell Laboratories, Norcross, Georgia.* COAXIAL CABLE; COMMUNICATION CABLES—in part.

Apker, Dr. L. *General Electric Research Laboratory, Schenectady, New York.* PHOTOEMISSION; PHOTOVOLTAIC EFFECT.

Applegate, Charles E. *Consulting Engineer, Weston, Massachusetts.* ELECTRICAL RESISTANCE; KELVIN BRIDGE; OHM'S LAW; RESISTANCE MEASUREMENT; other articles.

Arbuckle, Earl F., III. *Engineering Supervisor, WPIX, Inc., New York City.* TELEVISION CAMERA; TELEVISION TRANSMITTER.

Arimura, Dr. Kunitaka. *Consultant in Electronics, Washington, D.C.* YAGI-UDA ANTENNA.

B

Bagley, Dr. Brian G. *Bell Laboratories, Murray Hill, New Jersey.* AMORPHOUS SOLID.

Balabanian, Prof. Norman. *Department of Electrical and Computer Engineering, Syracuse University.* ELECTRIC FILTER.

Barna, Dr. Gabriel G. *Central Research Laboratories, Texas Instruments Inc., Dallas.* ELECTROCHROMIC DISPLAYS.

Barrett, Prof. Harrison H. *Optical Sciences Center, University of Arizona.* IMAGE PROCESSING.

Beams, Dr. Jesse. *Deceased; formerly, Department of Physics, University of Virginia.* AMMETER; CURRENT MEASUREMENT—both validated.

Belrose, Dr. John S. *Communications Research Centre, Department of Communications, Ottawa.* FREQUENCY-MODULATION RADIO.

Berger, Dr. France B. *Kearfott Division, Singer Company, Little Falls, New Jersey.* DOPPLER RADAR—in part.

Bergh, Dr. A. A. *Bell Telephone Laboratories, Murray Hill, New Jersey.* LIGHT-EMITTING DIODE.

Berglund, Neil. *Manager of Technology Development, Intel Corporation, Aloha, Oregon.* INTEGRATED CIRCUITS—in part.

Bernstein, Dr. Robert I. *Professor of Electrical Engineering and Associate Director, Electronics Research Laboratories, Columbia University.* ECHO BOX; MONOPULSE RADAR.

Bewley, Dr. Loyal V. *Dean (retired), College of Engineering, Lehigh University.* ELECTRIC TRANSIENT.

Birkhoff, Prof. Garrett. *Department of Mathematics, Harvard University.* BOOLEAN ALGEBRA.

Black, Dr. Harold S. *Communications Consultant, Summit, New Jersey.* AMPLITUDE MODULATION; CARRIER; ELECTRICAL SHIELDING; MULTIPLEXING; SIDEBAND; other articles.

Blatt, Prof. Frank J. *Department of Physics, Michigan State University.* HALL EFFECT.

Boast, Dr. Warren B. *Anson Marston Distinguished Professor Emeritus of Electrical Engineering, Iowa State University.* LIGHT PANEL.

Booth, Dr. Grayce M. *Manager, Special Projects, Marketing Support Operations, Honeywell Informations Systems, Inc., Phoenix,* DISTRIBUTED PROCESSING; MULTIPROCESSING.

Borrello, Sebastian R. *Central Research Laboratories, Texas Instruments Inc., Dallas.* PHOTOCONDUCTIVE CELL.

Breazeale, Prof. M. A. *Department of Physics and Astronomy, University of Tennessee.* TRANSDUCER.

Brown, Dr. Glenn H. *Professor of Chemistry and Director of the Liquid Crystal Institute, Kent State University.* LIQUID CRYSTALS.

Bube, Dr. Richard H. *Professor of Electrical Engineering, Department of Material Sciences, Stanford University.* PHOTOCONDUCTIVITY.

Burghard, Ron. *Intel Corporation, Aloha, Oregon.* INTEGRATED CIRCUITS—coauthored.

Bursky, Dave. *Senior Editor on Semiconductors, "Electronic Design," Hayden Publishing Company, Inc., Sunnyvale, California.* SEMICONDUCTOR MEMORIES.

C

Callaway, Prof. Joseph. *Department of Physics and Astronomy, Louisiana State University.* BAND THEORY OF SOLIDS; HOLES IN SOLIDS.

Cameron, Donald B. *Vice President of Technology, Battery Products Division, Union Carbide Corporation, New York City.* SOLION.

Carroll, Dr. John M. *Associate Professor of Computer Science, University of Western Ontario.* CONTINUOUS-WAVE RADAR; ELECTRONIC LISTENING DEVICES; MOVING-TARGET INDICATION.

Casey, Dr. H. C., Jr. *Department of Electrical Engineering, Duke University.* SEMICONDUCTOR HETEROSTRUCTURES—in part.

Chaffin, Dr. R. J. *Solid State Device Physics, Sandia National Laboratories, Albuquerque.* HIGH-TEMPERATURE ELECTRONICS.

Chapman, Dr. Richard A. *Central Research Laboratories, Texas Instruments Inc., Dallas.* PHOTOELECTRIC DEVICES.

Chaudhari, Praveen. *Thomas J. Watson Research Center, IBM. Yorktown Heights, New York.* MAGNETIC THIN FILMS.

Chin, Dr. Gilbert Y. *Bell Telephone Laboratories, Murray Hill, New Jersey.* FERRITE DEVICES.

Clarke, Prof. John. *Department of Physics, University of California, Berkeley.* SQUID; SUPERCONDUCTING DEVICES.

Cleland, J. W. *Solid State Division, Oak Ridge National Laboratory.* NEUTRON TRANSMUTATION DOPING.

Clemence, Dr. Gerald M. *Deceased; formerly, Observatory, Yale University.* QUARTZ CLOCK.

Cobine, Dr. James D. *Professor and Senior Scientist, Atmospheric Sciences Research Center, State University of New York at Albany.* GAS TUBE; THYRATRON; VOLTAGE-REGULATOR TUBE.

Cohn, Stanley H. *Department of Industrial Engineering, University of Toronto.* ELECTROMAGNETIC COMPATIBILITY.

Collier, R. J. *Bell Telephone Laboratories, Murray Hill, New Jersey.* MAGNETRON.

Collins, Dr. Dean R. *Director, CCD Technology Laboratory, Central Research Laboratories, Texas Instruments Inc., Dallas.* LIGHT AMPLIFIER.

Collins, Prof. J. H. *Department of Electrical Engineering, University of Texas, Arlington.* SURFACE ACOUSTIC-WAVE DEVICES.

Compton, Robert D. *Electro-Optical Systems Design, Milton S. Kiver Publications, Inc., Chicago.* OPTICAL ISOLATOR.

Conwell, Dr. Esther M. *Xerox Webster Research Center, Rochester, New York.* INTEGRATED OPTICS.

Corson, Almon J. *Consulting Engineer (retired), Instrument Department, General Electric Company, West Lynn, Massachusetts.* POTENTIOMETER (VOLTAGE METER); VOLTMETER.

Curdts, Edward B. *Consultant; Director of Engineering (retired), James G. Biddle Company, Philadelphia.* FREQUENCY MEASUREMENT—in part.

Cutler, Prof. C. C. *Professor of Applied Physics, Stanford University.* VIDEO TELEPHONE.

D

Date, Prof. H. *Acoustics Department, Kyushu Institute of Design, Fukuoka, Japan.* VIDEO DISK RECORDING.

Davis, Rowland F. *Products Planning Engineer (retired), American Telephone and Telegraph Company, New York City.* TELEPHONE.

de Boor, Dr. Carl. *Mathematical Research Center, University of Wisconsin.* NUMERICAL ANALYSIS.

Dekker, Dr. Adrianus J. *Professor of Solid State Physics, University of Groningen, Netherlands.* CONTACT POTENTIAL DIFFERENCE; ELECTRON EMISSION; SCHOTTKY EFFECT; THERMIONIC EMISSION; WORK FUNCTION (ELECTRONICS).

de Satnick, Steve. *Vice President, Operations and Engineering, KCET-TV, Los Angeles.* TELEVISION.

Dickieson, A. C. *Vice President (retired), Bell Telephone Laboratories, Sedonia, Arizona.* ELECTRICAL COMMUNICATIONS.

Duke, Dr. C. B. *Manager, Molecular and Organic Materials Area, Xerox Corporation, Rochester, New York.* SURFACE PHYSICS.

Duncan, Charles C. *International Communications Consultant, Manhasset, New York.* PRIVACY SYSTEMS (SCRAMBLING); TELEPHOTOGRAPHY.

E

El-Mansy, Youssef. *Intel Corporation, Aloha, Oregon.* INTEGRATED CIRCUITS—coauthored.

Engstrom, Dr. Ralph W. *Electro Optics and Devices, RCA Laboratories, Lancaster, Pennsylvania.* PHOTOMULTIPLIER.

Enslow, Prof. Philip H., Jr. *School of Information and Computer Science, Georgia Institute of Technology.* MULTIACCESS COMPUTER.

Esaki, Leo. *Thomas J. Watson Research Center, IBM, Yorktown Heights, New York.* TUNNELING IN SOLIDS.

F

Fan, Prof. H. Y. *Department of Physics, Purdue University.* ACCEPTOR ATOM; CONDUCTION BAND; DONOR ATOM; SEMICONDUCTOR; VALENCE BAND.

Fink, Donald G. *Director Emeritus, IEEE, Editor in Chief, "Electronics Engineers' Handbook," McGraw-Hill Book Company, New York City.* AUDION; ELECTRONICS; TELEVISION SCANNING.

Fiscarelli, A. *Transmission Performance, American Telephone and Telegraph Company, New York City.* CROSSTALK.

Foster, Dr. Mark G. *Department of Electrical Engineering, School of Engineering and Applied Science, University of Virginia.* HARMONIC ANALYZER; HIGH-FREQUENCY IMPEDANCE MEASUREMENTS; SPECTRUM ANALYZER.

G

Galler, Dr. Bernard A. *Computing Center, University of Michigan.* COMPUTER.

Galloway, Dr. William J. *Bolt, Beranek and Newman, Inc., Canoga Park, California.* FREQUENCY (WAVE MOTION); PHASE (PERIODIC PHENOMENA).

Galt, Dr. J. K. *Vice President, Sandia Laboratories, Albuquerque.* OPTICAL DETECTORS.

Gannon, Dr. J. D. *Department of Computer Science, University of Maryland.* PROGRAMMING LANGUAGES—coauthored.

Gewartowski, James W. *Bell Telephone Laboratories, Murray Hill, New Jersey.* BACKWARD-WAVE TUBE; TRAVELING-WAVE TUBE.

Gibbs, Hyatt M. *Optical Sciences Center, University of Arizona.* OPTICAL BISTABILITY.

Gifford, Richard P. *Deceased; formerly, General Electric Company, Lynchburg, Virginia.* MOBILE RADIO.

Gilbert, Dr. Edgar N. *Bell Telephone Laboratories, Murray Hill, New Jersey.* INFORMATION THEORY.

Ginzton, Prof. Edward L. *Chairman of the Board and formerly President, Varian Associates.* PULSE MODULATOR.

Glasford, Prof. Glenn M. *Department of Electrical Engineering, Syracuse University.* BLOCKING OSCILLATOR; CLAMPING CIRCUIT; COINCIDENCE AMPLIFIER; DELAY LINE; PULSE GENERATOR; other articles.

Gloge, Dr. Detlef C. *Crawford Hill Laboratory, Bell Telephone, Holmdel, New Jersey.* OPTICAL COMMUNICATIONS.

Golden, F. B. *Semiconductor Products Department, General Electric Company, Auburn, New York.* SEMICONDUCTOR RECTIFIER.

Gomer, Prof. Robert. *James Franck Institute, University of Chicago.* FIELD EMISSION.

Goodheart, Prof. Clarence F. *Department of Electrical Engineering, Union College.* CIRCUIT (ELECTRICITY); OPEN CIRCUIT; PARALLEL CIRCUIT; POTENTIOMETER (VARIABLE RESISTOR).

Goodstein, David H. *Director, Inter/Consult, Cambridge, Massachusetts.* ELECTRONIC PUBLISHING.

Gordon, Dr. J. P. *Bell Telephone Laboratories, Holmdel, New Jersey.* MASER—coauthored.

Gore, Edward W., Jr. *Vice President, International Market Requirements, Office Products Division, IBM, Franklin Lakes, New Jersey.* WORD PROCESSING.

Gränicher, Prof. H. *Laboratory of Solid State Physics, Swiss Federal Institute of Technology, Zurich.* BARIUM TITANATE; PIEZOELECTRICITY; ROCHELLE SALT.

Gregory, Dr. Bob L. *Sandia Laboratories, Albuquerque.* INTEGRATED CIRCUITS—in part.

Gross, Prof. Jonathan L. *Depertment of Mathematics, Columbia University.* GRAPH THEORY.

Gruenberger, Prof. Fred J. *Department of Computer Science, California State University.* DIGITAL COMPUTER PROGRAMMING.

Guidry, Dr. Mark R. *Fairchild Camera and Instrument Corporation. Mountain View, California.* CHARGE-COUPLED DEVICES.

H

Hailpern, Dr. Brent T. *IBM Research Laboratory, Yorktown Heights, New York.* CONCURRENT PROCESSING.

Halkias, Prof. Christos C. *Chair of Electronics, National Technical University, Athens, Greece.* BIAS OF TRANSISTORS; DIRECT-COUPLED AMPLIFIER; EMITTER FOLLOWER; LOAD LINE; PULSE TRANSFORMERS; other articles.

Harris, Prof. J. Donald. *U.S. Navy Submarine Medical Laboratory. Groton, Connecticut.* EARPHONES.

Harsh, M. Duffield, *Engineering Leader, Display Tube Product Development, RCA Electronic Components, Lancaster, Pennsylvania.* CATHODE-RAY TUBE.

Hartson, Prof. H. Rex. *Department of Computer Science, Virginia Polytechnic Institute and State University.* COMPUTER SECURITY.

Hayes-Roth, Dr. Frederick. *Executive Vice President, Technology, Teknowledge Inc. Palo Alto, California.* EXPERT SYSTEMS.

Helms, Harry L. *Professional and Reference Book Division, McGraw-Hill Book Company, New York City.* MICROCOMPUTER; MICROCOMPUTER DEVELOPMENT SYSTEM; OPERATING SYSTEM.

Herrell, Dr. Dennis J. *Thomas J. Watson Research Center, Yorktown Heights, New York.* SUPERCONDUCTING COMPUTERS.

Hill, Robert T. *U.S. Navy, Bowie, Maryland.* RADAR.

Hines, Dr. M. E. *Vice President, Microwave Associates, Inc., Burlington, Massachusetts.* PARAMETRIC AMPLIFIER.

Hodgin, David M. *Senior Member, Institute of Electrical and Electronics Engineers, Cedar Rapids, Iowa.* SINGLE SIDEBAND.

Hoffman, Dr. Alan J. *IBM Corporation, Yorktown Heights, New York.* LINEAR PROGRAMMING.

Holdeman, Dr. Louis B. *National Bureau of Standards.* JOSEPHSON EFFECT.

Holst, Per A. *Foxboro Company, Foxboro, Massachusetts.* ANALOG COMPUTER.

Holtje, Malcolm C. *General Radio Company, West Concord, Massachusetts.* DIGITAL VOLTMETER; VOLTAGE MEASUREMENT.

Horzepa, Joseph J. *Marketing Director, Visual Communications Services, American Telephone and Telegraph Company, Morristown, New Jersey.* CLOSED-CIRCUIT TELEVISION.

Hoxton, Prof. Llewellyn G. *Deceased; formerly, Professor Emeritus of Physics, University of Virginia.* JOULE'S LAW.

Huber, Dr. Robert J. *Electrical Engineering Department, University of Utah.* STATIC INDUCTION TRANSISTOR.

Hunter, Prof. Lloyd P. *Department of Electrical Engineering, University of Rochester.* CONTROLLED RECTIFIER; CRYOSAR; JUNCTION DIODE; TRANSISTOR; VARACTOR; other articles.

Huskey, Dr. Velma R. *Information Sciences, University of California, Santa Cruz.* ABACUS.

J

Jacobs, Prof. Stephen F. *Optical Sciences Center, University of Arizona.* LASER—coauthored.

Jen, Dr. C. K. *Supervisor, Microwave Physics Group, Applied Physics Laboratory, Johns Hopkins University.* CAVITY RESONATOR; DIRECTIONAL COUPLER; MICROWAVE; MICROWAVE OPTICS; WAVEGUIDE.

Joel, Amos E., Jr. *Bell Telephone Laboratories, Holmdel, New Jersey.* SWITCHING SYSTEMS (COMMUNICATIONS).

Johnson, Prof. Walter C. *Department of Electrical Engineering, Princeton University.* ELECTROMAGNETIC WAVE TRANSMISSION.

Jones, Dr. Edwin C., Jr. *Department of Electrical and Computer Engineering, Iowa State University.* THEVENIN'S THEOREM (ELECTRIC NETWORKS).

Jory, Howard R. *Varian Associates Palo Alto, California.* GYROTRON.

Juliussen, Dr. J. Egil. *Senior Member of Technical Staff, Texas Instruments Inc. Dallas.* MAGNETIC BUBBLE MEMORY.

K

Kaminow, Dr. Ivan P. *Bell Telephone Laboratories, Holmdel, New Jersey.* OPTICAL MODULATORS.

Kanal, Dr. Laveen N. *Department of Computer Science, University of Maryland.* CHARACTER RECOGNITION.

Kanzig, Prof. Werner. *Department of Physics, Massachusetts Institute of Technology.* FERROELECTRICS.

Kaufmann, R. H. *Consultant, Schenectady, New York.* ELECTRIC EQUIPMENT GROUNDING.

Kent, Dr. Ernest. *National Bureau of Standards.* ROBOTICS.

Kiehl, Dr. Richard A. *Solid State Device Physics, Division 5133, Sandia Laboratories, Albuquerque.* MICROWAVE SOLID-STATE DEVICES.

Kimbark, Dr. Edward W. *Head, Systems Analysis Group, Bonneville Power Administration, Portland, Oregon.* NEPER.

Kinch, Dr. Michael A. *Central Research Laboratories, Texas Instruments Inc., Dallas.* PHOTOELECTRICITY.

King, Dr. Roger. *Computer Science Department, University of Colorado.* ABSTRACT DATA TYPES.

Kinnard, Dr. Isaac F. *Decreased; formerly, Manager of Engineering, Instrument Department, General Electric Company, Lynn, Massachusetts.* ELECTRICAL MEASUREMENTS—in part; MULTIMETER; Q METER; VOLT-OHM-MILLIAMMETER; WAVEFORM DETERMINATION.

Klass, Philip J. *Senior Avionics Editor, "Aviation Week and Space Technology," McGraw-Hill Publications Company, Washington, D.C.* PASSIVE RADAR.

Klick, Dr. Clifford C. *Superintendent, Solid State Division, U.S. Naval Research Laboratory.* CATHODOLUMINESCENCE; ELECTROLUMINESCENCE—both in part.

Klock, Dr. Harold F. *Department of Electrical Engineering, Ohio University.* AUDIO AMPLIFIER; BRIDGE CIRCUIT; INTERMEDIATE-FREQUENCY AMPLIFIERS; PHASE INVERTER; PREAMPLIFIER; other articles.

Knowles, Hugh S. *President, Knowles Electronic, Inc., Franklin Park, Illinois.* LOUDSPEAKER.

Kotter, Dr. F. Ralph. *National Bureau of Standards.* CAPACITANCE MEASUREMENT.

L

Land, Dr. Cecil E. *Sandia Laboratories, Kirtland Air Force Base East, Albuquerque.* PHOTOFERROELECTRIC IMAGING.

Landes, Dr. Hugh S. *Department of Electrical Engineering, University of Virginia.* RADIO TRANSMITTER.

Lanford, Prof. William. *Physics Department, State University of New York at Albany.* SOFT FAILS.

Langenberg, Prof. D. N. *Department of Physics, University of Pennsylvania.* SUPERCONDUCTIVITY.

Laport, Edmund A. *Director, Communications Engineering, RCA, Princeton, New Jersey.* AMPLITUDE-MODULATION RADIO.

Lea, Dr. Wayne A. *Research Engineer, Speech Communications Research Laboratory, University of Southern California.* SPEECH RECOGNITION; VOICE RESPONSE.

Lee, L. K. *Assistant Director, Product Assurance, TRW Systems, Redondo Beach, California.* PRINTED CIRCUIT.

Lee, Prof. Samuel C. *Department of Electrical Engineering, University of Oklahoma.* MICROPROCESSOR.

Lehmer, Dr. Derrick H. *Department of Mathematics, University of California, Berkeley.* NUMBER SYSTEMS.

Le Page, Dr. Wilbur R. *Department of Electrical Engineering, Syracuse University.* CHOKE (ELECTRICITY); SUPPRESSOR; NOISE FILTER (RADIO).

Lesk, I. A. *Director, Central Research Laboratories, Motorola, Inc., Phoenix.* VARISTOR.

Lesso, Dr. William G. *Department of Mechanical Engineering, University of Texas, Austin.* OPERATIONS RESEARCH.

Lewis, Frank D. *James Millen Manufacturing Company. Malden, Massachusetts.* FREQUENCY COUNTER; FREQUENCY MEASUREMENT—in part; PIEZOELECTRIC CRYSTAL; WAVELENGTH MEASUREMENTS.

Lewis, Dr. Willard D. *Bell Telephone Laboratories, Murray Hill, New Jersey.* SWITCHING THEORY.

Lhermitte, Dr. Roger. *Professor of Physical Meteorology, University of Miami School of Marine and Atmospheric Science.* DOPPLER RADAR—in part.

Long, Dr. Stephen I. *Rockwell International Electronics Research Center, Thousand Oaks, California.* GALLIUM ARSENIDE INTEGRATED CIRCUITS.

Lyons, Prof. Walter. *Consultant in Telecommunications, Flushing, New York.* AUTOMATIC FREQUENCY CONTROL (AFC); AUTOMATIC GAIN CONTROL (AGC); RADIO RECEIVER; SIGNAL-TO-NOISE RATIO.

m

MacCarthy, Donnell D. *Manager, Voltage Regulator Engineering, Voltage Regulator Product Section, General Electric Company, Pittsfield, Massachusetts.* VOLTAGE REGULATOR.

McConnell, Ken R. *Manager of Engineering, Litton, Melville, New York.* FACSIMILE.

McFarlane, Robert. *Philips Laboratories, Briarcliff, New York.* DIGITAL OPTICAL DISK RECORDER.

McLeod, John H., Jr. *Consultant, La Jolla, California.* SIMULATION.

Mahoney, Daniel P. *Supervising Engineer, AT&T Long Lines, Somerset, New Jersey.* COMMUNICATION CABLES.

Maisel, Dr. Herbert. *Computer Science Program, Georgetown University.* DATA-PROCESSING SYSTEMS.

Manning, Dr. Kenneth V. *Professor Emeritus, Pennsylvania State University.* EDDY CURRENT; FARADAY'S LAW OF INDUCTION; INDUCTANCE; RELUCTANCE.

Markus, John. *Consultant, Sunnyvale, California.* ANTENNA; BREADBOARDING; COIL; ELECTRICAL LOADING; MICROWAVE REFLECTOMETER; STANDING-WAVE DETECTOR; other articles.

Matyas, Stephen M. *IBM Systems Communications Division, Kingston, New York.* CRYPTOGRAPHY.

Mellichamp, Prof. Duncan A. *Department of Chemical Engineering, University of California, Santa Barbara.* DIGITAL CONTROL.

Meszar, John. *Director (retired), Switching Systems Development, Bell Laboratories, Inc., New York City.* COUNTING CIRCUIT; SWITCHING CIRCUIT.

Meyer, Dr. Carl H. *Advisory Engineer, IBM Systems Communications Division, Kingston, New York.* CRYPTOGRAPHY—coauthored.

Meyer, Dr. John F. *Department of Computer and Communications Sciences, University of Michigan.* FAULT-TOLERANT SYSTEMS.

Miller, Barry. *Senior Editor, "Aviation Week and Space Technology," McGraw-Hill Publications Company, Los Angeles.* ELECTRONIC COUNTERMEASURES.

Miller, Dr. Glenn H. *Weapons Effects Division, Sandia Laboratories, Albuquerque.* ARC DISCHARGE; BREAKDOWN POTENTIAL; CATHODE RAYS; ELECTRICAL CONDUCTION IN GASES; SATURATION CURRENT; other articles.

Miller, John H. *Deceased; formerly, Vice President and Chief Engineer, Weston Instrument Division, Dayton, Ohio.* AMMETER; CURRENT MEASUREMENT—in part.

Millman, Dr. Jacob. *Department of Electrical Engineering, Columbia University.* PARASITIC OSCILLATION; SIGNAL GENERATOR.

Mills, Thomas B. *National Semiconductor Corporation, Santa Clara, California.* PHASE-LOCKED LOOPS.

Mitchell, Dr. Kim W. *Solar Energy Research Institute, Golden, Colorado.* SOLAR CELL—in part.

Moder, Dr. Joseph J. *Chairman, Department of Management Science, University of Miami.* PERT.

Mokhoff, Dr. Nicolas. *IEEE Spectrum, New York.* CONSUMER ELECTRONICS.

Moreno, Dr. Theodore. *Vice President, Information Systems Group, Varian Associates, Palo Alto, California.* MICROWAVE TRANSMISSION LINES.

Morrell, A. M. *Manager, Tube Development, Picture Tube Division, RCA Corporation, Lancaster, Pennsylvania.* COLOR TELEVISION.

Moyer, Dr. W. W. *Applied Research Laboratory, Pennsylvania State University.* CALCULATORS.

Myers, Ware. *Temple City, California.* SOFTWARE ENGINEERING.

n

Nelson, Prof. Raymond J. *Professor of Mathematics and Philosophy, Case Institute of Technology.* DATA REDUCTION.

Nelson, Dr. Richard B. *Chief Engineer, Varian Associates, Palo Alto, California.* KLYSTRON.

Nergaard, Dr. Leon S. *Director, Microwave Research Laboratory, RCA Laboratories, Princeton, New Jersey.* THERMIONIC TUBE.

Netravali, Arun N. *Bell Laboratories, Holmdel, New Jersey.* BANDWIDTH REDUCTION.

Neuhauser, Robert G. *Solid State Division, Electro Optics and Devices, RCA, Lancaster, Pennsylvania.* TELEVISION CAMERA TUBE.

Newell, Dr. Allen. *Department of Computer Science, Carnegie-Mellon University.* ARTIFICIAL INTELLIGENCE.

Nichols, Richard B. *Vice President, AT&T Long Lines, Morris Plains, New Jersey.* PRIVACY SYSTEMS (SCRAMBLING).

Niemoeller, Dr. Arthur F. *Professional Training, Research Laboratories, Central Institute for the Deaf, Saint Louis.* HEARING AID.

Nygren, Stephen. *Bell Telephone Laboratories, Reading, Pennsylvania.* SEMICONDUCTOR DIODE.

o

Olson, Dr. Harry F. *Formerly, Staff Vice President, Acoustical and Electromechanical Research, RCA Laboratories, Princeton, New Jersey.* AUDIO DELAYER; DISK RECORDING—in part; ELECTROACOUSTICS; FREQUENCY-RESPONSE EQUALIZATION; MICROPHONE; VOLUME CONTROL SYSTEMS.

Ovshinsky, Stanford R. *President, Energy Conversion Devices, Inc., Troy, Michigan.* GLASS SWITCH—coauthored.

P

Patel, Dr. C. K. N. *Bell Laboratories, Murray Hill, New Jersey*. PHOTOACOUSTIC SPECTROSCOPY.

Patrick, Norman W. *RCA Corporation, Lancaster, Pennsylvania*. STORAGE TUBE.

Paul, Prof. Igor. *Department of Mechanical Engineering, Massachusetts Institute of Technology*. DECISION THEORY; OPTIMIZATION.

Phister, Montgomery, Jr. *Consultant, Santa Fe*. DIGITAL COMPUTER.

Picraux, Dr. Samuel T. *Supervisor, Ion-Solid Interactions, Sandia Laboratories, Albuquerque*. ION IMPLANTATION—in part.

Preece, Dr. Carolyn M. *Bell Laboratories, Murray Hill, New Jersey*. ION IMPLANTATION—in part.

Pritchett, Prof. Wilson S. *Senior Project Engineer, Noller Control Systems, Inc., Richmond, California*. CAPACITOR; INDUCTOR; NONSINUSOIDAL WAVEFORM—all coauthored.

Pritsker, Prof. A. Alan B. *School of Industrial Engineering, Purdue University*. GERT.

Pullen, Dr. Keats A. *Ballistic Research Laboratories, Aberdeen Proving Ground, Maryland*. CIRCUIT (ELECTRONICS); TRANSFORMER—in part.

R

Ramberg, Dr. Edward G. *RCA Laboratories, Princeton, New Jersey*. ELECTRON LENS; ELECTRON MOTION IN VACUUM; LANGMUIR-CHILD LAW; SPACE CHARGE; other articles.

Ramey, Dr. Robert L. *Department of Electrical Engineering, University of Virginia*. DIRECT-CURRENT CIRCUIT THEORY; RESISTOR; TIME CONSTANT; WAVEMETER; other articles.

Reich, Prof. Herbert J. *(Retired) Department of Engineering and Applied Science, Yale University*. DIODE; ELECTRON TUBE; NEGATIVE-RESISTANCE CIRCUITS; VACUUM TUBE.

Reilly, Dr. Norman B. *Consultant, Computer Systems, Compata, Inc., Tarzana, California*. AUTOMATA THEORY.

Robb, D. D. *D. D. Robb and Associates, Consulting Engineers, Electric Power Systems, Salina, Kansas*. DIRECT CURRENT.

Robertson, Prof. Burtis L. *Professor of Electrical Engineering (retired), University of California, Berkeley*. ADMITTANCE; CAPACITOR—coauthored; ELECTRICAL IMPEDANCE; NONSINUSOIDAL WAVEFORM—coauthored; other articles.

Robinson, W. V. *Processor Technology Research Group, Bell Laboratories, Whippany, New Jersey*. LOGIC CIRCUIT—coauthored.

Rochkind, Mark M. *Bell Laboratories, Holmdel, New Jersey*. DATA COMMUNICATIONS.

Rockett, Frank H. *Engineering Consultant, Charlottesville, Virginia*. BREAKDOWN; CHOPPER; ELECTRICAL INSTABILITY; MICROWAVE TUBE; other articles.

Root, Dr. William L. *Department of Aerospace Engineering, University of Michigan*. ELECTRICAL NOISE; ELECTRICAL NOISE GENERATOR.

Rosenbaum, Prof. Fred J. *Department of Electrical Engineering, Washington University*. GYRATOR.

S

Samek, Michael J. *Director, Management Services, Celanese Corporation, New York City*. DATA-PROCESSING MANAGEMENT.

Schawlow, Prof. Arthur L. *Department of Physics, Stanford University*. LASER—coauthored.

Schelleng, J. C. *(Retired) Bell Telephone Laboratories*. CORNER REFLECTOR ANTENNA.

Schueler, Dr. Donald G. *Sandia Laboratories, Albuquerque*. SOLAR CELL—in part.

Schulman, Dr. James H. *U.S. Naval Research Laboratory* CATHODOLUMINESCENCE; ELECTROLUMINESCENCE—both coauthored.

Seager, Dr. C. H. *Sandia Laboratories, Albuquerque*. GRAIN BOUNDARIES.

Seifert, Dr. William W. *Professor of Electrical Engineering and Associate Director, Commodity Transportation and Economic Development Laboratory, Massachusetts Institute of Technology*. SLIDE RULE.

Sessler, Dr. Gerhard M. *Institut für Elektroakustik, Technische Hochschule, Darmstadt, West Germany*. ELECTRET TRANSDUCER.

Sheingold, Daniel H. *Manager of Technical Marketing, Analog Devices, Inc., Norwood, Massachusetts*. ANALOG-TO-DIGITAL CONVERTER; DIGITAL-TO-ANALOG CONVERTER.

Shively, R. R. *Supervisor, Processor Technology Research Group, Bell Laboratories, Whippany, New Jersey*. LOGIC CIRCUITS—coauthored.

Sibley, Dr. Edgar H. *Department of Information Systems Management, University of Maryland*. DATA-BASE MANAGEMENT SYSTEMS.

Siegman, Dr. A. E. *Stanford, California*. QUANTUM ELECTRONICS.

Sinclair, Dr. Donald B. *President (retired), General Radio Company, Concord, Massachusetts*. ELECTRICAL MEASUREMENTS—in part.

Singleton, John D. *Supervisor of Programming, Radio Division, National Broadcasting Company, New York City*. RADIO.

Sinnett, Chester M. *Deceased; formerly, RCA, Camden, New Jersey*. TELEVISION RECEIVER.

Sittner, Dr. W. R. *President, Electro Medical Systems, Inc*. PHOTODIODE; PHOTOTRANSISTOR.

Skilling, Prof. H. H. *Department of Electrical Engineering, Stanford University*. ALTERNATING CURRENT; COUPLED CIRCUITS; NETWORK THEORY; Y-DELTA TRANSFORMATIONS; other articles.

Skilling, James K. *General Radio Company, West Concord, Massachusetts*. SAMPLING VOLTMETER.

Smith, C. Price. *Manager, Power and Electro Optics Products, RCA Corporation, Lancaster, Pennsylvania*. PICTURE TUBE.

Smith, Fred W. *Western Union Telegraph Company, Mahwah, New Jersey*. TELETYPEWRITER.

Snow, William W. *Consulting Engineer, William W. Snow Associates, Inc., Woodside, New York*. EXCITATION; FREQUENCY DIVIDER.

Sohon, Dr. Harry. *Deceased; formerly, Moore School, University of Pennsylvania*. INDUCTANCE BRIDGE; INDUCTANCE MEASUREMENT; PHASE-ANGLE MEASUREMENT.

Sommer, Dr. Alfred H. *Thermo Electron Corporation, Waltham, Massachusetts*. SECONDARY EMISSION.

Soss, David A. *Chief Engineer, Instrumentation Research Laboratory, Electrical Engineering Department, University of Utah*. ELECTRONIC POWER SUPPLY.

Sterzer, Dr. Fred. *RCA Laboratories, Princeton, New Jersey*. MICROWAVE SOLID-STATE DEVICES.

Stevenson, Dr. Edward C. *Department of Electrical Engineering, School of Engineering and Applied Sciences, University of Virginia*. ELECTRICAL MEASUREMENTS; INDUCTANCE BRIDGE; MULTIMETER; WAVEFORM DETERMINATION; other articles—all validated.

Stewart, Dr. John W. *Department of Physics, University of Virginia*. CONDUCTION (ELECTRICITY); ELECTRIC CURRENT; IR DROP; JOULE'S LAW.

Stillman, Dr. Gregory E. *Department of Electrical Engineering, University of Illinois, Urbana*. PHOTOVOLTAIC CELL.

Sullivan, John. *Lee Allan Associates, Sunnyvale, California*. ELECTRIC UNINTERRUPTIBLE POWER SYSTEM.

Sutherland, J. R. *Power Transformer Department, General Electric Company, Pittsfield, Massachusetts.* TRANSFORMER—in part.

T

Tamura, Dr. Masahiko. *Acoustical Engineering Research Laboratory, Pioneer Electronic Corporation, Saitama, Japan.* HIGH-POLYMER TRANSDUCER.

Tang, Prof. K. Y. *Deceased; formerly, Department of Electrical Engineering, Ohio State University.* KIRCHHOFF'S LAWS OF ELECTRIC CIRCUITS.

Tannas, Lawrence E., Jr. *Aerojet Electro Systems Company, Azusa, California.* ELECTRONIC DISPLAY.

Tapia, Dr. Richard A. *Department of Mathematical Sciences, Rice University.* NONLINEAR PROGRAMMING.

Theis, Dr. Douglas J. *Computer Systems Department, Aerospace Corporation, Los Angeles.* COMPUTER STORAGE TECHNOLOGY.

Thiess, Helmut E. *Washington, D.C.* BIT.

Townes, Prof. Charles H. *Department of Physics, University of California, Berkeley.* MASER—coauthored.

Tuttle, Dr. W. Norris. *Consultant, Concord, Massachusetts.* CURRENT MEASUREMENT—in part; VACUUM-TUBE VOLTMETER.

U

Uhlig, Dr. Ronald P. *Bell Northern Research, Ottawa.* VIDEOTEXT AND TELETEXT.

V

Van Valkenburg, Dr. M. E. *Department of Electrical Engineering, University of Illinois, Urbana.* SWITCHED CAPACITOR.

Vollum, Dr. Howard. *Tektronix, Inc., Portland, Oregon.* OSCILLOSCOPE.

W

Waidelich, Dr. Donald L. *Department of Electrical Engineering, University of Missouri.* ELECTRONIC POWER SUPPLY—in part; RECTIFIER; RIPPLE VOLTAGE; WAVEFORM; other articles.

Waldron, Dr. Robert D. *Director, Research Enterprises, Scottsdale, Arizona.* ELECTRET; ELECTROSTRICTION.

Walsh, Dr. Walter M. *Bell Laboratories, Murray Hill, New Jersey.* CYCLOTRON RESONANCE EXPERIMENTS.

Ward, John E. *Electronics Systems Laboratory, Massachusetts Institute of Technology.* COMPUTER GRAPHICS.

Weaver, James L. *Lockheed Missiles and Space Company, Palo Alto, California.* PHOTOTUBE.

Weil, Robert T. *Deceased; formerly Dean, School of Engineering, Manhattan College.* KIRCHHOFF'S LAWS OF ELECTRIC CIRCUITS.

Weiser, Dr. M. D. *Department of Computer Science, University of Maryland.* PROGRAMMING LANGUAGES—coauthored.

Welch, Arthur A. *Engineering Consultant, Purcellville, Virginia.* CONDUCTANCE.

Wentworth, John W. *Director, Educational Development Engineering, RCA, Cherry Hill, New Jersey.* COLOR TELEVISION—in part; TELEVISION RECEIVER—validated.

Williams, Prof. Everard M. *Deceased; formerly, Department of Electrical Engineering, Carnegie-Mellon University.* TRANSMISSION LINES.

Williams, Perry F. *American Radio Relay League, Inc., Newington, Connecticut.* AMATEUR RADIO.

Williams, S. A. *Department of Physics, Iowa State University.* AMPLITUDE (WAVE MOTION).

Winch, Prof. Ralph P. *Department of Physics, Williams College.* CAPACITANCE; ELECTRIC CHARGE; ELECTROMOTIVE FORCE (EMF).

Wolf, Dr. P. *IBM Zurich Research Laboratory, Switzerland.* CRYOTRON.

Wolfe, Roger W. *Plainfield Semiconductor Operation, Burroughs Corporation, Plainfield, New Jersey.* NUMERICAL INDICATOR TUBE.

Wood, Dr. R. F. *Research Materials Program, Solid State Division, Oak Ridge National Laboratory.* NEUTRON TRANSMUTATION DOPING—coauthored.

Woodall, Dr. Jerry M. *IBM Watson Research Center, Yorktown Heights, New York.* SEMICONDUCTOR HETEROSTRUCTURES—in part.

Woodward, Dr. J. G. *David Sarnoff Research Center, RCA Laboratories, Princeton, New Jersey.* DISK RECORDING; MAGNETIC RECORDING.

Wright, Michael E. *National Semiconductor Corporation, Santa Clara, California.* DIGITAL COUNTER.

Y

Yamamoto, Dr. Takeo. *Acoustical Engineering Research Laboratory, Pioneer Electronic Corporation, Saitama, Japan.* HIGH-POLYMER TRANSDUCER.

Yang, Dr. Edward S. *Department of Electrical Engineering, Columbia University.* OSCILLATOR.

Yankee, Prof. Herbert W. *Mechanical Engineering Department, Worcester Polytechnic Institute.* COMPUTER-AIDED DESIGN AND MANUFACTURING.

Yelles, Marvin. *Formerly, Editor, "McGraw-Hill Encyclopedia of Science and Technology," McGraw-Hill Book Company, New York City.* OCTAL NUMBER SYSTEM.

Z

Ziegler, Dr. James. *Watson Research Laboratory, IBM, Yorktown Heights, New York.* SOFT FAILS—coauthored.

Zissis, Dr. George J. *Chief Scientist, Environmental Research Institute of Michigan, Ann Arbor.* INFRARED IMAGING DEVICES.

INDEX

INDEX

Asterisks indicate page references to article titles.

Abacus 1*
Abdank-Abakanowicz 43
Absorption wavemeter 922–923
Abstract boolean algebra 88
Abstract data types 1–4*
  classification 2
  data structures 2
  files and data bases 4
  lists 2–3
  primitive 2
  queues 2–3
  relation to algorithms 4
  sets 3–4
  stacks 2–3
  trees and graphs 3
ac *see* Alternating current
ac Josephson effect 436
ac potentiometer: Drysdale 634–635
  Tinsley-Gall 635
Accelerometer, high-polymer 405
Acceptor atom 4*
Acoustooptic optical modulators 577–578
A/D converter *see* Analog-to-digital converter
ADA (programming language) 163, 331–332, 660
  ADA Integrated Environment 332
  ADA Language System 331
Adaptive moving-target indication 530–531
Adjustable capacitors 98
Adjustable inductors 413
Adjustable resistors 710
Adleman, L. 181
Admittance 4*
  alternating-current circuit theory 13*
  conductance 164*
  susceptance 802*
ADT *see* Abstract data types
af amplifier *see* Audio-frequency amplifier
AFC *see* Automatic frequency control
AGC *see* Automatic gain control

Aggregate data types (programming) 649–650
  arrays 649–650
  records 650
Air capacitor 98
Algebraic coding theory (information theory) 416–417
Algebraic notation: calculator entry mode 92
ALGOL 657
ALGOL 68 656
Algorithms 4–6*
  artificial intelligence 5–6
  cryptographic 178–179
  Data Encryption Standard 154
  digital control 223–224
  hardware implementation 5
  programming languages 645–661*
  properties 5
  relation to abstract data types 4
  RSA 181–182
Alloy-junction transistors 443
Alphageometric display technology 894
Alphamosaic display technology 894
Alphanumeric electronic display 306
Alphaphotographic display technology 894
Alternating current 6–9*
  advantages 6
  circuit theory 9–17
  measurement 7, 186
  phase difference 7
  power factor 7–8
  reactance 696*
  rectifier 699–702*
  resistance measurement 705
  sinusoidal form 6–7
  three-phase system 8–9
  voltage measurement 906
Alternating-current circuit theory 9–17*
  admittance 4*
  alternating voltage and current 11–14
  circuits 117
  complex notation 10–11
  devices and models 14–15
  electric transients 271–272
  electrical impedance 278*
  examples 15–16

Alternating-current circuit theory —*cont.*
  nonsinusoidal waveform 551–553*
  phasors 9–10
  power 14
  resonance 710–712*
  shunting 747*
  susceptance 802*
  Thévenin's theorem 853–855*
  units 15
Alternating-current impedance bridge: inductance bridge 411–412*
AM *see* Amplitude modulation
Amateur radio 17–18*
  associations 18
  privileges 17
  public service 17
  regulations 17
  requirements 17
  technical developments 17–18
Ammeter 7, 18–21*
  direct-current measurement 186
  electrodynamic type 19
  permanent-magnet movable-coil type 19
  polarized-vane type 19
  resistance measurement 705–706
  soft-iron type 19–20
  taut-band type 19
  thermal type 20–21
Amorphous semiconductors 729
  glass switch 378–380*
Amorphous solid 21–22*
  glass switch 378–380*
  preparation 21
  semiconductors 21–22
  types 21
  uses 21
Ampere 185
Ampère, André Marie 185
Amplifier 22–30*
  amplitude modulation 41–42
  audio amplifier 74*
  audion 76*
  automatic gain control 79*
  backward-wave amplifier 80–81
  cascade 25–26
  cascode amplifier 100*
  coincidence amplifier 132*
  crossover network 177*
  differential 27–28

Amplifier—*cont.*
  direct-coupled amplifier 231–233*
  electrical instability 278*
  electrical noise generator 284*
  feedback 340–341
  frequency distortion 248
  gain 369–370
  hearing aid 394–395*
  intermediate-frequency amplifier 432*
  klystron 447–450
  laser 451
  linear circuits 424
  linearity 463
  maser 490–493*
  nonlinear 367–368
  operational 28–30
  parametric amplifier 593–594*
  paraphase 601
  parasitic oscillation 594
  power *see* Power amplifier
  preamplifier 636*
  push-pull amplifier 672–673*
  radio-frequency *see* Radio-frequency amplifier
  radio receiver 690–691
  regeneration 702*
  response 712*
  signal response 26
  tank circuit 825
  transistor 23–24
  traveling-wave tube 873
  undesirable conditions 26–27
  vacuum-tube 24–27
  video amplifier 889*
  voltage amplifier 900–905*
  volume control systems 915–916*
Amplifier-type voltmeter 887
Amplitude (wave motion) 30–31*
Amplitude modulation 31–34*, 275, 526–527
  detector *see* Amplitude-modulation detector
  microwave transmission 511
  mobile radio 524
  modulator 39
  modulator and demodulator 33–34
  radio *see* Amplitude-modulation radio
  radio transmitters 693–694
  single-sideband modulation 31–32

Amplitude modulation—cont.
uses in multiplexing 32–33
vestigial-sideband modulation 32
Amplitude-modulation detector 34–36*
diode detectors 34–35
regenerative detectors 36
square-law detectors 35–36
synchronous detectors 36
Amplitude-modulation radio 36–38*, 687
aviation and marine navigation aids 38
high-frequency (shortwave) 37
low-frequency (longwave) 37
medium-frequency 37
single-sideband hf telephony 38
telephony and telegraphy 37–38
television broadcasting 38
Amplitude modulator 33–34, 39–42*, 529
basic requirements 39
high-level modulation 39–41
low-level modulation 41–42
multiplier modulators 42
Amplitude resonance 710
Analog computer 42–57*, 144
applications 45
components 46–49
description and uses 44–46
digital equivalents 42–43
digital multiprocessor analog system 43
history 43–44
hybrid computers 55–57
inverse programming operations 50–52
linear computing units 46–48
multipliers 48–49
nonlinear computing elements 48
operation 55
operational amplifier accessories 47–48
operational amplifiers 28, 46–47
programming 49–55
programming for use as simulator 54–55
programming for use with calculus 53–54
programming symbols 50
programming to solve algebraic equations 52–53
representation of programming variables 50
signal-controlled programming 49
slide rule 752–753*
switched capacitor operations 805
types 42
unique features 45–46
Analog simulation 749
Analog switching systems (communications) 819
Analog-to-digital converter 57–59*
concepts and structure 57–58
data reduction 204
digital control 221
digital disk recording 246–247
hybrid computers 57

Analog-to-digital converter —cont.
techniques 58–59
AND circuit see AND gate
AND gate 211, 806
AND logic function 468
Anderson, P. W. 436
Anderson bridge 411
Angle modulation 59–60*, 527–528
frequency modulation see Frequency modulation
phase modulation 603–604*
Antenna 60–69*
AM radio transmitters 694
array antennas 68–69
bandwidth 64–65
corner reflector antenna 170*
direct-aperture type 66–67
efficiency 62
electrically small antennas 65
electromagnetic radiation from 289–290
electromagnetism 278
frequency-independent type 66
gain 370
impedance 63–64
microwave 511
monopulse radar 529
nonresonant antennas 65–66
polarization 62–63
radar systems, 676, 683–684
radiation mechanism 60–61
radiation pattern shape 61–62
radio receiver 693
radome 695–696*
reflector type 67
resonant antennas 65
thermal noise 283
transmitting for television 849–850
two-reflector antennas 67–68
Yagi-Uda antenna 930–931*
Antiferroelectric crystal 347–348
Antiresonance 70*
APL (programming language) 658–659
Appel, K. 384
Applicative programming languages 655
Arc discharge 70–71*
arc production 70
regions of an arc 70–71
Armstrong, E. H. 359
Armstrong, E. L. 275
Array antennas 68–69
Artificial intelligence 71–74*
algorithms 5–6
automata 78
examples 72–73
expert systems, 73, 334–337*
foundations 71–72
game programs 72
perception by computers 72–73
real-time expert systems 697
robotics 712–715*
scope and implications 73–74
voice response 895–899*
ASCR see Asymmetrical silicon-controlled rectifier
Assembler (microcomputer) 495
Assembler language (programming) 645–646
Astable blocking oscillator 87

Astable multivibrator 537
Astable negative-resistance circuits 541–542
Asymmetrical silicon-controlled rectifier 746
Asymmetrical triode thyristor 745
Asynchronous concurrent system 163
Atmospheric optical communications 572
Attenuator 74*
Audio amplifier 74*
automatic gain control 79
power amplifier (class B) 636
Audio delayer 74–76*
applications 76
digital system 75–76
loudspeaker-pipe-microphone system 74–75
magnetic tape recorder-reproducer system 75
Audio-frequency amplifier 25
see also Audio amplifier
Audio-frequency meters 353–357
audio-frequency bridge methods 355–356
direct-comparison methods 356
frequency counters 356–357
moving-coil meters 354
moving-iron meters 354–355
reed-type frequency meters 353–354
Audio magnetic recorders 482–484
cartridge 482–483
cassette 483
digital 483
reel-to-reel 483
sound quality 483
Audio transformers 861–862
distortion 862
rf and i-f transformers 862
Audion 76*
Audiotape cartridge 482–483
Audiotape cassette 483
Automata theory 77–78*
abstract theory 77
artificial intelligence 78
game theory 78
language theory 77
pattern recognition 78
self-organizing systems 78
structural theory 77–78
Turing machines 77
Automated tape library 162
Automatic frequency control (AFC) 79*
Automatic gain control (AGC) 79*
monochrome television receiver 843–844
Automatic volume control see Automatic gain control
Avalanche diodes 517–518
Axially symmetric electrostatic lenses 329–330

Backward-wave tube 79–81*
backward-wave amplifier 80–81
M-type backward-wave oscillator 80
O-type backward-wave oscillator 80
Balanced modulator 42
Ballistic galvanometer:
capacitance measurement 97
Band-pass filter 261
ladder filter 263
surface acoustic-wave 794–795
Band theory of solids 81–82*
effective mass 81–82
photoemission by semiconductors 615
transitions between states 82
Bandwidth reduction 82–83*
conditional replenishment 83
digital transmission 82
motion-compensated coding schemes 83
techniques 82–83
Bandwidth requirements (communications) 83–84*
Bardeen, John 327, 730, 783, 878
Bardeen-Cooper-Schrieffer theory 783, 789–790
Barium titanate 84–85*
piezoelectric crystal material 625
piezoelectricity 628
Barrier layer (semiconductor) 729
BASIC 657–658
Batch-oriented data communications 193
Battery: electronic power supply 317
uninterruptible power system 273
Battery charger: uninterruptible power system 273
Beam power tetrode 883
Beat-frequency oscillator see Heterodyne oscillator
Becquerel, A. C. 766
Bell, A. G. 607
Bellman, Richard 566
Berge, C. 386
BFL see Buffered FET logic
Bias of transistors 85–86*
collector-to-base bias 85
fixed-bias circuit 85
self-bias 85–86
Bidirectional clamp 125
Bidirectional diode thyristor 745
Bidirectional triode thyristor 745
Binary element see Bit
Binary number system 555–556
Binary relay circuit 171–172
Bipolar integrated circuits 421–424
linear circuits 424
semiconductor devices 424
Bipolar inverter circuit 421
Bipolar logic circuit 422
Bipolar microwave transistors 516
Bipolar RAM chips 157
Bipolar transistor 862
bistable multivibrator 535
clipping 127–128

Backus, John 656
Backward diode 734
Backward-wave amplifier 80–81

Bipolar transistor—*cont.*
 connections 867–868
 limiter circuit 461
 logic circuits 471
Bipotential electrostatic lenses
  329–330
Biquad filter 264
Bistable multivibrator: sym-
  metrical 534–535
  unsymmetrical 535–536
Bistable negative-resistance
  circuits 541
Bistable optical devices 569–
  571
 Fabry-Perot interferometers
  569–570
 properties 570–571
Bit 86*
Bit-rate compression *see*
  Bandwidth reduction
Bit-sliced microprocessors 508
Block chaining 182–183
Block ciphers 182–183
Block encryption 182
 encoders 414
Blocking oscillator 86–87*
 free-running astable type 87
 relaxation oscillator 702
 synchronized 87
Book-type ammeter 19
Boole, George 87
Boolean algebra 87–89*
 abstract relationships 88
 data types 648
 forms 89
 infinite relationships 88–89
 set-theoretic interpretation 87–
  88
Boolean rings 88
Bootstrap sawtooth generator 719
Boys, C. V. 43
Brattain, Walter 327
Braun, K. F. 299
Breadboarding 89*
Breakdown 89*
Breakdown diodes 231
Breakdown potential 89–90*
Bridge-circuit 90*
 audio-frequency meters 355–
  356
 capacitance measurement 95–
  97
 Kelvin bridge 445–446*
Bridge oscillator 586
Bridge rectifier circuits 701
Brooks deflectional dc
  potentiometer 634
Brush discharge 90*
Bubble memory chip 477
Buffered FET logic 371, 372
Bugs (eavesdropping) 314–315
Bulk store 155
Bush, Vannevar 43
Button-type potentiometer 633
Byte 86

C

C (programming language) 659–
  660
Cable television 129
Cache memory 155
CAD *see* Computer-aided design
  and manufacturing
CADUCEUS (expert system) 335

Calculators 90–93*
 electronic 90–92
 logic circuits 467–471*
 mechanical 90
 programmable 92
 special-purpose 92–93
Calibration 93*
CAM *see* Computer-aided design
  and manufacturing
Capacitance 93–94*
 alternating-current circuit
  theory 13
 body capacitance 94
 capacitor 98–100*
 guard ring 94
 properties of capacitors 93–94
Capacitance measurement 94–
  98*
 bridge methods 95–97
 capacitance standards 98
 distributed capacitance 97–98
 resonance method 95
 susceptance variation 95
 time-constant methods 97
Capacitor 98–100*
 air, gas, and vacuum types 98
 capacitance 93–94*
 capacitance measurement 94–
  98*
 charging and discharging 93
 classification 98
 diodes 231
 electrolytic types 99–100
 energy of charged capacitor 93
 geometrical types 93–94
 paper types 99
 solid-dielectric types 98–99
 switched capacitor 803–806*
 thick-film types 99
Carbon dioxide lasers 453
Carbon microphone 499
Carbon potentiometer 633
Carbon transducer: microphones
  497
Card punches 202
Card (punched) reader 202
Carey-Foster bridge 411–412
Carrier 100*
Carrier dc amplifier 232
Carson, John R. 359, 697, 752
Cartridge, audiotape 482–483
Cartridge disk 160
Cartridge magnetic-tape storage
  161
Cascade amplifier 25, 26
Cascode amplifier 100*
Casimir, H. B. G. 787
Cassette, audiotape 483
Cassette magnetic-tape storage
  161
Cathode lens 330
Cathode-ray tube 100–107*,
  879
 cathode rays 107*
 character-display types 106
 color 107, 134–135
 computer graphics 147–151*
 electron gun 101–103
 electronic display 304–305
 envelope 101
 flat tubes 106
 infrared image display 419
 linear sweep generation 802
 multigun tubes 106
 OCR scanner 110
 optically ported 106
 oscilloscope 587–593*

Cathode-ray tube—*cont.*
 phosphor screen 104–105
 photography 105–106
 picture tube 623–625*
 printing types 106–107
 radar displays 686
 rotating radial sweep
  generation 803
 special-purpose tubes 106–
  107
 television picture tube 844
Cathode-ray tube digital raster
  display: font 308
 pixels 308
Cathode rays 107*
Cathodoluminescence 107–108*
CATV *see* Cable television
Cavity resonator 108–109*
 cavity quality factor 108–109
 cavity resonance 108
 coaxial cavity magnetron 489
 forms 109
 klystron 449
 microwave wavemeter 511
Cayley, A. 385
CCD *see* Charge-coupled devices
Cellular mobile radio systems
  365
Central processing unit 155
 microprocessor 506–509*
Ceramic capacitor 99
Ceramic filters 264–265
Ceramic microphone 499
Ceramic transducer *see*
  Electrostrictive transducer
Chained block encryption 182
Character-display cathode-ray
  tubes 106
Character recognition 109–113*
 EDP systems 203
 optical *see* Optical character
  recognition
 pattern recognition 110
Characteristic curve 113*
Charge-coupled devices 113–
  116*
 control of charge motion 114
 input and output ports 115–
  116
 lifetime 115
 memory gap technology 158
 MOS structure 427
 soft fails 755–756
 television camera pickup
  devices 835
 television camera tube imager
  841–842
 transfer efficiency 114–115
Chemical lasers 454
Choke (electricity) 116*
Chopper 116*
Chopper amplifier 232
Chopper-stabilized dc amplifier
  232–233
Chopping 116*
Christie, S. H. 923
Church, A. 5
Cipher block chaining 183
Ciphertext 179
 stream ciphers 183
Circuit (electricity) 116–118*,
  164
 admittance 4*
 alternating-current circuit
  theory 9–17*
 alternating-current circuits
  117

Circuit (electricity)—*cont.*
 antiresonance 70
 bridge circuit 90*
 coupled circuits 117, 172–
  175*
 direct-current circuit theory
  233–235*
 direct-current circuits 117
 electric network 117
 electric transient 117, 267–
  272*
 electrical instability 278*
 electrical resistance 284*
 equivalent circuit 333*
 gyrator 386–389*
 inductance 410*
 inductance measurement 412–
  413*
 inductor 413*
 integrated circuit 118
 Kelvin bridge 445–446*
 Kirchhoff's laws of electric
  circuits 446–447*
 magnetic circuits 117
 negative-resistance circuits
  539–543*
 nonsinusoidal waveform, 117,
  551–553*
 Ohm's law 562*
 open circuit 118, 562*
 parallel circuit, 117, 593*
 parasitic oscillation 594–595*
 phase-angle measurement
  598–600*
 potentiometer (variable
  resistor) 632–633*
 Q (electricity) 673*
 resistor 709–710*
 resonance (alternating current)
  710–712*
 series circuit 117, 747*
 series-parallel circuits 117
 short circuit 118
 switching circuit 806–811*
 tank circuit 824–825*
 theory 116–117
 voltage-multiplier circuit 906–
  907*
Circuit (electronics) 118–124*
 bias of transistors 85–86*
 bipolar transistor circuits 535
 breadboarding 89*
 clamping circuit 124–126*
 clipping circuit 126–128*
 coincidence amplifier 132*
 comparator circuit 142–143*
 component ratings 122
 component specifications 120–
  124
 counting circuit 171–172*
 design 119–120
 direct-coupled amplifier 231–
  233*
 discriminator 236*
 electronic components 119
 electronic power supply 315–
  318*
 emitter follower 332–333*
 engineering design tests
  123
 environmental factors for
  operation 122
 failure tests 124
 feedback circuit 340–342*
 frequency modulator 367*
 frequency multiplier 367–
  368*

Circuit (electronics)—cont.
  frequency-response
    equalization 368–369*
  function generator 369
  functional tests 124
  gate circuit 375–376*
  IGFET circuits 535
  integrated see Integrated
    circuits
  integrated thermionic circuits
    406
  integrator circuits 804
  JFET circuits 535
  limiter circuit 460–461*
  logic circuits 467–471*
  logic gate multivibrator 538–
    539
  logical circuits 211–212
  optical isolator 579
  oscillator 583–587*
  phase inverter 601*
  phase-locked loops 601–603*
  phase modulator 604–607*
  power amplifier 635–636*
  printed circuit 636–645*
  production line tests 123–124
  pulse generator 660–661*
  push-pull amplifier 672–673*
  rectifier 699–702*
  relaxation oscillator 702–703*
  reliability 121–122, 124
  resistors 709
  scaling circuit 722*
  semiconductor rectifier 744
  soft fails 753–756*
  solion 766–767*
  specifications 121
  stabilization 122
  statistical test methods 124
  sweep generator 802–803*
  switching theory 822–824*
  tests 122–123
  transistor 862–867*
  trigger circuit 875–876*
  wave-shaping circuits 916–
    917
Circuit analyzer see Multimeter;
    Volt-ohm-milliammeter
Circuit switching (data
    communications) 194–195
Circular cylindrical cavity
    resonators 109
Circular sweep generator 802–
    803
Clamp see Clamping circuit
Clamping circuit 124–126*
  dc restorer 125
  keyed (synchronous) 125
  transmission gate 376
  triode 125–126
  voltage-amplitude-controlled
    125
Clipping circuit 126–128*
  diode 126–127
  limiter circuit 461
  transmission gate 376
  triode clipper 127–128
Closed-circuit television 128–
    130*
  applications 128
  cable television 129
  educational applications 128
  monitoring applications 128
  technical considerations 129–
    130
  theater television 128–129
  video conferencing 129

Closed-circuit television—cont.
  video telephone 129
CML see Current-mode logic
CMOS see Complementary metal-
    oxide semiconductor
Coaxial cable 130–131*
  communication cable systems
    138–141
  coupling with waveguide 920
  flexible 130
  loading 278
  rigid 131
  semirigid 130–131
  special coaxials 131
  types and uses 130–131
  waveguide 921
Coaxial cavity magnetron 489
Coaxial cavity resonators 109
Coaxial microwave transmission
    520–521
Coaxial resonators 109
Coaxial transmission lines 520
COBOL 657
Codetext 179
Coil 131–132*
Coincidence amplifier 132*
Cold-cathode diode 373
Cold-cathode gas tube 373–375
Cold-cathode triode 373–374
Collector modulation 39–41
Collector-to-base bias transistor
    circuit 85
Color cathode-ray tube 107
Color facsimile 339
Color television 132–135*
  cameras 132
  color-cathode-ray tubes 107
  color decoder 133–134
  color encoder 133
  electronic display 307
  encoding and decoding 132–
    134
  frequency interlace 133
  improvements 165–166
  picture tube 134–135, 624–
    625
  receiver see Color television
    receiver
  transmitting antenna 851
  use of primary colors 132
Color television receiver: color
    decoding circuits 846–847
  color kinescope and
    convergence circuits 847
  controls 847
  nature of color signal 845
  overall color receiver 846
Colpitts oscillator 585
Combinational switching circuit
    806–807
  switching theory 823–824
Communication cables 135–
    142*
  aluminum conductor cables
    137–138
  coaxial cable development
    140–141
  coaxial cable systems 138–
    141
  coaxial design characteristics
    140
  design 136–137
  gas pressure maintenance 138
  L-4 coaxial system 139
  L-5 coaxial system 139
  local and exchange area cable
    136

Communication cables—cont.
  special cables 136
  survivability of coaxial
    systems 141
  switchboard cable 136
  toll cable 136
  types and uses 136–138
Communications: electrical see
    Electrical communications
  electronics applications 328
  optical see Optical
    communications
Communications-jamming
    techniques 303
Community antenna television
    see Cable television
Commutable reverse-blocking
    triode thyristor 745
Compandor 178
Comparator circuit 142–143*
  analog computer programming
    49
  applications 143
  digital comparator 143
  timed circuit comparators
    143
  linear 142
Complementary
  semiconductor 471
  calculator storage 92
  logic circuit 471
Composition resistor 709
Compound semiconductor
    heterojunction devices
    406
CompuServe 324, 325
Computer 144–145*
  algorithm 4–6*
  analog see Analog computer
  artificial intelligence 71–74*
  cryotron 178*
  data communications 192
  data-processing systems 198–
    204*
  digital see Digital computer
  digital control 220–224*
  display electronic addressing
    308–309
  distributed processing 248
  embedded systems 330–332*
  expert systems 334–337*
  fault-tolerant systems 339–
    340*
  magnetic recorders 484
  microcomputer see
    Microcomputer
  microprocessor 506–509*
  multiaccess see Multiaccess
    computer
  multiprocessing 532–534*
  numerical analysis 557–560*
  operating system 562–563*
  portable 165
  programming languages 645–
    661*
  real-time systems 696–697*
  robotics 712–715*
  simulation 748–751*
  soft fails 753–756*
  software engineering 756–
    762*
  speech recognition 768–772*
  superconducting see
    Superconducting computers
  videotext and teletext 893–
    894*
  voice response 895–899*

Computer-aided design and
    manufacturing 145–147*
  CAD 145
  CAM 145
  numerical control 146–147
  programming languages 146
  vendor programs 146
Computer-compatible magnetic-
    tape units 161
Computer diodes 741
Computer graphics 147–151*
  applications 151
  graphical input 148
  graphical output 148–150
  home data retrieval 166–168
  input devices 150–151
  interactive input/output
    arrangements 150
  operating modes 147–148
  videotext and teletext 894
Computer security 151–154*
  control programs 563
  cryptographic controls 153–
    154
  cryptography 178–188*
  inference controls 153
  information flow controls
    153
  logical access controls 152–
    153
  physical and logical 152
  controls 153
Computer storage technology
    154–163*
  cartridge disks 160
  cassettes and cartridges 161–
    162
  charge-coupled devices 113–
    116*, 158
  digital optical disk recorder
    227–228*
  digital system 212
  disk file 485
  disk pack units 201–202
  electron-beam-accessed
    memories 158–159
  FET connections 868–869
  fixed-head disks 160
  floppy disks 160–161, 485
  function generator 369
  half-inch magnetic tape 161
  highly nonlinear B/H devices
    343
  logic circuits 211–212, 468
  magnetic bubble memory 159,
    477–478*
  magnetic tape units 161–162,
    202, 484–485
  main memory error checking
    and correction 157
  mass storage systems 162
  memory gap technologies
    158–159
  memory hierarchy 154–155
  memory organization 155
  MOSFET 865–866
  multiaccess system 532
  optical disks 161
  RAM chips 156–157
  ROMs, PROMs, and EPROMs
    157–158
  secondary memories 159–161
  semiconductor memories 739–
    741*
  soft fails 753–756*
  SQUID 773
  superconducting devices 782

Computer storage technology
 —cont.
 superconducting elements
  781–782
 Winchester technology 159–
  160
Computer word 86
Concurrent PASCAL 163
Concurrent processing 163–164*
 advantages and disadvantages
  163
 communication and
  synchronization 163–164
 specifying concurrency 163
Condenser see Capacitor
Condenser microphone see
 Electrostatic microphone
Conductance 164*
Conduction (electricity) 164*
Conduction band 164–165*
 valence band 888*
Connecting circuits 808
Constant-current dc
 potentiometer 633–634
Constant-current sawtooth
 generator 719
Constant-resistance dc
 potentiometer 634
Constant-voltage diodes 231
Consumer electronics 165–168*,
 328
 computers 165
 electronic games 165
 electronic publishing 322–326
 home data retrieval 166–168
 language translators 165
 television receivers 165–166
Contact barrier (semiconductor)
 730
Contact microphones 499–500
Contact noise 281
Contact potential difference 168*
Continuous-wave ion lasers 452
Continuous-wave radar 168–
 169*
 automobile safety 169
 detection of hostile targets 169
 laser radar system 169
 missile guidance 169
 surveillance of personnel 169
Control programs (operating
 systems) 562–563*
Control systems: digital control
 220–224*
 expert systems 334–337*
Controlled rectifier 169–170*
 silicon-controlled 701
Converse piezoelectric effect 627
Cooper, L. N. 783, 878
Cooper pairs 878
 Josephson effect 436–439*
Copper oxide rectifier 493
Corner reflector antenna 67,
 170*
Corona discharge 170–171*
 brush discharge 90*
Counter see Digital counter
Counting board see Abacus
Counting circuit 171–172*,
 810–811
 digital counter 225–226
 frequency counter 349–351*
 scaling circuit 722*
Counting-table abacus 1
Coupled circuits 117, 172–175*
 coefficient of coupling 173
 core loss 175

Coupled circuits—cont.
 equality of mutual inductance
  173
 equivalent circuits 173–174
 ideal transformers 173
 polarity 172
 steady-state equations 173
 transformation of impedance
  174–175
 transformers 174
 voltage equations 172–173
CPM see Critical path method
CPU see Central processing unit
Critical path method (CPM)
 175–176*
Crookes dark space 380
Crossed-field devices 489–490
Crossover network 177*
Crosstalk 177–178*
 remedies 177–178
 sources 177
 types 177
Cryosar 178*
Cryotron 178*
Cryptography 178–185*
 block ciphers 182–183
 cipher feedback 184–185
 cryptographic algorithms 178–
  179
 data encryption standard 180–
  181
 digital signatures 180
 privacy and authentication
  179–180
 RSA public-key algorithm
  181–182
 security controls 153
 stream ciphers 183–184
 strong algorithms 179
 unbreakable ciphers 179
Crystal filters 264
Crystal microphone 499
Crystal oscillator 586
Crystal transducer see
 Piezoelectric transducer
CSP (programming language) 163
Current see Electric current
Current measurement 185–186*
 ammeter 7, 18–21*
 direct and low-frequency
  currents 186
 high frequencies 186
 very small currents 186
Current-mode logic 471
CW radar see Continuous-wave
 radar
Cybernetics 74
Cyclotron resonance experiments
 187*
Cyclotron resonance maser see
 Gyrotron
Cylindrical capacitor 93–94

D

Dahlin's algorithm 224
Dantzig, George B. 462, 564
Dark current 187*
 phototube 622
Data base: files (abstract data
 types) 4
 inference controls 153
 on-line 321–322
 videotext and teletext 894–
  895

Data-base administrator 190
Data-base management systems
 187–191*
 access controls 152–153
 architecture 189–190
 cost of information systems
  188
 cryptography 178–185*
 data administration 190–191
 data dictionary 191
 data security 188–189
 design and implementation
  191
 integrity of data 189
 management control of data
  188–189
 modeling data 190
 organization of data 188
 real-time assistance 696
Data communications 191–195*
 applications 192–193
 communications processing
  193–194
 cryptography 178–185*
 data transmission 199
 digital filters 668
 digital transmission 667
 electronic publishing
  environment 319
 electronic switching systems
  821
 media conversion 193
 transmission 194–195
Data dictionary 191
Data Encryption Standard 154,
 180–181
Data flow systems 195–197*
 advantages and limitations
  197
 basic concepts 195–196
 comparison of conventional
  systems 196
 concurrent programs 163
 data dependence graphs 195–
  196
 data-driven and demand-
  driven execution 196
 static and dynamic
  architectures 196
Data-processing management
 197–198*
 demands of computer
  technology 198
 managerial control 198
 organization of people 197
 products of data processing
  197–198
Data-processing systems 198–
 204*
 character recognition 109–
  113*
 counting circuit 171–172*
 data-base management
  systems 187–191*
 data manipulation 199
 data-processing functions
  198–199
 data-processing management
  197–198*
 data recording 198–199
 data reduction 204*
 data retrieval 199
 data storage 199
 data transmission 199
 electronic see Electronic data-
  processing systems
 expert systems 334–337*

Data-processing systems
 —cont.
 multiaccess computer 531–
  532*
 punched-card systems 199–
  200
 report preparation 199
 switching circuit 806–811*
Data reduction 204*
Data retrieval, home 166
Data security: cryptography 178–
 185*
 data-base management
  systems 188–189
Data transmission 194,199
Data types (programming) 648–
 650
 aggregate types 649–650
 association with objects 651
 scalar types 648–649
 user-defined 656–660
Day's system (phase-
 discrimination multiplexing)
 32
DBA see Data-base administrator
DBMS see Data-base
 management systems
dc see Direct current
dc amplifier see Direct-coupled
 amplifier
dc Josephson effect 436
dc potentiometer: Brooks
 deflectional 634
 constant-current 633–634
 constant-resistance 634
dc restorer (clamp) 125
dc SQUID 779–781
DCFL see Direct-coupled FET
 logic
Deadbeat algorithm 224
de Broglie, L. 877
Debugging systems 335
Deception-jamming techniques
 303
Decimal number system 553–
 555
Decision making: optimization
 578–583*
Decision making under certainty
 206, 207
Decision making under risk
 206–208
Decision making under
 uncertainty 206, 208
Decision matrix 206–207
Decision theory 204–208*
 application and techniques
  206–208
 concepts and terminology
  204–206
 operations research 563–568*
 optimization 578–583*
Decision tree 205–206
Decoder: home data retrieval
 166–168
 set-top versus built-in 167
Decoding: information theory
 414–417
Decoupling filter 262
Decryption 154
Dedicated channels (data
 transmission) 194
Deflection-type frequency meters
 353
DeForest, Lee 76, 299, 327
Delay line 208–209*
 pulse generator control 661

Demodulator 209*
  phase-locked loops 601–603*
  see also Amplitude-modulation detector; Detector; Frequency-modulation detector
DENDRAL (expert system) 335
Depletion-mode FET 865
DES see Data Encryption Standard
Descriptive decision theory 204
Design systems 335
Destriau, G, 286
Detector 209–210*
  amplitude-modulation detector 34–36*
  detection fidelity 210
  frequency-modulation detector 361–363*
  microwaves 510–511
  phase-modulation detector 604*
  radar operation 676
  radio receiver 691
  types 210
Diagnosis systems 335
Dial channels (data transmission) 194
Dielectric-rod antenna 66
Dielectrics: barium titanate 84–85*
  capacitor 93, 98–100*
  cavity resonator 108–109*
  electret 257*
  electrostriction 330*
  ferroelectric properties 345–346
  piezoelectricity 626–631*
Differential amplifiers 27–28, 232, 904–905
Digital cassette transport 161
Digital comparator 143
Digital computer 144–145, 210–218*
  analog equivalents 42–43
  analog-to-digital converter 57–59*
  applications 144–145
  artificial intelligence 74
  capabilities 215–218
  cathode-ray-tube terminals 100
  codes 210
  counters 226
  counting circuit 171–172*
  digital system fundamentals 210–212
  efficiency 215–216
  electronic data-processing systems 200–204
  electronic versus mechanical computers 144
  fourth-generation systems 217
  harmonic analysis 394
  hybrid computers 55–57
  image processing 407–408
  industry growth 217–218
  logic circuits 211–212, 467–471*
  microcomputer see Microcomputer
  physical components 212
  programming see Digital computer programming
  semiconductor memories 739–741*

Digital computer—cont.
  stored program computer 144, 212–214
  switching circuit 806–811*
  switching gate 376
  system building blocks 212
  third-generation 216–217
Digital computer programming 218–220*
  addresses 218
  binary operation 218
  data flow systems 195–197*
  decimal operation 218
  flow charting 219–220
  instruction format 218–219
  languages 220
  notation 218
  number operations 218
  programming 219
  stored program computer 212–214
Digital control 220–224*
  analog control information 221–222
  computer/process interface 221–222
  control algorithms 223–224
  digital control information 221
  programming considerations 222
  real-time computing 222
Digital counter 212, 224–227*, 349–351
  applications 226
  specifications 226–227
Digital counting circuit 211
Digital delay line 209
Digital disk recording 246–247
Digital facsimile 339
Digital filter 265–266
Digital integrated circuits 422, 423
Digital multiprocessor analog computer 43
Digital optical disk recorder 227–228*
  complete recorder 228
  disk 227
  drive 227
  electronic subsystems 227–228
  error correction 228
  optical system 227
Digital recording, magnetic 481–484
Digital signatures (cryptography) 180
Digital simulation 749
Digital switching systems (communications) 819–821
Digital tape cartridge 161
Digital-to-analog converter 228–230*
  circuitry 230
  digital control 222
  digital disk recording 246–247
  uses 229–230
Digital voltmeter 230*
Dijkstrain, E. 385
Diode 230–231*, 881
  avalanche diodes 517–518
  capacitor diodes 231
  constant-voltage diodes 231
  junction diode 440–442*
  light-emitting diode 231, 457–459*
  light-sensitive diodes 231

Diode—cont.
  negative-resistance diodes 231
  point-contact diode 631*
  rectifier diodes 230–231
  semiconductor see Semiconductor diode
  semiconductor theory 731
  silicon diode 748*
  solion 766
  as switch for clipping 126–127
  tunnel diode 877*
  unijunction transistor 445
  varactor diodes 519
  voltage regulator circuit 908
  Zener diode 931*
Diode detectors 34–35
  conditions for high efficiency 34–35
  requirements for linearity 35
  special features 35
Diode function generator 48
Diode noise generator 284
Diode phase detector 602
Diode rectifier-amplifier voltmeter 887
Dipole antennas 65
Direct aperture antennas 66
Direct-coupled amplifier 25, 231–233*, 904
  carrier dc amplifier 232
  chopper-stabilized amplifiers 232–233
  dc amplifier 231
Direct-coupled FET logic 371, 372
Direct current 233*
  electronic power supply 316
  measurement 186
  rectifier 699–702*
  voltage measurement 905
Direct-current circuit theory 233–235*
  circuit response 235
  circuits 117
  classification 233–234
  current sources 234
  electric transients 269–271
  Ohm's law 562*
  physical laws of circuit analysis 234
  power 235
  series-parallel circuit 235
  shunting 747*
Direct-current vacuum-tube voltmeter 887
Direct piezoelectric effect 627
Direct-radiator speakers 474–475
Direct-view storage tubes 775–776
  display 151
Directional antenna: corner reflector antenna 170*
Directional coupler 235–236*
Discriminator 236*, 361
Disk file (memory storage) 485
Disk memories 159
  cartridge disks 160
  fixed-head disks 160
  floppy disks 160–161
  optical disks 161
  Winchester technology 159–160
Disk pack storage technology 201–202
Disk recording 236–247*
  commercial disk records 241

Disk recording—cont.
  digital recording 246–247
  distortion and noise 241–242
  distortion in reproduction 241–242
  frequency-response equalization 368–369*
  high-polymer phonograph cartridges 404–405
  monophonic system 236–241
  quadraphonic system 245–246
  stereophonic system 242–245
  surface noise 242
  video disk recording 889–892*
Disk-seal (lighthouse) tube 886
Disk storage technology: digital optical disk recorder 227–228*
Display storage tube 775–776
Distortion (electronic circuits) 247–248*
  amplitude distortion 247–248
  cross modulation 248
  crosstalk 177–178*
  frequency distortion 248
  phase distortion 248
  reduction by feedback 248
Distortion and noise meter: harmonic analysis 393
Distributed processing 248–250*
  advantages and disadvantages 163, 250
  distinguishing characteristics 248–249
  evolution 249
  forms 249–250
  hierarchical systems 249–250
  horizontal systems 250
  hybrid systems 250
Distributed RC filters 265
DMUC see Decision making under certainty
DMUR see Decision making under risk
DMUU see Decision making under uncertainty
Donor atom 251*
Doppler radar 251–253*
  airborne vehicular systems 251–252
  meteorological systems 252–253
Double-balanced phase detector 602
Double-base diode 866
Drift transistors 863
Drysdale ac polar potentiometer 634–635
Dual-slope A/D converters 59
Dual-trace oscilloscope 591
DuBridge, L. A. 614
DVST see Direct-view storage tube
Dynamic earphones 254
Dynamic-electrostatic earphone 254–255
Dynamic microphones: moving coil microphone 497–498
  moving conductor microphone 498
  ribbon-type 498
Dynamic programming: operations research 566

Dynamic random-access memory 156, 739
Dynamic transducer:
 microphones 496

EAROM *see* Electrically alterable read-only memories
Earphones 253–256*
 dynamic 254
 dynamic-electrostatic 254–255
 electrostatic 254
 hearing aid 394–395*
 magnetic 254
 miniature magnetic 254
 piezoelectric 255
 real-ear response 255
 realistic simulation 255–256
EBAM *see* Electron-beam-accessed memory
Eccles-Jordan circuit 540
ECD *see* Electrochromic displays
Echo box 256*
Eckert, J. P. 158
ECL *see* Emitter-coupled logic
ECM *see* Electronic countermeasures
Eddy current 256–257*
 causes 256–257
 laminations 257
 resistance measurement 706–707
Edison, Thomas 76, 299, 327, 564
EDP *see* Electronic data-processing systems
EEPROM *see* Electrically erasable programmable read-only memories
Einstein, Albert 613
Einstein photoelectric law 613–614
Electret 257*
Electret transducer 257–260*
 applications 259–260
 electromechanical transducers 259
 foil electret 257
 headphones 258–259
 microphones 257–258
Electric charge 260*
Electric circuit *see* Circuit (electricity)
Electric current 260–261*
 alternating current 6–9*
 chopping 116*
 conduction current 260
 direct current 233*
 displacement current 260–261
 eddy current 256–257*
 IR drop 436*
 measurement *see* Current measurement
 phasors 12
 reactance 696*
 rectifier 699–702*
 surging 802*
Electric filter 261–266*
 applications 262
 crystal and ceramic filters 264–265
 design 262
 digital filters 265–266

Electric filter—*cont.*
 distributed *RC* filters 265
 electronic power supply 317
 matched filters 265
 mechanical filters 265
 microwave filters 265
 noise filter (radio) 551*
 optimum filters 265
 piezoelectric crystals 626
 privacy system (scrambling) 645
 quartz resonator 631
 special filters 264–265
 surface acoustic-wave filters 794–795
 switched-capacitor filter 805–806
 switched filters 265
 types 261–262
Electric spark 266–267*
 mechanisms 266–267
 spark gap 767*
 theory 267
Electric transient 267–272*
 ac transients 271–272
 circuits 117
 dc transients 269–271
 surging 802*
Electric uninterruptible power system 272–274*
 subsystems 272–274
 types 272
Electrical communications 274–276*
 bandwidth requirements 83–84*
 communication cables 135–142*
 crosstalk 177–178*
 cryptography 178–185*
 data circuit grounding 314
 data communications 191–195*
 digital transmission 667
 electric filter 261–266*
 electrical noise 281–284*
 facsimile 337–339*
 frequency division multiplex 275
 frequency-modulation radio 363–367*
 information sources 274
 information theory 274, 413–417*
 modulation 526–528*
 multilink channels 274–275
 network configurations 274
 radio 686–688*
 single sideband 752
 switching systems 811–822*
 telephone 825–827*
 telephotography 827*
 television 830–833*
 time division multiplex 275
 transducers 274
 transmission lines 869–872*
 transmission media 274
 video telephone 892–893*
 voice response 895–899*
Electrical conduction *see* Conduction (electricity)
Electrical conduction in gases 276–278*
 arc discharge 70–71*
 basic effects 276–277
 breakdown potential 89–90*
 corona discharge 170–171*
 dark current 187*

Electrical conduction in gases —*cont.*
 diffusion of ions 277
 electric spark 266–267*
 free-charge removal 277
 glow discharge 380–381*
 ionization 435*
 mechanism of conduction 277–278
 motion of the charges 277
 phototube operation 620
 saturation current 718
 sources of free charges 276–277
 thyratron 855–856*
 Townsend discharge 857*
 voltage-regulator tube 911*
Electrical conductivity: photoconductivity 610–612*
Electrical impedance 278*
 admittance 4*
 alternating-current circuit theory 13
 high-frequency impedance measurements 395–402*
 impedance matching 409–410*
 impedance measurement 412–413
Electrical instability 278*
Electrical loading 278*
Electrical measurements 279–281*
 ac measurements 279
 accuracy 279
 circuit loading 280
 dc measurements 279
 field measurements 279–280
 frequency considerations 280
 harmonic analyzer 393
 high-frequency impedance measurements 395–402*
 inductance measurement 412–413*
 laboratory measurements 279
 measurement of parameters 280–281
 oscilloscope 592
 phase-angle measurement 598–600*
 potentiometer (voltage meter) 633–635*
 Q meter 673*
 resistance measurement 703–709*
 time dependence 280
 voltage measurement 905–906*
 Wheatstone bridge 923–925*
Electrical noise 281–284*
 electrical shielding 285*
 electronic equipment grounding 309–314*
 gas tubes 375
 mathematical analysis 282–283
 noise figure 284
 noise measurement 282
 shot noise 283–284
 signal-to-noise ratio 748*
 sources 281–282
 suppressor 793*
 thermal noise 283
 transmission lines 291–292
Electrical noise generator 284*
Electrical resistance 284*
 IR drop 436*

Electrical resistance—*cont.*
 resistance measurement 703–709*
 resistor 709–710*
Electrical shielding 285*
 electromagnetic shielding 285
 electrostatic shielding 285
 magnetostatic shielding 285
 shielded wires and cables 285
Electrical storage tubes 775
Electrically alterable read-only memories 158, 739, 740
Electrically erasable programmable read-only memories 739, 740
Electrically small antennas 65
Electroacoustic systems 285
Electroacoustic transducers *see* Electroacoustic systems
Electroacoustics 285*
Electrochromic displays 285–286*
 viologen displays 286
 WO_3 displays 286
Electrodynamic ac voltmeter 913–914
Electrodynamic ammeter 19
Electroluminescence 286–287*
 Destriau effect 286–287
 effects 287
 injection 287
 light-emitting diode 457–459*
 light panel 459–560*
 picture tube 624
Electrolytic capacitors 99–100
Electromagnetic compatibility 287–288*
 analysis and prediction 288
 design consideration 287–288
 interference potential 287
Electromagnetic shielding 285
Electromagnetic wave transmission 288–293*
 features of electromagnetic waves 288–289
 hollow waveguides 290–291
 propagation over the Earth 290
 radiation from antenna 289–290
 two-conductor transmission lines 291–292
Electromechanical high-polymer transducers 404–405
Electromechanical printers 203
Electromechanical transducers 259
Electromotive force (emf) 293*
 Faraday's law of induction 339*
 inductance 410*
 potentiometer (voltage meter) 633–635*
Electron-beam-accessed memory 158
Electron emission 293*
 field emission 348–349*
 secondary emission 723–725*
 thermionic emission 851–852*
 vacuum tube operation 879
Electron excitation 333
Electron lens 293*
Electron motion in vacuum 294–298*
 combined fields with axial symmetry 295–296

Electron motion in vacuum
—cont.
  effect of space charge 297–298
  magnetic fields 295
  quadrupole fields 296–297
  static electric fields 294–295
  time-varying fields 298
  vacuum tube operation 879
Electron optics 298–299*
  image-intensifier tubes 456
Electron tube 299*
  application 299
  audion 76*
  characteristics 299
  gas tube 372–375*
  history 299, 327
  magnetron 486–490*
  numerical indicator tube 560–561*
  phasitron 606–607
  phototube 620–622*
  storage tube 775–776*
  television camera tube 837–842*
  thermionic tube 852–853*
  vacuum tube 879–886*
Electronic ac voltmeter 915
Electronic analog computer 42–57
Electronic calculators 90–92
  display 91
  entry notation 92
  operand range 91–92
  speed 92
  storage registers 92
Electronic circuit see Circuit (electronics)
Electronic codebook mode 182
Electronic countermeasures 299–304*
  active 302–303
  categories 300
  reflectors 301–302
  surveillance systems 300
  systems 303
  trends 303–304
  warning receivers 300–301
Electronic data-processing systems 200–204
  card reader and punch 202
  character reader 203
  computer output microfilmers 204
  disk pack units 201–202
  displays 203
  magnetic tape units 202
  paper tape readers and punches 202–203
  printers 203
Electronic differential analyzer 53
Electronic digital counter 349–351
Electronic display 100, 304–309*
  calculators 91
  cathode-ray tube 106, 304–305
  color 307
  direct-view storage tubes 775–776
  display categories 305
  display electronic addressing 308–309
  EDP systems 203–204
  flat-panel displays 305

Electronic display—cont.
  font 308
  helmet-mounted and heads-up displays 306–307
  light-emitting diodes 458
  liquid crystals 465–466
  photoferroelectric imaging 616–618*
  picture tube 623–625*
  projection display 305–306
  radar operation 676
  radar system 686
  special-purpose displays 305–307
  technique 307–308
  three-dimensional imagery 306
  videotext and teletext 894
Electronic equipment grounding 309–314*
  data transmission 314
  ground gradients 311–312
  grounding and shielding practices 312–314
  low-noise environment 312
  noise coupling into signal circuit 309–311
  optical isolator 314
  thorough grounding 309
Electronic games 165
Electronic gate 806–807
Electronic listening devices 314–315*
  advanced devices 315
  bugs 314–315
  compromise of equipment 314
  CW radar surveillance 169
  defenses 315
Electronic phase-angle meter 598–599
Electronic power supply 315–319*
  battery 317
  direct-current 316
  filament or heater 316
  filters 317
  high-voltage 316–317
  switching regulators 317–319
  voltage-multiplier circuit 906–907*
  voltage regulator 907–909
Electronic publishing 319–326*
  case studies 324–326
  conduit versus content 320–321
  on-line data bases 321–322
  opportunities and obstacles 326
  production of electronic information products 320
  publishing environment 319–320
  video disk 324
  videotex 322–323
  videotext and teletext 894–895
Electronic stored-program digital computers 210
Electronic switch 326*
  phase-angle measurement 599–600
Electronic switching systems: pulse-code modulation 666–667
Electronic tablet: computer graphical input 151
Electronics 326–329*
  applications 328–329

Electronics—cont.
  consumer see Consumer electronics
  electroacoustics 285*
  high-temperature electronics 405–406*
  history 327–328
  industry growth 328
  quantum electronics 674*
Electrooptics: electrochromic displays 285–286*
  electronic display 304–309*
  infrared imaging device scanners 418
  liquid crystal display 466
  optical modulators 577
Electrophotoluminescence 287
Electrostatic earphone 254
Electrostatic lens 329–330*
  axially symmetric lenses 329–330
  lenses of plane symmetry 330
Electrostatic microphone 498–499
Electrostatic printers 203
Electrostatic shielding 285
Electrostatic speakers 473–475
Electrostatic transducer: microphones 496
Electrostatic voltmeter 912–913
Electrostriction 330*
Electrostrictive transducer: microphones 496–497
Embedded systems 330–332*
  ADA programming language 331
  algorithms 5
  VHSIC program 330–331
emf see Electromotive force
Emitter-coupled logic 471
Emitter-coupled phase inverter 601
Emitter follower 332–333*
  impedance matching 409
Emulator (microcomputer) 495
Encoding: information theory 414–417
Encryption 154
  cryptography 178–185*
End-fed monopole antenna 65
Energy measurement; watt-hour meter 7
Engan, H. 794
EPROM see Erasable programmable read-only memory
Equalizer, frequency-response 368–369
Equivalent circuit 333*
Erasable programmable read-only memory 158
Erdös, P. 386
Erlang, A. 564
Esaki, L. 518, 878
Esaki tunnel diode 734
Etched-well LED 458
Euler, L. 383
Excitation 333*
Excitation energy 333
Excitation potential 333–334*
Expert systems 334–337*
  accomplishments 334
  algorithms 6
  components 336–337
  distinguishing characteristics 334
  importance of expert knowledge 334

Expert systems—cont.
  real-time 697
  types 335–336
External photoelectric effect see Photoemission

F

Fabry-Perot interferometer: bistable optical devices 569–570
Facsimile 337–339*
  color 339
  digital 339
  direct recording 338
  frequency modulation 366
  modulation 338
  photographic recording 338–339
  scanning 337–338
  synchronization 339
  telephotography 827*
  transmission system 337
Far-end crosstalk 177
Faraday's law of induction 339*
Fault-tolerant systems 339–340*
  fault prevention and fault tolerance 340
  fault tolerance techniques 340
  faults and errors 340
Feedback circuit 340–342*
  amplifier feedback 340–341
  distortion reduction 248
  input and output impedances 341
  oscillator feedback 341
  parasitic oscillation 594
  positive and negative feedback 341
  regeneration 702
  servomechanism feedback 342
Ferrimagnetic garnets 342
Ferrite devices 342–344*
  applications 342–344
  chemistry and crystal structure 342
  highly nonlinear B/H devices 343
  linear B/H devices 342–343
  microwave devices 343–344
  nonlinear B/H devices 343
Ferroelectrics 344–348*
  antiferroelectric crystals 347–348
  applications 348
  barium titanate 84–85*
  classification 344
  crystal structure 347
  dielectric properties 345–346
  domains 344–345
  origin of phase transition 348
  photoferroelectric imaging 616–618*
  piezoelectric properties 346
  Rochelle salt 715–716*
FET see Field-effect transistor
Fidelity 348*
Field-effect transistor 862, 865–866
  clipping 127–128
  connections 868–869
  JFET 865
  keyed clamp 125–126
  limiter circuit 461
  monostable multivibrator 536

Field-effect transistor
—*cont.*
    static induction transistor
        773–775*
    VMOS and SOS 866
Field emission 348–349*
Field-emission vacuum tube 886
Field terminated diode 774
Filament power supply
    (electronics) 316
Film-type resistor 709–710
Filter *see* Electric filter
Finding circuit *see* Lockout
    circuit
Finite boolean algebra 88
First-generation computer 215–
    216
Fixed-bias transistor circuit 85
Fixed capacitors 98
Fixed-head disks 160
Fixed inductors 413
Flat cathode-ray tubes 106
Flat-panel electronic display 305
    font 308
    monochromatic color 307
Fleming, J. Ambrose 76, 299,
    327
Flexible coaxial cables 130
Flexible printed wiring 639
Flicker noise 281
Flip-flop 211–212, 807
    digital counter 225
    function generator 369
    logic gate multivibrator 538–
        539
    translating circuit 810
Floppy disk 160–161, 485
FM *see* Frequency modulation
FM-CW radar 169
Ford, G. 385
Ford, L. 385
FORTRAN 656–657
Foster-Seely discriminator 361
Foucault current *see* Eddy
    current
Fourth-generation digital
    computer 217
Fowler, R. 614
Free-electron laser 454
Free-electron theory of metals:
    field emission 348–349
Free-running astable blocking
    oscillator 87
Free-space optical
    communications 572
Frege, F. L. G. 5
Frequency (wave motion) 349*
Frequency counter 349–351*
    audio-frequency measurement
        356–357
Frequency divider 351–352*
    digital division 351
    subharmonic triggering 351–
        352
Frequency-division multiplexing
    32, 275, 532
Frequency-division multiplexing-
    frequency modulation 365–
    366
Frequency-independent antennas
    66
Frequency measurement 352–
    358*
    audio-frequency meters 353–
        357
    frequency counter 349–351*
    harmonic analyzer 393

Frequency measurement
—*cont.*
    high-frequency impedance
        measurements 395–402*
    primary frequency standard
        352–353
    radio-frequency measurements
        357
    secondary frequency standard
        353
    standards of time and
        frequency 352
Frequency modulation 39, 275,
    358–361*, 527
    advantages 359
    angle modulation 60
    detector *see* Frequency-
        modulation detector
    frequency modulator 360–
        361, 367*, 529
    instantaneous frequency 359
    magnetic recording 482
    microwave transmission 511
    noise advantage 359–360
    phase modulator 604–605
    preemphasis and deemphasis
        360
    principles 359
    production and protection
        360–361
    radio transmitters 694–695
    spectrum 359
    spectrum multiplication 360
    spectrum translation 360
Frequency-modulation
    broadcasting 363–364
    stereophonic broadcasting 364
    subscription service 364
Frequency-modulation detector
    361–363*
    discriminator 361–363
    integrating detector 363
    locked-oscillator detector 363
    phase-modulation detector
        604*
Frequency-modulation radio
    363–367*, 687
    facsimile 366
    FM broadcasting 363–364
    mobile radio 524
    mobile transmission 364–365
    quadraphonic sound 364
    radio relaying 365–366
    receivers 692
    telegraphy 366
    telemetry 366
Frequency modulator 360-361,
    367*, 529
    linearity and bandwidth 367
    varactor modulator 367
    voltage-controlled oscillators
    367
Frequency multiplier 367–368*
    nonlinear amplifier 367–368
    nonlinear coupler 368
    phase-locked loops 603
Frequency-response equalization
    368–369*
Frequency-shift keying 366
Frölich, H. 789
Frucht, G. R. 384
FTD *see* Field terminated diode
Fulkerson, D. 385
Full-wave diode rectifier 700
Function generator 369*
Fuse-programmable read-only
    memories 739–740

Fused-junction transistor *see*
    Alloy-junction transistors

Gain 369–370*
Gallium arsenide field-effect
    transistors 516
Gallium arsenide integrated
    circuits 370–372*
    designs 371–372
    gallium arsenide versus silicon
        371
Gallium arsenide MESFET 371
Gallium arsenide-phosphide LED
    457
Galvanoluminescence 287
Game theory: artificial
    intelligence 72
    automata 78
Gas capacitor 98
Gas-discharge lasers 453
Gas-discharge noise generator
    284
Gas-dynamic lasers 454
Gas-filled tubes 299
Gas masers 490–491
Gas phototube: principles of
    operation 620–621
    types of service 622
Gas tube 372–375*
    cold-cathode 373–375
    hot-cathode 374–375
    noise, oscillations, and surges
        375
    thyratron 855–856*
    volt-ampere characteristics
        373
    voltage-regulator tube 911*
Gate, electronic 806–807
Gate-array logic devices 469
Gate circuit 375–376*
    switching gate 376
    transmission gate 375–376
Gate turnoff device 746
General-purpose analog
    computer 42
General Radio admittance bridge
    400
General Radio rf bridge 398–
    399
Geometric programming:
    operations research 566
Geometrical electron optics 299
Germanium: amorphous 22
    semiconductor element 728
GERT 376–378*
    applications 378
    GERTE 376–377
    Q-GERT 377–378
GERTE 376–377
Giaever, I. 790, 878
Ginzburg, V. L. 789
Glass switch 378–380*
    amorphous switch materials
        379
    applications 379–380
    device characteristics 378–
        379
    glassy chalcogenides 22
    photographic film applications
        380
    read-only memories 379–380
    transistor applications 380

Glassy chalcogenides: amorphous
    semiconductors 22
Glow discharge 380–381*
Glow lamp 373
Gödel, K. 5
Goldberg, E. 44
Gorter, C. J. 787
Gradient microphone 500
    higher-order types 501–502
    unidirectional types 500–501
    velocity type 500
Grain boundaries
    (semiconductors) 381–383*
    atomic arrangements 382
    electronic structure 382
    transport of charge 382–383
Graph: abstract data type 3
Graph theory 383–386*
    applications 385–386
    definitions 384
    map coloring problems 384
    origin 383–384
    planarity 384–385
    variations 385
Gray tin: semiconductor element
    728
Grid-dip meter 923
Grid-dip oscillator 923
Grid-glow tube *see* Cold-cathode
    triode
Grid-rectification voltmeter 887
Grounded-base amplifier 904
Grounded-gate amplifier 904
Grounding, electronic equipment
    309–314*
Grown-junction transistors 443
GTO *see* Gate turnoff device
Gunn devices *see* Transferred-
    electron devices
Gyrator 263, 386–389*
    practical 387–389
    reciprocity 386–387
    theoretical 387
Gyrotron 389–391*
    basic characteristics 389–
        390
    capabilities 390–391

H

Haken, W. 384
Half-wave diode rectifier 699
Hall effect 391–392*
    ferromagnetic metals 392
    ionic crystals 392
    physical interpretation 391
    semiconductors 392, 728
    thermal side effects 391–392
"Ham" radio *see* Amateur radio
Harary, F. 385
Hard software: algorithms 5
Harmonic analyzer 393–394*
    digital computer resources 394
    heterodyne measurements 393
    high-speed analysis 393–394
    time compression analyzer 394
    tunable filters 393
    wattmeter measurement 393
    waveform determination 918
Harmonic oscillator *see*
    Sinusoidal oscillator
Harris, F. 564
Hartley, R. V. L. 86, 274, 414
Hartley oscillator 585
Harmann, C. S. 794

Headphone: electret 258–259
  high-polymer stereophonic 404
Heads-up electronic display
  306–307
Hearing aid 254, 394–395*
Heater power supply
  (electronics) 316
Heawood, P. J. 384
Helical-scan video tape recorder
  484
Helical whip antenna 65
Helium-neon lasers 453
Helmet-mounted electronic
  display 306–307
Helmholtz, H. 853
Helmholtz theorem see
  Thévenin's theorem
Henley, A. 284
Henry, Joseph 686
Hertz, Heinrich 299, 513, 686
Heterodyne frequency meters
  358
Heterodyne oscillator 587
Heterodyne principle 395*
  superheterodyne receiver 791*
Hewlett-Packard RX meter 399
Hierarchical distributed
  processing systems 249–
  250
High fidelity 348
High-frequency AM radio 37
High-frequency impedance
  measurements 395–402*
  automatic plotter 402
  general radio reflectometer
    402
  Hewlett-Packard network
    analyzer 402
  impedance plotter 402
  microwave null devices 399–
    400
  null methods 398-400
  parallel-resonance methods
    397.
  potentiometer method 396
  Q of tuned circuit or cavity
    398
  R_0/Q_0 of resonant cavity 398
  radio-frequency bridges 398–
    399
  reflection methods 401–
    402
  resonance methods 396–398
  resonant-rise method 397–398
  series-resonance methods
    396–397
  slotted line 401
  slotted section 401
  standing-wave methods 400–
    401
  vector impedance meter 396
  voltmeter-ammeter methods
    396
High-frequency voltmeter 717
High-polymer transducer 402–
  405*
  applications 405
  direct radiator loudspeakers
    403–404
  electromechanical transducers
    404–405
  microphones 404
  piezoelectric high polymers
    402–403
  stereophonic headphones 404
High-power, short-pulse lasers
  454–455

High-speed oscilloscopes 591–
  592
High-temperature electronics
  405–406*
  amorphous metal films 406
  compound semiconductor
    devices 405–406
  integrated thermionic circuits
    406
  silicon devices 405
High-voltage electric power
  supply 316–317
Higher-order gradient
  microphones 501–502
Highly nonlinear B/H devices
  343
Hilbert, D. 5
Hittorf dark space 380
Hogan, C. L. 388
Holes in solids 406–407*
  photoconductivity 611
Hollow waveguides 918
  attenuation 920–921
Honeycomb mass-storage system
  162
Hoperoft, J. 384
Horizontal distributed processing
  systems 250
Horn radiator antenna 66
Hot-cathode gas tubes 374–375
  noise, oscillations, and surges
    375
  phanotron tube 374–375
  thyratron tube 375
  tungar tube 374
Hot-wire ammeter 186
Hot-wire noise generator 284
Hull, A. W. 299
Human-machine systems 334
Hunting circuit see Lockout
  circuit
Hybrid computers 42, 55–57
Hybrid distributed processing
  systems 250
Hybrid integrated optical circuit
  431
Hybrid printed circuits 643–644
Hybrid simulation 749
Hyperbolic sweep generator 720,
  803

Ichbiah, Jean 660
Ideal transformer 173
i-f amplifier see Intermediate-
  frequency amplifier
IGFET see Insulated-gate field-
  effect transistor
Illiac multiprocessor system 533
Image intensifier see Light
  amplifier
Image isocon 839
Image orthicon 407*
  image section 837–838
  multiplier section 838–839
  scanning section 838
Image processing 407–409*
  global operation 408
  hardware 407–408
  local operations 408–409
  operation types 408–409
  point operations 408
Imaging devices, infrared 417–
  420

Immersion electrostatic lenses
  329–330
Impact avalanche and transit
  time diode 517–518, 741
Impact printer 203
IMPATT diode see Impact
  avalanche and transit time
  diode
Impedance see Electrical
  impedance 278
Impedance bridge, alternating-
  current 411
Impedance matching 409–410*
  emitter follower 409
  network 409
  transformers 409
Impedance measurement 412–
  413
  high-frequency 395–402
Impurity semiconductor 727
Independent-sideband radio
  receiver 693
Independent-sideband radio
  transmitter 695
Inductance 410*
  alternating-current circuit
    theory 12–13
  inductor 413*
  measurement see Inductance
    measurement
  mutual inductance 410
  self-inductance 410*
Inductance bridge 411–412*
  Anderson bridge 411
  Carey-Foster bridge 411–412
  general impedance bridge 411
  general inductance bridge 411
Inductance coil see Inductor
Inductance measurement 412–
  413*
  impedance measurement 412–
    413
  inductance bridge 411–412*
  inductance standards 412
Induction, Faraday's law of 339
Induction voltage regulator 910
Inductor 413*
  adjustable 413
  coil 131–132*
  fixed 413
  inductance 410*
  linear B/H devices 342–343
  variable 413
Infinite boolean algebra 88–89
Information systems see Data-
  processing systems
Information theory 78, 413–
  417*
  algebraic codes 416–417
  bit 86
  channel capacity 416
  communication systems 414
  electrical communication
    channel 274
  encoding and decoding 414–
    415
  information content of message
    415–416
Infrared imaging devices 417–
  420*
  characterization 419
  display 419
  scanning systems 418–419
Infrared masers 492
Injection electroluminescence
  287
Instruction systems 335

Instrument resistors 709
Instrumentation magnetic
  recorder 485
Insulated-gate field-effect
  transistor 535, 865
Integrated amplifier 905
Integrated circuits 118, 420–
  430*
  bipolar 421–424
  comparators 142–143
  digital systems 212
  fabrication 428–430
  fabrication processes 429
  fabrication requirements 429
  FET connections 868–869
  gallium arsenide see Gallium
    arsenide integrated circuits
  history of electronics 328
  integrated optical devices 428
  large-scale 155
  magnetic bubble memory 477–
    478*
  microcomputers 428
  microprocessor 506–509*
  MOS 424–428
  operational amplifiers 28
  printed 644
  reliability considerations 122
  semiconductor memories 739–
    741*
  soft fails 753–754
  thermionic 406
  types 420–428
Integrated optical circuit 431
Integrated optics 430–432*
  coupling of external light
    beams 431–432
  guided waves 431
  integrated circuit technology
    428
  lasers 432
  materials and fabrication 431
  modulators 432
Integrated oscillators 587
Integrated thermionic circuit 406
Integrating FM detector 363
Integrator circuits 804
Intelligible crosstalk 177
Interdigital transducers: surface
  acoustic-wave devices 793–
  794
Interlaced television scanning
  847
Intermediate-frequency amplifier
  25, 26, 432*
Interpretation systems 335
Intrinsic semiconductor 727
Inverter circuit 421–422
  FET connections 868–869
  MOS 426
IOC see Integrated optical circuit
Ion implantation 432–435*
  applications 433–435*
  insulators 434
  junction diode fabrication 441
  metals 434
  planar diffused epitaxial
    transistor 444
  process 433
  semiconductors 434
Ionization 435*
Ionization potential 435–436*
IR drop 436*
Iron-vane ammeter 19–20
ISB radio receiver see
  Independent-sideband radio
  receiver

ISB radio transmitter *see* Independent-sideband radio transmitter
ITC *see* Integrated thermionic circuit
Iverson, Kenneth 658
Ives, H. E. 892

J

JFET *see* Junction-gate field-effect transistor
Jobst, G. 299
Josephson, Brian D. 436, 791, 878
Josephson effect 436–439*, 878–879
  applications 439
  Josephson junction 438–439
  nature 436
  research superconductivity 791
  superconduction computers 776–779*
  theory 436–438
Josephson junction 879
  Josephson effect 436–439*
  SQUID 772–773*
Joule, J. P. 284, 439
Joule's law 439–440*
Junction diode 440–442*, 733
  avalanche diodes 517–518
  characteristics 442
  fabrication methods 440–441
  injection electroluminescence 287
  junction rectification 441
  optical properties 441–442
  photoconductive detector 612
  photodiode 612*
  photovoltaic cell 623*
  PIN diodes 518–519
  tunnel diode 877*
  varactor 888*
  Zener diode 931*
Junction-gate field-effect transistor 865
  bistable multivibrator 535
Junction transistor 23, 442–445*, 863
  alloy-junction transistors 443
  grown-junction transistors 443–444
  logic circuits 471
  mesa transistors 445
  planar diffused epitaxial transistors 444–445
  power transistors 445
  unijunction transistor 445

K

Kammerlingh Onnes, H. 783
Kawai, H. 402
Kell, R. D. 664
Kelvin, Lord 43
Kelvin bridge 445–446*
  errors 446
  sensitivity 445–446
Kelvin guard-ring capacitor 98
Kemeny, J. G. 657
Kerl, R. J. 608

Keyed clamp 125
Kinescope: aluminized screen 624
  construction 623–624
  deflection angle 623–624
  electromagnetic deflection 624
  electron gun 624
  focus 624
  glass envelope 624
Kirchhoff, G. 385
Kirchhoff's laws of electric circuits 446–447*
  current law 447
  voltage law 446–447
Kleene, S. C. 5
Klystron 284, 447–451*
  amplifier 449–450
  reflex oscillator 450–451
Kompfner, R. 299
Konigsberg bridge problem 383
Kuhn, H. 385
Kuratowski, K. 384
Kurtz, T. E. 657

L

Ladder filters 263
Lakin, K. M. 795
Lancaster, F. W. 564
Land mobile radio service *see* Mobile radio
Landau, L. D. 789
Langmuir, Irving 76
Langmuir-Child law 451*
Language translators 165
Large core storage 155
Large-scale integrated circuits: microcomputers 428
  optical devices 428
Laser 451–456*
  applications 455
  chemical 454
  comparison with other light sources 451–452
  free-electron 454
  free-space optical communications 572
  gas-discharge 453
  gas-dynamic 454
  high-power, short-pulse 454–455
  integrated optics device 432
  nuclear 454
  optical transmitters 573–574
  optically pumped 452–453
  photodissociation 454
  pulsed gas 453–454
  quantum electronics 674
  semiconductor 454
Laser printer 203
Laser video disk: business potential 324
Last-mask read-only memories 739
LC filter 263
LCD *see* Liquid-crystal display
Lecher wire wavemeter 922
LED *see* Light-emitting diode
Leibniz, G. W. 5
Lens, magnetic *see* Magnetic lens
Lewis, M. F. 795
Light amplification by stimulated emission of radiation *see* Laser

Light amplifier 456–457*
  applications 457
  image-intensifier tubes 456
  solid-state 457
  see also Laser
Light-emitting diode 231, 457–459*
  applications 458
  calculator displays 91
  fabrication methods 457–458
  indicators and displays 458
  optical emitter 575
  optical fiber transmission 459, 667
  optical transmitters 573–574
  optoisolators 458
Light-emitting diode-photodetector-amplifier optical isolators 576
Light-emitting diode-silicon optical isolators 576
Light panel 459–460*
Light pen: computer graphical input 148
Light-sensitive diodes 231
Lighthouse tube 886
Limiter circuit 460–461*
  diode limiters 460
  limiting by saturation 461
  triode limiters 461
  use of both limits 460–461
Line microphone 503
Line printer 203
Linear B/H devices 342–343
Linear circuits 424
Linear comparator 142
Linear programming 462–463*
  applications 462–463
  general theory 462
  graph theory 385
  methods of calculation 462
  operations research 565–567
Linear sweep generator 802
Linearity 463
Liquid-crystal display 91
Liquid crystals 463–466*
  applications 465–466
  calculator displays 91
  classification and structure 463–465
LISP (programming language) 655, 659
List: abstract data types 2–3
Load line 466–467*
Locked-oscillator FM detector 363
Lockout circuit 809
Loeb, L. B. 70
Logic circuits 469–471*
  boolean algebra 88
  combinational and sequential logic 468
  elements 211–212
  embodiment 468–469
  FET connections 868–869
  integrated circuits 420
  logic gate multivibrator 538–539
  operation 467
  SQUID 773
  superconducting devices 781–782
  switching circuit 806–811*
  switching gate 376
  technology 469–471
  types of logic functions 467–468

Logic gate *see* Switching circuit
Logic gate multivibrator 538
London, F. 788
London, H. 788
Long-wire antennas 65
Longwave AM radio 37
Loomis, L. 88
Loudspeaker 471–477*
  direct-radiator 474–475
  directional characteristics 472
  distortion 472
  efficiency 473
  electrical speaker impedance 472
  force factor 472–473
  high-polymer direct radiator 403–404
  physical performance characteristics 472–473
  placement 476–477
  power rating (input) 473
  pressure response 472
  pressure-response-frequency characteristic 472
  radiation-controlling structures 475–476
  radiation impedance 473–474
  requirements for speech and music 471–472
  response 712*
  systems 476
  types 473
Lovell, C. A. 44
Low-frequency AM radio 37
Low-pass filter 261, 262
Luminescence:
  cathodoluminescence 107–108*
  electroluminescence 286–287*
Lumped-circuit delay line 209
Luneberg lens antenna 66–67
Lyotropic liquid crystals 463

#

M-carcinotron *see* M-type backward-wave oscillator
M-type backward-wave oscillator 80
McCarthy, John 659
Magic tee junction 236
Magnetic bubble memory 159, 477–478*
  information retrieval 478
  ion-implanted materials 434
  loop organization 477–478
  principles of operation 477–478
  storage locations 477
  technology status 478
Magnetic circuits 117
Magnetic core memory 154, 155
Magnetic earphones 254
Magnetic-ink character recognition 109, 203
  stylized font characters 112
Magnetic lens 478–479*
Magnetic recording 479–486*
  audio delayer 75
  audio recorders 482–484
  digital 481–482
  FM recording 482
  frequency-response equalization 368–369*

Magnetic recording—*cont.*
  high-frequency bias 481
  instrumentation recorders 485
  magnetic heads 481
  noise and wavelength
    limitations 482
  playback process 481
  principles of recorder
    operation 480–482
  recorders for computers 484–
    485
  recording medium 480–481
  recording process 481
  tape duplication 485–486
  types of recording system
    482–486
  video recorders 484
Magnetic tape 480–481
Magnetic tape recorder 480
Magnetic tape storage 161, 202
  cassettes and cartridges 161–
    162
  half-inch tapes 161
Magnetic thin films 486*
  coercivity 486
  domain structure 486
  fabrication 486
  magnetic anisotropy 486
  magnetic bubble memory 477–
    478*
  magnetic order 486
Magnetic transducer:
    microphones 496
Magnetostatic shielding 285
Magnetron 486–490*
  coaxial cavity 489
  configuration 487
  crossed-field tubes 489
  electron bunching 488
  microwave circuit 488–489
  microwave generation 487–
    488
Main memory technology 154–
  158
  digital systems 212
  error checking and correction
    157
  RAM chips 156–157
  ROMs, PROMs, and EPROMs
    157–158
Marconi, Guglielmo 686
Marshall, F. G. 794
Maser 490–493*
  circuits 492–493
  gas masers 490–491
  optical and infrared 492
  solid-state 491–492
Mask-programmable read-only
    memories 739
Mass-storage systems 162
Matched filter 265
Mathematical modeling 748–749
Mathematical programming *see*
    Nonlinear programming
Matrix printers 203
Maxwell, James Clerk 509, 613,
    686
MDS *see* Microcomputer
    development system
Meacham, L. A. 630
Mechanical calculators 90
Mechanical electric filters 265
Mechanical rectifier 699
Medium-frequency AM radio 37
Meissner, W. 783
Meissner-Ochsenfeld effect 784
Memory *see* Computer storage
    technology

Memory gap 154
  technology 158–159
Memory tube *see* Storage tube
Menger, K. 385
MESA (programming language)
    163
Mesa transistors 445
MESFET *see* Metal-
    semiconductor field-effect
    transistor
Message-oriented data
    communications 192–193
Message switching (data
    communications) 195
Metal oxide semiconductor: logic
    device 470–471
  RAM devices 155, 157
  switched capacitor 803–806*
Metal oxide semiconductor field-
    effect transistor 371, 424,
    865–866
  connections 868–869
  high-temperature operation
    405
Metal oxide semiconductor
    integrated circuit 424–
    428
  complementary MOS 426
  n-channel MOS 426
  sampled-data devices 427–
    428
  silicon-gate MOS 426–427
Metallic-disk rectifier 493*
Mho (unit) 4
Mica capacitors 99
MICR *see* Magnetic-ink character
    recognition
Microcomputer 218, 494–495*
  digital control 221
  embedded systems 330–332*
  floppy disks 160–161
  integrated circuits 428
  logic network embodiment
    469
  operating systems 563
  personal 494–495
  RAM chips 157
  ROMs, PROMs, and EPROMs
    157–158
  single-board 494
  soft fails 753–754
Microcomputer development
    system 494, 495*
Microfilmers, computer output
    204
Microphone 496–506*
  calibration 504–506
  carbon transducer 497
  directional characteristic 504,
    505
  dynamic transducer 496
  electret 257
  electrical impedance 504
  electrical impedance
    characteristics 505
  electrostatic transducer
    496
  electrostrictive transducer
    496–497
  gradient microphone 500
  high-polymer 404
  magnetic transducer 496
  noise characteristic 504,
    506
  nonlinear distortion 504
  nonlinear distortion
    characteristic 505
  open-circuit response 504

Microphone—*cont.*
  performance characteristics
    503–504
  piezoelectric transducer 496
  pressure microphone 497–500
  response 712*
  response frequency
    characteristic 504–505
  transducers 496–497
  types 497–503
  wave microphone 502–503
Microprocessor 506–509*
  algorithms 5
  applications 509
  architecture and instruction
    506
  bit-sliced 508
  counters 226
  digital system design 508–509
  fundamentals of operation
    506–508
  microcomputer 494–495*
  reprogrammable memories 740
MicrOvonic File 380
Microwave 509–513*
  advantages 512
  applications 512–513
  circuit elements 509–511
  frequency and velocity 509
  frequency measurements 357
  generation 509
  propagation in space 511–512
  radar *see* Radar
  radio relaying 365–366
  reception 512
  transmission 511
Microwave amplification by
    stimulated emission of
    radiation *see* Maser
Microwave antenna 511
Microwave attenuator 510
Microwave circuit 509–511
  attenuators 510
  detectors 510–511
  phase changers 510
  probes 511
  waveguides 510
  wavemeters 511
Microwave detector 510–511
Microwave ferrite devices 343–
    344
Microwave filters 265
Microwave gallium arsenide
    field-effect transistors 516
Microwave optics 513–515*
  diffraction 514
  Faraday effect for microwaves
    514–515
  polarization 514
  rectilinear propagation 513
  reflection and refraction 513–
    514
Microwave oscillators 587
Microwave reflectometer 515*
Microwave solid-state devices
    515–519*
  active devices 515–518
  avalanche diodes 517–518
  GaAs field-effect transistors
    516
  passive devices 518–519
  PIN diodes 518–519
  point-contact diodes 518
  Schottky-barrier diodes 518
  transferred-electron (Gunn)
    devices 516–517
  tunnel diodes 518
  varactor diodes 519

Microwave transmission lines
    520–523*
  coaxial lines 520
  graphical solutions 522
  hollow-pipe waveguides 521
  impedance matching 522
  lines as circuit elements 522
  microwave structures 520–521
  striplines 521
  traveling and standing waves
    521–522
Microwave tube 79, 523–524*,
    879
  alternative designs 523–524
  effect of transit time 523
  effect of tube geometry 523
  gyrotron 389–391*
  traveling-wave tube 872–875*
Microwave waveguide 510
Microwave wavemeter 511, 923
Mid-frequency gain 900–901
Miller integrating circuit 720
Millikan, R. A. 614
Mixer 524*, 883
  radar operation 676
Mobile FM radio transmission
    364–365
  cellular systems 365
  radio paging service 365
Mobile radio 524–526*
  early development 524
  spectrum allocation and
    growth 524
  spectrum engineering 524–
    525
  technological trends 525–526
  uses 525
MODULA (programming
    language) 163
Modulation 526–528*
  amplitude *see* Amplitude
    modulation
  amplitude-modulation detector
    34–36*
  amplitude modulator 39–42*
  angle modulation 59–60*,
    527–528
  carrier 100*
  detector 209–210*
  engineering applications 528
  frequency *see* Frequency
    modulation
  modulator 528–529*
  multiple modulation 528
  phase *see* Phase modulation
  pulse *see* Pulse modulation
  sideband 747*
  single-sideband 31–32
  vestigial-sideband 32
Modulator 528–529*
  amplitude modulator 33–34,
    39–42*
  FM 360–361
  pulse modulator 668–671*
Monitoring systems 335
Monochromatic electronic
    display 307
Monochrome television receiver
    842–845
  antenna and transmission line
    842
  automatic gain control 843–
    844
  controls 844–845
  high-voltage supply 844
  horizontal deflection 844
  intermediate-frequency
    amplifier 842–843

Monochrome television receiver
—cont.
  picture tubes 844
  separation of video and audio
    843
  sweep systems 844
  sync separator circuits 844
  tuner 842
  vertical deflection 844
Monolithic integrated optical
  circuit 431
Monophonic disk recording 236–
  241
  cutters 237
  phonograph recorder 237
  record manufacture 238
  recording characteristics 237–
    238
  recording system 236–238
  sound-reproducing system
    238–241
Monopole antennas 64, 65
Monopulse radar 529*
Monostable multivibrator 536
Monostable negative-resistance
  circuits 542
Morse, Samuel F. B. 686
MOS see Metal oxide
  semiconductor
MOSFET see Metal oxide
  semiconductor field-effect
  transistor
Motorboating 278, 529–530*
Moving-coil frequency meter 354
Moving-coil microphone 497–
  498
Moving-coil speaker 473, 474
Moving-coil voltmeter 911–912
Moving-conductor microphone
  498
Moving-conductor speakers 473
Moving-iron frequency meter
  354–355
Moving-iron voltmeter 914–915
Moving-magnet voltmeter 912
Moving-target indication 530–
  531*
  adaptive 530–531
  charge-transfer devices 530
  measures of merit 530
  principles of operation 530
MTI see Moving-target indication
Müller, E. W. 349
Multiaccess computer 531–532*
  concurrent processing 163
  data communications 192
  data security 188–189
  software capabilities 532
  system components 531
  system operating requirements
    531–532
Multicomputer system 533
Multigrid vacuum tube 883
Multigun cathode-ray tube 106
Multilink communications
  channels 274–275
Multimeter 532*
  see also Volt-ohm-
    milliammeter
Multiple modulation 528
Multiplexing 532*
  amplitude modulation 32–33
  frequency-division 32
  phase-discrimination 32
  time-division 32
Multiplier: analog computer
  component 48–49
Multiplier modulator 42

Multiplier phototube see
  Photomultiplier
Multiprocessing 532–534*
  advantages 533
  conventional vs. data flow
    systems 197
  digital multiprocessor analog
    computer 43
  prospects 534
  types of system 533–534
Multiprogramming 532
Multiturn potentiometer 633
Multivibrator 534–539*
  astable 537
  logic gate 538–539
  monostable 536
  phase-locked loops 602
  relaxation oscillator 702
  symmetrical bistable 534–535
  triggering of 537
  unsymmetrical bistable
    circuits 535–536
Munkres, J. 385
Mutual inductance 410
Mutual-inductance frequency
  meter 354
MYCIN 73

n

n-channel MOSFET 426
NAND circuit 470
Natural resonance 710
Near-end crosstalk 177
Negative feedback 340, 341
  FM 360
Negative-resistance circuits 539–
  543*
  astable circuits 541–542
  basic circuits 539–540
  bistable circuits 541
  monostable circuits 542
  negative-resistance devices
    542–543
  sine-wave oscillators 542
  typical circuits 540–541
Negative-resistance diodes 231
Nematic liquid crystals 464
Neon lamp–photocell optical
  isolators 575–576
Neper 543*
Network analyzer: impedance
  measurements 402
Network theory 547–548*
  branch equations 544–545
  electric network 117
  elements of a network 543–
    544
  graph theory 385–386
  Kirchhoff's laws of electric
    circuits 446–447*
  loop equations 545–546
  node equations 546–547
  reciprocity principle 699
  superposition theorem 791–
    792*
  Thévenin's theorem 547–548,
    853–855*
  Y-delta transformations 929–
    930*
Neutron transmutation doping
  548–549*
  advantages 549
  doping of thin layers 549
  process 548–549
  solar cells 549

Nishizawa, J. 773
Nitrogen laser 454
Noise see Electrical noise
Noise filter (radio) 551*
Noise jamming 302–303
Noise-power measurements
  282
Noncrystalline solid see
  Amorphous solid
Nonintelligible crosstalk
  177
Nonlinear amplifier 367–368
Nonlinear B/H devices 343
Nonlinear potentiometer 632–
  633
Nonlinear programming 551–
  553*
  application 550
  computational methods 550–
    551
  general theory 550
  operations research 566
Nonresonant antennas 65–66
Nonsinusoidal waveform 551–
  553*
  circuits 117
  effect of even harmonics 552
  effect of odd harmonics 552
  even and odd functions 552
  examples 553
  Fourier series representation
    551–552
  power 553
  power factor 553
  rms value 552–553
  symmetry 552
Nonvolatile static random-access
  memory 739, 740
Norton amplifier 30
Norton's theorem 853
NOT logic function 467–468
Notarys-Mercereau microbridge
  438
npn transistor 424, 863
Nuclear lasers 454
Number systems 553–557*
  analog-to-digital converter 57–
    58
  binary system 555–556
  decimal system 553–555
  digital computer programming
    218
  digital counters 225
  digital system codes 211
  octal number system 556–
    557, 561*
Numerical analysis 557–560*
  differential equations 559–560
  interpolation and
    approximation 557–558
  solution of linear systems
    558–559
Numerical control: CAD/CAM
  146
Numerical indicator tube 560–
  561*
NV RAM see Nonvolatile static
  random-access memory
Nybble 86
Nyquist, H. 275, 414

o

O-carcinotron 80
O-type backward-wave oscillator
  80

Object-oriented programming
  languages 655–656
Ochsenfeld, R. 783
OCR see Optical character
  recognition
Octal number system 556–557,
  561*
Ohm 703
Ohm, Georg S. 385, 923
Ohmmeter 561–562*
Ohm's law 562*, 704
OMR see Optical mark reading
OMS see Ovonic Memory Switch
On-line data bases 321–322
Op amp see Operational amplifier
Open circuit 118, 562*
Open-wire transmission lines
  870
Operating system 155, 562–
  563*
  control programs 562–563
  digital control 220–224*
  microcomputer systems 563
  multiaccess computer software
    532
  processing programs 563
  security kernels 153
  time-sharing applications 563
Operational amplifier 28–30
  function in analog computer
    46–47
  history 44
  integrated circuit comparators
    142–143
  integrator 29–30
  Norton amplifier 30
  summing amplifier 29
  transmission gate 376
  voltage amplifier 28–29
Operations research 563–568*
  applications 568
  decision trees 568
  history 564
  linear program model 565–
    566
  Markov processes 567–568
  mathematical programming
    565–567
  methodology 564–565
  network models 566
  queueing theory 567
  simulation 567
  stochastic process 567–568
Optical bistability 569–571*
  all-optical systems 569
  bistable optical devices 569–
    571
  fundamental studies 571
Optical character recognition
  109–111, 203
  applications 111–113
  cursive writing 113
  different alphabets 113
  element design 111
  feature extraction and
    classification 111
  functional systems 110
  hand-printed characters 113
  postal address readers 112
  preprocessing 111
  scanning systems 110–111
  stylized characters 111–112
  transport systems 110
  typewritten and typeset
    characters 112–113
Optical communications 571–
  574*
  atmospheric 572

Optical communications—*cont.*
  free-space 572
  integrated optics 430–432*
  light-emitting diodes 459
  optical-fiber communication 572–573
  optical-fiber transmission systems 667
  optical receivers 574
  optical transmitters 573–574
Optical-coupled isolator *see* Optical isolator
Optical detectors 574–575*
  communications systems 574
  in optical isolators 575–576
Optical disk recorder, digital 227–228
Optical disks 161
Optical emitters 575
Optical fiber: communication systems 572–573
  transmission systems 667
Optical isolator 575–576*
  electronic equipment grounding 314
  light-emitting diode 458–459
  optical detectors 575–576
  optical emitters 575
Optical mark reading 109
Optical masers 492
Optical modulators 576–578*
  acoustooptic modulation and deflection 577–578
  electrooptic effect 576–577
  electrooptic intensity modulation 577
  integrated optics 432
  optical waveguide devices 578
Optically ported cathode-ray tubes 106
Optically pumped laser 452–453
Optics *see* Integrated optics; Microwave optics
Optimization 578–583*
  analytical approach 582
  application 579–580
  concepts and terminology 579–580
  criterion function 580–581
  functional constraints 581
  graphical analysis 581–582
  nonlinear programming 551–553*
  parameters 580
  problem definition 580
  problem formulation 580–583
  random search methods 583
  recursive methods 582–583
  regional constraints 581
  solution of problems 581–583
Optimum filters 265
Optoacoustics: photoacoustic spectroscopy 607–610*
Optocoupler *see* Optical isolator
Optoisolator *see* Optical isolator
OR gate 806
OR logic function 468
Oscillator 583–587*
  basic principles 583–584
  blocking oscillator 86–87*
  bridge 586
  bridge circuit 90
  crystal 586–587
  feedback 341
  function generator 369

Oscillator—*cont.*
  general form circuit 585
  heterodyne 587
  integrated 587
  klystron 450–451
  laser 451–456*
  M-type backward-wave 80
  magnetron 486–490*
  maser 490–493*
  microwave 587
  O-type backward-wave 80
  phase-locked loops 601–603*
  phase-shift 585–586
  piezoelectric crystals 626
  piezoelectric resonators 630
  quartz clock 674–675*
  relaxation *see* Relaxation oscillator
  surface acoustic-wave device 795
  very-high-frequency 585–586
  voltage-controlled 367, 369
Oscillograph: waveform determination 918
Oscilloscope 587–593*
  attenuators and gain controls 588–589
  cathode-ray tube 100, 588
  cathodoluminescence 107–108*
  current measurement 7
  differential input amplifiers 589–590
  distortion 590
  dual-trace 591
  electrical measurements 592
  high-speed 591–592
  horizontal sweep and synchronization 590–591
  linear sweep generation 802
  phase-angle measurement 599
  photography of oscilloscope patterns 592–593
  signal amplifier 588
  signal delay networks 588
  time-interval measurement 591
Oscilloscope tube 879
OTS *see* Ovonic Threshold Switch
Ovonic Memory Switch 378–380
  characteristics 378–379
  materials 379
  read-only memories 379–380
Ovonic Threshold Switch 378–380
  characteristics 378–379
  materials 379
  photographic films 380
  transistors 380
Oxide glass 21

Packet switching (data communications) 195
Paige, E. G. S. 794
PAM *see* Pulse-amplitude modulation
Paper capacitor 99
Paper tape punches 202–203
Paper tape readers 202–203
Parabolic reflector antennas 67
Parallel circuit 117, 595*
  dc 234

Parallel-plate capacitor 93
Parallel resonant circuit: tank circuit 824–825*
Parametric amplifier 593–594*
  advantages 594
  parametric effects in varistors 593–594
Paraphase amplifier 601
Parasitic oscillation 594–595*
Parkinson, D. B. 44
Partition noise 281
PASCAL 656, 659
Pass transistor *see* Transmission gate
Passive radar 595*
Patel, C. K. N. 608
PCM *see* Pulse-code modulation
PDM *see* Pulse-duration modulation
Pentode 883
Permanent-magnet movable-coil ammeter 18
Personal microcomputers 494–495
PERT 595–598*
  advantage 597–598
  background 595
  requirements 595–597
Phanotron tube 374–375
Phase (periodic phenomena) 598*
  phase inverter 601*
Phase-angle measurement 598–600*
  electronic phase-angle meter 598–599
  electronic switch 599–600
  oscilloscope methods 599
  phase-order indicators 600
  phase-relation indicators 600
  synchronizer or synchroscope 600
  three-voltmeter method 598
Phase-angle meter, electronic 598–599
Phase changer (microwave circuit) 510
Phase-discrimination multiplexing 32, 532
Phase inverter 601*
  paraphase amplifiers 601
  transformer inverter 601
Phase-locked loops 601–603*
  basic operation 602
  locked-oscillator detector 363
  phase detectors 602–603
  uses 603
  voltage-controlled oscillators 367, 602
Phase modulation 527–528, 603–604*
  advantages and applications 603
  detector 604
  fundamental properties 604
  microwave transmission 511
  noise response 604
Phase-modulation detector 604*
Phase modulator 529, 604–607*
  phasitron 606–607
  types 605–606
Phase-order indicators 600
Phase-relation indicators 600
Phase resonance 710
Phase-shift discriminator 361
Phase-shift oscillator 584–585

Phasitron 606–607
Phasors: alternating-current 9–10
  current 12
  voltage 12
Philbrick, G. A. 44
Phonograph cartridges, high-polymer 404–405
Phonograph record *see* Disk recording
Phosphor storage tube 776
Photoacoustic spectroscopy 607–610*
  condensed-phase spectroscopy 609–610
  gases 608–610
  methods of measuring absorption 607–608
Photoconductive cell 610*, 613
Photoconductivity 610–612*
  device forms 612
  photosensitivity 611
  spectral response of photoconductor 611–612
  speed of photoconductor response 612
  television camera tube 839–840
Photodiode 612*, 613
  optical-fiber transmission system 667
Photodissociation lasers 454
Photoelectric devices 613*
  photoconductive cell 610*
  photodiode 612*
  phototransistors 620*
  photovoltaic cell 623*
  solar cell 762–766*
Photoelectricity 613*
  photoelectric devices 613*
Photoemission 613–616*
  alkali halides 615–616
  compounds 616
  Einstein photoelectric law 613–614
  metals 614–615
  photoelectric applications 613
  phototube 620–622*
  semiconductors 615
Photoferroelectric imaging 616–618*
  intrinsic effect 617
  photosensitivity enhancement 617–618
Photoluminescence 287
Photomultiplier 618–620*
  applications 619–620
  detection limits 619
  dynode materials 619
  image orthicon 838–839
  operation and design 618–619
  photocathode materials 619
Phototransistor 613, 620*
Phototube 620–622*, 879
  dark current 622
  photocathode construction 622
  photocathode material 621–622
  photomultiplier 618–620*
  principles of operation 620–621
  sensitivity to incident light 621
  types of service 622
Photovoltaic cell 623*
Photovoltaic effect 623*

Picture tube 623–625*, 879
  aluminized screen 624
  color kinescope 624–625
  construction 623–624
  deflection angle 623–624
  electromagnetic deflection 624
  electron gun 624
  focus 624
  glass envelope 624
  monochrome television
    receiver 844
Picturephone 893
Pierce, G. W. 630
Piezoelectric crystal 625–626*
  applications 626, 629–631
  characteristics and
    manufacture 625–626
  crystal oscillators 586–587
  materials 625
Piezoelectric earphone 255
Piezoelectric resonator 628–629
  applications 629–631
Piezoelectric transducer:
    microphones 496
Piezoelectricity 626–631*
  applications 629–631
  electromechanical coupling
    627
  ferroelectric properties 346
  high polymers 402–403
  matrix formulation 627
  molecular theory 627–628
  necessary condition 627
  piezoelectric ceramics 628
  piezoelectric crystal 625–
    626*
  piezoelectric resonator 628–
    629
  quartz clock 674–675*
  Rochelle salt 715–716*
PIN diode 518–519, 741
Pippard, A. B. 788
Pixel 307–308
PL/1 (programming language)
    658
PLA see Programmable logic
    arrays
Planar diffused epitaxial
    transistors 444–445
Planck, M. 614
Plane sheet reflector antennas
    67
Planning systems 335
Plastic-film capacitors 99
Plate-circuit rectification meter
    886–887
Plate modulation 39–41
PLL see Phase-locked loops
Plotting boards, computer-
    controlled 148
PM see Phase modulation
PMOS devices 470
pn junction: junction transistor
    442–445*
  optical properties 441–442
  rectification in semiconductors
    441, 732
  semiconductor diode 732–
    735*
  silicon rectifier diode 741–742
pnp transistor 424
pnpn transistor: negative-
    resistance devices 542–543
Point contact: Josephson junction
    438
Point-contact diode 518, 631*,
    733

Point-contact transistor 631–
    632*
  relaxation oscillator 702
Polarized-vane ammeter 19
Polya, G. 385
Polyphase rectifier circuits 700
Polyvinylidene fluoride:
    piezoelectricity 402–403
Postdeflection acceleration
    cathode-ray tube 103
Postfix notation see Reverse
    polish notation
Positive feedback 341
Potential difference, contact 168
Potentiometer (variable resistor)
    632–633*
  construction 632–633
  use 632
Potentiometer (voltage meter)
    633–635*
  Brooks deflectional dc
    potentiometer 634
  constant-current dc
    potentiometer 633–634
  constant resistance dc
    potentiometer 634
  current measurement 186
  Drysdale ac polar
    potentiometer 634–635
  high-frequency impedance
    measurement 396
  Tinsley-Gall ac polar
    potentiometer 635
  voltage measurement 905–906
Power: alternating-current circuit
    theory 14
  measurement on wattmeter 7
Power amplifier 635–636*
  class A 635
  class AB 635–636
  class B 636
  class C 636
  push-pull amplifier 672–673*
Power resistors 709
Power-supply filter 262
Power transformer 858–861
  characteristics 859–860
  construction 858–859
  cooling 859
  overloads 861
  parallel operation 861
  phase transformation 861
  principle of operation 858
  taps 860–861
Power transistors 445
PPM see Pulse-position
    modulation
Preamplifier 636*
Prediction systems 335
Prescriptive decision theory 204
Pressure-gradient microphone
    500
Pressure microphone 497–500
  carbon type 499
  ceramic type 499
  contact type 499–500
  crystal type 499
  dynamic types 497–498
  electrostatic (condenser) type
    498–499
Primitive abstract data types 2
Printed circuit 636–645*
  applications 637–644
  engineering 637
  film deposition 644–645
  hybrid circuits 643–644
  integrated circuits 644

Printed circuit—cont.
  manufacturing 637
  manufacturing processes 644
  material-removal processes
    644
  mold and die processes 645
  photography 637
  printed wiring 637–639
  technology 637
  thick-film circuits 639–641
  thin-film circuits 641–643
Printed wiring 637–639
  connections 639
  flexible 639
  heat dissipation 639
  multilayer boards 637–639
  protective finish 639
Printer (computer) 203
Printing cathode-ray tube 106–
    107
Privacy systems (scrambling)
    645*
Processing programs (operating
    systems) 563
Program evaluation and review
    technique see PERT
Programmable calculators 92
Programmable logic arrays 469
Programmable read-only memory
    739–740
  chips 158
Programming: abstract data types
    1–4*
  algorithms 4–6*
  analog computer 49–55
  calculators 92
  computer control systems
    222–223
  data flow systems 195–197*
  digital computer see Digital
    computer programming
  embedded systems 330–332*
  languages see Programming
    languages
  linear see Linear programming
  nonlinear programming 551–
    553*
  operating system 562–563*
  ROMs, PROMs, and EPROMs
    157–158
  signal-controlled in analog
    computer 49
  software engineering 756–
    762*
Programming languages 645–
    661*
  abstract data types 2
  ADA 331, 660
  aggregate data types 649–650
  ALGOL 60 657
  APL 658–659
  applicative languages 655
  assignment statements 646–
    647
  association of data types with
    objects 651
  basic 657–658
  binding attributes to variables
    650–652
  C 659–660
  CAD/CAM 146
  COBOL 657
  compared with data flow
    systems 196–197
  compound statements 647
  concurrent programs 163
  conditional statements 647

Programming languages—cont.
  data types 648–650
  defining new data types 656
  digital computer programming
    220
  examples of different bindings
    653–654
  expert systems 336
  expressions 646
  extent of a variable 651–652
  FORTRAN 656–657
  functions 654–655
  GOTO statements 647–648
  LISP 659
  microcomputers 495
  name binding 653
  object-oriented languages
    655–656
  objects, variables, and
    identifiers 646
  parameters 652–654
  particular languages 656–660
  PASCAL 659
  PL/1 658
  procedure and function
    parameters 654–655
  procedures 648
  processing programs 563
  reference binding 653
  repetitive statements 647
  result binding 653
  scalar data types 648–649
  scope of an identifier 650–
    651
  side effects of functions 654
  SNOBOL 659
  software engineering 756–
    762*
  statements 646–648
  structure 646–648
  value binding 653
Projection electronic display
    305–306
PROM see Programmable read-
    only memory
PROSPECTOR (expert system)
    334
Proximity-effect microbridge 438
Proximity tubes 456
Pseudoanalog electronic display
    306
Public-key cryptographic system
    179–180
Public-key encryption 154
PUFF (expert system) 335
Pulse amplifier: video amplifier
    889*
Pulse-amplitude modulation 275,
    528, 663
  modulator scheme 668–669
  switching systems 819
Pulse-code modulation 275, 528,
    664–665
  applications 665–668
  coding and multiplexing
    665
  digital filters 668
  electronic switching systems
    666–667
  fiber optics 667–668
  global digital transmission
    667
  modulator scheme 670–671
  nationwide digital transmission
    667
  optical-fiber transmission
    systems 667

Pulse-code modulation—*cont.*
quantization 664–665
radio set AN/TRC 666
repeatering 665
sampling 664
switching systems 819–821
synchronization 666
telephone systems 666
video disk audio signal reproduction 892
Pulse compression filter 795–796
Pulse Doppler radar 252–253
Pulse-duration modulation 528, 663–664
modulator scheme 669–670
Pulse generator 660–661*
digitally controlled 661
pulse-forming networks 661
Pulse modulation 528, 661–668*
bandwidth reduction 82–83
basic concepts 662–663
microwave transmission 511
principles of operation 662–665
pulse-amplitude modulation 663
pulse-code modulation 664–665
pulse-code modulation applications 665–668
pulse-duration modulation 663–664
pulse modulator 668–671*
pulse-position modulation 664
radio transmission 687
time-division multiplex 275
Pulse modulator 668–671*
pulse-amplitude scheme 668–669
pulse-code scheme 670–671
pulse-duration scheme 669–670
pulse-position scheme 670
Pulse-position modulation 528, 664
modulator scheme 670
Pulse transformers 671–672*
Pulsed gas lasers 453–454
Punched-card data-processing equipment 199–200
Push-pull amplifier 672–673*
driver stages 672–673
operation 672
phase inverter 601*

Q

Q (electricity) 673*
coaxial cavity magnetron 489
Q meter 673*
Q-GERT 377–378
Q meter 673*
high-frequency impedance measurement 397
Quadraphonic disk recording: coding and matrixing system 245
compatibility with mono and stereo systems 246
discrete system 245–246
Quadruplex video tape recorders 484

Quality factor of circuit *see* Q (electricity)
Quantum electronics 674*
laser applications 674
nonlinear optical phenomena 674
stimulated emission and amplification devices 674
Quartz clock 674–675*
Quartz crystal resonator 626, 629–630
quartz clock 674–675*
Queue: abstract data types 2–3

R

Radar 512, 675–686*
antenna 683
automatic frequency control 79
carrier-frequency bands 677
clutter 679–680
composition of radar systems 681–686
continuous-wave radar 168–169*
demodulator 209*
detection process 680
display and control 686
Doppler radar 251–253*
echo box 256*
fundamentals of operation 676–677
historical development 675
hyperbolic sweep generation 803
kinds 675–676
monopulse radar 529*
moving-target indication 530–531*
noise 679
passive radar 595*
propagation 677–678
radio transmission method 687
radome 695–696*
range equation 677
receiver and signal processor 684–686
target characteristics 678–679
tracking 680–681
transmitter 682–683
Radio 686–688*
amateur radio 17–18*
amplitude modulation 687
amplitude-modulation radio 36–38*
amplitude modulator 39–42*
antenna 60–69*
bandwidth requirements 83–84*
code telegraphy 687
Federal Communications Commission regulation 688
frequency-division multiplex 275
frequency modulation 687
frequency-modulation radio *see* Frequency-modulation radio
frequency separation 688
frequency-shift transmission 687
methods of information transmission 687
mobile radio 524–526*

Radio—*cont.*
pulse transmission 687
radar 687
radio-frequency amplifier 688–690*
radio-frequency measurements 357
receiver *see* Radio receiver
technical history 686–687
transmitter *see* Radio transmitter
uses 687–688
Radio detection and ranging *see* Radar
Radio-frequency amplifier 25, 688–690*
basic circuit 688–689
capacitive feedback 689–690
cascode circuit 690
high-frequency limit 689
Radio-frequency bridges 398–399
Radio-frequency measurements 357–358
heterodyne frequency meters 358
microwave frequency measurements 357
transfer oscillator methods 357–358
Radio receiver 690–693*
amplification 690–691
antennas 693
automatic gain control 79*
demodulator 209*
detection 691
diode detectors 34–35
diversity reception 693
electrical noise generator 284*
FM receivers 692
heterodyne principle 395*
intermediate-frequency amplifier 432*
mixer 524
noise advantage with FM 359–360
noise filter (radio) 551*
radar operation 676
regenerative receiver 692
selectivity 690, 726*
single-sideband receivers 692–693
spectrum analyzer 767–768*
superheterodyne receiver 691–692, 791*
tuned amplifier 25–26
tuned-radio-frequency receiver 691
tuner 876
tuning 877*
types 691–692
Radio transformer: distortion 862
rf and i-f transformers 862
Radio transmitter 693–695*
amplitude-modulation transmitters 693–694
frequency-modulation transmitters 694–695
radar operation 676
single sideband/independent sideband 695
Radiometer 282
Radiometer-type frequency meter 355
Radiotelegraphy, amplitude-modulated 38

Radiotelephony, amplitude-modulated 37
Radome 695–696*
Ragazzini, J. R. 44
RAM *see* Random access memory
Raman-type masers 492
Randall, R. H. 44
Random-access memory 154, 155
capacity 739
chips 156–157
dynamic and static types 739
portable computer 165
Random-sampling voltmeter 717
Rare-earth ion lasers 452
Rayleigh, Lord 697, 793
RC-coupled amplifier 24, 900–902
high-frequency gain 901
interstage transfer function 901–902
low-frequency gain 901
passband 901
RC filter 263–264
distributed 265
RCT *see* Reverse-conducting thyristor
Reactance 696*
Reactance modulator: phase-locked loops 602
Read, W. T. 517
Read diode 517
Read-only memory 739–740
chips 158
electrically alterable 739, 740
electrically erasable programmable 739, 740
fuse programmable 739–740
language translators 165
mask programmable 739
Ovonic Memory Switch 379–380
ultraviolet erasable programmable 739, 740
Real-time systems 696–697*
artificial intelligence and expert systems 697
digital control 222
evolution 696–697
real-time simulation 697
types 696
Receiving tubes 879
Reciprocity principle 697–699*
electrical networks 699
electromagnetic systems 698
electrostatic systems 698
examples of reciprocal systems 697
gyrator 386–387
Rayleigh's theorem of reciprocity 697–698
Rectangular cavity resonators 109
Rectangular waveguide 918–920
designation 920
generalization to other TE waves 919
phase velocity 919
TM waves 919
Rectifier 699–702*
bridge rectifier circuits 701
controlled rectifier 169–170*, 701
current ratings 701–702
full-wave rectifier circuit 700

Rectifier—*cont.*
half-wave rectifier circuit 699
inverse voltage 701
metallic-disk rectifier 493*
parallel rectifiers 701
point-contact diodes 518
polyphase rectifier circuits 700
ripple voltage 712*
Schottky-barrier diodes 518
semiconductor *see* Semiconductor rectifier
varistor 888–889*
voltage-multiplier circuit 906–907*
Rectifier diode 230–231, 734
semiconductor diode 741–747*
Rectifier voltmeter 915
Rectigon tube *see* Tungar tube
Reed-type frequency meter 353–354
Reel-to-reel audiotape recorders 483
Reeves, H. A. 664
Reflector antennas 67
Reflector-type microphone 503
Relaxation oscillator: astable negative-resistance circuits 541–542
multivibrator 534–539*
Reflex oscillator: klystron 450–451
Regeneration 702*
Regenerative comparator 142
Regenerative detector 36
Regenerative radio receiver 692
Register circuits 810
Regulation 702*
Relaxation oscillator 702–703*
blocking oscillator 86–87*
Relay circuit 171
Relay connecting circuit 808
Relay lockout circuit 809
Relay selecting circuit 808
Relay translating circuit 809
Reluctance 703*
Repair systems 335
Reprogrammable semiconductor memory 740
Repulsion-vane ammeter 19
Resistance *see* Electrical resistance
Resistance decade 708–709
Resistance measurement: alternating-current resistance 705
applications 709
comparison by potential drop 706
eddy current method 706–707
extreme resistance 704
high resistance 704
inductive resistance 705
Kelvin bridge 446
low resistance 704
methods 705–709
ohmmeter 561–562*
resistance standards 707–709
thin-film resistance 704–705
time-constant method (high resistance) 706
time-constant method (low resistance) 706
voltmeter-ammeter method 705–706

Resistance measurement—*cont.*
Wheatstone bridge 923–925*
Resistor 709
adjustable 710
attenuator 74*
classification by construction 709–710
classification by use 709
composition 709
film-type 709–710
potentiometer (variable resistor) 632–633*
thermal noise 283
varistor 888–889*
wire-wound 710
Resistor-transistor logic gate 538
Resonance (alternating-current circuits) 710–712*
high-frequency impedance measurements 396–398
multiple resonance 712
series resonance 711
universal resonance curve 711–712
use 710–711
Resonant antennas 65
Resonant-cavity wavemeter 923
Resonant-reed-type frequency meters 353
Resonant-type frequency meter 354–355
Resonator: surface acoustic-wave device 795
Response 712*
Reverse-blocking diode thyristor 745
Reverse-blocking triode thyristor 745
Reverse-conducting thyristor 746
Reverse-conducting triode thyristor 745
Reverse polish notation: calculator entry mode 92
rf amplifier *see* Radio-frequency amplifier
rf SQUID 773, 780
Ribbon-type pressure microphone 498
Ribbon-type velocity microphone 500
Rigid coaxial cable 131
Ringel, G. 384
Ripple voltage 712*
Robotics 712–715*
emerging areas 715
logic circuits 467–471*
mechanical design of robots 712–713
real-time 696
robot control system 713–714
robot sensory system 714–715
Rochelle salt 715–716*
piezoelectric crystal material 625
Roentgen, W. C. 299
ROM *see* Read-only memory
Rotating radial sweep generator 803
Round, H. J. 287
Rowell, J. M. 436
RPN *see* Reverse polish notation
RSA algorithm 181–182
Russell, Bertrand 5
Russell, F. A. 44
Russell, J. B. 44

Saha, M. N. 70
Salamis marble tablet 1
Sampling gate *see* Transmission gate
Sampling voltmeter 716–717*
high-frequency 717
noise reduction techniques 717
random-sampling voltmeter 717
sampling technique 716–717
vector voltmeter 717
Saturation 717–718*
Saturation current 718*
Sawtooth current generation 721
Sawtooth voltage generation 718–721
Sawtooth-wave generator 718–722*
high-frequency limitations 721–722
Scalar data types (programming) 648–649
boolean 648
character constants 648–649
enumerated 649
integer constants 648
real numbers 649
Scaling circuit 722*
Schering bridge: capacitance measurement 95–96
Schmitt trigger circuit 536
Schottky, W. 299
Schottky barrier diode *see* Schottky silicon rectifier diode
Schottky barrier gate metal-semiconductor field-effect transistor 371
Schottky diode FET logic 371, 372
Schottky effect 722–723*
Schottky silicon rectifier diode 518, 733, 743–744
Schoty (abacus) 1
Schrieffer, J. R. 783, 878
SCR *see* Silicon-controlled rectifier
SDFL *see* Schottky diode FET logic
Second-generation digital computer 216
Secondary emission 723–725*
cathode-ray tube 104
cathodoluminescence 108
dependence on angle of primary beam 725
experimental yield curves 724–725
mechanism of the process 724
secondary energies 725
secondary yield 723–724
Secondary memory technology 154, 159–161
cartridge disks 160
fixed-head disks 160
floppy disks 160–161
optical disks 161
• Winchester technology 159–160
Seignette salt *see* Rochelle salt
Selectavision Video Disk Recording 890
Selecting circuit 808

Selectivity 726*
Selenium: semiconductor element 728
Selenium rectifier 493, 741
Self-bias transistor circuit 85–86
Self-inductance 410
Semiconductor 726–732*
acceptor atom 4
amorphous 21–22
amorphous metal films 406
band theory of solids 81–82*
barrier layer 729
breakdown 89
calculator storage 92
charge-coupled device 113–116*
conduction 164, 726–728
contact barrier 730
contact rectification 732
cryosar 178*
crystal growth 729
diffusion theory 731
diode theory 731
donor atom 251*
doping 729
electron distribution 726–727
elemental 728
field emission 348
frequency multiplier 368
gallium arsenide integrated circuits 370–372*
glass switch 378–380*
grain boundaries 381–383*
Hall effect 392, 728
high-temperature silicon devices 405
hole conduction 407
impurity 727–728
injection electroluminescence 287
integrated circuits 420–430*
intrinsic 727
ion implantation 434, 729
materials and their preparation 728–729
mobility of carriers 727
neutron transmutation doping 548–549*
photoconductivity 610–612*
photoemission 615
photovoltaic effect 623*
point-contact transistor 632
preparation of materials 729
random noise 281–282
rectification 729–732
rectification at *pn* junctions 732
saturation current 718
secondary emission 724–725
semiconducting compounds 728–729
single-carrier theory 730–731
surface barrier 729–730
surface electronics 732
thermionic emission 852
transistor action 863–865
transistor development 327
tunnel diode 877*, 878
tunneling theory 731
two-carrier theory 731–732
valence band 888*
zone refining 729
Semiconductor diode 732–735*
junction diode 440–442*
limiter circuit 460
microwave solid-state devices 515–519*

Semiconductor diode—*cont.*
  optical detector 574–575
  optical transmitters 573–574
  photodiode 612*
  with *pn* junctions 733–735
  without *pn* junctions 732–733
  solar cell 763
  varactor 231, 888*
  varistor 888–889*
  Zener diode 931*
Semiconductor heterostructures
  735–739*
  carrier and optical field
    confinement 735–736
  chemical vapor deposition 738
  chemistry 737
  fabrication 737
  heteroepitaxy 737
  high-temperature compound
    devices 405–406
  liquid-phase epitaxy 737–738
  modulation doping 737
  molecular-beam epitaxy 738
  quantum well effects 736
Semiconductor lasers 454
Semiconductor memories 739–
  741*
  fabrication technology 740–
    741
  nonvolatile memories 739–740
  read/write RAMs 739
Semiconductor rectifier 699,
  741–747*, 865
  ASCRs, RCTs, and GTOs
    746–747
  controlled rectifier 169–170*
  junction diode 440–442*
  light-emitting diode 457–459*
  metallic-disk rectifier 493*
  point-contact diode 631*
  rectifier circuits 744
  Schottky silicon rectifier diode
    743–744
  silicon-controlled rectifier
    744–746
  silicon rectifier diodes 741–
    743
  varistor 888–889*
Semirigid coaxial cables 130–
  131
Sensitivity 747*
Sequential relay switching circuit
  808
Sequential switching circuit
  807–808, 824
Series circuit 117, 747*
  ac transients 271–272
  conductance 164*
  dc 234
  dc transients 269–271
Series-parallel circuit 117
  dc 235
Servomechanism feedback 342
Set theory: boolean algebra 88
Sets: abstract data types 3–4
Shadow-mask color picture tube
  134–135
Shamir, A. 181
Shannon (unit) 86
Shannon, C. E. 74, 78, 86,
  274, 414
Shibata, J. 773
Shift register 810
Shockley, William 327, 773
Short circuit 118
Shortwave AM radio 37
Shot noise 281, 283–284

Shunt-compensated amplifier 889
Shunting 747*
Sideband 747*
  single *see* Single sideband
Signal generator 747–748*
Signal-to-noise ratio 748*
Silicon: amorphous 22
  bipolar inverter circuits 421
  compared with gallium
    arsenide for IC application
    371
  neutron transmutation doping
    548–549*
  semiconductor element 728
Silicon bipolar microwave
  transistor 516
Silicon-controlled rectifier 701,
  741, 744–746
Silicon diode 440, 748*
Silicon-gate MOS 426–427
Silicon intensifier target camera
  tube 456
Silicon intensifier television
  camera tube 840–841
Silicon MOSFET 405
Silicon on sapphire device 866
Silicon planar transistor 863
Silicon semiconductor rectifier
  diode 741–743
  Schottky 743–744
Simple aperture electrostatic
  lenses 329
Simulation 748–751*
  analog 749
  applications 750–751
  digital 749
  hybrid 749
  mathematical modeling 748–
    749
  real-time 697
  trends 749
Sine-wave oscillator: negative-
  resistance circuit 542
Single-board microcomputers
  494
Single-pulse relay counter 171
Single sideband 751–752*
  communications 752
  frequency stability
    requirements 751–752
  generation 751
  history 752
  mobile radio transmission 364
  modulation 31–32
  radio receivers 692–693
  radio transmitter 695
  reception 751
  telephony 38
Sinusoidal oscillator 583–587
SIT *see* Static induction
  transistor
Slant-track video tape recorder
  484
Slide-back voltmeter 887
Slide rule 752–753*
Slide-wire potentiometer 633
Small loop antenna 65
SMALL TALK (programming
  language) 655
Smectic liquid crystals 464–465
Smith, H. I. 794
Smith, P. H. 522
Smith, W. R. 794
SNOBOL (programming
  language) 659
Soft fails 753–756*
  ionization-induced 754–755

Soft fails—*cont.*
  rates for typical devices 755–
    756
Software: piracy 762
  portability 762
  see also Digital computer
    programming; Programming
    languages
Software engineering 756–762*
  design data representation
    759–760
  design of software 757
  development stages 756–758
  effectiveness 761
  functional specification 757
  methods 758–761
  motivation 756
  multiaccess computer 532
  program implementation 757–
    758
  program testing 758
  requirements definition 757
  software maintenance 758
  software management 761
  structured design 758–759
  structured programming 760–
    761
Solar cell 734
  amorphous silicon 22
  array applications 766
  arrays 765–766
  characteristics 763–765
  photovoltaic cell 623*
  principles of operation 763
  solar radiation 762–763
Solid-state masers 491
Solion 766–767*
Solion diode 766
Solion electrical readout
  integrator 766–767
Solion electroosmotic driver
  767
Solion pressure transducer 767
Soroban (abacus) 1
SOS device *see* Silicon on
  sapphire device
Sound-reproducing systems:
  fidelity 348*
  see also Disk recording
Space charge 767*
  effect on electron motion in
    vacuum 297–298
  Langmuir-Child law 451*
Space-division switching
  networks (communications)
  818
Spark gap 767*
Sparking potential *see*
  Breakdown potential
Speaker *see* Loudspeaker
Special-purpose analog computer
  42
Spectrum analyzer 767–768*
Speech recognition 768–772*
  accuracy 769
  artificial intelligence 73
  difficulty 768–769
  feature extraction 770
  higher-level linguistic
    components 771–772
  motivation for development
    768
  operation of speech
    recognizers 769–772
  pattern standardization and
    normalization 770–771
  prospects 772

Speech recognition—*cont.*
  word boundary detection 769–
    770
  word matching 771
Spherical capacitor 93
Sputtering 772*
  glow discharge 381
Square-law detectors 35–36
SQUID 772–773*, 879
  magnetometers 779–781
SSB *see* Single sideband
Stack: abstract data types 2–3
Stagger-tuned amplifier 25
Standing-wave detector 773*
Staples, E. J. 795
Static induction transistor 773–
  775*
  applications 774
  device physics 773–774
  fabrication 774–775
Static inverter: uninterruptible
  power system 273–274
Static random-access memory
  156–157, 739
Steering logic *see* Transmission
  gate
Step-recovery diodes 741
Step voltage regulator 909–910
Stereophonic disk recording:
  compatibility with
    monophonic systems 244–
    245
  recording system 243
  side thrust 244
  sound-reproducing system 243
  surface noise 244
  tracing distortion 243–244
  vertical tracking angle 244
Stereophonic FM broadcasting
  364
Stone, Marshall 88
Storage registers (calculators) 92
Storage tube 775–776*, 879
  direct-view 775–776
  display 775–776
  electrical 775
  operation 775
  phosphor 776
Stored program digital computer
  212–214
  characteristics 214
  instructions 213–214
  programming 218–220
Stream cipher 183–184
Striplines, microwave 521
Suan-pan (abacus) 1
Summing amplifier 29
Superconducting computers 776–
  779*
  advantages of Josephson
    technology 777–778
  disadvantages of Josephson
    technology 778–779
  limitations on cycle time 776–
    777
  performance 776
  progress and prospects 779
Superconducting devices 779–
  782*
  computer elements 781–782
  cryotron 178*
  electromagnetic radiation
    detectors 782
  Josephson effect 436–439*
  SQUID magnetometers 779–
    781
  standards 782

Superconducting quantum interference device see SQUID
Superconductivity 783–791*
  basic experimental properties 783–787
  electrical resistance of superconductors 784
  flux quantization 790
  Ginzburg-Landau theory 789
  high-frequency electromagnetic properties of superconductors 787
  isotope effect 787
  Josephson effects 878–879
  magnetic properties of superconductors 784–786
  microscopic (BCS) theory 789–790
  nonlocal theory 788–789
  pair tunneling 791
  quasiparticle tunneling 790–791
  superconducting computers 776–779*
  superconducting devices 779–782*
  theory 787–790
  thermal properties of superconductors 786–787
  tunnel junctions between metals 878
  two-fluid model 787–788
Superconductor: type I 784–785
  type II 785–786
Superheterodyne receiver 791*
  heterodyne principle 395*
  mixer 524
  radio receiver 691–692
  spectrum analyzer 767–768*
Superposition theorem (electric networks) 791–792*
  mathematical definition 792
  uses 792
Superregenerative detector 36
Suppressed carrier modulator 42
Suppression 792*
Suppressor 793*
Surface acoustic-wave devices 793–796*
  background 793
  bandpass filters 794–795
  oscillators 795
  pulse compression filters 795–796
  resonators 795
  transduction 793–794
Surface barrier (semiconductor) 729–730
Surface physics 796–801*
  data acquisition, analysis, and theory 797–801
  experimental apparatus 797
  LEED applications 799–800
  macroscopic models 799
  microscopic models 799
  physical principles of measurements 796–797
  quantum theory of surfaces 799
  surface atomic geometry 800–801
  surface atomic motion 801
  surface composition 800
  surface electronic structure 801

Surface physics—cont.
  surface preparation 797
  theoretical models for data analysis 799
Surging 802*
Susceptance 802*
Sweep generator 802–803*
  circular sweep generation 802–803
  hyperbolic sweep generation 803
  linear sweep generation 802
  rotating radial sweep generation 803
Switch see Electronic switch
Switchboard-type frequency meter 354
Switched capacitor 803–806*
  analog operations 805
  equivalent resistance 804–805
  filter design 805–806
  integrator circuits 804
  MOS structure 427
Switched channels (data transmission) 194
Switched filters 265
Switching circuit 806–811*
  analog computer programming 49
  clamping circuit 124–126*
  combinational circuits 806–807
  connecting circuits 808
  counting circuits 171, 810–811
  diode 231
  functional 808–811
  lockout circuits 809
  logic gate circuit 422
  register circuits 810
  selecting circuits 808
  sequential circuits 807–808
  switching gate 376
  translating circuits 809–810
Switching device: cryosar 178*
  cryotron 178*
Switching regulators, power supply 317–319
  applications 318–319
  configurations 318
  step-down 317–318
  step-up 318
Switching systems (communications) 811–822*
  coordinate switch systems 814–816
  counting circuit 171–172*
  data transmission 194–195
  electromechanical switches 814–816
  electronic switching 817–822
  fundamentals 811
  No. 1A electronic system 817–818
  numbering plan 812–813
  pulse-code modulation 666–667
  rotary switch systems 814
  service features 821
  signaling and control 811–812
  space-division networks 818
  switching circuit 806–811*
  switching connectives 813–814
  time-division networks 818–821
  XY system 814

Switching theory 822–824*
  combinational circuits 823–824
  sequential circuits 824
Symmetrical bistable multivibrator 534–535
  bipolar transistor circuits 535
  IGFET circuits 535
  JFET circuits 535
Synchronization 824*
Synchronized blocking oscillator 87
Synchronizer: phase-angle measurement 600
Synchronous clamp 125
Synchronous concurrent system 163
Synchronous detector 36
Synchroscope 824*
  see also Synchronizer

t

T network: capacitance measurement 96–97
Tancrell, R. H. 794
Tank circuit 824–825*
Tape recorder, magnetic 480
Tarjan, R. 384
Taut-band ammeter 19
TED see Transferred-electron device
Telecable 324–325
Telecommunication switching systems 811–822*
Telegraph: amplitude-modulated 38
  bandwidth requirements 83–84*
  block encoders 414
  frequency modulation 366
  radio transmission method 687
Telephone 825–827*
  accessory devices 826–827
  AM radio 37–38
  bandwidth requirement 83–84*
  communication cables 135–142*
  crosstalk 178
  data transmission 194
  electric network 826
  mobile 364–365
  privacy systems (scrambling) 645*
  pulse-code modulation systems 666
  radio relaying 365–366
  receiver 825–826
  single-sideband 38
  switching systems 811–822*
  transmitter 825
  video telephone 892–893*
  viewdata system 167
Telephotography 827*
Teleprinter see Teletypewriter
Teletel 322
Teletext see Videotext and teletext
Teletypewriter 827
  ASCII Code 829–830
  Baudot code 828
  encoding 414–415
  frequency-shift radio transmission 687

Teletypewriter—cont.
  operation 828–829
Television 830–833*
  AM broadcasting 38
  bandwidth 830–831
  bandwidth reduction 82–83*
  camera see Television camera
  cathode-ray tube 100
  cathodoluminescence 107–108*
  closed-circuit television 128–130*
  color see Color television
  frequency 831
  grades of service 832–833
  home data retrieval 166–168
  linearity 463
  mixer 524
  picture transmission 831–832
  picture tube 623–625*
  receiver see Television receiver
  recording programs 832
  scanning see Television scanning
  scrambling of signals 832
  sound transmission 831
  synchronization 824
  teletext and videotext 833
  video disk recording 889–892*
  video recorders 484
  videotex 322–323
  viewdata 167–168
Television camera 833–837*
  blanking, synchronizing, video processing, and control circuits 835–836
  color 132
  essential elements 834–836
  optical system 834
  photoconductive detector 612
  pickup device 834–835
  preamplifier 835
  scanning circuits 835
  tube see Television camera tube
  typical configurations 836–837
Television camera tube 837–842*, 879
  combined with image intensifier 456
  image isocon 839
  image orthicon 407*, 837–839
  photoconductive tubes 839–840
  silicon intensifier 840–841
  solid-state imagers 841–842
Television receiver 842–847*
  automatic frequency control 79
  cascode amplifier 100
  color 845–847
  color decoder 133–134
  color picture tube 134–135
  demodulator 209*
  fidelity 348
  heterodyne principle 395*
  improvements in color sets 165–166
  monochrome 842–845
  nonlinear B/H devices 343
  phase-locked loops 603
  tuned amplifier 25
  tuner 876

Television receiver—*cont.*
 tuning 877*
 video amplifier 889*
 videotext and teletext 893–894*
Television scanning 830, 847–848*
 color coding 848
 interlaced 847
 lines and frequency 847–848
 raster scan 802
 resolution 848
 scanning circuits 835
Television transmitter 848–851*
 designs 850–851
 power 849
 signal characteristics 848–849
 transmitting antennas 849–850
Tellegen, B. D. H. 299, 387
Tellurium: semiconductor element 728
Terasaki, T. 773
Tetrode 882–883
Thermal ammeter 20–21
Thermal matrix printer 203
Thermal noise 283
Thermionic emission 851–852*
 metals 852
 phanotron tube 374–375
 Richardson equation 851–852
 Schottky effect 722–723*
 semiconductors 852
 thermionic tube 852–853*
 work function 928–929*
Thermionic tube 852–853*, 879
 cathode materials 880–881
 electrode configurations 881–884
 structure and fabrication 879–880
Thermocouple meter 186
Thermometer *see* Thermal ammeter
Thermotropic liquid crystals 463
Thermovoltmeter 915
Thévenin, M. L. 853
Thévenin's theorem (electric networks) 853–855*
 examples 853–855
 proof 855
Thick-film capacitors 99
Thick-film printed circuits 639–641
 capacitances 641
 inductors 641
 protective coatings 641
 resistor networks 640–641
Thin-film microbridge 438
Thin-film printed circuits 641–643
 capacitances 643
 postdeposition treatments 643
 resistor networks 641–643
Thin films, magnetic *see* Magnetic thin films
Third-generation digital computer 216–217
Thomas, J. L. 707
Thompson-Lampard capacitor 98
Thomson, J. J. 70, 76, 107, 299
Thomson, William (Lord Kelvin) 43
Three-dimensional electronic display 306
Thyratron 375, 855–856*
 control action 855

Thyratron—*cont.*
 control characteristics 855–856
 ionization time 856
 ratings 856
Thyristor: ASCRs, RCTs, and GTOs 746–747
 silicon-controlled rectifier 744–746
 see also Controlled rectifier
Time compression harmonic analyzer 394
Time constant 856–857*
 resistance measurement 706
Time-division multiplexing 32, 275, 532
Time-division switching networks (communications) 818–821
 analog (PAM) systems 819
 digital (PCM) systems 819–821
Time-sharing system 216
 operating systems 563
Tinsley-Gall ac potentiometer 635
Top-loaded antenna 65
Townsend, J. S. 857
Townsend discharge 187, 857*
Transducer 857*
 earphone 253–256*
 electret transducer 257–260*
 electrical communications 274
 electromechanical 259
 high-polymer *see* High-polymer transducer
 microphone types 496–497
 piezoelectric crystals 626
 surface acoustic-wave devices 793–794
 telephone receiver 825–826
 telephone transmitter 825
Transfer switch: uninterruptible power system 274
Transferred-electron device 516–517, 542–543
Transformer 857–862*
 audio and radio-frequency transformers 861–862
 equivalent circuit 333
 ideal 173
 impedance matching 409
 phase-inverter 601
 power transformers 858–861
 pulse transformers 671–672*
 transformation of impedance 174–175
Transformer-coupled amplifier 24–25
Transformer-coupled circuits 174
Transistor 862–867*
 action 863–865
 amplifier *see* Transistor amplifier
 amplitude distortion 247–248
 bias of transistors 85–86*
 binary counter 172
 classification 863
 clipping 127–128
 connection *see* Transistor connection
 double-base diode 866
 equivalent circuit 333
 field-effect *see* Field-effect transistor
 gallium arsenide field-effect transistors 516
 gallium arsenide integrated circuits 370–372*

Transistor—*cont.*
 history of electronics 327–328
 integrated circuits 420
 junction transistor 442–445*
 linearity 463
 load line 466–467*
 manufacture 866–867
 mesa 445
 microwave solid-state devices 515–519*
 negative-resistance devices 542–543
 npn 424
 Ovonic Threshold Switch 380
 phototransistors 620*
 pnp 424
 point-contact transistor 631–632*
 power 445
 power switching 865
 silicon bipolar transistors 516
 static induction transistor 773–775*
 TTL 471
 voltage regulator circuit 908–909
Transistor amplifier 23–24
 base-modulated 40, 41
 basic principles 23
 cascode amplifier 100*
 collector-modulated 39
 potential distribution 23–24
Transistor connection 867–869*
 bipolar transistor 867–868
 field-effect transistors 868–869
Transistor direct-coupled amplifiers 231–232
Transistor-transistor logic 471
 gate 376
 NAND gate 471
Translating circuits 809–810
Translation-oriented data communications 192
Transmission gate 375–376, 470
Transmission lines 869–872*
 characteristic impedance 871
 circuit analysis 292
 circuit elements 872
 coaxial cable 130–131*, 870
 delay line 208–209*
 distortionless line 871
 electrical noise 291–292
 electromagnetic wave transmission 291–292
 high-frequency impedance measurements 400
 loading 278
 microwave transmission lines 520–523*
 neper 543*
 open-wire 870
 principal mode 870–871
 pulse transients 872
 sheathed cables 870
 standing waves 872
 tapered 872
 transmission-line equations 871–872
 traveling waves 870
TRAPATT diode *see* Trapped-plasma avalanche-triggered transit diode
Trapped-plasma avalanche-triggered transit diode 517–518
Traveling-wave antennas 65
Traveling-wave masers 493

Traveling-wave tube 872–875*
 backward-wave tube 79–81*
 beam collector 875
 electron gun 873
 focusing methods for the beam 873–874
 slow-wave circuit 874–875
Tree (abstract data type) 3
TRF receiver *see* Tuned-radio-frequency receiver
Trigger circuit 875–876*
 compared with pulse generator 661
 peaking (differentiating) circuits 875–876
 ringing circuits 876
Trimmer capacitor 98
Trimmer potentiometer 633
Triode 881–882
Triode clamp 125–126
TTL *see* Transistor-transistor logic
Tunable filter: harmonic analysis 393
Tuned amplifier 25, 902–903
 double-tuned 902–903
 intermediate-frequency amplifier 432*
 radio-frequency amplifier 688–690*
 single-tuned 902
Tuned-collector oscillator 585
Tuned-radio-frequency receiver 691
Tuner 876*
Tungar tube 374
Tungsten trioxide electrochromic displays 286
Tuning 877*
 tuner 876*
Tunnel diode 518, 734, 741, 877*, 878
 negative-resistance devices 542–543
Tunneling cryotron 178
Tunneling in solids 877–879*
 Josephson effect 878–879
 semiconductor theory 731
 tunnel diode 877*
 tunnel junctions between metals 878
Turan, P. 386
Turing, A. M. 5, 74
Turing machine 77
Tutte, W. 386
Twin-tee filter 262
Two-pulse relay counter 171–172
Two-pulse transistor counter 172
Two-reflector antennas 67–68
Two-way radio 17–18*
Type I superconductors 784–785
Type II superconductor 785–786

U

Uda, S. 931
Uhlenbeck, G. 385
Ultraviolet-erasable programmable read-only memory 739, 740
Unidirectional microphones 500–501
Unijunction transistor 445
 negative-resistance devices 542–543

Unijunction transistor—*cont.*
  relaxation-oscillator 703
  see also Double-base diode
Uninterruptible power systems 272–274
Unipotential electrostatic lenses 330
Untuned amplifier 25
UV EPROM *see* Ultraviolet erasable programmable read-only memory

Vacuum capacitor 98
Vacuum diode 881
Vacuum pentode 883
Vacuum phototube: principles of operation 620
  types of service 622
Vacuum tetrode 882–883
Vacuum triode 881–882
Vacuum tube 299, 879–886*
  amplifiers *see* Vacuum-tube amplifier
  audion 76*
  basic operation 879
  cathode materials 880–881
  classification 879
  electrode configurations 881–884
  equivalent circuit 333
  field-emission tubes 886
  frequency and gain limitations 885–886
  interelectrode capacitance 885
  lighthouse tube 886
  load line 466–467*
  magnetron 487
  microwave tube 523–524*
  random noise in electron-tube circuits 281
  saturation current 718
  shot noise 283–284
  structure and fabrication 879–880
  tube factors 884–885
  voltmeter *see* Vacuum-tube voltmeter
Vacuum-tube amplifier: basic principles 24
  classifications 24–26
  grid-modulated 40
  plate-modulated 39
Vacuum-tube dc amplifiers 232
Vacuum-tube rectifier 699
Vacuum-tube voltmeter 186, 886–888*
  amplifier-type meter 887
  diode rectifier-amplifier meter 887
  direct-current 887
  grid-rectification meter 887
  low-voltage limits 888
  plate-circuit rectification meter 886–887
  slide-back voltmeter 887
  types 886–887
  waveform error 888
Valence band 888*
Varactor 231, 519, 734, 888*
  frequency modulator 367
  multipliers 368
  parametric amplifiers 593
  phase-locked loops 602
Variable capacitance diode *see* Varactor

Variable capacitor 98
Variable inductor 413
Variable-resistance device: potentiometer (variable resistor) 632–633*
Varian, R. 299
Varian, S. 299
Varistor 888–889*
  grain-boundary properties 383
  parametric effects 593–594
Vector impedance meter 396
Vector voltmeter 717
Vectorgraphic electronic display 306
Velocity microphone 500
Vernam, G. S. 183
Vertical metal oxide semiconductor 866
Very-high-frequency oscillator 585–586
Very High Speed Integrated Circuit program 330
Very-large-scale integration: data flow systems 195–197*
Vestigial sideband modulation 32
VHSIC program *see* Very High Speed Integrated Circuit program
Video amplifier 889*
Video disk 161
  disk business 324
  publishing how-to literature 325
Video disk recording 889–892*
  reproduction of audio signals 892
  Selectavision Video Disk 890
  Video High Density Disk System 890–891
  Video Long Play System 891–892
Video electronic display 306
Video High Density Disk Recording 890–891
Video Long Play Disk Recording 891–892
Video tape recorders 484
  helical-scan 484
  quadruplex 484
Video telephone 129, 892–893*
  cost reduction approaches 893
  Picturephone 893
  video conferences 893
Video transformers 861–862
Videotex 322–323
Videotext and teletext 833, 893–894*
  display alternatives 894
  implementations 894
  services 894–895
  telecable 324–325
  teletext decoder 167–168
  videotex services 322–323
  viewdata compared with teletext 167–168
Vidicon tube 839
Viewdata *see* Videotext and teletext
Viologen electrochromic display 286
Virtual memory system 155
VMOS *see* Vertical metal oxide semiconductor 866
Voice response 895–899*
  assessment of systems 899

Voice response—*cont.*
  parametrized systems 896–897
  phonemic synthesizers 897–899
  reproduction of stored speech 895
  word concatenation systems 895–896
Volt-ohm-milliammeter 900*
  multimeter 532*
Voltage: measurement by voltmeter 7
  phasors 12
Voltage amplifier 28–29, 900–905*
  difference amplifiers 904–905
  direct-coupled amplifier 904
  grounded-gate (base) amplifier 903–904
  integrated amplifier 905
  preamplifier 636*
  RC-coupled amplifier 900–902
  source and emitter-follower stages 903
  tuned amplifier 902–903
Voltage-controlled oscillator: frequency modulator 367
  phase-locked loops 602
Voltage measurement 905–906*
  ac-dc transfer 906
  alternating current 906
  digital voltmeter 230*
  direct current 905
  potentiometer (voltage meter) 633–635*
  potentiometric measurement 905–906
  sampling measurement 906
  sampling voltmeter 716–717*
  volt-ohm-milliammeter 900*
  voltmeter 911–915*
Voltage-multiplier circuit 906–907*
Voltage regulator 907–910*
  diode regulator circuit 908
  electronic 907–909
  glow discharge 380
  induction voltage regulator 910
  other regulators 910
  power-system 909–910
  step voltage regulator 909–910
  transistor regulator circuits 908–909
Voltage-regulator tube 373, 911*
Voltmer, F. W. 793
Voltmeter 7, 911–915*
  classifications 911
  digital voltmeter 230*
  electrodynamic ac voltmeter 913–914
  electronic ac voltmeter 915
  electrostatic voltmeter 912–913
  moving-coil voltmeter 911–912
  moving-iron voltmeter 914–915
  moving-magnet voltmeter 912
  phase-angle measurement 598
  potentiometer (voltage meter) 633–635*
  rectifier voltmeter 915

Voltmeter—*cont.*
  resistance measurement 705–706
  sampling voltmeter 716–717*
  thermovoltmeter 915
  vacuum-tube voltmeter 886–888*
  voltage measurement 905–906*
Volume compressor 915
Volume control systems 915–916*
Volume expander 915–916
Volume limiter 915
von Neumann, John 74, 86

Watt-hour meter 7
Wattmeter 7
  harmonics measurement 393
Wave analyzers 393
Wave microphone 502–503
  line type 503
  reflector type 503
Wave motion: frequency 349*
Wave-shaping circuits 916–917*
  oscillator 583–587*
  pulse generator 660–661*
  pulse transformers 671–672*
  sawtooth-wave generator 718–722*
  trigger circuit 875–876*
Waveform 917*
Waveform determination 917–918*
Waveguide 918–921*
  acoustooptic modulator-deflector 578
  attenuation of hollow waveguides 920–921
  coaxial cable 921
  directional coupler 235–236*
  electromagnetic wave transmission in hollow waveguides 290–291
  high-frequency impedance measurements 400
  integrated optics 430–432*
  joints, bends, and junctions 920
  loading 278
  Maxwell's equations 918
  microwave 510
  microwave reflectometer 515*
  microwave transmission lines 521
  rectangular 918–920
  shapes 920
  transmission line modes 870
  transmission modes 918
Wavelength measurement 921–923*
  frequency measurement 922
  interferometer methods 922
  lecher wires 922
  standing-wave detector 773*
  tuned circuits 922–923
  wavemeter 923*
Wavemeter 923*
  standing-wave detector 773*
  wavelength measurement 921–923*
Wayne-Kerr rf bridge 399
Wehnelt, A. 299

Wheatstone, Sir Charles 923
Wheatstone bridge 923–925*
  accuracy 925
  capacitance measurement 95–97
  Kelvin bridge 445–446*
  sensitivity 924–925
White, R. M. 793
White-noise jamming 303
Whitehead, A. N. 5
Whitney, H. 386
Wien-bridge oscillator 90, 586
Wien frequency bridge 356

Wiener, Norbert 74, 86, 274
Williams tubes 158
Winchester technology (disk memory) 159–160
Wire-wound potentiometer 633
Wire-wound resistor 710
Wirth, Niklaus 659
Word definition 86
Word processing 925–928*
  application 926–928*
  equipment 926–928
  history and background 926
  input equipment 926

Word processing—*cont.*
  output equipment 927–928
  people 928
  procedures 928
  trends 928
Work function (electronics) 928–929*

Y-delta transformations 929–930*
  derivation 929

Y-delta transformations—*cont.*
  example 930
Yagi, H. 931
Yagi-Uda antenna 930–931*
Youngs, J. W. T. 384

Zener, C. 931
Zener diode 741, 931*
  voltage regulator circuit 908–909